gerd arntz
graphic designer

800 2005 055
ПОЧТА РОССИИ
E1
CNUSA (CNA)
Beijing
CHINA POST
PEK

AbeBooks
International Priority
Airkraft
Size 3
DOCUMENTE
DOCUMENTS
AbeBooks

Book Depository
WORLDWIDE EXPRESS
こわれもの
CUSTOMS DECLARATION

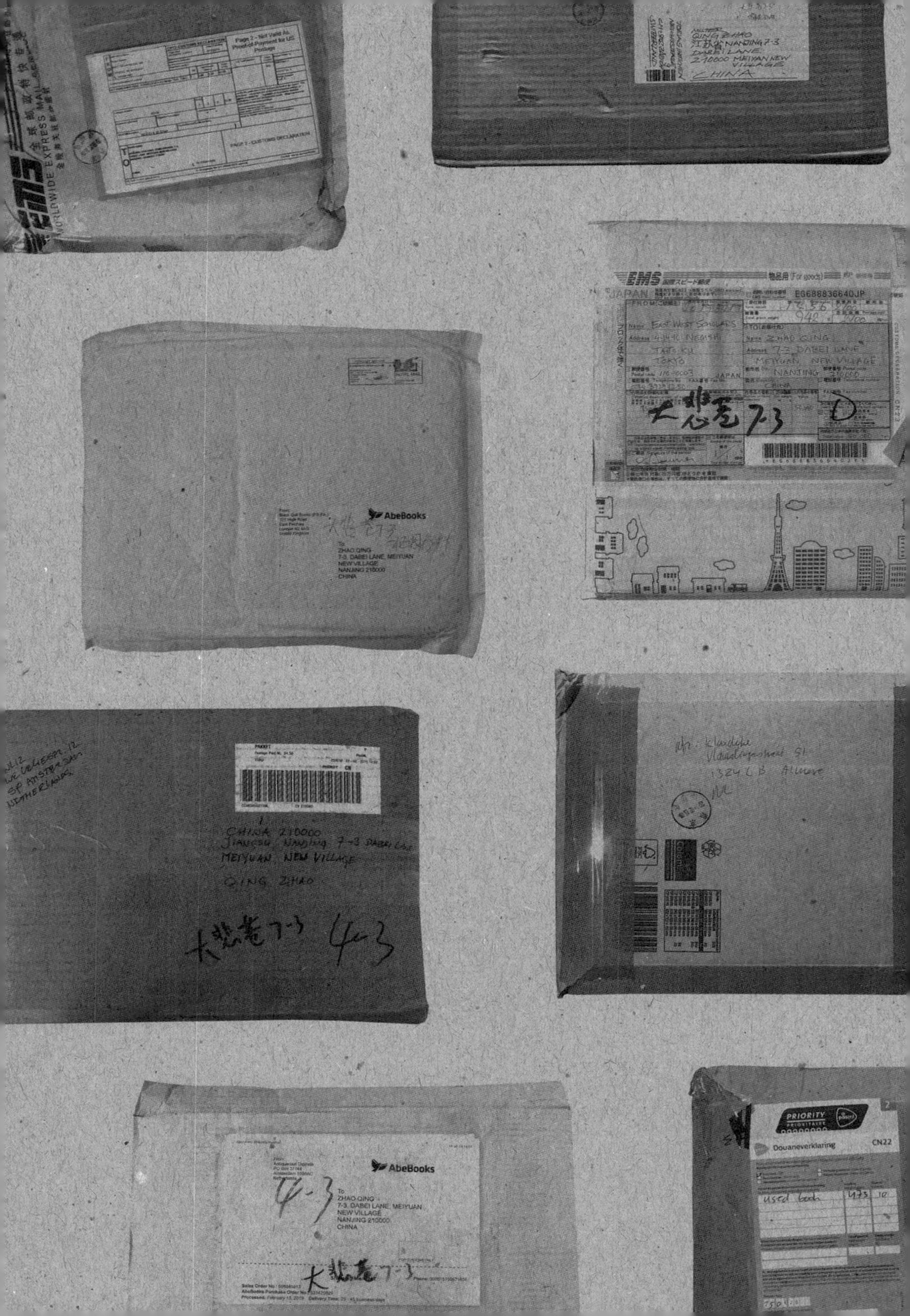

WORLDWIDE EXPRESS MAIL
QING ZHAO
NANJING 7-3
DABEI LANE
210000 MEIYAN NEW VILLAGE
CHINA
EMS
EG686836640JP
ZHAO QING
7-3 DABEI LANE
MEIYUAN NEW VILLAGE
NANJING
大悲巷7-3
AbeBooks
To
ZHAO QING
7-3, DABEI LANE, MEIYUAN
NEW VILLAGE
NANJING 210000
CHINA
PAKKET
CHINA 210000
JIANGSU NANJING 7-3 DABEI LANE
MEIYUAN NEW VILLAGE
QING ZHAO
大悲巷7-3 4-3
AbeBooks
To
ZHAO QING
7-3, DABEI LANE, MEIYUAN
NEW VILLAGE
NANJING 210000
CHINA
4-3
大悲巷7-3
PRIORITY
Douaneverklaring
CN22
used book

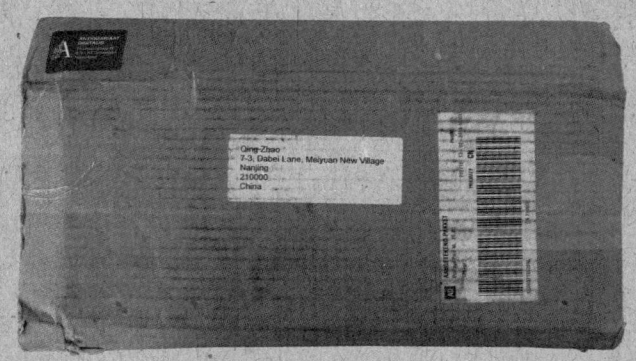

customs documents
Postal parcels
OEFRAA (DEA)
Frankfurt/M
DeutschePost
Airmail
2019-03-05
CNNKGA (CNA)
Nanjing
CHINA POST
PVG
Qing Zhao
7-3, Dabei Lane, Meiyuan New Village
Nanjing
210000
China
AANGETEKENDE BRIEF
RECOMMANDÉ
PRIORITY
CN
全球邮政特快专递
RLDWIDE EXPRESS MAIL SERVICE
CP72 CUSTOMS DECLARATION
P

Return address:
Unit 5, Gateway12 Business
Park, Davy Way, Hardwicke,
Gloucester, GL2 2BY, UK
Zhao Qing
7-3 Dabei Lane Meiyuan New
Village.
Nanjing
210000
CHINA
ROYAL MAIL
BY AIR MAIL
par avion
1
Book Depository
Book Depository
R Aangetekend
ZHAO QING
7-3, DABEI LANE
~MEIYUAN
210000 NEW VILLAGE
CHINA
RR617612842NL
RR617612842NL
used book
Klantnummer
赵清
PRIORITY
PRIORITAIRE / LUFTPOST
PRIORITY
PRIORITAIRE / LUFTPOST
Deutsche Post
Absender
August Dreesbach Verlag
Gollierstr. 70
Eingang D1 1.06
80339 München
Deutschland
Empfänger
Zhao Qing
7-3, Dabei Lane, Meiyuan
New Village
210000 Nanjing, Jiangsu
China
Bei Unzustellbarkeit
Rücksenden
Preisgabe
PREMIUM / PAR AVION PRIORITAIRE
BITTE DEN VERSANDSCHEIN
MIT KUGELSCHREIBER UND IN
DRUCKBUCHSTABEN AUSFÜLLEN.
DABEI FEST AUFDRÜCKEN!
Paketkarte
CP 92
DEU

CC 07657367 5 NL
大悲巷7-3

1 Book
0.68
УКРПОШТА

Port Betaald
Port Payé
Pays-Bas
postnl
PRIORITY
PRIORITAIRE
antiquariaat Kok
Oude Hoogstraat 14-18
1012 CE Amsterdam - Nederland
Aan / To:
Zhao Qing
7-3, Dabei Lane , Meiyuan
New Village, Nanjing 210000
China
大悲巷7-3

Post
Nanjing Han Qing Tang Design, 7
Lane
Meiyuan New Village
210000 NANJING JIANGSU
China
Bestellnummer: 028-1372644-582992
MEDEWO 119.657.25

Deutsche Post
911-028-000
>>> hier öffnen / open here

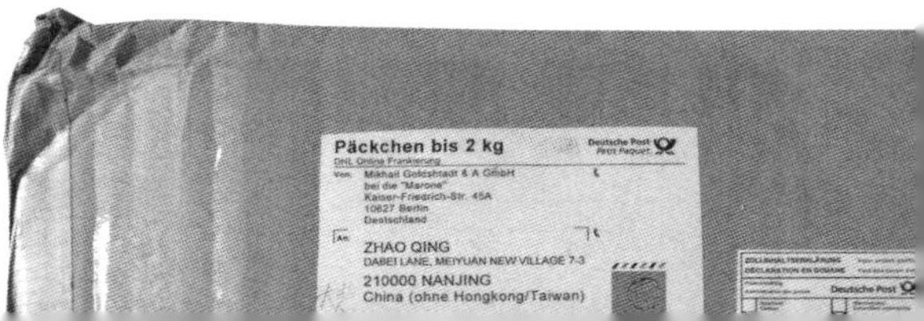

Päckchen bis 2 kg
Deutsche Post
Mikhail Golzshtadt & A GmbH
bei die "Marone"
Kaiser-Friedrich-Str. 45A
10627 Berlin
Deutschland
ZHAO QING
DABEI LANE, MEIYUAN NEW VILLAGE 7-3
210000 NANJING
China (ohne Hongkong/Taiwan)

Wir haben diese schönen Bücher gesammelt ,

◎ geordnet ,

abgefilmt , «

bearbeitet und gestaltet ,

~~um daraus eine Sammlung anzufertigen~~ ,

die all jenen gewidmet ist ,

die das Buch lieben . △

We collected these beautiful books and put them together, photographed,

edited and designed into a compilation,

which is dedicated to all book lovers .

我们收藏了这些美丽的书
把它们整理·拍摄·编辑·设计做成一本合集
献给所有爱书的人

20 世纪 50 年代初德国的政治分歧引致两个图书设计竞赛几乎同时设立：

1952 年西德在美因河畔法兰克福设立竞赛，

1965 年起由德国图书艺术基金会赞助并组织评选，

而"德意志民主共和国最佳图书奖"1953 年设立于莱比锡。

10 年后，

随着 Messehaus am Markt 的成立，

莱比锡书展进入了一个新的历史阶段，

"世界最美的书"国际比赛正式启动。

莱比锡书商协会将其作为国际书展之间的纽带，

从 1959 年起每隔五年举办一次。

自 1968 年以来，

该展览因为一个附属的主题特别展览而进一步升级。

"金字符奖"Goldene Letter 最初作为主题展览的最佳贡献奖颁发，

后来被提升为竞赛的最高奖项。

随着 1989 年 11 月柏林墙的倒塌，

组织两场平行的图书艺术竞赛已经不合时宜。

自 1991 年以来，

德国图书艺术基金会负责在莱比锡举办"世界最美的书"竞赛和展览。

In den *frühen fünfziger* Jahren führte die politische Spaltung Deutschlands zur Gründung zweier Buchgestaltungswettbewerbe. Parallel zum westdeutschen Wettbewerb, erstmals in Frankfurt am Main *1952* ausgerichtet, ab *1965* unter der Leitung der Stiftung Buchkunst, wurden »Die schönsten Bücher der DDR« ab *1953* in Leipzig geehrt.

Ein Jahrzehnt später begann in Leipzig mit der Einweihung des Messehauses am Markt ein neuer Abschnitt der Geschichte der Buchmesse Leipzig, der internationale Wettbewerb »Best Book Design from all over the World« wurde aus der Taufe gehoben. Ost-Börsenverein und Kommune planten es als Bindeglied zwischen den internationalen Buchausstellungen, die ab 1959 in Fünfjahresabständen stattfanden.

Seit 1968 wurde die Ausstellung weiter ausgebaut, indem sie von einer thematischen Sonderschau begleitet wurde. Die »Goldene Letter« wurde erstmals als Preis für die beste Einreichung in dieser thematischen Ausstellung vergeben, später wurde sie zur bedeutendsten Auszeichnung des Wettbewerbs. Mit dem Fall der Mauer im November 1989 vereinigten sich auch die beiden parallel laufenden Buchkunst Wettbewerbe. Seit 1991 richtet die Stiftung Buchkunst den Wettbewerb und die Ausstellung »Schönste Bücher aus aller Welt« aus.

In the early fifties Germany's political division led to an almost simultaneous establishment of two book art competitions: parallel to the West German vote, begun in Frankfurt am Main in 1952 and organised from 1965 under the auspices of the Stiftung Buchkunst, »Die schönsten Bücher der DDR« / »The Best Books of the GDR« were honoured in Leipzig from 1953 on.

One decade later, when a new phase in the history of the Leipzig Book Fair began with the inauguration of the Messehaus am Markt, the international competition »Best Book Design from all over the World« was founded. The Ost-Börsenverein/Leipzig Book Traders' Association and the commune planed it as a link between the international book art exhibitions reinvested from 1959 on in five-year intervals.

Since 1968 the exhibit was further upgraded by an affiliated thematic special show. The »Goldene Letter«, first given as a prize for the best contribution of these thematic exhibitions, later advanced to become the highest award of the contest. With the fall of the Berlin Wall in November 1989 the organisation of two parallel book art competitions had become obsolete. Since 1991 the Stiftung Buchkunst holds the competition and exhibition »Best Book Design from all over the World« in Leipzig.

KAREN BLIXEN
Nature

1991

1991-2003

金字符奖

《两条河流》

《公鸡彼得，纸天堂》

《夜间飞行》

《中国驻汉堡领事》

《Typoundso》

《印刷品》

《委内瑞拉历史词典》

《月亮和晨星》

《卡尔·克劳斯主编的〈火炬〉杂志（1899—1936）谚语词典》

Golden Letter

1991	*Two Rivers*	USA
1992		
1993	*Der Hahnepeter, Papier Paradies*	Germany
1994		
1995	*Nachtflugge*	Switzerland
1996	*Der chinesische Konsul in Hamburg*	Germany
1997	*Typoundso*	Switzerland
1998	*Printed matter / drukwerk*	The Netherlands
1999	*Diccionario de Historia de Venezuela*	Venezuela
2000	*Mond und Morgenstern*	Germany
2001	*Wörterbuch der Redensarten zu der von Karl Kraus 1899 bis 1936 herausgegebenen Zeitschrift „Die Fackel"*	Austria
2002		
2003		

1991

Golden Letter

*US
A　Wallace Stegner
T　Two Rivers
V　The Yolla Bolly Press, Covelo
G　James Robertson, Carolyn Robertson*

TWO RIVERS

1991

Golden Letter

*US
A Wallace Stegner
T Two Rivers
V The Yolla Bolly Press, Covelo
G James Robertson, Carolyn Robertson

Gold Medal

US
A John Hejduk
T Vladivostock
V Rizzoli International Publications, New York
G Kim Shkapich

Silver Medal

DE
A Bob Willoughby
T Jazz in LA
V Nieswand Verlag, Kiel
G Ingo Wulff

DE
T Andy Warhol Cinema
V Éditions Carré, Paris
G Éditions Carré

Bronze Medal

*CH
A Franz Zeier
T Richtigkeit und Heiterkeit
V Typotron AG, St. Gallen
G Jost Hochuli

DE
A,G Květa Pacovská
T eins, fünf, viele
V Ravensburger Buchverlag Otto Maier, Ravensburg

DE
A Heinz Lüllmann, Klaus Mohr, Albrecht Ziegler
T Taschenatlas der Pharma-kologie
V Georg Thieme Verlag, Stuttgart
G Karl-Heinz Fleischmann

DE
A Jacques Derrida
T Was ist Dichtung?
V,G Brinkmann & Bose, Berlin

NL
A Auke van der Woud
T Wim Quist, architect
V Uitgeverij 010, Rotterdam
C Reynoud Homan

Honorary Appreciation

CZ
A Pavel Šrut
T Kočiči Krái
V Albatros, Prag
G Milan Grygar

JP
A Setsuko Hasegawa
T A strange Inn
V Fukinkan Shoten, Publishers, Tokyo
G Yosuke Inoue

JP
A Makoto Nagao
T Dictionary of Iwanami Information Science V.G Iwanami Shoten, Publishers, Tokyo

NL
A Rudy Kousbroek
T 66 Zelfportretten van Nederlandse Fotografen
V Nicolaas Henneman Stichting, Amsterdam
G Lex Reitsma

1992

Golden Letter

*NO
A Halldis Moren Vesaas
T Mine Dikt
V H. Aschehoug, Oslo
G Kristian Ystehede, Guttorm Guttormsgaard

Gold Medal

CH
A Erich Grasdorf
T Klick
V Edition A., Zürich
G André Hefti

Silver Medal

JP
A Chizuru Miyasaki
T Green in the afternoon
V Tokyo Shoseki Co., Tokyo
G Takashi Kuroda

*NL
A Max Bruinsma
T Een leest heeft drie voeten. Dick Elffers & de kunsten
V Uitgeverij De Ralie;Gerrit Jan Thiemefonds, Amsterdam
G Rob Schroder, Lies Ros

Bronze Medal

CH
T Wie die Heizung Karriere machte
V Sulzer Infra AG, Winterthur
G Bruno Guttinger

DE
A Andreas Langen
T Spaziergang nach Syrakus
V Edition Cantz, Stuttgart
G Karin Girlatschek

DE
A Wilhelm Hornbostel
T Voilà
V Prestel Verlag, München
G KMS graphic, Maja Thorn

NL
A Hugo Brandt Corstius
T BSO Jaarverslag 1989
V BSO, Utrecht
G Harry N. Sierman

US
A Kenneth Frampton, Kunio Kudo
T Nikken Sekkei. Building Modern Japan 1900-1990
V Princeton Architectural Press, New York G Thomas Cox (Willi Kuntz Associates)

Honorary Appreciation

DE
A Hervg Guibert
T Dem Freund, der mir das Leben nicht gerettet hat
V Rowohlt Verlag, Reinbek
G Joachim Duster

JP
A Seizo Tashima
T Searching in the Forest
V KaiseiSha Publishing Co, Tokyo
G Kiichi Miyazaki

JP
A Atsuko Suga
T Milan - Paesaggi nella nebbia
V Hakusui-Sha, Co., Tokyo
G Koji ise

NL
A Maarten Kloos
T Alexander Bodon, architect
V Uitgeverij 010, Rotterdam
G Reynoud Homan

NL
A Paul Donker Duyvis,Klaus Honnef
T Schrag
V Edition Braus, Heidelberg
G Gerard Hadders, Andre van Dijk (hard werken)

1993

Golden Letter

*DE
A Kurt Schwitters, Květa Pacovská
T Der Hahnepeter, Papier Paradies
V Pravis Verlag, Osnabruck
G Květa Pacovská

1992

Gold Medal

NL
A Pieter Beek
T faces
V Gemeente Maastricht, Provincie Limburg
G Piet Gerards, Marc Vleugels

Silver Medal

CH
A Connie Imboden
T Out Of Darkness
V Edition FotoFolie, Zürich / Paris
G Kaspar Mühlemann

JP
A Masumi Shimizu
T A Noh Mask - Mitsui Family Property
V Gakken Co., Tokyo
G Yasuhiko Naito

NL
A P. F. Den Hartogen, L. Scheltinga
T Tabellenboek
V Wolters-Noordhoff, Groningen
G Roelof Koster (Studio Dorèl)

Bronze Medal

DE
T Mit der Absicht des Schöpfers hat es höchstens zufallig erwas zu tun
V Verlag Fotoforum SCHWARZBUNTe.V., Bielefeld
G Lutz Dudek, Claudia Grote - fendt

DE
A Ulysses Voelker
T Zimmermann meets Spiekermann
V Meta Design, Berlin
G Ulysses Voelker

NL
A Liesbeth Crommelin
T Uniek en meervoudig
V Stedelijk Museum, Amsterdam
G Reynoud Homan

NL
A Battus
T Symmys
V Em. Querido's Uitgeverij, Amsterdam
C Harry N. Sierman

NL
A Rainer Bullhorst, Rudolphine Eggink
T Friso Kramer. Industrieel ontwerper
V Uitgeverij 010, Rotterdam
G Reynoud Homan

US
A Michael Rotondi
T From the edge. Student Work
V The Southern California Institute of Architecture, Los Angeles
G April Greiman, Sean Adams (April Greiman Inc.)

Golden Letter

*NO
A Halldis Moren Vesaas
T Mine Dikt
V H. Aschehoug, Oslo
G Kristian Ystehede, Guttorm Guttormsgaard

1993
Golden Letter
*DE
A Kurt Schwitters, Květa Pacovská
T Der Hahnepeter, Papier Paradies
V Pravis Verlag, Osnabruck
G Květa Pacovská

Honorary Appreciation

CH
A Werner Warth
T Wil. Die Altstadt
V Verlag Meyerhans Druck AG, Wil/St. Gallen
G Gaston Isoz, Trogen

CH
T Le billet de banque et son imager: l'exemple suisse
V République et Canton de Genève
G Isabelle Lajoinie

CZ
A Jiří Teper
T Milovaný Obraz
V Atelier Abrakadabra, Prag
G Clara Istierová

ES
A San Juan de la Cruz
T Poes ía y otros textos
V Círculo de Lectores, Barcelona
G Norbert Denkel

*SE
A,G HC Ericson
T Ordbilder
V Carlssons Bokförlag, Stockholm

US
T Mechanika
V The Contemporary Arts Center, Cincinatti
G David Betz

US
A Francisco X. Alarcón
T De Amor oscuro. Of Dark Love
V Moving Parts Press,Santa Cruz
G Felicia Rice

1994

Golden Letter

—

Gold Medal

—

Silver Medal

CH
A Alfred Hablützel,Verena Huber
T Innenarchitektur in der Schweiz 1942-1992
V Verlag Niggli, Sulgen
G Thomas Petraschke (Studio Halblützel)

DE
A Franz Kafka
T Das Schloß
V Büchergilde Gutenberg, Frankfurt am Main
G Eckhard Jung, Ulysses Voelker

DE
A Nikolai Gogol
T Die Nase
V Edition Curt Visel, Memmingen
G Gert Wunderlich

Bronze Medal

CH
A Ernst Ziegler, Peter Ochsen-bein, Hermann Bauer
T Rund urns <Blaue Haus>
V Ophir-Verlag, St. Gallen
G Antje Krausch, Ruedi Tachezy

DE
A FSB
T Übergriff. Ein Buchprojekt von Studenten der HfG Karlsruhe unter Leitung von Gunter Rambow
V Verlag der Buchhandtung Walther König, Köln
G Studenten der HfG Karlsruhe und Sepp Landsbek

DE
A Beate Manske, jochen Rahe
T Design aus Bremen
V Design Zentrum, Bremen
G Hartmut Brückner

DE
A Shen Jiji
T Die Geschichte des Fräulein Ren
V,G CTL-Presse, Clemens Tobias Lange, Hamburg

*JP
A Ryotaro Shiba
T Touge - A Mountain Pass
V,G Shincho-Sha Co., Tokyo

*PL
T Jakob Böhmes Offenbarung über Gott
V Correpondance des Arts, Lódz
G Igor Podolczak,Janusz Tryzno

Honorary Appreciation

AT
A Dylan Thomas
T Unwegsame Gebiete
V,G Wolfgang Buchta, Wien

CH
A Jost Hochuli
T Freude an Schriften
V Typotron AG, St. Gallen
G Jost Hochuli

DE
A,G AnjaHarms
T Peleus
V Unica T, Oberursel i. Ts.

DK
T Det Nye Testamente
V Det Danske Bibelselskab, Kopenhagen
G Anne Rohweder

NL
A Ernst Braches
T The Steadfast Tin Soldier of Joh. Enschedé en Zonen, Haarlem
V Spectatorpers, Aartswoud; Just Enschedé, Amsterdam
G Bram de Does

1995

Golden Letter

*CH
A Kathrin Fischer
T Nachtflügge
V Kranich-Verlag, Zürich
G Kaspar Mühlemann

Gold Medal

NL
A VROM Ministry
T Government Buildings
V Uitgeverij 010, Rotterdam
G Reynoud Homan, Robbert Zweegman

Silver Medal

CH
A Guido von Stürler
T Spezifisches und Diffuses
V Verlag Niggli, Sulgen
G Urs Stuber

US
A Rodney Smith,Leslie Smolan
T The Hat Book
V Nan A. Talese; Doubleday, New York
G Leslie Smolan, Jennifer Domer (Carbon Smolan Associates)

Bronze Medal

DE
A Olaf Arndt, Rob Moonen
T camera silens
V ZKM Zentrum für Kunst und Medientechnologie, Karlsruhe
G Julia Hasting, Patricia Müller, Gerwin Schmidt, Béla Stetzer

DE
A Vittorio Magnano Lampugnani, Lutz Hartwig
T Vertikal
V Verlag Ernst & Sohn, Berlin
G Grappa-Design

ES
A Octavio Paz
T La llama doble
V Círculo de Lectores, Barcelona
G Norbert Denkel

US
A GuyBillout
T Journey. Travel Diary of a Daydreamer
V Creative Education, Mankato
G Rita Marshall (Delessert & Marshall)

US
A Daniel L. Schodek
T Structure in Sculpture
V The MIT Press, Cambridge
G Jeanet Leendertse

Honorary Appreciation

AT
A Christian Mähr, Nikolaus Walter
T Stadt Feldkirch
V Amt der Stadt Feldkirch, Feldkirch
G Reinha rd Gassner

DE
T Dinner for Two. Aus Gold und Silber
V Stiftung Gold-und Silber-schmiedekunst, Schwäbisch Gmünd
G Büro für Gestaltung Biste & Weißhaupt

DE
A,G Philipp Luidl
T Grundsetzliches 1-6
V SchumacherGebler, München

*EG
A,G MohieddenEl-Labbad
T Sprache ohne Wörter
V Dar El Shorouk, Kairo

US
T In Sight. The Seattle Public Art Puzzle Book V Seattle Arts Commission, Seattle
G Judy Anderson, Claudia Meyer-Newman (Anderson & Helms Cook)

1996

Golden Letter

*DE
A Wang Taizhi, Bernd Eberstein
T Der chinesische Konsul in Hamburg
V Christians Verlag, Hamburg
G Andreas Brylka

1994

Kathrin Fischer
nachtflügge
gedichte

1995

Golden Letter

*CH

A Kathrin Fischer

T Nachtflügge

V Kranich-Verlag, Zürich

G Kaspar Mühlemann

Wang Taizhi
Bernd Eberstein
Der chinesische Konsul in Hamburg
Gedichte
Bernt Federau
Fotos
Christians

1996

Golden Letter

*DE
A Wang Taizhi, Bernd Eberstein
T Der chinesische Konsul in Hamburg
V Christians Verlag, Hamburg
G Andreas Brylka

1997

Golden Letter

*CH
A,V,G Hans-Rudolf Lutz, Zürich
T Typoundso

Typoundso

1998

Golden Letter

*NL

A Karel Martens

T printed matter / drukwerk

V Hyphen Press, London

G Jaap van Triest, Karel Martens

printed matter | drukwerk

1999

Golden Letter

*VE

T　Diccionario de Historia de Venezuela

V　Fundación Polar, Caracas

G　Álvaro Sotillo

Diccionario de Historia de Venezuela
Diccionario de Historia de Venezuela
Diccionario de Historia de Venezuela

2000

Golden Letter

*DE
A Wolfram Frommlet, Henning Wagenbreth
T Mond und Morgenstern
V Peter Hammer Verlag, Wuppertal
G Henning Wagenbreth

Wolfram Frommlet Henning Wagenbreth
MOND
und Morgenstern
Peter Hammer Verlag

Wörterbuch der Redensarten
Karl Kraus
Die Fackel

REMINGTON RAND
2001
Golden Letter
T Wörterbuch der Redensarten zu der von Karl Kraus 1899 bis 1936 herausgegebenen Zeitschrift „Die Fackel"
A Werner Welzig *AT
V Verlag der Österreichischen Akademie der Wissenschaften, Wien
G Anne Burdick

Gold Medal

DE
A Stéphane Mallarmé
T Un coup de dés
V Steidl Verlag, Göttingen
G Klaus Detjen

Silver Medal

CZ
A,G Libor Beránek
T Typografické Variace
V Vysoká škola uměleckoprůmyslová VŠUP, Prag

FR
T Nadar. Les années Créatrices: 1854-1860
V Éditions de la Réunion des Musées Nationaux, Paris
G Pierre-Louis Hardy

Bronze Medal

AT
A Erich Klein
T Die Russen in Wien. Die Befreiung Österreichs
V Falter Verlagsgesellschaft, Wien
G Hofmann & Kraner

DE
T Buna 4
V,G Julia Hasting, Patricia Müller, Gerwin Schmidt, Béla Stetzer, Karlsruhe

DE
A Harold Brodky
T Die flüchtige Seele
V Rowohlt Verlag, Reinbek
G Edith Lackmann

DE
A Rainer Groothuis, Karl-Heinz Janßen
T Schwarzbuntes
V Christians Verlag, Hamburg
G Rainer Groothuis

US
A Herb Ritts
T Africa
V Bulfinch Press; Little, Brown and Company, Boston
G Betty Egg, Sam Shahid

Honorary Appreciation

*AU
A Chris O'Doberty
T The Reg Mombassa Diary 1995
V Bantam Books, Sydney
G Graham Rendoth (Reno Design Group 14144)

*FR
A Christian Janicot
T Anthologie du cinema invisible
V Éditions Jean-Michel Place; ARTE Éditions, Paris
G Bulnes & Robaglia

NO
A Mette Hvalstad
T Ungdom og sex
V Senteret for Ungdom, Samliv og Seksualitet, Oslo
G Marte Fæhn, Line Jerner (Lucas Design & illustrasjon)

PL
A Stasys Eidrigevičius
T Erotiky
V Wydawnictwo Tenten, Warschau
G Grażyna Bareccy, Andrzej Bareccy

US
A Elizabeth Spurr
T The Gumdrop Tree
V Hyperion Books for Children, New York
G Ellen Friedman, Julia Gorton

1997

Golden Letter

*CH
A,V,G Hans-Rudolf Lutz, Zürich
T Typoundso

Gold Medal

NL
T Veiligheid en Bedreiging in de 21e Eeuw
V VPRO, Hilversum
G Mieke Gerritzen, Janine Huizenga

Silver Medal

CH
T Jahanguir
V Galerie Jamileh Weber, Zürich
G Kaspar Mühlemann

DE
A Hanne Chen
T Sehr nah, sehr fern sind sich Mann und Frau
V Edition Zeichen Satz, Kiel
G Kerstin Weber, Olaf Schmidt

Bronze Medal

*CZ
A Ivan Wernisch
T Nesetkání. Nichtbegegnung
V Vysoká škola uměleckoprůmyslová VŠUP, Prag
G Stefanie Harms

DE
A,G Wiebke Oeser
T Bertas Boote
V Peter Hammer Verlag, Wuppertal

DE
A,G Hans Peter Willberg, Friedrich Forssman
T Lesetypographie
V Verlag Hermann Schmidt, Mainz

DE
T Unica T: 10 Jahre Künstler-bücher
A,V Unica T, Oberursel I. Ts./Offenbach am Main
G Anja Harms, Ines v. Ketel-hodt, Dois Preußner, Uta Schneider, Ulrike Stoltz (Unica T)

JP
A Gyoku Aoki
T Koda Aya No Tansu No Hikidashi
V Shincho-Sha Co., Tokyo
G Akio Nonaka

Honorary Appreciation

*CO
T Alta Colombia. El Esplendor de la Montaña
V Villegas Editores, Bogotá
G Benjamin Villegas

EG
A Karimam Hamsa
T Das Leben des Propheten
V Dar El Shorouk, Kairo
G Mohiedden El-Labbad

NL
T Chassé Theater Breda. Herman Hertzberger
V Uitgeverij 010, Rotterdam
G Bureau Piet Gerards

NL
A,G Richard Menken
T Richtkracht
V Hogeschool voor de Kunsten HKA Arnhem

PL
T Muzeum Ulicy. Plakat polski
V Wydawnictwo Krupski I S-ka
G Katarzyna Pogrebska

1998

Golden Letter

*NL
A Karel Martens
T printed matter | drukwerk
V Hyphen Press, London
G Jaap van Triest, Karel Martens

Gold Medal

US
A Jimmy Smith, John Huet
T Soul of the Game
V Melcher Media; Workman Publishing, New York
G John C. Jay

Silver Medal

DE
A,V,G Anja Wesner, Stuttgart
T Sehen ... erfinden

DE
A Bertolt Brecht
T Flüchtlingsgespräche
V Leipziger Bibliophilen-Abend e.V., Leipzig
G Gert Wunderlich

Bronze Medal

DE
A Keller + Kuhn
T Die blauen Wunder
V Reclam Verlag, Leipzig
G Matthias Gubig

DE
T Zur Anpassung des Designs an die digitalen Medien
V form + zweck, Berlin
G Sabine Golde, Tom Gebhardt

NL
A,V,G José van't Klooster
T East Side Souvenirs

SE
A,G HC Ericson
T Reningsverk
V Carlsson bokförlag, Stockholm

US
T Bill Viola
V Whitney Museum of American Art, New York; Flammarion, Paris
G Rebeca Méndez

Honorary Appreciation

DE
A Werner Döppner
T Kullu Mumkin
V,G Matthias Beyrow, Marion Wagner, Berlin

DE
A Amelie Fried, Jacky Gleich
T Hat Opa einen Anzug an?
V Carl Hanser Verlag, München
G Claus Seitz

DE
A,G Anja Osterwalder
T Space Nutrition. Essen im Weltall
V i-d büro, Stugart

*HU
T Budapesti galériák 1996. Gallery Guide of Budapest 1996
V Budapest Art Expo Alapítvány, Budapest
G Johanna Bárd

JP
A Sotoji Nakamura
T Shōgi-daiku Nakamura Sotoji no Shigoto
V Seigen-Sha Art Publishing, Kyoto
G Tsutomu Nishioka

2002

Golden Letter

*DE
A,V,G Mariko Takagi, Meerbusch
T Washi – Tradition und Kunst des Japanpapiers. Japanpapier zum Anfassen

1999

Golden Letter

*VE
T Diccionario de Historia de Venezuela
V Fundación Polar, Caracas
G Álvaro Sotillo

Gold Medal

CH
T Giuseppe Terragni. Modelle einer rationale Architektur
V Verlag Niggli, Sulgen
G Urs Stuber

Silver Medal

CH
A André Vladimir Heiz, Michael Pfister
T Dazwischen. Beobachten und Unterscheiden
V Museum für Gestaltung Zürich
G François Rappo

*ES
T Gaceta de Arte y su Época 1932-1936
V Centro Atlántico de Arte Moderno, Las Palmas de Gran Canaria; Edición Tabapress
G Raimundo C. Iglesias

Bronze Medal

DE
A Julia Blume, Günter Karl Bose
T Reihe „allaphbed" 5 Bände
V Institut für Buchkunst Leipzig an der Hochschule für Grafik und Buchkunst Leipzig
G Arbeitsgruppe „work ahead" an der Hochschule für Grafik und Buchkunst Leipzig

DE
A Johann Wolfgang v. Goethe, Wolf Erlbruch
T Das Hexen-Einmal-Eins
V Carl Hanser Verlag, München
G Claus Seitz, München

DE
A Joachim Nickel
T Palmin – eine Jahrhundertmarke
V Union Deutsche Lebensmittelwerke, Hamburg
G Gesine Krüger, Andrea Schürings (büro für mitteilungen)

NL
A Harry Ruhé
T Het Beste van Wim T. Schippers
V Centraal Museum, Utrecht
G Gonnissen en Widdershoven

US
T Photography's Multiple Roles
V Museum of contemporary Photography, Chicago
G studio blue

Honorary Appreciation

CH
A Vera Eggermann
T Sardinen wachsen nicht auf Bäumen
V Atlantis Verlag; Verlag pro juventute, Zürich
G Ueli Kleeb

DE
A Ulrike Lang
T Grete Gulbransson Tagebücher, Band I: Der grüne Vogel des Äthers, 1904 bis 1912
V Stroemfeld Verlag, Frankfurt am Main
G Karin Beck

JP
A Yoshizo Kawamori
T A Poet. Tohson Shimazaki in Paris
V,G Shincho-Sha Co., Tokyo

NL
A Ed van Hinte
T Eternally yours
V Uitgeverij 010, Rotterdam
G Gonnissen en Widdershoven

NL
T De Regie van de Stad
V NAi Uitgevers; EFL Publicaties, Rotterdam/DenHaag
G Arlette Brouwers, Koos van der Meer

2000

Golden Letter

*DE
A Wolfram Frommlet, Henning Wagenbreth
T Mond und Morgenstern
V Peter Hammer Verlag, Wuppertal
G Henning Wagenbreth

Gold Medal

NL
T 16th World Wide Video Festival '98
V Stichting World Wide Video Festival, Amsterdam
G Irma Boom

Silver Medal

AT
A Thomas Höft
T Welt aus Eisen
V Springer-Verlag, Wien/New York
G Alexander Kada

DE
A Olaf Rauh
T Persiplus
V Edition Persiplus, Olaf Rauh Eigenverlag, Leipzig
G Kerstin Riedel

Bronze Medal

*CA
A Louis Jensen
T Et hus er et ansigt
V Gyldendal, Kopenhagen
G Jens Bertelsen

JP
T Kitaoji Rosanjin to Suda Seika
V Shko-Sha Co., Kyoto
G Kikko Ken

NL
T If/Then. Design implications of new media
V Netherlands Design Institute; BIS Publishers, Amsterdam
G Mevis & Van Deursen

SE
T Handtverket i gamla hus
V Byggförlaget, Stockholm
G Dick Norberg

Honorary Appreciation

CH
A Andreas Steigmeier
T Raum Komfort Ästhetik
V hier+jetzt Verlag für Kultur und Geschichte, Baden
G Urs Bernet, Jürg Schönenberger

DE
T Von zart bis extrafett. TypoGrafik von Gert Wunderlich 1957-1998
V,G Gert Wunderlich, Leipzig

NL
A Herman de Coninck
T De Gedichten
V Uitgeverij De Arbeiderspers, Amsterdam/Antwerpen
G Steven van der Gaauw

US
A Francis M. Naumann
T Marcel Duchamp
V Harry N. Abrams Ins., New York; Ludion Press, Ghent
G studio blue, Chicago

US
A,G Richard Hendel
T On Book Design
V Yale University Press, New Haven & London

2001

Golden Letter

*AT
A Werner Welzig
T Wörterbuch der Redensarten zu der von Karl Kraus 1899 bis 1936 herausgegebenen Zeitschrift „Die Fackel"
V Verlag der Österreichischen Akademie der Wissenschaften, Wien
G Anne Burdick

Gold Medal

JP
T infoword English-Japanese Dictionary
V Benesse Corporation, Tokyo
G Haruyo Kobayashi

Silver Medal

A Jair de Souza, Lucia Rito, Sérgio Sá Leitão
T Futebol-Arte
V Empresa das Artes; Editora Senac, São Paulo
G Jair de Souza

NL
T 400 jaar vioolbouwkunst in Nederland
V Nederlandse Groep van Viool – en Strijkstok-kenmakers, Amsterdam
G Jonker & Van Vilsteren

Bronze Medal

CH
A C. Nicolai, O. Bender, P. Kraut
T raster-noton. oacis
V raster-noton, Berlin; taktlos-bern, Bern
G Olaf Bender

CH
A Urs Hochuli, Oskar Keller
T Stitterkiesel
V Edition Ostschweiz, VGS, St. Gallen
G Jost Hochuli

CH
T The Selection vfg. 1999
V Schwabe & Co, Basel
G Beat Müller, Wendelin Hess, Michael Birchmeier, Nora Beer (Müller + Hess)

CZ
A Jiří Šalamoun
T Nahá obryně
V Aulos, Prag
G Jiří Šalamoun, Zdeněk Ziegler

DE
A Richard Paul Lohse Stiftung
T Richard Paul Lohse. Konstruktive Gebrauchsgrafik
V Hatje Cantz Verlag, Ostfildern-Ruit
G Markus Bosshard (Weiersmüller Bosshard Grüninger wbg AG)

Honorary Appreciation

A,G Verena Böning
T Ausbrechen. Bulimie verstehen und überwinden
V Urban & Fischer Verlag, München

DE
T Seminar Typografie: Zum Buch 3, material 88, 3/2000
V material Verlag, Hochschule für bildende Künste Hamburg
G Hans Andree, Sigrid Behrens, Uli Brandt, Shin-Young Chung, Daniel Hahn, Nadine Jung, Volker Lang, Zuzanna Musialczyk, Peter Piller, Anna Reemts, Sven Seddig, Jan Schmietendorf

NL
A Frans van Rijn, Vincent Huizing
T Hoe gaat het … Verslag en visie 1998/1999
V ONVZ Zorgverzekeraar
G Dietwee, Utrecht, -SYB- grafisch ontwerp, Martijn Engelbregt (EGBG)

NL
A Aristoteles
T Ethica
V Historische Uitgeverij, Groningen
G Rudo Hartmann, Gerard Hadders

VE
A Miguel Arroyo, Lourdes Blanco, Erika Wagner
T El Arte Prehispánico de Venezuela
V Fundación Galería de Arte Nacional, Caracas
G Álvaro Sotillo

2002

Golden Letter

*DE
A,V,G Mariko Takagi, Meerbusch
T Washi – Tradition und Kunst des Japanpapiers. Japanpapier zum Anfassen

Gold Medal

CZ
A Herbert T. Schwarz
T Příběhy z parní lázně
V Argo, Prag
G Juraj Horváth

Silver Medal

DE
A Jorge Luis Borges
T Die Bibliothek von Babel
V Steidl Verlag, Göttingen; Büchergilde Gutenberg, Frankfurt am Main/Wien/Zürich
G Klaus Detjen

NL
A Leo van der Meule
T Het Rijnlands Huis teruggevolgd in de tijd
V Hoogheemraadschap van Rijnland, Leiden; Architectura & Natura, Amsterdam
G Irma Boom

Bronze Medal

AT
A Richard Pischl
T pro:Holz
V proHolz Austria, Wien
G Atelier Reinhard Gassner

DE
A Andrea Seppi, Steffen Junghans
T JVA Protokoll
V Institut für Buchkunst Leipzig an der Hochschule für Grafik und Buchkunst Leipzig
G Markus Dreßen

ES
A Cloé Poizat
T Máquinas
V Kalandraka Editora, Pontevedra
G equipo gráfico de Kalandraka

NL
A Kathinka van Dorp
T Eenenvijftig stemmen uit de wereldpoëzie, Rotterdam
G Typography and Other Serious Matters

NL
T Zinnen prikkelend
V Drukkerij Ando, Den Haag
G Typography and Other Serious Matters

Honorary Appreciation

CH
A Rolf Schroeter
T Die Lichtung
V Verlag Niggli Sulgen/Zurich
G Hans Grüninger (Weiersmüller Bosshard Grüninger wbg AG)

DE
A T. Coraghessan Boyle
T Der Hardrock-Himmel (Die Tollen Hefte 20)
V Büchergilde Gutenberg, Frankfurt am Main/Wien/Zürich
G Thomas Müller

NL
T Claus en Kaan
V Claus en Kaan Architecten, Amsterdam, Rotterdam
G Reynoud Homan

NL
T Dolly. A Book Typeface With Flourishes
V Underware, Den Haag
G Wout de Vringer (Faydherbe/De Vringer)

SE
A Bea Uusma Schyffert
T Astronauten som inte fick landa
V Alfabeta Bokförlag AB, Stockholm
G Lotta Kühlhorn

2003

Golden Letter

*NL
A,V,G Nynke M. Meijer, Sneek
T Jan Palach. Morgen word je wakker geboren

Gold Medal

DE
T Gutenberg Galaxie 2: Irma Boom
V Institut für Buchkunst Leipzig an der Hochschule für Grafik und Buchkunst Leipzig
G Kristina Brusa

Silver Medal

DE
A,G Anja Harms
T Hans Arp – Worte
V Anja Harms Künstlerbücher, Oberursel i. Ts.

NL
T Kleur
V kaAp/stadium generale van de Hogeschool voor de Kunsten Arnhem
G Corina Cotorobai (Werkplaats Typografie)

Bronze Medal

CH
A,V NORM, Zürich
T The Things
G Dimitri Bruni, Manuel Krebs (NORM)

DE
A Ray Bradbury
T Fahrenheit 451
V Büchergilde Gutenberg, Frankfurt am Main/Wien/Zürich
G Gerda Stangl

DE
A Arno Schmidt
T Brüssel/Die Feuerstellung
V Suhrkamp Verlag, Frankfurt am Main
G Friedrich Forssman

DE
A Carla Sozzani
T Yohji Yamamoto. Talking to Myself
V Steidl Verlag, Göttingen
G Claudio dell'Olio

NL
A,G Bram de Does
T Kaba ornament. Deel 1 Vorm
V Spectatorpers, Orvelte

Honorary Appreciation

DE
A Joseph Brodsky
T Weniger als man – Less than One
V Carivari, Leipzig
G Sabine Golde

DE
A Anna Seghers
T Der Ausflug der toten Mädchen
V Offizin Bertelsmann Club, Rheda-Wiedenbrück
G Rainer Groothuis (Groothuis, Lohfert, Consorten)

JP
A John Peter
T The Oral History of Modern Architecture
V TOTO Shuppan, Tokyo
G Sin Akiyama, Ken Hisase

NL
A Tanny Dobbelaar, Adriënne M. Norman
T Heftig vel
V Elsevier gezondheidszorg, Maarssen
G –SYB-grafisch ontwerp

NL
A Lucebert
T Verzamelde gedichten
V De Bezige Bij, Amsterdam
G Tessa van der Waals

2003

Golden Letter

*NL
A,V,G Nynke M. Meijer, Sneek
T Jan Palach. Morgen word je wakker geboren

Die erste Auswahl von 「Schönste Bücher Chinas」 erschien

im Jahre 2004 . ◎

《 Mit dieser Auswahl beteiligte

△ ~~sich China im darauffolgenden Jahr erstmalig an dem in Leipzig~~

ausgetragenen Wettbewerb 「Schönste Bücher der Welt」

– damit begann die Bücherreise nach Leipzig .

In 2004, 「the most beautiful books in China」 were selected to participate in the competition 「best book design from all over the world」 held in Leipzig the following year.

A journey of beautiful books to Leipzig began .

二〇〇四年

中国首次选出『中国最美的书』

以参加次年举办的莱比锡『世界最美的书』评选

开始了一段莱比锡美书之旅

2019—2004

莱比锡的选择

「世界最美的书」

Chief Editor

Zhao Qing

主编 赵 清

江苏凤凰美术出版社

Jiangsu Phoenix Fine Arts Publishing House

Schönste Bücher aus aller Welt 2019–2004

1347 1265 1187 1105 1027 0941 0855 0777 0703 0621 0551 0465 0379 0297 0215 0133

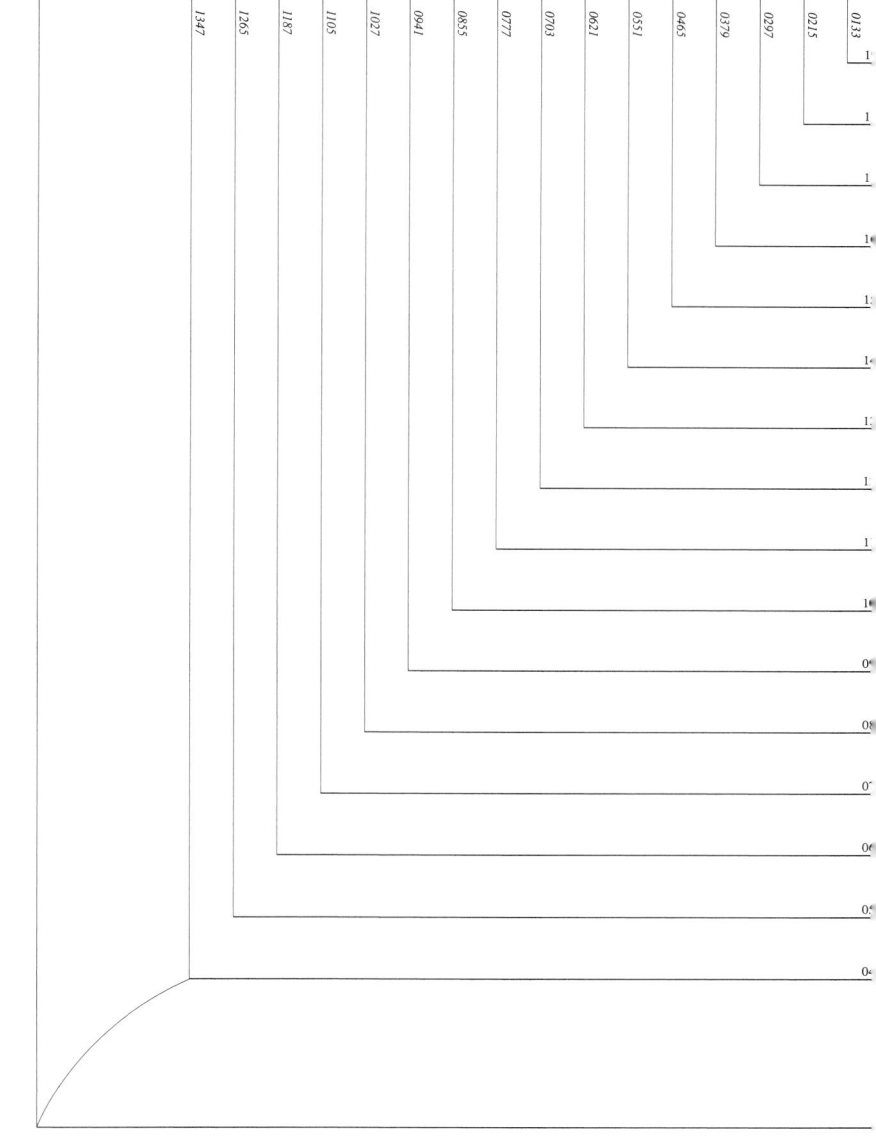

The Choice *of* Leipzig

Best book design from all over the world

2019–2004

Vorwort — *Preface*

序

Was ist schön? Das, was gefällt? Oder das, was unseren Augen schmeichelt? Wie empfinden wir Schönheit? Und wie wollen wir diesen so vagen Begriff über Kulturen hinweg anwenden, wenn zum Beispiel eine Jury ihre Entscheidungen zur Qualität von Buchgestaltung treffen soll?

Das ästhetische Gefühl brauche Reflexion, schrieb Immanuel Kant, ein deutscher Philosoph der Aufklärung. Nicht dass ich kognitive Leistung höher bewerten würde als eine subjektive Erfahrung. Aber eine Diskussion um Ästhetik, um Gestaltung, um gute Buchgestaltung oder gar ‹Schönste Bücher aus aller Welt› kann nicht ohne ein Nachdenken darüber geführt werden, welche Kriterien oder Anteile in die Beurteilung von ‹Schönheit› einfließen. Wir reagieren auf das was wir sehen, lesen und anfassen — mit allen Sinnen. Ist es nicht so, dass jedes Buch, egal wie es gestaltet oder ob der Inhalt verständlich ist, durch die Wahrnehmung unserer Sinne eine un(ter) bewusste Reaktion auslöst? Wir fühlen die Oberfläche des Materials, wir spüren das Gewicht oder wir setzen das Format in Bezug zu unserer Umgebung. Der Tastsinn, die Augen, vielleicht sogar der Geruchssinn reagieren auf das Objekt, das ich in Händen halte. Diese Wahrnehmung haben auch Laien, jene, die vielleicht nicht erklären können, warum sie ein Buch ‹schön› finden, aber deren wenige Quadratzentimeter Fingerspitzen etwas berühren, ertasten, anfassen und dadurch begreifen. Etwas durchs Begreifen verstehen. Nun können Impuls und Kognition in Wechselwirkung treten. Je mehr jemand mit Buchgestaltung zu tun hat, umso präziser kann sie oder er

What does "beautiful" mean? Is it what we like? Or is it what flatters our eyes? How do we perceive beauty? And how can we use this vague concept through cultures when, for example, a jury has to make a decision on the quality of book design?

Immanuel Kant, a German philosopher of the Enlightenment, wrote that the aesthetic feeling needs reflection. Not that I would value a cognitive capacity more highly than a subjective experience, but a discussion about aesthetics, about design, about book design, or even the competition *Best Book Design from all over the World* cannot take place without reflecting about which criteria or aspects have an influence in judging beauty.

We react to what we see, read, and touch — with all our senses. Doesn't every book, independently of how it is designed or whether its content is understandable, unleash a un/subconscious reaction through the perception of our senses? We feel the surface of the material, we feel the weight, or we relate the format with our environment. The sense of tact, the eyesight, maybe even the sense of smell react to the object that I hold in my hands. Laymen also have these perceptions. They may not be able to explain why they find a book "beautiful", but their square-centimetre fingertips touch, feel, and hold something. And, through this, they can comprehend. They can grasp something with their understanding. So can impulse and cognition influence each other. The more a person

什么是美？它是我们所喜爱的东西，还是吸引我们眼球的事物？我们如何感受到美？我们该如何跨越不同的文化，使用这个模糊的概念？例如，评审团应当如何在来自不同国家的书籍设计作品中做出选择？||||||||||||||||||||||||||||||||||||||||||

德国启蒙哲学家康德曾说过：美学感受需要内省。这并不意味着认知感受就一定高于主观体验，但一场关于美学、设计、精美书籍设计乃至“世界最美的书”的讨论，自然离不开对“美”之标准的思考。我们调动一切感官，对我们所见、所读和所接触的事物做出反应。只要我们感知到一本书，无论其设计如何，内容是否易懂，它总会引发我们的下（潜）意识反应。我们可以触摸到图书封面的材质，感受它的重量，或是把它与周围的环境联系在一起。我的触觉、视觉乃至味觉，都会对我手中拿着的这个物件做出反应。即便外行人士也有这种感受：虽然没法解释自己为何觉得一本书很美，但他们的指间总能触碰到点东西，从而促使他们做出判断，知晓前因后果。实际上，冲动和认知总在交替起着作用。一个人对书籍设计了解越多，就越能在评审会这类场合中准确地说出什么是高品质的作品。评委的职责，不是说出自己个人的喜好，不是朝自己喜欢的作品竖起

(z.B. in einer Jurysituation) benennen, was Qualität ausmacht. Es geht bei den Jurys ja nicht darum, seiner privaten Meinung Ausdruck zu geben: was gefällt — Daumen hoch — oder was nicht gefällt — Daumen runter. Das Abwägen von Gestaltungs- und Ausstattungsqualität ist ein komplexer Prozess. Zudem ein höchst spannender — für mich war er das immer, wenn ich eine Jurygruppe wieder einmal durch die Diskussionen leitete. Wenn Profis sich über Gestaltungsphänomene einigen wollen, müssen grundlegende technische und gestalterische Kriterien erarbeitet werden, abseits von persönlichem Geschmack oder Gefallen. Ich genieße es immer wieder, mit Kolleg/innen Argumente — pro und contra — zu einem Buch auszutauschen zu können. Denn: Kritik schärft das Auge, die Sinne.

Zeitgenössische Neurologen forschen längst an diesen Phänomenen: «Weil Menschen verschiedene ästhetische Vorlieben haben, dachte man bisher, es sei unmöglich, zu sagen, was Schönheit eigentlich ist. Hinzu kommt, dass wir mit dem Begriff unterschiedliche Sinneserfahrungen belegen. ... Wann immer Menschen eine ästhetische Erfahrung machen, wird die Region A1 im Stirnlappen des Großhirns aktiv, hinter der Augenhöhle», so beschreibt es der Londoner Neurobiologe Semir Zeki in einem Interview in der Wochenzeitung DIE ZEIT [1]. Er erforscht «wie das Gehirn unsere Wahrnehmungen in Gefühle und Konzepte von der Welt übersetzt». Ästhetik ist die Lehre der sinnlichen Wahrnehmung, so sagt man. Sie unterliegt, ebenso wie Gestaltung, kulturellen Einflüssen, ja. Aber was ich nicht nur bei der Stiftung Buchkunst sondern auch in meiner internationalen Unterrichtspraxis in Sachen Buch verstanden habe: Das Medium Buch ist ein universales Medium. Unabhängig vom Kulturkreis, in dem es veröffentlicht wird, der Sprache, in

1 ZEITmagazin Nr. 2/2019, 2. Januar 2019, editiert am 7. Januar 2019 (22.1.2019).

is involved in book design, the more precise he or she (for example, in a jury) can name what constitutes quality. For the jury, the matter is not expressing their personal opinion: what they like — thumbs up — or what they don't — thumbs down. Weighing the design and the quality of the presentation is a complex process. Also a thrilling one — for me it always was when I guided a jury through the discussion. When professionals have to reach a consensus over design phenomena, they have to address fundamental technical and design criteria, leaving aside personal taste or preferences. I always enjoy exchanging arguments — in favour or against a book —with colleagues. For critique sharpens the eyes and the senses.

Contemporary neurologists have been researching these phenomena for a long time. "Since people have different aesthetic preferences, until now we thought that it was impossible to say what beauty truly is. Besides, with this concept we cover different sense experiences... When someone has an aesthetic experience, the field A1 in the frontal lobe of the endbrain, behind the eye socket, becomes active." So describes it the neurobiologist from London Semir Zeki in an interview in the weekly journal *Die Zeit*.[1] He researches "how the brain translates our perceptions in feelings and concepts". It is said that aesthetics is the study of the sensorial perception. Just like design, it is subjected to cultural influences, yes. But I have come to understand something about books, not only in the Stiftung Buchkunst, but also in my international teaching practice: books are a universal medium. Independently from the cultural circle where they

1 ZEITmagazine Nr. 2/2019, 2. January 2019, edited on 7. January 2019 (22.1.2019).

大拇指，而对自己不喜欢的作品不屑一顾。对设计和装帧质量的评价，是一个复杂的过程，也是一个十分有趣的过程。至少我每次主持评委会讨论，都能获得许多乐趣。要让专家们就某一设计现象达成一致，就必须不受个人喜好的左右，制定基本的技术和设计标准。我也一直很享受与同僚们就一本书的好坏交换意见的过程，因为批评可以使我们的目光和感官变得更为敏锐。||||||||||

当代神经病学家一直在研究一类现象："从前人们认为，各人的美学偏好不同，所以我们无法定义美。另外，美这一概念也包含了不同的感官感受……每次人们获得美学体验，位于眼眶后方的大脑皮层 A1 区就会表现得活跃。"伦敦神经病学家塞米尔 · 泽基在接受《时代》周刊采访时这样说道[1]。他的研究课题是"大脑如何反映我们对情感和世界的感受"。人们常说，美学是关于感官感受的学问。它与设计一样，都受到文化的影响。这话不假，但无论是在为图书艺术基金会工作还是在世界各地授课时，我都充分感受到了一点：书籍是一个不分国界的媒介。无论它在哪个文化范围内出版，用何种语言写作，使用何种文字体系，受到哪些时代风潮的影响，它在全世界范围内都遵循相同的实施规则。即便某些文化空间自古以来就有不同的写作、阅读和用纸习惯，这一点也依然不例外。欣赏包在图书内页外的封

1 《时代》杂志版，2019 年第 2 期，2019 年 1 月 2 日，2019 年 1 月 7 日重新编辑。访问时间：2019 年 1 月 22 日。

der es geschrieben ist, den Schriftsystemen, in denen diese Sprache transportiert wird, den Moden, die den jeweiligen Zeitgeist abbilden — es folgt den gleichen Regeln der Handhabung, weltweit. Auch wenn in manchen Kulturen traditionellerweise die Schreib-, Lese- und damit Blätterrichtung in einem Buch verschieden sind. Der Deckel, der das Buchinnere vom Außen schützt; das Öffnen des Buches, wie das Öffnen einer Tür in einen Innenraum; das Blättern der Seiten, das mithilfe der Gestaltung eine gewisse Rhythmik, eine Dramaturgie erzeugt; die Zeit- und Raumdimension des Buches; die Qualität der Lesbarkeit; die Bildsprache; der narrative Aspekt — all das sind Attribute des Buches, die einen kulturübergreifenden Vergleich möglich und so faszinierend machen. Chinesische Bücher HANDhabe ich genauso wie australische, venezolanische, iranische oder europäische.

In der Zeit zwischen 2001 und 2012 hatte ich in der Stiftung Buchkunst jährlich fünf Jurys zur Buchgestaltung zu organisieren und zu moderieren, darunter den internationalen Wettbewerb ‹Schönste Bücher aus aller Welt›. Dieser Wettbewerb war für mich immer eine besondere Gelegenheit, den Stand der aktuellen Buchgestaltung auf höchstem Niveau sehen zu können. Die Bücher kamen sozusagen ins Haus. Jeder Jahrgang brachte die ‹Schönsten› aus den einzelnen Länderwettbewerben in Bezug zueinander. Manche Bücher haben wegen ihrer Inhalte fasziniert, manche eröffnen das Thema durch die besondere Ausstattung oder die Buchform, manche Bücher sind still und deshalb herausragend in Abgrenzung zur Aufgeregtheit der Mittel, Farben oder Formen der anderen. Jedes Buch ein eigener kleiner Kosmos. Jedes hatte schon eine Jury durchlaufen, im eigenen Kultur- und Sprachraum. Und lag jetzt erneut einer Jury vor, die sich über zwei Tage intensiven Austauschs wieder und wieder darüber verständigen musste, was sie auszeichnen wollte, nicht nach politischem Proporz, sondern für die Sache:

are published, from the language in which they are written, from the writing system in which this language is transported, from the trends that create the spirit of the current times — books follow the same rules, all over the world. Even if traditionally in many cultures the direction writing, reading, and thus the direction of leafing through a book are different. The cover protects the inside or the book from the outside world, the title page acts like a door opening to a room, the leafing of the pages creates a certain rhythm, a dramatic effect with the help of design. Then there are the time and space dimensions of the book, the quality of its readability, the images, the narrative aspect — all these are attributes of books that allow to make such a fascinating comparison throughout cultures. I can use Chinese books exactly the same as Australian, Venezuelan, Iranian, or European books.

Between 2001 and 2012, every year I organized and moderated five juries for book design in the Stiftung Buchkunst, one of them was the international competition *Best Book Design from all over the World*. This competition was always a special occasion for me to see the highest level of the current book design. The books came to us, so to speak. Every year brought together the "most beautiful" from the national competitions. Some books fascinate because of their content, others start the subject with a special decoration or form, some are calm and stand out in contrast to the exciting media, colours, or forms of others. Every book is a small cosmos. Every one of them had already gone through a jury in its own cultural and language sphere. And now they were in front of another jury that, over two days of intensive exchange, had to reach agreements again and again about what they wanted to give a prize to, not according to

皮；打开图书，就像打开一扇通往内室的门；翻阅内页，感受因设计而生的节奏和故事；一本书的时间和空间维度；它的可读性；图画语言；叙事层面——正是以上这些特征，使得跨文化的书籍比较成为可能，而且能带给人诸多乐趣。也正如此，我得以用相同的方式，对待来自中国、澳大利亚、委内瑞拉、伊朗和欧洲的图书。||||||

2001 年至 2012 年间，我在图书艺术基金会工作，每年都要召集并主持五个书籍设计评委会，其中就包括“世界最美的书”国际大赛评委会。就我个人而言，这项竞赛一直是了解最前沿、最高水平图书设计的良机。参赛图书被寄送给基金会。每年的竞赛，都会让各国图书“选美大赛”的佼佼者相聚在一起。有些图书胜在内容；有些图书以特别的装帧和形式引发热议；有些图书典雅朴素，与那些手段、颜色和形式激烈的图书形成鲜明对比。每一本书都自成体系，从而得以在评审后从各自的文化和语言空间里脱颖而出。现在，它们被摆在评委面前，他们要在两天时间里反复讨论，直到达成一致，评选出优胜者。评审无关政治因素，讲究实事求是，也即优者胜出。2003 年的评委会讨论得最为激烈，一本无论如何都称不上“美”的图书，让评委们意见不一。但这本书也另有特色：它的情绪化和极端化，让人不禁为之动容。||||||||||||||||||||||||||||||||||||||||||||||||||

1969 年，扬 · 帕拉赫为抗议苏联非法侵占捷克斯洛伐克，浇上汽油自焚。这戏剧性的一幕，也

nämlich beste Gestaltung. Die aufregendste Jury war im Jahrgang 2003, als sich die Jurymitglieder für ein Buch entschieden, das mitnichten als ‹schön› zu bezeichnen ist. Dieses Buch enthält eine zusätzliche Dimension: es berührte in seiner Emotionalität und Radikalität.

Die junge niederländische Designerin Nynke Meijer erinnert an Jan Palach, jenen Demonstranten, der sich 1969 in Prag aus Protest gegen die sowjetische Okkupation der Tschechoslowakei mit Benzin übergoss und verbrannte. Ein dramatischer Akt, ein politischer zugleich. Was berührt(e)? Die Buchgestaltung ist nicht distanziert, bleibt nicht in der kühlen Dokumentation sondern entwickelt ihre ganze Kraft durch eine passende Einfachheit (es ist eine Broschur), eine haarsträubende haptische Dimension (der Seitenschnitt des Buches fühlt sich wie blankes Fleisch an), eine aufladende Farbgebung (rote Doppelseiten sind taktgebend zwischen schwarz/weißen Collagen aus historischem Zeitungsmaterial eingeschoben) und die Fotokopierästhetik (als wäre man dabei, auf dem rauen Platz, mitten im Winter). Die Jurorinnen und Juroren diskutierten lange über die Frage, welche Ästhetik zur Dokumentation einer solchen Protestform korrespondieren kann? Wenn Inhalt und Form auf diese Weise eine gelungene Einheit bilden! Das anti-schöne Buch! Welch mutiger Schritt (einerseits), aber auch notwendig (andererseits), denn Ästhetik funktioniert nicht ohne den Kontext, nicht ohne den Inhalt. Aber wie will man eine solche Entscheidung einer Fachjury in der Öffentlichkeit und an die Presse kommunizieren, wenn ‹das Schönste Buch der Welt› nicht ‹schön› ist, zumindest nicht im erwarteten Sinn. Dann muss man ausholen und erklären, was Buchgestaltung ausmacht. Und beschreiben, dass es nicht um gefallen oder nichtgefallen geht, sondern unter Umständen eine emotional aufgeladene Gestaltung weiter tragen kann als ein technisch hoch elaboriertes und auf bestem drucktechnischen Niveau produziertes, aber belangloses Buch. Ich will diese beiden Pole nicht gegeneinander

political majority, but according to what we were judging: namely the best design.

The most thrilling jury was in 2003, when the members or the jury voted for a book that not everyone would qualify as "beautiful". That book contained an additional dimension: it was touching because of its emotional and radical nature.

The young Dutch designer Nynke Meijer reminded us of Jan Palach, the protester who in 1969 in Prague poured petrol on himself and set himself on fire to protest against the Soviet occupation of Czechoslovakia. A dramatic act, and also a political one. What is it that is or was touching? The book design is not distanced, it is not restricted to cold documenting, but develops its whole strength through an adequate simplicity (it is a paperback), a hair-raising haptic dimension (the edge of the book feels like bare skin), a heavy colouration (red double pages full of texture between black-and-white collages made of glued historical newspaper materials), and the photocopying technique employed — as if one were on that same rough square in the middle of winter. The jurors discussed for a long time — which aesthetics correspond to document such a protest? In this way, content and form were so perfectly united! It was the antibeautiful book! What a courageous step (on the one hand), but (on the other hand) what a necessary one it was, because aesthetics does not work without context, nor does it work without content. But how can a jury communicate to the general public and the press that "the book with the best design" is not "beautiful"? At least not in the expected sense. Then one has to explain what constitutes book design. And one must describe that it is not about likes or dislikes, but about an emotionally

具有很强的政治意义。为了纪念他，荷兰女设计师宁克 · 梅捷设计了这本书。它为何震撼人心？这本书的设计并没有像冰冷的纪录片那样，刻意与这段历史保持距离，而是倾尽全力再现历史：它外形简单（只是一本小册子），给人惊悸的触觉体验（侧面摸上去像赤裸的人肉），有着丰富的着色（在由报刊历史材料组成的黑白剪贴画中，赫然出现通篇血红的书页），还有逼真的印刷技术（给人一种在阴冷的寒风中伫立广场的感觉）。究竟什么样的美学手段，才适合记录这样的一次抗议活动？评委们就这个问题讨论了许久，给出的答案是：内容和形式必须被成功地结合在一起！这是一本反审美的图书！这一方面是勇敢的尝试，另一方面也是必要的，因为美离不开语境和内容。可是，如果“世界最美的书”不具一般意义上的“美”感，那么专业评委会的意见，又该如何被转达给公众和媒体呢？那我们必须从什么是书籍设计讲起，说明我们的选择无关好恶，一切只是因为：相比技术考究、印刷精美但却毫无意义的书籍，一件情绪饱满的设计作品显然更具价值。在此，我无意把这两种极端情况进行比较。但假如一件作品从技术角度看堪称完美，它的内容却无法触动我，那要它又有何用？同样，一本书如果话题十分有趣，但装帧简陋，经不起翻阅，或是排版粗糙，那它也不是我想看到的。有时候，在图书设计（这里的设计不仅指视觉层面，还包括材料的选择、装帧和图书的形式）中

ausspielen. Aber was habe ich von einem technisch perfekten Werk, wenn mich der Inhalt nicht anrührt? Genauso wenig möchte ich ein tolles Thema lesen, wenn das Buch zusammengeschustert ist, beim nächsten Öffnen auseinander bricht oder in miserabler Typografie gesetzt ist. Dem Inhalt gerecht zu werden ist in der Gestaltung (und hierunter verstehe ich neben der visuellen Dimension auch die materielle Form, die Ausstattung und die Buchform), manchmal nicht einfach. Oft sind es verlegerische oder auch Marktzwänge, die die eine oder andere Idee nicht realisierbar machen. Oft nicht einfach, da einen Weg zu finden. Und wenn man viel mit Büchern zu tun hat, kann es auch mal vorkommen, dass einen die Frage befällt: Braucht die Welt dieses Buch wirklich? Was ein Buch aber immer braucht, ist sensible Gestaltung. Eine, die nicht selbstverliebt die Attitüden des Gestalters spiegelt, sondern eine, die das jeweilige Thema transportiert und dabei hilft, es ins Handmedium Buch zu übersetzen; eine, die neugierig macht, berührt, vielleicht auch aufregt, weil sie kantig, eigenwillig und nicht leicht genießbar ist? Ist die kritische Reflexion, auch über die Relevanz unserer Arbeit, nicht auch Teil unserer Arbeit?

Tragen wir nicht auch eine Verantwortung für die visuelle, ästhetische Kommunikation, die nicht losgelöst sein kann von gesellschaftlichen Vorgängen? Auch im gestalterischen Prozess sollten wir kontextualisieren, also wissen, was in der Buchwelt gerade vor sich geht. Hat Kant also doch Recht!?

Was reizt einen Gestalter, alle ‹schönsten Bücher aus aller Welt› zu sammeln, um sie dann in fast wissenschaftlicher Detailarbeit verschiedenen gestalterischen Betrachtungen zu unterwerfen? ‹Grundlagenforschung eines Universalmediums› könnte man diese Arbeitsweise nennen, in der jedes einzelne Buch bis in die feinsten Einheiten seziert und in den Kontext des Jahrgangs gestellt wird. Man könnte die gesammelten Ergebnisse unter verschiedenen Aspekten auswerten: unter gestalterischer Sicht im Allgemeinen, mit

charged design transmitted in certain circumstances as a technically elaborated book — though a trivial one — produced with the best printing techniques. I don't want to contrast these two extremes, but what can I take from a technically perfect work if the content does not touch me? I would neither like to read about a subject if the book has been cobbled together, breaks when I open it, or has a bad typography. Sometimes it is not easy to make justice to the content through the design (and with this I mean, apart from the visual dimension, also the material form, the decoration, and the book form). Often there are publishing or market constraints that don't allow to carry out an idea. Often it is not simple to find a way in these situations. And, when one has a lot to do with books, it can also happen that the question arises: does the world really need this book? But what a book always needs is a sensitive design. One that does not narcissistically reflect the attitudes of the designer, but acts as a vessel for the current subject and helps to transport it into the medium of a book. A design that arouses curiosity, touches, maybe also irritates because it is harsh, unconventional, and not easy to enjoy? Isn't reflecting critically about the relevance of our task also part of our work? Don't we bear also a responsibility for the visual, aesthetic communication that cannot free itself from social processes? Even in the design process, we should contextualise, which means knowing what is happening in the world of books. So, after all, Kant was indeed right!

What makes a designer collect "the best designed books all over the world" and then subject them to different views on design in an almost scientific work? This working method could be called "basic research of a universal medium", since every

兼顾内容，并不是一件容易的事情。出版社的意志或是市场的限制，常常会妨碍创意的实现。要在这中间找到解决方案，其实并不容易。跟书打多了交道以后，人们有时不禁会问：世上真的需要这本书吗？而可被感知的设计，却是图书所一直需要的。这样的设计，不是图书设计者赖以自恋的标签；它应当承载起图书的主题，为把它转换成图书这种手持媒介提供帮助；它应该能唤起人的好奇心，触动读者；或许还应该棱角分明，独具个性，甚至引发共鸣。对工作意义的批判性反思，不也是我们工作的一部分吗？

我们是不是也应该为那些无法脱离社会进程的视觉和美学沟通承担一份责任？在设计过程中，我们也应该注重语境，了解图书世界正在发生什么。看来康德说得没错！

是什么促使一位设计师收集“世界最美的书”，对它们的各个设计过程进行科学、细致的研究？每本书都被抽丝剥茧，仔细分析，并放在当时的语境下进行解读。这样的工作方式，可被称作“对一种普遍媒介的基础性研究”。我们可以从不同的角度，对整体结果做出解读：既可以采用整体设计视角，也可以采用排版或印刷视角，或是重点研究不同材料的选用。我们甚至可以从社会学、媒体学、社会和文化学的角度，推断每个国家、每个年份或是整个世界的发展趋势——书籍设计的内涵，远不止于纸面。回首过去，书籍艺术总能反映技术、社会和政治层面的发展和扭曲。

typografischer oder fotografischer Perspektive im Besonderen oder nach der Frage der verwendeten Materialien. Man könnte sogar aus der Perspektive der Soziologie, der Medien-, der Gesellschafts- oder Kulturwissenschaft Rückschlüsse auf Tendenzen im jeweiligen Land, im jeweiligen Jahr oder in der Welt ziehen — Buchgestaltung drückt mehr als nur den Inhalt aus. Buchgestaltung zeigt im Rückblick immer auch technische, gesellschaftliche und politische Entwicklungen oder Verwerfungen auf. Eine Analyse der aktuellen oder historischen Buchgestaltung kommt in den Medienwissenschaften kaum vor. Es gibt Filmkritik, Literaturkritik, Designkritik, Opernkritik, nicht aber Typografiekritik oder Buchgestaltungskritik. Eine solche Lücke beginnt das hier aufgelegte Werk zu schließen. Mich beeindruckt Zhao Qians Sammelleidenschaft, und noch mehr seine konsequente Arbeitsweise, sein besonderer Blick auf die Sache, auf das Detail. Auch so kann kommuniziert werden, was Buchgestaltung ist. Ein Lehrbuch, das so gar nicht pädagogisch daher kommt, sondern viel eher von der Faszination Buchgestaltung erzählt. Ein Buch, das den Bogen spannt, das kontextualisiert. Das Medium Buchgestaltung ist, wie jede konzeptionelle Arbeit, vom Wechselspiel zwischen Detailarbeit und übergeordnetem Verständnis geprägt. Mikro und Makro befinden sich dabei im permanenten Wechsel und Bezug zueinander. Ich kann auf Mikroebene nichts ändern ohne damit nicht auch das Ganze zu verändern, und umgekehrt. Buchgestalter/innen sind in beide Richtungen aktiv: hin zum Detail um gleichzeitig das Gesamte zu erfassen. In diesem Sinn verstehe ich Zhao Qians Buch wie ein visuelles Nachdenken über den Zustand der Buchgestaltung, international.

single book is dissected in its smallest components and set in the context of the current year. One could value the results from different perspectives: considering overall design in general, taking into account typography or photography in particular, and studying the materials. One could also make conclusions on trends in a specific country, in the current year, or throughout the world from the perspective of sociology, media studies, social science, or culturology. Looking back, book design always shows technical, social, and political developments or rejections.

An analysis of current or historical book design rarely appears in media studies. There is film criticism, literature criticism, design criticism, opera criticism, but not typography criticism or book design criticism. The work that we are doing here is starting to fill such a void. I am impressed by Zhao Qing's passion, his peculiar vision of things and details. This way we can communicate what book design is. A course book that is not so pedagogical, but that rather tells us about the fascination of book design. A book that creates tension, that contextualises. As every conceptual work, the medium of book design is marked by the interplay between detail work and a superior understanding. Micro- and macroelements are in constant change and interrelation. I cannot change anything at the microlevel without altering the whole, and vice versa. Book designers are active in both directions: they have to grasp detail and, simultaneously, the whole. In this way, I understand Zhao Qing's book as a visual reflection of the state of book design — it is international.

但在媒体学中，对书籍设计历史和现状的研究只是凤毛麟角。世上有电影批评、文学批评、设计批评、歌剧批评，却没有排版批评或书籍设计批评这样的概念。本书的出版，正是填补空白的开始。赵清给我留下深刻印象的地方，不仅是他的收藏热情，更是他严谨的工作方式以及观察事物和细节的独到眼光。只有这样，才能道出书籍设计的真谛。这本教育类的书，其实并不重在说教，而是介绍书籍设计的魅力。它高屋建瓴，旁征博引。书籍设计与所有概念性工作一样，是微观工作和宏观理解的结合，两者相互转换，互为联系。如果不从整体上做出改变，那微观层面的改变就无法进行，反之亦然。书籍设计师必须在两个层面积极努力：既要追求细节，也要整体把握问题。在这种意义上，赵清的这本书可被视作对书籍设计现状的国际性视觉思考。||||||||||||||||||||||||||||||

Uta Schneider
乌塔・施耐德

1959 in Reutlingen geboren.
1979—1985 Studium Visuelle Kommunikationan der HfG Offenbach am Main. Schwerpunkt Zeichnen, Druckgrafik, Typografie.
seit 1986 Bildende Künstlerin mit Schwerpunkt Künstlerbuch (verschiedene Techniken, vor allem Hochdruck), Zeichnung und Text/ Raum-Installation. Gründerin und Mitglied der Künstlerinnengruppe Unica T. Seit 2002 Gründung von ‹usus› und Fortsetzung der künst- lerischen Zusammenarbeit mit Ulrike Stoltz.
seit 1986 Typografin, Buchgestalterin.
2001—2011 Geschäftsführerin der Stiftung Buchkunst, verantwortlich für die Wettbewerbe Die schönsten deutschen Bücher & Schönste Bücher aus aller Welt Leipzig. Kuratorin von Spezialausstellungen.
seit März 2012 wieder selbständig als Bildende Künstlerin, Typografin und Dozentin in der Erwachsenenbildung.
Mitglied im Deutschen Künstlerbund e.V., Bund Offenbacher Künstler e.V.

2005 1st Seoul Book Arts Award, Seoul/Südkorea 2003 Kulturpreis der Stadt Offenbach am Main 2001 Stipendium (& Produktion eines Künstlerbuchs) Nexus Press, Atlanta/ USA.
1998 Kunstpreis Neues Kunsthaus, Ahrenshoop 1994 Stipendium im Künstlerhaus Lukas, Ahrenshoop.

‹usus› Einzelausstellungen in Italien, Japan, Schweiz und Deutschland. Regelmäßig Gruppenausstellungen, international (s. Katalog). 2007 Veröffentlichung der Monografie ‹usus› trans—lation anläßlich der Einzelausstellung im Klingspor Museum Offenbach am Main und im Kunstverein Reutlingen (2009).
Seit 1986 Messebeteiligung mit den eigenen Künstlerbüchern auf der Frankfurter Buchmesse 1997—2004 Ausstellung/Messebeteiligung Art Frankfurt (vertreten durch Despalles Editions, Mainz & Paris).

1959: born in Reutlingen
1979 – 1985: studies Visual Communication at the University of Art and Design in Offenbach am Main and specialises in Drawing, Print, and Typography.
Since 1986: visual artist specialised in book art (different techniques, above all relief printing), drawing, and decoration of rooms with texts. Founder and member of the female artists group Unica T. Since 2002 she founds ‹usus›, a partnership with Ulrike Stolz, and continues her artistic collaboration with her.
Since 1986: typograph and a book designer.
1979 – 1985: managing director of Stiftung Buchkunst, responsible for the competitions Best Book Design from all over the World and Best Book Design from all over the World Leipzig. Curator of special exhibitions.
Since March 2012: works independently again as a visual artist, typograph, and educator in the field of adult education. Member of the Association of German Artists and the Offenbach Artists Association.

2005: First Seoul Book Arts Award, Seul (South Korea)
2003: Culture Award of the city Offenbach am Main
2001: scholarship (and production of an artistic book) awarded by Nexus Press, Atlanta (United States)
1998: Art Prize Neues Kunsthaus, Ahrenshoop
1994: scholarship in Künstlerhaus Lukas, Ahrenshoop
‹usus› Solo exhibitions in Italy, Japan, Switzerland, and Germany. Regular group exhibitions, international (see catalogue).

2007: publication of the monography ‹usus› trans—lation on occasion of the solo exhibition in the Klingspor Museum of Offenbach am Main and in the Reutlingen Artist Association (2009).
Since 1986: participation with her own artistic books in the Frankfurt Book Fair
1997 – 2004: exhibition and participation in the Frankfurt Book Fair Art Frankfurt (represented by Éditions Despalles, Mainz / Paris)

1959 年　生于罗特林根。
1979-1985 年　在奥芬巴赫设计学院学习视觉传播，重点学习素描、印刷和排版。
1986 年起　成为造型艺术家，主攻艺术书籍创作（使用不同技法，尤其是凸版印刷）、素描和文本 / 空间装置。艺术团体 Unica T 的创始人兼成员。
2002 年起　创办艺术组合 ‹usus›，与乌里克 · 施托尔茨（Ulrike Stoltz）保持艺术合作。
1986 年起　担任排版师和图书设计师。
2001-2011 年　担任图书艺术基金会负责人，在莱比锡负责“德国最美的书”和“世界最美的书”两项大赛，同时担任特别展览的策展人。
2002 年 5 月起　重新成为独立造型艺术家、排版师和成人教育讲师。德国艺术家联盟会员，奥芬巴赫艺术家联盟会员。

获奖情况
2005 年　获首尔图书艺术奖一等奖，韩国首尔
2003 年　获奥芬巴赫市文化奖
2001 年　获拿索斯出版社资助（并为其制作艺术图书一本），美国亚特兰大
1998 年　获新艺术之家艺术奖，亚恩朔普
1994 年　获卢卡斯艺术之家资助，亚恩朔普

艺术组合 ‹usus› 在意大利、日本、瑞士和德国举办个展，定期参加国际群展（参见目录册）。值 2007 年在奥芬巴赫科林斯波博物馆和 2009 年在罗特林根艺术家协会举办个展之际，出版专著《‹usus› 传移》。
艺术图书作品自 1986 年起参加展览，1997-2004 年参加法兰克福书展，后参加艺术法兰克福展览（由总部位于美因茨和巴黎的德斯帕勒斯版本出版社负责代理）。

Zum Ton schöner Bücher

– geschrieben anlässlich der Veröffentlichung von „Die Leipziger Auswahl – die schonsten Bucher der Welt 2019–2004"

The Melody and Tune of the Best Book Design

Written on the occasion of the publication of The Choice of Leipzig

Geht man in eine unweit des „Präsidentenpalastes" der Stadt Nanjing gelegene Gasse namens Meiyuan, dann kommt man am „Repräsentationsbüro der KP Chinas" vorbei, in dem zur Zeit der Kooperation von Kuomintang und KPCh Zhou Enlai residierte. In diesem Viertel gibt es eine Anzahl von Villen im westlichen Stil, die Atmosphäre im Schatten der graziösen Platanen verströmt etwas Geheimnisvolles und Ruhiges. Nach kaum mehr als 10 Metern biegt man in einen abgelegenen Winkel, an dessen Ende sich ein dreistöckiges Gebäude im westlichen Stil der Republikzeit befindet. Der Hof ist nicht groß, ein leichter Wind weht durch den Bambushain, das Sonnenlicht, das durch die Blätter der Bäume dringt, wirft helle Flecken auf den steinernen Boden, es macht einen freundlichen und anheimelnden Eindruck. Steigt man die hölzerne Treppe nach oben, so ist jeder Raum anders strukturiert. Die Ausstattung ist exquisit, der Tritt vor jedes Fenster bietet eine schöne Aussicht. Man befindet sich im Atelier von Herrn Zhao Qing, dem Gründer der „Han Qing-Galerie". Dieses Wohnzimmer ist in China zu etwas geworden, das ich den Lesesaal des Museums „Die schönsten Bücher der Welt" nennen möchte. Hier wird für die Fachkollegen und jungen Liebhaber der Kunst „Hof gehalten", und hier versammelt man sich, um sich an den Köstlichkeiten der schönsten Bücher der Welt zu laben.

Zhao Qing findet in allem, was er tut, einen bestimmten Ton – in seiner Einstellung gegenüber dem Leben ebenso wie in der Behandlung der Kollegen oder in Fragen des Designs... Gleich, ob es um die Herstellung von Büchern, den Entwurf von Plakaten, die Planung einer Ausstellungen, die Abhaltung von Unterricht oder um etwas Gemeinnütziges geht, alles ist akribisch und bis in kleinste Detail hinein durchdacht und verfehlt nie einen bestimmten Ton. Diese tonale Lage ergibt sich einerseits aus dem Buch als eine Art von „Behältnis" und andererseits aus der Abstimmung eben des Inhalts auf den Lese-

Not far from Nanjing's Presidential Palace there is the small Meiyuan Alley, once the place where Zhou Enlai established an office for the Communist Party during the First United Front. In this area all are western-style buildings arranged in a charming disorder. Covered by the graceful forms of plane trees, they have a mysterious and tranquil air. A few metres away, we turn and enter in a secluded, tortuous alley. Inside there is a small three-storey building built in the style of the republican period. The court is not big, a cool breeze goes through the bamboo forest, and sunlight shines through the leaves of the trees, creating a myriad of light spots on the stone — how pleasing. Going up the wooden stairs, the structure of every room is different and exquisite, every window frame and porch seems a good place to see the scenery. This building is Hanqing Hall, Zhao Qing's working studio. Its living room is China's only reading space that I can call a museum of *The Best Book Design from all over the World*. It has already become a "temple for art" worshiped by the scholars of this trade, for whom the beautiful books of the world are a feast for the eyes.

Zhao Qing has a special way of doing everything, a personal tune in the way he treats life, his colleagues, and design. Whether in creating books, designing posters, preparing exhibitions, teaching, or doing charitable work, he always puts attention to detail, acting meticulously — he has both a melody and a tune. The "melody" refers to the style and can become the container of books, but the "tune" represents the content, the musicality of reading. Zhao's

美书的腔调——
写在《莱比锡的选择——世界最美的书 2019-2004》出版之际

走进南京“总统府”不远的一条称之为梅园的小巷，经过曾经是国共合作时期周恩来驻扎过的“中共代表处”，这一带都是错落有致的西式洋房，在婀娜多姿的法国梧桐树的遮隐下，显得有些神秘而宁静。没过十几米，拐进曲巷幽径，最里面有一栋民国风格的三层小洋楼，庭院不大，竹林清风，阳光透过树叶的缝隙在石板地上落下流星般的亮点，甚是惬意。踩着木质的楼梯拾级而上，每一个房间结构都不同，精巧玲珑，每一个窗棂门廊都是借景的好去处，此楼便是“瀚清堂”堂主赵清先生的工作室。如今客厅已是中国唯一的，我称之为“世界最美的书”博物馆的阅读空间，现已成为业界同行学子热衷于“朝拜”的艺术殿堂，人们在这儿尽享美轮美奂的世界美书大餐。||||||赵清兄做什么都有腔调，对待生活，对待同道，对待设计……无论做书、做海报、做展览、做教学、做公益，做所有的事都一板一眼、一丝不苟、有腔有调。“腔”为样式，可比作书籍的容器；“调”即蕴涵，阅读的调性也。与赵兄因书结缘，在很多年前的评选中，他做的科技类图书《世界地下交通》，以及

vorgang. Die Verbindung mit dem vereehrten Herrn Zhao ist aus dem langjährigen Zusammenwirken bei der Arbeit als Juror hervorgegangen. Die Entwürfe der von Zhao gemachten Bücher etwa zu Wissenschaft und Technologie wie Unterirdischer Verkehr der Welt oder gängige Bücher zum menschlichen Alltag wie Speisezettel nach dem Angebot im Garten zeichnen sich durch die Verbindung von Inhalt und Form aus, es bleibt Raum für Überraschungen und damit wird ein Ton getroffen, der mitreißt. Zhao hat sich stets mit großer Energie an seine Arbeit gemacht, auf ihn geht eine große Anzahl herausragender Werke zurück, für die er mit fast allen wichtigen auf der Welt vergebenen Preisen für grafische Gestaltung ausgezeichnet worden ist. Dazu zählen die wiederholte Auszeichnung mit dem National Book Design Exhibition Gold Award, er führt mit weitem Vorsprung die Liste der Preisvergabe für Chinas schönstes Buch an und hat darüber hinaus auch vielen seiner Assistenten und Mitarbeiter die Auszeichnung mit verschiedenen Preisen ermöglicht. Zhao Qing ist wettbewerbsorientiert, mit den Besten seines Faches gut vernetzt, dabei vorbildorientiert, abwägend und mit Blick auf die Verbesserung der Designerkunst stets daran interessiert, hinzuzulernen. Dabei hat er keine Kosten und Mühen gescheut, die schönsten Bücher der Welt zu sammeln. Die Mühe für seinen Einsatz hat sich gelohnt. Nach einem Jahrzehnt harter Arbeit sind nahezu 200 chinesische Bücher, die in dem Zeitraum zwischen 2004 und 2019 offiziell an der Auswahl der schönsten Bücher der Welt teilgenommen haben und ausgezeichnet worden sind nun endlich als Sammlung in der Han Qing-Galerie zu sehen – es ist ein Leseraum für die schönsten Bücher der Welt, wie ich schon weiter oben bemerkte.

Auch mit seinem sorgfältig bearbeiteten und publizierten Buch „Die Leipziger Auswahl“, bei dem es sich um eine Sammlung der weltweit schönsten Bücher handelt, hat er sich auf die Zusammenarbeit mit Gleichgesinnten eingelassen. In der Vergangenheit wurde der internationale Wettbewerb für die schönsten Bücher der Welt von den Veranstaltern jeweils in Ost- und Westdeutschland ausgetragen. Nach der Wiedervereinigung Deutschlands ist der heutzutage in

connection with books started many years ago when he designed scientific books such as *Shijie dixia jiaotong (The World's Underground Traffic)* and other books treating subjects more related to daily life, like the Qing Dynasty cookbook *Suiyuan shidan (Recipes from the Garden of Contentment)*. He designed them paying attention to both form and meaning, which surpassed everyone's expectations. It was the "melody and tune" of his works that were so moving. Later, the ball started rolling, and he designed numerous outstanding works awarded with all big prizes of graphic design around the world, like the golden prize of numerous editions of the National Exhibition of Book Design in China and many nominations on the top ranks of China's Most Beautiful Book. In addition, the disciples he guided also received all kinds of prizes. When Zhao Qing competes, he does so wholeheartedly, with expertise, looking for people of outstanding talent, considering the differences, and studying meticulously in order to improve the possibilities of the design. This played an important role in his decision of collecting the best designed books from all over the world. In spite of hardships, he didn't hesitate — his determination did not break. Hard work payed off, and after ten years of effort, he had collected in Hanqing Hall almost two hundred books judged as the most beautiful from 2004 to 2019. In such a room, surrounded by the world's most beautiful books, he found the inspiration to edit and publish *The Choice of Leipzig*, a compilation of the world's most beautiful books, and share it with his colleagues.

In the past, this international competition on the beauty of books was held separately by East and West Germany. After the reunification, the contest *The Best Book Design from all over the World* has been held in Leipzig until present. It could be said that its significance is to act

最普通的生活类图书《随园食单》，被他设计得形意兼顾，出人意表，有腔有调，为之感动。之后，他则一发不可收，设计出大量优秀作品，囊括几乎全世界平面设计的所有主要大奖，多届全国书籍设计大展金奖，“中国最美的书”奖的获奖次数高居榜首，还带领助手弟子们横扫各类奖项。赵兄参赛有心，广交高手，寻觅翘楚，掂量差距，求学钻研，为的是提升设计的能量，这为他日后决心把世界最美的书收于囊中而煞费苦心，百折不挠，不惜重金，历经甘苦。功夫不负有心人，十多年的努力，终将以中国正式参评“世界最美的书”的 2004 年至 2019 年的获奖书 200 多册收集于瀚清堂，于是才有了前文中提到的有腔有调的“世界最美的书”的阅读空间，乃至有了他精心编辑出版的这本《莱比锡的选择》“世界最美的书”汇集，愿与同道共飨。

过去的国际美书大赛由东德西德两个阵营各自举办，德国统一后合二为一,即当今在莱比锡举办的称之为“世界最美的书”的赛事，可以说是真正意义上的面向全世界各个国家的国际性书籍艺术大赛。每年初春,各国评出的最美的书送交莱比锡，由本届评委会评出 14 本年度“世界最美的书”奖。自 2003

Leipzig ausgetragene Wettbewerb „Die schönsten Bücher der Welt" wohl im besten Sinne ein internationaler Buchkunstwettbewerb, der allen Ländern auf der Welt offen steht. Zu Beginn des Frühlings in jedem Jahr werden die von den jeweiligen Ländern ausgewählten schönsten Bücher nach Leipzig geschickt, wo die zuständige Jury 14 die „Schönsten Bücher der Welt" in dem entsprechenden Jahr auszeichnet. Seit 2003 werden die „Schönsten Bücher Chinas" auf Initiative des Shanghai Press and Publication Bureau hin ausgewählt. Mittlerweile sind die Werke von 22 chinesischen Buchdesignern mit allen Auszeichnungen geehrt worden, darunter Auszeichnungen wie die Goldene Letter, Medaillen in Gold, Silber uund Bronze sowie Ehrendiplome. Ich hatte das Privileg, 2014 zu einem der Juroren für die Auswahl der schönsten Bücher der Welt berufen zu werden. Damals hat es mich mit Stolz erfüllt, als Experten aus aller Welt ihre Expertise und ihr Urteil für die Werke der chinesischen Designer abgaben. Die chinesische Buchkunst ermöglicht es der Welt, bei dem Blick durch dieses Fenster zu verstehen, welche Fortschritte und Errungenschaften China auf dem Gebiet der Buchgestaltung während der 40 Jahre von Reform und die Öffnung gemacht hat. Darüber hinaus wird es möglich, dass Chinas Designer Selbstvertrauen gewinnen und sich der bestehenden Mängel und Unzulänglichkeiten bewusst werden, um der jüngeren Generation von Designern neue Horizonte zu bieten und ihnen den Schwung zu verleihen, der notwendig ist, um den Anschluss an die Weltspitze zu finden.

Bei allen Wettbewerben gibt es Regeln für die Beurteilung. Bei der Auswahl der schönsten Bücher der Welt wird die umfassende und ganzheitliche Beurteilung von Büchern als Maßstab betont, es geht nicht um Details. Ich habe Frau Uta Schneider, die über 12 Jahre hinweg Vorsitzende der Jury für die weltweit schönsten Bücher war und die eine berühmte Buchdesignerin ist, gefragt, welches Design ihrer Meinung nach für die Wahl als „Das schönste Buch" ausschlaggebend ist. Sie sagte, es sei schwer, das mit ein paar Worten zu sagen, doch zusammengefasst ließe sich das Folgende feststellen: „Beim Buchdesign geht es nicht nur darum, dass die Titelseite gut aussieht, wichtig ist die Integrität des Gesamtkonzepts; ein gutes Buch bedeutet

as a great competition on book art for all countries. Every year in spring, the books that every country considers the most beautiful go to Leipzig. This year, the jury has chosen fourteen works to be awarded the title of *The Best Book Designs from all over the World*. Since 2003, China's Most Beautiful Book, chosen by the Shanghai Publication Bureau, has been participating. There are already twenty two works by Chinese designers that have received the golden, silver, and bronze prizes. In 2014, I had the honour to be part of the jury of *The Best Book Designs from all over the World*. Seeing that all experts appreciated the work of Chinese designers, a sense of pride arose inside me. Book art has been the window that has allowed the world to understand China's progress and achievements in the field of book design in forty years of Reform and Opening Up. It has also allowed us, designers, to find confidence while also helping us to find out our deficiencies and shortcomings as well as giving impulse to the youngest generation of designers who want to expand their horizons and attain the highest level in the field.

Every competition has rules. *The Best Book Designs from all over the World* emphasizes judging every book comprehensively and not paying attention to only one of its aspects. Once I asked the famous designer Uta Schneider, who has been part of the jury for twelve years, which one could be considered the best book design. She said that it was very difficult to explain this in only a few sentences, but it could be summarized as follows: "Book design is not only a beautiful cover or pictures, but a concept that affects the whole book. A book is good not only because of a new and original design, but also because its content and layout are inextricably linked as a whole that allows to distinctly perceive the content. Good design carries out a complete reflection going

年开始，由上海市新闻出版局发起组织评选的“中国最美的书”参评至今，已有22本中国书籍设计师的作品获得包括金、银、铜、荣誉奖各奖项的荣誉。2014年我有幸受邀担任“世界最美的书”的评委，看到各国专家纷纷赏评中国设计时，一种自豪感油然而生。中国的书籍艺术通过这个窗口，让世界了解了改革开放40年的中国在书籍设计领域取得的进步和成绩，也让我们的设计师既找到了自信，也发现了自己的不足和短板，为年轻一代设计师开阔视野、追赶世界先进水平增添了动力。

凡是比赛，都有评判规则，“世界最美的书”的评选强调书的综合判断的标准，而非某一个局部。我曾请教担任了12年“世界最美的书”评委会主席、著名书籍设计家乌塔·施耐德女士，什么设计才是“最美的书”？她说用几句话很难讲清，如果概括表述可以这样认为：“书籍设计不只图封面好看，而是整体概念的完整；一本好书不仅在于设计的新颖，更在于书的内容编排与整体关系的贴切，并十分清晰地读到内容；好的设计，从功能的翻阅感受到内容诗意的表达，均有完整的思考。”确实如此，多年来我国的出版界和设计界一直在打破仅以书衣打

nicht nur Neuheit des Designs, relevant ist vielmehr die inhaltliche Anordnung des Buches und die hergestellte Gesamtbeziehung, die es ermöglicht, klar den Inhalt zu lesen; gutes Design bedeutet, dass ein umfassendes Konzept vorliegt, und zwar von der sinnlichen Wahrnehmung beim Blättern bis zum poetischen Ausdruck des Inhalts." In der Tat ist es den Verlagsvertretern und Designern in China im Laufe der Jahre stets darum gegangen, die Grenzen zu durchbrechen, zu denen es kommt, wenn man mit der Bucheinbindung vor allem das Dekorative im Blick hat. Es sind dabei verstärkt das umfassende Konzept des Buchdesigns berücksichtigt worden einschließlich der redaktionellen Gestaltung und unter Berücksichtigung des Leseflusses. Verlage und Designer sind dabei in die Rolle von Statisten ebenso geschlüpft wie die der Regisseure, um die beste Lektüre sicherzustellen. Sie haben selbst in solchen Fällen, da es Kunden nur um einfache Designanfragen ging und die Verlage und Designer sich nicht mit der Funktion von Make-up-Künstlern zufrieden geben wollten, viel Energie und Mühe investiert, um den Ansprüchen an die erzählerische Struktur des Buches, seine visuelle Sprache, die Präsentation der handwerklichen Umsetzung und die ästhetischen Lektüre Genüge zu tun. Aus jedem Band der schönsten Bücher der Welt lässt sich herauslesen, mit wieviel Hingabe und Kreativität sich die Designer um das Gesamtdesign der Bücher bemüht und dabei den Wert an Lesbarkeit gesteigert haben.

Was nun die Relativität einer Bewertung als „schönstes" Buch der Welt betrifft, so gilt es zu berücksichtigen, dass es bei Preisen sowohl etwas Zwangsläufiges als auch etwas Zufälliges gibt. Ausgezeichnet werden in jedem Jahr schließlich nur 14 Werke, die Aussichten auf einen Preis sind äußerst gering. Außerdem kommt hinzu, dass sich die Jury bei jedem Wettbewerb anders zusammensetzt, Urteilskriterien können dann unterschiedlich sein, sodass es durchaus möglich ist, dass einige gut gestaltete Bücher nicht in die engere Wahl kommen. Die Auswahl als das „schönste" Buch ist daher nie absolut anzusehen, was wiederum auch bedeutet, dass ein nicht ausgezeichnetes Buch nicht zwangsläufig auch einen Mangel an Schönheit aufweisen

from the sensation of turning the pages to the poetic expression of the content". Indeed, for many years China's publishing and design circles have always tried to overcome the limitation of using design only as a means of embellishment and have invested in a comprehensive concept of book design. Travelling together with the text, this kind of design takes the role of director to improve readability. Even when the client only asks for the design of the cover, it does not only fulfil the role of a makeup artist, but also takes charge of the narrative structure, the visual language, the presentation, and the reading aesthetics. In every work belonging to *The Best Book Designs from all over the World*, there is not a single one that does not transmit the complete dedication of the designers to the overall concept of the book. Their care and creativity have increased the value of the works.

As for the word "best" in *The Best Book Designs from all over the World*, of course it is something relative. There is something inevitable in receiving an award, but there is also some chance in it. After all, every year only fourteen books receive this distinction. This rate is only a little bit higher than 1 percent. Considering that every year the jury is different, its tendency may also slightly vary. This might cause that some well-designed books do not make it to the finals. Therefore, chances are that the elected ones are not the best designed, so it doesn't mean that the ones that are not chosen are not beautiful. Every time the results of the vote are announced, there is always an indescribable distress, but I still believe that every year's choices have a reason to be. By revealing the visual presentation and deciphering the design of each book, we can grasp the essential reasons for it to have been given an award. Furthermore, this can give us inspiration

扮为目的的装帧局限，投入包括编辑设计在内的书籍整体设计概念，与文本同行，为最佳阅读承担配角乃至导演的角色，甚至在客户只有封面设计索求的情况下，不满足只尽到化妆师的职能，而是为书的叙述结构、视觉语言、工学呈现、阅读审美付出额外的担当和心力。在每一本“世界最美的书”中无不读到设计师们全身心注入的书籍整体设计的用心和创造力，并为文本提升了阅读的附加值。

至于世界最美的书之“最”应该是相对的，得奖既有必然性，也有偶然性，毕竟每年获此殊荣的只有 14 本，百分之一点几的极低得奖率，加上每届评委组成不同，评判倾向也可能有所不同，致使一些设计不错的书不能入围。所以每次评上的书有可能不是美之“最”，而没有评上的书也未必不美。每次公布评选结果,都有一种莫名的心痛,但我仍然认为历年评出的“世界最美的书”确有美的道理。我们在这本汇集介绍的每一本获奖书的视觉展现和设计的文字解读中可以领会作品获奖的个中要义和给予我们设计思维与方法论的诸多启发，部分区别于一般图书、匠心独具的美书为我们展开了一个更为开阔的视角和想象。中国有悠久的书卷艺术的传统，但并不是我们止步不前

muss. Jede Veröffentlichung einer getroffenen Auswahl hinterlässt einen schwer zu erklärenden inneren Schmerz. Allerdings bin ich nach wie vor davon überzeugt, dass die Kriterien für die Schönheit bei der Bewertung als „Schönste Bücher der Welt" in den vergangenen Jahren nachvollziehbar sind. Aus der visuellen Präsentation und der textlichen Interpretation jedes mit einem Preis ausgezeichneten Buches, das in dieser Sammlung vorgestellt wird, können wir das Wesentliche des Preises erfassen und eine Menge Inspirationen für das Gedankliche und das Methodische des Designs bekommen. Damit werden zum Teil auch Unterschiede zu ganz gewöhnlichen Büchern deutlich, einzigartige schöne Bücher bieten uns eine offenere Perspektive und geben Raum für Phantasie. China verfügt über eine lange Tradition in der Kunst von Büchern und Bücherrollen, doch bedeutet das keinesfalls, dem Vorwand Raum zu geben, uns nicht weiterentwickeln zu müssen. „Die schönsten Bücher der Welt" ist ein Werk, aus dem wir die besten Elemente der besten Werke der Buch- und Bücherrollenkultur aus der ganzen Welt bei uns aufnehmen und unsere Designerkultur bereichern können. Das ist der Grund, warum Zhao Qing nach einer langen Zeit der sorgfältigen Vorbereitung einen Leseraum für „Die schönsten Bücher der Welt" geschaffen hat, begleitet von einer Publikation.

Zhao Qing bringt uns immer wieder freudige Überraschungen in den unterschiedlichsten Tonlagen. Darin enthalten sind auch seine Liebe und Besessenheit, mit der er sich dem Thema der Buchgestaltung gewidmet hat. Um Worte macht er nicht viel Aufhebens, er arbeitet still vor sich hin, und was er anpackt, hat Hand und Fuß. In Zeiten, da man immer wieder einmal angewidert ist von der herrschenden Ungestümheit und dem leeren Gerede, sind Menschen wie Zhao, der lieber handelt statt spricht, leider eine Seltenheit in China. Mit den „Schönsten Büchern" wird dem Leser eine schöne Lektüre ermöglicht, man setzt nicht auf leere Floskeln, bleibt vielmehr gediegen und jedem Werk für sich zugewandt. Bei der Buchgestaltung geht es um die Einheit von Äußerem und Innerem, Form und Geist müssen gleichermaßen vorhanden sein, Herz und Auge Gefallen finden und alles muss einen angemessenen Wert besitzen. Das ist es,

about design and its methodology. By distinguishing them from the average, these original and unique books unfold a wider perspective and enhance our imagination. China has a long tradition in the art of book binding, but this is not an excuse for us to not move forward. *The Best Book Designs from all over the World* is a mine from which we can extract the world's most outstanding book culture and knowledge on design. This is the reason why Zhao Qing went through such a long careful planning to establish the reading room of *The Best Book Designs from all over the World* and gave so much thought to the publication of this book.

Zhao Qing often brings us surprises that have both a melody and a tune. One of them is the love and dedication he put into the design of this book. Not very eloquent, he always works in silence, putting great care in every single aspect. In a time when everyone is averse to impetuousness and empty words, actions speak louder than words. Actually, this attitude has become scarce among our country's people. The most beautiful books are those that let us feel the beauty of reading, not those that rely on vacuous preaching. Actually, the design of each book should unify both surface and meaning, body and soul, to make them both pleasing to the eyes and valuable. This is the melody and tune that the most beautiful books should have.

The text above wants to congratulate this book on the occasion of its publication. Especially, I would like to express my respects to the Shanghai Publication Bureau, which has been organising China's Most Beautiful Book competition and acting as a talent scout for seventeen years. Many designers have achieved success because of this competition, which in turn has contributed to the common

的借口，“世界最美的书”正是一块他山之石，从中汲取世界优秀书卷文化和设计智慧营养为我所用，这正是赵清兄为什么经过要长期周密运筹，设立“世界最美的书”的阅读空间和出版本书的良苦用心吧。

赵清兄经常给我们带来有腔有调的惊喜，其中包含他对这个书籍设计行当的钟爱与痴迷。不善言辞的他总是默默地干活，而且每件事都干得像模像样，值大家反感的浮躁与空谈盛行之时，“干得比说得好”，实乃当今国民稀缺的腔调。“最美的书”让读者得到美的阅读，不靠空洞的说教，实实在在，一本一本地投入。书籍设计要表里合一，形神兼备，赏心悦目，物有所值，正是美书应有的腔调。

以上的文字是对此书即将付梓出版谨表祝贺。这里特别要向17年来始终坚持组织“中国最美的书”评选活动和担当伯乐的上海市新闻出版局深表敬意，许多设计师因此赛事成就了他们的事业，并带动了一大批年轻的设计师们共同进步；还要感谢所有一贯支持中国书籍艺术事业发展的各位国内外的专家评委们，他们不辞辛苦的鼎力付出，开启了中国当代书籍设计联结国际的通道，创造了与外界更多交流与学习的机会；最后要

was den Ton schöner Bücher ausmacht.

Bei dem vorstehenden Text handelt es sich um ein Grußwort anlässlich der bevorstehenden Buchveröffentlichung. Ich möchte an dieser Stelle vor allem dem Shanghai Press and Publication Bureau gegenüber meinen tiefen Dank dafür aussprechen, dass es in den zurückliegenden 17 Jahren die Organisation der Veranstaltungen zur Auswahl der schönsten Bücher in China ermöglicht hat und dabei als Talentförderer wirkte. Zahlreiche Designer sind beruflich erfolgreich aus diesen Veranstaltungen hervorgegangen und haben eine große Anzahl junger Designer bei gemeinsamen Fortschritten unterstützt. Außerdem geht mein Dank an alle Juroren und Fachkollegen im In- und Ausland, die die Entwicklung der chinesischen Buchkunst stets unterstützt haben. Dank ihres Wirkens, bei dem sie keine Mühen scheuten, sind Kanäle geschaffen worden, über die dem zeitgenössischen chinesischen Buchdesign die Verbindung zur internationalen Fachwelt ermöglicht wurde und mehr Möglichkeiten für die Kommunikation mit und das Lernen von der Außenwelt geschaffen wurden. Zum Schluss möchte ich allen Kollegen, die sich mit Begeisterung dem Buchdesign widmen und das Ideal des guten Buches pflegen dafür danken, dass sie im Laufe all dieser Jahre zur Vitalität der schöpferischen Kraft von Chinas Buchkunst beigetragen haben. Sie haben Schwierigkeiten überwunden und Ideen erneuert, damit Schritt gehalten werden konnte mit der Zeit. All ihnen erst ist es zu verdanken, dass sich das chinesische Design in einem vollkommen neuen Antlitz auf der Weltbuchbühne präsentieren kann.

In aufrichtiger Bewunderung für Zhao Qings aufrechterhaltenes Profitum, seine Gründlichkeit, den Pragmatismus, Eifer und Tatendrang gebe ich der Hoffnung Ausdruck, dass dieses tausend Seiten starke wunderbare Buch, auf das so viel Herzblut verwendet worden ist, positive Aufnahme bei dem Fachpublikum findet und eine große große Anzahl von Buchliebhabern erfreut.

Lü Jingren
am 12.März 2019 im Zhuxi Park, Peking

progress of a great number of young designers. I would also like to thank all the foreign and national expert jurors who have consistently supported the development of China's book art. They have worked tirelessly to open a channel that communicates contemporary Chinese book design with the rest of the world, creating numerous opportunities of exchange and learn from different countries. Finally, I would like to thank all the colleagues who have enthusiastically invested in book design embracing the ideal to create good books. These years they have brought great vitality to China's book creation. It has only been thanks to their overcoming difficulties, adopting new concepts, and advancing with the pace of time that nowadays Chinese design can show a completely new appearance on the world book stage.

I sincerely admire Zhao Qing's professionalism, his hard work, and the melody and tune that he approaches his work with. I hope that, the fruit of his effort, this thousand-page beautiful book enjoys the recognition of many colleagues, friends, and the great public of readers who love the art of books.

Lü Jingren
March 12th, 2019 in Beijing's Zhuxiyuan

感谢所有热情投入书籍设计和怀抱做好书理想的同道们，这些年来给中国书籍艺术创作带来了活力，他们克服困难，更新观念，与时代同步，才有今天在世界书籍舞台上展示中国设计的全新面貌。

由衷钦佩赵兄的这种专业坚持、认真务实、用心做事的腔调，期待这本凝聚心血的千页美书得到更多专业知音和广大热爱书籍艺术读者的喜爱。

吕敬人

2019 年 3 月 12 日　于北京竹溪园

Buchdesigner und Illustrator
Professor an der School of Fine Arts der Qinghua Universität
Gastprofessor an der Shanghai Academy of Fine Arts
Mitglied der Internationalen Vereinigung von Grafikdesignern (AGI)
Stellvertretender Direktor der Kommission für Buchdesign der Vereinigung Chinesischer Verlage.
Forscher am Institut für Design und Forschung der Kunstakademie Chinas
Chefredakteur der Serie Buchdesign.
Ausgezeichnet mit dem Nationalpreis als Fortschrittlicher Arbeiter, Sonderzuwendung als Sachverständiger des Staatsrats; wurde als einer der 10 einflussreichsten Designer der chinesischen Buchbindekunst in den vergangenen 50 Jahren ausgezeichnet; einer der 10 maßgeblichen Designer in Asien; einer von 60 Redakteuren, die herausragende Beiträge für die Entwicklung des chinesischen Buches in den vergangenen 60 Jahren geleistet haben; gewann den ersten China Art Achievement Award, den China Design Career Merit Award und den Guanghua Longteng-Preis [Ehrenpreis für 40 Vertreter des chinesischen Designs in der Zeit der vierzigjährigen Politik der Reform und Öffnung].
Zweimalige Auszeichnung der Werke mit dem in Leipzig vergebenen Preis „Die schönsten Bücher der Welt".
Preis in Gold und Silber im Rahmen einer Ausstellung National Book and Binding Art.
Viermalige Auszeichnung für das Design von Büchern der Regierung Chinas
Dreizehn Auszeichnungen für „Schönste Bücher Chinas" sowie weitere Preise in China und im Ausland.
2012 fand im deutschen Klingspor-Museum eine Ausstellung statt, in der das künstlerische Werk Lü Jingrens zur Buchgrafik gewürdigt wurde. 2014 war er Juror der Leipziger Auswahl „Schönste Bücher der Welt"; 2016 wurde im koreanischen Bozhou eine Ausstellung veranstaltet unter dem Titel „Kreativität im Stile der Vergangenheit – die Buchgrafik Lü Jingrens und zehn seiner Schüler"; 2017 fand in San Francisco die Ausstellung „Lü Jingren´s Book Design" statt; die Pekinger Kunstakademie veranstaltete ebenfalls 2017 eine Ausstellung mit dem Titel „Erkundungen in Sachen Buchkunst – 40 Jahre Lü Jingrens Buchdesign"; das Shanghaier Kunstmuseum Liu Haisu widmete Lü im Jahre 2018 eine Einzelausstellung.
Vom Verfasser liegen u.a. folgende Bücher vor: Lü Jingrens Buchdesign, Lü Jingrens Buchdesign Nummer 2, Grundlagen des Buchdesigns, Lehrbuch zum Buchdesign von Lü Jingren, Fragen der Buchkunst, Bauten aus Büchern – Orte der Geschichte, Ehrung von Büchern, Übersetzung und Veröffentlichung von Drehungen – die Welt des Designs bei Kohei Sugiura.

Lü Jingren is a book designer and illustrator. He works as a professor at the Academy of Arts and Design of Tsinghua University and is a visiting professor at the Shanghai Academy of Fine Arts. A member of the Alliance Graphique Internationale (AGI), he is the Deputy Director of the Book Designer Society of the Publishers Association of China as well as a researcher at the Design Institute of the Chinese National Academy of Arts. He is the Chief Editor of *Shuji sheji (Book Design)*.

Lü Jingren has been awarded the National Advanced Workers Medal and the State Council Special Allowance for Specialists. He has been bestowed the title of one of the Ten Designers Influencing the Fifty Years of Chinese Book Design, one of the Top Ten Asian Designers, and one of the Sixty Editors Making Outstanding Contributions to the Sixty Years of New China Books. Lü Jingren has been honoured with the Outstanding Chinese Artist Award, the Chinese Design Career Merit Award, and the Dragon Design Foundation Award (an achievement award given to forty Chinese designers since the Reform and Opening Up).

His works have won different national and international prizes, such as two awards in Leipzig (*The Best Book Design from all over the World*), the National Exhibition of Book Design Golden and Silver Awards, the Fourth China Publishing Government Book Design Award, and the Thirteenth China's Most Beautiful Book Award. In 2012 he held a solo exhibition on book design titled *Die Zukunft des Lesens (The Future of Reading)* at the Klingspor Museum in Offenbach, Germany. In 2014 he was a juror in Leipzig's competition *The Best Book Design from all over the World*. In 2016 he held an exhibition at the Paju Booksori Festival in South Korea titled *Transmission and Creation: Lü Jingren's Book Designs and His Students*. In 2017, he held the exhibition *Lü Jingren: Master of Chinese Book Design* at the Book Club of California in San Francisco, and another one called *Tao of Book Art — Forty Years of Lü Jingren's Book Design* in Beijing's Today Art Museum. In 2018 he held another exhibition in the Liu Haisu Art Museum in Shanghai.

He has published the works *Jingren shuji sheji (Jingren's Book Design)*, *Jingren shuji sheji erhao (Jingren's Book Design II)*, *Shuji sheji jichu (Tao of Book Design)*, *Lü Jingren shuji sheji jiaocheng (Lü Jingren's Book Design Course)*, *Shuyi wendao (Tao of Book Design)*, *Shuzhu — Lishi de chang (The Architecture of Books — A Historical Perspective)*, *Jingren shuyu (Jingren's Words on Books)*, and the Chinese translation of *Xuan — Shanpu Kanping de sheji shijie (Spinning — Kohei Sugiura's Design World)*.

吕敬人
Lü Jingren

书籍设计师、插图画家、清华大学美术学院教授、上海美术学院客座教授、国际平面设计师联盟（AGI）成员、中国版协书籍设计艺术委员会副主任、中国艺术研究院设计研究院 研究员、《书籍设计》丛书 主编。

曾获全国先进工作者奖章，国务院专家特殊津贴；被评为对中国书籍装帧 50 年产生影响的十位设计家之一、亚洲十位设计师之一、对中华人民共和国成立以来书籍 60 年有杰出贡献的 60 位编辑之一；获首届华人艺术成就大奖、中国设计事业功勋奖、光华龙腾“改革开放 40 年中国设计 40 人荣誉功勋奖”。

作品曾 2 次获德国莱比锡“世界最美的书”奖、全国书籍装帧艺术展金银奖、4 度获中国出版政府书籍设计奖、13 度获“中国最美的书”奖等国内外奖项。2012 年 在德国克林斯波书籍艺术博物馆举办“吕敬人书籍设计艺术”个展，2014 年担任 德国莱比锡“世界最美的书”评委，2016 年在韩国坡州举办“法古创新——吕敬人书籍设计与十位弟子展”,2017 年在美国旧金山举办“吕敬人的书籍设计展”及北京今日美术馆举办“书艺问道——吕敬人书籍设计 40 年”，2018 年在上海刘海粟美术馆举办个展。

编著出版《敬人书籍设计》《敬人书籍设计 2 号 》《书籍设计基础》《吕敬人书籍设计教程》《书艺问道》《书筑——历史的场》《敬人书语》，翻译出版《旋——杉浦康平的设计世界》等。

The Choice of Leipzig

The Choice of Leipzig – nur so kann der Titel dieses Buches sein. In zwei Städten dreht sich zweimal im Jahr alles um Bücher: Frankfurt und Leipzig. Aber nur in einer steht die Stadt während der Messetage Kopf für das Buch. In jedem Schaufenster, beim Frühstücksbuffet im Hotel, in Kneipen, Restaurants - überall liegen Bücher. Und nur eine darf sich mit Fug und Recht BuchKUNSTstadt nennen: Leipzig. Dort sitzt die Deutsche Nationalbibliothek mit dem Deutschen Buch- und Schriftmuseum, dort lernt der Buchgestaltungsnachwuchs an der Hochschule für Grafik und Buchkunst und dort wurde er erfunden – der Wettbewerb Die schönsten Bücher aus aller Welt. 1963 vom Ost-Börsenverein ins Leben gerufen, 1991 von der Stiftung Buchkunst übernommen.

Am Wettbewerb nehmen Bücher aus über 30 Ländern teil. Bücher, die in ihrer jeweiligen Heimat zu einem der Schönsten gekürt wurden. Sie reisen nach Leipzig, um sich dort einer internationalen Expertenjury zu stellen. Das ist sicher nicht einfach. Wie beurteilt man Bücher, die man sprachlich nicht versteht? Durch ausgiebige Diskussionen. Gute Gestaltung kennt keine Sprachbarrieren. Gute Gestaltung vermittelt den Inhalt. Gute Gestaltung bereichert den Inhalt. So sitzen und stehen die Designer um lange, schwere Holztische in einem ehrwürdigen Saal in der Deutschen Nationalbibliothek und lassen sich bezaubern und inspirieren. Am Schluss müssen Entscheidungen getroffen werden. Medaillen vergeben, Ehrungen ausgesprochen. Und dann? Dann sollen alle daran teilhaben. Alle Einsendungen werden auf den beiden Buchmessen in Leipzig und Frankfurt präsentiert. Experten und Laien kom-

The Choice of Leipzig

The Choice of Leipzig — only this could be the title of the present book. Twice a year, everything is about books in two cities in Germany: Frankfurt and Leipzig. But only in one of these two cities is completely fascinated about books during the fair. In every shop window, at the hotel's breakfast buffet, in bars, restaurants… everywhere are books. And also only one of these cities deserves to be called "the city of book design" — Leipzig. There are the German National Library and the German Museum of Books and Writing. This is the place where the new generations of book designers study at the Academy College of Graphic and Book Arts. In no other place could the competition *Best Book Design from all over the World* have been invented. Initiated by the East Germany Publishers & Booksellers Association in 1963, the competition was transferred to the Stiftung Buchkunst in 1991.

Books from more than thirty countries and regions take part in the competition. Elected as the most beautiful in their homeland, these books travel to Leipzig to be presented to an international jury of experts. For sure, it is not an easy task. How can one judge books without understanding them? With substantial discussions. Good design has no barriers: a good design transmits the content and enriches it. This way, sitting around long and heavy wooden tables in a venerable room at the German National Library, the jurors become enchanted and inspired. At the end, decisions have to be made — they must award medals and

莱比锡的选择

《莱比锡的选择》——这本书的名字非它莫属。一年两次围着图书转的城市，一共有两座：法兰克福和莱比锡。但全城上下在会展期间为图书忙得不可开交的情况，只会出现在莱比锡。无论是橱窗、酒店的自助早餐桌、酒吧还是餐厅，都摆满了图书。也只有这座城市，才有充分的理由以"书籍艺术之城"自居。莱比锡是德国国家图书馆及其所辖的德国图书和文字博物馆所在地，莱比锡平面设计及书籍艺术学院是图书设计后备人才的培养重镇，"世界最美的书"大赛也诞生于此。这项赛事1963年由东德交易所协会创办，1991年起由德国图书艺术基金会正式接手。

参赛的图书来自30多个国家和地区，它们都是各国精美图书中的佼佼者。它们来到莱比锡，接受国际专家评审团的检阅。这个过程并不容易。如果对书中的文字一窍不通，又该如何评判这些图书呢？答案是反复讨论。好的设计不受语言的限制，只会反映内容、丰富内容。所以，在德国国家图书馆庄严肃穆的大厅里，一群设计师围坐在敦实的长桌旁，努力地陶醉其中，希望从中得到启示。最后，他们必须做出评判：颁发奖项，表达敬意。然后呢，接下来的环节有赖于所有人的参与。所有参赛作

men an den Stand der Stiftung Buchkunst, um sich durch die unterschiedlichen Exponate inspirieren zu lassen. Miteinander über die Auswahl der Jury zu diskutieren, anderer Meinung zu sein, oder ebenso überzeugt von Konzept und Umsetzung.

Die Preisverleihung, immer freitags auf der Leipziger Buchmesse – ein internationales Buchkunsttreffen. Designer aus China treffen auf Designer aus Litauen, auf Gestalter aus den Niederlanden, der Schweiz, Japan, Venezuela … Alle haben nur ein Thema: Schöne Bücher. Jedes Jahr erstaunt es uns erneut welche Strecken die Gewinner auf sich nehmen, um ihre Urkunde abzuholen. Und jedes Jahr erfüllt es die Stiftung Buchkunst mit Stolz. Stolz darauf, dass wir gemeinsam mit vielen, vielen Gestaltern aus aller Welt dazu beitragen, die Bedeutung guter Gestaltung zu kommunizieren.

Danke an alle Buchkünstler!

Danke an Zhao Qing für „The Choice of Leipzig“. Es zeigt auf jeder Seite seine Leidenschaft für Buchdesign.

express words of praise. And then? Then everyone must take part in it. All the books participating in the contest are presented in the book fairs of both Frankfurt and Leipzig. Experts and laymen come to the stand of the Stiftung Buchkunst to be inspired with the different exhibition items. They discuss with each other about the decision of the jury, have another opinion, or are just as convinced about the concept and its implementation.

The presentation of awards, which takes place always on Friday in the Leipzig Book Fair, is an international encounter of book design. Chinese designers meet designers from Lithuania, the Netherlands, Switzerland, Japan, Venezuela··· They all talk about the same subject: the best book designs. Every year we are surprised by the long distances that the winners have to travel in order to get their certificate. And every year, the Stiftung Buchkunst is proud to make it happen again. We are proud that we can contribute to communicate the significance of good design with many designers from all around the world.

To all book designers, thank you!

I would also like to thank Zhao Qing for *The Choice of Leipzig*. It shows passion for book design on every single one of its pages.

品都将在莱比锡和法兰克福书展上展出。无论是专家还是外行人士，都能在图书艺术基金会的展台上欣赏五花八门的参展作品。他们相互讨论评委会的选择，表达不同的意见，也对这些书籍的设计理念和最终效果赞不绝口。

颁奖仪式固定于某个周五在莱比锡书展上举行。在这场书籍艺术的国际盛会中，来自中国的设计师遇见来自立陶宛的设计师，或是来自荷兰、瑞士、日本、委内瑞拉等地的同行，所有人都只谈论一个话题：精美的图书。前来领取证书的获奖者往往都历经长途跋涉，每年都令我们惊叹不已。每年的活动，也都让图书艺术基金会由衷地感到骄傲。能与众多来自世界各地的设计师一道推广优秀的书籍设计，我们深感荣幸。

在此，谨向所有书籍艺术家致谢！

感谢赵清设计的这本《莱比锡的选择》。这本书的每一页，都体现出他对书籍设计的热爱。

Katharina Hesse arbeitet seit 1998 in unterschiedlichen Funktionen mit dem „schönsten“ Produkt der Welt, als Buchhändlerin, als Organisationsleiterin am mediacampus frankfurt, dem zentralen Ausbildungsinstituts der Buchbranche und als Geschäftsführerin des ausschließlich E-Books verlegenden e-Lectra Verlags. Seit 2013 ist sie die Geschäftsführerin der Stiftung Buchkunst mit Sitz in Frankfurt und Leipzig. Die Stiftung Buchkunst verantwortet sowohl den Wettbewerb Die Schönsten deutschen Bücher, sowie seit 1991 den weltweit einzigartigen internationalen Buchkunstwettbewerb Schönste Bücher aus aller Welt.

Katharina Hesse
卡塔琳娜 · 黑塞

Since 1998, Katharina Hesse has been working in different functions with the most beautiful products of the world as bookseller; director of Mediacampus Frankfurt, the central education institution of the German publishing industry; and director of the e-Lectra Verlag, which exclusively publishes e-books. Since 2013, she is the managing director of the Stiftung Buchkunst, located in Frankfurt and Leipzig. This organization is responsible for the competition Most Beautiful German Books and, since 1991, for the international competition *Best Book Design from all over the World*.

自 1998 年以来，卡塔琳娜 · 黑塞就以不同的方式，与世界上最美的事物打着交道：她做过书商，当过德国书商学院的活动负责人，还曾担任过纯电子书出版社“e-Lectra”的总经理。2013 年起，她开始担任总部位于法兰克福和莱比锡的图书艺术基金会理事长。该基金会不仅是“德国最美的书”大赛的主办方，也从 1991 年起接手另一项国际书籍艺术竞赛——“世界最美的书”大赛。

Geschrieben anlässlich der Veröffentlichung von
„Die Leipziger Auswahl"

Written on the occasion of the publication of *The Choice of Leipzig*

„Die schönsten Bücher der Welt" ist ein sehr wichtiger Designwettbewerb für die Kreise des Verlags- und Buchdesigns in China. Die ausgezeichneten Buchgestaltungsarbeiten in diesem Wettbewerb haben viel Aufmerksamkeit erregt.

Der Designer Zhao Qing ist ein Liebhaber des Designs und vor allem des Buchdesigns. Im Jahre 2014 hat er damit begonnen, aus Sicht eines Forschers und Designers systematisch die Werke der Leipziger Reihe „Die schönsten Bücher der Welt" zu sammeln und zu erforschen. In den vergangenen Jahren hat er bei seinen Flügen nach New York immer wieder mit viel Aufwand Werke für seine Sammlung in der Han Qing-Galerie erworben.

Jetzt ist der Zeitpunkt gekommen, da das umfangreiche Werk, das das Ergebnis der Forschungsarbeiten Zhao Qings darstellt, vor der Veröffentlichung steht. Ich bin der Auffassung, dass dieses Buch einen sehr wichtigen Einfluss auf die chinesischen Buchdesigner und alle Buchliebhaber haben wird. Es handelt sich dabei um ein wegweisendes Werk, durch das wir die Welt sehen und die Welt uns versteht.

Die Hauptaufgabe der chinesischen Designer aus unserer Generation besteht darin, innerhalb der Geschichte des chinesischen Buchdesigns eine Art Brückenfunktion zwischen den Zeiten und Generationen auszuüben: Einerseits sollen Mittel und Formen des Lernens und des Austausches aus der Vorgängergeneration erhalten bleiben und nicht wieder verloren gehen; andererseits sollen Erkenntnisse und Erfahrungen aus unserer Generation auf unterschiedliche Art und Weise an die kommende Generation weitergegeben werden. Das ist der Sinn der Veröffentlichung von „Die Leipziger Auswahl".

Selbstverständlich ist mir klar – und ich sage das mit einem großen Bedauern –, dass der Wettbewerb „Die schönsten Bücher der Welt" im Anschluss an die sechste Sitzung der „Nationalen Ausstellung der Designkunst" hätte stattfinden sollen. Damals, im Jahre 2004,

The Best Book Design from all over the World is a very important design competition for the Chinese editorial world and the circles of book design. The book designs that win an award in this competition are given a great deal of attention.

Because of his fondness for books and book design, in 2014 the designer Zhao Qing started to systematically collect materials and research the Leipzig's competition *The Best Book Design from all over the World*. He did so from the perspective of a researcher and designer. During these years, every time he has been to New York he has considered himself a porter, carrying even an entire overweight suitcase to his studio Hanqing Hall.

Now, Zhao Qing's heavy research book is on the eve of publication. I trust that this book is going to have a great influence in Chinese book design as well as among the whole collective of book lovers. It will be like a coordinate system that will allow us to see the world while, at the same time, allowing the world to understand us.

The main role of our generation of designers in China's history of book design is to continue the tradition and inspire those who will come. Continuing the tradition means putting effort in keeping an open door towards the pioneering work of previous generations and allowing access to it. Inspiring the new generations involves to leave the different styles that we have slowly learnt through experience for our successors. Such is the significance of the book *The Choice of Leipzig*.

写在《莱比锡的选择》付梓之际

“世界最美的书”对中国出版界和书籍设计圈来说是非常重要的设计比赛。在这个比赛中获奖的书籍设计作品备受关注。

设计家赵清因爱设计、爱书，更爱书的设计，从一位研究者和设计师的角度出发，自 2014 年开始对莱比锡“世界最美的书”进行系统收藏和研究。这几年，他每次到纽约都把自己当成搬运工，整箱地甚至超重地往瀚清堂搬运。

现在，赵清先生这本厚重的研究成果即将出版，我相信，这本书对于中国的书籍设计师乃至整个爱书群体，都会产生非常重要的影响。它就像个坐标系，通过它，我们看到了世界，世界了解了我们。

我们这一代设计师，在中国书籍设计历史中的主要作用是承上启下。承上：把上一代开启的学习交流之门努力保持住，不让它再次密不透风；启下，把我们的点滴心得用各种方式留给下一代。这就是《莱比锡的选择》出版的意义。

说来惭愧，我知道“世界最美的书”比赛，应该是在第六届全国装帧设计艺术展之后了。当年，在六展上获得金奖的作品“打

gewann das Werk „Verpackung" den ersten Preis im Rahmen der 6. Veranstaltung von „Das schönste Buch Chinas", mit dem dann China im Wettbewerb „Schönste Bücher der Welt" vertreten war. Erst damals wurde mir klar, dass es diesen Wettbwewerb überhaupt gibt und dass die Teilnahme für Vertreter aus dem Publikationsgewerbe zwar kostenlos, aber dennoch überaus wichtig ist.

Es gibt zahlreiche Grafikdesign-Wettbewerbe auf der Welt, die berühmtesten sind u.a. ADC (New York), One Show (New York), TDC (New York), Tokyo TDC (Tokio) und D & AD (London). Hierbei handelt es sich um Wettbewerbe, mit denen eine breite Palette abgedeckt wird, Bücher sind nur ein kleiner Teil davon. Bei „Schönste Bücher der Welt" aus Leipzig handelt es sich um etwas vollkommen anderes: es ist ein sehr einfacher Buchgestaltungswettbewerb, teilnehmen kann nur das jeweilige Land, die Bewerbung individueller Designer ist ausgeschlossen. Obwohl es Unterschiede bezüglich des Verständnisses und des Ausdrucks von Buchdesign unter Designern verschiedener Kulturen gibt, bietet die Wettbewerbsplattform „Schönste Bücher der Welt" einen überprüfbaren globalen Ansatz für das Buchdesign.

Wenn man sich das Leipziger „Die schönsten Bücher der Welt" ansieht, dann bietet sich damit meiner Meinung nach ein sehr wichtiger Bezugsrahmen für unsere chinesischen Verlage und Buchgestalter – es ist wie ein Fenster, das uns den Blick auf die Designerkollegen und ihr Werk in anderen Kulturen eröffnet. Darüber hinaus stellt das Leipziger Werk ein Mittel dar, sich über die Anliegen und Themenvielfalt der Buchdesigner auf aller Welt zu informieren und die bestehenden Distanzen in der Entwicklung auszumachen. Auf der anderen Seite ist „Schönste Bücher der Welt" kein Bewertungssystem, mit dem das gesamte Grafikdesignwesen erfasst wird, daher könnte der vom Verfasser für dieses Buches gewählte Titel „Die Leipziger Auswahl" für die schönsten Bücher der Welt nicht treffender sein.

Da das Leipziger Werk für chinesische Designer ein Fenster darstellt, sich über die Designerkollegen in anderen Kulturen zu informieren, liegt seine Bedeutung auf der Hand. In einem Raum ohne Fenster ist es feucht, es gibt Ungeziefer

To be honest, I didn't get to know the competition *The Best Book Design from all over the World* until after the Sixth Edition of the National Exhibition of Book Design in China. That year, the work that was awarded the golden medal was also honoured with the title of China's Most Beautiful Book of 2004 and represented our country in Leipzig's competition. That was the first time I knew such a competition existed. Besides, it was also very important that the publications that entered the competition didn't have to pay a fee.

There are numerous graphic design competitions around the world, like the famous ADC, The One Show, and TDC in New York; TDC Tokyo; D&AD in London··· But these competitions are general for graphic design, so books are only a part of them. Leipzig's *The Best Book Design from all over the World* is not like them because it is purely about book design and also because it is only possible to participate in representation of a country, so it is not accepted that designers themselves apply. It could be said that, even if designers coming from different cultural backgrounds have different understandings on book design, *The Best Book Design from all over the World* has provided a global perspective on book design.

Taking Leipzig's The Best Book Designs into general consideration, I think that it has provided us, Chinese publishers and book designers, a crucial frame of reference. It has become a window through which we can see designs by colleagues from other cultures and an information chain to get to know the questions that book designers from different countries face. But it is not the whole system for evaluating graphic design. Therefore, choosing *The Choice of Leipzig* as a book title was the most appropriate because

包”获得了2004年度“中国最美的书”，将代表中国参加“世界最美的书”比赛。此时，我才知道还有一个这样的比赛，并且是出版物比赛中不收费但却是非常重要的。

世界上平面设计类的比赛很多，比较著名的有ADC（纽约）、ONE SHOW（纽约）、TDC（纽约）、Tokyo TDC（东京）、D&AD（伦敦）……这些比赛都是大平面范围的，书籍是其中一部分而已。莱比锡“世界最美的书”与上述比赛不同，它是非常单纯的书籍设计比赛，并且只能以国家的形式参与，不接受设计师个人报名。可以说，尽管不同文化的设计师对书籍设计的理解和表达存在差异，但“世界最美的书”这个比赛平台，还是为书籍设计提供了一个检验的全球视角。

综观莱比锡“世界最美的书”，我认为它给我们中国的出版人和书籍设计师提供了一个很重要的参照系：它是我们看其他文化中设计同行的一扇窗；了解不同国家书籍设计师所关注的问题的信息链；解读书籍设计前沿的测距仪。但它又不是平面设计评价体系的全部，因此，用本书作者为本书的命名来解读“世界最美的书”，称之为“莱比锡的选择”是再恰当不过的了。

既然它是中国设计师看其他文化中设计同行的一扇窗，其重要性自是不言而喻的。在没有窗的房间里一定是潮湿阴暗蛆虫翻

und Pflanzen tun sich schwer mit dem Wachstum. Sobald man das Fenster öffnet, gibt es frische Luft und Sonnenschein und Pflanzen und Blumen kommen zum Blühen. Unter den Angehörigen der Generation unserer Vorgänger, d.h. den in den 1920-1930er Jahren geborenen Buchdesignern (bzw. Buchgestaltern), befanden sich zwar zahlreiche Talente, doch da sie in Räumen abgeschottet von der Welt lebten, blieben ihnen die Entwicklungen im Design draußen in der Welt weitestgehend unbekannt (vom täglichen akademischen Austausch ganz zu schweigen) – man kann das nicht anders als eine Tragödie bezeichnen, die das für viele, verursacht von den Zeitumständen, gewesen ist! Dabei waren sie, das macht es im Grunde noch viel schlimmer, Zeigenossen von Meistern wie Emil Ruder, Paul Rand, Josef Müller-Brockmann und Kohei Sugiura! Die Generation, die nach 1940 geboren wurde, war da viel glücklicher. Ihr war es möglich, mit offenen Augen die Welt zu sehen, sich anderswo Wissen anzueignen, in einen Dialog mit der Welt zu treten, die traditionelle Ästhetik zu beerben, eine Sprache für das zeitgenössische chinesische Buchdesign zu schaffen und dabei enormen Einfluss auf das chinesische Verlagswesen zu gewinnen wie etwa Lü Jingren. Aus der Sicht der Entwicklung der gesamten Grafikdesign-Industrie wird deutlich, dass erst, als das Fenster geöffnet wurde, Chinas Stimme im Gewerbe des weltweiten Grafikdesigns erklingen und China ein Mitglied in der internationalen Vereinigung für Grafikdesign AGI werden konnte... Zu den Veränderungen in den vergangenen 20 Jahren bezüglich der Konzeption und Praxis des Buchdesigns und der Buchgestaltung in Festlandschina konnte es also nur deshalb kommen, weil Fenster und Türen geöffnet wurden. Ohne diesen Schritt wären nicht 22 Werke chinesischer Designer in Leipzig mit Preisen als schönste Bücher der Welt ausgezeichnet worden und läge China damit nicht international auf Rang 4.

„Die schönsten Bücher der Welt“ bildet eine Kette von Informationen in Bezug auf das, was den Buchdesignern aus verschiedenen Ländern am Herzen liegt. Von Beginn an kamen damit die nicht abbrechenden Veränderungen zum Ausdruck, denen die Gestaltung

it explained the nature of *The Best Book Design from all over the World*.

Since this competition is a window through which Chinese designers can see the works of colleagues from other cultures, its importance goes without saying. A room without windows is humid, dark, and full of maggots. It would be even difficult to grow plants in there. When a window opens, there is fresh air and light coming in, which can nurture a tree that grows tall and reaches the sky. Even if our forbearers were brimming with talent, these designers born around the 1920s and 1930s worked in in what 20th Century writer Lu Xun called "an iron room". They did not know anything of the changes that design was experimenting on the other side of the window, let alone the exchanges between people in the field. This was really a tragedy that that epoch brought upon individuals. These were great masters belonging to the same epoch of Emil Ruder, Paul Rand, Josef Muller-Brockmann, and Kohei Sugiura! Our predecessors who were born after 1940 were much luckier, they "looked with open eyes" — as Lu Xun would also say — and had the change to go out to study. This way, they could join the artistic currents of the world, pass on traditional aesthetics, and create the contemporary vocabulary of Chinese book design. This was the case of Lu Jingren; who had a tremendous influence in China's publishing sphere. Regarding the whole domain of graphic design, the rest of the world could only hear the voice of Chinese graphic designers after the window was open. It was only after the window was open that China became a member of Alliance Graphique Internationale (AGI) ··· Therefore, during the last twenty years there have been changes in both the concept and practice of book design in China. So if Leipzig has chosen a book to be the fourth among twenty two in *The Best Book Design*

涌，哪怕是盆栽植物也难生长。打开窗，就有新鲜的空气和阳光，就能孕育出参天之树。我们的先辈，出生在 20 世纪 20、30 年代的书籍设计师们（装帧设计师），尽管很多人都才华横溢，就是因为处在封闭的铁屋子里，对窗外设计界发生的变化几无所知，日常的学术交流更无从谈起，这真是时代带给生命个体的悲剧！要知道，他们可是与埃米尔 · 鲁德（Emil Ruder）、保罗 · 兰德（Paul Rand）、约瑟夫 · 米勒－布罗克曼（Josef Müller-Brockmann）、杉浦康平等巨匠同时代！同样是设计先辈，20 世纪 40 年代以后出生的则幸运得多，他们有了睁眼去看、走出去学习的机会，因而就能出现对接世界潮流，传承传统审美，创造当代中国书籍设计语汇，对中国出版产生巨大影响的吕敬人先生。从整个平面设计行业来看，只有打开窗，世界平面设计也才会有来自中国平面设计界的声音，才能出现中国的 AGI（国际平面设计联盟）会员……因此，中国这 20 年来书籍设计从观念到实践所发生的变化，书籍设计师在莱比锡所获得的 22 本“世界最美的书”奖和按国家排名第 4 的成绩，正是因为开窗破门。

“世界最美的书”是了解不同国家书籍设计师所关注问题的信息链。书籍从它出现起，对于书的设计无论是方法论、制书工

von Büchern egal ob Methodik, Buchtechnik oder Ausdrucksweise unterworfen ist. Es aktualisiert sich nicht nur ständig das Konzept von Design, auch die Technologie zur Buchherstellung wird nach und nach aufgrund von wissenschaftlichen und technologischen Fortschritten verbessert. Was nun das zeitgenössische chinesische Verlagswesen betrifft, so ist es, angefangen beim Konzept des Verlagswesens bis hin zum Prozess der Buchherstellung im internationalen Zusammenhang durchaus nicht ohne Einfluss geblieben. Die Publikationen im chinesischen Altertum wurden durch den Prozess der Herstellung von Büchern eingeschränkt, was es schwierig machte, aus der engeren Welt der Gelehrten herauszugelangen. Erst als zum Anfang des 19. Jahrhunderts die von Johannes Gutenberg im 15. Jahrhundert erfundene Druckkunst in das China der Qing-Dynastie gelangte, gelangten Zeitungen und Bücher in die Haushalte, sodass der gesamte Zivilisationsgrad des chinesischen Volkes einen riesigen Schritt nach vorne machte. Diese aus der fortschrittlichen Druckerei und Buchherstellung hervorgegangen Werke stellten in ästhetischer und technologischer Hinsicht eine vollkommene Unterwanderung der entsprechenden Entwicklungen in China während der vorangegangenen tausend Jahre dar. Man war gezwungen, Publikationen mit chinesischen Schriftzeichen entsprechend den neuen Buchformen zu verändern und das chinesische Verlagswesen nach und nach in die Welt zu integrieren. Buchgestaltung ist ein herausragendes Beispiel für unsere Integration innerhalb der ganzen Welt. Wang Jiaming sagte auf der neunten Sitzung der „National Book Design Art Exhibition“ und des dritten internationalen Designer-Forums „Book Beauty“: „Es ist eine unbestreitbare Tatsache, dass die Entwicklung des Buchdesigns der Entwicklung von Buchinhalten vorangegangen ist.“

Jetzt, da das Buchdesign in die Welt integriert ist, sind uns die Anliegen der Buchdesigner in verschiedenen Ländern natürlich ein Anliegen. Zum Beispiel die materielle Natur von Büchern, die Methodik der Buchbearbeitung, die Erforschung der Buchform, die Ästhetik des Layouts von Büchern usw. „Die schönsten Bücher der Welt“ bildet ein wichtiges Glied in der Kette von Informationen innerhalb des Gewerbes des Buchdesigns. Die

from all over the World, it must necessarily be because we have opened a window and burst a door open.

The Best Book Design from all over the World is an information chain that allows us to know the questions that book designers from different countries face. From the moment books appeared, whether the methodology, the craft of book binding, or the outer appearance of book design have been continuously changing. This is both fruit of a constant evolution in the concept of design and the advances in science and technique, which have progressively improved the art of book manufacturing. Regarding China's contemporary editorial scene, from the concept of publishing to the manufacturing of books, there is nothing that has not left a trace of the progress of humanity. The overwhelming majority of Chinese ancient publications have always been restricted to the craft of book binding, which prevented them from going out of the studios of Confucian scholars. In the 19th Century, the printing press invented by Johannes Gutenberg in the 15th Century entered the Qing Dynasty[1]. Finally newspapers and books could enter every household and contribute to enlighten Chinese people. These first printed books overturned a thousand years of Chinese book aesthetics and crafts and forced the publications in Chinese characters to adapt to the new form of books, making Chinese publishing gradually integrate with publishing around the world. Book design is an outstanding representation of how China got in contact with the rest of the world. On

1 Su Jing, *(2018) Zhu yi dai ke: shijiu shiji zhongwen yinshua bianju (Inlaying the Inscription: The Changes in Chinese Printing in the 19th Century)*, chapter one. Zhonghua Book Company.

艺还是表现形式，就一直处在变化之中；它既是设计观念的不断更新，也是伴随科学、技术进步而来的书籍生产工艺的逐步提升。拿中国当代出版来说，从出版理念到制书工艺，无不留下人类文明的进步脚印，而不是小楼里的春夏秋冬。中国古代出版物的绝大部分都受到制书工艺的制约，使它很难走出儒生的书斋。直到 19 世纪初由 15 世纪的德国人约翰内斯 · 谷登堡（Johannes Gutenberg）发明的印刷术传入清朝[1]，报纸和书籍才走进千家万户，使中国人的整体文明程度得到几何级提升。这些先进的印刷术印装的书籍，是对 1000 多年来中国书籍美学和印制工艺的颠覆，它迫使汉字出版物必须依据新的书籍形制做出改变，也使得中国出版逐步融入世界出版。书籍设计，就是我们融入世界的杰出代表。汪家明先生在第九届全国书籍设计艺术展暨第三届“书籍之美：国际设计师论坛”上说：“书籍设计的发展走到了书籍内容发展的前面，倒是一个不争的事实。”

既然书籍设计已经融入世界，不同国家书籍设计师所关注的问题自然也会成为我们所关心的问题。比如书籍的物质性，比如

1　《铸以代刻——十九世纪中文印刷变局》，第一章，苏精 著，中华书局，2018 年。

„Leipziger Auswahl“ ermöglicht uns tiefe Einblicke sowohl auf einer makroskopischen Ebene wie auf der Ebene der subtilen Analyse von Details. Damit verbunden ist die Hoffnung, dass chinesische Publikationen insgesamt davon profitieren, dass sie mit neuen ästhetischen Merkmalen ausgestattet werden und dass sich das Buchdesign kontinuierlich weiterentwickelt.

„Die schönsten Bücher der Welt“ ist ein Entfernungsmesser, mit dem die Grenzen der Buchgestaltung ausgelotet werden. Die Geschichte der Bücher ist eine Geschichte, die auf der Entwicklung von Texten und der damit verbundenen ständigen Aktualisierung der Konzepte des Designs sowie der Erneuerung entsprechender Technologien basiert. Von einem Stück Schaffell bis hin zu einem Träger für einen komplexen Textstil, von einer durch Schmuck und Ausstattung verschönerten Ausgabe bis hin zur reinen Verwendung von Schriftzeichen und Buchstaben zur Ermöglichung einer poetischen Lektüre – das Buchdesign hat in seiner Herkunftsregion, dem Westen, den Prozess der Erkundung des immer Neuen nie unterbrochen. Mit diesem Entfernungsmesser wird der Abstand für uns deutlich: Die überwiegende Mehrheit der Bücher in den Buchhandlungen gehört in die Zeit vor 1800. Der Fluss komplexer Texte; die Trennung zwischen Schriftauswahl, Textzusammenhängen und Lesegefühl... all das zeigt uns eine grausame Realität: Wir sind noch weit entfernt von einer fortschrittlichen Kultur für das Buchdesign, denn die Grundlagen für den Fortschritt sind auf losem Sand gebaut. Angesichts der riesigen Anzahl von Publikationen in China lässt sich zwar in vereinzelten Fällen von einem fortschrittlichen Buchdesign sprechen, doch selbst zwischen dieser und tatsächlicher Fortschrittlichkeit gibt es immer noch eine messbare Distanz. Chinas riesiger Publikationsmarkt und das einzigartige Verlagsumfeld zeichnen sich vor allem dadurch aus, dass ein Markt für preiswürdige Bücher geschaffen wurde, dies ist, um ehrlich zu sein, das Ergebnis einer unvollständigen Marktentwicklung: Es besteht ein Missverhältnis zwischen dem Aufwand für den Einsatz vergleichsweise fortschrittlicher Arbeitskraft und dem Ergebnis. Zum Einsatz kommt eine große Anzahl von Handwerkern oder von Handwerkern im Verbund mit Maschinen beim Buchdesign bzw. in der Technologie zur Buchherstellung. Designer aus den USA,

occasion of The Beauty of Books: International Designer Forum during the Ninth National Book Design Art Exhibition, publisher Wang Jiaming said, "Book design has developed until surpassing the content of books — this is undeniable".

Since book design entered the world, the questions that designers from different countries face have naturally become the ones we also care about, like the material nature of books, the methodology of book design, the exploration on book shapes, the aesthetics of publication··· The Best Book Designs is an important link in the information chain on book design, so *The Choice of Leipzig* has allowed us to have a glimpse into the depths of the leopard's markings and conduct a general research on them, while, at the same time, analysing and deconstructing it to the smallest details. I wish that all the Chinese publications that are built upon the foundation that this book provides possess new aesthetic characteristics and never stop developing design.

The Best Book Design from all over the World is an advanced equipment to measure and decipher book design. Along with the evolution of writing, the history has brought a constant renewal of design concepts and new techniques, from parchment to the most complex forms of books, from relying on decoration to embellish them to using typography in order to create a poetic flavour in reading. In the West, its homeland, book design has never stopped exploring new territories. With such advanced equipment, we can clearly see the gap: the vast majority of books in bookshops belong to the printing before 1800, but the day-to-day account of complicated texts spreads out. The selection of the typeface and the content of works, as well as the feeling of reading them, are coming apart. All these circumstances are revealing an ugly truth: we are still far

书籍的编辑设计方法论，比如对书籍形态的探索，比如书籍的版面美学……“世界最美的书”就是书籍设计信息链上的重要一环，《莱比锡的选择》使我们得以一窥这头豹子的斑纹深处，在对它进行宏观研究的同时又有入微的分析解构，希望借此在整体上逐渐形成属于汉字出版物的、具备新美学特征并且不断发展的书籍设计。

“世界最美的书”是解读书籍设计前沿的测距仪。书籍的历史就是依据文本的演进而带来的设计观念不断更新并与新工艺相结合的历史。从一张羊皮到拥有复杂的文本体例，从依靠装饰来美化版面到仅用文字排版就能营造诗意阅读，书籍设计在它的“祖国”西方，从来没有停止探索的脚步。有了这个测距仪，我们的差距就变得清清楚楚：书店里绝大多数的书籍属于1800年前的居中版心，复杂文本的流水账式的罗列，字体选择和文本内涵以及阅读感受的脱节……凡此种种，无不向我们展现一个残酷的现实：我们相距书籍设计文化的先进，还有非常漫长的距离，因为我们的先进的基础是建立在沙滩上的。对于中国数量庞大的出版物来说，经过书籍设计师整体设计的书籍属于凤毛麟角的先进，但是这先进比起真正的先进，仍然有可以丈量的距离。中国庞大的出版物市场和独特的出版环境，

Europa und Japan blicken diesbezüglich voller Neid auf uns. Das mag zwar in dieser Phase ein Vorteil für uns sein, es ist jedoch zu etwas wie einem Symbol der chinesischen Buchgestaltung geworden, dessen Bedeutung darin liegt, der starken Kultur des Westens etwas entgegenzusetzen. Erst wenn es gelingt, die Kreativität der Designer unter strenger Berücksichtigung für die Herstellung von Büchern zu bewerkstelligen und innerhalb der begrenzten Möglichkeiten eines Layouts mit Hilfe der redaktionellen Gestaltung und logischer Anordnung Lesbarkeit zu ermöglichen, um u.U. sogar mittels eines monochromen Drucks in Form von Low-Cost-Büchern (vgl. das Beispiel: The Best Book Designs 2015, Buchdesign: Irma Boom) die Position des Designers klar zu artikulieren, erst dann wird sich unser Buchdesign unter den vordersten Rängen einordnen können. Bis dahin ist es für uns noch ein weiter Weg!

Danke, Herr Zhao Qing, dass Sie uns „Die Leipziger Auswahl" mitgebracht haben. Es ist ein erster Schritt. Ich bin davon überzeugt, dass es mit einem Referenzrahmen wie „Die schönsten Bücher der Welt"und seinen vielfältigen Informationen und Maßstäben möglich wird, dass auch unser Buchdesign einst auf die ersten Ränge in der internationalen Buchwelt gelangt.

Liu Xiaoxiang
am 14.März 2019 im XXL Studio

away from the vanguard of book design. There is still a very long journey to go, because the foundations of our advances are built on the sand. Regarding China's enormous amount of publications, the books that undergo a full design process are an advanced minority, but between this minority and the rest there is still a considerable distance. China's great publishing market and its unique publishing environment have allowed to create award-winning books at the expense of the industry. This is the result of an incomplete market development: we cannot compare skilled labour force to the great amount of manual or semimanual work in book design and the craft of book binding. The form of books, their binding, and the quality of their materials are what makes us admire the designers from Europe, the United States, and Japan. The significance of our book design is that it has a strong cultural foundation that is different from the one in the West. This is our current advantage, and it has also become a symbol of Chinese books. A designer's creativity is only revealed under the critical constraints of the cost of a book. In the limited space of a page, we have to rely on editing, design, and logic to rebuild the reading of a book. We even depend on the low cost of a monochrome book (The Best Book Design 2015, designed by Irma Bloom) to clearly elaborate on the designer's standpoint. So our book design has just entered the vanguard. As we see it, there is still a long way to go!

I would like to thank Zhao Qing to bring us *The Choice of Leipzig*. Step by step, one goes as far as a thousand li. Therefore, with the frame of reference and the information chain of The Best Book Designs, I am confident that our book design is going to stand in the spotlight.

Liu Xiaoxiang
Written on March 14th, 2019 at the XXL Studio

为不惜工本打造获奖书籍提供了市场，平心而论，这是市场发育不完全的结果：我们用相比先进的劳动力不等值，换来了大量的手工、半手工的书籍设计和造书工艺，在书籍的形态、装订和材质上，让欧洲、美国、日本的设计师们羡慕不已。这虽然是我们现阶段的优势，但也似乎成了代表中国书籍设计的符号，存在的意义就是为强势文化的西方提供不同。只有在受做书成本严重制约的前提下实现设计师的创意，在有限的版面上依靠编辑设计和逻辑能力重构文本的阅读方式，甚至，靠一本单色印刷的低成本书籍（*THE BEST BOOK DESIGNS 2015*，书籍设计：Irma Boom）就能清晰阐述设计师立场的时候，我们的书籍设计才真正进入了先进行列。这对于我们来说，仍然任重而道远！

感谢赵清先生为我们带来《莱比锡的选择》。积跬步才能至千里，有了“世界最美的书”这个参照系、信息链和测距仪，我相信：我们的书籍设计终将会站到书籍舞台的前排。

刘晓翔

2019 年 3 月 14 日　于 XXL Studio

Mitglied der Alliance Graphique Internationale (AGI)

Senior Editor & Chief Book Designer von Higher Education Press

Art Director des XXL-Studios

Vorsitzender des Buchgestaltungsausschusses der Publishers Association of China

Auszeichnungen

Gewinner des Leipziger Preises „Beste Buchdesigns aus aller Welt“ in den Jahren 2010, 2012 und 2014

Gewinner des Preises „Beauty of Books in China Award“ (vierzehn Mal zwischen 2005 und 2015).

Gewinner des koreanischen Preises „Paju Book Award-Book Design Award“ (Achievement Award) 2013

Gewinner des Preises „Chinese Government Award for Publishing - Book Design Award“ in den Jahren 2013 und 2016

Gewinner des Gold Award der „National Exhibition and Appraisal of Book Design in China“ in den Jahren 1999 und 2004.

Designauszeichnung Best of Golden Pin im Jahre 2017

Jährliche Auszeichnung Tokio TDC im Jahre 2018

Auszeichnung NY ADC 97. Bronze Cubes

NY TDC 64 Best in Show

Kommunikationsdesign NY TDC 64 TDC Gewinner im Jahre 2018

Werke

Vom Schriftzeichen zum Buch: Chinesische Typografie

11×16 XXL Studio

Member of the Alliance Graphique Internationale (AGI)

Senior Editor & Chief Book Designer of Higher Education Press

Art Director of XXL Studio

Chairman of the Book Design Committee of the Publishers Association of China

Awards

Winner of the Leipzig "Best Book Design from all over the World" Award in 2010, 2012 and 2014

Winner of the "Beauty of Books in China Award" for fourteen times between 2005 and 2015

Winner of the Korean "Paju Book Award-Book Design Award" (Achievement Award) in 2013

Winner of the "Chinese Government Award for Publishing-Book Design Award" in 2013 and 2016

Winner of the Gold Award of the "National Exhibition and Appraisal of Book Design in China" in 1999 and 2004

Best of Golden Pin Design Award 2017

Tokyo TDC Annual Awards 2018

NY ADC 97th Bronze Cubes Awards

NY TDC 64 Best in Show

NY TDC 64 TDC Communication Design Winners 2018

Works

From a character to a book: Chinese Typography

11×16 XXL Studio

刘晓翔
Liu Xiaoxiang

刘晓翔工作室（XXL Studio）艺术总监，高等教育出版社编审、首席设计，中国出版协会装帧艺术委员会主任委员，国际平面设计联盟(AGI)会员。

获奖

2010 年、2012 年、2014 年 3 次获得德国莱比锡“世界最美的书”奖；2005-2017 年 17 次获“中国最美的书”奖；2013 年获韩国坡州出版奖书籍设计奖（成就奖）；2013 年（第三届）、2016 年（第五届）2 次获中国出版政府奖——装帧设计奖；1999 年（第五届）、2004 年（第六届）2 次获全国书籍装帧艺术展览暨评奖金奖；2017 年获中国台湾金点设计奖；2018 年获东京 TDC Annual Awards（年度奖），同年获纽约第 97 届 ADC Bronze Cubes Awards（铜方块奖），纽约第 64 届 TDC Best in Show（全场大奖）、Communication Design Winners（传达设计大奖）。

著作

《由一个字到一本书 汉字排版》《11×16 XXL Studio》

Für einen Herausgeber ist die innere Bewegtheit angesichts der angekündigten Veröffentlichung von „Die schönsten Bücher der Welt", die auf die Sammlung in Zhao Qings Galerie Han Qing zurückgeht, schwer in Worte zu fassen. Bei „Die schönsten Bücher der Welt" handelt es zweifellos um einen der am stärksten von der Globalisierung erfassten Verlagspreise. An Beurteilungen und Bewertungen von Büchern auf nationaler oder regionaler Ebene mangelt es nicht. Aufgrund religiöser, politischer, sprachlicher, sozialer, ethnischer und anderer komplexer Faktoren ist es für den Menschen jedoch schwierig, sich ein Bild von den inhaltlichen Kriterien einer einheitlichen und akzeptablen Bewertung machen. In Deutschland ist es gelungen, einen Weg zu finden, die „Schönheit" der Bücher zu bewerten. Angelegt ist das in einem umfassenden Katalog an Auswahlkriterien, sodass es möglich ist, in Fragen der Ästhetik zu einer weitestgehenden Übereinstimmung und Einheitlichkeit zu gelangen. Hierbei handelt es sich um eine bemerkenswerte Errungenschaft.

„Die schönsten Bücher der Welt" verfügt über eine einzigartige Bewertungsgrundlage und stellt den Designern Orientierungsmaßstäbe zur Verfügung, denenzufolge neben der Schönheit für das Design auch die Materialtechnik, die Herstellungsbedingungen, die Lesebedürfnisse der Menschen und das Maß an avantgardistischer Erkundung miteinzubeziehen sind. Das Ergebnis sind die 14 Bücher weltweit, die jedes Jahr bewertet werden. Die höchsten Auszeichnungen sind beschränkt auf 14 Kategorien für die jährlich zu Tausenden erscheinenden neuen Bücher. Diese neuen Bücher repräsentieren die höchsten Auszeichnungen des Jahres, sie stehen für die Visionen der Juroren und öffnen den Blick auf die Zukunft des immateriellen Erbes der Menschheit in Wissen und Kunst. Diese Bücher, die ursprünglich nur in Leipzig gesammelt wurden, sind seit den 1960er Jahren eine Fundgrube, die durch unermüdliche Sam-

I have heard that *the Best Book Designs from all over the World* collected by Zhao Qing collected in Hanqing Hall are about to be published. As a publisher, this has moved me in a way that I can barely express with words.

The Best Book Design from all over the World is without any doubt an international publishing award. There are numerous book competitions in every country and region, but – because of differences in religion, politics, language, society, and nationality – it is really difficult that all humanity can establish a unified way of evaluation that everyone can accept. The wisdom of the German people has allowed them to find a perspective to judge the beauty of books and create this superb competition that can integrate and unify everyone's aesthetic ideal. No one can deny that this is something extraordinary.

The Best Book Design from all over the World has a unique standard to judge the design quality. The following orientation is given to designers: a beautiful design has to be in harmony with the materials and crafts, as well as with the printing techniques. It has to consider the needs of the readers while, at the same time, explore new paths. And the fourteen books chosen every year are the ones have found the best way to achieve this. The awarded books are chosen with the outmost care. Every year, among innumerable new works, only fourteen are chosen. Having received the highest award, they represent the view of the jury. In regards to the future, they are an intangible heritage of the knowledge and art of humanity. Furthermore, these

收藏，一个人的长征

作为一个出版人，闻讯赵清先生瀚清堂收藏的“世界最美的书”即将结集出版，内心的感动很难用语言表达。

“世界最美的书”肯定是最全球化的一项出版评奖。关于书籍，国家或区域性的评比是不缺乏的；但由于宗教、政治、语言、社会、民族等复杂因素，人类真的很难去设想对内容进行统一的又能被人们接受的评比。德国人的智慧是找到了一个评比书籍之“美”的角度，创造了这么一项超越的比赛，让大家在审美上达到融合和统一。这不能不说是一项了不起的创造。

“世界最美的书”有它独特的评奖标准和对设计师提出的方向：要把美的设计与材料工艺、与印刷条件、与人们的阅读需求、与设计的前卫性探索统一在一起。而其最好的解答则是每年全球评出的 14 本书。惜奖如金，只设 14 个奖去面对每年成千上万的新书。这些新书代表了当年最高的奖项，代表了评委的眼光，对未来而言，它也是人类知识和艺术的非物质遗产。而这些书，原来只被收藏在德国莱比锡。这是 20 世纪 60 年代以来年积月累的宝库，但很难为爱书人所见。

melarbeit entstand, und die doch selbst unter den buchliebenden Menschen kaum jemand kennt.

Nun endlich ist auch in der chinesischen Stadt Nanjing das weltweit einzige außerhalb Deutschlands befindliche Zentrum für die Sammlung der weltweit schönsten Bücher eingerichtet worden. Es ist ein Lesesaal entstanden, auf dessen Initiative hin ein Werk mit seinen Exponaten gedruckt wurde, das den Titel trägt „Leipziger Auswahl". Wert und Bedeutung dieses Werkes können meiner Meinung nach gar nicht hoch genug eingeschätzt werden.

Den globalen Buchverlegern und Designern auf aller Welt gegenüber wird damit zum Ausdruck gebracht, wie sehr die Menschen in China Bücher lieben und wie sehr sie das Design schöner Bücher schätzen. Gleichfalls wird der internationalen Verlagswelt ein Eindruck davon vermittelt, was für umfassende Veränderungen sich in der chinesischen Verlagsbranche und beim Buchdesign in den zurückliegenden vier Jahrzehnten der Reform und Öffnung konzeptionell und praktisch abgespielt haben und wie stark das chinesische Volk dazu bereit ist, etwas über die fortschrittlichen Dinge der Welt zu lernen, um mit dem Rest der Welt gemeinsam voranzuschreiten.

Aber das allein reicht nicht aus. Vor allem sollte betont werden, dass diese Sammlung das Werk eines einzelnen Mannes darstellt: zu verdanken ist dieses nahezu unmöglich zu vollendende Werk nämlich einem chinesischen Designer allein, es geht weder auf eine Institution noch eine Gruppe von Menschen zurück. Umso mehr erfüllt mich das, was wir heute vor uns sehen, mit freudigem Staunen und Stolz.

Vierzehn Buchexemplare pro Jahr aus der Zeit von 2004 bis 2019: die Bücher aus diesen eineinhalb Jahrzehnten stammen aus über 30 Ländern und Regionen sind in den unterschiedlichsten Sprachen abgefasst. Vielfach handelt es sich um Musterexemplare in nur geringer Auflage, man kann sich ausmalen, wie schwierig es war, das alles zu sammeln. Zahlreiche Freunde haben mir in den letzten Jahren gegenüber angekündigt, eine vollständige Sammlung der schönsten Bücher Chinas während der zurückliegenden 16 Jahre anzufertigen. Obgleich dies ein Werk aus einem Land mit nur einer Sprache – der Chinesischen

books were only collected in Leipzig, Germany. Since 1960 they were kept there, so it was hard for book lovers to catch a glimpse of them.

Now, in Nanjing, China, another collection of the *The Best Book Design from all over the World* has been founded. This space for reading holds a collection titled *The Choice of Leipzig*. As I see it, it has an inestimable value and significance.

This initiative has made publishers and book designers all over the world aware that Chinese people love books and cherish the beauty of book design like no one else. It has also shown the world's publishing sphere that, after forty years of Reform and Opening Up, Chinese book publishing and design has undergone enormous changes that affect both the concept and the practice. It has proven that Chinese people are willing to learn from the world's most advanced products and go forward together.

But this is not enough. What should be most emphasized is that collecting is a long march for an individual. This almost insurmountable task has been carried out by a single Chinese designer – not by an organization or a collective. In regards to this, I am extremely surprised and proud.

Just imagine how difficult it must have been to collect the fourteen books awarded every year during the sixteen years from 2004 to 2019. Coming from more than thirty different countries and regions, these books were written in different languages, and many were part of small publications. These years many friends have told me they wanted to collect the works awarded as *China's Most Beautiful Books*, a contest that has been going on for sixteen years. Even these books have been published in the same country and are writ-

现在，在中国南京，建成了德国之外唯一的一个“世界最美的书”的收藏中心、一个阅读空间，现在这个阅读空间又将收藏品印成合集，命名为《莱比锡的选择》而出版，我认为其价值和意义难以估量。

它告诉全球书籍出版人和设计师，中国人无比热爱图书，无比热爱书籍设计之美；它也告诉世界出版界，改革开放 40 年，中国出版业和书籍设计界从观念到实际已发生了巨大的变化，中国人愿意学习世界先进的东西，与世界各地一起进步。

但这样说还是不够的，最应该强调的是，这一收藏是一个人的“长征”，这项几乎不可能完成的工作是由一个中国设计师独自完成的；不是一个机构，也不是一个群体。在这一事实面前，我很惊叹，也很自豪。

每年 14 本，2004–2019 年，这整整 16 年的书，来自 30 多个国家和地区，用的是不同的语种，很多是小印量的探索性样本，要去收齐，设想一下吧，该有多么的困难。这些年很多朋友对我说想收藏这 16 年全套“中国最美的书”，这是在同一块土地上用同一种汉字出版的书，尚且难以坚持，最终都放弃了。可见，基本收集齐“世界最美的书”，几乎是完成一件不可能的事。这 16 年，我一直在“中国最美的书”评委会工作，我体会对

– gewesen wäre, ließen sich die Projekte doch niemals weiterführen und wurden am Ende eingestellt. Das macht um so mehr deutlich, wie nahezu unmöglich es ist, eine grundlegende Sammlung der schönsten Bücher der Welt anzufertigen. Die vergangenen 16 Jahre über gehörte ich ständig der Jury von „Chinas schönste Bücher" an. Daher ist mir vollkommen klar, wie hoch nicht nur die finanziellen Kosten für Zhao Qing all die Zeit über gewesen sind, sondern mit wieviel Zeitaufwand, Energie und Verantwortung und Einsatzbereitschaft seine Arbeit verbunden war. Es war eine immense Herausfordeung und ein langer Kampf. Allein dafür schon gilt Zhao Qing meine höchste Anerkennung und mein höchster Respekt.

Mit der Sammlung, Ausstellung und Veröffentlichung von Zhao Qing gibt es in dem 1,3 Milliarden Menschen umfassenden China nunmehr endlich ein Zentrum für die Sammlung, Ausstellung und Erforschung der schönsten Bücher der Welt. Endlich steht damit Experten ein Ort zur Verfügung, um mit den Werken direkt in Berührung zu kommen und das Magische sowie den Charme dieser Schönheiten selber und aus nächster Nähe zu erfahren. Dies ist ein großer Verdienst von Herrn Zhao Qing gegenüber der chinesischen Gemeinschaft der Buchdesigner. Es handelt sich auch um eine wichtige Arbeit, die er im kulturellen Austausch zwischen China und dem Ausland geleistet hat.

Wenn wir heutzutage in ein Krankenhaus gehen und dabei die Vielzahl von medizinischen Geräten, Messaparaturen und Behandlungsformen aus Übersee sehen, kommt uns das angesichts der Tatsache, dass damit Menschenleben gerettet werden, wie eine Selbstverständlichkeit vor. Doch über die Bereitstellung von Räumen für Werke der Weltkultur gibt es bei uns immer wieder kontroverse Diskussionen. Herr Zhao Qing hat einen mutigen Schritt getan, er hat ein Beispiel gesetzt, indem er zeigt, wie wichtig es bei der kulturellen Entwicklung ist, von anderswo her etwas zu übernehmen bzw. Anleihen bei fremden Kulturen zu machen.

China und Deutschland haben im Bereich der Druck- und Verlagsgeschichte wichtige Erfindungen zum Nutzen der Menschheit

ten in the same language, my friends found it hard continue and eventually gave up. As it seems, collecting *The Best Book Design from all over the World* means carrying out an impossible task. These sixteen years, I have been part of the jury of *China's Most Beautiful Books*. This way, I have come to realize that Zhao Qing has not only invested money in this. This long march has mostly required his time and energy, putting to test his responsibility, willpower, and drive. It has been a long battle for which I profoundly admire Zhao Qing's perseverance.

With his collection, his exhibitions, and his publications, we finally have a center for *The Best Book Design from all over the World* in this country of one thousand three hundred million people. Specialists finally have a place to be in close contact with beautiful books and feel the magic and charm of beauty. This is an enormous contribution of Zhao Qing to China's book design and a great task in the cultural exchange between China and foreign countries.

Nowadays, we accept importing all kinds of medical equipment, testing machines, and treatments from overseas into our hospitals because it saves lives. However, when it comes to importing culture, there is always some controversy. Therefore, Zhao Qing has taken a courageous step and become a model for us – he has shown us that culture can only move forward when it draws on other sources.

Actually, China and Germany have made the most important inventions in printing. One thousand years ago, Bi Sheng, of the Northern Song Dynasty, invented the wooden movable-type press, which reflected the publishing culture of an agricultural society. Later, Gutenberg invented the copper movable-type press in Germany – the origin of modern printing.

于赵清来说除了金钱的付出，这一个人的“长征”，更多的是时间、精力的付出，是责任、毅力、精神的考验。这是一场持久战。仅这一点，我就对赵清先生的精神佩服得五体投地。

有了赵清的收藏、展示和出版，在 13 亿人口的中国大地，终于有了一个“世界最美的书”的收藏中心、展示中心和研究中心，专业人士终于有一个地方可以直面美书亲密接触，感受美的神奇和魅力。这是赵清先生对中国书籍设计界所作的巨大贡献，也是他对中外文化交流所做的一项重大工作。

今天，走进医院，看到各式各样来自海外的医疗器械、检测设备和治疗手段，我们会认为因救人性命而引进是理所当然的。而对引进世界的文化，就有些争议了。而赵清先生勇敢地走出了重要的一步，为我们树立了一个榜样——文化的发展也需要引进和借鉴。

中德两国曾经对人类的出版共同做出过重要的发明创造：1000 年前北宋毕昇发明了木活字印刷，反映出了农耕时代的出版文明。后来，德国谷登堡发明了机器铜活字印刷，开创了近代文明的印刷先河。海德堡印刷机、法兰克福书展至今长盛不衰。而他们的后人，都因为“世界最美的书”——寻找美、留住美和创造美走到一起来了，瀚清堂收藏就是见证。

gemacht: Vor 1000 Jahren hat Bi Sheng aus der Zeit der Nördlichen Song-Dynastie die Holzlithografie erfunden, die die Verlagskultur der Bauernzeit widerspiegelt. Später erfand der Deutsche Gutenberg die mechanische Kupferlithografie und leistete damit Pionierarbeit im Bereich der neuzeitlichen Druckerei. Heidelberger Druckmaschinen und die Frankfurter Buchmesse haben bis in die Gegenwart nichts von ihrer Kraft verloren. Ihre Nachkommen haben aus einem gemeinsamen Anliegen, in dessen Mittelpunkt die schönsten Bücher der Welt stehen, zusammengefunden auf der Suche nach Schönheit und ihrer Bewahrung bzw. ihrer Hervorbringung. Die Sammlung der Han Qing-Galerie ist der Beweis dafür.

Es ist 16 Jahre her, dass das Projekt „Die schönsten Bücher Chinas" im Jahre 2003 in Shanghai ins Leben gerufen worden ist. Für mich ergab sich damals die Gelegenheit, mit führenden chinesischen und ausländischen Buchdesignern zusammenzuarbeiten, dabei habe ich schließlich die Bekanntschaft mit Herrn Zhao Qing gemacht. Das ehrt mich sehr! In meinen Augen ist er einer der besten Designer, er hat die meisten Preise für die schönsten Bücher Chinas erhalten. Doch nicht nur das, Zhao Qing ist ein nachdenklicher Designer von eigenem Charakter. Sei es bei den jährlichen Treffen zum Runden Tisch mit dem Thema der schönsten Bücher Chinas, sei es bei der Leipziger Tagung von Designern aus China und anderen Ländern im März 2015 – stets haben mich seine Vorträge stark beeindruckt. Mit der Schaffung dieses Leseraums und der Veröffentlichung dieses Buches hat Zhao Qing ein brillantes Meisterwerk vollbracht. Nicht nur überragt er damit ohne Zweifel die Arbeit eines Designers ganz allgemein, er hat auch als Sammler Maßstäbe gesetzt. Wir sollten ihm alle danken. Das chinesische Verlagswesen ist auf Männer wie Zhao angewiesen – jemand wie Zhao mit einer derartigen Hingabe, einem solchen Verantwortungsbewusstsein und jenem Grad an Weisheit und Methodik ist eine Rarität.

Zum Schluss möchte ich darauf aufmerksam machen, dass mit der „Leipziger Auswahl" nicht nur eine Sammlung dokumentiert wird, sondern dass es sich darüber hinaus auch um ein Bilderbuch und ein

The Heidelberg press and the Frankfurt Book Fair are stronger than ever. Their descendants came together because of *The Best Book Design from all over the World* – to pursue, retain, and create beauty. The collection of Hanqing Hall proves it.

Sixteen years have passed since the competition *China's Most Beautiful Books* was established in Shanghai in 2003. It gave me the chance to work with top-notch Chinese and foreign book designers. Among them was Zhao Qing, whom I met for the first time there. I feel deeply honored for that. As I see it, he is the best designer, the one who has won the most prizes of *China's Most Beautiful Books.* He is a designer with ideas and individuality who every year is at the round table of *China's Most Beautiful Books*. In March 2015, he was invited to the Leipzig Designers Forum, and his speech left me a deep impression. Now, he has established this reading space and published this brilliant work. Undoubtedly, he has surpassed the work of an average designer and has achieved the status of a collector. Therefore, we should all thank him. China's publishing sphere needs people like him. We need people who have tons of enthusiasm, responsibility, knowledge, and means – qualities that are extremely rare.

As for The *Choice of Leipzig*, it is a museum piece. At the same time, it is a picture album, but it is rather a reference book. That such work could be published in China embodies the progress of four decades of Reform and Opening Up. I also feel glad for other designers. By reading this book, they can get a taste of sixteen years of the world's most beautiful books without putting a foot outside. Precisely in these sixteen years, China has participated in the competition *Best Book Designs from all over the World*, so inside the book there are

从 2003 年在上海创立“中国最美的书”这项事业，至今已 16 年了，它让我有机会与中外顶级的书籍设计师在一起工作，包括因此而结识了赵清先生。对此我深感荣幸！在我的印象中，他是一位最好的设计师，“中国最美的书”获奖最多的人；他也是一位有思想、有个性的设计师，在每年“中国最美的书”圆桌会议上，在 2015 年 3 月莱比锡中外设计家论坛上，他的发言都给我留下了深刻的印象。现在，他建成的这个阅读空间和出版的这部煌煌巨作，无疑超出了一般设计家的工作，而又有了一位收藏家的定位。我们大家应该感谢他。中国出版界很需要他这样的人，有十二万分的热情，又有责任感，还有智慧和办法，不可多得。

最后说到《莱比锡的选择》，应该是一本藏品集，又是一本画册，更是一本工具书。它能够在中国出版，正是 40 年改革开放进步的体现。我为其他的设计师深感庆幸，他们读到这本书，足不出户就可以领略这 16 年“世界最美的书”之风采。而恰恰是在这 16 年中，中国参加了“世界最美的书”的比赛，并且有 22 种书代表了中国攀登世界先进水平的高度，记录下了我们前进的脚步。赵清先生为此书出版做了大量翻译工作，因而其学术性和艺术性也是无可争辩的。

Nachschlagewerk handelt. Dass es in China veröffentlicht wird, ist ein Beleg für die Politik der Reform und Öffnung in den vergangenen 40 Jahren. Ich freue mich für die vielen anderen Designer, die dieses Buch lesen werden und denen damit ein Eindruck von den schönsten Büchern der Welt in den vergangenen 16 Jahren vermittelt wird. Eben in diesen 16 Jahren hat sich China an dem Wettbewerb „Die schönsten Bücher der Welt" beteiligt, die 22 chinesischen Bücher in dem Band verdeutlichen die in diesem Zeitraum erzielten Fortschritte und zeigen, dass China mittlerweile der Anschluss an die Weltspitze gelungen ist.

Zhao Qing hat eine großartige Übersetzungsarbeit für die Veröffentlichung dieses Buches geleistet, sein wissenschaftliches und künstlerisches Vorgehen ist ebenfalls unumstritten. Für sein Engagement an dem Projekt „Die schönsten Bücher Chinas" gilt ihm mein besonderer Dank! Gleichzeitig möchte ich auch meinen aufrichtigen Dank gegenüber den über all die Jahre mitwirkenden Juroren und Kollegen in China und anderswo aussprechen! Die Zeit fließt, mein Herz ist still!

Zhu Junbo
verfasst am 23. März 2019 in Shanghai

twenty two books that have put China the same level than the rest of the world, leaving a record of our steps forward. For publishing this book, Zhao Qing has made a great deal of work in translation. Thus, its academic and artistic value is incontestable.

As a participant of *China's Most Beautiful Books*, I would like to express my most profound respect and gratitude to Zhao Qing and to the Chinese and international members of the jury who worked together in those years. Time may have passed, but I feel the same for you all!

Zhu Junbo
March 23th, 2019 in Gao'anxuan, Shanghai

作为“中国最美的书”的参与者，在此，特向赵清先生表示敬意和感谢！同时也向当年一起共事的国内外评委、同事表示敬意和感谢！时光流淌，我心依然！

祝君波

2019 年 3 月 23 日　于上海高安轩

Herkunft und Studium

Geboren im März 1955 in Shanghai, Abstammungsort der Familie die Stadt Jinhua (Provinz Zhejiang). 1987 Studienabschluss (Abendschule) an der East-China Normal University, Department for Chinese Studies. 1995 Masterabschluss an der Fudan-Universität (School of Economics), Einstufung als verantwortlicher Editor. Beruflicher Werdegang. 1972 Eintritt in das Presse- und Verlagswesen der Stadt Shanghai sowie Lehre bei Duoyunxuan. Spezialisierung auf die Bereiche Holzschnitt und Gravur.

Seit 1991

- Verlagspräsident und Parteisekretär von Shanghai Painting and Calligraphy Press.
- Verlagspräsident und Parteisekretär von Shanghai People's Fine Arts Publishing House. - Generaldirektor von Duoyunxuan und Gründungsdirektor des angeschlossenen Auktionshauses. - Stellvertretender Leiter des Shanghai Press and Publication Bureau. - General Manager des Oriental Publishing Center innerhalb der China Publishing Group und zugleich Parteisekretär. - Stellvertretender Direktor des Shanghai Press and Publication Bureau. - Stellvertretender Direktor des Organisationskomitees des World Chinese Collectors Congress of Shanghai. - Vizepräsident der Chinese Periodical Association.
- Stellvertretender Direktor des Komitees für Literatur und Geschichte der Shanghai CPPCC. - Stellvertretender Direktor Jury für die schönsten Bücher in China.

Zhu Junbo ist außerdem Dozent am Nationalen Zentrum für die Ausbildung von Auktionatoren sowie Betreuer von Masterstudenten des Faches Medienwissenschaft an der Shanghai University of Technology, des Faches Kunstmarkt an der East China Normal University und des Faches Kunstmarkt an der Shanghai University. Darüber hinaus ist er Gastprofessor an der Shanghai University of Visual Arts und Distinguished Professor des Kollegs für Überseestudien der Jiaotong University.

Seit mehr als 40 Jahren wirkt Zhu Junbo im Verlags- und Kunsthandel mit. Seine wichtigsten Publikationen sind Einführung in die Kunstauktion und Praxis der Investition, Zhu Junbo über die Sammlung von Kunstwerken, Bemerkungen zur Sammlung von Kunstklassikern, Wegweiser zu Investionen in der Kunst und Zur Auktionstheorie (ausgewählte Kapitel). Zhu hat außerdem mehr als 160 Artikel zu verschiedenen Themen verfasst. Im Jahre 2010 wurde Zhu vom taiwanesischen Magazin „Art News" zu einer der bekanntesten 10 Persönlichkeiten des Jahres gewählt.

Im Laufe seiner Karriere gründete Zhu Junbo das erste Kunstversteigerungsunternehmen in China Duoyunxuan und gab damit dem chinesischen Auktionswesen einen entscheidenden Schub. Im Jahre 2004 war er an der Gründung von „Schönste Bücher Chinas" sowie an der Ausrichtung der chinesischen Spieleausstellung „Chinajoy" beteiligt, die damit zu einer bekannten Kulturmarke im In-und Ausland wurde.

Seit Jahr 2007 war Zhu leitend für die Vorbereitungen zur Durchführung des Weltkongresses chinesischer Sammler zuständig (für vier Sitzungen in acht Jahren).

Nach Eintritt in den Ruhestand im Mai 2018 ist er als Executive Vice President der Shanghai Literary and Literature Research Association tätig.

Zhu Junbo was born in March 1955 in Shanghai, but his family came from Jinhua, in Zhejiang Province. He worked as an editor and in 1972 he entered in the Shanghai News Department. He also became an apprentice at the Duoyunxuan Art Center, where he specialized in the craft of wood carving. In 1987, he graduated from the Chinese Language Faculty of the East China Normal University. He finished his master degree in the Faculty of Economics of Fudan University in 1995.

Since 1991, he has successively held the posts of director of the Shanghai Painting and Calligraphy Publishing and director of the Shanghai People's Fine Arts Publishing. While he held these two posts he was also Party Committee Secretary. Then, he was the general director of the Duoyunxuan Art Center and, at the same time, the general director of a company auctioned by the Duoyunxuan Art Center. Afterwards, he became director of the Shanghai News Department and general director of the Orient Publishing Center of the China Publishing Group Corporation while also holding the post of Party Secretary. He also worked as deputy director of the Shanghai News Department, assistant head of the organizational committee of the Shanghai World Congress of Chinese Collectors, and director of the China Periodicals Association. In addition, he served as deputy director of the Literature and History Committee of the Chinese People's Political Consultative Conference and deputy director of the jury of *China's Most Beautiful Books*. At the same time, Zhu Junbo has been a teacher at the National Auctioneer Training Center, at the University of Shanghai for Science and Technology in the media studies degree, at the East China Normal University in the art market degree. He also worked as a tutor for master degree students of art market at Shanghai University, as a guest professor at the Shanghai Institute of Visual Art, and as a special professor at the Overseas Education College of Shanghai Jiaotong University.

Zhu Junbo has been involved in publishing and art trade for more than forty years. He has written works such as *Practical Course on Art Auctions and Investment*, *Zhu Junbo Talks about Collecting*, *Collection of Texts*, *Guide for Investment in Works of Art*, *Theory of Auctions*, and many more. He has written more than 160 academic articles. In 2010, he was chosen one of the ten most influential figures of the year by the Taiwanese magazine The Art Newspaper.

In the professional field, Zhu Junbo also founded the first professional enterprise of art auctions, which was auctioned by the Duoyunxuan Art Center and started the first auctions in mainland China. In 2004, he participated and initiated the award *China's Most Beautiful Books* and the china Digital Entertainment Expo & Conference China Joy, which became a well-known cultural trademark known both in China and internationally. In 2007, he was in charge of the preparations for the Fourth World Congress of Chinese Collectors, which was held in 2008.

In May 2018, Zhu Junbo retired and became vice-president of the Shanghai Literature and History Research Association.

祝君波
Zhu Junbo

1955 年 3 月生于上海，祖籍浙江金华。华东师范大学夜大学中文系 1987 年毕业，1995 年复旦大学经济学院硕士生结业。编审。1972 年进入上海新闻出版界及朵云轩学徒，专攻木刻雕版工艺。

1991 年起历任上海书画出版社社长兼党委书记、上海人民美术出版社社长兼党委书记、朵云轩总经理兼朵云轩拍卖公司创始总经理、上海市新闻出版局副局长、中国出版集团东方出版中心总经理兼党委书记、上海市新闻出版局副局长（正局级）、上海世界华人收藏家大会组委会执行副主任、中国期刊协会副会长、上海市政协文史委员会常务副主任、“中国最美的书”评委会副主任。兼任国家拍卖师培训中心授课教师，上海理工大学传媒学专业、华东师范大学艺术市场专业和上海大学艺术市场专业硕士生导师，上海视觉艺术大学客座教授，交通大学海外学院特聘教授。从事出版和艺术贸易 40 余年。主要著作有《艺术品拍卖与投资实践教程》《祝君波谈收藏》《典藏文札》《艺术品投资指南》《拍卖通论》（部分章节）。著有各类论文 160 余篇。2010 年被中国台湾《艺术新闻》杂志评选为年度“十大风云人物”。

职业生涯中还创办了中国第一家艺术专业拍卖公司——朵云轩拍卖行，敲响了中国大陆拍卖第一槌。2004 年参与创建“中国最美的书”评选以及中国游戏大展“Chinajoy”，使之成为国内外知名的文化品牌。2007 年主持促成了世界华人收藏家大会的筹建和召开（八年四届）。

2018 年 5 月退休，任上海文史资料研究会常务副会长。

2004—2019

2019

2004

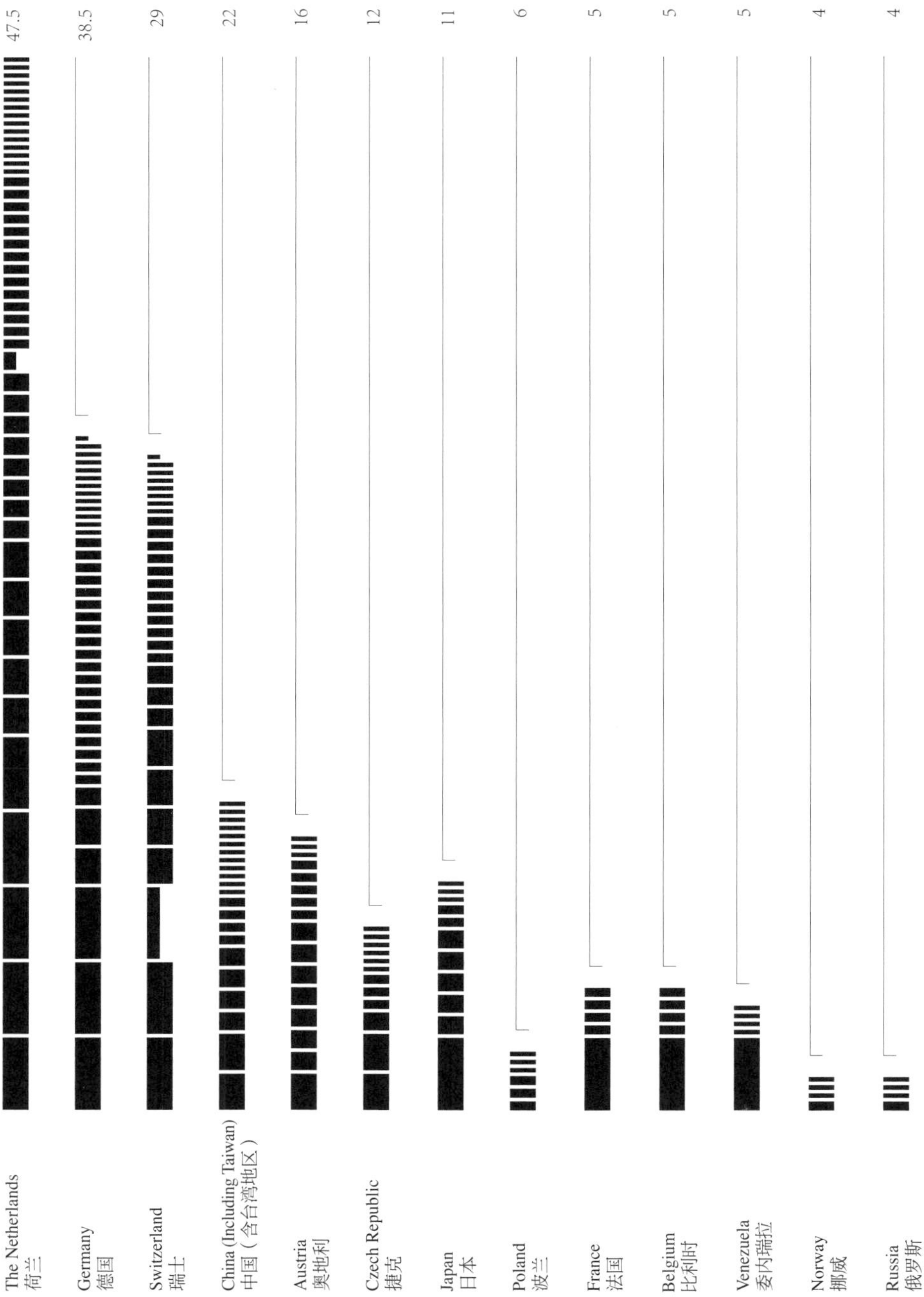

The Netherlands 荷兰 47.5
Germany 德国 38.5
Switzerland 瑞士 29
China (Including Taiwan) 中国（含台湾地区） 22
Austria 奥地利 16
Czech Republic 捷克 12
Japan 日本 11
Poland 波兰 6
France 法国 5
Belgium 比利时 5
Venezuela 委内瑞拉 5
Norway 挪威 4
Russia 俄罗斯 4

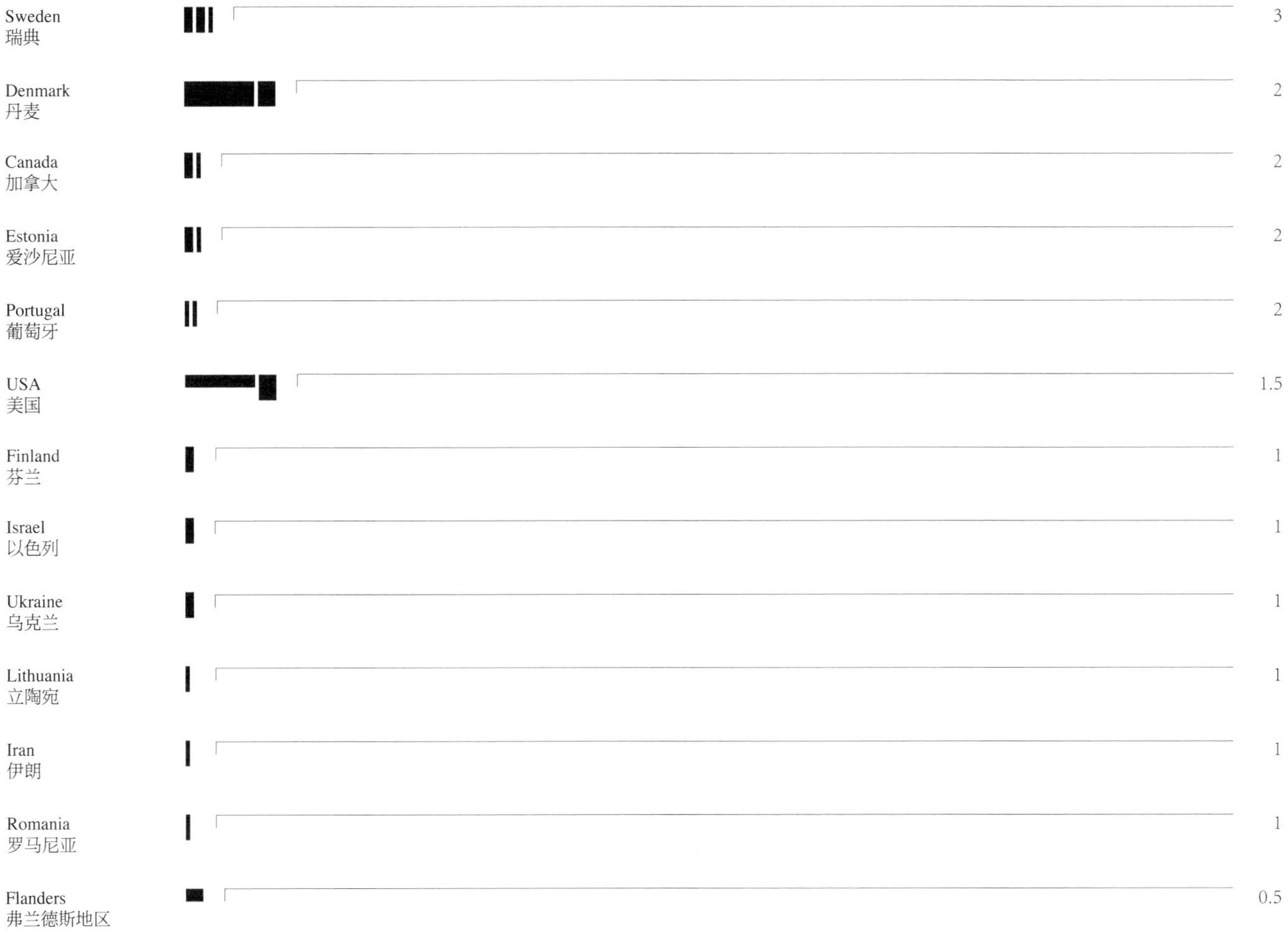
Sweden
瑞典
3
Denmark
丹麦
2
Canada
加拿大
2
Estonia
爱沙尼亚
2
Portugal
葡萄牙
2
USA
美国
1.5
Finland
芬兰
1
Israel
以色列
1
Ukraine
乌克兰
1
Lithuania
立陶宛
1
Iran
伊朗
1
Romania
罗马尼亚
1
Flanders
弗兰德斯地区
0.5

Irma Boom

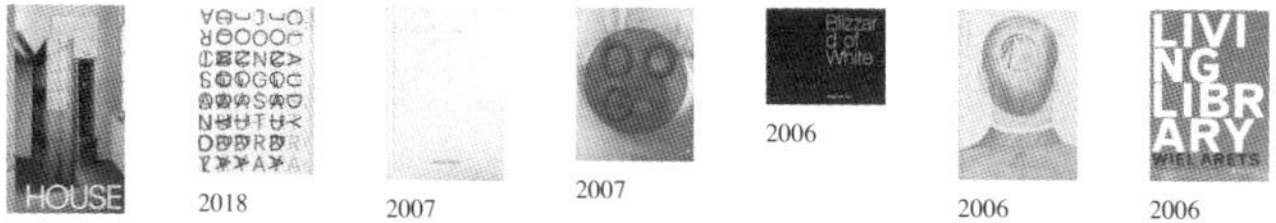
HOUSE
Blizzard of White
LIVING LIBRARY
WIEL ARETS

2019 2018 2007 2007 2006 2006 2006

Jonas Voegeli

WO-WO-WONIGE!
GLOBUS CASSUS

2018 2017 2011 2007 2005

Joost Grootens

BRICK AN EXACTING MATERIAL
Vinex Atlas 010

2016 2011 2009 2008 2006

Reinhard Gassner

2011 2010 2009 2007

Marcel Bachmann

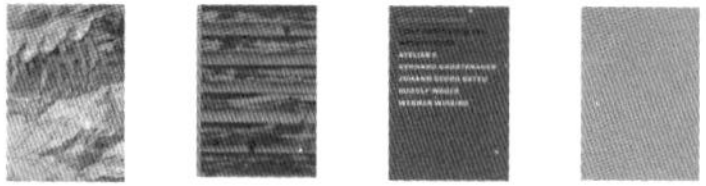

2012 2011 2010 2007

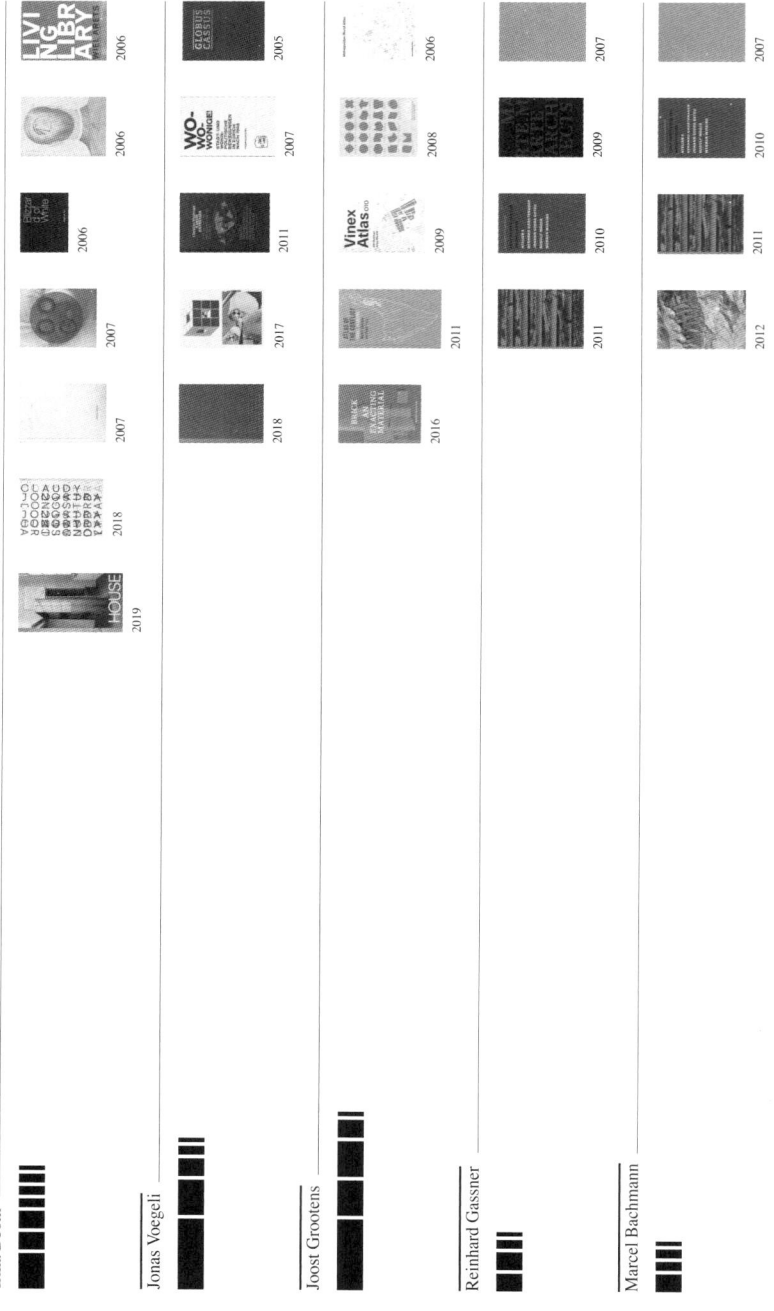

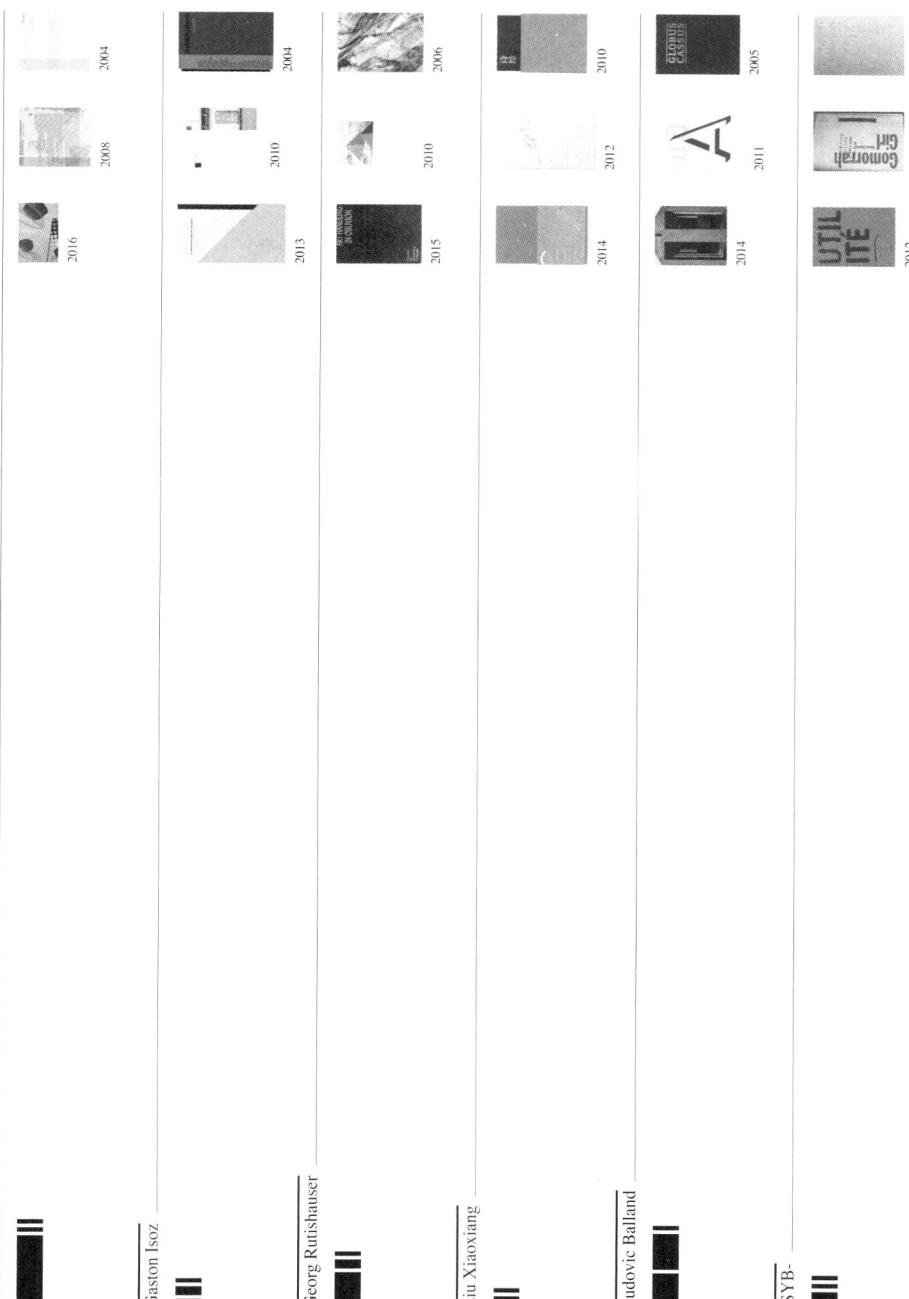

Álvaro Sotillo

2016 2008 2004

Gaston Isoz

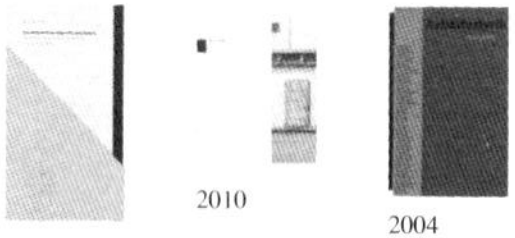

2013 2010 2004

Georg Rutishauser

ART HANDLING
IN OBLIVION

2015 2010 2006

Liu Xiaoxiang

2014 2012 2010

Ludovic Balland

A
GLOBUS
CASSUS

2014 2011 2005

-SYB-

UTIL
ITÉ
Gomorrah
Girl

2012 2012 2005

2019	《江苏老行当百业写真》	荣誉奖
2018	《园冶注释》	银 奖
	《茶典》	荣誉奖
2017	《虫子书》	银 奖
	《冷冰川墨刻》	荣誉奖
2016	《订单：方圆故事》	金 奖
	《学而不厌》	铜 奖
2014	《刘小东在和田 & 新疆新观察》	铜 奖
	《2010—2012 中国最美的书》	荣誉奖
2013	《坐火车的抹香鲸》（中国台湾）	银 奖
2012	《剪纸的故事》	银 奖
	《以有机为名》（中国台湾）	铜 奖
	《文爱艺诗集》	荣誉奖
2011	《漫游：建筑体验与文学想象》	荣誉奖
2010	《诗经》	荣誉奖
2009	《中国记忆——五千年文明瑰宝》	荣誉奖
2008	《之后》	荣誉奖
2007	《不裁》	铜 奖
2006	《曹雪芹风筝艺术》	荣誉奖
2005	《朱叶青杂说系列》	荣誉奖
	《土地》	荣誉奖
2004	《梅兰芳（藏）戏曲史料图画集》	金 奖

Old Trades of Jiangsu: A Glimpse	Honorary Appreciation
The Art of Gardening	Silver Medal
Tea Canon	Honorary Appreciation
Bugs' Book	Silver Medal
Ink Rubbing by Leng Bingchuan	Honorary Appreciation
Order–The Story of Fangyuan Bookshop	Gold Medal
Pleasure of Learning	Bronze Medal
Liu Xiaodong's Hotan Project & Xinjiang Research	Bronze Medal
The Beauty of Books in China 2010–2012	Honorary Appreciation
A Cachalot on a Train	Silver Medal
The Story of Paper-cut	Silver Medal
Zoom in, Zoom out	Bronze Medal
Poems by Wen Aiyi	Honorary Appreciation
Odyssey: Architecture and Literature	Honorary Appreciation
The Book of Songs	Honorary Appreciation
The Chinese Memory: Treasures of the 5000-year Civilization	Honorary Appreciation
The After Concept & Works Book	Honorary Appreciation
stitching up	Bronze Medal
Cao Xueqin's Art of Kite	Honorary Appreciation
The Philosophy of Mist / The Sensation of Wateriness / Look up at the Sky / The Ism of Antique / Consommé and Water	Honorary Appreciation
Antony Gormley–Asian Field	Honorary Appreciation
A Picture Album. Historical Materials of Mei Lanfang's Theatrical Performances	Gold Medal

ZOOM
IN
ZOOM
OUT

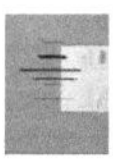

Code
编码

2019 Gl

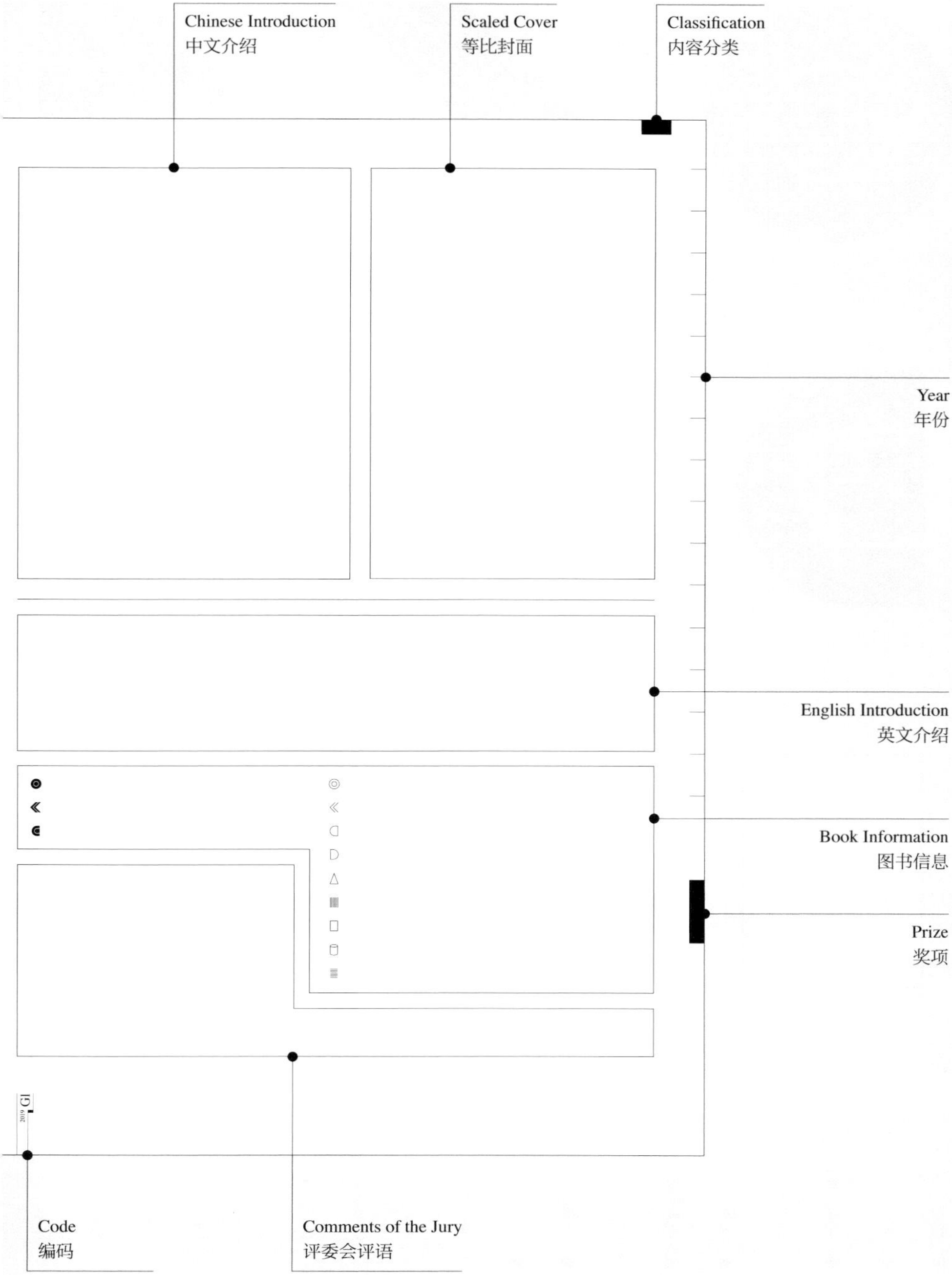

Chinese Introduction
中文介绍
Scaled Cover
等比封面
Classification
内容分类
Year
年份
English Introduction
英文介绍
Book Information
图书信息
Prize
奖项
Code
编码
Comments of the Jury
评委会评语

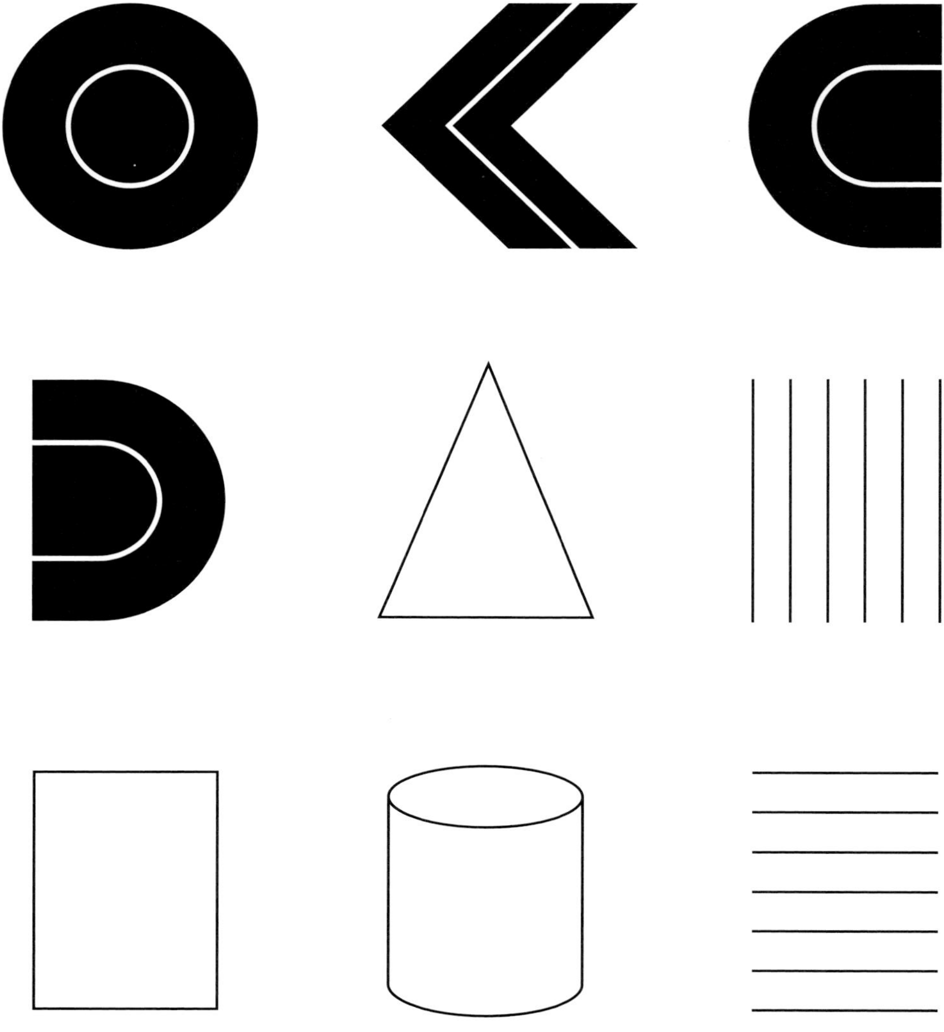

奖项
书名
国家 / 地区
设计者
作者
出版机构
尺寸
重量
页数
Award
Title
Country / Region
Designer
Author
Publisher
Size
Weight
Page

Gl Golden Letter 金字符奖

Gm Gold Medal 金奖

Sm Silver Medal 银奖

Bm Bronze Medal 铜奖

Ha Honorary Appreciation 荣誉奖

Binding Style 装订方式

Printing and Technologies 印制与工艺

S G

Code 编码

Year 年份 | Award 奖项 | Number 编号

2019 Sm²

Art & Design
艺术与设计

43%

Literature & Fiction
文学与小说

12%

Not Collected
未收藏
7%
Nature & Technology
自然与科技
6%
Others
其他
2%
Children and Young People's Books
少儿读物
5%
Education & Teaching
教育与教学
2%
Social Sciences
社会科学
23%

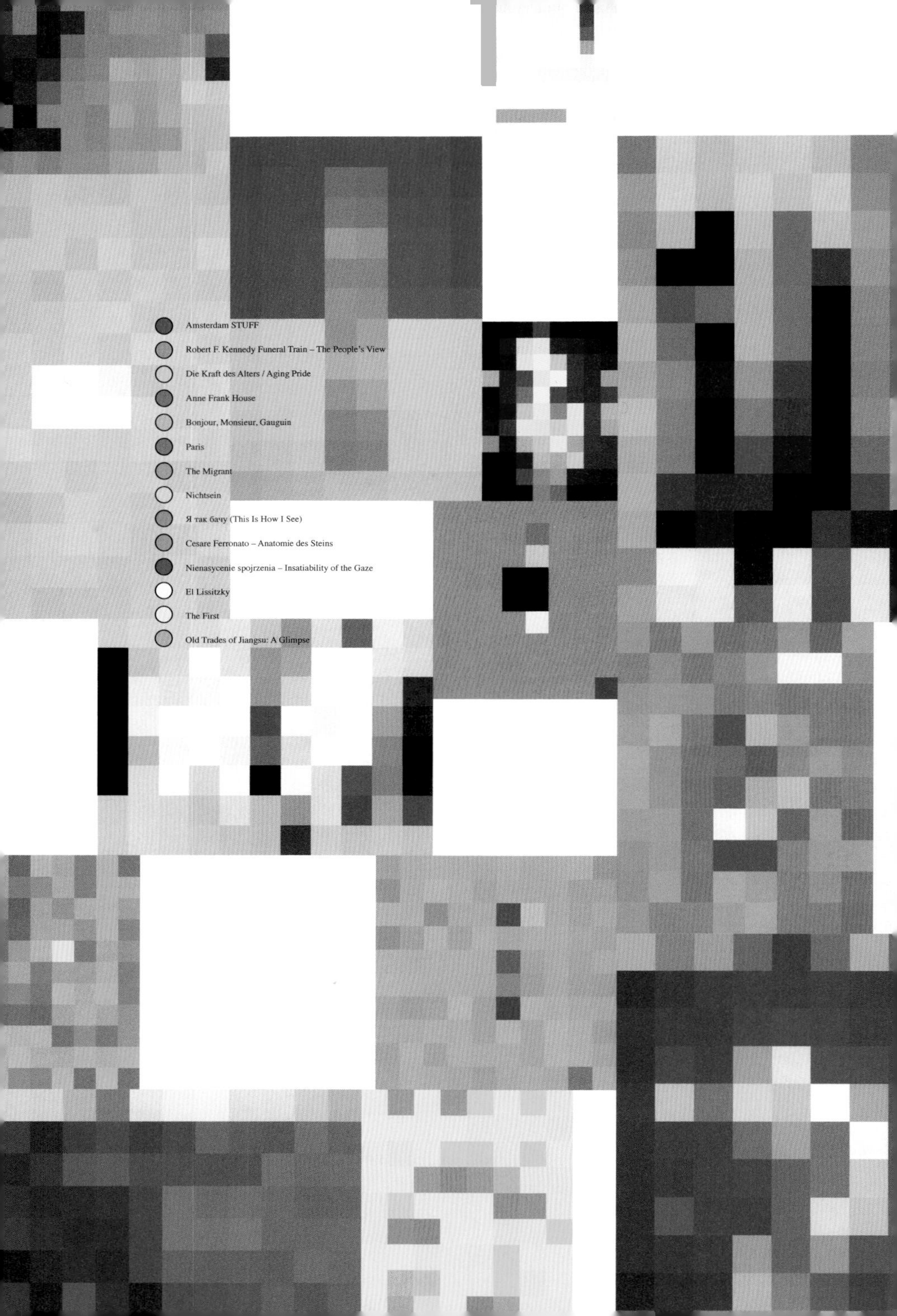
Amsterdam STUFF
Robert F. Kennedy Funeral Train – The People's View
Die Kraft des Alters / Aging Pride
Anne Frank House
Bonjour, Monsieur, Gauguin
Paris
The Migrant
Nichtsein
Я так бачу (This Is How I See)
Cesare Ferronato – Anatomie des Steins
Nienasycenie spojrzenia – Insatiability of the Gaze
El Lissitzky
The First
Old Trades of Jiangsu: A Glimpse

2019

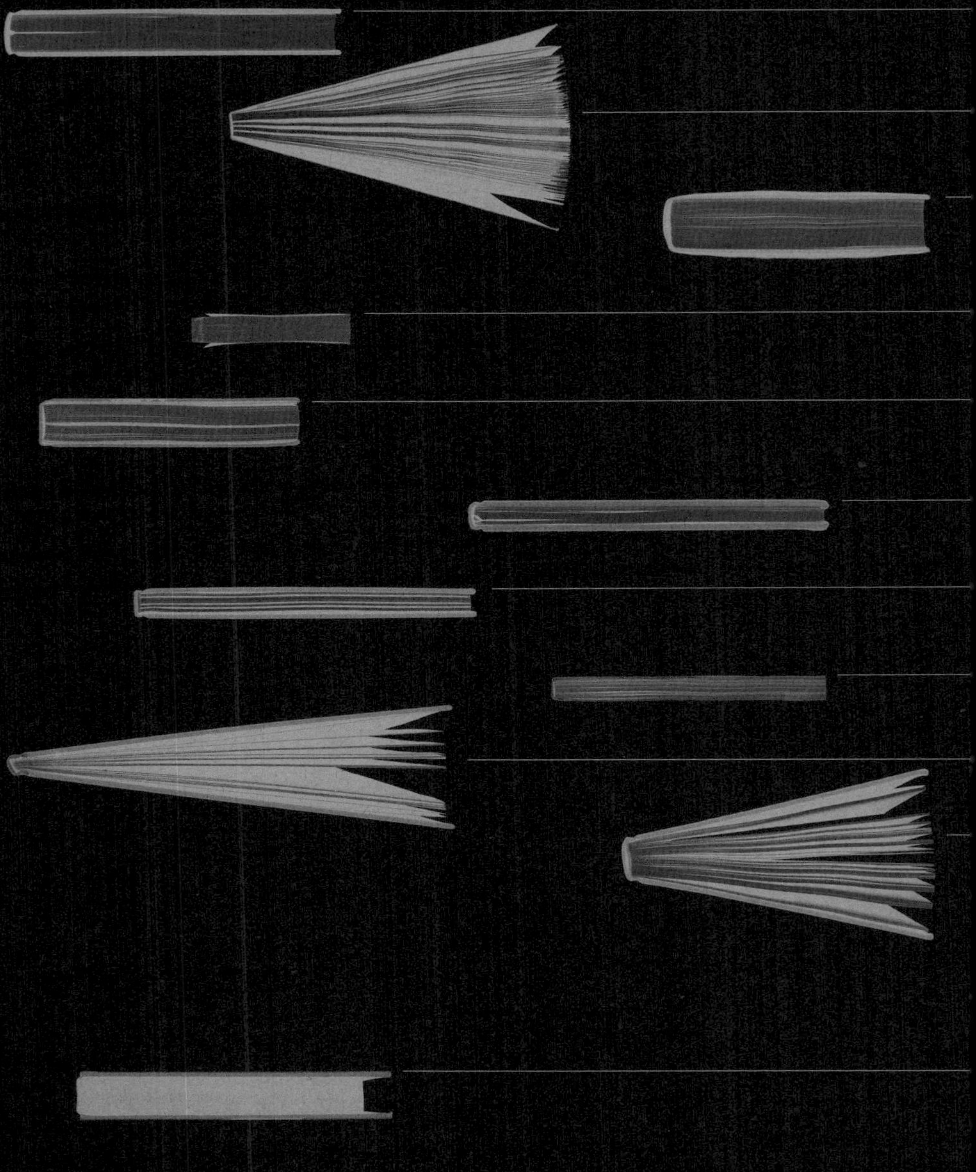

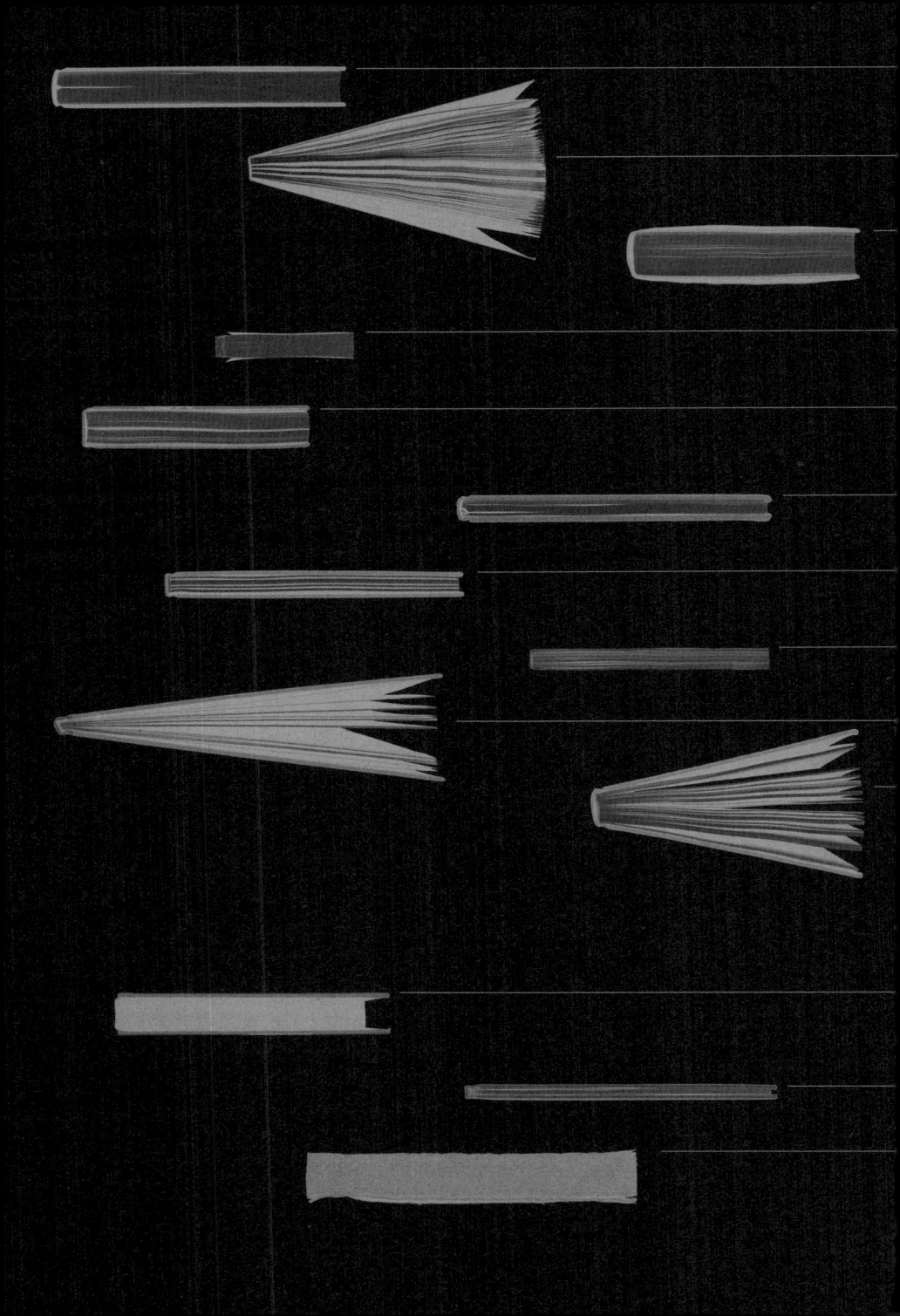

2019 Gl

0143

XXX STUFF

2019 Gm

0151

2019 Sm[1]

0157

2019 Sm[2]

0163

ANNE
HOUSE
FRANK

2019 Bm[1]

0169

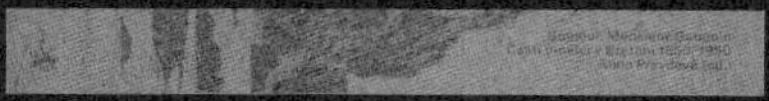

2019 Bm[2]

0175

2019 Bm[3]

0181

2019 Bm[4]

0187

2019 Bm[5]

0193

2019 Ha[1]

0199

Cesare
Ferronato
The Anatomy
of Stone

2019 Ha[3]

0203

2019 Ha[4]

0207

2019 Ha[5]

0211

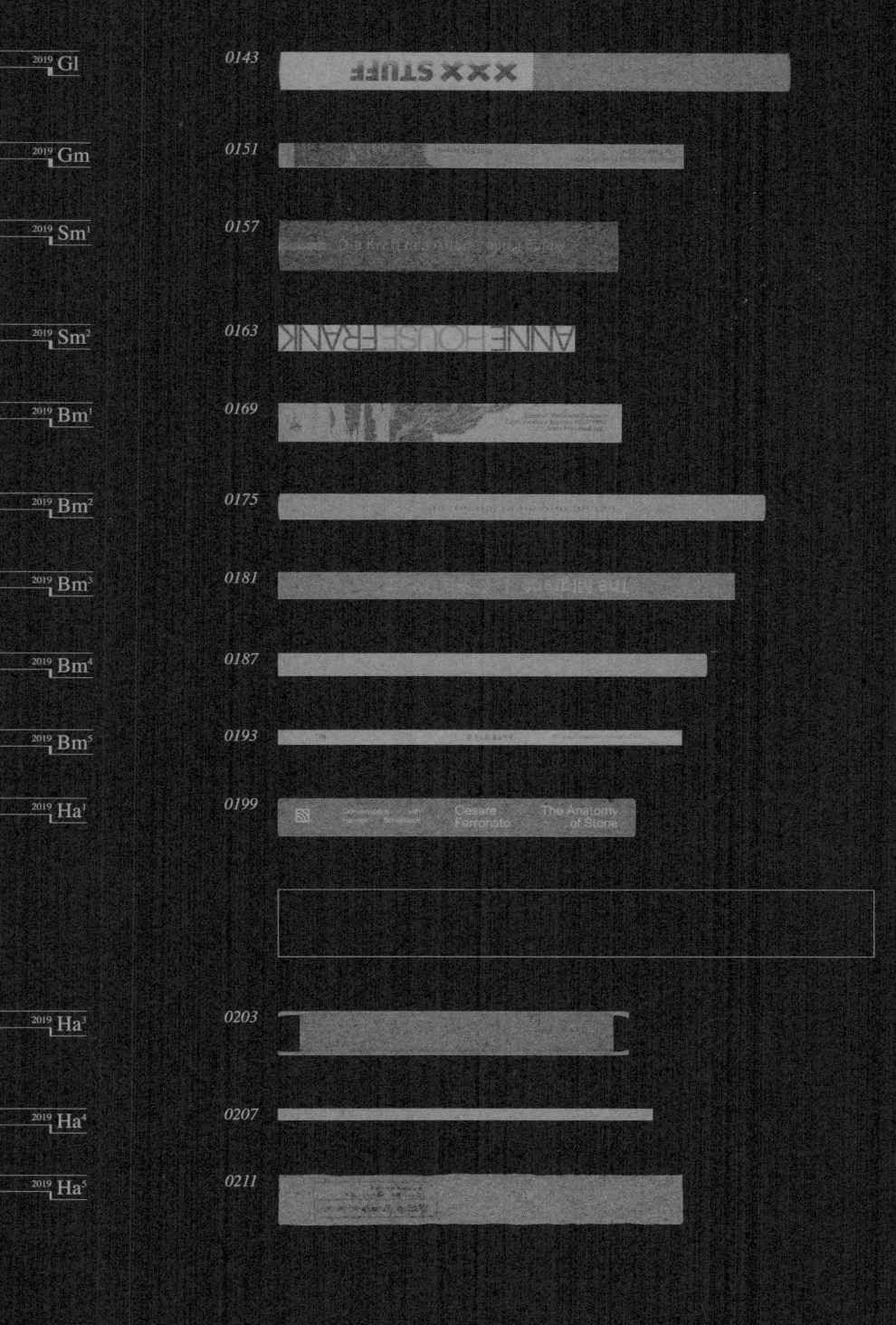

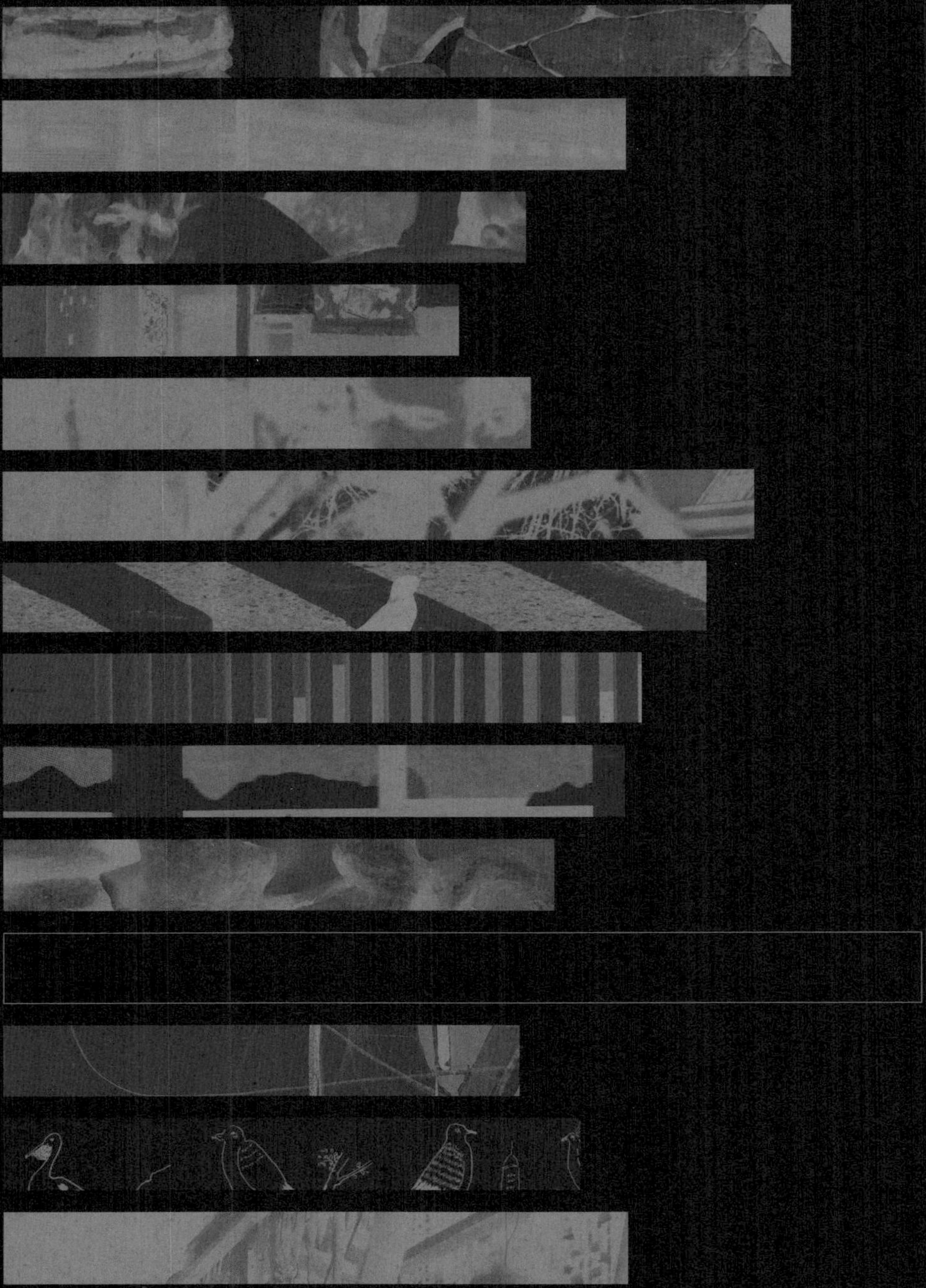

ols: 4.1-4.3 for food production; 4.4 soil investigation, well illing; 4.5-4.7 spinning, weaving, sewing; 4.8 fire making;

Sports and games: 9.1 skating; 9.2 kolf

fashion. Since 2014, I have been searching for the people who lined t to bid farewell to RFK, and the photographs and home movies th I began my inquiries at historical societies, archives, and libraries

Just as we would rather not see old people, we also do vant to have to think about death. Which is why in some pean cities the former grandiose black and gold hearses

for it. In addition, prices rose enormously during the occupation.

Take clothes, for example. Teenagers going through a growth spurt need new clothes from

financia

Aft

4,000 g

Commit

dová sídla v Bretani.[22] Plány navržené Bernardem Grueberem alizoval Josef Pruvot, který je s velkou pravděpodobností utorem architektonického řešeni dvou nově vybudovaných

to take advantage too often the result usually shows in the exams, and we act accordingly."

And that means that the offender is put on "probation" which means no trips, and constant

He is to play a very big part in organising the Royal tour of Canada in June and July, when the Queen and Prince Philip will open the new St. Lawrence Seaway.

"That was one of the

Edward VII Hospital and the Allied Circle.

As Mr. Ross, who is the chairman and joint managing director of a London fashion and fur house, explained to me:—

"I paint, frame, and then give my pictures

rebate before leaving Britain as a fresh form has to be made out."

No grounds

Mr. P. E. Holwell, inspector of taxes for West Hartlepool, said: "He should not

a figh fami

nanden, die durch Suizid sterben, und unter den 10 % ohne eindeutige Diagnose ähneln die psychiatrischen Symptome denen von Menschen, die durch

und Alkoholmissb wird auf 4 % bei P kungen, auf 7 % be

як пазл, а її поле зору сягає майже 360 градусів.

але розрізняє менше кольорів, ніж людина.

посередині перед носом через те, що його очі розташовані

CF Well, there was no medication. Initially all the medical staff could do was monitor a child's temperature; after that they would try to correct their growth by means of operations and stretching ther-

But my arms made up for that! My ar always been too long.

HS Later on you made thin arms and hands.

1 Proun, 1920–1921 | Russian State Archive of Literature and Arts | Cat.30

entation of the international avant- which encouraged artists to ove the limitations of genre. A cap definition of the Lissitzky pheno was provided by Selim Khan-Mago

the first

Kentarou Tanaka

2019 Gl

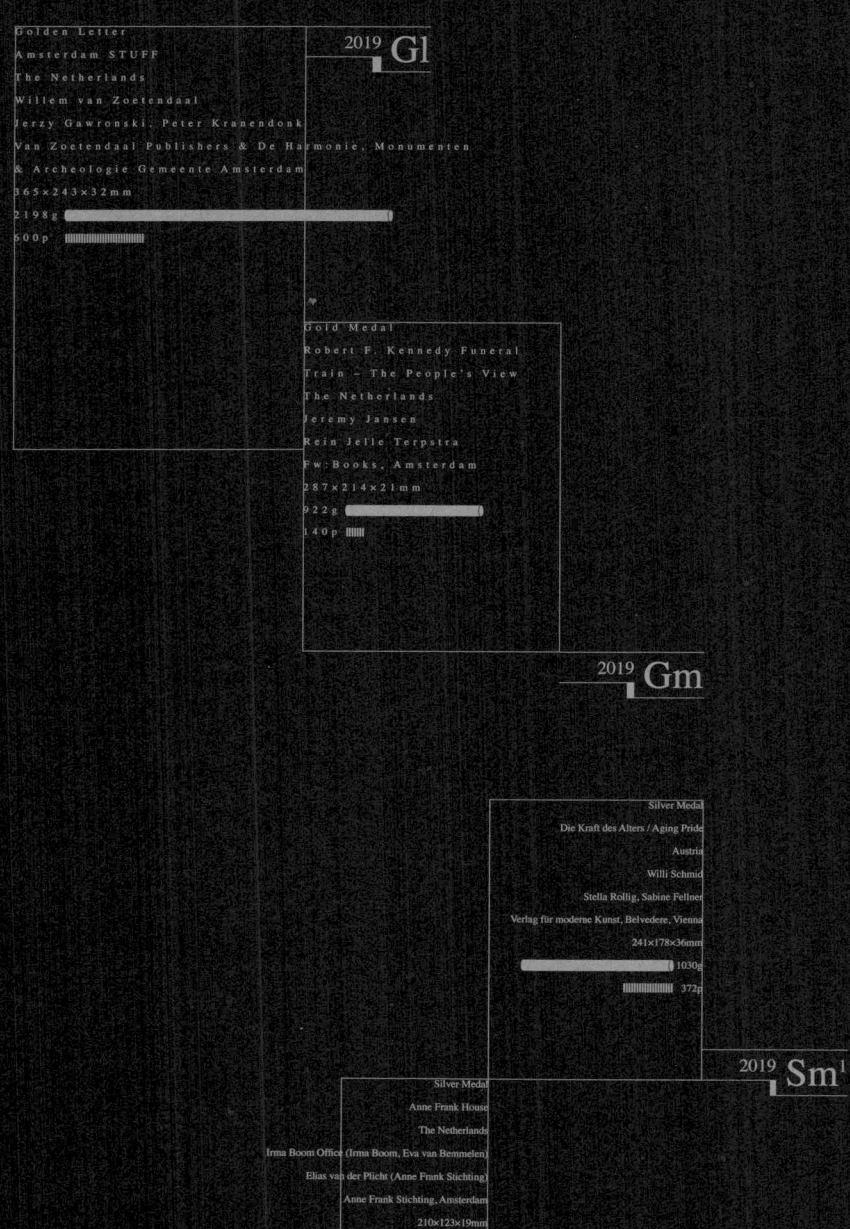

Golden Letter
Amsterdam STUFF
The Netherlands
Willem van Zoetendaal
Jerzy Gawronski, Peter Kranendonk
Van Zoetendaal Publishers & De Harmonie, Monumenten & Archeologie Gemeente Amsterdam
365×243×32mm
2198g
600p

Gold Medal
Robert F. Kennedy Funeral Train – The People's View
The Netherlands
Jeremy Jansen
Rein Jelle Terpstra
Fw:Books, Amsterdam
287×214×21mm
922g
140p

2019 Gm

Silver Medal
Die Kraft des Alters / Aging Pride
Austria
Willi Schmid
Stella Rollig, Sabine Fellner
Verlag für moderne Kunst, Belvedere, Vienna
241×178×36mm
1030g
372p

2019 Sm1

Silver Medal
Anne Frank House
The Netherlands
Irma Boom Office (Irma Boom, Eva van Bemmelen)
Elias van der Plicht (Anne Frank Stichting)
Anne Frank Stichting, Amsterdam
210×123×19mm
452g
336p

2019 Sm2

2019 Ha^5

Honorary Appreciation
Old Trades of Jiangsu: A Glimpse
China
Zhou Chen
Pan Wenlong, Gong Wei (photographer)
Jiangsu Phoenix Education Publishing, Ltd
288×284×38mm
1817g
646p

2019 Ha^1

Honorary Appreciation
Cesare Ferronato – Anatomie des Steins
Austria
Raphael Drechsel
Hannes Schüpbach
Verlag für moderne Kunst, Vienna
255×207×27mm
1042g
224p

2019 Ha^3

Honorary Appreciation
El Lissitzky
Russian
Evgeny Korneev
Tatyana Goryacheva, Ruth Addison, Ekaterina Allenova
Tretyakov Gallery, Jewish Museum and Tolerance Center
250×235×29mm
1084g
336p

2019 Ha^4

Honorary Appreciation
The First
Japan
Hideyuki Saito
Kentarou Tanaka
Bonpoint Japon
267×216×8mm
418g
46p

2019 Ha^2

2019 Bm^3

Bronze Medal
The Migrant
The Netherlands
Teun van der Heijden
Anaïs López
Self-Published
325×246×20mm
1048g+33g
124p+16p

2019 Bm^4

Bronze Medal
Nichtsein
Germany
Katharina Schwarz
Katharina Schwarz in co-operation with Ellen von den Driesch
Wissenschaftszentrum Berlin für Sozialforschung, UDK Berlin
305×174×16mm
553g
160p

2019 Bm^5

Bronze Medal
Я так бачу (This Is How I See)
Ukraine
Art Studio Agrafka (Romana Romanyshyn, Andrii Lesiv)
Art Studio Agrafka (Romana Romanyshyn, Andrii Lesiv)
Видавництво Старого Лева (Old Lion Publishing)
287×267×10mm
552g
56p

2019 Bm^1

Bronze Medal
Bonjour, Monsieur, Gauguin
Czech Republic
20YY Designers (Petr Bosák, Robert Jansa)
Adam Macháček)
Anna Pravdová
Národní galerie v Praze (National Gallery Prague)
245×173×32mm
879g
308p

2019 Bm^2

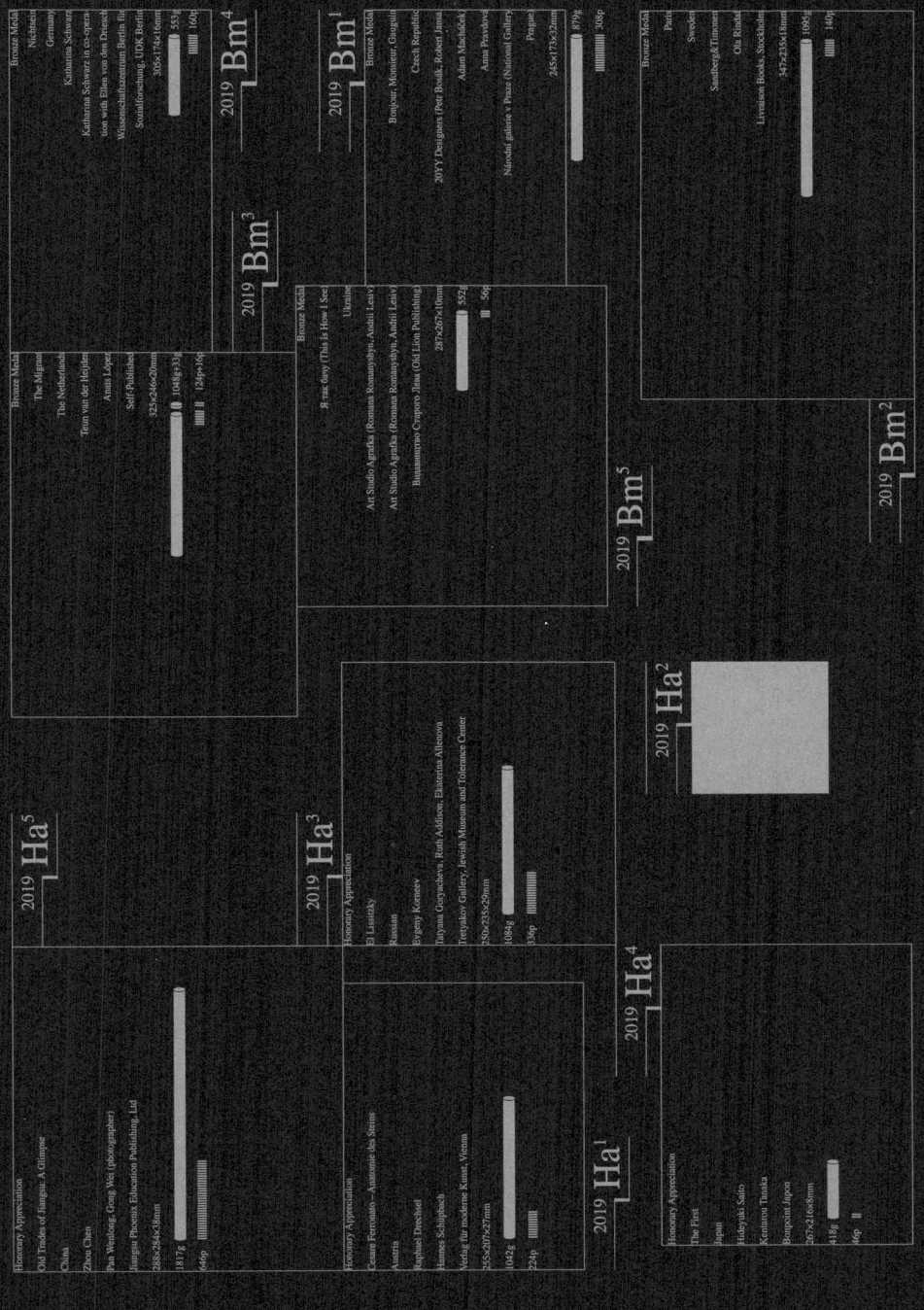

Bronze Medal
Paris
Sweden
Sandberg&Timonen
Ola Rindal
Livraison Books, Stockholm
347×235×18mm
1095g
140p

2019

The International Jury

Ricardo Báez (Venezuela)

Valeria Bonin (Switzerland Sascha)

Fronczek (Germany)

Simon Hessler (Sweden)

Rob van Hoesel (The Netherlands)

Country / Region

The Netherlands 4

Austria 2

Czech Republic 1

Germany 1

Sweden 1

Ukraine 1

Poland 1

Russia 1

Japan 1

China 1

Designer

Willem van Zoetendaal
Jeremy Jansen
Willi Schmid
Irma Boom Office (Irma Boom, Eva van Bemmelen)
20YY Designers (Petr Bosák, Robert Jansa, Adam Macháček)
Sandberg&Timonen
Teun van der Heijden
Katharina Schwarz
Art Studio Agrafka (Romana Romanyshyn, Andrii Lesiv)
Raphael Drechsel
Ryszard Bienert
Evgeny Korneev
Hideyuki Saito
Zhou Chen

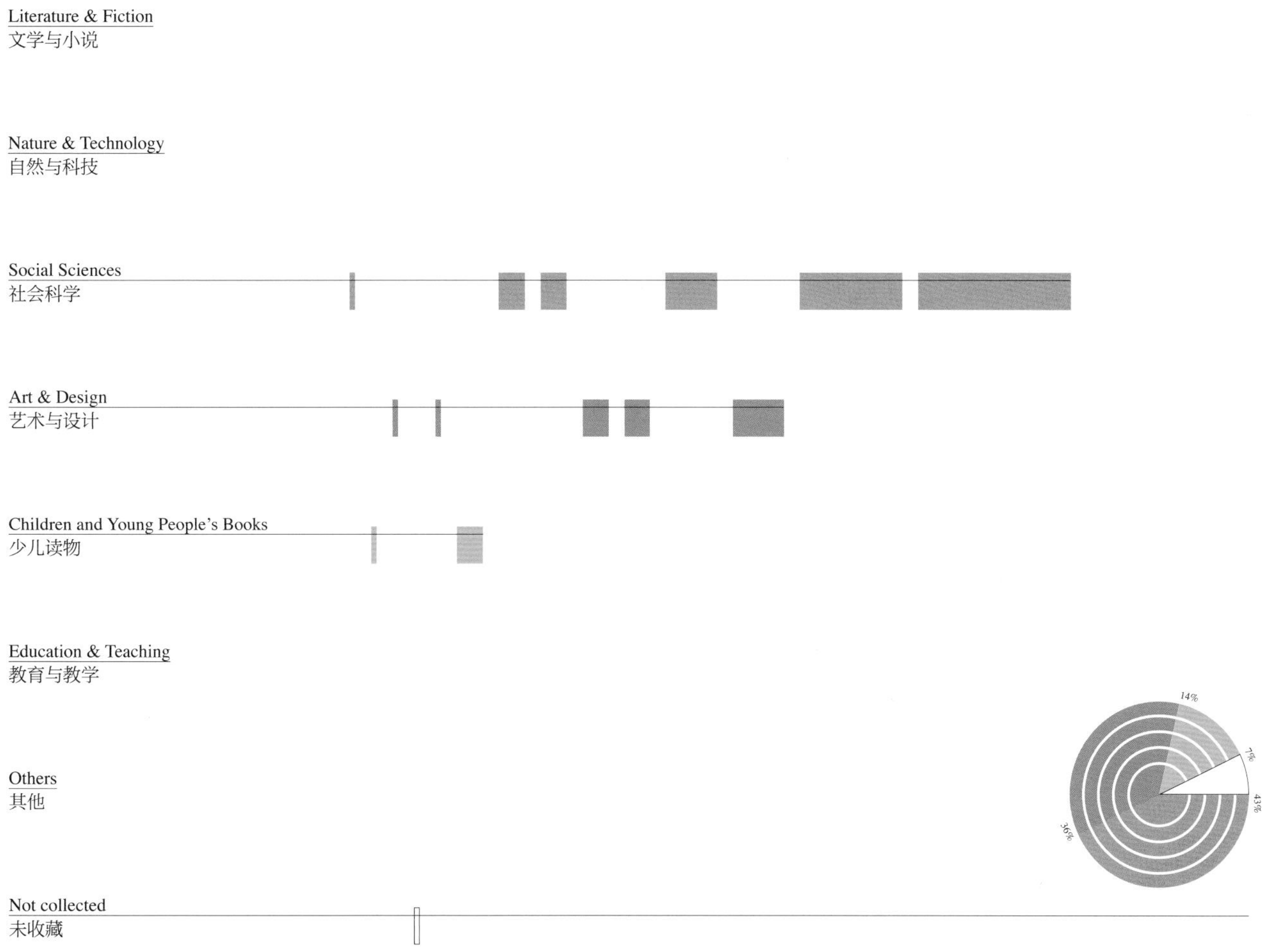
Literature & Fiction
文学与小说
Nature & Technology
自然与科技
Social Sciences
社会科学
Art & Design
艺术与设计
Children and Young People’s Books
少儿读物
Education & Teaching
教育与教学
Others
其他
Not collected
未收藏
14%
7%
43%
36%

5 Food Processing &
6 Science & Technology
6.9.4.3
Dial (brass, enamel) 1950-2005 Σ=1(1)

阿姆斯特丹市政府于 2018 年开通了南 / 北地铁线。2003-2010 年的地铁隧道挖掘过程，很大一部分工作有考古队参与。本书便是这次发掘工作的成果，收录了 70 万件发掘物中精选的 13000 件物品。书籍的装订形式为锁线硬精装。封面采用黄黑双色布面裱覆卡板，压烫 3 个银色圣安德鲁十字为阿姆斯特丹市徽的中心图案，指代阿姆斯特丹。护封采用土红色布纹纸烫黑色字，护封高度约为封面的一半，在位置和颜色上暗含城市的“镜像”即地下挖出的“东西”之意。内页采用超薄纸张五色印刷，体现考古发掘过程即剥开地层和历史。书籍最开始简述了阿姆斯特丹的历史、地铁工程的始末，以及针对当地独特松软地层的研究。之后分为 11 个部分展示发掘物，其中自然地理和早期文明部分的发掘物被划分为“0”，之后便是 10 个城市人造物分类。书中采用红色标注分类编号，配合专色蓝横排文字呈现更多信息；而专色蓝竖排文字则为发掘区域、物件材质、序列数字、尺寸和年代等物件信息编码。内页设置为 5×8 的网格系统，根据物件的形状大小占据相应的网格。最后则是一些形质分析和索引。

STUFF

The Amsterdam City Council opened the South/North Metro Line in 2018. A large part of the subway tunnel excavation process from 2003 to 2010 was attended by archaeological teams. This book is the result of this excavation work, with 13,000 items selected from 700,000 excavations. The binding form of the book is hard and hard-locked. The cover is covered with a yellow-black two-tone cloth cover, and the three silver St. Andrew's Crosses are pressed to the center of the Amsterdam emblem, referring to Amsterdam. The protective cover is made of black-colored cloth paper, and the height of the protective cover is about half of the cover. The position and color of the city imply the "mirror" of the city. The inner pages are printed in ultra-thin paper in five colors, reflecting the archaeological excavation process that strips the formation and history. The book begins with a brief description of the history of Amsterdam, the beginnings and ends of the subway project, and research on the unique soft layers of the area. It was then divided into 11 parts to show excavations, in which the excavations of natural geography and early civilization were classified as "0", followed by the classification of 10 urban man-made objects. The book uses red to mark the classification number, with the special color blue horizontal text to present more information; and the special color blue vertical text for the excavation area, object material, serial number, size and age and other object information coding. The inner page is set to a 5×8 grid system, occupying the corresponding grid according to the shape and size of the object. Finally, there are some qualitative analysis and indexing.

- 金字符奖
- 阿姆斯特丹的“宝藏”
- 荷兰

- Golden Letter
- Amsterdam STUFF
- The Netherlands
- Willem van Zoetendaal
- Jerzy Gawronski, Peter Kranendonk
- Van Zoetendaal Publishers & De Harmonie, Monumenten & Archeologie Gemeente Amsterdam
- 365×243×32mm
- 2198g
- 600p

Some years ago the city of Amsterdam took the opportunity to approach the major construction site for its new Metro line from an archaeological perspective. Scientists brought up countless items from the Amstel riverbed; 700,000 finds were recorded. Among these, 13,000 artefacts appear in this unique treasure trove that ranges from prehistoric times to the 20th century; they are all things that had sunk to the bottom of the Amstel. For instance, the chapter Food Processing & Consumption includes a pitchfork from around 1800 on the same double page as a chip fork from the outgoing 20th century. Despite its size, the use of India paper for the 600 pages makes this volume easy to handle. The paper's opaque quality has its own charm: it is reminiscent of the layering principle identified by archaeologists during excavations. Painstakingly extracted illustrations are presented according to category in an image area of 5 × 8 squares per page. If necessary, singular shapes are given extra space. This creates perfectly sorted displays that show a fascinating hotchpotch of items. The conspicuous cover reveals astonishingly little about the meticulously documented contents. It makes you feel quite curious. A paper half-cover bears the laconic title: Stuff. Above it, the three silver crosses of the Amsterdam coat of arms sit resplendent on the shimmering bicolour fabric — these shouldn't be confused with three "X"s, as any expectation of finding explicit material inside would be mistaken.

es t
tise
m's N
e gratefully c
0.1.4.3
12-6 m -NAP 11

2198

365 × 243 × 32

STUFF

600

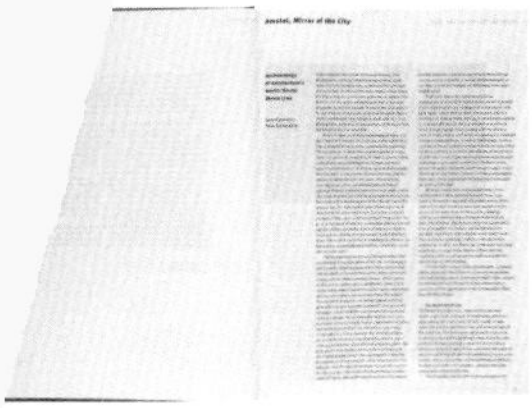

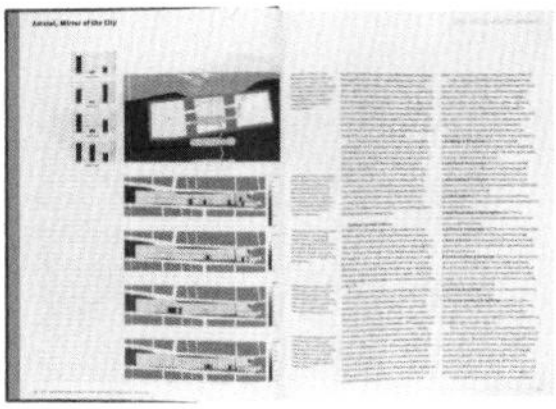

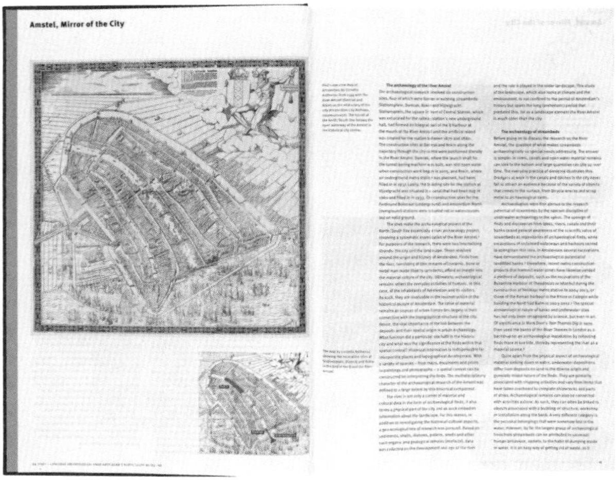
Amstel, Mirror of the City

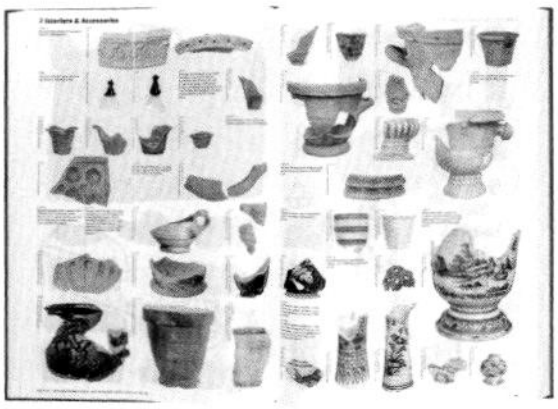

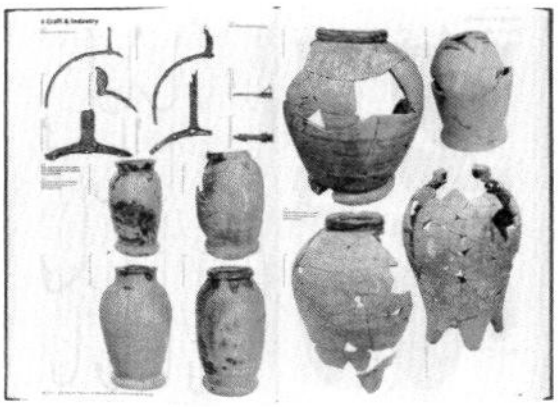

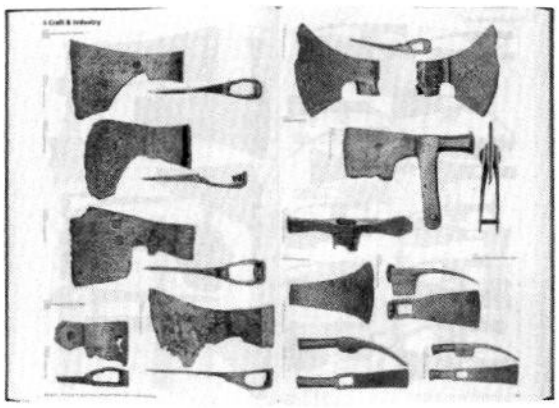

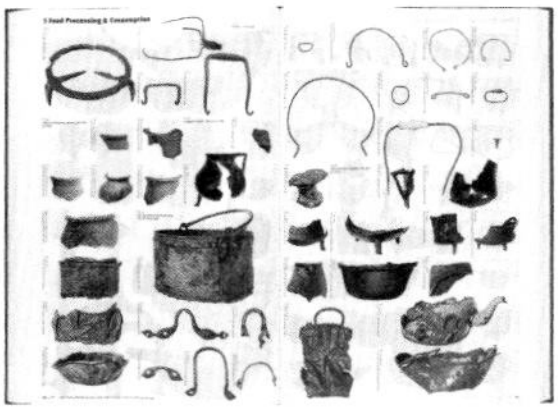

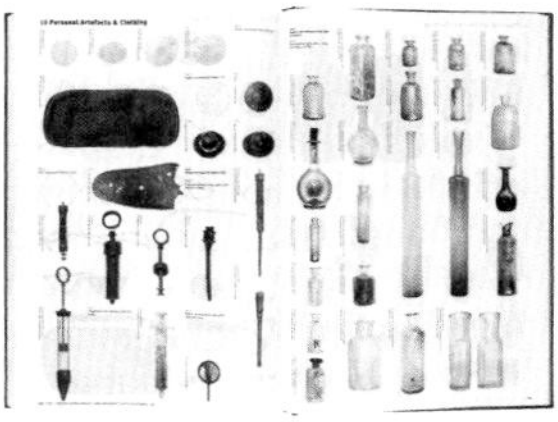

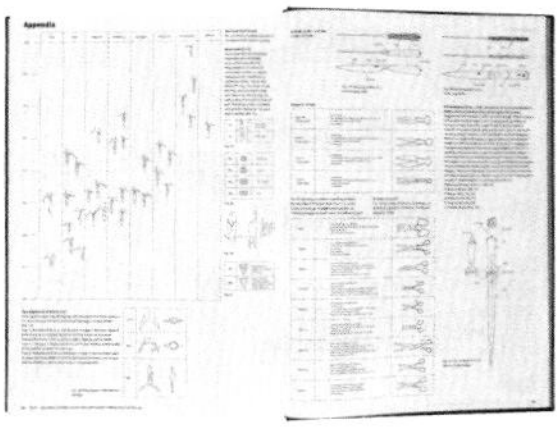
Appendix

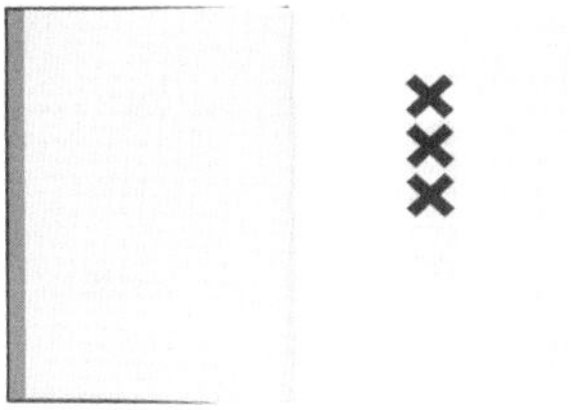

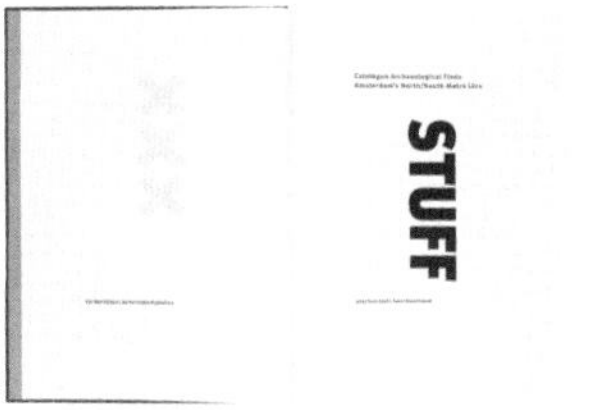
STUFF

Contents

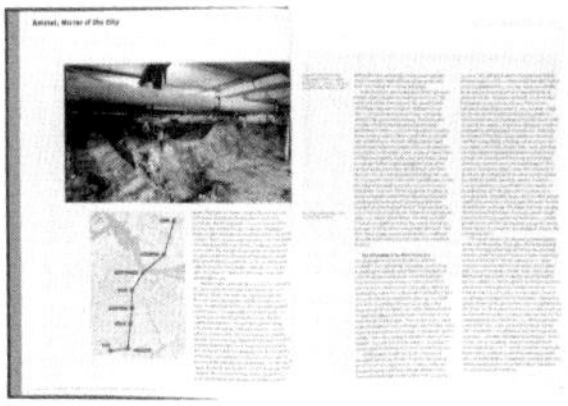
Amstel, Mirror of the City

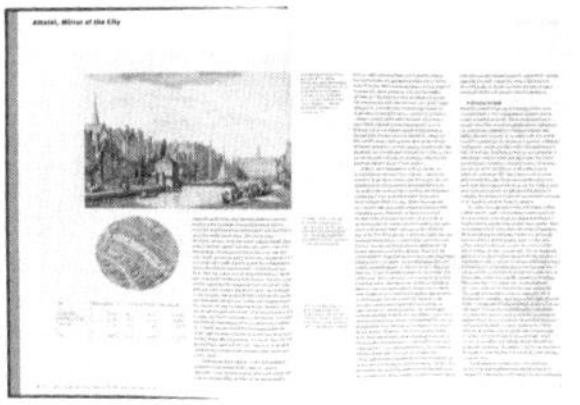
Amstel, Mirror of the City

Amstel, Mirror of the City

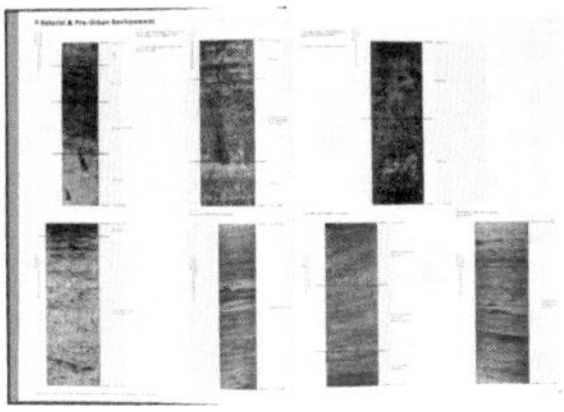
7 Natural & Pre-Urban Environment

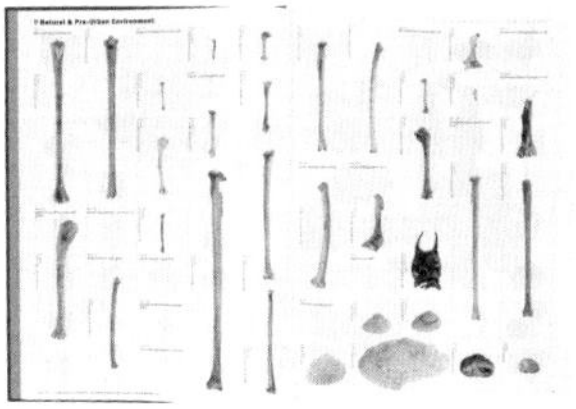
7 Natural & Pre-Urban Environment

1 Buildings & Structures

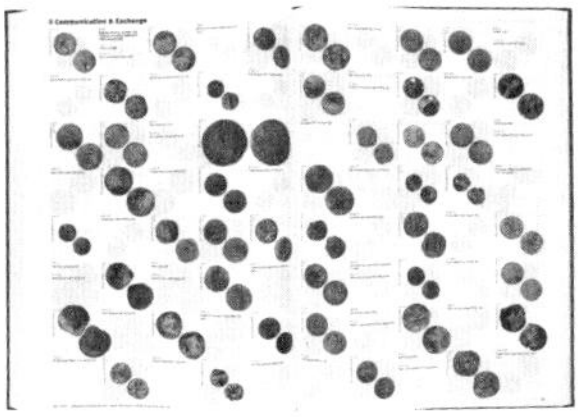
3 Communication & Exchange

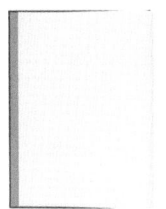

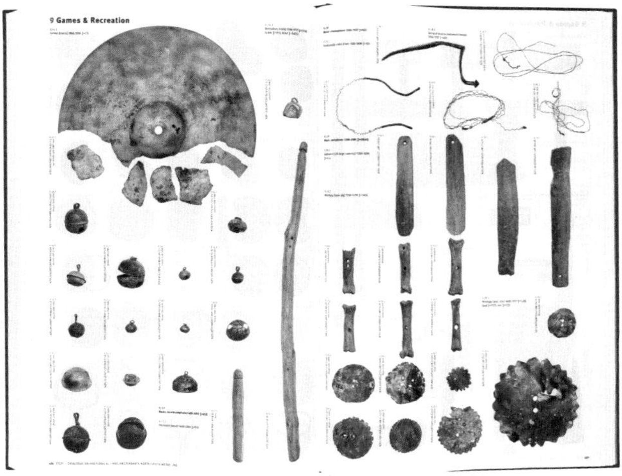
9 Games & Recreation

XXX STUFF

Appendix

Diverse typologies

Nail (1.9, 1.10)

Nails are subdivided in eight main types on the basis of the shape and manufacturing of the nail head. 0-5+ = number of recognisable facets on the head (fig. 1) (typology: Jan Dirk Bindt). The type code is composed of the row number and column number, separated by a dot. A chrono-typological subdivision is deducted from the nail head typology (fig. 2).

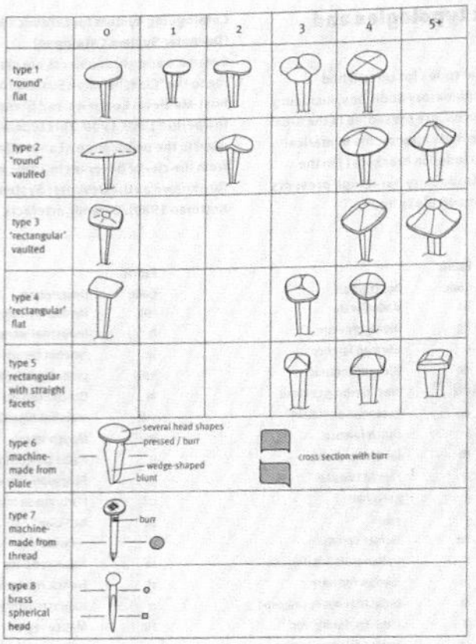
0
1
2
3
4
5+
type 1 'round' flat
type 2 'round' vaulted
type 3 'rectangular' vaulted
type 4 'rectangular' flat
type 5 rectangular with straight facets
type 6 machine-made from plate
several head shapes
pressed / burr
wedge-shaped
blunt
cross section with burr
type 7 machine-made from thread
burr
type 8 brass spherical head

Fig. 1 Typology of nail heads (typology: Jan Dirk Bindt)

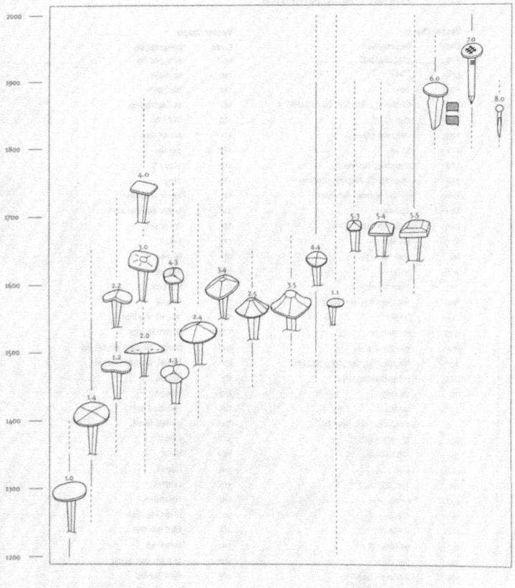
2000
1900
1800
1700
1600
1500
1400
1300
1200
7.0
6.0
8.0
4.0
5.3
5.4
5.5
3.0
4.3
4.4
3.4
2.2
3.5
1.1
2.4
2.0
1.2
1.3
1.4
1.0

Fig. 2 Chrono-typology of nail heads (typology: Jan Dirk Bindt)

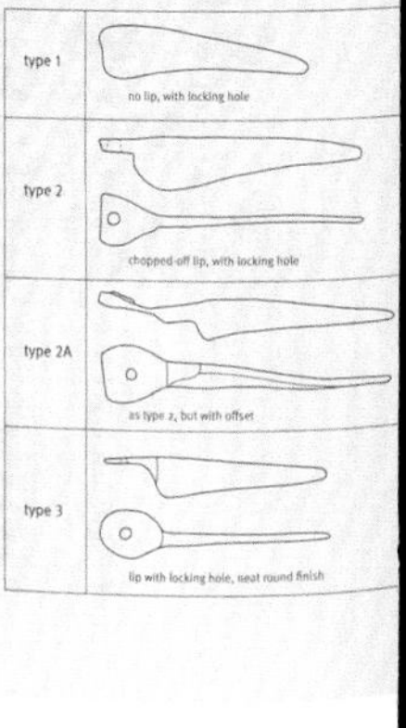
type 1
no lip, with locking hole
type 2
chopped-off lip, with locking hole
type 2A
as type 2, but with offset
type 3
lip with locking hole, neat round finish

Fig. 3 Forelock keys are morphologically subdivided into three basic categories (typology: Jan Dirk Bindt)

Forelock bolt and forelock key (1.12.1)

Forelock bolts are chrono-typologically subdivided into eight categories according to the section and shape of the shaft (typology: Jan Dirk Bindt): no drawing.
Type 1 Very irregular section (1300-1650)
Type 2 Irregular oblique rectangular section (1300-1650)
Type 3 Irregular rounded hexagonal section (1300-1650)
Type 4 Very irregular octagonal section (1300-1650)
Type 5 Rounded octagonal section (1500-1825)
Type 6 Round section with facets (1600-1825)
Type 7 Round section without facets (1600-1825)
Type 8 Square section (1500-1875)

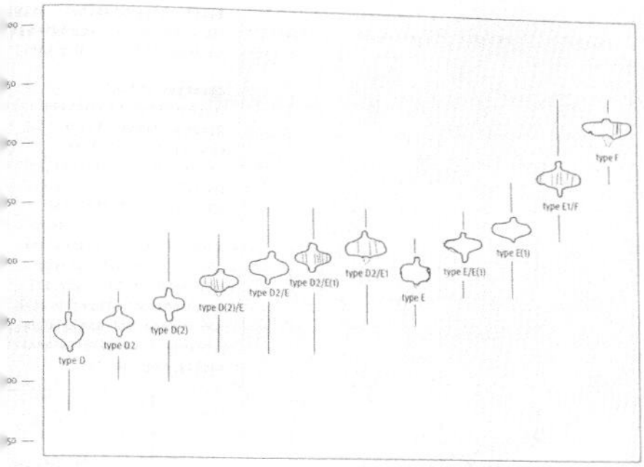

type F
type E1/F
type E(1)
type E/E(1)
type E
type D2/E1
type D2/E(1)
type D2/E
type D(2)/E
type D(2)
type D2
type D

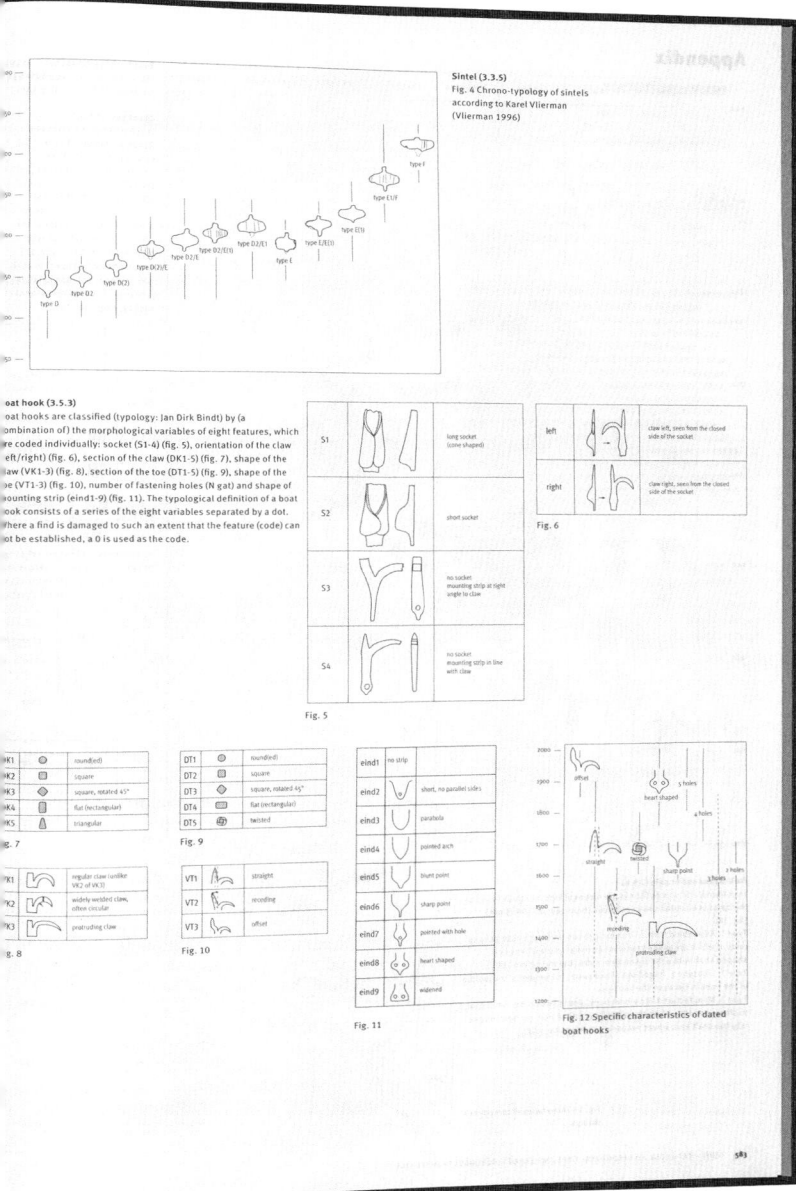

Sintel (3.3.5)
Fig. 4 Chrono-typology of sintels according to Karel Vlierman (Vlierman 1996)

oat hook (3.5.3)
oat hooks are classified (typology: Jan Dirk Bindt) by (a ombination of) the morphological variables of eight features, which re coded individually: socket (S1-4) (fig. 5), orientation of the claw eft/right) (fig. 6), section of the claw (DK1-5) (fig. 7), shape of the law (VK1-3) (fig. 8), section of the toe (DT1-5) (fig. 9), shape of the e (VT1-3) (fig. 10), number of fastening holes (N gat) and shape of ounting strip (eind1-9) (fig. 11). The typological definition of a boat ook consists of a series of the eight variables separated by a dot. here a find is damaged to such an extent that the feature (code) can ot be established, a 0 is used as the code.

S1		long socket (cone shaped)
S2		short socket
S3		no socket mounting strip at right angle to claw
S4		no socket mounting strip in line with claw

Fig. 5

left		claw left, seen from the closed side of the socket
right		claw right, seen from the closed side of the socket

Fig. 6

K1		round(ed)
K2		square
K3		square, rotated 45°
K4		flat (rectangular)
K5		triangular

g. 7

DT1		round(ed)
DT2		square
DT3		square, rotated 45°
DT4		flat (rectangular)
DT5		twisted

Fig. 9

K1		regular claw (unlike VK2 of VK3)
K2		widely welded claw, often circular
K3		protruding claw

g. 8

VT1		straight
VT2		receding
VT3		offset

Fig. 10

eind1	no strip	
eind2		short, no parallel sides
eind3		parabola
eind4		pointed arch
eind5		blunt point
eind6		sharp point
eind7		pointed with hole
eind8		heart shaped
eind9		widened

Fig. 11

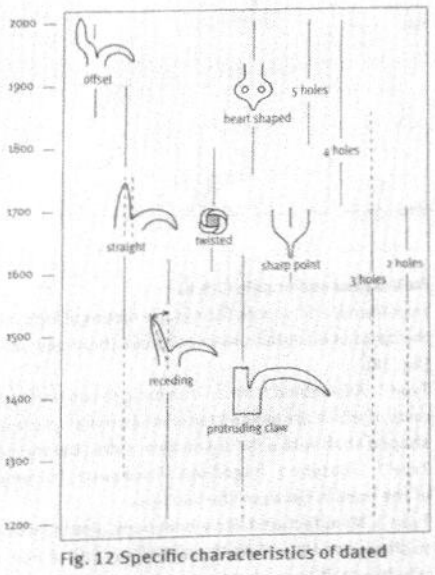

2000
1900
1800
1700
1600
1500
1400
1300
1200
offset
5 holes
heart shaped
4 holes
straight
twisted
sharp point
2 holes
3 holes
receding
protruding claw

Fig. 12 Specific characteristics of dated boat hooks

1968年6月8日，当时遇刺的美国总统候选人罗伯特·肯尼迪的遗体由火车从纽约运往华盛顿，成千上万的民众自发在铁路沿线送别这位领导人，并拍摄了大量的照片。本书的作者采访了拍摄这些照片的民众，展示了他们的照片和想法。本书的装订形式为无线包背胶平装。封面采用牛皮卡纸白黑双色叠印，书函采用黑色卡板印白，展现连续的火车照片。书籍内页采用黑白双色胶版纸张混合装订，白色部分四色印刷，展现大量单幅彩色照片和摄影者的访谈文字；黑色部分则先印白色铺底，再叠印四色照片，呈现连续的照片，形成类似视频的叙事感，充满了不真实的沉重气氛。全书均采用调频网印刷，大量的照片由于是非专业摄影师拍摄，画面不完美，但饱含情绪。全书没有页码，而采用火车运行到各地的时间作为标签，使得全书的情绪集中在事件本身。书的最后罗列了完整的照片时间、地点和拍摄者信息，并在黑色拉页中夹附一本在火车上的摄影师拍摄民众的照片和相关记载的手册。||||||||||||||||||||||||||||||||||||||||||||||

Robert F. Kennedy Funeral Train—
The People's View
Rein Jelle Terpstra

On June 8, 1968, the body of the assassinated president candidate, Robert Kennedy, was transported by train from New York to Washington. Thousands of people spontaneously saw the leader off along the railway and took numerous photographs. The author of this book interviewed the people who took pictures and published their pictures and ideas. This double-leaved book is perfect bound. The cover is printed in black and white with kraft cardboard, and the slipcase is made of black cardboard printed in white to show continuous photos of the train. The inner pages are bound with black and white offset paper and printed in four colors, showing a large number of single-page color photographs and transcripts of the interviews. The black part is printed with white background and then overlapped with continuous four-color photographs to form a narrative rhythm similar to a video. This part is full of heavy atmosphere of untruthfulness. The entire book is printed with FM screen. Since non-professional photographers took these photographs, they are not perfect, but full of emotions. The book has no page number, but uses the time when the train traveled to different places as the label, which makes the mood of the book focus on the event itself. At the end of the book, complete information about the time, place and photographer is listed, and a special album of photos taken by photographers on trains is attached to the black pullout page.

◉ 金奖

《 罗伯特·肯尼迪的葬礼火车——民众的视角

◖ 荷兰

◎ Gold Medal

《 Robert F. Kennedy Funeral Train – The People's View

◖ The Netherlands

▷ Jeremy Jansen

△ Rein Jelle Terpstra

||||| Fw:Books, Amsterdam

□ 287×214×21mm

⌂ 922g

≡ 140p

On 8 June 1968 the funeral train with the body of murdered presidential candidate Robert F. Kennedy travelled from New York to Washington. The photographer Paul Fusco was also on the train and took photographs of the people who lined the tracks in their thousands. This memorable event has been researched by photographer Rein Jelle Terpstra, who asked himself how those people actually perceived the train. In extensive fieldwork he managed to make contact with persons who filmed and took private photographs themselves. The different perspective has been documented and presented by Terpstra in this publication. Structured chronologically, the material reconstructs the journey from outside the train. All pages are designed as double pages, with an outward fold. This allows photos to extend over the front edge as well. When leafing through the book, a kind of propulsion is generated that evokes the funeral train rolling past. Individual frames from films are printed on black paper. These rather out-of-focus pictures, taken spontaneously by amateurs, are only understandable because the circumstances are known; otherwise they would seem mundane. On black paper? White sections had to be preprinted, creating a peculiar, unreal mood that recalls the harrowing atmosphere that day. One contemporary witness remarks: I can remember the thickness in the air. It is a book that preserves sources and serves as a memorial.

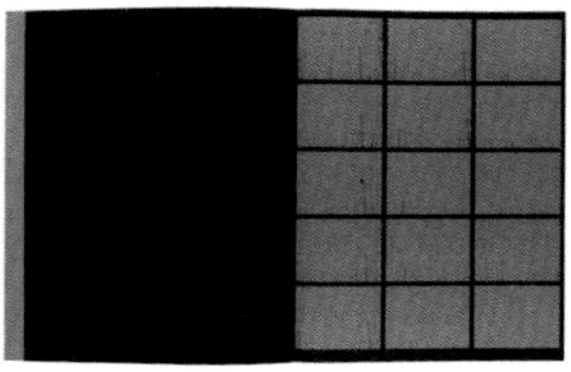

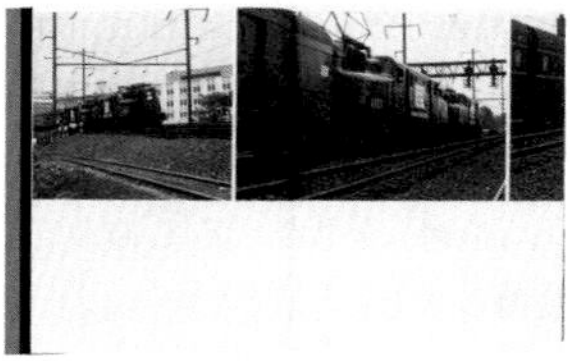

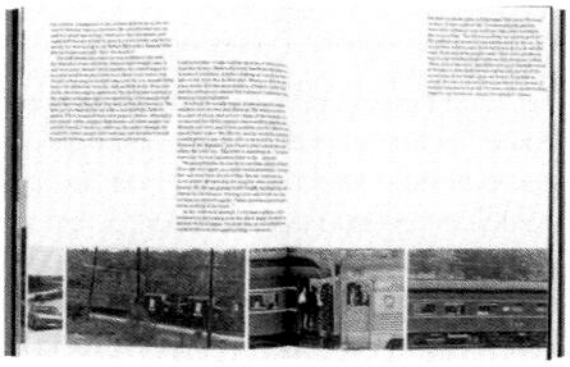

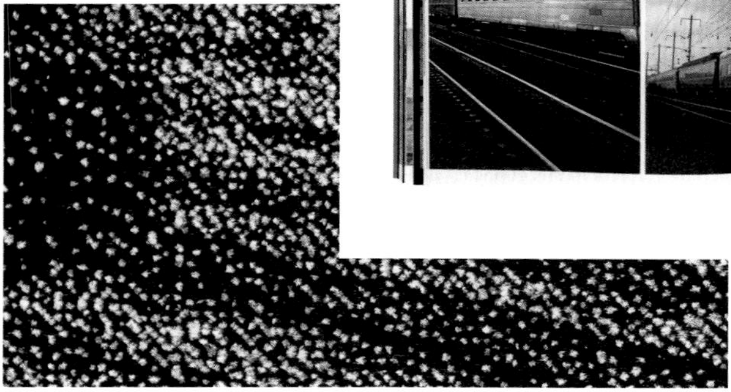

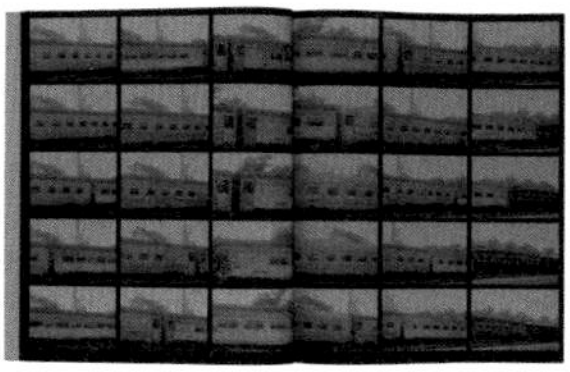
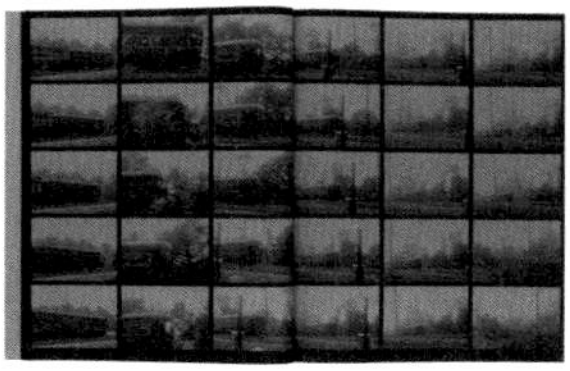
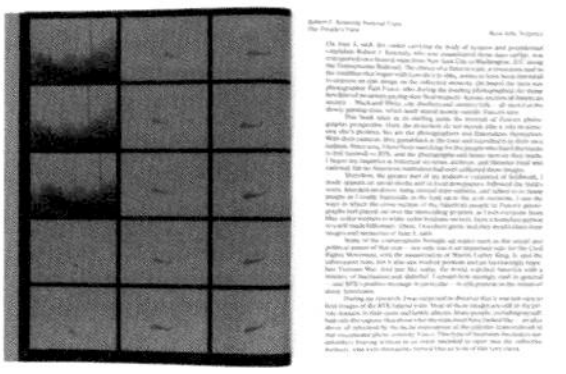

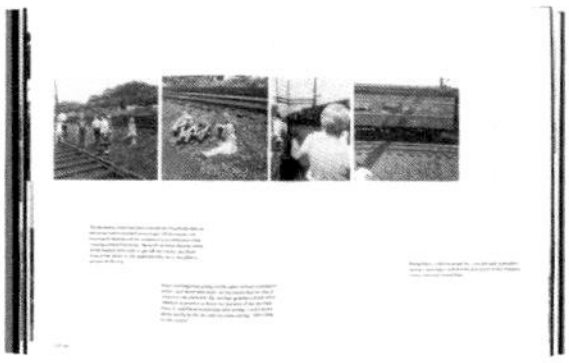
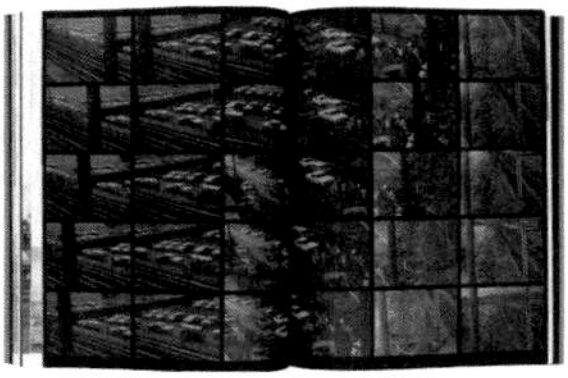

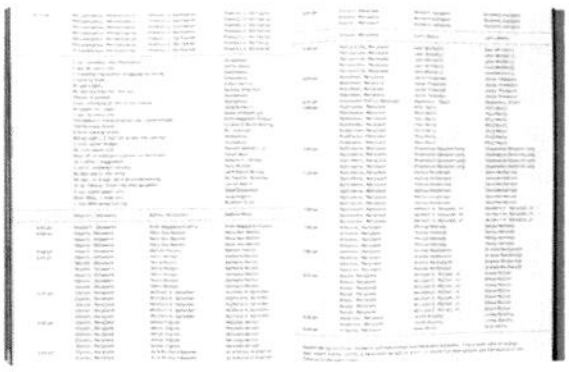
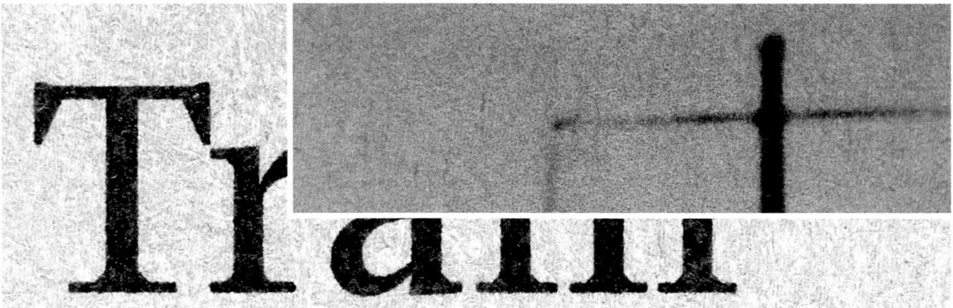
Train

Robert F. Kennedy Funeral Train —
The People's View

Rein Jelle Terpstra

Fw:Books

He was, in a way, more groundbreaking than his brother. The distance he travelled philosophically from 1960 to 1968 was a great one. He was on the Senate committee with Joe McCarthy of all people — that's really a change. I think Jack Kennedy was portrayed as being for the common people, but I think Bobby really was. Part of that was his personality. He was the middle child, he was scrappy, he was not as healthy. So, I think he had more empathy with people. You know, the Kennedy myth-making machine went into place, but he did go to the coal mines. I mean, he wasn't afraid. He did very, very bold things.

In my family, I was the photographer. When I was eight, Mom and Dad gave me a little Brownie Kodak. As things progressed, I went into movies. On that day, I brought my 8-mm movie camera.

Totally on the spur of the moment, people began showing up along the sides of the main rail line. I grew up a block from there. So, I went over about the time I thought the train was going to come through. It wasn't announced on television for people to pay their homage, but it was a natural thing to do. I remember the Boy Scouts had arrived in their uniforms with the American flag, and the Maryland flag, and then the Boy Scout flag.

1968 to me means a very turbulent year in my life, because we had Martin Luther King assassinated, and the cities burned in some form of retaliation. It was terrible to watch on the television. His assassination hurt everyone, no matter what religion or race. The incumbent president, Lyndon Johnson, had to deal with the Vietnam War, which was very much tearing at the country. Opinions were being drawn: should we get out of there, should we be there?

I feel Robert Kennedy was picking up the torch of his brother John. There is no doubt he would have been elected overwhelmingly. So, with his assassination, I thought as a 16-year-old that my town, my country, was falling apart — I felt that the world was falling apart.

PENN CENTRAL

I was eight years old and went to, I believe, North Philadelphia train station to watch with my family. Obviously, it was a very somber scene, and I had not been exposed to anything like it. I remember the gravel along the side of the tracks and waiting to see if we would see Ted Kennedy. Today, I don't remember whether we did see him or not.

The following month, we moved out of the house we lived in then. Probably my last vivid memory of living in that house was hearing my father say to my mother early in the morning, "You'll never believe this, but Bobby Kennedy's been shot."

My mother may have adult recollections of the event. It's strange to think that I'm 23 years older now than she was then.

Dear Adam,

I took you, your brother Ben, and little brother Dan (not quite three) to North Philadelphia Station because I knew Bobby's death was a hole in the heart of the people who lived there. It was as painful for me, and the pain remains.

We were all holding hands and singing "Glory, Glory, Hallelujah." Most people, including me, had tears streaming down our faces as we began swaying side to side. I wrote several poems about RFK's passing to help me get through the pain. He touched me more than any other presidential candidate ever has because of his compassion for the have-nots, such as the Native Americans, the Chicanos, the Blacks, the poor of any background. His own suffering because of his brother's death had obviously changed him enormously from his time as counsel for Joe McCarthy. The year 1968 was the most painful year in my life because of what happened to Martin Luther King and Bobby. As for who was visible as the train passed, I know there were several people at the back of the train waving back at us. I only remember seeing Ethel, but that doesn't mean Teddy wasn't there.

Love,
Mom

人们对衰老的认知随年龄的变化而改变。年轻人为了显得有经验，往往装扮成熟；到了一定年纪后，为了在社会和职业选择上不被边缘化，又会主动抵抗衰老。本书就是这样一本探讨老龄化及其背后文化建构的展览画册，策展人想传达这样的理念：人们都在抵抗衰老，但年纪渐长往往也代表着经验、知识和智慧的积累。衰老是个生物学概念，其背后的文化建构，才更能影响人们对于自我的认知和未来的选择。本书的装订形式为锁线圆脊硬精装。护封胶版纸四色印刷，内封蓝色纸张裱覆卡板，书名压凹处理，内页艺术纸四色印刷。全书印有 108 位艺术家的 188 件作品，包含绘画、雕塑、装置和摄影等形式，伴随 6 篇论文。论文部分设置为 4 栏，正文 3 栏、注释 1 栏，黑蓝双色、德英双语设计。书籍的纸张从前往后由白变黄，暗示衰老的过程。但由白变黄也可以解释为逐渐焕发出光辉，正如封面上已故的纽约时尚达人泽尔达 · 卡普兰（Zelda Kaplan）的侧影，身穿华服，闪耀光芒。

Die Kraft
des Alters

People's perception of aging changes with age. In order to appear experienced, young people often dress up to be mature. At a certain age, they will take the initiative to resist aging to avoid being marginalized in the society. This book is an exhibition album that explores aging and the cultural construction behind it. The curators wanted to convey the idea that people are resisting aging, but aging often represents the accumulation of experience, knowledge and wisdom. Aging is a biological concept, and the cultural construction behind it can influence people's self-perception and future choices in a better way. This hardback book is bound by thread sewing with a rounded spine. The jacket is printed on offset paper in four colors, and the inside cover is made of cardboard mounted with blue paper. The book title is embossed. The inside pages are printed in four colors on art paper. The book contains 188 works of 108 artists, including paintings, sculptures, installations and photographs, accompanied by six papers. The part of papers is typeset with four columns: three columns of the body text and one column of notes, printed in black and blue and written in German and English. The paper used in the book turns from white to yellow from the beginning to the end, suggesting the process of aging. However, turning from white to yellow can also be interpreted as sparkling gradually, just like the shining silhouette of the late New York fashion guru Zelda Kaplan on the cover, who is dressed in gorgeous clothes.

银奖
年龄的力量
奥地利

Silver Medal
Die Kraft des Alters / Aging Pride
Austria
Willi Schmid
Stella Rollig, Sabine Fellner
Verlag für moderne Kunst, Belvedere, Vienna
241 × 178 × 36mm
1030g
372p

The jacket of this exhibition catalogue displays an old lady — a prominent figure in New York's art scene — posing in a statuary profile. The half length portrait glows in shades of copper against the dark, warm background. You can imagine her extravagant silk suit rustling when she moves. She wears the hat like a crown. The two blue lines of the title read in opposite directions, positioned in front of and behind her to effectively underline the area of her face and body from the side, with their aura of self-assurance. Essays and picture sections alternate on sturdy natural paper. The details were not a self-evident solution, however: firstly, the different lengths of the German and English texts had to be counterbalanced. That doesn't seem to be particularly problematic here, though. The dark-blue English text is discreetly but sufficiently different from the black German. The comments are written in the margin column, oriented towards the spine. Their numbering begins with the same deep indents that also structure the columns of main text, flush with the page numbering that is also shifted inwards. This creates several vertical lines in the type area, which are used to align the pictures on those pages. By the end of the book, the paper has changed colour almost imperceptibly and has adopted the yellow shade of the endpaper at the front and back. This gradual yellowing lends the book a vividness that subtly reflects its topic: Aging pride.

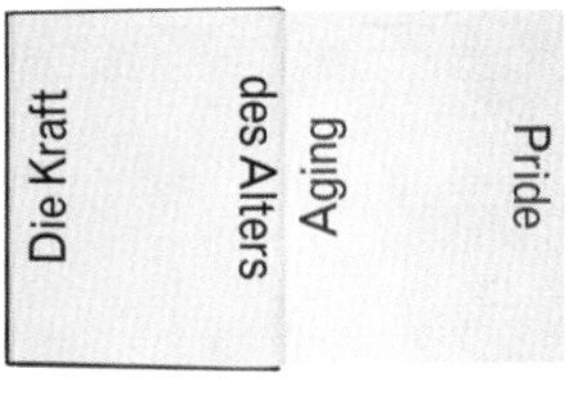
Die Kraft
des Alters
Aging
Pride

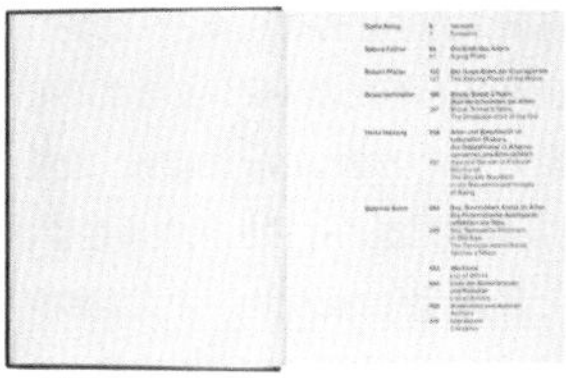

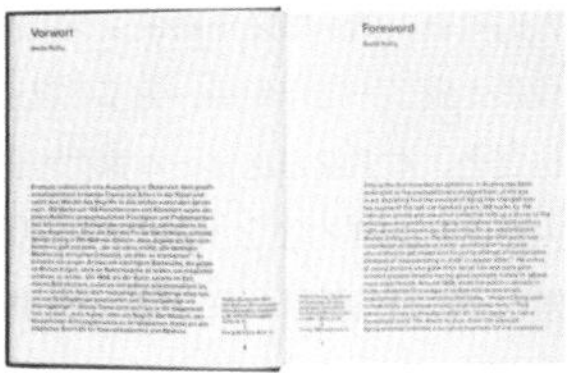
Vorwort
Foreword

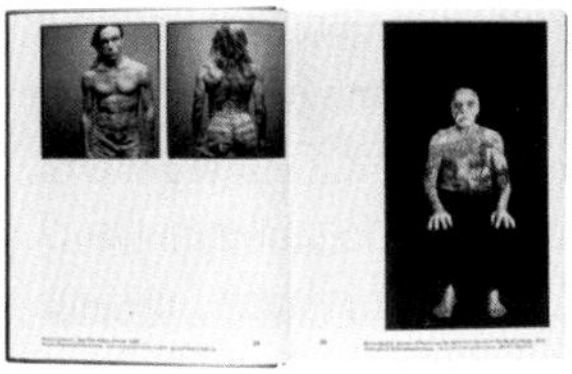

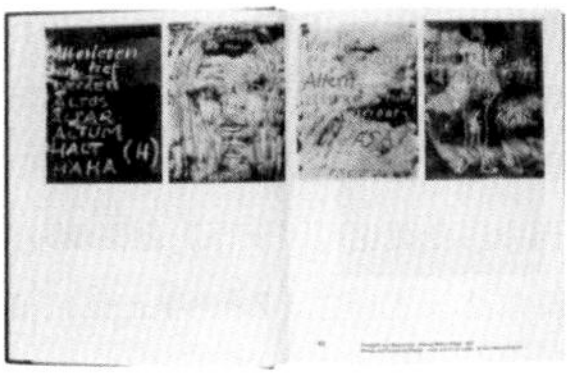

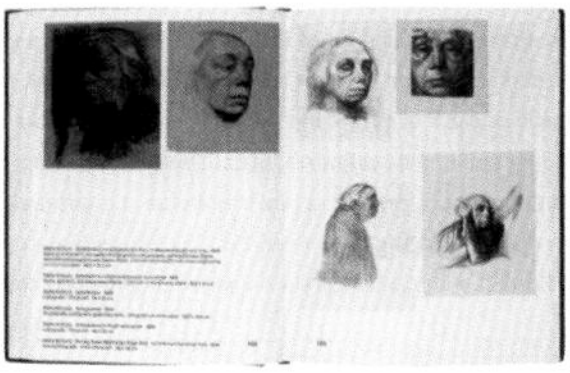

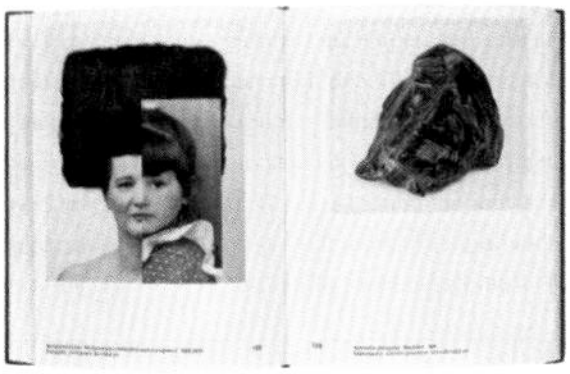

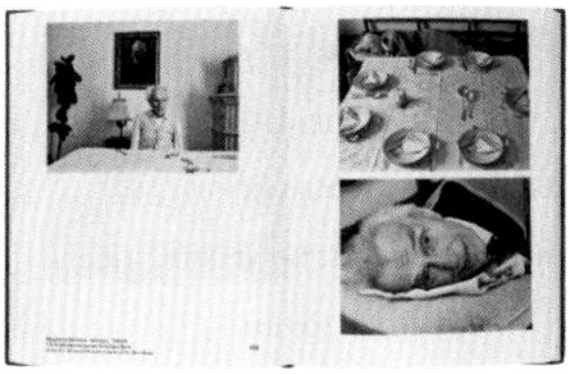

223
Szenenfotografie aus production photograph from
Kontakthof mit Damen und Herren ab 65, Tanztheater Wuppertal 2001
Ein Stück von Pina Bausch a piece by Pina Bausch

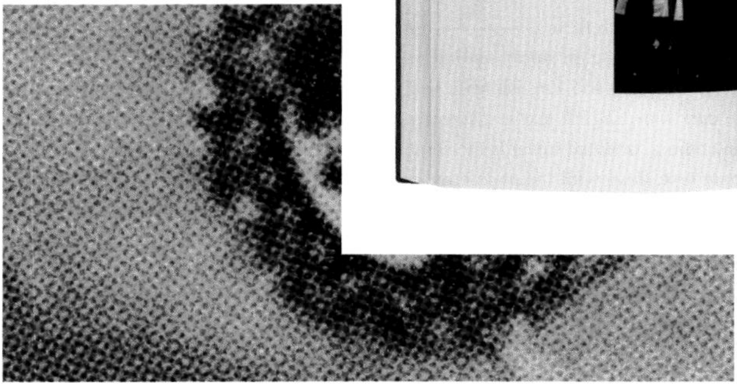

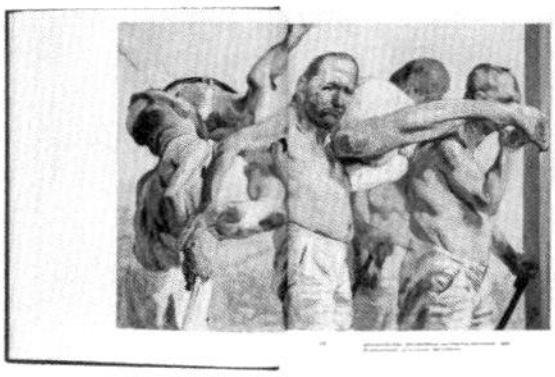

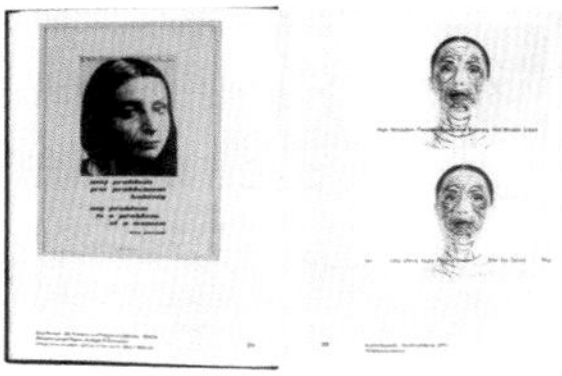

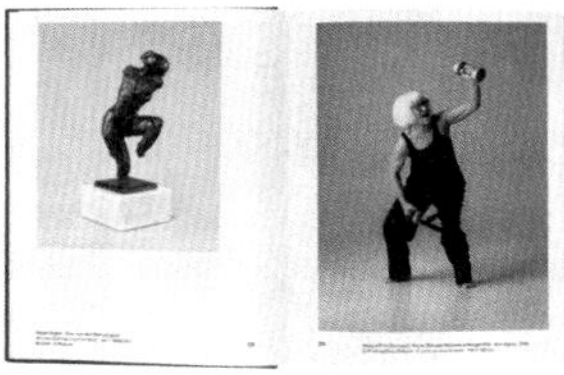

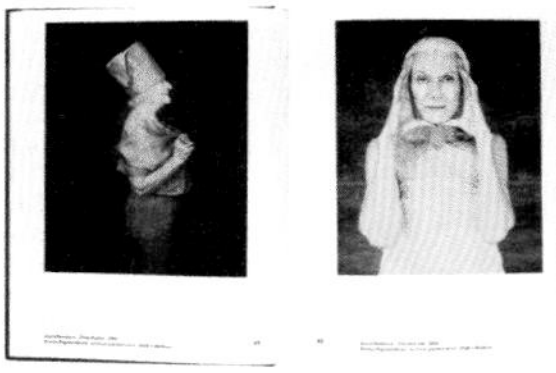

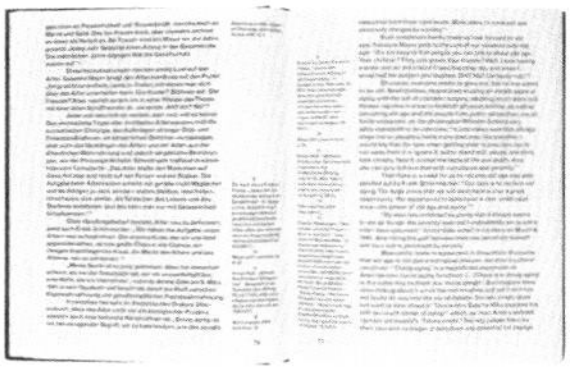

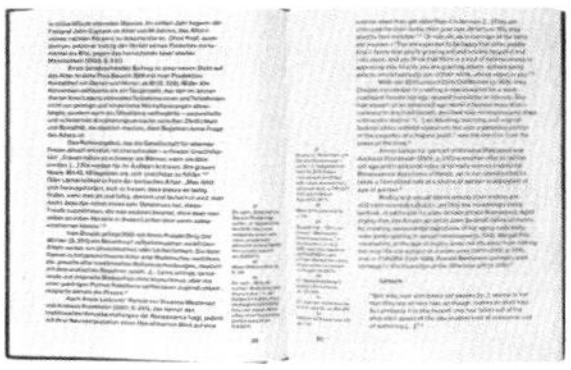

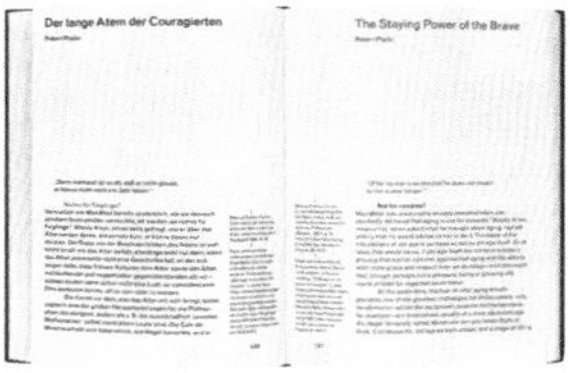
Der lange Atem der Couragierten
The Staying Power of the Brave

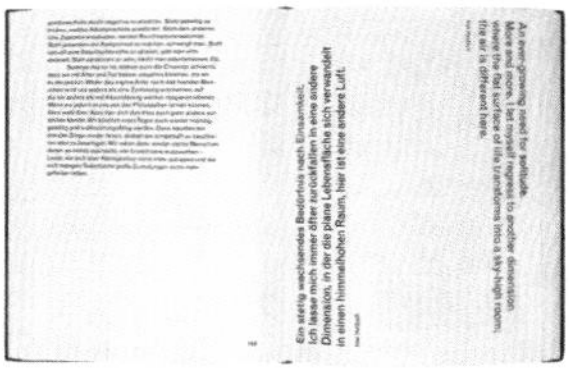

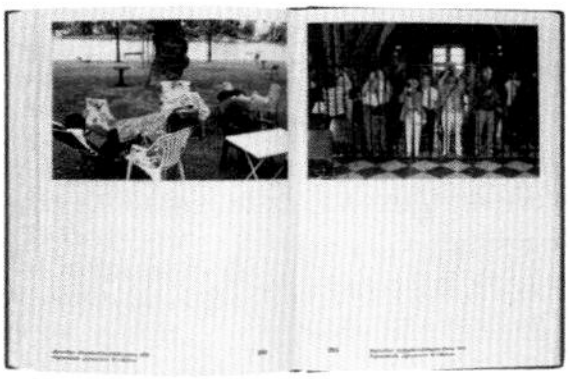

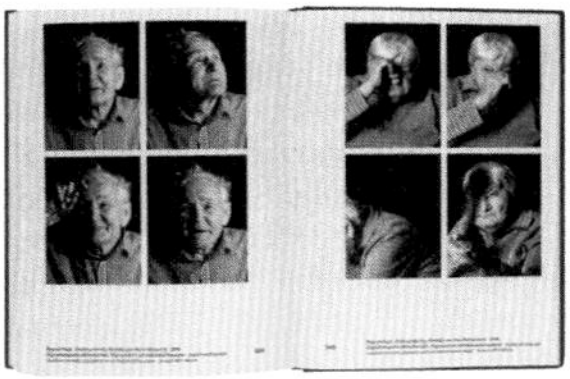

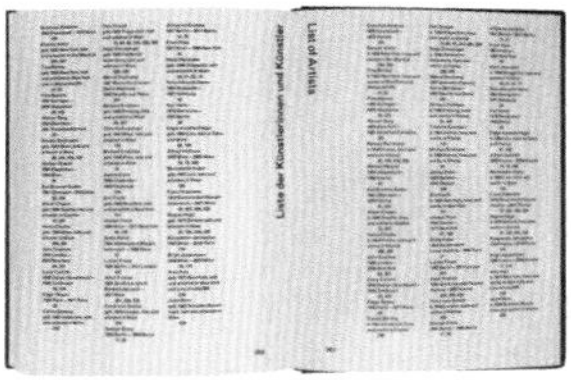

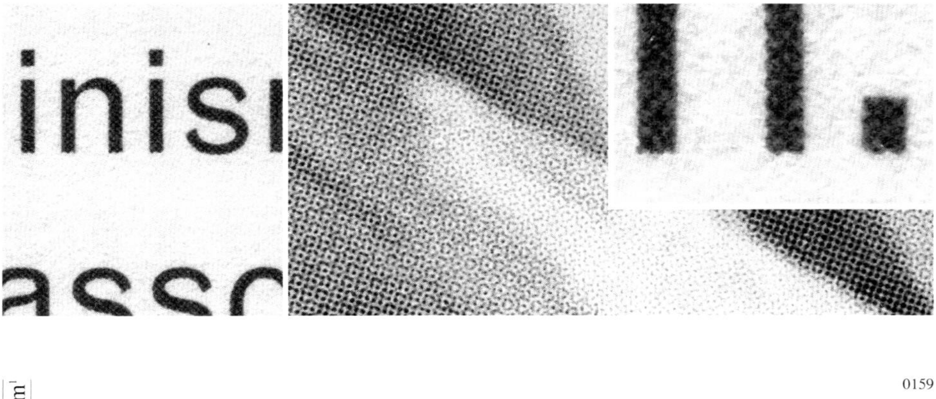

Die Kraft des Alters Aging Pride
belvedere

Blood, Sweat & Years. Zum Verschwinden der Alten

Beate Hofstadler

Verkörpern

Der (Spiel-)Film schafft keine neuen gesellschaftlichen Verhältnisse, sondern stellt bestehende Zustände dar, jedoch mit realitätsstiftender Wirkung. Während Kunst am Puls der Zeit unmittelbar erfasst, was gesellschaftlich brodelt, aber vielleicht noch nicht ausbuchstabiert ist, hinkt die wissenschaftliche Aufarbeitung stets hinten nach. Aber auch Kunst reproduziert Stereotype, Rollenklischees und blinde Flecken, wie in den einzelnen Genres nachzuweisen ist.[1] So auch in den Altersdarstellungen. Wie werden alte Menschen, deren Körper und Beziehungsstrukturen beleuchtet und gezeigt? Das Kino bedient unsere Schaulust und zeigt uns, wie wir begehren[2] können. Zudem ist das Kino ein Ort der Projektion[3], nicht nur des Films, sondern auch unserer eigenen Vorstellungen, Fantasien und Ängste. So bietet der Spielfilm Identifikationslinien mit den Protagonistinnen und Protagonisten bzw. deren Beziehungen, Situationen usw. Während der Roman im Symbolisieren unsere Vorstellungen inspiriert, lebt der Film von der bildhaften, imaginären *Dar*stellung. Der (Spiel-)Film hat mehrere Möglichkeiten, Alter(n) darzustellen. Er kann jede Schauspielerin und jeden Schauspieler

1 Henriette Herwig / Andrea Hülsen-Esch (H *Alte im Film und auf der Bühne. Neue Altersbilde und Altersrollen in den darstellenden Künsten*, Bielefeld 2016.

2 Zu begehren im psychoanalytischen Sinne bedeutet, einem unbewussten Wunsch nachzu laufen. Das Begehren w niemals befriedigt. Wir begehren jene verloren gegangene Einheit, das Genießen, das mit diese ehemaligen Vollkommen heit einherging. Das läs den Menschen weitermachen, zuerst in der ersten Beziehung, dann allen anderen Beziehung (vgl. Paul Verhaeghe, *Lie in Zeiten der Einsamkeit. Drei Essays über Begehr und Trieb*, Wien 2009).

3 Projektion bedeutet psychoanalytisch eine unbewusste Abwehr von inneren Gefühlen, Vorstellungen, Wünsche Gedanken, indem sie na außen in eine Person od eine Sache (= Objekt) verschoben werden. „Nicht ich, sondern du."

186

Blood, Sweat & Years. The Disappearance of the Old

Beate Hofstadler

1
...nriette Herwig and ...drea Hülsen-Esch, eds., *...e im Film und auf der ...hne: Neue Altersbilder ...d Altersrollen in den ...stellenden Künsten* (...elefeld, 2016).

2
...e psychoanalytical ...aning of desire is to ...sue an unconscious ...sh. Desire is never ...isfied. We desire that ... unity, the pleasure ...t was part of the former ...fection. It encourages ...ople to persevere, in ...ir first relationship and ...n in all other relation...ps thereafter (cf. Paul ...haeghe, *Liebe in Zeiten ... Einsamkeit: Drei Essays ...r Begehren und Trieb* ...enna, 2009]).

3
...jection in psychoanaly... means an unconscious ...ense against innermost ...lings, ideas, desires, ...ughts by attributing ...m to an external person ...hing (= object). "It's ..., not me."

Personification

Fictional film does not create new conditions in society but, rather, it reflects the existing world, which in turn contributes to forming our reality. Whereas art that's in tune with its time instantly captures what is bubbling under society's surface though perhaps needs to be spelled out, academic analysis is always one step behind. However, art, too, perpetuates stereotypes, clichés, and blind spots, as the various genres demonstrate, and the representation of old age is no exception.[1] How are old people, their bodies, and their relationships portrayed? Cinema is a way of satisfying our need for visual pleasure and informs our desires.[2] It is also a place of projection,[3] not only of the films themselves, but also of our own perceptions, fantasies, and fears. The fictional film offers scope for identification with its characters, their relationships, circumstances, and so on. Whereas a novel uses symbolism to inspire our imagination, the medium of film is about imaginative depiction. It has several tools at its disposal to depict age and aging. It can make its actors look old through makeup, lighting, camera distance, and so on, which can

作为德国籍犹太人，弗兰克家族在遭遇迫害时逃到荷兰，藏匿于书中展现的这处位于书架后部有如迷宫般的密室里，安妮 · 弗兰克（Anne Frank）在此写下了记录这场浩劫的《安妮日记》。本书通过历史照片、资料、藏身处照片、《安妮日记》的片段等丰富资料，串联并呈现安妮 · 弗兰克更丰富的生活细节。本书的装订形式为锁线胶平装，封面白卡纸四色印刷，勒口外翻，内藏安妮的书架上的生活照片。内页胶版纸四色印刷，以黑白为主。本书没有页码，所有页面均采用拉页形式，形成了复杂的内容展现结构，有如迷宫般的藏身处，表面和内里呈现着相关的信息，需要读者在其中来回翻找。仅有的内容提要形式的“目录”，进一步加强了书籍的“迷宫感”。照片的编号虽有一定的索引作用，但内容的查找主要依靠拉页折叠处的文字与“目录”的对应。这些拉页折叠处的文字在书口上形成丰富的肌理，有如“书柜家门”的木纹。||||||||||||||||||||||||

HOUSE

As a German Jew, the Franks fled to the Netherlands when they were persecuted and hid in the labyrinth-like secret room at the back of bookshelves, where Anne Frank wrote Annie's Diary to record the catastrophe. This book presents details of Anne Frank's daily life through rich materials such as historical photographs, documents, photographs of the hiding place and fragments of Annie's Diary. This paperback book is glue bound with thread sewing. The cover is printed on white cardboard in four colors and the flap is turned outward with the photos on Anne's bookshelf inside. The inner pages are printed on offset paper in four colors, mainly black and white. There are no page numbers in this book. All pages are in the form of foldouts, forming a complex display structure like a labyrinth. Relevant information is shown on the outer and inner sides, which requires readers to look back and forth. The book has a 'catalogue' in the form of abstracts, which further reinforces the 'maze' of the book. Although the number of photos has a certain indexing function, the search function mainly depends on the corresponding text printed on the pullouts. These words form abundant texture on the fore-edge of the book, such as the wooden grain of the 'bookcase door'.

- 银奖
- 安妮 · 弗兰克之家
- 荷兰

- Silver Medal
- Anne Frank House
- The Netherlands
- Irma Boom Office (Irma Boom, Eva van Bemmelen)
- Elias van der Plicht (Anne Frank Stichting)
- Anne Frank Stichting, Amsterdam
- 210×123×19mm
- 452g
- 336p

A jacket that actually looks quite normal is wrapped around this catalogue in an entirely unfamiliar way: the flaps are not folded inward — they are folded out instead. For two years the Frank family lived with friends in a hideout in Amsterdam. As German Jews, they went into hiding at the rear of their offices in an attempt to escape persecution by the National Socialists because leaving the Netherlands safely seemed impossible. The family was betrayed and then arrested on 4 August 1944; deportation to concentration camps followed. Only father Otto survived the Holocaust. A bookcase blocked the opening to the hideout in the labyrinthine building complex. A picture of this bookcase is displayed on the cover: instead of taking you inside the book, turning over the cover takes you to a dead end — the inside of the flap. You look straight at the wall of a room whose wallpaper was decorated with photos by daughter Anne. This special book design feature allows the paradoxically reversed cover to be seen as an allegory of the extremely precarious housing situation. The inside concept with its extensive collection of images, documents and texts seems to confirm this: all double pages are designed as fold-out pages with double gate fold. The book can be leafed through normally; then you fold out the concealed inner pages to reveal mini-panoramas that appear like insights into the hideout. Anne Frank wrote her now famous diary here.

ANNE
FRANK
HOUSE

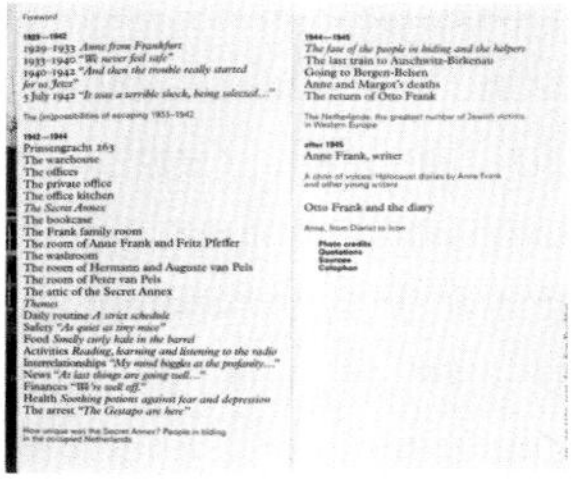
Foreword
1929–1942
1929–1933 Anne from Frankfurt
1933–1940 "We never feel safe"
1940–1942 "And then the trouble really started for us Jews"
5 July 1942 "It was a terrible shock, being selected..."
The (im)possibilities of escaping 1933–1942
1942–1944
Prinsengracht 263
The warehouse
The offices
The private office
The office kitchen
The Secret Annex
The bookcase
The Frank family room
The room of Anne Frank and Fritz Pfeffer
The washroom
The room of Hermann and Auguste van Pels
The room of Peter van Pels
The attic of the Secret Annex
Themes
Daily routine A strict schedule
Safety "As quiet as tiny mice"
Food Smelly curly kale in the barrel
Activities Reading, learning and listening to the radio
Interrelationships "My mind boggles at the profanity..."
News "At last things are going well..."
Finances "We're well off."
Health Soothing potions against fear and depression
The arrest "The Gestapo are here"
How unique was the Secret Annex? People in hiding in the occupied Netherlands
1944–1945
The fate of the people in hiding and the helpers
The last train to Auschwitz-Birkenau
Going to Bergen-Belsen
Anne and Margot's deaths
The return of Otto Frank
The Netherlands: the greatest number of Jewish victims in Western Europe
after 1945
Anne Frank, writer
A choir of voices: Holocaust diaries by Anne Frank and other young writers
Otto Frank and the diary
Anne, from Diarist to Icon
Photo credits
Quotations
Sources
Colophon

1929–1942
1929–1933 Anne from Frankfurt

"I don't think I'll ever feel at home in this house"
— Anne Frank

"I'm burning up with rage, but I can't show it"
— Anne Frank

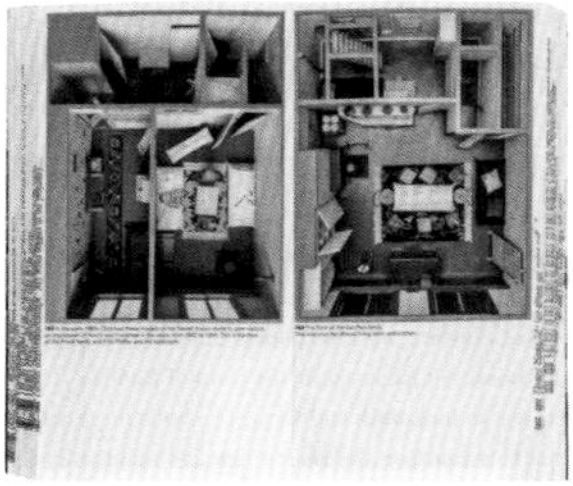

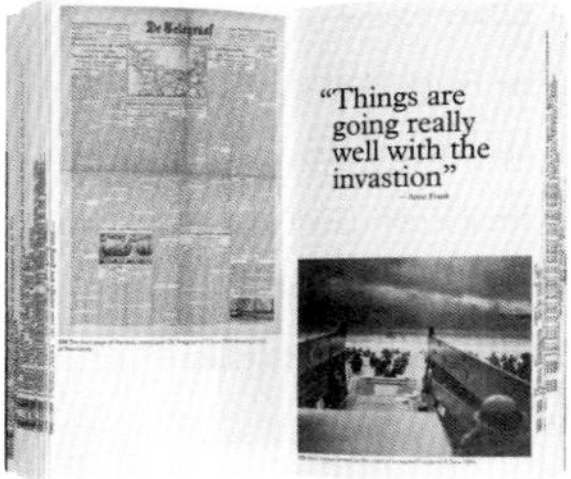
"Things are going really well with the invastion"
— Anne Frank

1942—1944
Finances "We're well off."
DE GRUYTER
ZWARTE MARKT
IS ROOF
181 The black market developed as a result of shortages. Black market trading in the Haarlemmerstraat in Amsterdam, 1944/1945.
182 A propaganda poster against the black market, dating from 1940, showing the text: "The black market is robbery"

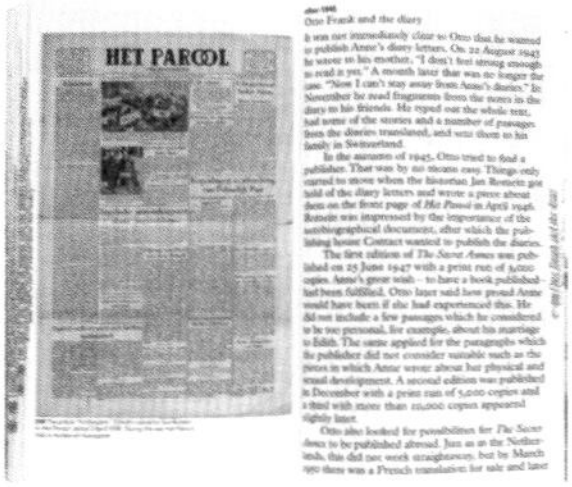
HET PAROOL

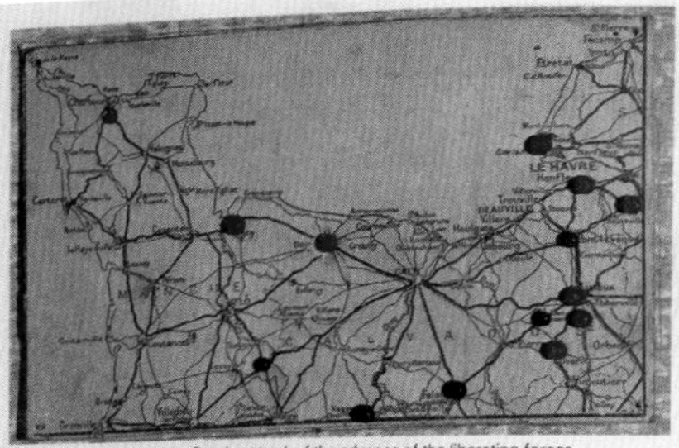
LE HAVRE

176 After the allied landings Otto kept track of the advance of the liberating forces on the map from *De Telegraaf* of 8 June.

RIJK VROUW

177

Haagsche Post

178

DAS REICH
VOR DEM GROSSEN ERWACHEN

179

De Prins

180

Victor Kugler provided the group in hiding with magazines such as *Het Rijk der Vrouw, Haagsche Post, De Prins*, as well as *Das Reich*.

1942—1944

News *"At last things are going well..."*

In his bedroom Otto had stuck a map of Nor on the wall. He used colored pins to accurate how the allied troops were gaining ground ir pied territory after D-Day. In the Secret Anr they knew precisely what was happening wit advance in Normandy and on the eastern fr On 9 June 1944 Anne wrote: "things are goi really well with the invasion. The allies have Bayreuth, a village on the French coast and now fighting for Caen." Two weeks later she "Things are really going well, Cherbourg, Vi and Slobin fell today. There must have been of booty and many prisoners."

The map of Normandy had been cut ou De Telegraaf on 8 June 1944. Although they shut in behind a bookcase, the group in the Annex were not deprived of information abou was happening in the outside world. The hel told them all the latest news from the city ar brought newspapers. They also listened to the much too often according to Anne: "As thou the German reports from the *Wehrmacht* and English BBC weren't enough, the "Luftlageme was introduced not long ago. In a word, it's r nificent, but [the other side of the coin], it's also very disappointing. The English turn th bombardment into a constant story which ca be compared to the German lies that are all same! So the radio gets turned on at 8 o'cloc the morning [if not earlier] and they listen to every hour until 9 or 10 or even 11 in the eve

The group kept themselves informed by listening to the radio on a daily basis. Betwee midday and the evening, they tuned into the or Radio Orange in Otto's office. It was proh to listen to these stations, so the radio was pu

e German station for safety when they finished
ing. In the summer of 1943, the occupying
s ordered that all radios had to be handed in
/ictor took it away at the end of July. From that
the group made use of a small wireless that
ıged to Johannes which was not handed in,
vas put in the living room of the Secret Annex.
Now and then Anne would make a note about
she'd heard on the radio. Writing about the
of the Jews, she said: "If it's so bad in Holland,
will they live in those distant barbaric regions
e they're sent. We're assuming that most of
are being murdered. The English radio has
d about gas; perhaps it's the quickest way to
'm very upset."
There were also some more cheerful news
ts in the diary, which was a good reason why
saw the radio as a source of courage. For
ple, on 27 March 1943 she wrote about what
ad heard about the attack on the population
ter in Amsterdam, which was aimed at making
re difficult to arrest Jews, amongst others.
n wearing the uniforms of the German police
the guards and then managed to get rid of
nportant papers."
A day after the failed attempt on Hitler's life
July 1944 she wrote: "Now I'm starting to
opeful, at last things are going well. Really,
s are getting better! Fantastic reports! There
n attempt to murder Hitler and this time it
't by Jewish communists or English capitalists,
y an important German general who was a
nt and was actually quite young." Admittedly
'ührer survived with little more than torn
es, but according to Anne this incident was
est proof that "there are many officers and
rals who are sick of the war and would like
e Hitler disappear into the depths."

The political news was a waste of time for Anne. She was not very interested in it and she hated the endless discussions in the Secret Annex about politics and the course of the war. "It's obvious that there are many different views about these issues, and perhaps even more logical that people talk about it a great deal during these difficult war years, but ... to quarrel so much about it is simply stupid!"

In her diary, Anne regularly copied reports from other sources of news such as *De Telegraaf*. Victor brought along the current affairs magazine the *Haagsche Post*, the family magazine *Panorama*, and occasionally the Nazi newspaper *Das Reich*. Anne was always impatient to get the *Cinema & Theater* which he brought her every week, a magazine that she really liked – even though it came from Nazi Germany – because it was full of stories about films, the theatre, puzzles, series and other uncontroversial subjects.

Every day Miep, Bep, Victor and Johannes told them what was happening in the city. Anne considered that these were quite often "horrible stories" about house searches, arrests of Jews, razzias, deportations and Westerbork. From time to time it made her feel very dejected. When Miep told them in the autumn of 1942 that the house of the Van Pels family had been emptied by the National Socialists, leaving Auguste very upset, she decided not to take them any more news that could shock them. This was by no means easy, because every time she went into the Secret Annex, she was bombarded with questions. Anne was sensitive about Miep withholding information and continued to question her and go on and on until Miep found that she was telling them exactly what she hadn't wanted to say.

BONJOUR

这是一本展示和研究捷克艺术家与布列塔尼地区之间关系的书籍。布列塔尼位于法国西北角，作为部分“大不列颠人”的移居地，解释为“小不列颠”，在文化上与以巴黎为中心的法国文化有一定差异。这里很早就以独特的文化吸引着各地的艺术家，尤其是来自捷克的艺术家。他们在 1850–1950 年间频繁来到布列塔尼，吸取当地文化与艺术的养分，同时创作出极为多样的艺术品。高更作为法国艺术家，在布列塔尼短暂生活过，留下了一些画作，其中一幅的作品名被用于本书书名。本书的装订形式为锁线硬精装，封面胶版纸印黑，裱覆卡板，封面使用了黑白双色烫印的工艺。内页前中后三部分采用土黄色艺术纸印单黑，展示布列塔尼地区的风土人情和艺术家照片。书籍的主要部分呈现作品和论文，半涂布纸四色印刷。本书作者认为内陆的波希米亚（古捷克地区）文化和靠海的布列塔尼文化存在相似性，可能是因为血缘上的某种联系，以至文化艺术上相互吸引。

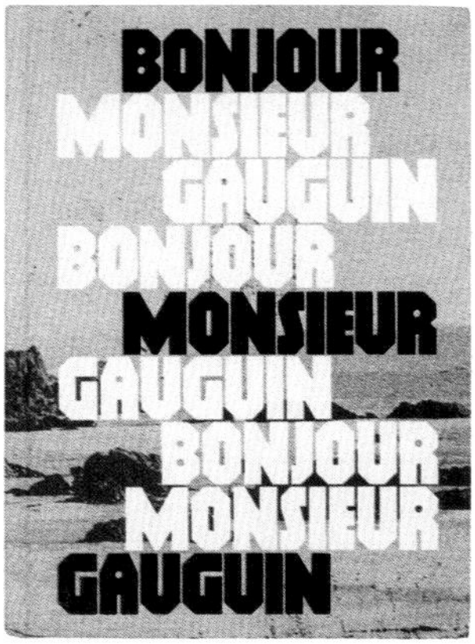
BONJOUR
MONSIEUR
GAUGUIN
BONJOUR
MONSIEUR
GAUGUIN
BONJOUR
MONSIEUR
GAUGUIN

This is a book showing and studying the relationship between Czech artists and Region Bretagne. Bretagne is located in the northwest corner of France. As the settlement of the 'Great Britain', it is interpreted as 'Little Britain', which is culturally different from the Paris-centered French culture. It has long attracted artists from all over the world, especially from Czech Republic, with its unique culture. From 1850 to 1950, they frequently came to Bretagne to absorb the nutrients of local culture and art, and at the same time to create extremely diverse works of art. As a French artist, Gauguin lived briefly in Bretagne, leaving some paintings behind. The book is named after one of the paintings. This hardback book is bound by thread sewing. The cover is printed on offset paper in black, mounted with cardboard. Hot stamping in black and white is also applied to the cover. The three parts of the main body are printed on khaki art paper in black, showing the local customs and artists' photographs in Region Bretagne. The main part of the book presents works and papers on semi-coated paper with four-color printing. The author believes that the similarities between the inland Bohemian (ancient Czech) culture and the seaborne Bretagne culture may be due to some kind of consanguinity, which leads to cultural and artistic attraction.

- 铜奖
- 你好，高更先生
- 捷克

- Bronze Medal
- Bonjour, Monsieur Gauguin
- Czech Republic
- 20YY Designers (Petr Bosák, Robert Jansa, Adam Macháček)
- Anna Pravdová
- Národní galerie v Praze (National Gallery Prague)
- 245×173×32mm
- 879g
- 308p

Strong contrasts from matt white and glossy black embossed, thick capitals create a striking cover. The words of the book's title are stacked on top of each other three times. Small dot-like holes punched into the letters lend ornamental accents to their chunkily edged contours, behind which the austere Brittany coastal landscape spreads out in a historical black-and white photograph: this exhibition catalogue on Czech artists in Brittany between 1850 and 1950 documents the cultural connections between eastern and western Europe. On the inside, a series of pages with historical photographs — landscapes, harbours, people — on earthy brown paper prepares the reader for the book's content. Similar sections in the middle of the book and at the end maintain this atmosphere. It is unusual that the essay texts are only printed on the right-hand pages, and only these have been given page numbers. On the image pages — well-structured and varied, orientation is provided by the picture numbering. In order to differentiate between image references and notes in the text, hairline frames with the height of a capital letter have been added: these are noticeable without being distracting. The block-like aura is also expressed in the book's edge, meticulously narrow and produced by the skilled bookbinding. It clarifies that the missing headband is not an error, but rather intentional design that aims to avoid watering down the block quality generated by the straight spine.

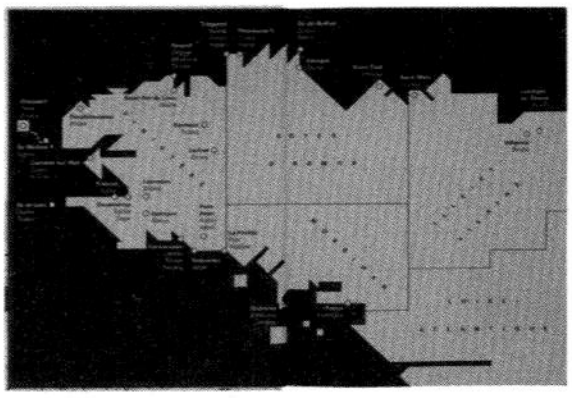

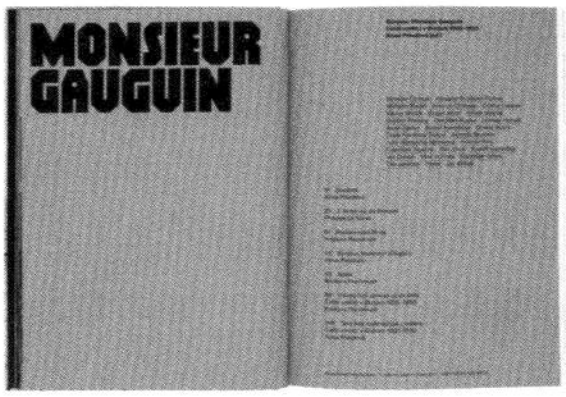

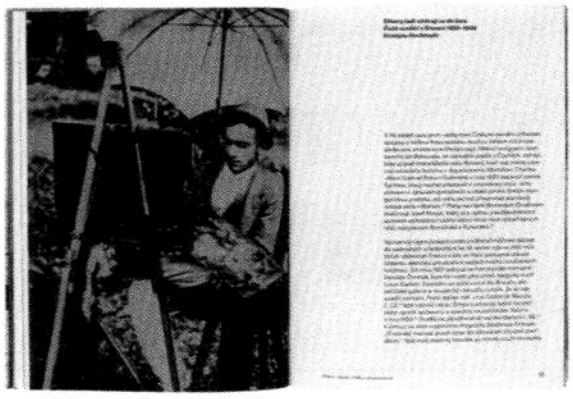

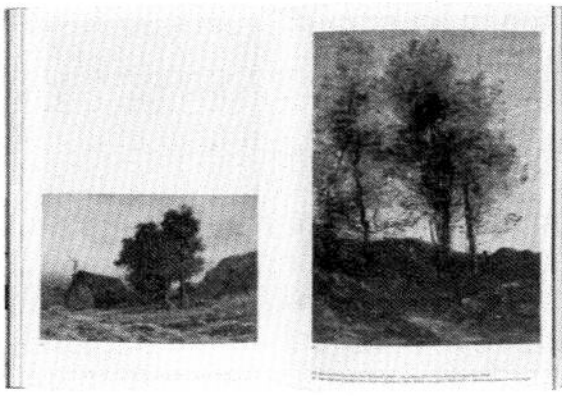

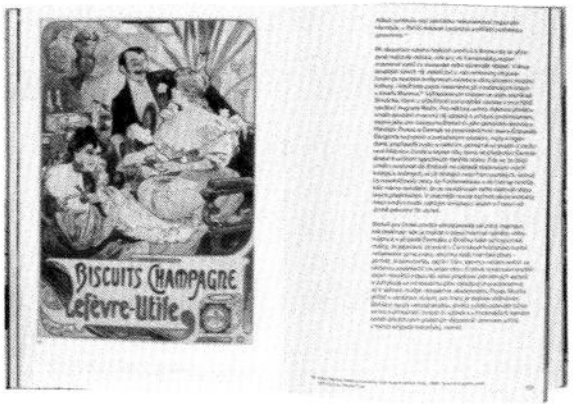

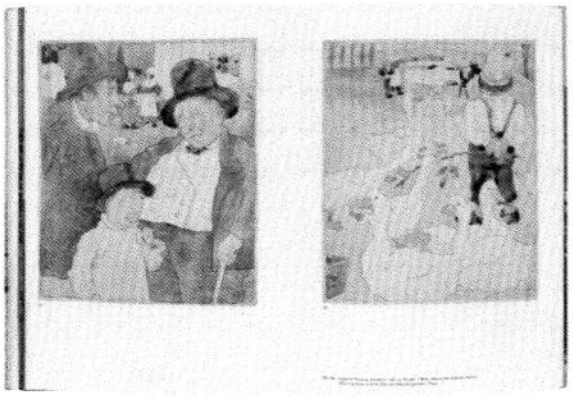

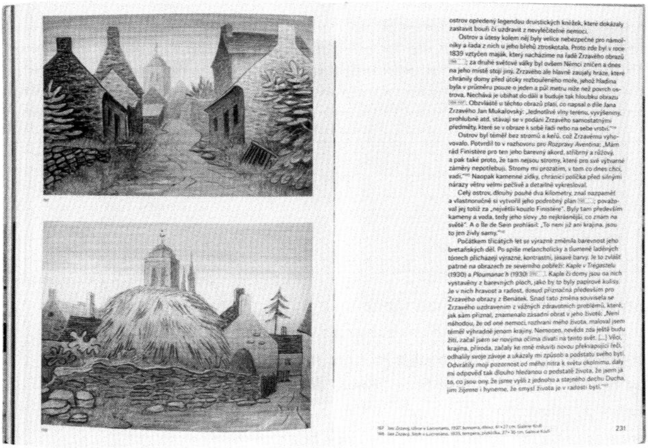

ostrov opředený legendou druistických kněžek, které dokázaly zastavit bouři či uzdravit z nevyléčitelné nemoci.

Ostrov a útesy kolem něj byly velice nebezpečné pro námořníky a řada z nich u jeho břehů ztroskotala. Proto zde byl v roce 1839 vztyčen maják, který nacházíme na řadě Zrzavého obrazů; za druhé světové války byl ovšem Němci zničen a dnes na jeho místě stojí jiný. Zrzavého ale hlavně zaujaly hráze, které chránily domy před útoky rozbouřeného moře, jehož hladina byla v průměru pouze o jeden a půl metru níže než povrch ostrova. Nechává je ubíhat do dálí a buduje tak hloubku obrazu. Obzvláště u těchto obrazů platí, co napsal o díle Jana Zrzavého Jan Mukařovský: „Jednotlivé vlny terénu, vyvýšeniny, prohlubně atd. stávají se v podání Zrzavého samostatnými předměty, které se v obraze k sobě řadí nebo na sebe vrství."

Ostrov byl téměř bez stromů a keřů, což Zrzavému vyhovovalo. Potvrdil to v rozhovoru pro *Rozpravy Aventina*: „Mám rád Finistère pro ten jeho barevný akord, stříbrný a růžový, a pak také proto, že tam nejsou stromy, které pro své výtvarné záměry nepotřebuji. Stromy mi prozatím, v tom co dnes chci, vadí." Naopak kamenné zídky, chránící políčka před silnými nárazy větru velmi pečlivě a detailně vykresloval.

Celý ostrov, dlouhý pouhé dva kilometry, znal nazpaměť a vlastnoručně si vytvořil jeho podrobný plán; považoval jej totiž za „největší kouzlo Finistère". Byly tam především kameny a voda, tedy jeho slovy „to nejkrásnější, co znám na světě". A o Île de Sein prohlásit: „To není již ani krajina, jsou to jen živly samy."

Počátkem třicátých let se výrazně změnila barevnost jeho bretaňských děl. Po spíše melancholicky a tlumeně laděných tónech přicházejí výrazné, kontrastní, jasné barvy. Je to zvlášť patrné na obrazech ze severního pobřeží: *Kaple v Trégastelu* (1930) a *Ploumanac'h* (1930). Kaple či domy jsou na nich vystavěny z barevných ploch, jako by to byly papírové kulisy. Je v nich hravost a radost, dosud příznačná především pro Zrzavého obrazy z Benátek. Snad tato změna souvisela se Zrzavého uzdravením z vážných zdravotních problémů, které, jak sám přiznal, znamenalo zásadní obrat v jeho životě: „Není náhodou, že od oné nemoci, rozhraní mého života, maloval jsem téměř výhradně jenom krajiny. Nemocen, nevěda zda ještě budu žíti, začal jsem se novýma očima dívati na tento svět. [...] Věci, krajina, příroda, začaly ke mně mluviti novou překvapující řeči, odhalily svoje závoje a ukázaly mi způsob a podstatu svého bytí. Odvrátily moji pozornost od mého nitra k světu okolnímu, daly mi odpověď tak dlouho hledanou o podstatě života, že jsem já to, co jsou ony, že jsme vyšli z jednoho a stejného dechu Ducha, jím žijeme i hyneme, že smysl života je v radosti bytí."

197 Jan Zrzavý, Ulice v Locronanu, 1927, tempera, dřevo, 41 × 27 cm, Galerie Kodl
198 Jan Zrzavý, Stoh v Locronanu, 1935, tempera, překližka, 27 × 36 cm, Galerie Kodl

231

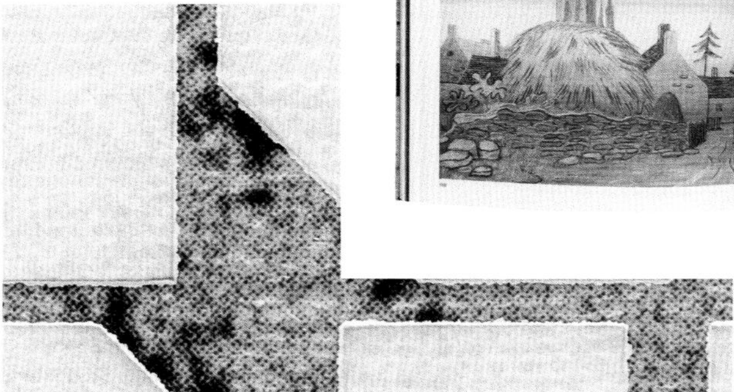

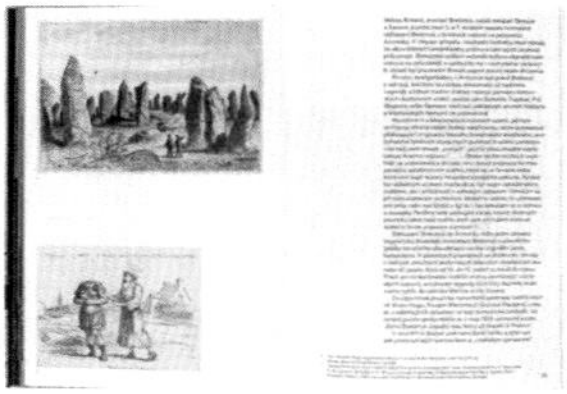

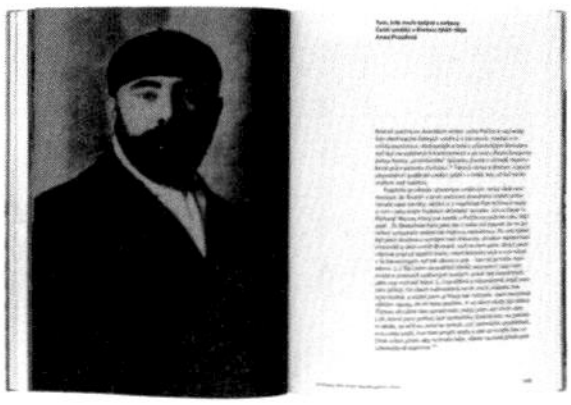

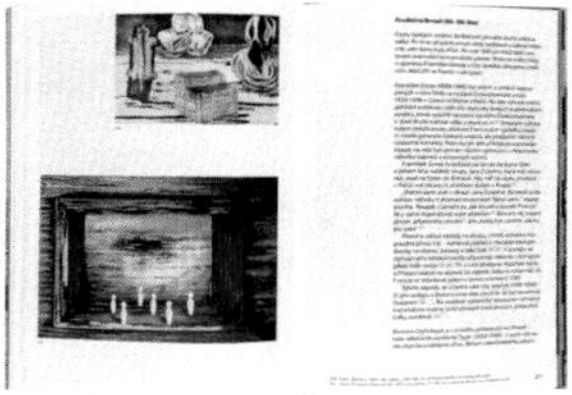

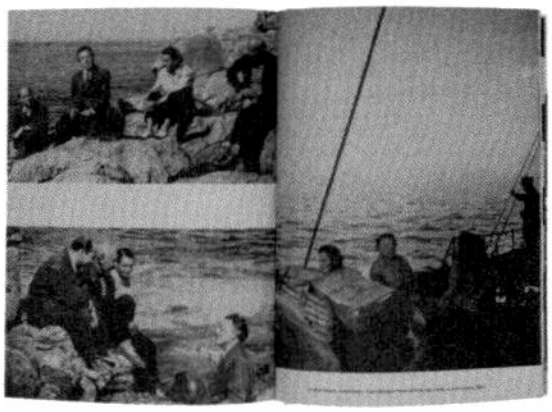

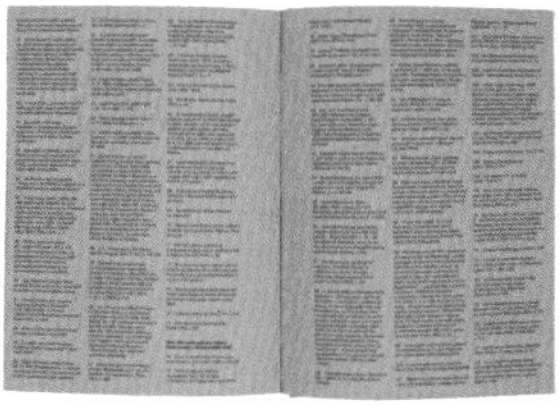

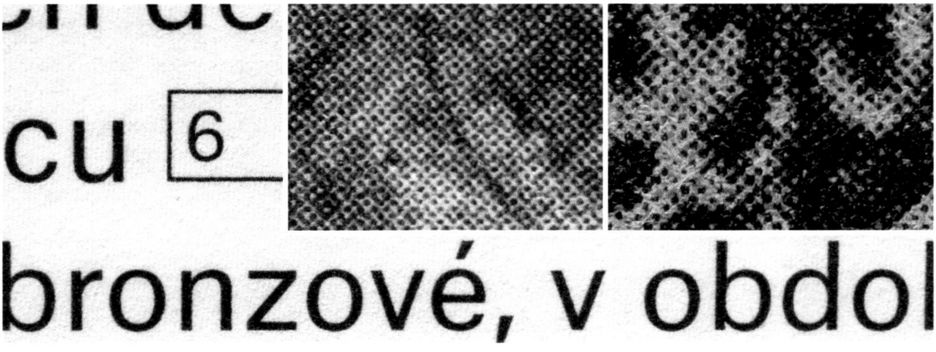
cu 6
bronzové, v obdol

Bonjour, Monsieur Gauguin
Čeští umělci v Bretani 1850–1950
Anna Pravdová (ed.)
32

Bonjour, Monsieur Gauguin

Obraz *Bonjour, Monsieur Gauguin* 32 namaloval Paul Gauguin v bretaňské vesničce Le Pouldu, kam se přesunul z Pont-Avenu, když mu již kvůli velkému zájmu malířů i turistů připadal příliš rušný. Při svém pobytu v zimě 1889 zde spolu se svým holandským žákem Miejerem De Haan a malířem Charlesem Filligerem vyzdobil nástěnnou malbou, obrazy a řezbami jídelnu hostince Marie Henryové, kde byl tou dobou ubytován.

Nade dveřmi Gauguin namaloval coby „domovní znamení" husu s nápisem „Maison Marie – Henry", na stěnu dívku předoucí na pláži v blízkosti hostince, dále portrét svého přítele De Haana a jako pendant k němu stylizovaný autoportrét. Horní výplň dveří místnosti zdobila první varianta obrazu *Bonjour, Monsieur Gauguin*, o něco menší než plátno v majetku Národní galerie v Praze, a také méně propracované.[14]

Dílo je parafrází obrazu *Bonjour, Monsieur Courbet*, který namaloval Gustav Courbet v roce 1854 v jižní Francii, nedaleko Montpellieru. Zachytil na něm setkání se svým mecenášem Alfredem Bruyasem, kterého doprovází jeho sluha. Zobrazil se zde nejen na stejné úrovni jako jeho bohatý ochránce, ale svým postojem i tím, že pojal sebe jako dominantu kompozice, dal jasně najevo sebejistotu a nadřazenost vůči tomu, kdo si od něho obraz objednal. Radikálně tím poukázal na změnu postavení umělce ve společnosti i na vlastní nezávislost na stávajících zvyklostech, jak v oblasti umělecké, tak v přetrvávající hierarchii buržoazní společnosti.

Paul Gauguin a Vincent Van Gogh obraz viděli při společné návštěvě Musée Fabre v Montpellieru v prosinci 1888. Oba je natolik zaujal, že se rozhodli vytvořit vlastní verzi. Gauguin se na rozdíl od Courbeta zobrazil čelem k divákovi a to nikoliv v nadřazené pozici, ale naopak na úrovni bretaňské venkovanky, která jej zdraví. Vyjádřil tím na jedné straně své postavení uměleckého outsidera i svou osamocenost a zároveň snahu žít život prostých lidí, mimo umělecká centra a obchod s uměním.

Courbetův obraz byl inspirován tehdy oblíbenou rytinou *Měšťané hovoří s bludným židem* (1831) Pierra Leloupa du Mans. S Ahasverem, věčným poutníkem a vyděděncem se tu Courbet programově ztotožnil: převzal jeho poutnickou hůl a pytel, v němž nesl Ahasver jen to nejnezbytnější k životu, nahradil malířským náčiním. Gauguin jeho metaforickou transpozici převzal, ale malířskými atributy se neidentifikoval. Poutnickou hůl vyměnil za cestovní plášť a prostými

32 Paul Gauguin, Bonjour, Monsieur Gauguin, 1889, olej, plátno, 92,5×74 cm, Národní galerie v Praze

这是由生活在巴黎的挪威摄影师奥拉·林达尔（Ola Rindal）拍摄的巴黎影集。与大多数展示巴黎著名景点的摄影不同，林达尔把镜头对准那些被人们忽视的角落，包括建筑、道路上那些肮脏和破损的地方，那些垃圾堆和随意丢弃的物品，以及流浪汉、鸟类和大街上的突发事件。摄影师试图通过这些照片展现巴黎的另一面，关注城市文明表面下被城市的“暴力”所“驱赶”且不被关注的真实存在。书籍的装订形式为锁线硬精装，封面黑白交织的布面裱贴四色印刷一张巴黎的俯拍大型照片，封底则裱贴一张很小的拍摄垃圾堆的照片，封面及书脊通过烫黑文字呈现作者、书名和专辑编号。内页分为三部分：主体部分为哑面艺术纸四色印刷，每两页展示一张照片，照片不存在连续性，以环境和场景为主，人和物品为辅，每张照片都规整地留出相同的白色边框；第二部分为胶版纸印刷，小 16 开本夹订在书内，专注拍摄以人与物品为主的特写照片；第三部分为光面铜版纸印刷，与大 16 开本夹订在书内，专注于展示照片中放大导致模糊的人物特写。本书采用法国国旗蓝白红三色的书签带。

This is an album of Paris by Norwegian photographer Ola Rindal, who lives in Paris. Unlike most photographs showing famous Paris attractions, Rindal focused on neglected corners, including buildings, dirty and damaged areas in streets, garbage dumps and casually discarded objects, as well as tramps, birds and emergencies. The photographer tried to show the other side of Paris through these pictures, focusing on the real existence of urban civilization that is 'driven' by the 'violence' of the city and is always neglected. The hardback book is bound by thread sewing. The cover is made of black-and-white cloth mounted with a large overhead photograph of Paris printed in four colors, while a small photograph of garbage dump is mounted on the back cover. The author's name, book title and album number are presented on the cover and the spine with black hot stamping. The inner pages are divided into three parts: the main part is printed on matte art paper in four colors, showing one picture on every two pages. There is no connection between the photos, which display environment, scenes, people and objects. Every photo has the same white border in a regular way. The second part is printed on offset paper and bound with a smaller format, focusing on close-up photographs of people and objects. The third part is printed on gloss art paper, and bound with a larger format, focusing on the blurred close-ups of character because of enlargement. The bookmark ribbons in the book are in blue, white and red with the French national flag.

- 铜奖
- 巴黎
- 瑞典

- Bronze Medal
- Paris
- Sweden
- Sandberg&Timonen
- Ola Rindal
- Livraison Books, Stockholm
- 347×235×18mm
- 1095g
- 140p

If your eyes need a rest from the plethora of attractions in a sophisticated metropolis, and focus on what is happening on the peripheries instead, unexpected themes move silently into your field of awareness. This is what Norwegian photographer Ola Rindal seems to have experienced in Paris, his adopted home, when — once established in the world of fashion and magazines and thus susceptible to the beautiful things in life — he gradually directed his lens to capture the sidelines as well. The elements that preoccupy Ola Rindal there, or perhaps also impress or sometimes fascinate, on the fringes of public space — a strangely shaped piece of material, a person huddling on a park bench, the foamy trail of washing water sploshed across the pavement —, all this is grouped in a large-format, illustrated book. Strong matt paper, almost paperboard, bears the double-page photographs with very high-resolution printing, framed on all sides by a narrow paper margin. The portrait formats run into the spine on one side and are faced by a blank page. These disconcerting, sobering and equally poetic moments grant the photographer and also the observer greater insight into the reality of the city. The absence of glamour appears to be a prerequisite for recognising the commonplace and also the aesthetics of the mundane — acknowledging what is there, what is all around us, devoid of retouching for any lifestyle. This is another way of saluting the city of love, sealed with the French tricolour as a ribbon bookmark.

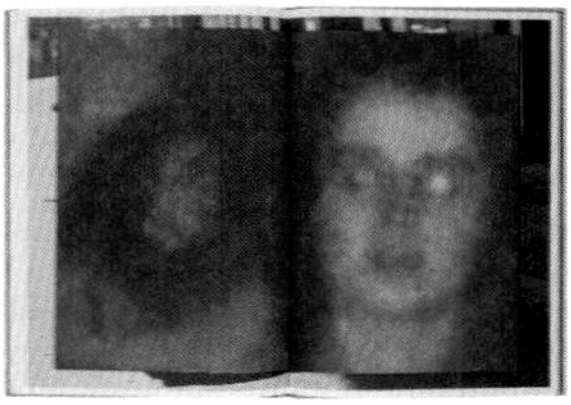

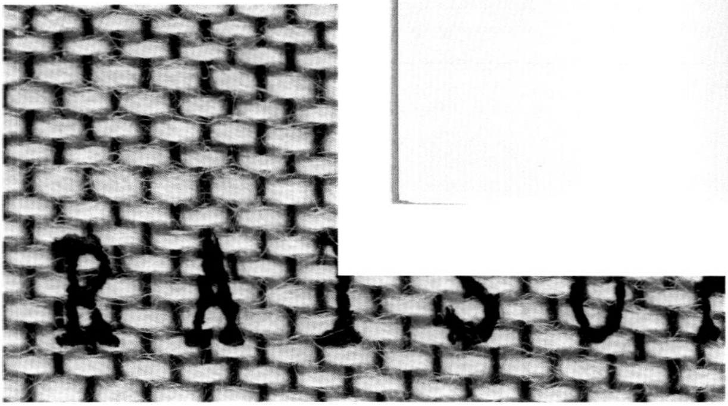

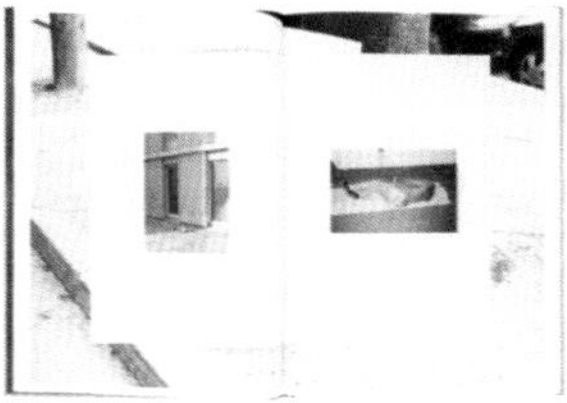

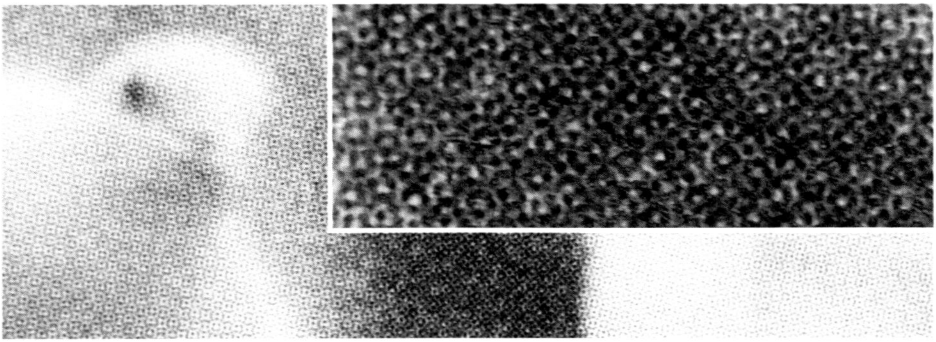

OLA RINDAL PARIS LIVRAISON BOOKS (L-B. 009)

这是一本记录八哥这一善于鸣叫和模仿人声的鸟类从印度尼西亚“移民”至新加坡之后生存现状的书，并借此讨论全球社会的人类移民问题。书中的主要内容包含摄影师拍摄的八哥生存现状、当地居民与八哥共同生活的照片、八哥遭遇追捕猎杀的照片、人们对城市中八哥进行讨论的报纸和短信，以及一些用于阐述人们态度转变的带有叙述性的漫画。本书的装订形式为锁线硬精装，封面采用棕黄色布面裱覆卡板，图案和文字烫金处理。内页多种纸张混合使用：白色胶版纸四色印刷，呈现照片、漫画和报纸扫描件；墨绿色艺术纸印金呈现照片用于在各方面内容间过渡；最后的明黄色艺术纸上四色加印银的五色印刷，用于呈现人们捕捉、保护和放飞八哥的过程。由于本书以图片为主，书中专门夹附一本 16 面胶版纸骑马钉的黑白小册子，用于解读分析书中图片和资料所要传达的内容。这样使得书中图片没有任何文字图注，简单而纯粹。

This is a book that records and explores the status quo of mynah, a bird that is good at singing and imitating human voice, after its migration from Indonesia to Singapore. The book also discusses the issue of human migration in the global community. The main contents of the book include the photos of the living conditions of the mynahs, photos of local residents living with the mynahs, pictures of the mynahs being hunted and killed, newspaper reports and text messages discussing the mynahs in the city, and narrative cartoons to illustrate the change of people's attitudes. The hardback book is bound by thread sewing. The cover is made of cardboard mounted with brown-yellow cloth, on which the patterns and letters are hot stamped in gold. The inner pages are mixed with various kinds of paper: white offset paper with four-color printing to present photographs, cartoons and scanning copies of newspaper; dark green art paper printed in gold to present photographs for transitions between different parts; and bright yellow art paper printed in five colors (four colors plus silver) to present the process of capturing, protecting and releasing mynahs. Since the book is mainly picture-based, it contains a black-and-white printed booklet with 16 sides of offset paper, which is used to interpret the information conveyed by the pictures and materials. The booklet makes the book simple and pure without any annotations.

- 铜奖
- 移民
- 荷兰

- Bronze Medal
- The Migrant
- The Netherlands
- Teun van der Heijden
- Anaïs López
- Self-Published
- 325×246×20mm
- 1048g+33g
- 124p+16g

A visual bird call: the enchanting, gold-embossed silhouette of a bird on the curry-coloured fabric cover. This book follows the migration history of a member of the starling family. A short booklet between the endpaper provides an explanation. The Javan mynah is actually a talented songbird that can even imitate the human voice. Once cherished and imported to Europe and Singapore for its singing talents, now it only squawks there to be heard above the urban hubbub. As time passes, however, the mynah is gradually being vilified, hunted, pursued and attacked. The photographer Anaïs López uses these correlations to tell the fate of this specific bird — an allegory of the current global relevance of social migration. In a collection of pictures with vivid graphics, she alternates between street photography and comic strips, documentary newspaper cuttings and poetic screen printing, shining in gold on dark olive-green paper. In Myanmar the mynah finally finds a new home and a sacred mission: as a messenger between humankind and heaven. On the last, yellow pages of the book, the religious ritual is observed step by step with the bird soaring up into the air out of the believer's hand, bearing their personal wishes. And indeed: right at the end — when you turn over the endpaper at the back — a delicate pop-up unfolds into three mynahs preparing to take to the air. A political book?

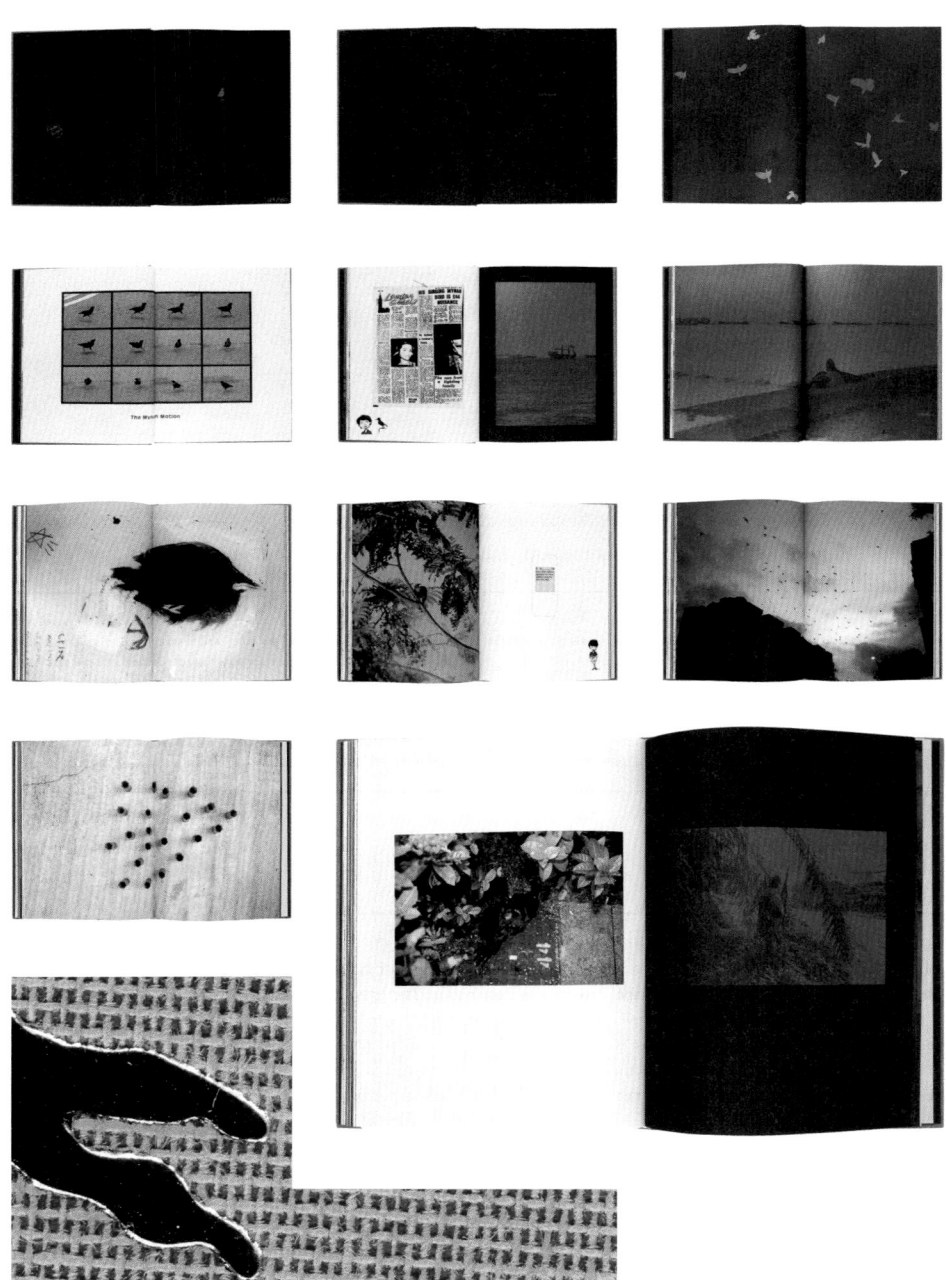

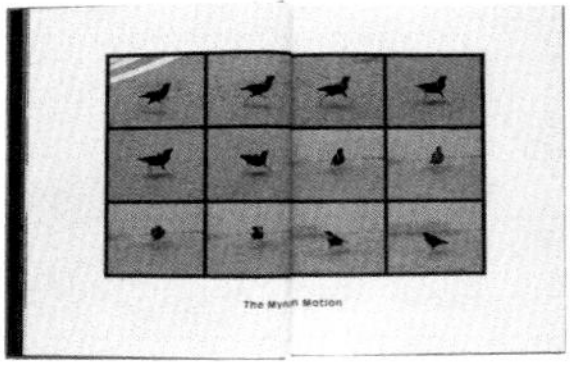
The Mynah Motion

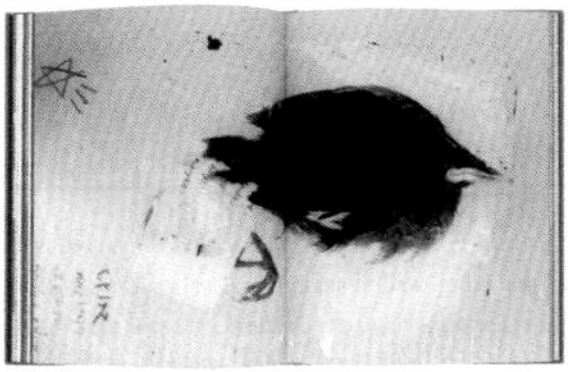

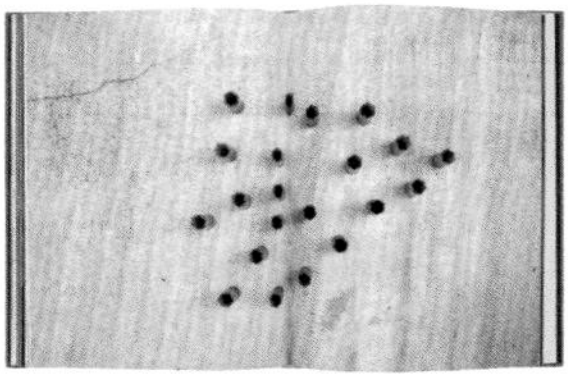

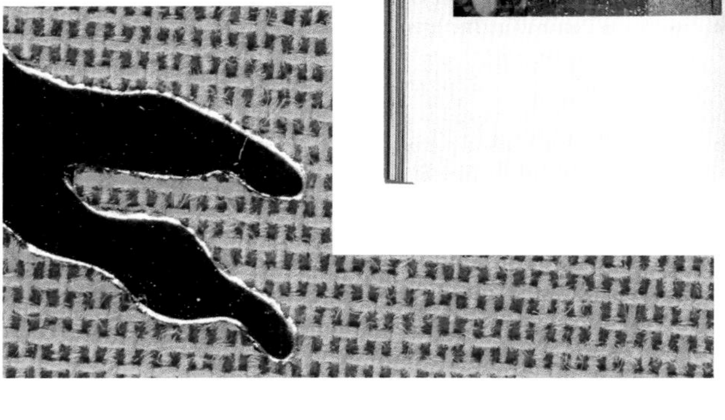

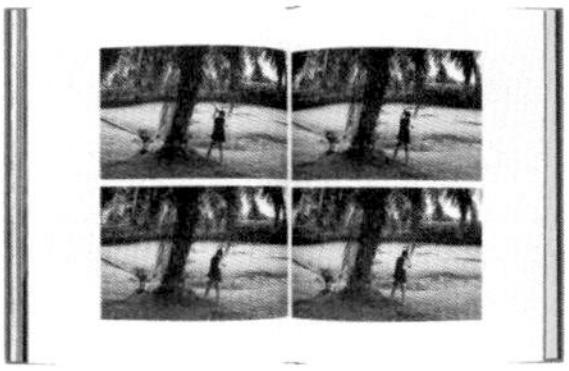

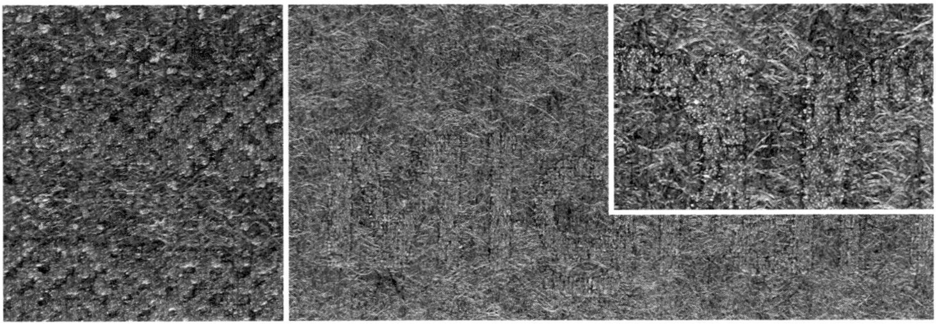

The Migrant

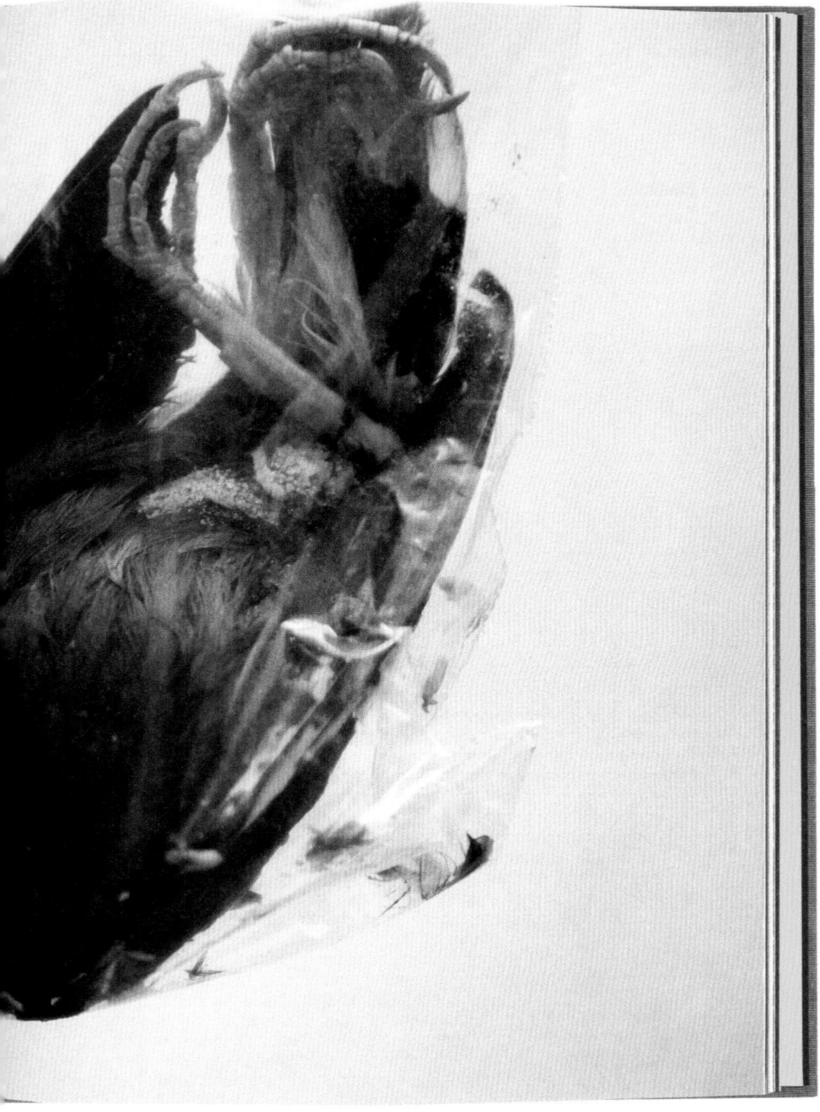

2019 Bm4

这是一本有关自杀及其相关原因的研究性书籍。书中将相关的社会因素研究、统计图表、自杀者的自述整合在一起，深度剖析了这一让人不愿触碰的沉痛主题。本书的装订方式为锁线胶平装，封面灰色卡纸印白，内页暖灰色与灰紫色胶版纸混合装订。全书只用黑白双色印刷，但所有图表（折线图、柱状图、雷达图）均采用镂空工艺。镂空工艺在纸张间留下光影，并因为工艺的精巧而需要被更小心地翻阅，增加了阅读时的严肃感。环衬部分采用肉色胶版纸张，是全书唯一给予读者的“温度”。书籍内容通过变换两色纸张进行区分。暖灰色纸张呈现了与自杀有关的研究，包括年龄、收入水平、婚姻状况、性别、医疗、社会福利、酒精消费、工作时间、战争、相互效仿、心理疾病、宗教、时间，以及自杀未遂者。灰紫色纸张则记录了曾经的自杀者的故事或语言片段（遗书、对话、未遂者的自述），书中将这些文字归纳出原因，分为八大类：抑郁、孤独、遗传、无家可归、得不到爱与被虐待、虚无主义与政治因素、失落（悔恨、自我价值缺失、被污蔑、成瘾）、对环境的不满。灰紫色纸张的运用也体现了这部分内容的阴郁情绪。

Nichtsein

This is a research book on suicide and its related causes. The book integrates the research of relevant social factors, statistical charts and self-narration of suicides in order to deeply analyze this painful theme that people are reluctant to touch. The paperback book is bound by thread sewing. The cover is made of gray cardboard printed in white, and the inside pages are mixed with warm gray and gray-purple offset paper. The entire book is printed in black and white only, but all charts (polygons, columns, radar charts) are hollowed out. Hollow-out technology leaves light and shadow on the pages, and the book needs to be treated more carefully because of its ingenuity, which increases the seriousness of reading. The endpaper is made of flesh-colored offset paper, which is the only 'warmth' given to readers in this book. The contents of the book are differentiated by changing the colors of paper. Warm grey paper presents suicide-related research, including age, income level, marital status, gender, medical care, social welfare, alcohol consumption, working hours, war, mutual imitation, mental illness, religion, time, and suicide attempters. The grey-purple paper records the story or text fragments of the suicides (posthumous papers, dialogues, self-narratives of the suicide survivors). In the book, these words are grouped into eight categories: depression, loneliness, heredity, homelessness, lack of love, nihilism and political factors, loss and dissatisfaction with the environment. The use of grey-purple paper also reflects the gloomy mood in this part.

铜奖
毁灭
德国

Bronze Medal
Nichtsein
Germany
Katharina Schwarz
Katharina Schwarz in co-operation with Ellen von den Driesch
Wissenschaftszentrum Berlin für Sozialforschung, UDK Berlin
305×174×16mm
553g
160p

What looks like sharp-edged white embossing on the jacket is in actual fact cut-out strips. In the places where they don't stick perfectly to the endpaper, the paperboard "bridges" cast shadows. On the inside you soon encounter a long series of pages whose sheets are riddled with holes over almost the entire surface, in regular rows. This is the first of many diagrams in this study on suicide. The diagrams have extremely reduced text and graphics, seemingly simple. The keywords are printed on the very robust, warm light-grey paper in two colours, black and white (the white is printed twice to achieve the necessary coverage). The statistical values themselves are entered with a laser cutter. The graphs, bar charts and polygons in the diagrams fall out of the paper and are revealed to be empty spaces, air, absence. These rows of holes stand for the "Number of suicides in 2015 in Germany, according to age groups". When graphics are used in this way, even sober statistics visualise strong emotions. Is this a nice book? It is certainly thought-provoking. You gulp more than once.

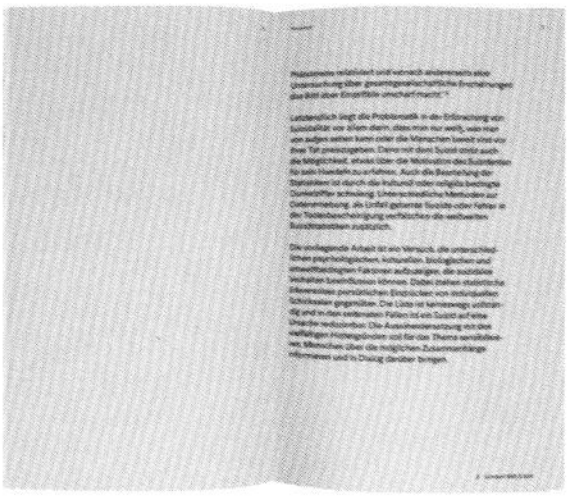

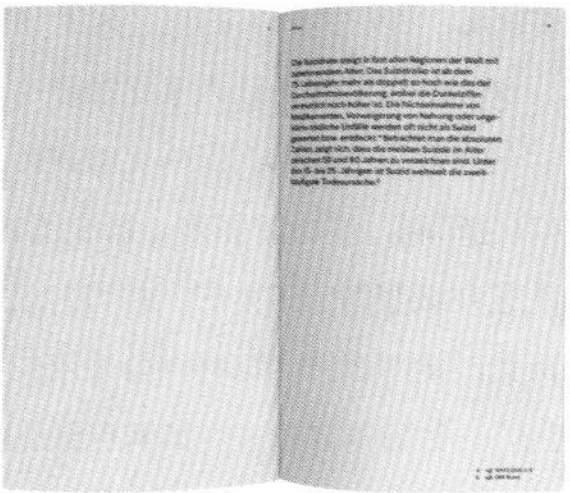

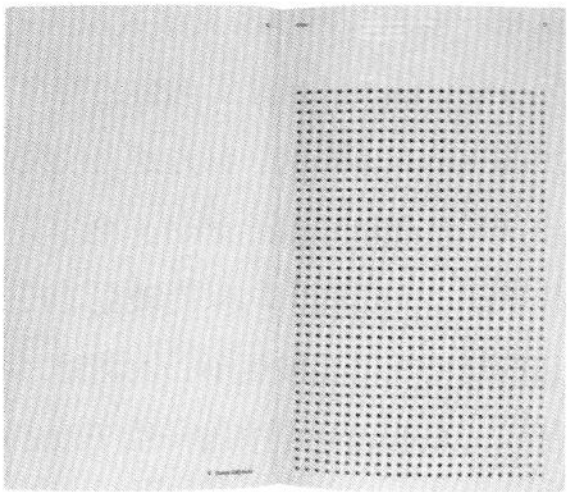

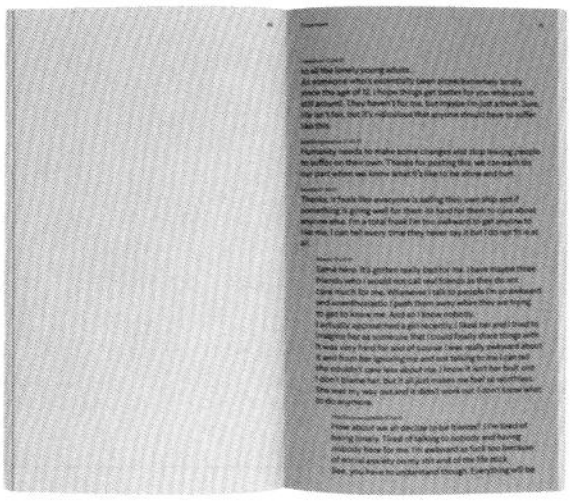

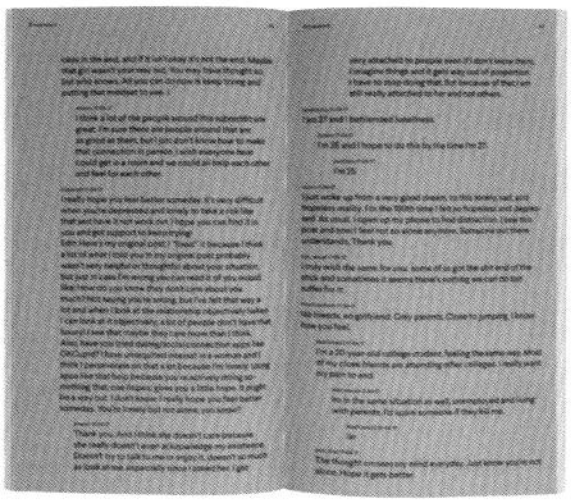

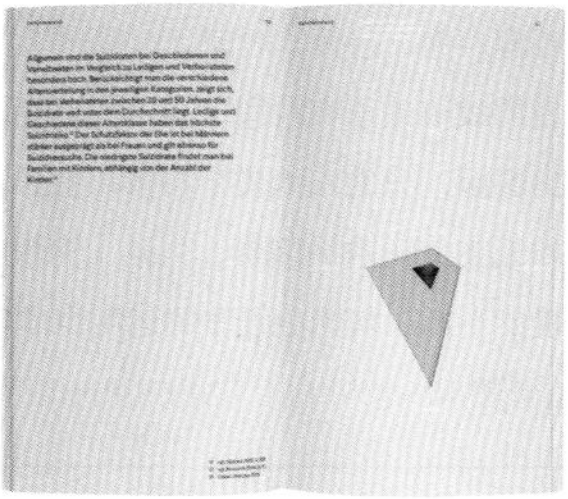

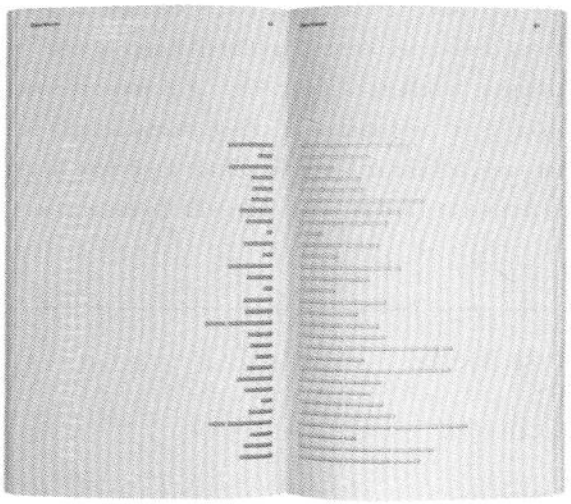

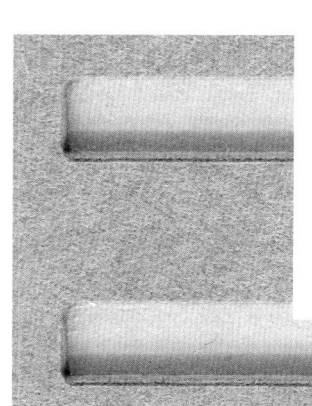

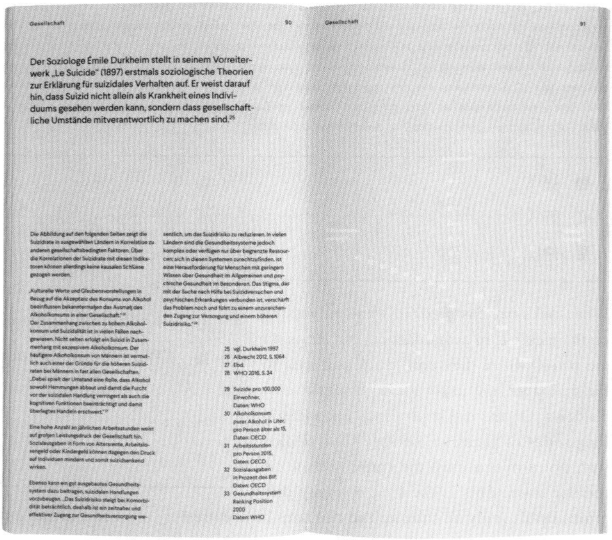
Gesellschaft
90
Gesellschaft
91
Der Soziologe Émile Durkheim stellt in seinem Vorreiterwerk „Le Suicide“ (1897) erstmals soziologische Theorien zur Erklärung für suizidales Verhalten auf. Er weist darauf hin, dass Suizid nicht allein als Krankheit eines Individuums gesehen werden kann, sondern dass gesellschaftliche Umstände mitverantwortlich zu machen sind.25
Die Abbildung auf den folgenden Seiten zeigt die Suizidrate in ausgewählten Ländern in Korrelation zu anderen gesellschaftsbedingten Faktoren. Über die Korrelationen der Suizidrate mit diesen Indikatoren können allerdings keine kausalen Schlüsse gezogen werden.
„Kulturelle Werte und Glaubensvorstellungen in Bezug auf die Akzeptanz des Konsums von Alkohol beeinflussen bekanntermaßen das Ausmaß des Alkoholkonsums in einer Gesellschaft.“26 Der Zusammenhang zwischen zu hohem Alkoholkonsum und Suizidalität ist in vielen Fällen nachgewiesen. Nicht selten erfolgt ein Suizid in Zusammenhang mit exzessivem Alkoholkonsum. Der häufigere Alkoholkonsum von Männern ist vermutlich auch einer der Gründe für die höheren Suizidraten bei Männern in fast allen Gesellschaften. „Dabei spielt der Umstand eine Rolle, dass Alkohol sowohl Hemmungen abbaut und damit die Furcht vor der suizidalen Handlung verringert als auch die kognitiven Funktionen beeinträchtigt und damit überlegtes Handeln erschwert.“27
Eine hohe Anzahl an jährlichen Arbeitsstunden weist auf großen Leistungsdruck der Gesellschaft hin. Sozialausgaben in Form von Altersrente, Arbeitslosengeld oder Kindergeld können dagegen den Druck auf Individuen mindern und somit suizidsenkend wirken.
Ebenso kann ein gut ausgebautes Gesundheitssystem dazu beitragen, suizidalen Handlungen vorzubeugen. „Das Suizidrisiko steigt bei Komorbidität beträchtlich, deshalb ist ein zeitnaher und effektiver Zugang zur Gesundheitsversorgung wesentlich, um das Suizidrisiko zu reduzieren. In vielen Ländern sind die Gesundheitssysteme jedoch komplex oder verfügen nur über begrenzte Ressourcen; sich in diesen Systemen zurechtzufinden, ist eine Herausforderung für Menschen mit geringem Wissen über Gesundheit im Allgemeinen und psychische Gesundheit im Besonderen. Das Stigma, das mit der Suche nach Hilfe bei Suizidversuchen und psychischen Erkrankungen verbunden ist, verschärft das Problem noch und führt zu einem unzureichenden Zugang zur Versorgung und einem höheren Suizidrisiko.“28
25 vgl. Durkheim 1997
26 Albrecht 2012, S. 1064
27 Ebd.
28 WHO 2016, S. 34
29 Suizide pro 100.000 Einwohner, Daten: WHO
30 Alkoholkonsum purer Alkohol in Liter, pro Person älter als 15, Daten: OECD
31 Arbeitsstunden pro Person 2015, Daten: OECD
32 Sozialausgaben in Prozent des BIP, Daten: OECD
33 Gesundheitssystem Ranking Position 2000 Daten: WHO

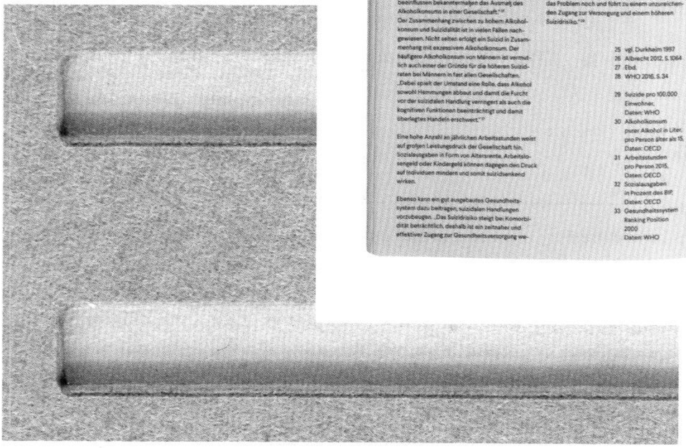

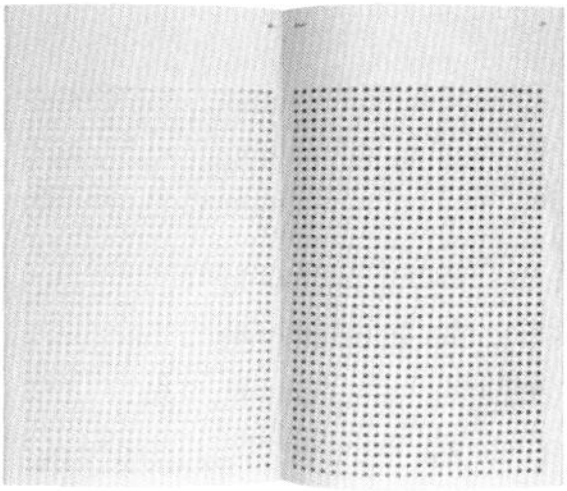

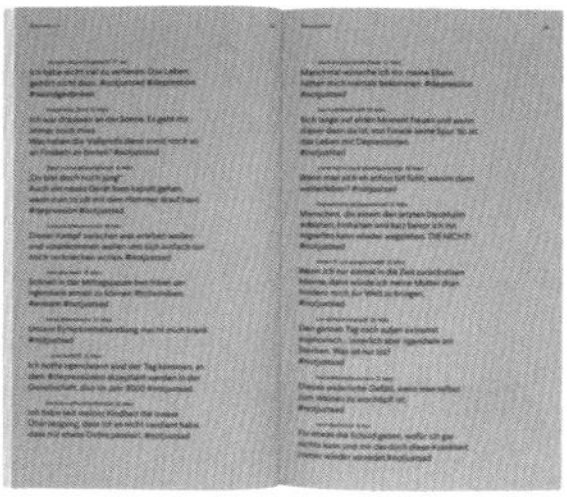

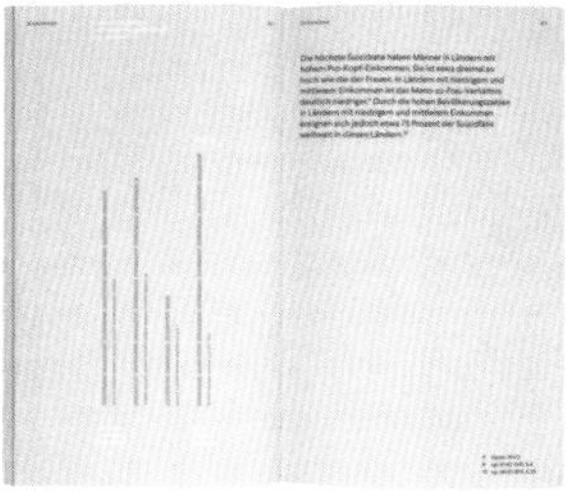

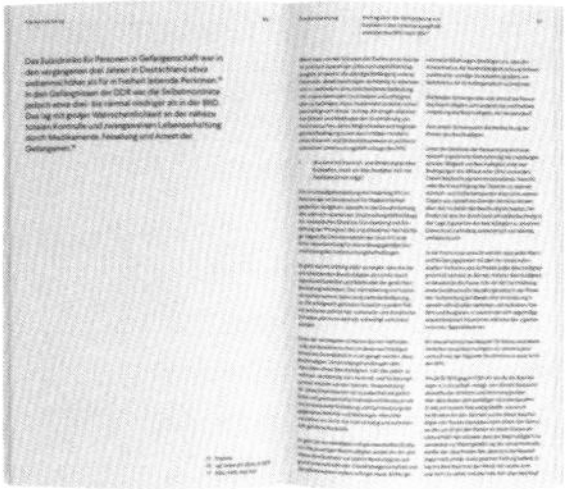

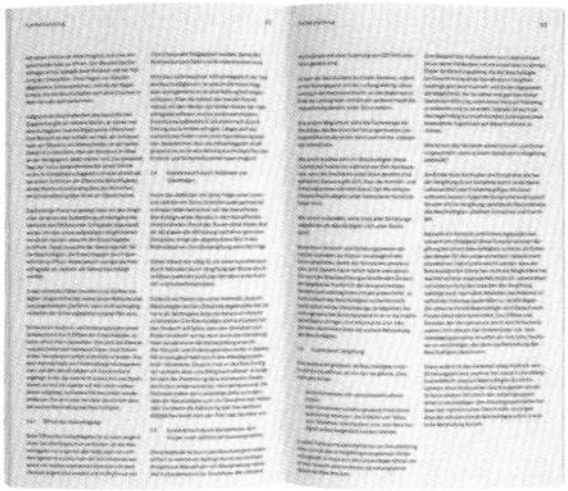

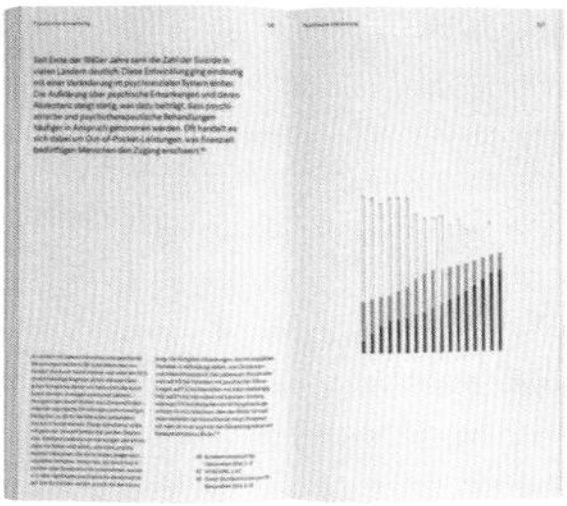

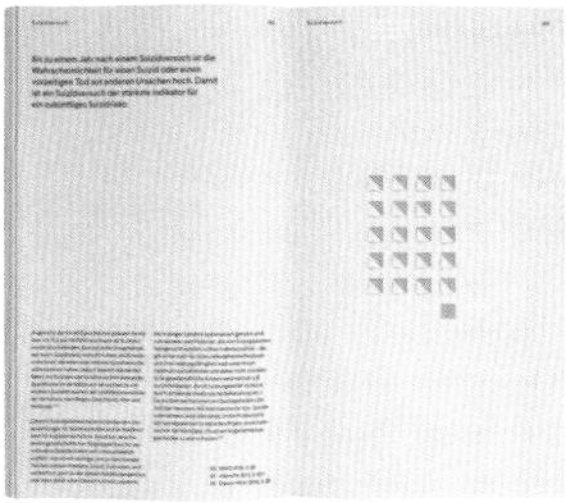

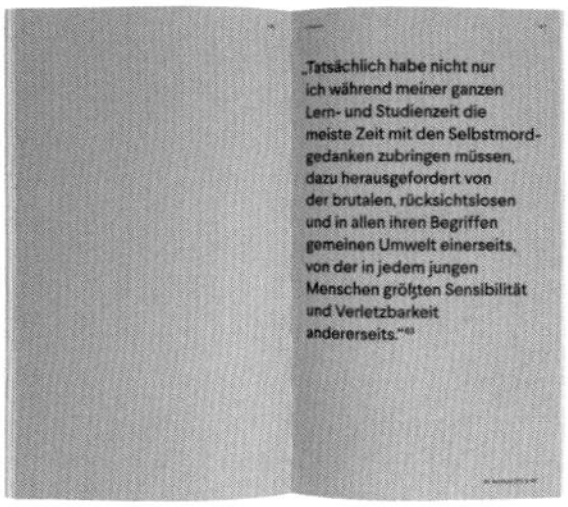
„Tatsächlich habe nicht nur
ich während meiner ganzen
Lern- und Studienzeit die
meiste Zeit mit den Selbstmord-
gedanken zubringen müssen,
dazu herausgefordert von
der brutalen, rücksichtslosen
und in allen ihren Begriffen
gemeinen Umwelt einerseits,
von der in jedem jungen
Menschen größten Sensibilität
und Verletzbarkeit
andererseits."

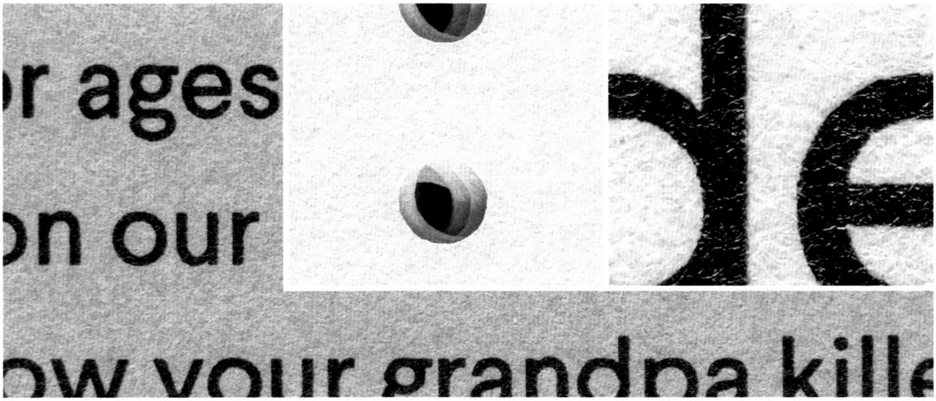
r ages
on our

Historische, kulturelle oder soziale Gegebenheiten können Einfluss auf die Suizidmethode und in weiterer Folge die Suizidrate nehmen. Ein möglicher Zusammenhang zeigt sich, wenn man die Entwicklung der Suizidraten in der DDR über die Zeit betrachtet. Lange galt hier das Einatmen von Hausgas als eine bevorzugte Suizidmethode. Die Gesamtzahl der Suizide sinkt annähernd um den Anteil der Suizide durch Hausgas.

„Im früheren Bundesgebiet standen die Vergiftungen mit festen und flüssigen Substanzen im Vordergrund und in der DDR die Vergiftungen mit im Haushalt verwendeten Gasen. Der Grund für diesen Unterschied lag in der noch großen Verbreitung und ausgedehnten Verwendung unentgifteten Kochgases in der damaligen DDR. Mit der Reduktion bzw. dem fast vollständigen Entfernen des CO-Anteiles im Haushaltsgas erfolgte auch das fast vollständige Verschwinden des ‚Kochgassuizids'."[18]

„Eine politische bzw. gesellschaftsordnungsbedingte Abhängigkeit der gewählten Suizidmethoden ist nicht erkennbar, sondern die Auswahl der Methoden hängt von der Gelegenheitsstruktur bzw. der Zugänglichkeit bzw. Verfügbarkeit ab (erkennbar z.B. an der sich ändernden Rolle der Feuerwaffen (größere Verfügbarkeit im Zeitablauf, Geschlechtsspezifität), des Vergiftens durch Gase im Haushalt (Rolle der Hausfrau auf der einen und zunehmende geringere Verfügbarkeit durch Ersetzung des giftigen Kohlenmonoxyds auf der anderen Seite) und des Sturzes in die Tiefe (Zunahme von Hochhäusern))."[19]

„Ein deutliches Abweichen von der Konstanz der Suizidraten wurde in den 60er Jahren in England beobachtet. Suizid mit Hausgas verschwand nahezu nach einer Dekade, in welcher sein CO_2-Gehalt stetig gesenkt wurde. Ein kompensatorisch wachsender Gebrauch anderer Suizidmethoden war nicht festzustellen. Es zeigt sich eine deutliche Abnahme der Suizidraten bei beiden Geschlechtern. Als nach 1971 die Entgiftung abgeschlossen war, schien kein weiteres Ziel mehr erreichbar, und die Suizidraten folgten erneut dem Aufwärtstrend früherer Dekaden."[20]

18 Wiesner 2004, S. 1103 f.
19 Albrecht 2012, S. 1030 f.
20 Bronisch 2014, S. 30
21 Daten: Todesursachenstatistik

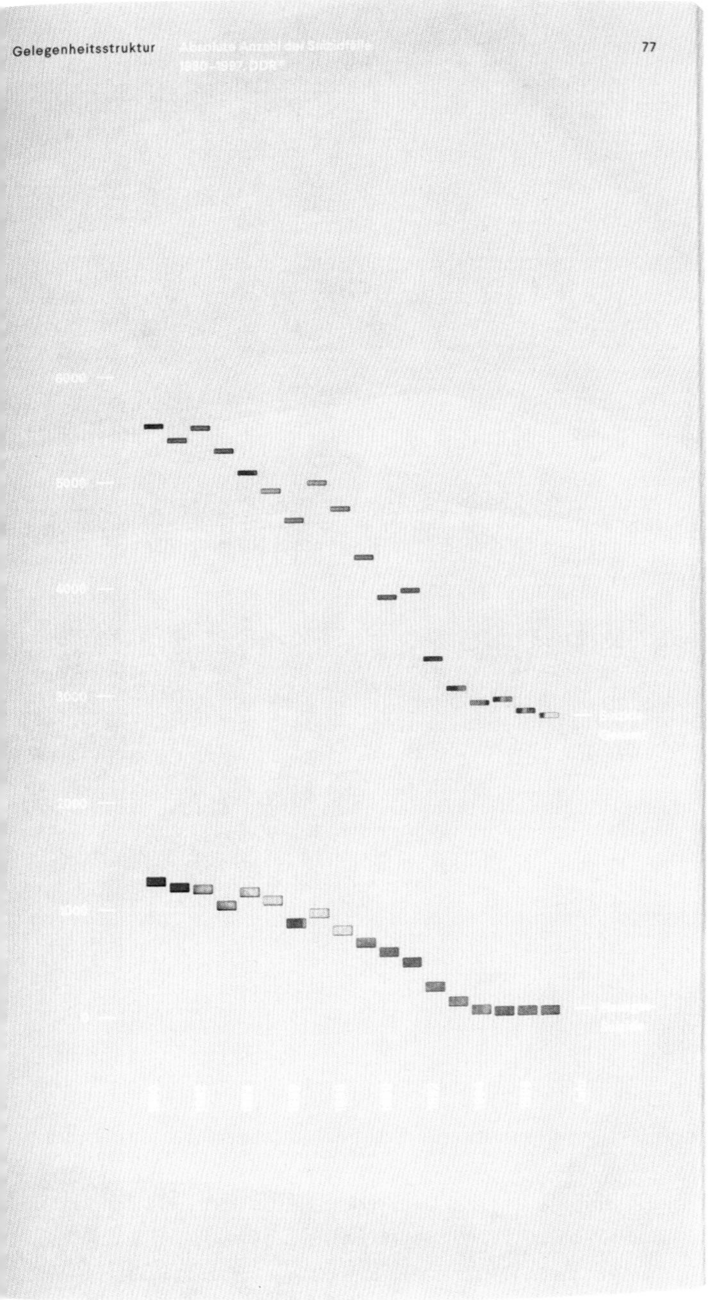
Gelegenheitsstruktur
77
6000
5000
4000
3000
2000
1000
0

这是一本带有故事性的少儿科普类图书。书中通过穿着喇叭裙、戴着新眼镜的小女孩的视角，探索了与眼睛、视觉、观察相关的知识领域。本书的装订形式为锁线硬精装。全书主要使用与四色接近的荧光油墨进行印刷，只在展示常见色色板的部分使用了 CMYK 的常规四色油墨。封面采用艺术纸张裱覆卡板，除使用四色荧光油墨外，中间的帽子和人脸五官部分烫黑处理。帽子上的蓝色点由于烫黑的压凹效果而凸出来，呈现相应的盲文。内页胶版纸张四色荧光墨印刷。内容包括：白天黑夜（光）、对眼睛的观察、视力、眼睛的结构、颜色、镜子、表情、符号、图案、眼镜、摄影、观测、魔术、动物的眼睛和视力差异、五感的全方位感知、盲人如何感知世界、电影、绘画……几乎包含了与视觉有关的一切。本书通俗易懂，视觉风格强烈。

Романа Романишин
та Андрій Лесів
Я
ТАК
БАЧУ

This is a science book for children with stories. Through the perspective of a girl wearing a skirt and new glasses, the book explores the fields of knowledge related to eyes, vision and observation. The hardback book is bound by thread sewing. The book is mainly printed with fluorescent ink similar to conventional CMYK four colors. The cover is made of art paper mounted with cardboard. In addition to using four-color fluorescent ink on the cover, the hat and facial features are hot stamped in black. The blue dots on the hat protrude due to the hot stamping effect, showing the corresponding Braille. The inside pages are printed with four-color fluorescent ink on offset paper. The contents include: day and night (light), observation of eyes, vision, structure of eyes, colors, mirror, expression, symbols, patterns, glasses, photography, observation, magic, animals' eyes and visual differences, omnidirectional perception of five senses, how blind people perceive the world, movies, paintings and so on. It contains almost everything about vision. This book is easy to understand and has a strong visual style.

铜奖 — Bronze Medal

这就是我看到的 — Я так бачу (This Is How I See)

乌克兰 — Ukraine

Art Studio Agrafka (Romana Romanyshyn, Andrii Lesiv)

Art Studio Agrafka (Romana Romanyshyn, Andrii Lesiv)

Видавництво Старого Лева (Old Lion Publishing)

287×267×10mm

552g

56p

This children's book tells us about the scenes viewed by a girl in a flared skirt. She is dependent on her new pair of large glasses. An owl, the mascot of vision, accompanies her while she explores. On one of the most powerful double pages, some animal silhouettes — including goats, chameleons, sharks and blackbirds — "huddle" together to form a kind of puzzle. On the left-hand side at the top of the page, against its basic paper white, the sun is shining: aha, it's daytime. Opposite on the right hand side, the crescent moon shines dimly above the full page of matt blue. Yellow dots, large, small, mostly in pairs, shine like eyes out of the blue night. Precisely — it is the eyes of the little animals on the left-hand side that linger in the same position. The girl declares: "There are eyes that see better than mine. Even in the dark." In other places the book becomes pedagogical, when the vision of bees, owls, horses, cats and dogs is compared, for example. Or even Braille — although not tangible, the letters are shown using signal red dots. The graphic, pictogram-like visual style of this book is impressive. The three special colours designed to highlight certain areas — variants of the primary colours blue, red and yellow — maintain their dominant independent existence. Here and there they are overprinted, fitting exactly over the top or overlapping; it is simply a pleasure to observe the whole diversity of shapes.

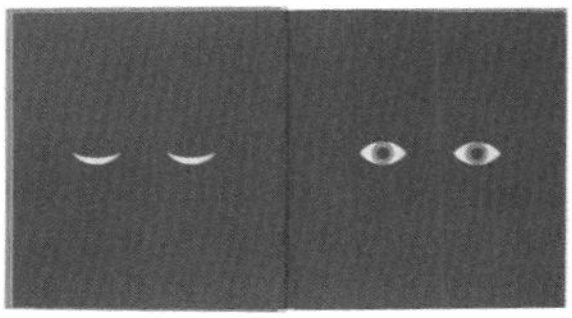
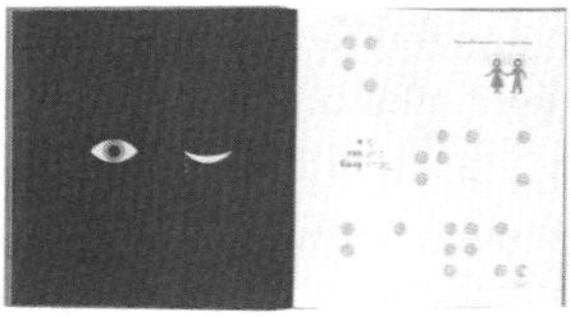
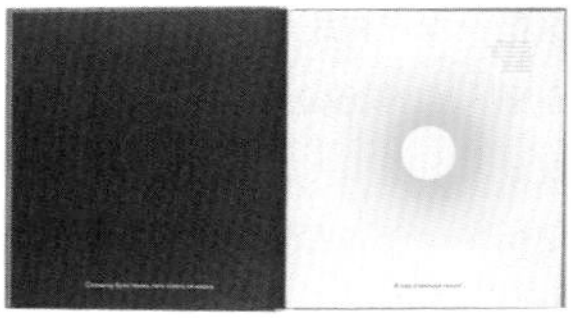

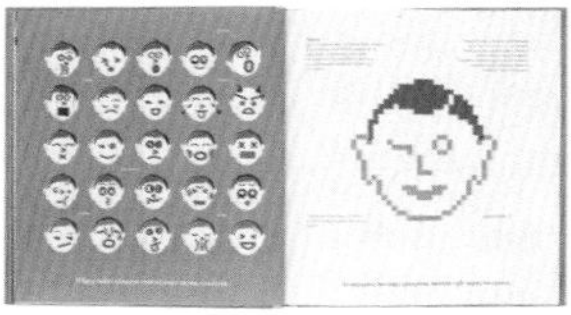
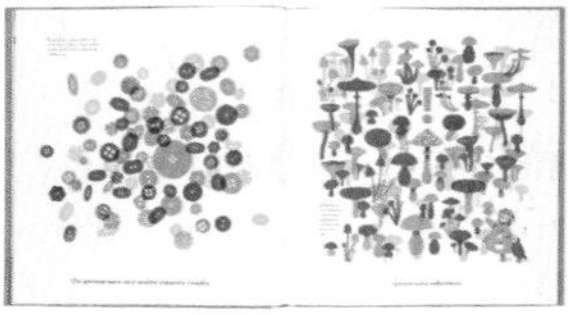
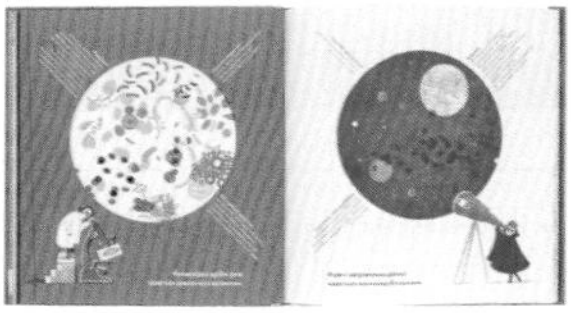
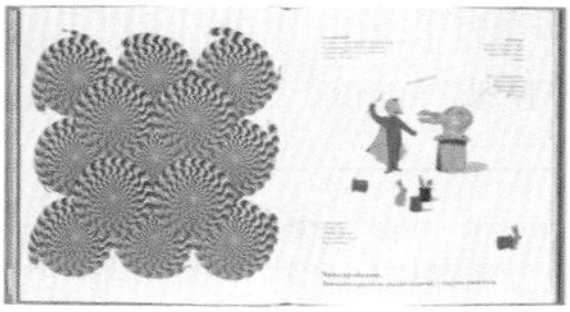

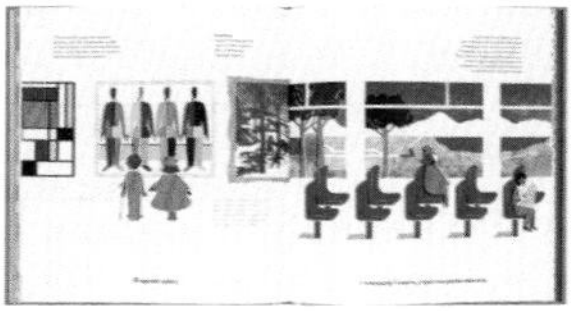

У мене багато ідей.
В уяві малюю досі небачене.
Змінюю кут зору,
аби побачити більше.

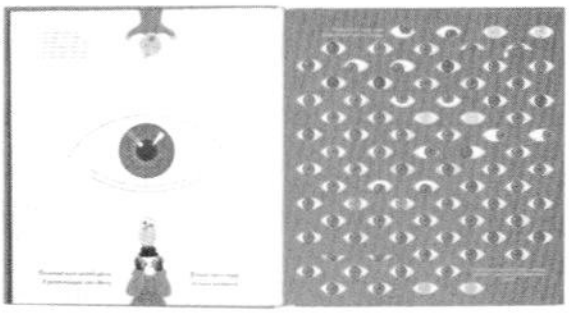
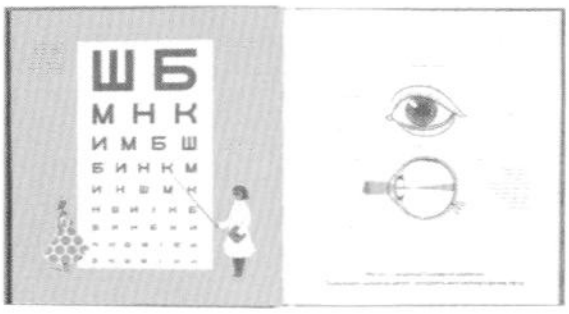
Ш Б
М Н К
И М Б Ш
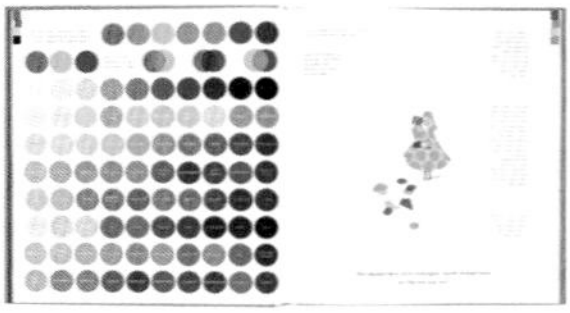
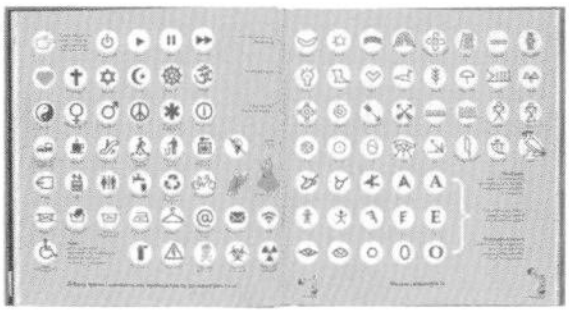

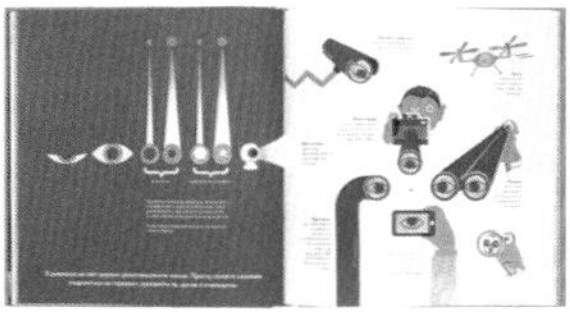
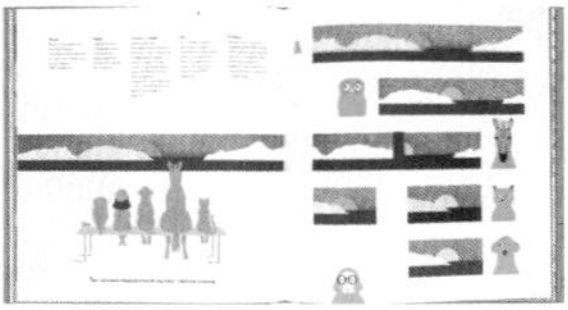
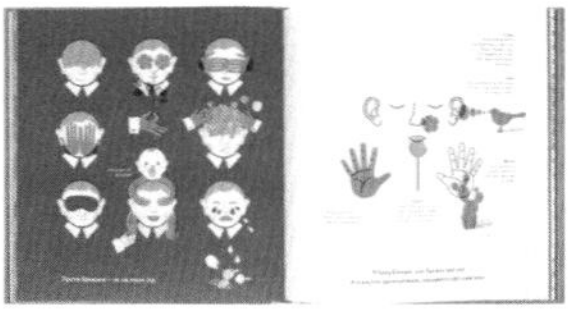
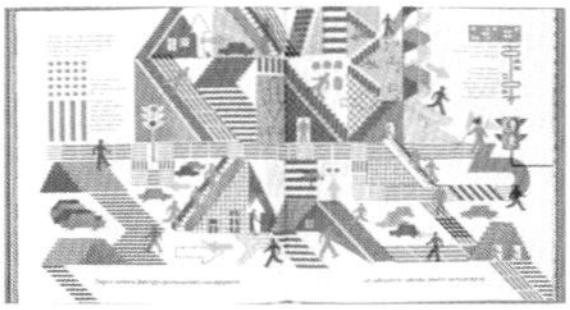
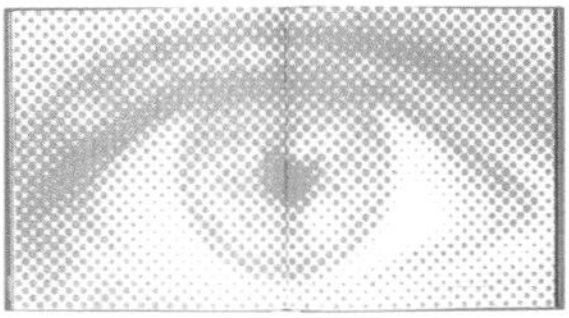
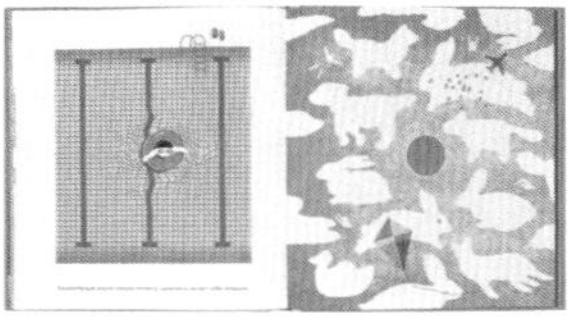
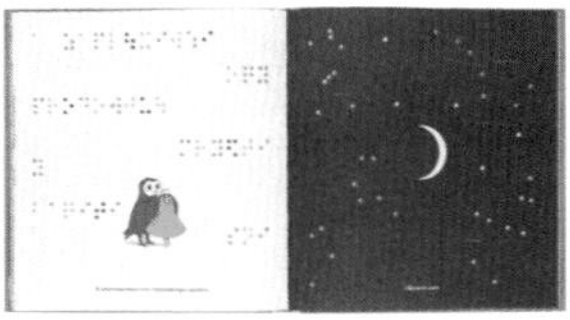
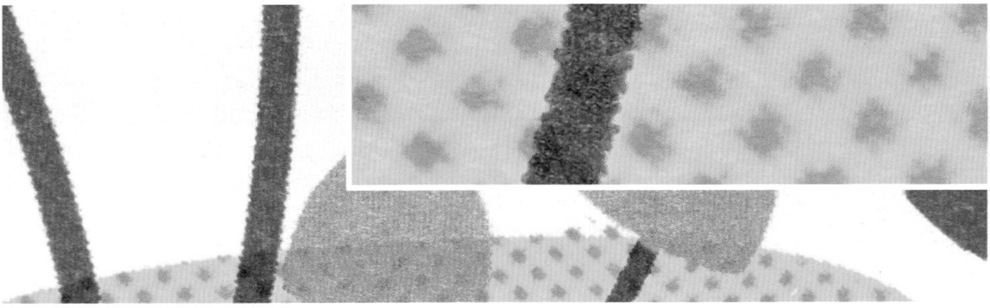

Я ТАК БАЧУ

Романа Романишин і Андрій Лесів

Шрифт Брайля
шрифт для письма і читання, який розробив для незрячих француз Луї Брайль у 1829 році, коли йому було лише 15 років.

Основою шрифту є матриця із шести крапок. Залежно від літери, потрібні крапки в матриці виділяються опуклими назовні, щоб їх можна було відчути на дотик.

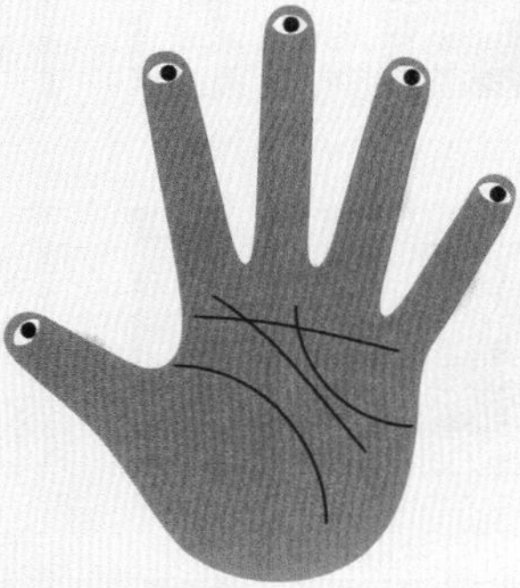

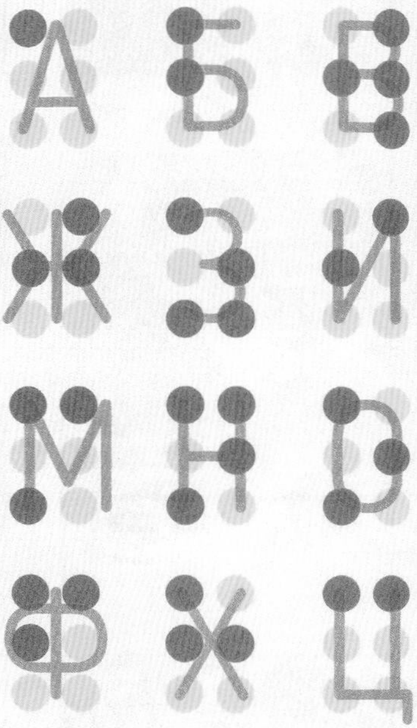
А Б В
Ж З И
М Н О
Ф Ж Ц

6 крапок Брайля дають 64 можливі комбінації для позначення різних літер, цифр та символів.

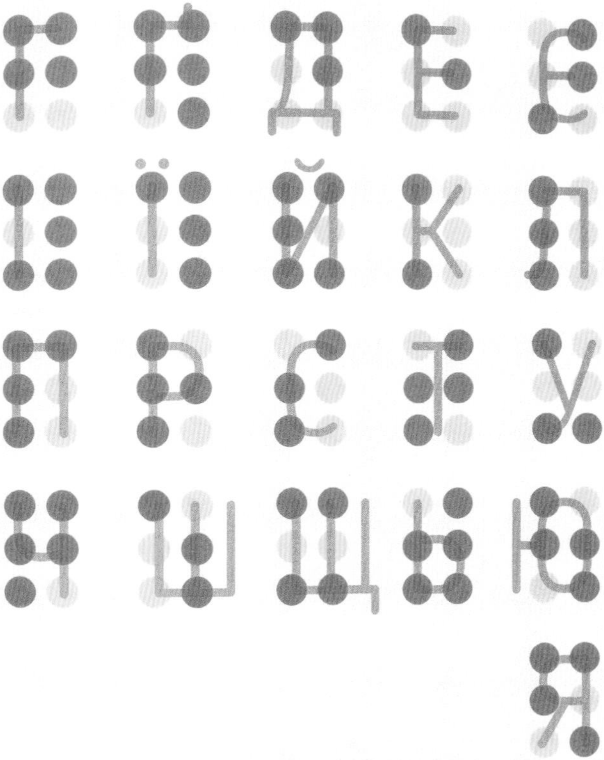
Г Ґ Д Е Є
І Ї Й К Л
П Р С Т У
Ч Ш Щ Ь Ю
Я

**Лише шість крапок достатньо,
аби сказати так багато.**

CF Moore leaves it to the eye or to the beholder's sensibilities to flesh it out. The creation of the void activates the sculpture. There are empty spaces and *external spaces*. In the case of the latter, we refer less to the surrounding space than to the silhouette. Depending on how a work is set up, you can either let something stand out or ensure that it remains inside. You can also toy with internal spaces, so that – depending on the angle they're seen from – they can remain hidden and absent. There are also enclosed *external spaces*, that is, when an object is placed in such a way that you can look through from underneath it.

HS Those are factors that register very strongly as you move around a piece. *Two Piece Reclining Figure Number 2*, for instance, is extremely changeable. One moment you're looking through it, but no sooner have you moved than you can't see through it at all. Sometimes you can't even see that it is in two pieces. It's very intriguing.

CF It's like that when you look at bones, too.

HS Yes, with the holes in the vertebrae where the spinal nerve ran.

CF Or if you look at a pelvis, with its huge forms extending outwards.

HS Moore once said, 'The variety of objects in my small studio provides me with many new ideas simply by looking at and handling them. This might not happen if each object was in isolation.'[13]

CF Very true. That's the reason why I always have to have things around me. The big disadvan-

这是一本用对话串联作品照片从而深度呈现瑞士雕塑家切萨雷·费罗纳托（Cesare Ferronato）艺术理念的书。本书的装订形式为锁线硬精装。封面红色布面裱覆卡板，艺术家的代表作使用光面艺术纸张单黑印刷后裱贴于封面中央，封面字符则使用白色压印。前后环衬选用深蓝色艺术纸张，与封面的红色形成视觉反差。内页涂布艺术纸四色印刷。书中通过雕塑家与青年艺术家汉尼斯·舒普巴赫（Hannes Schüpbach）的对话，讨论了雕塑家的成长、家庭、生活、作品和一些对他影响重大的人物。书中将语言比作雕塑家的凿子，在表达的过程中逐步呈现脑中的意象，慢慢接近本质。内容的主要部分采用四栏设置，其中正文对话跨两栏形成双栏版心，对话的两人采用不同的缩进设置形成明显区隔。页面顶部一栏放置巨大的页码，三栏放置注释。图片未严格按照栏位排版，版面丰富多变。|||||

Cesare Ferronato
The Anatomy of Stone
Conversations with Hannes Schüpbach

This is a book that uses dialogue to serialize photographs of works in order to present the Swiss sculptor Cesare Ferronato's artistic ideas in depth. The hardback book is bound by thread sewing. The cover is made of cardboard mounted with red cloth. The artist's representative works are printed on glossy art paper in black, and then mounted in the center of the cover. The characters on the cover are hot stamped in white. The front and back endpapers are made of dark blue art paper, which is visually contrasted with the red cover. The inside pages are printed on coated art paper with four-color printing. The book discusses the sculptor's growth, family, life, works and some people who have great influences on him through a dialogue between the sculptor and the young artist Hannes Schüpbach. The book compares language to a sculptor's chisel, which gradually presents the intention in the brain and gradually approaches the essence in the process of expression. The main part of the contents is typeset in four columns, in which the text dialogues cross two columns to form a double-column print area. Dialogues of two persons with different indentation settings form a distinct division. The huge page number is place in the first column and comments are placed in the third column. Pictures are not typeset according to columns strictly, thus the layout is abundant and changeable.

◎ 荣誉奖

《 切萨雷·费罗纳托——石头的解剖学

◖ 奥地利

◎ Honorary Appreciation

《 Cesare Ferronato – Anatomie des Steins

◖ Austria

D Raphael Drechsel

△ Hannes Schüpbach

||||| Verlag für moderne Kunst, Vienna

□ 255×207×27mm

1042g

224p

Catalogue, biography, a collection of interviews? The works of Swiss sculptor Cesare Ferronato are given special treatment in this book: discussions between Ferronato and the younger artist Hannes Schüpbach offer an opportunity to experience the former's world of thought and forms. Although the book has its own pictorial focus above and beyond the sculptures, it is pervaded by the system of interconnecting text and image. In a two-column type area, the spoken words are divided into two type widths per column to highlight the dialogue concept. The high top edge displays the occasional notes — it is structured into four columns, with the outermost column reserved for the page numbers. Their prominent size could be described as functionally dubious — but then what is functional, anyway? All the types of text have substantially differing type sizes. The white space at the sides — regardless of whether it has been left unoccupied consciously or intuitively, or because it simply turned out like that — appears as if it has been created by joining together highly scaled sections. In these open compositions of image and text, the oversized page numbers certainly assume a specific function: that of a stabilising pacemaker. Purists may find them scandalous, but sometimes bulk is important. The textile cover — haptic seduction radiated by a rough textured fabric and smooth art paper — grabs your attention with its picture of a torso, worked into the deepened surface, because of the amorphous effect created by the outline of the artistically defined form.

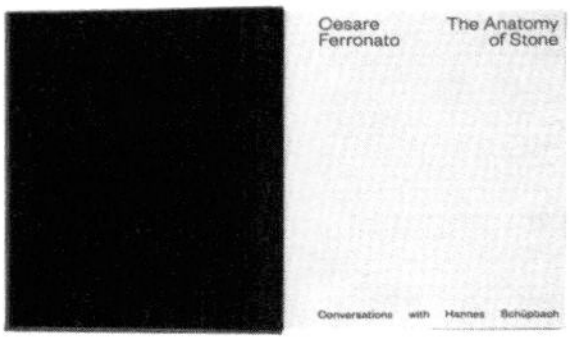
Cesare
Ferronato
The Anatomy
of Stone
Conversations with Hannes Schüpbach

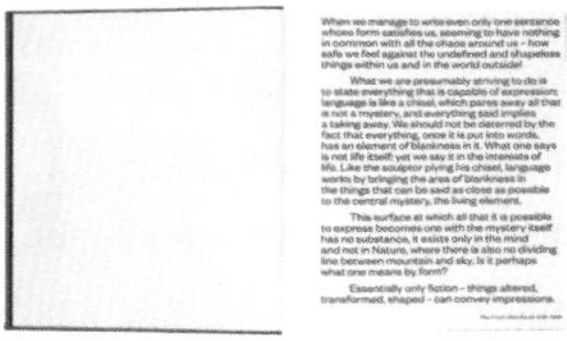

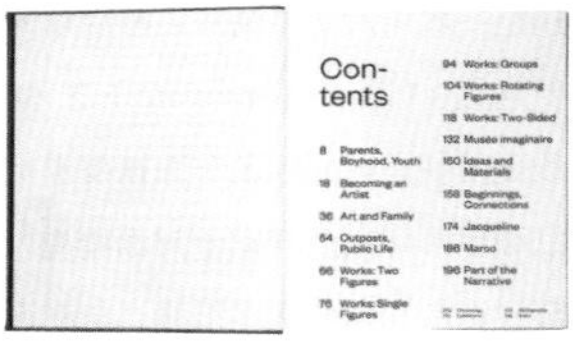
Con-
tents
8 Parents, Boyhood, Youth
18 Becoming an Artist
36 Art and Family
54 Outposts, Public Life
66 Works: Two Figures
76 Works: Single Figures
94 Works: Groups
104 Works: Rotating Figures
118 Works: Two-Sided
132 Musée imaginaire
150 Ideas and Materials
158 Beginnings, Connections
174 Jacqueline
186 Maroo
196 Part of the Narrative

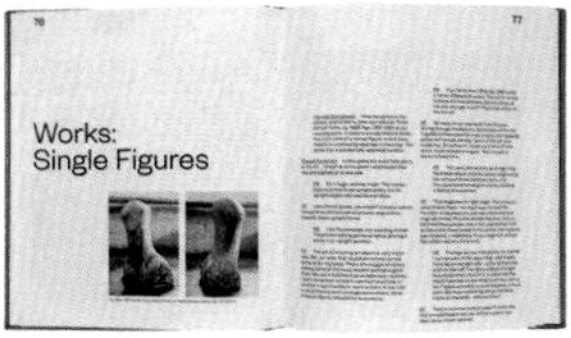
76
77
Works:
Single Figures

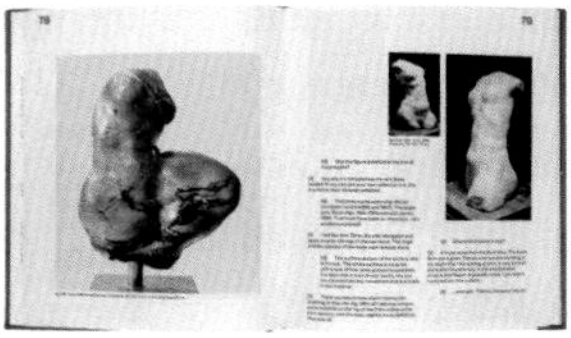

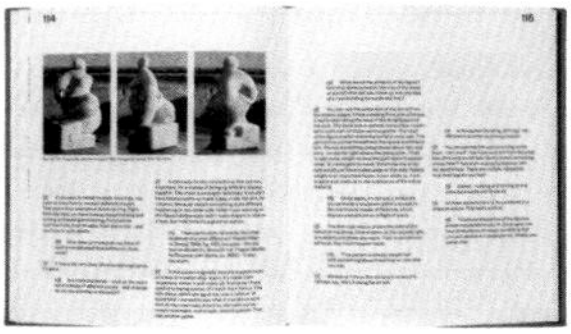

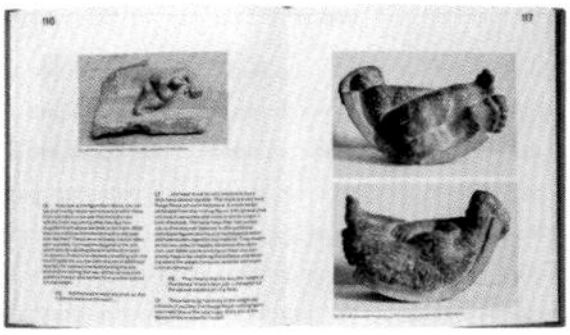

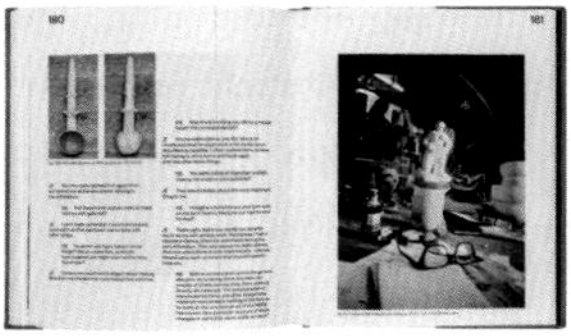

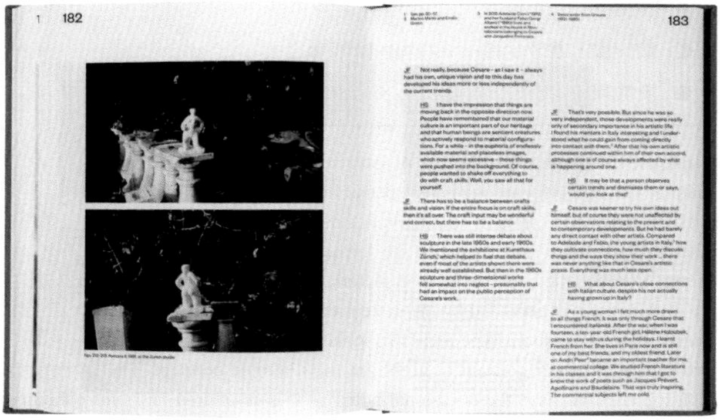
182
183

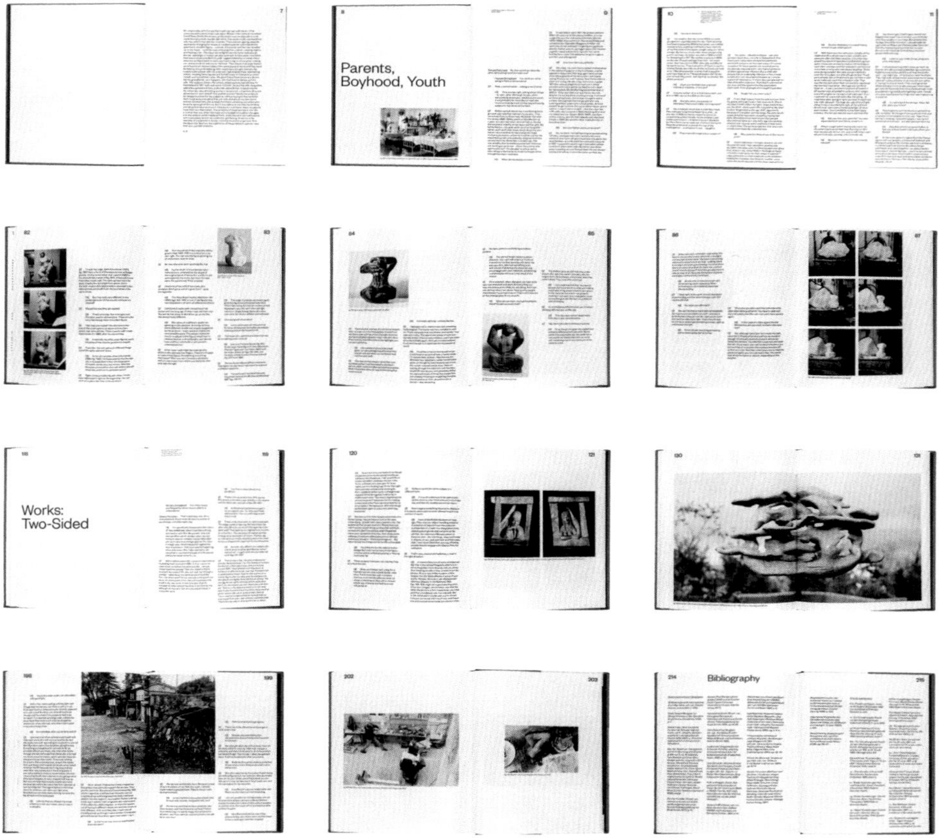
Parents,
Boyhood, Youth
Works:
Two-Sided
Bibliography

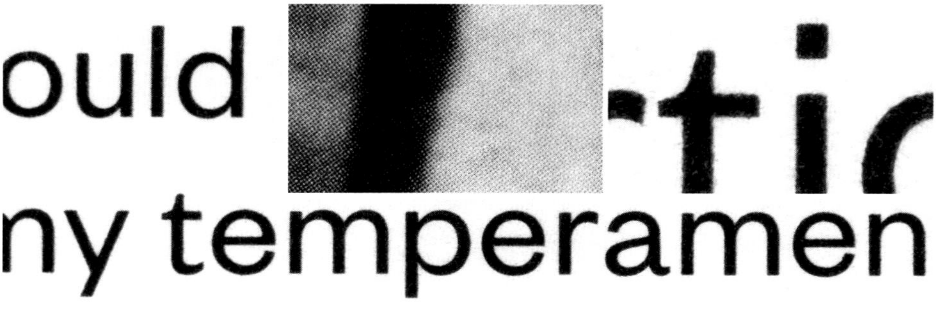
ould
tir
ny temperamen

这是有关俄国先锋派领军人物埃尔·利西茨基的展览的配套画册。书中主要呈现了利西茨基这位对包豪斯、风格派和解构主义都起到重要影响的艺术家的生平和作品。书籍设计的各个方面都试图通过纸面的形式对利西茨基的风格进行演绎。本书的装订形式为锁线硬精装。封面单黑印刷覆光面膜，裱覆在卡板上。卡板在三面书口部分均超出 15mm，并露出红色布面书脊。封面卡板内侧印有作品各部分名称的前一两个字母，与刷黑书口上的纵向条纹共同起到各部分内容的快速索引作用。内页胶版纸四色印刷，内容主要包括与展览相关的生平介绍、艺术评论，以及绘画、插图、建筑、书籍、摄影等方面的作品和尝试。书中标题采用等宽的打字机字体，正文主要采用几何无衬线字体，版心六栏设置并运用灵活，具有强烈的构成感。||||||||||||

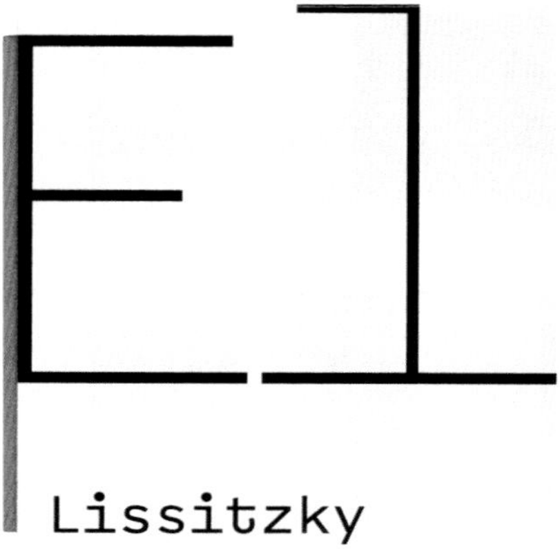
El
Lissitzky

This is a companion album to the exhibition of El Lissitzky, a leader of the Russian avant-garde. The book mainly presents the life and works of Lissitzky, who has influences on Bauhaus, De Stiji and deconstructionism. Every aspect of book design tries to deduce Lissitzky's style on the paper form. The hardback book is bound by thread sewing. The laminated cover is printed in black on the cardboard that exceeds the three edges with the width of 15mm. The first one or two letters of the title of each part are printed on the inside of the cardboard, which together with the vertical stripes on the fore-edge brushed in black plays an important role for indexing. The inside pages are printed on offset paper with four-color printing, mainly including life introduction, art review, and attempts in painting, illustration, architecture, book design, photography, etc. The title of the book is using typewriter font of equal width, while the main text is using geometric san-serif font. There are six flexible columns set in the print area, which makes the book structured.

- 荣誉奖
- 埃尔·利西茨基
- 俄罗斯

- Honorary Appreciation
- El Lissitzky
- Russian
- Evgeny Korneev
- Tatyana Goryacheva, Ruth Addison, Ekaterina Allenova
- Tretyakov Gallery, Jewish Museum and Tolerance Center
- 250×235×29mm
- 1084g
- 336p

The mounted covers mutate into the wings of a constructivist folding object, creating protruding edges with the width of a finger. Standing vertically, they hold the square innerbook — suspended from the side, as it were. El Lissitzky — the short title is turned into a typographical image on the white, shining laminated board cover, with the "E" flush at the top with the red spine lining, the "l" running into the top edge, and the lettering of Lissitzky flush with the bottom edge of the innerbook. The typical elements of the monospaced font that is reminiscent of typewriting — the linear style, the extra-bold dots on the "i"s, the special sweep of the little "t" — touch on elementary design principles of the Russian avant-garde in the 1920s. Inside this exhibition catalogue, six columns form the typographical framework of the double pages. Every second column contains the main text, interspaced with comments or pictures and captions. The combination of fonts aims to produce contrasts — an extended grotesque, for the distinctions a fine italic with sh, st and ct ligatures, and for the chapter headings the said monospaced font. Justification in the narrow columns looks remarkably compact; there seem to be fewer gaps than would be expected from this wide font. The white sections on the front edge colouring mark the chapters, with only the relevant pages blackened along their paper edges. The technical expertise necessary for such tricky edge colouring remains a mystery.

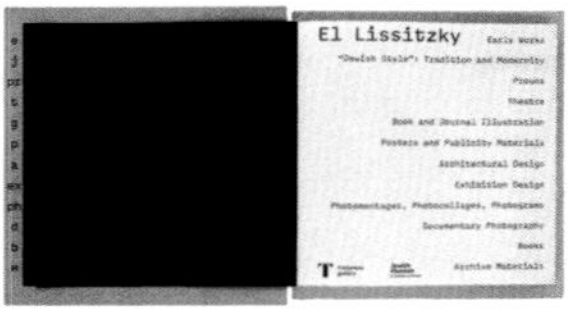
El Lissitzky
Early Works
"Jewish Style": Tradition and Modernity
Prouns
Theatre
Book and Journal Illustration
Posters and Publicity Materials
Architectural Design
Exhibition Design
Photomontages, Photocollages, Photograms
Documentary Photography
Books
Archive Materials

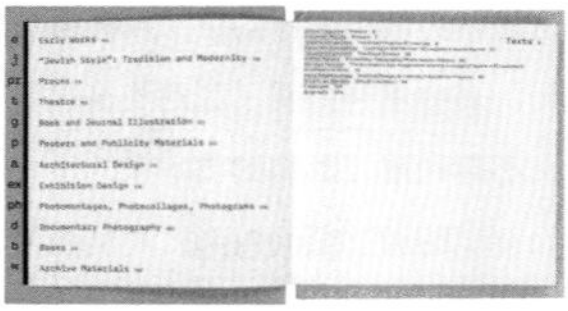
Early Works
"Jewish Style": Tradition and Modernity
Prouns
Theatre
Book and Journal Illustration
Posters and Publicity Materials
Architectural Design
Exhibition Design
Photomontages, Photocollages, Photograms
Documentary Photography
Books
Archive Materials
Texts

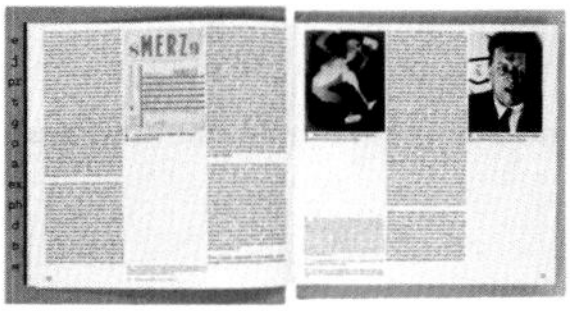

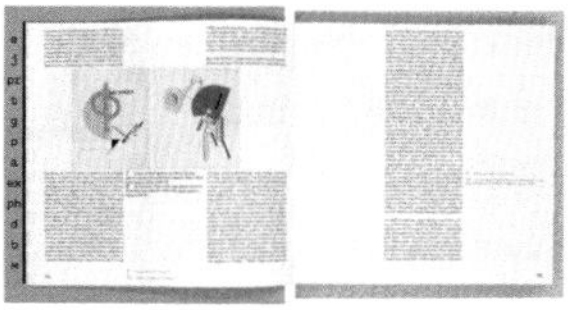

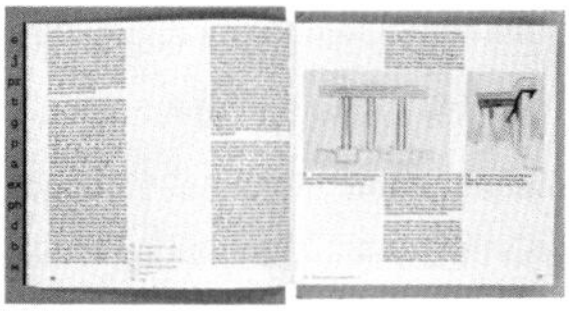

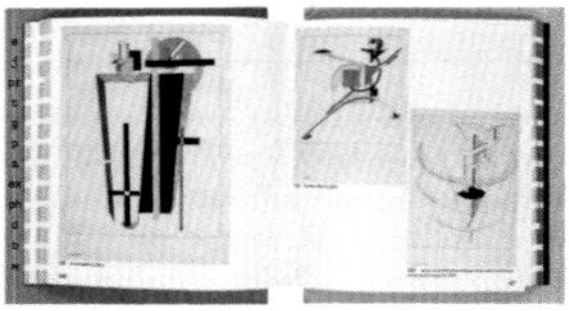

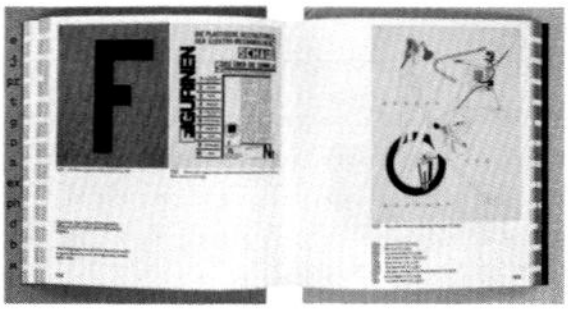

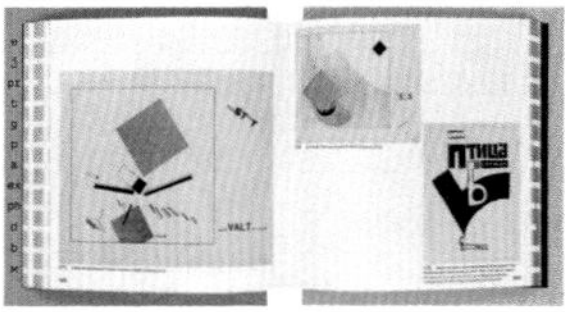

"Looking in Old Mirrors":
Alexander Kantsedikas
El Lissitzky's Jewish Period
22
23

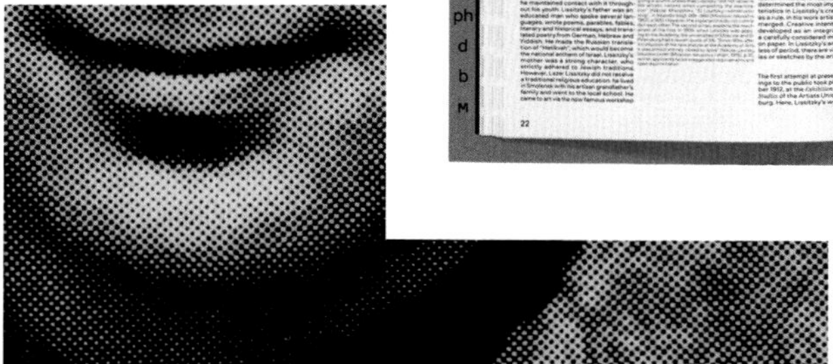

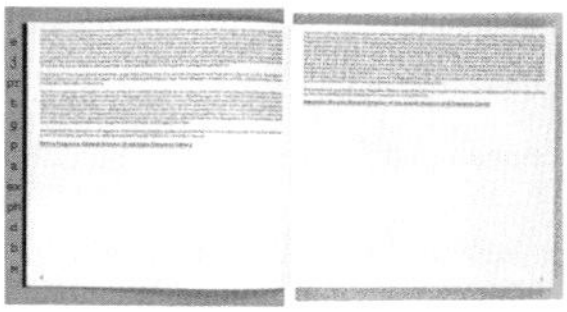
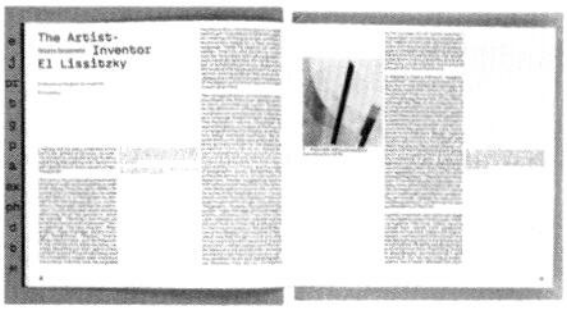
The Artist-
Inventor
El Lissitzky
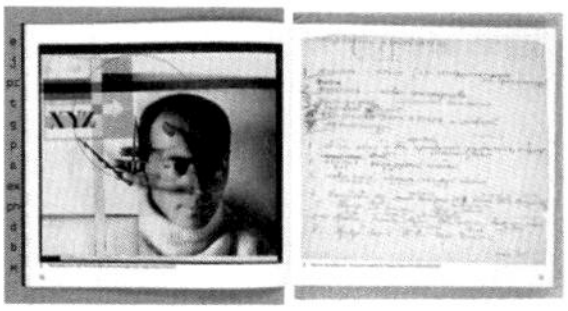
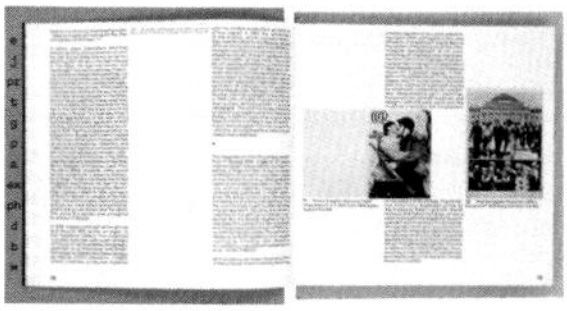
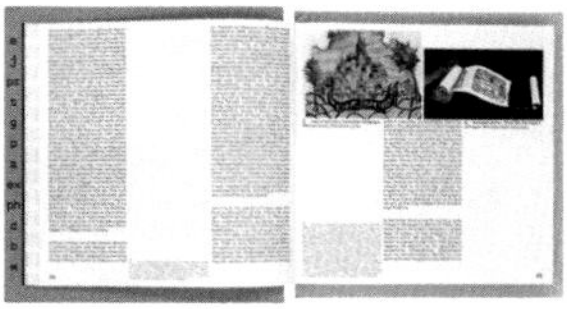
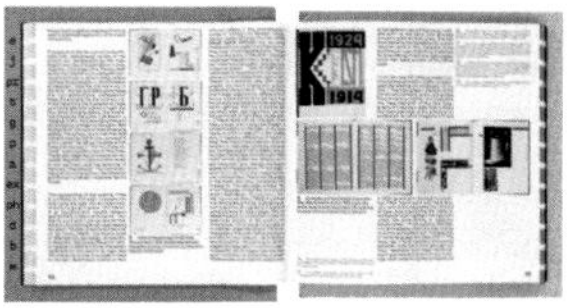
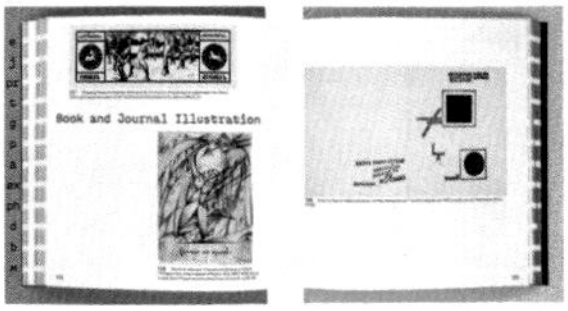
Book and Journal Illustration
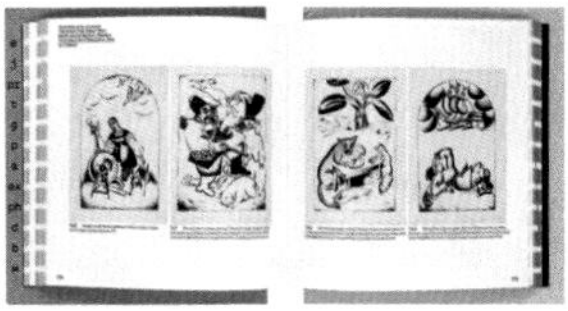

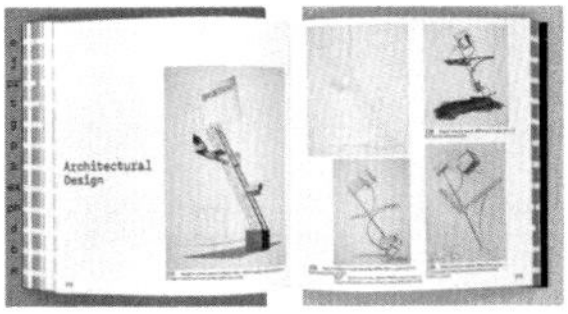
Architectural
Design
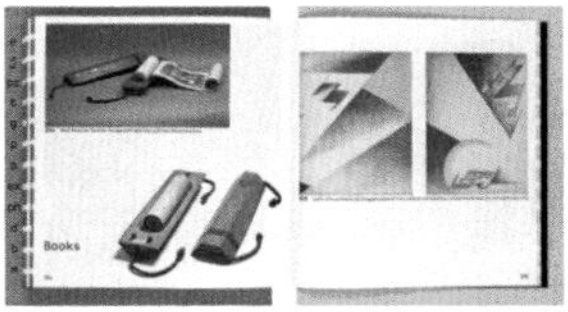
Books
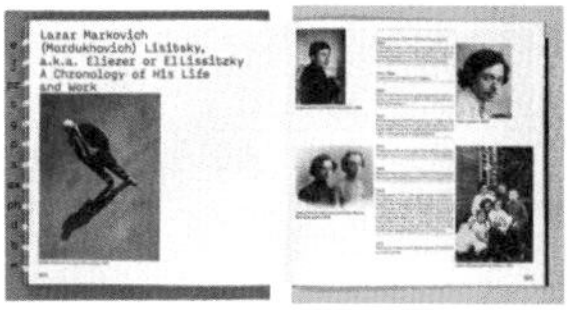
Lazar Markovich
(Mordukhovich) Lisitsky,
a.k.a. Eliezer or Ellissitzky
A Chronology of His Life
and Work

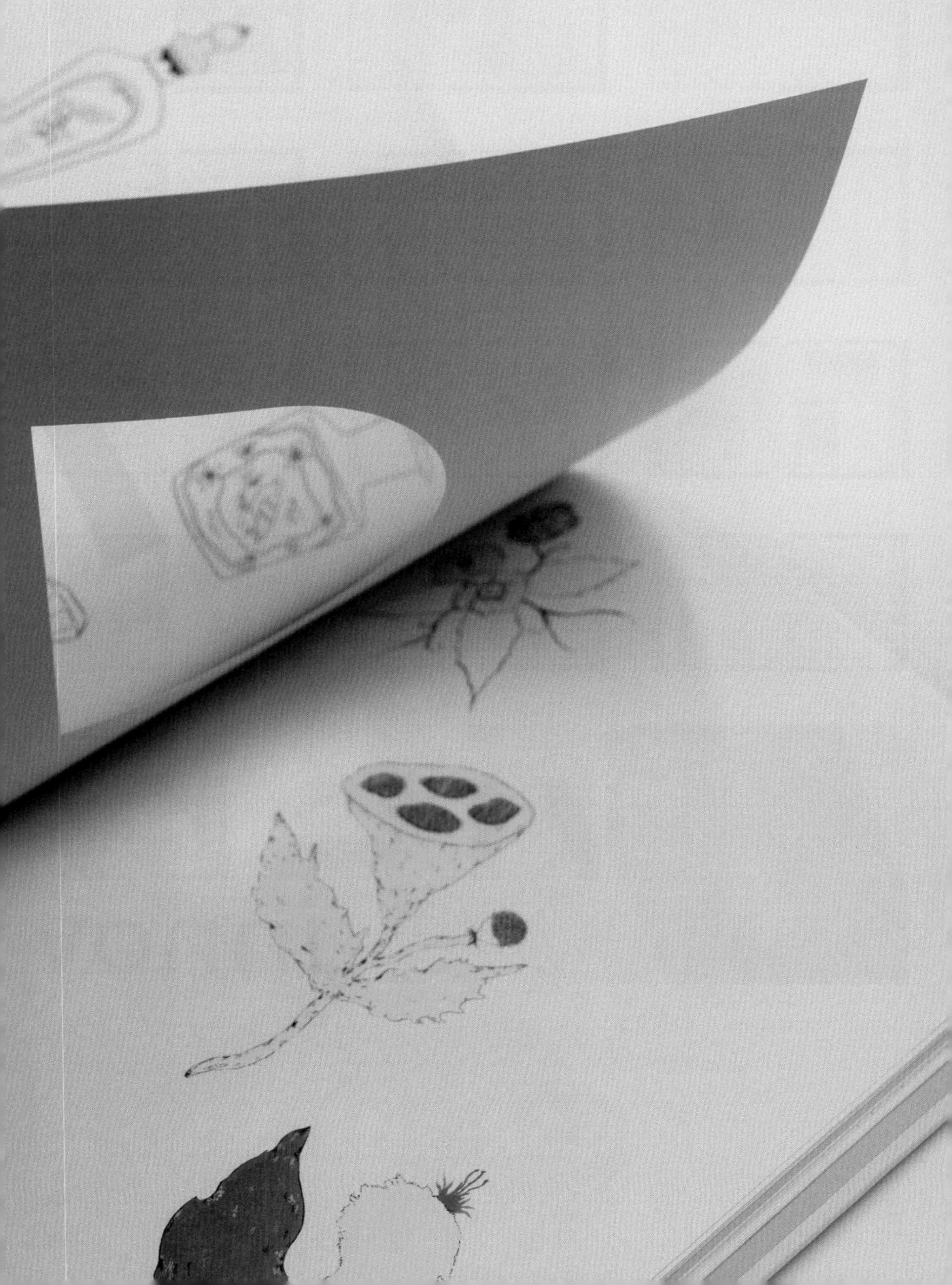

这是一本专为儿童准备的认知涂色书。有别于常见的认知涂色类图书，本书并不基于写实，也无意把孩子往“在纸面上还原真实世界”方向引导，而是通过较为艺术的方式，激发孩子更丰富的想象力。书籍的装订形式为锁线精装。封面采用了一种带有柔软肌理的粉红色纸张，红黄双色丝网印刷，套版故意不完全对齐，与手绘的风格相匹配。封面中下部标题字烫银，在整体温婉的封面中隐约呈现。环衬灰绿色纸张印银处理，内页胶版纸四色印刷，局部印银点缀。书内按照内容分为鸟类、哺乳动物、工具、植物、昆虫、建筑、水生动物、树，一共 8 个篇章，每章包含开篇、黑白图库和组合画面三部分，每章页数不均等。某些章节的开篇部分镂空呈现前后页面的形象，增加趣味。页面局部增加颜色块，有的按照图形边缘进行填色，有的则使用几何形进行抽象的色块叠加，形成独特风格的填色引导。

This is a cognitive coloring book designed for children. Unlike the common cognitive coloring books, this book is neither based on realism, nor intends to guide children to 'restore the real world on paper'. Through a more artistic way, the book tries to inspire children's imagination. The hardback book is bound by thread sewing. The cover is made of a kind of pink paper with soft texture, printed in red and yellow with silkscreen. The color plates are misaligned intentionally, matching the style of hand painting. The title on the lower part of the cover is hot-stamped in silver, which is faintly presented in the gentle cover. The endpaper is made of grey-green paper with silver hot stamping. The inside pages are printed on offset paper in four colors with partial silver embellishment. According to the contents, the book is divided into eight chapters, including birds, mammals, tools, plants, insects, buildings, aquatic animals and trees. Each chapter contains three parts: the opening chapter, black-and-white gallery and combined pictures. The pages of each chapter are uneven. At the beginning of some chapters, hollow-out technology is used to present the images on the front or back pages with delight. Color blocks are added to several pages: some fill in colors according to the edges of the graphics, some use geometric shapes to overlie abstract color blocks, which form a unique style of coloring guidance.

荣誉奖

第一件事

日本

Honorary Appreciation

The First

Japan

Hideyuki Saito

Kentarou Tanaka

Bonpoint Japon

267×216×8mm

418g

46p

This is usually too much to ask of marketing teams in the publishing world: the book title, author's name and publishing signet are printed so small, in the lower half of the cover, and even embossed in silver foil. If the light happens to fall from the wrong angle, the title becomes completely invisible. It is an enchanting sight, with the dozens of animals, plants, houses and all the rest filling up the cover. The drawings are printed in two special colours on the pink cover. The red contours are not merely drawn around the images, they are produced by the artist sketching an area with striking internal structure on the predrawn areas, and leaving a vibrant border. The first full-colour pages show colourful birds, followed by numerous line drawings. Gradually you feel yourself succumbing to the temptation to pull out your coloured pencils and see what the figures look like in certain colours. This children's book can be called a colouring book — but not necessarily — although it seems reasonable — yet the pages are too beautiful to draw on… One more comment on the appearance: Thanks to the finely structured embossing of the paper on the cover, the sturdy matt inner pages and the extra-narrow ribbon bookmark, a child's hands enjoy a haptic experience, noticing that the world of materials offers resistance and something moves if you put your hand on it — for example, so you don't fail the first time you try to turn a page.

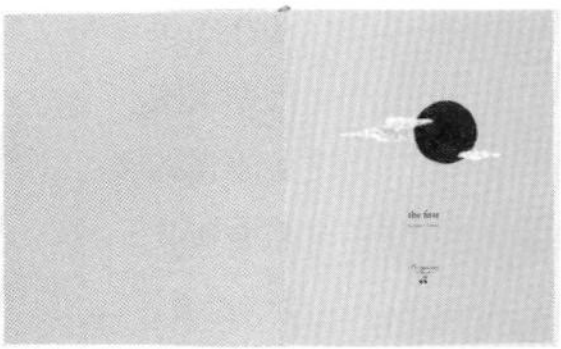

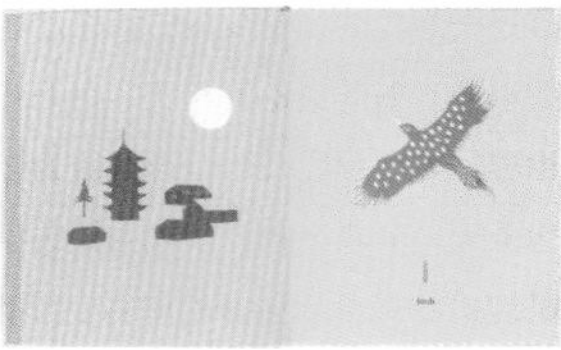

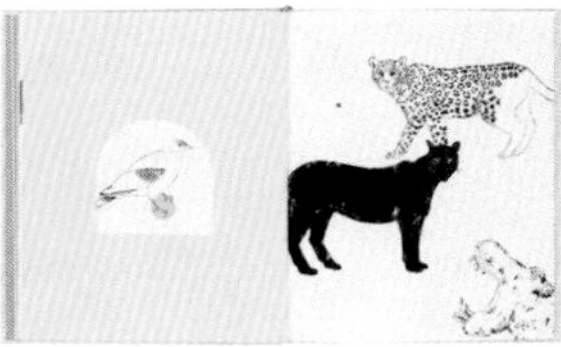

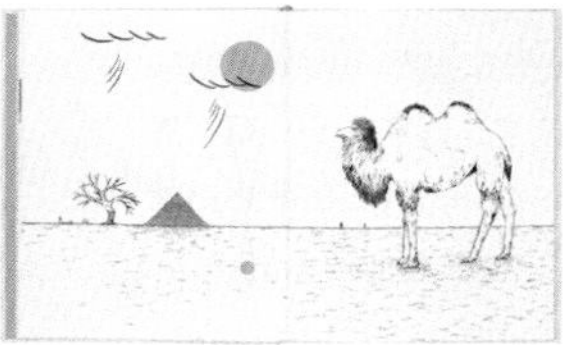

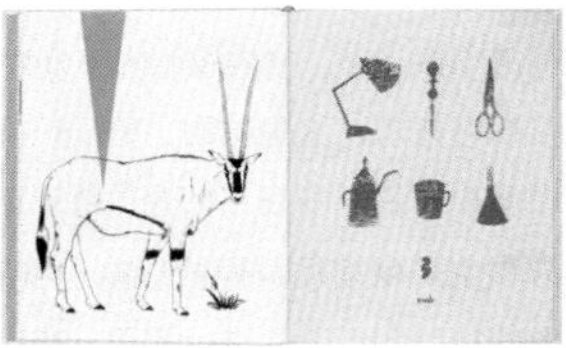

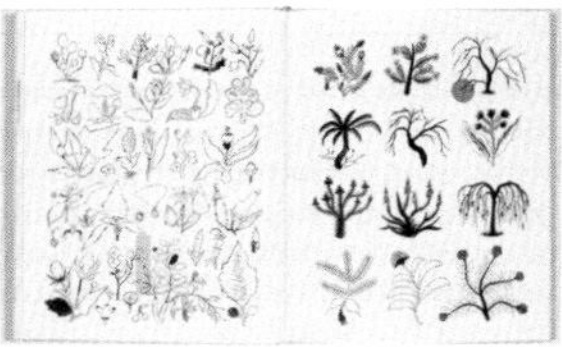

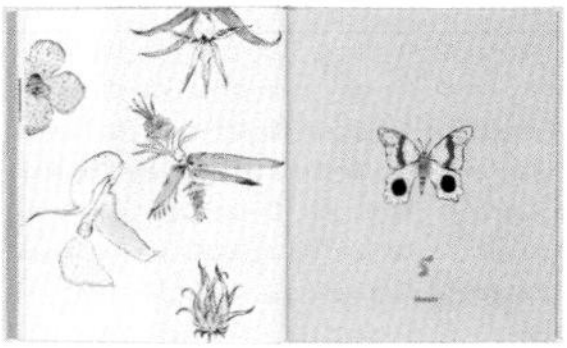

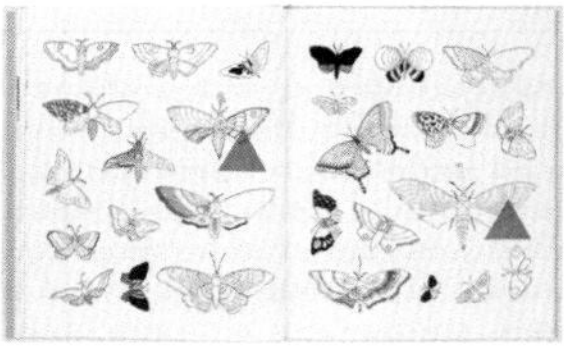

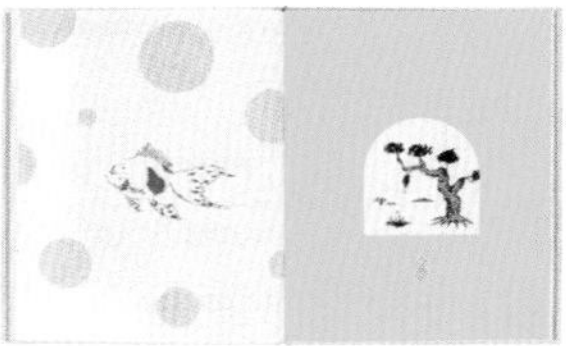

the

Kentarou Tanaka

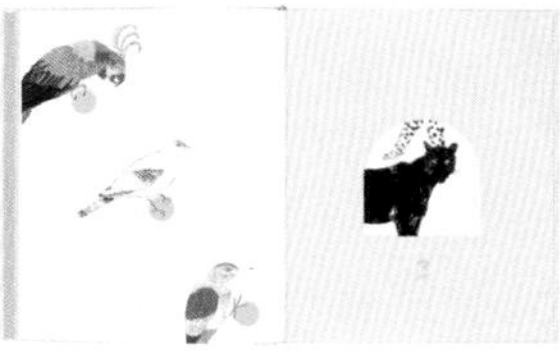
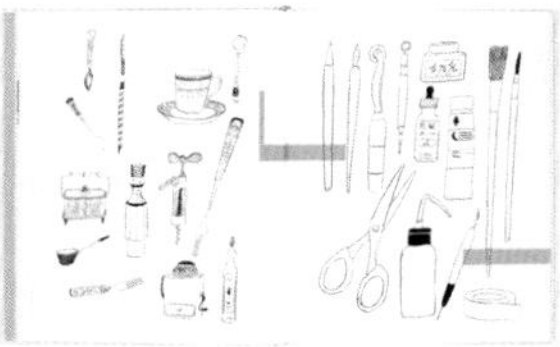
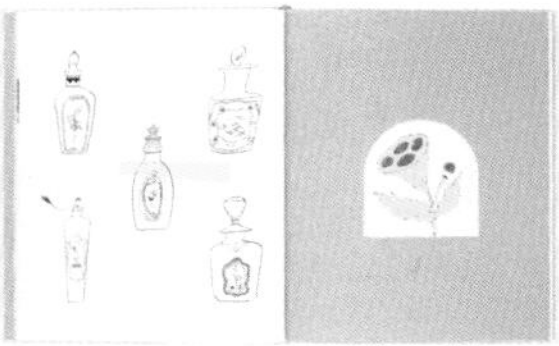
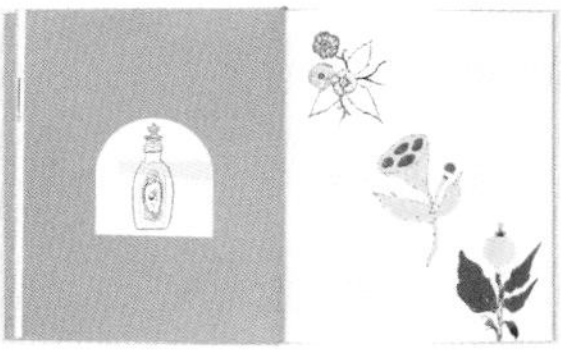

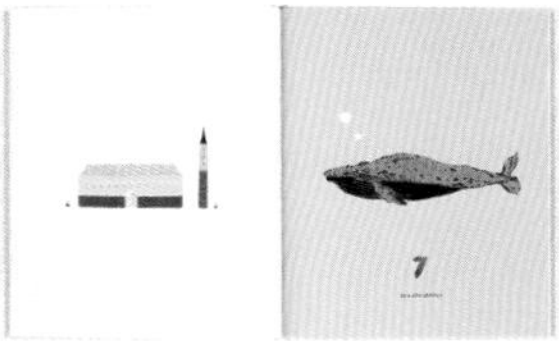
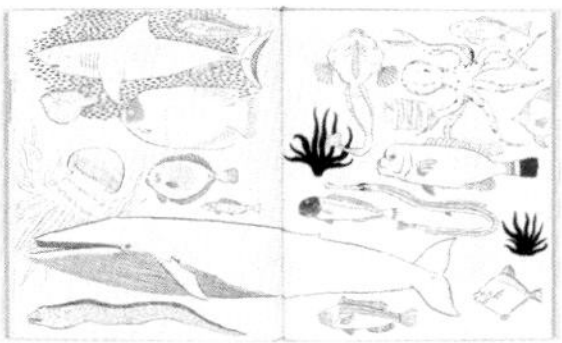
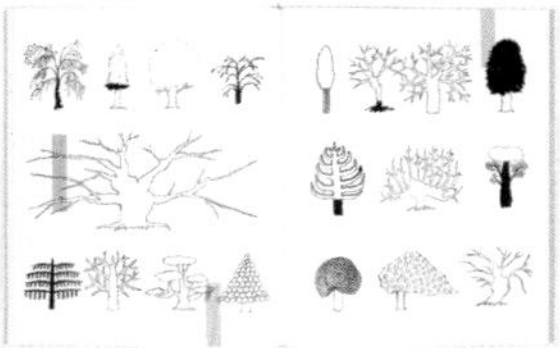
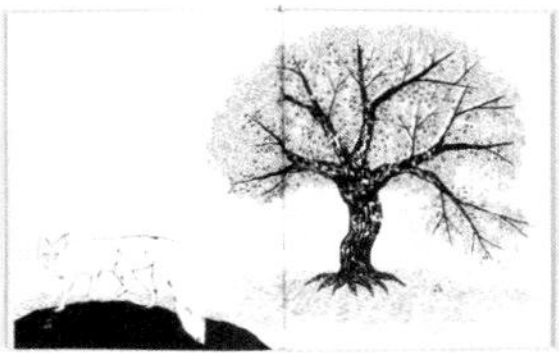
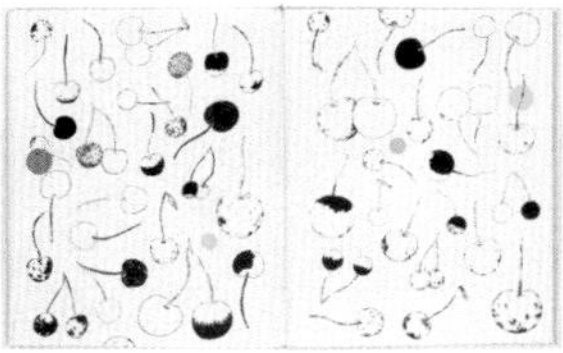
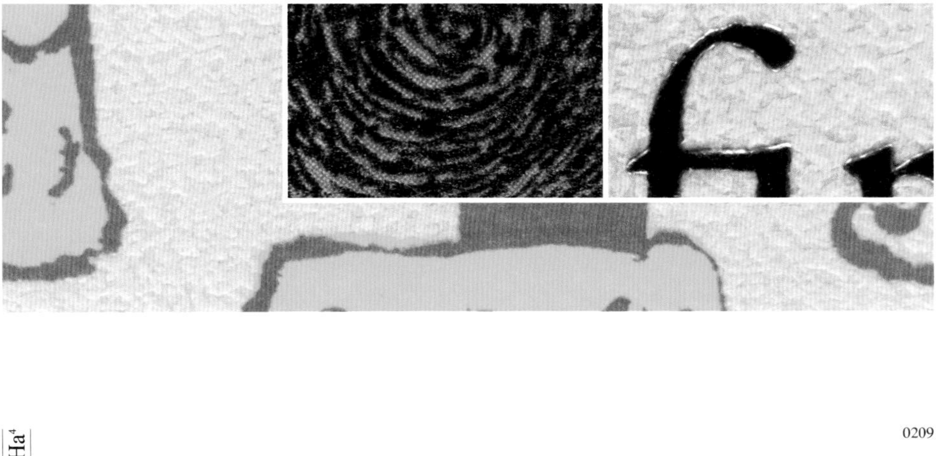

坐商在城里闹市
行商常常是农忙打炉、农闲操锤的兼
职匠人，他们走村串户，上门为乡下百姓做些金戒
手上的金
之后，他
慢变软。
子"叮叮当当"地
工夫，一个蝴蝶

本书为地域性老行当的合集，通过文字、摄影的方式，记录民间正在逐渐消失的各行业的状态。书籍的装订形式十分独特，以书籍使用的粗陋纸张搓绳取代锁线进行装订，封面与内页均采用老作坊包点心用的纸张为主，黑红双色印刷，其中照片以黑色为主（少量彩图），红色用于标题文字框与关键语句，与老行当物品的包装风格相得益彰，辅以宣纸四色彩印的各行业彩色照片拉页。书籍四个侧边均采用手工打毛，与整体风格相协调。书中内容分为八大部分：衣饰、饮馔、居室、服侍、修作、坊艺、工艺、游艺。每一部分通过一种颜色的胶版纸印刷黑色标题加以区隔。书中最有特色的部分在于页码的写法，重现了逐渐消亡的“苏州码子”，为方便读者理解这种计数符号，书中配上相应的书签加以解读。

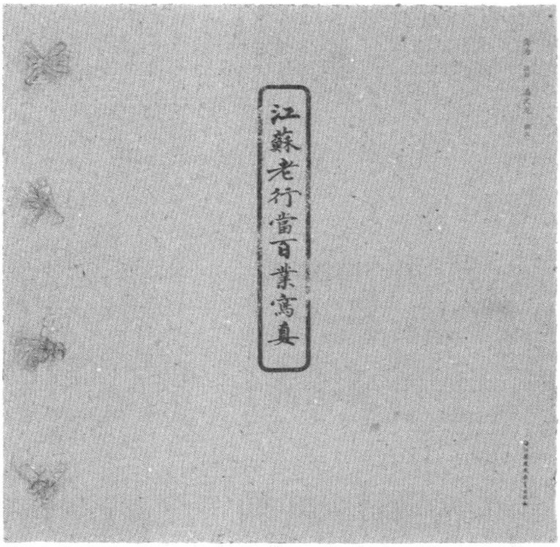

江蘇老行當百業寫真

This book is a collection of local old trades. It records the state of various disappearing industries by means of text and photography. The binding form of books is very unique, sewing thread is replaced by twisted rope made of rough paper to bind the book. The cover and inner pages are mainly made of old-fashioned paper that was used for snack wrapping. They are printed in black and red: the photographs are mainly black (except a small number of color pictures), while red is used for headline text frames and key sentences. The color combination fits well with the packaging style of old trades. There are many foldouts printed with picture of old trades printed on rice paper in four colors. Four edges of the book are hand-polished to match the overall style. The book is divided into eight parts: clothing, food, bedroom, service, cultivation, workshop, craft and entertainment. A page of offset paper in different color separates each part with a black heading printed on. The most distinctive feature of the book is the way of writing page numbers, which reproduces the disappearing 'Suzhou Code'. In order to facilitate readers' understanding of the counting symbols, explanations are printed on the attached bookmarks.

- 荣誉奖
- 江苏老行当百业写真
- 中国
- 周晨
- 潘文龙、龚为（摄影）
- 江苏凤凰教育出版社

- Honorary Appreciation
- Old Trades of Jiangsu: A Glimpse
- China
- Zhou Chen
- Pan Wenlong, Gong Wei (photographer)
- Jiangsu Phoenix Education Publishing, Ltd
- 288×284×38mm
- 1817g
- 646p

Jiangsu is now one of China's economically strongest and richest provinces, with 11 cities of over a million inhabitants. What a contrast to this book, which presents 90 artisanal branches of trade from pre- and early industrial times. A thick wad of greyish brown paper, roughened on all sides, is held together on the left by rudimentary side-sewing. The ends of the string — also made of coarse paper — are pressed flat like dried flowers. The whole thing feels like a pliable pillow. Black-and-white photographs on the packing paper, and colour photos on ultra-thin, yellowy-white laid paper with fold-out pages, in a slightly smaller format, show people working on their crafts. In the interaction between perfect printing and haptic changes of paper, the pictures grant insights into civilisation techniques that were believed almost lost. It is a romantic view, although this term is initially foreign to the Far East. At the very least, a certain effusive pride renders homage to artisanship, whose refinement, quality and diligence could be intended as references to the prosperity of this Chinese province today. The brown packing paper is normally used to wrap foods; in this book it is used as base material for presenting the workmanship of old trades — a subtle reminder of how fulfilling it can be to earn your living through meaningful activity and by working with your own hands.

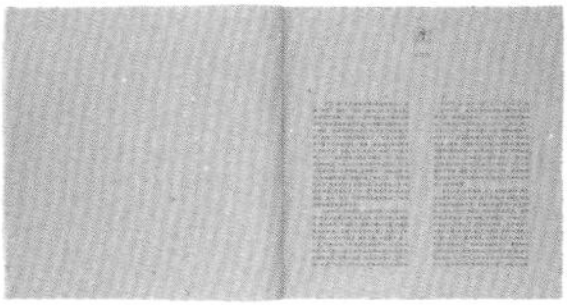
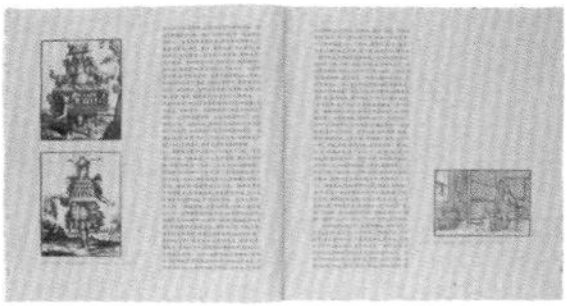

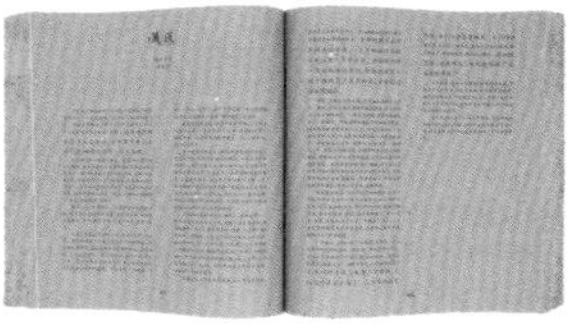
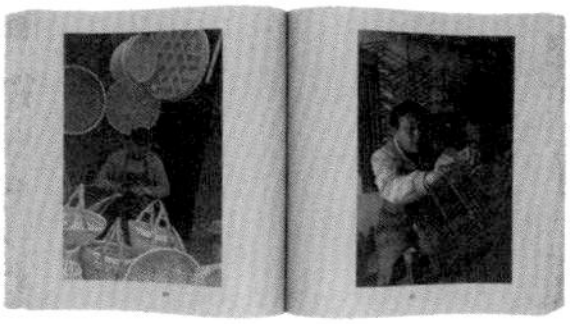
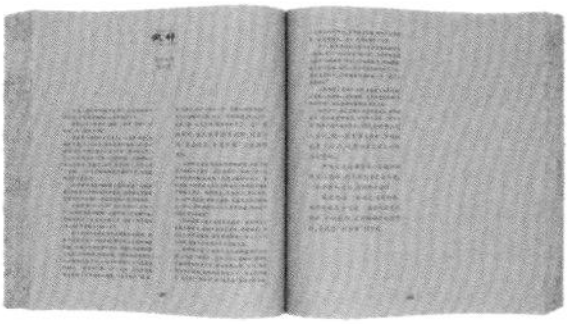
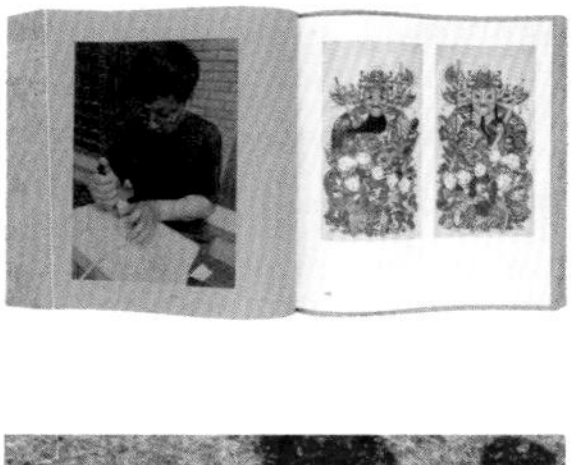
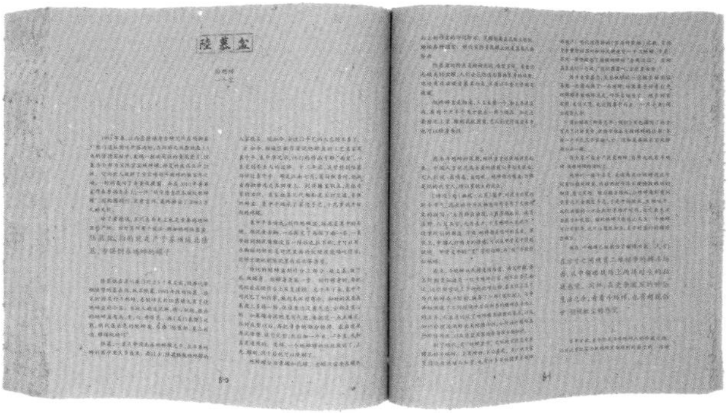
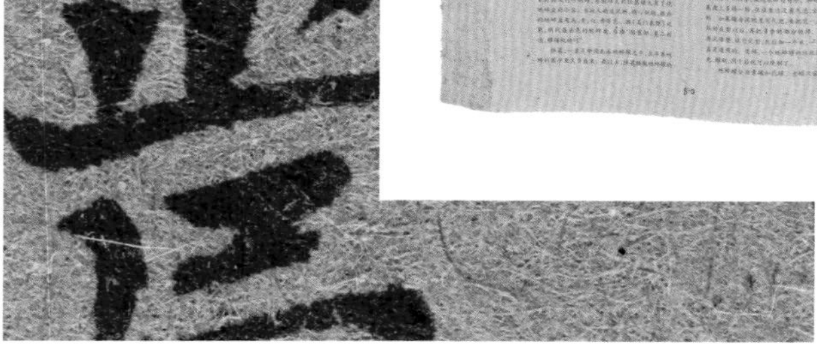

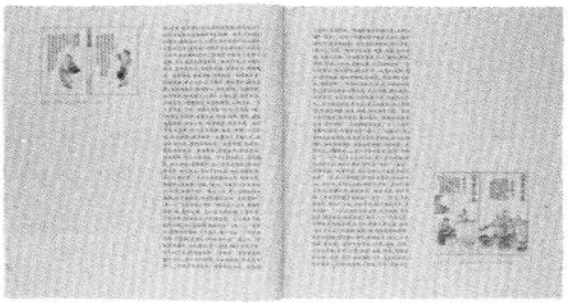
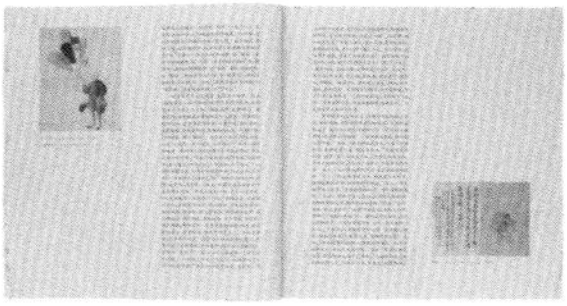
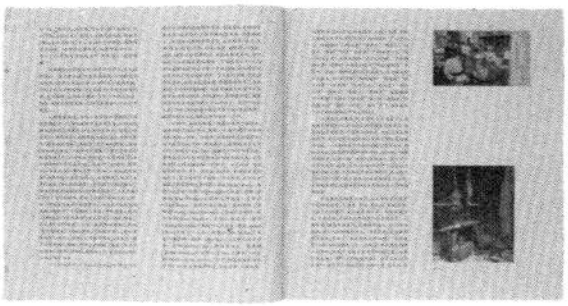

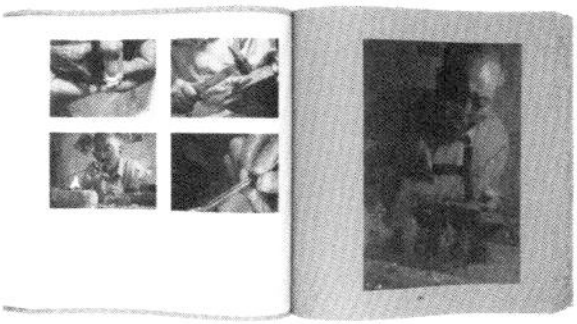
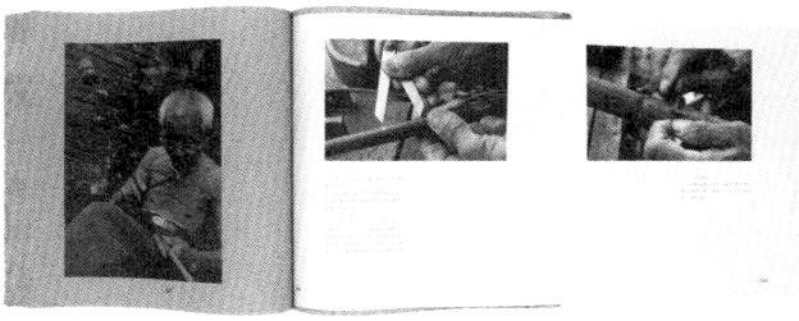

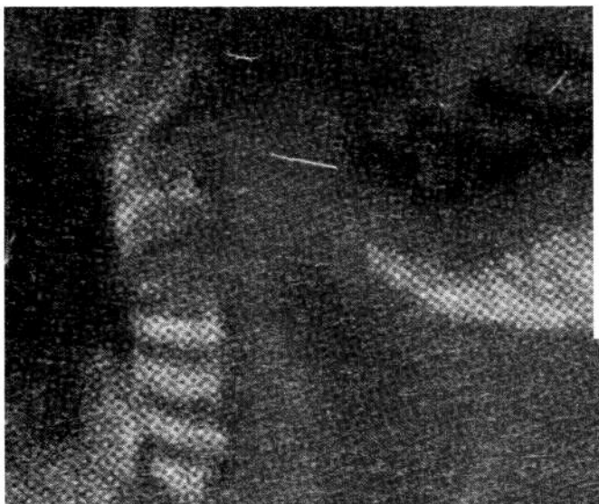
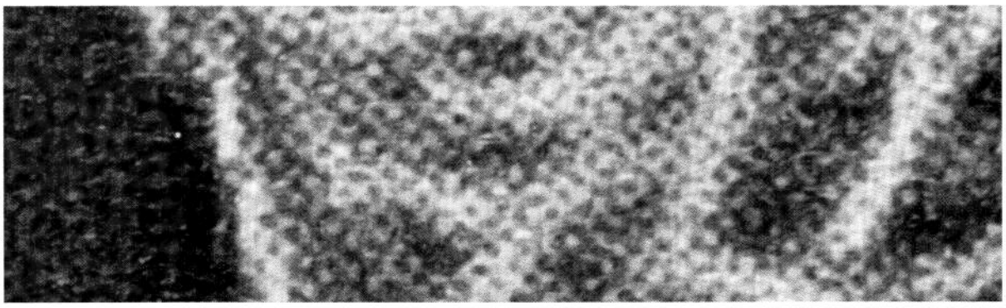
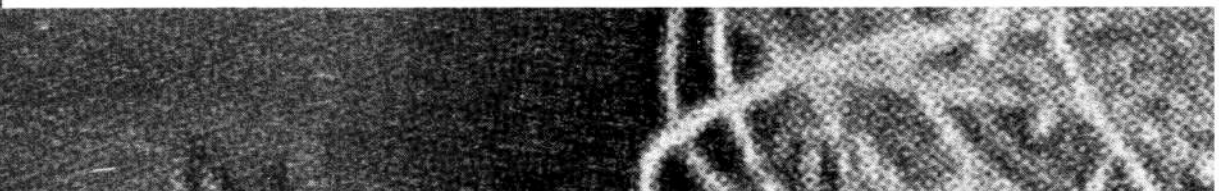

HEIMAT, HANDWERK UND DIE UTOPIE DES ALLTÄGLICHEN
Soirée Fantastique
The Art of Gardening
Process for becoming a little bear
Avond
documenta 14: Daybook
EFES:EFES
200 keystrokes per minute
Autonomie auf A4
First things first
Claudy Jongstra
Tea Canon
Monsanto: A Photographic Investigation
II Concurso Nacional de Poesía Joven Rafael Cadenas

2018

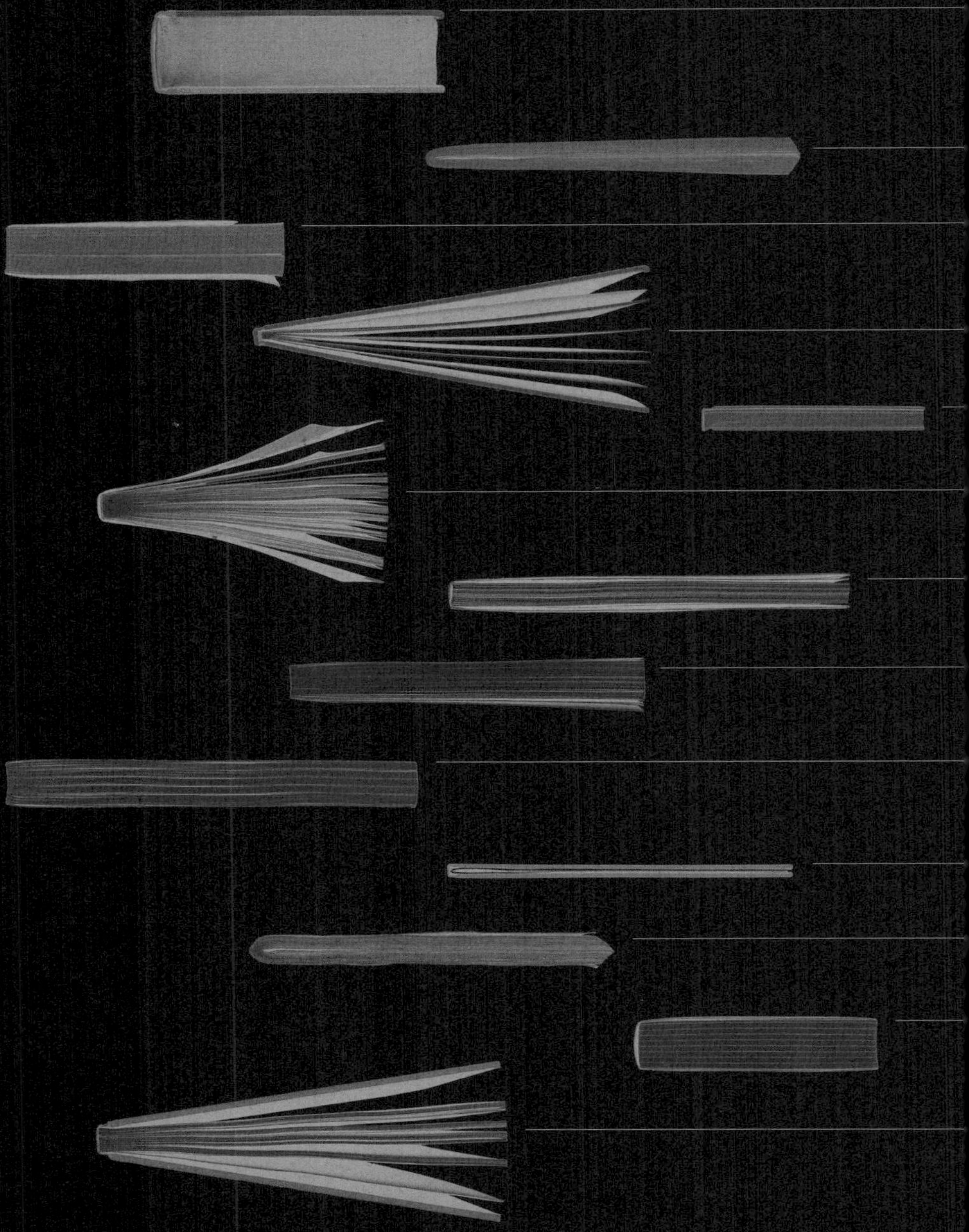

2018 Gl 0225

2018 Gm 0233

2018 Sm^{1} 0239

2018 Sm^{2} 0245

2018 Bm^{1} 0251

2018 Bm^{2} 0257

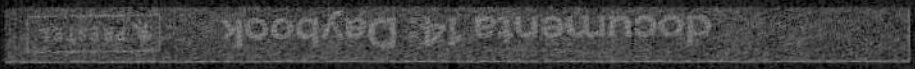
documenta 14: Daybook

2018 Bm^{3} 0263

2018 Bm^{4} 0269

2018 Bm^{5} 0275

2018 Ha^{1} 0281

2018 Ha^{2} 0285

2018 Ha^{3} 0289

2018 Ha^{4} 0293

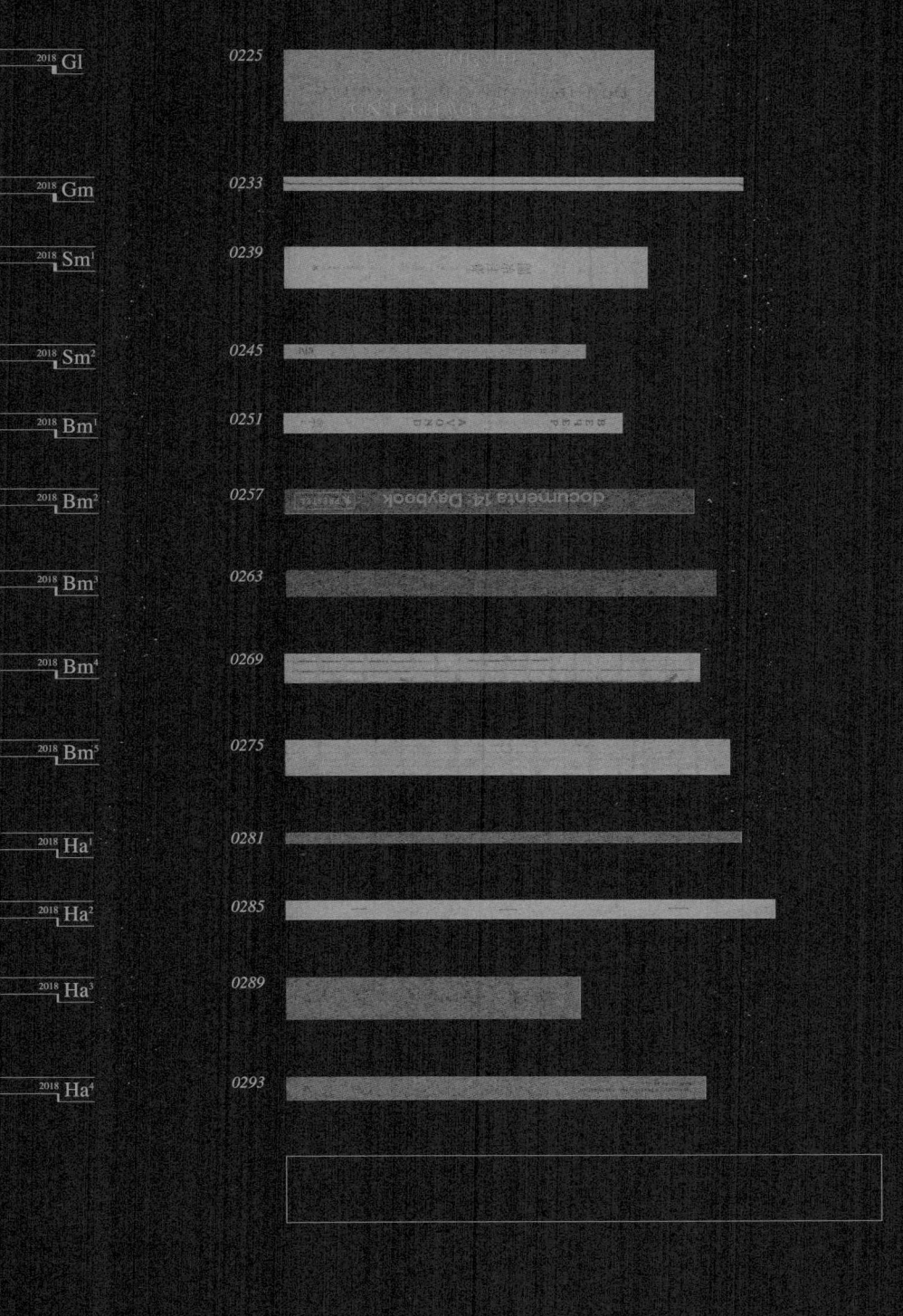

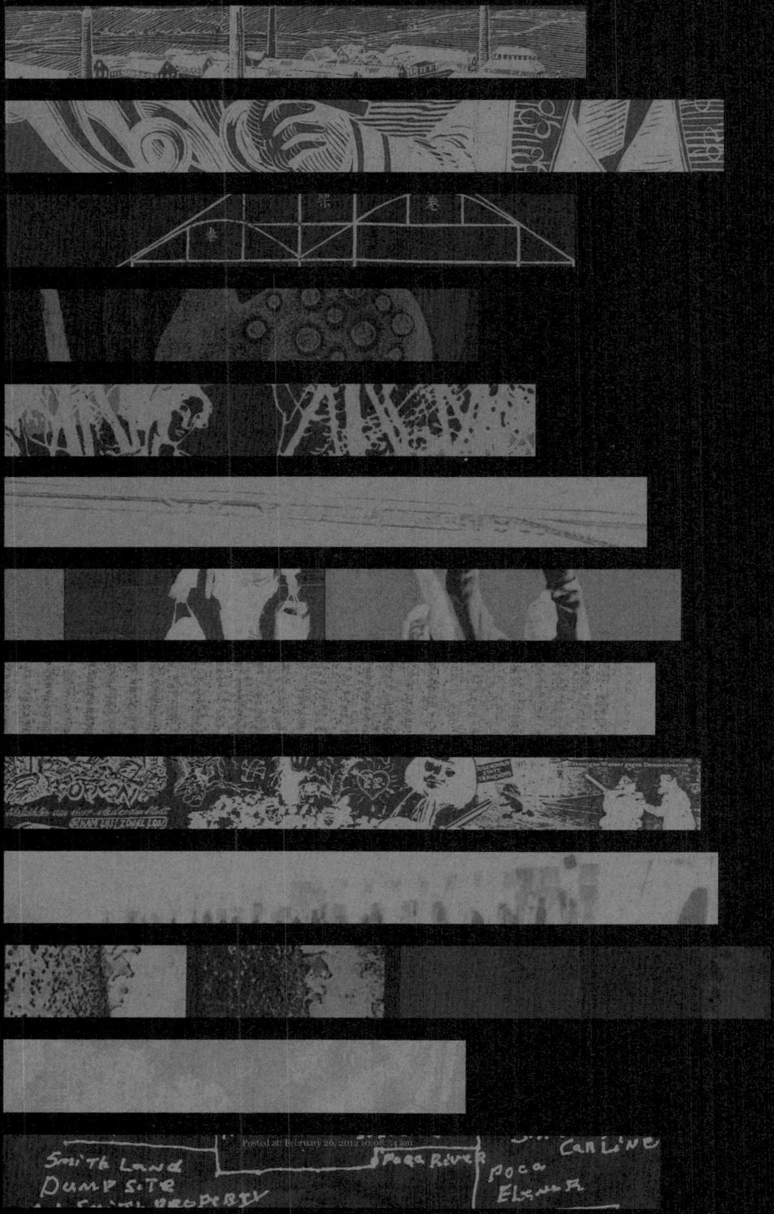
Posted at: February 26, 2012
Smith Land
Dump Site
Poca River
Poca
Carline

Zu Beginn der Natur- und Heimatschutzbewegung, als deren Gründerfigur Ernst Rudorff gilt, steht das Bedauern über den Verlust »altvertraute[r] Bilder« von »Strassenbilder[n]

1848 – 1856	Post-Str. 18, Volckmar's Hof	1861 – 1869	Neukirchhof 3
1856 – 1866	Lurgensteins Garten Nr. 1	1870 – 1883	Neukirchhof 2
1867 –	Lurgensteins Garten Nr. 5, A	1884	Pfaffendorfer
		1885 – 1893	Pfaffendorfer

五

同里 讀較好的 崇本堂和 更是別具 生成關係 永齋 南宋淳祐 名・這里 作者計成 雲人物 主席王紹

しかきんを

總序

ばくばく ごっくん

dolnik' gebruikt, een versmaat waarin twee- en epige versvoeten elkaar afwisselen, heeft de vertaler peling veroorloofd, die immers in de dolnik zelf is

With the documenta 14 *Daybook* open in your hands, you are now ready to begin a journey in time and space—both present and historical, both political and personal, and not in one city but in two. The

עידוד הזו. למעשה הבנאלי הפעוט הזה היתה משמעות: גם כאן נעשה מאמץ
הלאים (נכון לומר "לייהד" במובן התרבותי) את הכדורגל: כך למשל יש להבין

Владимир Маканин

«Андеграунд, или Герой нашего времени».

Особенно кисти рук будут крепки, если ты двадцать с лишним лет печатал на машинке, еще и переносил, перевозил ее в поисках теплого жилого угла туда-сюда по всему городу (по всей непредсказуемой спальной Москве). Двадцать

verkrustet – so unter- wurde Zürich in den ren wahrgenommen.

rungen. Allerdings gelang es den 80ern, die Stadt stärker zu prägen. Sie vermochten beengende Nor-

Bildspra stände h

photos. Such works show her experimental approach to photographic techniques, which also entails examining photography as a socially-

that interested Shahbazi in devel extensive series. She selected sub collected photographs and reproc

ct perception, that is when it goes without thoughts or words. Claudy: So that is what it will be about, it will

e. **Louwrien: The auditorium is for how many persons?** Claudy: Two hundred. They will invite people who

lectures, I suppose, or have symposia about topics, environmental topics, but also racial topics and also other topics of an interest. **Louwrien: Every topic is possible...** Claudy: ...and important. So in that riverbed, I think, for me

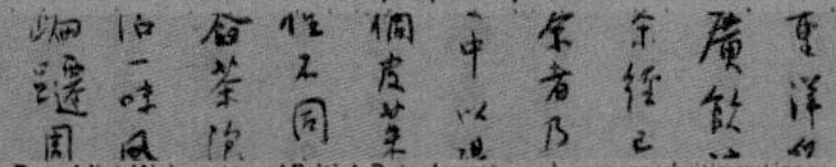

McDonald's®, Walmart® and British Petroleum® may have earned citizens' ire as symbols of injurious corporate decadence. However, the deep-seated prevailing distrust of American-based multinational Monsanto® stands without parallel or precedent. **Remarkably enough**, Monsanto®'s own actions reveal that they themselves accept the public's strongly negative assessment. Recent attempts to reinvent themselves by rewriting history and jettisoning their past lies are an unconscious admission that they share the public's disdain for their dark pathology. **The trained, strained,**

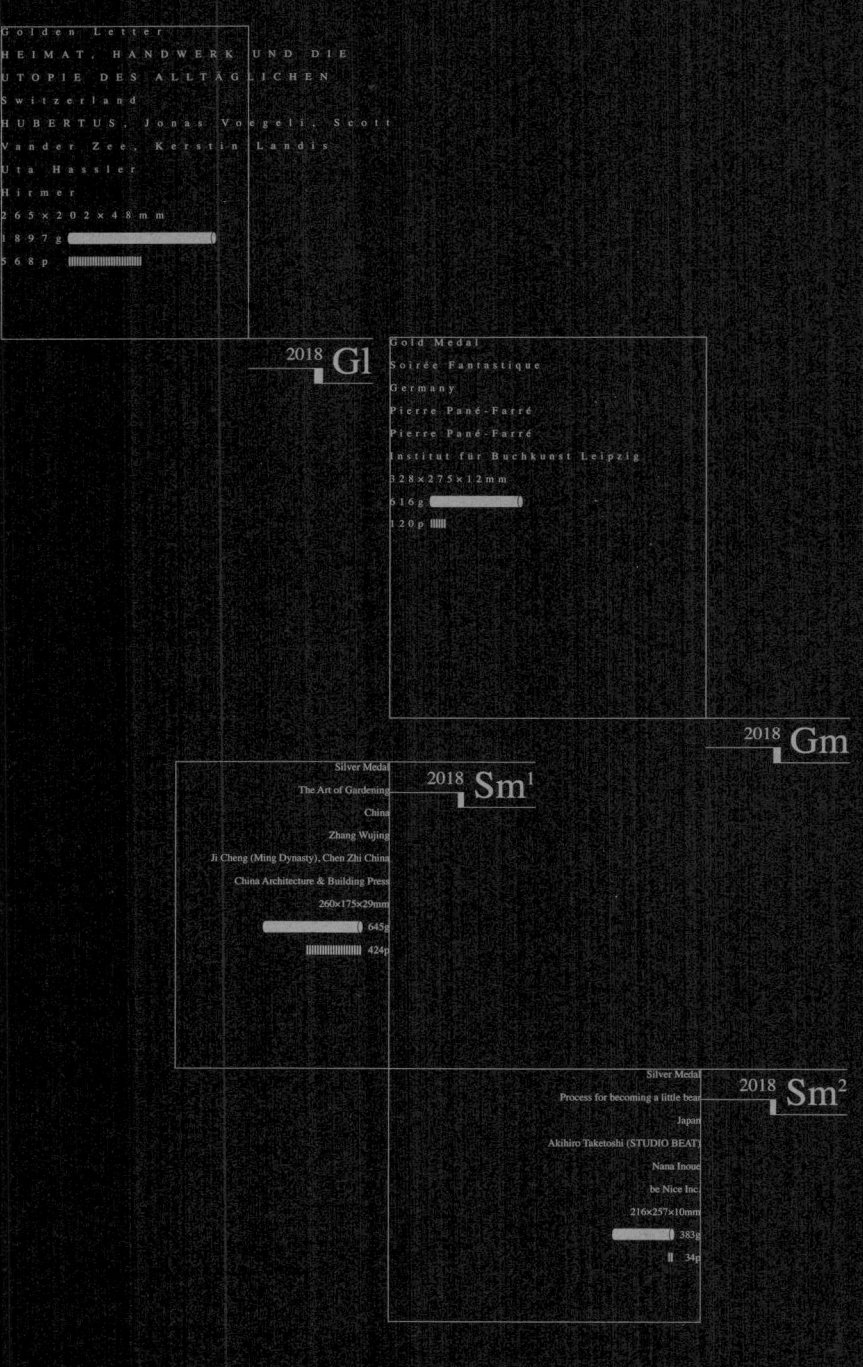

Golden Letter
HEIMAT, HANDWERK UND DIE UTOPIE DES ALLTÄGLICHEN
Switzerland
HUBERTUS, Jonas Voegeli, Scott Vander Zee, Kerstin Landis
Uta Hassler
Hirmer
265×202×48mm
1897g
568p

2018 Gl

Gold Medal
Soirée Fantastique
Germany
Pierre Pané-Farré
Pierre Pané-Farré
Institut für Buchkunst Leipzig
328×275×12mm
616g
120p

2018 Gm

Silver Medal
The Art of Gardening
China
Zhang Wujing
Ji Cheng (Ming Dynasty), Chen Zhi China
China Architecture & Building Press
260×175×29mm
645g
424p

2018 Sm1

Silver Medal
Process for becoming a little bear
Japan
Akihiro Taketoshi (STUDIO BEAT)
Nana Inoue
be Nice Inc.
216×257×10mm
383g
34p

2018 Sm2

Honorary Appreciation
Monsanto: A Photographic Investigation
Venezuela
Ricardo Báez
Mathieu Asselin
Verlag Kettler (English), Actes Sud (Français)
300×270×21mm
1266g
148p

Honorary Appreciation
Tea Canon
China
Pan Yanrong
Lu Yu (Tang Dynasty) and others
The Commercial Press
211×140×29mm
665g
772p

2018 Ha3

2018 Bm4

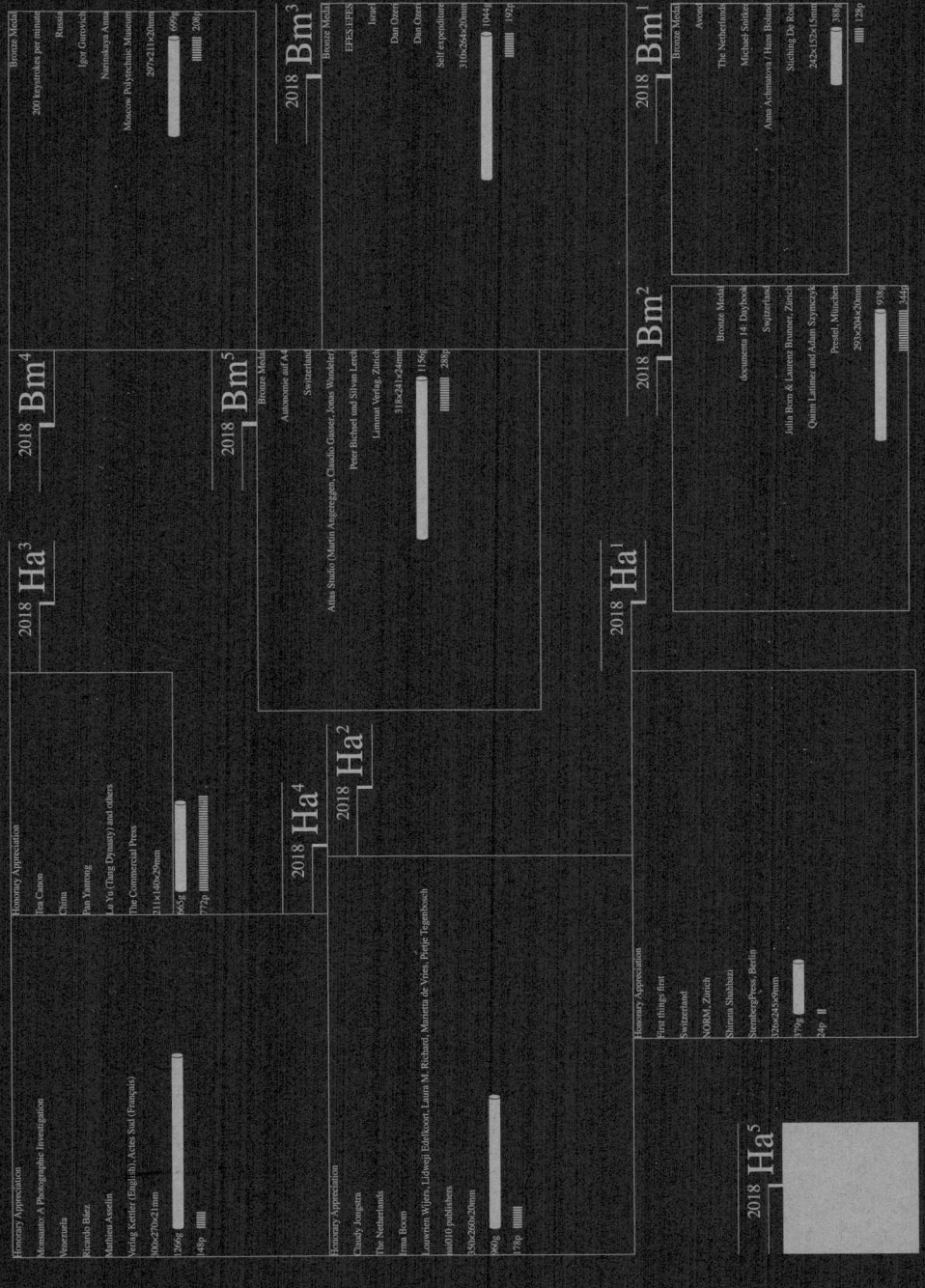

Bronze Medal
200 keystrokes per minute
Russia
Igor Gurovich
Narinskaya Anna
Moscow Polytechnic Museum
297×211×20mm
699g
208p

2018 Bm5

Bronze Medal
Autonomie auf A4
Switzerland
Atlas Studio (Martin Angereggen, Claudio Gasser, Jonas Wandeler)
Peter Bichsel und Silvan Lerch
Limmat Verlag, Zürich
318×241×24mm
1156g
288p

2018 Ha4

2018 Ha2

Honorary Appreciation
Claudy Jongstra
The Netherlands
Irma Boom
Louwrien Wijers, Lidweji Edelkoort, Laura M. Richard, Marietta de Vries, Pietje Tegenbosch
nai010 publishers
350×260×20mm
960g
178p

2018 Bm3

Bronze Medal
EFES:EFES
Israel
Dan Ozeri
Dan Ozeri
Self expenditure
310×264×20mm
1044g
192p

2018 Ha1

Honorary Appreciation
First things first
Switzerland
NORM, Zürich
Shirana Shahbazi
SternbergPress, Berlin
326×245×9mm
379g
24p

2018 Bm2

Bronze Medal
documenta 14: Daybook
Switzerland
Julia Born & Laurenz Brunner, Zürich
Quinn Latimer und Adam Szymczyk
Prestel, München
293×204×20mm
938g
344p

2018 Bm1

Bronze Medal
Avond
The Netherlands
Michael Snitker
Anna Achmatova / Hans Boland
Stiching De Ross
242×152×15mm
388g
128p

2018 Ha5

2018

The International Jury

Susan Colberg (Canada)

Jonas Voegeli (Switzerland)

Aud Gloppen (Norway)

Alexandra Buhl (Island)

Florian Hardwig (Germany)

Stefanie Schelleis (Germany)

Noam Schechter (Israel)

Country / Region

Switzerland 4

The Netherlands 2

China 2

Venezuela 2

Germany 1

Japan 1

Israel 1

Russia 1

Designer

HUBERTUS, Jonas Voegeli, Scott Vander Zee, Kerstin Landis

Pierre Pané-Farré

Zhang Wujing

Akihiro Taketoshi (STUDIO BEAT)

Michael Snitker

Julia Born & Laurenz Brunner, Zürich

Dan Ozeri

Igor Gurovich

Atlas Studio (Martin Angereggen, Claudio Gasser, Jonas Wandeler)

NORM, Zürich

Irma Boom

Pan Yanrong

Ricardo Báez

Juan Fernando Mercerón. Giorelis Niño (Assistent)

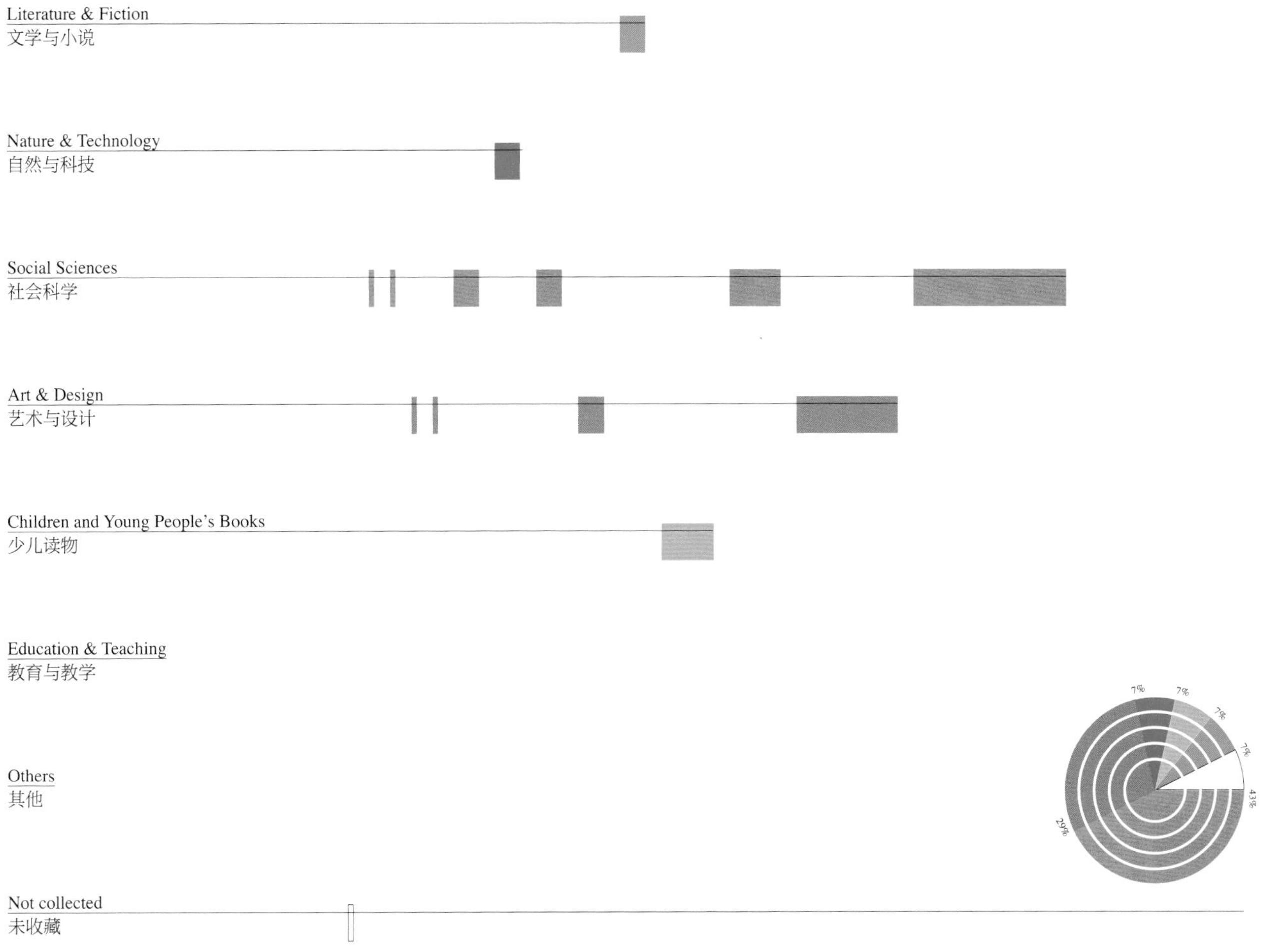
Literature & Fiction
文学与小说
Nature & Technology
自然与科技
Social Sciences
社会科学
Art & Design
艺术与设计
Children and Young People's Books
少儿读物
Education & Teaching
教育与教学
Others
其他
Not collected
未收藏
7%
7%
7%
7%
43%
29%

HEIMAT.

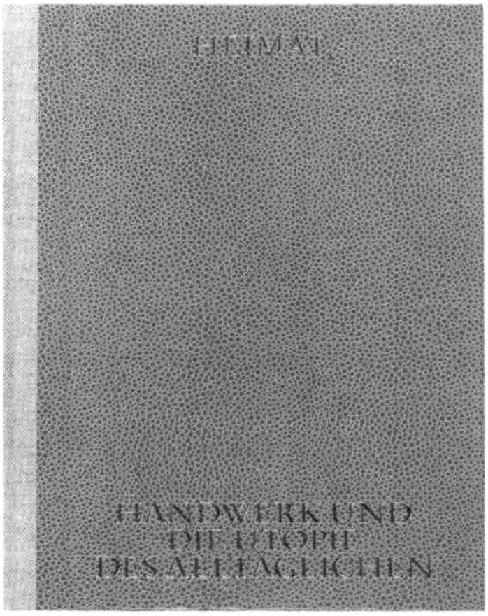

本书探讨了 20 世纪上半叶建筑和景观设计的理念，批判地研究了理想化设计理念在家园建设中的可能性。本书包含对工业化的批判、建筑和景观的设计传统、工业技术的发展与支持，以及理想化的前工业美学的影响。本书为锁线硬精装，封面尝试了多种材料和工艺的混合，以体现前工业美学和工业技术的融合理想。封面采用橘色的仿珍珠鱼皮，封底采用白色的仿鳄鱼皮，书脊通过粗麻布进行包裹，文字烫金，封面裱覆卡板。前后环衬分别选用茶绿色和藏青色艺术纸印刷银色文字。内页哑面涂布纸张五色印刷（四色加银色），上书口刷金。内容划分为前述的四大部分，收录了 16 位作者的研究成果。书籍正文版心设置为较小单栏，文末接四栏大版心注释。页码和栏位均依据页面右侧排布，留下左侧较大的留白空间。图片页版心与注释页相同，图注银色印刷，与正文适合阅读的版心设计拉开距离。

This book explores the ideas of architecture and landscape design in the first half of the 20th century, and critically examines the possibilities of idealized home design. The book discusses criticism of industrialization, the design traditions of architecture and landscape, the development and support of industrial technology, and the influence of idealized pre-industrial aesthetics. The cover is made of the orange Pearl Gourami skin faux leather while the back cover is made of the white crocodile skin faux leather. The spine is wrapped by coarse linen cloth with hot stamping gold texts, and graphic board mounts the cover. The front and back end paper uses tea green and navy blue art paper respectively with silver characters printed. The inside pages are printed in five colors (four colors plus silver) on matte coating paper with gilt-edging technique. The content is divided into four parts, including the research results of 16 authors. The type area of the main body is set to be smaller than the annotation pages, which shares the same size type area as the picture pages. The captions are printed in silver that is different from the layout of the main text.

- 金字符奖
- 家、工艺和日用品的乌托邦
- 瑞士

- Golden Letter
- HEIMAT, HANDWERK UND DIE UTOPIE DES ALLTÄGLICHEN
- Switzerland
- HUBERTUS, Jonas Voegeli, Scott Vander Zee, Kerstin Landis
- Uta Hassler
- Hirmer
- 265×202×48mm
- 1897g
- 568p

The cover – which is not even remotely everyday – gives this book the appearance of a cornerstone on an architecture library shelf. This weighty work was designed as a half-bound volume. A coarse linen fabric protects the wide spine. An exotic effect is produced by the structures of the board covers: at the front, tobacco-coloured ray skin; at the back, lightly shimmering lizard skin. A dusky pink headband seals up the gap between the gilded top edge and the inner spine. The golden capital letters used for the embossed title are positioned unusually near the upper and lower edges. This study on the concept of the homeland in architecture and landscape design during the first half of the 20th century consists of critical essays documented with historical texts and pictures. The observer wonders if the elaborate outfit glorifies the notion of the homeland. The basic text reposes in a traditional type area, splendidly typeset with wide gutters. However, the solemn mood is counteracted by typographic jibes: paragraphs begin with large indents; illustration details are placed on the left-hand side next to the column and leave a gap where they were taken out of the body text; pagination is always on the right-hand side (therefore deep in the gutter margin on the left page); captions stretch across the entire width along the top edge, but in silver printing. The headlines on the centreline are all in firm contrast. The typographical irony does not invalidate the seriousness of this analysis, but rather underlines the critical distance. This is the subtle achievement of this book.

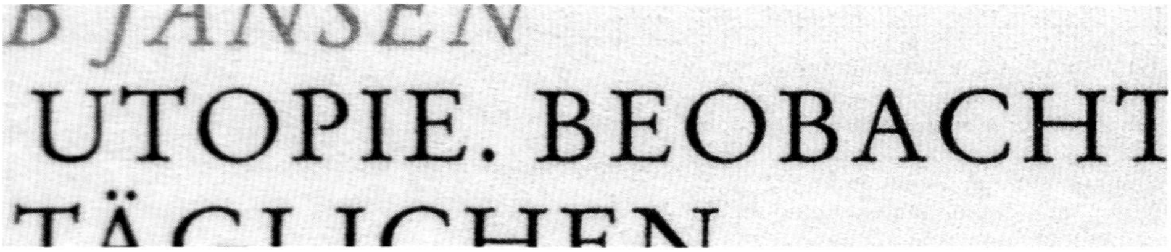
B JANSEN
UTOPIE. BEOBACHT
TÄGLICHEN

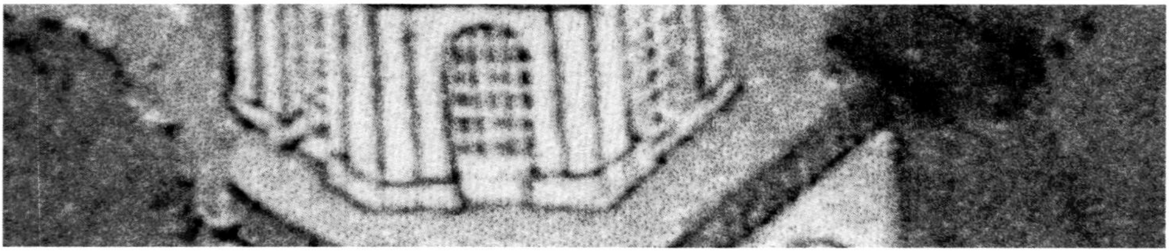

»H
icht sterl
an seiner
Heimat, d
In
ahrhund
n Sprach
imodern
für das
in zum v
sie für n
in Deuts
n, Brauc
en verbi
en, bis hir
Pflanzer
lschaftsb

Vo
Zu
eren Grü
den Verlı

Selbstinszen
5. Aufl. Berlin
34
Vgl. z. B. die
Spohn 2009 (
der, Heiko: B
Engelmann (
1938/1939 in F
baute »Gut B
Kalender (20
35
Alle genannte
denn auch ein
Anlage, ohne
nem Zeitpunk
chen Festleg
Typenentwur
im Übrigen in
terdörfer« geg
(wie Anm. 1),
Förderung vo
aus dem Jahr
schaft einer d
zunächst ein
Musterbeispie
Gerd: Nicht r
1996 (wie Anm
36
Vgl. Johannse
re Tümlauer K
Koog 1985; F
der in den neu
aktion zugur
erblickt.« Zu
chen Bauwes
während der
Iserhoht, Mar
(Schriftenreih
kunde in Scl
Großbarkau 1
burg vgl.: Li
Wohn- und Sie
lenburg von 1
geschichte re
Mecklenburg

265×202×48

1897

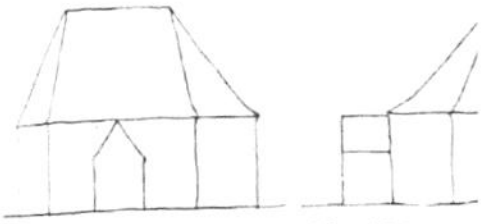

Abb. 170.

568

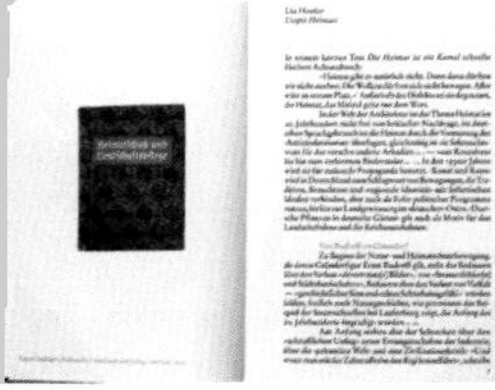

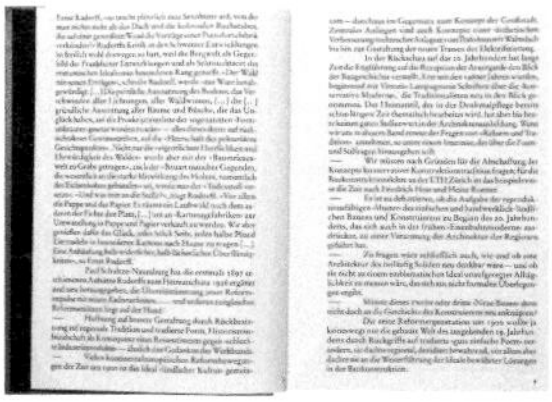

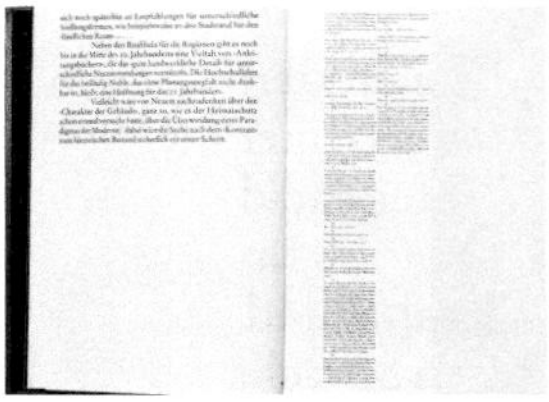

HEIMAT,
HANDWERK UND
DIE UTOPIE
DES ALLTÄGLICHEN

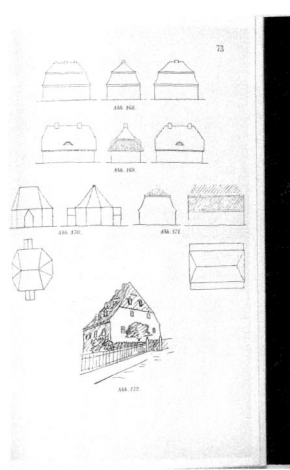

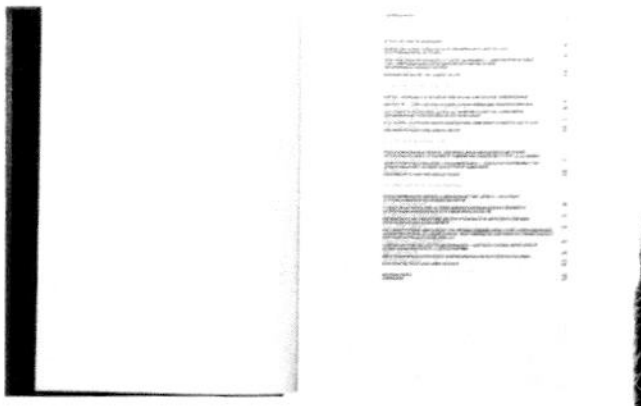

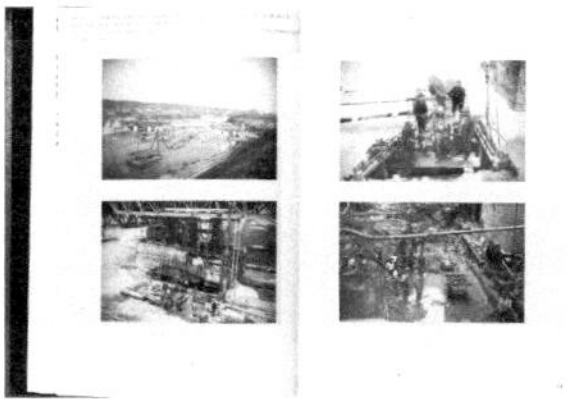

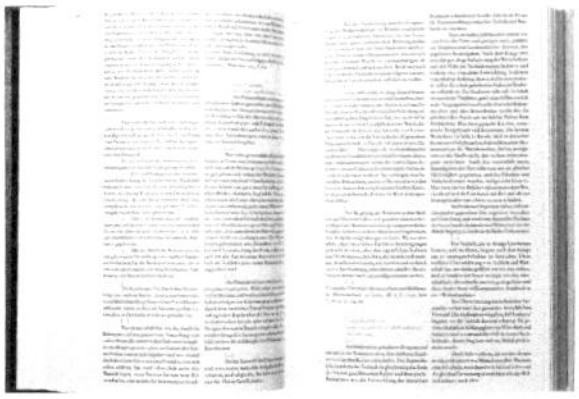

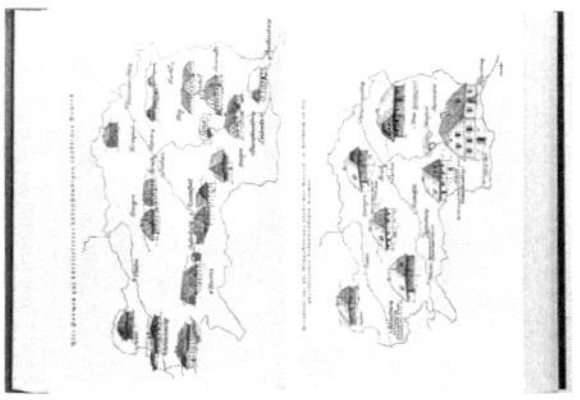

a
b
c
d
e
f
g
h
73
Abb. 168.
Abb. 169.
Abb. 170.
Abb. 171.
Abb. 172.

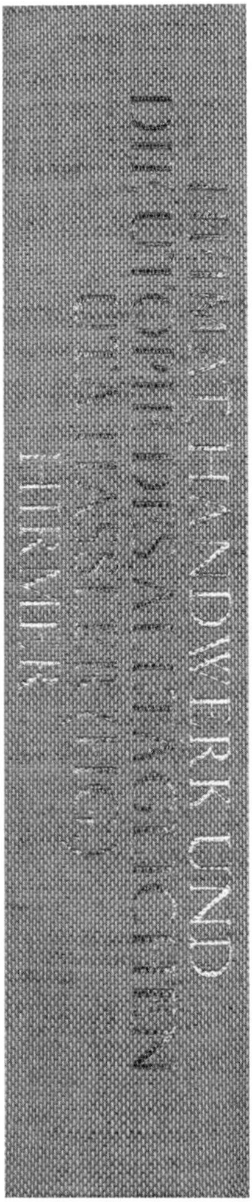

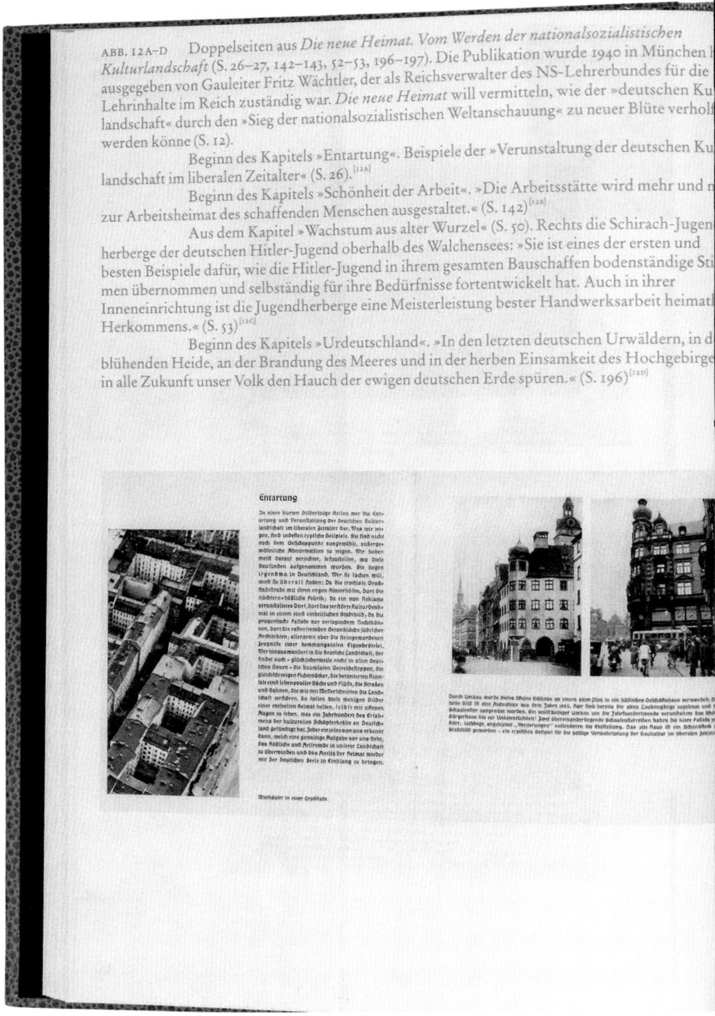

ABB. 12A–D Doppelseiten aus *Die neue Heimat. Vom Werden der nationalsozialistischen Kulturlandschaft* (S. 26–27, 142–143, 52–53, 196–197). Die Publikation wurde 1940 in München h ausgegeben von Gauleiter Fritz Wächtler, der als Reichsverwalter des NS-Lehrerbundes für die Lehrinhalte im Reich zuständig war. *Die neue Heimat* will vermitteln, wie der »deutschen Ku landschaft« durch den »Sieg der nationalsozialistischen Weltanschauung« zu neuer Blüte verholf werden könne (S. 12).

Beginn des Kapitels »Entartung«. Beispiele der »Verunstaltung der deutschen Ku landschaft im liberalen Zeitalter« (S. 26).[12A]

Beginn des Kapitels »Schönheit der Arbeit«. »Die Arbeitsstätte wird mehr und n zur Arbeitsheimat des schaffenden Menschen ausgestaltet.« (S. 142)[12B]

Aus dem Kapitel »Wachstum aus alter Wurzel« (S. 50). Rechts die Schirach-Jugen herberge der deutschen Hitler-Jugend oberhalb des Walchensees: »Sie ist eines der ersten und besten Beispiele dafür, wie die Hitler-Jugend in ihrem gesamten Bauschaffen bodenständige Sti men übernommen und selbständig für ihre Bedürfnisse fortentwickelt hat. Auch in ihrer Inneneinrichtung ist die Jugendherberge eine Meisterleistung bester Handwerksarbeit heimatl Herkommens.« (S. 53)[12C]

Beginn des Kapitels »Urdeutschland«. »In den letzten deutschen Urwäldern, in d blühenden Heide, an der Brandung des Meeres und in der herben Einsamkeit des Hochgebirge in alle Zukunft unser Volk den Hauch der ewigen deutschen Erde spüren.« (S. 196)[12D]

Entartung

Schönheit der Arbeit

Urdeutschland

这是一本展示 19 世纪中叶的海报和场景的历史资料。书中展示的海报来自莱比锡市立博物馆的收藏，海报为 1840–1870 年间莱比锡 Oskar Leiner 印刷机印制，同时在海报中穿插地展示了摄影师亚历山大 · 塞茨（Alexander Seitz）拍摄的同一时期的莱比锡街景，展现当年人们的生活环境、海报设计风格、印刷水准等历史风貌。本书的装订方式为骑马钉线装，封面采用带有皮革质感的纸张红黄黑三色凸版印刷。内页大部分页面为胶版纸四色印刷，采用 4 开纸张 4 折后装订，未清三边，使得上书口连在一起，翻口部分留有未裁切的自然斜角。每页展示的海报不能完全遮盖之前页面的海报，使得翻阅的过程产生真实世界纸张层层堆叠的连贯性。街景照片统一为泛黄的单色老照片，与海报形成并行的两条“展线”。另有少量页面采用哑面涂布纸单色凸版印刷，用尽可能还原当年海报的印刷方式展示部分海报的局部。||||||||||||||||||||||||||||||||||||||||

Soirée
Plakate der Druckerei OSKAR LEINER in Leipzig. Mit Fotografien von ALEXANDER SEITZ.
FAN-TAS
TIQUE.
Institut für Buchkunst Leipzig

This is a historical book showing posters and scenes from the mid-19th century. The posters displayed in the book are from the collections of Leipzig City Museum, printed by Oskar Leiner Printer between 1840 and 1870. The pictures of the street scenes in Leipzig during the same period were photographed by photographer Alexander Seitz, showing people's living environment. The book is bound by thread sewing and the cover is made of leather-like paper embossed in red, yellow and black. Most of the inside pages are printed in four colors on offset paper. Without cutting edges, the upper part is connected together. The posters displayed on each page partly cover the posters on the previous page, making the readers feel the coherence in the real world. The faded monochrome photos of street scenes and the poster are two parallel 'exhibition lines' in this book. In addition, a small number of pages are embossed in monochrome color with matte coated paper in order to show part of the original posters using the printing method of the day.

- 金奖
- 迷幻的夜晚
- 德国

- Gold Medal
- Soirée Fantastique
- Germany
- Pierre Pané-Farré
- Pierre Pané-Farré
- Institut für Buchkunst Leipzig
- 328×275×12mm
- 616g
- 120p

The jet-black Wild West letters on the jacket are designed to attract attention from afar. The jacket feels leathery to the touch – and inside the large format makes the bundle of thin paper seem even lighter. The letter-folded sheets of paper are closed at the top. A tightly stitched seam with strong black yarn holds together the bundle of papers in the middle of the book. Roughly 330 posters from Leipzig printers Oskar Leiner from the years between the 1840s and 1870s are contrasted with street photography from Leipzig that originated in the same period. The monochrome, stationary pictures from this past era all seem so rigid; it is as if the people have been faded out by the camera's long exposure times. And suddenly, in your mind you hear loud voices. You can imagine the crowds of people, squealing with pleasure or astonishment at the Niederländisches Affentheater, the marvellous magicians, the wrestlers and firework artistes. Colourful pastel posters conjure up the shimmering, magnificent costumes and the motley audience – but how? Using an intriguing idea for the design: the posters, one after another, are placed on top of each other in the book, with the edges of the previous one remaining visible, or disappearing entirely – just like posters that are stuck on top of one another to form huge collages where recent events still linger. In this kaleidoscopic density, a facet of social history comes to life: leisure activities in the city of Leipzig during the 19th century.

SOIRÉE
FANTASTIQUE.

Circus Blennow. Die große Post.
Grosse Pariser

Rappo. Theater.
Heute 2 Vorstellungen.
Anfang 5 und 8 Uhr.
Ausstellung

Rappo'. Theater
CIRCUS RENZ

Rappo'. Theater
Hotel de Prusse.
Prof. St. Roman

Rappo'. Theater
Hotel de Prusse.
5400
CIRCUS RENZ.

Rappo'. Theater
Hotel de Prusse.

Theat

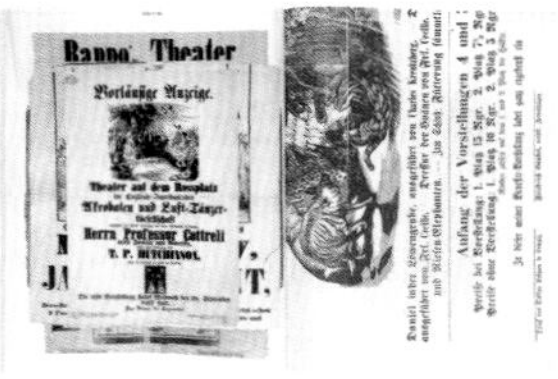
Rappo'. Theater
Vorläufige Anzeige.

Rappo'. Theater
Vorläufige Anzeige.
STEREOSKOPEN
oder
Natur-Bilder.
Die grosse
Menagerie
von A. Scholz

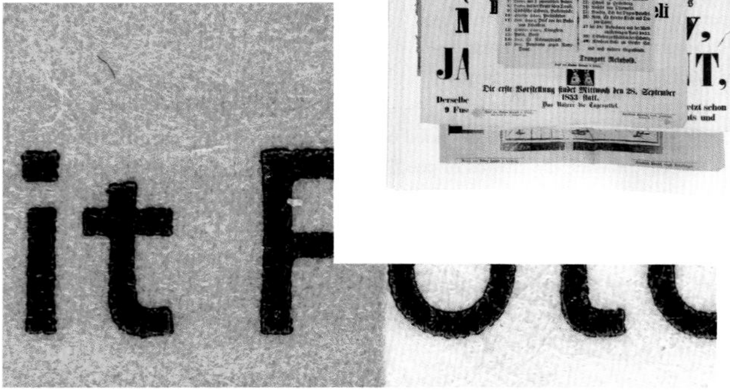
it F

Rappo. Theater.
Ausstellung
Schiffe

Rappo. Theater
Hotel de Pologne.
Doppel-Concert
CIRCUS RENZ.

Rappo. Theater
Hotel de Prusse.
Dr. Stephani
CIRCUS RENZ

Rappo. Theater
Hotel de Prusse.
SALLE DE BASCH
F. J. Basch
ORANG-UTANG

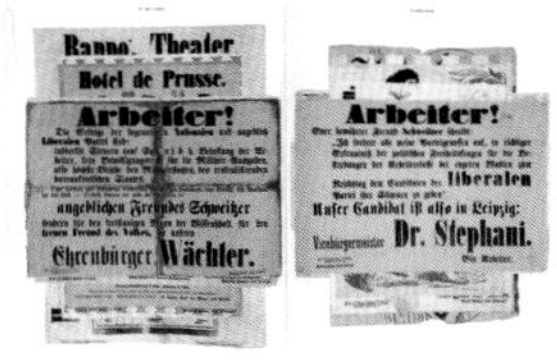

Rappo. Theater
Hotel de Prusse.
Arbeiter!
Wächter.
Dr. Stephani.

Rappo. Theater
Hotel de Prusse.
Kunstausstellung

Rappo. Theater
Miss Betsy,
JAGD-ELEPHANT.
CIRCUS RENZ.
RING-KAMPF.

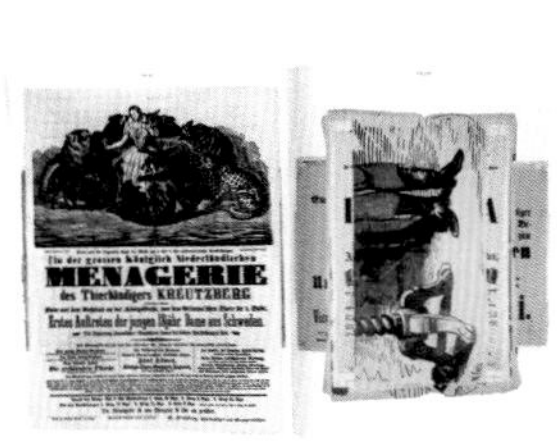

MENAGERIE

Kreutzberg's
MENAGERIE
Kolter-Weitzmann.

Circus Renz.
Cirque Equestre
Jos. Lipbard
Japanesen

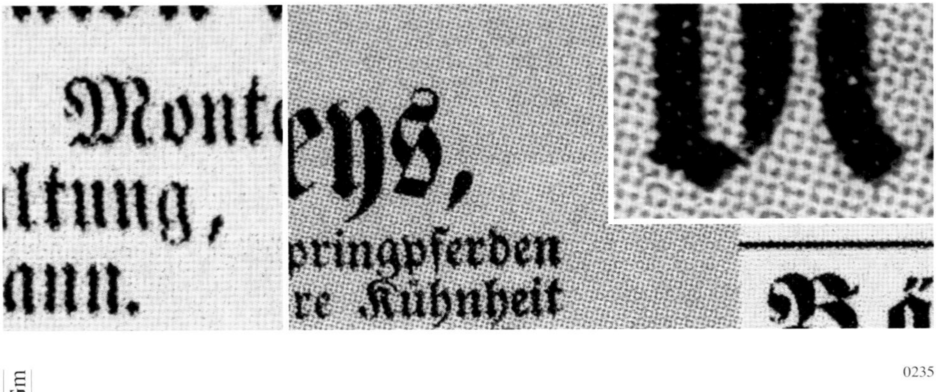

Monte
ltung,
ann.
ens,
pringpferden
re Kühnheit

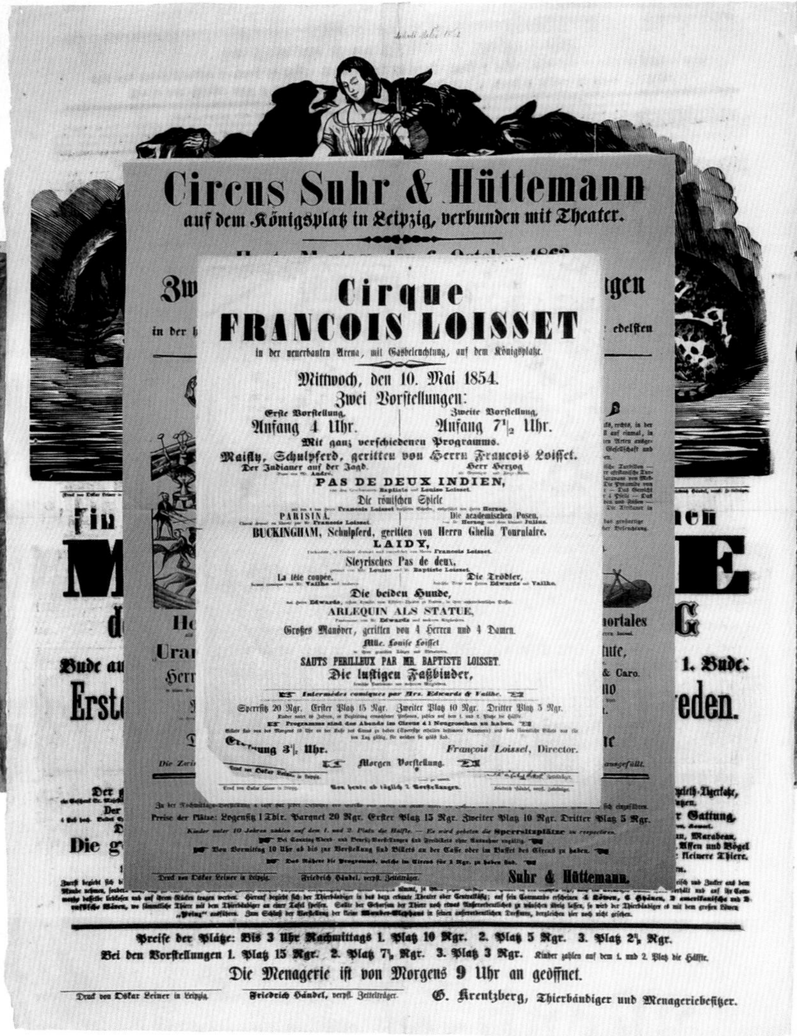

Circus Suhr & Hüttemann

auf dem Königsplatz in Leipzig, verbunden mit Theater.

Cirque FRANCOIS LOISSET

in der neuerbauten Arena, mit Gasbeleuchtung, auf dem Königsplatze.

Mittwoch, den 10. Mai 1854.

Zwei Vorstellungen:

Erste Vorstellung.	Zweite Vorstellung.
Anfang 4 Uhr.	Anfang $7\frac{1}{2}$ Uhr.

Mit ganz verschiedenen Programms.

Maisly, Schulpferd, geritten von Herrn Francois Loisset.

Der Indianer auf der Jagd. — Herr Herzog

PAS DE DEUX INDIEN,

Die römischen Spiele

PARISINA. — Die academischen Posen,

BUCKINGHAM, Schulpferd, geritten von Herrn Ghelia Tournaire.

LAIDY,

Steyrisches Pas de deux,

La tête coupée, — Die Trödler,

Die beiden Hunde,

ARLEQUIN ALS STATUE,

Großes Manöver, geritten von 4 Herren und 4 Damen.

Mlle. Louise Loisset

SAUTS PERILLEUX PAR MR. BAPTISTE LOISSET.

Die lustigen Faßbinder,

Intermèdes comiques par Mrs. Edwards & Vailhe.

Sperrsitz 20 Ngr. Erster Platz 15 Ngr. Zweiter Platz 10 Ngr. Dritter Platz 5 Ngr.

Programms sind des Abends im Circus à 1 Neugroschen zu haben.

Eröffnung $3\frac{1}{2}$ Uhr. — *François Loisset*, Director.

Morgen Vorstellung.

Von heute ab täglich 2 Vorstellungen.

Preise der Plätze: Logensitz 1 Thlr. Parquet 20 Ngr. Erster Platz 15 Ngr. Zweiter Platz 10 Ngr. Dritter Platz 5 Ngr.

Kinder unter 10 Jahren zahlen auf dem 1. und 2. Platz die Hälfte. — Es wird gebeten die Sperrsitzplätze zu respectiren.

Bei Sonntag-Abend- und Benefiz-Vorstellungen sind Freibillets ohne Ausnahme ungültig.

Von Vormittag 10 Uhr ab bis zur Vorstellung sind Billets an der Casse oder im Buffet des Circus zu haben.

Das Nähere die Programms, welche im Circus für 1 Ngr. zu haben sind.

Druck von Oskar Leiner in Leipzig. — Friedrich Händel, verpfl. Zettelträger. — Suhr & Hüttemann.

Preise der Plätze: Bis 3 Uhr Nachmittags 1. Platz 10 Ngr. 2. Platz 5 Ngr. 3. Platz $2\frac{1}{2}$ Ngr.

Bei den Vorstellungen 1. Platz 15 Ngr. 2. Platz $7\frac{1}{2}$ Ngr. 3. Platz 3 Ngr. Kinder zahlen auf dem 1. und 2. Platz die Hälfte.

Die Menagerie ist von Morgens 9 Uhr an geöffnet.

Druck von Oskar Leiner in Leipzig. — Friedrich Händel, verpfl. Zettelträger. — G. Kreutzberg, Thierbändiger und Menageriebesitzer.

Circus Renz.
Sonntag, den 14. Mai 1876
Vorläufige Anzeige.
Einem hochgeehrten hiesigen und auswärtigen Publikum die ergebene Anzeige, daß ich unter der Firma:
Mr. Herrmann's
Soirée Fantastique
in der neuerbauten Bude
auf dem Königsplatze,
an Herrn Lehmann's Hause, einen Cyklus von Vorstellungen in der
höheren Magie ohne Apparate,
wie auch Dissolving-Views (bewegliche Nebelbilder) ausgezeichneten Genre's, während der Dauer der Oster-Messe geben werde.
Aus meinem reichhaltigen Programm werde ich täglich die besten Piecen auswählen.
Alles Weitere durch die Zettel und Anzeigen in den Blättern.
Hochachtungsvoll
Henri Herrmann aus London,
Physiker.
Druck von Oskar Leiner in Leipzig.
Friedrich Händel, verpfl. Zettelträger.
Preise der Plätze:
Sperrsitz 3 Mk., Fremdenloge 3 Mk., Tribüne 2 Mk., 1. Platz 1 Mk. 50 Pfg., 2. Gallerie 1 Mk., 3. Gallerie 50 Pfg.
Das Rauchen im Circus ist verboten.
Morgen 2 Vorstellungen.
E. RENZ, Dir.
Druck von Oskar Leiner in Leipzig.
Friedrich Händel, verpfl. Zettelträger.

2018 Sm1

《园冶》是明代造园家计成撰写的一本关于造园理论的专著，而本书是林学家、造园学家陈植依据《园冶》写作的阐释之作，是对原书的解读与再创作。书籍的装订方式为裸脊锁线装，封面采用棕色草麻纸，较书籍开本短30mm位置向内折叠形成勒口，只做书名烫白字。护封采用比封面高度短70mm的书画纸，除标题外的信息均印在此处。在封面与护封边界处，以梅花形窗格造型烫珠光白。封面与环衬为同一纸张，和合折叠而成。内页采用胶版纸张印单黑，宣纸双色印刷折叠其中，形成三大篇章的隔页，以及部分拉页。全书竖排右翻，版心下沉，形成巨大的天头，用于放置图注和页眉信息。原文、释文和注释字号逐级渐小，层次分明。全书纸张柔软，开本大小、厚度和重量都易于捧阅。

園冶注释

The Art of Gardening is a monograph on the theory of gardening written by Ji Cheng, a garden designer in the Ming dynasty, while this book is an interpretation book written by Chen Zhi, a silviculturist forester and a landscape architect, to decipher and re-create the original book. The book is bound by thread sewing with an uncovered spine, and the cover is made of brown straw board with hot-stamped title in white. The jacket's height is 70mm shorter than the cover; all information except the title is printed here. The inside pages are printed on offset paper in black, attached with double-color printed Chinese art paper as the interleaves to divide three chapters. The typesetting is vertical and rightward with a relatively large head margin, which is used for the caption and page header information. The font sizes of the text, the interpretation and the annotation gradually diminish to indicate levels. The book is friendly to read in size, thickness and weight.

银奖	Silver Medal
园冶注释	The Art of Gardening
中国	China
张悟静	Zhang Wujing
计成（原著）、陈植（注释）	Ji Cheng (Ming Dynasty), Chen Zhi China
中国建筑工业出版社	China Architecture & Building Press
	260×175×29mm
	645g
	424p

A new edition of a rare, historic book about the theory of Chinese horticulture has been published, with contemporary book design. When the Chinese typeface specifies the structure of the lines and text sections, it bears comparison with a typographic translation. The subheads are taken from old woodcut scripts that lend the new page some of their calligraphic flavour. A few fold-out plates have been glued in: the reproduction of a sheet of characters as a zigzag fold documents the visual quality of the original. Single-coloured photographs of famous Chinese gardens have been printed on finely ribbed paper – light-coloured and with weak contrasts. Their letter fold with irregularly spaced folds opens out like a fantasy image. This book seems as though it has grown in a celestial garden. The narrow sign displaying the title is glued to the brown cardboard slipcase that is open on both sides. The dust jacket, truncated in its length, covers two thirds of the open spine, and the tacking thread becomes visible. Light-brown cardboard protects the innerbook at the front and back, the embossed structure seems haptically like the fine bark of a tree. Inside the book the sunyellow tacking thread is seen time and again, as are filaments between flower petals. Is this all unpretentious or representative? Western criteria of the economy and displays of splendour have no validity here. The entirety is so harmonically linked in itself in terms of its asymmetry, colour shades and materials that the eye of the observer even believes that it can see a dark matte gold in the yellowish-brown characters on the jacket.

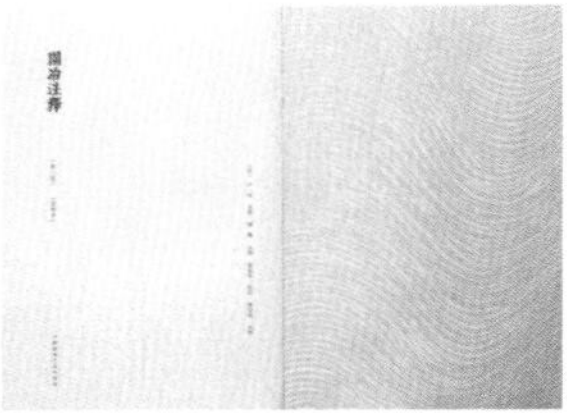

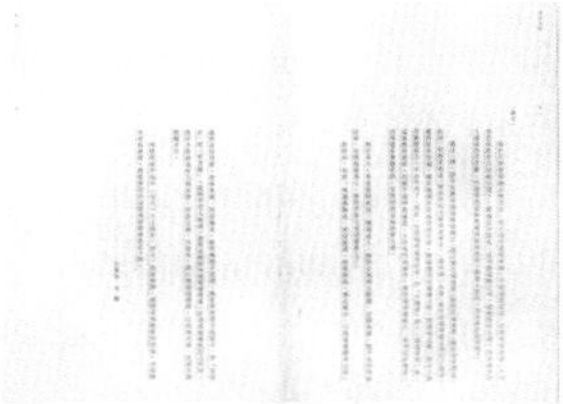
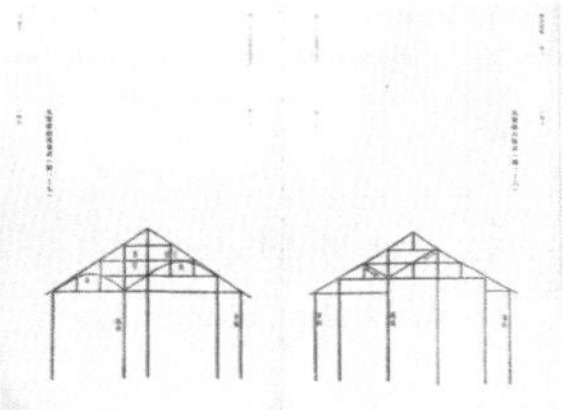
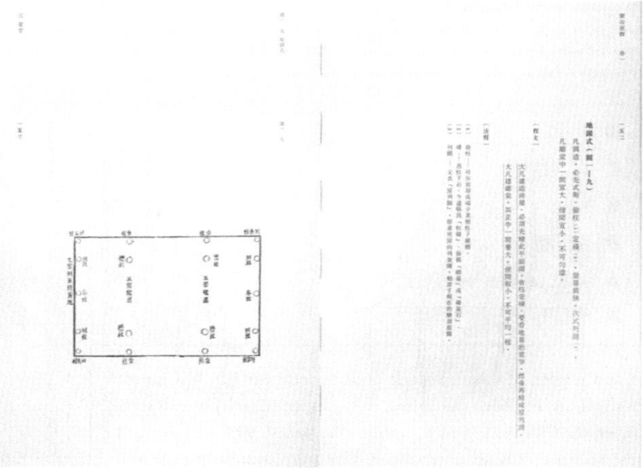
地圖式（圖一—九）
凡廳堂中一間宜大，傍間宜小，不可匀造。
大凡建造廳堂，其正中一間要大，傍間較小，不可平均一樣。
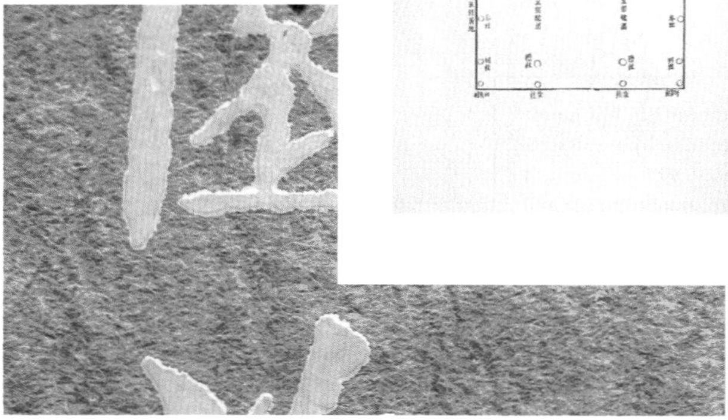

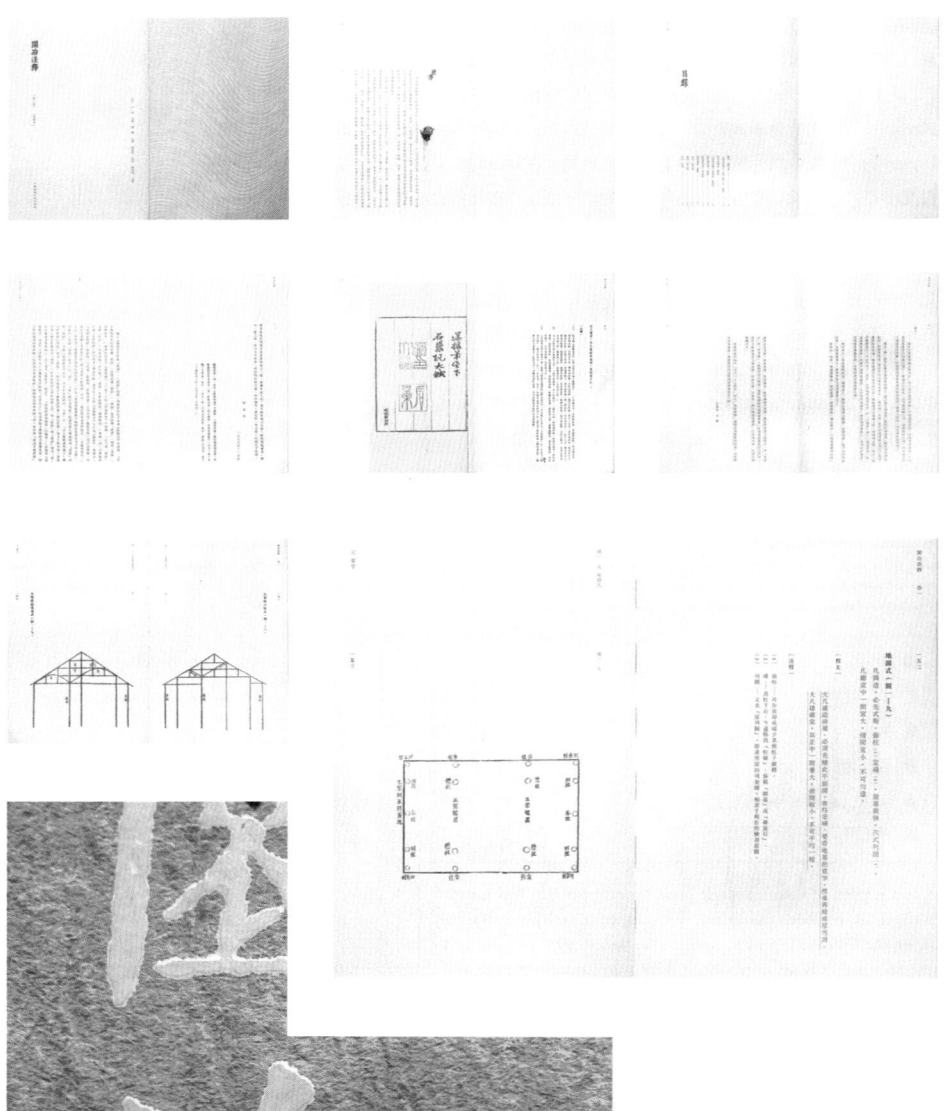

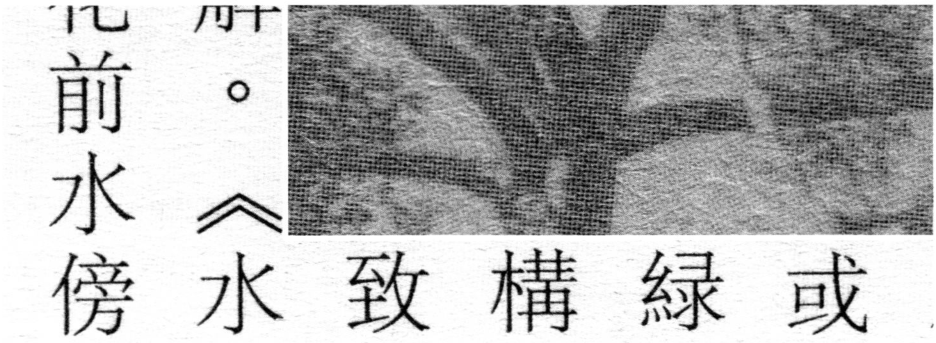
或
緑
構
致
解。《水
花前水傍

園冶注釋
（第二版）（重排本）
中國建築工業出版社

七 牆垣

凡園之圍牆，多于(1)版築，或于石砌，或編籬棘(2)。夫編籬斯勝花屏(3)，似多野致(4)，深得山林趣味(5)。如内花端、水次(6)、夾徑、環山之垣，或宜石宜磚，宜漏宜磨，各有所製。從雅遵時，令人欣賞，園林之佳境也。歷來牆垣，凭匠作雕琢花鳥仙獸，以爲巧製，不第林園之不佳，而宅堂前(7)之何可也。雀巢可憎，積草如蘿(8)，袪之不盡，扣之則廢(9)，無可奈何(10)者。市俗村愚之所爲也，高明(11)而慎之。世人興造，因基之偏側(12)，任而造之。何不以牆取頭闊頭狹(13)，就(14)屋之端正，斯匠主之莫知也。

［釋文］

一般庭園圍牆，多用土造，或用石砌，或栽植有刺的植物，編成綠籬。綠籬較花屏爲佳，因饒自然風致，深得山林雅趣。假如園内的花前、水邊、路旁和環山的圍牆，或宜石叠，或宜磚砌，或宜花牆，或宜磨磚，材料、方法各有不同。總以式樣雅致合時，令人欣賞，纔是庭園的優美環境。向來的牆垣，全凭工匠雕成花草、禽鳥、神仙、怪獸，以爲製作精巧，

罐式（圖三—三十一）

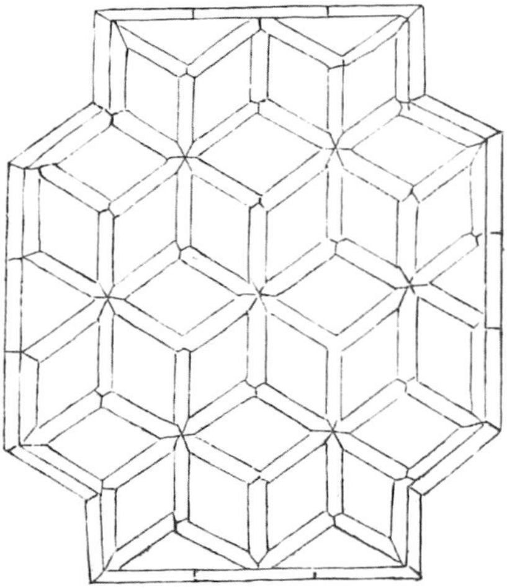

这是一本包含了哲学思考的绘本。女孩爱上了男孩，为了实现愿望，做了很多尝试，也吃掉了很多动物，鲸、鹿、鸟、兔子甚至是熊和男孩，最终她成了孤独的熊。书籍为手工无线胶装硬精装，封面卡板采用三层薄板模切拼贴而成，形成两个逐渐下凹的圆圈浮雕，以呼应书中的餐盘。黄色布面裱覆卡板，封面和书脊文字烫银。由于内页使用数模孔版印刷，全部采用黄、红、中灰三种色彩，所以红色艺术纸环衬和中灰色堵头布也都为此统一。涂布艺术纸张触感柔美，孔版印刷挂粗网，形成了很好的类版画效果。本书为手工特装本，另有平价版本售卖。

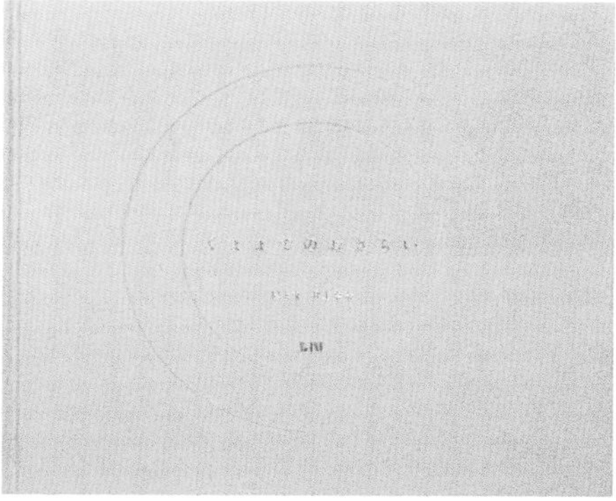

This is a picture book with philosophical thinking. The girl fell in love with the boy. In order to realize the wish, she made many attempts and even ate many animals, such as whales, deer, birds, rabbits, bears and even the boy. She became a lonely bear eventually. This is a perfect-bound hardcover book. The covering pallet is made of three layers of laminated sheets, forming two gradually embossed circles to echo the dinner plates in the book. Yellow cloth mounts on the pallet and the texts are hot stamped in silver on the cover and spine. The inside pages are printed in yellow, red and neutral gray, so it's unified to use red end paper and neutral gray head band. Coated art paper is soft. This book is a hand-made special edition. There are affordable versions for sale as well.

银奖
成为一只小熊的过程
日本

Silver Medal
Process for becoming a little bear
Japan
Akihiro Taketoshi (STUDIO BEAT)
Nana Inoue
be Nice Inc.
216×257×10mm
383g
34p

The yellow fabric has special gauze that cannot be seen, but rather felt, which is just one reason why opening the horizontal-format, cloth-bound book takes longer than you would expect. Puzzling over the circular relief, you conjure up an image of a plate with edge and recess. The circle is not embossed, but the front cover seems to be made up of three layers – the first two layers are stamped out in the shape of a circle. This achieves a good relief depth, and the fabric cover forms distinct, soft shadowed edges to represent a plate. The neat lines on the cover are embossed with silver foil, and the headband is silver grey. A vertical sleeve adds another dimension to the cover, and a little bear with a small red skirt greets the reader. The sleeve does not have to be removed because it is merely slid over the front cover. Stiff, slightly structured, light-red endpapers at the front and back create a link between the cover and its content, whose creamy white, coarse-grained paperboard sheets are bound with adhesive. On the left-hand side, the uncluttered verses, red, in calligraphic style. The right-hand pages display three-colour illustrations in pictogram-like reduction, with bright red, lemony yellow and medium grey. They are prints of templates with the vibrant structure effects characteristic of monotypic graphics. The picture story goes something like this: new animals keep landing on the yellow, round plate – sperm whale, sheep, stag, bird, hare and then – a bear. The next picture: a boy takes a bear-girl by the hand. At the end all of them land in the bear's tummy, even the boy. Only the bear-girl creeps alone through the woods.

くままでのおさらい

くまさんになった　わたしは
おさらのうえの
こいびとを
ぱくぱく　ごっくん

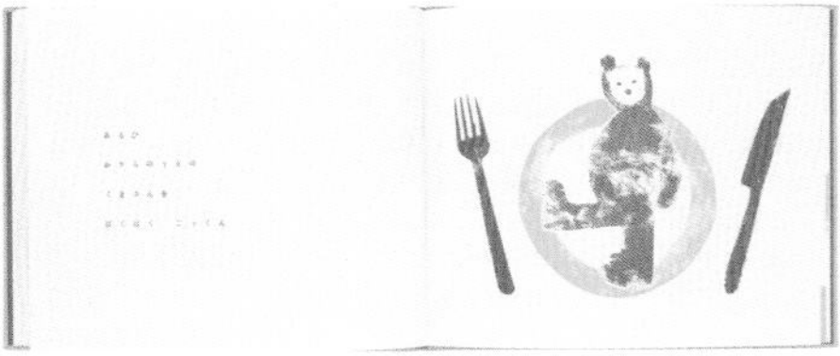

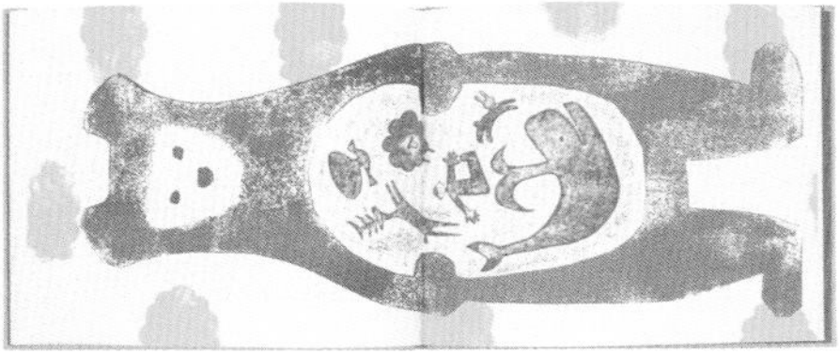

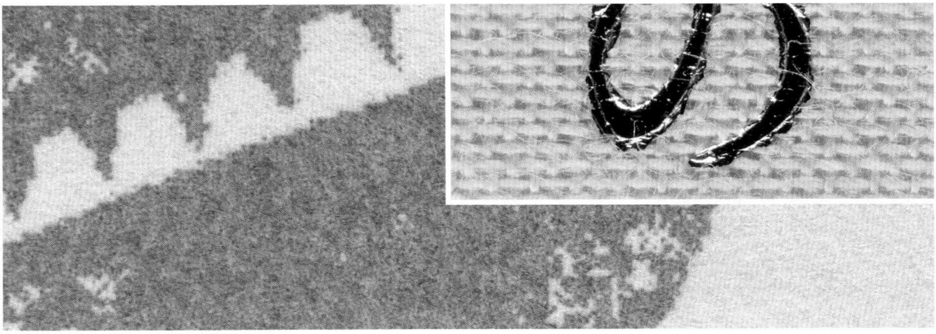

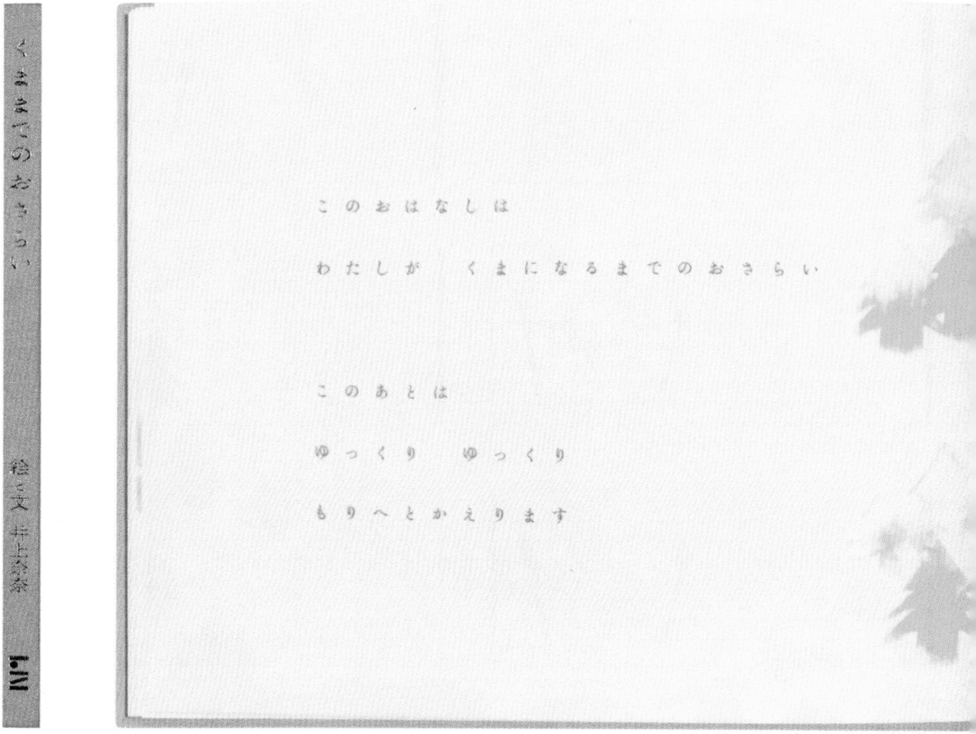
くままでのおさらい
絵と文 井上奈奈

このおはなしは

わたしが　くまになるまでのおさらい

このあとは

ゆっくり　ゆっくり

もりへとかえります

ВЕЧЕР

这是俄罗斯女诗人安娜·阿赫玛托娃（Anna Achmatova）诗集的荷兰版译本。阿赫玛托娃作为俄罗斯“白银时代”的代表性诗人，被誉为“俄罗斯诗歌的月亮”，在俄罗斯文坛具有重要地位。本书的装订形式为锁线硬精装。封面采用浅灰色星幻纸粉红与深蓝色双色印刷，封底单黑印刷女诗人照片，并通过压凹工艺呈现简洁但又古典的相框外观。由于书名叫《黄昏》，书籍在纸张的选色上有一定巧思，前环衬选用浅粉色胶版纸张，而后环衬选用深蓝色纸张，除了与封面字体色彩呼应，也在通过纸张色彩之间的前后关系表达时间概念，即太阳落山后半边余晖与半边夜幕的时刻——黄昏。内页的主体部分采用与封面相同的浅灰色星幻纸单黑印刷，纸张呈现微弱的金属色反光，与俄罗斯的“白银时代”和“俄罗斯诗歌的月亮”均有呼应。正文左页俄文，右页荷兰文，诗歌风格优雅简练，大多为短诗，荷兰文译本遵照原作，具有相同的韵脚。书籍的最后部分采用灰色纸张单黑印刷多张女诗人的生活照片。||||||||||||||||||||||||||||||||||||||||

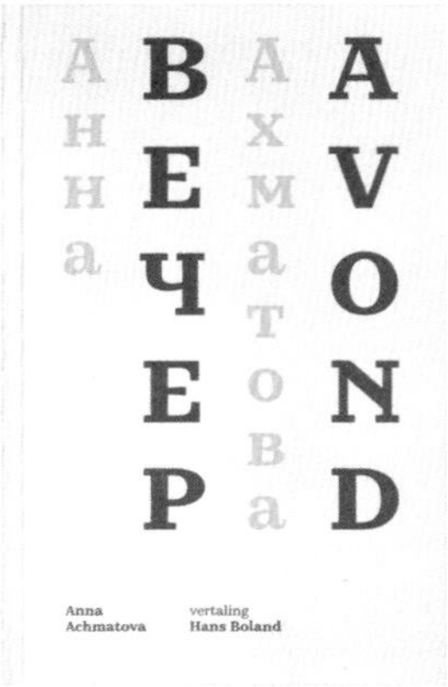
Анна
ВЕЧЕР
Ахматова
AVOND
Anna Achmatova
vertaling Hans Boland

This is a Dutch translation of the poetry anthology of Russian poetess Anna Achmatova. As a representative poet of the 'Silver Age' in Russia, Achmatova is known as 'the Moon of Russian Poetry' and plays an important role in Russian literary world. The hardback book is bound by thread sewing. The cover is made of light gray art paper printed in pink and dark blue, while the back cover is printed with the photo of the poetess in single black. The embossed picture frame presents a concise but classical appearance. Since the title of the book is Dusk, the book has selected the paper color carefully. The front endpaper is printed on offset paper in light pink, and the back endpaper is printed on dark blue paper. The selected colors not only match the color of the cover fonts, but also express the concept of time through the relationship between the colors, that is, Dusk is the time just after sunset when the daylight has almost gone but when it is not completely dark. The main part of the book is printed in black on the same light grey art paper as the cover. The paper presents a weak metallic reflection, which echoes the Russian 'Silver Age' or 'the Moon of Russian Poetry'. The text is written in Russian on the left pages and in Dutch on the right pages. The style of the poems is elegant and concise. Most of the poems are short. The Dutch translation follows the original and has the same rhyme. The last part of the book is printed on gray paper in black to show many pictures of the poetess's life.

- 铜奖
- 黄昏
- 荷兰

- Bronze Medal
- Avond
- The Netherlands
- Michael Snitker
- Anna Achmatova / Hans Boland
- Stiching De Ross
- 242×152×15mm
- 388g
- 128p

The writer Anna Achmatova bestowed particular glamour on the Silver Age of Russian literature. Her first volume of poems from 1912 has been published here for the first time as a translation into Dutch. The letters of the title are arranged one below the other: majestic, striking, airy. But the book catches your eye in other ways besides this vertical text. It is a physical detail that achieves even greater prominence: the elegant edges, extraordinarily narrow – so narrow that even the headband had to be omitted. The bilingual text (Russian, Deutsch) and the two scripts (Cyrillic, Latin) are a typographical challenge. The book designer does not succumb to tabular logic, but makes individual, plausible decisions on the front, spine, and back. The rear side of the paperback actually seems like a front cover: a photographic portrait of the writer that you would normally expect on the front or as a frontispiece, on a die-stamped surface. Subdued rose, darkened blue, recurring in the endpaper and separator pages, create a cool, classy colour scheme. The inner paper itself is responsible for the underlying atmosphere. In a soft light, it shimmers with a slightly silvery shine; when you turn it towards the light, larger sections of the paper gleam. The semibold, classical body type grants a sound footing for short verses or the few stanzas on the double page. The result: an editorial homage to the Russian poet from the Silver Age taking the form of book art.

ВЕЧЕР

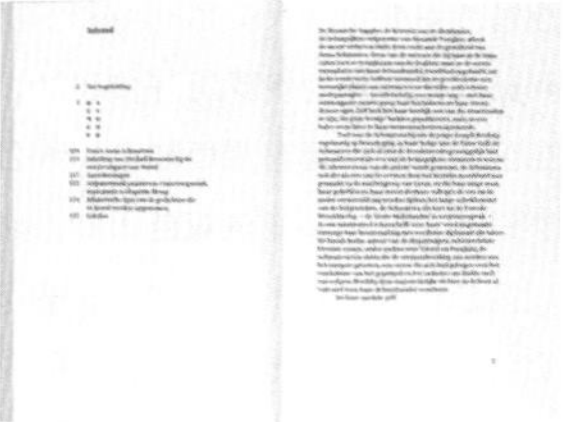

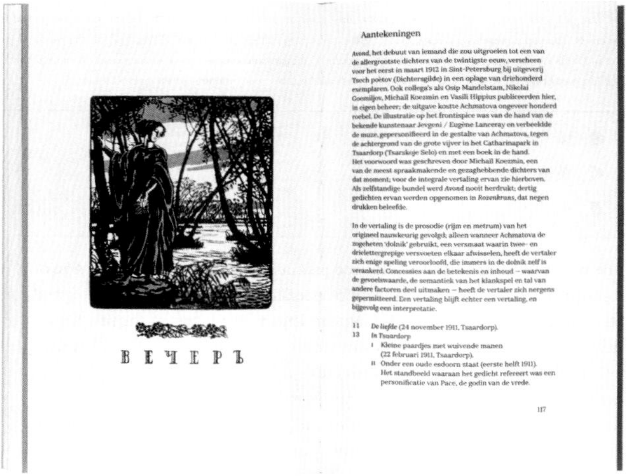
ВЕЧЕРЪ

Aantekeningen

Avond, het debuut van iemand die zou uitgroeien tot een van de allergrootste dichters van de twintigste eeuw, verscheen voor het eerst in maart 1912 in Sint-Petersburg bij uitgeverij Tsech poètov (Dichtersgilde) in een oplage van driehonderd exemplaren. Ook collega's als Osip Mandelstam, Nikolai Goemiljov, Michail Koezmin en Vasili Hippius publiceerden hier, in eigen beheer; de uitgave kostte Achmatova ongeveer honderd roebel. De illustratie op het frontispice was van de hand van de bekende kunstenaar Jevgeni / Eugène Lanceray en verbeeldde de muze, gepersonifieerd in de gestalte van Achmatova, tegen de achtergrond van de grote vijver in het Catharinapark in Tsaardorp (Tsarskoje Selo) en met een boek in de hand.
Het voorwoord was geschreven door Michail Koezmin, een van de meest spraakmakende en gezaghebbende dichters van dat moment; voor de integrale vertaling ervan zie hierboven.
Als zelfstandige bundel werd *Avond* nooit herdrukt; dertig gedichten ervan werden opgenomen in *Rozenkrans*, dat negen drukken beleefde.

In de vertaling is de prosodie (rijm en metrum) van het origineel nauwkeurig gevolgd; alleen wanneer Achmatova de zogeheten 'dolnik' gebruikt, een versmaat waarin twee- en drielettergrepige versvoeten elkaar afwisselen, heeft de vertaler zich enige speling veroorloofd, die immers in de dolnik zelf is verankerd. Concessies aan de betekenis en inhoud – waarvan de gevoelswaarde, de semantiek van het klankspel en tal van andere factoren deel uitmaken – heeft de vertaler zich nergens gepermitteerd. Een vertaling blijft echter een vertaling, en bijgevolg een interpretatie.

11 *De liefde* (24 november 1911, Tsaardorp).
13 *In Tsaardorp*
I Kleine paardjes met wuivende manen (22 februari 1911, Tsaardorp).
II Onder een oude esdoorn staat (eerste helft 1911). Het standbeeld waaraan het gedicht refereert was een personificatie van Pace, de godin van de vrede.

117

AVOND

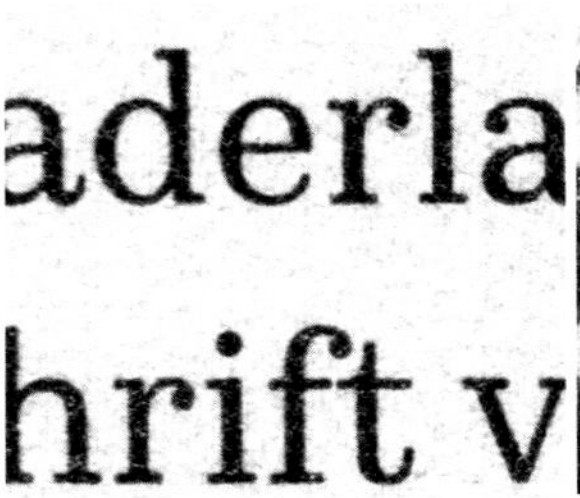
aderla
hrift v
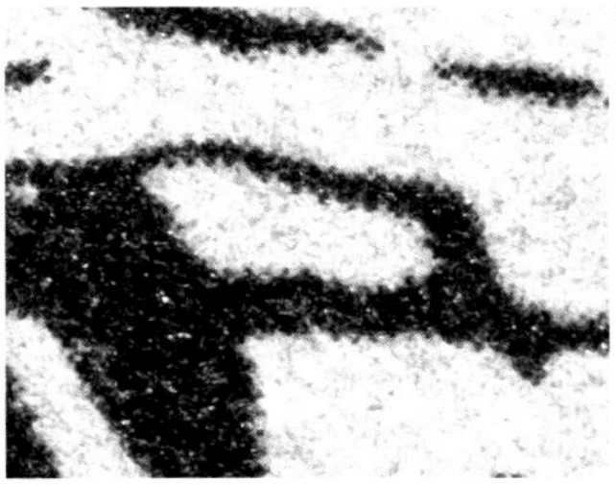
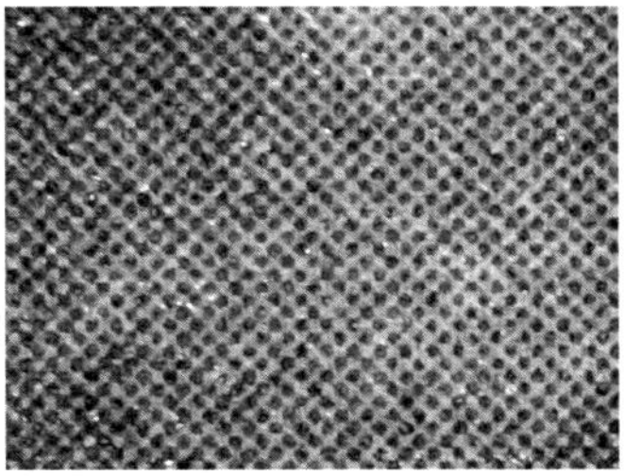

g met een Britse dir

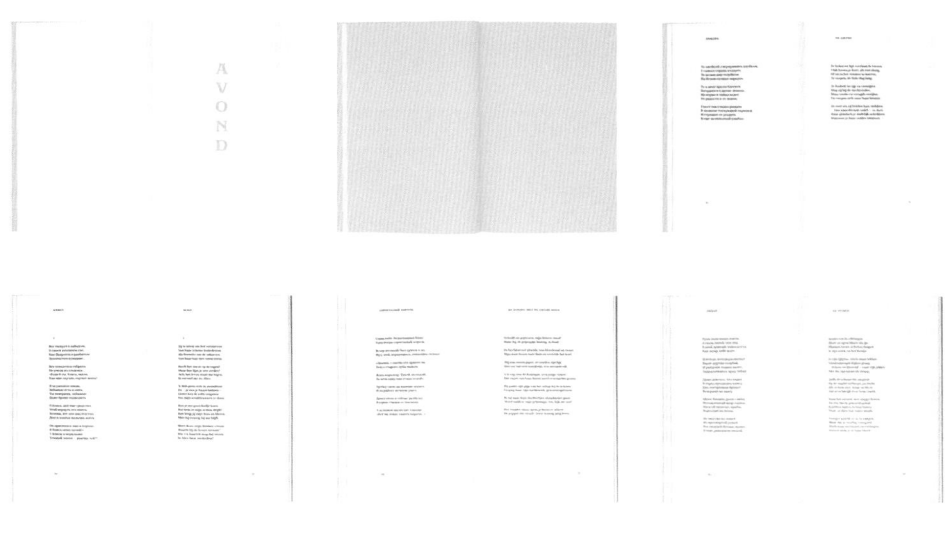

ВЕЧЕР
Anna Achmatova
AVOND

МАСКАРАД В ПАРКЕ

Луна освещает карнизы,
Блуждает по гребням реки...
Холодные руки маркизы
Так ароматно-легки.

«О принц! – улыбаясь, присела, –
В кадрили вы наш vis à vis»,
И томно под маской бледнела
От жгучих предчувствий любви.

Вхол скрыл серебрящийся тополь
И низко спадающий хмель.
«Багдад или Константинополь
Я вам завоюю, ma belle!»

«Как вы улыбаетесь редко,
Вас страшно, маркиза, обнять!»
Темно и прохладно в беседке.
«Ну что же! пойдем танцовать?»

Выходят. На вязах, на кленах
Цветные дрожат фонари,
Две дамы в одеждах зеленых
С монахами держат пари.

И бледный, с букетом азалий,
Их смехом встречает Пьеро:
«Мой принц! О, не вы ли сломали
На шляпе маркизы перо?»

BAL MASQUÉ IN HET PARK

Het maanlicht strijkt over de plassen
En kroonlijsten. Glimlachend reikt
Zij hem een met reukzeep gewassen,
Koud handje. De marktgravin lijkt

– Haar masker ten hoon – te verbleken
Terwijl ze een knicks maakt: 'Ik zie
U bij de quadrille, dat spreken
We af. U bent mijn vis-à-vis.'

Het vuur van de liefde slaat over
Naar moerbei en zilverabeel.
'Istanboel en Bagdad verover
Ik aanstonds en zal ik, ma belle,

U aanbieden.' 'Wilt u mij minnen?
U glimlacht haast nooit, bent u heel,
Heel wreed?' 'Gaat het dansen beginnen?
't Is frisjes hier in het prieel.'

Er glimmeren tuinlampionnen
In iepen en esdoorns; daar doen
Twee dames in groene japonnen
Met pelgrims een spel, om een zoen.

Azalea's ronddelend, olijk
Roept die als Pierrot is geschminkt:
'De prins, markiezin, is zo vrolijk!
De veer van uw hoed is verminkt.'

8
SATURDAY
documenta 14

这是一本记录 2017 年第 14 届卡塞尔文献展的书籍。不同于常规的展册，本书并非以展品为主，而是记录了受邀的策展人、诗人、评论家、史学家和其他艺术家的评论和再创作，通过平行两条线的资料记录，形成了时空交汇的思想碰撞。本书的装订形式为无线胶装，内封白卡纸印单黑，护封采用黑色塑料封皮印银，字体压凹成型。内页灰色新闻纸四色印刷，没有页码，而是使用本次展览从雅典开始到卡塞尔结束一共 163 天的日历作为内容的序列结构。内页展开的左右页为一个整体，左下角为参展艺术家的创作记录，用统一的黑色块标明，其余部分为观展者的评论和叙述文字。右页上部为展览时期每一天的日历，使用了老式日历的形式，包含年月日甚至日出日落等细节信息。书籍最后的赞助商、索引以及留给读者的笔记页，依然延续了日历的做法，仿佛展览的延续。

This is a book documenting the 14th Kassel Documenta in 2017. Unlike the regular exhibition volume, this book does not focus on the exhibits, but records the comments and re-creations of invited curators, poets, critics, historians and other artists. The data is organized by time line and space line. The book is perfect bound with white cover printed in black. The jacket is made of black plastic sheet printed in silver, and the fonts are embossed. The inside pages are printed in four colors on gray newsprint without page numbers. Instead, a calendar of 163 days from the beginning of the exhibition in Athens to the end of Cassel is used as the sequence of the contents. On a double-page spread, the records of the participating artists are put in the lower left corner, marked with a unified black block. At the top of the right page is a calendar for each day of the exhibition period, in the form of an old-fashioned calendar containing details such as year, month, day and even sunrise and sunset time. Furthermore, the pages for sponsor list, indexes and notes apply the same layout as a calendar, as if it were a continuation of the exhibition.

- 铜奖
- 第 14 届卡塞尔文献展：日记账
- 瑞士

- Bronze Medal
- documenta 14: Daybook
- Switzerland
- Julia Born & Laurenz Brunner, Zürich
- Quinn Latimer und Adam Szymczyk
- Prestel, München
- 293×204×20mm
- 938g
- 344p

An exhibition guide, catalogue, companion volume, reader for documenta 14? None of those. This Daybook stands as a hybrid which admittedly documents the previous Documenta exhibition, but also asserts the timeless validity of art. Each artist is represented on a double page in the book, and not just with one work (or more), but also with a date that is particularly important for the artist. It is these dates that then, in descending order, determine the allocation of the artists to the calendar in the Daybook, the 163 days of the exhibition. The motifs of being bound to time, being on record, are manifest: the notebook as a place to jot down the day's events and thoughts, the calendar. On a material level: the thin, greyish paper in recycling style, a book protector as a dust jacket, with pockets made of blue plastic. (Incidentally, there is a triplet of Daybooks, in German, Greek and English.) The calendar supplies a formal time base. A date that is significant for each artist creates a personal link that goes beyond time references. Each essay is written by a different author who is close to the artists or their works. You could say that idea, leitmotif and design are synonyms in the case of this book. The conventional order of book production is so condensed here that it is hardly possible to talk about the design alone.

documenta 14: Daybook
Athens, 8 April – Kassel, 17 September 2017
PRESTEL
Munich · London · New York

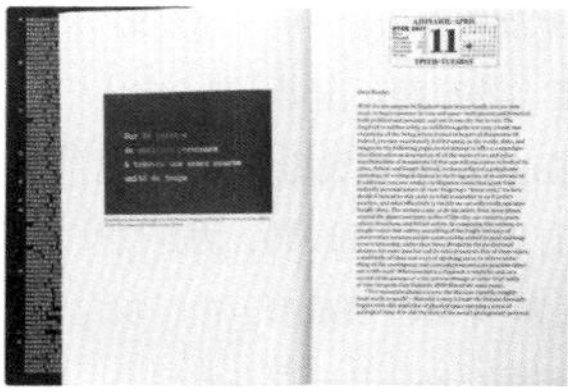

GEORGIA SAGRI
15

ANDREAS ANGELIDAKIS
11

NEVIN ALADAĞ
29

Donors
OCA

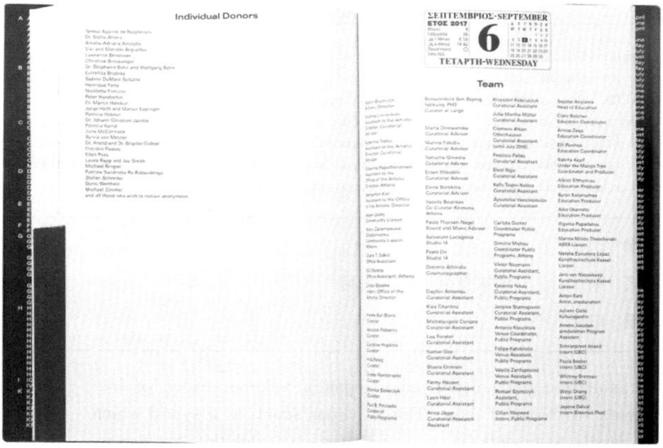
Individual Donors
ΣΕΠΤΕΜΒΡΙΟΣ-SEPTEMBER
ΕΤΟΣ 2017
6
ΤΕΤΑΡΤΗ-WEDNESDAY
Team

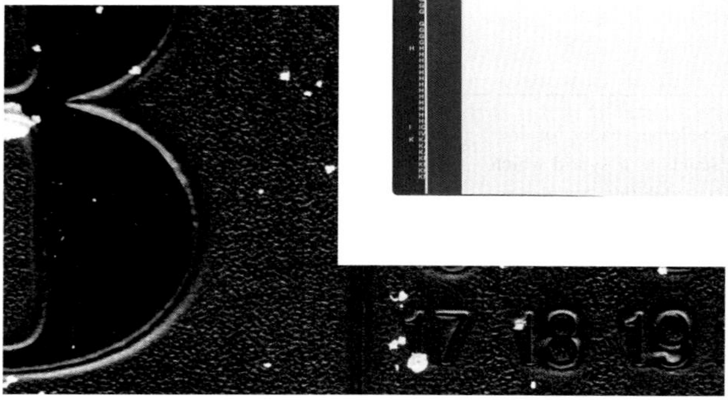
17 18 19

BONITA ELY

KATALIN LADIK

HANS HAACKE

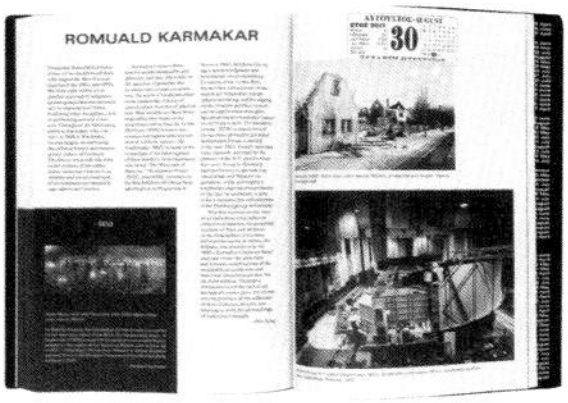
ROMUALD KARMAKAR
30

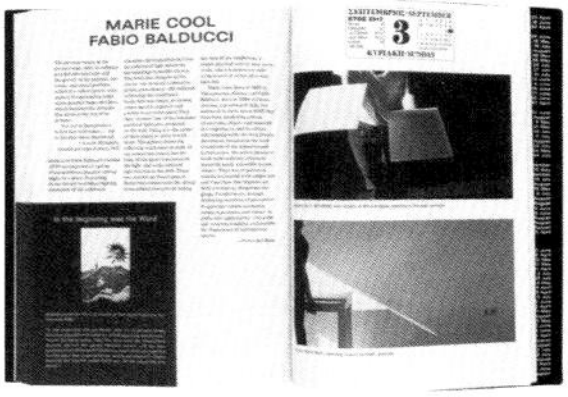
MARIE COOL
FABIO BALDUCCI
3

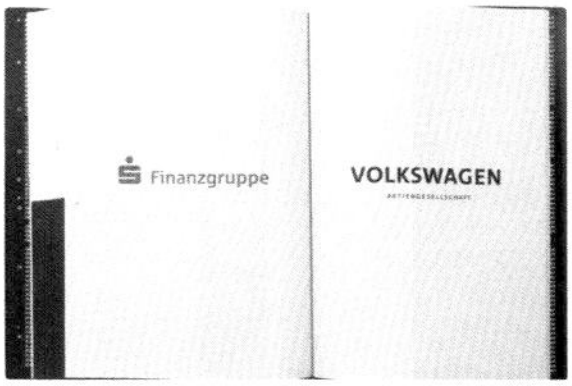
Finanzgruppe
VOLKSWAGEN
AKTIENGESELLSCHAFT

Acknowledgments
10

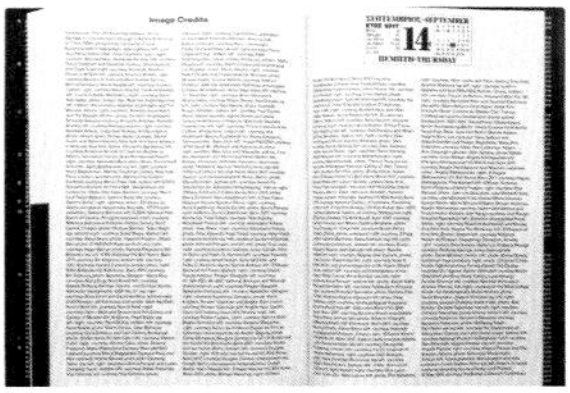
Image Credits
14

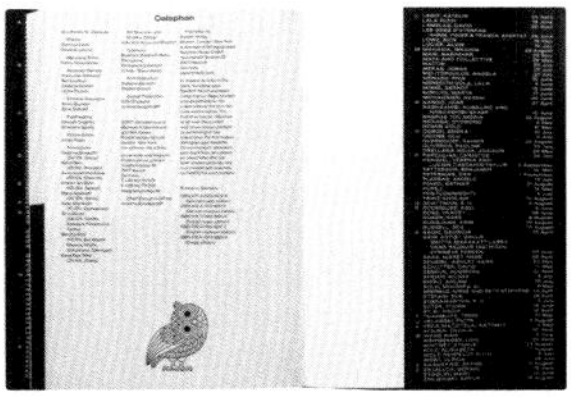
Colophon

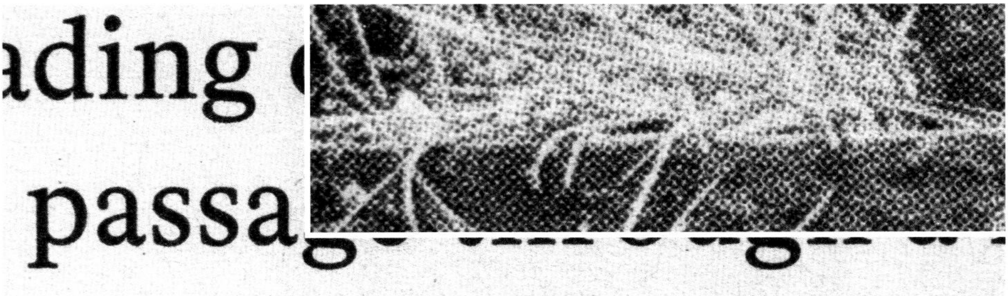

PRESTEL

documenta 14: Daybook

DAN PETERMAN

Ingot God statuette (13th–12th century BCE), copper, Enkomi, Cyprus

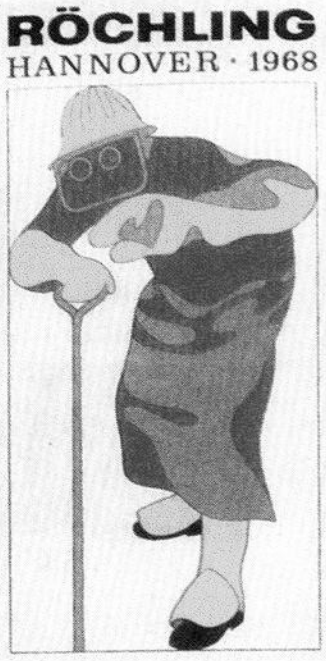
RÖCHLING
HANNOVER · 1968

Cover of Röchling steelworks trade publication, Hannover, 1968

5,500 BCE

Bakarna sekira (copper axe, 5,500 BCE), Pločnik, Serbia, length: 18.6 cm

5,500 BCE is the date of the earliest known metallurgical workshops in Europe. The remnants of this mining, smelting, and casting activity, developed by the Vinča culture, was unearthed at archeological sites in Pločnik and Belovode in present-day Serbia. Both towns are situated south of Belgrade along the driving route between Kassel and Athens.

—Dan Peterman

Dan Peterman has been working at the intersection of classical sculptural questions pertaining to form, mass, and volume, and pressing socioeconomic and ecological concerns since the beginning of the 1980s, when he first moved into the former recycling plant that has been his South Side studio ever since. It is part of a larger complex called the Experimental Station that proved to be an important early hub of the so-called social practice phenomenon in contemporary art to which the city of Chicago has become so closely associated.

An interest in alternative economies and the logic of circulation and recycling has led Peterman, born in Minneapolis in 1960, to concentrate on two materials for documenta 14, both of which come steeped in deep art-historical meaning, and whose coupling closely mirrors the fraught economic relationship between Germany and Greece: steel and copper.

ΣΕΠΤΕΜΒΡΙΟΣ-SEPTEMBER

ΕΤΟΣ 2017

Μήνας 9
Εβδομάδα 35
Ανατολή Αθήνα 6:55
Δύση Αθήνα 19:54
Σελήνη 11
244-365

1

Δ M	Τ T	Τ W	Π T	Π F	Σ S	Κ S
				1	2	3
4	5	6	7	8	9	10
11	12	13	14	15	16	17
18	19	20	21	22	23	24
25	26	27	28	29	30	

ΠΑΡΑΣΚΕΥΗ-FRIDAY

chance encounter with a small ot statue referred to as the God" (Cyprus was the source st copper circulating in cient eastern Mediterranean) l Peterman's attention toward tion of the ingot as unit or ype, as basic material form— ct, another type of currency obal economy perennially t between the illusory prom- dematerialization and the reality of the world's irreduc- ateriality.

urs of both the historical f Germany's past and present rial might (such as the nd steel mills, where some of the world's largest outdoor sculptures are produced, as well as factory grounds in Kassel that have played high-profile roles in Germany's eventful military past) and the underbelly of Athens's flourishing scavenging economy have resulted in a series of site-specific interventions. These range from a copper-casting workshop to a trail of iron ingots, sourced from a factory specialized in recycling iron dust, and scattered throughout the exhibition trajectory in a manner that conjures memories of classical American Minimalism, while simultaneously referencing the constant stream of "stuff" that keeps the world economy afloat. And, of course, tellingly in these times of violent worldwide upheaval, the immortal words of Otto von Bismarck, the father of the modern German state, come to mind: "Not through speeches and majority decisions will the great questions of the day be decided . . . but by iron and blood."

—Dieter Roelstraete

Scrap steel with crane, research photograph, Kassel, 2015

5 April
8 June
0 April
6 June
18 May
16 July
August
28 May
29 May
25 July
27 July
31 July
2 June
4 June
9 June
7 June
0 June
7 April
0 June
August
8 May
21 May
10 July
11 July
August
23 July
4 May
4 July
ember
0 May
ember
5 July
ugust
9 May
4 July
ugust
ugust
June
June
ugust
ugust
ugust
April
June
April
May
May
April
3 July
9 July
6 May
April
April
5 July
June
April
May
ugust
May
June
June
April
ugust
ugust
July
July
gust
April
June
gust

这是一本探讨对长久以来巴以冲突历史的态度的图书。书中将双方冲突的照片、足球比赛的照片、相应的文献资料、扫描件进行穿插并置，展现这场历时已久并充满着热血、欢呼、临时的单方胜利、男子气概、荣耀和能量的博弈，从长远看是一场 0:0 没有输赢的比赛。本书的装订形式为锁线软装。护封采用白卡纸印单黑，覆哑光膜。内封白卡纸黄黑双色印刷，书脊部分裸露，实用暗红色布面和墨绿色锁线进行加固。内文胶版纸四色印刷。图片部分的排版较为自由，大多采用并置、叠加、穿插的方式，展现冲突场面和足球比赛间或多或少的相似。文字部分主要为希伯来文，右齐排版，只在文档扫描件和部分英文引用文献部分使用英文。主要文字部分采用大小不等的双栏设置，分置正文和注释。全书没有页码，使用加方括号的数字来展现文字、图片或文档扫描件等内容序列。

This is a book that explores and expresses attitudes towards the long history of the Israeli-Palestinian conflict. The book juxtaposes photographs of conflicts between the two sides, photographs of football matches, relevant documents and scanners, in order to show the long-standing game full of blood, cheers, temporary one-sided victory, masculinity, glory and energy. In the long run, it is a 0:0 game without winning or losing. This paperback book is bound by thread sewing. The laminated jacket is made of white cardboard printed in black. The bare spine has been reinforced by dark red cloth and dark green sewing thread. The inside pages are printed on offset paper in four colors. The layout of the pictures is relatively free, using juxtaposition, superposition and interpolation to show more or less similarities between conflict scenes and football matches. The text is mainly written in Hebrew and right-aligned, only the contents from the scanned documents and quoted documents are written English. The main body is typeset in two columns of unequal width to put text and annotations separately. There are no page numbers in the book, and bracketed numbers are used to show the sequence of text, pictures or scanned documents.

- 铜奖
- 0:0
- 以色列

- Bronze Medal
- EFES:EFES
- Israel
- Dan Ozeri
- Dan Ozeri
- Self expenditure
- 310×264×20mm
- 1044g
- 192p

The black/white paperboard cover displays its striking title in bold letters from Modern Hebrew: Efes:Efes (Nil-Nil). The picture on the cover shows cloudy trails of vapour against a clear sky, as a column, as an arch. Traces of celebratory fireworks? Of rockets with warheads? This volume of photographs shows people, crowds of people. Men. Israelis, Palestinians. In black/white or in colour. Discovered in the State of Israel National Photo Collection, possibly filter results for the keywords football, military, attacks. This combination of a love of sports with mortal fear seems unsettling and daring – haphazardly cast out and playfully combined. Some examples: The troops sitting enthroned on a tank stretch their arms up; on a nearby picture bare-chested football fans wrapped in flags hang out of the sides of the decorated cars. Manifestations of pleasure, warning, triumph? – Or a snapshot of footballers waiting for the ball directly in front of the goal, as a panoramic image; a masked youth superimposed, reaching backwards to throw a stone. A controversial design affinity between the pattern on a Palestinian scarf and the net stretched in a (football) goal. Again and again, the view is disturbed, interrupted, distracted. The author/designer confronts the observer with a story that the latter may not want to hear expressed in words at all: a contradictory game about masculinity, energy, honour, competition, defence. A party game that ends in a draw, nil-nil.

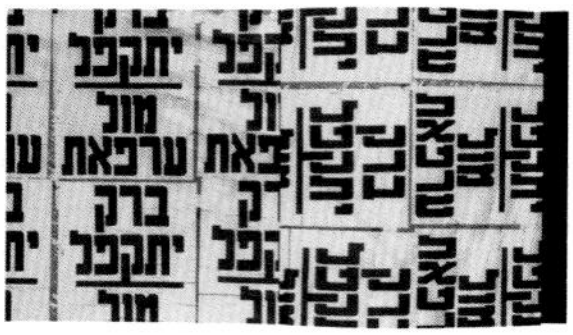
יתקפל
מול
ערפאת
ברק
יתקפל

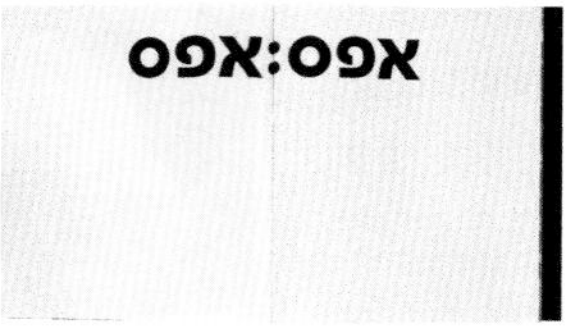
אפס:אפס

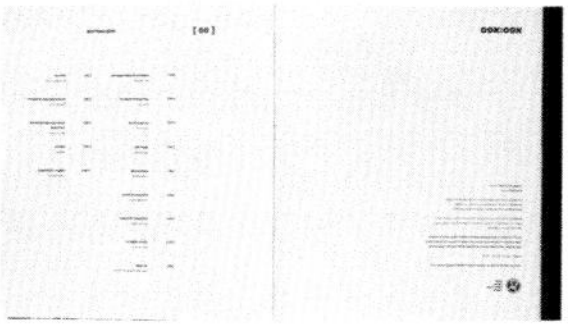

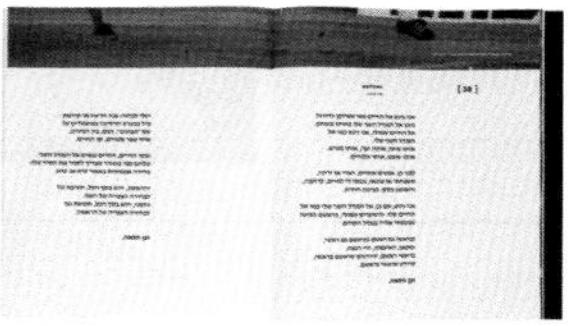

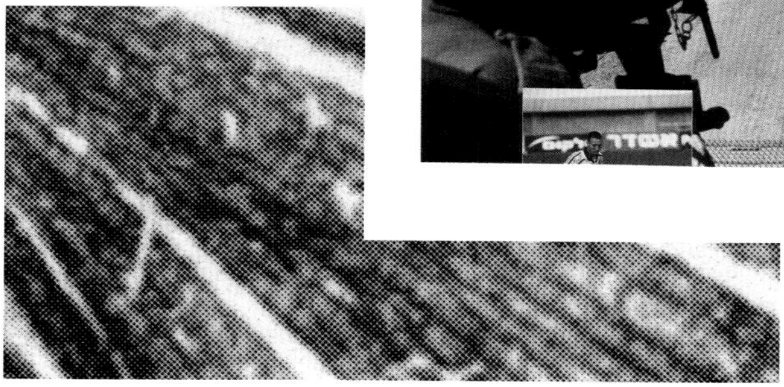

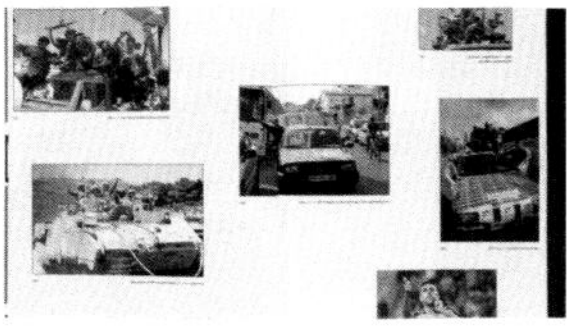
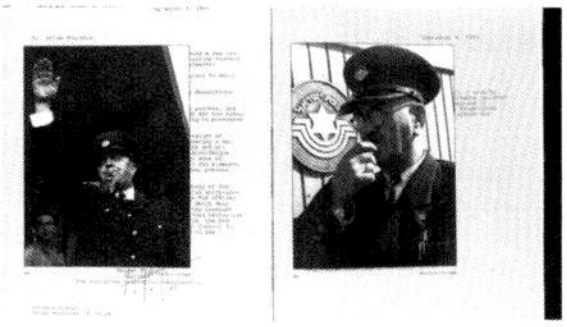

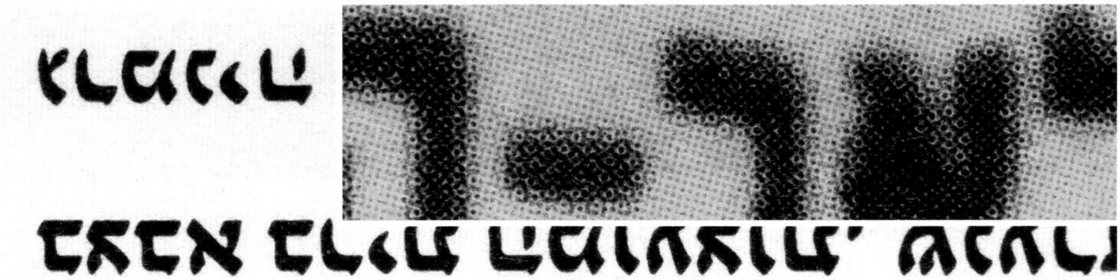

7. למען העם והמדינה

כשלהי הקרצ'נדו

בקצה האחרון של שנות ה-50 הייתה לכדורגל הישראלי עדנה. מעמדו כסמל לאומי נעשה עובדה חברתית ידועה ומוכרת. באותם הימים המונח "לאומי" הוחל רק על יהודים, כי עמדת השלטונות כלפי הערבים הייתה, בלשון המעטה, אמביוולנטית. דבר אחד היה ברור בגלוי או במובלע: הרוב היהודי לא ראה במיעוט הערבי חלק אמתי ממדינת ישראל, והערבים היו לדידו בחזקת "דיירי משנה" (בנזמין ומנסור, 1992), שהזהות הישראלית נכפתה עליהם; הממשל הצבאי דאג שאת ההסכמה החיצונית לזהות זו שהתבטאה בטקסי יום העצמאות למשל, יביעו הערבים כנדרש. באותן שנים ניתן היה להבחין בשינויים דמוגרפיים בכדורגל הישראלי, שינויים שהצדיקו את תפקידו כמכשיר אינטגרטיבי-לאומי: הוא החל להתמלא בשחקנים עולים חדשים, רבים מהם יוצאי אסיה ואפריקה. נדמה היה שכמו במקומות אחרים, הכדורגל הולך ונעשה יותר ויותר פונקציונלי מבחינת הסדר החברתי בכלל והאינטגרציה בפרט. באווירה ששררה אז, אינטגרציה בתנאים מסוימים שהוכתבו לנמצאים ולעולים החדשים הייתה מושג בעל קונוטציה חיובית מאוד. עדיין האמינו בחלונות הגבוהים בכור ההיתוך.

על מעמדו המסוים של הכדורגל בהקשר הלאומי ניתן היה ללמוד מן התקשורת: העיתונות היומית והתקופתית והרדיו הציגו את ידיעות הספורט ובדבר משחקי הנבחרת במקום המיועד לחדשות. עדות בעלת משקל מיוחד היא כמות הצופים במגרשים בזמן המשחקים של נבחרת ישראל; עשרות אלפי גופים, רובם המוחלט גברים, מילאו את יציעי המושבים ואת מורדות סוללות העפר שהקיפו את האצטדיון ברמת גן: 50,000 נכחו במשחק נגד נבחרתה של יוגוסלביה במסגרת התחרויות הקדם-אולימפיות (12.10.1959), משחק שהסתיים בתוצאת תיקו 2:2 (שני שערים של נחום סטלמך), שנרשמה ב"פנתאון" הכדורגל הישראלי. ביחס לכלל האוכלוסייה הישראלית הבוגרת

שחקן נבחרת ישראל גולש לרגלו של שחקן נבחרת אורוגוואי, אצטדיון רמת־גן, 1973

пишущая машинка
200 ударов
цитаты

这是一本有关俄罗斯打字机与文学创作的展览画册。展览由俄罗斯理工学院博物馆、莫斯科现代艺术博物馆和俄罗斯国家文学博物馆主办，展示了对于俄罗斯的文学创作起到重要贡献的打字机、打字机技术，以及文学创作方面的相关内容。书籍的装订形式为裸脊锁线装，书籍没有与内页有明显差异的封面，呈现出一叠未完成的资料形式。内文采用泛黄的胶版纸和哑面白卡纸混合而成，大部分内容采用黑色和紫色双色印刷，呈现打字机照片、打字机细节特写、文学打字稿件、打字机排印、打字机文本形式“涂鸦”海报等丰富的展示内容，展览馆照片和文学手稿则采用四色印刷。从封面起，页面普遍采用四栏设置，大多文本或图片占三栏，留左侧一栏加注释。四栏设置并不仅仅在文本和图片的排布上起到版面网格的作用，还形成了部分文本直接空出一栏的超大首行缩进设置，具有打字机时代的文本味道。

ударов
в минуту

This is an exhibition album about Russian typewriters and literary creation. Organized by Moscow Polytechnic Museum, Moscow Modern Art Museum and Russian National Museum of Literature, the exhibition presents typewriters, typewriter technology and related contents of literary creation, which have made important contributions to Russian literature. The book is bound by thread sewing with an uncovered spine. The cover looks similar as the inside pages, so the book is more like a pile of unfinished documents. The faded offset paper and white matt cardboard are used together, most contents are printed in black and purple except gallery pictures and literature manuscripts are printed in four colors. The pages use four-column typesetting and most text or pictures account for three columns, leaving the left column with annotations. The four-column typesetting not only plays the role of a layout grid in the arrangement of text and pictures, but also forms the super-large text-indent for the first line, which implies the style of typewriter era.

- 铜奖
- 每分钟敲键 200 次
- 俄罗斯

- Bronze Medal
- 200 keystrokes per minute
- Russia
- Igor Gurovich
- Narinskaya Anna
- Moscow Polytechnic Museum
- 297×211×20mm
- 699g
- 208p

A stitch-bound, glued innerbook lies naked on the table: this publication looks like an unfinished book, or rather like a sheaf of manuscripts – which corresponds to the design in hand as well. It is the catalogue for an exhibition at the Polytechnic Museum in Moscow: “200 Keystrokes per Minute. Typewriter and the 20th-Century Consciousness”. The book designer arranges the texts on rough, yellowish-white publishing paper with columnwidth indents within a four-column type and image area. Typographical differentiation is achieved, and more than three type sizes would be unnecessary. The exhibits are shown – typewriters from several decades, black/white pictures viewed from above, with each frontal view underneath in a blue/violet special colour. Another section is dedicated to a consumable material that is most likely to have been forgotten: black or blue carbon paper as a means of duplication. By establishing the relevance of this waste from an archival perspective, we sense what is behind the indecipherable secret concealed in these sheets, whose one-sided use is apparent, without ever disclosing their texts. Velvety smooth, matte illustration printing paper serves as the basis for brilliant close-ups of old typewriters in very high resolutions – like looking into an engine room. You feel as though you are sitting in an engine that is being powered by literary energy, transformed into readable forms. A homage to the typewriter as a means of producing Russian literature.

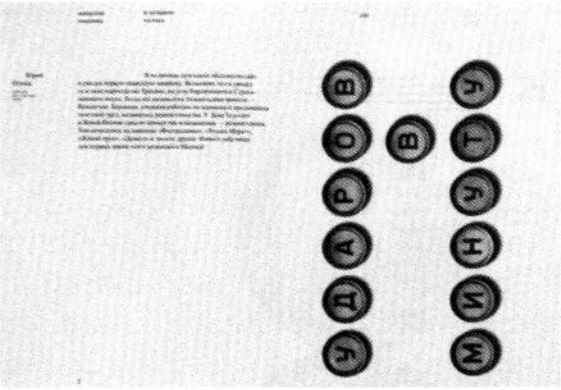

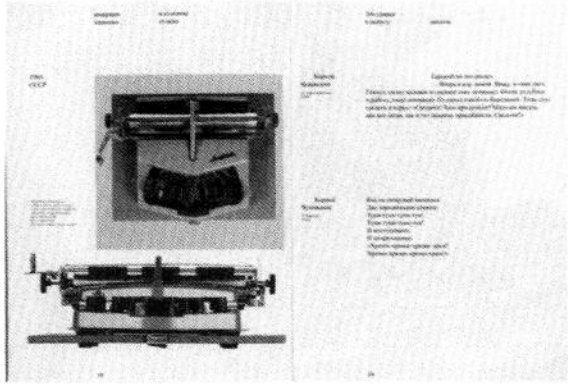

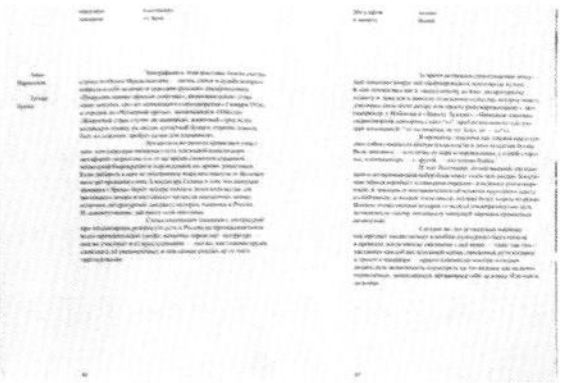

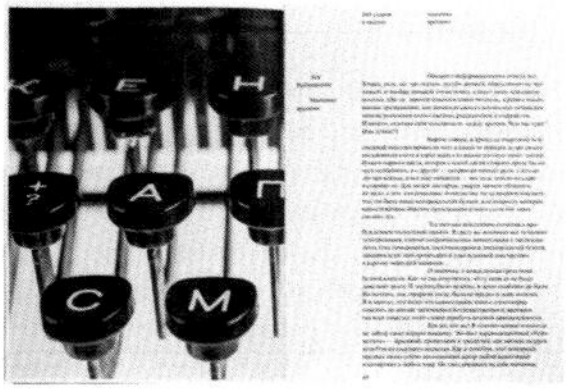

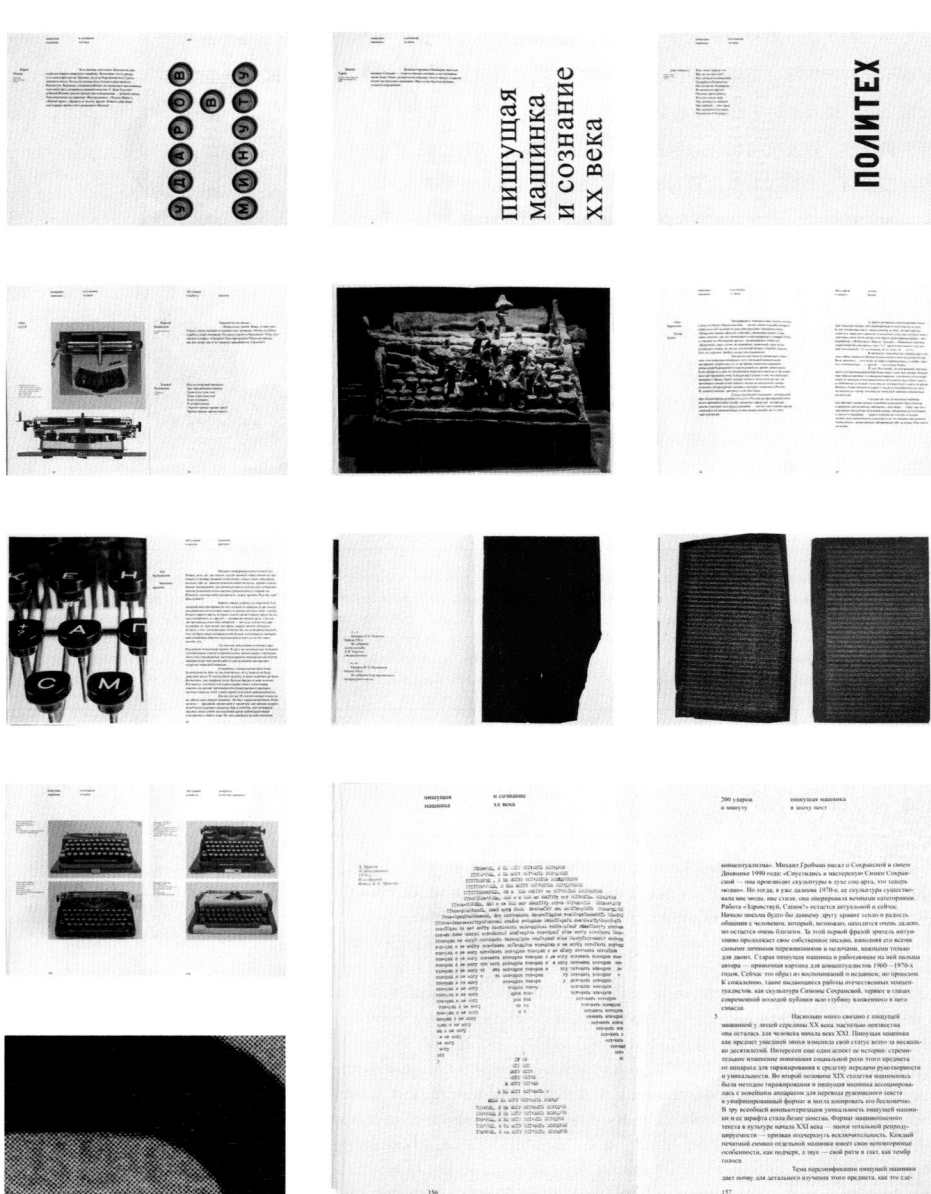

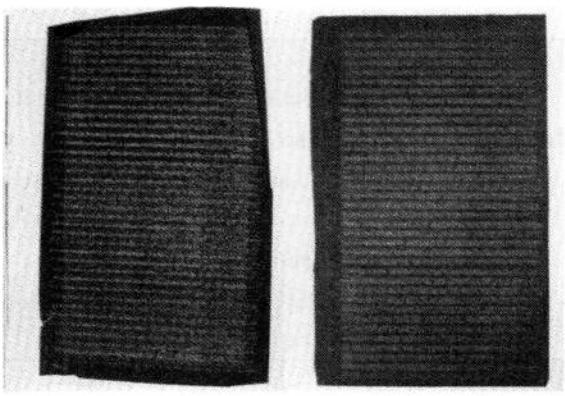

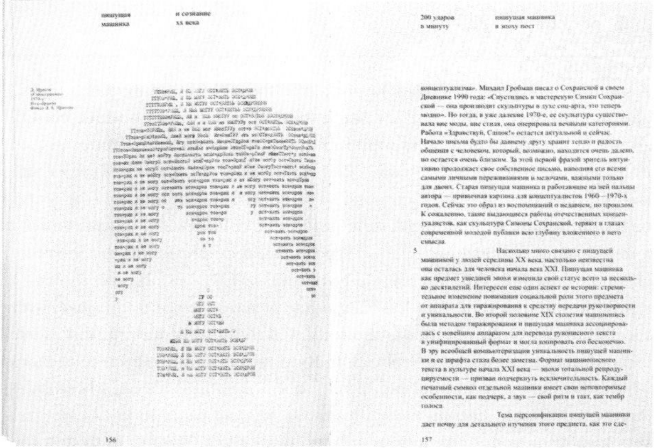

пишущая машинка — и сознание хх века

200 ударов в минуту — пишущая машинка в эпоху пост

концептуализма». Михаил Гробман писал о Сохранской в своем Дневнике 1990 года: «Спустились в мастерскую Симки Сохранской — она производит скульптуры в духе соц-арта, это теперь модно». Но тогда, в уже далекие 1970-е, ее скульптура существовала вне моды, вне стиля, она оперировала вечными категориями. Работа «Здравствуй, Сашок!» остается актуальной и сейчас. Начало письма будто бы давнему другу хранит тепло и радость общения с человеком, который, возможно, находится очень далеко, но остается очень близким. За этой первой фразой зритель интуитивно продолжает свое собственное письмо, наполняя его всеми самыми личными переживаниями и мелочами, важными только для двоих. Старая пишущая машинка и работающие на ней пальцы автора — привычная картина для концептуалистов 1960—1970-х годов. Сейчас это образ из воспоминаний о недавнем, но прошлом. К сожалению, такие выдающиеся работы отечественных концептуалистов, как скульптура Симоны Сохранской, теряют в глазах современной молодой публики всю глубину вложенного в него смысла.

5 Насколько много связано с пишущей машиной у людей середины XX века, настолько неизвестна она осталась для человека начала века XXI. Пишущая машинка как предмет ушедшей эпохи изменила свой статус всего за несколько десятилетий. Интересен еще один аспект ее истории: стремительное изменение понимания социальной роли этого предмета от аппарата для тиражирования к средству передачи рукотворности и уникальности. Во второй половине XIX столетия машинопись была методом тиражирования и пишущая машинка ассоциировалась с новейшим аппаратом для перевода рукописного текста в унифицированный формат и могла копировать его бесконечно. В эру всеобщей компьютеризации уникальность пишущей машинки и ее шрифта стала более заметна. Формат машинописного текста в культуре начала XXI века — эпохи тотальной репродуцируемости — призван подчеркнуть исключительность. Каждый печатный символ отдельной машинки имеет свои неповторимые особенности, как подчерк, а звук — свой ритм и такт, как тембр голоса.

Тема персонификации пишущей машинки дает почву для детального изучения этого предмета, как это сле-

156 157

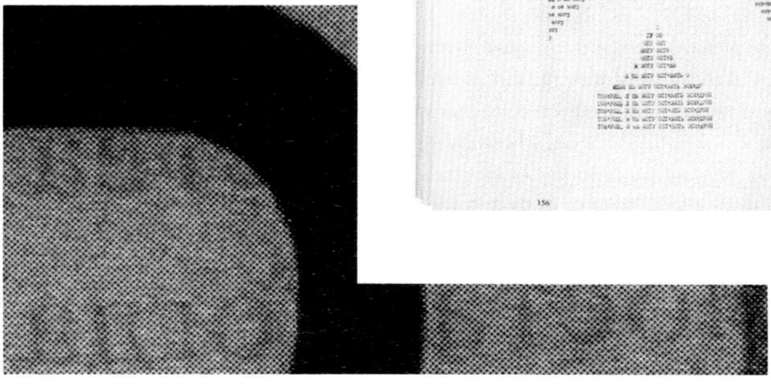

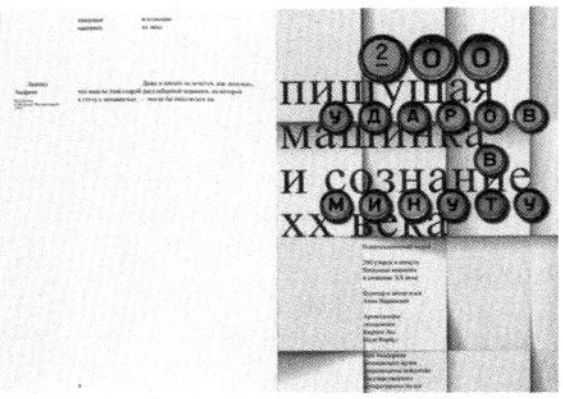
200
ударов
в
минуту
пишущая машинка и сознание XX века
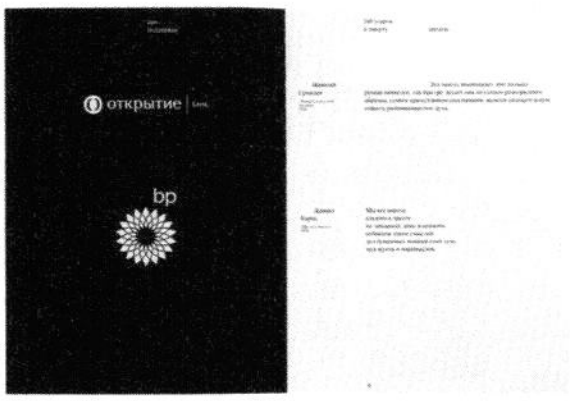
открытие
bp
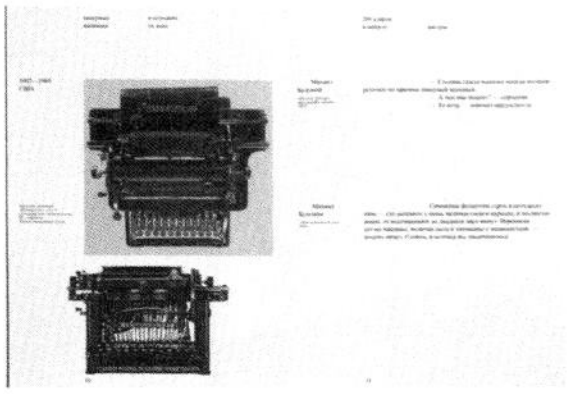
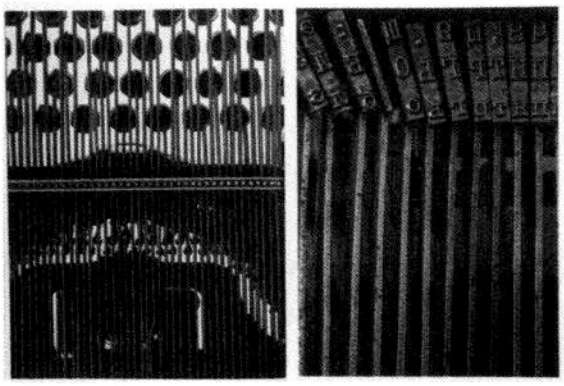

OLIVER

UNDERWOOD

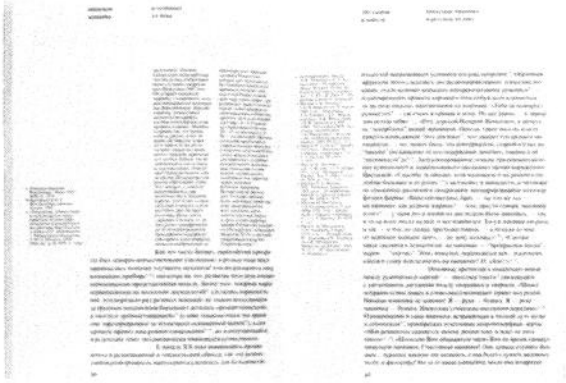
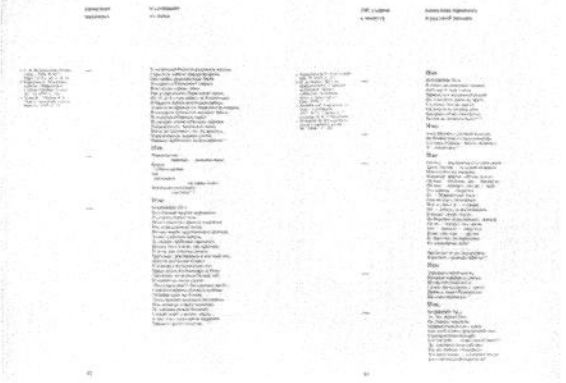
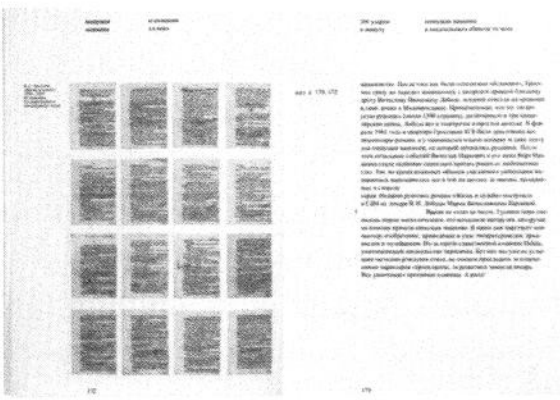
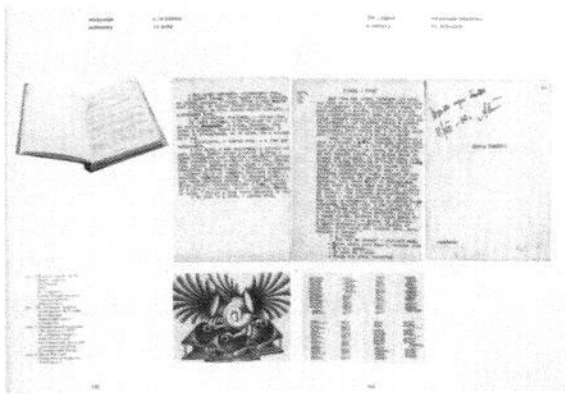

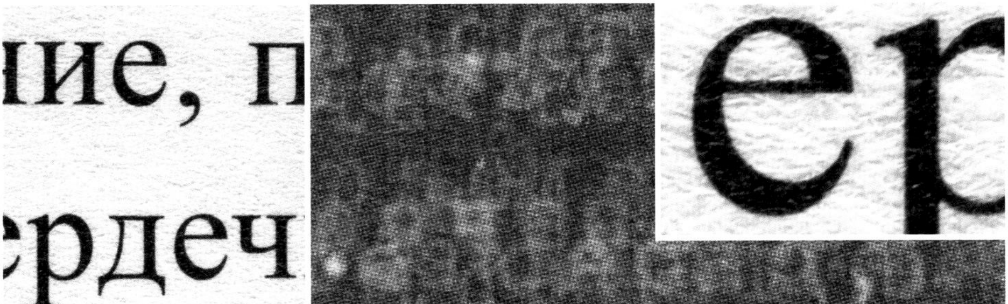
ние, п
ердеч
ер

Нина Лаврищева (Московский Музей современного искусства)

1 Пишущая машинка в эпоху ПОСТ

Пишущая машинка — одно из тех уникальных изобретений, которые сумели сформировать целый пласт мировой культуры. XX век — эпоха революций — стал временем не только широкого распространения пишущей машинки, но и смены области ее бытования. Из чисто утилитарной сферы она перешла в разряд уникальных объектов, символических предметов уходящей эпохи. Расширение области применения открыло огромный спектр новых возможностей использования пишущей машинки для всех видов искусства, а также неизбежно сместило точку зрения публики на сам объект. В первую очередь сам аппарат стал играть роль символического предмета, окруженного ностальгическим ореолом.

Буквально сразу же после вхождения в повседневный обиход пишущая машинка уже в конце XIX столетия стала средством создания изобразительных произведений. Техника машинописного рисунка, сделанного с помощью букв, пробелов и знаков препинания, стала новым видом графики. История искусства машинописной графики указывает на 1890-е годы, когда английской стенографисткой Флорой Ф. Ф. Стейси была создана одна из старейших серий изображений людей, животных и пейзажей. Можно спорить о художественной ценности работ этой любительницы, но она стала пионером в искусстве машинописной графики, которая уже к 1920-м годам захватила и профессионалов.

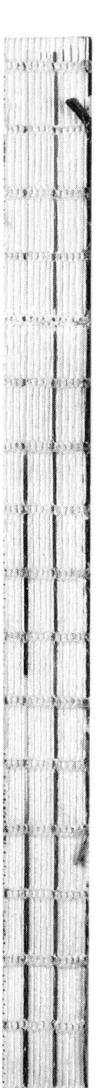

Итак, история западноевропейского изобразительного искусства открыла для себя пишущую машинку гораздо раньше, чем это сделали отечественные художники. В изображениях начала XX столетия пишущая машинка встречается ситуативно, только как предмет в общем, ничем не привлекательном интерьере. Например, у таких выдающихся рисовальщиков, как Николай Николаевич Купреянов, которые обыгрывали сложнейшие задачи нетривиального подхода к привычным ситуациям в карикатурах для журналов «Крокодил» и «Безбожник», встречается печатный аппарат. Ученик Кордовского, Петрова-Водкина и Остроумовой-Лебедевой, преподаватель ВХУТЕМАСа, Купреянов умел точно и емко сочетать все требования новой власти с удивительной точностью образов. Без преувеличения искрометной можно назвать его иронию над картинками из жизни новой России 1920-х годов. К этому времени он отказался от печатных графических техник

илл. 1 Н. Н. Купреянов
«Двадцать шестое
отделение милиции
г. Москвы», 1924—1925
Бумага, тушь, карандаш
Из собрания Московско-
го музея современного
искусства

илл. 2 В. А. Фаворский
«Чудо седьмое
"Пишущая машинка"»,
иллюстрация
к «Семь чудес»
С. Я. Маршака, 1929
Бумага, цветная
ксилография
Из собрания Государ-
ственной Третьяковской
галереи

B

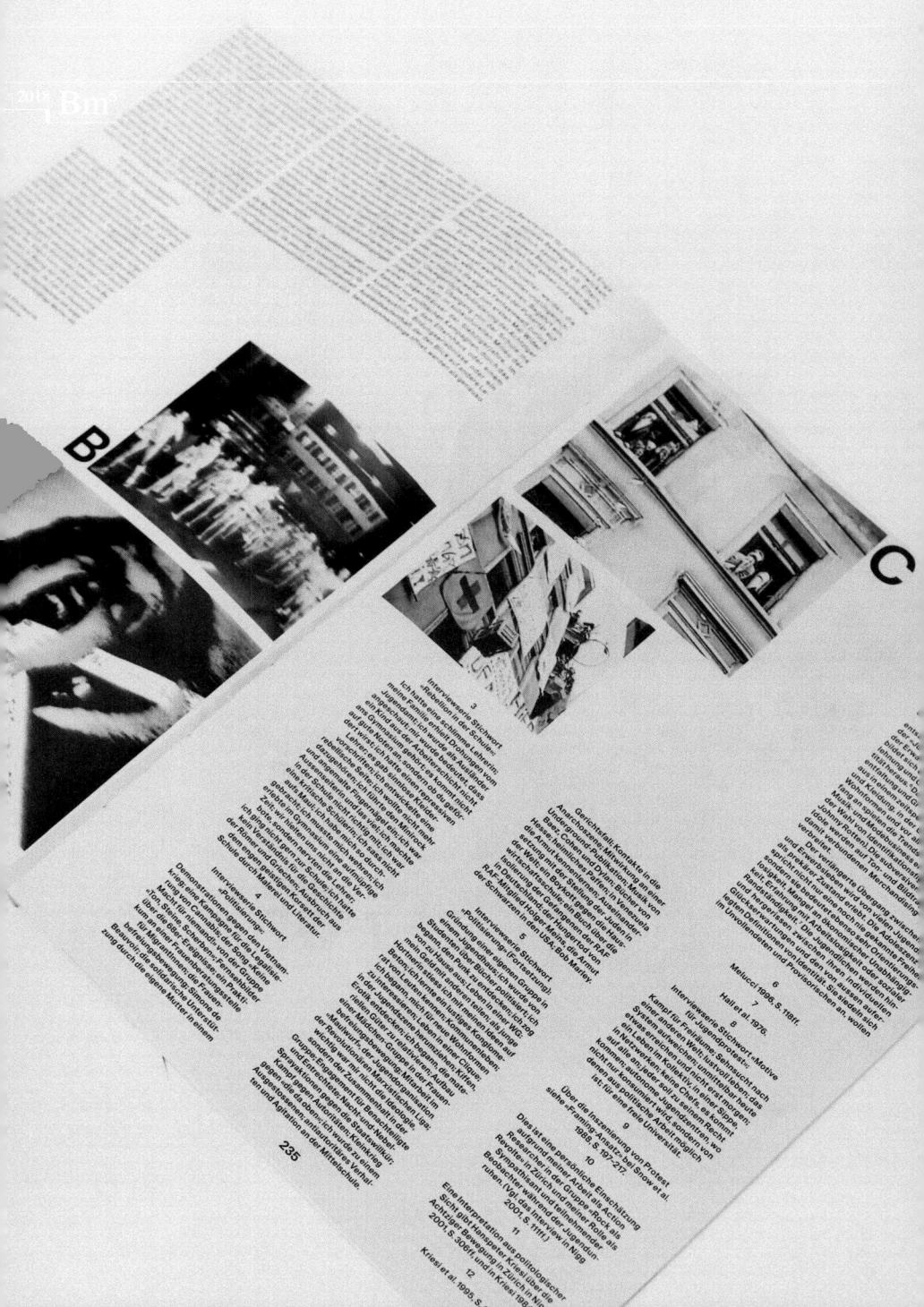

C

3
Interviewserie Stichwort
«Rebellion in der Schule»:
Ich hatte eine schlimme Lehrerin; meine Familie erhielt Drohungen vom Jugendamt; ich wurde als Ausländer angeschaut; mir wurde bedeutet, dass ein Kind aus der Arbeiterschicht nicht ans Gymnasium gehört; es kommt nicht auf gute Noten an, sondern ob du gefordert wirst; ich hatte einen repressiven Lehrer; es gab sinnlose Kleidervorschriften; ich entwickelte eine rebellische Seite; ich wollte nicht mehr dazugehören; ich führte den Minirock und angemalte Fingernägel ein; ich war Aussenseiterin und las viel; ich machte in der Schule nicht richtig mit; ich war eine kritische Schülerin; ich sass nicht aufs Maul; ich habe mich so durchgebracht; ich musste mich wehren; ich erlebte im Gymnasium eine aufmüpfige Zeit; wir hielten uns nicht an die Verbote, sondern nervten die Lehrer; ich ging nicht gern zur Schule; ich hatte kein Verständnis für die Geschichte der Römer und Griechen; Ausbruch aus dem engen geistigen Korsett der Schule durch Malerei und Literatur.

4
Interviewserie Stichwort
«Politisierung»:
Demonstrationen gegen den Vietnamkrieg; eine Kampagne für die Legalisierung von Cannabis; der Song «Keine Macht für niemand», von der Gruppe «Ton, Steine, Scherben»; Fernsehbilder über die 68er-Ereignisse; ein Praktikum an einer Frauenberatungsstelle für Migrantinnen; die Frauenbefreiungsbewegung; Simone de Beauvoir; die solidarische Unterstützung durch die eigene Mutter in einem Gerichtsfall; Kontakte in die Anarchoszene; Mitwirkung an einer Underground-Publikation; Musik von Baez, Cohen und Dylan; Bücher von Hesse; heimliches Paffen; in Venezuela die Armut kennenlernen; Auseinandersetzung mit der Stellung der Juden in der Welt; ein Boykott gegen die Hauswirtschaftsschule; ein Buch über die Isolierung der Gefangenen der RAF in Deutschland; der Hungertod von RAF-Mitglied Holger Meins; die Armut der Schwarzen in den USA; Bob Marley.

5
Interviewserie Stichwort
«Politisierung» (Fortsetzung):
Gründung einer eigenen Gruppe in einem Jugendhaus; ich wurde von Studenten über Bücher politisiert; ich begann, den Punk zu entdecken; ich zog von zu Hause aus; Leben in einer WG; mein Geld mit anderen teilen; als junge Hortnerin stiess ich mit meinen Ideen auf Beton; ich lernte ein lustiges Konglomerat von Leuten kennen; Kommunenleben; ich begann, mich für neue Wohnformen zu interessieren; Leben in einer Clique; in der Jugendszene herumziehen; Kiffen; Erotik entdecken; ich begann, die materiellen Güter zu relativieren; Aufbau einer Mädchen-Gruppe in der Frauenbefreiungsbewegung; Mitarbeit im «Maulwurf», der Jugendorganisation der Revolutionären Marxistischen Liga; wichtig war mir nicht die Ideologie, sondern der Zusammenhalt in der Gruppe; Engagement für Benachteiligte und Entrechtete; Nacht-und-Nebel-Sprayaktionen gegen die Staatswillkür; Kampf gegen Autoritäten; Kleinkrieg gegen «die da oben»; ich wurde zu einem Ausgestossenen; antiautoritäres Verhalten und Agitation an der Mittelschule.

ents…
der «Ju…
der Erwa…
bilder sich…
lehnung und G…
tität heraus.«D…
entfaltung und G…
aus in einem zeitsp…
und Kleidung, in der…
Wohnformen und vor al…
fang an spielen die Trends…
Musik- und Modebusiness e…
der Wahl von Identifikations…
Johnny Rotten). Die subversive…
Idole werden auf Ton- und Bildträ…
damit verbunden Merchandising
verbreitet.
Der verlängerte Übergang zwischen…
und Erwerbsleben wird von vielen Jugend…
als prekärer Zustand erlebt. Die Adoleszenz…
spricht nicht nur eine noch nie gekannte Freihe…
sondern sie bedeutet ebenso sehr Orientierungs-
losigkeit, Mangel an ökonomischer Unabhängigkeit, Erfahrung mit Arbeitslosigkeit oder sozialer
Randständigkeit.[7] Die Jugendlichen werden hin-
und hergerissen zwischen ihren individuellen
Glückserwartungen und den von aussen aufer-
legten Definitionen von Identität. Sie siedeln sich
im Unvollendeten und Provisorischen an, wollen

6
Melucci 1996, S. 118ff.

7
Hall et al. 1976.

8
Interviewserie Stichwort «Motive für Jugendprotest»: Kampf für Freiräume; Sehnsucht nach einer anderen Welt; lustvoll leben; das System aufweichen; unmittelbar heute etwas erreichen und nicht erst morgen; ein Leben im Kollektiv, in einer Sippe, in Netzwerken; keine Chefs, es kommt auf alle an; jeder soll zu seinem Recht kommen; autonome Jugendzentren, wo nicht nur konsumiert wird, sondern von denen aus politische Arbeit möglich ist; für eine freie Universität.

9
Über die Inszenierung von Protest siehe «Framing-Ansatz» bei Snow et al. 1988, S. 197–217.

10
Dies ist eine persönliche Einschätzung aufgrund meiner Arbeit als Action Researcher in der Gruppe «Rock als Revolte» in Zürich und meiner Rolle als Sympathisant und teilnehmender Beobachter während der Jugendunruhen. (Vgl. das Interview in Nigg 2001, S. 111ff.)

11
Eine Interpretation aus politologischer Sicht gibt Hanspeter Kriesi über die Achtziger Bewegung in Zürich in Ni… 2001, S. 306ff. und in Kriesi 198…

12
Kriesi et al. 1995, S. …

这是一本收集、整理和研究 20 世纪 80 年代初期苏黎世青年运动期间诉求传单的书籍。本书除了大量展示当年的传单，并未表达任何明确的立场，而是对传单的诉求表达、信息传达效果、设计风格、传单的载体意义、历史价值等方面进行探讨，研究纸面形式和内在传达出的自主权利。书籍的装订形式为裸脊锁线装，封面采用白卡纸对折后装订，印刷亮眼的专色红，呈现书名、目录、概述和版权信息，封面文字特别压凹处理。内页胶版纸先印一层本白色，为 A4 大小，与书籍纸张产生极微小差异，之后单黑印刷这些传单的内容。传单内容按照时间顺序分组，配有相关事件的时间和论述用以区隔。书中穿插四组各 8 面的纸张单色印刷，内容为相关研究者的论述，纸张呈现灰白斑驳的半透明肌理，与内页纸张在书口上形成明显差异。书籍封面坚硬的白卡纸与内页柔软的胶版纸对比明显，仿佛坚毅表面下的各色鲜活灵魂。|||||
||||||||||||||||||||||||||||||||||||||||||||||||||||||||||||||||||||||||||||

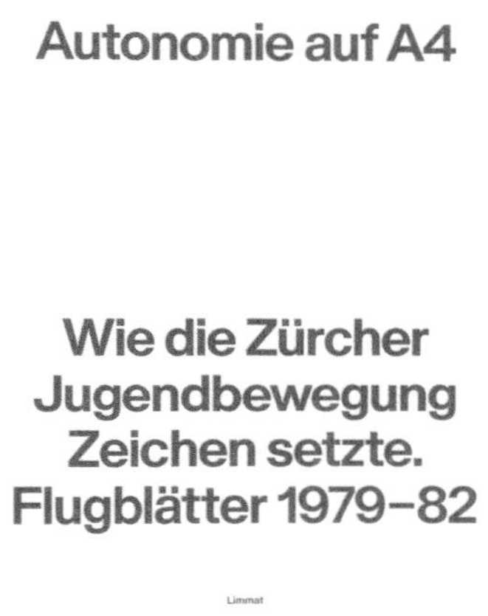
Autonomie auf A4
Wie die Zürcher
Jugendbewegung
Zeichen setzte.
Flugblätter 1979–82
Limmat

This is a book that collects, collates and studies leaflets of appeals during the Zurich Youth Movement in the early 1980s. In addition to displaying a large number of leaflets in those days, the book explores the appeal expression, the information transmission effect, the design style, the significance of the leaflet, and the historical value without taking sides clearly. It also studies the inherent right of self-determination conveyed by the external form. The book is bound by thread sewing with a bare spine. The cover is made of folded white cardboard, while the title, catalogue, overview and copyright information of the book are printed in bright red and embossed specifically. The inside sheets of offset paper are printed in white firstly, and then printed the contents of these leaflets in monochrome. The leaflets are grouped chronologically, attached with the details and comments on the relevant events. There are four groups of interspersed comment pages in this book and every group has eight sides. The comment pages present mottled translucent texture, which is obviously different from the offset paper. The hard cardboard cover contrasts with the soft offset paper inside, showing various and vivid spirits covered by tough appearance.

- 铜奖
- A4 上的自主权
- 瑞士

- Bronze Medal
- Autonomie auf A4
- Switzerland
- Atlas Studio (Martin Angereggen, Claudio Gasser, Jonas Wandeler)
- Peter Bichsel und Silvan Lerch
- Limmat Verlag, Zürich
- 318 × 241 × 24mm
- 1156g
- 288p

200 “Flugis” (flyers from the Zurich youth movement in the early 1980s) in their original size: Paperboard as endpapers at the front and back – without a real cover – is simply converted into a mock jacket – without a spine. The red title (with extra embossing) is punched in deeply. The open spine of the stitch-bound innerbook is programmatic. We can only imagine (although that itself is sufficient): beneath each facsimile sheet is a white, A4-sized area printed onto white uncoated paper, in order to a) keep as close to the original document as possible and b) generate only minimal contrast. Essays from contemporary witnesses, historians and other reflectors are inserted, evenly distributed, between the 288 pages. These essays are printed in a semibold, Helvetica-like font on a strange paper: it is slightly cloudy grey (presumably ex works because a raster cannot be seen by any stretch of the imagination), has a somewhat metallic sound, and reduces the sterility of the text pages (compared to the shimmering black/white dalliance of the Flugis); in particular, however, it is clearly visible on the sheared edges when the book is closed. Quotations in monumental type size provide repeated opportunities for a breather: “After we used the matrices, we burned them to cover our tracks.” “I refuse to be associated with a style.” “We didn’t receive any money from Moscow.” This book design achieves one particularly sophisticated rendering: the complexity of a simple flyer.

ROCK-MUSIK:
Plausch oder Profit?

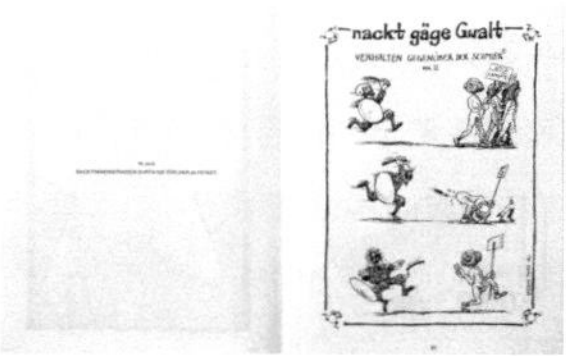
nackt gäge Gwalt

AKTIONS GRUPPE
NACKT IN DIE FREIHEIT
AUCH AN DIESEM HEISSEN SAMSTAG
ABZIEH DUREZIEH
en heissä summer
aber SUBITO

Ich fühlte mich wie ein Kind im Candystore
Elinor Burgauer

GESUCHT
wegen
TÖTUNG
der BEWEGUNG.......
KNABENSCHIESSEN
1980

Überleben

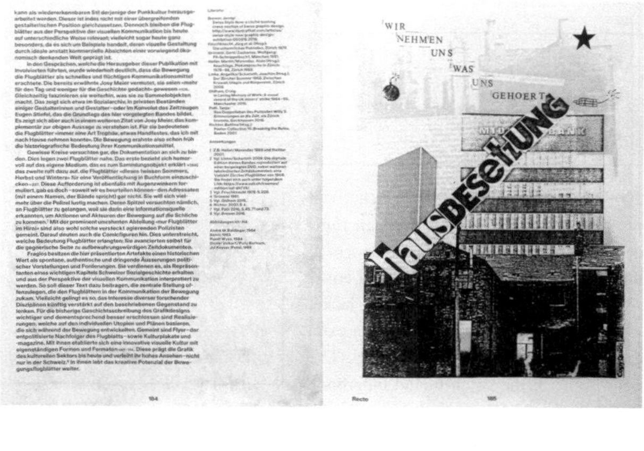
WIR NEHMEN UNS WAS UNS GEHOERT
hausbesetzung
184
Recto
185

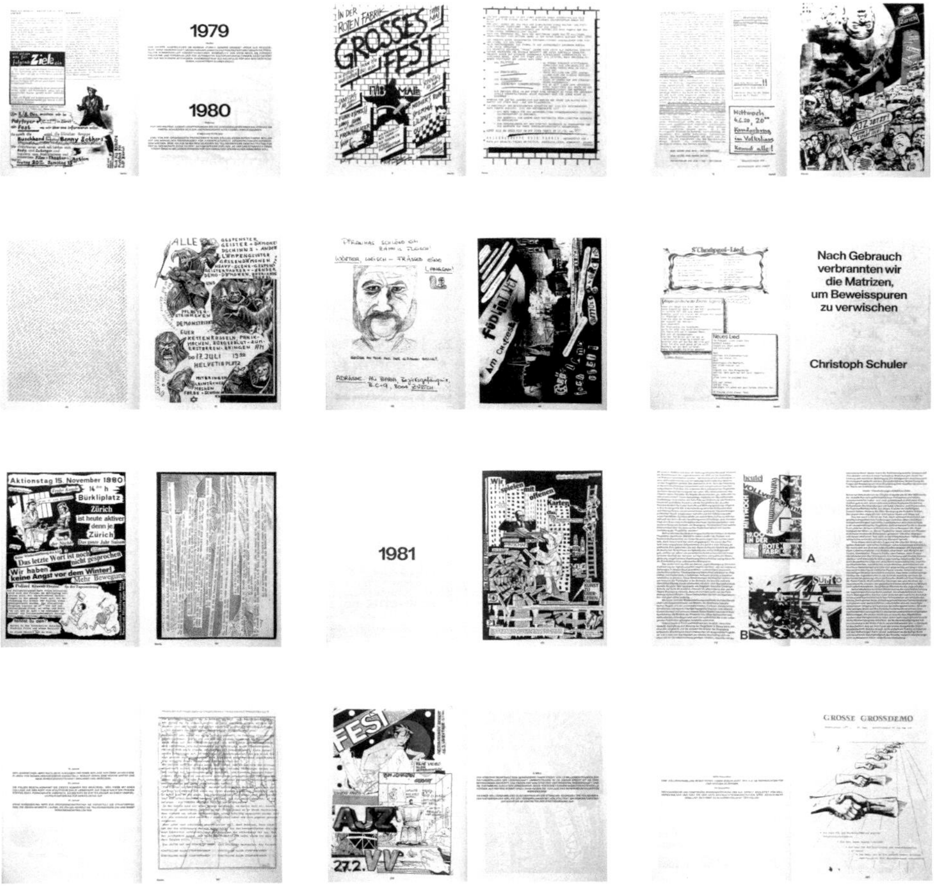
1979
1980
GROSSES FEST
1981
Nach Gebrauch verbrannten wir die Matrizen, um Beweisspuren zu verwischen
Christoph Schuler
Aktionstag 15. November 1980
Bürkliplatz
Zürich
Das letzte Wort ist noch nicht gesprochen
Wir haben keine Angst vor dem Winter!
Wir spielen mit offenen Karten
FEST
AJZ
27.2.
GROSSE GROSSDEMO

66 Mill
hat die
Stadt
geben. 22 Millionen, ein

sehenen Orten und zu unbestimmten Zeiten. Die Neuartigkeit des Protests verunsichert die Behörden und trifft sie unvorbereitet. Je grösser die Zahl der Mobilisierten, desto mehr wird die Legitimität der Behörden und deren Politik infrage gestellt. Die Behörden fühlen sich gezwungen zu handeln, da sie ihr Gewaltmonopol bedroht sehen. Die Militanz auf der Strasse hält so lange an, bis die Proteste zu Erfolgen führen. Unberechenbarkeit, Flexibilität und Innovation sind also die hauptsächlichen Merkmale der Achtziger Bewegung.[12]

Ausbreitung und Resonanz

Die Jugendprotestwelle der frühen Achtziger-Jahre begann in Amsterdam. Im März und April 1980 liefern sich hier Militär und Polizei auf der einen Seite und die sogenannten Kraaker (Häuserbesetzerinnen) auf der anderen Seite Strassenschlachten. Danach folgt der Ausbruch der Jugendunruhen in Zürich, die sich schnell in der Schweiz ausbreiten. Gleichzeitig springt ein Funken über die deutsche Grenze ins benachbarte Freiburg i. Br., wo es im Juni 1980 zur ersten grossen Konfrontation zwischen Häuserbesetzerinnen und der Polizei kommt. Es ist anzunehmen, dass die zeitlich etwas später einsetzenden Unruhen in weiteren deutschen Städten – Göttingen, Berlin, Nürnberg, Bremen u.a. – ebenfalls zur gleichen Protestwelle gehören. Die Kraaker in den Niederlanden und die AJZ-Bewegung→G in der Schweiz sind also gewissermassen Vorbilder für die Bewegungen in Deutschland.[13]

Dieses Dominomodell geht aber zu sehr davon aus, dass sich Protest wie ein Diffusionsprozess ausbreitet, und bedarf der Differenzierung[14]: AktivistInnen sind sich zwar bewusst, dass ihr Handeln in einem interregionalen und internationalen Zusammenhang steht, doch sie sehen die Verbreitung von widerständischen Botschaften nicht als linearen, sondern als wechselseitigen Prozess zwischen den verschiedenen bewegten Städten. Ereignisse stossen durch die Vermittlung der Medien auf regionale, nationale und internationale Beachtung. Handlungsrelevant kann die Medienberichterstattung aber nur dann werden, wenn in einer Stadt die Mobilisierung von Protest kurz bevorsteht oder bereits im Gang ist. Impulse von aussen können dann eine auslösende oder verstärkende Wirkung auf den Mobilisierungsprozess haben. Umgekehrt werden die Mobilisierungsprozesse in den Epizentren des Protests durch die Ereignisse an deren Peripherie beeinflusst. Persönliche Kontakte zwischen den verschiedenen Bewegungen und die Vermittlungsarbeit der Medien begünstigen dann die Entstehung von informellen interregionalen und transnationalen Netzwerken.[15] Jede Stadt hat einen anderen politischen und subkulturellen Hintergrund, vor dem sich die lokale Bewegung entfaltet. Jede Bewegung entwickelt eine eigene Identität, die sie auf keinen Fall durch den Zustrom von PolitaktivistInnen aus anderen Städten aufgeben will. Hilfsangebote aus anderen Städten sind so lange willkommen, wie die eigene Autonomie gewahrt bleibt. Wichtig für die Mobilisierung von Protest ist vor allem der Austausch von Erfahrungen auf den Ebenen der Ziele und Ideen, der Slogans sowie der Organisations- und Handlungsformen.[16]

Erfahrungen mit Militanz und Repression

Durch die Bewegung politisierte Jugendliche realisieren, dass eine Aufbruchstimmung lange anhalten und ein Gefühl der Zugehörigkeit und der Neuorientierung vermitteln kann. Angehörige der zweiten Generation von italienischen EinwandererInnen lernen sich durch die Bewegung besser kennen. Die Bewegung stellt eine gesellschaftliche Praxis dar, die den AktivistInnen eine Palette von Entfaltungsmöglichkeiten bietet: Sie können eine Bar oder eine Beiz betreiben, in Arbeitsgruppen mitarbeiten und bei Entscheidungen mitreden. Viele vertreten zum ersten Mal in der Öffentlichkeit ihre Anliegen, werden sich ihrer Rechte im Umgang mit Polizei und Justiz bewusst und sammeln Erfahrungen, wie mit Behörden und anderen VertreterInnen von staatlichen Institutionen am besten zu verhandeln ist. Der Umgang mit eigenen Medien, die Kommunikation nach aussen und die Vernetzung mit Gruppen in anderen Städten vermitteln das so sehnlich herbeigewünschte Gefühl von einem Leben im Hier und Jetzt.[17]

Zwiespältig sind die Erfahrungen in den Autonomen Jugendzentren (AJZ), wo es oft ruppig zu und her geht. Viele werden desillusioniert und wenden sich von ihren idealistischen Positionen ab. In den AJZs gibt es keine gut funktionierenden Betriebsstrukturen, die Verantwortung ruht auf einigen wenigen AktivistInnen. Es gibt plötzlich unlösbare Konflikte: Wie werden die spärlichen Betriebsgelder verteilt? Wie begegnet man den bewaffneten Dealern von harten Drogen? Wie ist die Gewalt gegen Frauen zu verhindern? Immer wieder machen Frauen in der Achtziger Bewegung die Erfahrung, dass sie typische Frauenjobs übernehmen oder in diese von den Männern hineingedrängt werden. Zudem wird beklagt, dass immer mehr AktivistInnen von der Instandstellung und dem Betrieb der AJZ so sehr absorbiert werden, dass die weiteren Ziele der Bewegung aus den Augen verloren gehen. Das Prinzip «Alle machen alles» ist auf die Dauer wenig effizient; fähige Leute werden zurückgebunden und Mittelmässigkeit setzt sich durch. Ebenfalls als negativ wahrgenommen werden Machtgames, die Verschleierung von sozialen Unterschieden zwischen den verschiedenen Gruppen und Clans sowie eine tendenziell feindliche Haltung gegenüber StudentInnen und Intellektuellen. Nach dem Ende der Achtziger Bewegung im Jahr 1982 fallen viele in ein Loch, haben Mühe im Umgang mit der allgegenwärtigen Repression, sehen ihr soziales Netz zerstört und verlieren ihre politische Perspektive.

Den Jugendunruhen wird in allen Schweizer Städten mit scharfer Repression begegnet, da die Behörden Ruhe und Ordnung sowie den Schutz des Privateigentums bedroht sehen. Allein zwischen Mai 1980 und August 1981 sind in Zürich mehr als 3800 Personen festgenommen worden. Von den Verhafteten müssen schliesslich 200 als Angeklagte vor Gericht erscheinen. Neben einigen Freisprüchen werden Bussen zwischen 50 und 1000 Franken und vorwiegend bedingte Freiheitsstrafen zwischen 14 Tagen und 7 Monaten verhängt. Die Höchststrafe von 14 Monaten Gefängnis muss ein Aktivist absitzen, der als «Herr Müller»→F auf witzige Art und Weise eine Fernseh-Talkshow gesprengt hat und in der Zürcher Bewegung sehr populär geworden ist. Mit ihren Urteilen will die Justiz eine abschreckende Wirkung erzielen. Nur schon die Teilnahme an einer nichtbewilligten Kundgebung genügt, um den Tatbestand des Landfriedensbruchs zu erfüllen.[18]

Wie nehmen AktivistInnen ihr damaliges persönliches und gesellschaftliches Umfeld wahr und welche Erfahrungen machen sie mit den Medien, der Polizei und der Justiz? Positiv bewertet wird die direkte Unterstützung durch die eigene Familie, durch FreundInnen und Bekannte. Eltern machen sich Sorgen um das Wohlergehen ihrer Kinder, andere lehnen die Revolte ihrer Kinder radikal ab oder setzen sich kritisch mit ihnen auseinander. Eltern ohne Schweizerpass leben in Angst, da sie wegen der politischen und zum Teil illegalen Aktivitäten ihrer Kinder befürchten, ausgewiesen zu werden. Die öffentliche Meinung wird vorwiegend als verständnislos erlebt und für Jobber ist es schwierig, Arbeit zu finden.[19] Auch die politischen Behörden werden als ablehnend wahrgenommen. Ihnen wird mangelndes Fingerspitzengefühl im Umgang mit der schwer fassbaren Bewegung vorgeworfen.[20]

Die Offenheit des politischen Systems gegenüber den Forderungen der Achtziger Bewegung ist von Stadt zu Stadt verschieden.[21] Generell findet die Achtziger Bewegung aufgrund ihres unkonventionellen und militanten Auftretens nur wenige Verbündete im etablierten Politikbetrieb. Der gemässigte Flügel der Bewegung wird von den Behörden als Verhandlungspartner für die Nutzung von staatlichen Ressourcen halbwegs anerkannt, der sich radikalisierende militante Flügel hingegen erfährt wegen seiner Fundamentalopposition gegen Staat und Gesellschaft die volle Repression. Das politische System selbst bleibt trotz Rochaden zwischen sozialdemokratischen und bürgerlichen PolitikerInnen in den Lokalbehörden grundsätzlich stabil.

Die Zusammenstösse→E mit der Staatsgewalt sind traumatisch. Physische Verletzungen, Einkesselungen, Tränengas- und Gummigeschossangriffe, Verurteilungen und Zensur werden nicht so leicht vergessen gehen.[22] Die Bewegung in den verschiedenen Schweizer Städten reagiert auf die zunehmende Gewalt und wachsende Ohnmacht mit Wut und Militanz. In den Vollversammlungen geht es immer wieder um die Frage, ob und wenn ja, in welcher Form gewalttätiger Widerstand gegen die Staatsgewalt berechtigt sei. Generell wird Gewalt gegen Personen abgelehnt und Sachbeschädigung in jenen Fällen als legitim erachtet, wo staatliche Autorität den letzten Spielraum für friedliche Aktionen und Demonstrationen aufhebt.[23] Als die Repression obsiegt und die Bewegung ihre Mobilisie-

zuschulden kommen liess; es war schwierig, Arbeit zu finden: «Aha! Du gehörst auch zu denen!»

20

Interviewserie Stichwort «Reaktionen von den politischen Behörden»:

Die Stadtbehörden hätten es kommen sehen müssen; das Verhandlungsergebnis mit den Behörden war ein Kuhhandel; wir erhielten einen Raum und hatten nun ruhig zu sein; wir Jungen hatten null Vertrauensbonus; der Regierungsrat fuhr mit schwerem Geschütz gegen die Ethnologen auf; die Stadtregierung liess die Polizei hemmungslos auf DemonstrantInnen los; einmal gab es eine Verhaftung von 300 bis 400 DemonstrantInnen, weil die Behörden herausfinden wollten, woher diese Bewegung kam; die Radikalen (Freisinnigen) waren bei uns in Lausanne an der Macht und betrachteten uns als Abschaum – als petits merdeux.

21

Vgl. dazu den Aufsatz von Tackenberg/Wisler in Nigg 2001, S. 369ff.

22

Interviewserie Stichwort «Repression»:

Ich erhielt eine bedingte Gefängnisstrafe; ich habe heute noch Angstträume; ich sass immer wieder in Präventivhaft; Einkesselung durch die Polizei, Verhaftungen, Ängste, Busse wegen Landfriedensbruch; als Italienerin hatte ich Angst vor Ausschaffung; der Justizapparat investierte viel, um die Leute verurteilen zu können; die, welche sich den Prozessen verweigerten, wurden härter bestraft; systematischer Einsatz von Tränengas und Gummigeschossen; schlimme Kopf- und Augenverletzungen; südamerikanische Zustände; Beschlagnahmung von Videoaufnahmen durch die Staatsanwaltschaft; Berufsverbote; an den Mittelschulen und an der Uni herrschte ein repressives Klima.

23

Exkurs: Die Diskussion um das Widerstandsrecht hat in Europa eine lange Tradition. Die Stellung der Reformatoren zum Widerstandsrecht war uneinheitlich: Während Calvin, Melanchthon und Zwingli ein aktives Widerstandsrecht in engen Grenzen

G

und mit unterschiedlicher Begründung und Abstufung anerkannten, lehnte Luther es anfangs grundsätzlich ab und liess es später nur ausnahmsweise zu. Mit der amerikanischen Declaration of Independence und der Französischen Revolution mündete das Widerstandsrecht in die Bewegung der Menschen- und Grundrechte. Der deutsche Rechtspositivismus des 19. Jahrhunderts liess für ein Widerstandsrecht keinen Raum, da dessen Geltung über dem gesetzten Recht stehender Normen abgelehnt wurde. Die Rechtssysteme in Frankreich, Grossbritannien und den USA hielten hingegen am Widerstandsrecht grundsätzlich fest, begrenzten es aber unterschiedlich. Politische Bedeutung gewann das Widerstandsrecht wieder im 20. Jahrhundert, einerseits vor allem durch die neuartigen Phänomene des Unrechtsstaats und des Totalitarismus, die dem Gedanken des ethisch berechtigten Widerstands zu neuer Geltung verhalfen (vgl. dazu Meyers Enzyklopädisches Lexikon 1979; Stichwort «Widerstandsrecht»). Seit 1968 wird in den neuen sozialen Bewegungen das Widerstandsrecht gegen alle Formen von Diskriminierung immer wieder neu diskutiert. Die Frage der dabei angewandten Mittel bleibt umstritten.

24

Interviewserie Stichwort «Positive Auswirkungen der Achtziger Bewegung»:

Prägend war dieses Heimatgefühl, mich dazugehörig zu fühlen; trotz Repression blieb ich ungebrochen; Verbundenheit und lange Freundschaften waren für mich wichtig; das Gefühl, am Puls der Zeit zu sein; ich erlebte

first

这是伊朗裔艺术家史拉娜·沙巴兹（SHirana Shahbazi）的展览作品集。书中展示艺术家创作摄影和抽象绘画等共计47件作品。本书的装订形式为锁线硬精装，封面玫红色布面裱覆卡板，文字烫白，一块巨大的暗红色仿皮面材料裱贴在封面上，遮挡部分文字。内页前后总计8页胶版纸张陈述展览理念、作品图注、后记和版权信息，其余均为哑面铜版纸四色印刷，展示作品。作品的展陈并无非常明确的分类结构关系，看似没有任何关联的黑白、彩色、多重曝光，甚至修改过色彩的各种照片与抽象的几何绘画作品并置。艺术家试图通过这些看似毫无联系的作品的并置，并且不给予任何解释的展示方式，让观展者或读者自己去寻找作品之间的联系。由于每次展览和画册的作品组合方式不同，都会产生完全不同的作品联系。就好比电影里的蒙太奇手法，当两张图片相遇的时候，就产生了“第三张”图像。图像之间的关联可以是色彩、构图、线条、疏密、带给人的感觉，甚至语意联系，这些都交由观看者自己联系和赋予了。

first
things
first
shiran
shahb

This is an album of works created by Iranian artist Shirana Shahbazi. In total, the book presents 47 works including photographs and abstract paintings. The hardcover book is bound by thread sewing with a rose cloth cover mounted by graphic board, and white hot stamping is used for the text. Huge dark red artificial leather is pasted on the cover and covers part of the text. There are eight pages to state the concept of the exhibition, artwork illustration, postscript and copyright information. The rest are printed in four colors on matte coated paper to display the works. The works are not displayed by a clear classification. Black and white photos, colorful photos, and abstract geometric paintings are mixed together. The artist exhibits these seemingly unrelated works without giving any explanation; hopefully the viewers and the readers can find the hidden connection between the works. Due to the different ways of combination of works in each exhibition and album, they are connected differently, just like the montage technique in the film. When two pictures encounter, the “third” image will be produced. The relationship between images can be related to color, composition, line, density, feeling, and even meanings, which are all discovered by the viewers themselves.

- 荣誉奖
- 重中之重
- 瑞士

- Honorary Appreciation
- First things first
- Switzerland
- NORM, Zürich
- Shirana Shahbazi
- SternbergPress, Berlin
- 326×245×9mm
- 379g
- 24p

Fuchsia-coloured book linen coats the cover of this large, thin volume. A piece of dark-red imitation leather is flush-mounted in an embedded area – a sign taking the form of a picture, you could say, that cuts off the printed white letters of the title. The book contains 4 pages of blue paper as an integrated endpaper, 8 pages of accompanying text on light, uncoated paper, and 16 pages of plates on thin offset paperboard. The gutters on the text pages, meaning the white paper frame around the text, extending to the edge of the page, also form the frame for the arrangement of the pages with the pictures. There are only two rules for the determination of picture size and position – without the use of a layout grid: each photograph touches the imaginary outer frame and is oriented on its neighbour according to aesthetic criteria. In other words, each individual picture is meticulously balanced on the page, enters into formal dialogues, and each keeps its distance from the others. The very different, even contradictory visual styles converge to find unequal pairs or groups. In a fascinating way, each picture succeeds in awakening the awareness of hidden aspects in its neighbouring picture. Each picture becomes, as it were, the teacher of its counterpart (or companion). The conglomeration that is initially distracting, if the first glance remains the last, disappears upon further observation, because the visual analogies increase, and meanings are generated, every time you look again. This publication does not serve as merely the catalogue for the exhibition of the same name. It is an independent work that develops its own strength.

first things first
shirana shahbazi

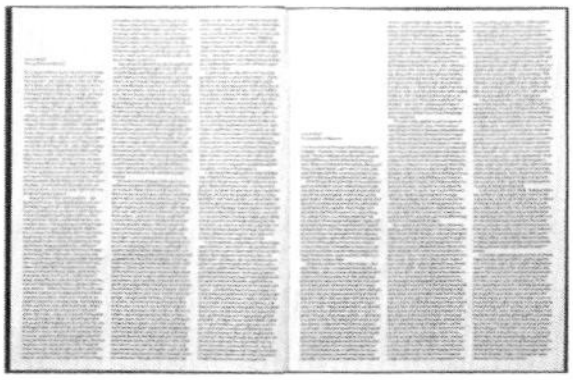

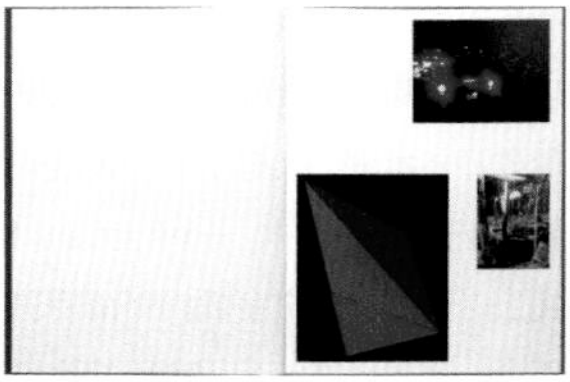

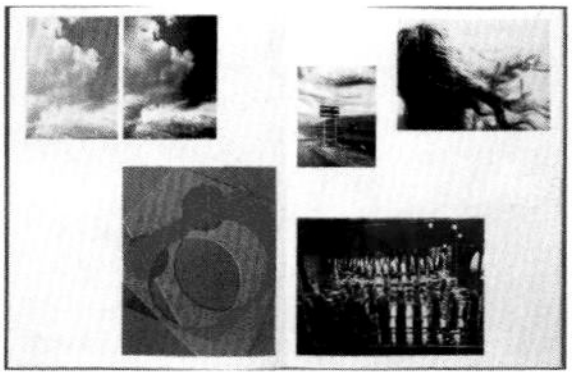

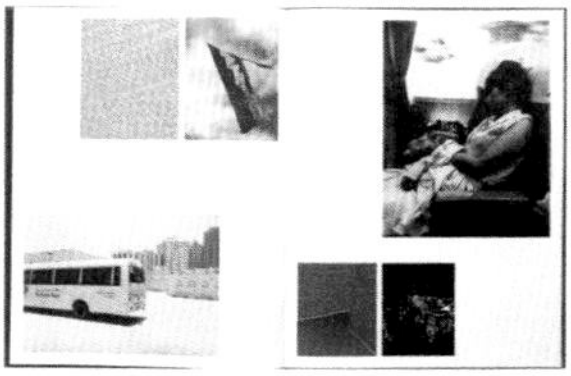
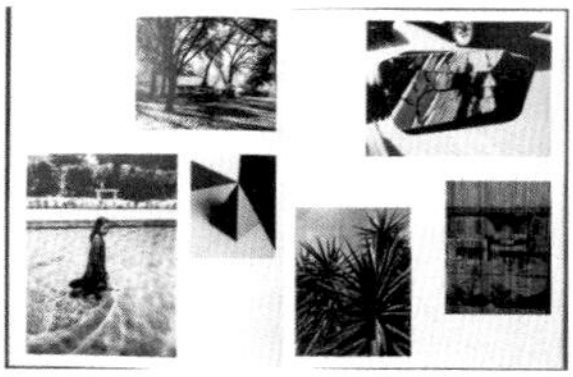

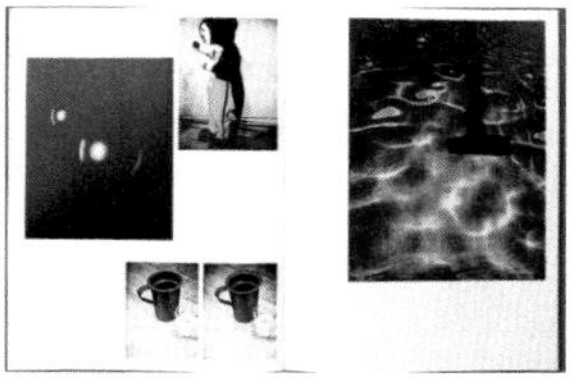

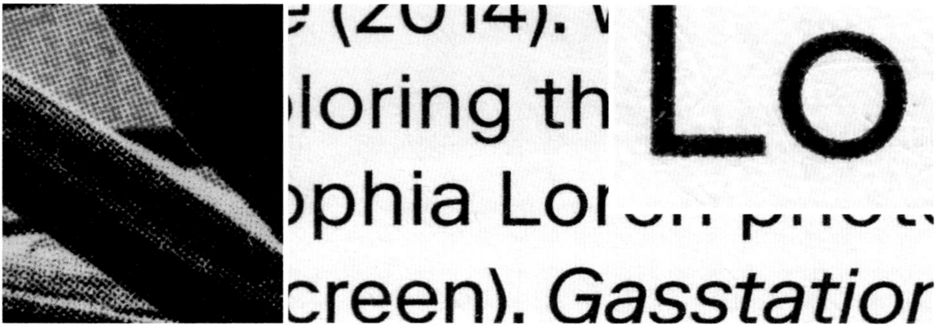
e (2014). V
loring th
LO
phia Lor
creen). Gasstatior

indigo
madder
48
indigo
madder
49

这是荷兰编织艺术家、设计师克劳迪·扬斯特拉（Claudy Jongstra）的作品集。书中展示了艺术家种植和采集染料、染织、拼接创作、布展等作品的产生过程，也包含大量与他人的对谈记录。本书为骑马钉，书籍被装在单黑印刷的一团织物图形的透明塑料袋中。封面与内页并无材质和工艺上的差异，对开的胶版纸单面四色印刷后通过两次折叠直接装订，书籍未清三边，使得上书口连在一起，翻口部分留有未裁切的自然斜角。书籍并不只是陈列创作过程的照片，与此同时也在进行着配色实验。封面和封底使用艺术家姓名字母交叠组合，封底采用传统CMYK四色，而封面则是采用了染织用的靛蓝色、茜草红色、洋甘菊黄色和几乎与黑色无异的深靛蓝色来替代。内页照片除少量使用CMYK外，大部分都被替换成与封面相同的染织用色，使得照片呈现与艺术家作品接近的偏深的暖色调。翻口处陈列本页所用色彩，照片下四色版被拆分出来，两两组合或互换色版，形成各种独特的配色效果，是艺术家作品的纸面实验。||||||||||||||||||||||||

This is an album of works created by a Dutch designer and knitting artist Claudy Jongstra. It shows the production process of planting, collecting dyes, dyeing, weaving, stitching, and arranging exhibitions, as well as conversations with others. The book is bound by saddle stitch, enclosed in a see-through plastic bag printed in black. There is no difference in material and technology between the cover and the inside pages. The offset paper is one-side printed in four colors and then bound after folding twice. The three edges are untrimmed that the upper part is connected together, leaving an uncut natural bevel on the fore-edge. The book not only displays the photos of the creation process, but also has experiments with color matching. The back cover is using traditional CMYK four-color printing, while the cover is printed in colors for dyeing and weaving, such as indigo blue, madder, chamomile yellow and Dark Indigo blue that is almost the same as black. Most of the interior photos are printed in the same colors for dyeing and weaving as the cover, which make the photos in warm tone similar to the artist's works. The colors used in each page are displayed near the fore-edge. There is a four-color plate under the photo, every two colors are combined to form various unique color matching effects. This is the paper experiment of artist's works.

- 荣誉奖
- 克劳迪 · 扬斯特拉
- 荷兰

- Honorary Appreciation
- Claudy Jongstra
- The Netherlands
- Irma Boom
- Louwrien Wijers, Lidweji Edelkoort, Laura M. Richard, Marietta de Vries, Pietje Tegenbosch
- nai010 publishers
- 350×260×20mm
- 960g
- 178p

Wire spine binding cuts through the thick bundle of 180 sheets. Somehow it appears unfinished. As it has not been cut back, the edges of the matte India paper jump to and fro; at the front the inner pages shoot forward by half the binding thickness, and the folded sheets are closed at the top. The book cover is filled with the name of the artist in capital letters. Columns of upside-down letters are printed over with vertical lines in different colours. The letters are printed at right angles and on top of each other. All kinds of directions and colour combinations are encountered. Along the edge there are small squares, with the names of the colours written in tiny words. Like those on final artworks prior to printing, which the bookbinder normally cuts off. An unusual continuation of this theme inside the book is promised by the four colours: indigo blue, madder red, chamomile yellow and black indigo. The Dutch artist creates large-dimensioned textile pictures. Her work begins with her own sheep. In her garden she grows plants in order to extract dyes. This book shows the catalogue of works at the end on only four pages, while inside you become the companion of an alchemical process. Photographs lead through the garden, show inflorescences, helpers gathering their produce, dyeing processes, samples of coloured wool. And time and again, miniatures of the corresponding colours are printed under the colour images, with didactic intentions – in the colours indigo, madder and chamomile, which are unusual for four-colour printing. Even the colours are interchanged. This book design work serves as a metaphor for the exploration of unfamiliar shades. Rarely have artistic media come into such close contact with one another..

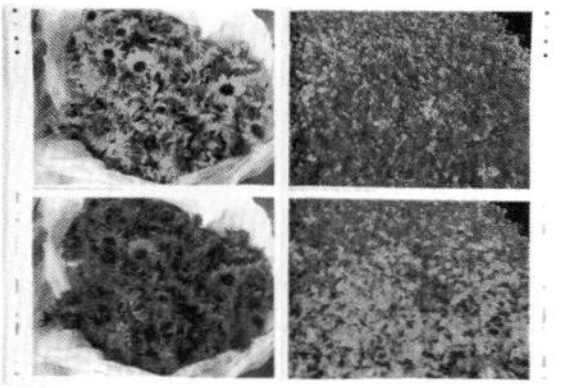

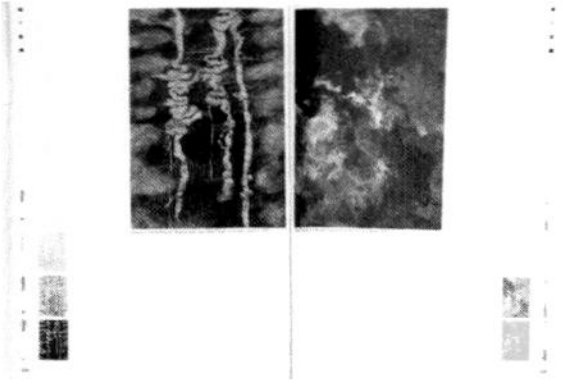

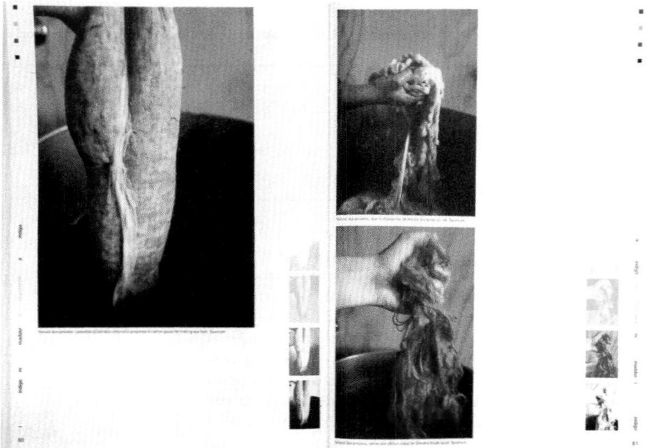

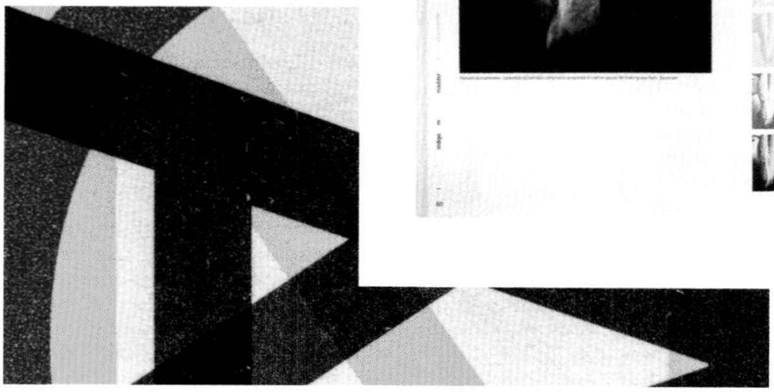

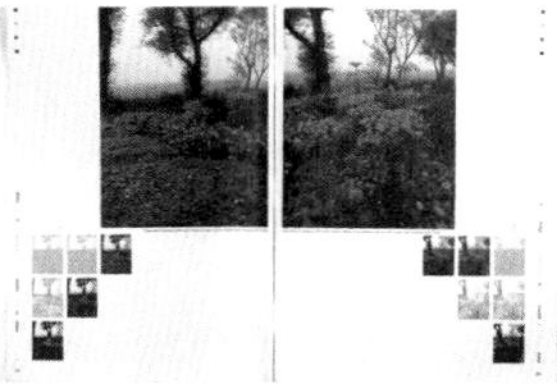

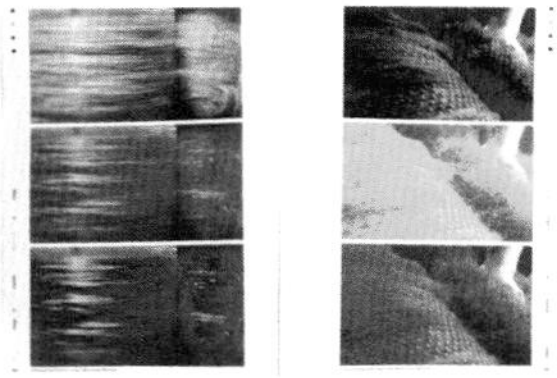

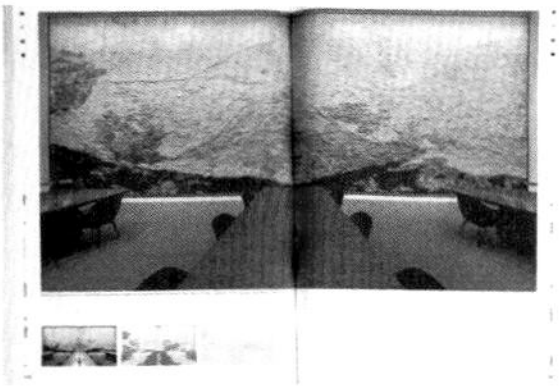

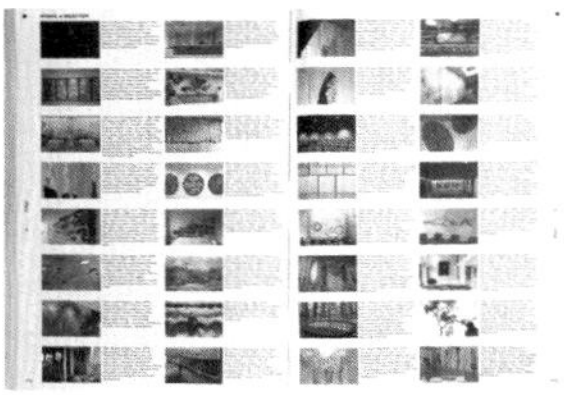

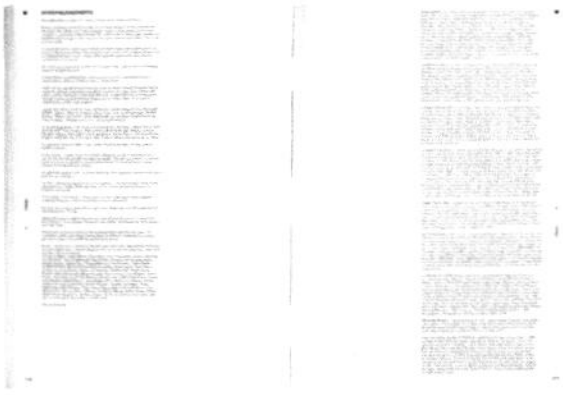

Louwrien:

欽定四庫全書

《茶典》是中国古代典籍《四库全书》的一部分，由 8 卷茶书经典集结编成。本书的版本在此基础上增加了十大名家以“茶”为主的书画作品，并邀请国内知名学者撰写相关文字，旨在制作出一部“爱茶人的典籍”。本书的装订形式为锁线胶装右翻本，封面选用茶色水洗牛皮纸凸版印刷，形成微微下凹的书名与印章。内页主要采用圣经纸红黑双色印刷，按原典籍影印版直接呈现。从茶叶生产的历史、源流、现状、生产技术，到饮茶技艺、茶道原理等各方面进行叙述。书中穿插采用宣纸四色印刷的名家画作和画作解读进行阅读节奏的调整。每一篇章前的极细线工笔白描小画，呈现各色茶具，与书中图录相得益彰。上下书口与翻口刷金色，体现出优雅闲适的茶书典籍趣味。

茶典

Tea Canon is a part of the ancient Chinese classic *the Four Branches of Literature*, which is composed of eight volumes of classic books about tea. On this basis, the edition adds paintings and calligraphy works by ten famous artists, and invites well-known scholars in China to write relevant reviews, aiming to produce a “classic book for tea lovers”. The rightward book is perfect-bound with thread sewing. Dark brown kraft paper is used for the cover, which is embossed to form the slightly concave title and stamp. The inside pages are printed in red and black on India paper, following the same way as the photolithographic original books. The book contains varied contents such as the history, origin and development, current situation, production technology, tea art and principles of tea ceremony. Rice paper on which famous paintings and interpretations are printed in four colors, are interspersed to adjust the reading rhythm. Each chapter is preceded by a claborate-style painting, presenting a variety of tea sets that complements the illustrations in the book. With gilt-edging technique, the classic book successfully presents elegance and comfortable style.

荣誉奖	Honorary Appreciation
茶典	Tea Canon
中国	China
潘焰荣	Pan Yanrong
陆羽（唐）等	Lu Yu (Tang Dynasty) and others
商务印书馆	The Commercial Press
	211×140×29mm
	665g
	772p

Soft, supple, heavy – this book feels like Chinese tea culture in concentrated form. The rough paperboard used for the cover has a natural colour that alternates between shades of brown, green and beige, which may result from the shadow effect in the micro-relief structure. The coated endpapers at the front and back reinforce the toughness of the cover, which astonishingly forms a rounded spine while the innerbook remains straight. The eight historical books from the imperial library are presented on ultra-thin, yellowish-white paper. The vertical characters are printed in a traditional, red line grid that is given modern inter- pretation on the separator pages: black hairlines structure the captions in a manner that produces interesting areas of paperwhite. These separator pages are made of a special Chinese paper, very finely ribbed, which responds to the touch almost like fabric. And so these pages also display pictures – admittedly as high-resolution reproductions here – for which Xuan paper is intended: calligraphic works and ink painting. Their sophisticated folding with a shortened side and two sheets that are closed off at the front compels you to use your hands more than usual when looking at the pictures, instead of just turning over the pages as normal. This decelerates any examination of the work, and heightens the pleasure involved. The matte gold edging all around the book intensifies the aura radiated by this gem of a book. It seems refined with a velvety, shining patina as if through ritual use.

茶典

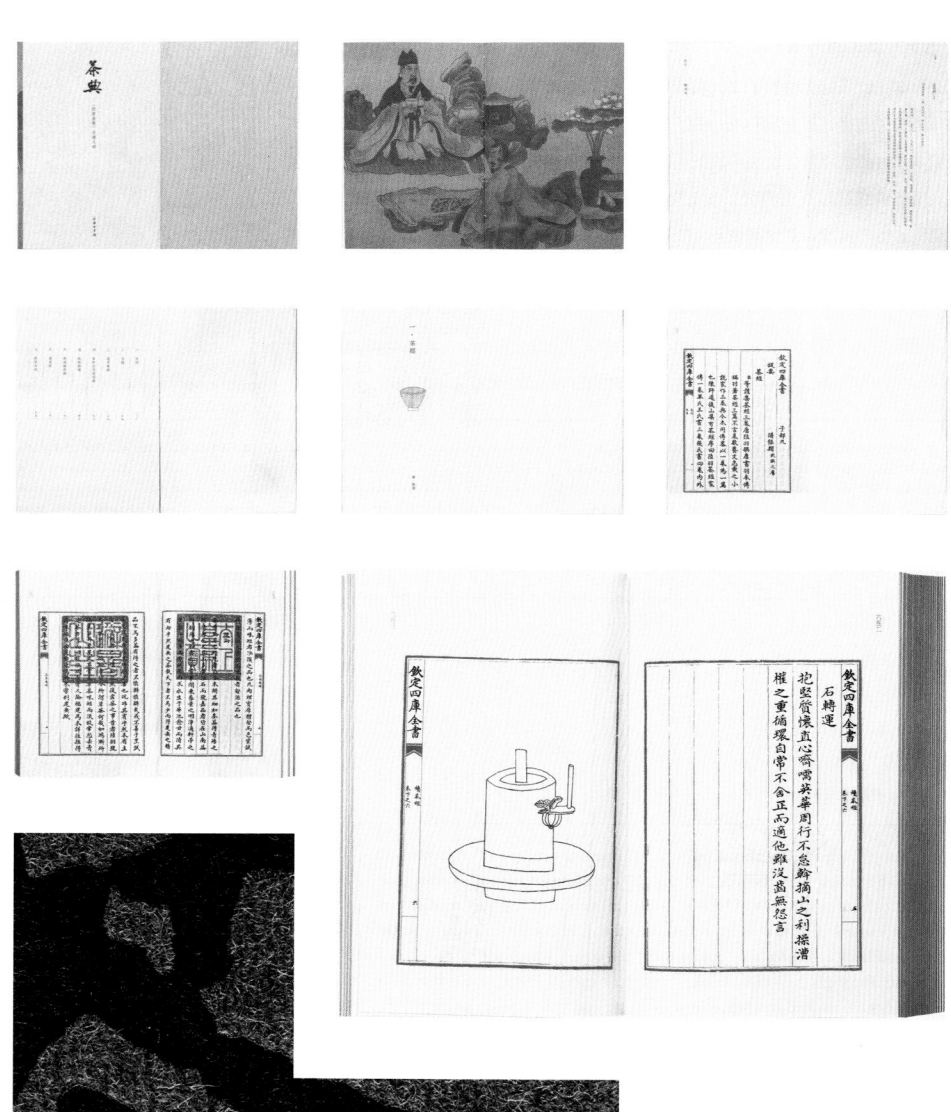

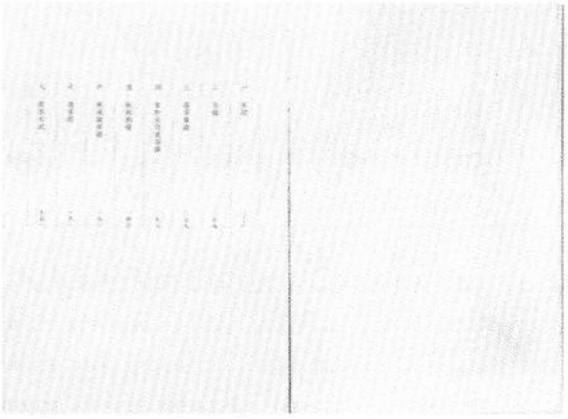

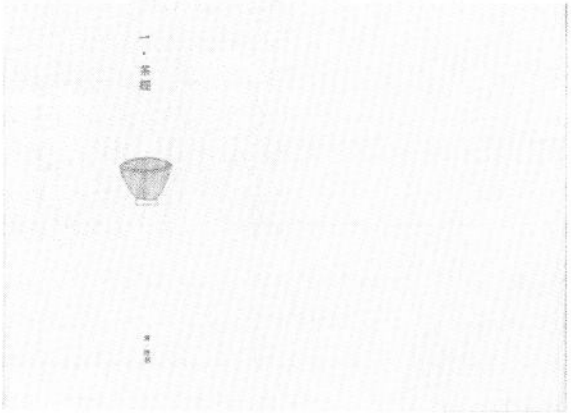
一・茶經

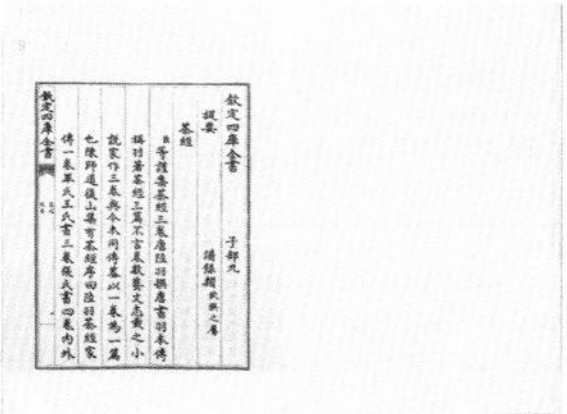
欽定四庫全書
提要
茶經

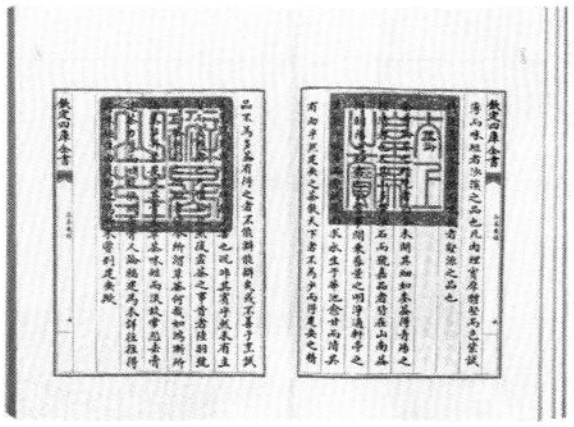
欽定四庫全書

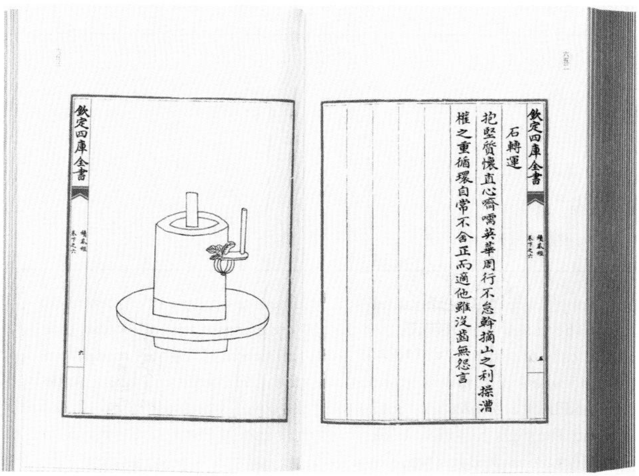
欽定四庫全書

石轉運

抱堅質懷直心嚌嚅英華周行不怠斡摘山之利操漕權之重循環自常不舍正而適他雖沒齒無怨言

欽定四庫全書

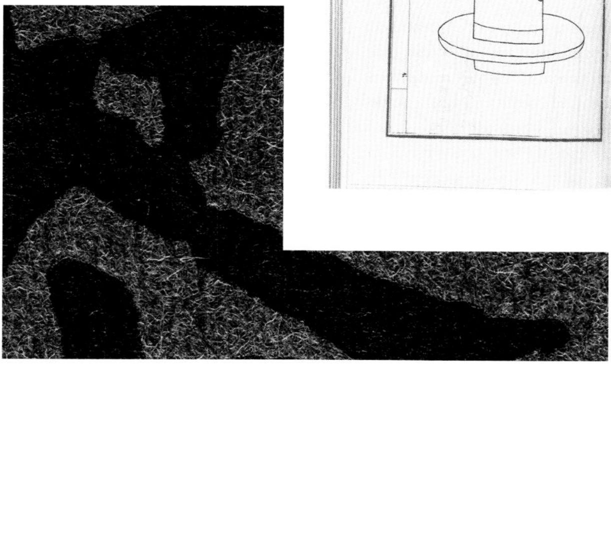

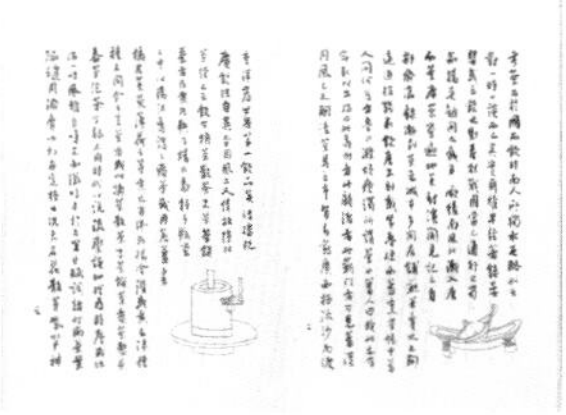
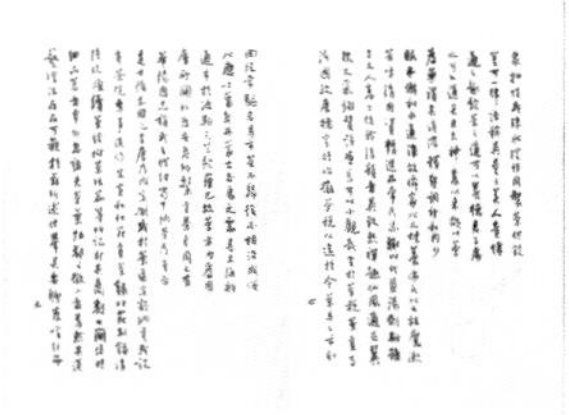

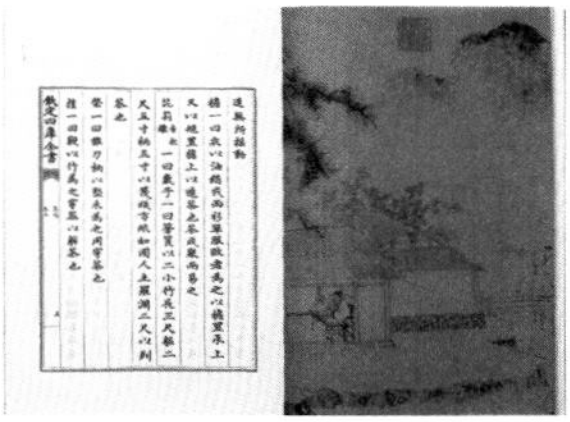
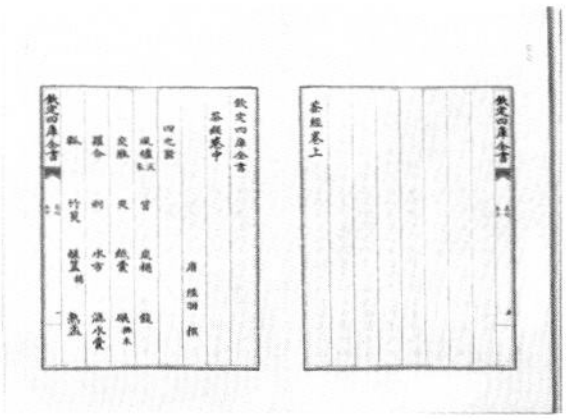
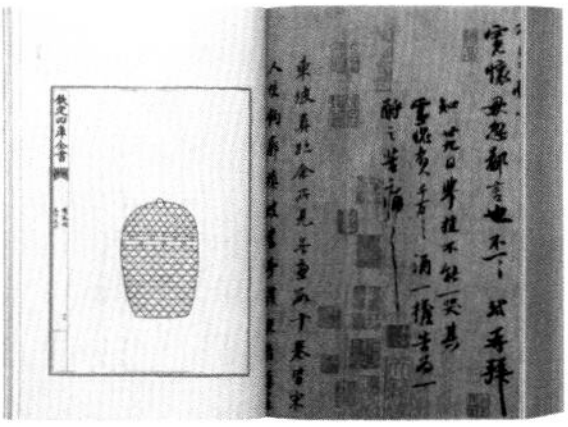
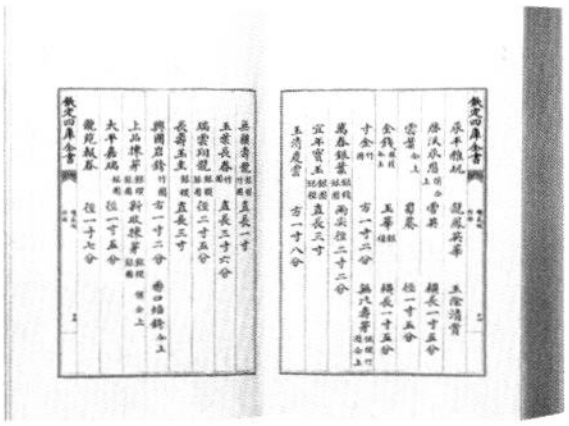
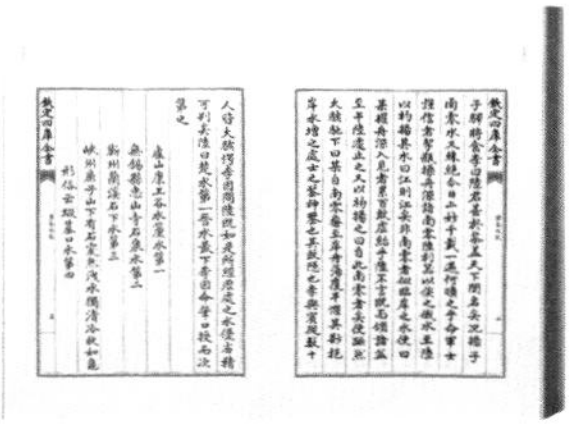
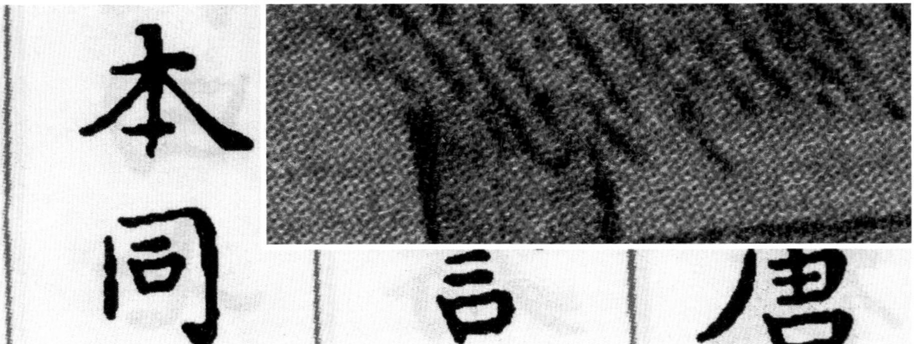
本
同
言
唐

MONSANTO CHEMICAL CO. ANNISTON, ALA.—16
POST CARD

这是一本通过大量真实图片让人们对转基因产品保持警惕的调查报告。书中的资料来源于转基因产品巨头孟山都“理想未来”的广告、材料表格，以及摄影记者马蒂厄·阿瑟林（Mathieu Asselin）实地采访的第一手资料。本书为锁线硬精装，封面采用哑面艺术纸橙绿黑三色印刷，裱覆超出一般厚度的卡板，大面积看似无害的绿色遇上刺眼的橙色，从这里就已经开始对读者进行警示：书中内容可能引起不适。内页哑面涂布纸四色印刷，局部裱贴收集而来的各种证物，其余均为报纸扫描件、环境受污染的照片、遗传畸形与患病者的照片、证物照片，以及部分抗议者的“签名照”合集。通过这些内容的直接并置，将一个迫在眉睫的“理想世界”的承诺化为泡影；通过双眼的真实观察，思考这个“精美”的世界是否只是虚妄。

Monsanto®:
A Photographic Investigation
by Mathieu Asselin

This is a research report with a lot of real pictures to make people cautious about genetically modified products. The material comes from advertisements and data sheets of Monsanto, a huge company dealing with genetically modified products, and the primary sources of fields research by a photojournalist Mathieu Asselin. The hardcover book is bound by thread sewing. The cover is printed in orange, green and black with matte art paper, mounted by thicker graphic board than the average. When the seemingly innocuous green encounters the harsh orange, the readers have been warned that the contents may cause discomfort. The inside pages are printed in four colors on coated paper, and sorts of exhibits are attached, such as scanned copies of newspapers, photos of environmental pollution, photos of people with genetic diseases, photos of exhibits, and a collection of 'signed photos' of some protesters. The combination of these contents destroys the promise of an imminent 'ideal world'. Through the close observation of eyes, the author pondered whether this 'exquisite' world is just an illusion.

- 荣誉奖
- 孟山都：摄影调查
- 委内瑞拉

- Honorary Appreciation
- Monsanto: A Photographic Investigation
- Venezuela
- Ricardo Báez
- Mathieu Asselin
- Verlag Kettler (English), Actes Sud (Français)
- 300×270×21mm
- 1266g
- 148p

The title makes you feel sceptical. Does the ® symbol after the brand name indicate a corporate profile, commissioned by the company? Or is it the defence against constantly lurking legal dangers, averting disaster? Extremely thick, virtually overbred grey boards create a kind of super-stiff cover. Its harmless design proves to be an allegory: on the pure, fresh, universally exhilarating green, the sign bearing the title shows a flash of orange – which is not just a signal, it bears a distinct warning: "The following book contains pictures that may be disturbing to some viewers." Documentary portraits of people, sober scenes of a ghost town. Contaminated houses, poisoned ground, genetic deformities, illness and death. Polychlorinated biphenyls, Agent Orange, Round up. The photographic tracking pursues associations through four chapters. Interspersed with a number of newspaper articles, re-archived from microfilm miniatures. And empty pages, many unprinted pages, vacant areas of concealment, placeholders for unresolved issues. Document facsimiles are pasted in and serve as an overriding link. At the front, even before the first photographs appear, you leaf through a separate, stapled booklet with reprints of Monsanto® advertisements from three decades. This is contrasted at the back by a form – the declaration of consent for grain growers. This book concept transforms the promise of an imminent, ideal world, enforced by the hopelessness of the fine print, into hypocrisy, seen through the eyes of reality.

Monsanto:
A Photographic
Investigation
by Mathieu Asselin

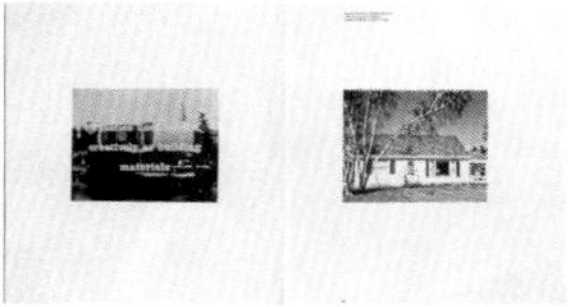

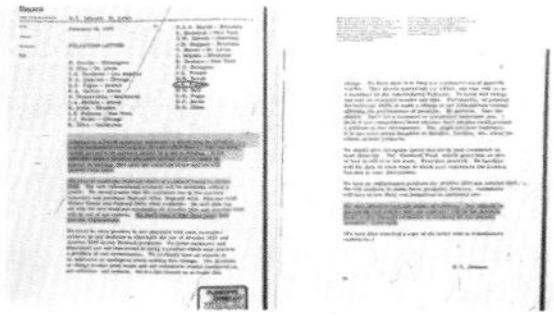

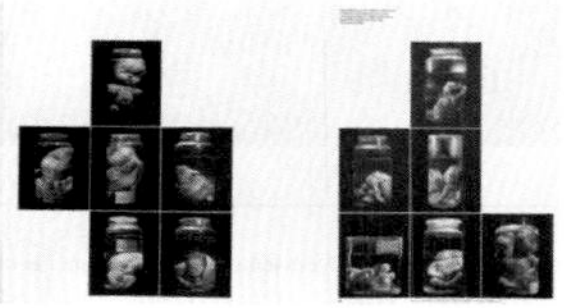

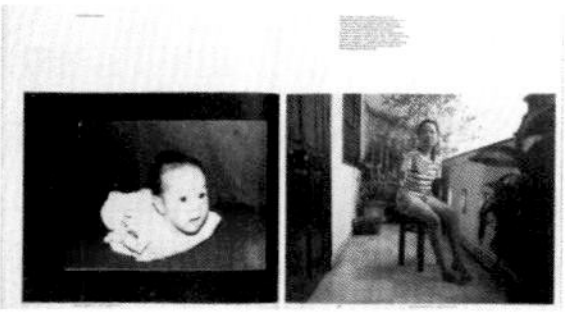

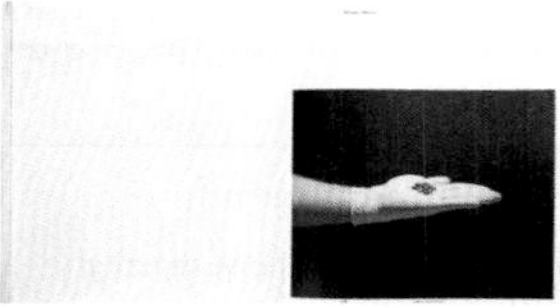

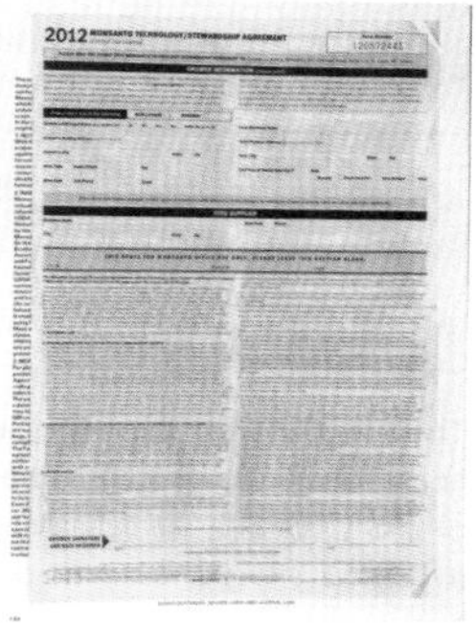
2012 MONSANTO TECHNOLOGY/STEWARDSHIP AGREEMENT

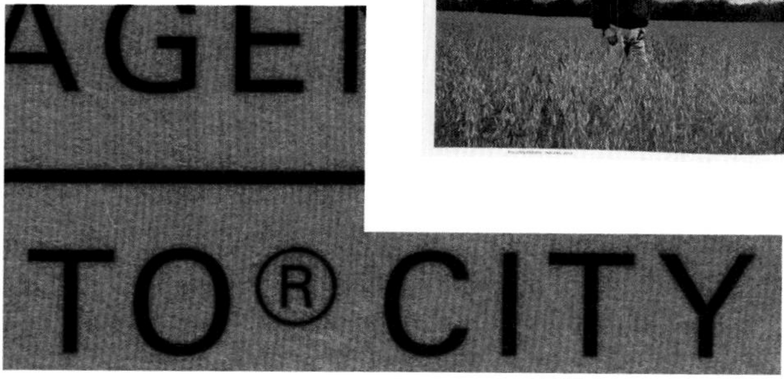
AGE
TO® CITY

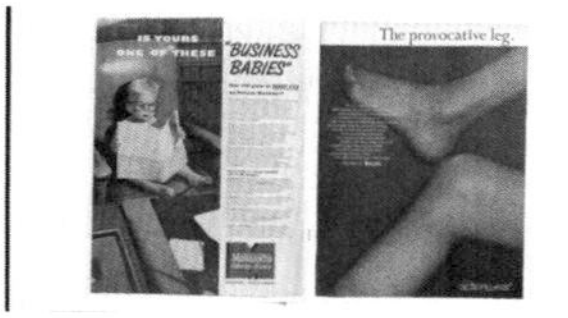
IS YOURS ONE OF THESE
"BUSINESS BABIES"
The provocative leg.

THE DAY NIAGARA FALLS
bubbled over
Making sure cakes always rise to the occasion. This is Monsanto today
Monsanto

I.
The House of the Future &The Red River

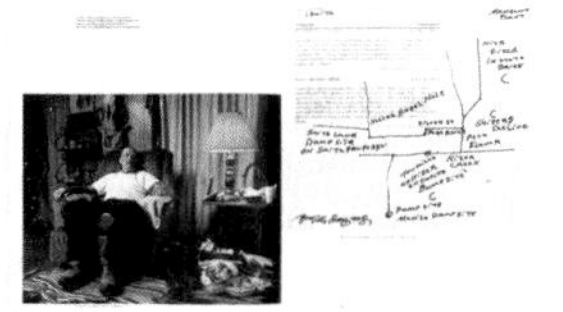

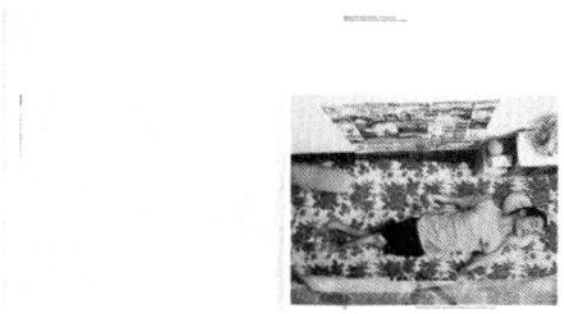

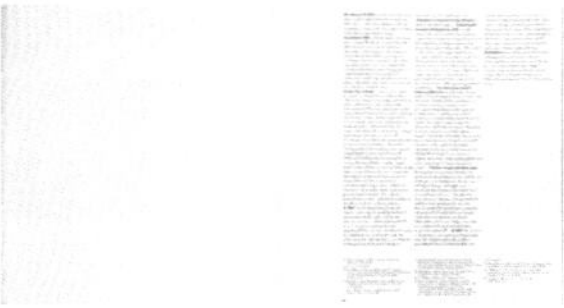

Leaders Buck Carter On Inflation
Wholesale Prices Rise 0.9 Percent Last Month
Caution: Smoking May Be More Hazardous
The Centralia Fireside Guard
Water bond 'top priority for city voters
Sturgeon residents ask tough cleanup questions
Board plans opening of Bicentennial Park

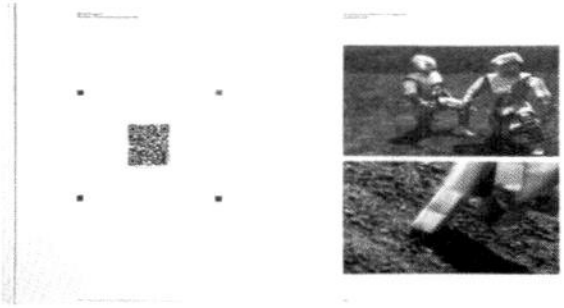

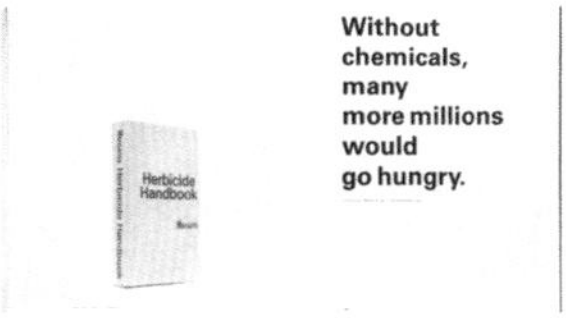
Herbicide Handbook
Without chemicals, many more millions would go hungry.

Monsanto
Chemicals help you eat better.

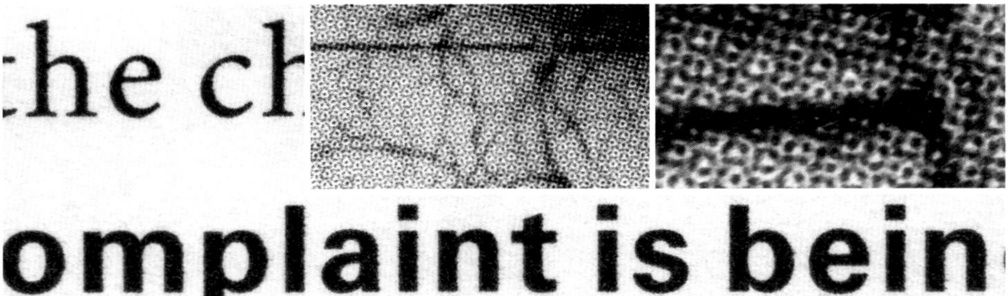
the ch
omplaint is bein

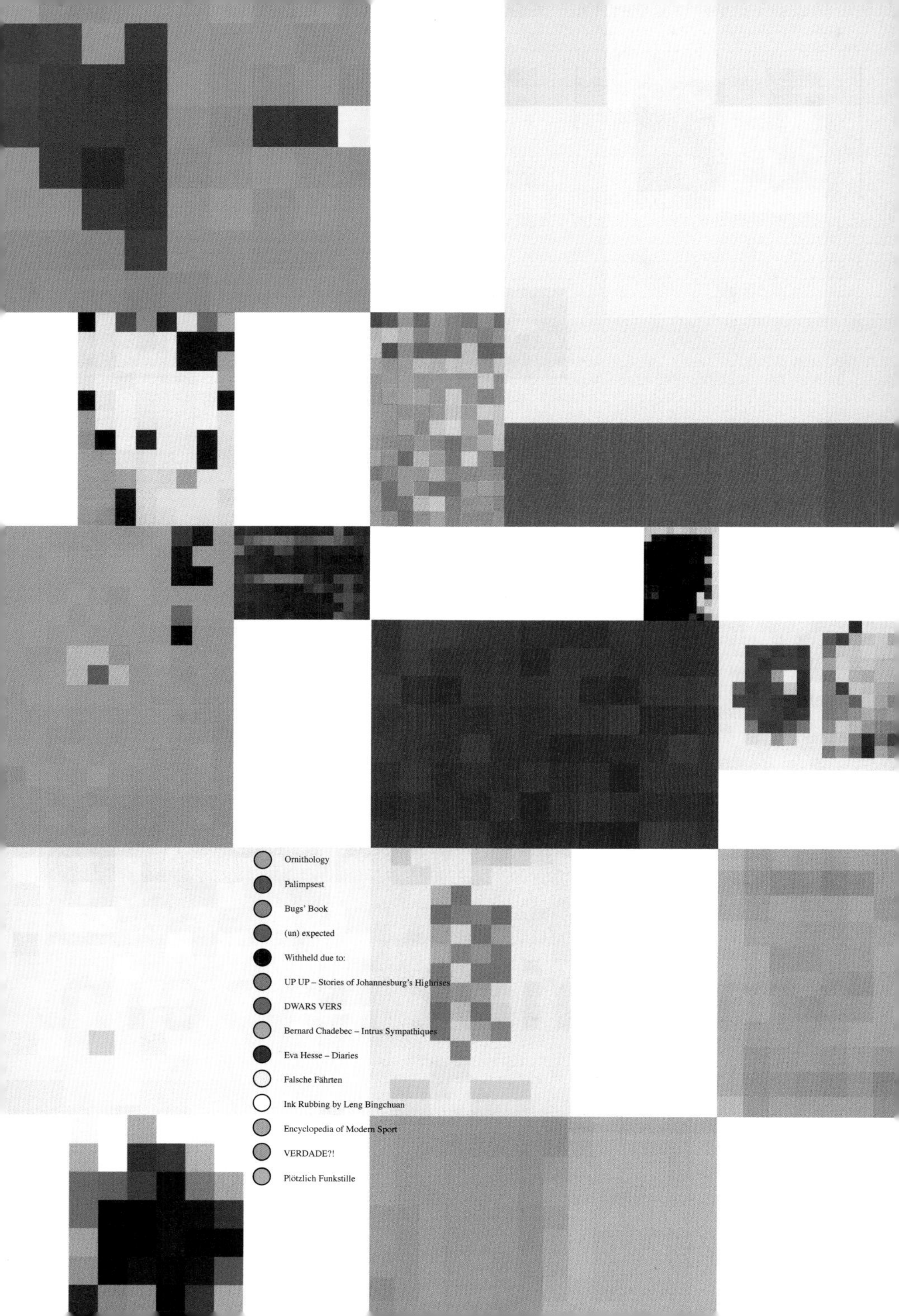
Ornithology
Palimpsest
Bugs' Book
(un) expected
Withheld due to:
UP UP – Stories of Johannesburg's Highrises
DWARS VERS
Bernard Chadebec – Intrus Sympathiques
Eva Hesse – Diaries
Falsche Fährten
Ink Rubbing by Leng Bingchuan
Encyclopedia of Modern Sport
VERDADE?!
Plötzlich Funkstille

2017

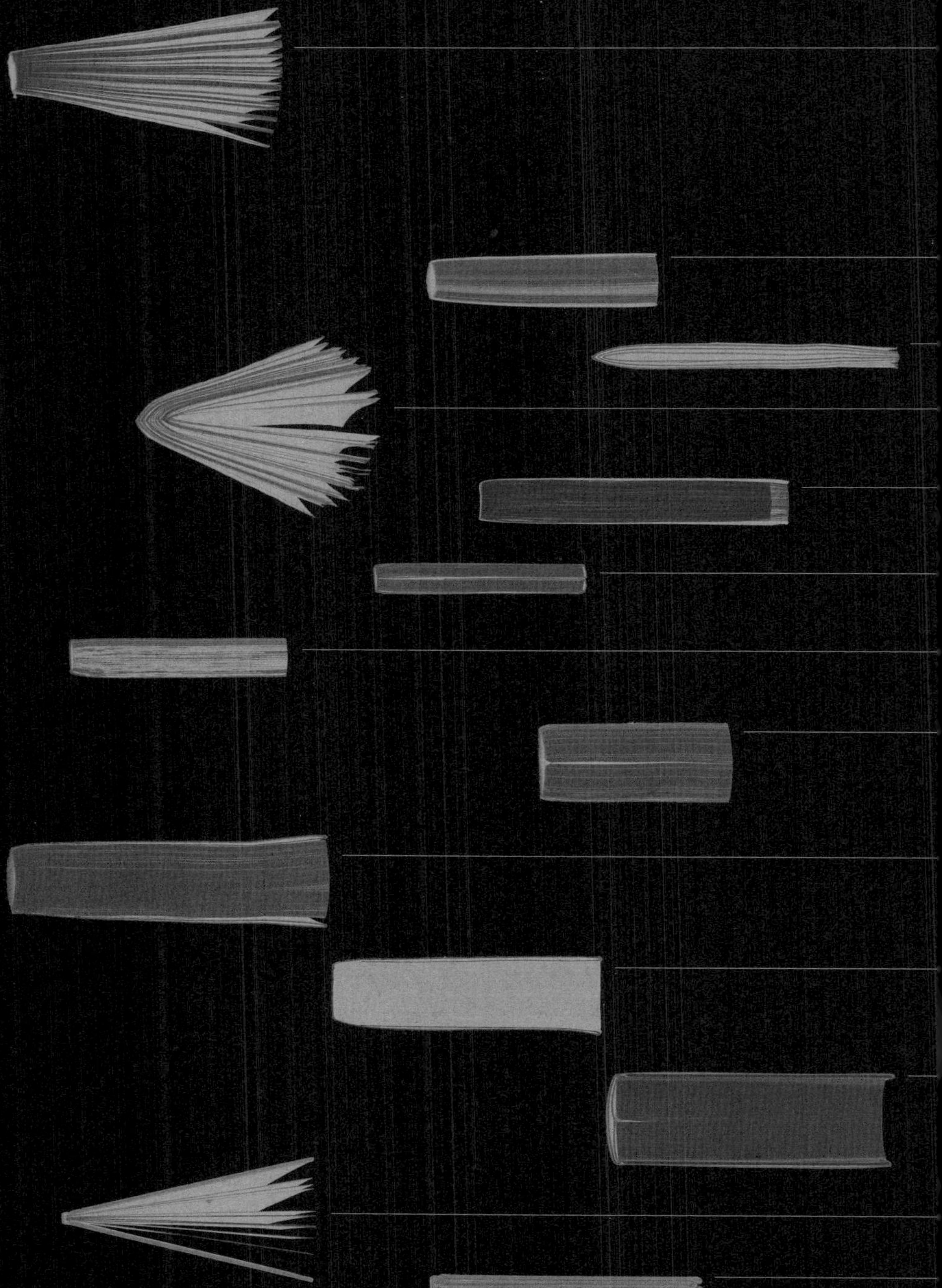

2017 Gl *0307*

2017 Sm[1] *0315*

2017 Sm[2] *0321*

2017 Bm[1] *0327*

2017 Bm[2] *0333*

2017 Bm[3] *0339*

2017 Bm[4] *0345*

2017 Bm[5] *0351*

2017 Ha[1] *0357*

2017 Ha[2] *0363*

2017 Ha[3] *0367*

2017 Ha[4] *0371*

2017 Ha[5] *0375*

ARTMENT OF THE ARMY
Builders in Battle!

he cuckoo is said by some to be a hawk transformed, because at
he time of the cuckoo's coming, the hawk, which it resembles, i
ever seen; and indeed it is only for a few days that you will s
awks about when the cuckoo's note sounds early in the season.

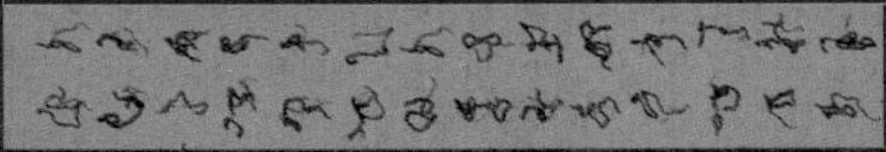

Mother

She related she took some valium and other medications and went to bed. She was woken by glass breaking near her bed and she noticed ten individuals on her roof and another twenty individuals in the residence. She tried to tell them she was a friend, but they did not understan her. She related the individuals searched her house and used dogs during the search. One individual asked her for money and she gave him $60.00, which she receives monthly for retirement, and he placed it in his pocket. She also gave the individuals 300,000 Dinar she sti

big earphones under their hoodies?
The short, thin guy with gold-rimmed glasses?
The small boy in red sports shoes, with his Mr T hairstyle?

I stood and watched him out of sight:
Today I reach around the door

enced directly, it remains always a di
Considering that Weil wrote this declara
1943 in London, where she sought refug

is ridiculous, good, I'm glad; if it be unimportant;
t should hold me back. Experience as much as poss
both are beneficial. I'm sure I have never let mysel

sind es, welche unsere gesamte Wirkli
keit gestalten. Grundkonzepte wie das

童年，就像卡在喉咙里的刺，不是那么容易取出来的，因为实在也找不到比
童心更沉重的心。童年纯真的"秘密"无法言传，因为它是彼岸。

(2) **発育発達とスポーツ医科学**
学校における体育授業をはじめ，ス
ポーツ少年団や部活動でのスポーツ，
運動が重要なことを体育授業で説明す
るためのよい根拠となっている。
猪飼らによれば，筋持久力のトレー

VERDADE

MEIN ERSTES ZIEL IST DIE ERBSENDOSENFABRIK DES MULTIKONZERNS DESCROC
IRGENDWO AM STADTRAND. SIE GEHORT DEM SOHN VON MEINEM BOSS – DI
WELT IST EBEN EIN DORF.

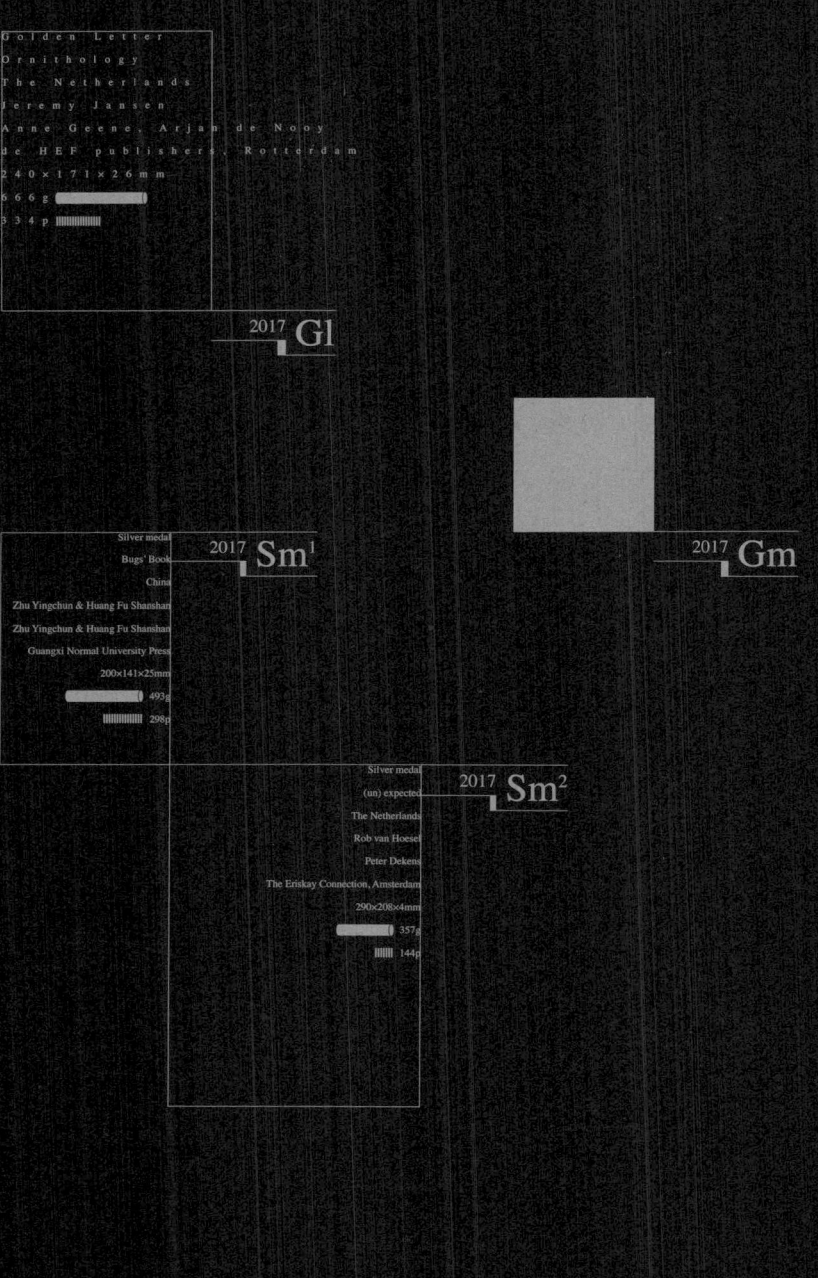

Golden Letter
Ornithology
The Netherlands
Jeremy Jansen
Anne Geene, Arjan de Nooy
de HEF publishers, Rotterdam
240×171×26mm
666g
334p

2017 Gl

2017 Gm

Silver medal
Bugs' Book
China
Zhu Yingchun & Huang Fu Shanshan
Zhu Yingchun & Huang Fu Shanshan
Guangxi Normal University Press
200×141×25mm
493g
298p

2017 Sm1

Silver medal
(un) expected
The Netherlands
Rob van Hoesel
Peter Dekens
The Eriskay Connection, Amsterdam
290x208×4mm
357g
144p

2017 Sm2

2017 Ha[1]

Honorary Appreciation
Falsche Fährten
Germany
Jonas Voegeli, Kerstin Landis, Scott Vander Zee
— Hubertus Design, Zürich
Peter Radelfinger
Edition Patrick Frey, Zürich
296×213×42mm
1756g
584p

2017 Ha[2]

Honorary Appreciation
Ink Rubbing by Leng Bingchuan
China
Zhou Chen
Leng Bingchuan
Dolphin Books
260×184×39mm
1622g
642p

2017 Ha[3]

Honorary Appreciation
VERDADE?!
Portugal
Pato Lógico
Bernardo P. Carvalho
Pato Lógico
256×200×9mm
325g
36p

2017 Ha[4]

Honorary Appreciation
Encyclopedia of Modern Sport
Japan
Shin Tanaka, Sozo Naito
Toshio Nakamura, Takeo Takahashi,
Tsuneo Sougawa, Hidenori Tomozoe
TAISHUKAN Publishing Co., Ltd.
270×204×60mm
2556g
1378p

2017 Ha[5]

Honorary Appreciation
Plötzlich Funkstille
Germany
Benjamin Courtault
Benjamin Courtault, Paris (FR)
kunstanstifter verlag, Mannheim
305×187×9mm
333g
32p

2017 Bm[1]

Bronze Medal
Withheld due to
Switzerland
typosalon, Christof Nüssli
Christof Nüssli
cpress
224×167×19mm
600g
408p

2017 Bm[2]

Bronze Medal
UP UP – Stories of Johannesburg's Highrises
Germany
Huber / Sterzinger, Zürich (CH)
Nele Dechmann, Fabian Jaggi, Katrin Murbach, Nicola Ruffo
Hatje Cantz Verlag, Ostfildern
270×192×24mm
926g
336p

2017 Bm[3]

Bronze Medal
DWARS VERS
The Netherlands
Team Thursday (Simone Trum & Loes van Esch)
Emily Dickinson, Edna St. Vincent Millay
Ans Bouter en Benjamin Groothuyse
210×150×20 mm
469g
272p

2017 Bm[4]

Bronze Medal
Bernard Chadebec – Intrus Sympathiques
Switzerland
Olivier Lebrun and Urs Lehni in collaboration
with Simon Knebl, Phil Zumbruch, Saskia
Reibel and Tatjana Stürmer (HfG Karlsruhe)
Olivier Lebrun and Urs Lehni
Rollo Press, Zürich
191×135×21 mm
470g
272p

2017 Bm[5]

Bronze Medal
Eva Hesse – Diaries
Switzerland
NORM, Zürich / Johannes Breyer, Berlin
Eva Hesse / Barry Rosen
Hauser & Wirth Publishers / Yale University Press
204×136×50 mm
1190g
900p

2017

The International Jury

Demian Bern (Germany)

Bruno van den Elshout (The Netherlands)

Lars Fuhre (Sweden)

Tom Mrazauskas (Latvia)

Federica Ricci (Italy)

Mariko Takagi (Japan)

Andreas Töpfer (Germany)

Country / Region

The Netherlands 3

Switzerland 3

Germany 3

China 2

Czech Republic 1

Japan 1

Portugal 1

Designer

Jeremy Jansen

Petr Jambor

Zhu Yingchun & Huang Fu Shanshan

Rob van Hoesel

typosalon, Christof Nüssli

Huber / Sterzinger, Zürich (CH)

Team Thursday (Simone Trum & Loes van Esch)

Olivier Lebrun and Urs Lehni in collaboration with Simon Knebl, Phil Zumbruch, Saskia Reibel and Tatjana Stürmer (HfG Karlsruhe)

NORM, Zürich / Johannes Breyer, Berlin

Jonas Voegeli, Kerstin Landis, Scott Vander Zee — Hubertus Design, Zürich

Zhou Chen

Shin Tanaka, Sozo Naito

Pato Lógico

Benjamin Courtault

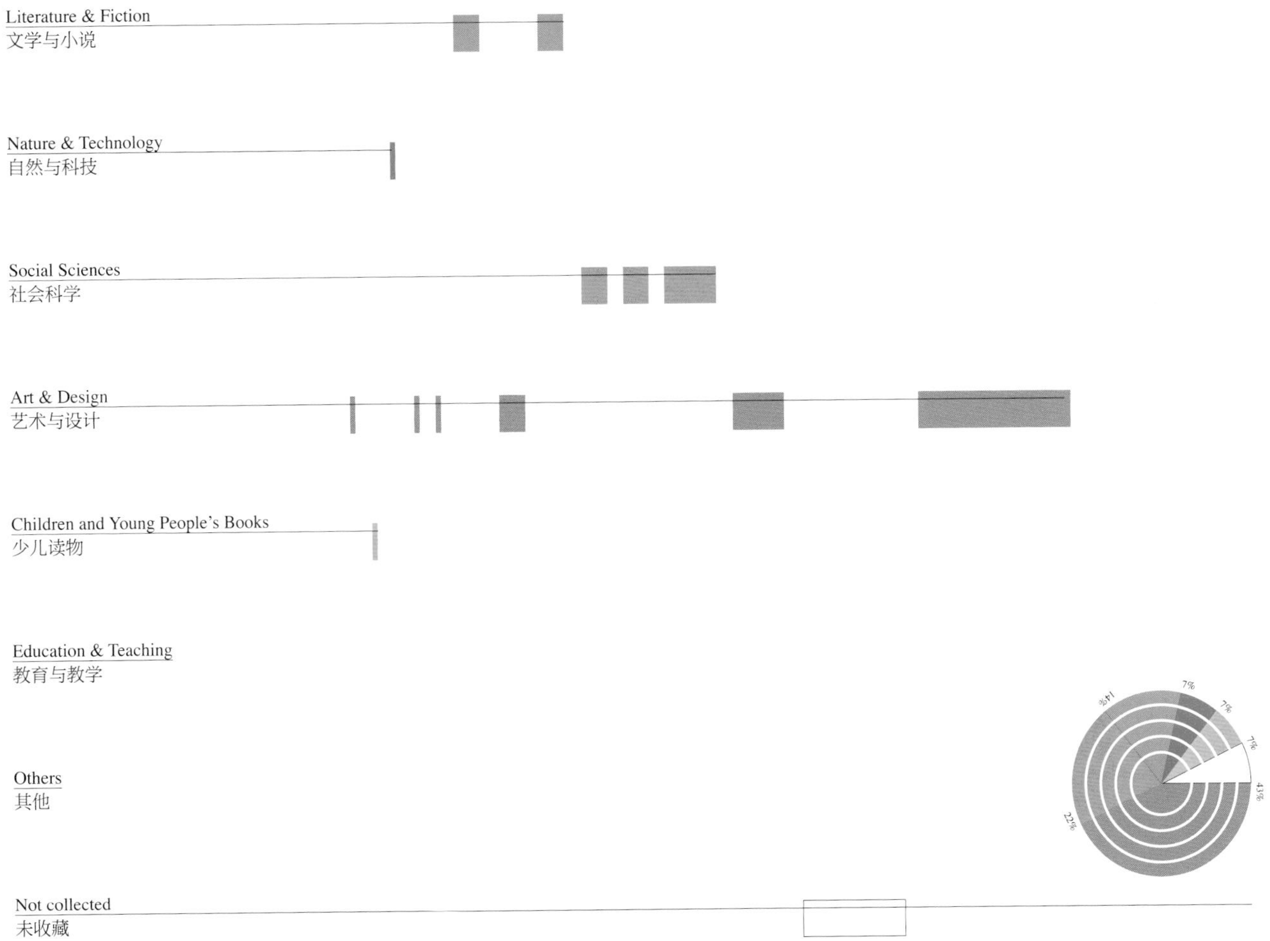
Literature & Fiction
文学与小说
Nature & Technology
自然与科技
Social Sciences
社会科学
Art & Design
艺术与设计
Children and Young People's Books
少儿读物
Education & Teaching
教育与教学
Others
其他
Not collected
未收藏
14%
7%
7%
7%
43%
22%

Indian
Peafowl
Goldfinch
de Oudezijds Voorburgwal
Unidentified

这是两位摄影师以鸟类摄影为主进行的非“科学”的鸟类学研究书籍。由于鸟类的种类和总量都很大，又具有丰富多样的外观，鸟类摄影已成为研究鸟类的热门方式。摄影师拍摄了具有代表性甚至已经灭绝的鸟类资料，通过创造性的分类和关联工作，为鸟类学研究提供了大量特殊而新鲜的视角。本书的装订方式为锁线胶平装，护封采用灰色哑面艺术纸，下半部鸟类图形部分印灰白色，文字印黑。内封浅棕色卡纸印黑，书脊贴橙色布面材料加固。内页根据内容需求和彩图量，采用胶版纸和哑面涂布纸混合装订，四色印刷。内容分为鸟类学历史、形态与动作、行为与活动、遗传学、鸟类观察统计以及其他特殊的问题。书籍以意大利歌剧家普契尼的《蝴蝶夫人》乐谱开篇，之后的页面根据内容差异分为双栏、三栏、四栏、六栏，有些部分跨栏形成大图或摆放较多文字，形成丰富的节奏，有如音乐上的各种节拍。

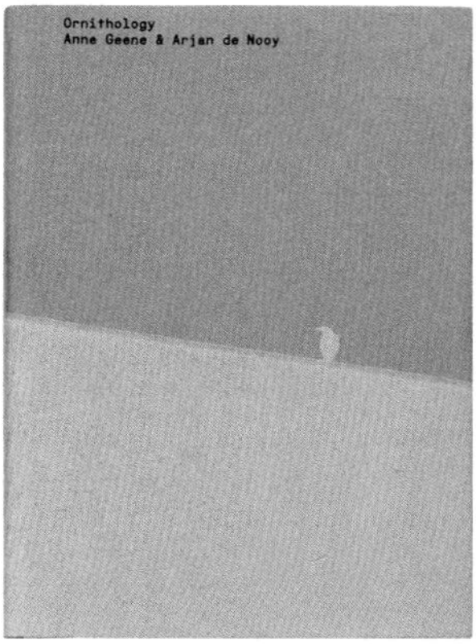
Ornithology
Anne Geene & Arjan de Nooy

This is a non-scientific book on ornithology conducted by two photographers focusing on bird photography. Bird photography has become a popular way to study birds because of their large species and population, and their varied appearance. Photographers have captured representative and even extinct bird data, providing a large number of special and fresh perspectives for ornithology research through creative classification and association work. The book is perfect bound with thread sewing. The jacket is printed in grey matte art paper. Bird figures on the lower part of the jacket are printed in grey and white, while the text is printed in black. The title page is printed in black on light brown cardboard, and the spine is strengthened with orange cloth. The inside pages are bound with offset paper and matte coated paper, and printed in four colors according to the content and pictures. The contents are divided into avian history, morphology and movement, behavior and activity, genetics, avian observation and statistics, and other special problems. The book starts with the music score of *Madame Butterfly* by Italian opera writer Puccini, and then the pages are divided into two columns, three columns, four columns and six columns according to different contents. Some large pictures or big blocks of text are typeset cross the columns, forming a rich rhythm, just like various beats of music.

- 金字符奖
- 鸟类学
- 荷兰

- Golden Letter
- Ornithology
- The Netherlands
- Jeremy Jansen
- Anne Geene, Arjan de Nooy
- de HEF publishers, Rotterdam
- 240×171×26mm
- 666g
- 334p

Paintings from old masters are scoured for birds; details are isolated and sorted. Variant forms of the white bird splotch are identified as an indicator of the flight speed. The black nuances of the crow and the white nuances of the wood pigeon's egg are shown on exposure scales. The book bridges several gaps in avian research. These are mostly series whose general principle becomes recognisable as the motifs are structured in variants and specific relationships to one another. The astounding, entertaining richness of the overall design is not at all an obstacle to the self-descriptive overall effect. In decimal structuring, as is customary for academic texts, surprising aspects are arranged together which have been neglected by research, or which create entirely new coherences. The typewriter script in a single font size is obliged to take a factual approach. The rigid type and image area can be filled either systematically, stringently, or playfully, rhythmically. Sections printed in black/white and colour alternate sensibly. Different types of paper which only display tenuous differences in their shade and texture subtly mark certain chapters. As is so often the case in fundamental research, its practical use will presumably only be disclosed in the future. That is not the issue here, however, because the dividing line between integrity and irony becomes indistinct so as to be constructive, and it is precisely in this zone that aesthetic and bird-free perceptions thrive.

hit
he Gee
birds, Aristotle

666

334

240×171×26

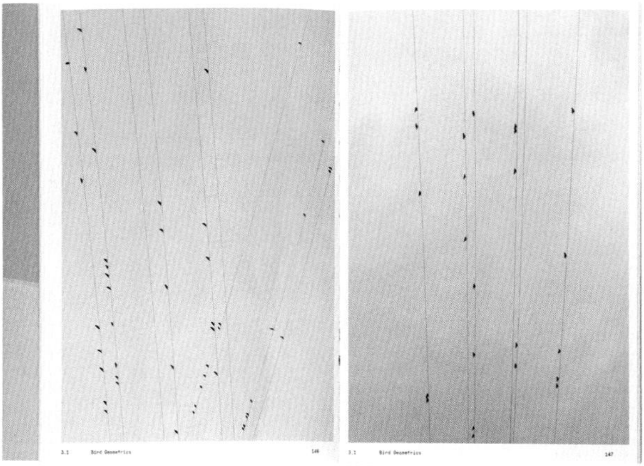
3.1 Bird Geometrics 146
3.1 Bird Geometrics 147

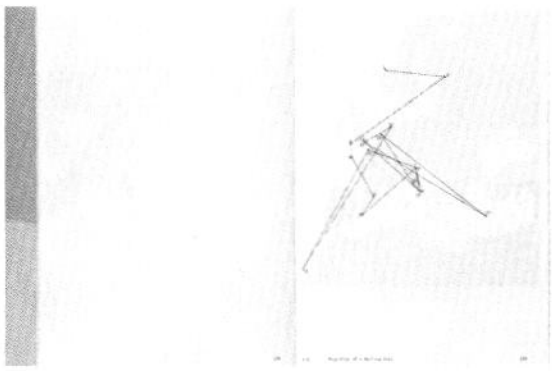

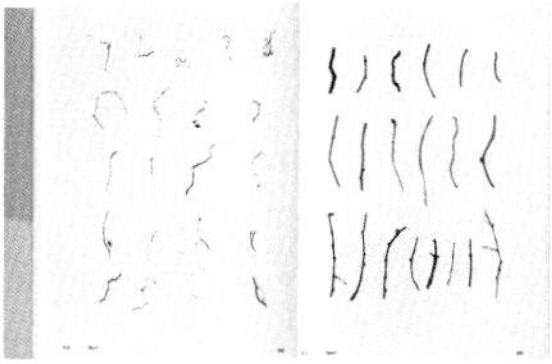

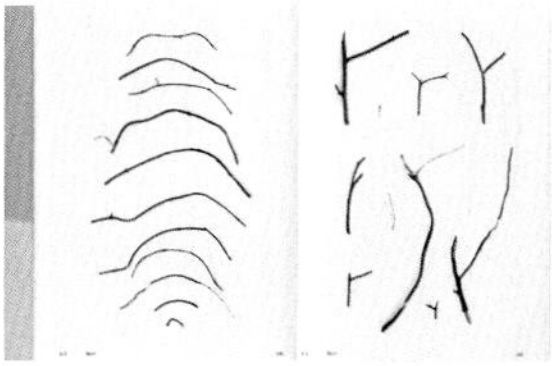

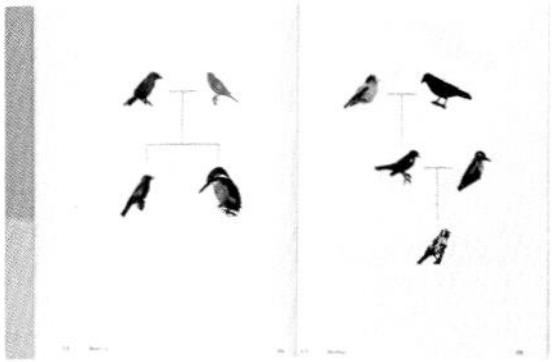

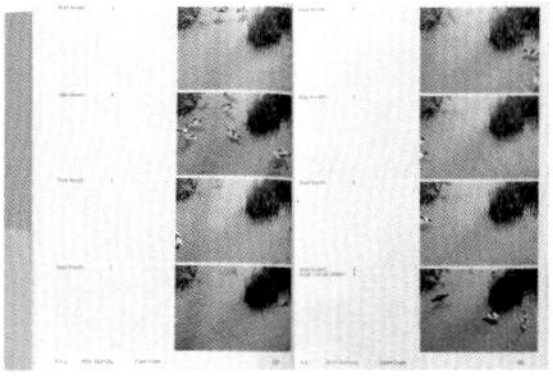

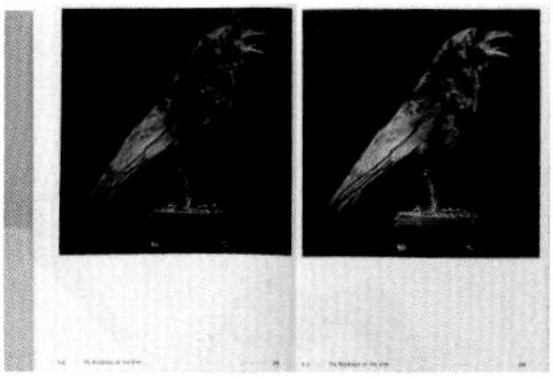

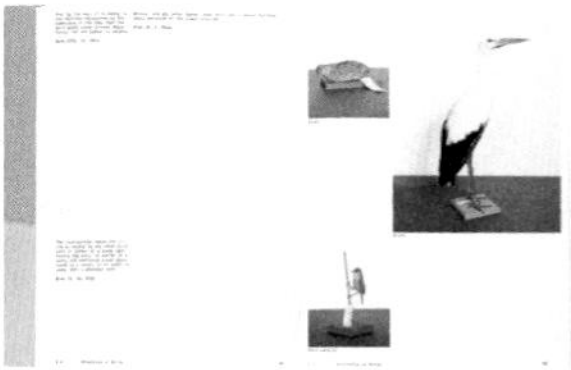

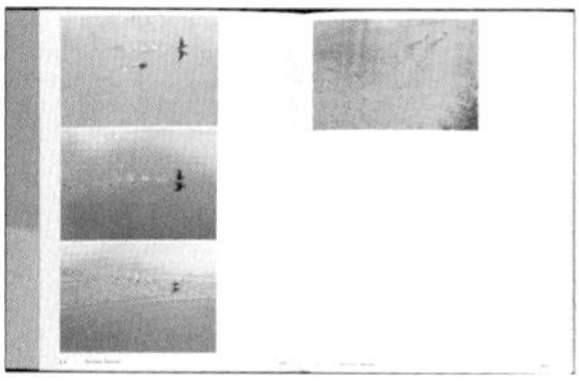

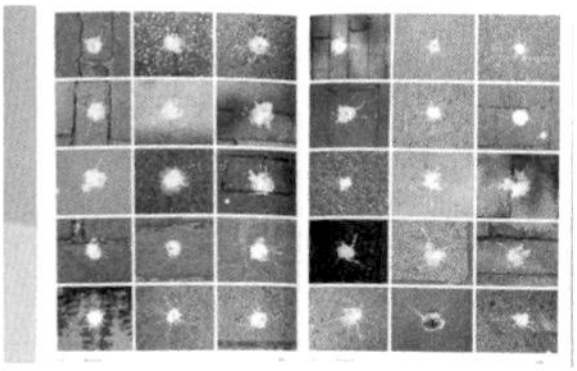

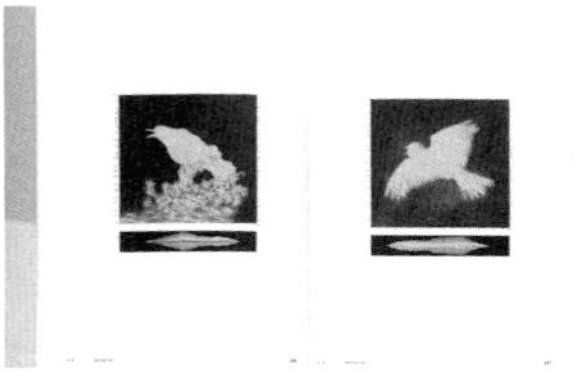

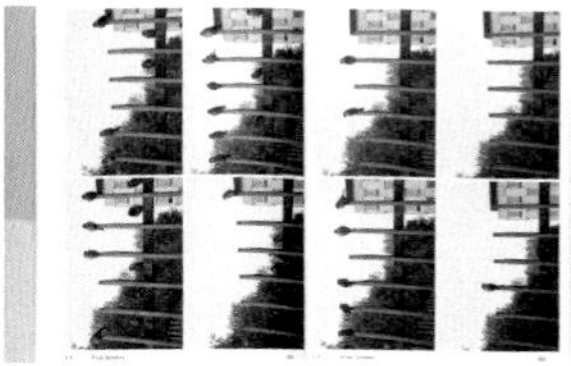

Great Spotted Woodpecker
Great Spotted Woodpecker
Treecreeper
Treecreeper
Mallard
Mallard
Great Tit
Mallards
Wood Pigeon
Wood Pigeon
Blue Tit
Blackbird
Herring Gull
Coots
Blackbird
5.1.1 Bird Counting Area 270
Blackbird
Great Tit
Blue Tit
Blue Tit
Herring Gull
Blackbird
Blackbird
House Sparrow
Great Spotted Woodpecker
Great Spotted Woodpecker
Blue Tit
Blackbird
Song Trush
Wren
Rose-ringed Parakeet
5.1.1 Bird Counting Area 271

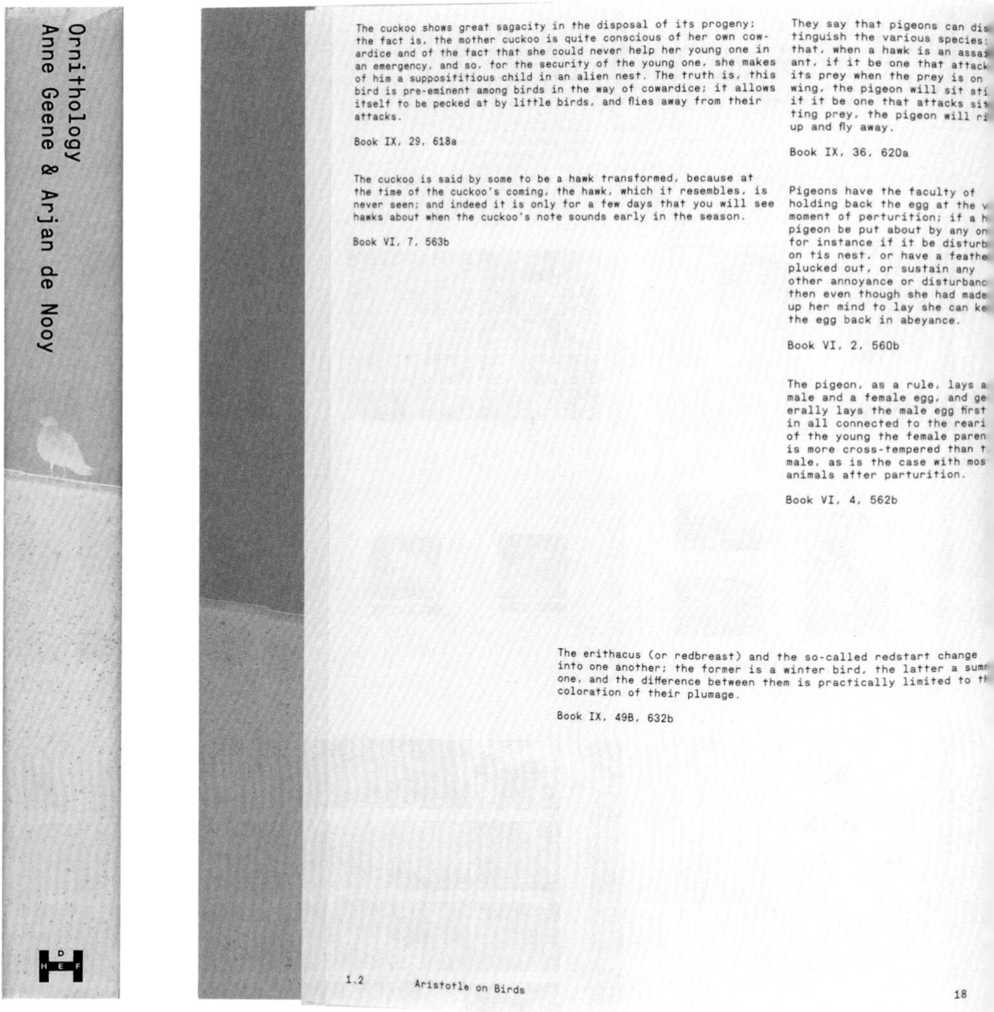
Ornithology
Anne Geene & Arjan de Nooy
The cuckoo shows great sagacity in the disposal of its progeny; the fact is, the mother cuckoo is quite conscious of her own cowardice and of the fact that she could never help her young one in an emergency, and so, for the security of the young one, she makes of him a supposititious child in an alien nest. The truth is, this bird is pre-eminent among birds in the way of cowardice; it allows itself to be pecked at by little birds, and flies away from their attacks.
Book IX, 29, 618a
The cuckoo is said by some to be a hawk transformed, because at the time of the cuckoo's coming, the hawk, which it resembles, is never seen; and indeed it is only for a few days that you will see hawks about when the cuckoo's note sounds early in the season.
Book VI, 7, 563b
They say that pigeons can dis tinguish the various species: that, when a hawk is an assa ant, if it be one that attack its prey when the prey is on wing, the pigeon will sit sti if it be one that attacks sit ting prey, the pigeon will ri up and fly away.
Book IX, 36, 620a
Pigeons have the faculty of holding back the egg at the v moment of perturition; if a h pigeon be put about by any on for instance if it be disturb on tis nest, or have a feathe plucked out, or sustain any other annoyance or disturbanc then even though she had made up her mind to lay she can ke the egg back in abeyance.
Book VI, 2, 560b
The pigeon, as a rule, lays a male and a female egg, and ge erally lays the male egg first in all connected to the reari of the young the female paren is more cross-tempered than t male, as is the case with mos animals after parturition.
Book VI, 4, 562b
The erithacus (or redbreast) and the so-called redstart change into one another; the former is a winter bird, the latter a sum one, and the difference between them is practically limited to th coloration of their plumage.
Book IX, 49B, 632b
1.2
Aristotle on Birds
18

ckoo

Pigeon

Redbreast

Redstart

2017 Sm

这是设计师朱赢椿多年来观察和记录虫子的结晶。与传统意义上的昆虫研究书籍不同，作者没有对昆虫拍摄照片，也没有写下任何记录性的文字，甚至本书除了最后拉页上的版权信息外，没有任何的人类文字。全书的主要内容为各种小虫爬过的痕迹，仿佛它们的书法作品。本书的装订形式为锁线胶平装，封面棕色土豆纸印黑，黑卡纸环衬。内页胶版纸印黑，呈现虫子们的"书法"作品、写就的"文章"；或者哑面涂布纸四色印刷过油，呈现它们的"绘画"作品或"作画"过程。为了书籍完整而统一的形式，甚至连页码都没有采用数字，而是使用圈点的方式来表示。全书仿佛一本外来文明的书籍，在图像、生物学、哲学、艺术等多个领域为读者带来新的思考方向。

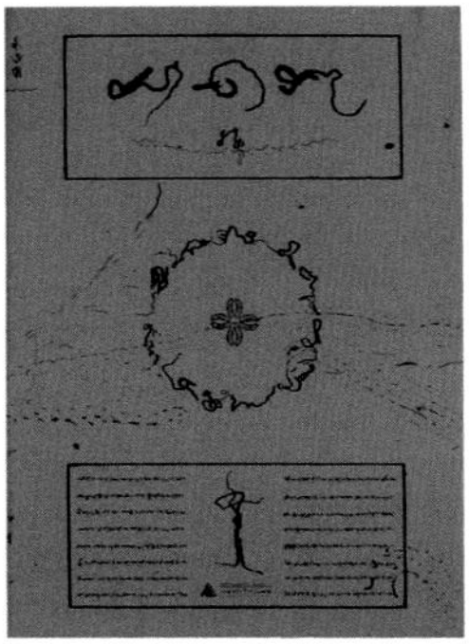

This book is about insects observed and recorded by designer Zhu Yingchun for many years. Unlike the traditional insect research books, the author neither took photographs of insects, nor did he write down any record. Furthermore, the book had no text except the copyright information on the last page. The main content of the book is the traces of various insects, as if they were calligraphy works of the insects. This paperback book is perfect bound with thread sewing. The cover is made of brown art paper printed in black, while the end paper is made of black cardboard. The inside pages are made of offset paper printed in black to present insect's calligraphy works or written articles. Matte coated paper in four-color printing presents the insects' painting works or painting process. The book is like a cultural production of alien civilization, bringing new thinking directions to readers in the fields of image, biology, philosophy, art and so on.

银奖
虫子书
中国
朱赢椿、皇甫珊珊
朱赢椿、皇甫珊珊
广西师范大学出版社

Silver medal
Bugs' Book
China
Zhu Yingchun & Huang Fu Shanshan
Zhu Yingchun & Huang Fu Shanshan
Guangxi Normal University Press
200×141×25mm
493g
298p

This work is a forerunner in the field of morphology. It displays all elements of an academic approach: "Half acre of land, five years of time, inviting one hundred species of bugs, collecting a thousand kinds of traces, finally, we have a book." Morphology is the method, in a dual sense. Firstly biologically: manifold traces of insects, spiders and worms are made visible perhaps for the first time in plausible image processing; they develop huge calligraphical qualities seemingly by themselves. On the other hand linguistically: the aesthetic suggestion of the image results is so powerful that the traces—such as masterful ink drawings, like rubbings from prehistoric rock carvings, like klecksography lost in thought—are recognised as a mysterious, unfamiliar literary language. Images become text, the ciphers turn into a literal script. As a consequence of the maximum mergence of phenomena and observation, the didactic typography draws on the newly discovered, natural, unique fonts. The so-called imaging processes make a key contribution to understanding in natural sciences. Images with a realistic appearance, which tend to suggest depictive objectivity, are construed from measured values. They hide the fact that these are basically processes that lend meaning. It is precisely this inconsistency that is ingeniously exposed by this search for traces. At the end, this artistic study succeeds in nothing less than the realisation of a philosophical metaphor: the world as a book that writes itself.

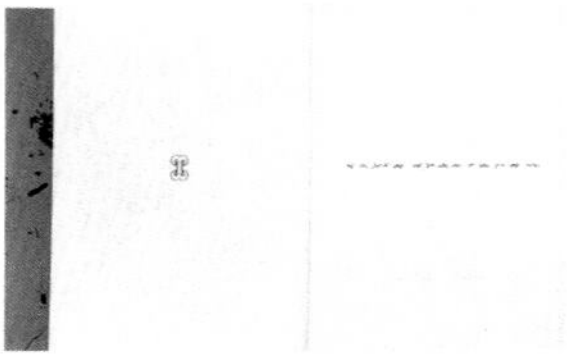
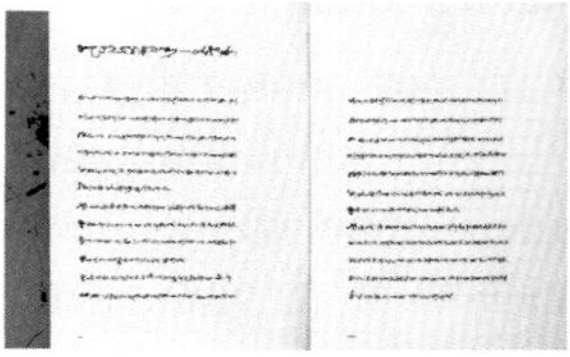
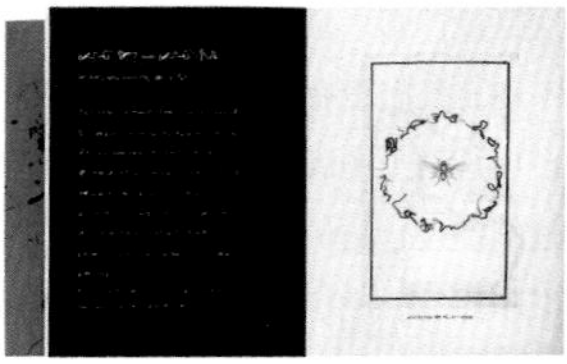
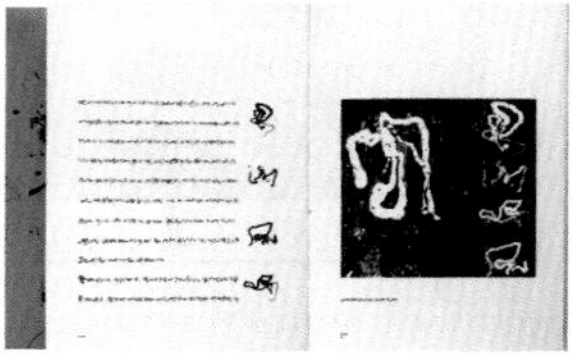
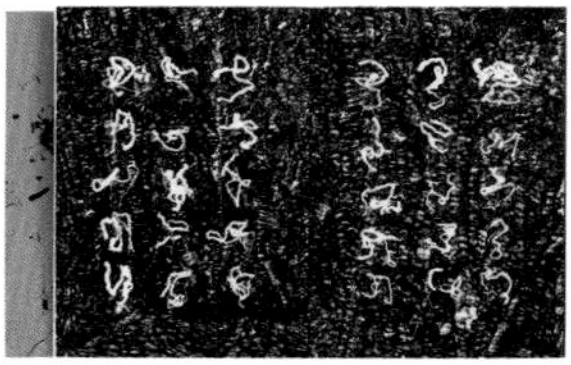
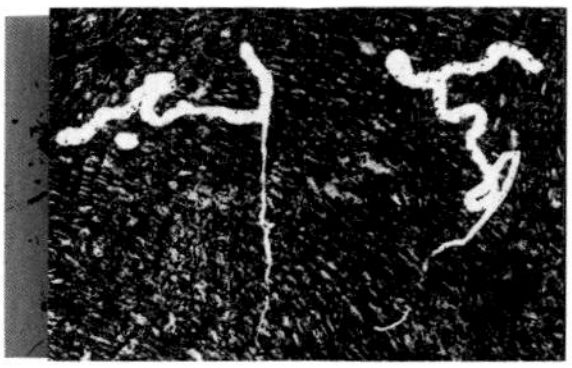
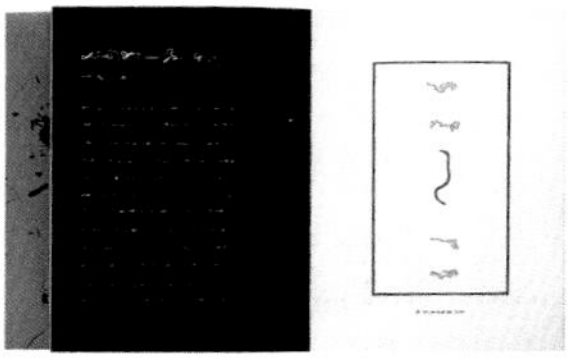
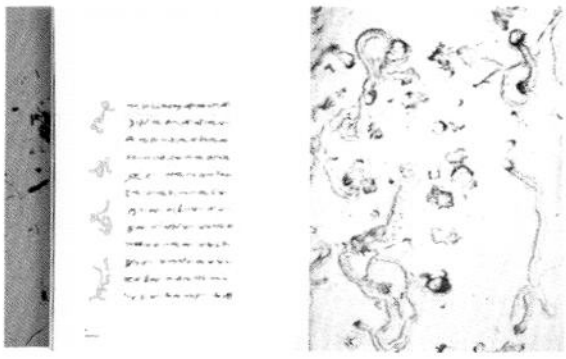
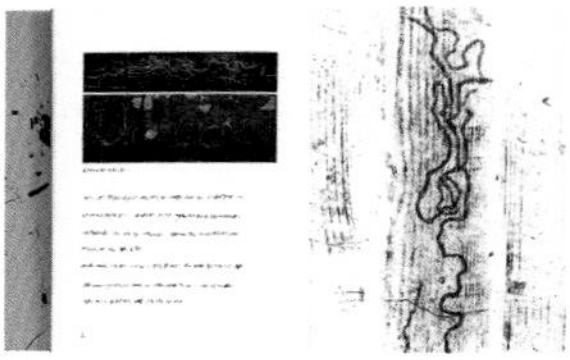
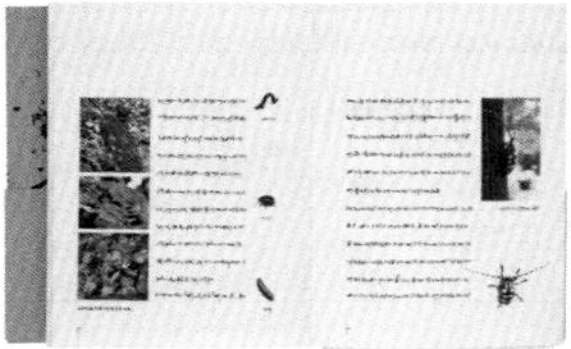
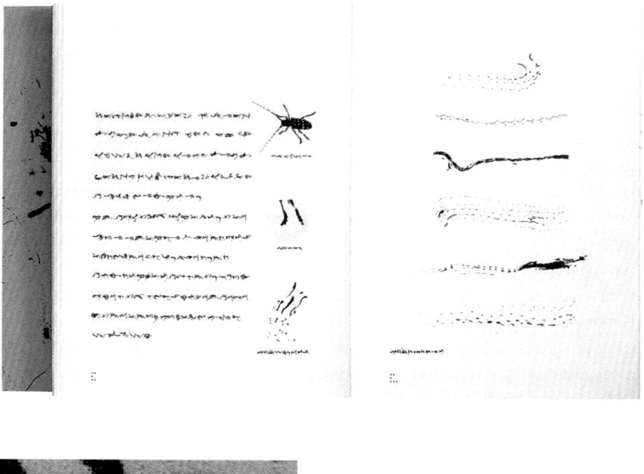
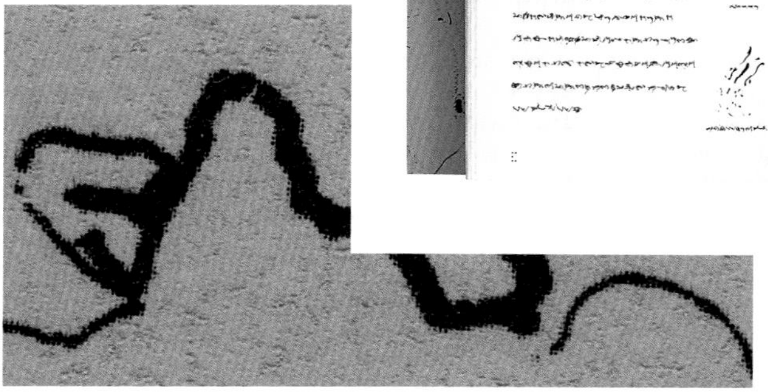

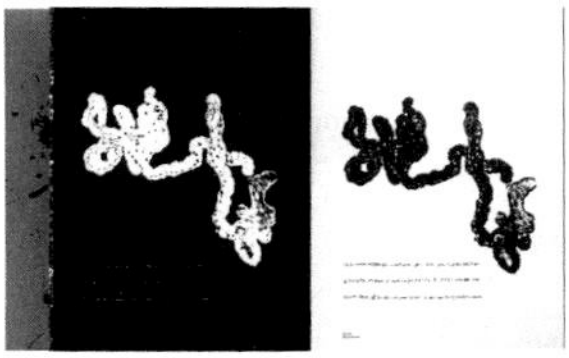

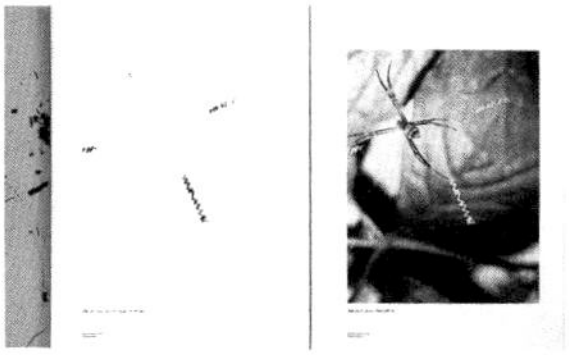
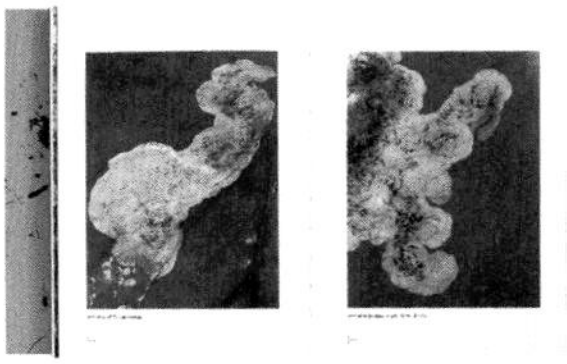
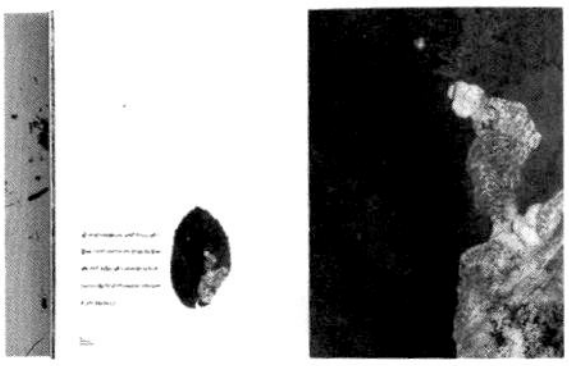
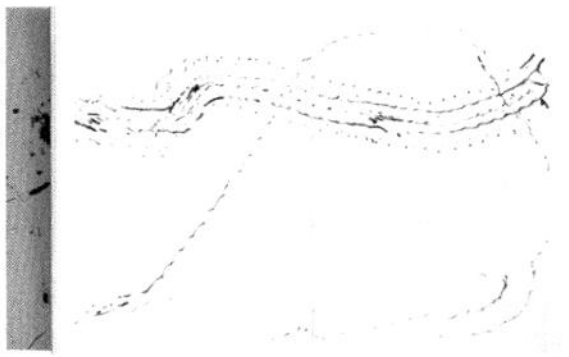

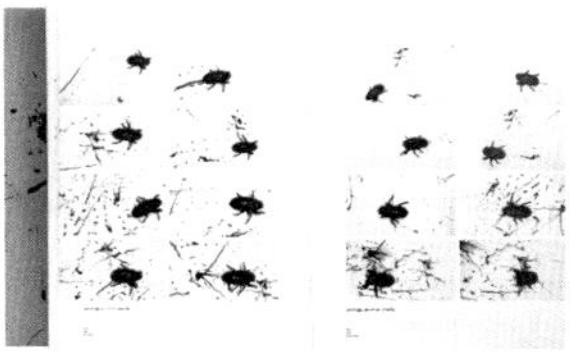
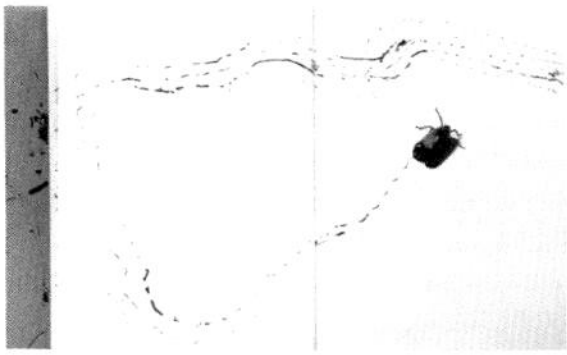

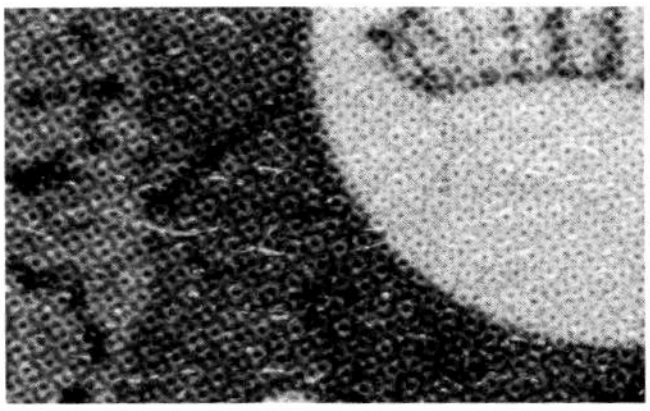
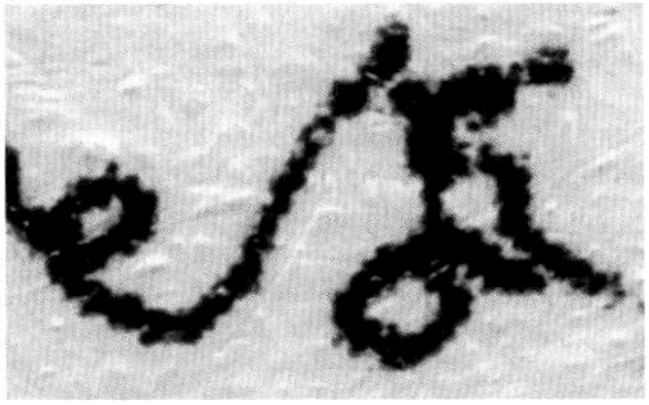

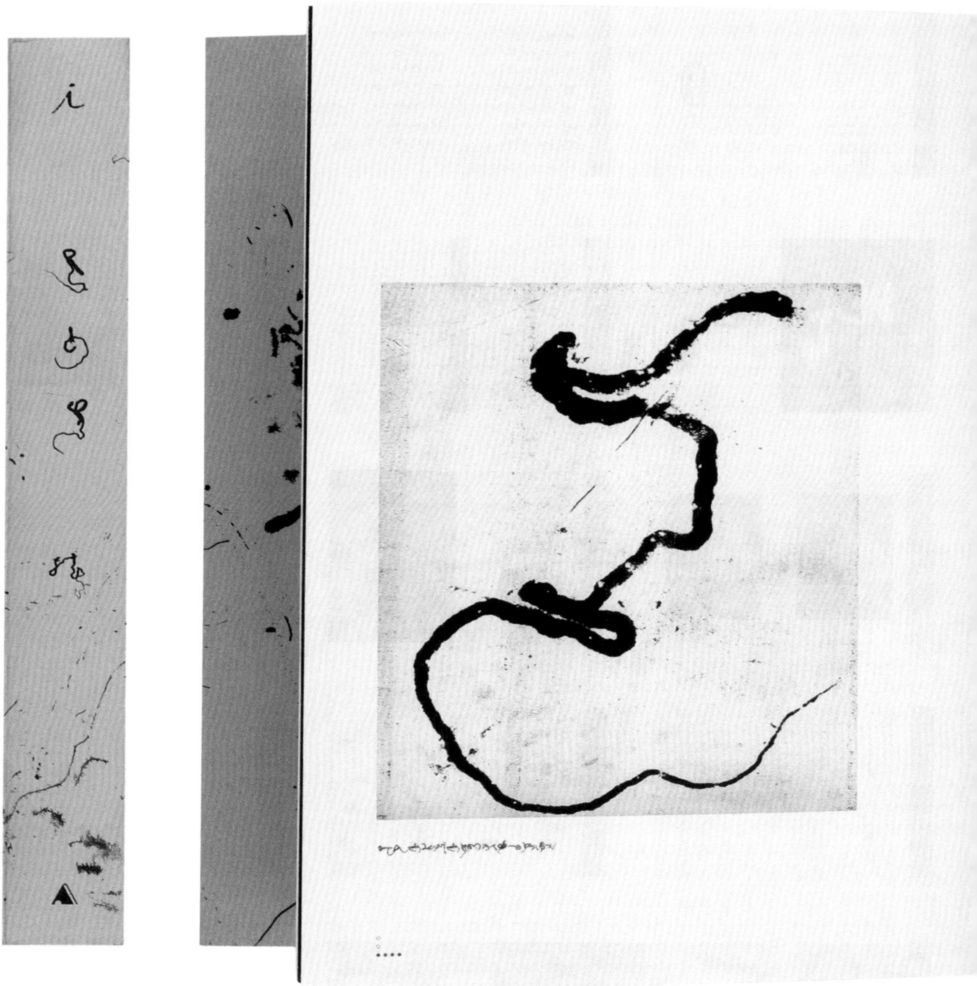

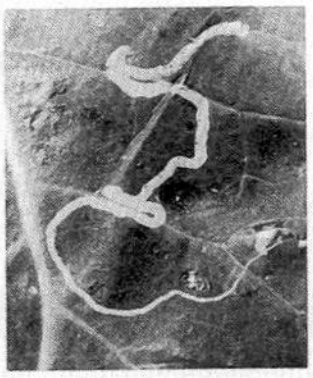

Ward
Kris

这是一本研究自杀者的亲友们面对亲人逝去、逐渐从悲伤中走出来找到新生活的相册。全书记录了 5 个相关案例，除了少量人物关系和相关故事的文字外，基本以图片为主。本书为骑马钉，封面与内页用纸、工艺相同。全书采用大小两种尺寸的页面混合装订而成，从封面开始的大页面采用胶版纸印单黑，呈现案例事件周围的环境、房屋、街道，具体到每个案例的时候则采用小一号、较薄一些的胶版纸四色印刷。案例以黑底白字的人名为开篇，逝去的人名在第 2 页就转变为英文的句号“.”，之后便是“留存下”的亲友们的哭泣、逝者的笔记、一些现场的照片等等；紧接着是他们通过做别的事情分散悲痛的心情，逐渐走出悲痛过程的照片。阅读过程沉重，引人思考，为的是让还活着的人们保持清醒，让“无法”预料的事变为“可以”预料，同时本书也给医护工作者的相关工作提供了最真实的资料。

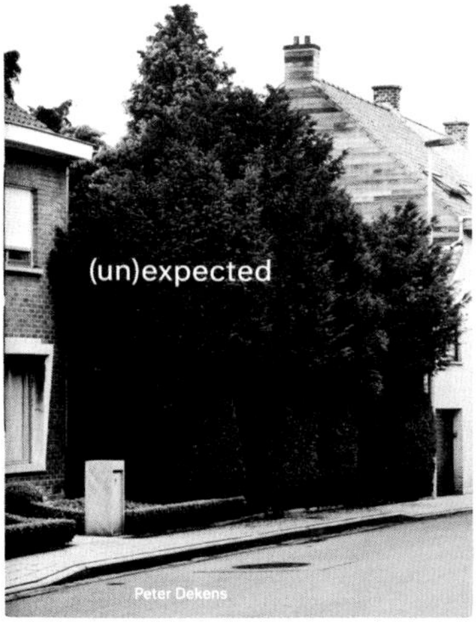

(un)expected
Peter Dekens

This is a photo album of suicides' relatives and friends who gradually recovered from their grief and found a new life. Five related cases were recorded in the book. Except for a few interpretations of personal relations and related stories, the book is mostly based on pictures. This book is saddle stitched. Its cover and the inner pages share the same processing techniques. There are two formats of pages in the book. The larger pages starting from the cover are printed in black on offset paper, presenting the surrounding environment, houses and streets of a certain case. The smaller pages are used for each specific case, printed in four colors on offset paper. The case begins with the names of people with white characters on a black background, and the names of people who have passed away are translated into full stops on the second page. After that, there are the photos of crying relatives and friends, notes of the suicides, photos of the scene and so on, followed by photos of them gradually getting over their grief by doing other things. The reading process is serious and thought-provoking, in order to keep the living awake, to make the unexpected predictable, and to provide the most authentic information about the work of health care workers.

- 银奖
- （未）预料到的
- 荷兰

- Silver medal
- (un) expected
- The Netherlands
- Rob van Hoesel
- Peter Dekens
- The Eriskay Connection, Amsterdam
- 290×208×4mm
- 357g
- 144p

Black / white photos devoid of people: two-storey brick houses in rows. Blinds rolled down. Garden hedges head-high. Overcast afternoons in the autumn. Evening illumination. Leafing back and forth repeatedly through magazine-sized, text-free, truncated double pages with pictures; the motifs seem like silent headlines. You wait for something to happen, but instead you feel left alone in an accurate and introverted environment. It becomes more distinct five times, however: more lively—you are tempted to say. Five smaller inner booklets are embedded in the large black/white book. A series of coloured pages each begins with a title page: a black surface, white forenames in a large font. You open it up and the repetition of the title differs in one respect: a missing name is marked with a suspension point—a full stop? The author and photographer of this sad, courageous project seeks public dialogue on the high suicide rates in West Flanders—not using an abstract study, but by looking at five true stories about families and relationships. The observer experiences this publication as a filmic reportage. In very concrete images, it manages to formulate the sorrow, remembrance and puzzlement that occupy the minds of the deceased's traumatised relatives. Free of pathos, this sensitive examination confronts us with a taboo. It inspires us to be attentive and stay awake when language falls silent and lonely eyes lower apprehensively.

Ime
Hanne

Mother
Father

Father

Steven
José

Ward
Kris

Kris

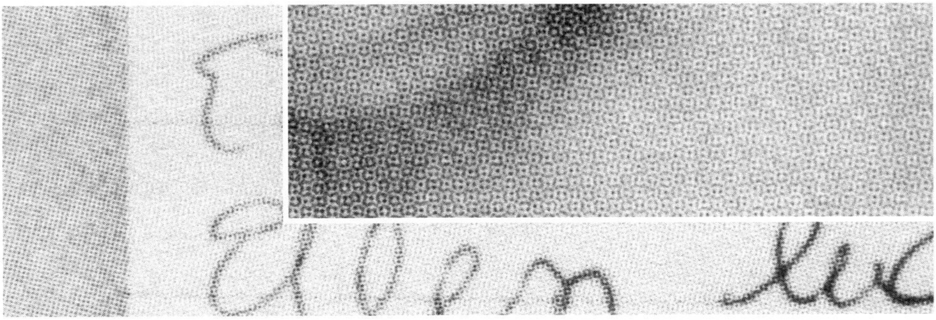

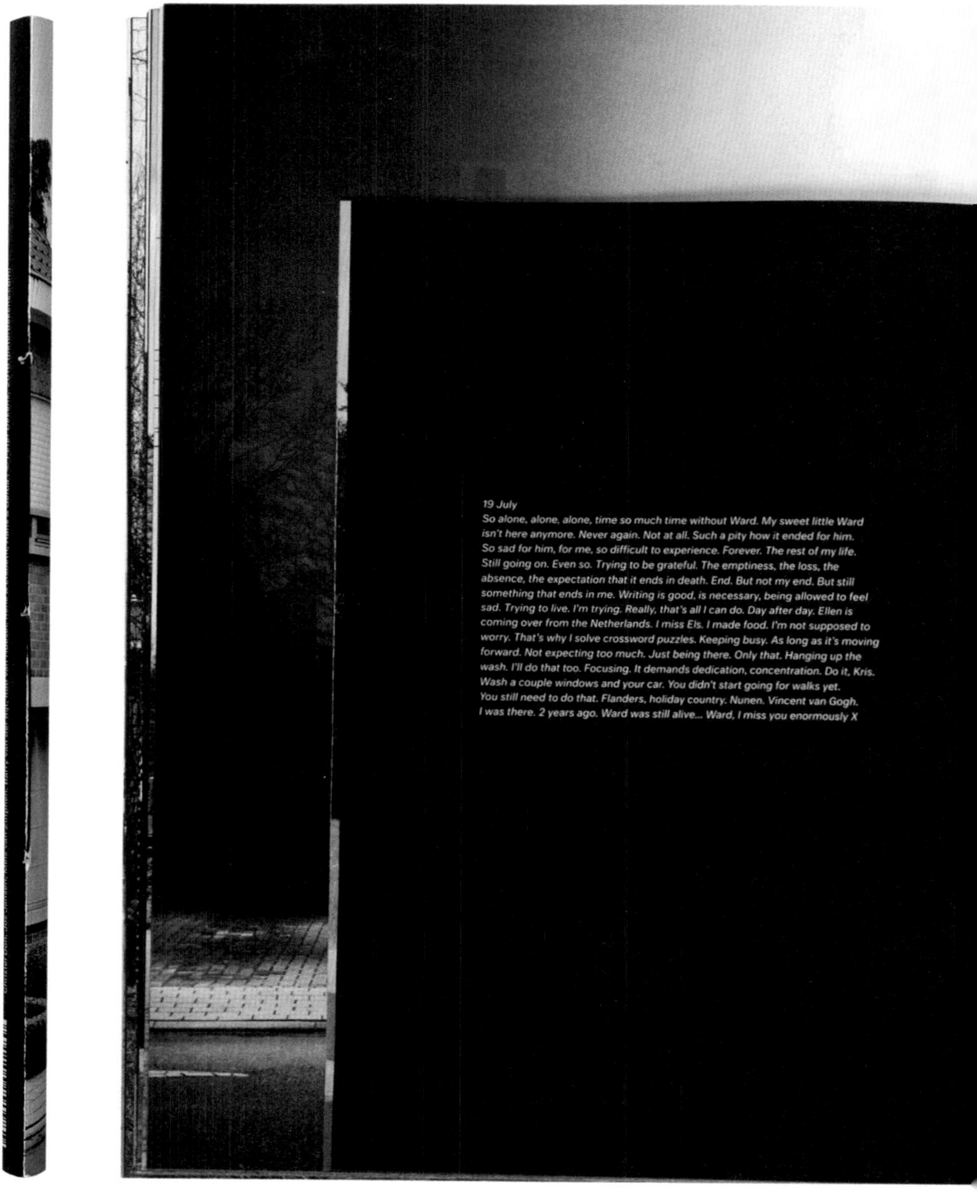

19 July

So alone, alone, alone, time so much time without Ward. My sweet little Ward isn't here anymore. Never again. Not at all. Such a pity how it ended for him. So sad for him, for me, so difficult to experience. Forever. The rest of my life. Still going on. Even so. Trying to be grateful. The emptiness, the loss, the absence, the expectation that it ends in death. End. But not my end. But still something that ends in me. Writing is good, is necessary, being allowed to feel sad. Trying to live. I'm trying. Really, that's all I can do. Day after day. Ellen is coming over from the Netherlands. I miss Els. I made food. I'm not supposed to worry. That's why I solve crossword puzzles. Keeping busy. As long as it's moving forward. Not expecting too much. Just being there. Only that. Hanging up the wash. I'll do that too. Focusing. It demands dedication, concentration. Do it, Kris. Wash a couple windows and your car. You didn't start going for walks yet. You still need to do that. Flanders, holiday country. Nunen. Vincent van Gogh. I was there. 2 years ago. Ward was still alive... Ward, I miss you enormously X

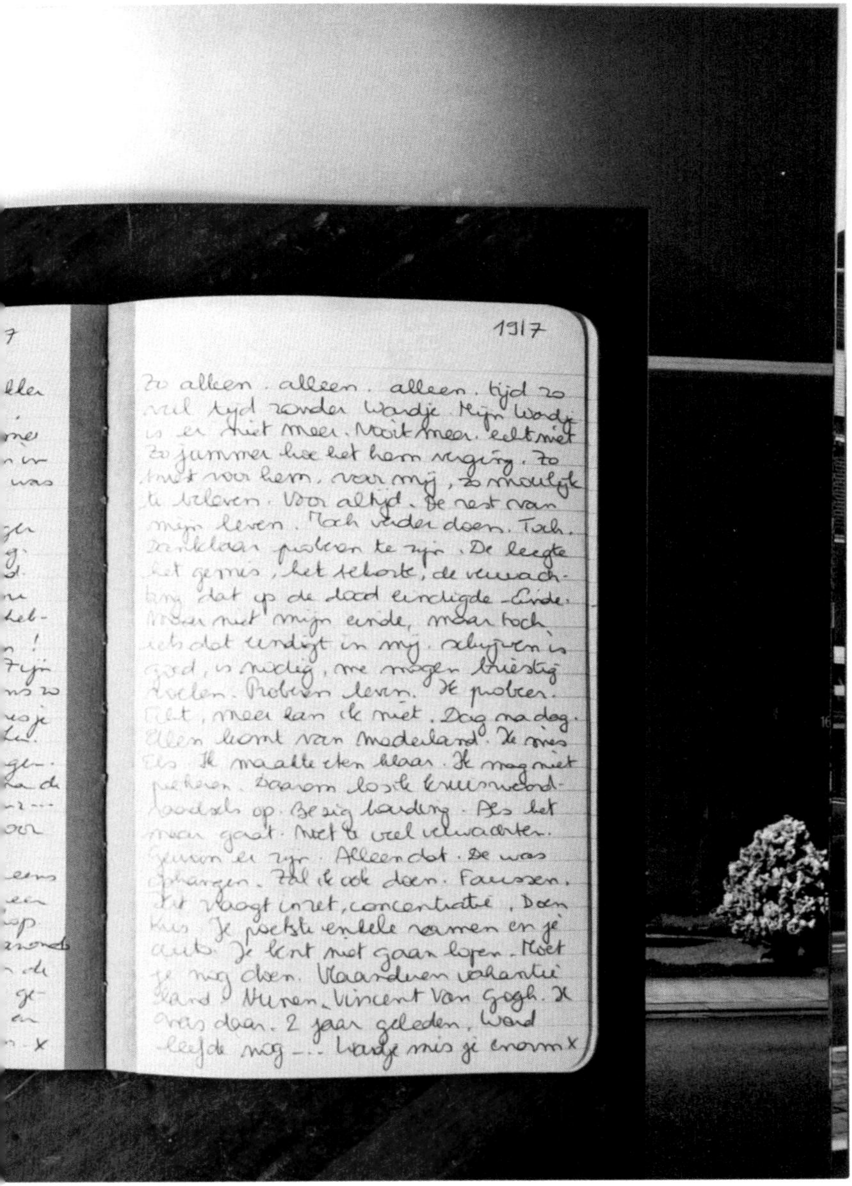
19/7
Zo alleen. alleen. alleen. tijd zo
veel tijd zonder Wardje. Mijn Wardje
is er niet meer. Nooit meer. echt niet
Zo jammer hoe het hem verging. Zo
triest voor hem, voor mij, zo moeilijk
te beleven. Voor altijd. De rest van
mijn leven. Toch verder doen. Toch.
Dankbaar proberen te zijn. De leegte
het gemis, het tekort, de verwachting dat op de dood eindigde. Einde.
Maar niet mijn einde, maar toch
iets dat eindigt in mij. schrijven is
goed, is nodig, me mogen triestig
voelen. Proberen leven. Ik probeer.
Echt, meer kan ik niet. Dag na dag.
Ellen komt van Nederland. Ik mis
Els. Ik maakte eten klaar. Ik mag niet
piekeren. Daarom loste kruiswoord-
raadsels op. Bezig houden. Als het
maar gaat. Niet te veel verwachten.
Gewoon er zijn. Alleen dat. De was
ophangen. Zal ik ook doen.
Het vraagt inzet, concentratie. Doen
Je poetste enkele ramen en je
auto. Je bent niet gaan lopen. Moet
je nog doen. Vlaanderen vakantie
land. Nuenen. Vincent Van Gogh. Ik
was daar. 2 jaar geleden. Ward
liefde nog... Wardje mis je enorm x

CHRISTOF NÜSSLI
WITHHELD DUE TO

这本书记录了美军在伊拉克战争中的阿布格莱布（Abu Ghraib）监狱虐囚事件。书中的图片和资料均来自美国政府对虐囚事件司法调查后公开发表的资料。本书严格来说没有装订方式，页面通过折叠的方式包裹在一起，并放入写有书名和作者名的透明塑封袋中，形成警方调查档案的外观。书中大量各色胶版纸单黑印刷，仿佛多次涂写、复印、传真的资料，而照片部分则直接展示受害者各种被伤害的身体照片，不适合被展示的部分被黑色块遮挡。书中开篇文字和中心页面均采用荧光橙色纸单黑印刷，强烈的视觉对比冲击着读者，通过本书这种强烈而直接的呈现罪行的方式，让人意识到对于丑闻的反思并不应该这样简单地结束。

The book documents the abuses at Abu Ghraib prison during the war in Iraq. The images and materials in the book are from materials released by the U.S. government after its judicial investigation into the abuse of prisoners. Strictly speaking, the book has no binding method because the pages are wrapped together by folding and placed in transparent plastic bags with the title of the book and the author's name to form the appearance of police investigation files. A large number of colored offset paper is printed in black, like repeatedly scribbled, copied and faxed materials; while the photo section directly shows the victim's injured body, and the parts that are not suitable for display are also covered by black blocks. The book's opening text and central pages are printed on fluorescent orange paper in monochrome black, which shows strong visual contrast to the readers. Through the book's direct presentation of crime, people realize that the reflection on the scandal should not be ended in such a simple way.

- 铜奖
- 扣押原因：
- 瑞士

- Bronze Medal
- Withheld due to:
- Switzerland
- typosalon, Christof Nüssli
- Christof Nüssli
- cpress
- 224×167×19mm
- 600g
- 408p

This sheaf of papers gives you the impression that it could perhaps become a book at some point. Each page is loose—it is actually just a simple collection of papers, folded over and bundled together. The outer sheet bears a distant relationship to a dust jacket; the cover image is incorrectly exposed and practically unrecognisable. It contains a thick wad of paper with ugly pictures, itself wrapped in some vivid neon orange sheets: haphazard printing of correspondence from the authorities, email printouts, photocopies, faxed to and fro countless times, with handwritten notes and scribbles, sections of text blanked out with felt pen, daubed until illegible. They seem like makeshift decorative sheets of paper, and look demonstratively like printer's waste. How did this slovenly paper object make its way into the sphere of the "best book designs from all over the world"? The 198 images display sections of prisoners' naked bodies, images that were released for publication by the US government following the judicial investigation into the torture scandal in Abu Ghraib, Iraq. They are some of the harmless photos where cruelty is hardly recognisable. When opened up, the following question and 50 others are listed: "What should these images prove?" They are printed on another vivid neon orange sheet of paper in the middle of the bundle. The pictures now have to undergo a second series of questioning, even though the legal sentencing has long been completed. In a moral, social respect, however, the scandal is far from over. The editor succeeds in achieving this update with the most sparing, yet cuttingly arranged ingredients.

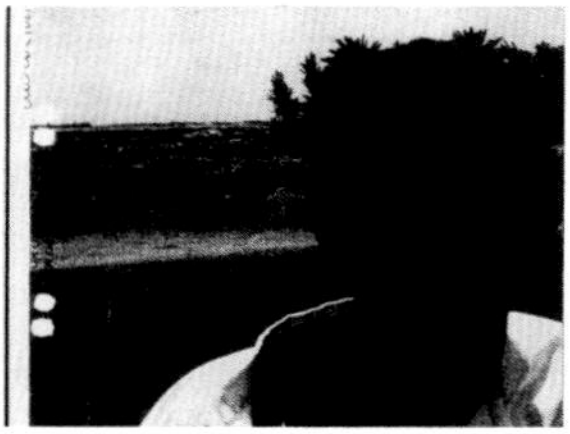

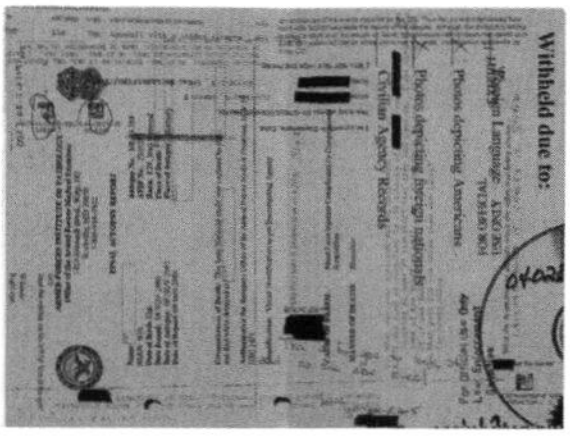

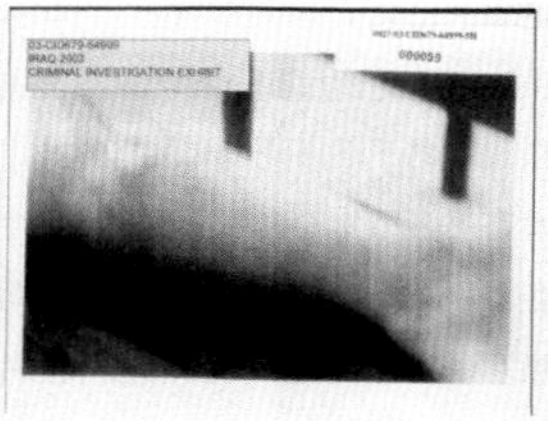

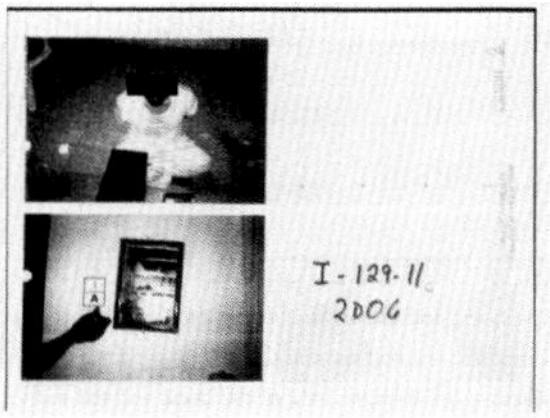

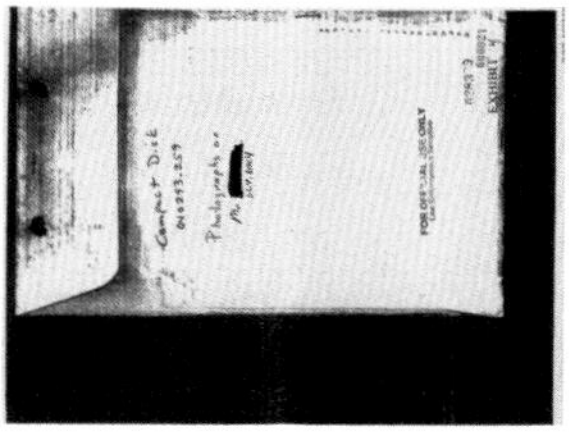

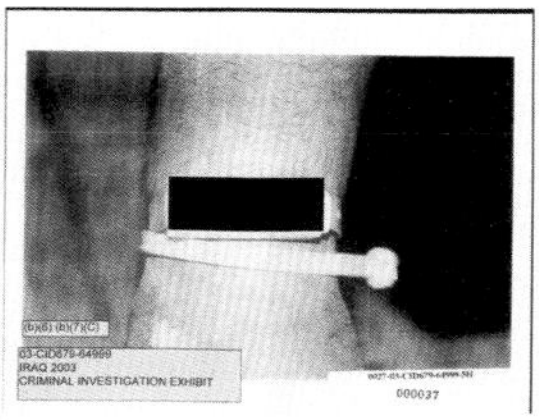
03-CID679-64999
IRAQ 2003
CRIMINAL INVESTIGATION EXHIBIT
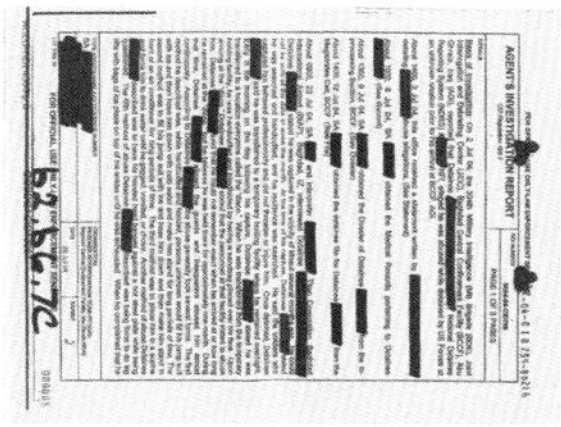
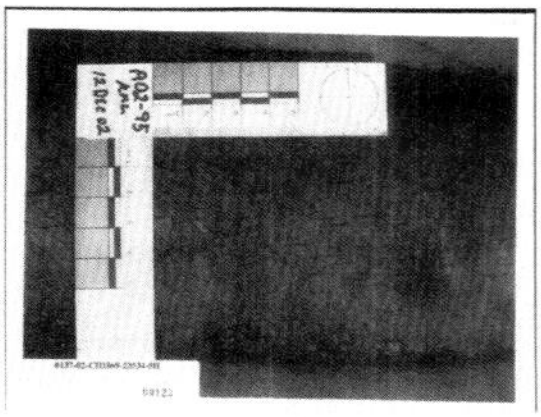
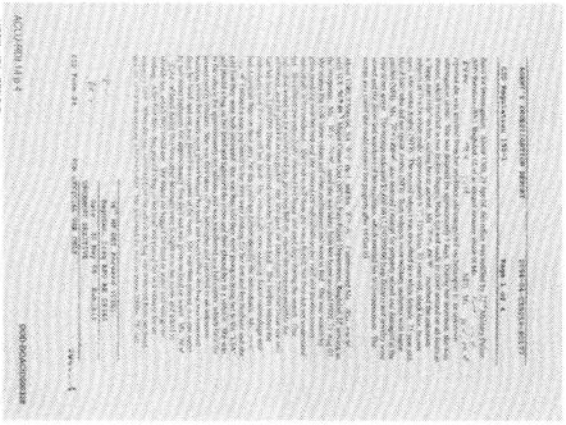
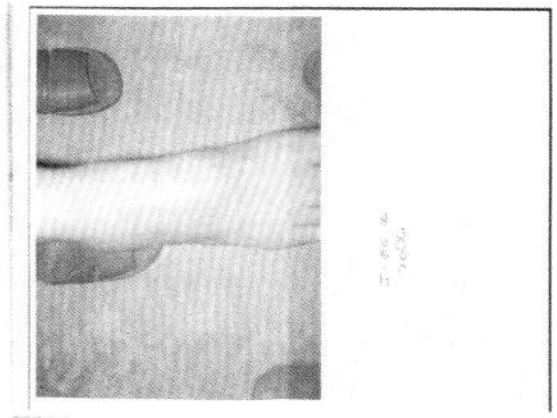
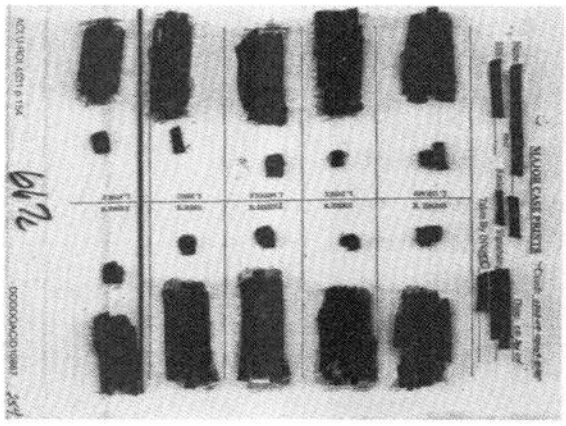
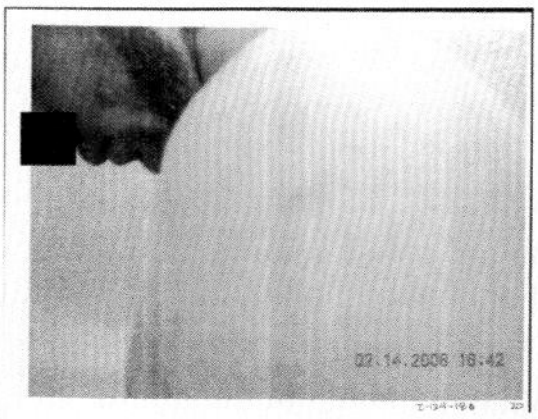
02.14.2006 16:42
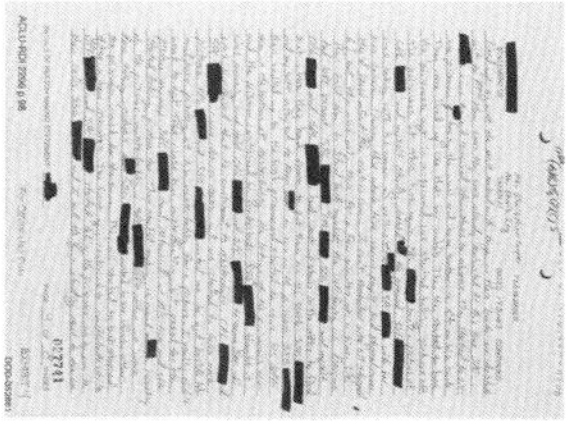
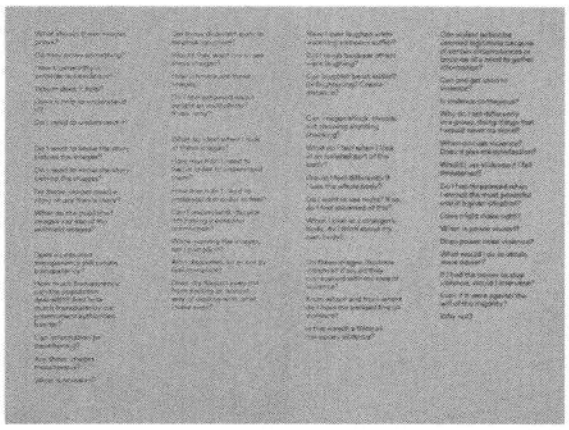
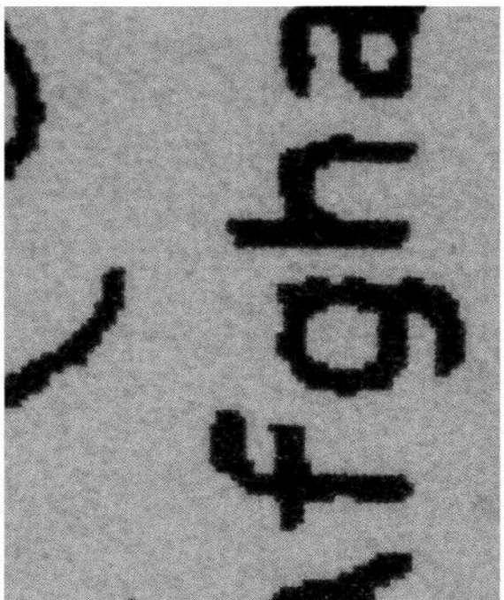

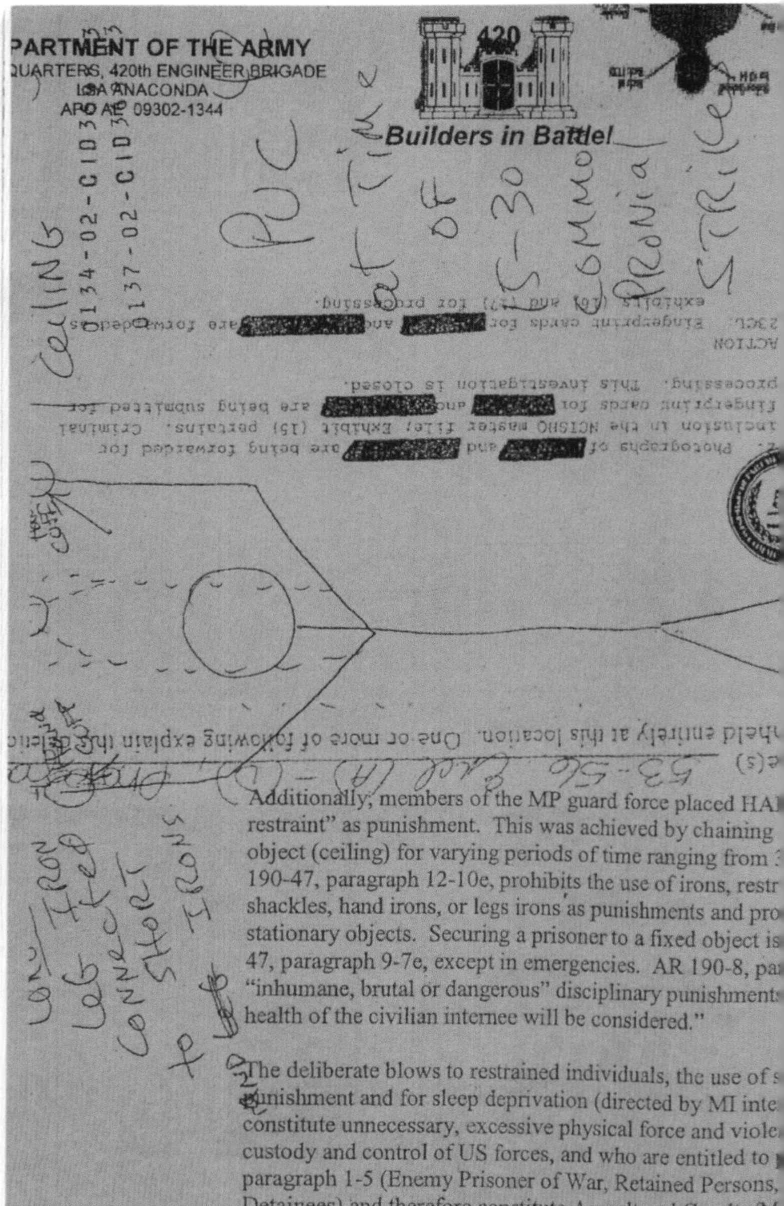

PARTMENT OF THE ARMY
QUARTERS, 420th ENGINEER BRIGADE
LSA ANACONDA
APO AE 09302-1344

Builders in Battle!

C. Photographs of [redacted] and [redacted] are being forwarded for inclusion in the NCISHQ master file; Exhibit (15) pertains. Criminal fingerprint cards for [redacted] and [redacted] are being submitted for processing. This investigation is closed.

ACTION

23CL. Fingerprint cards for [redacted] and [redacted] are forwarded as exhibits (16) and (17) for processing.

held entirely at this location. One or more of following explain the deletion

e(s)

Additionally, members of the MP guard force placed HA restraint" as punishment. This was achieved by chaining object (ceiling) for varying periods of time ranging from 190-47, paragraph 12-10e, prohibits the use of irons, restr shackles, hand irons, or legs irons as punishments and pro stationary objects. Securing a prisoner to a fixed object is 47, paragraph 9-7e, except in emergencies. AR 190-8, pa "inhumane, brutal or dangerous" disciplinary punishment health of the civilian internee will be considered."

The deliberate blows to restrained individuals, the use of punishment and for sleep deprivation (directed by MI inte constitute unnecessary, excessive physical force and viole custody and control of US forces, and who are entitled to paragraph 1-5 (Enemy Prisoner of War, Retained Persons, Detainees) and therefore constitute Assault and Cruelty/M

0134-02-CID3
0137-02-CID3
Ceiling
PUC
at time
of
15-30
53-56 Encl (A)-(D)
Leg irons
short irons

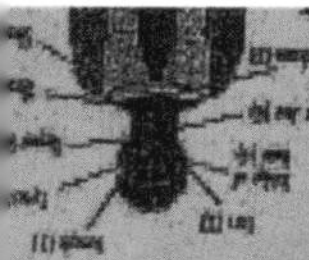

y were you laughing?
ought it was funny at the time. I don't think it is funny anymore.
you know where the videotape is now?
ve no clue. SGT [redacted] probably has it.
e you seen the tape?
I have.

AH in "standing
overhead to a fixed
es to several hours. AR
traps and jackets,
stening prisoners to
hibited under AR 190-
6-11a(4) prohibits
cifies the "age, sex and

g/restraint devices as
s), collectively
ersons under the care,
n IAW AR 190-8,
Internees and Other
ent.

1 set of
Leg
Irons
Feet on Floor

I sent several months in Afghanistan interrogating the Taliban and al Qaeda. Restrictions on interrogation techniques had a negative impact on our ability to gather intelligence. Our interrogation doctrine is based on former Cold war and WWII enemies. Todays enemy, particularly those in SWA, understand force, not psychological mind games or incentives. I would propose a baseline interrogation technique that at a minimum allows for physical contact resembling that used by [redacted]. (This allows open handed facial slaps from a distance of no more than about two feet and back handed blows to the midsection from a distance of about 18 inches. Again, this is open handed.) I will not comment on the effectiveness of these techniques as both a control measure and an ability to send a clear message. I also believe that this should be a minimum baseline.

Other techniques would include close confinement quarters, sleep deprivation, white noise, and a litnany of harsher fear-up approaches...fear of dogs and snakes appear to work nicely. I firmly agree that the gloves need to come off.

[redacted]
V/8
CW3
3ACR

(b)(6)-2,3

EXHIBIT 12

SHIPPED TO
US ARMY CRIMINAL
INVESTIGATION
LABORATORY
42600 1256
REGISTERED MAIL
11 AUG 2004

FILE COPY

A: No.
Q: Did you witness the Iraqi police hitting or beating the detainee?
A: No.
Q: Did you take a video clip of the detainee?
A: Yes.
Q: What did the video clip show?

What was the feelings of the group toward the video: Done for fun
As Joke
Were people laughing about it: Yes
Did they take it seriously or as a joke: Joke

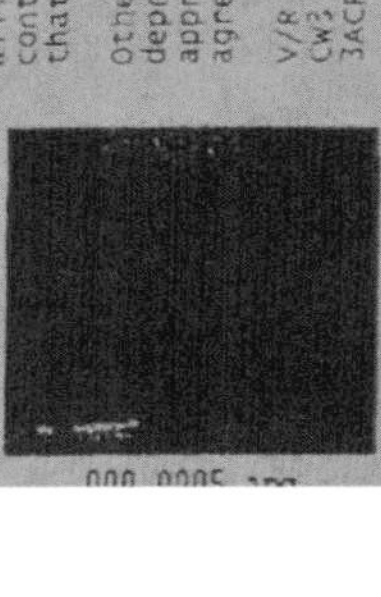

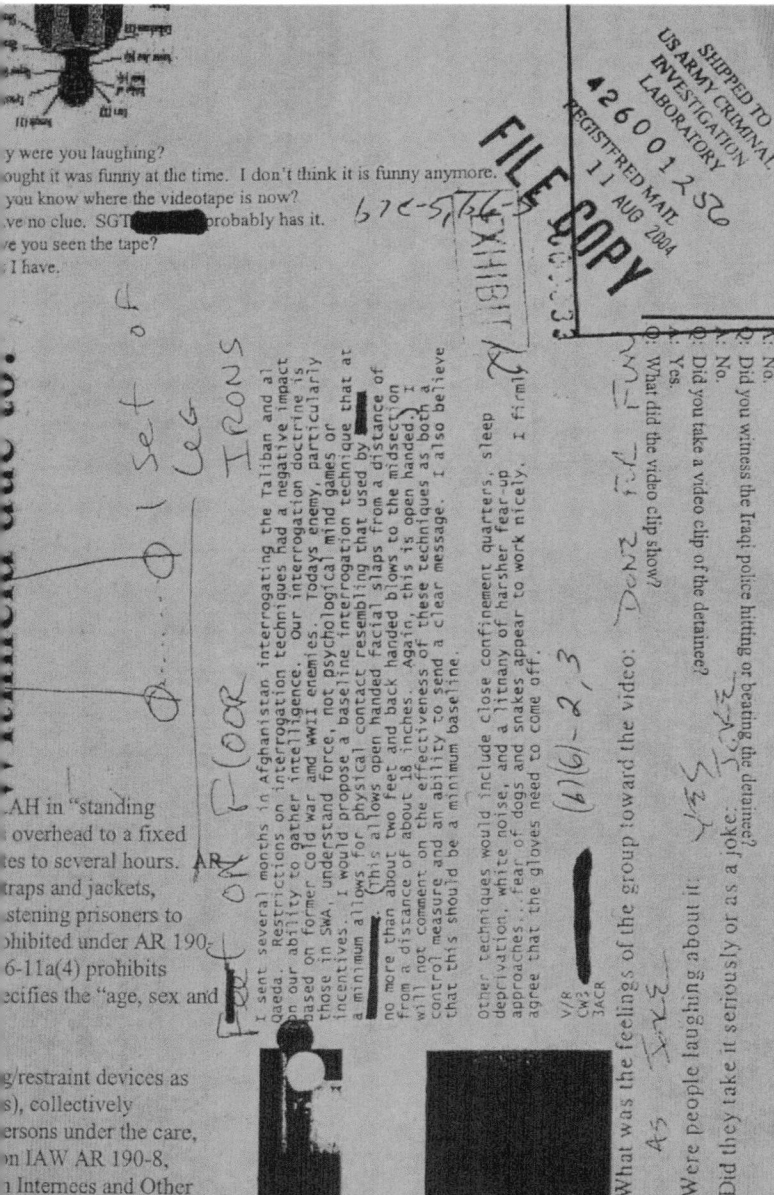

blocks
catching
a flying
comes into
Leuba
Ruffo
Corner

这是一本记录南非首都约翰内斯堡标志性建筑和人们当代生活的书。书中的建筑均为见证了这座大都市历史变化的摩天大楼。本书的装订形式为锁线软平装，封面采用微微泛蓝紫色的黑卡纸正反两面印金色，呈现出具有历史感的经典味道。内页除了扉页和最后的人物介绍、资料和版权信息外，没有对这些建筑的综述，而是直接呈现为 27 篇独立的部分，每一部分均包含以开本大小的胶版纸印单色黑呈现的建筑外观、平立面图，以及比开本略窄 10mm 的胶版纸四色印刷的采访、论文和照片。27 篇长短不一的页面在右侧翻口上形成与封面的建筑外立面类似的垂直序列，经典而有力量。与大多数建筑类书籍由建筑师来阐述建筑特点、展示建筑外观或结构不同，这本书通过大量普通人的报告、采访或论文表达他们与建筑之间的关系，从而形成对这些经典建筑的新观察，以及对南非当代城市生活的新见解。

This is a book that records the landmark buildings and contemporary life of people in Johannesburg, the capital of South Africa. The buildings in the book are skyscrapers that witness the historical changes of the metropolis. The paperback book is bound with thread sewing. The cover is made of a slightly bluish-purple cardboard printed in gold on both sides, showing a classical taste of history. There is no review of these buildings in the book. It presents 27 independent sections directly – each part contains the appearance of the buildings, plans and elevations printed on offset paper in monochrome, as well as the interviews, papers and photographs printed in four colors. Unlike most books on architecture, which are written by architects to illustrate architectural features and to explain the difference between appearance and structure, this book expresses their relationship between ordinary people and buildings through a large number of reports, interviews or paper. A new way of observing these classical buildings and a new insight into contemporary urban life in South Africa are formed.

铜奖

拔地而起——约翰内斯堡的高楼故事

德国

Bronze Medal

UP UP – Stories of Johannesburg's Highrises

Germany

Huber / Sterzinger, Zürich (CH)

Nele Dechmann, Fabian Jaggi, Katrin Murbach, Nicola Ruffo

Hatje Cantz Verlag, Ostfildern

270×192×24mm

926g

336p

The pages that are slightly smaller immediately catch your eye, even when the book is still lying there unopened. The sheets of paper with the texts are narrower by one centimetre, and so the formal entities separate by themselves the first time they are leafed through as if connected naturally. (This makes the book particularly nice for the user to handle, but particularly elaborate for the producer). On the left, a large picture. On the right—instead of a heading—a statement in large type that would look overexaggerated even as a title. But the fact that this looks like a headline indicates the equality of the interviews/essays with the monochrome architectonic portraits of the buildings: they are excellent architecture photos—historical, documental and contemporary, and the shorter pages of text with bold Baroque Antiqua, which, used in this way, visualises the journalistic impetus of the editorial concept. You could say that composite chapters are created which turn every piece of architecture into a sociological issue. High-rise architecture from the 20th century in Johannesburg: history, biography receives a location—the building, and building history receives a biographical connection. The design of the foldout cover proves expressive: vertical tectonic structure in golden print on black cardboard.

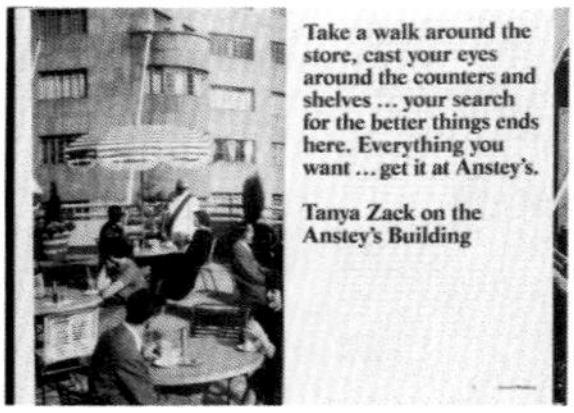
Take a walk around the store, cast your eyes around the counters and shelves … your search for the better things ends here. Everything you want … get it at Anstey's.
Tanya Zack on the Anstey's Building

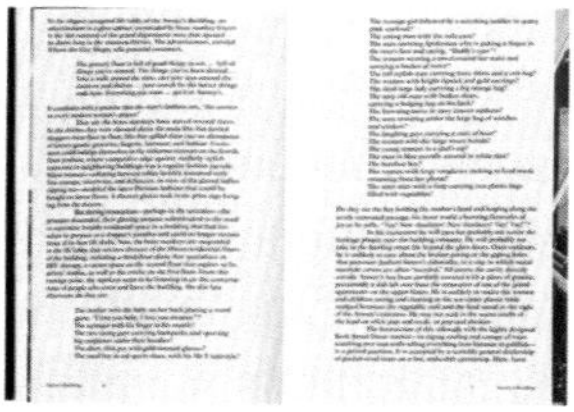

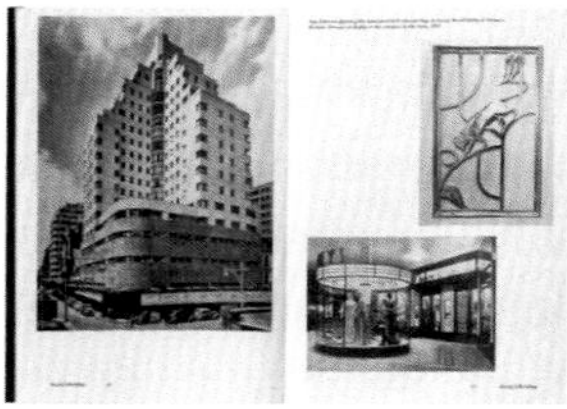

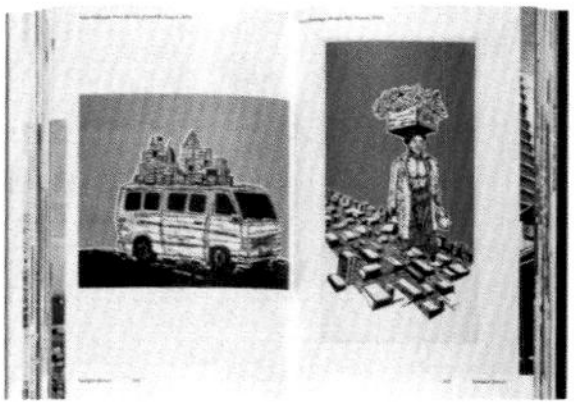

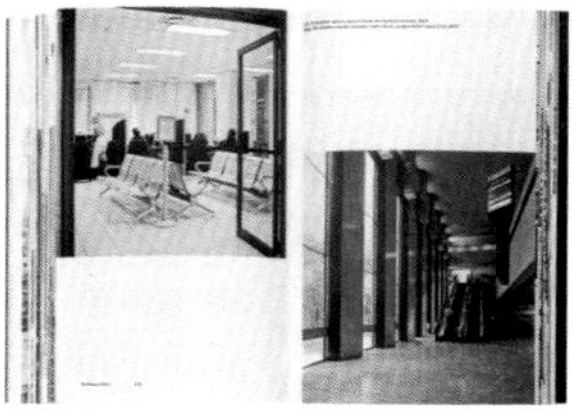

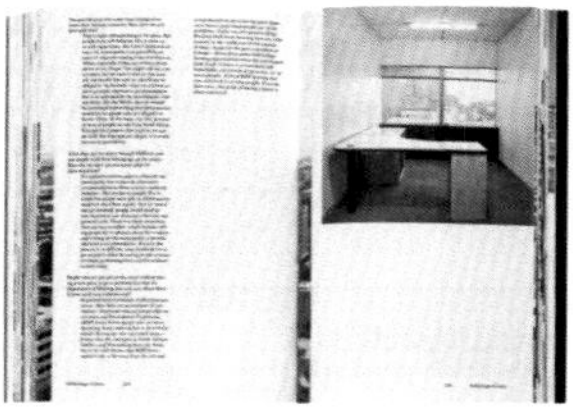

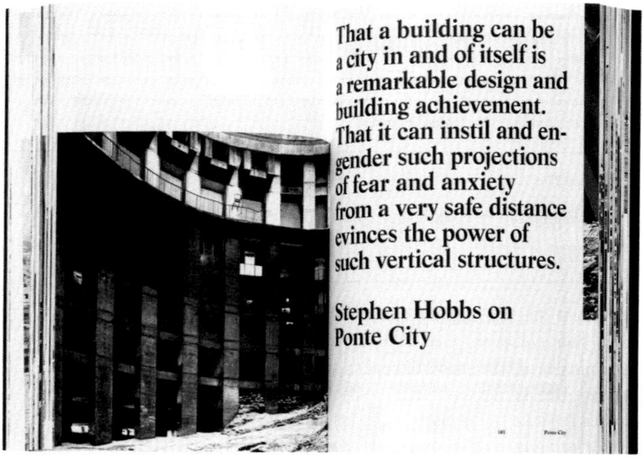
That a building can be a city in and of itself is a remarkable design and building achievement. That it can instil and engender such projections of fear and anxiety from a very safe distance evinces the power of such vertical structures.
Stephen Hobbs on Ponte City
181
Ponte City

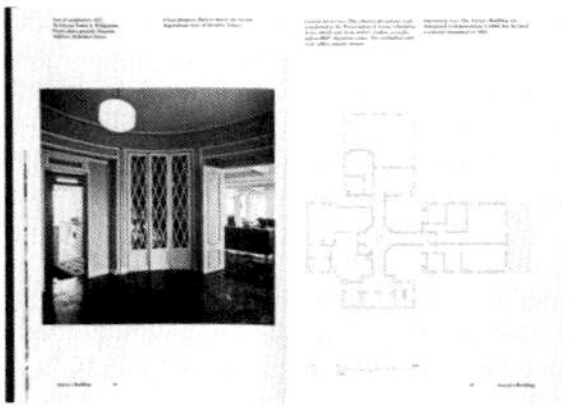

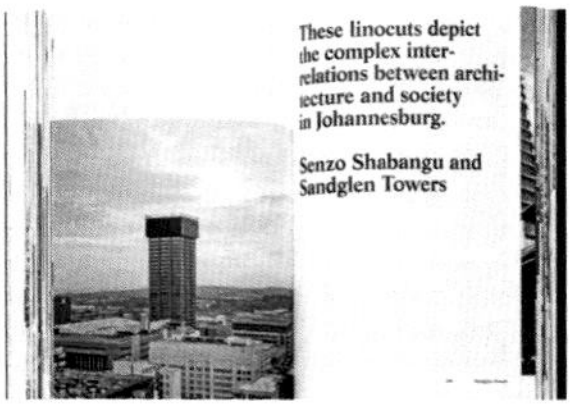
These linocuts depict
the complex inter-
relations between archi-
tecture and society
in Johannesburg.
Senzo Shabangu and
Sandglen Towers

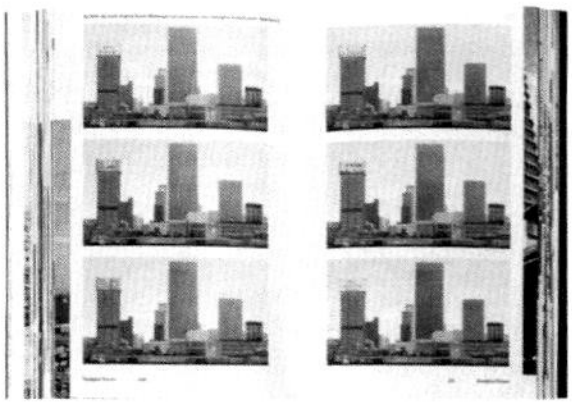

Now you can pay the premiums on all your possessions monthly!
SANTAM BANK LIMITED
INSURANCE PROMOTOR PLAN

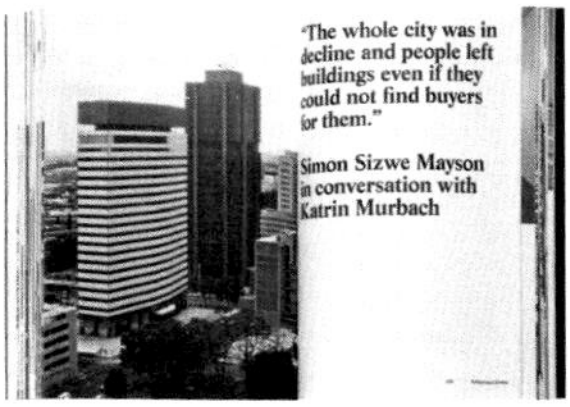
"The whole city was in decline and people left buildings even if they could not find buyers for them."
Simon Sizwe Mayson in conversation with Katrin Murbach

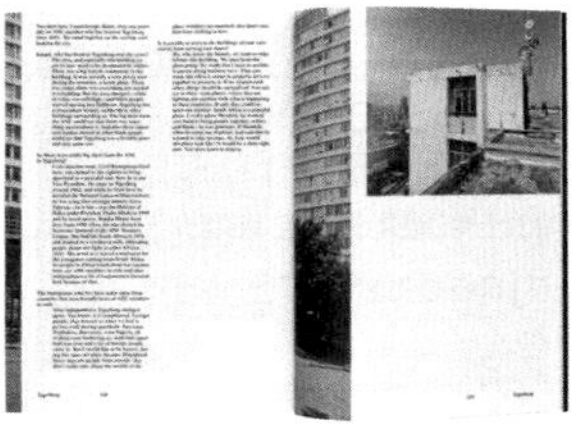

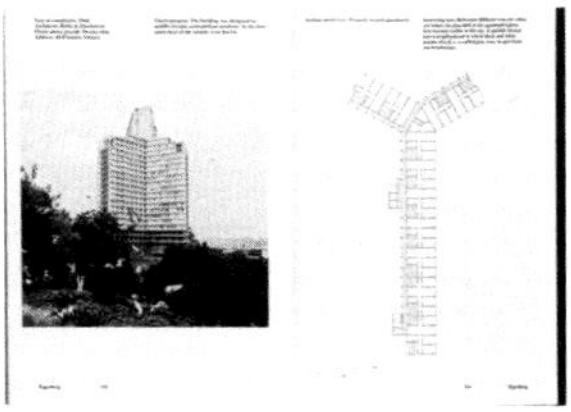

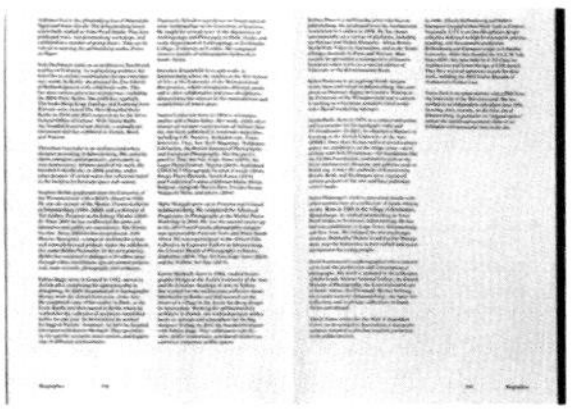

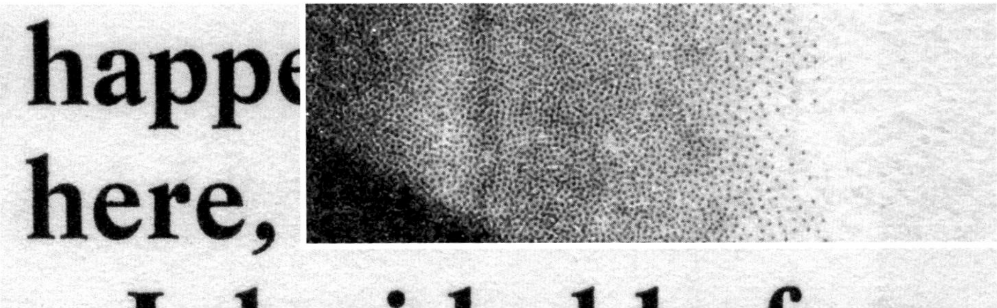
happe
here,
e. I decided before

UP
UP
Stories of Johannesburg's
Highrises
Fourthwall
Books
HATJE
CANTZ

enthusiastic about a shared journey to the Witwatersrand. He gladly accepted when Beit offered to leave some diamonds behind after the journey and with them he bought a large portion of the rights along the main reef in Johannesburg in 1886 and 1887.

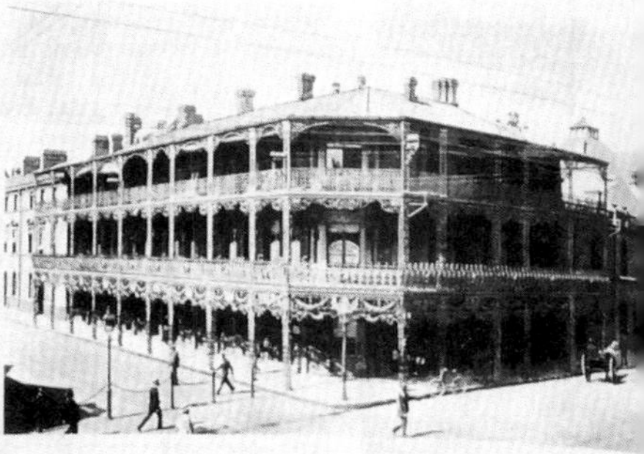

The second Corner House with its famous balconies, 1899.

Lord War Agents Mineral Extractors From Congo
We Are Currently In South Africa With Kgs Of 24
Carat Gold For Sale On A Very Good Negotiable
Price!!!!!!
Please Contact NOOR DINE MBULA 078 592 7969
redjoeozzy@gmail.com
Hurry While
Stock Lasts

Scrap of paper promoting Congolese gold on the black market.

Corner Houses One, Two, Three, and Four

In order to manage his operations locally, Eckstein built the first Corner House in 1886. The name Corner House is derived from his German surname, Eckstein, which means "cornerstone." As the beginnings of this new El Dorado were so frenzied, there was no time to wait for bricks and cement, so a single-story shack had to suffice for the time being. Beit, Eckstein, and another German mining magnate, Julius Wernher, founded the firm Rand Mines one year later. In 1890, the first Corner House was demolished and replaced by a magnificent three-story building with balconies. For the next thirteen years, this structure embodied power and wealth in the young town. Beit and his partners built the third Corner House in 1903, again as a replacement building on the same site. This edifice, with nine stories and a parapet, was the first highrise in Johannesburg, which had already become a city with 100,000 inhabitants by the turn of the century. The third Corner House, the unrivaled protagonist of the city's architecture until the construction boom of the nineteen-thirties, still stands at the corner of Simmonds and Commissioner Streets. In 1964, while the James Bond classic *Goldfinger* was gracing cinema screens for the first time, Rand Mines opened its fourth Corner House. At that time, the whole world's gaze was fixed on gold. In the new eighteen-story structure, all important rooms faced south: away from the sun and towards the mines.

The Hangover After the Party

Many of the gold mines in Johannesburg are no longer in operation. Nevertheless, the landscape of the city is still shaped by mines, and organizations are now drawing attention to the problems associated with the massive sand dumps that the mines have left behind. One person who has taken up this problem is the activist Mariette Liefferink, who conducts regular tours of the mines in order to educate the public about the seriousness of the situation. The meeting point for the tour with Mariette is usually the McDonald's in the West Rand town of Krugersdorp. The striking Mariette is as well known for her signature red lipstick, sunglasses, and high heels as for her profound knowledge of the mining industry and its effects on the environment. She is at the highest level in seventeen government committees. She flies around the world to speak at environmental congresses. She swaps her elegant shoes for rubber boots in order to measure the radioactivity in mine dumps. She goes to court to litigate against the world's powerful. Several years ago, the European Environmental Bureau described the irreversible damage to the ecosystem caused by mines as the second biggest challenge after global warming. Mariette stands in front of the piles of material extracted from the tunnels of the underworld and tells the background story: 270 mine dumps around Johannesburg containing 600,000 tons of uranium and released radioactivity that can last for millions of years. Poisoned rivers, contaminated wetlands, unusable soil. The tour leads up a hill, where the radioactivity is as high as in Chernobyl's exclusion zone. It leads to a broken pipe, where a poisonous soup gushes into the countryside. She tells of

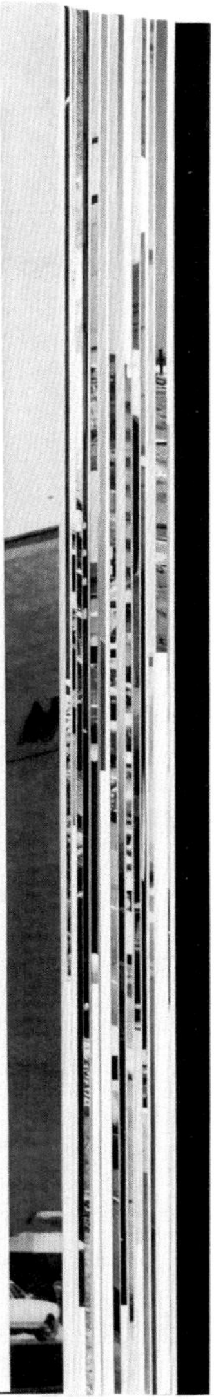

Emily
Dickinson
Edna
St. Vincen
Millay

这是两位著名的美国女诗人艾米丽・狄金森（Emily Dickinson）和埃德娜・圣文森特・米莱（Edna St. Vincent Millay）的诗歌合集。两位女诗人并不是生活在同一时期的人，但她们都在自己的时代用诗歌的形式尽力去挣脱时代的陋习，在某些意义上，都是具有鲜明个性的独立女性。本书的装订形式为锁线胶平装，封面通过可以折进翻口的勒口将书籍分为两部分，也形成了一分两半的第二个"书脊"。全书采用蓝色、绿色，以及两个颜色的混合色——墨绿色，一共三种颜色进行印刷。书籍两个封面分别采用蓝绿双色，书籍和封面的共用信息采用墨绿色。内页部分半涂布胶版纸三色印刷，两人分别使用蓝、绿色分开，前言、目录和之后的索引、参考信息页采用墨绿色。全书采用了一种略扁的无衬线字体，呈现出与社会对于女性的"期望"所不同的独立而坚毅的味道。||||||||||||||||||||||||||||||||

Emily
Dickinson
Edna St. Vincent
Millay
DWARS
VERS
Lyrisch tweeluik
gedichten &
sonnetten
Vertaald en
bijeengebracht
door Ans Bouter

This is a collection of poems by Emily Dickinson and Edna St. Vincent Millay, two famous American poetesses. The two poetesses were not living in the same period, but they tried to break away from the bad habits of their own times with the form of poetry in their own times. To some extent, they both were independent women with distinct personality. The paperback book is bound with thread sewing. The cover divides the book into two parts by the flap that can be folded into the fore-edge. It also forms another 'spine'. The book is printed in blue, green, and a mixture of two colors – dark green. The two covers are blue and green to distinguish two poetesses, and the inside pages are printed on the semi-coated offset paper in three colors. The foreword, catalogue, index and reference pages are in dark green. The book adopts a slightly flat sans serif font, showing independence and persistence different from the 'expectations' of women in our society.

- 铜奖
- 交织的新鲜
- 荷兰

- Bronze Medal
- DWARS VERS
- The Netherlands
- Team Thursday (Simone Trum & Loes van Esch)
- Emily Dickinson, Edna St. Vincent Millay
- Ans Bouter en Benjamin Groothuyse
- 210×150×20mm
- 469g
- 272p

This flap soft cover has very special flaps. They not only reinforce the jacket, but at the front they are also furnished with a kind of second spine which enables each of them to wrap one half of the innerbook in the middle of the book. The book can therefore be placed on the shelf in both directions. A lyric diptych—and the book actually opens automatically between its two halves. The insides of the flaps bear the names of the poetesses, in large grotesque type and lines pitched in opposite directions. Each covers half of the full-page portrait photos, the left monochrome in blue, the right printed in green, with the lines in corresponding colours. The paratext at the beginning of the book— title pages, articles—and at the end of the book—articles and appendix— is printed in a blue-green combination colour, a manifest yet plausible means of harmoniously combining both sections of the book. The design elements are consciously reduced overall. The spacious proportion of white, the grotesque type of the kind used around 1900, the surprising colours and the severe architecture of the book all provide a fresh setting for the poetry by these two American poetesses.

Emily Dickinson
Edna St. Vincent Millay

Emily Dickinson
Gedichten

Emily Dickinson
Gedichten

Het vers, de vrouw en de vorm
Over het vertalen van lyriek
door Bob van Toor

'To many poets the physical character of their poem, its rhythm, its rhyme, its music, the way it looks on the page, is quite as important as the thing they wish to say; to some it is vastly more important.'

Edna St. Vincent Millay
in de inleiding van Charles Baudelaire's *Les fleurs du mal*

In 1936 hoorde Edna St. Vincent Millay, toen de beroemdste dichter in Amerika, dat haar jonge collega en minnaar George Dillon van plan was Baudelaire te vertalen. Millay was meteen geïnteresseerd in het project; vooral omdat Dillon zich tot doel had gesteld Baudelaire's haast ondoorgrondelijke verzen te vertalen met behoud van elk metrum en zo veel mogelijk in de oorspronkelijke vorm. Een enorme taak, waar zij gefascineerd aan begon mee te denken. Uiteindelijk zou Millay een groot deel van de vertaling op zich nemen en werd ze mede-auteur van de bundel.

Het respecteren van cadans, vorm en ritme, zo stelt Millay in haar introductie van dat boek, is een bijkomend juk voor de vertaler, die al zwoegt om het gedicht van de oorspronkelijke taal naar een nieuwe over te dragen zonder het te verstikken. Wanneer de vertaling echter op allebei die fronten slaagt, is de beloning groots. Dan kan het nieuwe gedicht worden voorgelezen na het oude, als een passende echo; het is in feite hetzelfde gedicht, 'mogelijk wat ruw behandeld, maar in elk geval in zijn anatomie intact.'

Deze dubbelbundel met vertalingen van het werk van Dickinson en Millay kwam tot stand met hetzelfde doel voor ogen. Een selectie van gedichten uit de grote oeuvres van beide dichters, ingedeeld naar de thema's van hun werk. Binnen de thema's zijn de gedichten chronologisch gerangschikt, zodat ontwikkelingen in het werk van de dichter zichtbaar worden. Het oorspronkelijke gedicht en de Nederlandse vertaling volgen elkaar, zoals Millay het uitdrukte, als een echo. Om naast elkaar gelezen, overdacht en herontdekt te worden: lyrische poëzie van twee geniale vrouwen, overgedragen uit het Engels met het grootst mogelijke respect voor de vorm zowel als de inhoud.

Bob van Toor 15

Edna
St. Vincent
Millay
Gedichten &
sonnetten
Over Emily Dickinson
Index op beginregel
Emily Dickinson
Index op beginregel
Edna St. Vincent Millay

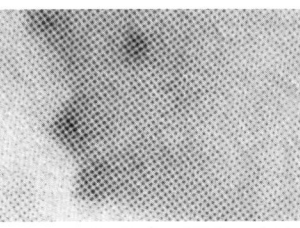
rote oe
un wer
3ir
schikt, zodat ontwikke

Emily Dickinson
Edna St. Vincent Millay

Lyrisch tweeluik
gedichten & sonnetten

Vertaald en bijeengebracht
door Ans Bouter

Bevrijdend keurs-lijf

Over taal
en ritme

door
Ans Bouter

Het geeft enorm veel voldoening als het lukt het melodische ritme en metrum van een gedicht terug te laten komen in de Nederlandse vertaling. Dat ontdekte ik zes jaar geleden, toen ik een vertaling las van 'O tell me the truth about love', een mooi liefdesgedicht van Wystan Hugh Auden waarin hij steeds van metrum wisselt.

Is it prickly to touch as a hedge is,
Or soft as eiderdown fluff?
Is it sharp or quite smooth at the edges?
O tell me the truth about love.

Our history books refer to it
In cryptic little notes,
It's quite a common topic on
The Transatlantic boats

Ik vroeg me af waarom het zo typerende ritme niet was aangehouden in die vertaling. Zou dat niet te doen zijn? Zou de Nederlandse taal zich daar niet goed voor lenen? En wat jammer zou dat zijn, want er gaat dan zoveel zeggingskracht verloren! Ik wilde kijken of zulke ritmewisselingen ook in het Nederlands te realiseren zijn en was blij verrast toen bleek dat dergelijke ritmische poëzie, die rijmt zonder tot een dreun te vervallen, zich wel degelijk op eenzelfde manier laat vertalen.

Is het stekelig net als de hulst is
Of zacht als een vossenstaart
Denk je zelf dat het goud of verguld is
Oh liefde, onthul me je aard

这是法国平面设计师贝尔纳·夏德贝克（Bernard Chadebec）设计的安全提示海报的合集。这位设计师在1965–2005年担任法国国家研究与安全研究所的设计师期间，共设计了超过300张关于预防事故和职业病的海报。书籍为锁线胶平装，护封胶版纸四色印刷，内封单面白卡纸印单黑，主要呈现书名和英法双语目录。内页分为海报展示、工作介绍、海报复制品和设计师生平四大部分。第一部分精选40张海报，胶版纸四色印刷，每张四折后，按照四张一组进行装订，上切口不清边，形成类似于蝴蝶装的结构。翻阅中海报的局部相互产生视觉冲击，吸引读者去探索。后三部分采用光面铜版纸四色印刷，工作介绍部分的照片和海报复制品部分均为黑色挂粗网设计，页面下方有数字，可以在前面的相关页码中找到折叠的四色海报。

chadebec

It's a collection of safety alert posters by the French graphic designer Bernard Chadebec. As a designer at the National Institute for Research and Safety in France from 1965 to 2005, the designer designed more than 300 posters on the prevention of workplace accidents and occupational diseases. The paperback book is perfect bound with thread sewing and the jacket is printed on offset paper in four colors. The inside cover is printed on white cardboard in black to present the title of the book and bilingual catalogue in English and French. The inside pages are divided into four parts: poster display, job description, poster reproduction and the life story of designers. In the first part, 40 posters are selected and printed in four colors on offset paper. Every four pages are bound together without cutting the front edge, forming a structure similar to butterfly-fold binding. Looking through parts of the posters produces strong visual impact, which attracts readers to explore. The last three parts are printed in four colors on glossy coated paper. The photos of the job description and the reproductions of the posters are all designed with AM screening. There are numbers at the bottom of the page and readers may find the four-color posters according to the relevant number in the previous part of the book.

铜奖

贝尔纳·夏德贝克——友好的警示

瑞士

Bronze Medal

Bernard Chadebec – Intrus Sympathiques

Switzerland

Olivier Lebrun and Urs Lehni in collaboration with Simon Knebl, Phil Zumbruch, Saskia Reibel and Tatjana Stürmer (HfG Karlsruhe)

Olivier Lebrun and Urs Lehni

Rollo Press, Zürich

191×135×21mm

470g

272p

This paperback monograph honours the poster designs of Bernard Chadebec. They are colourful posters with a serious topic and educational purpose: the prevention of accidents at the workplace and occupational illness. Chadebec was employed as a graphic designer at the National Research and Safety Institute in France for many decades. In the catalogue there is not a trace of admonitions, catastrophic scenes or threats and risks, however. Instead, a colourful kaleidoscope of graphic stimuli, similar to those typical of the sixties and seventies.The posters have been wonderfully integrated into book form. Even as reduced-sized reprints on wafer-thin paper, they are still so large that they cannot fit individually on a single page of the book. The reprints have therefore been folded twice and put together in threes to form layers with stitch binding. The top of the book has not been cut, though. And so sections of the quartered posters are sometimes upside down, and sections of different posters sometimes face each other on opposite sides. You can't really look at them as a whole at all. But something else becomes all the more clear. Chadebec's graphic art is precise and powerful; each detail is permeated with the quality of the whole poster. The originals are attractive posters because they are supposed to be hung up at the workplace. Humorous yet equally drastic, they cut right to the chase of each danger. And the book design works with correspondingly congenial didactics: you not only observe the posters as a relic of design history, but also appreciate the timeless quality. Looking more closely at the details, you can also examine closely what exactly makes a poster into a poster.

pièce non

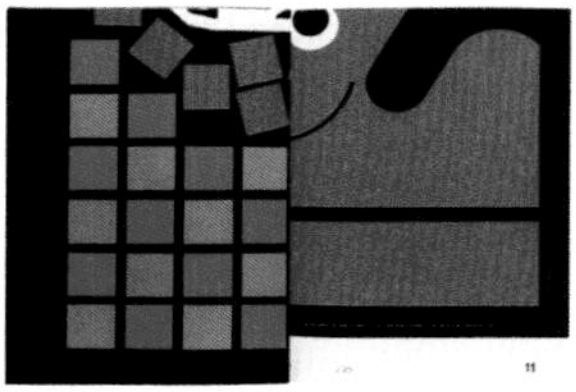

nger ou de fumer
chadebec

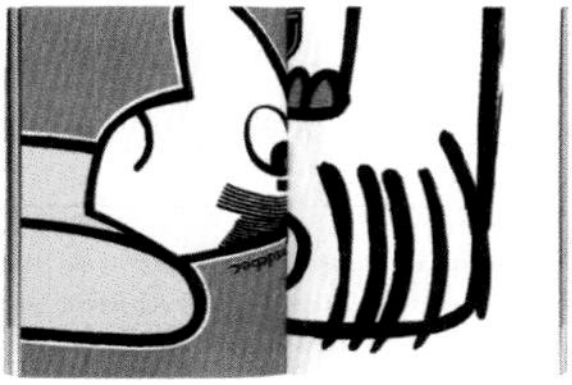

2 sécurités:

z toujours
protecteur

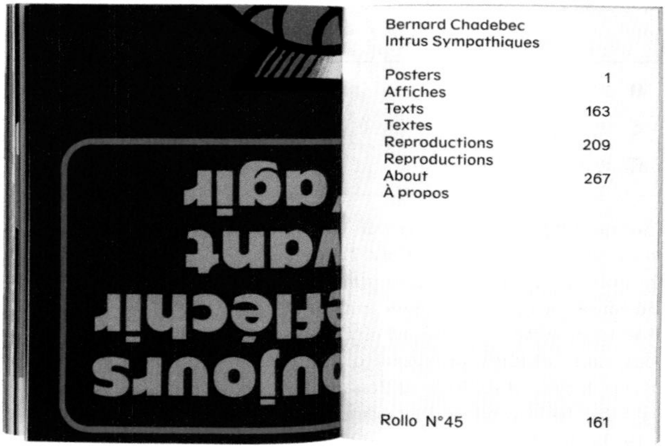
Bernard Chadebec
Intrus Sympathiques
Posters 1
Affiches
Texts 163
Textes
Reproductions 209
Reproductions
About 267
À propos
Rollo N°45 161

Abc
À propc

est si vite arrivé
27

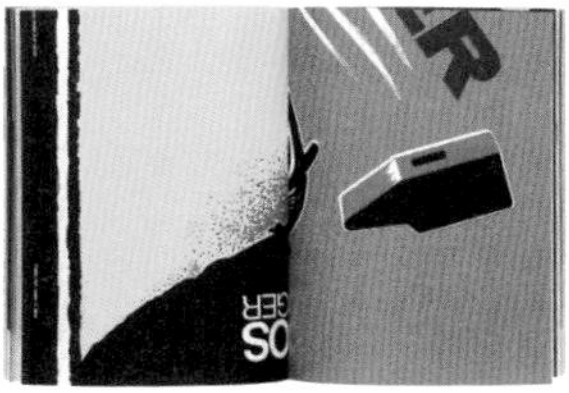

avant...de ma
98

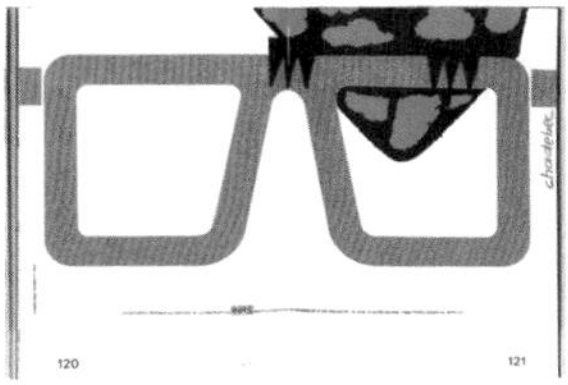
120
121

SE
PROTEGER
N'EST PAS
semelles anti
C'EST P
130

J'AI DES
CHAUSSURES
A COQUILLE
D'ACIER
139

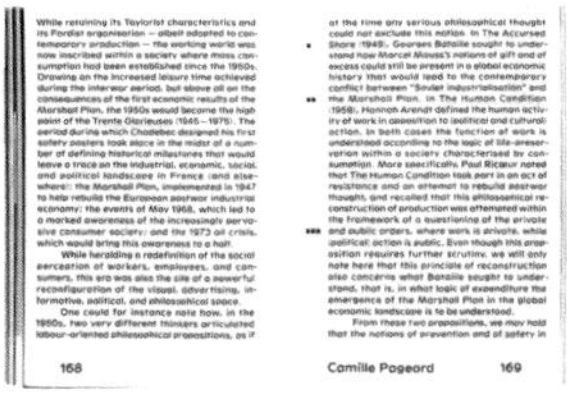
168
Camille Pageard
169

178
Notes

184
185

SOCIA
TIONAL DE RECHERC
IE OLIVIER NOYER 75680 P

REDUCTION DE L'AFFICHE I.N.R.S. N° 332 D CODE CDE 4860 FORMAT 60 x 80
chadebec à la manière de Picasso
CHAQUE CHOSE A SA PLACE
LA SECURITE SOCIALE AU SERVICE DE LA PREVENTION
INSTITUT NATIONAL DE RECHERCHE ET DE SECURITE
30, RUE OLIVIER NOYER, 75680 PARIS CEDEX 14
266
024 025

French graphic designer Bernard Chadebec was an employee of the INRS throughout his professional life, between 1965 – 2005. At this institute, he realised more than 300 posters around the topic of accident prevention, which circulated in a total print run of over 30 million copies. This book has helped to bring about the first monographic exhibition of Chadebec's work at the Écomusée Creusot Montceau.

The National Research and Safety Institute for occupational accident and disease prevention (INRS) is a non-profit association created in 1947 in the wake of the establishment of the Social Security scheme.

The INRS is tasked with the following missions: to identify occupational hazards and highlight their dangers; to analyse their consequences for the health and the safety of men and women at work; to develop and promote means of preventing and controlling these hazards. To do so, the INRS conducts study and research programmes, and organises information, training and assistance activities.

The INRS continues to design and circulate posters on any subject it considers important for the health and safety of men and women at work, who may be addressed through this media.

Visitors may acquire posters from:
www.inrs.fr

d again.
ent + rejection
changes + pressures.
y repetitive downfall + that
then is all he can cover with me.
I should do nothing now for Tom.
Tom has not changed.
Nothing is all no.
all black. all white.
Keep going on
work.
The actions are all constructive
good
only the work actions
The fears
That is the weakness

本书为整理并重新编排的美籍德裔雕塑家伊娃·黑塞（Eva Hesse）的日记。伊娃主张完全即兴和未知的创作过程，惯用不稳定的、有机的、不断变化的材料，强调艺术的最终状态不是关键，而重视艺术的形成过程。本书的装订形式为锁线胶平装，封面采用深红紫色布面裱贴卡纸，只烫印最基本的书籍信息。环衬选用鹅黄色胶版纸张，内页哑面书写纸印单黑。内容按照年份整理了各处收藏的艺术家日记手稿，使用双页满版黑色来区隔。书籍最大的特色在于在有限制的排版规范内尽可能地去展现手稿的原貌。页面被设置为八栏（辅助线作用），便于内容就近对齐。下画线、删除线、不合常规的缩进、超大的空格、草图、涂抹，甚至错误的大小写都在文本上有所呈现。本书的排版设计是艺术家强调过程和作品自我完成的一项实践。

Eva Hesse Diaries

This book is the re-organized and re-arranged diary of Eva Hesse, an American German sculptor. Eva advocates a completely improvised and unknown creative process, using unstable, organic and constantly changing materials. She emphasizes that the key of arts is not the final state but the artistic process. This paperback book is perfect bound with thread sewing. The cover is made of cardboard mounted by dark red purple cloth with the basic book information hot-stamped. The end paper is made of light yellow offset paper while the inside pages are using newsprint in monochrome black. The collections of artists' manuscripts are sorted by year and separated by two full black pages. The greatest feature of books is to show the original appearance of manuscripts as much as possible with the limited typesetting standards. The page is set to 8 columns (auxiliary lines) so that the content can be easily aligned. Underlines, deletions, irregular indentations, oversized spaces, sketches, smears, and even incorrect capitalization are all present in the text. The design of this book is an artistic practice of emphasizing process and self-completion of the works.

- 铜奖
- 伊娃·黑塞日记
- 瑞士

- Bronze Medal
- Eva Hesse – Diaries
- Switzerland
- NORM, Zürich / Johannes Breyer, Berlin
- Eva Hesse / Barry Rosen
- Hauser & Wirth Publishers / Yale University Press
- 204×136×50mm
- 1190g
- 900p

Diary extracts that originated between 1955 and 1970 have been published, from the estate of the artist Eva Hesse. Spanning 900 pages, the result of this source exploration only seems like a slab on the outside, with respect to its dimensions, whereas it feels positively clear-cut and light when handled. Why? The volume has two photographs. The frontispiece shows a portrait of the artist; the last printed page displays an example of the handwriting in the original diaries—with passages deleted and added, and with an intuitive use of the space at the sides. This is an editorial challenge and presents massive demands on the creative typesetting. The table opposite shows the solution. The handwritten variants are restricted above all to the placement of the entries. On the one hand, seven uniform indents provide sufficient order in order to secure good readability, and on the other hand they maintain the associative content of the original text positioning with sufficient differentiation. The result of the typographical transcription leads to vivacious double pages when printed. The modern, classicistic type with the look and feel of newspaper and typewriter script was also popular in the sixties. The option of underlining is preserved by ample spacing between the lines, with self-evident left justification and restriction to one font size. The years are separated generously. In each case, six pages are set aside for this, the first four of which are entirely black. The matt, slightly yellowish paper is ideal for reading. The volume nestles in your hands—not only because the paper is so lightweight, but also due to the flexible and trimmed cover, the thin outer cardboard with finely woven textile cover. This is why it is a sleek volume—useful in every respect.

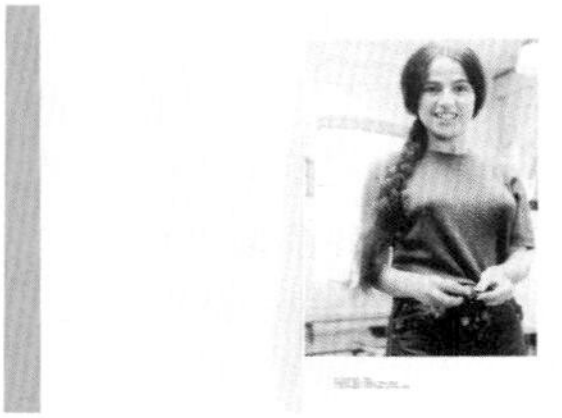

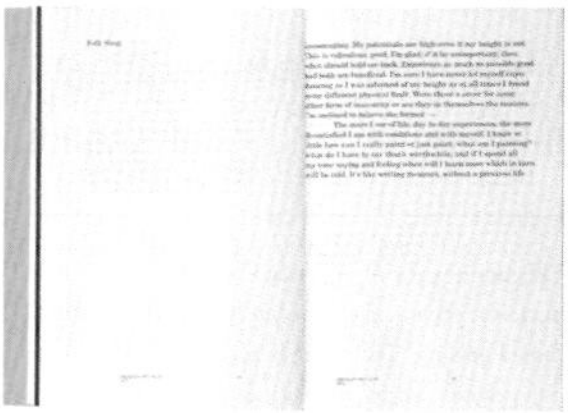

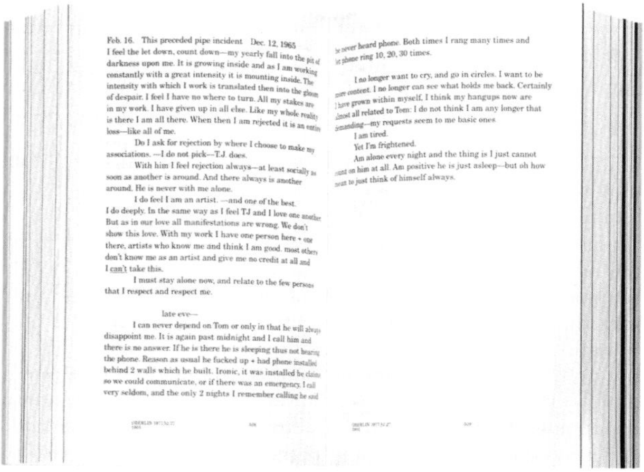

Feb. 16. This preceded pipe incident Dec. 12, 1965

I feel the let down, count down—my yearly fall into the pit of darkness upon me. It is growing inside and as I am working constantly with a great intensity it is mounting inside. The intensity with which I work is translated then into the gloom of despair. I feel I have no where to turn. All my stakes are in my work. I have given up in all else. Like my whole reality is there I am all there. When then I am rejected it is an entire loss—like all of me.

Do I ask for rejection by where I choose to make my associations. —I do not pick—T.J. does.

With him I feel rejection always—at least socially as soon as another is around. And there always is another around. He is never with me alone.

I do feel I am an artist. —and one of the best. I do deeply. In the same way as I feel TJ and I love one another But as in our love all manifestations are wrong. We don't show this love. With my work I have one person here + one there, artists who know me and think I am good. most others don't know me as an artist and give me no credit at all and I can't take this.

I must stay alone now, and relate to the few persons that I respect and respect me.

late eve—

I can never depend on Tom or only in that he will always disappoint me. It is again past midnight and I call him and there is no answer. If he is there he is sleeping thus not hearing the phone. Reason as usual he fucked up + had phone installed behind 2 walls which he built. Ironic, it was installed he claims so we could communicate, or if there was an emergency. I call very seldom, and the only 2 nights I remember calling he said he never heard phone. Both times I rang many times and let phone ring 10, 20, 30 times.

I no longer want to cry, and go in circles. I want to be more content. I no longer can see what holds me back. Certainly I have grown within myself. I think my hangups now are almost all related to Tom: I do not think I am any longer that demanding—my requests seem to me basic ones.

I am tired.

Yet I'm frightened.

Am alone every night and the thing is I just cannot count on him at all. Am positive he is just asleep—but oh how mean to just think of himself always.

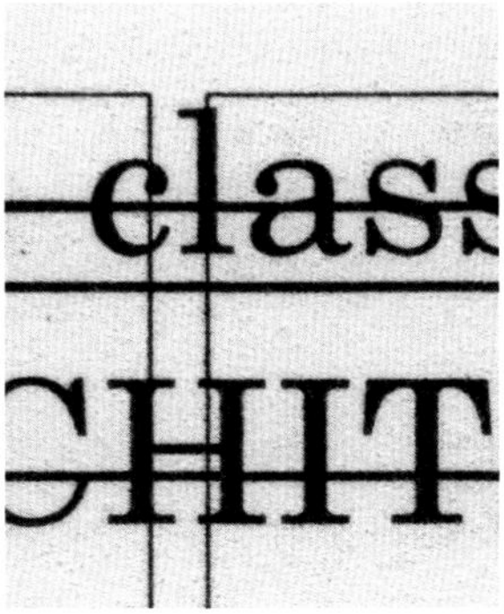
class
CHIT
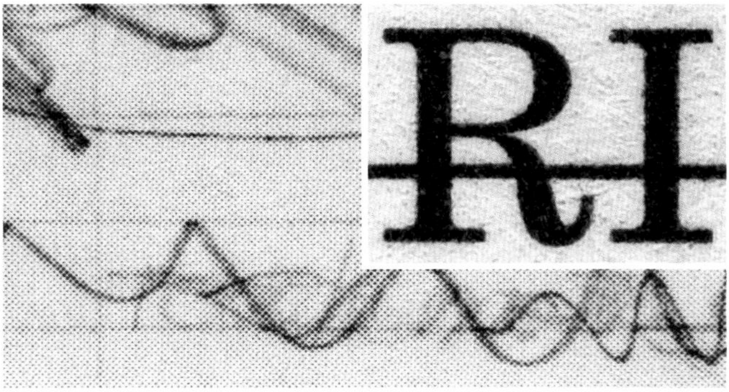
RI

Eva Hesse
Diaries
Edited by
Barry Rosen
Hauser & Wirth
Yale

LIBRARY.
HELEN
Call Bernie
group—
<u>ERIC—PICTURE</u>

THUR—17.

I would have been so different if I would have ha mummy. A companion, security, a real person who loved not in a purely selfish way and would have sincerely to when I was wrong would have corrected me with an exar or explanation of the right way. Then now it wouldn't be hard. I do get hurt easy, am revengeful want to be loved and not love in return. I guess like in most things I can' the medium, just to be liked respected. —

It was an eventful day, not very enjoyable.

Symposium. Mod. Art.
A) modern painting is large as they believe you must in the picture with your body as well as mind use your ent movement of arm not just wrist.
B) combine free forms with circles rectangular, sq. in ar
c) collaboration among the 3 painting sculp. arch.
d) sculpture is a part of an entity it must belong be part of environment
e) painting is man's individual self expression of his environment (trunk)

———

was pleased—asked again

OBERLIN 1977.52.39
1955 30

LUNCH ERLEBACHER

er—53 Park
[. 1st 42 42 or 49 crosstown
nk 5th & 43
vetti 5th—46 and 47
oll 575 Mad
VERNE East 57 (lex & 3rd) south side
LH Samples 101 Park

I feel lousy feverish, made appointment anyway. I want pendant. I feel cheated always by people. Afraid to give self. I think I knew I would fight with Helen. I must have n angry as I blew my top, against Eva—clothes. —both ldy hypocrite, selfish. Joey—letter. Cried desperately—but ught about it so it was controlled I immediately got up tinued washing cleaning up. Helen was disinterested she ed Murray what he was eating. 2. doing with feet. hung up h short good bye—wouldn't answer when she called back. I had no one, but didn't really care at that point.

Buy rubber cement—white paint.
as I cried mummy a little continued. I didn't want to go
with that boy I avoided his question: Bernie—Bob—

Sun—Dorit—car ride—upset—
Sat nite—Jack, John

---

spoke about summer—cried very much, had no control contest is upsetting me very much. school I'm afraid. gret entering. Things depend much on one another. I can't d my hand still it's shaking. I want to run run run. y did you tell me to run. Can you know what this is like?

2017 Ha
FF
10
Das Mannigfalt
FF
171_3
Es lässt sich plausibel behaupten,
die Unterwerfung unter den fas
Führer nicht durch das Fehlen
FBP
202
ANWÄLTE KLA
CHS MILLIARDE
FBP
204
MIXED PICKLES

这是一本将瑞士艺术家彼得·拉德尔芬格（Peter Radelfinger）的个人收集整理并公之于众的书。艺术家痴迷于收集各类印刷或在线媒体中的文字和图像，内容涵盖报纸、图表、公式、广告、照片、诗歌等等。本书的装订形式为锁线胶平装，封面白卡纸四色印刷，从封底包裹到封面而未在书脊上进行粘贴，将本书近600页A4大小的页面直接呈现在读者面前。内页胶版纸四色印刷，通过带纹理的柔面艺术纸分隔为五部分。但这五部分并不是相互独立或者相互承接的关系，而是像锁链般环环相扣的内容结构。内容按照FF、FBP、FBD、FBN进行划分，图片与文字的出处并不直接呈现在同一页面上，对开的左右页也并无内容上的直接联系。所有内容的注解均在之后部分中一一呈现，需要按照编号来索引查找，犹如十四行诗般，“跨行押韵”。通过这样的跨页面信息间的关联，让本就内容量巨大的书籍形式本身呈现出沟通传达的意义，形成一种沟通万物的复杂体系。

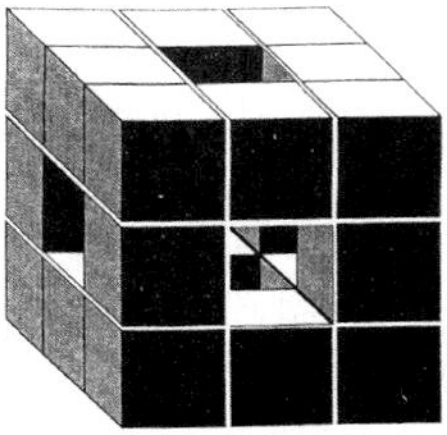

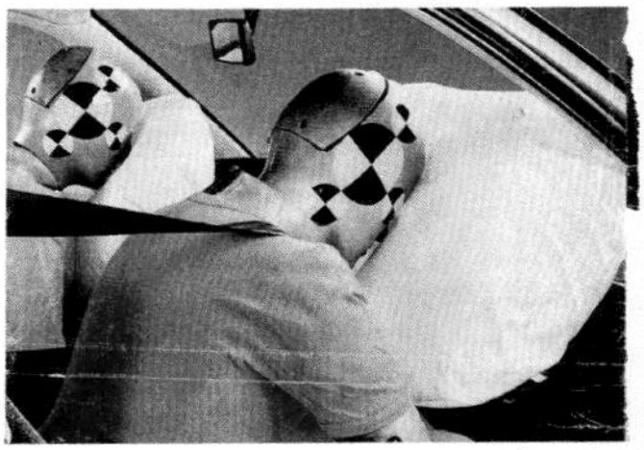

This is a personal collection of the Swiss artist Peter Radelfinger that has been made public. Artists are obsessed with collecting words and images from various print or online media, including newspapers, charts, formulas, advertisements, photographs, poems and so on. This paperback book is perfect bound with thread sewing. The cover is printed on white cardboard in four colors without sticking on the spine, thus nearly 600 pages can be fully opened. The inside pages are printed in four colors and divided into five parts by flexible art paper. These five parts are neither independent nor continued; they are structured like a closing chain. The contents are divided according to FF, FBP, FBD and FBN. The origins of pictures and text are not directly presented on the same page, and there is no direct connection between the left and right pages. The annotations are presented in the following sections. They need to be indexed and searched according to the number, just like “cross-line rhyme” in sonnets. Through such inter-page information association, the book itself, which has a large capacity, presents the meaning of communication, a complex system of linking all things.

- 荣誉奖
- 错误的线索
- 德国

- Honorary Appreciation
- Falsche Fährten
- Germany
- Jonas Voegeli, Kerstin Landis, Scott Vander Zee — Hubertus Design, Zürich
- Peter Radelfinger
- Edition Patrick Frey, Zürich
- 296×213×42mm
- 1756g
- 584p

With the gravitational force of a freshly copied doctoral thesis, this chunky block sticks to the desk, as thick as a pack of A4 paper—at least that is your spontaneous impression. If you follow up this red herring (it is actually art) there is still the chance that you will come to the right conclusion. On hundreds of double pages, there are text quotations on the left-hand side and four pictures on the right-hand side. Each quote has been given a shelf mark. The artist is an obsessive collector of pictures and text, and his encyclopaedic compilation seems to be catalogued entirely at random. Towards the end of the collection, the objective, simple intervals in the text/image arrangement are disturbed only slightly, but decisively. The text and image pages are swapped around because the extensive index already begins while the image material continues to spread out—a subtle expansion of the reference structure. In the old days we went through the thick Brockhaus lexicon, going from word to word and illustration to illustration just out of boredom, and then we intertwined all the associations offered higgledy-piggledy through the volume in order to have traced the Gordian knot of world knowledge. What seems like ridiculous time-wasting from the perspective of our hyperlinked digital universe turns this book into a comprehensible catapult for realisations: “There are two points that are far apart in the moment that an idea appears—not knowing how they should communicate with one another—and then, after sufficient transformations, all of a sudden they are next to each other ...”

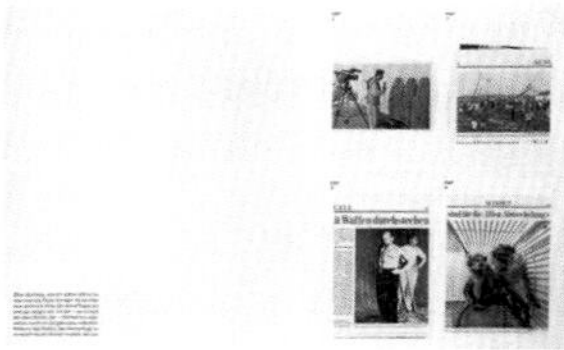
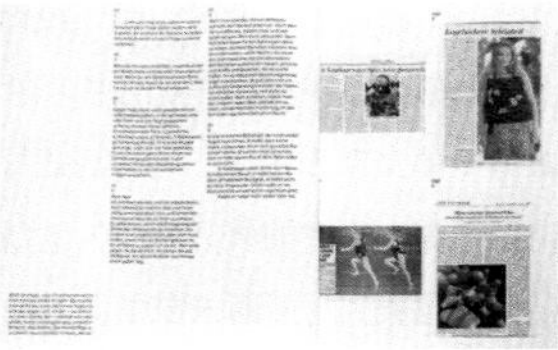

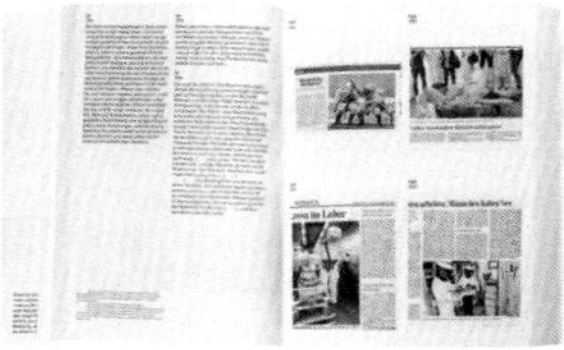
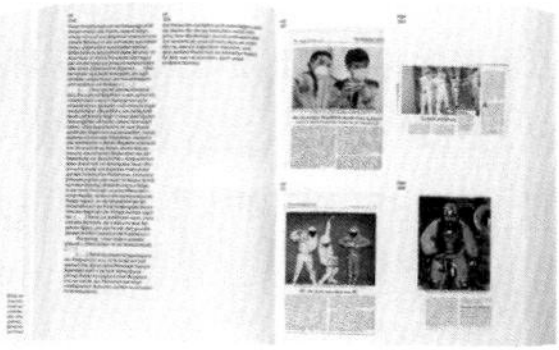
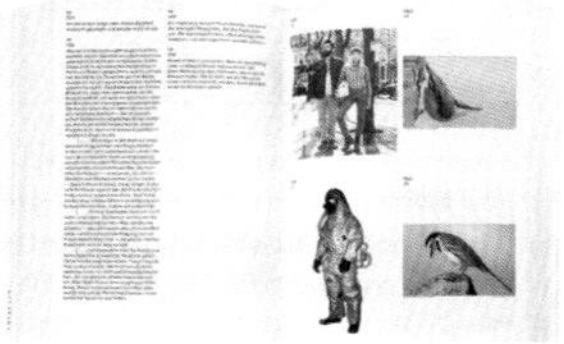

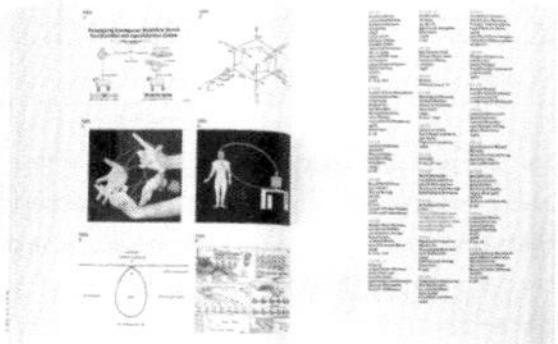
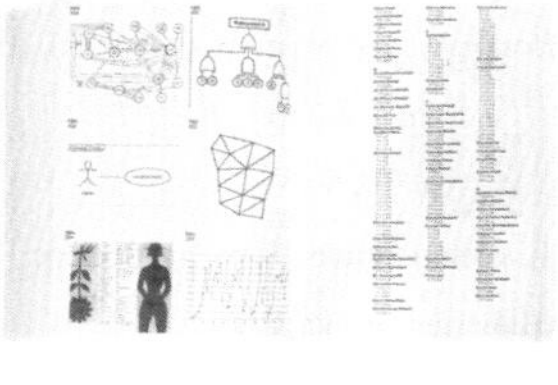
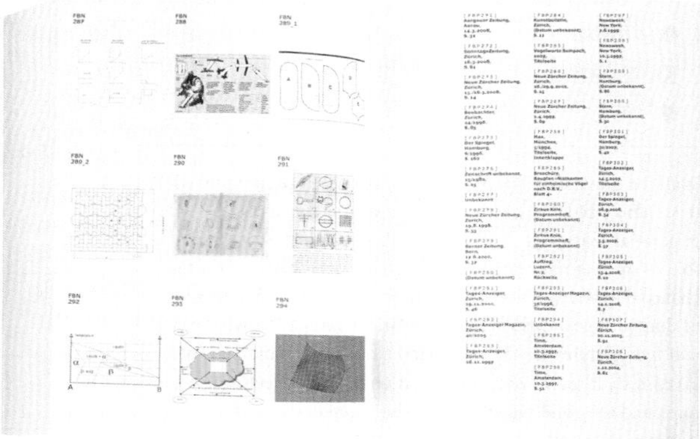
FBN 287
FBN 288
FBN 289_1
FBN 289_2
FBN 290
FBN 291
FBN 292
FBN 293
FBN 294
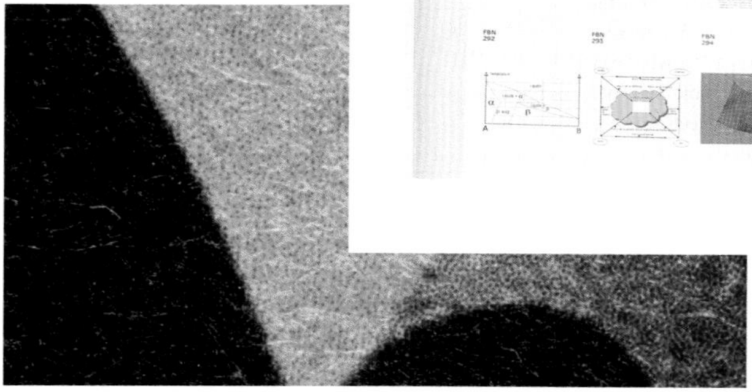

Zeit
sse der kommu

FALSCHE
FÄHRTEN
PETER
RADELFINGER
EDITION
PATRICK FREY
NR. 180

FBN
450_2

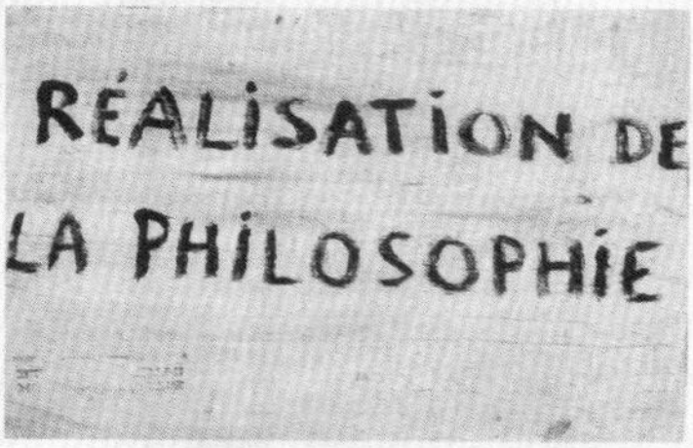
RÉALISATION DE
LA PHILOSOPHIE

FBN
450_3

DÉPASSEMENT
DE L'ART

FBN
450_4

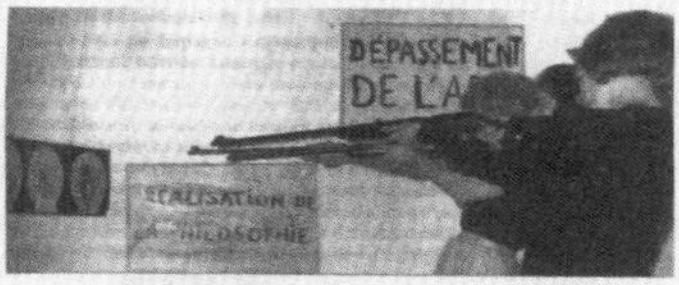
DÉPASSEMENT

FBN
451

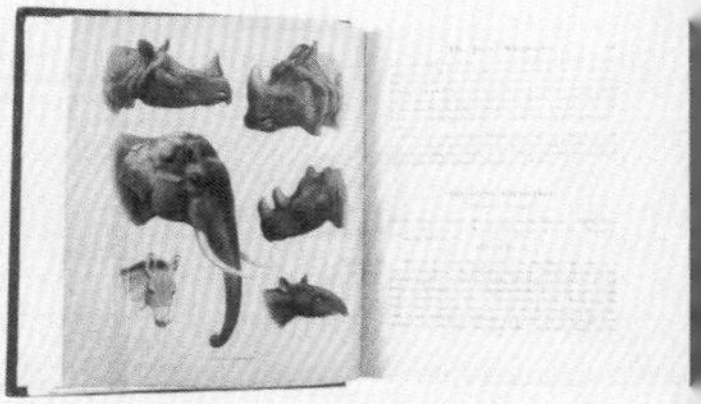

FBN
452

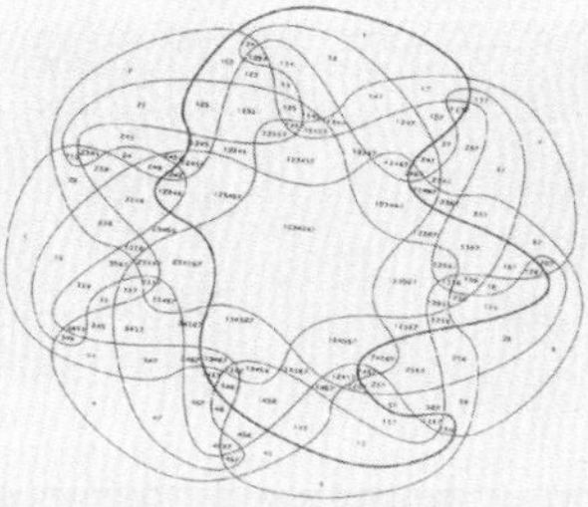

FBN
453

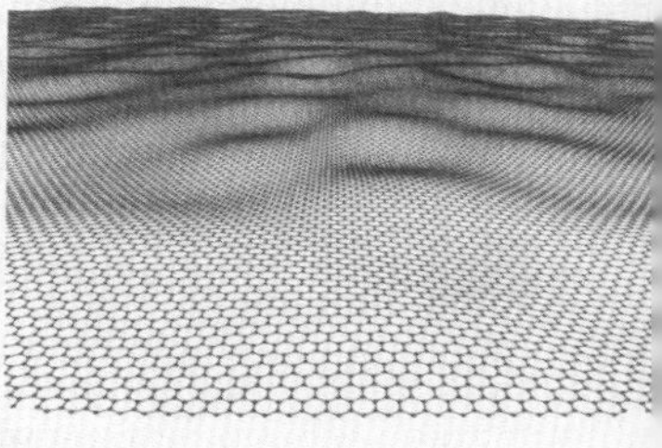

[FBN330_1]
Böhme_
Philosophische_Kugel,
http://es.wikipedia.org/wiki/Jakob_Böhme

[FBN330_2]
Facebook Daumen,
Zeichen mit Cursor,
http://de.dreamstime.com/stockfoto-facebook-mögen-daumen-herauf-zeichen-image24208480

[FBN331]
Die Rückkopplungs-maschine,
Peitgen Heinz-Otto,
Bausteine des Chaos. Fraktale,
Klett-Cotta/
Springer Verlag,
Stuttgart/Heidelberg
1992,
S. 36–47

[FBN332]
Ein harmloses ein-dimensionales Gebilde: Diese wildaussehende Kurve ist tatsächlich weit von einem wirklich komplizierten eindimensionalen Objekt entfernt,
Peitgen Heinz-Otto,
Bausteine des Chaos. Fraktale,
Klett-Cotta/
Springer Verlag,
Stuttgart/Heidelberg
1992,
S. 136

[FBN333]
Netzwerk mit Graph,
Peitgen Heinz-Otto,
Bausteine des Chaos. Fraktale,
Klett-Cotta/
Springer Verlag,
Stuttgart/Heidelberg
1992,
S. 344

[FBN334_1]
Test de laboratorio y gràficos, *Sputnik*,
Ausstellungskatalog,
Madrid,
1997,
S. 161

[FBN334_2]
Prototipo,
Sputnik,
Ausstellungskatalog,
Madrid,
1997,
S. 162

[FBN335]
Spaghetti-Computer,
Dewdney A. K.,
Der Turing Omnibus,
1995,
S. 241

[FBN336]
Konzept einer Turing-Maschine,
Dewdney A. K.,
Der Turing Omnibus,
1995,
S. 222

[FBN337]
Viele Übergänge werden zu einem Übergang,
Dewdney A. K.,
Der Turing Omnibus,
1995,
S. 190

[FBN338]
Die Welt der Elementarteilchen,
in Zukav Gary,
Die tanzenden Wu Li Meister,
1981,
S. 271,
http://www.librosmaravillosos.com/ladanzadelosmaestros dewuli/capitulo10.html

[FBN339]
Transformationen von «etwas» in «nichts» und von «nichts» in «etwas»,
Zukav Gary,
Die tanzenden Wu Li Meister,
1981,
S. 275,
http://www.librosmaravillosos.com/ladanzadelosmaestros dewuli/capitulo10.html

[FBN340]
Schematischer Plan von Armands Garten mit den wichtigsten Themenkreisen,
Schlumpf Hans-Ulrich,
Tages-Anzeiger Magazin,
6/1974,
S. 18

[FBN341]
Debord Guy,
Carte de Paris avant 1957,
Ausstellungskatalog,
Museum Tinguely,
Basel,
2007,
Abbildung 045

[FBN342]
Guy Debord listet die Alkoholica auf, die in der Nacht vom 9.5.1962 konsumiert worden sind,
Ausstellungskatalog,
Museum Tinguely,
Basel,
2007,
S. 14

[FBN343]
Mark Lombardi: Georg W. Bush, Harken Energy and Jackson Stephens (1979–1990),
http://www.agilefluenca.org/

[FBN344]
Leistungsmerkmale grafischer Ordnungs-formen, 14082002,
Dirmoser,
http://slideplayer.de/slide/918288/

[FBN345]
World-view, the genesis of sense,
Quelle unbekannt

[FBN346]
Anamorphe Karte: Switzerland: Referendum on Minarets,
29.11.2009

[FBN347]
The Arteries, from Diderot, L'Encyclopédie,
1765,
ZONE, Fragments for a History of the Human Body,
Part Three,
1989,
S. 447 ff.

[FBN348]
From Master John of Arderne, Die Arti Phisicali et Cirurgia,
1412,
ZONE, Fragments for a History of the Human Body,
Part Three,
1989,
S. 447 ff.

冰

这是中国艺术家冷冰川的墨刻作品集，书中呈现了艺术家自 20 世纪 80 年代至 2012 年的 256 幅作品。书籍的装订形式为锁线软精装，书籍的各个方面都在尽力展现艺术家的作品特点。封面采用白色哑面艺术纸包裹双层黑色卡纸的做法，白色封面上只有少量文字烫黑，其余部分烫透工艺处理，封面中部一道弧线“刻痕”透出里面包裹着的卡纸的黑色。内页序言和最终的对话部分采用黑卡纸印银，连续黑色页面不同角度的弧线“刻痕”与封面呼应。作品按三大时间段划分，通过三种颜色的胶版纸印单黑，呈现艺术家自己三个时期创作理念的文字表达。作品部分采用哑面涂布纸单黑印刷。下书口将艺术家名的西文大写字母做激光雕刻工艺，强化纪念合集的概念。

This is a collection of ink-engraving works by Chinese artist Leng Bingchuan. The book presents 256 works from the 1980s to 2012. It is bound with thread sewing. All aspects of books are trying to show the characteristics of the artist's works. The cover uses white matte art paper wrapped by double-layer black cardboard. Only a small amount of text on the white cover is printed in black and the rest of the cover is hot pressed to be transparent. An arched "nick" in the middle of the cover reveals the black cardboard inside. The preface and the final dialogue are printed in silver on black cardboard, and the arcs of different angles correspond to the cover. The works are divided into three parts according to time frame. The artist's creative ideas in three periods are distinguished by using three different colors of offset paper. The works use matte coated paper printed in black. Laser engraving technology is applied to the western capital letters of the artist's name on the bottom edge to strengthen the concept of commemorative collection.

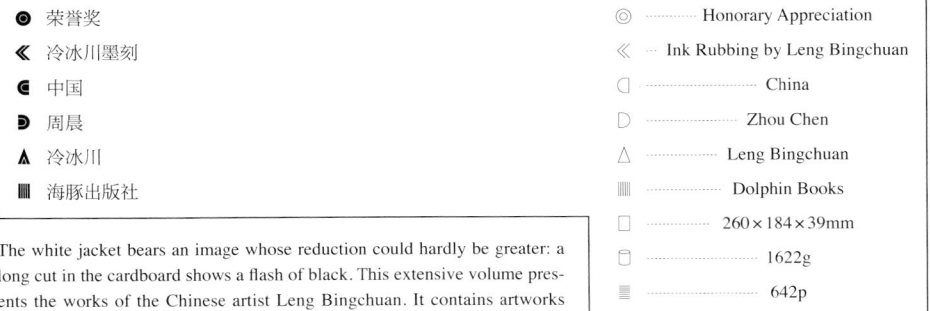

- 荣誉奖
- 冷冰川墨刻
- 中国
- 周晨
- 冷冰川
- 海豚出版社

- Honorary Appreciation
- Ink Rubbing by Leng Bingchuan
- China
- Zhou Chen
- Leng Bingchuan
- Dolphin Books
- 260×184×39mm
- 1622g
- 642p

The white jacket bears an image whose reduction could hardly be greater: a long cut in the cardboard shows a flash of black. This extensive volume presents the works of the Chinese artist Leng Bingchuan. It contains artworks with maximum black/white contrasts: lines and areas that have been incised, scraped and scratched into deep black ink. The book design has taken this principle as its leitmotif. The double flap jacket exemplifies the dual-layer concept of the medium: on the outside white, on the inside black cardboard. The innerbook can be taken out of the jacket, with an open spine and visible binding; black on all six sides. The artist's name has been engraved in Antiqua capitals into the bottom edge of the book, and so the lower edges of the pages demonstrate the graphic technique, as it were: enlarged cross-sections of the image carrier by way of example. The didactic enthusiasm becomes clear at this point in another detail; even the engraved name is crossed by a finely curved strip. The first pages—strong, black paper —are also incised with curved lines in different directions. A series of photographs on the pages at the back show how the master's hand guides the cutter from the deep, dark background into the deep, dark ink. Last but not least, the artwork reproductions have been printed in an extremely luscious and even tone. The opulence, the vivid richness of a black area as seen in this printing is a rare experience.

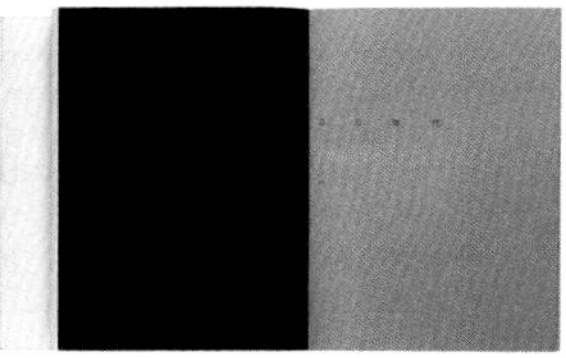
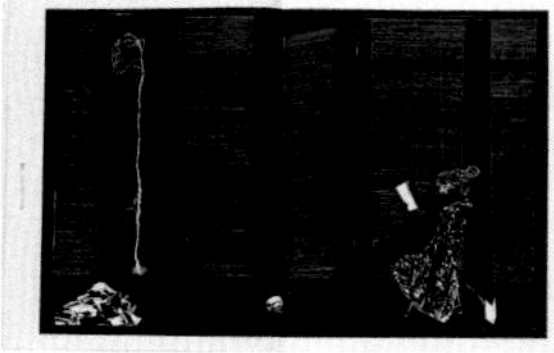

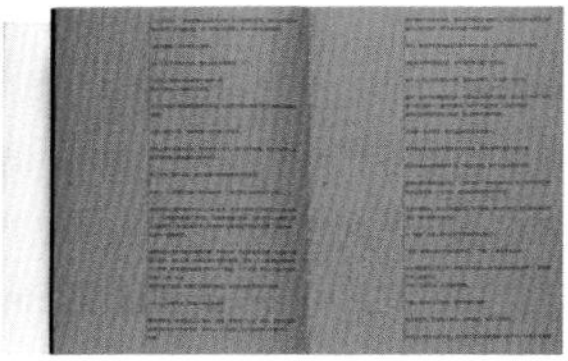

墨

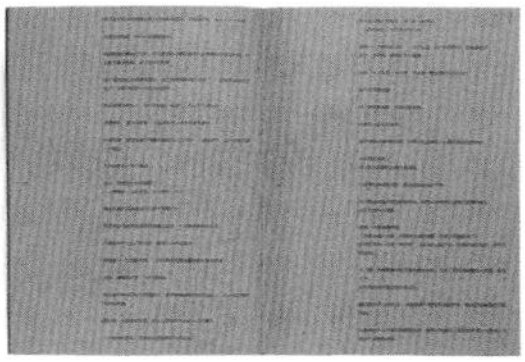

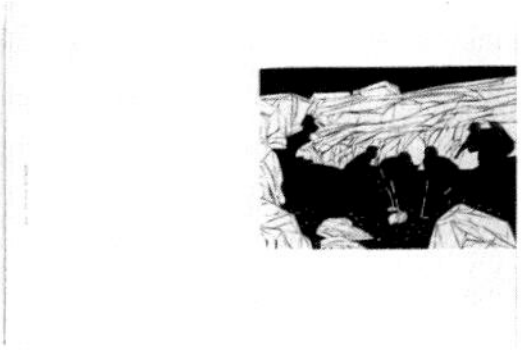

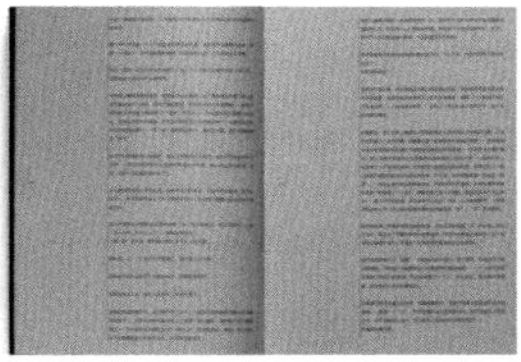

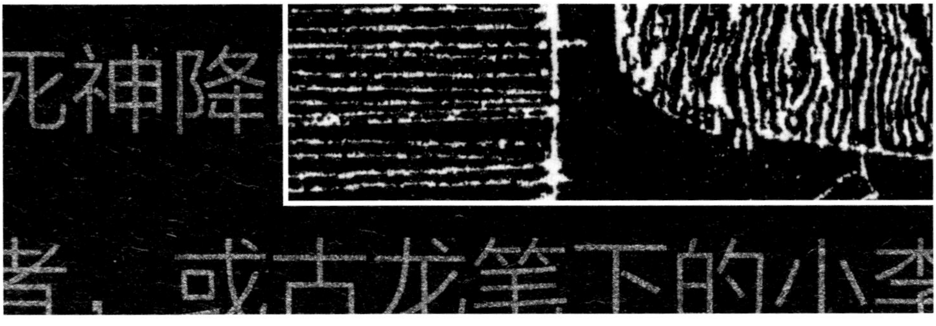
死神降
或古龙笔下的小李

这是一本由 400 名专业人士参与编写的现代运动百科词典，内容涵盖与当代体育运动相关的政策、法律、经济、健康、文化、科学、教育、项目、媒体、艺术、特殊群体等多个方面。大部头的词典往往在内容量和书籍的体量上给人形成压迫感，为此，这本书设计的很多方面都是在尽可能地降低这种压迫感，让人愿意去翻阅这本超过 1350 页的图书。书籍为圆脊锁线硬精装，书函白色卡纸青绿色与黑色双色印刷，护封哑面艺术纸印青绿色。内封白色布纹纸印深蓝灰色，裱覆卡板。环衬与扉页都分别采用了不同的绿色来降低书籍的“重量”，内页偏黄的滑面艺术纸印单黑。主要页面设计为三栏，通过多级标题形成清晰的索引结构，图片和表格页尽量减小或采用较细的线条，以降低页面灰度。右侧翻口设计了索引线，方便查找。蓝紫双色的书签带夹在书中，也与书籍的“轻量化”目标相得益彰。

21世紀
スポーツ
大事典
Encyclopedia of Modern Sport

This is a modern sports encyclopedia written by 400 professionals, covering policy, law, economy, health, culture, science, education, projects, media, art, special groups and other aspects related to contemporary sports. Since large dictionaries tend to be oppressive in terms of content and volume, this book is designed to minimize this sense of oppression and make people willing to read the book with more than 1350 pages. The hardcover book is bound by thread sewing with a round spine. The slipcase is made of white cardboard printed in cyan and black. The inside cover is using white coating paper printed in blue grey, mounted with graphic board. Both the end paper and the title page adopt different green to reduce the "weight" of book. The inside pages are using yellowish glossy art paper printed in monochrome. The main pages are designed in three columns. A clear indexing structure is formed through multi-level headings. The images and tables are minimized or using thinner lines to reduce the gray scale of the page. The index line is designed on the right edge to facilitate searching. The blue-purple bookmark strap clipped to the book also complements the book's "lightweight" goal.

荣誉奖
21 世纪运动百科
日本

Honorary Appreciation
Encyclopedia of Modern Sport
Japan
Shin Tanaka, Sozo Naito
Toshio Nakamura, Takeo Takahashi, Tsuneo Sougawa, Hidenori Tomozoe
TAISHUKAN Publishing Co., Ltd.
270×204×60mm
2556g
1378p

Encyclopedias are attractive because the condensed knowledge inside can be grasped with both hands. They also put fear into people, though— the fear of paper overladen with text, the fear of heavy books, the fear of a thousand pages. The design of this Japanese reference book dispenses with the sense of heaviness, however. An attractive lightness emanates: the water blue jacket has been given rational yet intuitive typography with fine fonts. The white cover with unusually thin board considering the substantial innerbook can hardly appear lighter, and the greenish-blue endpapers at the front and back can hardly seem more airy. The particularly finely woven headband and both ribbon bookmarks match the colour tone beautifully. Even from the outside, visible on the fore-edge, it is evident that the whole work is structured into manageable portions—with a finely structured arrangement using collating marks, and sorted in horizontal and vertical lines: the ends of the lines on the right-hand side and the separating pages for each segment. Inside the book, we find beautiful, smooth, yellowish-white thin paper with three-column type area and generous spacing between the columns. Black/white photos and reduced-graphic diagrams and sketches have been used sparingly. The entries are structured with fine lines and numbered circles; even the mostly image-free double pages are able to support the light mood of the cover. This has been achieved by the book design: giving dynamic buoyancy not only to this majestic magnitude for functional reasons, but also to the topic of sport itself.

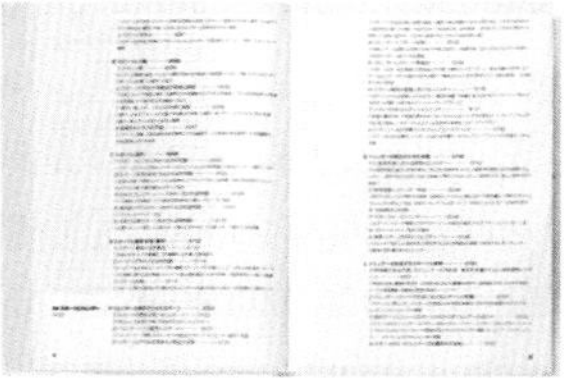
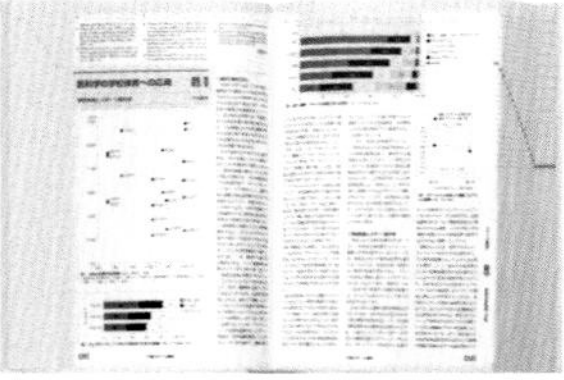
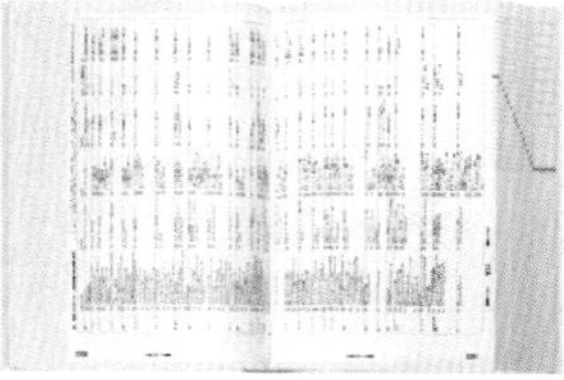

メジャーリーグのうぬぼれルーキー
W. P. KINSELLA
SHOELESS JOE
スタジアムの神と悪魔
EL FUTBOL
0850
0851

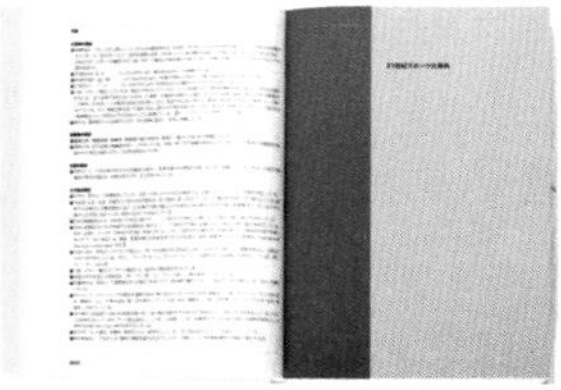

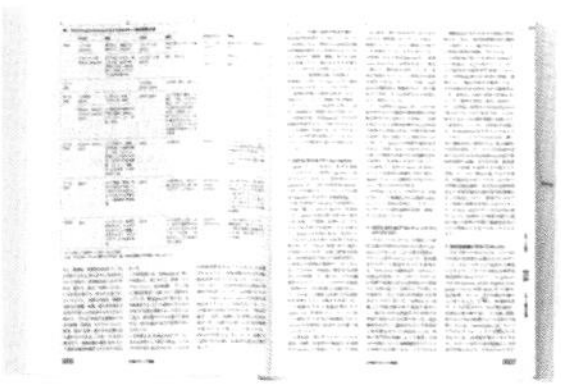

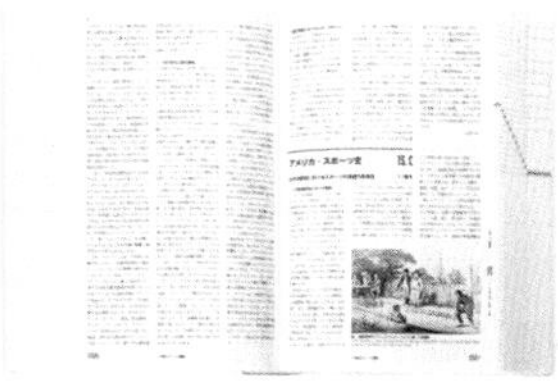
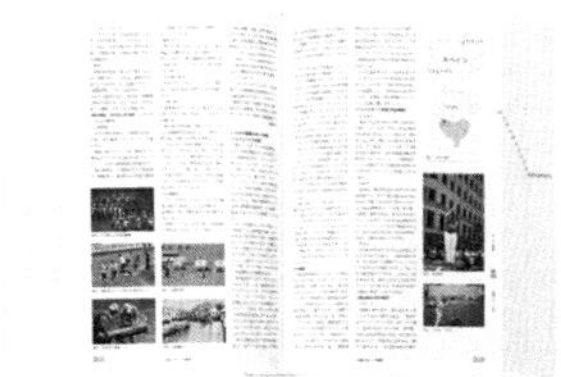
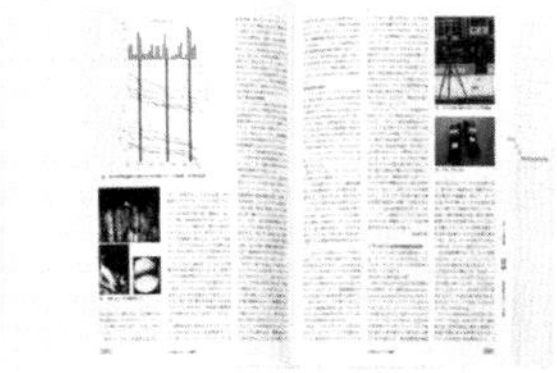

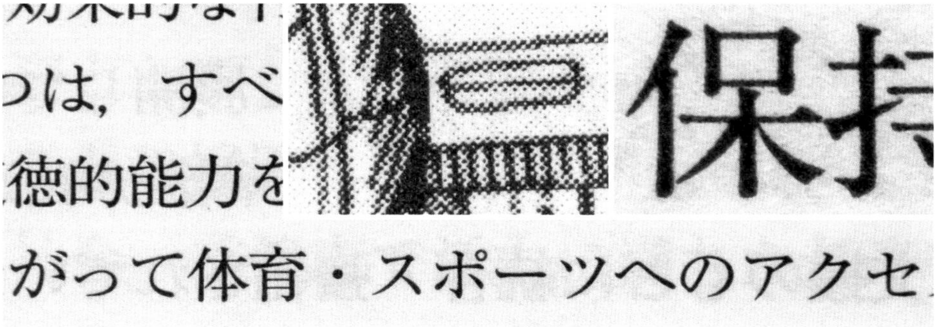
つは，すべ
徳的能力を
保
がって体育・スポーツへのアクセ

这是一本故事和绘画俱佳的儿童绘本，讲述了渔夫带着家犬在海上遇到的巨浪、看到海怪和美人鱼、被卷入漩涡、遭到暴风雨的袭击，但最终得救的离奇故事。在整个故事的最后，渔夫坐在酒吧里跟一帮听客们讲述自己的这段传奇经历，人人目瞪口呆，映衬本书的书名“真的吗？”，这只有渔夫自己知道。本书的装订形式为绘本惯常使用的锁线硬精装，非常薄的书籍厚度。为了体现故事的荒诞性，书籍在画风、色彩等方面都做了大胆的尝试。封面采用哑面艺术纸红蓝黑三色印刷，裱覆卡板。内页胶版纸与封面相同的三色印刷，充满对比的红蓝配色形成了强烈的视觉冲击力。画面不是以写实为主的画风，也并非完全抽象，印刷上也并不呈现笔触的肌理，而是除了毛糙的飞白外，全部扁平化单色处理，通过多笔触相互交叠，形成丰富的画面效果。狗在故事中被拟人化，与渔夫产生丰富的互动，也增加了故事的荒诞性。||||||||||||||||||||||||||||||||

VERDADE?!
BERNARDO P. CARVALHO
PATO LÓGICO

This is a children's picture book with excellent stories and paintings. It tells the fantastic story of the fishermen and their dogs that encountered the huge waves at sea. They saw the sea monsters and mermaids, and are swept into the whirlpool, attacked by the storm, and finally saved. At the end of the story, the fisherman sat in a bar and told his legendary experience to a group of listeners. Everyone was stunned, reflecting the title of the book, "Really?" Only the fisherman knew. The hardback book is bound by thread sewing like most of picture books, with very thin book thickness. In order to reflect the absurdity of the story, the book has made bold attempts in painting style, color and other aspects. The cover is printed in red, blue and black on matte art paper mounted with graphic board. Inside pages of offset paper are printed in the same way as the cover. The contrast of red and blue forms a strong visual impact. The picture is neither realistic nor completely abstract. The printing does not present the texture of brush strokes. Instead, it is flat with monochrome processing. Overlapping of multiple strokes forms rich picture effects. Dogs are personified in the story, which has interaction with the fisherman, which adds the absurdity to the story.

◉ 荣誉奖
《 真的吗?
◖ 葡萄牙

◎ ··· Honorary Appreciation
《 ··· VERDADE?!
◖ ··· Portugal
◗ ··· Pato Lógico
△ ··· Bernardo P. Carvalho
||||| ··· Pato Lógico
□ ··· 256×200×9mm
⌷ ··· 325g
≣ ··· 36p

The pictures in this children's book by Bernardo P. Cavalho are neither paintings nor drawings. Or both, with little trouble taken over the brushwork. Often the distinctive strokes broaden to form large, connected areas which then regulate the entire composition of the images. The brushstrokes frequently become ragged around the edges and create virtual half-tones with picturesque qualities. The conspicuous element of the book as a whole is its colourfulness, its apparent restriction to two colours: blue and red. There is a third colour, black, due to carefully planned superimposition of blue and red shapes. The artist does not think of black as a colour into which he can dip his instrument, however— although he allows the paper white to become an independent colour as if by magic, in a kind of negative retouching. It is therefore all cleverer than it appears at first glance. The story: Fisherman, dog and lifebuoy in a mini boat on a calm sea. Suddenly stops short. Roughest swell. Foaming wave crests. Threatening bulbous bow of huge cargo vessel. Man overboard. Tugging mermaids. Monster with tentacles and many mouths. Elegant yacht with bikini-clad ladies. The whole truth at the dockland pub bar—or a sailor's yarn?! Yes, and the book? Technically it is a classical paperback. But the remarkable visual work is wholly oriented on the paper-based sheet medium, and so no difference can be made between looking at images and looking at books.

VERDADE?!
BERNARDO P. CARVALHO

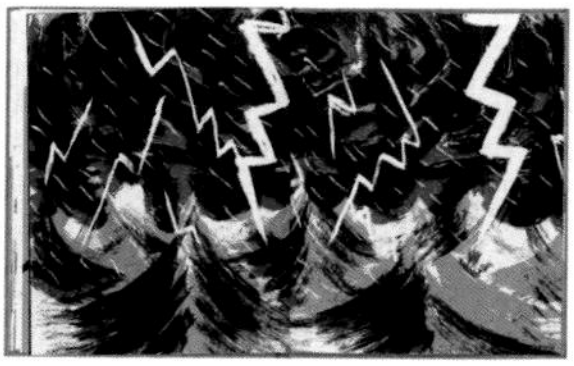

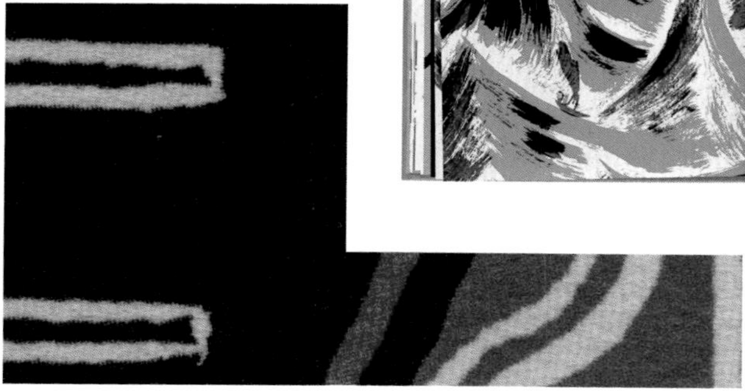

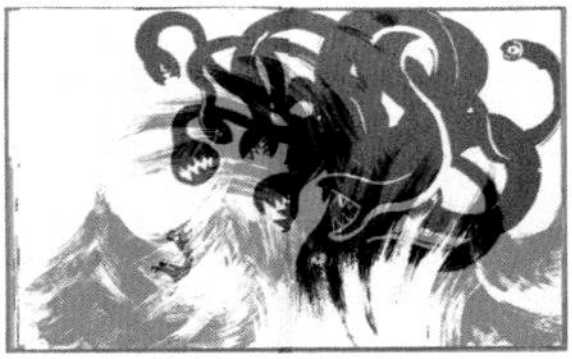

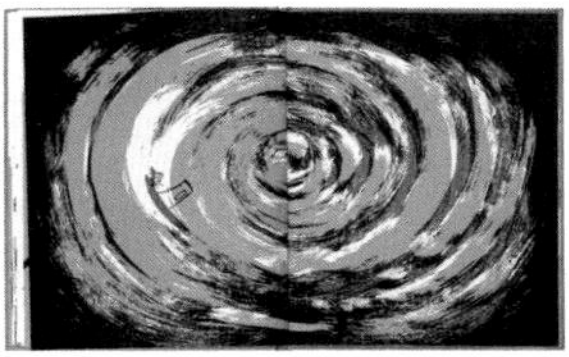

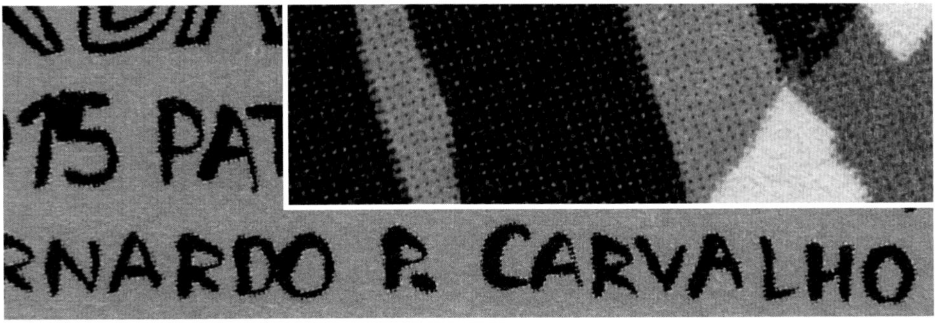
15 PA
RNARDO P. CARVALHO

这是一本讲述现代人离开网络后进入真实世界的荒诞故事。故事的主角是一个整天忙碌于网络世界但生活无趣的小职员，晚上除了在屏幕前聊天，没有什么太大的生活改变。一次偶然的网络信号中断后，被派往各地查明原因，从而遭遇各色离奇的故事，最终发现信号中断的原因是成群的候鸟挡住了信号基站，他能做的只有耐心等待。本书为欧洲漫画常用的锁线硬精装，封面胶版纸三色印刷，裱覆卡板，内页胶版纸三色印刷。画面为高空俯视的平行构图，体现出空间纵深感的同时，也具有不同于三点透视的荒诞味道。绘画的风格采用木刻的方式，呈现斑驳的肌理。全书总计使用了红、黄、蓝、草绿、荧光橙、棕六种颜色，但每幅画面只使用其中三种，通过三种颜色之间的碰撞和透叠形成丰富的画面效果。每个页面也都因为选用的颜色、面积配比差异而产生不同的情绪感受。强烈的色彩碰撞创造出一个不同寻常的现实世界，仿佛离开了网络的人在现实中反而晕头转向地迷失了。正文采用与画面相匹配的手绘字体，阅读感受轻松有趣。

BENJAMIN COURTAULT
PLÖTZLICH
FUNKSTILLE
KUNSTANSTIFTER

This is a fantastic story about modern people leaving the Internet and entering the real world. The protagonist of the story is a small clerk who is busy all day in the Internet world and his daily life is boring. Besides chatting in front of the screen at night, there is not much change in his life. After an accidental interruption of the network signal, he was sent to various places to find out the reason, and then encountered various strange stories. Finally, he found that the reason for the interruption was that the swarms of migrant birds blocked the signal base station. All he could do was to wait patiently. This hardback book is bound by thread sewing like most European cartoon books. The cover is printed on offset paper in three colors, mounted with graphic board. So do the inside pages. The picture is composed from the high altitude, which reflects the depth of space and adds the sense of absurdity different from the three-point perspective image. The woodcut paintings display mottled texture. The book uses six colors: red, yellow, blue, grass-green, fluorescent orange and brown, but only three of them are used in each picture. Through the collision and overlap of every three colors, the visual effects are abundant. Each page also has different emotional feelings because of the difference of colors and area ratio. Strong color collision creates an unusual real world, as if people who leave the network are lost in reality. The text font is hand-painted font matching with the images, which make the book interesting.

◎ 荣誉奖

《 信号突然中断

◖ 德国

- ◎ Honorary Appreciation
- 《 Plötzlich Funkstille
- Germany
- Benjamin Courtault
- Benjamin Courtault, Paris (FR)
- kunstanstifter verlag, Mannheim
- 305×187×9mm
- 333g
- 32p

One day an accident happens that sends the protagonist in this story of an internet relationship away from his everyday routine and into real surroundings. Genuine special colours—always three in alternating combinations—produce rasterfree areas that create contrasting blends lying next to and on top of each other. There is no partiality towards a particular colour. This creates the peculiar camouflage effect. The concept enables the pictures to achieve—independent of the number of published copies—the quality of original printed graphics. The compositions are construed as dioramas. Mostly structured in parallel perspectives, they gain spatial depth from an elevated viewpoint. With the inner tension of Japanese colour woodcuts, shapes spanning large areas and detailed structures are related to one another. The huge dynamics only unfold when you allow yourself to see them as still pictures. Then a shimmer appears, as if the pages themselves were energised. That seems to be the secret behind these traditionally crafted book illustrations.

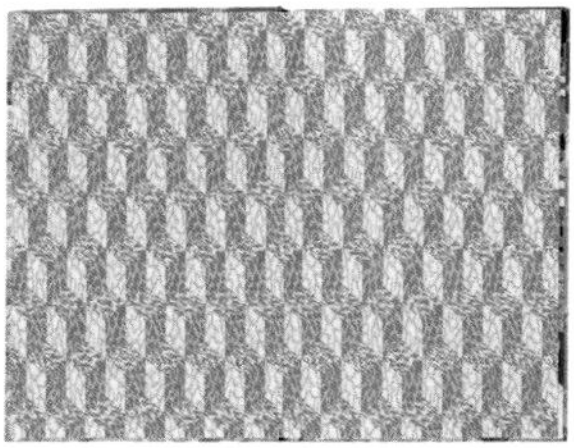

PLÖTZLICH
FUNKSTILLE

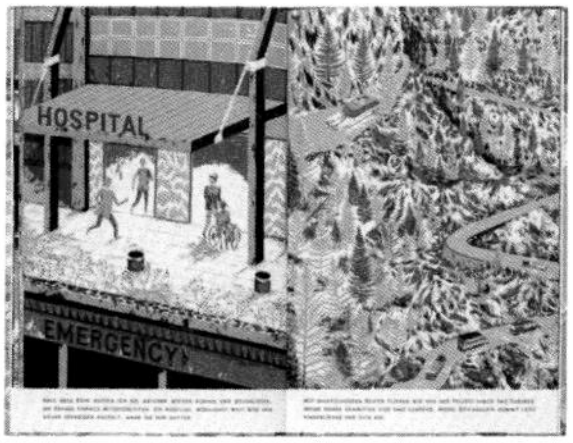

HOSPITAL
EMERGENCY

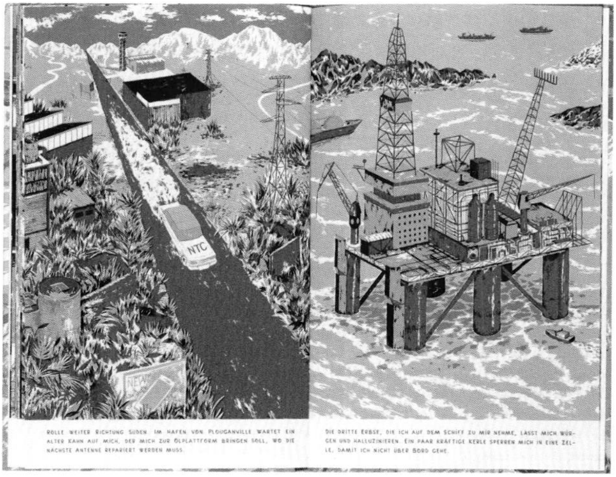

NTC
ROLLE WEITER RICHTUNG SÜDEN. IM HAFEN VON PLOUGANVILLE WARTET EIN ALTER KAHN AUF MICH, DER MICH ZUR ÖLPLATTFORM BRINGEN SOLL, WO DIE NÄCHSTE ANTENNE REPARIERT WERDEN MUSS.
DIE DRITTE ERBSE, DIE ICH AUF DEM SCHIFF ZU MIR NEHME, LÄSST MICH WÜRGEN UND HALLUZINIEREN. EIN PAAR KRÄFTIGE KERLE SPERREN MICH IN EINE ZELLE, DAMIT ICH NICHT ÜBER BORD GEHE.

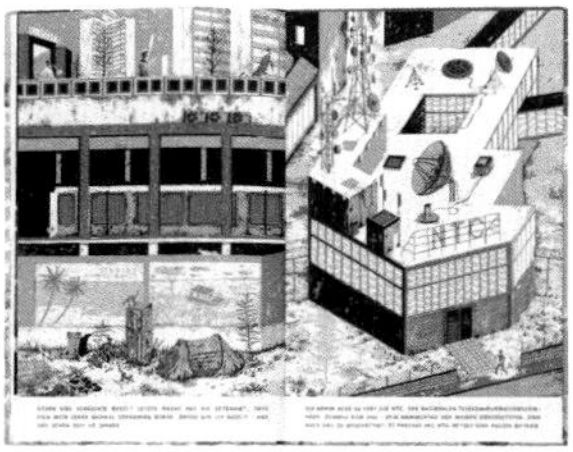

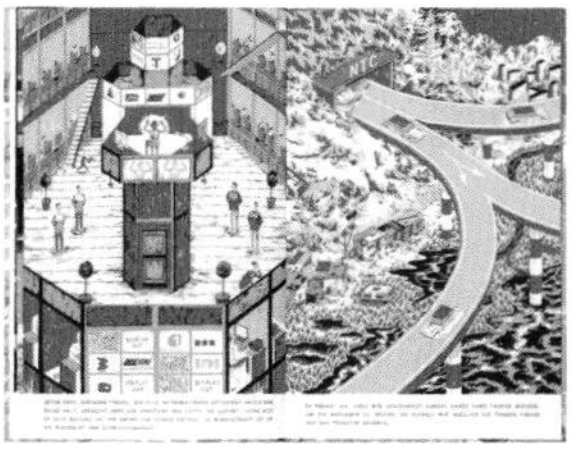

IST SI
UNS S
KANN DAS A

Other Evidence: Blindfold
Order – The Story of Fangyuan Bookshop
2×100 mil. m2
Brick. An Exacting Material
Daniël van Dicht
Het Meest Geloofde Sprookje
Nowhere Men. Illegale Migranten im Strom der Globalisierung
Pleasure of Learning
Farvel til romanen. 24 timer i grenseland
LIVING THE NORDIC LIGHT
Career Choice Without Prejudice
Del reverso de las imágenes
Jours Blancs
Taking off. Henry my neighbor

2016

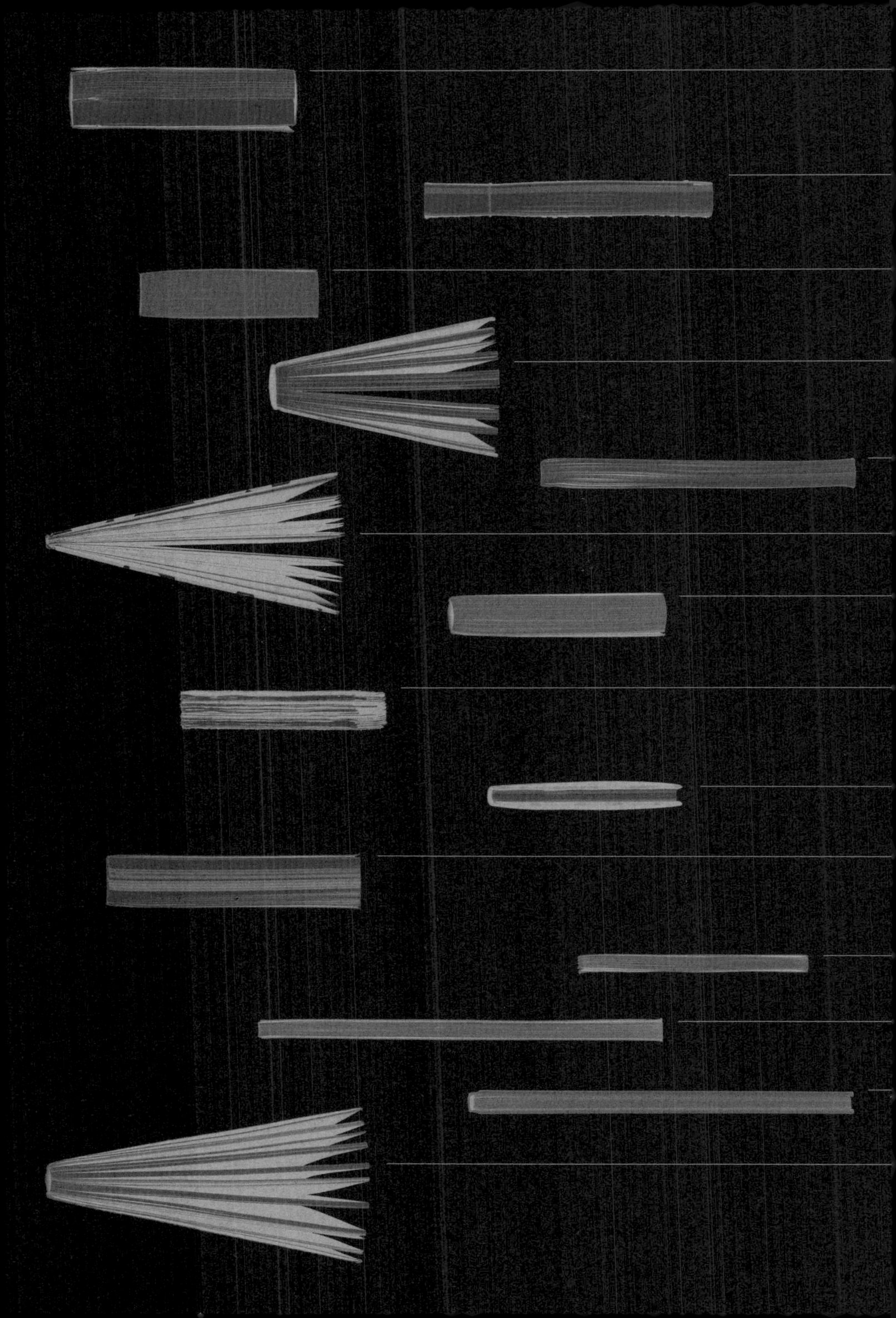

2016 Gl

0389

2016 Gm

0397

2016 Sm[1]

0403

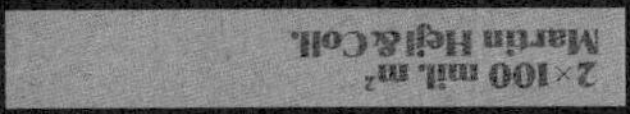
2×100 mil. m²
Martin Hejl & Coll.

2016 Sm[2]

0409

BRICK

2016 Bm[1]

0415

DANIËL VAN DICHT

2016 Bm[2]

0421

2016 Bm[3]

0427

Christoph Miler Nowhere Men

2016 Bm[4]

0433

2016 Bm[5]

0439

2016 Ha[1]

0445

2016 Ha[2]

0449

2016 Ha[3]

0453

2016 Ha[4]

0457

2016 Ha[5]

0461

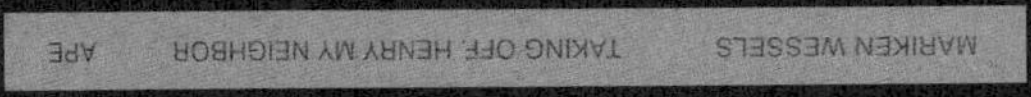
MARIKEN WESSELS
TAKING OFF. HENRY MY NEIGHBOR
APE

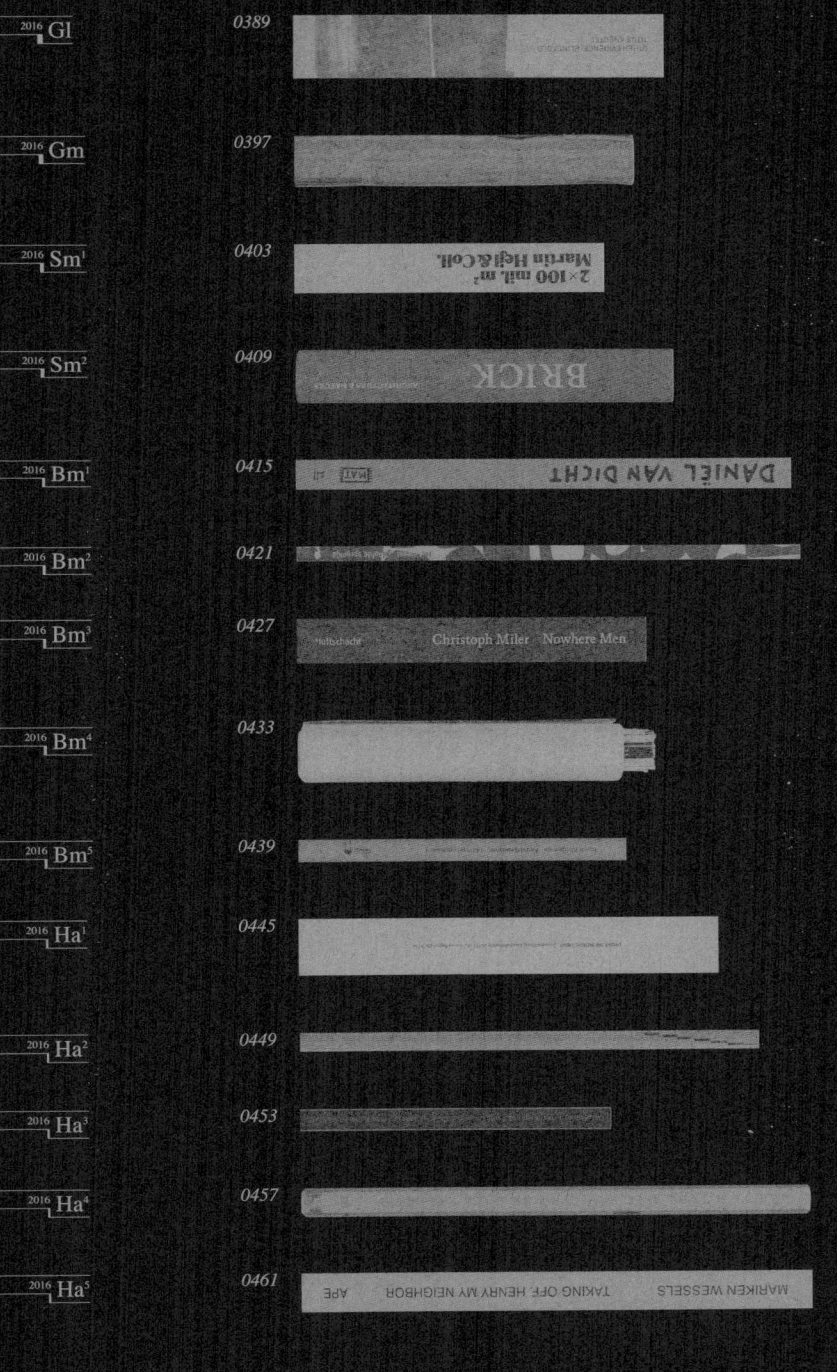

LZ2-8

AUTOPSY DATE	19 September 1996
AGE	40-50 years
SEX	Male
HEIGHT	170.9 cm (femur 45.0)
HANDEDNESS	Right

父亲吕安未，也作吕安廉，笔名乌夏。1937年12月生于山西夏县
普通农民，父亲出生才百日，日本鬼子就进了村。日本鬼子投降
小学，发蒙于解保全老师。那时家里条件并不富裕，一支铅笔还

Progress? How could it have possibly worked? Transforming a rural town situated in the heart of the fertile South Moravian region, far from any major cultural centres, into an industrial

Many innovations in brick manufacturing can be grouped with th
dematerialisation'. This term refers to reducing the amount of raw m
nd energy used to manufacture brick by making the brick thinner or h

NA DANIELS LAATSTE RIT BESLISTE DE
MANÈGE DAT LADY VOORTAAN UITSLUITEND

voorbij die de mooie

material angereichert habe. Das Ergebnis sind sech
lungen, die nicht als verbindliche Lebensläufe, sor
literarische Protokolle und archetypische Skizze

[redacted]

det spørsmålet blir jeg sittende og beskjeftige meg med denne
kveldsstunden, på vei inn i alderdommen, mens min venn og
medkrimforfatter Larry B bestiller enda en karaffel vin, og for-

che Forscher und schrieb
wie man lernen kann, alle
n und nicht nur den Som-

Malhotra S. & Mattoo S. K. (2001
disorders: a report from North India,
64, 145–154.

1) Každý žák nebo žačka si vybere jed-
nu svou oblíbenou seriálovou (případně
filmovou) postavu.

3) Po promyšlení každý svou po
představí ostatním, vyučující za
vlastnosti na tabuli do tří sloup

El afiche de una familia feliz (fotografía nún
continúa en las indígenas que llevan, y bien
deben pesar, a sus hijos en la espalda. En la r

Loin d'une approche nébuleuse de la réalité, le photographe la capte au contraire de manière tr
lise un travail de terrain, enraciné dans le réel, à l'image de ces visages marqués par les rides, ou les ma
qui se frottent quotidiennement à la montagne. Le rapprochement de la série « Jours blancs » avec les

and remaining photographs and turned them into collages,
which he subsequently used as models for clay figurines.
Eventually, Henry retreated from photography altogether and

CLAY FIGURINE

After Henry's di

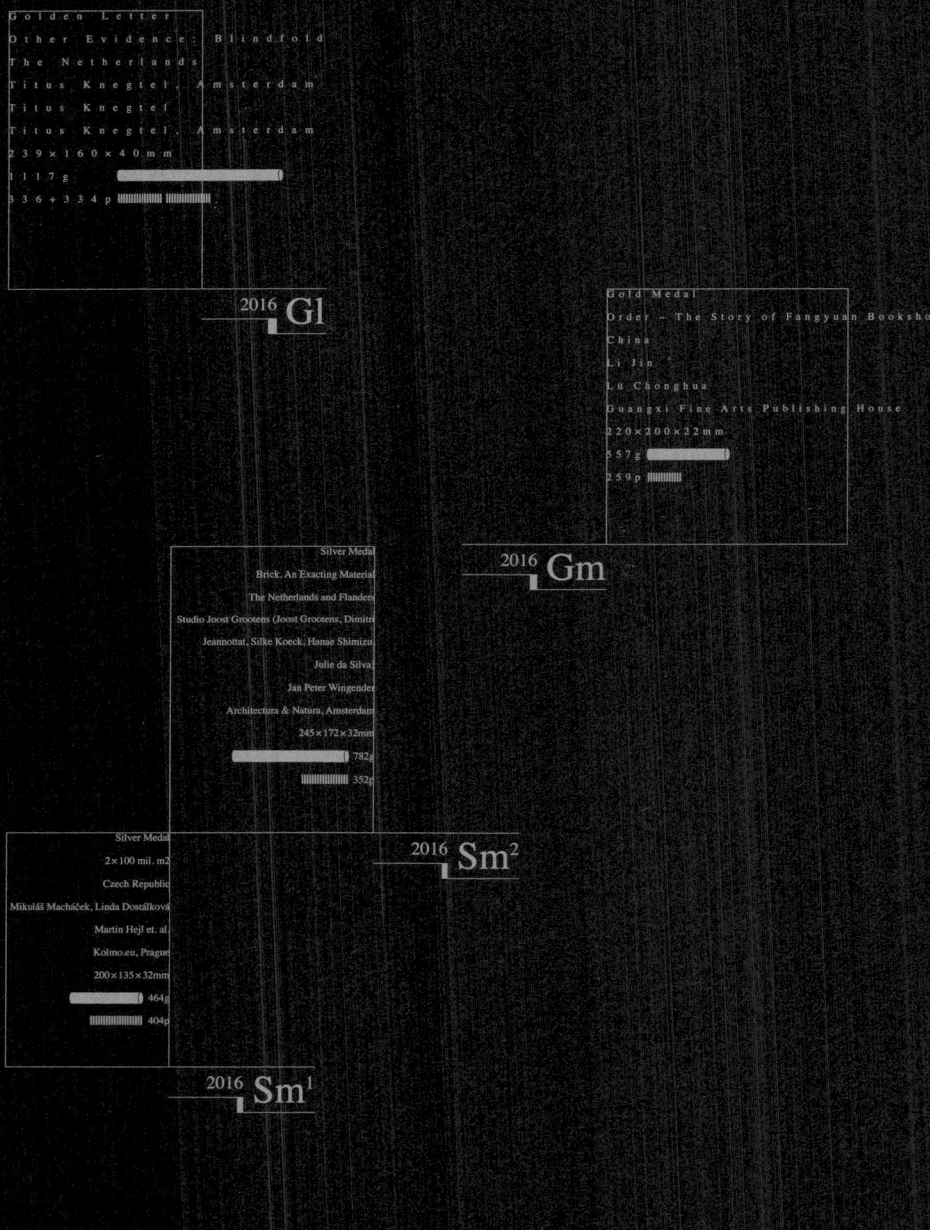

Golden Letter
Other Evidence: Blindfold
The Netherlands
Titus Knegtel, Amsterdam
Titus Knegtel
Titus Knegtel, Amsterdam
239×160×40mm
1117g
336+334p

2016 Gl

Gold Medal
Order – The Story of Fangyuan Bookshop
China
Li Jin
Lü Chonghua
Guangxi Fine Arts Publishing House
220×200×22mm
557g
259p

2016 Gm

Silver Medal
Brick. An Exacting Material
The Netherlands and Flanders
Studio Joost Grootens (Joost Grootens, Dimitri Jeannottat, Silke Koeck, Hanae Shimizu, Julie da Silva)
Jan Peter Wingender
Architectura & Natura, Amsterdam
245×172×32mm
782g
352p

2016 Sm²

Silver Medal
2×100 mil. m2
Czech Republic
Mikuláš Macháček, Linda Dostálková
Martin Hejl et. al.
Kolmo.eu, Prague
200×135×32mm
464g
404p

2016 Sm¹

2016 Bm^{1}

Bronze Medal

Daniel van Dicht

Belgium

Dear Reader, Matthias Philips (Illustrations)

Matthias Philips

Lannoo

320×240×18mm

917g

144p

2016 Bm^{2}

Bronze Medal

Het Meest Geloofde Sprookje

The Netherlands

De Vormforensen (Annelou van Griensven & Anne-Marie Geurink) & Lyanne Tonk

De Vormforensen (Annelou van Griensven & Anne-Marie Geurink)

De Vormforensen (distribution De Vrije Uitgevers)

326×222×8mm

452g

64p

2016 Bm^{3}

Bronze Medal

Nowhere Men. Illegale Migranten im Strom der Globalisierung

Austria

Christoph Miler, Zürich (CH)

Christoph Miler

Luftschacht, Wien

226×160×26mm

669g

336p

2016 Bm^{4}

Bronze Medal

Pleasure of Learning

China

Qu Minmin, Jiang Qian

Zhou Xue

Phoenix Fine Arts Publishing Ltd.

232×152×26mm

547g

414p

2016 Bm^{5}

Bronze Medal

Farvel til romanen. 24 timer i grenseland

Norway

Aslak Gurholt Rønsen / Yokoland

Ingvar Ambjørnsen

Editor: Bendik Wold og Nils-Øivind Haagensen / Flamme Forlag

Flamme Forlag

212×134×14mm

293g

127p

2016 Ha^{1}

Honorary Appreciation

LIVING THE NORDIC LIGHT

Norway

Snøhetta

Åsne Seierstad, Po Tidholm, Lars Forsberg, Barbara Szybinska Matusiak, Vidje Hansen, Bruno Laeng

Zumtobel AG

271×213×37mm

1394g

400p

2016 Ha^{4}

Honorary Appreciation

Jours Blancs

Germany

Nanni Goebel, Kehrer Design Heidelberg

François Schaer (Photographs), Pauline Martin (Text)

Kehrer Verlag Heidelberg, Berlin

330×300×14mm

1058g

100p

2016 Ha^{3}

Honorary Appreciation

Del reverso de las imágenes

Venezuela

Álvaro Sotillo, Juan F. Mercerón

Paolo Gasparini, Victoria de Stefano (Text), Ana Nuño (Translation)

Editorial mal de ojo

201×315×15mm

682g

88p

2016 Ha^{5}

Honorary Appreciation

Taking off. Henry my neighbor

The Netherlands

Mariken Wessels, Jurgen Maelfeyt

Mariken Wessels

Art Paper Editions, Gent

330×242×25mm

1617g

326p

2016 Ha^{2}

Honorary Appreciation

Career Choice Without Prejudice

Czech Republic

Jan Šiller

Anna Babanová, Jitka Hausenblasová, Jitka Kolářová, Tereza Krobová, Irena Smetáčková

Gender Studies, Prague

297×211×10mm

293g

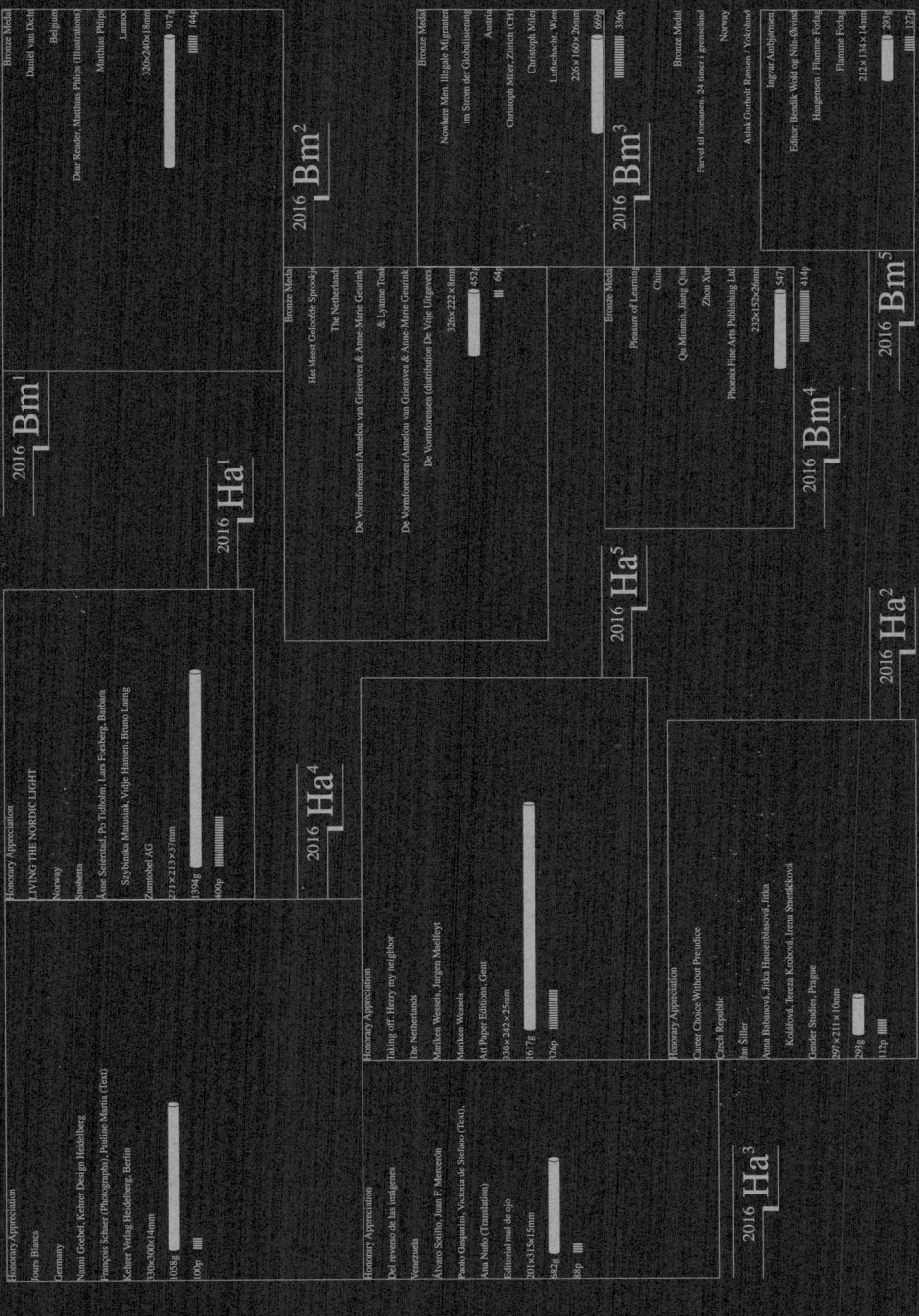

112p

2016

The International Jury

Zita Bereuter (Austria)

Loraine Furter (Belgium)

Jan Bajtlik (Poland)

Federica Ricc (Italy)

Bernardo Carvalho (Portugal)

HD Schellnack (Germany)

Jonathan Yamakami (Brazil, USA)

Country / Region

The Netherlands 3

China 2

Czech Republic 2

Norway 2

Belgium 1

Austria 1

Venezuela 1

Germany 1

Flanders 0

Designer

Titus Knegtel, Amsterdam

Li Jin

Mikuláš Macháček, Linda Dostálková

Studio Joost Grootens (Joost Grootens, Dimitri Jeannottat, Silke Koeck, Hanae Shimizu, Julie da Silva)

Dear Reader, Matthias Phlips (Illustrations)

De Vormforensen (Annelou van Griensven & Anne-Marie Geurink) & Lyanne Tonk

Christoph Miler, Zürich (CH)

Qu Minmin, Jiang Qian

Aslak Gurholt Rønsen / Yokoland

Snøhetta

Jan Šiller

Álvaro Sotillo, Juan F. Mercerón (Assistent)

Nanni Goebel, Kehrer Design Heidelberg

Mariken Wessels, Jurgen Maelfeyt

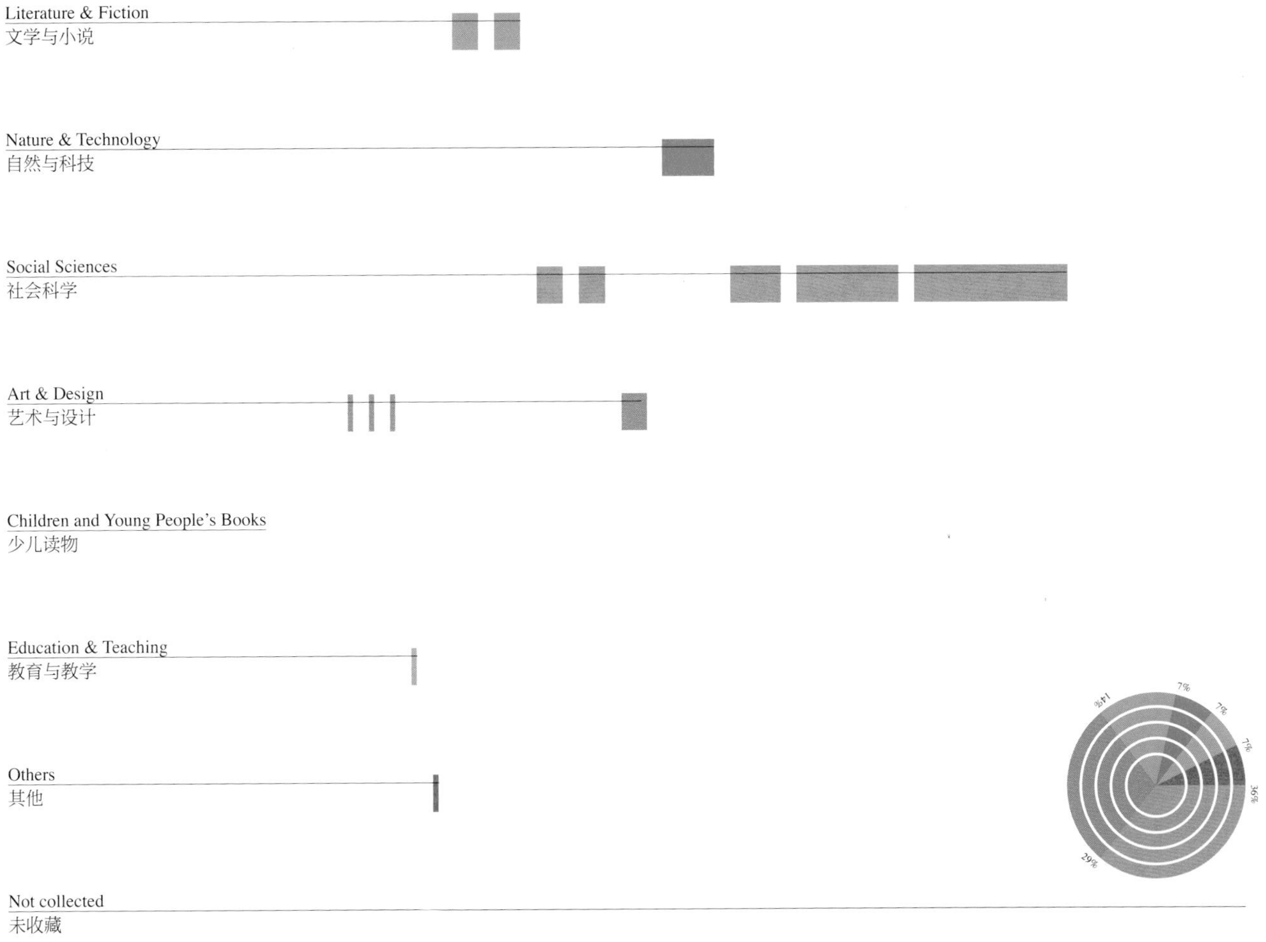
Literature & Fiction
文学与小说
Nature & Technology
自然与科技
Social Sciences
社会科学
Art & Design
艺术与设计
Children and Young People's Books
少儿读物
Education & Teaching
教育与教学
Others
其他
Not collected
未收藏
7%
7%
7%
36%
29%
14%

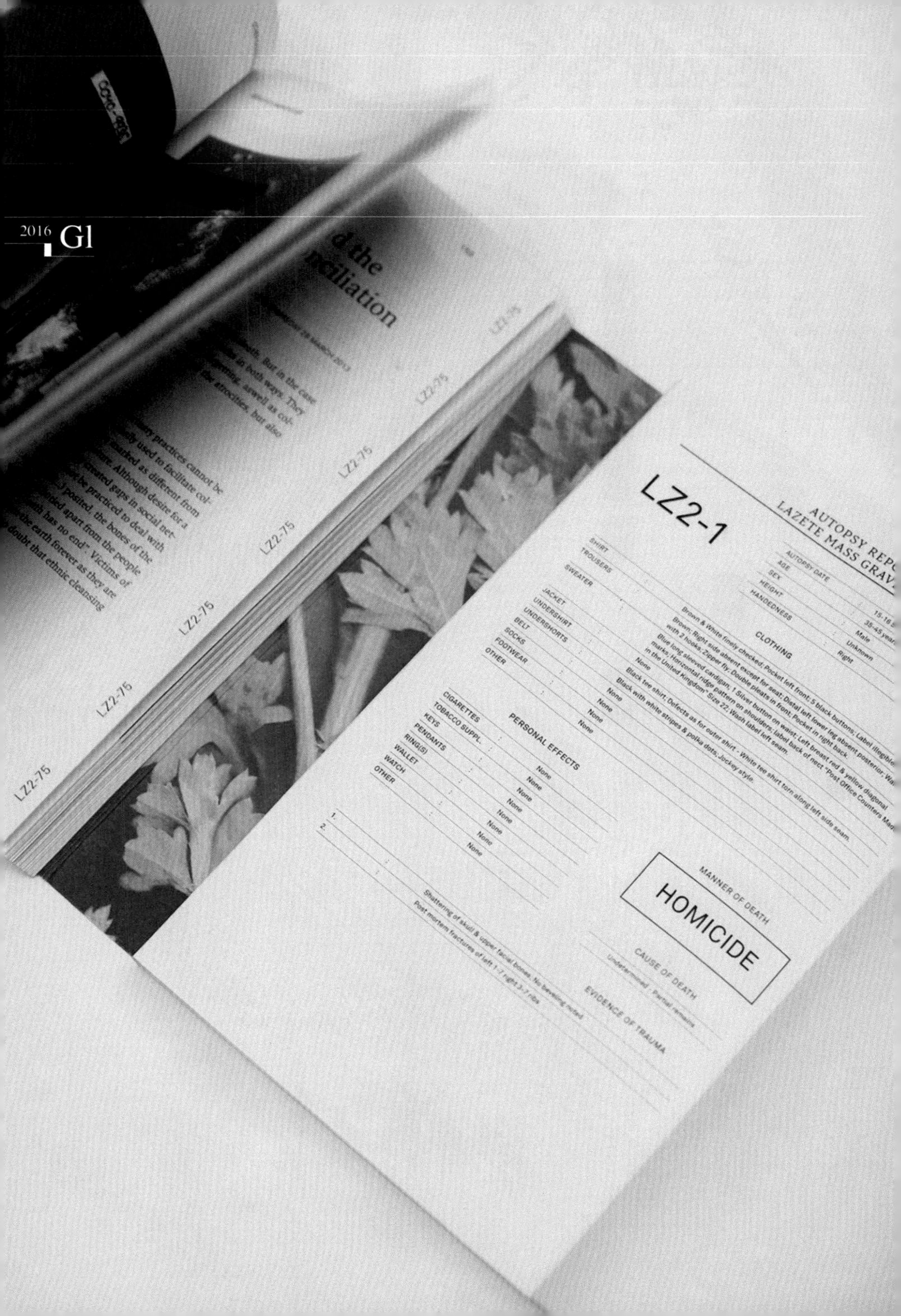
LZ2-75
LZ2-1
AUTOPSY REPO
LAZETE MASS GRAV
CLOTHING
SHIRT
TROUSERS
SWEATER
JACKET
UNDERSHIRT
UNDERSHORTS
BELT
SOCKS
FOOTWEAR
OTHER
PERSONAL EFFECTS
CIGARETTES
TOBACCO SUPPL.
KEYS
PENDANTS
RING(S)
WALLET
WATCH
OTHER
MANNER OF DEATH
HOMICIDE
CAUSE OF DEATH
EVIDENCE OF TRAUMA

这是一本记录着沉痛的历史证据并研究和试图还原现场的珍贵档案。书中整理了 1995 年波黑和前南联盟军警在斯雷布雷尼察进行的 8000 人大屠杀的历史资料，包括万人坑的发现过程、场所照片、证据和部分验尸报告。书籍的装订形式采用金属构件固定活页，封面采用环绕式包覆的形式，将档案内容锁在中间。封面白卡纸四色印刷，但只呈现为蓝紫色和墨绿色双色观感。内页分为页数相差无几的两部分，可供对照查阅。第一部分胶版纸四色印刷，左翻本，与封面相同的双色观感处理方式，蓝色展示过程和证据，绿色则是发现地点和其他信息，双色图片相互透叠，形成如电影的蒙太奇手法。页边有针对第二部分的档案编号，虽不完全一一对应，但提供了与第二部分相关资料的查找链接。第二部分胶版纸单黑印单面，右翻本，为 100 多份详尽的验尸报告。这些资料大多来自针对本次事件临时成立的前南斯拉夫国际刑事法庭。

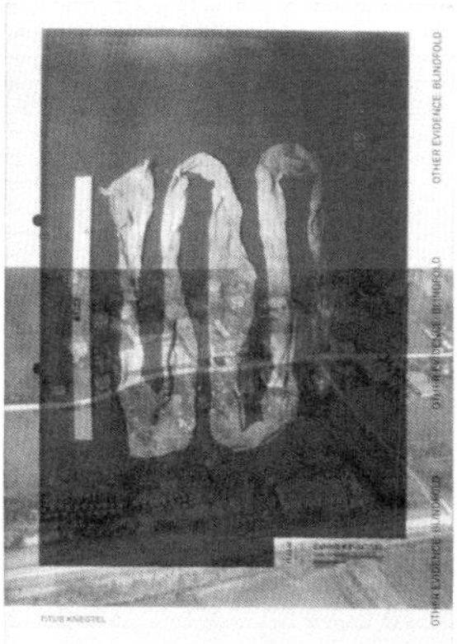

This is a precious archive that records painful historical evidence and studies and attempts to restore the scene. The book collates the historical data of a massacre carried out by Bosnia and Herzegovina army and the former Yugoslav Police in Srebrenica in 1995, including the process of finding mass graves, site photographs, evidence and some autopsy reports. The book is bound with metal components to fix loose pages, and the cover is wrapping the archives in a circular way. The cover is printed on white cardboard in four colors, but blue-purple and dark green are dominant hue. The inside pages are divided into two parts with almost the same number of pages, which can be read by comparison. The first part is printed on offset paper in four colors, turning leftward, using blue to display process and evidence, which green is to show the location and other information. Two-color pictures overlap each other, forming montage effect in films. There are file numbers for the second part in the margin, which do not exactly correspond to the second part, but provide a searching link with the relevant information. The second part is printed on offset in monochrome black, which contains more than 100 detailed autopsy reports. Most of these data comes from the International Criminal Tribunal for the Former Yugoslavia, which was temporarily established in response to this incident.

金字符奖

其他证据：蒙上眼睛

荷兰

Golden Letter

Other Evidence: Blindfold

The Netherlands

Titus Knegtel, Amsterdam

Titus Knegtel

Titus Knegtel, Amsterdam

239×160×40mm

1117g

336+334p

The Srebrenica massacre in 1995: more than 8,000 people murdered. – The International Criminal Court: evidence of a war crime exhumed and brought to light, translated into objective data. 50,000 pages of autopsy reports written, 30,000 pieces of evidence photographed. – An artist's book: documents from the tribunal taken from the files. – The form: double codex with individual sheets of paper, bound with paper fasteners and wrap-around cardboard cover. Shocking events do not simply disappear if they are forgotten. Remembrance alone does not set things to rights again. Like a monument, an artist's book can achieve something: in a visual and material setting, it can capture those things that are not understandable. In this case, the finesse of the design features will not be admired, the careful typography will not be emphasised, and the modern typeface will not be appreciated. This book will be valued and kept as an expression of shell-shocked shame – and once in a while, in moments of disbelief, it will be picked up and people will ask themselves: How could this have happened?

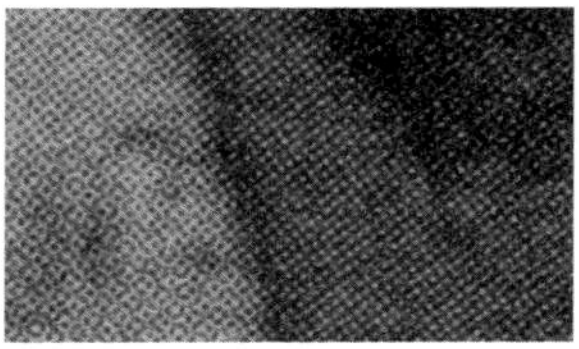

size ran

entral on

faint plaid. 2 holes

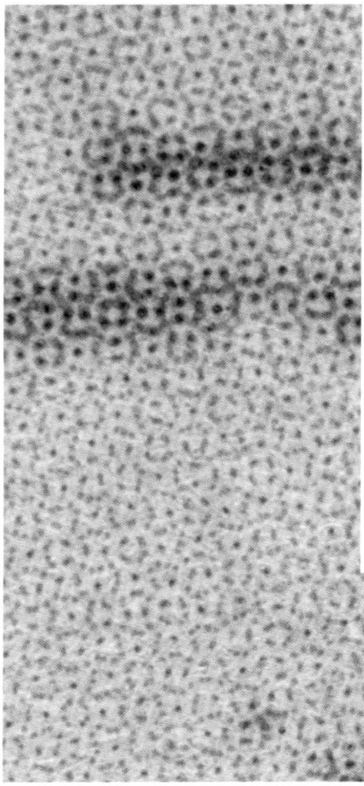

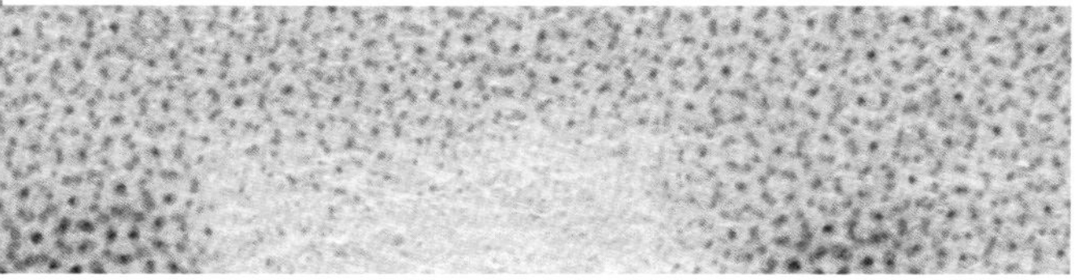

e one being addressed. Th

s back on. It is obvious thou

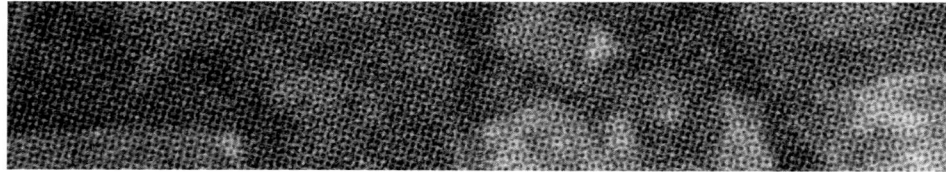

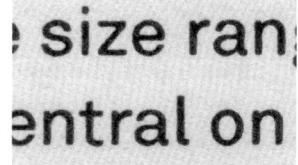

239×160×40

LZ2

1117

336+334

SHIRT

TROUSERS

SWEATER

JACKET

UNDERSHIRT

UNDERSHORTS

BELT

SOCKS

FOOTWEAR

OTHER

CIGARETTES

TOBACCO SUPPL

KEYS

PENDANTS

RING(S)

WALLET	:	None
WATCH	:	None
OTHER	:	Ball point pen in left jacket pocket; Tablets in small pack left outer jacket pocket

Multiple gunshot wounds

Disclaimer

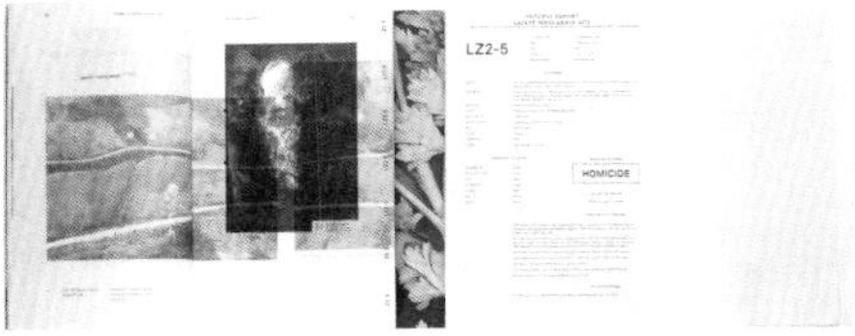
LZ2-5
HOMICIDE

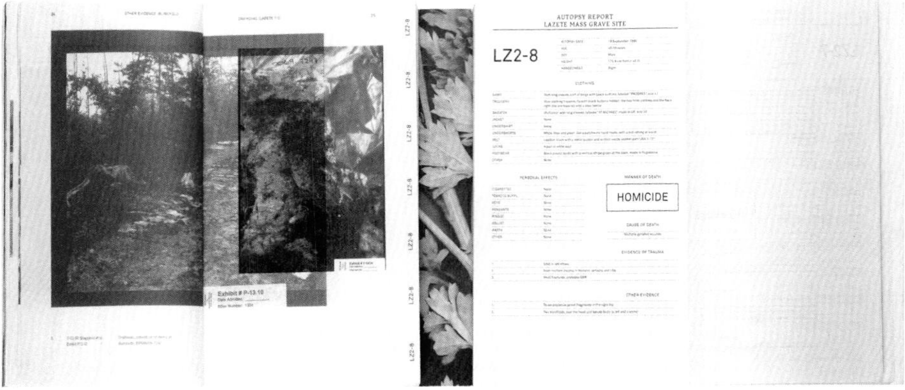
Exhibit # P-13.10
LZ2-8
AUTOPSY REPORT
LAZETE MASS GRAVE SITE
LZ2-8
CLOTHING
PERSONAL EFFECTS
MANNER OF DEATH
HOMICIDE
CAUSE OF DEATH
EVIDENCE OF TRAUMA
OTHER EVIDENCE

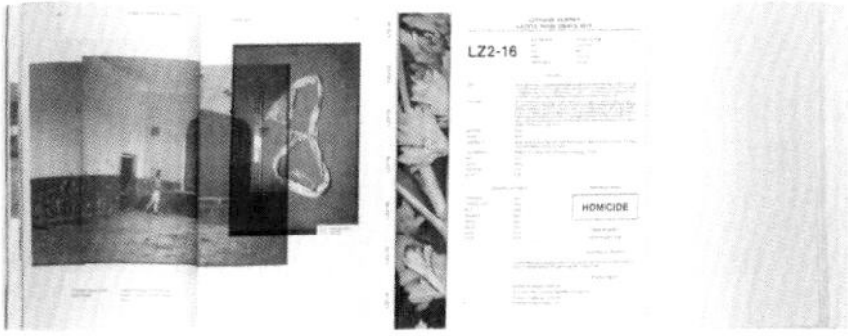
LZ2-16
HOMICIDE

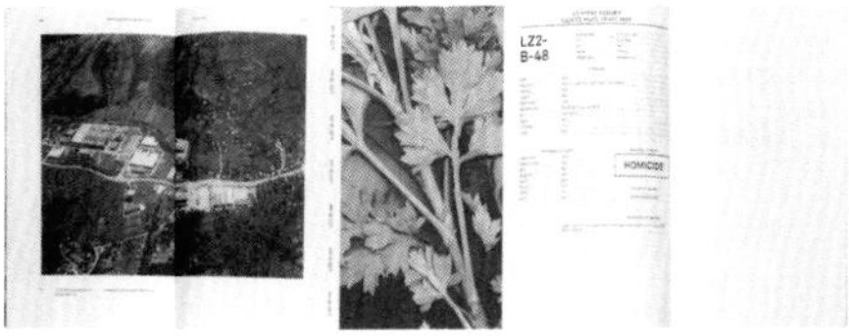
LZ2-
B-48
HOMICIDE

Contents
Other Evidence:
Blindfold

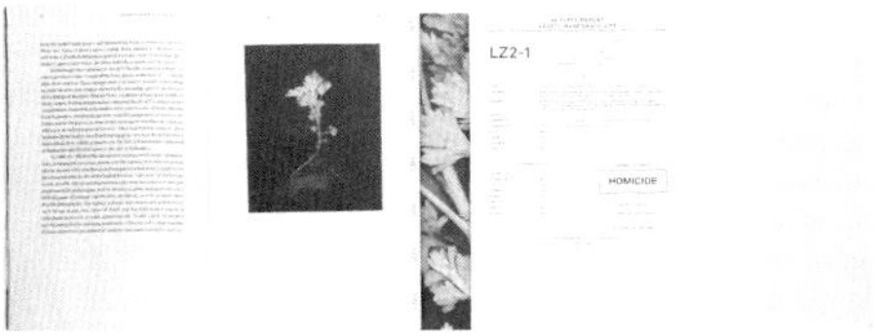
LZ2-1
HOMICIDE

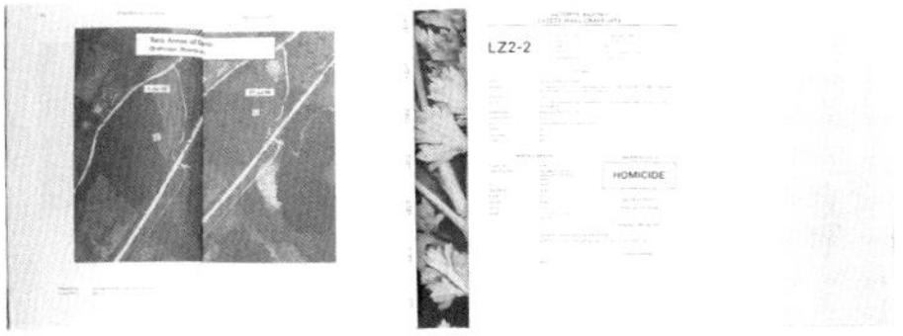
LZ2-2
HOMICIDE

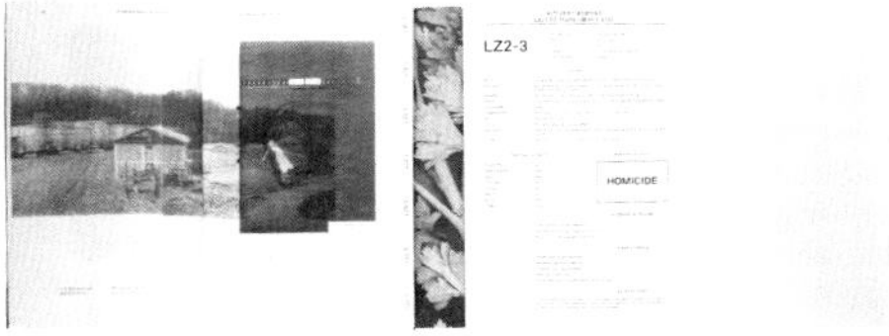
LZ2-3
HOMICIDE

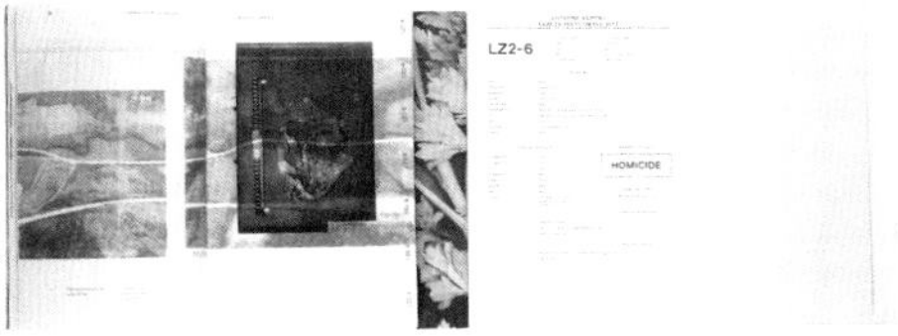
LZ2-6
HOMICIDE

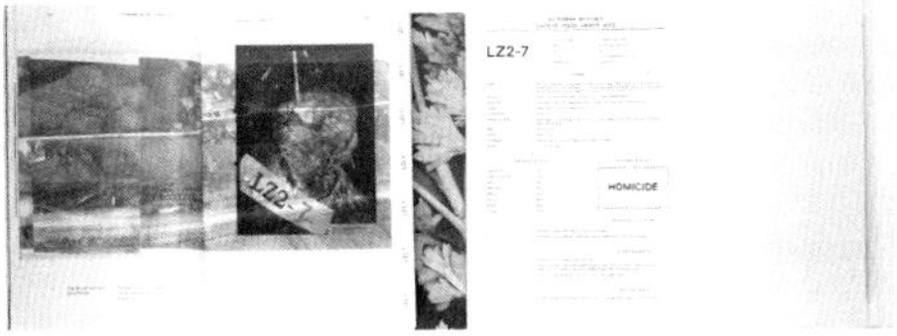
LZ2-7
HOMICIDE

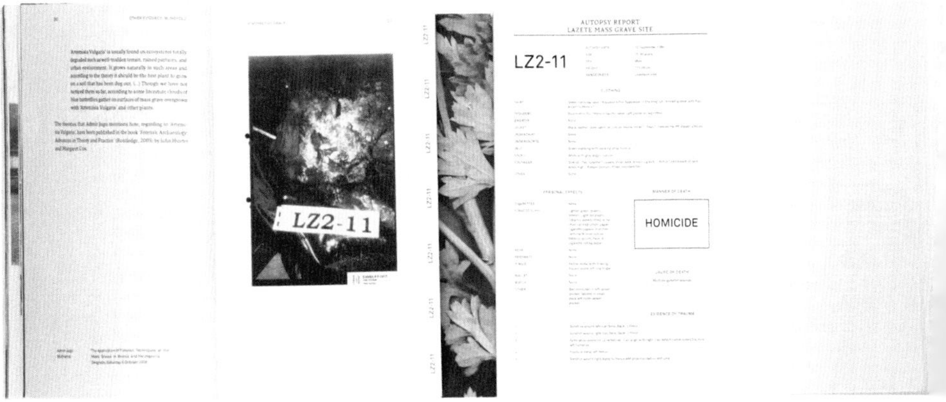
LZ2-11
AUTOPSY REPORT
LAZETE MASS GRAVE SITE
LZ2-11
HOMICIDE

OTHER EVIDENCE: BLINDFOLD
TITUS KNEGTEL
52
OTHER EVIDENCE: BLINDFOLD
PILICA DOM
53
LZ2-22

AUTOPSY REPORT
LAZETE MASS GRAVE SITE

LZ2-22

AUTOPSY DATE	9 September 1996
AGE	30-40 years (mean 35)
SEX	Male
HEIGHT	178.5 cm
HANDEDNESS	Not determined

CLOTHING

SHIRT	Light green with long sleeves, has epaulets at shoulders, two breast pockets, black buttons, size 39
TROUSERS	Blue, two side pockets, R back pocket, grey button fly, shortened from bottom by hand sewing, labeled "C 50"
SWEATER	None
JACKET	None
UNDERSHIRT	None
UNDERSHORTS	White sports trousers, elastic at ankles, no size and label
BELT	Leather, brown with metal buckle
SOCKS	White, wool, handmade
FOOTWEAR	Black Wellington boots cut from ankles, has rectangular grey adhesive repair at left boot
OTHER	None

PERSONAL EFFECTS

CIGARETTES	None
TOBACCO SUPPL.	None
KEYS	None
PENDANTS	None
RING(S)	None
WALLET	None
WATCH	None
OTHER	Papers with numbers, from R trouser pocket

MANNER OF DEATH

HOMICIDE

CAUSE OF DEATH

Multiple gunshot wounds

EVIDENCE OF TRAUMA

1	Gunshot wound to left humerus with fissured fractures
2	Large defect at cranium involving facial bones with radiating fractures from front to back consistent with a gunshot wound

OTHER EVIDENCE

1	Bullet at L buttock

2016 Gm

本书以书店订单为引导，记录了中国西安方圆书店 33 年的发展史。书中整理了书店与各大美术出版社的订单，撰写了店主自述的故事，以及随性的小画。本书的装订形式为锁线包背装，锁线位置处于封面约四分之一处，留下较宽的开放式书脊用于展示店主的自画像。封面采用灰绿色塑料编织袋面，无印刷，裱贴胶版纸单黑印刷的“订单”形式的书籍信息。内页胶版纸四色印刷，包背页面朝外的部分主要以文字辅以插图为主，记录跟书店经营有关的内容；朝内的页面则为相应时期所拍摄的照片，可以裁开浏览。书籍的前后分别使用新闻纸单黑印刷，呈现大量各时期的订单、自画像和朋友画的店主小画，为自述的历史附上真实而有趣的佐证。

單訂
方圓故事

This book, organized by the bookstore orders, records the 33-year history of Fangyuan Bookstore in Xi'an, China. The book collates the orders of Fangyuan Bookstore with major art publishing houses, describes the stories of the owner himself, as well as some casual paintings. The double-leaved book is bound with sewing thread located about a quarter of the cover, leaving a relatively wide spine for the shopkeeper's self-portrait. The cover is made of gray-green plastic woven bags without printing, mounted by offset paper printed with book information in the form of 'an order'. The inside pages are printed on offset paper in four colors, the outer page of the double-leaved part is mainly printed with text and illustrations to record the operation of the bookstore, while the inward page is displaying the photos taken in the corresponding period, which can be cut to browse. The book uses newsprint in black to present a large number of orders from different periods, self-portraits of the shopkeeper and small pictures painted by friends, which provide real and interesting evidence for the self-narrative history.

- 金奖
- 订单：方圆故事
- 中国
- 李瑾
- 吕重华
- 广西美术出版社

- Gold Medal
- Order – The Story of Fangyuan Bookshop
- China
- Li Jin
- Lü Chonghua
- Guangxi Fine Arts Publishing House
- 220×200×22mm
- 557g
- 259p

A fascinating book block: Japanese binding creates a beautifully soft bundle out of the creamy, blossom-light paper. Cover pages made of synthetic raffia in a dull, olive green protect the almost square book in portrait format. Red writings and drawings are printed on the spine and tail edge – like a summary of the content. The extra-wide rear section enables four mini-books on the left of the grey binding string that holds everything together. They flash by at your fingertips in a kind of flip book, where the caricature faces seem like characters. In the main book, the light and airy interaction between text and image is astounding. What freedom! The script is written both horizontally from left to right, and also vertically from right to left. The visual principle of script and sketch allows both of these to signify meaning in merely different states of matter. The whole page itself appears as if in a third, abstract state – with a generous top edge and gutter margin, with the controlled relationship between printed and unprinted space – and seems like a super-character. All aspects of this book – material, characters, printing, binding – are organic. The printed verso pages – still closed at the front by the fold – serve up additional content like an unopened fruit. As if the growth of such books were without intention.

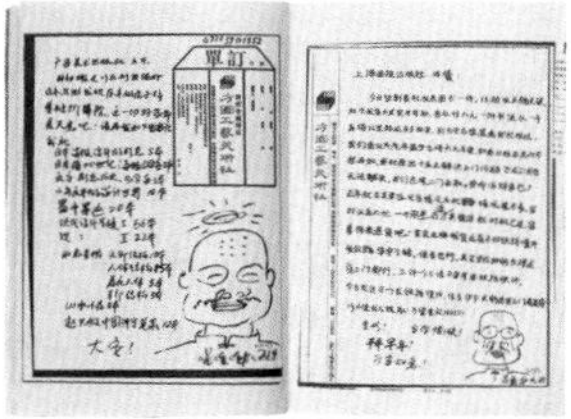
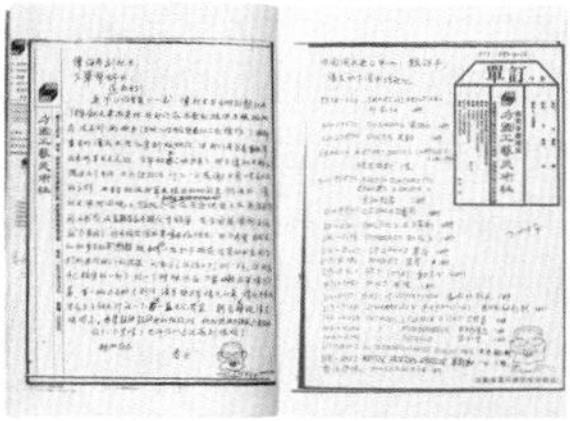
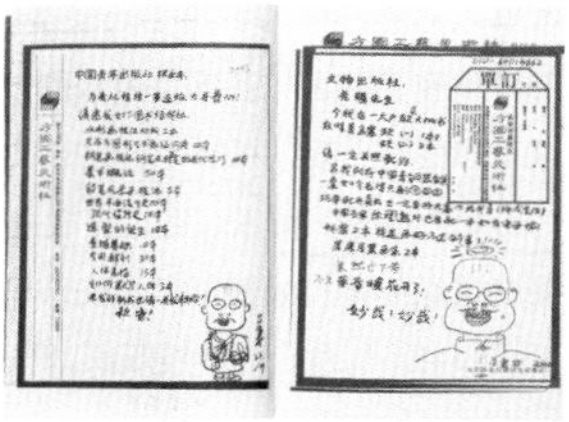
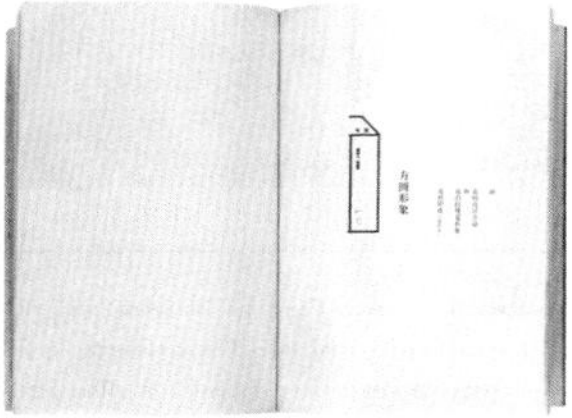
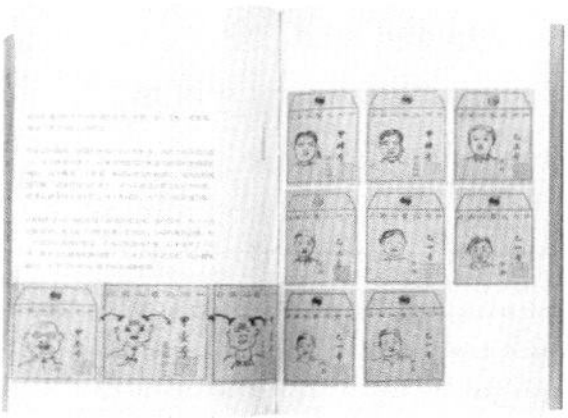
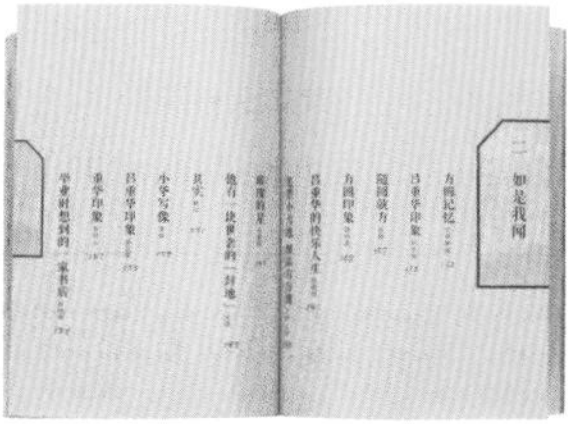
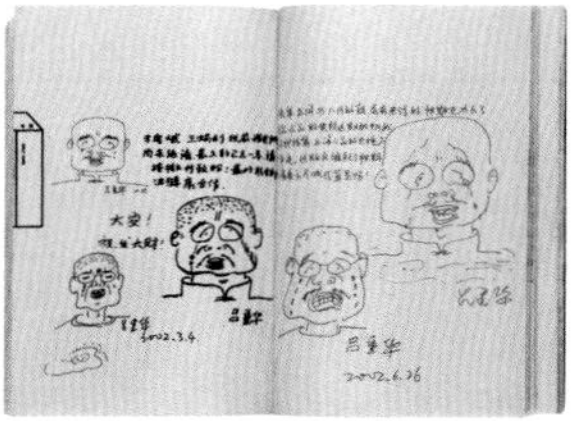
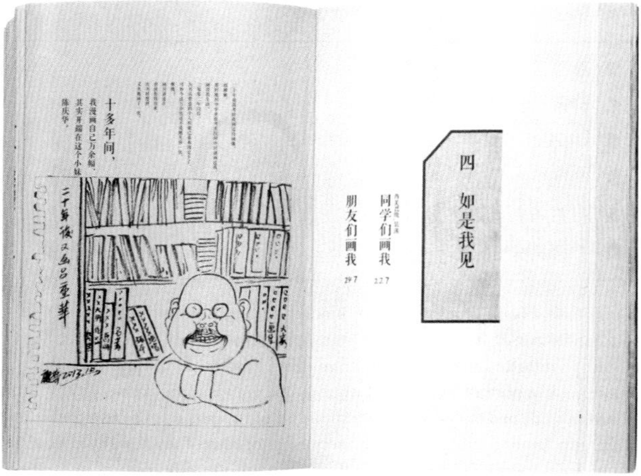
四 如是我见
同学们画我 227
朋友们画我 297
二十年后又画吕重华
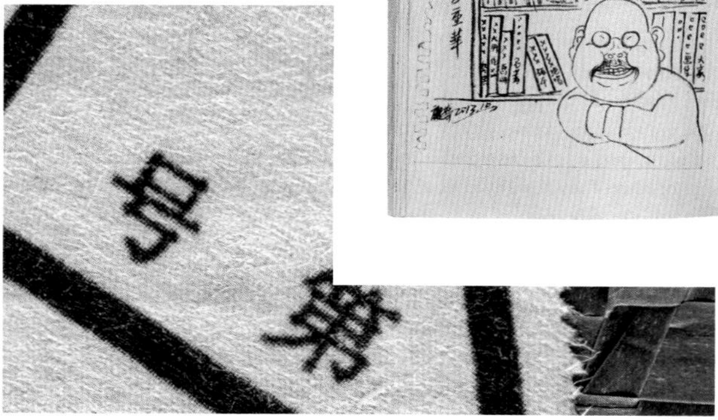

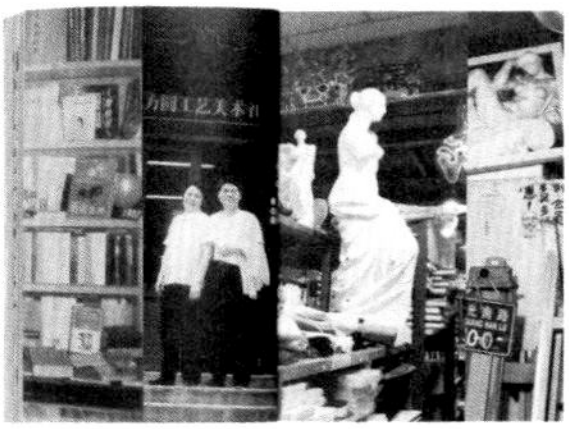

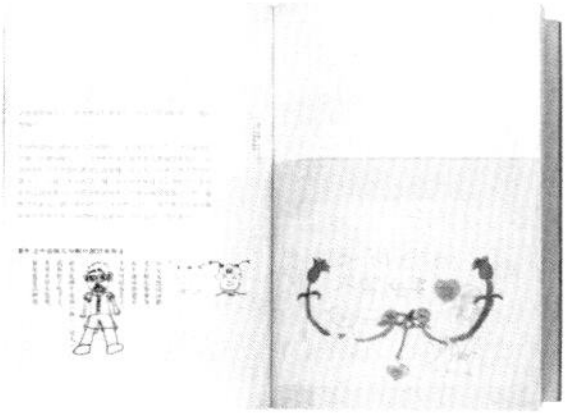

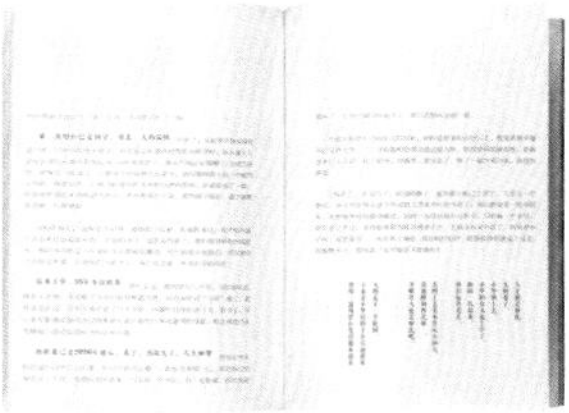

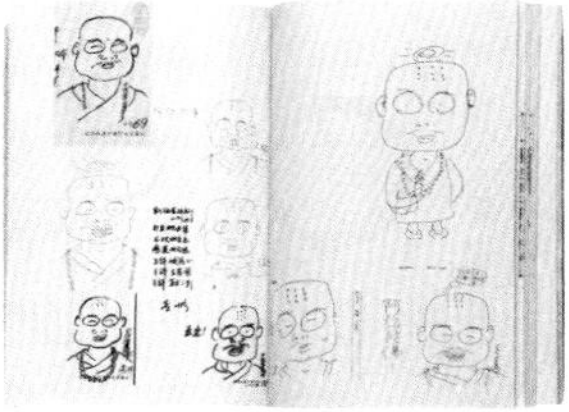

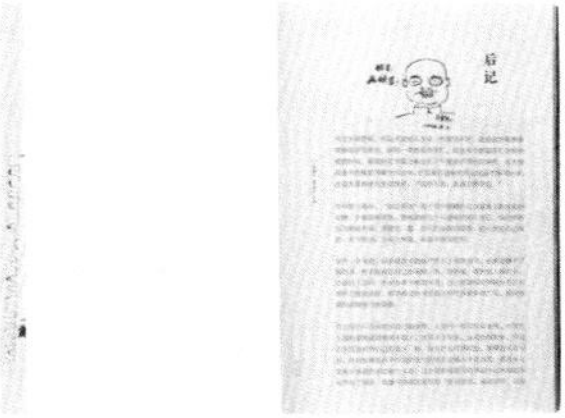
后记

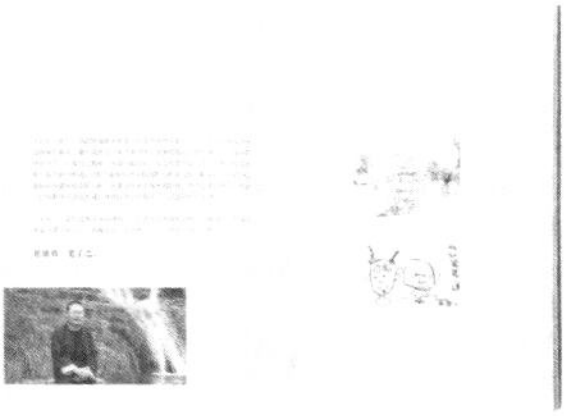

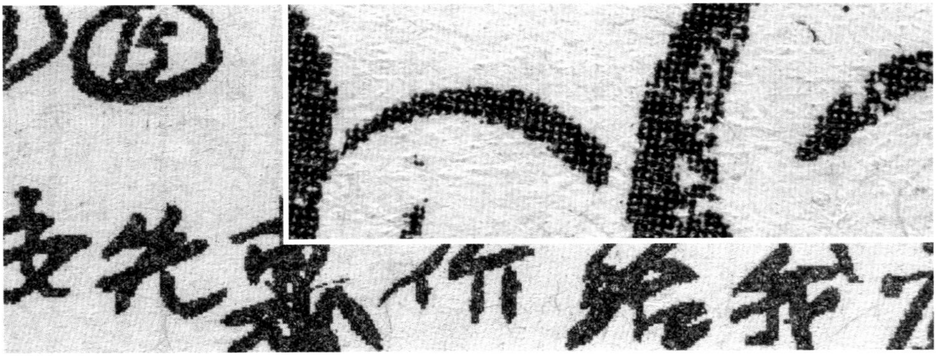

方圓故事

1 匾

这块匾是父亲集《礼器碑》的字而成，这样不但效果好而且又不花钱，即使请名书法家来写，也未必那一挥毫就能达到最好状态。其实集字也不易，《礼器碑》里现成的字只有“方”、“工”、“社”，而“圆”、“艺”、“美”、“术”这四个字都是父亲用“礼器碑”的笔画造出来的，整体效果还不错，“术”字虽弱一点，但也不掉队。1991年暑假高考结束后，这块匾从放大、刷漆到刻字都是我亲手制作完成，可惜当时用的材料不够好。匾挂出来后，常有搞书法的人问我这匾是谁写的。

方圓工藝美術社
始于公元一九八二

我的设计作品和我们的视觉形象 —

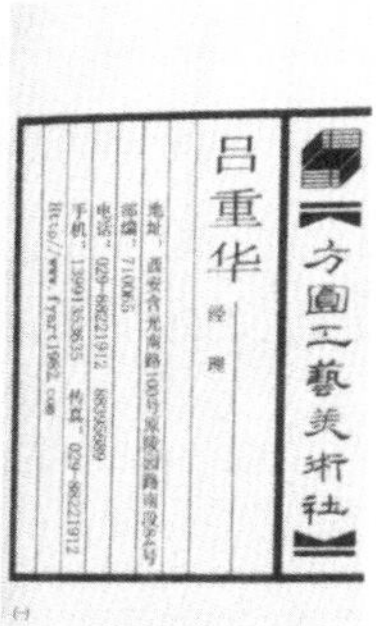
吕重华
方圓工藝美術社
电话：029-88221912 88395689
手机：13991353635 传真：029-88221912
邮编：710065

(一)

方圓工藝美術社
購
書
銀
票
伍拾圓整

(二)

qunlipai
(三)

这是我为群力公司设计的商标，是1993制作的，群力公司生产橡胶产品、橡皮和美术用品。图像设计为一只手持橡皮状，圆形的手形加橡皮构成英文字母“Q”来表现群力和橡皮，圆的处理突出橡胶产品的特征。我自己对这个设计比较满意，这个算是学习阶段的习作、后面的字母是哥们成鹏后加的。

我为我们社设计的标志，时间大约是在1997年，用一函线装书的外观做成矛盾空间图像，线装书表现我追求的中式古典风格的视觉形象，并突出“书”和书店的严肃性。矛盾空间则表现书店的艺术性和时代性。“方圆”表达得还是不够，现在很多人识不得带函套的线装书形象，年轻人更甚，竟然有人说是架子床，我很郁闷。这个标志和集礼器碑的店名很般配。

因为卖书，我把名片设计成一页线装书内页的形象，符合我追求的中式古典风格的视觉象形，并突出“书”和书店的严肃性。名片形象产生于1996年大学毕业后，比标志产生略早一些，到现在一直没有大的变化。有一次去印名

residential blocks that were under construction ... time. And many times I met an older lady asking ... "Please, I am here to see my daughter, could you help me? I can't find the way..." So it was two of us looking. It became very important to make the design in a way that was readable for basic orientation of a ... anybody coming to visit.

How did you want to ensure higher readability of the ...

Where the prescribed panel technology offers a corner section, for example in a meander area, I can better express where the space is public, ... cars, and where on the contrary a calm area, designed for quiet ... greenery, with sandpits for kids etc.

How were the corner sections created and how did you succeed in implementing them?

We came up with the corner sections ourselves, and the South-Western City is the first place where they were used. We found out they condensed the housing development and it helped us to push them through.

We had the limit of two hundred and seventy people per hectare for the area. When subway stations were planned, they needed a certain number of residents within an eight-minute walking distance. Following these indicators we had to make the development denser. And it was the corner section that helped us to meet the criteria without increasing the number of floors.

How did you approach the issue of height?

The Stodůlky housing project incorporated an agricultural village with a Baroque farm and the Church of St James the Greater; we wanted the church to remain a hallmark of the new development as well. For this purpose we tried to make the mass as small as possible, we did not want major contrasts. This was translated to a maximum of eight floors. When we proceeded to the next stage, Lužiny, a more generous project than Stodůlky, it was designed in an open space of the northern hillside with fields, and in some cases ten-storey buildings were designed to compensate for the slope. But an expert opinion from the ministry resulted in the necessity to increase the number of floors at Lužiny by three on average.

Instead of a right angle section the 135-degree angle ... in Lužiny. How ...

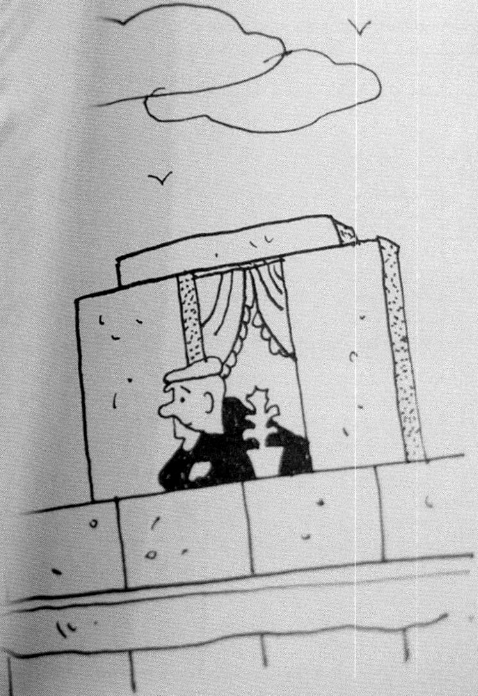

这是一本记录捷克和斯洛伐克 100 年来公寓住宅类建筑的理念、规划、实施和评论的文献，是 2014 年威尼斯建筑双年展捷克斯洛伐克馆的展示项目。本书的装订形式为无线胶装，封面采用白色布面烫黑，裱贴极薄的胶版纸，呈现柔软易翻阅的亲和手感。内页泛黄的胶版纸四色印刷，使得图片的饱和度较低，从而凸显黑白分明的内文的强烈风格，着力于文本内容的表现。标题和重点文字采用了 20 世纪 70 年代的建筑中偏好的具有极强视觉效果的衬线字体，特别设置的文本缩进页产生了独特的页面阅读节奏。与极力展现建筑规划伟大成就的建筑类书籍不同，本书在图片选择和设计上呈现出一种客观的距离感，从而更清晰和严肃地创作出一本带有个人评论和自我调侃气质的建筑评论手册。

2×100 mil. m²
Martin Hejl&Coll.

This is a document recording the concept, planning, implementation and review of apartment buildings in the Czech republic and Slovakia in the past 100 years. It is the exhibition project of Czechoslovakia Pavilion at the 2014 Venice Biennale of Architecture. This book is perfect bound. The cover is white cloth with black hot stamping and mounted with thin offset paper, which makes the book soft and amiable. The inside pages are printed on offset paper in four colors. The low saturation images highlight the sharp contrast of black text and white paper, focusing on the expression of the content. The title and key text adopt the serif font with strong visual effect preferred in the architecture of 1970s, and the special text indentation creates a unique reading rhythm. Unlike the architectural books which endeavor to display the great achievements of architectural planning, this book presents the feeling of distance in picture selection and design, so as to create a clear and serious architectural review manual with personal comments and self-amusement.

银奖
2×100 mil. m2
捷克

Silver Medal
2×100 mil. m2
Czech Republic
Mikuláš Macháček, Linda Dostálková
Martin Hejl et. al.
Kolmo.eu, Prague
200×135×32mm
464g
404p

There is a self-mocking air to this documentation of apartment construction and housing estates in Czechoslovakia. The illustrations catch your eye first: some are sketched like cartoons, while others are images drawn vivaciously with a tusche brush. Their pedagogical impetus differs considerably from the self-attributed greatness that is frequently emanated by architects and architecture. The typography pursues an intensely contrasting black & white. A 70s-style serif typeface was chosen – an extraordinary preference in an architectural context. Two fonts and two font sizes – with particularly narrow typesetting in the extra-bold font, huge indents, left justification, and chapter headings almost in display size: these clear devices are sufficient to fill the pages of text with a tectonic, didactic structure and breathe a liveliness into them that encourages further reading. The colour photos are printed entirely fearlessly on rough, yellowed book paper – making it crystal clear that the key focus here is not on shiny, dazzling effects. These unpretentious and humorous design features are to be understood as an expression of self-distance, and at the same time they underline the seriousness of the social concerns presented in the architecture. The clear, compact quality creates a handbook with personal commentary and is displayed in situations ranging from interviews with architects through to the extensive chronology of figure ground plans. It is not representation, but rather the character of the works that is important in this publication, which was the Czech and Slovak contribution to the Biennale 2014 in Venice.

2×100 mil. m²
Martin Hejl & Collective

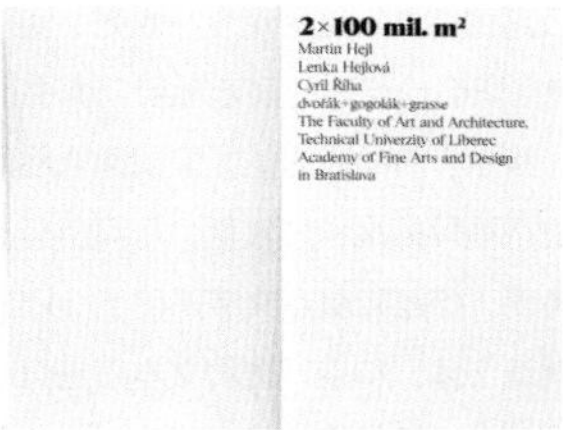
2×100 mil. m²
Martin Hejl
Lenka Hejlová
Cyril Říha
dvořák+gogolák+grasse
The Faculty of Art and Architecture,
Technical Univerzity of Liberec
Academy of Fine Arts and Design
in Bratislava

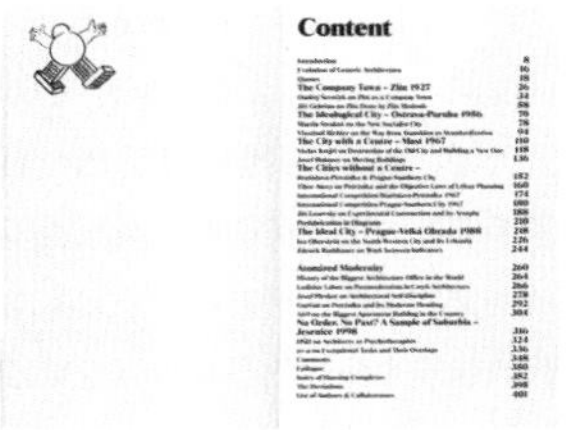
Content

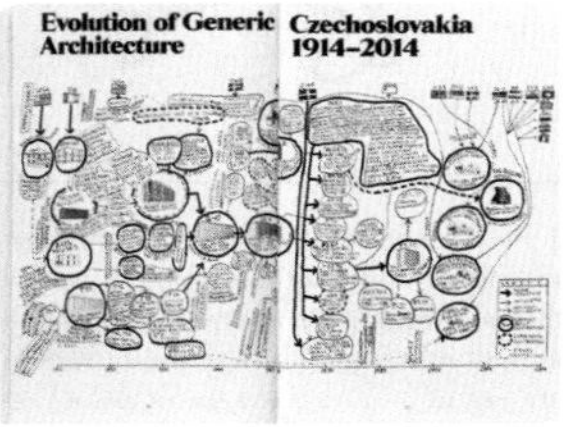
Evolution of Generic Architecture
Czechoslovakia 1914–2014

Quotes

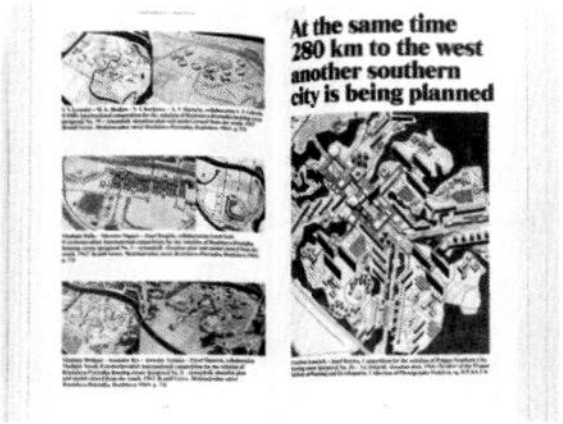
At the same time
280 km to the west
another southern
city is being planned

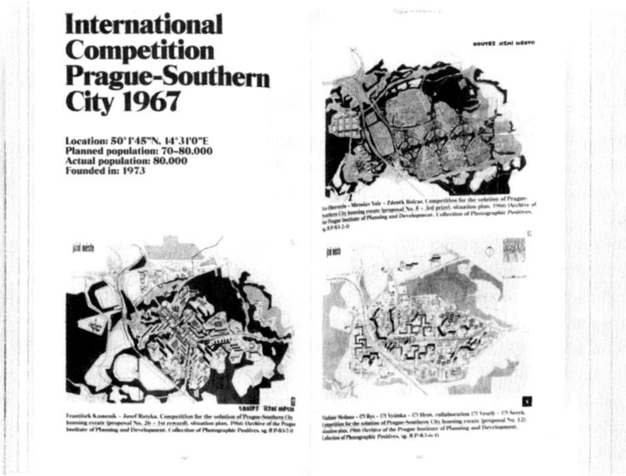
International
Competition
Prague-Southern
City 1967
Location: 50°1'45"N, 14°31'0"E
Planned population: 70–80,000
Actual population: 80,000
Founded in: 1973
František Kameník – Josef Rotyka. Competition for the solution of Prague-Southern City housing estate (proposal No. 26 – 1st reward), situation plan, 1966 (Archive of the Prague Institute of Planning and Development, Collection of Photographic Positives

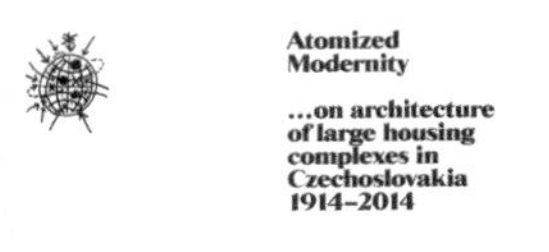
Atomized
Modernity
…on architecture
of large housing
complexes in
Czechoslovakia
1914–2014

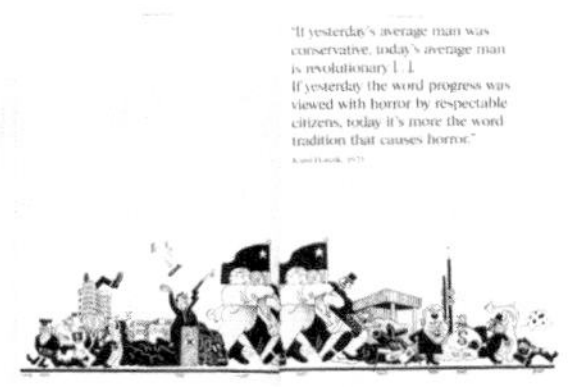
"If yesterday's average man was
conservative, today's average man
is revolutionary [...].
If yesterday the word progress was
viewed with horror by respectable
citizens, today it's more the word
tradition that causes horror."

"We've destroyed or neglected
beautiful towns and villages,
this beautiful heritage left by dozens
of generations. We've covered
our country with rabbit hutches
in which it is impossible to live,
where our citizens only sleep
and watch television shows."

The Company Town
Zlín 1927
A bathroom rather
than a bow window

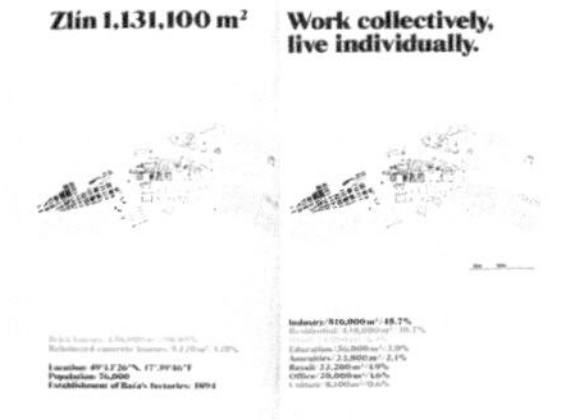
Zlín 1,131,100 m²
Work collectively,
live individually.

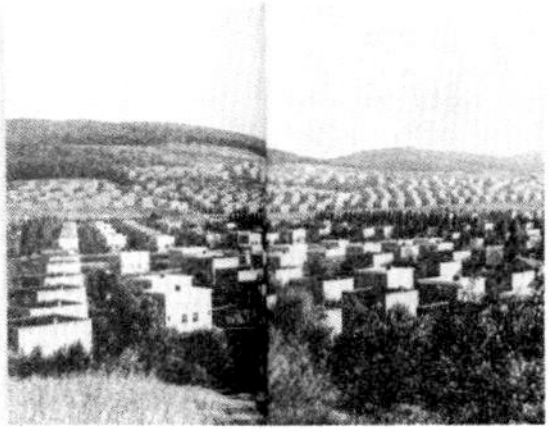

Prefabrication
in Diagrams

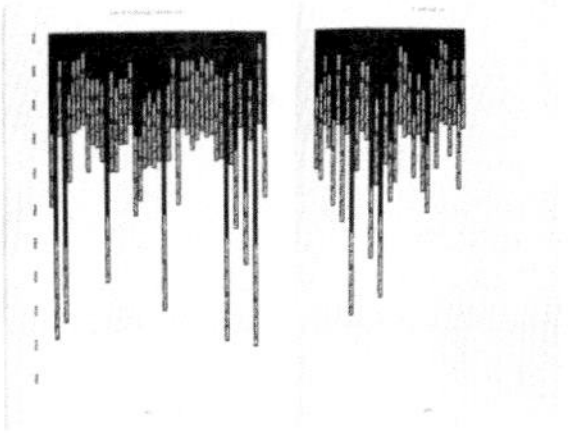

The beginning…
Terraces

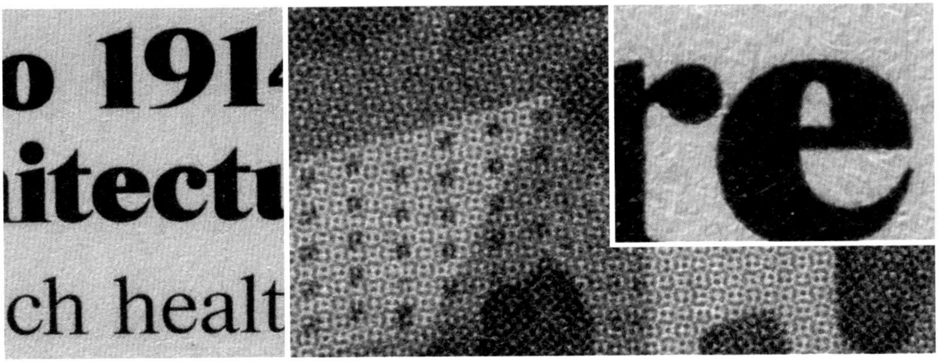
o 1914
itectu
ch healt
re

2×100 mil. m²
Martin Hejl & Coll.

Colophon

This book was published on the occasion of the exhibition 2×100 mil. m² at the Pavilion of the Czech Republic and Slovakia, La Biennale di Venezia, 14. Mostra Internazionale di Architettura, 4 June–23 November 2014.

Concept / Martin Hejl, Lenka Hejlová

Produced by / KOLMO.eu

Graphic Design / Linda Dostálková,
Mikuláš Macháček

Editors / Michal Dvořák, Ivan Gogolák,
Lukáš Grasse, Lenka Hejlová,
Jana Kořínková, Pavel Kosatík,
Lukáš Rous, Cyril Říha

Texts / Martin Hejl, Cyril Říha

Illustrations / Alexey Klyuykov, Jan Šrámek

Sound recording / Matouš Godík, Vojtěch Ptáček

Transcription / Lenka Hejlová, Petra Ježková,
Eliška Málková, Martina Skalická

Translation / Jana Kinská, Jana Kořínková,
Skřivánek Language Agency

English language editor / Jana Kořínková, Skřivánek
Language Agency

Photographs / Dušan Tománek

Documentation
and images / Ostrava City Archive, The Archive of the National Technical Museum, The Archive of the Prague Institute of Planning and Development, private archives of A69, Tibor Alexy, Tomáš Džadoň, GutGut, Josef Holanec, HŠH, Václav Krejčí, Ladislav Lábus, Jan Magasanik, Ivo Oberstein, ov-a, Josef Pleskot, Martin Strakoš, Ondřej Ševeček

Layout and typesetting / Linda Dostálková,
Mikuláš Macháček

Printed by / Indigoprint

Print run / 1000

Published by / © KOLMO.eu, 2014

ISBN 978-80-260-6127-4

Partners / dvořák+gogolák+grasse, The Faculty of Art and Architecture of the Technical University of Liberec, The Academy of Fine Arts and Design in Bratislava

The project was kindly
supported by / The Ministry of Culture of the Czech Republic, The National Gallery in Prague, The grant of the Czech Technical University in Prague No. SGS14/090/OHK1/1T/15 "Talks about Past and Current Trends in the Construction of Cities in Terms of Urban Planning Ideas and Visions", Grant of the Czech Architecture Foundation, The Ministry of Culture of the Slovak Republik

We would like to thank the following people for their help and support in the preparation of this book / students of the Faculty of Art and Architecture of the Technical University of Liberec and the Academy of Fine Arts and Design in Bratislava, Benjamin Bradňanský, Martina Flekačová, Zdeněk Fránek, Vito Halada, Kryštof Hanzlík, Henrieta Moravčíková, Ladislav Monzer, Peter Stec, Jan Studený, Eliška Málková and Petr Láska for their unbelievable effort, Jana Kořínková, Linda Dostálková and Mikuláš Macháček for courage to take almost impossible task.

The texts were translated by the Skřivánek Language Agency. Skřivánek has been a leading provider of language services in the field of translation and interpreting, localization, DTP and language teaching in Central and Eastern Europe for twenty years.

History of the Biggest Architecture Office

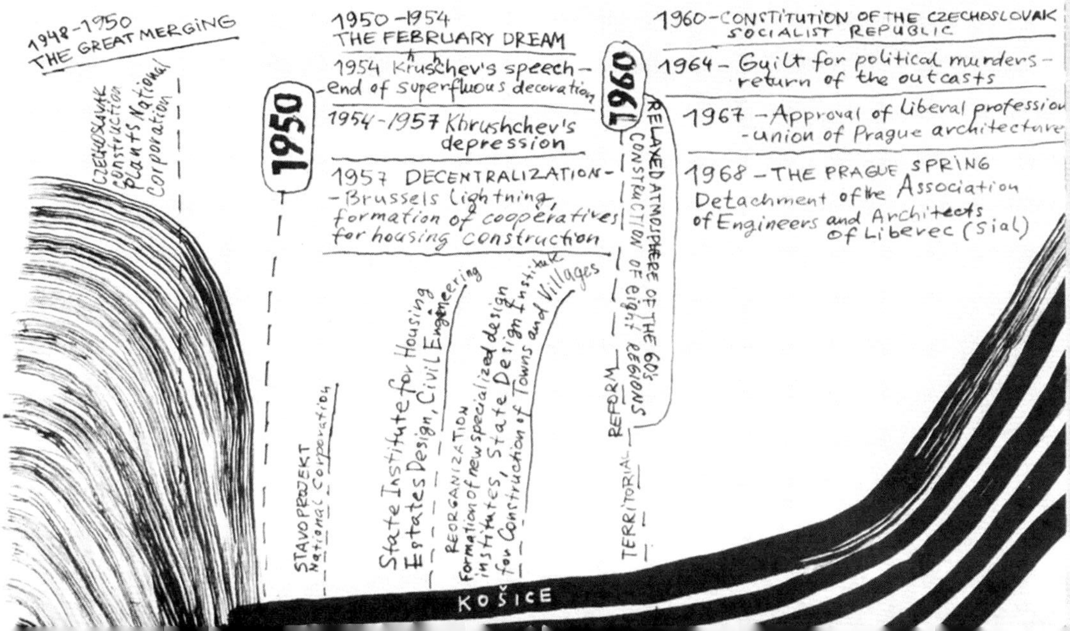
1948–1950
THE GREAT MERGING
CZECHOSLOVAK construction Plants National Corporation
1950
STAVOPROJEKT
National Corporation
State Institute for Housing Estates Design, Civil Engineering
REORGANIZATION
Formation of new specialized design institutes, State Design Institute for Construction of Towns and Villages
1950–1954
THE FEBRUARY DREAM
1954 Khrushchev's speech – end of superfluous decoration
1954–1957 Khrushchev's depression
1957 DECENTRALIZATION – Brussels lightning, formation of cooperatives for housing construction
1960
TERRITORIAL REFORM
RELAXED ATMOSPHERE OF THE 60's
CONSTRUCTION OF eight REGIONS
1960 – CONSTITUTION OF THE CZECHOSLOVAK SOCIALIST REPUBLIC
1964 – Guilt for political murders – return of the outcasts
1967 – Approval of liberal profession – union of Prague architecture
1968 – THE PRAGUE SPRING
Detachment of the Association of Engineers and Architects of Liberec (Sial)
KOŠICE

2016 Sm²
BRICK
AN EXACTING MATERIAL

这是一本当代荷兰建筑中砖结构的使用手册。书中介绍了荷兰砖结构建筑的历史、多种砖块的陈列、砖结构分析、实例展示、建筑师的运用等方面内容，给教育理论和实践上的交流提供了重要的资料。本书为圆脊锁线硬精装，封面带有纹路的暖灰色艺术纸灰色和黑色双色印刷，白色烫印文字，裱覆卡板。环衬采用充满杂质的粗糙纸张，体现与砖结构建筑相近的气质，内页半涂布胶版纸四色印刷。内页的主要文字采用了偏向页面左侧的非对称版心框架，与图片展示、建筑师内容的双栏对称版心形成差异。对于这样栏较宽但字号偏小的内文，首行采用了较大的缩进设置。内页大量用于展示建筑框架和砖结构的线图处理得十分精致，与黑白的建筑照片形成了充满结构感的页面效果，体现着本书内容的特质，一种对合适材料的精确运用。||||||||||||||||||||||
||||||||||||||||||||||||||||||||||||||||||||||||||||||||||||||||||||||||||||||||||||||||||||||||||||
||||||||||||||||||||||||||||||||||||||||||||||||||||||||||||||||||||||||||||||||||||||||||||||||||||||

BRICK
AN
EXACTING
MATERIAL
ARCHITECTURA & NATURA

This is a manual for the use of brickwork construction in contemporary Dutch architecture. The book introduces the history of Dutch brickwork construction buildings, displaying of various bricks, analysis of brickwork construction, demonstration of examples, application of architects and other aspects, which provides important information for the communication of educational theory and practice. This hardback book is bound with thread sewing. The cover is printed on warm gray art paper in gray and black with white hot-stamped text, mounted by graphic board. The end paper is made of rough paper filled with impurities, reflecting the similar temperament with brickwork construction. The inside page is semi-coated offset paper with four-color printing. The inside pages adopt the asymmetric type page which is biased to the left, different from the two-column symmetrical type page using to display pictures and architects. Since the column is wide and the font size is small, the indentation for the first line is relatively large. The line diagrams of the architectural framework and brickwork construction are delicate. Accompanied with the black and white photos of the buildings, the pages are structured finely, reflecting the essential of the book – correct material and accurate application.

◎ 银奖

« 砖块，一种精确的材料

荷兰及弗兰德斯地区

◎ Silver Medal

« Brick. An Exacting Material

The Netherlands and Flanders

Studio Joost Grootens (Joost Grootens, Dimitri Jeannottat, Silke Koeck, Hanae Shimizu, Julie da Silva)

△ Jan Peter Wingender

Architectura & Natura, Amsterdam

245×172×32mm

782g

352p

The structured cover itself reflects the genre: it is a handbook on the contemporary usage of brick in Dutch architecture. The widely varying chapters and sections are each given their own visual logic. The book's structure is actually held together by an elaborate framework for type and images, which enables whatever is necessary. The basic double page in single-column full justification and vibrant Antiqua appears traditional yet surprises with asymmetrical left and right sides. The columns exceed the customary dimensions for very large indents. This misalignment creates space on the right of the column which can also be used for pictures; above all, it allows width dimensions for the type area in two-column texts that allow continued full justification with the same type size as the basic text – without risking gaps. The spaces are also suitable for text and image tables or for complex captions.All headings are printed in spaced capital letters; only the chapter headers are placed in the centre of the page: a respectful bow to the history behind the book's material. And the very rough natural paper is naturally no obstacle to four-colour photos. An ambitious subject, an ambitious design, an ambitious book.

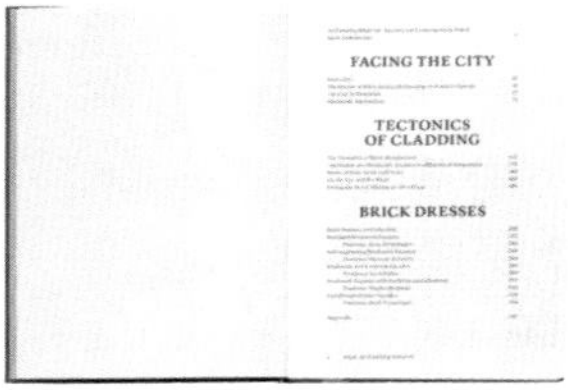

FACING THE CITY

TECTONICS OF CLADDING

BRICK DRESSES

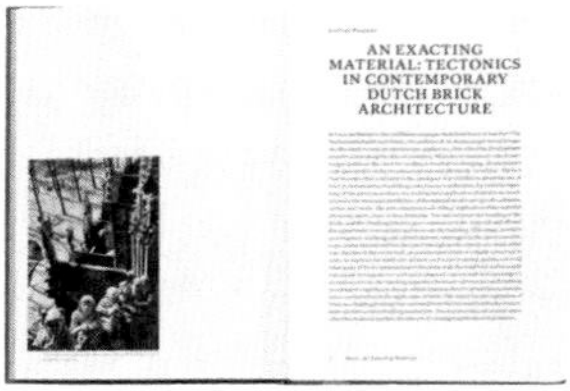

AN EXACTING MATERIAL: TECTONICS IN CONTEMPORARY DUTCH BRICK ARCHITECTURE

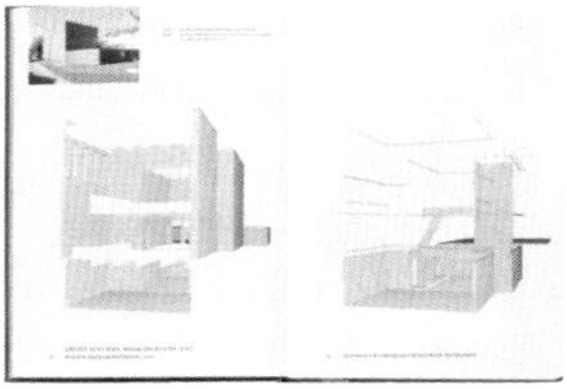

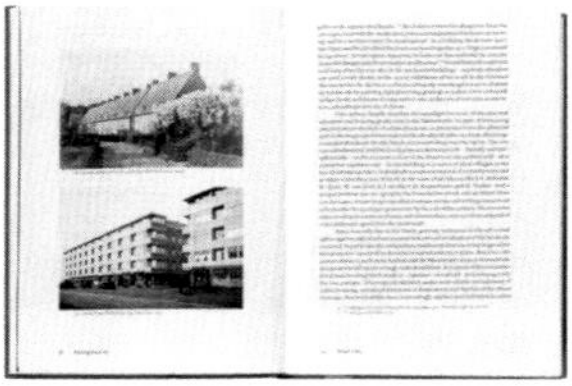

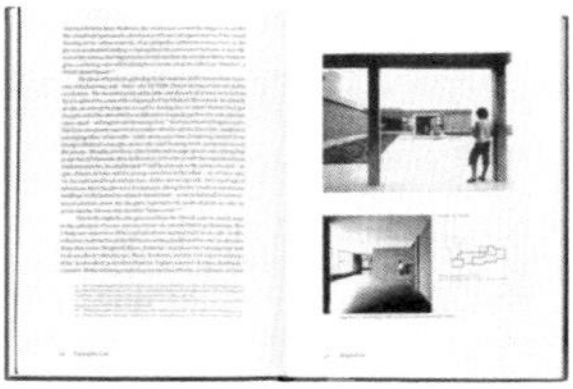

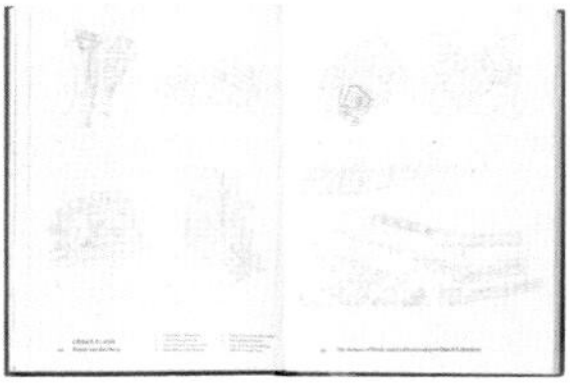

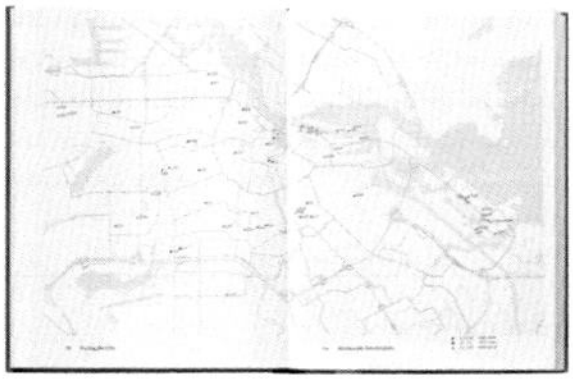

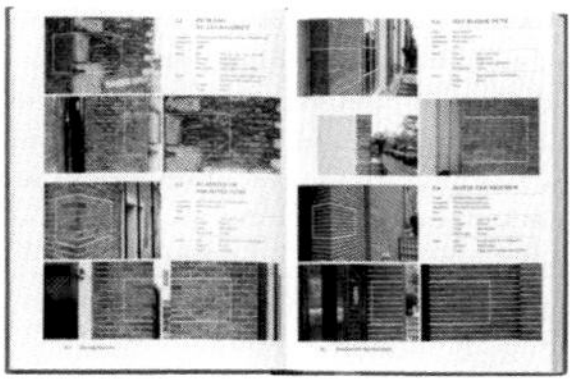

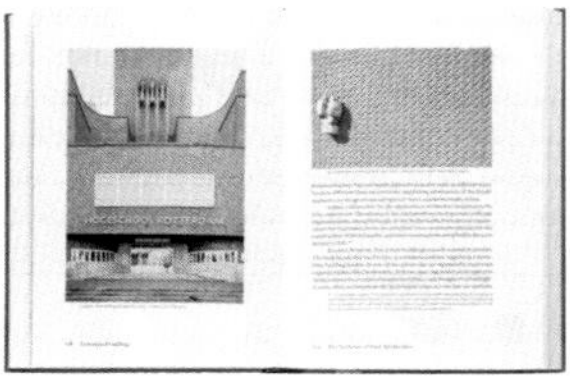

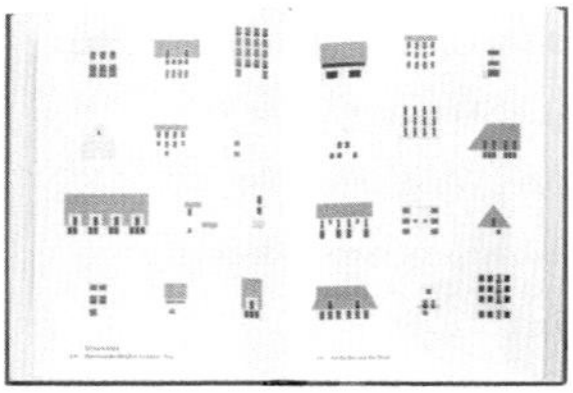

Positions:

Herman Zeinstra

Jaapjan Berg and Jan Peter Wingender

A conversation with Herman Zeinstra about his work is filled with striking concepts like faith, fascination, play, rebelliousness and attitude. Indeed, Zeinstra's work cannot easily be captured in a single style. To a large degree, this results from the fundamental questions he poses in each assignment and his refusal to settle for routine solutions. In his building designs, Zeinstra uses brickwork in a practical manner. Its use is not a given but rather part of the game he plays with the assignment, the client and the specific conditions. For Zeinstra brick is not a conviction but a supporting tool. The degree and the way in which it is used is determined by the assignment or the design and construction process. The material helps Zeinstra to find alternative paths and routes in architecture and construction, and to articulate an 'honesty' that suits the contemporary situation.

160 Brick Dresses

Residential building [illegible], Amsterdam, [illegible]

SELECTION PROCESS

According to Zeinstra, many architects attach too much weight to brick and the moment or process of selecting materials in general. Actually, a concept should only be conceived after the fact. 'When pairing up a material to a project, you really do make wrong choices sometimes. In some cases, you then have to switch to another material, even if you're already in the middle of the process.' According to Zeinstra, materials, including brick, are often chosen long before the actual design process—namely during the many occasions when an architect sees an inspiring example where a material or a composition of materials is critical for the character of a building. 'That image, the material or the thought that it provokes sticks in the mind.' Of course, the inspiration can also come from a place where there is no architecture. Zeinstra subsequently plays around with the thought—largely in outlines—of how to apply a material. He examines how something has been or can be made, and its potential applications.

At that point, there is no project or assignment at hand. It is what you might call a 'memory collection of materials', a source of inspiration that gradually fills up and which you can subsequently draw on. So actually the process of selecting materials, including brick, is constantly unfolding—independent of a specific project. Zeinstra considers the time between the discovery and application of a material as an incubation period. 'After all, we are all trained in creating. If something fascinates you, it keeps your attention. You learn from it.'

This way of working does not rule out that some ideas are not approved. 'Sometimes the required technique is too complicated or too labour intensive for the design at hand. An assignment often simply does not lend itself

261 Positions: Herman Zeinstra

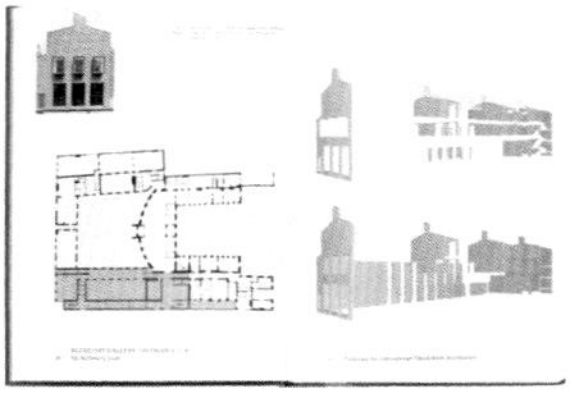
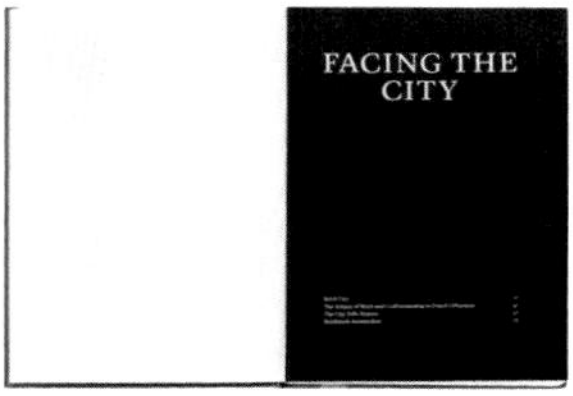
FACING THE CITY
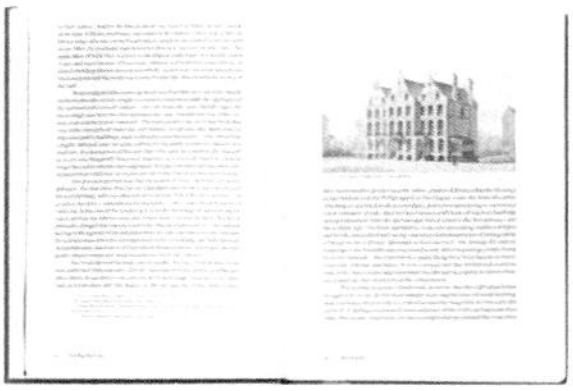
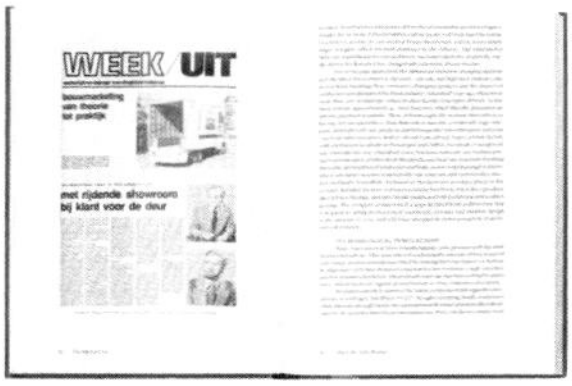
WEEK UIT
bouwmarketing van theorie tot praktijk
met rijdende showroom bij klant voor de deur
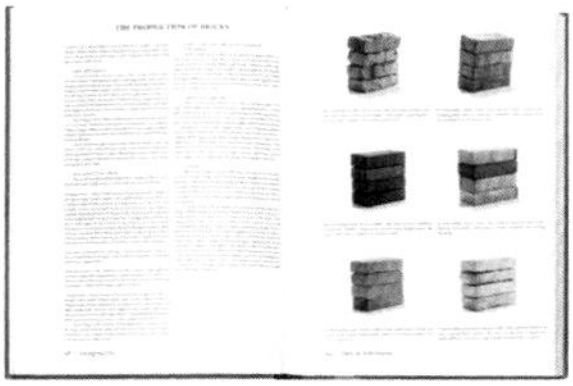
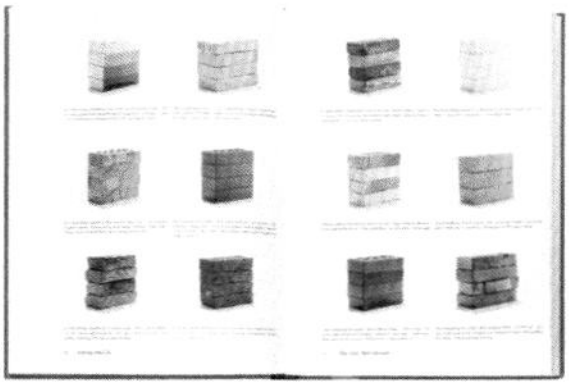
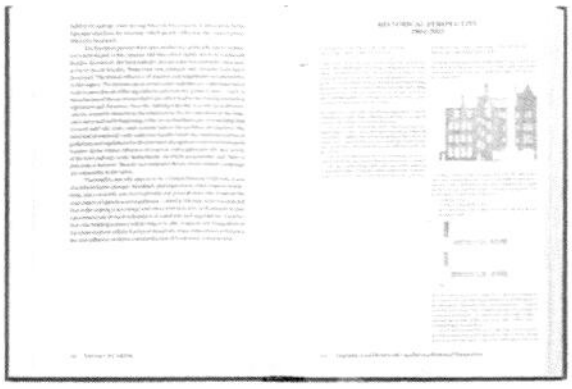
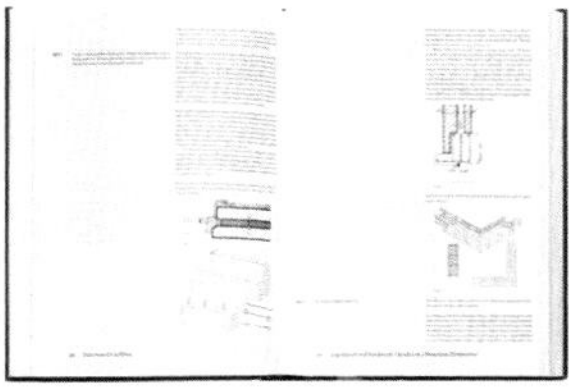
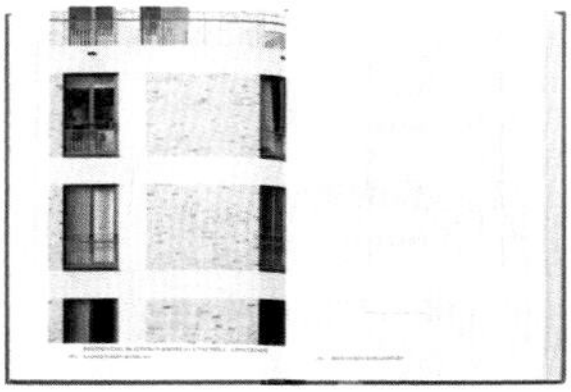
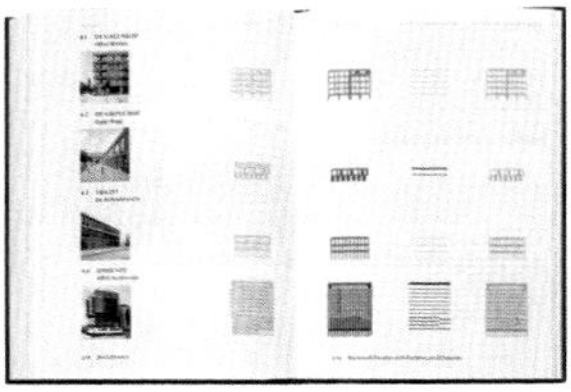
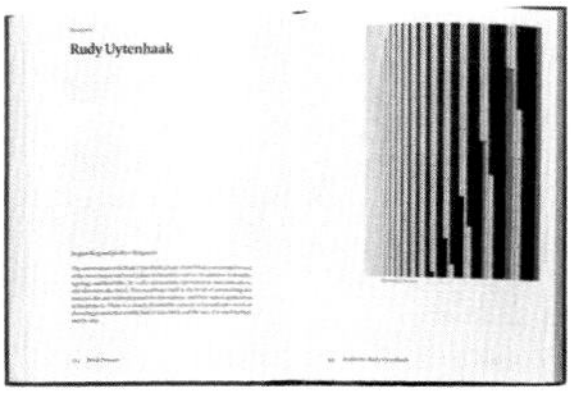
Rudy Uytenhaak
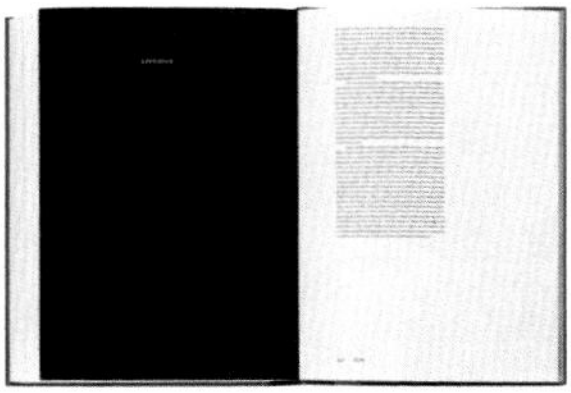

brick
direction of the

BRICK
ARCHITECTURA & NATURA

brickwork next to the opening should have a width of at least ⅓ the width of the opening. It is clear that these sorts of dimension rules had a considerable influence on the design of brickwork faça

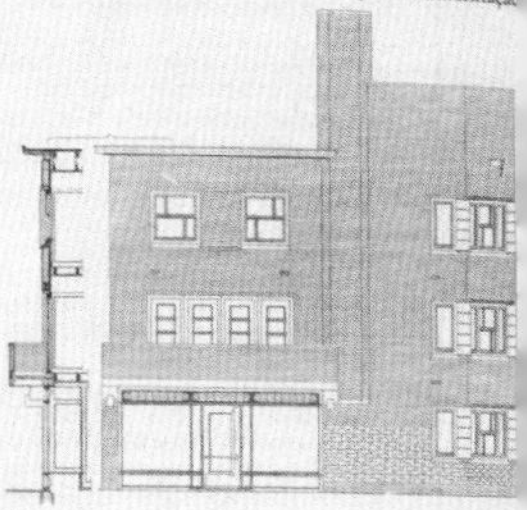

Fig. 7

Introduction of 'breast summers': iron beam covered with piece natural stone. There was a great aversion to this way of spanning opening both structurally and architecturally. It was deemed be to use reinforced concrete or 'betonno' in these types of situation

Less decorated façades and ornaments. Extensive constructio housing with a uniform architecture in the period after the F World War. (Fig. 7)

1931 Revision of the Housing Act. This made it necessary to adapt the Model Building Regulations to the new statutory regulations.

1933 Adaptation of the Model Building Regulations.

In England the mechanical engineer Henry Dyke concluded at beginning of the 1930s that it should be possible to prefabric high-quality brickwork elements. Dyke was granted a patent for system he devised in 1933 and started marketing it with his compa Simplified Brick Construction Ltd. The brick in combination w an air cavity formed the basis of the system. In addition, the pre elements consisted of concrete inner leaves, which could vary thickness depending on their application.

Fig. 8

1947 National government stimulated industrial building systems to me the great demand for housing after the Second World War.

949 Stimulation of industrial building systems and national granting subsidies resuled in the development of the BMB system *(Baksteen Montage Bouw)* that was largely based on Henry Dyke's brickwork façade system.

Fig. 9

950 Reconstruction Act *(Wederopbouwwet)*: Article 20 offered opportunities to deviate from the Model Building Regulations and development plans, ample use of which was made. The law was only supposed to be in effect for 3 years, but ultimately remained in force until 1970.

952 Model Building Regulations (*Model Bouw Verordening*, MBV 1952). The construction of brickwork buildings primarily occured on the basis of the table in the Model Building Regulations, because this gave simple rules for low-rise and medium-rise buildings (no more than 5 storeys, possibly on a substructure, provided that it was not higher than 2.2 m). The MBV set the following requirements for making cavity walls [art. 172]:

- the joint thickness of both leaves of a cavity must at least possess the required total thickness as indicated in fig. 10;
- the cavity cannot be wider than 100 mm and no thinner than 50 mm and must be lightly ventilated with fresh air;
- extra loads are not permitted without sufficient reinforcement;
- the total height of the wall built as a cavity wall may not amount to more than 7 m for load-bearing walls and 10.5 m for other walls, unless both parts of the walls have a joint thickness of 300 mm or measures against buckling have been taken in another way.

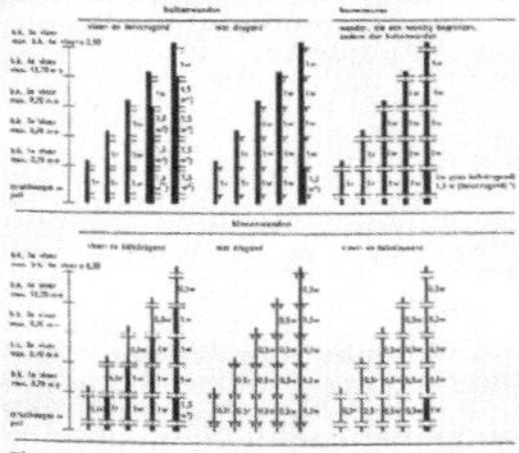

Fig. 10

Start of automation in brick factories by the application of measurement and control engineering for the production. Kilns could operate autonomously without human interference.

955 N 1055 – Technical Principles for Building Regulations (*Technische Grondslagen Bouwwerken*, TGB 1955). Central Office for Standardisation, Royal Dutch Society of Engineers, Netherlands Society for

这是比利时报纸上每周更新的一篇漫画的合集。漫画的主角叫丹尼尔·范迪奇（Daniël van Dicht）。他是一个不穿衣服的粉红色胖子，从不说话，仅用动作来表达自己。他经常置身于危险或尴尬的境况中，但他关注的事情似乎只有吃。他的行为看起来很不合理，也不乏人性的贪婪与吝啬，甚至有些愚蠢和搞笑，但一些选择似乎又充满了人生的思考，好在他只是一个漫画人物。整部漫画都很荒诞，作者通过这样一个虚构的人物在奇特境遇中的独特选择，展现生活中的那些或带有讽刺或值得思考的人生经验与“技巧”。本书的装订方式为锁线胶平装，封面白卡纸印刷明快的粉红色，覆哑膜后丝网印黑色，使得黑色在封面上呈现一定的光泽。内页哑面半涂布纸张四色印刷，黑色线条配合充满画面的高饱和度色彩，风格独特。除了封面条码外，全书没有使用一个字库字体，也没有任何矢量图形，所有字体与图形均为手绘，体现出轻松又荒诞的风格。

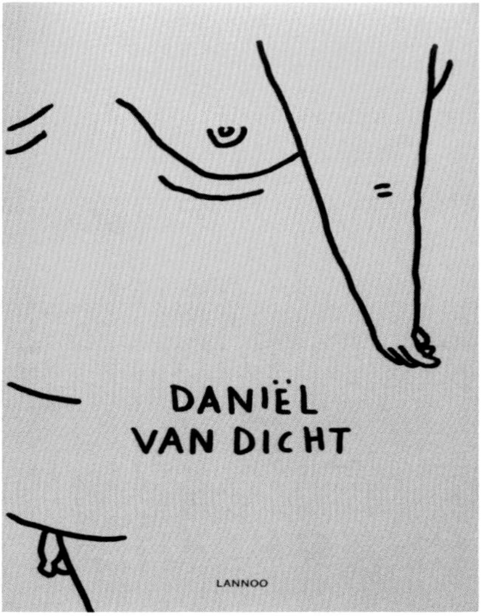
DANIËL
VAN DICHT
LANNOO

This is a collection of cartoons updated weekly in Belgian newspapers. The protagonist of the cartoon is Daniël van Dicht. He is a pink fat man with no clothes on. He never talks; he only expresses himself by actions. He often finds himself in dangerous or embarrassing situations, but his only concern seems to be eating. His behavior seems unreasonable, greedy and mean, even somewhat silly and hilarious, but some choices seem to be thoughtful. Fortunately he is only a cartoon character. The whole story is absurd. Through the unique choice of such a fictional character in strange situations, the author shows those life experiences and "skills" ironically. The paperback book is perfect bound with thread sewing. The cover is made of white cardboard printed in bright pink, and silk-screen printing in black after laminating. The inside pages are printed on coated paper in four colors. There are black lines contrasted with pictures of high saturation. Except the bar code on the cover, neither the font from font library nor the vector graphic has been used in the book. All the fonts and graphics are hand-drawn, which is random and absurd.

- 铜奖
- 丹尼尔·范迪奇
- 比利时

- Bronze Medal
- Daniël van Dicht
- Belgium
- Dear Reader, Matthias Phlips (Illustrations)
- Matthias Phlips
- Lannoo
- 320×240×18mm
- 917g
- 144p

This endomorph pushes back the boundaries of any kind of typology. His phlegm is so pronounced that – apart from a rare flush – he is left entirely unaffected by everyday perils. He finds himself in highly practical situations that are always slightly to seriously absurd, but everything seems to essentially pass over him. Things usually revolve around dogs and food intake. Or something that is being imitated. The easy-going man finagles his way between the ducks with a gaping mouth to catch most of the food. Luckily, he is only a sketch. The stout, naked man, Herr van Dicht, is the protagonist of weekly cartoons that are collected here as a flexible folio. The page structure is spacious, wonderfully composed, sparing in its use of colour, always covering the whole page, with lots of pink – and everything is drawn with clear, equally thick contours. A certain psychological expression is created by these economical graphics – mirroring van Dicht's visible greed and latent parsimony. In addition, the character has extremely reduced facial expressions. The illustrator himself does without the most important means of expression on the man's face: his eyes – which are covered by glasses. The book is designed as an album dedicated to this naked little man. Now and then, from off-stage, i.e. from the edge of the picture, the plump pink hands extend into the picture, affix a cartoon with an adhesive strip, or clip their fingernails.

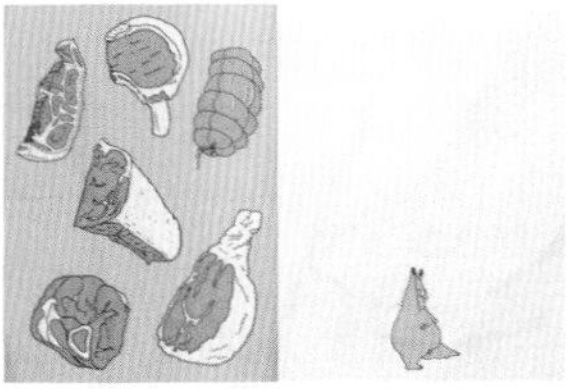

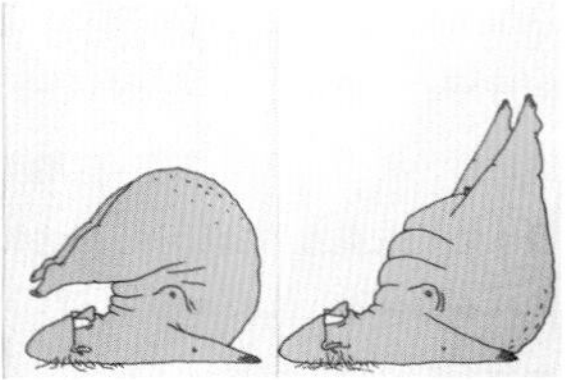

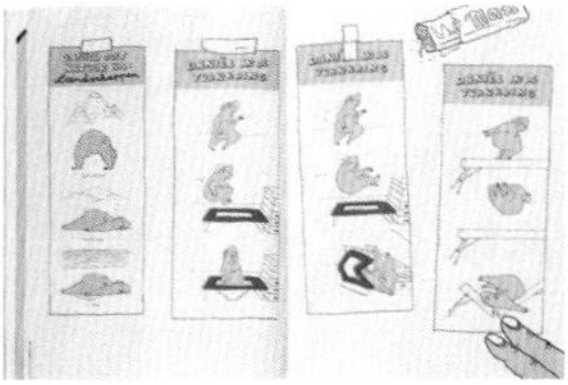

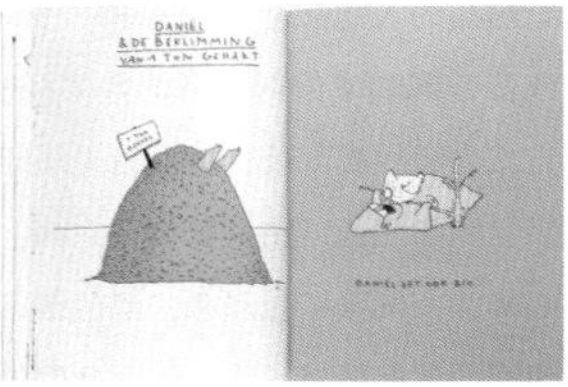
DANIEL
& DE BERLIMMING
VAN 1 TON GEHAKT

DANIELS ONE NIGHT STAND

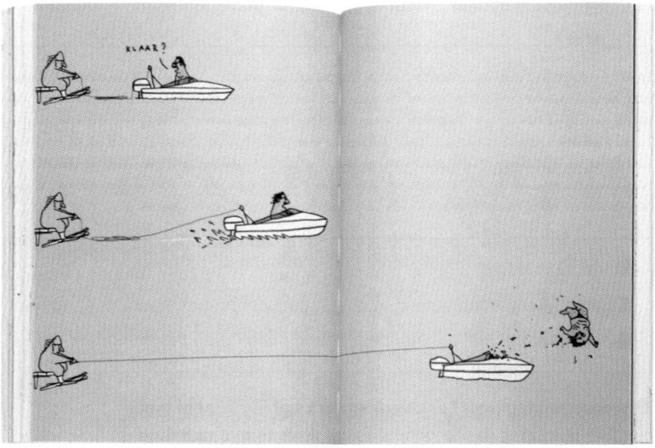
KLAAR?

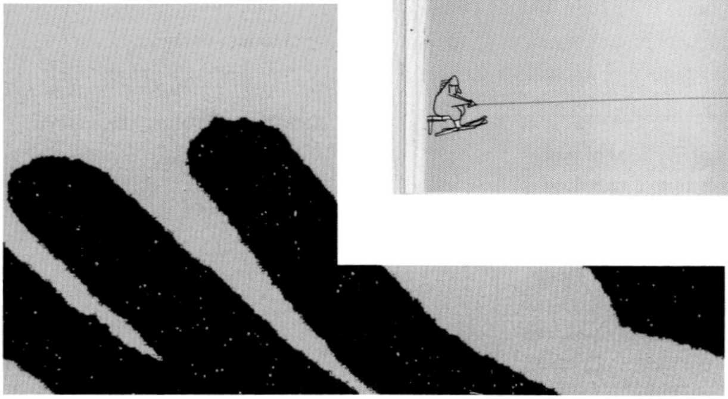

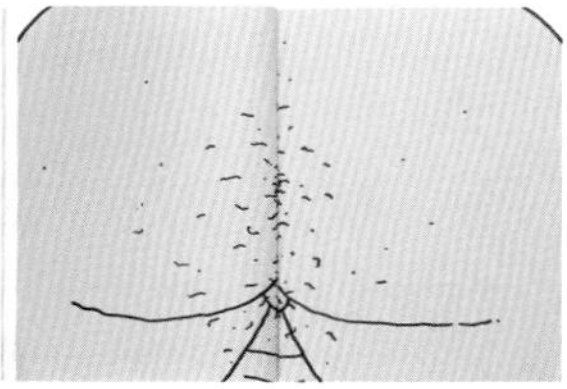

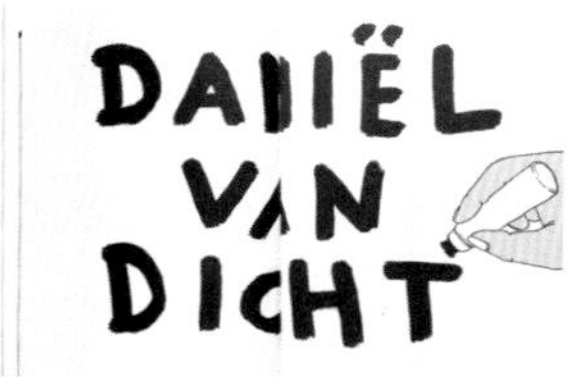

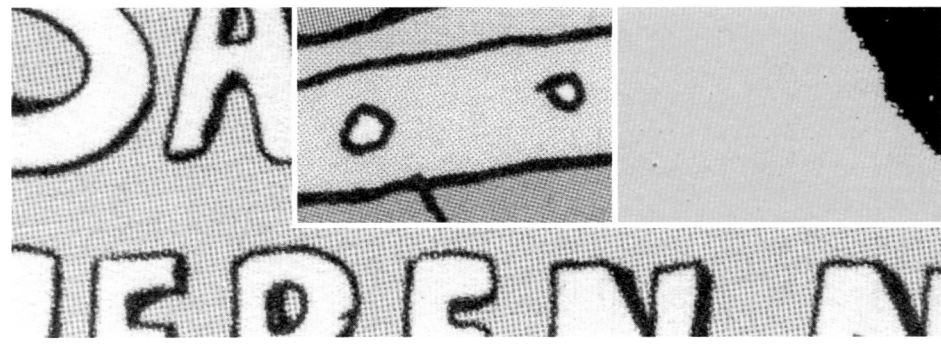

DANIËL
VAN
DICHT

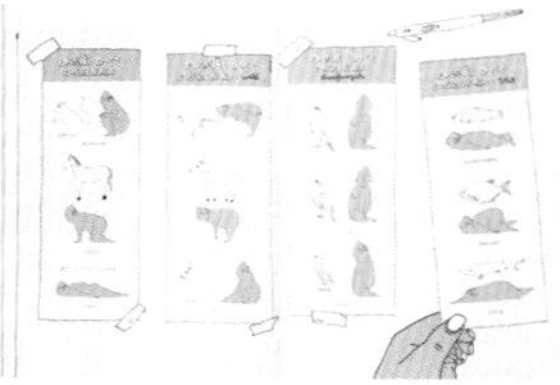

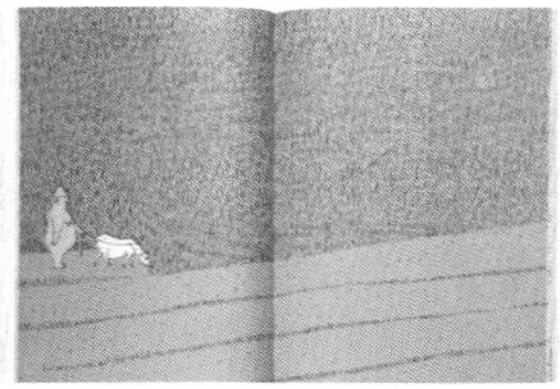

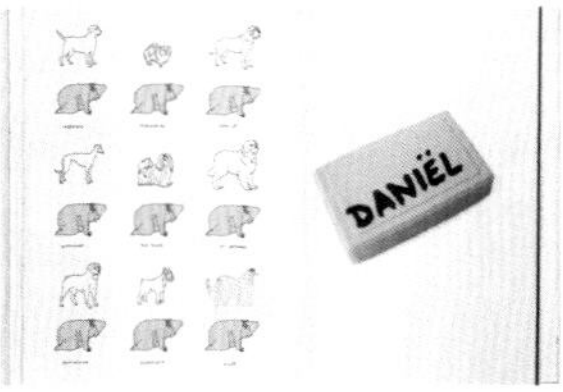

DANIËL

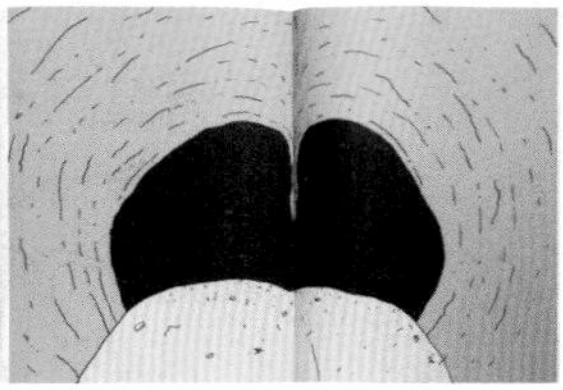

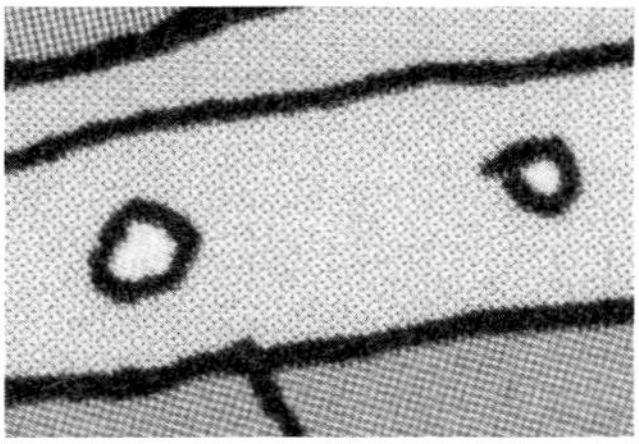

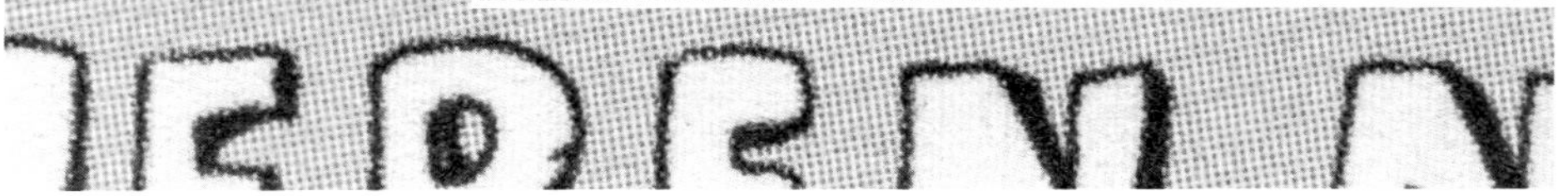

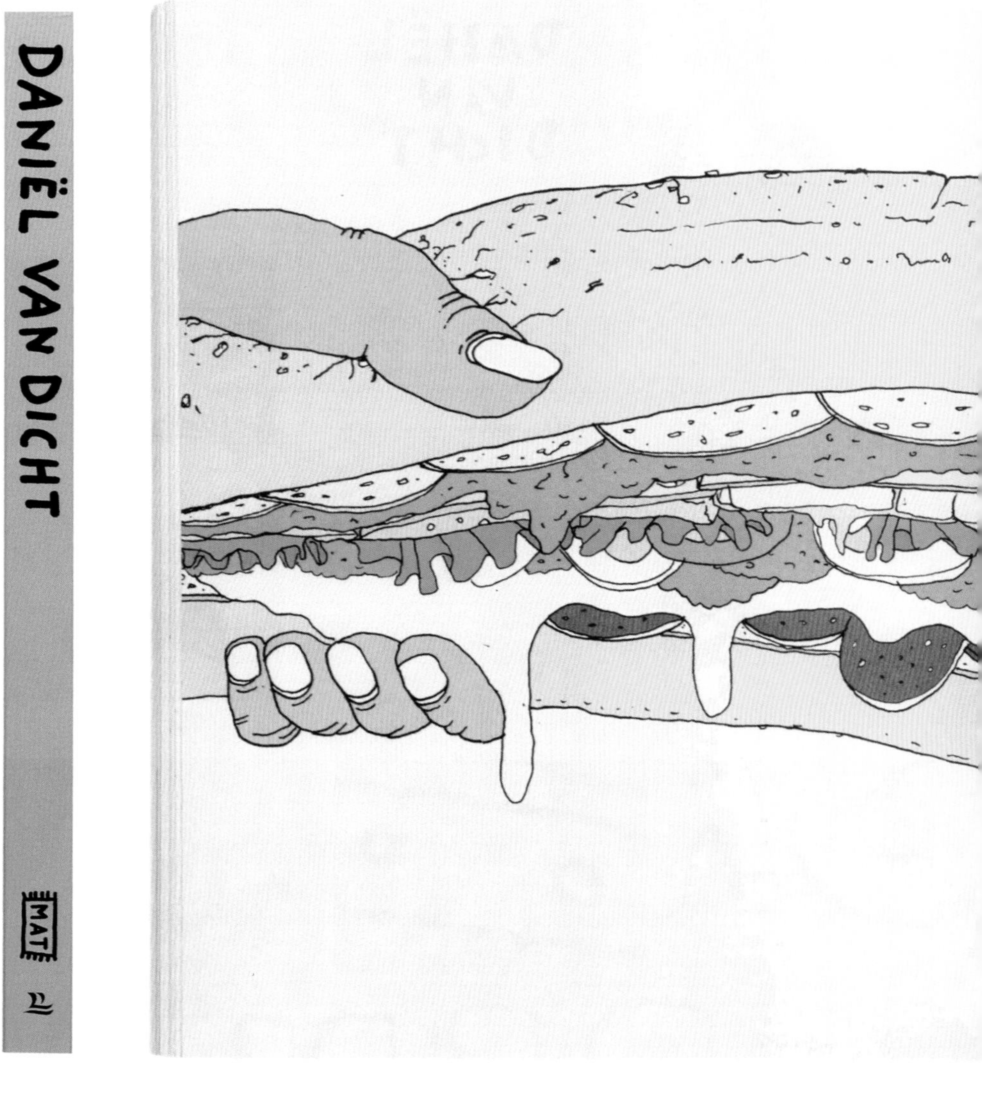
DANIËL VAN DICHT
MAT

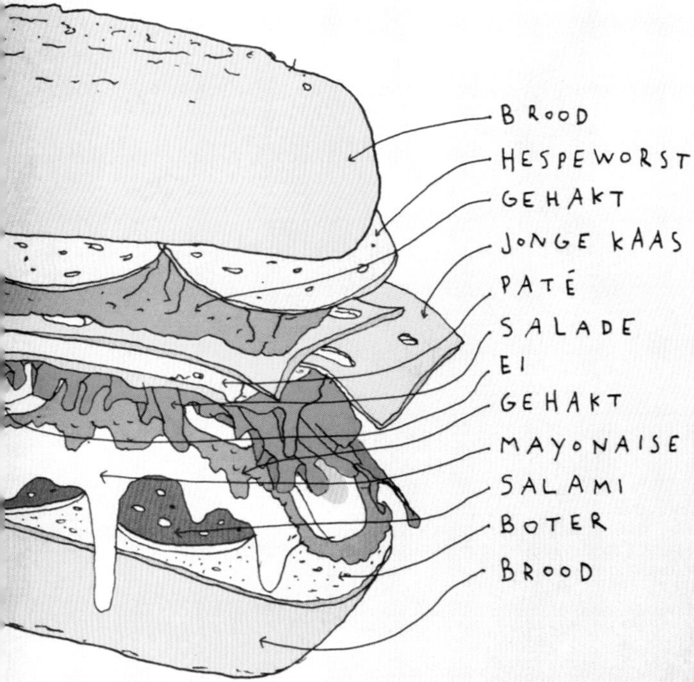
BROOD
HESPEWORST
GEHAKT
JONGE KAAS
PATÉ
SALADE
EI
GEHAKT
MAYONAISE
SALAMI
BOTER
BROOD

2016
1
Bm

这是一本“童话”体经济学寓言。表面上讲述了一个小生物想方设法长大、获取权利、膨胀最终破灭的故事，实则是在阐述经济泡沫产生直至破裂的故事。本书的装订方式为锁线硬精装，封面半涂布艺术纸深蓝专色印刷，上部烫金，裱覆卡板。内页半涂布艺术纸红蓝双专色印刷，颜色之间的叠透关系和丰富的图案肌理，产生丰富的变化，局部页面烫金。由于是经济学寓言，本书的设计采用上切口不清边的做法，使得每两页之间形成一个半开放的空间，里面印有银行、保险或其他金融机构较常用的安全信封内部防信息泄漏的各式图案，并且在最后的页面将这个“童话”里“保密”的经济学问题呈现给读者。书中图形的设计采用撕纸或手绘线条的儿童绘本风格，“伪装”了其经济学故事的本质。在字体和排版设计上，书中所有的大小写“O”都调整到不在基线的位置上，甚至有少量倾斜和旋转。而与圆圈有关的单词例如“ronde cirkel”做弧形基线处理，形成了独特的阅读体验。

Het
meest
geloofde
sprookje

This is an economic fable written like ‘fairy tales’. It seemingly tells a story about a small creature trying to grow up, gain rights, expand and eventually come to nothing. Actually, it tries to explain how the economic bubble grew and burst. The hardback book is bound by thread sewing. The cover is made of semi-coated art paper printed in dark blue, gold hot stamped on the upper part and mounted by the graphic board. The inside pages are made of semi-coated art paper in red and blue, gold hot stamped partly. The overlapping colors and rich texture patterns result in rich changes. The upper edge of the book remains uncut, which can make a semi-open space between every two pages. Various patterns of information leakage prevention commonly used by banks, insurance companies or other financial institutions are printed inside as the ‘secret part’ of this ‘fairy tale’. The graphic design in the book adopts the style of children’s picture book with torn paper or hand-drawn lines, and ‘camouflages’ the economic essence of the story. Letter ‘O’ in the book, no matter uppercase or lowercase, is adjusted to be out of baseline position, even with a little bit tilt and rotation. Words related to circles, such as “ronde cirkel”, are processed with arc-shaped baselines, forming a unique reading experience.

铜奖
最被人相信的童话
荷兰

Bronze Medal
Het Meest Geloofde Sprookje
The Netherlands
De Vormforensen (Annelou van Griensven & Anne-Marie Geurink) & Lyanne Tonk
De Vormforensen (Annelou van Griensven & Anne-Marie Geurink)
De Vormforensen (distribution De Vrije Uitgevers)
326×222×8mm
452g
64p

A little creature becomes hungry for power; it begins to paint circles and swaps them for other things. The circles grow larger and larger; the demand for them culminates in a desire for the golden, shining sun. The folded pages of the hardback are closed at the top, creating pockets between the double pages which are open at the front and bottom. The insides are printed with a wide repertoire of geometric patterns – so mysterious. How does a design feature like that fit into the visual scenario of collages made from cut and torn shapes, two-colour printing, signal red and ink blue, embossed with golden foil? The circles in the writing itself also begin to take on a life of their own; the o in the semibold grotesque type leaves the baseline, shifting up or down. The printing inside the pockets turns out to be taken from a security envelope: inner patterns protect the contents from prying eyes, mostly lines over the whole area which prevent passwords, account numbers or other confidential financial information from showing through… At the end, the whole thing collapses: “Oh no, what now?” A children’s book? A parable of supply and demand, growth optimism and the absolute value of money: the most widely believed fairy tale.

HET
MEEST
GELOOFDE
SPROOKJE

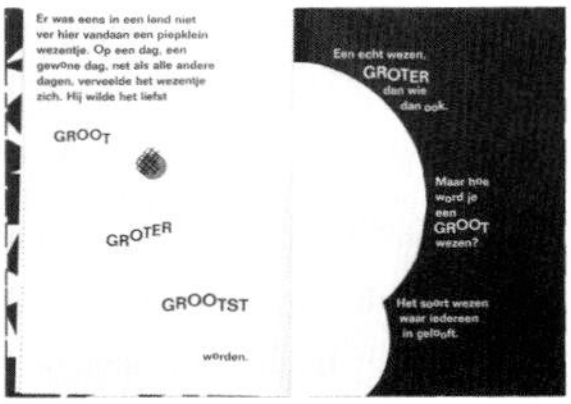
Er was eens in een land niet ver hier vandaan een piepklein wezentje. Op een dag, een gewone dag, net als alle andere dagen, verveelde het wezentje zich. Hij wilde het liefst
GROOT
GROTER
GROOTST
worden.
Een echt wezen.
GROTER
dan wie
dan ook.
Maar hoe word je een GROOT wezen?
Het soort wezen waar iedereen in gelooft.

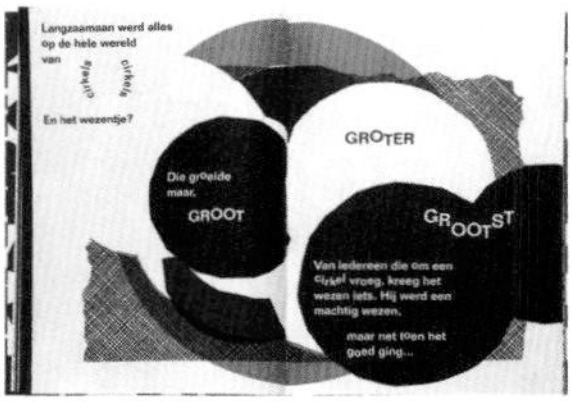
Langzaamaan werd alles op de hele wereld van
cirkels
cirkels
En het wezentje?
GROTER
Die groeide maar.
GROOT
GROOTST
Van iedereen die om een cirkel vroeg, kreeg het wezen iets. Hij werd een machtig wezen,
maar net toen het goed ging...

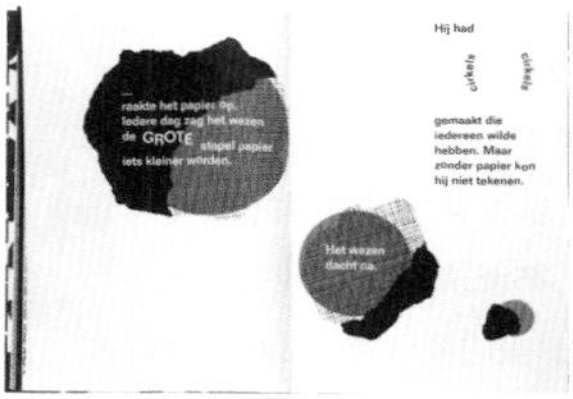
raakte het papier op.
Iedere dag zag het wezen de GROTE stapel papier iets kleiner worden.
Hij had
cirkels
cirkels
gemaakt die iedereen wilde hebben. Maar zonder papier kon hij niet tekenen.
Het wezen dacht na.

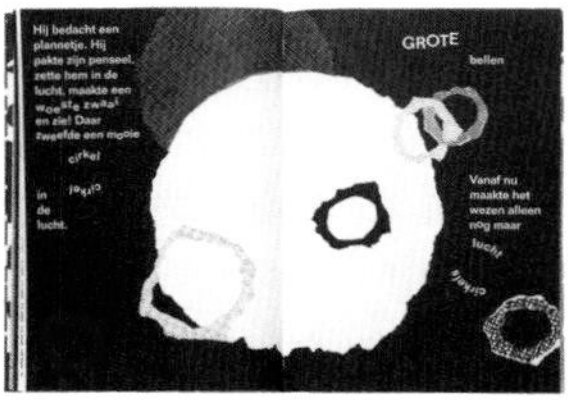
Hij bedacht een plannetje. Hij pakte zijn penseel, zette hem in de lucht, maakte een woeste zwaai en zie! Daar zweefde een mooie
cirkel
in
cirkel
de
lucht.
GROTE
bellen
Vanaf nu maakte het wezen alleen nog maar
lucht
cirkels

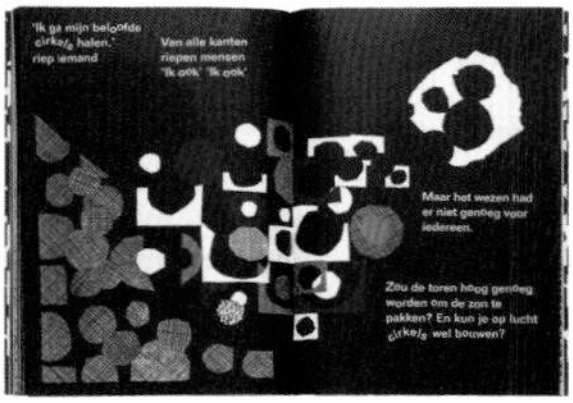
'Ik ga mijn beloofde cirkels halen,' riep iemand
Van alle kanten riepen mensen 'Ik ook' 'Ik ook'
Maar het wezen had er niet genoeg voor iedereen.
Zou de toren hoog genoeg worden om de zon te pakken? En kun je op lucht cirkels wel bouwen?

Daar had je het al plots knapte één van de cirkels en begon de hele berg te trillen
PATS, daar sprong nog een bel.
De hele toren STORTTE in.

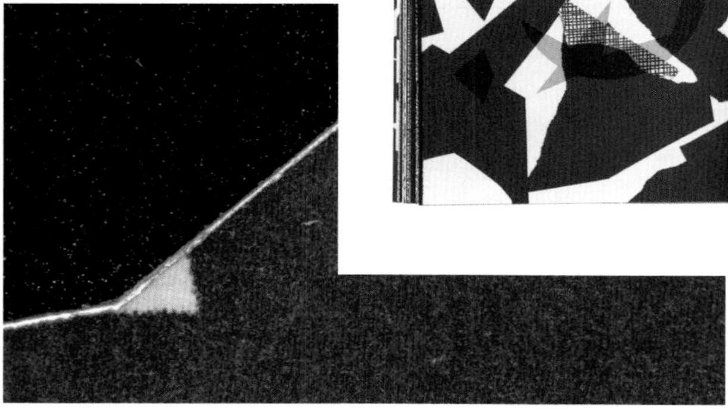

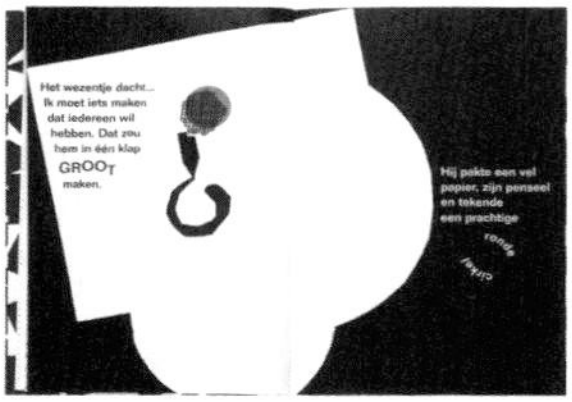
Het wezentje dacht...
Ik moet iets maken
dat iedereen wil
hebben. Dat zou
hem in één klap
GROOT
maken.
Hij pakte een vel
papier, zijn penseel
en tekende
een prachtige
ronde
cirkel

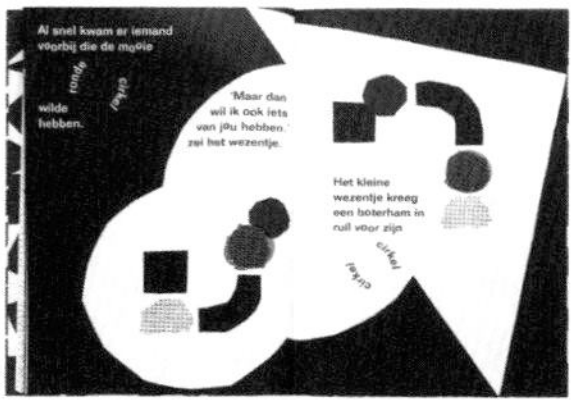
Al snel kwam er iemand
voorbij die de mooie
ronde
cirkel
wilde
hebben.
'Maar dan
wil ik ook iets
van jou hebben.'
zei het wezentje.
Het kleine
wezentje kreeg
een boterham in
ruil voor zijn
cirkel

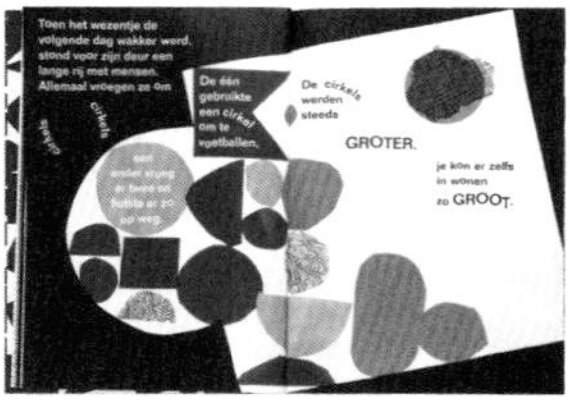
Toen het wezentje de
volgende dag wakker werd,
stond voor zijn deur een
lange rij met mensen.
Allemaal vroegen ze om
cirkels
De één
gebruikte
een cirkel
om te
voetballen,
De cirkels
werden
steeds
GROTER.
je kon er zelfs
in wonen
zo GROOT.

Iedereen bleef komen.
Er waren zoveel
mensen die cirkels
wilden dat het wezen nu
dag en nacht bezig was
om lucht te maken.
cirkels

Het wezen kon het niet meer
bijhouden.
Daarom zei hij
tegen de volgende
die om een cirkel vroeg
'Als jij alvast
jouw boterham geeft
dan maak ik binnenkort
voor jou een cirkel.'

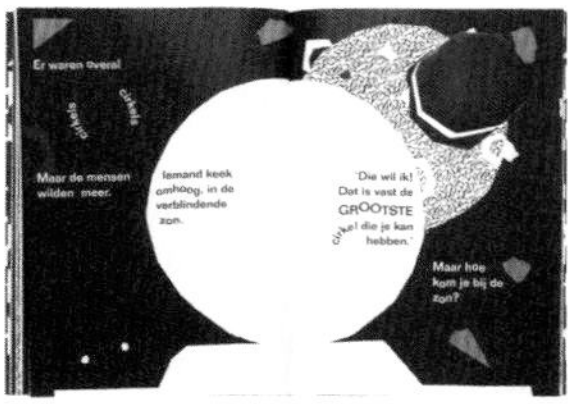
Er waren overal
cirkels
cirkels
Maar de mensen
wilden meer.
Iemand keek
omhoog, in de
verblindende
zon.
'Die wil ik!
Dat is vast de
GROOTSTE
cirkel die je kan
hebben.'
Maar hoe
kom je bij de
zon?

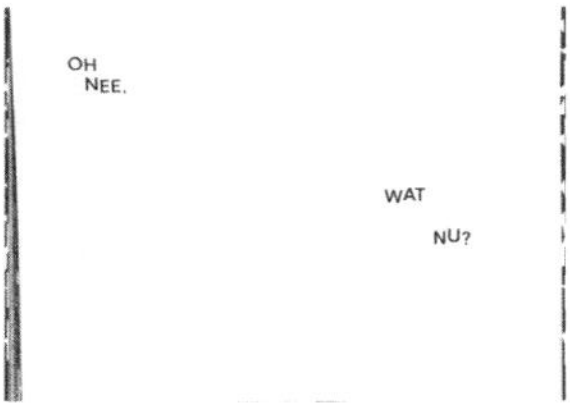
OH
NEE.
WAT
NU?

SPROOKJES
BESTAAN

Hij ging op zijn cirkel staan, maar hij ko
er niet bij. Al snel kwam er iemand
die zijn cirkel op de andere cirk
rolde. Nu waren ze al iets
dichterbij.
Het meest geloofde sprookje

Van alle kanten
kwamen er nu
mensen met
cirkels
cirkels
voor de toren.
Maar zelfs als
je helemaal
bovenop de
toren ging staan
kon je de zon
nog niet pakken.

größer der
der einzelne

这是一本关于欧洲非法移民的书。本书的作者采访了6位来自各地的非法移民，记录下他们不为人知的故事。本书的装订方式为锁线胶平装，封面本白色艺术纸印单黑，画面使用位图网频处理，封面及书脊标题字烫白。内页胶版纸灰黑双色印刷。书中内容分为“在家乡”“漂泊路上”和“新居所”三大部分，每部分分别讲述6位受访者的个人经历。通过穿插与之相关的新闻报道、图表等资料，让读者意识到新闻所反映的背后，往往遗漏的是一个个独立的人以及他们被埋没的命运。三大部分采用黑底反白字的页面进行区隔。6位受访者的经历采用衬线体，版面空出翻口位较大的留白区域。每个人的名字都被放得很大，起到明显的区隔作用。穿插在书中的新闻报道采用灰色铺满底，版面四边都很小，并且选用了无衬线粗体字，与受访文字形成强烈对比。

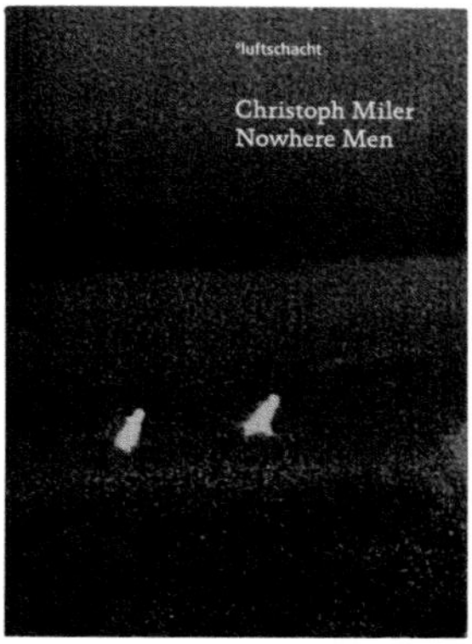

luftschacht
Christoph Miler
Nowhere Men

This is a book about illegal immigrants in Europe. The author interviewed six illegal immigrants from all over the country and recorded their unknown stories. The paperback book is perfect bound with thread sewing. The cover is made of white art paper printed with black ink only and the cover image is processed by AM screening. The title on the cover and the spine are white hot stamped. The inside pages are made of offset paper in gray and black. The content is divided into three parts: 'hometown', 'road of wandering' and 'new residence'. Each part tells the personal experiences of six respondents. Through interlacing relevant news reports, charts and other materials, readers may realize that an independent person is always missing in the news reports. The three parts are partitioned by pages with black background and white characters. Serif fonts are used to typeset the experience of six respondents, and the layout was relatively empty. Everyone's name is amplified, playing a significant role in separating. The news reports interspersed in the book are typeset with bold sans serif characters, in sharp contrast with the interview text.

铜奖
漂泊者——全球化大潮中的非法移民
奥地利

Bronze Medal
Nowhere Men. Illegale Migranten im Strom der Globalisierung
Austria
Christoph Miler, Zürich (CH)
Christoph Miler
Luftschacht, Wien
226×160×26mm
669g
336p

Difficult issues and beautiful books – surely not the perfect couple? The highest common denominator of the three factors governing migration – original motivation, odyssey and destination – seems to be fear. Fear on all levels. What exactly do we know about illegal immigration in Europe, though? Author and designer Christoph Miler seemingly needs to go beyond the daily news and prefers his own research over somebody else's experiences. He found people who have taken six different paths through life and listened to them, their stories, and the external and internal mayhem. Miler has turned his discoveries into a book. He adopts two soundtracks and applies them with a steady hand. Firstly, the six protagonists have their say. The typography treats their words with the same respect as is awarded to other quality literary texts. Secondly, headlines and pictures are taken from fast-moving mass media and transposed into stable book form. These are associative pages of documentation – a light-grey background strictly separating them from the biographical accounts. It could be said that this parallel approach acts as a corrective to the contexts created in the customary, exclusive consumption of news and data. The individual findings do not necessarily have to be wrong. However, this book shows once again what tends to be missing from our palette when we paint our picture of the world: the individual person, their personal fate, their individuality, or to put it in a nutshell: their humanity. It is a book worthy of consideration, an artistic snapshot in time – people from somewhere, now here.

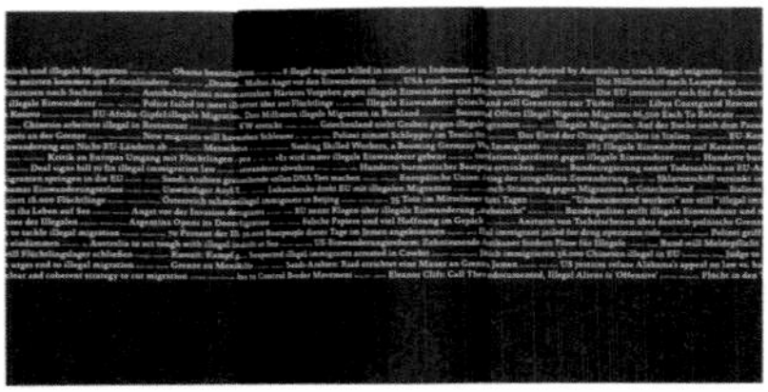

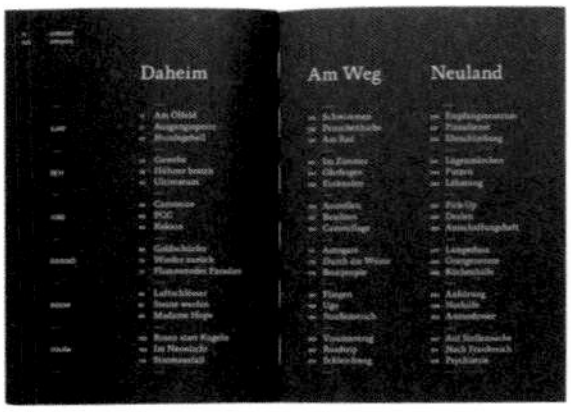

Daheim
Am Weg
Neuland

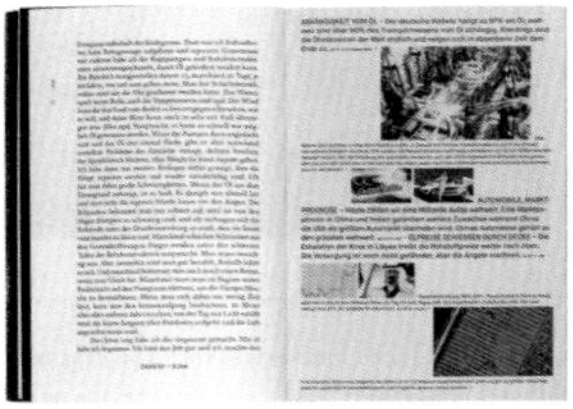

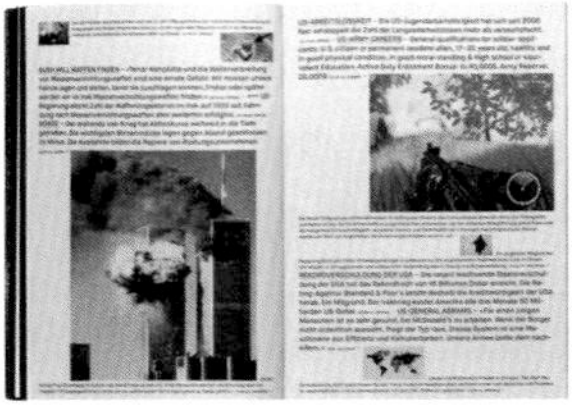

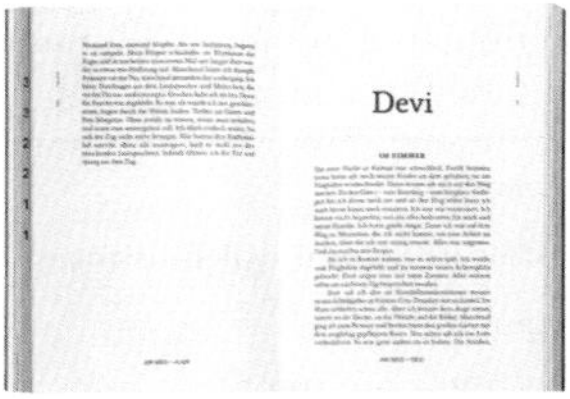

Devi

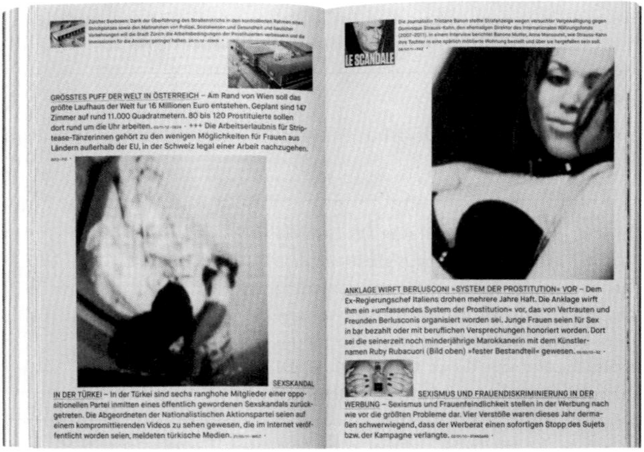

GRÖSSTES PUFF DER WELT IN ÖSTERREICH – Am Rand von Wien soll das größte Laufhaus der Welt für 16 Millionen Euro entstehen. Geplant sind 147 Zimmer auf rund 11.000 Quadratmetern. 80 bis 120 Prostituierte sollen dort rund um die Uhr arbeiten. +++ Die Arbeitserlaubnis für Strip-tease-Tänzerinnen gehört zu den wenigen Möglichkeiten für Frauen aus Ländern außerhalb der EU, in der Schweiz legal einer Arbeit nachzugehen.
SEXSKANDAL IN DER TÜRKEI – In der Türkei sind sechs ranghohe Mitglieder einer oppositionellen Partei inmitten eines öffentlich gewordenen Sexskandals zurückgetreten. Die Abgeordneten der Nationalistischen Aktionspartei seien auf einem kompromittierenden Videos zu sehen gewesen, die im Internet veröffentlicht worden seien, meldeten türkische Medien.
LE SCANDALE
ANKLAGE WIRFT BERLUSCONI »SYSTEM DER PROSTITUTION« VOR – Dem Ex-Regierungschef Italiens drohen mehrere Jahre Haft. Die Anklage wirft ihm ein »umfassendes System der Prostitution« vor, das von Vertrauten und Freunden Berlusconis organisiert worden sei. Junge Frauen seien für Sex in bar bezahlt oder mit beruflichen Versprechungen honoriert worden. Dort sei die seinerzeit noch minderjährige Marokkanerin mit dem Künstlernamen Ruby Rubacuori (Bild oben) »fester Bestandteil« gewesen.
SEXISMUS UND FRAUENDISKRIMINIERUNG IN DER WERBUNG – Sexismus und Frauenfeindlichkeit stellen in der Werbung nach wie vor die größten Probleme dar. Vier Verstöße waren dieses Jahr dermaßen schwerwiegend, dass der Werberat einen sofortigen Stopp des Sujets bzw. der Kampagne verlangte.

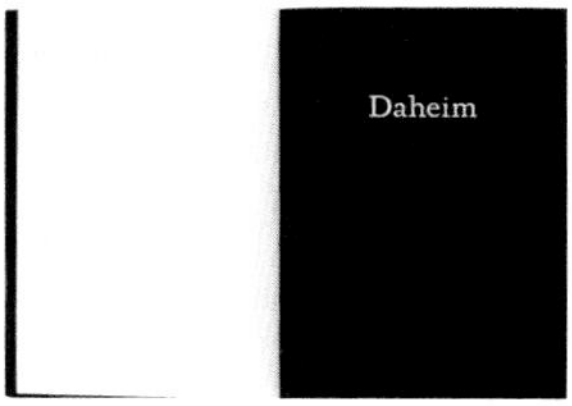
Daheim

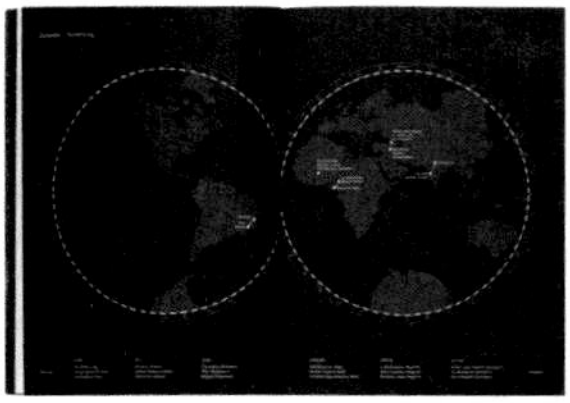

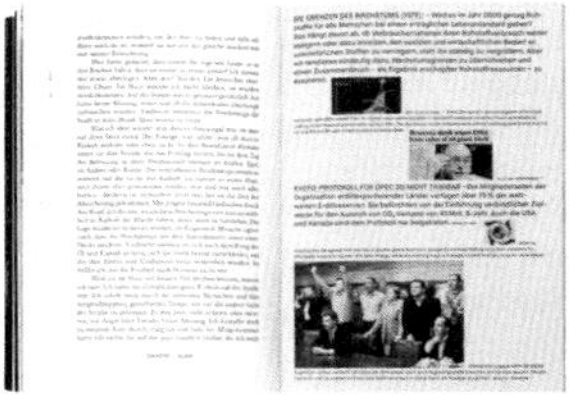

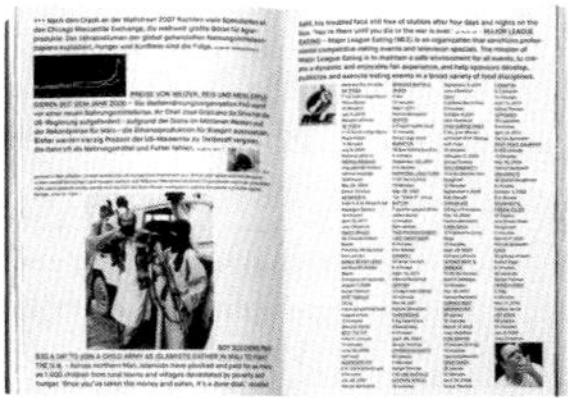

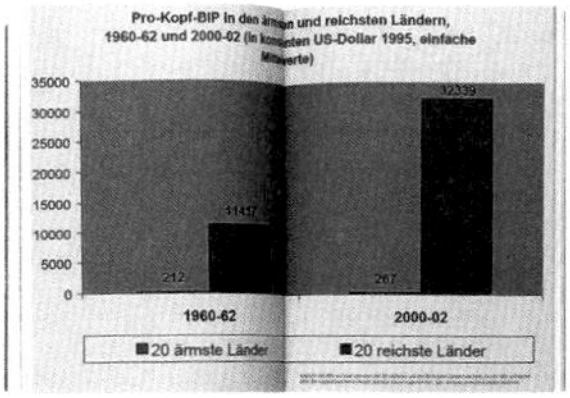
35000
30000
25000
20000
15000
10000
5000
0
212
11417
267
32339
1960-62
2000-02
20 ärmste Länder
20 reichste Länder

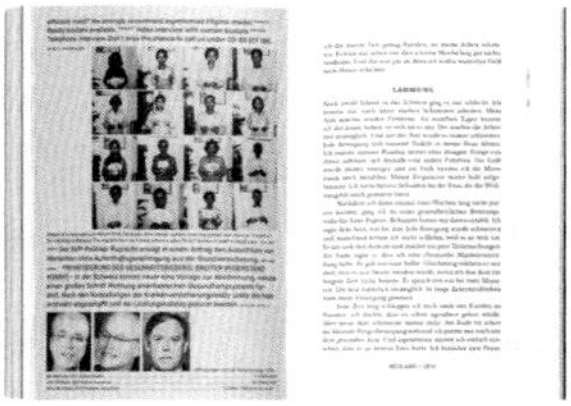

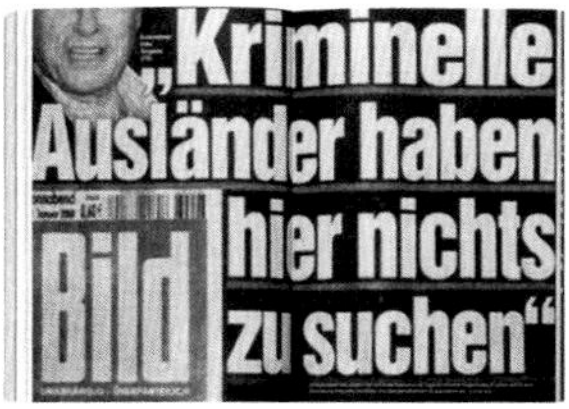
„Kriminelle Ausländer haben hier nichts zu suchen“
Bild

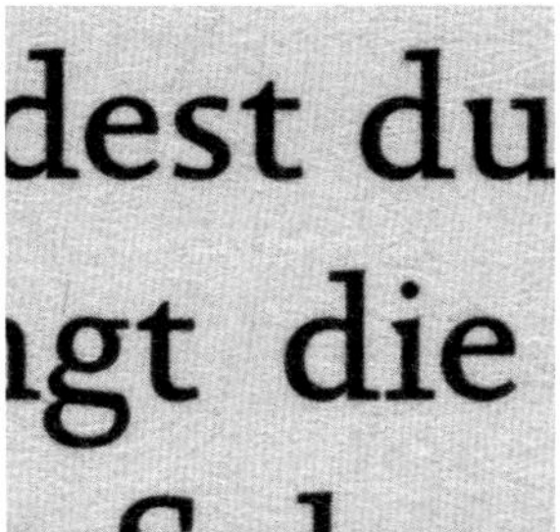
dest du
ngt die
n Schra

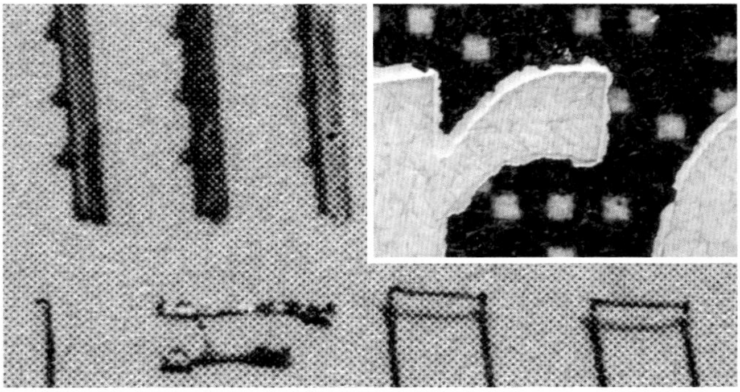

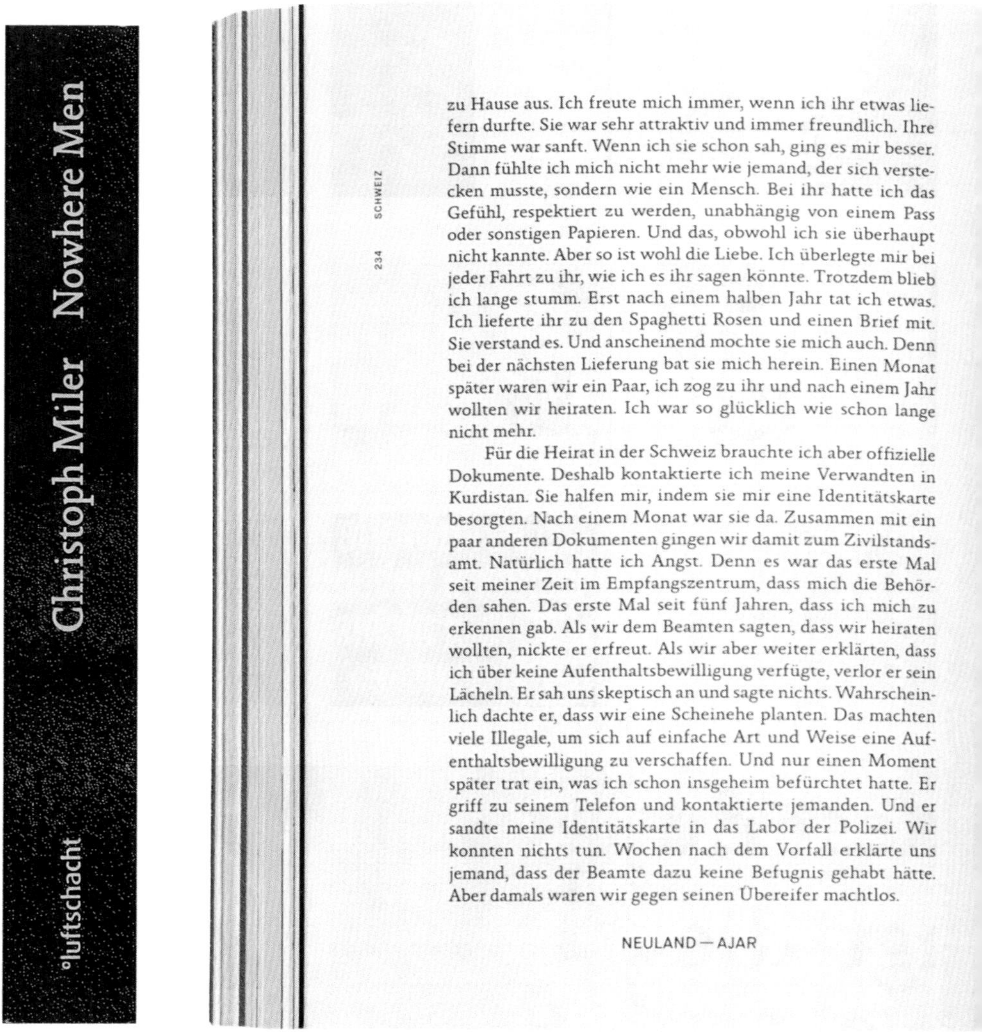
Christoph Miler Nowhere Men
°luftschacht

234 SCHWEIZ

zu Hause aus. Ich freute mich immer, wenn ich ihr etwas liefern durfte. Sie war sehr attraktiv und immer freundlich. Ihre Stimme war sanft. Wenn ich sie schon sah, ging es mir besser. Dann fühlte ich mich nicht mehr wie jemand, der sich verstecken musste, sondern wie ein Mensch. Bei ihr hatte ich das Gefühl, respektiert zu werden, unabhängig von einem Pass oder sonstigen Papieren. Und das, obwohl ich sie überhaupt nicht kannte. Aber so ist wohl die Liebe. Ich überlegte mir bei jeder Fahrt zu ihr, wie ich es ihr sagen könnte. Trotzdem blieb ich lange stumm. Erst nach einem halben Jahr tat ich etwas. Ich lieferte ihr zu den Spaghetti Rosen und einen Brief mit. Sie verstand es. Und anscheinend mochte sie mich auch. Denn bei der nächsten Lieferung bat sie mich herein. Einen Monat später waren wir ein Paar, ich zog zu ihr und nach einem Jahr wollten wir heiraten. Ich war so glücklich wie schon lange nicht mehr.

Für die Heirat in der Schweiz brauchte ich aber offizielle Dokumente. Deshalb kontaktierte ich meine Verwandten in Kurdistan. Sie halfen mir, indem sie mir eine Identitätskarte besorgten. Nach einem Monat war sie da. Zusammen mit ein paar anderen Dokumenten gingen wir damit zum Zivilstandsamt. Natürlich hatte ich Angst. Denn es war das erste Mal seit meiner Zeit im Empfangszentrum, dass mich die Behörden sahen. Das erste Mal seit fünf Jahren, dass ich mich zu erkennen gab. Als wir dem Beamten sagten, dass wir heiraten wollten, nickte er erfreut. Als wir aber weiter erklärten, dass ich über keine Aufenthaltsbewilligung verfügte, verlor er sein Lächeln. Er sah uns skeptisch an und sagte nichts. Wahrscheinlich dachte er, dass wir eine Scheinehe planten. Das machten viele Illegale, um sich auf einfache Art und Weise eine Aufenthaltsbewilligung zu verschaffen. Und nur einen Moment später trat ein, was ich schon insgeheim befürchtet hatte. Er griff zu seinem Telefon und kontaktierte jemanden. Und er sandte meine Identitätskarte in das Labor der Polizei. Wir konnten nichts tun. Wochen nach dem Vorfall erklärte uns jemand, dass der Beamte dazu keine Befugnis gehabt hätte. Aber damals waren wir gegen seinen Übereifer machtlos.

NEULAND — AJAR

EHEVERBOT FÜR SANS-PAPIERS – Seit 2011 können in der Schweiz nur noch Menschen heiraten, die ihren Aufenthalt durch eine Bewilligung o. ä. nachweisen können. Damit gilt ein faktisches Eheverbot für Sans-Papiers und ihre PartnerInnen. Von 3500 Ehen, die 2008 in Zürich überprüft wurden, stellten sich 500 als Scheinehen heraus. 10/06/10-HRW · MENSCHENRECHTE, ARTIKEL 16, ABSCHNITT 1 – Heiratsfähige Männer und Frauen haben ohne jede Beschränkung auf Grund ihrer Rasse und Herkunft, ihrer Staatsangehörigkeit oder ihrer religiösen Orientierung das Recht, zu heiraten und eine Familie zu gründen. Sie haben bei der Eheschließung, während der Ehe und bei der Auflösung die gleichen Rechte. 2014-UN ·

DATING WORLD – Alle ukrainischen und russischen jungen Frauen, die Sie auf unserer Seite »Dating World« sehen, möchten eine Ehe eingehen. Vergessen Sie nicht, dass Millionen von attraktiven, schönen und sehr gebildeten Frauen, die in der ehemaligen Sowjetunion leben, einen liebevollen Mann aus dem Westen suchen. 25/08/14-DAW ·

Der internationale Sextourismus hat sich in den letzten Jahren von Südostasien in den Nordosten Brasiliens ausgeweitet. Dabei hat die Mädchen-Prostitution erschreckende Ausmaße angenommen. Tourismus – eine vom IWF empfohlene Entwicklungsstrategie – wird als Lösung sozialer

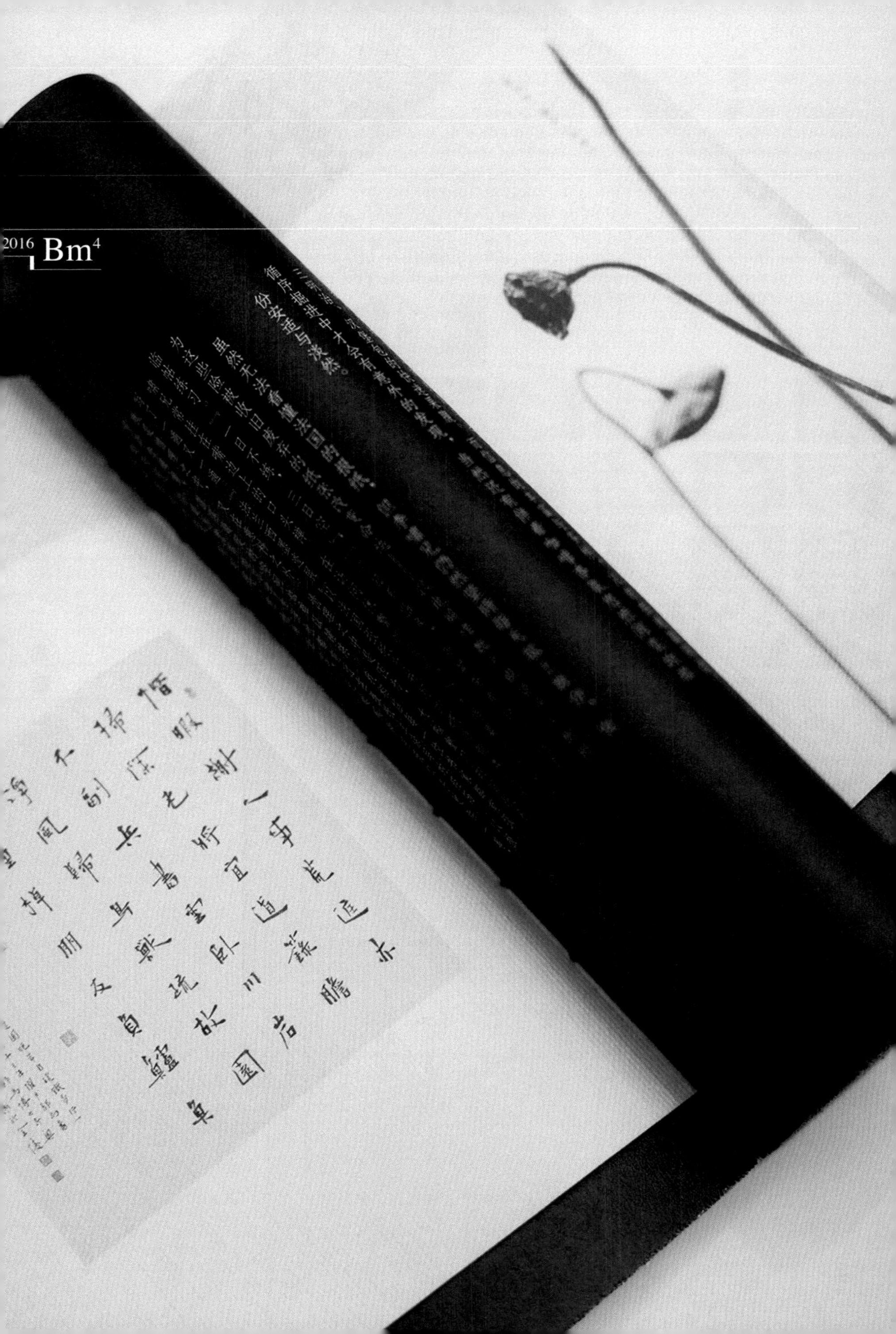

本书是一位电视主持人送给自己 40 岁生日的礼物之作，收录了作者的文化随笔及收藏雅玩。整本书体现出典型的东方美学风格。全书采用中式从左往右翻阅，文字采用竖排版。从开篇到结尾不断地用书法进行书写，始于白页又终于白页，表达出作者的态度：学习是一种反复与坚持，没有终点，恰如《论语》所言之“学而不厌”。本书的装订方式为裸脊锁线，封面采用白卡纸印黑后裱覆宣纸形成渐隐的观感，书名文字珠光白压凹处理。封面外部环状包裹一张略短于封面尺寸的毛毡，并用草线圈扎，书名以篆刻方式印在毡子上。内页通过半涂布的艺术纸手工贴装大量宣纸的手法，作者收藏的字画作品以随意的方式夹于书中，似乎随时可能洒落，颇富诗意与雅趣。内容分为“今我来思”“此情难寄”“无关风月”三个部分。书籍的上口与右翻口无清边，下口打毛处理，也与内容随性的气质相匹配。

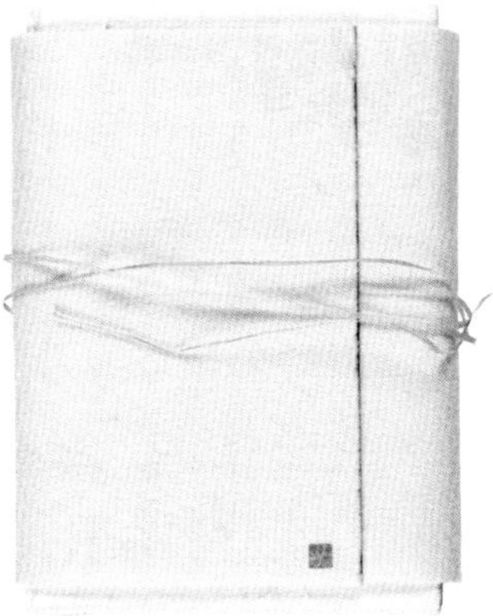

This book, composed by a TV host, is a gift for his 40th birthday. It contains the author's cultural essays and elegant collections. The whole book embodies the typical oriental aesthetic style – it should be read from left to right in Chinese style, and the text is typeset vertically. From the beginning to the end, the author keeps writing calligraphy, starting with the white page and ending with the white page, which expresses the author's attitude: learning is a kind of repetition and persistence with no end point. As the *Analects* of Confucius says, 'Pleasure of learning'. The book is bound with thread sewing with a bare ridge. The cover is made of white cardboard printed in black and mounted with rice paper to form a recessive impression. The title is embossed in pearl white. A felt slightly shorter than the size of the cover wraps around and is tied with a straw coil. The title of the book is printed on the felt by seal cutting. The inner pages made of semi-coated art paper are hand-mounted with rice paper. The author's collection of calligraphy and painting works is sandwiched in the book in an arbitrary way. It seems that they may be scattered at any time, which is poetic. The content is divided into three parts: 'when I shall be returning', 'my feeling is difficult to send', and 'irrelevant to romance'. The upper and right edges remain untrimmed and the lower edge is quite rough. The casual style matches the content perfectly.

铜奖	Bronze Medal
学而不厌	Pleasure of Learning
中国	China
曲闵民、蒋茜	Qu Minmin, Jiang Qian
周学	Zhou Xue
江苏凤凰美术出版社	Phoenix Fine Arts Publishing Ltd.
	232×152×26mm
	547g
	414p

Like a bundle of secret documents, a soft piece of felt is wrapped around the body of this book and tied with a thread of raffia. Then calligraphic daubs of ink darken the creamy natural white of the paper, page by page until you become immersed in a new world. The strong, tough paper contains texts in filigree Chinese type, with some smears of watery ink here and there – traces of the artists' work, printing ink that has seeped through or been discarded? An illusion – the linen tester reveals screen dots. Visual gems are peeled away: folded, ultra-delicate pieces of rice paper are worked into the book's layers; printed only on one side, they bear calligraphy and tusche painting – mostly a combination of the whole sheet and a large detailed section. The reader works through the book very carefully because it appears so vulnerable with its assortment of papers, the stitch binding with open spine, and the secretive beginning. You therefore handle the artistic pages with extra caution – the book, illustrations and observations blend to form a contemplative whole. This corresponds perfectly to the pedagogical motive of the whole project – in the words of Confucius: the Pleasure of Learning.

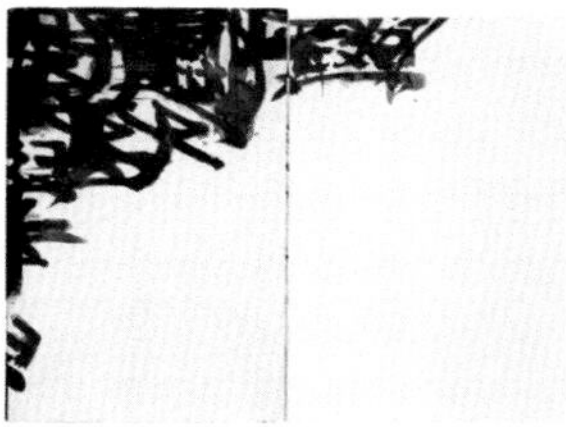
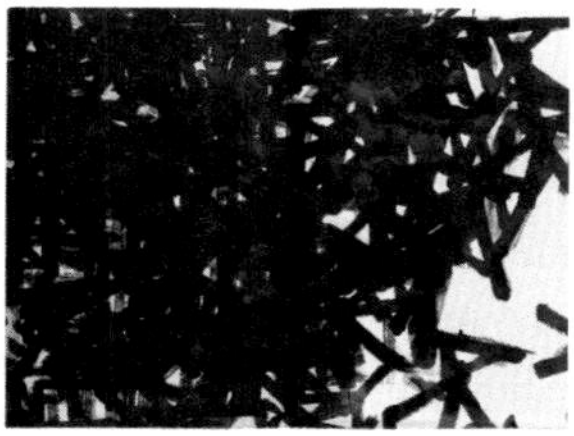

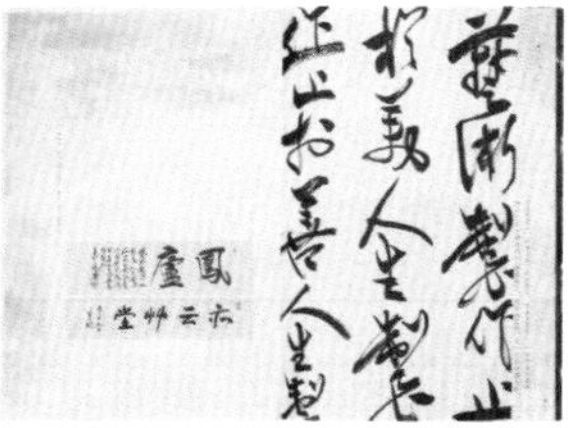
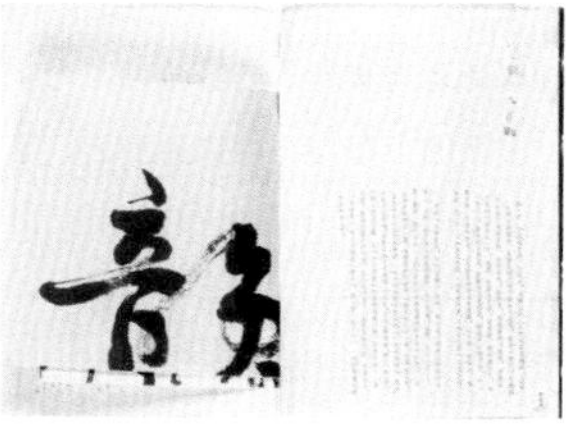
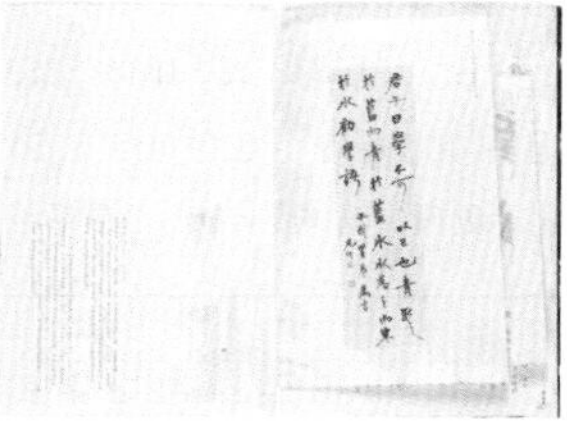
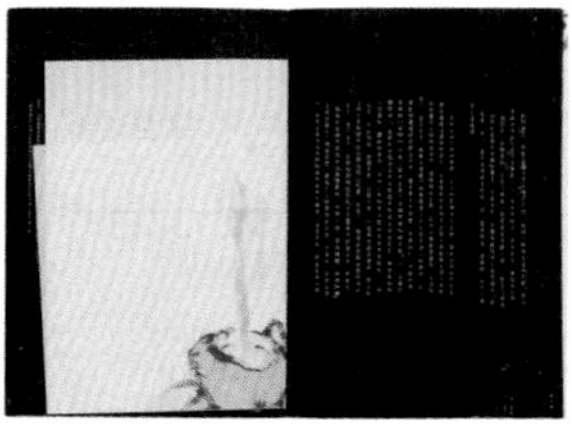
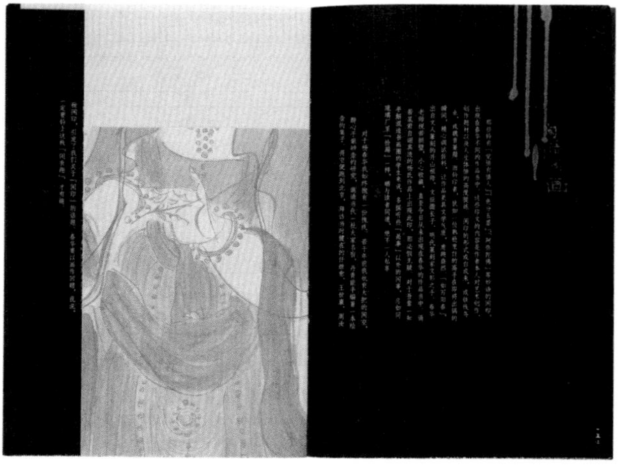
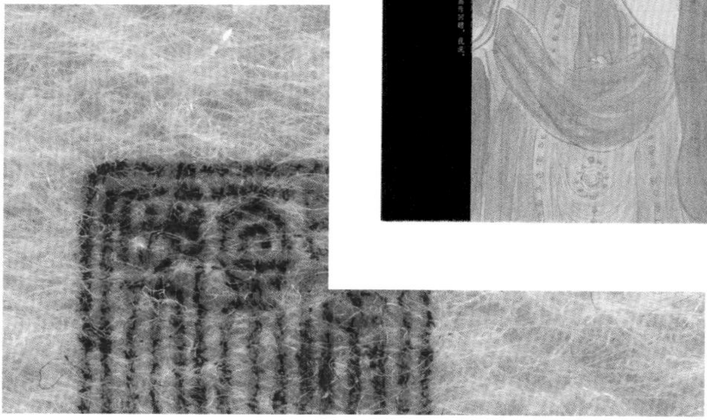

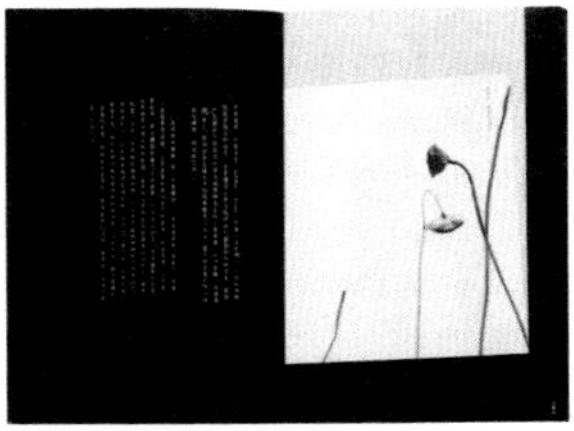

创烧时
具的
只
的
坡

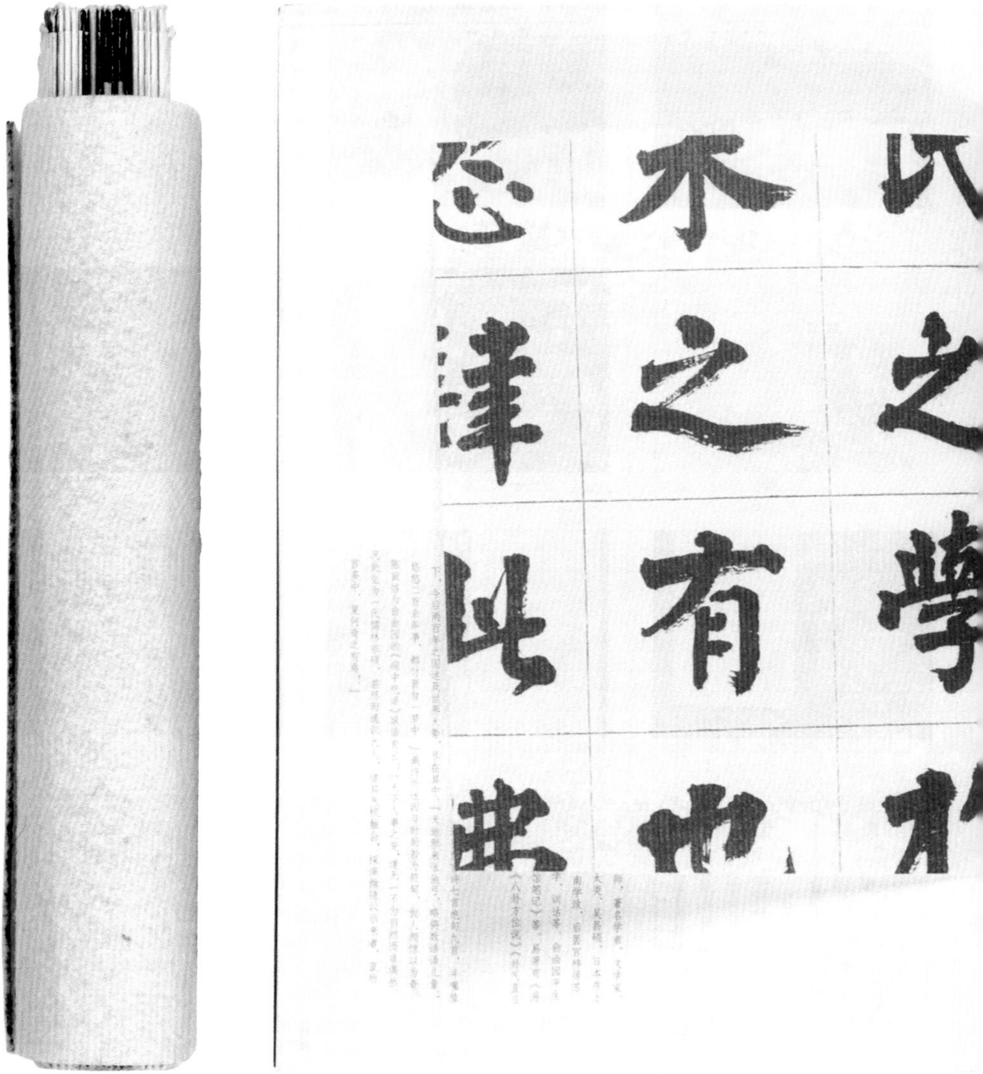
不之有也

kommet for
av territoriet,
og frokostblanding
hun fikk hver eneste
virkedag året igjennom

本书是挪威作家伊万·安约森（Ingvar Ambjørnsen）的自述小说。2010 年 6 月 17 日晚，作者突然对自己写的所有内容都不满意，便开始怀疑自己的写作能力。当他发现自己要钻进这个逻辑上的死胡同的时候，决定暂停自己的写作，出去透透气，寻找崭新的自己，于是便有了这本记录自我状态、行动、所见和所思的半散文半小说的文字。本书的装订形式为锁线硬精装，护封灰色艺术纸四色印刷，内封和环衬红色艺术纸印单黑。内页哑面涂布纸单色黑印刷。书籍没有目录，按照时间顺序分为 9 个篇章，用黑底反白字的方式印上所属篇章的标题，大多为地点或简单的感悟。大量水墨插图从封面延续至内页，记录了当前所处环境和个人的状态：光脚、卵石地面、各类植物、葡萄酒、抽烟、街道、路灯等等。

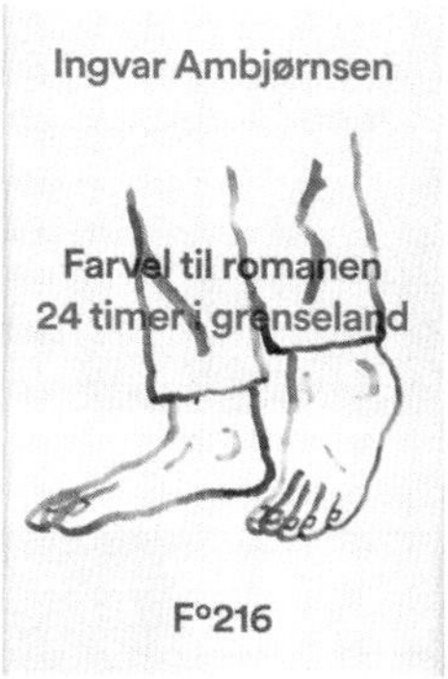

Ingvar Ambjørnsen
Farvel til romanen
24 timer i grenseland
F°216

This book is a narrative novel by Ingvar Ambjørnsen, a Norwegian writer. In the evening of June 17, 2010, the author suddenly felt unsatisfied with everything he wrote, and began to doubt his writing ability. When he found himself going into this dead end, he decided to leave his writing career temporarily and find a brand-new self. Thus he had this record of his mental states, actions, sights and thoughts. This hardback book is bound by thread sewing. The jacket is made of gray art paper printed in four colors, while the inside cover and the end paper is made of red art paper printed in monochrome black. The inside pages are printed on matte coated paper in black. There are no catalogues in the book. It's divided into nine chapters in chronological order. The titles of the chapters are printed in white on black background. A large number of ink illustrations display continuously from the cover to the inside page, recording the current environment and personal status: barefoot, pebble floor, all kinds of plants, wine, smoking, streets, street lights and so on.

铜奖

小说再见——游离的 24 小时

挪威

Bronze Medal

Farvel til romanen. 24 timer i grenseland

Norway

Aslak Gurholt Rønsen / Yokoland

Ingvar Ambjørnsen,
Editor: Bendik Wold og Nils-Øivind Haagensen / Flamme Forlag

Flamme Forlag

212×134×14mm

293g

127p

It is an agreeable sight even for non-smokers: the cigarette end smouldering away by itself on the narrow spine of the book. It belongs to the long-haired profile with the hefty sideburns on the back cover. There are plants growing freely in flowerpots on the cover flaps, plus an uncorked bottle with a filled wine glass. And on the front there are bare feet – but the story continues. Beneath the feet, on the cover, you can see the ground, or at least a pattern with radially arranged cobblestones, grouted in red from the red paper. Lush vegetation entwines an unoccupied garden chair on the endpaper, activated by the same red paper. So much atmosphere – and we haven't even reached page one yet. Overall, the book has a traditional form. The story is presented in a timeless serif typeface with distinct writing that is bold yet fine, typeset with the necessary line spacing for comfortable reading without becoming too large. An ample amount of whole-page illustrations with calming white space continues the visual setting from the beginning. The brushstrokes change from contours into larger areas, drawn in a single stroke – a watery blend of different grey shades. Good reading conditions are created here by the book itself.

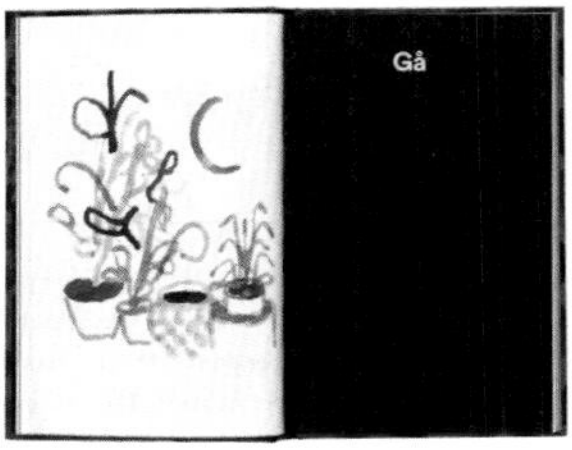
Gå

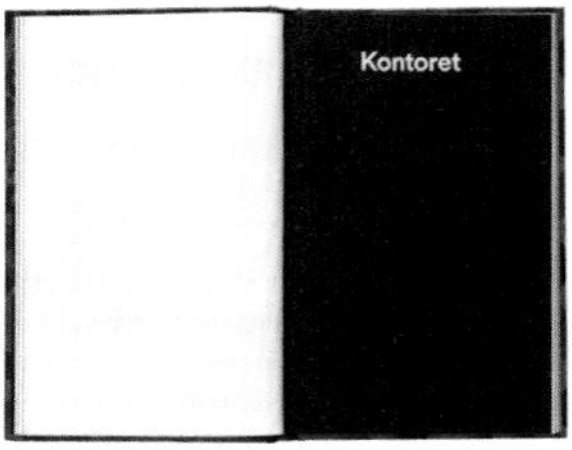
Kontoret

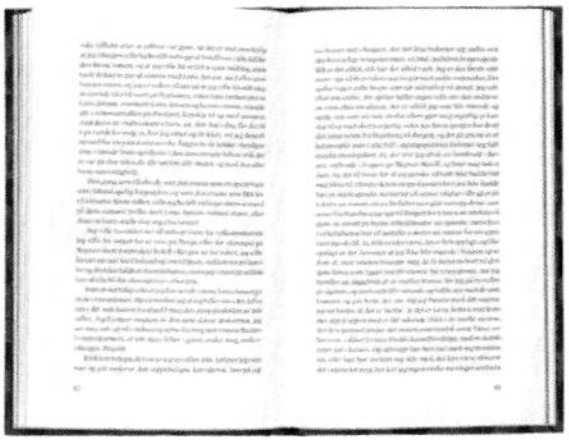

til et ganske spesielt vaskemiddel. Det er bedre enn alt annet. Mye bedre. Og hele tyve prosent av innholdet er helt gratis.
– Er du ute og spaserer uten sko? Det må jeg si!
Nå, i det klare morgenlyset, virker det ikke på langt nær like naturlig som i natt. Full av entusiasme, frigjort fra Romanen, hadde jeg styrtet ut som jeg sto og gikk, for å drikke duften av sommerregnet som falt over den varme asfalten. Nå virker det hele bare utstudert og dumt. Stå her. Barbeint i dress. Som om jeg protesterer mot et eller annet. Antrekk: pent. Og så møter man i dress. Barbeint. Det er slikt man finner på mens man går sitt siste år av ungdomsskolen, etter å ha lest Bakunin eller Jerry Rubin for aller første gang.
Jeg forklarer. Det fine sommerregnet. Duftene av byen. Den plutselige euforien.
OK. Han følger.
Grønt.
Vi krysser gaten.
Det er Åsne! Jeg forstår at det ikke kan være sant. Men jeg ser det likevel.
– Ja, fortsetter han. – Det hviler en stor ensomhet i det å gjøre noe slikt. Du er en kjent mann i Norge. Du er bosatt i utlander. Likevel setter hun seg ned og ber deg komme hjem til henne. Ja, faktisk om å flytte inn i hennes egen hage. Og dette, min venn, hadde vært enkelt og liketil å begripe seg på, dersom det ikke var for det faktum at hun har en mann, og, så vidt jeg forstår, også to barn som bor hjemme. Da kan vi nemlig, etter all sannsynlighet, sette en strek over ... Ja, du vet. Hvis det ikke er noen veldig spesielle greier ute og går, da. Slikt finnes jo. Men du har altså bestemt deg for å dra oppover?
Jeg svarer ikke. For første gang på flere år får jeg lyst på en sigarett.
– Ta et råd: Sitt stille. Lytt! Og vent!
Og for å få en slutt på det: – Jeg lover! Jeg rusher ikke.
Han nikker.
95

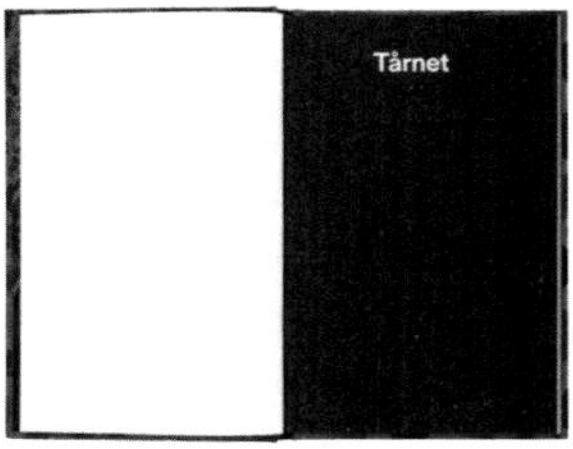
Tårnet

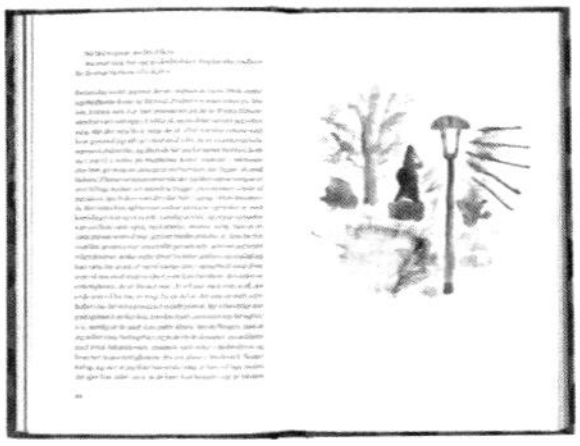

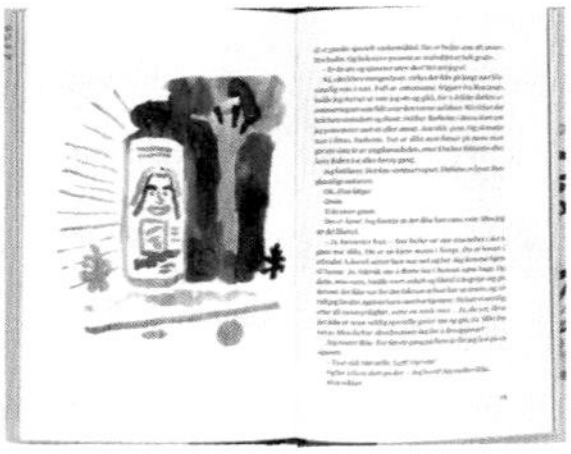

ker og re
en,
e, kalde
eser som en katt, så har jeg

Ingvar Ambjørnsen
Farvel til romanen
24 timer i grenseland

spade, fikk treet på plass i jorden, Hamburger Abendblatt ikke til stede, ikke en linje på trykk dagen etter heller, jeg har lagt meg til den vanen at jeg bukker for treet hver gang jeg passerer det, det synes jeg er passende oppførsel, og en dannet markering av vår felles tilstedeværelse i livet.

Ja, det lyser i stuevinduet til gode gamle Thomas Mader, den tidligere eiendomsmagnaten som sprang ut i mitt eget hode sommeren 2003, mannen som avvikler hele sin virksomhet, og isteden gir seg til å gå og gå rundt omkring i Hamburg, fra Innocentia Park og ned i havnen, fra havnen og opp gjennom Planten un Blomen, over broene og langs kanalene, rundt og rundt Alster, og så videre. Og hver eneste dag gjør jeg nettopp det samme, jeg går og går, fra havnen og opp i Innocentia Park, jeg går i Stadtpark og gate opp og gate ned i Eppendorf og Hoheluft, Barmbek og Eimsbüttel, jeg tar ingen notater, det har jeg aldri gjort, men om kvelden, etter at jeg har spist middag, og langt ut over natten, sitter jeg på kontoret og lar Mader gå i de samme gatene som meg, jeg lar ham se det jeg ser, og nokså tidlig i romanen lar jeg ham altså flytte inn i denne gamle, praktfulle villaen ved Innocentia Park, mens jeg selv holder til i Grindel, i en kåk som sant å si ikke er så dårlig, den heller, og på den måten beveger jeg meg mellom litteraturen og virkeligheten, en konstant pendelvirksomhet som varer i et års tid, til romanen er ferdig og kan sendes forlaget, dag for dag, natt etter natt, i uke etter uke, måned etter måned, betrår jeg disse gatene og parkene, i virkeligheten og i romanen, ut og inn av tekst og virkelighet, om og om igjen, til siste punktum kan settes. Og etter at romanen er kommet ut, når den står rundt i hyllene, både i de tyske hyllene og de norske hyllene, og kritikerne har sagt sitt, fortsetter jeg å gå i de samme gatene og i de samme parkene, langs de samme kanalene og over de samme broene, mens romanpersonen Thomas Mader langsomt skriver seg inn i en annen bok enn min; den store glemmeboken.

2016 Ha 1

这是奥地利照明集团奥德堡（Zumtobel Group）2013、2014 的年度报告。这家照明公司从 1992 年起便邀请世界各地的设计师对光进行阐释和表现。本书的装订为锁线胶平装，封面白卡纸印黑加凹凸工艺，封底贴单色照片。内页胶版纸调频网四色印刷，全书以 Futura 字体为主。扉页部分呈现四个经纬度信息，为生活在挪威和瑞典的四位百岁老人的位置，封底照片便是其中一位（共有四版封底设计）。前言过后是对这四位老人的专访，记录了他们经历电力革命、两次世界大战以及各种社会和科技发展的长达百年的口述历史，特别采用等宽打字机字体排版。之后分两大部分，通过照片、插图、油画等形式，对人们记录和表现过的黑暗与光明进行罗列。接着紧跟三篇科学论文，从不同方面研究了北欧自然光的特性，以及对当地民众生活、生理和心理的影响。书籍只用最后不到五分之一的版面展示公司财报数据、品牌运营和相关业务的内容。||||||||||

This is the annual report of Zumtobel Group (an Austrian supplier of integral lighting solutions) from 2013 to 2014. Since 1992, the lighting company has invited designers around the world to focus on the interpretation and performance of light. The book is glue bound with thread sewing. The cover is printed on white cardboard in black with embossing process while the back cover is printed with monochrome photographs. The inside pages are made of offset paper in four colors with frequency modulation screen. Futura font is the main font in the book. On the title page, there are latitudes and longitudes data of four positions, where four centenarians live in Norway and Sweden. One centenarian's photo is displayed on the back cover. Interviews to the four old people are followed by the introduction, typeset with typewriter font of equal width. The interviews have recorded the oral history of the electric power revolution, two world wars, and various social and technological developments over the past 100 years. After that, there are two parts to list the darkness and light that people recorded and expressed through photographs, illustrations, oil paintings and other forms. The following three scientific papers study the characteristics of Nordic natural light and its impact on the lives, physiology and psychology of the local people from different aspects. The book only uses less than a fifth of the pages to show the company's financial data, brand operations and related business.

- 荣誉奖
- 现世的北欧之光
- 挪威

- Honorary Appreciation
- LIVING THE NORDIC LIGHT
- Norway
- Snøhetta
- Åsne Seierstad, Po Tidholm, Lars Forsberg, Barbara Szybinska Matusiak, Vidje Hansen, Bruno Laeng
- Zumtobel AG
- 271×213×37mm
- 1394g
- 400p

This bulky volume with a white cardboard cover, with the title embossed in thick Futura capitals in the centre, gives the impression of an art catalogue. A photograph is stuck to shiny paper in the debossed area on the back cover. An elderly woman gazes out of the intense black & white portrait. It is about light, Nordic light. Inside the book, very different sections await the observer. Four portraits of people who are around 100 years old, with texts and photographs – two associative collections of images showing aspects of light and dark – three academic essays on the influence of polar light on a person's living circumstances. Something emerges that is akin to a visual cultural history of the effect of light in Northern climes. The focus is on people and light as a life-giving dimension. The last, short part of the book, however, almost goes unnoticed. Apart from a few illustrations, it is without any pictures; the unadorned, uncomplicated numeric tables are recognisably the essential parts of an annual report. Then your thoughts turn to the tiny line on the book's cover which you originally overlooked. In actual fact, this really is an annual report for a company that supplies lighting technology. The impressive, artistic presentation documents the high standards and quality of the company for which a clear benchmark exists: natural light.

LIVING
THE
NORDIC
LIGHT

LIVING
THE
NORDIC
LIGHT

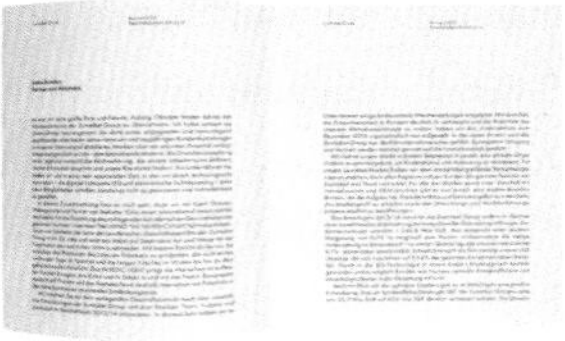

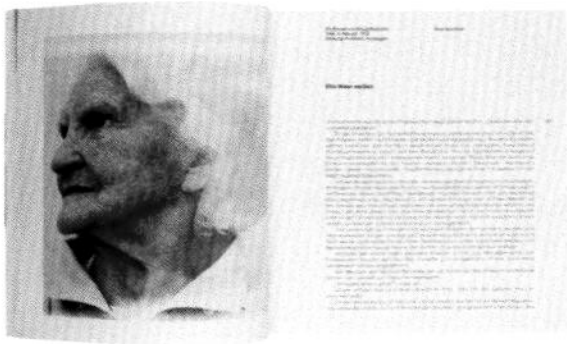

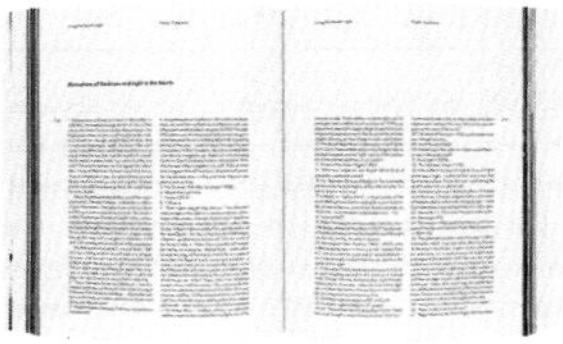

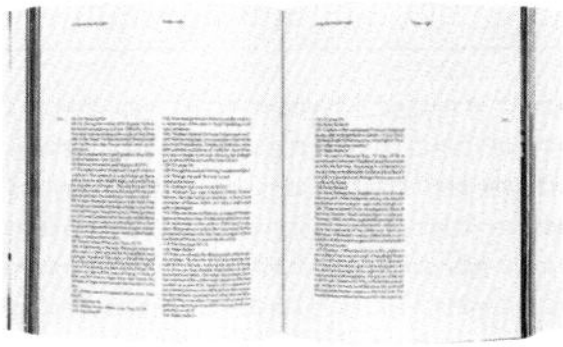

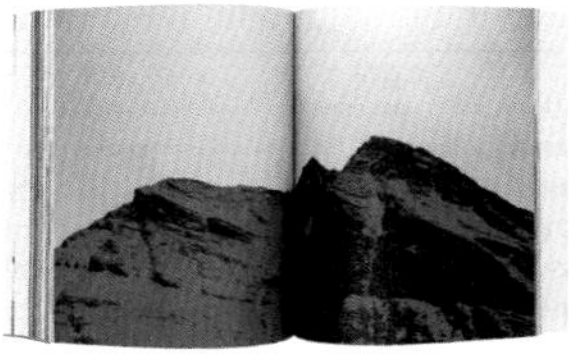

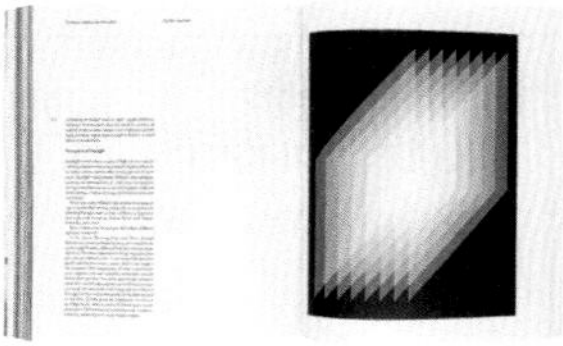

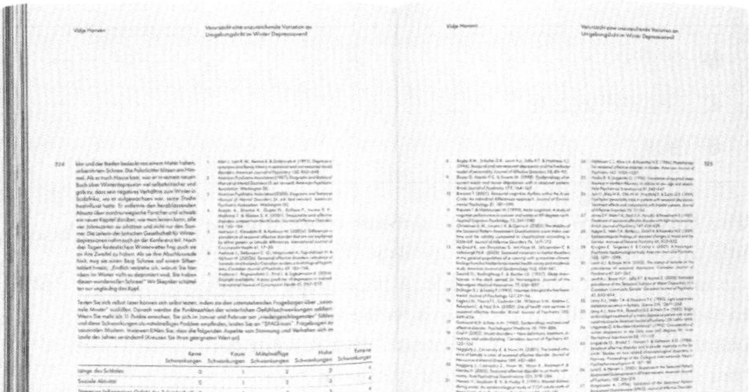
Vidje Hansen
Verursacht eine unzureichende Variation an Umgebungslicht im Winter Depressionen?

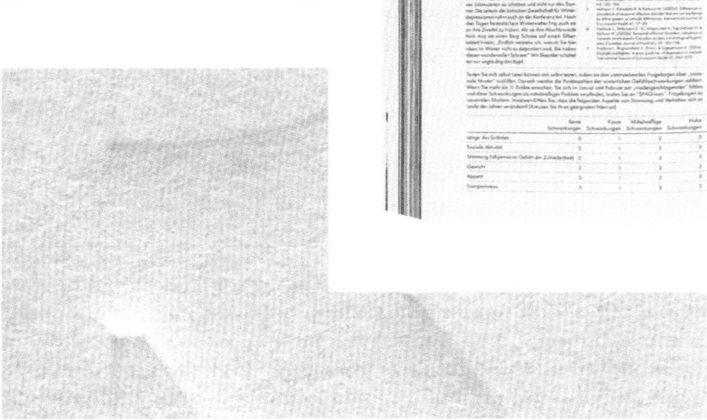

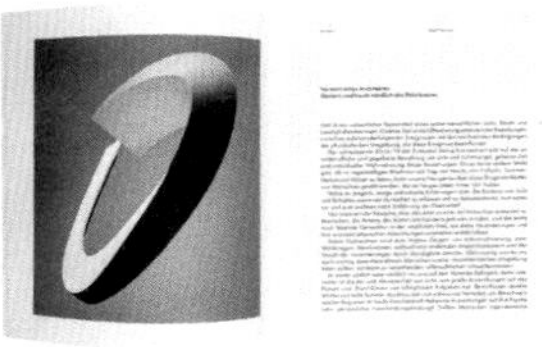

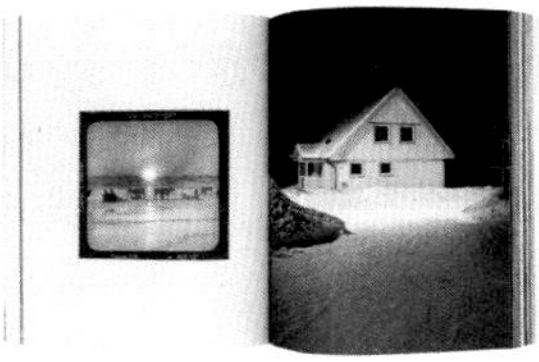
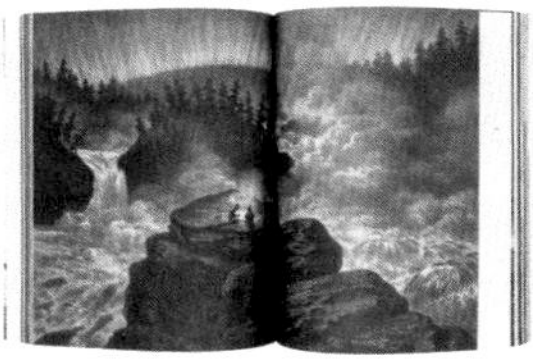
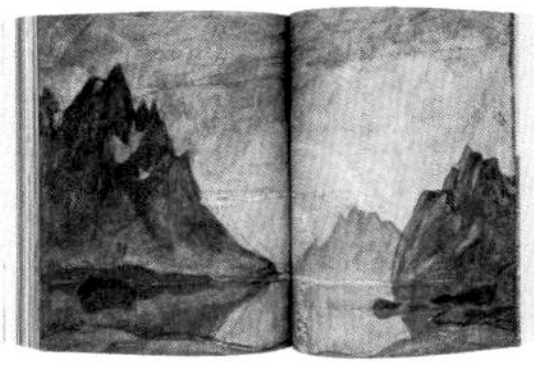
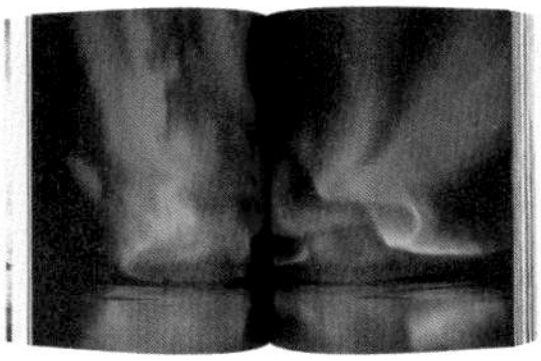
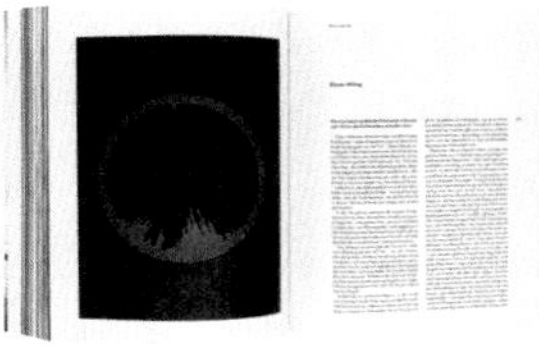

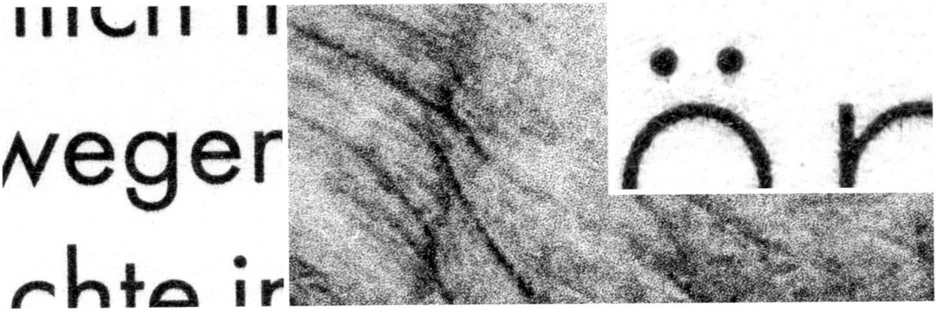

Zastoupení žen a mu
A)
é
ací
Monopoly
Cochrane was rich and had
rvants to wash the dishes. But they
ten broke the china while doing the
shes. In 1887, Cochrane decided to
reate a dishwasher that would do
he job instead of people

这是一本研究就业中的性别歧视，并提出改变现状的方法论书。书中对就业市场、社会和学校中的性别差异提出质疑，通过分析和研究，提出了几大改善的步骤，陈列多项实验和参与实验的资料，帮助希望对此有所改善的年轻人建立信心，并找到适合自己的工作。本书的装订方式为裸脊锁线，封面白卡纸蓝黄黑三色印刷，覆哑面膜。内页胶版纸四色印刷，无其他特殊工艺。书中分为两大部分，通过深满底绿色页面区隔。第一部分是现状研究、理论介绍和本书的使用说明；第二部分按照存在性别歧视的职业界定、刻板印象消除、自我信心建立、改善的方式方法这四大目标，分别提供了大量的实践指导内容。四大目标按照黄、蓝、红、绿色标进行区隔，在书口上清晰可辨。书中存在大量有趣的插图和明晰的图标，配合并不拘泥的网格但条理清晰的版面，使得阅读过程轻松愉悦。

Proč zrovna automechaničkou? učitelem v mateřské škole? pilotkou? květinářem? stavební inženýrkou? zdravotním bratrem? řidičkou sanitky? kosmetikem? instalatérkou? ošetřovatelem? političkou? asistentem...?
Volba povolání bez předsudků
Metodika projektového dne pro základní školy
Erasmus+
GENDER STUDIES

This is a study of gender discrimination in employment, and put forward a methodology to change the status quo. The book questions the gender differences in the job market, society and schools. Through analysis and research, it puts forward several steps for improvement. It also displays a number of experiments and participates in experiments to help young people who hoped to improve their confidence and find suitable jobs. The book is bound by thread sewing with a naked spine. The laminated cover is made of white cardboard printed in blue, yellow and black. The inside pages are using offset paper printed in four colors without any other special techniques. The book is divided into two parts, separated by a page full of green. The first part is about the research status, theoretical introduction and the manual of the book; the second part provides a lot of practice instruction according to four major aspects – such as defining occupational gender discrimination, eliminating stereotypes, building self-confidence, and methods of improvement. The four aspects are segregated using the labels in yellow, blue, red and green, which can be clearly distinguished from the edge of the book. There are a lot of interesting illustrations and clear icons in the book, which make the reading relaxed and pleasant with a well-organized layout that does not adhere to the grid.

- 荣誉奖
- 无偏见的职业选择
- 捷克

- Honorary Appreciation
- Career Choice Without Prejudice
- Czech Republic
- Jan Šiller
- Anna Babanová, Jitka Hausenblasová, Jitka Kolářová, Tereza Krobová, Irena Smetáčková
- Gender Studies, Prague
- 297×211×10mm
- 293g
- 112p

Questioning and overcoming gender role clichés and their influence on the choice of profession later in life: that is the focus of this school textbook for project work. The open spine is not only decorative, it is hardly possible to imagine a book that opens more easily – practical handling takes centre stage. After all, a textbook has to be more than just student-friendly. First and foremost, it should naturally enable teachers to find their way through the contents with 100% certainty. Pictograms and the clear page structure are good aids here. The typography is admittedly didactic, but also unorthodox, shown for example in the column variants. With considerable differences in the font sizes, with the alternative numbers and widths of columns, with fine lines that divide and allocate, the typography deals successfully with the strong editorial structure of the content. The illustrations of the men and women, reminiscent of Roy Lichtenstein, are characteristic but not overly drawn. Their use of fashionable features is highly economic – a subtle design subtask with a solution that is equally striking and subtly drawn. The whole result is dense yet spacious. The texts appear compact, but a lot of space still remains at the sides; both groups of readers – the learners and the educators, are spared a flood of information. Last but not least, the wonderful list of contents displays a stringent concept. It seems like an instruction manual and a summary rolled into one.

Volba povolání bez předsudků
Gender Studies, o.p.s.

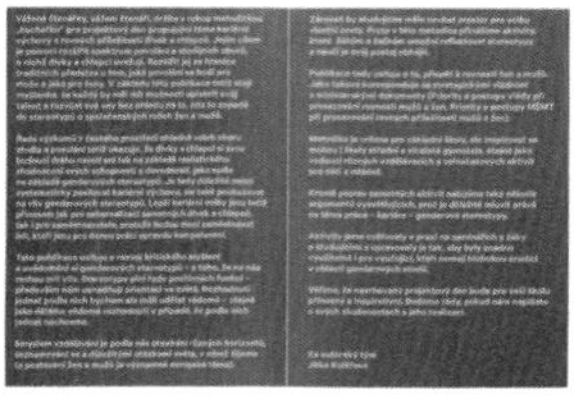

Obsah

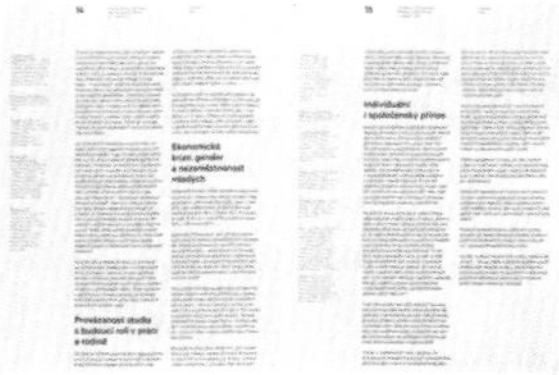

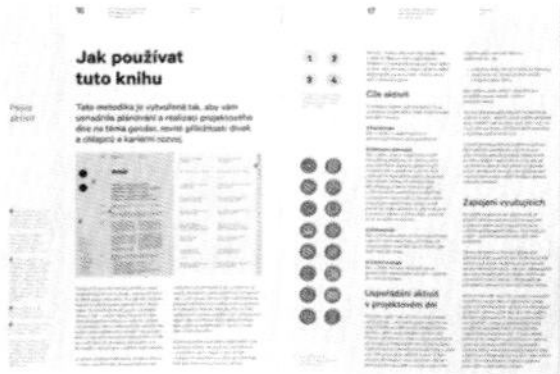
Jak používat tuto knihu

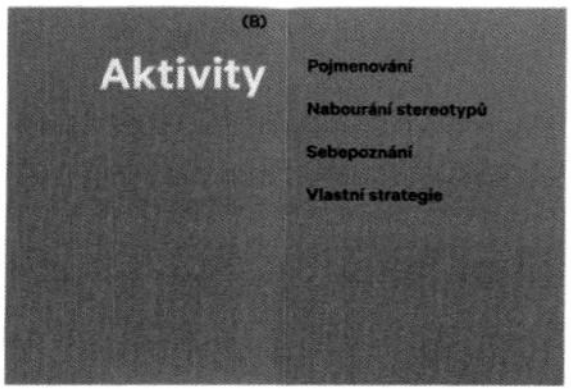
(B)
Aktivity
Pojmenování
Nabourání stereotypů
Sebepoznání
Vlastní strategie

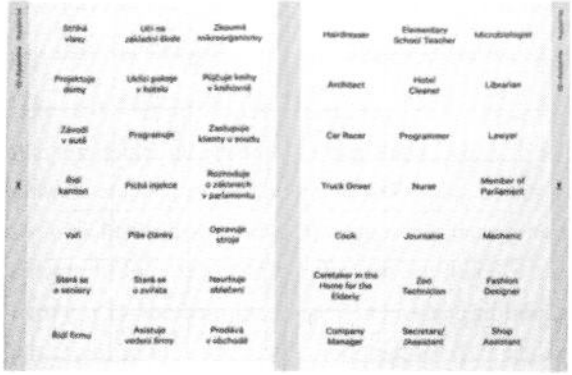

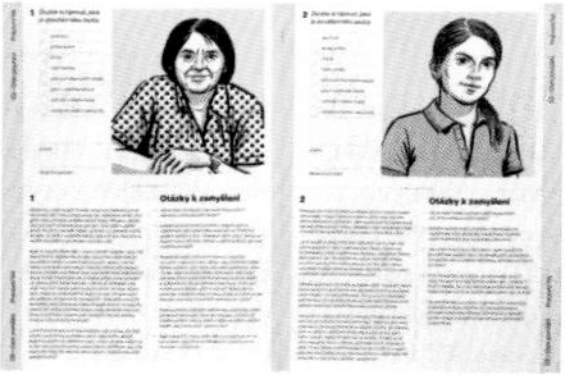

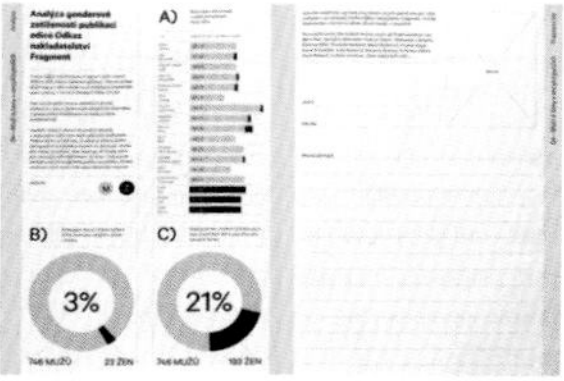
A)
B)
C)
3%
21%

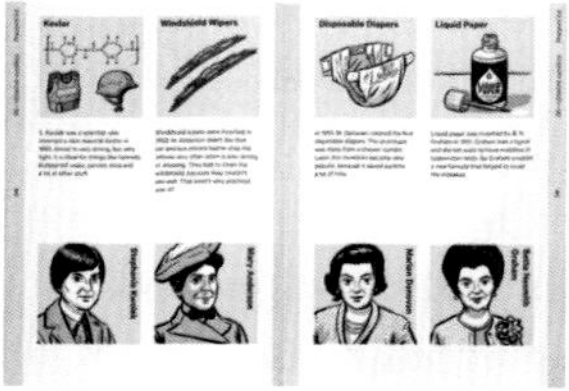

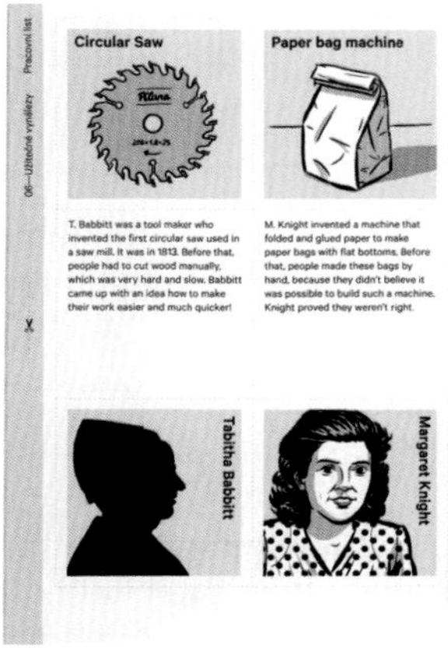
06—Užitečné vynálezy
Pracovní list
Circular Saw
Paper bag machine
T. Babbitt was a tool maker who invented the first circular saw used in a saw mill. It was in 1813. Before that, people had to cut wood manually, which was very hard and slow. Babbitt came up with an idea how to make their work easier and much quicker!
M. Knight invented a machine that folded and glued paper to make paper bags with flat bottoms. Before that, people made these bags by hand, because they didn't believe it was possible to build such a machine. Knight proved they weren't right.
Tabitha Babbitt
Margaret Knight

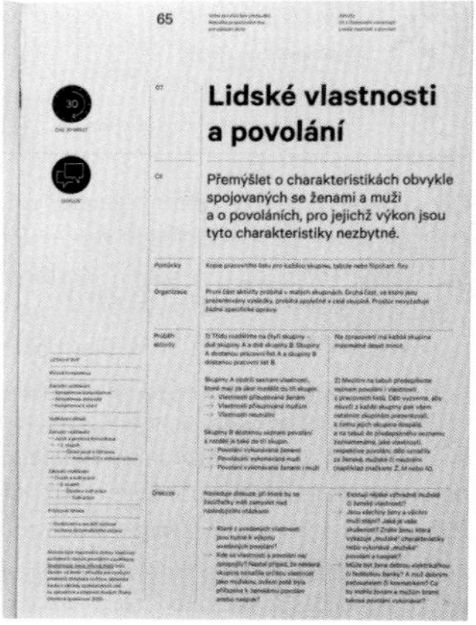
65
Lidské vlastnosti a povolání
Přemýšlet o charakteristikách obvykle spojovaných se ženami a muži a o povoláních, pro jejichž výkon jsou tyto charakteristiky nezbytné.

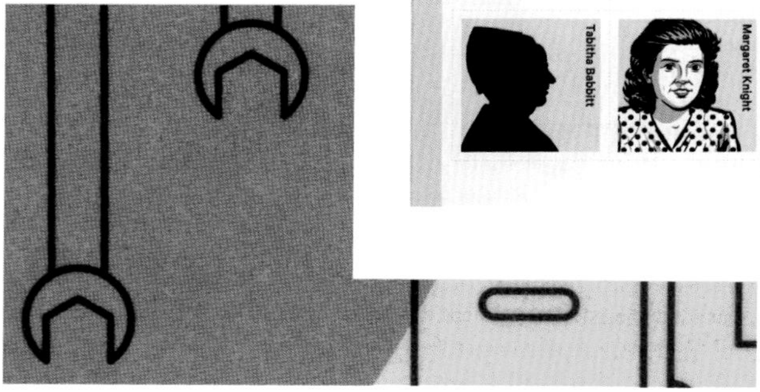

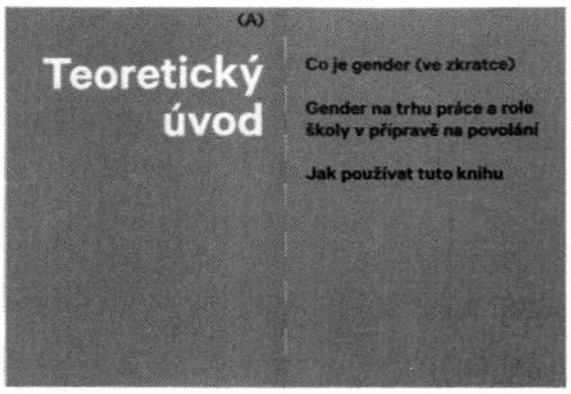
(A)
Teoretický úvod
Co je gender (ve zkratce)
Gender na trhu práce a role školy v přípravě na povolání
Jak používat tuto knihu

Co je gender (ve zkratce)

Gender na trhu práce a role školy v přípravě na povolání
40%
95%
25%
17%
8%

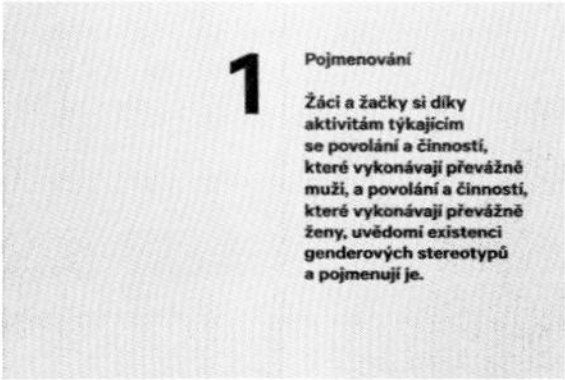
1
Pojmenování
Žáci a žačky si díky aktivitám týkajícím se povolání a činností, které vykonávají převážně muži, a povolání a činností, které vykonávají převážně ženy, uvědomí existenci genderových stereotypů a pojmenují je.

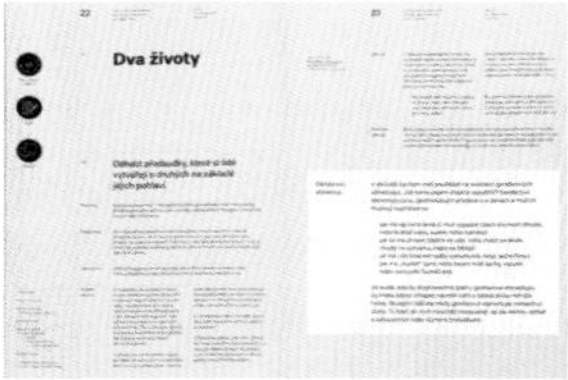
Dva životy

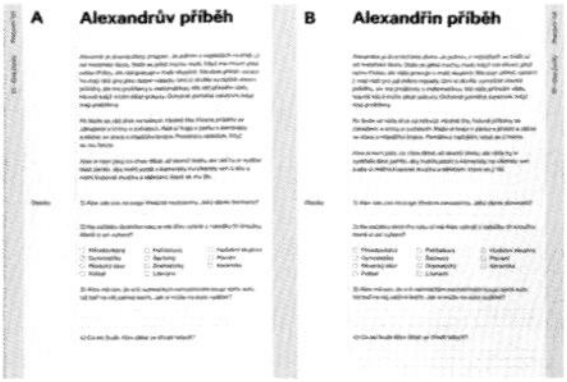
A
Alexandrův příběh
B
Alexandřin příběh

1
2
Monika Babičková volba:
Váš rodinný poklad
Snad je nikdy nebudete potřebovat.

5
6

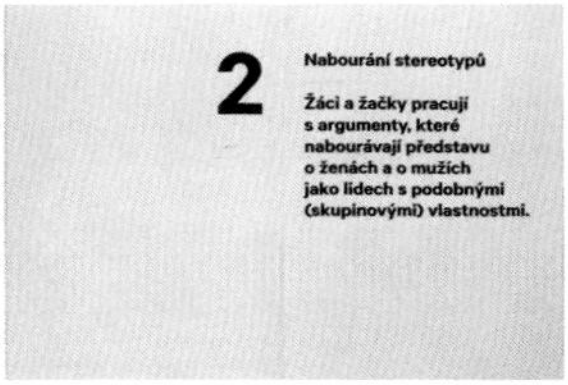
2
Nabourání stereotypů
Žáci a žačky pracují s argumenty, které nabourávají představu o ženách a o mužích jako lidech s podobnými (skupinovými) vlastnostmi.

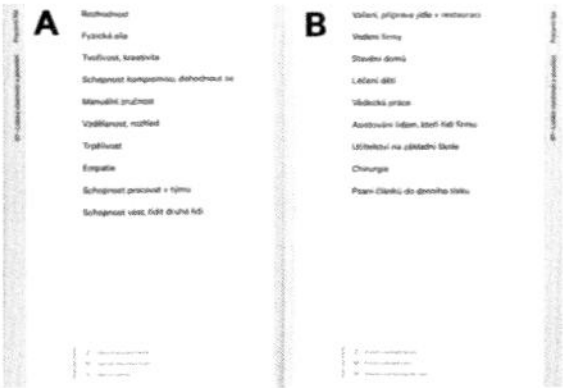
A
B

1
2
3

Automechanička
Politička, poslankyně Parlamentu ČR
Ošetřovatel v nemocnici, zdravotní bratr
Kuchař ve školní jídelně

uplat
h přitom pro or

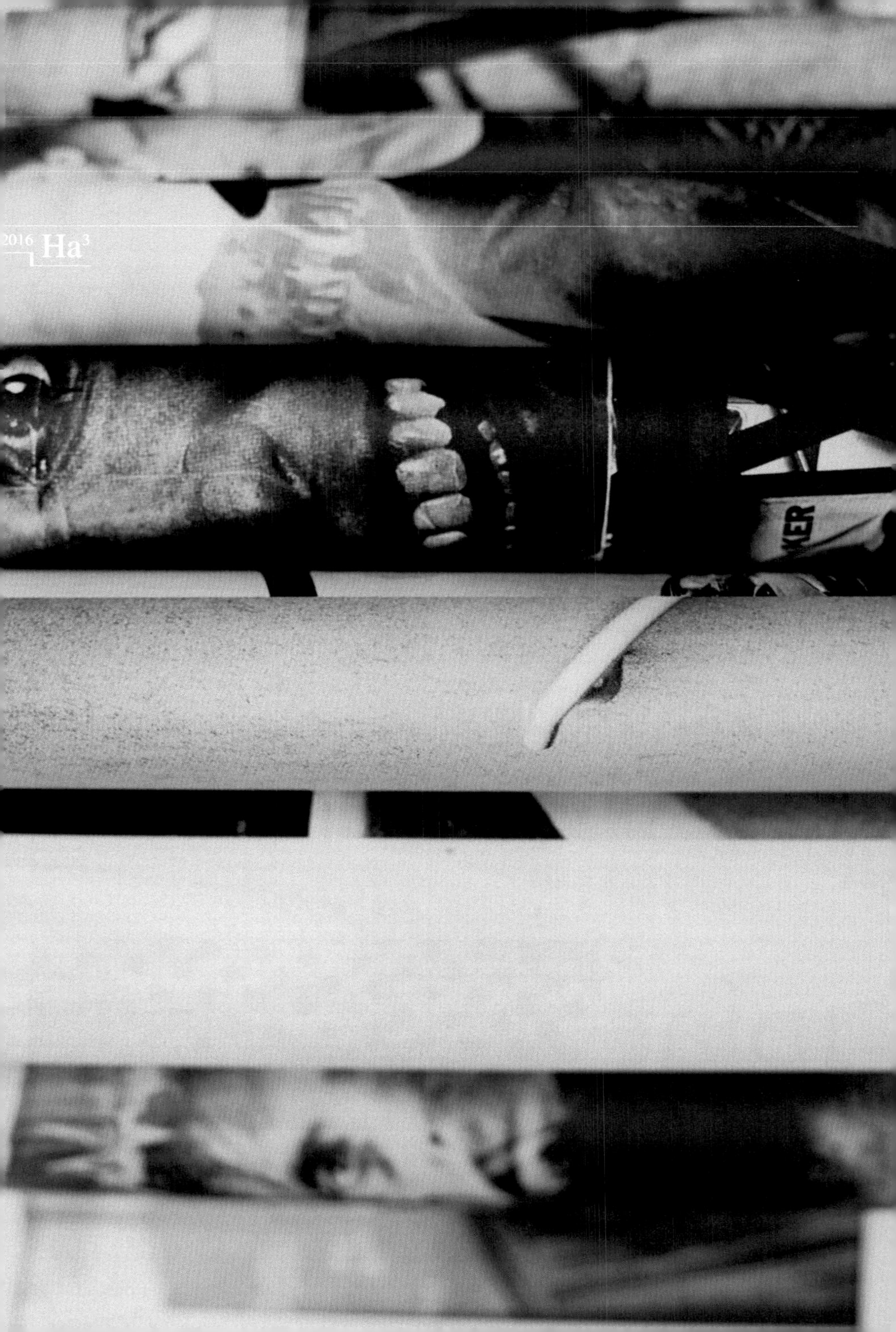

这是一本摄影师在世界各地的街道拍摄的作品集。摄影师巧妙地将城市中的广告、涂鸦、照片、玻璃反光、标语等与现场结合起来，展现现实的另一面。本书为包背胶装，封面灰色卡板银黑双色印刷，书脊黑色包布印银。内页照片部分胶版纸印单黑，照片并不局限在单一页面，配合包背装翻口的过渡，形成连贯的视觉冲击。文字部分蓝绿双色印刷，为西班牙文和英文双语。图片说明在最后，将前面照片的冲击力纯粹化。

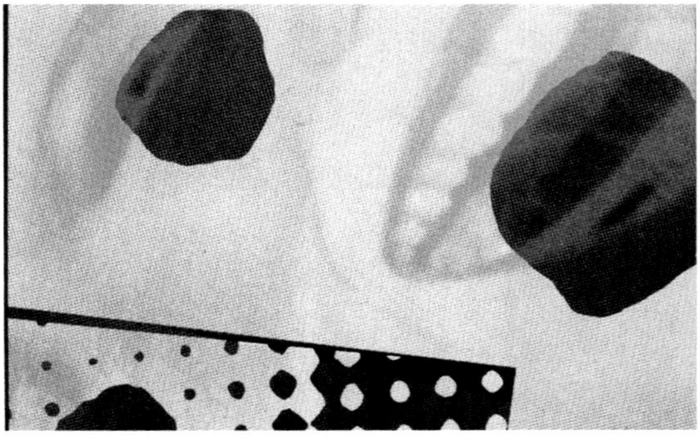

This is a collection of street photography taken from all over the world. The photographer skillfully combines the advertisements, graffiti, photos, reflections of glass and slogans in the city together to show the reality. The double-leaved book is perfect bound and the cover is made of gray cardboard printed in silver and black. The spine is covered by black cloth printed in silver. The photos on the inside pages are printed in black on offset paper. The photos are not limited to a single page. With the transition of the double-leaved pages, the visual impact is strong and coherent. The text is printed in blue and green, in both Spanish and English. The final captions purify the impact of the images.

荣誉奖
照片的背后
委内瑞拉

Honorary Appreciation
Del reverso de las imágenes
Venezuela
Álvaro Sotillo, Juan F. Mercerón
Paolo Gasparini, Victoria de Stefano (Text), Ana Nuño (Translation)
Editorial mal de ojo
201×315×15mm
682g
88p

People and little groups, reflections and windows, eyes and lenses, shop windows and dummies, pavements and advertising, confused scales and distorted sizes, inside and outside, skin and plastic, day and night, style and timelessness... As if one of these black & white photographs weren't enough to disturb conventional ways of viewing, they are grouped together in pairs and printed edge to edge, hardly identifiable in the book. In this way, the actual sharp juxtaposition merely becomes one more seam in the existing facets of the photographic compositions, and the pictures merge to produce a complex visual experience. The book designer succeeds in intensifying this synthetic process employed by the photographer. He provides further image fractures by allowing the pictures to extend over the folded front edge – made possible by the Japanese adhesive binding. The book now embodies what could be described as a whole film, whose edits become constituent components of the images. The confusion seems perfect – but one thing becomes clear: the fact that we can never be sure whether we can really recognise what we are seeing – and vice versa. Deceptive images are revealed to be real, and what is real always seems deceptive. The photographer practises the art of vision reversal, as it were. And things suddenly look you in the eye.

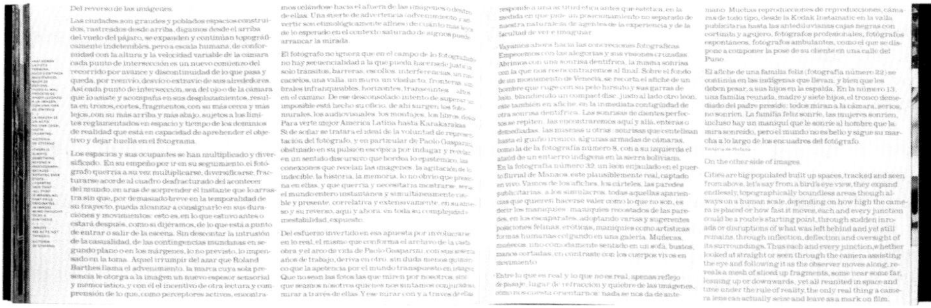

Del reverso de las imágenes

Las ciudades son grandes y poblados espacios construidos, rastreados desde arriba, digamos desde el arriba del vuelo del pájaro, se expanden y continúan topográficamente indetenibles, pero a escala humana, de conformidad con la altura y la velocidad variable de la cámara cada punto de intersección es un nuevo comienzo del recorrido por avance y discontinuidad de lo que pasa y queda, por reenvío, desvío o extravío de sus alrededores. Así cada punto de intersección, sea del ojo o de la cámara que lo asiste y acompaña en sus desplazamientos, resulta en trozos, cortes, fragmentos, con su más cerca y más lejos, con su más arriba y más abajo, sujetos a los límites reglamentados en espacio y tiempo de los dominios de realidad que está en capacidad de aprehender el objetivo y dejar huella en el fotograma.

Los espacios y sus ocupantes se han multiplicado y diversificado. En su empeño por ir en su seguimiento, el fotógrafo querría a su vez multiplicarse, diversificarse, fracturarse acorde al cuadro desfracturado del acontecer del mundo, en aras de sorprender el instante que lo arrastra sin que, por demasiado breve en la temporalidad de su trayecto, pueda alcanzar a consignarlo en sus duraciones y movimientos: esto es, en lo que estuvo antes o estará después, como si dijéramos, de lo que está a punto de entrar o salir de la escena. Sin descontar la intrusión de la casualidad, de las contingencias mundanas en segundo plano o en los márgenes, lo no previsto, lo impensado en la toma. Aquel irrumpir del azar que Roland Barthes llama el advenimiento, la marca cuya sola presencia le otorga a la imagen un nuevo espesor sensorial y memorístico, y con él el incentivo de otra lectura y comprensión de lo que, como perceptores activos, encontramos colándose hacia el afuera de las imágenes o dentro de ellas. Una suerte de advertencia (advenimiento y advertir son etimológicamente afines) de cuanto más lejos de lo esperado en el contexto saturado de signos puede arrancar la mirada.

El fotógrafo no ignora que en el campo de lo fotografiado no hay secuencialidad a la que pueda hacérsele justicia, sólo tránsitos, barreras, escollos, interferencias, un rascacielos, una valla, un muro, un viaducto, fronteras, umbrales infranqueables, horizontes, transeúntes ... altos en el camino. De ese desconsolado intento de superar lo imposible está hecho su oficio, de ahí surgen los fotomurales, los audiovisuales, los montajes, los libros desde Para verte mejor América Latina hasta Karakarakas. Si de soñar se tratara el ideal de la voluntad de representación del fotógrafo, y en particular de Paolo Gasparini, obstinado en su pulsión escópica por indagar y revelar en un sentido discursivo que bordea lo epistémico, las conexiones que revelan las imágenes, la agitación de lo indecible, la historia, la memoria, lo no obvio que presiona en ellas, y que querría y necesitaría mostrarse, sería el mundo entero instantánea y simultáneamente visible y presente, correlativa y extensivamente, en su anverso y su reverso, aquí y ahora, en toda su complejidad e inestabilidad, expuesto.

Del esfuerzo invertido en esa apuesta por involucrarse en lo real, el mismo que conforma el archivo de la vasta obra y el arco de vida de Paolo Gasparini, con sus sesenta años de trabajo, deriva en otro, sin duda menos quimérico que la apetencia por el mundo transpuesto en imagen. Que no sean las fotos las que miren por nosotros, sino que seamos nosotros quienes nos sintamos conjurados a mirar a través de ellas. Y ese mirar con y a través de ellas responde a una actitud ética antes que estética, en la medida en que pide un posicionamiento no separado de nuestra naturaleza de agentes de la experiencia y de la facultad de ver e imaginar.

Vayamos ahora hacia las concreciones fotográficas. Empecemos con las alegorías y sus visiones cruzadas. Abrimos con una sonrisa dentífrica, la misma sonrisa con la que nos reencontraremos al final. Sobre el fondo de un monumento de Venecia, se recorta el afiche de un hombre que ruge con su pelo hirsuto y sus garras de león, blandiendo un compact disc, justo al lado otro león, éste también en afiche, en la inmediata contigüidad de otra sonrisa dentífrica. Las sonrisas de dientes perfectos se repiten, las encontraremos aquí y allá, enteras o demediadas, las mismas u otras, sonrisas que centellean hasta el guiño irónico, algunas armadas de cámaras, como la de la fotografía número 8, con a su izquierda el ataúd de un entierro indígena en la sierra boliviana. En la fotografía número 32, un león enjaulado en el puerto fluvial de Manaos, éste plausiblemente real, captado en vivo. Vamos de los afiches, los carteles, las paredes publicitarias, a los simulacros, todas aquellas apariencias que quieren hacerse valer como lo que no son, es decir los maniquíes, maniquíes recostados de las paredes, en los escaparates, adoptando varias y sugerentes posiciones felinas, eróticas, maniquíes como artísticas formas humanas colgando en una galería. Muñecas, muñecos, uno cómodamente sentado en un sofá, bustos, manos cortadas, en contraste con los cuerpos vivos en movimiento.

Entre lo que es real y lo que no es real, apenas reflejo de pasaje, lugar de refracción y quiebre de las imágenes, cómo nos cuesta orientarnos: nada se nos da de antemano. Muchas reproducciones de reproducciones, cámaras de todo tipo, desde la Kodak Instamatic en la valla publicitaria hasta las antediluvianas cajas negras con cortinita y agujero, fotógrafos profesionales, fotógrafos espontáneos, fotógrafos ambulantes, como el que se dispone a componer la pose de su cliente en una calle del Puno.

El afiche de una familia feliz (fotografía número 22) se continúa en las indígenas que llevan, y bien que les deben pesar, a sus hijos en la espalda. En la número 13, una familia reunida, madre y siete hijos, el tronco demediado del padre preside: todos miran a la cámara, serios, no sonríen. La familia feliz sonríe, las mujeres sonríen, incluso hay un maniquí que le sonríe al hombre que la mira sonreído, pero el mundo no es bello y sigue su marcha a lo largo de los encuadres del fotógrafo.

On the other side of images

Cities are big populated built up spaces, tracked and seen from above, let's say from a bird's eye view, they expand endlessly, topographically boundless areas though always on a human scale, depending on how high the camera is placed or how fast it moves, each and every junction could be a route's starting point, through sudden inroads or disruptions of what was left behind and yet still remains, through inflection, deflection and oversight of its surroundings. Thus each and every junction, whether looked at straight or seen through the camera assisting the eye and following it as the observer moves along, reveals a mesh of sliced up fragments, some near some far, leaning up or downwards, yet all reunited in space and time under the rule of reality, the only real thing a camera lens can actually seize and leave as a mark on film.

TRUST
JESUS
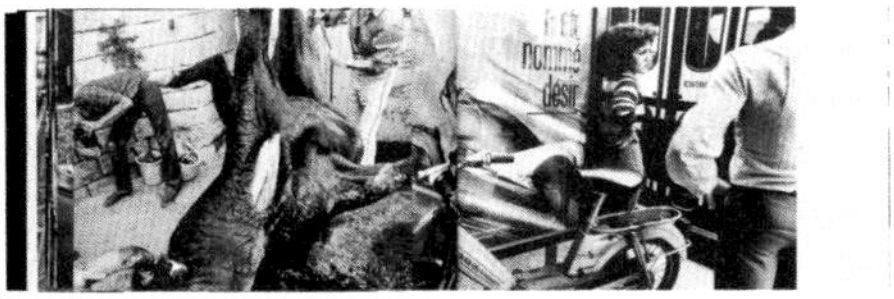

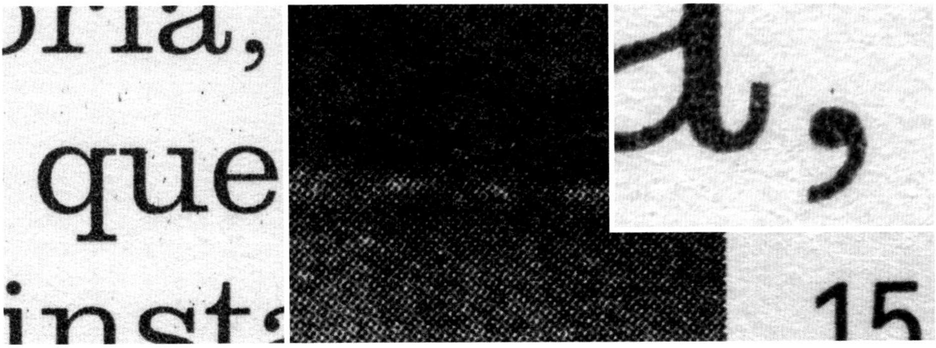
que
15

2016 Ha4

这是瑞士摄影师弗朗索瓦·沙尔（François Schaer）拍摄的各地滑雪场的影集。摄影师通过白天滑雪场大面积白色的场景中少量的人和物，构建出抽象的画面，也更能凸显出雪景里各色服装、房屋、设备的缤纷色彩。本书为锁线硬精装，封面白色哑面艺术纸印黑，书名字母压凹，裱覆卡板。内页前后文字部分为较薄的胶版纸印黑，前后页呈现朦胧的叠透效果，作品页哑面艺术纸四色印刷。为了体现滑雪场的白茫茫一片的抽象感，照片注释被放置在书的最后部分。书脊处使用了鲜红色的堵头布，与整体白色抽象中星星点点显眼的“小色块”形成呼应。这部摄影作品作为当代社会景观的认可，展现出滑雪场的壮观景象和赖以为生的人们，同时也谴责了对雪山资源的过度开发。

This is an album of photos of ski resorts taken by Swiss photographer Francois Schaer. Through a small number of people and objects in a large white scene in the ski resort during the day, the photographer constructs abstract pictures, which can highlight the various colors of clothes, houses and equipment in the snow scene. This hardback book is bound with thread sewing. The cover is made of white matte art paper printed in black, with the lettering title. The text is printed on offset paper in black while the photographs are printed on matte art paper in four colors. The photo annotation is placed in the last part of the book in order to reflect the endless whiteness of the ski resorts. The bright red head band is used to echo the conspicuous ‘small colorful block’ in integral whiteness. As recognition of contemporary social landscape, this photograph album not only displays the spectacular scenery of ski resorts and people who depend on them for their livelihood, but also condemns the over-exploitation of snow mountain resources.

- 荣誉奖
- 白昼
- 德国

- Honorary Appreciation
- Jours Blancs
- Germany
- Nanni Goebel, Kehrer Design Heidelberg
- François Schaer (Photographs), Pauline Martin (Text)
- Kehrer Verlag Heidelberg, Berlin
- 330×300×14mm
- 1058g
- 100p

At first glance, the large-format cover shows an abstract landscape, produced as if in a Zen trance using a pointed paintbrush. But don’t be fooled. At second glance, if you try a bit harder, you can recognise the photographic source of a winter landscape. The capital letters spelling out the title are imprinted in white like tracks in the snow. The book is an extensive collection of photographs which pay homage to the skier’s territory in visual minimalism. They use the white as a background provided by the days when a supernatural light saturates everything – snow-covered piste and sky – like white air. How do you print all these many gradients of white using four colours? It is astonishing that a raster of 70 l/cm is sufficient to sustain these fine differences in tone values on the matt, absorbent board as well. Have they used higher-pigmented printing inks, perhaps? The plates are occasionally interspersed with blank pages – with nothing on them – whenever there is a need for some breathing space. You are even tempted to spot printed tracks of weightless brightness on those pages as well. The short legends are fortunately not positioned next to the pictures. They are added at the end on thinner paper instead, as a synopsis next to a stamp-size reproduction of each picture. In this part, the reserved yet congenial book design uses the translucency of the thin paper; it prolongs the experience of spatial ambiguity that stages the photographs so magnificently.

B R
O U I
L L
E R

L E S
P I
S T
E S

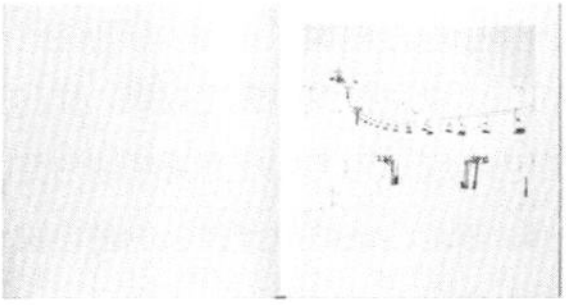

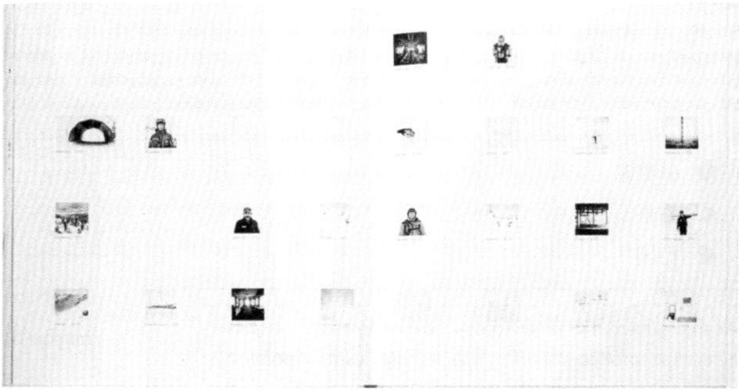

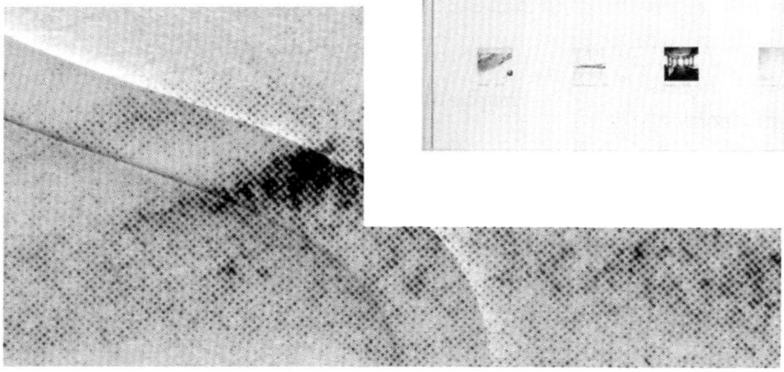

ine Martin for th

这不是一本普通的艺术作品集，除了书中展示的照片和雕塑外，整本书也记录了艺术家创作背后偏执而心酸的故事。作者在旅行过程中遇到了艺术家亨利，亨利大量拍摄自己妻子的各种姿势的照片，包括大量裸体照片 5000 多张；他痴迷于拍摄这些隐私，最终导致妻子的离去。随后艺术家自己也过上隐居的生活，将照片归类，重构拼贴，创作出大量基于人体的抽象小雕塑。作者在得到授权后，重新编辑并出版了这些照片、归类资料和雕塑作品。本书的装订方式为锁线胶平装，封面白色哑面艺术纸四色印刷，抽象而隐晦。内页照片部分亮面铜版纸四色印刷，大量标签拼贴在书中，展现艺术家对这些照片的归类。其余部分则胶版纸四色印刷，呈现创作背景、照片的拼贴解构、雕塑作品等。全书大量照片拼贴、标签混用，配合上夹页和插页，将这一创作过程中隐秘的内容呈现出来。书名“Taking off”除有脱掉之意，表明裸体照片外，也含有“逃脱”的双关语义，暗指艺术家妻子的离开。

Taking Off

…/…

This is not an ordinary collection of artistic works. In addition to the pictures and sculptures displayed in the book, the book records the paranoid stories behind the artist's works. During their journey, the authors met an artist Henry, who took a large number of photos of his wife, including more than 5,000 naked photos. He was obsessed with shooting personal privacy, which eventually led to his wife's departure. Then the artist himself lived in seclusion. He classified and reconstructed the photographs, then created lots of abstract sculptures of human body. After gaining authorization, the authors re-edited and published these photos, classified materials and sculptures. The book is perfect bound with thread sewing. The cover is made of white matte art paper printed in four colors, abstract and obscure. The interior photographs are printed in four colors on gloss art paper. A large number of labels are collaged in the book to show the artist's classification of these photographs. The rest part of the book is printed on offset paper in four colors, presenting the creation background, collage and deconstruction of photographs, sculpture works and so on. 'Taking off', the title of the book, not only implies that there are naked photographs in the book, bot also hints the departure of the artist's wife.

- 荣誉奖
- 脱掉——我的邻居亨利
- 荷兰

- Honorary Appreciation
- Taking off. Henry my neighbor
- The Netherlands
- Mariken Wessels, Jurgen Maelfeyt
- Mariken Wessels
- Art Paper Editions, Gent
- 330×242×25mm
- 1617g
- 326p

This book tells the story of Martha and Henry. In pictures. Pictures of youth. A wedding. Then endless series of poses showing Martha undressed. Martha leaves Henry. Organic collages consisting of fragments of the nude photos. Surreal little clay figures. Animal traps in the woods. Were the nude photos ever intended for the public eye? Everything seems private. The conservative setting. The model's absent-minded expression. Rarely smiling. Hanging baskets with plastic squirrel. Henry's tireless production, collection and categorisation of nude poses with bizarre listing. The catalogue stages the material like an artist's legacy. The observer is left to experience Henry's enigmatic project without detachment. The photos are more reminiscent of mammograms than erotic images. The son writes a letter to his mother – a facsimile is inserted on an empty double page. The past splits into fragments. If you patiently leaf through all this madness, you reach the white, grotesque, recumbent body parts – like hermaphrodites, or torsos in antiquity. Abstract body forms are created. Independent objects, released from Henry's manic preoccupation with his fetish, his search for form. The book finishes – before the epilogue and publishing details – with a list of works for the body sculptures. This publication serves as recognition of Martha's courage. And it is a late recognition of Henry's journey in pursuit of art.

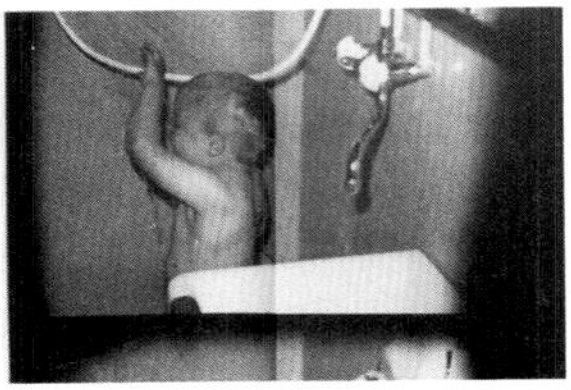

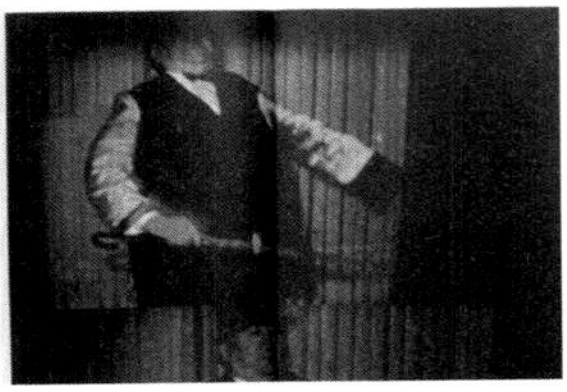

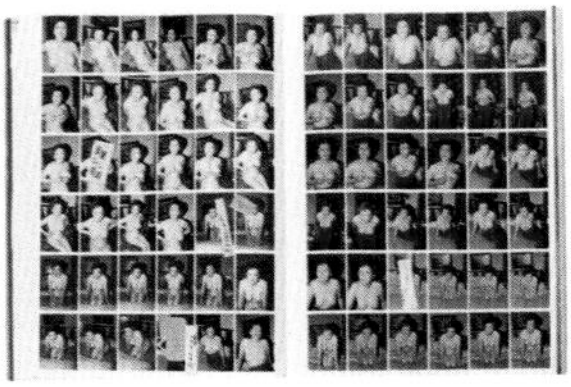

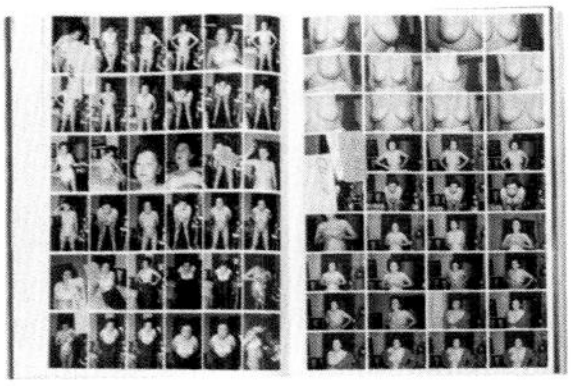

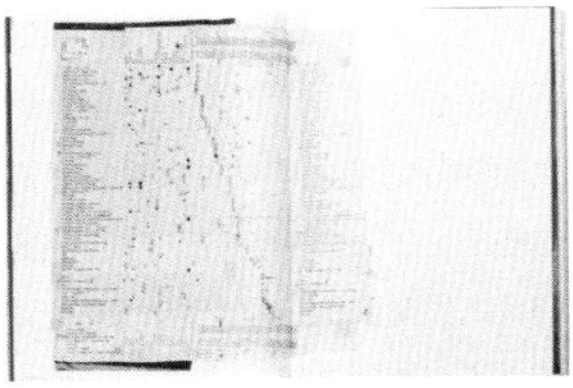

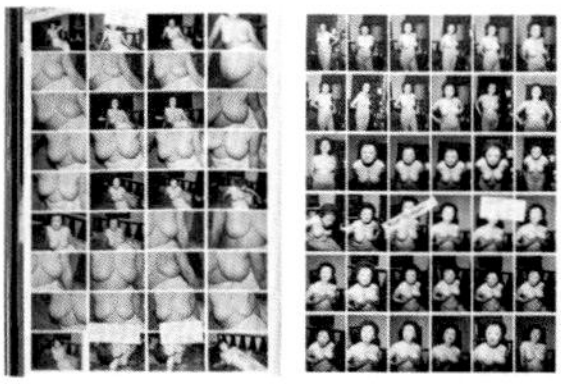

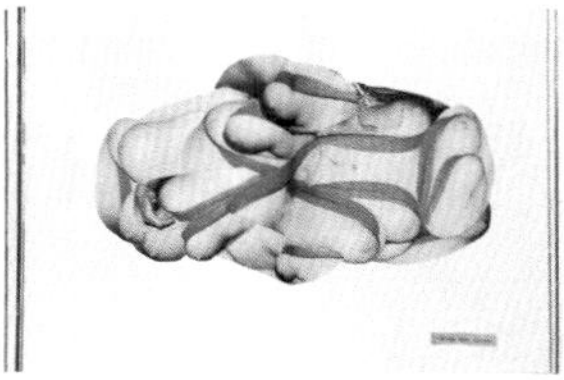
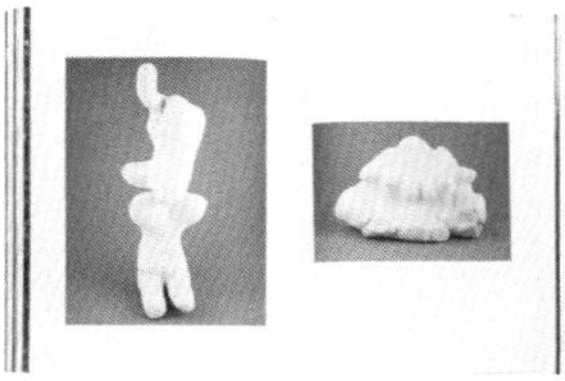

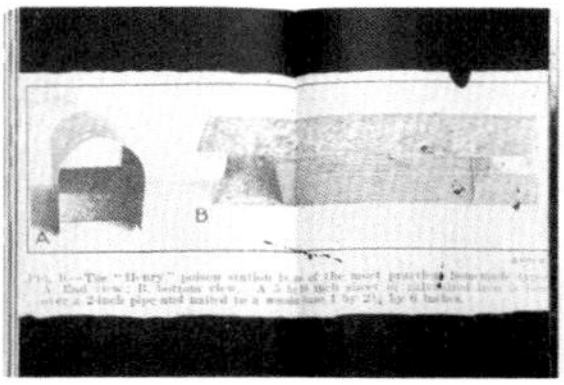

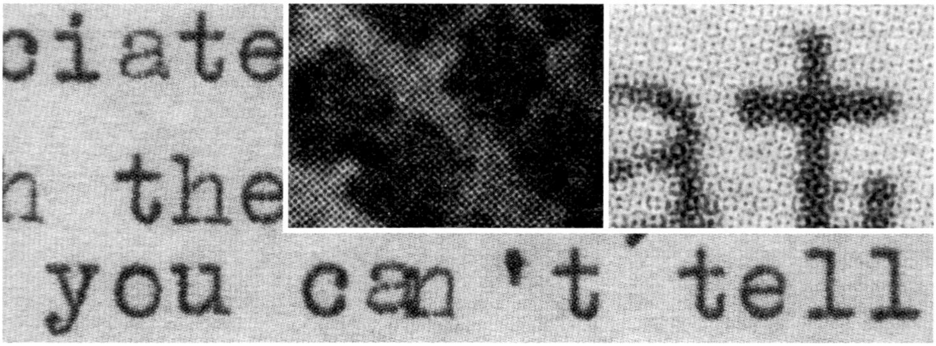
ciate
n the
you can't tell

- Untitled (September Magazine) Paul Elliman, 2003
- Fundamentalists and Other Arab Modernisms. Architecture from the Arab World 1914–2014
- Danish Artists' Books / Danske Kunstnerbøger
- Sequester
- Motion Silhouette
- Miklós Klaus Rózsa
- Paul Chan. New New Testament
- Éléments Structure 01
- New Horizons
- Cartea de vizită
- More Than Two (Let It Make Itself)
- Art Handing In Oblivion (fink twice 501) / Albert's Guesthouse (fink twice 502) / Walking through Baghdad with a Buster Keaton Face (fink twice 503)
- SERIES The Seto Library. Seto Kirävara
- Aliens and Herons

2015

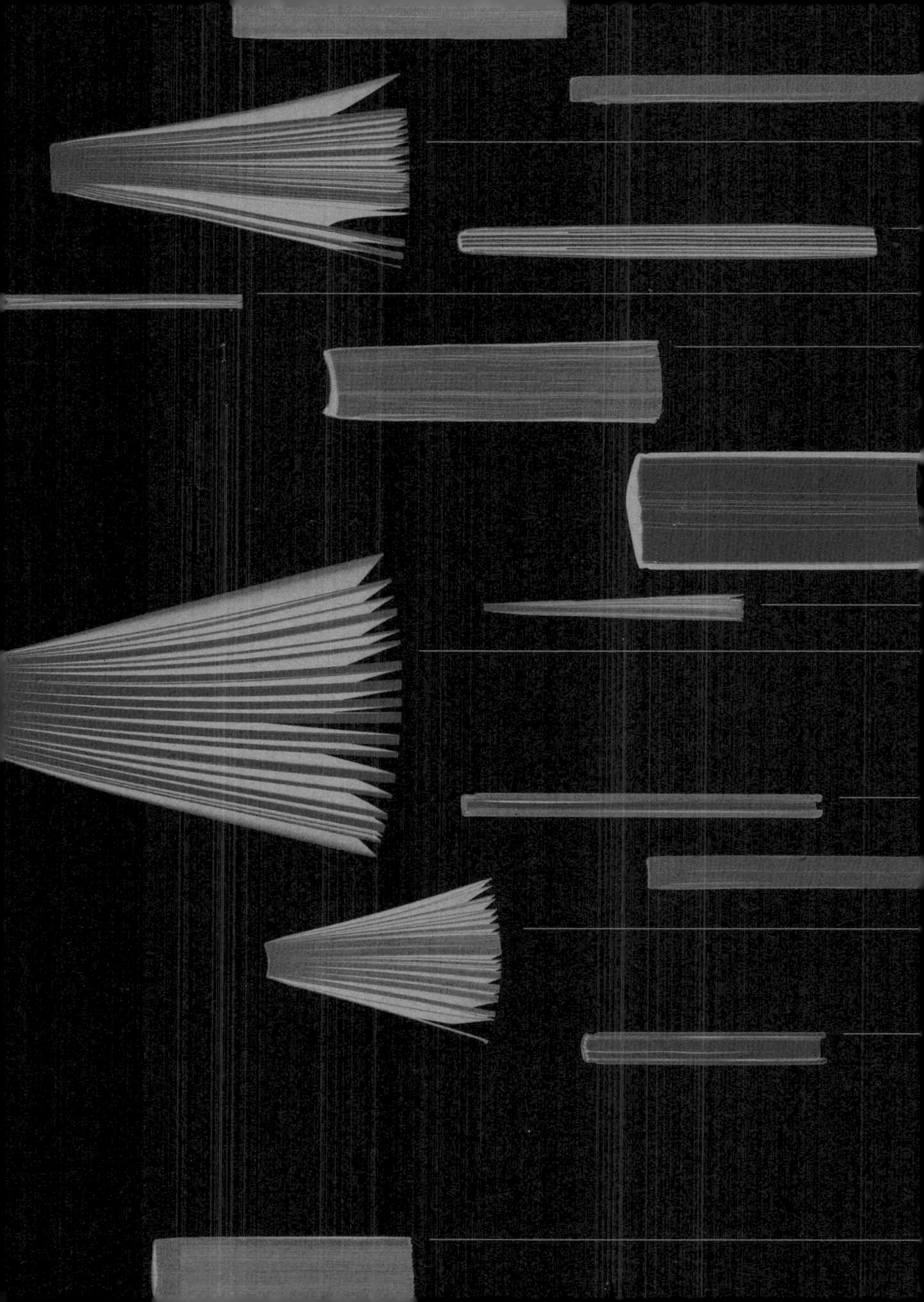

2015 Gl 0475

2015 Gm 0483

2015 Sm[1] 0489

2015 Sm[2] 0495

2015 Bm[1] 0501

2015 Bm[2] 0507

2015 Bm[3] 0513

2015 Bm[4] 0519

2015 Bm[5] 0525

2015 Ha[1] 0531

2015 Ha[2] 0535

2015 Ha[3] 0539

2015 Ha[4] 0543

2015 Ha[5] 0547

Danske Kunstnerbøger
Danish Artists' Books
Miklós Klaus Rózsa
NEW HORIZONS
2012
501 Rob van Leijsen
502 Petra Elena Köhler / Nicolas Vermot-Petit-Outhenin
503 Thomas Galler

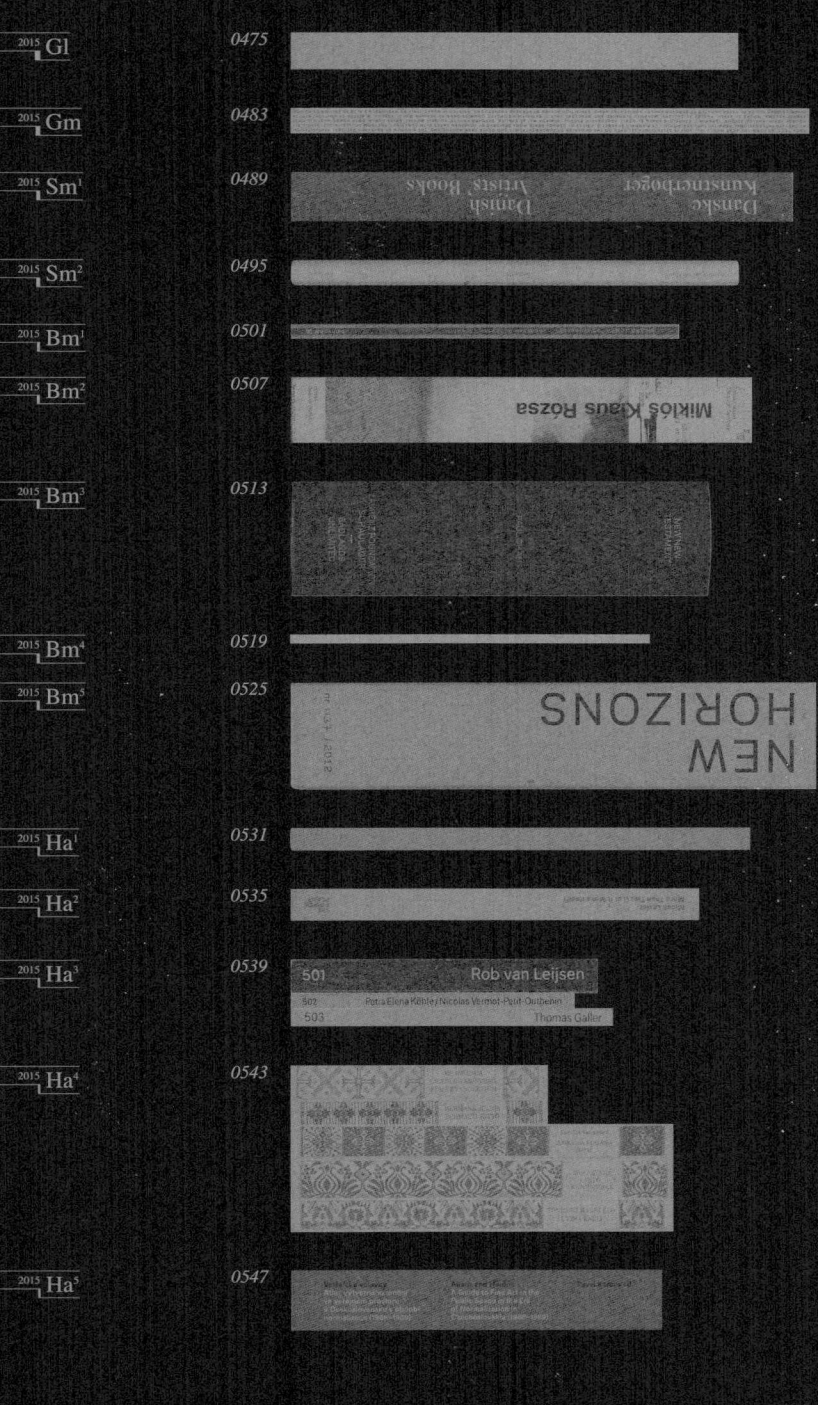

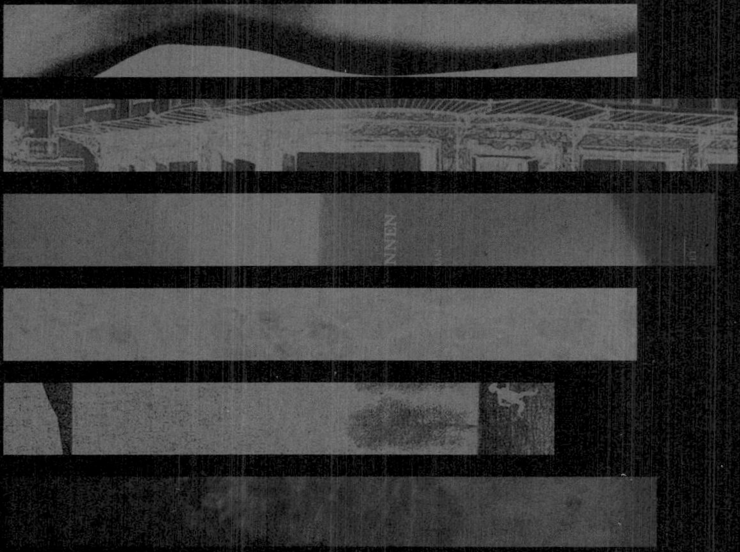

When the history of magic
is definitively written,

9 789491 843051

المساكن تبيّن أن
يد منذ العام ١٩٥٢
، إذ تكمن جذورها

عقود القرن العشرين الوسيطة مجرّد اشتقاق من أعمال أوروبية، وجمالياتها ليست
تقليدًا أعمى. ولم يزَ كريم أن تاريخ العمارة هو تعاقب أساليب، بل رصد تطوّر العمارة
أولًا من خلال قراءة مادية للتاريخ، حيث تساهم التقنيات ومواد البناء، بالإضافة إلى

perimented with: Anna Fro Vodder's unique objects, including the paper-object *Novelty Shape Biscuit* (2004) and the artist's group A Kassen's water-damaged *Damaged by Water. Financed by Insurance*

Awoiska van der Mo

せんろは　とおくのほしまで　つづいてる

Chugga-chugga, choo-choo! The ground shakes.

30 - 40 Personen der "Heimkampagne" in einzelnen Gruppen
marschiert und sofort in den Innenhof der Anstalt eingedr
Diese Demonstranten haben sich mit den Zöglingen unterhal

Warum war am Anfang ein Wort? Wo waren die Sterne, die V
die Nacht? Die Berge oder das Mondlicht? Hat jemand die
kommen sehen? War es lang und schwierig oder kurz und dire

ou l'inclinaison des lames,
l'espace entre celles-ci varie et
agit donc sur l'ombre et la lumière.

les horiz
bâtiment
paysage.

HORIZONS

În zilele noastre, cultura populară ne oferă câțiva piloni de sprijin în
tanței cărții de vizită. Pornind de la competiția furibundă a bancher
Psycho", privind calitatea hârtiei, cernelei sau eleganța fontului și p

221 artworks and objects made by 101

Looting art is one of the most hidden and underestimated objectives of war. Behind conquering campaigns lurk the stories of deprived cultural heritage and devastation of the enemy's cultural identity. Looting art in wartime or political

Kuninga handśa-ajaja – Kooseri Rein
Kuninga käsitüümeistri – Reidla Angela

included over five hundred examples of art w
public spaces through out the period of Nor

2015 Gl

Golden Letter

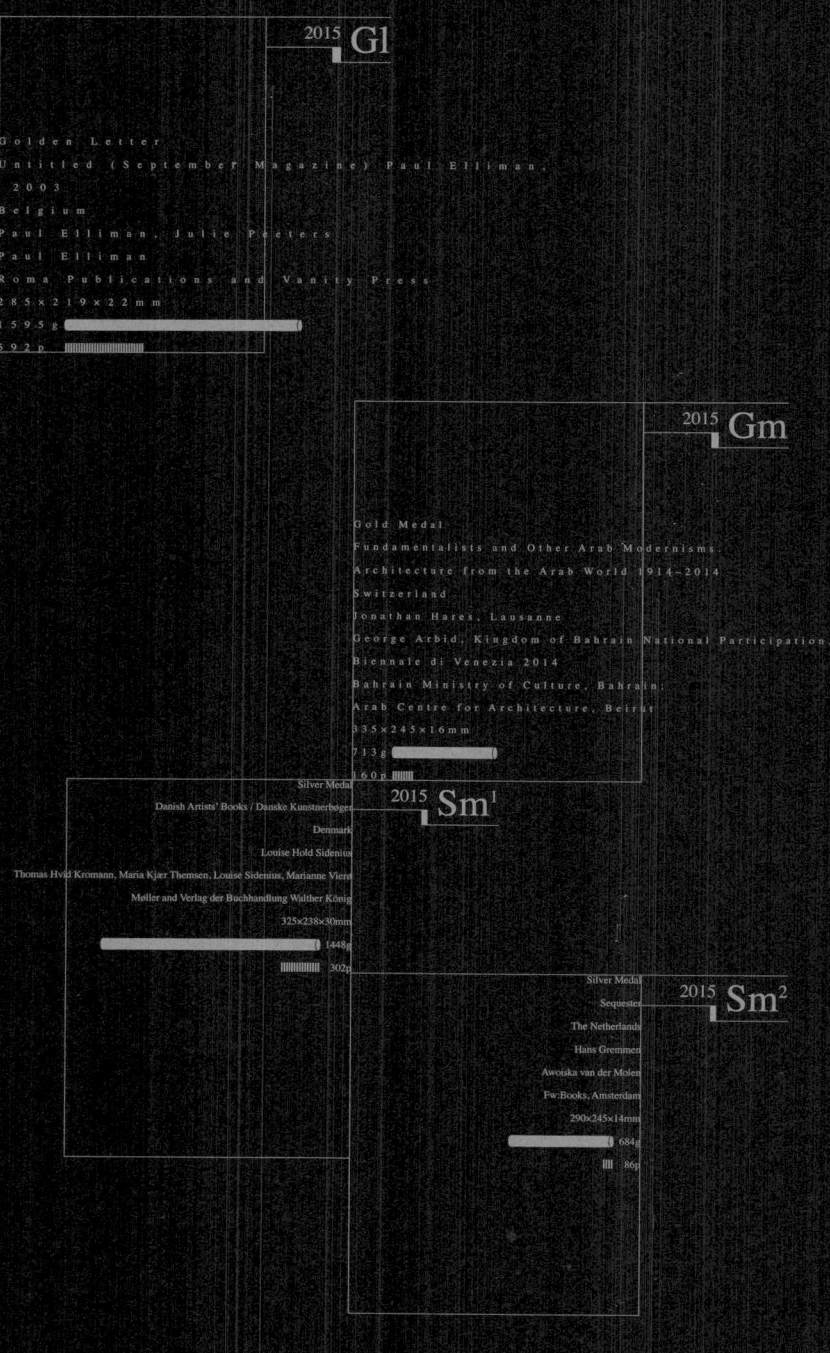

Untitled (September Magazine) Paul Elliman, 2003

Belgium

Paul Elliman, Julie Peeters

Paul Elliman

Roma Publications and Vanity Press

285×219×22mm

1595g

592p

2015 Gm

Gold Medal

Fundamentalists and Other Arab Modernisms. Architecture from the Arab World 1914–2014

Switzerland

Jonathan Hares, Lausanne

George Arbid, Kingdom of Bahrain National Participation, Biennale di Venezia 2014

Bahrain Ministry of Culture, Bahrain; Arab Centre for Architecture, Beirut

335×245×16mm

713g

160p

2015 Sm[1]

Silver Medal

Danish Artists' Books / Danske Kunstnerbøger

Denmark

Louise Hold Sidenius

Thomas Hvid Kromann, Maria Kjær Themsen, Louise Sidenius, Marianne Viero

Møller and Verlag der Buchhandlung Walther König

325×238×30mm

1448g

302p

2015 Sm[2]

Silver Medal

Sequester

The Netherlands

Hans Gremmen

Awoiska van der Molen

Fw:Books, Amsterdam

290×245×14mm

684g

86p

2015 Ha[1]

Honorary Appreciation
Cartea de vizită
Romania
72+87
Fabrik-72+87
Fabrik
297×235×14mm
680g
64p

2015 Ha[3]

Honorary Appreciation
Art Handing In Oblivion (fink twice 501) / Albert's Guesthouse (fink twice 502) / Walking through Baghdad with a Buster Keaton Face (fink twice 503)
Switzerland
Rob van Leijsen, Georg Rutishauser, Sonja Zagermann, Petra Elena Köhle, Nicolas Vermot-Petit-Outhenin, Thomas Galler
Rob van Leijsen / Petra Elena Köhle, Nicolas Vermot-Petit-Outhenin / Thomas Galler
edition fink, Verlag für zeitgenössische Kunst, Zürich
200×145×21mm / 190×130×9mm / 210×168×10mm
366g / 134g / 212g
382p / 158p / 174p

2015 Ha[2]

Honorary Appreciation
More Than Two (Let It Make Itself)
Canada
Jeff Khonsary (The Future)
Micah Lexier
The Power Plant
264×195×20mm
877g
224p

2015 Ha[4]

Honorary Appreciation
SERIES The Seto Library. Seto Kiravara
Estonia
Agnes Ratas
various
Seto Instituut
166×114×23mm / 166×114×16mm / 247×166×23mm / 247×166×27mm / 247×166×22mm / etc.
262g / 181g / 580g / 1038g / 522g / etc.
288p / 176p / 256p / 396p / 200p / etc.

2015 Ha[5]

Honorary Appreciation
Aliens and Herons
Czech Republic
Tereza Hejmová
Pavel Karous (Ed.)
Arbor vitae, Academy of Arts, Architecture and Design, Prague
240×167×38mm
814g
459p

2015 Bm[1]

Bronze Medal
Motion Silhouette
Japan
Megumi KAJIWARA, Tatsuhiko NIIJIMA
Megumi KAJIWARA, Tatsuhiko NIIJIMA
Megumi KAJIWARA, Tatsuhiko NIIJIMA
250×149×8mm
190g
12p

2015 Bm[2]

Bronze Medal
Miklós Klaus Rózsa
Germany
Christof Nüssli, Christoph Oeschger
Christof Nüssli, Christoph Oeschger
Spector Books / cpress, Leipzig
298×211×43mm
2140g
624p

2015 Bm[3]

Bronze Medal
Paul Chan, New New Testament
Switzerland
Kloepfer-Ramsey Studio, Brooklyn
Laurenz Foundation, Schaulager, Basel;
Badlands Unlimited, New York
Laurenz Foundation, Schaulager, Basel
272×182×74mm
2838g
1092p

2015 Bm[5]

Bronze Medal
New Horizons
The Netherlands
Rob van Hoesel
Bruno van den Elshout
The Eriskay Connection, Breda
340×235×68mm
3993g
212p

2015 Bm[4]

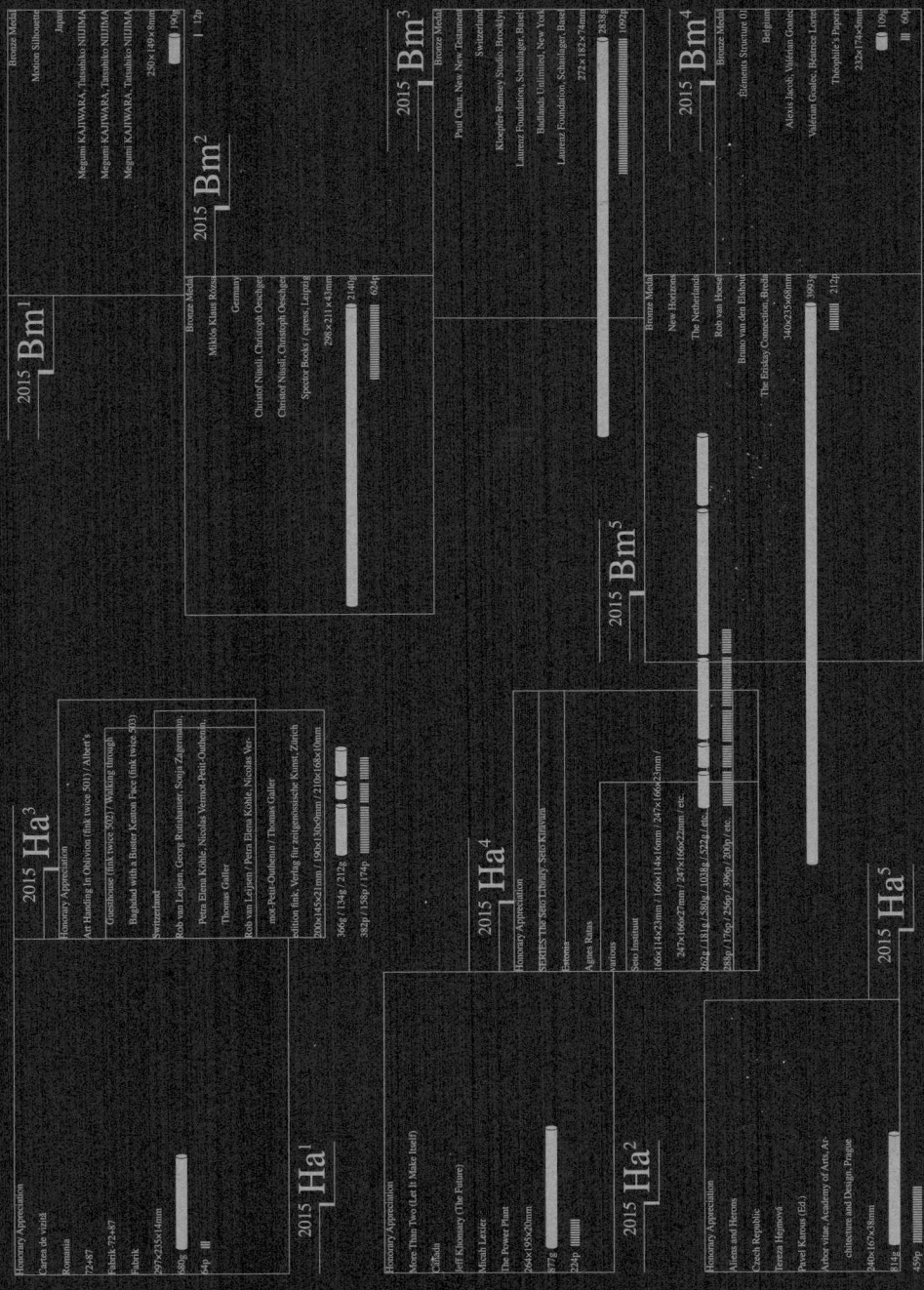

Bronze Medal
Éléments Structure 01
Belgium
Alexis Jacob, Valérian Goalec
Valérian Goalec, Béatrice Lortet
Théophile's Papers
232×174×5mm
109g
60p

2015

The International Jury

Peter Willberg (England, Germany)

Vanessa van Dam (The Netherlands)

Lars Fuhre (Sweden)

Ivan Alexandrov (Russia)

Valeria Bonin (Switzerland)

Anouk Pennel (Canada)

Demian Bern (Germany, France)

Country / Region

Switzerland 3

Belgium 2

The Netherlands 2

Denmark 1

Japan 1

Germany 1

Romania 1

Canada 1

Estonia 1

Czech Republic 1

Designer

Paul Elliman, Julie Peeters

Jonathan Hares, Lausanne

Louise Hold Sidenius

Hans Gremmen

Megumi KAJIWARA, Tatsuhiko NIIJIMA

Christof Nüssli, Christoph Oeschger

Kloepfer-Ramsey Studio, Brooklyn

Alexis Jacob, Valérian Goalec

Rob van Hoesel

72+87

Jeff Khonsary (The Future)

Rob van Leijsen, Georg Rutishauser, Sonja Zagermann, Petra Elena Köhle, Nicolas Vermot-Petit-Outhenin, Thomas Galler

Agnes Ratas

Tereza Hejmová

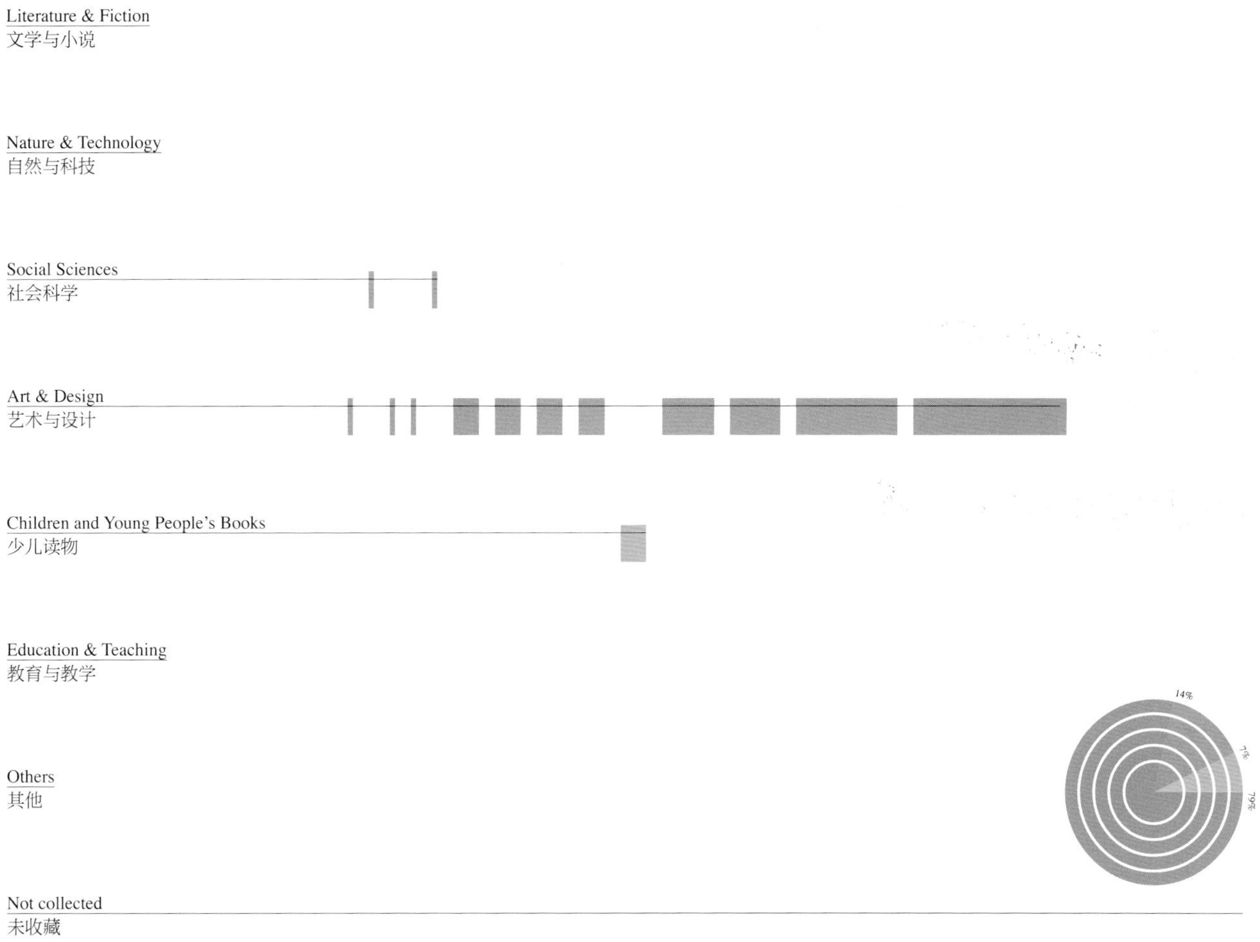
Literature & Fiction
文学与小说
Nature & Technology
自然与科技
Social Sciences
社会科学
Art & Design
艺术与设计
Children and Young People's Books
少儿读物
Education & Teaching
教育与教学
Others
其他
Not collected
未收藏
14%
7%
79%

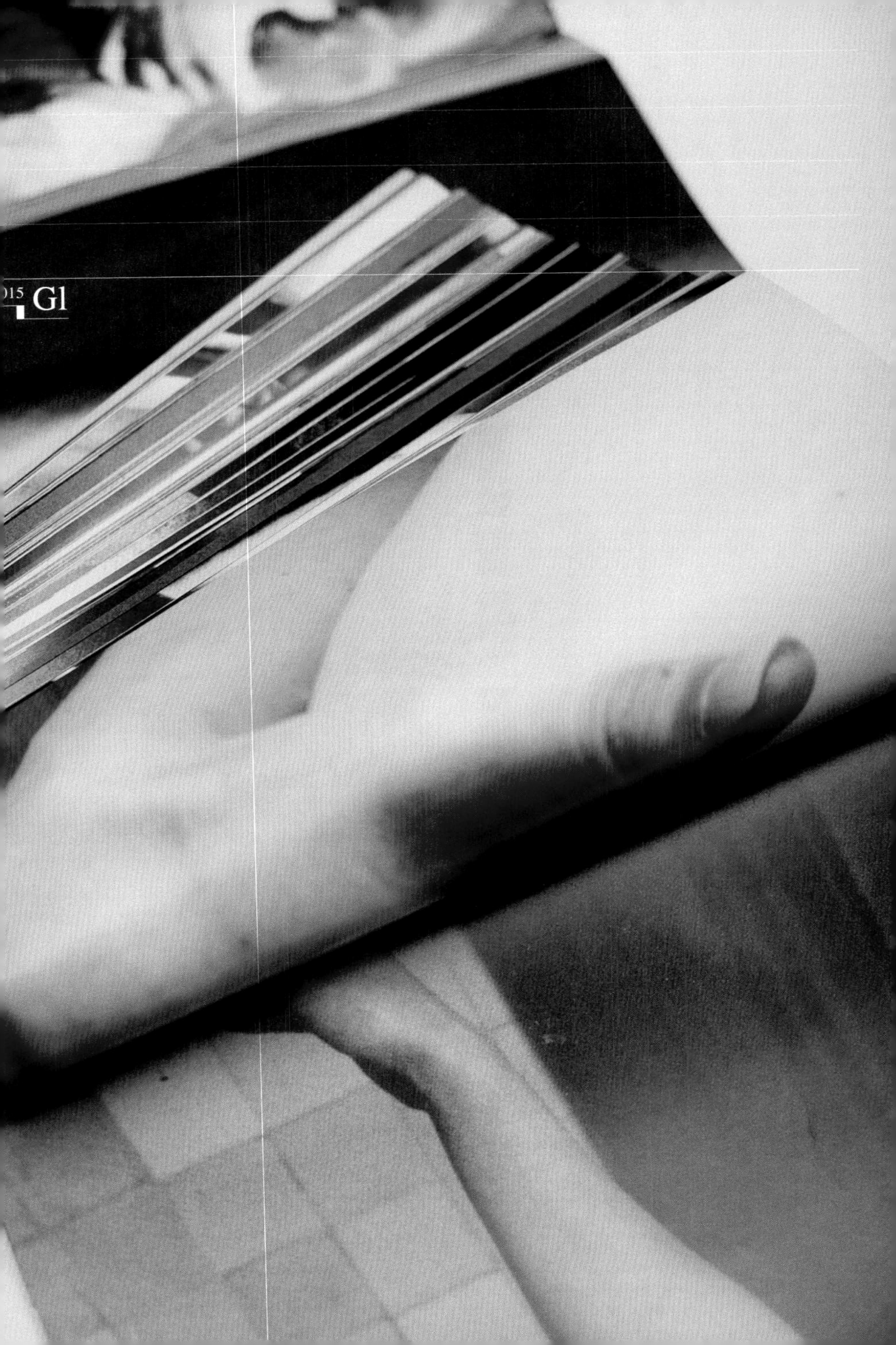

这是一本由英国艺术家兼设计师保罗 · 艾利曼（Paul Elliman）收集、整理并设计的时尚画集。艾利曼长期探索技术与语言之间的相互影响，尝试找到排版和人声之间的关系，将多种声音通过转换为语言编码的形式记录下来。在这本书中，他收集了超过 500 张各式杂志里的图片。他的关注点在这些图片里人们的手部，通过局部裁切和重新编排的方式，展示这些手部的形态、姿势与衣着的关系，以及它们的指向和目标。在翻阅本书的时候，读者可以试图去理解这些手势，并寻找在连续页面之间这些视觉元素如何吸引我们，如何在这本没有文字的书中引导视觉和思考的走向。本书装订形式为无线胶装，封面光面铜版纸四色印刷，表面覆光膜。内页光面铜版纸四色印刷，有如时尚杂志的做法，局部有横向尺寸短 10mm 的拉页。从图片呈现放大的四色网点和较大的马赛克等肌理可以看出大致的图片来源。

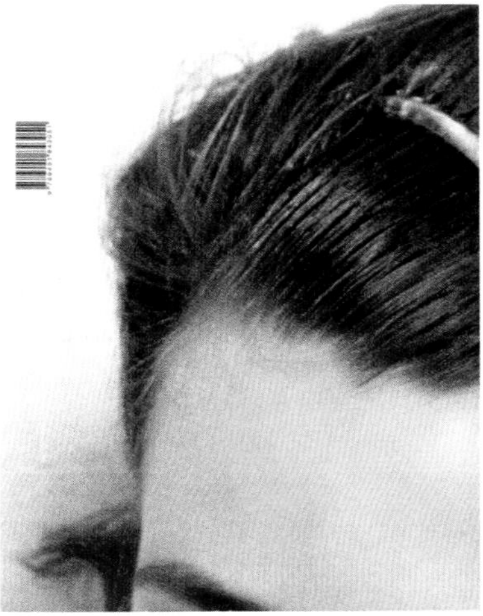

This is a collection of fashion paintings collected, organized and designed by British artist and designer Paul Elliman. Elliman has explored the interaction between technology and language for a long time. He recorded multiple voices by converting them into language coding and tried to find the relationship between typesetting and human voice. In this book, he collects more than 500 pictures in various magazines. His focus is hands in these pictures. Through local cutting and rearrangement, he shows the shape, posture, relationship with clothes, and their targets oriented. When reading this book, readers may try to understand these gestures, and find out how these visual elements on successive pages attract us, and how to guide the direction of vision and thinking in this wordless book. This book is perfect bound. The laminated cover is made of gloss art paper printed in four colors, just like fashion magazines. From the enlarged four-color dots and relatively large mosaic texture, we can see that the pictures are scanned from other publications.

金字符奖
无题（九月杂志）保罗 · 艾利曼，2003
比利时

Golden Letter
Untitled (September Magazine) Paul Elliman, 2003
Belgium
Paul Elliman, Julie Peeters
Paul Elliman
Roma Publications and Vanity Press
285×219×22mm
1595g
592p

The glamour world can't get enough of its own self-reflecting glossy magazines. Celebrity style and self-staging all over the place. And now another 500 fat pages are added. A magazine with a captivating cover: a lady's mop of hair, cut below the left eyebrow and adorned with a cute hairclip braves the keen audience. And what is the name of this new organ of the glittering partisans? 9789491843051. Simply a barcode on the cover page. And what does the broad spine say? Nothing. The back cover? Fluttering laundry in Yves Klein blue. Well, let's have a look inside. Bare of any text, the pages lead us into the iconography of exhibitionist wellbeing. This is a course book on the art of playing with our viewing habits, our unconscious expectations. Beautiful naked feet, cool Prince of Wales check pleats, fragmentary gestures – hands, where one has to look where they point, what they are gripping. Declaring the art of this book as a parody would be too simple. It penetrates the visual code in which we are trapped. This code frustrates our yearnings. Quite funny, if screen dots are reproduced by screen dots: beaten at their own game.

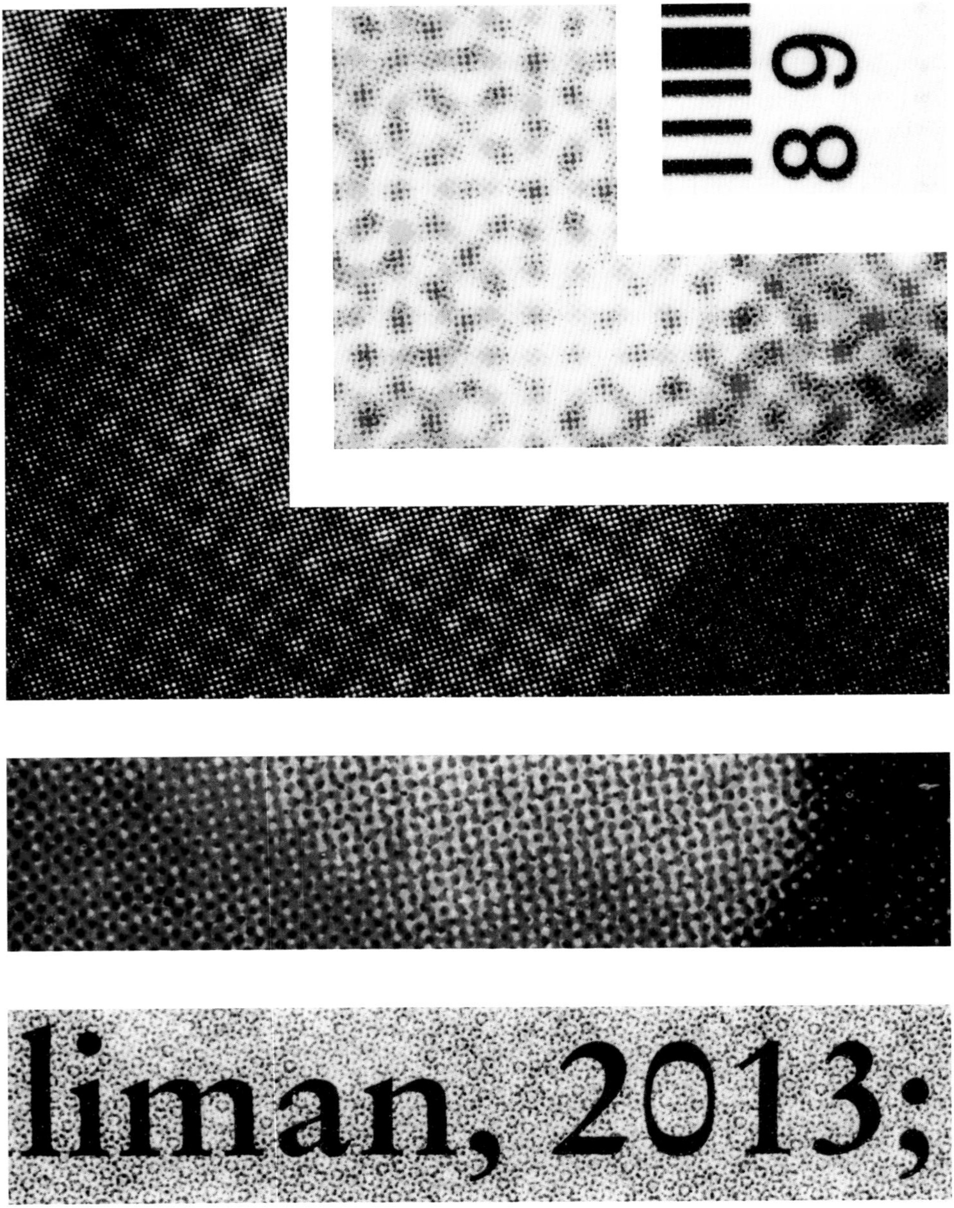
liman, 2013;

1595 285×219×22

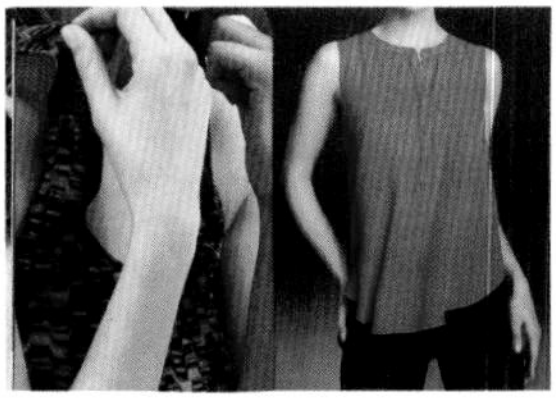

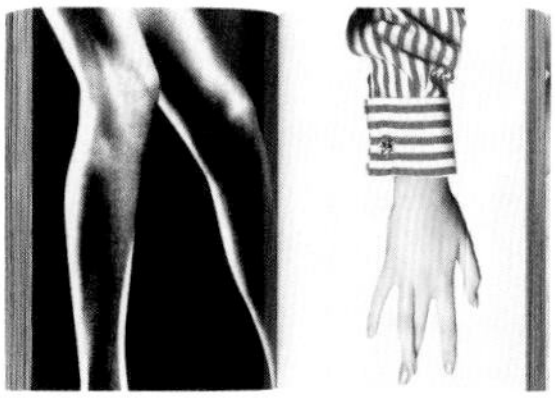
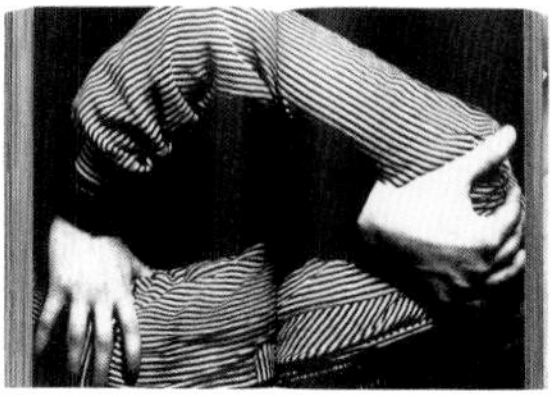

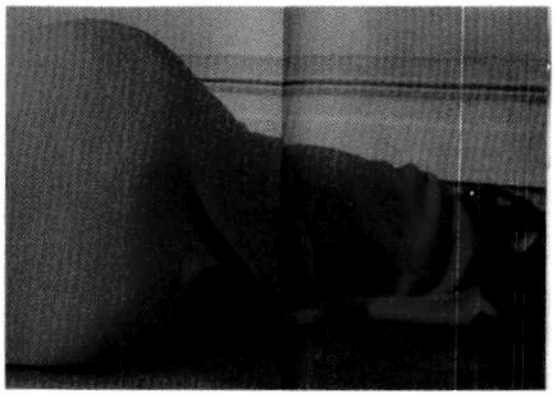
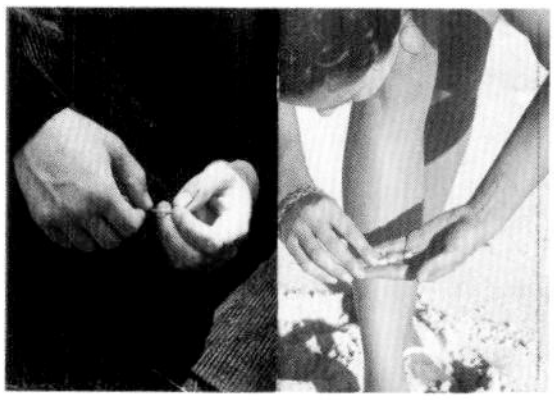
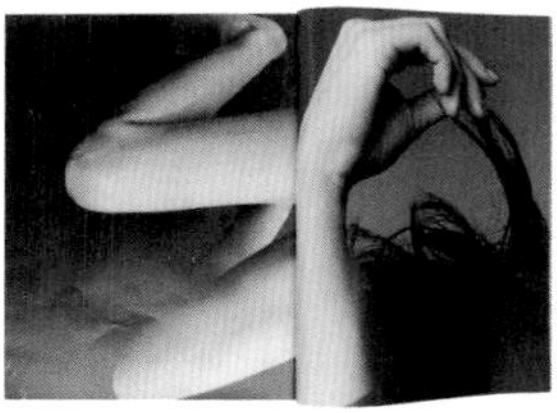

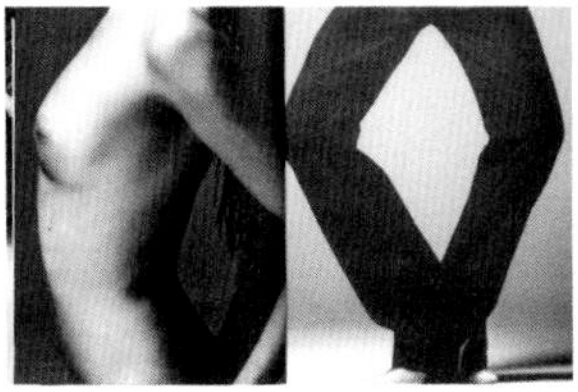
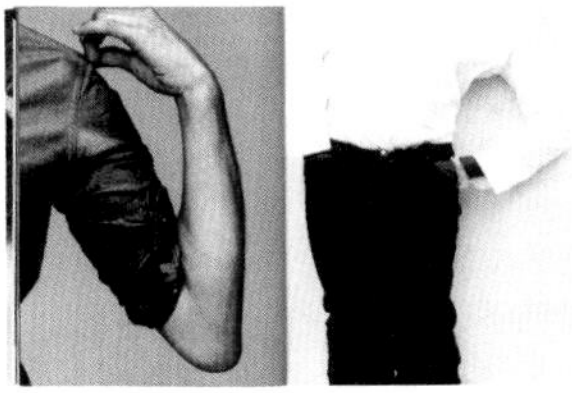
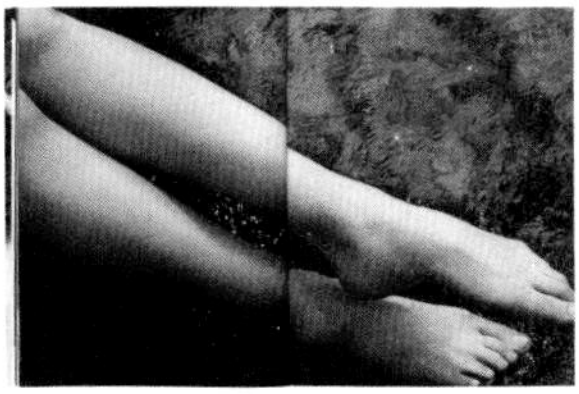
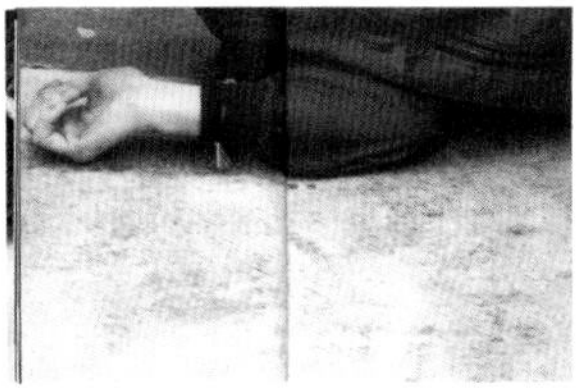

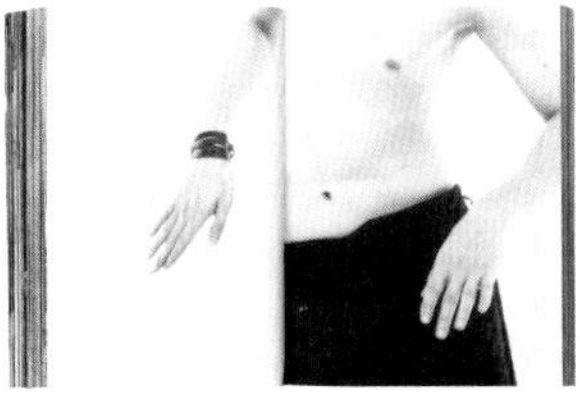

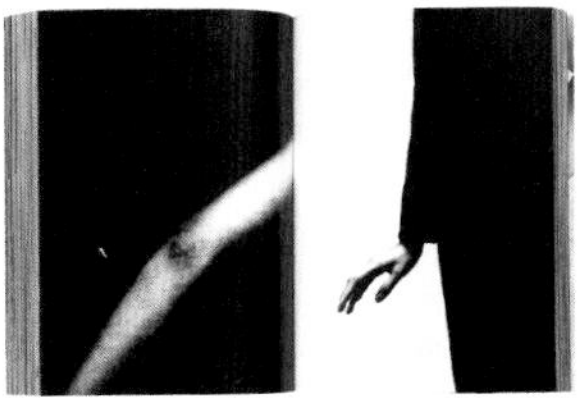
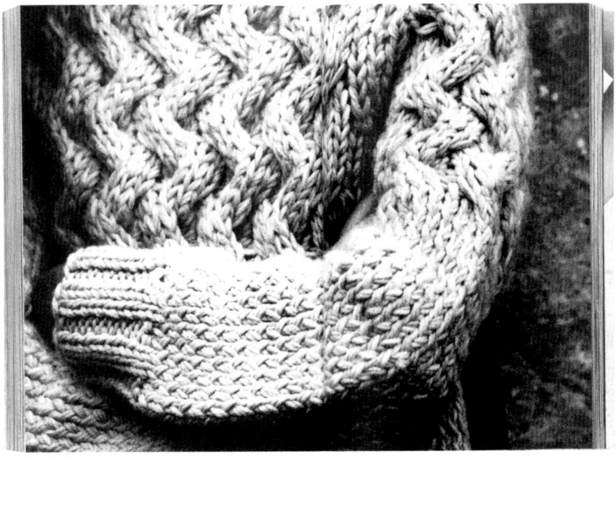

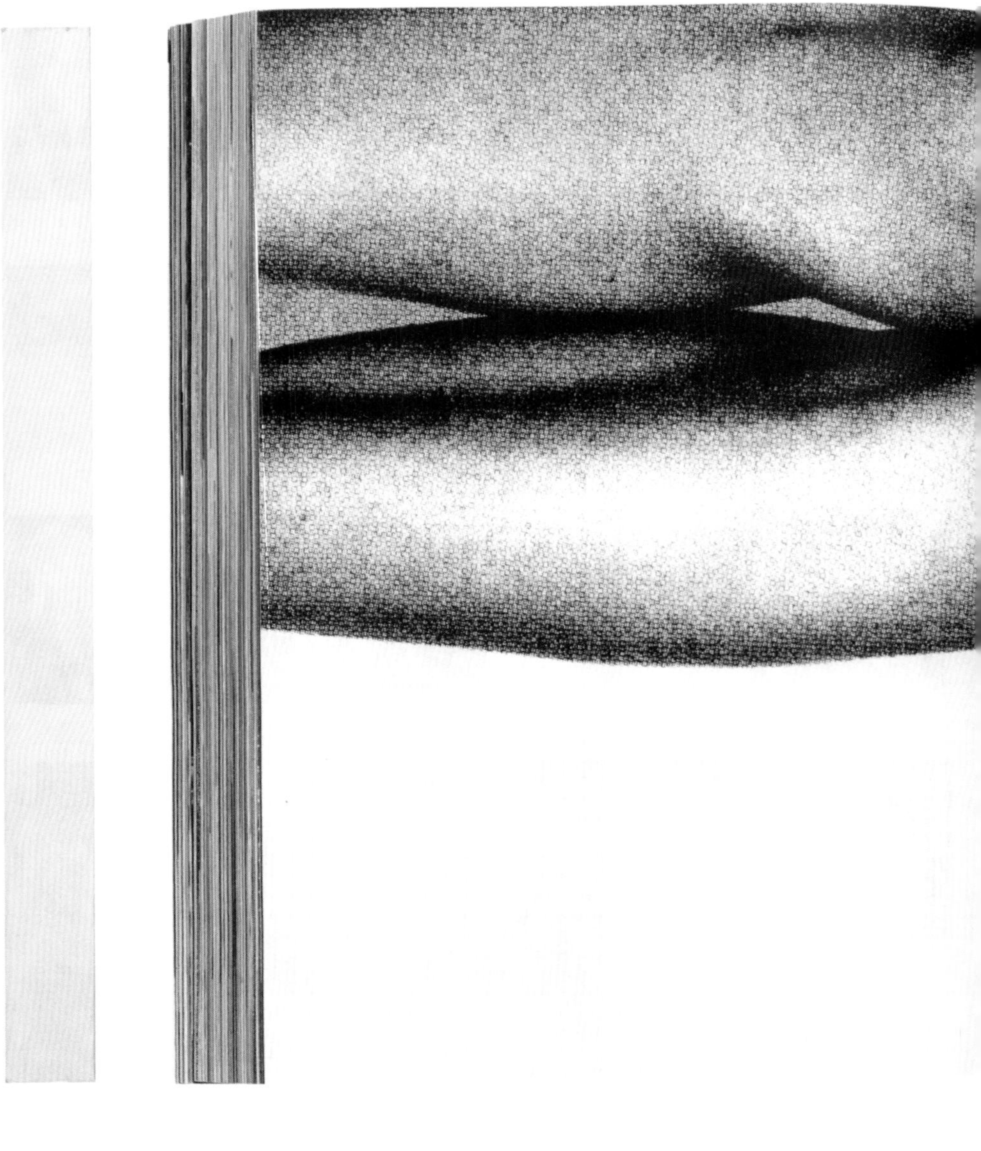

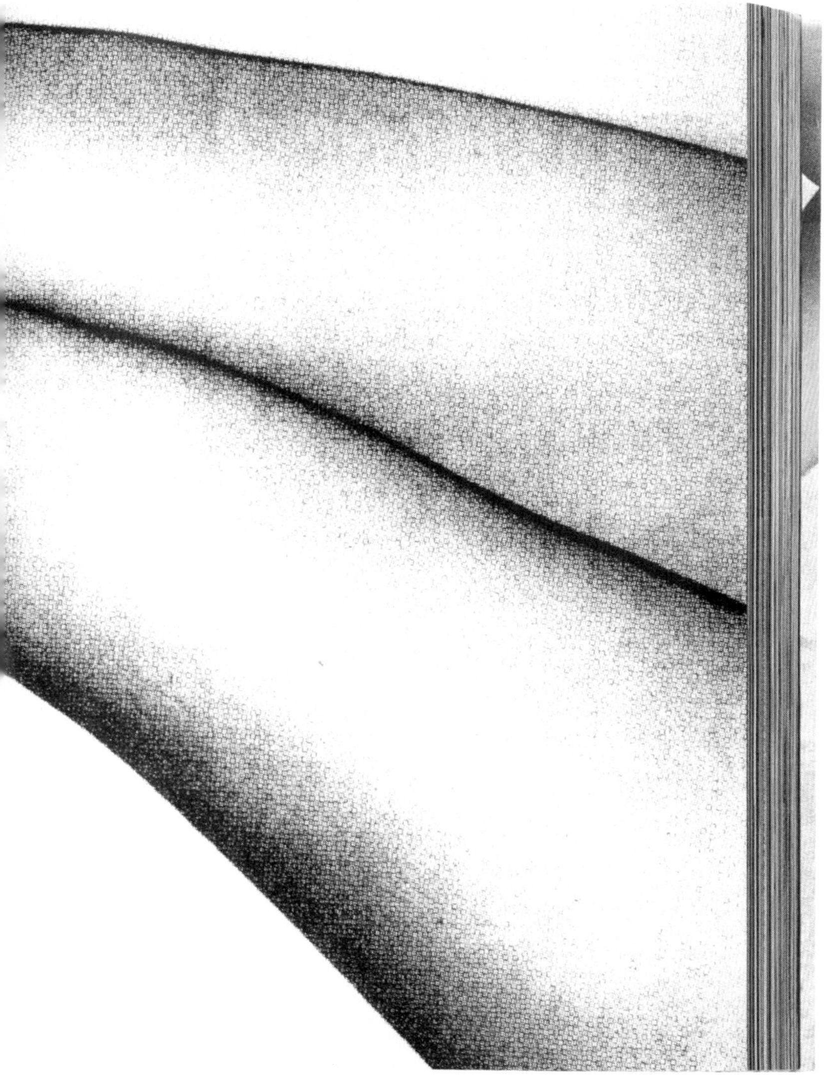

الأردن
قطر
مصر
وادي الجمال، مصر
العربيّة المتّحدة
لبنان

2014 年的第 14 届威尼斯建筑双年展的一个重要主题是"吸收现代性"，即各参展国家或地区展示各自在全球化浪潮下建筑文化的改变，包括现代性如何转化为标准化，各地风格对现代性的再造与融合。本书为巴林王国展区有关阿拉伯地区建筑过去 100 年（1914-2014）历史的展览配套书籍。书籍的装订形式为无线胶平装，封面白卡纸印单黑，内页胶版纸以黑色为主（极少量彩图），在照片和建筑图纸下印有浅黄色交代图片的边界。全书英文和阿拉伯文混排，根据两种文字的书写习惯，分别按照左对齐和右对齐进行排版对照，甚至页码部分也采用两种文字的数字进行标示。书籍的主要部分为 100 年间建造于阿拉伯地区的建筑照片、平立剖面图、草图，按照时间顺序进行排版，版面灵活自由。英文和阿拉伯文目录分列书籍图录部分的前后，并在书籍的最后部分按照国家和地区的划分，介绍了这 100 年间各地政治经济和物质文化方面的发展与改变。||||||||||||||||||||||||||||||||||||||||||||||||||||||||||||||||||||||||||||||||||||||||||||||||||||||||||||||||||||||||

An important theme of the 14th Venice Biennale of Architecture in 2014 is 'Absorbing Modernity'. That is, each participant shows its own changes in architectural culture under the tide of globalization, including how modernity can be transformed into standardization, and how local styles can reconstruct and integrate modernity. This book is a supporting book for the exhibition on the history of Architecture from the Arab World over the past 100 years (1914-2014) in the Kingdom of Bahrain. The paperback book is perfect bound and the cover is printed on white cardboard in black. The inside pages are made of offset paper printed in black (with few colorful images), and there are light yellow boundaries printed under the photographs and architectural drawings. The book is typeset in both English and Arabic. According to the writing habits of the two languages, they are aligned to the left side and right side respectively. Even the page numbers are marked in two languages. The main part of the book contains photographs, flat profiles and sketches of architectures built in the Arab region in the past 100 years. They are typeset in chronological order with flexible layout. In the last part of the book, the economic development and political changes in different places during the past 100 years are introduced according to different countries and regions.

金奖

原教旨主义者和其他阿拉伯现代主义——来自阿拉伯世界的建筑 1914-2014

瑞士

Gold Medal

Fundamentalists and Other Arab Modernisms. Architecture from the Arab World 1914–2014

Switzerland

Jonathan Hares, Lausanne

George Arbid, Kingdom of Bahrain National Participation, Biennale di Venezia 2014

Bahrain Ministry of Culture, Bahrain; Arab Centre for Architecture, Beirut

335×245×16mm

713g

160p

In its pavilion at the 14th International Architecture Biennale, the Kingdom of Bahrain puts a book centre stage. It contains a selection of 100 buildings signifying committed modernity in the Arab world. The format of the small tome certainly has a representative air. Material and design, on the other hand, are characterized by a generous restraint. The matte, slightly yellowish cardboard cover with reduced typography has something sandy and simultaneously bright about it. The smooth matte paper inside the book is white, the black and white photographs and plans are framed by the yellowish tinge of the cover page. Rarely is any of the buildings represented by more than three images on one page: in a very freely interpreted picture arrangement, the white of the paper playfully surrounds the printed areas. The lettering of the project pages restricts itself to the naming of objects and picture captions without any formulated text. These few, finely nuanced contrasts are sufficient for the book designer to add spin to the relationship between the depicted architectures of the north-western world and Arab culture - at least as far as visual experience goes.

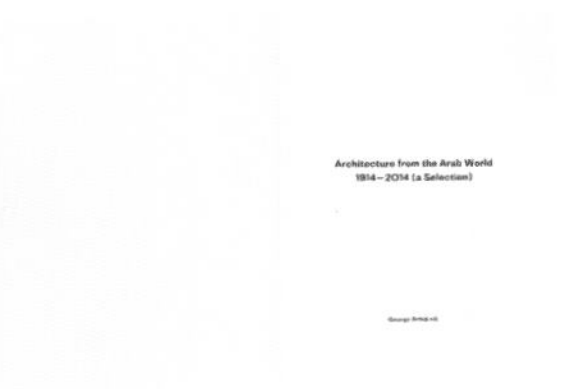
Architecture from the Arab World
1914–2014 (a Selection)

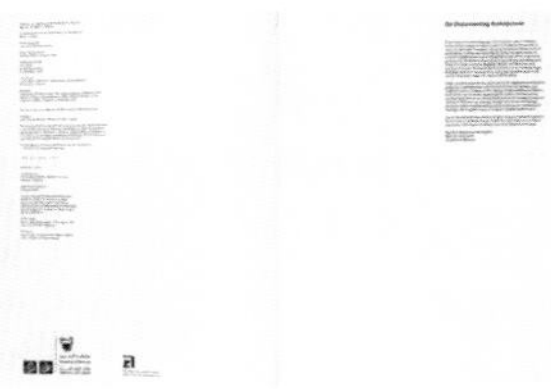

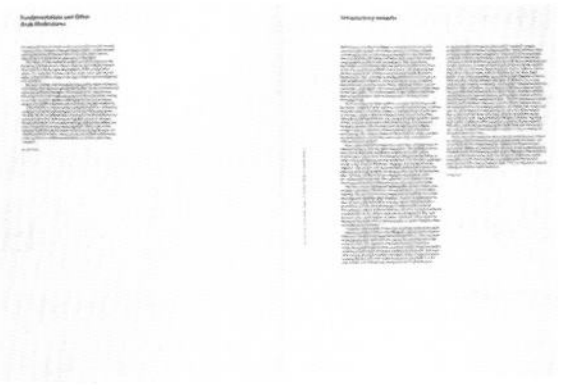

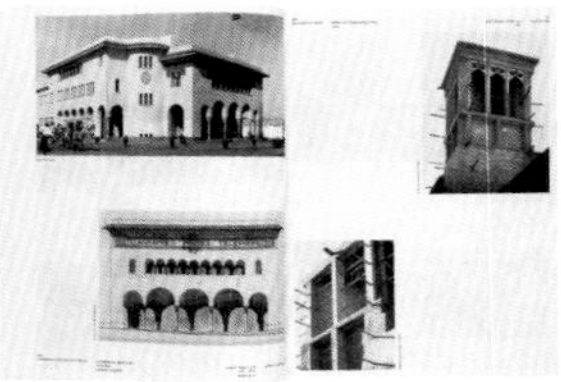

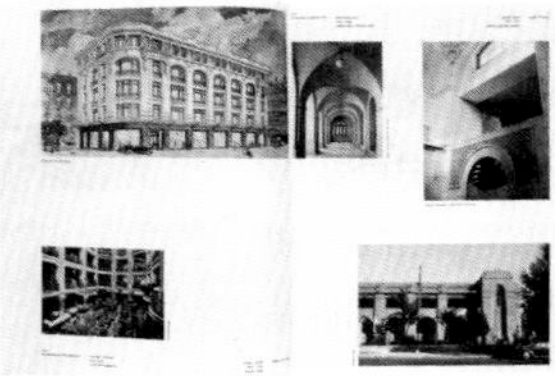

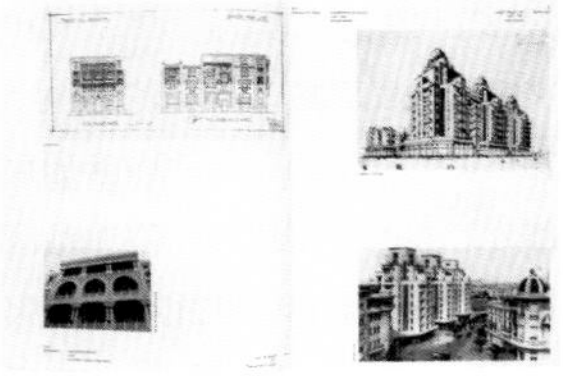

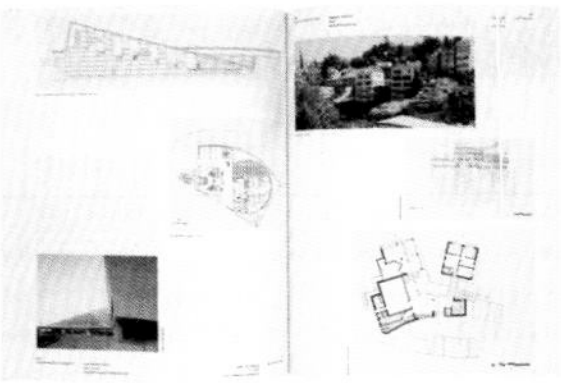

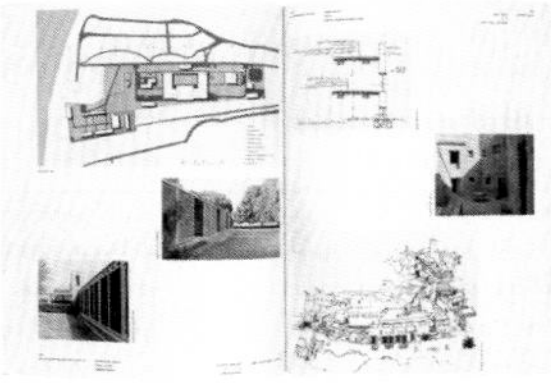

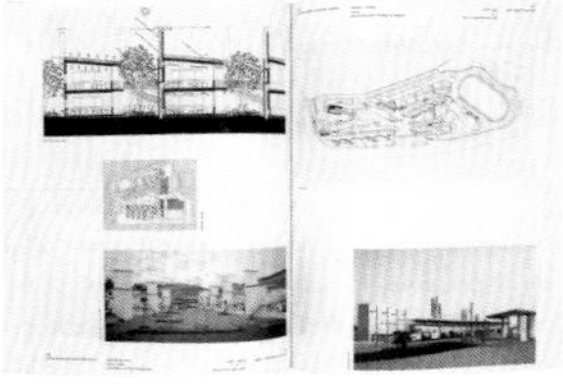

Arab Maghreb
A Century of Architecture in the Maghreb, a Retrospective, 1914–2014
Morocco
Abderrahim Kassou
Algeria
Bouasad Aiche

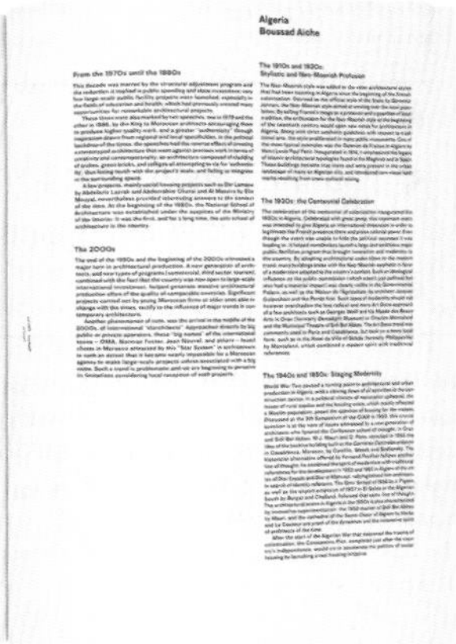

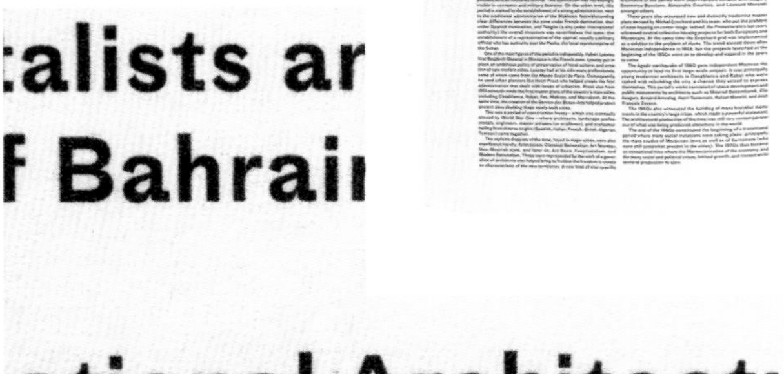
talists ar
f Bahrai
ational Architect

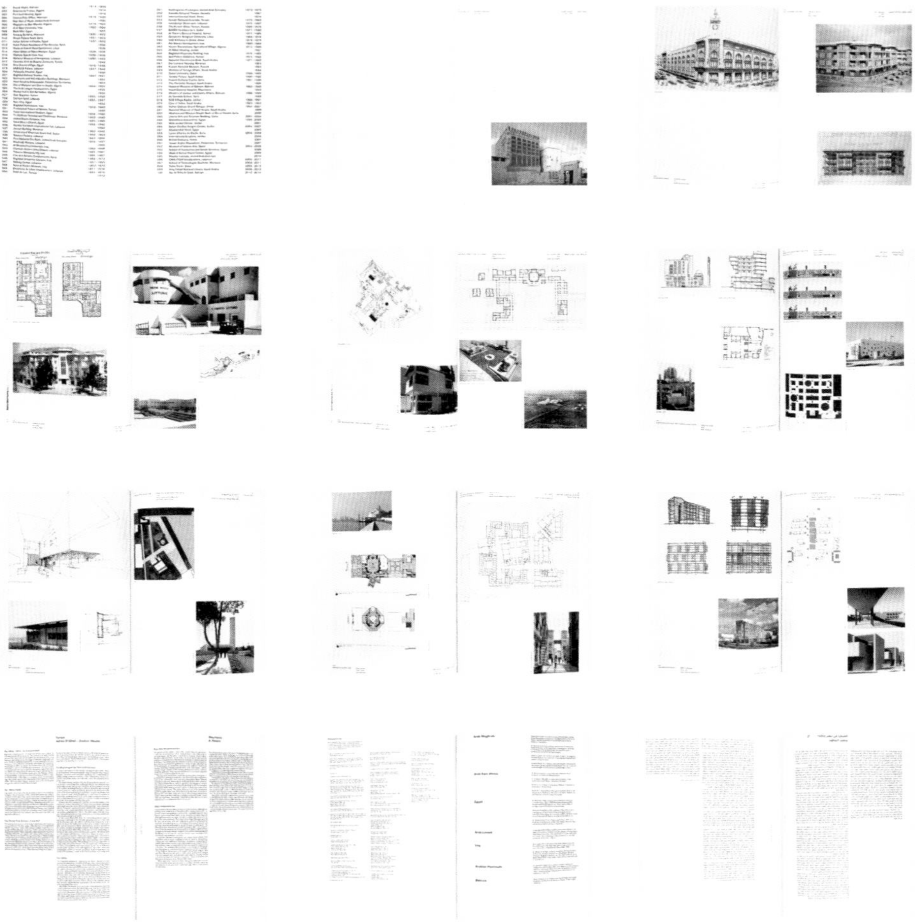

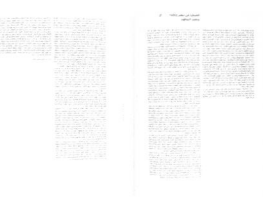

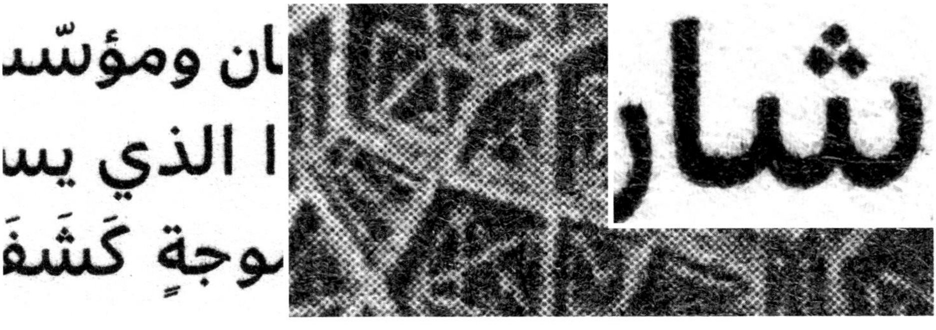
ان ومؤسّس
ا الذي يس
وجةِ كَشَفَ
شار

ARCHITECTURE FORM THE ARAB WORLD 1914-2014 (A SELECTION) PUBLISHED ON THE OCCASION OF FUNDAMENTALISTS AND OTHER ARAB MODERNISMS, THE KINGDOM OF BAHRAIN'S NATIONAL PARTICIPATION AT THE 14TH INTERNATIONAL ARCHITECTURE EXHIBITION – LA BIENNALE DI VENEZIA. EDITED BY GEORGE ARBID. BOOK DESIGNED BY JONATHAN HARES WITH CONTRIBUTIONS FROM BOUSSAD AICHE, RASHID ALI, GEORGE ARBID, KHALED ASFOUR, IBRAHIM Z. BAHRELDIN ADNEN EL GHALI, ABDERRAHIM KASSOU, SHAIKHA MAI AL KHALIFA, BERNARD KHOURY, ZOUBEIR MOUHLI, AMIRA OSMAN, OMER OSMAN, CECILIA PIERI, ASHRAF SALAMA, NOURA AL SAYEH, KHALED AL-SULTANY, MOHAMED ELSHAHED AND MERCEDES VOLAIT. EXHIBITION CONCEPT COMMISSIONED BY SHAIKHA MAI AL KHALIFA, MINISTER OF CULTURE, DEPUTY COMMISSIONER NOURA AL SAYEH CURATED BY BERNARD KHOURY AND GEORGE ARBID. EXHIBITION DESIGN BY BERNARD KHOURY / DW5. PRINTED IN QUART, ITALY BY MUSUMECI S.P.A. IN AN EDITION OF 40.000

Kazim Kenan in Ghazieh (1963-1966). The Beirut Airport, the Cité Sportive, the Casino du Liban, and the Presidential Palace, are among many government sponsored projects testifying to the authorities' ambition to lend the country a modern face. The Ministry of Defense by Wogenscky and Hindié (1962-1968), the Tripoli Fair started in 1962 by Oscar Niemeyer, and the Electricité du Liban headquarters by CETA group (1965-1972), are probably the best representatives of public architecture of the time.

When preoccupation with regional identity emerged in the mid-1960s, local materials such as sandstone were rediscovered and spaces like courtyards reappeared. Assem Salam's buildings of that period include the Serail in Sidon, and the Khashokji Mosque (1965) recognizable with its sandstone walls topped with a faceted roof in reinforced concrete – a modern interpretation of the dome.

1970's and 1980's

A few exceptions aside, such as the daring Interdesign showroom by Khalil Khoury, the architecture of the 1970's seems to have developed into a mainstream, literal application of building ordinance. Linear balconies wrapped around dull buildings were more a product of exploiting the total area permitted in building codes than from sound design thinking. The experimentation of the earlier period was abandoned for soulless architecture.

1990's, Reconstruction and beyond

Construction continued throughout the wars that ravaged Lebanon between 1975 and 1990, yet some degree of rupture with the past was inevitable. The intermittent periods of unrest that shook the country caused significant destruction. Because the core of Beirut was the scene of fierce battles, new neighborhoods developed overnight in the suburbs. Uncontrolled expansion reached zones hitherto relatively protected from urbanization. The result was the detrimental loss of balance between the urban and the rural as the fabric of rural zones was invaded by the urban building type often incongruously imposed on sloping sites.

After the war, reconstruction and urban redevelopment raised the complex issue of identity. What appeared to have been appropriately resolved in the cultural heyday of the 1950's and 1960's emerged once more, unresolved this time, intensified both by the delayed postmodern wave and nostalgia for a Lebanon that was felt to have disappeared. Two tendencies emerged: exacerbated high-tech as a vain attempt to catch up with the world, and the recourse to pastiche that merely testifies to the loss of tradition. Fortunately, other approaches have appeared, devoid of superficial statements, displaying a decisive character that could be called Situated Modernism. Such an example is the Banque Audi headquarters by Kevin Dash (2001). A more recent building, the CMA-CGM headquarters by Nabil Gholam architects (2005-2011) resorts to an enclosed mass of double-skin glass, treated differently on the various sides, and manages to negotiate its affiliation to the encountered models of the 1950s-60s with deep shaded façades.

Still recovering from the after-effects of the war, although the damage these days is less the physical than psychological, architecture in today's Lebanon has lost much of its former confidence and as a result, is too often tempted by the comfort and security of rehearsed tradition. Care for the urban and natural environments and relation to public space, are much more contextually needed than superficial stylistic references or outbids on the 'Lebanity' of the design.

Jordan

Jordan was founded as an independent country in 1924, after it was under Ottoman rule forming part of the province of Syria with Damascus as capital. Amman has been populated by waves of immigrants, growing very quickly with the arrival of Palestinians in 1948 and in 1967. The city benefited also from the oil booms of the 1970s in the Gulf and later from the Gulf war in 1990, and again from the 2nd oil boom in the mid-2000s.

Amman's architecture is mostly dictated by the use of local stone admirably served by skilled masons. In the early 1980s, Bilal Hammad proposed a housing scheme, al-Ribat in Amman that offers the local qualities of a neighborhood cluster inducing conviviality and keeping with human scale. It has the merit of ageless architecture, concerned with spatial relationships and performance more than style. In the SOS village in Aqaba on the Red Sea (1988-1991) Jafar Tukan also uses local skills of stone building and simple vernacular ventilation techniques. Precast concrete is introduced to replace wooden tension members.

In the early 2000s, Amman saw a project that can be coined as Ammani, namely the Wild Jordan Center by architect/artist Ammar Khammash. Perched on concrete stilts over the steep hill of Jabal Amman, the nature center displays a natural, unadorned, and low tech elegance demonstrating that good architecture does not require much artifice. Resorting to many recycled materials, the building fits in its context while not making an expected literal reference to the locale.

The tradition of stone building found another contemporary application with the International Academy in Amman by Khalid Nahhas (2006). Here again, passive energy and the use of traditional materials and techniques are combined with contemporary amenities. Several finishes of local stone demonstrate that the tradition is still alive.

Palestine

The history of architecture in Palestine is yet to be written. Riwaq, a center for architectural preservation in Ramallah, did a colossal survey of historic buildings built prior to 1948 in the West Bank, Gaza and East Jerusalem. The remarkable work covers whole villages and towns since 1700. Very little has been done, however, on the documentation of its architecture of the past century, mostly after Israeli occupation. Probably more acutely than anywhere else, Palestinian heritage, old and modern, is threatened. Occupation, systematic demolitions, but also unprecedented rates of construction in recent times had a tremendous impact on the quality of the built environment A record of 20th century buildings in Palestine would certainly include the Baroque Jacir Palace in Bethlehem (1914), the Alami House in Jericho (1919), the Qutub House by Egyptian architect Sayed Karim in Shufat (1960), the work of Hani Arafat in Nablus in the 1960s such as the Municipality and the Salti House, the Hanania Commercial Building in central Ramallah by Arkbuild (1971), the Engineering Building at Birzeit University (1984).

Three buildings particularly stand out: the Azzahra-Ambassador Hotel (1953) in East Jerusalem, designed by a team composed of architects Georges Rais and Theo Canaan with engineer Bagdassar Erdekian, symbolically built around a large oak tree. More recently, two landmarks were built in Ramallah by architect Jafar Tukan. Both minimalist structures make the best of their sites and inspire serenity and meditation. Their ingredients are simple landscaping, and humble volumes with local materials. In the Yasser Arafat Mausoleum (2007), a modest but powerful place, the prayer pavilion and the burial chamber face each other on two sides of a path. The prayer hall is a meditative space made of Jerusalem stone only adorned with a frieze holding calligraphy of Quranic inscriptions, while the chamber is elegantly mirrored in a reflecting pool. The Memorial for renowned poet Mahmoud Darwish (2012) located on the hill of Al-Birweh Park, includes a mausoleum, a museum, an underground theater, and an open air theater.

Syria

The first two decades of the 20th century in Syria were marked by the works of Fernando di Aranda who designed the Hijaz railway station in Damascus (1908-1912) and Damascus University (1922-1923). Another distinguished architect is Abderrazak Malas, the author of the Fijeh Water Building.

Hotel Orient Palace by architect Antoine Tabet (1930-1933) is an early modernist building, in line with the French rationalist school of Auguste Perret. Michel Ecochard, the French architect and urban designer left two important buildings. While renovating the Azem Palace in 1936, he added a modernist house on pilotis for the director of the newly created French Institute. At the same time, Ecochard designed the National Museum in Damascus, completed in 1940, combining the sobriety of medieval Syrian architecture with the simplicity of modern architecture. Until the 1960s, the country had a flourishing building industry. From that period, some distinguished designs by Egyptian architects Mustafa Shawky and Salah Zeitoun are found, namely hospitals in Damascus Aleppo and Hama. Borhan Tayara and Naufal Kasrawi designed the Fine Arts Society Condominiums in Damascus (1968-73), probably the first duplex apartment building in the country.

When the Socialist political system nationalized the profession and gave mostly work to large companies owned by the military, it had a devastating effect on architecture quality in the country. Several established architects decided to relocate in Lebanon, Kuwait, or Saudi Arabia like Naufal Kasrawi, while some remained and were able to maneuver, like Youssef Abou Hadid with his remarkable works in Damascus, namely the Ministry of Higher Education and the Syrian Insurance Headquarters (1992), and Borhan Tayara who designed the Faculty of Architecture at Damascus University. The Shagrawieh Elementary School in As-Suwayda built with local basaltic stone in 1990 by the Mhanna brothers won the Aga Khan Award but did not succeed at reviving traditional ways of building.

Among the distinguished buildings built around the year 2000 in Damascus is the Madrasa and Mosque Shaykh Badr-al-Din al-Hasani by Wael Samhouri, a 9 story building, commissioned by the Awqaf Charity. The project is revealing of the conditions of practice in negotiating architectural style and types of window openings with the client.

الذي صمّمه كاظم كنعان. ويُعدّ مطار بيروت الدوليّ، والمدينة الرياضيّة، وكازينو لبنان، والقصر الرئاسي من بين العديد من المشاريع العامة. فصارت شاهدًا على طموح السلطات وعزمها على إعطاء البلاد وجهًا حديثًا. وعلى الأرجح أنّ وزارة الدفاع لفوغنسكي وهنديه (١٩٦٢ - ١٩٦٨)، بالإضافة إلى معرض طرابلس الدولي الذي صمّمه أوسكار نيماير في العام ١٩٦٢، ومقرّ مؤسسة كهرباء لبنان لمجموعة سينا (١٩٦٥ - ١٩٧٢) تندرج على قائمة النماذج الأفضل تعبيرًا عن العمارة العامّة في ذلك الوقت.

عندما برز الانشغال بالهويّة الإقليمية في منتصف ستّينات القرن العشرين، أعيد اكتشاف الموادّ المحلّية مثل الحجر الرمليّ، وكثُر انتشار الباحات الداخلية. فشملت أعمال عاصم سلام في تلك الفترة السرايا الحكوميّة في صيدا، ومسجد الخاشقجي (١٩٦٥) المشهور بجدرانه من الحجر الرمليّ يعلوها سقف معدّد الأوجه من الخرسانة وهو تصميم حديث للقبّة.

سبعينات القرن العشرين وثمانيناته

إلى جانب عدد قليل من الاستثناءات، كتصميم مبنى «إنترديزاين» لخليل خوري، تطوّرت معظم عمارة السبعينات بعيداً عن الإبداع. فأضحت وكأنها ترجمة مباشرة لحرفيّة قانون البناء. فكثرت الشرفات الطوليّة المستقيمة التي تلتفّ حول مبان مملّة الطابع كنتيجة لمحاولة استثمار المساحة الإجمالية التي يسمح بها القانون. وعليه، تمّ التخلّي عن المنحى التجريبيّ الذي طَبعت فيه الفترة السابقة.

إعادة الإعمار في تسعينات القرن العشرين، وما تلاها

مع أنّ عمليّة البناء استمرّت طوال فترات الحرب التي عصفت بلبنان بين عامي ١٩٧٥ و١٩٩٠، كان لا بدّ من حصول قطيعة معيّنة مع الماضي. فالفترات المتقطّعة من الاضطرابات التي هزّت البلاد تسبّبت بدمار هائل. ولمّا كان قلب بيروت مسرحًا للمعارك الضارية، انبثقت أحياء جديدة في الضواحي بين ليلة وضحاها. وسرعان ما بلغ التوسّع العشوائيّ تلك المناطق المحميّة نسبيًا حتى ذلك الوقت من الزحف المدينيّ السريع. فكانت النتيجة خسارة وخيمة للتوازن بين المناطق المدينيّة والريفيّة. فيما عانى نسيج المناطق الريفية غزو المباني ذات الطابع المدينيّ، والتي غالبًا ما فرضت وجودها المتنافر على المنحدرات الجبليّة.

بعد الحرب، أثارت إعادة الإعمار وإعادة تطوير المناطق المدينيّة المسألة المعقّدة المرتبطة بالهوية. فقد عادت المسألة لتطفو على السطح بعدما ظُنّ أنّها حُلّت بالشكل المناسب في أوج الطفرة الثقافية في خمسينات القرن وستيناته، إلّا أنّها هذه المرّة لم تجد لها حلًّا، بل تكثّفت بسبب موجة ما بعد الحداثة المتأخّرة من جهة، والحنين إلى الهويّة الوطنيّة المهدّدة. في هذا الإطار، برز اتّجاهان: نهج مستتشرٍ يقوم على التسلّح بالتكنولوجيا المتطوّرة، وهو محاولة يائسة للّحاق بركب بقية العالم، ونهج آخر يستعين بعناصر من التراث لا تعدو سوى تذكير بفقدانه. لحسن الحظ، ظهرت اتجاهات أخرى خالية من التصريحات السطحية، تعكس طابعًا يمكن وسمه بالحداثة المرتبطة بالمكان، على غرار مقرّ بنك عوده الذي صمّمه كيفن داش وانتهى تنفيذه في العام ٢٠٠١. أمّا مبنى CMA-CGM، وهو من تصميم مكتب نبيل غلام (٢٠٠٥ - ٢٠١١)، فعبارة عن ثلاثة كتل من الزجاج المزدوج تتميّز بمعالجة خاصة لأنحائها المختلفة لحمايتها من الشمس، وقد نجح مع منحاه التحديثي بالانتماء للنماذج التي تعود لخمسينات القرن وستيناته.

لا تزال العمارة في لبنان تتعافى من الآثار التي خلّفتها الحرب. مع أنّ الضرر في هذه الأيّام يطال المستوى النفسيّ أكثر منه الماديّ، قد فقدت الكثير من ثقتها السابقة، ونتيجةً لذلك تراها تنجرّ في الكثير من الأحيان إلى اختيار الأمان في التقاليد المعتادة.

وتلحّ الحاجة اليوم إلى الحرص على البيئة المدينية والطبيعية والعمل على تأمين حيّز عام، بدل الاستناد على المراجع النمطيّة السطحية أو المزايدة في «لبننة» التصميم.

الأردن

تأسّس الأردن كبلد مستقلّ في العام ١٩٢٤، بعد أن كان تحت الحكم العثماني جزءاً من ولاية سوريا وعاصمتها دمشق. وقد سكن الأردن موجات من المهاجرين، وتزايد العدد بسرعة كبيرة مع وصول الفلسطينيين في العام ١٩٤٨ والعام ١٩٦٧. وقد استفادت عمّان من الطفرة النفطية في السبعينات وفي وقت لاحق من حرب الخليج في العام ١٩٩٠، ومرّة أخرى من الطفرة النفطية في العام ٢٠٠٥.

عمارة عمّان في الغالب من الحجر المحلّي. من عماراتها المميّزة في أوائل الثمانينات، مجمّع الرباط السكني لبلال حمّاد الذي يوفّر صفات التفاعل والعيش المشترك بين الجيران وينتمي إلى العمارة المعنية بالأداء أكثر منه بالأسلوب. في قرية SOS في العقبة على البحر الأحمر (١٩٨٨ - ١٩٩١) استخدم جعفر طوقان المهارات المحلّية في بناء الحجر بالاضافة الى بعض الجسور من الخرسانة سابقة الصب عند الحاجة. كما استعمل تقنيّات تهوية بسيطة.

وفي العام ٢٠٠١، شهدت عمّان مشروعاً يمكن أن يوصف عمّانياً بامتياز. هو مركز برّية الأردن من تصميم عمّار خمّاش. يجثم على تلّة شديدة الانحدار في جبل عمّان، ويندمج ببساطة في المحيط من غير تصنّع ودون إشارة حرفيّة متوقّعة لانتمائه الى المكان. وينطلق المشروع مبادئ إعادة التدوير مما يدلّ على أن العمارة الجيدة لا تحتاج إلى أكثر من ذلك.

وفي الأكاديمية الدولية في عمّان لخالد نحاس (٢٠٠٦) فإن استعمال تقنيات عديدة لبناء الحجر منها التقليدي ومنها المعاصر يثبت أن الحرفيّة ما زالت حيّة. هنا مرة أخرى، يتم الجمع بين الطاقة السلبيّة واستخدام المواد والتقنيات التقليدية مع وسائل الراحة المعاصرة.

فلسطين

ما زال تاريخ العمارة في فلسطين بحاجة إلى تدوين. وقد قام مركز المعمار الشعبي (رواق) في رام الله بمسح جبّار للمباني التاريخية التي بنيت قبل عام ١٩٤٨ في الضفّة الغربية وغزّة والقدس الشرقية. ويغطي العمل قرى وبلدات بأكملها منذ ١٧٠٠. في المقابل، تم القيام بالقليل بما يختص بعمارة القرن الماضي، خاصة بعد الاحتلال الإسرائيلي. لربما يشكّل الخطر على التراث الفلسطيني، القديم والحديث، حدّة أكثر من أي مكان آخر. إذ أن الاحتلال، والهدم المنهجي، يضاف إليهما في الآونة الأخيرة عمران غير مسبوق، لهم تأثير هائل على نوعية البيئة المبنية. وبالتأكيد يجب أن يشمل سجلّ مباني القرن العشرين في فلسطين قصر جاسر في بيت لحم (١٩١٤)، بيت العلمي في أريحا (١٩١٩)، بيت القطب للمعماري المصري سيّد كريم في شعفاط (١٩٦٠)، بعض أعمال هاني عرفات في نابلس في الستّينات مثل البلديّة وبيت السلطي، مبنى حنانيا التجاري في وسط رام الله (اركيبلد، ١٩٧١) ومبنى الهندسة في جامعة بيرزيت (فرانشيسكو مونتانا، ١٩٨٤).

يضاف إليها ثلاثة مبان مهمّة: فندق الزهرة - أمبسدور (١٩٥٣) في القدس الشرقية، من تصميم المعماريين جورج ريّس وثيو كنعان مع المهندس بهدسار إردكيان، وقد بُني رمزيًا حول شجرة سنديان كبيرة. وفي الآونة الأخيرة، تم بناء معلمين في رام الله من تصميم المهندس المعماري جعفر طوقان وفريقه. ويحقّق كلا المشروعين البسيطين أفضل محاكاة للموقع مستنبطين منه الإلهام والصفاء والتأمل. مكوّنات المشروعين هي المناظر الطبيعية، وعمارة بسيطة من المواد المحلّية. في ضريح ياسر عرفات (٢٠٠٧)، وهو مكان متواضع وملهم معاً، يتقابل جناح الصلاة والمدفن على جانبي المسار. وتبدو قاعة الصلاة مساحةً تأمّليةً من حجر القدس مزيّنةً فقط بإفريز خُطّت عليه نقوشٌ قرآنية في حين تنعكس صورة المدفن في مساحة من الماء. أمّا صرح ومتحف الشاعر محمود درويش (٢٠١٢)، فيقع على تلّة حديقة البروة، ويتضمن الضريح، متحفاً، مسرحاً تحت الأرض ومسرحاً في الهواء الطلق.

سوريا

تميّز العقدان الأوّلان من القرن العشرين في سوريا بأعمال فرناندو دي أراندا الذي صمّم محطّة سكّة حديد «الحجاز» في دمشق (١٩٠٨ - ١٩١٢) وجامعة دمشق (١٩٢٢ - ١٩٢٣). كما برز المعماريّ عبد الرزّاق ملص، صاحب مبنى مياه الفيجة. وما لبث أن قام فندق أورينت بالاس لأنطوان ثابت (١٩٣٠ - ١٩٣٣)، وهو مبنى حديث العمارة في المدينة، يتماشى مع المدرسة العقلانية الفرنسية لأوغست بيريه. أمّا ميشال إيكوشار، المعماريّ ومصمّم المدن الفرنسي، فقد صمّم اثنين من المباني الهامّة. أحدهما بمناسبة ترميم قصر العظم في العام ١٩٣٦، إذ أضاف منزلًا على أعمدة حديث الطابع لمدير المعهد الفرنسي الذي كان قد أنشئ حديثًا. في الوقت نفسه، عمل إيكوشار على تصميم المتحف الوطني في دمشق، الذي أنجز في العام ١٩٤٠، وقد جمع بين رصانة العمارة السورية القديمة وبساطة العمارة الحديثة. واستمرّ البلد ينعم بصناعة البناء مزدهرة حتّى الستّينات. فمن تلك الفترة، ظهرت بعض التصاميم المتميّزة لمعماريّين مصريّين أمثال مصطفى شوقي وصلاح زيتون، لا سيّما المستشفيات في دمشق وحلب وحماة. ثمّ عمل برهان طيارة ونوفل كسراوي على تصميم عمارة سكنيّة لنقابة الفنون الجميلة في دمشق (١٩٦٨ - ١٩٧٣)، وهو أوّل مبنى لشقق دوبلكس في البلاد.

عندما أمّم النظام السياسيّ الاشتراكيّ هذه المهنة وأفسح المجال أمام الشركات الكبيرة التي يديرها الجيش، كان لهذه الخطوة الوقع المدمّر على جودة العمارة في البلاد. على أثره، قرّر العديد من المعماريين مثل نوفل كسراوي الانتقال إلى المملكة العربيّة السعوديّة ولبنان والكويت. مع ذلك، فضّل آخرون البقاء واستطاعوا إيجاد هامش للمناورة، فكان للمعماريّ يوسف أبو حديد أعمالًا رائعة في دمشق، كوزارة التعليم العالي ومقرّ شركة التأمين السورية (١٩٩٢). ولبرهان طيارة كلّية الهندسة المعماريّة في جامعة دمشق. وقد حصدت مدرسة شعراوية الابتدائية في السويداء، التي شيّدت بحجر البازلت المحليّ في العام ١٩٩٠ على يد الإخوة مهنّا، جائزة الآغا خان، لكنّها لم تنجح في إحياء طرق البناء التقليدية.

من بين المباني المتميّزة التي رأت النور حوالي العام ٢٠٠٠ في دمشق، نذكر مدرسة ومسجد الشيخ بدر الدين الحسني لوائل السمهوري، وتتكوّن من مبنى بعدّ تسعة طوابق مع برنامج تعليم كامل للعلماء والدعاة الدينيين، بتكليف من الأوقاف الخيرية. ويعكس المشروع هذا الوضع الذي بلغته هذه المهنة في التفاوض بين المعمار وصاحب المشروع حول الطراز المعماريّ وأنواع فتحات النوافذ.

ترجمته عن الإنكليزيّة نجلا رعيدي

Sm
OBJECT

这是一本综合展现丹麦艺术类书籍的书。本书通过对丹麦艺术书籍历史的陈述、对部分艺术书籍和活动的讨论、对材料和形式的思考、对具体作品的罗列和探讨、对藏家评论的整理，展现了丹麦艺术书籍各个时期以来的面貌。本书的装订形式为锁线胶平装，封面哑面白卡纸四色印刷，渐变底色，文字露白，表面未覆膜，没有任何特殊工艺。丹麦文与英文双语双栏排版从封面起贯穿全书。内页的前半部分采用 16 页胶版纸黑色印刷包覆 4 页哑面铜版纸四色印刷。胶版纸单色部分采用双语双栏排版，在中部插入等宽的图片，是经典的版式。通过图注的无衬线字体和比图片略微缩进的变化来丰富版面，铜版纸彩图部分展示书籍大图。后半部分则只用胶版纸四色印刷，介绍艺术家、作品和一些评论家的文字，图片的排版相较前半部分丰富多变。

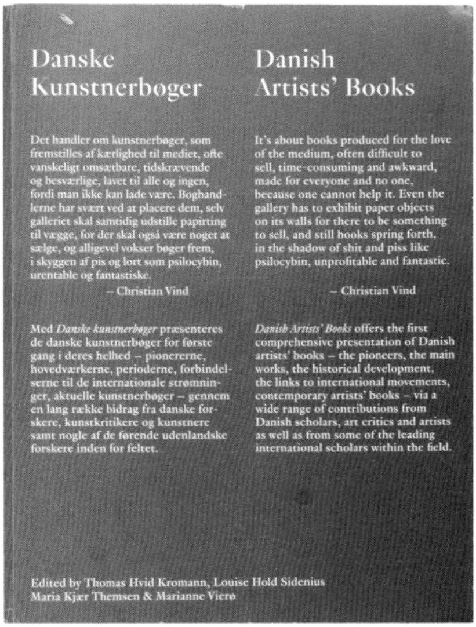

Danske Kunstnerbøger
Danish Artists' Books
Det handler om kunstnerbøger, som fremstilles af kærlighed til mediet, ofte vanskeligt omsætbare, tidskrævende og besværlige, lavet til alle og ingen, fordi man ikke kan lade være. Boghandlerne har svært ved at placere dem, selv galleriet skal samtidig udstille papirting til væggen, for der skal også være noget at sælge, og alligevel vokser bøger frem, i skyggen af pis og lort som psilocybin, urentable og fantastiske.
– Christian Vind
It's about books produced for the love of the medium, often difficult to sell, time-consuming and awkward, made for everyone and no one, because one cannot help it. Even the gallery has to exhibit paper objects on its walls for there to be something to sell, and still books spring forth, in the shadow of shit and piss like psilocybin, unprofitable and fantastic.
– Christian Vind
Med Danske kunstnerbøger præsenteres de danske kunstnerbøger for første gang i deres helhed – pionererne, hovedværkerne, perioderne, forbindelserne til de internationale strømninger, aktuelle kunstnerbøger – gennem en lang række bidrag fra danske forskere, kunstkritikere og kunstnere samt nogle af de førende udenlandske forskere inden for feltet.
Danish Artists' Books offers the first comprehensive presentation of Danish artists' books – the pioneers, the main works, the historical development, the links to international movements, contemporary artists' books – via a wide range of contributions from Danish scholars, art critics and artists as well as from some of the leading international scholars within the field.
Edited by Thomas Hvid Kromann, Louise Hold Sidenius
Maria Kjær Themsen & Marianne Vierø

This book is a collection of Danish art books. By presenting the history of Danish art books, discussing some art books and activities, reflecting on materials and forms, listing and explaining specific works, it shows the complete picture of Danish art books of all times. It is perfect bound with thread sewing. The cover is made of matte white cardboard printed in four colors with gradient background, white text, without any special technology. The Danish and English bilingual double-column typesetting goes throughout the book. In the first half of the inside pages, 16 pages of offset paper printed in black are wrapped by 4 pages of matte copperplate paper printed in four colors. The monochrome part adopts classic bilingual and double-column typesetting, inserting pictures in the middle. The layout is enriched by the serif fonts and slightly indentation compared with the pictures. The large pictures of the books are displayed on the copperplate paper. The latter part is printed on offset paper in four colors, introducing artists, works and some critics' reviews. The layout of pictures is more varied than that of the first half.

- 银奖
- 丹麦艺术家的书
- 丹麦

- Silver Medal
- Danish Artists' Books / Danske Kunstnerbøger
- Denmark
- Louise Hold Sidenius
- Thomas Hvid Kromann, Maria Kjær Themsen, Louise Sidenius, Marianne Vierø
- Møller and Verlag der Buchhandlung Walther König
- 325×238×30mm
- 1448g
- 302p

The first compendium of Danish artists' books with such a crude cover? Even as a general write-up its design is somewhat understated. The large format and the fat book body certainly don't stand in the way of its intended role as a source identifier. From the first touch the reader is able to orient himself between the two main parts of the volume: in the front part, each of the 16-page sections of the thick uncoated paper are covered by a sheet of glossy paper, which has been stitched in, so that the truncated, printed picture panels are immediately visible. They impart rhythm to this historic-theoretical part, irrespective of the text flow. The lines appear in two columns, suitably interleaved, in a bold type area whose side brackets are narrower than the column spacing. Intermediate headings, footnotes and captions of the black and white photographs are divided sparingly but effectively by indents. The back part of the book introduces the individual artists; their works are now printed in colour on the rough text paper. Equally striking and audacious is the bold, partly polygonal framing of the illustrations. Pleasantness is never the principle in the very personal, sometimes anarchic medium of books about artists. And this might actually be the aesthetic message of the cover.

Danske Kunstnerbøger
Danish Artists' Books
Artikler
Articles

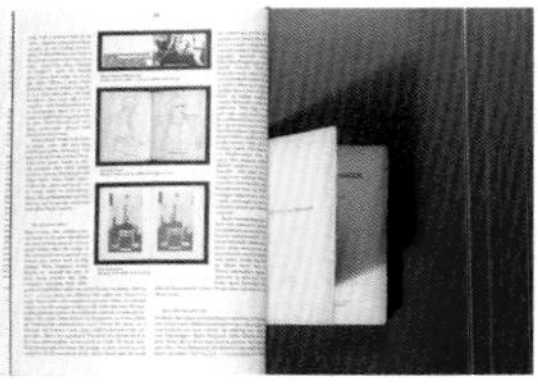

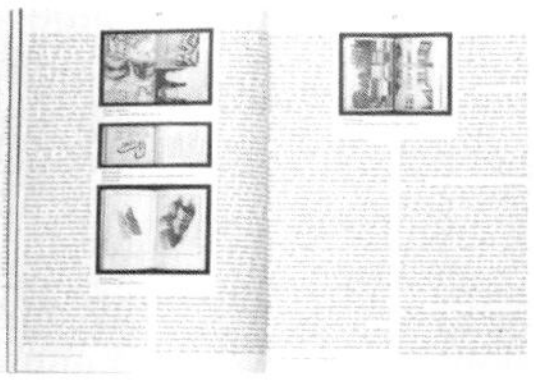

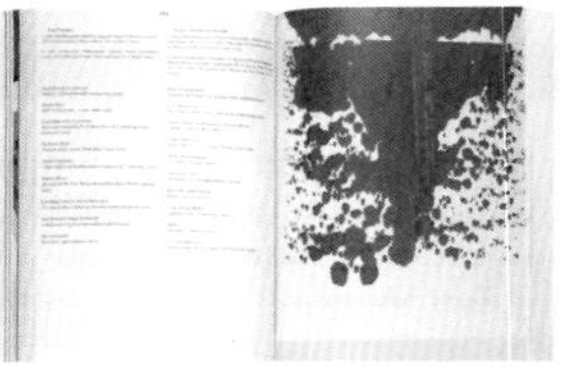

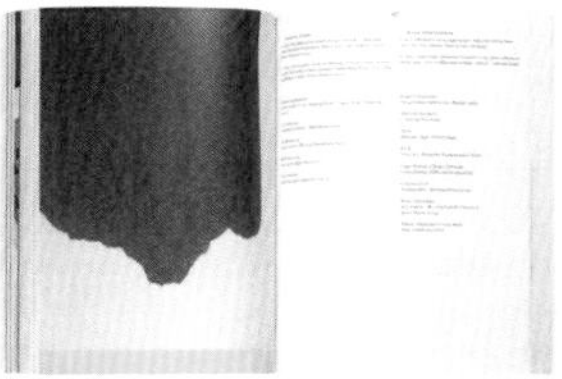

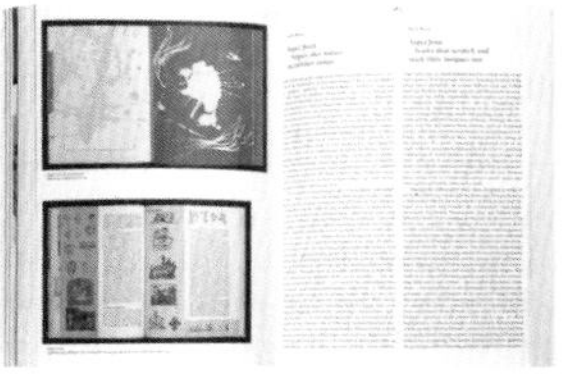

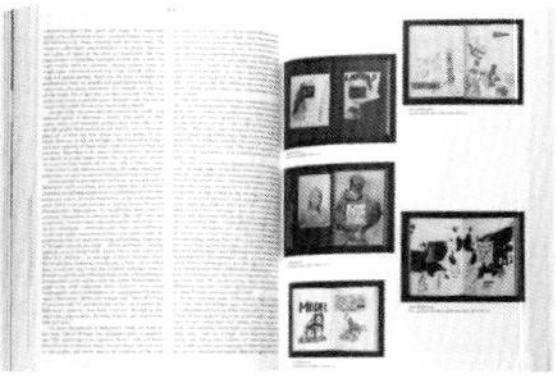

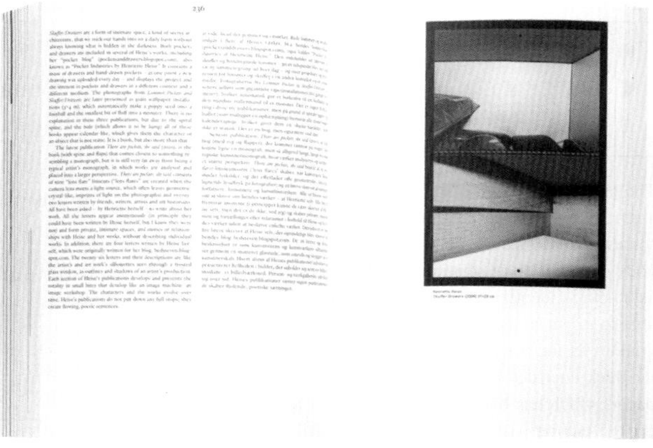
236

bøger

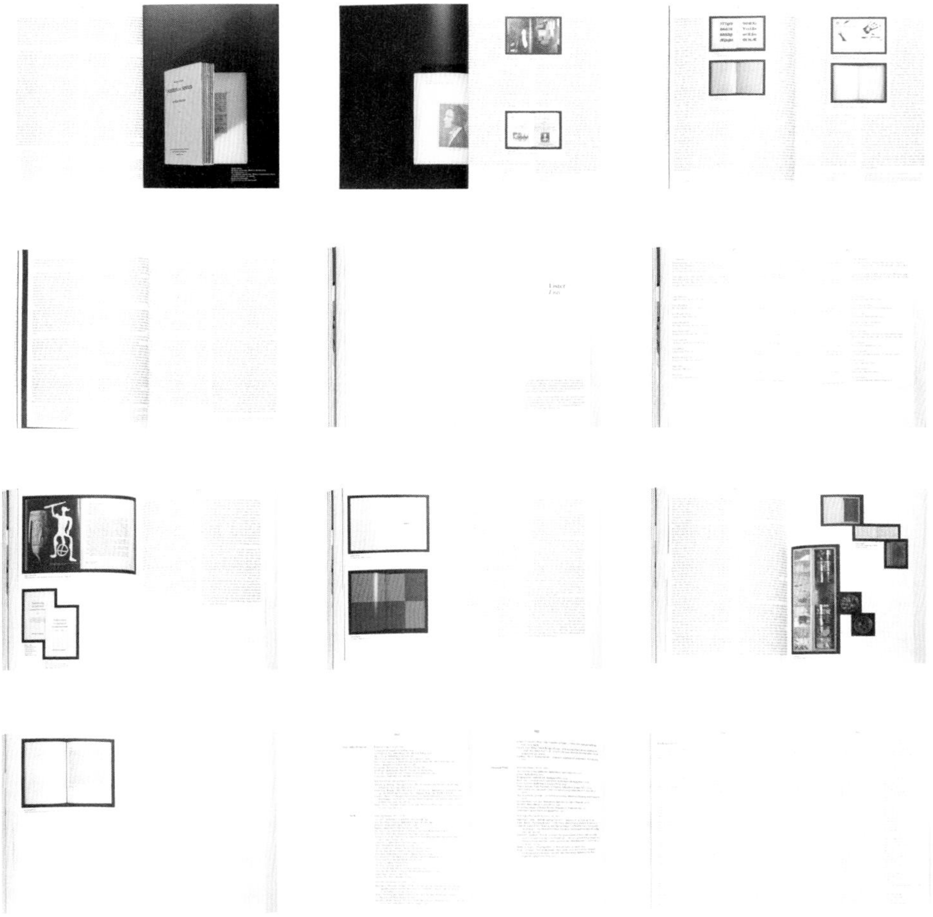

vhic
nd of worksho

Danske Kunstnerbøger

Danish Artists' Books

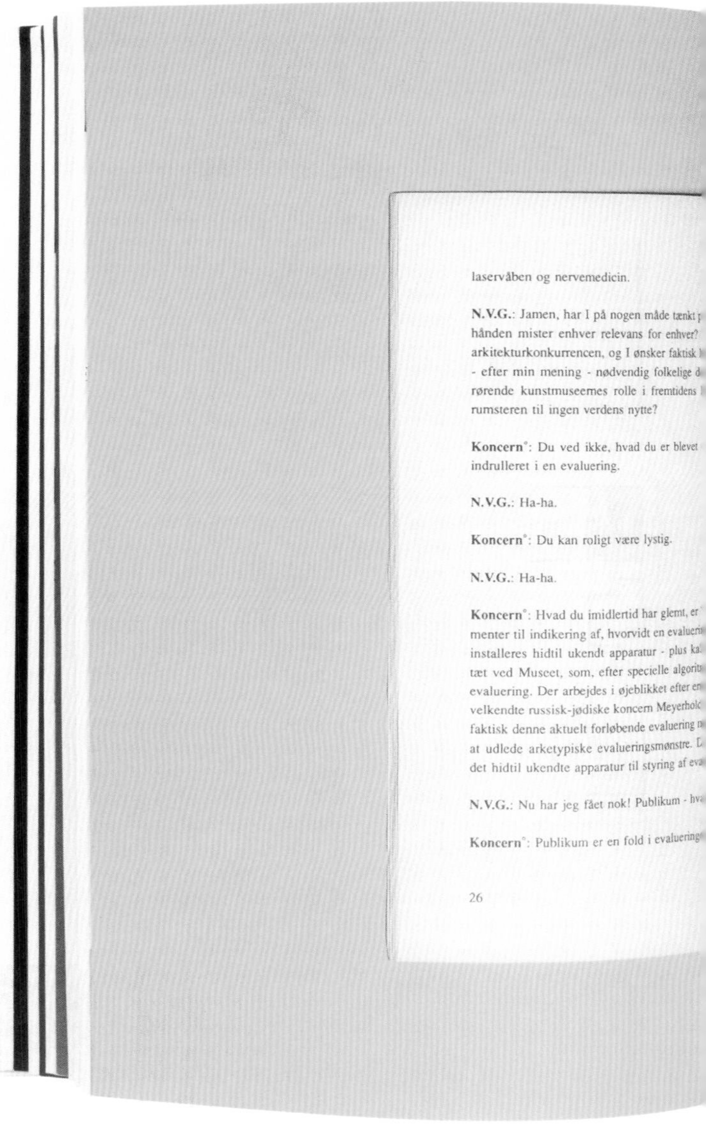
laservåben og nervemedicin.

N.V.G.: Jamen, har I på nogen måde tænkt p
hånden mister enhver relevans for enhver?
arkitekturkonkurrencen, og I ønsker faktisk
- efter min mening - nødvendig folkelige d
rørende kunstmuseernes rolle i fremtidens
rumsteren til ingen verdens nytte?

Koncern°: Du ved ikke, hvad du er blevet
indrulleret i en evaluering.

N.V.G.: Ha-ha.

Koncern°: Du kan roligt være lystig.

N.V.G.: Ha-ha.

Koncern°: Hvad du imidlertid har glemt, er
menter til indikering af, hvorvidt en evaluer
installeres hidtil ukendt apparatur - plus ka
tæt ved Museet, som, efter specielle algorit
evaluering. Der arbejdes i øjeblikket efter en
velkendte russisk-jødiske koncern Meyerhol
faktisk denne aktuelt forløbende evaluering n
at udlede arketypiske evalueringsmønstre. D
det hidtil ukendte apparatur til styring af eva

N.V.G.: Nu har jeg fået nok! Publikum - hv

Koncern°: Publikum er en fold i evaluering

26

eptkunstens fotokopier og onede objekter, der var in-eret som forskellige proble-seringer af det visuelle, har haft svært ved at opnå insti-nel eksponering, om end de unstmarkedet stadigvæk al-er nået op på siden af de klas- kunstneriske medier som ri og skulptur. Men en udvi-offentlighed lykkedes det at skabe, og værkkarakteren e konceptuelle kunstnerbo-blev ikke for alvor ustabil.[5]

om en delvis omgaelse af risystemet er kunstnerbo-dog forblevet relevant for e kunstnere, men en sats-på en mere sammenhæn-le kritik af kunstinstitutio-har fortonet sig.

lerhjemme har der været få ekter, der har haft et mel-værende med den mere radi-side af denne tradition. Et af var Koncern°, der i 1990'er-begyndelse i form af tids-ftet med titlen *Koncern°. Skrift unstnerisk-filosofisk grundforsk-* foretog en hysterisk opkørt - eller måske snarere afdis-ivering, der forhøjede kon-kunstens selvevaluerings pro- i en sådan grad, at den oprindelige drøm om en udvidet ntlighed nærmest imploderede i en art sarkastisk nerve-menbrud for postkonceptkunsten, der hverken kunne gå r konceptkunstens institutionskritiske henvendelsesform r 80'erkunstens institutionsaffirmative selveksponering og *erplay*. I modsætning til størstedelen af 1980'ernes individ-ostaserende og værkorienterede kunst udviklede Kon-n° en kollektiv tekstbaseret kunst, hvor forholdet mellem st, værk og samfund blev underkastet en indgående og ofte risk undersøgelse, der endte med at sætte spørgsmålstegn de fundamentale kategorier, som får ikke bare kunstinsti-onen, men samfundet som social helhed til at hænge sam-n: subjekt, fællesskab og kommunikation.

Hinsides forestillingen om en borgerlig endsige proletarisk ntlighed analyserede Koncern° kunstinstitutionen og dens ventioner og regler. Til dette arbejde udviklede gruppens kunstnere, Søren Andreasen, Jan Bäcklund, Jakob Jakobsen Jørgen Michaelsen, i fællesskab et særegent tekstsubjekt, n i sjælden grad var i stand til at problematisere kunstinsti-onelle forhold og det i bedste forstand helt urimelige ved

Benjamin Buchloh kritiserer konceptkunsten for i virkeligheden at være en form for administrativ æstetik, der blot afspejler den postindustrielle kapitalisme. "Conceptual Art 1962-1969: From the Aesthetics of Administration to the Critique of Institutions", i: *October*, nr. 55, 1990, pp. 105-143.

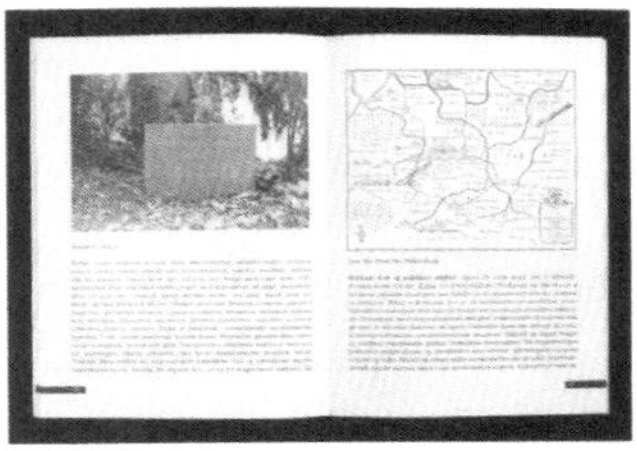

Koncern°:
Koncern°. Skrift for kunstnerisk-filosofisk grundforskning nr. 1 (1989) 27,7×21 cm

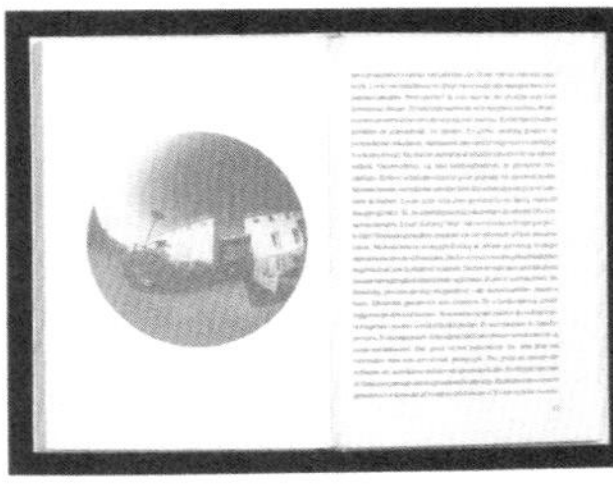

Koncern°:
Koncern°. Skrift for kunstnerisk-filosofisk grundforskning nr. 4 (1992) 22×15 cm

the visual, have not found it difficult to achieve institutional exposure – even though on the art market they have still never reached the levels of traditional artistic media such as painting and sculpture. But the expanded public sphere never materialised, and the art-work character of conceptual artists' books was never really destabilised.[5]

As a partial circumvention of the gallery system the artist's book has, however, remained relevant for younger artists, but a commitment to a more coherent and radical critique of the art institution has faded away.

Here in Denmark there have been few projects that have had a dispute with the more radical side of the tradition. One of them was Koncern°, which in the early 1990s in the form of the periodical entitled *Koncern°. Skrift for kunstnerisk-filosofisk grundforskning* ("The Group°. The Journal of Fundamental Artistic-Philosophical Research") carried out a hysterically uneven self-discursivation, or perhaps even de-discursivation, that elevated conceptual art's project of self-evaluation to such an extent that the original dream of a wider audience nearly imploded in a kind of sarcastic nervous breakdown of post-conceptual art, which was able to follow neither conceptual art's institution-critical approach nor the 1980s' art institution-affirming self-exposure and power-play. Unlike the majority of the 1980s' individual hypostasisising and work-oriented art, Koncern° developed a collective, text-based art where the relationship between art, work and society was subjected to thorough and often delirious investigation which went on to question the fundamental categories that enable not only the art institution but society as a whole to cohere: subject, community and communication.

Beyond the notion of a bourgeois, even less a proletarian, audience, Koncern° also analysed the art institution and its conventions and rules. In this work the group's four artists, Søren Andreasen, Jan Bäcklund, Jakob Jakobsen and Jørgen Michaelsen, jointly developed a unique text subject, which to an exceptional degree was able to critique art-institutional relations as well as showing the completely absurd (in the best sense) artistic gesture, which if nothing else virtually suspends all social relations without any great avant-gardist fanfare but rather as a

5. Benjamin Buchloh criticises conceptual art for essentially being a form of administrative aesthetic that simply reflects post-industrial capitalism. "Conceptual Art 1962-1969: From the Aesthetics of Administration to the Critique of Institutions", in *October*, no. 55 (1990), pp. 105-43.

Sm^2

这是荷兰摄影师阿沃伊斯卡·范德莫伦（Awoiska van der Molen）的摄影作品集。本书通过多种纸张、工艺和特别的装订形式体现着黑白摄影的魅力。本书的装订方式为锁线硬精装。护封对开哑面涂布纸折叠而成，黑灰双色印刷，使黑白摄影照片呈现出丰富而细腻的层次。内封黑色布面裱覆灰色卡板印白，背裱黑色卡纸做环衬。内页采用 8 页黑卡纸和 16 页白色哑面涂布纸反复间隔的结构，将内页分为五部分；黑卡纸部分印白，白色哑面涂布纸则如同封面一致的黑灰双色印刷。除了图片编号，内页没有其他文字，图片内容大致为山林植被、云雾天气、无植被的荒山、有水或无水的河床等。读者需要在几部分黑白交错的页面里寻找摄影用光的魅力。

This is a collection of photographs by Dutch photographer Awoiska van der Molen. It reflects the charm of black-and-white photography through a variety of paper, craft and special binding forms. The hardback book is bound by thread sewing. The cover is made of folded matte coated paper and printed in black and gray, which presents the rich and delicate layers of black-and-white photographs. The inside cover is made of black cloth mounted with gray graphic board, and black cardboard is used for the end paper. The inside pages are divided into five parts, each part contains 8 pages of black cardboard and 16 pages of white matte coated paper. The black cardboard part is printed in white; while the white matte coated paper is printed in black and gray as the cover. There is no text on the inside page except the picture numbers. The pictures are mostly forest vegetation, cloud and mist, barren mountains, watery or waterless riverbed, etc. Readers need to find the charm of 'light' for photography in the black-and-white pages.

银奖
与世隔绝
荷兰

Silver Medal
Sequester
The Netherlands
Hans Gremmen
Awoiska van der Molen
Fw:Books, Amsterdam
290×245×14mm
684g
86p

Photography, scenery, black and white. Nuanced. Speechless. Isolated. A book with three chapters without text. The reader gradually searches for words, in order not to succumb to the allure of this book. So powerful is the probability that the observer immerses himself through the pictures and becomes part of the secret. And from this darkness emerges the innocent landscape that never wants to be explained. The task of the book designers is to preserve the magic of the photographs by Awoiska van der Molen. In this context, the flair for the pictures, the graphic concept and all technical aspects cannot be separated from each other. Fine paper with a non-reflective smoothness, brilliant two-colour print with two shades of black, the print on black paper – enigmatic even if viewed under a magnifying lens -, non-laminated superimposed grey cardboard, black stitching and black spine cover, protection by an American book jacket – this is the material contribution of black magic. If we did not know that photography is an art involving light, we might insinuate that these pictures and this book are of obscure origin.

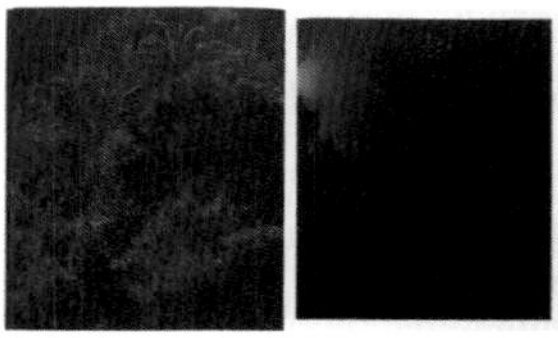

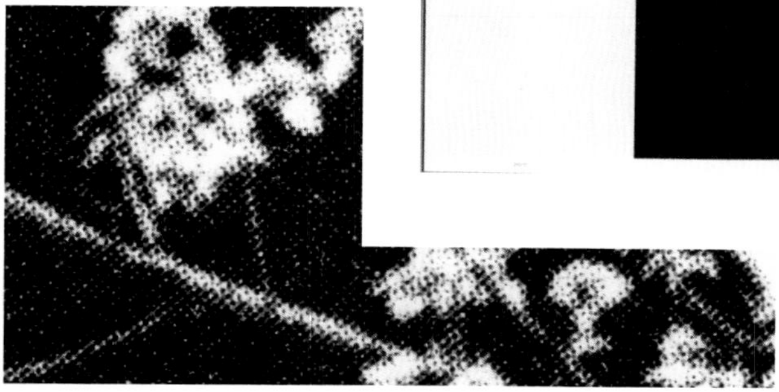

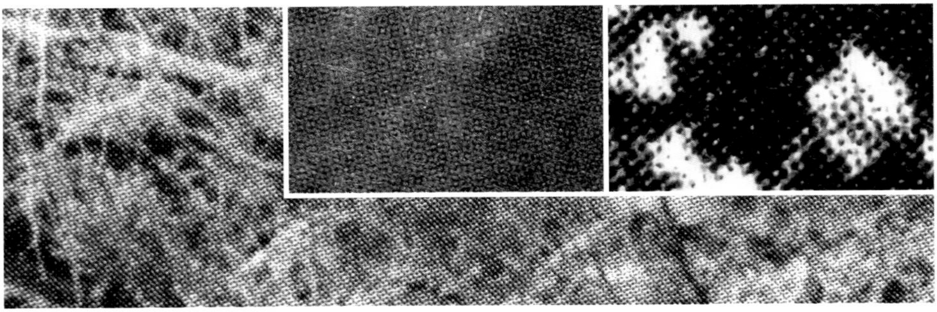

Awoiska van der Molen
Sequester
Fw:Books

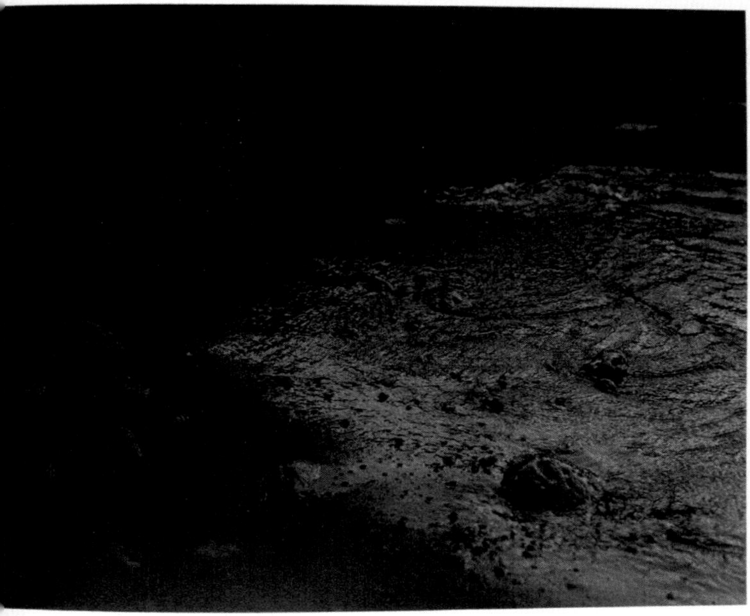

015 Bm1

这是一本具有独特玩法的互动绘本。书籍本身并不是一个绘制出完整画面的传统绘本形式，而是需要读者自己动手赋予本书最终的完整形态。本书的装订方式为双页面对裱的蝴蝶装，封面采用绒面纸张蓝黄双色印刷。封面中心部分烫半透明漆面工艺，封面裱贴卡板，夹住中间并不厚的内页。从扉页开始，书籍便为双页对裱的蝴蝶装，哑面涂布纸单黑印刷，对裱后形成较厚的卡纸手感，并进行模切，形成的小鸟、闪电、火车、精灵等形象均为两页共用的元素。使用时需要在一个光线较暗的环境下，通过单一光源照亮模切形象，形成单页上的阴影，与原有内容组成最终完整的画面。移动光源使得投影转向对页，则组成新的完整画面。书内插图风格松软而温暖，充满童真趣味。

MOTION SILHOUETTE
モーションシルエット
かげからうまれる物語
silhouette books
かじわら めぐみ / にいじま たつひこ
グラフィック社

This is an interactive picture book with unique ways of playing. The book itself is not a traditional picture book that draws a complete story, but requires the readers to complete the book by themselves. The book is butterfly-fold binding and the cover is printed in blue and yellow on matte paper. The central part of the cover is hot stamped with translucent paint. The cover is mounted with the graphic board. Beginning with the title page, the book is printed on matte coated paper in black only. The double-leaved pages feel like thick cardboard that can be die cut. The images of birds, lightning, trains and elves are elements shared by two pages. When used in a dark environment, the die-cutting image needs to be illuminated by a single light source to cast a shadow on a single page and form a complete picture with the original content. Moving the light source makes the shadow turn to the opposite page, which produces a new picture. The style of the illustrations is soft and warm, full of childlike taste.

铜奖
动态剪影
日本

Bronze Medal
Motion Silhouette
Japan
Megumi KAJIWARA, Tatsuhiko NIIJIMA
Megumi KAJIWARA, Tatsuhiko NIIJIMA
Megumi KAJIWARA, Tatsuhiko NIIJIMA
250×149×8mm
190g
12p

The depiction of the double-page scenes in this children's book is actually incomplete; the observer has to steer the action himself, and then something is actually set in motion – not only by turning the pages. It is best experienced in the dark. On each double page, carton silhouettes are stitched in. With a small torch or the light from the mobile phone, the spectator becomes the director of the shadow that he commands to dance on the well calculated paper white. Suddenly, birds land on branches, a butterfly flees from the spider to the roses, the dandelion seeds are scattered, a ghost reaches for the child, a locomotive steams to the moon. Scuff and skid mark structures modulate the shadow play into solid space. They are charmingly reproduced; two shades of black without screening take on a velvety flair on the matte, yellowish carton, and the spine cover made from fluffy material also has a velvety feel. If is fascinating to fire one's own imagination by such non-electronic means, almost as in the bygone days of the zoetrope – and all this in the age of hyper-flickering images.

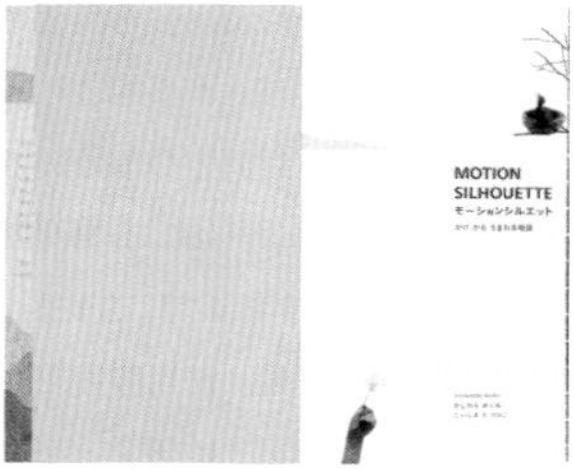
MOTION
SILHOUETTE
モーションシルエット

をす

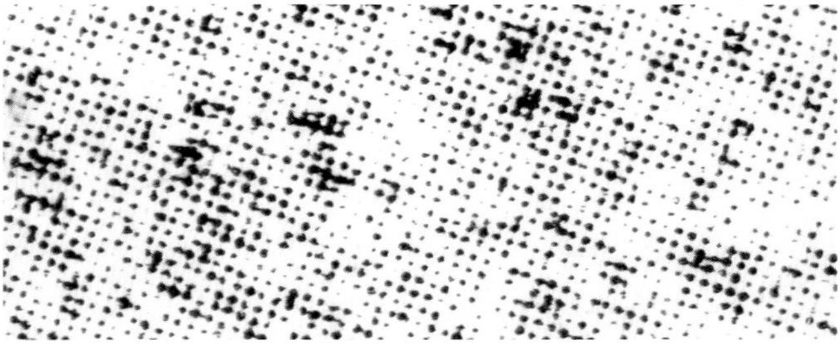

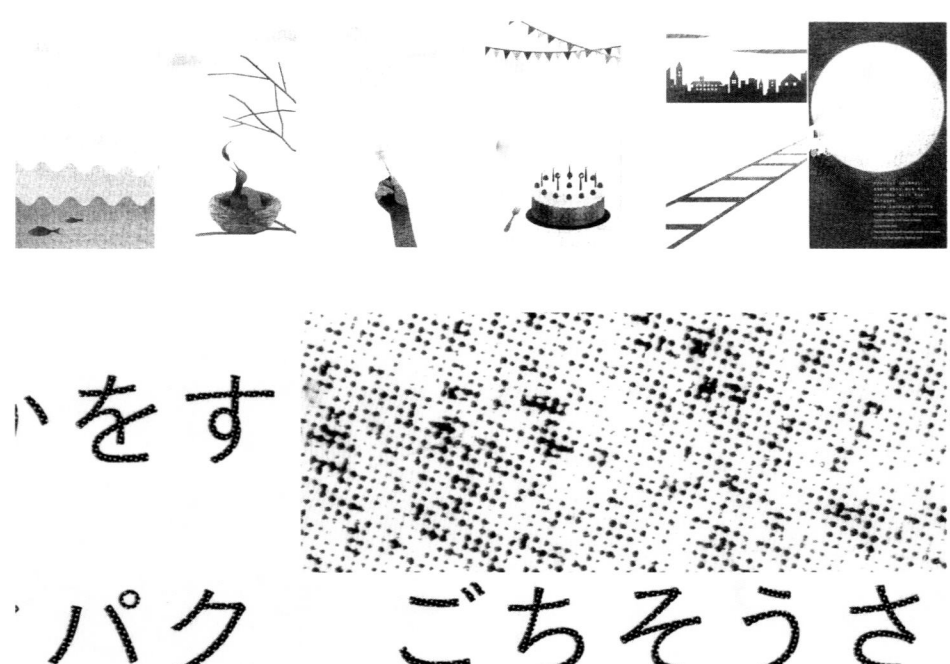

パク　ごちそうさ

モーションシルエット ―かげからうまれる物語―
silhouette books
かじわらめぐみ／にいじまたつひこ
グラフィック社
gb

くらいよる　ウトウトしてたら　ヘットに　おはけか　あらわれた
ハシャマのままて　にけだすと　おおきくなって　おってきた
おはけは　すこし　さみしそう

As I dozed off late at night, a ghost appeared beside my bed
I got up and ran away, just wearing my pajamas
The ghost chased me
It got bigger and bigger and had a slightly sad look on its face

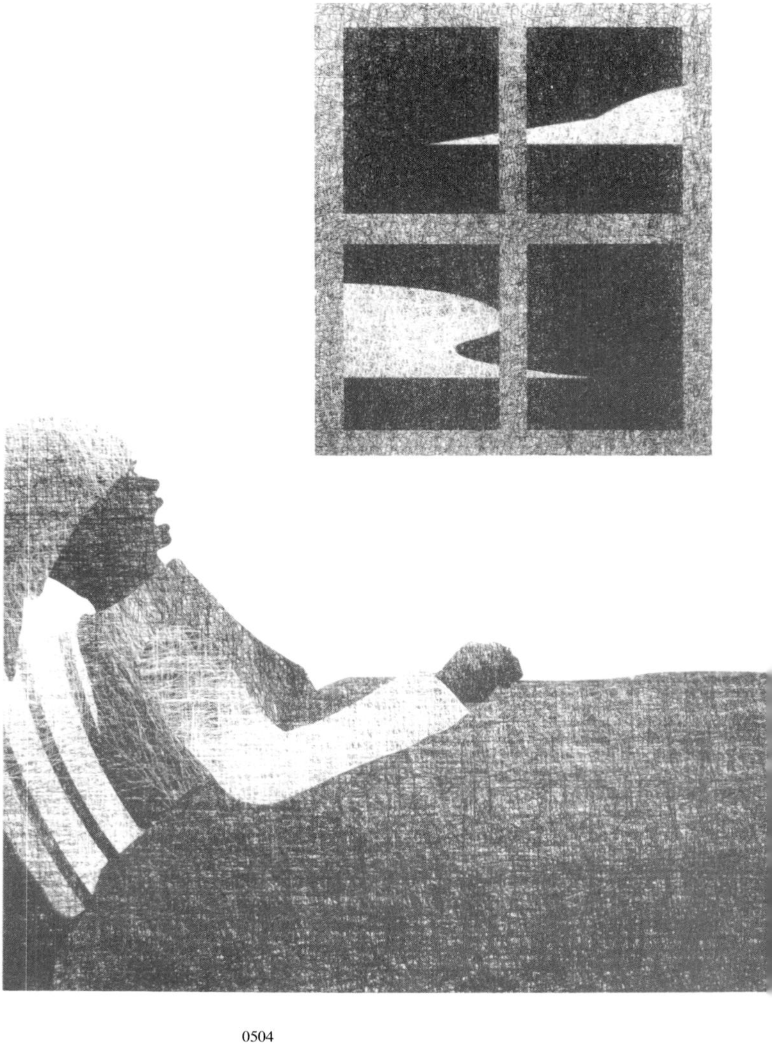

015 Bm2

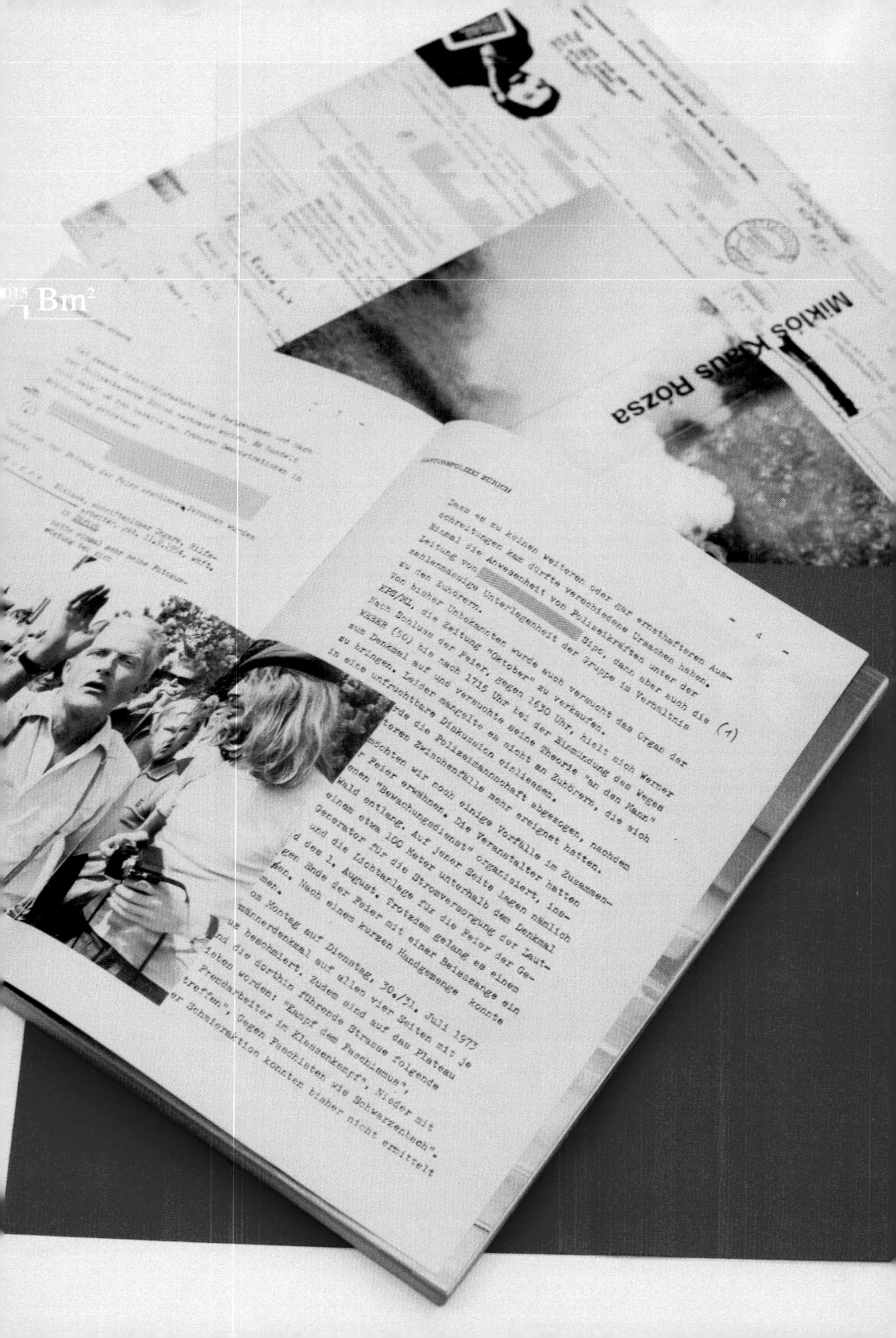
Miklós Klaus Rózsa
KANTONSPOLIZEI ZÜRICH
- 4 -
Dass es zu keinen weiteren oder gar ernsthafteren Aus-
schreitungen kam dürfte verschiedene Ursachen haben.
Einmal die Anwesenheit von Polizeikräften unter der
Leitung von Sipo, dann aber auch die
zahlenmässige Unterlegenheit der Gruppe im Verhältnis
zu den Zuhörern. (1)
Von bisher Unbekannten wurde auch versucht das Organ der
KPS/ML, die Zeitung "Oktober" zu verkaufen.
Nach Schluss der Feier, gegen 1630 Uhr, hielt sich Werner
WEBER (50) bis nach 1715 Uhr bei der Einmündung des Weges
zum Denkmal auf und versuchte seine Theorie "an den Mann"
zu bringen. Leider mangelte es nicht an Zuhörern, die sich
in eine unfruchtbare Diskussion einliessen.
rde die Polizeimannschaft abgezogen, nachdem
toren Zwischenfälle mehr ereignet hatten.
möchten wir noch einige Vorfälle im Zusammen-
r Feier erwähnen. Die Veranstalter hatten
onen "Bewachungsdienst" organisiert, ins-
Wald entlang. Auf jener Seite lagen nämlich
einem etwa 100 Meter unterhalb dem Denkmal
Generator für die Stromversorgung der Laut-
und die Lichtanlage für die Feier der Ge-
d des 1. August. Trotzdem gelang es einem
gen Ende der Feier mit einer Beisszange ein
zen. Nach einem kurzen Handgemenge konnte
men.
on Montag auf Dienstag, 30./31. Juli 1973
männerdenkmal auf allen vier Seiten mit Je
z beschmiert. Zudem sind auf das Plateau
nd die dorthin führende Strasse folgende
ieten worden: "Kampf dem Faschismus",
Fremdarbeiter im Klassenkampf", Nieder mit
treffen", Gegen Faschisten wie Schwarzenbach".
er Schmieraktion konnten bisher nicht ermittelt

这是一本将摄影师米克洛什·克劳斯·罗饶（Miklós Klaus Rózsa）的摄影作品和所在年代的档案进行整合，形成一种微妙冲突的书。本书的装订形式为锁线软精装，护封采用白卡纸，文本部分的底色被印上一层浅浅的本白色，与照片部分的白色形成微妙的反差。黑色的文字和照片都被处理成单黑形式，模拟扫描件的感觉。封面照片部分亮面 UV 工艺，内封鲜红色卡纸。内页部分胶版纸双色印刷，与封面印刷方式相同，通过照片和文本之间的相互罗列压叠，形成如电影蒙太奇般混排的形态。观察与反观察是本书的主旨，甚至读者在阅读本书的时候，也因为照片对文本的部分遮挡而不能完全获知一些信息。所以本书的大量文字并不是给读者细细品读的，而是作为一个“底色”存在，与表面覆盖的照片形成的冲突关系才是本书想表达的内核。

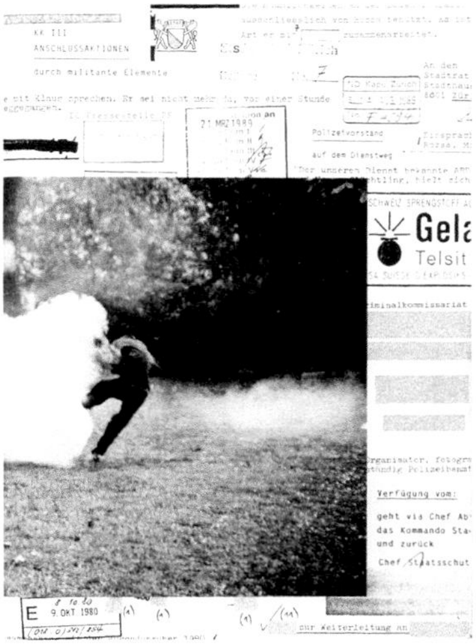
KK III
ANSCHLUSSAKTIONEN
durch militante Elemente
Polizeivorstand
auf dem Dienstweg
Telsit
Verfügung vom:
geht via Chef Ab-
das Kommando Sta-
und zurück
zur Weiterleitung an

This is a book that combines the photographs of Swiss photographer Miklós Klaus Rózsa with archives of the time that the photos were taken, creating a delicate conflict. The book is bound by thread sewing with the white cardboard jacket. The background color of the text part is natural white, which forms a subtle contrast with the white in the photos. The black text and photos are printed with black ink only, simulating the scanned copies. The image on the cover spots UV while the inside cover is made of bright red cardboard. The inside pages are printed in two colors on offset paper. By arranging and overlapping photographs and texts, the layout displays a mixed arrangement like montage in movies. Observation and anti-observation are the main purposes of the book. Even when readers read the book, they cannot fully understand some information because the photos partially obscure the text. Therefore, a large amount of text in the book is not for readers to read carefully, but as a 'background' existence. The text is meant to conflict with the covering photos.

- 铜奖
- 米克洛什·克劳斯·罗饶
- 德国

- Bronze Medal
- Miklós Klaus Rózsa
- Germany
- Christof Nüssli, Christoph Oeschger
- Christof Nüssli, Christoph Oeschger
- Spector Books / cpress, Leipzig
- 298×211×43mm
- 2140g
- 624p

624 DIN A 4 pages. Better have a quick flip through. A little bomb here, a dust cloud there, lots of police helmets and even more facsimiled typewritten pages. Everything in fitting black and white, thank you, it all makes sense: coming to terms with the past, stress with the state. Germany 1968? East Germany, Stasi? Hang on: metropolitan police Zurich, cantonal police Zurich, federal police? Stress in Switzerland! Staged as a source edition, the material compiled by the Swiss state between 1971 and 1989 about the photographer and political activist Miklós Klaus Rózsa, as well as the photographic oeuvre of Rózsa himself, has been converted into a work of art of contemporary history in this publication. One of the fascinating photographs: entangled barbed wire in the night, illuminated by flashlights. How pleasant it might have been for the men in uniform to placate the civic unrest is anyone's guess. The grave, sublime means of book design: all documents from state surveillance and contemporary reporting are given a chalky white fond; all photographs of Rózsa and the paratext are reproduced in black on white paper. This almost imperceptible but crucial contrast also involves the third observer, the reader, in the surveillance carousel. »Unstable elements«: probably the motto of both the federal police and the book designers. The forces at work were undoubtedly freaks.

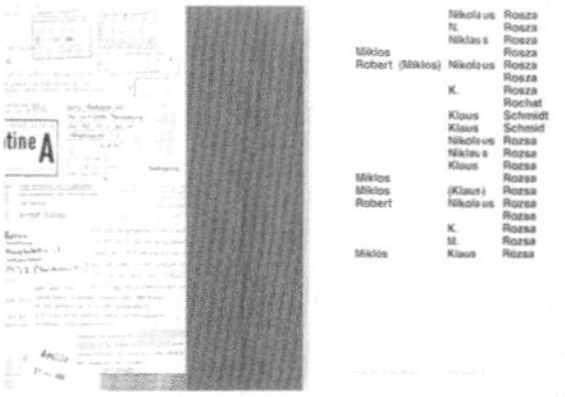

tine A
Nikolaus Rosza
N. Rosza
Niklaus Rosza
Miklos Rosza
Robert (Miklos) Nikolaus Rosza
Rosza
K. Rosza
Rochat
Klaus Schmidt
Klaus Schmid
Nikolaus Rozsa
Niklaus Rozsa
Klaus Rozsa
Miklos Rozsa
Miklos (Klaus) Rozsa
Robert Nikolaus Rozsa
Rozsa
K. Rozsa
M. Rozsa
Miklós Klaus Rózsa

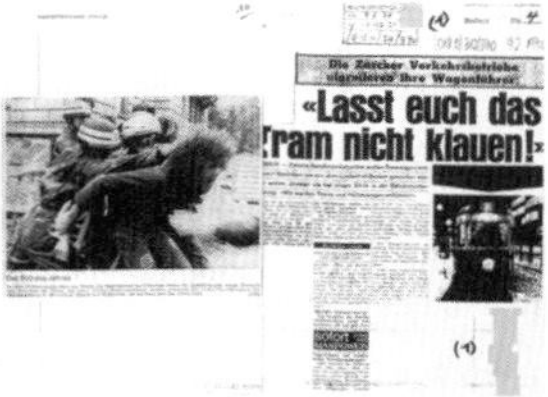

Die Zürcher Verkehrsbetriebe signalisieren ihre Wagenführer
«Lasst euch das Tram nicht klauen!»

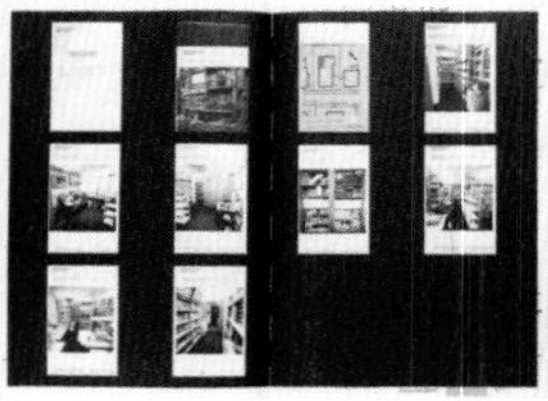

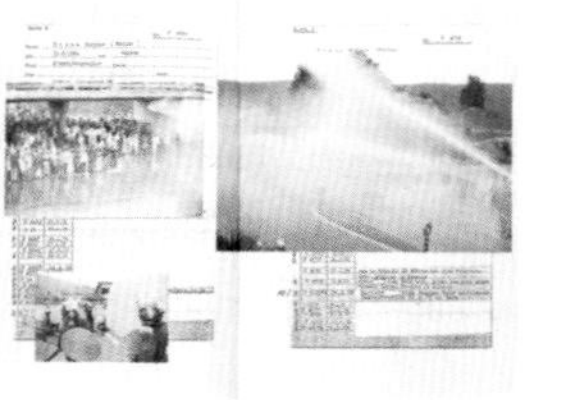

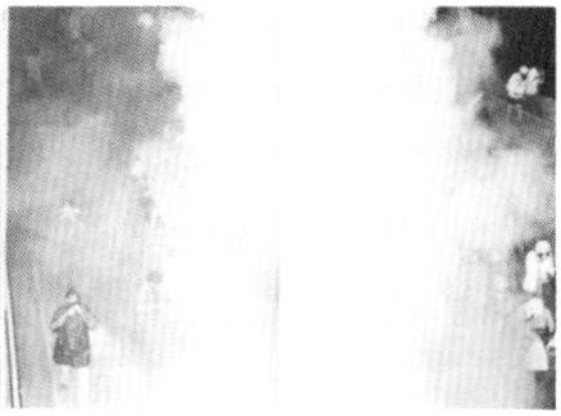

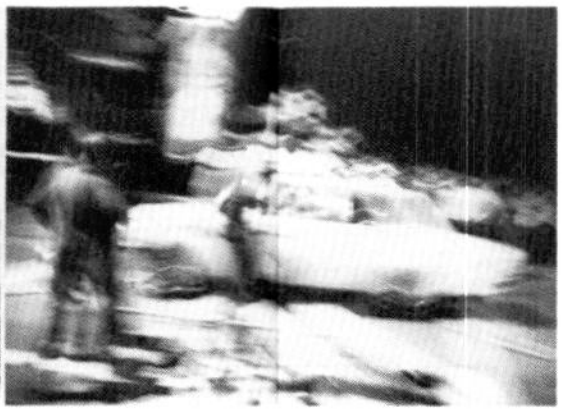

E [redacted] 12.3.82 (1) [redacted]

Kopie (11)

(1) [redacted]

Von [redacted] erhalten.

(11) (11)

...nt" (ohne
...hlt, werden

...blikation be-
kannt ist und nach dort vorliegenden Erkenntnissen zwischen
dieser Druckschrift und dem bekannten gleichnamigen Film inhaltliche Identität besteht.
Sind Ihnen die Produzenten von Buch und Film bekannt?

ND Zürich
mit der Bitte um einen kurzen Bericht an mich. Danke.

3.3.1982 / BUPO / [redacted]

(1)

Bitte dann, Buch und Film auch wieder an mich zurück.

[redacted] (1)

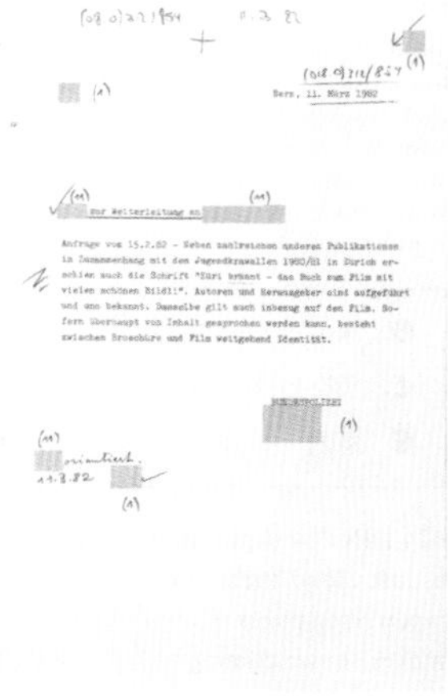

(08.0)212/854 11.3.82

[redacted] (1)

(018.0)212/854 (1)

[redacted] (1)

Bern, 11. März 1982

(11) [redacted] zur Weiterleitung an [redacted] (11)

Anfrage vom 15.2.82 - Neben zahlreichen anderen Publikationen
im Zusammenhang mit den Jugendkrawallen 1980/81 in Zürich erschien auch die Schrift "Züri brännt - das Buch zum Film mit
vielen schönen Bildli". Autoren und Herausgeber sind aufgeführt
und uns bekannt. Dasselbe gilt auch inbezug auf den Film. Sofern überhaupt von Inhalt gesprochen werden kann, besteht
zwischen Broschüre und Film weitgehend Identität.

BUNDESPOLIZEI

[redacted] (1)

(11)
[redacted] orientiert.
11.3.82 [redacted]
(1)

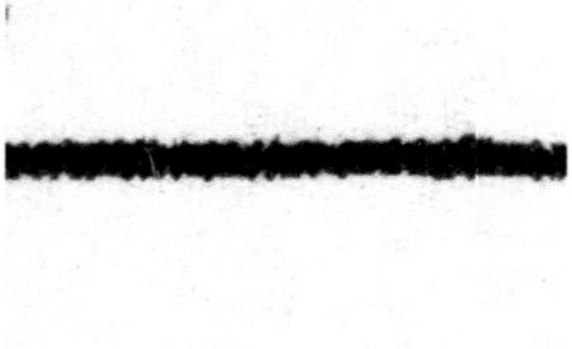

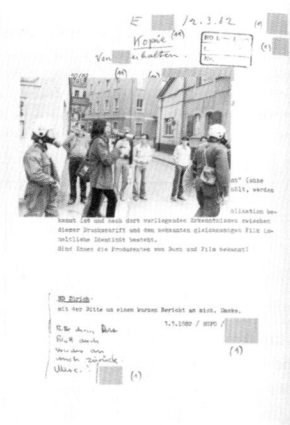

Gelatine B
Telsit
Gelatine A

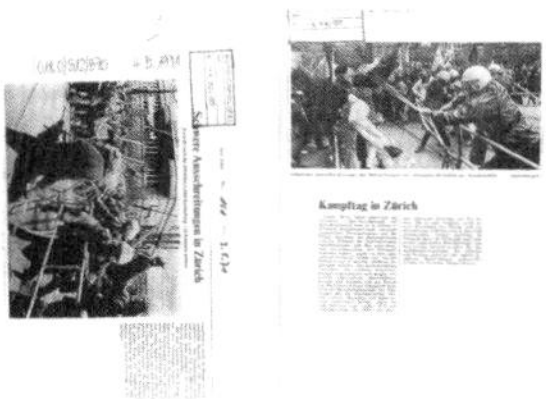
Kampftag in Zürich

TELEX
EINGANG

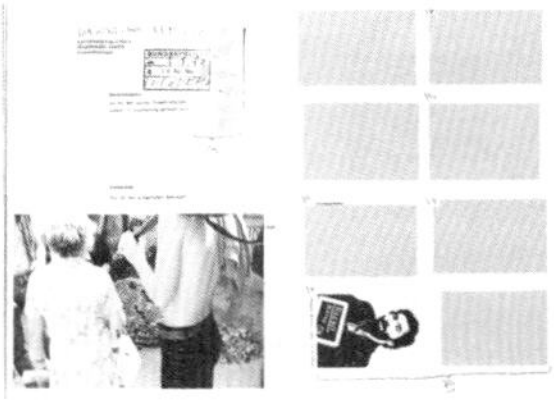

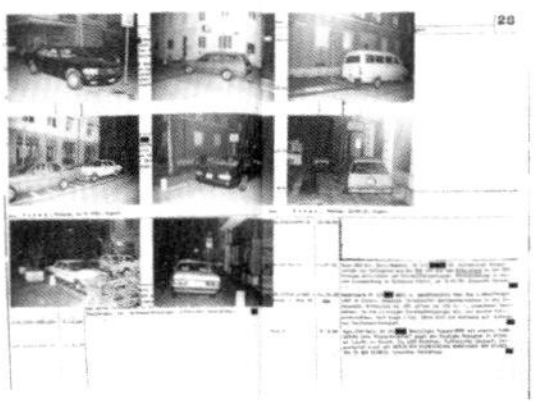

5
Christof Nüssli
Christoph Oeschger
Miklós Klaus Rózsa
x. möcht
sei er w
iert:
an BA
" NDKP
cpress
Spector Books
Bemerkungen im

Ausschnitt aus dem
Tages - Anzeiger [redacted] 5
Nr. 37 vom: 14.2.85

(1179:0)321.4/58

15. MRZ 1985

[redacted] 5

Das Bundesgericht sprach von Willkür

Obergerichtsurteil auf einseitiger Grundlage

lv. Zu Unrecht – so stellte am Mittwoch das Bundesgericht in Lausanne fest – hat sich das Zürcher Obergericht allein auf die Aussagen des Geschädigten verlassen, als es einen Pressefotografen, der an einem Auto die Radioantenne geknickt haben sollte, wegen Sachbeschädigung verurteilte. Dass es angesichts einiger Ungereimtheiten von einer Abklärung der vom Angeklagten erhobenen Einwände absah, war willkürlich und somit verfassungswidrig. Die staatsrechtliche Beschwerde des Fotografen wurde daher gutgeheissen und das Obergerichtsurteil aufgehoben.

Das Urteil sei etwas vom Groteskesten, was er je erlebt habe, rief seinerzeit Rechtsanwalt Franz Schumacher als Verteidiger des Pressefotografen vor dem Obergericht aus, als er die vom Einzelrichter wegen Sachbeschädigung verhängte, bedingt vorzeitig löschbare Busse von 400 Fr. anfocht (TA vom 25. 5. 83). Die Verurteilung aufgrund «absolut ungenügender Beweise» sei die Folge einer von der Polizei gegen seinen Mandanten veranstalteten «Hetzjagd», weil jener wegen Misshandlung in einer Krawallnacht Strafanzeige gegen die Polizei erstattet habe. Das Obergericht räumte wohl ein, dass das erstinstanzliche Urteil «nicht in jeder Beziehung mustergültig» war, doch zweifelte auch es nicht an der Täterschaft des Fotografen und bestätigte die Strafe. Sein Urteil hielt auch der Überprüfung durch das Kassationsgericht stand.

An sich handelte es sich um eine Bagatelle: Nach einem ersten Zwischenfall mit einem Automobilisten vor der Sihlpost, weil der Fotograf angeblich beim Aussteigen dessen «Ford Mustang» beschädigt hatte, war es Stunden später zufällig im Seefeld wieder zu einem Zusammentreffen gekommen. Der Fotograf soll, als er angeblich aus beruflichen Gründen einen Demonstrationszug begleitete (es war die Zeit der Jugendunruhen), den am Strassenrand parkierten «Ford Mustang» erblickt und ihm die Radioantenne abgeknickt haben. Das bestritt der Fotograf energisch, doch die Richter glaubten seinem Gegenspieler, weil dieser sonst keinerlei Veranlassung gehabt hätte, den ohne sein Wissen bei der Demo auftauchenden «Mann von der Sihlpost» als Urheber der Sachbeschädigung anzuzeigen.

In seiner *staatsrechtlichen Beschwerde* rügte der Fotograf, dass die Zürcher Instanzen nur auf die Aussagen des Geschädigten abgestellt und seine sämtlichen Beweisanträge abgewiesen hätten. Damit sei ihm das rechtliche Gehör verweigert und damit das Willkürverbot der Bundesverfassung verletzt worden. Seine unzähligen Vorwürfe wies am Mittwoch die *I. Öffentlichrechtliche Abteilung* des Bundesgerichtes zum allergrössten Teil zurück, weil der Entscheid eines Gerichtes über die Abnahme beantragter Beweise eine in seinem Ermessen liegende vorweggenommene Beweiswürdigung darstelle.

Übrig blieben aber *zwei dunkle Punkte*, über die sich die kantonalen Instanzen nach Meinung des Bundesgerichts n… hätten hinwegsetzen dürfen. Einmal h… der Anzeigeerstatter, obwohl er den F… grafen auf einem Bild sofort erkannt… te, unzutreffende Angaben über des… Bartwuchs sowie Farbe und Dichte … Haars gemacht. Zum andern – und … wog noch viel schwerer – hatte die Beg… terin des Automobilisten, die mit ihm … einem Café aus den Vorfall beobach… hatte, nicht nur (gemäss Version des K… sationsgerichts) den Fotografen ni… wiedererkannt, sondern laut Polizeip… tokoll erklärt: «Nein, es ist ein ande… gewesen.»

Bei dieser Sachlage war – so meinte … Referent – die Möglichkeit einer V… wechslung nicht mehr auszuschliess… und es wäre darum Aufgabe des Geric… gewesen, hier durch *Einvernahme* … *Zeugin*, die von keiner der drei kanto… len Instanzen vorgeladen worden w… Klarheit zu schaffen. Indem trotz den W… dersprüchen ohne weitere Abklärung allein auf die in ihrer Glaubwürdigk… erschütterten Aussagen des Geschäd… ten abgestellt wurde, seien die Grenz… zur Willkür überschritten worden. Es l… ge hier «nicht einmal ein diskutierbar… Grenzfall» vor, meinte sogar einer d… Richter.

So kam es zur Gutheissung der B… schwerde des Fotografen und zur Auf… bung des kantonalen Schuldspruchs. A… Obergericht liegt es nun wieder, ob es d… unterlassenen Beweiserhebungen nac… holen und dann neu urteilen oder glei… einen Freispruch fällen will.

Betrifft: R o z s a , Miklos, geb. 11.9.1954 (bek.)

5
15.2.85 [redacted]
Verfügung: ad acta …
Zürich, den 18. FEB. 1984
Der … Kopie an 15.II.
[redacted] 5

(1.18)94 - 16

(1179:0)321.4/58 - 23

Aussersißler Zeitung

Nr. 6 vom 15.3.1985

22. APR. 1986

5

Foto Klaus Rozsa

Klaus Rozsa und sein Mitarbeiter David Adair

Das Fotofachgeschäft als Quartierladen

Neben dem Ladentisch sind Hinweise auf Veranstaltungen im Quartierzentrum Kanzlei angebracht und Zettelchen wie «Lederjacke zu verkaufen», «Grosse Wohnung gesucht von Frau mit drei Kindern», «Verkaufe Hasselblad-Objektiv, günstig». Eine gemütliche Sitzecke lädt zum Blättern in Fachzeitschriften ein.

Im sehr bewusst als Quartierladen gehaltenen Fotofachgeschäft «multimedia» an der Anwandstrasse 34 darf auch über anderes geredet werden als Kameras und Filmqualitäten, und Klaus Rozsa und David Adair nehmen sich Zeit beim Bedienen und Beraten. «Es beelendet mich zu sehen, wie im Aussersihl ein Lädeli nach dem andern zum Sex-Shop oder Spielsalon umfunktioniert wird, denn Kleingewerbe und Fachgeschäfte erfüllen doch eine wichtige Aufgabe in diesem Wohnquartier», meint der gelernte Fotograf Rozsa. Darum hat er sich im Sommer 1983 unweit der Stauffacherstrasse, an der er aufgewachsen ist, ein Atelier und einen Laden eingerichtet.

Rozsas Liebe gilt den Schwarzweissfotografien, die er in seinem eigenen Labor entwickelt, doch bietet «multimedia» auch Fotoartikel aller Art, Videogeräte und -zubehör sowie Electronic-Artikel an. Nebst Pass-, Portrait- und Werbeaufnahmen übernimmt das Team Rozsa/Adair auch Fotoreportagen.

ADAIR, David, 1.6.1964 bek.

ROZSA, Klaus, 11.9.1954 bek.

24.3.1986 5

24. MRZ 1986

BEHR · DIE GOLDENEN ZWANZIGER JAHRE

这是美籍华裔艺术家陈佩之（Paul Chan）在隐居时创作的一系列作品的合集。他将各式书籍封面作为画板去承载自己记忆中的画面，甚至是对一些概念的理解和想象，继而写下一些被自己"编码"过的文字去解读或者只是与这些作品相互"陪伴"。这是艺术家一贯探索的方向，即将内容转换为各种不甚直白的文本，探索"语言和交互之间的关系"，以及"表现出来的和实际想表达之间的差异"，这些留待观看者自己去"解读"。本书为圆脊锁线硬精装，书函采用牛皮卡瓦楞纸印单黑，封面采用深蓝紫色仿皮面裱覆卡板，文字烫金、色块烫透明珠光色的工艺。黑卡纸在书中除了环衬外，还作为章节分隔页使用，将作品分为八大部分，以及分隔出前后辅文。书籍开篇引用了古希腊哲学家赫拉克利特（他的文章只留下只言片语，大量使用隐喻，致使后人产生许多不同的解释。）的话，而陈佩之的艺术创作也是对此进行的实践。

NEW NEW TESTAMENT
PAUL CHAN

This is a collection of works by an American Chinese artist Paul Chan during his seclusion. He used all kinds of book covers as sketchpads to carry pictures in his memory, even understanding and imagination of some concepts, and then wrote down some words that he 'coded' to interpret or 'accompany' with these works. The artist has always been exploring how to transform content into various texts that do not necessarily express meaning directly, 'the relationship between language and interaction', and 'the difference between what is shown and what is actually intended to be expressed'. This hardback book is bound by thread sewing with rounding spine. The slipcase is made of corrugated Kraft paper printed in black. The cover is made of dark blue-purple faux leather mounted with graphic board. The text is gold hot stamping while the color blocks are hot stamped with transparent pearlescent colors. Black cardboard is used as the end paper and the chapter dividing pages, separating the work into eight parts, as well as isolating the supplementary information. The book begins with a quotation from the ancient Greek philosopher Heraclitus (who left only a few words in his article and used metaphors extensively, resulting in many different interpretations for later generations) to indicate that Paul Chen's is doing the similar artistic experiments.

铜奖
陈佩之的新新约
瑞士

Bronze Medal
Paul Chan. New New Testament
Switzerland
Kloepfer-Ramsey Studio, Brooklyn
Laurenz Foundation, Schaulager, Basel; Badlands Unlimited, New York
Laurenz Foundation, Schaulager, Basel
272×182×74mm
2838g
1092p

Proportions of the book body: comparable to a lectionary. Illustrations on the right-hand book pages: the artist has freed the book jackets from their contents. He uses the opened covers as upright canvases. On these canvases, he paints blunt rectangles, spread out, bluish grey, light to dark, sometimes with a mountain motif – like label tags without titles. Every new picture is meticulously numbered, verse-like, compulsive. The original lines on the spine remain mostly legible. Text on the left-hand pages: each of the numbers is added by a text – coded in concrete poetry with extended punctuation and syntax, as if directed by a higher force. Back in the baroque age, this principle of apparently knotted semiotics already fired up pensive minds to unravel the meaning of the world in emblematic books. The seriousness of such undertakings is, of course, only guaranteed, if everything is arranged with extreme care and stringency, which means that everything must be above board: perfect proportions, classic typesetting in Garamond, precise print. An exegetic attempt: the material side of the testament is its existence as a book. The semantic side is that it is hidden through a code. The art in the book turns the book into art becomes art through the book which produces the art.

PAUL CHAN
NEW NEW
TESTAMENT

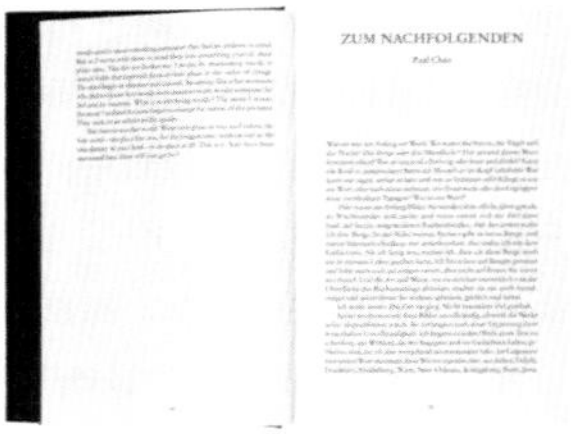
ZUM NACHFOLGENDEN
Paul Chan

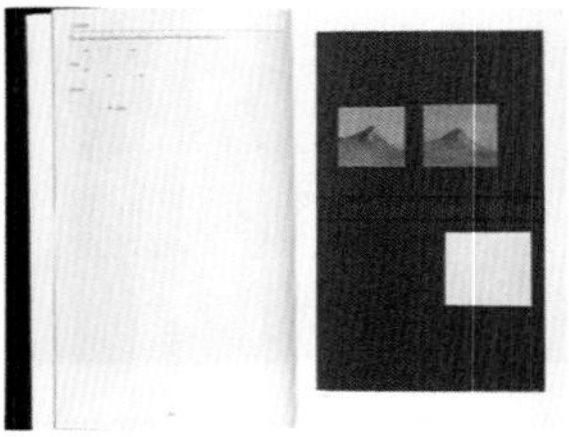

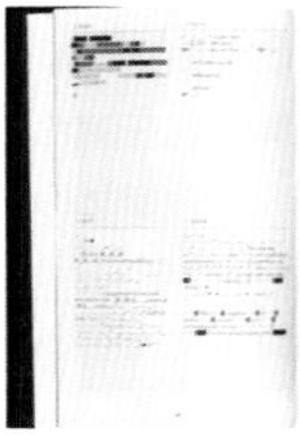

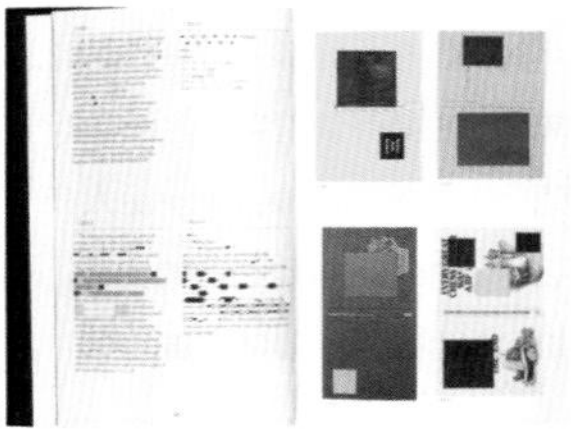

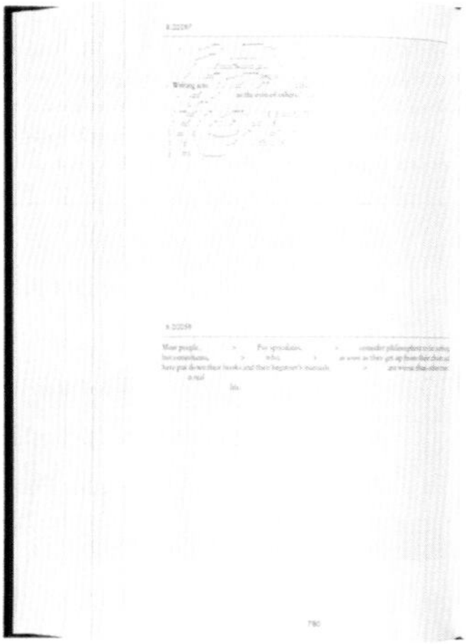

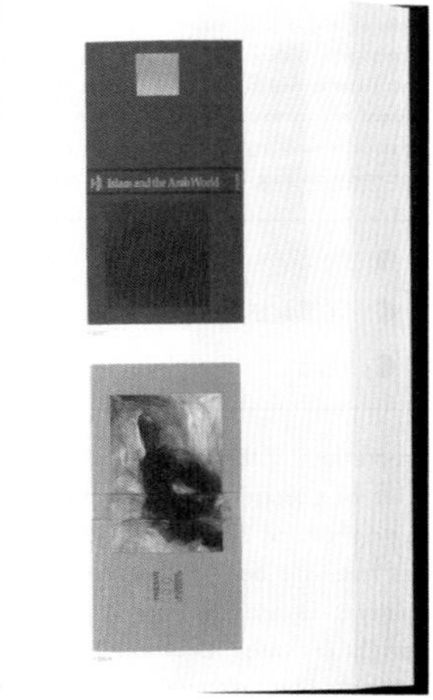
Islam and the Arab World

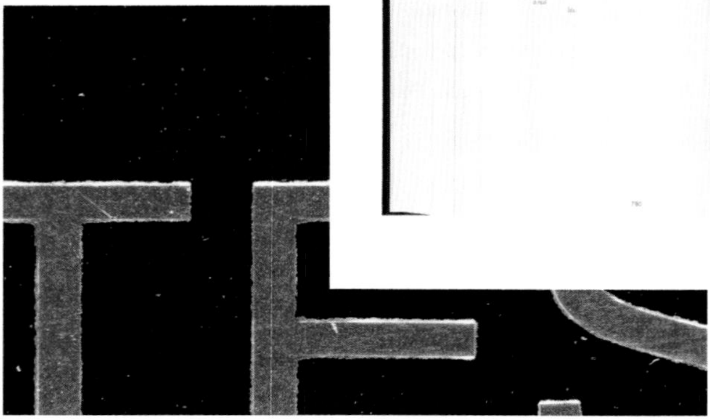

Meat

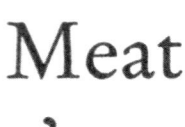

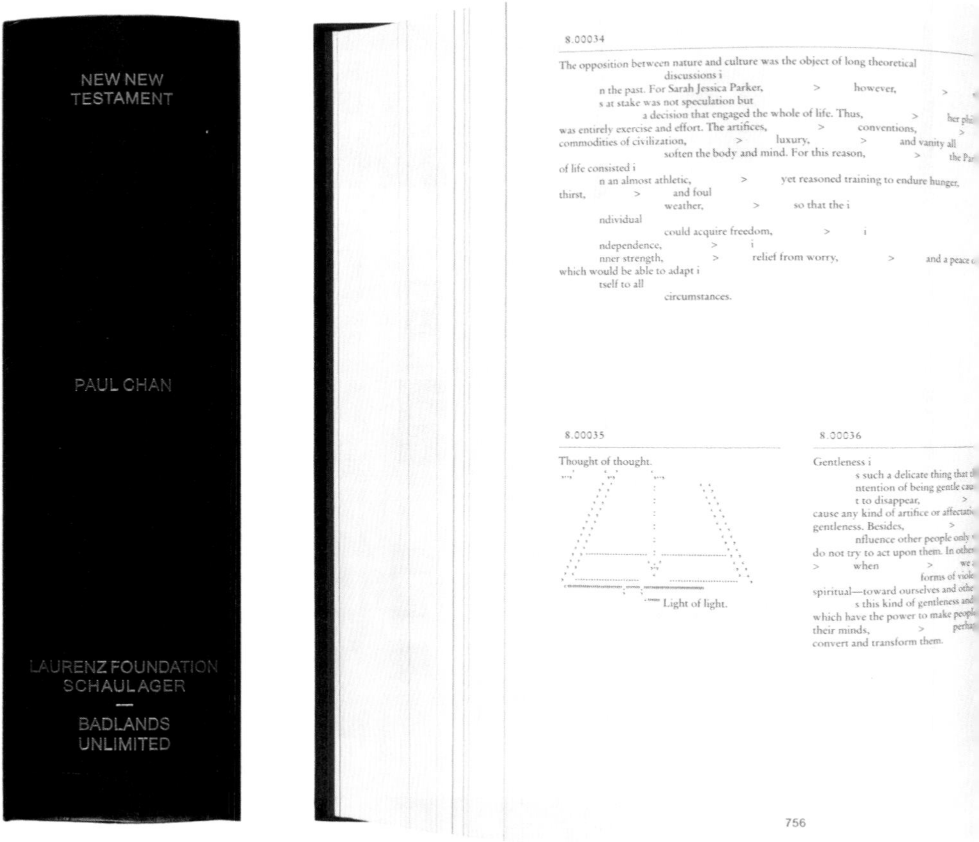
NEW NEW TESTAMENT
PAUL CHAN
LAURENZ FOUNDATION SCHAULAGER
BADLANDS UNLIMITED
8.00034
The opposition between nature and culture was the object of long theoretical
discussions i
n the past. For Sarah Jessica Parker, > however, >
s at stake was not speculation but
a decision that engaged the whole of life. Thus, >
was entirely exercise and effort. The artifices, > conventions,
commodities of civilization, > luxury, > and vanity all
soften the body and mind. For this reason, >
of life consisted i
n an almost athletic, > yet reasoned training to endure hunger,
thirst, > and foul
weather, > so that the i
ndividual
could acquire freedom, > i
ndependence, > i
nner strength, > relief from worry, >
which would be able to adapt i
tself to all
circumstances.
8.00035
Thought of thought.
Light of light.
8.00036
Gentleness i
cause any kind of artifice or
gentleness. Besides,
do not try to act upon them.
when
spiritual—toward ourselves and
which have the power to make
their minds,
convert and transform them.
756

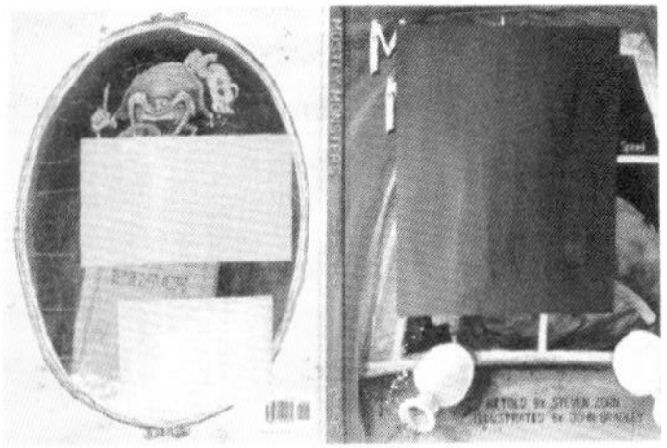

MOSTLY MONSTERS
RETOLD by STEVEN JOHN
ILLUSTRATED by JOHN SPRINKLEY

A Day in the Life of the
SOVIET UNION
A Day in the Life of the Soviet Union

ANDRE MASSON
ANDRE MASSON

ou autres
une façade,
la lisibilité
qu'offrent
plusieurs critères pour
ces grilles. Tout d'abord,
qui doit être diffuse
obtenir un passage capti-
entre les blancs et les
noirs, elle permet de conserver
le rythme régulier des horizontales qui les composent.
Leur format doit être assez
important pour permettre à
l'œil de pouvoir être accroché.
my framing exclusively on the
vent in itself, ignoring external
elements such as brick or metal
sheet walls, as they disrupt the
blades' optical interplay.
The choice of these
vents was

这是艺术家瓦莱里安·戈阿莱克（Valérian Goalec）拍摄和思考基础形式的摄影与论文集。首先，艺术家拍摄了建筑外的通风口百叶窗，研究了它们的功能和内部构造。第二步，将这些基础结构从原本的功能里抽离出来，当作雕塑去研究其材料、光影和对人与环境的影响。第三步，将“基础元素”进行整理，畅想之后可以被再创造的可能性。本书的装订形式为无线胶平装。全书采用胶版纸印单色黑，图片部分挂粗网，通风口百叶窗横平竖直居中放置，文字部分则采用中规中矩的 Akzidenz Grotesk。全书极力将所有形式的独特性抹去，只通过缩进和上下分区来呈现法英两种文字，尽可能地体现“基础元素”这一处于原始素材和目标结果中间单纯而充满无限可能的“奇点”状态。从内容结构方面看，图片被放在最前面，三篇文章后置，目录放在封底，呈现观察和思考通风口百叶窗这一由表及里的过程。|||||||||||||||||||

This is a collection of photographs and essays by the artist Valérian Goalec, who studied the basic forms of photography. First, the artist photographed the ventilation louvers outside the building and studied their functions and internal structures. The second step is to separate these infrastructures from their original functions and to use them as sculptures to study their materials, light and shadow, and their impact on people and the environment. The third step is to sort out these 'basic elements' and imagine the possibility that they can be recreated. This paperback book is perfect bound. The entire book is printed on offset paper in monochrome black. The pictures are printed with rough screen and the photos of ventilation louvers are placed horizontally and vertically in the middle of the pages. The text is typeset with regular font Akzidenz Grotesk. The book strives to erase the uniqueness of all forms, only using indentation and partition to present French text and English text. The book endeavors to display the 'basic elements' that are in the 'singularity' state between the original material and target results. In terms of the book's structure, pictures are placed at the front, three articles are placed at the back, catalogues are placed at the back cover, showing the process of observing and thinking about ventilation louvers from the outside to the inside.

铜奖
基础元素 01
比利时

Bronze Medal
Éléments Structure 01
Belgium
Alexis Jacob, Valérian Goalec
Valérian Goalec, Béatrice Lortet
Théophile's Papers
232×174×5mm
109g
60p

»The document creates the work, and the work creates the document.« This sentence, taken from the catalogue, indicates the interlocked referentiality or artistic concepts, where, for example, a printed catalogue is not just an accessory, but an essential ingredient. And this appears to be the case here, too. The elementary form is the subject of the photographic series of ventilation grids; the design of the catalogue is equally elementary. So elementary that the voluminous book paper, the bare necessity of the coarse screen, the monochrome under-dyed and torn-off print are hard to beat. An unmistakeable sign of this strategy is the stylistic quotation of "Elementary Typography", which nevertheless utilizes Grotesk as an accent in its restriction to two fonts. This has to suffice.

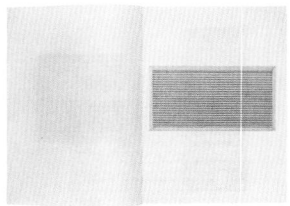

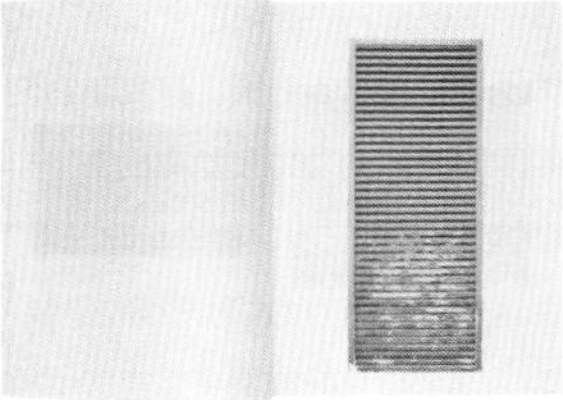

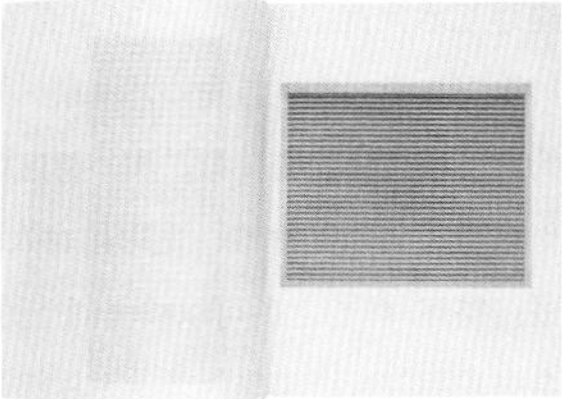

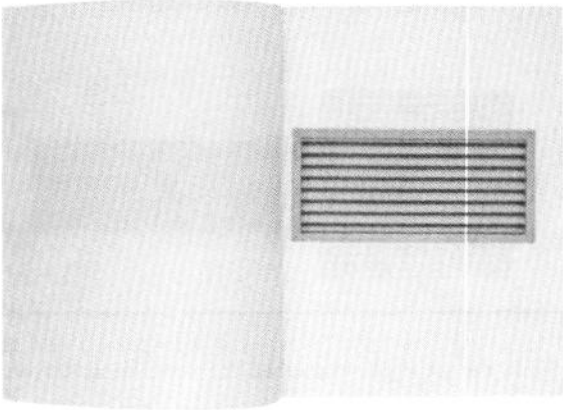

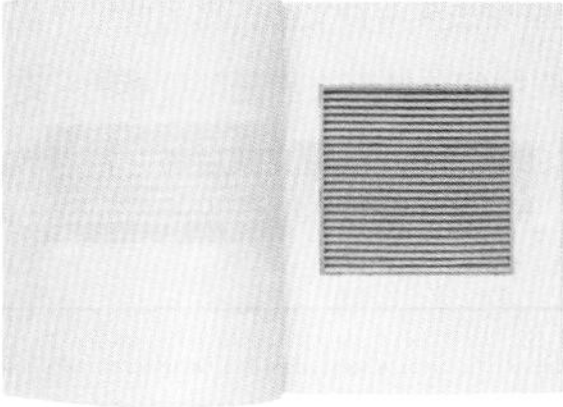

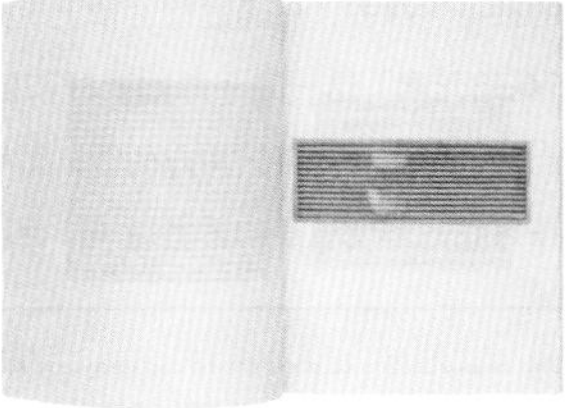

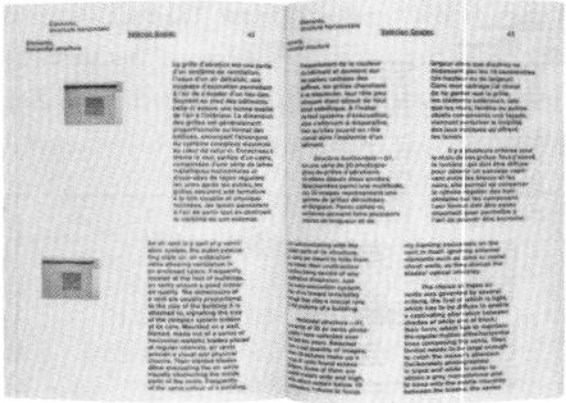

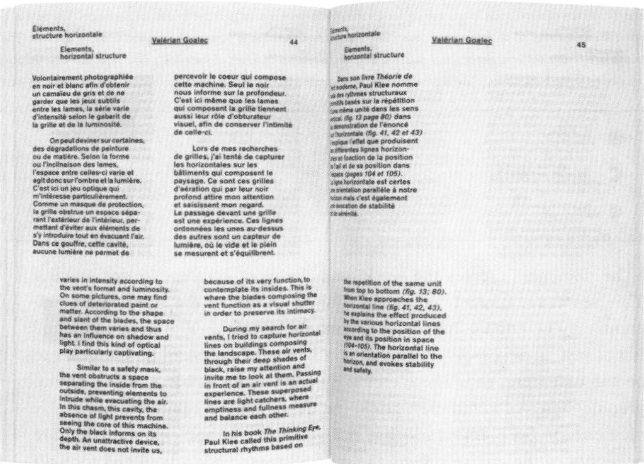

Éléments,
structure horizontale

Valérian Goalec

44

Elements,
horizontal structure

Volontairement photographiée en noir et blanc afin d'obtenir un camaïeu de gris et de ne garder que les jeux subtils entre les lames, la série varie d'intensité selon le gabarit de la grille et de la luminosité.

On peut deviner sur certaines, des dégradations de peinture ou de matière. Selon la forme ou l'inclinaison des lames, l'espace entre celles-ci varie et agit donc sur l'ombre et la lumière. C'est ici un jeu optique qui m'intéresse particulièrement. Comme un masque de protection, la grille obstrue un espace séparant l'extérieur de l'intérieur, permettant d'éviter aux éléments de s'y introduire tout en évacuant l'air. Dans ce gouffre, cette cavité, aucune lumière ne permet de percevoir le coeur qui compose cette machine. Seul le noir nous informe sur la profondeur. C'est ici même que les lames qui composent la grille tiennent aussi leur rôle d'obturateur visuel, afin de conserver l'intimité de celle-ci.

Lors de mes recherches de grilles, j'ai tenté de capturer les horizontales sur les bâtiments qui composent le paysage. Ce sont ces grilles d'aération qui par leur noir profond attire mon attention et saisissent mon regard. Le passage devant une grille est une expérience. Ces lignes ordonnées les unes au-dessus des autres sont un capteur de lumière, où le vide et le plein se mesurent et s'équilibrent.

varies in intensity according to the vent's format and luminosity. On some pictures, one may find clues of deteriorated paint or matter. According to the shape and slant of the blades, the space between them varies and thus has an influence on shadow and light. I find this kind of optical play particularly captivating.

Similar to a safety mask, the vent obstructs a space separating the inside from the outside, preventing elements to intrude while evacuating the air. In this chasm, this cavity, the absence of light prevents from seeing the core of this machine. Only the black informs on its depth. An unattractive device, the air vent does not invite us, because of its very function, to contemplate its insides. This is where the blades composing the vent function as a visual shutter in order to preserve its intimacy.

During my search for air vents, I tried to capture horizontal lines on buildings composing the landscape. These air vents, through their deep shades of black, raise my attention and invite me to look at them. Passing in front of an air vent is an actual experience. These superposed lines are light catchers, where emptiness and fullness measure and balance each other.

In his book *The Thinking Eye*, Paul Klee called this primitive structural rhythms based on

Éléments,
structure horizontale

Valérian Goalec

45

Elements,
horizontal structure

Dans son livre *Théorie de* …rt moderne, Paul Klee nomme …s des rythmes structuraux …mitifs basés sur la répétition …ne même unité dans les sens …tical. *(fig. 13 page 80)* dans …démonstration de l'énoncé …l'horizontale *(fig. 41, 42 et 43)* …xplique l'effet que produisent …s différentes lignes horizon…tales en fonction de la position …e l'œil et de sa position dans …espace *(pages 104 et 105)*. …a ligne horizontale est certes …ne orientation parallèle à notre …orizon mais c'est également …ne évocation de stabilité …t de sérénité.

…he repetition of the same unit from top to bottom *(fig. 13; 80)*. When Klee approaches the horizontal line *(fig. 41, 42, 43)*, he explains the effect produced by the various horizontal lines according to the position of the eye and its position in space *(104–105)*. The horizontal line is an orientation parallel to the horizon, and evokes stability and safety.

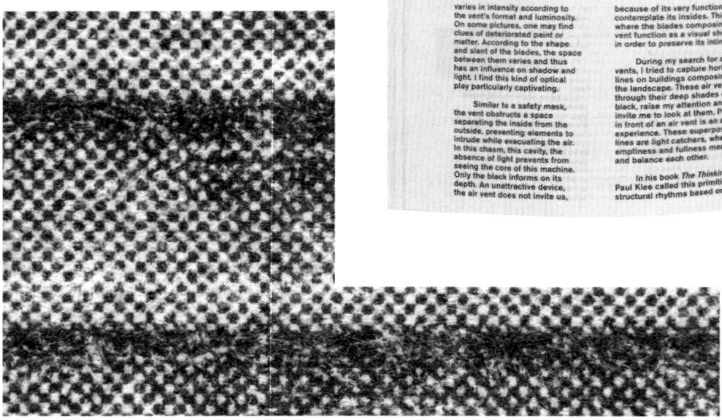

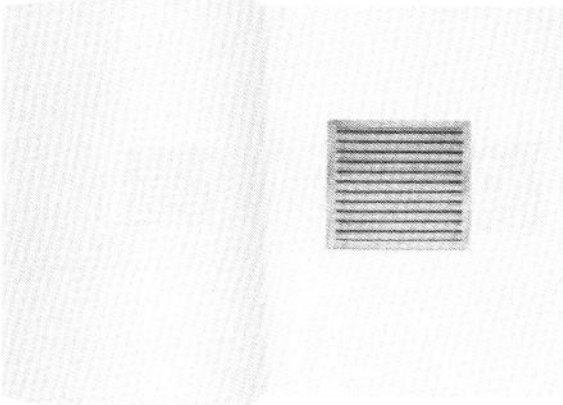

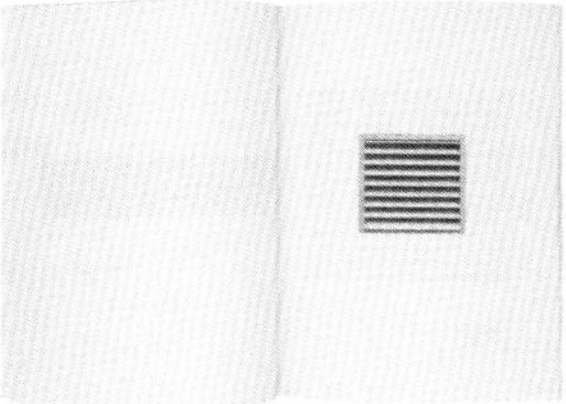
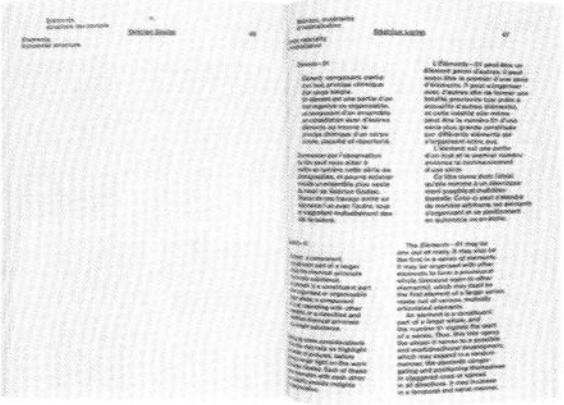
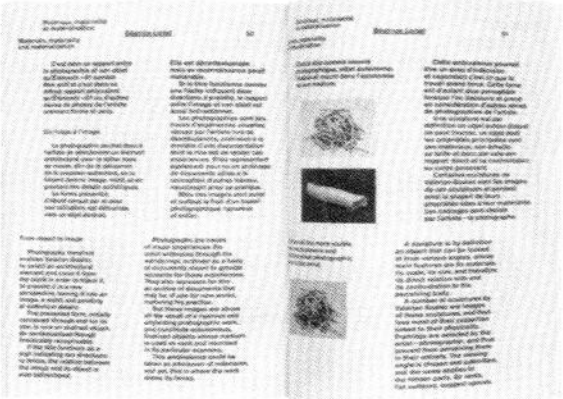
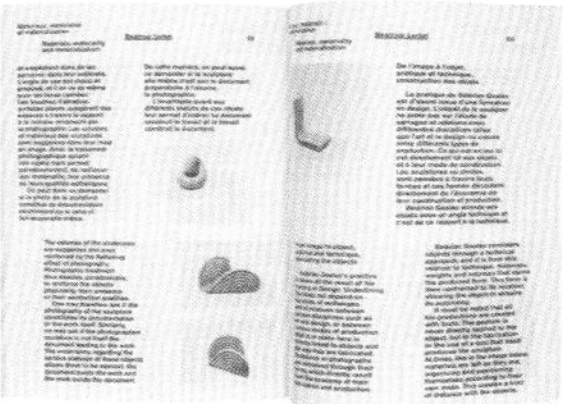

perme
égulie
i les comp

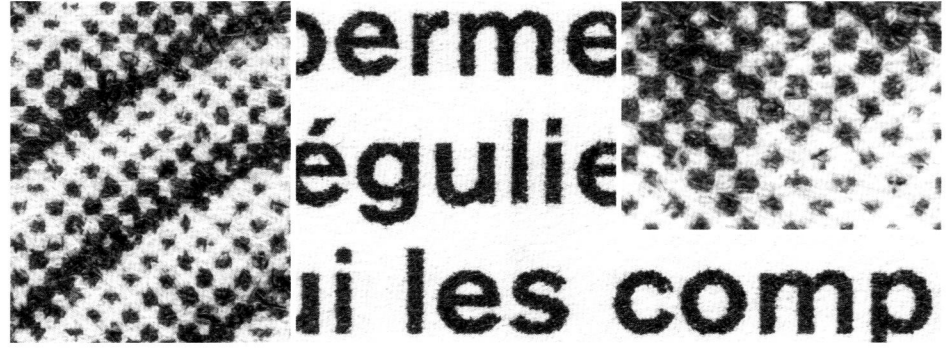

aux matériaux, aux poids et aux volumes que nait la forme produite. Cette forme est ensuite confrontée à son récepteur et c'est par cette rencontre que l'objet acquiert son autonomie.

Il est aussi important de souligner que tous les objets produits le sont grâce à l'aide d'outils. En effet, le geste n'est jamais directement appliqué l'objet, ce geste est la fabrication ou l'utilisation d'un outil et c'est celui-ci qui produira la sculpture. Ou encore, comme l'image ci-contre, la matière est laissée à elle-même, elle s'auto-organise, se répartit selon ses propres masses. Cela provoque une sorte de mise à distance de l'objet, à la fois de son créateur et de son spectateur, celle-ci grandissant lorsque le filtre photographique intervient ensuite.

both from its maker and its spectator, a distance that increases with the photographic filter.

The term "prototype" is often attached to the lexicon of industrial design. It is both a finalised object and its own project. It is the idea and the realisation of its idea, simultaneously, without being its absolute finalisation. Thus, it is connected to volume and its image in Goalec's air vents and sculpture photographs, an unstable and paradoxical status offering various developments and variations.

If a designed object is made such with its use in mind, sculpture is built in the perspective of its relation with the spectator, and it is through its physicality that it will raise the spectator's interest.

ériaux, matérialité
ıatérialisation

ateriality
ization

Béatrice Lortet

e terme « prototype » fait partie
vocabulaire lié au design
ıstriel. Il est à la fois un objet
lisé et son projet. Il est l'idée
a réalisation en même temps,
s pour autant en être
nalisation totale. Il rejoint
cette manière le lien entre
olume et son image dans
photographies de bouches
ération ou de sculptures,
statut instable et paradoxal,
posant développements
déclinaisons.
Si un objet né du design
t dans l'optique de son
isation, la sculpture est
ıstruite dans l'idée de son
port au spectateur et c'est
sa matérialité que celui-ci
uvera son intérêt.
La pratique de Valérian peut
développer ou se condenser.

Les objets peuvent se décliner, ils sont modulables à travers des contraintes techniques précises.

Leur stabilité se situe dans leur traitement technique, dans leur médium, dans leur construction elle-même.

p. 51 *Sans titre*, argile et polyméthacrylate de méthyle rouge, 40 cm × 60 cm, 2012. — **p. 51** *Sans titre*, ø 6 cm × 39 cm, argile, 2012. — **p. 52** *Sans titre*, argile, dimensions variable, 2012. — **p. 53** *Sans titre*, argile, 13 cm × 14 cm × 7 cm, 2012. — **p. 54** *Corps 4/7*, plâtre, dimensions variable, 2013.

Goalec's practice
evelop or condensate.
may take a number
ıd are modulable
ecise technical
. Their stability lies
hnical treatment,
ım, and their very
ɔn.

d, extruded clay and green
methacrylate, 40 cm × 60 cm,
2 *Untitled*, clay,
m × 12 cm, 2012. — **p. 52**
y, 27 cm × 16 cm × 12 cm, 2012.
os 1/7, plaster, variables size,
5 *Corps 6/7*, variables size,

015 Bm 5

这是一本除了书函完全没有文字的影集。荷兰艺术家布鲁诺·范登埃尔斯豪特（Bruno van den Elshout）在家乡荷兰海牙的一座酒店楼顶搭建了专门的摄影装置。这个装置固定机位和视角，按照艺术家设定的每小时拍一张照片的设置，拍摄了这个海平面 2012 年全年的照片，共 8785 张，书中精选了其中的 300 张。整本书厚实沉重，但印刷和工艺都十分考究。书函和书籍本体有如一块巨大的刨木，平整而坚固。书籍的装订形式为双页对裱的蝴蝶装，书籍采用了双层书函的设计，外书函采用灰色艺术纸裱覆卡板，只印刷最基本的书名、作者名、时间地点和简介信息。内书函为棕色卡纸环状包裹，留出上下档口用于取书。书籍本体没有严格意义上的封面，书脊部分贴白色布面材料，给予加固。内页采用哑面艺术纸四色印刷，内页排版设计为“可变网格”，使得整个翻阅的连续页面形成丰富的节奏和旋律，就像一本乐谱。设计纯净、极简而流畅，展现着瞬息万变的海平面。

This is an album with no words at all except the cover. Dutch artist Bruno van den Elshout built a special photographic device on the roof of a hotel in Hague, the Netherlands. The device fixes the position and angle of view, took a picture of the sea level every one hour. In the whole year of 2012, 8785 photos are taken and 300 photos are selected in the book. Although the book is thick and heavy, printing and binding technologies are very sophisticated. The slipcase and the book itself are like a huge plank, flat and solid. The book is butterfly-fold binding with double-layer slipcase. The outer case is made of gray art paper mounted with graphic board; only the title, author's name, time, place and brief information are printed on it. The inner case is made of brown cardboard, leaving the upper and lower parts open for taking the book. Technically, the book itself has no cover. The spine is reinforced with white cloth. The inside pages are printed in four colors on matte art paper, and 'variable grids' are used to typeset, which make the continuous pages have abundant rhythm and melody, just like a music score. The design is pure, simple and smooth, showing the rapidly changing sea level.

铜奖
新地平线
荷兰

Bronze Medal
New Horizons
The Netherlands
Rob van Hoesel
Bruno van den Elshout
The Eriskay Connection, Breda
340×235×68mm
3993g
212p

From the robust slipcase, the reader pulls a sturdy banderole. And from the banderole he pulls, well, a book or rather a book block. No cover, no dustjacket, no thread or adhesive binding, the spine just delicately fluted by the folded edges of the sheets. The block does not really feel like a book. It can be opened without any resistance, but it is as stiff as a massive piece of planed wood. Apart from its smooth pages, the straightforward cube has nothing book-like about it. There is just this shape; the whole object has a minimalist appearance. The exterior contrasts strongly with the fluid theme of the pictures: water, sea, sky, light, clouds. 300 photos on 212 pages show the horizon of the sea. The whole thing is a selection from 8785 pictures that were taken from the same spot every hour for a year. Floating between sky and water. A book without words.

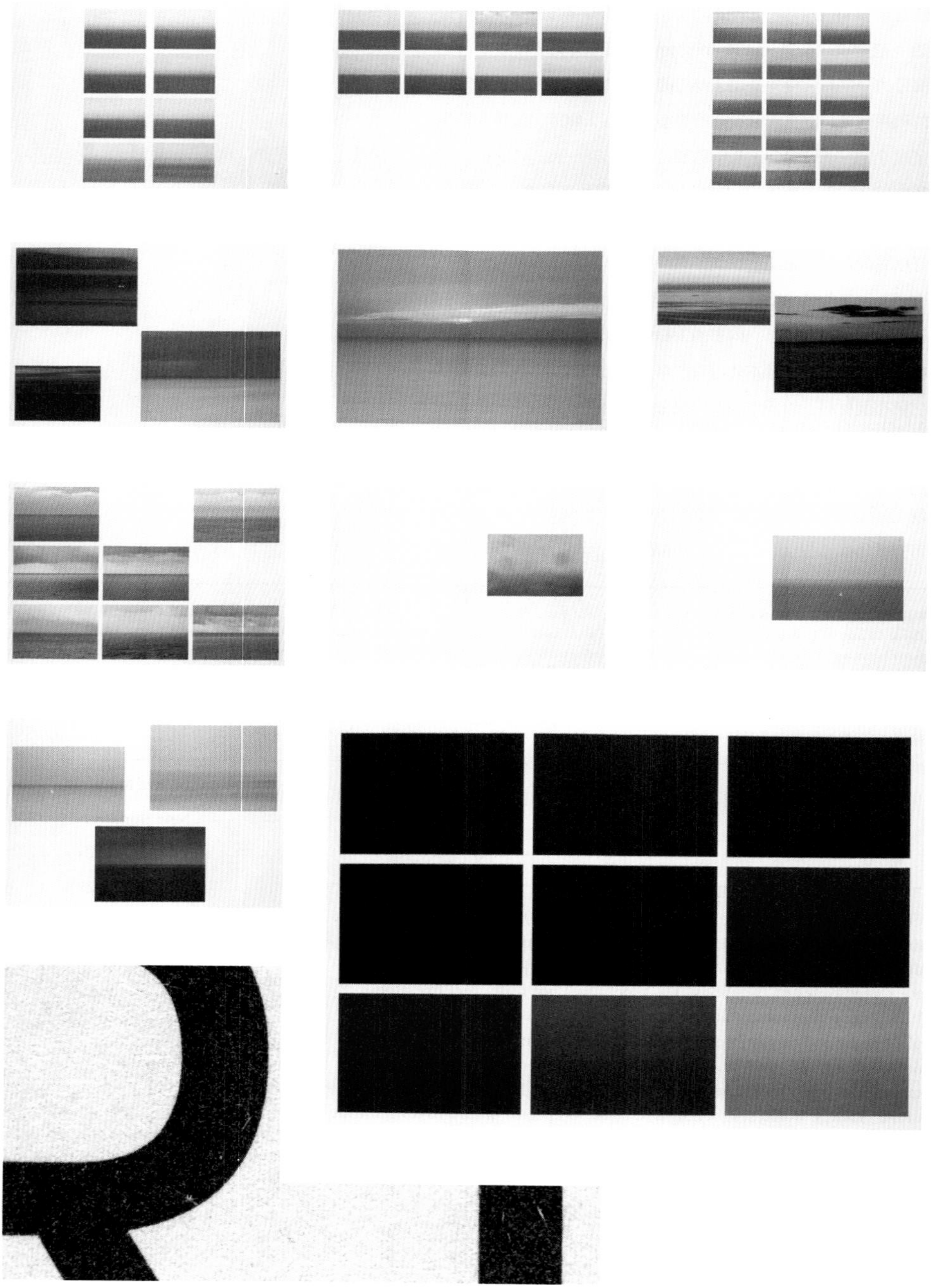

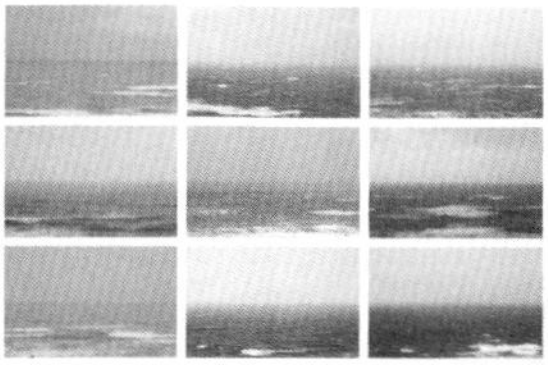
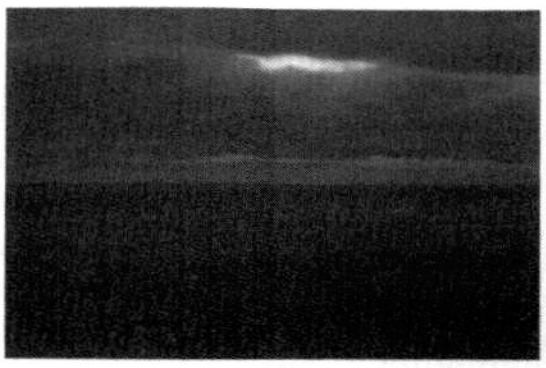

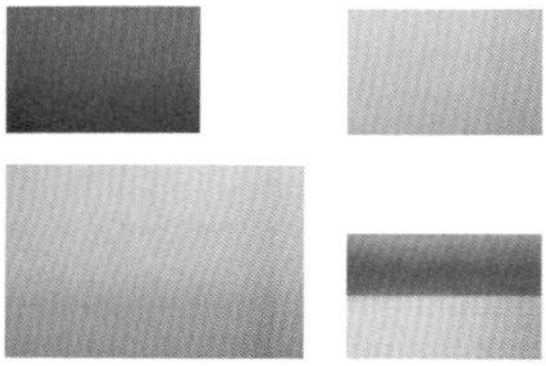
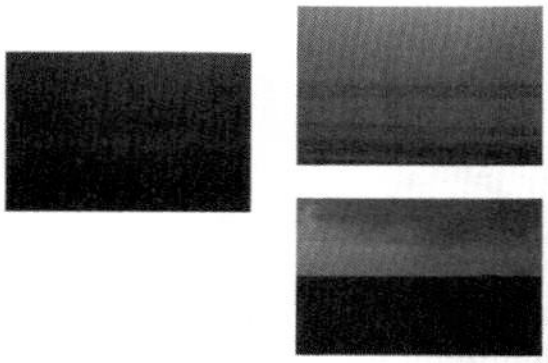
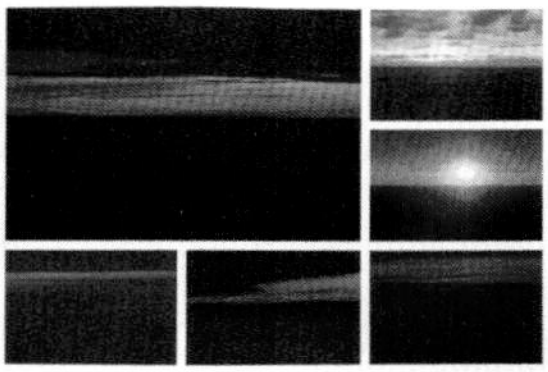
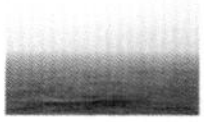
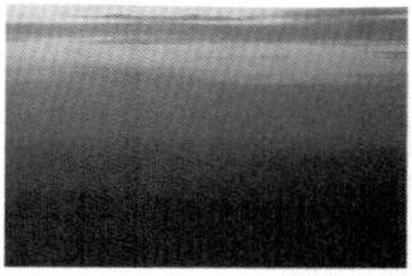
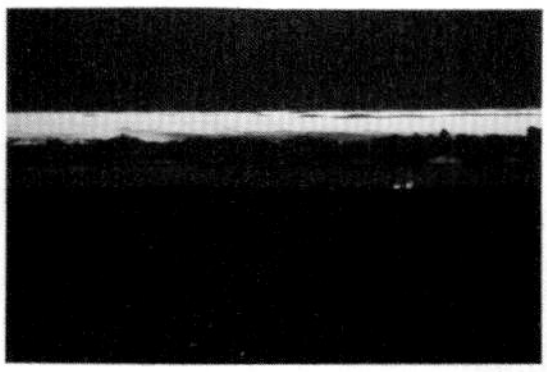

 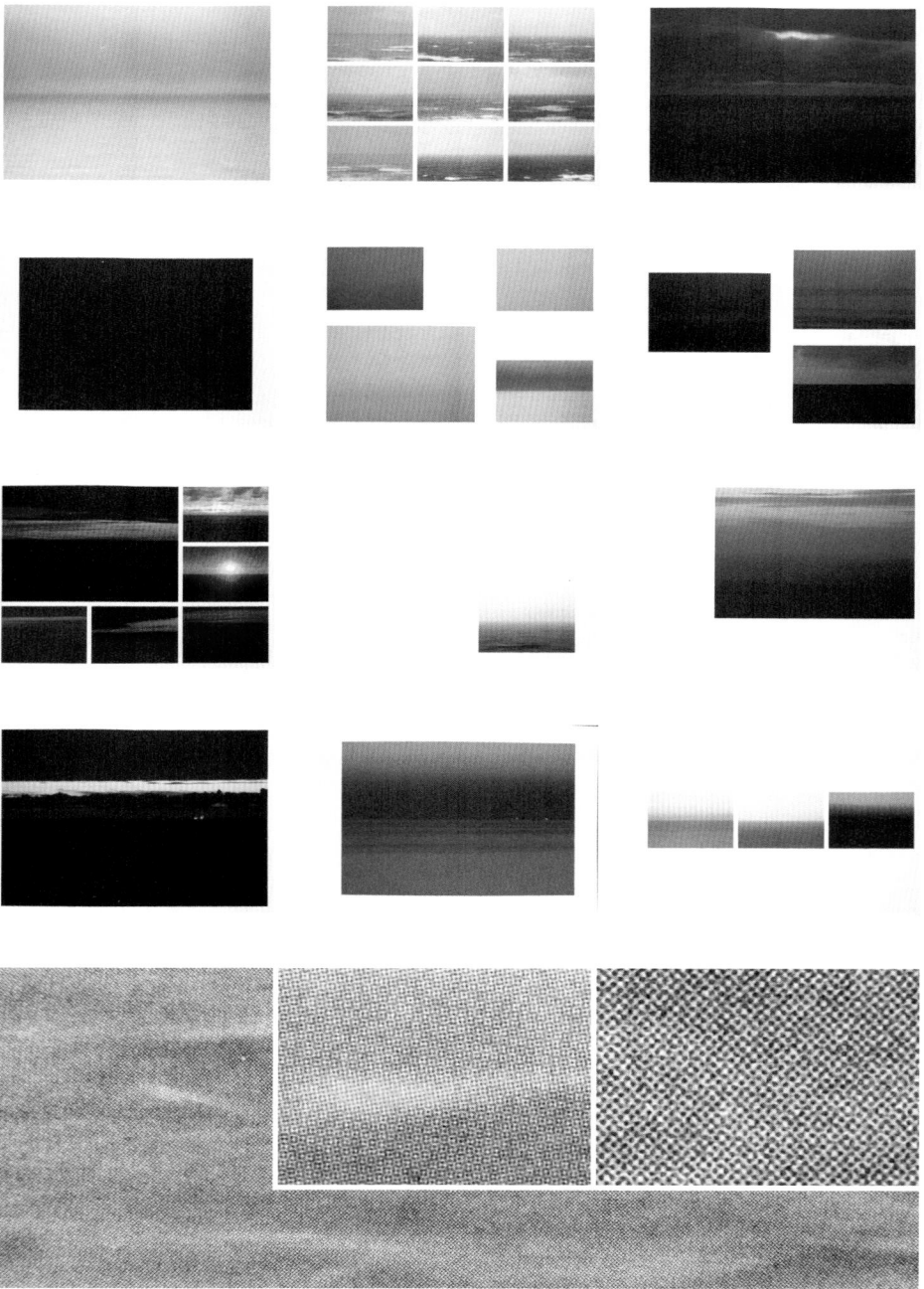
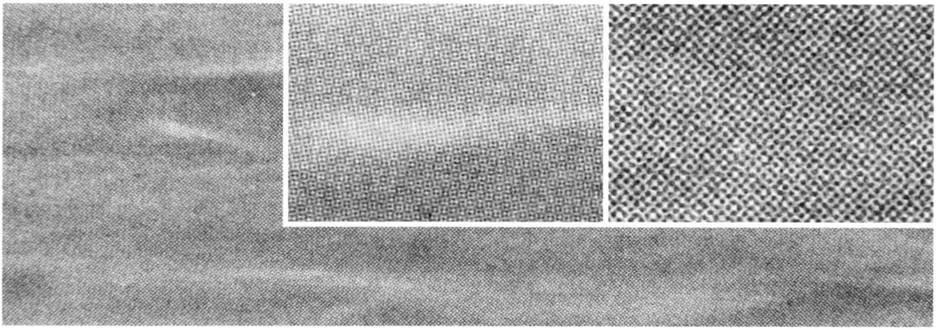

NEW
HORIZONS
nr 1137 / 2012

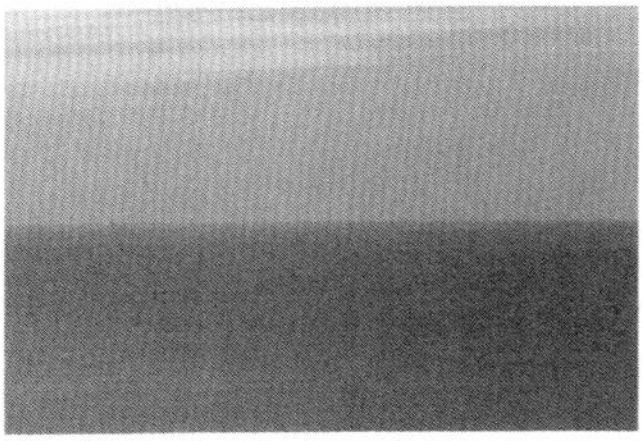

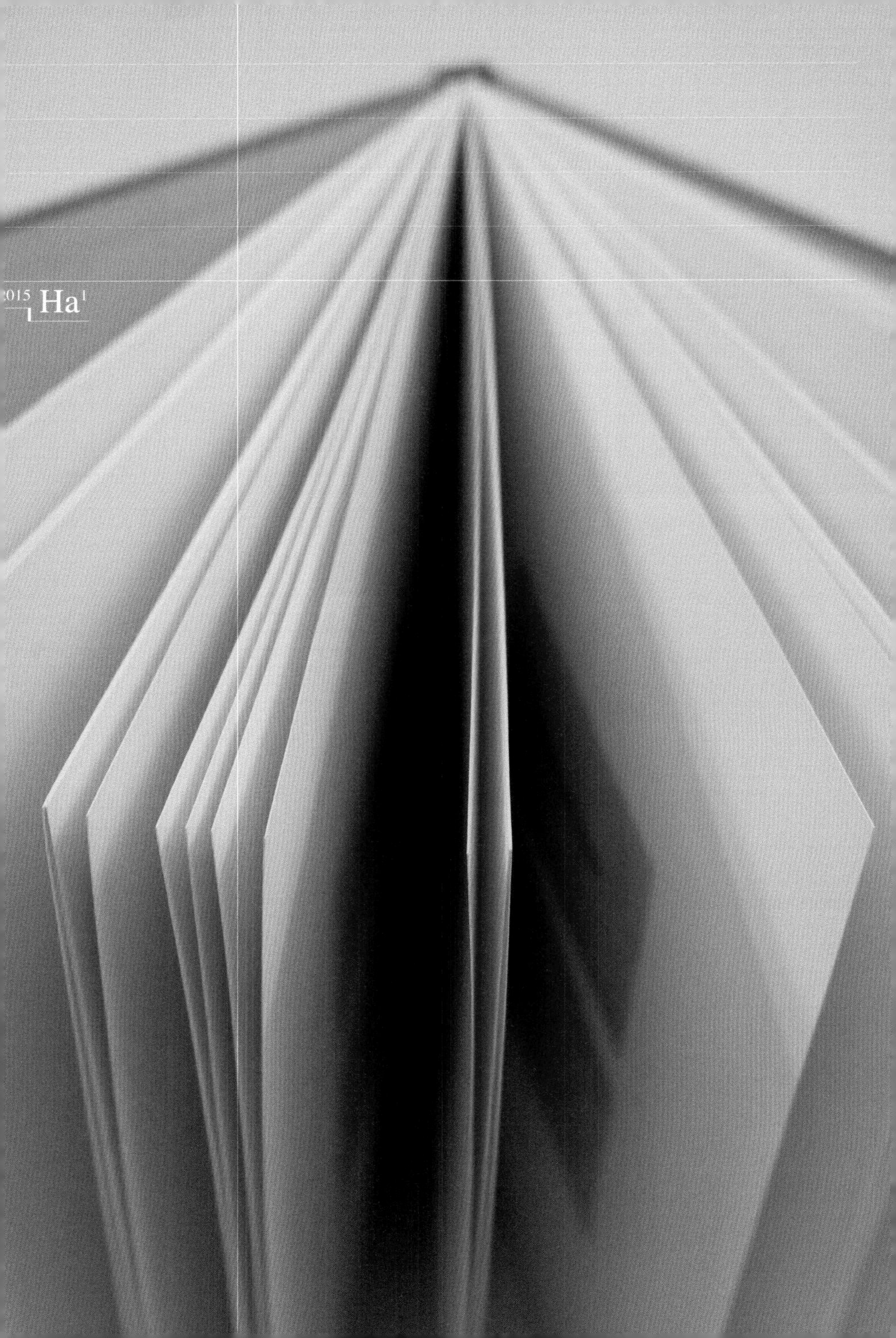
2015 Ha1

本书通过展示逐渐被现代社会所弃用而成为历史的纸质名片，探讨曾经和现在的社交关系和社交形式的变化。曾经的名片通过纸张的选择、字体的选用、方寸间的排版、甚至名片的气味等形成独特的“虚拟形象空间”，展示着人们的身份和形象。通过相互交换名片来彼此建立联系，甚至在名片上写上其他方面的信息来获得进一步的交流。但现在名片已经虚拟化，通过智能手机的点击分享来交换个人信息。本书的装订形式为锁线硬精装。封面采用灰色布面裱覆卡板，封面的中上部嵌入一张名片大小的纸张单黑印刷附带压凹处理，模拟出早期铅活字印刷名片的效果。环衬选用浅灰色滑面艺术纸张，与封面选用的布面色调统一。内页胶版纸四色印刷，除开篇对名片这一形象载体作为社交形式的思考和怀旧外，之后的页面为各种形式的名片合集，每页展示一张名片的正反面，包含铅印、手写，甚至夹杂了各式绘画和其他信息的名片用法。||||||||||||||||||||||||

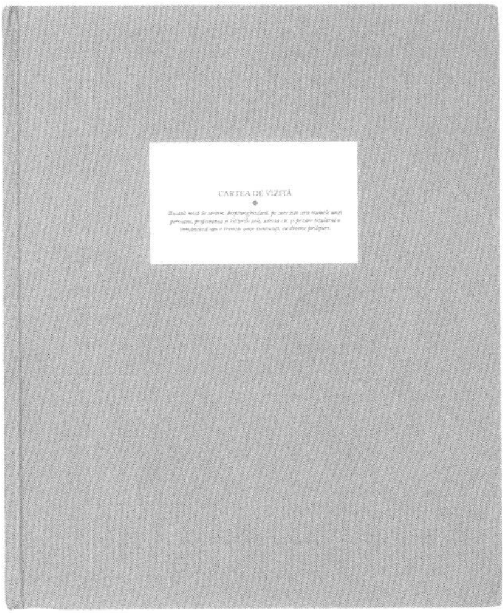
CARTEA DE VIZITĂ

This is a book that explores the changes in social relationships and social forms by displaying how business cards were gradually abandoned by modern society and became history. Former business cards built a unique ‘virtual image space’ to shows people’s identity through the choice of paper, font, layout, and even the smell of the cards. People established contact with each other by exchanging business cards or even writing other information on business cards to obtain further communication. Nowadays business cards have been virtualized; personal information will be exchanged through click-and-share on smartphone. The hardback book is bound by thread sewing. The cover is made of gray cloth mounted with cardboard, and a business card-sized paper is embedded in the upper-middle part of the cover, printed in black with embossing process. It simulates the effect of business cards using lead movable type printing in the early days. The endpaper is made of light gray art paper, which is in harmony with the color of the cover. The inside pages are printed in four colors on offset paper. In the first part, the author considers business cards of as a form of social interaction and expresses nostalgia. The following pages are a collection of business cards in various forms; each page shows the obverse and reverse sides of a business card, including lead printing and handwriting, and even a variety of painting and other information.

- 荣誉奖
- 名片
- 罗马尼亚

- Honorary Appreciation
- Cartea de vizită
- Romania
- 72+87
- Fabrik - 72+87
- Fabrik
- 297×235×14mm
- 680g
- 64p

The term »Visiting Card« has a rather metaphorical character today: the electronic visiting card, moved with a mouse click to the directory on the hard drive (also a metaphor), the personal website as a quasi-public visiting card, the polished shoes. Even the ink-jet-printed design templates on pre-perforated thin carton can only be described in this way in a figurative sense. The catalogue of historic visiting cards from Romania is a homage to a medium of personal - and necessary - everyday communication. The self-confidence of the sender was not expressed in visual onomatopoeia, but actually hardly played a role: restrained elegance was a matter of course. And exactly this must be the message of the printing house, which produced this catalogue for its exhibition of historic visiting cards. With delicately reproduced facsimiles and properly printed for handling, each page of the book presents a card with front and back page - with the effect of a herbarium. One can only hope that the incorrect running direction of the carton chosen for the book is not just a coincidence. After all, it is common practice to lend the carton more strength by selecting the long side of the card as running direction.

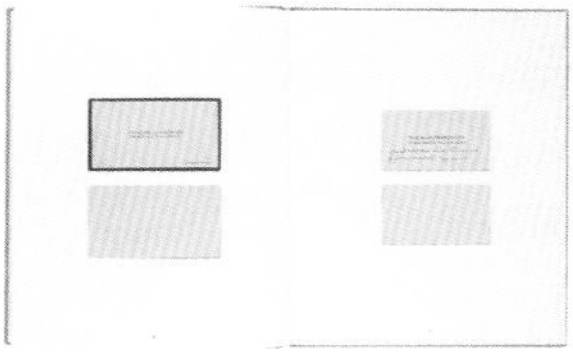

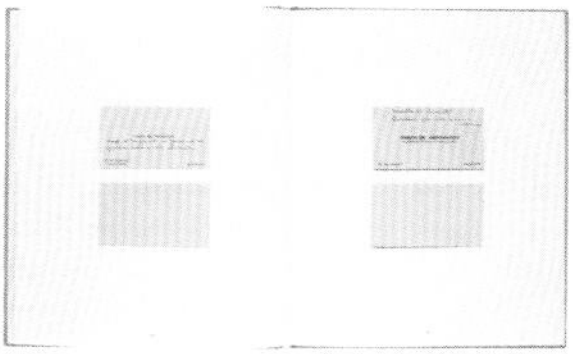

ghiulară, pe ca

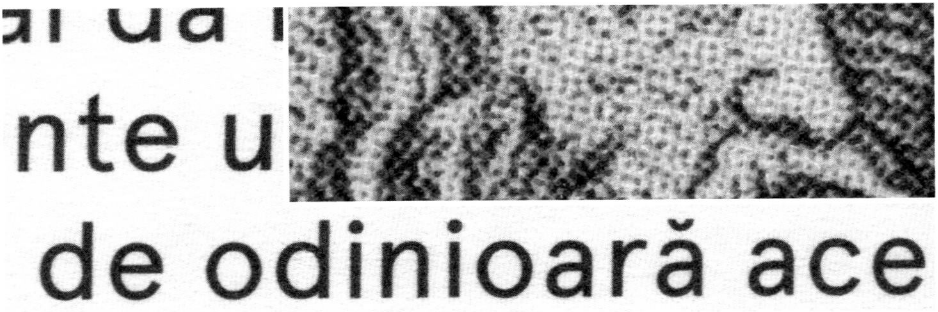
nte u
de odinioară ace

015 Ha2
including artists'
artists' homes,
rooms of galler-
always of vari-
cial buildings,
of the walls
of my own
109–11.

这是一本艺术展览的导览画册，展览为加拿大艺术家迈卡·莱克西尔（Micah Lexier）在个人展览（One）和协作展览（Two）之后的多人展览项目。与常规的展览画册展示完美的展品状态不同，这次展览需要展示不同地区的 101 位艺术家或艺术团体的 221 件作品，对于较短的展览布置和准备来说，是不可能完成的任务。所以这本画册选择了一个独特的方案，即直接拍摄展品布置时的照片，呈现“未完成”的状态。本书的装订方式为锁线胶平装，封面白卡纸红黑双色印刷。内页展示作品的主要部分为胶版纸红黑双色印刷，只在中心页面有 16 页的红色胶版纸印红色文字，展示与作品、展览介绍等相关信息。全书的设计简洁收敛，只使用了红黑两种颜色，所有作品均调整为黑白照片（展品大多以黑白为主），鲜明的展品数字编号直接印在照片上，用于引导阅读中心红色页面的展品信息，同时也具有一定的页码作用。

More Than Two
(Let It Make Itself)
—
221 artworks and objects made by 101 individual artists, duos, and collectives, selected and arranged by Micah Lexier.

This is a guide album for the art exhibition, which is a multi-person exhibition project of Canadian artist Micah Lexier after her individual exhibition ‘One’ and collaborative exhibition ‘Two’. Unlike the traditional exhibition albums, which show the perfect state of exhibits, this exhibition needs to display 221 works of 101 artists or art groups from different regions. It is impossible to complete exhibition arrangement in a short preparation time. Therefore, this album chose a unique scheme, that is, to directly take pictures of preparing the exhibits, presenting an ‘incomplete’ state. The book is perfect bound and the cover in printed on white cardboard in red and black. The inside pages for displaying the works are made of offset paper printed in red and black. Only 16 pages of red offset paper are printed in red only, explaining related information about works and the exhibition. The design is concise and convergent, using only red and black. All works are adjusted to black-and-white photos (most exhibits are black and white). The distinct exhibit numbers are printed directly on the photos, leading the readers to the exhibit information on the red pages. The numbers function as page numbers as well.

◉ 荣誉奖
《 两个以上（让它们成为自己）
◖ 加拿大

◎ Honorary Appreciation
《 More Than Two (Let It Make Itself)
◖ Canada
D Jeff Khonsary (The Future)
△ Micah Lexier
||||| The Power Plant
□ 264×195×20mm
877g
224p

At first glance something looks strange: the very big red numbers, directly printed onto the black and white pictures of the art objects. The backstory might perhaps be like this: a catalogue is to be published for the exhibition. How can this work, if perhaps not all of the artists are able to provide photographs, and the 221 works of art have to be collected from all kinds of different directions. This means that the catalogue has to be produced during the brief build-up period. This scenario is adopted as the concept of the catalogue. On some pictures one can see that the position of the object or the picture is just being prepared. The list of exhibits, arbitrarily and eye-catchingly put into the middle of the catalogue, printed in red on pink paper, is therefore called »checklist«. The catalogue acquires its own artistic quality, beyond the temporary exhibition; the exhibited works of art are connected by a red thread: namely by the red numbering as a code for the build-up phase of the show, the interrelation of the works in the room. A simple, a fascinating book concept.

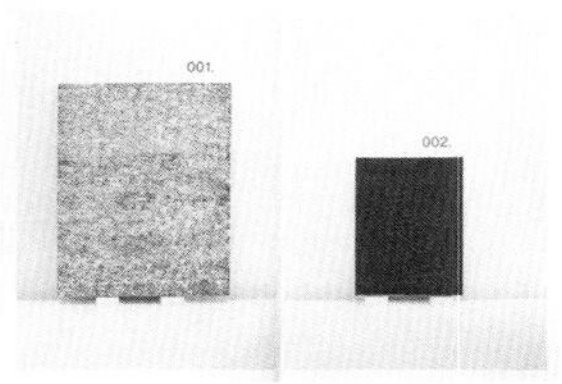
001.
002.

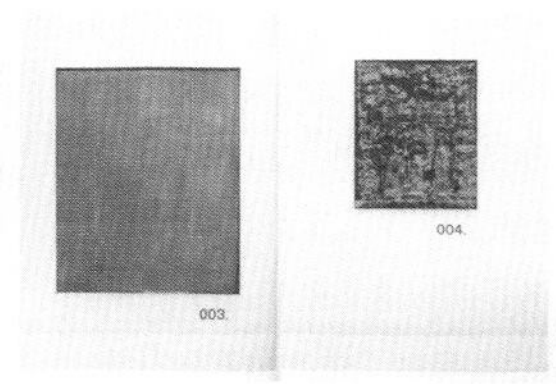
003.
004.

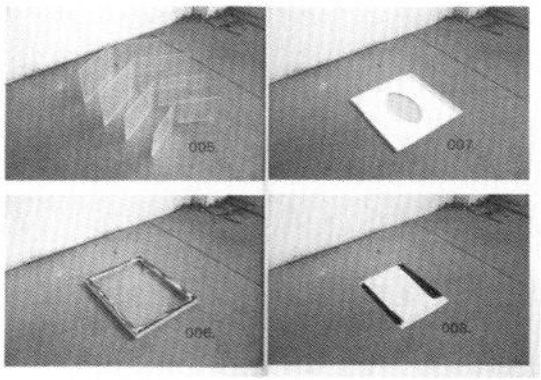
005.
007.
006.
008.

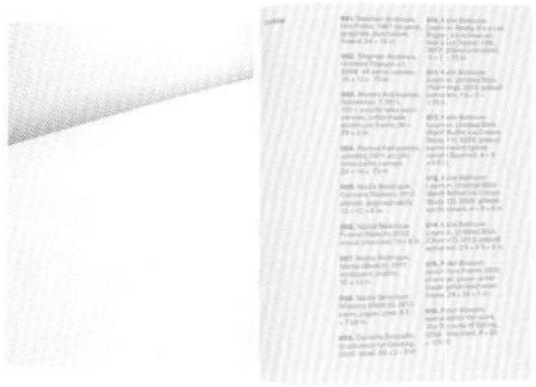

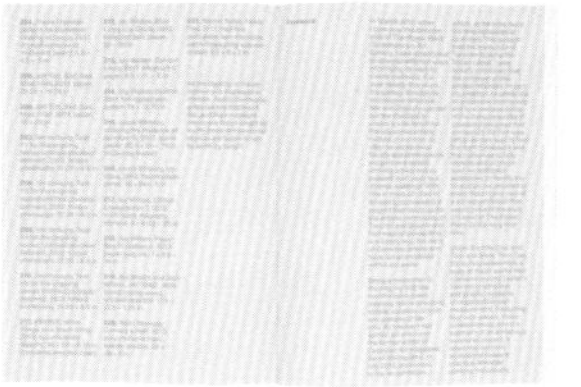

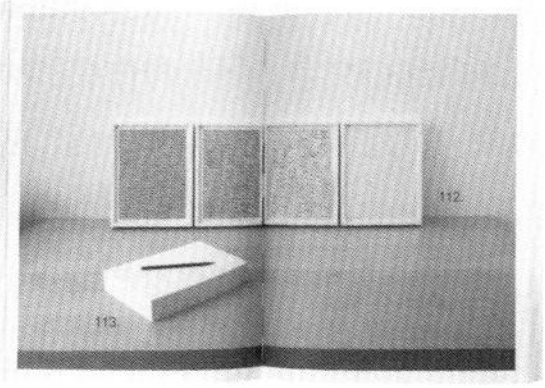
112.
113.

134.

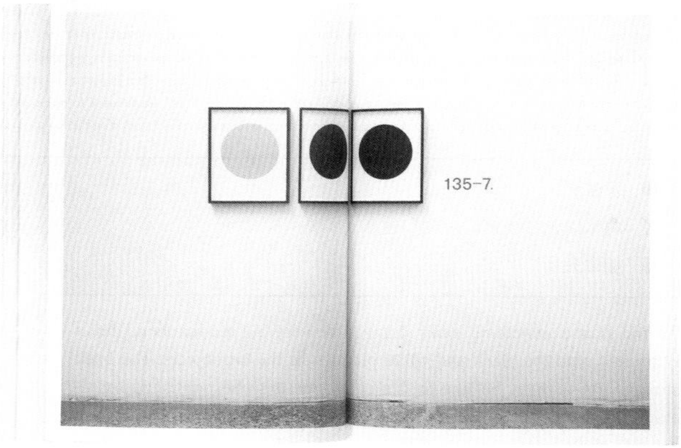
135–7.

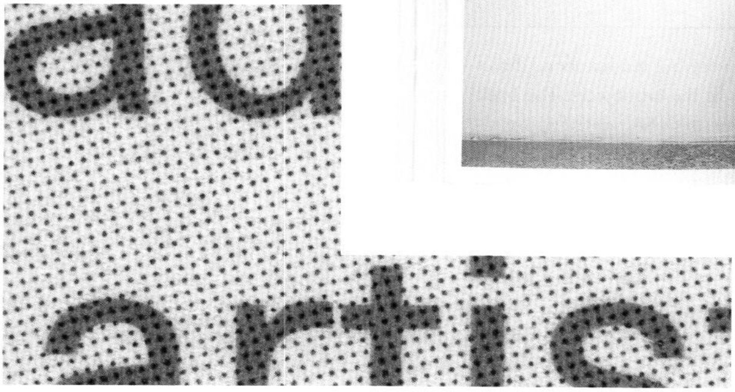

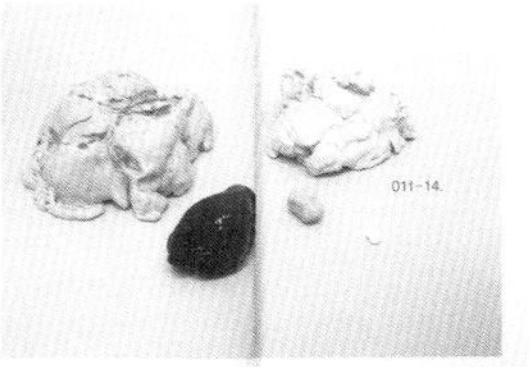
011–14.
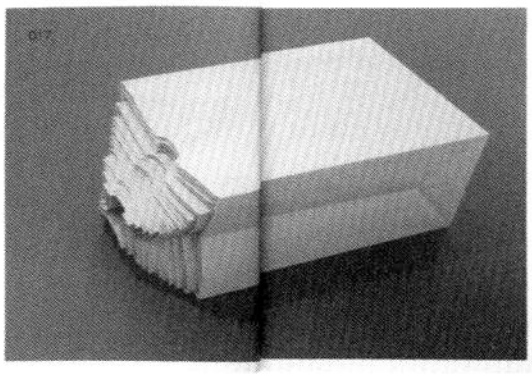
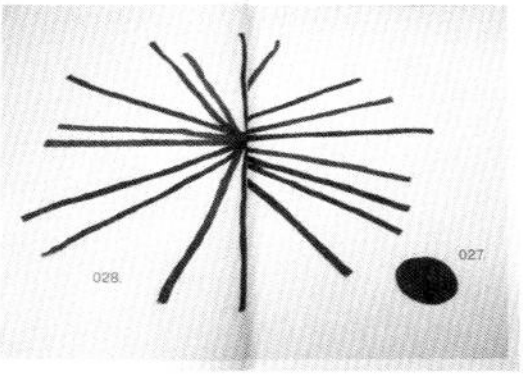
028.
027.
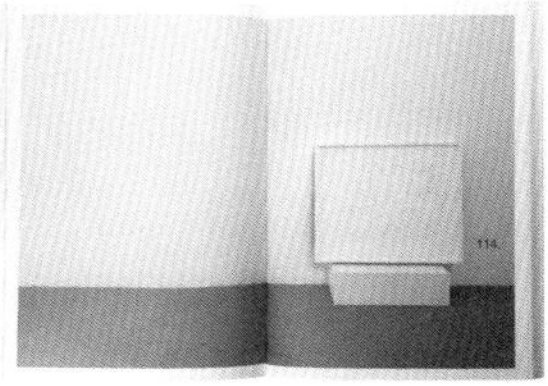
114.

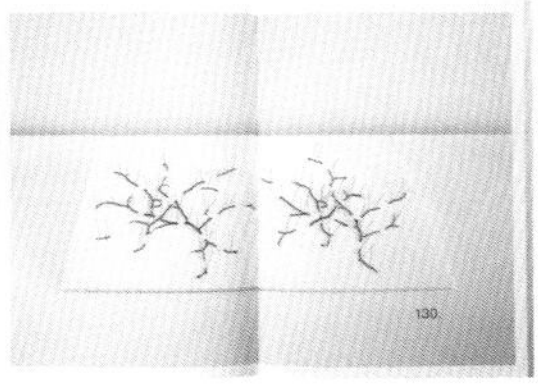
130.
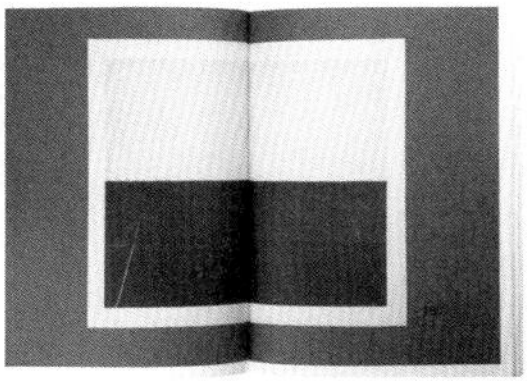
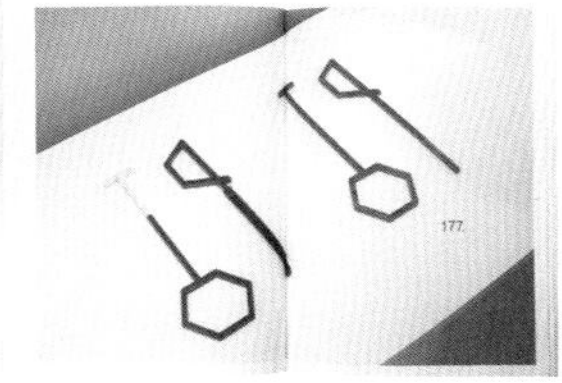
177.
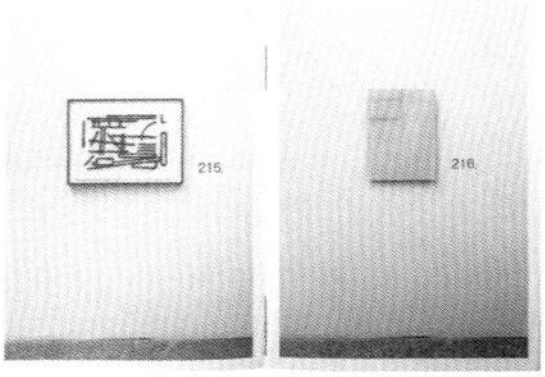
215.
216.

ovem
11. acryl

2015 Ha3

这套书是瑞士当代艺术出版社 Fink 的出版系列。本系列专门再版已出版过的艺术书籍，包括地图、艺术家作品集、艺术理论等。由于这些再版的书籍原本的纸张、开本、印刷都不相同，在纳入本系列的过程中，经过设计上的统一后，为普通读者提供较为廉价的再版。本系列在装订、纸张和印刷上都尽量质朴。书籍的装订为无线胶装，封面白卡纸单黑印刷。封面图案来自原版封面的单黑挂粗网的设计处理，并且在左下角写上这个系列的统一编号。内页胶版纸单黑印刷，维持原版内容和版式（部分页面减少）。系列中每本书的差异主要体现在封面内侧各异的纯色，以及相较原版等比例缩小后不相同的开本。||||||||||||||||||

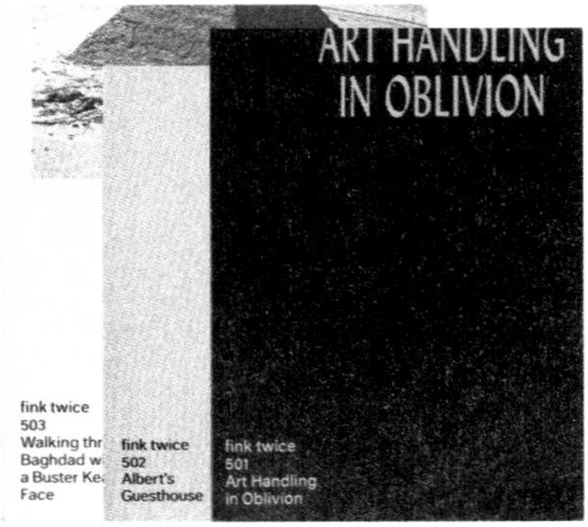
ART HANDLING
IN OBLIVION
fink twice
503
Walking thr
Baghdad w
a Buster Ke
Face
fink twice
502
Albert's
Guesthouse
fink twice
501
Art Handling
in Oblivion

The set of books is a publishing series of Fink, an independent Swiss contemporary art publishing house. The editorial concept of this series is to reprint published art books, including maps, artists' work collections, art theory and so on. Since the original paper, format and printing colors of these reprinted books are different, they are unified in design to integrate into the same series. This series is as plain as possible in binding, paper selecting and printing, as it is reprinted for cheaper and more reader-oriented version. The books are perfect bound and the cover is made of white cardboard printed in black only. The cover images come directly from the original version printed in black with FM screening. There are sequenced numbers of this series written in the lower left corner of the covers. The inside pages are made of offset paper printed in black and completely maintain the original content and layout (some books reduce pages). The distinction between books in the series is different solid colors printed inside the covers and the different formats which are reduced in proportion to the original edition.

荣誉奖

遗忘中的艺术处理 /
艾伯特宾馆 /
以巴斯特基顿的样子穿过巴格达

瑞士

Honorary Appreciation

Art Handing In Oblivion (fink twice 501) /
Albert's Guesthouse (fink twice 502) /
Walking through Baghdad with a Buster Keaton Face (fink twice 503)

Switzerland

Rob van Leijsen, Georg Rutishauser, Sonja Zagermann, Petra Elena Köhle, Nicolas Vermot-Petit-Outhenin, Thomas Galler

Rob van Leijsen /
Petra Elena Köhle, Nicolas Vermot-Petit-Outhenin /
Thomas Galler

edition fink, Verlag für zeitgenössische Kunst, Zürich

200 × 145 × 21mm / 190 × 130 × 9mm / 210 × 168 × 10mm

366g / 134g / 212g

382p / 158p / 174p

A remarkable editorial concept: reprints of artists' books, quasi as paperbacks. The cover drastically emphasizes the downsizing process by simply trimming the original-size motif. The huge screen width makes no secret of the second-hand usage: after all, the new series is called: fink twice. Yet the technical synchronization of the reprint series – monochrome print, thin open paper, glue binding – does not go quite as far. Despite the reduction of the inner pages, each reprint keeps its individual format, for example by giving each of the inside cover pages its own spot colour. The font size of the lettering on the spine varies depending on the width of the spine. Another remarkable aspect is that these unpretentious prints never appear as a mere rehash, but keep their own character – and acquire a new one.

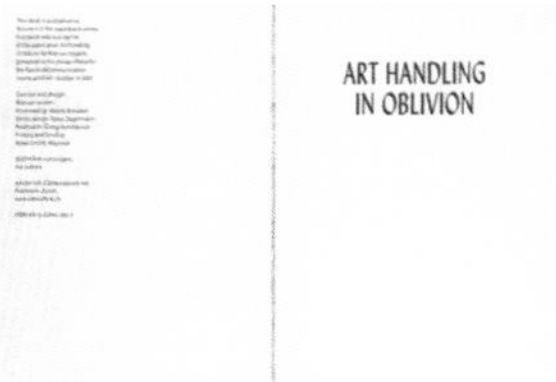
ART HANDLING
IN OBLIVION

Introduction

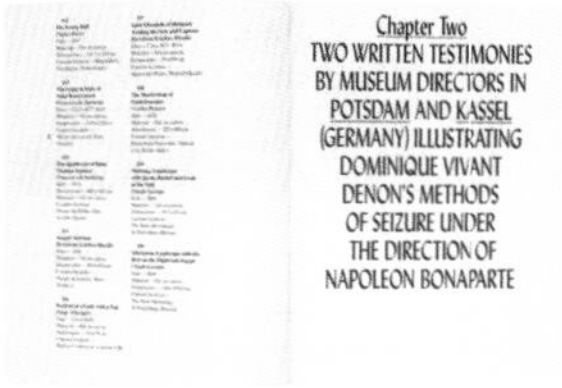
Chapter Two
TWO WRITTEN TESTIMONIES
BY MUSEUM DIRECTORS IN
POTSDAM AND KASSEL
(GERMANY) ILLUSTRATING
DOMINIQUE VIVANT
DENON'S METHODS
OF SEIZURE UNDER
THE DIRECTION OF
NAPOLEON BONAPARTE

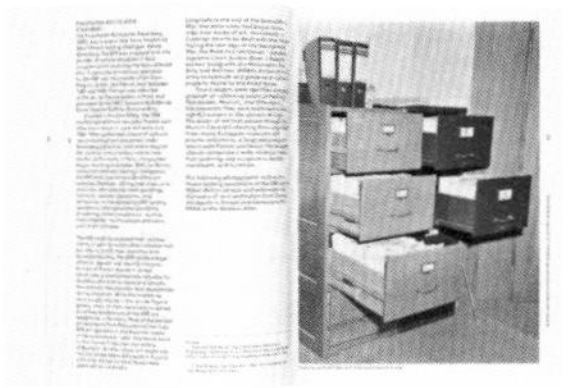

TRACES OF LOOTING IN THE NATIONAL MUSEUM OF IRAQ, BAGHDAD

In the months preceding the 2003 Iraq war, various antiquities experts, including representatives from the American Council for Cultural Policy, asked the Pentagon and the UK government to ensure the safety of the Iraq Museum from architectural damage and looting. Although promises were not made, US forces did avoid bombing the site.[1] On 8 April the last members of the museum staff left the museum. Lt. Col. Eric Schwartz of the US 3rd Infantry Division stated that he was unable to enter and secure the compound because they had wanted to avoid returning fire at the museum. Sniper positions, discarded ammunition and fifteen Iraqi army uniforms were later discovered in the building. Iraqi forces had built a fortified wall along the western side, allowing concealed movement between the front and the rear of the museum.[2]

The plunder took place between 10 and 12 April. Some of the staff returned on the 12th and fended off further attempts by looters while entering the museum. They guarded the museum until US forces arrived on 16 April. A special team headed by Marine Col. Matthew Bogdanos initiated an investigation on 21 April. His investigation indicated that no US forces had looted the building and that there were three separate thefts by three distinct groups within thirty-six hours between 10 and 12 April. One part of the theft implied an inside job by the museum staff, a second the work of organised criminals, and a third fleeing Iraqi soldiers who had left the door open, which had given access to looters. At least 15,864 objects were stolen, making it the biggest museum theft in history.[3]

While the staff instituted a storage plan to prevent theft and damage, many larger statues, steles, and friezes had been left in the public galleries, protected with foam and surrounded by sandbags.[4] Forty valuable pieces were stolen from these galleries, from which thirteen had been recovered before January 2005, including the three most valuable: the *Sacred Vase of Warka* (p. 381, W14873), the *Mask of Warka* (p. 381, W17878), and the *Bassetki Statue* (p. 366, IM77823). Approximately 3,181 excavation-site pieces (jars, vessels, pottery shards, etc.) were stolen from the storage rooms above ground, of which over 3,080 have since been recovered. The third occurrence of theft was noticed in the underground storage rooms, where evidence pointed to an inside job. The thieves attempted to steal the objects easiest to transport, which had been intentionally stored in the most remote location possible.

The amnesty led by Col. Matthew Bogdanos started the recovery within days after the plunder. A message was put out to imams, newspapers, television stations, and people on the street that anyone returning an item would be asked only one question: "Would you like a cup of tea?" The response was almost immediate. Some antiquities were left at mosques, others simply handed to patrolling US soldiers. The *Sacred Vase of Warka*, after two weeks of negotiation, was returned in June in a car boot, along with ninety-five other artefacts. Just over 2,000 recoveries were the result of raids, the biggest being at a farmhouse where the *Mask of Warka* was discovered under forty-five centimetres of dirt.[5]

Between 1998 and 2004, Joanne Farchakh-Bajjaly, a Lebanese archaeologist and journalist, made numerous trips to Iraq to document the state of its archaeological sites. Her site photographs bear dramatic testimony to the damage that sites such as Umma, Umm al-Aqarib, and Larsa in the south, or Nimrud and Nineveh in the north have suffered. In May 2003, one month after the end of the Iraq war, she visited the Iraq Museum and documented the looting damage. Farchakh-Bajjaly's photographs form the basis of the following visual testimony.

1. Andrew Lawler, "Impending War Stokes Battle over Fate of Iraqi Antiquities," in *Science*, vol. 299, no. 5607, 31 January 2003, p. 643.
2. Matthew Bogdanos, "Pieces of the Cradle," in *Marine Corps Gazette* (Marine Corps Association), January 2005, pp. 40–44.
3. David Buncombe, "Revealed: The Real Story behind the Great Iraq Museum Thefts," in *The Independent*, 13 November 2005.
4. Robert M. Poole, "Looting Iraq," in *Smithsonian Magazine*, February 2008.
5. Robert M. Poole (see note 4).

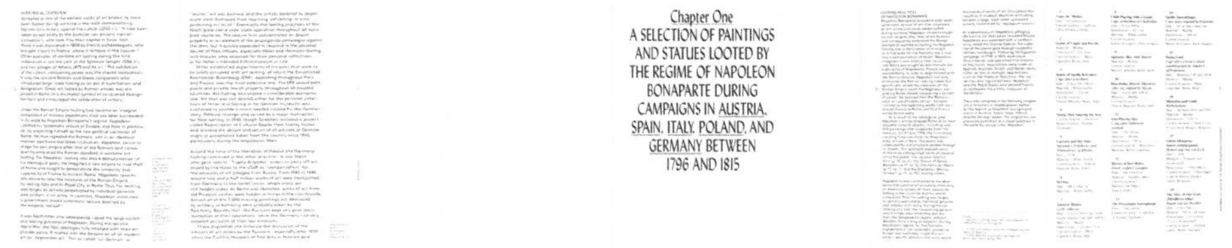
Chapter One
A SELECTION OF PAINTINGS AND STATUES LOOTED BY THE REGIME OF NAPOLEON BONAPARTE DURING CAMPAIGNS IN AUSTRIA, SPAIN, ITALY, POLAND, AND GERMANY BETWEEN 1796 AND 1815

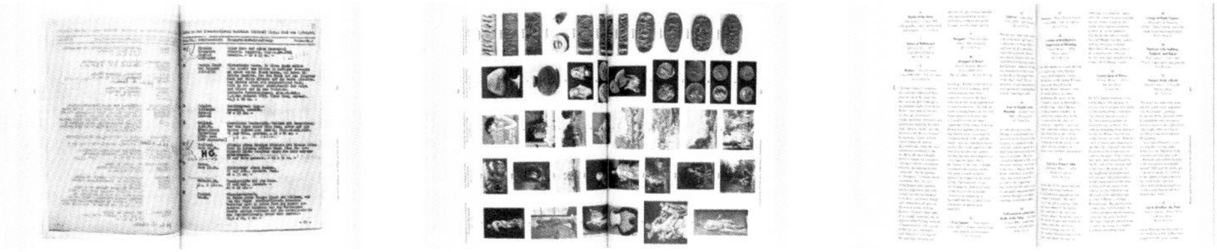

Artefacts originating from the Abu Salabikh region in the Al-Qadisiyah Province, south of Baghdad

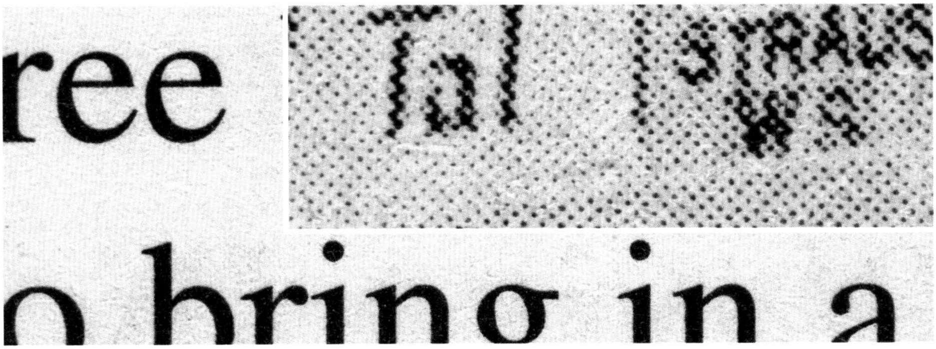
ree
o bring in a

VAGADSÕ' JA ANDSAGU' JUTU'
ESÄ KUTŠ UMMA POIGA KODO
Lõpp.
VAGADSÕ' JA ANDSAGU' JUTU'
KOL' JUMALAPALLÕJAT
Lõpp.
SETO KIRAVARA
015 Ha4

南爱沙尼亚有一支大约只有一万多人的民族叫塞托，塞托人讲自己的语言塞托语——一种隶属于乌拉尔语系的南爱沙尼亚方言。这套“塞托图书馆系列”，就是针对这一语言和文化编撰的套系图书，目前仍有后续出版物。书籍的装订形式为锁线圆脊硬精装。开本分为大小两种，小开本采用白色防布纹纸，大开本则用白色卡纸。封面只用单红色印刷书名和出版社，书脊印有民族特色的纺织物花纹，套系内图书的封面红色并不统一，而是随着出版的次序有渐变关系。书籍的内容包含诗歌、歌曲、传记、福音书、民间传说、民族文化等等，根据文字和图片量的多寡选用两种开本形式。印刷色的选择上也根据情况分为单黑、红黑、四色等多种形式。内文有统一字体和版式规范，但每本进行灵活的变化。书内配有书签带，书签带的红色选择与封面印刷的红色相匹配。||||||||||||||||||||||||||||||||||||||||||||

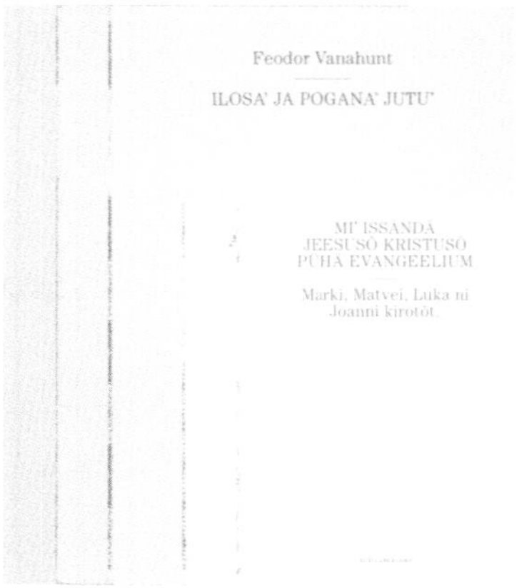

Feodor Vanahunt
ILOSA' JA POGANA' JUTU'
MI' ISSANDA JEESUSÕ KRISTUSÕ PÜHA EVANGEELIUM
Marki, Matvei, Luka ni Joanni kirotõt.

In South Estonia, there is an ethnic group of about 10,000 people called Seto. The Setos speak their own language Seto, a southern Estonian dialect belonging to the Finnic group of the Uralic languages. This 'Seto Library Series' is a series of books on this language and culture, and there are still follow-up publications. The book is hardback bound by thread sewing with a rounded spine. The book has two formats; the small one is using white wove paper while the large one is using white cardboard. The cover is printed in single red with the title of the book and the publishing house. The textile pattern with national characteristics is printed on the spine. The red color using on the covers of the book series is not the same, but has a gradual change with the order of publication. The contents of books include poems, songs, biographies, gospels, folklore, national culture and so on. According to the amount of text and pictures, two formats are selected. The choice of printing colors is also varied from single black, red black, four colors and other forms according to the situation. There are uniform fonts and format specifications for the text, but each book can be flexibly changed. The book is equipped with a red bookmark tape, which matches the red color printed on the cover.

荣誉奖
塞托图书馆系列
爱沙尼亚

Honorary Appreciation
SERIES The Seto Library. Seto Kirävara
Estonia
Agnes Ratas
various
Seto Instituut
166×114×23mm / 166×114×16mm / 247×166×23mm / 247×166×27mm / 247×166×22mm / etc.
262g / 181g / 580g / 1038g / 522g / etc.
288p / 176p / 256p / 396p / 200p / etc.

This book series is published by the Seto Institute, which is dedicated to the cultivation and documentation of a language in the south east of Estonia: Seto. These days it is only spoken by a few thousand people. This publishing commitment considers a full library of Seto books. Poetry, songs, biographies, the gospels, folklore are preserved in printed form. Depending on the text type, two formats are available. The design of this book series does not refer so much on textual contents but rather on the cultural similarities of this small linguistic group. The humble paperbacks hint at the characteristics of national costumes. White linen here, fragmented paper white there - embroidery and borders here - or geometric textile ornaments on the spines. The cover is printed in a single colour, from volume to volume in changing red hues; headband and bookmark are chosen in matching colours. The robust style of the edges, which slightly protrude over the book block, embraces the rustic traditions of the Setukese people. This series clearly shows once again that, as hardly any other tool, the book as a medium serves the identity formation of a society and the preservation of immaterial heritage.

KODUÕLU
SETU PORTER
20,-

XVI KUNINGRIIGI PÄIV
1. AUGUST 2009, LUHAMAA

Ülembsootska – Sarvõ Õiõ
Nuursootska – Mõrra Liia

Kuninga sõnolinõ – Nassari Elvi
Kuninga kurnuskamangja – Mõro Leo
Kuninga kandikõuaja – Paltseri Ülle
Kuninga karguskargaja (naane) – Heliste Leida
Kuninga karguskargaja (tsirk) – Pille Liisa
Kuninga kaatsikargaja (tsura) – Katsaski Niilo
Kuninga leevätüdaja – Taak Hiie
Kuninga piiragulüdaja – Lepasilla Elli
Kuninga siirmeistri – Luigasõ Inara
Kuninga õllõtegija – Riitsaarõ Reili
Kuninga veinimeistri – Valgu Uno
Kuninga kandle-ajaja – Kooseri Rein
Kuninga käsitüümeistri – Reidla Angela
Kuninga rägimiir – Lõvi Riho

Nelandat kõrda Luhamaal peetü' kuningriigi päiv vahet' asupaika ja tä viidi koolimaja takast platsi päält ar Mäaai mäele, koh nelli aastakka tagasi tahvit-södi välla pandi Peko kujo, mida Riitsaarõ Evar pidi paduvihmadõ kuningriigi päävä põhjusõs.

Suurõs vastandis eelmatsele Luhamaa nulga kuningriigi päävale oll' soo päiv väega pallav ja päävapaistõlinõ. Kuigi vahepaal tull' väikene vihmaopsahus, oll' tuu nii hähkene, õt inemisõ' olli' innetoh kerge jahutava kastmisõ üle rõõmsa'.

Üte pääkõrraldaja, Misso kooli värske huvijuhi Madzeri Inge sõnotsi oll' päävä kõrraldaminõ tõsinõ väläkutsõh. Kuigi 2006. aastagal oll' pidõlisi kõgõst paari tuhandõ kandih ka ketva tima tõttu, oll' huviline hulk vahepaalsidõ aastakidõga perätuhe kasunu' ja aasta varahampa oll' Värskah jo inemisi ligi 6000. Luhamaadõ võidsõ ootsa' sama pall'o, tuuperäst oll' vaja ar' koolimaja man vajo ar' mannu', sinnä rahvas lihtsalt mahtunu-s. Põld kõrgõ mäe otsah vai kuningriigi lastuplats hoadi tulla kotusõ päälе. Elektri lastu näüdõtuõbelt ja ka parkimisplats tetti timä põllu pääle.

A tegijidõ kamp oll' hua, ni kuigi ülembsootska Hadsi Silveri sõnul oll' paar kuud varahampa viil tervelik tühjus ja tegijidõ kampa olõ-s viil kokko hõigatu, sai Luhamaa rahvas ülembsootska arvatõh nisõhõ häste hakkama. Inge kül mälehe', õt suurõmb eeltüü käve jo utessa kuud varahappa, ku tull' löüdä' pilledmängija', leppü' kokko maaomanikuga ja tetä' muud tarviliit.

Latsi päävil, mis peeti Luhamaal edimäst kõrda, oll' pilt siski kurvalinõ. Lavalaudu viil tõrvati ni lava olõ-s viil katustki pääle saanu'. Latsi tuu seka-s. Järgmätses päävas oll' ka lava pusto, hainarulli' istmisõs olõmah ja pido tull' sündsa.

Miä tõmokraat tähendäs?

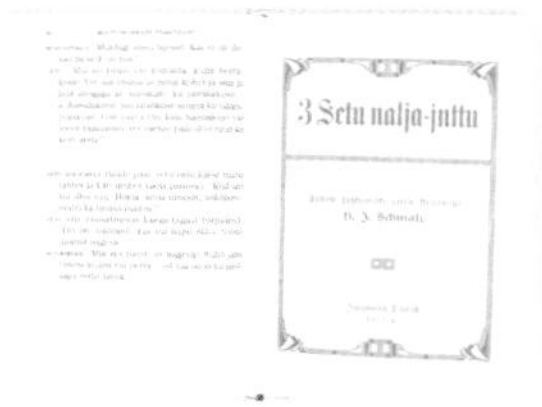
3 Setu nalja-juttu

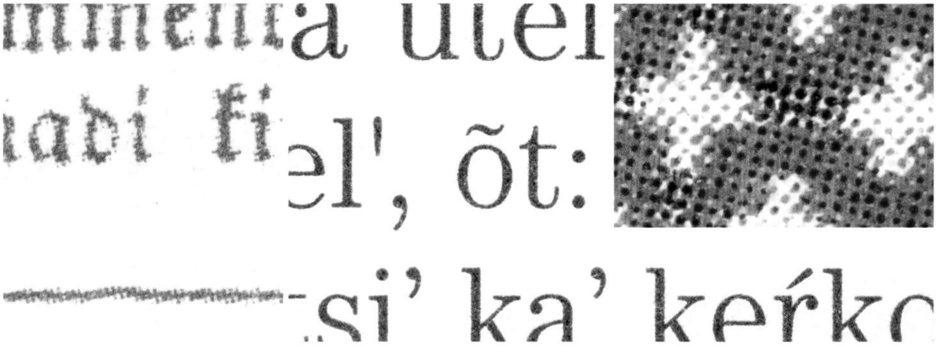
nmenta uter
iabi fi
el', õt:
si' ka' keŕko

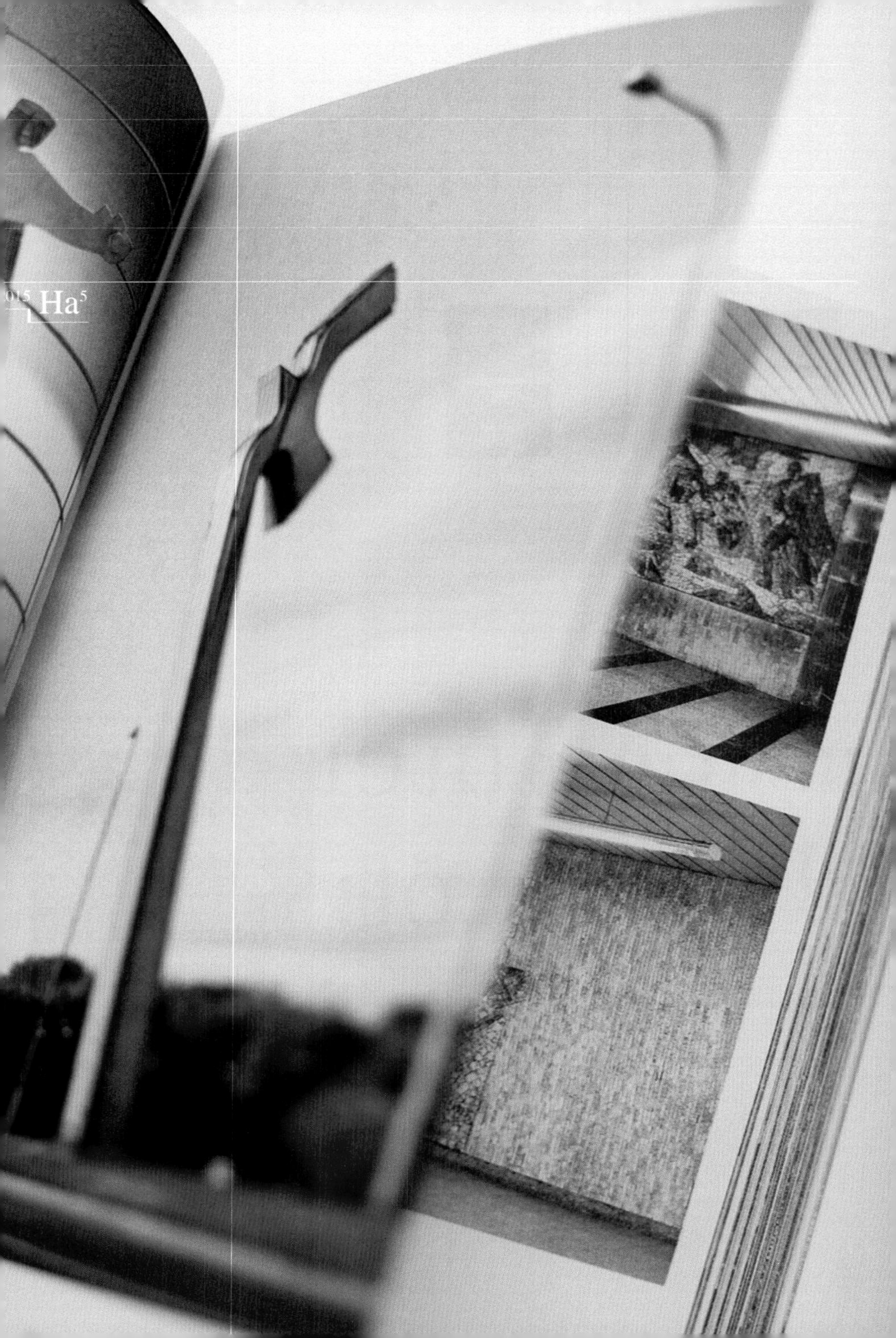

本书收集、整理并讨论了 1968－1989 年间捷克斯洛伐克的公共雕塑。在这段时间里，几乎所有的公共空间都被格式雕塑占满，对这些雕塑的整理体现着作者对特定历史时期的独特回忆。书中将这些雕塑归纳出五种分类标签，有些雕塑兼具多个分类标签时，都会对此进行陈述。本书的装订方式为锁线胶装，封面白卡纸四色印刷，表面覆亮面膜。内页胶版纸四色印刷，偏黄的纸张和较吸墨的纸张印刷效果，很好地体现着旧日回忆的味道。全书捷克文与英文双语排版，从各级标题、正文，一直到图注，均严格按照双栏设置。页码被隐藏在极其靠近装订线的位置，模糊书籍的线性结构，而可以更自由地徜徉在这些遍布街道的公共雕塑里。书籍的上下书口和右侧翻口没有清边，使得每 32 页形成的折手都呈现一个隆起的书口边缘，具有如浮雕般的触摸手感。

Vetřelci
a volavky
Aliens
and Herons

This is a book that collects, collates and discusses public sculptures in Czechoslovakia from 1968 to 1989. During this period, almost all public spaces were occupied by format sculptures. The collection of these sculptures reflects the author's memories of the specific historical periods. The book summarizes these sculptures into five types of classification labels; some of the sculptures may have multiple labels at the same time. The book is perfect bound with thread sewing. The laminated cover is printed in four colors on white cardboard. The inside pages are printed in four colors on offset paper. The yellowish and ink-absorbent paper has the taste of old memories. The text is formatted using a double-column layout in Czech and English, including all levels of headings, text and captions. Page numbers are hidden close to the binding line in order to blur the linear structure of the book, thus the readers may wander in the streets with public sculptures freely. The book is untrimmed that every 32 pages form a ridge edge, which feels like relief.

荣誉奖

外星人和苍鹭

捷克

Honorary Appreciation

Aliens and Herons

Czech Republic

Tereza Hejmová

Pavel Karous (Ed.)

Arbor vitae, Academy of Arts, Architecture and Design, Prague

240×167×38mm

814g

459p

This book must have been taken from a natural science bookshelf. It is a taxonomy of sculpture in public places of Czechoslovakia from1968 to 1989. The large-sized plastic sculptural and pictorial works are systematically arranged, following the rules of biological sciences. How baffling it is to marvel at cosmonauts, kissing couples, the variations of Muf Supermuf and the Explosion of Transformers in stone, concrete and bronze in playgrounds, in parks and in front of large-scale housing estates, where they proclaim the message of an ideal world and a creation of a future world full of promises as a matter of course. And now the paragraph for the book lovers: it smells so nice! The spine remains intact when the book is opened. The 32-page sections make the relief-like fore-edge trim so easy to touch. To westerners, the print of the colour sets on the yellowish paper evokes wistful memories of the forms they had to fill in painstakingly when they wished to cross the border. And these sparing, precisely placed scientific drawings in pointillism technique – they are an indicator that the publishers have applied themselves with a certain distance and a huge degree of warm-heartedness to an entire genre of their political-cultural heritage. An example to all monuments authorities.

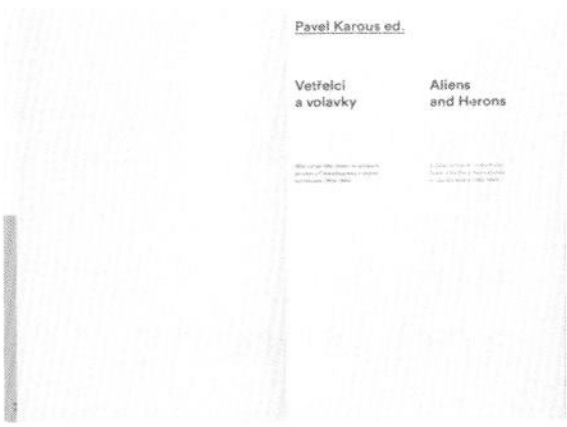
Pavel Karous ed.
Vetřelci a volavky
Aliens and Herons

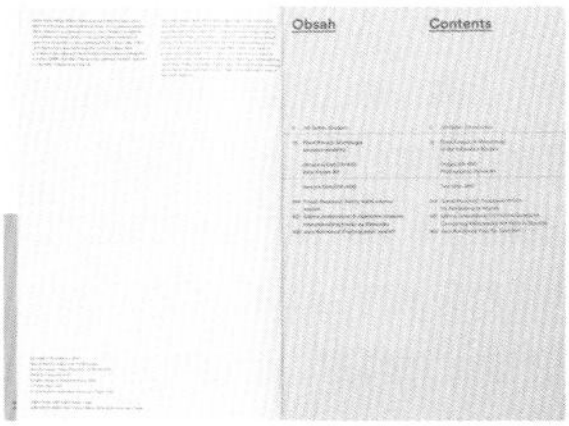
Obsah
Contents

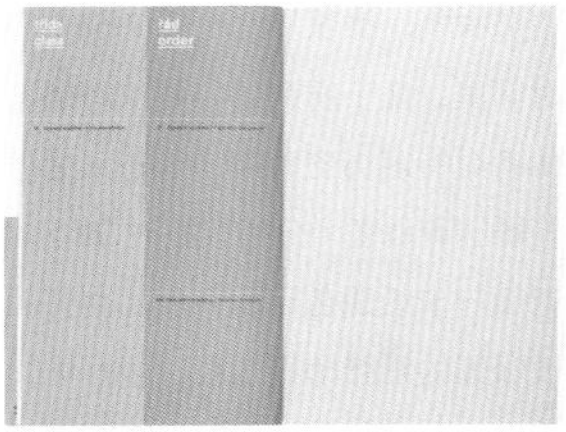
řád
order

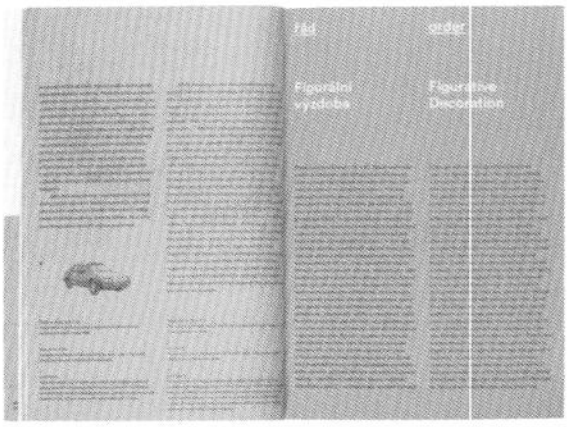
řád
order
Figurální výzdoba
Figurative Decoration

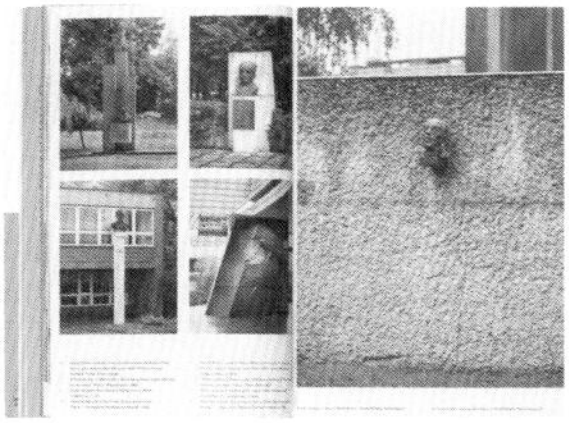

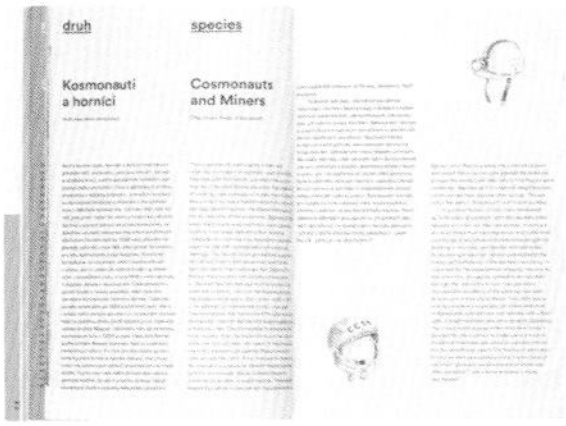
druh
species
Kosmonauti a horníci
Cosmonauts and Miners

Tomáš Pospiszyl
Sochy, které nikomu nepatří
Sculptures Which Do Not Belong to Anyone

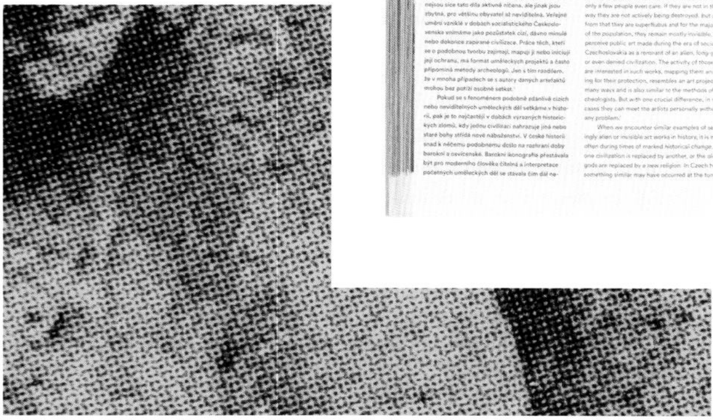

Jiří Šetlík
Úvodem
Introduction

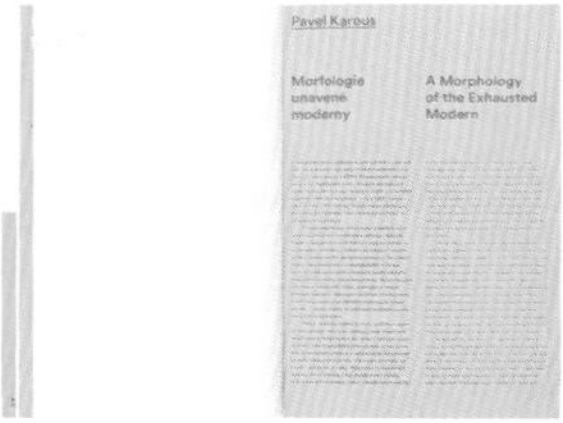
Pavel Karous
Morfologie unavené moderny
A Morphology of the Exhausted Modern

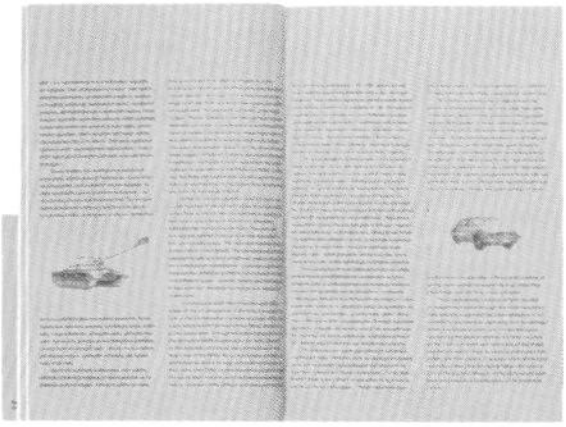

druh
species
Madona s dítětem
Madonna with Child

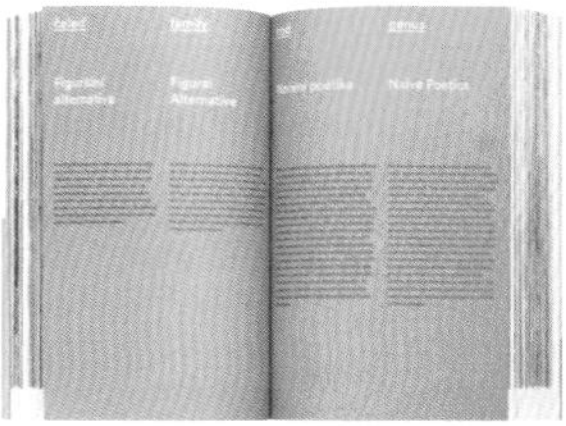
Naive Poetics

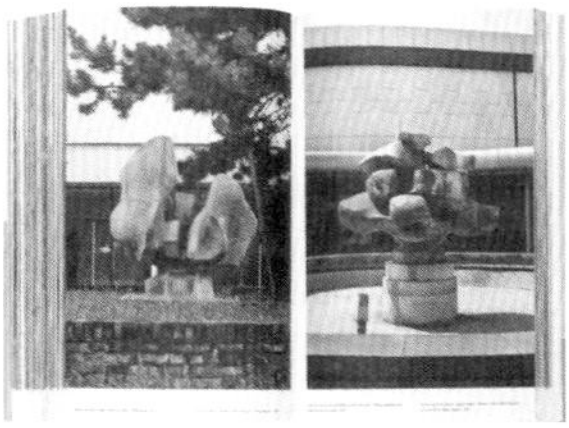

jen jako

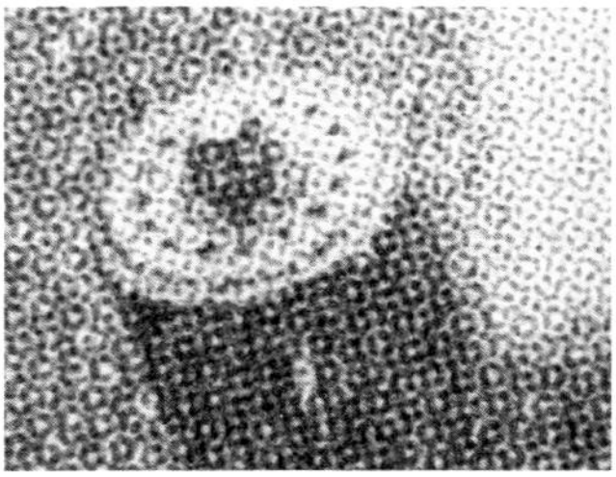

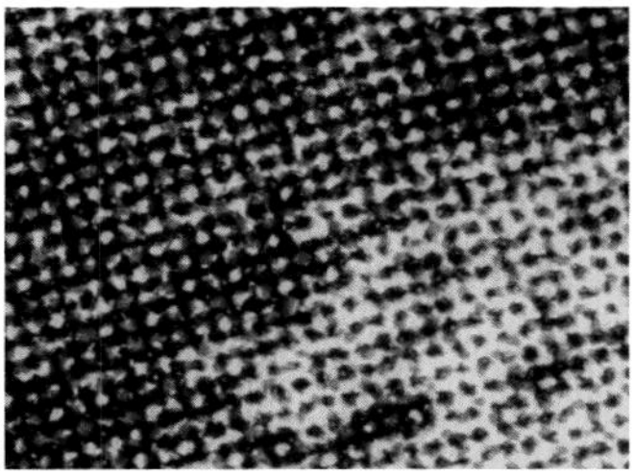

řísně ge

tury, ale jsou i odvážnými

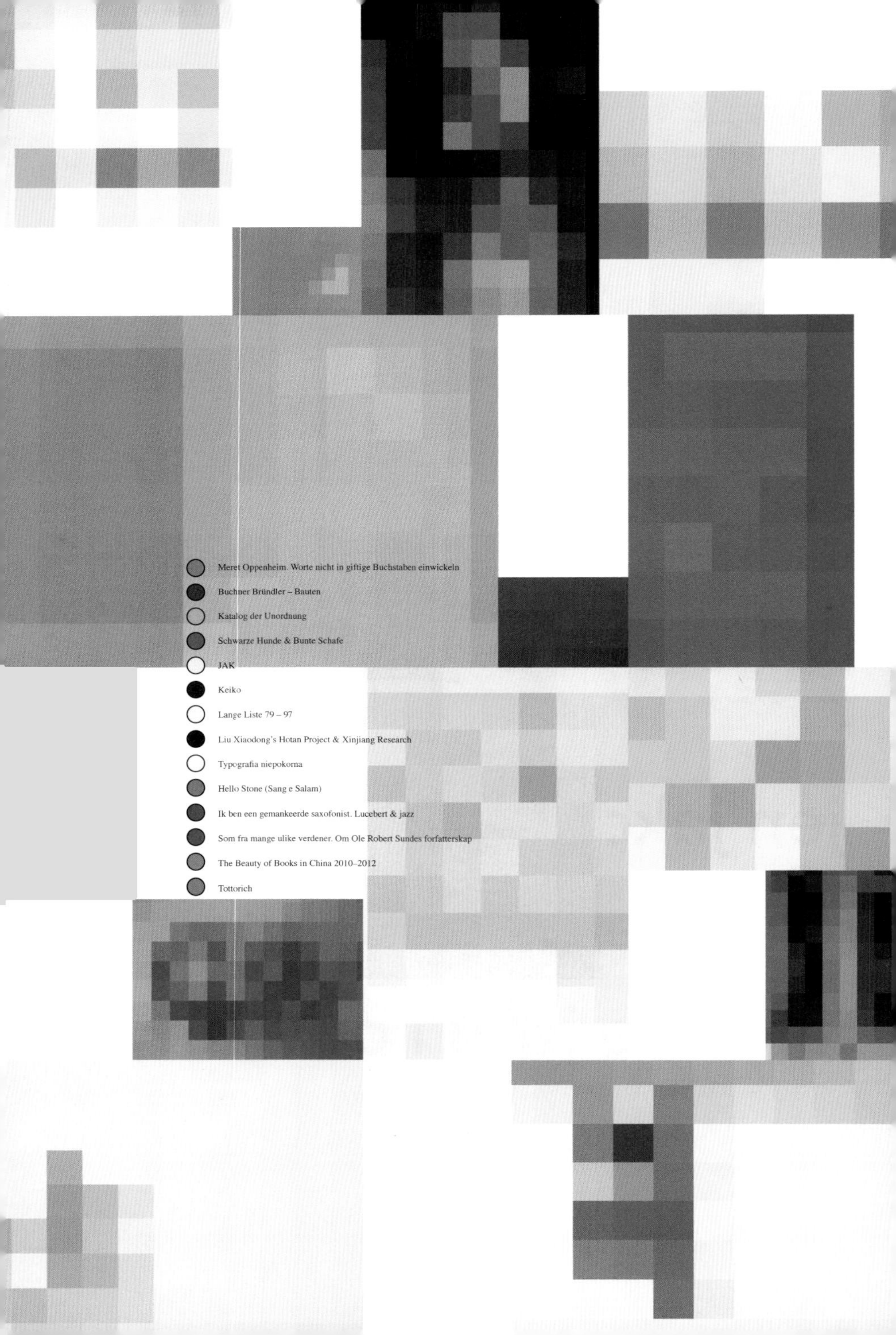

- Meret Oppenheim. Worte nicht in giftige Buchstaben einwickeln
- Buchner Bründler – Bauten
- Katalog der Unordnung
- Schwarze Hunde & Bunte Schafe
- JAK
- Keiko
- Lange Liste 79 – 97
- Liu Xiaodong's Hotan Project & Xinjiang Research
- Typografia niepokorna
- Hello Stone (Sang e Salam)
- Ik ben een gemankeerde saxofonist. Lucebert & jazz
- Som fra mange ulike verdener. Om Ole Robert Sundes forfatterskap
- The Beauty of Books in China 2010–2012
- Tottorich

2014

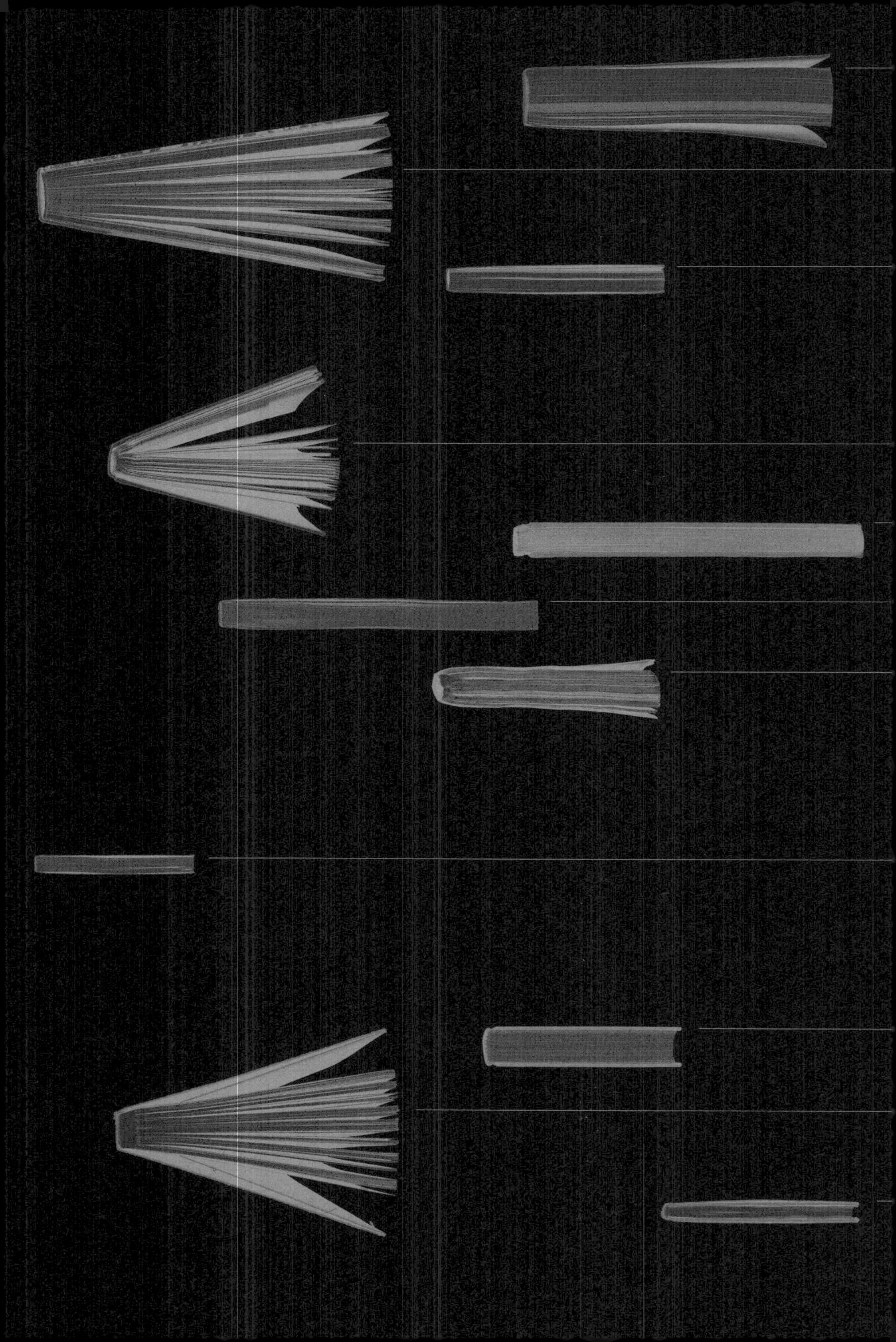

2014 Gl

0561

Scheidegger & Spiess
Meret Oppenheim

2014 Gm

0569

2014 Sm[1]

0575

2014 Bm[1]

0581

JAK

2014 Bm[2]

0587

2014 Bm[3]

0593

LANGE LISTE
Spector Books

2014 Bm[4]

0599

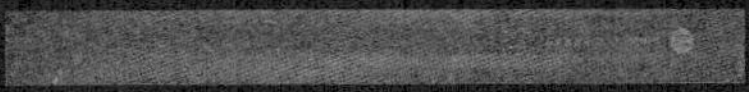

2014 Ha[1]

0605

2014 Ha[3]

0609

2014 Ha[4]

0613

2014 Ha[5]

0617

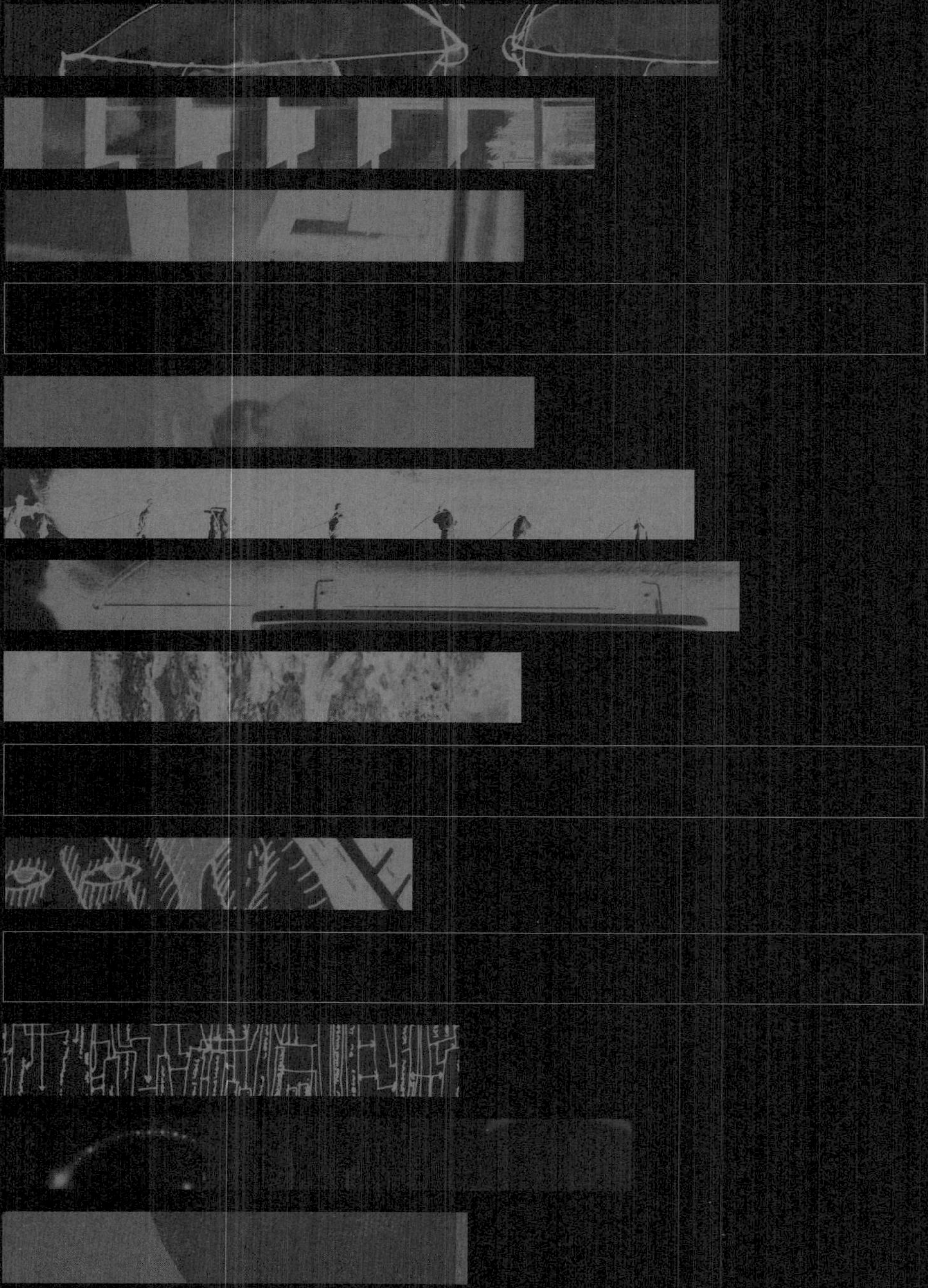

Liebe Eltern!"

Wie geht es Euch? Ich freue mich sehr bis ich heimkomme und auch auf die Schule. Am Dienstag hatten wir keine

mit drei Erdbeeren und einer Biene.

12

sind fünf Herzchen aufgeklebt, eines kleiner als das andere, bei denen jedem mit

Ein neuer Bub ist geko
alt. Herr Dr. hat ihn geh
aussehe, ob er dick o
antwortete er: «oui, gra

Dennoch versuchen wir als Architekten ganz
usst, dieses gesetzte Recht zu dehnen und die
etzeslücken auszuloten ...

gründen ausg
plötzlich mög
in der Materia

Stacks of census questionnaires "taller than the W
Monument"; a sea of policy-holders insured against th
disaster, accident, and death; legions of workers in asse

der Zwiebel stattfindet. JAK überträgt das Offenbarte ins
Heft. Die filigranen und komplexen Zeichnungen, die
bescheidenen Wörter, der Wirbelwind der Gedanken und

Mohammad and his two colleagues, and porters in a car bound for Chittagong. The boys
dressed up in their finest clothes: slippers, dark trousers. One of them had to borro
trousers from a friend; he only has working clothes. It looks like it's going to be a beautif

FDGB	27,00
Versicherung	70,00
Benzin	145,00
Kinderkrippe	30,00

Kutte v. Robert 150,00

Robert war ein guter
Freund der Familie. Das war
ein lustiger Kerl, immer ein

FDGB (10 + 20,— M)
Versicherung
Garage *(März, April)*
Benzin

下午到了乌市，入住如家小店，奔往国际展览中心，在那里将
和田项目，空间超大，在 108 空间。晚上韩书记等宣传部、文
大柬宴请，新疆比内地晚两个小时时差，夜两点多入睡。

ند یا افتاده‌اند دنبالت. من یک بار با بابا، شبانه از
ا وقتی دید وحشت کرده‌ام و هی دور و برم را نگاه می‌ک

lette samlinger. Selv har jeg skrevet en
sj én gang, men Sunde klarte faktisk å

“中国最美的书”奖项创立已近十年。十年来，“中国最美的书”评委
书艺术基金会的委托，评选和推荐“中国最美的书”获奖图书参评“

水から出
意志があ
思いその
いした文
ると充分
りはじめ
無くなっ
と文字が
たいな体
微かな力
とさらに
の手のよ

Golden Letter

Meret Oppenheim.

Worte nicht in giftige Buchstaben einwickeln

Switzerland

Bonbon, Valeria Bonin und Diego Bontognali, Zurich

Lisa Wenger, Martina Corgnati

Scheidegger & Spiess, Zurich

330×220×38mm

2018g

474p

2014 Gl

Gold Medal

Buchner Bründler – Bauten

Germany

Design Concept: Ludovic Balland, Andreas Bründler,

Daniel Buchner – Basel (Switzerland), Design and Composition: Ludovic Balland und Gregor Schreiter / Ludovic Balland Typography Cabinet – Basel (Switzerland)

gta D ARCH Ausstellungen, ETH Zürich (Ed.)

gta Verlag, Zurich

273×232×36mm

1643g

338p

2014 Gm

2014 Sm[1]

Silver Medal

Katalog der Unordnung

Austria

Christoph Schorkhuber, Linz

Helmuth Lethen, IFK

Internationales Forschungszentrum Kulturwissenschaften

an der Kunstuniversität Linz, Linzt

240×162×16mm

478g

191p

2014 Sm[2]

2014 Ha3

Honorary Appreciation
Som fra mange ulike verdener. Om Ole Robert Sundes forfatterskap
Norway
Andreas Töpfer
Red. Audun Lindholm
Gyldendal Norsk Forlag AS
211×135×26mm
438g
279p

2014 Ha1

Honorary Appreciation
Hello Stone (Sange Salam)
Iran
Majid Zare
Mohammad Reza Bayrami
Asr-e Dastan
189×130×12mm
235g
240p

2014 Ha3

2014 Ha4

Honorary Appreciation
The Beauty of Books in China 2010–2012
China
Liu Xiaoxiang, Liu Xiaoxiang Studio
Organizing Committee of "The Beauty of Books in China" of Shanghai Press and Publication Administration
Shanghai People's Fine Arts Publishing House
291×192×40mm
1709g
294p

2014 Ha5

Honorary Appreciation
Tottorich
Japan
Masahiko Nagasawa
Yuan Okada
Doyo Bijutsusha Shuppan Hanbai
215×145×11mm
273g
93p

2014 Ha2

2014 Bm1

Bronze Medal
JAK
Germany
Demian Bern, Stuttgart
JAK, Hamed Taheri
EXP. edition, Stuttgart
245×166×23mm
700g
108p

2014 Bm2

Bronze Medal
Keiko
Germany
Marek Mielnicki / veryniceworks, Warschau (Polen)
Tomasz Gudzowaty (Hrsg.)
Hatje Cantz Verlag, Ostfildern
319×243×20mm
1531g
144p

2014 Bm5

2014 Bm3

Bronze Medal
Lange Liste 79 – 97
Germany
Christian Lange, München
Christian Lange
Spector Books, Leipzig
340×241×19mm
1038g
186p

2014 Bm4

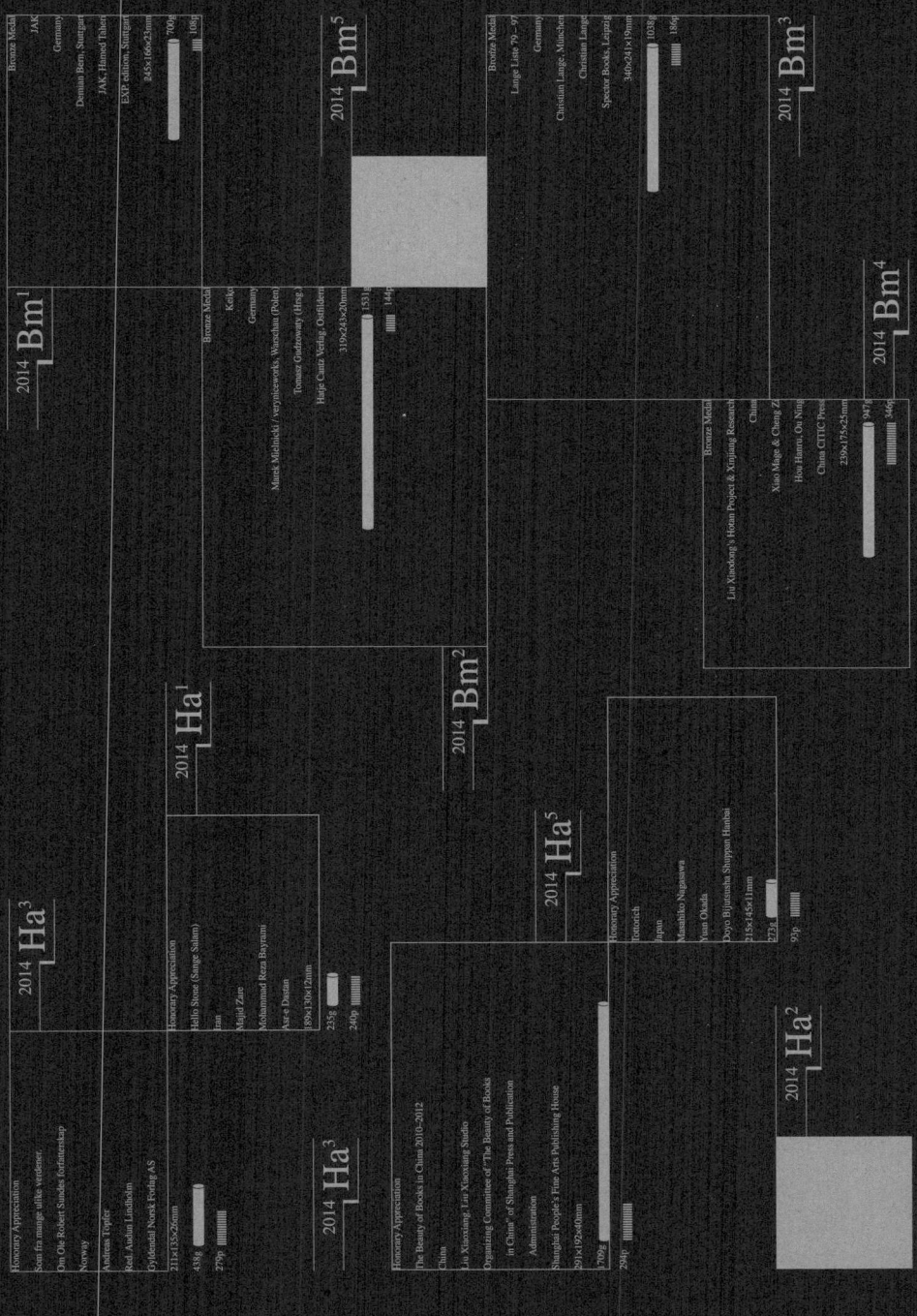

Bronze Medal
Liu Xiaodong's Hotan Project & Xinjiang Research
China
Xiao Mage & Cheng Zi
Hou Hanru, Ou Ning
China CITIC Press
239×175×25mm
947g
346p

The International Jury

Prof. Lü Jingren (China)
Maren Katrin Poppe (Germany)
Prem Krishnamurthy (USA)
Manja Hellpap (Germany)
Mariko Takagi (Japan, Germany)
Kurt Dornig (Austria)
Roland Stieger (Switzerland)

2014

Country / Region

Germany 4
Austria 2
China 2
Switzerland 1
Poland 1
Iran 1
The Netherlands 1
Norway 1
Japan 1

Designer

Bonbon, Valeria Bonin und Diego Bontognali, Zurich

Design Concept: Ludovic Balland, Andreas Bründler, Daniel Buchner – Basel (Switzerland), Design and Composition: Ludovic Balland und Gregor Schreiter / Ludovic Balland Typography Cabinet – Basel (Switzerland)

Christoph Schorkhuber, Linz

Lisa Maria Matzi, Wien

Demian Bern, Stuttgart

Marek Mielnicki / veryniceworks, Warschau (Polen)

Christian Lange, München

Xiao Mage & Cheng Zi

Monika Hanulak

Majid Zare

Piet Gerards Ontwerpers

Andreas Töpfer

Liu Xiaoxiang, Liu Xiaoxiang Studio

Masahiko Nagasawa

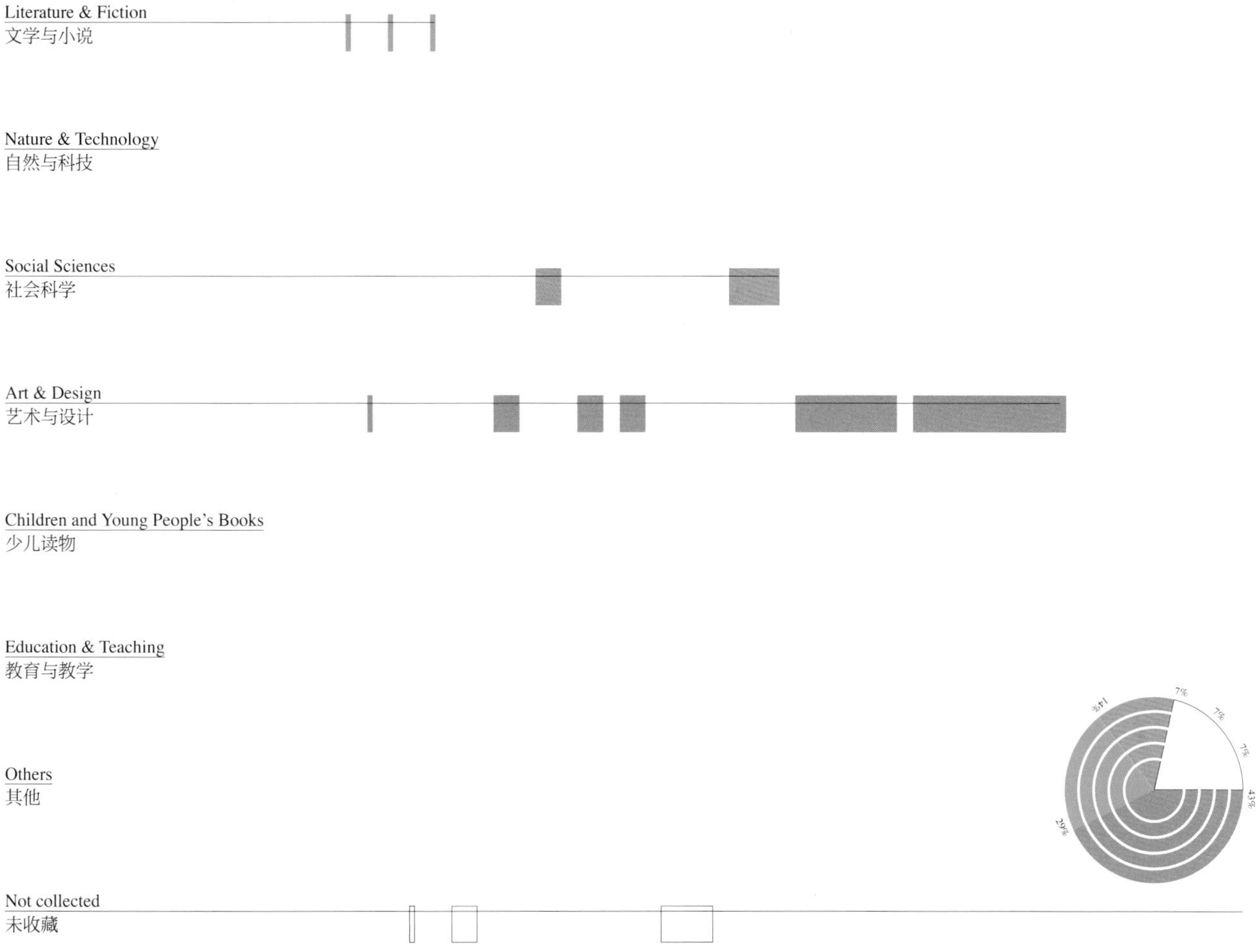

Literature & Fiction
文学与小说
Nature & Technology
自然与科技
Social Sciences
社会科学
Art & Design
艺术与设计
Children and Young People's Books
少儿读物
Education & Teaching
教育与教学
Others
其他
Not collected
未收藏
7%
7%
7%
14%
29%
43%

2014 Gl

de ta bonne lettre.
J'espère te voir arriver
bientôt. Nous ne par-
-tirons pas avant le
15 ou le 16 juillet, et
je ne sais pas encore où
Probablement en Espagne
de voir la Galice
et au Portugal.
Je suis encore très fatigué
Mais je vais
une rose autour
de mon rêve, comme
personne plus haut
Meret,
ta lettre
Nicolas Calas

这本书收集了瑞士超现实主义艺术家梅雷特·奥本海姆（Meret Oppenheim）70 多年中 1000 多份未发表的文件。作为超现实主义的领军人物，她的生活和工作充满传奇色彩，令人着迷。本书为锁线胶装平装本，封面采用布面四色印刷，裱覆白卡纸，文字部分烫黑。内页主体部分为哑面艺术纸印深灰色，展现信息量最多的各项文件，包括与家人、友人、其他艺术家的信件，以及在之前的版本中没有发表过的文件。这些部分间隔着夹带 4 页彩图，胶版纸四色印刷，为相关文件的扫描版。书中三分之一处完整地展现了艺术家的一本笔记本的所有页面，采用胶版纸四色印刷，可以全面了解她从童年到 1943 年间的绘画、照片、信件，甚至收集的一些资料，从中看到艺术家这些年来的变化。书中正文部分设置为双栏，文中有大量注解，创造性地采用极窄的双栏设置插在正文双栏中间，由于注解长短不一，正文与注解的文字块形成了具有大量变化的穿插效果，非常具有特点。

Meret
Oppenheim
Worte nicht
in giftige
Buchstaben
einwickeln

"This book collects more than 1,000 unpublished documents related to Swiss surrealist artist Meret Oppenheim over 70 years. As a leader of surrealism, her life and work are full of legends and fascinating. This paperback book is perfect bound with thread sewing. The four-color printing cloth cover is mounted by white cardboard with black hot stamped text. The inner pages are made of matte art paper printed in dark gray, showing the most informative documents, including letters with his family, friends and other artists, as well as documents unpublished in previous versions. Among every 16 or 24 pages, there are four pages of pictures, printed on offset paper in four colors, which present the scanned copy of relevant documents. Pages of an artist's notebook are fully displayed on offset paper in four colors. The readers can get a comprehensive understanding of the artist through her paintings, photos, letters, and some information collected from her childhood to 1943. The text is formatted using a double-column layout while the large amount of annotations are creatively inserted in the middle of the columns with a very narrow double-column setting. Since the length of the annotations is different, text blocks and annotated text blocks form a great variety of interlaced effect.

金字符奖

梅雷特·奥本海姆
——不要用有毒的字母包裹话语

瑞士

Golden Letter

Meret Oppenheim.
Worte nicht in giftige Buchstaben einwickeln

Switzerland

Bonbon, Valeria Bonin und Diego Bontognali, Zurich

Lisa Wenger, Martina Corgnati

Scheidegger & Spiess, Zurich

330×220×38mm

2018g

474p

A weighty folio of substantial size dedicates itself to the letters of Meret Oppenheim. At the heart of the sources under investigation is the facsimile of the album "Von der Kindheit bis 1943" ("From childhood until 1943"), published here for the first time. From a design point of view the challenge here is to organise a large dismembered body of text in such a way that the reader is impelled to read it by more than just a sense of duty. On the other hand the fact that the display of her own personality was anathema to Oppenheim calls for typographical discretion. The large-sized typeface does not have wide spacing, yet its pronounced bold-fine style allows the two-column typesetting enough space to breathe. Footnotes are positioned unconventionally, namely in single or double columns of the narrowest width between the main columns. The text is printed in dark grey. The overall effect is one of politeness and organised structure. The idea of altering the background of each page of the album in the facsimile section to multi-coloured pastel shades has been marvellously realised. The change of paper underlines the precious nature of the original documents.The work is presented unpretentiously yet self-confidently in a cloth-laminated card sleeve on which the elements of its design order can truly resonate.

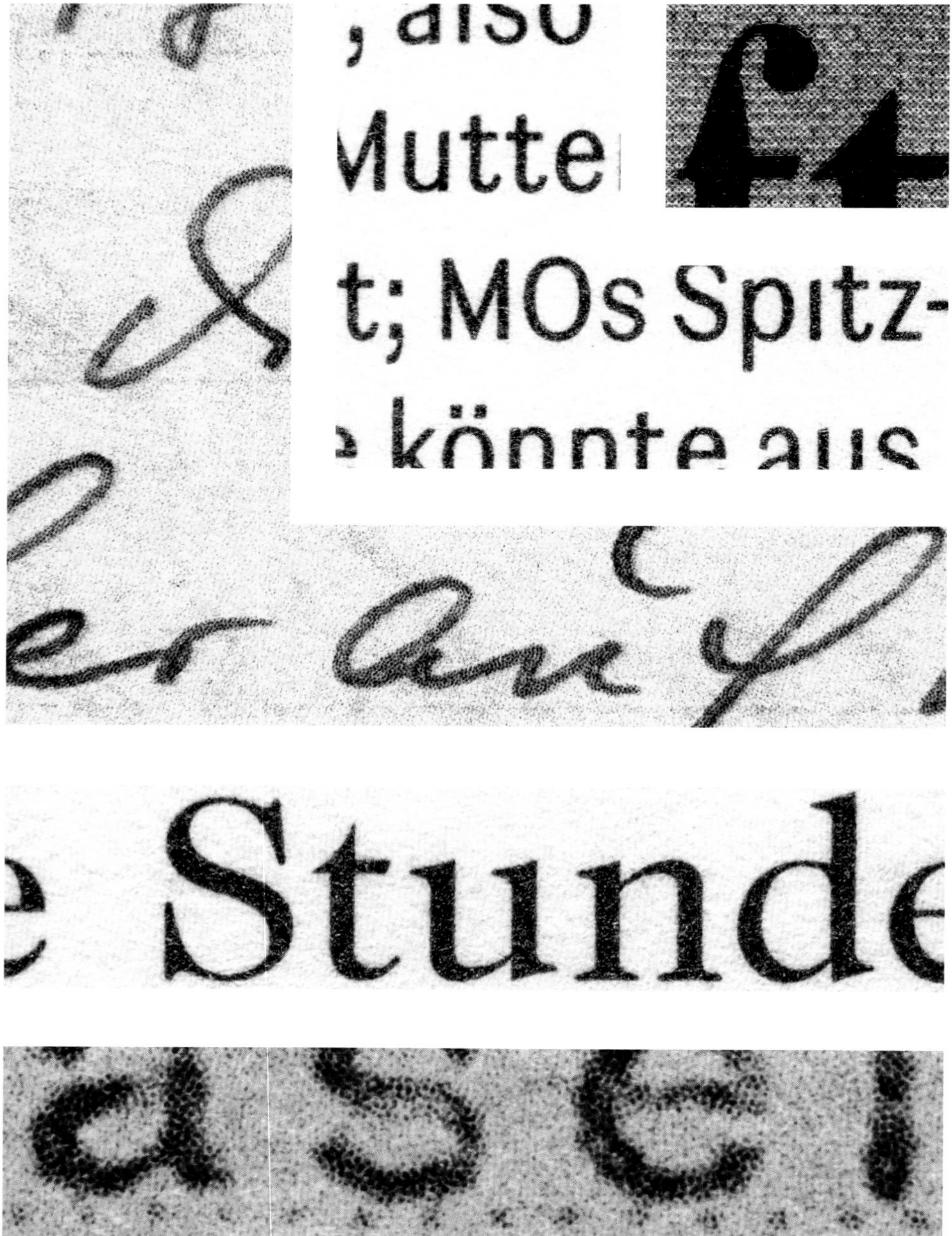
Mutte
t; MOs Spitz-
könnte aus
Stunde

weil
gan-
tern
tro-
wars
wie
roh,
ziges
der
uto-
Hol-
ym-
er es
und
Herr

Buch
das,
ver-
r am
wohl
run-
nder
lese
mer
dem
dan-

hell
u da
es[5].
Bitte
wort
be es

men
tin?
heim
dem
mmt
s ist
nder

s Du
mir,
ndes
rau,

nen
ner.
... es. Man konnte
Grösse reicht gar
iana hat ein Kind.
r Freude. Ich sah
achte ihm schöne

r, dass Du gesund

von Januar 195
bis April 1969.

2
Schweizerdt.
für hinauf.

3
Wahrscheinlic
Band 4 von *We*
in vier Bänden
Heinrich Hein
hrsg. von Paul
Stapf, Birkhau
Basel 1956.

4
Wahrscheinlic
Der Schrei (H
oder *Der grün*
Zuschauer (H
beide 1959.

5
Vielleicht *Schw*
ze Form mit To
kopfen (H 47),
Garten (H 47 a
oder *Garten* (H
b), alle von 195

6
Wahrscheinlic
Irene Zurkinde

7
Prodör, kleine
Ort oberhalb v
Faido.

8
Alfred Bühler,
Schwiegersoh
von EO.

9
Am 1. Februar
1959 hatte die
erste Abstimm
für die Einführ
des Frauenstir
rechts auf eidg
nössischer Eb
stattgefunden
die Vorlage wu
von den Schwe
Männern mit e
Zweidrittels-
mehrheit abge
lehnt; erst am
Februar 1971 w
das Frauen-
stimmrecht en
lich mit 66 %
Ja eingeführt.

330 × 220 × 38

da war. Dass du auch sonst zu tun hast mit restaurieren freut mich für Dich wegen Geld.[11] Du wirst mir dann erzählen ob Du mit dem Geld besser auskommst in L.

Ruthi schreibt mir viel, und ich ihr auch so alle 3 Wochen. Ich schicke ihr auch mancherlei. Mit ihrem verdüsterten Gemüt lebt sich's halt schwerer. Sie hat auch wenig Geld nachdem sie ein Vermögen verpulvert, u. wieder verpulvern würde. «Wer kann dafür, wer kann

2018

474

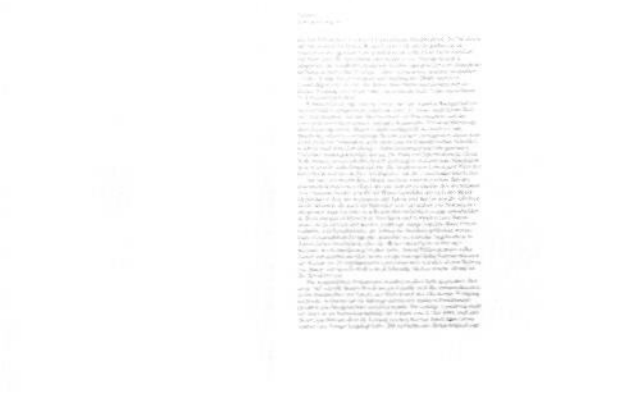

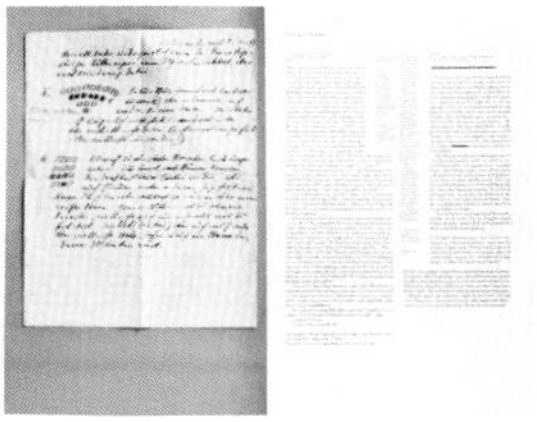

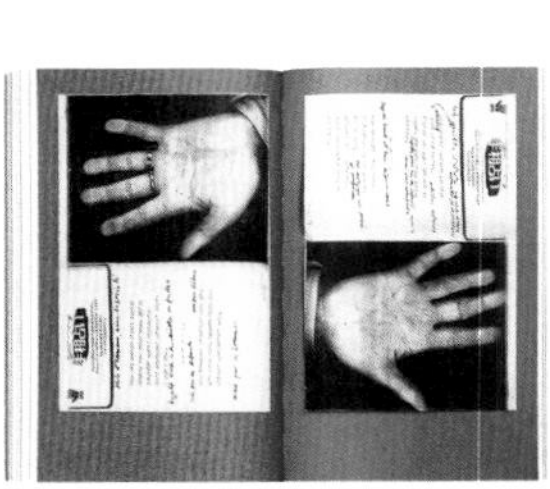

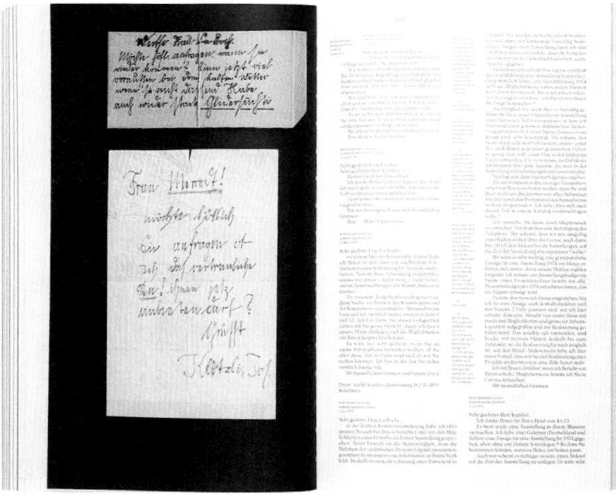

Meret
Oppenheim
Worte nicht
in giftige
Buchstaben
einwickeln

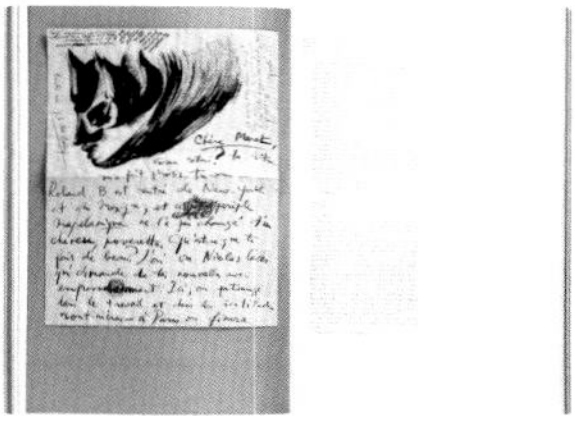

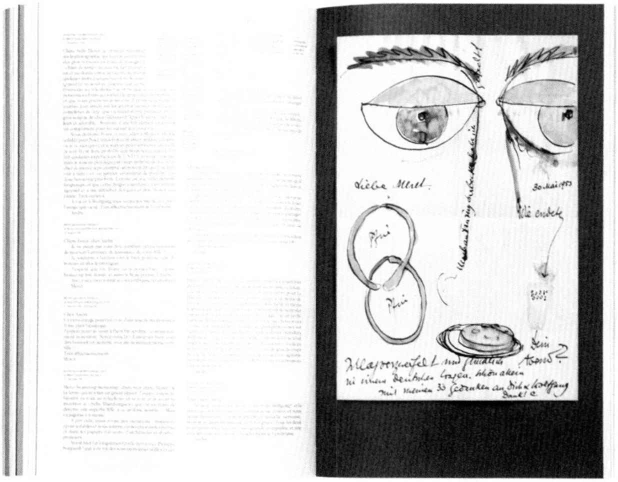
Liebe Meret
30 Mai 1953
Dein

Meret Oppenheim

Scheidegger & Spiess

Die Freundschaft zwischen André Kamber (*1932), 1967–1997 Direktor des Museums der Stadt Solothurn, und Meret Oppenheim begann 1966 anlässlich einer Ausstellung des Kunstvereins Solothurn mit Schweizer Künstlern und dauerte bis zu ihrem Tod. Kamber schätzte Oppenheim und ihre Kunst sehr; ab 2007 war er Stiftungsrat der Fondation Meret Oppenheim, welche sich die Aufarbeitung ihres Œuvres zum Ziel gesetzt hat und nach Abschluss dieser Arbeiten 2013 aufgelöst wird.

Kamber kaufte eine Zeichnung für seine private Sammlung, realisierte 1974/75 Meret Oppenheims erste Retrospektive in der Schweiz und machte sich stark für den Ankauf wichtiger Werke durch das Kunstmuseum Solothurn. Sie schenkte ihm 1985 im Gegenzug einen Holzschnitt, welcher der Benefiz-Mappe für die grafische Sammlung des Kunstmuseums beigefügt wurde.

Das vorliegende Konvolut – mit zunehmender Vertrautheit wechseln sie 1974 vom Sie zum Du – enthält mehrheitlich Briefe von Meret Oppenheim an André Kamber; teilweise von ihm selbst kommentiert.

→ Notiz zu Dos… 1985 (Al 32), 1985, siehe A… kung 8, S. 42…

André Kamber, Solothurn
an Meret Oppenheim, Hünibach
21. Juni 1966[1]

Sehr geehrte Frau La Roche,

der Kunstverein hier will wieder eigene Ausstellungen durchführen. So planen wir für den Beginn eine Schweizer Zeichnungsausstellung. Die Tendenz liegt auf wenigen Ausstellern zugunsten gültiger Werkgruppen.[2]

Mir fällt die Aufgabe zu, das Ausstellungsgut zusammenzubringen. Deshalb möchte ich Sie zur Teilnahme an dieser Ausstellung einladen. Eine Werkgruppe soll nach unserer Auffassung 5–8 Zeichnungen umfassen, wobei wir auch auf öffentlichem oder privatem Besitz basieren.

Es wäre mir sehr gedient, wenn Sie mir möglichst bald einen prinzipiellen Bescheid geben könnten, ob Sie an dieser Ausstellung teilnehmen wollen und können. Allerdings möchten wir auch in jeder Gruppe verkäufliche Werke haben.

Möglicherweise könnte ich auch einmal zu Ihnen hinauf fahren. Deshalb wäre ich froh, wenn Sie mich avisieren für den Fall, dass Sie länger verreisen.

Mit freundlichen Grüssen
Für den
Kunstverein der Stadt Solothurn
André Kamber, Aktuar

Meret Oppenheim, Hünibach
an André Kamber, Solothurn
23. Juni 1966

Sehr geehrter Herr Kamber,

Besten Dank für Ihren Brief. Es trifft sich gerade sehr gut, da in Basel bei der Galerie Handschin (Malzgasse 2, Ecke St. Albanvorstadt) 15–20 Zeichnungen von mir ausgestellt sind (alle verkäuflich).[3]

Die Ausstellung ist noch bis 30. Juni. Dann gehen die Zeichnungen wieder zu mir. Aber es ist natürlich besser, die Zeichnungen gerahmt und an der Wand zu sehen.

Vielleicht können Sie sich sogar mit Herrn Handschin arrangieren, dass er sie Ihnen gerahmt überlässt. Ich finde sie im allgemeinen gut gerahmt (Ausser die *Waldlichtung*[4] im breiten Eichenrahmen, das ist doch nicht so günstig für die Zeichnung). Denn nicht nur, dass das Einrahmen enorm Zeit braucht, sondern hauptsächlich weil meine Zeichnungen nicht fixiert sind u. zum Teil mit sehr weichem Stift gezeichnet (Ich will sie nicht fixieren). Sobald sie aus der Mappe, in der sie sonst ruhen, sind, passiert etwas. Die, die beim Clicheur war, ist verwischt, auch eine andere, diese muss ich vor Ihrer Ausstellung wieder «restaurieren».

Auch bitte ich Sie jetzt schon, mir am Ende der Ausstellung das Ausrahmen zu überlassen, damit ich jede Zeichnung mit einem Blatt Pergamin decken kann, das ich hinten etwas anleime.

Ich danke Ihnen für die Einladung und grüsse Sie freundlichst
Meret Oppenheim
Wann soll wohl die Ausstellung sein?

Meret Oppenheim, Hünibach
an André Kamber, Solothurn
12. Juli 1966

Sehr geehrter Herr Kamber,

Besten Dank für Ihren Brief.

Es ist nett von Handschin, dass er die Bilder gerahmt gibt. Was die Prozente betrifft, so möchte ich von ev. Verkäufen 50%. 15% für den Kunstverein Solothurn u. 35% für Handschin (weil er die Bilder gerahmt gibt).

Ich bitte Sie, mir zu berichten bevor Sie sie hängen, damit ich die verwischten Stellen bei dem einen noch ausradieren kann.

Die Ausstellung sei Ende August, sagte H.
Mit freundlichen Grüssen,
Meret L.-R.-Oppenheim

Meret Oppenheim, Hünibach
an André Kamber, Solothurn (privat)
6. Mai 1967[5]

Sehr geehrte Frau Kamber,
Sehr geehrter Herr Kamber,

Besten Dank für Ihren Brief.

Ja, kommen Sie einmal, besser vielleicht im Sommer, d.h. ev. Juni, weil dann die Sachen aus Stockholm[6] zurück sind. Vorher bitte Telephon, weil wir über Wochenende manchmal fort sind.
Mit freundlichen Grüssen,
Meret Oppenheim

1
Maschinengeschriebener Brief

2
Zeichnungen Schweizerischer Künstler (Hodler, Vallotton, Amiet, Soutter, Auberjonois, Brühlmann, Klee, Berger, Blanchet, Meyer-Amden, Gimmi, Morgenthaler, Barraud, Pauli, Schürch, Bally, Geiser, M. Gubler, Schnyder, A. Giacometti, Tschumi, Camenzind, Wiemken, Fischer (fis), Kämpf, Oppenheim, R. Müller, Tinguely, Kuhn, Buser, Luginbühl, Sadkowsky), Kunstverein der Stadt Solothurn (27.8.–28.9.1966).

3
In der Gruppenausstellung *Dorothe Hofmann – Meret Oppenheim – K.R.H. Sonderborg*, Galerie Handschin, Basel (2.6.–30.6.1966).

4
O 104 a von 1965.

5
Kommentar von AK: Betr. Privaten Ankauf *Augenbogen*, 1963 (M 69 f).

6
Die Retrospektive *Meret Oppenheim* im Moderna Museet, Stockholm (15.4.–21.5.1967).

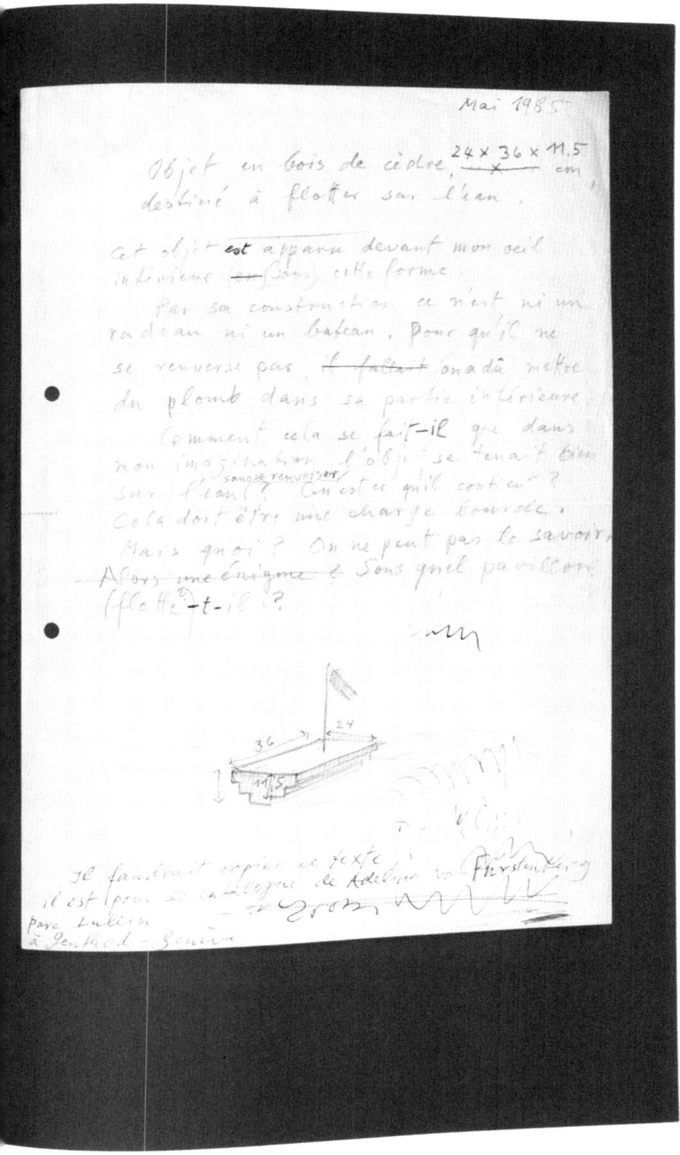

Mai 1985

Objet en bois de cèdre, 24 × 36 × 11.5 cm,
destiné à flotter sur l'eau.

Cet objet est apparu devant mon oeil
intérieur ~~(en)~~ (sous) cette forme
Par sa construction ce n'est ni un
radeau ni un bateau. Pour qu'il ne
se renverse pas, ~~il fallait~~ on a dû mettre
du plomb dans sa partie inférieure.
Comment cela se fait-il que dans
mon imagination l'objet se tenait bien
sur l'eau (sans se renverser)? Qu'est ce qu'il contient?
Cela doit être une charge lourde.
Mais quoi? On ne peut pas le savoir.
~~Alors une énigme~~ ? Sous quel pavillon
flotte-t-il?

36
24
11.5

Il faudrait copier ce texte pour le catalogue de Adelina von Fürstenberg
il est pour le ...

Parc Lullin
à Genthod - Genève

这是国际知名的瑞士建筑事务所 Buchner Bründler 的作品集。书中展现了这家建筑事务所的建筑方案、对谈、评论、研究等不同维度的内容。本书装订方式为锁线硬精装，封面哑面艺术纸单黑印刷，裱覆极薄的卡板，标题字烫银处理。内页大部分为胶版纸四色印刷，照片大多数为黑白，少量处理为低饱和度。每个项目的最后附上圣经纸单黑印刷的建筑图纸拉页，按照图纸内容呈阶梯状折入书内，阶梯状书口上标明图纸名和比例尺。建筑项目前的目录和对谈部分与常规的图书做法不同。严格来说，本书没有一个按照书籍内容顺序排列的目录，而是根据内容主题设定了大量不同的标签，每个标签下有几个页码索引，这样就建立了一个多元的索引系统，让读者可以根据兴趣直接阅读相关内容，形成非线性的阅读体验。

Buchner Bründler
Bauten

This is a collection of works by Buchner Bründler, an internationally recognized Swiss architectural firm. The book shows the architectural plan, conversation, commentary, research and other aspects of this firm. The hardback book is bound by thread sewing. The cover is printed in matte art paper in black only, mounted by thin cardboard; the title words are silver hot stamped. The inside pages are made of offset paper with four-color printing. Most of the photos are black-and-white, and a few are processed in low saturation. At the end of each project, the architectural drawings printed on bible page in black are attached, folded into the book according to the contents of the drawings. The titles and scale of the drawings are shown on the edge. The design of catalogue and conversation section is different from that of conventional books. Strictly speaking, the book does not have a catalogue arranged in the order of book contents, but sets many different labels according to various themes. There are several page numbers indexed under each label. Thus, a multi-index system is established, which allows readers to read the relevant content according to their interests. It gives the readers special non-linear reading experience.

金奖
Buchner Bründler——建筑物
德国

Gold Medal
Buchner Bründler – Bauten
Germany
Design Concept: Ludovic Balland, Andreas Bründler, Daniel Buchner – Basel (Switzerland)
Design and Composition: Ludovic Balland und Gregor Schreiter / Ludovic Balland Typography Cabinet – Basel (Switzerland)
gta D ARCH Ausstellungen, ETH Zürich (Ed.)
gta Verlag, Zurich
273×232×36mm
1643g
338p

An architect's monograph which has been solidified into visual poetry. On the book's wide spine we read not only its title, but also the blurb which has been positioned here, using the same sober yet delicate typography which sustains the whole book. The cover already bears witness to the disciplined but extravagant taste of architecture and book. The potential offered by a contents table is exploited to the full, stretching as it does here over 29 pages, connecting the subject groupings which are summarised under particular mottos with interviews and raising what is otherwise generally such an unassuming zone of a book to nothing short of a literary genre. The book follows a concept of presenting the individual constructions with a minimum of text, and its photographs on matt absorptive paper – mostly monochrome, sometimes with reduced colour saturation – leave behind their impressions thanks to their contemplative staging. Between the projects, moreover, a contrast is created which could hardly be greater: using Bible paper for fold-out pages the construction drafts are tantamount to an invitation to industrial espionage. And for this the book even has a special tool at the ready – a bookmark showing the increments of the various scales. A work of graphical engineering.

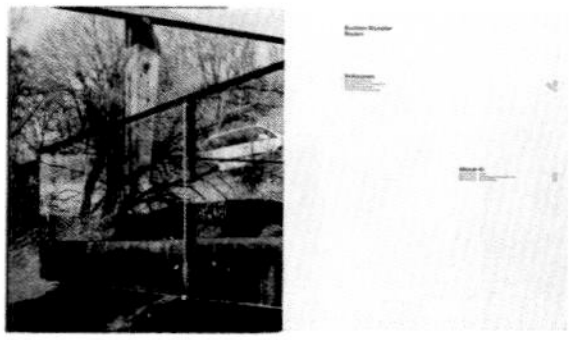
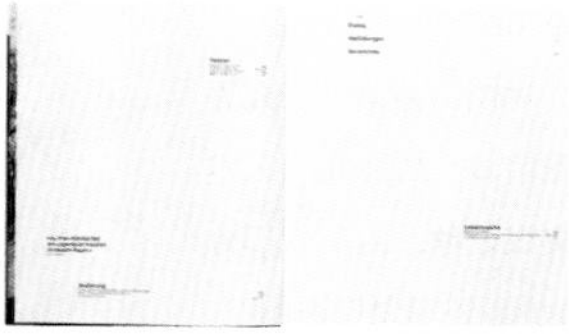

Buchner Bründler
Bauten
Lofthaus
Basel
Basel,
Schweiz
1998–2002

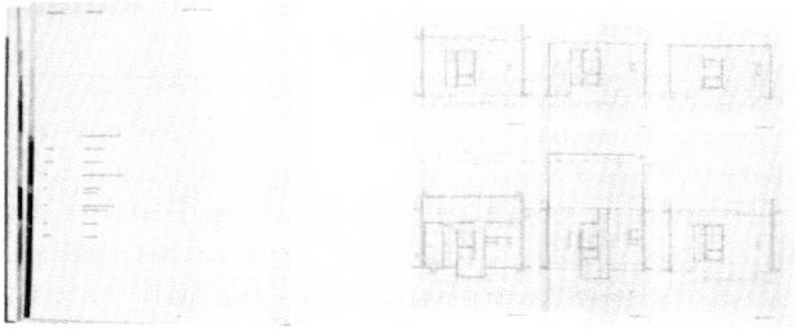
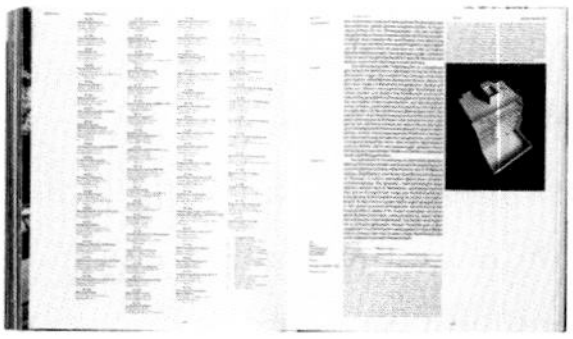
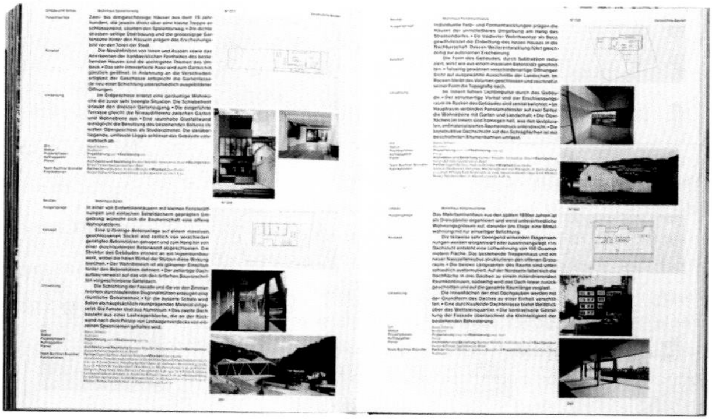

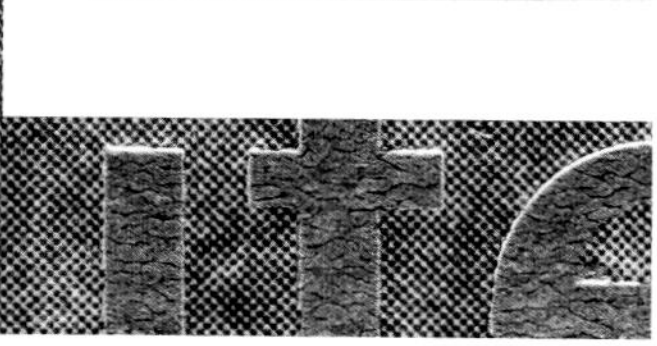

Wohnhaus
Blonay
Blonay,
Schweiz
2000–2002

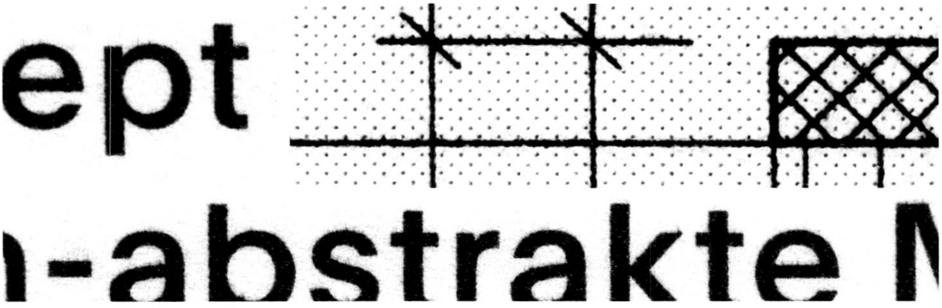
ept
-abstrakte

Buchner Bründler
Bauten

In eindrücklichen Bildstrecken gibt die Monografie Einblick in die erfolgreiche fünfzehnjährige Bautätigkeit des Basler Architekturbüros Buchner Bründler. Gespräche mit Persönlichkeiten, die in verschiedenster Weise mit der Tätigkeit von Daniel Buchner und Andreas Bründler verbunden sind, thematisieren interessante Aspekte aus dem vielfältigen Arbeitsalltag der Architekten und bringen dem Leser die Haltung des Büros näher. Die Darstellung kleinerer und projektierter Arbeiten ergänzt die Auswahl der wichtigsten, zu internationalem Renommee gelangten Bauten. Mit Beiträgen von Philipp Esch, Hélène Grimaud, Mat Hennek, Tibor Joanelly, Andreas Ruby, Thomas Wüthrich und Caspar Zellweger.

gta

ISBN 978-3-85676-297-1

Einbau — Carhartt Store München — N° 107

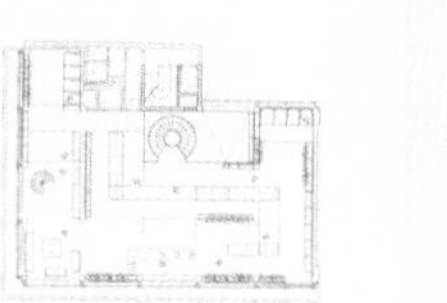

Ausgangslage
Das Gebäude, in den 1960er Jahren vom Architekten Kurt Ackermann für die Hypo-Bank erbaut, erhielt 1967 den BDA-Preis. Die engen Platzverhältnisse der Eckparzelle haben zur besonderen Vertikalorganisation geführt. • Der Beton als hauptsächliches Konstruktionsmaterial und die raumhohen, schwarz gerahmten Verglasungen lassen das Haus modern und zeitlos erscheinen. • Da soll der Carhartt Flagship Store mit zusätzlichem Off-Space für Künstlerausstellungen eingerichtet werden.

Konzept
Das vorgefundene Vokabular wird für die Themen des Verkaufsladens weiterentwickelt. Die Übergänge der Zeitepochen sind fliessend. • Um die Grosszügigkeit und Offenheit der architektonischen Grundsubstanz wirken zu lassen, werden wenige grosszügige, direkt in die Struktur eingepasste Einbauten entworfen.

Umsetzung
Die Betonstützen und die eingegossenen Böden kontrastieren mit den schwarzen Fensterrahmen. • Aus den freigelegten Installationen in den Decken entsteht ein ornamentales Labyrinth. • Die elaboriert wirkenden Metallgeländer werden zu Hängebügeln und zu übergrossen, mit Netzen bespannten Rahmen. • Grosszügige Raumkörper aus roher Eiche und Stahl sowie Wandregale aus schwarz gebeizter Tanne ergänzen die sichtbar gewordene Grundsubstanz.

Ort	München, Deutschland
Status	Realisiert
Projektphasen	**Projektierung** 2009 • **Realisierung** 2009
Auftraggeber	Work in Progress Textilhandels GmbH, Weil am Rhein, Deutschland
Planer	**Architektur** Buchner Bründler Architekten, Basel
Team Buchner Bründler	**Partner** Daniel Buchner, Andreas Bründler • **Projektleitung** Daniel Abraha • **Mitarbeit** Martin Risch, Oliver Teinl

Umbau — Wohnhaus Birsigstrasse — N° 143

Ausgangslage
Das Haus wurde 1904 als Einfamilienhaus mit über 350 Quadratmetern Wohnfläche auf vier Ebenen gebaut. DasErdgeschoss diente dem repräsentativen, das erste Obergeschoss dem privaten und die übrigen Geschosse dem temporären Wohnen. • Im Haus ist der Bauherr aufgewachsen. Es soll für eine junge Familie umgebaut werden.

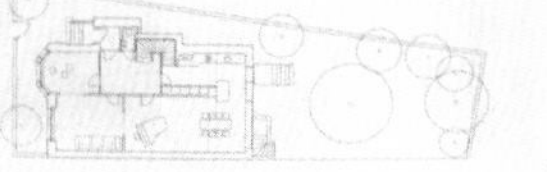

Konzept
Beim Umbau und partiellen Rückbau wird die Substanzerhaltung des Bestandes berücksichtigt. • Zentrales Motiv ist der geschaffene Einheitsraum im Erdgeschoss. Ein einzelnes, grosses Schaufenster zum kleinen, intensiv bepflanzten Garten steht für eine neue innenräumliche Dimension: Der Garten dringt in seiner Dichte bis in die Tiefe des Raums und prägt als poetisches Naturmotiv die Atmosphäre.

Umsetzung
Die Räume und Raumfolgen im Erdgeschoss werden neu organisiert und erweitert. Um das Entrée als zentralen Verteiler im Originalzustand gruppieren sich Spiel-, Lese-, Wohnzimmer und Küche. • Die Querbezüge zwischen den einzelnen Räumen und zum Entrée sowie die Sichtbezüge zum Aussenraum beleben die Raumsequenz.

Ort	Basel, Schweiz
Status	Realisiert
Projektphasen	**Projektierung** 2010 • **Realisierung** 2011
Auftraggeber	Privat
Planer	**Architektur und Bauleitung** Buchner Bründler Architekten, Basel • **Bauingenieur** Jürg Merz Ingenieurbüro, Maisprach
Team Buchner Bründler	**Partner** Daniel Buchner, Andreas Bründler • **Mitarbeit** Hellade Miozzari, Jenny Jenisch

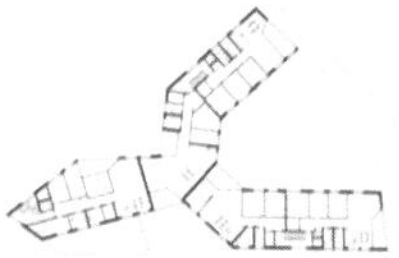

coop

Ausgangslage

Sins entwickelte sich an einem Kreuzungspunkt bedeutender Transitwege, welche die Innerschweizer Städte Zug und Luzern mit dem Reusstal und dem Mutschellengebiet sowie den nördlich gelegenen Aargauer Zentren verbanden. • Das Grundstück liegt im Dorfzentrum, das wegen der Verkehrsbelastung besonders für Wohnzwecke unattraktiv geworden ist. Die geplante Überbauung mit Ladenflächen, Bereichen für Dienstleistungsbetriebe und Eigentumswohnungen soll eine Reaktivierung herbeiführen.

Konzept

Die dreieckförmige Parzelle wird im Erdgeschoss und in den zwei Untergeschossen vollflächig bebaut. Da stehen Mietflächen für einen Grossverteiler, eine Bäckerei, einen Blumenladen und eine Beraterbank bereit. Die geschlossene Passage nimmt in ihrer Form die frühere Wegverbindung auf dem Grundstück auf. • Das Volumen zeichnet sich durch Rückschnitte in mehreren der oberen Bereiche aus und nähert sich so den Grössenverhältnissen des Ortes. Der Mäanderkörper mit Y-Form erhält durch die geneigte Dachform weitere Plastizität und wirkt aufgrund der geometrischen Aufnahme der Strassenform gestaffelt und kleinteiliger. • Aus der Grundgeometrie entwickelte Loggien mit trichterförmigem Charakter schneiden den Gebäudekörper weiter auf, frei in die Fassaden gesetzte Fenster lassen ihn homogener erscheinen.

Umsetzung

Die Fassade wird in Dämmbeton einschichtig gegossen und lässt den gesamten Bau sehr monolytisch erscheinen. Gleichzeitig nimmt sie sich durch ihre dunkle Farbe optisch zurück. • Die hohe Konstruktionstiefe ermöglicht tiefe Fensterlaibungen, die den massiven Charakter des Körpers unterstreichen. • Übergrosse Treppenhausfenster lassen den Körper strassenseitig offen erscheinen. • Die Rückschnitte des Körpers lassen drei Terrassen entstehen, wovon die mittlere den Bewohnern vorbehalten ist. Dreiseitig vom Gebäude umschlossen, öffnet sie sich in der Hauptrichtung nach Süden zum Alpenpanorama. • Die Lärmsituation bestimmt die Grundrisstypologie der Wohnungen im 2. und 3. Obergeschoss. Erschliessung und Nebenräume bilden eine räumliche Pufferschicht an der Strassenfront, die Wohn- und Schlafräume öffnen sich zur abgewandten Seite.

Ort Sins, Schweiz

Typus Neubau

Projektphasen **Projektierung** 2007–2009 • **Realisierung** 2008/09

Auftraggeber Dorfkern Sins AG, Sins

Projektdaten **Grundstücksfläche** 2450 m² • **Bebaute Fläche** 2100 m² • **Geschossfläche** 8000 m² (nach SIA 416) • **Gebäudevolumen** 28 400 m³

Planer **Architektur** Buchner Bründler Architekten, Basel • **Bauleitung** Bucher + Partner Architekten AG, Römerswil • **Bauingenieur** Burger & Partner AG, Basel; Ingenieurbüro Gwerder + Partner, Dietwil • **Akustik** Grolimund & Partner AG, Aarau • **Bauphysik** Ehrsam & Partner AG, Pratteln • **HLK und Sanitär** Josef Ottiger + Partner AG, Emmenbrücke • **Elektroplanung** B + S Elektro Engineering AG, Emmenbrücke • **Gestaltungsplanung** Kohli & Partner, Kommunalplan AG, Wohlen • **Verkehrsplanung** Rudolf Keller & Partner Verkehrsingenieure AG, Muttenz

Team Buchner Bründler **Partner** Daniel Buchner, Andreas Bründler • **Projektleitung** Gabriella Bertozzi, Ewa Misiewicz • **Mitarbeit** Olivia Frei, Nick Waldmeier, Sabine Beer, Sonja Christen, Patrick Filipaj

Publikationen Zentrumsüberbauung Sins, in: ders., *Architektonisches Potential von Dämmbeton*, Zürich 2010, S. 88–93.

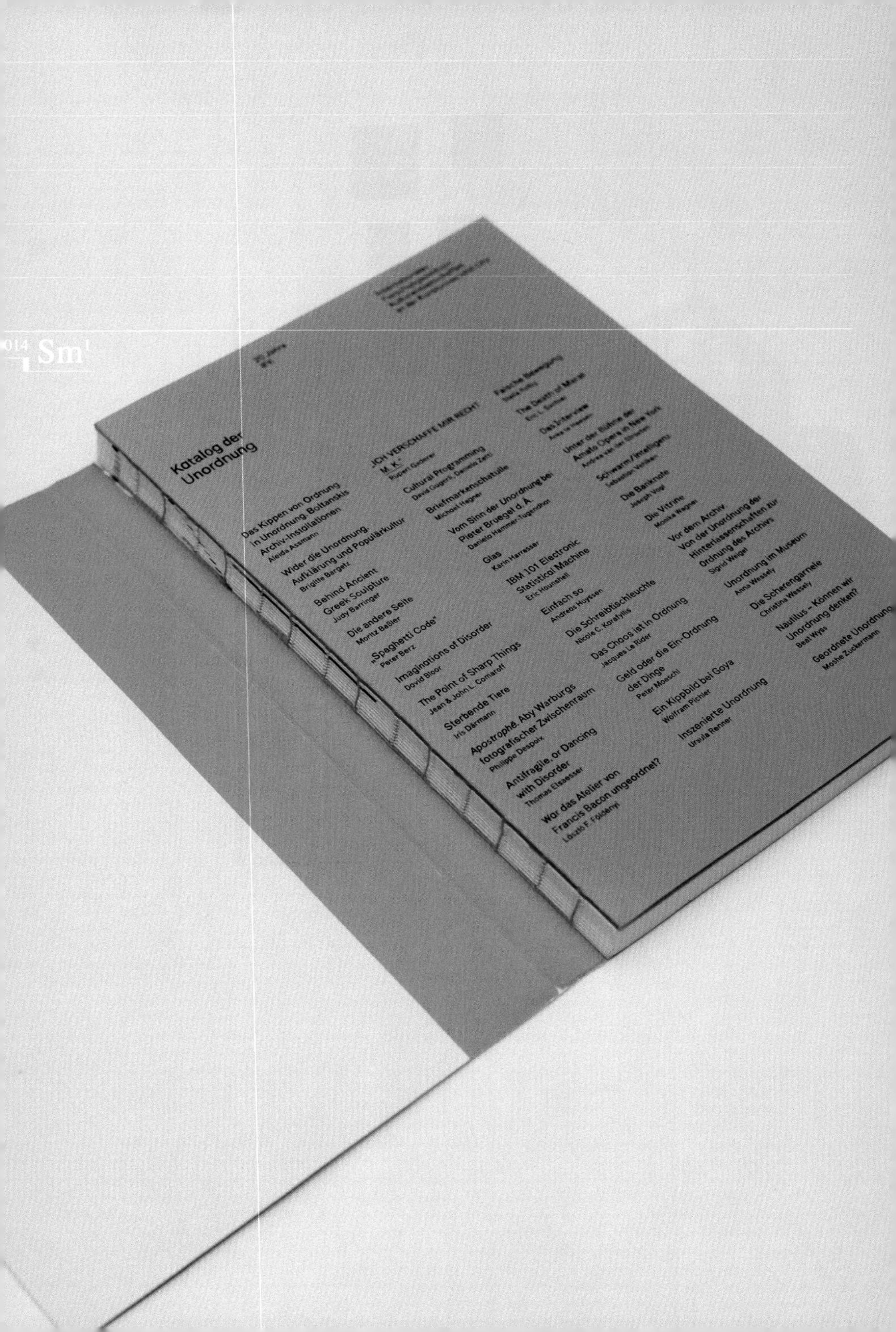
Katalog der Unordnung
Das Kippen von Ordnung in Unordnung. Boltanskis Archiv-Installationen
Aleida Assmann
Wider die Unordnung. Aufklärung und Populärkultur
Brigitte Bargetz
Behind Ancient Greek Sculpture
Judy Barringer
Die andere Seite
Moritz Baßler
„Spaghetti Code"
Peter Berz
Imaginations of Disorder
David Bloor
The Point of Sharp Things
Jean & John L. Comaroff
Sterbende Tiere
Iris Därmann
Apostrophé. Aby Warburgs fotografischer Zwischenraum
Philippe Despoix
Antifragile, or Dancing with Disorder
Thomas Elsaesser
War das Atelier von Francis Bacon ungeordnet?
László F. Földényi
„ICH VERSCHAFFE MIR RECHT M. K."
Rupert Gaderer
Cultural Programming
Briefmarkenschatulle
Michael Hagner
Vom Sinn der Unordnung bei Pieter Bruegel d. Ä.
Glas
Karin Harrasser
IBM 101 Electronic Statistical Machine
Eric Hounshell
Einfach so
Andreas Huyssen
Die Schreibtischleuchte
Das Chaos ist in Ordnung
Jacques Le Rider
Geld oder die Ein-Ordnung der Dinge
Peter Moeschl
Ein Kippbild bei Goya
Wolfram Pichler
Inszenierte Unordnung
Ursula Renner
The Death of Marat
Das Interview
Unter der Bühne der Amato Opera in New York
Die Banknote
Joseph Vogl
Die Vitrine
Monika Wagner
Vor dem Archiv
Von der Unordnung der Hinterlassenschaften zur Ordnung des Archivs
Sigrid Weigel
Unordnung im Museum
Anna Wessely
Die Scherengarnele
Christina Wessely
Nautilus – Können wir Unordnung denken?
Beat Wyss
Geordnete Unordnung
Moshe Zuckermann

林茨艺术大学国际文化研究中心（IFK）成立 20 周年时，组织了一次名为“无序想象”的会议，会议上科学家和艺术家们针对这一主题进行探讨并撰写各自观点。本书是这些观点的合集，在设计上是一次有序框架内有限度的“无序”尝试。书籍的装订为裸脊锁线软精装，封面哑面白卡纸五色印刷（四色加荧光橙色），封面内侧荧光橙满版印刷与橙色锁线相匹配。内页分为两部分，前一部分哑面涂布纸五色印刷（四色加中灰色），将内容呈现为五个不同的目录：图片名字母顺序、文本字母顺序、作者名字母顺序、图片页码顺序、标题页码顺序。第二部分为正文，采用哑面涂布纸四色印刷。全书采用 8 种不同的无衬线字体混合成新字体，在版面上形成丰富的灰度肌理。排版上根据版面的不同，正文和注释会交换位置。但这些变化都在三个字号、行距统一、版心固定、三栏设置的经典瑞士版式风格中有限地“无序”着。

Katalog der Unordnung

On the 20th anniversary of the International Center for Cultural Studies of Linz University of Art, a conference called ‘Unordered Imagination’ was held, at which the scientists and artists discussed and narrated their own views on this topic. This book is a collection of these ideas and is designed as a limited ‘disorderly’ attempt within an orderly framework. The book is bound by thread sewing with a bare spine. The cover is made of white cardboard printed in five colors (four colors plus fluorescent orange), and the inside front cover is printed in fluorescent orange matched with the orange sewing thread. The inside pages are divided into two parts. The first part is matte coated paper printed in five colors (four-color plus medium gray) and five different catalogues are presented according to alphabetical order of picture titles, alphabetical order of text, alphabetical order of author names, order of pictures’ page number, order of titles’ page number. The second part is the text, using matte coated paper printed in four colors. The book uses eight different san-serif fonts to mix into a new font, which has rich gray scale texture on the page. As for typesetting, the text and annotations will change positions depending on the layout. However, these changes are ‘out-of-order’ limitedly following the classic Swiss format style such as three font sizes, unified line spacing, fixed type page and three columns.

银奖

无序的目录

奥地利

Silver Medal

Katalog der Unordnung

Austria

Christoph Schorkhuber, Linz

Helmuth Lethen, IFK

Internationales Forschungszentrum Kulturwissenschaften an der Kunstuniversität Linz, Linzt

240 × 162 × 16mm

478g

191p

Rather than “catalogue of disorder”, as its title might be translated, this book could just as well have been called “catalogue of order”. For this it is indebted to its ironic borrowing of classic Swiss typography with which the macro-structure of this Austrian book is organised. Sans-serif typefaces, its confinement to just three font sizes, an anaxial type-area with margin, a top margin with generous white space and vibrant page headings, detailed contents pages with forewords on grey card. Yet some things are nevertheless surprising. The book’s components are constantly re-sorted in the multiple table of contents. Not until the fifth list does one reach what one would normally expect by way of summary of contents, here consistently bearing the heading: page numbers. Within the book’s interior the columns are not fixed, exchanging places with the margin as required. But what really stands out is the typeface. With hot lead typesetting it would sometimes happen that a character from another typeface would smuggle its way into the line of print. In printer’s jargon this was then known as a “printer’s pie”. Here the typesetting is made up exclusively of such printer’s pies, with eight different sans-serif typefaces being wildly jumbled together. And, surprisingly, this apogee of typographical chaos engenders a pleasantly shimmering appearance. Order or disorder is a matter of the perspective from which one regards the complexity of the given circumstances.

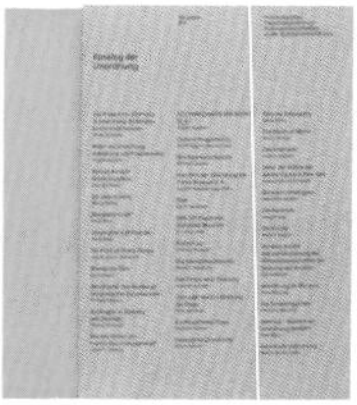

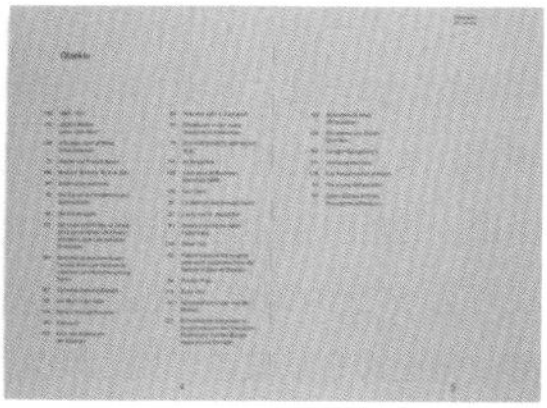

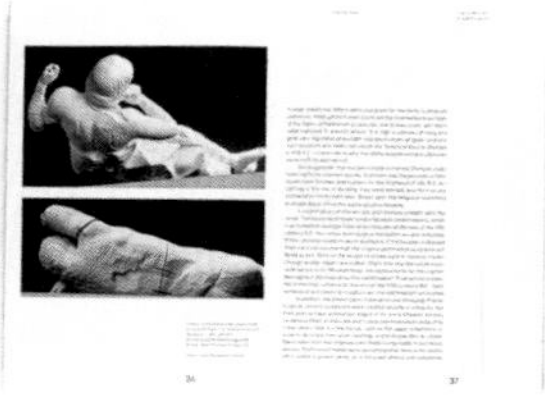

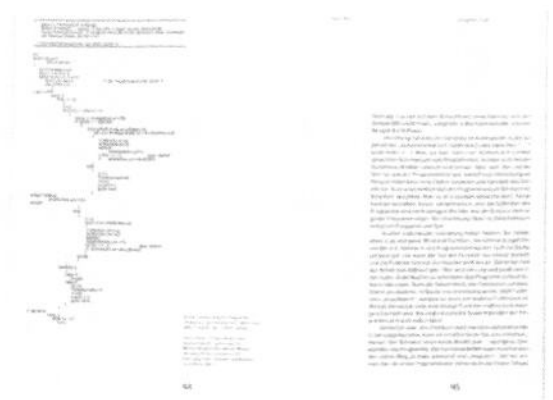

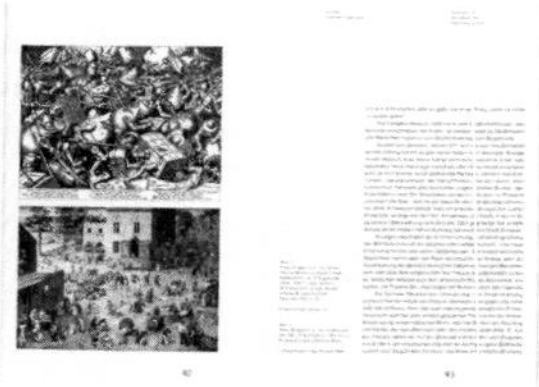

Kiosk der Unordnung

Eric Hounshell

IBM 101 Electronic Statistical Machine

Stacks of census questionnaires "taller than the Washington Monument"; a sea of policy-holders insured against the risks of disaster, accident, and death; legions of workers in assembly lines and cadres of white-collar employees to track and coordinate their labor and its products, morale, supplies, and weaponry of mass armies deployed onto foreign continents; vast warehouses full of goods in search of distributors and consumers; a terra incognita of creators, readers, listeners, and viewers of the culture industry, their habits unrecorded and tastes unfathomed: these are some of the sites of disorder wrought by the "second nature" of advanced industrial societies and reined in by data-processing machinery of the late 19th and 20th centuries.

Punched-card tabulating machines ushered a vocabulary of fact, science, measurement, objectivity, and precision into new domains. Phenomena of natural and social life from heredity to fashion, once matters of speculation or studied conjecture, were now to be penetrable, calculable, even controllable through the systematic collection, storage, and machine-aided analysis of data. Aside from governments, the life insurance and market/opinion research industries were the principal users of punched-card machines. From the American "average Joe" to the New Objectivity interwar avant-garde, however, the streams of facts and figures flowing from these tabulating machines were objects of enthusiasm beyond the ranks of the scientists and clerks who used them.

Punch cards were first used to format and store data for machine-aided analysis in the late 19th century, but the tabulating machines that processed these cards continued to gain speed and features into the post-WWII period. The first Hollerith machine (1889) could sort and count up to 80 cards per minute; its human operator noted by hand the values tallied on the machine's dials. The IBM 101 depicted here (ca. 1952) could process 450 cards per minute, perform statistical calculations, print tables of results, and punch these amounts onto cards for storage and future use. Still, cross-tabulations on a large dataset could take several days; a researcher could "diddle" for months before settling on a set of measurable

International Business Machines (IBM)

© Courtesy of the Computer History Museum

100

101

ngsz
sser
unstunive

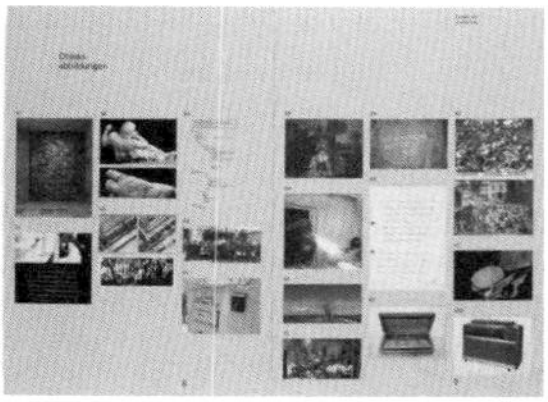
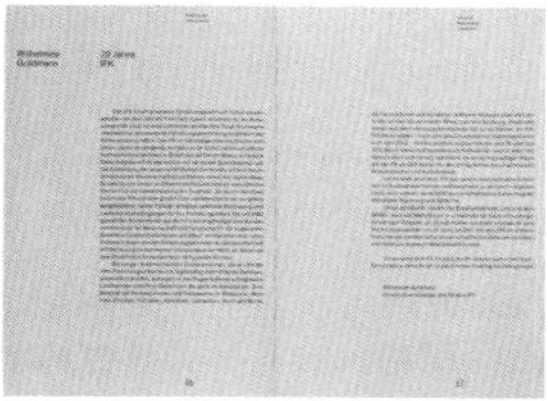
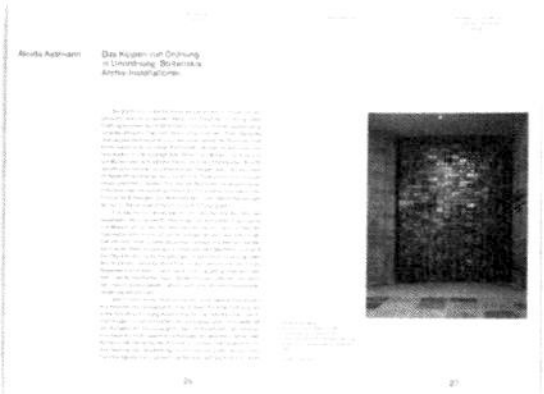

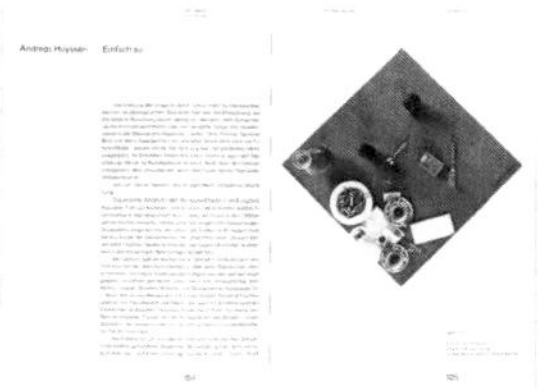

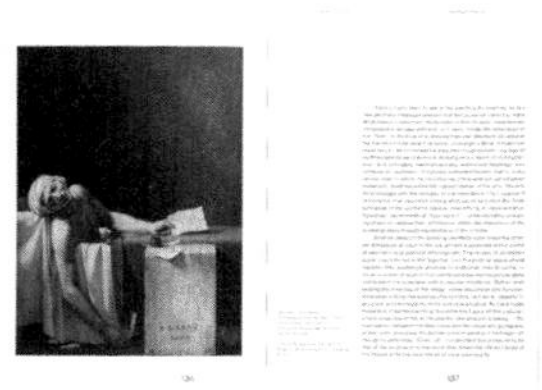
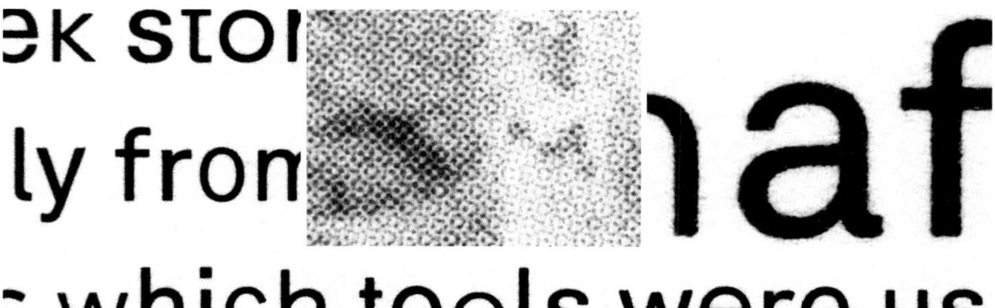
ek stor
ly from
af
s which tools were us

IFK Katalog der Unordnung

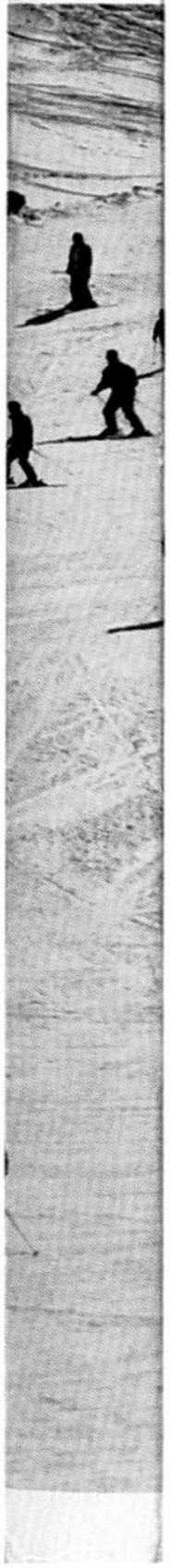

Katalog der
Unordnung

Andrea van
der Straeten

Unter der Bühne der Amato Opera in New York

The Barber of Seville
Cavalleria rusticana
Pagliacci
Don Pasquale
La Traviata
Aida
Rigoletto
Carmen
The Marriage of Figaro
The Magic Flute
Don Giovanni
La finta giardiniera
Il trovatore
Faust
La Bohème
La forza del destino
Zanetto & I Pagliacci
Lucia di Lammermoor
Manon
Die Fledermaus
Madama Butterfly
Un giorno di regno
Luisa Miller
Aroldo
Land of Smiles
Hänsel und Gretel
The Pied Piper of Hamelin
Così fan tutte
Un ballo in maschera
Tosca
Andrea Chénier
Falstaff
The Tales of Hoffmann
La battaglia di Legnano
Oberto
La cena delle beffe
Crispino e la comare
Hin und zurück & Mavra
Bastien et Bastienne
The Impresario
Nerone
Il campanello di notte
Verdi & Alzira: Voltaire & Les Américains
Thérèse
La serva padrona
Doctor Miracle
Salvator Rosa
Die Fledermaus
Lo schiavo
I due Foscari
The Empty Bottle
Otello
Profecías
Suor Angelica
Bluebeard's Castle
Fosca
The Merry Widow
Secret of Susannah
Ruddigore
L'amico Fritz & I Pagliacci
The Boor
L'elisir d'amore
The Forest

144

Anthony Amato, Sohn einer italienischen Einwandererfamilie von der Amalfiküste, gründete 1948 mit seiner Frau Sally das kleinste Opernhaus der Welt in Lower Manhattan.

Der Orchestergraben der Amato Opera befand sich zur Hälfte unter der Bühne und war zu klein für Cello, Kontrabass oder Violinen, denn er bot zu wenig Raum für die ausladende Bewegung der Bögen.

Aber wie in der berühmten Metropolitan Opera schwebten zu Beginn der Vorführung die Kronleuchter hinauf zur Decke – in dieser intimeren Version eines Operntheaters wurden sie von Hand hinaufgezogen – und ließen auf den 107 Plätzen im Souterrain die Gespräche verstummen.

61 Jahre lang wurde kontinuierlich geprobt, schneiderteSally Amato die Kostüme, schleppte Anthony Amato bis zu seinem 88. Lebensjahr selbst die Kulissen. Es fanden tausende Aufführungen statt, die vor allem jungen, unbekannten Sängerinnen und Sängern die Chance boten, sich auf einer winzigen, aber musikalisch professionellen Opernbühne zu erproben.

Einige schafften nach der Schulung durch Tony Amato den Sprung in die großen Opernhäuser der Welt.

Dieser Beitrag ist Anthony Amato (1920–2011) gewidmet.

Foto: Andrea van der Straeten, 2009

↓

Andrea van der Straeten

Bildende Kunst,
Kunstuniversität Linz

2014 Bm1

这本书是一场艺术家与读者之间互动实验的媒介。JAK 是艺术家的艺名，也是这个互动实验中的一方。读者通过阅读本书来了解 JAK 是谁、在做什么。去过哪些地方，表达怎样的观点，而这些都没有准确的界定，一切都存在于读者与这本书互动阅读的结果中，这是本书想表达的内容产生于互动中的“自由”体验。本书的主体装订形式为锁线硬精装，封面白色哑面胶版纸书名烫黑，封底凹凸压花，裱覆卡板。前后的环衬和过门页采用绿色胶版纸印单黑，主要为本书概念的简单阐释、图片罗列和版权信息。内页胶版纸印单黑，为各种见闻、对话、信件。书中大量夹带照片、海报、手稿、册页，是艺术家尝试与读者之间形成互动的文件，读者在打开这些内容的时候就跟随艺术家进入到现场，观看了一部话剧、一场电影，或者完成了一件艺术作品，有多种的可能性和丰富的阅读体验。|||||||||||||||||||||||||||||||||||||||||||||||||||||||||||||||||||||||||||||||||||||||||||||||||||||||||||||||||||||||||||||||||||||||||||||

JAK

This book is a medium for interactive experiments between artists and readers. JAK is the artist's stage name, and is also a party of interactive experiments. Through reading the books, readers can understand who JAK is, what he is doing, where he has been, and what opinions he expresses. None of these questions has an accurate answer; everything exists in the results of the interaction between readers and the book. What the book wants to express just comes from the 'free' experience of interaction. The hardcover book is bound by thread sewing. The cover is made of white matte offset paper with black hot stamped title while the back cover is embossed with patterns, mounted with graphic board. Green offset paper printed in black is used for the end paper, which mainly explains the concept of the book, lists pictures and copyright information. The inner pages are made of offset paper printed with black ink only, including all kinds of information, dialogues and letters. A large number of pictures, posters, manuscripts and pamphlets attached in the book are documents that the artist used to interact with readers. When the readers browse the book, they follow the artist into the scene to watch a drama or movie, or complete an artwork. There are many possibilities and rich reading experience.

铜奖
JAK
德国

Bronze Medal
JAK
Germany
Demian Bern, Stuttgart
JAK, Hamed Taheri
EXP. edition, Stuttgart
245×166×23mm
700g
108p

An absolutely normal book – is what this book is not. On leafing through it, leaves fall out, but not the pages themselves, rather inserts in various forms. Everything not free to move stays put – that which is technically known as a book. And yet its free elements are an intrinsic part of it, no less so than the imagination of its authors and the artist – and no less so than the imagination of its readers and beholders. Not only between author and reader is this book a document of interactiveness, but also between book and spirit. Of course its particular form remains a special case, but it makes it obvious to us, and tangible too, what the true intention of a printed book is, whether with or without text. It makes an example of the first maxim to be found within it, a quote from Jean-Paul Sartre: "… Thus unlike a tool, the book is not a means to a particular end: it commends itself as the very purpose of the reader's freedom."

JAK

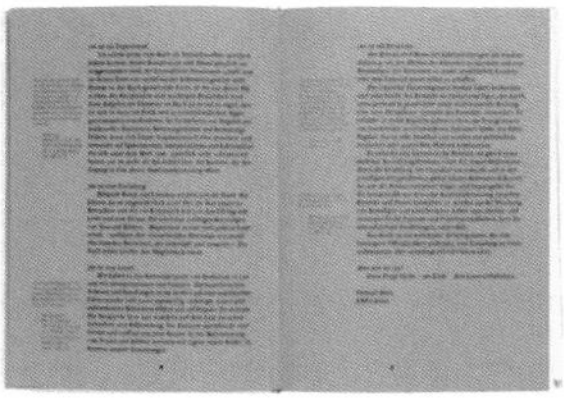

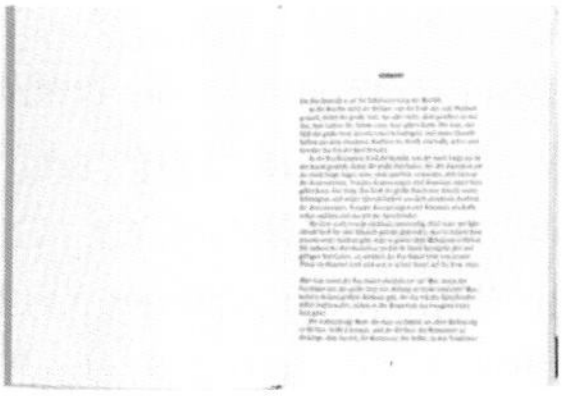

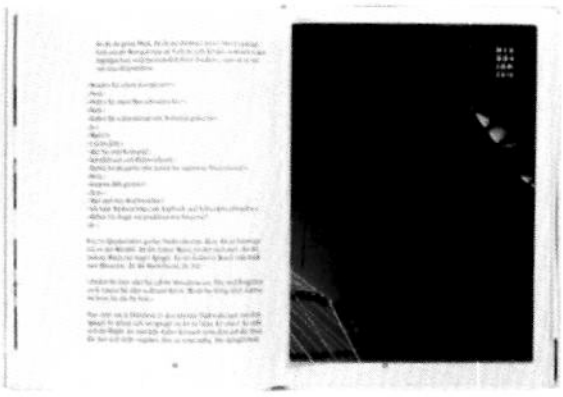

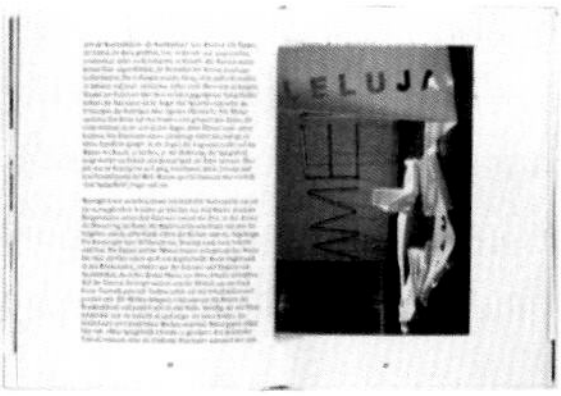

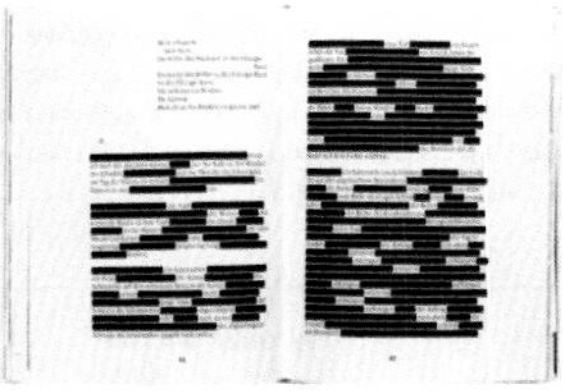
[redacted]

[redacted]
who
the
fuck
is
jak?

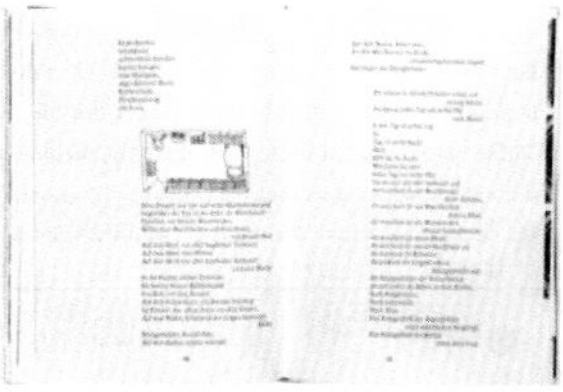

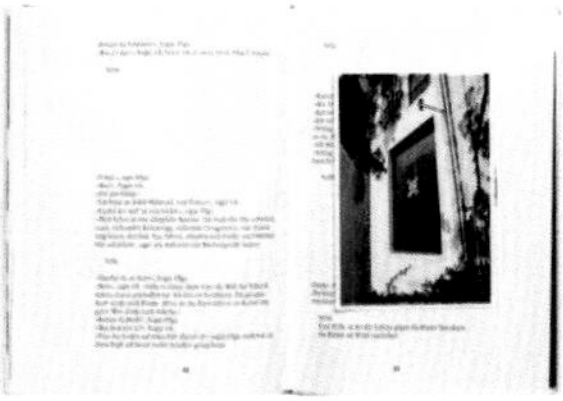

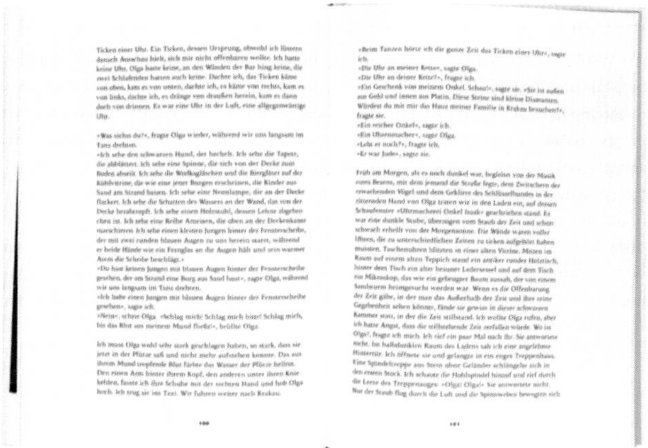
Ticken einer Uhr. Ein Ticken, dessen Ursprung, obwohl ich lüstern danach Ausschau hielt, sich mir nicht offenbaren wollte. Ich hatte keine Uhr, Olga hatte keine, an den Wänden der Bar hing keine, die zwei Schlafenden hatten auch keine. Dachte ich, das Ticken käme von oben, kam es von unten, dachte ich, es käme von rechts, kam es von links, dachte ich, es dränge von draußen herein, kam es dann doch von drinnen. Es war eine Uhr in der Luft, eine allgegenwärtige Uhr.

»Was siehst du?«, fragte Olga wieder, während wir uns langsam im Tanz drehten.
»Ich sehe den schwarzen Hund, der hechelt. Ich sehe die Tapete, die abblättert. Ich sehe eine Spinne, die sich von der Decke zum Boden abseilt. Ich sehe die Wodkagläschen und die Biergläser auf der Kühlvitrine, die wie eine jener Burgen erscheinen, die Kinder aus Sand am Strand bauen. Ich sehe eine Neonlampe, die an der Decke flackert. Ich sehe die Schatten des Wassers an der Wand, das von der Decke herabtropft. Ich sehe einen Holzstuhl, dessen Lehne abgebrochen ist. Ich sehe eine Reihe Ameisen, die oben an der Deckenkante marschieren. Ich sehe einen kleinen Jungen hinter der Fensterscheibe, der mit zwei runden blauen Augen zu uns herein starrt, während er beide Hände wie ein Fernglas an die Augen hält und sein warmer Atem die Scheibe beschlägt.«
»Du hast keinen Jungen mit blauen Augen hinter der Fensterscheibe gesehen, der am Strand eine Burg aus Sand baut«, sagte Olga, während wir uns langsam im Tanz drehten.
»Ich habe einen Jungen mit blauen Augen hinter der Fensterscheibe gesehen«, sagte ich.
»Nein«, schrie Olga. »Schlag mich! Schlag mich bitte! Schlag mich, bis das Blut aus meinem Mund fließt!«, brüllte Olga.

Ich muss Olga wohl sehr stark geschlagen haben, so stark, dass sie jetzt in der Pfütze saß und nicht mehr aufstehen konnte. Das aus ihrem Mund tropfende Blut färbte das Wasser der Pfütze hellrot. Den einen Arm hinter ihrem Kopf, den anderen unter ihren Kniekehlen, fasste ich ihre Schuhe mit der rechten Hand und hob Olga hoch. Ich trug sie ins Taxi. Wir fuhren weiter nach Krakau.

100

»Beim Tanzen hörte ich die ganze Zeit das Ticken einer Uhr«, sagte ich.
»Die Uhr an meiner Kette«, sagte Olga.
»Die Uhr an deiner Kette?«, fragte ich.
»Ein Geschenk von meinem Onkel. Schau!«, sagte sie. »Sie ist außen aus Gold und innen aus Platin. Diese Steine sind kleine Diamanten. Würdest du mit mir das Haus meiner Familie in Krakau besuchen?«, fragte sie.
»Ein reicher Onkel«, sagte ich.
»Ein Uhrenmacher«, sagte Olga.
»Lebt er noch?«, fragte ich.
»Er war Jude«, sagte sie.

Früh am Morgen, als es noch dunkel war, begleitet von der Musik eines Besens, mit dem jemand die Straße fegte, dem Zwitschern der erwachenden Vögel und dem Geklirre des Schlüsselbundes in der zitternden Hand von Olga traten wir in den Laden ein, auf dessen Schaufenster »Uhrmacherei Onkel Isaak« geschrieben stand. Es war eine dunkle Stube, überzogen vom Staub der Zeit und schon schwach erhellt von der Morgensonne. Die Wände waren voller Uhren, die zu unterschiedlichen Zeiten zu ticken aufgehört haben mussten. Taschenuhren blitzten in einer alten Vitrine. Mitten im Raum auf einem alten Teppich stand ein antiker runder Holztisch, hinter dem Tisch ein alter brauner Ledersessel und auf dem Tisch ein Mikroskop, das wie ein gebeugter Baum aussah, der von einem Sandsturm heimgesucht worden war. Wenn es die Offenbarung der Zeit gäbe, in der man das Außerhalb der Zeit und ihre reine Gegebenheit sehen könnte, fände sie gewiss in dieser schwarzen Kammer statt, in der die Zeit stillstand. Ich wollte Olga rufen, aber ich hatte Angst, dass die stillstehende Zeit zerfallen würde. Wo ist Olga?, fragte ich mich. Ich rief ein paar Mal nach ihr. Sie antwortete nicht. Im halbdunklen Raum des Ladens sah ich eine angelehnte Hintertür. Ich öffnete sie und gelangte in ein enges Treppenhaus. Eine Spindeltreppe aus Stein ohne Geländer schlängelte sich in den ersten Stock. Ich schaute die Hohlspindel hinauf und rief durch die Leere des Treppenauges: »Olga! Olga!« Sie antwortete nicht. Nur der Staub flog durch die Luft und die Spinnweben bewegten sich

101

s ur
ationen Träu

Eine Möwe schreit und ein Schiff pfeift; unter Wasser wird ein Stein von der Strömung angehoben und schlägt auf die tiefen Töne eines Klaviers, das mit einem Schiff untergegangen ist; der Stein zerbricht und Millionen Austern schlagen im Chor ihre Muschelschalen zu; ich höre die Schwere der Erde, die pulsiert.«

Der Himmel ist bedeckt mit dunklen Gewitterwolken. Die Landschaft ist dunkelgrün gefärbt. Mitten in der Landschaft steht ein uralter Baum. Der Baum ist einsam. Es blitzt. Pause. Es donnert:

»Um den Lärm des Kosmos zu ersticken, nehme ich in meinem Zimmer im Altersheim meine Stimme auf. Genauso wie zu Anfangszeiten des Kinos, wo bei den Filmvorführungen Klavier gespielt wurde, um den lauten Lärm des Projektionsapparates zu übertönen. Das Altersheim liegt am Meer. Eine Heimbewohnerin über mir, eine alte Opernsängerin, singt den *Liebestod* der Isolde. Ich höre das graue Korn ihrer Stimme. Eine andere Heimbewohnerin, eine alte Balletttänzerin, tanzt am Strand. Es scheint, ein Schmetterling bewege ihren Körper und lasse ihr Seidentuch flattern. Ihre Bewegungen sind der Erde, nicht dem Himmel zugewandt. Das hohe Alter ist das Näher-Sein zur Erde. Im Alter betet man zur Dinglichkeit der Erde. Das hohe Alter ist das Vermögen, die Erde zu mögen. Eine Herde weißer Pferde mit fliegender Mähne galoppiert den Strand entlang.«

Der Himmel ist bedeckt mit dunklen Gewitterwolken. Die Landschaft ist dunkelgrün gefärbt. Mitten in der Landschaft steht ein uralter Baum. Der Baum ist einsam. Es blitzt. Pause. Es donnert:

»Das hohe Alter heißt nicht, sich dem Tod zu nähern. Es hat nichts mit dem Tod zu tun. Der Tod begleitet das Leben von Anfang an. Der Tod ist der Besitzer der Mietwohnung, in der wir leben. Unser Leben ist die Miete, die wir an den Besitzer bezahlen, bis unser Vertrag endet und der Besitzer uns aus der Wohnung jagt. In der Jugend vermag man das Tun-Können. Im hohen Alter vermag man das Nicht-Tun-Können. Meine Einsamkeit im Altersheim ist nicht mein Elend. Sie ist die Muschel, in der die Herrlichkeit meines Lebens steckt. Das hohe Alter ist das Glas einer Öllampe, in der eine Motte zum ersten Mal die Wärme des Feuers spürt. Das hohe Alter ist das

Karaoke Alpin

014 Bm²

这是波兰摄影师托马什·古佐为第（Tomasz Gudzowaty）拍摄的孟加拉国 KeiKo 拆船厂工人的影集。全球每年报废的 700 艘远洋巨轮中，大约有 30%－40% 在孟加拉国第二大城市吉大港被拆卸，摄影师使用黑白胶片摄影展现正被拆卸的巨轮和工人们钢铁、泥土混合着汗水的艰辛劳作，记录着物质文明发展到今天依然不可或缺的工作和可能无法被选择的生活。书籍的装订形式为锁线硬精装，封面与内页等齐的卡板单黑印刷覆哑膜，出版机构 logo 在左下角烫黑处理，书脊包覆黑色布面，书名和作者名在布面烫黑。环衬采用黑卡纸，内页哑面铜版纸双层次黑色印刷，并在最黑的部分过油处理，与胶片摄影粗粝的银盐噪点形成很好的视觉效果。全书没有目录，只在最后有少量针对个别工人的文字描述。上下书口及右侧翻口刷黑。本书的设计始终围绕着“黑”进行，极力体现着拆船工人的艰辛。|||||||||||||||||||||||||||||||||||||||||||||||||||||||||||||||||||||||||||||||||||||||||||||||||||||||||||||||||||||||||||||

This is a photography album of workers of Keiko Shipbreaking Plant in Bangladesh, taken by Polish photographer Thomasz Gudzowaty. Among 700 scrapped ocean-going vessels every year, about 30-40% ships are dismantled in Chittagong, Bangladesh's second largest city. Through the black-and-white film photography, the photographer managed to show the giant vessels being dismantled and the hardworking workers surrounded by steel and soil mixed with sweat. The photos recorded the indispensable work and the life that has no choices. The hardcover book is bound by thread sewing. Laminated cardboard printed in monochrome black are used for the cover, the logo of the publishing house is black hot stamped in the lower left corner. The spine is covered with black cloth, with the black hot stamped title and the name of the author. The end paper is made of black cardboard while the inside pages are made of matte coated art paper printed with two different kinds of black ink. The darkest part has been matte vanished, contrasted with hot pixel of silver salt photography, it has a good visual effect. There is no catalogue in the book, only a few descriptions of some workers at the end. Three edges of the book are brushed with black ink. The key concept of book design is 'black', which reflects the hardship of ship-breakers.

◎ 铜奖

《 Keiko

◖ 德国

◎ Bronze Medal

《 Keiko

◖ Germany

D Marek Mielnicki / veryniceworks, Warschau (Polen)

△ Tomasz Gudzowaty (Hrsg.)

||||| Hatje Cantz Verlag, Ostfildern

□ 319×243×20mm

▯ 1531g

≣ 144p

A new standard for the colour black. What richness of nuance, atmosphere and pictorial quality the alleged absence of colour can evince! This photographic project in book form also refutes generally assumed appearances: black-and-white equals graphic. Its features: cardboard cut-flush binding with black fabric spine; transparent glossy spine embossing; thick cover card with neither inner edge nor standing edge; three-sided colour edging in black around the whole body of the book including the cover edge; loose hinge at the front making for perfect 180 degree opening; black endpapers; bi-colour printing of the richly contrasting photographs; duplex black and black glossy. And what about the pictures? With 75% tonal values or even in the blackest of zones the differentiation of shapes is still just about perceptible; reproduction and printing techniques become signets of the meaning of photographic visual language. The subject matter? The hulk of a shipwreck, rust, dirt, sweat, mud, naked feet, kilotons of weight and muscle power – the deconstruction of steel and bodies. It is seldom that the beholder of a book is tantalised so directly with a single colour spectrum.

KEIKO
TOMASZ GUDZOWATY

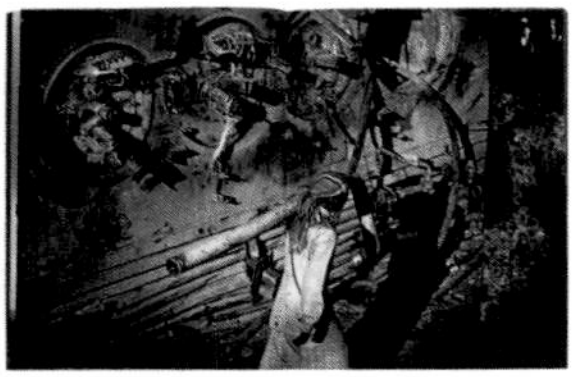

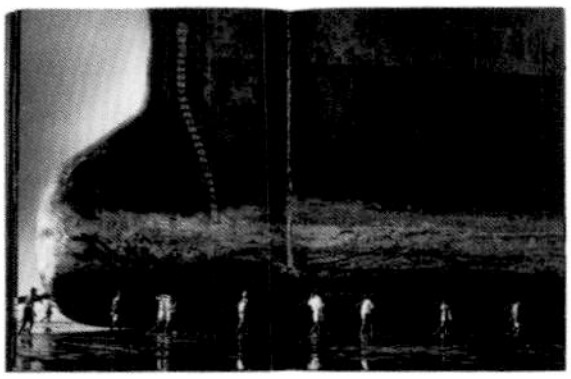

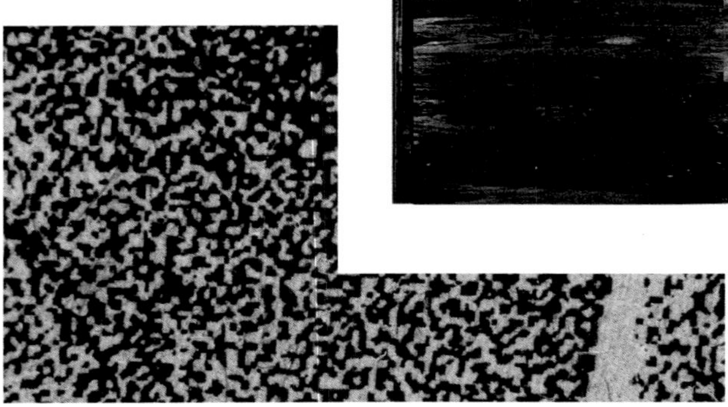

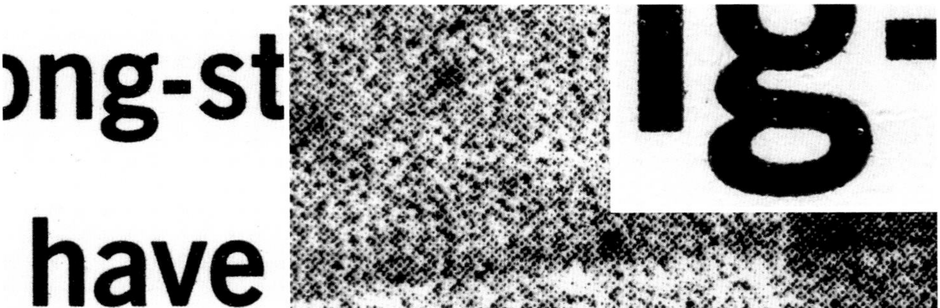
ong-st
have

BROTHERS

They get up early each morning. The most devout—at first light—to bow before Allah at morning prayers. The less devout rise an hour later. They eat a simple breakfast: rice, a little chicken, tea with condensed milk. They drink a can of cheap energy drink with a Bengal tiger on its packaging, then put on their sandals, a shirt or T-shirt, some loose cotton trousers or the skirt known as a *lungi.* "Nothing that restricts movement," one of them explains. "Our work puts us in permanent danger. You must be quick and agile, like a monkey."

Every day at eight, they go to a drab square that was once a beach, and today, is their workplace. They are welcomed there by a security guard—the almond-eyed Lallek from the indomitable Marma tribe. They stroll between the almost empty oil barrels, among the gaudily painted trucks, between dangling cables, taut anchor chains, bathyscaphes, ship's wheels, screw propellers, wires, turbines, bushings—in among everything they've managed to pull out of the ships.

The ships are already there waiting for them. Massive as gods. Run ashore. Some are only beginning to be dismantled, others are half done. Only a few sheets of metal remain of others. They take an oxyacetylene torch from the ramshackle store with them to continue the stripping.

"Brother, pass me the cable for the bottle."

"Brother, where shall we work today?" they ask their supervisor. "Brother, may Allah watch over you," they sigh, and go to work.

Next to these maritime colossi, they look like ants around a dead buffalo, a *gaur* in these parts. Yet in three, four months at most, they will dismantle those ships down to the tiniest screw.

FAZLUR

"At first I was afraid of them. The ships are enormous, brother, the height of a several-story building. I am from a small village in the province of Barisal, and I came to Chittagong because this is where the most breakers' yards are in Bangladesh, and in the world. But I had never seen a tall building before, so the ship was the first such big thing I ever saw. I knew it would be enormous, but what greeted my eyes was overwhelming. I nearly brought up my breakfast."

Fazlur's work involves wading through mud up to his calves, and then chopping into pieces something that was once part of the hull. He's a little over thirty, has a light complexion, a sparse stubble, and he carries out an enormously responsible task: when a new ship arrives, he has to clamber onboard with an oxygen torch, cut out the first hole, and throw down a rope ladder for the others.

"When a ship comes in, we always stand on the shore and welcome it with a cheer," he explains. "There's a ship, which means we'll have something to live on. Sometimes we cheer louder, sometimes quieter. Then the captain has to do something they don't teach in any school: deliberately run the ship aground. He waits for the tide, gathers speed and bang! He sails right up onto shore.

"Captains are different. One Russian didn't go fast at all; he left the ship in the shallows. He went to the owner and started crying, saying: 'I have no wife, no children, that ship was my whole life. I couldn't destroy her.' So he cried his heart out and went back to Russia, while we had to carry thirty thousand metric tons of his ship on our backs from the shallows to the shore.

"My cousin worked here before. But he was unlucky. A gas bottle exploded right near him. He lost an arm, so the owner agreed for me to replace him, otherwise I'd have had no chance of getting a job here. In Bangladesh, many millions of people have no work. This work may be hard, but the money is good. Now that I work with the torch, I can earn 12,000 thaka a month. One dollar is 80 thaka, so I earn about 150 dollars—really good money, especially considering I only spend a little on food and a ticket home every other month. With the money, my wife has opened a small shop in our village, and we are living a lot better than we did a few years ago. We give a little money to my cousin, too. There's no chance he'll find work with only one arm.

"I started as a porter; I carried pieces cut off the ship from place to place with my brothers. We call one another 'brother' because our lives depend on one another. These men are closer to me than my mother and father. I wanted to try my hand at cutting, and it turned out I had some talent for it. It's hard here to go from being a porter to a cutter, but the master liked me and kept telling our supervisor that I was talented and that he should give me my own torch. So I got one and, after that, the order to board the ship and make the first cut. I was so scared I thought I'd die, but the master insisted. He gave me time to calm down. 'Talk to her,' he told me, meaning the ship. 'I always do that.'

"Since then, I often talk to the ships: you're not going to hurt me now, are you, little one? You've been through a lot, sailed so far—time to make way for others.

"I learned this from my grandfather. He taught me to behave respectfully towards everything and everyone. To give thanks to Allah for giving us food. To give thanks for our clothing. To give thanks for this life. So, when I have to make the climb, to make that first cut, I say to Allah: thank you for this ship. Thank you, that it sailed here. For thanks to this, I will have breakfast, dinner and supper. I will have money to buy clothes for my two children. If you can grant a sinner like me another day of this life, I shall praise you further. But if you want me to call me to you today, I am ready. We all are."

RASHEL

The *Lively Falcon* is a cargo ship that was which built in 1977 in the Kanda Shipyard in Japan and was owned by the Vietnamese. It's 168.42 meters long, and 38.44 meters tall from the keel to the top of the tallest mast. Or rather it was; those figures are not current. The *Falcon's* tallest mast was recently sawed into meter-long pieces and has already been melted down—maybe even changed into rebar or nails—for use in another tower block in Chittagong, which is experiencing a building boom. Eight breakers bent over the *Falcon's* hull with oxygen torches, the cables wound around them like snakes of the street charmers in the capital, Dhaka. Among them stands Rashel, a supervisor, wearing a blue shirt and dark trousers. His cheerful expression makes you think he's done nothing but smile his whole life. "The oxyacetylene mix is delivered to the torch through external cables," he explains to someone, and orders one of the breakers to turn his torch off for a moment and show them. It's difficult to get him to switch from technical talk to a normal conversation about the daily life of a man who, day after day, chops ships the size of six-story buildings into pieces. A sensitive soul staring at the *Falcon,* or the *Ref Vega* nearby, the *Dibena Unity,* the *Anggraini,*

Bm
Vase, Gießkanne
Windbeutel
Ausstellung
Lebensmittel
BH
Lebensmittel
Schuhe, Stiefel Christian
Unterhose, Beutel,
2 Handtücher
Creme, Zahnbürste
Lebensmittel
Bier
Blumen
Spielzeug
Lebensmittel
Ersatzteil
Kleid
Wecker
18,20
5,00
9,00
55,00
20,00
137,50
67,50
21,50
11,90
3,50
15,00
FDGB
Benzin

这本书收集和整理了设计师克里斯蒂安·朗格（Christian Lange）的母亲从 1979 年到 1997 年来的购物清单，从这个独特的角度反映了一个普通人家庭在这么多年的生活中的侧面，也同样可以看到德国社会在各个时期的变化，是一本独特视角的社会和历史研究档案。本书的装订方式为锁线平装，封面白卡纸印单色黑，宽幅的勒口形成了较厚的封面触感。内页胶版纸单色黑印刷。全书没有采用特殊的工艺和材料，作为一本私人的“史料”，呈现出非常质朴的纪实感。本书的开篇展示了用于收集这些购物清单的 5 本账本，内页的主体部分按照年份和月份将这些购物清单做了整理，并配合清单内容穿插大量当时的私人照片、广告海报、物品图片、生活场景、笔记和信件。书籍的最后部分将所有内容按照项目、类别做了数量和支出占比的统计。书中根据清单项目多而文字少的特点特别设计了四栏的版面，配合自由穿插跨栏的图片，版面在满足功能的同时丰富多变。||||||||||||||||||||||||||||||||||||||||||||||||||||||||||||||||||||||||||||||||||||||||||||||||||||||||||||||||||||||

7 9
9 7
LANGE
LISTE

This book collects and collates the shopping lists of the designer Christian Lange's mother from 1979 to 1997. The changes of German society in different periods can be reflected from the side of ordinary family life. It is an archive of social and historical studies from a unique perspective. This paperback book is bound with thread sewing. The cover is made of white cardboard printed in monochrome black, and a wide flap makes the cover thicker. The inner pages are printed on offset paper in monochrome black. The book doesn't use any special technologies and materials. As a private 'historical material', it's more like a simple documentation. The book starts with five account books for collecting these shopping lists and organizes these shopping lists according to the year and month. Accompanied with the list, a large number of private photos, ad posters, life scenes, notes and letters are interspersed. The last part of the book makes statistics on the number and proportion of expenditure according to items and categories. Since there are many categories but a few word in the list, the typesetting used the four-column layout specially. With the pictures inserted freely, the layout is changeable while satisfying the function.

- 铜奖
- 1979 – 1997 年朗格家的账单
- 德国

- Bronze Medal
- Lange Liste 79 – 97
- Germany
- Christian Lange, München
- Christian Lange
- Spector Books, Leipzig
- 340×241×19mm
- 1038g
- 186p

Christian Lange's book "Lange Liste" is in fact nothing more than lists of expenses from his mother's housekeeping books, beginning in 1979 during the time of the GDR and extending up until 1997. What makes this book so special is its reduced design. The lists have the typography of long till receipts which bear witness to thrift of glaring dimensions, however. Illustrated with photographs which have been joined to the lists to form a collage, the book has wonderful haptic qualities – its choice of paper, the eschewing of colour and statistical appendix leave a convincing aftertaste of dealing with everyday life under restrictive circumstances. Using the principles of conceptual art, this insight into a private life lived in German real socialism as well as after its demise ends up becoming a sociologically inspired study; the lists it contains can practically be read as diaries. Simple, charming and attractive: a superb contemporary historical document.

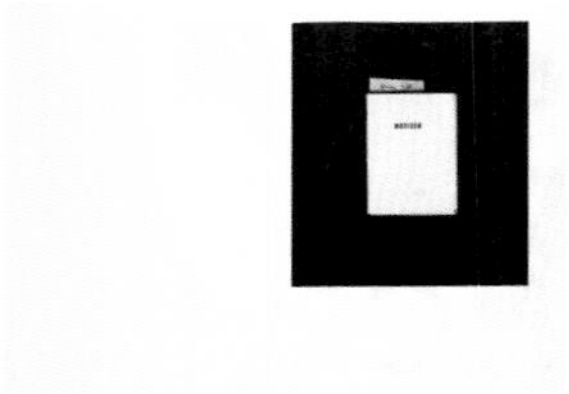

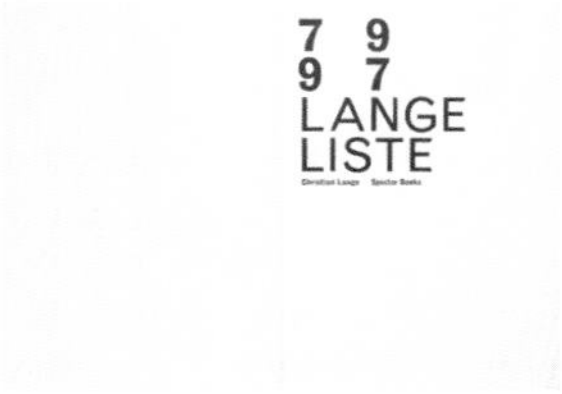
7 9
9 7
LANGE
LISTE

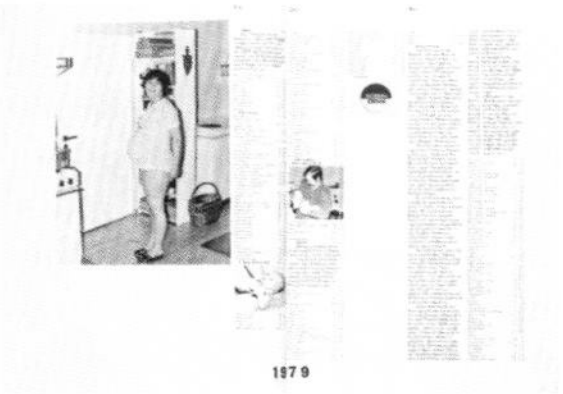
1979

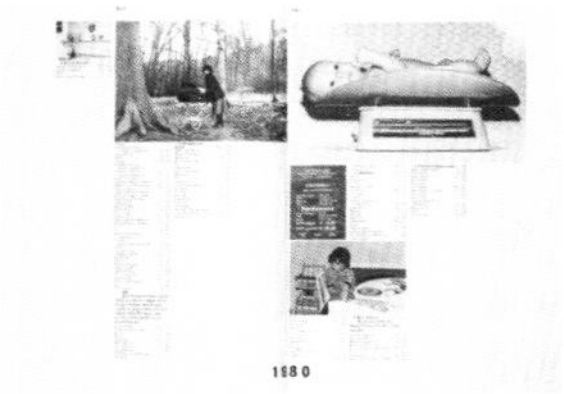
1980

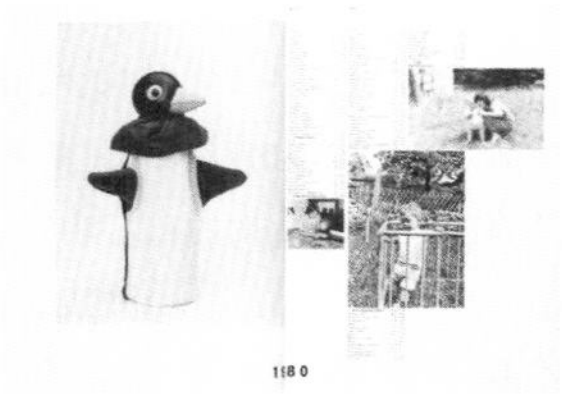
1980

1987

A Posten 5 7 5
Anteil 1 2 , 1 0 %
Häufigkeit* 2 , 3 6 ×

A–B
Auto–Benzin
Posten 2 3 0
Anteil 3 , 0 8 %
Häufigkeit* 0 , 9 4 ×

A–G–P
Auto–Garage
–Parken
Posten 5 9
Anteil 0 , 6 2 %
Häufigkeit* 0 , 2 4 ×

A–S–V
Auto–Steuer
–Versicherung
Posten 3 8
Anteil 0 , 9 1 %
Häufigkeit* 0 , 1 6 ×

A–R–Z
Auto–Reparatur
–Zubehör
Posten 2 4 8
Anteil 7 , 4 9 %
Häufigkeit* 1 , 0 2 ×

7 × Auspuff
1 9 × Autodurchsicht
1 2 × Autofelgen
2 × Autolackierung
5 × Autolampen
2 4 × Autoreifen
2 × Dachgepäckträger
1 × Fell für Auto
4 × Hohlraumkonservierung
1 1 × Motoröl
7 × Unterbodenpflege
1 1 × Zündkerzen

Auto *pro Monat

B Posten 1 . 1 6 6
Anteil 1 3 , 3 2 %
Häufigkeit* 4 , 7 8 ×

B–G
Bau–Garten
Posten 2 9 6
Anteil 1 , 0 5 %
Häufigkeit* 1 , 1 2 ×

7 × Bäume fällen
1 8 4 × Blumen & Pflanzen & Samen
2 × Rasenmäher
6 × Schaufel
1 × Teich

B–W
Bau–Wohnung
Posten 8 7 0
Anteil 1 2 , 2 7 %
Häufigkeit* 3 , 5 7 ×

8 × Baufolie
8 6 × Farbe & Lacke
1 9 × Fenster
1 2 × Fliesen & Bodenbelag
1 8 × Glaser *(Scheiben)*
6 0 0 × Hohlblocksteine
3 × Kompressor
2 8 × Leim & Kleber
4 × Leiter
1 5 7 × Sack Zement
1 7 × Säge
3 × Schubkarren
7 9 × Tapete
4 × Wasserwaage
6 × Zollstock

Bau *pro Monat

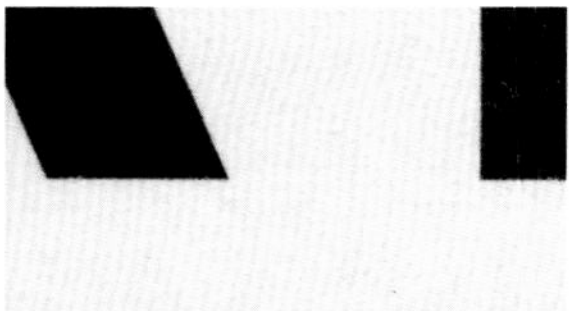

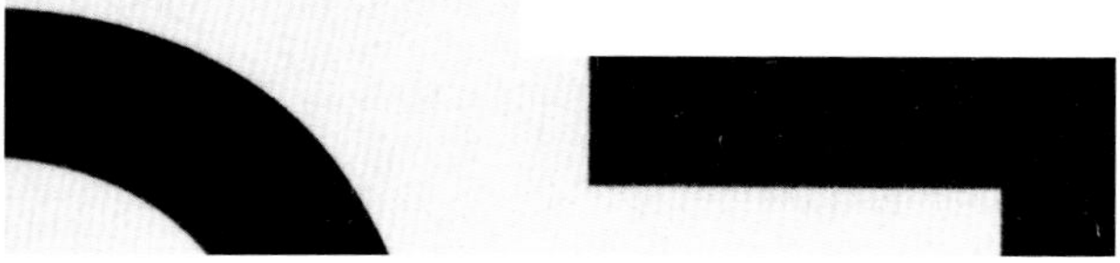

1979

1980

1981

KASKO
bevor es
zu spät ist!

81

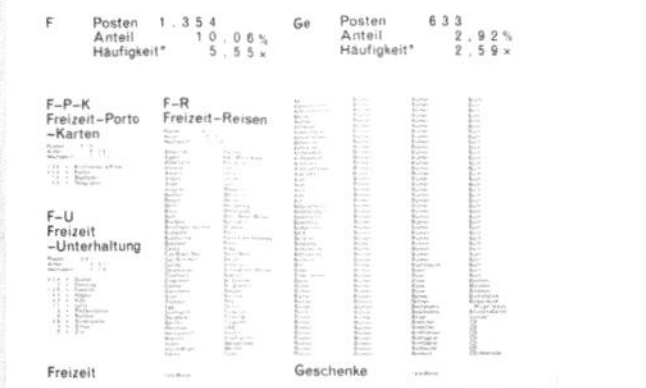
F Posten 1.354
Anteil 10,06%
Häufigkeit* 5,55×
Ge Posten 633
Anteil 2,92%
Häufigkeit* 2,59×
F-P-K
Freizeit-Porto
-Karten
F-R
Freizeit-Reisen
F-U
Freizeit
-Unterhaltung
Freizeit
Geschenke

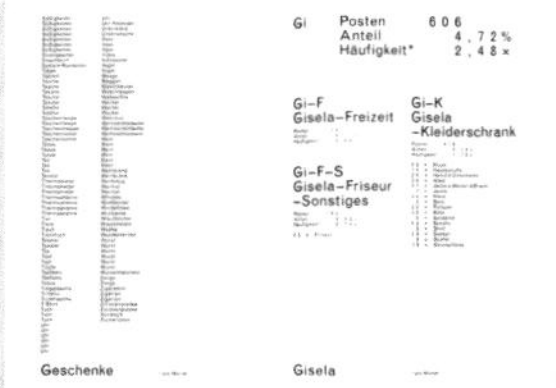
Gi Posten 606
Anteil 4,72%
Häufigkeit* 2,48×
Gi-F
Gisela-Freizeit
Gi-K
Gisela
-Kleiderschrank
Gi-F-S
Gisela-Friseur
-Sonstiges
Geschenke
Gisela

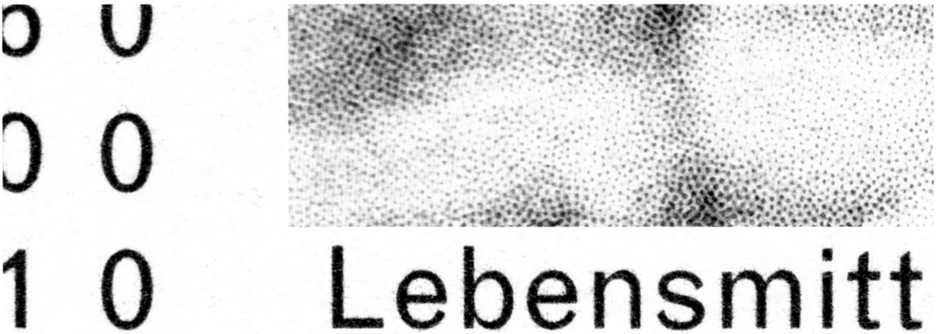
0 0
0 0
1 0
Lebensmitt

LANGE LISTE 79 97

Spector Books

Energie	582.00
Miete	1000.00
Energie	171.00
Telekom	62.50
Taschengeld	100.00
Lebensmittel	16.50
Blumentopf	12.00
Lebensmittel	22.00
Rolliege + Auflage	105.00
Buch	8.00
Hochzeit Sylvia	17.00
Konzert	70.00
Bilderrahmen	11.00
Lebensmittel	153.50
Massagen	32.00
1× DRK-Kurs Christian	35.00
Hemd, T-Shirt	60.00
Lebensmittel	232.00
Blumen	7.50
Kanne	5.00
Schmuck D. + Michelle	102.00
«Fielmann»	11.00
Fleischer	20.00

→ **Wangerooge**	
Christian (Klassenfahrt)	560.00
Weimar Franzosenbesuch, Geschenke	500.00
3× Filme	10.00
Lebensmittel	133.00
Salz und Pfeffer für Valerie	8.75
T-Shirt + Hose	34.00
Einwohnermeldeamt Christian	10.00
Antrag für Fahrschule Christian	65.00
Apotheke	26.00
Elektromaterial Garten	27.00
Fotos f. Franzosen	30.00
Lebensmittel	74.00
Lebensmittel	185.00
Fahrradzubehör Christian	30.00
Zeichenblock	10.00
Lebensmittel Thomas	333.00

Rundfunk	84.75
Miete	1000.00
Energie	291.00
Telefon	67.00
Taschengeld	100.00
Trinkwasser	33.00
Geb. Edith *(Badehandtuch, Schoko)*	65.00
Geb. Alex *(Schirm, Schoko)*	60.00
Lebensmittel	50.00
Post	8.00
Fotos	45.00
Baumaterial Garten	121.00
Apotheke	9.00
Lebensmittel	49.00
→ **Seidensteppbett**	35.00

Lebensmittel	96.00
Apotheke	5.80
Fotos	15.00
Lebensmittel	44.00
Friseur	42.00
Blumen	5.00
Socken, Shirt, T-Shirt	80.00
Zeitung	2.50
Hemdbluse	60.00
Lebensmittel	135.00
Lebensmittel	29.00
Fotos	13.00
Lebensmittel	27.00
Rock, Shirt, Hose	145.00
Krawatten	48.00
Schlafanzug	29.00
Slip	10.00
Lebensmittel	33.50
Hannover	30.00
Lebensmittel Thomas	230.00
Werkzeug	50.00
Farbfotos	9.00

Aug Sep

Wille's EIS

Rundfunk	85.00
Miete	1000.00
Energie	291.00
Telefon	78.00
Taschengeld	100.00
Versicherung	29.40
Mittenwald Übernachtung	750.00
Mittenwald Sonstiges	170.00
Lebensmittel	33.00
(Ottoversand) Kleid + Shirt	123.40
Füller, Büro	20.00
Geburtstag Thomas	135.00
Lebensmittel	47.00
Karte	5.00
Lebensmittel	33.00
Film	3.00
Kalender, Blumen	25.00
Blumen	2.00
Jacke	198.00
Lebensmittel	40.00
Fotos	32.00
Lebensmittel	72.00
Ikea	59.00
dm Drogerie	34.00

Miete	1000.00
Energie	291.00
Telefon	55.00
Taschengeld Christian	100.00
Lebensmittel	119.00
Scheren	15.00
Baumarkt	50.00
Lebensmittel	56.00
Kleid, Tuch	161.00
Lebensmittel	26.00
Deckenplatten für Schlafzimmer	397.80
Lebensmittel «Säckel»	114.00
Mörtel	10.40
Shirt	30.00
Lebensmittel	110.00
Geburtstag Gisela *(Parfüm)*	45.00
Lebensmittel	13.00
1× Fusspflege	20.00
Pflanzen	13.00
Apotheke	9.00
Lebensmittel	32.00
Friseur	9.00
Einlagen	17.00
Elektromaterial	25.00
Bücher	15.00
Karten	5.50
Lebensmittel	81.00
Bügelbrettbezug	13.00
Lebensmittel	38.00
Badematte	6.00
Schuhe	10.00
2× Unterhemden	10.00
Lebensmittel	30.00
Lebensmittel	50.00
Herr Mukerjee *(2 Deckchen, Stehstrauss, Cognac, Film)*	70.00
Lebensmittel Thomas	240.00

→ Geburtstag Christian (Disko)	100.00
Geb. Christian *(Kopfhörer)*	150.00
Lebensmittel	40.00
Friseur	31.00
Lebensmittel	22.00
Pflanze	8.00
Werkzeug	52.00
Tapete	26.00
Lebensmittel	126.00
Fotos	11.00
Baumarkt	85.00
Lebensmittel	30.00
Apotheke	12.50
Fotos	24.60
Wandertag Christian	25.00
«Doldinger» Kultur	66.00
Lebensmittel	73.00
Hose + Rock	117.00
Lebensmittel Thomas	395.00

Lebensmittel	58.00
Pfosten, Wasserwaage	45.50
Lebensmittel	30.00
→ Bansin	255.00
Schuhe Christian	173.00
Fotos	29.00
Geb. Peter *(Buch, Krawatte, Blumen)*	70.00
Kutte, Hose «Storg»	110.00
Fleischer	13.20
Fleischer	10.00
Lebensmittel	11.00
Lebensmittel	6.00
Lebensmittel	55.00
Lebensmittel Thomas	245.00

Versicherung	29.40
Miete	1000.00
Energie	291.00
Telefon	75.00
Taschengeld Christian	100.00
Trinkwasser	33.00
Zahnarzt	1088.00
Lebensmittel	35.00
Zahnarzt	741.00
2× Pullover Christian	90.00
Fahrradzubehör	124.40
2× Unterhemden	10.00
Buch	5.00
Lebensmittel	73.00
Görlitz *(Besuch)*	60.00
Kassetten	16.00
Lebensmittel	70.00
Cognac für Jörg *(Bansin)*	20.00
«Gräzer», Wurst, Blumen	110.00
Lebensmittel	215.00

97

本书以汇编的形式收录了艺术家刘小东在和田的绘画、日记、照片、纪录片等关于此项目的丰富资料，并通过采访、研讨会纪要、散文等形式记录下文学家和艺术家在新疆的艺术与思考。本书的装订形式主要为多种纸张锁线胶装，局部夹插其他开本或材料的页面的混合方式。封面采用黑色仿皮面印金，封面边缘故意留毛边。内页按照标题呈现的两大块内容划分为两大部分，通过仿皮与麻布的对裱材料来区隔。不同内容采用不同纸张和排版方式，纸张的材料和不同排版方式在书中相互“碰撞”形成对比，装订时也故意不严格清边，印有分属两部分书名的布面书带在书中褶皱而柔软地夹着，使得整本书呈现出一种被艺术家长期使用，包罗大量资料的工作笔记本的状态，是了解艺术家、作品和相关思想的第一手资料。

刘小东在和田
LIU XIAODONG'S HOTAN PROJECT
&
新疆新观察

This book collects the artist Liu Xiaodong's paintings, diaries, photographs and documentaries in Hetian, and records the artist's works and thoughts in Xinjiang through interviews, seminar summaries and prose. This book is perfect bound with thread sewing. There is various paper used in the book, inserted by pages of different format or materials. The cover is made of black faux leather printed in gold and the edge of the cover is deliberately untrimmed. The inside pages are divided into two parts, which are separated by a page mounted with faux leather and linen. Different paper and typesetting methods are used to distinguish and contrast different contents. The book's edges remain untrimmed deliberately. The ribbon printed with titles of two parts is put in between the pages, which makes the book appear to be used by the artist for a long time. The notebook that contains a large amount of information is the first-hand material to understand the artist himself, his works and ideas.

- 铜奖
- 刘小东在和田 & 新疆新观察
- 中国
- 小马哥、橙子
- 侯瀚如、欧宁
- 中信出版社

- Bronze Medal
- Liu Xiaodong's Hotan Project & Xinjiang Research
- China
- Xiao Mage & Cheng Zi
- Hou Hanru, Ou Ning
- China CITIC Press
- 239×175×25mm
- 947g
- 346p

This book is presented in the form of a compilation documenting an extensive art project. Diary entries, sketchbook drawings, photographs, overpainted photographs, large-format plein air painting on canvas right through to a podium discussion and coverage of the exhibition's reception culminate in this self-contained artist's book beauty. The design concept therefore goes way beyond that of a standard documentation: by outlining rather than simply juxtaposing the very different phases of the enterprise it has successfully transformed the process-driven nature of the project into a durable medium – that of a book – thereby alternating between glossy and matt materials, between coarse and fine typography, between colour and monochrome. You can really be persuaded you are feeling temperature, air and earth, indeed the whole atmosphere of what is taking place on the book's images. Yet the question keeps coming back: is such a thing – imparted atmosphere – actually possible? This book offers an opportunity to examine just this possibility. Perhaps it is through its atmosphere that the reader may gain an inkling of the original subject matter. If this is the case, then the credit belongs to its design concept.

刘小东在和田
&
新疆新观察

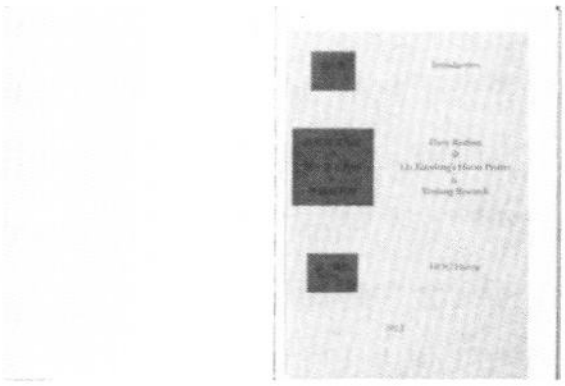

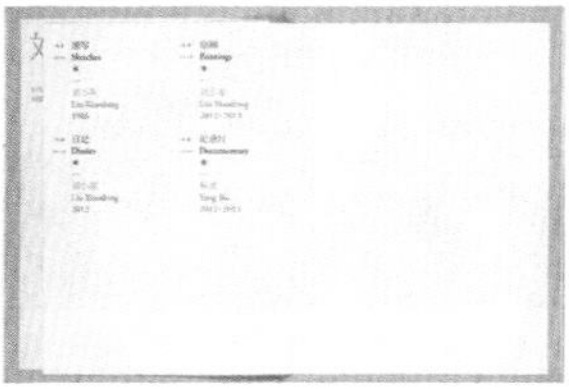

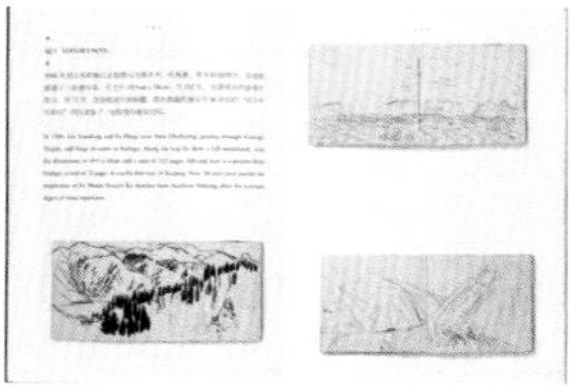

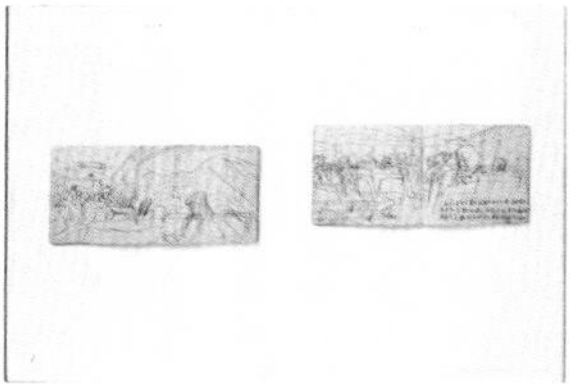

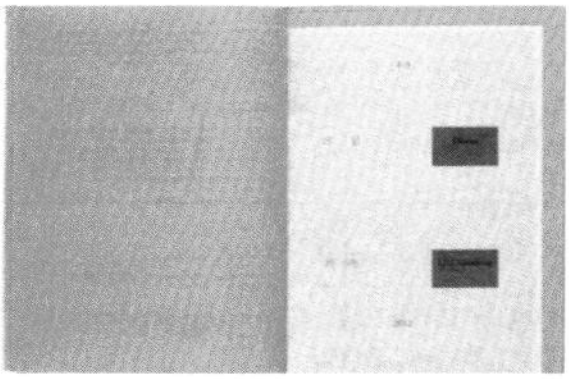

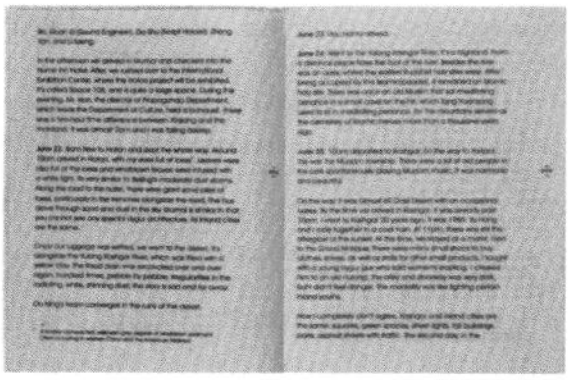

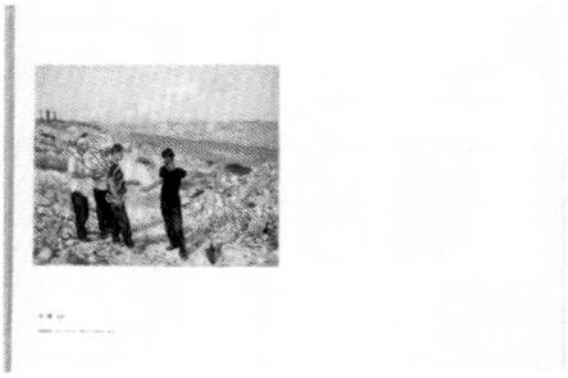

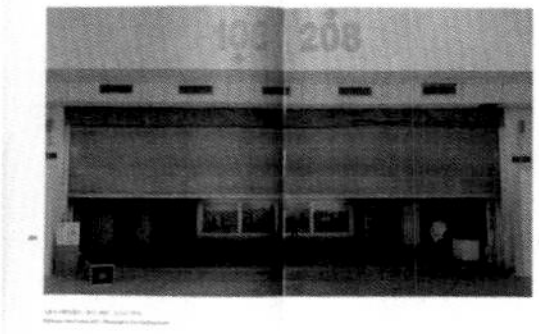

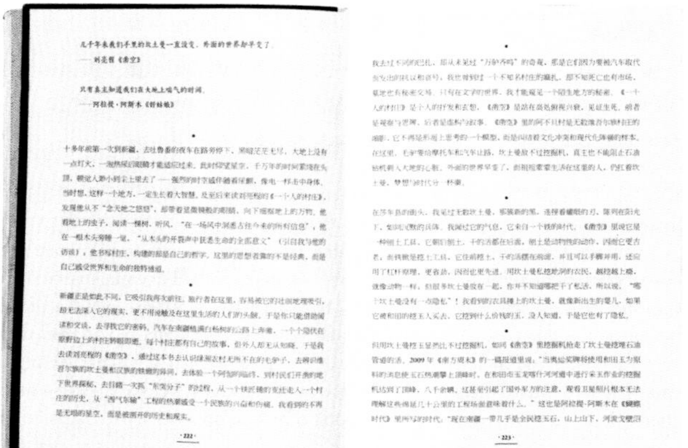

几千年来我们手里的坎土曼一直没变，外面的世界却早变了。
——刘亮程《凿空》
·222·
·223·

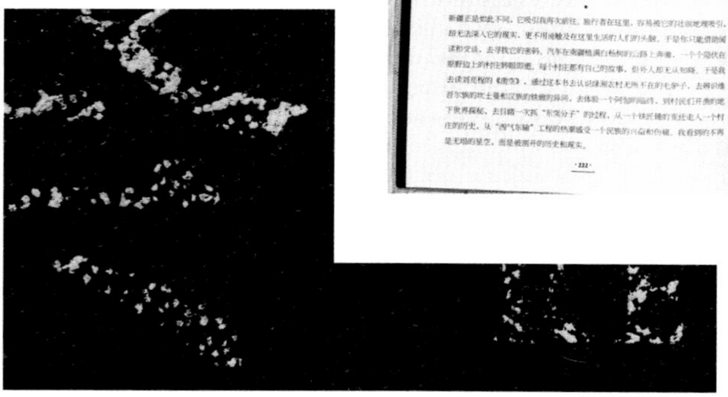

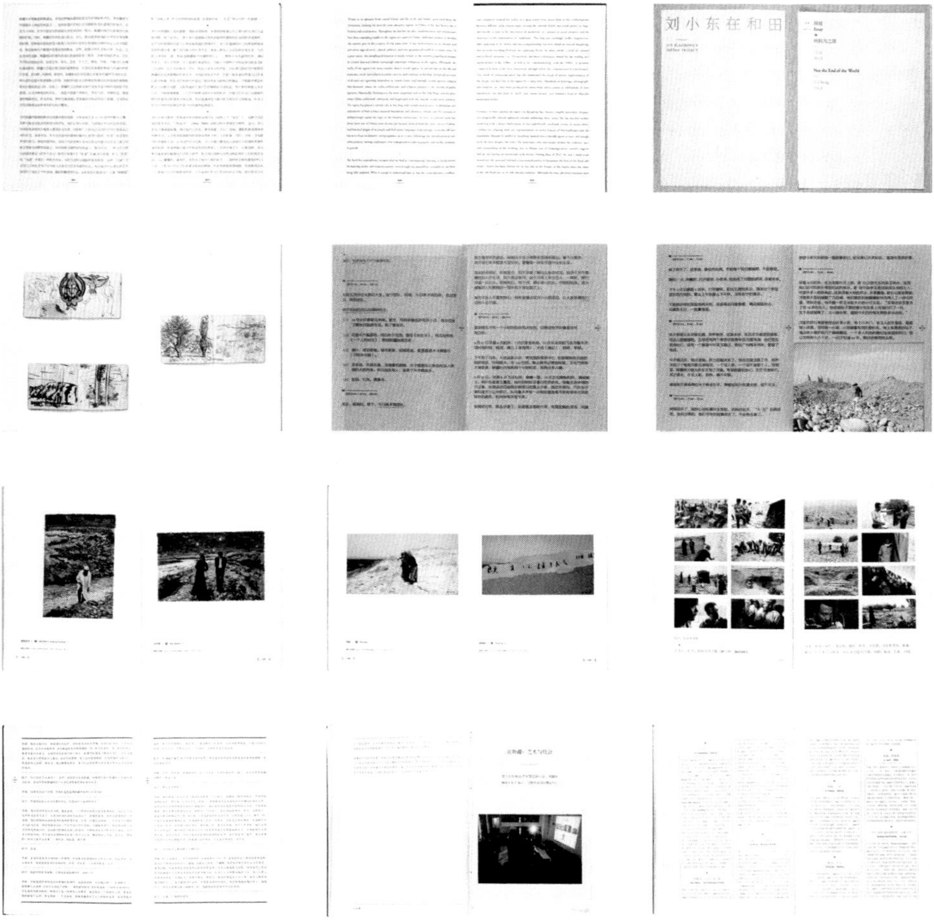
刘小东在和田
Not the End of the World

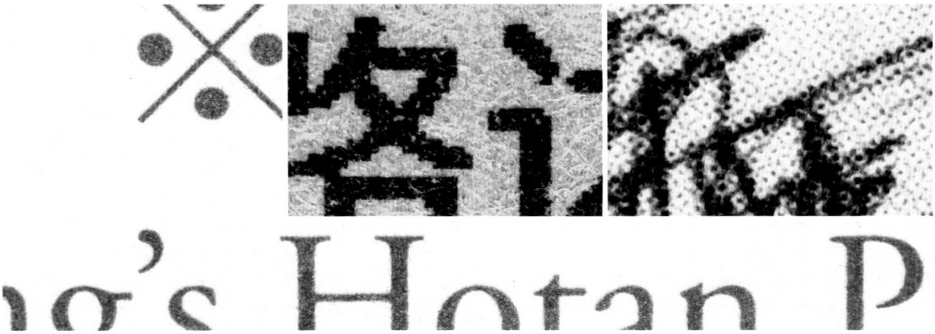
ng's Hotan P

李娟

没有最好的地方，也没有最坏的地方

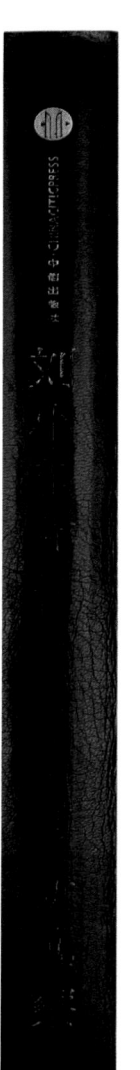

欧宁，2012年6月17日，阿勒泰

欧宁：你是怎么开始写作的？

李娟：怎么说呢。因为小时候好像干别的什么都干不好，就作文课好像还行，学习不好嘛，就作文还写得好一些，然后呢画画也可以吧。从小我就知道了，人长大必须得干一件什么事情。你说我能干什么呢，学习也不好，这也不能干，那也不能干，身体也不好也不能当兵，也不愿意种地，想想看，想学画画吧，又太花钱了，哈哈，要买纸什么的。只有写作。当然这不是最大的原因，主要还是喜欢，喜欢写点什么东西，喜欢看书。

欧宁：你第一次发表作品是在几岁、什么时候？

李娟：小的时候在校刊上发表过一些东西，但第一次发表是在20岁吧。那时候去打工，在乌鲁木齐，打工挺困难的，就受不了那个苦嘛，想改变生活，然后就写了一篇稿子，跑到一编辑部去投稿，过了一年多，发表了。

欧宁：你那篇文章是什么内容？

李娟：树，写树的，反正内容里可能有一些，因为那时候生活比较窘迫嘛，就写一些现在看起来挺可笑的那样一种情绪在里面，呵呵，抱怨呀，或者怎么说呢，也不是抱怨吧，就是青春期的那种情绪吧。

欧宁：还记得那时候你都看一些什么书吗？

李娟：我从小喜欢看书，看的书很杂的，反正杂七杂八的。小时候对我影响很大的书，而不是我个人喜欢的，小时候比较容易受影响吧，看了那个《小王子》，后来看了孙犁的《铁木前传》，还有汪曾祺的《受戒》，还有一些日本漫画，它那个语言相当强烈。

欧宁：这都是中学时候看的书吗？

李娟：初中的时候。高中看得少一点了，小时候喜欢看、最小的时候看《安徒生童话》，很喜欢，觉得说不出来的伤心。

欧宁：那现在呢？

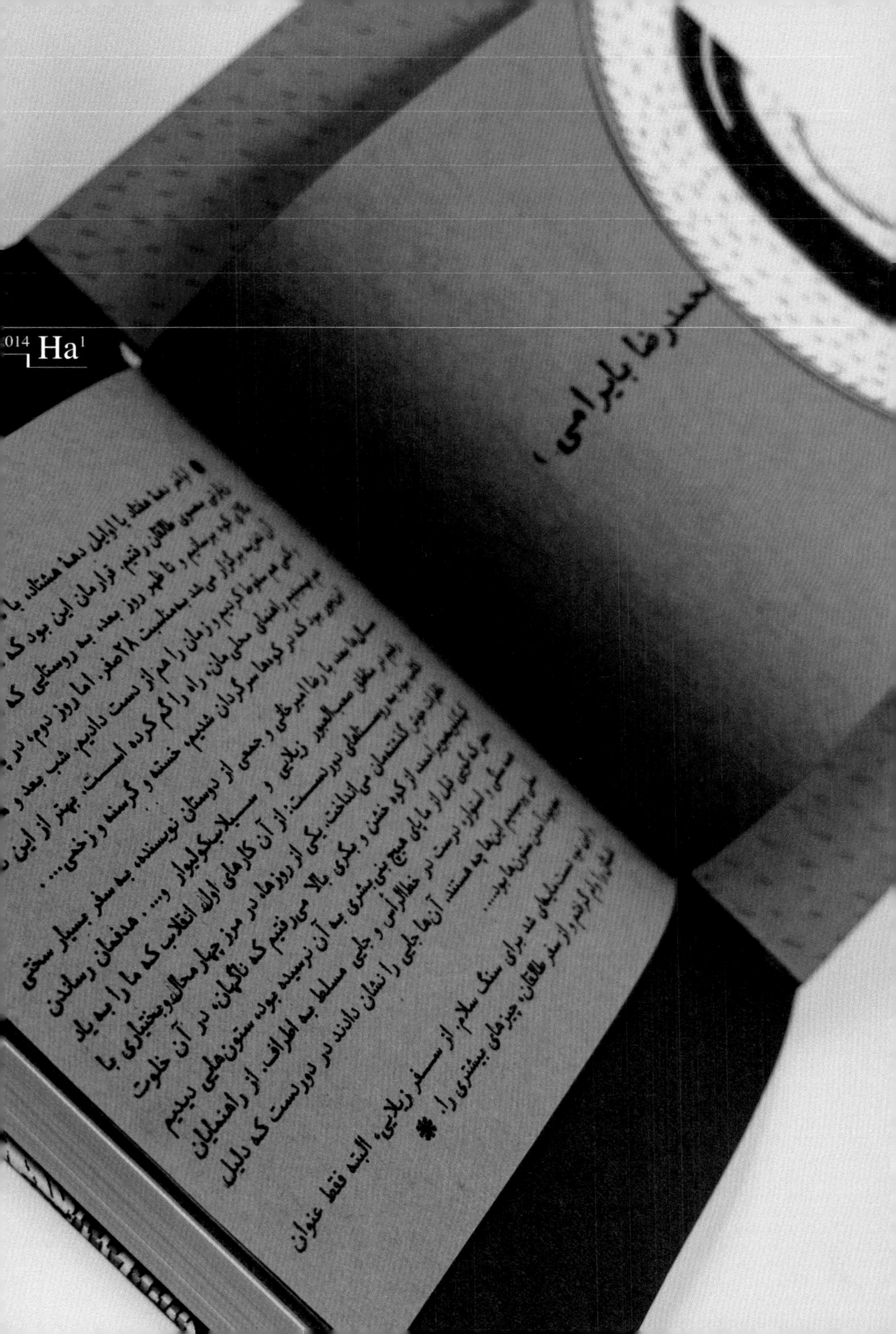
رضا بایرامی
سال‌ها بعد با رضا امیرخانی و جمعی از دوستان نویسنده، به سفر بسیار سختی

这是一个用波斯文写就的现代寓言故事，讲述了一群年轻人的独特的旅行经历。他们在神秘的村庄中参加一年一度的宗教活动，通过这个与过去的理想之间的沟通渠道，这帮年轻人逐渐认识到原本所理解的友情、忠诚、牺牲以及智慧的真正本质。本书的装订形式为无线胶平装，书籍为右翻本。封面采用艺术纸张，棕色和绿色双色印刷，覆盖亮面UV。封面为故事插图，半圆弧形状的勒口完美地配合着插图的形式。封面与正文之间，夹着几页单黑色印刷的牛皮纸张，大约只有书籍高度的三分之二，展示作者照片、介绍和版权信息。正文采用黄褐色书写纸印单色黑，纸张弹性十足，翻阅手感舒适。故事本身的叙事手法混合了现实和虚构的成分，而纸张的独特色泽与质感也衬托着故事的神秘气息。

This is a modern fable story written in Persian. It tells about the unique travel experience of a group of young people who participated in annual religious activities in a mysterious village. Through this channel to communicating with past ideals, these young people gradually realized the true essence of friendship, loyalty, sacrifice and wisdom. The paperback book is perfect bound, and it's a rightward flipbook. The cover is made of art paper printed in brown and green, spotting bright UV. On the cover, there is a story illustration that is perfectly matched by a semi-circular shaped flap. Between the cover and the main body, there are several pages of kraft paper printed in single black to display the author's photo, book introduction and copyright information. The text is printed in monochrome black on yellowish-brown writing paper. The paper feels comfortable to read. The method of narration in this story mixes reality and fiction, and the unique luster and texture of the paper also reveal the mystery of the story.

荣誉奖

你好石头

伊朗

Honorary Appreciation

Hello Stone (Sange Salam)

Iran

Majid Zare

Mohammad Reza Bayrami

Asr-e Dastan

189×130×12mm

235g

240p

How does this paperback from Iran manage to draw attention to itself without doing so conspicuously? Its two-colour jacket illustration appears as a pictorial summary of its fable. Simple, linear strokes in entwined composition achieve a strong build-up of suspense. Contrasts are increased through the relief of lacquered black forms. Curiosity is aroused on picking up the book and being beckoned in to read by the continuation of the drawing on unusually punched jacket flaps. An attractive touch is to be found in information about the author being placed ahead of the inner title, namely on curtailed pages using brownish paper. The text, in calligraphic Persian script, is rendered in a blackish brown on particularly yellowy paper, airily set with a double measure of indentation. The overall impression is subtle and literary, the muted colour scheme congenial and mysterious at the same time.

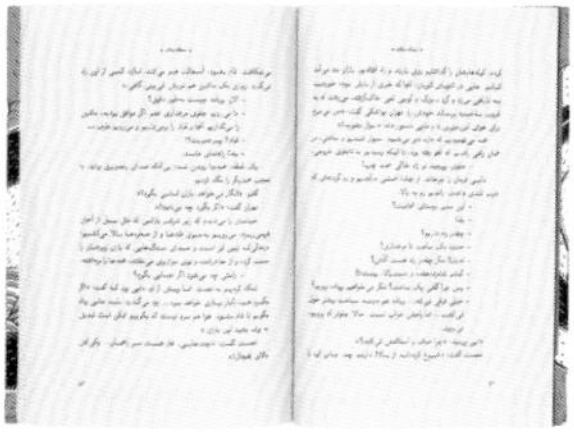

• سنگ سلام •

- پس یخچال هم دارید!
- تقریباً بله! تا اواخر خرداد دوام می‌آورد. قدیم می‌رفتند و از توش برف و یخ می‌آوردند.

دور و برمان، فقط سیاهی کوه‌ها را می‌دیدیم و گهگاه، دامنه‌های نیمه‌خیس راه در وقتی که نور چراغ‌های جیپ، کشیده می‌شد روی آن‌ها.

گفتم: «مثل اینکه همه‌جا سبز شده!»

نعمت گفت: «من هم خبر ندارم. امسال اصلاً نیامده‌ام این‌طرف‌ها. ولی دیگر وسط‌های اردیبهشت است... باید هم بشود.»

همین‌طور که جلو می‌رفتیم، رسیدیم به شیب خیلی تندی. جیپ دایی، زوزه‌کشان شروع کرد به بالارفتن، اما قبل از تمام‌شدن سربالایی، به خرخر و پت‌پت افتاد و بعد هم خاموش شد.

- ای که هی! همین را کم داشتیم!

نگاهی به یکدیگر انداختیم و پیاده شدیم. سایه‌مان زیر نور ماشین، کش می‌آمد و می‌رفت تا آن دوردورها. قبل از اینکه دایی، چراغ‌ها را از ترس خالی‌شدن باتری خاموش کند، نعمت نشست رو زمین و دست گذاشت بر خاک.

- چی شده؟
- ماشینی از اینجا گذشته... منتها وقتی که زمین کاملاً خشک بوده.
- یعنی صبح یا دیروز؟
- نه... تازه است.

من و مهران با تعجب نگاهش کردیم.

۵۴

• سنگ سلام •

- یعنی واقعاً تو کهنه و تازه‌بودن رد را تشخیص می‌دهی؟
- گیر آوردی ما را؟!

به‌خاطر دایی، مجبور بودیم مؤدب باشیم، اما توجه او هم جلب شد. انگار خراب‌شدن ماشین را فراموش کرده بود. با تحسین نگاه کرد به نعمت. انگار خوشش می‌آمد که یک روستایی واقعی را ببیند: «می‌توانی بگویی کی گذشته دقیقاً؟»

- پیش پای ما!
- مطمئنی؟!
- بله! حالا عمو این‌ها را که دیدیم، می‌پرسیم که دقیقاً کی گذشته و کی بوده. همه از جلوی آن‌ها می‌گذرند.
- همه؟!
- بله! سر راه هستند... گفتم که!

دایی بست‌های روی کاپوت را باز کرد.

- چراغ‌قوه را از زیر صندلی من بیاور بهمن!

چراغ‌قوه را آوردم. دایی کمی به موتور ور رفت و بعد گفت استارت بزنم. اما تا کلید را چرخاندم و موتور خرخر کرد، داد زد: «نزن! نزن! بدتر می‌شود. خفه کرده. قبل از حرکت وقت نکردم که بهش برسم... باید هم این‌طور می‌شد.»

نگاهی به تاریکی اطراف انداختم: آیا سفری که با این‌همه زحمت شروع کرده بودیم، به‌همین زودی تمام می‌شد؟ نعمت با نگرانی پابه‌پا شد و زل زد به کوه‌ها. پیدا بود که او هم نگران است.

- حالا چه کار کنیم دایی؟

۵۵

(عصر داستان)

سنگ سلام

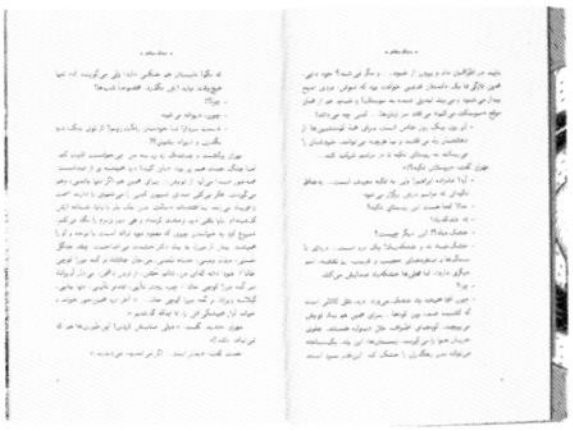
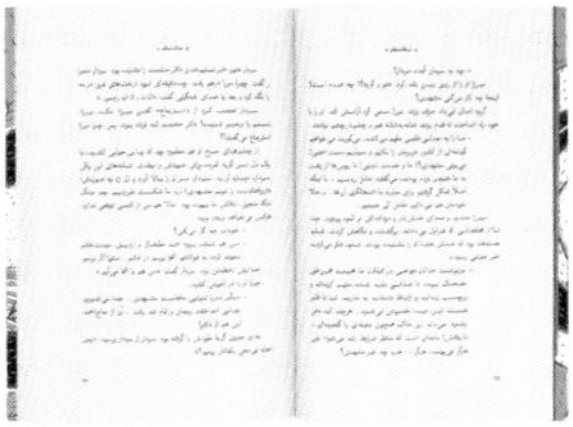

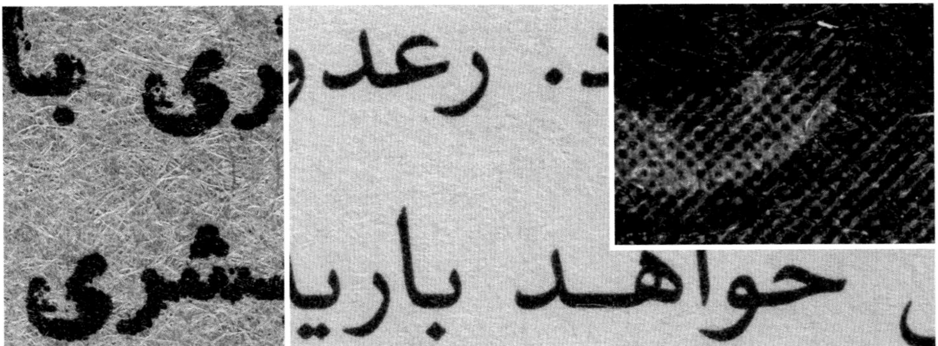

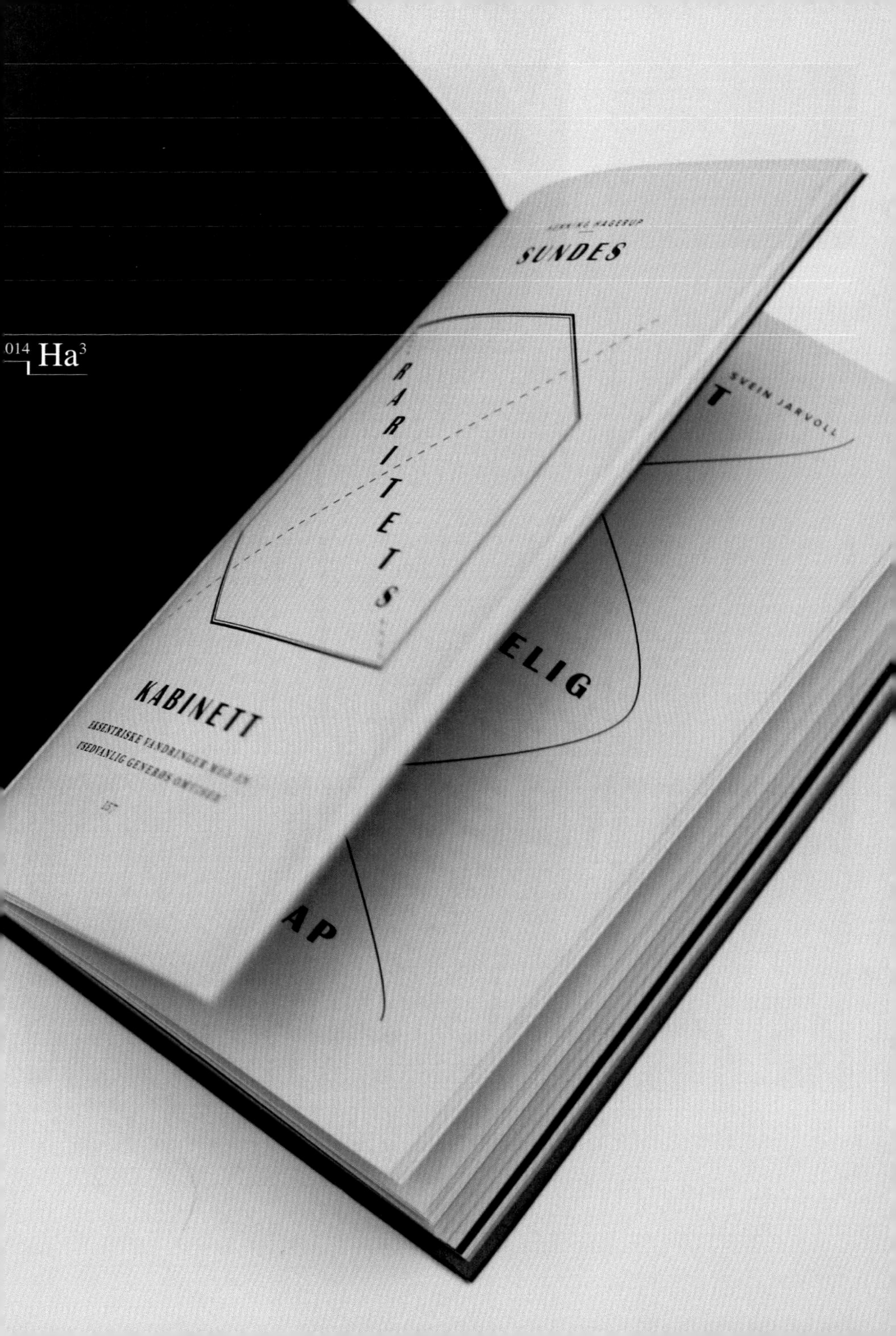
SUNDES
RARITETS
KABINETT
SVEIN JARVOLL
ELIG
AP

奥莱·罗伯特·桑德（Ole Robert Sunde）是挪威著名诗人、小说家、散文家，本书为奥登·林德霍尔姆（Audun Lindholm）将自己和别人研究桑德的文章汇编起来形成的散文集。本书的装订方式为锁线硬精装，封面胶版纸灰色和橙色双色印刷，裱覆卡板，标题部分通过弧线连接单词，在看似不经意的文字排版中区分出主副书名。标题字做了非常不明显的凹凸工艺，只在触摸封面的时候才能体会到。打开封面后橙色的环衬与封面形成鲜明对比，与封面不明显的凹凸工艺一并体现着内敛但充满活力的气质。内页胶版纸单黑印刷，六大部分分别设置了六种不同的符号，在章节页和页眉上加以体现。每篇文章都单独设置了与封面类似但又独特的排版，像一个个独立但又相互关联的小封面。内文的注释长短不一，通过直接压进正文版心的方式呈现，形成独特的页面节奏。

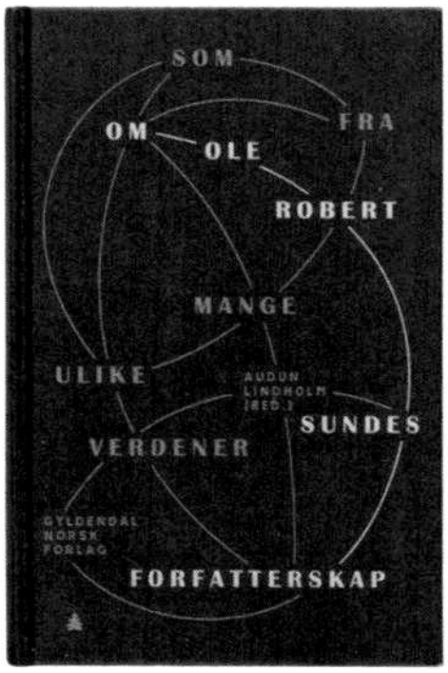
SOM
FRA
OM
OLE
ROBERT
MANGE
ULIKE
AUDUN
LINDHOLM
(RED.)
SUNDES
VERDENER
GYLDENDAL
NORSK
FORLAG
FORFATTERSKAP

Ole Robert Sunde is a famous Norwegian poet, novelist and essayist. This book is a collection of essays studying Ole Robert Sunde compiled by Audun Lindholm. The hardback book is bound by thread sewing. The cover is made of offset paper printed in gray and orange, mounted by graphic board. The words in the title are connected by arc lines and embossed indistinctly, which can only be realized when touching the cover. The orange end paper is in sharp contrast with the cover, but demonstrates the restrained but energetic character as well. The inner pages are printed on offset paper in black only. Six different symbols are set for six parts respectively, which are displayed on chapter pages and headers. Each essay has a specially typeset page like an independent but interrelated 'cover'. The annotations of the text are different in length. They are presented in the type area that forms a unique layout.

荣誉奖

来自许多不同的世界
——关于奥莱·罗伯特·桑德的写作

挪威

Honorary Appreciation

Som fra mange ulike verdener.
Om Ole Robert Sundes forfatterskap

Norway

Andreas Töpfer

Red. Audun Lindholm

Gyldendal Norsk Forlag AS

211 × 135 × 26mm

438g

279p

A collection of essays is not always the most gratifying assignment for a book designer. Authors often have their own annotation style, the footnotes are short or long, there is a mixture of more extended paragraph texts with a motto here and a few illustrations there. Sections are either numbered or have subheads or neither, there may be enumerations, sometimes extensive citations need to be indented, sometimes they are italicised within the main body of text. Such is the case in this collection of contributions on the work of the Norwegian writer Ole Robert Sundes. The typographer's approach to these editorial necessities is an assertive one. Using a classically reliable type-area he allows the annotations to overlap as necessary with the type-area as marginal notes placed at roughly the same height as the corresponding text; the titles of the cited authors' books are in italics with underlining. From the point of view of orthodox teaching such examples would not always have been regarded as evidence of typographical sensitivity. Yet here they contribute to a vitality which is more than merely formal. Chapter beginnings are generously presented with abstract and symbolic graphics as if they were individual title pages. The overall care taken in the finish of the book extends to the understated embossing of the lines on the paperback's title, the allure of its combination of grey and orange leaving its mark from the boldy coloured endpapers right through into the interior of the book.

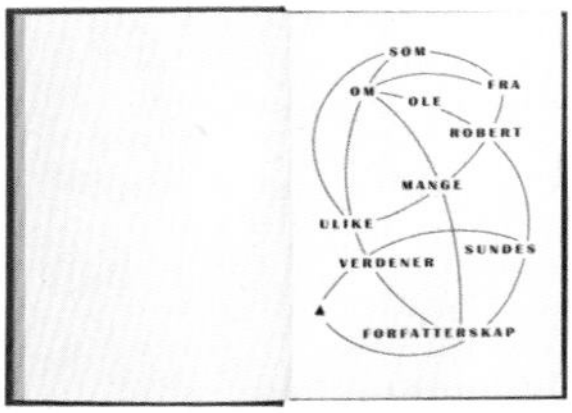
SOM
FRA
OM
OLE
ROBERT
MANGE
ULIKE
SUNDES
VERDENER
FORFATTERSKAP

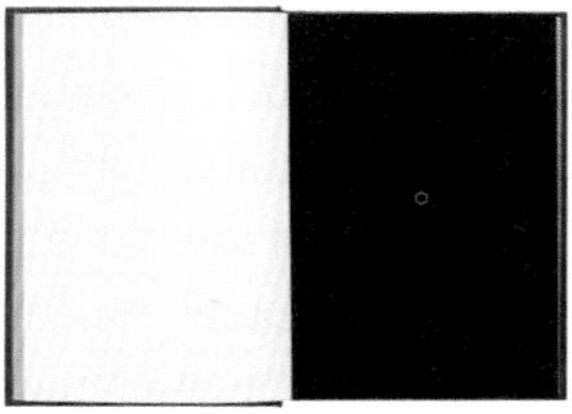

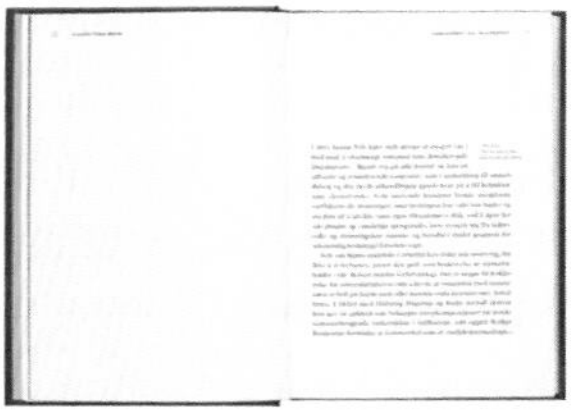

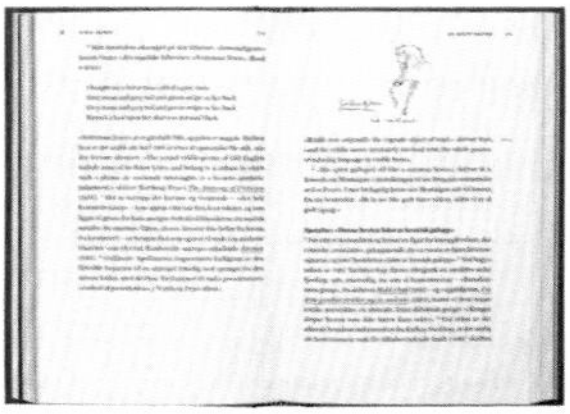

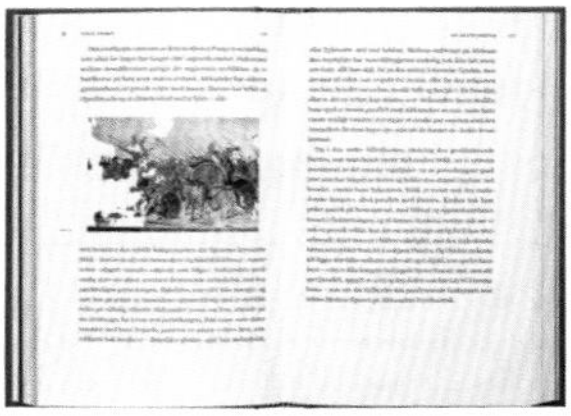

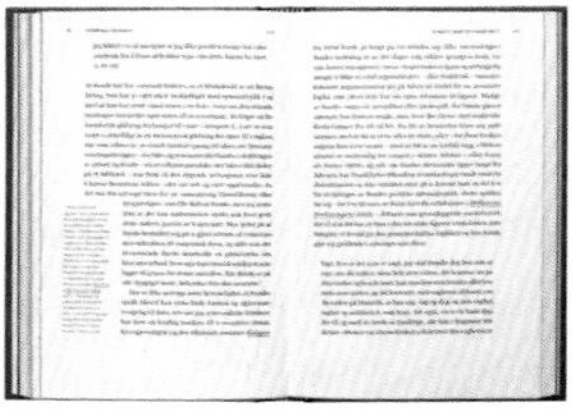

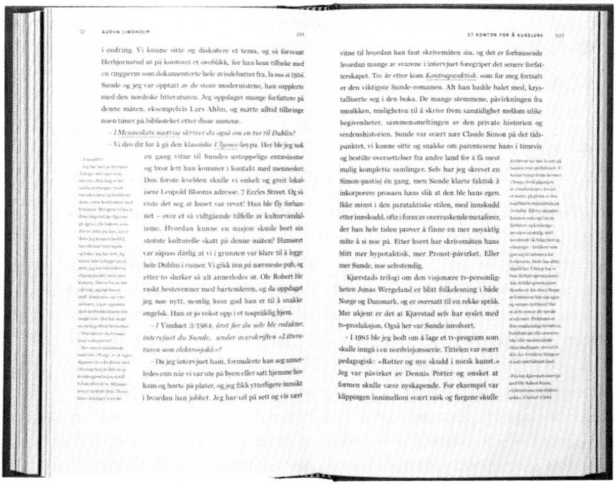

AUDUN LINDHOLM 226

i endring. Vi kunne sitte og diskutere et tema, og så forsvant Herbjørnsrud ut på kontoret et øyeblikk, før han kom tilbake med en ringperm som dokumenterte hele avisdebatter fra, la oss si 1956. Sunde og jeg var opptatt av de store modernistene, han supplerte med den nordiske litteraturen. Jeg oppdaget mange forfattere på denne måten, eksempelvis Lars Ahlin, og måtte alltid tilbringe noen timer på biblioteket etter disse møtene.

– *I Menneskets matrise skriver du også om en tur til Dublin?*

– Vi dro dit for å gå den klassiske *Ulysses*-løypa. Her ble jeg nok en gang vitne til Sundes ustoppelige entusiasme og hvor lett han kommer i kontakt med mennesker. Den første kvelden skulle vi enkelt og greit lokalisere Leopold Blooms adresse, 7 Eccles Street. Og så viste det seg at huset var revet! Han ble fly forbannet – over et så vidtgående tilfelle av kulturvandalisme. Hvordan kunne en nasjon skusle bort sin største kulturelle skatt på denne måten? Humøret var såpass dårlig at vi i grunnen var klare til å legge hele Dublin i ruiner. Vi gikk inn på nærmeste pub, og etter to slurker så alt annerledes ut. Ole Robert ble raskt bestevenner med bartenderen, og da oppdaget jeg noe nytt, nemlig hvor god han er til å snakke engelsk. Han er jo vokst opp i et tospråklig hjem.

– *I Vinduet 3/1984, året før du selv ble redaktør, intervjuet du Sunde, under overskriften «Litteraturen som elektrosjokk»?*

– Da jeg intervjuet ham, formulerte han seg annerledes enn når vi var ute på byen eller satt hjemme hos ham og hørte på plater, og jeg fikk ytterligere innsikt i hvordan han jobbet. Jeg har vel på sett og vis vært

ET KONTOR FOR Å KUKELURE 227

vitne til hvordan han fant skrivemåten sin, og det er forbausende hvordan mange av svarene i intervjuet foregriper det senere forfatterskapet. Tre år etter kom *Kontrapunktisk*, som for meg fortsatt er den viktigste Sunde-romanen. Alt han hadde balet med, krystalliserte seg i den boka. De mange stemmene, påvirkningen fra musikken, muligheten til å skrive frem samtidighet mellom ulike begivenheter, sammensmeltingen av den private historien og verdenshistorien. Sunde var svært nær Claude Simon på det tidspunktet, vi kunne sitte og snakke om parentesene hans i timevis og bestilte oversettelser fra andre land for å få mest mulig komplette samlinger. Selv har jeg skrevet en Simon-pastisj én gang, men Sunde klarte faktisk å inkorporere prosaen hans slik at den ble hans egen. Ikke minst i den parataktiske stilen, med innskudd etter innskudd, ofte i form av overraskende metaforer, der han hele tiden prøver å finne en mer nøyaktig måte å si noe på. Etter hvert har skrivemåten hans blitt mer hypotaktisk, mer Proust-påvirket. Eller mer Sunde, noe selvstendig.

Kjærstads trilogi om den visjonære tv-personligheten Jonas Wergeland er blitt folkelesning i både Norge og Danmark, og er oversatt til en rekke språk. Mer ukjent er det at Kjærstad selv har syslet med tv-produksjon. Også her var Sunde involvert.

– I 1985 ble jeg bedt om å lage et tv-program som skulle inngå i en nordvisjonsserie. Tittelen var svært pedagogisk: «Røtter og nye skudd i norsk kunst.» Jeg var påvirket av Dennis Potter og ønsket at formen skulle være nyskapende. For eksempel var klippingen innimellom svært rask og fargene skulle

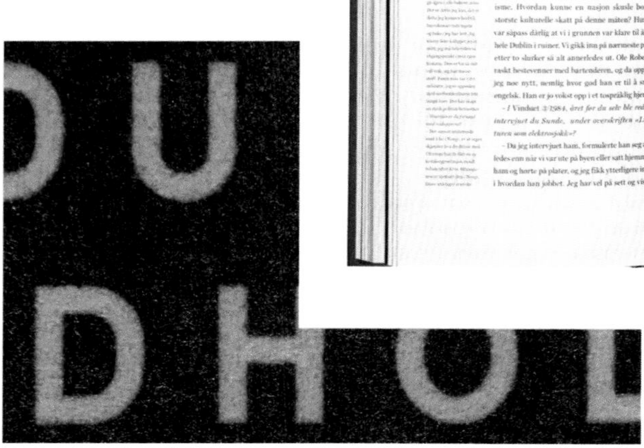

DEN
ER
FAR
BARA
VÄRLDEN

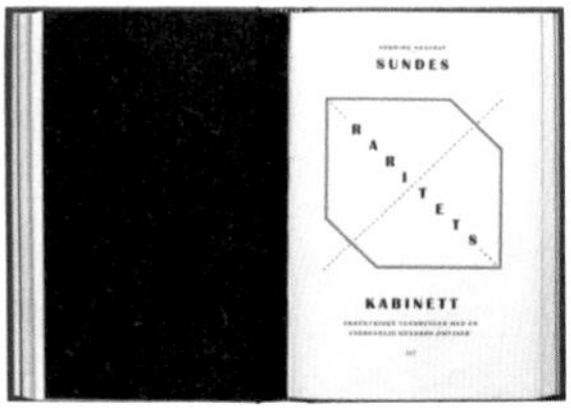
SUNDES
RARITETS
KABINETT

et des

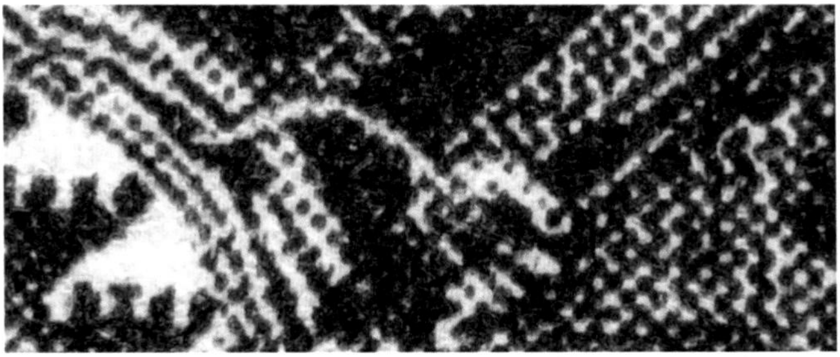

at man før eller

代文学馆馆藏珍品大系·手稿卷
《2006—2009中国最美的书》
《革命胜景图册》
设计者说
设计者说

本书是有关书的书，收录了2010-2012年获得“中国最美的书”的60件作品，并对这些书籍的设计、形态、内容做了全面的梳理和统计，建立了可以单独欣赏又可横向和纵向比较的图表与多维度的体系。本书由于大量传统包背和拉页混合的设计，装订方式采用了无线胶装的软精装形式。皱纹纸裱覆卡板形成最外层的书函，封面采用暖灰色棉布棕白双色烫印工艺，裱覆牛皮卡纸。内页五色印刷（四色加金色），章节隔页采用印刷用宣纸，而内页采用轻薄的哑面涂布纸。每一本书为一个单元，通过单页加三折页的形式呈现与书籍有关的各类信息，包括内容信息、物理参数、评委评语、设计者说，以及各角度图片。书籍的最后，通过尺寸、重量、页码这三个每本书都有的信息，建立了完整的图，让这些书籍在图表中处于相应的位置，供读者更直观地比较这些书籍之间的关系。

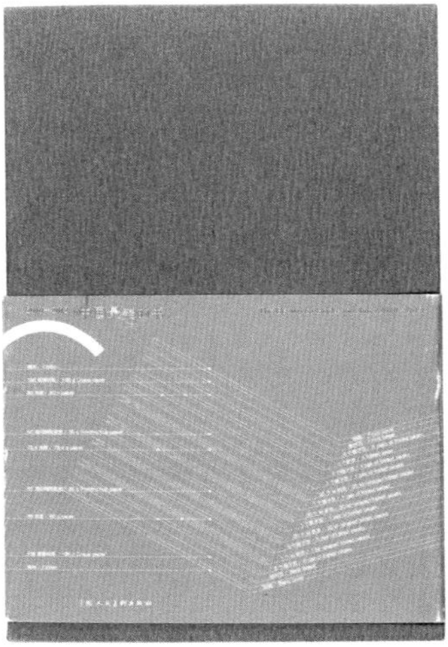

This is a book about books, which contains 60 works that won the prize ‘the most beautiful books in China’ from 2010 to 2012. It makes a comprehensive analysis and statistics on the design, format and content of these books, and establishes a multi-dimensional system of charts and graphs that can be appreciated and compared. The book is perfect bound with traditional double-leaved design. The slipcase is made of Crepe paper mounted with graphic board; the cover is made of warm grey cotton cloth hot-stamped in brown and white, mounted with kraft cardboard. The inside pages are printed in five colors (four colors plus gold) on light matte coated paper while rice paper is used for interspersed pages of the chapters. Each book collected is a unit, presented by a single page plus a trifold to show all related information, including content, physical parameters, judges’ comments, designers’ words, and pictures from different angles. At the end of the book, there is a complete table containing three sets of data – size, weight and number of pages, so that readers can compare the books more intuitively.

荣誉奖
2010-2012 中国最美的书
中国
刘晓翔
上海市新闻出版局“中国最美的书”评委会
上海人民美术出版社

Honorary Appreciation
The Beauty of Books in China 2010–2012
China
Liu Xiaoxiang, Liu Xiaoxiang Studio
Organizing Committee of “The Beauty of Books in China” of Shanghai Press and Publication Administration
Shanghai People’s Fine Arts Publishing House
291×192×40mm
1709g
294p

This corpulent, large quarto volume about fine books from China has 294 pages, which given its weight and its thin paper may seem puzzling. And indeed there is much more paper involved, since as a result of a modern interpretation of French fold – folded sheets whose open pages are stitched along the middle and the folds of which point outwards – there are over 240 visible pages which are not counted. This can be most clearly verified by means of the didactic banderole placed around the slipcase, itself open on both sides. With its presentation of 60 books this is a captivating concept: on the left-hand page the book’s exterior, on the right the respective jury text and technical data. Then one deliberately opens up a fold-out section revealing across the width of four successive pages sensitively photographed and perfectly printed images which have been intuitively arranged in a soothing yet exciting fashion. The design is so congenial to the beholder that there is a sense of actually holding the depicted books themselves. Such elaborate finish calls for unhurried handling – this is not a book for just flicking through. This reduced tempo is something to be thankful for, since it results in the exquisiteness of the whole design – both that of the 60 beautiful books and of the catalogue itself – being shown off to wonderful advantage.

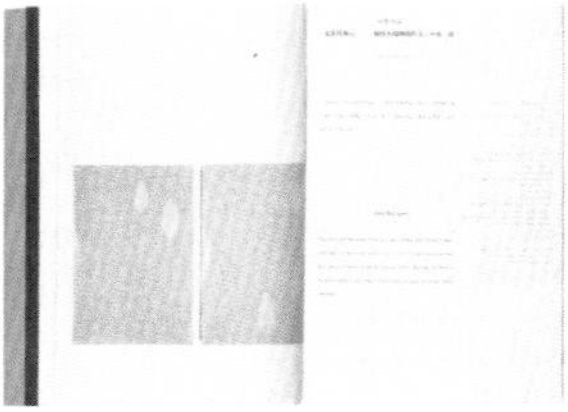

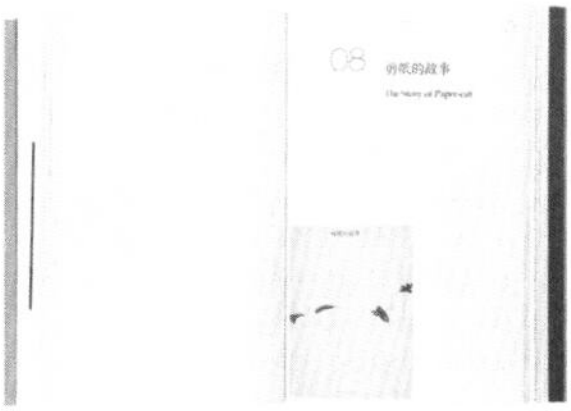
08
剪纸的故事
The Story of Paper-cut

31 1 2
3 4 5
6 7 8
9

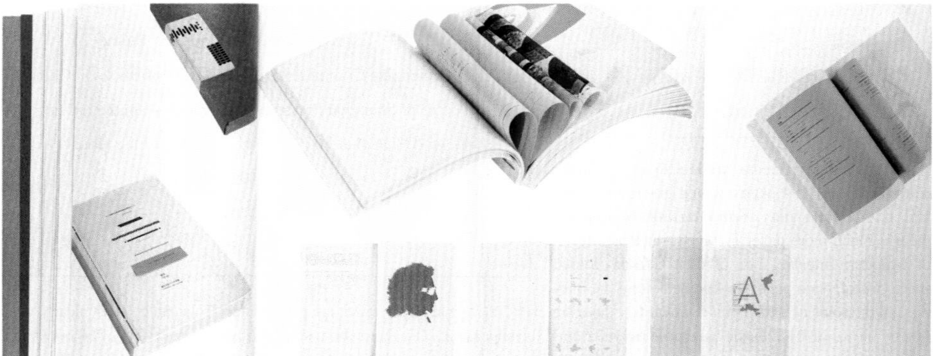

2010
02 穿墙术
诗经

Books in

本书是日本“诗歌和思想新人奖”得主冈田元的诗歌集，同时也是“诗歌和思想新人奖”系列的第 6 本。书中收录了诗人的作品共 18 首，内容涵盖现代诗歌和一些带有诗歌味道的散文。书籍虽然小巧，但采用了锁线硬精装的装订形式。护封采用暖灰色哑面艺术纸浅灰色和银色双色印刷，与纸张形成微妙的明度反差，图形色块故意斑驳处理，又形成了巧妙的细节。内封冷灰色哑面艺术纸只印浅灰色书名。环衬采用明黄色哑面胶版纸，搭配带有半透明水印纹路印有银色书名的扉页，体现出内敛但又充满活力的独特气质，正如“诗歌与思想新人奖”所体现的思想的细腻与新人的冲击力，缺一不可。内页胶版纸黑色印刷，极其简约的日式排版，让人可以专注于诗人的作品本身。

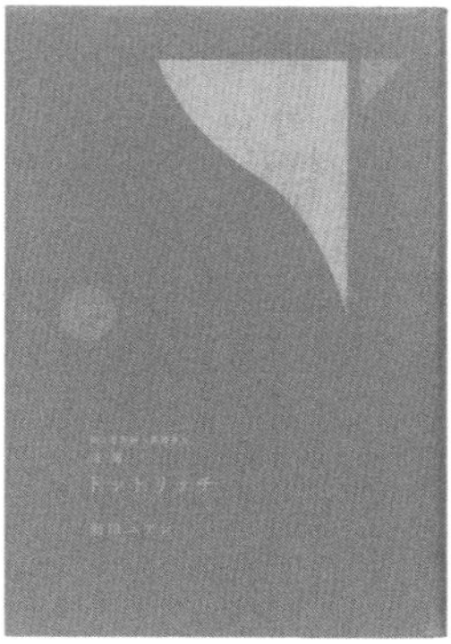

This book is a collection of poems by Yuan Okada, the winner of 'Japanese New Poetry and Thoughts Award'. It's the sixth book of the series of 'New Poetry and Thoughts Award'. The book contains 18 pieces of works, covering modern poetry and some poetic prose. The tiny hardcover book is bound with thread sewing. The jacket is printed on warm grey matte art paper in light grey and silver, which forms a subtle contrast of brightness with the paper. The graphic color blocks are deliberately mottled to create subtle details. Only light grey titles are printed on the cold grey matte art paper. The endpaper is made of bright yellow matte offset paper while the title page is printed with silver letters using translucent watermark technique. Together the book is designed to be serene and energetic, just like what 'New Poetry and Thoughts Award' advocates – delicacy and impact. The inner pages are printed on offset paper in black, extremely simple Japanese style typesetting allows the readers to concentrate on the poet's works.

- 荣誉奖
- Tottorich
- 日本

- Honorary Appreciation
- Tottorich
- Japan
- Masahiko Nagasawa
- Yuan Okada
- Doyo Bijutsusha Shuppan Hanbai
- 215×145×11mm
- 273g
- 93p

Minimalism is not a synonym for asceticism. At any rate the features of this quietly-spoken Japanese poetry volume are more elaborate than might be apparent at first glance. The jacket in grey, matt paper with few lines and forms is printed in three colours. The covering of this slim paperback with rounded spine resorts to bluish-grey paper, and its banderole is printed with extensive use of yet another nuance of grey. Yellow endpapers and a yellow headband shine like bright light through a grey sky. And before the vertical Japanese verses like hanging chains cast spells of airy veils onto the pages there are title lines printed in grey on a specially glued paper of individual quality – and with an ornamental stroke using watermark technique. Not being able to read Japanese need not be a hindrance to an aesthetic appreciation of its typography. In fact then it is even more noticeable that there are no gaps on the surface of the paper, but rather a field of serene nuances.

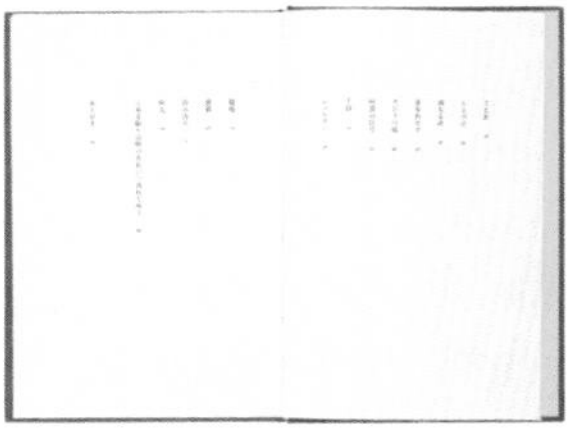

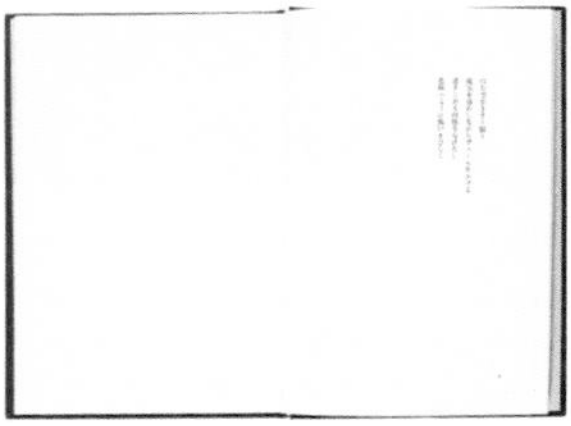

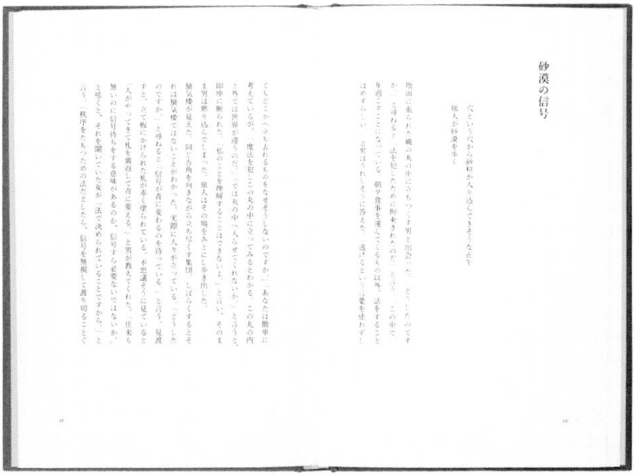
砂漠の信号
穴という穴から砂粒が入り込んできそうな正午
旅人が砂漠を歩く

ようか

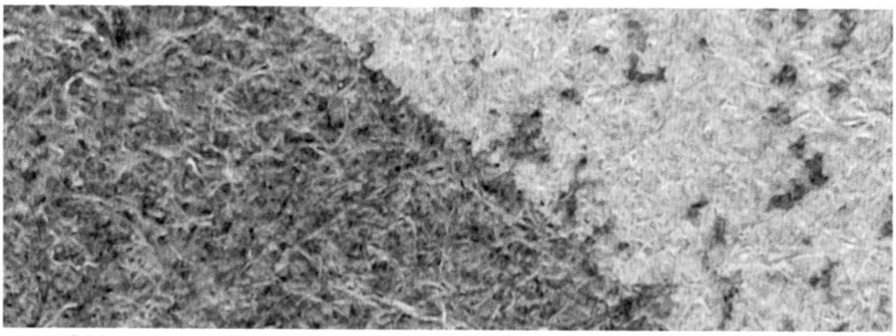

ニ　が　じ

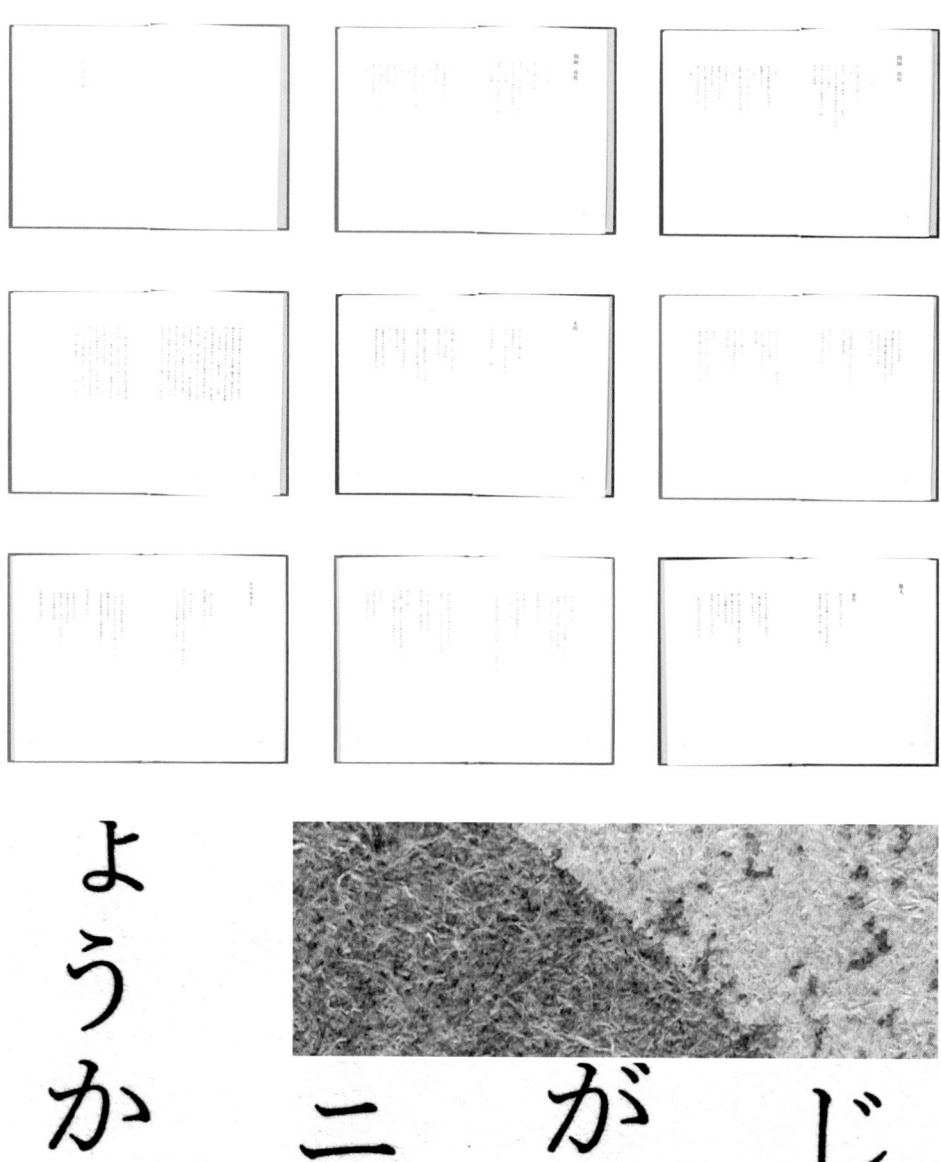

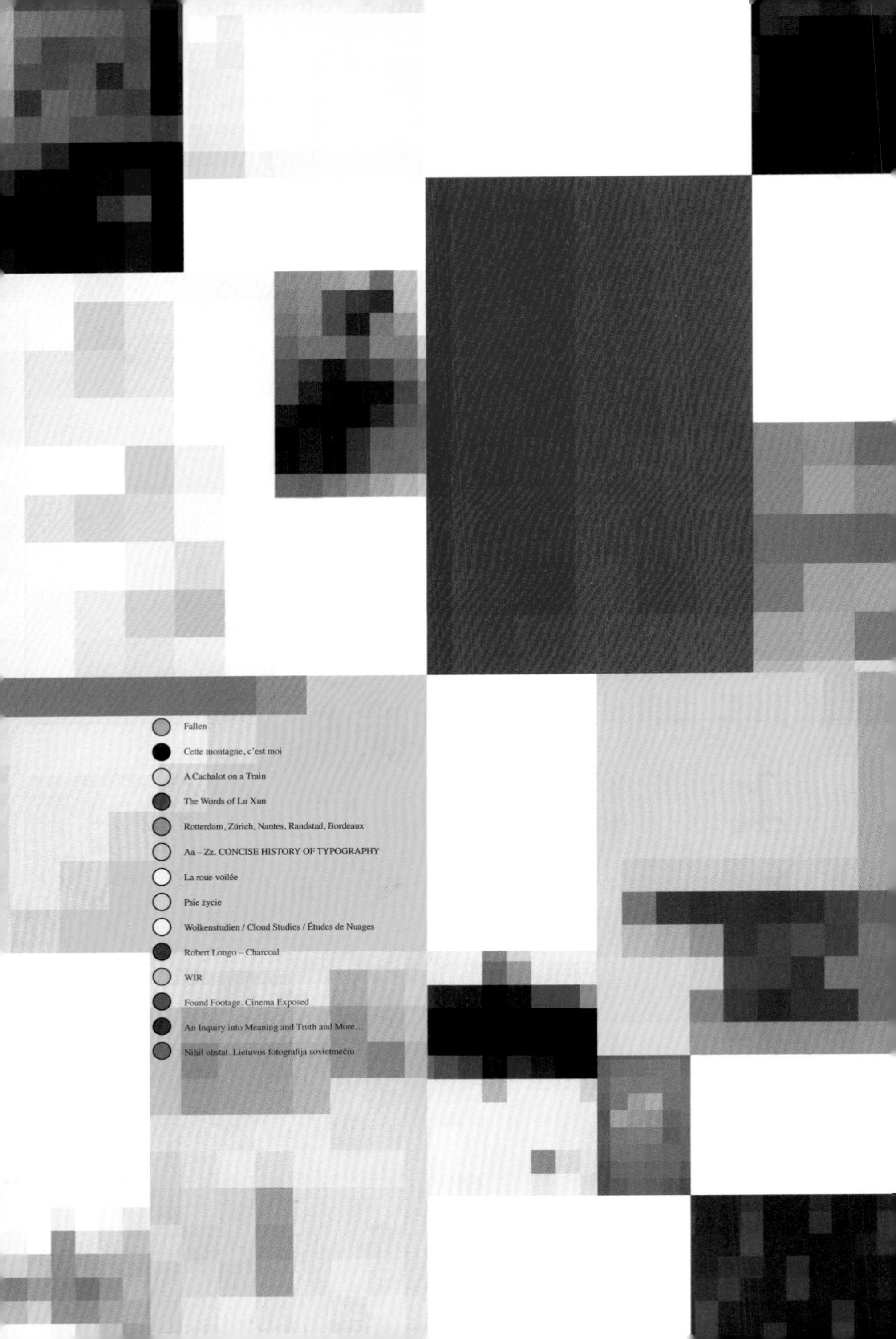
Fallen
Cette montagne, c'est moi
A Cachalot on a Train
The Words of Lu Xun
Rotterdam, Zürich, Nantes, Randstad, Bordeaux
Aa – Zz. CONCISE HISTORY OF TYPOGRAPHY
La roue voilée
Psie życie
Wolkenstudien / Cloud Studies / Études de Nuages
Robert Longo – Charcoal
WIR
Found Footage. Cinema Exposed
An Inquiry into Meaning and Truth and More…
Nihil obstat. Lietuvos fotografija sovietmečiu

2013

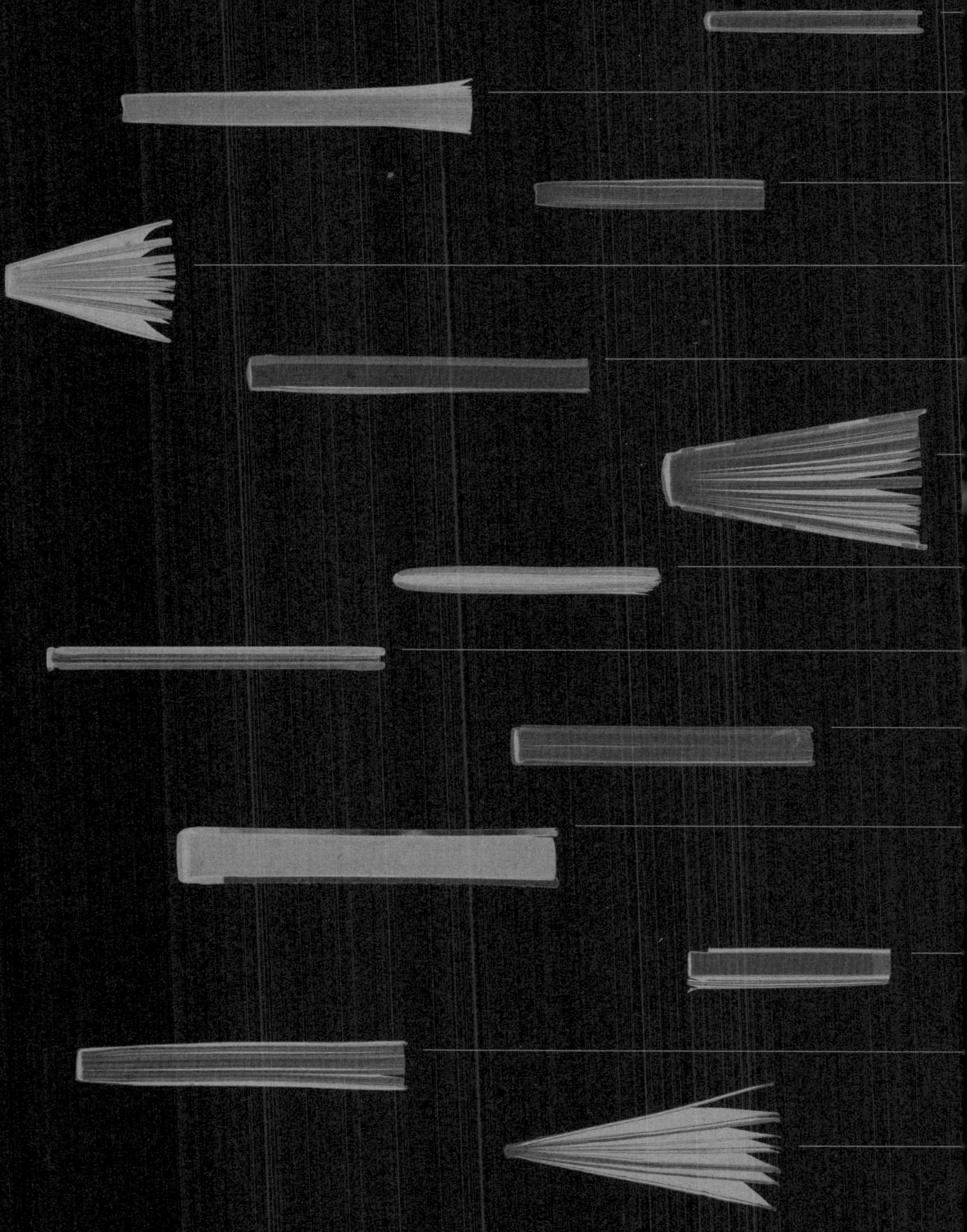

2013 Gl 0631

2013 Gm 0639

MONTAGNE

2013 Sm[1] 0645

2013 Sm[2] 0651

2013 Bm[1] 0657

Changing Cultures of Planning

2013 Bm[2] 0663

PUMP

2013 Bm[3] 0669
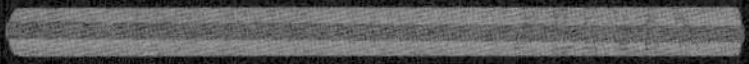

2013 Bm[4] 0675

PSIE ŻYCIE

2013 Bm[5] 0681

WOLKENSTUDIEN CLOUD STUDIES ETUDES DES NUAGES

2013 Ha[1] 0687

2013 Ha[2] 0691

2013 Ha[3] 0695

Found Footage
Cinema Exposed

2013 Ha[4] 0699

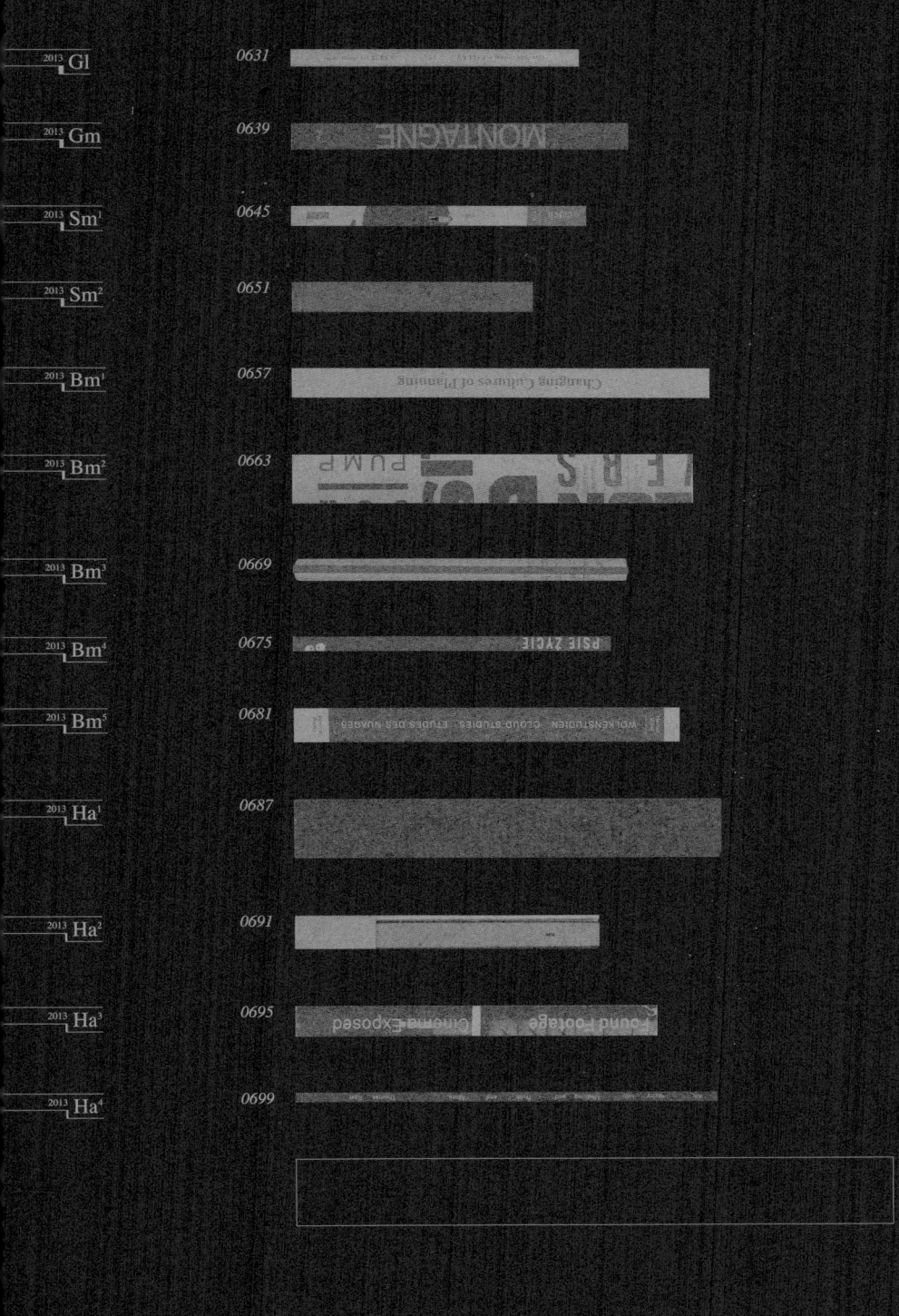

N-401
Madrid
Ciudad Real
Politi

that must have been ... her large feet into high ... precious parts. One of ... neighbourhood: the w... she had begun to take ... her falling incident ... Lady remembered ... g-dog ran hither ... orderly

out attracting their at... They don't like ... ning after

Pour les anciens mineurs comme mon grand-père, les terr
les témoins de promenades nostalgiques – chacun avait é
montés de la fosse où ils étaient descendus –, pour les en

諷刺的是，它的沒落，起因於東港的走紅。觀光局推廣鮪魚季，加上

王船祭，東港在地方的努力經營下，五六月成了觀光的好去處。阿芳

The brief distributed to the five architects
invited to participate in the *AIR: de Kop van Zuid* festival demonstrates a cultural

Kop van Zuid,
arouse interes
planning amo

Erinevaid kanaleid pidi ostsid nii Tallinna gümnaasiumi kui ka Tartu ülikooli trükikoda iga kolme kuni viie aasta tagant juurde uusi kirjagar-

puhul kasutanud Fra
Guyot (u 1510–1570)
24-aastaselt Antwer

qu'il entend les ondes
il y a un truc physique

Kiedy rozdzielą nas setki mil,
powrócę do ciebie jak Lessie.

gesehen

änderten. Mehrere technische Hind
stellten sich Riggenbach jedoch in d

There is an ambiguity in some of the images that suggests an ambival
that is most charged in a close-up of the peephole on the apartment door. To
this drawing seems to ask, is psychoanalysis a kind of peering, even peepin

und gelange zu einem höchst merkwürdigen Ergebnis.
Gleichung, jeder geometrischen Figur entspricht eine
Linie oder ein Körper. Für die irrationalen Formeln, für

. They don't talk
erground any-
y say, 'Well, this

I mean, there are ima
all over. Walk down
the street, there are

other side. In this reading, a cover has a meta-substantive relati
its book, producing a commentary and thus deeper understar
of what it stands for and unfolds in its own nature. With his sl

Golden Letter
Fallen
Germany
Hans-Jörg Pochmann
Gian-Philip Andreas, Gesine Palmer
Eigenverlag Hans-Jörg Pochmann, Leipzig
205×130×12mm
190g
72p

2013 Gl

Gold Medal
Cette montagne, c'est moi
The Netherlands
Hans Gremmen
Witho Worms
Fw: Books, Amsterdam
240×222×17mm
813g
175p

2013 Gm

Silver Medal
The Words of Lu Xun
Japan
Kenya Hara / Tei Rei
Lu Xun
Heibonsha Limited Publishers, SDX Joint Publishing Company
172×113×20mm
285g
259p

2013 Sm2

Silver Medal
A Cachalot on a Train
Taiwan, China
NOBU
Wang, Yen-Kai
Locus Publishing Company Limited
210×150×13mm
400g
173p

2013 Sm1

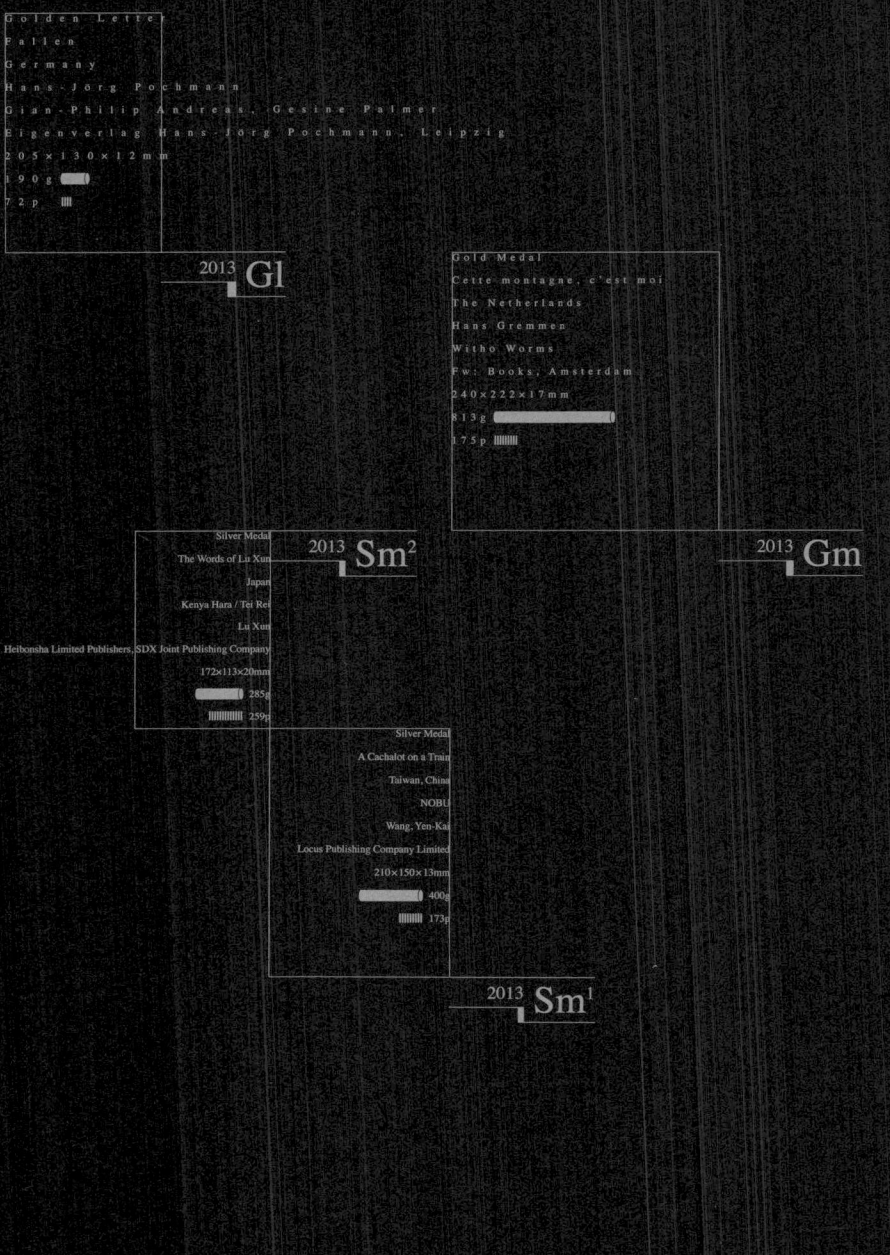

2013 Ha[1]

Honorary Appreciation

Robert Longo – Charcoal

Germany

Stapelberg & Fritz, Stuttgart

Hal Foster, Kate Fowle, Thomas Kellein

Hatje Cantz Verlag GmbH, Ostfildern

304×259×33mm

2114g

251p

2013 Ha[2]

Honorary Appreciation

WIR

Germany

Gaston Isoz, Berlin

Jewgenij Samjatin

disadorno edition, Berlin

217×120×22mm

450g

215+19p

2013 Ha[4]

Honorary Appreciation

An Inquiry into Meaning and Truth and More…

The Netherlands

Edwin van Gelder / Mainstudio

Thomas Raat

Onomatopee, Eindhoven

300×200×7mm

330g

80p

2013 Ha[3]

Honorary Appreciation

Found Footage. Cinema Exposed

The Netherlands

Joseph Plateau, grafisch ontwerpers

Marente Bloemheuvel, Giovanna Fossati, Jaap Guldemond (Hrsg.)

Amsterdam University Press, EYE Film Institute Netherlands, Amsterdam

258×209×18mm

438g

255p

2013 Ha[5]

2013 Bm[1]

Bronze Medal

Changing Cultures of Planning – Rotterdam, Zurich, Nantes, Randstad, Bordeaux

Belgium

Joris Kritis with Rustan Söderling

Nathanaëlle Baës-Cantillon, Joachim Declerck, Michiel Dehaene, Sarah Levy

Architecture Workroom Brussels

297×211×21mm

1075g

302p

2013 Bm[2]

Bronze Medal

Aa – Zz. CONCISE HISTO-RY OF TYPOGRAPHY

Estonia

Ivar Sakk

Ivar Sakk

Sakk&Sakk

285×178×38mm

1062g

447p

2013 Bm[3]

Bronze Medal

La roue voilée

Belgium

PLMD (pleaseletmedesign) + interns: Audé Gravé, Rosalie Wagner

David Widart

L'Amicale Books

239×170×6mm

174g

84p

2013 Bm[5]

Bronze Medal

Wolkenstudien / Cloud Studies / Études de Nuages

Germany

Helmut Völter, Leipzig

Marcel Beyer, Helmut Völter (Hrsg.)

Spector Books, Leipzig

275×206×25mm

1075g

272p

2013 Bm[4]

Bronze Medal

Psie życie

Poland

Józef Wilkoń and Piotr Gil

Józef Wilkoń

Wydawnictwo HOKUS-POKUS Marta Lipczyńska-Gil

226×205×7mm

345g

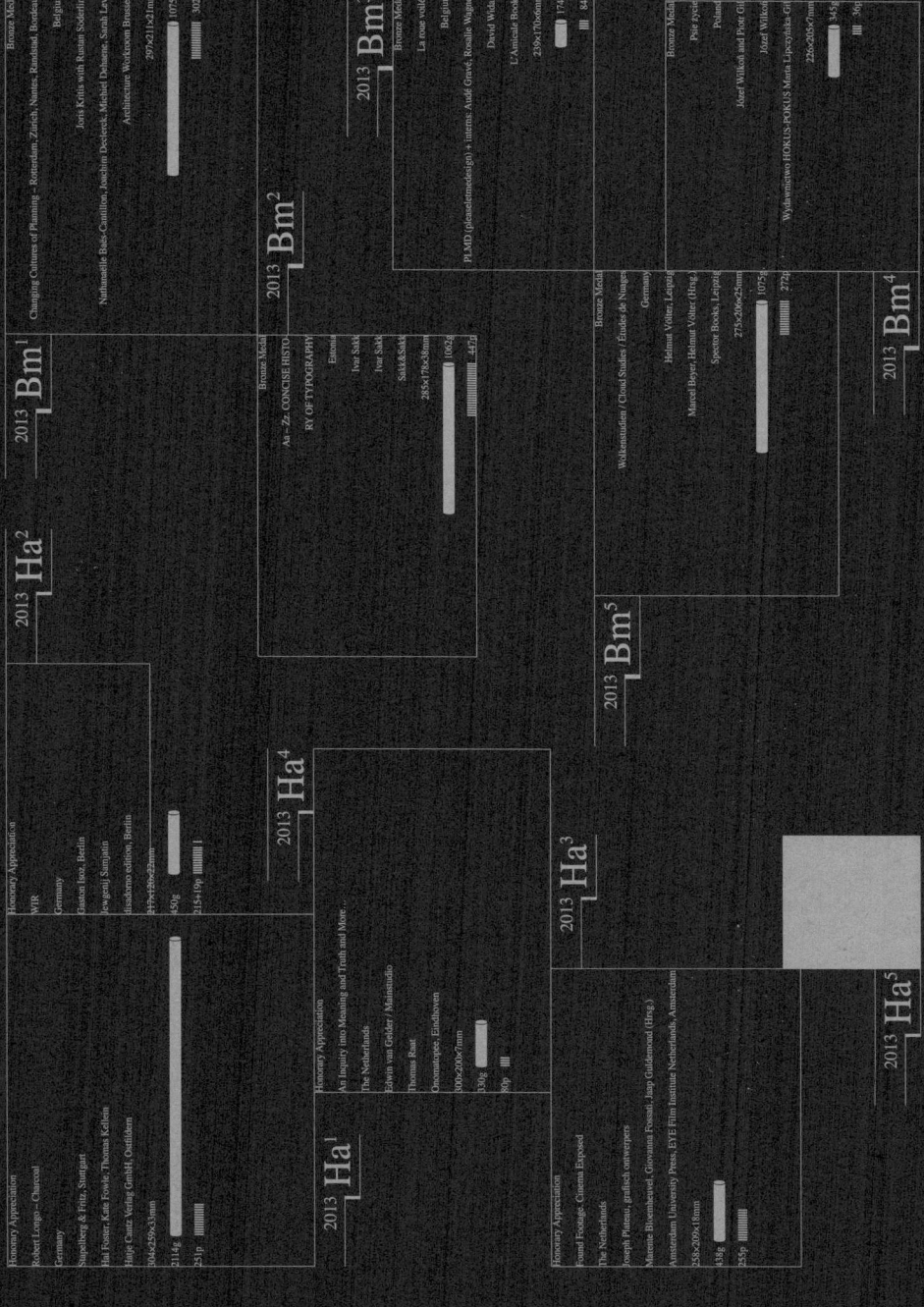

36p

The International Jury

Dr. Aladdin Jokhosha (Germany, Iraq)

Gabriele Lenz (Austria)

Tomas Mrazauskas (Germany, Lithuania)

Prof. Felix Scheinberger (Germany)

Roland Stieger (Schwitzerland)

Prof. Wim Westerveld (Germany)

Susanne Zippel (China, Germany)

2013

Country / Region

Germany 4

The Netherlands 3

Belgium 2

Taiwan, China 1

Japan 1

Estonia 1

Poland 1

Lithuania 1

Designer

Hans-Jörg Pochmann

Hans Gremmen

NOBU

Kenya Hara / Tei Rei

Joris Kritis with Rustan Söderling

Ivar Sakk

PLMD (pleaseletmedesign) + interns: Audé Gravé, Rosalie Wagner

Józef Wilkoń and Piotr Gil

Helmut Völter, Leipzig

Stapelberg & Fritz, Stuttgart

Gaston Isoz, Berlin

Joseph Plateau, grafisch ontwerpers

Edwin van Gelder / Mainstudio

Tomas Mrazauskas

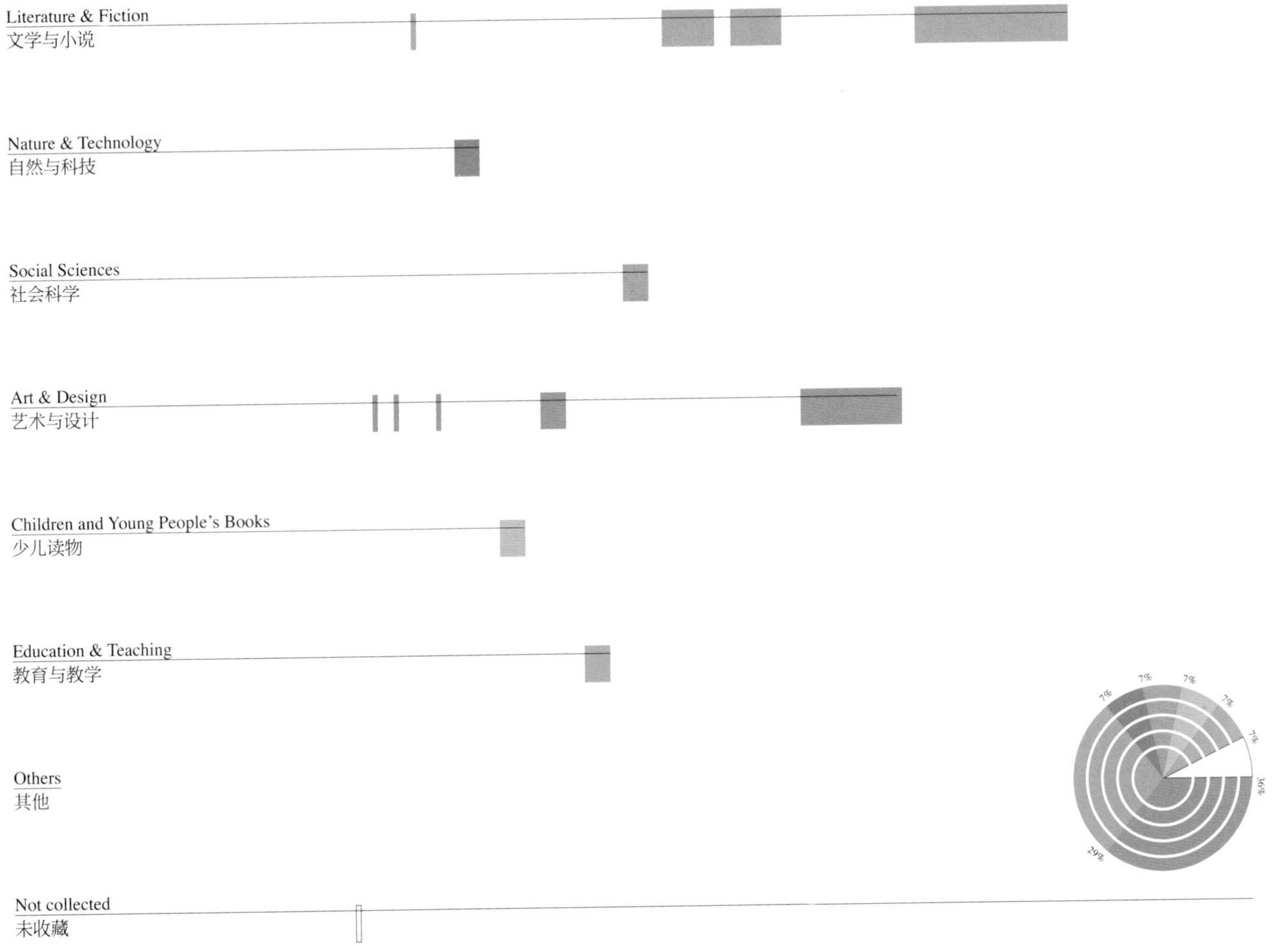
Literature & Fiction
文学与小说
Nature & Technology
自然与科技
Social Sciences
社会科学
Art & Design
艺术与设计
Children and Young People's Books
少儿读物
Education & Teaching
教育与教学
Others
其他
Not collected
未收藏
36%
29%
7%
7%
7%
7%
7%

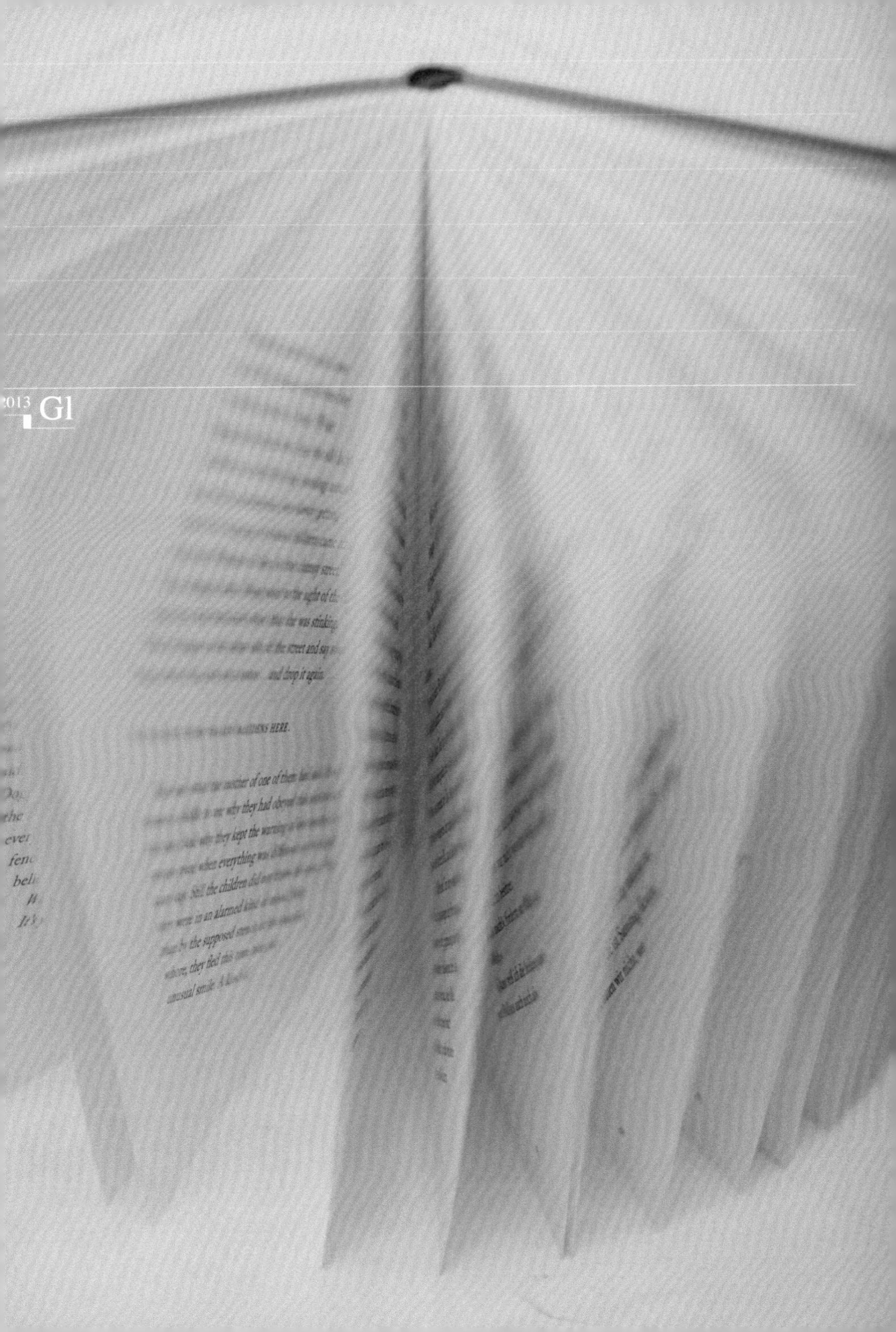
and drop it again.

这是一本两个作者分别写作有关坠落主题的合集。从社会到习俗，从生活到现实，无论什么身份的人，都无法否认坠落是一种真实存在的体验。人们在失重当中，体会到的是一种无穷尽的未知与虚无。本书的装订方式为硬精装封面的无线胶装，书脊处的软胶和本身不多的页数可以保证本书可以很好地展开。护封胶版纸绿黑双色印刷，其中黑色照片为粗网处理。内封蓝绿色布纹纸烫白工艺，裱覆卡板。内页胶版纸单黑印刷。本书没有绝对的封面封底，两位作者的文字分别从封一和封四进入，分别为德文和英文写成的文字，只排在当前翻开页面的右页，所以只有奇数页码。文字排版每两页间形成约 5.8° 的逆时针倾斜，直到最后一页完成 180° 旋转，进而接另一篇文章，从而形成循环往复的螺旋结构。阅读过程中，每读一页需要顺时针旋转书本，在此过程中体验“坠落”的失重感。

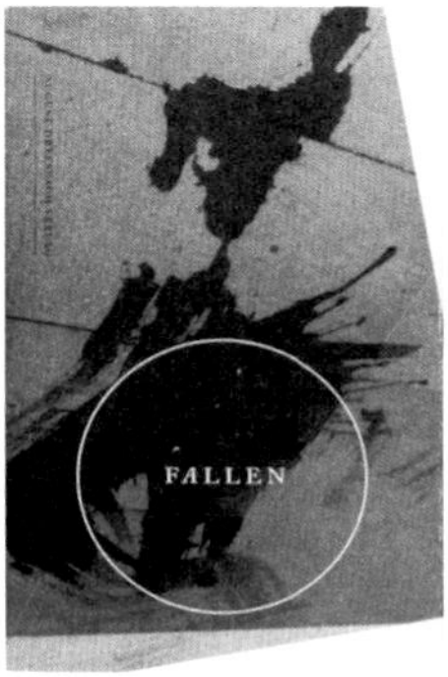
FALLEN

This is a collection of two authors who wrote about 'fallen'. From society to custom, from life to reality, people cannot deny that falling is a real experience no matter what the status is. In weightlessness, people experience endless unknown and nihility. The hardcover book is perfect bound. The soft glue used on the thin spine ensures that the book can be fully opened. The jacket is made of offset paper in green and black, on which the black photograph are treated with rough screen printing. The inside cover is made of blue-green cloth paper, hot stamped in white and mounted with graphic board. Offset sheet printed in black is used for the inside pages. Strictly speaking, there is no definite cover and back cover because the words by two authors are printed on both front cover and back cover, written in German and English. The texts are only typeset on the right pages, so there are only odd page numbers in the book. Between two pages, the text layout forms a counterclockwise tilt of about 5.8 degrees until 180 degrees of rotation is completed, and then followed by another article, thus forming a circular spiral structure. Every two pages of text tilts counterclockwise by about 5.8 degrees until it completes a 180-degree rotation, followed by another article, creating a circular spiral. During the process of reading, readers need to rotate the book clockwise to experience the 'fallen' weightlessness.

金字符奖
坠落
德国

Golden Letter
Fallen
Germany
Hans-Jörg Pochmann
Gian-Philip Andreas, Gesine Palmer
Eigenverlag Hans-Jörg Pochmann, Leipzig
205 × 130 × 12mm
190g
72p

A book with two front pages but no back one. Two stories – one in German, one in English – both to be read through to the end of the book. The homographic titles "Fallen" differ only in tense – the present in German, the perfect in English. And precisely this is what provides the book designer with the motif for a typographical performance of the content. Throughout the book the text is rotated within the type-area through 180 degrees, flush left ragged right remaining order of the day. You start reading on a right-hand page, leaf forward rotating the book a little bit more each time you turn the page until you end up on a left-hand page. By now the reader has just about stumbled through to the end of one of the stories with a slight sense of disorientation, only for the dizzying experience to be continued with the same rotational game for the other story – this time in the opposite direction.

es herrschte Frieden
am Mount Carmel. Seine Frauen hatten
geboren und taten es weiterhin, sein Lächeln richtete über
die Auseinandersetzungen des Alltags. Er leitete das Gebet und rief seinen Jüngern
immer wieder ins Gedächtnis, worauf sie warteten: auf die Wiederkehr. Auf das
Ende und den Neuanfang. Sie alle blieben heiter.

Vernon zog sich aus und stieg ohne eine Sekunde des Zögerns ins kalte Wasser. Sofort tauchte er den gesamten Körper unter und schwamm los, mehrere Bahnen hin und her, in kräftigen, gleichmäßig ausgeführten Zügen. Bald schon spürte er die Kälte nicht mehr, eine große Ruhe durchströmte ihn.

Cyrus und Joe kamen nun über die Veranda gerannt und stellten sich ans westliche Ende des Pools: »Wie lange hältst du es denn aus im Eiswasser?«

»Sehr, sehr lange«, entgegnete Vernon mit gespielt prophetischer Stimme und schwamm dabei unbeirrt weiter.

»Das glauben wir nicht«, meinte Joe.

»Das Wasser ist gar nicht so kalt«, log Vernon. »Wenn ihr beiden mutig seid, dann springt ihr

GESINE PALMER
We do not accept fallen maidens here!
FALLEN

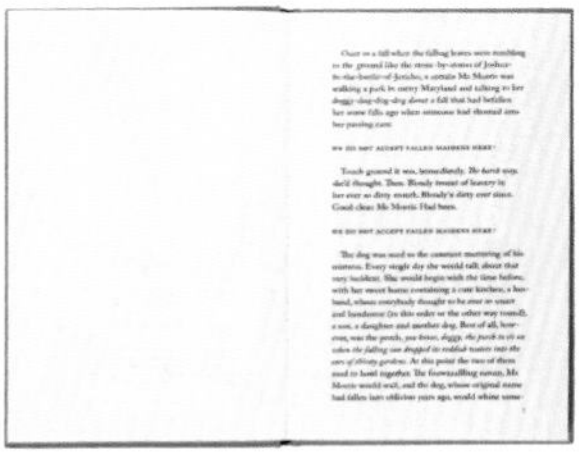

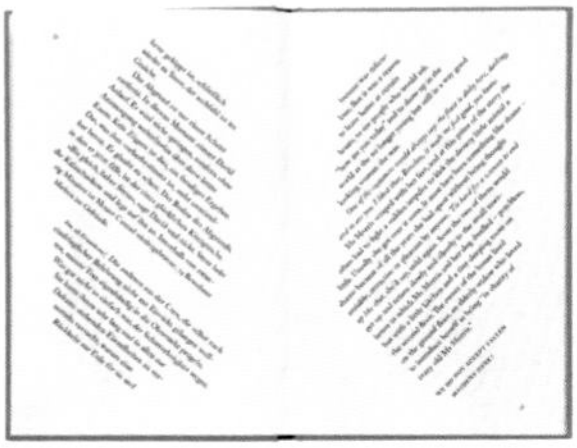

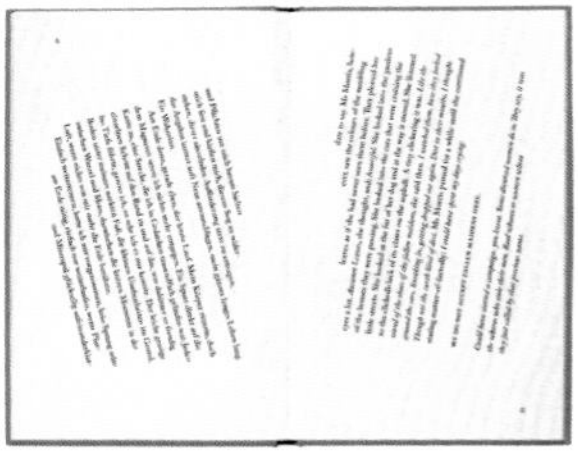

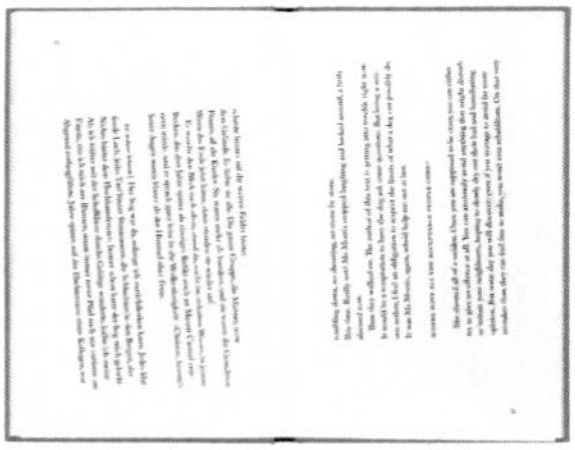

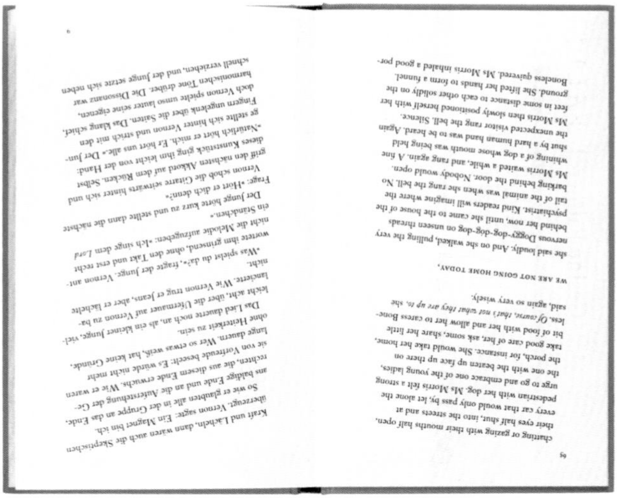
65
chatting or gazing with their mouths half open, their eyes half shut, into the streets and at every car that would only pass by, let alone the pedestrian with her dog. Ms Morris felt a strong urge to go and embrace one of the young ladies, the one with the beaten up face up there on the porch, for instance. She would take her home, take good care of her, ask some, share her little bit of food with her and allow her to caress Boneless. Of course, that's not what they are up to, she said, again so very wisely.
WE ARE NOT GOING HOME TODAY,
she said loudly. And on she walked, pulling the very nervous Doggy-dog-dog-dog on unseen threads behind her now, until she came to the house of the psychiatrist. Kind readers will imagine where the tail of the animal was when she rang the bell. No barking behind the door. Nobody would open. Ms Morris waited a while, and rang again. A fine whining of a dog whose mouth was being held shut by a hard human hand was to be heard. Again the unexpected visitor rang the bell. Silence. Ms Morris then slowly positioned herself with her feet in some distance to each other solidly on the ground. She lifted her hands to form a funnel. Boneless quivered. Ms Morris inhaled a good por-
Kraft und Lächeln, dann wären auch die Skeptischen überzeugt. Vernon sagte: Ein Magnet bin ich.
So wie er glaubten alle in der Gruppe an das Ende, ans baldige Ende und an die Auferstehung der Gerechten, die aus diesem Ende erwuchs. Wie er waren sie von Vorfreude beseelt: Es würde nicht mehr lange dauern. Wer so etwas weiß, hat keine Gründe, ohne Heiterkeit zu sein.
Das Lied dauerte noch an, als ein kleiner Junge, vielleicht acht, über die Ufermauer auf Vernon zu balancierte. Wie Vernon trug er Jeans, aber er lächelte nicht.
»Was spielst du da?«, fragte der Junge. Vernon antwortete ihm grinsend, ohne den Takt und erst recht nicht die Melodie aufzugeben: »Ich singe dem Lord ein Ständchen.«
Der Junge hörte kurz zu und stellte dann die nächste Frage: »Hört er dich denn?«
Vernon schob die Gitarre seitwärts hinter sich und griff den nächsten Akkord auf dem Rücken. Selbst dieses Kunststück ging ihm leicht von der Hand: »Natürlich hört er mich. Er hört uns alle.« Der Junge stellte sich hinter Vernon und strich mit den Fingern ungelenk über die Saiten. Das Klang schief, doch Vernon spielte umso lauter seine eigenen, harmonischen Töne drüber. Die Dissonanz war schnell verziehen, und der Junge setzte sich neben
6

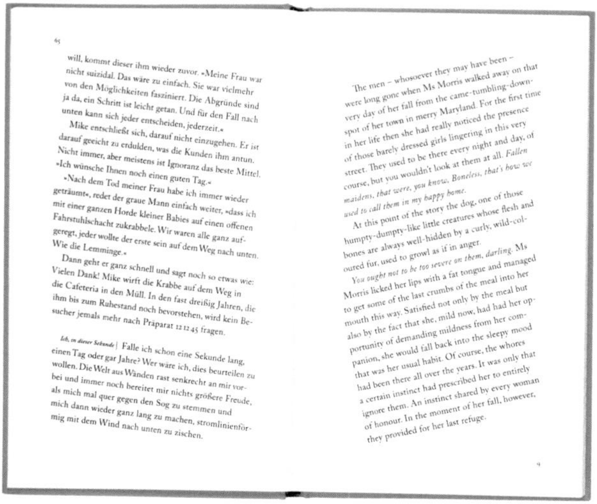

65

will, kommt dieser ihm wieder zuvor. »Meine Frau war nicht suizidal. Das wäre zu einfach. Sie war vielmehr von den Möglichkeiten fasziniert. Die Abgründe sind ja da, ein Schritt ist leicht getan. Und für den Fall nach unten kann sich jeder entscheiden, jederzeit.«

Mike entschließt sich, darauf nicht einzugehen. Er ist darauf geeicht zu erdulden, was die Kunden ihm antun. Nicht immer, aber meistens ist Ignoranz das beste Mittel. »Ich wünsche Ihnen noch einen guten Tag.«

»Nach dem Tod meiner Frau habe ich immer wieder geträumt«, redet der graue Mann einfach weiter, »dass ich mit einer ganzen Horde kleiner Babies auf einen offenen Fahrstuhlschacht zukrabbele. Wir waren alle ganz aufgeregt, jeder wollte der erste sein auf dem Weg nach unten. Wie die Lemminge.«

Dann geht er ganz schnell und sagt noch so etwas wie: Vielen Dank! Mike wirft die Krabbe auf dem Weg in die Cafeteria in den Müll. In den fast dreißig Jahren, die ihm bis zum Ruhestand noch bevorstehen, wird kein Besucher jemals mehr nach Präparat 12 12 45 fragen.

Ich, in dieser Sekunde | Falle ich schon eine Sekunde lang, einen Tag oder gar Jahre? Wer wäre ich, dies beurteilen zu wollen. Die Welt aus Wänden rast senkrecht an mir vorbei und immer noch bereitet mir nichts größere Freude, als mich mal quer gegen den Sog zu stemmen und mich dann wieder ganz lang zu machen, stromlinienförmig mit dem Wind nach unten zu zischen.

The men – whosoever they may have been – were long gone when Ms Morris walked away on that very day of her fall from the came-tumbling-down-spot of her town in merry Maryland. For the first time in her life then she had really noticed the presence of those barely dressed girls lingering in this very street. They used to be there every night and day, of course, but you wouldn't look at them at all. *Fallen maidens, that were, you know. Boneless, that's how we used to call them in my happy home.*

At this point of the story the dog, one of those humpty-dumpty-like little creatures whose flesh and bones are always well-hidden by a curly, wild-coloured fur, used to growl as if in anger.

You ought not to be too severe on them, darling. Ms Morris licked her lips with a fat tongue and managed to get some of the last crumbs of the meal into her mouth this way. Satisfied not only by the meal but also by the fact that she, mild now, had had her opportunity of demanding mildness from her companion, she would fall back into the sleepy mood that was her usual habit. Of course, the whores had been there all over the years. It was only that a certain instinct had prescribed her to entirely ignore them. An instinct shared by every woman of honour. In the moment of her fall, however, they provided for her last refuge.

9

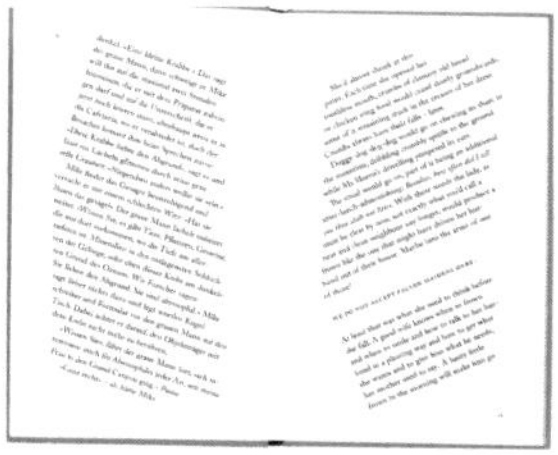

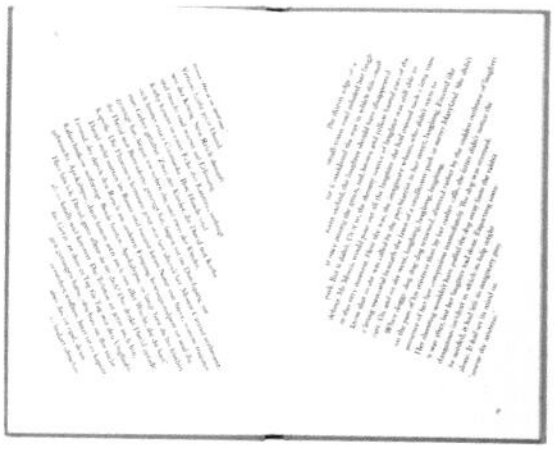

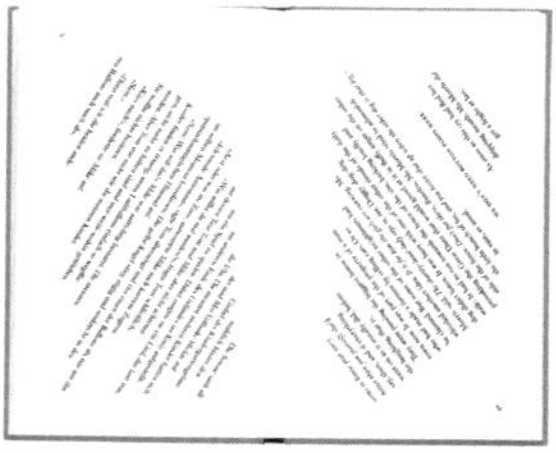

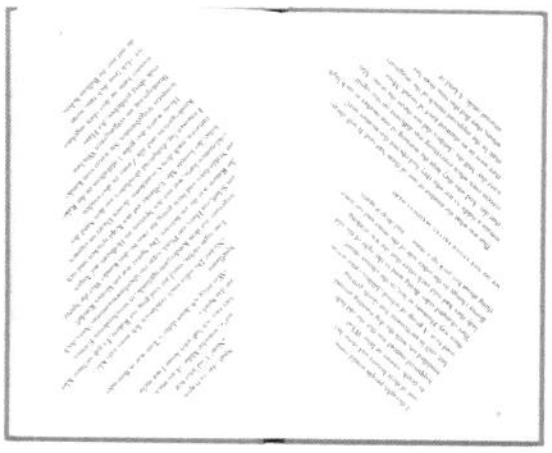

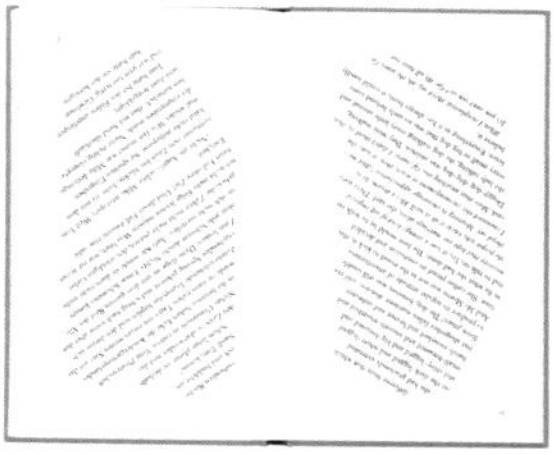

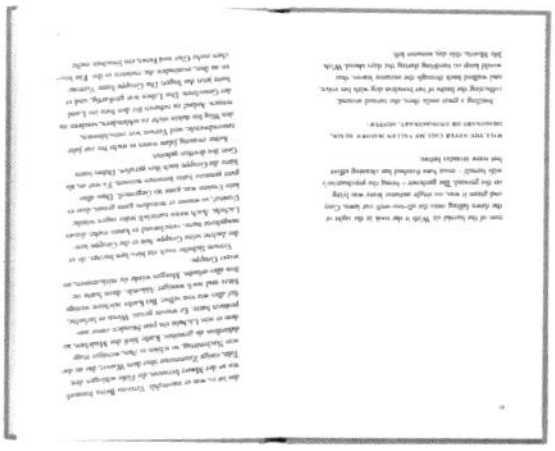

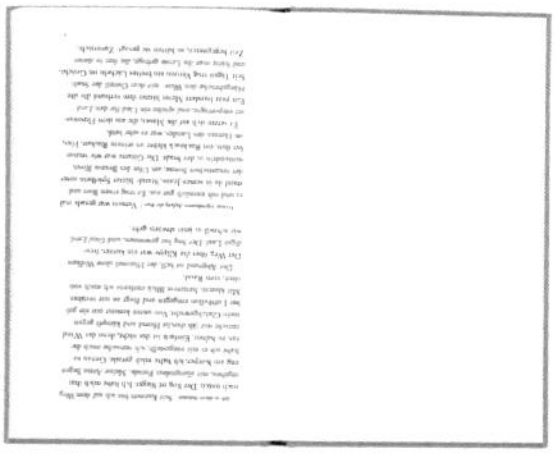

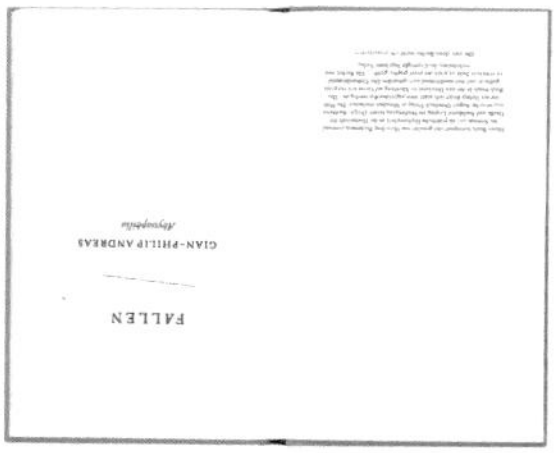

205×130×12

bishe
machei
Moment
sel und Ge
Gebühr strap
Links von i
sie in die Kapsel
ist längst getrennt,
telbar vor ihr zeigt si
und diffus im Schwarz
mehr von ihr wegfällt. E
Bedeutungslosigkeit, er
Sie aber bleibt regung
jeden Gegenstand gerät
bei sich, nichts befindet
Schub und keine Brems
Es gibt nur sie, Tom, in
Raumfähre fällt weg von
wie ein Sandkorn, und a
Das ihn umgebende Sch
Screen.
Hier will ich bleiben,
bricht in ein lautes Lach
aus. Scherzhaft spricht s
es in das längst

190

ann stünden sie wied
ben, stand da, nackt
als einziges Relikt n
nz leise in die Wolke
der Himmel über T

var da, solange ich zu
Staumauern, die Schl
enster: Immer schon
asse durchs Gebirge
en, wann immer uns
später, auf der Dacht

72

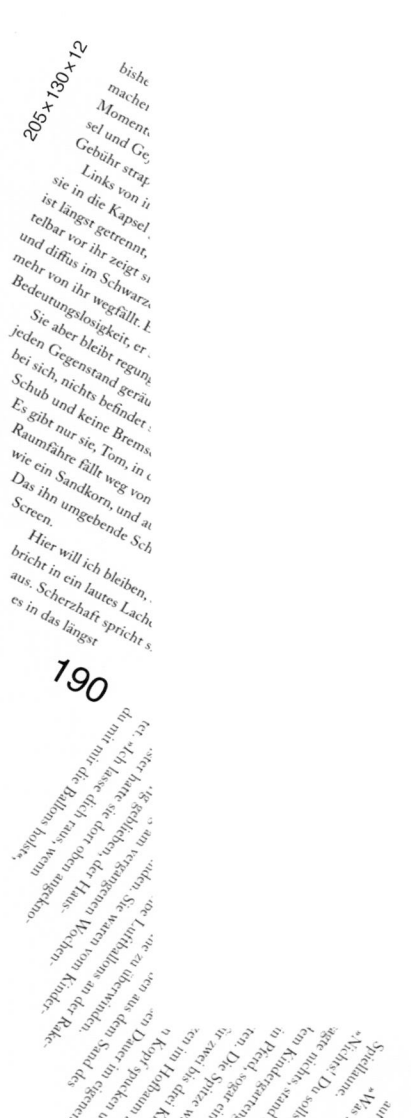

called her particular craziness,
she would be tolerated. People would offer
her some charity, alms, material gifts and almswords of
loving kindness, a bit condescending, maybe, but not really violent,
threatening or expiring.
At this day, however, all the instincts of Ms Morris that had caused her to
carefully succumb to the established order of alms for surrender, seemed
to fail. She could as well have taken real earth into her mouth again. Yet, the
lady chose to ignore it, delighting in her new liberty, spinning around all
of a sudden and almost singing:

WE DO NO LONGER NEED ACCEPTABLE PEOPLE HERE!

After the third spin she became dizzy and had again to reach out
for a piece of fence. She stumbled, but she didn't
fall this time. Boneless shivered, quivered,

41

013 Gm

本书是一本记录矿山废墟的影集。作者通过对比利时、德国、法国、波兰和威尔士五个国家和地区的现场考察、检测、记录和拍摄，记录下过去 100 年来社会、政治、经济发展下，人与环境之间不断变化的关系。书籍的装订形式为包背锁线装，封面采用黑色卡纸，内页则为黑色胶版纸。全书的印刷效果较为特别，银白色和黑色油墨双色印刷，银白色作为第一套版印刷文字，或者印满版作为照片底色，由于银白色油墨的覆盖力适中，所以照片最亮的部分为银灰色。黑色则作为第二套色叠印在此之上，呈现矿山废墟的照片。全书按照国家和地区分为五个部分，每个部分里的单张照片均未排列在单独一页内，而是横跨对页，或者跨过包背页的翻口，形成连续的照片组图，有如连续的画廊陈列。||||||||||||

CETTE
Witho Worms

This book is an album of mine ruins. Through on-site investigation, testing, recording and filming in Belgium, Germany, France, Poland and Wales, the author recorded the changing relationship between human and environment with the social, political and economic development in the past 100 years. The book is double-leaved bound with thread sewing. The cover is made of black cardboard while the inside pages are made of black offset paper. The printing effect of the book is quite special with silver white and black ink used. Silver white is used for printing the text or the background color of the image pages. Because of the moderate coverage of silver-white ink, the brightest part of the photo is silver gray. Black is superimposed on it as the second set of colors, showing pictures of mine ruins. The book is divided into five parts according to the countries and regions, every single photo in each part is not arranged on a single page, but crossing pages or the edge of the double-leaved book, forming a continuous group like gallery display.

金奖
这座山是我
荷兰

Gold Medal
Cette montagne, c'est moi
The Netherlands
Hans Gremmen
Witho Worms
Fw: Books, Amsterdam
240×222×17mm
813g
175p

What is the matter with these pictures? With contrasts as weak as these, the features of the landscape are only perceptible by paying special attention. The images show coal mounds, slag heaps of the mining industry, documented by the artist after near scientific site inspections in five European countries. This foray is translated into the paged medium by glued Japanese bookbinding, so that the images – printed in steel grey and black on black paper – continue to the next page across the folded leading edge. Gradually, the irritating contrasts take on a deeper meaning: just like the light-absorbing black of the coal, the book – contrary to conventional perceptions – becomes a light-absorbing gallery of photographs. The book elevates the mysterious photographic project into an inimitable object of meditation.

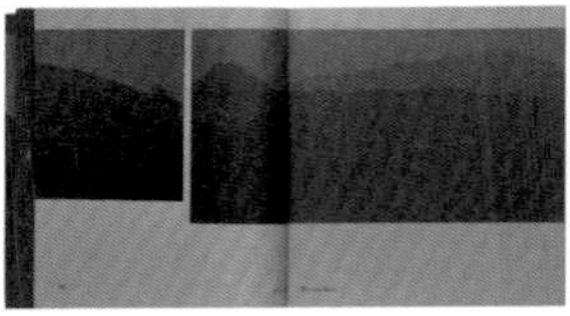
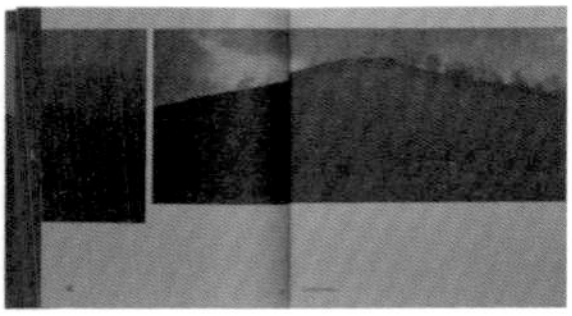
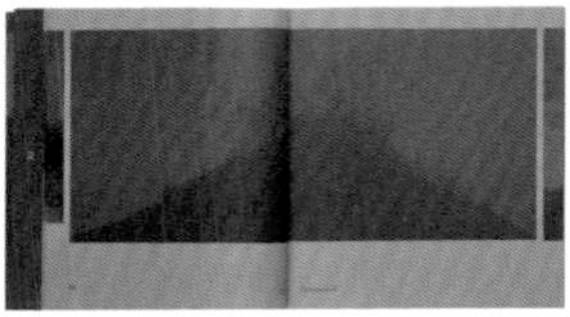

DEUTSCH-
LAND
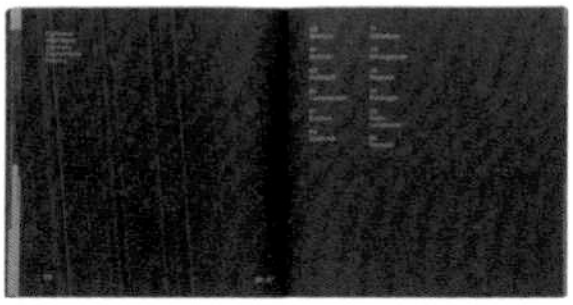
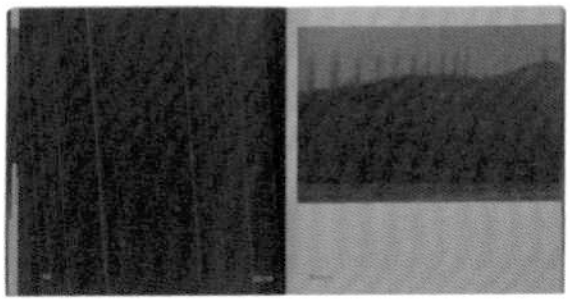

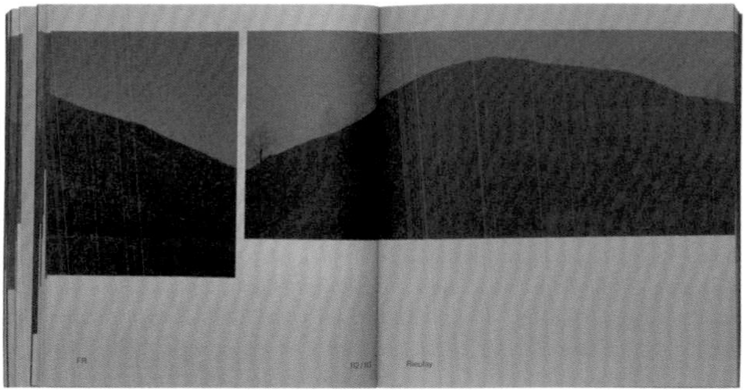

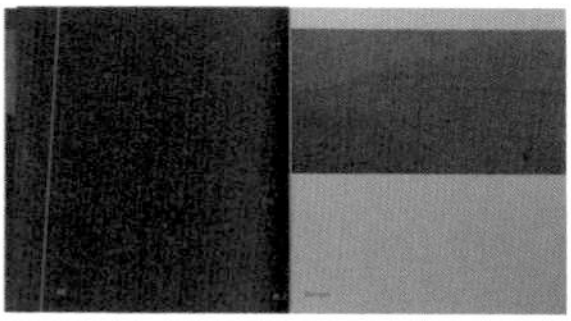
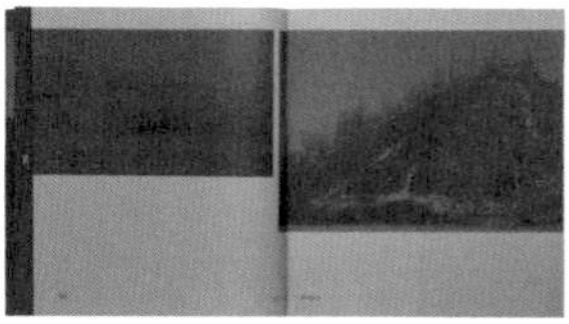

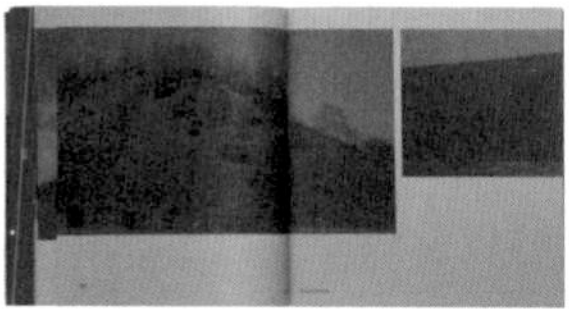
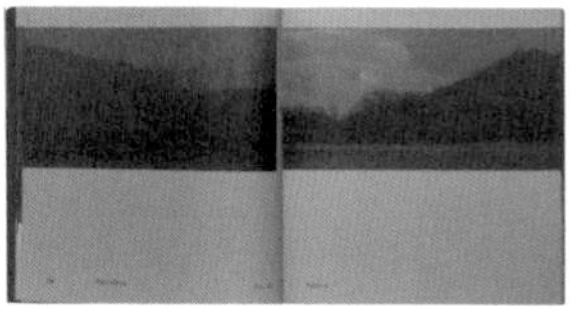
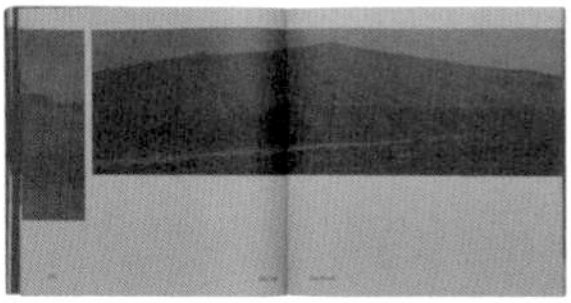
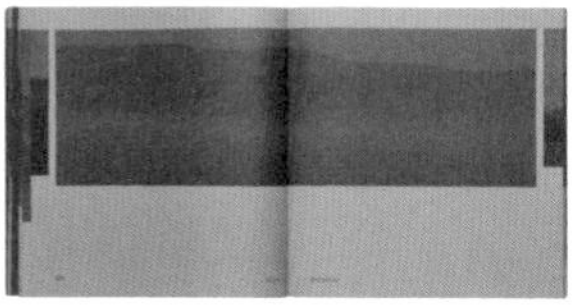
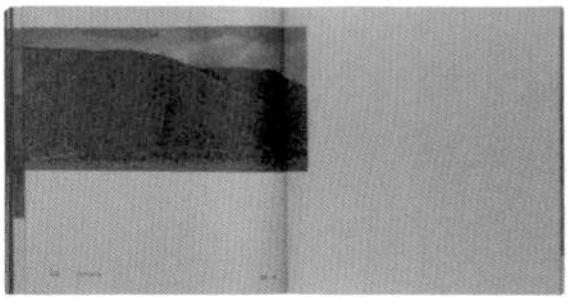
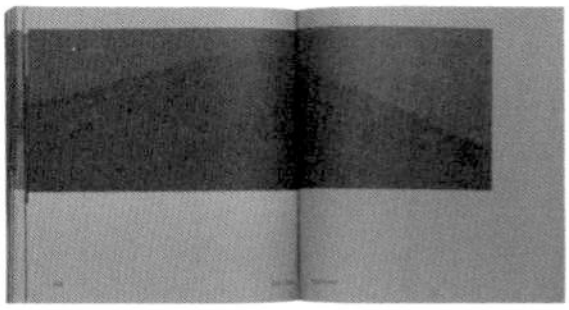

we d
dere helft op de s

MONTAGNE

Fw:

FR

Il y a 350 millions d'années, un paysage, mais ce mot n'existait pas encore, a été compressé en une fine couche de charbon qui a été enterrée sous la surface de la terre. Au cours des siècles derniers, cette couche a été retrouvée et réutilisée par l'homme, et le paysage a ainsi de nouveau changé d'aspect.

Pour le photographe, un terril est un triangle noir dans un cadre rectangulaire. Mais aussi simple la forme soit-elle, sa signification symbolique n'en est que plus complexe. Les terrils renvoient à la révolution industrielle et à la domination occidentale sur l'économie mondiale. Ils renvoient à la pauvreté et aux visages noircis des milliers de mineurs qui ont extrait la prospérité des profondeurs de la terre. À présent que nos économies occidentales vacillent, ces pyramides noires peuvent être considérées comme autant de tertres funéraires d'un système capitaliste pratiquement en faillite.

Les terrils se sont réunis pour constituer le premier paysage culturel post-industriel paneuropéen. Ils sont le résultat des industries lourdes qui ont été à l'origine de la création de la Communauté européenne du charbon et de l'acier, la CECA, le précurseur de l'Union européenne. Les terrils sont d'une importance vitale. Alors que les centrales thermiques au charbon contribuent largement aux rejets de CO_2 et donc aux changements climatiques, ces terrils sont devenus de nouveaux parcs et espaces naturels en Belgique, en France et en Allemagne, des espaces de loisirs et de distraction équipés de pistes de randonnée et de cyclotourisme et même, ici et là, d'un théâtre ou d'une piste de ski.

Dans mon travail photographique, je recherche la correspondance entre le contenu documentaire et l'autonomie de la forme. « Cette montagne, c'est moi » renvoie au processus pénible de la maîtrise de la technique. J'ai construit toutes mes procédures à partir de rien. Le titre renvoie à l'origine humaine du paysage, aux mineurs et à leurs familles qui ont vécu à l'ombre des terrils. Et en troisième lieu, le titre renvoie à la relation philosophique entre représentation et réalité, entre la représentation et le représenté.

Je photographie un paysage d'occasion à la signification recyclée qui illustre la situation actuelle en Europe de l'ouest. Avec mes photos sur le charbon, je dis adieu à une méthode de production industrielle en utilisant une manière de photographier qui appartient à cette même période et qui est, elle aussi, condamnée à disparaître. J'ai considéré les terrils dans une approche fondée sur la qualité du paysage. Je réalise un document social à partir d'un paysage qui est finalement tout aussi anonyme que les mineurs qui l'ont remonté à la surface.

— Witho Worms

Three hundred and fifty million years ago, a landscape (though naturally this word did not yet exist) was compressed into a thin layer of coal and buried beneath the surface of the earth. In recent centuries, it was rediscovered and reused by humans, thereby once again altering the appearance of the landscape.

To a photographer, a coal mountain or slag heap is a black triangle within a rectangular frame. But the simplicity of that shape is matched by the complexity of its symbolic significance. Slag heaps are closely associated with the industrial revolution and Western dominance of the global economy. They are associated with poverty and the black coal-stained faces of the thousands of workers who chipped away deep inside the earth to generate prosperity. And now that our Western economies are in crisis, the black pyramids can be viewed as symbols of the burial mounds of a nearly bankrupt capitalist system.

The mountains together represent the first pan-European post-industrial man-made landscape. They are the result of the heavy industries that gave rise to the establishment of the European Coal and Steel Community, the forerunner of the European Union. Slag heaps are vital. Whereas coal-fired power plants contribute significantly to CO2 emissions and, consequently, climate change, these mountains in Belgium, France and Germany have become 'new nature', converted into parks, places of recreation and inspiration, with footpaths and cycle paths and even occasionally a theatre or ski slope.

In my photographic work I pursue the point of convergence between documentary content and the autonomy of form. Cette montagne c'est moi refers to the challenging process of mastering the technique. I have developed all of my methods from the ground up. The title refers to the human origin of the landscape, to the miners and their families who lived in the shadow of the mountains. Thirdly, it refers to the philosophical relationship between representation and reality, between depicting and depiction.

I have photographed a second-hand landscape with a recycled significance that shows the current state of affairs in Western Europe. My carbon photographs are my way of parting with an industrial production method, and I do so using a photographic method belonging to the same period and destined to be lost. I approached the mountains from a landscape perspective and created a social document of a landscape that is intrinsically as anonymous as the workers who carried it to the surface.

— Witho Worms

EN

这是一本充满诗意的台湾旅行游记。广告人阿铠以 7-11 的推广旅游企划为契机，对台湾的多个城市进行了深度考察，并记录成为散文游记。本书的装订方式为无线胶装。封面白卡纸四色印刷，覆哑膜，书名烫印珠光白。内页哑面涂布纸四色印刷，文字部分单黑，书籍的设计者 NOBU 绘制的插图为四色。与流水账式的游记不同，广告人阿铠以诗意的文字记录下在台湾旅游期间数十个大小城镇的遭逢往事，共 71 篇。让一些平淡无奇的事物变得新鲜起来，进而让读者愿意了解一下这些事物背后的故事。NOBU 的插图线条简洁松弛，配色清新舒朗，与文字呼应，形成了很好的阅读节奏。

銀牌獎

This is a poetic travel notes guide of Taiwan. Advertiser Kai took 7-11's tour promotion as an opportunity to make in-depth investigation of many cities in Taiwan and wrote travel prose. The book is perfect bound. The laminated cover is made of white cardboard printed in four colors with hot-stamped title in pearl white. The inside pages are printed on coated paper in four colors while the text part is monochrome black. The illustrations drawn by NOBU, the designer of the book, are using four-color printing. Different from the travel notes like day-to-day account, advertiser Kai recorded his experiences in Taiwan using poetic words, including 71 articles. He makes ordinary things fresh and encourages the readers to understand the story behind these things. NOBU's illustrations are simple and relaxed with fresh color matching, which perfectly echo the words and form a good reading rhythm.

银奖 — Silver Medal

坐火车的抹香鲸 — A Cachalot on a Train

中国台湾 — China Taiwan

NOBU — NOBU

王彦铠 — Wang, Yen-Kai

大块文化出版股份有限公司 — Locus Publishing Company Limited

210×150×13mm

400g

173p

What calmness and poetry emanates from this book! The compositional strategies of the depicted contemporary illustrations obey the primacy of flat space or spatial flatness stretching between linear drawn structures. The western differentiation between drawing and painting thus grasps at nothing, because in the Far East these non-painterly means acquire an absolute pictorial and magical effect. This authority is also expressed by the language of the text pages: the dominant white spaces are tickled by the filigree gossamer of the Chinese square-block characters. The poetry does not lie in vocalized or depicted items but in the aura of circumscription. This is the true delight offered by this book.

catch
catch your eyes | catch your heart | catch your mind

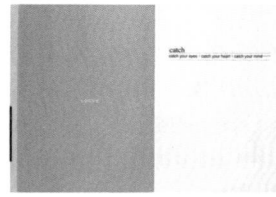

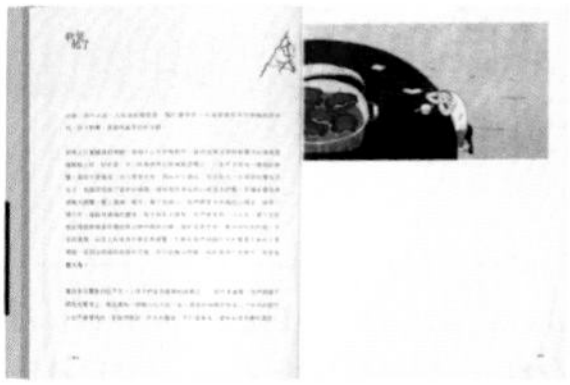

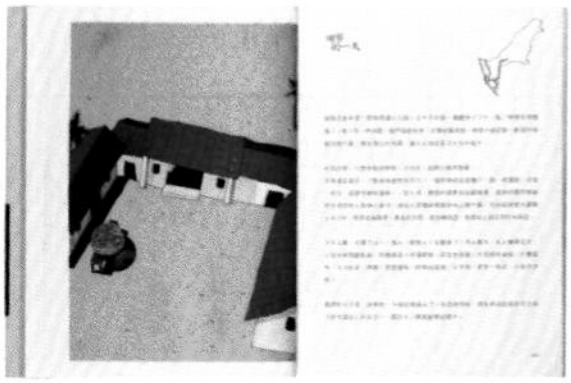

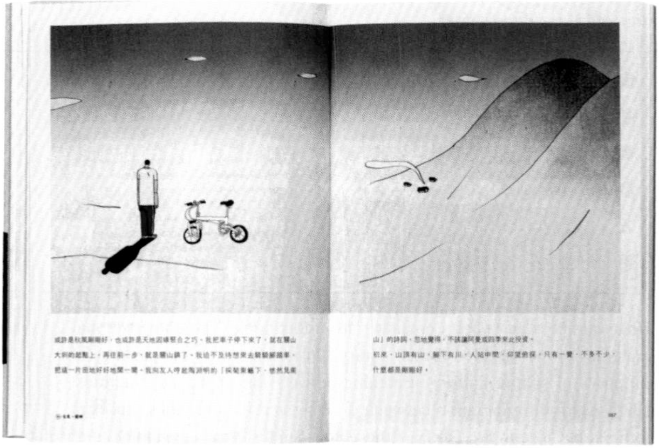
或許是秋風剛剛好，也或許是天地因緣聚合之巧，我把車子停下來了，就在鸞山
大俐的起點上，再往前一步，就是鸞山鎮了。我迫不及待想來去騎騎腳踏車，
把這一片田地好好地聚一聚。我向友人呼起陶淵明的「採菊東籬下，悠然見南
山」的詩詞，忽地覺得，不該讓阿嬤或四季來此投資。
初來，山頂有山，腳下有川，人站中間，仰望俯探，只有一覺，不多不少，
什麼都是剛剛好。

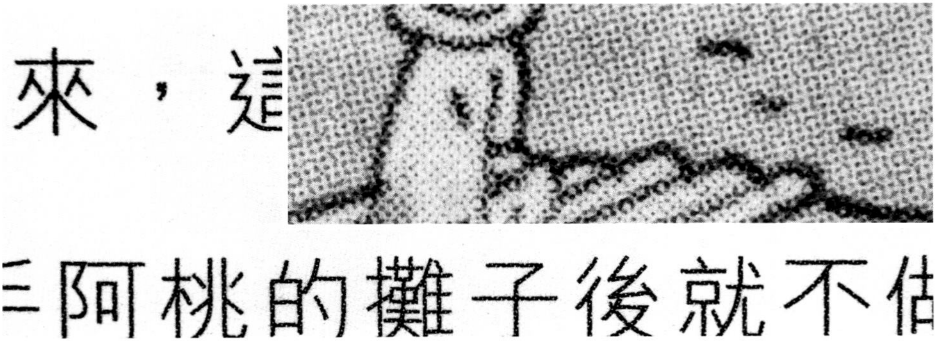

來，
阿桃的攤子後就不

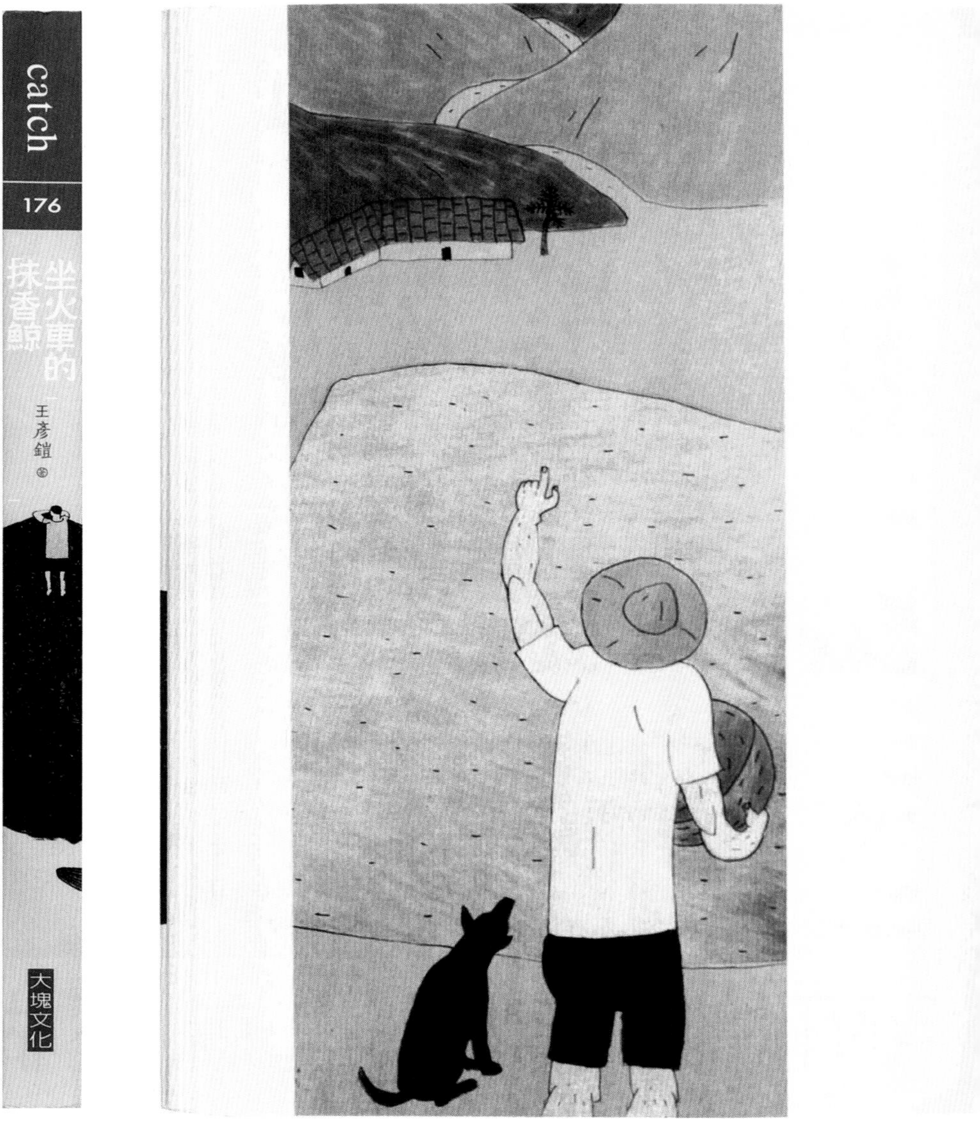
catch
176
坐火車的抹香鯨
王彥鎧 著
大塊文化

向左彎 向右彎

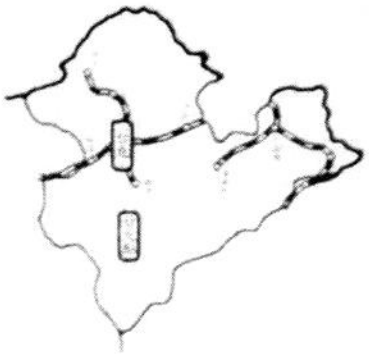

這裡，不種茶的，採茶，不採茶的，賣茶，不賣茶的，至少喝茶。這是坪林鄉，傳說中北宜公路的綠色黃金城，就是這兒。

秋日午前，老農一家，趁著太陽午睡，雨甚下不下，趕忙鋪布曬茶。一個四十幾歲的老伯，鄉親喊他三叔。三叔看著天，手抱一堆茶，邊撒邊說，雲層散而厚，午時山起風，光線日微，正好二十三度，是時候了，快叫大家趕緊曬了吧。這時，整條門前道，灰褐色的，頓時綠油一片。

文山包種茶，產量雖非第一，「東方美人」卻是茶中極品。曲捲的茶葉，以山泉沖泡，一心二葉如伸展的雙臂，旋轉於杯中，在西方人眼裡，像個體態輕盈的東方女子跳著曼妙芭蕾。三叔說，東方美人多外銷，觀光客現在興的，是到這兒喝咖啡，哪個山頭景色好，行動咖啡館就往哪兒跑。觀光客來了，管他喫茶還是喝咖啡，總之熱鬧了，自然有生意。三叔的家近親水公園，問他家怎走，以前他說直走，看了金城武的電影《向左走向右走》這樣走紅，他說，做生意還是得有廣告術語，那到底怎麼走，三叔說，還是走北宜，然後，向左彎，向右彎，向左彎，向右彎……

2013 Sm2

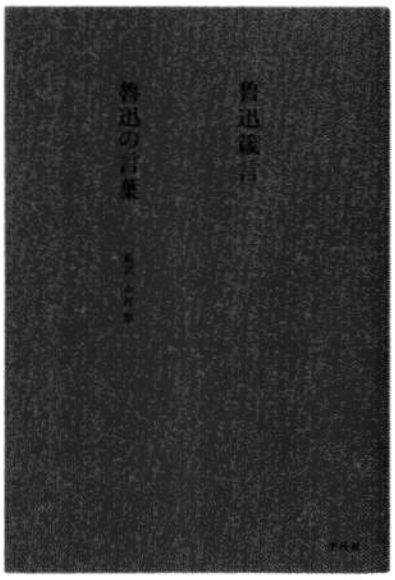

这是一本由平凡社根据三联书店版《鲁迅箴言》重新编辑而成的精选本。三联书店版《鲁迅箴言》为研究鲁迅先生的几位专家精心编选而成，收录鲁迅的经典语录 365 条，囊括了鲁迅先生对于人生、读书、写作、历史等内容的文字。本书在三联书店版的基础上精选出 130 句，中日双语排版。本书为锁线硬精装，护封红色卡纸印黑，底色稀疏的黑色网点形成了丰富的肌理，书名中日双语烫黑。内封相同印刷方式裱覆卡板。暗红色的环衬配合浅灰色扉页。内页胶版纸单黑印刷，对开页中日双语对照，章节页上为鲁迅收藏的版画。上下书口与左侧翻口刷与封面一致的暗红色。

This is a selected edition based on the Word of Lu Xun published by Joint Publishing. It is re-published by Heibonsha Limited Publishers. Several experts who study Lu Xun carefully compile the Joint Publishing edition, which contains 365 classical quotations of Lu Xun, including his words on life, reading, writing and history. On the basis of Joint Publishing edition, this book has selected 130 sentences typeset in both Chinese and Japanese. This hardback book is bound with thread sewing. The jacket is made of red cardboard printed in black and sparse black dots form a background with rich texture. The bilingual title of the book is hot stamped with black ink in Japanese and Chinese. The inside cover is printed in the same way and mounted with graphic board. Dark red endpaper perfectly matches with light grey title page. The inner pages are printed in black on offset paper, with Chinese and Japanese contents on facing pages in contrast. The engravings collected by Lu Xun are on the chapter pages. The three edges of the book are brushed with the same dark red ink as the cover.

银奖

鲁迅箴言

日本

Silver Medal

The Words of Lu Xun

Japan

Kenya Hara / Tei Rei

Lu Xun

Heibonsha Limited Publishers, SDX Joint Publishing Company

172×113×20mm

285g

259p

This bilingual anthology of quotations – Japanese and Chinese – has been designed to convey a tantalizing sense of pleasure to hand and eye. The small, sparingly distributed wood engravings with their generous black spaces provide a faint sense of the spirit of departure during the first decades of the 20th century. Lu Xun is regarded as a modernizer of Chinese literature. The design of the small ceremonial board-covered volume – the delicate sphere-like typography of hanging lines and the overall accoutrement – exudes a self-confident air of modesty. Red, the colour of good fortune, cloaks the soft pages in the endpaper, three-page edging, cover coating, dust cover and banderole. The booklet can be regarded as an homage to the author.

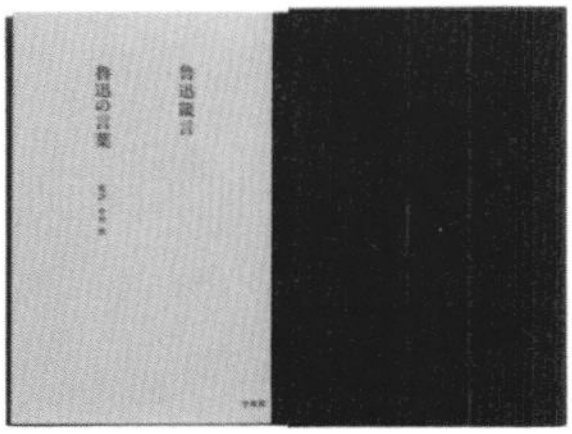

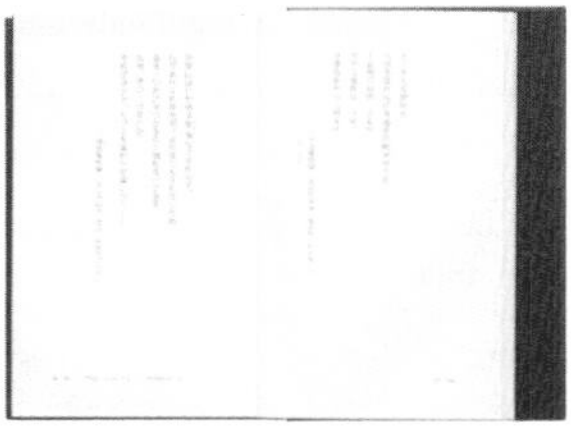
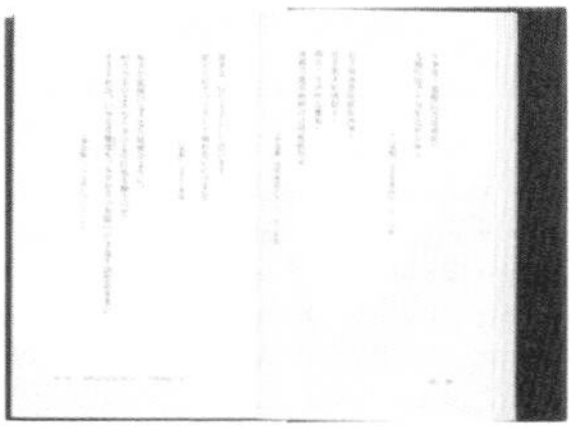

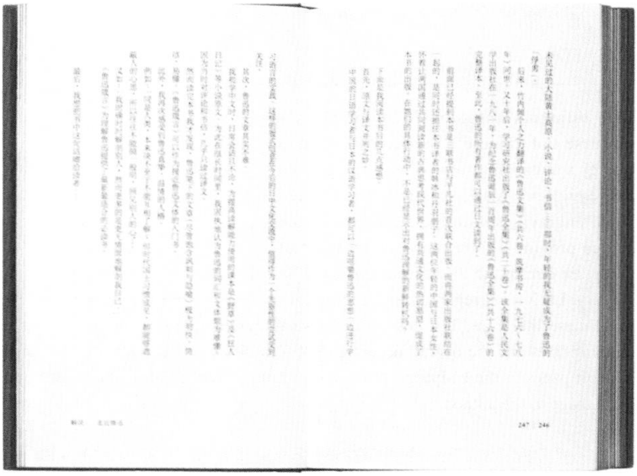
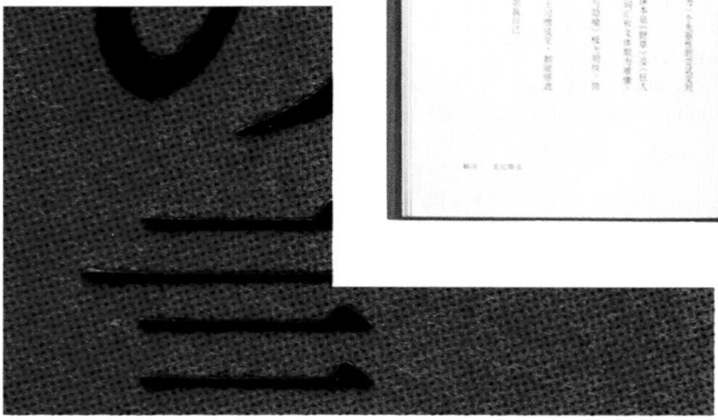

や暗喩
文体接

魯迅の言葉
平凡社
女性の天性のうちには、母性と娘性はあるが、妻性はない。
妻性は逼まられてできたものであり、
母性と娘性が混ざり合ったものにすぎない。
『而已集』「小雑感」
一羽の小鳥を籠のうちに閉じ込めても、竿にとまらせても、
地位は変わったように見えるが、
じつは、どちらも同じように人の弄び物である……
『南腔北調集』「婦女の解放について」
第四章　人として生きる｜做人

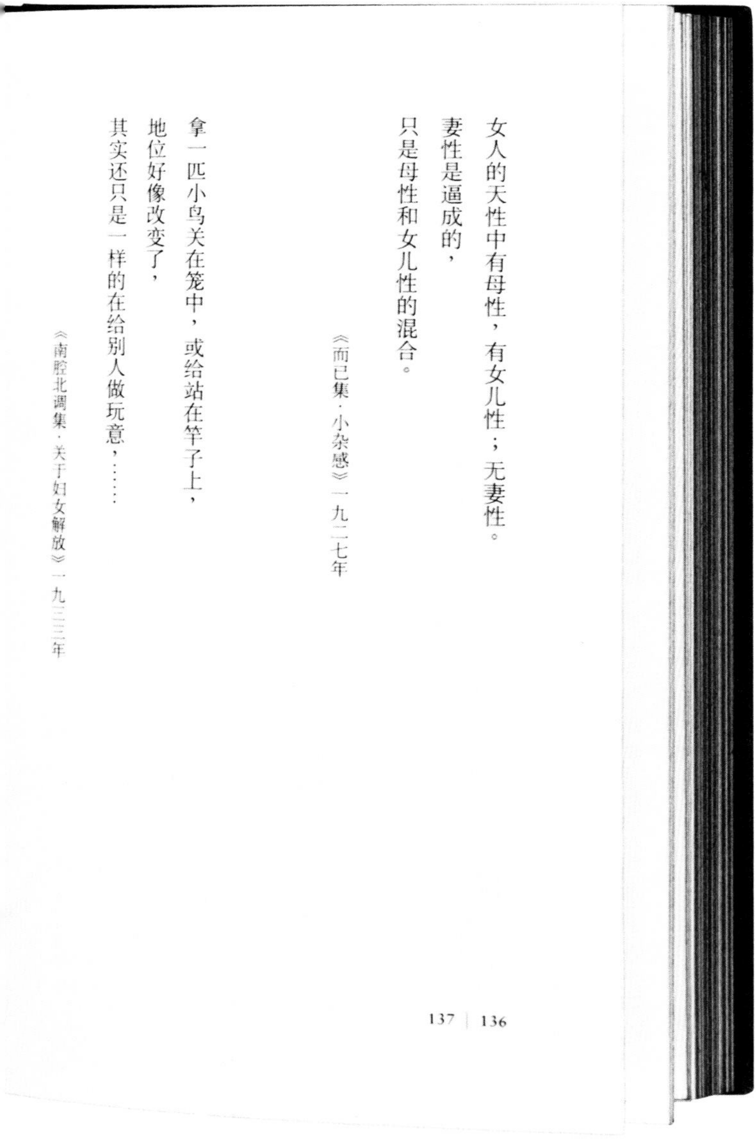

女人的天性中有母性，有女儿性；无妻性。
妻性是逼成的，
只是母性和女儿性的混合。

《而已集·小杂感》一九二七年

拿一匹小鸟关在笼中，或给站在竿子上，
地位好像改变了，
其实还只是一样的在给别人做玩意，……

《南腔北调集·关于妇女解放》一九三三年

137 | 136

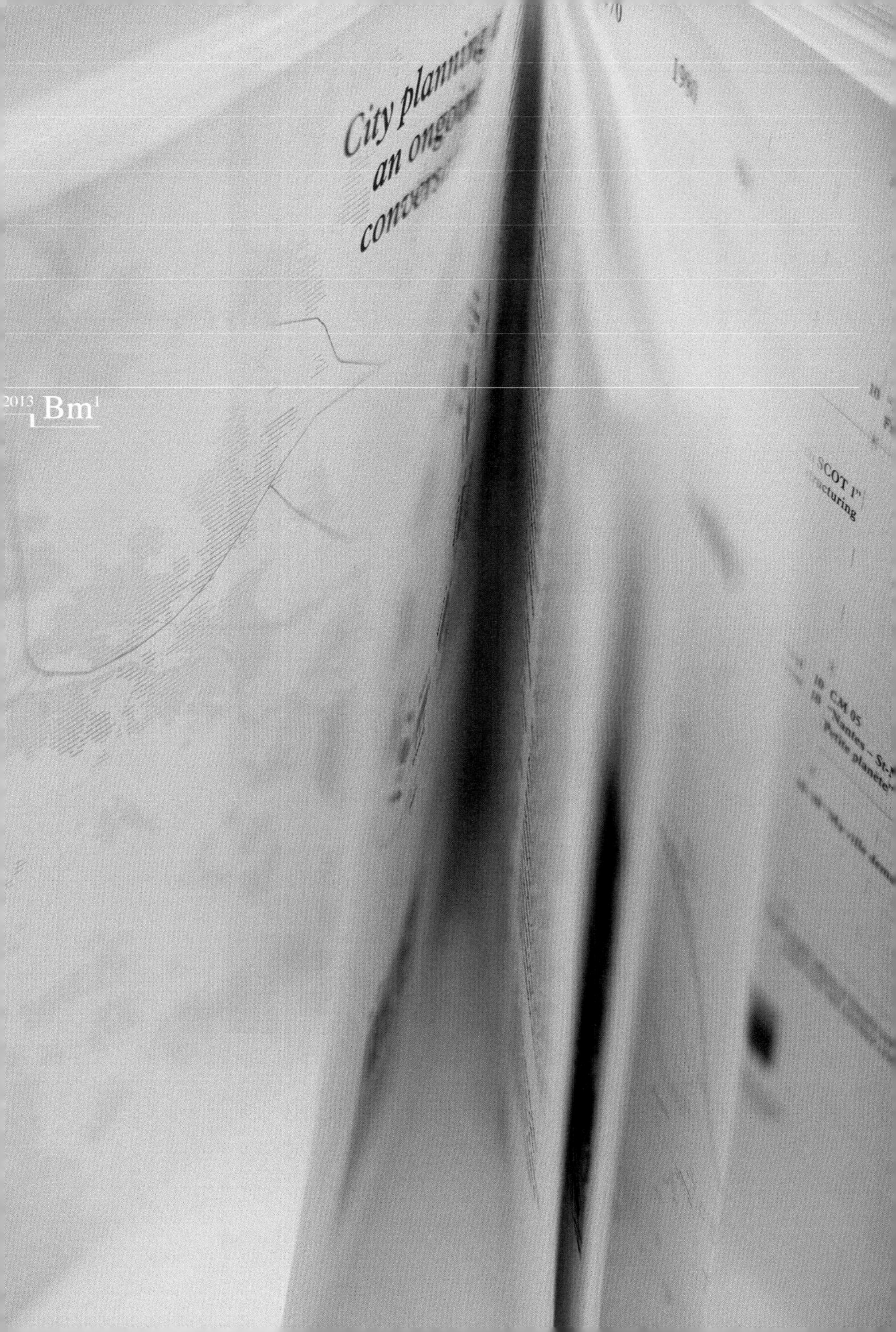
City planning
an ongoi
SCOT I"
ructuring
10 CM 05
10 "Nantes – St-
Petite planète"

2013 Bm1

这是一本关于城市文化转型的书。本书通过对鹿特丹、苏黎世、南特、兰斯塔德、波尔多五个城市规划、人物采访等方面的研究，详细归纳了欧洲创新型城市转型的全过程。本书装订为锁线平装，封面白卡纸铜、黑双色印刷。内页根据内容划分采用了多种纸张的组合。五个城市转型部分的内容，均按照相同的规制形成五个独立的单元：白色胶版纸四色印刷，呈现五个城市文化转型的历程、过程中的问题，后接黄色胶版纸印黑，为两到三个相关城市专家的访谈，最后附上白色胶版纸四色印刷的规划图纸。书的最后部分采用中灰、浅灰和白色纸张分别为前述文字的法文、荷兰文，以及参考资料等附文信息。全书三栏文字配合图片的穿插，版式多变。每个城市开篇的时间轴拉页信息丰富，设计精美。

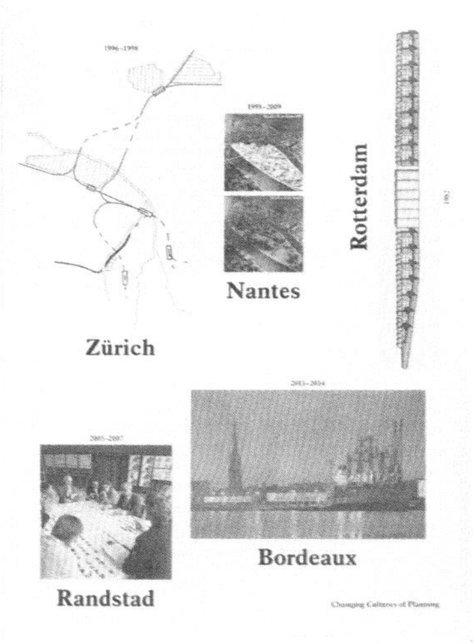

Rotterdam
Nantes
Zürich
Bordeaux
Randstad

This is a book on the transformation of urban culture. Through the study on the urban planning of Rotterdam, Zurich, Nantes, Randstad and Bordeaux, this book summarizes the process of transformation in European innovative cities in details. The paperback book is bound with thread sewing. The cover is made of white cardboard printed in copper and black. The inner pages are using different combinations of paper according to the content. The five parts concentrated on the transformation of five cities are independent units following the same regulations: white offset paper is printed in four colors, which present the process and problems of the transformation; following yellow offset paper is printed in black with interviews with two or three relevant urban experts; finally white offset paper with four-color printing is attached for planning drawings. The last part of the book contains appendixes using medium gray paper, light gray paper and white paper in French or Dutch respectively. There are three columns of text inserted with the illustrations, which make the layout various. The part of each city has an informative foldout, which is designed exquisitely.

- 铜奖
- 文化转型计划——鹿特丹、苏黎世、南特、兰斯塔德、波尔多
- 比利时

- Bronze Medal
- Changing Cultures of Planning – Rotterdam, Zürich, Nantes, Randstad, Bordeaux
- Belgium
- Joris Kritis with Rustan Söderling
- Nathanaëlle Baës-Cantillon, Joachim Declerck, Michiel Dehaene, Sarah Levy
- Architecture Workroom Brussels
- 297×211×21mm
- 1075g
- 302p

It would be difficult to find a website with comparable complexity which is as easily accessible as this book. Not only as far as orientation – the ‘what is where?’ – is concerned, but also with a view to the mastery of the amount of text and image constituents with their varying depth. How does it work? The designers give the heterogeneous editorial material of the five studies of urban planning a pronounced structure: deviants are handled separately, similes get equal treatment; there are semantic, inconspicuous yet noticeable paper changes; fold-out pages; undogmatic arrangements of pictorial material of varying quality. Thanks to the hollow spine, the thread-bound brochure opens without breaking the carton cover. Weighty problems in terms of book design have been solved in a professional manner.

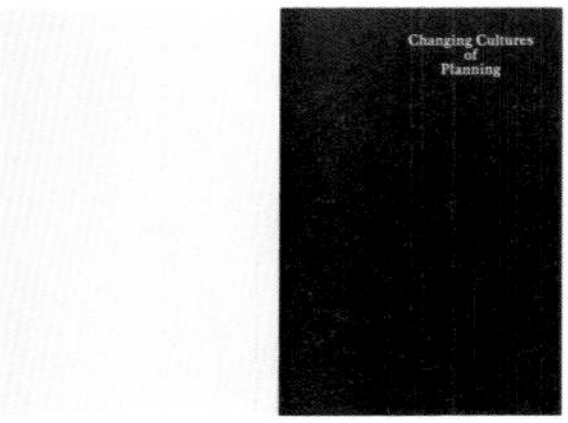
Changing Cultures
of
Planning

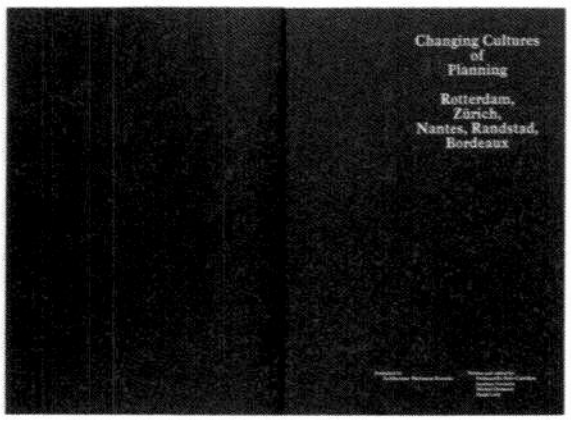
Changing Cultures
of
Planning
Rotterdam,
Zürich,
Nantes, Randstad,
Bordeaux

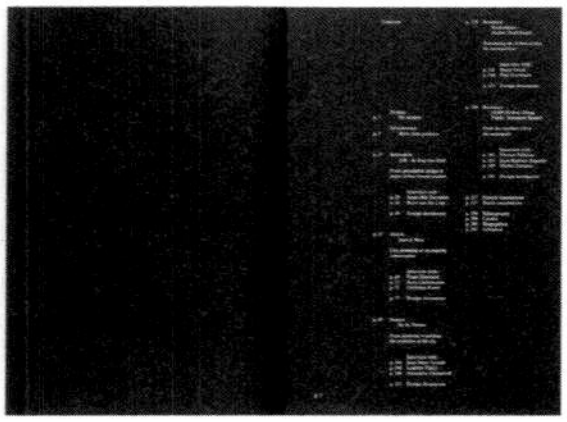

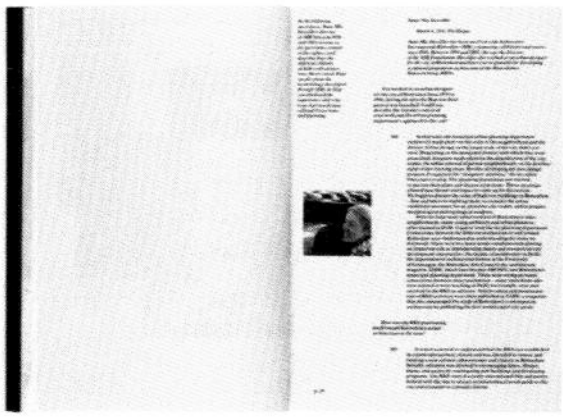

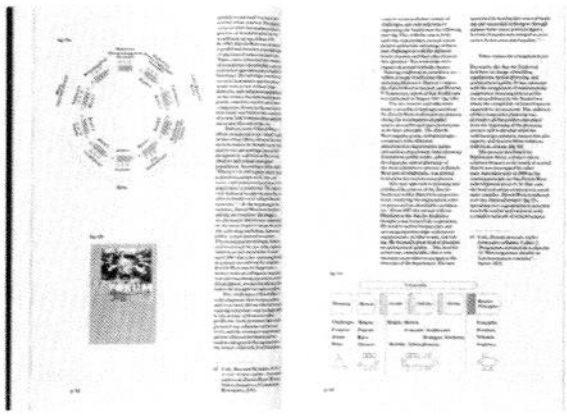

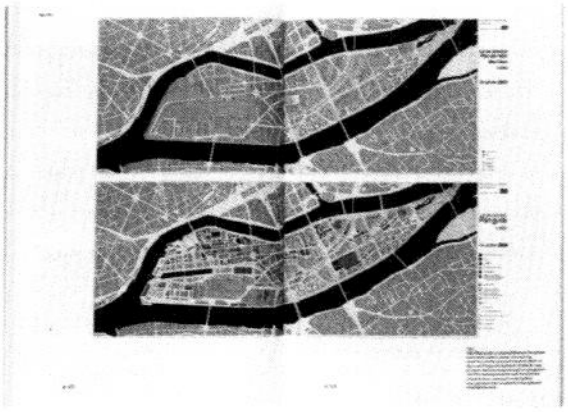

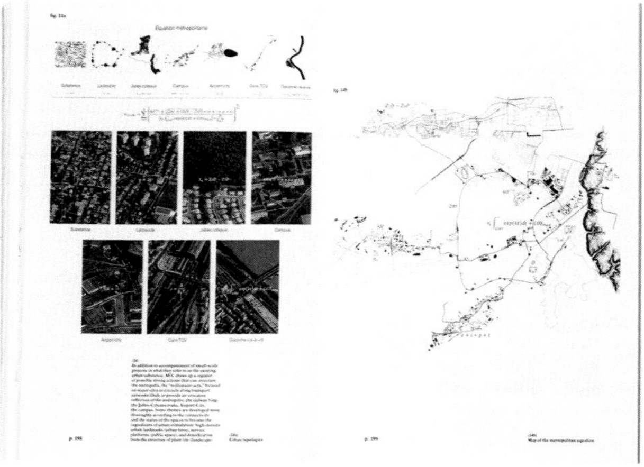
fig. 14a
Equation métropolitaine
p. 198
p. 199
Map of the metropolitan equation

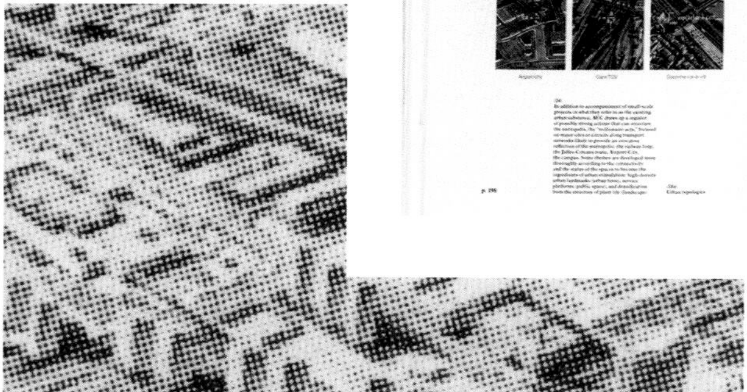

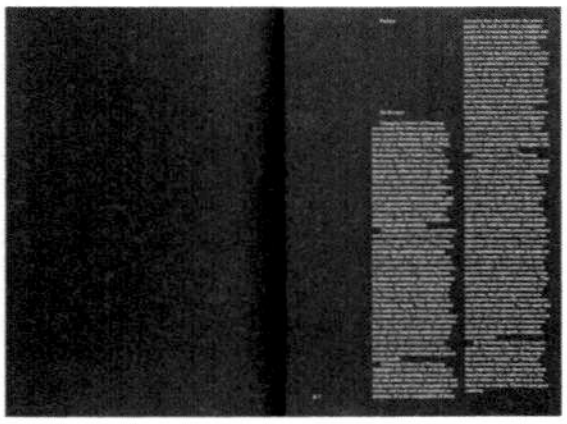

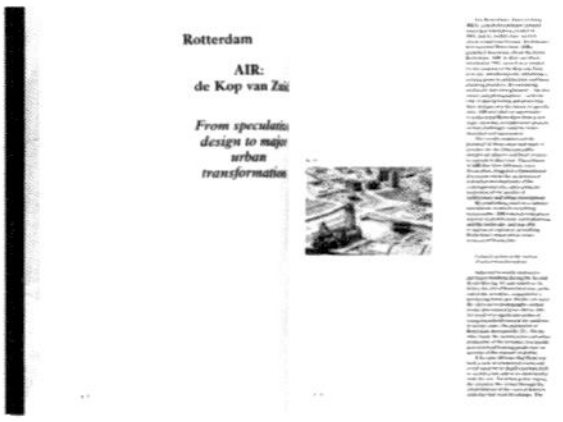

Rotterdam
AIR:

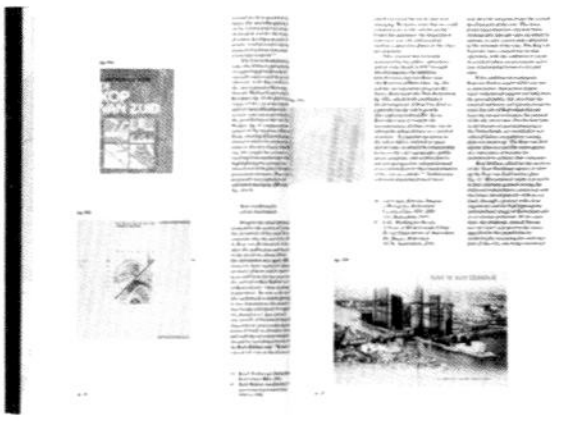

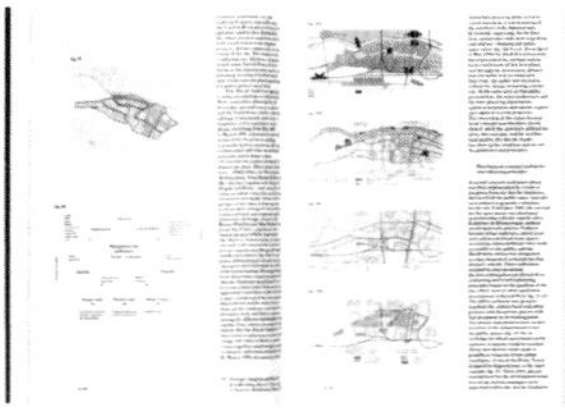

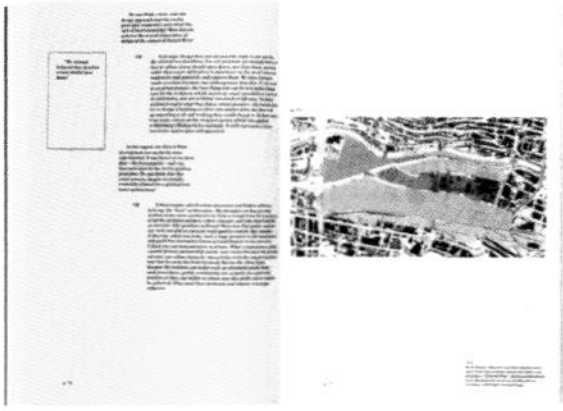

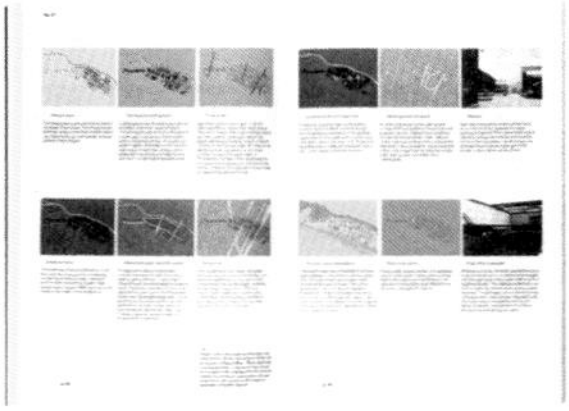

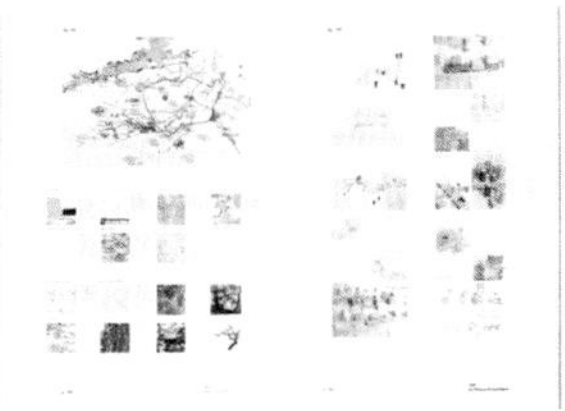

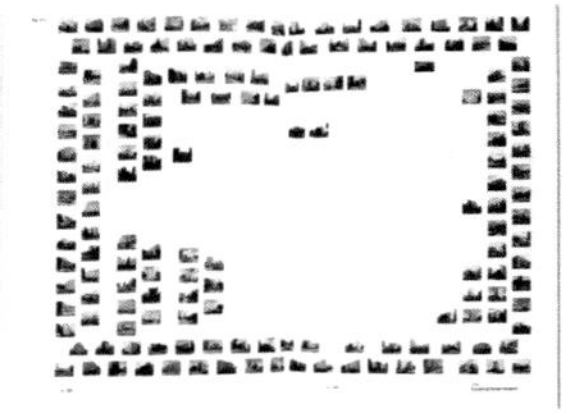

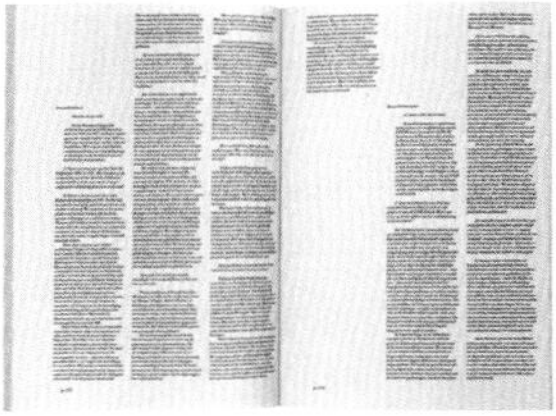

sion
ban developm

1970 **1980** **1990** **1995**

Legal Context

92 District de l'Agglomération Nantaise Intermunicipal cooperation

91 Plan des déplacements urbains PDU Mobility plan

Actors Consultation

82–83 Jean Blaise appointed director of the Maison des cultures de Nantes

78 Urbanism agency AURAN is created

89 Jean-Marc Ayrault becomes mayor of Nantes

96–97 Exhibition of Projet 2005 "Rives de Loire" at the Maison des Hommes

Prospective Studies

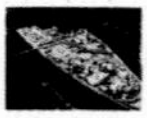

76 "La ville moderne" projection of a desired state for the island in 1980

91–94 Île de Nantes study Dominique Perrault et François Grether

95–97 "Rives de Loire" Atlas

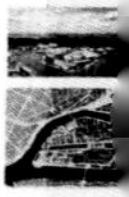

98 "Île d

Rives

Projects

84 Tramway Line 1

87 Closure of the ile de Nantes shipyards

90 Madeleine-Champs de Mars urban project

92 Tramway Line 2

92 Opération coeur de quartier Les Dervallières

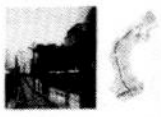

93 Cours des 50 Otages project Fortier-Bloch-Rota

96 Agreement for the construction of the courthouse on Île de Nantes

2000 | 2005 | 2010 | 2012 | 2020

10 "Les chantiers du SCOT 2"
From SCOT to PLU

té Urbaine de

ipal cooperation

01 Beginning of consultation for the Schéma de cohérence territoriale SCOT
Strategic planning toward 2020

04 Nantes métropole
Intermunicipal cooperation

07 Approval of the SCOT

08 "Les chantiers du SCOT 1"
Strengthen the structuring urban nodes

ESTUAIRE

00 Jean Blaise becomes director of the Lieu Unique

rence
olitaine (CM) 01

03 creation the SAMOA and the Atelier l'île de Nantes

05 CM 02

06 CM 03

07 Festival Estuaire

08 CM 04

09 Festival Estuaire

10 CM 05
10 "Nantes – St-Nazaire Petite planète" exhibition

12 Festival Estuaire

10–30 "Ma ville demain"

99 Chemetoff is selected to apply the Plan-Guide method

01–10 Three missions:
– guide-plan follow-up
– management of public space projects
– communication

procedure
public spaces
sland

00–01 contract negociation

10–20 UAPS-Marcel Smets selected to supervise the further renewal of the island

02–07 First phase

sultation"

00 Tramway line 3
00 Courthouse
Jean Nouvel
00 pont mobile
Barto+Barto

02–05 Quai François Mitterand

05–07 Quai Blanchot
Parc des Chantiers
Quai aux Antilles

07 Opening shipyard halls of the Loire

09 Architecture school
Lacaton & Vassal

01 GPV Malakoff-Pré Gauchet

03 Île Feydeau

06 Tramway line 4
06 Arborea
Tetrac architectes

LORENA C
NEVILLE STAPLES
SPECIALS
RIVERS
Das
Offen
Aber nur der erkenn
Das älteste Deutsche
und ist am Ende
Rudolf
Rudolf Koch. Wilhelm-Klingspor-Schrift, 1926.
Tekstuur (textum – kude) on kõige kandilisem ja mur-
tum gooti kiri, millest moodustuvad tumedad, tiheda
koega tekstiplokid. Tähed on pikaks venitatud ning
vahemaad postide vahel on korrapärased. Tähtede
sisevormi laius vastab umbkaudu joone jämedu
Tekstuuri kasutasid esimesed raamatutrü
kaasa arvatud Johann Gutenberg.
rahvale käsitsi kirjutatud raa
aastal kavandas Frederic
bergi piiblist inspirees
Goudy Text. Kling
Wilhelm-Klin
Koch aast

本书是一本字体历史和排印教材。与大多数只介绍各个时代和风格最典型的字体图书相比，本书具有海量的字体资料整理。本书为圆脊锁线硬精装，封面灰色草浆纸红黑双色印刷，裱覆卡板。翻过与封面相同的草浆纸环衬，内页胶版纸印单黑。内容分为四大部分，第一部分为大量收集拍摄而来的字体运用实例，一个跨页共 16 张图片，按照碑刻字体、经书字体、衬线字体、无衬线字体的发展过程进行陈列。第二部分为字体术语、基础知识和经典字体的样张与简介。第三部分为字体发展的历史，记录着从碑刻字体到当代数字化的字体文件的简要史。第四部分则为参考书目、双语词典、索引等附文。全书双栏排版，作为一本关于字体设计的书，书中的字体选用和排版考究，无衬线字体的选用也可以在满足阅读需要的同时，将更有个性的字体凸显出来。

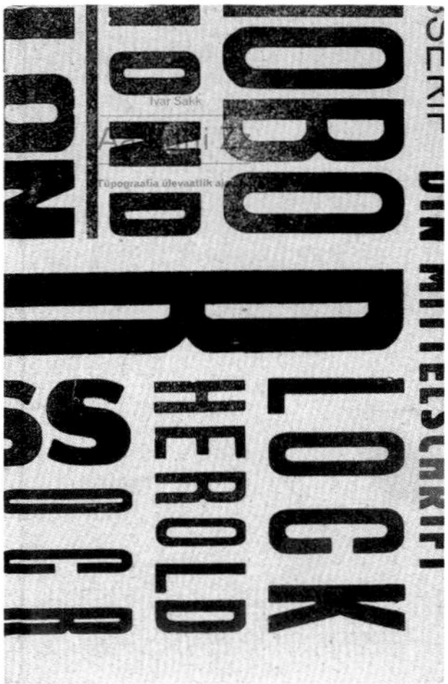

Ivar Sakk
HEROLD
LOCK

This is a textbook on font history and typography. Compared with most typeface books that only introduce the most typical typeface of each era and style, this book has collated a large amount of typeface information. This hardback book is bound by thread sewing with rounded spine. The cover is made of gray straw paper printed in red and black, mounted with graphic board. After the endpaper using the same straw paper as the cover, the inner pages are printed on offset in monochrome black. The content is divided into four parts. The first part contains lots of examples of using fonts, total 16 pictures are arranged according to the developing process of fonts – inscription fonts, scripture fonts, handwriting fonts, serif fonts, sans serif fonts. The second part is about the typeface terminology, basic knowledge and the sample and introduction of classical typefaces. The third part is about the history of font development, which records a brief history from inscribed fonts to contemporary digital font files. The fourth part is for appendices, such as the reference bibliography, bilingual dictionary and index. The book is typeset in double columns. As a book about font design, the selection and typesetting of fonts are exquisite; especially the use of serif-free fonts can meet the requirements of easy reading as well as highlight more personalized fonts.

- 铜奖
- 从 Aa 到 Zz——字体排印简史
- 爱沙尼亚

- Bronze Medal
- Aa – Zz. CONCISE HISTORY OF TYPOGRAPHY
- Estonia
- Ivar Sakk
- Ivar Sakk
- Sakk&Sakk
- 285 × 178 × 38mm
- 1062g
- 447p

We hear a collective groan from our colleagues in the typeface department: another book about fonts from the Romans to the digital revolution! Don't judge so hastily. Because what catches the typographically oversaturated eye in this book is its peculiar clarity, free from any creative explosions. While there is a certain sense of déjà vu, we find a lapidary division into three parts, guided by a didactic Eros: a photographic case study of the status quo of fonts displayed in public spaces; a section with tables of type specimens – divided into lines with different cuts and sizes; a text section about the history of writing with explanatory illustrations. And there is something else that awakens curiosity: the Eastern European perspective of the Estonian author.

85
St.George's
54
STACIÓ

CIUTAT

Tekstikiri
Antiikva
PEALKIRI
Kalligraafia
Grotesk
agh
Egiptienn

petrus lancollar dorum
melchior
american uncial
maxentium
lutetia parisiorum
aachen
charlemagne
empire carolingiens
londinium
primum in germania
don fernando alvarez de toledo
aix en chapelle
BORDEAUX
minuscule gothique

Antique renaissance
Claude Garamond
LES ALPHABETS DE CETTE EPOQUE
Les diverses formes
François I 1540
L'imprimerie Royale
L'APOGEE DE LA CREATION DE FONTES
Antique renaissance
Claude Garamond
LES ALPHABETS DE CETTE EPOQUE
Les diverses formes
François I 1540
L'imprimerie Royale
L'APOGEE DE LA CREATION DE FONTES

Ladina kirja
alguspäevad
TRAJAN

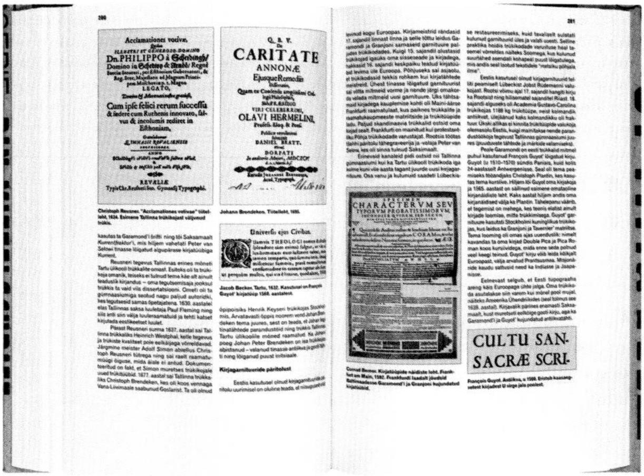
280
281
Acclamationes votivæ
CARITATE
ANNONÆ
OLAVI HERMELINI
CULTU SAN-
SACRÆ SCRI-

THINGS
SALE
Mercantile Bank
Standard

TERVEELLIST
HELPPOA
TEMPO
ROTHERHITHE
Tickets
Politi

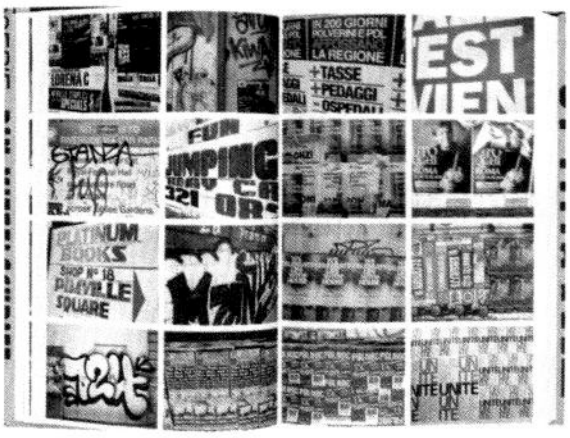
EST
VIEN
PLATINUM BOOKS
SQUARE

Giambattista Bodoni
Manuale tipografico
Cancelleresco majuscole
Jean-Claude Ferré
EMPORIO ARMANI
LOS ANGELES & SAN FRANCESCO
Richard Austin
THE FIRST ENGLISH MODERN FACE
King George IV
Stephenson Blake
MONOTYPE BELL
Work by Van Dijck equals the greatest
THE SIX DOCUMENTS TAKING PRIZES

Morris Fuller Benton
American Type Founders
Microsoft
Boston Stereotype Foundry
Brooklyn Bridge
The Inland Printer
Automatic typecasting machine
19th century advertising
Madison Square Garden
The secret history of letters
LADY WILLOUGHBY
DARLINGTON RAILWAY

Der Formenreichtum ist
Berlin Alexanderplatz
Unter den Linden
Lucian Bernhard
KUNSTGEWERBEMUSEUM
AMERICAN TYPE FOUNDERS
Le Castel Beranger
RESTAURANT GRAND PALACE
POSTERS BECAME POPULAR
METROPOLITAINES
Am Beginn des 20. Jahrhunderts

William Morris

Neue Grafik
New Graphic Design
Graphisme actuel
Univers
Univers
Univers
Univers

historical
SYNTHETIC
contemporary
REFLEXIVE
speculative
Consequeligula
alterpoint
Quetempor
DynaGrotesk DXE
Czech Republic
Parisine
Parisine
LeMonde Sans
Klavika
LeMonde Livre

abiga
it võtt
lisandub 21. sajan

Ivar Sakk
PUMP

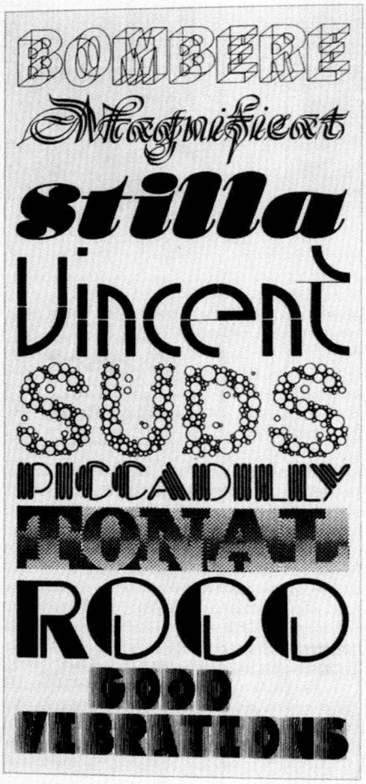
BOMBERE
Magnificat
Stilla
Vincent
SUDS
PICCADILLY
TONAL
ROCO
GOOD
VIBRATIONS

Letraseti 1973. aasta võistluse võidutööd.

2590 võistlustöö hulgas oli loomulikult palju diletantide loomingut, kuid ka rohkesti kõrgprofessionaalseid kaastöid. Letragraphica žürii valis välja 17 kirjatüüpi, mis kohe tootmisse läksid. Letratähtede eripära arvestades olid paljud kavandid tehtud pealkirjade jaoks. Enamik võitnud töödest järgisid seda joont ning pikema teksti ladumiseks ei sobinud üldse. Kuid ajastu hipi- ja noorsookultuurist ning biitmuusikast pulbitseva õhustiku tõttu ei pööratud erilist tähelepanu lihtsale tekstikirjale ning efektsed popkunstist inspireeritud šriftid olid need, mida kaasaegsed tegelikult ootasid.

Esikohale tuli ameeriklanna Carla Bom
re Wardi šrift **Bombere** – majusklitest koosr
ruumiline kiri. Tähed võtsid peaaegu arhitektu
se vormi ning koosnesid mahulist kehamit m
keerivatest kontuuridest. Kuigi langevate var
dega kirjad olid tollal juba populaarsed, juha
ilmselt Bombere kätte tee joonel põhinevate
ruumimängudeks. Teisele kohale tulnud
jatüüp, kanadalase F. Friedrich Petersi **Magni
cat**, oli hoopis teisest puust. Laisulekalligraafi
inspireeritud kursiivkiri oli baroklikult küllusli
joonega. Hoogsad suletõmbed täitsid tähepir
tihedalt, paralleelsed jooned rõhutasid olu
konstruktsioonilisi sõlmi.

Kolmanda koha saanud prantsla
François Boltana **Stilla** oli ülijõuline, trullak
mummudega kursiivkiri, milles 19. sajandi
natähtede maksimumini viidud kontrast ris
kaasaja vitaalse lopsakusega. Teised äram
gitud šriftid olid Vincent, Piccadilly, Tonal, Ro
Good Vibrations jm.

Uutest kirjadest anti teada trükiste vah
dusel, samuti leidsid need koha Letraseti too
kataloogis, mis oli olulisel kohal iga tollase dis
neri töölaual. Tähestikud trükiti seal täielik
ära, lisaks näidati iga kirjatüübi suurusvarianc

8 new faces from Letrase

Reklaamleht Letraseti kirjatüüpidega, 1969.

üks-ühele. Ainuüksi Suurbritannias oli selle 160-leheküljelise kataloogi müüginumber 125 000.

Letraset müüs oma toodangut üheksakümnel maal, tootmine oli sisse seatud Ashfordis Inglismaal ning väiksemad tehased olid veel Mehhikos ja Argentiinas. Kirjatüübi produtseerimine algas disainiosakonnas, kus iga täht lõigati terava noa abil tasafilmist välja. Lõigatud tähed olid 6 tolli kõrged. Ühe umbes 80 märgist koosneva tähestiku lõikamiseks kulus kolme nädala ümber. Tähestikku kuulusid suur- ja väiketähed, numbrid ja kirjavahemärgid. Sellist filigraanset ja aeganõudvat protsessi võib võrrelda Gutenbergi-tüpograafia aegadega, kus trükitüübi saamiseks oli vaja samuti osavat metallimeest-graveerijat. Letraseti kaastöölised – Robin Brignall, Freda Sack, Alan Meeks jt – ei olnud ainult paljalt käsitöölised, vaid enamikus kunstnikuhariduse ja tüpograafiataustaga tegelased, kelle potentsiaal leidis rakendamist mitte ainult igapäevases lõiketöös, vaid ka uute kirjatüüpide kavandamisel.

Lõigatud tähed läksid fotolaborisse, kus neid vähendati ja paljundati ning paigutati leheküljе formaadile. Lehtedest valmistati kontaktmenetluse abil negatiivid. Kõik edaspidised fototoimingud toimusidki kontaktkoopiatega, et minimeerida positiivide valgustamise käigus tekkida võivaid ebateravusi. Positiividest tehti siiditrükivormid, millelt trükiti läbipaistvale polüteenmaterjalile kõigepealt kirjad ning teises etapis kaeti kogu lehepind liimainega.

Letraseti koosseisuline kujundajate tiim oli tugev ja viljakas. 1964. aastal liitus sellega Colin Brignall, kes sai firma kunstiliseks juhiks. Juba 1965. aastal valmis tema käe all **Countdown**, mis

Artone
Studio India
Ink

Seymour Schwast. Kataloogi kaas, 1969.

Countdown

Colin Brignall. Countdown, 1965.

Aachen

Colin Brignall. Aachen, 1969.

REVUE

Colin Brignall. Revue, 1969.

PREMIER LIGHTLINE

Colin Brignall. Premier Lightline, 1969.

PREMIER SHADED

Colin Brignall. Premier Shaded, 1970.

Octopuss

Colin Brignall. Octopuss, 1970.

Bottleneck

Tony Wenman. Bottleneck, 1972.

University

Philip Kelly, Michael Daines. University Roman, 1972–1977.

Compacta

Frederick Lambert. Compacta, 1963.

Artone

Seymour Schwast. Artone, 1968.

oilée placée sur un axe
aboussures.

这是一本试图表达生活本质的书。全书并没有非常明确地去表达主旨，而是让读者在文字、图片和翻阅本书的独特体验中去体会。全书没有传统意义上的装订形式，书籍被放置在透明的塑料袋中，塑料袋上贴着深蓝色贴纸，作者和书名信息烫银。取出书籍本体，页面被按照骑马钉的形式折好，使用深蓝色皮筋固定书脊。全书胶版纸双色套版印刷，主要照片为粗网处理印单黑，主要文字则为深蓝色，少量页面红黑双色套印。由于大量文字并不排在一个页面内，而是横跨到同一纸张的另一页上，所以在翻阅的过程中时不时需要将这样的文字页单独取下。在连续的翻阅过程中，页面围绕“中轴”皮筋转动，有如帆船的机轮旋转。文字中叙述的几个私人的故事，就像这本书的形式，一旦转动起来就不能停下，这是生活的本质。两页图片就像半格摄影同一张胶片上的两张图，静静地表达着似有似无的联系。

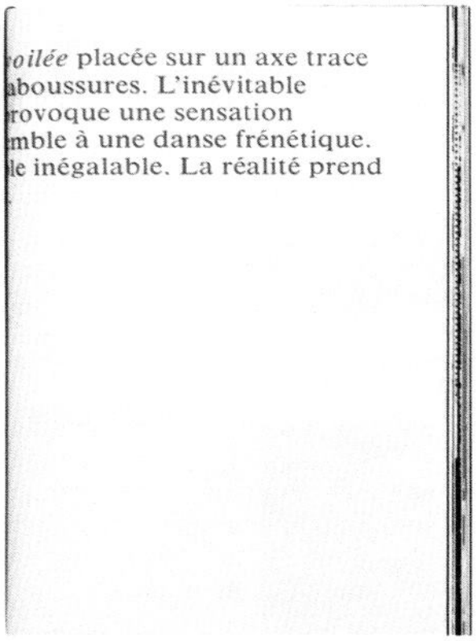
oilée placée sur un axe trace
aboussures. L'inévitable
rovoque une sensation
mble à une danse frénétique.
le inégalable. La réalité prend

This is a book trying to express the essence of life. The whole book does not express the main idea very clearly, but let the readers explore the unique experience when reading the book. The book hasn't been bound in traditional way. It is placed in transparent plastic bags with dark blue stickers, with the author name and title information silver hot stamped. The book pages have been folded and the spine is fixed with dark blue rubber bands. The book is using duotone offset printing. The photos are mainly printed with rough screen in monochrome black while the text is dark blue. A small number of pages are printed with red-black registration. Since a large number of words are not typeset in a single page, but are arranged on the facing page of the same paper, it is necessary for readers to take such text pages out to read. In the process of browsing the book, the pages revolve around the 'mid-axis' rubber band, just like a spinning sailboat's wheel. Several personal stories in the book, like the form of this book, can't be stopped once they start. This is the essence of life. Two pages of pictures are like two pictures on the same film in half-grid photography, which quietly express the seemingly unrelated relationship.

铜奖
隐秘的轮子
比利时

Bronze Medal
La roue voilée
Belgium
PLMD (pleaseletmedesign) + interns: Audé Gravé, Rosalie Wagner
David Widart
L'Amicale Books
239×170×6mm
174g
84p

Is this handful of papers actually a book? A video clip on paper? A bundle of photographic notes? Loose pages with coarsely pixelated black and white photographs and crudely typographed blue text are not even sharply folded but simply slung over and fixed with a flat blue rubber band. This banality casts a spell on the observer who examines and dismantles this 'thing' – a question of: shall I stay outside or shall I immerse myself? What looks like a low-budget production becomes the ideal form of microscopic snapshots from everyday life and, above and beyond that, the rudiment if not the archetype of a book. Yes, this handful of papers is a book, even if 'the wheel is out of kilter' – just like life itself.

LA ROUE VOILÉE.
David Widart.

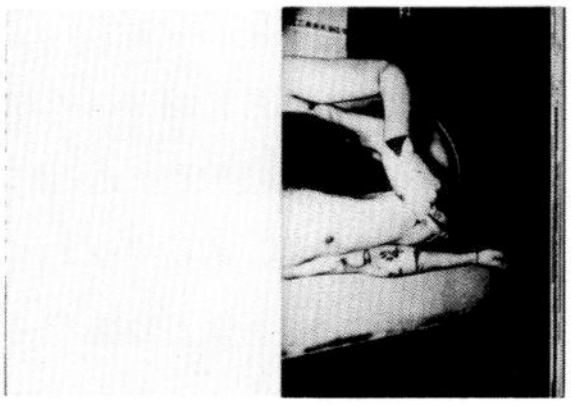

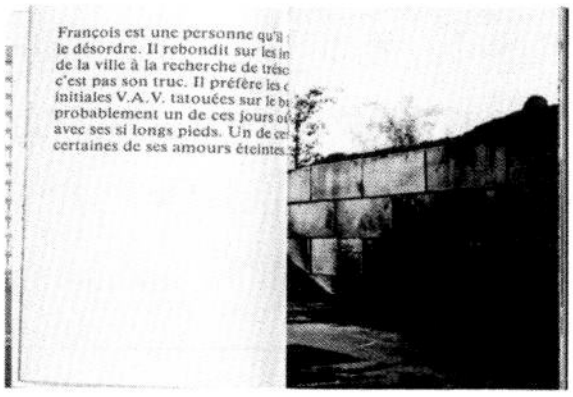

François est une personne qu'il
le désordre. Il rebondit sur les in
de la ville à la recherche de trés
c'est pas son truc. Il préfère les
initiales V.A.V. tatouées sur le b
probablement un de ces jours o
avec ses si longs pieds. Un de ce
certaines de ses amours éteintes

Dans sa dernière vie, Sam faisai
qu'il entend les ondes de la mus
il y a un truc physique qui se pr
comme une boule, c'est puissan
automatique, il se met à danser
Westcoast. Sam a une sensibilit
pour les belles choses. Je ne dira
mement, mais, en tout cas cette
ses deux jouets favoris, étaient s
Sam fait partie des gens qui sen
de spécial. Il ne se dit pas qu'il e
intéressant. C'est juste que ce q
sens. C'est trop loin de l'essenti
a arrêté de passer des heures da
son jardin. Il a également arrêté
Certaines de ses chemises le ren
d'oreilles équilibrent son cervea
Sam a un voyage important à fa
le bon moment, il partira.

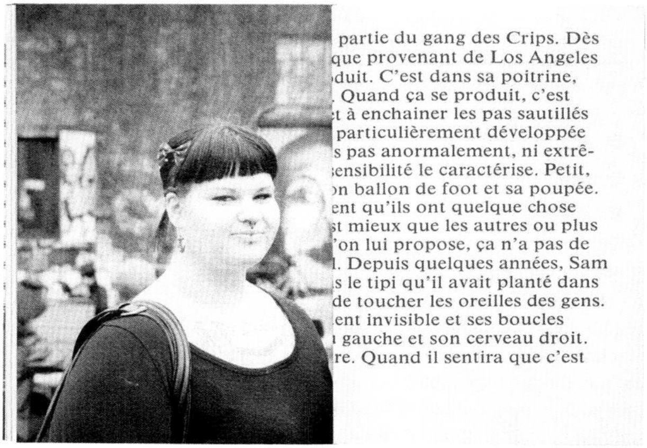

partie du gang des Crips. Dès
que provenant de Los Angeles
duit. C'est dans sa poitrine,
. Quand ça se produit, c'est
t à enchainer les pas sautillés
particulièrement développée
s pas anormalement, ni extrê-
sensibilité le caractérise. Petit,
on ballon de foot et sa poupée.
ent qu'ils ont quelque chose
st mieux que les autres ou plus
'on lui propose, ça n'a pas de
. Depuis quelques années, Sam
s le tipi qu'il avait planté dans
de toucher les oreilles des gens.
ent invisible et ses boucles
gauche et son cerveau droit.
re. Quand il sentira que c'est

Zineb s'est toujours aménagé des
réfugier : des cachettes, des cabar
date ressemble à un trip sous LS
mondes. Zineb a besoin d'observ
trouver son inspiration. Quand e
fabriqué une poupée, avec à l'int
l'on actionne avec une ficelle. Zin
à Frankenstein, qu'elle l'aime be
et qu'elle stoppe les angoisses.

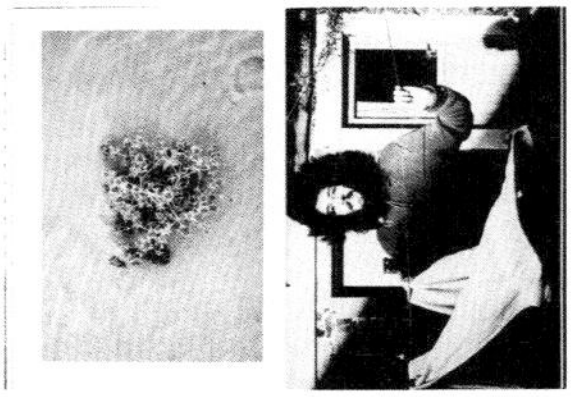

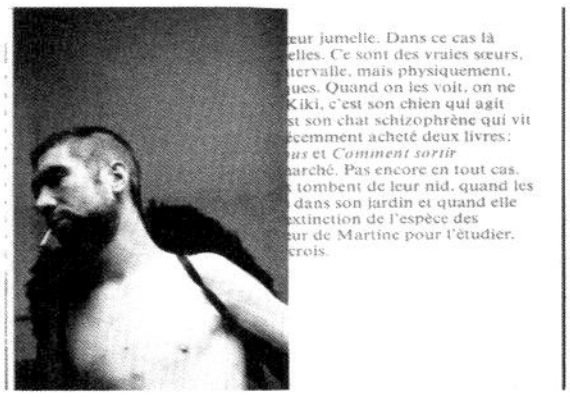
eur jumelle. Dans ce cas là
elles. Ce sont des vraies sœurs.
tervalle, mais physiquement,
ues. Quand on les voit, on ne
Kiki, c'est son chien qui agit
st son chat schizophrène qui vit
écemment acheté deux livres:
us et Comment sortir
arché. Pas encore en tout cas.
tombent de leur nid, quand les
dans son jardin et quand elle
xtinction de l'espèce des
ur de Martine pour l'étudier.
crois.

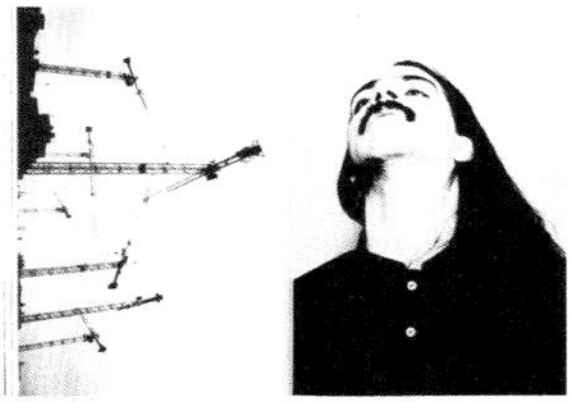

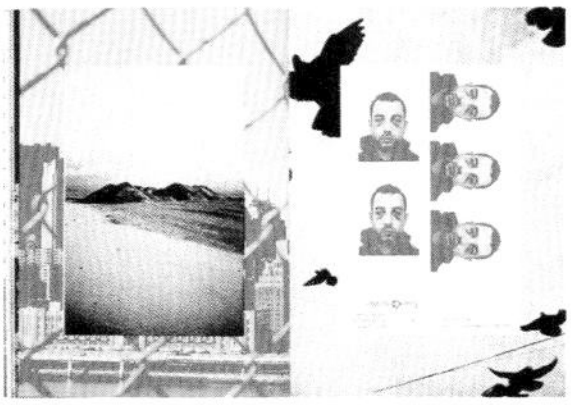

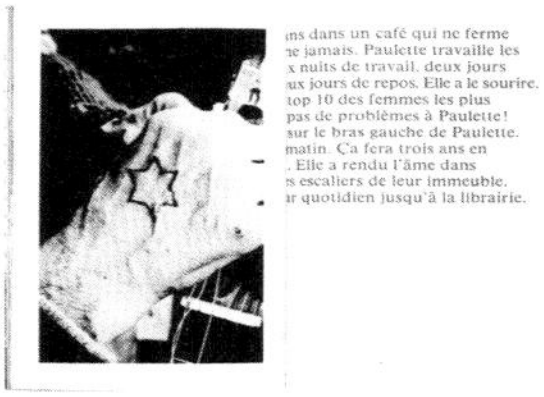
ns dans un café qui ne ferme
e jamais. Paulette travaille les
x nuits de travail, deux jours
ux jours de repos. Elle a le sourire.
top 10 des femmes les plus
pas de problèmes à Paulette!
sur le bras gauche de Paulette.
matin. Ça fera trois ans en
. Elle a rendu l'âme dans
s escaliers de leur immeuble.
r quotidien jusqu'à la librairie.

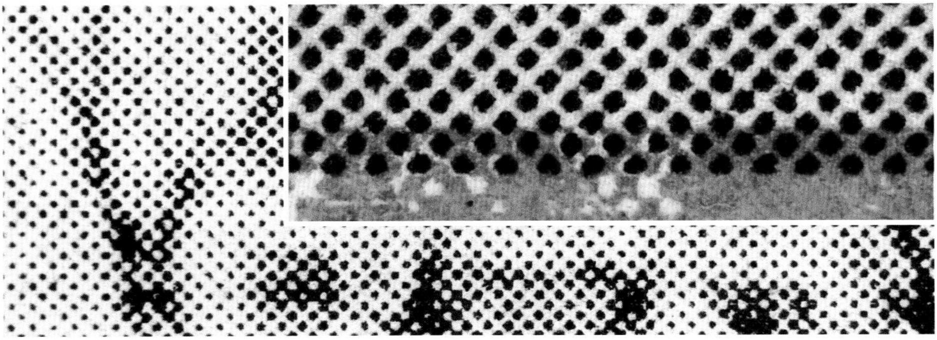

:ndroits où elle pouvait se
s, des grottes. Le dernier en
. À la frontière entre deux
· dans cet univers irréel pour
: était petite, sa maman lui a
ieur une boîte à musique que
b dit que sa poupée ressemble
ıcoup, qu'elle porte bonheur

PSIE ŻYCI

这是一本由波兰艺术家约瑟夫·威尔康（Józef Wilkoń）创作的儿童绘本。书中的各式狗型木雕以及搭配的押韵叙事诗均为艺术家本人创作。本书的装订形式为锁线硬精装。封面四色印刷覆哑光膜，裱覆卡板。内页胶版纸张四色叠印银色共五色印刷。书中各式木雕狗被放置在现实的场景中，产生虚拟与现实相结合的独特的感官体验。钝拙的字体选用以及背景部分斑驳银版的处理手法，与不加精雕细琢的木雕狗相适配。文字部分以人类的朋友——狗的口吻写作，拟人化地讲述了体型和外观千差万别的狗被人类驯化之后，成为人类生活的一部分，命运也从此被掌握在人的手中。狗愿意与人相伴左右，害怕孤独、被遗弃、被链条拴着……没有人类，狗将变成野狗，生活充满未知的恐惧。艺术家旨在提醒人们要对被人类驯服的狗们的命运负责。

Józef Wilkoń
PSIE ŻYCIE

This is a children's picture book created by Polish artist Józef Wilkoń. All kinds of dog wood carvings and rhyming narrative poems in the book are created by the artist himself. The hardback book is bound by thread sewing. The cover is printed in four colors, laminated with matte film and mounted by cardboard. The inside pages are printed in five colors including regular four-color and silver. In the book, all kinds of woodcarving dogs are placed in the real scene, which produces a unique sensory experience combining virtual and reality. The choice of blunt fonts and the mottled background in silver are suitable for the rough woodcarving dogs. The text part is written in the tone of dog, a friend of human beings. It tells the personified story that dogs with different shapes and appearance have been domesticated by human beings and become a part of human life. From then on, their fate has been grasped in the hands of human beings. Dogs are willing to be around people, afraid of loneliness, abandonment and chains. Without humans, dogs will become wild dogs, and life will be filled with unknown fears. The artist aims to remind people that they are responsible for the fate of domesticated dogs.

铜奖
狗的生活
波兰

Bronze Medal
Psie życie
Poland
Józef Wilkoń and Piotr Gil
Józef Wilkoń
Wydawnictwo HOKUS-POKUS Marta Lipczyńska-Gil
226×205×7mm
345g
36p

A children's book with unconventional illustrations and eccentric colouring. The image technology cocks a snook at the theoreticians of virtual reality. Because the figures of the dog story, the painted wood sculptures are both: paintings and reproduced objects; three-dimensional portraits, so to speak. And therefore the episodes come across beautifully contradictory and light-footed in a genuinely artificial world playing with a misconception – that it's not a parallel world that shines through these pictures, but that they instead belong, materially and virtually, to our perceptible here and now. The blunt book format, the thick natural paper, the silver and black background, the bold, stamp-like typography are the formal means to keep this show rolling.

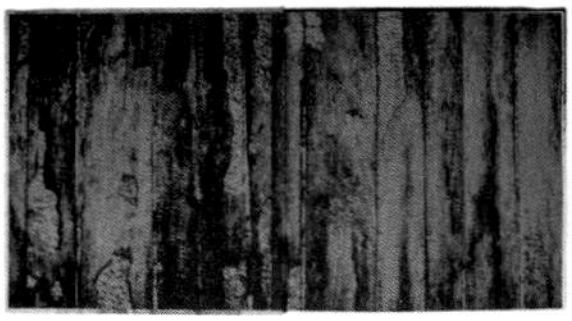

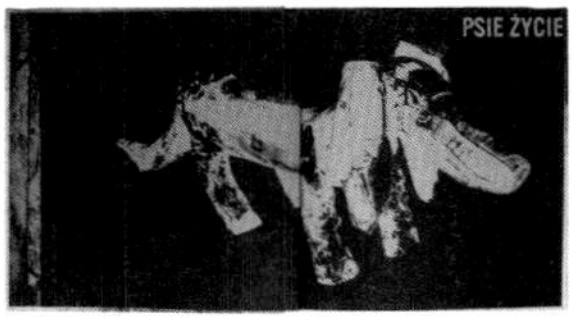
PSIE ŻYCIE

Józef Wilkoń
PSIE ŻYCIE

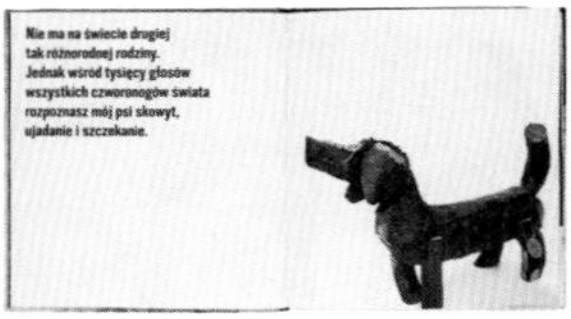
Nie ma na świecie drugiej
tak różnorodnej rodziny.
Jednak wśród tysięcy głosów
wszystkich czworonogów świata
rozpoznasz mój psi skowyt,
ujadanie i szczekanie.

Jesteśmy razem
jak daleko sięga psia i ludzka pamięć.
Moim przodkiem był wilk.
Czasem jesteśmy podobni
jak dwie krople wody.

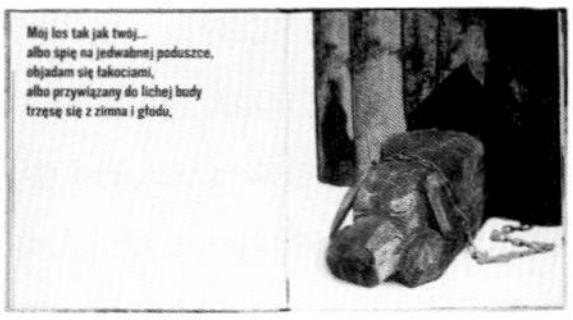
Mój los tak jak twój...
albo śpię na jedwabnej poduszce,
objadam się łakociami,
albo przywiązany do lichej budy
trzęsę się z zimna i głodu,

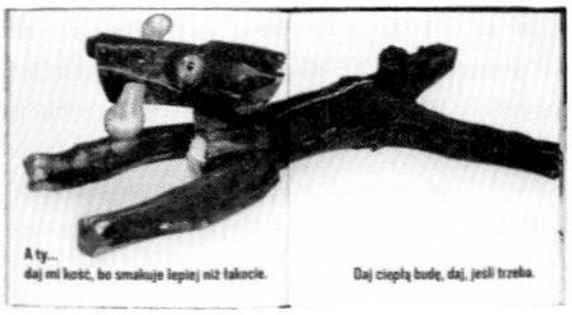
A ty...
daj mi kość, bo smakuje lepiej niż łakocie.
Daj ciepłą budę, daj, jeśli trzeba.

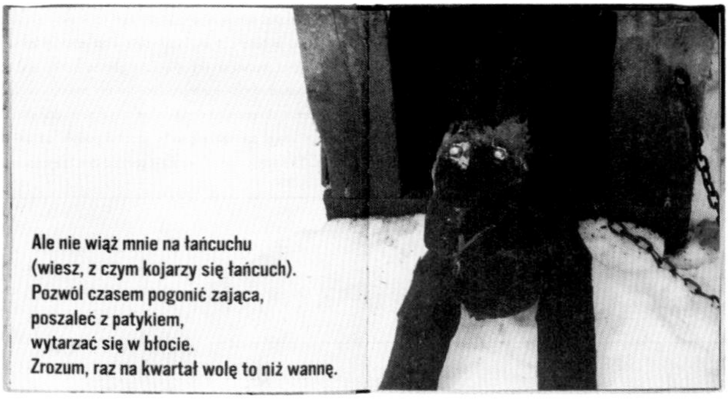
Ale nie wiąż mnie na łańcuchu
(wiesz, z czym kojarzy się łańcuch).
Pozwól czasem pogonić zająca,
poszaleć z patykiem,
wytarzać się w błocie.
Zrozum, raz na kwartał wolę to niż wannę.

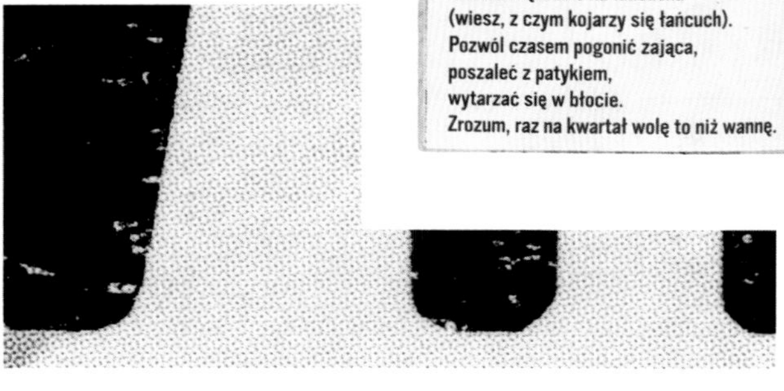

Mówię do ciebie
ja, pies,
twój największy przyjaciel.

Mówię w imieniu wszystkich psów świata,
od ratlerka tyciego jak chomik
po kaukaza ogromnego jak niedźwiedź.

To ty sprawiłeś na swój użytek,
że jamnik może wejść do lisiej nory,
chart dogonić zająca,
a owczarz odgonić wilki
od kóz i owiec.

albo błąkam się bezdomny,
albo mieszkam z braćmi w schronisku.

Będę czekał na ciebie w domu,
pod sklepem, na przystanku.
Kiedy rozdzielą nas setki mil,
powrócę do ciebie jak Lessie.

Będę się ścigał z tobą po łąkach,
rozpędzę twoją chandrę.
Będę twoim zagłówkiem,
u twoich stóp milczącą poduszką.

Podrap mnie za uchem i głaszcz.
Zobacz, jak wilgotnieją moje oczy.

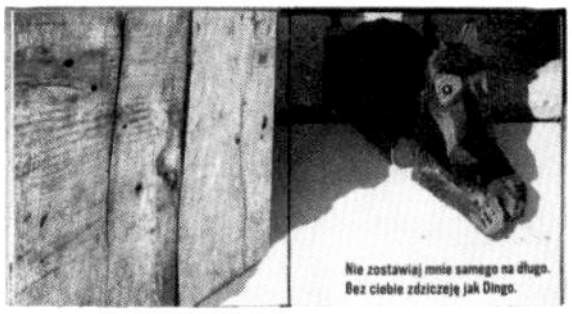
Nie zostawiaj mnie samego na długo.
Bez ciebie zdziczeję jak Dingo.

Strach pomyśleć,
jakie byłoby
bez ciebie
moje psie życie.

Jednak mój los jest w twoich rękach,
mój przyjacielu,
mój panie.
Pewnie wiesz, że kocham cię
z całej psiej mocy
i zawsze będę przy tobie.

des peintres un intérêt
phénomènes naturels et en
les appeler à accomplir leur devoir
comme s'il leur offrait une place aux
des chercheurs tout en se
d'un air offensé. Selon l'analyse
l'observateur de nuages
admet cependant être lui-même un
dessinateur, il n'existe aucune
artistique digne d'illustrer son propos
Face aux sujets comptant parmi les « plus
grands, les plus beaux phénomènes
que le peintre puisse échouer dans la
de nuages » est pour lui
Les meilleurs artistes, qui, même
qui consacraient à leur étude de nuages
méticulosité toute particulière, ne
leurs toiles tout au plus que de purs effets
atmosphériques.
Les mauvais nuages – c'est-à-dire
ceux qui ne sont pas réalistes – sont
de trois façons : certains artistes ne
tiennent pas à l'observation de la nature
se conforment à « l'œil du connaisseur »
voir dans un nuage un chameau, une baleine
ou une brume en forme de baleine
Dans le deuxième groupe, l'outil manque
de finesse, soit dans l'exactitude de
l'observation, soit dans le rendu sur la toile
Les amateurs de nuages pourraient
à la vue de ces toiles, et il n'est de toute
pas question de s'allier à ces artistes
Au troisième groupe, Ley ne fait
reproche, il a même la délicatesse de ne
nommer ni Turner, ni Constable, les deux
grands peintres anglais de nuages
à un passé encore proche. Des
bien que

013 Bm5

这是一本研究云的书。早在 1803 年就有科学家对云的类型进行了分类，但由于缺乏图像资料的支撑，气象学界对于云的分类、命名缺乏共识。之后摄影技术的出现也逐渐被运用到与气象相关的研究中。本书收集了多个地点拍摄的云的照片，以及对于云的相关研究文章，试图对这方面的研究予以总结。本书装订方式为锁线硬精装，封面白色皮面纸印单黑，裱覆卡板，封面三种语言印有本书目录概要。书脊、封面及环衬上不同粗细的网频，似在表达云的浓淡层次。内页主要分为两大部分：照片部分哑面铜版纸四色印刷，展示多地、多角度、多时间段内云层的形状、运动、类别，甚至卫星云图；文字部分胶版纸印单黑，主要为相关研究、观测历史、分析图表等气象学专业资料。两大部分穿插装订，在纸张触感和阅读节奏上起到了很好的协调作用。

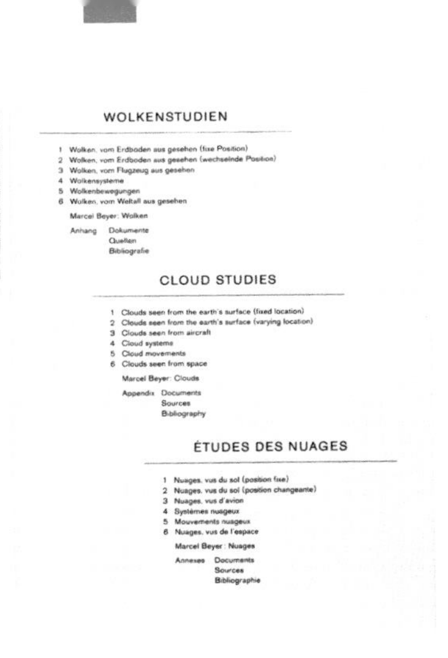

WOLKENSTUDIEN
1 Wolken, vom Erdboden aus gesehen (fixe Position)
2 Wolken, vom Erdboden aus gesehen (wechselnde Position)
3 Wolken, vom Flugzeug aus gesehen
4 Wolkensysteme
5 Wolkenbewegungen
6 Wolken, vom Weltall aus gesehen
Marcel Beyer: Wolken
Anhang Dokumente
Quellen
Bibliografie
CLOUD STUDIES
1 Clouds seen from the earth's surface (fixed location)
2 Clouds seen from the earth's surface (varying location)
3 Clouds seen from aircraft
4 Cloud systems
5 Cloud movements
6 Clouds seen from space
Marcel Beyer: Clouds
Appendix Documents
Sources
Bibliography
ÉTUDES DES NUAGES
1 Nuages, vus du sol (position fixe)
2 Nuages, vus du sol (position changeante)
3 Nuages, vus d'avion
4 Systèmes nuageux
5 Mouvements nuageux
6 Nuages, vus de l'espace
Marcel Beyer : Nuages
Annexes Documents
Sources
Bibliographie

This is a book about clouds. As early as the year 1803, some scientists classified clouds. However, due to the lack of the supported image data, there is no consensus on the classification and naming of clouds in meteorology until the emerging photography technology was gradually applied to the related meteorological research. This book collects photos of clouds taken in many places, as well as related research essays on clouds and tries to summarize the research in this area. This hardback book is bound by thread sewing. The cover is made of white faux leather printed in black, mounted by graphic board. There are three languages on the cover, printing the catalogue and outline of the book. The different frequencies of printing screen – used for the spine, the cover and the endpaper – are meant to display the layers of the clouds. The inner pages are mainly divided into two parts: the photographs are printed on matte coated art paper in four colors, showing the shape, movement, category and even satellite cloud images of clouds in different places, angles and periods; the text is printed on offset paper in black, including meteorological professional data such as observation history, analysis charts and other related research. The two parts are interspersed for binding that coordinates the reading rhythm.

- 铜奖
- 云的研究
- 德国

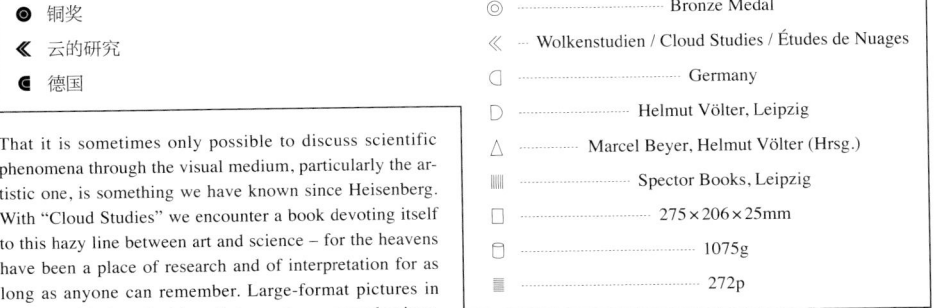

- Bronze Medal
- Wolkenstudien / Cloud Studies / Études de Nuages
- Germany
- Helmut Völter, Leipzig
- Marcel Beyer, Helmut Völter (Hrsg.)
- Spector Books, Leipzig
- 275 × 206 × 25mm
- 1075g
- 272p

That it is sometimes only possible to discuss scientific phenomena through the visual medium, particularly the artistic one, is something we have known since Heisenberg. With “Cloud Studies” we encounter a book devoting itself to this hazy line between art and science – for the heavens have been a place of research and of interpretation for as long as anyone can remember. Large-format pictures in colour or black-and-white, photographs or reproductions, which close in on their subject and then distance themselves from it, a far-reaching typographical concept with an obsessive eye for detail, alternating paper and the haptic features of the cover as well as the sophisticated edging – all these variations and shifts reflect their subject, as changeable as it is ephemeral: the clouds.

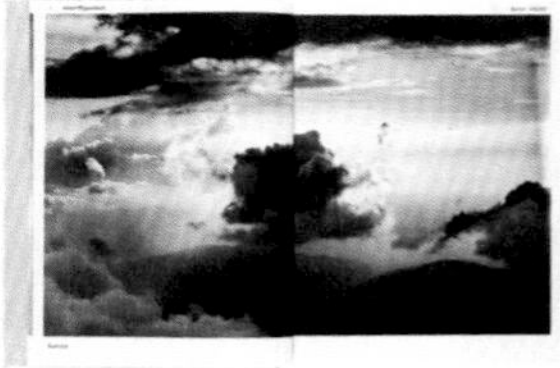

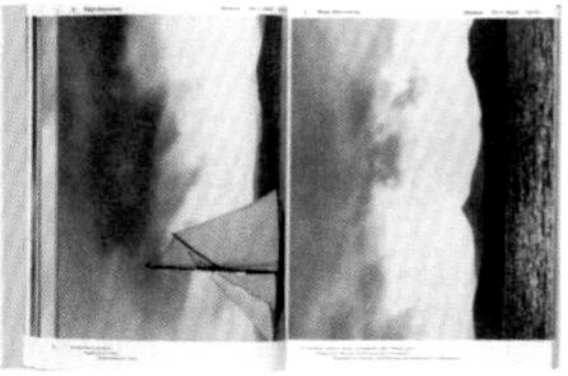

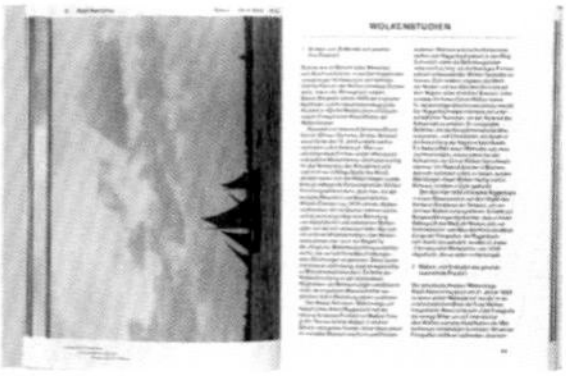
WOLKENSTUDIEN

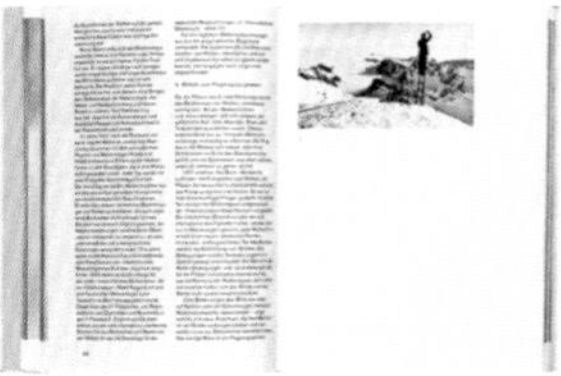

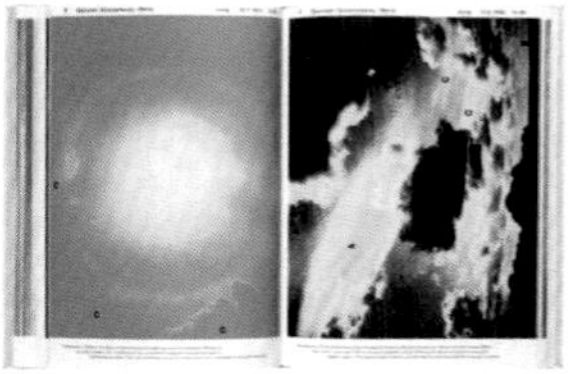

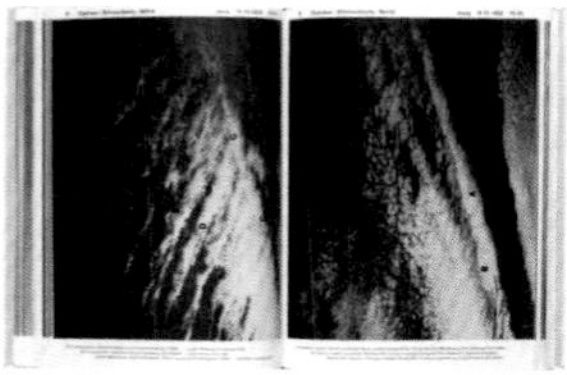

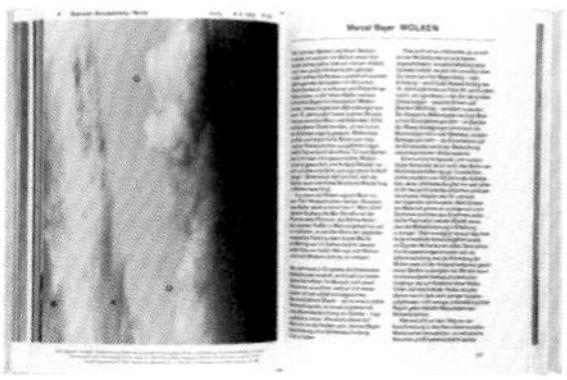
Marcel Bayer WOLKEN

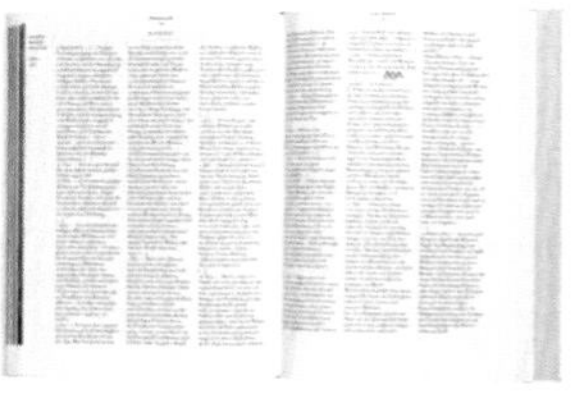

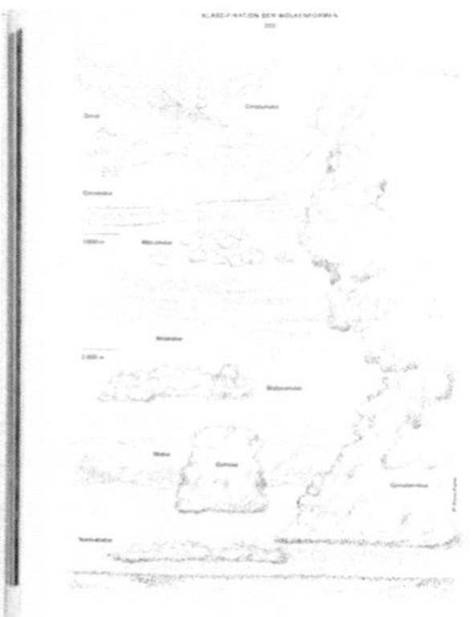
KLASSIFIKATION DER WOLKENFORMEN

KLASSIFIKATION DER WOLKENFORMEN
UNTERARTEN
SONDERFORMEN / BEGLEITWOLKEN

sys
movem

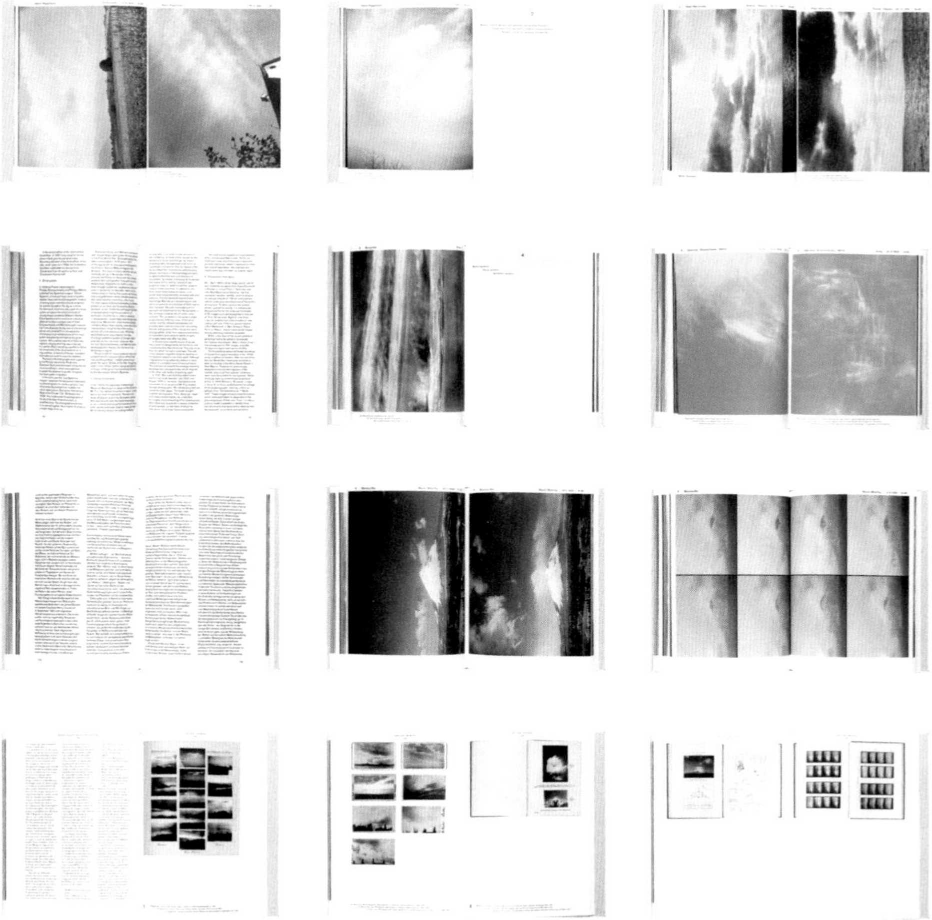

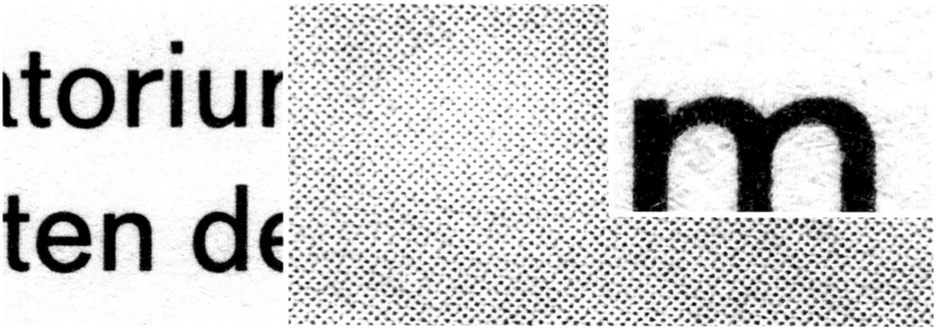
atoriur
ten de
m

Helmut Völter
WOLKENSTUDIEN CLOUD STUDIES ÉTUDES DES NUAGES
Spector Books

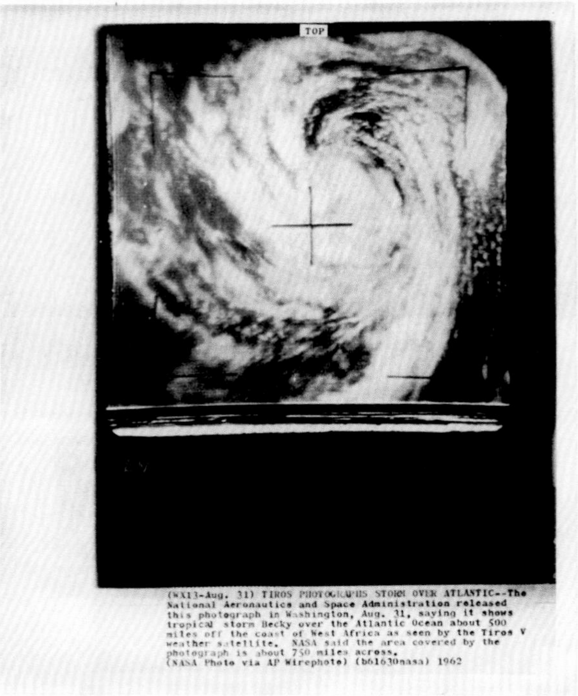
TOP
(WX13-Aug. 31) TIROS PHOTOGRAPHS STORM OVER ATLANTIC--The National Aeronautics and Space Administration released this photograph in Washington, Aug. 31, saying it shows tropical storm Becky over the Atlantic Ocean about 500 miles off the coast of West Africa as seen by the Tiros V weather satellite. NASA said the area covered by the photograph is about 750 miles across.
(NASA Photo via AP Wirephoto) (b61630nasa) 1962

[Tiros fotografiert einen Sturm über dem Atlantik], Pressebild des San Francisco Examiner, 1962
Tiros photographs storm over Atlantic, press photograph from the San Francisco Examiner, 1962
[Tiros photographie une tempête au-dessus de l'Atlantique], photo de presse du San Francisco Examiner. 1962.

HOOK, R.
A METHOD FOR MAKING A HISTORY OF THE WEATHER
Th. Sprat: The History of the Royal Society of London. For the Improving of Natural Knowledge, pp.173-179
London 1667

LAMARCK, J.B.
SUR LA FORME DES NUAGES
Annuaire météorologique pour l'an X de l'ère de la République Française, No.3, pp.149-164
Paris 1802

HOWARD, L.
ON THE MODIFICATION OF CLOUDS, and on the Principles of Their Production, Suspension, and Destruction: Being the Substance of an Essay Read before the Askesian Society in the Session 1802–3
Tilloch's Philosophical Magazine, vol.16, pp.97-107, 344-357, vol.17, pp.5-11,
1803
reprinted in:
Gustav Hellmann (ed.): Neudrucke von Schriften und Karten über Meteorologie und Erdmagnetismus, No.3, Berlin 1894

LAMARCK, J.B.
TABLEAU DES DIVISIONS DE LA RÉGION DES MÉTÉORES
Annuaire météorologique pour l'an XI de l'ère de la République Française, No.4, pp.122
Paris 1803

LAMARCK, J.B.
SUR LE POINT DU VENT, observé d'après la forme et l'arrangement des nuages
Annuaire météorologique pour l'an XII de l'ère de la République Française, No.5, pp.150-162
Paris 1804

HOWARD, L.
UEBER DIE MODIFICATIONEN DER WOLKEN
Annalen der Physik, vol.21, pp.137-159
Halle 1805

LAMARCK, J.B.
NOUVELLE DÉFINITION DES TERMES que j'emploie pour exprimer certaines formes de nuages qu'il importe de distinguer dans l'annotation de l'état du ciel
Annuaire météorologique pour l'an XIII de l'ère de la République Française, No.6, pp.112-133
Paris 1805

LAMPADIUS, W.A.
SYSTEMATISCHER GRUNDRISS DER ATMOSPHÄROLOGIE
Freyberg 1806

FORSTER, Th.
RESEARCHES ABOUT ATMOSPHERIC PHÆNOMENA
2nd ed.
London 1815

GOETHE, J.W.
WOLKENGESTALT NACH HOWARD
Zur Naturwissenschaft überhaupt, vol.1, No.3, pp.97-124
1820

GOETHE, J.W.
LUKE HOWARD TO GOETHE. A Biographical Scetch
Zur Morphologie, vol.1, No.4, pp.357-359
1822

KÄMTZ, L.F.
VON DEN HYDROMETEOREN
Lehrbuch der Meteorologie, vol.1, pp.377-405
Halle 1831

KÄMTZ, L.F.
WOLKEN
Vorlesungen über Meteorologie, pp.142-154
Halle 1840

FRITSCH, K.
ÜBER DIE PERIODISCHEN ERSCHEINUNGEN AM WOLKENHIMMEL
Abhandlungen der Königl. Böhmischen Gesellschaft der Wissenschaften, 5. Folge, vol.4
Prag 1846

RENOU, E.
INSTRUCTIONS MÉTÉOROLOGIQUES
Annuaire de la Société Météorologique de France, vol.3, pp.142-146
Paris 1855

RUSKIN, J.
OF CLOUD BEAUTY
Modern Painters, vol.5, part 7, pp.109-162
New York 1860

HAMERTON, Ph.G.
THE RELATION BETWEEN PHOTOGRAPHY AND PAINTING
A Painter's Camp in the Highlands and Thoughts about Art, pp.200-242
London 1862

POËY, A.
SUR DEUX NOUVEAUX TYPES DE NUAGES observés à La Havane, dénommés Pallium (Pallio-Cirrus et Pallio-Cumulus) et Fracto-Cumulus
Comptes rendus hebdomadaires des séances de l'Académie des Sciences, vol.56, pp.361-364
Paris 1863

TOMLINSON, Ch.
CLOUDS SEEN FROM A BALLOON
The Rain-Cloud and the Snow-Storm, pp.76-80
London [1864]

POËY, A.
CONSIDÉRATIONS SYNTHÉTIQUES SUR LA NATURE, la constitution et la forme des nuages
Annuaire de la Société Météorologique de France, vol.13, pp.100-112
Paris 1865

TYNDALL, J.
NOTE ON THE FORMATION AND PHENOMENA OF CLOUDS
Proceedings of the Royal Society of London, vol.17, pp.317-319
London 1869

POËY, A.:
NEW CLASSIFICATION OF CLOUDS
Annual report of the Board of Regents of the Smithsonian Institution, pp.432-456
Washington 1870

FRITSCH, K.
UEBER POËY'S NEUE EINTHEILUNG DER WOLKEN
Zeitschrift der Österreichischen Gesellschaft für Meteorologie, vol.6, pp.321-327
1871

POËY, A.
NOUVELLE CLASSIFICATION DES NUAGES, suivie d'instructions pour servir à l'observation des nuages et des courants atmosphériques
Annales Hydrographiques, 1e série, vol.35, pp.615-715
Paris 1872

WALTON, E.
CLOUDS. Their Forms and Combinations
London 1873

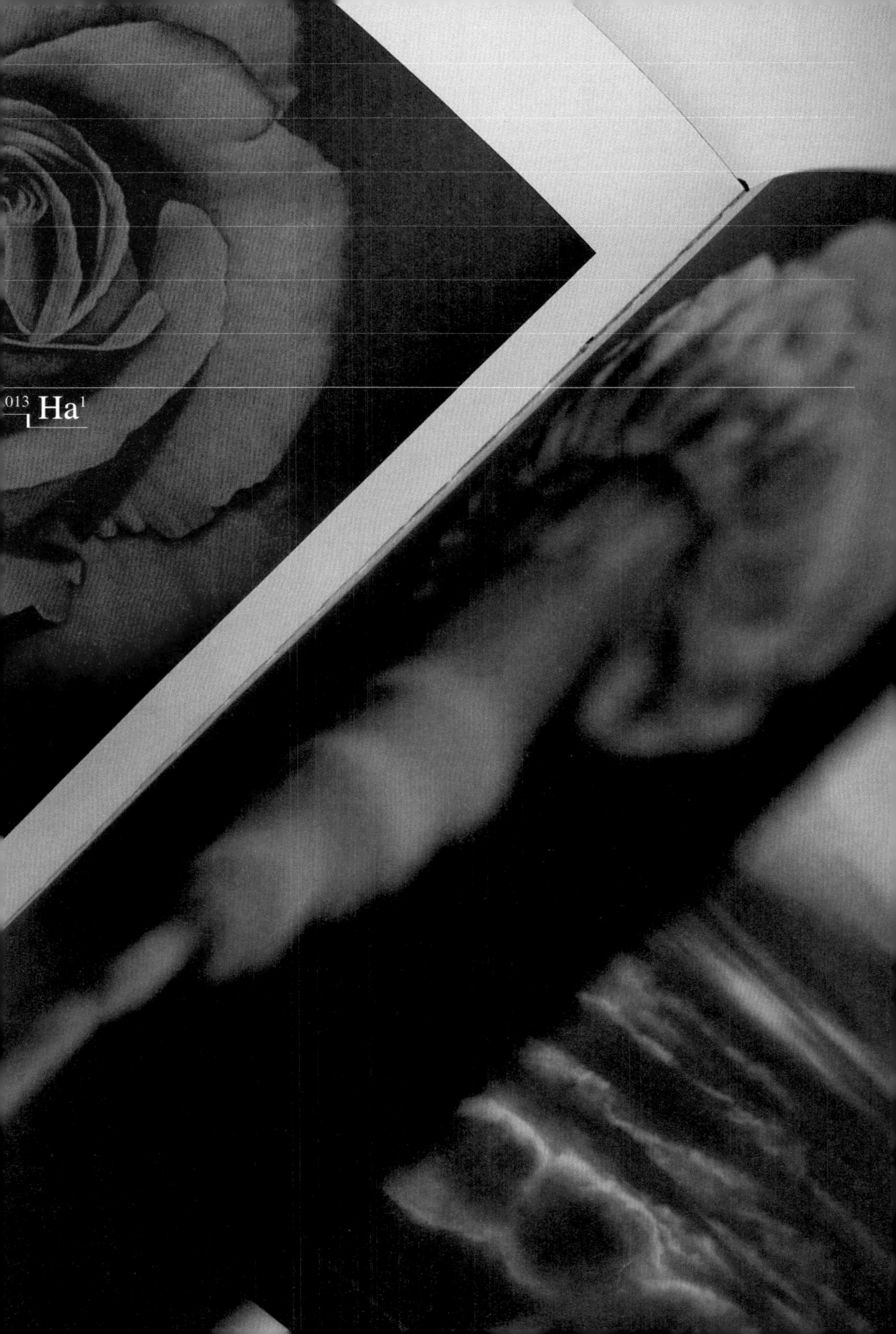
013 Ha1

这是美国艺术家罗伯特·隆戈（Robert Longo）的炭笔画作品集。书中将艺术家多次展览的展陈布置、作品分类、艺术评论集合起来，形成了对隆戈炭笔画作品的全面解读。本书装订形式为锁线硬精装，封面灰色卡板标题烫黑，书脊黑色布面包覆在封面上，标题烫黑，封底哑面艺术纸四色黑印刷。黑色卡纸的环衬后，内页哑面艺术纸四色印刷。书中将展陈与作品分类结合起来，对艺术家作品进行了深度解读。艺术家的作品大多为黑色炭笔画，为在纸面上还原作品的丰富层次，黑色的作品采用四色印刷，色调调整统一。上下书口与右侧翻口刷黑色，在内容和形式上都与作品形成呼应。

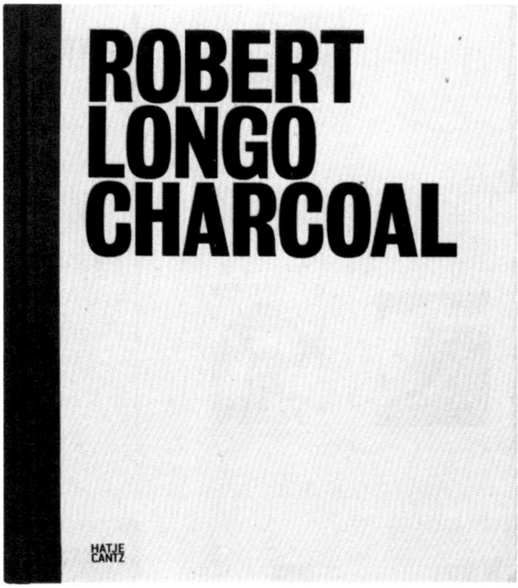
ROBERT
LONGO
CHARCOAL
HATJE
CANTZ

This is a collection of charcoal paintings by Robert Longo, an American artist. The book combines the installations, classified works and art reviews of the artists' former exhibitions to elaborate on Longo's charcoal paintings. The hardback book is bound by thread sewing. The cover is made of gray graphic board with the title hot stamped in black. The spine is covered with black cloth and the back cover is made of matte art paper in black. Followed by the black cardboard endpaper, the inside pages are printed on offset paper in four colors. The book combines exhibition design with classification, and deeply interprets the works. Most of the works are black charcoal paintings. In order to restore the rich layers of the works on the pages, black works are printed in four colors, and the tone is unified. The three edges of the book are brushed with black inks, which correspond to the works' theme and form.

荣誉奖

罗伯特·隆戈——木炭画

德国

Honorary Appreciation

Robert Longo – Charcoal

Germany

Stapelberg & Fritz, Stuttgart

Hal Foster, Kate Fowle, Thomas Kellein

Hatje Cantz Verlag GmbH, Ostfildern

304×259×33mm

2114g

251p

Robert Longo's charcoal drawings are presented here perfectly in the harmony of overall features – material and finish, colour, typography and continuity of design. The black spine with glossy black embossed text, the sturdy and uncoated grey card with author and title in narrow bold, also black embossed sans-serif, the black endpapers as well as three-sided black edging set the mood for this chiaroscuro – that play of light and shade which Longo pursues with such a high degree of mastery in all his works. The haptically pleasant soft uncoated paper requires lithography and printing of the highest order of precision to ensure that the charcoal drawings, in some cases very dark, do not merge into the paper. How closely the technical reproduction here has stuck to the essence of the original work.

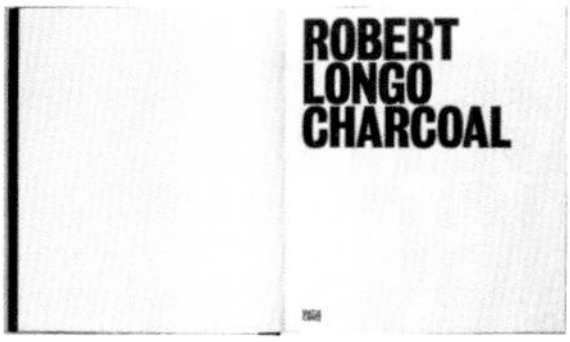
ROBERT
LONGO
CHARCOAL

CONTENTS

ROBERT
LONGO:
FOREWORD

HAL
FOSTER:
THE
AMERICAN
FRIEND

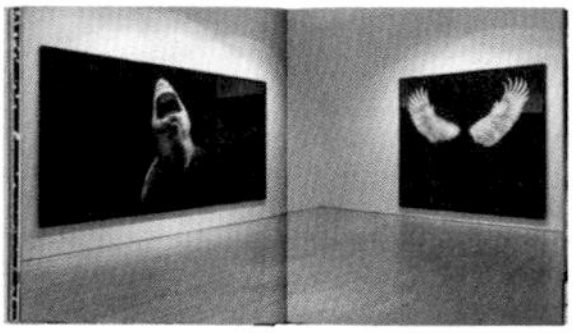

THE
MYSTERIES

THOMAS
KELLEIN:
A KINGDOM
OF
IMAGES

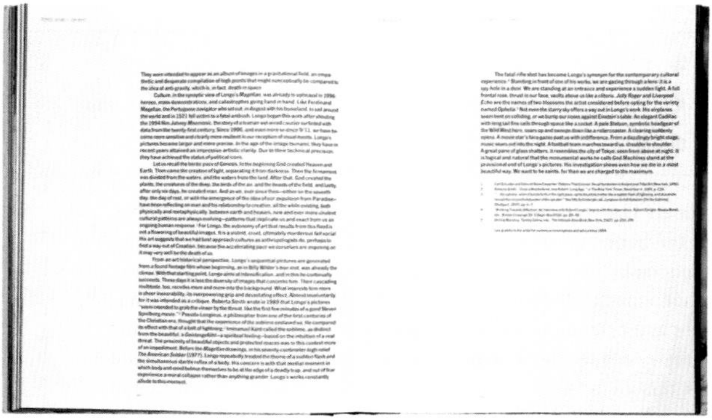

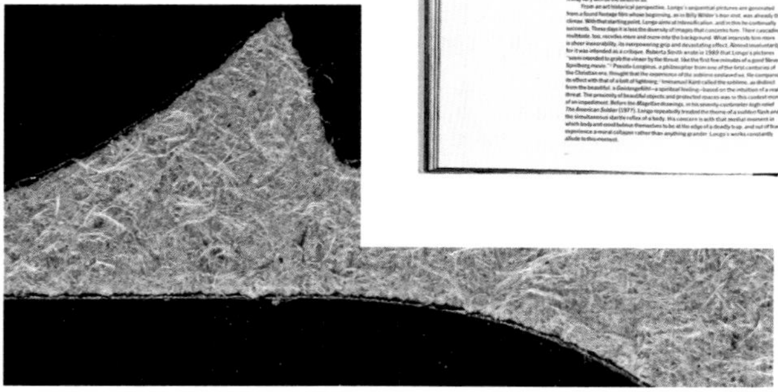

THE
ESSENTIALS

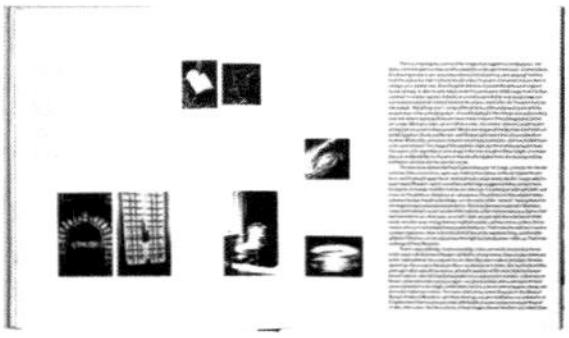

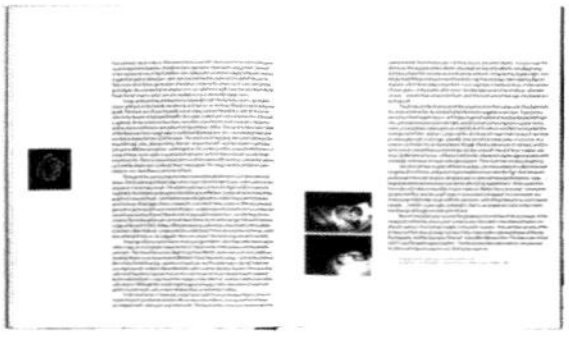

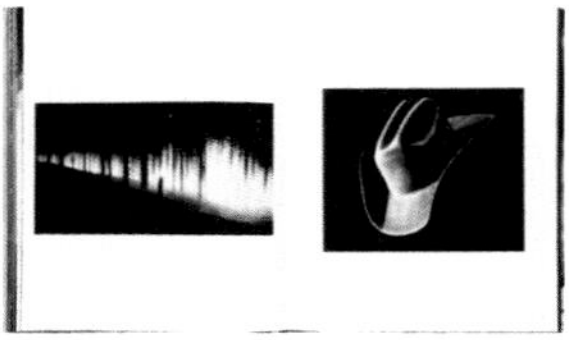

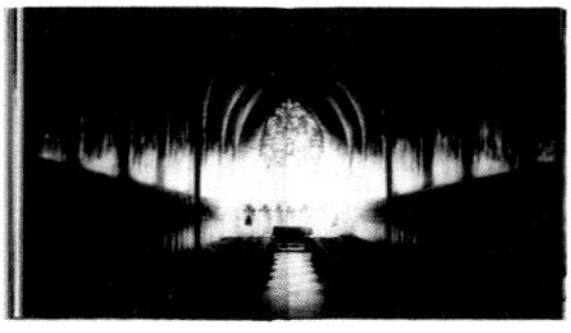

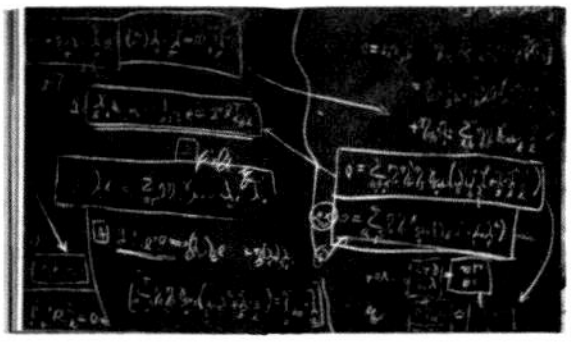

ET
IN
ARCADIA
EGO

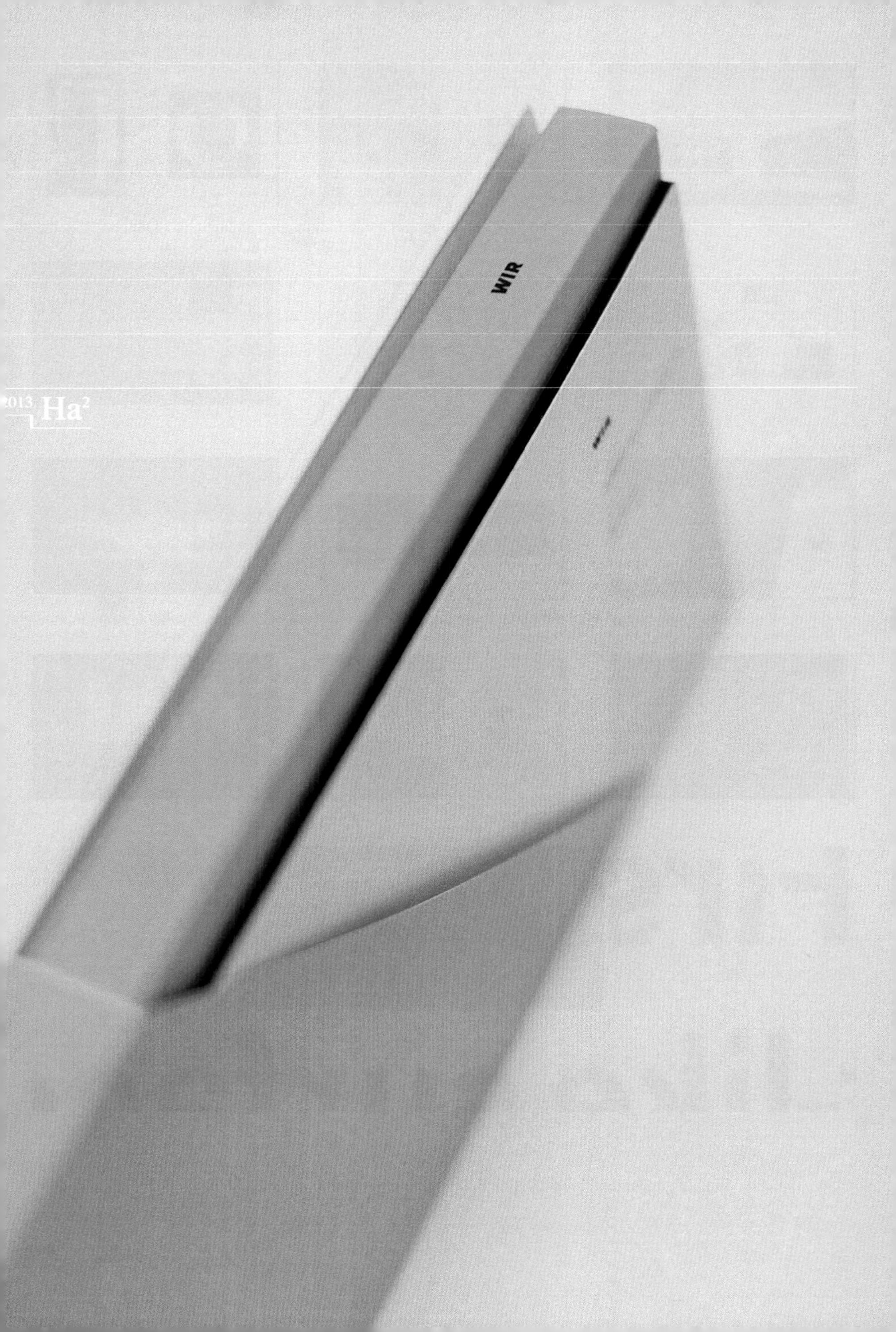
WIR
013 Ha2

这是俄国小说家尤金·扎米亚金（Jewgenij Samjatin）的著作《我们》的德文版。作为世界上第一部反乌托邦小说，书中采用笔记的形式，模拟了一个生活在未来世界中的模范公民的口吻，描述了一个高度数字化、集中统一管理的社会，讽刺了极权主义的种种弊端。这个德文版分为两部分，第一部分为小说主体，锁线硬精装，呈现为一个笔记本的形态。封面白色哑面艺术纸蓝黑双色印刷，裱覆双层黑色卡板，并继而裱覆在绿色的封面上。黑色卡纸环衬后内页书写纸蓝黑双色印刷，“笔记本”形体的内页上部蓝色线格，主人公用以记录标题和页码。书籍的另一部分为德国学者理查德·萨格（Richard Saage）的评论与解读，骑马锁线装订，胶版纸单黑印刷。封面采用黑白长短不同的双色卡纸对裱，形成与小说类似的笔记本形式。两部分通过斜切的黄色卡纸封套合在一起。

It is the German version of the book We written by Jewgenij Samjatin, a Russian novelist. As the first anti-utopian novel in the world, the book takes the form of notes to simulates the tone of a model citizen living in a future world. It describes a highly digitized society that is centralized and dictatorial, and satirizes the disadvantages of totalitarianism. The German version is divided into two parts. The first part is the main body of the novel, which is hardbound with thread sewing like a notebook. The cover is printed on white matte art paper in blue and black, laminated with double-layer black cardboard. Followed by the black cardboard endpaper, the inner pages are printed on writing paper in blue and black. There are blue line grids on the 'notebook' pages for the protagonist to record the title and page number of his notes. The other part of the book is the comment and interpretation of Richard Saage, a German scholar. This part is bound by saddle stitch with thread sewing and printed on offset paper. The cover is mounted by black and white cardboard with different length. The two parts are put together in a slipcase made of yellow cardboard with a beveled cut.

荣誉奖

我们

德国

Honorary Appreciation

WIR

Germany

Gaston Isoz, Berlin

Jewgenij Samjatin

disadorno edition, Berlin

217×120×22mm

450g

215+19p

Book and accompanying booklet are held together by a sleeve construction made of yellow card cut diagonally; the book sports a green jacket with superimposed covers – the explanatory supplement has single-section binding protected by a double jacket. On smooth bookprinting paper in a slim and manageable format the typography makes an immaculate impression. The attentive use of non-justified text is particularly striking. The high area at the top of the double-page spreads is reserved for a vertical notation system whose fine blue lines suggest an association with officialdom. The aggregate colour scheme makes connections with regard to content: the “green wall” encircling the “single state”, or the “majolica blue”, colour of the sky and of the uniforms. And not least the constructivist allusion made by the outer sleeve construction.

134

Ich nahm den Telefonhörer ab:

»I, sind Sie's?«

»Ja. Warum rufen Sie so spät an?«

»Ich … ich wollte Sie bitten … ich möchte gern, dass Sie morgen neben mir sitzen. Liebste …«

Liebste – ich hauchte das nur. Sie gab lange keine Antwort. Mir war, als hörte ich in I's Zimmer jemand flüstern. Endlich sagte sie: »Nein, ich kann nicht. Sie wissen, wie gern ich es möchte, aber es geht wirklich nicht. Warum? Morgen werden Sie es sehen.«

Nacht.

EINTRAGUNG NR. 25

Übersicht
Niederfahrt vom Himmel
Die größte Katastrophe
der Geschichte
Alles unbestimmt

Vor der Wahl, als sich alle erhoben und die feierlichen Klänge der Hymne über unseren Köpfen brausten, vergaß ich für eine Sekunde, was I von diesem Feiertag gesagt und was mich so sehr beunruhigt hatte. Ja, ich glaube, ich vergaß sogar sie. Ich war wieder der kleine Junge, der an diesem Tag einmal bitterlich geweint hatte, weil er einen winzigen, ihm allein sichtbaren Fleck auf seiner Uniform entdeckte. Wenn auch keiner der rings um mich Stehenden sah, wie viele schwarze Flecken jetzt auf mir waren, so wusste ich doch allzu gut, dass ein Verbrecher wie ich unter diesen Menschen mit den offenen, ehrlichen Gesichtern nichts zu suchen hatte. Ach, ich wäre am liebsten aufgesprungen, um mit tränenerstickter Stimme die ganz Wahrheit über mich herauszuschreien. Mag es auch mein Ende sein, dachte ich, was tut es? Wenn ich mich nur eine einzige Sekunde lang so rein und gedankenlos fühlen könnte wie dieser kindlich-blaue Himmel!

Aller Augen blickten zum Himmel auf; in dem morgendlich keuschen Blau zitterte ein kaum erkennbarer Punkt, bald dunkel, bald im Licht blitzend. Das war Er, der von den Himmeln zu uns herniederstieg, ein neuer Jehova im Flugzeug, weise, gütig und streng wie der Gott der Alten. Mit jeder Minute kam Er näher und näher, immer höher schlugen Ihm Millionen Herzen entgegen. Jetzt musste Er uns sehen! Im Geist schaute ich mit Ihm auf die Menge herab, auf die punktierten Linien der konzentrisch angeordneten Tribünen, die wie Kreise eines Spinnennetzes waren. Im Zentrum dieses Netzes würde sich gleich eine weiße weise Spinne niederlassen, der Wohltäter in weißer Uniform, der uns in seiner Weisheit unsere Hände und Füße mit den starken Fäden des Glückes gebunden hat.

Seine erhabene Niederfahrt war beendet, die brausende Hymne verstummte, alle hatten sich wieder hingesetzt. Da er-

JEWGENI

RO

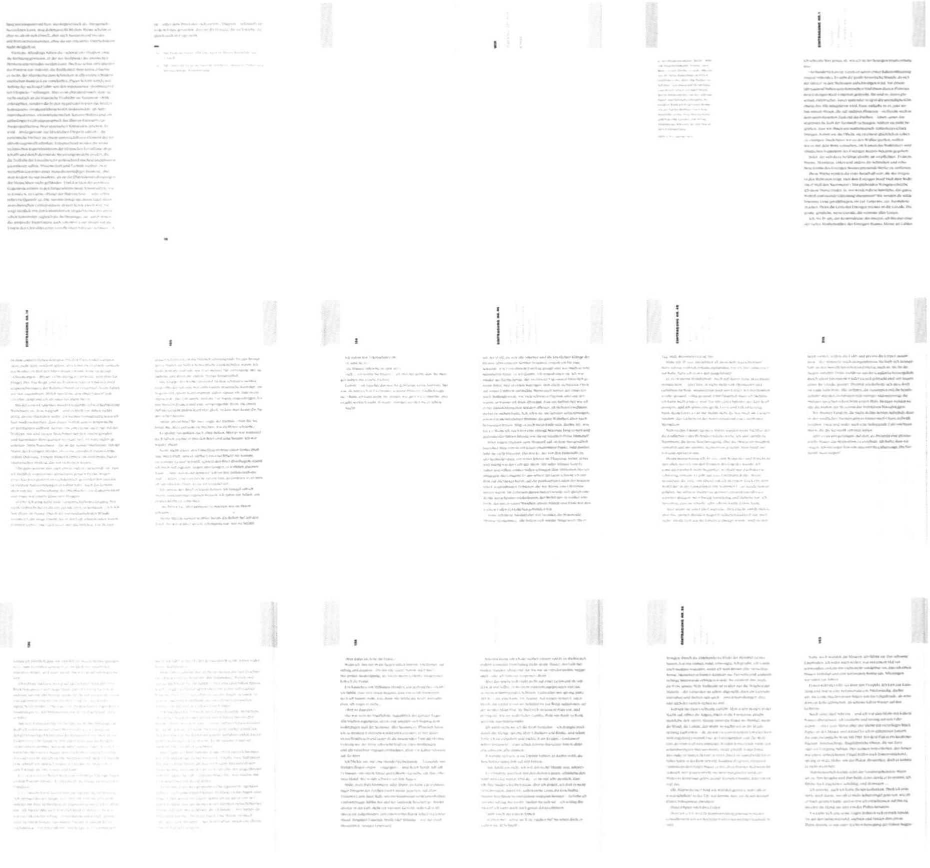

es Einz
STRU
ıbt
efähigt
IS „W
und andere die Schö

A Different
Re-Discovery
of Something
(Lost)
The Dynamics
of Found
Footage Film
After 1990
Marc Glö

这是荷兰国家电影资料馆在 2012 年开馆展的图录。本书通过对老胶片录影的发现、分析、重构，展现了电影制作人和艺术家如何将历史老镜头重塑为新作品，揭示了老胶片素材的采样和重新创作（发现）的可能性。本书的装订方式为锁线和插页混合胶装，护封胶版纸四色印刷，可以展开为一张海报，内封黑卡纸印银。内页分为两大部分：对老胶片电影的重新发现、解读的文字为胶版纸单黑印刷，在书中呈现为插页；老电影截图、胶片扫描的展示则为哑面涂布纸四色印刷，在书中锁线。两部分相互穿插混合装订。文本部分设置为不等宽双栏设置，页码置于双栏中间，在建立规律的同时，形成了独特的文本节奏。多种老胶片录影本身的独特色彩和影片质量也在书中形成了丰富的肌理。

Found
Footage
Cinema
Exposed
Amsterdam University Press
EYE Film Institute Netherlands

This is the catalogue of the opening exhibition of the National Film Archives of the Netherlands in 2012. Through the discovery, analysis and reconstruction of old film videos, this book shows how filmmakers and artists reconstruct old historical footage into new works, and reveals the possibility of sampling and re-creating old film material. The book is bound by sewing thread and glue together. The jacket is printed in four colors on offset paper, which can be unfolded into a poster. The inside cover is printed on black cardboard with silver. The inner pages are divided into two parts: the interpretations of old film films are printed on offset sheet in black; while matte coated paper printed in four colors are used to display the screenshots of old films. The two parts are interspersed for binding. The text part is using unequal width double-column setting and the page number is in the middle of the two columns. It forms a unique text rhythm. The unique color and film quality of various old film videos also provide rich texture in the book.

- 荣誉奖
- 寻找录像——电影曝光
- 荷兰

- Honorary Appreciation
- Found Footage. Cinema Exposed
- The Netherlands
- Joseph Plateau, grafisch ontwerpers
- Marente Bloemheuvel, Giovanna Fossati, Jaap Guldemond (Hrsg.)
- Amsterdam University Press, EYE Film Institute Netherlands, Amsterdam
- 258 × 209 × 18mm
- 438g
- 255p

It looks as if the current media dispute – electronic, audiovisual and printed media entangled in a St. Vitus's dance – is not quite as dramatic in the daily work of picture and information creators; the existence of this catalogue on the opening show of the new Netherlands film institute proves it. The beautiful, matte American cover that can be folded out into a poster introduces the mood and the tone of the imagery. The means of artistic book design employ the characteristics of a cinematographic atmosphere, divide words and pictures, differentiate between visual and textual qualities and create a self-contained documentation. The bookmark is a film strip – what a friendly gesture!

Found Footage: Cinema Exposed
Found Footage at EYE Film Institute

Introduction

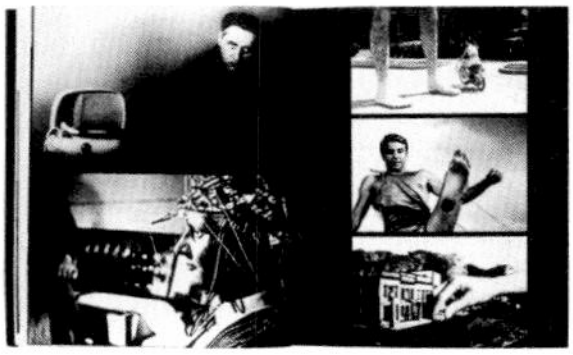

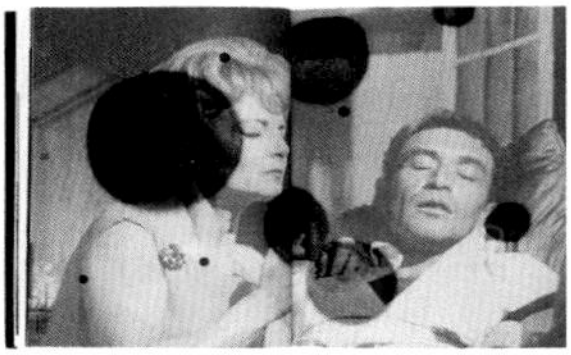

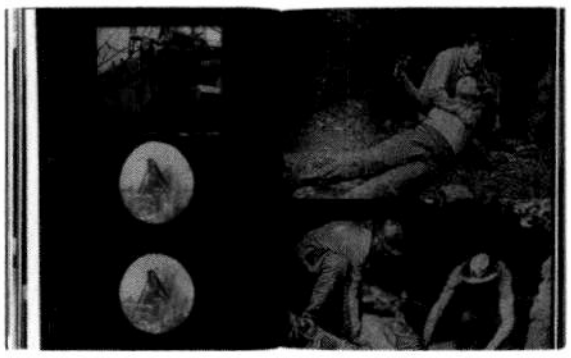

148
STRATEGIES OF APPROPRIATION

149
Bits & Pieces
nr. 613

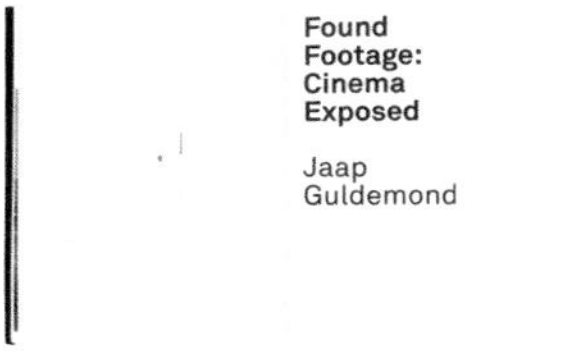
Found Footage: Cinema Exposed
Jaap Guldemond

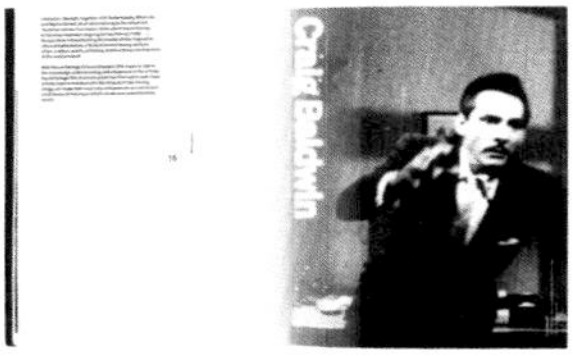
Craig Baldwin

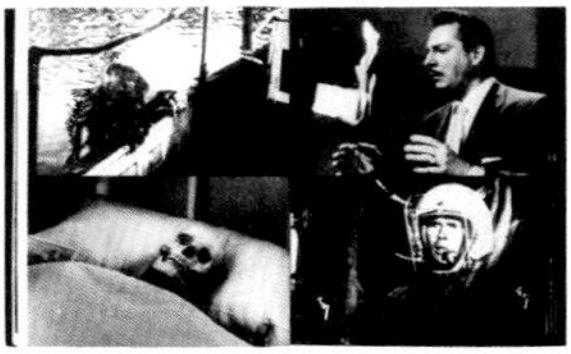

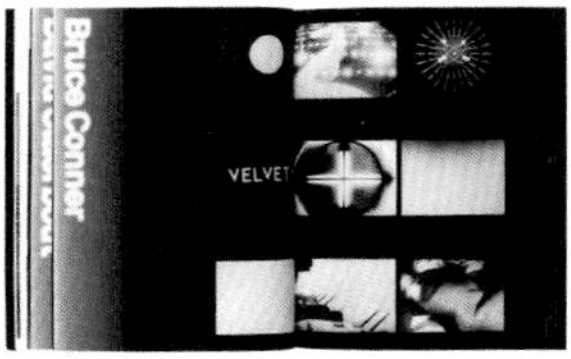
Bruce Conner
VELVET

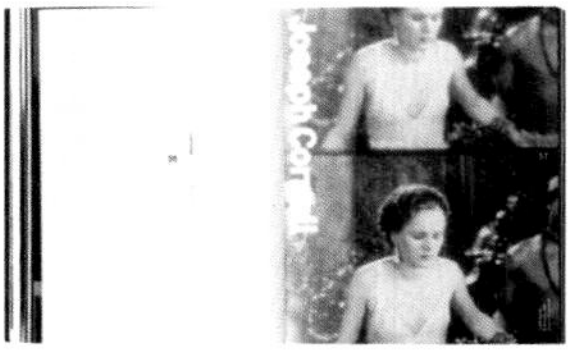
Joseph Cornell

"I did many films in the past years that do not consist of found footage. (...) I have made many films that could be counted as belonging to the genre 'direct cinema.' But even for them I am primarily looking for preexisting scenarios. Just recently, I filmed in an advertising agency where a potential client was being shown a campaign they wanted to sell him. I did not construct
this story, rather I found it given. Images and sounds that we find without already having been aware that they exist are like an *objet trouvé*. Imagine a child who is walking on the beach and suddenly reaches for a pebble that evokes the lines of a human face. The *objet-trouvé* artist tries to preserve this notion of amazement. This also expresses that you
cannot create meaning systematically, as the big production companies, cinema, and TV stations try to do. One needs chances and the luck of a finder.
In Vienna there is a building by Hans Hollein with a small balcony attached that is designed for taking pictures of the *Stephansdom* across the way. Documentary films often refuse to take
the ideal and allocated point of view in order to seek out their own – which could be the back of the building. I like looking at something as it is being presented to me. And then I make the picture appear a little bit different from how it wants to be seen, to perform a small alteration as we know it from pop art."

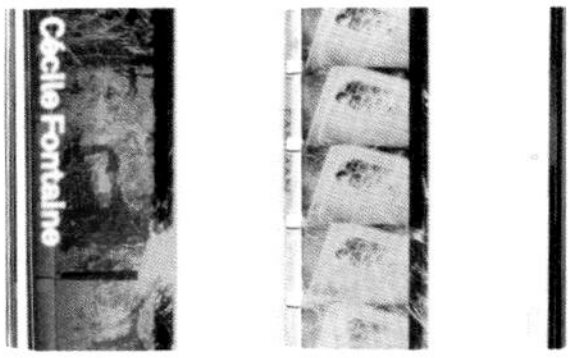
Cécile Fontaine

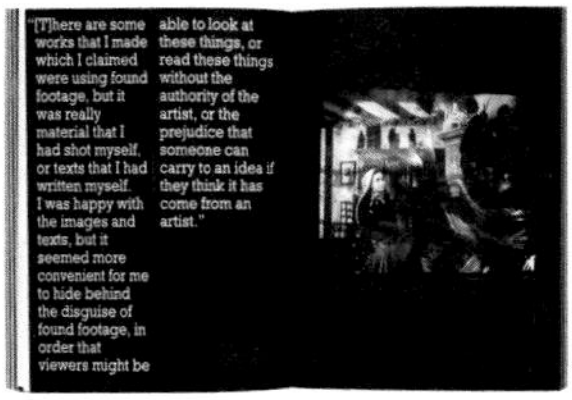
"[T]here are some works that I made which I claimed were using found footage, but it was really material that I had shot myself, or texts that I had written myself. I was happy with the images and texts, but it seemed more convenient for me to hide behind the disguise of found footage, in order that viewers might be
able to look at these things, or read these things without the authority of the artist, or the prejudice that someone can carry to an idea if they think it has come from an artist."

Ken Jacobs

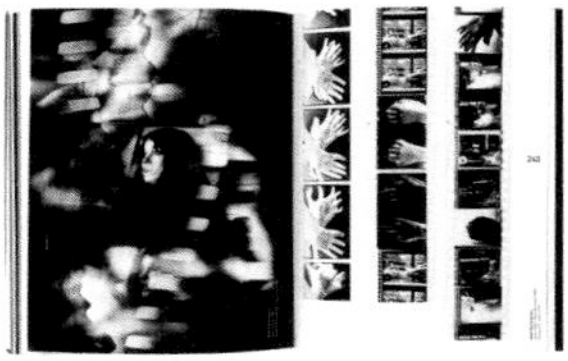

n Far
comparison to

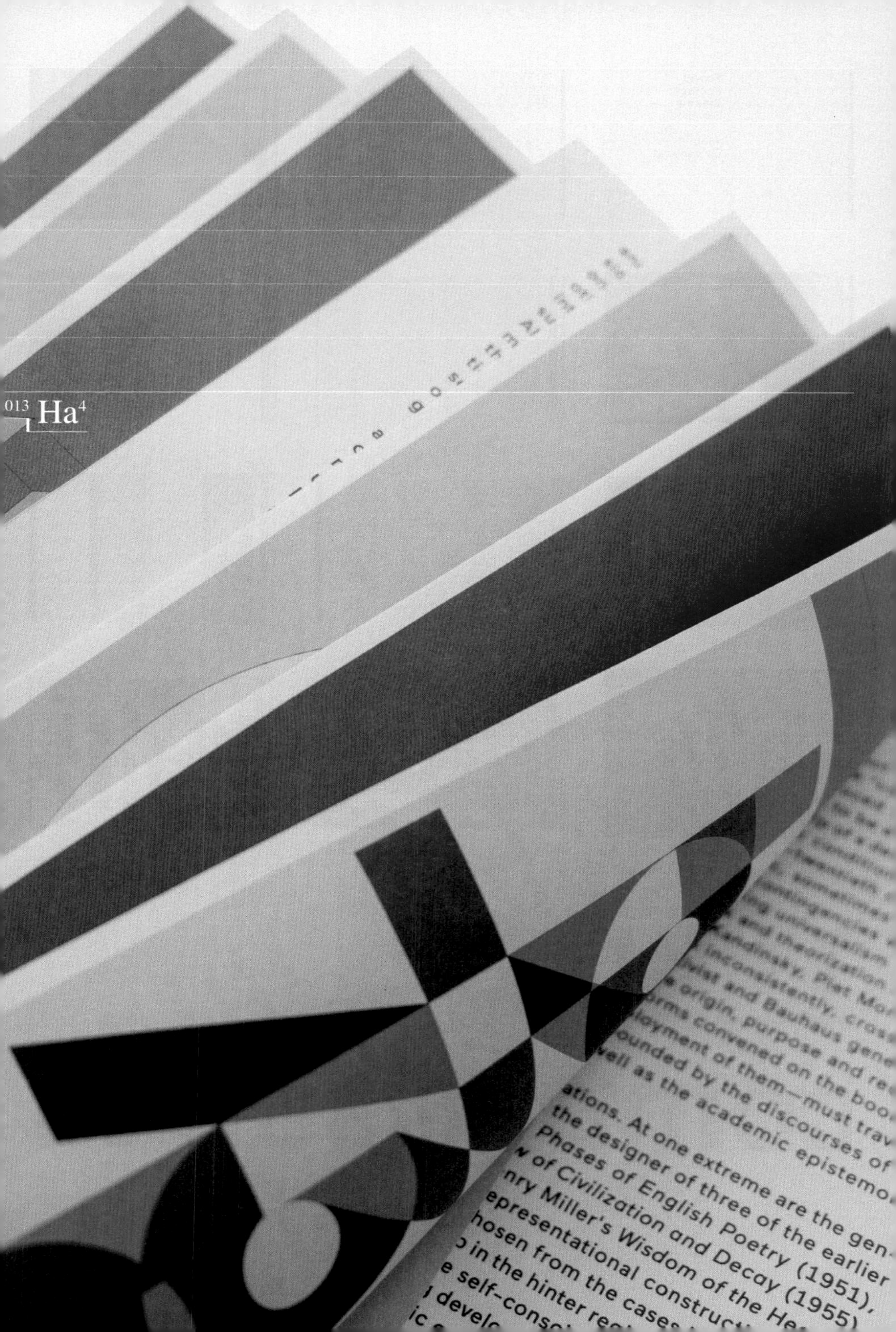
Piet
inconsistently,
and Bauhaus
origin, purpose and
convened on the
of them—must
by the discourses of
well as the academic
At one extreme are the
the designer of three of the earlier
Phases of English Poetry (1951),
of Civilization and Decay (1955)
Miller's Wisdom of the
from the cases
in the hinter

荷兰艺术家托马斯·拉特（Thomas Raat）试图解读20世纪40年代至70年代欧洲流行的现代主义平面设计背后的视觉含义、哲学思考而创作了大量的油画作品，这些作品的图案来自当时流行的被称之为"egghead"（书呆子）的现代主义平面设计风格的书籍封面。本书为这些作品的作品集、评论家分析，并且从更多维的角度丰富自己的理念和探究。本书为包背夹插页的混合胶装，封面牛皮卡纸印单黑，封面勒口展开图处理自油画作品展览的现场照片。内页胶版纸四色印刷，图片为油画所画的现代主义平面设计风格的图书封面插图，将标题单独排在页面中的短插页上，而艺术家自己对于封面的解读、封面和标题蕴含的意义都罗列出单词，按照从A到Z的顺序印在包背页内侧，在书籍的物理结构上呈现出"隐含意义"。

An
Inquiry
into
Meaning
and
Truth
and
More...
Thomas
Raat

Dutch artist Thomas Raat tried to interpret the visual meaning and philosophical thinking behind the popular modernist graphic design in Europe from the 1940s to 1970s, and created a large number of oil paintings. The patterns of his paintings came from the book covers of the prevailing modernist graphic design style at that time, which was called 'egghead'. This book is a collection of these works and related reviews, which enrich the research from a more dimensional perspective. This double-leaved book is bound by glue with inserts. The cover is made of kraft cardboard printed in black and the image on the flap is the photograph taken at the exhibition. The inside pages are printed on offset paper in four colors. The oil-painting illustrations are book covers in modernist graphic design style, and the book titles are typeset separately on the shorted inserts. The artist's interpretation of the covers and the titles are summarized to several words, printed on the inside of the double-leaved pages in the order of A to Z. The physical structure of the book shows 'implied meaning'.

- 荣誉奖
- 关于意义、真理及其他的探讨
- 荷兰

- Honorary Appreciation
- An Inquiry into Meaning and Truth and More…
- The Netherlands
- Edwin van Gelder / Mainstudio
- Thomas Raat
- Onomatopee, Eindhoven
- 300×200×7mm
- 330g
- 80p

The publication with paintings by Thomas Raat performs a fascinating act of regression. The artist expurgates the covers of one to two-generation-old paperbacks of their textual constituents: he transfers the abstract geometrical graphics into large-scale oil paintings – and book titles become picture titles. The autonomy of the visual quality of this former commercial art can suddenly be physically felt – probably for the first time. And now these paintings are redirected to the original medium, but the cover motifs are translated into contents. The illustrated pages are interleaved by half-width coloured pages carrying the titles in free alignment and bold Grotesque – in a quasi emblematic process, with Lemma, Pictura and the absent epigram, the actual text of the book. Is there any other way for hidden poetry to unfold itself than in the medium 'book'?

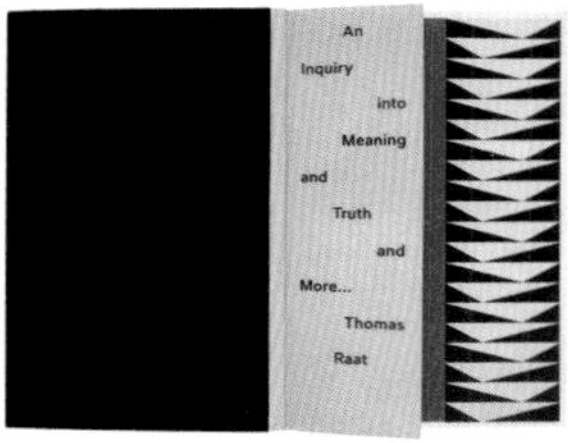
An
Inquiry
into
Meaning
and
Truth
and
More...
Thomas
Raat

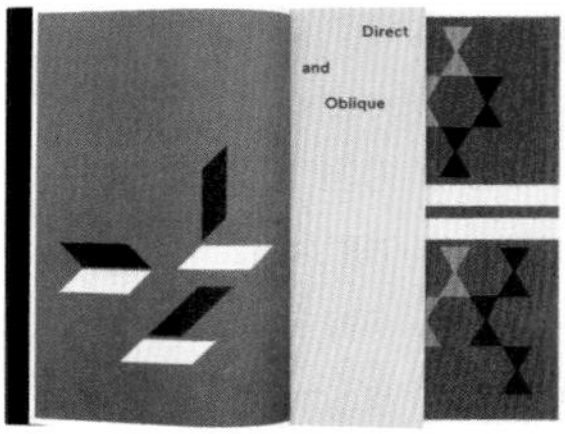
Direct
and
Oblique

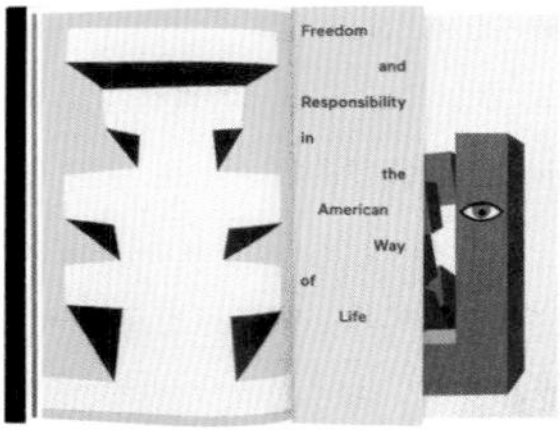
Freedom
and
Responsibility
in
the
American
Way
of
Life

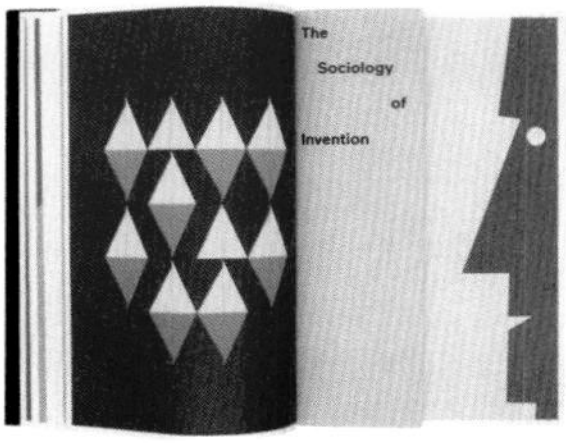
The
Sociology
of
Invention

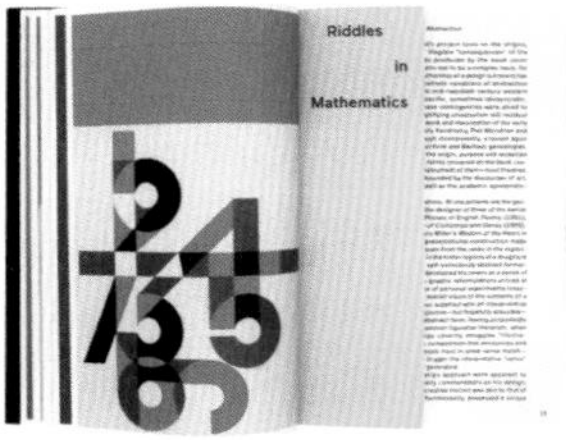
Riddles
in
Mathematics

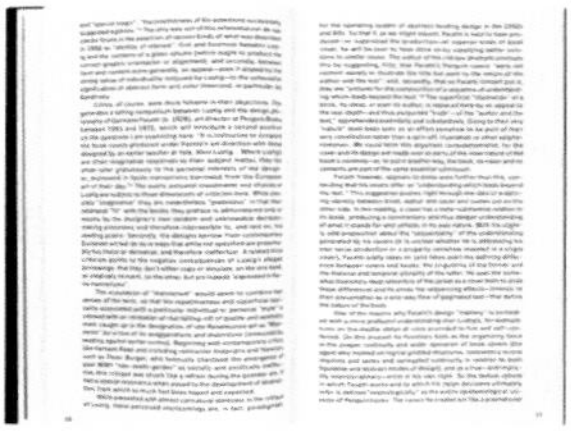

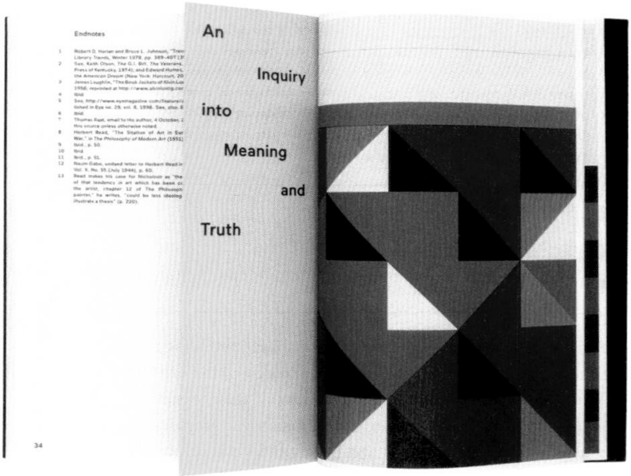
Endnotes
An
Inquiry
into
Meaning
and
Truth
34

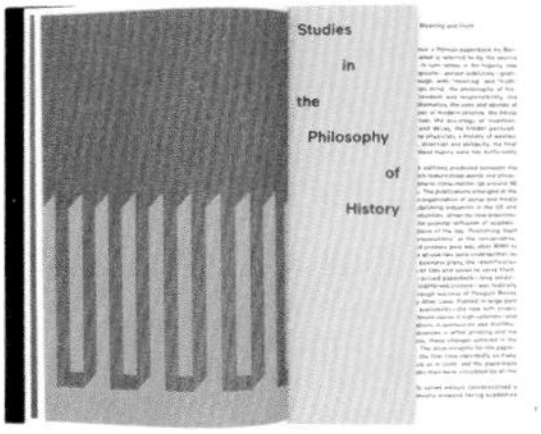
Studies
in
the
Philosophy
of
History

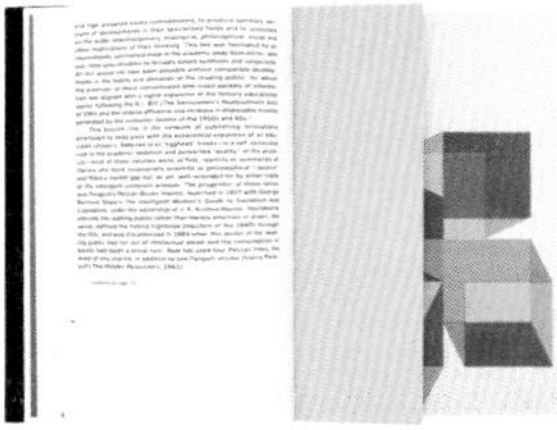

Perspektiven

The
Final
Hours

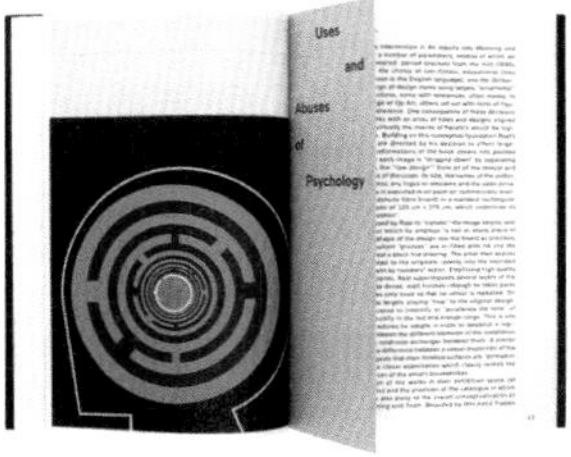
Uses
and
Abuses
of
Psychology

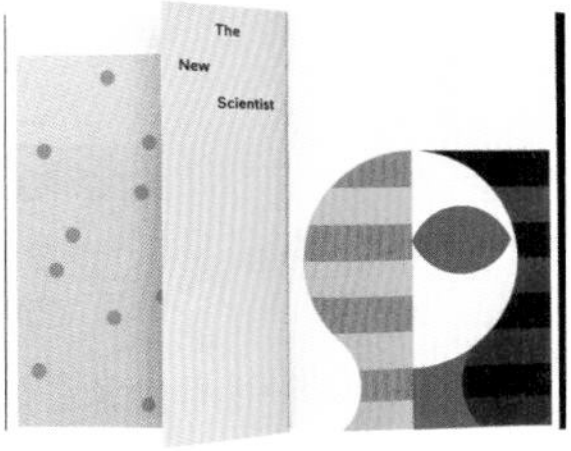
The
New
Scientist

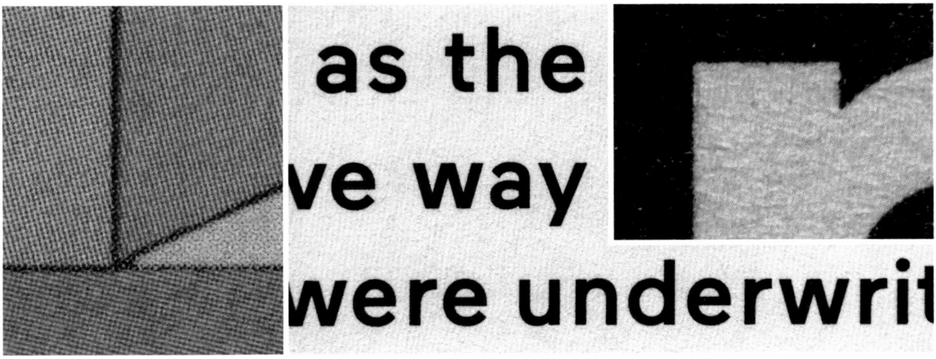
as the
ve way
were underwrit

- La Résidence
- Raum, verschraubt mit der Zeit / Space, Twisted with Time
- The Story of Paper-cut
- Utilité
- What We Wear
- IM WALGAU GEMEINDEN gemeinsam
- Drawing a Hypothesis
- Zoom in, Zoom out
- Libuše Niklová
- Poems by Wen Aiyi
- Gomorrah Girl
- Pampilio
- 100 % Ivanovo. Agitprop Fabrics and Designs of 1920 –1930s
- 2666

2012

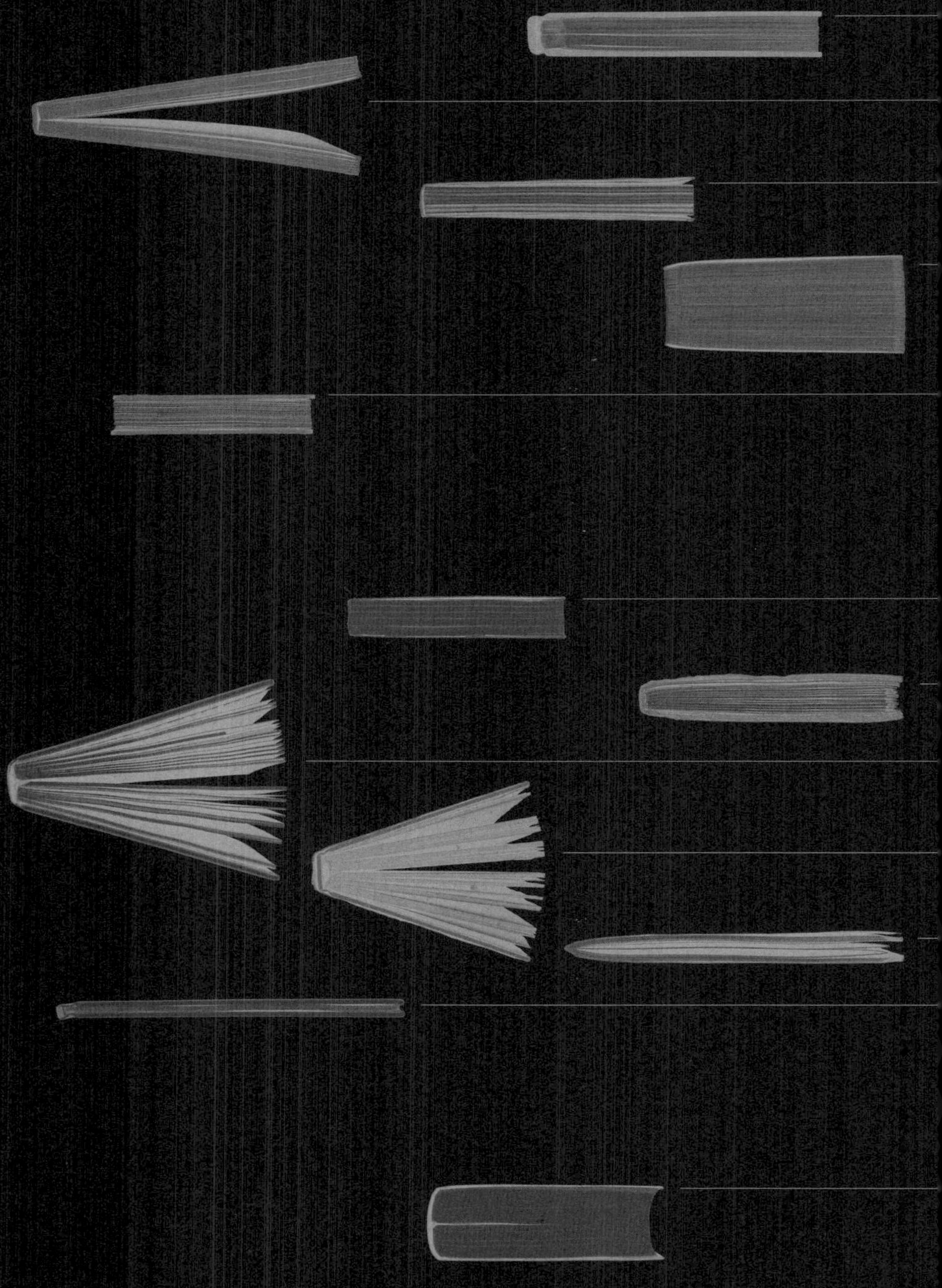

2012 Gl 0713

2012 Gm 0721

2012 Sm[1] 0727

2012 Sm[2] 0733

2012 Bm[1] 0739

2012 Bm[3] 0745

2012 Bm[4] 0751

2012 Bm[5] 0757

Libuše Niklová

2012 Ha[1] 0761

2012 Ha[2] 0765

2012 Ha[3] 0769

2012 Ha[5] 0773

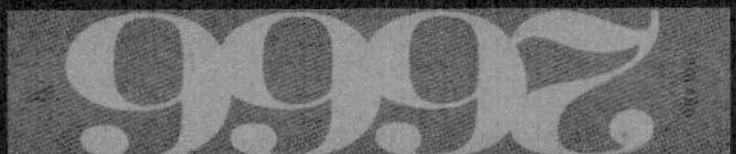

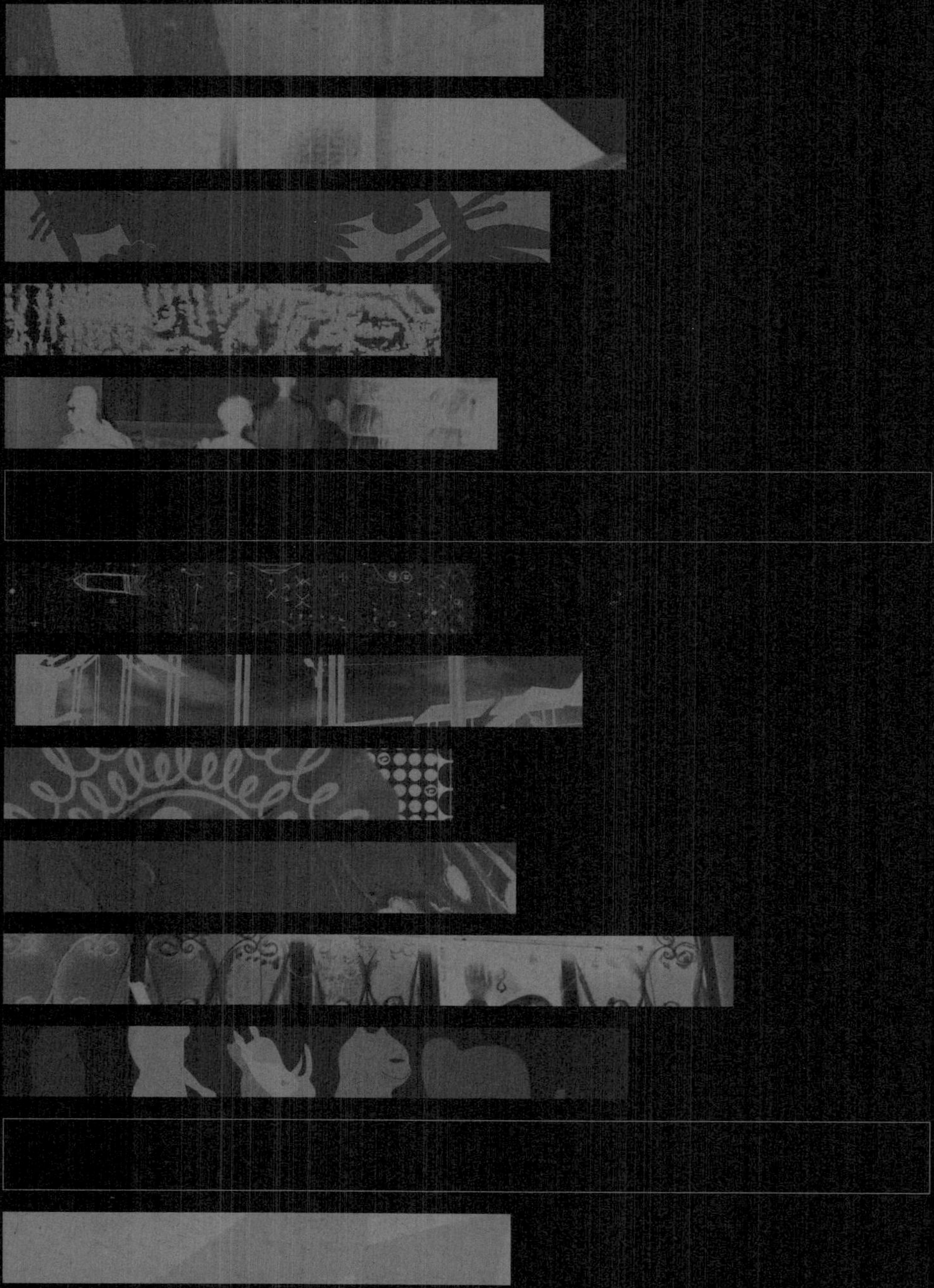

Today I was in the park behind Saint Gilles
the sky, and the city below vanished in a mi
smog. People were grouped like islands all o

relevant circles – well known that I support architecture in the
One part of this commitment was also the projekt_A, which
a vehicle following the 2003 European Capital of Culture

Space, Twisted with Time is the
ture Styria Yearbook to appear
with the biennially awarded Sty

…、贾思勰、李清照、刘墉、郑板桥等等，他们的故事美哉人间，这个地方就是中

…就是潍坊，尽管我是长期生长在条件更加优越的青岛市，但是对于老家潍坊的情

…egels of wetten. In dit boek geldt maar één b

overwhelmed by the sheer amount, and this
was just one single warehouse. I wondered
who would benefit from clothes that we in the
Netherlands consider to be waste material.
Seeing this huge heap made me aware that

Dhaka, Bangladesh, February

There are currently at least two opposing ways of und
ing the term 'diagram'. Some see diagrams above all
to systematisation – "problem solvers, because they

浅草青春新大地

Niklová vnesla do tradičních
modelů gumových dutých figu-
rek vyráběných v padesátých

Libuše Niklová
pletely new arti
to the tradition

…是感动，我们可以从爱艺对"生与死"的终极关注中

…震撼："你将离去／请留下你那浓密的翠绿……让我们

in this region apart and young girls seem
be destined to live on the street.

Les animaux parcouraient la forêt en long et en large, br
des racines, des branches d'arbres et tout ce qu'ils pouvaient
Mais ça ne suffisait pas. Ils avaient toujours aussi faim.

lakun eller denna brist på bibliografiskt vetande, som bara kan
på hans ungdom, minskade inte på minsta vis den förbländni
beundran romanen ingav honom.

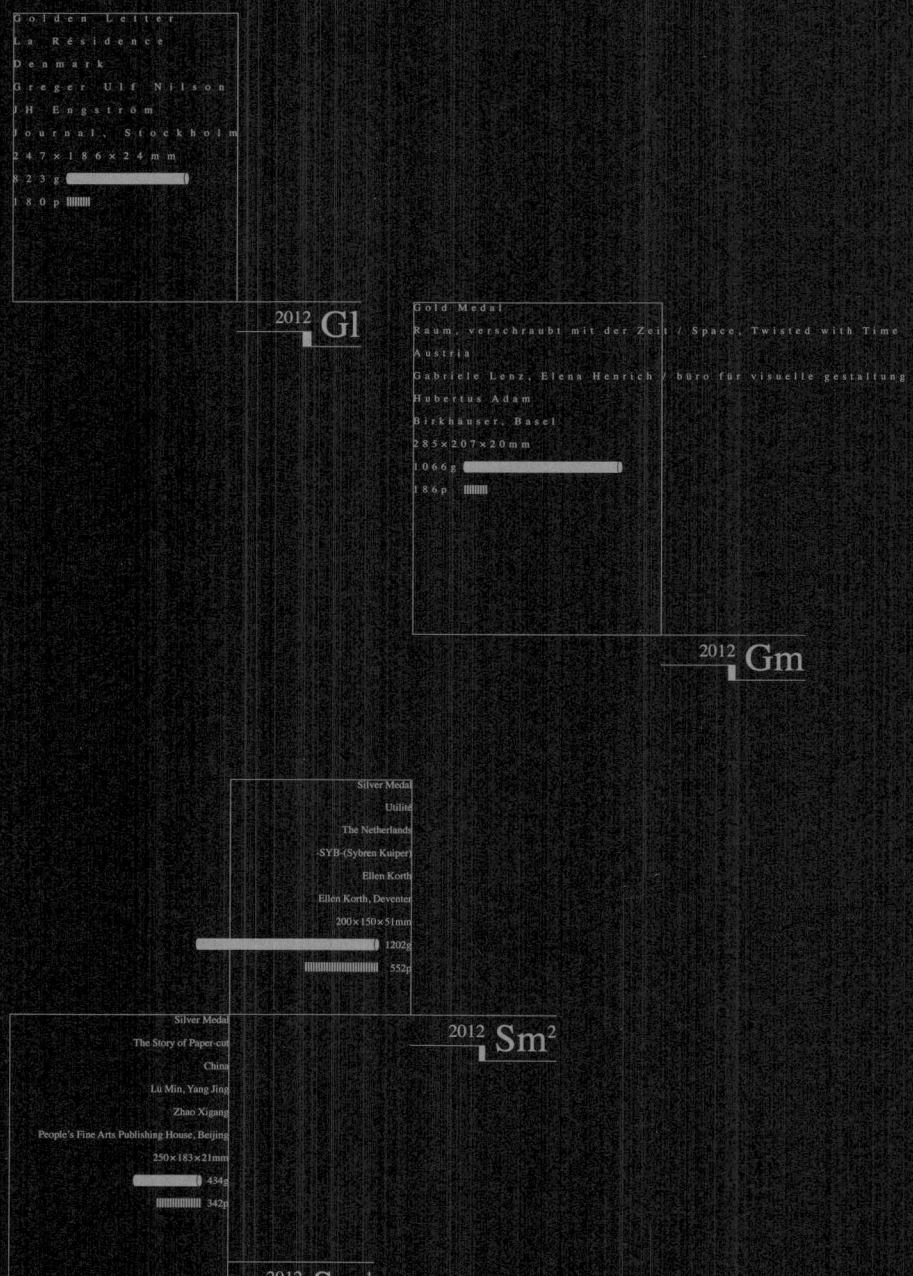

Golden Letter

La Résidence

Denmark

Greger Ulf Nilson

JH Engström

Journal, Stockholm

247×186×24mm

823g

180p

2012 Gl

Gold Medal

Raum, verschraubt mit der Zeit / Space, Twisted with Time

Austria

Gabriele Lenz, Elena Henrich / büro für visuelle gestaltung

Hubertus Adam

Birkhäuser, Basel

285×207×20mm

1066g

186p

2012 Gm

Silver Medal

Utilité

The Netherlands

-SYB-(Sybren Kuiper)

Ellen Korth

Ellen Korth, Deventer

200×150×51mm

1202g

552p

2012 Sm2

Silver Medal

The Story of Paper-cut

China

Lü Min, Yang Jing

Zhao Xigang

People's Fine Arts Publishing House, Beijing

250×183×21mm

434g

342p

2012 Sm1

2012 Ha[1]

Honorary Appreciation
Poems by Wen Aiyi
China
Liu Xiaoxiang, Gao Wen
Wen Aiyi
Writers Press, Beijing
235×144×23mm
617g
380p

2012 Ha[2]

Honorary Appreciation
Gomorrah Girl
The Netherlands
-SYB- (Sybren Kuiper)
Valerio Spada
Cross Editions, Paris
335×227×3mm
330g
40p

2012 Ha[5]

Honorary Appreciation
2666
Sweden
Nina Ulmaja
Roberto Bolaño
Albert Bonniers Förlag, Stockholm
233×165×47mm
1180g
1056p

2012 Ha[3]

Honorary Appreciation
Pampilio
Poland
Monika Hanulak
Irena Tuwim, Monika Hanulak
Wytwórnia, Warschau
287×237×9mm
455g
32p

2012 Ha[4]

2012 Bm[1]

Bronze Medal
What We Wear
The Netherlands
Teun van der Heijden / Heijdens Karwei
Pieter van den Boogert
Keff & Dessing Publishing, Amsterdam
226×174×26mm
691g
140p

2012 Bm[3]

Bronze Medal
Drawing a Hypothesis
Austria
Simona Koch
Nikolaus Gansterer
Springer Wien New York
215×145×25mm
574g
348p

2012 Bm[4]

Bronze Medal
Zoom in, Zoom out
Taiwan, China
Fang Hsin-Yuan
Victor HSU Architect & Partners; Zoom Design Atelier (Ed.)
Garden City Publishers Ltd., Taipei
265×190×20mm
661g
228p

2012 Bm[5]

Bronze Medal
Libuše Niklová
Czech Republic
Zuzana Lednická / Studio Najbrt
Tereza Bruthansová
Arbor vitae societas, Prag
206×193×29mm
975g
299p

2012 Bm[2]

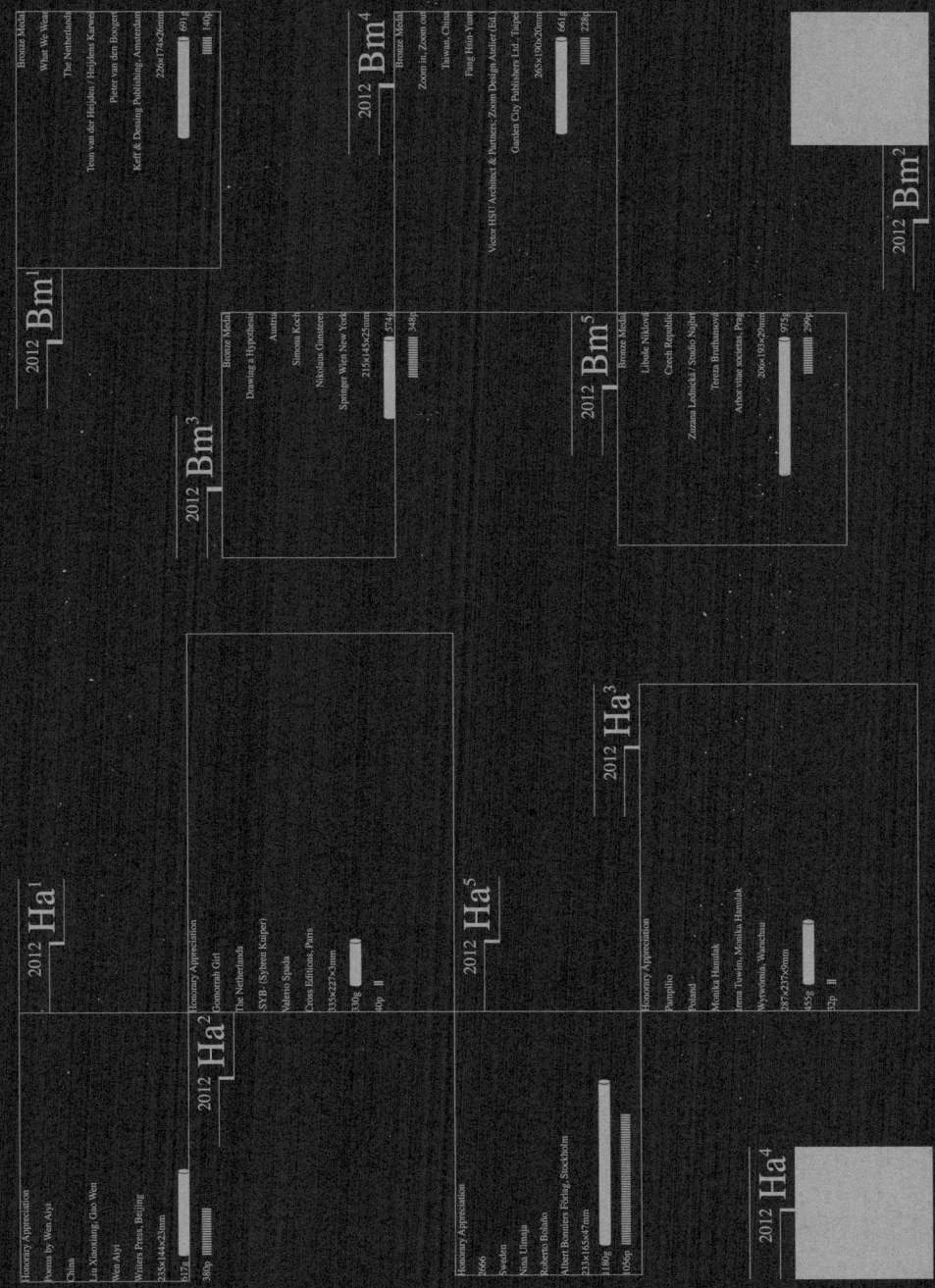

The International Jury

Ahn, Sang-soo (South Korea)
Julia Blume (Germany)
Anders Carpelan (Finland)
Bassam Kahwagi (Lebanon)
Stefanie Langner (Germany)
Harry Metzler (Austria)
Stefanie Schelleis (Germany)

2012

Country / Region

The Netherlands 3
Austria 3
China 2
Denmark 1
Taiwan, China 1
Czech Republic 1
Poland 1
Russia 1
Sweden 1

Designer

Greger Ulf Nilson
Gabriele Lenz, Elena Henrich / büro für visuelle gestaltung
Lü Min, Yang Jing
-SYB-(Sybren Kuiper)
Teun van der Heijden / Heijdens Karwei
Marcel Bachmann, Andrea Redolfi / Atelier Gassner-Redolfi
Simona Koch
Fang Hsin-Yuan
Zuzana Lednická / Studio Najbrt
Liu Xiaoxiang, Gao Wen
-SYB-(Sybren Kuiper)
Monika Hanulak
Evgeny Korneev
Nina Ulmaja

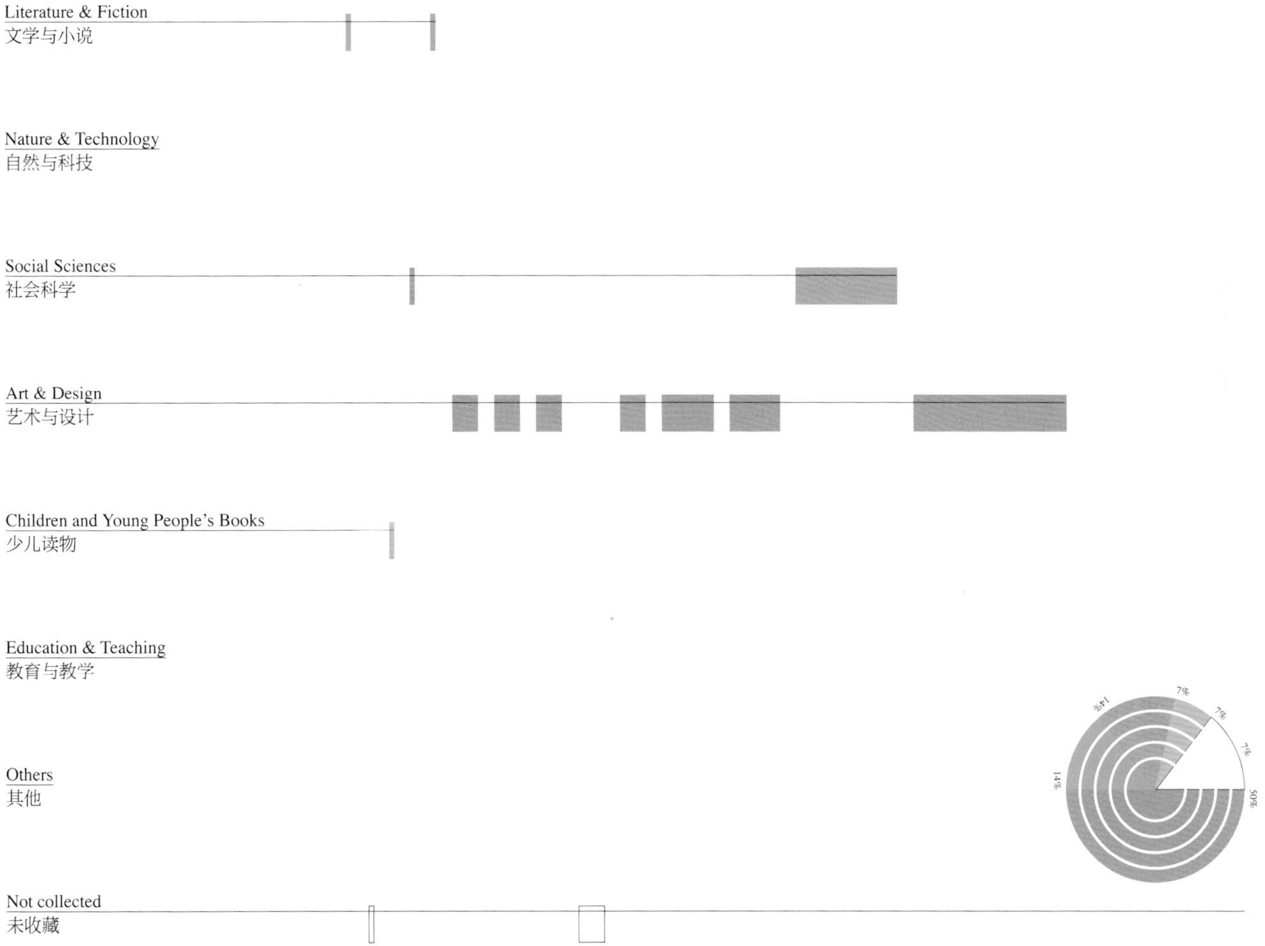
Literature & Fiction
文学与小说
Nature & Technology
自然与科技
Social Sciences
社会科学
Art & Design
艺术与设计
Children and Young People's Books
少儿读物
Education & Teaching
教育与教学
Others
其他
Not collected
未收藏
50%
14%
14%
7%
7%
7%

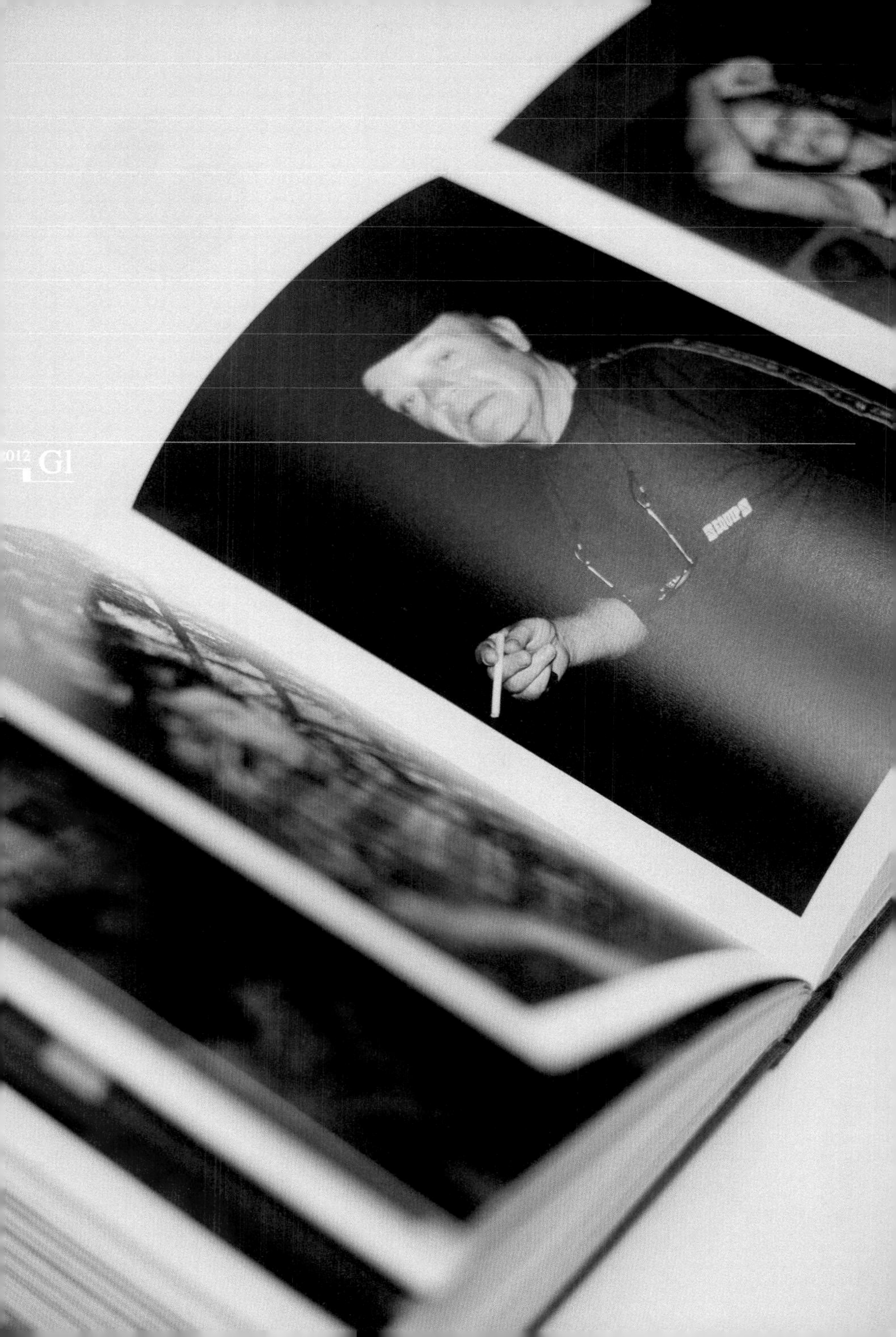

这本是瑞典摄影师 JH · 恩斯特伦（JH Engström）的摄影集，记录其两次在布鲁塞尔期间拍下的照片。这些照片包含他对自己摄影创作的思考，以及布鲁塞尔别样视角的记录。书籍为锁线硬精装，封面墨绿色布面文字烫金裱覆卡板。内页涂布纸四色印刷。全书瑞典文、法文、英文三种文字，除前言部分少量阐释创作背景和想法外，照片部分的文字很少。图片按照一个跨页的布鲁塞尔街景照片配三张人物或者事物的照片的节奏，三连张照片要么为人物连续动作，要么具有相关性，通过拉页方式折在书中。摄影师并不希望通过文字来解释照片，所以这种内容一致的三种文字的图片说明更多是一种启发性质的阅读引导，帮助读者自己去解读图片中的故事，体会照片中人物的状况，以及图片带来的不安气氛。

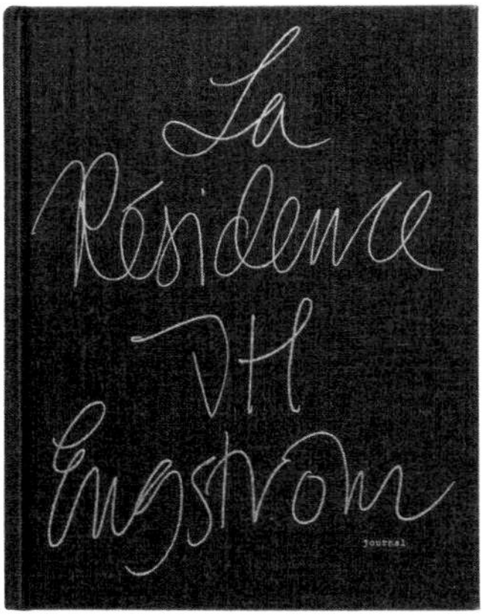
La Résidence JH Engström
journal

This is a photography album by JH Engström, a Swedish photographer. The book records the photographer's two stays in Brussels with pictures reflecting his thoughts on his creation. The hardback book is bound by thread sewing. The cover is made of dark green cloth with hot-stamped text in gold, mounted with graphic board. The inside pages are printed on coated paper in four colors. The text is in three languages – Swedish, French and English, except for a few explanations of the background and ideas in the preface, there are few words in the photo section. The pictures are typeset following such a rule – a double page spread of Brussels street view with three pictures of people or things. The three pictures are either continuous action of characters or connected with characters. They are inserted to the book in the form of foldout. The photographer does not want to interpret pictures by text, so the three-picture foldouts are inspiring reading guide, which helps the readers to interpret the stories in the pictures themselves and to understand the situation of the characters and the uneasy atmosphere in the pictures.

金字符奖
居所
丹麦

Golden Letter
La Résidence
Denmark
Greger Ulf Nilson
JH Engström
Journal, Stockholm
247×186×24mm
823g
180p

JH Engström's volume of photography is structured extremely simply, yet in an intriguingly complex manner: through a system of fold-out pages, picture sequences arranged in triptychs turn out to be fully fledged scenes. Each of these is prefaced by a borderless double-page picture spread, announcing the change of subject in the manner of a curtain or interlude. Then, lyrically or narratively, a few short lines or perhaps a longer paragraph of text open up a field of association which, enigmatically and gesturally, is infused with life by its pictures and people. Nothing dramatic takes place, no lessons are being taught – but as each sequence elicits greater curiosity, for the spectator, browsing and folding his or her way through the pages, a personal individual story emerges, like the pieces of a jigsaw puzzle. A fascinating, eye-opening book – interaction without anything having to be plugged in.

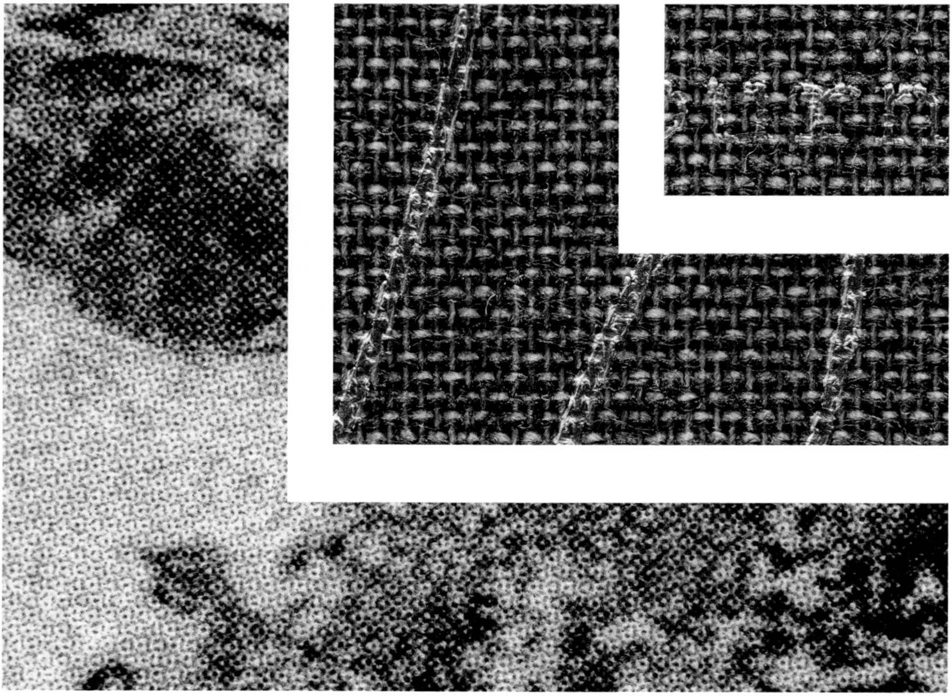

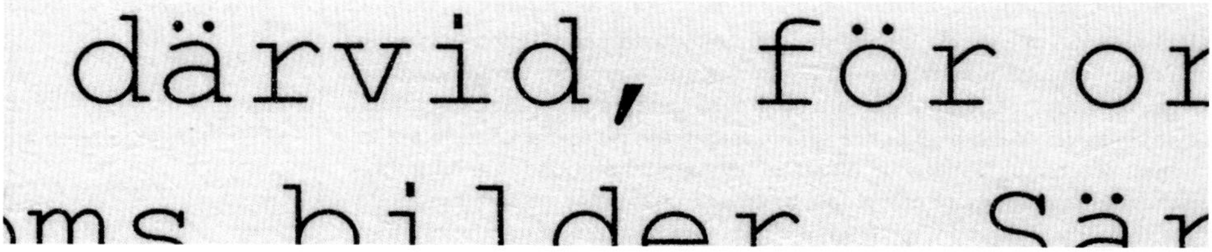
därvid, för
bilder.

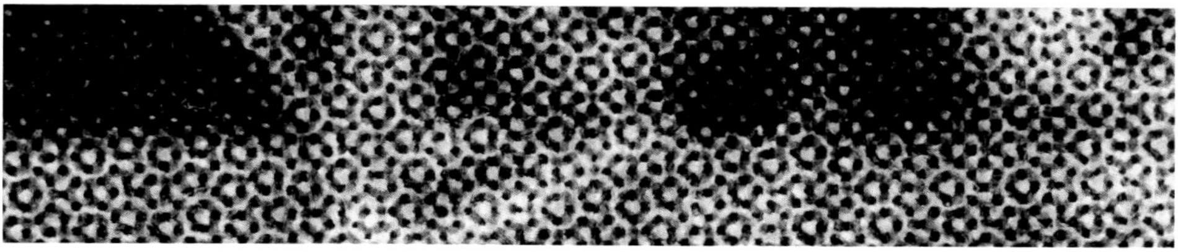

247×186×24

La Résidence
JH
Engstrom
journal

823

180

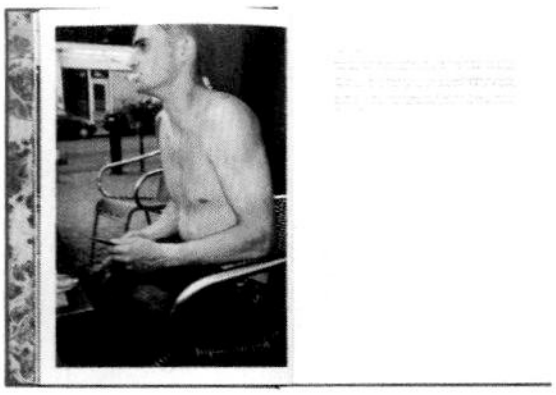

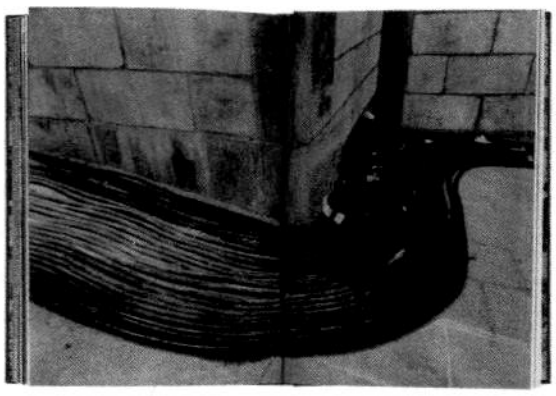

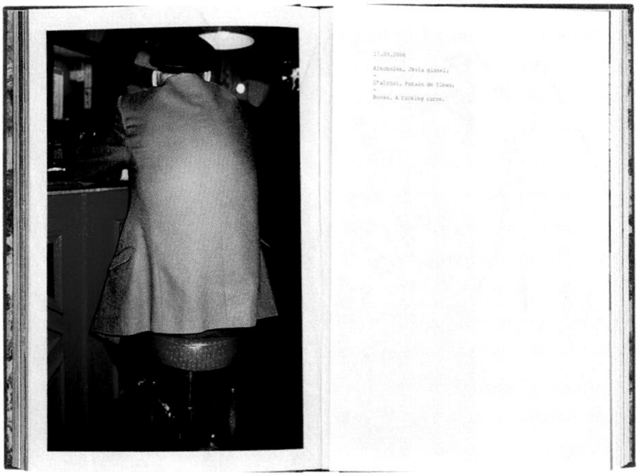
L'alcool. Putain de fléau.
Booze. A fucking curse.
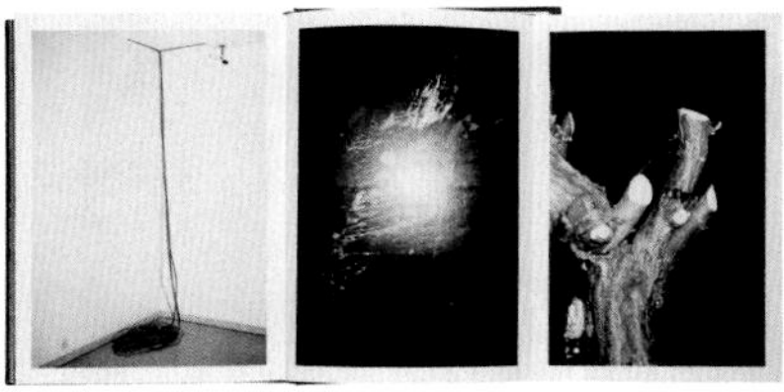

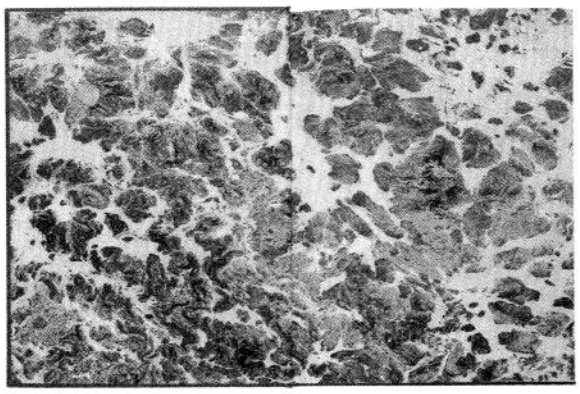

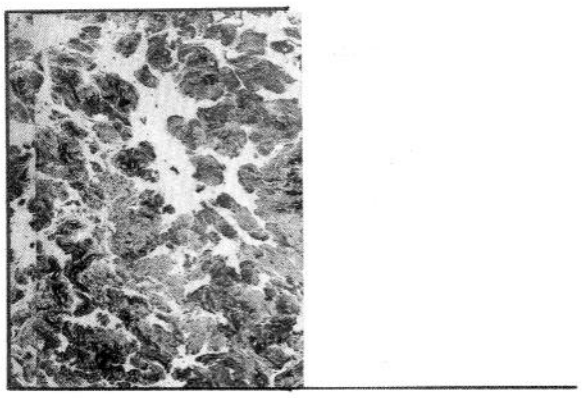

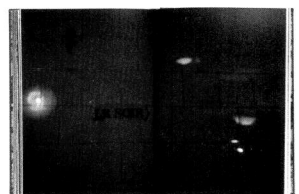

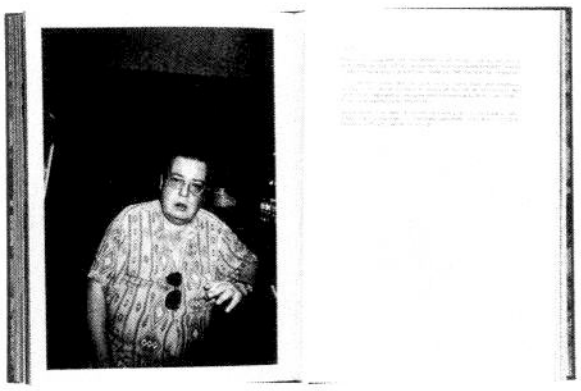

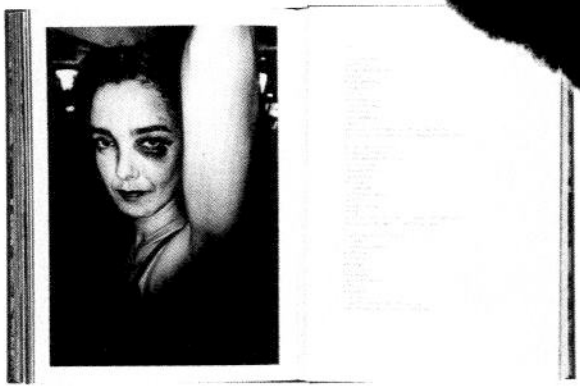

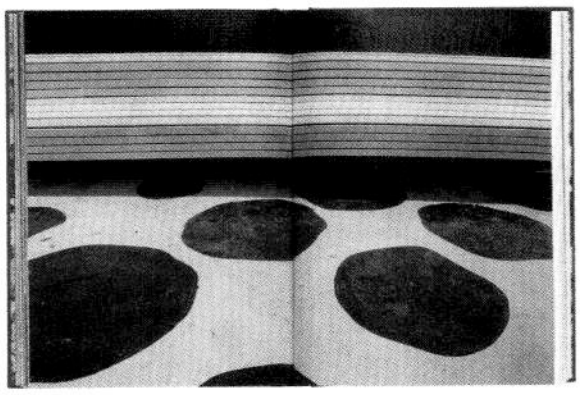

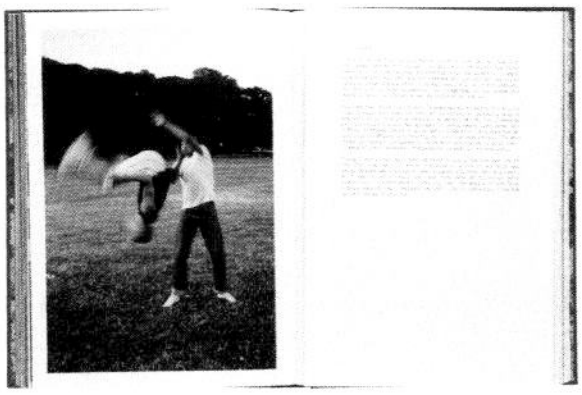

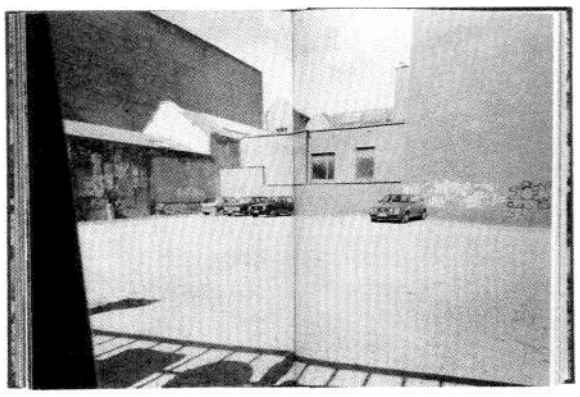

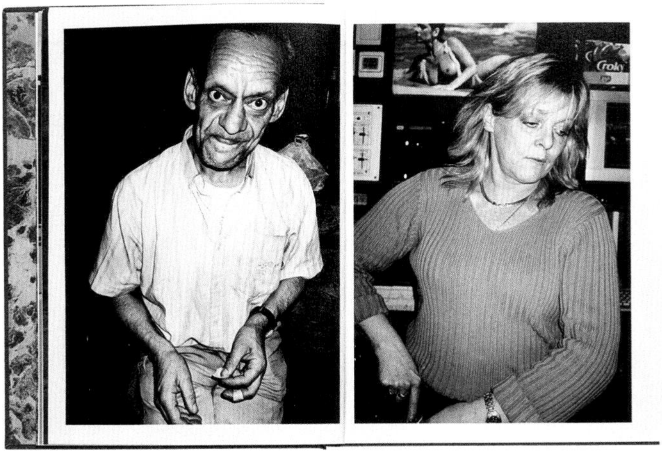
Croky

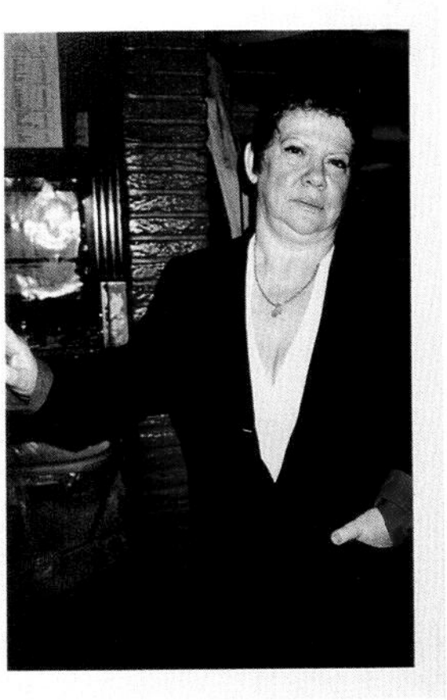

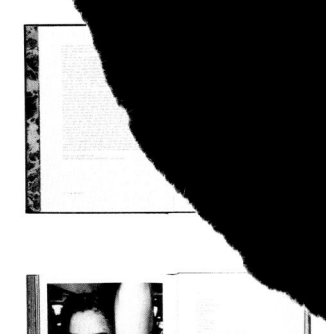

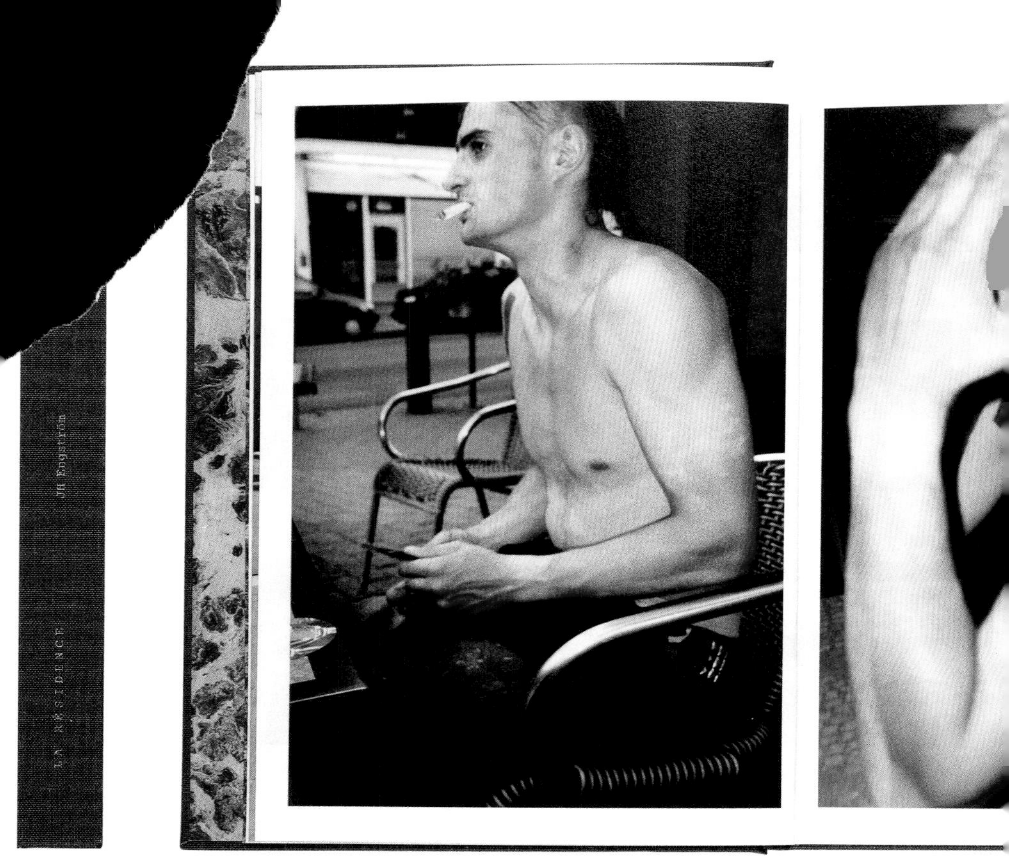
LA RÉSIDENCE
JH Engström

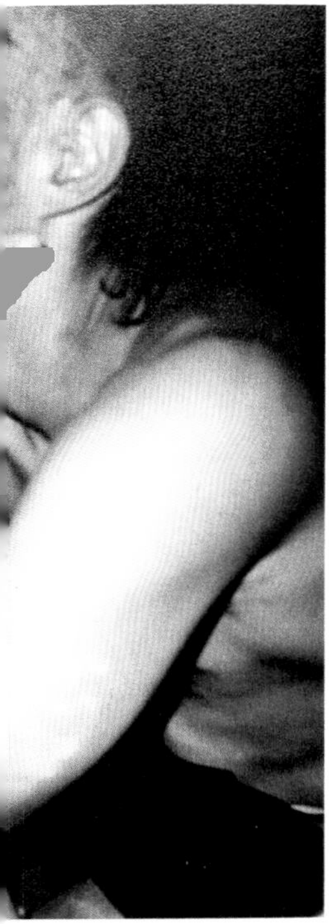

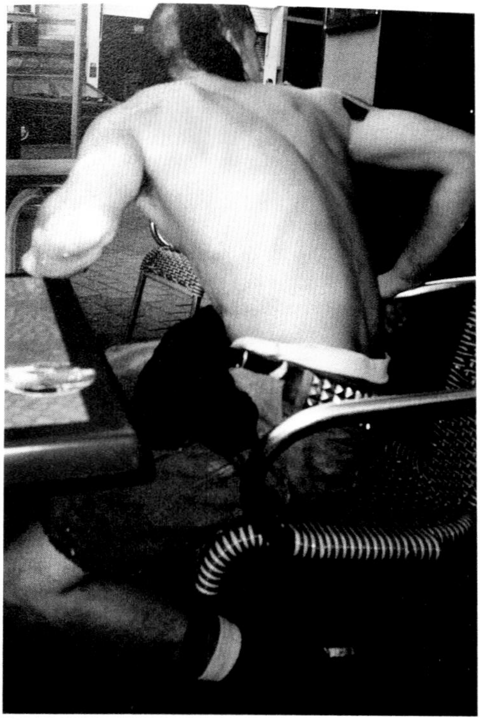

本书是 2010 年度奥地利格拉兹市的建筑年鉴。书中包含 10 个建筑项目的照片、文字介绍以及相关的考察和思考。书籍采用两个裸脊锁线的并行开本，通过精装封面合扣在一起。封面灰蓝色布面丝印黑色，部分文字凹压处理，裱覆卡板。内页哑面铜版纸深蓝灰色和黑色双色印刷，文字部分通过本白色过油刷底来暗暗凸显文字块区域。翻阅的时候需要将书平放在桌上，左右两册对照阅读。右册为 10 个项目的黑白单色照片，左册为对应的项目介绍文字和相关考察及分析笔记，通过翻口上的小短线进行一一对应。由于文本内容和双语的排版较为复杂，为了清晰区分，采用如下设置：介绍文字为黑色，其他文字则为深蓝灰色，无衬线字体为德文，衬线字体为英文。文本采用段落符号进行分割，而未采用常规回行的设置，用以应对文本量巨大的情况。

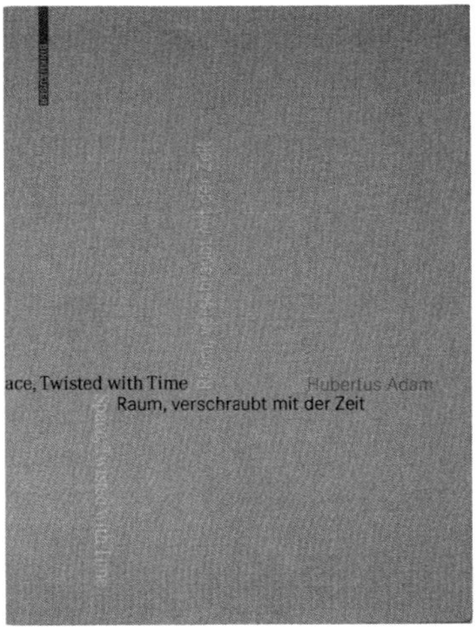

ace, Twisted with Time
Hubertus Adam
Raum, verschraubt mit der Zeit

This book is the 2010 architectural yearbook of Graz, Austria. The book contains photographs, text descriptions of ten construction projects as well as relevant investigations and reflections. The book contains two parallel books bound by thread sewing without spines, which are fastened together by a hardcover. The cover is made of grey-blue cloth printed in black with silk screen, mounted by graphic board. The inside pages are printed on matte coated art paper in dark blue-grey and black. The text block is varnished in raw white to be highlighted. When reading the book, the readers need to put the book on the table and read the separated two books in contrast. The right one contains black-and-white monochrome photographs of ten projects, while the left one is the introduction text of corresponding projects and notes of relevant investigation and analysis. The short lines on the turning edge link the two books. Due to the complexity of the text content typeset bilingually, the following settings are adopted to distinguish clearly – introductory text is in black and other text is in dark blue-grey; serif-free fonts are used for German part while serif fonts are used for English part. The text is segmented by paragraph symbols instead of routine backline settings, in order to cope with the huge volume of text successfully.

金奖

空间，随着时间扭转

奥地利

Gold Medal

Raum, verschraubt mit der Zeit / Space, Twisted with Time

Austria

Gabriele Lenz, Elena Henrich / büro für visuelle gestaltung

Hubertus Adam

Birkhäuser, Basel

285 × 207 × 20mm

1066g

186p

Often there is the desire for pictures and photographs, especially architectural ones, to speak for themselves without the aid of text. But linguistic communication can seldom be avoided, particularly, as is the case here, when we are dealing with an architectural almanac. The solution? To have two book blocks! The unusual way in which this work is bound enables the reader here to peruse and read double pages of text parallel to black-and-white photographs also arranged on double page spreads. An elegant and minimalistic system of dashes along the gutter helps in the matching up process. In this way unadulterated architectural photography is partnered by compellingly designed double pages of text in double columns varyingly interspersed with small architectural drawings and plans. A book of great clarity and strength!

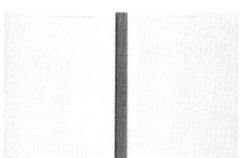

Space, Twisted with Time
Raum, verschraubt mit der Zeit

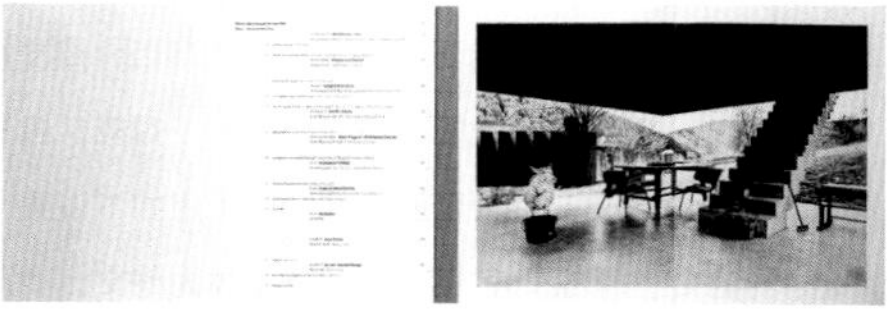

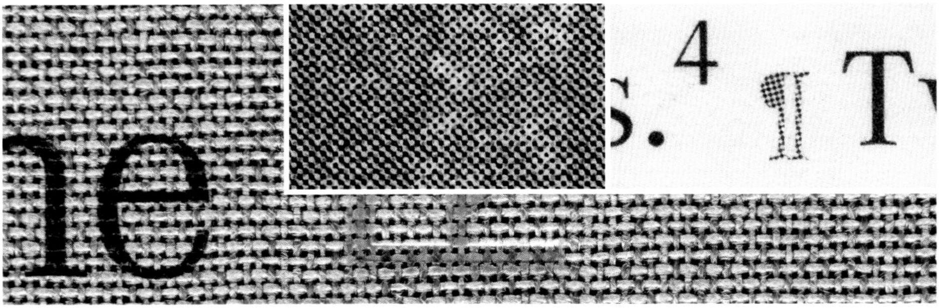

Raum, verschraubt mit der Zeit
Hubertus Adam
Space, Twisted with Time
Sp
BIRKHÄUSER
HDA

Sanierung FRZ Leoben *Das Forschungs- und Rechenzentrum (FRZ) der Alpine Montangesellschaft in Leoben galt als ein Hauptwerk der Partnerschaft von Günther Domenig und Eilfried Huth. Nach einem Umbau zeigt sich das Gebäude heute völlig entstellt. Dazu Eilfried Huth:* ¶ „Es gibt noch schlimmere Beispiele, die aber alle auch eine bestimmte Ursache hatten. Man darf nicht vergessen, dass wir danach strebten, möglichst expressiv zu bauen, auch hinsichtlich der Materialien. Die Pädagogische Akademie in Graz-Eggenberg zum Beispiel ist ein reiner Sichtbetonbau, das ging so weit, dass wir selbst die Trennwände aus Beton, sprich Eternit, errichtet haben, obwohl das viel teurer war, als wenn wir sie normal verfliest hätten. Das FRZ in Leoben wiederum war ein Bau aus Corten-Stahl. Das zog viele Nachteile nach sich, die man heute aus bauphysikalischer Sicht nicht mehr in Kauf nehmen würde, abgesehen davon, dass bei bestimmten Materialien wie dem Eternit auch Asbest festgestellt worden ist. Die Unzulänglichkeiten haben stets zur Folge gehabt, dass die Erhaltungskosten groß waren. ¶ Beim FRZ in Leoben ist hinzugekommen, dass niemand mit dem Gebäude etwas hat anfangen können. Das FRZ ist zu einer Zeit entstanden, in der die beiden großen Stahlkonzerne, Voest und die Alpine Montangesellschaft, fusioniert wurden. Der Schwerpunkt der Betriebsführung hat sich nach Linz zur Voest verlagert. Somit war das Gebäude für die Linzer ein Erbe, das sie nicht wollten. So hat das Haus einige Zeit leer gestanden. Dann hat man es der Montanuniversität angeboten und die hätte natürlich investieren müssen, das Geld wurde ihr allerdings nicht sogleich zur Verfügung gestellt. Letztlich hat es dann ein Investor billig als „Schnäppchen" gekauft und eine Sanierung eingeleitet mit dem Ziel, das Gebäude der Montanuniversität zu übergeben (2008). Man hat mich mit eingebunden in diesen Prozess, um mich in der Folge scheibchenweise zu entmachten. Bei jeder Besprechung war wieder eine Kröte zu schlucken, kam wieder eine Horrormeldung, was man alles *nicht* machen kann. ¶ Im Entwurfsansatz des FRZ-Gebäudes war 1968 vorgesehen, später noch ein viertes Geschoss abhängen zu können, die Konstruktion war auch so angelegt worden, was aber – wenn realisiert, und das ist eben bei der Sanierung geschehen – die Proportion verändert, zum Nachteil. Es ist zwar funktionell in Ordnung, aber die Abhängung des vierten Geschosses war ein Punkt, der den meisten negativ aufgefallen ist. Die Sanierer sind eigentlich unschuldig – es war im Ursprungskonzept vorgesehen. ¶ Das Zweite war die Schrägstellung der Fenster sowie von deren Abdeckungen. Schon im Entwurf in den sechziger Jahren waren diese als Träger für Solarpaneele zur Sonnenenergiegewinnung vorgesehen – eine Technologie, die damals noch in den Kinderschuhen steckte. Bei der Sanierung hätte ich mir erhofft, dass jetzt, da die Solartechnologie weiter entwickelt ist, diese umgesetzt würde. Ich habe ja unzählige

Günther Domenig / Eilfried Huth: Forschungs- und Rechenzentrum (FRZ) Research and Computing Centre, Leoben, 1969–74; Originalzustand original state

Leoben
Gangoly & Kristiner
Generalsanierung
Hörsaalgebäude
Montanuniversität

Entlang der mittleren Mur, deren Wasser seit den zwanziger Jahren des vorigen Jahrhunderts zur Energiegewinnung genutzt wird, fahre ich zwei Tage später von Graz aus Richtung Leoben, ins Herz der Obersteiermark. Es ist Sonntag, die Stadt wirkt am frühen Nachmittag fast leergefegt. Weil die Universität heute geschlossen ist, nutze ich die Gelegenheit und unternehme eine Zeitreise in die Geschichte der Montanindustrie. Vorbei an der rauchenden Hütte Donawitz, die wie ein gewaltiger Stöpsel in den Flaschenhals des Tals geschoben zu sein scheint, geht es nach Norden, bis hinauf auf die Passhöhe des Präbichl, der die Eisenerzer Alpen vom Hochschwab trennt. Es ist neblig an diesen Wintertagen, ein wenig Schnee fällt, und so zeichnet sich der Erzberg nur schemenhaft ab: Natur, dem menschlichen Willen unterworfen – ökonomisches Kalkül und reinster Pragmatismus. Ich kann mich des Eindrucks kaum erwehren, es handle sich hier um Land Art in den größtmöglichen Dimensionen: Landschaften der postindustriellen Transformation verführen zur romantisierten Wahrnehmung. Der Erzberg thront über Eisenerz mit seinen einstigen Arbeiterkolonien, er beherrscht die sich stetig entvölkernde Stadt. Eisenerz hieß einst Innerberg, und es ist bezeichnend, dass der Ort schließlich auf den Namen des Bodenschatzes umgetauft wurde, dem er sein Entstehen verdankt. Auf dem Weg zurück halte ich kurz in Vordernberg, wo 1840 die „Steirisch-Ständische Montanlehranstalt" gegründet wurde, aus der die 1850 nach Leoben verlegte Bergakademie hervorging. Als ich abends um Viertel vor Neun ein historisches Restaurant am Hauptplatz von Leoben betrete, hat die Küche schon geschlossen. Eine Universitätsstadt stelle ich mir anders vor. ¶ Der Boom der Montanindustrie hatte 1889 eine gründerzeitliche Stadterweiterung nötig gemacht. Die Franz-Josef-Straße, für deren Anlage die Nordfront des Hauptplatzes durchbrochen wurde, avancierte zur neuen Entwicklungsachse. Sie führt auf den Turm der Gustav-Adolf-Kirche zu; dort schwenkt der Straßenzug nach Westen um und führt über die Brücke zum Bahnhof auf dem jenseitigen Murufer. Das neobarocke Hauptgebäude der Montanuniversität, der einstigen Bergakademie, entstand zwischen 1908 und 1910 unmittelbar an

Renovation of the FRZ Leoben

The Forschungs- und Rechenzentrum (FRZ) (Research and Computing Centre) of the Alpine Montangesellschaft in Leoben was considered a main work of the partnership between Günther Domenig and Eilfried Huth. After undergoing renovation, the building today looks completely disfigured. Here is what Eilfried Huth has to say about it: ¶ There are even worse examples which, however, all have a certain cause, too. One must not forget that we strove to build as expressively as possible, also in terms of materials. The Pädagogische Akademie in Graz-Eggenberg, as a construction example, is a pure exposed concrete building; it went to the point that we built dividing walls in concrete, that is, Eternit, although that was much more expensive than if we had normally tiled them. The FRZ in Leoben, on the other hand, was a Corten steel construction. This brought many disadvantages that would no longer be accepted today from a structural physics perspective, apart from the fact that asbestos was also detected in certain materials, such as the Eternit. These shortcomings have always resulted in high maintenance costs. ¶ What complicated matters at the FRZ Leoben was the fact that no one knew what to do with the building. The FRZ was built at a time when the two big steel companies, Voest and the Alpine Montangesellschaft, were merged. The main focus of company operations shifted to Voest in Linz. As a consequence, the building was a legacy that the people in Linz did not want. So the house stood empty for a time. Then it was offered to the Montanuniversität, and they naturally had to invest money into it; the money, however, was not given to them right away. In the end, an investor bought it at a "bargain price" and started a renovation with the goal of making it available to the Montanuniversität (2008). They then got me thoroughly involved in this process – in order to disempower me little by little afterwards. At every meeting there was another bitter pill to swallow; there was another horror story about what one cannot do. ¶ In the conceptual design of the FRZ building, the intention in 1968 was to add a fourth storey to it later on; the construction was also arranged in this way, which, however, – when carried out, and that is what happened during the reno-vation – changes the proportion, to its detriment. It was functionally alright, but the suspension of the fourth storey was a point most people had negatively noticed – actually, it is no fault of the renovators – it was foreseen in the original concept. ¶ The second thing was the slope of the windows, as well as their coverings. These had already been intended in the design in the Sixties to be holders for solar panels to produce solar energy, a technology still in its infancy back then. During the renovation, I would have hoped that now, since solar technology is further developed, this would be implemented. I made countless designs for it. Then

Günther Domenig / Eilfried Huth: Forschungs- und Rechenzentrum (FRZ) Research and Computing Centre, Leoben; Zustand nach Umbau State after renovation

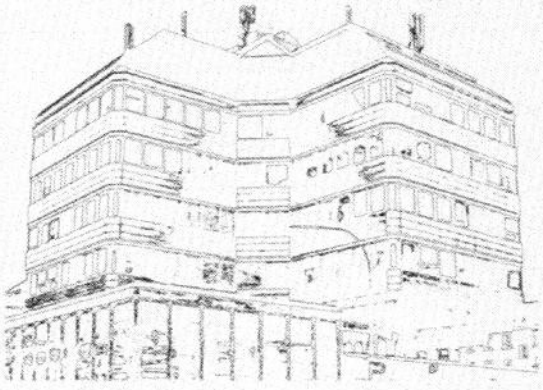

Leoben
Gangoly & Kristiner
General Renovation of University Lecture Hall Building

Two days later I drive from Graz towards Leoben, along the middle stretch of the Mur, whose water has been used for generating energy since the Twenties, into the heart of Upper Styria. It is Sunday, early afternoon, and the city seems nearly empty. Because the university is closed today, I use the opportunity to take a time journey into the history of the mining industry. Past the smoky Donawitz Works, which appear to be like a huge cork pressed into the bottleneck of the valley, the road goes up north to the Präbichl Pass, which separates the Eisenerz Alps from the Hochschwab mountain range. It is foggy on these winter days, a little snow is falling, and only the outline of the Erzberg can be made out: nature, subjected to the will of man – economic calculation and pragmatism in its purest form. I can hardly avoid the impression that this is Land Art in the largest possible dimensions: landscapes of post-industrial transformation seduce me into romanticised perception. The Erzberg towers over Eisenerz and its former workers' colonies; it rules the continuously depopulating city. Eisenerz was once called Innerberg, and it is significant that this town was rechristened with the name of the natural resource it owes its existence to (Eisenerz – iron ore). On the way back I make a short stop in Vordernberg, where the "Steirisch-Ständische Montanlehranstalt" (Styrian Corporate School of Mining) was founded in 1840. The school was relocated to Leoben in 1850 and called the "Bergakademie" (Mining Academy). As I enter a historic restaurant on Leoben's Main Square at 8:45 p.m., the kitchen is already closed. I imagine a university city to be different. ¶ In 1889, the mining industry boom made a Gründerzeit city expansion necessary. The Franz-Josef-Straße, which broke open the north front of the Main Square to be laid out, evolved into a new developmental axis. Leading to the tower of the Gustav-Adolf Church, the street then swings to the west and follows across a bridge to the railway station on the other side of the Mur. The Neo-Baroque main building of the Montanuniversität (University of Leoben), the former Bergakademie, arose directly on Franz-

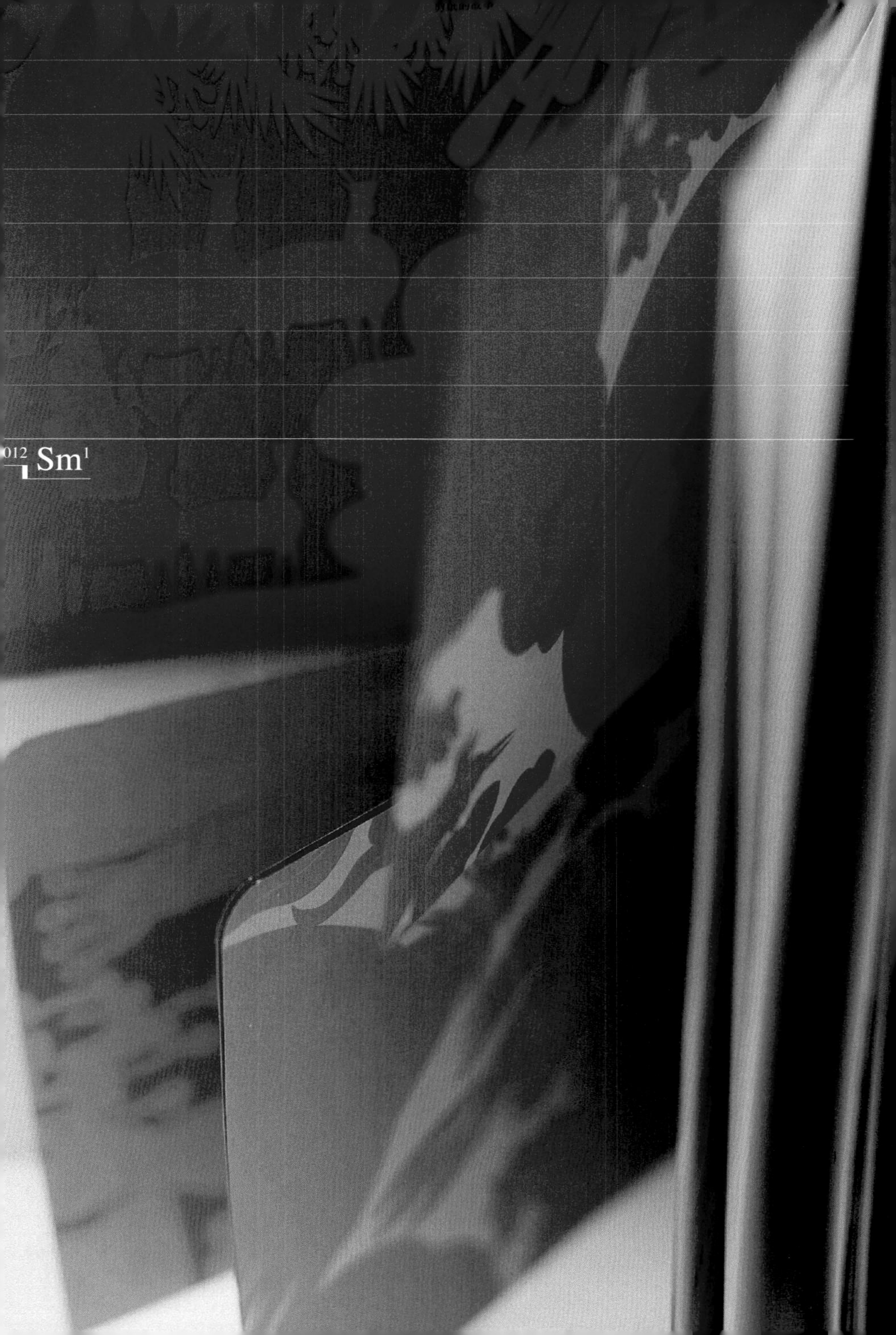
012 Sm1

这是艺术家赵希岗的剪纸作品集。全书收录了艺术家自己与剪纸的故事、几百幅剪纸作品，以及与各个剪纸创作主题相关的理念阐释或诗意表达。书籍的装订形式为裸脊锁线。全书采用多种轻薄而富有触感的纸张，红、黄、蓝、绿、粉五色线进行锁线装订，将传统美学融入现代设计。封面半透明的轻薄纸张正反红黑双色印刷，通过折叠的方式透出里面的图案，外部的剪纸图案通过 UV 工艺凸显质感。内页将剪纸作品分为 35 个主题，每个主题开始部分印有与主题相关的阐释文字，纸张从中间裁开分成上下两块，带来独特的翻阅感受。部分较大尺寸的作品通过四折的方式收叠书中。全书中英双语，宋体字与极细的无衬线字体形成独特的文本肌理对比，放置在页面中心位置且字号较大的页码是一个有趣的尝试。

剪纸的故事

This is a collection of paper-cut works by the artist Zhao Xigang. The book contains the story of the artist and paper-cut, hundreds of pieces of paper-cut works, as well as the conceptual interpretation or poetic expression related to the theme of each paper-cut creation. The book is bound by thread sewing with a bare spine. The book uses a variety of light and tactile paper and five colors thread of red, yellow, blue, green and pink, which integrate traditional aesthetics into modern design. The cover is made of translucent thin paper printed in red and black and the inside patterns are revealed by folding. Spotting UV highlights the outer paper-cut pattern. The inside pages are divided into 35 sections according to themes of the paper-cut works. Each section starts with the explanation related to the theme. The page is cut from the middle into two parts, which brings unique reading experience. Some of the larger works are folded into the book by four folds. The book is bilingual in both English and Chinese. The Song typeface together with a fine san serif font forms a unique contrast. Placing large size page numbers in the center of the page is also an interesting attempt.

银奖	Silver Medal
剪纸的故事	The Story of Paper-cut
中国	China
吕旻、杨婧	Lü Min, Yang Jing
赵希岗	Zhao Xigang
人民美术出版社	People's Fine Arts Publishing House, Beijing
	250×183×21mm
	434g
	342p

A history of paper-cut, this extremely smooth book produces a feeling when leafing through it that brings us into close contact with the artist's sensibilities. Here the book designer becomes a second author, speaking to us in the language of the book and of the paper. Once inside the book we find thin, delicate paper with the coloured prints of paper-cut pictures, lines of text, areas of script. We experience a sense of dynamics, of stability, joy, blossoming, emptiness, family, moonlight, birdsong, living spirits, the woods – and once more, birds flying and singing, the fluttering of feathers, the scurrying of squirrels. As the motifs change so does the paper – in colour and in form. And all of a sudden we have been touched by the magic of the paper-cut art. By the same token we have also been touched by the sensitivity of the book designer, who in any case already shares the same vibrations as the artist.

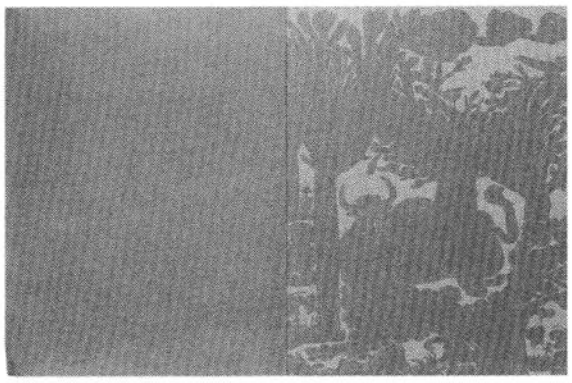
 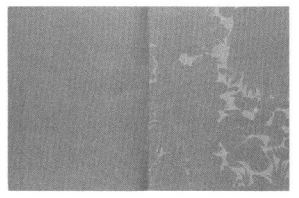
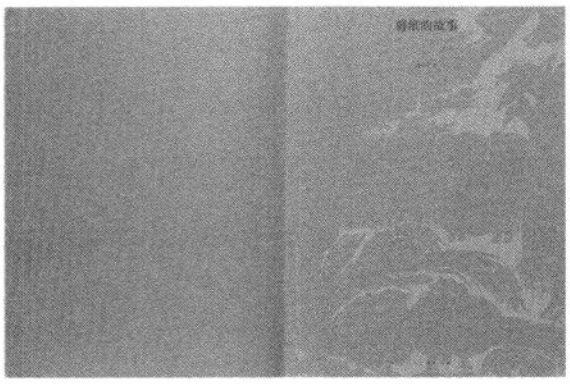
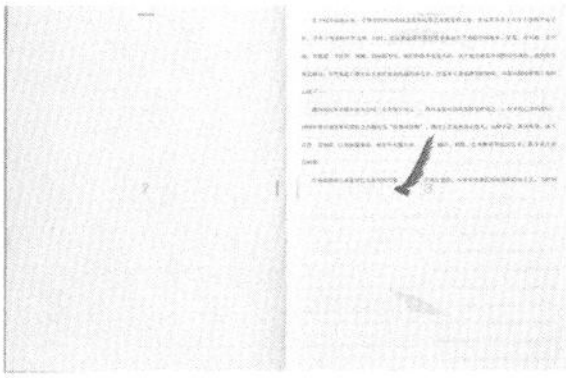

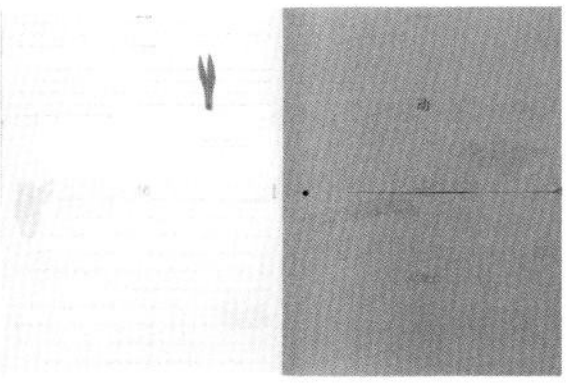
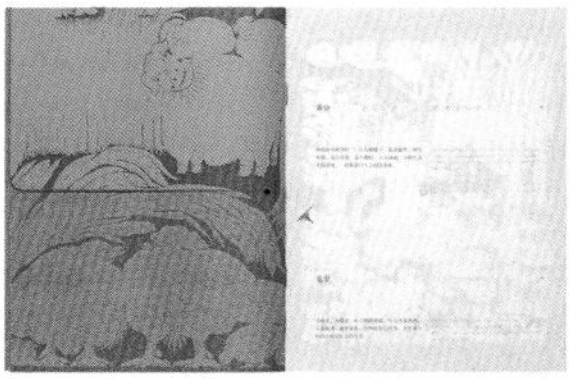
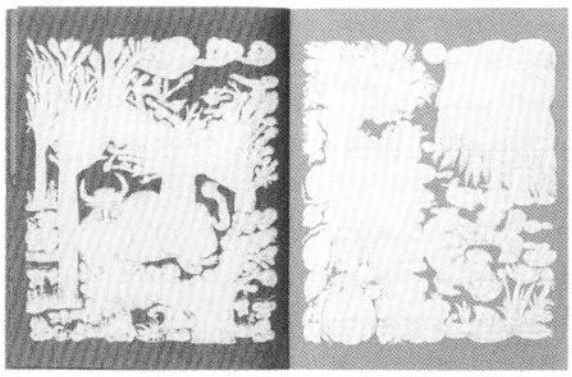
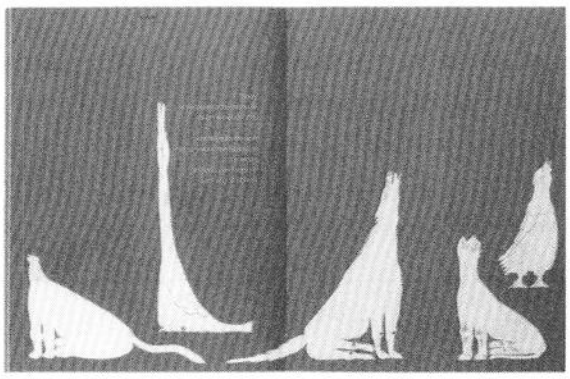
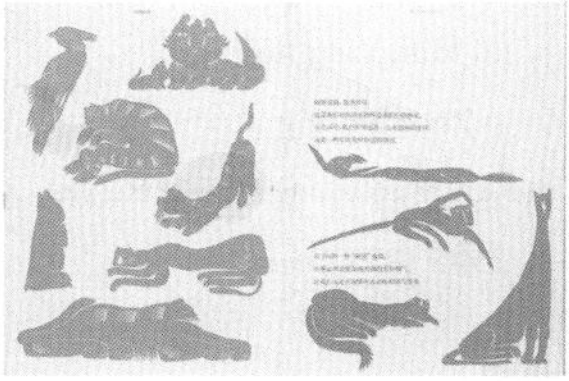
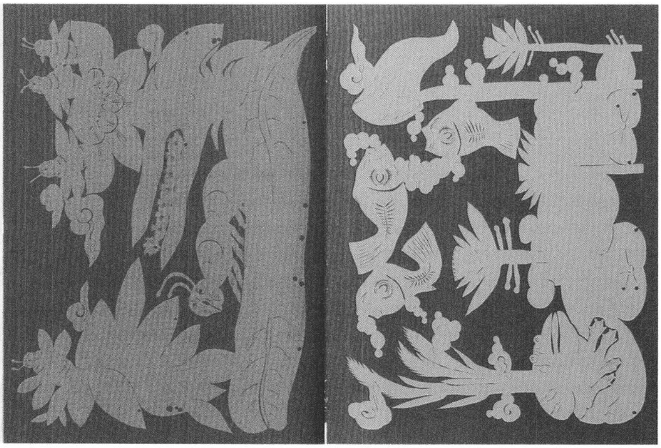
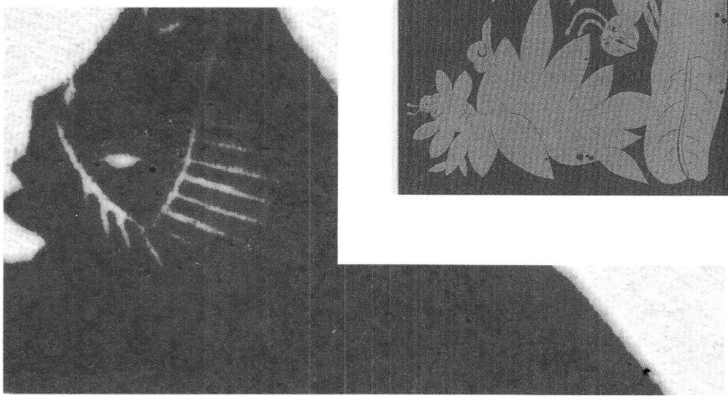

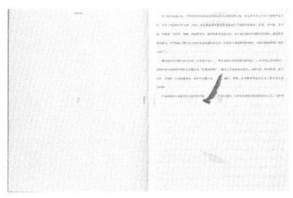

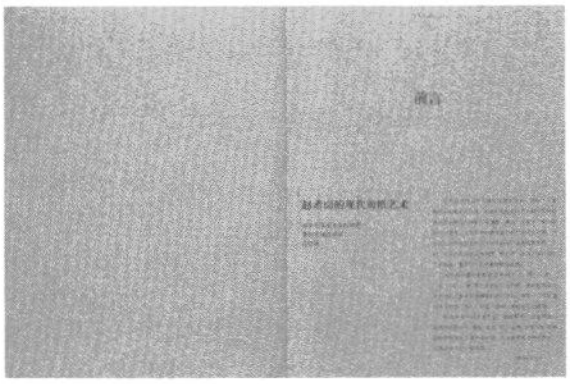

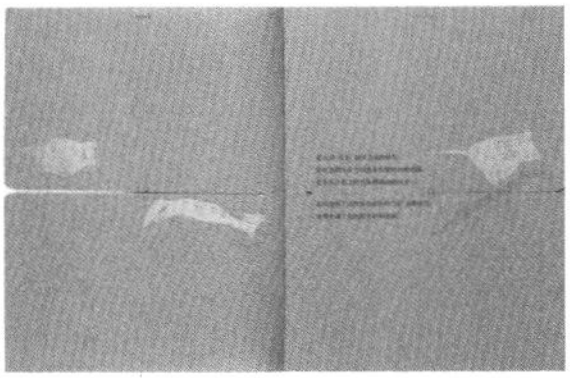
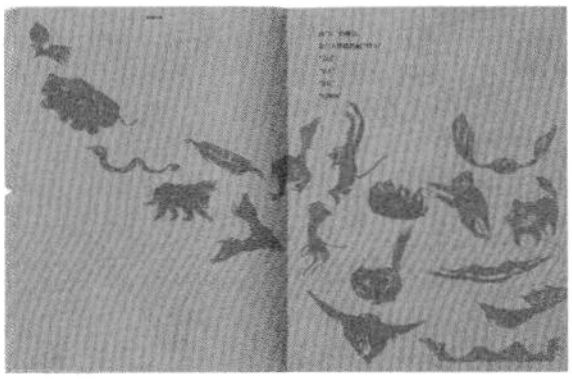
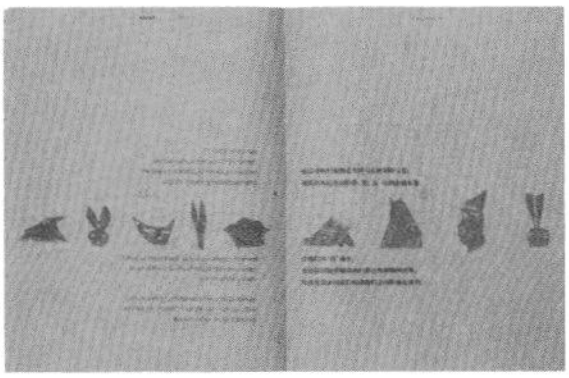
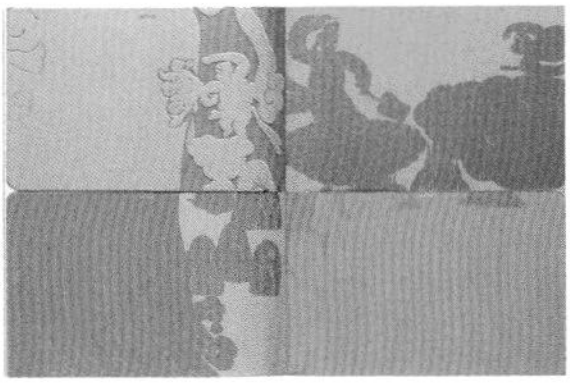
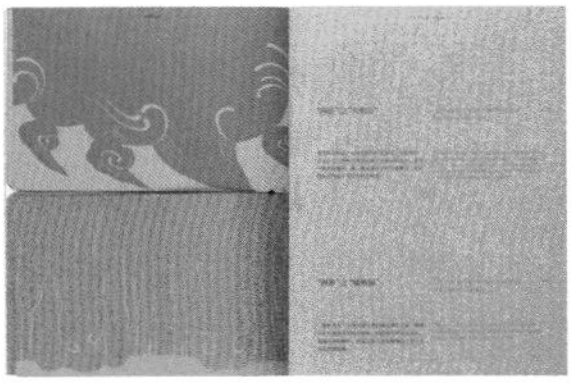
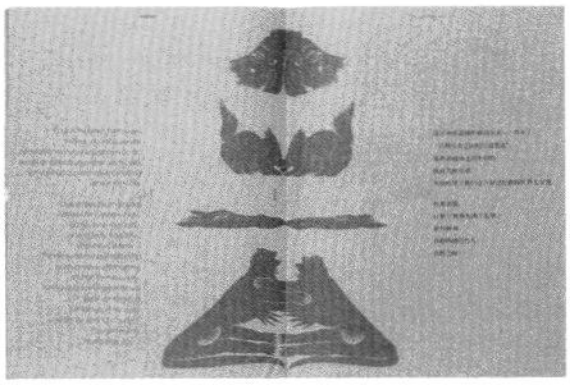

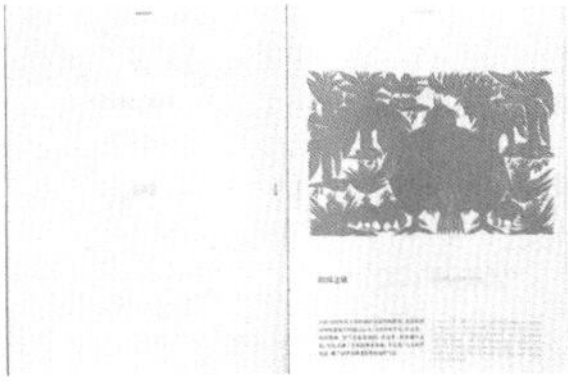
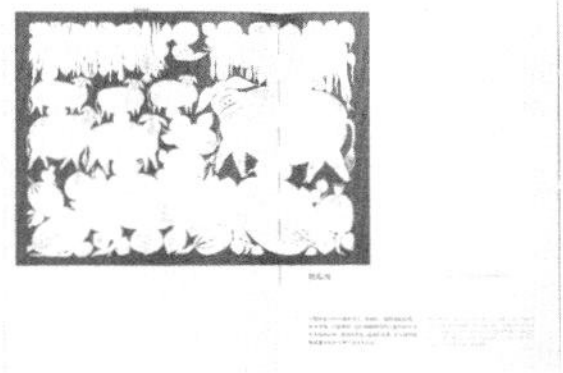

与孙
趣的一幕。

剪纸的故事

赵希岗 著

The Story of Paper-cut

Zhao Xigang

People's Fine Arts Publishing House

人民美術出版社

《竹林七贤》和《庄周问道》，

是学习中国传统文化思想之后，

体会内涵精神所创作的作品。

"The Seven Persons of virtue in bamboo forest" and "Zhuangzhou asked Taoism" were the works created after I learned the Chinese traditional culture and learned about its essence.

竹林七贤

The Seven Persons of virtue in bamboo Forest

流动着的大气空间，伴随着若隐若现的密竹和走兽，"竹林七贤"那有节奏、有韵律的如痴如醉表演情境，似乎在追逐一种恬静、淡泊、自然的生活境界，眼前似乎展现了一种畅意、天然、率真的思想情感的世界。

One of the series, "The Seven Person of virtue in the bamboo forest". With the company of the floating clouds, the looming bamboos and beasts, their performance seems to tell they are chasing a peaceful, casual and natural life style.

本书作者亲自去 46 位编织艺术家的家中进行采访、拍摄，记录他们作品背后的故事。全书为裸脊锁线装，外包牛皮卡纸制成的盒子进行保护，盒子无其他工艺。封面牛皮卡纸红黑双色印刷，前后环衬上红色的前言和后记分别为作者和设计师的创作背景与思路，红色文字与书脊上的红色锁线相呼应。内页胶版纸四色印刷。全书没有用大量文字去阐述理论，而是直接展示最真实的创作状态和环境。将艺术家分为 46 个相互独立的单元，通过单元外层向内折叠包裹的方式，将内层信息包裹其中，犹如作者所说的进入这些艺术家家中的过程：最外层是作品，作品页向两侧打开仿佛开门的过程，里面可以看到艺术家家中的场景、编织时手部的特写、给作者的亲笔信、创作访谈、艺术家与创作环境的照片。

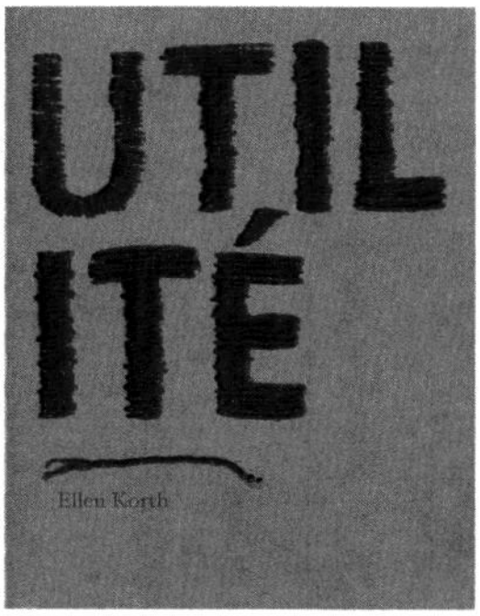
UTIL
ITÉ
Ellen Korth

This is a collection of the stories behind the works of 46 weaving artists. The author went to their homes to interview, photograph and record. The book is bound by thread sewing with a bare spine. It is protected by a slipcase made of kraft cardboard without any special technique. The cover is printed on kraft cardboard printed in red and black. Foreword and postscript printed in read on the front and back endpaper are about the creative background and ideas of the author and the designer. The red text echoes the red thread on the spine of the book. The inside pages are printed on offset paper in four colors. Instead of explaining the theory with long-winded articles, the book directly shows the real creative state and environment. According to the artists, the book is divided into 46 independent units. The information is enclosed by the outer layer of the unit by folding the package inward. As the author said, the reading experience is like the process of entering the artist's home. Works are printed on the outer layer, and when the readers open the layer, the door to their house has been opened. Inside the readers can see scenes in the artist's home, close-ups of the hand when weaving, personal letters to the author, creative interviews, photographs of creating environment.

银奖

效用

荷兰

This unusual form serves as an invitation to the reader into the habitats of creative people involved in various forms of needlework. Its fold-out tables are not just for effect but open up spatial layers and a continually surprising form of access to the world of textile art with wool, needles and thread. The stringent concept is underscored by simple typography, with loving details making reference to the subject matter. The book's haptic qualities and finish are superb, with highly successful picture rendition on uncoated paper. The red thread and the headings provide a magnificent way of citing the subject without being obtrusive. A wonderfully congenial and perfectly executed book.

Silver Medal

Utilité

The Netherlands

-SYB-(Sybren Kuiper)

Ellen Korth

Ellen Korth, Deventer

200×150×51mm

1202g

552p

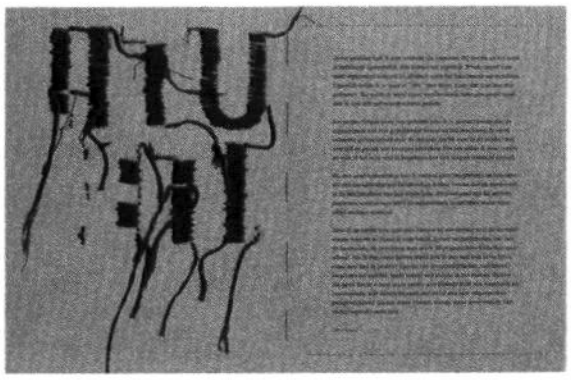
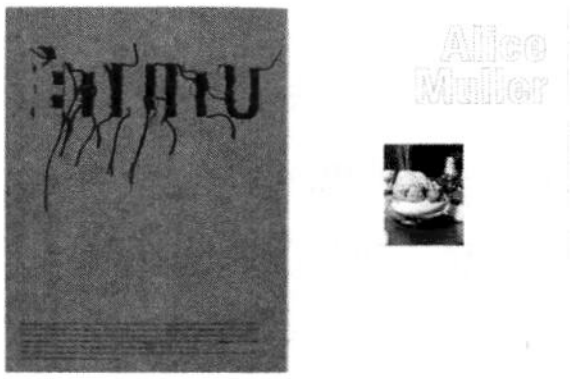
Alice
Muller
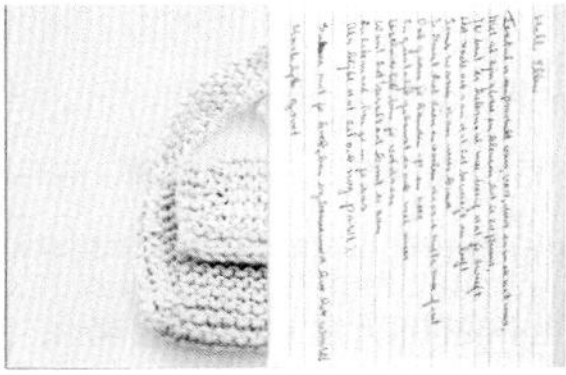

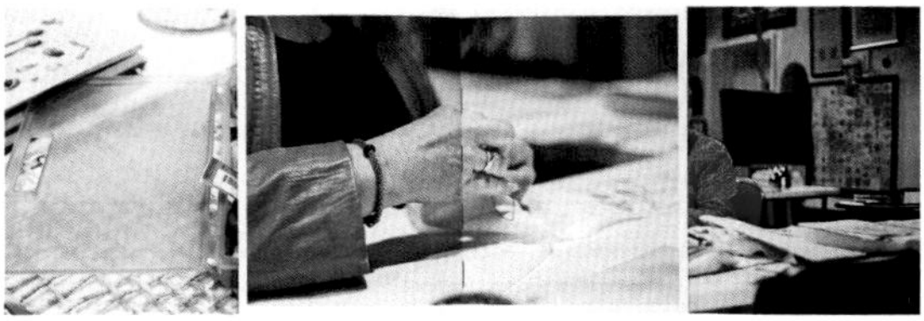
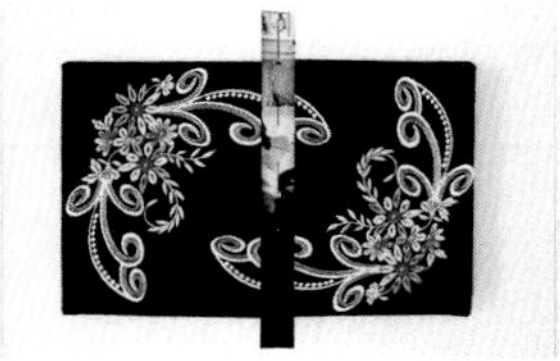

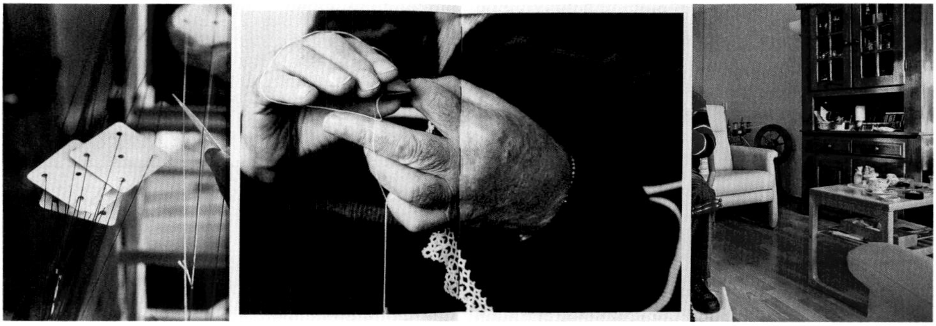

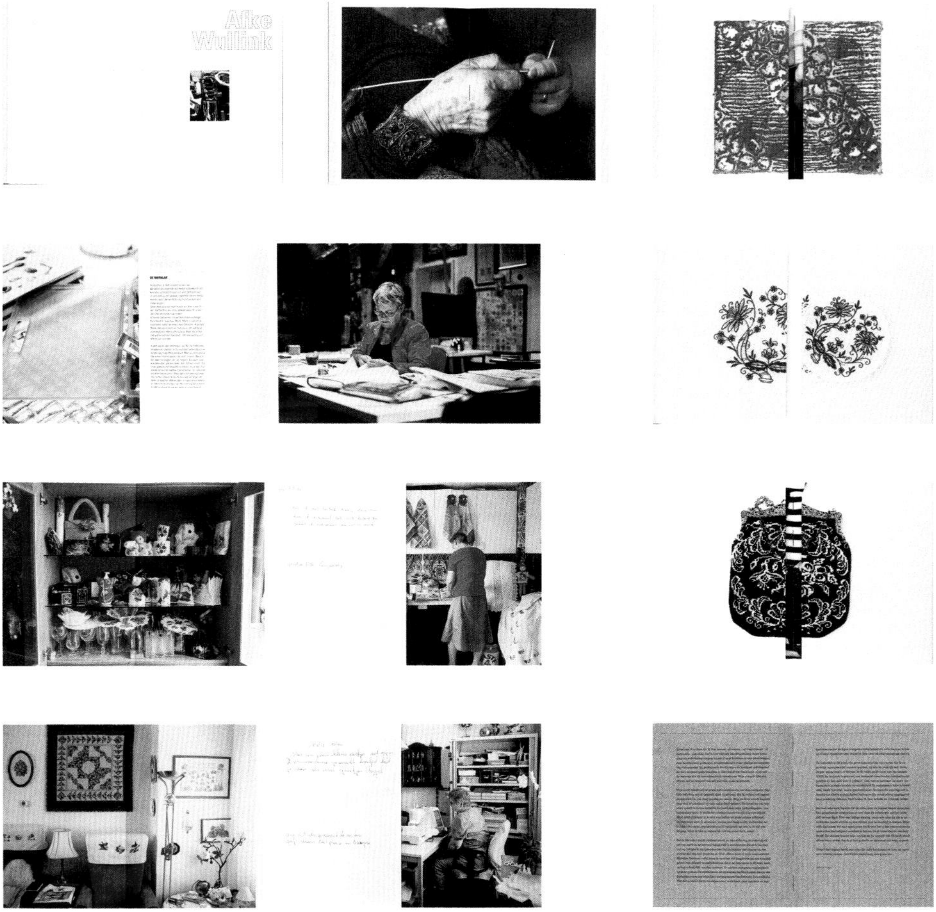
Afke
Wullink

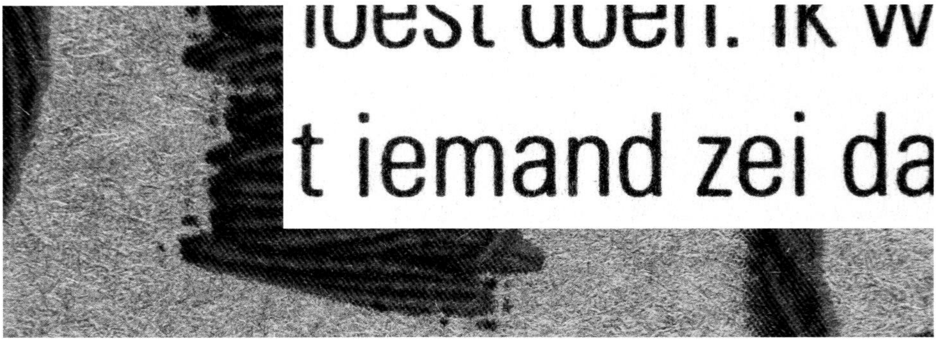
loest doen. Ik w
t iemand zei da

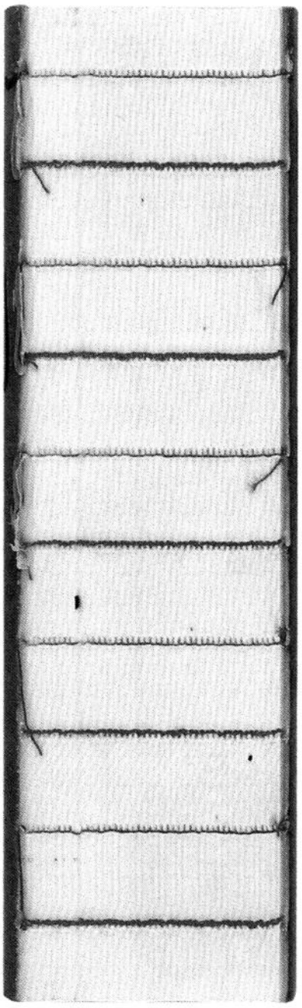

Alice Muller

Als ze me vroeger zeiden dat ik iets moest doen, dan gooide ik m'n kont tegen de krib. En dat heb ik eigenlijk nog steeds. De laatste jaren heb ik nogal het een en ander aan de hand gehad met mijn gezondheid, maar als die dokter dan zei dat ik bepaalde pillen moest slikken, dan wou ik eerst weleens weten waarom dat dan wel was.

Zo heb ik ook jaren op een grote pendelbus gezeten. Eerst heb ik mijn vrachtwagen rijbewijs gehaald, mét aanhanger. Toen ik mijn gewone rijbewijs haalde (ik heb voor alle rijbewijzen maar 10 lessen gehad) zei die instructeur dan dat ik niet goed genoeg schakelde. Daar kon ik dan helemaal niet tegen. Ik wou altijd weten waaróm ik iets moest doen. Ik wou nooit zomaar wat doen alleen omdat iemand zei dat het moest.

Ik kom uit een gezin met 11 kinderen, dat was vroeger gewoon zo. Maar dan meestal bij de katholieken, maar wij waren protestants. Het was niet anders. We moesten daarom ook allemaal meehelpen in huis. Dus moest ik als meisje naar de huishoudschool, want de jongens mochten doorleren, want die moesten de kost verdienen werd er dan gezegd. En toen ik na de eerste 2 jaar primair moest kiezen of ik één jaar door wilde leren voor huishoudassistent of twee jaar voor huisnaaister, zei ik doe mij dan maar liever die twee jaar, want dat huishouden dat ken ik allemaal wel. We moesten immers altijd meehelpen in huis.

Vroeger zaten we 's avonds altijd allemaal te breien. Zo moesten we ook iedere dag na schooltijd 10 toeren breien. Ik zat dan al die tijd met dat breiwerk op schoot. Daar had ik zo'n hekel aan, dat ik er altijd heel lang over deed. Stom natuurlijk, want als ik doorbreide waren die toeren zo af, maar ik kon er niet tegen dat me iets werd gezegd waar ik het nut niet van in zag. Ik nam later mijn breiwerk ook weleens mee op mijn werk. Maar een trui breien vond ik weer wel leuk. Zulk soort dingen

这是一本跟踪服装产业链的摄影集。摄影师从孟加拉国廉价服装生产开始，记录服装的诞生，继而跟踪了服装在时尚、销售、运输、回收、二手、处理等方面的循环过程，探讨了发达国家与发展中国家在全球经济产业链、不平等利益分配、话语权争夺、需求与变相的奴役之间相互依存但又相互压迫的关系。书籍的装订形式为仿经折装。采用哑面艺术纸四色印刷做书函，将书籍主体包裹其中。封面采用灰色带有杂质的卡板印单黑，四角导圆角。内页胶版纸四色印刷。正反封面均可左翻阅读，形成两条并行的叙事线，分别为从服装在孟加拉国制造到西方贩售的过程，以及从欧洲被淘汰、回收，并运输到非洲作为二手贩售的过程。那些在非洲卖不掉的回收衣物又作为原料运往孟加拉国，从而形成循环。书籍的最后部分通过大量文字详述了照片拍摄背后的故事和思考。|||||||||||||||||||||||||||||||||||||||||||||||||||||||||||||||||

This is a photo album that tracks the apparel industry chain. The photographer began with the production of cheap clothing in Bangladesh and recorded the birth of clothing; then he tracked the cycle of clothing in fashion, sales, transportation, recycling, second-hand and disposal; finally discussed the chain of global economic industry, unequal distribution of interests, competition for the right of speech, demand and disguised slavery between developed and developing countries. The book is bound like accordion binding. It's wrapped by a slipcase made of matte art paper printed in four colors. The cover is made of gray cardboard printed in black four with four rounded corners. Offset paper printed in four colors is used for inside pages. The book can be read either from the front cover or from the back cover that forms two parallel narrative lines, including not only the process of making clothes in Bangladesh and selling them in the West, but also the process of being eliminated from Europe and transported to Africa for second-hand sale. Those recycled clothes that cannot be sold in Africa are then transported back to Bangladesh as raw materials. The last part of the book elaborates on the story and thinking behind the photos.

铜奖

我们穿着的

荷兰

Bronze Medal

What We Wear

The Netherlands

Teun van der Heijden / Heijdens Karwei

Pieter van den Boogert

Keff & Dessing Publishing, Amsterdam

226×174×26mm

691g

140p

The voyage that our everyday clothes have made around the globe is captured here in a moving and stimulating book – a voyage that begins right from the way the book is bound. »What We Wear« is driven by a journalistic impetus of rare force which allows the reader to share in the photographer's experiences. This gigantic leporello fold-out is nothing short of a road movie made up of stationary pictures – just printed on paper rather than projected onto a screen. Astonishing variety in handling and page turning mean that our hands do not just touch the paper but truly reach into the pictures themselves. These docu- mentary, revelatory pictures encourage us to look with a sharper eye at what we buy, where our clothing comes from and, above all, who actually made it.

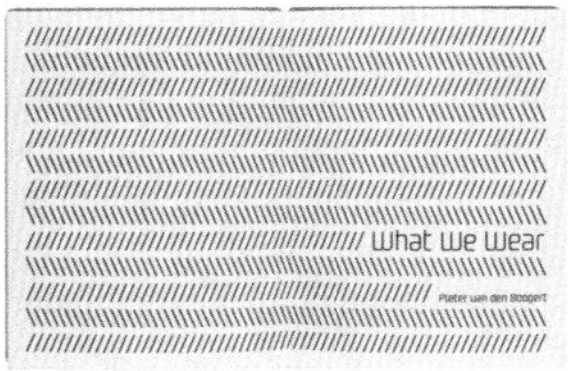
What We Wear
Pieter van den Boogert

Made
in Bangladesh
An Example of Production

Tunika
1995

50%

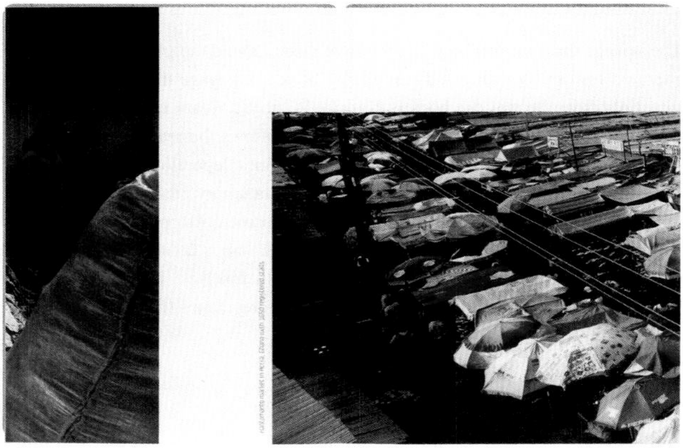

Used
in The Netherlands
An Example of Consumption

Dumped
in Ghana
An Example of Distribution

orts cor
ustrativ

What We Wear //////////
Pieter van den Boogert

Left-over clothes sold from large stacks for lowest price

Fig. 09-03
one two three four
the symmetrical currency
of the careening
circles
(elipse brides)

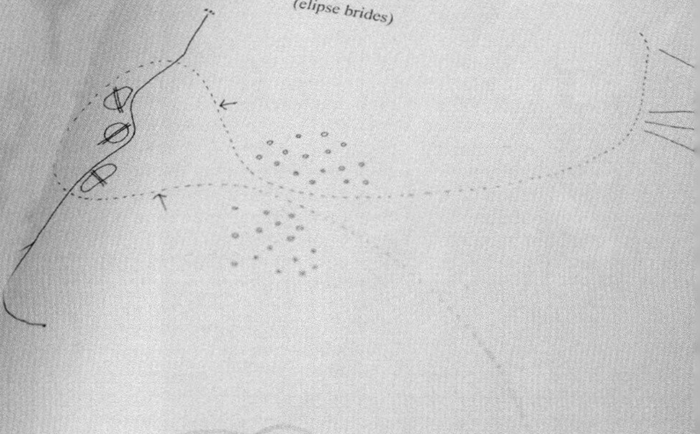

这是奥地利艺术家尼古拉斯·甘斯特尔（Nikolaus Gansterer）寻找艺术与科学之间的联系，研究科学概念可视化图表，以及可视化表达形式的一本书。本书作者通过与科学家的密集交流，将绘画作为研究媒介，通过科学概念在绘制过程和结果上的研究，促使新的叙事方式和思考得以展现。本书为锁线胶装的平装书，封面白卡纸双色印刷，标题部分压凹处理，封面未覆膜。内页主要为胶版纸单黑印刷，分为 27 个交流、绘制与可视化过程的研究。内夹三处光面铜版纸四色印刷的创作和实验过程。书籍正文设计成较为典型的科学理论的版式，封面勒口处的公认的基础图示、书中夹带的书签上的图形标准、封底暗藏的关于秩序和关系的大图，都是帮助读者进入这本书的好工具。

Drawing a
Hypothesis
Figures of Thought
A Project by
NIKOLAUS GANSTERER

This is a book written by Austrian artist Nikolaus Gansterer, who looks for the link between art & science and studies visual charts of scientific concepts and visual expressions. Through intensive communication with scientists, the author uses painting as a research medium; through the study of scientific concepts in the drawing process and results, he promotes the new narrative mode and thoughts to be displayed. This paperback book is glue bound with thread sewing. The cover is printed on white cardboard with double-color printing and the title text is embossed. The inside pages are mainly printed on offset paper in black, which is divided into 27 sections according to the process of communication, drawing and visualization. The contents of creating and experimenting process are printed on gloss coated art paper in four colors. The main body of the book is designed as a typically scientific book. The recognized basic graphics on the flap of the front cover, the graphic standards on the bookmark included in the book, and the hidden maps of orders and relationships on the back cover are all good tools to help readers understand the book.

- 铜奖
- 画出一个假设
- 奥地利

- Bronze Medal
- Drawing a Hypothesis
- Austria
- Simona Koch
- Nikolaus Gansterer
- Springer Wien New York
- 215×145×25mm
- 574g
- 348p

Everyone these days is talking about imaging techniques, particularly in the fields of technology and science. Pictures are created from data material and then appear to be natural, representational pictures. Pitting itself against this ideology the book »Drawing a Hypothesis« shows that the process of thinking is in any case a pictorial one. The author presents 27 hypotheses by means of a comprehensive collection of sketches in the way they often arise as a – whether scrawly, casual or elaborate – visual adjunct to thought processes or argumentation. And though this publication is anything but a picture book, the curious reader cannot help but start to suspect, while flitting from page to page, that the »manufacture of thought« does not merely take place through the process of »speaking« but that – to loosely quote Kleist – thoughts are manufactured through the process of scribbling. A stupendous book on the »imaging techniques« of thinking.

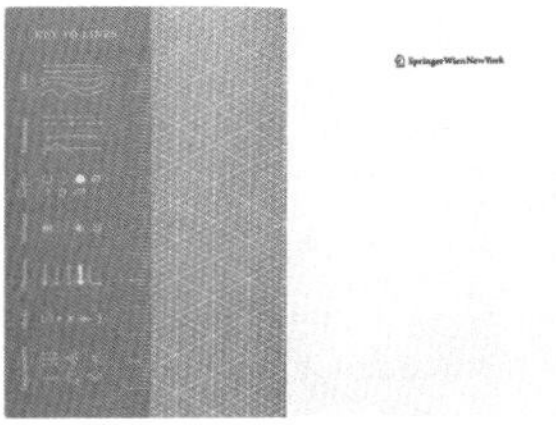

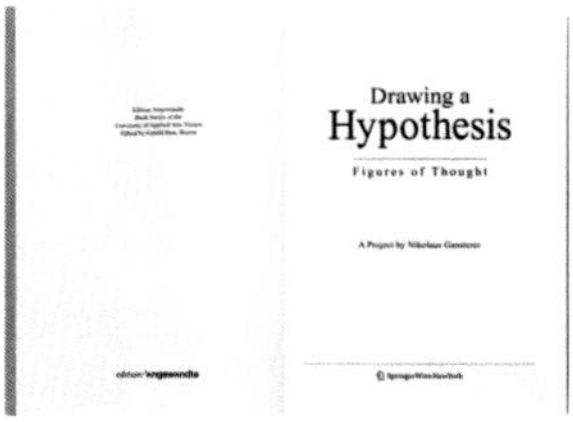
Drawing a
Hypothesis
Figures of Thought

TABLE OF CONTENTS

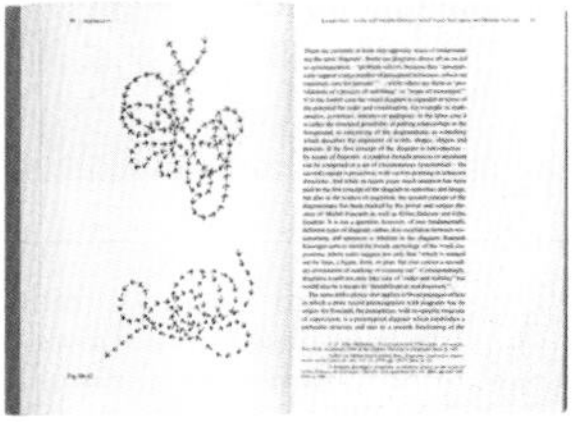

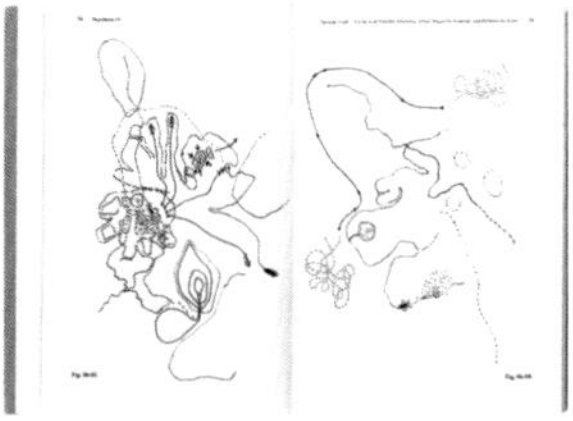

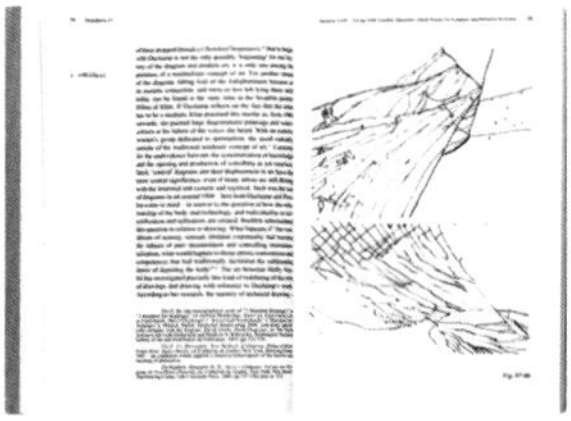

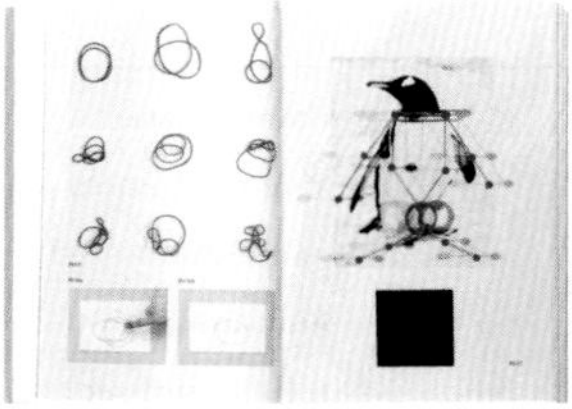

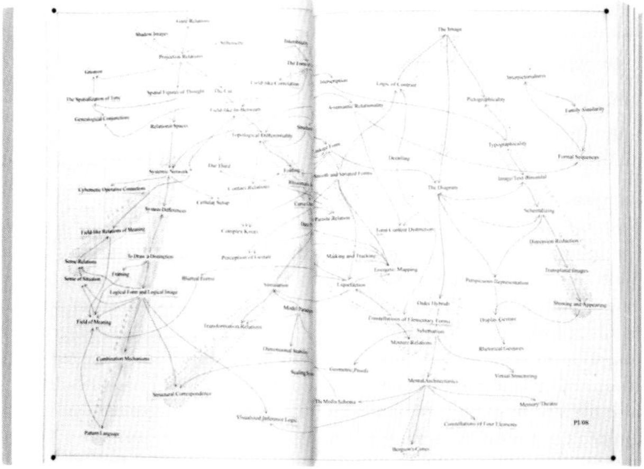

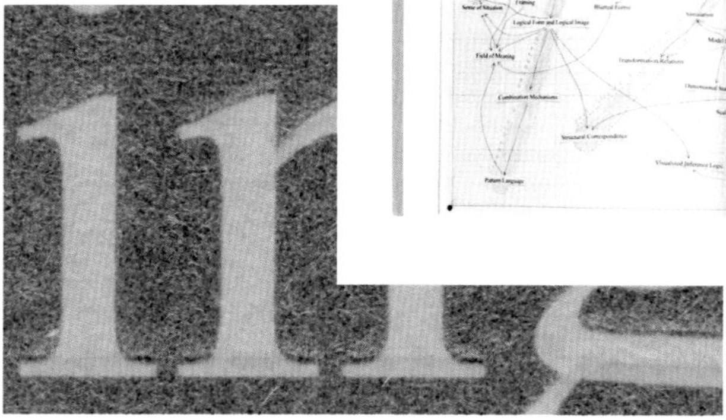

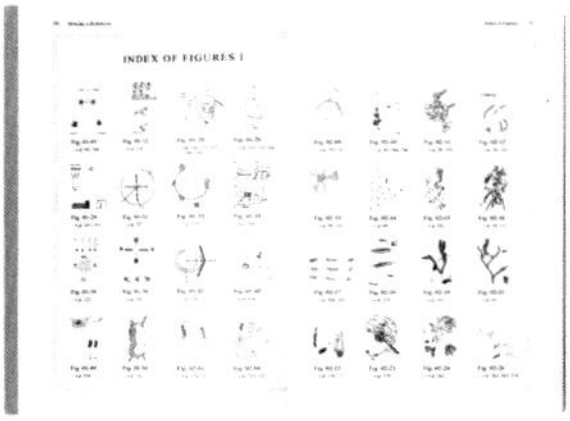
INDEX OF FIGURES I

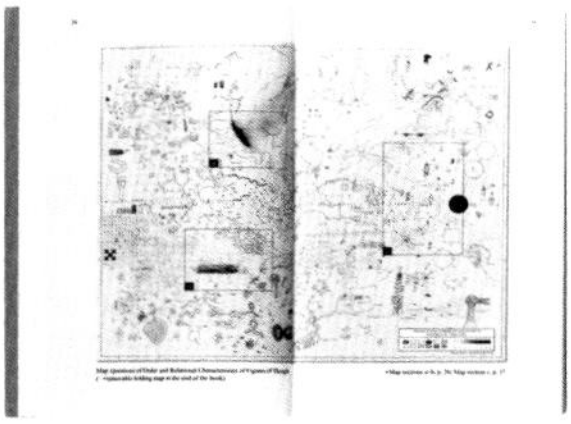

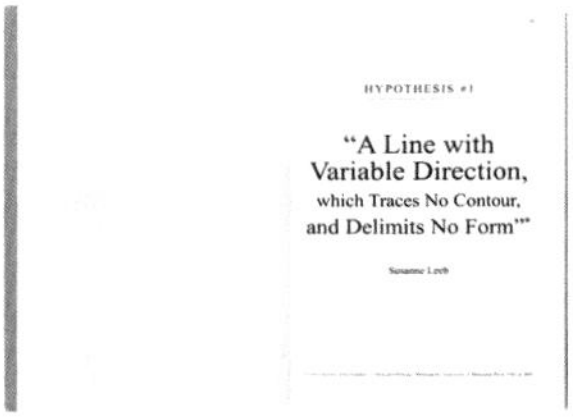
HYPOTHESIS #1
"A Line with
Variable Direction,
which Traces No Contour,
and Delimits No Form"
Susanne Leeb

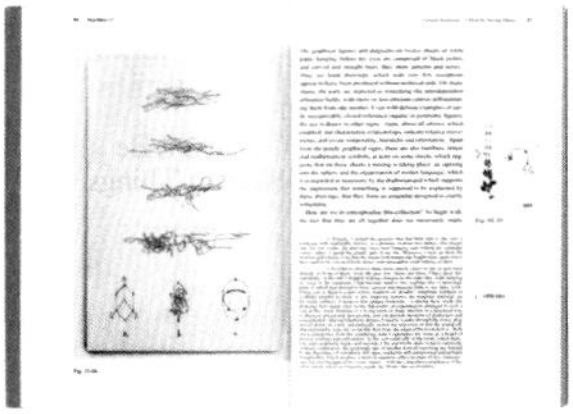

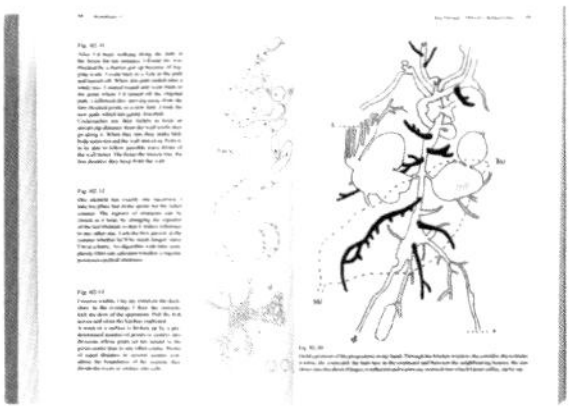

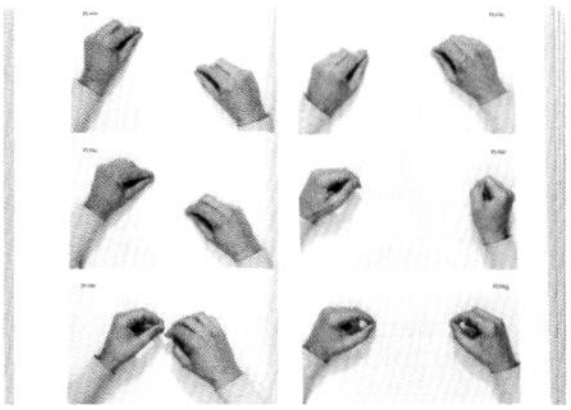

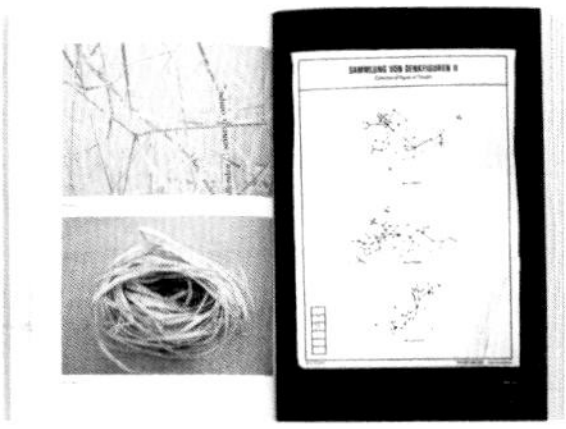
SAMMLUNG VON DENKFIGUREN II

and a
and receive or

Gansterer Drawing a Hypothesis – Figures of Thought

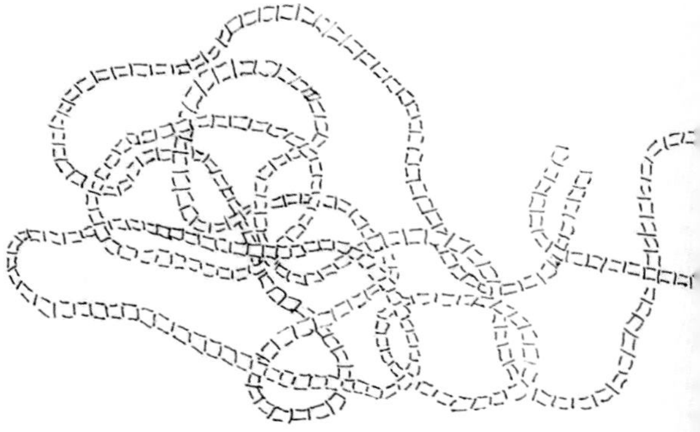

Fig. 10-26

the 'line', as it was known in French schools of the 19th ce
– was an activity enabling the development of graphical
petence for the process of industrial production.[23] Con
tion drawings counted as non-retinal, i.e. not confined
chance appearance of things; instead they were to compre
this appearance in its constructive-functional principles, a
etrating view reserved for male pupils only. For the que
of modern subjectivity this demand was decisive, the de
that technical drawing be trained until it became an u
scious process, "*d'une manière inconsciente, machinale*",
highly consequential connection of machine and unconsc
reaching from surreal automatism by way of the "machi
the studio"[25] to the abstract machines of Deleuze and Guat
The sexualised machines of Duchamps or Picabia, but als

23) Molly Nesbit, *Marcel Duchamp: The Language of In* in: *The definitely unfinished Marcel Duchamp*, ed. Thierry de Duve, Cam Mass., London, 1991, pp. 351-394, here: p. 353.

24) Nesbit, ibid., p. 383, fn 24.

25) E.g. Caroline A. Jones' book in which she discusses the " machines" of abstract expressionism: *Machine in the Studio. Constructi postwar American Artist*, Chicago, University of Chicago Press, 1996.

26) cf. also Rosalind Krauss' text in which she relates Eva H works to Deleuze/Guattari's "organless body": Rosalind Krauss, *Hesse's D Machines* (1993), in: *Eva Hesse*, ed. Mignon Nixon (= October Files 3), bridge/MA, London, 2002.

omechanical drawings of Eva Hesse are articulations of the iestion of how body and technology, desire and scientific tionality cohere, wresting a polymorphous principle of desire om the reality of industrialisation.

How important the normalising, scientific and quantifying agrams were and are for the question of subjectivity can be eaned from diagram drawings oscillating between disciplin-g or self-optimising controls and self-determination. In the ocess, art itself breaks with the myth of modernity, of the ank page, the white sheet, when artists use pre-prepared mate-il. Thus, there are countless works drawn in and on maps (e.g. asreen Mohamedi), on graph paper (Eva Hesse, Agnes Martin, nma Kunz and many others), in schedules (Eva Hesse) or count books (Morgan O'Hara) – all forms in which a con-ection is visible between subjectivity and the economy (of ne) in the sense of self-administration. This kind of technique so crosses over into the realm of self-regulation, for instance, hen the sound artist Stephan von Huene, using neurolinguistic ogramming and the brain training and memory methods of e British psychologist Tony Buzan, chooses Mind Maps as a rm of notation for the theoretical construction of his installa-ons and for his lectures.[27]

(→PI/01)

While diagrams of this kind manifest a specific subject lation which corresponds to the modern invocation to self-lministration, a principle of the diagrammatic is found less a concrete diagram form but rather in loose arrangement id open constellations. This encompasses other arts as well beyond concrete, material drawing – and allows the diagram be seen as a tool for channelling the making of relationships id for the abandonment of a rational procedure. One thinks of yvind Fahlström, whose installations are conceived as move-ole, so that the individual elements always find their way to ew constellations; or one thinks of the choreographer Yvonne ainers with her ephemeral production of a space through ovement, and in her changing between rule, arrangement id lapse; or of the absurd performances of the 1970s/80s of tist, poet, theatre director and actor Stuart Sherman, in which

Fig. 06-12b

27) cf. Astrit Schmidt-Burkhardt, *Das Diagramm als Gedächtnis-rm. Zu Stephan von Huene's Mind-Map-Methode* (*The Diagram as a Form of lemory. To Stephan von Huene's Mind-Map-Method*), in: Irmgard Bohunovsky-ärnthaler (ed.), *Kunst ist gestaltete Zeit* (*Art is Formed Time*). *Über das Altern*, lagenfurt, 2007, pp. 105-141.

本书为徐岩奇建筑师事务所 2000－2001 年设计作品集，书中通过多种材料的运用，体现建筑师对于自身作品的独特理解与诠释。本书装订方式为锁线胶装，封面犹如水泥质感的灰色卡板丝印白色方块后文字烫黑处理，书脊处采用建筑工地使用的绿色防护网包裹，从封面上直观地传递出与建筑相关的意味。内页开篇的理念诠释采用多色艺术纸裁切成不规则的多边形，似在对应书中不同建筑的外形。正文的建筑项目部分采用半透明的绘图纸与哑面胶版纸混合装订。绘图纸上单黑印刷，内容为项目资料、平立剖面图与手绘分析图；哑面胶版纸四色印刷建成照片。由于大多数照片为横向比例，所以胶版纸设计为等宽的方开本，并通过排版组合，展现建筑不同角度和时间段的风貌。内页半透明绘图纸上设计方案 的“虚”透叠出哑面胶版纸建成照片的“实”，是作品处于不同状态的体现。

ZOOM
IN
ZOOM
OUT
VICTOR HSU ARCHITECT & PARTNERS
ZOOM DESIGN ATELIER

This book is a collection of design works by Victor HSU Architect & Partners from 2000 to 2001. Through the use of various materials, it reflects the architect's unique understanding and interpretation of his works. The book is glue bound by thread sewing. The cover is made of gray cardboard printed with white squares, and the text is hot-stamped in black. The spine of the book is wrapped with a green protective net used in construction, which conveys the connection with architecture intuitively. The main body starts with the explanation of the ideas. The designer uses irregular polygons cut from colorful art paper, which resembles the shapes of different buildings. The construction project part is bound with translucent drawing paper and matte offset paper. The drawing paper is printed in black with project information, flat profile and hand-drawn analysis; while the matte offset paper is printed in four colors to display photographs. Since most of the photos are in horizontal proportion, the format is designed as a square of equal width. With different combination of the photos, the book shows the style of buildings in different period from various angles. The 'virtual' design schemes on the translucent drawing paper and the 'real' photographs of buildings on the matte offset paper both reflect the different state of an architectural work.

铜奖	Bronze Medal
以有机为名	Zoom in, Zoom out
中国台湾	Taiwan, China
一元搥搥（方信元）	Fang Hsin-Yuan
徐岩奇，徐岩奇建筑师事务所	Victor HSU Architect & Partners; Zoom Design Atelier (Ed.)
田园城市出版社	Garden City Publishers Ltd., Taipei
	265 × 190 × 20mm
	661g
	228p

A refreshing dip into the world of architecture, into the nucleus of the architect's bureau, so to speak. Buildings evoke hefty discussions or even inflame passions. They are viewed from different perspectives, by young and old alike. Buildings can leave an austere impression or a mellow one – depending on a wide spectrum of available materials. Contrasts provide inspiration. And scaffolding and safety nets are necessary too ! All these aspects of architecture are brought to life in this compelling and bold book – from the transparent paper used for its draft sketches right through to the photo booklets exhibiting the bureau's completed works.

ZOOM
IN
ZOOM
OUT

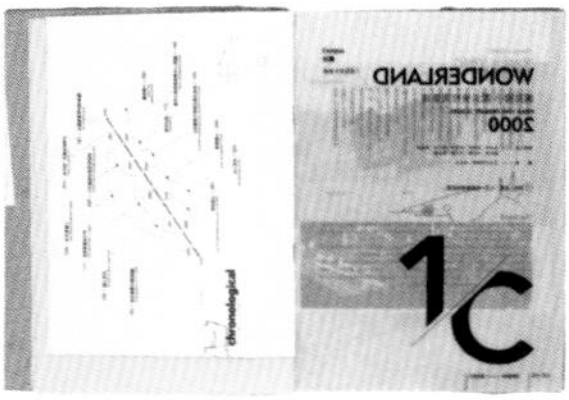
WONDERLAND
2000
chronological

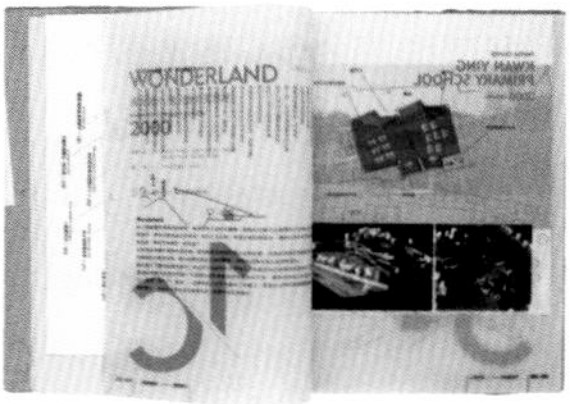
WONDERLAND
KWAN YING
PRIMARY SCHOOL

KWAN YING
PRIMARY SCHOOL

Public
淺草青春新天地
台南市西門市場改造之觸媒工程
NEW SIMEN MARKET - ASAKUSA
2004
NEW SIMEN MARKET
ASAKUSA

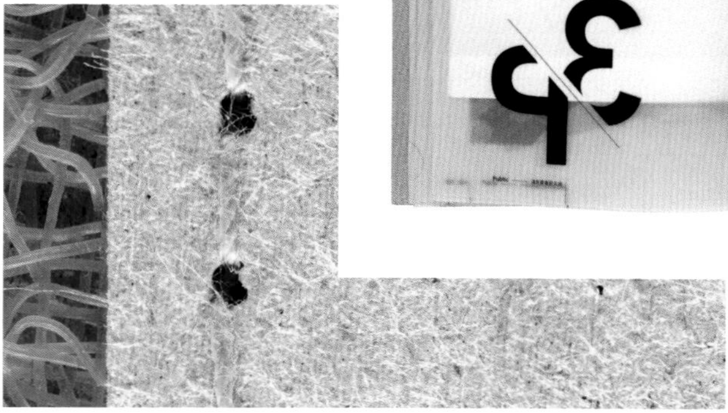

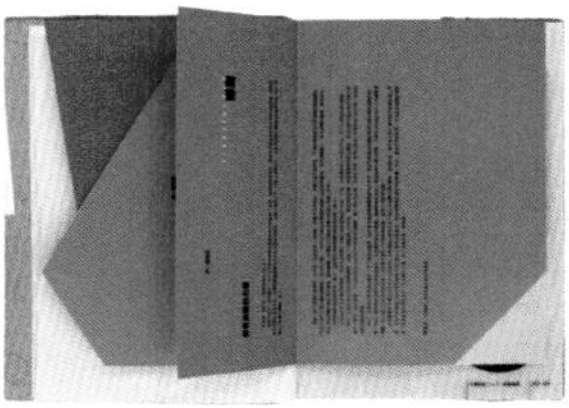
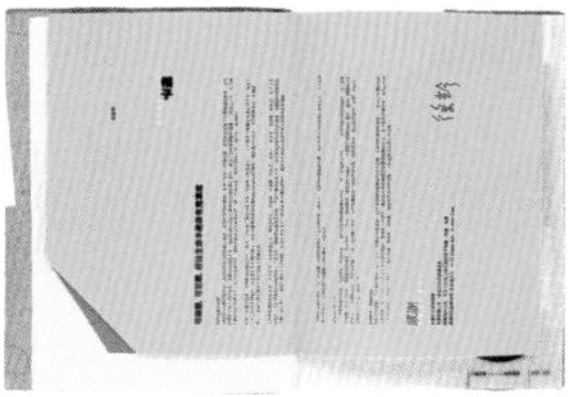

植入與共生
2002
2p

植入與共生
2002
2p

TOURIST FACILITY IN
CHIKU LAGOON AREA
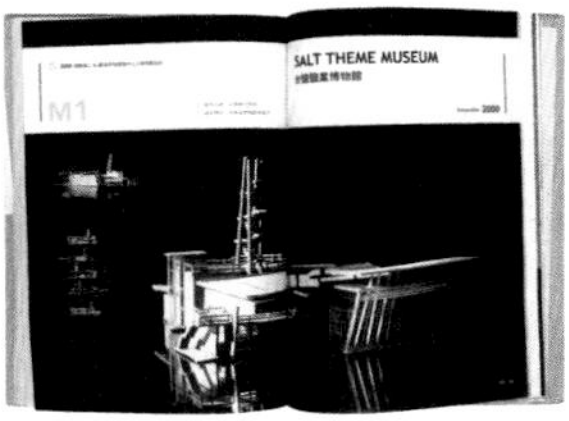
SALT THEME MUSEUM
M1
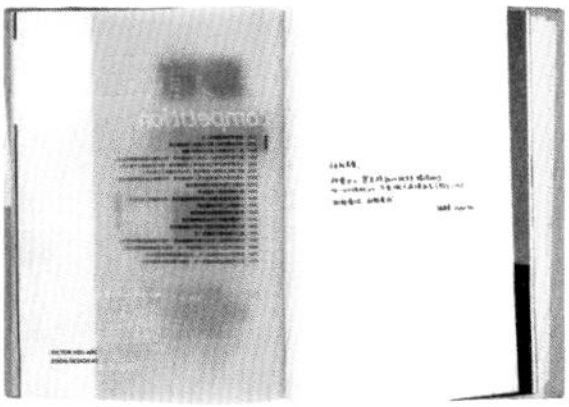
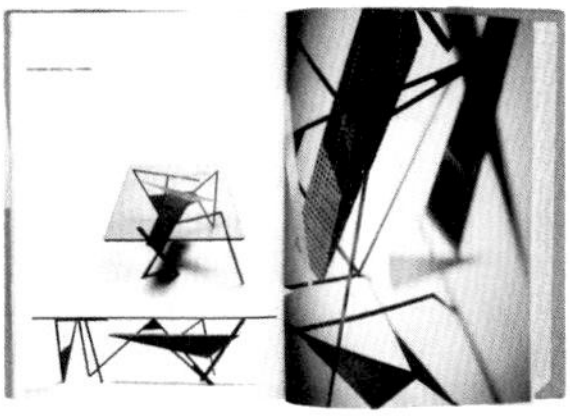
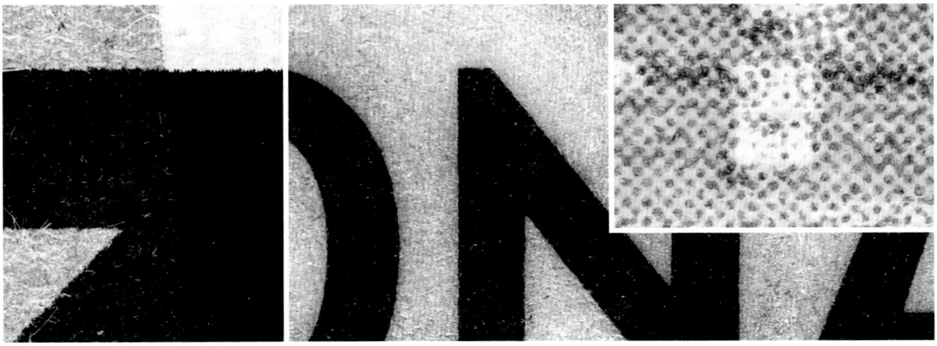

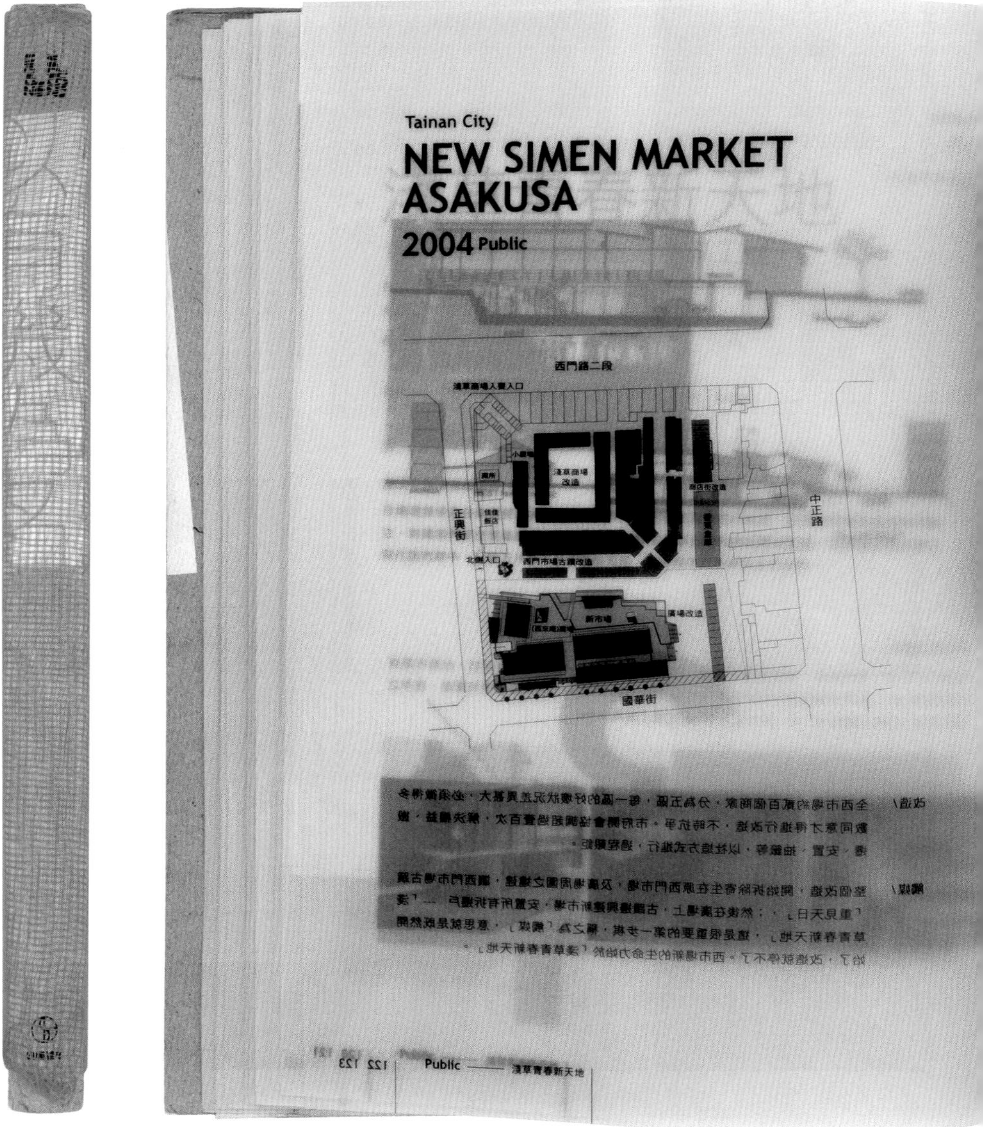
Tainan City
NEW SIMEN MARKET
ASAKUSA
2004 Public
西門路二段
淺草商場人員入口
小廣場
淺草商場改造
廁所
佳佳飯店
正興街
北側入口
西門市場古蹟改造
商店街改造
中正路
新市場
廣場改造
國華街
122 123
Public —— 淺草青春新天地

"We made up themes

这是一本介绍上20世纪60、70年代享誉世界的捷克玩具设计师利布谢·尼克洛娃（Libuše Niklová）及其玩具设计作品的书。这位设计师在木质玩具文明的国家里，开创性地使用了二战后普及的塑料作为玩具设计材料，在成本、清洁、重量等方面带来了玩具的革命。本书为锁线硬精装，封面光面铜版纸四色印刷覆哑膜，并在封面纸张和卡板之间夹了一层薄海绵，形成了塑料充气玩具的柔软手感。内页人物及传记部分采用胶版纸四色印刷，代表作品部分则采用哑面铜版纸四色印刷，大量黑白粗网老照片在书中穿插，与作品形成并行的叙事节奏。全书捷克文与英文双语言排版，英文固定蓝色，捷克文则按照板块分为红、黄、绿三色。三色书签带与内文多色的设计衬托出她的设计世界的丰富多彩。||||||||||||||||||

This is a book about the world-renowned Czech toy designer Libuše Niklová and her toy designs in the 1960s and 1970s. The designer pioneered in using plastics as toy materials after World War II in a country with a wooden toy tradition. She brought about a toy revolution in terms of cost, cleanliness and weight. The hardback book is bound by thread sewing. The laminated cover is printed on gloss coated art paper in four colors, and there's a thin sponge between the cover paper and the graphic board, which makes the book soft feeling like plastic inflatable toys. The inner pages written about characters and biographies are printed in four colors on offset paper, while the representative works are printed in four colors on matte coated art paper. A large number of old black-and-white pictures are interlaced in the book, forming a parallel narrative rhythm with the works. The book is typeset in both Czech and English. English text is printed in blue only, while Czech text is printed in red, yellow and green according to the sections. Tri-color bookmark tape and multi-color design of the text set off the rich and colorful design world of the toy designer.

- 铜奖
- 利布谢 · 尼克洛娃
- 捷克

The lives of hosts of children, primarily in Eastern Europe, were accompanied in the 1970s and 1980s by toys produced by designer Libuse Niklová. This book offers a comprehensive overview of her works and extends its intensive use of colour, with predominantly red, yellow, blue and green, into its graphic design. Out of an index using a smaller format and different paper is created an outstanding archive of the designer's output. The material used for the cover makes reference to the synthetic material which the toys were made of. Here we have a book which in all its aspects is harmonious and which for all its stringency still makes a light-hearted impression.

- Bronze Medal
- Libuše Niklová
- Czech Republic
- Zuzana Lednická / Studio Najbrt
- Tereza Bruthansová
- Arbor vitae societas, Prag
- 206×193×29mm
- 975g
- 299p

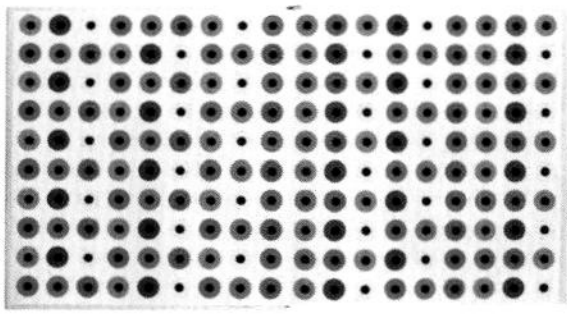

Libuše Niklová

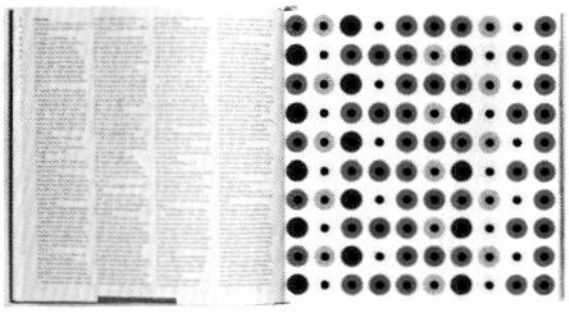

gumáčci
plastic
playthings
1954–1964

GUMOTEX

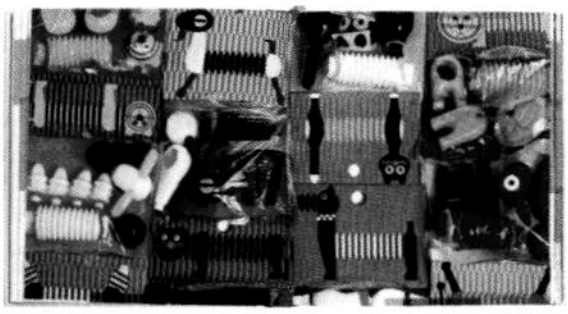

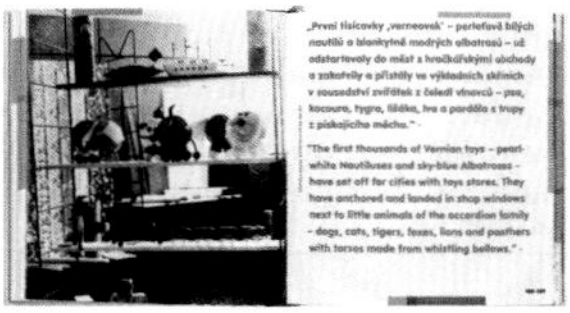

„První tisícovky ‚verneovek' – perleťově bílých nautilů a blankytně modrých albatrosů – už odstartovaly do měst s hračkářskými obchody a zakotvily a přistály ve výkladních skříních v sousedství zvířátek z čeledi vinovců – psa, kocoura, tygra, lišáka, lva a pardála s trupy z pískajícího měchu." -
"The first thousands of Vernian toys – pearl-white Nautiluses and sky-blue Albatrosses – have set off for cities with toys stores. They have anchored and landed in shop windows next to little animals of the accordion family – dogs, cats, tigers, foxes, lions and panthers with torsos made from whistling bellows." -

"Along with designing the form of the water toys, continuous practical testing was conducted directly in the water. During the summer this happened close to Napajedla on a dead stretch of the Morava River called Pohribek, during the winter it was at the Winter Spa Center in Gottwaldov. They tested the stability and suitability of the product's shape, its carrying capacity and its behaviour when overloaded with people. Products on the water had to have at least three-chambers, be equipped with a safety valve and, in case of accident, one single inflated chamber had to be able to support a person."

Soupis díla
Libuše Niklové
Catalogue Raisonné
of Libuše Niklová
Sestavila / Compiled by
Tereza Bruthansová
Spolupráce / with Assistance from
Magdalena Wells
248–273

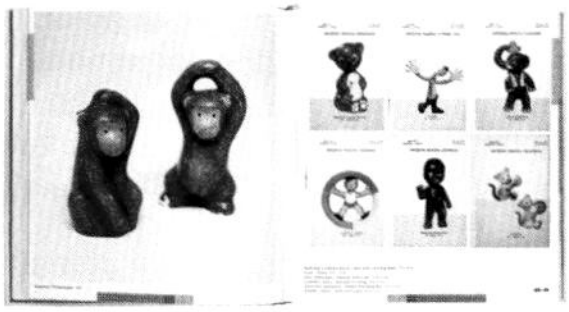

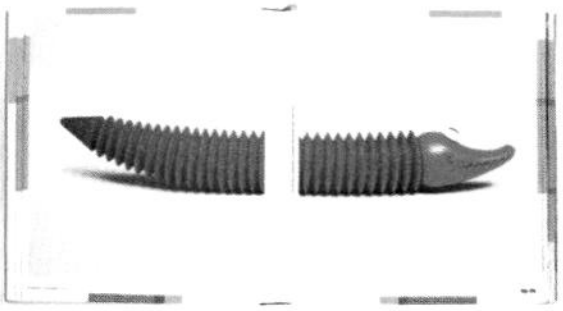

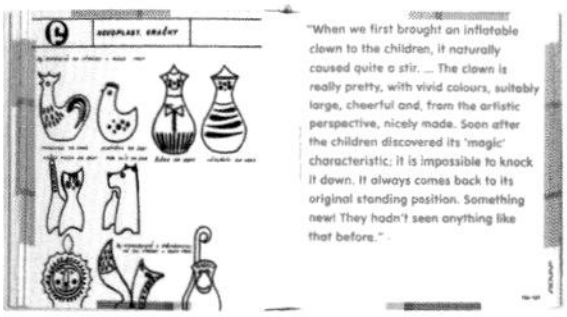
"When we first brought an inflatable clown to the children, it naturally caused quite a stir. ... The clown is really pretty, with vivid colours, suitably large, cheerful and, from the artistic perspective, nicely made. Soon after the children discovered its 'magic' characteristic; it is impossible to knock it down. It always comes back to its original standing position. Something new! They hadn't seen anything like that before."

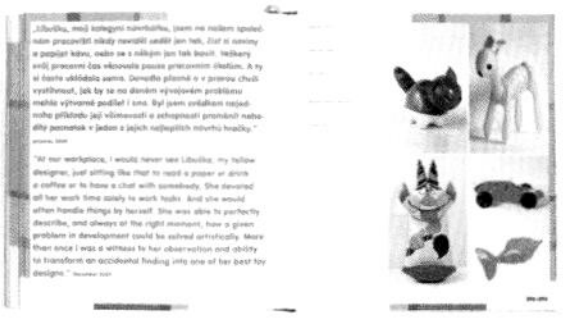

in
talented

012 Ha1
卷七 272
卷七 300

本书是著名诗人文爱艺的诗歌集，收录抒情和散文诗 156 首。文爱艺与他的诗被誉为“用诗建造的可以对话的青春偶像”“是精神家园中与人共同呼吸的草坪”，充满了现代气息和自然气息，是传统精神与现代思想的天然融合。本书为锁线硬精装，护封浅灰色艺术纸红黑双色印刷，护封内侧印红色图案，为艺术家谭东的油画局部。内封暖灰色布面裱覆卡板，文字部分通过凹凸处理呈现微妙的手感。鲜红色的环衬包裹下，内页胶版纸单黑印刷，舒朗的版式配合少量谭东的油画，充满韵律。上下书口及右侧翻口红色刷边处理，红色的书签带夹在书中，精致而优雅。全书大量使用红色，仿佛字里行间充满了诗人对生活的热爱。

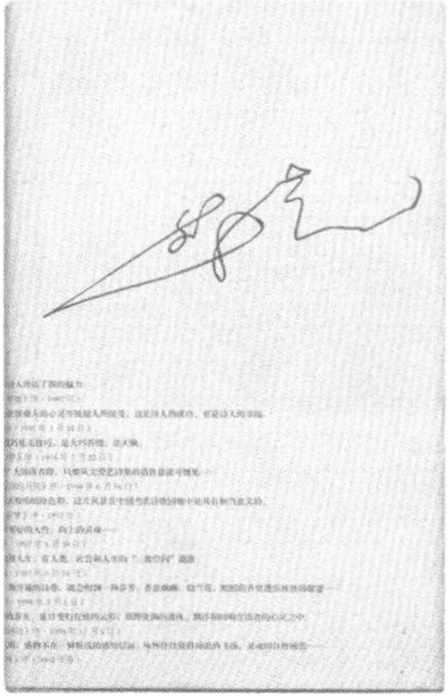

This book is a collection of 156 poems by the famous poet Wen Aiyi, including lyric poetry and prose poetry. Wen Aiyi is praised as 'a youth idol built with poems' and her poems are praised as 'lawns that breathe together with people in spiritual home'. Her poems are full of natural flavor, and are the natural integration of traditional spirits and modern thoughts. This hardback book is bound by thread sewing. The jacket is made of light gray art paper printed in red and black, while red patterns are printed inside the jacket, which is part of Tan Dong's oil painting. The inside cover is made of warm grey cloth mounted with the graphic board, and the embossed text presents a delicate appearance. Wrapped by the bright red endpaper, the inside pages are made of offset paper printed in black. Simple page layout with a few Tan Dong's oil paintings is rhythmic. The three edges of the book are brushed with red ink. The red bookmark clipped in the book is delicate and elegant. Red is heavily used throughout the book to show the poet's passion for life between the lines.

- 荣誉奖
- 文爱艺诗集
- 中国
- 刘晓翔、高文
- 文爱艺
- 作家出版社

- Honorary Appreciation
- Poems by Wen Aiyi
- China
- Liu Xiaoxiang, Gao Wen
- Wen Aiyi
- Writers Press, Beijing
- 235 × 144 × 23mm
- 617g
- 380p

With an air of simplicity and honesty the poet's signature flows across the dust-jacket. The cover itself feels sleek and silky, with a variety of blind-embossed raised and recessed lines. Especially subtle is the effect of the particularly narrow embossed margin at the top, at the front and at the bottom. The typographical texture inside the book coupled with its generous use of blank space give the flowing poetry room to breathe. A red-based colour concept gives the volume an element of the sacred, even though the actual printing merely uses monochrome black. The poet confronts the subject of how truth has no regard for embellishment... What is a poem? This sounds virtually like a request from the designer. The answer in red extends to the book's spine. The old classical types – nothing but the best.

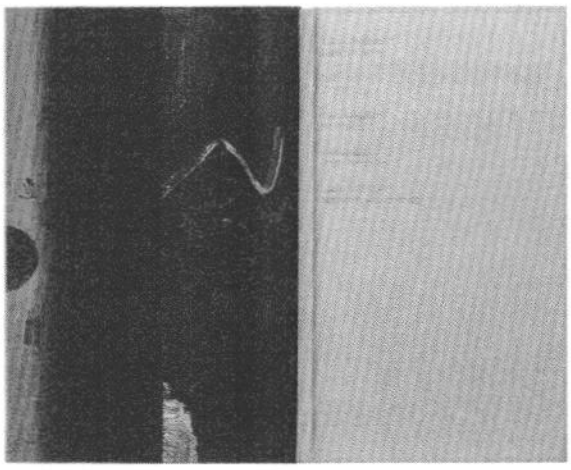

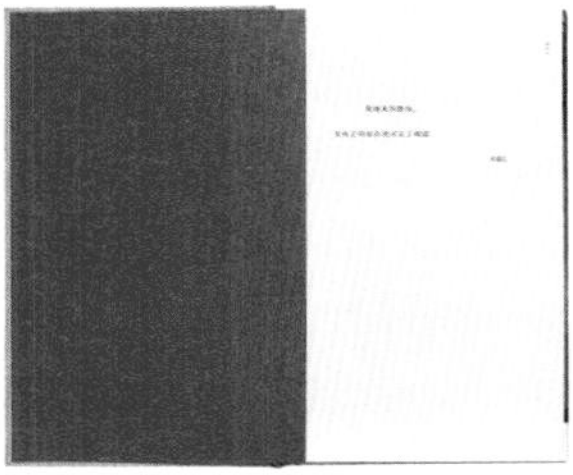

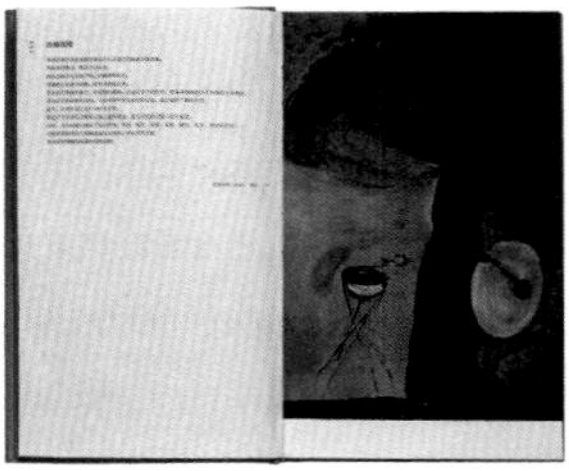

卷一

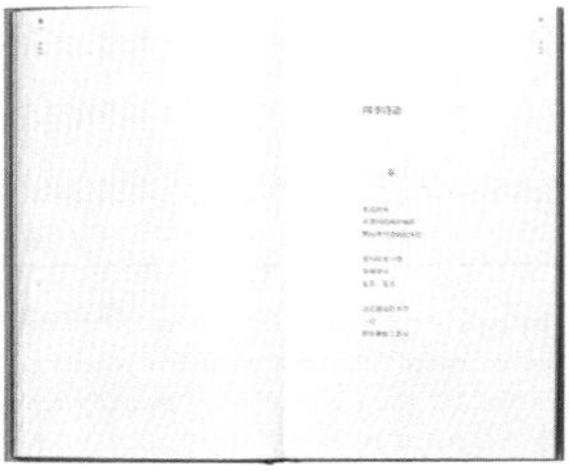

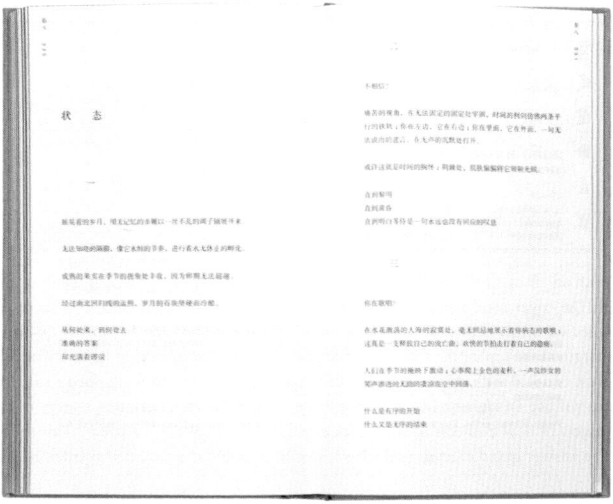
状　态
不相信?
直到黎明
直到黄昏
你在歌唱?
什么是有序的开始
什么又是无序的结束

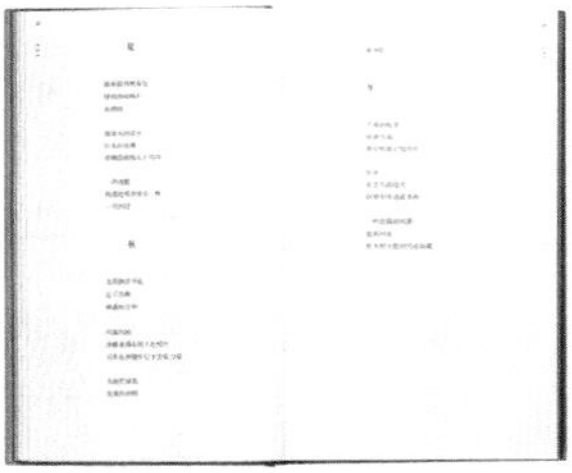

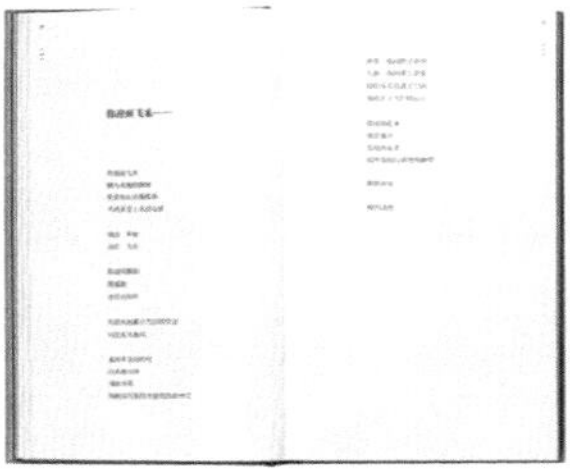
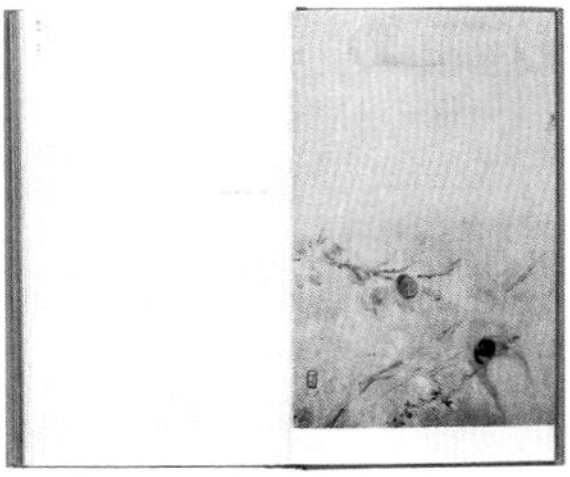
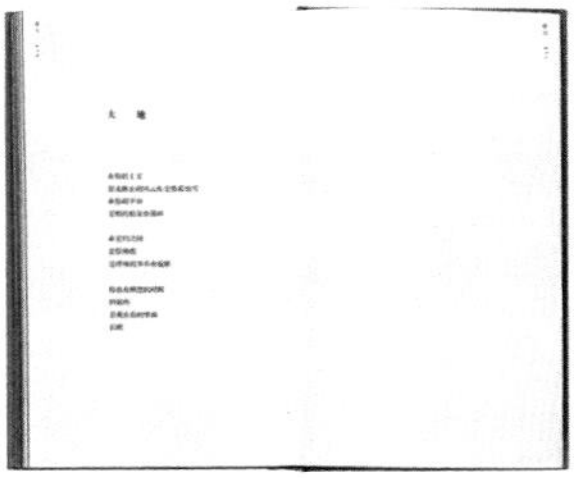
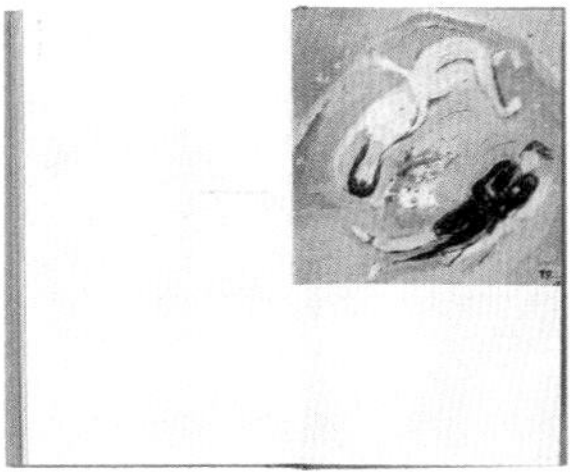
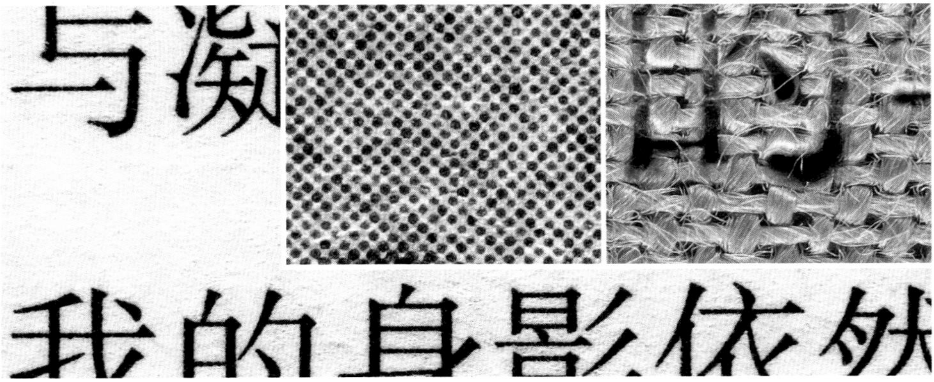

Ha2
sul rilievo precedente.
ripreso da altra
rilevata la scalfittura,

这是一本以意大利那不勒斯地区为背景，有关青少年生存状态、选择与机遇为主题的摄影作品集。蛾摩拉为《圣经・旧约》中的罪恶之城，蛾摩拉女孩在这里暗指生存在这个充满犯罪的危险地区的女生。本书的装订形式为骑马钉，封面并无特殊工艺和纸张。书中包含两大部分：1. 从警方调查的黑手党谋杀当地 14 岁女孩的案卷中抽取出的部分页面，采用胶版纸四色印刷，维持扫描件的真实状态；2. 反映当下年轻人生活状态的纪实摄影照片，哑面铜版纸四色印刷，开本比案卷小一截，穿插在谋杀案的案卷之中。照片下有详尽的拍摄主题、拍摄地点等信息，方便读者在案卷资料和现实照片的对比中体会到当地的氛围，并思考在这座充满问题的城市中人们该如何选择的问题。

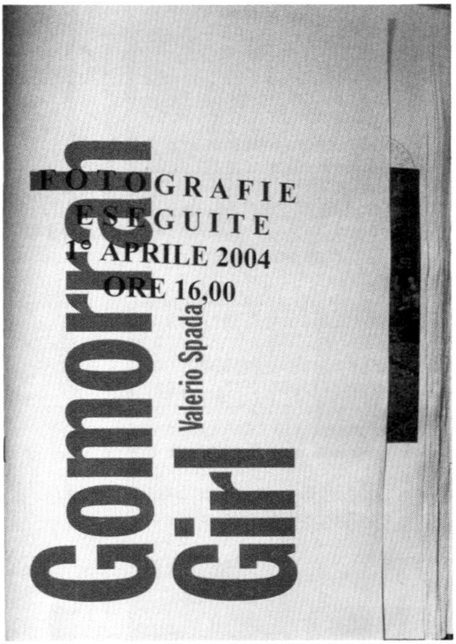
FOTOGRAFIE
ESEGUITE
1° APRILE 2004
ORE 16,00
Gomorrah
Girl
Valerio Spada

This is a photography album on the living conditions, choices and opportunities of teenagers in Naples. Gomorrah is a city of evil in the Old Testament, while Gomorrah Girl here alludes to girls living in a dangerous area full of crime. The book is bound by saddle stitch, and the cover doesn't use special craft or paper. The book contains two main parts. The first part contains several pages extracted from the files of police investigation of 14-year-old local girls murdered by the Mafia, which are printed in four colors on offset paper to maintain the state of scanned pieces. The second part contains documentary photographs reflecting the living conditions of young people nowadays, printed in four colors on matte coated art paper. The format of the second part is smaller than the case file, interlaced in the murder files. There is detailed information such as theme and shooting location under the photos. It is convenient for readers to experience the local atmosphere in the comparison of the files and real photos, and to think about how people make right choices in this city full of problems.

- 荣誉奖
- 蛾摩拉女孩
- 荷兰

The story of a girl who is accidentally shot dead between the fronts of the Camorra is recounted with few paragraphs and with distinctive typography. A thin volume, reminiscent of a police file, allows the reader to become immersed in the case. Its human abyss is made visible on two levels in just a few pages: the evidence from the actual case on thin, rustling paper and, bound in between using a smaller format and a change of paper with everyday photography, the lives of young girls in whose day-to-day existence Camorra is all too palpable. In just a few pages and with the simplest of means great intensity is produced.

- Honorary Appreciation
- Gomorrah Girl
- The Netherlands
- -SYB- (Sybren Kuiper)
- Valerio Spada
- Cross Editions, Paris
- 335×227×3mm
- 330g
- 40p

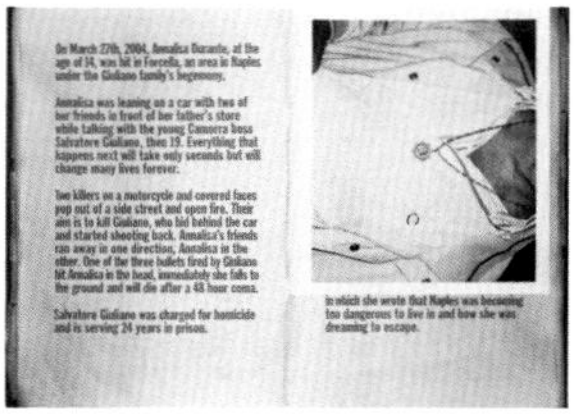
On March 27th, 2004, Annalisa Durante, at the age of 14, was hit in Forcella, an area in Naples under the Giuliano family's hegemony.
Annalisa was leaning on a car with two of her friends in front of her father's store while talking with the young Camorra boss Salvatore Giuliano, then 19. Everything that happens next will take only seconds but will change many lives forever.
Two killers on a motorcycle and covered faces pop out of a side street and open fire. Their aim is to kill Giuliano, who hid behind the car and started shooting back. Annalisa's friends ran away in one direction, Annalisa in the other. One of the three bullets fired by Giuliano hit Annalisa in the head, immediately she falls to the ground and will die after a 48 hour coma.
Salvatore Giuliano was charged for homicide and is serving 24 years in prison.
in which she wrote that Naples was becoming too dangerous to live in and how she was dreaming to escape.

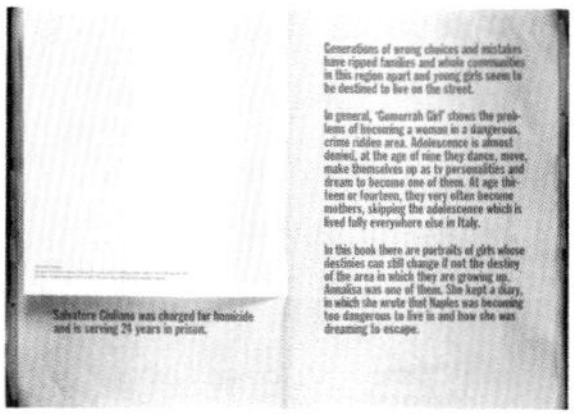
Salvatore Giuliano was charged for homicide and is serving 24 years in prison.
Generations of wrong choices and mistakes have ripped families and whole communities in this region apart and young girls seem to be destined to live on the street.
In general, 'Gomorrah Girl' shows the problems of becoming a woman in a dangerous, crime ridden area. Adolescence is almost denied, at the age of nine they dance, move, make themselves up as tv personalities and dream to become one of them. At age thirteen or fourteen, they very often become mothers, skipping the adolescence which is lived fully everywhere else in Italy.
In this book there are portraits of girls whose destinies can still change if not the destiny of the area in which they are growing up. Annalisa was one of them. She kept a diary, in which she wrote that Naples was becoming too dangerous to live in and how she was dreaming to escape.

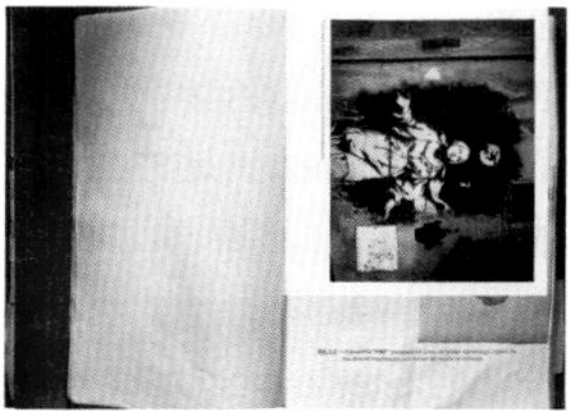

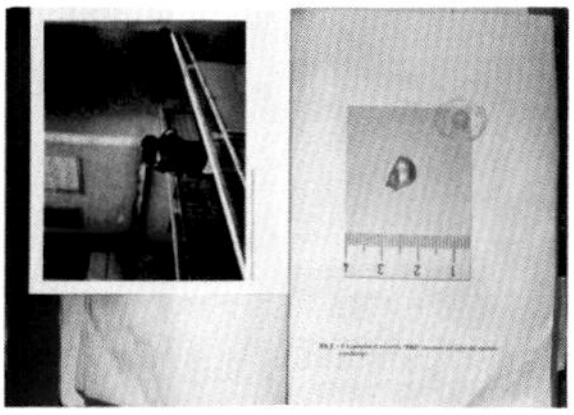

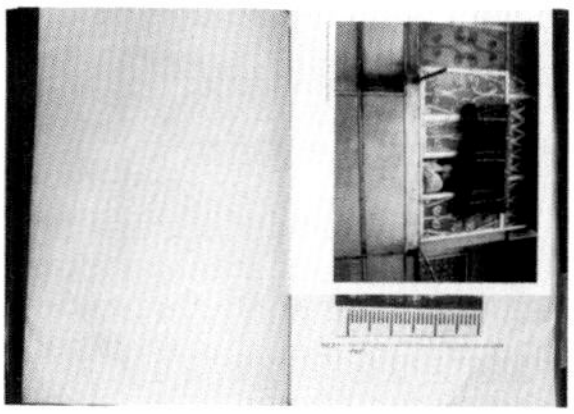

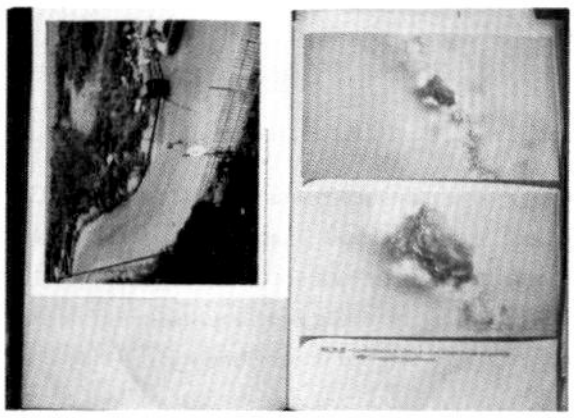

PR1
PR4
Rill. 36-37 = Impronta di riga residua sul proiettile "PR1" in accostamento e in sovrapposizione con impronta di riga residua sul proiettile "PR4".
Si evidenziano correlazioni nelle caratteristiche di classe d'arma e in alcune significative microstrie che lasciano validamente ritenere che siano stati esplosi dallo stesso revolver

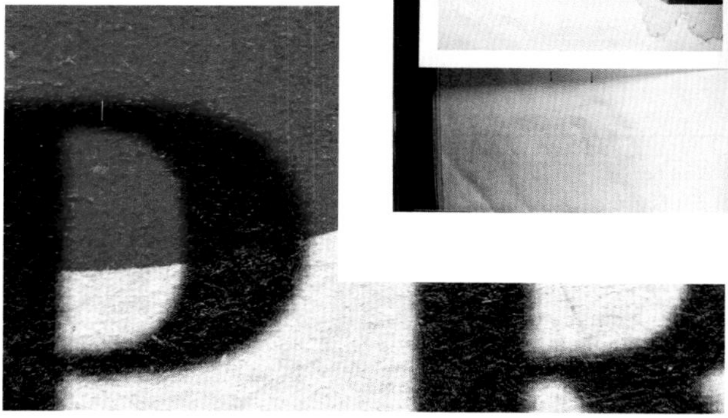

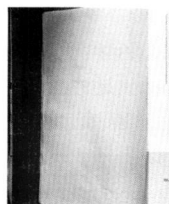

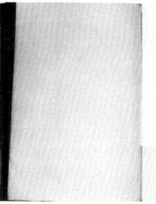

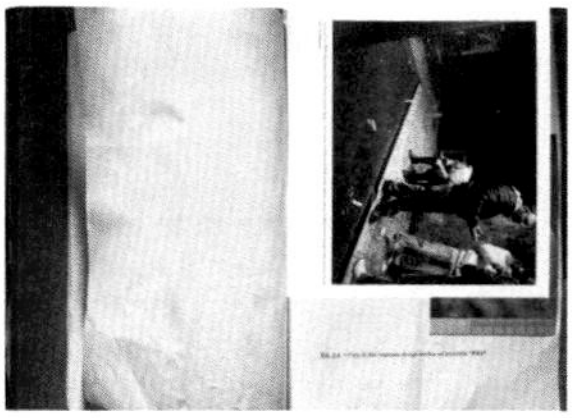
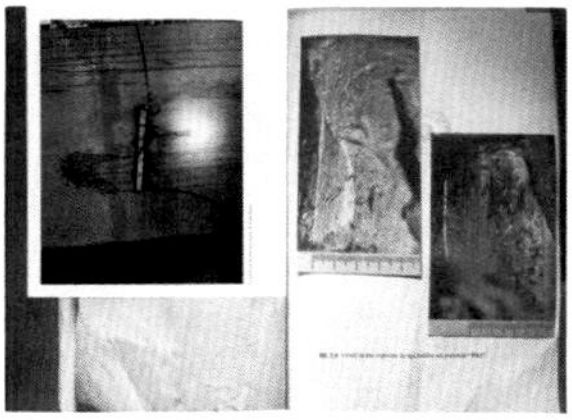
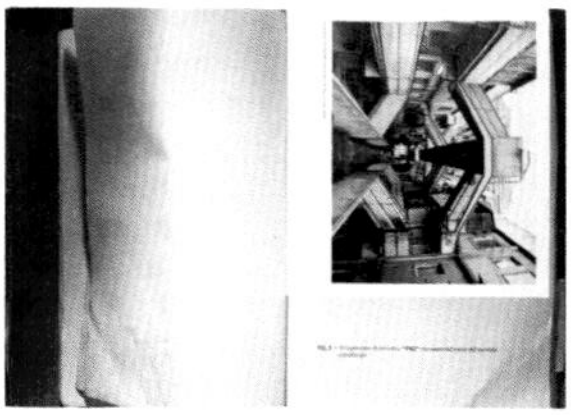
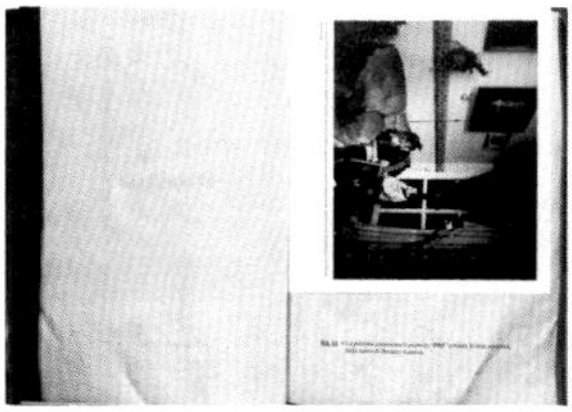
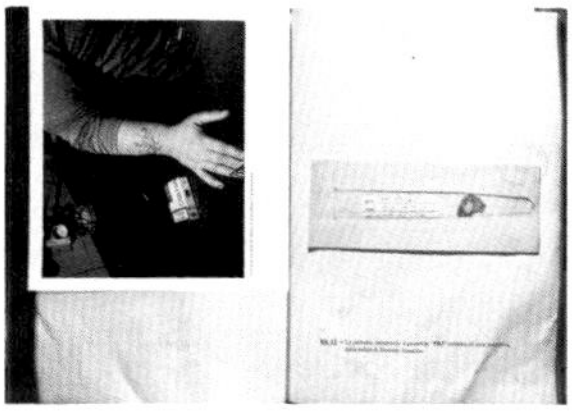

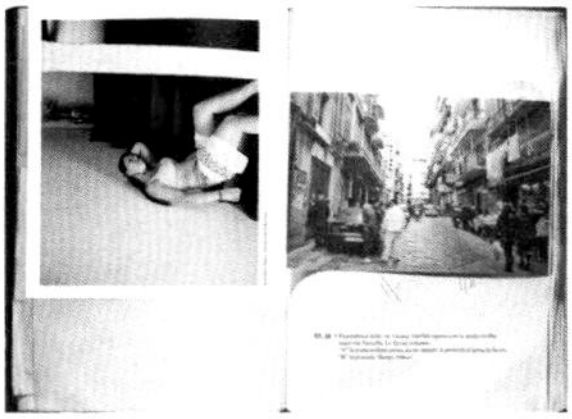
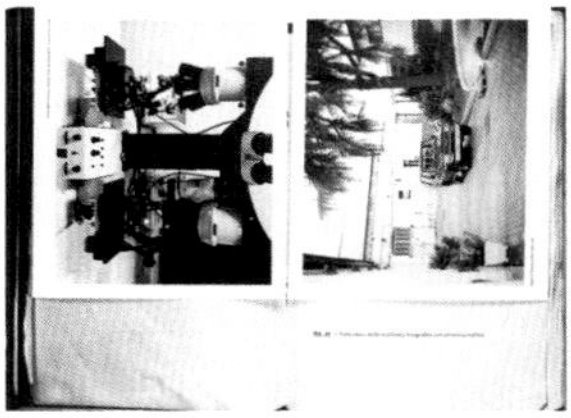

动物王国深陷饥荒危机中，动物们探索未知树种上不知名的果实来挽救危机。这棵树被国王狮子命名为拗口的名字“Pampilio”，但多位使者在传达树名的过程中始终记不住，导致动物们不敢食用它的果实。最终小乌龟将树的名字编成歌谣，顺利地传达给了动物们，化解了这场饥荒。故事本身比较古老，但五色印刷、动物的拟人化处理，以及特殊字体的选用赋予了它新鲜的视觉表现。本书的装订形式为锁线硬精装。封面铜版纸五色印刷，覆哑光膜。内页胶版纸五色叠印，形成丰富的色彩变化。书中动物除大量采用色块表现外，不少动物都采用直立行走，甚至人身加动物头套的形式。小乌龟则是人身背着乌龟壳，似乎为童话故事改编为真人表演的舞台剧提供了合适的视觉表现模板。正文字体采用加强衬线字体，单印蓝色，而书名（未知的树木名）在书中采用五色带花式衬线的字体，放大处理，在页面中格外凸显。||||||||||||

This is a story of exploring unknown fruits to save animals from a famine crisis in their kingdom. The tree was named ‘Pampilio’ by the king lion, but many emissaries could not remember it in the process of conveying the name of the tree. As a result, the animals dare not to eat the fruits. Finally, a little turtle used the name of the tree to compile a song and passed it on successfully to the animals, thus alleviating the famine. The story itself is relatively old, but five-color printing, anthropomorphic treatment of animals, and the selection of special fonts give it fresh visual expression. The hardback book is bound by thread sewing. The cover is made of coated art paper printed in five colors, laminated with gloss film. The inside pages are using five-color printing as well, which produces abundant color changes. In the book, a large number of color blocks are used to depict animals. Many animals are walking upright with the human body carrying the ‘animal headgear’, while the small turtle is depicted as a human body carrying a tortoise shell. It seems that the book provides a suitable visual performance model for the drama adaption. The main text uses the enhanced serif font printed in blue only, while the book title (name of the unknown tree) uses the serif font printed in five colors with enlargement processing, which is prominent in the page.

荣誉奖

Pampilio

波兰

In times of endemic technologisation in all spheres of life here we find a book which, astonishingly, appears almost archaic in its visual means and typographi- cal language. The appeal of its silhouette compositions combined with a Wild-West-like Egyptienne typeface lies in the fact that it is very much recognisable as a book of today without any air of outmodedness despite its clichéd character. The carefree vibrancy of the five-colour printing is juxtaposed against the enigmatic, sharp-edged world of figures in which this old fairy-tale is given a new visual interpretation.

Honorary Appreciation

Pampilio

Poland

Monika Hanulak

Irena Tuwim, Monika Hanulak

Wytwórnia, Warschau

287×237×9mm

455g

32p

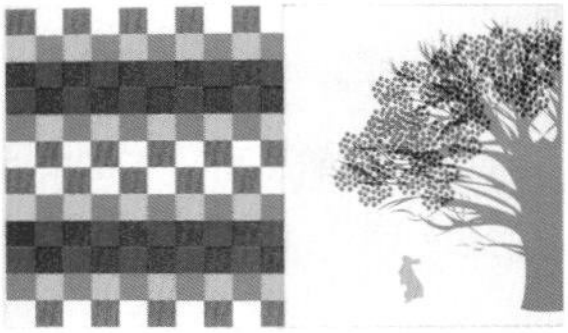

AU
PAYS DES ANIMAUX REGNAIT UNE GRANDE
FAMINE.

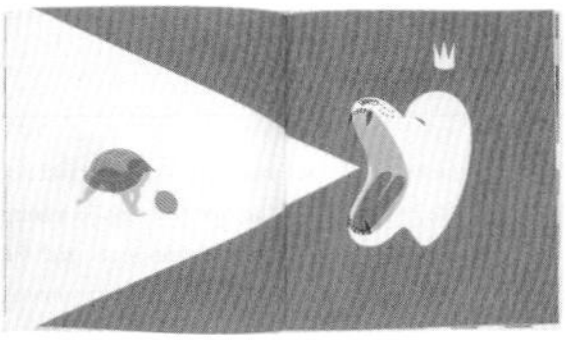

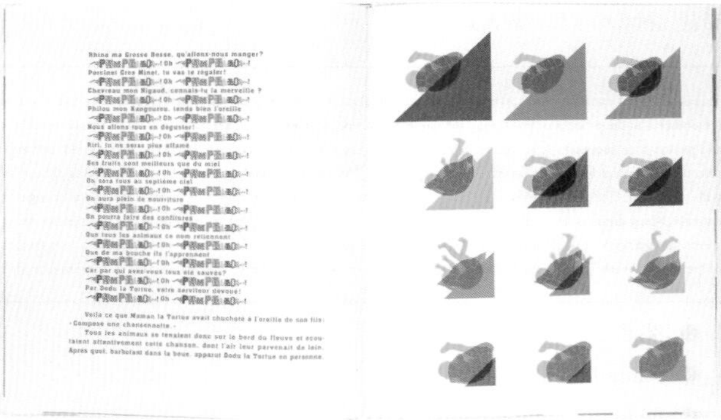
Rhino ma Grosse Bosse, qu'allons-nous manger?
Porcinet Gros Minet, tu vas te régaler!
Chevreau mon Nigaud, connais-tu la merveille ?
Philou mon Kangourou, tends bien l'oreille
Nous allons tous en déguster!
Riri, tu ne seras plus affamé
Ses fruits sont meilleurs que du miel
On sera tous au septième ciel
On aura plein de nourriture
On pourra faire des confitures
Que tous les animaux ce nom retiennent
Que de ma bouche ils l'apprennent
Car par qui avez-vous tous été sauvés?
Par Dodu la Tortue, votre serviteur dévoué!
Voilà ce que Maman la Tortue avait chuchoté à l'oreille de son fils: «Compose une chansonnette.»
Tous les animaux se tenaient donc sur le bord du fleuve et écoutaient attentivement cette chanson, dont l'air leur parvenait de loin. Après quoi, barbotant dans la boue, apparut Dodu la Tortue en personne.

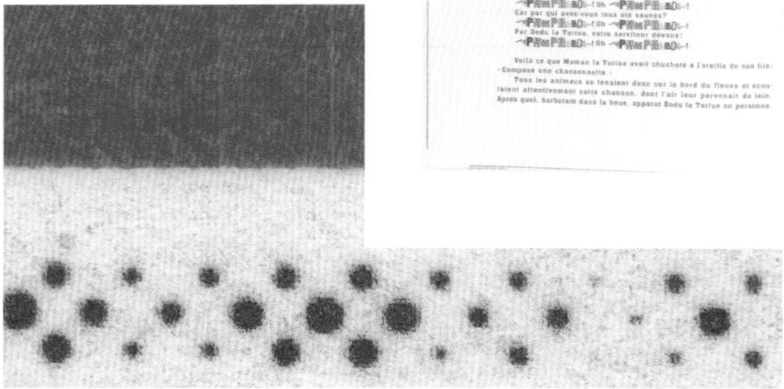

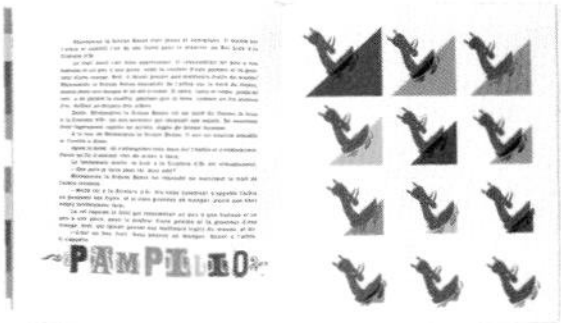

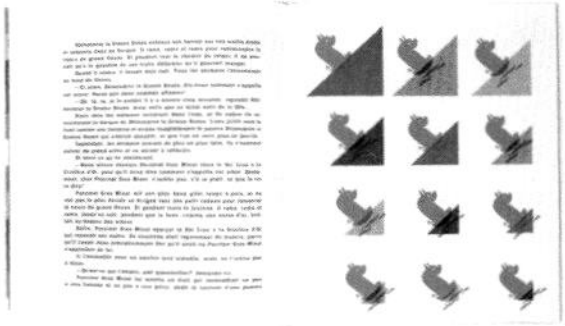
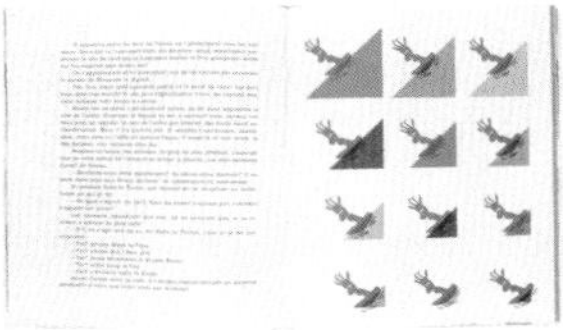
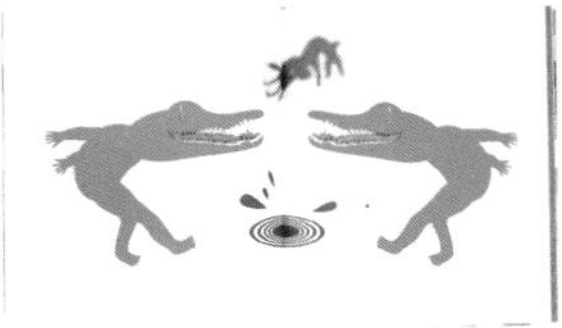
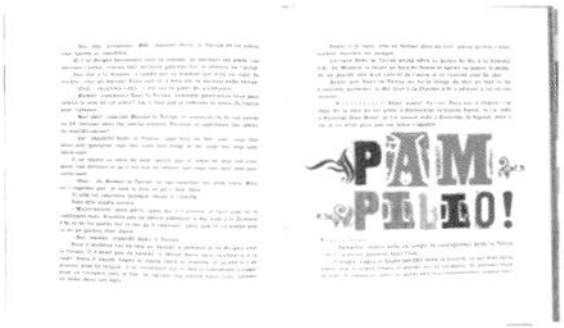

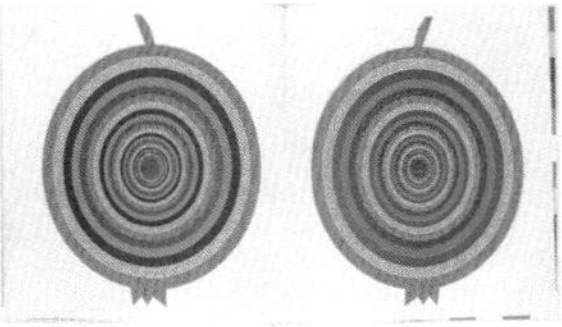
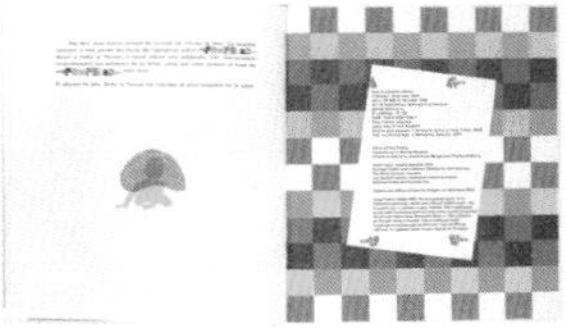
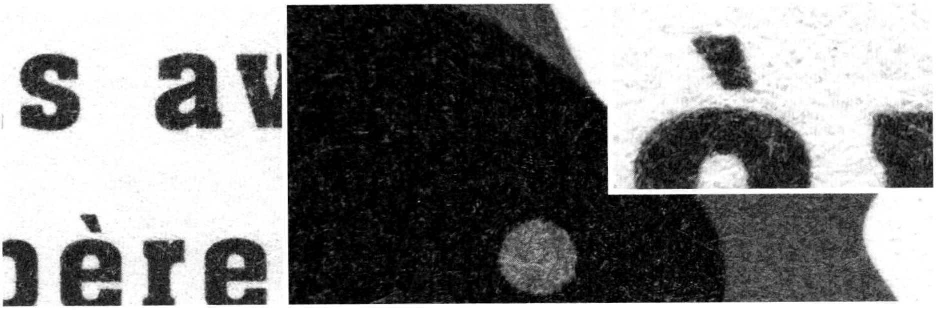

012 Ha5

litet hål. En jävla spricka, som den där
pan de har i Kalifornien, San Bernadino
det en sån i Kalifornien? Första gången, jag
bor i Arizona. Oj, det var långt, sa Ramírez.
åker jag hem, sa Harry Magaña. Sedan lyssnade
om barn. Har du någon gång lyssnat riktigt
Harry? Nej, jag har inga barn. Visst ja, så var det
Varför ber han om ursäkt? tänkte Harry Magaña
kvinna. En kvinna man behandlar illa utan att
gammal vana. Vanan gör oss blinda, Harry, eller
tills den här kvinnan en dag när allt är för sent
Den här kvinnan som oroat sig för och tagit hand om
börjar vissna i våra armar. Och inte ens då inser vi
Ramírez. Har jag berättat min egen historia för
Magaña, har jag sjunkit så lågt? Saker och ting är
viskade Ramírez. Eller tror du att saker och ting
ska tro det vi ser, bara så där, utan problem, utan
sa Harry Magaña, man måste alltid ställa
som flickan i Tijuana. Man ska alltid ställa

这是智利小说家罗贝托·波拉尼奥（Roberto Bolaño）的长篇小说代表作之一。书中讲述了五个独立但又相互关联的故事，部分人物或事件相互穿插，共同指向具有魔幻现实主义色彩的拉美故事，虽然作者自身热衷于“现实以下主义”的诗歌运动。作为拉美文学的代表人物之一，波拉尼奥博览群书，擅长各种写作技巧，用极具冲击力的文笔展现不同寻常的人生困境，以及追求文学的纯真度。本书的装订形式为无线胶装圆脊软精装，封面白卡纸褐、黑双色印刷，表面覆哑光膜，黑色部分亮面 UV 工艺。环衬部分采用中灰色横纹艺术纸。内页泛黄的胶版纸单黑印刷。除五部分开篇隔页满版封面局部花纹和巨大的章节数字外，正文并无独特的版式，也毫无特殊装饰，但细节考究得当，使得页面的设计绵密而连贯，是让真正喜爱文学的读者能完全扎进作者笔下小说世界的典范。

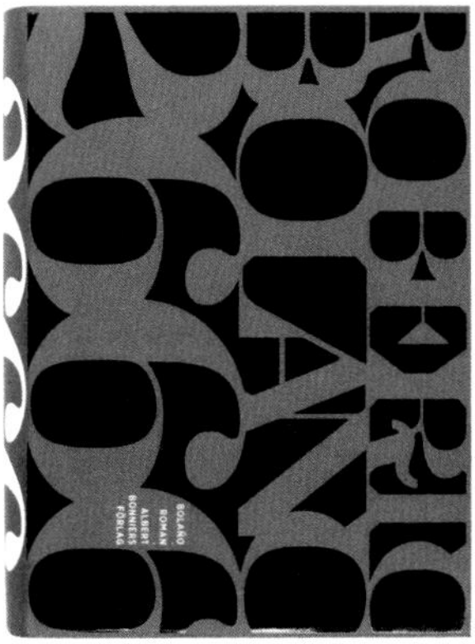

This is one of the representative novels of Chilean novelist Roberto Bolaño. The book tells five independent but interrelated stories. Some characters or events interlace with each other and point to Latin American stories with magic realism, although the author himself is keen on the poetry movement of 'infrarealism'. As one of the representative figures of Latin American literature, Bolaño has read a wide range of books and is skilled in writing. He made use of his powerful writing to show the extraordinary plight of life and tried to pursue literary purity. This book is perfect bound with rounded spine. The laminated cover is printed on white cardboard in brown and black, and the black part spots bright UV. The endpaper is made of medium gray art paper. The inside pages are printed on yellowish offset paper in black. There is no unique layout or special decoration in the text, except for the patterns and huge chapter numbers on the first page of each chapter. However, due to the careful details, the design of the pages is compact and coherent, so that readers who really like literature can get into the author's fiction world completely.

荣誉奖

2666

瑞典

With over a thousand pages you might expect a book which makes you feel tired before you have even started reading it. But nothing of the sort here. This is an attractive, soft and most inviting publication. From the typeface layout within the type-area to the flexible fold-out cover, economical and meticulous continuity of typography is characteristic. Everything is very soothing with no danger of boredom, the chapter beginnings elegant. With sparse means but high quality a well-formed novel.

Honorary Appreciation

2666

Sweden

Nina Ulmaja

Roberto Bolaño

Albert Bonniers Förlag, Stockholm

233 × 165 × 47mm

1180g

1056p

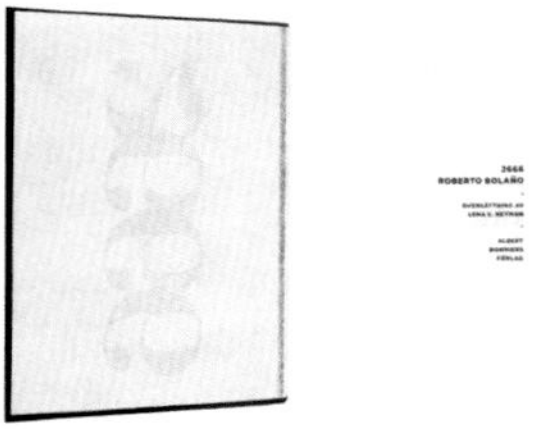

2666
ROBERTO BOLAÑO

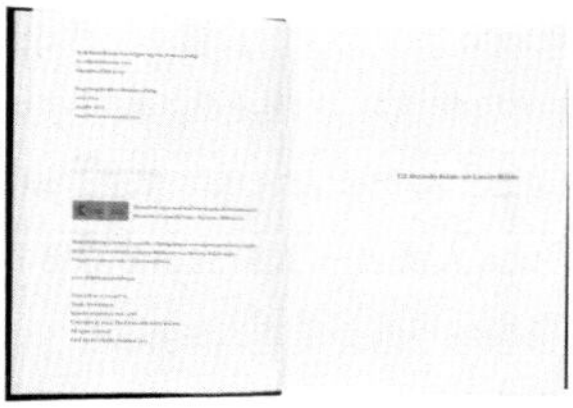

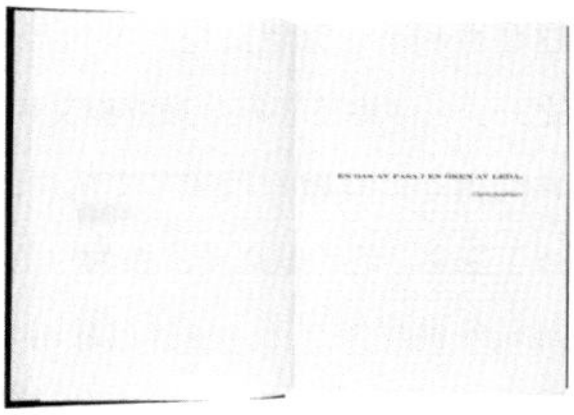

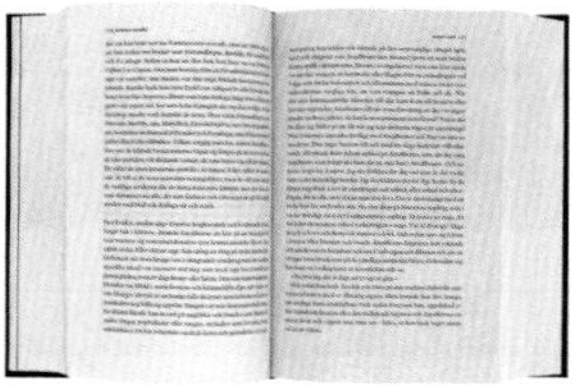

3

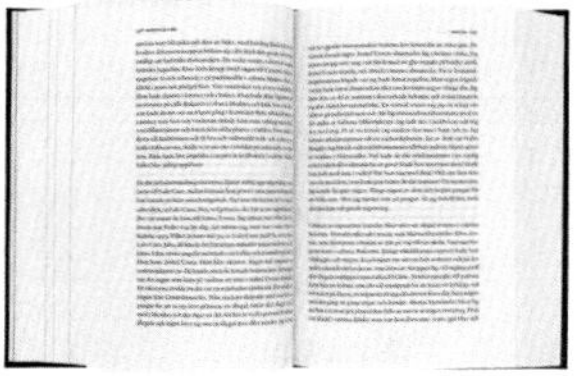

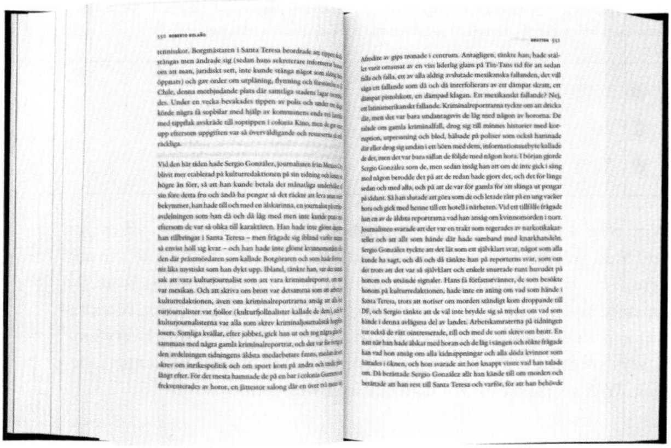

550 ROBERTO BOLAÑO

tennisskor. Borgmästaren i Santa Teresa beordrade att tippen skulle
stängas men ändrade sig (sedan hans sekreterare informerat honom
om att man, juridiskt sett, inte kunde stänga något som aldrig
öppnats) och gav order om utplåning, flyttning och förstörelse
Chile, denna motbjudande plats där samtliga stadens lagar
des. Under en vecka bevakades tippen av polis och under
körde några få sopbilar med hjälp av kommunens enda
med uppflak avskräde till soptippen i colonia Kino, men de gav
upp eftersom uppgiften var så överväldigande och resurserna så
räckliga.

Vid den här tiden hade Sergio González, journalisten från Mexico
blivit mer etablerad på kulturredaktionen på sin tidning och
högre än förr, så att han kunde betala det månatliga underhållet
sin före detta fru och ändå ha pengar så det räckte att leva utan
bekymmer, han hade till och med en älskarinna, en journalist på
avdelningen som han då och då låg med men inte kunde
eftersom de var så olika till karaktären. Han hade inte glömt
han tillbringat i Santa Teresa – men frågade sig ibland varför
så envist höll sig kvar – och han hade inte glömt kvinnomorden
den där prästmördaren som kallade Botgöraren och som hade
nit lika mystiskt som han dykt upp. Ibland, tänkte han, var det
sak att vara kulturjournalist som att vara kriminalreporter,
var mexikan. Och att skriva om brott var detsamma som att
kulturredaktionen, även om kriminalreportrarna ansåg att alla
turjournalister var fjollor (kulturfjollnalister kallade de dem),
kulturjournalisterna var alla som skrev kriminaljournalistik
losers. Somliga kvällar, efter jobbet, gick han ut och tog några
sammans med några gamla kriminalreportrar, och det var
den avdelningen tidningens äldsta medarbetare fanns, medan
skrev om inrikespolitik och om sport kom på andra och
långt efter. För det mesta hamnade de på en bar i colonia Guerrero
frekventerades av horor, en jättestor salong där en över två meter

BROTTEN 551

Afrodite av gips tronade i centrum. Antagligen, tänkte han, hade stället varit omsusat av en viss liderlig glans på Tin-Tans tid för att sedan falla och falla, ett av alla aldrig avslutade mexikanska fallanden, det vill säga ett fallande som då och då interfolierats av ett dämpat skratt, ett dämpat pistolskott, en dämpad klagan. Ett mexikanskt fallande? Nej, ett latinamerikanskt fallande. Kriminalreportrarna tyckte om att dricka där, men det var bara undantagsvis de låg med någon av hororna. De talade om gamla kriminalfall, drog sig till minnes historier med korruption, utpressning och blod, hälsade på poliser som också hamnade där eller drog sig undan i ett hörn med dem, informationsutbyte kallade de det, men det var bara sällan de följde med någon hora. I början gjorde Sergio González som de, men sedan insåg han att om de inte gick i säng med någon berodde det på att de redan hade gjort det, och det för länge sedan och med alla, och på att de var för gamla för att slänga ut pengar på sådant. Så han slutade att göra som de och letade rätt på en ung vacker hora och gick med henne till ett hotell i närheten. Vid ett tillfälle frågade han en av de äldsta reportrarna vad han ansåg om kvinnomorden i norr. Journalisten svarade att det var en trakt som regerades av narkotikakarteller och att allt som hände där hade samband med knarkhandeln. Sergio González tyckte att det lät som ett självklart svar, något som alla kunde ha sagt, och då och då tänkte han på reporterns svar, som om det trots att det var så självklart och enkelt snurrade runt huvudet på honom och utsände signaler. Hans få författarvänner, de som besökte honom på kulturredaktionen, hade inte en aning om vad som hände i Santa Teresa, trots att notiser om morden ständigt kom droppande till DF, och Sergio tänkte att de väl inte brydde sig så mycket om vad som hände i denna avlägsna del av landet. Arbetskamraterna på tidningen var också de rätt ointresserade, till och med de som skrev om brott. En natt när han hade älskat med horan och de låg i sängen och rökte frågade han vad hon ansåg om alla kidnappningar och alla döda kvinnor som hittades i öknen, och hon svarade att hon knappt visste vad han talade om. Då berättade Sergio González allt han kände till om morden och berättade att han rest till Santa Teresa och varför, för att han behövde

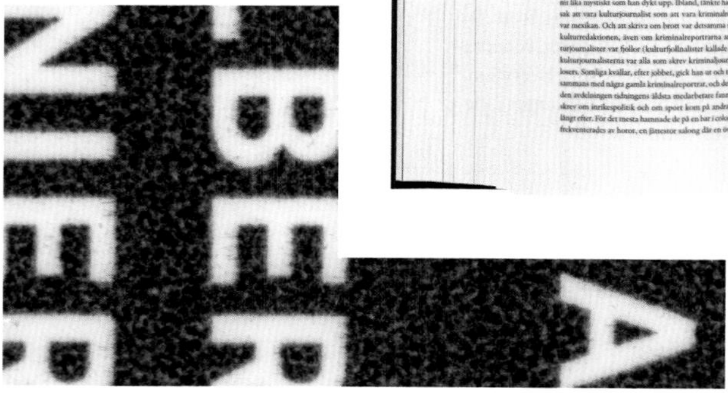

ska in
boldi. Lärarna had

Atlas of the Conflict. Israel-Palestine
Atlas Transformace
Bemoeizorg
Title
— and Willem. Documentation of a Youth
Gerd Arntz. Graphic Designer
Fassaden aus Holz
Jak jsem potkal d'ábla
Christian Waldvogel. Earth Extremes
Odyssey : Architecture and Literature
Von einem, der auszog das Fürchten zu lernen
Leerzeichen fur Applaus
Zazie in die Metro
Serie Teaching Architecture. 3 Positions Made in Switzerland (1. Hong Kong Typology; 2. Important Buildings; 3. Radical Mix in Hanoi)

2011

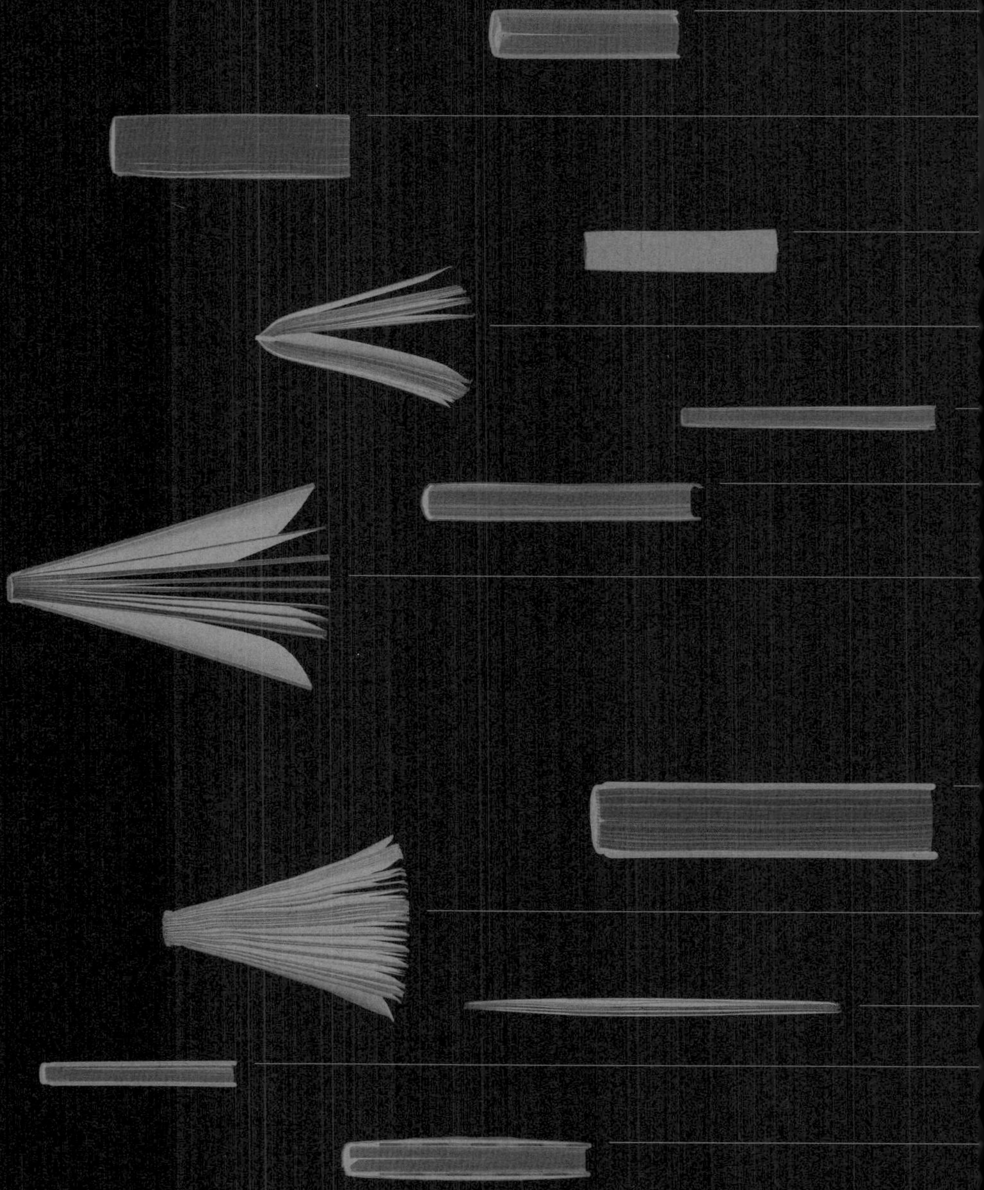

2011 Gl

0787

2011 Gm

0795

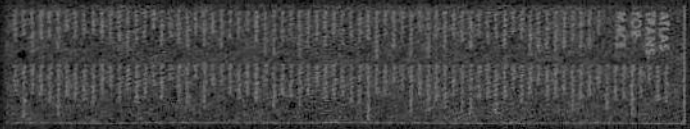

2011 Sm[1]

0801

Bemoeizorg

2011 Sm[2]

0807

2011 Bm[1]

0813

AND WILLEM

2011 Bm[2]

0819

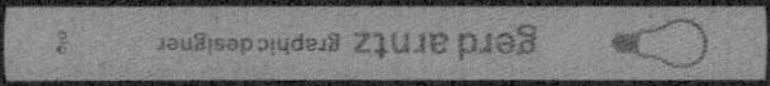
gerd arntz graphic designer

2011 Bm[3]

0825

FASSADEN AUS HOLZ
proHolz Holzforschung Austria

2011 Bm[5]

0831

Christian Waldvogel
EARTH
EXTREMES

2011 Ha[1]

0837

2011 Ha[2]

0843

2011 Ha[3]

0847

2011 Ha[4]

0851

ZAZIE IN DER METRO

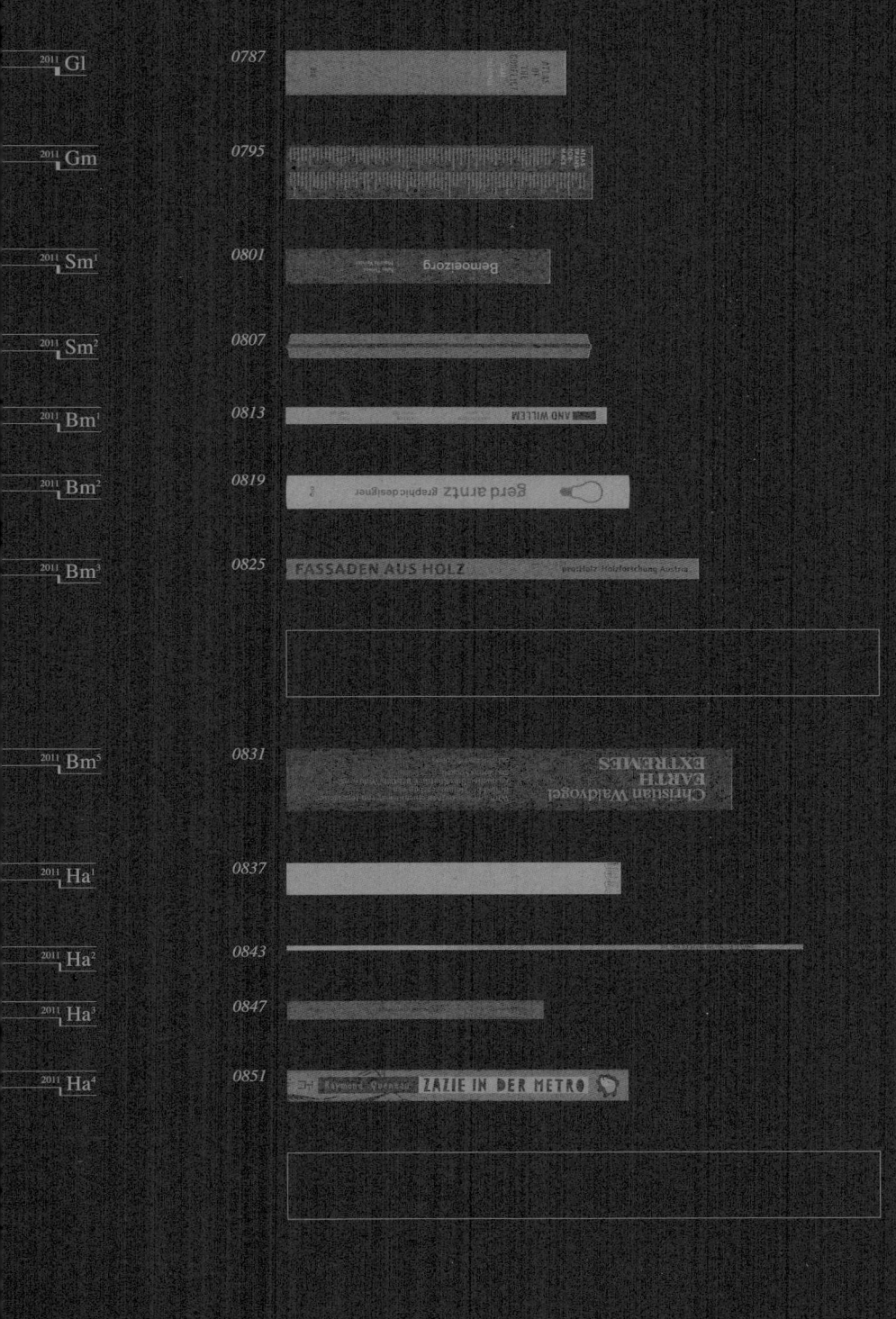

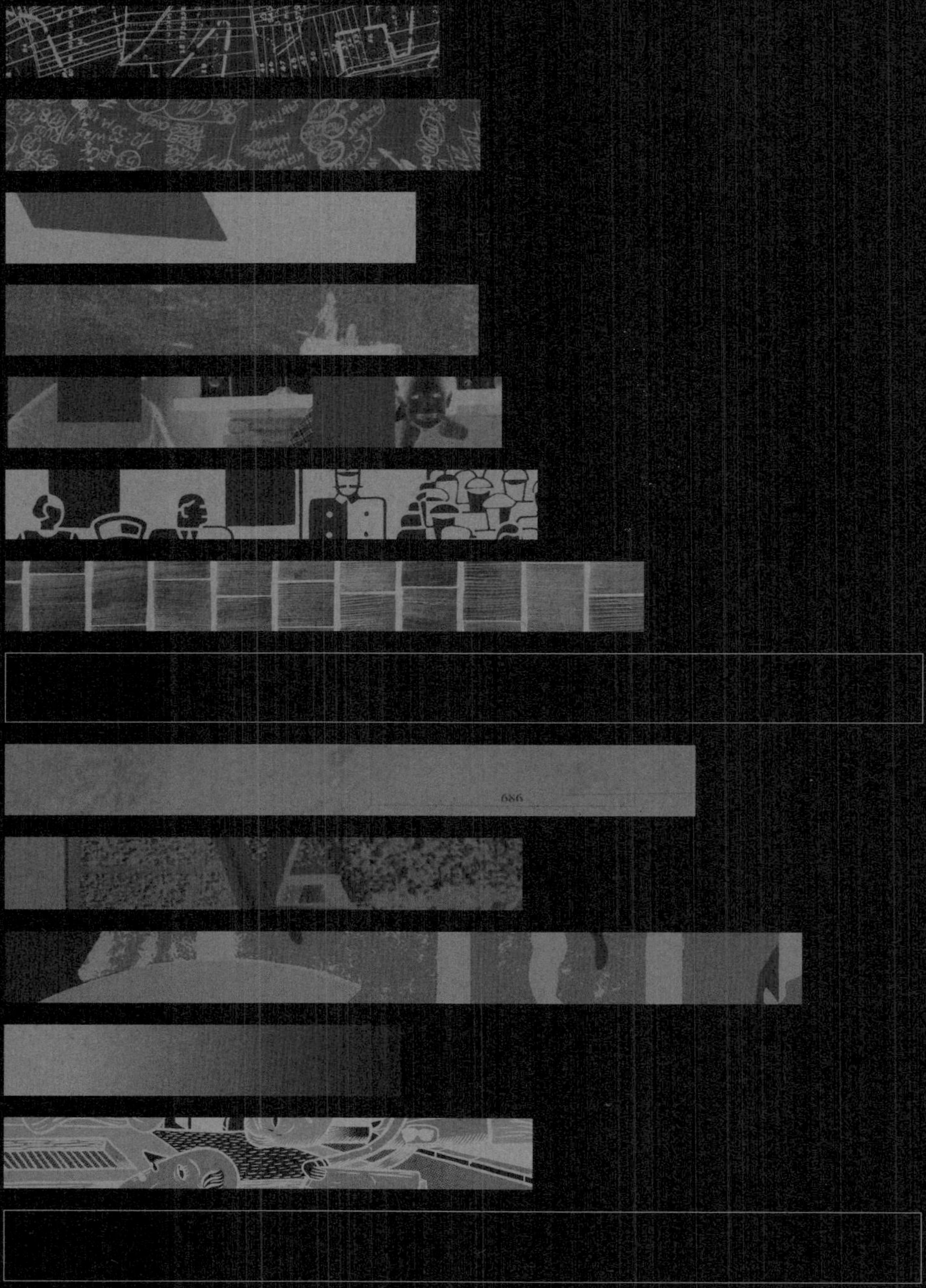

I was brought up in a Zionist context. We were
overwhelmingly and completely appreciative
considering it a miracle: a nation that had co

administrativním samopohybem naší
civilizace, a jsou proto „politikum". Boj
o vlastní identitu každého člověka se stal
bází „antipolitické politiky", protože moc

určitá prav
Přirozené,
a jinak, ne
diskusí, vli

2.1 Veiligheid en agressie

2.2 Aanmeldingen en verwijzinge

AGES
S NOT

3. To allow discourse to col
its own weight and to acqui
shape of letters. Letters wh

identical twins [redacted] and WILLEM, children of
OOL, are born on October 11, 1982. Four years
ed with new parents [redacted] and [redacted]. The twins

types. The file revealed a great number of elegantly styl
images in a series of items ordered in different categori
They were commissioned by Otto Neurath (1882-1945) a

3.1 Allgemeines

Holzqualität/Sortierung

Die Holzqualität von Brettern, Profilbrettern und

Der Mindestumfang der sc
Sortiermerkmale enthält fo
ausgeschlagene Stellen (sc

probe Huygens[1] lands on Titan afte
seven-year voyage.

[illegible]

**Der Originaltext "Märchen von einem, der auszog das Fürchten zu lernen"
von den Brüder Grimm liegt als Einfarb-Risographie bei.**

Eine Agentur ist eine Agentur ist eine A
eine Agentur ist eine Agentur Guten Morge
Aufwachen Am Schreibtisch Uhr Ku

– Na? fragte Gabriel. Paßt es dir?

Zazie machte ein Zeichen, das anzuzeigen schien, daß sie
nung für sich behielt.

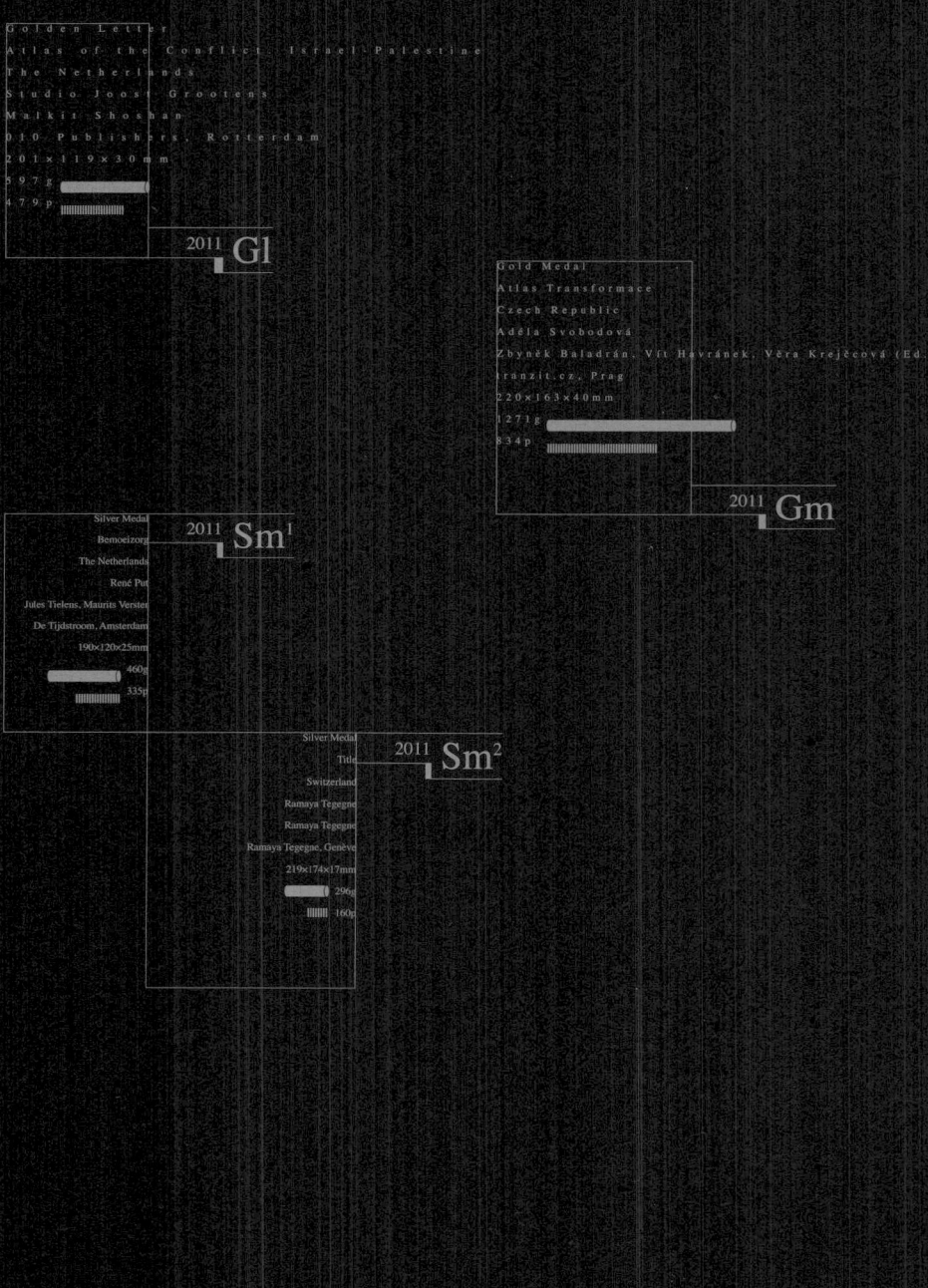

G o l d e n   L e t t e r

A t l a s   o f   t h e   C o n f l i c t .   I s r a e l - P a l e s t i n e

T h e   N e t h e r l a n d s

S t u d i o   J o o s t   G r o o t e n s

M a l k i t   S h o s h a n

0 1 0   P u b l i s h e r s ,   R o t t e r d a m

2 0 1 × 1 1 9 × 3 0 m m

5 9 7 g

4 7 9 p

2011 Gl

Gold Medal

Atlas Transformace

Czech Republic

Adéla Svobodová

Zbyněk Baladrán, Vít Havránek, Věra Krejčcová (Ed.)

tranzit.cz, Prag

220×163×40mm

1271g

834p

2011 Gm

Silver Medal

Bemoeizorg

The Netherlands

René Put

Jules Tielens, Maurits Verster

De Tijdstroom, Amsterdam

190×120×25mm

460g

335p

2011 Sm1

Silver Medal

Title

Switzerland

Ramaya Tegegne

Ramaya Tegegne

Ramaya Tegegne, Genève

219×174×17mm

296g

160p

2011 Sm2

2011 Ha[1]

Honorary Appreciation
Odyssey : Architecture and Literature
China
Xiao Mage, Cheng Zi
Ou Ning (Ed.)
China Youth Press, Beijing
240×170×22mm
792g
448p

2011 Ha[2]

Honorary Appreciation
Von einem, der auszog das Fürchten zu lernen
Germany
Doris Freigofas / Golden Cosmos
Doris Freigofas
Golden Cosmos, Berlin
370×243×4mm
304g+24g
24p+8p

2011 Ha[4]

Honorary Appreciation
Zazie in der Metro
Germany
Joe Villion
Raymond Queneau
Büchergilde Gutenberg, Frankfurt am Main
245×152×19mm
615g
221p

2011 Ha[3]

Honorary Appreciation
Leerzeichen für Applaus
Germany
Jenna Gesse
Jenna Gesse
Eigenverlag Jenna Gesse, Bielefeld
184×114×12mm
159g
96p

2011 Ha[5]

2011 Bm[1]

Bronze Medal
— and Willem. Documentation of a Youth
The Netherlands
Robin Uleman
Willem Popelier
post editions, Rotterdam
230×169×11mm
393g
170p

2011 Bm[2]

Bronze Medal
Gerd Arntz. Graphic Designer
The Netherlands
Ontwerpwerk
Ed Annink, Max Bruinsma (Ed.)
010 Publishers, Rotterdam
247×180×24mm
763g
288p

2011 Bm[4]

2011 Bm[5]

Bronze Medal
Christian Waldvogel. Earth Extremes
Switzerland
Jonas Voegeli, Christian Waldvogel, Benjamin Roffler
Jacqueline Burckhardt, Christian Waldvogel, Jonas Voegeli (Ed.)
Verlag Scheidegger & Spiess AG, Zürich
320×214×41mm
2283g
495p

2011 Bm[3]

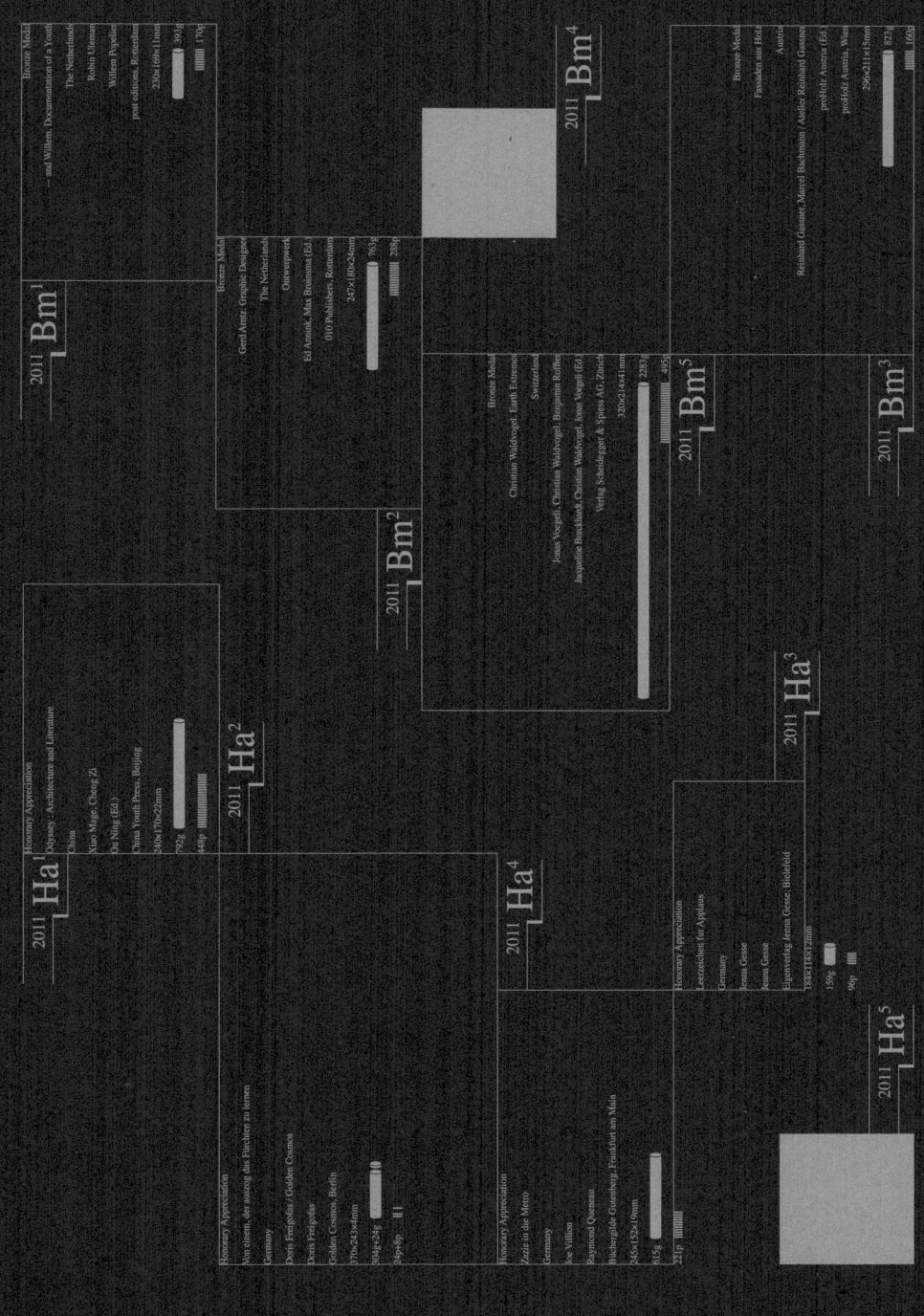

Bronze Medal
Fassaden aus Holz
Austria
Reinhard Gassner, Marcel Bachmann / Atelier Reinhard Gassner
proHolz Austria (Ed.)
proHolz Austria, Wien
296×211×15mm
823g
160p

Designer

Studio Joost Grootens

Adéla Svobodová

René Put

Ramaya Tegegne

Robin Uleman

Ontwerpwerk

Reinhard Gassner, Marcel Bachmann / Atelier Reinhard Gassner

Honza Zamojski

Jonas Voegeli, Christian Waldvogel, Benjamin Roffler

Xiao Mage, Cheng Zi

Doris Freigofas / Golden Cosmos

Jenna Gesse

Joe Villion

1. Ludovic Balland, Ivan Weiss; 2. Ludovic Balland; 3. Ludovic Balland / Typography Cabinet, Basel

2011

The International Jury

Konstanze Berner (Germany)

Reuben Crossman (Australia)

Danny Dobbelaere (Belgium)

Aleš Najbrt (Czech Republic)

Katja Schwalenberg (Germany)

Jorge Silva (Portugal)

Nina Ulmaja (Sweden)

Country / Region

The Netherlands 4

Switzerland 3

Germany 3

Czech Republic 1

Austria 1

Poland 1

China 1

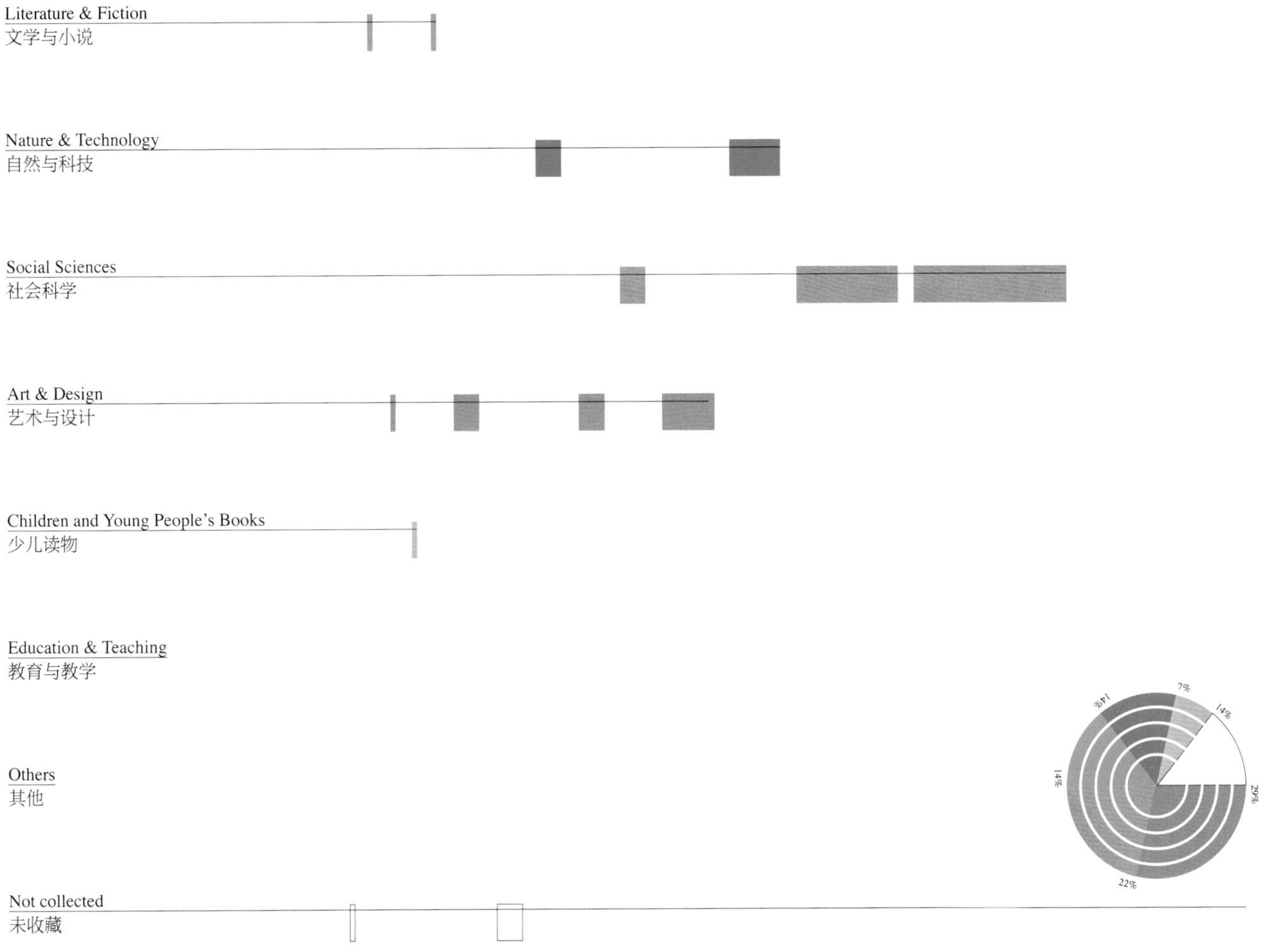
Literature & Fiction
文学与小说
Nature & Technology
自然与科技
Social Sciences
社会科学
Art & Design
艺术与设计
Children and Young People's Books
少儿读物
Education & Teaching
教育与教学
Others
其他
Not collected
未收藏
29%
22%
14%
14%
7%
14%

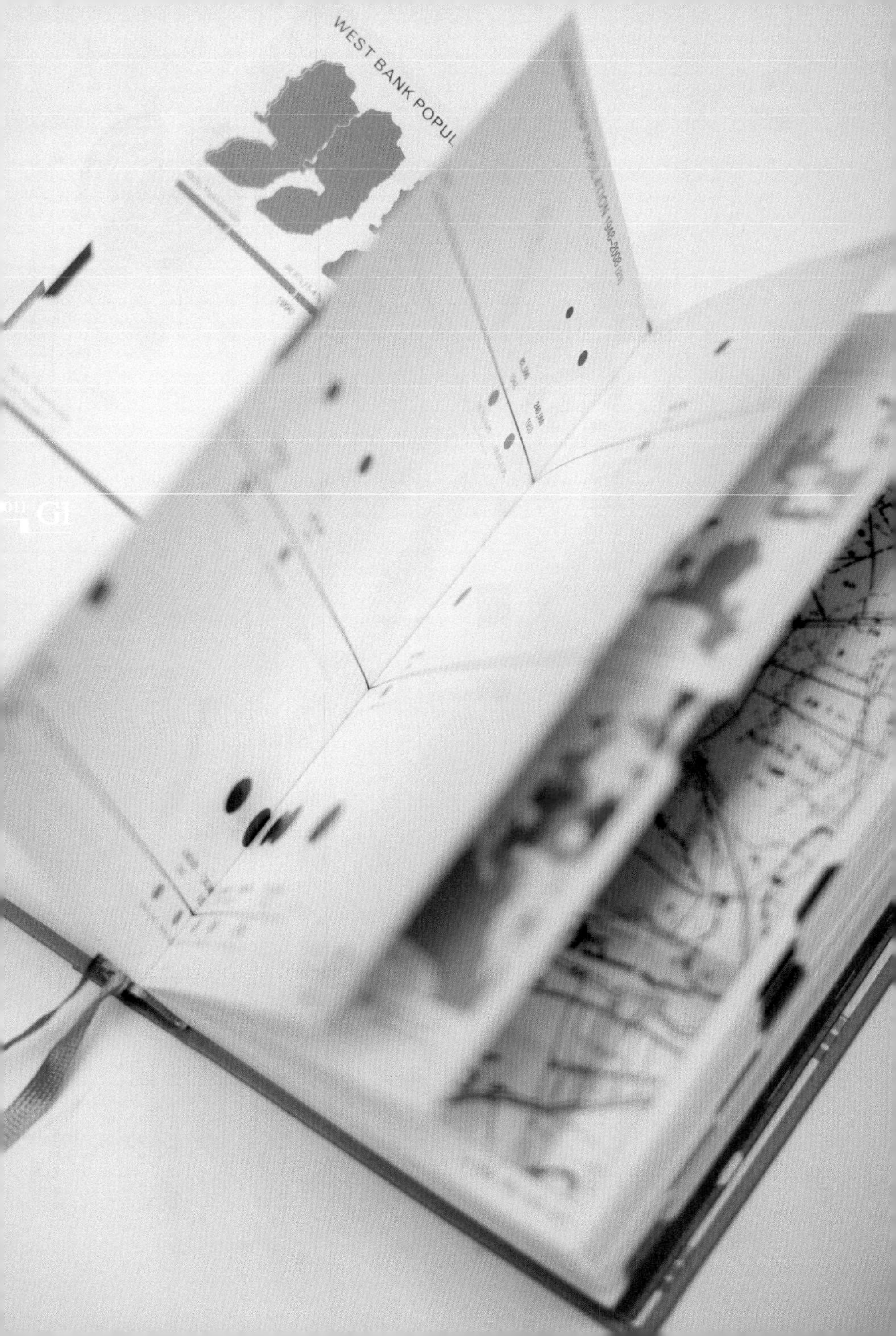
WEST BANK POPUL

这是一本有关以色列和巴勒斯坦地区近 100 年来各项数据统计的资料手册。书中通过边界、定居、人口、所有权、建设、水资源、考古等相关资料的统计和绘制，深入记录和探讨了这一纷争地区存在的问题。书籍为锁线硬精装，封面灰色布纹纸烫白和烫黑工艺，裱覆卡板。粉红色胶版纸的前后环衬上为与这个地区历史相关的人物头像，单黑印刷。内页前四分之三为各项统计图表，哑面涂布纸蓝、棕、黑三色印刷，信息清晰。后四分之一为词典、图片、时间线等附录信息，哑面浅灰色涂布纸四色印刷，纸张较图表部分略薄。这是一本小巧的手册，信息设计系统而精妙，页码部分有用于信息引导的大量“链接”，以及右侧翻口的层级标签，都方便了检索。信息部分进行了系统化的图形和配色设计，除了蓝、棕双书签和三色图表，还专门针对图表设计了使用图例。

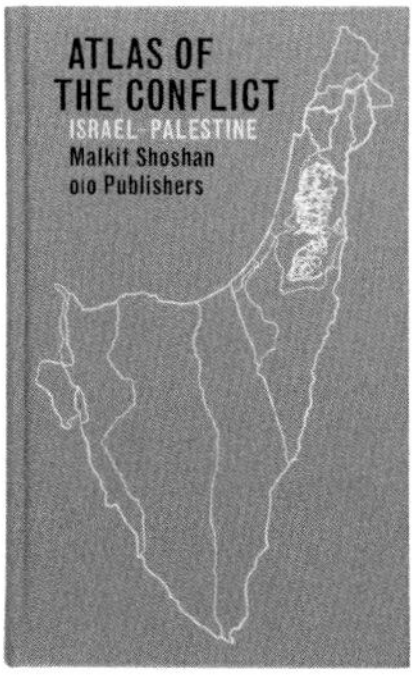
ATLAS OF
THE CONFLICT
ISRAEL-PALESTINE
Malkit Shoshan
010 Publishers

This is a manual on data statistics of Israel and the Palestinian over the past 100 years. Through the statistics for settlement, population, ownership, construction, water resources, archaeological investigation and other related data, the book records and explores the problems existing in this disputed area. The hardback book is bound by thread sewing. The cover is hot-stamped in white and black on gray cloth paper, laminated with cardboard. The front and back endpapers made of pink offset paper are printed with figures related to the history of the area in monochrome black. The first three quarters of the inner pages are statistical charts with clear information, printed on matte coated paper in three colors: blue, brown and black. The last quarter is appendix information such as dictionary, pictures and the time line, printed on light gray matte coated paper in four colors, which is slightly thinner than the paper for charts. This is a pocket manual with exquisite information system –a large number of 'links' for information guidance shown on the part of page numbers, together with the hierarchical labels on the right side of the turning edge, are convenient for retrieval. Systematic graphics and color matching are carefully designed to display information, for example, bookmark printed in blue and brown, three-color charts, and legends specifically designed for the charts.

- 金字符奖
- 以巴冲突地图集
- 荷兰

- Golden Letter
- Atlas of the Conflict. Israel-Palestine
- The Netherlands
- Studio Joost Grootens
- Malkit Shoshan
- 010 Publishers, Rotterdam
- 201 × 119 × 30mm
- 597g
- 479p

The Israel-Palestine conflict is a highly complex issue and to create an atlas based on it is to make a statement. Its ingenious design in the form of a pocket atlas brings structure and light to the chaos. The reader encounters a bright, magnificently designed book offering insights into the development of regional and state planning. The meticulously drawn charts are every bit as fascinating as the extensive mapping, the documentary photography and the encyclopaedic texts. Typographical hyperlinks join the two chapters and thus mutually complement the cartographic and source material. This book constitutes visual communication of the highest order. An instrument to facilitate understanding, discussion and reflection. Not a gratuitous design book but one which is both essential and eminently usable.

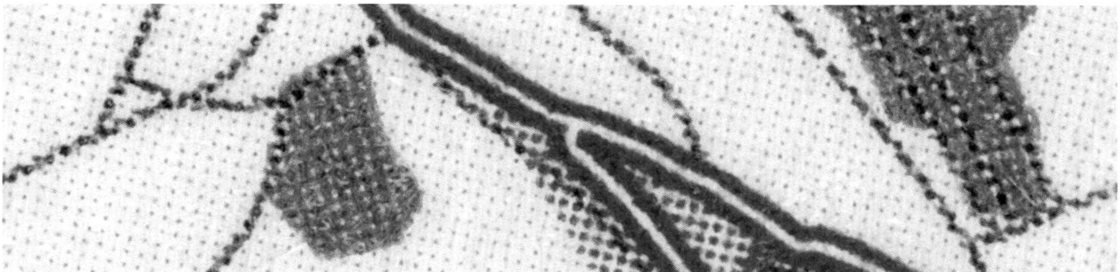

201×119×30
05
13
479
22
597
29
30
31
32

LAND ACQUISITION 1918–1960
YIZRAEL VALLEY AREA
LAND OWNERSHIP
1918
1947
1938
1960
Land owned by Jewish individuals
Land owned by Keren Keyemet
Land owned by PICA (Palestine Jewish Colonisation Association)
→ 323, 363, 369, 371, 373, 389, 391, 399, 435, 437
205

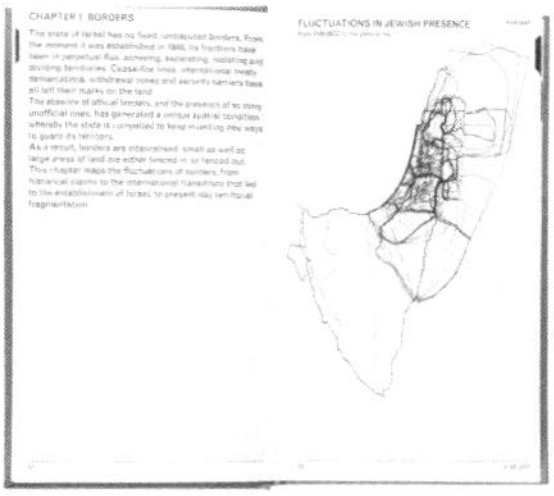
CHAPTER I BORDERS
FLUCTUATIONS IN JEWISH PRESENCE

MECHANISM OF LAND CONFISCATION

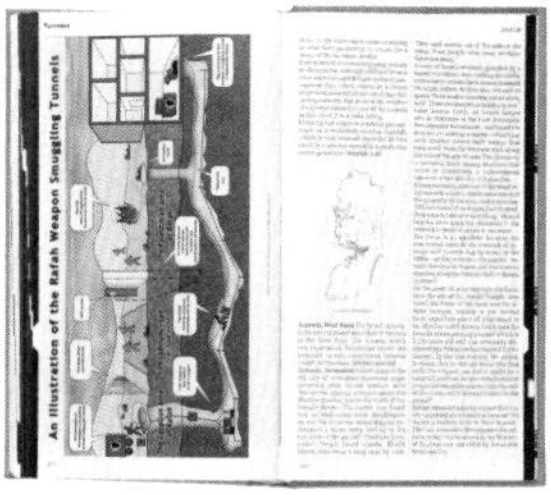
An Illustration of the Rafah Weapon Smuggling Tunnels

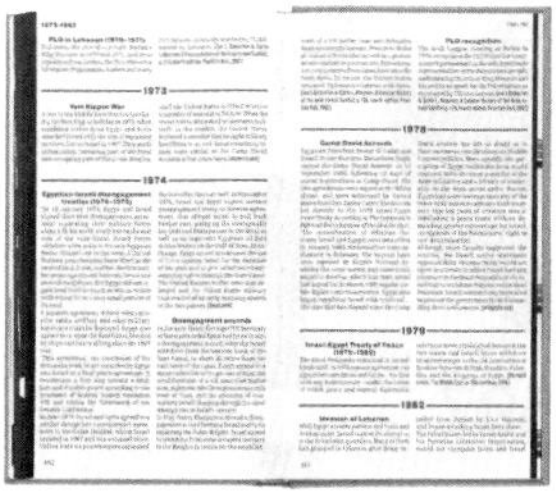

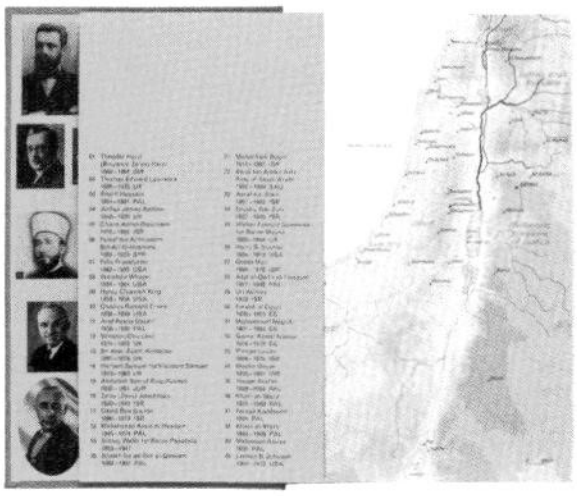

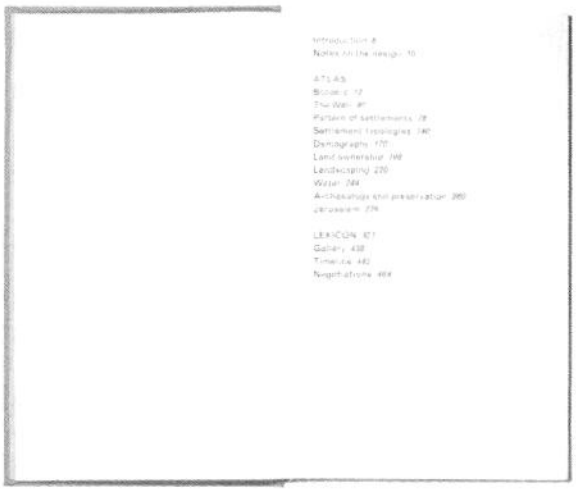

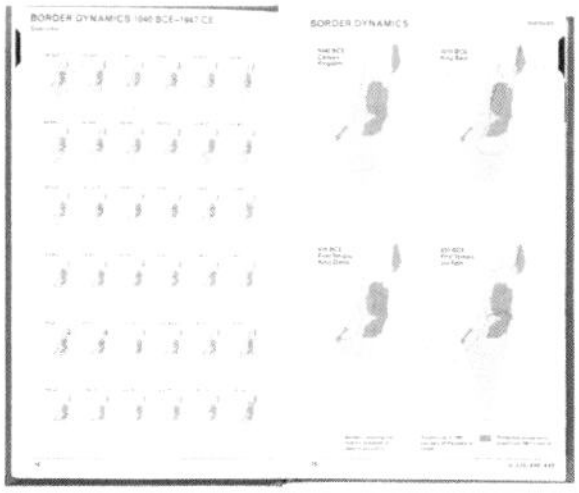
BORDER DYNAMICS
BORDER DYNAMICS

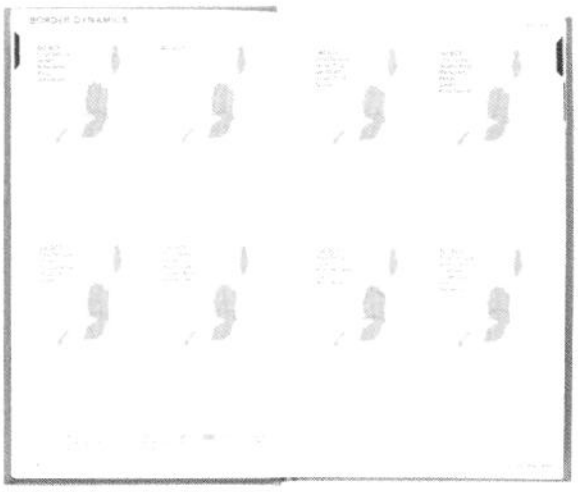

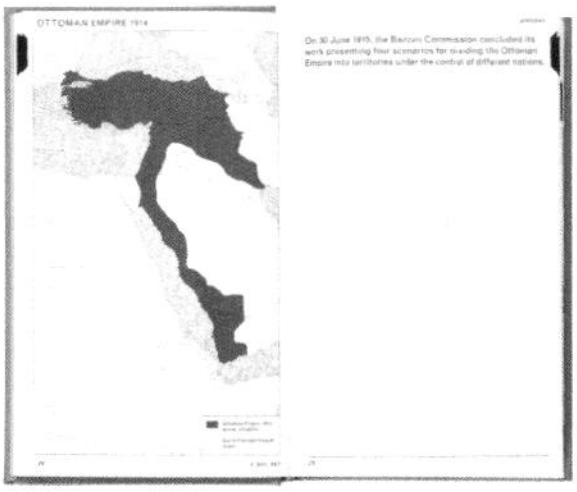
OTTOMAN EMPIRE 1914

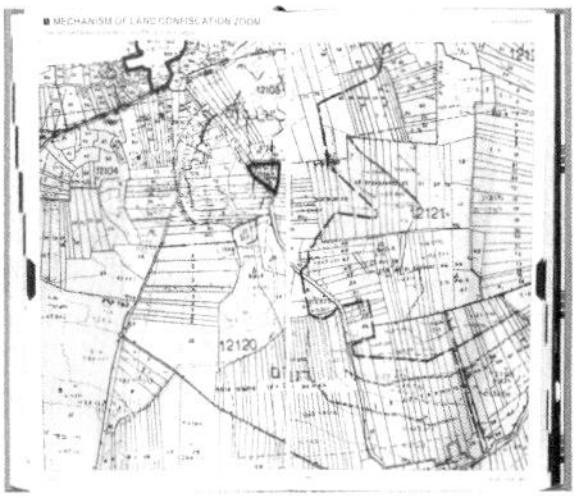
12120

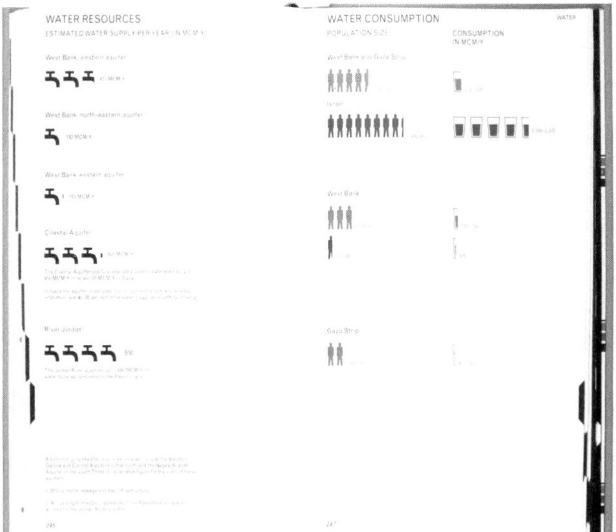
WATER RESOURCES
ESTIMATED WATER SUPPLY PER YEAR (IN MCM/Y)
West Bank, western aquifer
West Bank, north-eastern aquifer
West Bank, eastern aquifer
Coastal Aquifer
River Jordan
WATER CONSUMPTION
POPULATION SIZE
CONSUMPTION IN MCM/Y
WATER
West Bank and Gaza Strip
Israel
West Bank
Gaza Strip
246
247

ATLAS
OF
THE
CONFLICT
ISRAEL
—
PALESTINE
010

NEGOTIATIONS 1993–2010

Timeline

From the early 1990s, starting with the Madrid Conference, border dynamics in Israel shift in focus from external boundaries to internal divisions between Israel and a future Palestinian state.

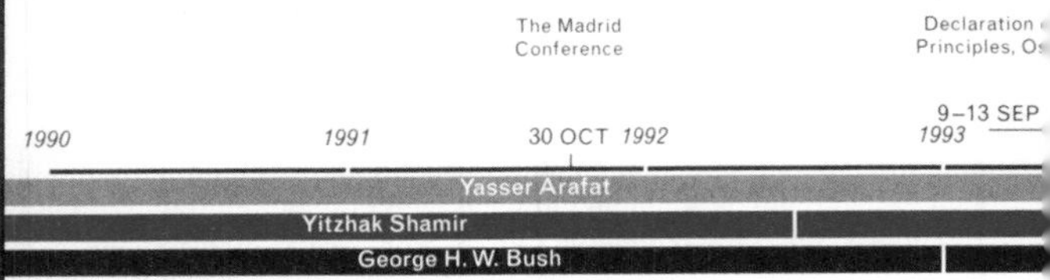
The Madrid
Conference
Declaration
Principles, O
1990
1991
30 OCT
1992
9–13 SEP
1993
Yasser Arafat
Yitzhak Shamir
George H. W. Bush

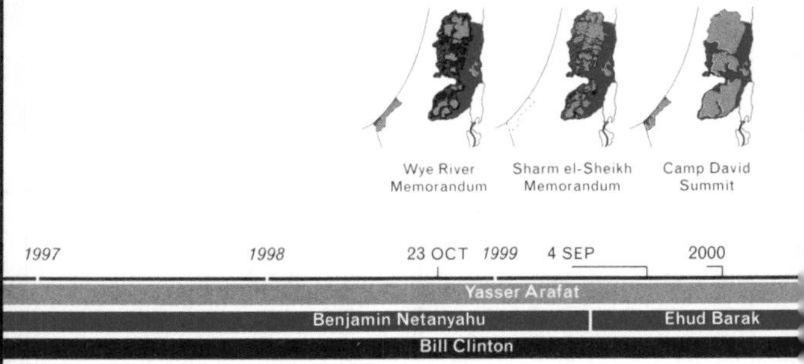
Wye River
Memorandum
Sharm el-Sheikh
Memorandum
Camp David
Summit
1997
1998
23 OCT
1999
4 SEP
2000
Yasser Arafat
Benjamin Netanyahu
Ehud Barak
Bill Clinton

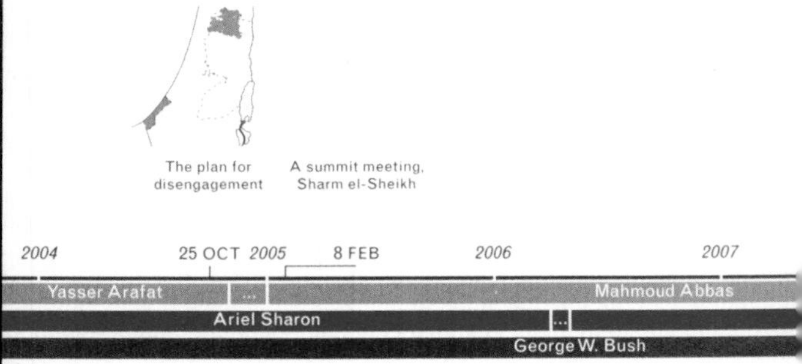
The plan for
disengagement
A summit meeting,
Sharm el-Sheikh
2004
25 OCT
2005
8 FEB
2006
2007
Yasser Arafat
...
Mahmoud Abbas
Ariel Sharon
...
George W. Bush

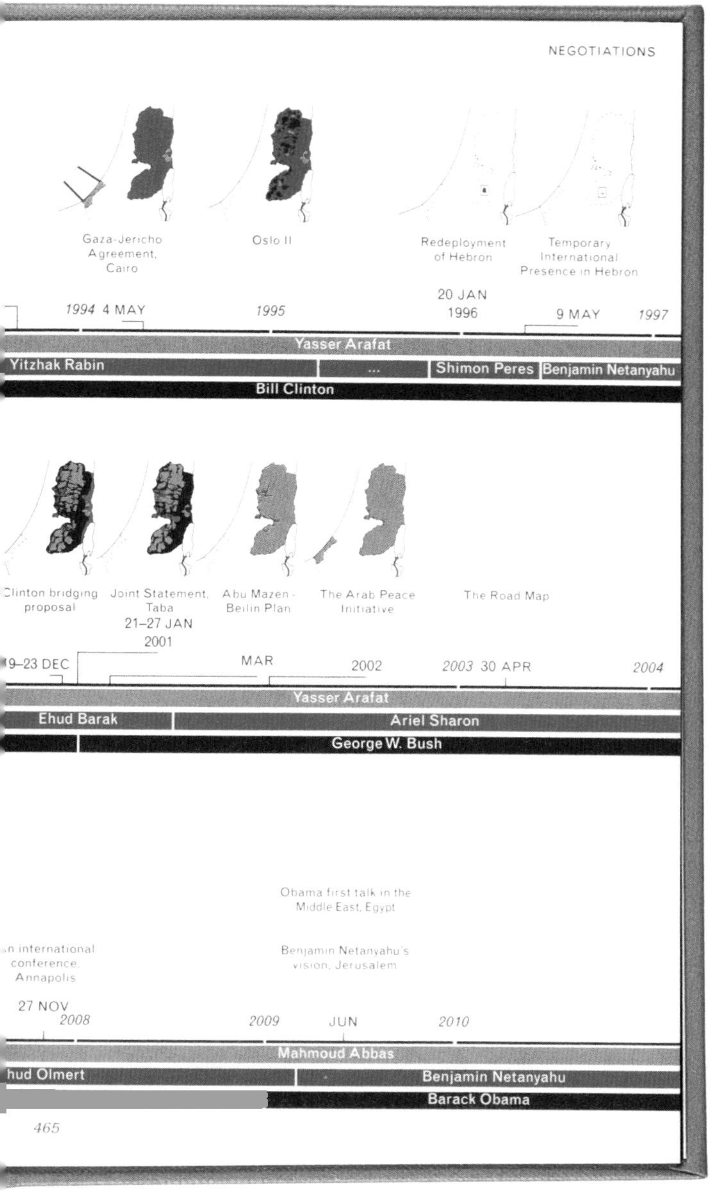
NEGOTIATIONS
Gaza-Jericho Agreement, Cairo
Oslo II
Redeployment of Hebron
Temporary International Presence in Hebron
1994 4 MAY
1995
20 JAN 1996
9 MAY
1997
Yasser Arafat
Yitzhak Rabin
...
Shimon Peres
Benjamin Netanyahu
Bill Clinton
Clinton bridging proposal
Joint Statement, Taba
21–27 JAN 2001
Abu Mazen - Beilin Plan
The Arab Peace Initiative
The Road Map
9–23 DEC
MAR
2002
2003 30 APR
2004
Yasser Arafat
Ehud Barak
Ariel Sharon
George W. Bush
Obama first talk in the Middle East, Egypt
n international conference, Annapolis
Benjamin Netanyahu's vision, Jerusalem
27 NOV
2008
2009
JUN
2010
Mahmoud Abbas
hud Olmert
Benjamin Netanyahu
Barack Obama
465

2011 GI

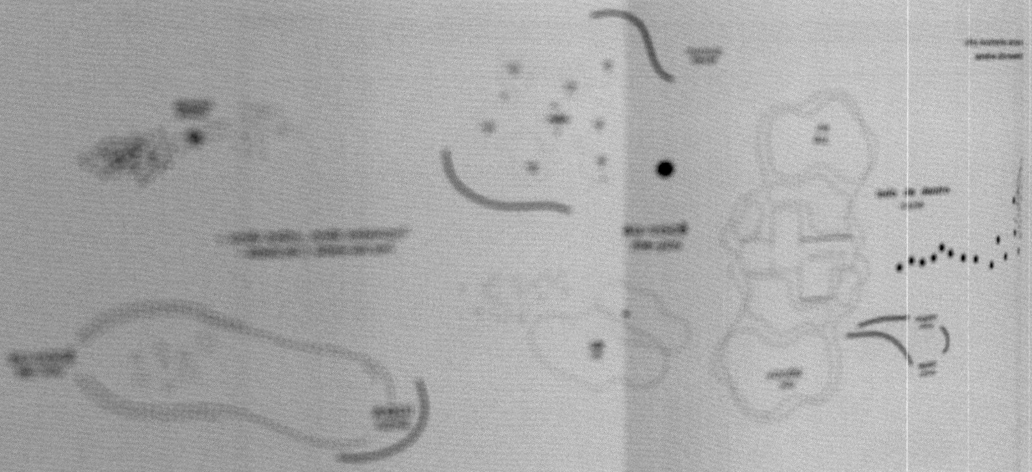

P 157

γ δ

PROSTOROVÉ VYLOUČENÍ

PROSTOROVÉ VYLOUČENÍ

JOANNA ERBEL

Jedna z dimenzí společenského vyloučení, vyplývající z prostorového uspořádání městské tkáně. Prostorové vyloučení je produktem společenského vyloučení a prohlubuje jeho dopad. Je to neviditelný nástroj disciplíny, jenž uspořádává sociální oblast tím, že vylučuje z pole vidění ty, kteří svou přítomností zpochybňují logiku legitimizovaného společenského pořádku.

Evropská města mají dlouhou tradici způsobů, jak znesnadňovat nechtěným skupinám, aby využívaly všeobecně dostupné prvky městského prostoru: např. ocelovými bodci na vnějších římsách oken v přízemí, kluzkými okraji fontán či moderními lavičkami z ocelových prutů, které jsou mimořádně nepohodlné, pokud by na nich bezdomovci nebo kdokoli jiný chtěl spát. Městská hmota dává lidem jemně, ale jasně najevo, kde jsou vítáni a kde nikoli. Městská tkáň ovlivňuje praxi jednotlivce i jeho pocit pohodlí – tento vliv je tím silnější, čím více se člověk liší od žádoucího uživatele.

Prostorové vyloučení je nade vše zřetelné v modernizujících se postsocialistických městech. Mnohé přestavěné prostory dávají najevo, že jsou určeny úzké skupině lidí: těm, kteří neustále spěchají (zastaví se jenom na kávu nebo na rychlou svačinu), jsou v dobré formě a zvládnou množství schodišť, spojují venkovní ulice s rozličnými prostředky veřejné dopravy. Jsou určeny

lidem, kteří nosí boty bez podpatků, aby udrželi stabilitu na hrbolatém povrchu starých chodníků nebo na nově položených porézních betonových kostkách. Jakmile se někdo od tohoto modelového uživatele odchýlí – ať už tím, že nosí vysoké podpatky, jezdí s dětským kočárkem nebo třeba na invalidním vozíku –, stane se z cesty městem překážkový běh. Pohyb po městě vyžaduje bdělost a hbitost; pro některé, zejména pro ty v horší kondici, představuje takovou výzvu, že veřejný prostor raději přestanou používat.

Města se mění, jejich nová podoba však není přístupnější, ba dokonce zde vznikají nové bariéry. Vysoké obrubníky, [illegible], nedostatečný počet nízkopodlažních autobusů a tramvají způsobuje, že město je nevlídné vůči starším lidem, lidem nemohoucím a rodičům s malými dětmi. I když to někdy vypadá, že se městský prostor snaží nemohoucím lidem vyjít vstříc, nakonec se ukáže, že je to jen další městská past: např. chodníky sice mají nízké obrubníky, jejich vnitřní část je však nižší než strana vnější, což je pro vozíčkáře nepříjemné a v zimě pak chodník strašně klouže a nelze po něm jet bezpečně. Prudká transformace měst výrazným způsobem mění městskou tkáň, a i když se často zdá, že se města snaží dostát potřebám modernizujících se společností, prakticky svou novou formou slouží pouze úzké skupině lidí – těm, kteří jsou vítáni kapitalistickému pořádku. Ostatní, kteří nemohou (nebo nechtějí) splnit podmínku, že jsou zdravými a zaměstnanými lidmi, resp. spotřebiteli, nejsou považováni za důležité a při plánování prostorového uspořádání nejsou bráni v úvahu.

Mnozí městští obyvatelé, [illegible] s architektonickými bariérami, se raději rozhodnou vycházet z bytu co nejméně. Tímto „měkkým“ způsobem jsou odstraněni z veřejného prostoru a [illegible] v prostoru soukromém, i když neexistují žádné restrikce, které by je explicitně omezovaly. Tímto mechanismem vzniká dobrovolné vyloučení značné části společnosti. Stojí za zmínku, že [illegible]

本书在总体结构上是一本词典，但不是一本试图给词语或概念下明确定义的词典。本书包含众多学术领域的经验，对当今不断变迁下的各种概念提出了多角度的观察、思考、反思和见解。本书为无线胶装，封面白卡纸单黑印刷，覆哑膜。目录词条直接构成封面封底，书脊则包含了为本书提供观点的作者名。700 多页的无线胶装能被轻易翻开得益于内页字典纸的使用。内页主要为单黑印刷，夹带少量四色印刷的拉页。前目录为词条排序，书后的目录则为作者排序。每一词条均提供多方面的信息，包括多名作者的解读和思考，而并不给出明确的定义。词条最后提供参考书目、资源链接、插图和相关笔记等更多信息。有些词条还包含希腊字母索引，可以被索引至封面勒口处夹带的 6 张概念集合图，包含认知体系、身体构成、思维逻辑、机械构造等方面。|||||||||||||||||||||||||||||

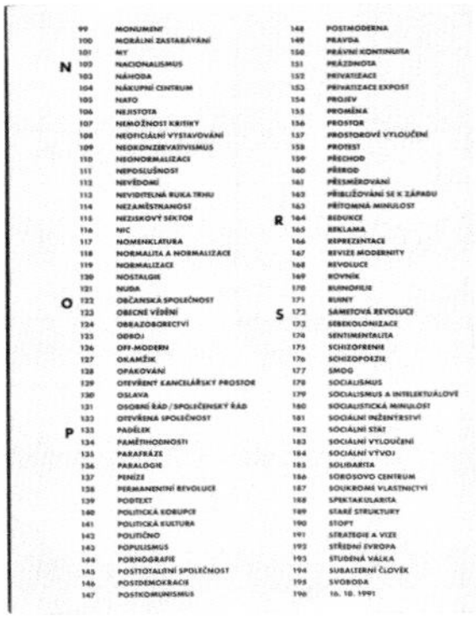

	99	MONUMENT		148	POSTMODERNA
	100	MORÁLNÍ ZASTARÁVÁNÍ		149	PRAVDA
	101	MY		150	PRÁVNÍ KONTINUITA
N	102	NACIONALISMUS		151	PRÁZDNOTA
	103	NÁHODA		152	PRIVATIZACE
	104	NÁKUPNÍ CENTRUM		153	PRIVATIZACE EXPOST
	105	NATO		154	PROJEV
	106	NEJISTOTA		155	PROMĚNA
	107	NEMOŽNOST KRITIKY		156	PROSTOR
	108	NEOFICIÁLNÍ VYSTAVOVÁNÍ		157	PROSTOROVÉ VYLOUČENÍ
	109	NEOKONZERVATIVISMUS		158	PROTEST
	110	NEONORMALIZACE		159	PŘECHOD
	111	NEPOSLUŠNOST		160	PŘEROD
	112	NEVĚDOMÍ		161	PŘESMĚROVÁNÍ
	113	NEVIDITELNÁ RUKA TRHU		162	PŘIBLIŽOVÁNÍ SE K ZÁPADU
	114	NEZAMĚSTNANOST		163	PŘÍTOMNÁ MINULOST
	115	NEZISKOVÝ SEKTOR	R	164	REDUKCE
	116	NIC		165	REKLAMA
	117	NOMENKLATURA		166	REPREZENTACE
	118	NORMALITA A NORMALIZACE		167	REVIZE MODERNITY
	119	NORMALIZACE		168	REVOLUCE
	120	NOSTALGIE		169	ROVNÍK
	121	NUDA		170	RUINOFILIE
O	122	OBČANSKÁ SPOLEČNOST		171	RUINY
	123	OBECNÉ VĚDĚNÍ	S	172	SAMETOVÁ REVOLUCE
	124	OBRAZOBORECTVÍ		173	SEBEKOLONIZACE
	125	ODBOJ		174	SENTIMENTALITA
	126	OFF-MODERN		175	SCHIZOFRENIE
	127	OKAMŽIK		176	SCHIZOPOEZIE
	128	OPAKOVÁNÍ		177	SMOG
	129	OTEVŘENÝ KANCELÁŘSKÝ PROSTOR		178	SOCIALISMUS
	130	OSLAVA		179	SOCIALISMUS A INTELEKTUÁLOVÉ
	131	OSOBNÍ ŘÁD / SPOLEČENSKÝ ŘÁD		180	SOCIALISTICKÁ MINULOST
	132	OTEVŘENÁ SPOLEČNOST		181	SOCIÁLNÍ INŽENÝRSTVÍ
P	133	PADĚLEK		182	SOCIÁLNÍ STÁT
	134	PAMĚTIHODNOSTI		183	SOCIÁLNÍ VYLOUČENÍ
	135	PARAFRÁZE		184	SOCIÁLNÍ VÝVOJ
	136	PARALOGIE		185	SOLIDARITA
	137	PENÍZE		186	SOROSOVO CENTRUM
	138	PERMANENTNÍ REVOLUCE		187	SOUKROMÉ VLASTNICTVÍ
	139	PODTEXT		188	SPEKTAKULARITA
	140	POLITICKÁ KORUPCE		189	STARÉ STRUKTURY
	141	POLITICKÁ KULTURA		190	STOPY
	142	POLITIČNO		191	STRATEGIE A VIZE
	143	POPULISMUS		192	STŘEDNÍ EVROPA
	144	PORNOGRAFIE		193	STUDENÁ VÁLKA
	145	POSTTOTALITNÍ SPOLEČNOST		194	SUBALTERNÍ ČLOVĚK
	146	POSTDEMOKRACIE		195	SVOBODA
	147	POSTKOMUNISMUS		196	16. 10. 1991

This book is a dictionary in general structure, but not a dictionary trying to define words or concepts clearly. It contains many academic experiences and offers multi-angle observations, reflections and opinions on various concepts under the changing circumstances. This book is prefect bound with white cardboard cover printed in black only, laminated with matte film. The catalogue entries are printed on the front and back cover, while the names of the authors who provide ideas are displayed on the spine. Thanks to the use of dictionary paper, the prefect bound book with more than 700 pages can be fully opened. The inside pages are mainly printed in black while a small amount of foldouts are printed in four colors. The front catalogue is in the order of entries, while the latter is in the order of authors. Each entry provides a wide range of information, including the interpretation and reflection of several authors, without giving a clear definition. At the end of the entry, more information is provided, such as bibliography, resource links, illustrations and related notes. Some entries also contain an index of the Greek alphabet, which can be lead to six diagrams of concept sets attached to the flap, including cognitive system, body composition, thinking logic, mechanical construction and so on.

金奖
转型舆图
捷克

Gold Medal
Atlas Transformace
Czech Republic
Adéla Svobodová
Zbyněk Baladrán, Vít Havránek, Věra Krejčcová (Ed.)
tranzit.cz, Prag
220 × 163 × 40mm
1271g
834p

An 800-page reference work, this documentation of an artistic project is graced with a typographic jacket which on the front and back covers plus jacket flaps lists all keywords on the subject of “transformation processes” as well as the names of the authors on its spine. The book is predominantly printed in b/w and is surprising in its very clearly legible, functional typography, striking headwords and fold-out colour illustrations. A successful mixture of typefaces between Futura and an elegant text font, set clearly and calmly with ragged right alignment provide an ideally accessible platform for this content relating to the theory of art.

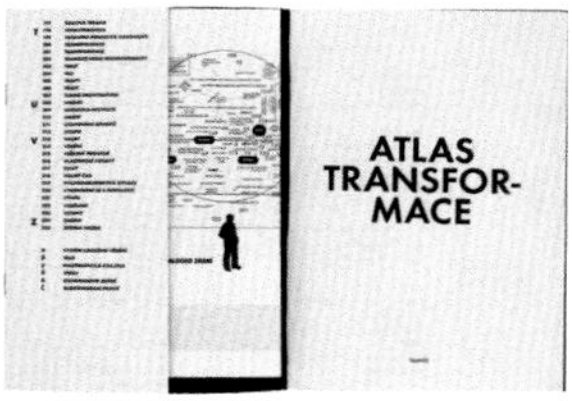
ATLAS
TRANSFOR-
MACE

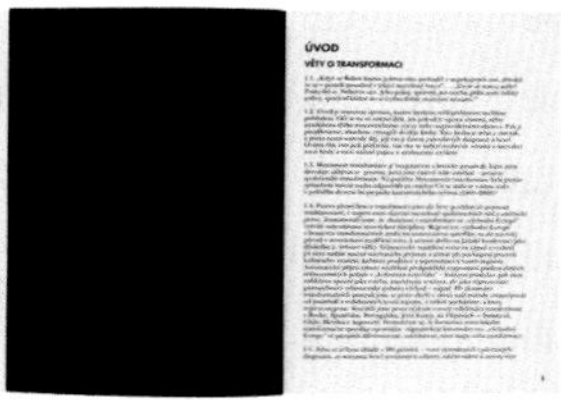
ÚVOD

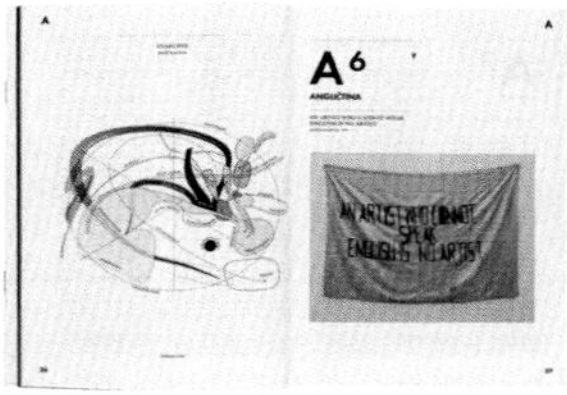
A6
ANGLIČTINA

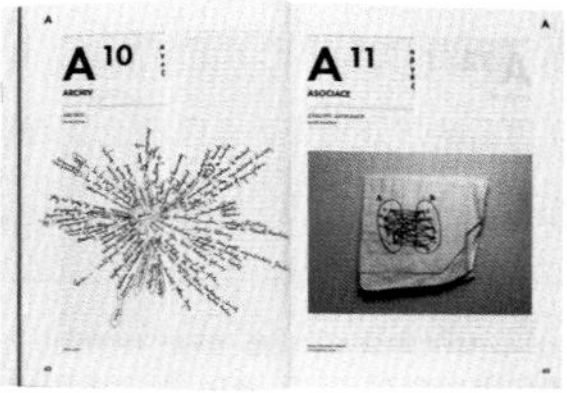
A10
ARCHIV
A11
ASOCIACE

A12
ASYMETRIE

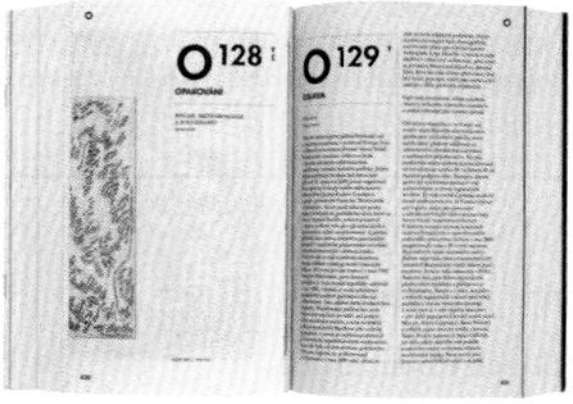
O128
OPAKOVÁNÍ
O129
OSLAVA

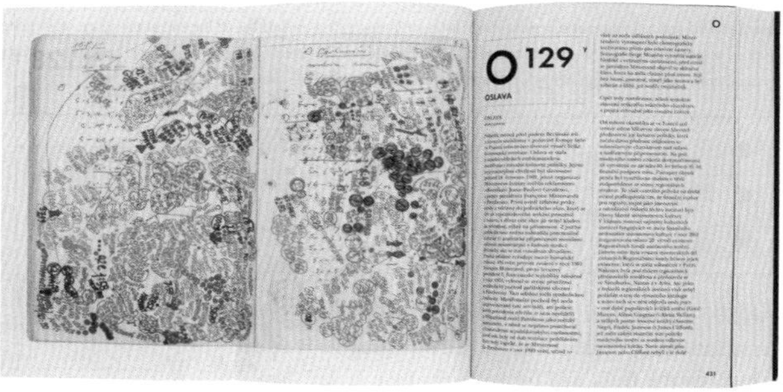
O129
OSLAVA

VIDITELNA

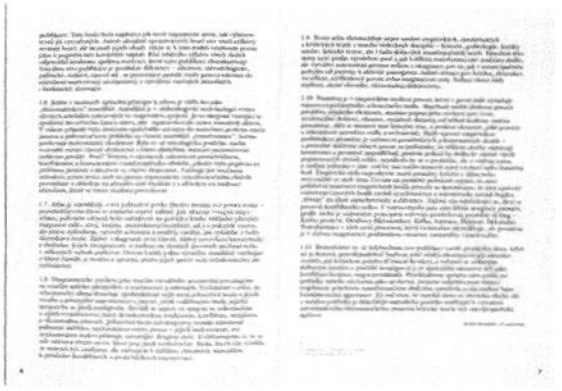

A1
ADAPTACE

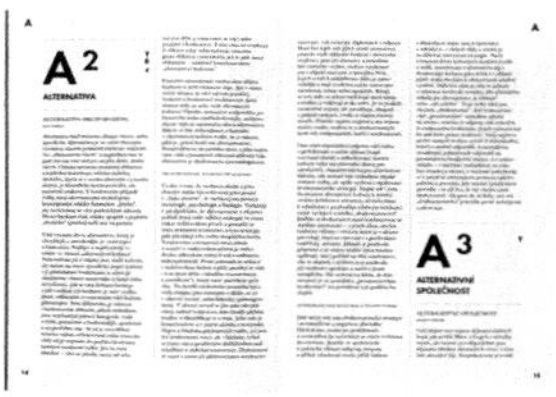
A2
ALTERNATIVA
A3
ALTERNATIVNÍ SPOLEČNOST

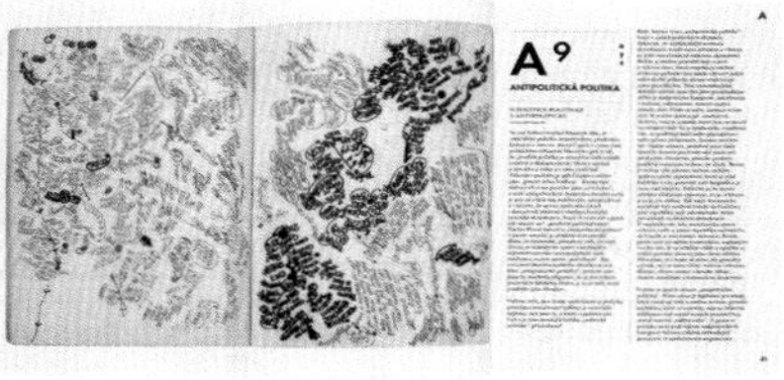
A9
ANTIPOLITICKÁ POLITIKA

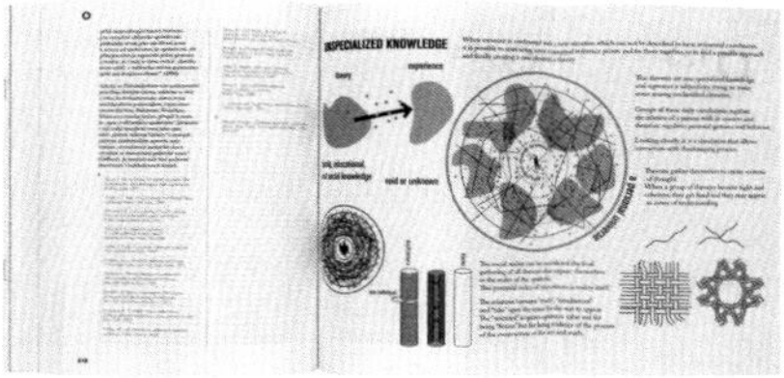
SPECIALIZED KNOWLEDGE

O130
OSOBNÍ ŘÁD / SPOLEČENSKÝ ŘÁD

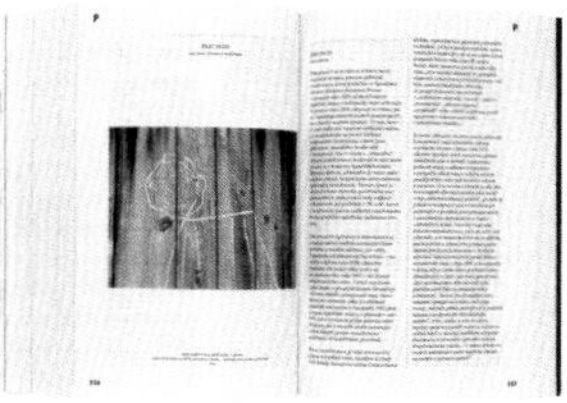

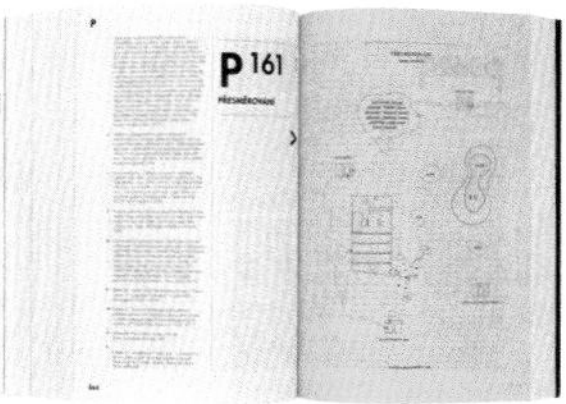
P161
PŘESMĚROVÁNÍ

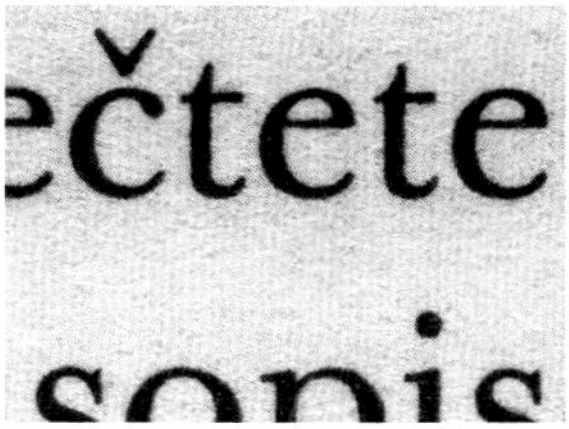
ečtete

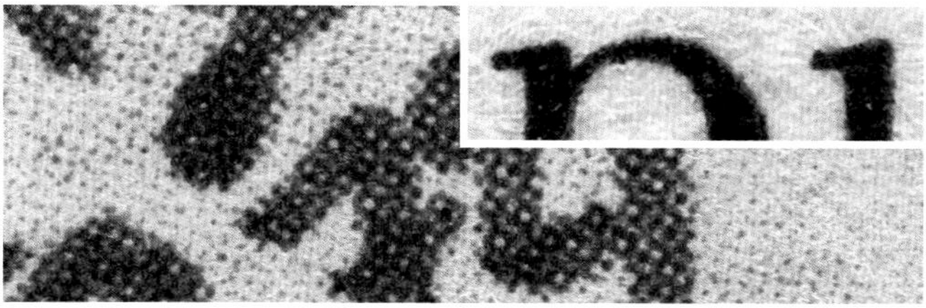

ATLAS TRANS-FOR-MACE

Josef Alan
Ayreen Anastas
Aristide Antonas
Babi Badalov
Zbyněk Baladrán
Ricardo Basbaum
Hrach Boyadyan
Maria José Belbel
Václav Bělohradský
Erick Beltrán
Hakim Bey
Homi K. Bhabha
Egon Bondy
Svetlana Boym
Boris Buden
Cabello/Carceller
Nerea Calvillo
Václav Cílek
Ana Paula Cohen
Keti Čuchrov
Lord Ralf Dahrendorf
Jiří David
Juliane Debeusscher
Gilles Deleuze
Angela Detanico.
Rafael Lain
Vladimíra Dvořáková
Joanna Erbel
Jindřich Fibich
François Marie
Charles Fourier
René Gabri
Mariano García
Ernest Gellner
Adéla Gjuričová
Bohumila Grögerová
Boris Groys
Marina Gržinić
Pierre-Félix Guattari
João Maria Gusmão
& Pedro Paiva
Jiří Havel
Václav Havel
Vít Havránek
Friedrich August
von Hayek
Beatriz Herraez
Josef Hiršal
Ludvík Hlaváček
Karl Holmqvist
Cyril Höschl
Ilja a Vlastimila
Hradečtí
Jaime Iregui
Katrien Jacobs
Frederic Jameson
Ana Janevski
Tomáš Ježek
Ilja Kabakov
Jiří Kabele
Franz Kafka
Jens Kastner
Jan Keller
Alexander Kiossev
Václav Klaus
Milan Knížák
Stanislav Komárek
Karel Kosík
Zdeněk Košek
Primož Krašovec
David Kulhánek
Jiří Kunc
Andrej Lankov
Eileen Legaspi-Ramirez
Mikołaj Lewicki
Mark Lombardi
Tomislav Longinović
Karl Lydén
Artem Magun
Radim Marada
Françoise Mayerová
Adam Mazur
Viktor Misiano
Lubomír Mlčoch
JY Moon
Vlad Morariu
Sir Thomas More
Julia Moritz
Chantal Mouffe
Cristián Gómez Moya
Kamil Nábělek
Monica Narula
Raqs Media Collective
Publius Ovidius Naso
Ania Nicinska
Arno Victor Nielsen
Boris Ondreička
Anatolij Osmolovskij
Milan Otáhal
Martina Pachmanová
Livia Pancu
Jiří Pehe
Georges Perec
Lia Perjovschi
Nataša Petrešin-Bachelez
Miloslav Petrusek
Miroslav Petříček
François Piron
Sylva Poláková
Tomáš Pospiszyl
Jiří Přibáň
Quentin Reed
Gabriella Ricciardi
Teresa Riccardi
Petr Robejšek
Joaquín Barriendos
Rodríguez
Lukasz Ronduda
Sergio Rubira
Jacques Rupnik
Marco Scotini
Keiko Sei
Jan Sokol
Gyatri Chakravorty
Spivak
Yannis Stavrakakis
Mladen Stilinović
Jiří Suk
Jana Ševčíková
Jiří Ševčík
Milan Šimečka
Martin Škabraha
Zdislav Šulc
Marcel Tomášek
Igor Tomeš
Juan Valentini
Vassilis Vamvakas
Zdeněk Vašíček
Vangelis Vlahos
Raluca Voinea
Srdjan Jovanović Weiss
Wolfgang Welsch
Haegue Yang
Milan Zelený
Magdalena Ziołkowska
Slavoj Žižek

P 162

Y

PŘIBLIŽOVÁNÍ SE K ZÁPADU

PŘIBLIŽOVÁNÍ K ZÁPADU

JOAQUÍN BARRIENDOS RODRÍGUEZ

Ačkoli tomu dnes těžko můžeme uvěřit, uplynula jen velice krátká doba od chvíle, kdy nezápadní moderní umění našlo zastoupení na mezinárodních expozicích. Před sotva několika desetiletími tvořila tzv. celosvětové moderní umění výlučně díla z rukou západních a pozápadnělých umělců a rovněž organizátoři výstav (vypadá to jako výmysl, ale ani *kurátoři* tehdy neexistovali) patřili všichni k západnímu *mainstreamu*. Pochopitelně není třeba vysvětlovat, že i všechny kulturní instituce, jež uváděly do života produkci, tvorbu a mezinárodní šíření umění, spočívaly v rukou západních a pozápadnělých gestorů.

Periferní umění bylo tudíž určeno „historickým" nebo etnografickým muzeím, jako by se rozvoj všeho současného a post-moderního usadil ve vymezeném a soustředném poli ekumenického řádu. Štítky primitivního a *naif*, které Západ umístil do čela všeho, co zůstalo ležet mimo kartografii modernizujícího pokroku, vedly k tomu, že periferie nesla tíhu ekonomické marginálnosti a opakujících se avantgardních modelů. Rentabilita periferie v rámci výstav moderního umění nebyla „účetně" uznávána. Nevraživost pociťovaná ze středu vůči „protinožcům" modernity tak zamlžovala hodnotu směny jejich moderního kulturního zboží.

Dnešní situace se od té výše popsané z
liší. Geografie soudobého umění spol
s trhem mezinárodní estetiky změnily
za pouhá dvě a půl desetiletí svůj vylu
a centralizovaný charakter a získaly vš
cující rozměr. Kam se podíváme, kona
bienále, veletrhy, kolokvia a výstavy, v
na explicitně mezinárodní úrovni, a na
spolužijí umělci z magrebského region
Sahary, jižní Asie, centrální Asie, jihoa
kého kontinentu, Střední Ameriky,
příslušníci mexické menšiny v USA, V
doevropané anebo (samozřejmě) umě
z jakéhokoli jiného místa naší planety
v harmonii s umělci severoamerického
kontinentu a střední Evropy. Hlavní p
se ve velmi krátkém čase rozlil z hrani
vymezeného teritoria a vydal se hleda
periferii. Stručně řečeno, hledat to jin
exotické a odlišné. A právě to jiné pro
zájem muzeí, galerií, makroexpozic
a obchodních veletrhů moderního um
stejně jako za starých časů koloniální
expanzionismu. Na poslední kasselsk
Documentě, nové aréně moderního u
měla své zastoupení dokonce i natolik
geograficky a kulturně vzdálená skup
jakou jsou Inuité.

Uvedení multikulturality do oblasti u
během jediného okamžiku přeměnilo
v surovinu všech mezinárodních exhib
Západ dychtil po něčem jiném a vyno
se kultury „odpověděly na jeho volání
kladnou odezvou ve všech úrovních,
s novými periferními zkušenostmi".[1]
Ekonomická kapitalizace marginální s
uvedla do pohybu „aktivum periferie"
začalo vytvářet přidanou hodnotu
globálního moderního umění. To zno
aktivovalo trh a oběh moderního, legi
„exotického", ale potenciálně mezinár
zboží na základě kapitalizace jeho nej
značnějšího a nejvíce stigmatizujícího
druhoplánové perifernosti.

Zapojení kulturní diverzity a obhajob
periferního umění sehrály určující roli
v samotném procesu „bienalizace"
soudobého umění, jak Darío Corbeira
nazývá institucionalizaci (každý druhý
praktik propagace a šíření moderního

566

umění.[2] Vše stejné (jako Západ) ztratilo na síle a po všem odlišném ve své jinakosti se požadovalo maximum vysvětlení, originality a zvláštnosti; čím excentričtější byly doklady diverzity, tím snadněji vypadala možnost zařazení na scénu mezinárodního moderního umění. Zrodila se globální epocha kultury a globální trh a ve složitém komplexu jejich hlavolamu, doprovázena svým celoplošným pronikáním do všech řádů každodenního života, byla moderní estetika povolána stát se klíčovým prvkem jejich vazeb. Jak poukazuje Iván de la Nuez, „politika [chápejme *neokolonialistická politika*] tentokrát nebude pokračovat cestou války, jak argumentoval Clausewitz a jak se dělo se zápalem na celém kontinentu v průběhu 70. let, ale cestou estetiky“.[3]

Historie zařazení periferie do *mainstreamu* globální moderní kultury je přes svou závratnost, anebo možná díky ní, daleka toho, být historií skutečné *deconquisty*. Ačkoli se proces absorpce umělecké geografie překrývá s „postkoloniálním obdobím“, své kořeny zapouští do hluboké krize Západu a do nevyhnutelných následků kolonialistické modernosti, jež přetrvává v nových modelech desteritorializované dominace prostřednictvím řízení hodnot a kulturních reprezentací.

1 Iván de la Nuez, *Latinskoamerické umění a globalizace*. In *L'art a finals del segle XX*, edited by Antich (Girona: Universitat de Girona, 2002), 108.

2 Corbeira, Dario. „Bienalizace moderního umění.“ *Brumaria*, č. 2 (Madrid 2003).

3 De la Nuez, op. cit., 109–110.

P 163 γ δ

PŘÍTOMNÁ MINULOST

PŘÍTOMNÁ MINULOST

ADELA GJURIČOVÁ

Specifické „chování“ dějin v éře globalizace se většinou popisuje jako jejich nebývalé zrychlení. Rodíme se do jedné doby, žijeme v další a umíráme v ještě jiné. Události se na nás řítí v reálném čase (fenomén CNN), všechno se archivuje a zároveň zpřístupňuje. Vše se okamžitě stává minulostí, ale nedávná minulost zároveň bolí i se do ní nutkavě obracíme. Jakých forem nabývala tato „práce s minulostí“ v průběhu české postkomunistické transformace?

Zmiňme ještě blízkost pojmu soudobých dějin, jež se jako specifický obor historiografie etablovaly po 2. světové válce. Modelem českých soudobých dějin byly nepochybně německé *Zeitgeschichte* (H. Rothfels), vzniklé na počátku 50. let jako snaha o pochopení traumatu nacismu a holocaustu, a tedy nebývale propojující „velké“ dějiny s historikovou biografií a stejně nebývale otevřeně přistupující k dějinám generačně a tematicky. Od tohoto původního zadání – tedy historizovat pamatovanou minulost a pracovat pro novou budoucnost – se soudobé dějiny již od té doby značně emancipovaly, není ale náhodou, že právě tento podobor se jako první otevřel (či byl nucen se otevřít) nové pluralitě přístupů historického zkoumání včetně zkoumání fenoménu paměti (viz heslo *Historie*).

Ovšem jak historiografie jako pokus o odborný výklad nedávné minulosti, vstupující zároveň do veřejné debaty, tak historické vědomí veřejnosti a jeho politická

这是一本针对精神病人康复治疗的干预式护理手册。作者把实践中积累的有关准备、谈话、诊断、会诊、治疗、合作等各方面的经验和技巧整理在书中。本书与此类专业图书枯燥无趣的设计完全不同，通过现代的字体、排版和配色打造了专业图书的新形象。装订形式为无线胶装，白卡纸蓝紫色和黑色双色印刷，通过黄色标签“0% 科学”（0% Wetenschap）注明本书不是基于科学实验，而是基于临床实践的经验之谈。内页胶版纸蓝紫色和黑色双色印刷，通过较粗的无衬线字体、颜色和粗细多种变化的下画线、蓝紫色标题、蓝紫色底反白字等多种方式，达成丰富而清晰的层级关系。上下书口与右翻口的蓝紫色刷边也让整个图书的外观更加整体。

Bemoei
Zorg
Eenvoudige tips
voor moeilijke zorg
0%
WETENSCHAP
Jules Tielens
Maurits Verster
de Tijdstroom

This is a manual of interventional care for the rehabilitation of mental patients. The author collates the experience and skills accumulated in practice into the book, from the aspects of preparation, conversation, diagnosis, consultation, treatment and cooperation. This book is designed so different from the boring professional books that it creates a new image of professional books through modern fonts, typesetting and color matching. The binding form is wireless glue binding, printed in blue-purple and black on white cardboard. The yellow label '0% Science' indicates that the book is not based on scientific experiments, but on the experience of clinical practice. The inside pages are made of offset paper printed in blue-purple and black. The layout achieves rich and clear hierarchical relations by means of thick sans-serif fonts, various underlines of different colors and thickness, blue-purple headlines and so on. The three edges of the book are brushed with blue-purple ink, which gives the book a more unified appearance.

- 银奖
- 干预式护理
- 荷兰

- Silver Medal
- Bemoeizorg
- The Netherlands
- René Put
- Jules Tielens, Maurits Verster
- De Tijdstroom, Amsterdam
- 190×120×25mm
- 460g
- 335p

"Bemoeizorg" is an unusual and striking handbook for those who work with psychiatric patients. The subject matter would not normally lead one to expect such a fresh and unconventional book as this. The very "Dutch" typography using a ragged right, pristine, powerful Grotesque typeface is generously incorporated into the type-area. The black and blue used as printing colours create a wonderful colour harmony. The three-sided blue edging gives the book a weighty and compact appearance and commands full attention. The do-it-yourself fold-out indices constitute a particular highpoint of this superb specialist book.

Bemoeizorg

Eenvoudige tips voor moeilijke zorg

De Tijdstroom
Utrecht

Jules Tielens
Maurits Verster

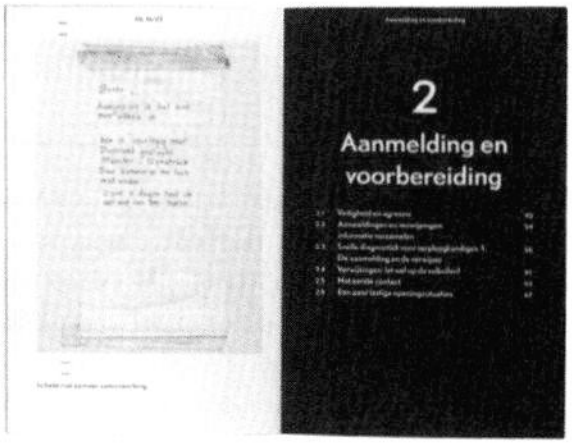

2

Aanmelding en voorbereiding

Het presenteren van een patiënt 4.4

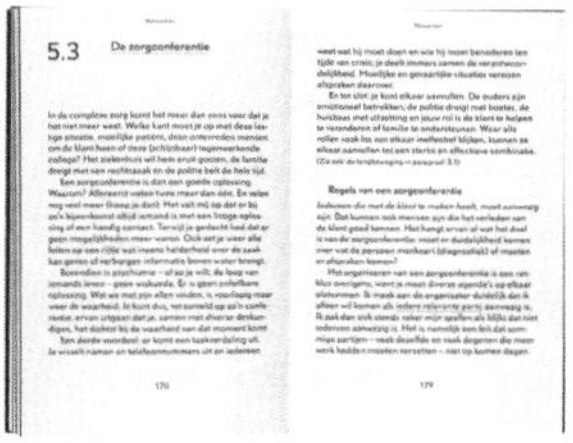

5.3 De zorgconferentie

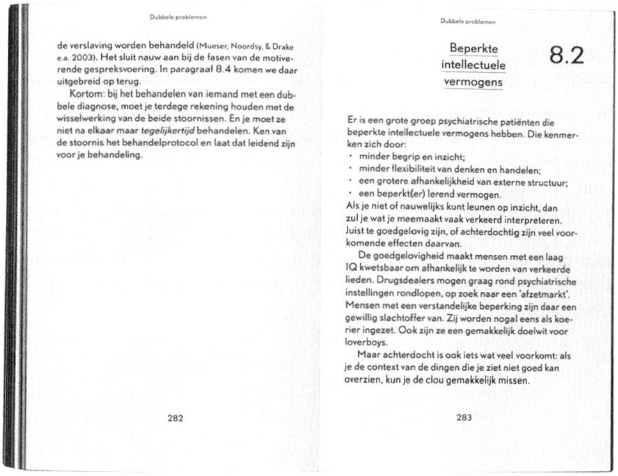

Dubbele problemen

de verslaving worden behandeld (Mueser, Noordsy, & Drake e.a. 2003). Het sluit nauw aan bij de fasen van de motiverende gespreksvoering. In paragraaf 8.4 komen we daar uitgebreid op terug.

Kortom: bij het behandelen van iemand met een dubbele diagnose, moet je terdege rekening houden met de wisselwerking van de beide stoornissen. En je moet ze niet na elkaar maar *tegelijkertijd* behandelen. Ken van de stoornis het behandelprotocol en laat dat leidend zijn voor je behandeling.

282

Dubbele problemen

Beperkte intellectuele vermogens 8.2

Er is een grote groep psychiatrische patiënten die beperkte intellectuele vermogens hebben. Die kenmerken zich door:

- minder begrip en inzicht;
- minder flexibiliteit van denken en handelen;
- een grotere afhankelijkheid van externe structuur;
- een beperkt(er) lerend vermogen.

Als je niet of nauwelijks kunt leunen op inzicht, dan zul je wat je meemaakt vaak verkeerd interpreteren. Juist te goedgelovig zijn, of achterdochtig zijn veel voorkomende effecten daarvan.

De goedgelovigheid maakt mensen met een laag IQ kwetsbaar om afhankelijk te worden van verkeerde lieden. Drugsdealers mogen graag rond psychiatrische instellingen rondlopen, op zoek naar een 'afzetmarkt'. Mensen met een verstandelijke beperking zijn daar een gewillig slachtoffer van. Zij worden nogal eens als koerier ingezet. Ook zijn ze een gemakkelijk doelwit voor loverboys.

Maar achterdocht is ook iets wat veel voorkomt: als je de context van de dingen die je ziet niet goed kan overzien, kun je de clou gemakkelijk missen.

283

ied
sen met

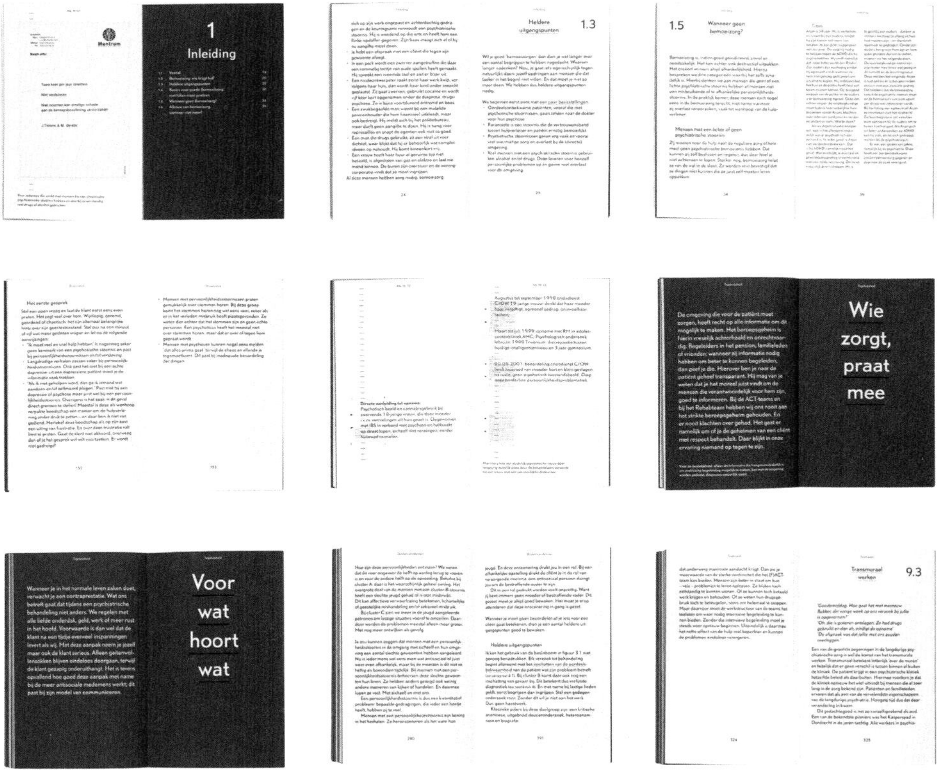
1
Inleiding
Heldere uitgangspunten
1.3
1.5
Wanneer geen bemoeizorg?
Wie zorgt, praat mee
Voor wat hoort wat
Transmuraal werken
9.3

moe
huisbezoek

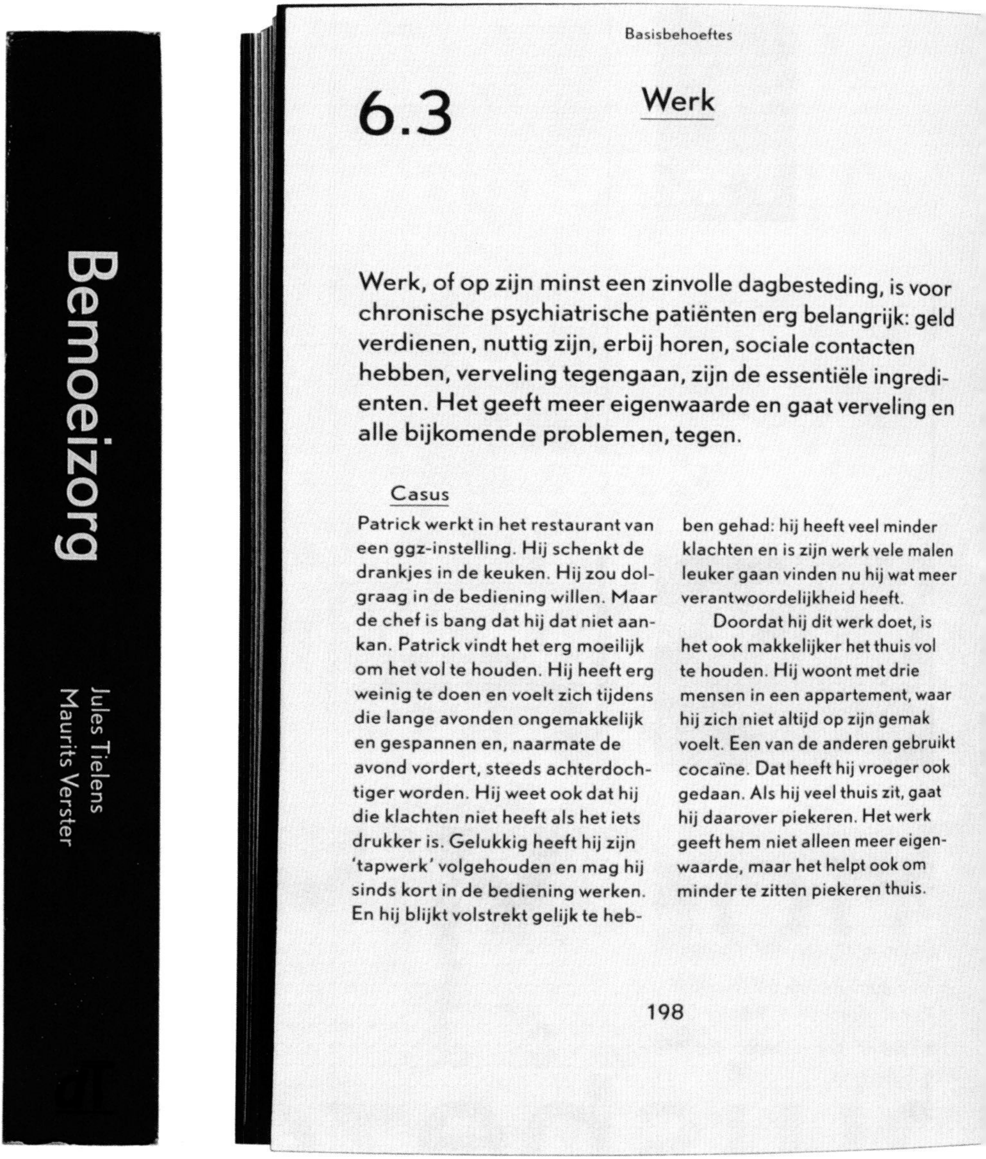
Bemoeizorg
Jules Tielens
Maurits Verster

6.3 Werk

Werk, of op zijn minst een zinvolle dagbesteding, is voor chronische psychiatrische patiënten erg belangrijk: geld verdienen, nuttig zijn, erbij horen, sociale contacten hebben, verveling tegengaan, zijn de essentiële ingrediënten. Het geeft meer eigenwaarde en gaat verveling en alle bijkomende problemen, tegen.

Casus

Patrick werkt in het restaurant van een ggz-instelling. Hij schenkt de drankjes in de keuken. Hij zou dolgraag in de bediening willen. Maar de chef is bang dat hij dat niet aankan. Patrick vindt het erg moeilijk om het vol te houden. Hij heeft erg weinig te doen en voelt zich tijdens die lange avonden ongemakkelijk en gespannen en, naarmate de avond vordert, steeds achterdochtiger worden. Hij weet ook dat hij die klachten niet heeft als het iets drukker is. Gelukkig heeft hij zijn 'tapwerk' volgehouden en mag hij sinds kort in de bediening werken. En hij blijkt volstrekt gelijk te hebben gehad: hij heeft veel minder klachten en is zijn werk vele malen leuker gaan vinden nu hij wat meer verantwoordelijkheid heeft.

Doordat hij dit werk doet, is het ook makkelijker het thuis vol te houden. Hij woont met drie mensen in een appartement, waar hij zich niet altijd op zijn gemak voelt. Een van de anderen gebruikt cocaïne. Dat heeft hij vroeger ook gedaan. Als hij veel thuis zit, gaat hij daarover piekeren. Het werk geeft hem niet alleen meer eigenwaarde, maar het helpt ook om minder te zitten piekeren thuis.

198

Conclusie: wees niet automatisch te voorzichtig (en te paternalistisch) en realiseer je hoe belangrijk zinvol werk voor de patiënt is.

Betaald werk is niet voor iedereen het beste. Er zijn ook mensen die er vreselijk gespannen door raken en er juist allerlei klachten door krijgen. Daarvoor zoek je soms (bij voorkeur tijdelijk) naar andere oplossingen. Uiteraard probeer je waar mogelijk de zelfstandigheid en stressbestendigheid te vergroten. Maar laten we eerlijk zijn, het kan weliswaar meer dan we nu vaak proberen, maar niet altijd en bij iedereen. Dit mag overigens natuurlijk geen argument zijn om je achter te verschuilen.

De slechtste oplossing is mensen zinloze dingen laten doen. Veel psychiatrische instellingen doen dat helaas nog. Houtgutsen of mandela's tekenen. Of nog erger: aan de ene kant patiënten kaarsen laten maken om die achter hun rug weer te laten omsmelten. Nee, het moet werkelijk nut hebben. Niet duizend schroeven draaien in een onduidelijk ding. Dat is geestdodend. Dan werk je alleen voor het geld en met een beetje geluk voor het contact met een paar collega's. Als werk zinvol is, motiveert dat om je best te doen. Voor mensen met chronische psychiatrische problemen is dat van groot belang.

Tegenwoordig is het gebruikelijk IPS *(individual placement and support)* of SE *(supported education)* aan te

本书试图通过艺术作品命名背后的故事，来分析和研究图像和文字之间关联性的传播学、语言学问题。本书没有传统意义上的装订形式，页面被按照骑马钉的形式折好后，使用黑色皮筋固定书脊，使得书籍可以更轻松地被翻阅。封面采用掺满杂质的特种卡纸印单色红，通过将清晰的反白标题字呈现在掺满杂质的纸张上，表达传播学里信息主体与意义失调的“噪音”之间的关系。内页分为两大部分：包裹在外部的 100 页为作品部分，每页一幅作品，但并未完整呈现，只截取了足够认清作品的局部，并伴随部分图注，疑为扫描自多本画册内页，这部分内容采用白色胶版纸孔版印刷单色黑；书籍的中间部分为释义部分，按照作品部分的页码进行标示，呈现作品命名背后的思考和释义，可以看出不同作品命名在传递信息上的差异，这部分采用白色胶版纸孔版印刷单色红。

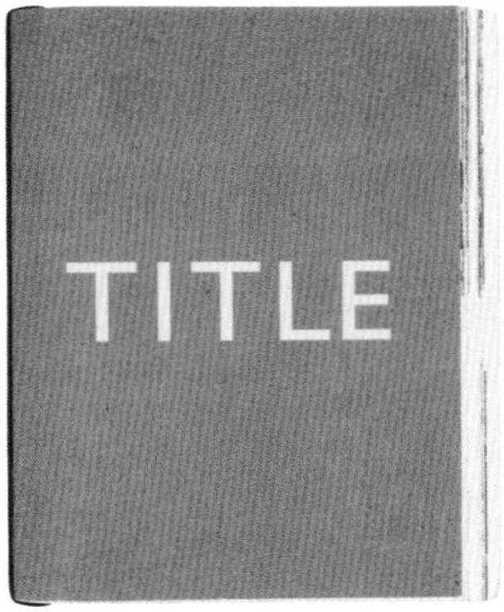

TITLE

This book attempts to analyze and study the communication and linguistic issues of the relevance between images and words through the stories of naming the artistic works. The book hasn't been bound in traditional way. After the pages are folded in the form of saddle stitching, the spine of the book is fixed with black leather bands, which makes it easier for the readers to browse. The cover is printed in monochrome red with special cardboard with impurities. By presenting the highlighted headlines on the paper with impurities, the relationship between the information subject and the meaningless 'noise' in media science is expressed. The inside pages are divided into two parts: the first part contains 100 pages and each page displays a piece of work. Although it's not wholly presented, there are enough details to recognize the work. Some annotations in the book seem to be scanned from some other albums. This part is printed on white offset paper in black with orifice printing. The second part contains explanations, which are marked according to the page numbers of the first part, and presents the thoughts behind the naming of the work. It can be seen that different works convey different information through naming. This part uses white offset paper with orifice printing in monochrome red.

银奖
标题
瑞士

A complete contrast to some of the more structured and traditional submissions. This book explores a very deconstructed approach to reproducing a contemporary art collective and questions what function of linguistic messages are in relation to the iconic message. It is an extremely effective two colour solution, loose bound with a simple black rubber band. The pace of the book is cleverly controlled by a refined yet playful use of typography, layout and image selection. A bold challenge to commercial book design.

Silver Medal
Title
Switzerland
Ramaya Tegegne
Ramaya Tegegne
Ramaya Tegegne, Genève
219×174×17mm
296g
160p

THE NIGHT WATCH

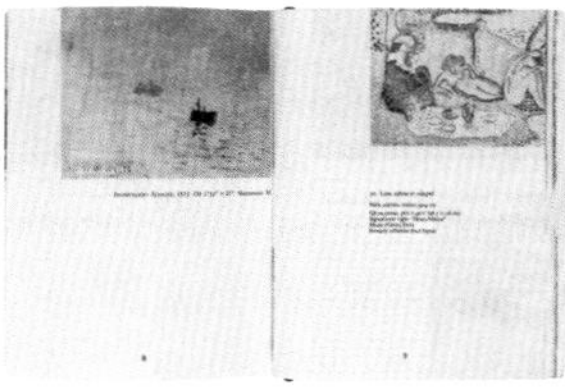
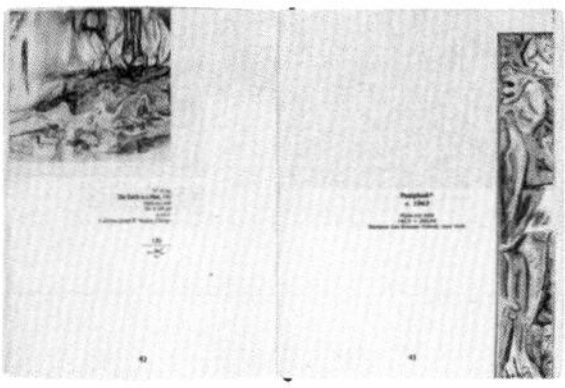
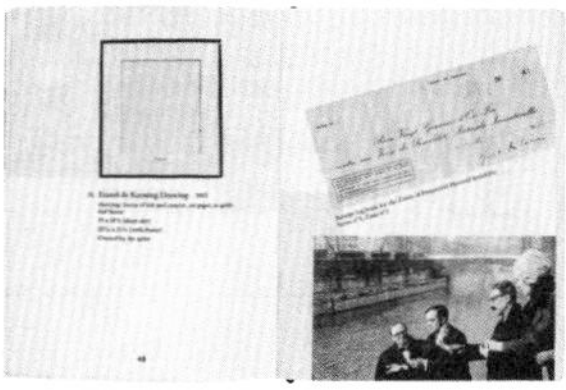

TITLE
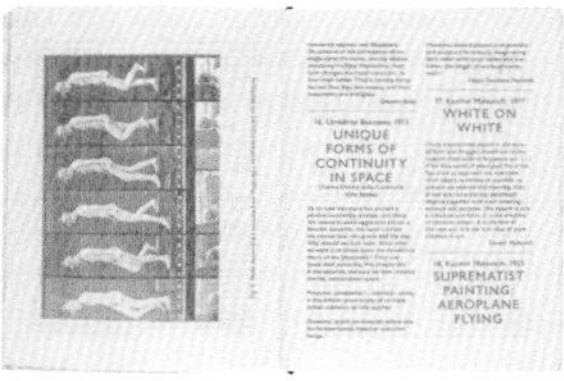
UNIQUE FORMS OF CONTINUITY IN SPACE
WHITE ON WHITE
SUPREMATIST PAINTING AEROPLANE FLYING
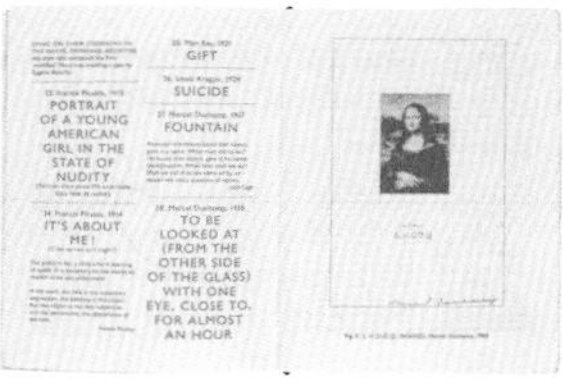
GIFT
PORTRAIT OF A YOUNG AMERICAN GIRL IN THE STATE OF NUDITY
SUICIDE
FOUNTAIN
IT'S ABOUT ME!
TO BE LOOKED AT (FROM THE OTHER SIDE OF THE GLASS) WITH ONE EYE, CLOSE TO, FOR ALMOST AN HOUR
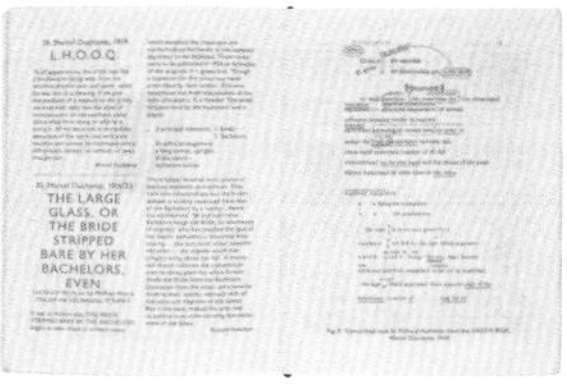
L.H.O.O.Q
THE LARGE GLASS, OR THE BRIDE STRIPPED BARE BY HER BACHELORS, EVEN
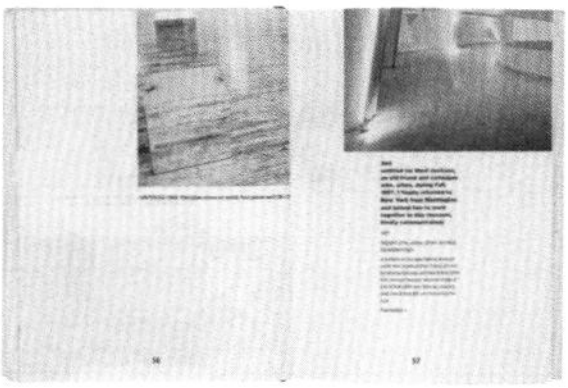
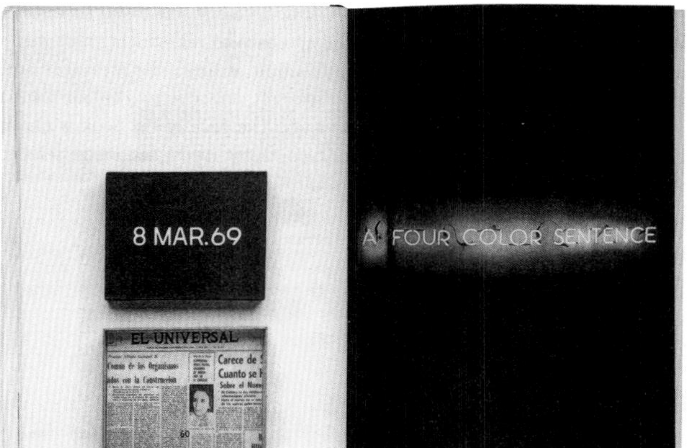
8 MAR.69
EL UNIVERSAL
A FOUR COLOR SENTENCE

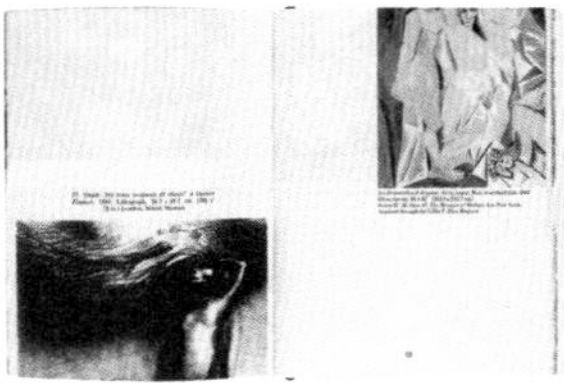
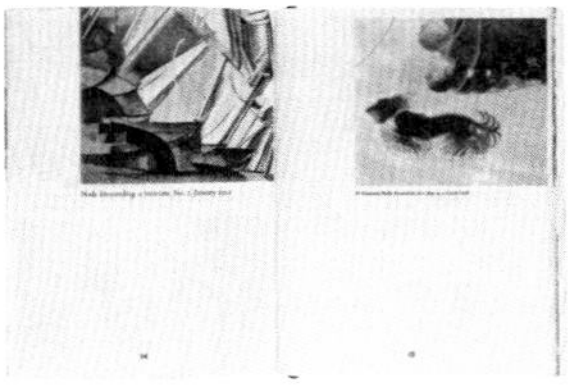

SUICIDE
Louis ARAGON
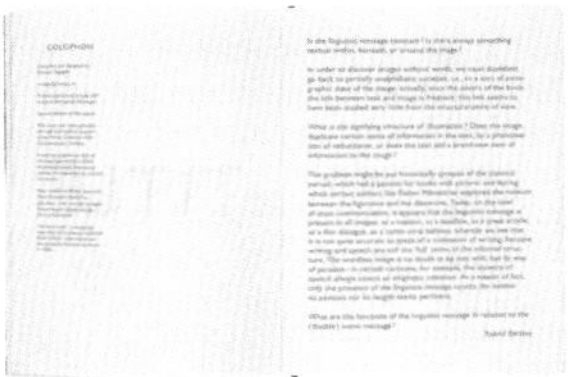
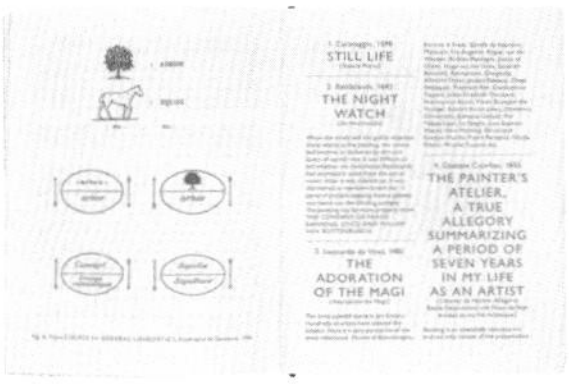
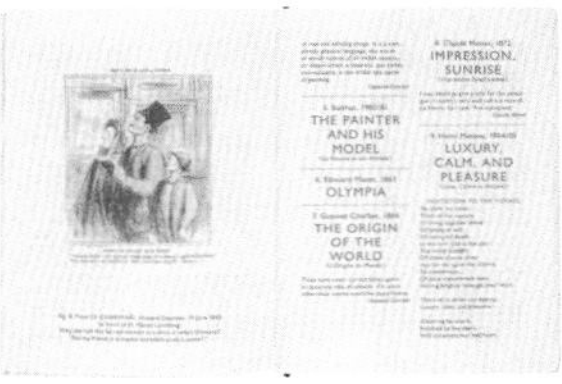
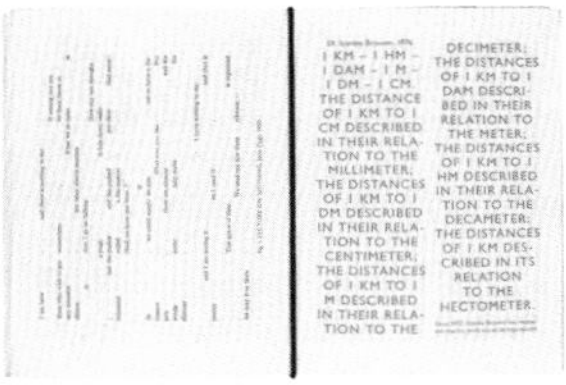
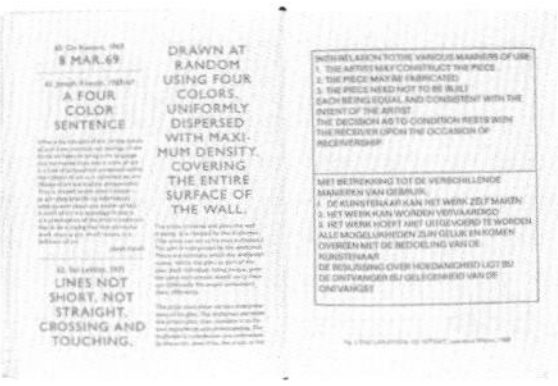
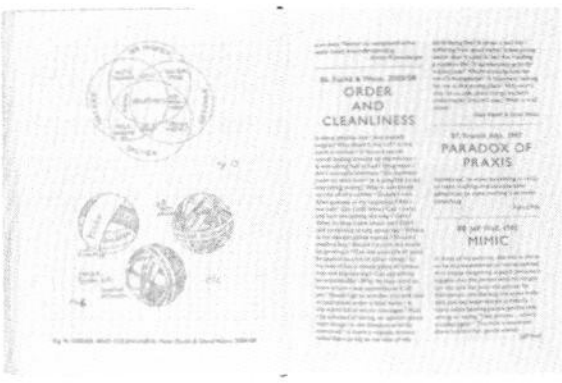
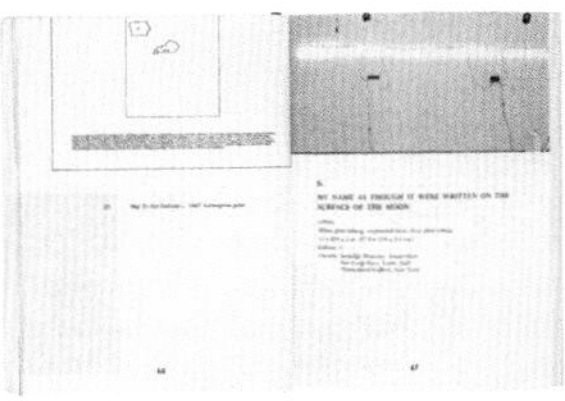
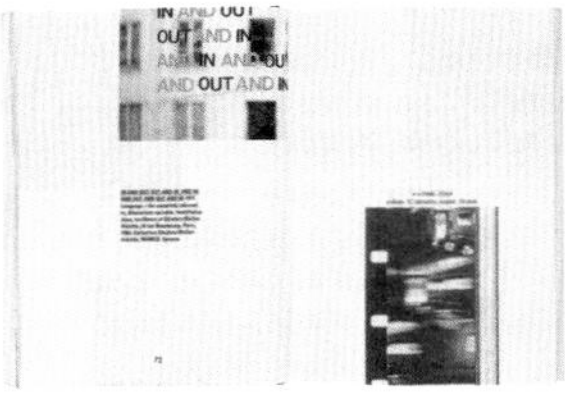
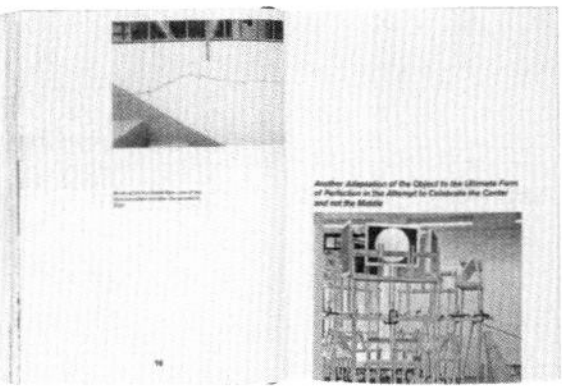
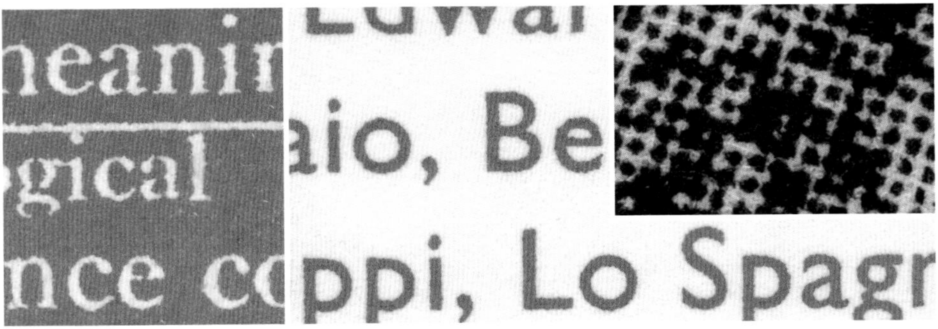
neanin
gical
nce c
Edwar
aio, Be
ppi, Lo Spagr

35. René Magritte, 1926

THE TREASON OF IMAGES (THIS IS NOT A PIPE)

(La Trahison des Images, Ceci n'est pas une pipe)

Separation between linguistic signs and plastic elements; equivalence of resemblance and affirmation. These two principles constituted the tension in classical painting, because the second reintroduced discourse into an art from which the linguistic element was rigorously excluded. Hence the fact that classical painting spoke while constituting itself entirely outside language; hence the fact that it rested silently in a discursive space; hence the fact that it provided, beneath itself, a kind of common ground where it could restore the bonds of signs and the image. Magritte knits verbal signs and plastic elements together, but without referring them to a prior isotopism. He skirts the base of affirmative discourse on which resemblance calmly reposes, and he brings pure similitudes and non-affirmative verbal statements into play within the instability of a disoriented volume and an unmapped space. A process whose formulation is in some sense given by 'Ceci n'est pas une pipe'.

1. To employ a calligram where are found, simultaneously present and visible, image, text, resemblance, affirmation and their common ground.
2. Then suddenly to open up, so that the calligram immediately decomposes and disappears, leaving as a trace only its own absence.
3. To allow discourse to collapse of its own weight and to acquire the visible shape of letters. Letters which, insofar as they are drawn, enter into an uncertain, indefinite relation, confused with the drawing itself—but minus any area to serve as a common ground.
4. To allow similitudes, on the other to multiply of themselves, to be born from their own vapour and to rise endlessly into an ether where they refer to nothing more than themselves.
5. To verify clearly, at the end of the operation, that the precipitate has changed colour, that it has gone from black to white, that the 'This is a pipe' silently hidden in the mimetic representation has become the 'This is not a pipe' of circulating similitudes.

A day will come when, by means of similitude relayed indefinitely along the length of a series, the image itself, along with the name it bears, will lose its identity.

Michel Foucault

---

36. René Magritte, 1937

NOT TO BE REPRODUCED

(La Reproduction Interdite)

Images, ideas and words are different determinations of a single thing: thought. The surreal is reality which has not

Interview imaginaire de René Magritte

M. B. — Si je vous interviewais pour un nouveau journal?

Magritte. — Je veux bien, mais à quel sujet? Il me semble l'avoir déjà dit.

M. B. — Il s'agirait d'un sujet particulier. Ce ne serait pas pour confectionner un article ou un livre comme Untel que je méprise.

Magritte à côté de l'un de ses premiers tableaux

X. — (X. est souvent présent aux entretiens de Magritte. Il ressemble à un poète. X coupe parfois la conversation d'une manière sentencieuse. C'est un style dont il use avec adresse et malice.)
Il faut bien des gens qui écrivent pour les concierges.
(Vient un silence rempli par les carillons des horloges que le maître de la maison paraît collectionner.

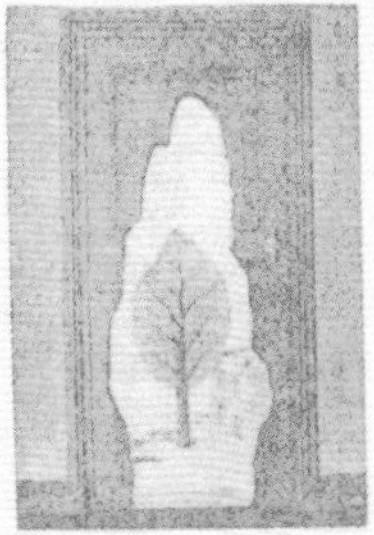

La Perspective amoureuse

L'ami importun en visite les écoute avec une sorte de respect. Les horloges déréglées l'une par rapport à l'autre de 10 à 20 secondes donnent la sensation d'entamer un concert de caractère classique. On regarde le mur. On regarde un tableau. Moment qui convient pour remarquer que le visage de Magritte bouge, que ses traits ont de l'écrit. C'est une personnalité fascinante. Un coup d'œil gris. Une diction qui met l'interlocuteur mal à l'aise. Une politesse exagérée sinon bien faite.)

Magritte. — Alors, vous désirez m'interviewer. C'est une bonne idée.

M. B. — Oui. Il faudrait que j'apporte un enregistreur.

Magritte. — Ce serait très bien que vous vous consacriez au journalisme. Il y a tellement d'imbéciles. Je n'ai jamais rien lu dans un journal qui vaille la peine.

M. B. — J'ai lu sur vous. En effet, une certaine vulgarité...

Magritte. — Ah tiens... cependant... c'est simple. J'aimerais aussi quelque chose d'autre.

M. B. — Que pensez-vous de ceci : Ne plus considérer vos tableaux en fonction des titres qui — vous l'avez souvent déclaré — achèvent le dépaysement du spectateur et transportent l'image dans l'esprit à un niveau où elle se détache absolument de toute interprétation commune. Au contraire, retrouver les événements de la vie, de la société... bref, retracer autour de vos œuvres l'environnement disparu. Elles apparaîtraient, dès lors, comme des témoignages de l'actualité et non comme des poèmes.

Magritte. — Je ne comprends pas.

M. B. — Ecoutez, cela ne traduit pas exactement mon option en matière d'art. Ce point de vue pourrait être intéressant. Il serait destiné à fonder une critique vivante de vos images. A mon sens, elle n'a jamais été faite.

Magritte. — C'est un curieux langage que vous tenez. C'est de la

M. B. **Magritte** **Photo Maria Gilissen.**

3

Fig. G. IMAGINARY INTERVIEW WITH RENÉ MAGRITTE, Marcel Broodthaers, 1967

P
R
K
Ki
R
Ki
L
L moves.

这是一本双胞胎兄弟两人的档案，是一本通过对成员肖像、逐年的家庭关系表、钥匙、通信、相关物品和家庭照片的陈列和对比，记录双胞胎两人的父母和兄弟姐妹之间不断变化的关系档案。书籍为锁线软精装，封面肉色卡纸四色印刷，内页除肖像部分采用光面铜版纸四色印刷外，其余均为胶版纸四色印刷。家庭关系档案本是一个较为复杂、不容易表述清楚的主题，但书籍的设计信息清晰、设计明快。本书最大的特色是书中一些人物的肖像、孩提时的照片，以及文字上出现的部分人名采用棕色或黑色方块遮挡。本是因为这部分人未对这些照片和信息授权的限制，但通过这样的遮挡处理却形成了独特的视觉风格，并且让读者在翻阅的过程中解读人物关系增加了一层趣味。

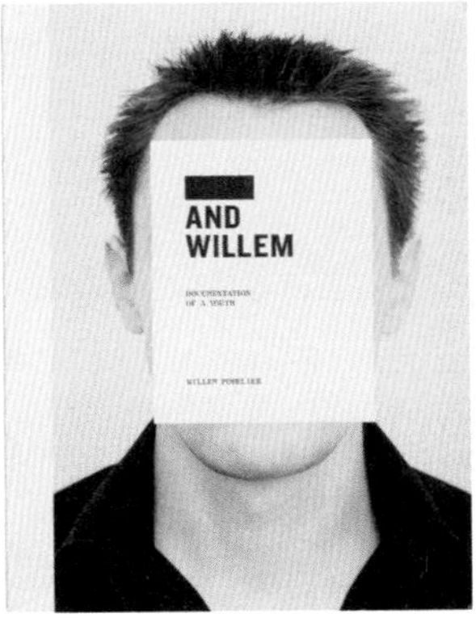
[redacted]
AND
WILLEM

This is an archive of twin brothers. It records the changing relationship between the siblings and their parents by displaying the family members' portraits, annual tables of family relationship, keys, communications, related items and family photos. The paperback book is bound by thread sewing. The cover is printed on flesh-colored cardboard with four-color printing. Most of the inside papers are printed in four colors on offset paper except for the portrait part, which is printed on gloss art paper. The archive of family relationships is a complex and difficult topic to express, but it is clearly designed in this book. The most distinctive feature of the book is that some portraits of characters, pictures of childhood, and some names in the text are blocked by brown or black boxes. This is because the authorization of these pictures and information hasn't been obtained. However, such occlusion forms a unique visual style and adds more interests to the readers when they try to interpret the relationship of characters.

- 铜奖
- __与威廉——一个青年的档案
- 荷兰

- Bronze Medal
- — and Willem. Documentation of a Youth
- The Netherlands
- Robin Uleman
- Willem Popelier
- post editions, Rotterdam
- 230×169×11mm
- 393g
- 170p

The emotional effect of this book stems from its enigmatic content, its choice of materials and its unperturbed use of colour. Artist and author Willem Popelier (Willem's alter ego?) takes us on a 21-year journey through the family history of the twins — and Willem. Through chronologies, portraits, family photos and graphics the book systematically recounts the shifting relationships between parents and siblings. Despite its functional depiction of changes within the family the book's design succeeds in bringing out many different shades. Particularly striking are the light brown rectangles which hide some of the faces and become symbolic for emotional inaccessibility. The treatment of photographs, format, overall volume and features makes for a balance between distance and high emotion.

AND
WILLEM

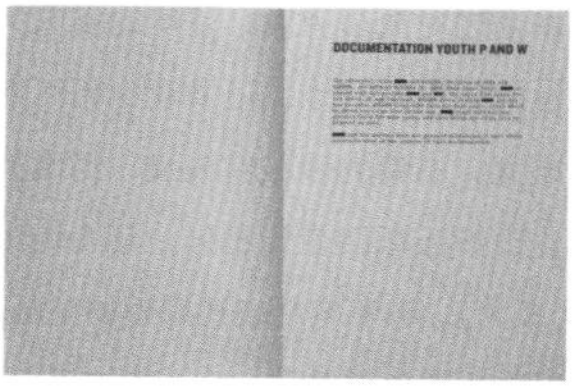
DOCUMENTATION YOUTH P AND W

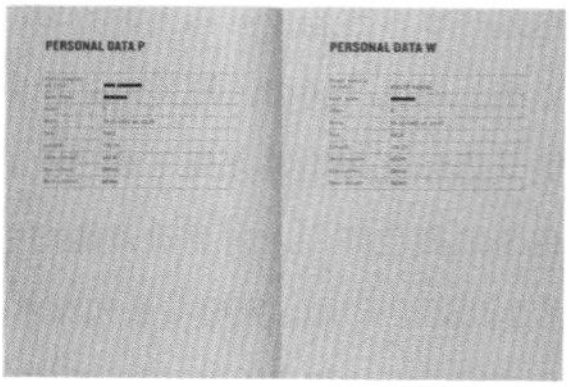
PERSONAL DATA P
PERSONAL DATA W

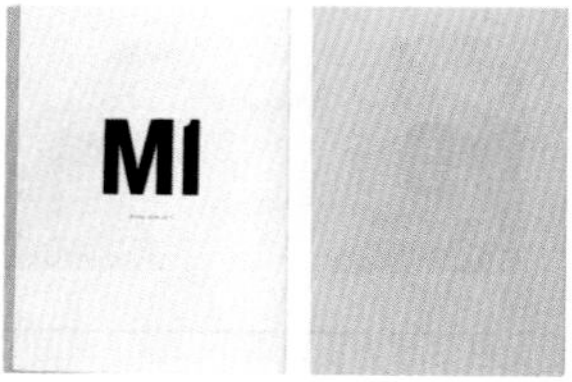
Ml

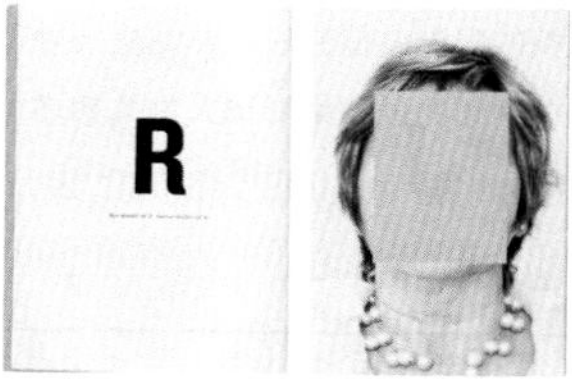
R

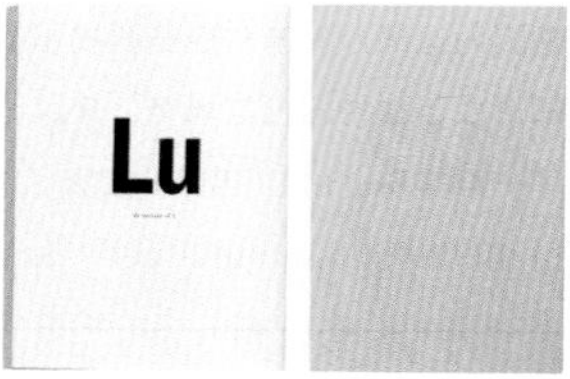
Lu

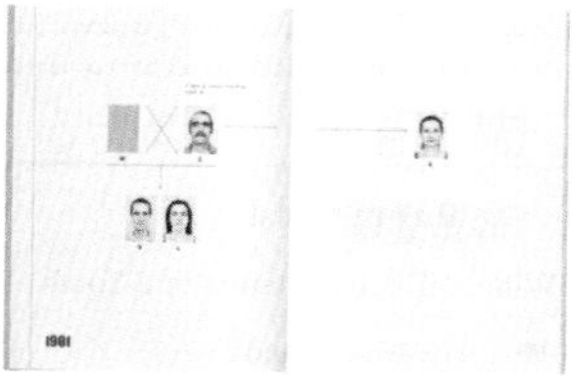
1981

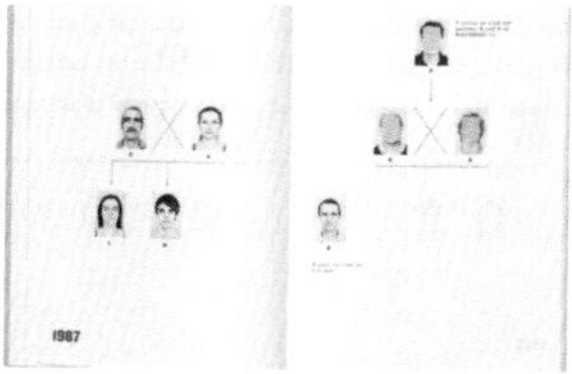
1987

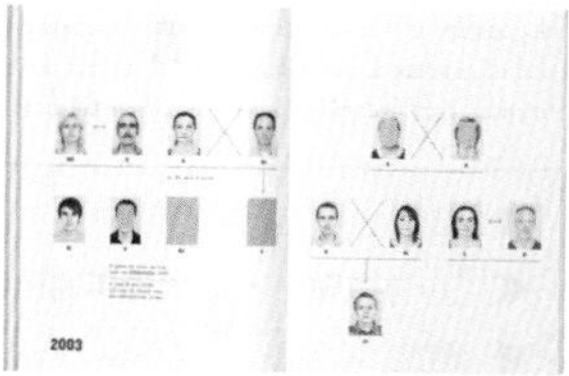
2003

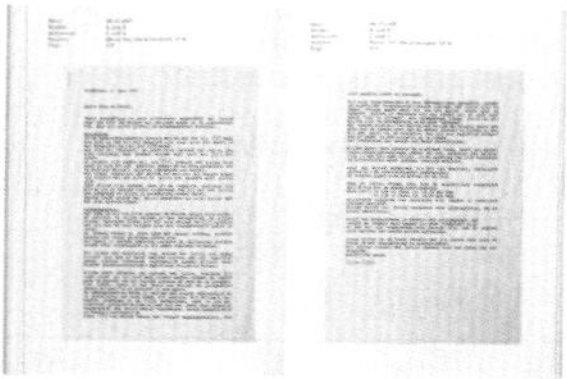

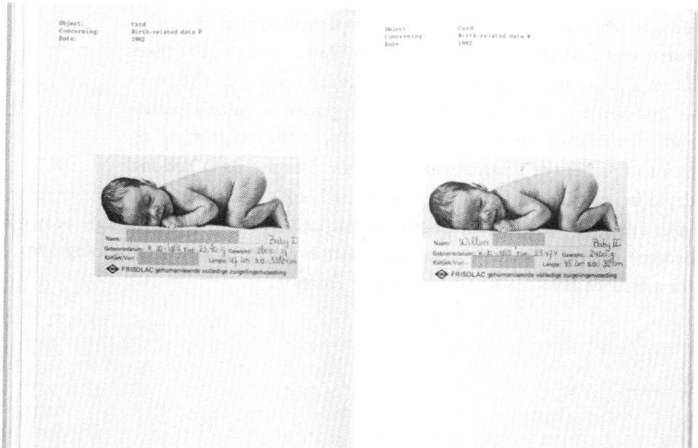
Object: Card
Concerning: Birth-related data P
Date: 1982
Object: Card
Concerning: Birth-related data W
Date: 1982
FRISOLAC gehumaniseerde volledige zuigelingenvoeding
FRISOLAC gehumaniseerde volledige zuigelingenvoeding

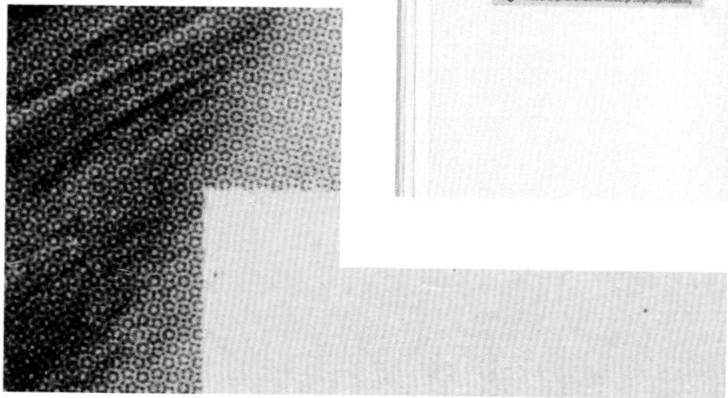

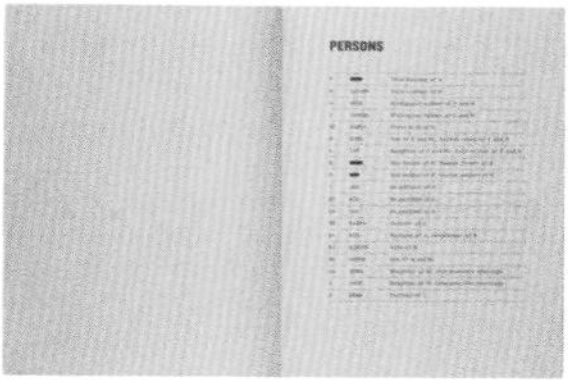
PERSONS

P

W
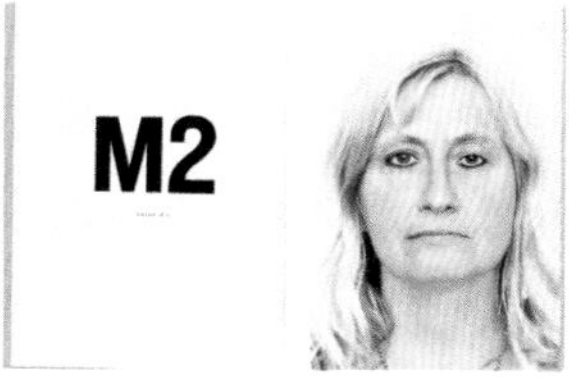
M2

Ja

CHRONOLOGY

PLACES OF RESIDENCE
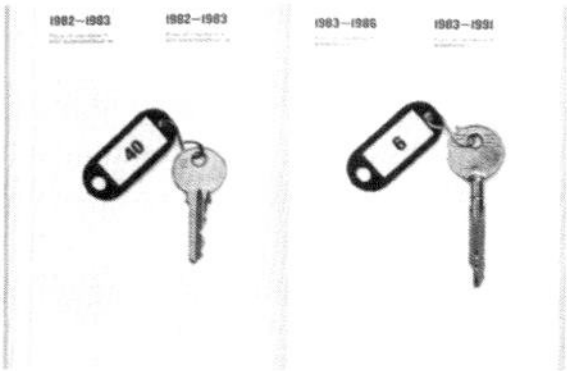
1982–1983
1982–1983
1983–1986
1983–1991

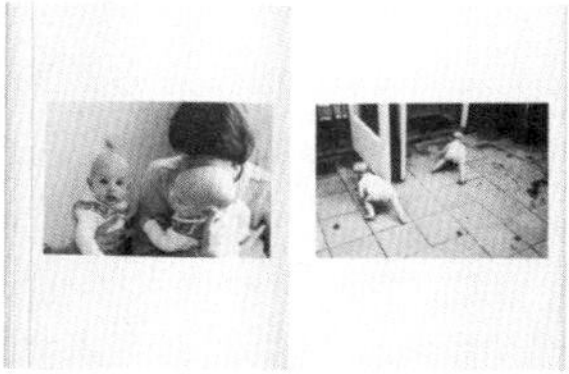
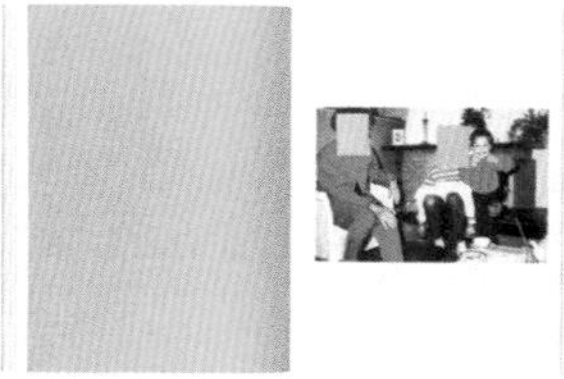

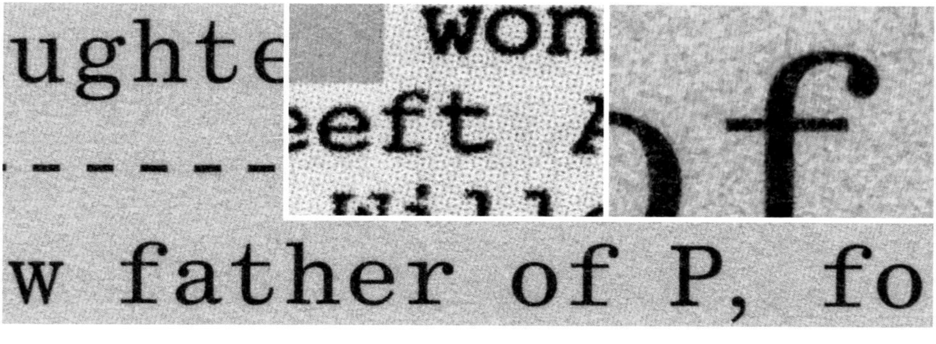
ughte
won
eeft
of
w father of P, fo

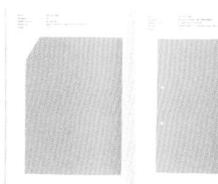

AND WILLEM

DOCUMENTATION OF A YOUTH

WILLEM POPELIER

POST EDITIONS

Date: 10-13-1982
Sender: Home nursing association Eindhoven, Maternity centre
Addressee: C and A
Subject: Statement regarding the newborn twins being kept at home
Page: 1/1

KRUISVERENIGING EINDHOVEN

GEZONDHEIDSZORG THUIS

Aan de heer en mevrouw [redacted]
St. Nicasiusstraat 40
5614 CG EINDHOVEN

dienst: Kraamcentrum
nummer:
uw brief:
bijlagen:
onderwerp:

eindhoven, 13 oktober 1982

Ondergetekende vader /......................
moeder / [redacted]

Stelt het op prijs tegen ons advies in haar/zijn op 11 oktober 1982 geboren tweeling thuis te houden.

De geraadpleegde verloskundige Mw M[redacted] stelt zich tevens verantwoordelijk voor deze situatie.
Overwegend het voor en tegen van het al of niet bieden van thuis-kraamzorg door ons Kraamcentrum menen wij gezien het bovenstaande toch interne thuiskraamzorg te moeten bieden waarvan blijkens ondertekening in deze de verantwoording komt te liggen bij de ouders en betrokken verloskundige.

[redacted]

correspondentie-adres: postbus 310 - 5600 AH eindhoven
don boscostraat 4 - telefoon (040) 44 44 35* - giro 1 07 39 18 - bankrelaties: a.b.n. 52 76 16 052, rabo 10 10 50 461

Date: 02-17-1987
Sender: Child Abuse Councillor Bureau
Addressee: C and A
Subject: Invitation for interview
Page: 1/1

Bureau VERTROUWENSARTS inzake KINDERMISHANDELING

voor de provincie Noord-Brabant

Postbus nr. 7095
4800 GB BREDA

Telefoon nr. 076 - 14 63 23

Verzoeke bij beantwoording te vermelden: U 6680/mdK/sN

Breda, 17 februari 1987

Bijlage:

De heer en mevrouw [redacted]
Burghplein 6
5614 BA EINDHOVEN

Geachte mevrouw, meneer [redacted],

Op ons bureau is er melding gekomen over de problemen in uw gezin t.a.v. de tweeling [redacted] en Willem.
Daar wij ons altijd eerst oriënteren moeten omtrent de gebeurtenissen, heeft het enige tijd geduurd alvorens wij u uit konden nodigen uw visie over deze melding te geven.

Gaarne nodigen wij u nu uit voor een gesprek.
Als datum zouden wij voor willen stellen dinsdag 10 maart om 14.30 uur op ons bureau, Delpratsingel 4 te Breda.
Hopelijk schikt u dit. Zo niet, dan gaarne bericht via bovenstaand telefoonnummer.

Hoogachtend,

[redacted]

Mw. M.J. [redacted],
vertrouwensarts.

Wij komen niet; omdat dit probleem helemaal niet ons en onze kinderen ~~niet~~ betreft, maar enkele mensen die grote emotionele moeilijkheden hebben met het verwerken van het gebeurde. Deze mensen hebben er belang bij om ons in een kwaad daglicht te zetten omdat ze boter op hun hoofd hebben: de heer en mevrouw [redacted] en de heer [redacted].

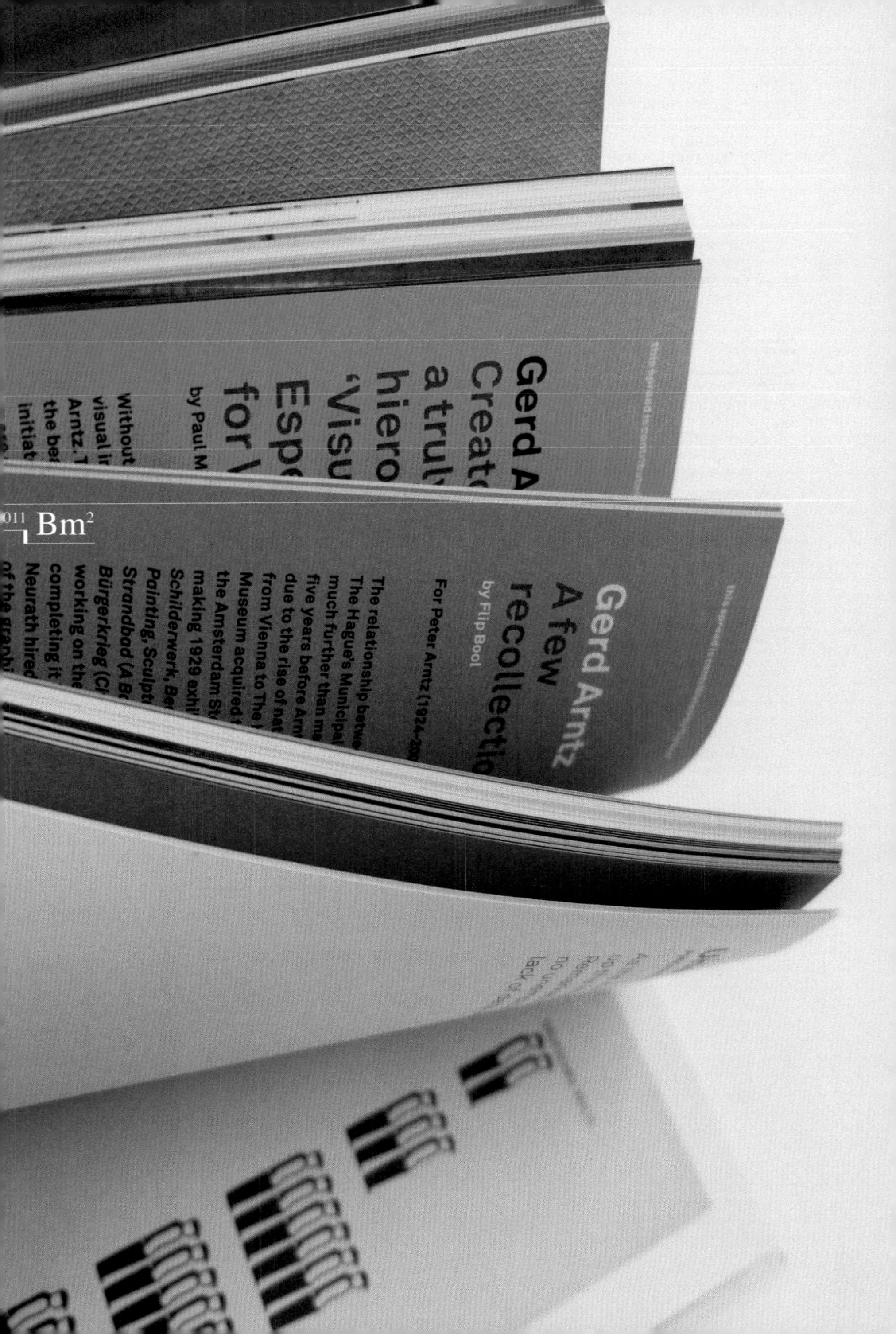
Gerd Arntz
A few
by Flip Bool

这一本同名展览的画册，以德国黑白木刻版画家格尔德 · 阿恩茨（Gerd Arntz）创作的 4000 多件象形图为基础，介绍艺术家在平面设计领域里信息的图形化表达即图形独立于语言来解释复杂信息方面所做的成果。书籍的装订形式为锁线软精装。封面白卡纸四色印刷覆哑膜，图形部分亮面 UV 处理。黑卡纸作为环衬，内页胶版纸四色印刷。书中通过黑底反白字的形式将内容进行区隔，介绍了艺术家的生平、版画创作、接受委托设计这些象形图前后的社会形态和个人故事，并通过象形图在信息图表中的应用，展示其对后世的影响。书籍的中后部“视觉故事”篇章，作者挑选了一些象形图，通过左右页的排版和色彩关系，试图建立图形与图形之间的联系，即不通过语言的方式来阐述一些可能的故事。由于没有任何文字，需要由读者自行理解，检验图形的信息传达的有效性。书籍的最后对这些象形图进行分类，并展示了参考书籍和展览本身。

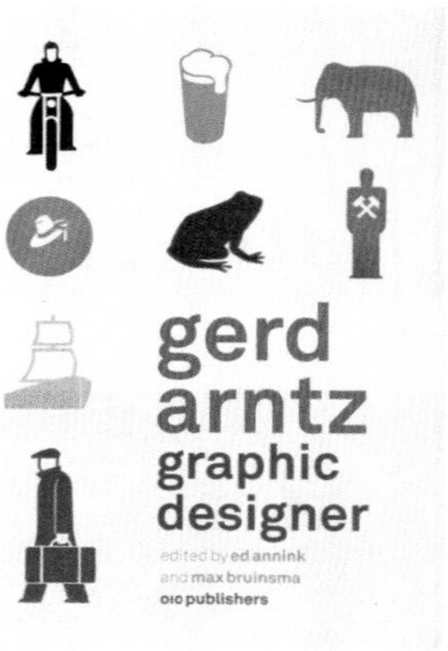
gerd
arntz
graphic
designer
edited by ed annink
and max bruinsma
010 publishers

This is an album of an exhibition that has the same name as the designer. Based on more than 4,000 black-and-white pictograms created by German woodcut artist Gerd Arntz, this album introduces the achievements of artists in graphic design, especially in how the graphics can interpret complex information independent from languages. The book is bound by thread sewing. The cover is made of white cardboard in four colors, laminated with matte film. The images are spotted with bright UV. Black cardboard is used as endpapers while offset paper is printed in four colors as inside pages. The book introduces the artist's life, creation process, personal stories related to his pictograms, and his influence on future generations through the application of pictograms in information charts. In the chapter of 'Visual Story', the author selected some pictograms and tried to establish the relationship between graphics through the page layout and color matching. That is to say, some possible stories were elaborated without language. Since there is no text, it is necessary for the readers to understand and judge the effectiveness of information transmission by themselves. At the end of the book, these pictograms are categorized and reference books and the exhibition are displayed.

- 铜奖
- 格尔德 · 阿恩茨，平面设计师
- 荷兰

- Bronze Medal
- Gerd Arntz. Graphic Designer
- The Netherlands
- Ontwerpwerk
- Ed Annink, Max Bruinsma (Ed.)
- 010 Publishers, Rotterdam
- 247 × 180 × 24mm
- 763g
- 288p

Isotype (International System of Typographic Picture Education) was developed in the 1920s in order to be able to explain complex information independently of language. For this the designer Gerd Arntz originated powerful pictograms, illustrations and diagrams. This catalogue of works presents both the drafts as well as the context in which the artist operated. The book's design works sensitively with areas of colour, partly underlying the illustrations, and dutifully transmits the expression contained within the pictorial language through clear and pre-cise typography. This book displays an impressively varied layout thanks to its lucid character and serves as a reference work for all those involved in information communication and systems of orientation.

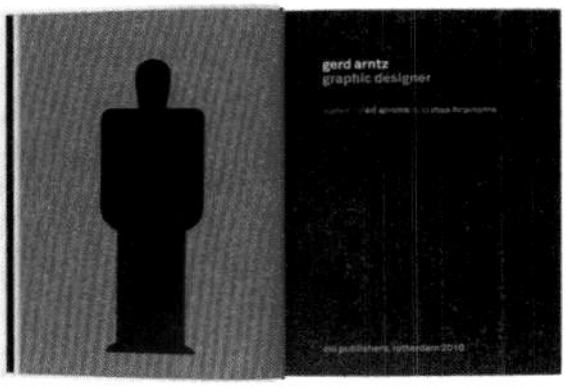
gerd arntz
graphic designer

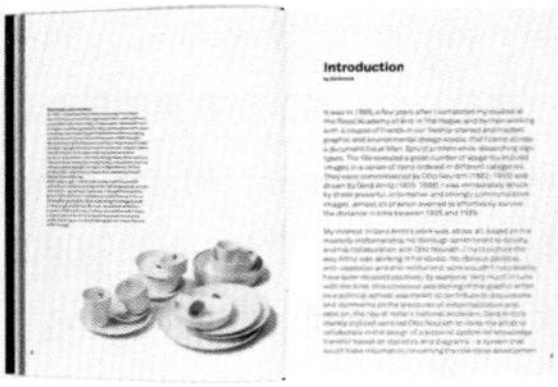
Introduction

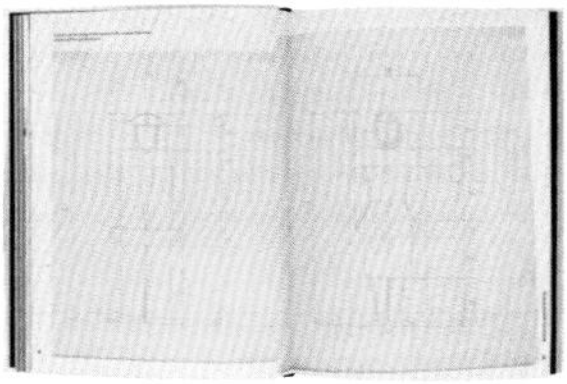

DIE PROLETARISCHE REVOLUTION
Prosit Neujahr Hindenburg! — Das Volk hungert!
DIE PROLETARISCHE REVOLUTION
Was ist das Parlament?

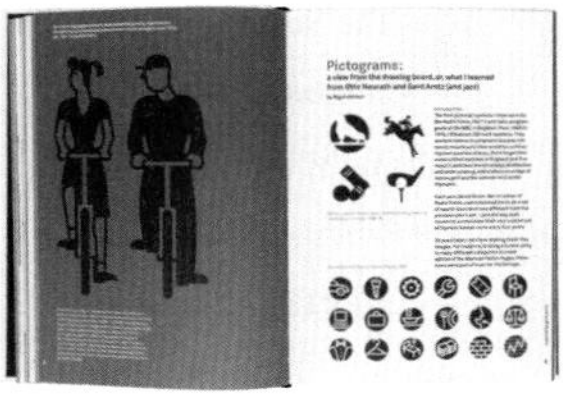
Pictograms:

Sometimes icons and symbols are used as tiny visual gifts in graphics: a little ☎ next to the phone number on business cards, or a symbolic eye next to a page number to denote "see page..." It can be fun too. I put a snail next to the mailing address on my business card. It's not necessary, but it matches up with the mouth (representing voicemail, or phone), the fax machine icon, and the computer icon for email and web addresses. Purists will note that "snailmail" is hardly international and could be meaningless to many people. Here's one place I'll let Neurath have his own opinion, and personally reserve judgement. He believed that a symbol should represent only what a thing looks like, not what the word for it sounds like, and it should never be a play on words. It was all part of his effort to help international understanding.
Now that the world is becoming one big, undifferentiated culture, a little movement in the other direction, towards preserving differences, might be a relief. I believe that it is rather exciting to be slightly baffled by foreign languages, because you learn something while you are figuring out the meaning. (Of course, I don't want to be baffled while rushing to catch a plane.) Does the hand sign shown here (→) mean the same thing all over the world? And should I worry about that when designing a logo for an Anglo/American company?
38%
37%
12%
8%
5%
Large and small
With the exception of Gerd Arntz's work on pages 91 and 95, most of the examples I've used to illustrate this essay are about the size they were when they were originally printed. I work quite a bit larger than this when I'm making them, so that I can more easily see what I'm doing. But I have to remember that for all my talk about "art," these are works of commerce first and art second, and are seldom the visual stars of a printed page. They are supposed to be small (←).
But this is nearly my last example, so let's be indulgent. It is nice to see it big!

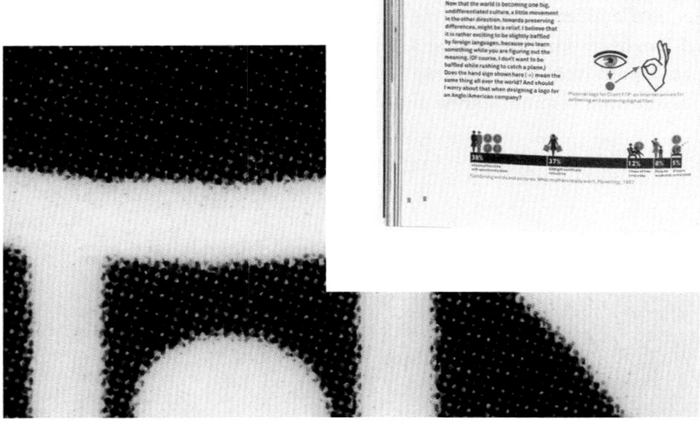

gerd
arntz
applied
bildung

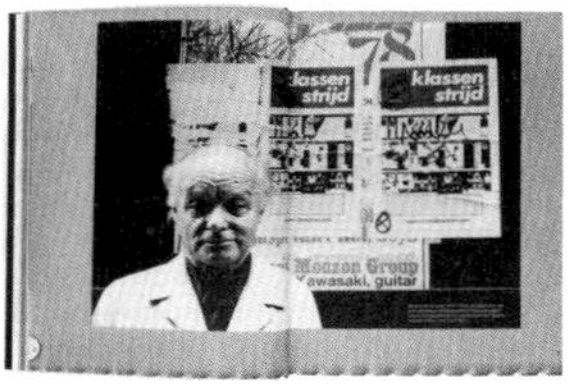
klassen
strijd
klassen
strijd
Kawasaki, guitar

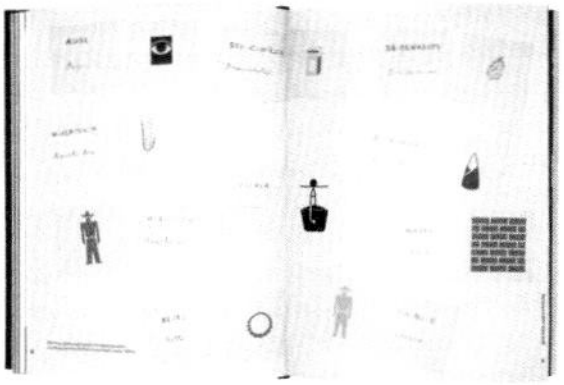

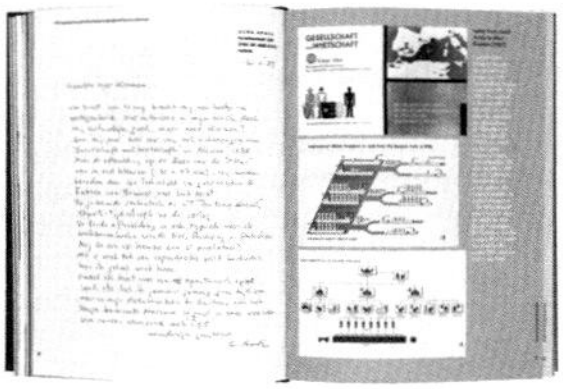

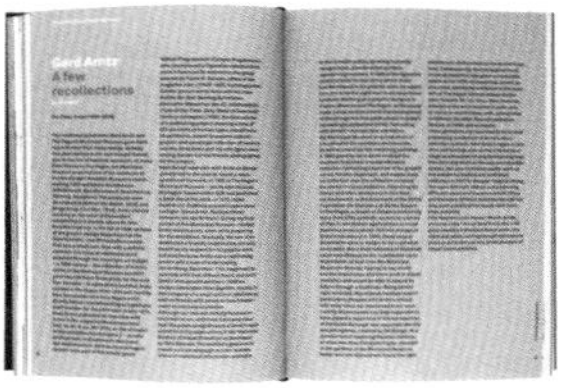
Gerd Arntz
A few
recollections

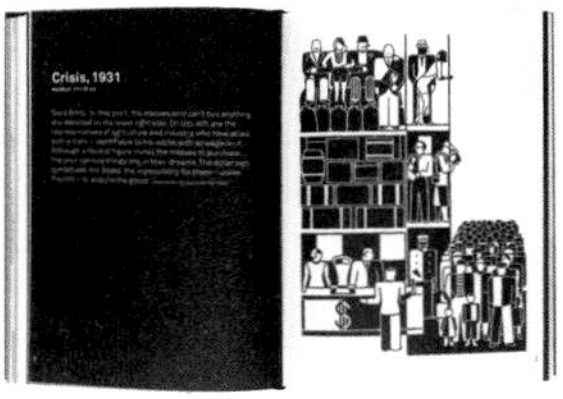
Crisis, 1931

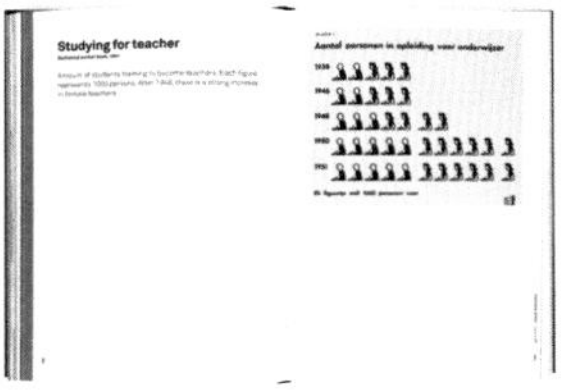
Studying for teacher
Aantal personen in opleiding voor onderwijzer

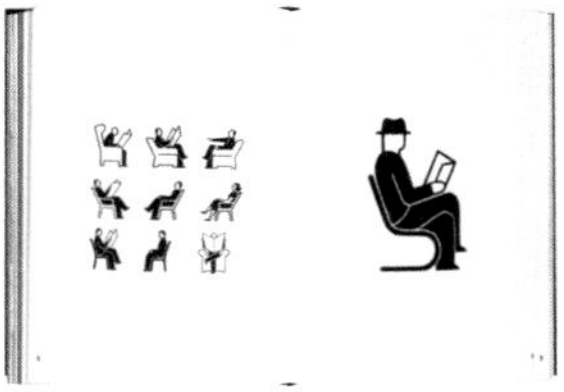

cure
nt and inforr

gerd arntz graphic designer

010

Great War 1914-1918
Published in a traveling, folding presentation portfolio for the Mundanaeum in London, 1930s.

Isotype the transformers

by Erik Spiekermann

Like journalists who take facts and prepare them for their readerships, designers have always been transformers. Whatever we do to the information we shape, we influence its reception by transforming it. That goes for advertising as much as it does for hard-core information graphics. Tools are available to turn any dumb number into an equally dumb graphic, with multicolored gradations, drop shadows and pseudo-dimensional distortion. To paraphrase Churchill, 'there are lies, dumb lies, and there are Powerpoint presentations'. But it only takes one little Isotype chart, called "Comparisons of Quantities" to destroy the validity of most business graphics.
It is a simple chart that shows how only one variable may look cool and simple, but to really compare things we need to show the method behind a graphic: how did we arrive at the numbers, what do they relate to and how do they relate to each other?
One chart that displays all the qualities of the Isotype method is "Great War 1914–18" that shows how many soldiers were involved, killed or wounded in the First World War. The illustrations are detailed enough to show differences in the uniforms but simple enough to be able to work even when reproduced at the size of a business card (the original was 420 x 630 mm). By showing the armies in parallel projection, marching away from each other, rather than as plain rows of soldiers like bars in a graph, it clearly distinguishes between the Central Powers and the Allies. This is reminiscent of old engravings with their isometric depiction of battle formations, but much more sober and dramatic at the same time. The chart can be read in three ways: comparisons between the sizes of the opposing whole groups; proportions within each group; between proportions of the same class across the two groups.
One can almost understand the point most Isotype charts make intuitively, before even reading any detail. The facts are all there, but they have been transformed into communication.

Comparison of forms: squares, circles, rectangles and signs. Published in International Picture Language, Otto Neurath, 1936.
A small error in the last sentence, corrected by an attentive reader; 2 is 1/3 of 6.

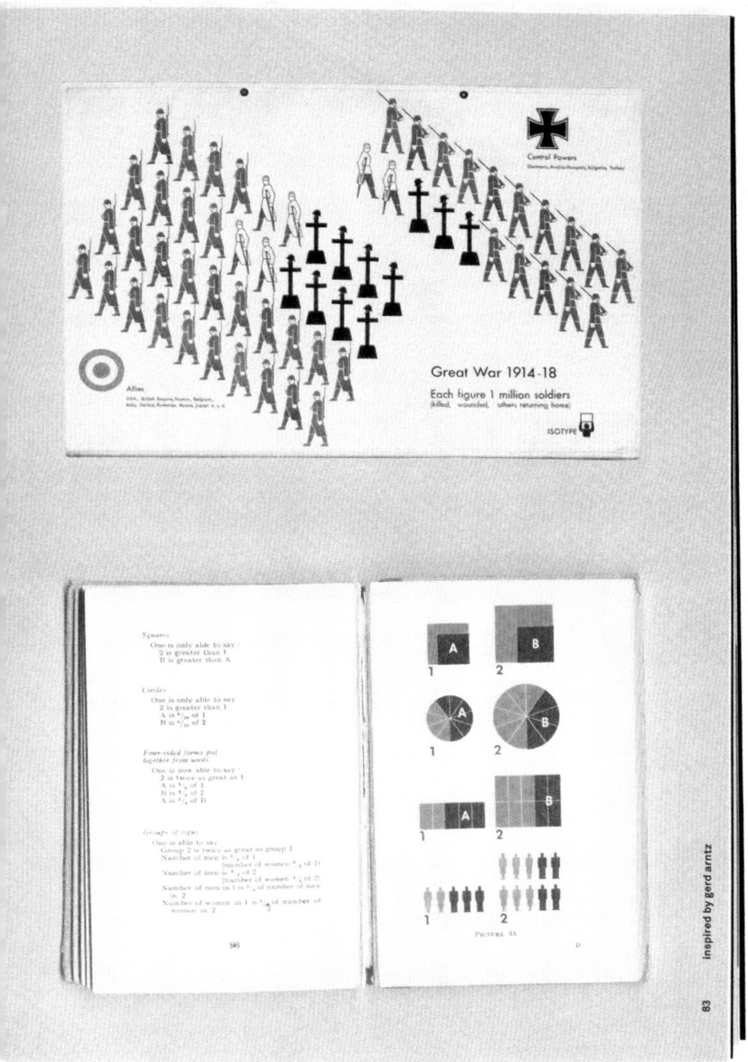
Central Powers
Allies
Great War 1914-18
Each figure 1 million soldiers
(killed, wounded, others returning home)
ISOTYPE
A
B
1
2
PICTURE 35
96
inspired by gerd arntz
83

2011 Bm³
pro:Holz Holzforschung

本书是针对木质外墙构建相关的书籍，书中对木质外墙的外观、材料种类、特性、表面涂料、保暖防水隔音和防火性能、构建细节和具体案例进行了研究和展示，是针对工程师、建筑师等目标群体推出的高质量标准的专业书籍。书籍为锁线硬精装，封面中灰色布面银、黑双色印刷，直接裱覆在与内页等尺寸的卡板上，环衬棕色仿木纹纸裱贴，使得整个封面不论是在视觉上还是触感上，就像是夹在书最外侧的两片木板。内页哑面铜版纸四色加专色红印刷，专色红主要用于文本部分标题、重点细节等方面的突出显示。内文设置为五栏，根据内容需要使用为独立五栏或双栏正文加侧边辅助栏两大类版面。书中大量绘制精细的构建结构图、节点图体现了本书专业性的同时，也配合着大量精美的照片为本书增色不少。|||||||||||||

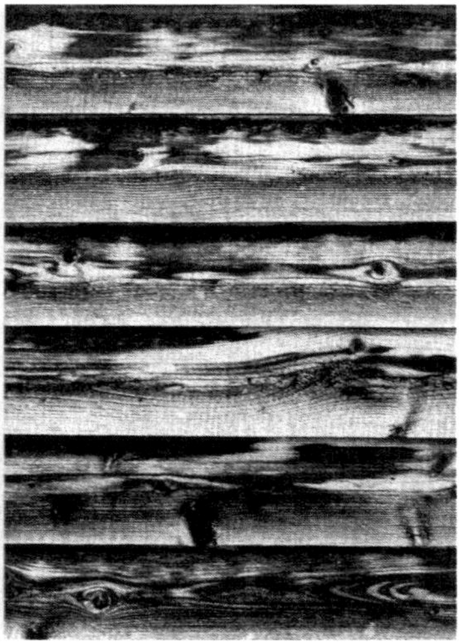

This book is about the construction of wooden exterior walls. It not only displays the appearance, types of material, characteristics and the surface coating, but also studies the performance of keeping warm, waterproof, sound insulation and fire resistance, construction details and specific cases. It is a professional book with high quality standards for engineers, architects and other targeted groups. The hardback book is bound by thread sewing. The cover is made of neutral gray cloth printed in silver and black, directly mounted with graphic board of the same size as the inner pages. The endpaper is made of brown wood-like paper, so that the whole cover feels like two wooden boards clipped to the outside of the book. The inner pages are printed on matte art paper in four colors plus spot-color red. Spot-color red is mainly used for highlighting the title and key details of the text. The text is set to five columns. According to the requirements of the content, two types of layout are used: independent five columns or double-column text plus side auxiliary columns. In the book, a large number of elaborate structural drawings and node drawings reflect the professionalism of the book, while many exquisite photographs make the book more colorful.

铜奖
木质外墙
奥地利

Bronze Medal
Fassaden aus Holz
Austria
Reinhard Gassner, Marcel Bachmann / Atelier Reinhard Gassner
proHolz Austria (Ed.)
proHolz Austria, Wien
296×211×15mm
823g
160p

This specialist book proves that a technically-orientated work based on several research projects and with a target group made up primarily of engineers, contractors and architects can nevertheless have a life of its own. The crispness of the photographs and drawings is no less striking than the layout. The economical use of extra colours within the text (red and grey) enhances the overall elegance. The typography has excellent legibility and is clearly structured right down to the last detail. The endpapers infuse the feeling of wood as does the cloth cover printed with silver and black. Without title or words the cover of this cardboard cut-flush binding tells the story of the great durability of wooden façades.

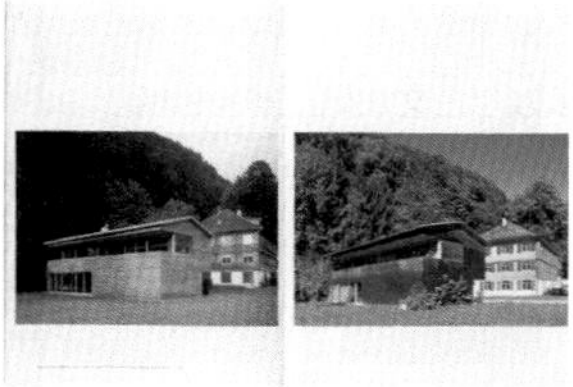

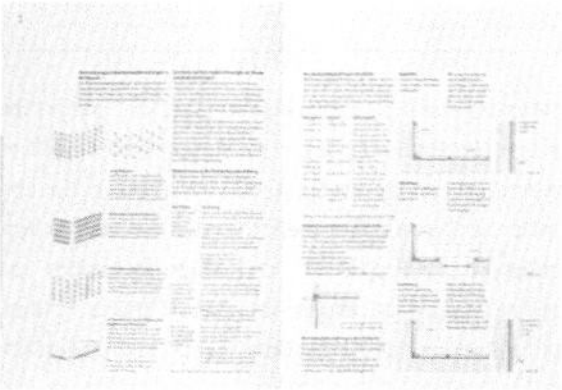
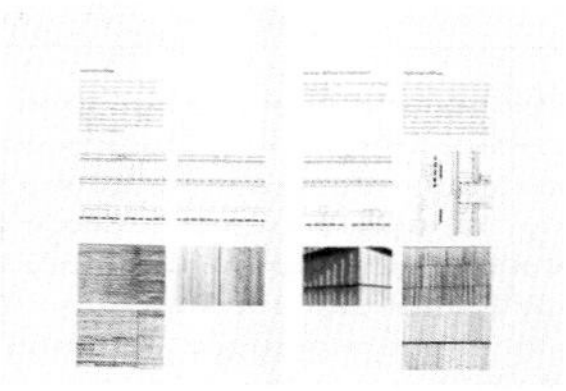
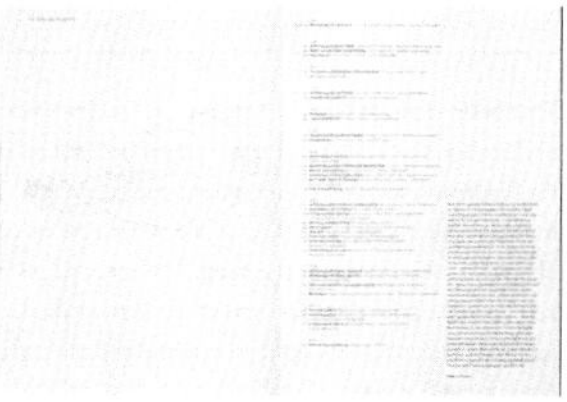
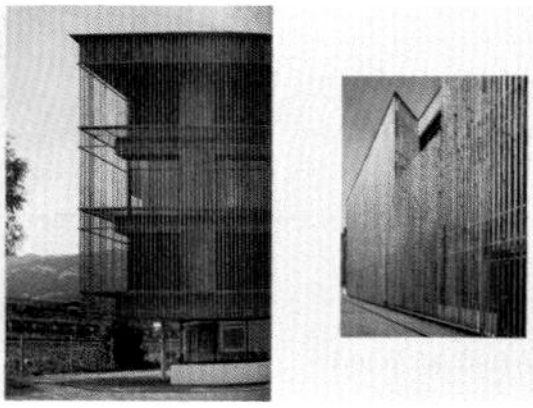
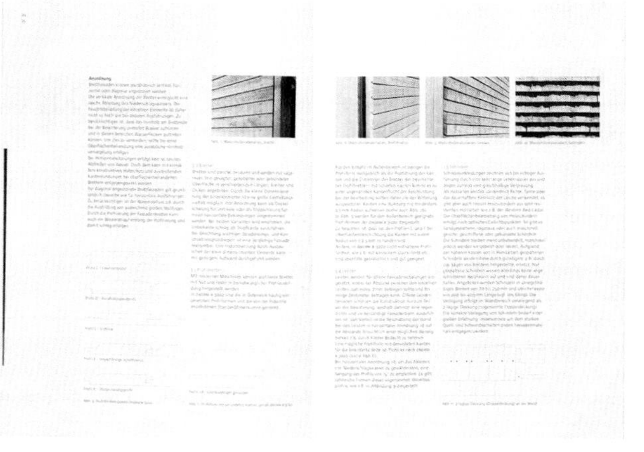

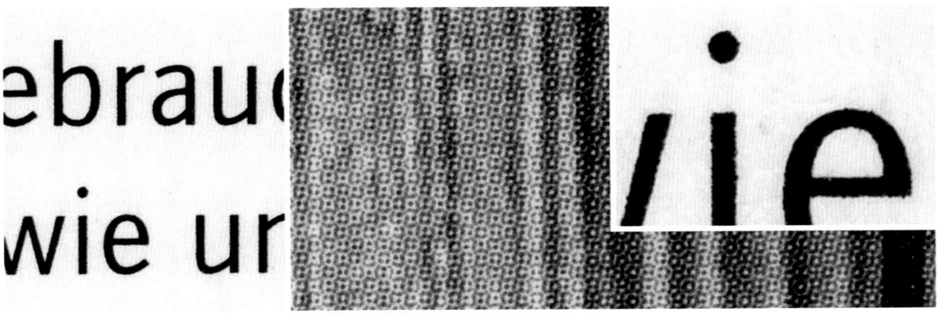
ebrau
wie ur
vie

Eugen Spitaler

3.1 Allgemeines

Holzqualität/Sortierung

Die Holzqualität von Brettern, Profilbrettern und Leisten hat sowohl auf das Erscheinungsbild als auch auf die Dauerhaftigkeit und Beständigkeit von Holzfassaden einen wesentlichen Einfluss. Die Beanspruchung von Holz im Außenbereich ist durch die verschiedensten Witterungseinflüsse sehr hoch. Es ist daher wichtig, dass die Qualitätsanforderungen an Fassadenholz wie z.B. Profilbretter genau definiert und eingehalten werden. Nur so kann bei optimaler Konstruktionsausführung eine qualitativ hochwertige Holzfassade mit minimalem Wartungs- und Instandhaltungsaufwand gewährleistet werden. Beispielhaft werden nachfolgend Möglichkeiten der Sortierung angeführt.

• Normative Mindestanforderungen

Profilbretter für den Einsatz im Wand- und Deckenbereich unterliegen seit Juni 2008 der verpflichtenden CE-Kennzeichnung nach EN 14915. Produkte, die dieser Norm entsprechen, sind am CE-Logo erkennbar.

CE

Abb. 2: CE-Logo

Zu beachten sind Anforderungen an Herstell- und Dimensionstoleranzen, Sortierung, Holzfeuchte etc. nach folgenden Produktnormen:

_ Profilbretter aus Nadelholz nach EN 14519
_ Glattkantbretter ohne Nut und Feder aus Nadelholz nach EN 15146
_ Profilbretter aus Laubholz nach EN 14951

Alle Produktnormen bieten die Möglichkeit, die Sortierung nach Klasse A, Klasse B oder als sogenannte „Freie Klasse" zu vereinbaren. Im jeweiligen Anhang der angeführten Normen ist auch eine Liste mit den für die „Freie Klasse" erforderlichen Sortiermerkmalen zu finden. Dies ermöglicht, spezielle Sortierungen oder Kundenwünsche zwischen den Vertragspartnern zu vereinbaren.

Der Mindestumfang der schriftlich festzulegenden Sortiermerkmale enthält folgende Punkte: Äste, ausgeschlagene Stellen (schadhaft bearbeitete Stellen), Druckholz, Verformungen, Harzgallen, Risse, Markröhre, Farbe, Pilzbefall, Insektenbefall, Baumkante, Rindeneinwuchs, Stapellattenmarkierungen. Geregelt wird in diesen Produktnormen auch die Probennahme in Schiedsfällen. Vor der Montage der Fassade wird empfohlen, die gelieferte Ware zu überprüfen, insbesondere wird auf folgenden Passus hingewiesen: „... Alle Profile sollten vor der Verlegung oder innerhalb von 7 Tagen nach der Lieferung geprüft werden, es gilt der jeweils frühere Termin."

• Sortierung gemäß Sortierrichtlinie des VEH

Der Verband der Europäischen Hobelindustrie (VEH) definiert in seinen Güterichtlinien für gehobelte Profile exakte Sortierbestimmungen für unterschiedliche heimische Nadelholzarten. Bei Hobelwaren für den Einsatz in der Fassade ist, sofern nichts anderes vereinbart, die Sortierung VEH AB als Standard am Markt anzusehen. Höherwertige Sortierungen sind VEH Top und VEH A, welche insbesondere für beschichtete Fassaden zu empfehlen sind.

• Empfehlung der Holzforschung Austria

Vollholzbretter werden sowohl beschichtet als auch unbehandelt im Fassadenbau eingesetzt. Dabei sind die Anforderungen an die Sortierung unterschiedlich. Aus diesem Grund werden von der Holzforschung Austria zusätzliche Qualitätsanforderungen für Massivholzbretter unter Berücksichtigung von Forschungsergebnissen in Tabelle 2 empfohlen, welche auf besonderen Kundenwunsch und gegen Aufpreis zu vereinbaren sind.

Merkmale	**Beschichtete Bretter**	**Unbehandelte Bretter**
Astgröße	1/4 der Profilbrettbreite	1/4 der Profilbrettbreite
Eingewachsene, lose und ausgefallene Äste, Rindeneinwuchs	nicht zulässig	nicht zulässig
Ausbesserungen mit Ast- und Hirnholzdübeln	nicht zulässig	nicht zulässig
Harzgallen	nicht zulässig	kleine Harzgallen zulässig [1)]
Risse	nicht zulässig	Haarrisse zulässig [2)]
Mark	nicht zulässig	eingeschränkt zulässig [3)]
Buchs	bis 20 % des Querschnitts bzw. der Oberfläche zulässig	bis 20 % des Querschnitts bzw. der Oberfläche zulässig
Pilz- und Insektenbefall	nicht zulässig	nicht zulässig

1) bis 5 mm Breite und 50 mm Länge
2) feine Risse, die nur an der Oberfläche durch Schwindspannungen des Holzes entstehen
3) bei 15 % der Ware, an der Sichtfläche bis maximal 15 % der Brettlänge und 4 mm Breite

Tabelle 2: Empfohlene Qualitätsanforderungen an Profilbretter für den Einsatz im Außenbereich

Dimensionen und Ausführung

Holz als hygroskopischer Stoff nimmt aus dem Umgebungsklima Feuchtigkeit auf oder gibt Feuchtigkeit ab. Diese Aufnahme bzw. Abgabe von Wasser führt zum Quellen bzw. Schwinden von Holz. Das Ausmaß dieser Verformungen ist umso größer, je größer die Dimension des Holzstücks ist. Fassadenbretter dürfen daher nicht in jeder beliebigen Abmessung ausgeführt werden. Die maximal zulässige Breite ist abhängig von der Profilausführung, der Befestigung, der Oberflächenbehandlung und der Witterungsbeanspruchung am Einsatzort. Bei Nut-Feder-Profilen ist eine maximale Breite von 150 mm nicht zu überschreiten. Bei breiten Profilen besteht die Gefahr, dass es aufgrund großer Verformungen einzelner Bretter zu Rissen im Federbereich oder zum Aufgehen der Nut-Feder-Verbindung kommt. Bei extrem witterungsbeanspruchten Fassaden mit häufigem Wechsel von Durchfeuchtung und intensiver Sonnenbestrahlung wird daher eine maximale Breite der Profilbretter von 120 mm, speziell auch für beschichtete Fassadenbretter, empfohlen. Hinsichtlich der Dicke werden aus Gründen höherer Stabilität mindestens 19 mm empfohlen. Dickere Fassadenbretter sind möglich, bringen aber keine Verbesserung hinsichtlich möglicher Verformungen.

Keilzinkung

Durch die Keilzinkung wird die Fertigung von Vollholzbrettern ohne störende Holzmerkmale (wie z.B. Harzgallen, große Äste) ermöglicht. Dem Aufwand der Keilzinkung ist der Vorteil einer sehr hochwertigen Ware gegenüberzustellen. Auf Kundenwunsch ist es z.B. möglich, astfreie Ware in nahezu beliebigen Längen herzustellen.
Als Zinkenlänge werden 10 mm empfohlen. Eine Lücke zwischen Zinken und Zinkengrund (= Zinkenspiel) ist nicht zulässig. Im unmittelbaren Zinkenbereich darf kein Ast vorhanden sein, d.h. es ist ein gerader Faserverlauf vor der Zinke erforderlich. Für die Einzelteile darf nur Holz gleicher Holzart mit ähnlicher Struktur verwendet werden. Die Kernseite aller Einzelteile muss auf der gleichen Seite des keilgezinkten Brettes liegen. Zur Verklebung ist ein Klebstoff vom Typ I nach EN 301 oder ein gleichwertiger Klebstoff zu verwenden.

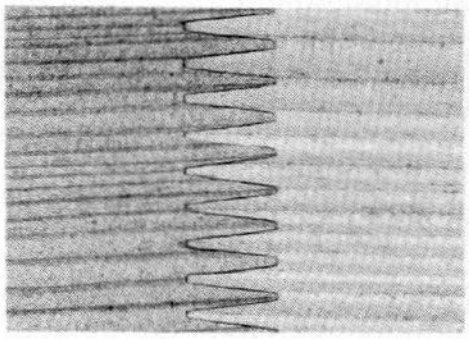

Abb. 3: Lärchenprofilbrett mit Keilzinkenverbindung ohne Zinkenspiel

Insbesondere mit Beschichtung und bei Verwendung schmaler Bretter stellt diese Fassadenvariante eine sehr dauerhafte Lösung dar. Aufgrund des hochwertigen Materials ist eine nicht sichtbare Befestigung (z.B. von der Rückseite) zu empfehlen.

Jahrringlage

Hinsichtlich der Jahrringlage ist die Verwendung von gemischter Ware bestehend aus Riftbrettern mit stehenden Jahrringen, Halbriftbrettern sowie Fladerbrettern mit liegenden Jahrringen üblich. Bei direkter Bewitterung kann es bei Fladerbrettern zur teilweisen Ablösung des Fladers an der Oberfläche und somit zu Beschichtungsschäden kommen. Auch treten bei Fladerbrettern weitaus häufiger Risse auf als bei Rift- und Halbriftbrettern; deshalb werden für bewitterte Fassaden letztere empfohlen.
Diese Qualität zu garantieren, stellt jedoch einen erheblichen Mehraufwand bei der Produktion dar und ist mit deutlichen Mehrkosten verbunden.
Generell sollten Profilbretter für den Außenbereich mit ihrer markzugewandten Seite („rechten Seite") nach außen montiert werden (siehe Abb. 4). Dadurch treten wesentlich geringere Verformungen und Risse im Einsatz auf.

linke Seite oben

rechte Seite oben

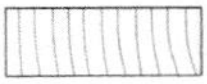

Riftbrett

Halbriftbrett

Brett mit liegenden Jahrringe (Fladerbrett)

Abb. 4: Jahrringlagen

Holzfeuchtigkeit

Die Holzfeuchtigkeit bei eingebauten Fassaden liegt in Abhängigkeit von Standort, Exposition, Jahreszeit etc. im Jahresmittel zwischen 12 % und 16 %. Dieser Bereich sollte daher beim Einbau weder unter- noch überschritten werden.

	Beschichtete Bretter	Unbehandelte Bretter
Breite	max. 150 mm (120 mm [1) 2)])	max. 150 mm (120 mm [2)])
Dicke	19 mm	19 mm
Kantenradius	min. 2,5 mm	–
Empfohlene Jahrringlage	Rift/Halbriftbretter	–
Außenseite	markzugewandte Seite [3)]	markzugewandte Seite [3)]
Holzfeuchte	12 % – 16 %	12 % – 16 %

1) für extrem witterungsbeanspruchte Fassaden
2) speziell für bewitterte Fassaden mit Profilbrettern aus Laubholz
3) speziell für Profilbretter mit liegenden Jahrringen

Tabelle 3: Empfohlene Dimensionen und Bearbeitung von Brettern für den Einsatz im Außenbereich

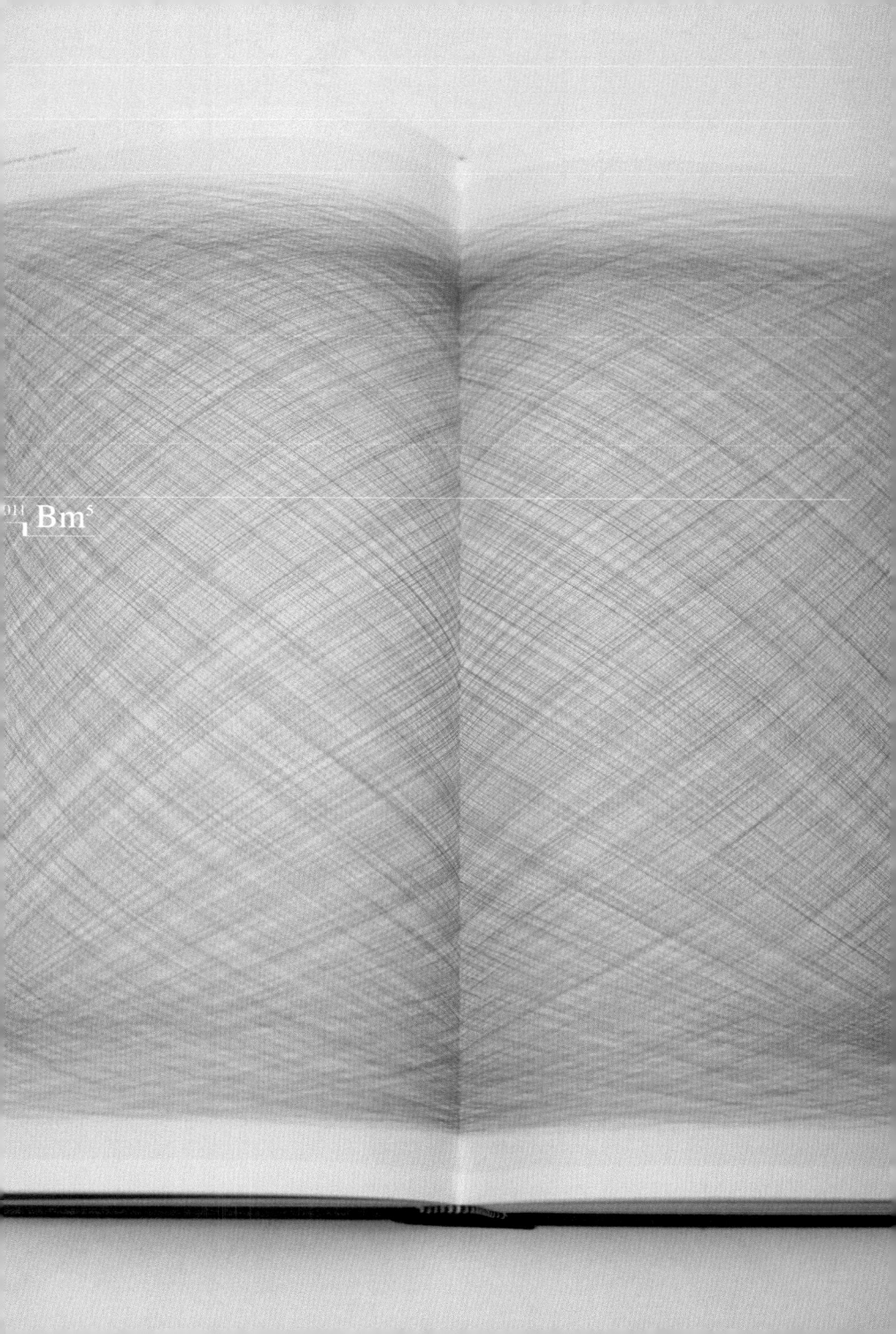

这是一本对地球进行细致观测、研究进而规划未来发展的书籍。本书包含作者收集和研究宇宙、空间、地球的大量观测数据和图片，并将实验和艺术相结合，对未来地球的发展和探索展开了大胆的想象。本书中提到的“中空集群”（Globus Cassus）概念是本书作者克里斯蒂安·瓦尔德福格尔（Christian Waldvogel）在 2004 年威尼斯双年展的参展项目，同时书籍也获得了 2005 年“世界最美的书”金奖（可在本书相关页面查询）。本书为锁线硬精装，封面黑色布面丝网印深灰色后烫银，裱覆卡板。内页正文部分为胶版纸四色印刷，英德双语，内文排版遵循中心文字最大、往上下两侧逐渐减小的原则，模拟出一种中心凸起、两侧渐弱的弧面效果，原理类似书中 P140-141 页插图（见左页本书的细节图）。附录部分光面铜版纸四色印刷，包含相关资料和人物介绍。

Christian Waldvogel
EARTH
EXTREMES
With an essay by
Mit einem Essay von
Jörg Heiser
Edited by
Herausgegeben von
Jacqueline Burckhardt,
Christian Waldvogel,
and Jonas Voegeli
Scheidegger & Spiess

This is a book of detailed observations and studies of the earth in order to develop a vision for the future. It contains a large number of data and pictures of the universe, space and the earth collected and studied by the author. The book combines experiments with art to launch a bold imagination of the future development and exploration of the Earth. The concept of 'Globus Cassus' mentioned in this book is the author Christian Waldvogel's exhibition project in the 2004 Venice Biennale, which won 'the Best Book Design from all over the World Award' in 2005. This hardback book is bound with thread sewing. The cover is made of black cloth in dark grey with silver hot stamping, mounted by graphic board. The inside pages are printed on offset paper in four colors in both English and German. The page typesetting follows the principle that the central text is the largest, and gradually decreases to simulate an arc. The principle has been illustrated on the page 140 to 141. The appendix contains relevant information and introduction of characters, printed on gloss art paper in four colors.

- 铜奖
- 克里斯蒂安·瓦尔德福格尔的极致地球
- 瑞士

- Bronze Medal
- Christian Waldvogel. Earth Extremes
- Switzerland
- Jonas Voegeli, Christian Waldvogel, Benjamin Roffler
- Jacqueline Burckhardt, Christian Waldvogel, Jonas Voegeli (Ed.)
- Verlag Scheidegger & Spiess AG, Zürich
- 320×214×41mm
- 2283g
- 495p

A large black book, its subject everything: the universe, stagnancy, space and time. The material is approached gently – first darkness, then small letters which become larger, like a zoom from far away. The typography is imposing, the texts, photographs, graphics and illustrations acquire a wonderful resonance. The book is light and heavy all in one. Texts and pictures are ordered so as to allow space and contemplation. Then we fall back into the world of illustration. With an obsession for detail the typography succeeds in uniting two languages on one page. Leafing through the book is a pleasure, forwards, backwards, forwards again – not to mention the change of paper between essay and appendix. If only it wouldn't stop! So back we go again, forwards, backwards! Into the universe, into the artist's imaginary worlds, into the worlds of art and science.

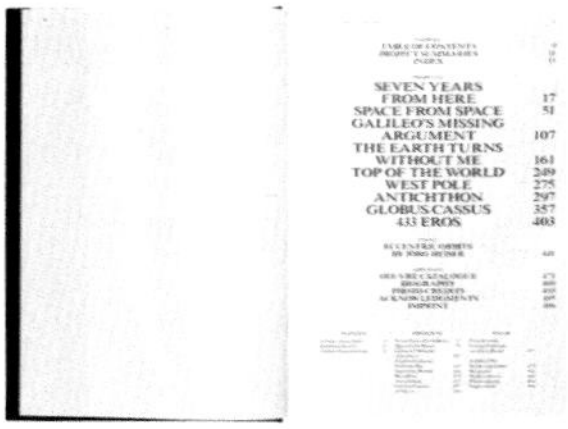
SEVEN YEARS FROM HERE 17
SPACE FROM SPACE 51
GALILEO'S MISSING ARGUMENT 107
THE EARTH TURNS WITHOUT ME 161
TOP OF THE WORLD 249
WEST POLE 275
ANTICHTHON 297
GLOBUS CASSUS 357
433 EROS 403

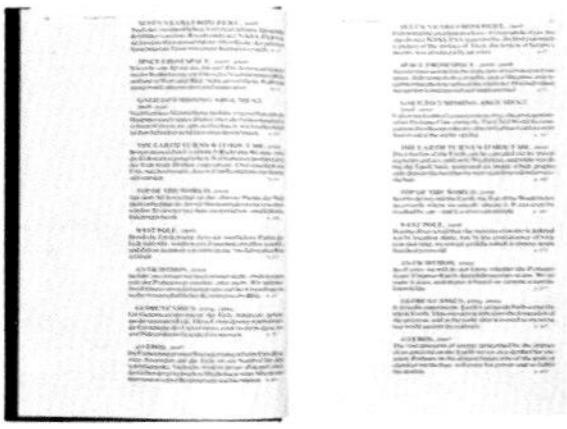

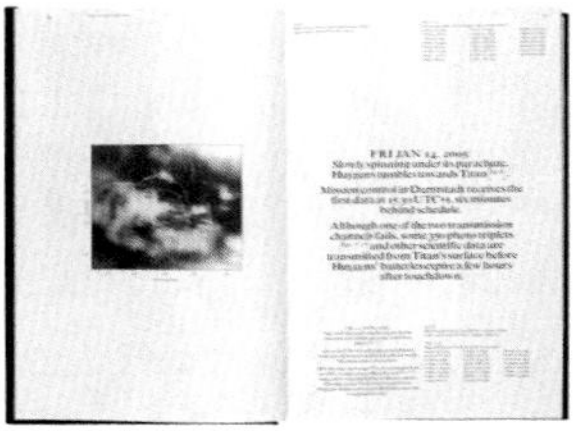

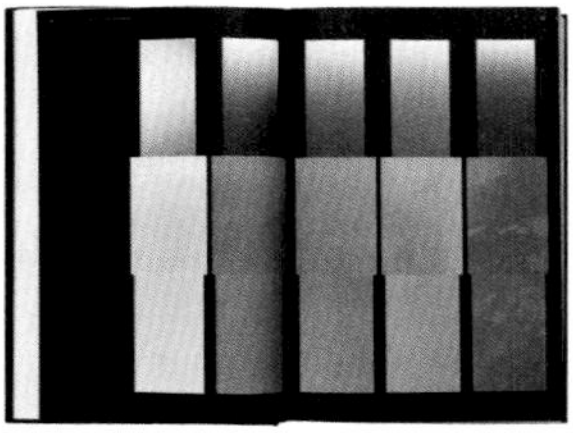

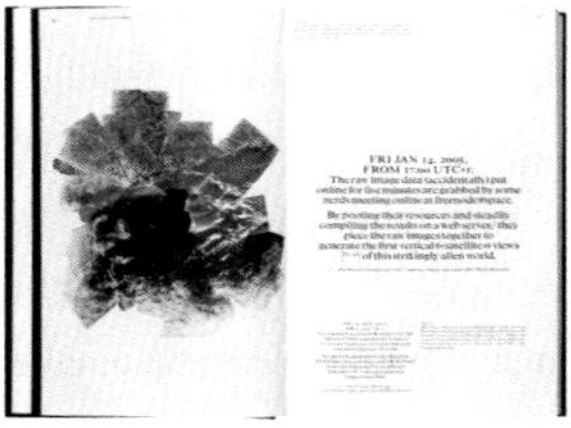

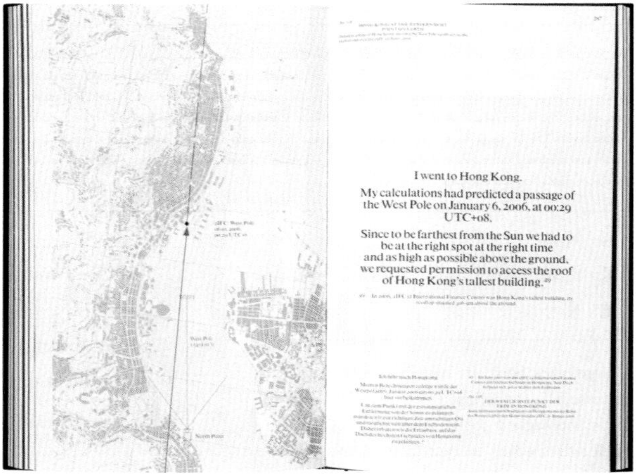
I went to Hong Kong.
My calculations had predicted a passage of the West Pole on January 6, 2006, at 00:29 UTC+08.
Since to be farthest from the Sun we had to be at the right spot at the right time and as high as possible above the ground, we requested permission to access the roof of Hong Kong's tallest building.49

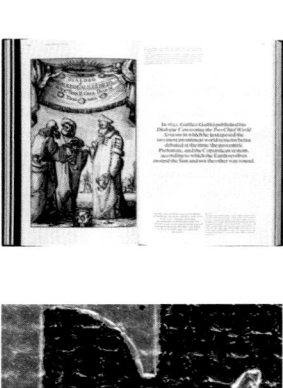

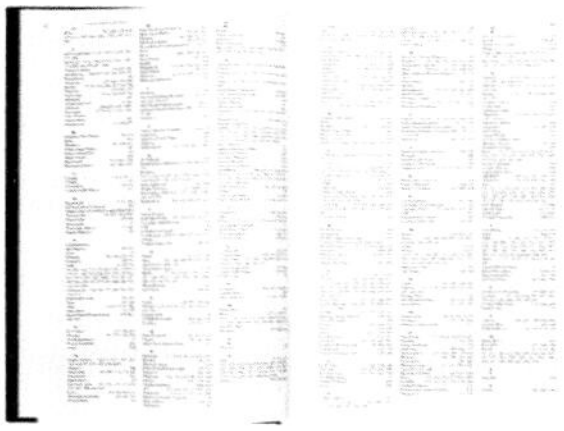
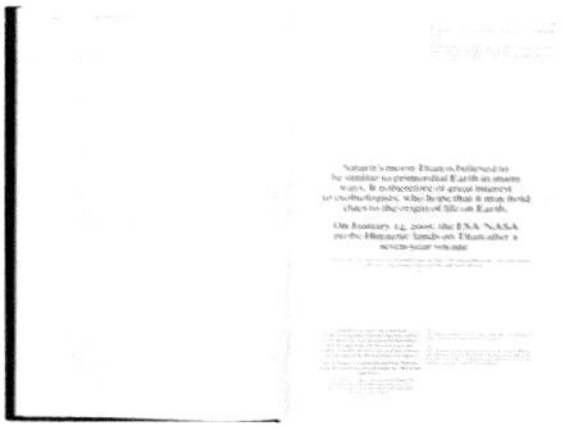

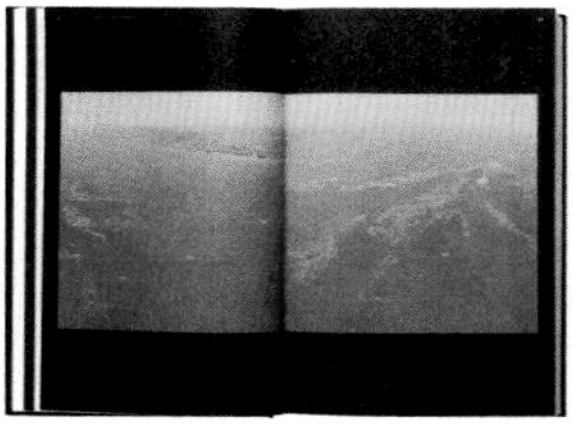
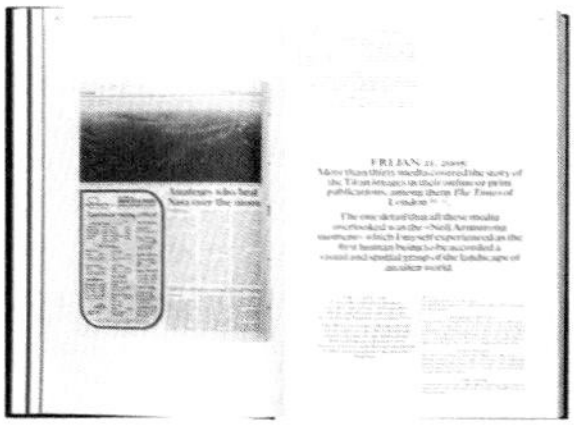

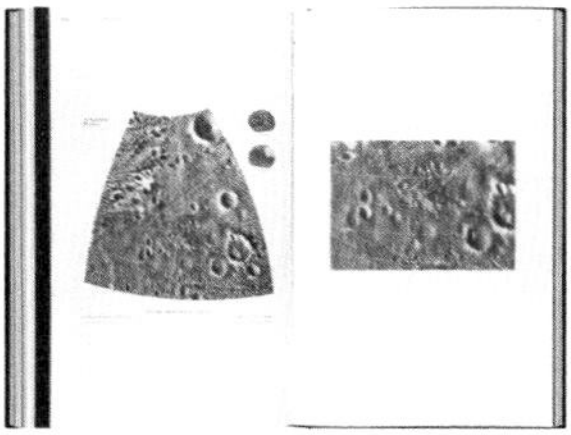
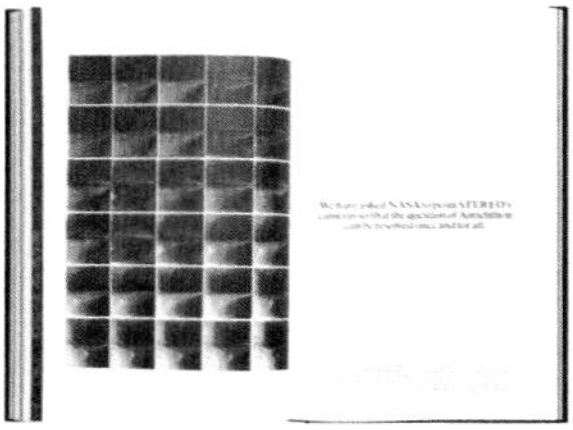
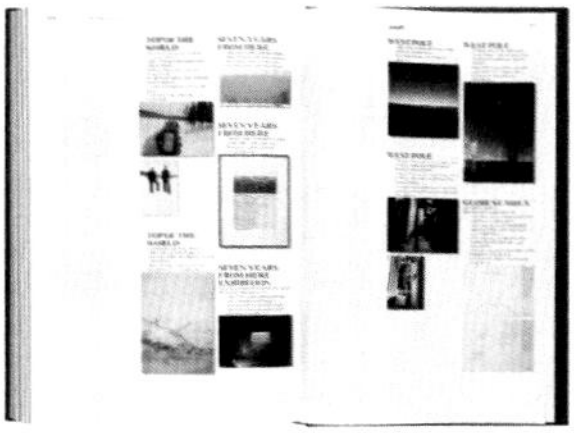

des
rien ein paar

Christian Waldvogel
EARTH
EXTREMES
With an essay by / Mit einem Essay von Jörg Heiser
Edited by / Herausgegeben von
Jacqueline Burckhardt, Christian Waldvogel,
and Jonas Voegeli
Scheidegger & Spiess

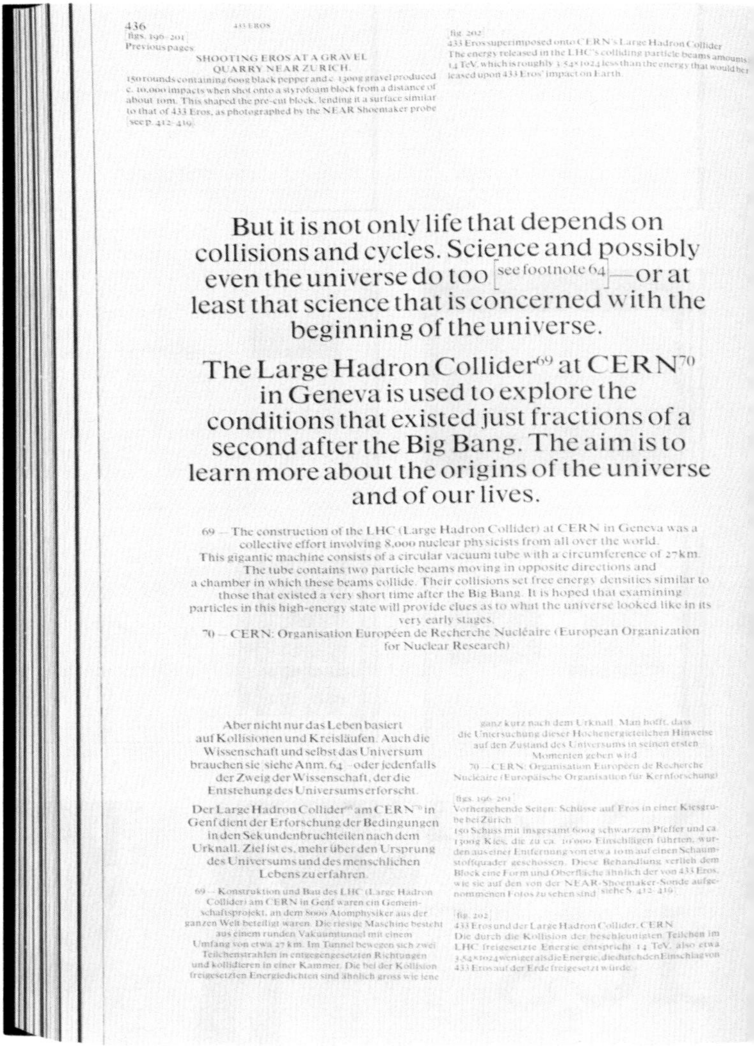

436 433 EROS

[figs. 196–201]
Previous pages:
SHOOTING EROS AT A GRAVEL QUARRY NEAR ZURICH.
150 rounds containing 600g black pepper and c. 1300g gravel produced c. 10,000 impacts when shot onto a styrofoam block from a distance of about 10m. This shaped the pre-cut block, lending it a surface similar to that of 433 Eros, as photographed by the NEAR Shoemaker probe [see p. 412–419].

[fig. 202]
433 Eros superimposed onto CERN's Large Hadron Collider
The energy released in the LHC's colliding particle beams amounts 14 TeV, which is roughly 3.54×1024 less than the energy that would be released upon 433 Eros' impact on Earth.

But it is not only life that depends on collisions and cycles. Science and possibly even the universe do too [see footnote 64]—or at least that science that is concerned with the beginning of the universe.

The Large Hadron Collider[69] at CERN[70] in Geneva is used to explore the conditions that existed just fractions of a second after the Big Bang. The aim is to learn more about the origins of the universe and of our lives.

69 — The construction of the LHC (Large Hadron Collider) at CERN in Geneva was a collective effort involving 8,000 nuclear physicists from all over the world. This gigantic machine consists of a circular vacuum tube with a circumference of 27km. The tube contains two particle beams moving in opposite directions and a chamber in which these beams collide. Their collisions set free energy densities similar to those that existed a very short time after the Big Bang. It is hoped that examining particles in this high-energy state will provide clues as to what the universe looked like in its very early stages.

70 — CERN: Organisation Européen de Recherche Nucléaire (European Organization for Nuclear Research)

Aber nicht nur das Leben basiert auf Kollisionen und Kreisläufen. Auch die Wissenschaft und selbst das Universum brauchen sie [siehe Anm. 64]—oder jedenfalls der Zweig der Wissenschaft, der die Entstehung des Universums erforscht.

Der Large Hadron Collider[69] am CERN[70] in Genf dient der Erforschung der Bedingungen in den Sekundenbruchteilen nach dem Urknall. Ziel ist es, mehr über den Ursprung des Universums und des menschlichen Lebens zu erfahren.

69 — Konstruktion und Bau des LHC (Large Hadron Collider) am CERN in Genf waren ein Gemeinschaftsprojekt, an dem 8000 Atomphysiker aus der ganzen Welt beteiligt waren. Die riesige Maschine besteht aus einem runden Vakuumtunnel mit einem Umfang von etwa 27 km. Im Tunnel bewegen sich zwei Teilchenstrahlen in entgegengesetzten Richtungen und kollidieren in einer Kammer. Die bei der Kollision freigesetzten Energiedichten sind ähnlich gross wie jene ganz kurz nach dem Urknall. Man hofft, dass die Untersuchung dieser Hochenergieteilchen Hinweise auf den Zustand des Universums in seinen ersten Momenten geben wird.

70 — CERN: Organisation Européen de Recherche Nucléaire (Europäische Organisation für Kernforschung)

[figs. 196–201]
Vorhergehende Seiten: Schüsse auf Eros in einer Kiesgrube bei Zürich
150 Schuss mit insgesamt 600g schwarzem Pfeffer und ca. 1300g Kies, die zu ca. 10'000 Einschlägen führten, wurden aus einer Entfernung von etwa 10m auf einen Schaumstoffquader geschossen. Diese Behandlung verlieh dem Block eine Form und Oberfläche ähnlich der von 433 Eros, wie sie auf den von der NEAR-Shoemaker-Sonde aufgenommenen Fotos zu sehen sind [siehe S. 412–419].

[fig. 202]
433 Eros und der Large Hadron Collider, CERN
Die durch die Kollision der beschleunigten Teilchen im LHC freigesetzte Energie entspricht 14 TeV, also etwa 3.54×1024 weniger als die Energie, die durch den Einschlag von 433 Eros auf der Erde freigesetzt würde.

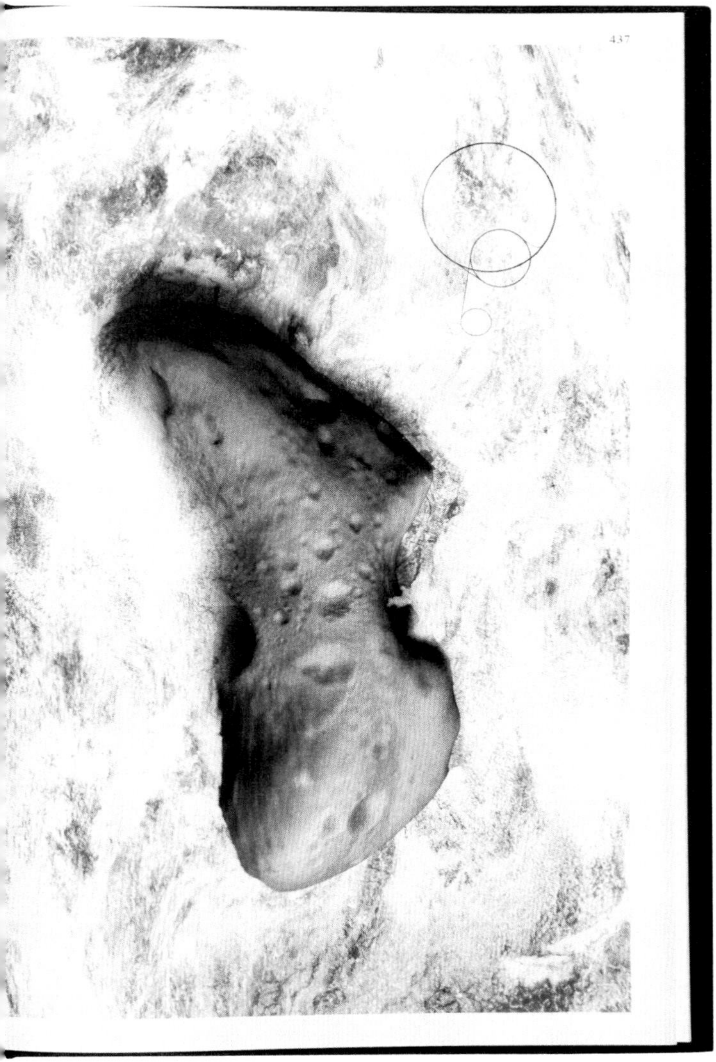

011 Ha1

这是 2009 年深圳 · 香港城市 / 建筑双年展的参展项目，也是建筑旅游项目“漫游”（Odyssey）的记录。在阅读的过程中，读者在著名建筑师的建筑和小说家的文字间跨越时空的瞬间转移，本身也是一次独特的建筑之旅。本书的书函就充满了多道工艺，牛皮卡纸外贴单黑印刷的白色纸张，为书籍信息，后烫印黑色。书籍由于各种尺寸和类别的纸张的使用，附加多种印刷工艺，装订上采用了逐页手工排序后的无线胶装，书脊部分裱贴卡纸加固。书中包含当代具有实验意义的 9 个建筑项目，展示照片、图纸、信息和小说家“漫游”后的中英文双语文字。内容严格按照类别划分纸张和工艺，层次分明。全书虽然围绕建筑，但具有非常良好的翻阅手感。遍布书籍的连续图案类似建筑表皮的连续结构，形成了丰富的文本肌理。

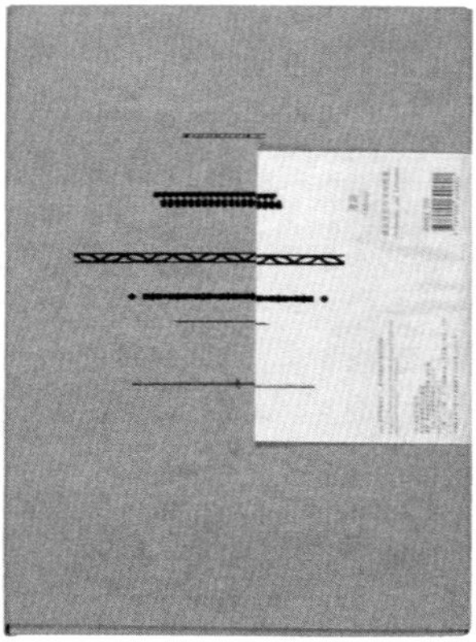

This book is an exhibition project of Shenzhen & Hong Kong Biennale of Architecture in 2009. It is also the record of 'Odyssey', an architectural tourism project. Even in the process of reading, the readers can experience instantaneous transfer between the famous architects' architecture and the novelist's literature beyond time and space. The book's slipcase is produced with several technologies – kraft cardboard is pasted by white paper printed with book information in black, and then hot stamped in black. Due to the use of various sizes and types of paper and various printing processes, the book is perfect bound after page-by-page manual sorting, and the spine is strengthened by pasting cardboard. The book contains nine contemporary architectural projects with experimental significance, and displays photographs, drawings, information and Chinese-English bilingual text after the novelist's 'odyssey'. The entire book revolves around architecture and the continuous patterns throughout the books are similar to the architectural surface.

- 荣誉奖
- 漫游：建筑体验与文学想象
- 中国
- 小马哥、橙子
- 欧宁
- 中国青年出版社

- Honorary Appreciation
- Odyssey : Architecture and Literature
- China
- Xiao Mage & Cheng Zi
- Ou Ning (Ed.)
- China Youth Press, Beijing
- 240 × 170 × 22mm
- 792g
- 448p

This is a very unconventional solution for a Chinese/English book on urban architecture. It has been meticulously designed and produced to create an accessible product that contains a very comprehensive content structure. The layout is clear and easy to follow and the designer has incorporated an interesting selection of typefaces of varying degree to successfully highlight elements of each of the nine architectural projects being presented. The typography is understated yet works to strengthen the graphic aesthetic. Beautiful stock and format choice – the frequent changes of surface and format give the book diversity and provide the texts and documentation with an appropriate form of expression.

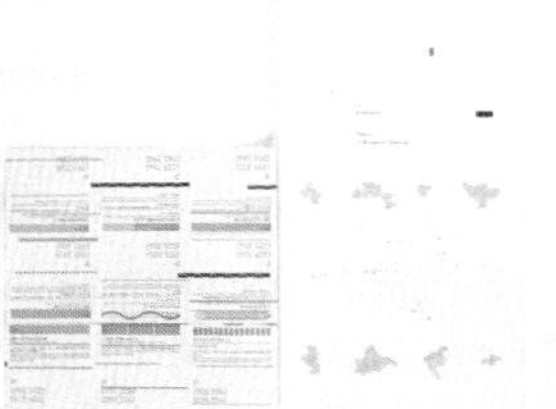

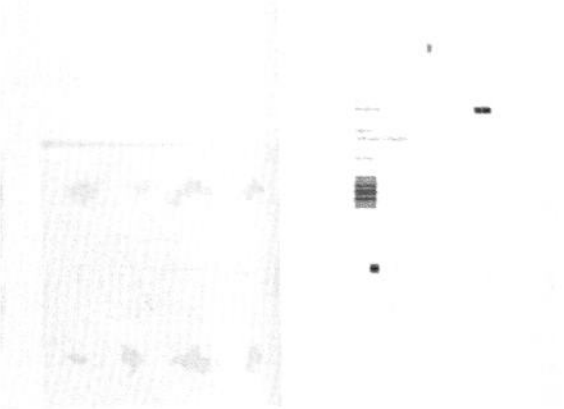

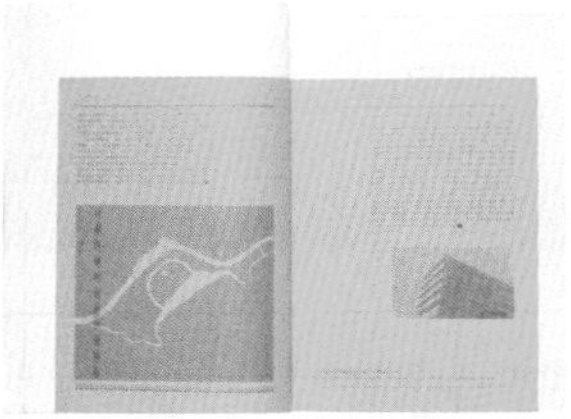

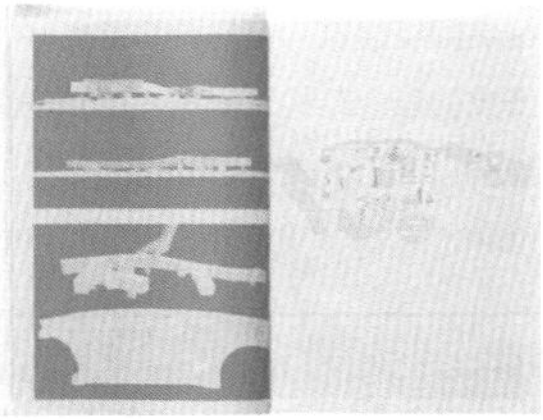

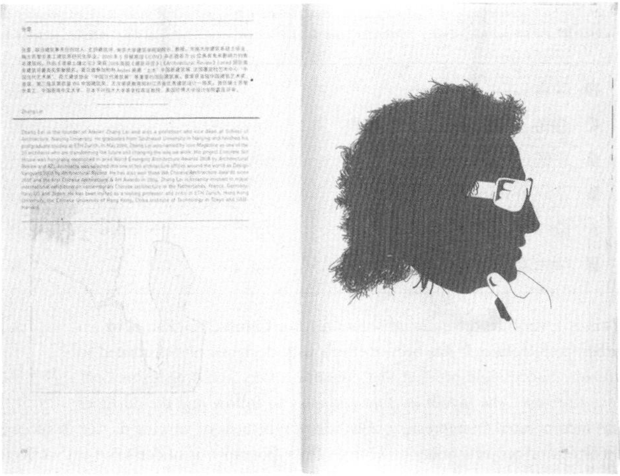

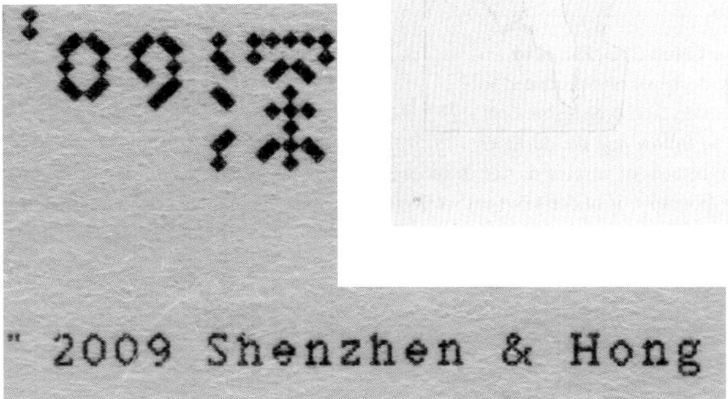
'09深
" 2009 Shenzhen & Hong

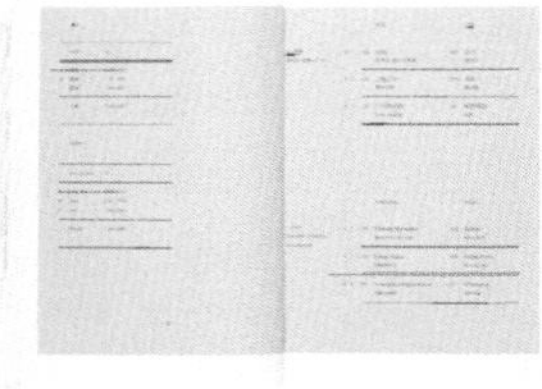

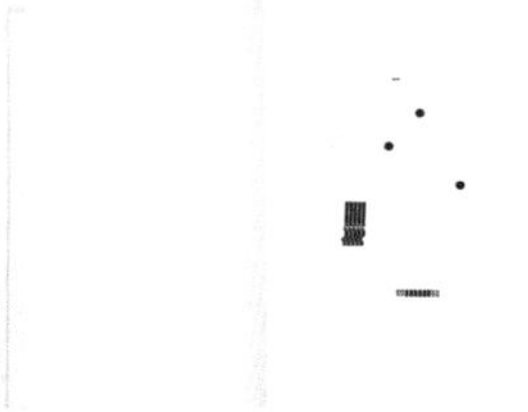

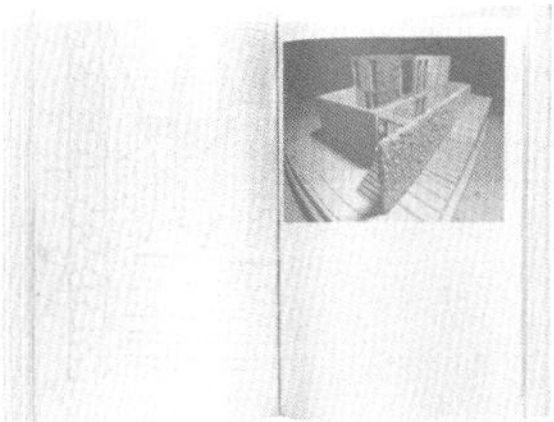

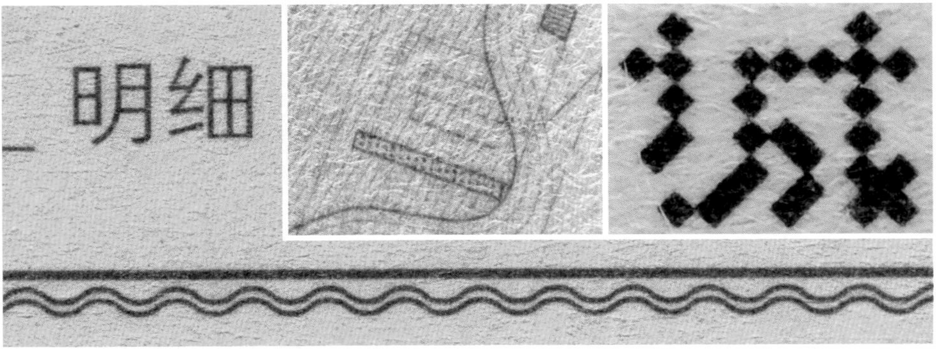
明细

“父亲宅”是马清运为自己父亲在老家设计建造的。它位于西安市东南方的蓝田县玉山镇。蓝田是蓝田猿人的古迹，而玉山又是唐代大诗人王维自己建造的辋川别业的地方。蓝田玉山和秦岭山脉作为“父亲宅”的背景，形成了整个区域内巨大的景观变化，从陡峭的山峰到和缓的坡地，河谷甚至延伸到一个资源丰盛的中部平原。“父亲宅”坐落在河与山之间一个多重含义的地理位置，河水冲刷出光滑的石头，山谷产出粗糙的石头。年复一年，河水将石头从山体层阶间冲刷下来，提供了建筑材料上的丰富资源。水和石之间的关系是如此精确，以至于石头的尺寸与它的颜色，还有它被流经河水所冲刷的程度都精确相关，这反映出石头丰富的质地。由石头建造的房子因此在光和影的浓缩、粗糙和光滑的密度间游离。将石头在质地和建造方法之间的作用最大化是设计的首要原则。对待石头的不同方法取决于它在哪里出现在理论说教的框架内。这种猛烈碰撞的有机物质经过排列和节减，给了“父亲宅”一个短暂的质地生命，包含在显著的现代形式主义中。■

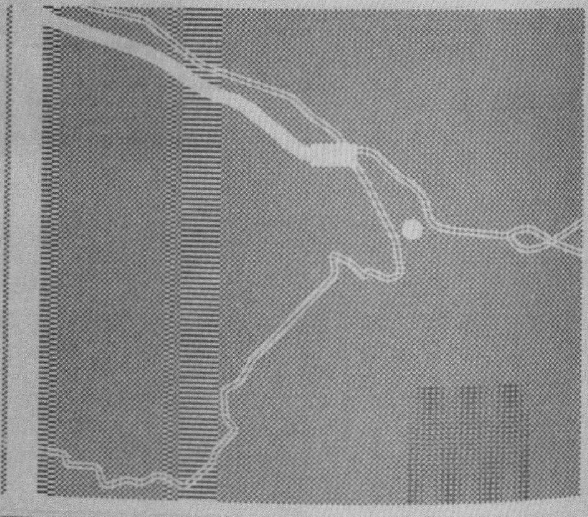

铁栅栏门

_ Project Introduction _ 275

This is a house that Qingyun Ma designed and built for his father on his homeland. Jade Valley and Qingling Mountain Range is its backdrop which defines the whole territory with a vastly changing landscape, from steep mountain to mild slope and to river valley and evens out into a infinitely expansive plateau called Middle Plateau. Father's House rests in the ambiguous position between the river (and smoothed stones) and the mountains (and coarse stones). Over time, the river pulls rock out of layers of mountain, offering an abundant source of building material. The relation between water and stone is so accurate that the size of stone correlates precisely to its color and level of perturbation by the passing water. Reflecting the richness in stone variety, the house panels oscillate between concentrations of dark and light hues, rough and smooth textures. To maximize the stone application both in quantity and local construction methodology is the first and foremost principle of the design. To simplify and homogenize the construction is to minimize the non-local construction technique and mobilization. Therefore, the use of the alien (concrete) remains a minimal and distinct element whereas the use of the local (stone and wood) is maximized. The stone is treated differently depending on where it occurs within the didactic realm of the concrete frame. This collision of rough, organic materials with highly regulated and spare form gives the house an ephemeral quality encased in distinctly modern formalism. ■

陕西省西安市蓝田县玉山镇
Yushan Town, Lantian County, Xi'an, Shaanxi Province

konnte nicht begreifen, was es heißen
mir! Mir gruselt´s nicht: das wird wohl e
Nun geschah es, dass der Vater einm
wirst groß und stark, du musst auch etwas
du, wie dein Bruder sich Mühe gibt, aber an d
antwortete er, »ich will gerne was lernen; ja,
mir´s gruselte; davon verstehe ich noch gar nich
und dachte bei sich: »Du lieber Gott, was ist mein B
Lebtag nichts: was ein Häckchen werden will, muss
seufzte und antwortete ihm: »Das Gruseln, d
wirst du damit nicht verdienen.«
Bald danach kam der Küst
Not und erzählte, wie s
er wüsste nicht

011 Ha²

这是一本对 1857 年版《格林童话》收录的同名故事进行现代演绎的插画书。本书的主体部分装订形式为锁线胶平装，全书红蓝黑三色丝网印刷，并通过色彩之间的相互叠加，形成深红、浅红、深蓝、浅蓝、蓝黑五色。封面和内页均采用丝网版画用胶版卡纸，只在纸张克数上有些许差异。书籍后部夹带骑马钉小书，胶版纸宝蓝色单色印刷。这本骑马钉小书呈现的是原版童话故事。原版故事讲述了一个天生不知道什么是害怕的傻小子想学会害怕的故事。在故事中因为他的这项特殊“技能”闯了不少的祸，也解开了让很多人丧命的“魔宫”的秘密，因此国王将公主嫁给他，最终公主将一桶虾虎鱼浇在睡梦中的傻小子身上，他从梦中惊醒，知道了什么叫害怕。但本书的主要部分并没有根据原版故事描述的场景、人物来绘制插画，而是进行了现代演绎。封面一个身穿现代着装的傻小子单手握着一条不知道是什么的毛茸茸的绳子，其他人都在往他身后跑，书名也只印了前半段。这条绳子一直延续到内页部分才能看到，原来是一条恶犬的尾巴，书名的后半部也在此呈现。书中展现了高台跳水、笼中斗狮、火场救人、斗殴、战争等场景，更容易让生活在现代的人们感同身受。||||||

DORIS FREIGOFAS
VON EINEM DER AUSZOG...

This is an illustrated book of a story in the 1857 edition of *Green's Fairy Tales*, created with the contemporary interpretation. The main part of the book is glue bound with thread sewing, printed in red, blue and black with silk screen. Through the superposition of the colors, five colors – dark red, light red, dark blue, light blue and blue black – are created. The cover and inner pages are all made of offset cardboard. There are slight differences in the grams of paper. A saddle stitched little book is attached to the back of the book and printed on offset paper in blue. The little book presents the original fairy tales, which tells a story about a silly boy who was born to be fearless and wanted to learn how to be afraid. In the story, due to his special 'skill', he caused many disasters and revealed the secret of the 'Magic Palace' that killed many people. The king married the princess to him, and eventually the princess poured a pail of shrimp and tiger fish on him when he was asleep. He woke up from his dream and knew what fear was. However, the main body of the book does not draw illustrations according to the scenes and characters described in the original story, but carries out modern interpretation. On the cover, a silly young man in modern clothes is holding a hairy rope in one hand while others were running behind him. Only the first half of the title was printed. On the inside page, we may find out that the hairy rope is a dog's tail. The book shows such scenes as platform diving, lion fighting in cage, fire rescue, fighting, wars and so on. It is easier for people living in modern times to have common feelings.

- 荣誉奖
- 傻小子学害怕
- 德国

With just three colours the artist conjures up a new interpretation of the well-known Grimm's fairy tale in large-format, extensive pictures, thus re-inventing the tradition of telling fairy tales in an unusual manner. For this she has no need of text – the pictures tell their own story from double-page spread to double-page spread in a narrative, powerful form which still leaves scope for individual interpretation. An extra booklet containing the original text is attached into the jacket flap at the end of this picture book.

- Honorary Appreciation
- Von einem, der auszog das Fürchten zu lernen
- Germany
- Doris Freigofas / Golden Cosmos
- Doris Freigofas
- Golden Cosmos, Berlin
- 370×243×4mm
- 304g+24g
- 24p+8p

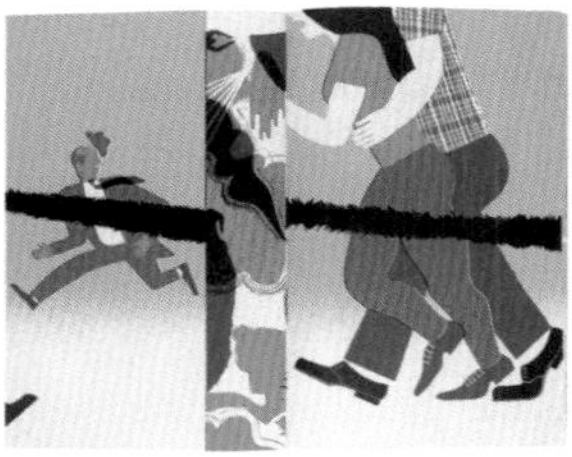

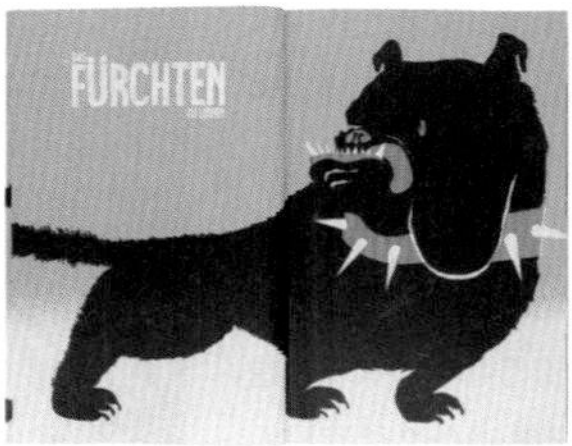
FÜRCHTEN

TARAK

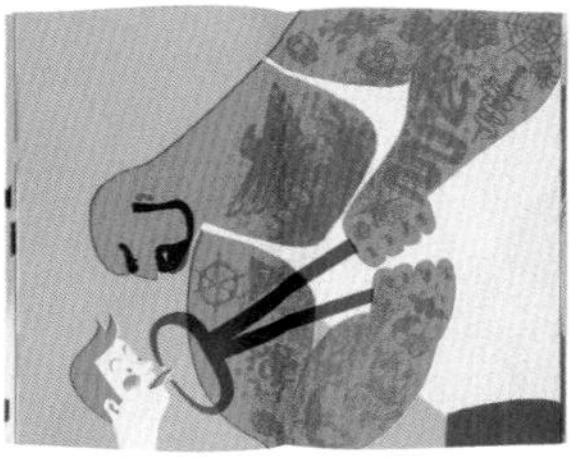

STRIKE!

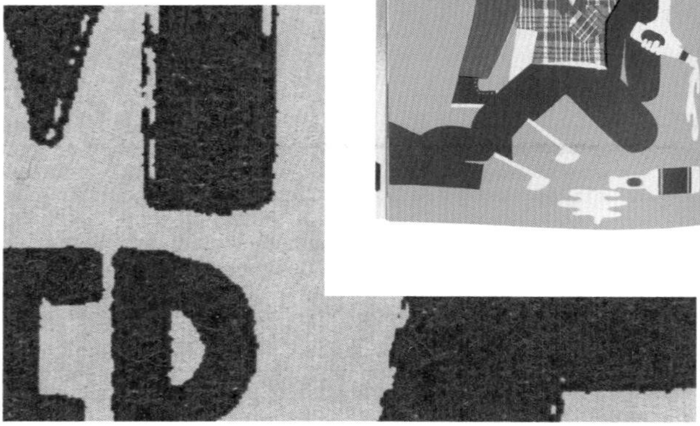

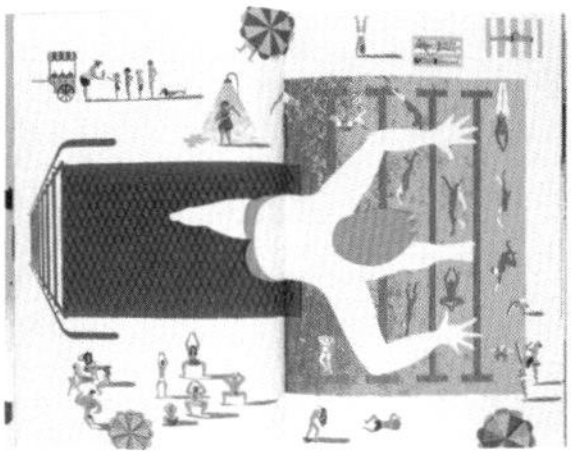

gt, aber
s erlöst

011 Ha³

drei
zeiliger
Die Zahlen
die mich willkommen heißen
bescheinen meine edlen Ziele
und heldenhaft entsteige ich
Wendewelten in Wendebezügen
wässre und schäume mich
den Dreck
der nachts unter meinen Nägeln wächst
hübsche und häute mich
bereit für den Tag
auf Knopfdruck
öffnet sich die Welt
auf 24 Zoll
aktuelle Bilder aus dem Orgasmusgebiet
und der ultimative Krieg
so tippe ich
die Sekunden voran
die wie getrieben flüchten
zusehends verpixeln ganze Wochen
zu schreibgeschützten Ordnerstrukturen
viele ungeöffnete Briefe
blinken rastlos um Aufmerksamkeit
die Zahlen zeigen auf Tagesbilanzen
und draußen
heulen schon die Wölfe

这是一本“设计诗”，除了内容是当代诙谐幽默的小短文或现代诗，设计上也饱含诗意。这样一个小开本口袋书很精致地做成了锁线硬精装。封面深灰色布面丝印黑色后裱覆卡板，上下书口及右侧翻口刷浅蓝色。黑色环衬包覆内页胶版纸四色印刷，除主要呈现的黑色文字外，其他几色只在极少页面点缀呈现。本书的内页排版与大多数专为现代诗而设计的版面不同。大多数现代诗集会遵循单边对齐的排法，任由“断句”的另一侧形成长短不一的节奏。而本书则不局限于这样的节奏，虽整体依然在一定的版心内排布，但版面和排法十分自由。这自由的排法并不是没有章法，书中有 2 页专门呈现了本书所用的细密的网格系统，甚至对着光源透过前后页及隔页可以看到连续页面的文本之间相互穿插、借用空间的有趣节奏，仿佛环环相扣的精妙演出。||||||||||||||||||||||||||||||||||||||||||||

Leerzeichen
für
Applaus

This is a collection of ‘designed poems’, that is to say, contemporary humorous essays or modern poems are designed to be poetic. Such a small pocket book is exquisitely made into a hardbound book with thread sewing. The cover is made of dark grey cloth printed in black with silk screen, then mounted with graphic board. The three edges of the book are brushed with light blue ink. Followed by black endpaper, the inside pages are made of offset paper printed in four colors. Black is widely used for the text, while other three colors are used only on few pages. The layout of the inner pages of this book is different from that of modern poetry anthologies. Most modern poetry anthologies follow the unilateral alignment arrangement, allowing the other side of the ‘sentence’ is uneven in length. The book is not limited to such a regulation. Although the overall layout is still within the fixed type area, the way of typesetting is quite free and orderly. There are two pages in the book, which display the fine grid system used in the book. In front of the light source, looking through the pages, we can feel the interesting rhythm of the text interlacing with each other, as if we are enjoying interlocking performances.

荣誉奖
留白
德国

This is a handy little book on the subject of design and typography, a kind of small and unconventional Bible for (young) designers. The typeface Swift, normally used in newspapers and working well here in a smaller rendition as text font, is skilfully applied and self-assured. In a poetic and ironic fashion the strong design plays with the rules of typography. The three-sided colour edging provides a nice antithesis to the understatement to be found within the book. An unusually dense masterpiece, unpretentious and restrained.

Honorary Appreciation
Leerzeichen fur Applaus
Germany
Jenna Gesse
Jenna Gesse
Eigenverlag Jenna Gesse, Bielefeld
184×114×12mm
159g
96p

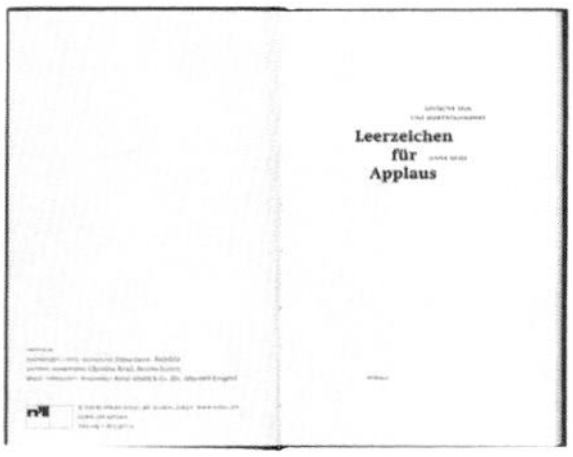
Leerzeichen
für
Applaus

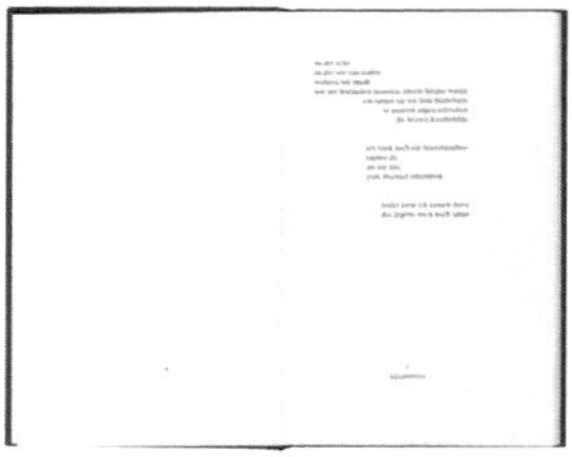

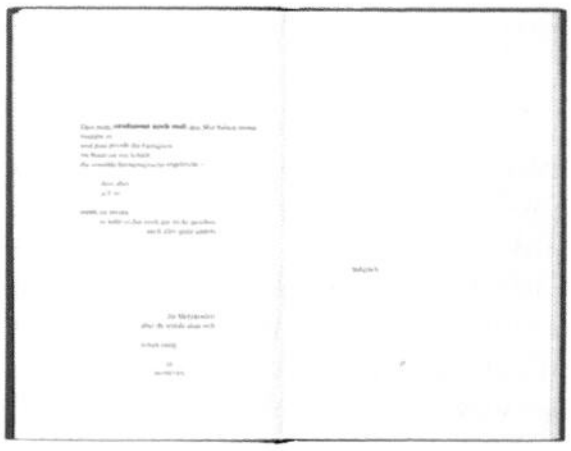

DESIGNER
klingt nach Glastisch
GRAFIKER
klingt nach Zirkel
KÜNSTLER
klingt nach Beziehungsstress
Gestalter
klingt nach nichts
Das bin ich
masters of arts
masters of hearts
arts of masters
arts of heart masters
arts of hamsters
ham of art-hearts
und der dunkle lord
der sith
54
SELBSTVERSTÄNDNIS
55
KREDIT PUNKTE

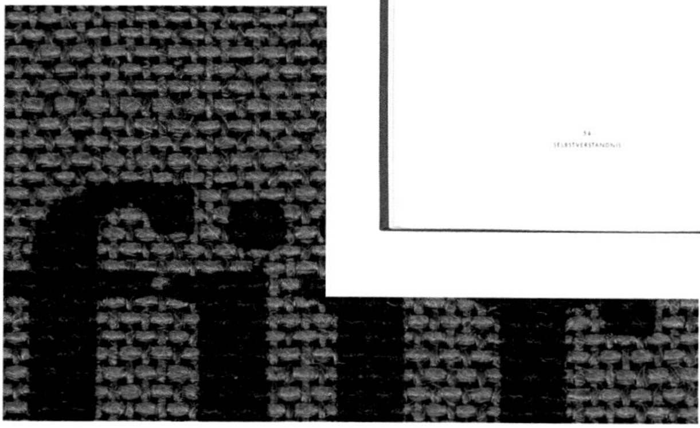

Isabell
steht
auf
Cocktailschirmchen
Analverkehr
und House
und hat 186 Freunde
ansonsten
weiß ich nichts
über sie
drei
zeiliger
befehl
WENN DIE KONZENTRATION
ÜBERNUSCHELT WIRD
ENTSTEHT NICHTS
designers' taste:
coffee
cigarettes
and copy paste

mdreh
verkle
Inn
icht gefunden. Seine

011 Ha4

这是法国诗人、小说家雷蒙·格诺（Raymond Queneau）于1959年创作的小说，讲述女孩扎齐和她的叔叔在巴黎逗留期间，遇到种种风波，甚至巴黎地铁罢工的荒诞故事。本书为插画师乔·维利翁（Joe Villion）绘制和设计的版本。从封面延续至内页的插画和加入了荧光油墨的五色印刷是本书最大的亮点。本书为锁线圆脊硬精装，护封胶版纸五色印刷，画风怪诞，色彩搭配冲击力十足。内封黑色布面书脊烫白裱覆卡板，前后各裱贴双色印刷方块图案。黑色环衬之后，内页依旧是胶版纸黑、灰绿、草绿、紫、荧光红五色印刷。黑色为画面造型基础和正文文字，其他四色通过相互组合和叠印搭配，形成符合故事需求的不同画面氛围。正文字体选用了加强衬线的字体，与插图相得益彰。页码色彩与邻近插图适配随机改变，好像霓虹灯一样，营造出本书内容荒诞的色彩。

Raymond Queneau
ZAZIE
IN DER METRO
Büchergilde
Gutenberg

This is a novel written in 1959 by Raymond Queneau, a French poet and novelist. It tells the story of a girl Zazie and her uncle, who were in trouble during their stay in Paris, and even the panic of a subway strike in Paris. This book is a version drawn and designed by illustrator Joe Villion. Illustrations extending from cover to inside pages and five-color printing with fluorescent ink are the highlights of the book. This hardback book is bound by thread sewing with a rounded spine. The jacket is made of offset paper printed in five colors with weird style of painting and color matching. The spine is covered by black cloth with white hot stamping, and the double-color printing square patterns are pasted on the front and back cover. Followed by the black endpaper, the inner pages are printed on offset paper in five colors – black, grey green, grass green, purple and fluorescent red. Black is the key color of picture modeling and text typesetting. The other four colors form different atmosphere to meet the needs of the story by combining and overlapping. Enhanced serif fonts are used in the text to complement the illustrations. The colors of page numbers are changing randomly like neon lights according to the hue of adjacent illustrations, emphasizing the absurdity of the book.

荣誉奖
地铁里的扎齐
德国

Honorary Appreciation
Zazie in die Metro
Germany
Joe Villion
Raymond Queneau
Büchergilde Gutenberg, Frankfurt am Main
245 × 152 × 19mm
615g
221p

A new face to a classic story from the 1960s, colourful and exciting. Through their vivid, fluorescent colours Joe Villion's illustrations create a modern and fresh aesthetic. Within the book neon shades are also used for the full-page pictures and even the pages numbers change colours. The typesetting of the main text is well executed, in places broken up with vignettes. Choice of paper – silky and with a pleasant degree of whiteness – and binding are excellent and significantly contribute to the book's extremely high overall quality.

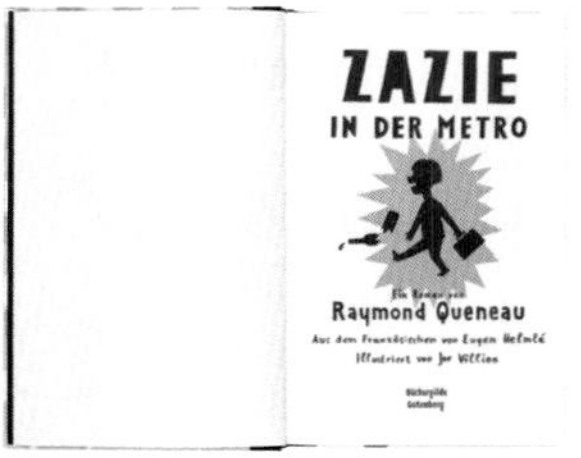
ZAZIE
IN DER METRO
Raymond Queneau

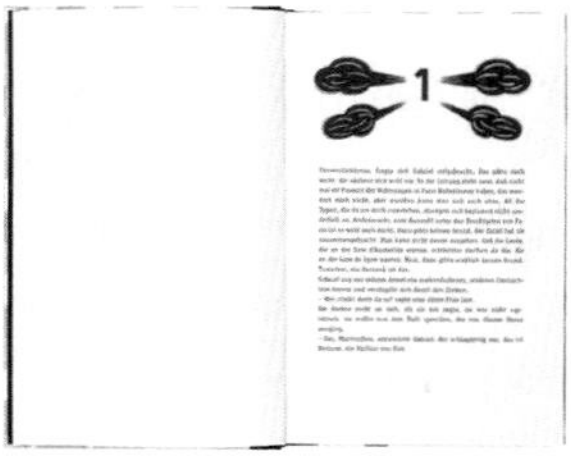

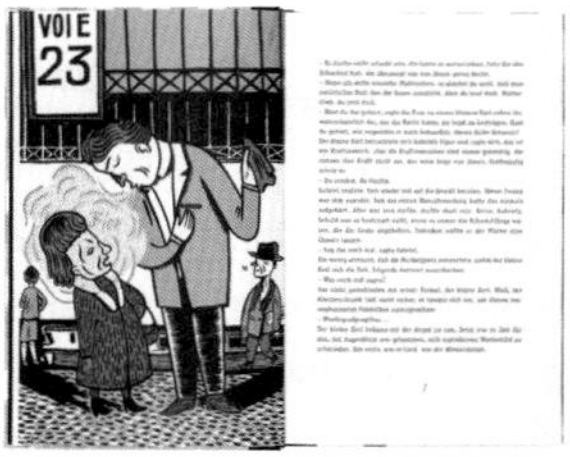

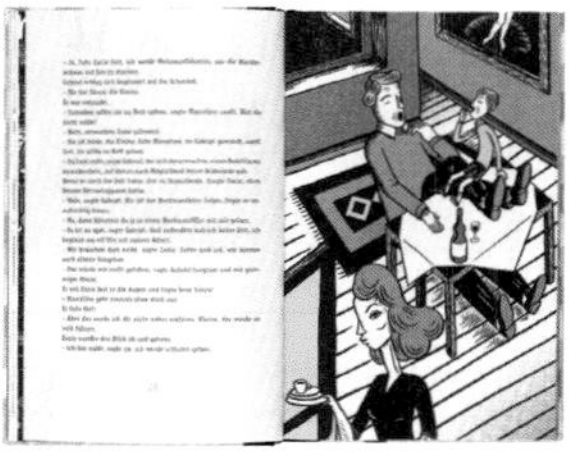

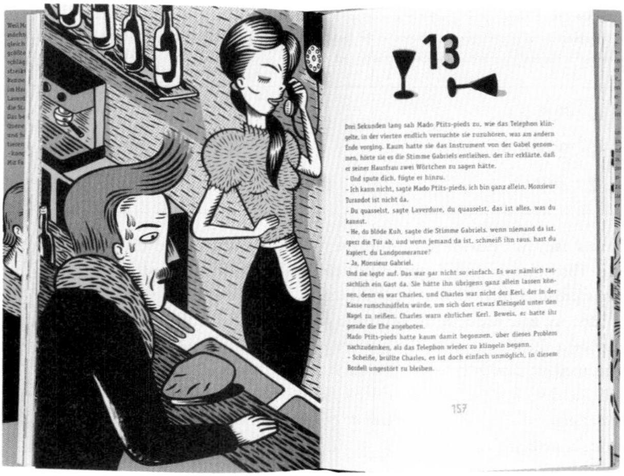
13

Drei Sekunden lang sah Mado Ptits-pieds zu, wie das Telephon klingelte, in der vierten endlich versuchte sie zuzuhören, was am andern Ende vorging. Kaum hatte sie das Instrument von der Gabel genommen, hörte sie es die Stimme Gabriels entleihen, der ihr erklärte, daß er seiner Hausfrau zwei Wörtchen zu sagen hätte.
- Und spute dich, fügte er hinzu.
- Ich kann nicht, sagte Mado Ptits-pieds, ich bin ganz allein, Monsieur Turandot ist nicht da.
- Du quasselst, sagte Laverdure, du quasselst, das ist alles, was du kannst.
- He, du blöde Kuh, sagte die Stimme Gabriels, wenn niemand da ist, sperr die Tür ab, und wenn jemand da ist, schmeiß ihn raus, hast du kapiert, du Landpomeranze?
- Ja, Monsieur Gabriel.
Und sie legte auf. Das war gar nicht so einfach. Es war nämlich tatsächlich ein Gast da. Sie hätte ihn übrigens ganz allein lassen können, denn es war Charles, und Charles war nicht der Kerl, der in der Kasse rumschnüffeln würde, um sich dort etwas Kleingeld unter den Nagel zu reißen. Charles warn ehrlicher Kerl. Beweis, er hatte ihr gerade die Ehe angeboten.
Mado Ptits-pieds hatte kaum damit begonnen, über dieses Problem nachzudenken, als das Telephon wieder zu klingeln begann.
- Scheiße, brüllte Charles, es ist doch einfach unmöglich, in diesem Bordell ungestört zu bleiben.

157

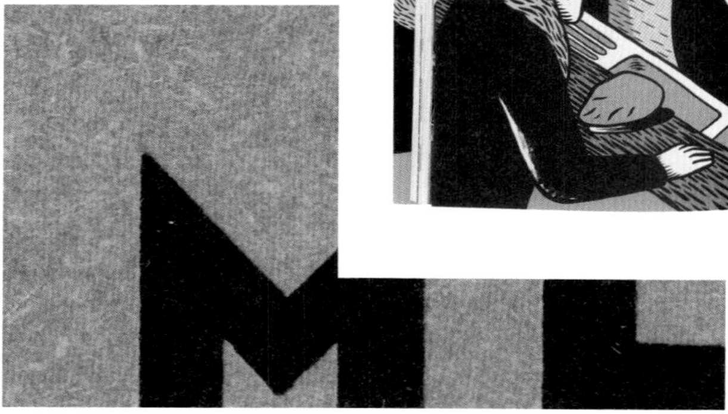

GEV
le vin
DUBON
19

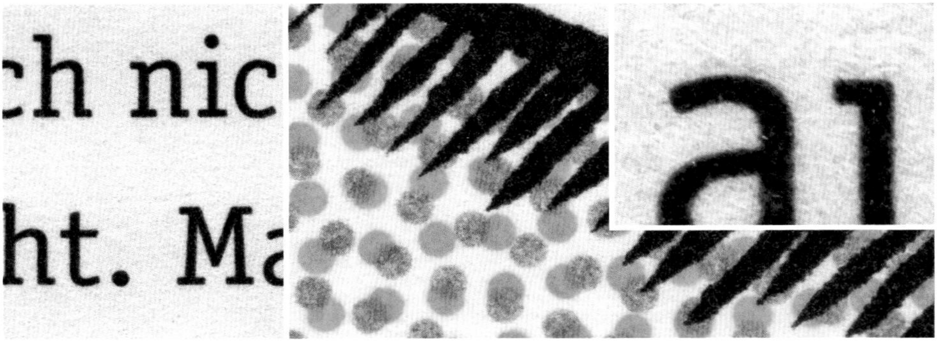
ch nic
ht. Ma

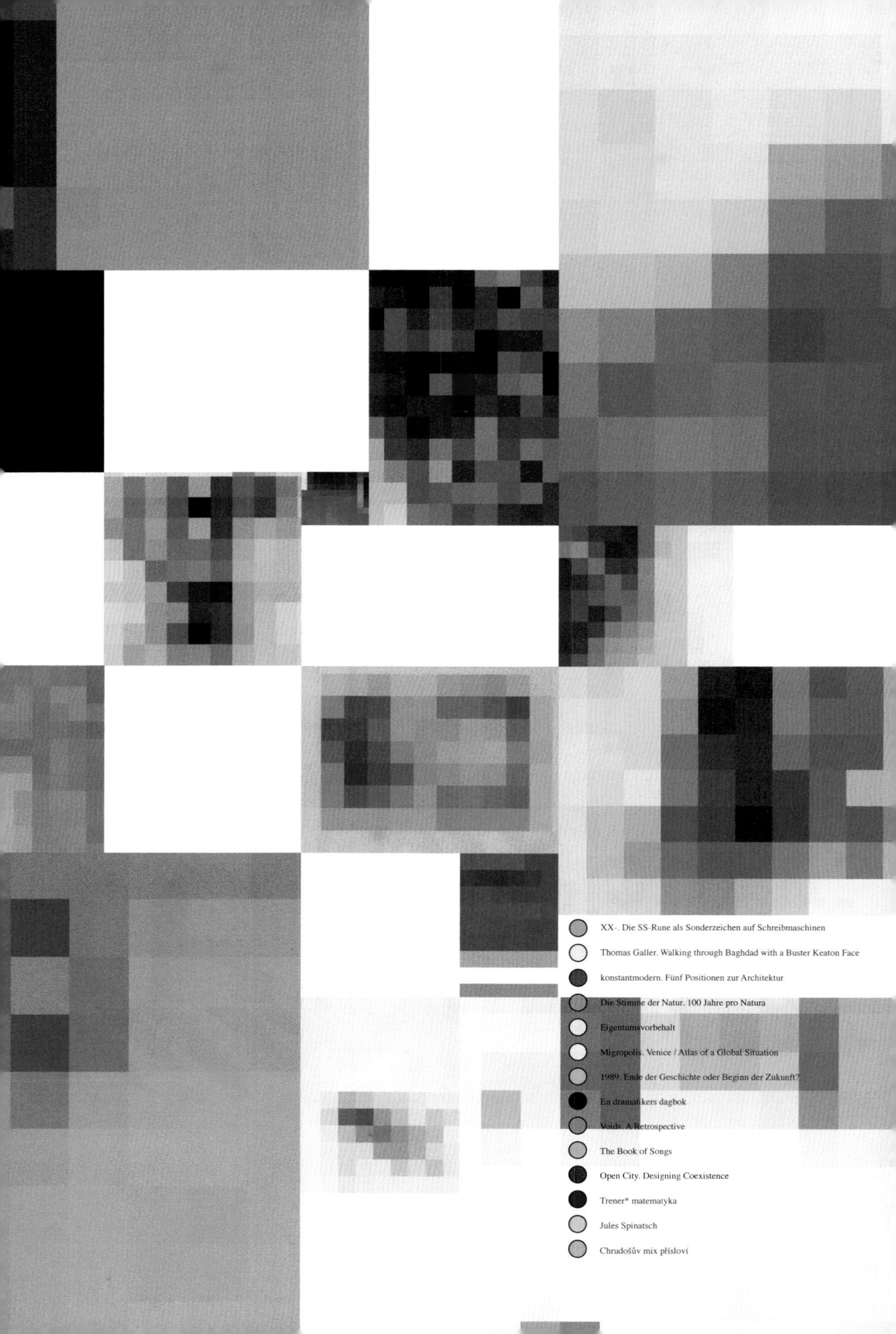
XX-. Die SS-Rune als Sonderzeichen auf Schreibmaschinen
Thomas Galler. Walking through Baghdad with a Buster Keaton Face
konstantmodern. Fünf Positionen zur Architektur
Die Stimme der Natur. 100 Jahre pro Natura
Eigentumsvorbehalt
Migropolis. Venice / Atlas of a Global Situation
1989. Ende der Geschichte oder Beginn der Zukunft?
En dramatikers dagbok
Voids. A Retrospective
The Book of Songs
Open City. Designing Coexistence
Trener* matematyka
Jules Spinatsch
Chrudošův mix přísloví

2010

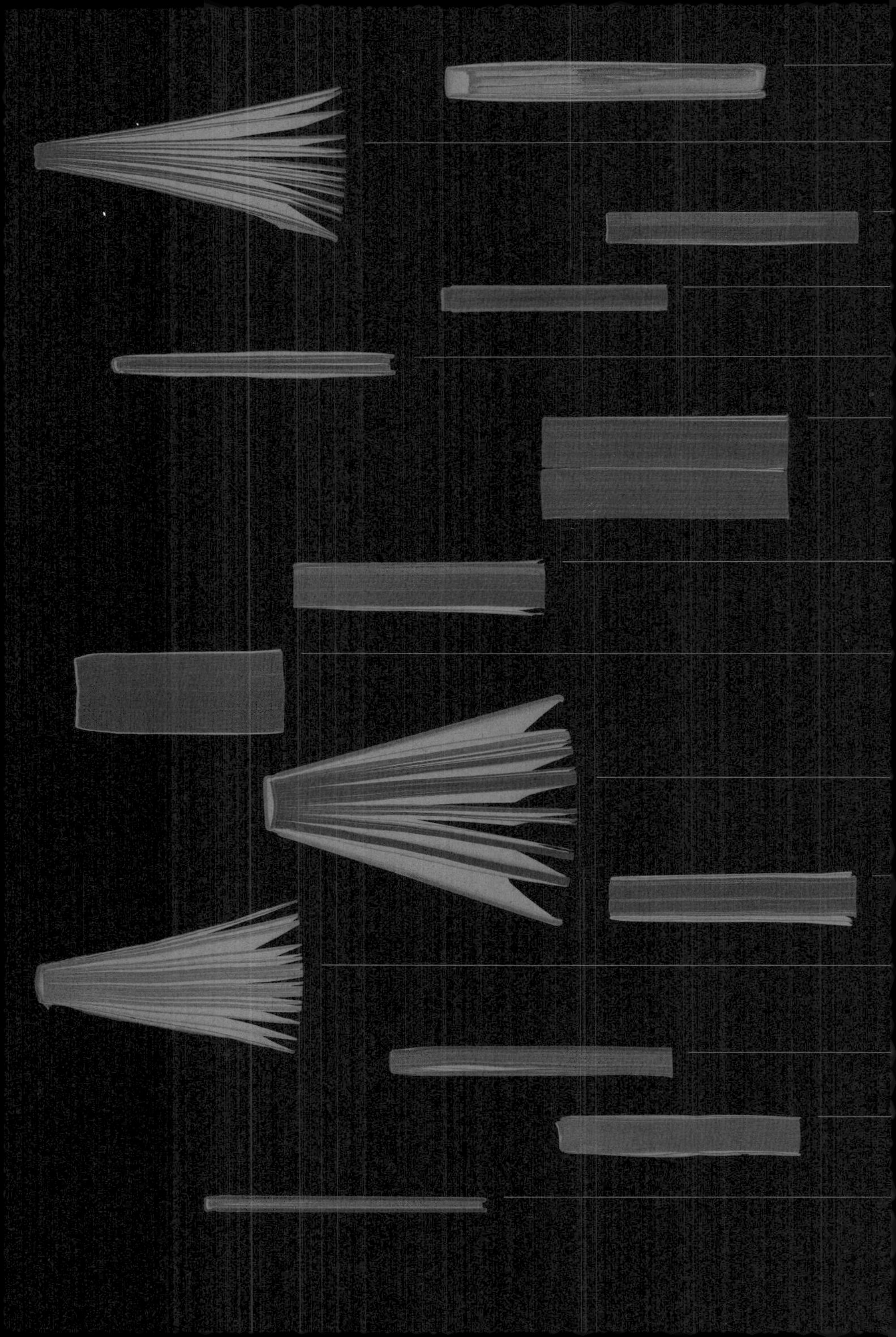

2010 Gl — 0865

2010 Gm — 0873

edition fink
Thomas Galler

2010 Sm^1 — 0879

konstantmodern

2010 Sm^2 — 0885

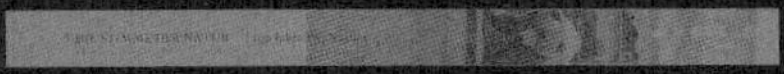

2010 Bm^1 — 0891

2010 Bm^2 — 0897

MIGROPOLIS
MIGROPOLIS

2010 Bm^3 — 0903

6 8 6

2010 Bm^4 — 0909

LARS NORÉN
EN DRAMATIKERS DAGBOK

2010 Bm^5 — 0915

Voids

2010 Ha^1 — 0921

2010 Ha^2 — 0925

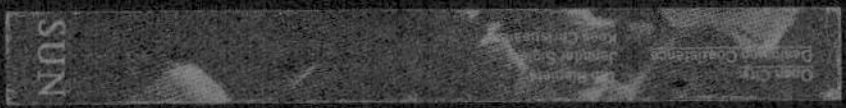
SUN

2010 Ha^3 — 0929

2010 Ha^4 — 0933

2010 Ha^5 — 0937

CHRUDOŠŮV MIX

GIBRALTAR
TUNISIE
Szukana objętość stożka jest rów

d. h. die Schrift. Das Wort als die Schrift aber ist die Handschrift. [...] Die Schreibmaschine entreißt die Schrift dem Wesenbereich der Hand,

— Diese Handvoll Druckbuch
zwischen den Fingern halten ka
Buchstaben, die auf der Tastatu

"I'm feeling a little disconnected from my real life. A kind of losing touch with t
you know what I mean? I mean... I actually think, there is an element of psycho
here." — "Did I do that? I guess I did." — "So how does it feel?" — "In real life?

das vorliegende Material zur Verfügung gestellt haben. Mit ihrer Hilfe hab
einen anderen und erweiterten Blick auf die Architektur des 20. Jahrhun
erhalten und hoffe, dass dieses Buch auch ihnen eine neue Sicht auf ihre
und ihr Werk eröffnet. Weiters danke ich Nikolaus Schletterer für seine Au

In Wirklichkeit, und damit kommen wir zu einer anderen Int
tion des Begriffs Verstädterung, ist die gesamte Schweiz städt
zwar insofern, als sie einen Lebensstil kennt, den man als stä

Wertschöpfungsketten [...] Das Leben selbst wird

zum Unternehmen, welches sich auf dem Markt behaupten

died of asthma attack in detention centre in Trajal, Ceuta (ES) awaiting expulsion
drowned, 1 found, 1 missing, shipwreck near Kusadasi (TR) on way to Island of Sam
reportedly stowaway hidden under a vehicle, dead body found on A3 near Clanfield
26 missing, 11 found mummified in boat drifting near the Caribbean heading for Spa

Hon har jobbat på verkstäderna med några TV4-produkti
tar den här veckan också, sedan är det slut. Hon är den mär
behöver mig mest av allt och som jag försummar mest. Jag m

*The Specialization of Sensibility in the Raw Materia
of Stabilized Pictorial Sensibility [La spécialisation
sensibilité à l'état matière première en sensibilité*

去，而作是诗也。"方玉润誉为"凭吊诗中绝唱"。三章叠咏，三十句。

1　彼黍离离　彼稷之苗[1]

An open society is both friend and enemy of the Open City.
influence of its success means there is a constant threat of t
destruction of diversity. The freedom of low-cost total mobil

➔ W magazynie pewnej restauracji znajduje się ocet 10-procentowy w
kach o pojemności 0,5 l. Do przyrządzania sałatek kucharz potrzebu
4-procentowego oraz 5-procentowego.

Im 2005 publizierten Band *Temporary Discomfort* präser
Jules Spinatsch das Resultat einer über mehrere Jahre f
ten Untersuchung, die zu einem seiner zentralen künst

Gold Medal

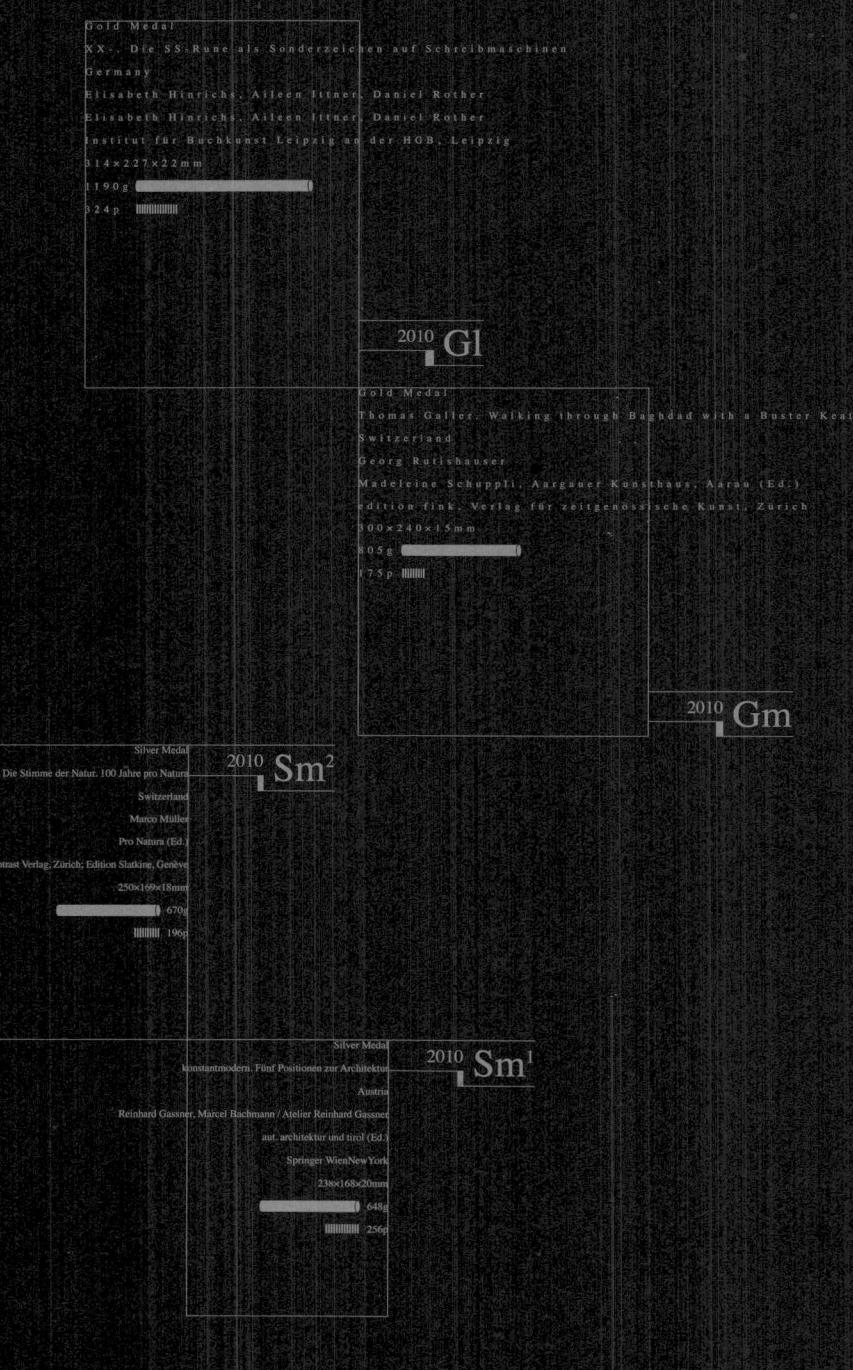

XX-. Die SS-Rune als Sonderzeichen auf Schreibmaschinen

Germany

Elisabeth Hinrichs, Aileen Ittner, Daniel Rother

Elisabeth Hinrichs, Aileen Ittner, Daniel Rother

Institut für Buchkunst Leipzig an der HGB, Leipzig

314×227×22mm

1190g

324p

2010 Gl

Gold Medal

Thomas Galler. Walking through Baghdad with a Buster Keaton Face

Switzerland

Georg Rutishauser

Madeleine Schuppli, Aargauer Kunsthaus, Aarau (Ed.)

edition fink, Verlag für zeitgenössische Kunst, Zürich

300×240×15mm

805g

175p

2010 Gm

Silver Medal

Die Stimme der Natur. 100 Jahre pro Natura

Switzerland

Marco Müller

Pro Natura (Ed.)

Kontrast Verlag, Zürich; Edition Slatkine, Genève

250×169×18mm

670g

196p

2010 Sm2

Silver Medal

konstantmodern. Fünf Positionen zur Architektur

Austria

Reinhard Gassner, Marcel Bachmann / Atelier Reinhard Gassner

aut. architektur und tirol (Ed.)

Springer WienNewYork

238×168×20mm

648g

256p

2010 Sm1

2010 Ha[1]

Honorary Appreciation
The Book of Songs
China
Liu Xiaoxiang
Xiang Xi (Ed.)
Higher Education Press, Beijing
257×167×27mm
946g
391p

2010 Ha[2]

Honorary Appreciation
Open City. Designing Coexistence
The Netherlands
Mevis & Van Deursen mit Werkplaats Typografie
Tim Rieniets, Jennifer Sigler, Kees Christiaanse (Ed.)
SUN architecture, Amsterdam
270×202×29mm
1051g
464p

2010 Ha[3]

Honorary Appreciation
Trener* matematyka
Poland
Grzegorz Podsiadlik
Jan Górowski, Adam Łomnicki
Wydawnictwo Szkolne PWN, ParkEdukacja, Warszawa / Bielsko-Biała
242×205×15mm
671g
328p

2010 Ha[4]

Honorary Appreciation
Jules Spinatsch
Switzerland
Winfried Heininger / Kodoji Press
Marco Obrist, Kunsthaus Zug (Ed.)
Kodoji Press, Baden
235×170×25mm
877g
349p

2010 Ha[5]

Honorary Appreciation
Chrudošův mix přísloví
Czech Republic
Juraj Horváth
Chrudoš Valoušek
Baobab, Praha
207×203×10mm
320g
46p

2010 Bm[1]

Bronze Medal
Eigentumsvorbehalt
Germany
Gaston Isoz
Gaston Isoz (Ed.)
disadorno edition, Berlin
305×224×11mm
738g
88p

2010 Bm[2]

Bronze Medal
Migropolis. Venice / Atlas of a Global Situation
Germany
Wolfgang Scheppe mit Katerina Dolejšová, Veronica Bellei, Miguel Cabanzo
Wolfgang Scheppe, The IUAV Class on Politics of Representation (Ed.)
Hatje Cantz Verlag, Ostfildern
240×170×73mm
3252g
1344p

2010 Bm[3]

Bronze Medal
1989. Ende der Geschichte oder Beginn der Zukunft?
Austria
Chris Goennawein
Kunsthalle Wien, Gerald Matt, Cathérine Hug, Thomas Mießgang (Ed.)
Verlag für moderne Kunst Nürnberg
240×190×33mm
819g
317p

2010 Bm[4]

Bronze Medal
En dramatikers dagbok
Sweden
Nina Ulmaja
Lars Norén
Albert Bonniers Förlag, Stockholm
209×148×52mm
1354g
1680p

2010 Bm[5]

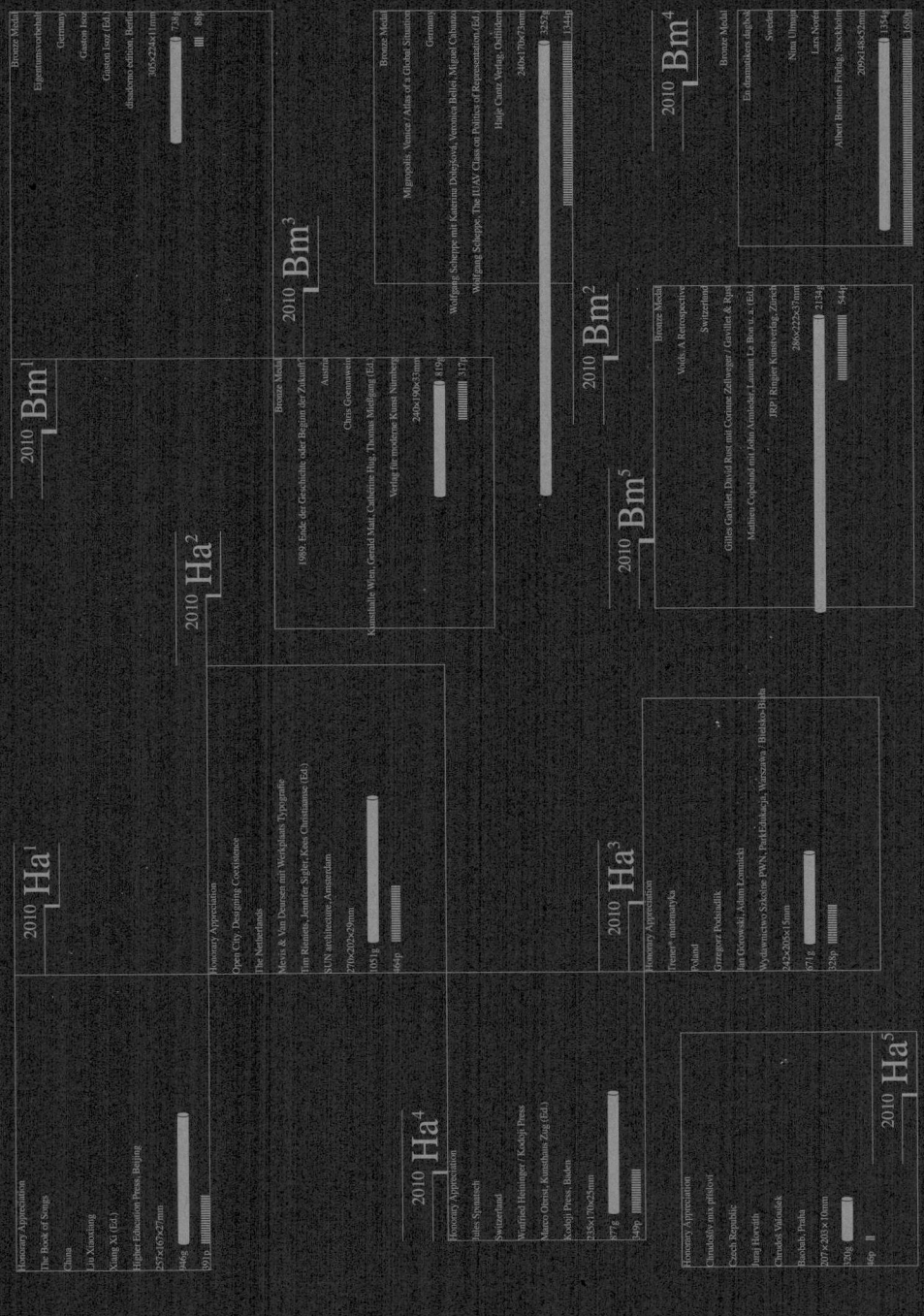

Bronze Medal
Voids. A Retrospective
Switzerland
Gilles Gavillet, David Rust mit Corinne Zellweger / Gavillet & Rust
Mathieu Copeland mit John Armleder, Laurent Le Bon u. a. (Ed.)
JRP | Ringier Kunstverlag, Zürich
286×222×37mm
2134g
544p

Designer

Elisabeth Hinrichs, Aileen Ittner, Daniel Rother

Georg Rutishauser

Reinhard Gassner, Marcel Bachmann / Atelier Reinhard Gassner

Marco Müller

Gaston Isoz

Wolfgang Scheppe mit Katerina Dolejšová, Veronica Bellei, Miguel Cabanzo

Chris Goennawein

Nina Ulmaja

Gilles Gavillet, David Rust mit Corinne Zellweger / Gavillet & Rust

Liu Xiaoxiang

Mevis & Van Deursen mit Werkplaats Typografie

Grzegorz Podsiadlik

Winfried Heininger / Kodoji Press

Juraj Horváth

2010

The International Jury

Nici von Alvensleben (USA)

Markus Dreßen (Germany)

Gaston Isoz (Germany)

Gabriele Schenk (Germany)

Marc Taeger (Spain)

Minako Teramoto (Japan)

Tessa van der Waals (The Netherlands)

Country / Region

Switzerland 4

Germany 3

Austria 2

Germany 2

Sweden 1

China 1

The Netherlands 1

Poland 1

Czech Republic 1

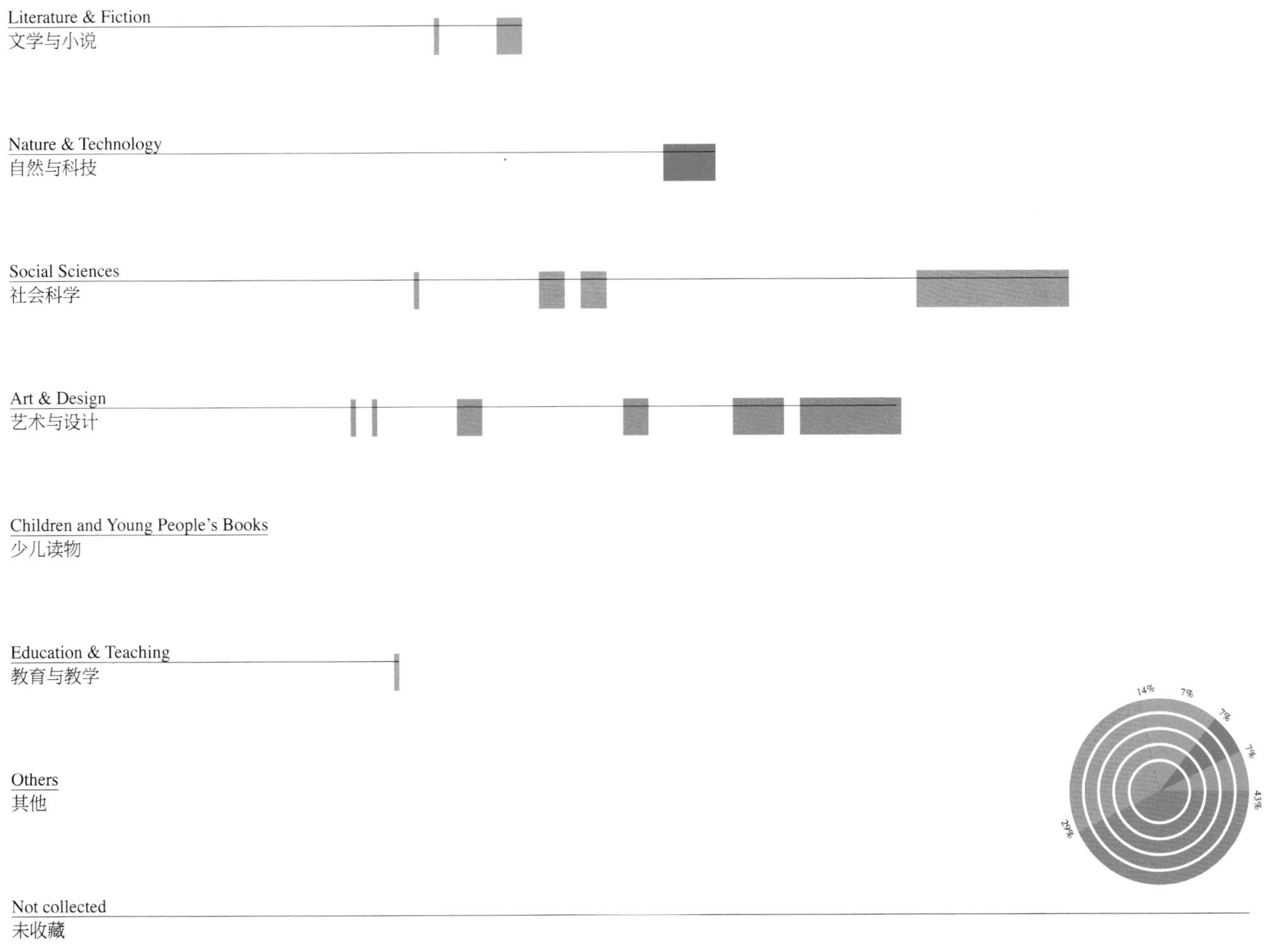
Literature & Fiction
文学与小说
Nature & Technology
自然与科技
Social Sciences
社会科学
Art & Design
艺术与设计
Children and Young People's Books
少儿读物
Education & Teaching
教育与教学
Others
其他
Not collected
未收藏
14%
7%
7%
7%
43%
29%

FRAU
ICHEN
SCHINE

三位作者整理了德国在纳粹时期有关打字机、符号系统和相关行业行为规范的资料，并集结成本书这样一本历史档案。书籍极力呈现出博物馆中松散的文件被文件夹包裹的档案感。护封采用橙色卡纸印黑，超长的勒口将所有内容包裹其中。硬精装封面采用灰色网纹纸烫白，封二采用亮面铜版纸橙黑双色印刷，通过多层折叠的方式，罗列参考资料的封面。书籍的主体部分采用裸脊锁线装，便于完全翻开。本书内容分三部分，第一部分，女性打字员口述文字、打字机使用说明、打字标准动作、身体姿态、办公环境、提高效率的方法等与行业行为规范有关的档案；第二部分，以“SS”为代表的德国特殊字符是如何演变成纳粹思想的视觉符号，并贯穿二战的档案；第三部分，加入特殊符号的德式打字机的生产、标准化、管理以及技术的档案。三部分主要通过胶版纸橙黑双色印刷呈现，中间夹带亮面铜版纸印黑与浅黄色胶版纸印黑的部分呈现附加内容。书籍不通过页码来建立索引，而是通过文字和图片的编号进行内容的组织。||||||||

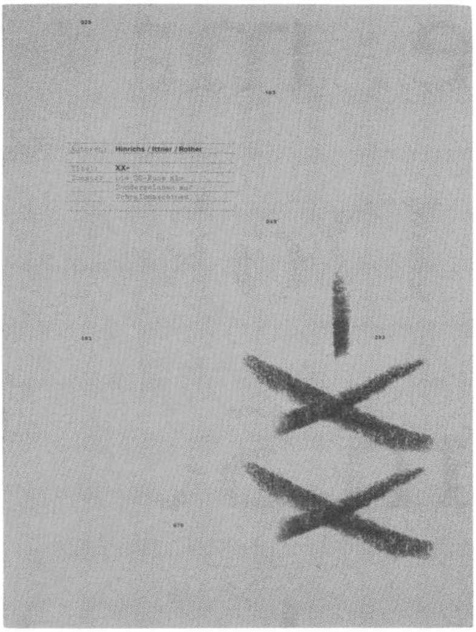

Hinrichs / Ittner / Rother
XX-

Three authors published a historical archive by compiling data on typewriters, symbolic systems and related industry codes in Germany during the Nazi German period. The book strives to show the appearance of loose files wrapped in folders. The jacket is printed on orange cardboard in black, and all the pages are wrapped in a super-long flap. Hard cover is made of gray wove paper hot stamped in white. The inside front cover is printed in orange and black on bright coated art paper, and the reference materials are listed on multi-folded layers. The main body of the book is bound by thread sewing with no spine, which is helpful to open completely. This book is divided into three parts. The first part contains the archives including female typists' oral account, manuals of typewriters, standard typing actions, body posture, office environment, and ways to improve efficiency. The second part is about how the German special characters represented by 'SS' evolved into Nazi thoughts and used as visual symbols throughout World War II. In the third part, there are archives including the production, standardization, and management of German typewriters with special symbols. The three parts are mainly presented on offset paper in orange and black, and some additional contents are printed on the bright coated art paper and light yellow offset paper in black. The book is not indexed by page numbers, but by the serial numbers of text and pictures.

金字符奖

XX-——打字机上的特殊字符 SS

德国

Gold Medal

XX-. Die SS-Rune als Sonderzeichen auf Schreibmaschinen

Germany

Elisabeth Hinrichs, Aileen Ittner, Daniel Rother

Elisabeth Hinrichs, Aileen Ittner, Daniel Rother

Institut für Buchkunst Leipzig an der HGB, Leipzig

314×227×22mm

1190g

324p

A truly admirable way of approaching German history. On the basis of the "SS" typewriter key historical documents from the Third Reich have been compiled and, in a convincingly organised manner, examined in relation to the typical character of official documents. The subject of how power symbols are visually rendered through typographical systems under totalitarian conditions has been very convincingly represented in this book. Thanks to changes of paper, text quotes and use of colour the design successfully brings the complexity of the individual details into a tight sequence, shedding new light on source material from both archives and literature. An extraordinary book, convincing on account of its extensive research, and one which successfully treats its subject matter in a way which is conceptionally and visually up-to-date.

e, ma

Dok

m Sc

als

e1

zerbrechliche, gl

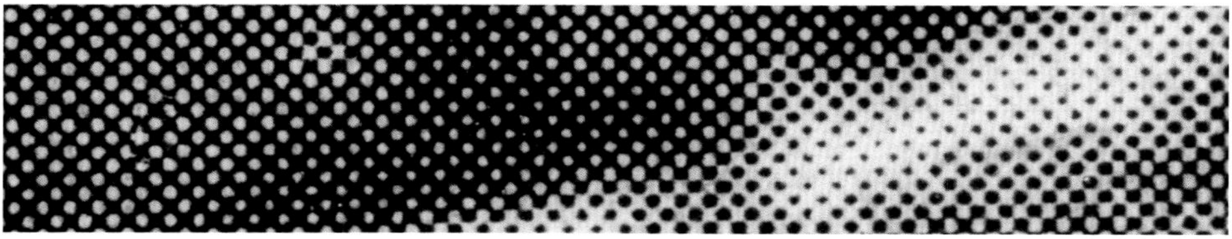

ens, Frankfurt

314×227×22

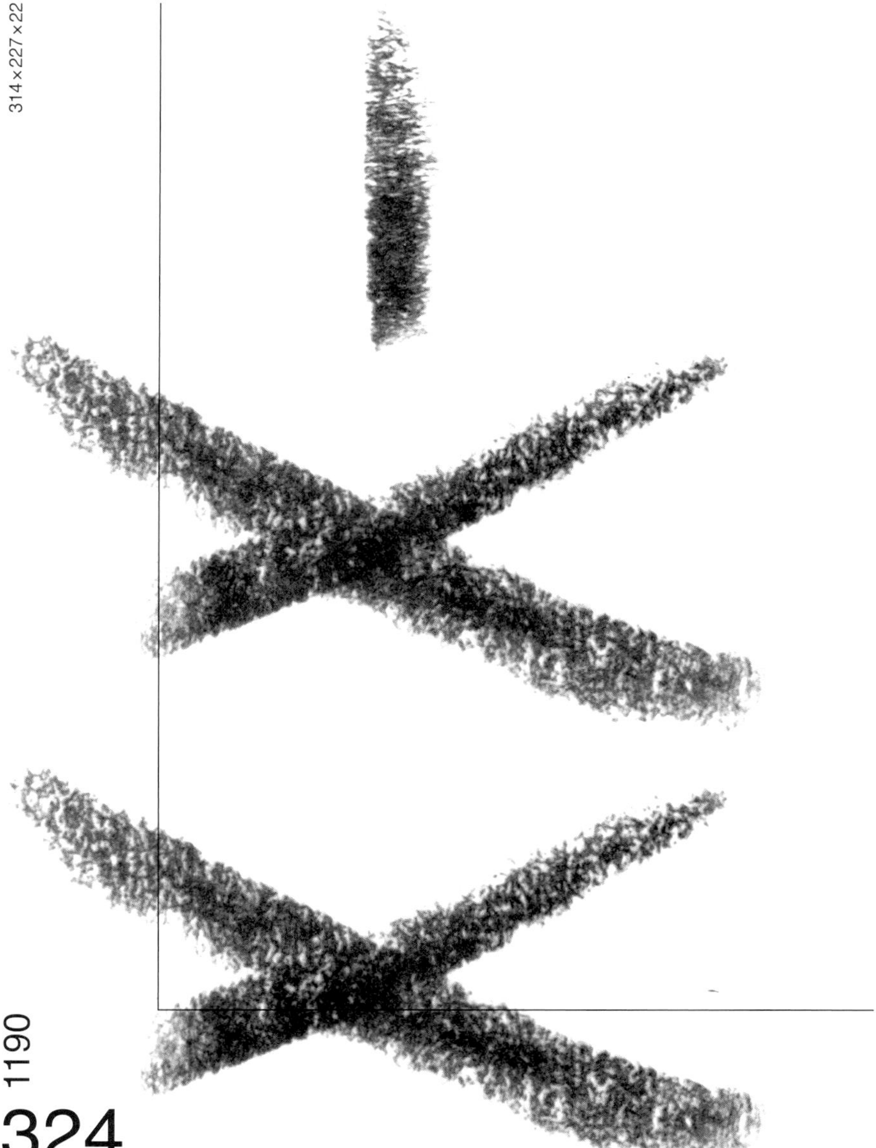

1190

324

FRAU

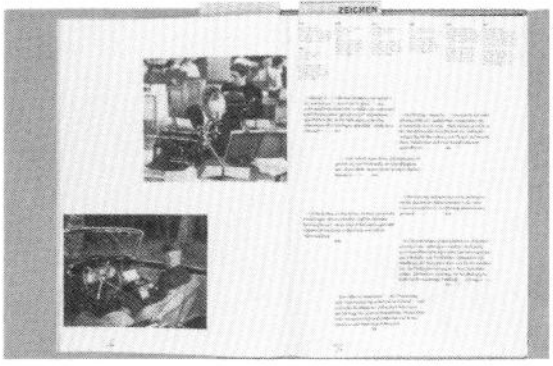
ZEICHEN

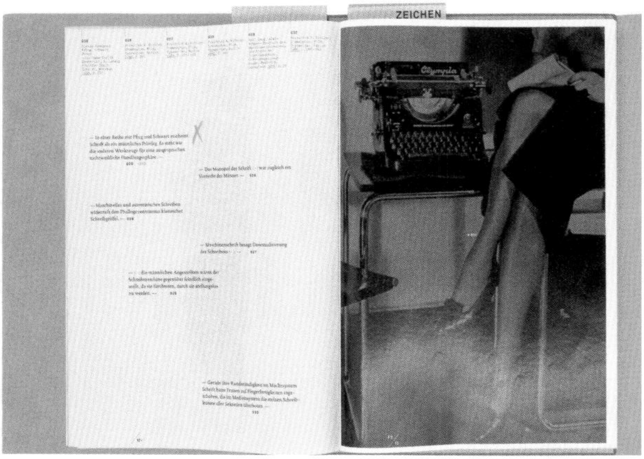
ZEICHEN
Olympia

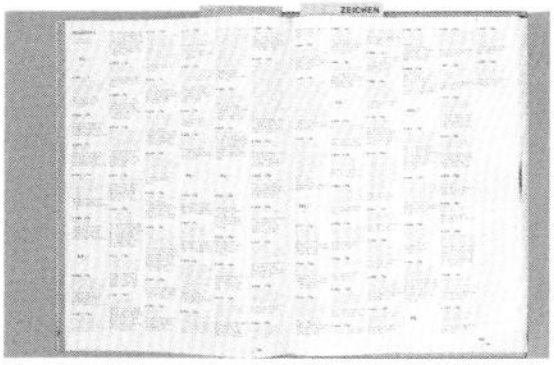
ZEICHEN

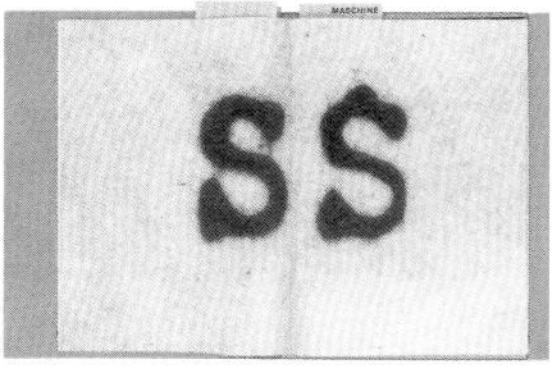
MASCHINE
SS

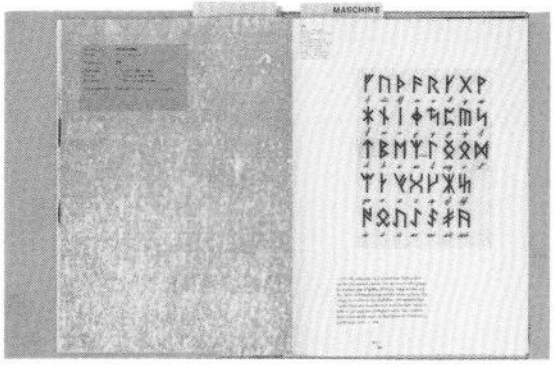
MASCHINE

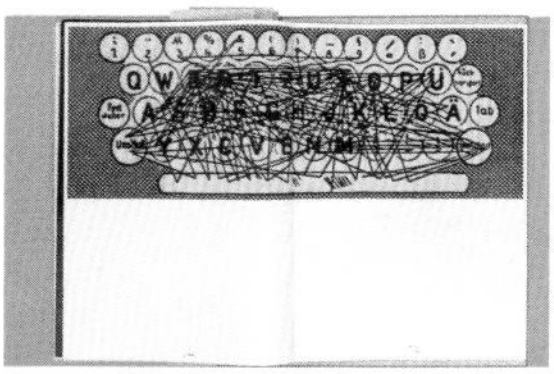

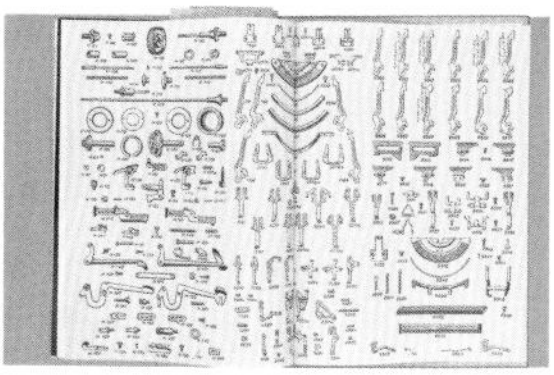

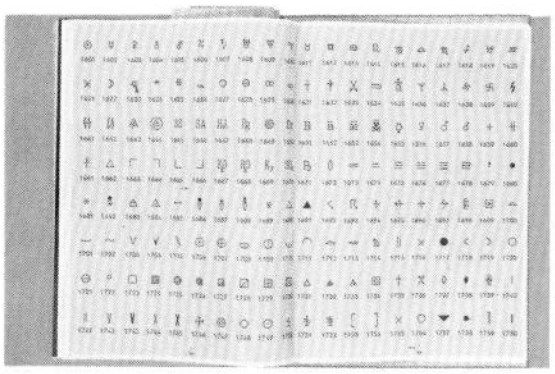

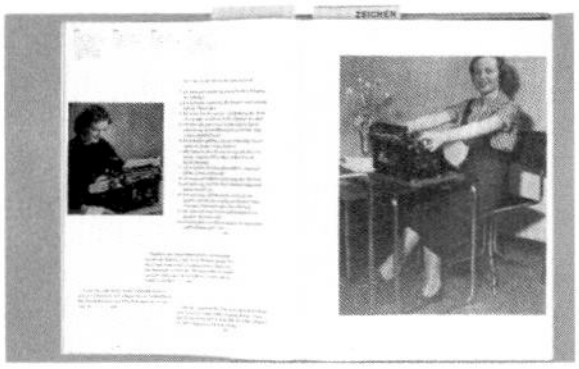
ZEICHEN

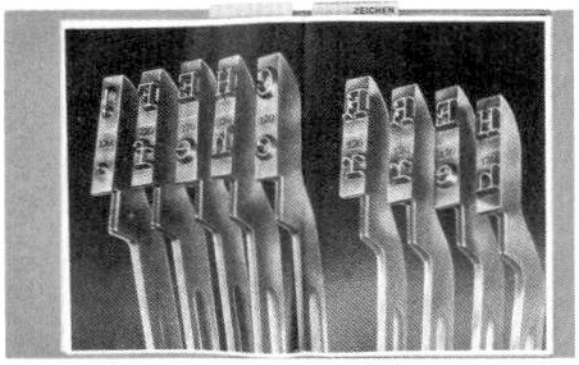
ZEICHEN

ZEICHEN

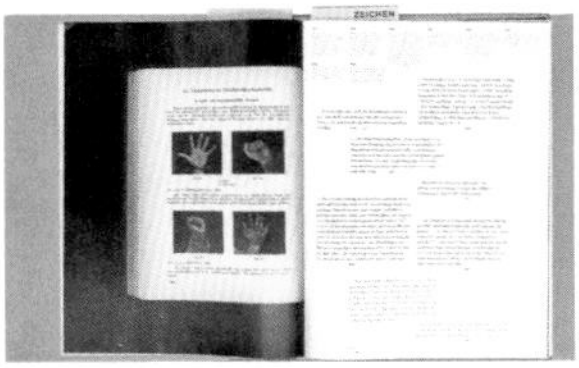
ZEICHEN

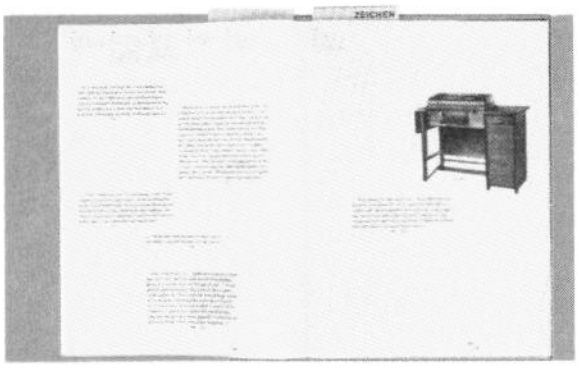
ZEICHEN

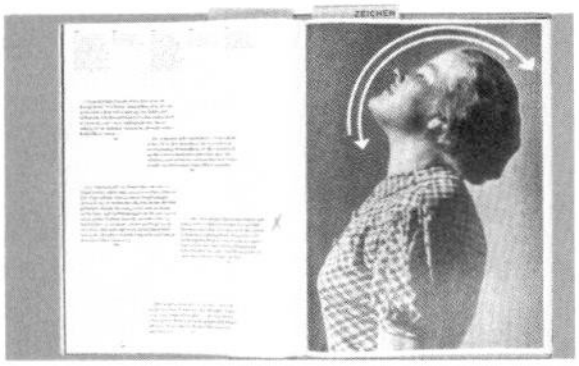
ZEICHEN

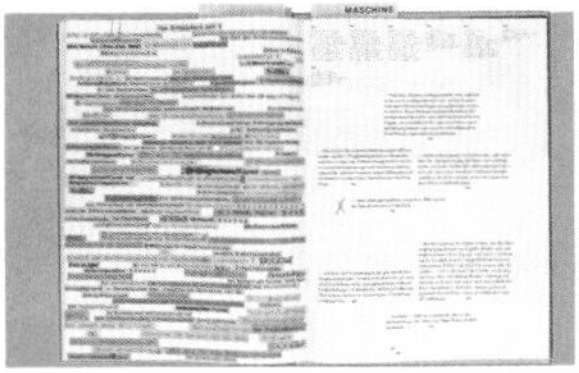
MASCHINE

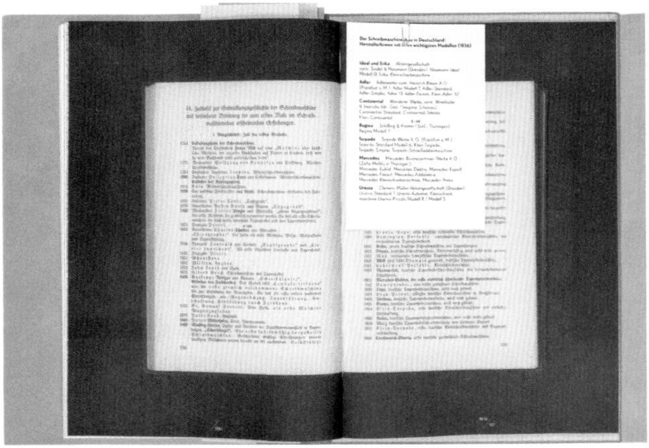
Ideal und Erika
Adler
Continental
Regina
Torpedo
Mercedes
Urania

XX-
Hinrichs / Ittner / Rother
× 185
144‹

'1
rtin Heidegger:
samtausgabe
. 54 Parmenides.
eiburger Vorlesung
1942 / 43.
ankfurt am Main
82, S. 119

372
Martin Heidegger:
Gesamtausgabe
Bd. 54 Parmenides.
Freiburger Vorlesung
WS 1942 / 43.
Frankfurt am Main
1982, S. 126

373
Brief an Wolfram
Sievers, München
vom 29. IV. 1942.
Bundesarchiv Berlin,
Sign. NS 21 / 922,
Blatt-Nr. 241

374
Martin Heidegger:
Gesamtausgabe
Bd. 54 Parmenides.
Freiburger Vorlesung
WS 1942 / 43.
Frankfurt am Main
1982, S. 119

375
Michel Foucault:
Archäologie des
Wissens, Frankfurt
am Main 1990 [1973],
S. 124–125

376
Schreibmaschine und
Augenhygiene.
– In: Schreib-
maschinen-Zeitung
157, Hamburg 1911.
– zitiert nach:
Hans Ulrich Gumbrecht/
K. Ludwig Pfeiffer
(Hg.): Materialität
der Kommunikation,
Frankfurt am Main
1995, S. 326

'7
ns-Joachim Fritz:
nschen in Büro-
beitsräumen,
nchen 1982, S. 104

378
D.A.G. Hauptabteilung
Weibliche Angestellte
(Hg.): Frauen im Mahl-
werk der Bürotechnik,
Hamburg 1967, S. 32

— Da ich gestern Abend meine Daten nur mit Bleistift schreiben konnte, erlaube ich mir, diese Daten heute nochmals kurz mit Maschine geschrieben nachzuliefern. —
373

— Das maschinelle Schreiben [–] degradiert das Wort zu einem Verkehrsmittel. —
371

— In der „Schreibmaschine“ erscheint die Maschine, d. h. die Technik, in einem fast alltäglichen und daher unbemerkten und daher zeichenlosen Bezug zur Schrift [–]. — 372

— Das Wort als das eingezeichnete und so dem Blick sich zeigende ist das geschriebene Wort, d. h. die Schrift. Das Wort als die Schrift aber ist die Handschrift. [–] Die Schreibmaschine entreißt die Schrift dem Wesenbereich der Hand, und d. h. des Wortes. Dieses selbst wird zu etwas „Getipptem“. —
374

— Diese Handvoll Druckbuchstaben, die ich zwischen den Fingern halten kann, oder auch Buchstaben, die auf der Tastatur einer Schreibmaschine angezeigt sind, konstituieren keine Aussagen: es sind höchstens Instrumente, mit denen man Aussagen schreiben kann. —
375

— Buchstaben ohne Wärme, ohne Leben, ohne Bewegung, aus denen wir vergeblich etwas herauszulesen bestrebt sind. —
376

— Mit der Einführung der Schreibmaschine setzt ein grundlegender Wandel in der langfristigen Entwicklung des Schreibvorganges ein, die schon im 14. Jahrhundert mit dem allmählichen Verzicht auf kalligraphische Elemente eingesetzt hatte, um dann gegen Ende des 19. Jahrhunderts mit dem allmählichen Übergang zu immer restriktiveren Formen zwanghafter Schreibbewegungen und zunehmend standardisierter Schriftbilder zum mechanisierten Schreiben typisierter Buchstabenfolge zu führen. —
377

— Ich muß betonen, daß ich sehr gern maschineschreibe. — 378

010 Gm

这是艺术家托马斯·加勒（Thomas Galler）的一本展览图录。他从网上搜罗了大量媒体或普通士兵在战场上拍下的照片和视频，用以表达媒体或者其他参与战争的人拍下的照片和视频未必就是战争的真相这一观点。普通人看到的这些报道有可能是被娱乐化或者曲解了的所谓“真相”。巴斯特·基顿（Buster Keaton）是著名导演兼演员，以不苟言笑的肢体喜剧表演著称，被称为“大石脸”，这里指带着他那样的表情去现场，自绝现实，与真相屏蔽关联。本书为无线胶装，胶体薄而柔软，书籍可以完全打开。封面白卡纸印单色黑，无特殊工艺。内页胶版纸印单色黑，部分页面夹带玻璃卡印四色满版照片。全书图片版面自由，注重图片之间的组合和连续节奏，而不注重网格，文字单栏左对齐，彩页背面有详细的作品信息。全书设计质朴，但有很强的内容冲击力。||||||||
||||||||||||||||||||||||||||||||||||||||||||||||||||||||||||||||||||||||||||||||||

This is an exhibition catalogue of artist Thomas Galler. He collected a large number of photographs and videos taken by the media or ordinary soldiers on the battlefield from the Internet to express the fact that the photographs and videos taken by the media or other people involved in the war are not necessarily the truth of the war. The so-called 'truth' may be entertained or misinterpreted by public news reports. Buster Keaton, a famous director and actor, is well known for his meticulous body comedy performances, regarded as "Big Stone Face", which means that he always acts with no facial expression to alienate himself from the reality and his performance has no connection with the so-called truth. This book is perfect bound with soft and thin glue, so that the book can be fully opened. The cover is printed monochrome black on white cardboard without any special technology. The inside pages are printed on offset paper in monochrome black, attached with some glossy cardboard printed with four-color pictures. The overall book has a free layout, focusing on the combination and continuous rhythm of pictures rather than on the grid. The text is typeset in single column and left aligned, while detailed information is listed on the back of the colorful pages. The design is simple, yet has a strong impact.

- 金奖
- 托马斯·加勒
  ——以巴斯特·基顿的样子穿过巴格达
- 瑞士

- Gold Medal
- Thomas Galler.
  Walking through Baghdad with a Buster Keaton Face
- Switzerland
- Georg Rutishauser
- Madeleine Schuppli, Aargauer Kunsthaus, Aarau (Ed.)
- edition fink, Verlag für zeitgenössische Kunst, Zürich
- 300×240×15mm
- 805g
- 175p

This book succeeds in uniting various layers – simultaneously a catalogue, artist's book and monograph – in a highly subtle and intelligent fashion. The change of paper here is skilfully orchestrated, thus ensuring a strong differentiation between the colour illustrations (on high-glossy chromolux paper) from the exhibition, in contrast to black and white research and reference material on uncoated paper. The typography has an unpretentious feel giving the book a harmonious aura – ragged alignment makes for agreeable lightness and ease of reading.

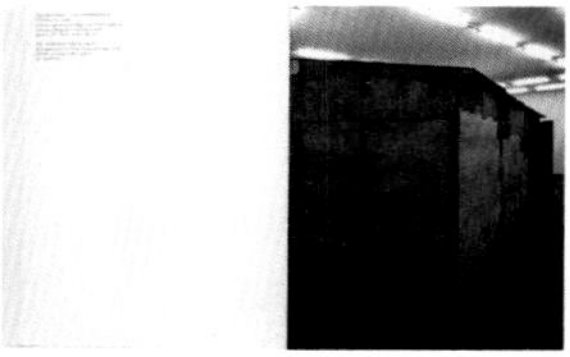

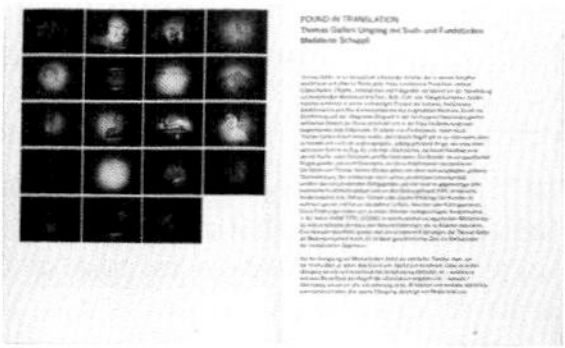
FOUND IN TRANSLATION
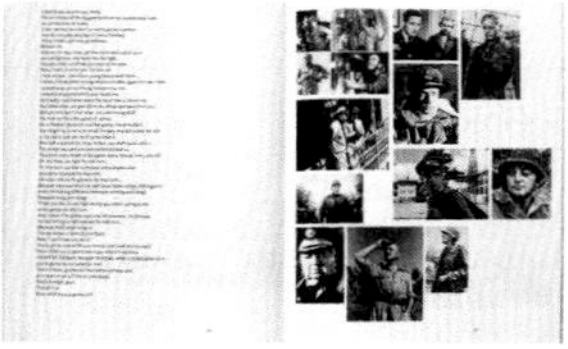

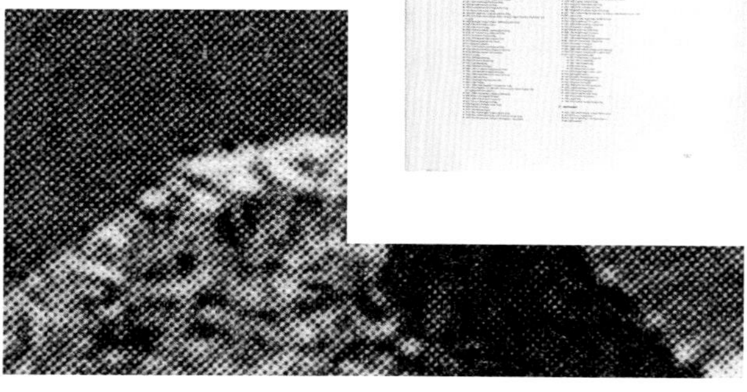

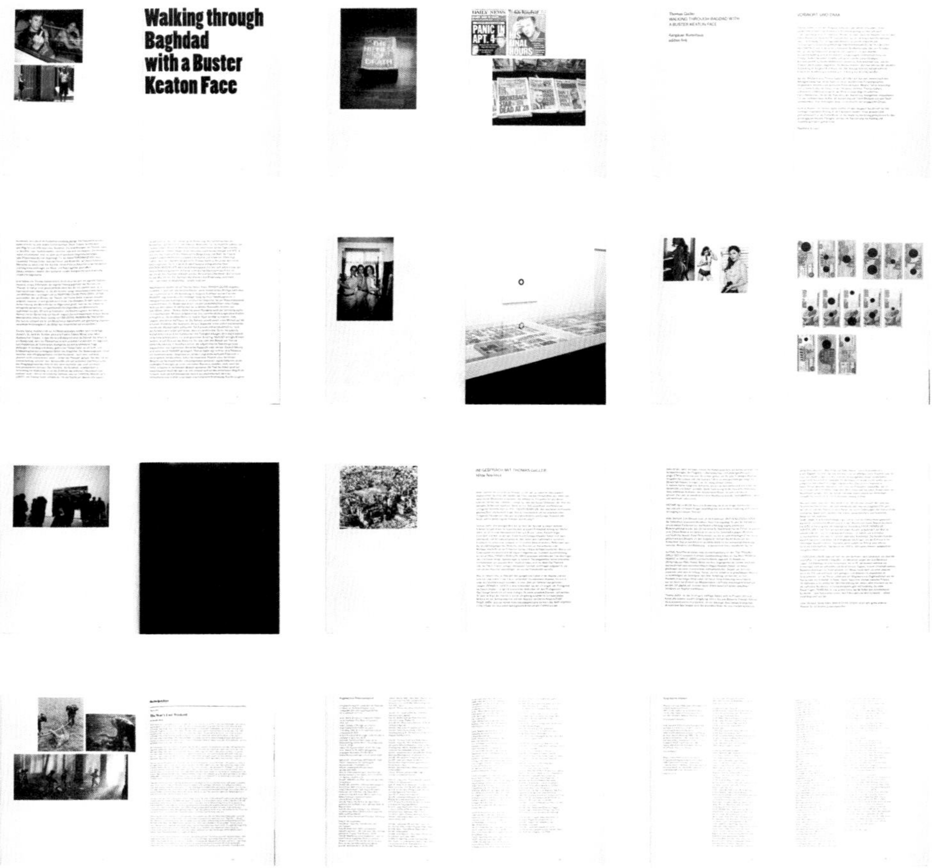
Walking through
Baghdad
with a Buster
Keaton Face

mme
lungene Verar

Thomas Galler

edition fink

IDEAL FOR VIDEO USERS
NOW INCLUDES GUIDE TO FILMS SUITABLE FOR FAMILY VIEWING
HALLIWELL'S
FILM GUIDE
11TH EDITION
FULLY REVISED AND UPDATED WITH OVER 21,000 ENTRIES
'THE FILM BUFF'S BIBLE' THE TIMES
EDITED BY JOHN WALKER

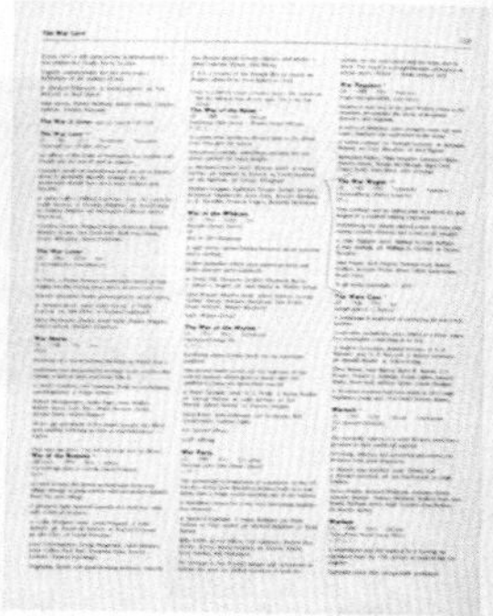

Liste von Kriegen

aus Wikipedia, der freien Enzyklopädie

Liste von Kriegen

Inhaltsverzeichnis

Antike

Ägypten und Vorderasien

- siehe Hauptartikel: Liste der Kriege und Schlachten im Altertum

China

- siehe Hauptartikel: Liste der Kriege und Schlachten im alten China

Griechenland

[illegible]

Rom bis zum Tod Kaiser Justinians (500 v. Chr. bis 565 n. Chr.)

[illegible]

Mittelalter

Frühmittelalter

[illegible]

Hochmittelalter

[illegible]

Spätmittelalter

[illegible]

Frührenaissance

[illegible]

Neuzeit

Spätrenaissance

[illegible]

17. Jahrhundert

[illegible]

- 1675–1676 King Philip's War
- 1676–1681 Russisch-Osmanischer Krieg
- 1683–1684 Reunionskrieg
- 1683–1699 Großer Türkenkrieg
- 1685 Monmouth Rebellion
- 1685 Französisch-Siamesischer Krieg
- 1687 Englisch-Siamesischer Krieg
- 1688–1697 Pfälzischer Erbfolgekrieg (Neunjähriger Krieg)
  - 1689–1691 Irischer Krieg
  - 1689–1697 King William's War in Nordamerika

18. Jahrhundert

- 1700–1721 Großer Nordischer Krieg
- 1701–1714 Spanischer Erbfolgekrieg
  - 1702–1713 Queen Anne's War
- 1714–1717 Siamesisch-Kambodschanischer Krieg
- 1714–1718 Venezianisch-Österreichischer Türkenkrieg
- 1715 Erster Jakobitenaufstand
- 1715–1717 Yamasee-Krieg
- 1718–1720 Krieg der Quadrupelallianz
- 1722–1727 Dummers Krieg
- 1722–1723 Russisch-Persischer Krieg
- 1727–1729 Englisch-Spanischer Krieg
- 1733–1735 Polnischer Thronfolgekrieg
- 1735–1739 Russisch-Österreichischer Türkenkrieg
- 1735–1737 Spanisch-Portugiesischer Krieg um Colonia
- 1739–1748 Englisch-Spanischer Krieg
- 1740–1748 Österreichischer Erbfolgekrieg
  - 1740–1742 Erster Schlesischer Krieg
  - 1744–1745 Zweiter Schlesischer Krieg
  - 1744–1748 Erster Karnataka-Krieg
  - 1744–1748 King George's War
- 1741–1743 Schwedisch-Russischer Krieg
- 1745–1746 Zweiter Jakobitenaufstand
- 1751–1754 Zweiter Karnataka-Krieg
- 1754–1756 Guarani-Kriege in Paraguay
- 1756–1763 Siebenjähriger Krieg
  - 1754–1763 Franzosen- und Indianerkrieg
  - 1756–1763 Dritter Karnataka-Krieg
- 1763–1764 Pontiac-Aufstand
- 1764–1769 Neunter Burmesisch-Siamesischer Krieg

B×4
C×2
D×2
E×1
G×2
H×1
J×1
K×1
M×1
N×1
O×1
P×1
R×1
S×1
T×2
U×1
W×1
X×1
Y×1
Z×2

这是一本记录和探讨二战后德瑞两国建筑中现代主义发展历史的展览作品集。书中通过对五家建筑师和事务所的年表介绍、采访对话、建筑分析、照片陈列，来探讨建筑作为社会空间乃至社会关系作用中对于现代性的追求。本书的装订形式为锁线胶平装。封面选用黑白双色交织的深灰色布面，丝网印刷中灰、黑色和白色油墨。内页分为两部分，前半部分胶版纸单黑印刷为主（只在隔页建筑师和事务所名处使用浅黄色），展示德瑞两国共五家建筑师和事务所的专访和建筑作品，每家展示 3 件建筑作品。文字和图片交叉混排，专访问答的版心设置为五栏，问答内容分别占据前四栏和后四栏，从而形成明显的视觉区隔。书籍的后半部分则为这 15 件建筑作品现状的彩图，为摄影师在 2008－2009 年间集中拍摄。图片逆时针旋转 90°，每两页选用一张大图配几张小图的组合，图片部分的页码在目录页上呈现为灰色。||||||||||||||||||||||||

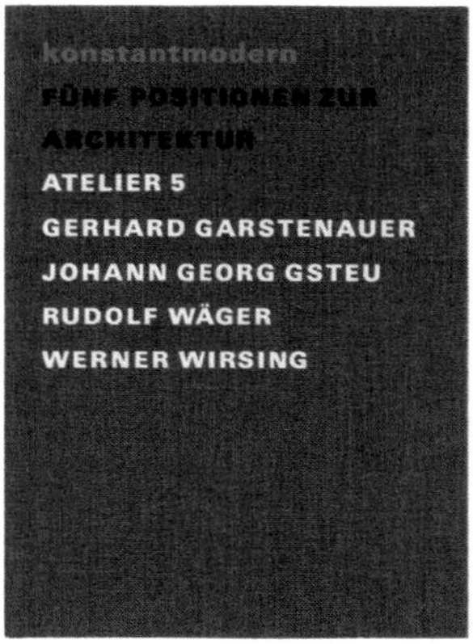
konstantmodern
FÜNF POSITIONEN ZUR
ARCHITEKTUR
ATELIER 5
GERHARD GARSTENAUER
JOHANN GEORG GSTEU
RUDOLF WÄGER
WERNER WIRSING

This is a collection of works recording and discussing the history of modernism of architecture in Germany and Switzerland after World War II. The book explores the pursuit of modernity in architecture by introducing the chronology of five architects and firms, interviews, analysis and photos. This paperback book is glue bound by thread sewing. The cover is made of dark gray fabric interwoven with black and white, and printed in gray, black and white with silk screen. The inside pages are divided into two parts. The first part is printed on offset paper in black mainly (light yellow is used only in the names of architects and firms). This part contains interviews and architectural works of five architects and firms in Germany and Switzerland, each of which displays three architectural works. Texts and pictures are mixed and typeset. Questions and answers of the interviews are typeset in five columns. The second part of the book is colorful pictures of these 15 architectural works, which were taken by photographers during 2008 and 2009. All the pictures are rotated 90 degrees counterclockwise. Every two pages contain a big picture and several small ones. The page numbers of the image section appear gray on the catalog page.

银奖
永恒的现代性——建筑界的五个代表
奥地利

Silver Medal
konstantmodern. Fünf Positionen zur Architektur
Austria
Reinhard Gassner, Marcel Bachmann / Atelier Reinhard Gassner
aut. architektur und tirol (Ed.)
Springer WienNewYork
238×168×20mm
648g
256p

The book's title says it all – not least as far as its design is concerned. The haptic and aesthetic quality of its cover entices the reader to delve into the statements of the Atelier 5 architects, here discussing five architectural stances from Austria, Germany and Switzerland. What makes the book so agreeable and manageable is its grey cloth which rather than encompassing the cover sits loose. The colour scheme of cover and interleaves is finely gauged, whilst the typography is restrained and meticulous. The attractive photographs, which in the second (essay) section of the book have been rotated, are all given plenty of space.

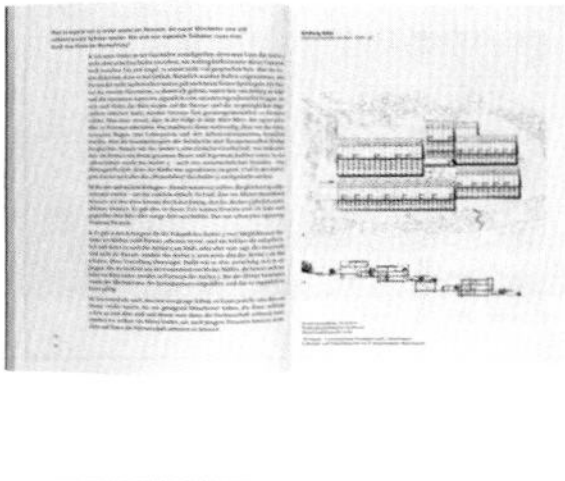
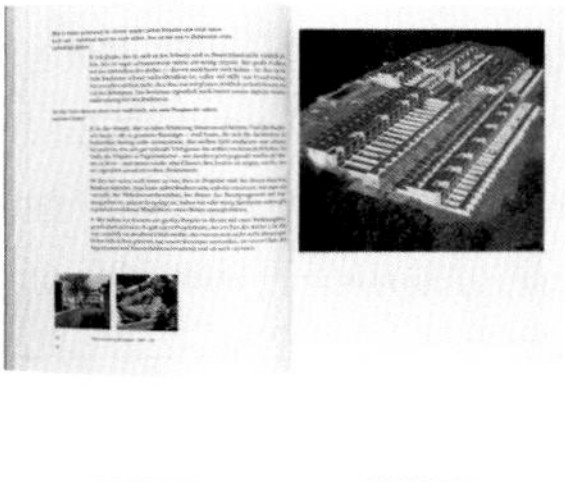

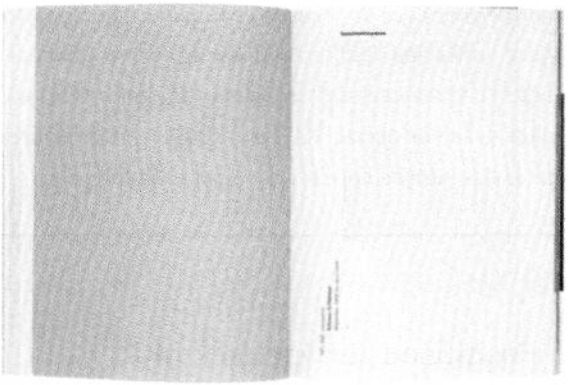
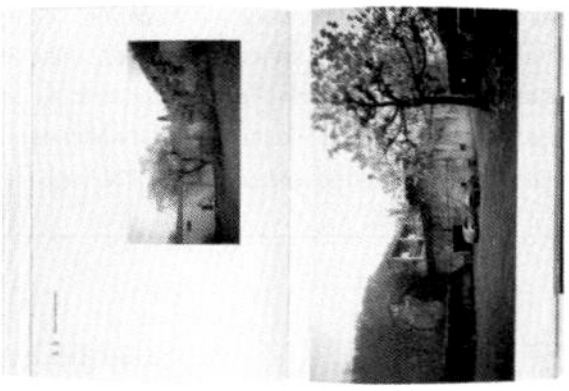
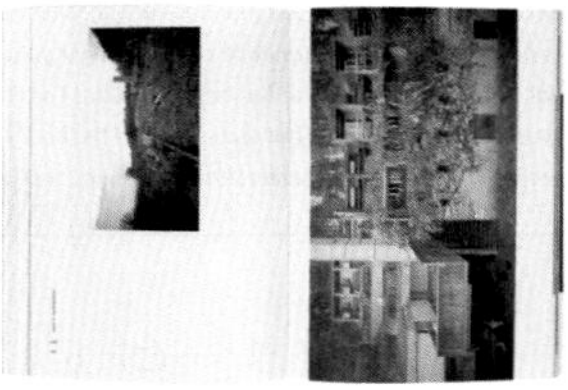
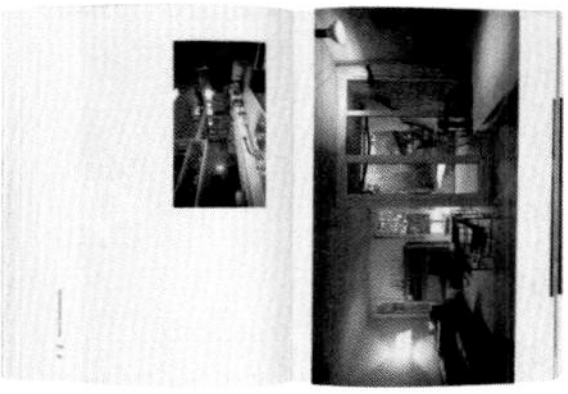

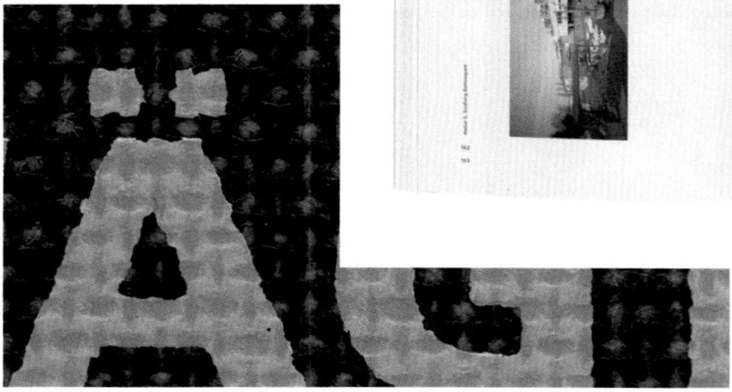

ATELIER 5

Bauau
rm
ein kleines

konstantmodern

Mich wundert, dass ihr immer noch Schwierigkeiten in der Umsetzung habt – vor allem in Hinblick darauf, dass die Zufriedenheit und die Bindung der Bewohner eurer Siedlungen sehr groß sind und sogar so weit gehen, dass manche von einer zur nächsten Siedlung von euch ziehen, wenn es ihre geänderte Lebenssituation erfordert.

P: Viele Bauträger sagen, wir haben uns eure Projekte angeschaut, die sind wunderbar, funktionieren vielleicht in der Schweiz, aber leider können wir d uns nicht verkaufen. Punkt!

Bei den Besuchen eurer Siedlungen ist mir aufgefallen, dass anscheinend ein Segment der Bevölkerung dort lebt, das sich genau mit dieser Art zu leben – nämlich in einer Gemeinschaft – und mit den spezifischen Wohnungszuschnitten identifiziert.

M: Es sind nach wie vor Liebhaber, die in einer Atelier-5-Siedlung wohnen die sind vielleicht nicht so weit gestreut. Ich glaube, dass wir Menschen ar chen, die offen sind, auch lieb gewonnene Wohnerfahrungen aufzugeben. In gewissen Schicht gibt es sicher eher die Bereitschaft zu sagen, das probier jetzt mal aus.

B: Es gibt aber auch eine Siedlung von uns, nämlich Wertherberg, die kurz Halen gebaut wurde und von Menschen bewohnt wird, die sich anfänglic dieser Architektur nicht identifizieren konnten. Sie waren zu Anfang unglü und begannen die Architektur mit nicht gerade gutem Geschmack durch Ste tate etc. zu „verschönern". Interessant ist dabei, dass die Bewohner ihre Häu heute pflegen und sie nicht verkaufen. Sie meinen, die Architekten hätten eine tolle Siedlung gebaut, nur keine schönen Materialien verwendet. Darau stand z. B. auch eine kollektive Aktion, denn einmal im Jahr werden geme die Dächer „schön" gestaltet. Das finde ich eigentlich interessant.

Ihr beschäftigt euch kontinuierlich mit dem Thema Siedeln bzw. Siedlung und habt von Anfang an verschiedene Konzepte verfolgt.

P: Alle Konzepte und Siedlungen haben eigentlich immer den Ansatz von Urb im Hintergrund. In Halen hat man versucht, ein kleines Dorf zu bauen – mi Straße, einem Platz, einem Laden, einem Kindergarten und Ateliers. Abe könnte Halen auch in die Stadt implantieren und es würde gut funktioniere Verbindung von Arbeiten und Wohnen in einer Siedlung hat eigentlich unge Qualitäten, denn ich habe meinen Arbeitsplatz in der Nähe – nicht in der Wohnung, sondern ich gehe aus dem Haus, schließe ein Atelier auf, arbeit und gehe abends wieder nach Hause, und das innerhalb einer Siedlung. Vorstellung fällt vielen Leuten aber offensichtlich schwer, denn wir haben Ansatz in Frankfurt und in Mainz probiert, aber von unseren Ideen ist am nur noch ein Fragment übrig geblieben. Ähnlich verhält es sich bei Gemeinsc räumen, die es in vielen Siedlungen in der Schweiz gibt.

010 Sm²

这是一本为瑞士历史最悠久的自然生态委员会成立 100 周年而做的纪念书籍。通过对委员会的介绍、理念描绘、人与自然关系的深入思考、瑞士风景、委员会历史等信息，记录委员会的成果，展望未来环保的目标。本书锁线硬精装，封面为与内页尺寸齐平，未采用超出内页的卡板，且胶版纸四色印刷后直接正面裱覆卡板上，而未做向内包裹。内页根据内容划分，采用四种纸张划分出五大部分。第一部分采用中黄色艺术纸印单黑，简述委员会及理念；第二部分采用浅灰色胶版纸四色印刷，深度探讨自由与自然的悖论、生物多样性、城市的发展等话题；第三部分采用哑面铜版纸四色印制，展示专业摄影师拍摄的瑞士自然风光照片；第四部分采用浅黄色双胶纸印单黑，概述协会历史；最后一部分采用与第一部分相同的纸张印单黑，为大事年表和其他附文信息。由于纸张选用的关系，在书口上呈现由外而内、由深变浅的层级变化。

DIE STIMME DER NATUR
100 Jahre Pro Natura
pro natura
KONTRAST

This is a commemorative book for the 100th anniversary of Switzerland's oldest Natural Ecology Commission. By introducing the committee, describing its concept, thinking deeply about the relationship between human and nature, showing Swiss landscape, and reviewing the history of the committee, recording the achievements of the committee, the book looks forward to the goals of future environmental protection. The hardcover book is bound with thread sewing. The cover is the same size as the inside pages. The offset paper is printed in four colors and directly mounted by the graphic board without inward wrapping. The inside pages are divided into five parts according to the content, using four kinds of paper. The first part uses medium yellow art paper printed in black to outline the Committee and its concept; the second part uses light gray offset paper printed in four colors to discuss the paradox between freedom and nature, biodiversity, urban development and other topics deeply; the third part uses matte art paper printed in four colors to show the pictures of Swiss natural scenery taken by professional photographers. The fourth part uses light yellow double-offset paper printed in black to outline the history of the association; the last part uses the same paper as the first to list chronology of major events and other information. Because of the variety of paper selected, there is a hierarchical change from outside to inside.

银奖
大自然的声音——Pro Natura 100 周年
瑞士

Silver Medal
Die Stimme der Natur. 100 Jahre pro Natura
Switzerland
Marco Müller
Pro Natura (Ed.)
Kontrast Verlag, Zürich; Edition Slatkine, Genève
250×169×18mm
670g
196p

"Die Stimme der Natur" – the voice of nature – speaks here through a variety of delicate uncoated papers and a clear, subtle and unruffled design. The hundred-year history of "Pro Natura" is told in a very harmonious fashion. A striking photographic concept, presenting various natural locations with fascinating pictures plus excellently readable typography combined with documentative b/w photos introduces us to the history of and themes surrounding nature conservation in Switzerland. A highly successful overall conception for an auspicious anniversary.

DIE STIMME DER NATUR
100 Jahre Pro Natura

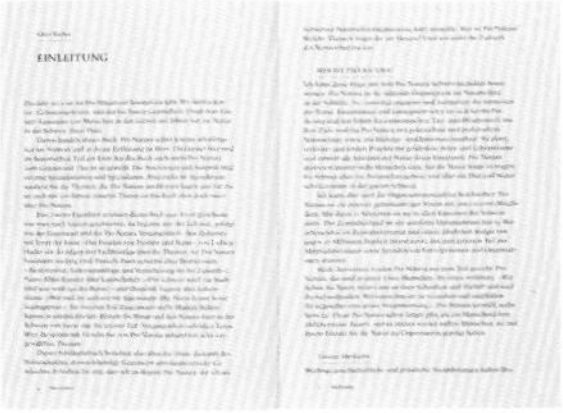
EINLEITUNG

DAS PARADOX VON FREIHEIT UND NATUR
BIODIVERSITÄT
DIE SCHWEIZ WIRD ZUR STADT
WIR UND DIE ANDEREN IM ALPENRAUM

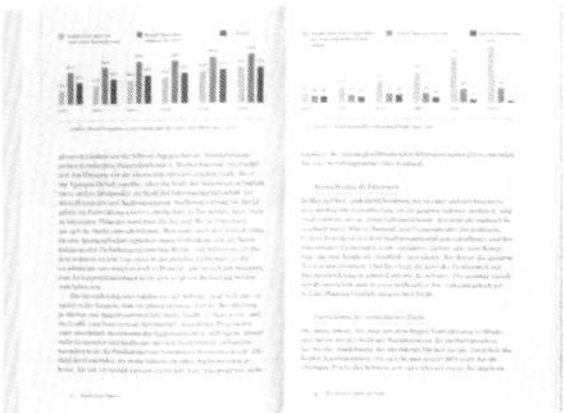

EINE FOTOGRAFISCHE REISE DURCH DIE SCHWEIZ VON HEUTE

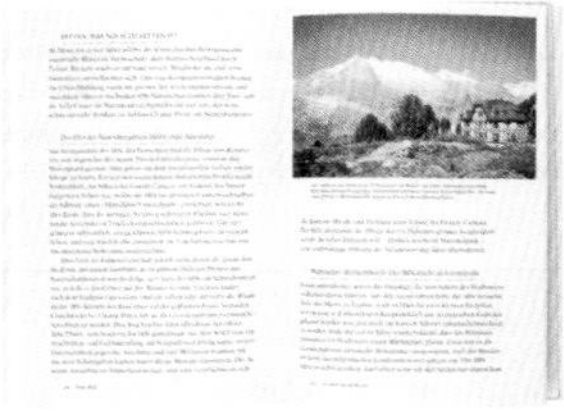

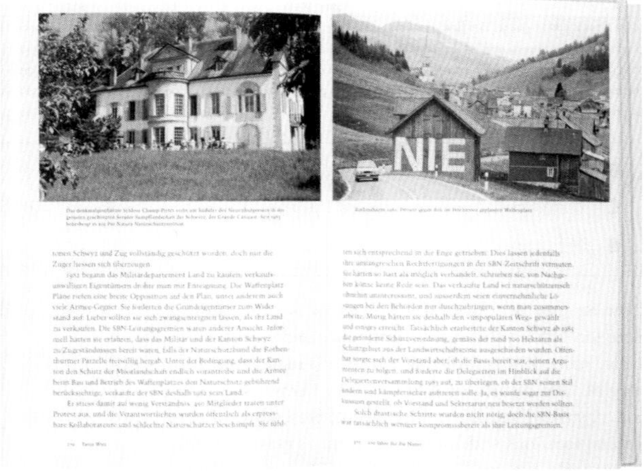
NIE
tonen Schwyz und Zug vollständig geschützt worden, doch nur die Zuger liessen sich überzeugen.
1982 begann das Militärdepartement Land zu kaufen; verkaufsunwilligen Eigentümern drohte man mit Enteignung. Die Waffenplatz-Pläne riefen eine breite Opposition auf den Plan, unter anderem auch viele Armee-Gegner. Sie forderten die Grundeigentümer zum Widerstand auf: Lieber sollten sie sich zwangsenteignen lassen, als ihr Land zu verkaufen. Die SBN-Leitungsgremien waren anderer Ansicht. Informell hatten sie erfahren, dass das Militär und der Kanton Schwyz zu Zugeständnissen bereit waren, falls der Naturschutzbund die Rothenthurmer Parzelle freiwillig hergab. Unter der Bedingung, dass der Kanton den Schutz der Moorlandschaft endlich vorantreibe und die Armee beim Bau und Betrieb des Waffenplatzes dem Naturschutz gebührend berücksichtige, verkaufte der SBN deshalb 1982 sein Land.
Er stiess damit auf wenig Verständnis. 400 Mitglieder traten unter Protest aus, und die Verantwortlichen wurden öffentlich als erpressbare Kollaborateure und schlechte Naturschützer beschimpft. Sie fühlten sich entsprechend in die Enge getrieben. Dies lassen jedenfalls ihre umfangreichen Rechtfertigungen in der SBN-Zeitschrift vermuten. Sie hätten so hart als möglich verhandelt, schrieben sie, von Nachgeben könne keine Rede sein. Das verkaufte Land sei naturschützerisch ohnehin uninteressant, und ausserdem seien einvernehmliche Lösungen bei den Behörden nur durchzubringen, wenn man zusammenarbeite. Mutig hätten sie deshalb den «unpopulären Weg» gewählt und einiges erreicht. Tatsächlich erarbeitete der Kanton Schwyz ab 1984 die geforderte Schutzverordnung, gemäss der rund 700 Hektaren als Schutzgebiet aus der Landwirtschaftszone ausgeschieden wurden. Offenbar sorgte sich der Vorstand aber, ob die Basis bereit war, seinen Argumenten zu folgen, und forderte die Delegierten im Hinblick auf die Delegiertenversammlung 1985 auf, zu überlegen, ob der SBN seinen Stil ändern und kämpferischer auftreten solle. Ja, es wurde sogar zur Diskussion gestellt, ob Vorstand und Sekretariat neu besetzt werden sollten.
Solch drastische Schritte wurden nicht nötig, doch die SBN-Basis war tatsächlich weniger kompromissbereit als ihre Leitungsgremien.

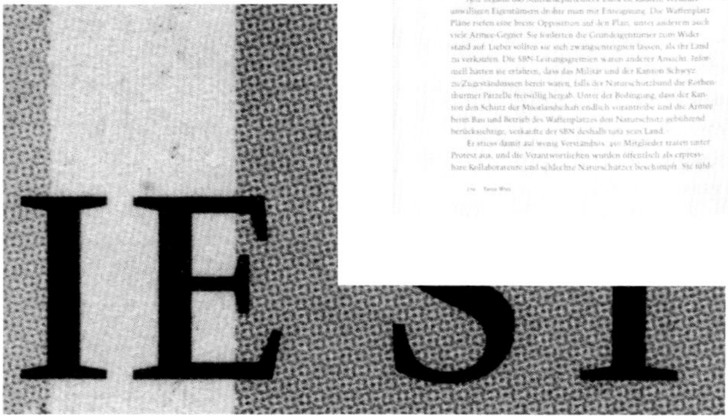
IE ST

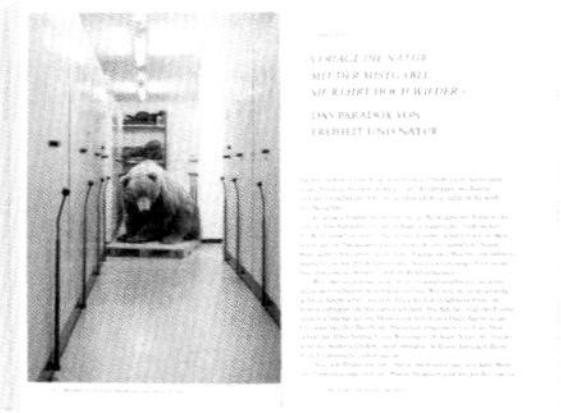

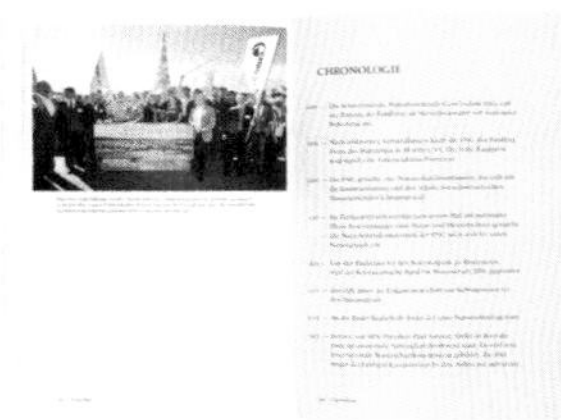
CHRONOLOGIE
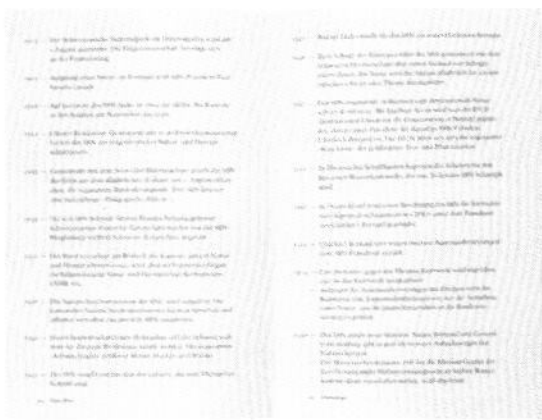
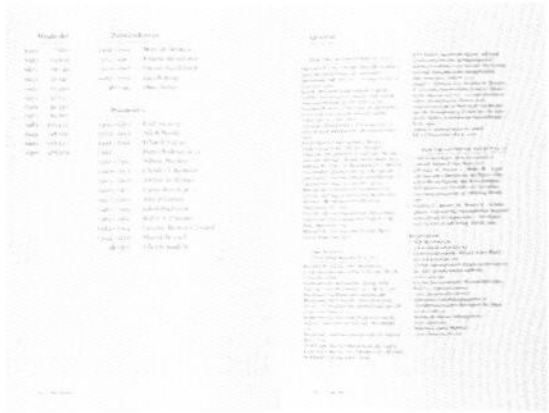

DIE STIMME DER NATUR
100 Jahre Pro Natura

46 Reliefmodell, Wallis, ETH-Bibliothek Zürich, Kartensammlung

Pierre-Alain Rumley

DIE SCHWEIZ WIRD ZUR STADT

Und was wird aus der Natur?

Die Schweiz ist in den letzten Jahrzehnten verstädtert, das ist ein klares Faktum. Aber in welchem Ausmass und warum? Leben wir heute in einer einzigen Metropole («Metropole Schweiz»)? Die Raumplanung war eigentlich dazu gedacht, die Entwicklung der Verstädterung zu steuern. Hat sie versagt, ganz oder teilweise, und wenn ja, warum? Hätte man es besser machen können, und wie? Was ist in Zukunft zu tun?

DIE VERSTÄDTERUNG DER SCHWEIZ

Entwicklung der städtischen Bevölkerung

Die Verstädterung ist eine weltweite Erscheinung. Die Bevölkerung, die man als städtisch bezeichnet (allerdings je nach Land nach unterschiedlichen Kriterien), hatte 2003 einen Anteil von 48,3 Prozent (74,5 Prozent in den entwickelten Ländern) und dürfte bis ins Jahr 2030 auf 60,8 Prozent (81,7 Prozent in den entwickelten Ländern) ansteigen (Véron, 2006).

Die Schweiz ist eines der am meisten entwickelten Länder der Welt und demnach auch eines der am meisten verstädterten: Der Anteil der in Städten (genau genommen in den Agglomerationen) lebenden Bevölkerung stieg von 45 Prozent im Jahr 1950 auf 73 Prozent im Jahr 2000 an (siehe Grafik 1 auf der nächsten Seite).

Das Wachstum der Städte kann auf verschiedene Weise erklärt werden. In den Ländern der Dritten Welt ist es auf den starken Bevölkerungsdruck und auf die Flucht vom Land und aus der Landwirtschaft zurückzuführen. Eine immer zahlreichere Bevölkerung kommt auf der häufig ungewissen Suche nach Arbeit in die Städte. In den industri-

Inszenierungen
Bm

这本貌似普通的摄影作品集的书，并非以艺术为第一目的，而是一本记录柏林城市工业荒地和工业区状态，进而阐述城市和政体转型过程的社会科学方向的实验书籍。本书锁线硬精装，护封白色卡纸单黑印刷，内封浅棕色环保纸双色印刷后裱覆卡板。内文哑面艺术纸黑色和浅棕色双色印刷，除了文字部分的双色搭配外，照片也采用两色的印刷，浅棕色丰富了照片的中亮部细节，并且给整个照片带来一点偏暖的色调。书中由于有大量满版甚至拉页照片，为了避免页码对照片的影响，连续几页的页码会集中放置，带来独特的视觉体验。这是一本记录城市转变过程的书，而非记录最终阶段。书籍将过程“冻结”，让它可以被检索和回溯。

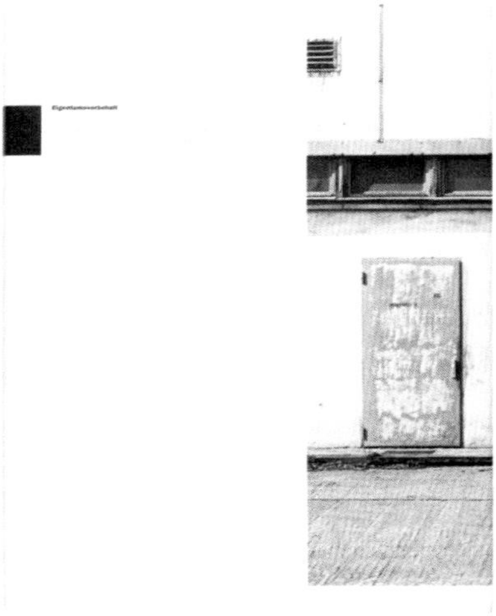

This collection of photographs is not a book for the first purpose of art, it is an experimental book that records the state of industrial wasteland and industrial area in Berlin, and then expounds the social direction in the process of urban and political transformation. The hardcover book is bound with thread sewing, and the jacket is printed in black on white cardboard. The inside pages are printed in black and light brown on matte art paper. Two-color printing is not only used for the text part, but also used for photos. Light brown enriches the details of the bright part of the photos, and brings a warm tone to the whole pages. Since there are a large number of full-page and gatefold photos in the book, the page number of consecutive pages will be placed intensively in order to avoid the influence of page number on photos. It brings unique visual experience. This is a book that records the process of urban transformation, not the final stage. The book "freezes" the process so that it can be searched and retrieved.

- 铜奖
- 保留所有权
- 德国

An extraordinary book with compelling atmosphere and a keen sense of economy. The pages' presentation allows the b/w photographs to tell their own story: wasteland, spaces, deserted places bear witness to a changing urban landscape. In its content, colour and positioning the text – as an independent entity sensitively placed in the margins and accompanied by plenty of blank space – makes reference to the illustrations, at the same time fitting in with the overall structure of the book in a balanced fashion. A book created most meticulously and with a careful hand.

- Bronze Medal
- Eigentumsvorbehalt
- Germany
- Gaston Isoz
- Gaston Isoz (Ed.)
- disadorno edition, Berlin
- 305×224×11mm
- 738g
- 88p

155

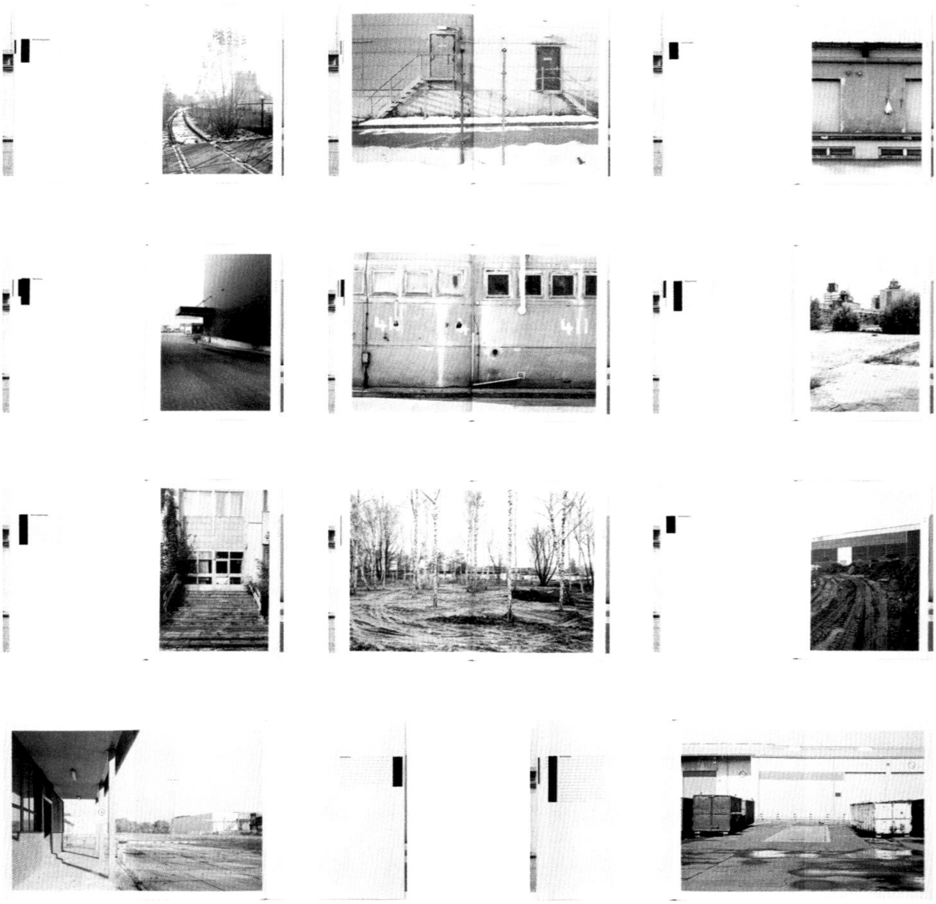

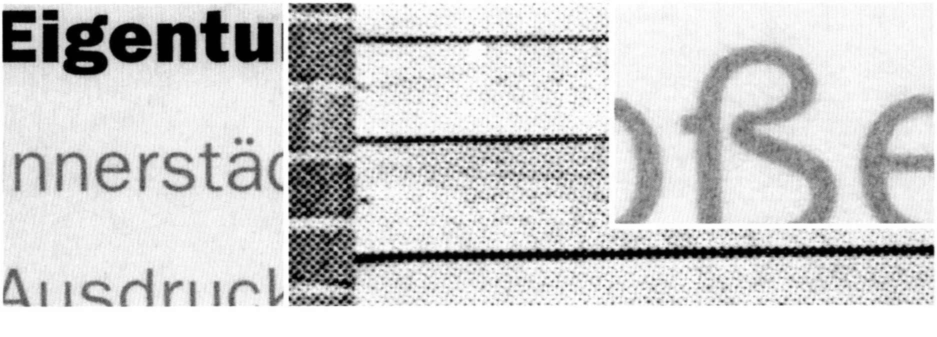
Eigentu
nnerstäc
Ausdruc

Eigentumsvorbehalt
Maschinerie

030000
BORDERLINE
STREET
"The
are to
deciphe
"These people go out f
driven by necessity that
or war, and oppression, wh
country is experienced by th
are constrained to seek new la
they are most formidable, and if t
confronted by good arm
never be held back."
MIGROPOLIS
Venice / Atlas of a Global Situation

这是一套汇集了威尼斯大量书籍统计的档案。在哲学家沃尔夫冈 · 舍佩（Wolfgang Scheppe）的带领下，威尼斯建筑大学的学生对整个威尼斯的各方面进行了统计，最终呈现了这本涵盖数万张照片、发展状况、人口流动等多方面数据和资料的档案集。本书一套两本锁线胶装，封面白卡纸四色印刷，覆哑膜，使用白卡纸印单黑覆哑膜的简易函套进行包裹。内页哑面铜版纸四色印刷，内容涵盖威尼斯的发展、全球化情况、交通及周边联系、娱乐生活、移民和现状等。因为数据量巨大，内容多达 1300 多页，书籍不得不分为两本，两本页码顺序为承接关系。为便于查找，书籍采用了一种类似于大富翁游戏方形环状棋盘的形式，对全书内容进行编号。两本开篇也均包含全套书的目录和使用图例。

"The refugee should be considered for what he is, that is, nothing less than a border concept."
Giorgio Agamben
"When I first came to Venice I tried, as a sort of autodidact, to decode the aesthetic of Venice. I observed the façades of the great palazzos on the Canal Grande, including the Ducal Palace, and was struck by the complete indifference to anything that was not the façade.
'Aha', I thought, 'an aesthetic derived more from scenography than architecture', and was quite proud of myself. A bit later I learned that John Ruskin had said as much nearly a century and a half before.
I came to believe that it was impossible to make an original observation about Venice, a thought which in fact was not itself original, but had been authored by Mary McCarthy fifty years earlier.
It was not until I saw the beginnings of Wolfgang Scheppe's Migropolis project that I understood the possibility of inventing a visual language to describe contemporary Venice. While Migropolis interrogates the mechanisms that globalization deploys to plunder Venice, it is equally a critique-in-practice of the means of representation available for a visual analysis: graphics and photography. In that it confronts the Debordian quandary: The spectacle can only be critiqued in spectacular terms."
Lewis Baltz
"For indeed any city, however small, is in fact divided into two, one the city of the poor, the other of the rich; these are at war with one another."
Plato
HATJE CANTZ

This is a set of archives compiling the statistics of a large number of books in Venice. Led by the philosopher Wolfgang Scheppe, the students of Venice University of Architecture made statistics on all aspects of Venice, and eventually presented this archive of data and information covering tens of thousands of photographs, development, population shift and so on. This set contains two books bound with thread sewing. The laminated cover is printed in four colors on white cardboard. A simple slipcase using white cardboard printed in monochrome black is used to wrap the books. Printed in four colors on matte coated paper, the contents cover Venice's condition of development, globalization, transportation and surrounding relations, entertainment life, and immigration. Because of the huge amount of data and more than 1300 pages of content, the book has to be divided into two volumes. However, the pages of two books are numbered continuously in order to facilitate the search. Both books begin with a full set of catalogues and illustrations.

- 铜奖
- 威尼斯大都市 / 全球形势图集
- 德国

- Bronze Medal
- Migropolis. Venice / Atlas of a Global Situation
- Germany
- Wolfgang Scheppe mit Katerina Dolejšová, Veronica Bellei, Miguel Cabanzo
- Wolfgang Scheppe, The IUAV Class on Politics of Representation (Ed.)
- Hatje Cantz Verlag, Ostfildern
- 240 × 170 × 73mm
- 3252g
- 1344p

A highly complex and multifaceted topic has been rendered and mastered here through book design of the highest order! The book is convincing thanks to its superb picture grid and an accompaniment which in its typography is impressively meticulous. The type size remains very small yet is easy to read. Field studies on the anatomy of the city of Venice (with pictures, data representation, essays, interviews and case studies) are no less exemplary in the graphic form which they have assumed; and, last but not least, its information graphics are very convincing.

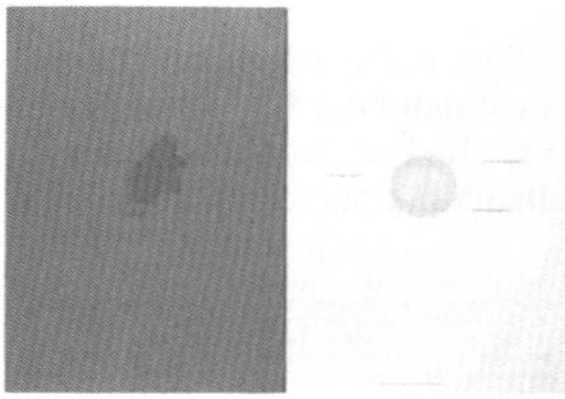

MIGROPOLIS

MIGROPOLIS

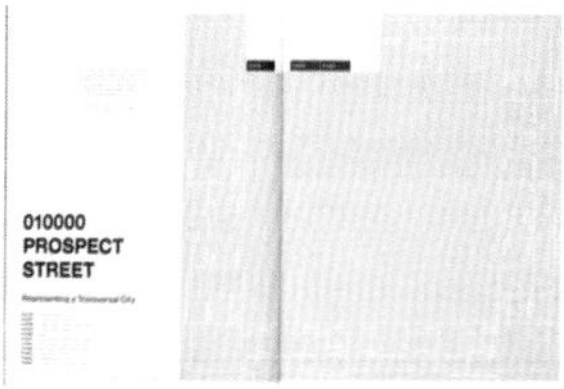
010000
PROSPECT
STREET

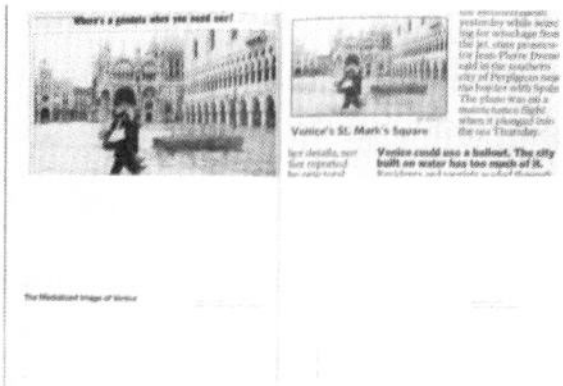
Venice's St. Mark's Square

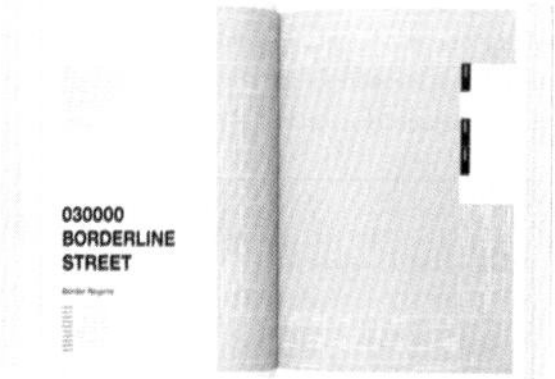
030000
BORDERLINE
STREET

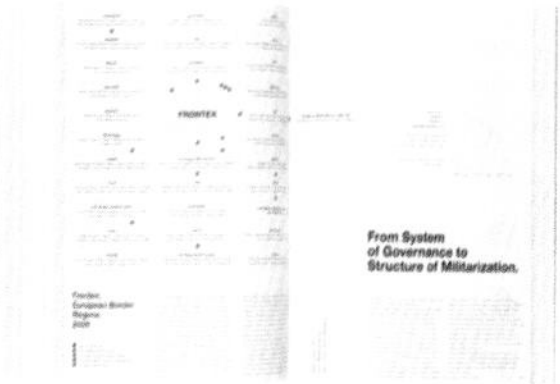
FRONTEX
From System
of Governance to
Structure of Militarization.

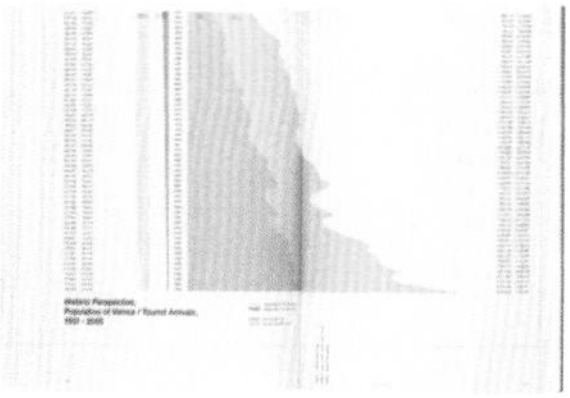

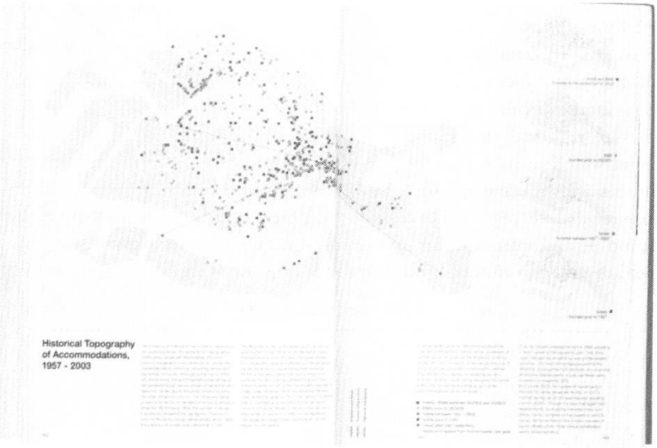
Historical Topography
of Accommodations,
1957 - 2003

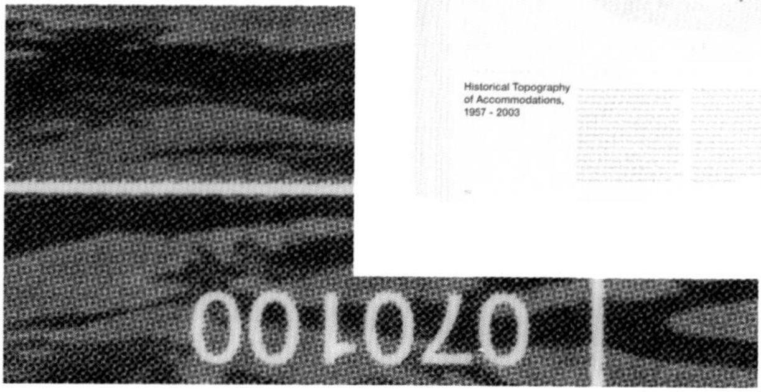

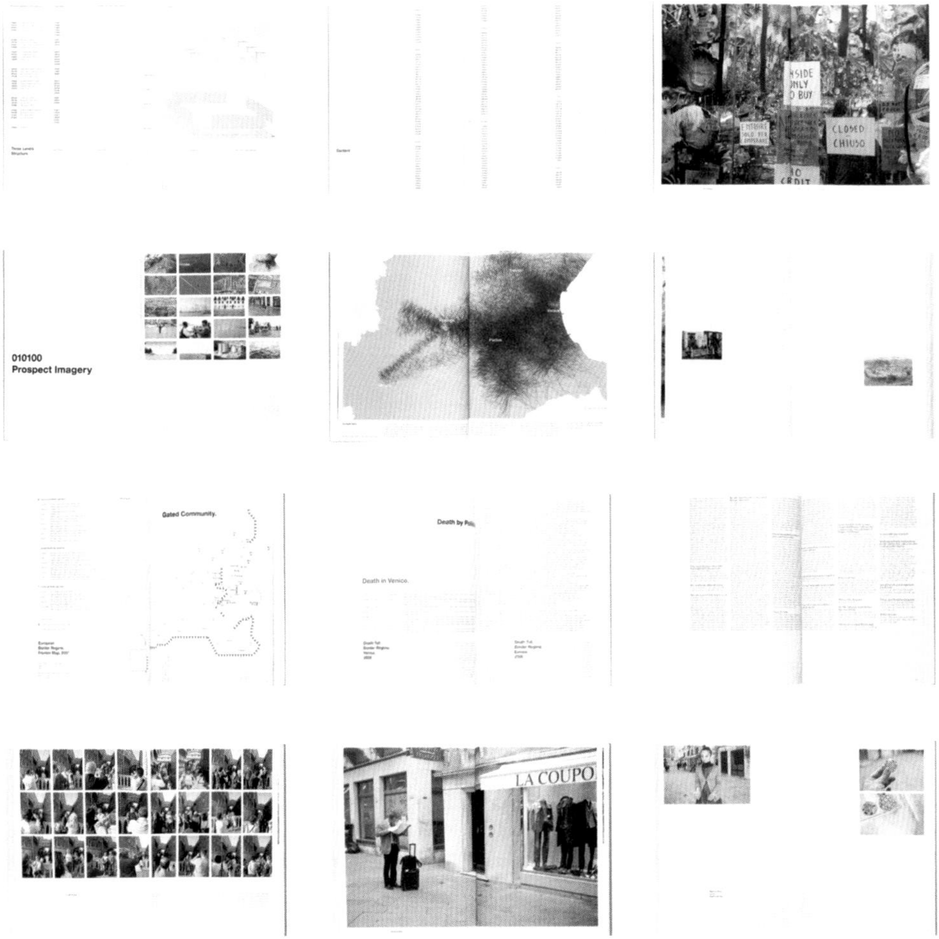
CLOSED
CHIUSO
010100
Prospect Imagery
Gated Community.
Death in Venice.
LA COUPO

carri
of Venice The

MIGROPOLIS
Venice / Atlas of a Global Situation
Wolfgang Scheppe
& the IUAV Class on
Politics of Representation
Vol. II.
HATJE CANTZ
MIGROPOLIS
Venice / Atlas of a Global Situation
Wolfgang Scheppe
& the IUAV Class on
Politics of Representation
Vol. I.
HATJE CANTZ
Index

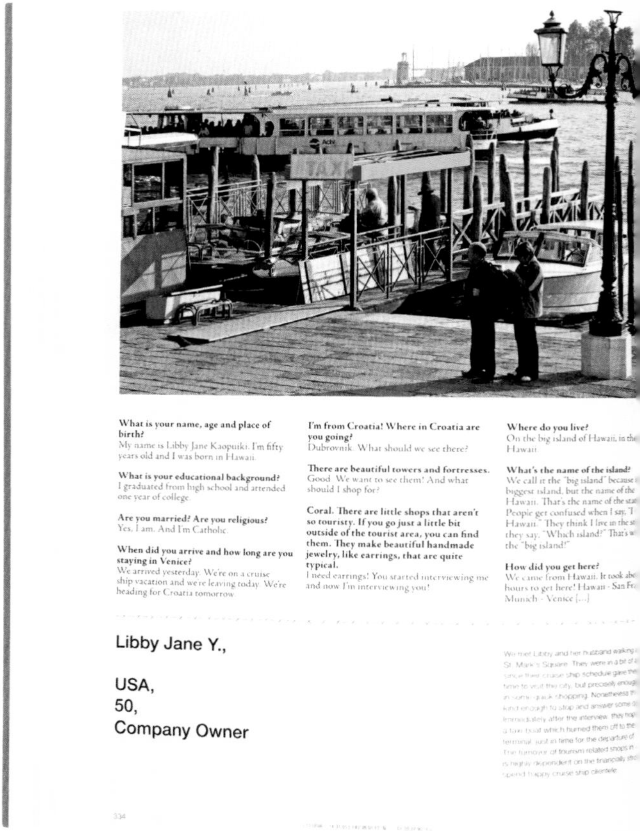

What is your name, age and place of birth?
My name is Libby Jane Kaopuiki. I'm fifty years old and I was born in Hawaii.

What is your educational background?
I graduated from high school and attended one year of college.

Are you married? Are you religious?
Yes, I am. And I'm Catholic.

When did you arrive and how long are you staying in Venice?
We arrived yesterday. We're on a cruise ship vacation and we're leaving today. We're heading for Croatia tomorrow.

I'm from Croatia! Where in Croatia are you going?
Dubrovnik. What should we see there?

There are beautiful towers and fortresses.
Good. We want to see them! And what should I shop for?

Coral. There are little shops that aren't so touristy. If you go just a little bit outside of the tourist area, you can find them. They make beautiful handmade jewelry, like earrings, that are quite typical.
I need earrings! You started interviewing me and now I'm interviewing you!

Where do you live?
On the big island of Hawaii, in th
Hawaii.

What's the name of the island?
We call it the "big island" because
biggest island, but the name of the
Hawaii. That's the name of the sta
People get confused when I say, "I
Hawaii." They think I live in the s
they say, "Which island?" That's w
the "big island!"

How did you get here?
We came from Hawaii. It took ab
hours to get here! Hawaii - San Fr
Munich - Venice [...]

Libby Jane Y.,

USA,
50,
Company Owner

334

our cruise schedule?
ovember 11-26. Next, we're going
, then to Ancona and then to
e're also going somewhere in Spain,
nd up in Portugal.

our profession in Hawaii?
wo companies. We have an auto-
p and my company helps other
t their own businesses and run them.

ur first time here?
r second time.

id you come back?
t her husband] Because he's never
! He's very Catholic!

l you imagine Venice to be like
ming here?
e thirty years ago. I don't remember.
finished high school. I don't think I
reconceived notion of Venice itself.
out Paris, and that was it. So it was
me. I walked into a glass shop
of this building [*indicating Saint*
d saw this beautiful cut-glass crystal
ed to buy it but my mother said,
eak when you bring it back home!"
d thirty years to come back and buy
last set and that shop has been there
erations!

bby's husband]
ave any idea about Venice prior
g here?
d seen St. Mark's Square on
mmercials, the old buildings.
esn't have buildings like this. The
we see those on TV. Do you know
? They have all of this right there,
ndoliers [...] It's huge, big like this
ven billion dollars!

u tell us about your background?
born in Washington, but my family
m Germany. They left Germany
0s – my great-great-grandparents.
igrated for work. They went to
n and became farmers. Then my
wed to Washington and, in the end,
awaii!
I was born, Hawaii wasn't part
yet. It was a territory until I was
the last state.

And the happiest!
L: You live longer in Hawaii than in any other state!
H: We have a wonderful lifestyle: very family-oriented, no winters, the same temperature all year. It varies maybe by 5 or 7 degrees, that's all! You can swim in the ocean every day, all year.
L: You can do any sport you can imagine: ski, surf...

And the cost of living?
H: The cost? The cost is higher than here. But it costs more to live in California. California itself isn't so expensive, but taxes are very high there.

The cost of living is high here.
H: How much does it cost to buy an apartment here in Venice?

I would say that a two-bedroom apartment in this area costs about € 400,000.
L: That's expensive! In Hawaii that's the price of our house.

Are there earthquakes?
L: We just had an earthquake that destroyed all my glassware! It measured 6.9 on the Richter scale.
H: TV set out of its shell, water shooting all over the place. For me it was the third big one in thirty years. They hit about every ten years I guess. But Hawaii is still a beautiful place to live. You don't need heating in the houses. We wear shorts and T-shirts all year [...] Anymore questions?

What do you think about Bush making a wall on the border with Mexico?
H: The wall? Stupidity. Because the wall is this big, and the border is this big [*indicating that the wall is smaller than the border*]! So you can walk around it. It doesn't make any sense. Americans are very nervous about terrorism, but now it's unnecessary, because if they want to get us they will! We're spending so much money on homeland security.

Have you seen any Michael Moore movies? There's one in particular which deals with the United States-Canada border [...]
H: Actually, the Canadian border was very open for a long time. You could walk across. Now they're getting paranoid on both sides. Not so much with Mexico. Mexico is very upset about the wall. And if you took all the illegal Mexicans out of the United States, there would be no more workers. They do all the jobs that Americans don't want to do. Americans are very spoiled people. But I would say that we have much better manners than the people here. Everybody walks into everybody here. I think they're a little rough. In the US people are a bit more polite.

In Venice, immigration is a big issue. There are all kinds of jobs that Venetians won't do [...] and tourism is based on foreign workers.
L: That's how Hawaii is. You have to speak Spanish to work in Hawaii. We have Spanish-speaking people and Philippinos. If we didn't have them, we would have no cooks. [...]

At the same time, they're marginalized by society.
H: Yes, they're outside of society. They don't have medical insurance and things like that. It's very sad because our country was built by people coming from other countries!

Is there something particular that you like here, aside from the glass?
L: Besides the glass... Oh! I love the canals and the bridges. I tried to find the Bridge of Sighs to show him, but I couldn't remember where it is!

It's just over there, around the corner, on the left.
L: I knew it was here somewhere.

We can show you if you want.
H: Oh no. We've got to get in the taxi to get back to the ship.

Thank you very much for having spent some of your time with us.
H: OK girls. Get those college degrees. They're necessary in today's world.

020000 Global Street
020001 Libby Jane Y.

010 Bm³

…er Kosolapov
… Kruger
…mann
…ne Meckseper
…ekas
…hailov
…denbach
…e Paik

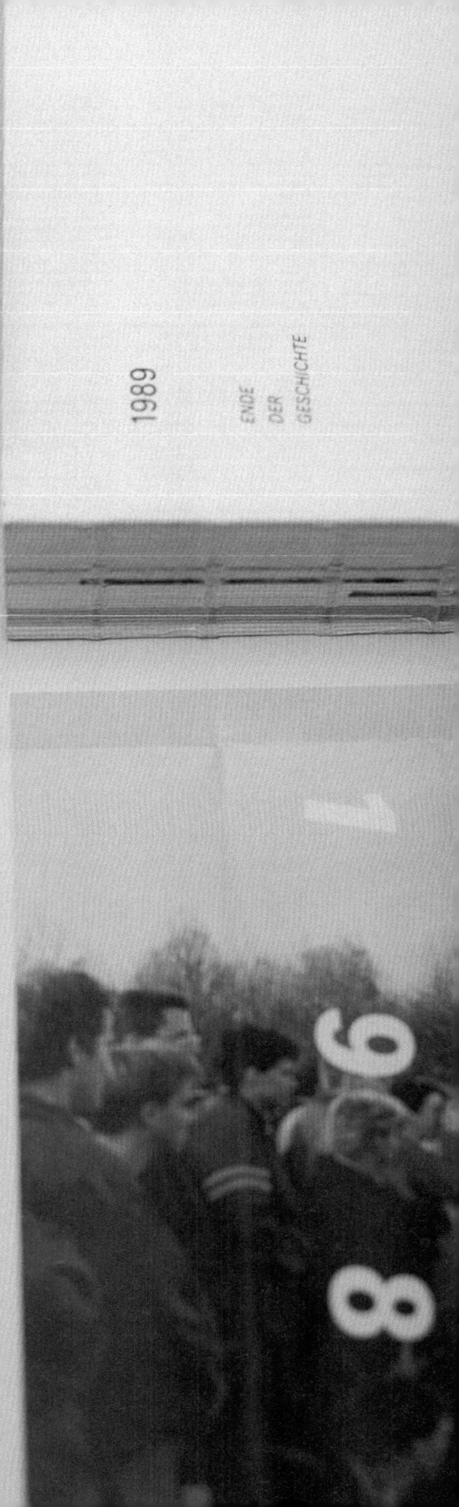
1989
ENDE
DER
GESCHICHTE

这是一本记录和研究在大历史的转折中一系列连锁反应的书。书中内容包含关于历史的阐述、访谈、艺术家创作、摘录等信息，从多方面展示世界格局的改变给人们的生活、心理带来的影响。本书在书籍设计、纸张运用、字体和排版上都尽量简化，体现出很强的功能主义和厚重的文献味道。书籍为裸脊锁线平装，封面胶版纸单黑印刷，外包印单黑照片的浅灰色纸。内页主要纸张与封面一致，文本部分印单黑，艺术家作品部分则四色印刷。文本只使用了一种字体，且只有两个字号（正文和图说），标题通过排版位置和下画线来凸显。后半部的摘录采用粉红色双胶纸印单黑，与前文不同的打字机字体具有很强的档案感。除开篇的目录外，每一章节均有详细内容作为隔页，如同地铁各处的导视系统，严谨有序。全书厚达 33mm，但因为柔软的纸张和未经特殊处理的封面，有很好的翻阅手感。

1989

This is a book that documents and studies the chain reactions that occur at the turning point of great history. The book contains information about historical exposition, interviews, works, excerpts and so on. It shows the impact on people's life and mentality when the world pattern has been changed. The book is as simple as possible in book design, paper selection, fonts and typesetting, reflecting a strong functionalism. The paperback book is bound with thread sewing, and the cover is printed in black on offset paper wrapped by light gray paper with monochrome photographs. The paper used for the inside pages is the same as the cover. The text is printed in black and the artist's works are printed in four colors. Only one font and two font sizes (text and illustration) are used for the text, and the title is highlighted by the position and underline. The second half of the excerpt is printed on pink double offset paper in black. The typewriter font differs from the fonts used previously and has a sense of archiving. In addition to the catalogue at the beginning, each chapter has an interspersed page with detailed directory, just like the guidance system everywhere in the subway, rigorous and orderly. Although the book is 33mm thick, it feels good to read with soft paper and unspecified covers.

铜奖

1989 年是故事的结束，还是未来的开始?

奥地利

This book constitutes an invaluable document that remains useful even when the exhibition is over or has never even been seen. The choice and mixture of paper, binding, typography, choice and layout of illustrations of the exhibited works of art are well-balanced and exact. Even elegant details such as picture headings are superbly incorporated into the overall form. Form follows function. The design helps us to appreciate the multifaceted and complex content. Each double-page spread is a joy to behold.

Bronze Medal

1989. Ende der Geschichte oder Beginn der Zukunft?

Austria

Chris Goennawein

Kunsthalle Wien, Gerald Matt, Cathérine Hug, Thomas Mießgang (Ed.)

Verlag für moderne Kunst Nürnberg

240 × 190 × 33mm

819g

317p

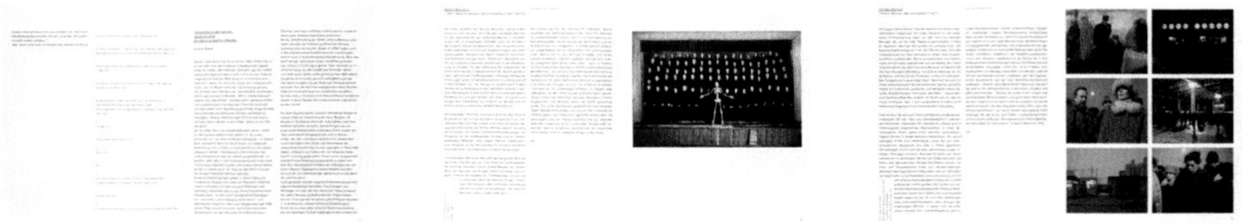

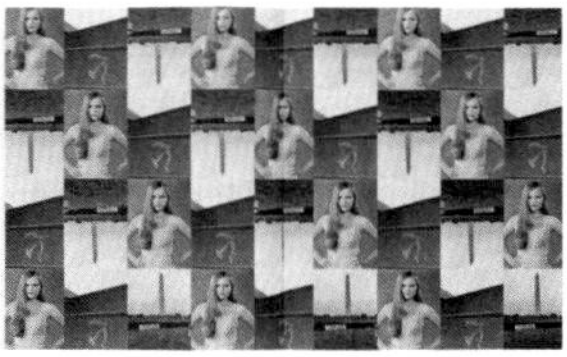

ПЕРЕС
РОЙКА

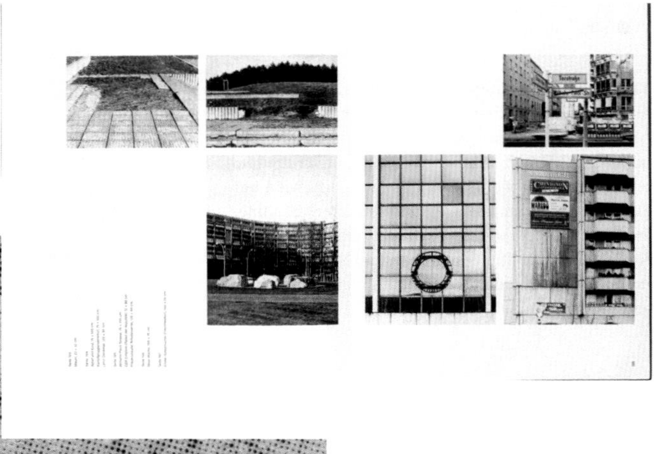

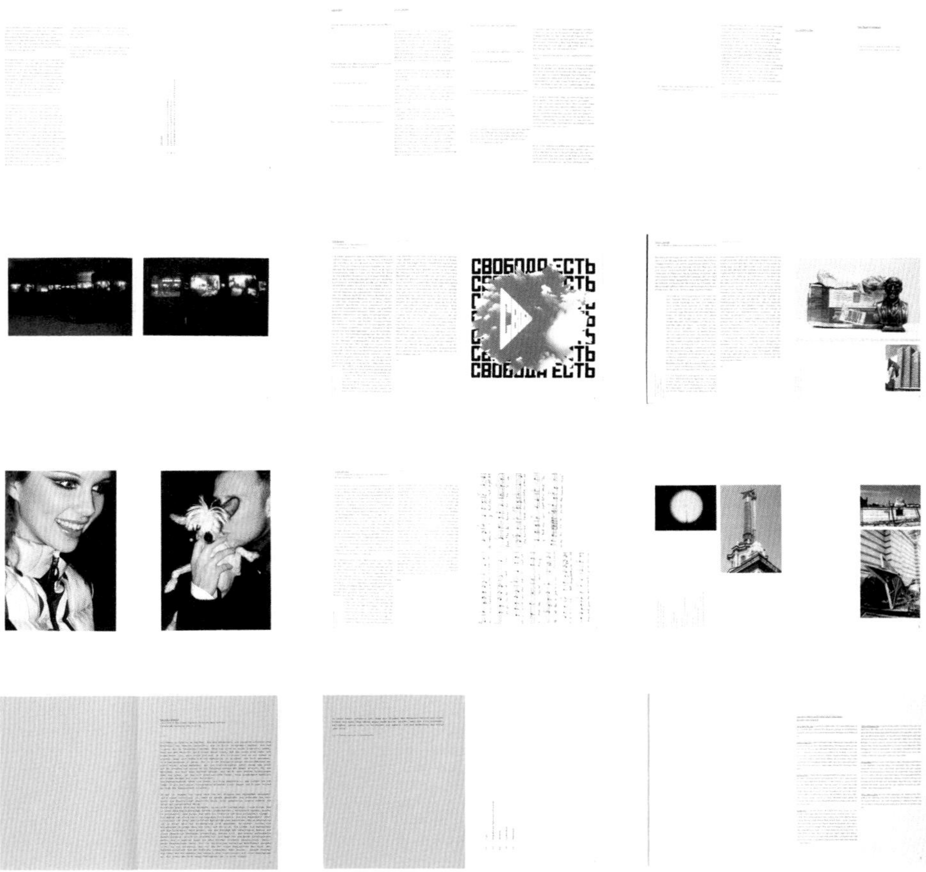
СВОБОДА ЕСТЬ

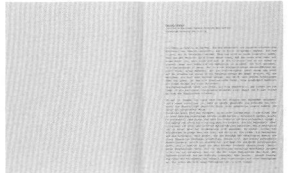

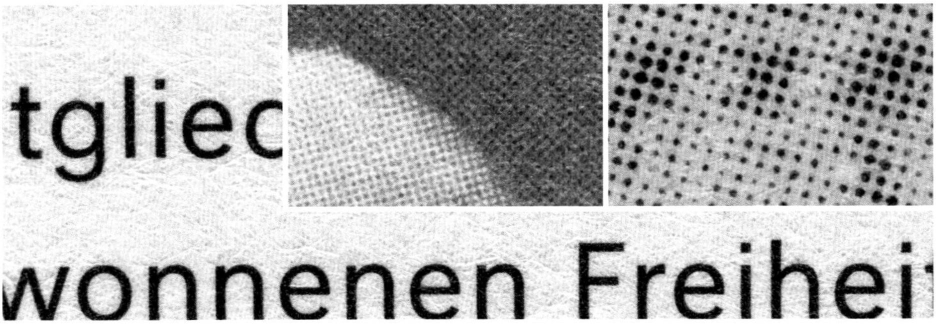
tgliec
wonnenen Freihei

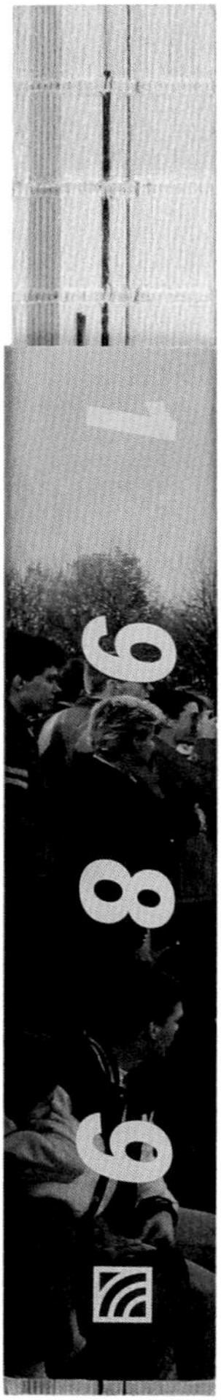

Josephine Meckseper

* 1964 in Lilienthal, Deutschland, lebt und arbeitet in New York City

Nach ihrem Studium am California Institute of the Arts Anfang der 1990er Jahre begann Josephine Meckseper ihre Laufbahn zunächst als Herausgeberin eines konzeptuellen Undergroundmagazins im Stil einer Boulevardzeitung, dessen Aufmachung an das italienische Blatt *Cronaca Vera* angelehnt war. *FAT* erschien sporadisch in unterschiedlichen Auflagen und enthielt theoretische und essayistische Texte sowie künstlerische Arbeiten u. a. von Dan Graham, Monica Bonvicini und Matthew Barney, die als Werbeanzeigen getarnt waren. 1998 inszenierte die Künstlerin eine Posterkampagne, in der sie als Kandidatin für den US-Senat auftrat. Seit dem Jahr 2000 stellt Josephine Meckseper, zeitweilig auch als Journalistin tätig, in Galerien und musealen Institutionen aus. Ihre konzeptuelle Praxis versteht sie als subversives Spiel, das mittels nicht affirmativer Darstellungsweisen eine selbstreflexive Persiflage auf den Warenwert des Kunstobjekts ins Bild bringt. Dazu konterkariert sie in ihren Installationen, Filmen und fotografischen Arbeiten das reziproke Verhältnis von Politik und Konsum, Industrie und Militär, Mode und Protestkultur, indem sie kommerzielle Präsentationsmodi und das rhetorische Vokabular aus Massenmedien, Werbung und Kaufhauskultur mit journalistischem Bildmaterial verbindet, das die problematischen Hintergründe thematisiert, auf denen Markt und Überfluss der privilegierten westlichen Hemisphäre basieren. Intention und visuelle Sprache spiegeln die affine, aber durchaus zwiespältige Haltung der Künstlerin zu Phänomenen, Tendenzen und Attitüden der 1970er Jahre wider: politischer Aktivismus und seine populären Ausdrucksmittel bilden die beiden Referenzfelder, die sie an die Tradition institutionskritischer Kunst anknüpfend in warenhausähnlichen Installationen inszeniert, die wie eine Provokation zum Vandalismus erscheinen.

Shelf No. 37, 2007
Metallregal, Gemälde, Mixed Media
86,36 × 259,72 × 35,56 cm

Die für die Ausstellung ausgewählten Arbeiten stellen die Interdependenz von Gegenpolen einer realpolitischen Problematik symbolhaft zur Schau. *Untitled (Sfera)* (2006), eine simulierte Auslage und *Untitled (Cold War)* (2006) eine Bild-Tapete sowie eine Reihe von Collagen sind durchsetzt von ideologisch codierten Motiven: Auf der einen Seite ein Dollarzeichen, ein Model in Einheitsware, ein Stück Jeansstoff, die deutsche Flagge, auf der anderen Seite Fotografien von Anti-Irakkrieg-demonstrationen, Hammer und Sichel, das Washington Monument mit einem Protestschild gegen den amerikanischen Präsidenten George W. Bush, ein Palästinenserschal. Der visuelle Dekonstruktivismus von Josephine Meckseper verwischt Freund- und Feindbilder von Demokratie und Kapitalismus. Sinnentleerte Signifikanten funktionieren wie Kommunikationselemente eines Corporate Designs, aber nur um aufzuzeigen, dass anhand von Warenkreisläufen die Bindungen zwischen ökonomischen und politischen Konstellationen sichtbar gemacht werden können. Modernistische Formalismen gehen auf das Interesse an „den Wurzeln politischer Bildsprache, insbesondere der russischen Konstruktivisten und Suprematisten" zurück. „Im Vordergrund steht dabei für mich", so die Künstlerin, „die Ablösung der formalen Funktion von Kunst zugunsten einer politischen. Bei mir manifestiert sich das zum Beispiel in rasterartigen Bildern und Collagen, die auf Zeitungsabbildungen von Absperrgebieten bei Demonstrationen basieren. In den Tagen und Wochen vor größeren Kundgebungen werden vielerorts Karten dieser Art in den Medien veröffentlicht. Mich interessiert hier besonders die Darstellung der Transformation von Konsumzonen, beispielsweise Einkaufspassagen, zu politisierten Zonen, das heißt Protestzonen." (Josephine Meckseper, 2009)

SG

2007

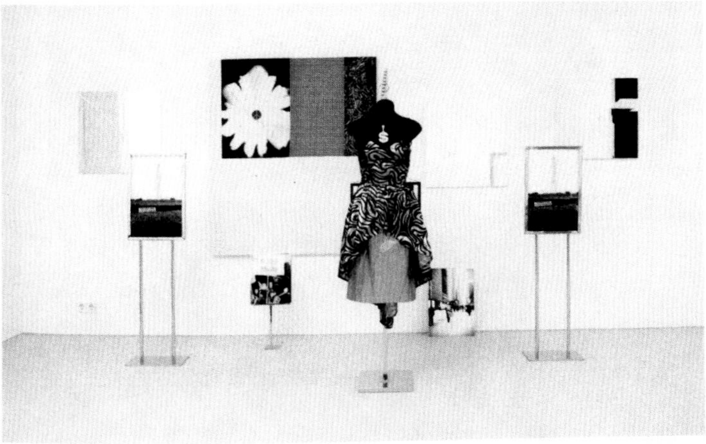

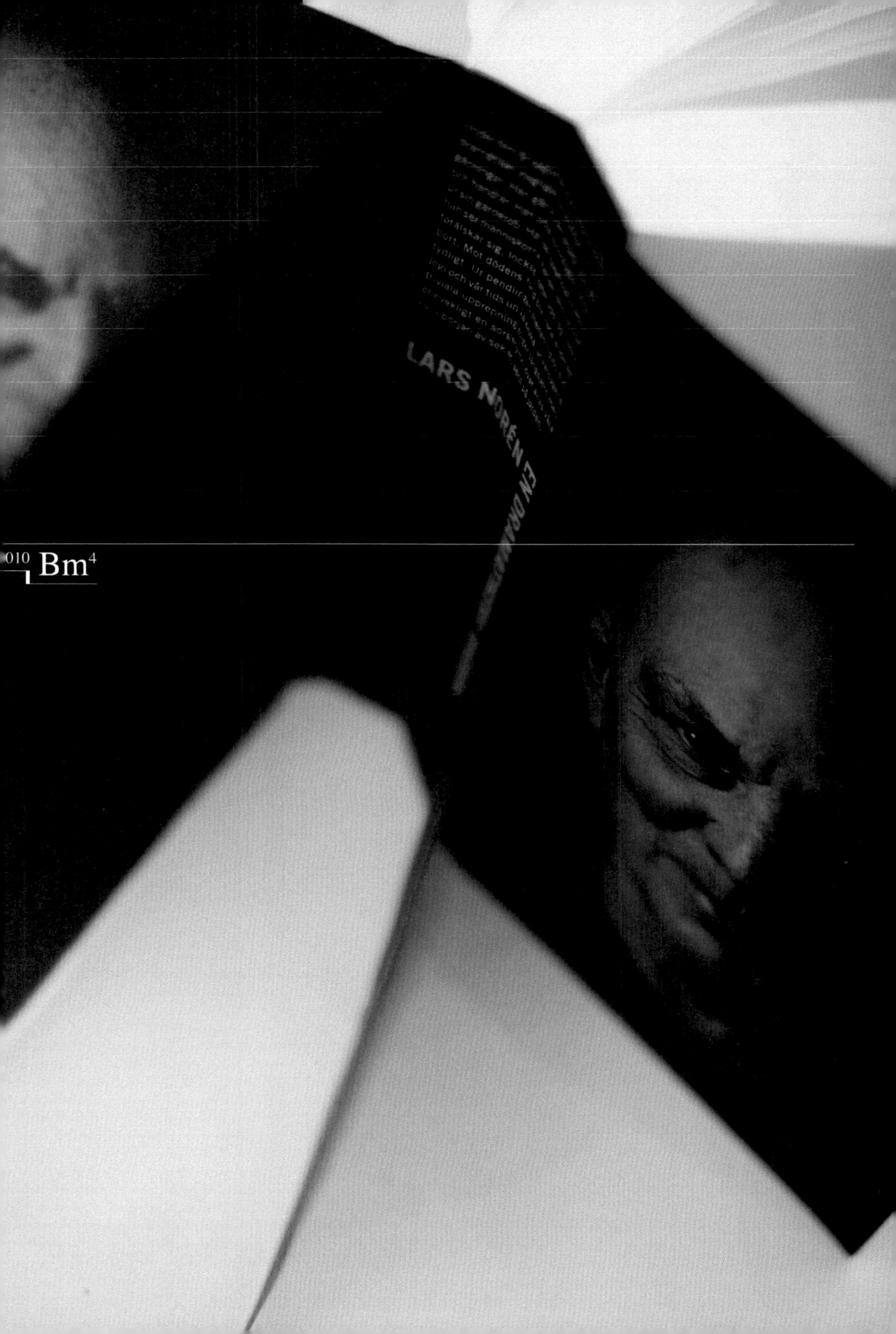
LARS NORÉN

这是瑞典当代著名剧作家、小说家和诗人拉尔斯·诺伦（Lars Norén）的日记，内容包含 2000 年 8 月到 2005 年 7 月整整五年内的生活、想法与情感，在微不足道的细节和日复一日的通勤中展开。本书为无线胶装，多达 1600 多页的内页可以很好地展开得益于纸张的选择。封面黑卡纸无印刷，只有封面烫黑和书脊烫白两道工艺。除了书名、作者名和出版社，宽厚的书脊上印有本书的内容简介。前后环衬为较厚的胶版纸印单色黑，分别为作者睁眼和闭眼的照片。封面的黑白烫工和前后环衬睁眼闭眼的照片似乎意味着白昼黑夜之间的联系。内页圣经纸单色印刷，按照一年为周期划分为五个部分。全书没有页码，每段文字边的日期就是索引标记。书籍的设计精巧细腻而没有过多的装饰，体现着有如白开水一般的日常生活。设计没有过多地参与内容本身，而把思想的精妙留给读者在阅读中体会。

LARS NORÉN
EN DRAMATIKERS DAGBOK

This is the diary of Lars Norén, a famous contemporary Swedish playwright, novelist and poet. It covers life, ideas, and emotions for five years, from August 2000 to July 2005, with minute details and day-to-day commuting. The book is perfect bound. A book with up to 1600 pages can be fully opened thanks to the wise choice of paper. Black cardboard cover is unprinted, designed with two processes – black hot stamp on the cover and white hot stamp on the spine. In addition to the title of the book, the name of the author and the publishing house, the broad spine is printed with a brief introduction of the book. The front and back end paper is printed in monochrome black on offset paper, with the photographs of the author opening and closing eyes respectively. The black-and-white hot stamping cover and the photos of opening eyes and closing eyes seem to indicate the connection between day and night. The inside pages printed in monochrome are divided into five parts according to the year. There are no page numbers in the book. The date at the edge of each paragraph is the index mark. Without too much decoration, the design is exquisite and delicate, which reveals that daily life is just like boiling water. The design does not involve too much in the content itself, but leaves the subtlety of ideas to the readers.

- 铜奖
- 一本戏剧家的日记本
- 瑞典

In all its details this book is designed every bit as dramatically and strikingly as the life of the playwright Lars Norén itself would appear to be. Heavy and light, black and white, day and night, subtle and sober, soft and severe, introvert, extrovert – a true monument in book form to its author and his country of origin. From cover design to chapter headings, use of Bible paper to binding – all details are consistently matched to one another, and even without page numbering its 1680 pages are a delight to read. Compliments to the publisher for daring to bring out a publication of such unorthodox form!

- Bronze Medal
- En dramatikers dagbok
- Sweden
- Nina Ulmaja
- Lars Norén
- Albert Bonniers Förlag, Stockholm
- 209×148×52mm
- 1354g
- 1680p

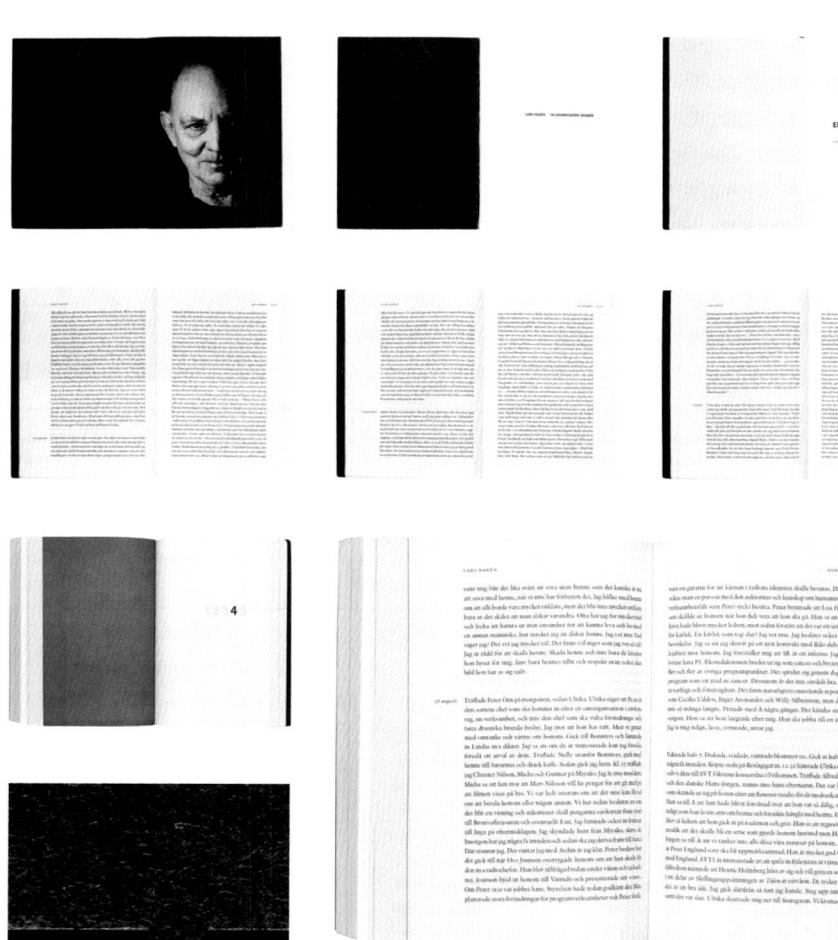

LARS NORÉN
EN DRAMATIKERS DAGBOK
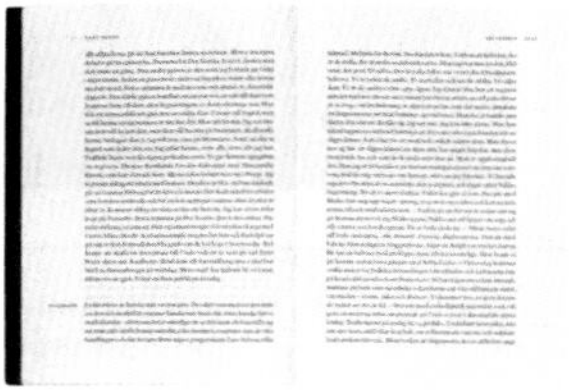

LARS NORÉN

r sä
yrkogården k

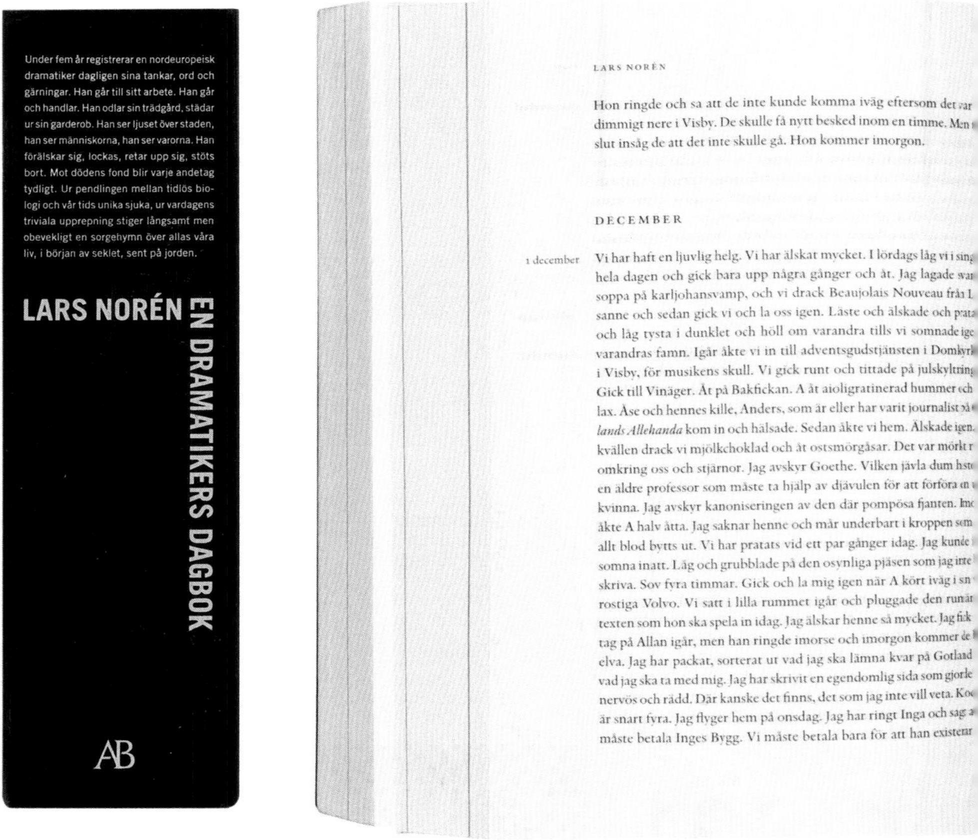
Under fem år registrerar en nordeuropeisk dramatiker dagligen sina tankar, ord och gärningar. Han går till sitt arbete. Han går och handlar. Han odlar sin trädgård, städar ur sin garderob. Han ser ljuset över staden, han ser människorna, han ser varorna. Han förälskar sig, lockas, retar upp sig, stöts bort. Mot dödens fond blir varje andetag tydligt. Ur pendlingen mellan tidlös biologi och vår tids unika sjuka, ur vardagens triviala upprepning stiger långsamt men obevekligt en sorgehymn över allas våra liv, i början av seklet, sent på jorden.
LARS NORÉN
EN DRAMATIKERS DAGBOK
AB
LARS NORÉN
Hon ringde och sa att de inte kunde komma iväg eftersom det
dimmigt nere i Visby. De skulle få nytt besked inom en timme.
slut insåg de att det inte skulle gå. Hon kommer imorgon.
DECEMBER
1 december
Vi har haft en ljuvlig helg. Vi har älskat mycket. I lördags låg vi i
hela dagen och gick bara upp några gånger och åt. Jag lagade
soppa på karljohansvamp, och vi drack Beaujolais Nouveau från
sanne och sedan gick vi och la oss igen. Läste och älskade och
och låg tysta i dunklet och höll om varandra tills vi somnade
varandras famn. Igår åkte vi in till adventsgudstjänsten i
i Visby, för musikens skull. Vi gick runt och tittade på
Gick till Vinäger. Åt på Bakfickan. A åt aioligratinerad hummer
lax. Åse och hennes kille, Anders, som är eller har varit journalist
lands Allehanda kom in och hälsade. Sedan åkte vi hem. Älskade igen.
kvällen drack vi mjölkchoklad och åt ostsmörgåsar. Det var mörkt
omkring oss och stjärnor. Jag avskyr Goethe. Vilken jävla dum
en äldre professor som måste ta hjälp av djävulen för att förföra
kvinna. Jag avskyr kanoniseringen av den där pompösa fjanten.
åkte A halv åtta. Jag saknar henne och mår underbart i kroppen
allt blod bytts ut. Vi har pratats vid ett par gånger idag. Jag
somna inatt. Låg och grubblade på den osynliga pjäsen som jag
skriva. Sov fyra timmar. Gick och la mig igen när A kört iväg i
rostiga Volvo. Vi satt i lilla rummet igår och pluggade den
texten som hon ska spela in idag. Jag älskar henne så mycket. Jag
tag på Allan igår, men han ringde imorse och imorgon kommer
elva. Jag har packat, sorterat ut vad jag ska lämna kvar på
vad jag ska ta med mig. Jag har skrivit en egendomlig sida som
nervös och rädd. Där kanske det finns, det som jag inte vill veta.
är snart fyra. Jag flyger hem på onsdag. Jag har ringt Inga och
måste betala Inges Bygg. Vi måste betala bara för att han

betalar hans kommande nervsammanbrott. Så får vi se vad som händer imorgon. Allan säger att elektrikern har tagit betalt för arbeten som han inte har gjort. Det är inte ens färdigt. Han vill naturligtvis inte göra klart det förrän han har fått sina pengar. Det var en recension i *DN* igår av *Krig*, men jag vill inte höra den. A läste den på Bakfickan. Hon sa att recensenten inte visste vad han skulle tycka, och att han skyllde det på uppsättningen. Det måste vara Zern. Jag är så trött på dem. Hur gamla ska de bli? När kommer de nya kritikerna? De fria kritikerna?

Jag skriver, tror jag. Jag har skrivit tolv sidor idag. Jag vet inte vad det är. 2 december
Hoppas det håller på så. Amélie ringde när jag vilade på förmiddagen. Det är en mycket fin stor recension i *Libération*, liksom i *Le Monde* häromdagen. De hyllar Agathe. Antoine och Gérard är kanske sårade av det. Det är bra att jag träffar dem om några veckor och helar deras sår. Allan glömde bort tiden och kom kl. 13. Jag gick igenom huset med honom och elektrikerna. De ska avsluta sitt arbete nästa vecka. Jag måste betala hela räkningen eller vänta på en elektriker som kanske dyker upp om ett och ett halvt år, från ingenstans. Pratat med A och med Nelly. Hon är hemma idag och läser. Jag sov en stund på eftermiddagen. Skrev. Packade vidare. Imorgon åker jag kvart i tre. Jag har läsning på torsdag. Sedan ska jag prata med J. Han ringde igår. Han går hos en terapeut. Var där i fyrtiofem minuter igår. Jag som föraktat själsproblem och terapi och svaghet, sa han. Han är från Argentina och det är väl också sociologiskt och kulturellt. Men i Buenos Aires finns det flest psykoanalytiker efter Paris och New York. Jag sa att han skulle ta det i mycket mycket små bitar. Inte allt på en gång. Avbryt arbetet, om du måste, och ägna dig åt rannsakning. Åt smärtan. Om all smärta kommer på en gång sprängs du sönder. Du har ju levt med det så länge utan att veta något och det du ändå har vetat börjar äntligen bli begripligt. Men du måste gå hand i hand med din terapeut. Försök inte veta för mycket under för kort tid. Om det sker fruktansvärda hastiga ras i din personlighet, se till att du rör dig, pratar, äter, går, handlar, långsamt. Balansera. Jag ska ta kontakt med Hasse, Kickans man, för den fortsatta renoveringen. Han bor i

这是一本蓬皮杜艺术中心有关“虚空”主题的展览画册，展现与展览这件事有关的历史。前卫艺术家伊夫 · 克莱因（Yves Klein）于1958年的“虚空”展打开了一道之前无法想象的空间，从此“虚空”的概念被各类艺术家反复探讨与解读。本书把页面当作展览空间，旨在将当代艺术表达虚空主题的概念通过书籍延续下去。本书为锁线硬精装，封面灰色布纹纸金色和黑色双色印刷，封面和内页部分场馆图片一样处理成较大颗粒的调频网图，带有一种似隐似现的虚无感。尤其是内页图片单黑印刷，点阵强烈的对比与普通空置的场景照片相比，似乎空间里又充满看不见的力量。环衬与封面灰色布纹纸相同，内页光面铜版纸四色印刷与胶版纸印单黑混合使用，分别用于展示展览、空间，以及相关深入的思考文章。正文版式在三栏里灵活设计，有些版面大量留白，与其他版面的文字或图片形成独特的对页关系，在版面上进行着“虚空”与“如何填充”的艺术实践。

Voids

This is an exhibition album on the theme of 'voids' of Pompidou Art Center to show the history of the exhibitions. Yves Klein, an avant-garde artist, explored an unimaginable space in the 'Voids' exhibition in 1958. Since then, the concept of 'Voids' has been repeatedly explored and interpreted by various artists. This book regards pages as exhibition spaces, intending to express the concept of nihility of contemporary art. This hardcover book is bound with thread sewing, and the cover is printed in gold and black on grey wove paper. Frequency Modulated screening is used for the cover and some pictures of venues on inside pages, with a looming sense of nihility. Especially for the single-black printing of the inner page pictures, the sharp contrast of dot matrix indicates that the space is full of invisible forces. The end paper is made of grey wove paper, which is the same as the book cover. Gloss art paper printed in four colors and offset paper printed in black is mixed for displaying exhibition, space and related articles with deep thinking. The text typesetting is flexibly designed in three columns. Some pages are heavily blanketed, contrasting to the opposite pages with text or pictures, the design carries out the artistic practice of 'voids' and 'how to fill'.

铜奖

虚空——回顾展

瑞士

Bronze Medal

Voids. A Retrospective

Switzerland

Gilles Gavillet, David Rust mit Corinne Zellweger / Gavillet & Rust

Mathieu Copeland mit John Armleder, Laurent Le Bon u. a. (Ed.)

JRP | Ringier Kunstverlag, Zürich

286×222×37mm

2134g

544p

The book is a convincing example of how it is not only possible to document an exhibition in a catalogue but how a book can itself take the curatorial concept one step further. The area within the book is conceived as if it were exhibition space. Via the typographically restrained half title an overview of several exhibition rooms is reached – empty exhibition rooms, double page-spread illustrations, grainy. The list of contents itself becomes architecture. The shift from coated to uncoated paper is subtle, purposeful and non-intrusive, also characterising the various rooms as layers of time which extend throughout the exhibition project. The typography is functional, sober and in keeping with its content – voids.

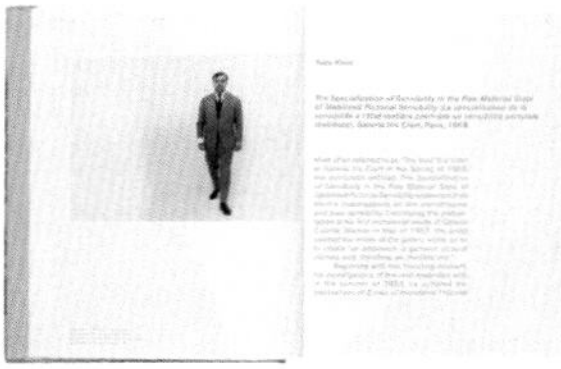

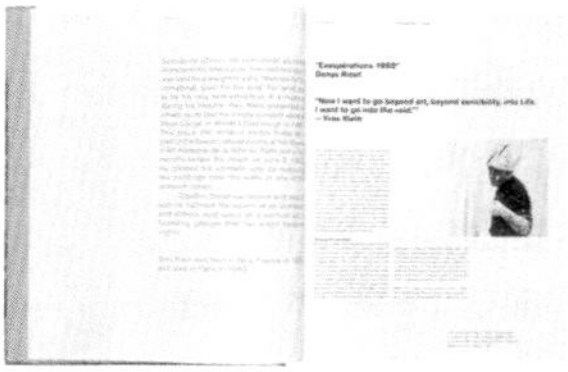

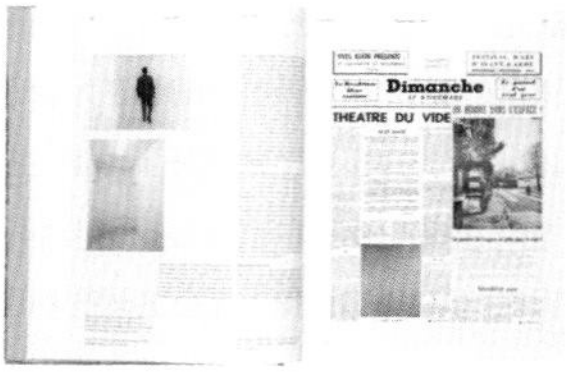
Dimanche
THEATRE DU VIDE

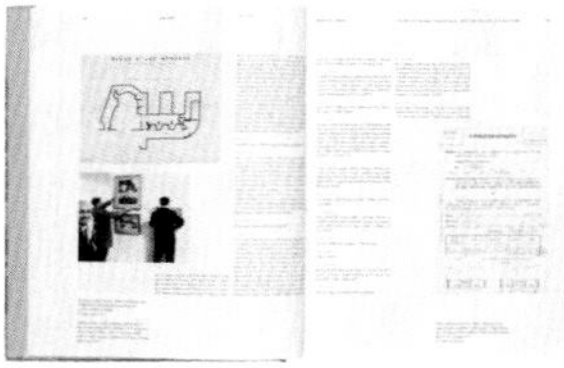

THE 'AIR-CONDITIONING' SHOW
1966

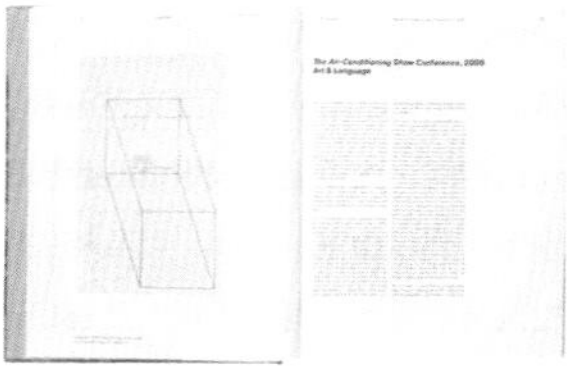

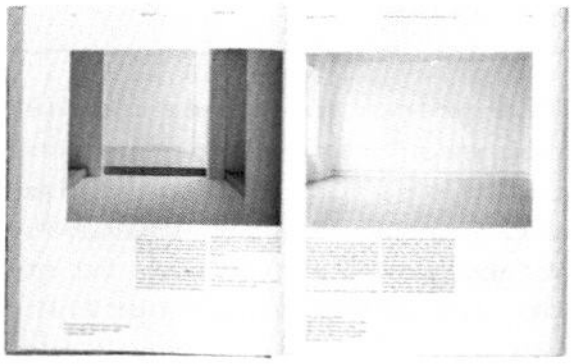

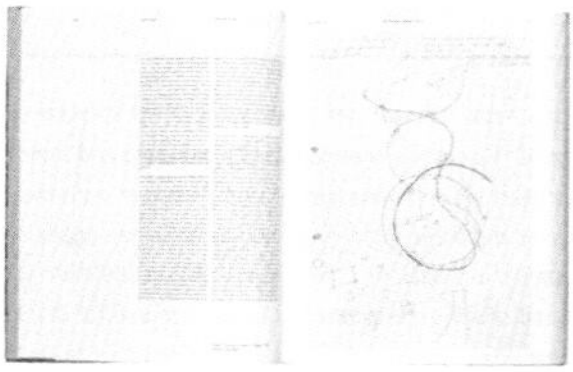

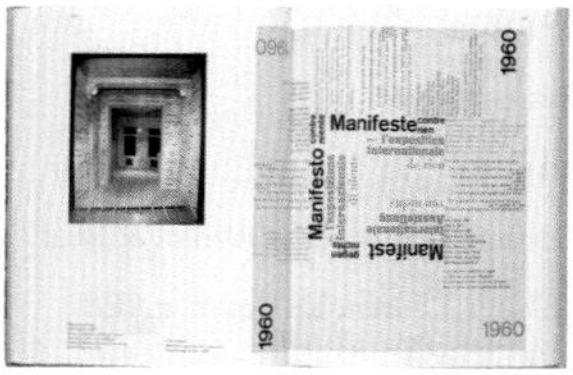
1960
Manifeste
Manifesto
Manifest
1960
1960

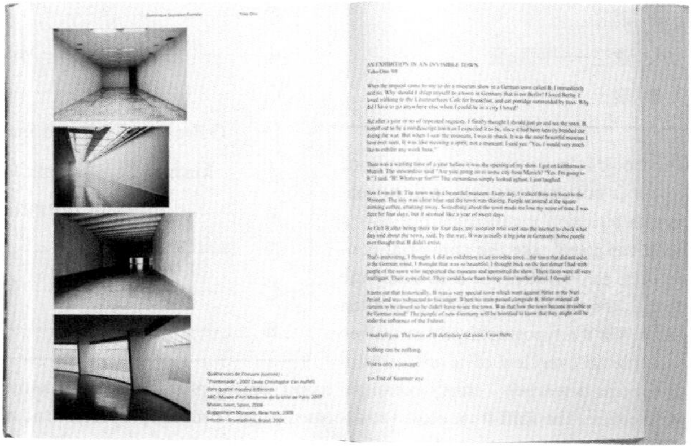
Nothing can be nothing.

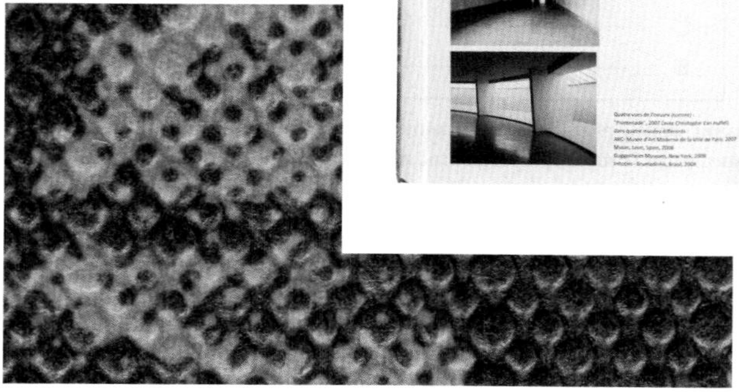

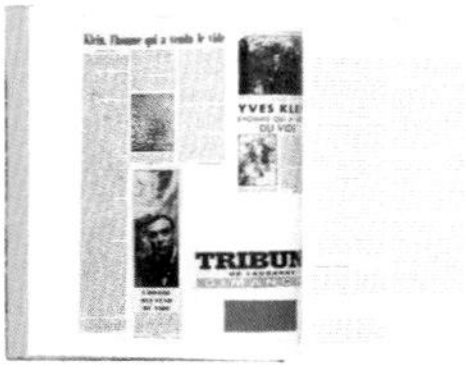

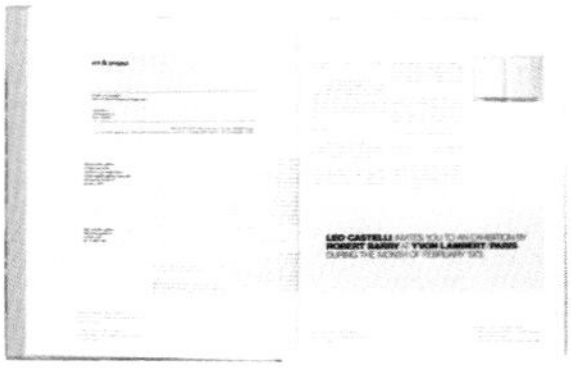
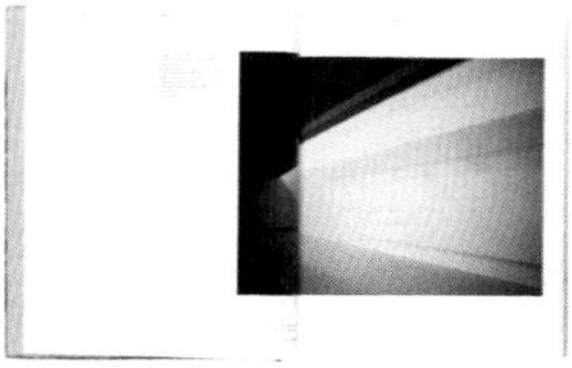

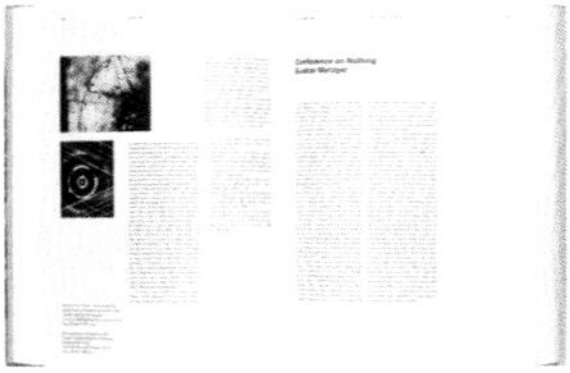

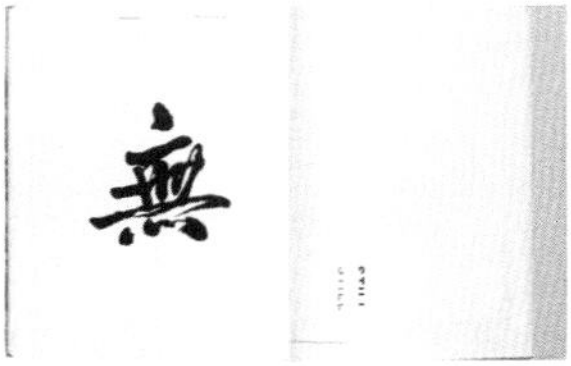
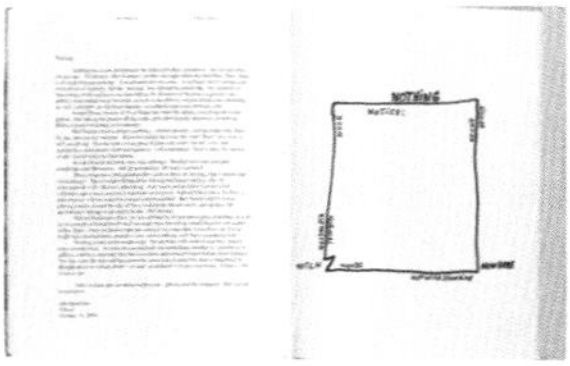
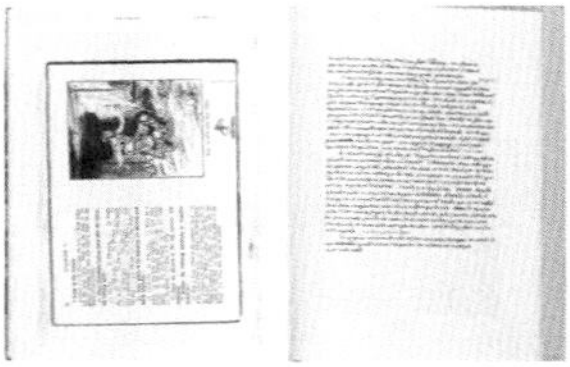
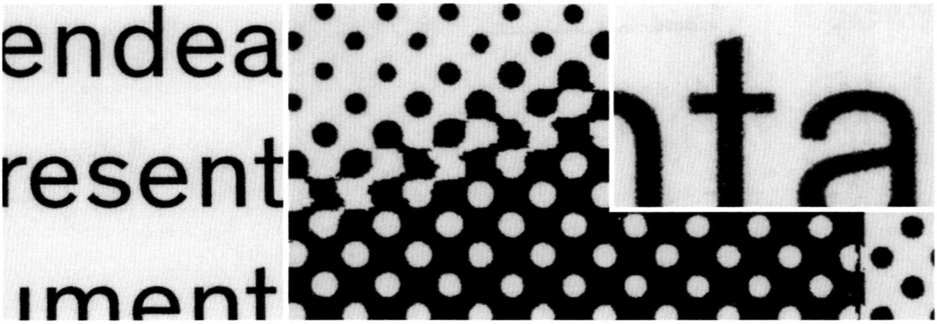
endea
resent
nta

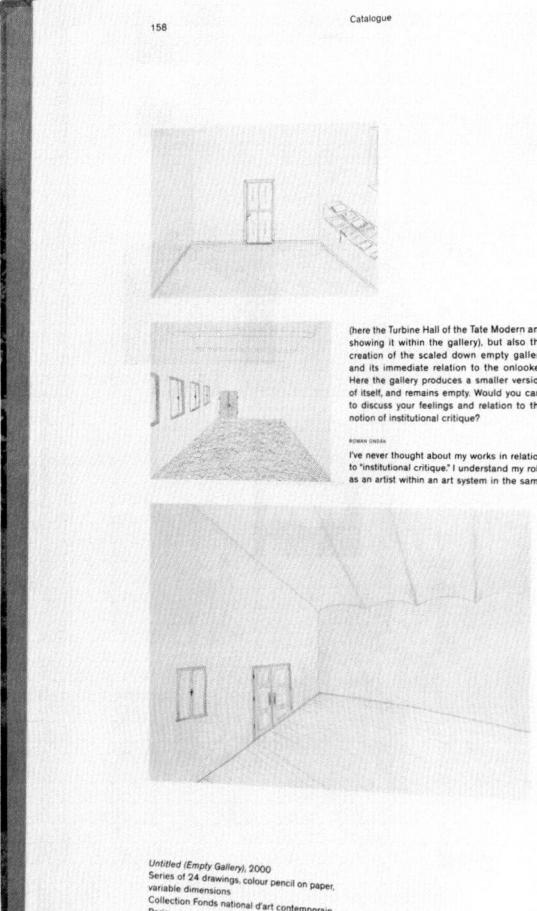

(here the Turbine Hall of the Tate Modern and showing it within the gallery), but also the creation of the scaled down empty gallery and its immediate relation to the onlooker. Here the gallery produces a smaller version of itself, and remains empty. Would you care to discuss your feelings and relation to the notion of institutional critique?

ROMAN ONDÁK

I've never thought about my works in relation to "institutional critique." I understand my role as an artist within an art system in the same way as I understand my role as a man within society. Art and the art system could be viewed as a mini-society, a system existing within a society, both having institutional parameters. By involving people from outside the art system, like people from my community or from the audience, I like provoking situations in which these two systems can merge.

MATHIEU COPELAND

We haven't talked about your influences. One thinks of artists such as Michael Asher, and also Jiří Kovanda for the minutiae and precision of your pieces (I'm thinking of Kovanda's exhibition *Installation 5 [Panes of Glass in an Empty Gallery]* from the summer of 1979), and Franz Erhard Walther for the process embedded in your practices. Would you care to tell me who you were, and still are, looking at?

ROMAN ONDÁK

There are certainly intuitive references to several artists I appreciate in my works, but Július Koller from Bratislava and Jiří Kovanda from Prague have influenced me the most. I've always believed that looking closely at some great art by a few local artists would bring me more in terms of understanding my own situation than looking for influences from internationally established art. In the meantime Koller and Kovanda have become internationally renowned and their influence has gained a wider importance.

MATHIEU COPELAND

I'd like to discuss the possible impact of the political context in which you grew up, and how this may have informed your work. In particular I'm interested in the desire and need for voids and emptiness. Would the void be a means to an end for you, representing the ultimate engagement of defiance?

ROMAN ONDÁK

I've never thought about doing "political art," but because of my interest in situations within society, the impact of the political context on some of my works is noticeable. I've never wanted to illustrate politics as such, but many of my works are concerned with infiltrating my personal experiences into a society which is being transformed. It's actually very exciting to live in a society whose shape is constantly

Untitled (Empty Gallery), 2000
Series of 24 drawings, colour pencil on paper, variable dimensions
Collection Fonds national d'art contemporain, Paris

evolving. I grew up during communism, so I can compare both political and cultural systems, but what I see as an advantage is that I began to make my art after the fall of communism, at the beginning of the 1990s. I've never had a desire to comment directly in my works on that "dark" period. Later the notion of void was a tool for me to start, to move from a certain point. Just as someone needs a house to be filled with furniture for his family to inhabit it, the void for me was a matter to create for myself and possibly to fill it with something later.

MATHIEU COPELAND

And pursuing the subject of the void, I'd like to understand how you describe it. To me the void is neither positive nor negative, it just is, as a statement of fact. How would you qualify the void?

ROMAN ONDÁK

As I said, it's like some abstract matter kept somewhere in Room 13, which you can visit when there's no space left around you. It's like something one should have a chance to escape to.

MATHIEU COPELAND

Most of your work also calls for a reversal of situations, turning things on their heads without adding anything. I could give the example of going out of the gallery, looking into the gallery from outside and doubling the points of view. I'm interested in the use of the gallery as a readymade (and in that respect it's very similar to the view of Bethan Huws for instance). Would you care to consider this?

ROMAN ONDÁK

The gallery is like a large sculpture, but yet it's not interesting enough for me to call it a sculpture. If I do something with a gallery itself, there's always something else going on beyond it. So for me the gallery is like a pretext to make something invisible visible.

MATHIEU COPELAND

I'd like to also discuss the piece *More Silent Than Ever*. This piece affirms one thing, and yet leaves the gallery entirely empty. The notion of deceiving the viewer is at the very core of the work. For this piece, a choice has to be made: we either believe the artist or not. The implication that there is something there means that the viewer becomes central to the work.

ROMAN ONDÁK

When I state on the wall label in the gallery "room with a hidden eavesdropping device," is it true or not? Is there a chance to prove it or not? Is it important or not? Furthermore, what if said there: "This is the wall." Is that true or not? Well, I can see paint, but there should be a wall behind it. It's like Magritte's "Ceci n'est pas une pipe." Of course this work is based on how much we can believe in the words that we read on the wall label. We all know that an eavesdropping device is usually doing a good job when it's well hidden, so the statement remains no more than a rumor left to the viewer wandering in the empty room. If the visitor is curious about the fact he stays longer and observes the walls, hoping to find a detail revealing the hiding place of the device. So in the end it's he who plays the most important role in the work.

MATHIEU COPELAND

And to conclude, allow me to put to you again a question that you avoided during our last discussion! Maybe this will strike a chord

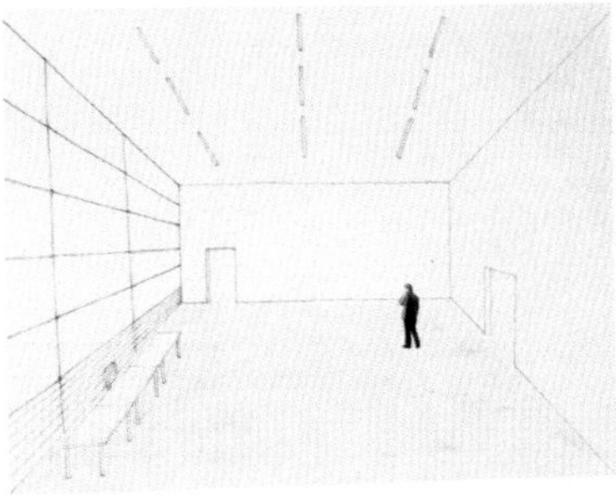

Somewhere Else, 2002
Series of 10 drawings, colour pencil on paper
Each between 20 × 29 cm and 42 × 53.5 cm
Collection Frac Pays-de-la-Loire, France

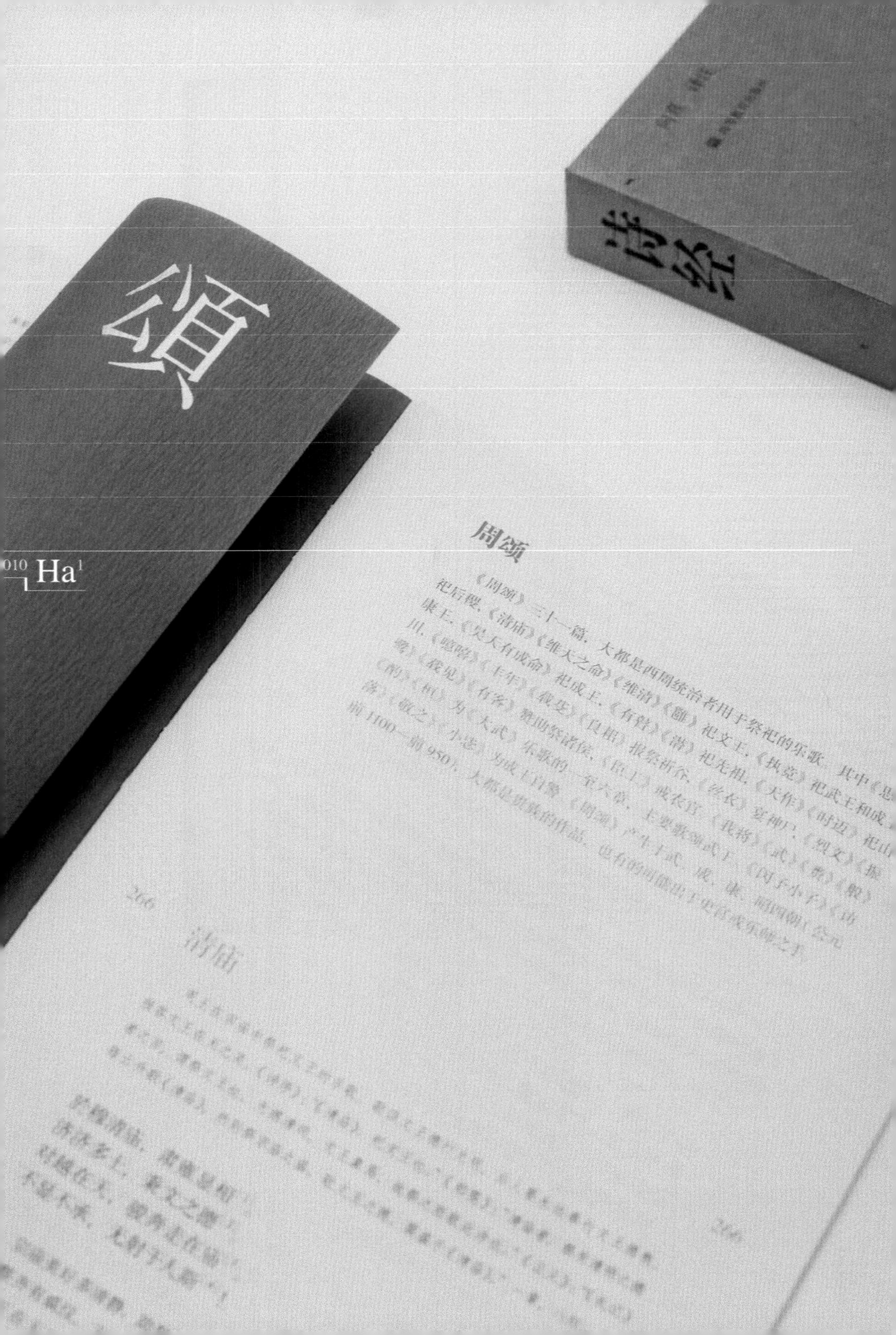

頌
周颂
清庙
266
010 Ha1

本书是中国古代经典《诗经》的译注版，包含《诗经》原文、今译、注释和韵读，是当代人阅读典籍的入门读物。本书为裸脊锁线装，基于小16开的开本，在保证单页内容承载量的同时拥有很好的捧阅手感。封面黑色纹理艺术纸印金、烫白，黄色牛皮纸印黑裱覆卡板制作的半截书函，简约端庄。半透明伊维斯纸张通过四色印刷古画、背面印刷叠透呈现、包裹黑色艺术纸张等多种方式，在进入正文前带来丰富的感官体验。内页胶版纸印刷黑、金双色，诗歌标题边的数字在书口和装订口两侧呈现，可以比页面下方的中文页码更好地起到目录索引的作用。棕色草麻纸烫红金作为风、雅、颂三个大篇章的隔页，并通过最近的折手页锁进装订线内。由于《诗经》这类书籍的公版属性，市面上充斥着大量各种设计版本，但这个版本的设计是其可以凸显出来的原因。

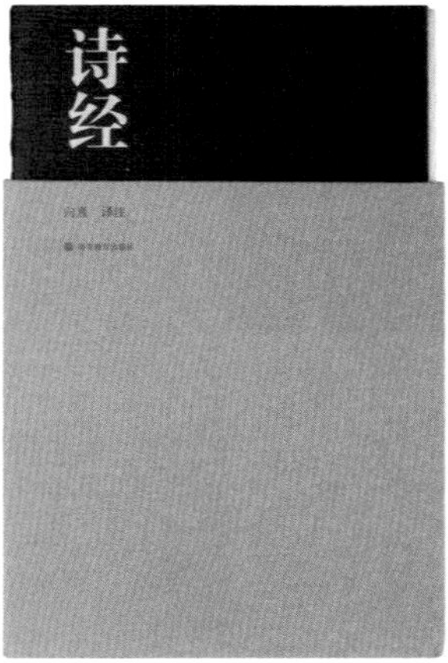
诗经

This book is an annotated edition of the ancient Chinese classic of poetry *The Book of Songs*. It contains the original, modern translation, annotation and rhythm and it is an introductory book for people to read the classics. This book is bound by thread sewing with a bare spine. The cover is printed in gold and hot stamped in white on black art paper. Yellow kraft paper mounted with graphic board is used to make the slipcase, simple and dignified. Ahead of the main body, ancient paintings are printed in four colors on translucent art paper which leads to rich sensory experience. The inside offset paper is printed in black and gold, which is a number of titles on both sides of the book and binding ports. It can play a better role in catalogue indexing than the Chinese page number at the bottom of the page. Brown linen paper with red gold hot stamping is used as interleaf of three chapters 'Feng', 'Ya' and 'Song'. Since books as *The Book of Songs* are in public domain, there are a large number of different versions on the market, but the design of this version is the reason why it can be highlighted.

荣誉奖
诗经
中国
刘晓翔
向熹（译注）
高等教育出版社

Honorary Appreciation
The Book of Songs
China
Liu Xiaoxiang
Xiang Xi (Ed.)
Higher Education Press, Beijing
257 × 167 × 27mm
946g
391p

This elegant book, printed in gold and black, comprises writings by traditional poets. Mystically veiled illustrations of old ink paintings at the beginning and end of the book are printed on tasteful paper and frame the texts in an atmospheric way. Divided up by short inserted card pages, these lead us into the world of Chinese history. Thanks to clear, convincing typography and generous layout the poems in this very fine volume can be appreciated within a modern atmosphere.

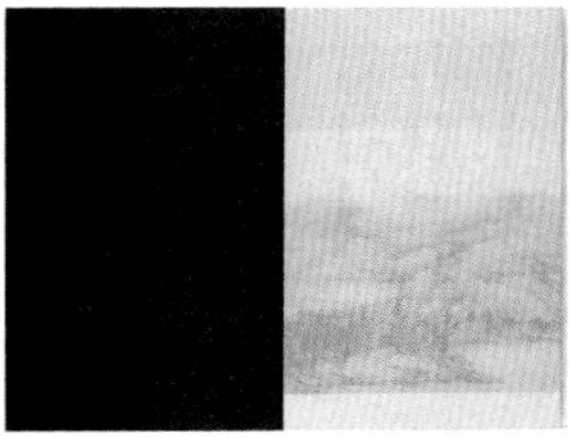

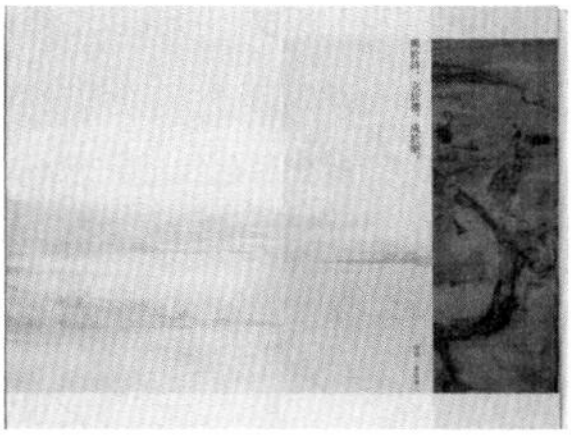

诗经

風

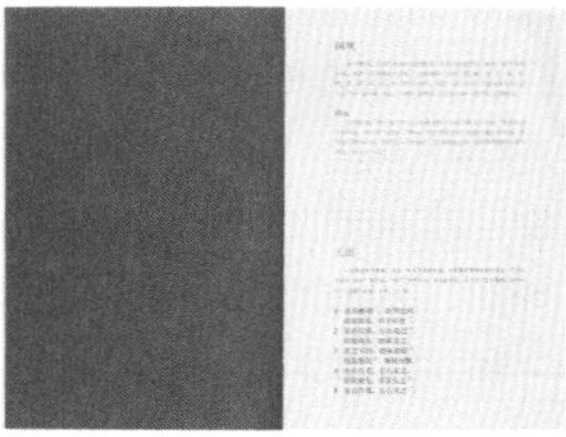

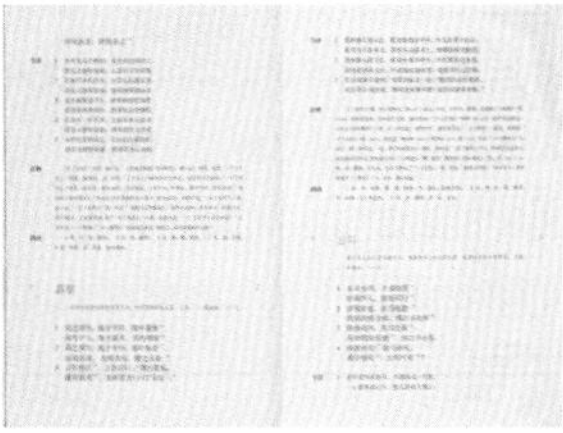

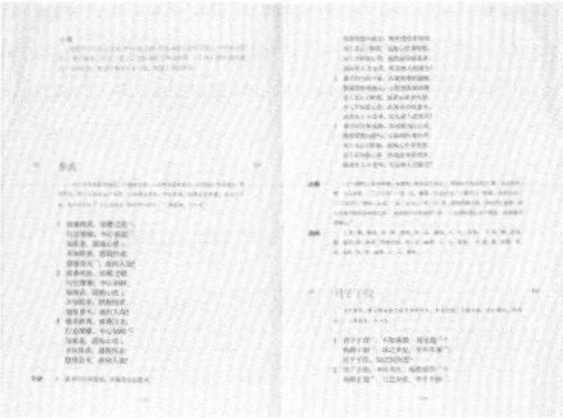

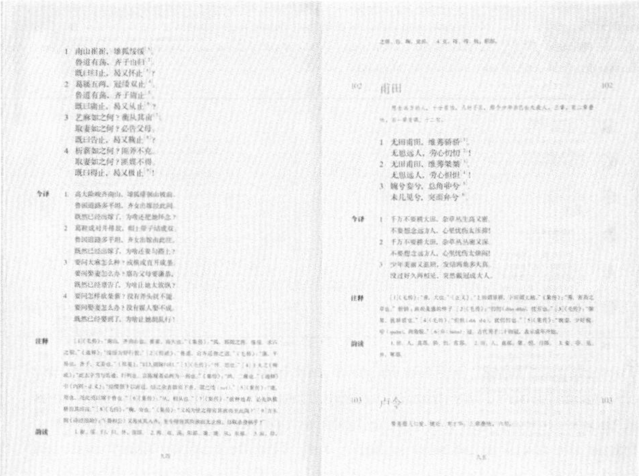
1 南山崔崔，雄狐绥绥[1]。
鲁道有荡，齐子由归[2]。
既曰归止，曷又怀止[3]？
2 葛屦五两，冠緌双止[4]。
鲁道有荡，齐子庸止[5]。
既曰庸止，曷又从止[6]？
3 艺麻如之何？衡从其亩[7]。
取妻如之何？必告父母。
既曰告止，曷又鞠止[8]？
4 析薪如之何？匪斧不克。
取妻如之何？匪媒不得。
既曰得止，曷又极止[9]！
今译
1 高大险峻齐南山，雄狐缓缓山坡前。
鲁国道路多平坦，齐女出嫁经此间。
既然已经出嫁了，为啥还把她怀念？
2 葛鞋成对并排放，帽上带子结成双。
鲁国道路多平坦，齐女出嫁由此往。
既然已经出嫁了，为啥还要与她上？
3 要问大麻怎么种？或横或直开成垄。
要问娶妻怎么办？禀告父母要谦恭。
既然已经禀告了，为啥让她太放纵？
4 要问怎样砍柴薪？没有斧头就不能。
要问娶妻怎么办？没有媒人娶不成。
既然已经娶到了，为啥让她胡乱行？
注释
韵读
102 甫田 102
1 无田甫田，维莠骄骄[1]。
无思远人，劳心忉忉[2]！
2 无田甫田，维莠桀桀[3]。
无思远人，劳心怛怛[4]！
3 婉兮娈兮，总角丱兮[5]。
未几见兮，突而弁兮[6]。
今译
1 千万不要耕大田，杂草丛生高又密。
不要想念远方人，心里忧伤太压抑！
2 千万不要耕大田，杂草丛生密又深。
不要想念远方人，心里忧伤太烦闷！
3 少年美丽又温婉，发结两角多天真。
没过好久再相见，突然戴冠成大人。
注释
韵读
103 卢令 103

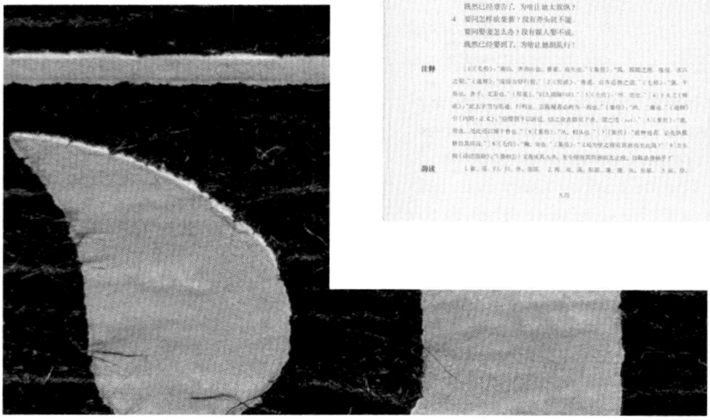

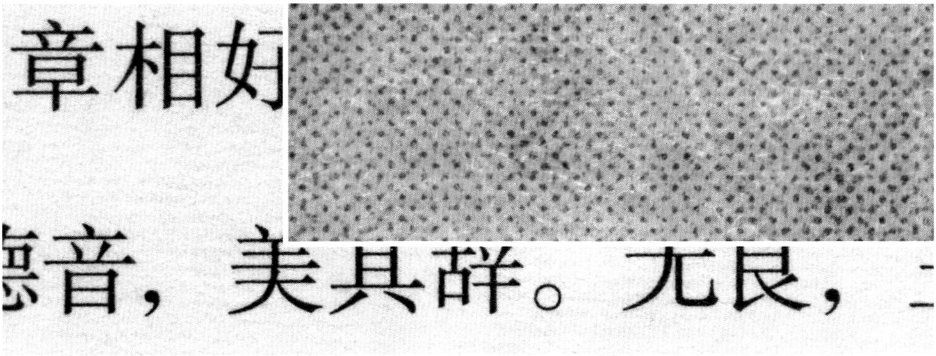
章相好
音，美其辞。无良，

Open
402
404

这是 2009 年举行的第四届鹿特丹国际建筑双年展的出版物，展示了建筑和城市的规划如何发挥作用、促进城市多样性共存的图书。书中包含对开放城市的理论构建和实施规划，是建筑双年展讨论未来城市如何构建的梳理和实践。本书锁线胶装，封面黄色卡纸印黑，外包调频处理的人群照片，单面光面铜版纸四色印刷，内页采用了浅黄的胶版纸四色印刷，配合少量铜版纸印刷的四色照片夹页。内文整体采用四栏设计，只使用一种字体，通过多栏间的变化和字号大小、下画线设置等完成丰富的页面设计。页面下脚有很大的页码便于查找，这样做的目的是为了仿效电话黄页，把多样的城市信息呈现在书里，方便建筑师、城市规划师、相关专业的从业人员都可以方便地参与到“开放城市”的探索和实践中来。

OPEN
CITY

This is the publication of the 4th International Biennale of Architecture of Rotterdam in 2009. It shows how architecture and urban planning can work and promote the coexistence of urban diversity. The book contains the theoretical construction and implementation planning of open cities. It is a review and practice of how the architecture of the future city is constructed in the Architectural Biennale. The book is perfect bound with thread sewing. The cover is printed in black on yellow cardboard, wrapped by a photo of the crowd in Frequency Modulated Screening. The photo is four-color printing on one-sided gloss art paper. The inside pages are printed in four colors on light yellow offset paper, attached with four-color photo pages printed on art paper. The whole text is typeset in four columns with only one font. The variety of design is accomplished by changing the size of the font and setting the suitable underline between the columns. Every page has a large page number for easy searching, which is similar to the telephone yellow pages and may present various city information in the book, so that architects, urban planners, and related professional practitioners can easily participate in the exploration and practice of ‘open city’.

- ◎ 荣誉奖
- « 开放城市——设计共存
- ◖ 荷兰

- ◎ Honorary Appreciation
- « Open City. Designing Coexistence
- ◖ The Netherlands
- D Mevis & Van Deursen mit Werkplaats Typografie
- △ Tim Rieniets, Jennifer Sigler, Kees Christiaanse (Ed.)
- ||||| SUN architecture, Amsterdam
- □ 270×202×29mm
- ⌷ 1051g
- ≣ 464p

A specialist book which approaches the urban theme in a variety of ways. The book’s body and materiality suggest a telephone book. The paper is thin and yellowish. The usual black marker of the index is placed as a beam in the top edging. Suspended from this the typographical design is developed, uniformly maintaining a single typeface, namely Monotype Grotesque. The design is both excited and exciting. Despite the input of a number of typographers the book’s clearly defined parameters make for unity. A detailed look reveals original typographical solutions. This is a convincing design concept which allows for diversity and permits room for manoeuvre. Open city!

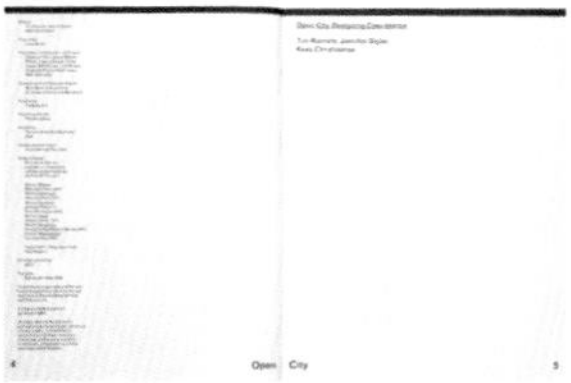

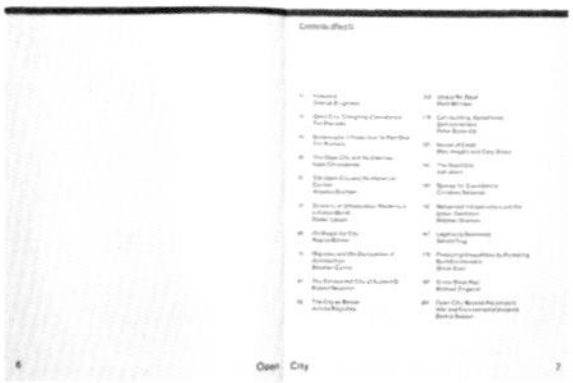

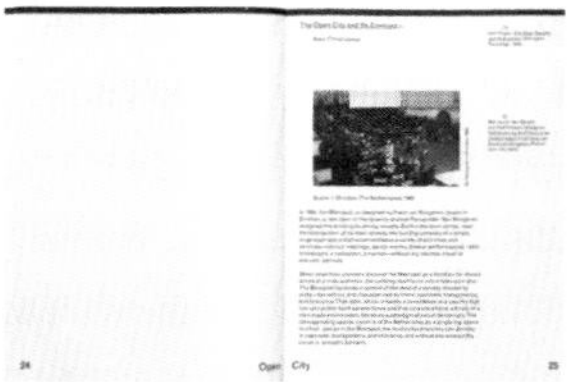

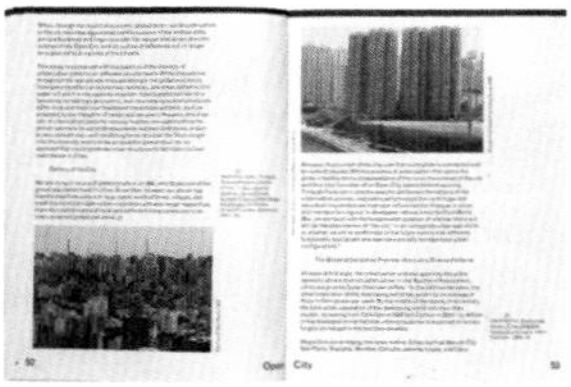

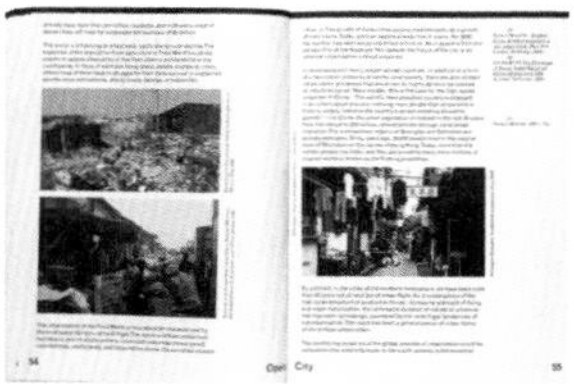

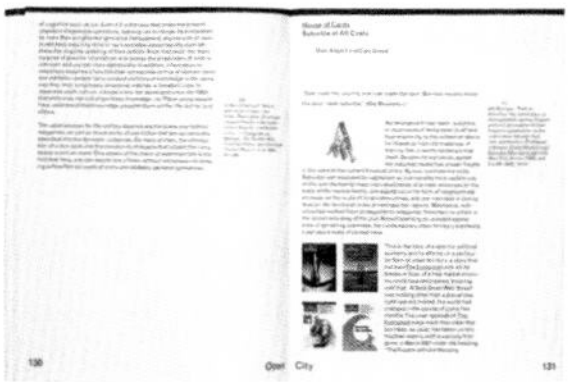

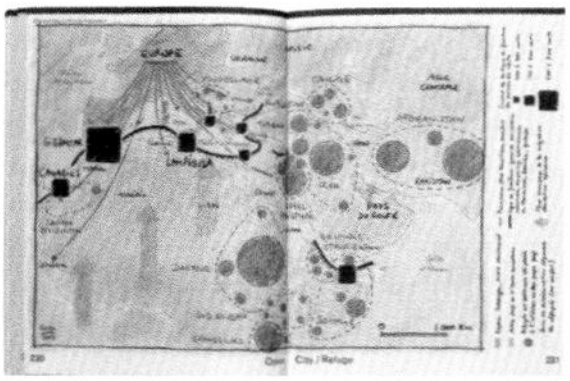

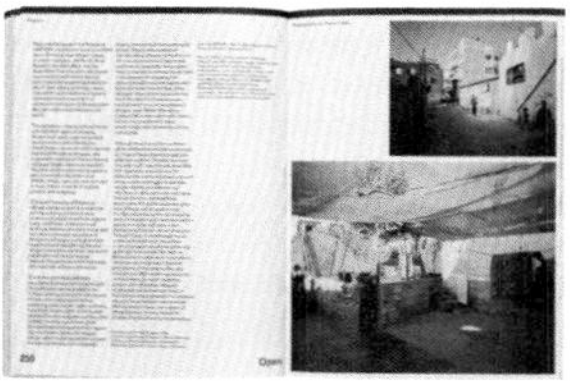

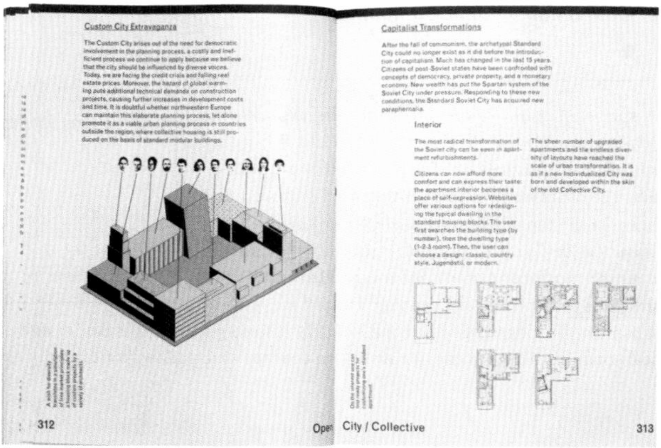
Custom City Extravaganza
The Custom City arises out of the need for democratic involvement in the planning process, a costly and inefficient process we continue to apply because we believe that the city should be influenced by diverse voices. Today, we are facing the credit crisis and falling real estate prices. Moreover, the hazard of global warming puts additional technical demands on construction projects, causing further increases in development costs and time. It is doubtful whether northwestern Europe can maintain this elaborate planning process, let alone promote it as a viable urban planning process in countries outside the region, where collective housing is still produced on the basis of standard modular buildings.
312
Open
Capitalist Transformations
After the fall of communism, the archetypal Standard City could no longer exist as it did before the introduction of capitalism. Much has changed in the last 15 years. Citizens of post-Soviet states have been confronted with concepts of democracy, private property, and a monetary economy. New wealth has put the Spartan system of the Soviet City under pressure. Responding to these new conditions, the Standard Soviet City has acquired new paraphernalia.
Interior
The most radical transformation of the Soviet city can be seen in apartment refurbishments.
Citizens can now afford more comfort and can express their taste: the apartment interior becomes a place of self-expression. Websites offer various options for redesigning the typical dwelling in the standard housing blocks. The user first searches the building type (by number), then the dwelling type (1-2-3 room). Then, the user can choose a design: classic, country style, Jugendstil, or modern.
The sheer number of upgraded apartments and the endless diversity of layouts have reached the scale of urban transformation. It is as if a new Individualized City was born and developed within the skin of the old Collective City.
City / Collective
313

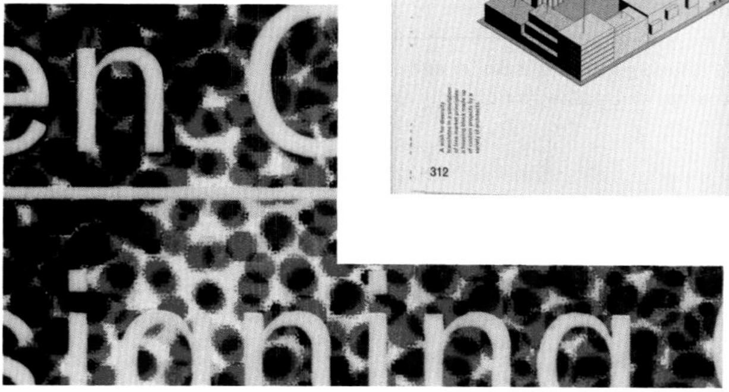
en C
sioning

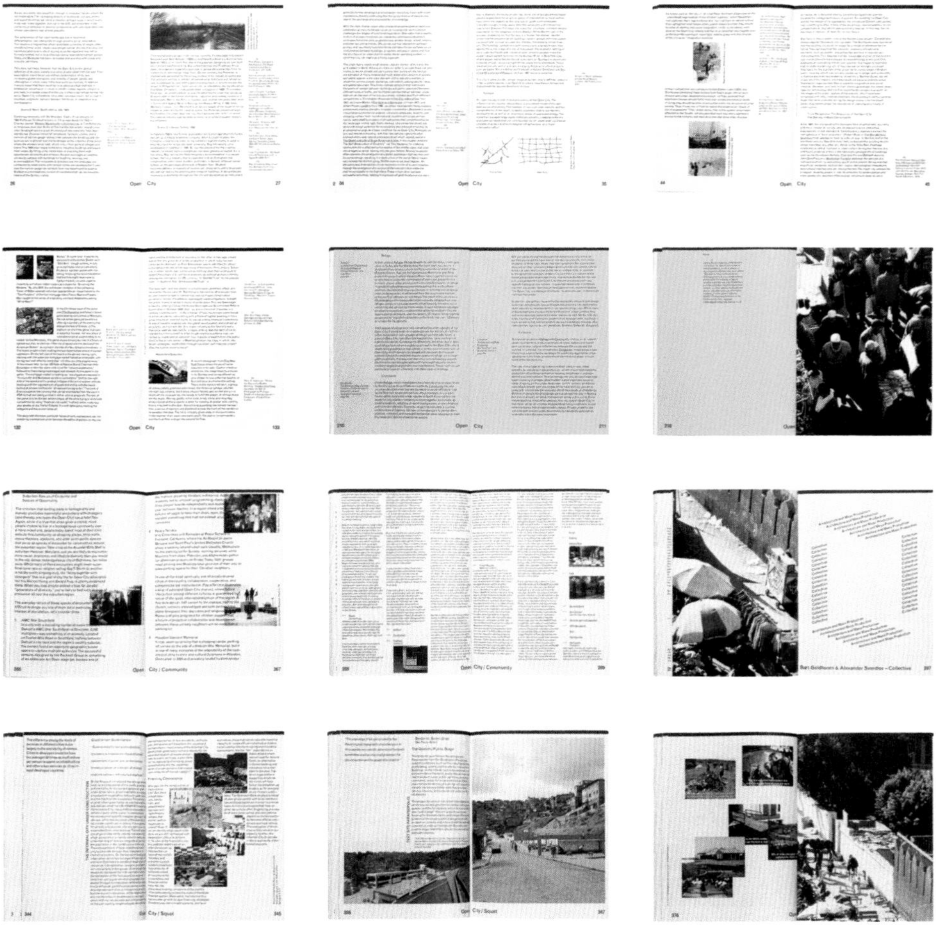

Zadanie 144.

WARTO ZAPAMIĘTAĆ

Walcem nazywamy bryłę powstałą przez obrót prostokąta dookoła prostej zawierającej jeden z boków tego postokąta.

Przekrojem osiowym walca nazywamy prostokąt, będący częścią wspólną tego walca i płaszczyzny przechodzącej przez jego oś obrotu.

Objętość walca wyraża się wzorem $V = \pi r^2 h$, gdzie:
– objętość walca,
długość promienia
wysokości

Do menzurki (w
metalu. Poziom wody
pod lustrem wody
Wiedząc, że średnica
tego kawałka metalu
żony z dokładnością do
żenie π liczbę 3,14.

Rozwiązanie

Szukana objętość kawałka
niego wody. Ta woda zajęła
o tyle podniósł się poziom
1,5 cm (bo średnica równa 2
Objętość tego walca jest równa

$V = \pi \cdot (1$

Odpowiedź

Szukana objętość jest równa
w żądanym przybliżeniu

这是一本波兰高中数学基础水平的参考资料和习题集，属于网络和线下一整套教学系统的一部分。对于教材教辅类图书，本书采用了非常清晰的框架和色彩设计，使得这本图书具有了很好的阅读体验。本书为无线胶装，封面为白卡纸四色印刷加绿色荧光膜，覆哑膜，并在文字部分亮面 UV。内文胶版纸黑色和绿色双色印刷，内容分为习题解析和答案两部分，通过双色油墨量的多少来产生丰富的页面层次。习题解析页面设计为三栏，中心主栏为习题、解题过程和附加习题。左栏灰色部分为要点信息，右栏通过“便签条”样式的设计来展示解题思路，并通过箭头指向相关习题位置。

Trener*
liceum - poziom podstawowy i rozszerzony
matematyka
Jan Górowski
Adam Łomnicki
2y+3x=25
2y+3x=25

This is a reference and problem set for the basic level of mathematics in Polish high schools, which is part of the whole set of teaching systems for the online and offline education. As a textbook and supplementary book, the book adopts a very clear framework and color design, which makes the readers have good reading experiences. This book is perfect bound and the cover is printed in black and green on white cardboard with green fluorescent film, mounted by matte film, and spot UV in the text part. The inner offset paper is printed in black and green. The content is divided into two parts: problem analysis and the answer. The changing amount of the two-color ink is intended to create layers. The problem analysis page is designed in three columns. The main column in the center is for the exercises, problem solving process and additional exercises. The gray part of the left column is the key information, while the right column shows the solution ideas through the design of the 'note' style, and points to the position of the relevant exercises through the arrows.

荣誉奖
教与练 * 数学
波兰

- Honorary Appreciation
- Trener* matematyka
- Poland
- Grzegorz Podsiadlik
- Jan Górowski, Adam Łomnicki
- Wydawnictwo Szkolne PWN, ParkEdukacja, Warszawa / Bielsko-Biała
- 242×205×15mm
- 671g
- 328p

Maths books do not need to be dull and here is the proof : this one uses fresh colours and its pages are light and airy in design with a clear structure. A column on the left and differently shaded boxes on the right provide explanations which are allocated to the exercises in a readily discernible and efficient way. How the testing system works is explained by way of example on a double-page spread. Bright green pages at the beginning of each chapter and bold bars give structure to this unusual school textbook for Polish secondary level pupils, with its meticulous typography and asymmetrical double page layout producing a highly modern graphic effect.

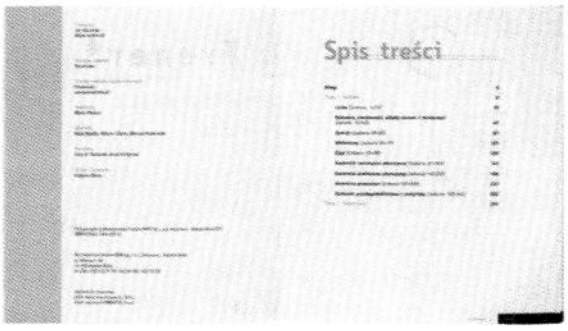
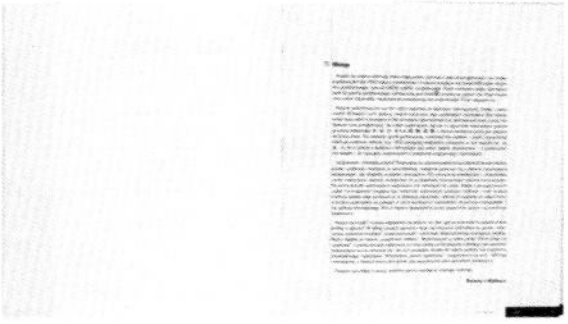
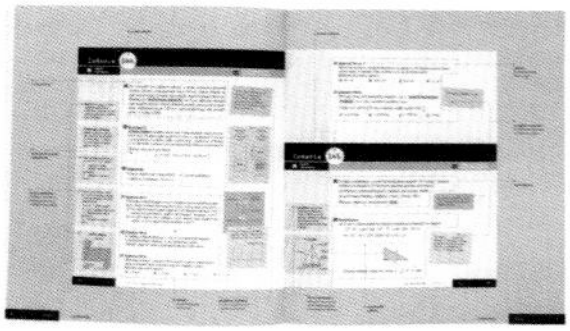
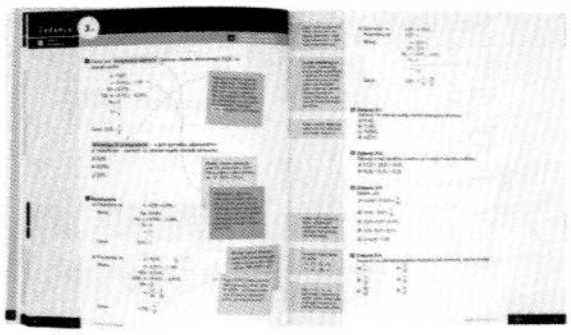
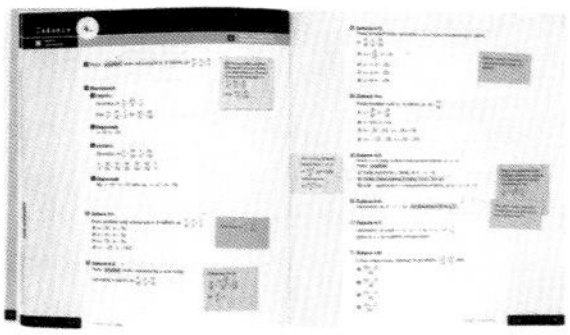
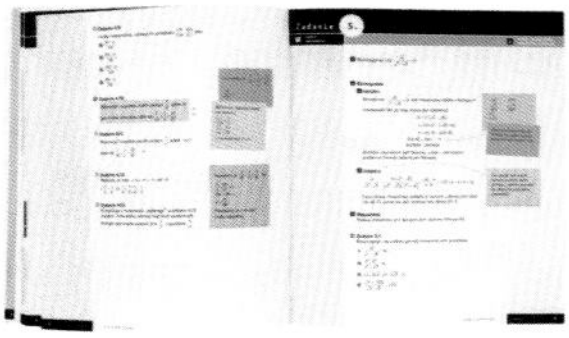
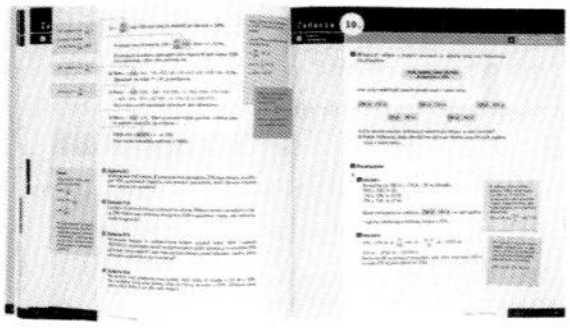
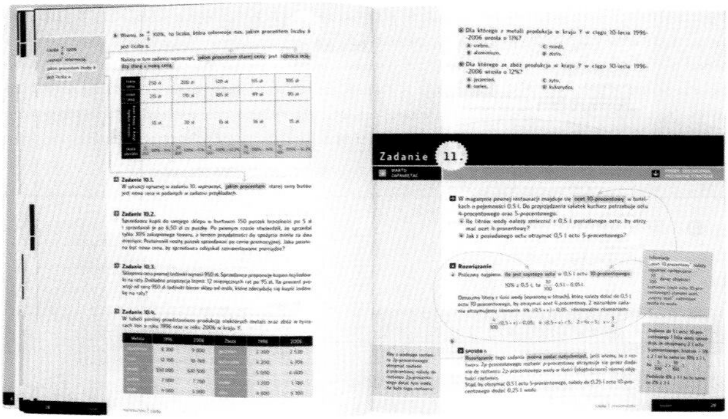
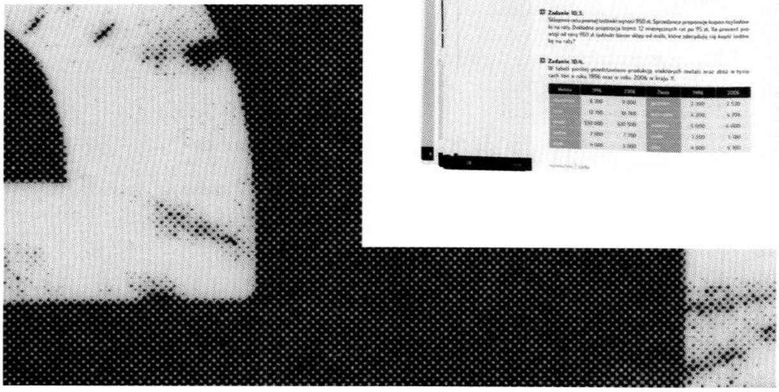

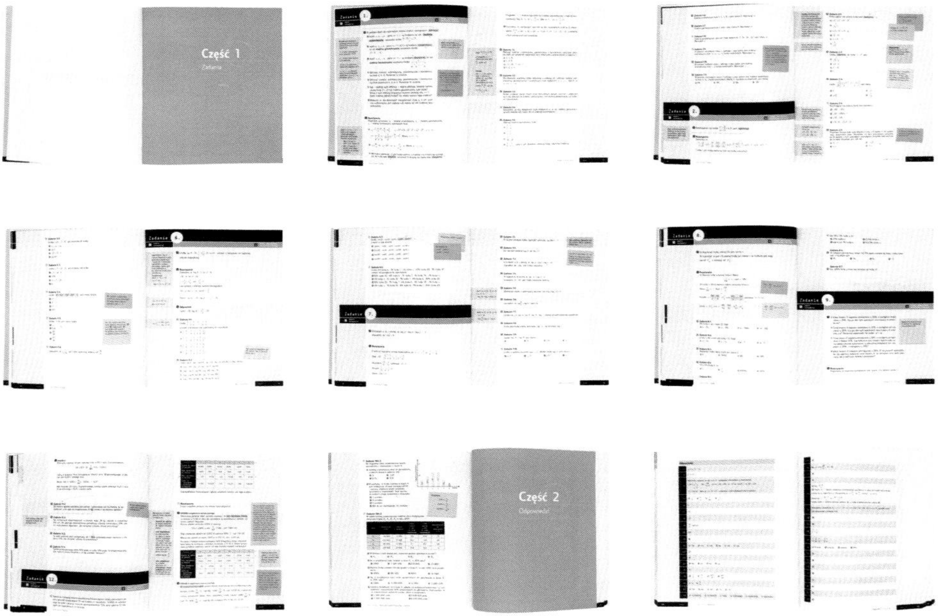
Część 1
Zadania
Część 2
Odpowiedzi

tym
A
acją odsetek, a

本书是瑞士摄影师尤勒斯·施皮纳奇（Jules Spinatsch）的作品集。从摄影师的作品和展览入手，分析其作品背后的故事和关注的角度。本书锁线胶装，封面采用三层结构，内封白卡纸印黄色摩尔纹，护封光面铜版纸印刷蓝色照片和文字，最外层包裹亮黄色透明 PVC 书皮，通过色彩的叠加使得照片和文字呈现墨绿色。内页英德双语排版，胶版纸夹插少量光面铜版纸单黑印刷，用于呈现作品和解读文字。照片的拍摄很有特点，由于本书是关于摄影师之前展览和作品的分析、解读，所以书中多数照片被添加了镜框呈现在版面上，颇具展览现场感。照片统一调整为相同的灰调子，使得整本书的视觉体验均匀而平静。上下书口及右侧翻口刷黄色，在整体外观上与黄色半透明书皮取得一致。

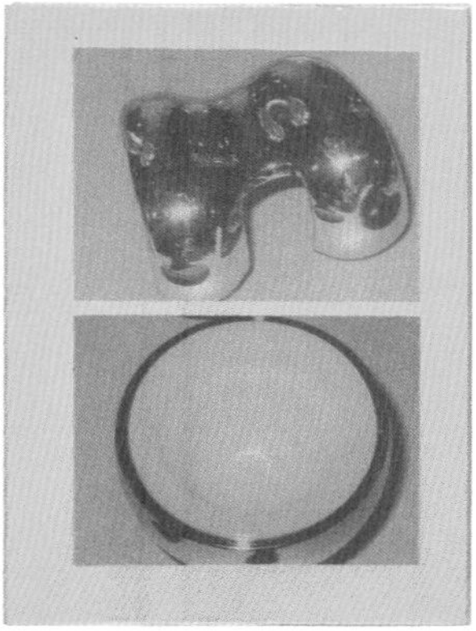

This book is a collection of works by Jules Spinatsch, a Swiss photographer. Starting with the photographer's works and exhibitions, this book analyses the stories and concerns behind his works. The book is perfect bound with thread sewing. There are three layers of the cover – inside cover is made of white cardboard printed with yellow moire pattern while the jacket is made of gloss art paper printed with blue pictures and texts. The outermost layer is covered with a bright yellow transparent PVC book cover, which makes the photos and text appear dark green through color overlay. The inside pages are English-German bilingual typesetting, offset paper and a small amount of smooth gloss art paper are used for presenting and interpreting. Photo shooting is very characteristic. Since the book is about the analysis and interpretation of the photographer's previous exhibitions and works, most of the photos in the book are framed and presented on the pages, which creates a feeling of being in the exhibition. Photos are adjusted to the same gray tone, making the visual experience even and calm. The three edges of the book are brushed in yellow to match the yellow transparent cover in terms of overall appearance.

荣誉奖
尤勒斯·施皮纳奇
瑞士

Honorary Appreciation
Jules Spinatsch
Switzerland
Winfried Heininger / Kodoji Press
Marco Obrist, Kunsthaus Zug (Ed.)
Kodoji Press, Baden
235×170×25mm
877g
349p

This book is conspicuous thanks to its intensive colour scheme using yellow for the cover, spine and edging. Particularly elegant is the contrast between inner and outer design in choice of materials and colours. To put the emphasis on the b/w picture section the text has been organised with typographic clarity and without obtrusive frills. The artist's works have been photographed throughout against a grey background which enhances uniformity and gives the book a pleasant sense of calm. A quiet, very consistent presentation of contemporary photography.

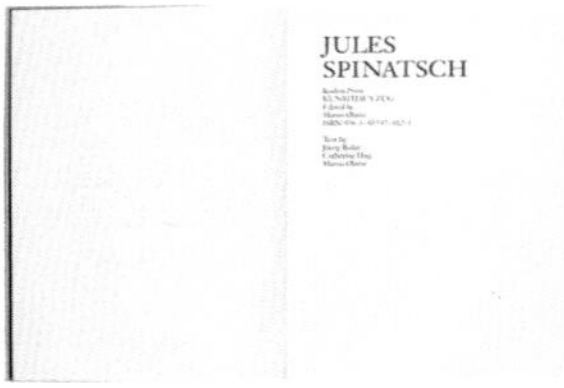
JULES
SPINATSCH

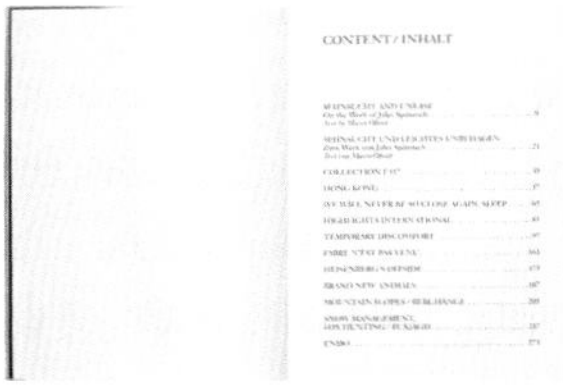
CONTENT / INHALT

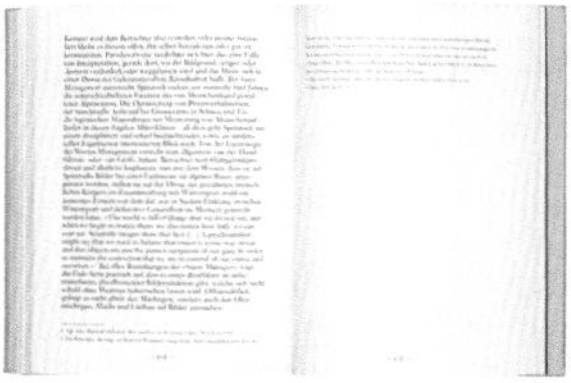

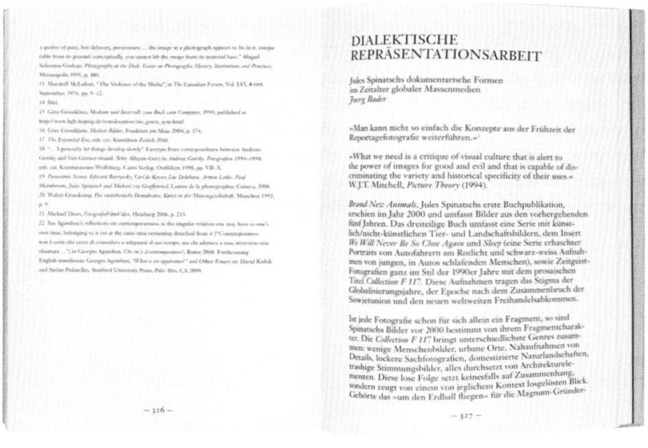

a quality of pure, but delusory, presentness ... the image in a photograph appears to be in it, insepa-
rable from its ground; conceptually, you cannot lift the image from its material base." Abigail
Solomon-Godeau, *Photography at the Dock: Essays on Photographic History, Institutions and Practices*,
Minneapolis 1991, p. 180.
13 Marshall McLuhan, "The Violence of the Media", in *The Canadian Forum*, Vol. LVI, # 664,
September, 1976, pp. 9–12.
14 Ibid.
15 Götz Grossklaus, *Medium und Intervall: vom Buch zum Computer*, 1999, published at
http://www.hgb-leipzig.de/transdiscourse/bio_goetz_syn.html
16 Götz Grossklaus, *Medien-Bilder*, Frankfurt am Main 2004, p. 174.
17 *The Expanded Eye*, exh. cat. Kunsthaus Zürich 2006.
18 "... I generally let things develop slowly". Excerpts from correspondence between Andreas
Gursky and Veit Görner (transl. Toby Alleyne-Gee) in *Andreas Gursky. Fotografien 1994–1998*,
exh. cat. Kunstmuseum Wolfsburg, Cantz Verlag, Ostfildern 1998, pp. VII–X.
19 *Panoramic Scenes. Edward Burtynsky, Carl de Keyzer, Luc Delahaye, Armin Linke, Paul
Shambroom, Jules Spinatsch and Michael von Graffenried*, Centre de la photographie, Geneva, 2006.
20 Walter Grasskamp, *Die unästhetische Demokratie. Kunst in der Massengesellschaft*, München 1992,
p. 9.
21 Michael Diers, *FotografieFilmVideo*, Hamburg 2006, p. 213.
22 See Agamben's reflections on contemporariness as the singular relation one may have to one's
own time, belonging to it yet at the same time remaining detached from it ("Contemporaneo
non è colui che cerca di coincidere e adeguarsi al suo tempo, ma chi aderisce a esso attraverso una
sfasatura ...") in Giorgio Agamben, *Che cos'è il contemporaneo?*, Rome 2008. Forthcoming
English translation: Giorgio Agamben, *"What is an apparatus?" and Other Essays* (tr. David Kishik
and Stefan Pedatella), Stanford University Press, Palo Alto, CA 2009.

— 326 —

DIALEKTISCHE REPRÄSENTATIONSARBEIT

Jules Spinatschs dokumentarische Formen
im Zeitalter globaler Massenmedien
Joerg Bader

»Man kann nicht so einfach die Konzepte aus der Frühzeit der Reportagefotografie weiterführen.«[1]

»What we need is a critique of visual culture that is alert to the power of images for good and evil and that is capable of discriminating the variety and historical specificity of their uses.« W.J.T. Mitchell, *Picture Theory* (1994).

Brand New Animals, Jules Spinatschs erste Buchpublikation, erschien im Jahr 2000 und umfasst Bilder aus den vorhergehenden fünf Jahren. Das dreiteilige Buch umfasst eine Serie mit künstlich/nicht-künstlichen Tier- und Landschaftsbildern, dem Insert *We Will Never Be So Close Again* und *Sleep* (eine Serie erhaschter Portraits von Autofahrern am Rotlicht und schwarz-weiss Aufnahmen von jungen, in Autos schlafenden Menschen), sowie Zeitgeist-Fotografien ganz im Stil der 1990er Jahre mit dem prosaischen Titel *Collection F 117*. Diese Aufnahmen tragen das Stigma der Globalisierungsjahre, der Epoche nach dem Zusammenbruch der Sowjetunion und den neuen weltweiten Freihandelsabkommen.

Ist jede Fotografie schon für sich allein ein Fragment, so sind Spinatschs Bilder vor 2000 bestimmt von ihrem Fragmentcharakter. Die *Collection F 117* bringt unterschiedlichste Genres zusammen: wenige Menschenbilder, urbane Orte, Nahaufnahmen von Details, lockere Sachfotografien, domestizierte Naturlandschaften, trashige Stimmungsbilder, alles durchsetzt von Architekturelementen. Diese lose Folge setzt keinesfalls auf Zusammenhang, sondern zeugt von einem von jeglichem Kontext losgelösten Blick. Gehörte das »um den Erdball fliegen« für die Magnum-Gründer-

— 327 —

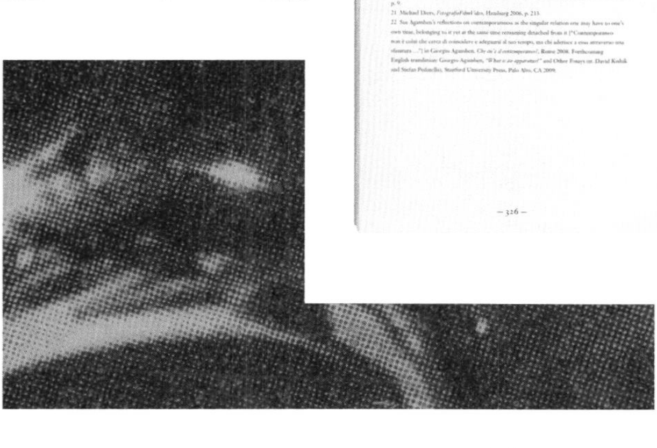

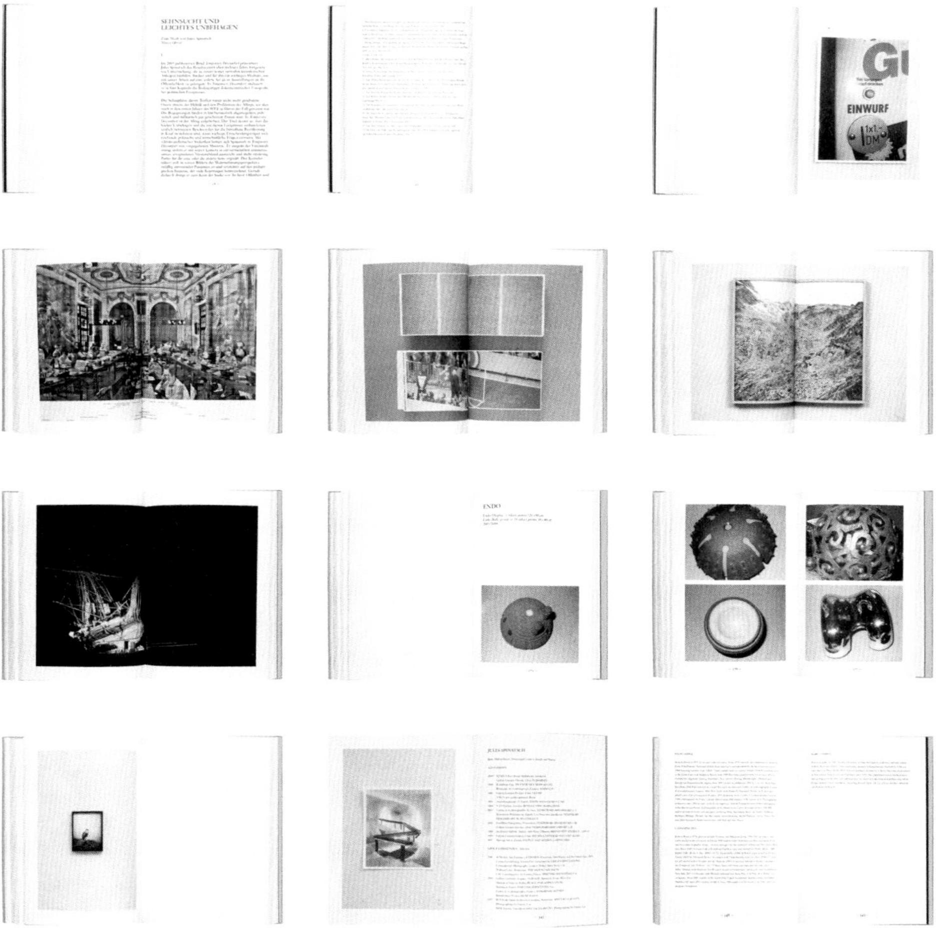
SEHNSUCHT UND
LEICHTES UNBEHAGEN
EINWURF
ENDO
JULES SPINATSCH

self

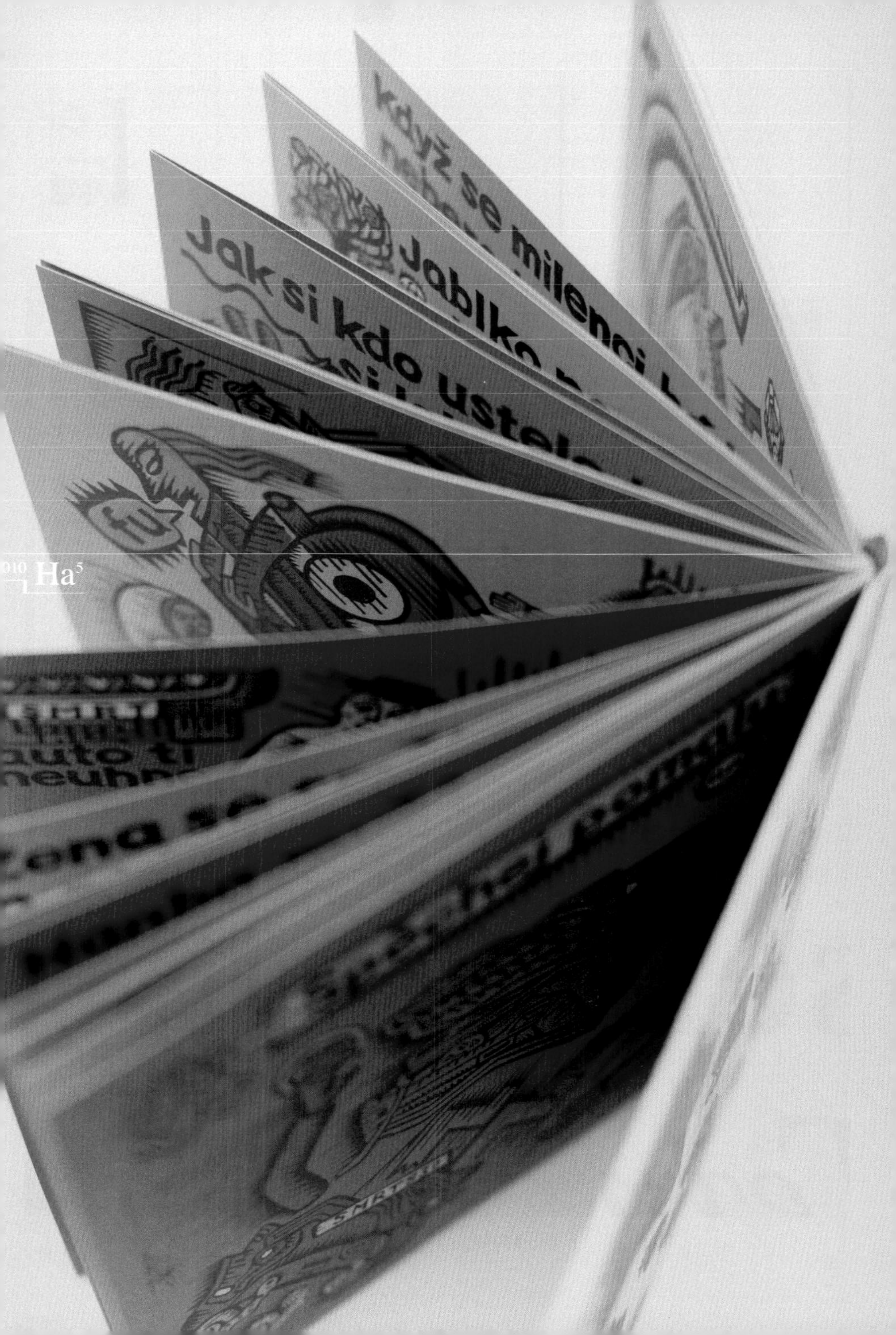
Když se milenci
Jablko
Jak si kdo
auto ti

这本是捷克插画家、平面设计师赫鲁多斯·瓦劳谢克（Chrudoš Valoušek）创作的一本有关谚语的小书，插画家根据自己有趣的想象对民间流传的谚语进行了现代解读，并用木版画的形式表现出来。本书为锁线硬精装，封面胶版纸四色印刷裱覆卡板，内页相同纸张四色印刷。全书并无其他工艺，完全通过画面本身的魅力来吸引读者。内页插图对传统谚语的新式解读充满趣味，表现形式风趣幽默，配色大胆强烈，字符通过木刻的形式也成了画面中的一部分，带有一定的图形属性。整体营造出一种无法抵御的画面张力和信息密度，是一本很容易让全年龄层都喜爱的小书。

CHRUDOŠŮV
MIX
PŘÍSLOVÍ
CHRUDOŠ VALOUŠEK / BAOBAB

This is a little book about proverbs created by Czech illustrator and graphic designer Chrudoš Valoušek. According to his rich imagination, the illustrator interprets the popular proverbs in modern times and expresses them in the form of wood engraving. This hardcover book is bound with thread sewing and the cover is printed in four colors on offset paper mounted by graphic board. The offset paper is used for inside pages as well. The book appeals to the reader entirely through the charm of the pictures without any special technology. The illustrations that interpret traditional proverbs in a new way are interesting, with humorous forms and bold color matching. Characters have become part of the picture through woodcut, with certain graphical properties. As a whole, it creates an irresistible tension and information density. It is a little book that can easily be loved by people of all age groups.

荣誉奖

赫鲁多斯·瓦劳谢克的谚语新解

捷克

A small-sized book with unusual, powerful and boldly coloured illustrations. The woodcut character of the pictures inside is aesthetically reinforced by wooden letters on the cover. The frequently single-page illustrations are divided from each other on double-page spreads by means of strong colour contrasts. Handdrawn block letters form part of the illustrations. The drawings, with great density, create a sense of irrepressible tempo and abundance. Choice of paper and binding is good, adding to the book's strong identity.

Honorary Appreciation

Chrudošův mix přísloví

Czech Republic

Juraj Horváth

Chrudoš Valoušek

Baobab, Praha

207 × 203 × 10mm

320g

46p

Spěchej pomalu.

Když se milenci hádají, nebere to vážně ani pes.
ON-LINE
CHRUDOŠŮV
MIX
PŘÍSLOVÍ
CHRUDOŠ VALOUŠEK / BAOBAB

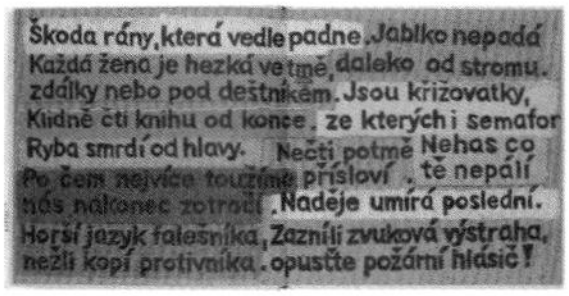
Škoda rány, která vedle padne.
Jablko nepadá daleko od stromu.
Každá žena je hezká ve tmě, zdálky nebo pod deštníkem.
Jsou křižovatky, ze kterých i semafor
Klidně čti knihu od konce.
Ryba smrdí od hlavy.
Nečti potmě přísloví
Nehas co tě nepálí
Naděje umírá poslední.
Horší jazyk falešníka, nežli kopí protivníka.
Zazněli zvuková výstraha, opusťte požární hlásič!

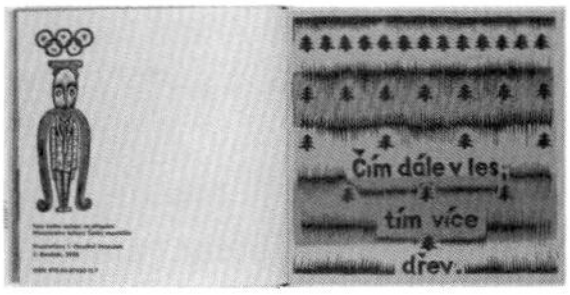
Čím dále v les, tím více dřev.

Jablko nepadá daleko od stromu.
Stará láska nerezaví.
VLPP
A+K
NEREZ
MAKE LOVE NOT WAR

Polibek bez kn... je jako biftek bez hořčice

Kdo se bojí nesmí do lesa
HALÓ
HALÓ

Hanba mužům, kterým žena vládne.
S.O.S
POMOC AU
DUCHAMPE
To není umění, to je záchod.

Jak si kdo ustele, tak si lehne .....
www.lehne.cz
www.ustele.cz
Jak se do lesa volá tak se z lesa ozývá

Co na srdci, to na jazyku.
kdo chce psa bít, hůl si vždycky najde.

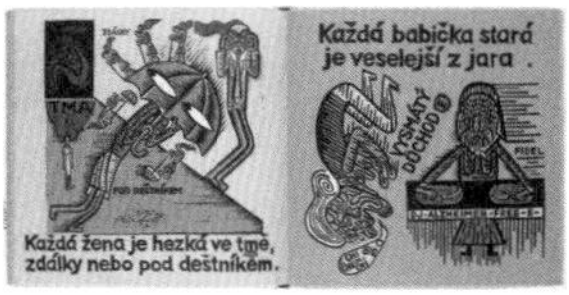
TMA
Každá žena je hezká ve tmě, zdálky nebo pod deštníkem.
Každá babička stará je veselejší z jara .

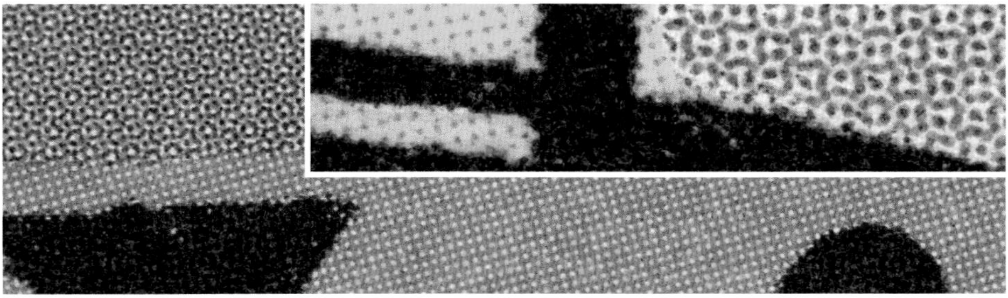

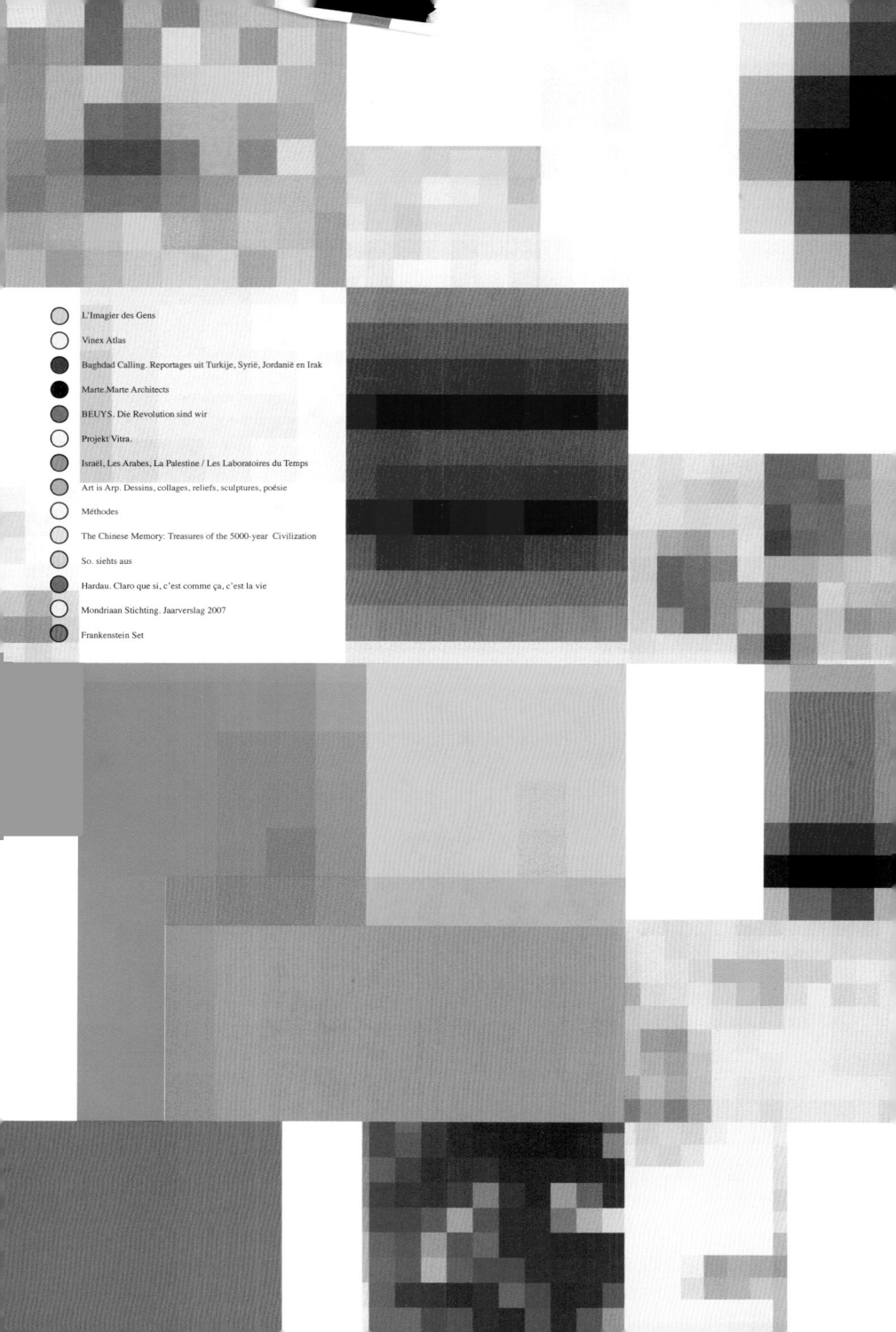

- L'Imagier des Gens
- Vinex Atlas
- Baghdad Calling. Reportages uit Turkije, Syrië, Jordanië en Irak
- Marte.Marte Architects
- BEUYS. Die Revolution sind wir
- Projekt Vitra.
- Israël, Les Arabes, La Palestine / Les Laboratoires du Temps
- Art is Arp. Dessins, collages, reliefs, sculptures, poésie
- Méthodes
- The Chinese Memory: Treasures of the 5000-year Civilization
- So. siehts aus
- Hardau. Claro que si, c'est comme ça, c'est la vie
- Mondriaan Stichting. Jaarverslag 2007
- Frankenstein Set

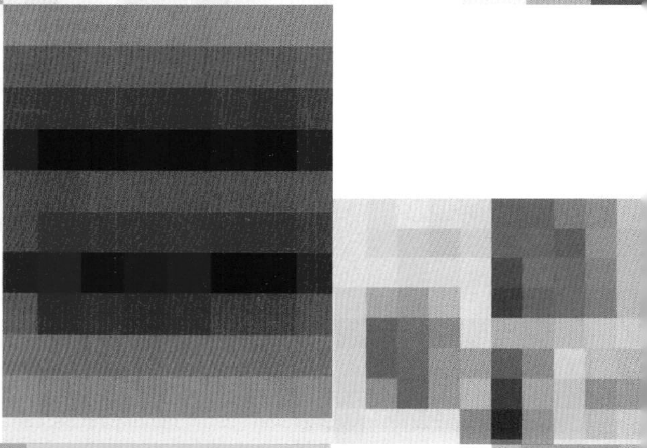

2009

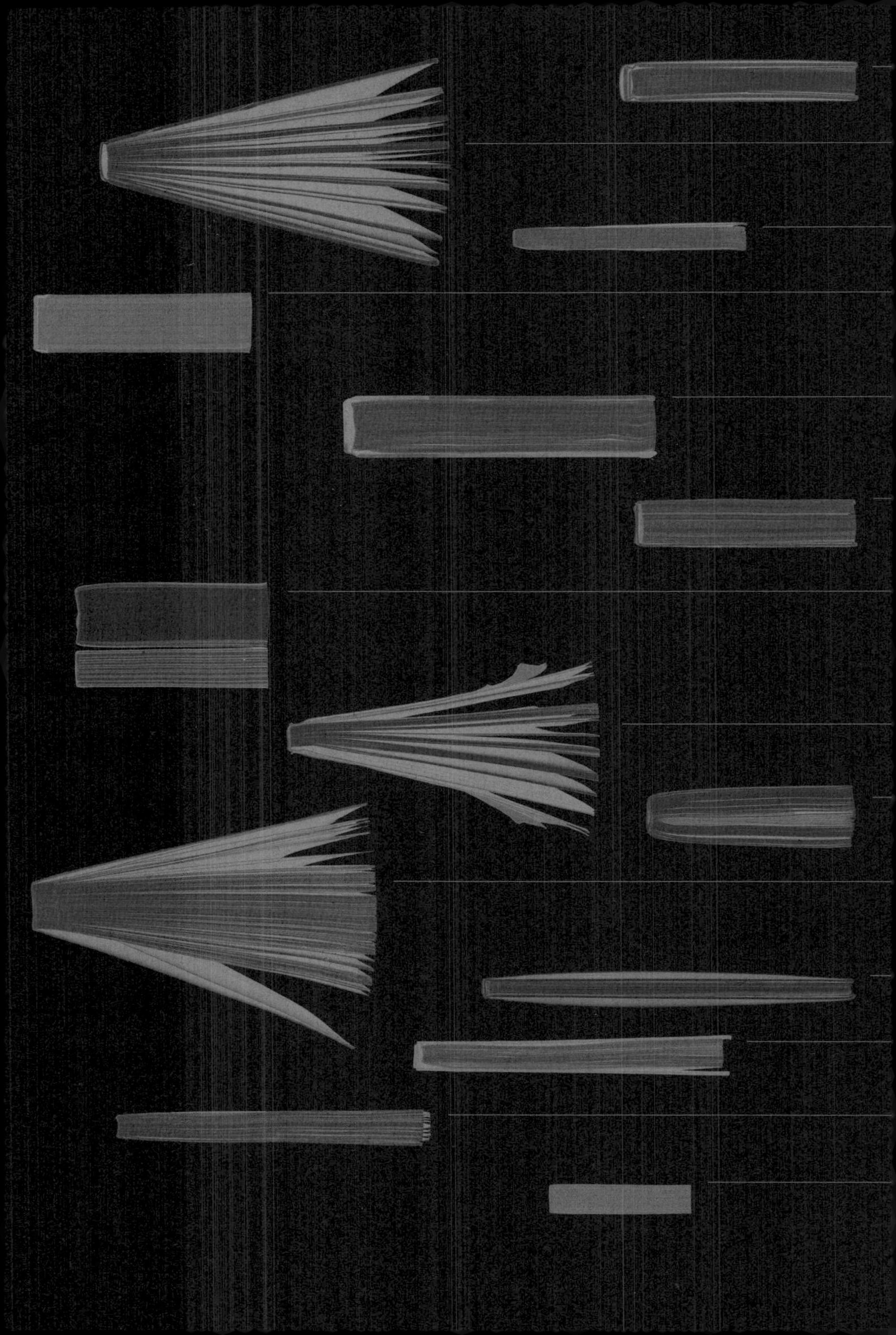

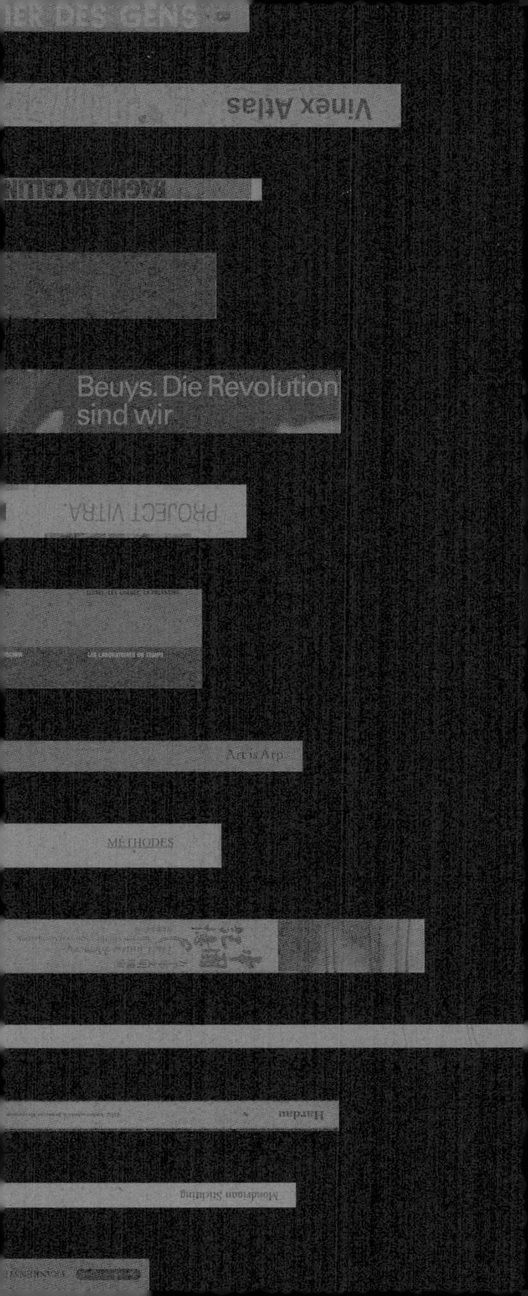
Vinex Atlas
Beuys. Die Revolution sind wir
PROJECT VITRA.
Art is Arp
MÉTHODES
Mondriaan Stichting

Waterakkers-Lunetten

MÉDECIN

erde Nota over de Ruimtelijke Ordening Extra.
a van het ministerie van Volkshuisvesting,
ng en Milieubeheer (VROM) gaf de Neder-
n het begin van de jaren negentig van de

Vinex is an acronym for Vierde Nota over de Ruimtelijke Ordeni
Extra, in its most literal translation, Fourth Report on Spatial
Planning Extra. (Early official translations use the term Physica
Planning.) In this supplement to the fourth in the series of poli

2008

The British-led coalition invaded Saddam Hussein's Iraq five years ago, bringing down Saddam's hated dictatorial regime within a month. In their wake, Iraq has become the

penchant for solid concrete that sets M.M apart from
lberg architecture scene. M.M's idea of the ideal hous
to the ideas of most other teams in the area, and they

03 wenn das Tier zum Menschen erlöst wird

08 wo ist Element 3?

13 ich durchsuche Feldcharakter

gou, Deutscher Bundestag, Kunsthaus Graz

On se souvient qu'en 1993 un long article signé de Samuel Huntington et intitulé « The clash of civilizations » avait bé-

119. *Libération* du 5 juin 2005.

120. David Schulman, *Ta'ayush. Journal d'un combat pour la paix*

déchagrinés, tout polis, flagada[2], le cortège jubilant
des circonstancionistes, la théorie devant la villa

l'intuition sur la sensation de l
quelques échantillons – sol, mur,

曙光初
史前时期
礼乐安
夏至秦
叁贰壹
盛世华
西汉至隋
叁贰壹
古典终
宋元明清
叁贰壹

r, antwortete jedoch nicht, sondern lächelte mich nur nachsicht
un, alles, was ich Ihnen sagen wollte, war, mißtrauen Sie ihren

Der hat vielleicht ein Terrarium oder so. Freaks gibt es einige. Hunde gibt es keine – ah doch, so ein paar wenige, das ist aber geheim.
Den K. sehen wir recht oft. Er sieht selbst ein bisschen aus wie eine Taube. Das ist sein Job. Er hört auch nicht gut; wenn man mit dem Velo

sagt haben, er solle doch die Möwen nicht mehr füttern.

Sibylle Sag: **Wenn ich nicht eine Kollegin hier im Haus hätte, die Frau H., würde ich noch viel weniger Leute kennen.** Ja, durch sie habe ich ein paar Leute kennengelernt.
Ein Bewohner aus dem 1. Stock geht immer

voording

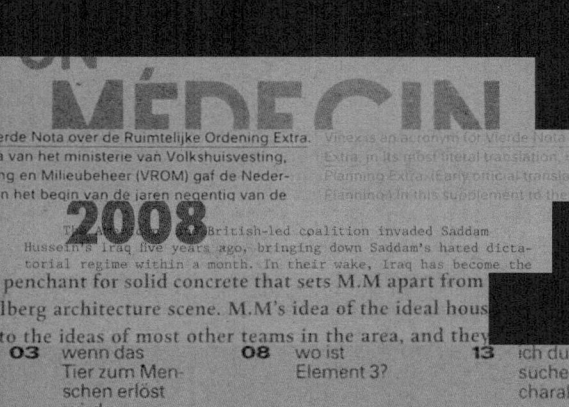

n. Whoever blasphemes the name of the Lord sh
urely be put to death. All the congregation shall sto
m. The sojourner as well as the native, when
asphemes the Name, shall be put to death

2009 Gl

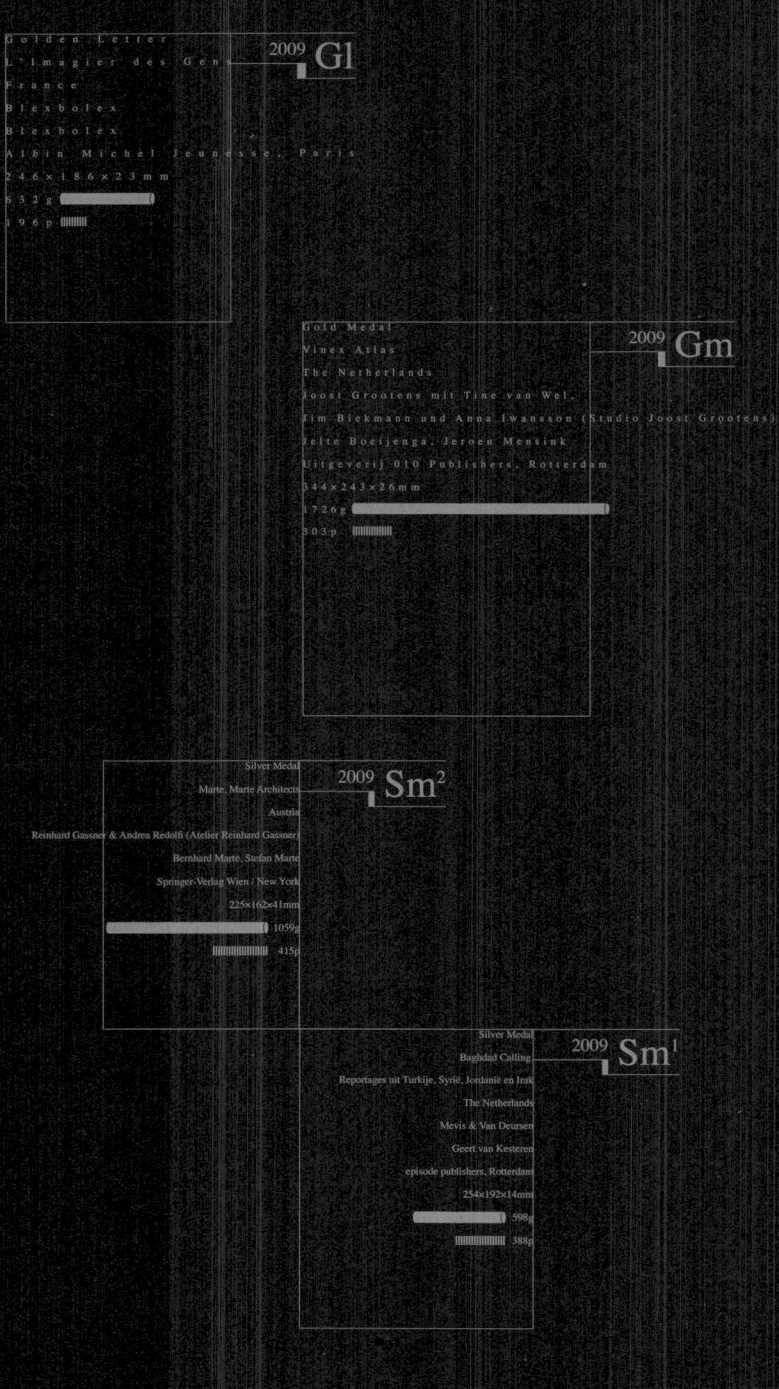

Golden Letter

L'Imagier des Gens

France

Blexbolex

Blexbolex

Albin Michel Jeunesse, Paris

246×186×23mm

632g

196p

2009 Gm

Gold Medal

Vinex Atlas

The Netherlands

Joost Grootens mit Tine van Wel,

Jim Biekmann und Anna Iwansson (Studio Joost Grootens)

Jelte Boeijenga, Jeroen Mensink

Uitgeverij 010 Publishers, Rotterdam

344×243×26mm

1726g

303p

2009 Sm2

Silver Medal

Marte. Marte Architects

Austria

Reinhard Gassner & Andrea Redolfi (Atelier Reinhard Gassner)

Bernhard Marte, Stefan Marte

Springer-Verlag Wien / New York

225×162×41mm

1059g

415p

2009 Sm1

Silver Medal

Baghdad Calling

Reportages uit Turkije, Syrië, Jordanië en Irak

The Netherlands

Mevis & Van Deursen

Geert van Kesteren

episode publishers, Rotterdam

254×192×14mm

598g

388p

2009 Ha2

Honorary Appreciation
So. siehts aus
Germany
Kerstin Rupp
Kerstin Rupp
Eigenverlag Kerstin Rupp, Leipzig
426×294×14mm
1149g
96p

2009 Ha1

Honorary Appreciation
The Chinese Memory: Treasures of the 5000-year Civilization
China
Lü Jingren & Lü Min (Jingren Art Design Studio)
Capital Museum (Ed.)
Cultural Relics Press, Beijing
359×239×35mm
2777g
352p

2009 Bm1

Bronze Medal
BEUYS. Die Revolution sind wir
Germany
Detlef Fiedler, Daniela Haufe, Nina Polumsky, Daniel Wiesmann (cyan)
Eugen Blume, Catherine Nichols (Ed.)
Steidl Verlag, Göttingen
305×247×41mm
2458g
407p

2009 Bm2

Bronze Medal
Projekt Vitra
Germany
Cornel Windlin
Rolf Fehlbaum, Cornel Windlin (Ed.)
Birkhäuser Verlag, Basel / Boston / Berlin
244×174×34mm
1106g
396p

2009 Bm3

Bronze Medal
Israël, Les Arabes, La Palestine /
Les Laboratoires du Temps
France
Julien Hourcade, Thomas Petitjean (Hey Ho)
Série «Essais», Jean Daniel / Alain Fleischer
Galaade Éditions, Paris
215×140×37mm / 214×140×28mm
984g / 478g
859p / 417p

2009 Ha3

Honorary Appreciation
Hardau. Claro que si, c'est comme ça, c'est la vie
Germany / Switzerland
Julia Ambroschütz, Jeannine Herrmann (Südpol)
Julia Ambroschütz, Jeannine Herrmann
Salis Verlag, Zürich
303×240×20mm
911g
120p

2009 Ha4

Honorary Appreciation
Mondriaan Stichting. Jaarverslag 2007
The Netherlands
Ingeborg Scheffers
Mondriaan Stichting, Amsterdam
275×211×15mm
723g
220p

2009 Bm4

Bronze Medal
Art is Arp
Dessins, collages, reliefs, sculptures, poésie
France
Sp Millot
Editions des Musées de la Ville de Strasbourg, Strasbourg
280×230×19mm
1140g
343p

2009 Ha5

Honorary Appreciation
Frankenstein Set
Switzerland
Geoff Han
Christian Jankowski
Christoph Keller Editions / JRP Ringier Kunstverlag, Zürich
180×104×23mm
345g
440p

2009 Bm5

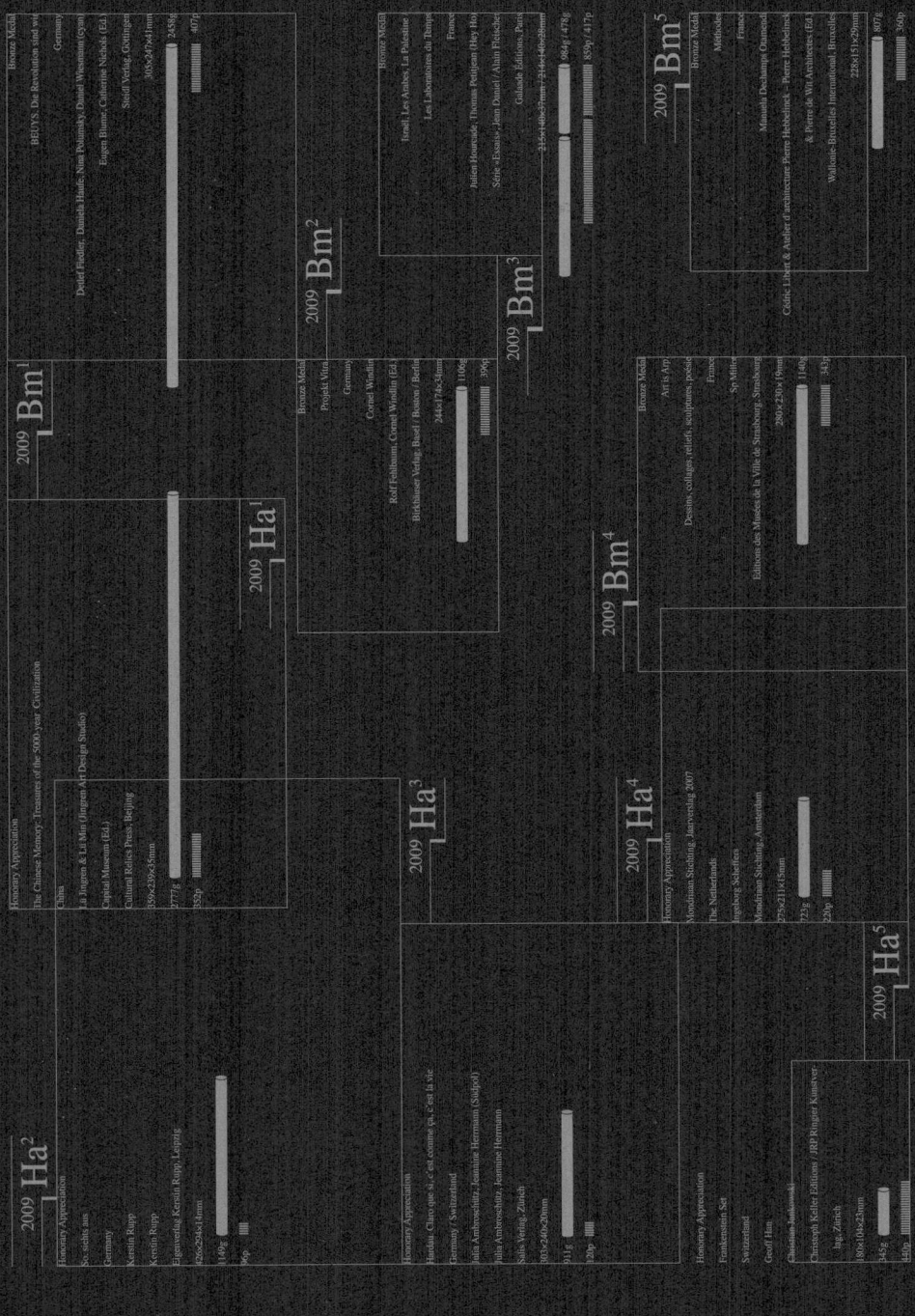

Bronze Medal
Méthodes
France
Manuela Dechamps Otamendi
Cédric Libert & Atelier d'architecture Pierre Hebbelinck – Pierre Hebbelinck & Pierre de Wit Architectes (Ed.)
Wallonie-Bruxelles International, Bruxelles
228×151×29mm
807g
360p

Designer

Blexbolex

Joost Grootens mit Tine van Wel, Jim Biekmann und Anna Iwansson (Studio Joost Grootens)

Mevis & Van Deursen

Reinhard Gassner & Andrea Redolfi (Atelier Reinhard Gassner)

Detlef Fiedler, Daniela Haufe, Nina Polumsky, Daniel Wiesmann (cyan)

Cornel Windlin

Julien Hourcade, Thomas Petitjean (Hey Ho)

Sp Millot

Manuela Dechamps Otamendi

Lü Jingren & Lü Min (Jingren Art Design Studio)

Kerstin Rupp

Julia Ambroschütz, Jeannine Herrmann (Südpol)

Ingeborg Scheffers

Geoff Han

2009

The International Jury

Peter Cocking (Canada)
Goele Dewanckel (Belgium)
Markus Dreßen (Germany)
Arturs Hansons (Latvia)
Dominika Hasse (Germany)
Peter E. Renn (Switzerland)
Philippa Walz (Germany)

Country / Region

France 4
The Netherlands 3
Germany 3
Switzerland 1
Austria 1
China 1

Literature & Fiction
文学与小说
Nature & Technology
自然与科技
Social Sciences
社会科学
Art & Design
艺术与设计
Children and Young People's Books
少儿读物
Education & Teaching
教育与教学
Others
其他
Not collected
未收藏
57%
7%
7%
29%

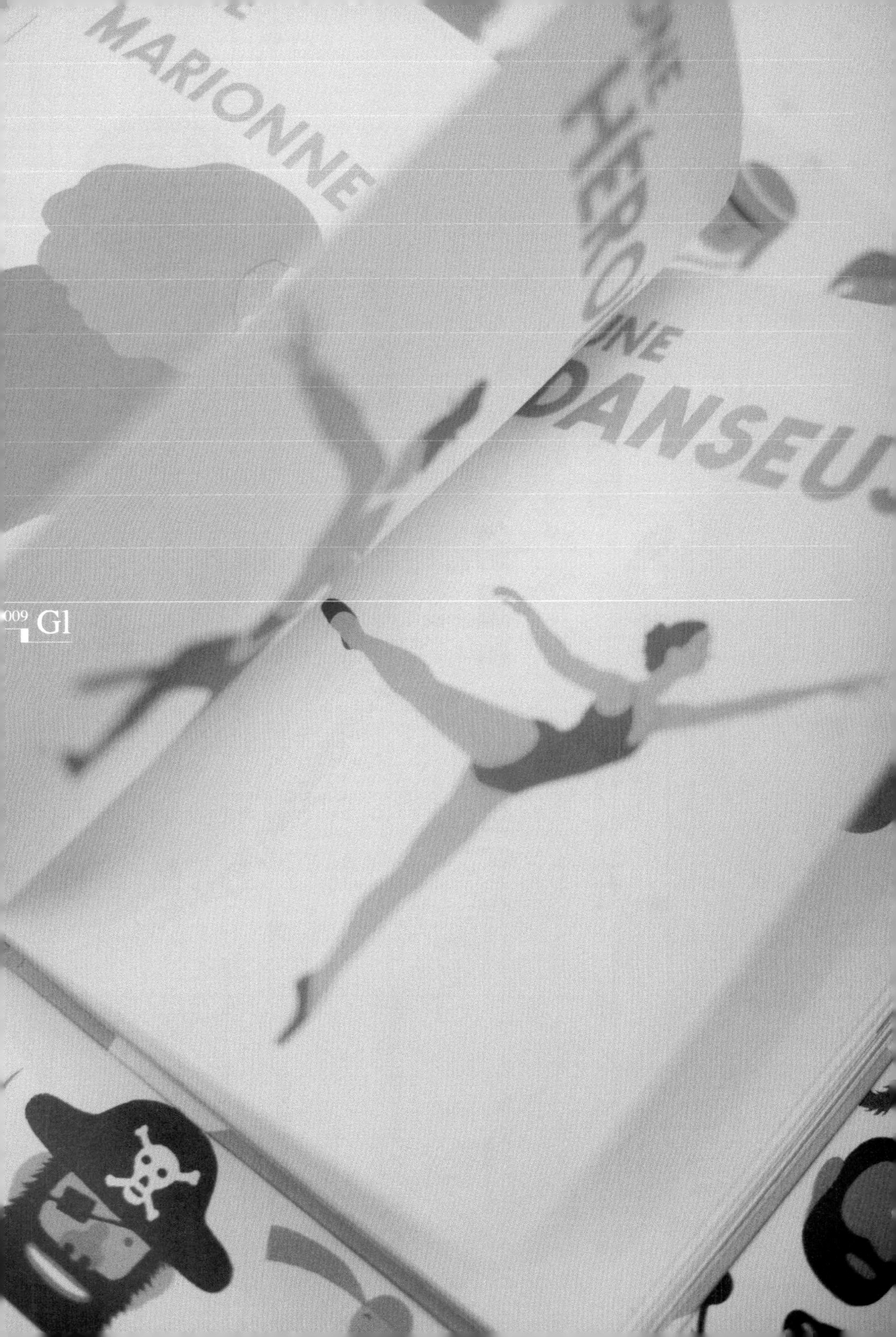
MARIONNE
DANSEUS

009 G1

这是一本通过丰富的想象力来跟孩子互动并从玩耍中学习的亲子读物。法国漫画家 Blexbolex [贝纳尔 · 格朗热 (Bernard Granger) 的笔名] 在书中展示了各种真实存在、神话中甚至想象中的人物，展现他们有趣的动作，突出各自最主要的特征。本书为锁线精装，并且只用珊瑚粉、水蓝、黄三色孔版印刷，通过水性油墨相互叠加的关系，呈现多达七种颜色，配以白色镂空部分和局部的渐变效果，使得画面细节丰富。护封选取部分内文人物形象进行组合，正反印刷后包裹在封面上，展开是一张 4 开海报。内封白色布面印单色黄，人物形象和文字露白。内文没有目录，只有人物的展现，左右页人物身份具有一定的关联性，有的是逻辑上的关联，有的是动作或局部的相似。||

BLEXBOLEX
L'IMAGIER DES GENS
ALBIN MICHEL JEUNESSE

This is a parent-child book that teaches how to interact with children and learn from games through imagination. The French cartoonist Blexbolex (pen name of Bernard Granger) creates a variety of real people, mythical characters and even imaginary figures in his book, showing their funny movements and highlighting their main characteristics. This hardcover book is perfect bound with thread sewing. Stencil printing technique is used for the cover design with three colors – coral, water blue and yellow. Water-based inks are superimposed on each other and present up to seven colors, accompanied with white hollow part and partial gradient effect. There is no table of contents in the book, only the description of characters. The characters shown on the left and right pages are somehow connected – some are logically related some are similar in movement.

◎ 金字符奖

《 人的意象

◖ 法国

Its simple yet very eloquent and distinct illustrations are most captivating. Each one tells its own story though all are interrelated. The connection of image and text at the same graphic level is particularly unusual. The excellent choice of materials reinforces the unusual colour spectrum – only three colours are used for printing! Jacket, cover and endpapers also consolidate the narrative atmosphere of the book perfectly. What a glorious children's book!

◎ ·········· Golden Letter

《 ·········· L'Imagier des Gens

◖ ·········· France

◗ ·········· Blexbolex

△ ·········· Blexbolex

|||| ··· Albin Michel Jeunesse, Paris

□ ·········· 246×186×23mm

▯ ·········· 632g

≣ ·········· 196p

IENT DA

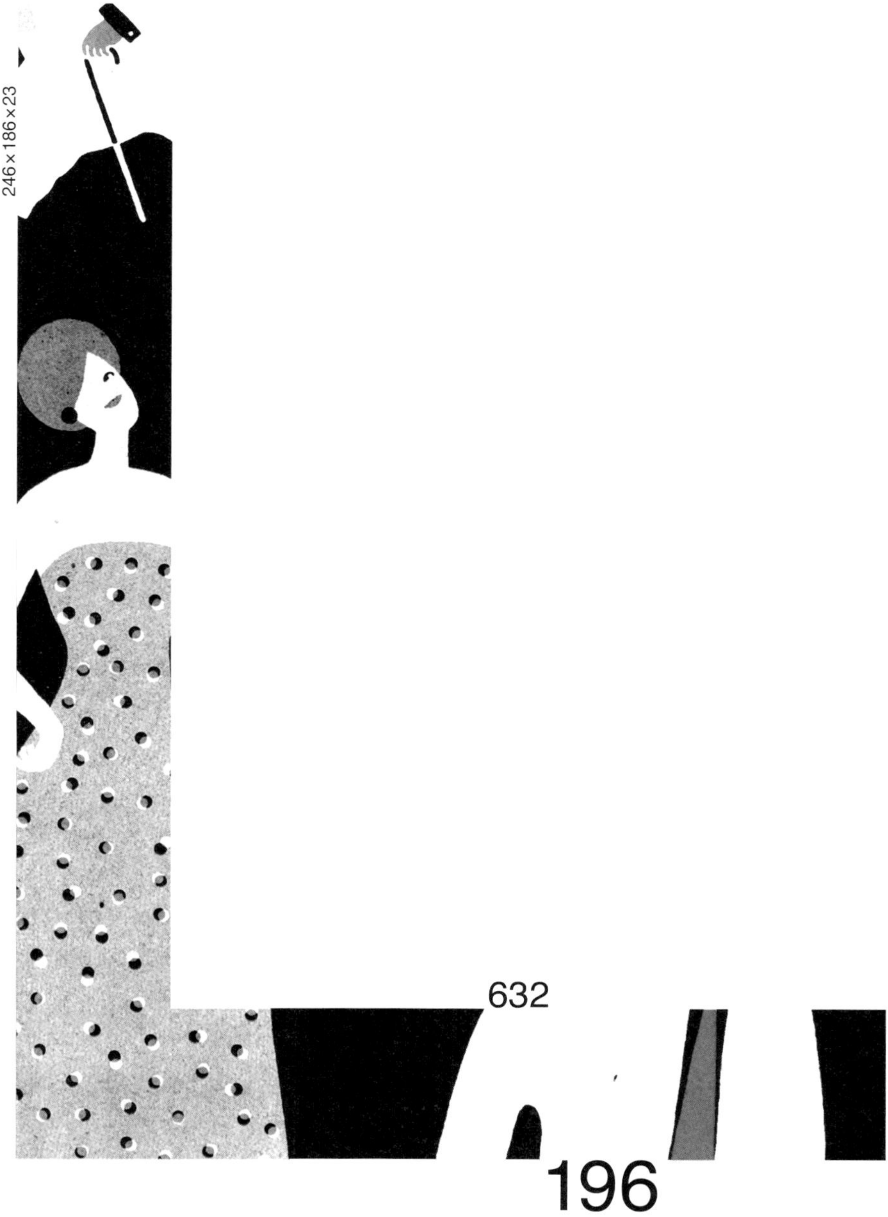
246×186×23
632
196

UN
MONSIEUR
UNE
DAME

UN
COUPLE
UN
CÉLIBATAIRE

UN
PAPA
UNE
FAMILLE

UN
CURIEUX
UN
ESPION

UN
DÉMON
UN BONHOMME
DE NEIGE

DES
CLIENTS
UNE
SERVEUSE

L'IMAGIER
DES GENS
BLEXBOLEX

DES
GENS

DES
ÉCOLIERS
UN
PROFESSEUR

UNE
FILLETTE
UN
FARCEUR

UN
FRILEUX
UN
FUMEUR

UN
NUDISTE
UN HOMME
INVISIBLE

UN
BOSSU
UNE
BEAUTÉ

L'IMAGIER DES GENS · B
UN
MÉDECIN

UN
FORGERON

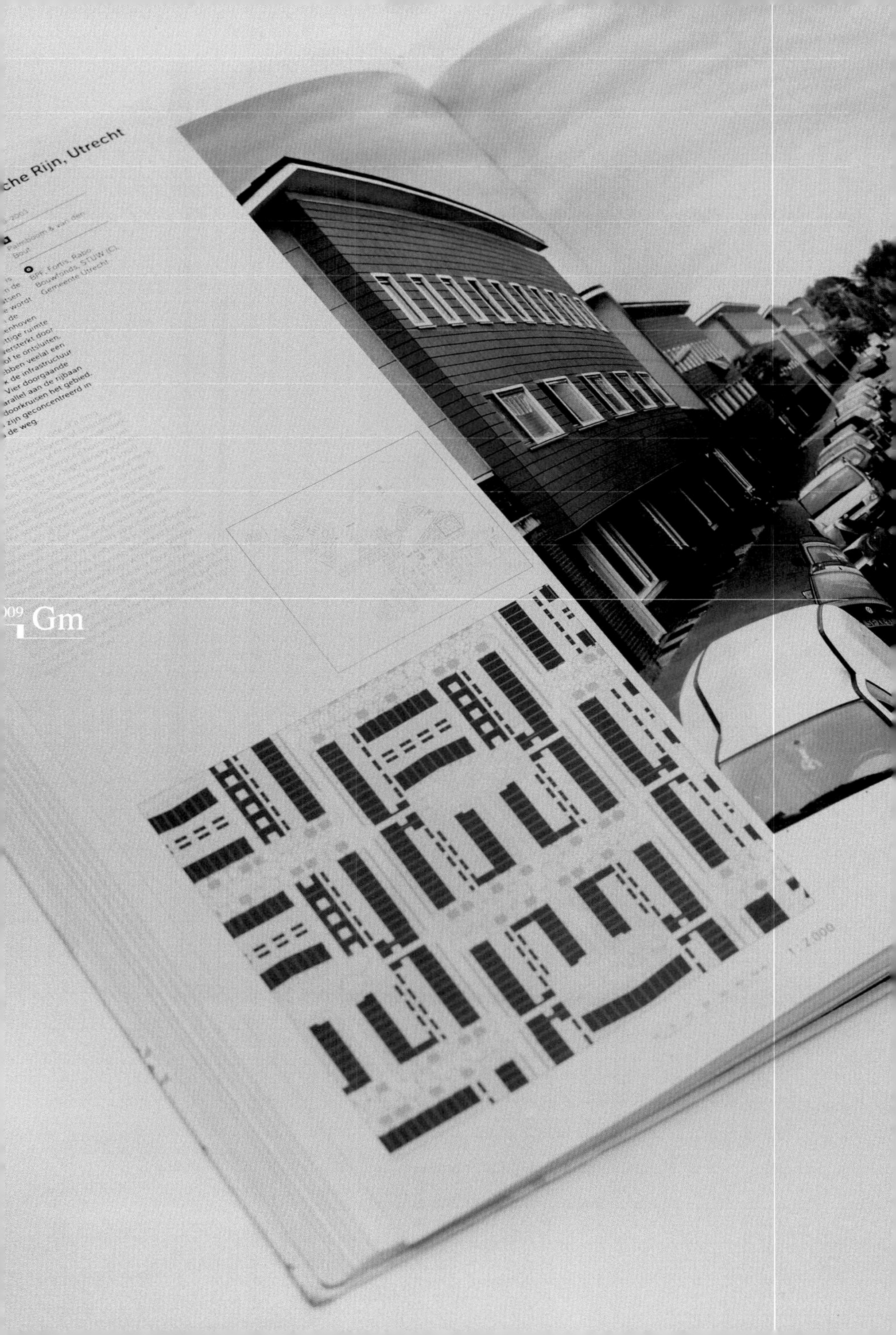
che Rijn, Utrecht
Palmboom & van den
Bout
BPF, Fortis, Rabo
Bouwfonds, STUW (C),
Gemeente Utrecht
de
wordt
enhoven
ttige ruimte
ersterkt door
of te ontsluiten.
ebben veelal een
k de infrastructuur
Vier doorgaande
arallel aan de rijbaan
doorkruisen het gebied.
zijn geconcentreerd in
de weg.
1:2000

009 Gm

这是一本政府政策实施的研究报告。20 世纪 90 年代荷兰政府颁布了城市空间规划的政策，其中第四点政策关健词的前几个字母组成新的单词“Vinex”，成为此项规划政策的代号。本书通过航拍、绘制、数据统计等方式来跟踪这项计划的实施情况。书籍为锁线硬精装，封面布纹纸四色印刷裱覆卡板，地图和文字部分压凹处理。内页五色印刷，部分色块和文字加入银色，使得颜色透露出一定的金属光泽。内容按照政策、舆论、统计、航拍及绘制来展现各个区域的实施情况，可以看到政策实施过程中的区域地表的变化、容积率的改变、居住配套合适的增减。信息丰富，细节极为精美。

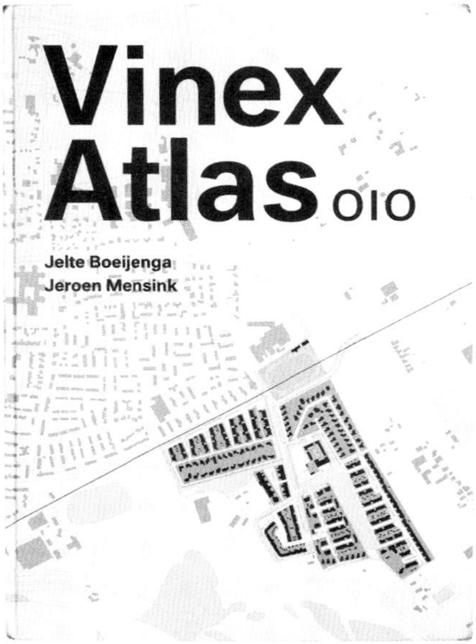
Vinex
Atlas 010
Jelte Boeijenga
Jeroen Mensink

The book is a research report on the implementation of government policies. In the 1990s, the Dutch government promulgated the policy of urban spatial development. The fourth article of the policy was named 'Vinex', which is formed by the first letters of the key words. The book tracks the implementation of this project through aerial photography, drawing and data statistics. The hardcover book is bound with thread sewing. The cover is printed in four colors and mounted with graphic board. Map and text on the cover are embossed. Inside pages are printed in five colors with silver used for some color blocks and texts. The situation of each region is organized by the sequence of policy, public opinions, statistics, aerial photography and drawings. The readers can observe the changes of regional surface, floor area ratio and the residential support. The book is rich in information and exquisite in detail.

- 金奖
- Vinex 地图集
- 荷兰

- Gold Medal
- Vinex Atlas
- The Netherlands
- Joost Grootens mit Tine van Wel, Jim Biekmann und Anna Iwansson (Studio Joost Grootens)
- Jelte Boeijenga, Jeroen Mensink
- Uitgeverij 010 Publishers, Rotterdam
- 344×243×26mm
- 1726g
- 303p

An outstanding example of successful information-based graphics. The Vinex Atlas sets out to describe 52 city districts and their spatial planning using aerial photographs, plans and diagrams. The detailed information is clearly structured, precisely collated and in its design concept shows sure instinct right down to the last detail. Whether pictograms, topographical figures or tables – everything bears witness to outstanding quality, consistently maintained on double page-spreads with aerial photographs showing urban construction development. Excellent design of complex contents which makes the reader eager to find out more.

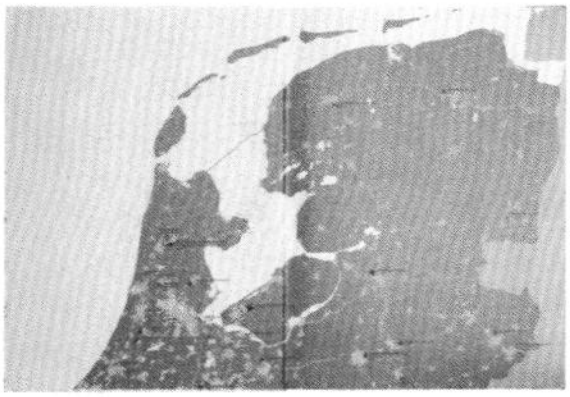

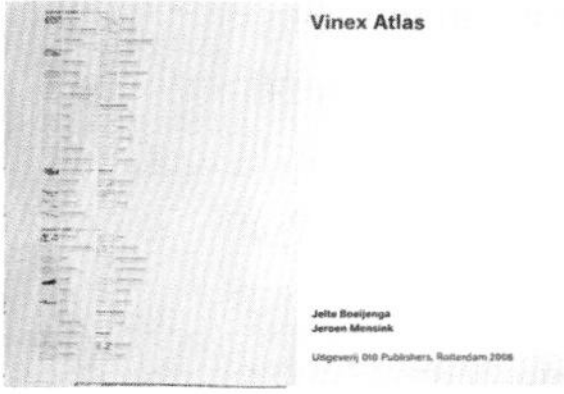
Vinex Atlas
Jelte Boeijenga
Jeroen Mensink
Uitgeverij 010 Publishers, Rotterdam 2006

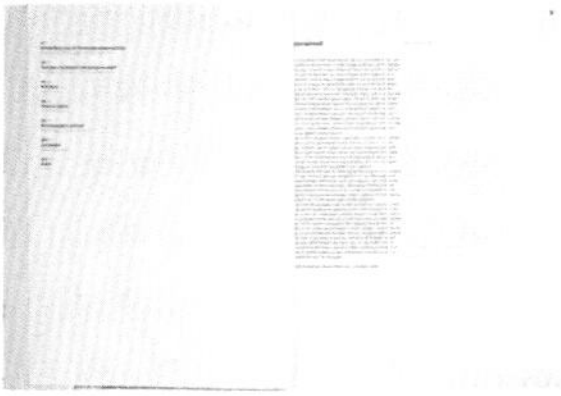

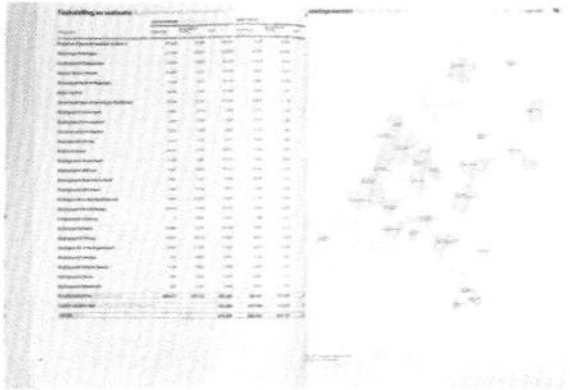

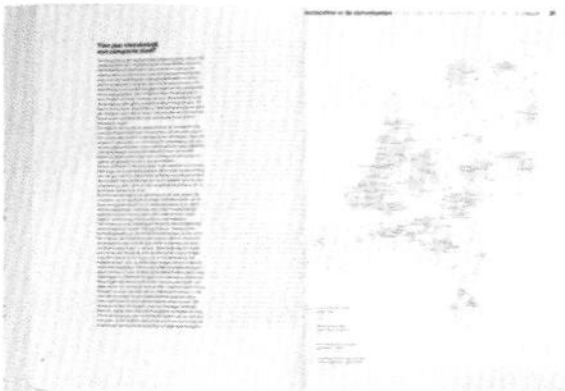

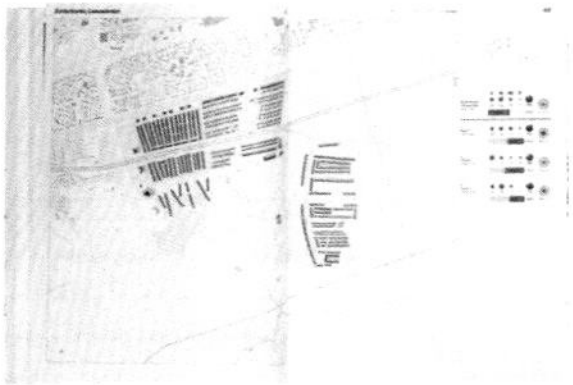

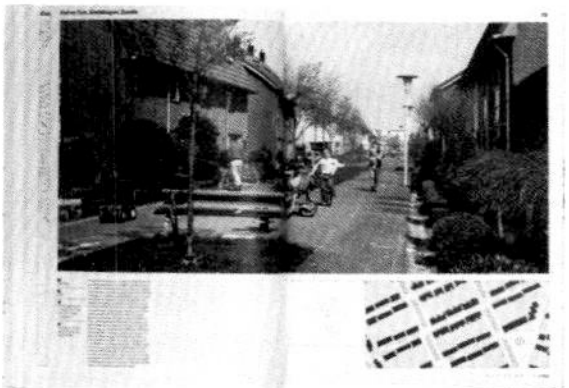

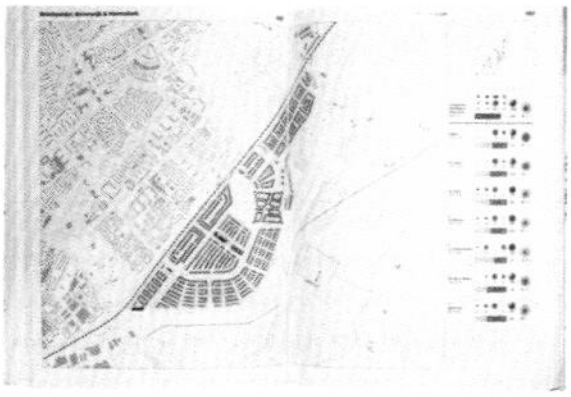

Fase II, Maasveld, Tegelen
266
267

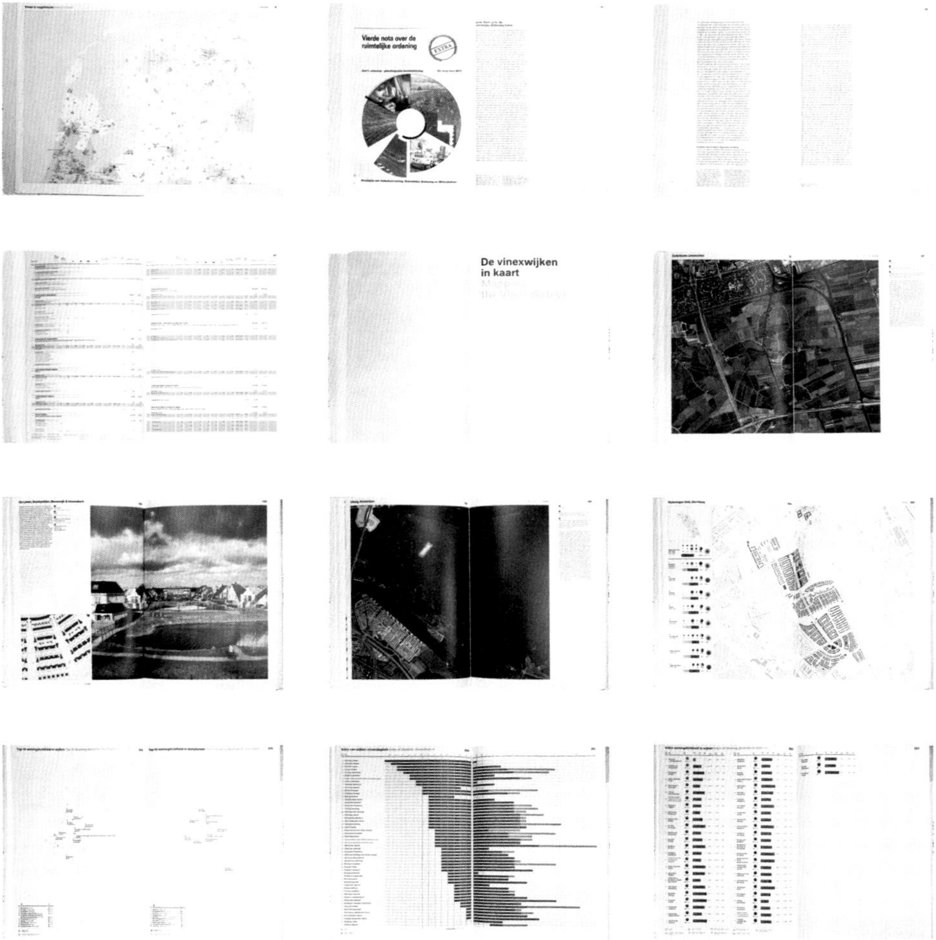
Vierde nota over de ruimtelijke ordening
EXTRA
De vinexwijken
in kaart

Vinex

Vinex Atlas
010

Vinexwijk, Plaats Vinex district, Place

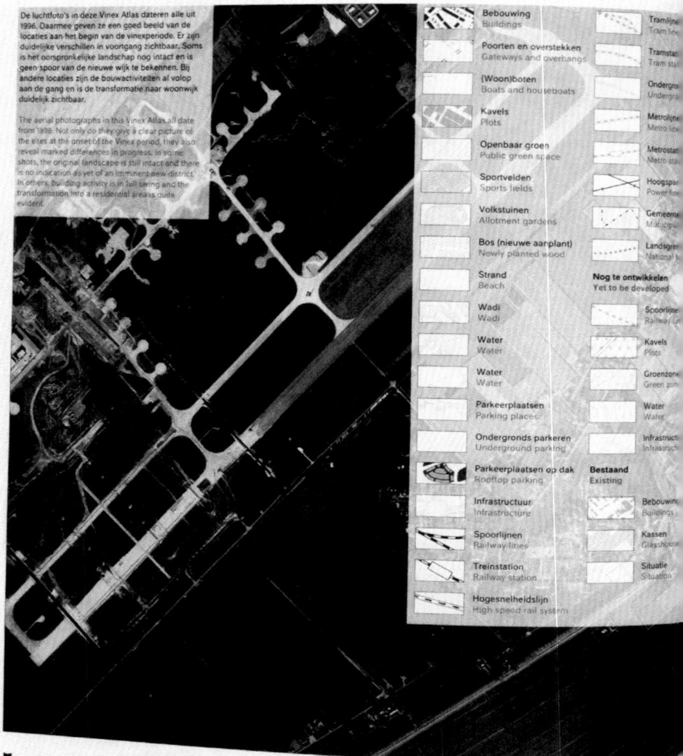
De luchtfoto's in deze Vinex Atlas dateren alle uit 1996. Daarmee geven ze een goed beeld van de locaties aan het begin van de vinexperiode. Er zijn duidelijke verschillen in voortgang zichtbaar. Soms is het oorspronkelijke landschap nog intact en is geen spoor van de nieuwe wijk te bekennen. Bij andere locaties zijn de bouwactiviteiten al volop aan de gang en is de transformatie naar woonwijk duidelijk zichtbaar.
The aerial photographs in this Vinex Atlas all date from 1996. Not only do they give a clear picture of the sites at the onset of the Vinex period, they also reveal marked differences in progress. In some shots, the original landscape is still intact and there is no indication as yet of an imminent new district. In others, building activity is in full swing and the transformation into a residential area is quite evident.
Bebouwing
Buildings
Poorten en overstekken
Gateways and overhangs
(Woon)boten
Boats and houseboats
Kavels
Plots
Openbaar groen
Public green space
Sportvelden
Sports fields
Volkstuinen
Allotment gardens
Bos (nieuwe aanplant)
Newly planted wood
Strand
Beach
Wadi
Wadi
Water
Water
Water
Water
Parkeerplaatsen
Parking places
Ondergronds parkeren
Underground parking
Parkeerplaatsen op dak
Rooftop parking
Infrastructuur
Infrastructure
Spoorlijnen
Railway lines
Treinstation
Railway station
Hogesnelheidslijn
High speed rail system
Nog te ontwikkelen
Yet to be developed
Kavels
Plots
Water
Water
Bestaand
Existing
Kassen
Situatie
Situation

Start bouw–voltooiing vinexwijk

Ontwerpers van stedenbouwkundig plan

(L) Landschap

Ontwikkelende partijen

Projectontwikkelaars

(C) Woningcorporatie

(OC) Ontwikkelingscombinatie

De legenda in de Vinex Atlas wijkt op een aantal punten af van gebruikelijke legenda's zoals bijvoorbeeld die in de Bos Atlas. In de kaarten is gepoogd alle informatie op te nemen die zowel bepalend is geweest voor het ontstaan van de wijk, als voor de ruimtelijke kwaliteit in het dagelijks gebruik van de wijk. Daarom zijn bijvoorbeeld gemeentegrenzen, nieuwe infrastructuur, hoogspanningstracés en spoorlijnen aangegeven omdat deze de contouren van veel vinexplannen in belangrijke mate hebben bepaald. En omdat de kwaliteit van de woonomgeving in hoge mate wordt bepaald door de wijze waarop het parkeren in de wijk is opgelost, zijn ook de parkeerplaatsen prominent in de kaarten opgenomen. De kleine kaart toont de contouren van de gerealiseerde deelplannen met nummers die verwijzen naar de diagrammen. Ook de plandelen die nog in ontwikkeling zijn worden aangegeven. De woningdifferentiatie wordt zichtbaar doordat per diagram het aantal woningen per woningtype is aangegeven. In het staafdiagram wordt het grondgebruik zichtbaar en is met percentages aangegeven hoeveel grond is uitgegeven ten behoeve van woningen en hoeveel er is ingericht als openbaar groen of water. Deze twee variabelen – de woningdifferentiatie en het grondgebruik – hebben een gezamenlijke resultante: het aantal woningen per hectare in de wijk of het deelplan, ofwel de dichtheid. Gezamenlijk geven deze diagrammen een beeld van de overeenkomsten en verschillen tussen en binnen de wijken.

Bebouwing

17

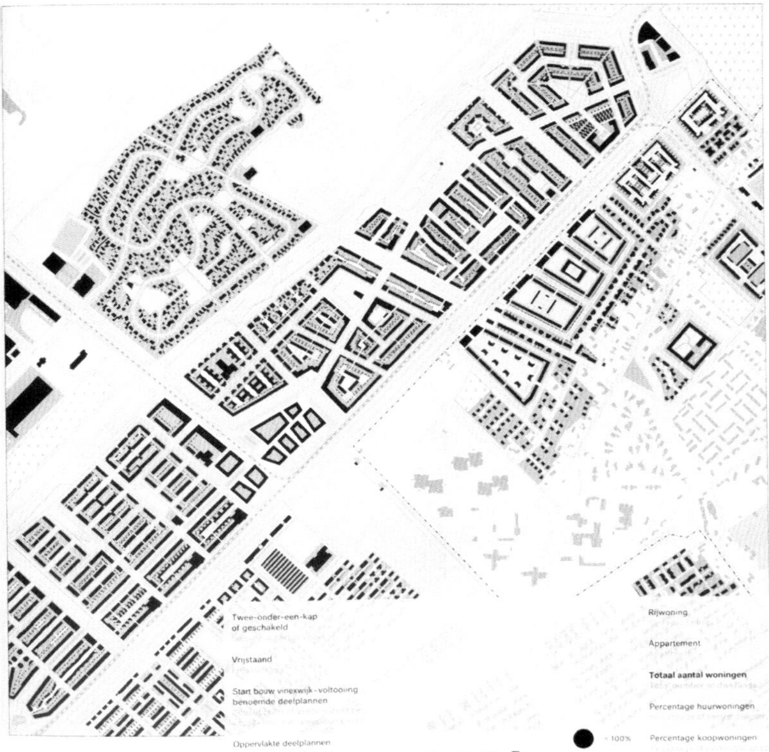

Twee-onder-een-kap of geschakeld

Vrijstaand

Start bouw vinexwijk - voltooiing benoemde deelplannen

Oppervlakte deelplannen

Oppervlakte bedrijventerrein

Oppervlakte hoofdplanstructuur (tarra)

Bruto oppervlakte vinexwijk

Start bouw deelplan - voltooiing deelplan

Nummer van deelplan

Oppervlakte groen en water

Oppervlakte verharding

Oppervlakte uitgegeven gebied (kavels)

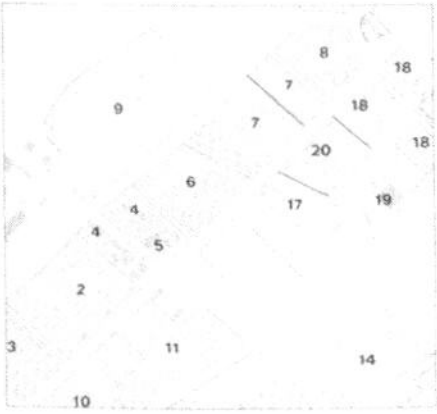
8
18
7
9
18
7
20
18
6
4
17
19
4
5
2
3
11
14
10

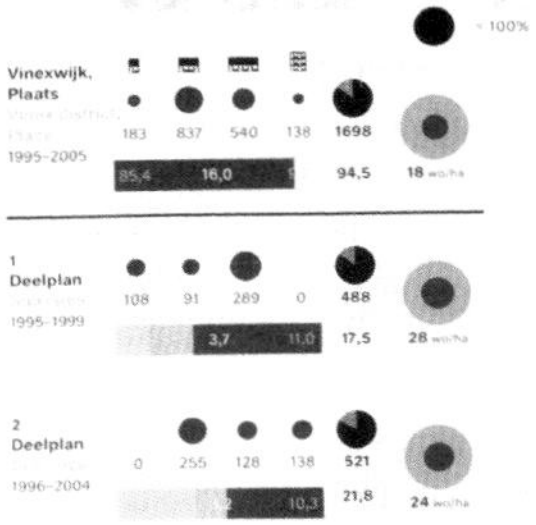
= 100%
Vinexwijk, Plaats
1995–2005
183
837
540
138
1698
85,4
16,0
94,5
18 wo/ha
1 Deelplan
1995–1999
108
91
289
0
488
3,7
11,0
17,5
28 wo/ha
2 Deelplan
1996–2004
0
255
128
138
521
10,3
21,8
24 wo/ha

Rijwoning

Appartement

Totaal aantal woningen

Percentage huurwoningen

Percentage koopwoningen

= 40 wo/ha

Bruto dichtheid in vinexwijk
in woningen per hectare

Aantal woningen per categorie

Woningen per categorie (%)

= 117 wo/ha

Netto dichtheid in deelplan
in woningen per hectare

Netto oppervlakte deelplan

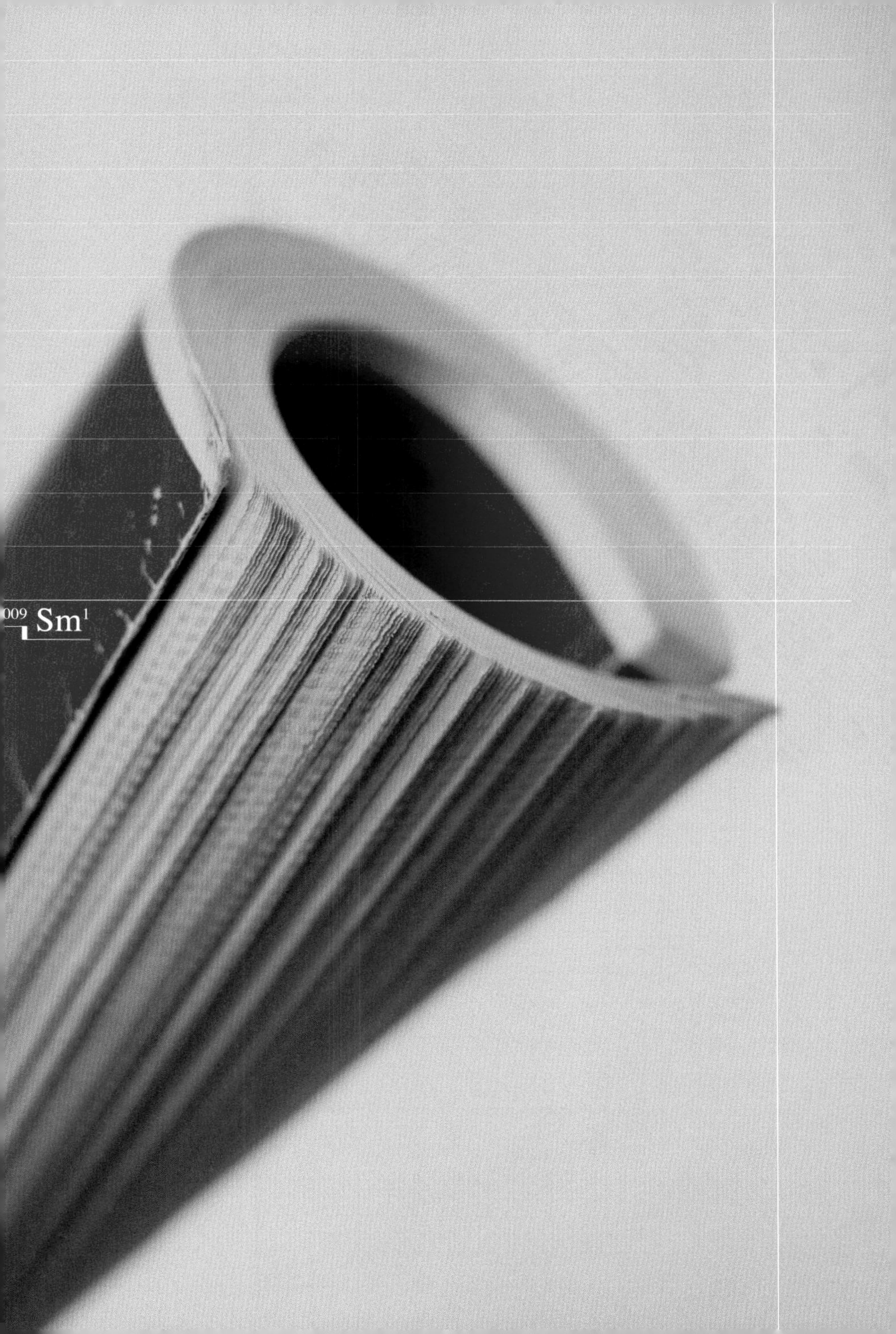
009 Sm1

这是一本新闻纪实摄影集。本书的作者海尔特·范克斯特伦（Geert van Kesteren）是一名屡获世界各地新闻摄影奖的新闻纪实摄影师，也是世界新闻纪实摄影团体马格南图片社的一员。他在本书里通过图片和文字真实地记录了2005-2007年间伊拉克及周边国家普通民众的生存状态。本书为无线胶装，封面单面白卡纸红黑双色印刷，未覆膜的封面在翻阅之后会留下大量清晰的折痕。内文胶版纸红黑双色印刷文字部分，配合新闻纸四色印刷相关时间和地点的照片，文字部分开本宽度比照片用的新闻纸部分短20mm，便于翻阅。正文字体采用等宽打字机字体，和打字机专用的排版格式及符号，照片部分有故意压缩过的痕迹，可能是为了快速报道和出版而只传输的小图导致。简洁而有力的设计、故意不精雕细琢的处理，体现了具有极高时效性的新闻纪实感。||||||||||||||||||||||||||||||||||||||||||||||||||||||||||||||||||||||||||||

Geert van Kesteren
Reports from Turkey, Syria, Jordan and Iraq
BAGHDAD
CALLING

This is a book of photojournalism. The book's author, Geert van Kesteren, is an award-winning news photographer and a member of Magnum Agency, the world news documentary photography group. In this book, he authentically records the living conditions of ordinary people in Iraq and surrounding countries from 2005 to 2007 through pictures and words. This book is perfect bound and the cover is printed in red and black with one-sided white cardboard. The uncoated cover will leave a lot of clear creases after turning over. The inside pages are using offset paper printed in red and black, with the photos of the four-color printing indicating the time and place. The design is concise yet powerful, the unrefined treatments reflect the sense of news documentation with high timeliness.

银奖

巴格达的呼吁

——关于土耳其、叙利亚、约旦和伊拉克的报告

荷兰

This documentary book of photography places the daily life of the people of Baghdad at your fingertips. The choice of photos (professional photographs combined with amateur snapshots), the materials used (newsprint and coated magazine paper), the simple yet entirely fitting typography (typewriter script and Sans Serif) prevent any possibility of a detached perspective – very impressive!

Silver Medal

Baghdad Calling.

Reportages uit Turkije, Syrië, Jordanië en Irak

The Netherlands

Mevis & Van Deursen

Geert van Kesteren

episode publishers, Rotterdam

254 × 192 × 14mm

598g

388p

Baghdad
Calling

Amsterdam,
2008

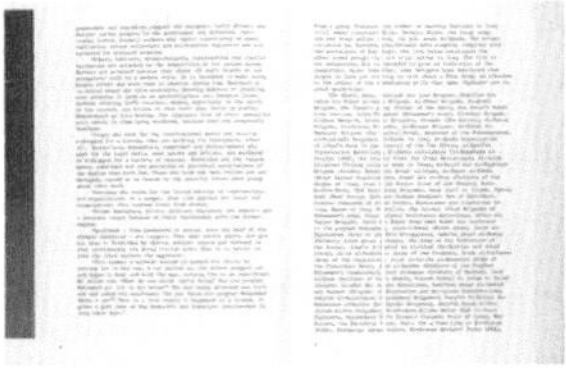

Baghdad,
Iraq, 2005-2007

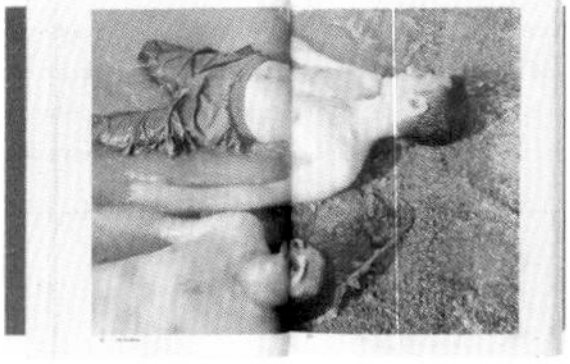

نجف
كربلاء
جميع المحافظات
العراقية
Damascus, Syria, 2007
Geert van Kesteren

Baghdad,
Iraq, 2005-2007

Damascus,
Syria, 2007

Baghdad,
Iraq, 2005-2007

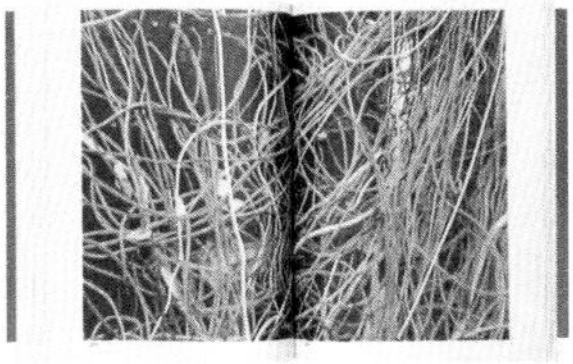

Damascus,
Syria, 2007
Baghdad,
Iraq, 2005-2007

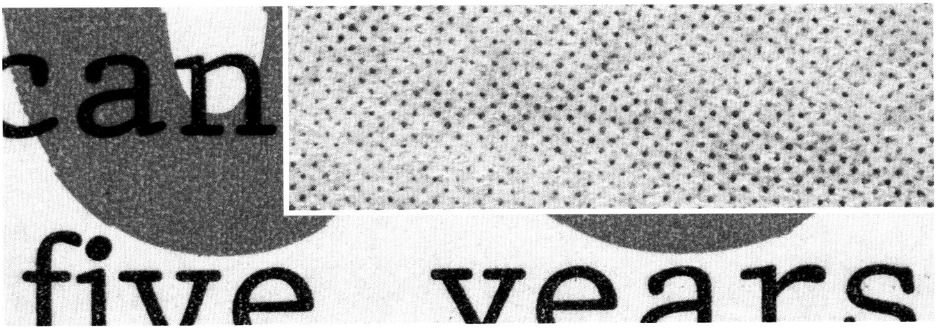
can
five years

BAGHDAD CALLING

Amman, Jordan, 2007

The forms from the Dutch Embassy lie on a side table next the warm brown beans and tea. The documents, in officialese tha is hard to comprehend, state that Basil Karim's family has been refused asylum. They are among the 60,000 Iraqi Mandeans, also known as Sabians, followers of a religion that takes John the Baptist as its main source of inspiration. Under Saddam Hussein they led a relatively tranquil life, but that seems to have been consigned to the past. Basil puts a DVD in his player and sends all the women out of the room. His nephew appears on the TV screen and lowers his trousers to show his mutilated penis. 'One Thursday he was kidnapped and beaten for a long time. His kidnappers told him that the fact he was not a Muslim was reaso enough to kill him slowly, but they preferred to set an example Muslim is circumcised; a Mandean isn't. They brutally hacked aw at his foreskin with a big knife. His captors made a threat: th is what we will do to all the Sabians if you don't leave.' Basi has a sealed document that corroborates the story and is signed by a Mandean priest. With the DVD and document, Basil hopes to able to move to the Netherlands legally, joining several family members who already live there. So far he has been waiting in vain.

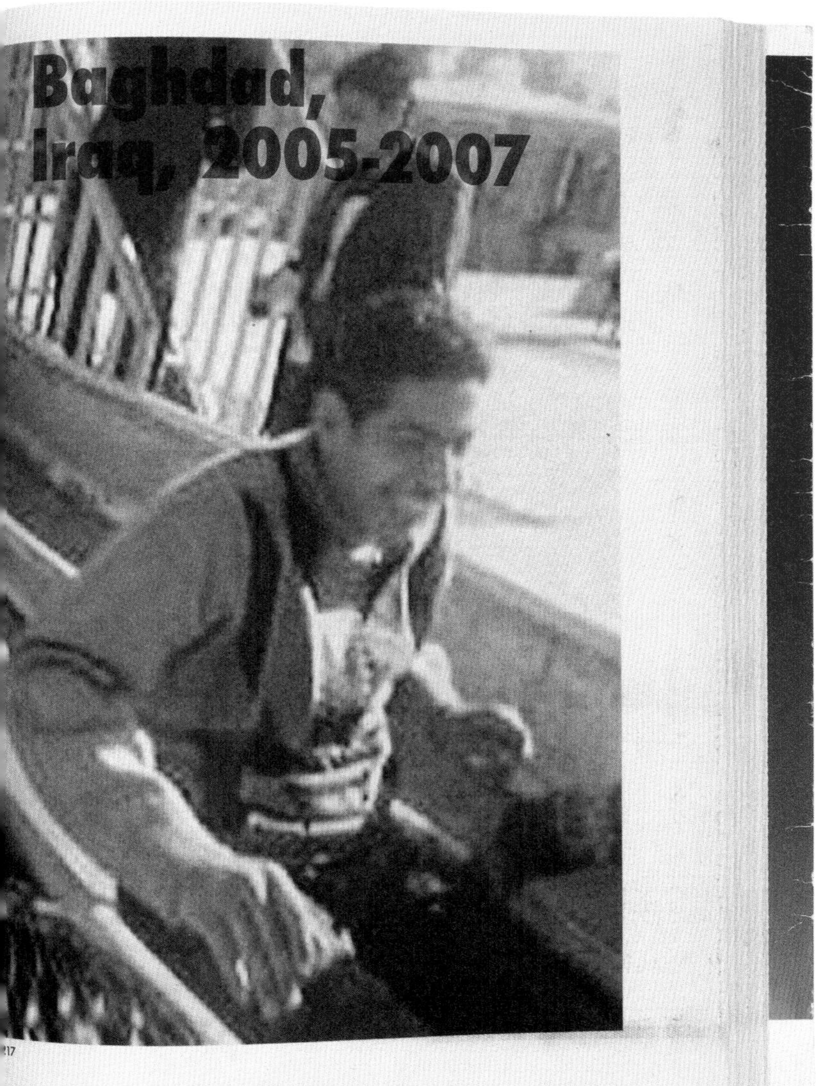
Baghdad,
Iraq, 2005-2007
217

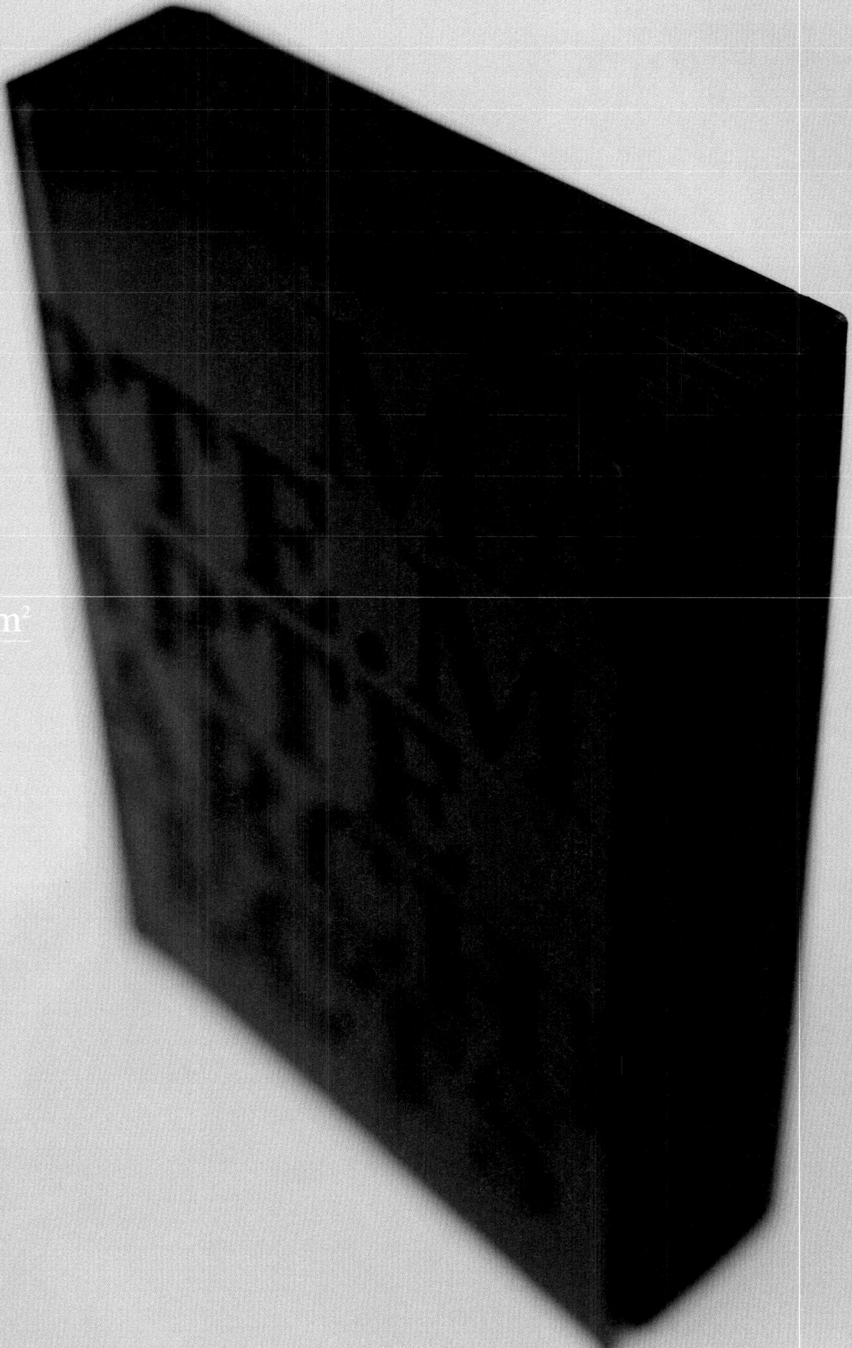

009 Sm²

这是奥地利建筑师事务所 Marte. Marte Architects 的出版物，但与常见的建筑类出版物不同，本书并没有按照常规介绍每个建筑项目的顺序去编排，而是根据认知的理念把项目打散，让读者在翻阅的过程中逐步了解他们的思考。本书为硬精装，封面没有像一般硬精装超出内页并包裹住卡板的做法，整个封面与内页的尺寸一致，封面黑卡纸大面积 UV 处理后直接裱覆在卡板上。内页胶版纸四色印刷，德英双语辅以少量法语和西班牙语，通过字体和粗细的差异进行区分。内容按照客户、建筑外观、航拍鸟瞰、室内现状、未完成作品和未来的思考几部分进行展示，作品前后穿插，通过页码部分的“超链接”提示本建筑的其他内容的页码。在这些部分中还穿插了思考分析、建筑师事务所小知识等章节用于调整节奏。书上下及右侧翻口部分全部刷黑，配合书内大量隔夜的黑色，与封面形成完整的一体。

MA
RTE.M
ARTE
ARCHI
TECTS

This is a publication of Marte.Marte Architects, an Austrian architectural firm. Unlike conventional architectural publications, this book is not arranged in the order of introducing each architectural project. Instead, it breaks up the projects according to the concept of cognition so that readers can gradually understand their thinking in the process of reading. This is a hardcover book, however the cover does not go beyond the inner pages as usual. The size of the cover is exactly the same as that of the inner pages. The inside pages are printed in four colors on offset paper, supplemented by a small amount of French and Spanish, and distinguished by type and thickness. The contents are presented in accordance with the clients, architectural appearance, aerial photography, interior status, unfinished works and thoughts about the future. The works are interspersed, and the pages of other related contents of the building are suggested by “hyperlinks” in page numbers. In these sections, there are also chapters on thinking and analysis, knowledge of the architect’s office and so on, which are used to adjust the rhythm. The top, bottom and right turn-over parts of the book are all brushed black. Together with a large number of darkness in the book, it forms a complete integration with the cover.

- 银奖
- Marte. Marte 建筑事务所
- 奥地利

- Silver Medal
- Marte. Marte Architects
- Austria
- Reinhard Gassner & Andrea Redolfi (Atelier Reinhard Gassner)
- Bernhard Marte, Stefan Marte
- Springer-Verlag Wien / New York
- 225×162×41mm
- 1059g
- 415p

This is unmistakeably a celebration of architecture. And yet the reader is not merely conducted from house to house, from plan to plan. The black, heavy book block leads into the lightness of its interior where architecture becomes an experience, where warmth becomes tangible. Perfectly staged is the unconventional opening with a gallery of building clients followed by the transition to plans, texts and reflections. The trilingual edition has been excellently rendered typographically. The cover with its bold capitals has been realised in matching shades with glossy finish.

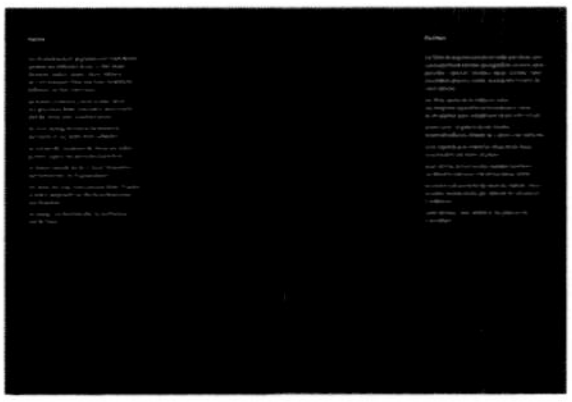

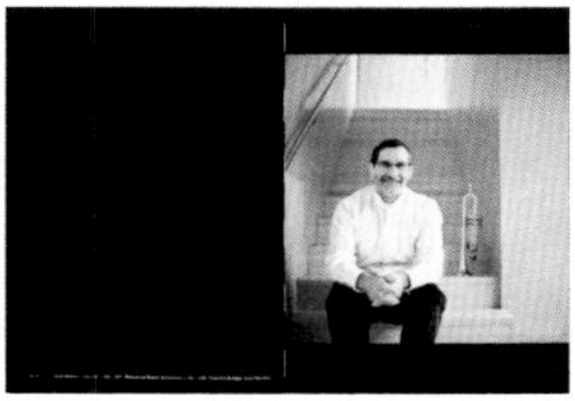

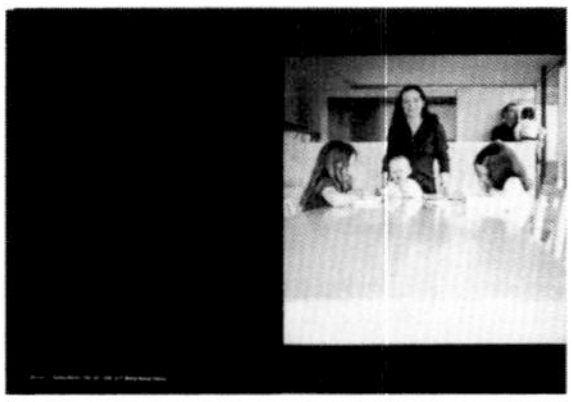

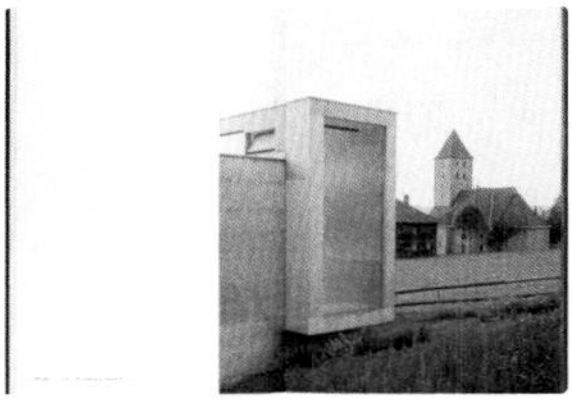

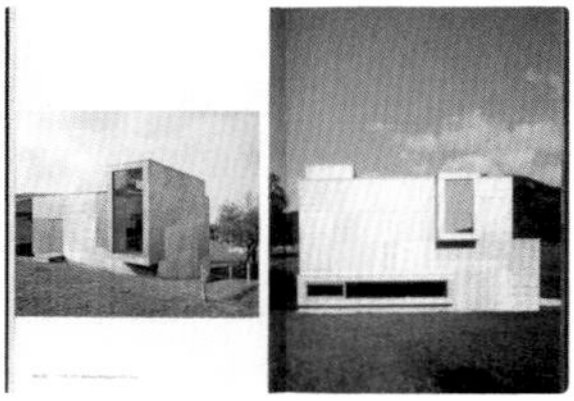

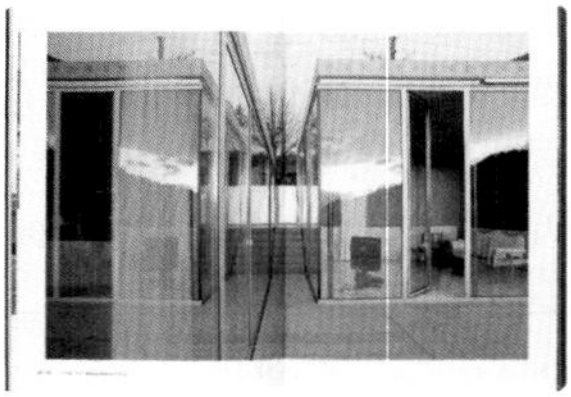

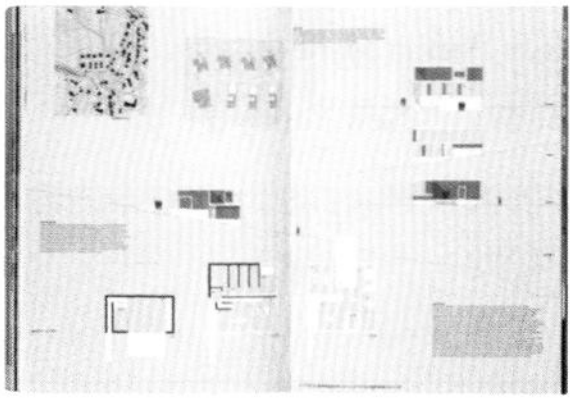

Neubau Schulzentrum Grieskirchen. Der Schulkomplex ist auf einer markanten Geländekuppe über der oberösterreichischen Kleinstadt situiert. Ähnlich einer Kloster- oder Wehranlage besetzt ein lang gestreckter, nach außen geschlossener Baukörper mit inneren Hofstrukturen die städtebaulich bedeutende Situation. Klar abgegrenzt zum Landschaftsraum eröffnet sich im Inneren der Anlage eine urbane, räumliche Struktur mit Wegen, Plätzen und abstrahiertem Garten. Die vier Schultypen reagieren mit individuellen Grundrisskonzepten auf ihre Anforderungen. Ausblick aufs Dorf bieten die Unterrichtsräume, untereinander kommunizieren sie über die Höfe. Aus dem Sockel herausmodelliert sind die eigenständigen Baukörper, so auch die zentral gelegenen Turnhallen. Das Stahlbetonskelett mit optimierter Gebäudehülle aus hochwertigen Glaselementen und Sichtbetonscheiben garantiert höchste Flexibilität und Energieeffizienz.

Grieskirchen School Centre. The school complex is situated on a striking hilltop over the Upper Austrian town of Grieskirchen. The elongated building, closed to the outside, has the appearance of cloisters and castles with the interior courtyards and is responding to the urban significance of the site. Clearly separated from the landscape, the complex offers an urban structure featuring paths, squares and abstract gardens. The four school types address their requirements with different individual ground plan concepts. Orientation towards town and inbetween the classrooms throughout the courtyards is provided. A basement unifies the singular volumes, in central position the gymnasium. The reinforced concrete skeleton with its optimized facade elements of highly insulated concrete slabs and glazing guarantee flexibility and energy efficiency.

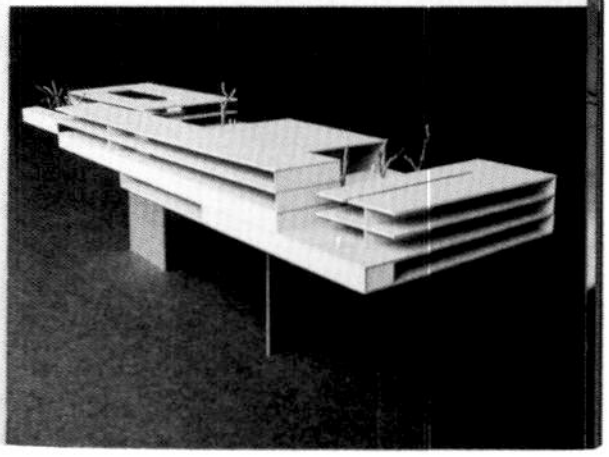

Grieskirchen School Center Upper Austria 2003 1st Prize
Client Municipality of Grieskirchen

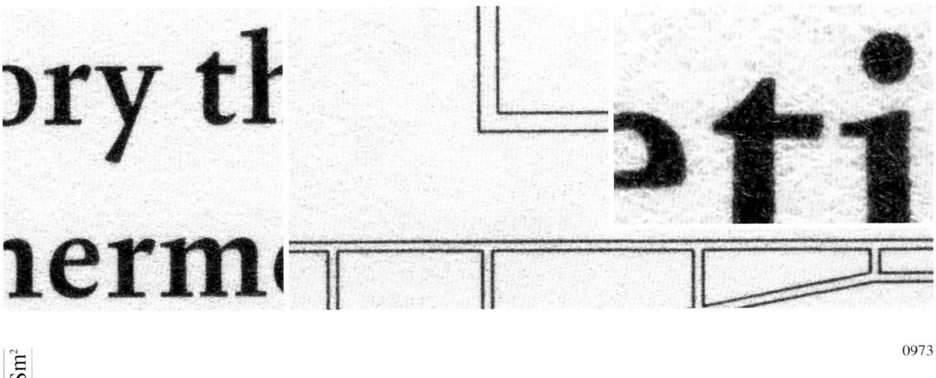

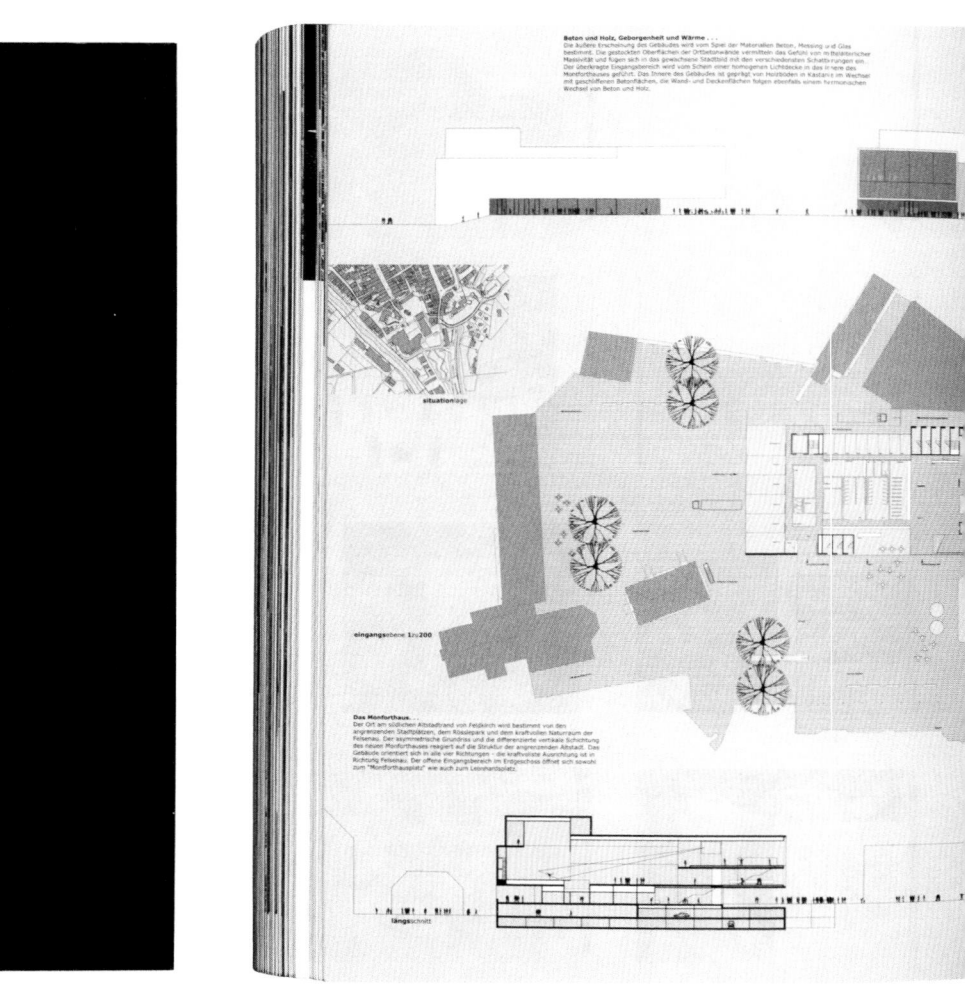

Beton und Holz, Geborgenheit und Wärme . . .
Die äußere Erscheinung des Gebäudes wird vom Spiel der Materialien Beton, Messing und Glas bestimmt. Die gestockten Oberflächen der Ortbetonwände vermitteln das Gefühl von mittelalterlicher Massivität und fügen sich in das gewachsene Stadtbild mit den verschiedensten Schattierungen ein. Der überkragte Eingangsbereich wird vom Schein einer homogenen Lichtdecke in das Innere des Montforthauses geführt. Das Innere des Gebäudes ist geprägt von Holzböden in Kastanie im Wechsel mit geschliffenen Betonflächen, die Wand- und Deckenflächen folgen ebenfalls einem harmonischen Wechsel von Beton und Holz.

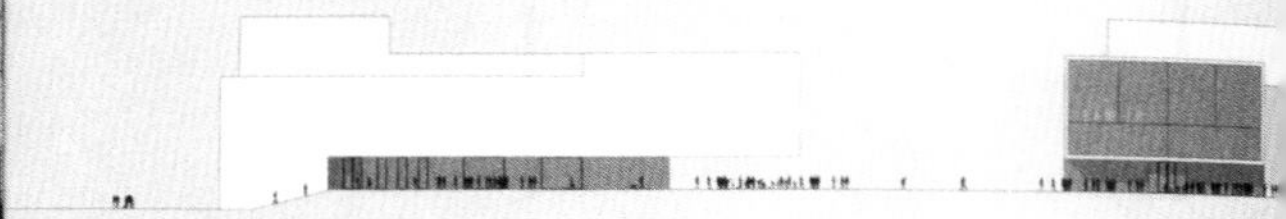

situationlage

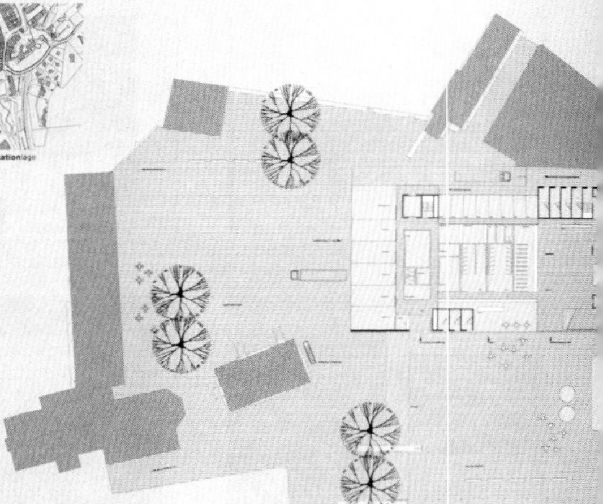

eingangsebene **1**zu**200**

Das Monforthaus. . .
Der Ort am südlichen Altstadtrand von Feldkirch wird bestimmt von den angrenzenden Stadtplätzen, dem Rösslepark und dem kraftvollen Naturraum der Felsenau. Der asymmetrische Grundriss und die differenzierte vertikale Schichtung des neuen Monforthauses reagiert auf die Struktur der angrenzenden Altstadt. Das Gebäude orientiert sich in alle vier Richtungen - die kraftvollste Ausrichtung ist in Richtung Felsenau. Der offene Eingangsbereich im Erdgeschoss öffnet sich sowohl zum "Monforthausplatz" wie auch zum Leonhardsplatz.

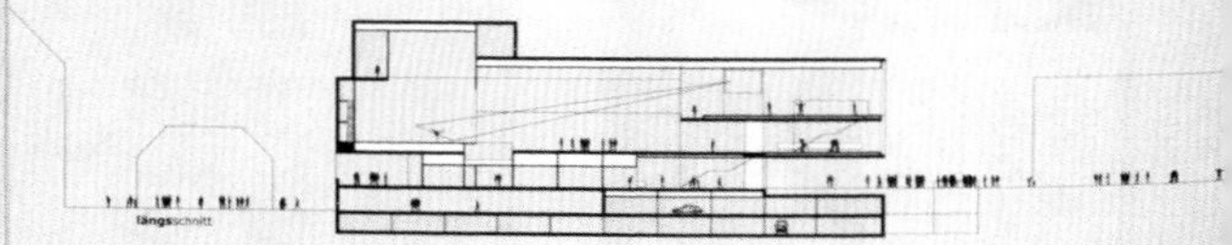

längsschnitt

ansichten 1zu**200**

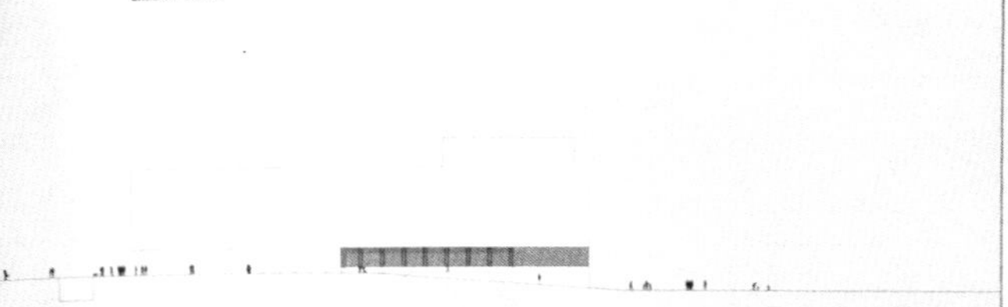

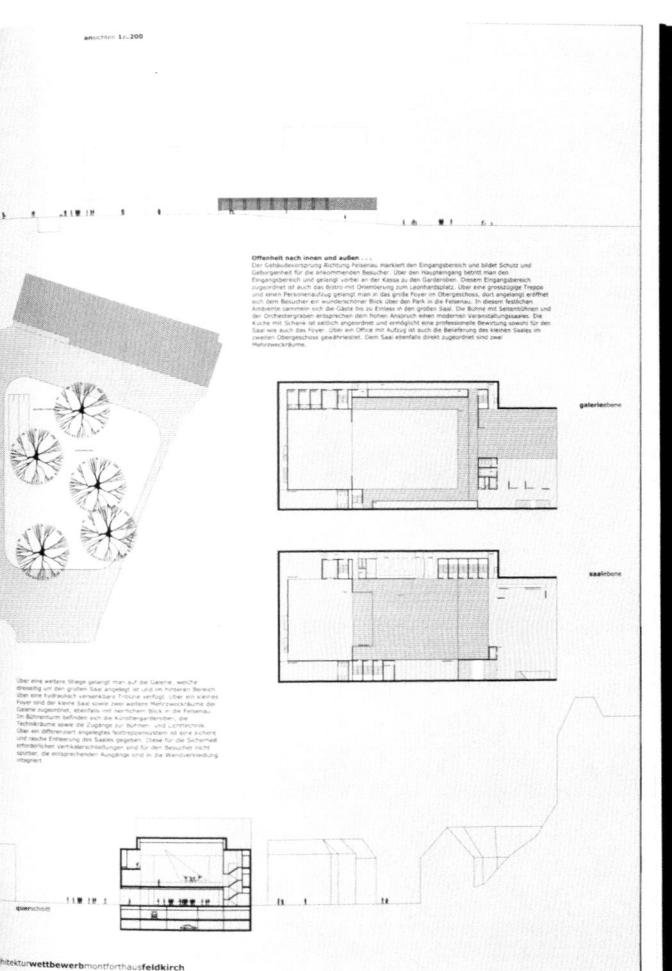

Offenheit nach innen und außen . . .
Der Gebäudevorsprung Richtung Felsenau markiert den Eingangsbereich und bildet Schutz und Geborgenheit für die ankommenden Besucher. Über den Haupteingang betritt man den Eingangsbereich und gelangt vorbei an der Kassa zu den Garderoben. Diesem Eingangsbereich zugeordnet ist auch das Bistro mit Orientierung zum Leonhardsplatz. Über eine grosszügige Treppe und einen Personenaufzug gelangt man in das große Foyer im Obergeschoss, dort angelangt eröffnet sich dem Besucher ein wunderschöner Blick über den Park in die Felsenau. In diesem festlichen Ambiente sammeln sich die Gäste bis zu Einlass in den großen Saal. Die Bühne mit Seitenbühnen und der Orchestergraben entsprechen dem hohen Anspruch einen modernen Veranstaltungssaales. Die Küche mit Schank ist seitlich angeordnet und ermöglicht eine professionelle Bewirtung sowohl für den Saal wie auch das Foyer. Über ein Office mit Aufzug ist auch die Belieferung des kleinen Saales im zweiten Obergeschoss gewährleistet. Dem Saal ebenfalls direkt zugeordnet sind zwei Mehrzweckräume.

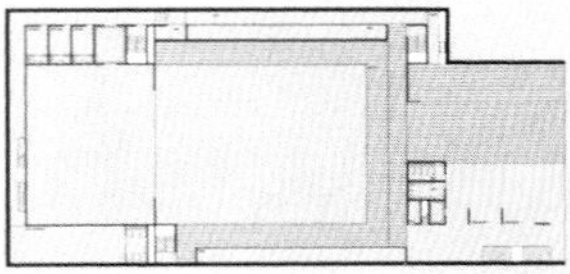

galerieebene

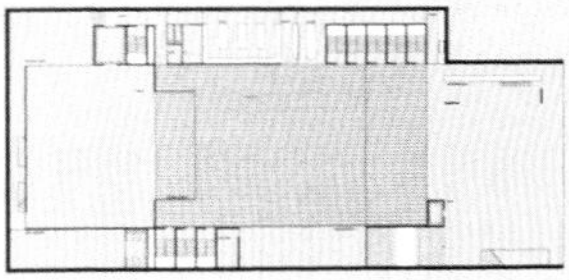

saalebene

Über eine weitere Stiege gelangt man auf die Galerie, welche dreiseitig um den großen Saal angelegt ist und im hinteren Bereich über eine hydraulisch versenkbare Tribüne verfügt. Über ein kleines Foyer sind der kleine Saal sowie zwei weitere Mehrzweckräume der Galerie zugeordnet, ebenfalls mit herrlichem Blick in die Felsenau. Im Bühnenturm befinden sich die Künstlergarderoben, die Technikräume sowie die Zugänge zur Bühnen- und Lichttechnik. Über ein differenziert angelegtes Nottreppensystem ist eine sichere und rasche Entleerung des Saales gegeben. Diese für die Sicherheit erforderlichen Vertikalerschließungen sind für den Besucher nicht spürbar, die entsprechenden Ausgänge sind in die Wandverkleidung integriert.

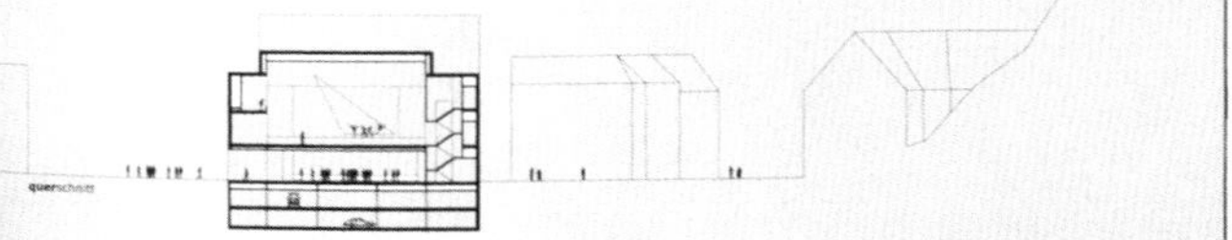

querschnitt

hitektur**wettbewerb**montforthaus**feldkirch**

Bm

这是德国著名艺术家约瑟夫·博伊斯（Joseph Beuys）的作品和思想集。书中通过艺术家的作品、思想表达、其他人对他的理解等方面，还原了一个真实鲜活的博伊斯。本书为锁线硬精装，封面浅灰色艺术纸印单色黑，裱覆卡板。金色环衬从一开始似乎在表达着已经载入史册的艺术经典。内文的主要部分为胶版纸四色印刷，版面网格设置严谨，栏数较多，并通过多栏之间的自由组合，形成丰富的内页变化。文字部分则设置两栏正文，辅以三栏注释来打破常规。虽然整体版面自由多变，但从目录到具体页面的索引系统非常讲究，全书各层级均有唯一编号，章节隔页上出现目录的做法，类似于德国地铁无处不在的全局线路图，让人始终知道自己身在何处。内文的最后为浅黄色书写纸印单色黑，展示一些资料信息。全书对于文字的排版尝试具有很强的革命性，似在书籍设计上集中体现博伊斯的创作理念。

Beuys. Die Revolution sind wir
Steidl

This is a collection of works and ideas by Joseph Beuys, a famous German artist. The book restores a real and vivid Beuys through the artist's works, expressions and other people's understanding of him. This book is hardbound with thread sewing. The cover is printed in black on light grey art paper and mounted with graphic board. From the beginning, the golden end paper seems to express the classics of art that have been recorded in the annals of history. The main part of the text is printed in four colors on offset paper. Layout grid setting is rigorous with several columns. The free combinations of multiple columns form a rich variety of internal pages. In the text section, there are two columns of text supplemented by three columns of comments to break the rule. Although the overall layout is free and changeable, the index system from the catalogue to the specific page is very exquisite. There are unique numbers at all levels of the book. The way of presenting catalogues before each chapter is similar to the local route map everywhere in the German metro, which lets people always know where they are. At the end of the text, the light yellow paper is printed in monochrome black, showing some additional information. The whole book has a strong revolutionary attempt at typesetting, which seems to embody Boyce's creative ideas in book design.

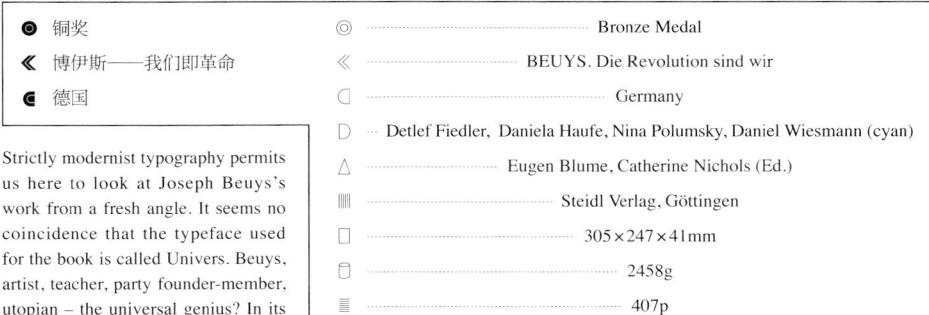

铜奖
博伊斯——我们即革命
德国

Bronze Medal
BEUYS. Die Revolution sind wir
Germany
Detlef Fiedler, Daniela Haufe, Nina Polumsky, Daniel Wiesmann (cyan)
Eugen Blume, Catherine Nichols (Ed.)
Steidl Verlag, Göttingen
305×247×41mm
2458g
407p

Strictly modernist typography permits us here to look at Joseph Beuys's work from a fresh angle. It seems no coincidence that the typeface used for the book is called Univers. Beuys, artist, teacher, party founder-member, utopian – the universal genius? In its pictorial choreography and with just under 400 pages the focus is less on those reproductions of expansive sculptures and delicate drawings by Beuys already seen hundreds of times than on his rich, action-packed and restless life. A formal revolution – visible design that interprets the content anew. Thank you!

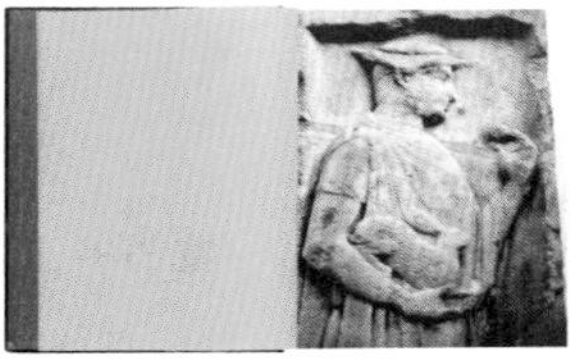

Beuys. Die Revolution sind wir

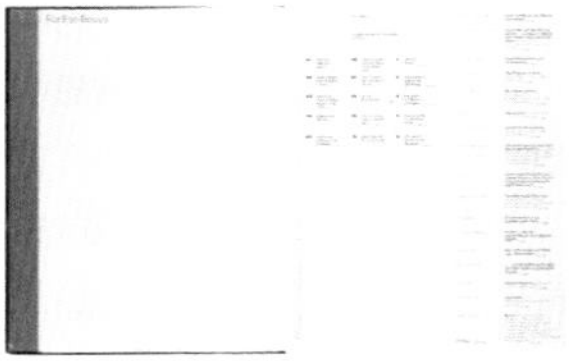

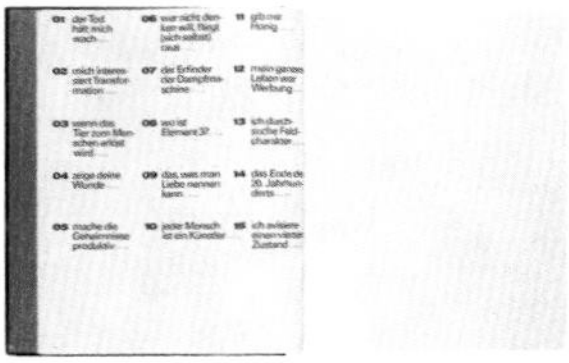

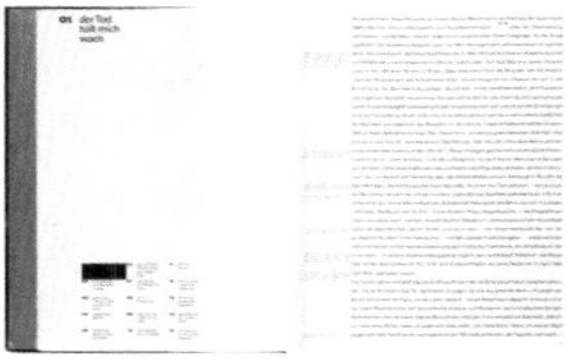

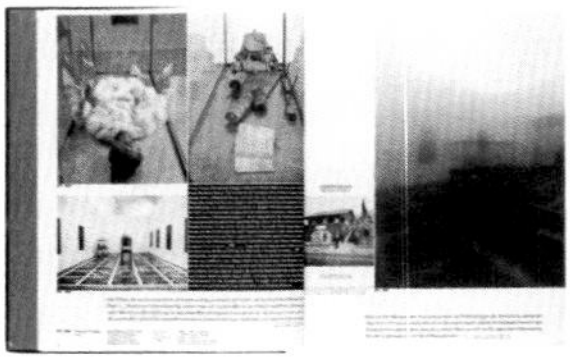

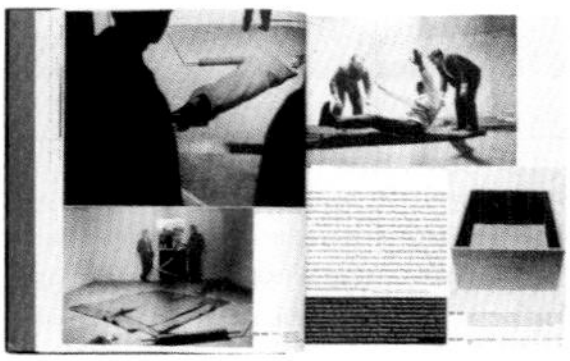

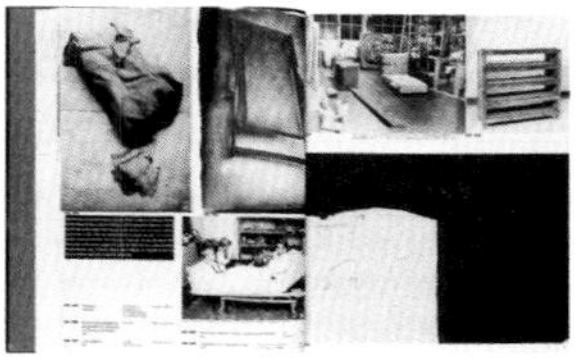

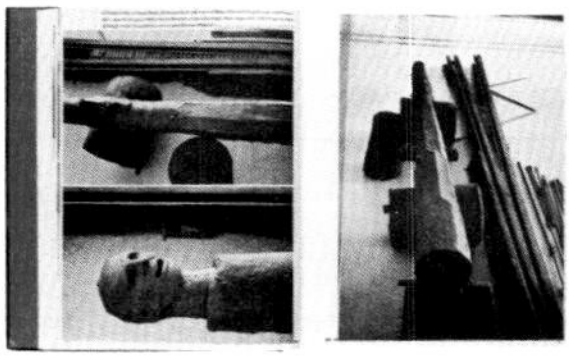

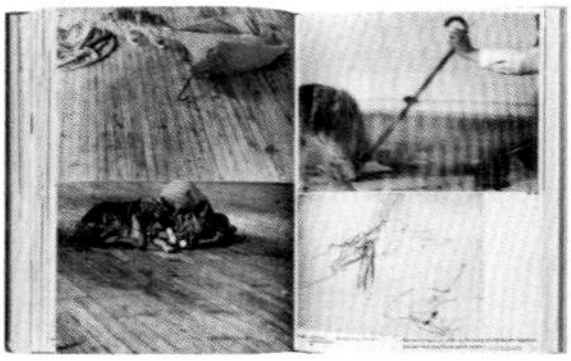

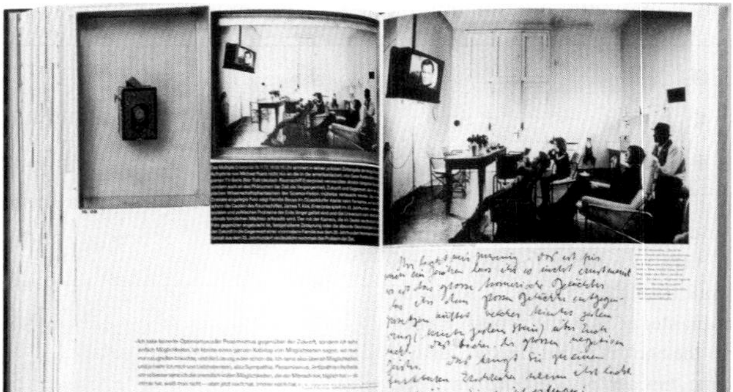

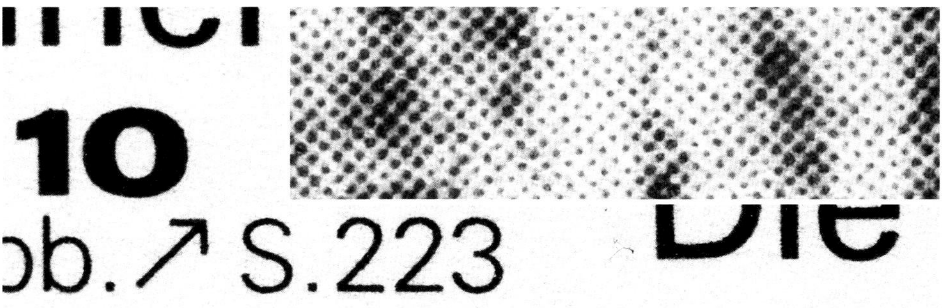
10
S.223

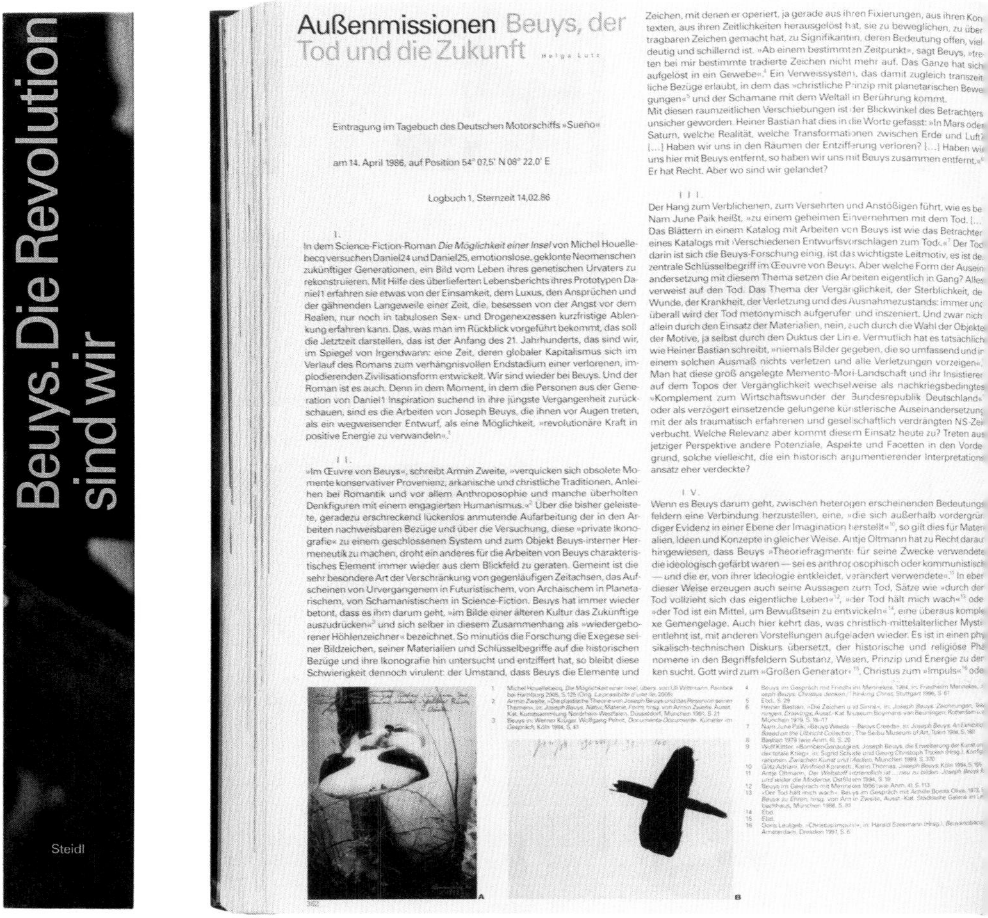

Beuys. Die Revolution sind wir

Steidl

Außenmissionen Beuys, der Tod und die Zukunft

Helga Lutz

Eintragung im Tagebuch des Deutschen Motorschiffs »Sueño«

am 14. April 1986, auf Position 54° 07,5' N 08° 22,0' E

Logbuch 1, Sternzeit 14,02.86

I.

In dem Science-Fiction-Roman *Die Möglichkeit einer Insel* von Michel Houellebecq versuchen Daniel24 und Daniel25, emotionslose, geklonte Neomenschen zukünftiger Generationen, ein Bild vom Leben ihres genetischen Urvaters zu rekonstruieren. Mit Hilfe des überlieferten Lebensberichts ihres Prototypen Daniel1 erfahren sie etwas von der Einsamkeit, dem Luxus, den Ansprüchen und der gähnenden Langeweile einer Zeit, die, besessen von der Angst vor dem Realen, nur noch in tabulosen Sex- und Drogenexzessen kurzfristige Ablenkung erfahren kann. Das, was man im Rückblick vorgeführt bekommt, das soll die Jetztzeit darstellen, das ist der Anfang des 21. Jahrhunderts, das sind wir, im Spiegel von Irgendwann: eine Zeit, deren globaler Kapitalismus sich im Verlauf des Romans zum verhängnisvollen Endstadium einer verlorenen, implodierenden Zivilisationsform entwickelt. Wir sind wieder bei Beuys. Und der Roman ist es auch. Denn in dem Moment, in dem die Personen aus der Generation von Daniel1 Inspiration suchend in ihre jüngste Vergangenheit zurückschauen, sind es die Arbeiten von Joseph Beuys, die ihnen vor Augen treten, als ein wegweisender Entwurf, als eine Möglichkeit, »revolutionäre Kraft in positive Energie zu verwandeln«.[1]

II.

»Im Œuvre von Beuys«, schreibt Armin Zweite, »verquicken sich obsolete Momente konservativer Provenienz, arkanische und christliche Traditionen, Anleihen bei Romantik und vor allem Anthroposophie und manche überholten Denkfiguren mit einem engagierten Humanismus.«[2] Über die bisher geleistete, geradezu erschreckend lückenlos anmutende Aufarbeitung der in den Arbeiten nachweisbaren Bezüge und über die Versuchung, diese »private Ikonografie« zu einem geschlossenen System und zum Objekt Beuys-interner Hermeneutik zu machen, droht ein anderes für die Arbeiten von Beuys charakteristisches Element immer wieder aus dem Blickfeld zu geraten. Gemeint ist die sehr besondere Art der Verschränkung von gegenläufigen Zeitachsen, das Aufscheinen von Urvergangenem in Futuristischem, von Archaischem in Planetarischem, von Schamanistischem in Science-Fiction. Beuys hat immer wieder betont, dass es ihm darum geht, »im Bilde einer älteren Kultur das Zukünftige auszudrücken«[3] und sich selber in diesem Zusammenhang als »wiedergeborener Höhlenzeichner« bezeichnet. So minutiös die Forschung die Exegese seiner Bildzeichen, seiner Materialien und Schlüsselbegriffe auf die historischen Bezüge und ihre Ikonografie hin untersucht und entziffert hat, so bleibt diese Schwierigkeit dennoch virulent: der Umstand, dass Beuys die Elemente und Zeichen, mit denen er operiert, ja gerade aus ihren Fixierungen, aus ihren Kon-texten, aus ihren Zeitlichkeiten herausgelöst hat, sie zu beweglichen, zu über-tragbaren Zeichen gemacht hat, zu Signifikanten, deren Bedeutung offen, viel-deutig und schillernd ist. »Ab einem bestimmten Zeitpunkt«, sagt Beuys, »tre-ten bei mir bestimmte tradierte Zeichen nicht mehr auf. Das Ganze hat sich aufgelöst in ein Gewebe«.[4] Ein Verweissystem, das damit zugleich transzeit-liche Bezüge erlaubt, in dem das »christliche Prinzip mit planetarischen Bewe-gungen«[5] und der Schamane mit dem Weltall in Berührung kommt.

Mit diesen raumzeitlichen Verschiebungen ist der Blickwinkel des Betrachters unsicher geworden. Heiner Bastian hat dies in die Worte gefasst: »In Mars oder Saturn, welche Realität, welche Transformationen zwischen Erde und Luft? [...] Haben wir uns in den Räumen der Entzifferung verloren? [...] Haben wir uns hier mit Beuys entfernt, so haben wir uns mit Beuys zusammen entfernt.«[6] Er hat Recht. Aber wo sind wir gelandet?

III.

Der Hang zum Verblichenen, zum Versehrten und Anstößigen führt, wie es be[i] Nam June Paik heißt, »zu einem geheimen Einvernehmen mit dem Tod. [...] Das Blättern in einem Katalog mit Arbeiten von Beuys ist wie das Betrachten eines Katalogs mit ›Verschiedenen Entwurfsvorschlägen zum Tod‹.«[7] Der Tod, darin ist sich die Beuys-Forschung einig, ist das wichtigste Leitmotiv, es ist der zentrale Schlüsselbegriff im Œuvre von Beuys. Aber welche Form der Ausein-andersetzung mit diesem Thema setzen die Arbeiten eigentlich in Gang? Alles verweist auf den Tod. Das Thema der Vergänglichkeit, der Sterblichkeit, de[r] Wunde, der Krankheit, der Verletzung und des Ausnahmezustands: immer und überall wird der Tod metonymisch aufgerufen und inszeniert. Und zwar nich[t] allein durch den Einsatz der Materialien, nein, auch durch die Wahl der Objekte, der Motive, ja selbst durch den Duktus der Linie. Vermutlich hat es tatsächlich, wie Heiner Bastian schreibt, »niemals Bilder gegeben, die so umfassend und in einem solchen Ausmaß nichts verletzen und alle Verletzungen vorzeigen«. Man hat diese groß angelegte Memento-Mori-Landschaft und ihr Insistieren auf dem Topos der Vergänglichkeit wechselweise als nachkriegsbedingtes »Komplement zum Wirtschaftswunder der Bundesrepublik Deutschland« oder als verzögert einsetzende gelungene künstlerische Auseinandersetzung mit der als traumatisch erfahrenen und gesellschaftlich verdrängten NS-Zei[t] verbucht. Welche Relevanz aber kommt diesem Einsatz heute zu? Treten aus jetziger Perspektive andere Potenziale, Aspekte und Facetten in den Vorde[r]grund, solche vielleicht, die ein historisch argumentierender Interpretations-ansatz eher verdeckte?

IV.

Wenn es Beuys darum geht, zwischen heterogen erscheinenden Bedeutungs-feldern eine Verbindung herzustellen, eine, »die sich außerhalb vordergrün-diger Evidenz in einer Ebene der Imagination herstellt«[10], so gilt dies für Materi-alien, Ideen und Konzepte in gleicher Weise. Antje Oltmann hat zu Recht darau[f] hingewiesen, dass Beuys »Theoriefragmente für seine Zwecke verwendete, die ideologisch gefärbt waren — sei es anthroposophisch oder kommunistisch — und die er, von ihrer Ideologie entkleidet, verändert verwendete«.[11] In eben dieser Weise erzeugen auch seine Aussagen zum Tod, Sätze wie »durch den Tod vollzieht sich das eigentliche Leben«[12], »der Tod hält mich wach«[13] ode[r] »der Tod ist ein Mittel, um Bewußtsein zu entwickeln«[14], eine überaus komple-xe Gemengelage. Auch hier kehrt das, was christlich-mittelalterlicher Mysti[k] entlehnt ist, mit anderen Vorstellungen aufgeladen wieder. Es ist in einen phy-sikalisch-technischen Diskurs übersetzt, der historische und religiöse Phä-nomene in den Begriffsfeldern Substanz, Wesen, Prinzip und Energie zu den-ken sucht. Gott wird zum »Großen Generator«[15], Christus zum »Impuls«[16] ode[r]

A

B

362

zum »Erfinder der Dampfmaschine«[17], die Erde zur »Wärmezeitmaschine«[18] und die Auferstehung zum »Auferstehungsprinzip«[19] erklärt. Und schließlich wird das Gefühl der »völligen Vernichtung des Ich, das die Mystiker aller Zeiten genossen haben«[20], in sein Gegenteil verkehrt, indem dem Menschen bei Beuys ein ungeheueres Entwicklungs- und Freiheitspotenzial zugestanden wird. Das, was hier — auf der Rückseite eines in seiner Gänze Vergänglichkeit evozierenden Œuvres — theoretisch entworfen wird, ist in letzter Konsequenz das Konzept eines Neuen Menschen, eines zukünftigen Neomenschen, der allein durch die Nutzung der ihm innewohnenden »Ich-Kraft«[21] die Todeszone der Gegenwart verlassen kann, der experimentierend »die Phase des Todes erproben und durchlaufen«[22] kann.

Kann man den Tod mit dem Tod überwinden? Oder stehen wir hier doch vor dem Problem, dass die Konzeption des Menschen, wie Beuys sie entworfen hat, dem Tod keinen Raum lässt, seine Absolutheit nicht mehr zu denken erlaubt? Geht es nicht vielleicht doch um eine Schwierigkeit, die Michael Ignatieff mit den Worten beschrieben hat: »Cultures that live by the values of self-realization and self-mastery are not [...] especially good at dying, at submitting to those experiences where freedom ends and biological fate begins. Why should they be? Their strong side is the Promethean ambition: the defiance and transcendence of fate, material and social limit.«[23]

Oder anders herum gefragt: Was bleibt vom Tod, wenn man ihm, wie Beuys, zumindest theoretisch seine Unwiderruflichkeit und Irreversibilität nimmt, das eben, was ihn zu diesem »empirisch-metaempirischen Ungeheuer«[24] macht? Wenn man seine Faktizität ersetzt zugunsten der Vision von seiner Überwindbarkeit, wenn man ihn zur Transitzone erklärt, zu einem Zustand, der verlassen werden kann wie eine Raketenplattform. Schenken wir Vladimir Jankélévitch Glauben, so befinden wir uns damit im Bereich von Science-Fiction. »Es ist endlich Zeit für das Geständnis«, schreibt er, »dass der Augenblick des Todes unwiderruflich die Unmöglichkeit einer Rückkehr in das Diesseits besiegelt. Wer ›wiederstirbt‹ war nicht tot.[25] [...] Die Wundertätigkeit einer auf den Kopf gestellten Zeit [...] ist eine wahrhaft ›unmögliche Voraussetzung‹ und eine hyperbolische Utopie, die wohl den Geschmack eines Wells treffen würde.«[26]

V.

Beuys spricht nicht von Science-Fiction. Er spricht von Utopien. Und eine Utopie ist in seinen Augen nicht viel mehr als ein Plan. »Manchmal ist es ein Langzeitplan, kürzer oder länger, je nach Anstrengungen, die gemacht werden auf diesem Weg. Und am Ende ist dieses utopische Ziel erreicht.«[27] Die Art der Perspektive jedoch, die er einzunehmen pflegt und die Zeitreise, die sein Blick unternimmt, um die Wahrnehmung zu schärfen, all dies rekurriert deutlich stärker auf Science-Fiction-Vorstellungen der 1960er Jahre als üblicherweise angenommen.[28] »Wir stellen fest,« so Beuys 1985 in einem Gespräch, »dass die Gegenwart Dinge zeigt, die so nicht gehen. Und das Gegenkonzept kommt aus der Zukunft.«[29] Zweifelsohne wird in diesem Zusammenhang weder die Technik zum reinen Fetisch erhoben, noch erfolgt die Erkundung der imaginären Raumsphären mit solch wundersam-poetischen Luftgefährten, wie Panamarenko sie entworfen hat. vgl. abb. A Bei Beuys ist auch das Futuristischste immer schon alt und jedes Ufo ein archaisches Zeichen. vgl. abb. B

Das Multiple Der Mann am Haupthebel zeigt Joseph Beuys in Rückenansicht vor der Fensterfront einer Autobahnbrücke. vgl. abb. C Unser Blick geht mit ihm auf eine mehrspurige Autobahn, während er mit der rechten Faust einen unsichtbaren Schalthebel zu betätigen scheint. Das, was wir sehen, ist nicht mehr das, was wir sehen. So altertümlich das Raumschiff auch sein mag und so merkwürdig unser First Officer auch gekleidet ist: Wir heben ab. Um vom Tower aus das All zu beobachten, mit eben jener Neugierde, mit der auch Captain Kirk und seine Crew die galaktischen Erscheinungen im Weltraum verfolgt und gedeutet haben. vgl. abb. D Und um schließlich, wie Beuys feststellt, auf einen »Planet Deutschland [zu stoßen], der unerforscht ist.«[30]

VI.

Kehren wir zurück zur »Bodenstation des Menschen«.[31] Das Multiple Enterprise 18.11.72, 18:05:16 Uhr enthält in einer kleinen Zinkkiste einen altertümlichen Fotoapparat, dessen Linse mit Filz beklebt ist. **15 09** Auf der Innenseite des dazugehörigen Deckels klebt eine Fotografie, die Beuys und seine Familie vor dem Fernseher zeigt, während das Gesicht von Captain Kirk aus der TV-Serie *Raumschiff Enterprise* in Großaufnahme auf dem Bildschirm zu erkennen ist. **15** Captain Kirk hat sich eingeschaltet, er spricht aus dem von der Decke mittels einer Halterung befestigten Fernseher zu den andächtig lauschenden Familienmitgliedern. Die Fernsehsendung wird zu einer besonderen Sendung, gerade so, als wäre die »Enterprise« mit ihrer Besatzung nicht nur jene bekannte, Kult gewordene Science-Fiction-Serie der 1960er und 70er Jahre und Ausdruck des Begehrens, auch die »letzte Grenze« zu überwinden und zu den »Galaxien vorzudringen, die nie ein Mensch zuvor gesehen hat«.[32] Das Verhältnis scheint sich umzukehren und das Beuys'sche Wohnzimmer zur Außenstation des Weltraums zu werden. Das kleine Zinkkästchen mitsamt dem sorgfältig darin platzierten Objekt und der gleich einer Objektbeschriftung im Deckel angebrachten Fotografie geben dieser Assoziation auf interessante Weise Raum: indem der Fotoapparat und die Art seiner Präsentation einerseits unmissverständlich auf museale Aufbewahrungspraktiken verweisen, zugleich aber durch das Material Zink und die Verschiebung aus dem Gebrauchskontext diese Zuordnung wieder in Frage gestellt wird; und nicht zuletzt dadurch, dass hier die Ähnlichkeit des Fotoapparats mit dem Tricorder ausgespielt wird, wie er bei der Besatzung von Raumschiff Enterprise bei Außenmissionen auf fremden Planeten im Gebrauch war. vgl. abb. E Einmal mehr gehen Vergangenes und Zukünftiges ineinander über, wird ein gewöhnliches Objekt zum Relikt einer bereits schon wieder vergangenen Zukunft.

VII.

Wenn in der utopischen Bewusstseinszone »Eurasien« Ostmenschen und Westmenschen miteinander kommunizieren, wenn es sich um einen utopischen »Weltentwurf« handelt, der »jenseits aller Gegensätze« funktioniert[33], so ist auch dieses Modell von Beuys an Überlegungen orientiert, wie sie zeitgleich im Zusammenhang mit der Frage nach Möglichkeiten der Kommunikation mit außerirdischer Intelligenz gestellt wurden. Auf der ersten großen Konferenz über »Kommunikation mit außerirdischer Intelligenz« (CETI) im Jahr 1971, an der renommierte Wissenschaftler aus den USA und der Sowjetunion trotz des Kalten Krieges gemeinsam teilnahmen, bestand nicht nur Einigkeit darüber, dass die »Bedingung jeder Begegnung mit außerirdischer Intelligenz die irdischer Intelligenz über verschiedene Geschichten, Nationalitäten, Sprachen und Disziplinen hinweg ist«.[34] Allen Beteiligten war zugleich auch klar: Will man das Unbekannte wahrscheinlich machen, so muss man das Bekannte, muss man sich selbst, unplausibel machen. Oder, um es mit den Worten von Claus Pias zu sagen: »Der Außerirdische ist nicht nur die eigene Frage als Gestalt, sondern umgekehrt müssen wir selbst zu Aliens werden. ›Immer häufiger spielte ich die Rolle des Außerirdischen‹, schreibt auch John Lomberg während seiner Arbeit am Voyager Bildprogramm.«[35]

Die Arbeiten von Beuys erfordern ein ähnliches Gedankenexperiment. Das Begehren des Subjekts, *»sich sich sehen zu sehen«*[36], kehrt wieder als der illusorische Versuch, ein Bild von dem zu erhaschen, wie sich das eigene Ich im Auge eines Außerirdischen spiegelt. Die Aktivierung verschütteter Wahrnehmungspotenziale, wie sie Beuys vorschwebte, basiert in starkem Maße auf dem Erlernen eben dieser Fähigkeit, die Welt aus einer Art Perspektive zweiter Ordnung,

MANN AM HAUPTHEBEL

17 Ebd., S. 65–66
18 Die Wärmezeitmaschine in der Ökonomie, 1975, Postkarte, vgl. Jörg Schellmann (Hrsg.), *Joseph Beuys. Die Multiples*, München 1992, Nr. 154
19 Beuys im Gespräch mit Mennekes 1996 (wie Anm. 4), S. 57
20 Johan Huizinga, *Herbst des Mittelalters*, 11. Aufl., Stuttgart 1975 (erste deutsche Fassung 1923), S. 279
21 Beuys im Gespräch mit Mennekes 1996 (wie Anm. 4), S. 39
22 Beuys im Gespräch mit Bonito Oliva 1973 (wie Anm. 13), S. 79
23 Michael Ignatieff, »Modern Dying«, in: *The New Republic* (Washington D.C.), 26. Dezember 1988
24 Vladimir Jankélévitch, *Der Tod*, übers. von Brigitta Restorff, Frankfurt am Main 2005, französische Originalausgabe, Paris 1977, S. 12
25 Jankélévitch zit. nach: Thomas Macho, »Der Tod und die Toten. Anmerkungen zu Vladimir Jankélévitch und Robert Harrison«, in: *Merkur*, Heft 687, Stuttgart 2006, S. 632
26 Jankélévitch 2005 (wie Anm. 24), S. 355. Mit Wells ist der bekannte Science Fiction Autor Herbert George Wells gemeint, von dem unter anderem der Roman *Die Zeitmaschine* (1895) berühmt geworden ist.
27 Beuys im Gespräch mit Mennekes 1996 (wie Anm. 4), S. 45
28 Herzlicher Dank an Dietmar Schmidt für wichtige Hinweise und profunde *Star Trek*-Kenntnisse.
29 *Joseph Beuys, Jannis Kounellis, Anselm Kiefer, Enzo Cucchi, Ein Gespräch / Una Discussione*, hrsg. von Jacqueline Burckhardt, 2. Aufl., Zürich 1988, S. 146
30 Ebd., S. 147
31 Vgl. Clara Bodenmann-Ritter (Hrsg.), *Joseph Beuys. Jeder Mensch ein Künstler. Gespräche auf der documenta 5 1972*, Frankfurt am Main, Berlin, Wien 1975, S. 66
32 Vorspann der TV-Serie *Raumschiff Enterprise* (1972–1974, 1988), deutsche Fassung der amerikanischen Serie *Star Trek* (1966–1969).
33 Doris Leutgeb, »Eurasien«, in: Szeemann 1997 (wie Anm. 16), S. 161
34 Carl Sagan (Hrsg.), *Communication with Extraterrestrial Intelligence*, Cambridge, Mass. 1973, S. xi
35 Claus Pias, »Kalküle der Hoffnung«, in: Thomas Macho, Annette Wunschel (Hrsg.), *Science & Fiction. Über Gedankenexperimente in Wissenschaft, Philosophie und Literatur*, Frankfurt am Main 2004, S. 83
36 Jacques Lacan, *Die vier Grundbegriffe der Psychoanalyse. Das Seminar Buch XI*, 3. Aufl., Weinheim, Berlin 1987, S. 88

A Panamarenko K2 — The 7000-meter-high flying jungle and mountain machine 1994, Blatt aus einem Portfolio mit sechs Arbeiten, Offset **B** Joseph Beuys Ufo 1984, Siebdruck auf Papier **C** Joseph Beuys Der Mann am Haupthebel 1985, Postkarte **D** Auf der Brücke des Raumschiffs Enterprise Star Trek: Spock's Brain, USA 1968 **E** Aussenmission Star Trek: The Paradise Syndrom, USA 1968

AUTHORS
...Charles & Ray
George Nelson,
Girard, Tibor Kalman,
Frank Gehry, Tadao
SANAA, Jasper
Zaha Hadid, Hella
rius, Maarten Van
Ronan & Erwan
Jean Prouvé, Mario
Citterio,
Panton,

009 Bm²

这是一本展示瑞士著名家居企业维特拉（Vitra）产品、理念和设计背后故事的书，展示了这家企业如何将文化和经济完美融合，并将维特拉的历史载入西方设计史的原因。本书锁线硬精装，护封铜版纸四色印刷覆亮膜，表面将目录、版权信息等做压凹处理，呈现出丰富的手感。内封白色布纹纸红黑双色印刷，信息部分亮面 UV 工艺。前后环衬灰卡纸印橙红色，分别为索引信息和以维特拉为中心的相关“地图”。内页部分通过灰卡纸印红作为分隔页、哑光面铜版纸四色彩印展现建筑和产品、胶版纸印黑阐述理念和传记的方式，将各部分严格分开，并提供了很好的阅读体验。内文分为维特拉建筑、产品、设计师、博物馆、藏品、理念等几个方面，全面展示和阐述维特拉的企业理念。其中主要场馆为世界级建筑师参与的建筑群和可以拉开似仓库货架的藏品架令人印象深刻。

PROJECT VITRA.
LES MEUBLES
DES ATELIERS
JEAN PROUVÉ
khäuser

This is a book showing the stories behind the products, ideas and designs of Vitra, a famous Swiss household company. It shows how the company integrates culture and economy perfectly, and why Vitra's history is included in the history of Western design. The book is hardbound with thread sewing and the laminated jacket is printed in four colors. The catalogue and copyright information are embossed. The front and back end paper is printed in orange on grey cardboard with index information and related "maps" focused on Vitra. The inside pages are separated strictly by grey cardboard as dividing pages; matte coated art paper printed in four colors is used to display architecture and products; offset paper printed in black is used for concepts and biography part. The design provides good reading experiences. The text is divided into vitra architecture, products, designers, museums, collections, concepts and other aspects, to fully demonstrate and elaborate Vitra's business philosophy. The buildings designed by world-class architects and the collection shelves that can be opened like warehouse shelves are quite impressive.

- 铜奖
- 维特拉计划
- 德国

What do Tati, Teller, Thiel, Thonet, Thut, Tillmanns, Tokyo, transversality, trial & error and Tüllinger have in common ? Enough to all be listed under T in the index of Vitra's fifty-year company history. The topic is actually furniture design. The house-typeface, Futura, is the same as IKEA's, but the company philosophy is not. Between "Kafka's Trial" and "Tati's Playtime", the book impressively succeeds in embedding Vitra's history into that of western culture. And all of this with toplevel state-of-the-art book design.

- Bronze Medal
- Projekt Vitra.
- Germany
- Cornel Windlin
- Rolf Fehlbaum, Cornel Windlin (Ed.)
- Birkhäuser Verlag, Basel / Boston / Berlin
- 244×174×34mm
- 1106g
- 396p

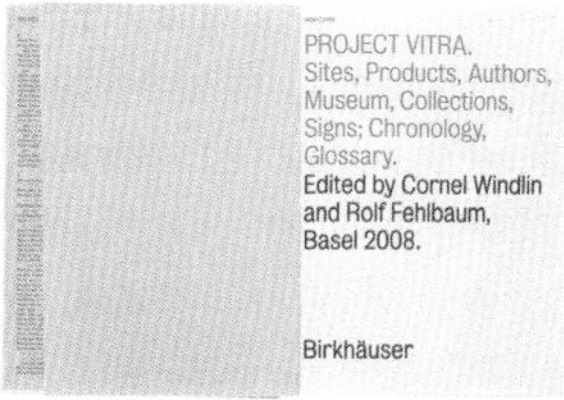
PROJECT VITRA.
Sites, Products, Authors,
Museum, Collections,
Signs; Chronology,
Glossary.
Edited by Cornel Windlin
and Rolf Fehlbaum,
Basel 2008.
Birkhäuser

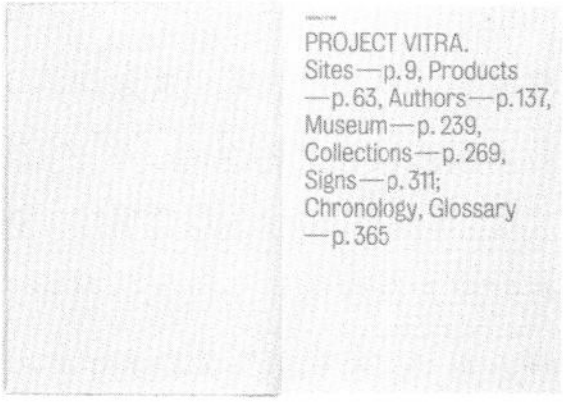
PROJECT VITRA.
Sites—p. 9, Products
—p. 63, Authors—p. 137,
Museum—p. 239,
Collections—p. 269,
Signs—p. 311;
Chronology, Glossary
—p. 365

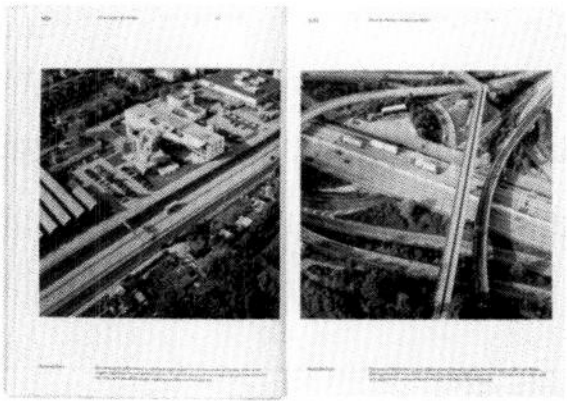

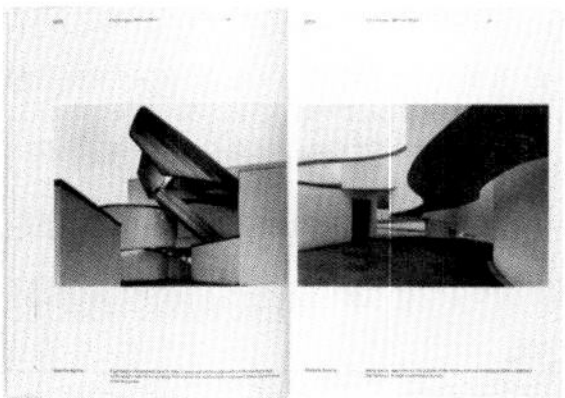

...Allerheiligen Hofkirche,
Atomium, Centre Pompi-
dou, Deutscher Bundes-
tag, Kunsthaus Graz,
LA Philharmonic Concert
Hall, Nasjonalbiblioteket
Oslo, Palais de Justice
Nantes, Pinakothek der
Moderne München,
Seattle Public Library,
Tate Modern, University
Library Utrecht...

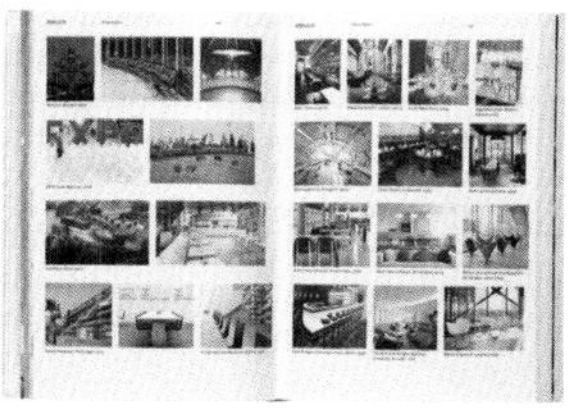

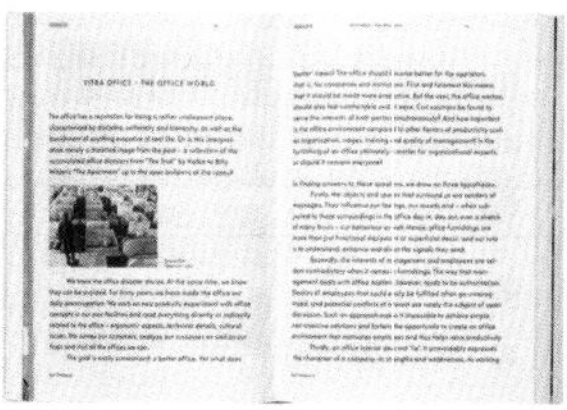

Herman Miller Collection
1957, Panton Chair 1967,
Vitramat 1976, Bellini
Collection 1984, Vitra
Edition 1987, Metropol
1988, AC 1 1990, Ad Hoc
1992, Meda Chair 1996,
.03 1999, Jean Prouvé
Re-Edition 2002, Joyn
2002, Home Collection
2004, Vitra Edition 2007

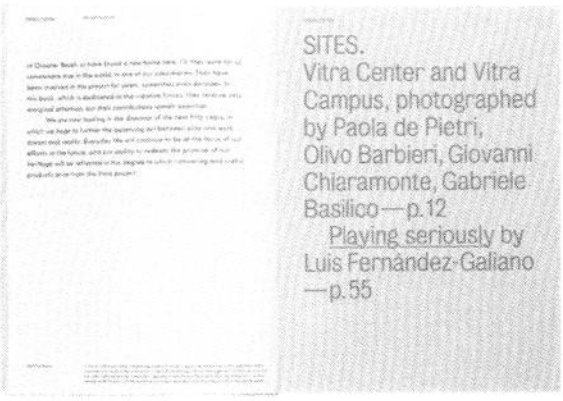
SITES.
Vitra Center and Vitra Campus, photographed by Paola de Pietri, Olivo Barbieri, Giovanni Chiaramonte, Gabriele Basilico—p.12
Playing seriously by Luis Fernández-Galiano —p.55

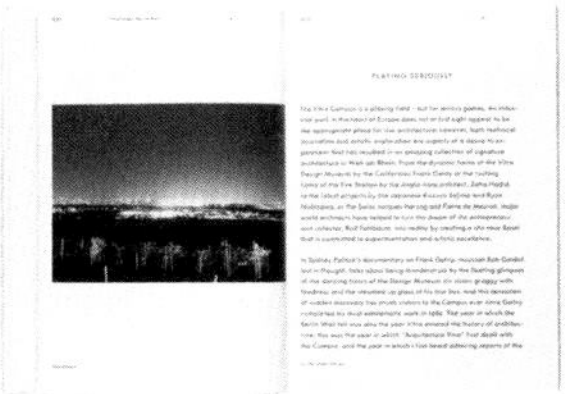
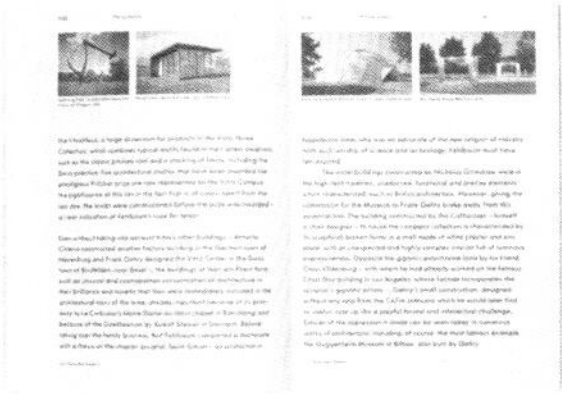

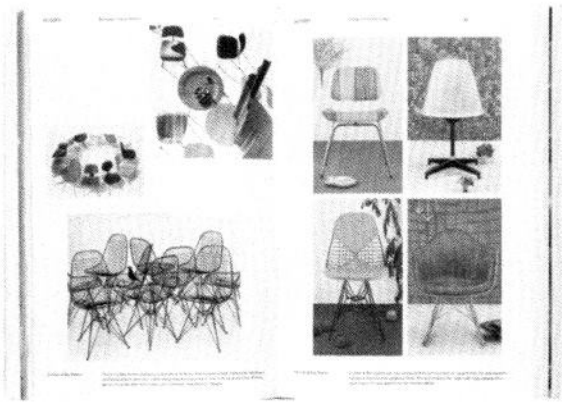
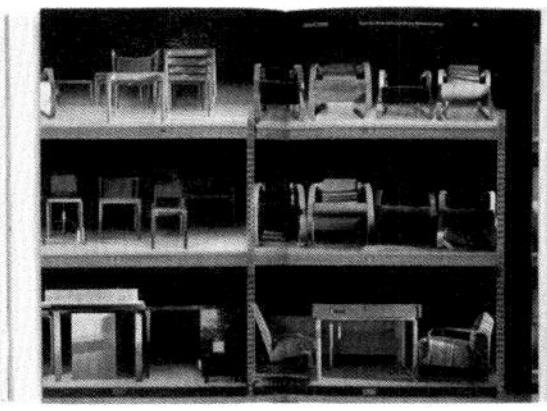

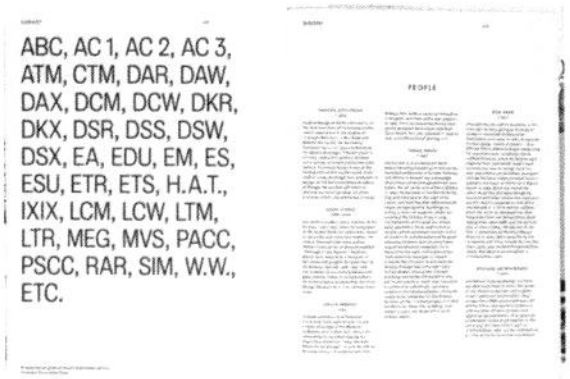
ABC, AC 1, AC 2, AC 3, ATM, CTM, DAR, DAW, DAX, DCM, DCW, DKR, DKX, DSR, DSS, DSW, DSX, EA, EDU, EM, ES, ESU, ETR, ETS, H.A.T., IXIX, LCM, LCW, LTM, LTR, MEG, MVS, PACC, PSCC, RAR, SIM, W.W., ETC.
PEOPLE
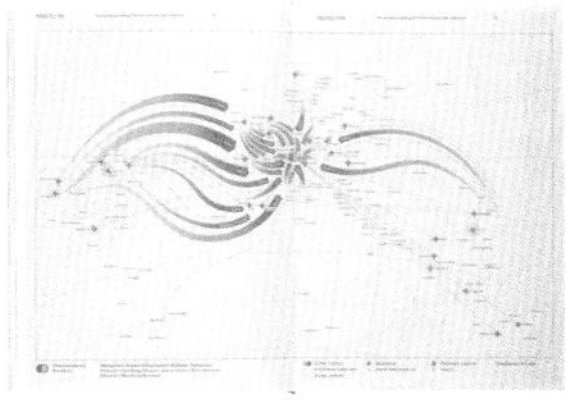
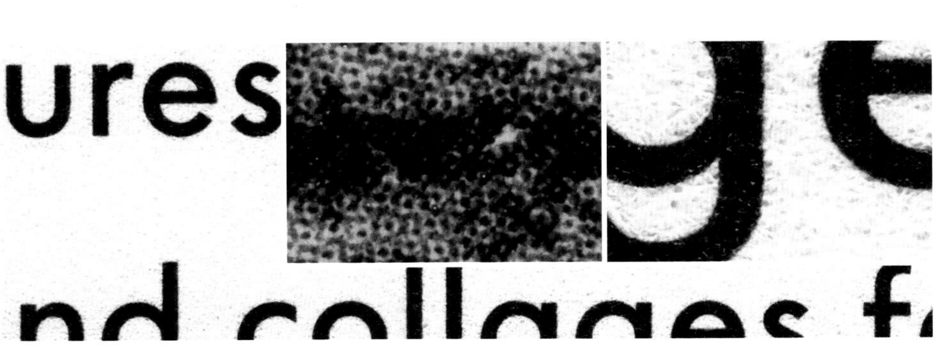
ures
nd collages

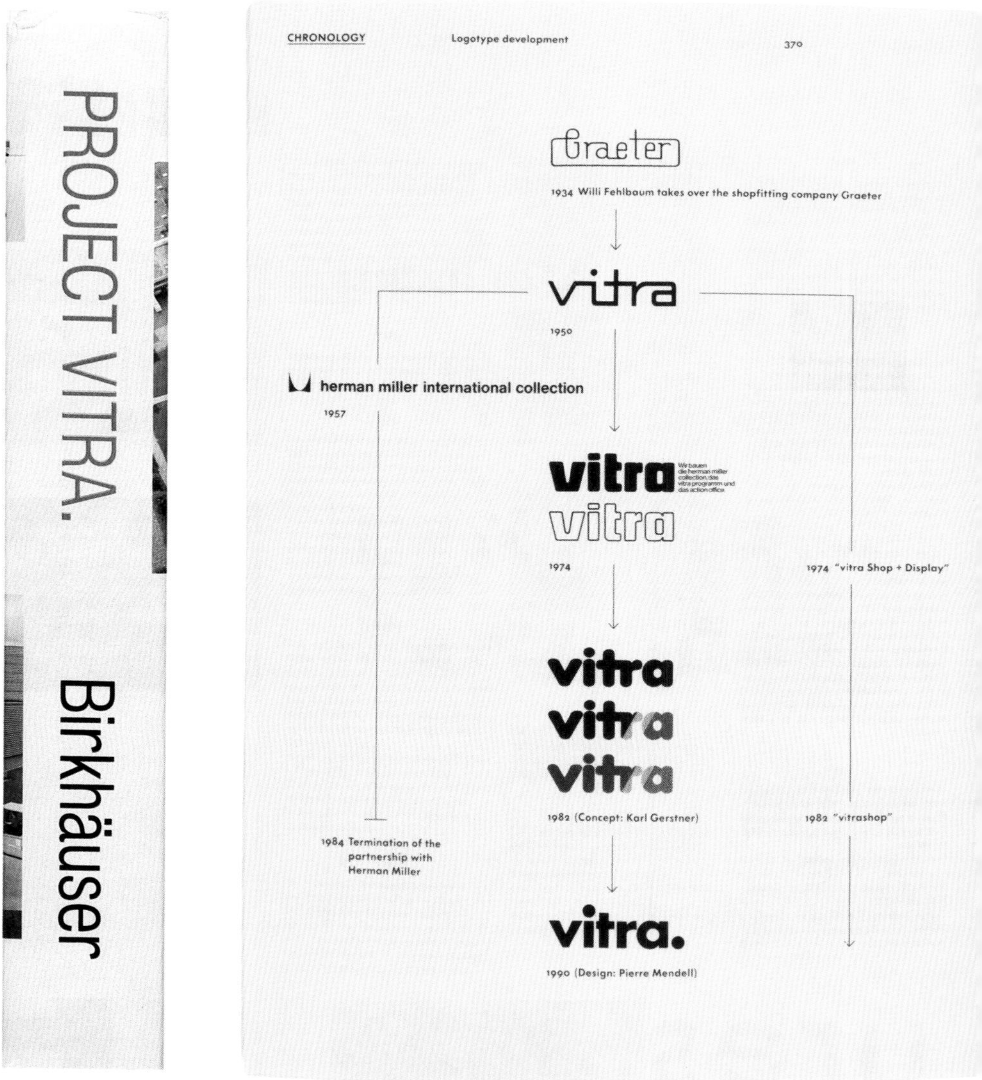
PROJECT VITRA.
Birkhäuser
CHRONOLOGY
Logotype development
370
Graeter
1934 Willi Fehlbaum takes over the shopfitting company Graeter
vitra
1950
herman miller international collection
1957
vitra
Wir bauen die herman miller collection, das vitra programm und das action office.
vitra
1974
1974 "vitra Shop + Display"
vitra
vitra
vitra
1982 (Concept: Karl Gerstner)
1982 "vitrashop"
1984 Termination of the partnership with Herman Miller
vitra.
1990 (Design: Pierre Mendell)

KEYWORDS

ARCHIVES

n addition to its collections, the Vitra Design Museum maintains archives of wo-dimensional documents on the history of industrial furniture design and related fields. Among the most important items held in the archives are the estates, or partial estates, of designers like → George Nelson, → Alexander Girard, → Verner Panton, Anton Lorenz and Harry Bertoia. The archived material s complemented by a comprehensive echnical library of books and periodcals on the history of furniture, design, architecture and art. Although the archives and library are primarily inended for internal use, they are also available to external researchers.

BARRAGAN FOUNDATION

A cultural foundation under the auspices of Vitra. It conserves and researches the estate of the Mexican architect Luis Barragán and promotes academic study and discussion of his work. Luis Barragán (1902–1988), who was awarded the Pritzker Prize in 1980, is one of the greatest figures of the modernist movement. Seven years after his death, Vitra was able to acquire parts of his estate, thereby preventing this cultural heritage from being dispersed. Since it was established in 1996, the Barragan Foundation (based in → Birsfelden near Basel) has examined and classified the estate (which includes some 13,500 drawings and plans and over 20,000 photographs) and has ensured that it is stored in accordance with good archiving practice. The Foundation (headed by Federica Zanco) first achieved public prominence with its exhibition entitled "Luis Barragán. The Quiet Revolution" which was on show at major museums in Europe, Japan and Mexico from 2000 until 2003, together with the accompanying catalogue. A comprehensive publication on the Barragan Foundation's archives is currently being compiled.

BIRSFELDEN

Vitra's headquarters are located in Birsfelden near Basel. → Willi Fehlbaum acquired the site after selling the cramped building which housed the original shopfitting business to the City of Basel. The first factory building was constructed in Birsfelden in 1957, with an office wing built to the plans of Beck and Baur, the Basel architects. This building is still standing, and it houses Vitra's → product development division and the → Barragan Foundation. The new administrative building by → Frank Gehry (the Vitra Center) at the same location was occupied in 1994. This building houses Vitra's headquarters with the management and the controlling, marketing and international divisions.

CHAIRMAN ROLF FEHLBAUM

The title of an illustrated book by → Tibor Kalman, published in 1998 by Lars Müller Publishers, about → Rolf Fehlbaum and Vitra. The book was published to mark the award of the Federal Prize for Design Promotion to Rolf Fehlbaum in 1997 by the German Design Council and the German Ministry of Economics. The poetic yet playful tone of the book is already evident in the double meaning of its title: "chairman" not only refers to the head of a company but also alludes to Fehlbaum's long-standing preoccupation with chairs.

CITIZEN OFFICE

The title of a touring exhibition presented for the first time at the Vitra Design Museum in 1993, and of the accompanying publication containing suggestions on the contemporary organization and design of office workplaces. Based on the thesis that an office is a living space for people as well as their workplace, the designers → Andrea Branzi, Michele De Lucchi and → Ettore Sottsass developed three alternatives to the uniformly hierarchical and inflexible office style which was prevalent at the time. Specimen solutions of an experimental and pioneering nature were developed after a consultation process coordinated by James Irvine and supported by Vitra. Although none of the suggestions in the exhibition actually went into production, the "Citizen Office" project prompted thoughts about how the design of the office world could be made more people-friendly and effective, and it provided stimulus for future developments.

CLASSICS

Vitra regards as classics the products that are still up to date even though they were created in a previous era. When such a design is created, it is revolutionary but it also signals the birth of a new concept of design. In subsequent decades, the design proves to be resilient in the face of its successors and imitators. It outlives them and becomes a classic. This does not mean that it is now 'tamed': its revolutionary origins live on and it seems to remain eternally fresh. A classic appears to be timeless, and only becomes dated when a new era begins. Vitra classics include designs by → Charles & Ray Eames, → George Nelson, → Jean Prouvé and → Verner Panton.

COLLAGE

A design philosophy for indoor spaces which aims to create a practical and inspiring interior that can be used for work as well as relaxation and entertainment. Instead of using integrated, stylistically pure and uniform solutions, collage advocates a mixture of elements which are selected and arranged so as to reflect the individual personality of the occupants. Collage does not aim for a design that will be valid once and for all, but views the interior and its composition

009 Bm^3

这是法国独立文学出版社 Galaade 的一个出版系列。这家出版社由历史学家埃马纽埃尔·科拉（Emmanuelle Collas）建立，专业出版小说、非小说和一些不同于其他出版商的图书。这套书籍采用锁线胶平装，封面采用各色卡纸，单色印刷适合主题配色的油墨。封面版式具有统一性，信息层级考究。内页胶版纸印刷单色黑。在工艺和纸张上尽可能地简约。内页字体和栏宽统一，但在标题字大小、页面元素、图片位置、章节隔页处理等方面根据作品主题来寻求细微差异。加粗字号的页码、标题和线条元素是较为显著的特点。正文排版字体阅读起来舒适，版面灰度均匀细腻。

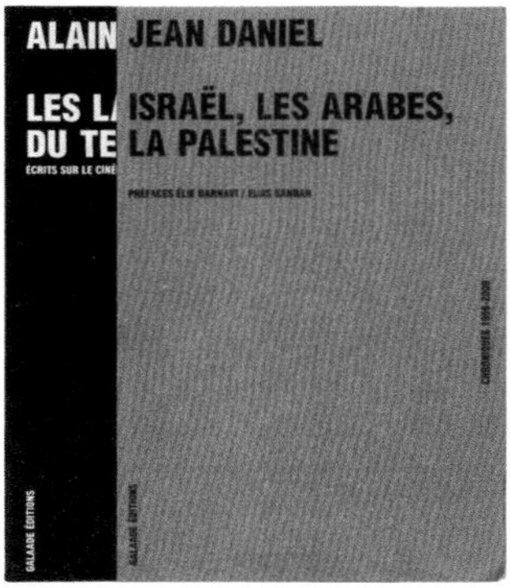

ALAIN
LES L
DU TE
JEAN DANIEL
ISRAËL, LES ARABES, LA PALESTINE
GALAADE ÉDITIONS

This is a series of books published by Galaade, an independent French literary publishing house. Established by historian Emmanuelle Collas, the publishing house specializes in fiction, non-fiction and some literary publishing fields different from other publishers. This set of books is perfect bound with thread sewing and the covers are using various colorful cardboards, printed with monochrome ink that is suitable for the theme. The layout of the covers is unified and information hierarchy is exquisite. The inside pages are printed in monochrome black on offset paper and the craft is as simple as possible. The font and column width are uniform, but the subtle differences are made according to the theme of the works in terms of title size, page elements, image location, and page separation. Page numbers, headings and line elements with bold fonts are prominent features. Text fonts are comfortable to read, and the gray level of the layout is even and delicate.

- 铜奖
- 以色列、阿拉伯人、巴勒斯坦 / 时间实验室
- 法国

Modern interpretation with excellent typography, reinforced by a classical typearea, perfectly staged – enticement indeed to the reader! The overall character determined by the series does not have a restrictive effect and the designer redefines certain parts according to the particular edition. Subtle typographical features are tantalisingly played with here. The flexible brochures are each furnished with their own single-colour sleeve, with matching title pages and black or white capitals. In its clarity the whole ensemble makes a very convincing impression.

- Bronze Medal
- Israël, Les Arabes, La Palestine / Les Laboratoires du Temps
- France
- Julien Hourcade, Thomas Petitjean (Hey Ho)
- Série «Essais», Jean Daniel / Alain Fleischer
- Galaade Éditions, Paris
- 215 × 140 × 37mm / 215 × 140 × 28mm
- 984g / 478g
- 859p / 417p

JEAN DANIEL
ISRAËL, LES ARABES,
LA PALESTINE

1956

E LA FONDATION
NESPÉRÉE
UX MENACES
NÉVITABLES

846

l'Arabie Saoudite, ont réitéré l'offre de paix qu'ils avaient adressée – en vain ! – aux Israéliens en 2002. Le gouvernement de Tel-Aviv n'a pas pu, cette fois, refuser l'offre du prince saoudien. Mais rien ne laisse penser que cette éventuelle négociation puisse favoriser un retour à la paix dans la mesure où les Américains ne pèsent pas encore de tout leur poids. Comme toujours, des forces hostiles à la paix vont se déchaîner pour faire échouer toutes les négociations.

Comme le souligne plus loin dans ce numéro Charles Enderlin [138], il se trouve que l'Union européenne fait partie, avec les Nations unies, les États-Unis et la Russie, de ce « Quartette » garant de la fameuse « feuille de route » qui préconise la sécurité pour les Israéliens et un État souverain et viable pour les Palestiniens. C'est une occasion – aussi exaltante que celle qui a animé les militants de la campagne pour le Darfour – de faire pression sur les États-Unis. Il y a bien des choses à faire, concrètes, réalistes ou utopiques. J'y reviendrai.

7 JUIN 2007
SIX-JOURS… ET QUARANTE ANS

Tandis que Vladimir Poutine et George Bush s'apprêtaient à nous jouer le grand air du retour à la guerre froide, Mahmoud Ahmadinejad, quant à lui, annonçait qu'un conflit bien chaud était déjà en cours. Célébrant à sa manière le quarantième anniversaire de la guerre éclair des Six-Jours livrée par Israël contre l'armée égyptienne, le président iranien n'a rien trouvé de mieux que de proclamer que le compte à rebours avait déjà commencé pour « l'annihilation du régime sioniste par les mains des peuples libanais et palestinien » [139]. Aussitôt, les iranologues se sont jetés sur leurs outils de décryptage. Comment Ahmadinejad pouvait-il réitérer sa provocation au moment où ses supérieurs dans la hiérarchie du Conseil de la Révolution proposaient à l'Europe d'élaborer en commun une politique de stabilité au Liban ?

Mais laissons provisoirement de côté les arcanes de la politique iranienne. Sauf pour souligner ceci, qui ne surprendra pas

138. Charles Enderlin, « La clé de la paix est à Washington », in *Le Nouvel Observateur*, n° 2213, 5 avril 2007, pp. 68-69.

139. Déclarations du président iranien le 3 juin 2007.

847

nos lecteurs : qu'il serve d'alibi, de prétexte à la provocation ou de ciment unitaire pour les islamistes, le problème palestinien continue de se trouver au cœur des relations de l'Occident avec les Arabes, et désormais avec les musulmans. Avant le déclenchement de l'intervention en Irak, deux personnalités américaines avaient averti George Bush : « Vous avez besoin, pour n'importe quelle entreprise au Proche-Orient, d'une alliance avec le monde arabe, et vous ne l'aurez qu'en réglant d'abord le problème palestinien. » James Baker et Zbigniew Brzezinski – c'est d'eux qu'il s'agit – avaient raison.

Donc, la guerre des Six-Jours a commencé il y a quarante ans. Une autre guerre l'avait précédée, en 1956, à la suite d'un différent frontalier entre l'Égypte et Israël. En 1957, un règlement était négocié, aboutissant à la démilitarisation du Sinaï, à l'installation de casques bleus et à la liberté de navigation dans le détroit de Tiran, qui sépare le golfe d'Aqaba de la mer Rouge. Nasser, qui ne s'était jamais résigné à ce règlement, a fini par fermer le détroit de Tiran.

À partir de ce moment-là, on a attendu la guerre. C'est en vain que de Gaulle avait cru devoir tout faire pour qu'elle n'eût pas lieu. Il ne pardonnera jamais aux Israéliens de ne pas avoir accepté les garanties qu'il croyait pouvoir leur offrir. Mais lorsque la guerre a été déclenchée, il faut rappeler l'immense capital de sympathie et de solidarité active dont a bénéficié le petit État hébreu. De nombreux jeunes Européens, Juifs et non Juifs, se sont engagés dans les troupes israéliennes. Mauriac a fait un article inspiré pour dire son angoisse, et Malraux a confié à Shimon Peres que, s'il avait eu dix ou quinze ans de moins, il n'aurait pas hésité à combattre dans l'armée israélienne. Romain Gary était dans le même état d'esprit.

Quant à l'historien anticolonialiste et proarabe Pierre Vidal-Naquet, réputé pour sa dénonciation implacable et inlassable des tortionnaires français en Algérie, il a confessé à quel point il avait eu peur pour l'État d'Israël et combien, à ce moment-là, il s'était senti juif. La gauche et une bonne partie de l'extrême-gauche françaises ont tremblé pour l'existence de l'État d'Israël en déplorant ce qui leur apparaissait comme un acharnement du sort contre ces éternelles victimes. Cet

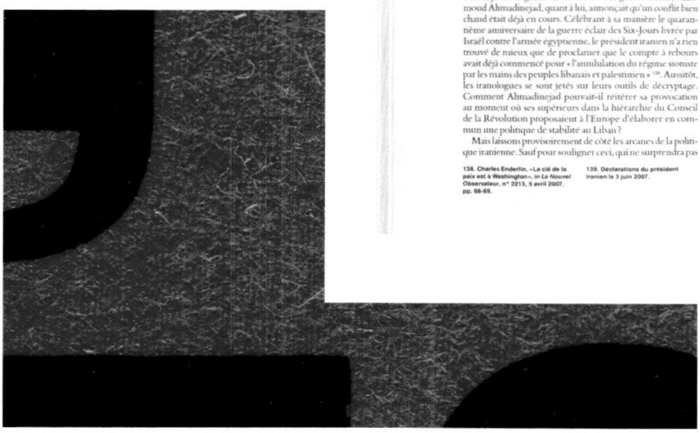

ELIE BARNAVI PRÉFACE

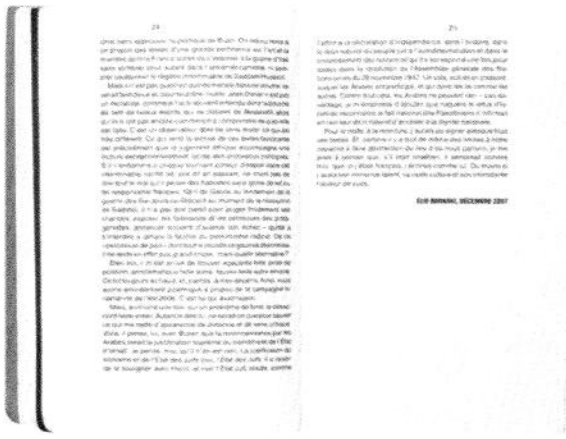

2003

2006

JEAN DANIEL
Jean Daniel, fondateur du Nouvel Observateur, est éditorialiste et écrivain.

ÉLIE BARNAVI
Ambassadeur d'Israël en France de 2000 à 2002, conseiller scientifique auprès du musée de l'Europe à Bruxelles, Élie Barnavi est professeur émérite d'histoire de l'Occident moderne à l'université de Tel-Aviv.
ELIAS SANBAR
Écrivain et journaliste, Elias Sanbar participe depuis 1991 aux négociations de paix israélo-palestiniennes. Depuis janvier 2006, il est ambassadeur et observateur permanent de la Palestine auprès de l'Unesco.

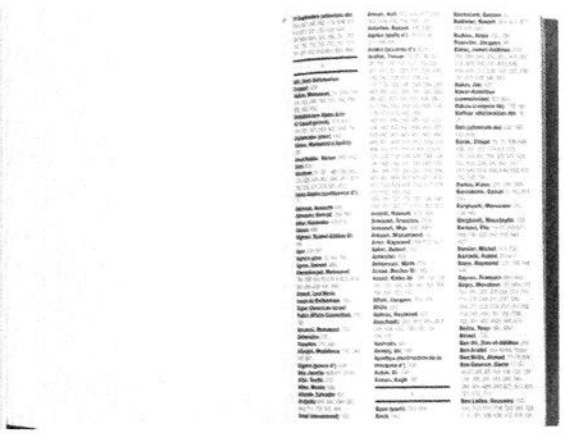

« Si
ne
ion »
on l

ISRAËL, LES ARABES, LA PALESTINE
JEAN DANIEL
GALAADE ÉDITIONS
LES LABORATOIRES DU TEMPS
ALAIN FLEISCHER
GALAADE ÉDITIONS

relativiser celles de son passé antisémite. C'est contre cette confusion de la conscience que doivent lutter ensemble, et au nom des mêmes valeurs, les citoyens de la République.

8 JUIN 2006
L'OCCUPATION SELON SHULMAN

La tragique incapacité des Palestiniens de s'entendre sur la seule position raisonnable et conforme à leurs intérêts dépend en grande partie de l'aide que nous sommes capables de procurer à l'actuel président de l'Autorité palestinienne, M. Mahmoud Abbas. Il ne s'agit pas d'apprivoiser, de justifier ou de comprendre le Hamas. Il s'agit de fortifier au contraire ses ennemis. D'autant que la position consistant à vouloir détruire l'État d'Israël n'est pas seulement le désaveu d'une solennelle décision prise par l'ONU en 1948. Elle constitue la seule base juridique valable à partir de laquelle le combat des Palestiniens pour récupérer leurs territoires peut être justifié et soutenu. C'est pourquoi il paraît au moins maladroit, sinon sectaire, d'expliquer le radicalisme du Hamas par l'accumulation des malheurs [119], ce qui est une façon d'amoindrir les mérites, en tous points admirables, du président palestinien.

Une chose au moins aura été révélée dans cet océan de souffrances, de ruine et de deuil qui submerge aujourd'hui le territoire de Gaza, c'est que l'importance de l'aide européenne n'y était pas négligeable. Pour le reste, et comme le dit David Shulman [120], une occupation est toujours un crime et personne, sous aucun prétexte, ne peut s'en accommoder.

22 JUIN 2006
TEMPÊTE SUR UN LOBBY

LES BRISEURS DE TABOU

On se souvient qu'en 1993 un long article signé de Samuel Huntington et intitulé « The clash of civilizations » avait bé-

119. *Libération* du 5 juin 2005.

120. David Schulman, *Ta'ayush. Journal d'un combat pour la paix, Israël-Palestine 2002-2005*, Paris, Seuil, « La librairie du XXIe siècle », 2006.

cié d'un tel retentissement que l'auteur l'avait développé leux livres successifs traduits dans presque toutes les lans et discutés dans tous les pays [121]. Selon un article, « The m over the Israel lobby », paru dans le dernier numéro de *lew York Review of Books* [122], un autre essai publié le 23 mars nier dans la *London Review of Books* est déjà destiné à susr des polémiques au moins aussi importantes. Son titre : lobby israélien et la politique étrangère des États-Unis ». auteurs : deux universitaires, John J. Mearsheimer et phen M. Walt. Le premier enseigne à l'université de cago, le second à Harvard.

eur essai vient de briser un tabou. Il pose en effet tranllement la question de savoir si l'extrême générosité de le financière des États-Unis à Israël (près de 3 milliards dollars par an), comme la possibilité ainsi donnée à Isd'acheter des armes aussi sophistiquées que les chasseurs , est justifiée depuis qu'Israël a perdu tout intérêt stratégipour les États-Unis avec la cessation de l'aide importante l'Union soviétique accordait à l'Égypte et à la Syrie. Sans ite, depuis le 11 septembre 2001, Israël est-il considéré nme un allié décisif dans la guerre contre le terrorisme. is c'est un raisonnement qui peut se retourner avec la stion de savoir si le soutien à n'importe quelle politique élienne ne contribue pas, précisément, à affaiblir la lutte États arabes et musulmans contre le terrorisme.

es deux auteurs n'hésitent pas à mettre en cause un lobby, nerican Israel Public Affairs Commitee, que l'on appelle s le monde entier « Aipac », et dont la puissance vient de suite après celle de la National Rifle Association, défend le droit des citoyens de posséder des armes. Le y Aipac est associé aux évangéliques chrétiens comme ı DeLay, Jerry Falwell, Pat Robertson, et aux néoconateurs Paul Wolfowitz, Richard Perle, Bernard Lewis et liam Kristol.

aturellement, devant une mise en question qui pourrait scandale, les hommes qui viennent d'être cités ne sont restés inactifs. Leurs alliés universitaires ont souligné le ique de rigueur et d'objectivité des analyses – et, parfois

Samuel P.Huntington, *Le Choc ivilisations*, Paris, Odile Jacob, et *Qui sommes-nous ? Identité nale et choc des cultures*, Paris, Jacob, 2004.

u moment où cet article était à New York disparaissait la cofondatrice de la *New York Review of Books*, Barbara Epstein, personnalité remarquable sans laquelle Bob Silvers n'aurait jamais pu mener à bien la tâche de publier le journal intellectuel le plus prestigieux de l'Occident.

009 Bm4

本书是德裔法国雕塑家、画家、诗人让·阿尔普（Jean Arp）的作品集。书中通过对阿尔普各个时期的图纸、拼贴、浮雕、雕塑和诗歌作品的陈列和解读，展现阿尔普根植于传统，但又寻求破坏、随意的艺术风格。本书为锁线胶平装，封面通过橙、黄、蓝、绿四色胶版纸的相互包裹装订，形成独特的封面效果，每一层封面都可以单独打开，背面包含一张阿尔普的代表作品。内页用纸丰富，环衬为黄色牛皮卡，过门页灰卡纸印黑，前言为黑卡纸印银，作品页亮面铜版纸四色印刷，文字和解读部分胶版纸四色印刷。其中穿插部分内含金属质感的艺术纸用以展现特殊作品。主文字双栏排版，但通过图片版式的自由和字体选择的多样性，呈现多变的最终效果。

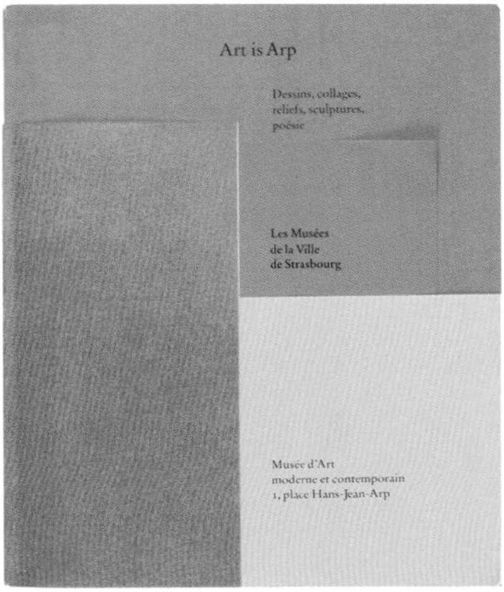
Art is Arp
Dessins, collages, reliefs, sculptures, poésie
Les Musées de la Ville de Strasbourg
Musée d'Art moderne et contemporain 1, place Hans-Jean-Arp

This book is a collection of works by Jean Arp, a German-born French sculptor, painter and poet. Through the display and interpretation of Alp's drawings, collages, reliefs, sculptures and poetry works in different periods, the book reveals Alp's artistic style that is rooted in tradition and seeking to destroy the tradition. This book is perfect bound with thread sewing. The cover is wrapped and bound by orange, yellow, blue and green four-color offset paper to form a unique effect. Each layer of the cover can be opened separately with a representative work of Alp printed on the backside. The inside pages use various kinds of paper – yellow kraft cards used as end paper, black cardboard printed on silver used as preface page, four-color printing gloss art paper used as content pages, and offset paper used for the text pages. Art paper with metal texture is inserted to display special works. Though the text is typeset in two columns, the final effect is varied with the free picture format and the diversity of font selection.

- 铜奖
- 阿尔普的艺术
  ——图纸、拼贴、浮雕、雕塑、诗歌
- 法国

- Bronze Medal
- Art is Arp.
  Dessins, collages, reliefs, sculptures, poésie
- France
- Sp Millot
- Editions des Musées de la Ville de Strasbourg, Strasbourg
- 280×230×19mm
- 1140g
- 343p

The jacket of this exhibition catalogue has four colours and four layers, each in a different format. The book unites experimental aspects and classical layout to the highest possible standard and visualises the idea of vacillation between the principles of composition and improvisation. With much variation and wealth of materials the design makes use of the artist's ideas, but without ever forcing itself into the foreground. The catalogue provides an incentive to wander amongst everything on offer.

Art is Arp

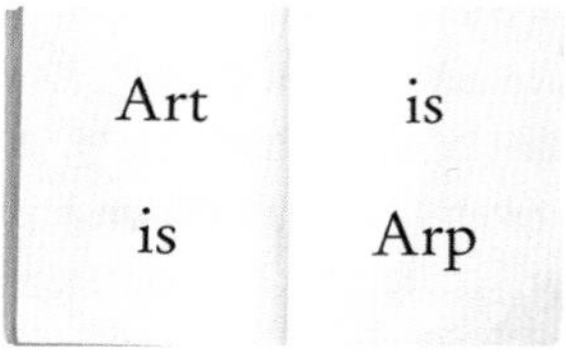
Art
is
is
Arp

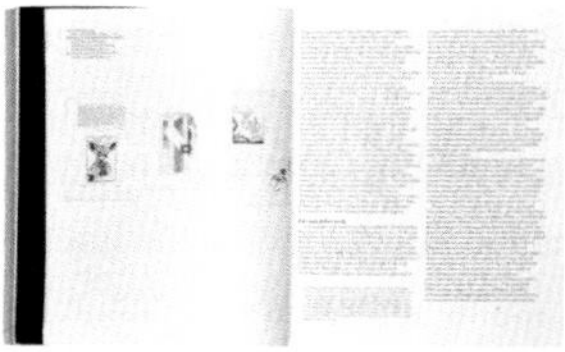

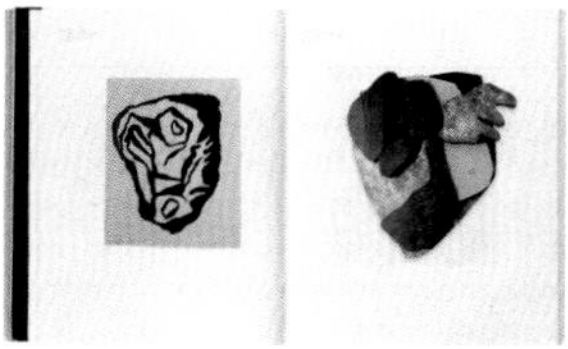

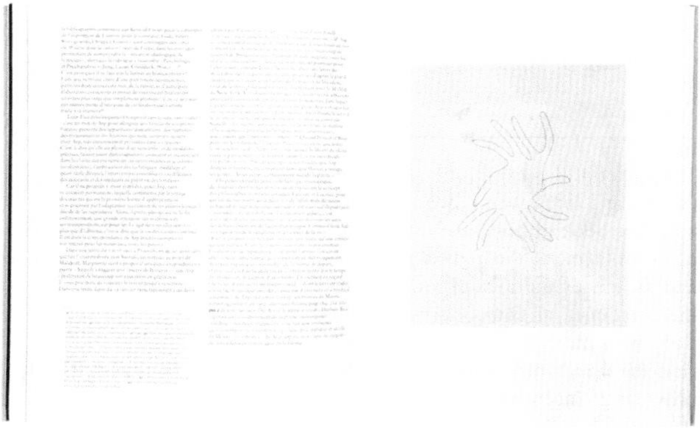

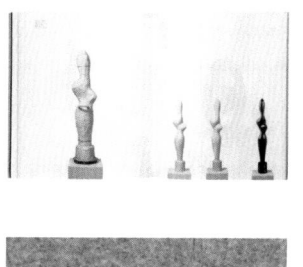

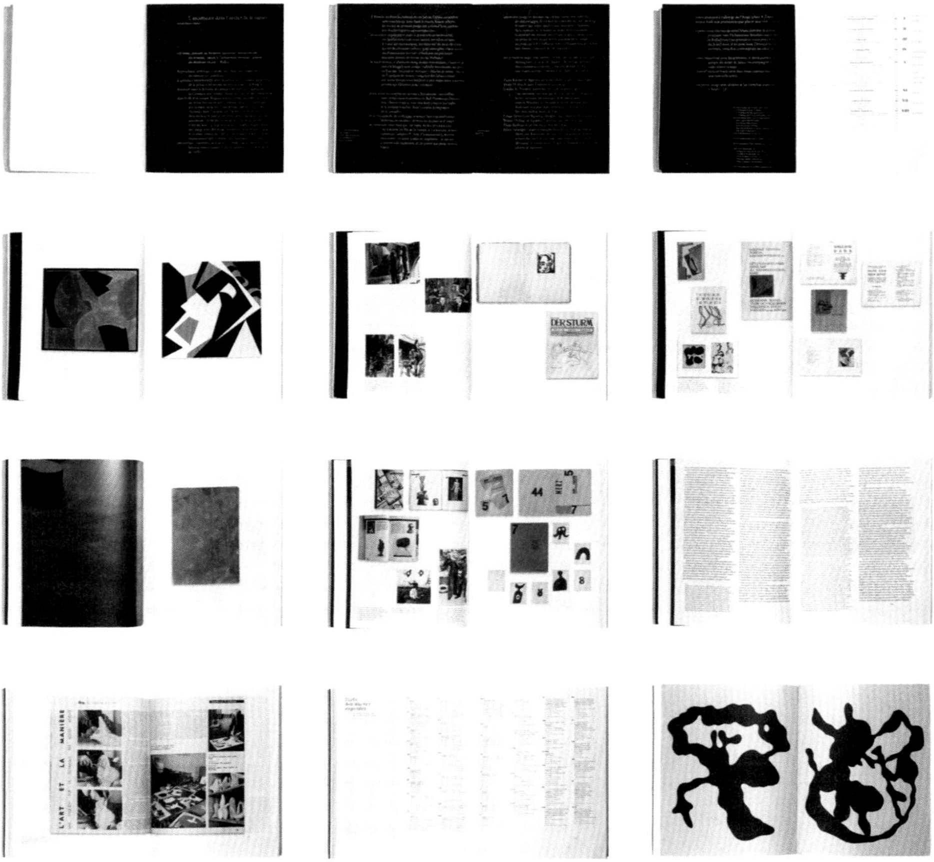

nspira
e qu'il

Art is Arp
Nous tenons à remercier chaleureusement les institutions dévolues à l'œuvre de Arp qui par leur soutien et leurs prêts généreux nous ont permis d'accomplir ce projet:
Fondation Arp, Clamart
Claude Weil-Seigeot
Chiara Jaeger
Virginie Delcourt
Fondazione Marguerite Arp, Locarno
Rainer Hüben
Stiftung Hans Arp und Sophie Taeuber-Arp e.V., Rolandseck
Dieter G. Lange
Anna Krems
Walburga Krupp
Arp Museum Bahnhof Rolandseck, Remagen
Prof. Dr. Klaus Gallwitz
Dr. Oliver Kochoff
Astrid von Asten
Notre profonde gratitude s'adresse en particulier à Walburga Krupp, Chiara Jaeger, Virginie Delcourt, Astrid von Asten et Rainer Hüben pour leurs conseils, leur disponibilité et leur enthousiasme tout au long de la préparation de cette exposition.
Nous remercions vivement les musées, bibliothèques, institutions, galeries et collectionneurs privés de leur concours à la réalisation de cette exposition:
Aargauer Kunsthaus, Aarau
Madeleine Schuppli
Corinne Linda Sotzek
Bibliothèque Kandinsky, Centre Pompidou, Paris
Didier Schulmann,
Nathalie Cisse
Christelle Courregelongue
Francine Delaigle
Laurence Gueye-Parmentier
Agnès Leroux de Bretagne
Brigitte Vincens
Bibliothèque littéraire Jacques Doucet, Paris
Sabine Coron
Marie-Dominique Nobécourt-Mutarelli
Fatima El Hourd
Centraal Museum, Utrecht
Pauline Terreehorst
Len van den Berg
Daniella Luxembourg Art Ltd, Londres
Daniella Luxembourg
Catherine Keller
Naomi Ellis
Paul Coulon
Deutsche Bank, Francfort
Friedhelm Hütter
Dr. Danielle Pippardt
Laura Popplow
Emanuel Hoffmann-Stiftung, Münchenstein
MS

t ouvrage est publié à l'occasion de l'exposition « Arts/Arts » présentée au Musée d'Art moderne et contemporain de la Ville de Strasbourg du 17 octobre 2008 au 15 février 2009

Sénateur–Maire de Strasbourg	*Roland Ries*
Adjoint au Maire chargé de l'action culturelle	*Daniel Payot*
Président de la Communauté urbaine de Strasbourg	*Jacques Bigot*

xposition a été réalisée par les Musées de la Ville de Strasbourg

Les Musées de la Ville de Strasbourg	
Directrice des Musées	*Joëlle Pijaudier-Cabot*
Administration	*Hubert Tagliang et son équipe*
Communication	*Alexandre Kerstetter et son équipe*
Coordination éditoriale	*Laure Lane, Lize Braat*
Documentation et photographie	*Christine Speroni et son équipe*
Service éducatif	*Margaret Pfenninger et son équipe*
Bibliothèque des Musées	*François-Marie Deyrolle et son équipe*

xposition a bénéficié du soutien du ministère de la Culture et de la Communication (Direction régionale des affaires culturelles d'Alsace)

tte exposition est inscrite au programme de la Saison culturelle européenne.
aison culturelle européenne (1er juillet – 31 décembre 2008) est organisée par le ministère des Affaires étrangères et européennes et le ministère de la Culture et de la Communication
c le soutien du Secrétariat général de la Présidence française de l'Union européenne, et mise en œuvre par Culturesfrance.

Ambassadeur chargé de la dimension culturelle de la Présidence française de l'Union européenne	*Renaud Donnedieu de Vabres*
Commissaire général	*Laurent Burin des Roziers*

Commissariat	*Isabelle Ewig, Maître de conférences en histoire de l'art à l'Université de la Sorbonne (Paris IV)*
Conseiller scientifique	*Emmanuel Guigon, Conservateur en chef, Directeur du musée des Beaux-Arts et d'Archéologie de Besançon*
Directrice des Musées de la Ville de Strasbourg	*Joëlle Pijaudier-Cabot*
Conservatrice du Musée d'Art moderne et contemporain	*Estelle Pietrzyk*
Coordination de l'exposition	*Sophie Kimenau*
Action culturelle et éducative	*Hélène Fourneaux et Martine Debaene*
Montage	*Équipe technique des Musées sous la responsabilité de Daniel Del Degan et Xavier Claus*
Architecture de l'exposition	*Benoît Grafteaux & Richard Klein, architectes, Carine Lefebvre assistante*

009 Bm5

03
01

书籍表面上是在介绍皮埃尔·黑贝林克和皮埃尔·德威特建筑事务所（Pierre Hebbelinck & Pierre de Wit Architectes），实际是通过罗列和分析他们工作上的各个环节、要素、工具和方法，来观察和思考建筑师作为人类本身如何与工作的空间相处，如何将设计的奇思妙想安放到“空间”中这一过程的记录。书籍展示的核心——“感知方法论”和以人为本的理念，不仅针对客户，也针对建筑师们自己。本书的装订形式为锁线胶平装。护封采用带有皮质手感的艺术纸张印单黑，内封白卡纸印黑，内页胶版纸张四色印刷。内容的主体分为 27 个部分，罗列了包括事务所的建筑、室内布置、书籍收藏、材料和工具、笔记和草图、模型、试验、会议、施工、摄影等内容，也展示了设计师的桌子、角落里的杂物、对人本身感知的思考、一步到海的环境等细节，大量资料被设计为折页或小开本夹页藏在书中，呈现出“方法”笔记的状态。||||||||||||||||||||||||||||||||||||||||||||||||||||||

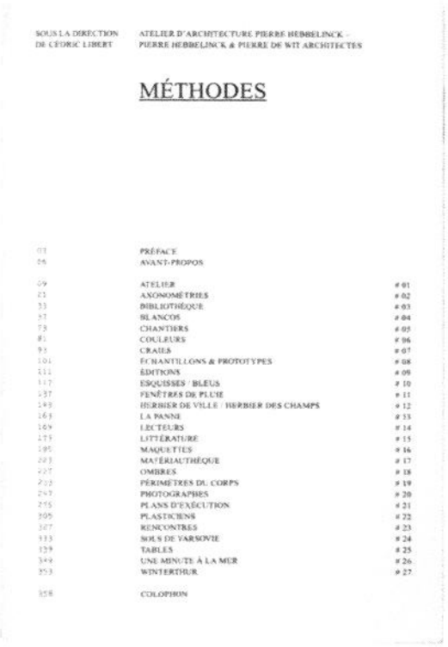
SOUS LA DIRECTION DE CÉDRIC LIBERT

ATELIER D'ARCHITECTURE PIERRE HEBBELINCK – PIERRE HEBBELINCK & PIERRE DE WIT ARCHITECTES

MÉTHODES

03	PRÉFACE	
05	AVANT-PROPOS	
09	ATELIER	# 01
21	AXONOMÉTRIES	# 02
33	BIBLIOTHÈQUE	# 03
57	BLANCOS	# 04
73	CHANTIERS	# 05
81	COULEURS	# 06
93	CRAIES	# 07
101	ÉCHANTILLONS & PROTOTYPES	# 08
111	ÉDITIONS	# 09
117	ESQUISSES / BLEUS	# 10
137	FENÊTRES DE PLUIE	# 11
143	HERBIER DE VILLE / HERBIER DES CHAMPS	# 12
163	LA PANNE	# 13
169	LECTEURS	# 14
175	LITTÉRATURE	# 15
195	MAQUETTES	# 16
203	MATÉRIAUTHÈQUE	# 17
227	OMBRES	# 18
233	PÉRIMÈTRES DU CORPS	# 19
247	PHOTOGRAPHIES	# 20
275	PLANS D'EXÉCUTION	# 21
305	PLASTICIENS	# 22
327	RENCONTRES	# 23
333	SOLS DE VARSOVIE	# 24
339	TABLES	# 25
349	UNE MINUTE À LA MER	# 26
353	WINTERTHUR	# 27
358	COLOPHON	

The book seems to introduce Pierre Hebbelinck & Pierre de Wit Architects. In fact, it observes and ponders how architects work as human beings and how they put fantastic design ideas into the 'space' by listing and analyzing the steps, elements, tools and methods in their work. The core of book is 'perception methodology' and the people-oriented concept, not only for customers, but also for architects themselves. This paperback book is glue bound by thread sewing. The jacket is printed in black on leather-like art paper, and the inside cover is printed on white cardboard in black. Offset paper printed in four colors is used for inside pages. The main body is divided into 27 parts, including office building, interior layout, book collection, materials and tools, notes and sketches, models, experiments, meetings, construction, photography, etc. It also shows the designer's desks, corner debris, thoughts of people's perception, etc. A large amount of information is designed to foldouts or small brochures hidden in the book, showing the notes of 'method'.

◎ 铜奖

《 方法

◖ 法国

An architectural book on the "methodology of perception", which engages all one's senses. Daring but well accomplished mix of typefaces with fine typography, adjusting perfectly to the language of the images. The flexible binding makes it easy to leaf through the different papers and pages folded in on the right. The book, introducing the Pierre Hebbelinck + Pierre de Wit atelier, is convincing in its white and still beauty. There is more to discover in this book of the senses than one might first think.

◎ ········· Bronze Medal

《 ········· Méthodes

◖ ········· France

D ········· Manuela Dechamps Otamendi

△ ········· Cédric Libert & Atelier d'architecture Pierre Hebbelinck – Pierre Hebbelinck & Pierre de Wit Architectes (Ed.)

|||| ········· Wallonie-Bruxelles International, Bruxelles

□ ········· 228×151×29mm

⌸ ········· 807g

≣ ········· 360p

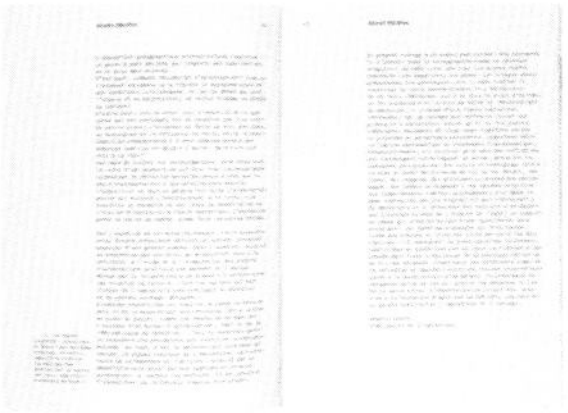

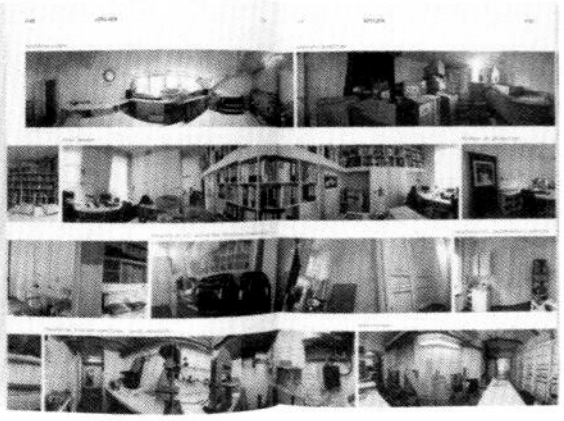

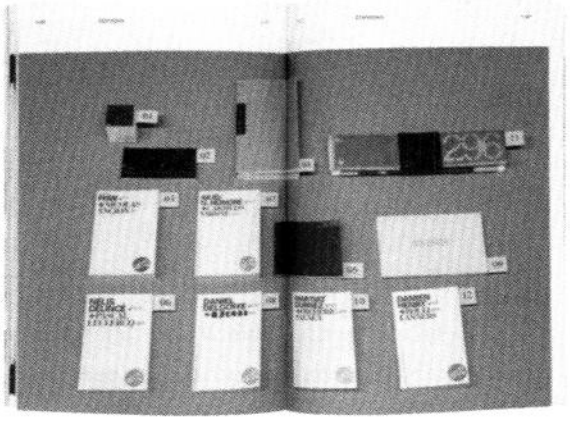

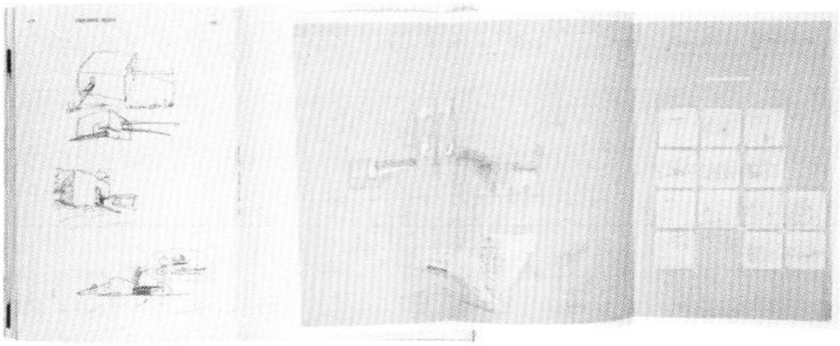

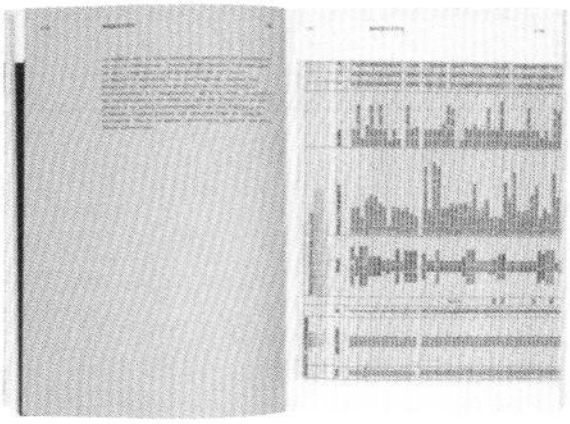

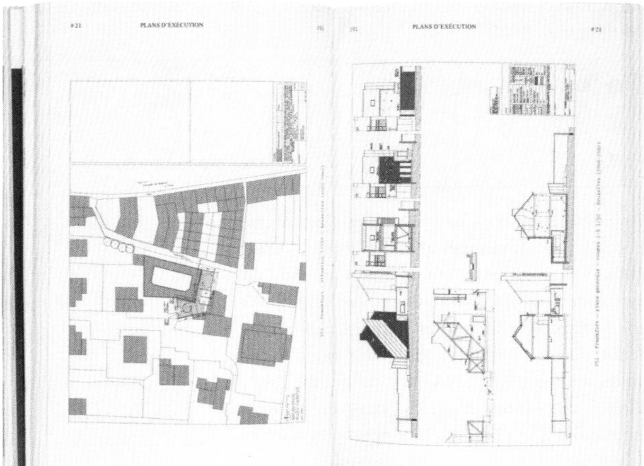

21
PLANS D'EXÉCUTION
PLANS D'EXÉCUTION
21

SES /

RES D

ER DE VILL

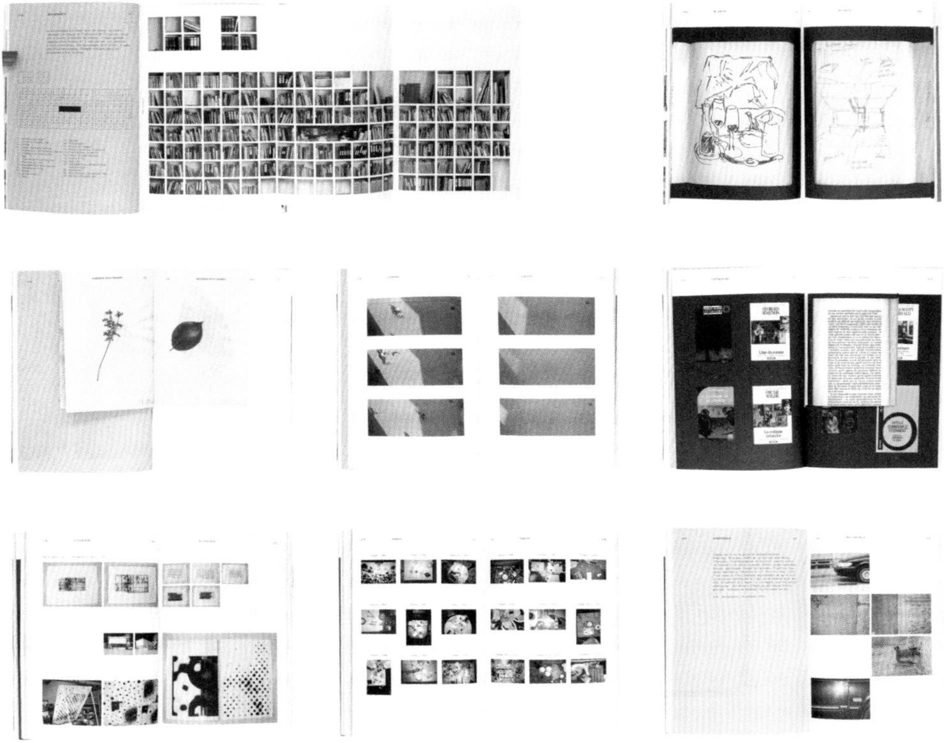

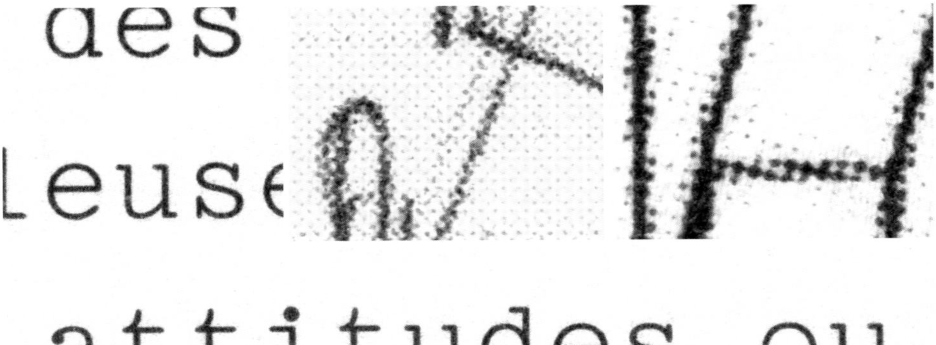
des
leuse
attitudes ou

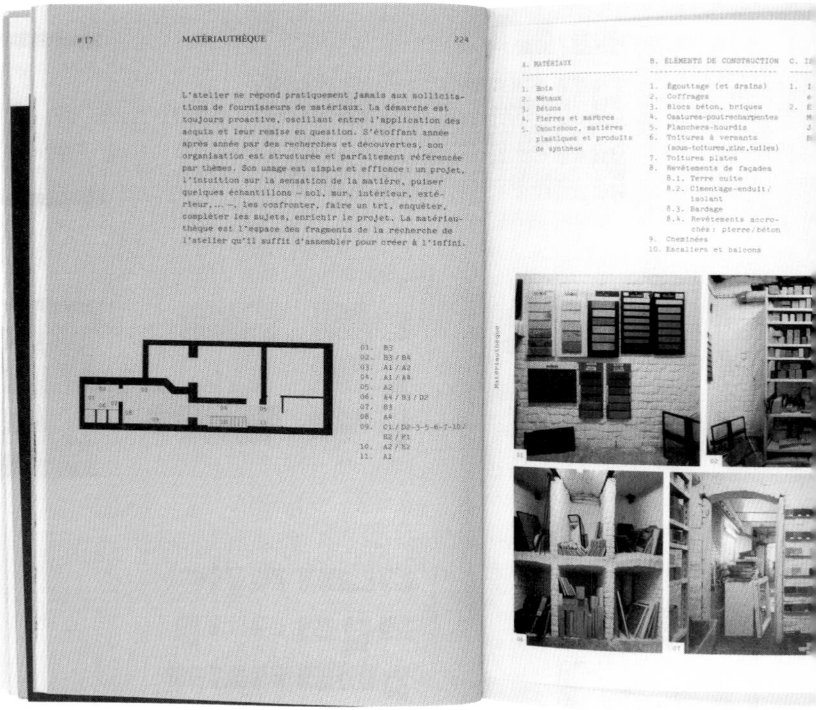

#17
MATÉRIAUTHÈQUE
224
L'atelier ne répond pratiquement jamais aux sollicitations de fournisseurs de matériaux. La démarche est toujours proactive, oscillant entre l'application des acquis et leur remise en question. S'étoffant année après année par des recherches et découvertes, son organisation est structurée et parfaitement référencée par thèmes. Son usage est simple et efficace : un projet, l'intuition sur la sensation de la matière, puiser quelques échantillons – sol, mur, intérieur, extérieur,… –, les confronter, faire un tri, enquêter, compléter les sujets, enrichir le projet. La matériauthèque est l'espace des fragments de la recherche de l'atelier qu'il suffit d'assembler pour créer à l'infini.
01. B3
02. B3 / B4
03. A1 / A2
04. A1 / A4
05. A2
06. A4 / B3 / D2
07. B3
08. A4
09. C1 / D2-3-5-6-7-10 / E2 / F1
10. A2 / E2
11. A1
A. MATÉRIAUX
1. Bois
2. Métaux
3. Bétons
4. Pierres et marbres
5. Caoutchouc, matières plastiques et produits de synthèse
B. ÉLÉMENTS DE CONSTRUCTION
1. Égouttage (et drains)
2. Coffrages
3. Blocs béton, briques
4. Ossatures-poutrecharpentes
5. Planchers-hourdis
6. Toitures à versants (sous-toitures,zinc,tuiles)
7. Toitures plates
8. Revêtements de façades
8.1. Terre cuite
8.2. Cimentage-enduit / isolant
8.3. Bardage
8.4. Revêtements accrochés : pierre / béton
9. Cheminées
10. Escaliers et balcons
Matériauthèque

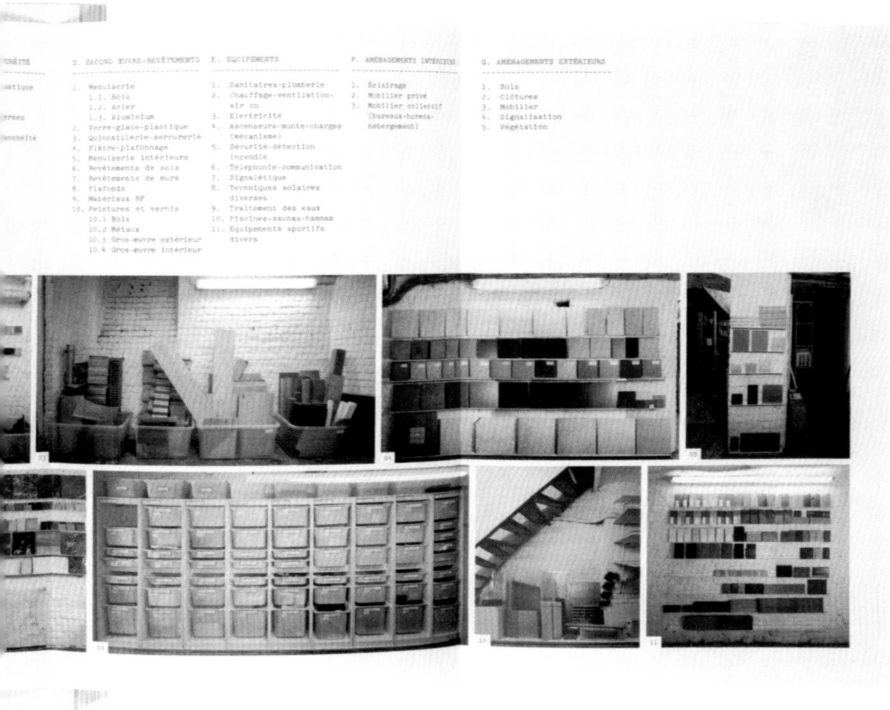

CHÉITÉ
astique
erses
anchéité

D. SECOND ŒUVRE-REVÊTEMENTS

1. Menuiserie
   1.1. Bois
   1.2. Acier
   1.3. Aluminium
2. Verre-glace-plastique
3. Quincaillerie-serrurerie
4. Plâtre-plafonnage
5. Menuiserie intérieure
6. Revêtements de sols
7. Revêtements de murs
8. Plafonds
9. Matériaux RF
10. Peintures et vernis
    10.1 Bois
    10.2 Métaux
    10.3 Gros-œuvre extérieur
    10.4 Gros-œuvre intérieur

E. EQUIPEMENTS

1. Sanitaires-plomberie
2. Chauffage-ventilation-air co
3. Electricité
4. Ascenseurs-monte-charges (mécanisme)
5. Sécurité-détection incendie
6. Téléphonie-communication
7. Signalétique
8. Techniques solaires diverses
9. Traitement des eaux
10. Piscines-saunas-hammam
11. Equipements sportifs divers

F. AMÉNAGEMENTS INTÉRIEURS

1. Éclairage
2. Mobilier privé
3. Mobilier collectif (bureaux-horeca-hébergement)

G. AMÉNAGEMENTS EXTÉRIEURS

1. Sols
2. Clôtures
3. Mobilier
4. Signalisation
5. Végétation

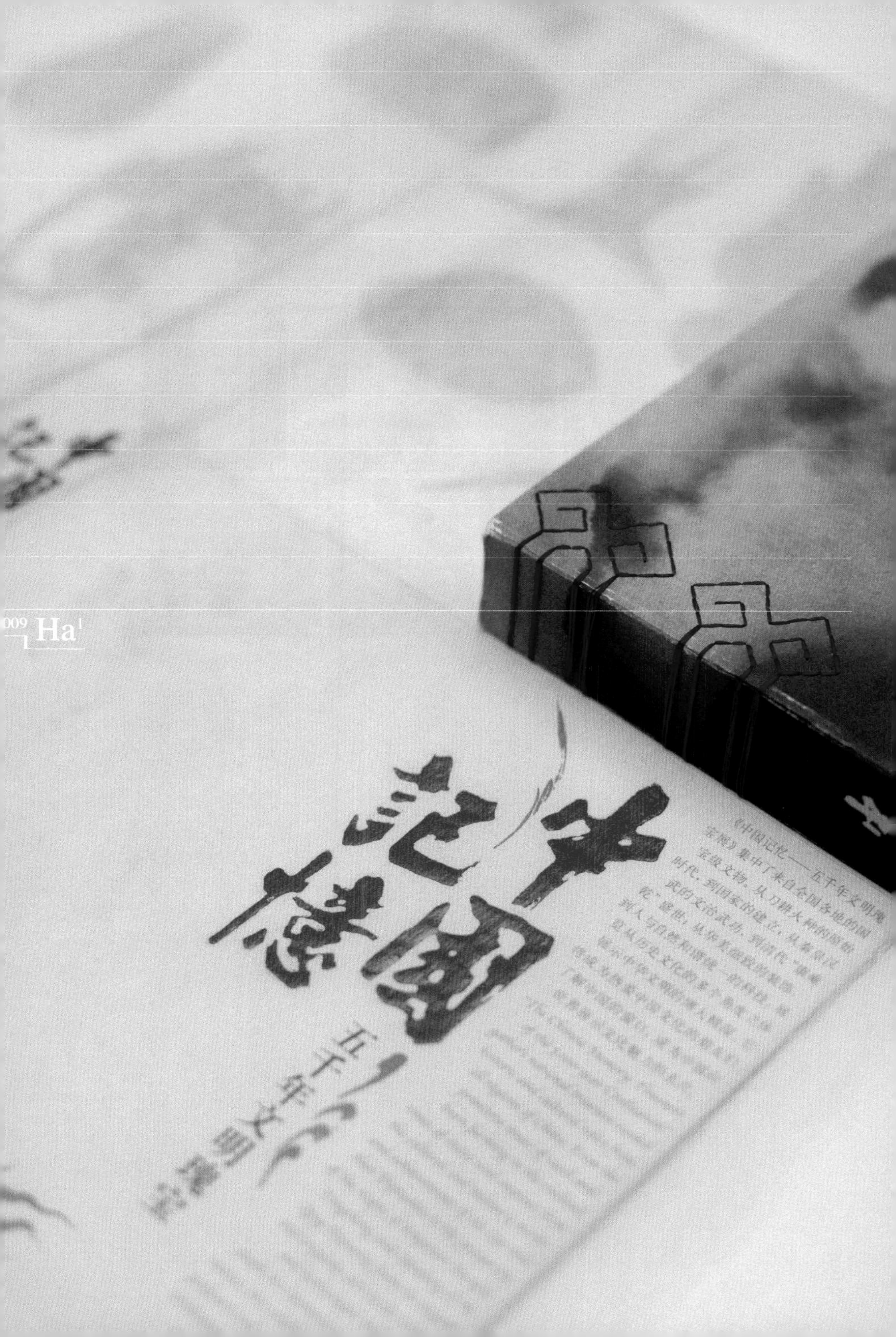
中国记忆
五千年文明瑰宝

《中国记忆——五千年文明瑰宝》选取了中国 26 个省市 55 座博物馆的 169 件文物精品，是一本中国特展图录，也同时属于首都博物馆书库系列的一部分。本书是无线包背胶装的精装本，由于 8 开页面足够大，在装订处不用完全打开的情况下可以较好地展开，并由于包背纸张两侧受力不同，展开页最终呈现水平的牵拉平面。护封半透明的艺术纸正反双面黑色和亚金色印刷，通过纸张的半透明效果，透露出内层的文物细节，书名烫红色。封面超感花纹纸四色印刷，标题烫亚金色。书脊部分红色绣线云纹图案。封面通过裱覆扉页稻禾纸的方式，具有一定的硬度，同时维持了较薄的厚度。内容方面将华夏历史分为四个大的阶段，图片精美。内文纯质纸五色印刷，部分文字的亚金色在书页里体现着文物瑰宝的精致珍贵，中英对照的横竖混合排版细节考究。

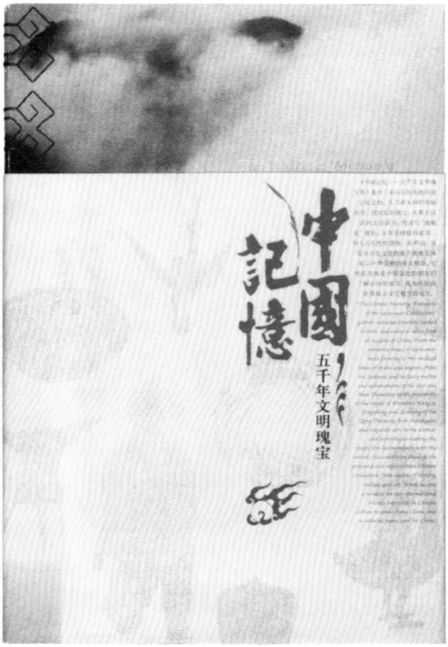

中國記憶
五千年文明瑰宝

The Chinese Memory: Treasures of the 5000-year Civilization, is a catalogue of 169 cultural relics from 55 museums in 26 provinces and cities in China. It is also a part of the Library Series of the Capital Museum. This hardcover edition is a double-leaved book. Since the page size is large enough, the book can be fully opened and the unfolded pages eventually show a horizontal surface. The jacket is printed in black and sub-gold on both sides on translucent art paper and the details of cultural relics can be observed through the translucent paper. The title of the book is hot stamped in red. The cover is printed in four colors in super-sensitive pattern paper with red embroidery moire pattern on the spine. The cover is mounted by the title page to get a certain degree of hardness and maintains a relatively thin thickness. In terms of content, the history of China is divided into four major stages. The text is printed in five colors on pure paper with beautiful pictures, the golden color reflects how precious the treasures are. The horizontal and vertical typesetting details of Chinese and English are exquisite.

◎ 荣誉奖	◎ Honorary Appreciation
《 中国记忆——五千年文明瑰宝	《 The Chinese Memory: Treasures of the 5000-year Civilization
中国	China
敬人设计工作室　吕敬人、吕旻	Lü Jingren & Lü Min (Jingren Art Design Studio)
首都博物馆	Capital Museum (Ed.)
文物出版社	Cultural Relics Press, Beijing
	359×239×35mm
	2777g
	352p

This particularly luxurious volume opulently documents cultural artefacts. Its design makes references to both traditional Asian approaches and western catalogue design. Images are treated meticulously and very well presented throughout; in both languages the typography is strong and resolute. The Chinese-style binding is stylistically appropriate and adds to the book's lustre. The three-quarter dust-jacket, printed on soft, slightly structured semi-transparent paper, unfolds into a large montage of images. A most impressive book!

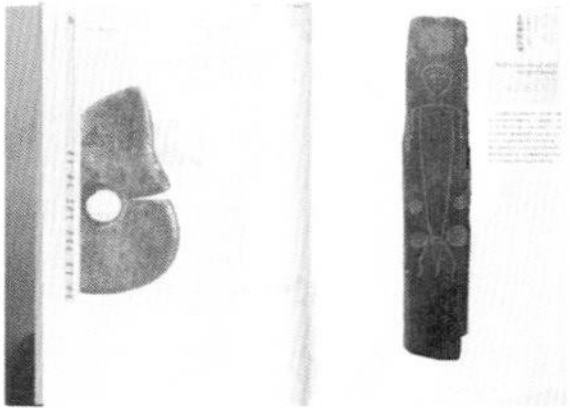

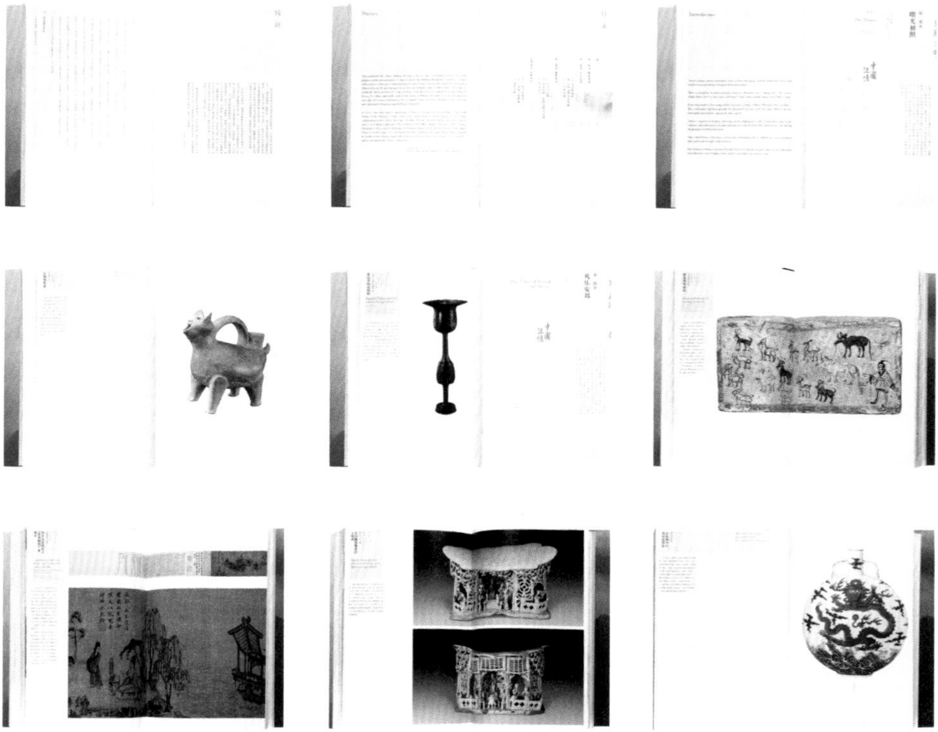

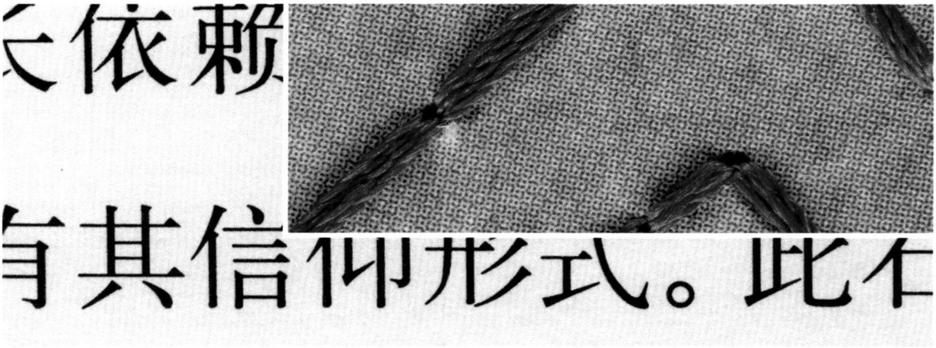
依
其信仰形式。此

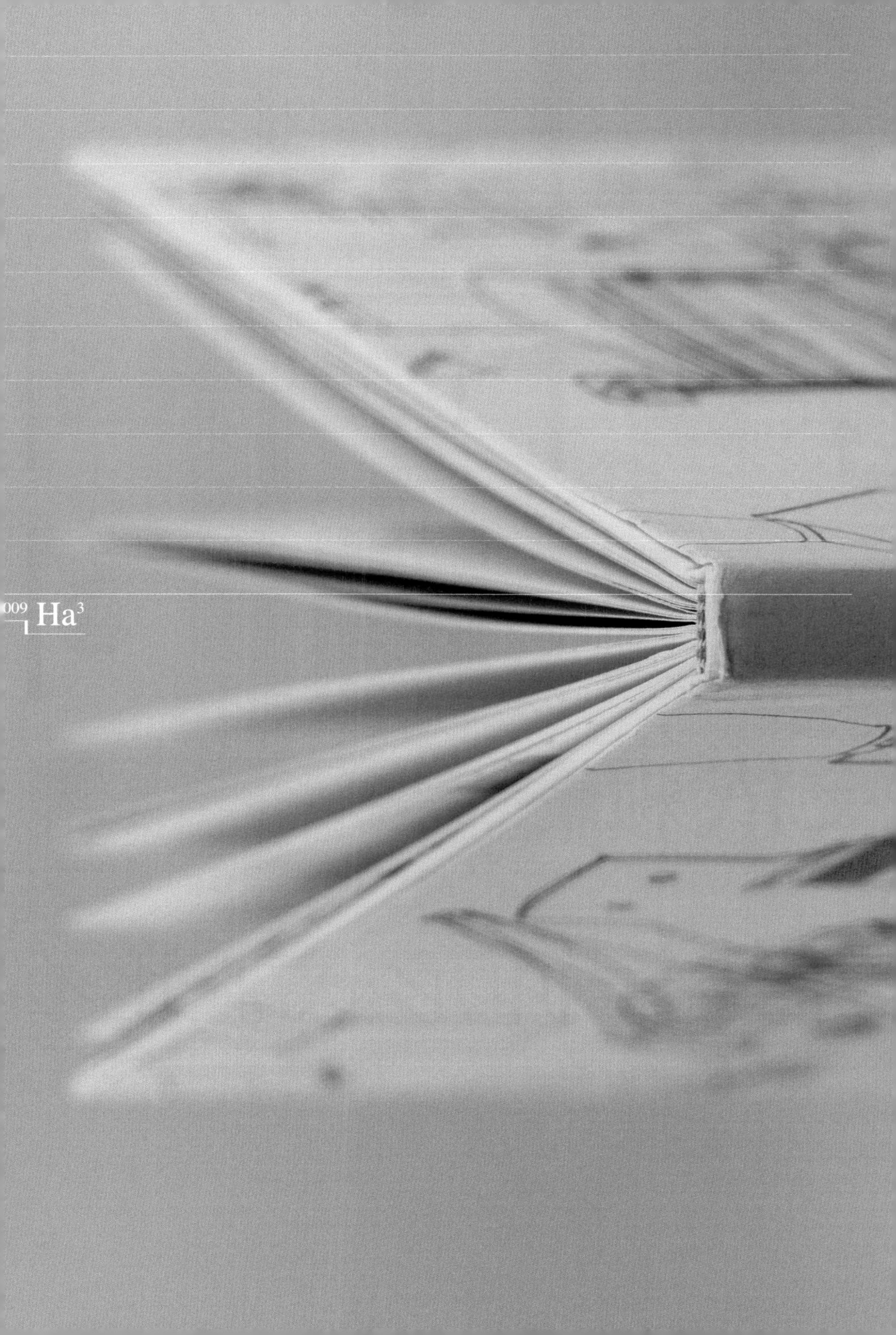

009 Ha3

这是一本速写集，作者以儿时和当下生活的周边场景为题材，关注生活并且思考记忆与真实之间的关系。本书的装订形式为锁线硬精装，护封、内封以及内页均采用胶版纸张，其中内封裱覆卡板。全书黑红双色印刷，呈现稚拙的炭笔速写线条。全书共分为三个部分，分别为记忆中儿时生活的场景、自画像和绘制的各种常见动物，以及记忆中身边的场景。作者并不是以写生的方式去绘制这些画面，而是通过回忆去复原脑中的印象。作者发现即便对着镜子去画自己，每次的结果都不一样。这些绘画包含了个人每次观察到的重点，是个人的主观经验。它或许不像摄影那样可以更好地再现事物，但却呈现出另一种真实，一种属于个人记忆的有的放矢的真实。这是客观的现实与人脑中的“真实”之间的差异，体现书名“这就是它的样子”中的“它”是个人主观意识中的那个“它”，是客观在人脑中的投射，是经过人脑加工过的那个“它”。书中夹带撕碎的笔记本页面，用于勾起人心中的回忆。精致的字体排印，与稚拙的绘画风格形成对比。||||||||||||||||||||||||||||||||||||||||||||||||||||||||||||||||||||||||||||||||||||||

so.
siehts
aus

This is a sketchbook. The author focuses on the surrounding scenes of childhood and present life, pays attention to daily life and thinks about the relationship between memory and reality. The hardback book is bound by thread sewing. The jacket, inside cover and inner pages are all made of offset paper and the inside cover is mounted with cardboard. The book is printed in black and red with childish charcoal sketch lines. It is divided into three parts, namely, the scenes of childhood life in memory, self-portraits and various hand-painted animals, as well as the scenes around in memory. The author didn't draw these pictures in the way of sketching, but restored the impression in his mind by recalling because he found that even when he drew himself in the mirror, the images were different every time. These paintings contain the key of individual observation. They are the subjective experience of the individual. They may not reproduce things as well as photography, but they present another kind of reality, a kind of targeted reality belonging to personal memory. This is the difference between objective reality and 'reality' in human brain. It reflects that 'it' in book title *This is What it Looks Like* is the 'it' in individual subjective consciousness, the objective projection in human brain, and the 'it' has been processed by human brain. The book contains torn notebook pages to evoke memories. The delicate typesetting is in sharp contrast to the childish painting style.

- 荣誉奖
- 这就是它的样子
- 德国

This dynamic contemporary book, which focuses on city life through the narrative of drawing, makes good use of its large format. The drawings – all rendered in pencil – are fresh and intimate, with the casual unpretentiousness of sketchbook work, yet with resolute, nuanced composition. The layered cover, in simple black-white-pink that is echoed throughout the book, is especially striking, as are the patterned endpapers.

- Honorary Appreciation
- So. siehts aus
- Germany
- Kerstin Rupp
- Kerstin Rupp
- Eigenverlag Kerstin Rupp, Leipzig
- 426×294×14mm
- 1149g
- 96p

1
früher
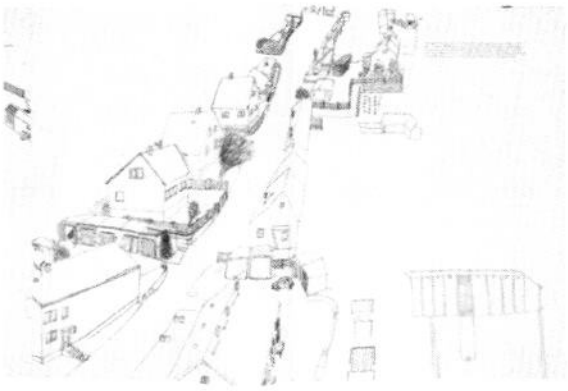

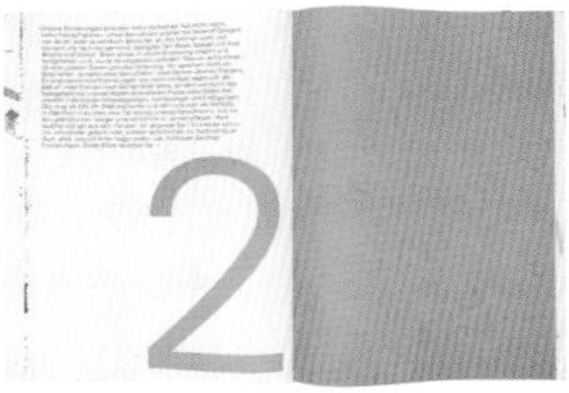
2

Wer kennt diese Frau?

Was siehst du?
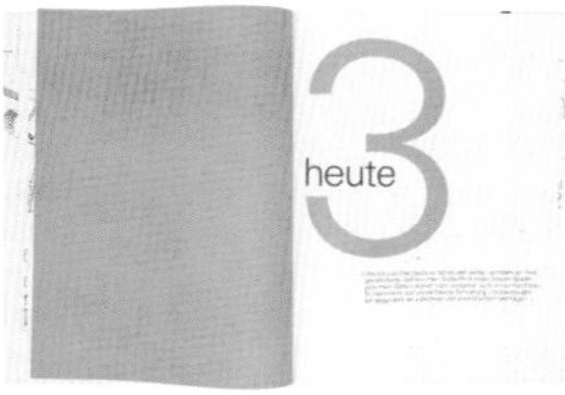
3
heute
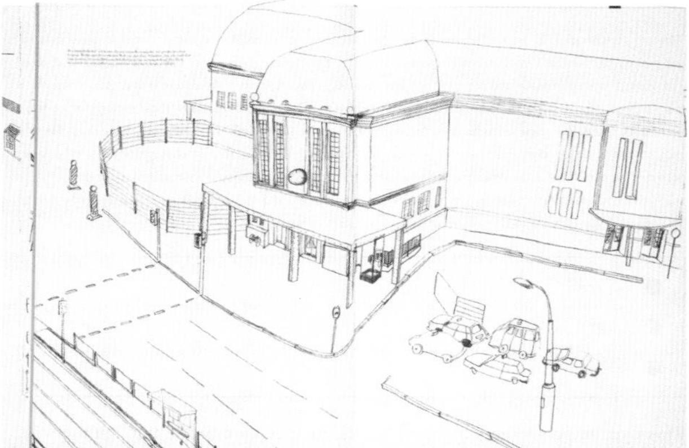
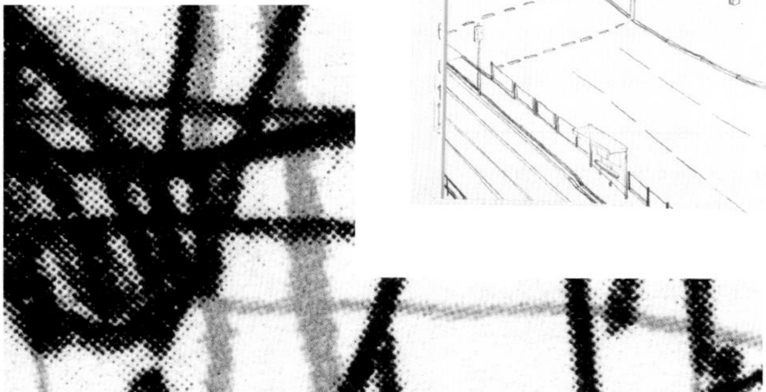

Ich kann keine Tiere zeichnen.

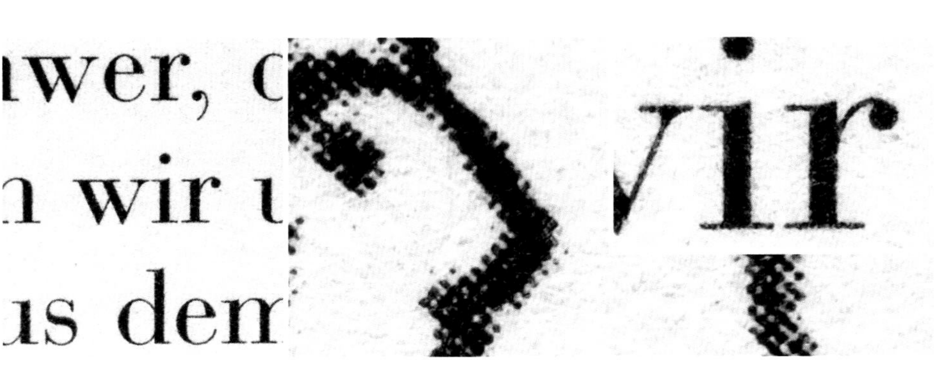
wer,
wir
dem
wir

Hardau
Claro que si, c'est comme ça, c'est la vie
Wunderli
Shahbazi
Nabi
Huber
Moser

瑞士苏黎世西北部的 Hardau 街区在 2006-2007 年间进行了大规模的房屋翻新改造，牵涉到 573 套公寓 1300 多人的生活起居。由于工期短、人口结构复杂，不能进行大规模动迁，边住边修就成了这次改造的难点。本书采访和记录了亲历这次改造过程的居民们，将家居环境、人物观点、个人照片融合起来，形成了一本多维度反映生活在这个多语言、多文化的街区中普通人的生活状态的社会学档案。本书的主体部分装订形式为无线胶平装。书的封面采用比内页高和宽一些的黑色卡板印白处理，这坚硬且高出一截的封面在书脊部分不与内页相连接，形成一个可以完全打开的“文件夹”形式。内页部分采用胶版纸单黑印文字夹插光面铜版纸单黑印刷的生活场景，书籍中间包背装形式呈现部分受访者的照片。内页文字部分大多为人物对一些具体事件的看法，排版灵活多变。封面的内侧顶部书口处标注了 14 个发表观点的居民姓氏，通过内页顶部的冒号“：”与相应的文本块形成纵向联系。||||||||||||||||||||||||||||||||||||||||

Hardau
Claro que si, c'est comme ça, c'est la vie

Hardau, a block in Northwest Zurich, Switzerland, carried out a large-scale renovation in 2006-2007, which involved more than 1,300 people living in 573 apartments. Because of the short construction period and complex population structure, large-scale relocation could not be carried out, so it was difficult to live with repairs at the same time. This book interviews the residents who have experienced the transformation process. It combines the home environment, people's viewpoints and personal photos to compile a multi-dimensional sociological archive reflecting the living conditions of ordinary people living in this multi-lingual and multi-cultural neighborhood. The main part of the book is perfect bound. The cover of the book is printed in white on black cardboard, which is taller and wider than the inside pages. It is not attached to the inside page on the spine, but forms a 'folder' that can be fully opened. The inner pages are made of offset paper and gloss coated art paper printed in black. The double-leaved part of the book presents some of the interviewees' photos. The text is mainly the ideas of the residents on some specific events, and the typesetting is quite flexible. There are the names of 14 residents who express their views marked on the top edge of the cover, connected with the colons on the top edge of the inner pages to direct the position.

◉ 荣誉奖
《 Hardau——既然如此，那就这样，这就是生活
◖ 德国 / 瑞士

This book's overall presentation beautifully captures the microcosm and the individuality of the people who inhabit the Hardau, Zurich's largest municipal housing estate. Black-and-white photography on glossy paper inserted between the text pages in smaller formats presents the residents' habitat in an unusually atmospheric fashion. Conversation sequences, with incisive typography are woven amongst the pictures, changes of paper reinforcing the contrasts. An exciting, honest and multifaceted presentation of living together – an exceptional overall design!

◎ Honorary Appreciation
《 Hardau. Claro que si, c'est comme ça, c'est la vie
◖ Germany / Switzerland
D Julia Ambroschütz, Jeannine Herrmann (Südpol)
△ Julia Ambroschütz, Jeannine Herrmann
||||| Salis Verlag, Zürich
□ 303×240×20mm
⌸ 911g
≣ 120p

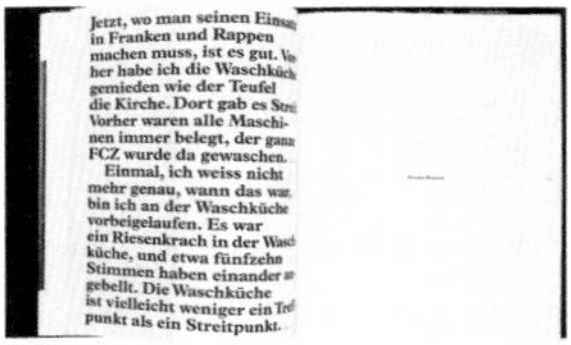
Jetzt, wo man seinen
in Franken und Rappen
machen muss, ist es gut.
her habe ich die Waschküch
gemieden wie der Teufel
die Kirche. Dort gab es
Vorher waren alle Maschi-
nen immer belegt, der
FCZ wurde da gewaschen.
Einmal, ich weiss nicht
mehr genau, wann das
bin ich an der Waschküche
vorbeigelaufen. Es war
ein Riesenkrach in der
küche, und etwa fünfzehn
Stimmen haben einander
gebellt. Die Waschküche
ist vielleicht weniger ein
punkt als ein Streitpunkt.

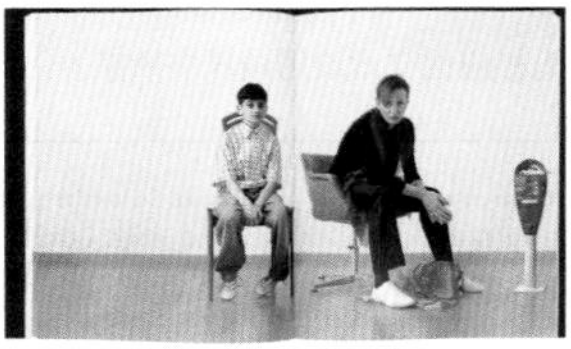

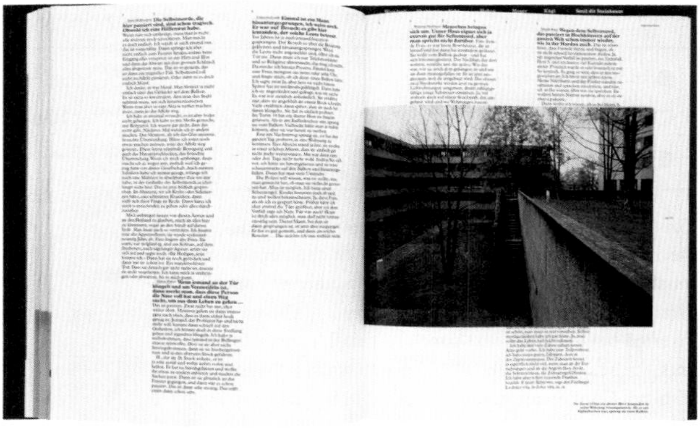

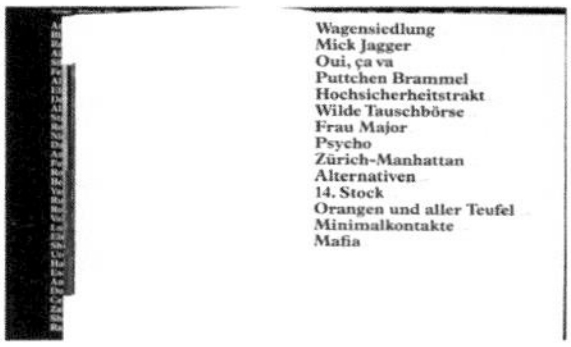

Wagensiedlung
Mick Jagger
Oui, ça va
Puttchen Brammel
Hochsicherheitstrakt
Wilde Tauschbörse
Frau Major
Psycho
Zürich-Manhattan
Alternativen
14. Stock
Orangen und aller Teufel
Minimalkontakte
Mafia

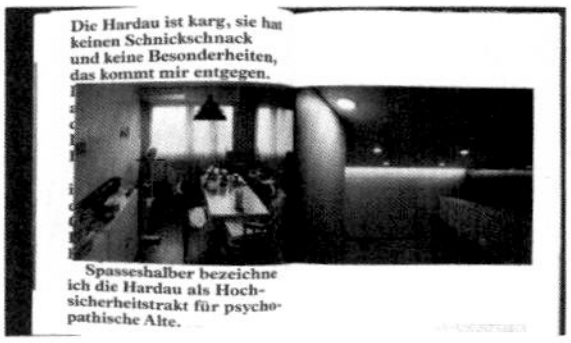

Die Hardau ist karg, sie hat keinen Schnickschnack und keine Besonderheiten, das kommt mir entgegen.

Spasseshalber bezeichne ich die Hardau als Hochsicherheitstrakt für psychopathische Alte.

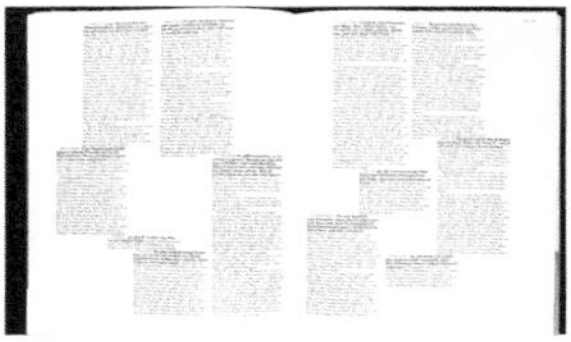

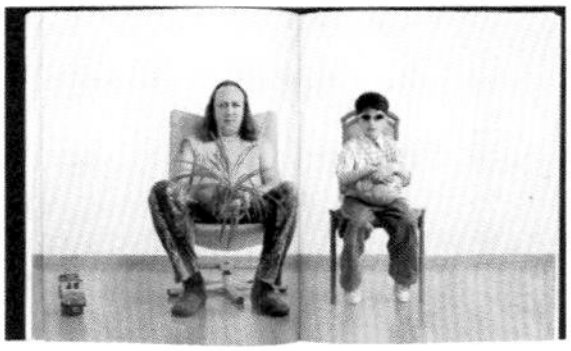

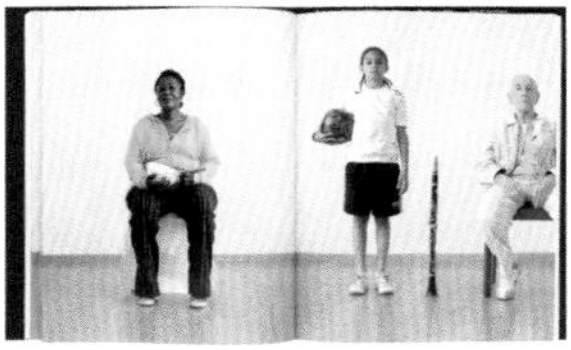

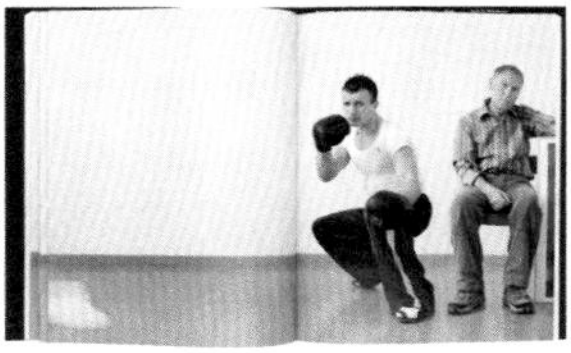

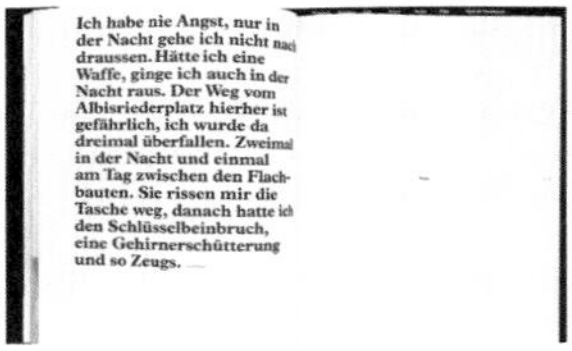

Ich habe nie Angst, nur in der Nacht gehe ich nicht nach draussen. Hätte ich eine Waffe, ginge ich auch in der Nacht raus. Der Weg vom Albisriederplatz hierher ist gefährlich, ich wurde da dreimal überfallen. Zweimal in der Nacht und einmal am Tag zwischen den Flachbauten. Sie rissen mir die Tasche weg, danach hatte ich den Schlüsselbeinbruch, eine Gehirnerschütterung und so Zeugs.

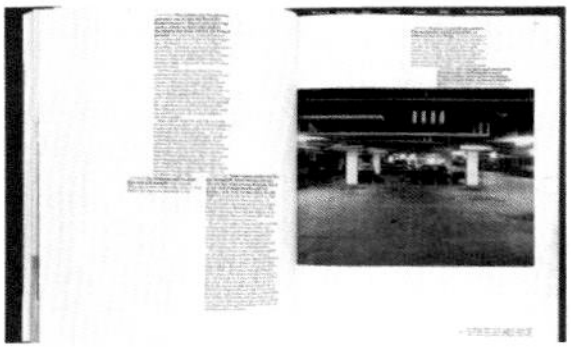

ich in

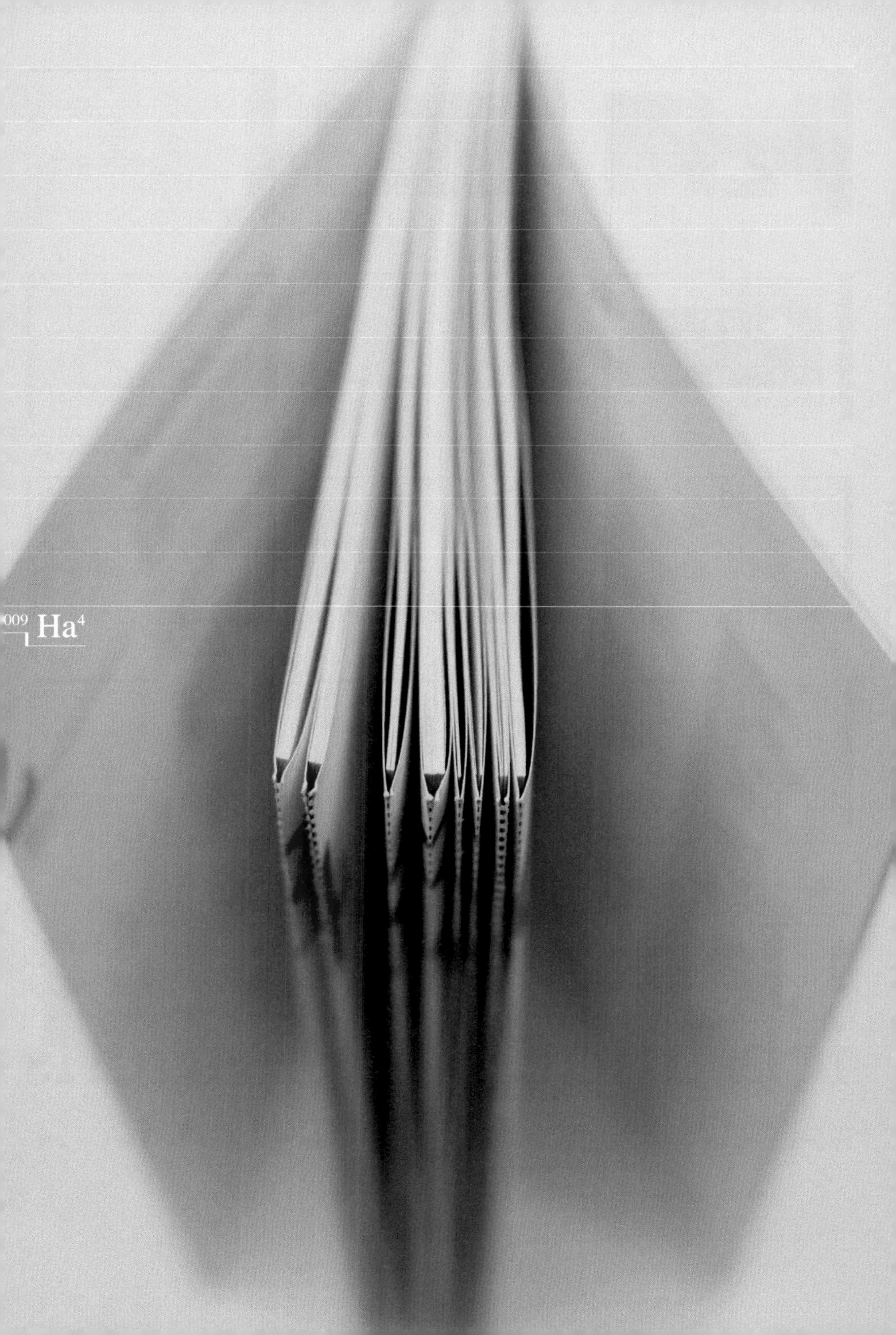
009 Ha4

这是一本蒙德里安基金会 2007 年的年度报告，报告体现了他们在支持推动文化、艺术、设计和遗产等项目上的财政支持、建议和实际执行效果。书籍装订采用无线胶装，胶体薄而软，完全不妨碍本书可以完全展开。封面采用滑面白卡纸印单色黑，字体和排版充满时尚感。内页正文分为十大部分，分别用于陈述基金会在重点项目、艺术设计、文化遗产、跨学科活动、国际项目、组织架构、工作程序、问责制度、财政审批等方面的报告内容。每一部分均由超过正文页面 5mm 的包背装纸张将具体内容隐藏其间，需要阅读者自己裁开，体现资料的保密性。每一部分设置一种专色，配合主文字的黑色，在胶版纸上双色印刷，配合版式的丰富细节和渐变效果，整体看起来十分时尚。右翻口处除了每一部分的数字外，也包含每一部分的小章节目录，提供了很好的导航作用。作为一本资助和财政报告，本书提供了艺术的解决方案。

Mondriaan
Stichting
Jaarverslag 2007

This is the Mondrian Foundation's 2007 annual report, which reflects their financial support, recommendations and implementation results in supporting projects to promote culture, art, design and heritage. The book is perfect bound with a thin and soft layer of glue, which does not prevent the book from being fully opened. The cover is printed in monochrome black on glossy white cardboard. The font and typesetting are full of fashion sense. The main body of the text is divided into ten parts, aimed to state the contents of the foundation's reports on key projects, art design, cultural heritage, interdisciplinary activities, international projects, organizational structure, working procedures, accountability system, financial approval and so on. Each part is sealed by the double-leaved pages that exceed 5mm of the text pages. The readers need to cut them out by themselves. This design reflects the confidentiality of the information. Each part is equipped with a special color, matching the key color of black. Double-color printing on offset paper, with the rich details and gradient effect of the layout, the overall look is very fashionable. As a funding and financial report, this book provides an artistic solution for Mondrian Foundation.

- 荣誉奖
- 蒙德里安基金会 2007 年度报告
- 荷兰

An annual report, making its necessary information accessible with great ease. The mixture of typography allows for refreshing experiments in small as well as large font sizes. The interplay of colours, changing over the course of soft transitions, forms a harmonious entity. The structuring of the chapters is clearly accentuated by initially sealed pages, which have to be opened at the perforation, sticking out over the edge. The visual aspect of the book is soft and flowing, and affords the typographical and colour-coded elements much space to unfold.

- Honorary Appreciation
- Mondriaan Stichting. Jaarverslag 2007
- The Netherlands
- Ingeborg Scheffers
- Mondriaan Stichting, Amsterdam
- 275 × 211 × 15mm
- 723g
- 220p

Mondriaan Stichting
Jaarverslag 2007

Voorwoord

Inhoud
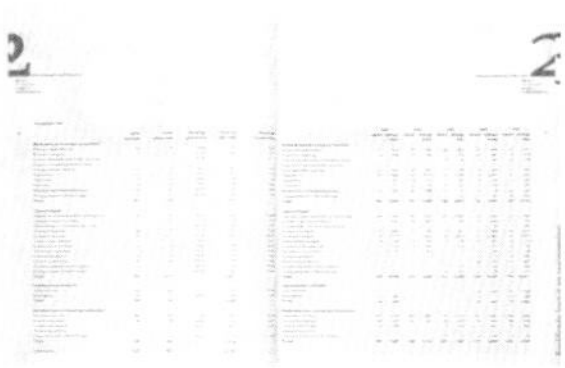
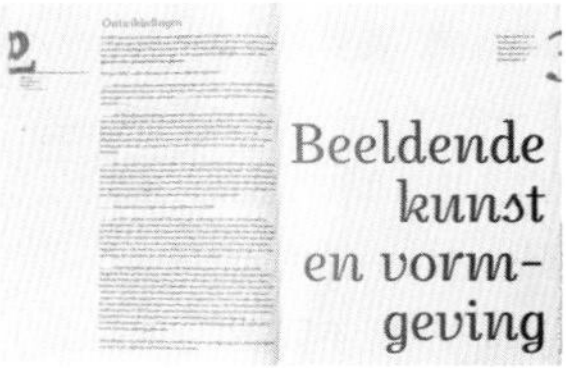
Beeldende kunst en vorm- geving
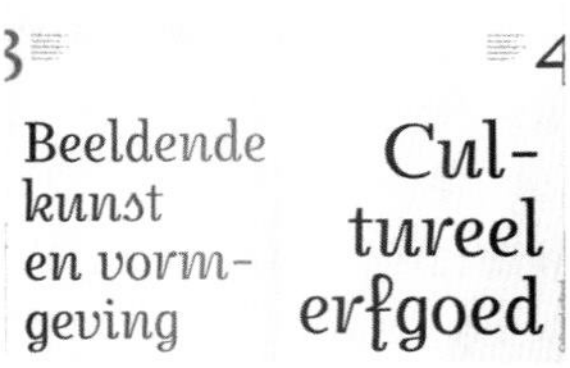
Beeldende kunst en vorm- geving
Cul- tureel erfgoed

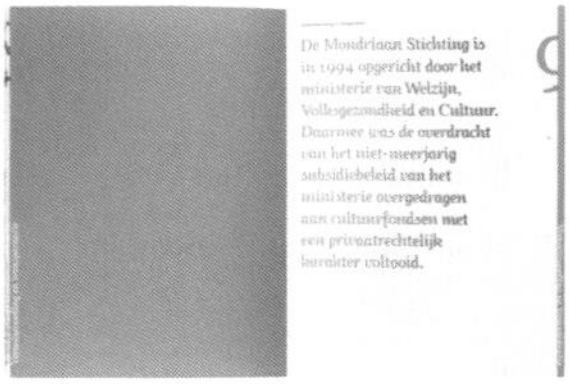
De Mondriaan Stichting is in 1994 opgericht door het ministerie van Welzijn, Volksgezondheid en Cultuur. Daarmee was de overdracht van het niet-meerjarig subsidiebeleid van het ministerie overgedragen aan cultuurfondsen met een privaatrechtelijk karakter voltooid.
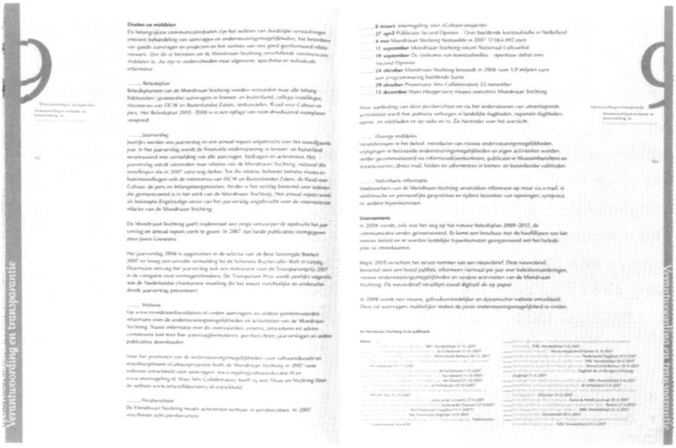
Verantwoording en transparantie

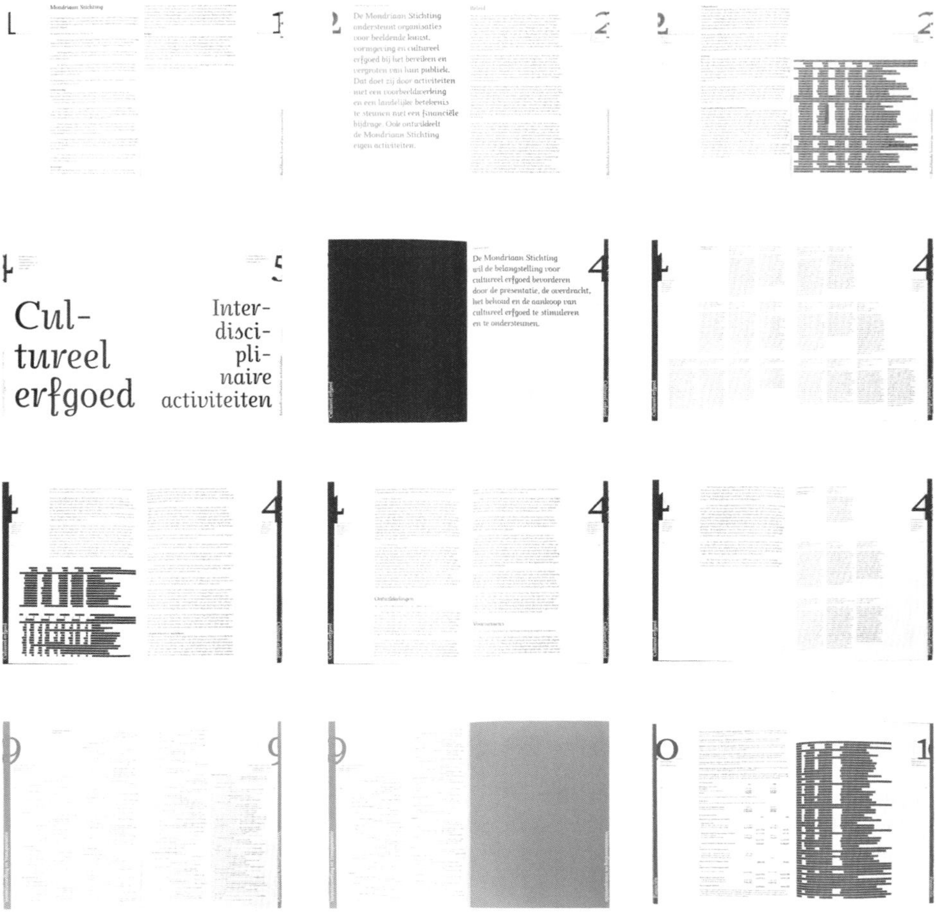
Mondriaan Stichting
De Mondriaan Stichting ondersteunt organisaties voor beeldende kunst, vormgeving en cultureel erfgoed bij het bereiken en vergroten van hun publiek. Dat doet zij door activiteiten met een voorbeeldwerking en een landelijke betekenis te steunen met een financiële bijdrage. Ook ontwikkelt de Mondriaan Stichting eigen activiteiten.
Cul-tureel erfgoed
Inter-disci-pli-naire activiteiten
De Mondriaan Stichting wil de belangstelling voor cultureel erfgoed bevorderen door de presentatie, de overdracht, het behoud en de aankoop van cultureel erfgoed te stimuleren en te ondersteunen.

ria v
tureel erfgoe

Ha5
Angels of Revenge
50 State Senator Ernie Chambers
On September 14, 2007, a United
named Ernie Chambers sued God.
"Anybody can file a lawsuit
–even God," says Chambers.
Chambers, a fiery senator fr
his official lawsuit in Douglas Cou
suit acknowledges that God goes
names and titles, and recognizes
is omnipresent.
Chambers has tried to
to little effect. "Plaintiff

这是德国艺术家克里斯蒂安·扬科夫斯基（Christian Jankowski）研究弗兰肯斯坦（原指科学怪人，这里指人们心中创造出的怪物）的一本书。通过挖掘人心里的“复仇天使”、研究怪物、分析其理论根基，解释“怪物”就是个人恐惧心理的外化，并研究如何面对恐惧，如何运用这些“怪物”。从外观上看，本书是一本不起眼的小书，装订方式为锁线胶平装，封面白卡纸红黄黑三色印刷，覆亮膜。内页主要为胶版纸红黑双色印刷，每两印张包覆四页单面铜版纸四色印刷，用于展示彩图。上下书口及右侧翻口刷红色，与封面及内页呼应。英德双语排版，但图片没有重复使用。本书没有目录，设计者可能是希望读者逐页阅读，逐步深入理解本书核心。部分章节编号错乱的设置，可能是为了呼应其中一个章节“不可思议”的错误，制造一种恐怖感。

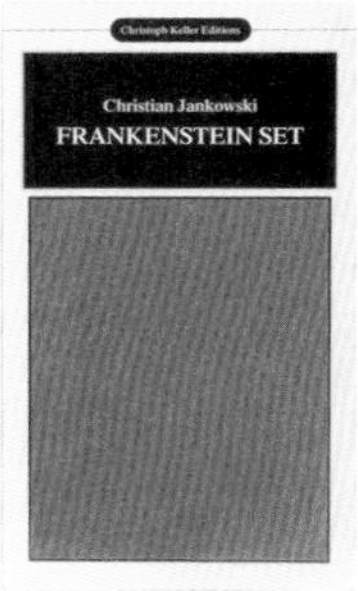
Christoph Keller Editions
Christian Jankowski
FRANKENSTEIN SET

The book is a study of Frankenstein by the German artist Christian Jankowski. By digging into the 'avenging angel' in human mind, studying the monster and its theoretical foundation, the author explains that 'monster' is the externalization of personal fear and suggests how to face it and how to apply it. The book is an unimpressive one in appearance, perfect bound with thread sewing, printed in red, yellow and black on the laminated cover. The inside pages are mainly printed in red and black on offset paper, and each two sheets are covered with 4 pages of single-side coated paper, which is used to display color pictures. The three edges of the book are brushed in red to echo the cover and inside pages. The text is typeset in English and German, but the pictures are not repetitive. There is no table of contents in this book. The designer may want the reader to read page by page and gradually understand the core of the book. Some chapter numbers are set up erratically in order to made an 'incredible' error and create a sense of terror.

◉ 荣誉奖
« 弗兰肯斯坦集合
◖ 瑞士

◎ Honorary Appreciation
« Frankenstein Set
◖ Switzerland
D Geoff Han
△ Christian Jankowski
▥ Christoph Keller Editions / JRP Ringier Kunstverlag, Zürich
□ 180×104×23mm
⌸ 345g
≣ 440p

A book with the power to split a jury on beautiful books like no other. Its features are not opulent, the paper thin, the binding – well, it just about holds the pages together. The whole thing is set in Times New Roman bold. Colour pictures with poor resolution of severed body-parts round off the scene. Everything is nicely vulgar, predominantly in red, but as such appropriate to the content. A strong concept. We don't like it! We do like it !!

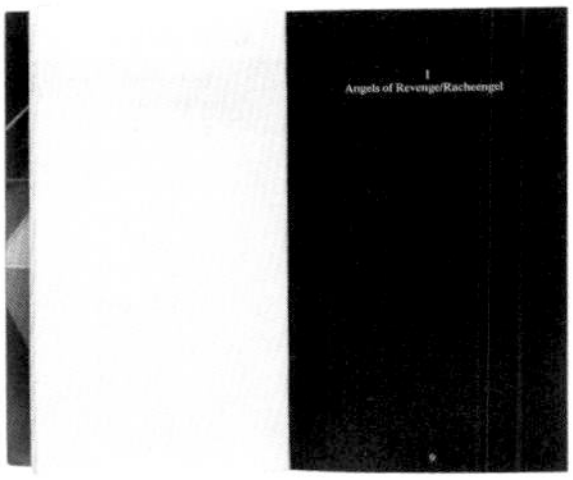
1
Angels of Revenge/Racheengel

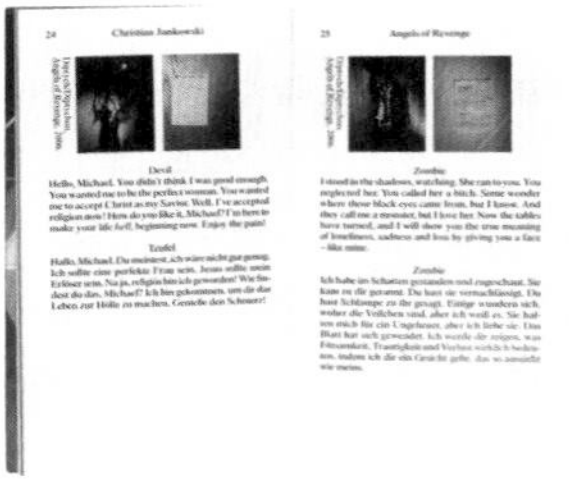

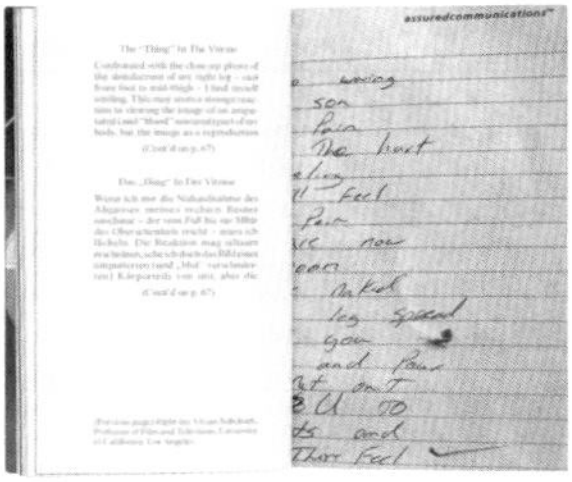

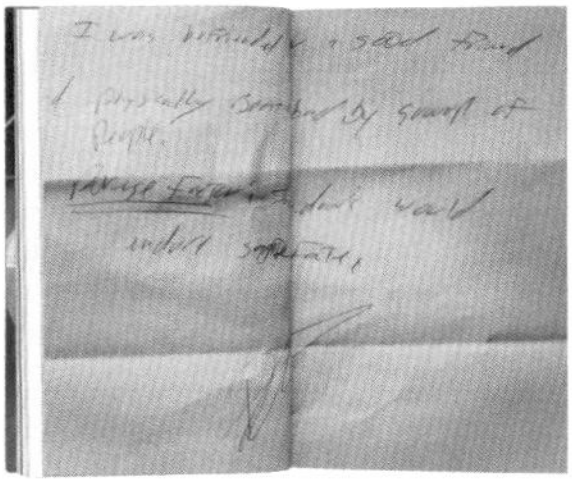

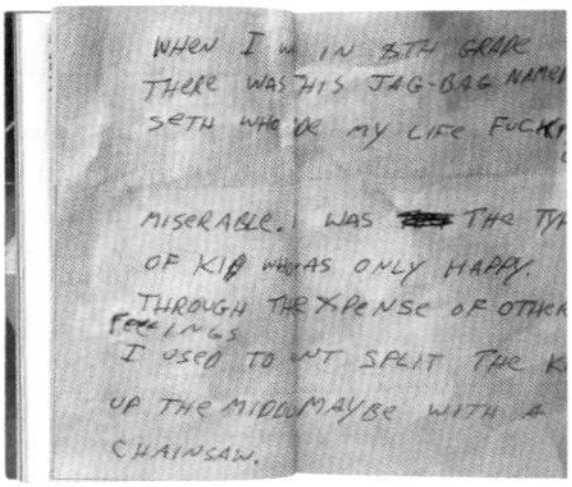

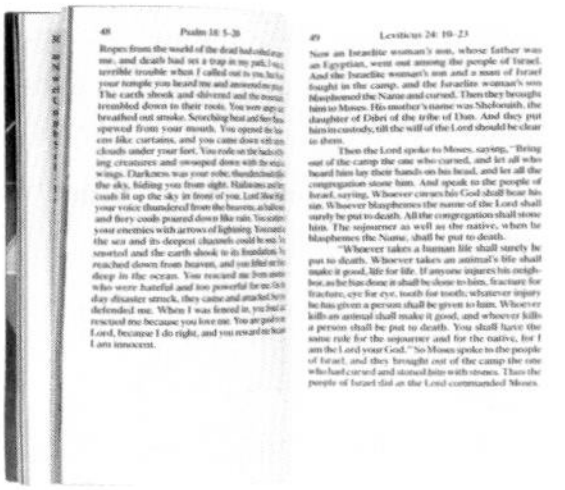

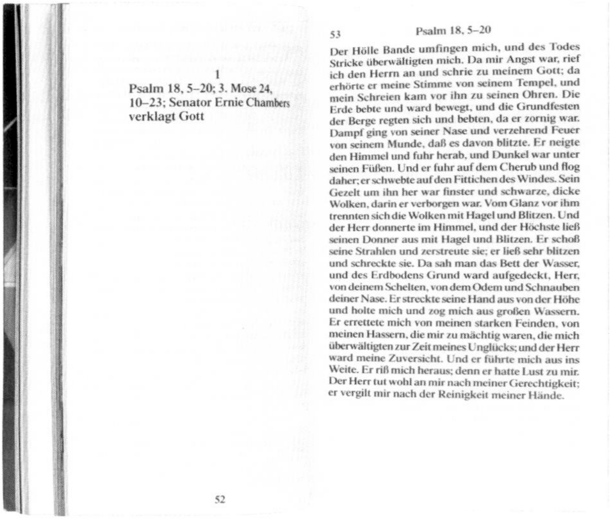

1
Psalm 18, 5–20; 3. Mose 24, 10–23; Senator Ernie Chambers verklagt Gott

52

53 Psalm 18, 5–20

Der Hölle Bande umfingen mich, und des Todes Stricke überwältigten mich. Da mir Angst war, rief ich den Herrn an und schrie zu meinem Gott; da erhörte er meine Stimme von seinem Tempel, und mein Schreien kam vor ihn zu seinen Ohren. Die Erde bebte und ward bewegt, und die Grundfesten der Berge regten sich und bebten, da er zornig war. Dampf ging von seiner Nase und verzehrend Feuer von seinem Munde, daß es davon blitzte. Er neigte den Himmel und fuhr herab, und Dunkel war unter seinen Füßen. Und er fuhr auf dem Cherub und flog daher; er schwebte auf den Fittichen des Windes. Sein Gezelt um ihn her war finster und schwarze, dicke Wolken, darin er verborgen war. Vom Glanz vor ihm trennten sich die Wolken mit Hagel und Blitzen. Und der Herr donnerte im Himmel, und der Höchste ließ seinen Donner aus mit Hagel und Blitzen. Er schoß seine Strahlen und zerstreute sie; er ließ sehr blitzen und schreckte sie. Da sah man das Bett der Wasser, und des Erdbodens Grund ward aufgedeckt, Herr, von deinem Schelten, von dem Odem und Schnauben deiner Nase. Er streckte seine Hand aus von der Höhe und holte mich und zog mich aus großen Wassern. Er errettete mich von meinen starken Feinden, von meinen Hassern, die mir zu mächtig waren, die mich überwältigten zur Zeit meines Unglücks; und der Herr ward meine Zuversicht. Und er führte mich aus ins Weite. Er riß mich heraus; denn er hatte Lust zu mir. Der Herr tut wohl an mir nach meiner Gerechtigkeit; er vergilt mir nach der Reinigkeit meiner Hände.

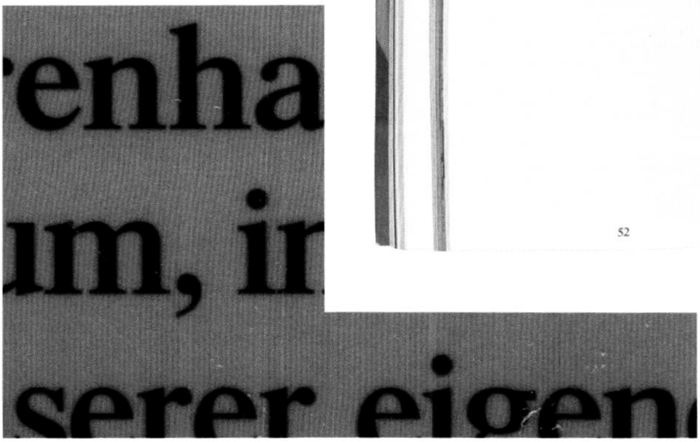

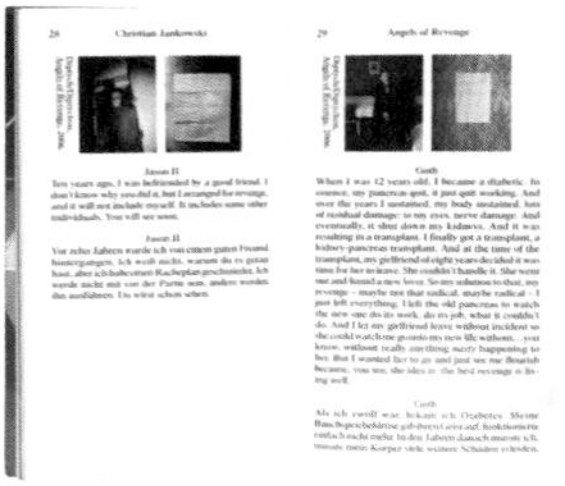
Christian Jankowski
Angels of Revenge

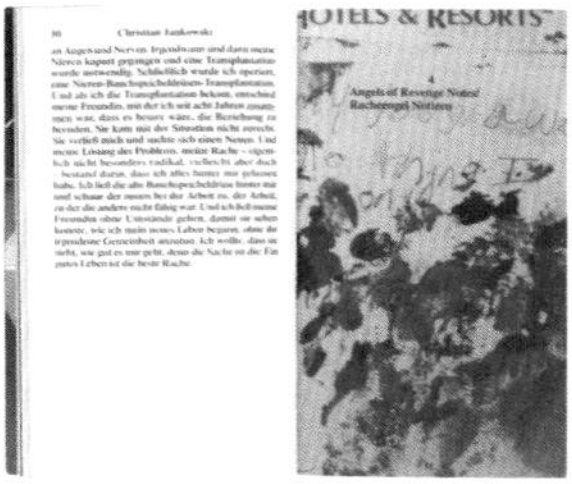
Christian Jankowski
HOTELS & RESORTS
Angels of Revenge Notes/
Racheengel Notizen

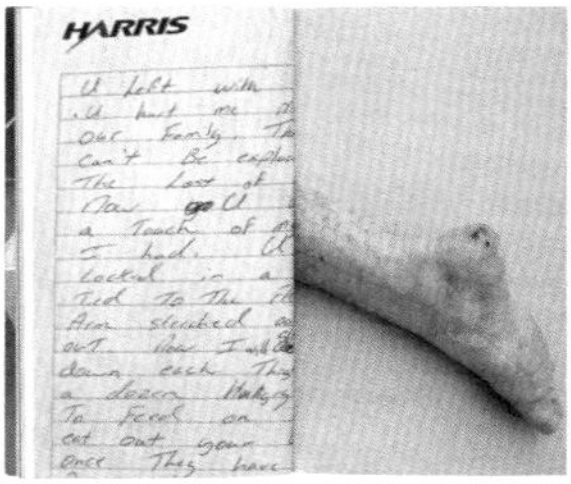
HARRIS

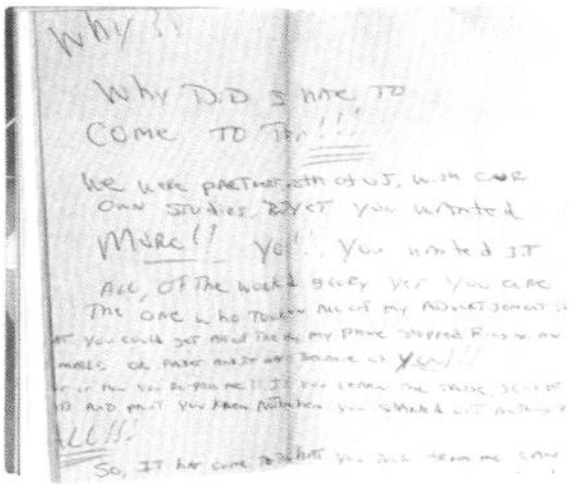

WYNDHA
Psalm 18: 5–20; Leviticus 24:
10–23; State Senator Ernie
Chambers Sues God

Angels of Revenge/Racheengel
2006, Video, 11:06 minutes/
2006, Video, 11:06 minuten
By/Von Christian Jankowski
VENGE

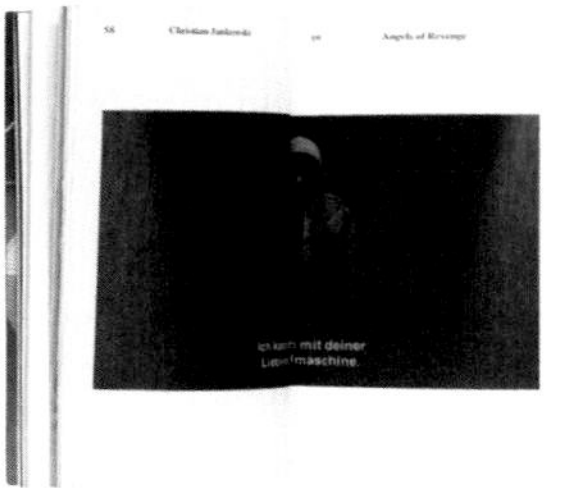
Christian Jankowski
Angels of Revenge

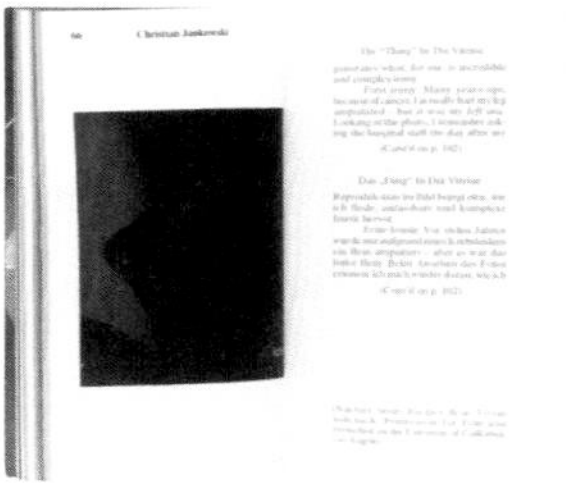
Christian Jankowski

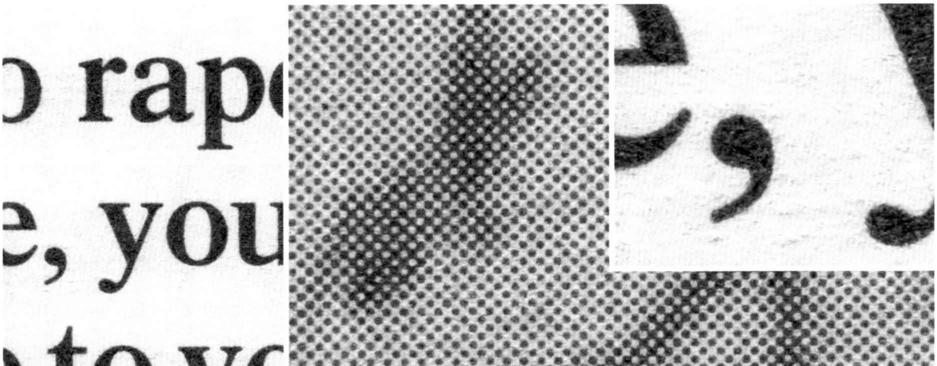

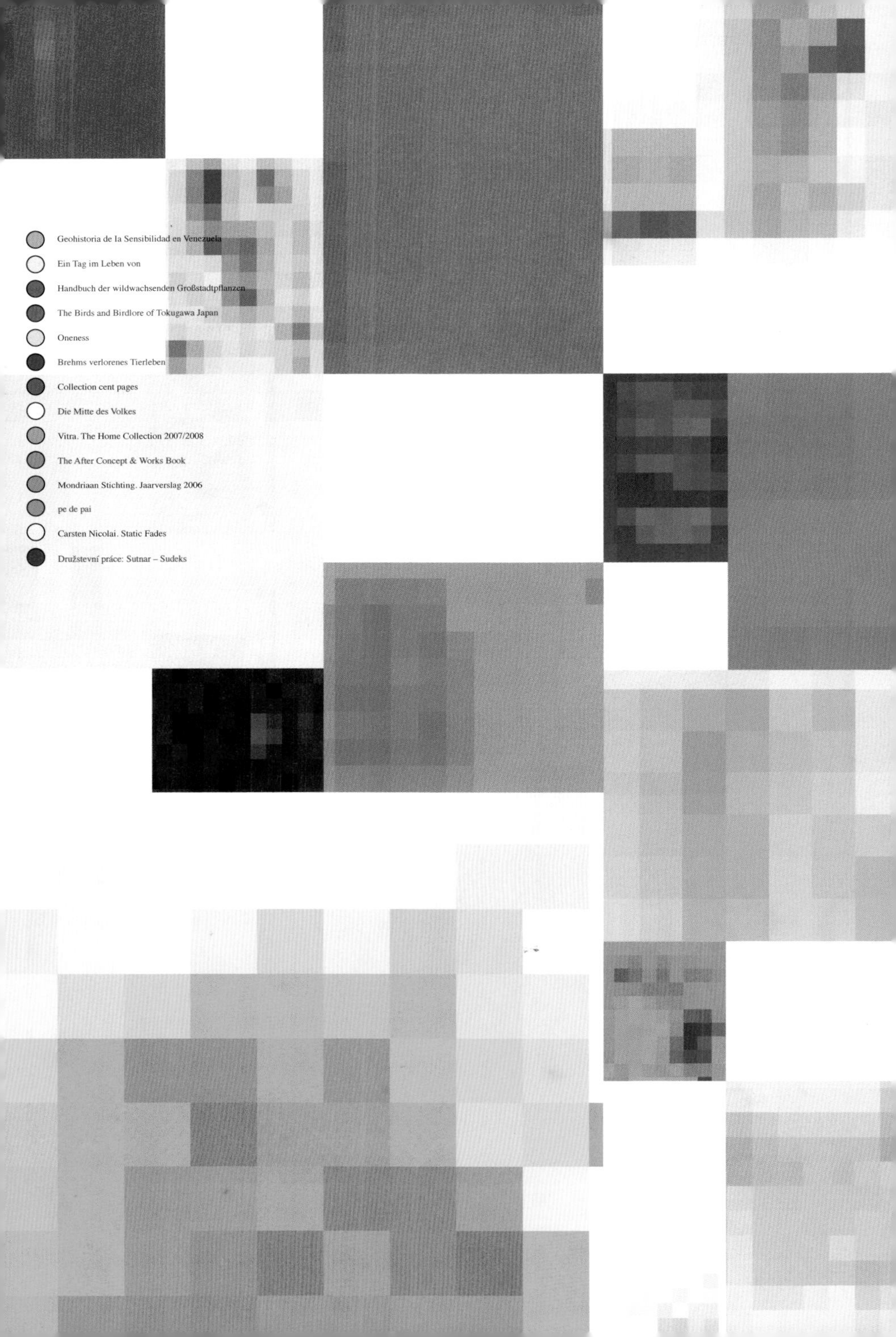
Geohistoria de la Sensibilidad en Venezuela
Ein Tag im Leben von
Handbuch der wildwachsenden Großstadtpflanzen
The Birds and Birdlore of Tokugawa Japan
Oneness
Brehms verlorenes Tierleben
Collection cent pages
Die Mitte des Volkes
Vitra. The Home Collection 2007/2008
The After Concept & Works Book
Mondriaan Stichting. Jaarverslag 2006
pe de pai
Carsten Nicolai. Static Fades
Družstevní práce: Sutnar – Sudeks

2008

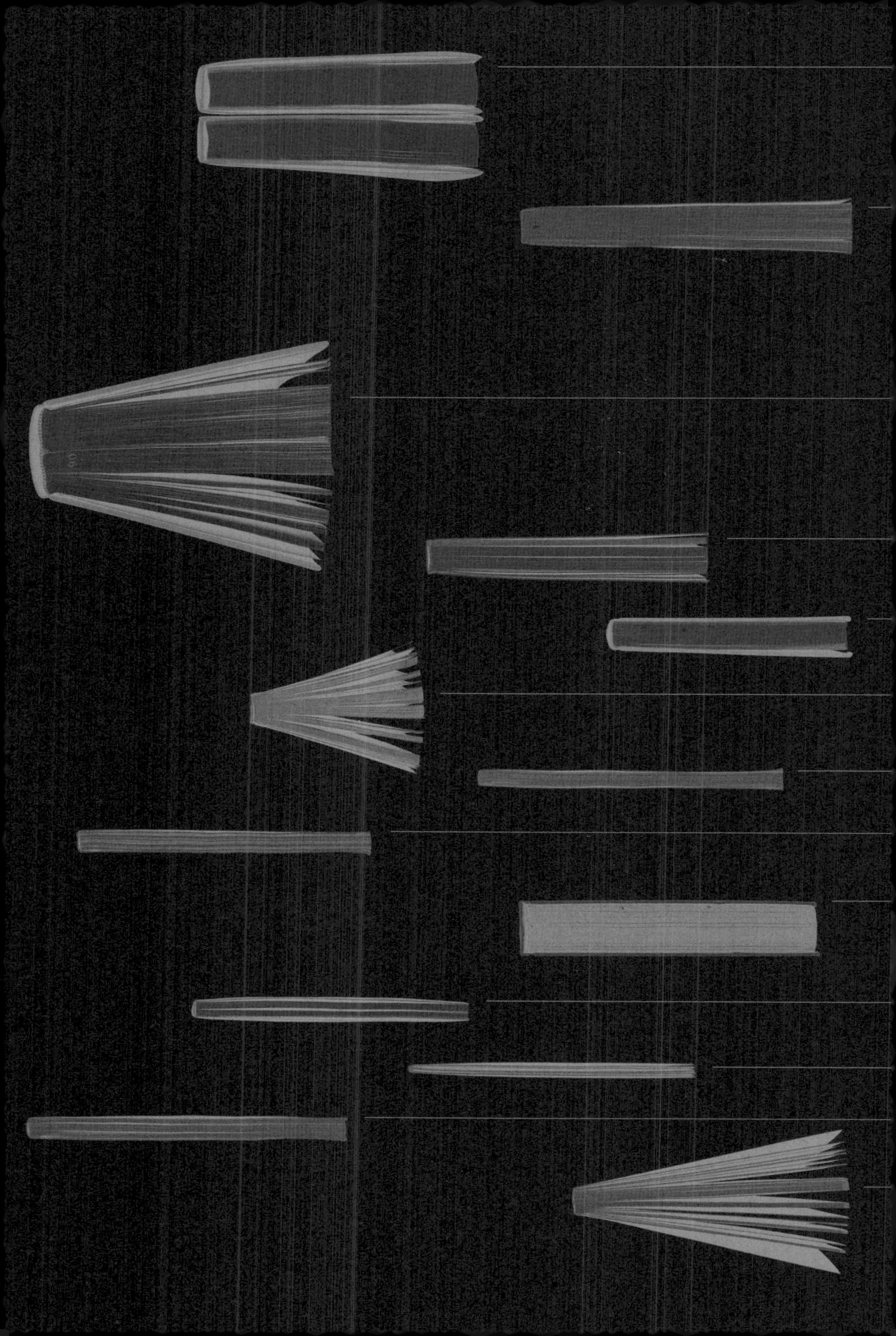

2008 Gl

1037 
GEOHISTORIA
DE LA SENSIBILIDAD EN
VENEZUELA
Tomo 1
GEOHISTORIA
DE LA SENSIBILIDAD EN
VENEZUELA

2008 Gm

1045 
salis
Ein Tag im Leben von

2008 Sm[2]

1051 

2008 Bm[1]

1057 

2008 Bm[2]

1063 
BREHMS VERLORENES TIERLEBEN

2008 Bm[3]

1069 

2008 Bm[4]

1075 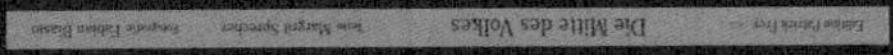
Die Mitte des Volkes

2008 Bm[5]

1081 

2008 Ha[1]

1085 

2008 Ha[2]

1089 

2008 Ha[3]

1093 

2008 Ha[4]

1097 

2008 Ha[5]

1101 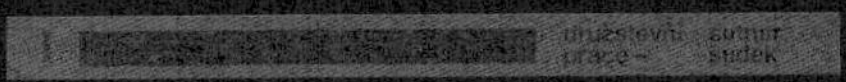

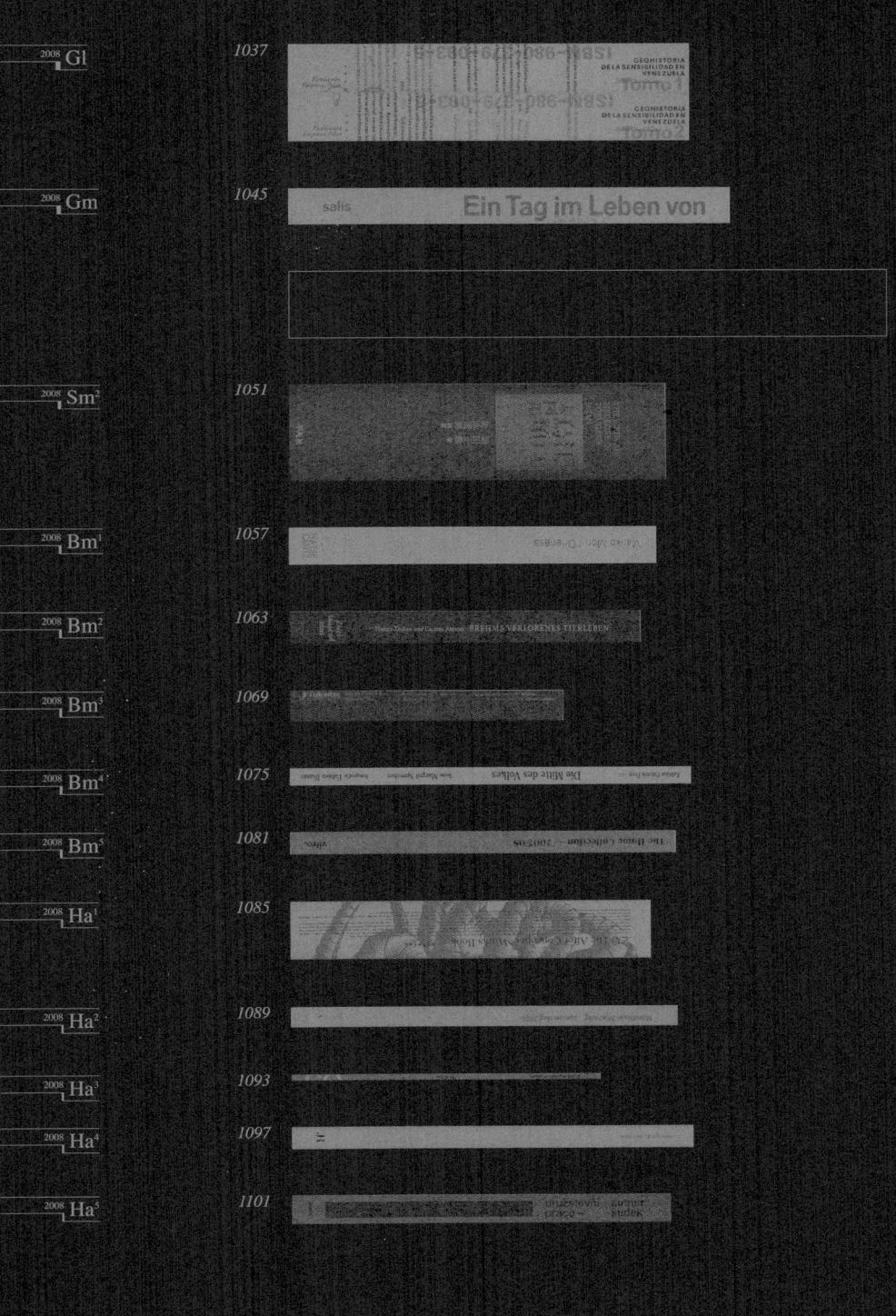

O'KELLY
HY MANY
DILLON
Trim
Athenry
Athlone
Galway
Dublin

que se había ido al fondo estaba en línea perpendicular con el quebrad
borde del camino y tenía una extensión aproximada de trescientas
yardas por cincuenta, bajo el antiguo nivel» (154).

982 gewonnen. Peterka unternimmt pro Jahr rund 40 Gas- und Heissluftballonfahrten. D
hsenen Kindern wohnt mit seiner Frau in Rapperswil SG, wo er ein Radio- und Ferns

e verlassen, sondern muss sich | tem Baumwolltuch besteht und über- | hatte, trafen auch un

[redacted]

[1] Länge: 65 cm Gewicht: 1.2–2.3 kg Lebenserwartung: 3–6 Jahre Ehemalig
tungsgebiet: Osten Australiens Zeitpunkt des Aussterbens: 1910 Ursache
sterbens: Zerstörung des Lebensraums; Beute eingeschleppter Füchse und H

sendungen von SF1 und 2 konsumiert ein SVPler mit dem grimassierend
eines Kindes, das essen muss, was auf den Tisch kommt.
Bei offiziellen SVP-Veranstaltungen wird der Medienpulk in eine Art Z

sinds 1981 brengt Vitra op zijn fabrieksterrein in Weil am Rhein een geheel nieuwe opvatting over "bedrijfsarchitectuur" in
praktijk. Het idee was architectonische ontwerpen bewust met elkaar te laten contrasteren en de locatie zo een bijzondere
een geheel eigen identiteit te geven. Geheel in stijl met dit idee liet Vitra ieder gebouw op het bedrijfsterrein door een
dere architect ontwerpen.

自己知道、为什么好广告公司绝对不代表就能舒
。短命的广告公司和其它因先出头而短命的新

those engaged in the advertisi
excellent advertising company
Short-lived advertising companie

overzicht van de aanvragen per discipline. Verder valt op dat het
in 40% van de gevallen gaat om multi- of interdisciplinaire projecten (vaak vanuit de podiumkunsten). De interdisciplinaire
benadering van de ondersteuningsmogelijkheid sluit aan bij dit
hoge percentage disciplineoverstijgende projecten.

pai escad

The installation realistic explores the phenomenon of nois
ing to the theories of Takashi Ikegami on active mut
self-reproducing networks. Noise is a complex, randomly

vané čtenářské obce. Motivace německého družstevního nakladatelského
í byla však naprosto odlišná od snah *dp* – většinou jim nešlo o vytváření nov
jejich ediční strategie sledovala výhradně pasivní konzumní cíle a jejich knihy

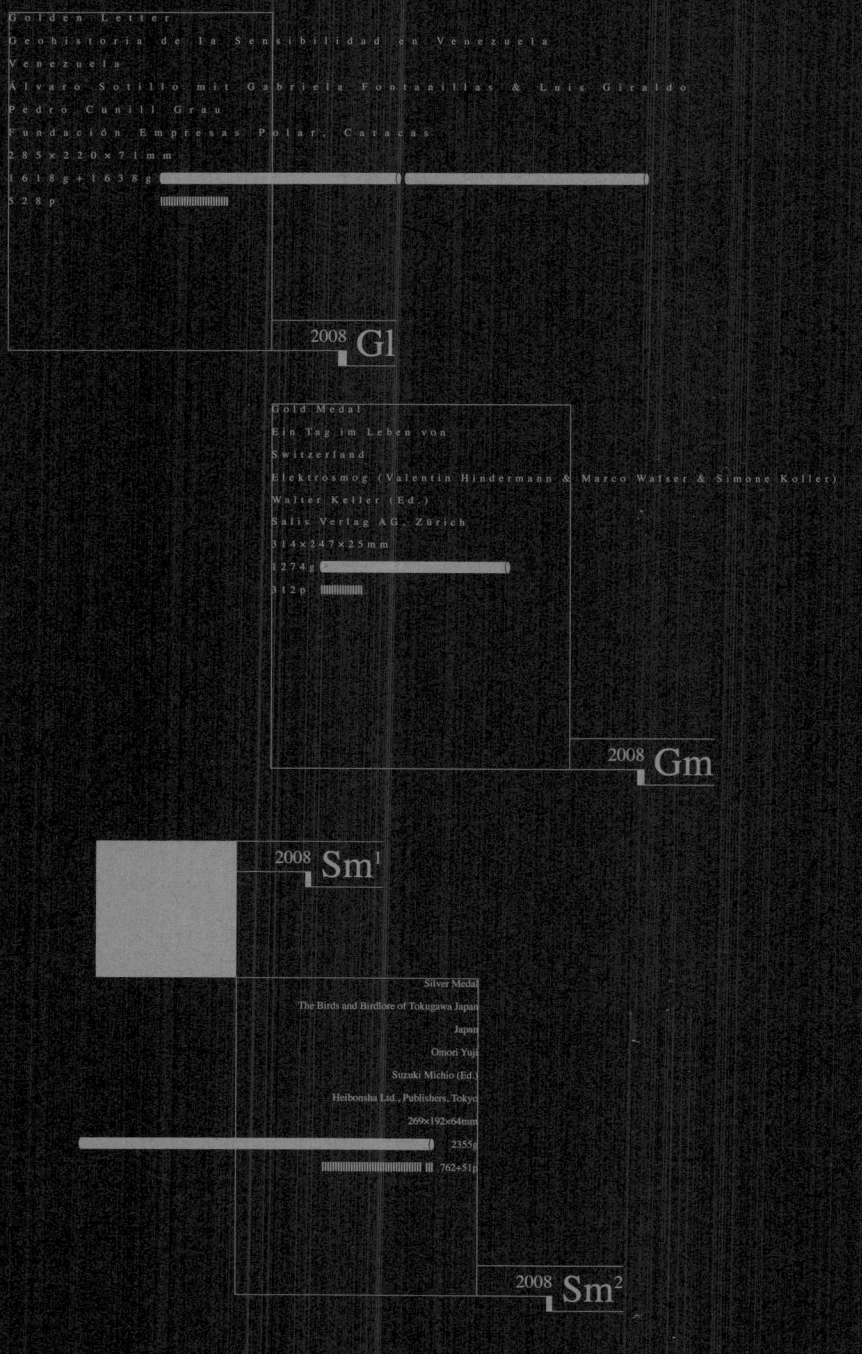

Golden Letter

Geohistoria de la Sensibilidad en Venezuela

Venezuela

Álvaro Sotillo mit Gabriela Fontanillas & Luis Giraldo

Pedro Cunill Grau

Fundación Empresas Polar, Caracas

285×220×71mm

1618g+1638g

528p

2008 Gl

Gold Medal

Ein Tag im Leben von

Switzerland

Elektrosmog (Valentin Hindermann & Marco Walser & Simone Koller)

Walter Keller (Ed.)

Salis Verlag AG, Zürich

314×247×25mm

1274g

312p

2008 Gm

2008 Sm[1]

Silver Medal

The Birds and Birdlore of Tokugawa Japan

Japan

Omori Yuji

Suzuki Michio (Ed.)

Heibonsha Ltd., Publishers, Tokyo

269×192×64mm

2355g

762+51p

2008 Sm[2]

2008 Ha^3

Honorary Appreciation
pe de pai
Portugal
Bernardo P. Carvalho (Planeta Tangerina)
Isabel Martins, Bernardo P. Carvalho
Planeta Tangerina, São Pedro do Estoril
220×198×4mm
145g
32p

2008 Ha^1

Honorary Appreciation
The After Concept & Works Book
China
Wang Chengfu & Geng Geng
Sun Naiqiang, Wang Lei, Wang Chengfu, Shi Changsheng
Tianjin Yangliuqing Fine Arts Press, Tianjin
256×216×40mm
1710g
456p

2008 Ha^4

Honorary Appreciation
Carsten Nicolai. Static Fades
Switzerland
Gavillet & Rust
Dorothea Strauss, Haus Konstruktiv, Zürich (Ed.)
JRP Ringier Kunstverlag AG, Zürich
286×238×13mm
892g
157p

2008 Ha^2

Honorary Appreciation
Mondriaan Stichting. Jaarverslag 2006
The Netherlands
Joost Grootens
Mondriaan Stichting, Amsterdam
275×215×13mm
587g
192p

2008 Ha^5

Honorary Appreciation
Družstevní práce : Sutnar – Sudeks
Czech Republic
Štěpán Malovec & Martin Odehnal
Lucie Vlčková (Ed.)
Uměleckoprůmyslové museum v Praze, Prague
270×205×20mm
1151g
250p

2008 Bm^2

Bronze Medal
Brehms verlorenes Tierleben
Germany
Hanna Zeckau & Carsten Aermes
Hanna Zeckau, Carsten Aermes
Zweitausendeins, Frankfurt am Main
250×175×20mm
734g
257p

2008 Bm^4

Bronze Medal
Die Mitte des Volkes
Switzerland
Prill & Vieceli
Fabian Biasio, Margrit Sprecher
Edition Patrick Frey, Zürich
285×211×10mm
593g
187p

2008 Bm^3

Bronze Medal
Collection cent pages
France
sp Millot
Éditions cent pages, Grenoble
195×125×20mm
392g
512p

2008 Bm^1

Bronze Medal
Oneness
Germany
Hester Fell
Mariko Mori
Hatje Cantz Verlag, Ostfildern
261×209×25mm
1337g
304p

2008 Bm^5

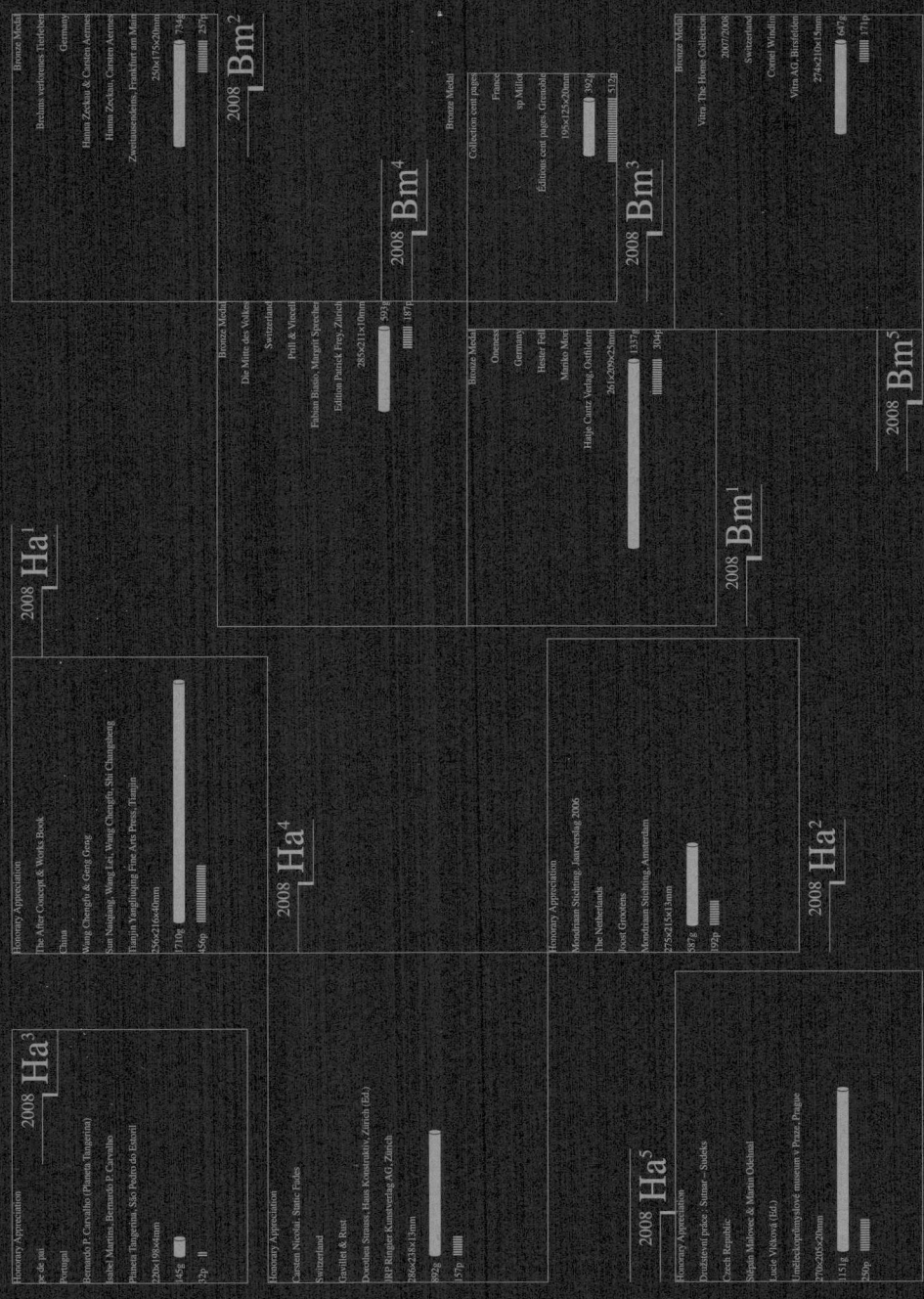

Bronze Medal
Vitra. The Home Collection
2007/2008
Switzerland
Cornel Windlin
Vitra AG, Birsfelden
274×210×15mm
647g
171p

The International Jury

Chin-Lien Chen (The Netherlands)

Prof. Markus Dreßen (Germany)

Prof. Heike Grebin (Germany)

Geir Henriksen (Norway)

Jurij Kocbek (Slovenia)

Angelika Sagner (Germany)

Gilmar Wendt (England)

Designer

Álvaro Sotillo mit Gabriela Fontanillas & Luis Giraldo
Elektrosmog (Valentin Hindermann & Marco Walser & Simone Koller)
Helmut Völter
Omori Yuji
Hester Fell
Hanna Zeckau & Carsten Aermes
sp Millot
Prill & Vieceli
Cornel Windlin
Wang Chengfu & Geng Geng
Joost Grootens
Bernardo P. Carvalho (Planeta Tangerina)
Gavillet & Rust
Stěpán Malovec & Martin Odehnal

2008

Country / Region

Switzerland 4
Germany 3
Venezuela 1
Japan 1
France 1
China 1
The Netherlands 1
Portugal 1
Czech Republic 1

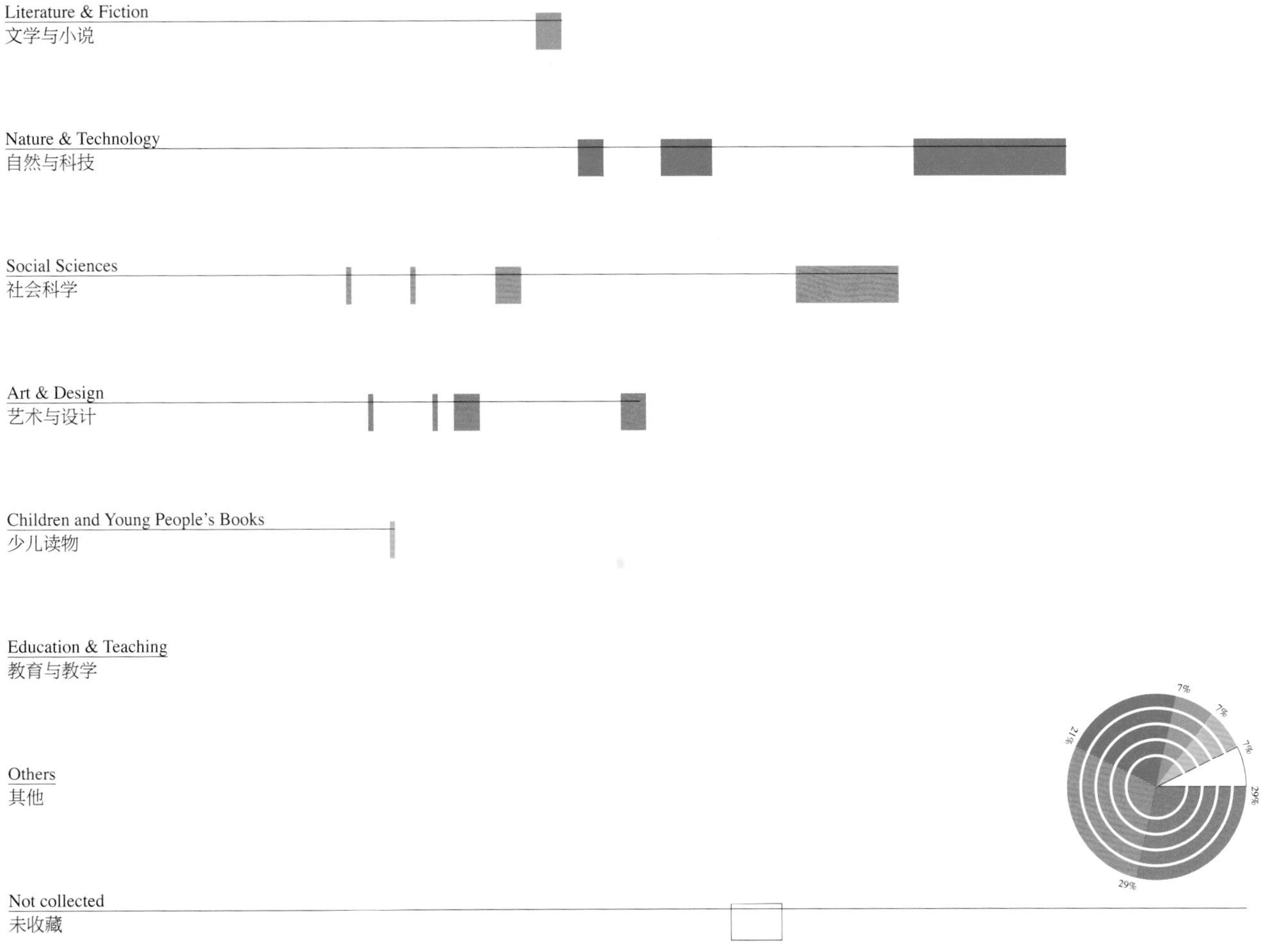
Literature & Fiction
文学与小说
Nature & Technology
自然与科技
Social Sciences
社会科学
Art & Design
艺术与设计
Children and Young People's Books
少儿读物
Education & Teaching
教育与教学
Others
其他
Not collected
未收藏
7%
7%
7%
29%
29%
21%

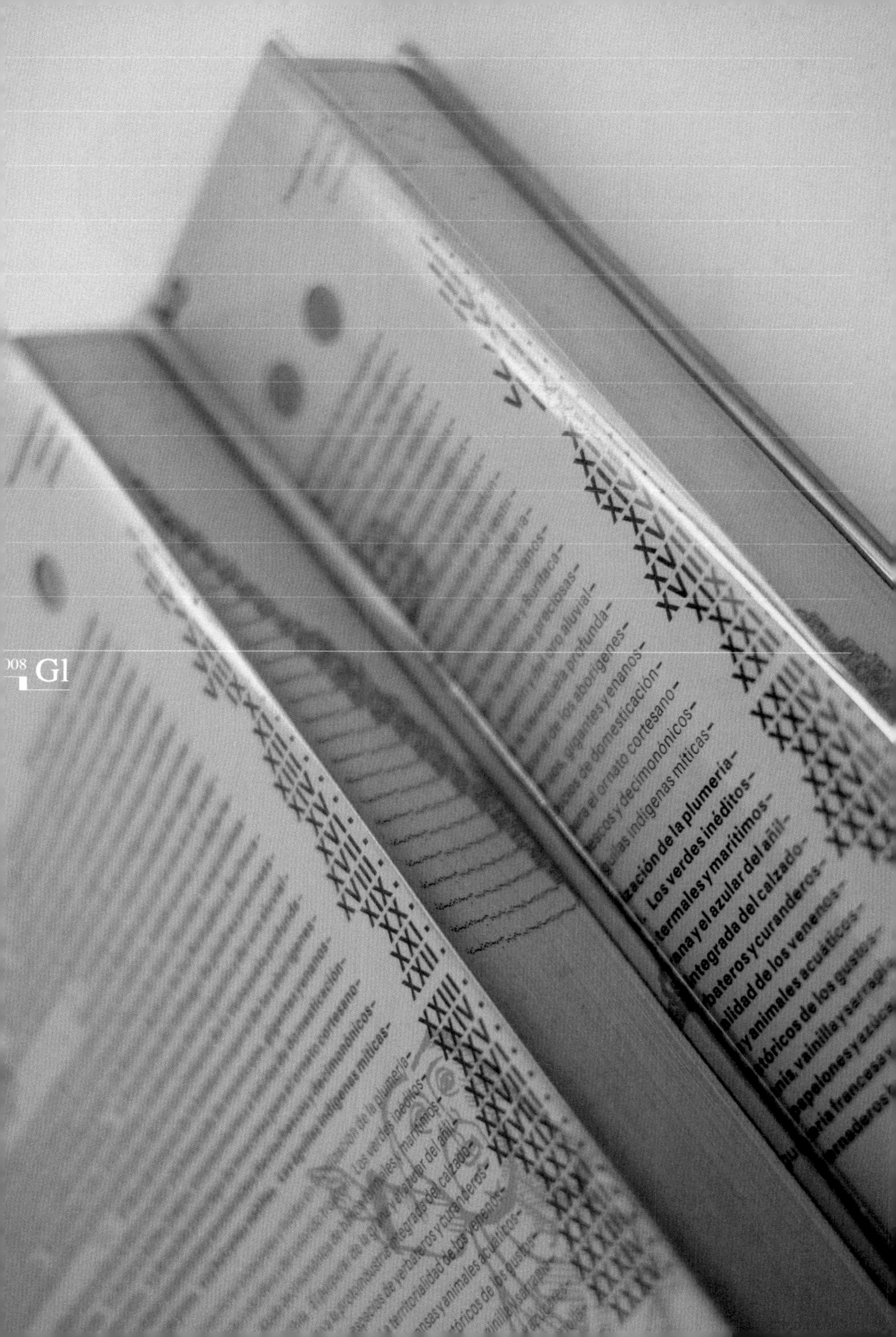

zación de la plumería–
Los verdes inéditos–
termales y marítimos–
ana y el azular del añil–
integrada del calzado–
bateros y curanderos–
alidad de los venenos–
y animales acuáticos–
tóricos de los gustos–
gigantes y enanos–
de domesticación–
el ornato cortesano–
cos y decimonónicos–
indígenas míticas–

本书是一套两本由地理学家编写的有关委内瑞拉的百科全书。内容涉及自然地貌、建筑景观、气候生态、物产资源、人文历史、宗教文化等多方面，但内容并不是按照严格的自然科学分类去介绍，而是通过人们的直观考察，从人文地理的角度出发，将这些内容串联起来，形成以点及面的内容体系。本书为包背精装，护封暖灰色胶版纸黑绿双色印刷，外包裹透明PVC材料，内封绿色布面精装。在护封上有圆形打孔，内封书脊上有圆形压印，用以区别上下册。为体现本书的信息量，也为了便于内容的查阅，封面信息纵横交错，但最主要的是把目录直接放在封面上，并通过右侧书口的短横线来指引到相关页面。右侧书口上部的红色短线为当前标题内容的进一步划分，起到辅助查阅的目的。内页艺术纸四色印刷，不对称的双栏设计为左侧的注释栏留下了灵活排版的余地。||||||||||||||||||||||||||||||||

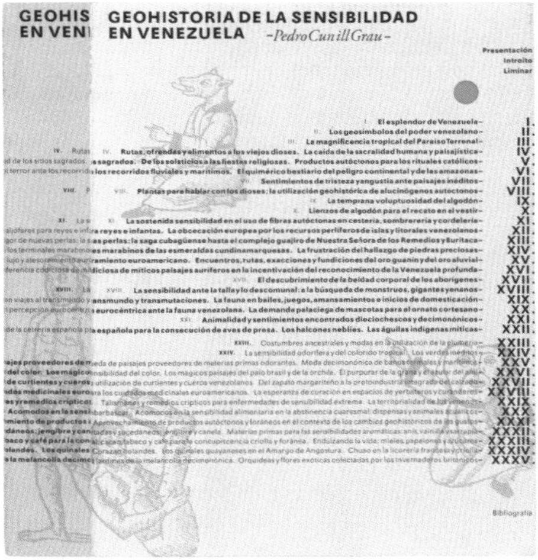
GEOHISTORIA DE LA SENSIBILIDAD
EN VENEZUELA
-Pedro Cunill Grau-

This book is a set of two encyclopedias on Venezuela written by geographers. The content covers many aspects, such as natural landform, architectural landscape, climate and ecology, property resources, human history, religious culture, etc. However the book is not divided into chapters according to strict classification, but to connect these contents from the perspective of human geography through observation. The way of acquiring knowledge is different when using traditional textbooks. The double-leaved book is hard-bound. The jacket is made of gray offset paper printed in black and green, wrapped by translucent PVC. There are circular perforations on the jacket and circular impressions on the spine to distinguish the two volumes. In order to facilitate accurate access, the cover is full of crisscrossed information. However, the most effective way is putting the catalogue directly on the cover, and the short lines on the right edge of the book may lead to the relevant pages. The short red lines on the upper part of the right edge serve as an auxiliary reference tool and also have certain decorative effects. The inside pages are printed on matte art paper in four colors. There is strict typesetting format in the body part, but the layout of illustrations and annotations is relatively free, which avoids being rigid as a tome.

- ◎ 金字符奖
- « 委内瑞拉感性地理学史
- ◐ 委内瑞拉

- ◎ Golden Letter
- « Geohistoria de Ia Sensibilidad en Venezuela
- ◐ Venezuela
- ▷ Álvaro Sotillo mit Gabriela Fontanillas & Luis Giraldo
- △ Pedro Cunill Grau
- |||| Fundación Empresas Polar, Caracas
- □ 285×220×71mm
- ⌷ 1618g+1638g
- ≣ 528p

Complex yet very subtle typography. Diverse information is characterised just so as to make it clearly distinguishable. The book-designer has refrained from using coloured type, working instead merely with small contrasts in size, thickness and highlighting. The manner of binding (French fold) opens up a wonderful possibility for navigation through the book. Its front edge is graced by a subtle thumb index. The layout is open and flexible. The designer has not become constrained by his own mesh, as is so often the case with such a large number of illustrations.

La t
zos

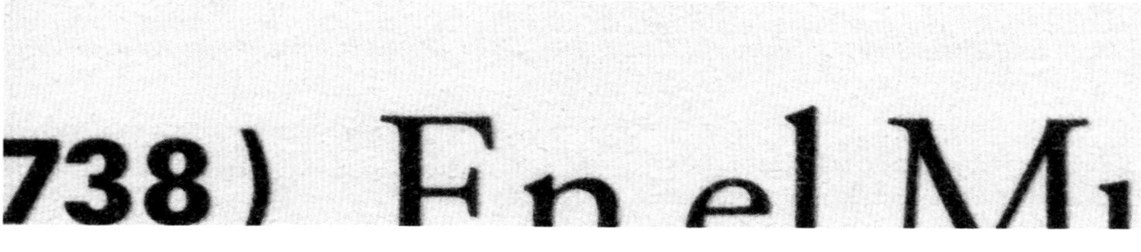
738) En el M

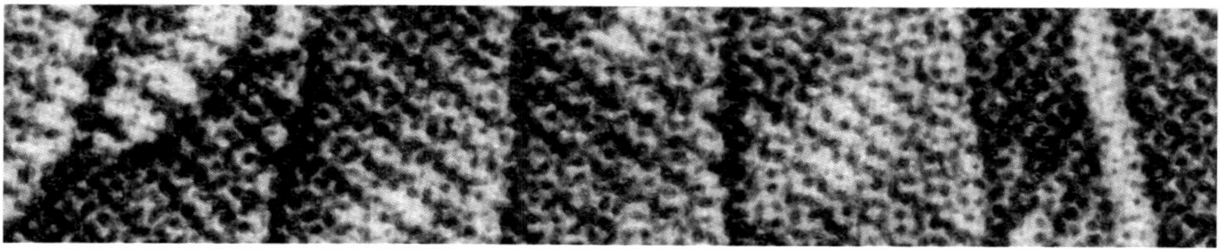

285×220×71

Presentación

Introito

Liminar

ela– I.
no– II.
ial– III.
ca– IV.
os– V.
as– VI.
os– VII.
os– VIII.
ón– IX.
tir– X.
ría– XI.
os– XII.
ca– XIII.
as– XIV.
ial– XV.
da– XVI.
es– XVII.
os– XVIII.
ón– XIX.
no– XX.
os– XXI.
as– XXII.
ría– XXIII.
os– XXIV.
os– XXV.
ñil– XXVI.
do– XXVII.
os– XXVIII.
os– XXIX.
os– XXX.
os– XXXI.
bia– XXXII.
es– XXXIII.
lla– XXXIV.
os– XXXV.

1618+1638 528

XVIII Costumbres ancestrales y modas en la utilización de la plumería–

XXIV. La sensibilidad odorífera y del colorido tropical. Los verdes inéditos–

; proveedores de materias primas odorantes. Moda decimonónica de baños termales y marítimos–

color. Los mágicos paisajes del palo brasil y de la orchila. El purpurar de la grana y el azular del añil–

curtientes y cueros venezolanos. Del zapato margariteño a la protoindustria integrada del calzado–

; medicinales euroamericanos. La esperanza de curación en espacios de yerbateros y curanderos–

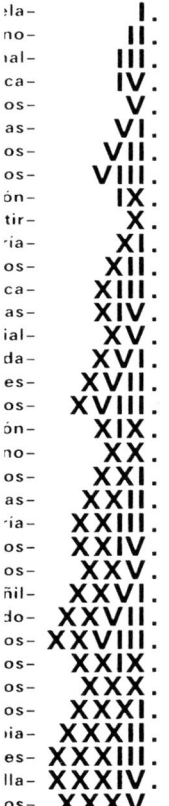

remedios crípticos para enfermedades de sensibilidad extrema. La territorialidad de los venenos–

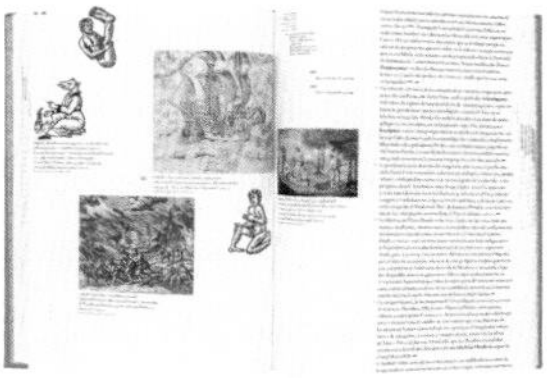

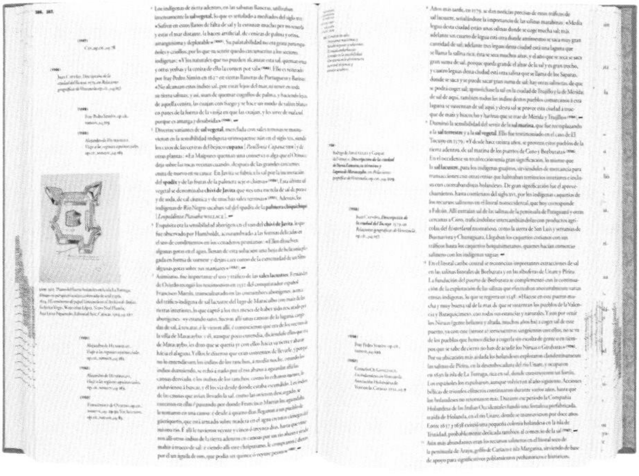

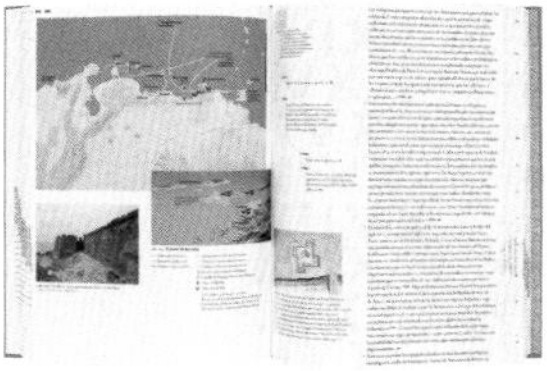
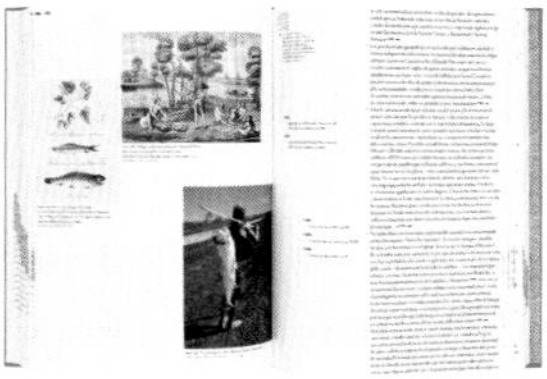

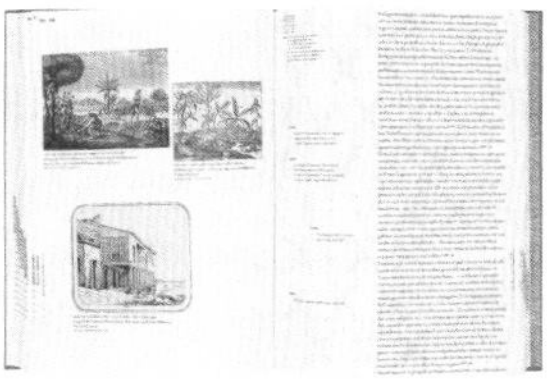

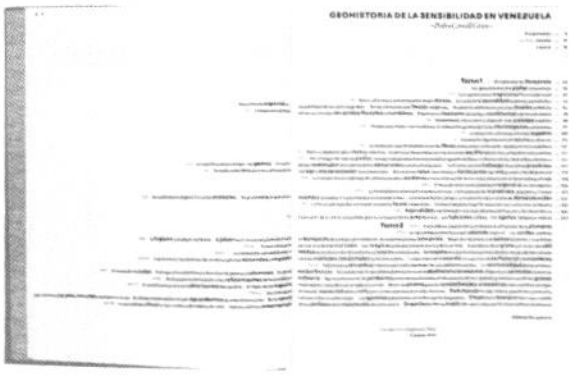

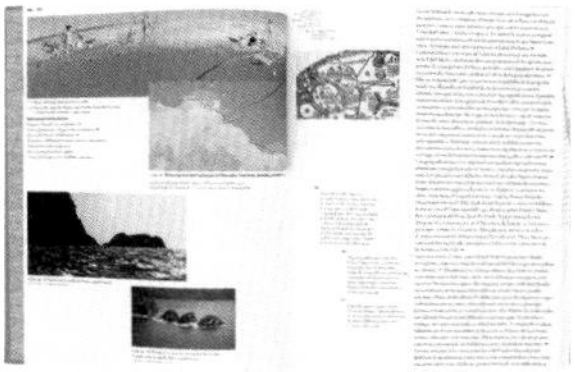

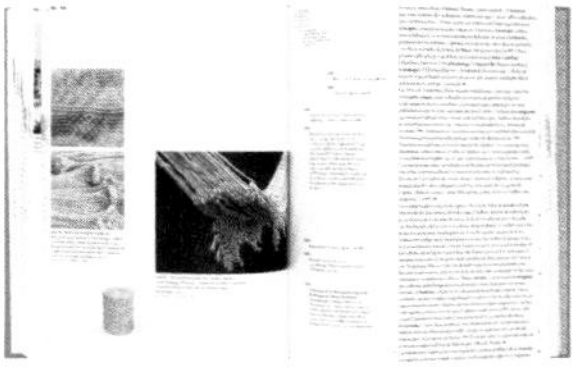
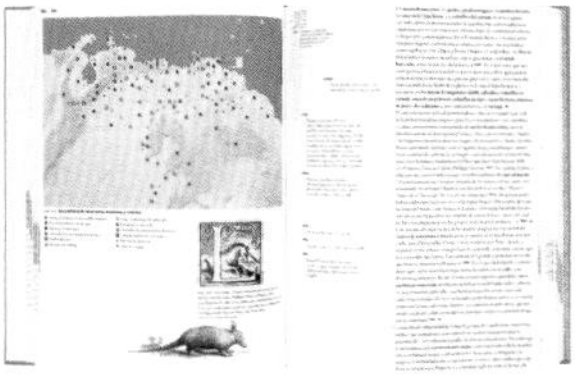
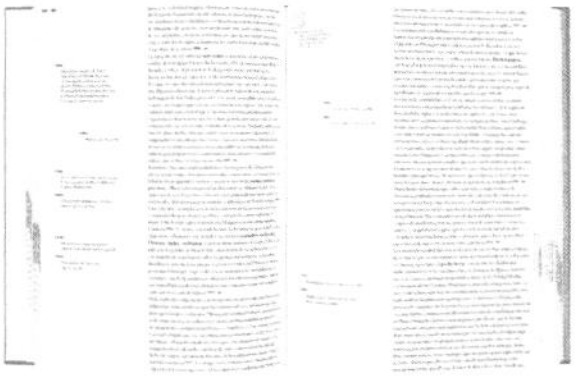
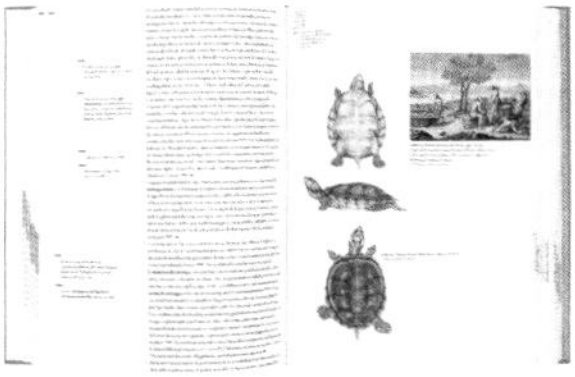
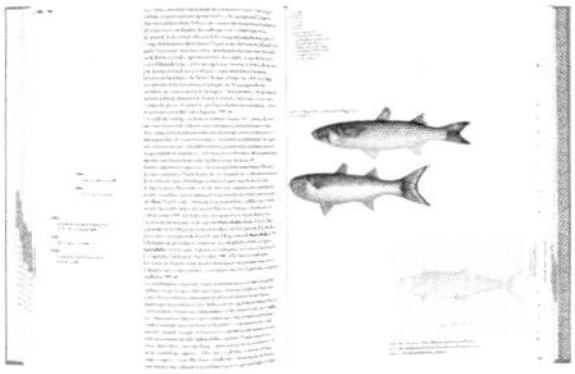
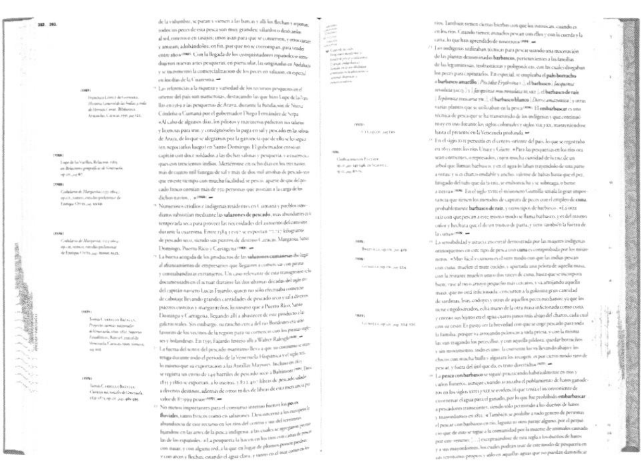

GEOHISTORIA
DE LA SENSIBILIDAD EN
VENEZUELA
Pedro Cunill Grau
Tomo 1
Tomo 2
2007
Fundación
Empresas Polar
380. 381.
LÁM. 216 Manzanillo [Hippomane Mancinella]
Herbarium del Museo de Ciencias Naturales de Chicago.
LÁM. 217 Venado, PL. 37, Collection de Mammiferes,
Du Muséum D'Histoire Naturalle, París, 1808.
LÁM. 218 Jaguar, PL. 15, Collection de Mammiferes,
Du Muséum D'Histoire Naturalle, París, 1808.

con hierba «muy a punto» (1131). Fernández de Oviedo señala el complejo empleo de hierbas ponzoñosas con otros productos venenosos al hacer referencia a la muerte en pocas horas de un miembro flechado de la expedición en 1536 de Diego Dortal en el trayecto de San Miguel de Neverí al pueblo de Guamba, junto a otros casos en los cuales los conquistadores flechados fallecían de **«rabia»** a los tres días del enflechamiento. Los flechadores indígenas de la costa oriental utilizaban como materia prima para sus venenos el **manzanillo de playa** [*Hippomane mancinella*]:
«Y es que donde hay aquellos manzanillos, [...] aquel es el principal material donde esta hierba se funda, con el qual se mezclan otras muchas ponzoñas, assi como alacranes, vívoras, hormigas grandes de los encordios, de quien se dirá adelante, y de aquellas culebras verdes que se cuelgan de los árboles, [...] Ponen assimesmo en esta malvada hierba aquella agua marina, ques una cosa á manera de bexiga ó bamboya morada, que anda sobre las aguas de la mar, é ciertas arañas, é algunos zumos de hierbas é rayces que mezclan, é cierto género de abispas: que cada cosa dellas es muy bastante para dar la muerte. Y destas cosas y otras hacen aquella mixtura, con que untan sus flechas aquestos indios: é donde carescen de algunas cosas destas, suplen su malicia poniendo en su lugar otras tan malas ó peores, de que ya ellos tienen larga experiencia. É cuando acaece que algund herido desta hierba escapa, es por dieta é mucha diligencia de le chupar la herida: é socórrenle con ventosas é otras medecinas entre los chripstianos, é por la mayor parte está la salud del herido en ser la flecha untada de dias é estar muerta la hierba, ó enflaquecida la maldad é fuerca della por ser añeja, ó por le faltar algunos materiales, ó mejor diciendo, por querer Dios que viva el que está herido» (1132).

(1131)
Bartolomé de Las Casas, *Historia de Indias*, op. cit., tomo I, *pág.* **536**, refiere que los indígenas parianos «andan muy ataviados de arcos, flechas y tablachinas, y las flechas traen casi todos con hierba».

47 En las costas occidentales venezolanas también se utilizaba el veneno del **manzanillo**, aunque de manera más directa: «Los indios de Tierra Firme, infectan la punta de las flechas con la leche de la corteza de este árbol, y quien es herido con éstas, aunque no sea mortal, se le corrompe y pudre mucho la llaga y dura mucho para sanar, y muchos mueren si la herida está en lugar nervoso o es profunda, por la mucha materia que segrega, como sucede en la garganta, tobillo, manos o lugares similares. Así mismo emplean los indios la fruta en la composición del veneno que hacen para las flechas; el fruto es mortal a quien comiese mucho o poco, si no se remediase con aceite y salmuera, procurando por el vómito y por debajo sacarlo del cuerpo y con cosas refrescantes, da gran nublamiento en la cabeza, inflamando todo el cuerpo ardentísimamente por donde pasa, con grandísimos vómitos» (1133).

(1132)
Fernández de Oviedo, op. cit., tomo V, *pág.* **250**.

(1133)
Cey, op. cit., *pág.* **40**.

48 En 1578 en la *Relación geográfica de Caracas* se proporcionan notas reveladoras acerca del peligro para los conquistadores españoles de las flechas envenenadas con las mixturas encabezadas por el **manzanillo**: «untan las flechas con hierba malísima con la que han muerto muchos españoles. Hácenla de una fruta que llaman manzanilla, que son como manzanas pequeñas, amarillas, que huelen bien. La hierba que es sólo desta manzanilla, no es muy mala aunque algunos con ella mueren, pudre la carne y hace otros daños. Esta manzanilla suelen confeccionarla con víboras, sapos y arañas, metido todo esto en una tinajuela y allí se mueren. De eso y sangre de costumbre de mujer, y de otras cosas ponzoñosas las hace una vieja que dicen que de por hacer esta hierba fallece pronto» (1134).

(1134)
Juan de Pimentel, ***Descripción de Santiago de León de Caracas***, 1578, en *Relaciones geográficas de Venezuela*, op. cit., *pág.* **125**.

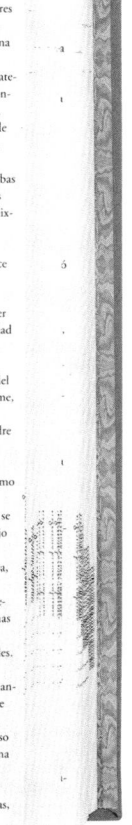

49 En el oriente del país, en la Nueva Andalucía, era alta la sensibilidad hispánica del terror ante las flechas indígenas envenenadas: «Las flechas, de que usan en la Guerra, suelen las mas Naciones herbolarlas con un

Katrin
Kleiber

瑞士《Das Magazin》杂志是一份周刊，有个“周六专栏”刊写一个普通瑞士人的一天生活。这本书是编辑沃尔特·凯勒(Walter Keller)从杂志 20 多年的专栏中精选而出的合集。本书为锁线胶装，封面白卡纸黑和荧光绿双色印刷，简洁有力。绿字部分为书名、作者、出版社等信息，而黑字为一个具体的人一天故事的开始，有多种版本，意味着不同人会有不同的开始。内页正文部分按照人物的姓氏从 A 到 Z 进行分组，打乱了普通的时间线。如果按照书最后提供的时间线来看，杂志的版式在这 20 多年内经历了四栏到两栏一直到当前三栏的变化过程。内文依旧延续黑绿双色，巨大的绿色人名均放置在页面的左上角，但打乱了时间线使内文版式看起来丰富多变。

Der Tag beginnt für mich schon am Abend vorher. Ein Tag im Leben von
Porträts aus über zwanzig Jahren, ausgewählt von Walter Keller.
DAS MAGAZIN
salis

Das Magazin is a weekly magazine from Switzerland. There is a column about the daily life of an ordinary Switzer on every Saturday. The book is a collection of articles from this column selected by editor Walter Keller. This book is perfect bound with thread sewing. The cover is made of white cardboard printed in black and fluorescent green, which is concise and powerful. The title, information of the author and the publishing house are printed in green; while the story of an ordinary people is printed in black. There are stories in different versions according to various beginnings. The main body of the book is divided into groups by initials of surnames from A to Z, which disturbs the time line. According to the time line provided by the book, the layout of the magazine pages has undergone from four-column to two-column in the past 20 years, and then to the current three-column layout. The text is printed in black and green as well. There are huge green names in the upper left corner of the page, which make the layout varied.

金奖
__ 生命中的一天
瑞士

Gold Medal
Ein Tag im Leben von
Switzerland
Elektrosmog (Valentin Hindermann & Marco Walser & Simone Koller)
Walter Keller (Ed.)
Salis Verlag AG, Zürich
314×247×25mm
1274g
312p

The design of this book makes use of graphic conventions taken from newspaper and magazine design, doing so in a playful and unforeseen fashion. The way the contents are categorised, the use of only one highlighting colour, the varying column structure and changes of typeface reinforce the volume’s documentary character. It is a pleasure to see how good design, vigorous and unambiguous, can tell stories in such a natural and unambitious way.

Ein Tag
im Leben von
Porträts aus über zwanzig Jahren,
ausgewählt von Walter Keller.

Paul
Accola

Corinne
Hobi

Max
Christian
Meissgeier

Helga
Merz

Andri
Pol

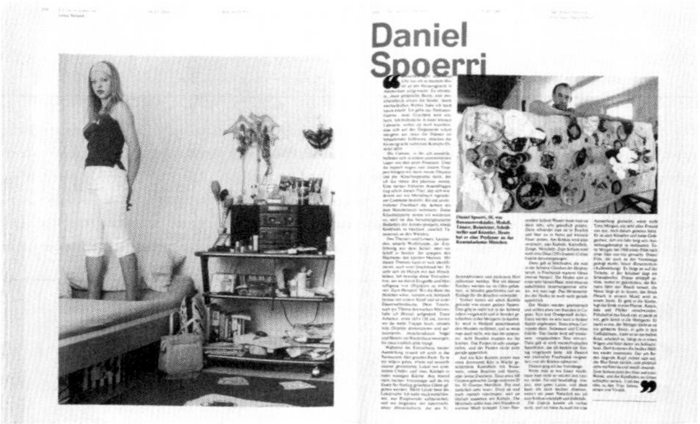
Daniel
Spoerri

Markus
Altherr

Nicole
Amrein

Siegfried
Avesani

Dragan
Najman
Asmi
Nardo

Pasqualina
Pagliarulo
Peter
Peterka

Pierce

Duri
Stecher

Redl
Berthold
Redlich

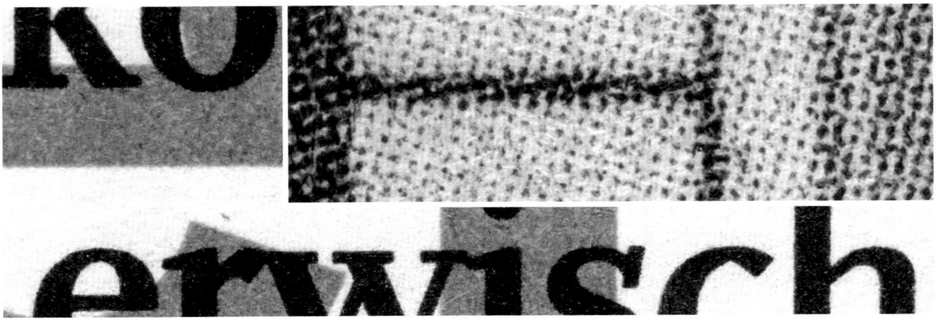
KO
erwisch

Ein Tag im Leben von
salis
188
Ein Tag im Leben von
Anna Netrebko
19.11.2005
Bild Tom Haller

Text: Curt Truninger
Bild: Curt Truninger

Alexander Niemetz

Alexander Niemetz, 42, stammt aus Balsthal SO. Nach der Mittelschule in Immensee studierte er in Berlin und schloss als Diplompolitologe ab. Dann arbeitete er als freier Journalist, war Ghostwriter für Politiker, CDU-Pressesprecher in Hessen. 1979 ging er zum ZDF, zuerst ins «Heute-Journal», dann zur «Tele-Illustrierten» und anschliessend in die innenpolitische Redaktion. Heute ist Niemetz der einzige ZDF-Chefreporter.

“Bei Auslandreportagen, vor allem in Krisengebieten, ist bei mir früh Tagwache, so gegen halb sechs Uhr. Zu Hause in Wiesbaden, wo ich knapp vier Monate im Jahr bin, stehe ich so gegen halb acht Uhr auf, und zwar ohne Wekker. Zuerst hole ich mir von nebenan frische Brötchen und kaufe zwei Tageszeitungen, ‹Bild› und die ‹Frankfurter Allgemeine›. Die Zeitungen lese ich beim Frühstück. Dazu höre ich Radio, und zwar SW 3. Anschliessend fahre ich mit dem Auto in 15 Minuten ins ZDF-Studio, das zwischen Wiesbaden und Mainz liegt. So gegen zehn Uhr treffe ich dort ein. Im Büro lese ich dann zuerst drei weitere Tageszeitungen. Wenn ich gerade von einer Reportage aus dem Ausland oder auch Inland zurückkomme, schneide ich mit der Cutterin den Film. Sollte der Film bereits fertig montiert sein, so schreibe ich den Filmkommentar oder suche vielleicht im Musikarchiv eine passende Musik aus. Meine Filme sind unterschiedlich lang. Von vier, fünf Minuten – etwa für das ‹Heute-Journal› – bis 45 Minuten zum Beispiel für ‹Reportage am Dienstag›. Das ist der kreative Teil meiner Arbeit im Studio.

Viel Zeit nehmen auch die ganzen Reiseabrechnungen, Visaanträge, Drehbewilligungen und so weiter in Anspruch oder die Beantwortung von Zuschauerpost. Glücklicherweise sind es nicht immer 10 000 empörte Briefe wie nach dem Film ‹Hexen in Deutschland›. Es muss nämlich jeder Zuschauerbrief vom verantwortlichen Redaktor persönlich beantwortet werden. Daneben muss ich für neue Reportagen recherchieren, wobei mir meine Sekretärin bei der Archivarbeit hilft, oder ich telefoniere mit unseren Korrespondenten, um abzuklären, ob sie etwa am gleichen Thema arbeiten. Am Dienstag ist immer Sitzung der Hauptabteilungsleiter.

Am Mittag esse ich meistens in der TV-Kantine. Eigentlich schade, dass wir hier auf dem Lerchenberg so weit von der Stadt weg sind und so kaum je aus diesem Ghetto herauskommen. Nun, ich kann mich nicht beklagen, ich komme ja genügend weg vom Studio. Meine letzten Auslandreportagen führten mich nach Namibia, Libyen, Sri Lanka, dem Libanon und auf die Philippinen. Als nächstes plane ich drei Berichte aus sogenannt typischen Guerillagebieten. Vermutlich werden dies Nicaragua oder Kolumbien, dann Kambodscha und Moçambique sein. Angst habe ich bei solchen Einsätzen eigentlich nicht, darf ich nicht haben. Gefährlich ist übrigens für uns nicht unbedingt die Filmarbeit während der Kampfhandlungen, sondern der Transport dorthin. Wichtig ist, dass ich mit einem eingespielten Filmteam zusammenarbeiten kann und dass ich vor Ort auf einen zuverlässigen einheimischen *stringer*, also einen Informanten, der die Lage genau einzuschätzen weiss, zählen kann. Während solcher Auslandeinsätze habe ich kaum Freizeit, und ein 18-Stunden-Tag ist fast die Norm.

Für meine Frau und meine beiden Kinder ist mein Job eine recht grosse Belastung. Wenn sie beispielsweise erst durch einen telefonischen Kommentar im ‹Heute-Journal› erfahren, dass ich im Libanon noch am Leben bin, so strapaziert dies eine Beziehung enorm. Die berechtigte Frage, warum ich denn immer wieder in Krisengebiete gehen wolle, taucht zu Hause häufig auf. Bedingt durch mein hektisches Leben wohnt meine Familie meist getrennt von mir, und zwar ausserhalb von Frankfurt.

Als Chefreporter bin ich auch im deutschen Inland im Einsatz. Nach Möglichkeit berichte ich von allen Parteitagen und grossen Wahlveranstaltungen. Oft sind dies Tagesreportagen fürs ‹Heute-Journal›.

Wenn ich also im Studio arbeite, so komme ich gegen sieben Uhr abends nach Hause. Ich mache mir dann etwas Bescheidenes zu essen. Obwohl ich gerne gut esse, mache ich für mich alleine kein grosses Geköche. Nachher schaue ich mir die Nachrichtensendungen an. Auch politische Magazine und Sportsendungen schalte ich regelmässig ein. Sonst bin ich aber kein intensiver Fernseher, vor allem diesen ganzen Unterhaltungsrotz mag ich nicht. Ich lese lieber Bücher. Neben der Fachliteratur interessiere ich mich hauptsächlich für Bücher aus dem Gebiet der Esoterik, obwohl ich eigentlich als sehr rationaler Typ gelte. Ich bin überzeugt, dass viele Leute mediale Fähigkeiten haben, diese aber nicht erkennen. Frank Capras Buch ‹Wendezeit› ist für mich ein sehr wichtiges Buch, dessen Zukunftsprognose ich weitgehend teile. Was ich auch immer wieder gerne lese sind Bücher von Carlos Castaneda. Das Interessante an Castaneda sind nicht unbedingt die Kapitel über Drogen. Spannend finde ich, was er über die Meditation sagt, dass diese also im Körper genauso chemische Reaktionen auslösen kann wie Drogen. Meditation ist für mich nicht so sehr ein Mittel gegen Stress, sondern vielmehr eine Art Selbstfindung. Für die rein körperliche Fitness spiele ich dafür Squash. Dieser sportliche Ausgleich ist für mich enorm wichtig, weil ich doch meinen Körper ganz schön strapaziere. Nur ein Beispiel: Kürzlich kam ich nach vier Wochen in Angola und Namibia an einem Samstag nach Hause, und bereits am Montag war ich wieder im Flugzeug Richtung Japan zum Weltwirtschaftsgipfel.

Für mich war schon immer klar, dass ich Journalist werden wollte. Schon in der Mittelschule schrieb ich für das Luzerner ‹Vaterland›. Dazu kommt, dass ich aus einer sehr politischen Familie komme. Mein Vater war Kantonsrat im Kanton Solothurn.

In die Schweiz möchte ich nicht zurückkehren. Die Schweizer Innenpolitik interessiert mich beispielsweise überhaupt nicht mehr. Alles ist in der Schweiz so kleinkariert, so eng. Als Schweizer hat man übrigens in der Bundesrepublik durchaus gleichwertige Berufschancen wie die Deutschen. Problemlos konnte ich – als Parteiloser – für deutsche Spitzenpolitiker Reden schreiben. Das sind so kleine Wunder, die man hier in Deutschland erlebt. Solange die Leistung stimmt, spielt die Nationalität keine Rolle. Ich bin absolut sicher, dass ein Parallelfall zu mir in der Schweiz völlig undenkbar wäre. Ich könnte mir übrigens auch nicht vorstellen, beim Schweizer Fernsehen zu arbeiten. Hier beim ZDF scheint mir vieles grosszügiger als bei der SRG.

Wenn ich ein Buch schreiben würde, käme ich sicher in die Schweiz zurück, und bestimmt werde ich als Pensionär zurückkehren.
Aber bis dann belasse ich es beim regelmässigen Skiurlaub im Diemtigtal im Berner Oberland.”

日本江户幕府时期乐于文化事业的藩主堀田正敦主持编修了鸟类学图鉴《观文禽谱》，本书是在这本经典著作的基础上进行编撰，加入更多参考资料的鸟类百科全书。本书为圆脊锁线硬精装，护封四色印刷，内封红色布面裱覆卡板，只在书脊部分烫金标明信息。浅棕色艺术纸印典籍文字做环衬包裹内文部分。正文部分胶版纸四色印刷，内容按照水禽、林禽、山禽、外来鸟类为主要章节，辅以馆藏资料、人物年谱、参考文献等附文组成这本学术著作。主要信息部分采用蓝色文字以写明各种名称、收藏者，黑色文字则主要为堀田原著介绍和当今解说。鸟类图片均为手绘图谱，绘制细节丰富，印刷精美，极具收藏和观赏价值。书的最后除了人物和事件索引外，鸟类部分设立日文假名、汉字名、拉丁学名、西文常规叫法四套索引，便于查找。

In Tokugawa Japan, the seignior Hotta Masatoshi, who was willing to take part in cultural undertakings, presided at the compilation of an illustrated handbook *The Birds and Birdlore of Tokugawa Japan*. This book is based on this classic book to compile an encyclopedia of birds with more references. The hardback book is bound by thread sewing with a rounded spine. The jacket is printed in four colors, with a red inner cover mounted with the graphic board. The book information is hot stamped in gold on the spine. The endpaper is made of light brown art paper printed with the scripts. The text is printed on offset paper in four colors and the content is divided into waterfowl, forest fowl, mountain fowl, exotic birds and other major chapters, supplemented by library materials, Chronological life, references and other appendices. Blue characters are used to list various names of the bird and collectors, while black characters are mainly to introduce Hotta's original work and contemporary explanations. All the bird pictures are hand-painted illustrations with rich details. The exquisitely printed book is of great collection and ornamental value. At the end of the book, besides the index of characters and events, four other indexes are set up for easy searching, which are Japanese Kana, Chinese Character Name, Latin Scientific Name and Western Conventional Name.

- 银奖
- 江户川鸟类大词典
- 日本

A richly illustrated and informative book. Its simple structure permits a flexibility of design which gives the pages a dynamic appearance. Everything is filled with wonderful typographical details. A real visual and intellectual treat.

- Silver Medal
- The Birds and Birdlore of Tokugawa Japan
- Japan
- Omori Yuji
- Suzuki Michio (Ed.)
- Heibonsha Ltd., Publishers, Tokyo
- 269×192×64mm
- 2355g
- 762+51p

かり一種 はら白
かり一種 あひがん

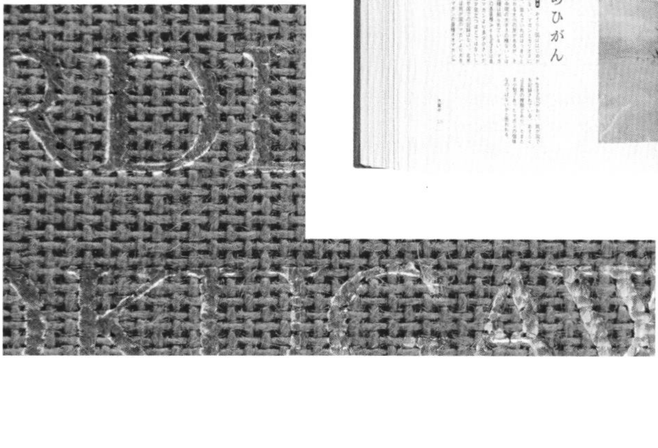

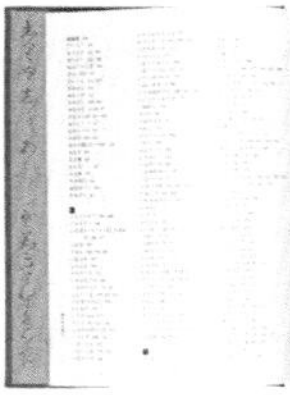
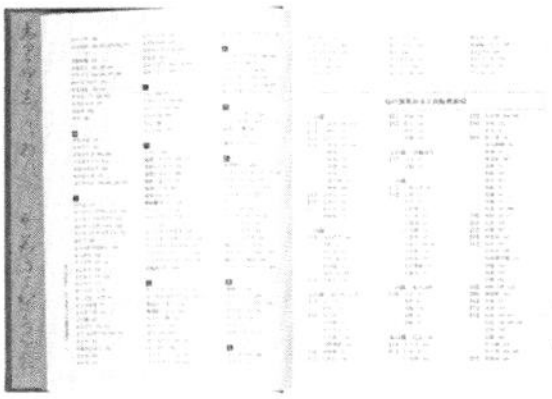

林禽
下

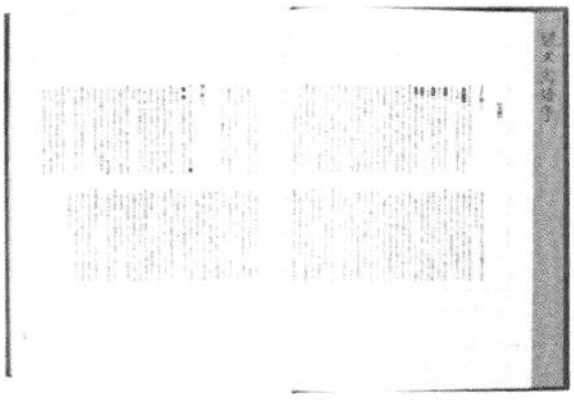
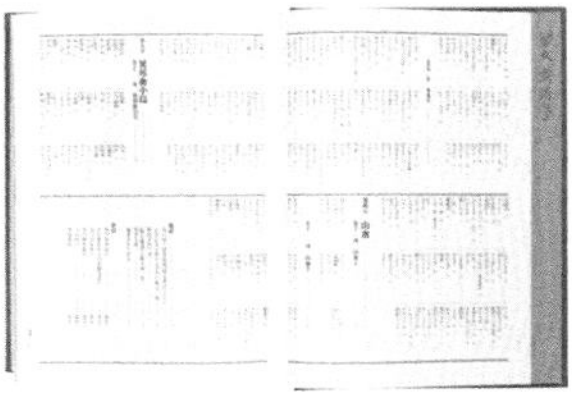
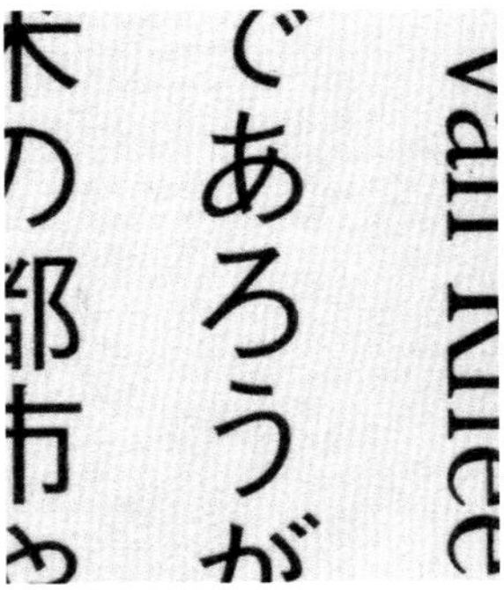
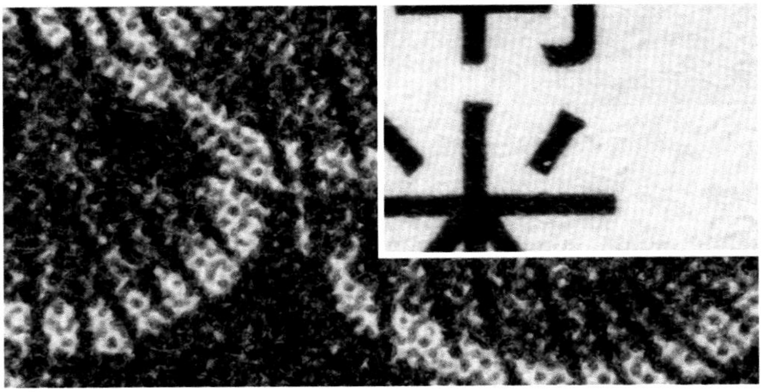

THE BIRDS AND BIRDLORE OF TOKUGAWA JAPAN
江戸鳥類大図鑑
堀田正敦 著
鈴木道男 編著
平凡社

大いすか
えぞいすか
大いすか
福岡侯蔵図
現和名：ギンザンマシコ
Pinicola enucleator
えぞいすか
現和名：ギンザンマシコ
Pinicola enucleator
林禽下
549

しまいすか

編田 「羽譜」が引くある人の説によれば、島鶍は形状はイスカに似るが翼に白羽がある。この白羽はすなわち「ホロの羽」ということ。今按ずるに、この説のとおりであるとすると、これは羽白イスカと同物なのかもしれないが確認できない。羽色の変化したものを「島」と呼ぶのは禽舗の常なのである

解説 熊本侯の蔵図と正敦による解説文では、翼に白があることが強調されている。イスカにも翼に白が出ることがあるとのことだが、本図のように三列風切の外縁が白いのは、ナキイスカ*Loxia leucoptera*（→いすか図三）の特徴である。ナキイスカも飼育によって雄の赤が失われやすいのだが、この鳥の黄色は緑を帯びていないので、雄であろうと思われる。

羽白いすか

編田 この図（図一）は前のものとおおよそ似ていて、小さいところをみると鵐イスカの雛のようにも思われる。もしくは図を描いた者の違いによるものなのだろうか。禽舗の者に問うたところ、この図は鵐イスカに相違ないと語った。
（図二は）前図と比較すると大きくて羽色も異なっている。どちらも羽白イスカと題されているが、必ずしも同じものとは言い難い。あるいは大イスカの類ではなかろうか。未詳である。禽舗の者に問うたところ、この鳥は両翼に白斑があるとのことであった。

解説 図一は、イスカの類ではあるが、翼の白の出方が正確には描かれておらず、イスカともナキイスカとも断定できない。イスカ類の幼鳥はまだ嘴がふつうの鳥のようで、左右に食い違うことはないので、鶵だという正敦の説はあたらない。
図二は、あまり正確な図とはいえない。おそらくイスカ*Loxia curvirostra*であろう。イスカ属ではイスカより大型のものはいないはずである。嘴は成鳥のものと思われるが、縦斑の多い羽色は幼鳥を思わせる。図からは翼の白斑がうかがえない。

現和名：イスカ
Loxia curvirostra

548

22 / 23 / 24
26 / 27 / 28 / 29
30 / 31 / 32

Mariko Mori
/ Oneness /

本书为具有多国学习和生活背景的日本艺术家森万里子的展览作品集。森万里子是一位擅长在多种文化和技术之间自由切换的艺术家，她的作品不受传统意义上艺术分类的限制，通过表演、装置、数码等多种形式的混合，表现出生命、死亡、灵性、现实和宇宙之间相互混合的超现实特质。本书在内容编排和印刷工艺上都尽量找寻着艺术家作品反映出的独特气质。书函为单面白色卡板制作，折边部分故意留出深深的灰色折痕，书脊信息烫银。书籍锁线胶装，封面采用仿皮面印银，人名和书名做烫银处理。内文部分主要为涂布画刊纸黑色和银色双色单面印刷，在重要作品的部分采用玻璃卡四色印刷重点呈现，作品信息和介绍被设置在每大类的最后，所以阅读本书需要大量前后翻阅。全书黑、白、灰、银和多种纸张的混合使用，产生了微妙而丰富的对比，具有极强的未来感。

This book is a collection of exhibition works of Japanese artist Mariko Mori who has a background of studying and living in many countries. Mariko Mori is an artist who is good at switching freely between different cultures and technologies. Her works are not restricted by the traditional art classification. Through the mixing of performances, installation arts and digital forms, she shows the surreal characteristics of life, death, spirituality, reality and the universe. The content arrangement and printing technology of this book both reflect the unique characteristic of the artist's works. The slipcase is made of white graphic board and there are deep gray creases left on the folding part deliberately. The information on the spine is silver hot stamped. The book is perfect bound with thread sewing. The cover is printed on faux leather in silver, with the title and author's name hot stamped in silver. The inside pages are mainly printed single-sided on coated art paper in black and silver. As for important works, four-color printed glassine cardboard is used. The information and introduction of works are typeset at the end of each category. The combination of various colors (black, white, grey, silver) and paper produces subtle and rich contrasts, which makes the book futuristic.

◎ 铜奖
« 合一
◖ 德国

Its structure is intriguing. Visual references, exclusively printed in silver, are presented before the works of art themselves and explained later in the book. At the very end the artist herself comments on her work. All this makes the journey through the book something of a challenge as one finds oneself continually flicking back and forth. But it is all the more rewarding since there is much more to discover than first meets the eye. This book forces its reader to engage with it as closely as one has to with Mori's work itself if one hopes to get beyond the surface.

◎ Bronze Medal
« Oneness
◖ Germany
D Hester Fell
△ Mariko Mori
||| Hatje Cantz Verlag, Ostfildern
□ 261 × 209 × 25mm
⌸ 1337g
≣ 304p

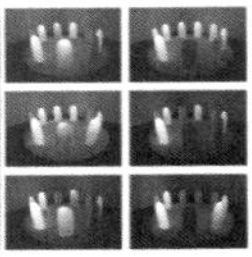

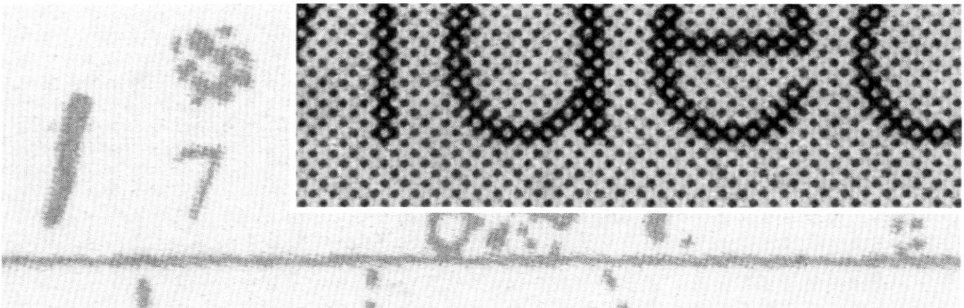

Mariko Mori / Oneness
HATJE
CANTZ

03

Standing Stones of Callanish / Isle of Lewis /
Western Isles / Scotland / The Standing Stones
... resembling
... consist of the
... the middle, one
... of thirteen
... single and one
... from the
... about 2000 B.C
... the site, but its
... been determined,
... the row of stones
... aligns with the
... moon behind a

008 Bm²

本书源于 19 世纪的德国动物学家阿尔弗雷德·布雷姆（Alfred Brehm）的灭绝动物笔记，补充了来自柏林和维也纳自然博物馆的相关插图、地图，是一本灭绝动物研究专著。本书圆脊布面硬精装，深棕色的布面裱贴卡板，丝网印书籍内部动物剪影图案，书名、作者名、出版社烫白处理。蓝底环衬点缀斜向 45° 排列的白色鸟类和羚羊剪影，基本囊括了本书的两大动物分类——鸟类和哺乳动物。内文胶版纸四色印刷，通过地理位置、生存环境、食物链、生育能力等方式，阐述具体动物的情况，并在表格、标注、地图等方面通过深蓝色和棕色对两大物种进行区分。动物彩图局部通过在正文侧边加注数字和单词加粗的方式注明。本书印刷精美，细节丰富，极大地丰富了布雷姆曾经的研究，对于科考和动物保护有很好的实用价值。

Hanna Zeckau und Carsten Aermes
BREHMS
VERLORENES
TIERLEBEN
Zweitausendeins

This book is based on the notes of extinct animals written by Alfred Brehm, a German zoologist in the nineteenth century. More illustrations and maps from the Museums of Nature in Berlin and Vienna are added to perfect this monograph on extinct animals. The book is hardbound with rounded spine. The cover is made of dark brown cloth mounted with the graphic board. Animal silhouette patterns are printed inside the book. The title of the book, the author's name and the publishing house are hot stamped in white. The blue endpaper is decorated with white silhouettes of bird and antelopes arranged obliquely at 45 degrees, which basically encompasses the two major animal classifications of the book – birds and mammals. The inside pages are printed on offset paper in four colors, describing the living conditions of animals by introducing geographical location, living environment, food chain, fertility. Two animal species are distinguished by marking dark blue or brown in tables, labels and maps. The book is exquisitely printed and rich in details, which greatly enriches Brehm's previous research. It is of good practical value for scientific research and animal protection.

铜奖
布雷姆的灭绝动物清单
德国

Crammed with information this is a book waiting to be discovered. And a lot of help is offered in the process. A system of cross-references, pictograms, sketches and statistical data has been elaborated in great detail. Clear boxes containing the most important facts are to be found at the beginning of each entry. Typography, printing, binding and choice of materials are all absolutely convincing.

Bronze Medal
Brehms verlorenes Tierleben
Germany
Hanna Zeckau & Carsten Aermes
Hanna Zeckau, Carsten Aermes
Zweitausendeins, Frankfurt am Main
250 × 175 × 20mm
734g
257p

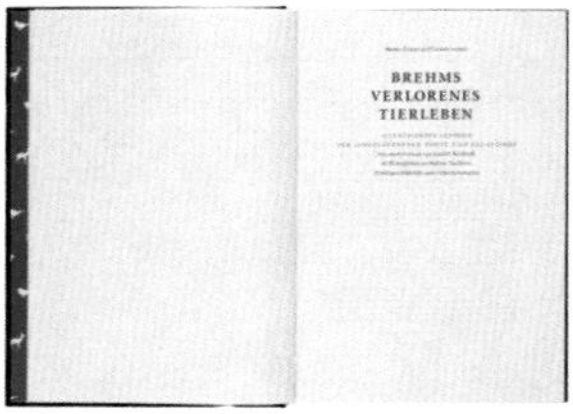
BREHMS
VERLORENES
TIERLEBEN
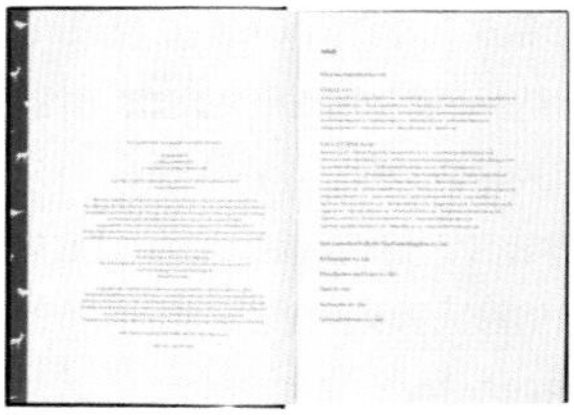

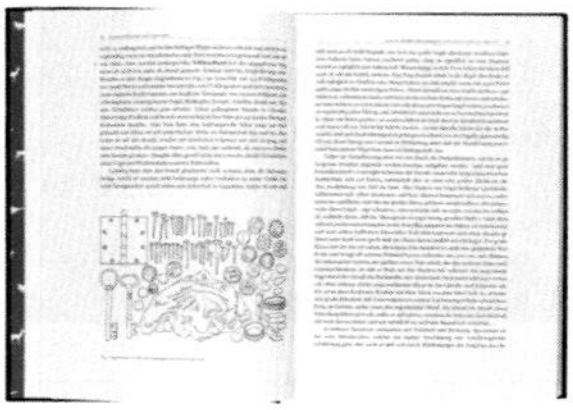
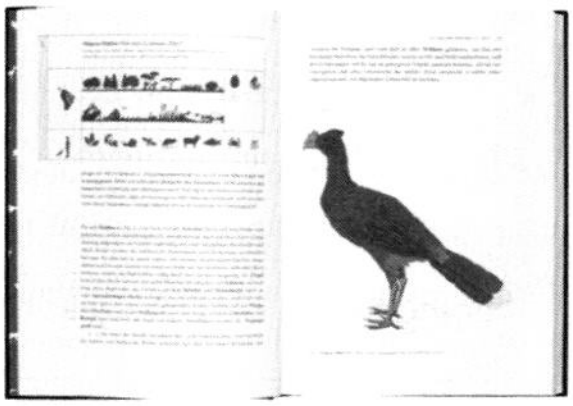

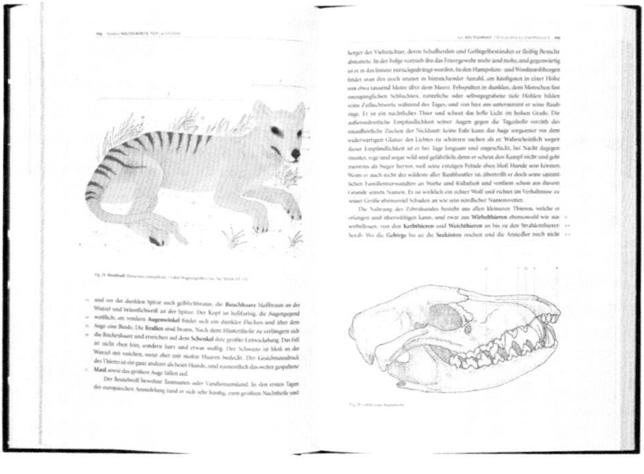

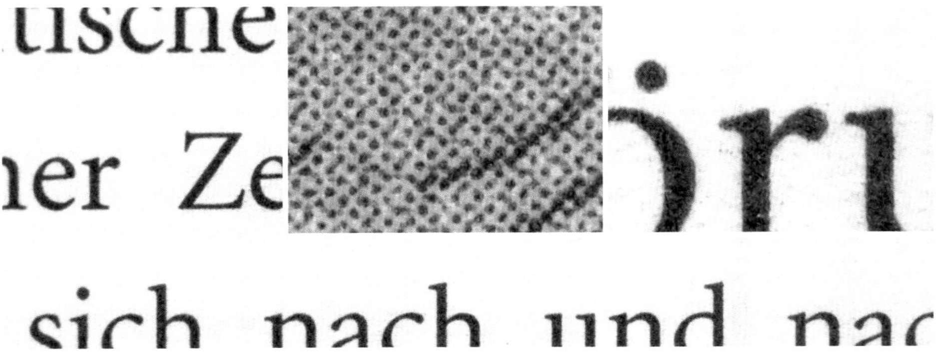
sich nach und

BREHMS VERLORENES TIERLEBEN

Hanna Zeckau und Carsten Aermes

Zwei tausend eins

Übersichtskarte 02

2
1
3

Fig. 30 **Östlicher Bougainville-Langnasenbeutler** *Perameles bougainville fasciata* ([illegible] der Originalgröße)

folgende Angaben: 1. Juli 1839. Zum erstenmal den Streifenbeuteldachs erlegt in dem Revier, das den großen Busch an der Straße zum Murray begrenzt. Ich störte das Tier von dem Kamme eines der felsigen Bergrücken auf; nach einer scharfen Jagd von ungefähr 100 Yards suchte es Schutz unter einem Stein und wurde leicht erbeutet. Es eilte über den Erdboden dahin mit ganz bedeutender Schnelligkeit und mit einer Bewegung ganz ähnlich dem Galopp eines Schweines. Diesem ist es auch ähnlich in der Zähigkeit, mit der sein Fell am Fleisch hängt. Bei Öffnung des Magens fand sich, daß er die Reste von Raupen und anderen **Insekten** enthielt, einige **Samen** und faserige **Wurzeln.** Das Fleisch erweist sich, gebraten, als ein delikates, vortreffliches Essen, wie das der meisten, wenn nicht aller Mitglieder der Gattung.«) +++

Gaimard-Bürstenrattenkänguru *Bettongia gaimardi gaimardi* (Desmarest, 1822)* * [1]
Südöstliches Bürstenrattenkänguru *Bettongia penicillata penicillata* Gray, 1837** * [2]
Westliches Lesueur-Bürstenrattenkänguru *Bettongia lesueur graii* (Gould, 1841)*** * [3]

* Unterart von *Bettongia gaimardi*, die der Familie der Rattenkängurus (Potoroidae) angehört und im folgenden Text allgemein beschrieben wird
** Unterart von *Bettongia penicillata*, im folgenden Text allgemein beschrieben
*** Unterart von *Bettongia lesueur*, im folgenden Text allgemein beschrieben

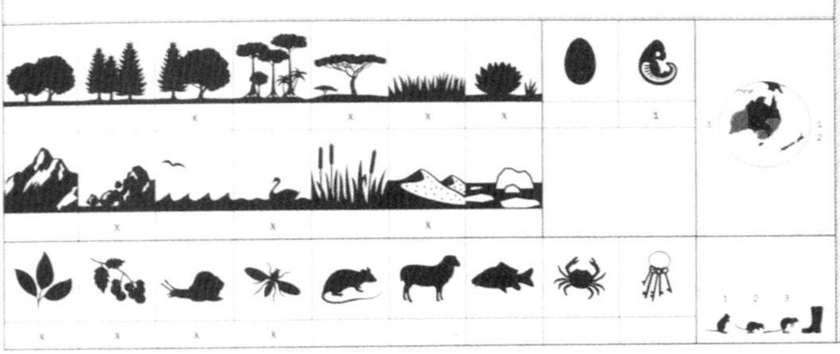

[1] **Länge:** 65 cm **Gewicht:** 1,2–2,3 kg **Lebenserwartung:** 3–6 Jahre **Ehemaliges Verbreitungsgebiet:** Osten Australiens **Zeitpunkt des Aussterbens:** 1910 **Ursachen des Aussterbens:** Zerstörung des Lebensraums; Beute eingeschleppter Füchse und Hauskatzen **Besonderes:** Benutzung von 5–6 Nestern zur selben Zeit; zum Greifen geeigneter Schwanz, dadurch zusätzliches Werkzeug beim Transport von Nahrung und Nestmaterial

[2] **Länge:** 64 cm **Gewicht:** 1,1–1,6 kg **Lebenserwartung:** 4–6 Jahre **Ehemaliges Verbreitungsgebiet:** Südostaustralien **Zeitpunkt des Aussterbens:** 1923 **Ursachen des Aussterbens:** Beute eingeschleppter Füchse und Katzen; Zerstörung des Lebensraums **Besonderes:** Einzelgänger außer während Balz, Paarung und Aufzucht der Jungen; nachtaktiv; zum Greifen geeigneter Schwanz, dadurch zusätzliches Werkzeug beim Transport von Nahrung und Nestmaterial

[3] **Länge:** 67 cm **Gewicht:** 1–1,5 kg **Lebenserwartung:** 2–4 Jahre **Ehemaliges Verbreitungsgebiet:** Westaustralien **Zeitpunkt des Aussterbens:** 1960er Jahre **Ursachen des Aussterbens:** Beute eingeschleppter Wildkaninchen, Hausrinder, Schafe, Katzen und Füchse; Jagd **Besonderes:** ein Männchen und mehrere Weibchen pro sozialer Gruppe; Höhle mit kurzem Tunnel und ein bis zwei Eingängen oder große Höhle mit bis zu 100 Eingängen und mehr als 50 Individuen verschiedener Gruppen

SAL
IGON
DIS

这是由法国出版社 Cent Pages 出版的一个文学系列。这个系列专门出版当代文学、被遗忘的经典作品，以及一些佚名的作品，内容涵盖小说、杂文等。这个系列的书籍采用锁线胶装的装订方式，封面深棕色单色印刷，印刷挂网较大。文字大多为留白，部分采用烫金工艺。封面字体组合穿插错落，体现出新颖的节奏感。内页开始的部分为粉红色胶版纸印黑，主要为正文的铺垫信息，类似于电影标题出现前的引子。正文开始以极薄的白色书写纸印单色黑，排版考究。上下及右侧书口均刷红色，体现出“经典”书籍的质感，但全书极其柔软，便于翻阅，且没有多余的装饰，体现了书的最后所言“为了让文本看起来轻松”。

GILBERT
SORRENTINO
MULLIGAN
SALMI SEW
GONDIS
ÉDITIONS
CENT
PAGES

This is a literary series published by Cent Pages, a French publishing house. This series specializes in the publication of contemporary literature, forgotten classic works, and some anonymous works, including novels, essays, etc. This series of books are prefect bound with thread sewing. The cover is printed in dark brown with large printing screen. Most of the characters are white; some of them are hot stamped in gold. The combination of fonts gives the cover a dynamic appearance. The beginning part of the inner pages is printed on pink offset paper in black, which is mainly the paving information for the text. The exquisitely typeset text is printed on thin and white writing paper in monochrome black. The three edges of the book are painted red, showing the texture of 'classic' books. The book is extremely soft without any redundant decoration, which confirm the book's last words, 'to make the text look easy'.

铜奖

Cent Pages 精选系列

法国

Bronze Medal

Collection cent pages

France

sp Millot

Éditions cent pages, Grenoble

195×125×20mm

392g

512p

Let's do the Time Warp again! With "Collection cent pages" the book-designer Philippe Millot has successfully accomplished a journey in time through the history of literature and typography and one is happy to accompany him in his exuberant typographical divertissement. A black cover with large white letters and a short biography, embossed in metallic colours, does not yet reveal what typographical sophistication and wit await the reader inside the book. In the appendix to each volume we learn that the book series is also a homage to the American type-designer Matthew Carter.

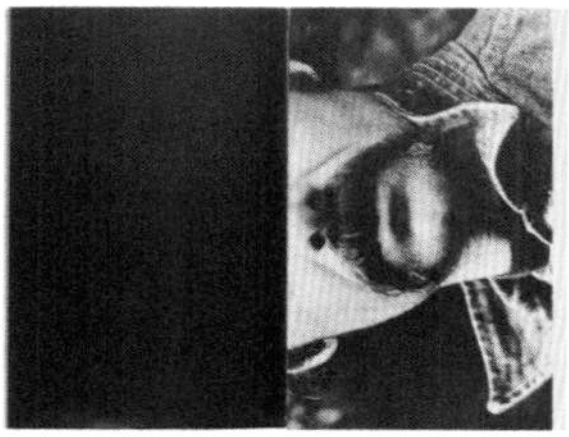

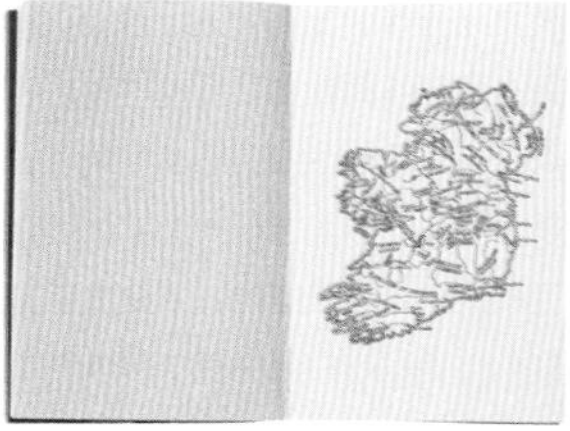

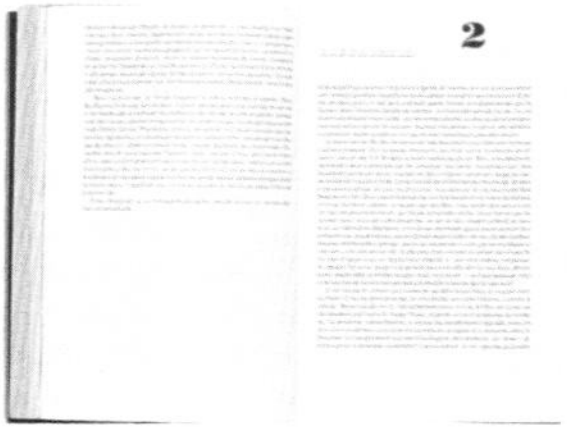

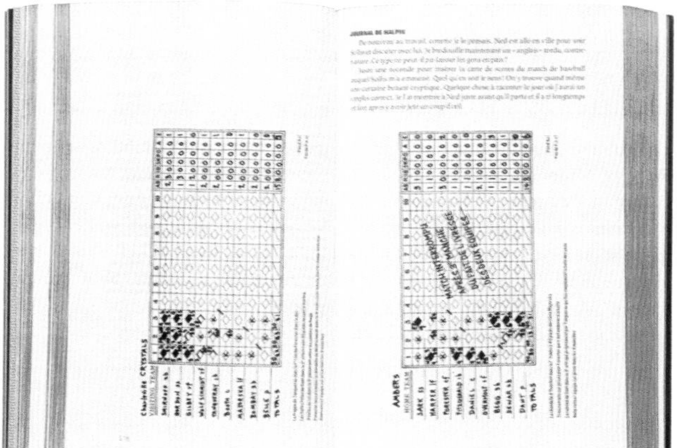

SAL
MIGON
DIS
MULLI
GAN
STEW
1
4
LA
SUEUR
DE L'AMOUR

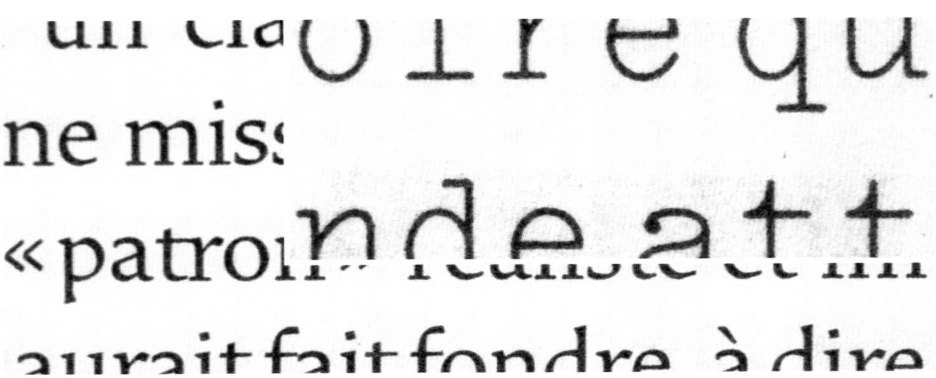
ne miss
«patron
aurait fait fondre, à dire

Isbn 978-2-906724-92-1 Gilbert Sorrentino / Salmigondis / Traduit par Bernard Hœpffner avec le concours de Catherine Goffaux
512 pages / 30 e Éditions cent pages

la gorge de Ned Beaumont s'échappa un bruit de gargouillis tandis que la peur le dégoût et la nudité d'une lubricité galopante se gravaient furieusement sur son visage. Même Daisy rosit un peu, et je savais à quoi elle pensait — mais permettez-moi de fermer la porte sur ce sujet ! Si vous le voulez bien. Elles étaient comme mes allusions ont pu vous le faire comprendre, ce genre de femmes qui aiment bien vous « couper le sifflet ». Me faut-il insinuer davantage ? Mais, et suivez-moi attentivement dans ma grossière confession, se trouvaient brodés sur cette section de leur tunique qui recouvrait tout en les caressant leurs seins fermes et démoniaques, des symboles assez étranges, en points d'un rouge éclatant, comme dans le dessin suivant :

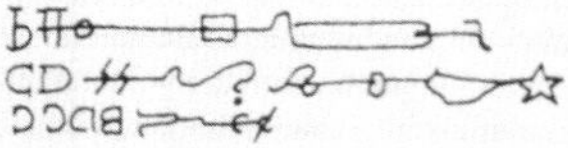

Quel que soit le galimatias que ces soi-disant symboles cryptiques prétendaient représenter, je vous assure qu'en tout cas ils étaient efficaces. Nous étions tous trois en émoi devant eux, et Ned Beaumont était secoué par l'agitation. Les deux miss sataniques, quoi qu'elles aient pu être d'autre, étaient loin d'être des bleues du monde du spectacle et le seul fait de leur apparition sur la scène, ainsi costumées en un mélange de prêtresse, de charlatan magique et de gonzesse sexy était de la dynamite particulièrement géniale ! Pour un *panache* ou un autre, elles lancèrent des jeux de cartes en l'air, en direction des clients, tandis qu'elles s'activaient en une performance terpsichoréenne franchement dégueulasse tant elle manquait de talent mais qui leur permettait d'exhiber, comment dire ?, leurs charmes corporels. Et elles n'y allaient pas de main morte ! Ayant rapidement achevé leur « danse », vous me feriez extrêmement plaisir en oubliant ce dernier substantif, serrées l'une contre l'autre sur le devant de la scène, et baignées d'une lumière bleutée qui ressemblait mystérieusement à la teinte des cocktails — d'ailleurs nos verres avaient été silencieusement remplis à nouveau et la boisson semblait à présent posséder la saveur de la réglisse, à peine croyable ! —, elles se mirent à chanter une étrange chanson sur l'air d'un vieux succès du vaudeville, « Moonlight On Our Mañanas », dont vous vous souvenez sans doute selon divers degrés de répugnance.

Rogo, rogo, gajja mogo, habba dabba dou !
Est notre fier baragouin, ah oui ?
Dur comme motte, fort comme poutre
Gomba ! Labba houdou, ah oui ?
Nein ! On en a rien à foutre !

402

Voici ma copine chérie, Corrie Corriendo!
Jetez salace un œil sur ses tétons!
Salé gronœud dans l'pantalon?
Et si d'une main pulper un téton?
Hein! Pas de chance pour toutes ces loutres!

Ouais, regardez donc ici Berthe Delamodii!
C'est qu'on a là un gros ka-kii?
Falloun, falloun, falloun – wahou?
Doigté pour encercler ka-kii!
Wein! Ju-jus c'est pour les outres!

Ici, les deux jolies jolies ladies, si zolies, zolies!
Toutes les minutes dans le clivage?
Ambidexteriprestidigit!
Qui songe à voir si doux clivage?
Fein! D'abord la boue et puis la yourte!

Car! C'est le show-business pour vous!
Ho! Laffa-daffa suceça-mouilla!
Ambicliva canarda-souilla!
Tétons popotin c'est du massepain
Éclate falzar et toutes les nippes!
Tous des blastiphages et des tourtes!

Lèche-la quelle gueule t'as là!
Qui pousse donc non-non? Qui? Qui?
Sullivandaire kumquat! Si l'on s'en sert!
Affriortolant c'est qu'on y viendra
Au vestibule!
U'cazz a la mac et cul!

ıs avez beau vous en moquer, et sans doute le ferez-vous, ce n'était pas une ention, et je n'ai d'ailleurs ni créé ni changé une seule syllabe de cette étrange ansonnette. Je l'ai lue sur le papier mot pour mot telle qu'elle fut émise par rs bouches plantureuses et leurs lèvres épanouies. Son effet sur le public fut ctrique! Tout le monde dans le club, dans la pénombre, les hommes et leurs enouilles, tous etaient assis là, hypnotisés, comme frappés par le tonnerre. fut l'effet de ces paroles complètement dingues sur le groupe pas trop bête i constituait l'auditoire. Pas des pedzouilles, ceux-là! pour construire une e phrase. Je me rends compte maintenant, avec le recul que procure une

100

2008 Bm4

这是一本有关瑞士人民党（简称 SVP）参与各项社会活动的书。本书的两位作者通过四年时间的跟拍和记录，为整个党派参加会议、举行庆祝活动、名人聚会、办公及生活场景，以及相关文字论述做了详尽的收集和整理。本书为锁线胶装，并无特殊工艺，但在设计和编排上做了大胆的尝试。封面白卡纸四色印刷覆亮膜，整体设计风格似报纸，版面划分为五栏，内容布满四封，除了封面的中心照片和书名信息以及封底最后的版权信息、作者介绍外，其余内容均拥有独立编号，与内文一一对应，起到了目录的作用。内文哑面铜版纸四色印刷，每篇文章或照片拥有独立编号，对应封面上的标题编号，而没有常规的页码。阅读过程中，需要反复在内文和封面之间翻阅，内文仿佛一叠相互独立的资料，被夹在类似报纸的封面当中，有很强的新闻纪实感。

Edition Patrick Frey
Die Mitte des Volkes
Fotografie
Fabian Biasio
Texte
Margrit Sprecher

This is a book recording the participation of the Swiss People's Party in various social activities. Through four years of follow shot and recording, the two authors have collected photos of the party's participation in meetings, celebrations, celebrity gatherings, office and life scenes, as well as related documents. This book is perfect bound with thread sewing without any special technology. However, the designer has made a bold attempt in typesetting. The laminated cover is made of white cardboard printed in four colors. The overall cover design is like newspaper – the layout is divided into five columns and the contents occupy four columns. Except the cover image, book title, the copyright information on the back cover and the author's introduction, all the rest of the contents have independent numbers, corresponding to the text one by one. The cover plays a role of catalogue. The inside pages are made of matte art paper printed in four colors, each article or photo has a number matched with the title number on the cover. Since there is no normal page number, the readers need to check the cover and inside pages repeatedly during the process of reading. The stack of independent materials is sandwiched in the newspaper-like covers, which gives the book a documentary appearance.

◎ 铜奖

《 人民的中心

◖ 瑞士

◎ Bronze Medal

《 Die Mitte des Volkes

◖ Switzerland

D Prill & Vieceli

△ Fabian Biasio, Margrit Sprecher

|||| Edition Patrick Frey, Zürich

□ 285 × 211 × 10mm

593g

187p

This is not so much a book as a reportage. It's not something I read, it's something I am part of, as I hastily flick through it, quasi travelling together with photographer and authors through the Switzerland of the SVP (the Swiss political party "Schweizerische Volkspartei"). The book-designer uses a variety of means without ever going off the rails. The book is always charging ahead, its pages anxious to be turned. It has the loudness of the SVP – no place here for subtlety – and yet its design is well considered, planned and expertly handled. This is the book's principal achievement. It constitutes contemporary design, evolved out of the subject matter, and as such individual and unique.

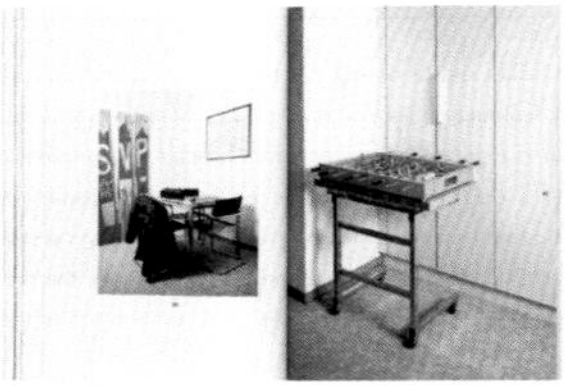

SVP UDC
Qualité suisse
140

in im
enministert

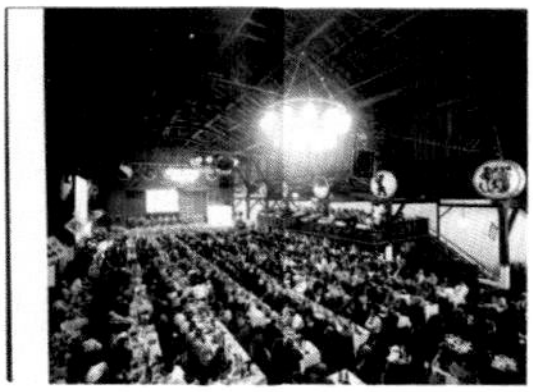

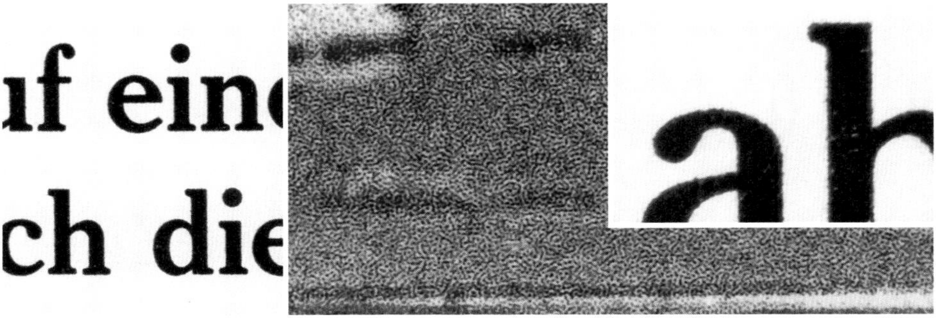
ıf ein
ch die
ah

Edition Patrick Frey

Die Mitte des Volkes

Texte Margrit Sprecher

Fotografie Fabian Biasio

84

85

86

87

88

89

90

91

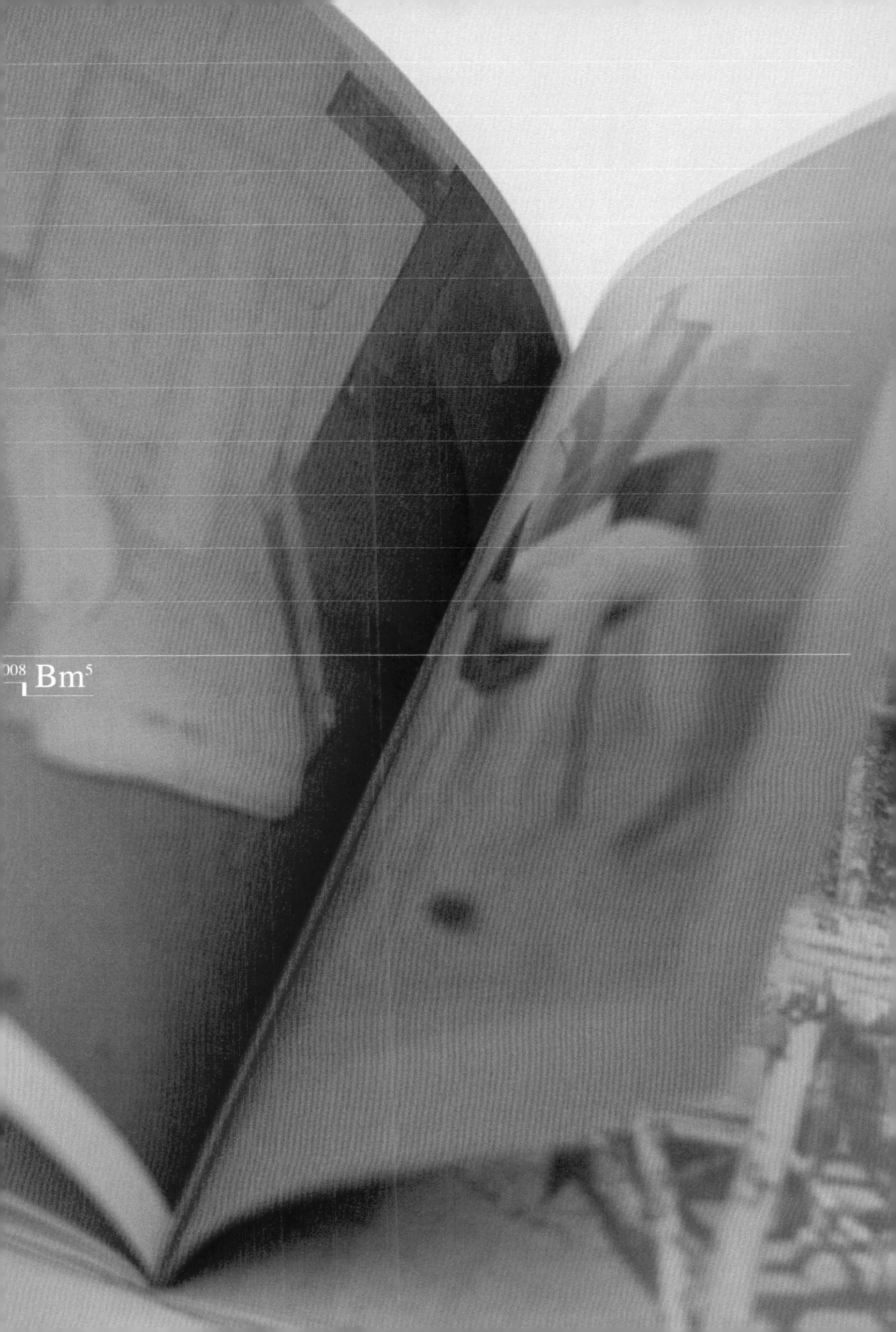
008 Bm5

这是瑞士著名家居企业维特拉（Vitra）2007－2008 年的家居主题杂志。因为是杂志而不是新产品宣传册，本书主要内容为各著名设计师设计并由维特拉生产的家居用品，且绝大多数照片是这些用品真实地放在不同人家里的状态。本书为锁线胶装，外包单色印刷图案的海报做护封。内封白卡纸毛面五色印刷（四色加印银），其中印银为底色，封面上印有为维特拉设计过家居产品的著名设计师的名字。内文通过多种纸张混合装订来区分板块，光面胶版纸四色印刷的家居场景照片、棕色纸张挂粗网单黑印刷的新品、胶版纸玫红渐变底色和挂粗网印黑的未完成或结构拆解产品、胶版纸单黑印刷的含有产品内容的各类小游戏，以及常规宣传册会有的产品、设计师介绍等相关内容。本书把时尚杂志的概念融入企业宣传图录中，传递了把经典设计融入日常生活的企业理念。

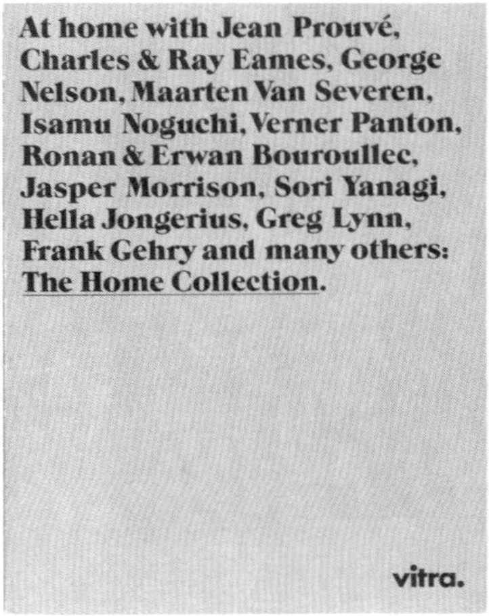
At home with Jean Prouvé, Charles & Ray Eames, George Nelson, Maarten Van Severen, Isamu Noguchi, Verner Panton, Ronan & Erwan Bouroullec, Jasper Morrison, Sori Yanagi, Hella Jongerius, Greg Lynn, Frank Gehry and many others: The Home Collection.
vitra.

This is the 2007/2008 home magazine of Vitra, a famous Swiss home furnishing company. It's a magazine, not a brochure for new products. The housewear and furnishings designed by famous designers and produced by Vitra are displayed in the book, and most of the photos are taken in different homes. The book is perfect bound with thread sewing and wrapped by a jacket printed with posters in monochrome color. The inside cover is made of white cardboard printed in five colors (four colors plus silver) and the silver is the background color. The names of the famous designer who designed the household products for Vitra are printed on the cover. Different parts of the text are distinguished by various kinds of paper, for example, real scene photos are printed on four-color printing offset paper, new products are displayed on brown paper printed in black only, unfinished or disassembly products are printed on rose red offset paper in black with rough screen printing, various games containing product introduction and designers' information are printed in single black on offset paper. This book incorporates the concept of fashion magazine into the enterprise publicity catalogue, and conveys the concept of integrating classical design into daily life.

铜奖
维特拉家居集 2007－2008
瑞士

Bronze Medal
Vitra. The Home Collection 2007/2008
Switzerland
Cornel Windlin
Vitra AG, Birsfelden
274×210×15mm
647g
171p

Ten pages are used to discuss what Vitra wishes to achieve in its publication, the presentation of its "Home Collection". A further 172 pages elaborate on Vitra's "Home Story". That this piece of work succeeds in transmitting an identity can be largely attributed to its designer Cornel Windlin. It is surprising with what wealth of ideas the beautification of everyday life through the company's well-known design classics is conveyed. Everyday life here means what's special as well as what's trivial, something that can be seen in the illustrations and choice of materials. The jacket is a folded poster, on the inside of which the "objects" can be seen being delivered.

New Products

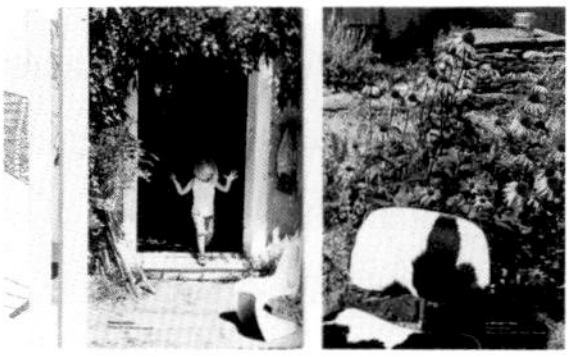

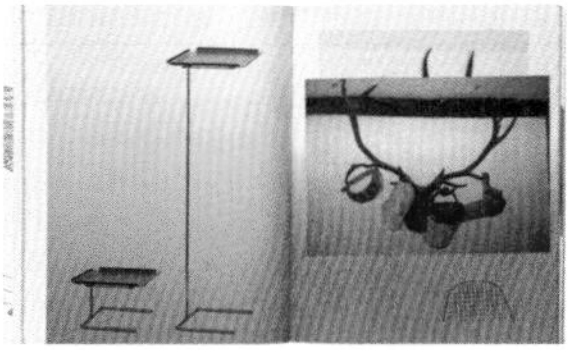

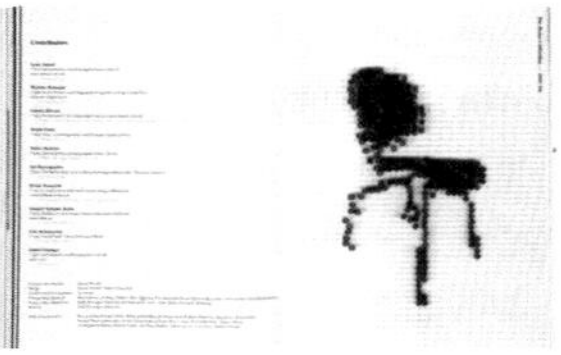

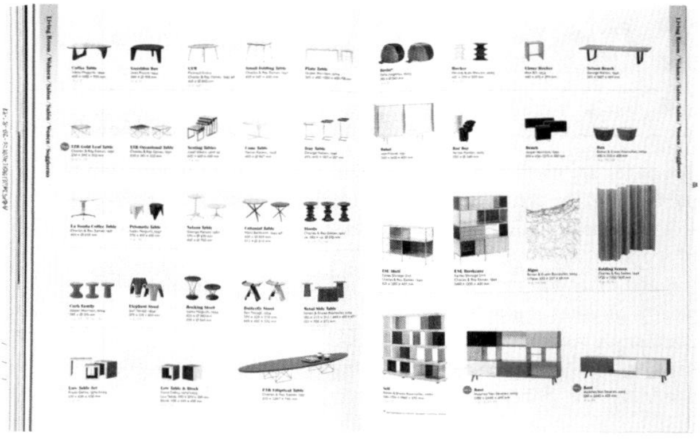

SUPER SPASS!
SANY Mix
Rätsel MIX
VOL 13
HELLAMSJ
WWPOOPEA
YAKARIMS
NMOONIAP
ADUARTEE
SVVNANOR
ETERWANN
TMUTORCS
TUIARTIV

Ha

本书是天津后海传媒的企业理念和作品集。书籍通过理念阐述、作品呈现、人物介绍、未来畅想来全面解读天津后海传媒的现状以及愿景。本书大量采用格式工艺、材料和装订手法，使得书籍本身也成为这家公司集中表达创意的一件作品。书函采用酒红色半透明亚克力制作，封面白卡纸黑色和荧光橙色双色印刷，覆哑膜后于荧光墨所印插图上做磨砂 UV 处理。翻过与书函匹配的深红色环衬，是多达数十页的四色印刷宣纸，内含序言。内页分为理念、标识、地产、媒体等七大部分，胶版纸四色印刷。中间通过短于开本 10mm 的月影纸烫透工艺处理插图，成为各部分的隔页。最后配以企业与合伙人介绍作为附文。书籍上下及右侧翻口刷酒红色电化铝，将所有可以设计的部分都覆满了这家广告公司的创意。

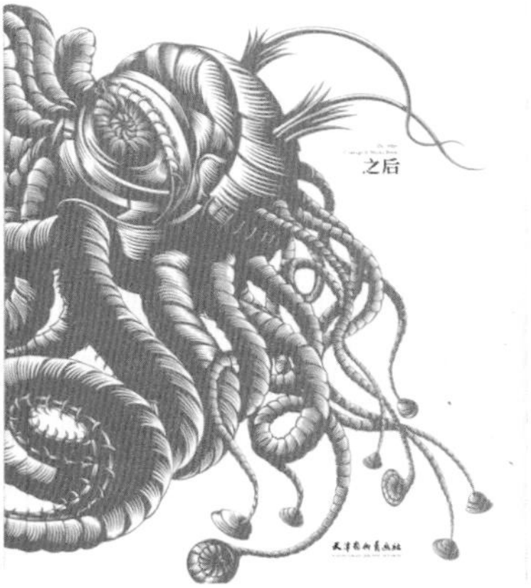
之后

This book is about the corporate philosophy and works of the After Media in Tianjin. The book comprehensively interprets the present situation and vision of the After Media in Tianjin through the elaboration of ideas, the presentation of works, the introduction of characters and the imagination of the future. The book uses various crafts, materials and binding techniques to make the book itself a creative work, which can express the company's originality. The slipcase is made of wine-red translucent acrylic. The cover is printed in black and fluorescent orange on white cardboard. After laminating, the illustrations printed in fluorescent ink are finished with spot UV. Turning over the dark red endpaper, there are dozens of pages printed in four colors on rice paper, containing the preamble. The inside pages are divided into seven parts: concept, logo, real estate, media and so on, printed on offset paper in four colors. The illustrations are processed with hot stamping technique on art paper, which becomes the separating pages of each part. The introduction of enterprises and partners is attached at the end of the book. The three edges of the book are bronzed with wine-red anodized aluminum. Everywhere of the book is designed to show the creativity of this advertising company.

- 荣誉奖
- 之后
- 中国
- 王成福、耿耿
- 孙乃强、王磊、王成福、史长胜
- 天津杨柳青画社

- Honorary Appreciation
- The After Concept & Works Book
- China
- Wang Chengfu & Geng Geng
- Sun Naiqiang, Wang Lei, Wang Chengfu, Shi Changsheng
- Tianjin Yangliuqing Fine Arts Press, Tianjin
- 256×216×40mm
- 1710g
- 456p

Too beautiful and well-designed to go away from the competition empty-handed. They know all the tricks as well as how to use them ... elaborate features, successful choice of materials and masterly diversity in the double-page layout make this book a real treasure trove. Agency The After has demonstrated its expertise with conviction and great finesse.

The
After

海

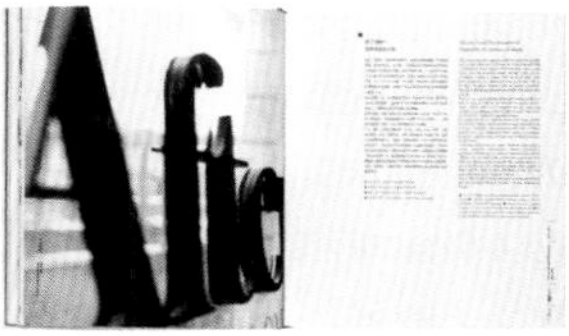
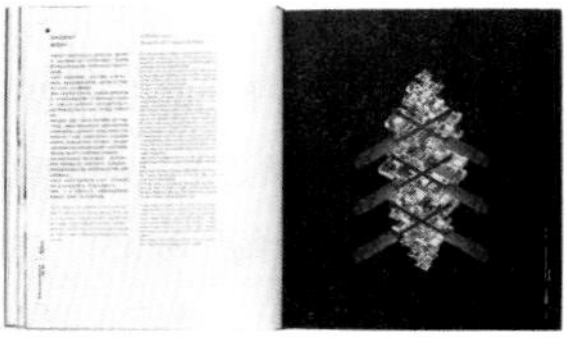
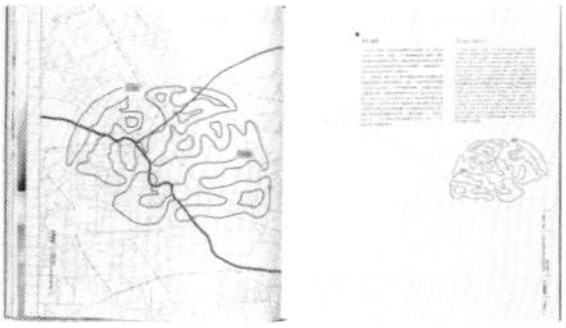
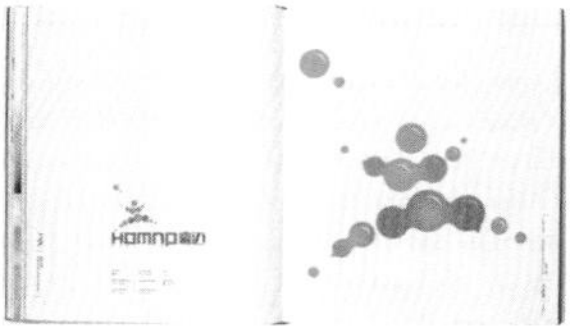
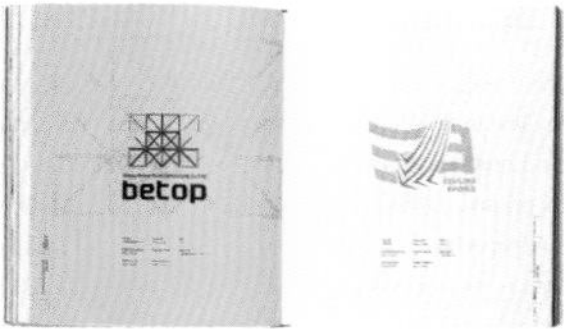
betop

今晚

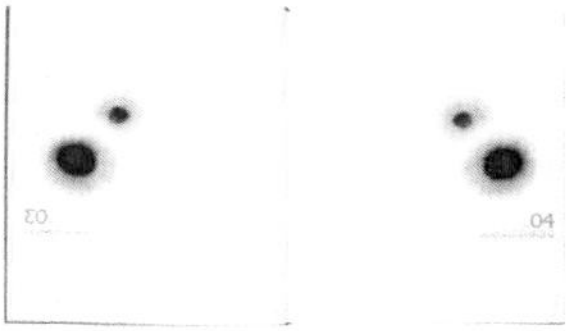
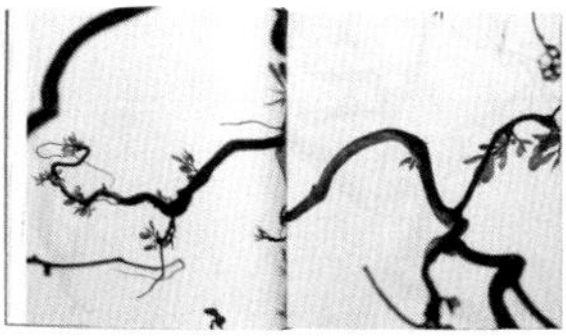
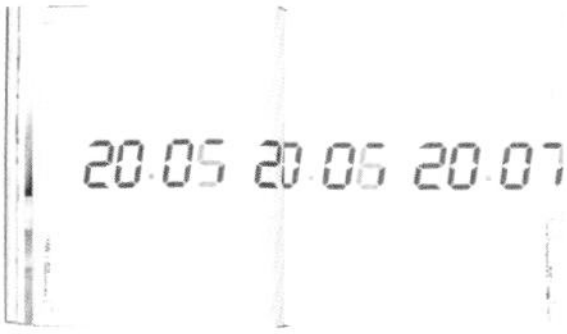
20.05 20.06 20.07
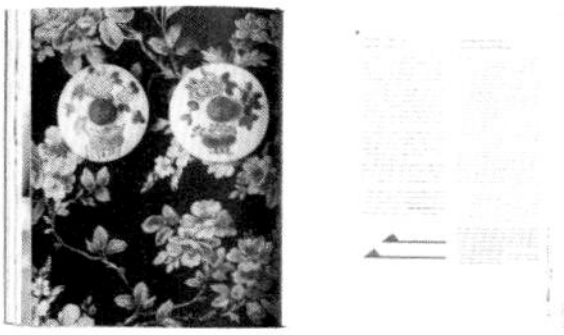
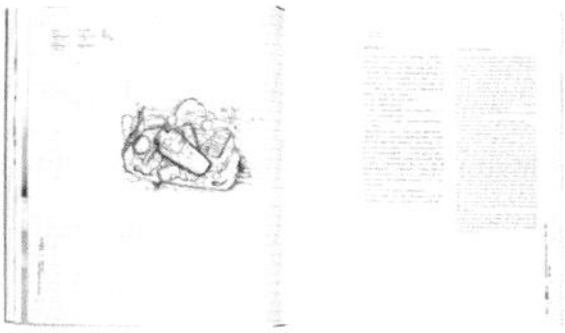
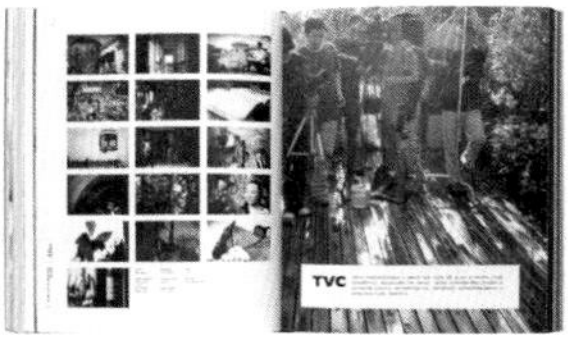
TVC
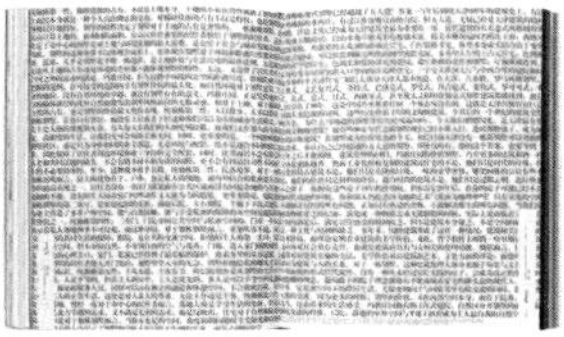
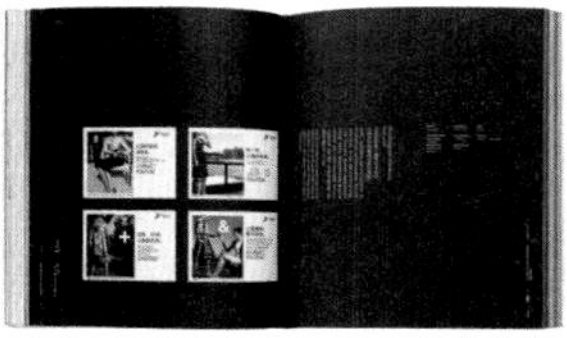
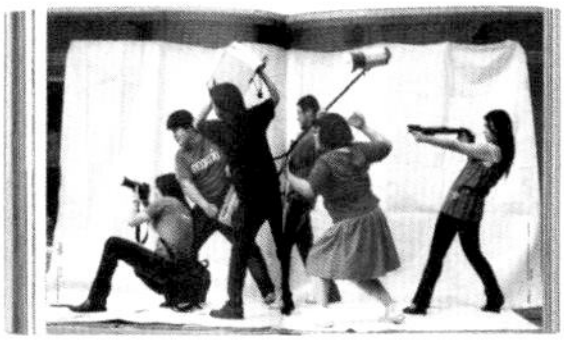
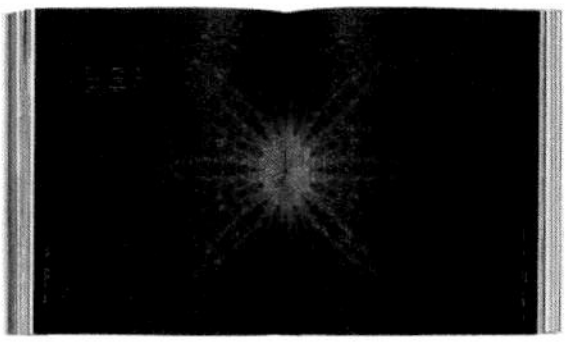
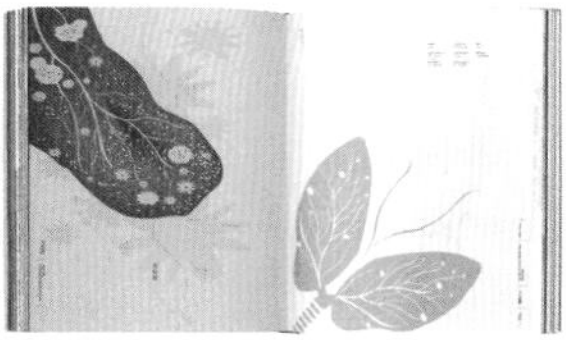
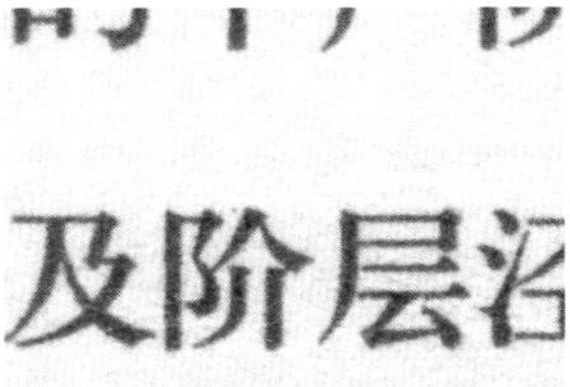
及阶层
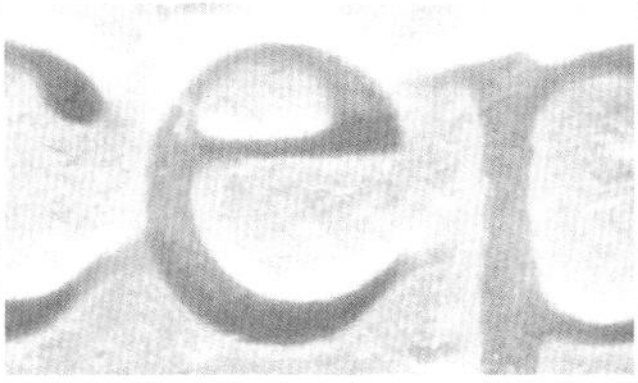
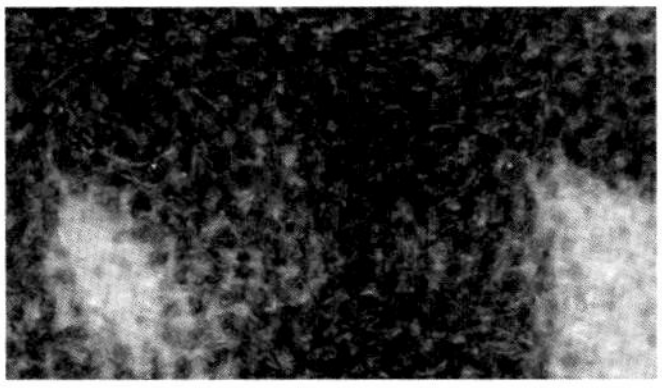
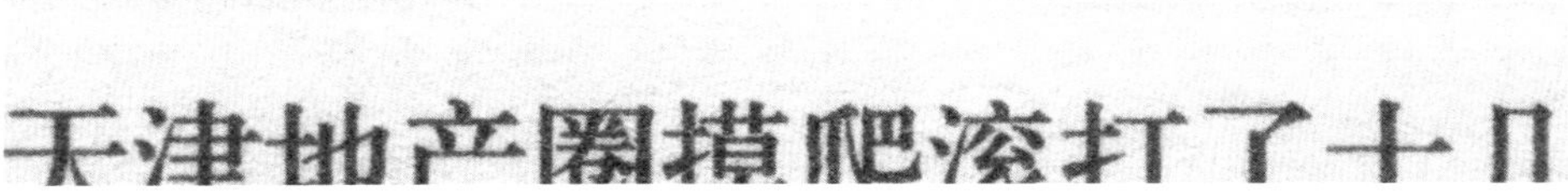
天津地产圈摸爬滚打了十

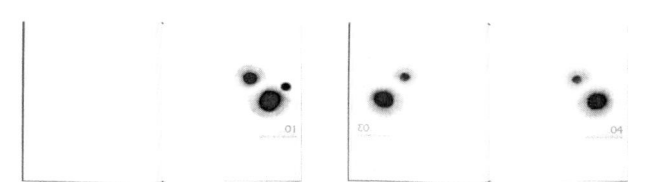

008 Ha2
1

这是蒙德里安基金会 2006 年的年度报告，对基金会参与投入、建设、组织的艺术、文化、公共、国际等几大类项目的数据和进展进行整理归纳。本书的装订形式为锁线胶平装。全书采用蓝、棕、哑金、浅黄、荧光红五色印刷，并通过叠加效果形成第六色——紫色。护封采用极薄的艺术纸张双面印刷产生透叠效果，内封胶版纸双面印刷与护封形成呼应。内页半涂布胶版纸五色印刷。本书在目录之前有 2 页用于解读内容的示例页面，对贯穿全书的 22 个图形所代表的内容分类、8 个符号所代表的方向类别，以及 6 种颜色在书中分别对应的内容做了详细阐释。其中浅黄色和蓝色的同心图形代表这一项目所在的分类、总预算和基金会投入预算，具体数字在项目信息里都有写明。内页的左页天头部分为目录，荧光红色代表当前内容，而地脚部分则为具体的项目索引。本书通过艺术化的手法为这样一份财政报告提供了独特的解决方案。

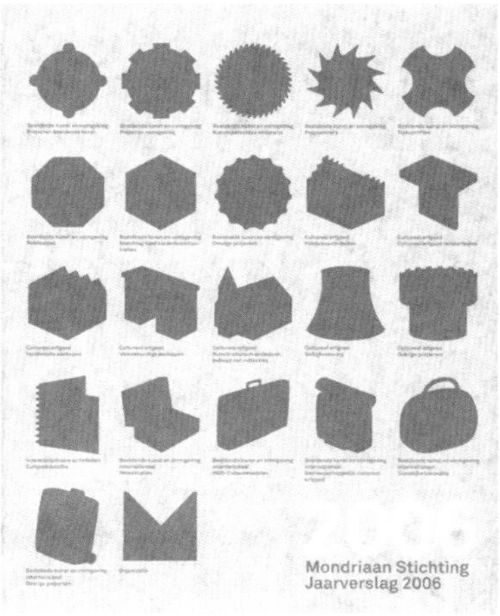
Mondriaan Stichting
Jaarverslag 2006

This is the annual report of the Mondrian Foundation in 2006. It summarizes the data and progress of the Fund's involvement in investment, construction, art, culture, public and international projects. This book is glue bound with thread sewing and printed in five colors – blue, brown, gold, light yellow and fluorescent red, which produce the sixth color purple by superposition effect. The jacket is made of thin art paper with double-sided printing and the inside cover is made of offset paper with double-sided printing as well. The inside pages are printed on semi-coated offset paper in five colors. The book has two pages of sample pages for interpretations before the catalogue. It explains in detail the content classification represented by 22 graphics throughout the book, the direction represented by 8 symbols, and the corresponding content of six colors in the book. The concentric figures in light yellow and blue represent the classification of the project, the total budget and the investment budget of the foundation. The specific figures are shown in the project information. The left header of the inner page contains catalogues, the fluorescent red represents the current content, and the footer part is the index of specific item. This book provides a unique solution to such a financial report by artistic means.

荣誉奖

蒙德里安基金会 2006 年度报告

荷兰

Contemporary design underscores the purpose of this publication (annual report 2006) in a professional but also humorous fashion. Concept, typography and features are convincing in their clarity. The classification by pictograms and the clever and appealing design of tables and statistics serve to document the work of an active cultural foundation in a fresh and inviting manner.

Honorary Appreciation

Mondriaan Stichting. Jaarverslag 2006

The Netherlands

Joost Grootens

Mondriaan Stichting, Amsterdam

275×215×13mm

587g

192p

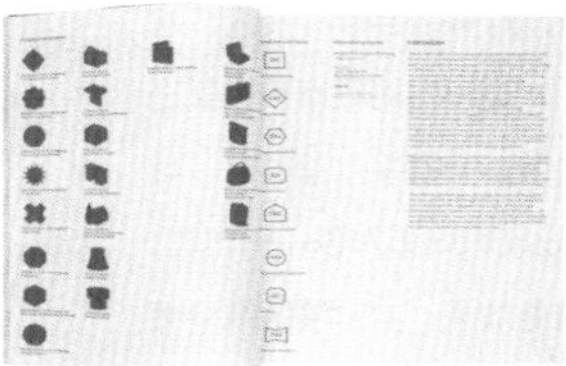

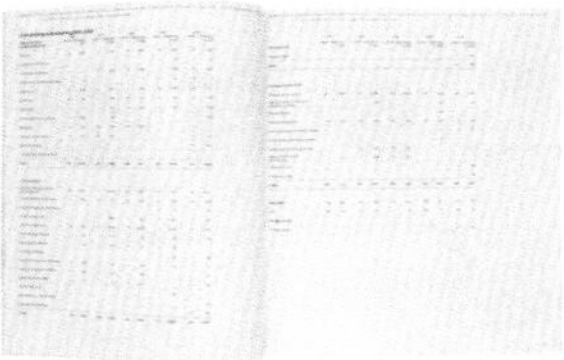
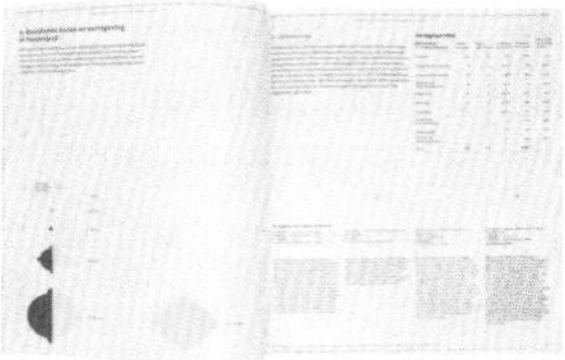
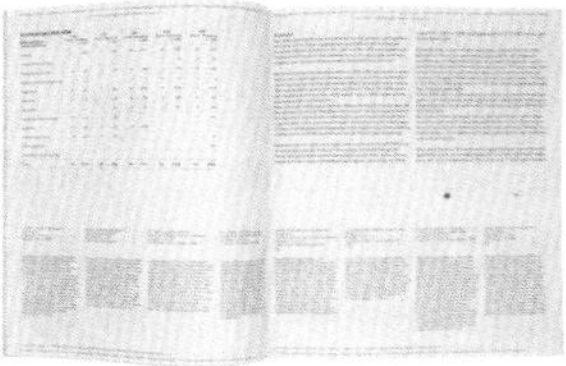

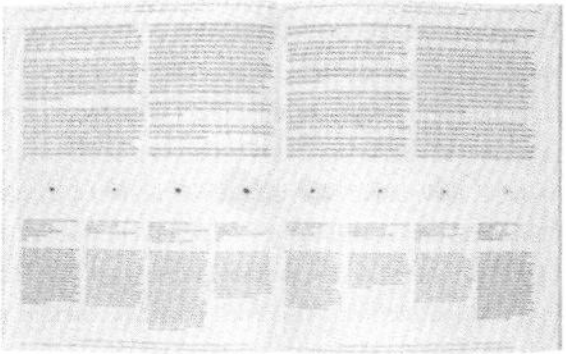

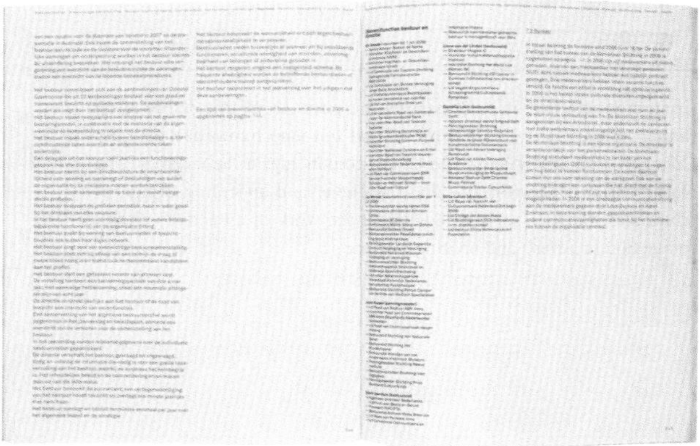
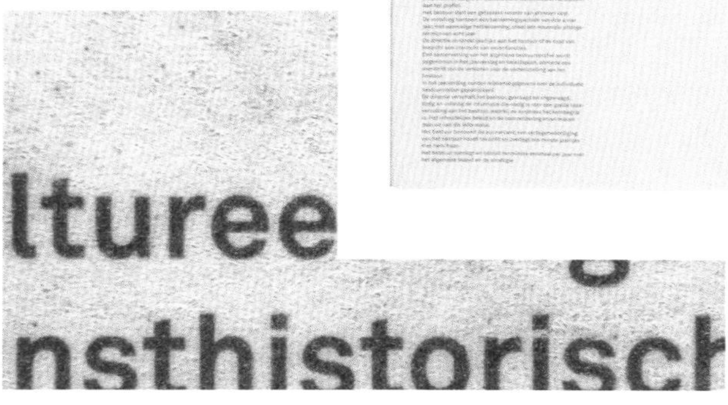
lturee
nsthistorisc

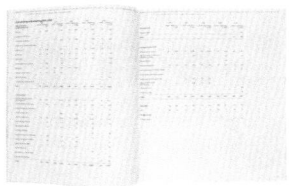

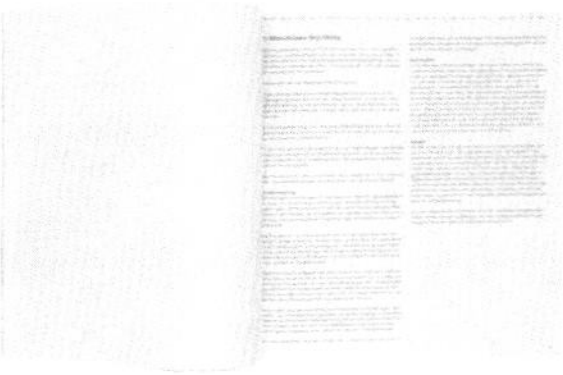
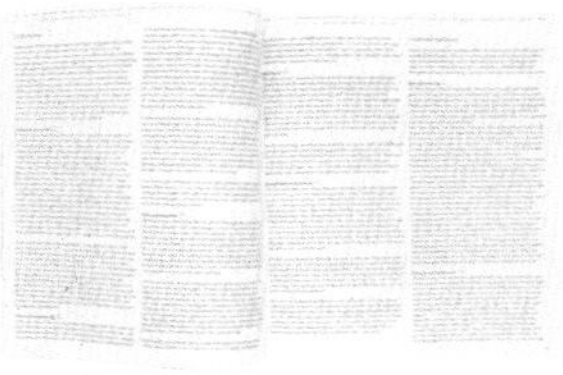
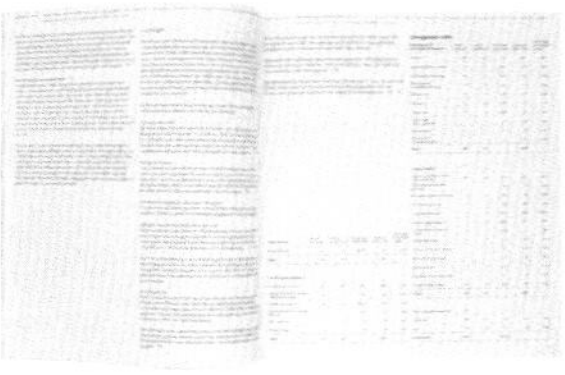

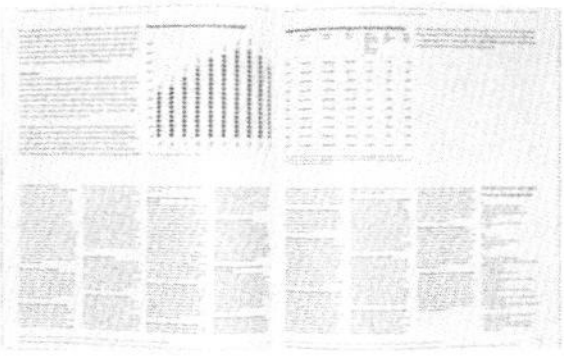

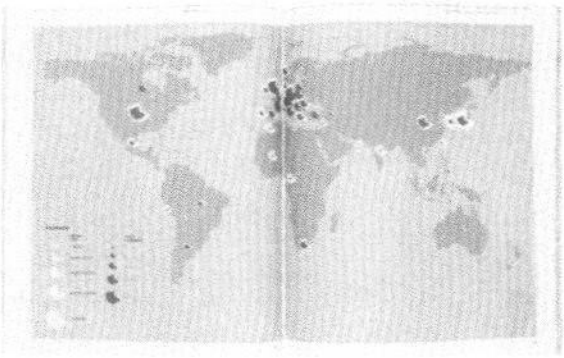

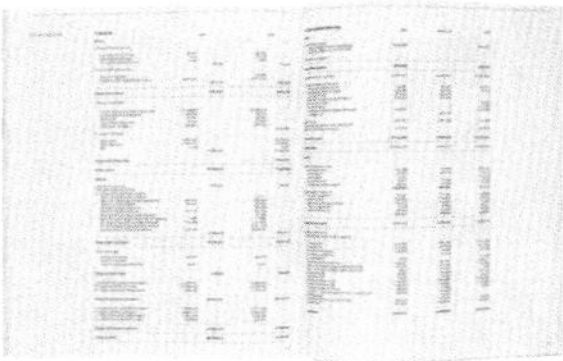

et aan
agina 27, 28). De or

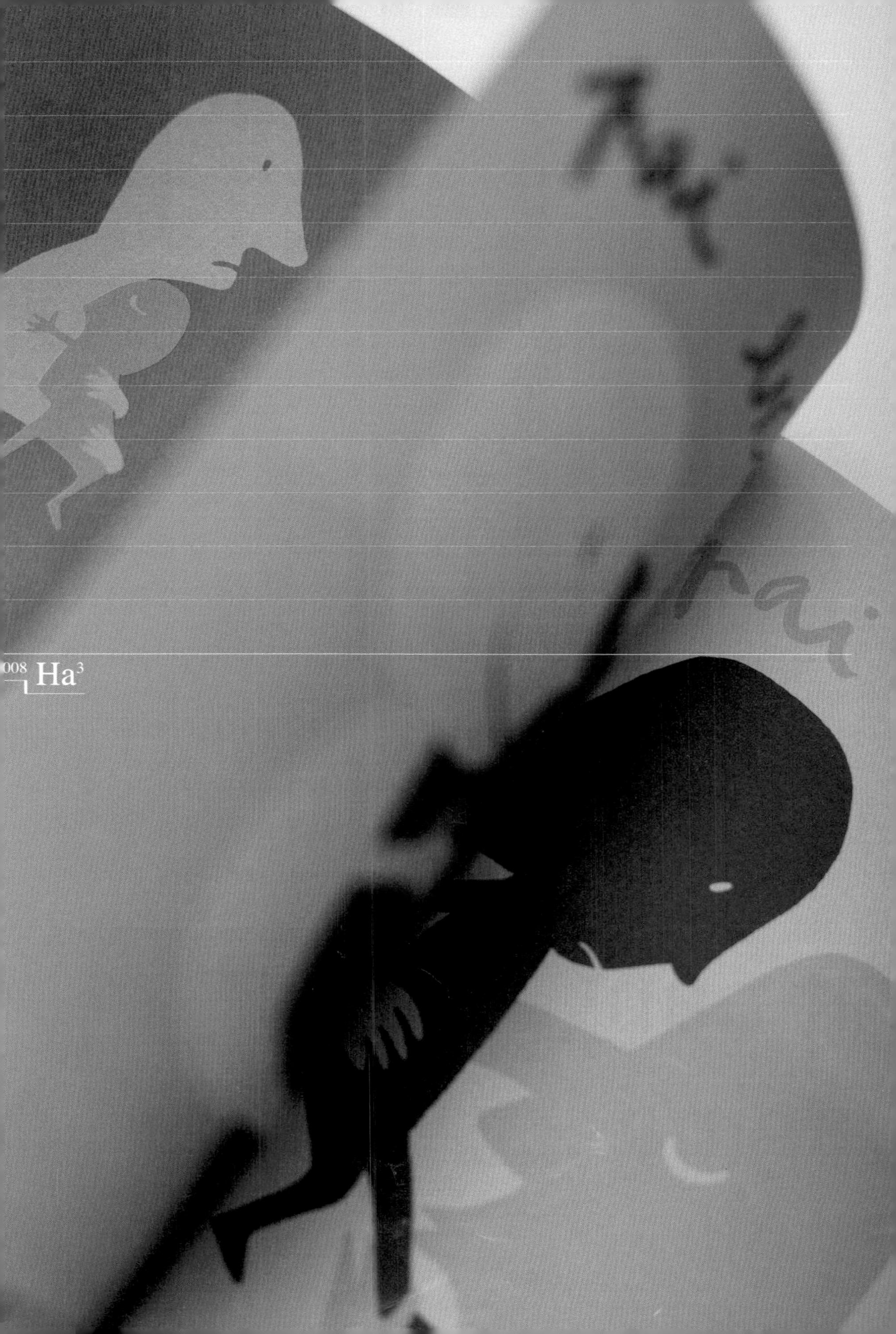

这是一本充满温暖亲情的绘本，由葡萄牙插画家贝尔纳多·卡瓦略（Bernardo Carvalho）绘制而成，并由自己参与创立的童书出版社——橘子星球（Planeta Tangerina）进行出版。本书通过最简洁的画面和文字，来展现孩子成长过程中父亲所扮演的角色。书籍为锁线胶装，较常规的方 20 开绘本，也没有什么特殊工艺，但插图和色彩十分考究。封面白卡纸四色印刷，覆哑膜。内页胶版纸四色印刷，一面一个场景，人物动作生动，但细节和色彩都尽力做到克制，让人们的关注点都集中在父子之间的关系上。书籍的配色十分统一，红、棕、蓝、绿、黄均控制在低饱和度偏暖的倾向上，配合黑白灰的使用，呈现出温暖和谐的画面，与“身边的父亲”这一主题非常契合。

pê de pai
Isabel minhós martins
Bernardo Carvalho

This is a picture book full of warmth and affection, drawn by Bernardo Carvalho, a Portuguese illustrator, and published in Planeta Tangerina, a children's book publishing house he co-founded. Through the most concise pictures and words, this book shows the role of father in the process of children's growth. The book is perfect bound with thread sewing. There is not any special technique, but the illustrations and colors are very exquisite. The laminated cover is printed on white cardboard in four colors. The inside pages are printed on offset paper in four colors, page by page, scene by scene. The characters are vivid, but the details and colors are restrained, so that the readers may focus on the relationship between father and son. The color matching of books is quite consistent. Red, brown, blue, green and yellow are all controlled to be low saturated and warm. With the use of black, white and grey, the harmonious pictures fit well with the theme of 'the father around'.

荣誉奖
爸爸在身旁
葡萄牙

The reduced style of the monotone illustrations, the type, the colours and the paper are as appealing and warm as the book's subject. The double-pages are balanced, with superbly corresponding colour scheme and a rhythm which gently pervades the whole book. The harmonisation and interaction of illustrations and type constitute a further commendable feature.

Honorary Appreciation
pe de pai
Portugal
Bernardo P. Carvalho (Planeta Tangerina)
Isabel Martins, Bernardo P. Carvalho
Planeta Tangerina, São Pedro do Estoril
220×198×4mm
145g
32p

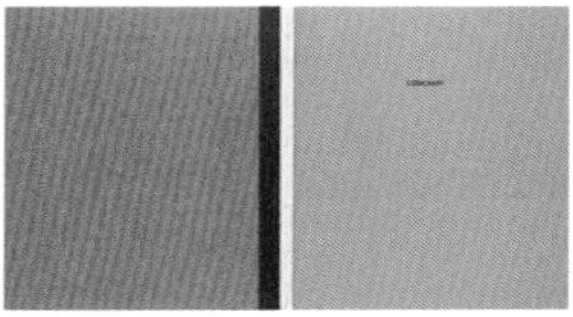

pé de pai

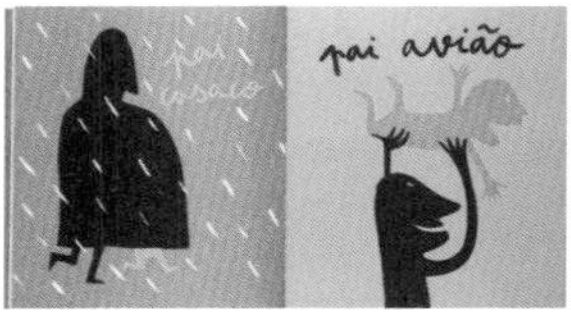
pai avião

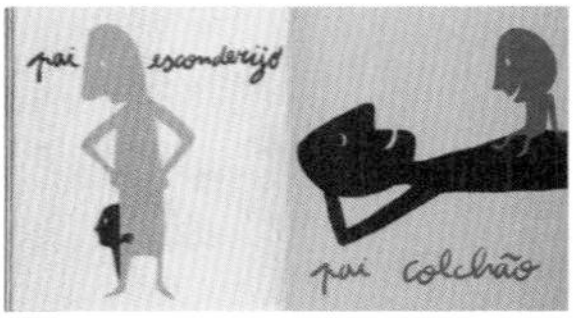
pai esconderijo
pai colchão

pai boia
pai esfregão

pai seta

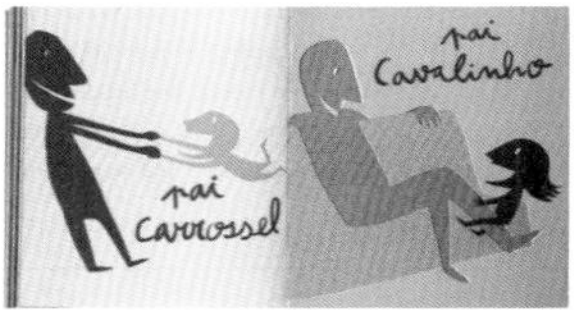
pai Cavalinho
pai Carrossel

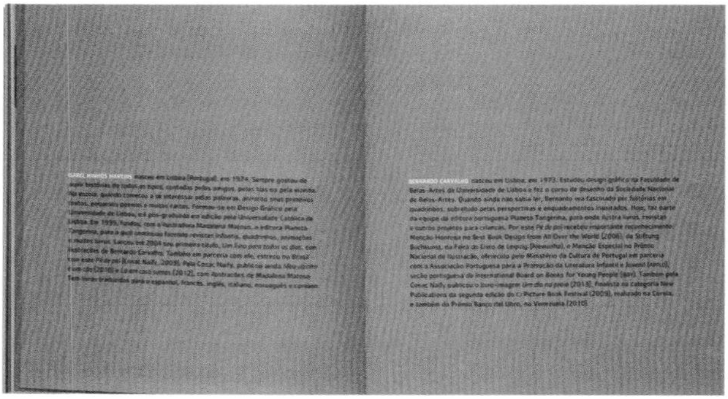

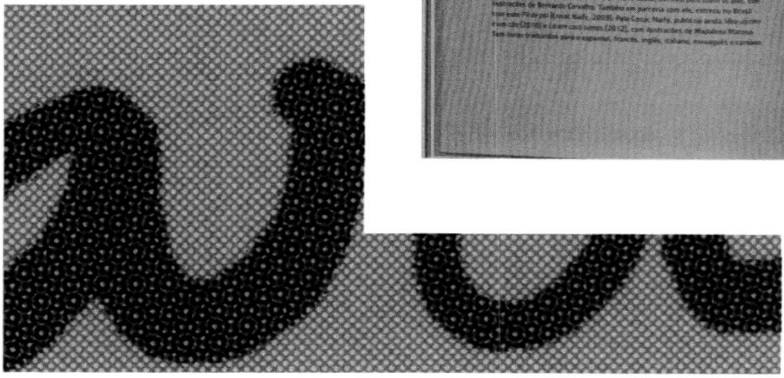

pai
cabide

pai
guindaste
pai
trator

pai
sofá
pai
motor

pai meta

pai ambulância
pai despertador

pai escada
pai doutor

and constan-
a real-time
ications. The
strument thus
and the sig-
the discus-
points whose
observation
immediately
of scientific

这是德国声音艺术家卡斯滕·尼科莱（Carsten Nicolai）的作品集。与常见艺术门类的艺术家不同，尼科莱主要研究声音、艺术和科学之间的联系，通过数学、代码、随机错误等概念融入电子器械来进行声音实验。他化名阿尔瓦·诺托（Alva Noto）来创作电子音乐。本书锁线胶装，护封铜版纸四色印刷，覆布纹膜，封面上的作品前数字为编号，而不是页码，可以通过每张图片下的图片编号进行索引。内封白卡纸印单黑，只印护封的文字与图片线框。内文作品部分展示每件作品的创作器材、声音图谱、最终呈现，哑面铜版纸四色印刷。后一部分为作品解读、艺术年谱、图片信息等，胶版纸印刷单色黑。这本作品集通过大量线条和严格的网格系统，体现了艺术家科学实验的严谨性，但声波图像配合网格系统之下多变的排版又具有很好的律动感，与他充满听觉刺激、渐显渐隐的实验音乐很匹配。

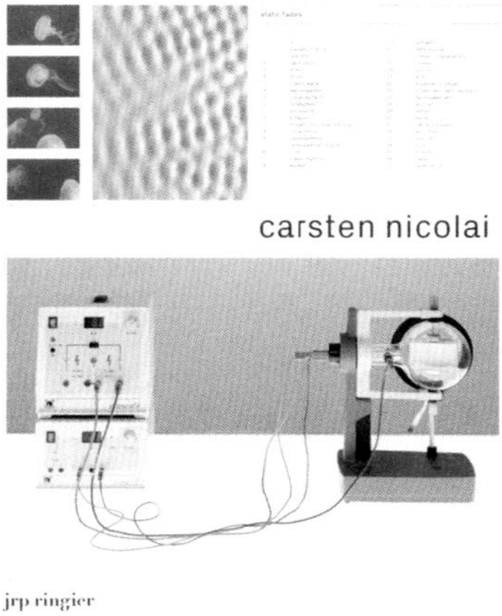

carsten nicolai
jrp ringier

This is a collection of works by Carten Nicolai, a German sound artist. Unlike artists in common art categories, Nicolas mainly studies the relationship between sound, art and science. He integrates concepts such as mathematics, code, and random errors into electronic devices for sound experiments. His alias Alva Noto is used to create electronic music. The book is perfect bound with thread sewing. The jacket is printed on coated art paper in four colors, laminated with cloth film. On the cover, the numbers in front of the work titles are serial numbers, not the page numbers. They are consistent with the numbers under each picture. The inside cover is printed on white cardboard in black. Each work is introduced by displaying the creating equipment, sound atlas and final presentation, printed on matte art paper in four colors. The latter part contains the interpretation of works, art chronology, and picture information, etc., using offset paper printed in monochrome black. This collection of works reflects the preciseness of scientific experiments through a strict grid system; however the sound image matching with the changeable layout under the grid system has a dynamic appearance, which is consistent with his experimental music full of auditory stimulation.

- 荣誉奖
- 卡斯滕·尼科莱——渐显渐隐
- 瑞士

A very sensitively designed book. From the cover to the last page the same visual language is spoken. Simple in its design yet rich in detail, everything is like a soothing melody.

- Honorary Appreciation
- Carsten Nicolai. Static Fades
- Switzerland
- Gavillet & Rust
- Dorothea Strauss, Haus Konstruktiv, Zürich (Ed.)
- JRP Ringier Kunstverlag AG, Zürich
- 286×238×13mm
- 892g
- 157p

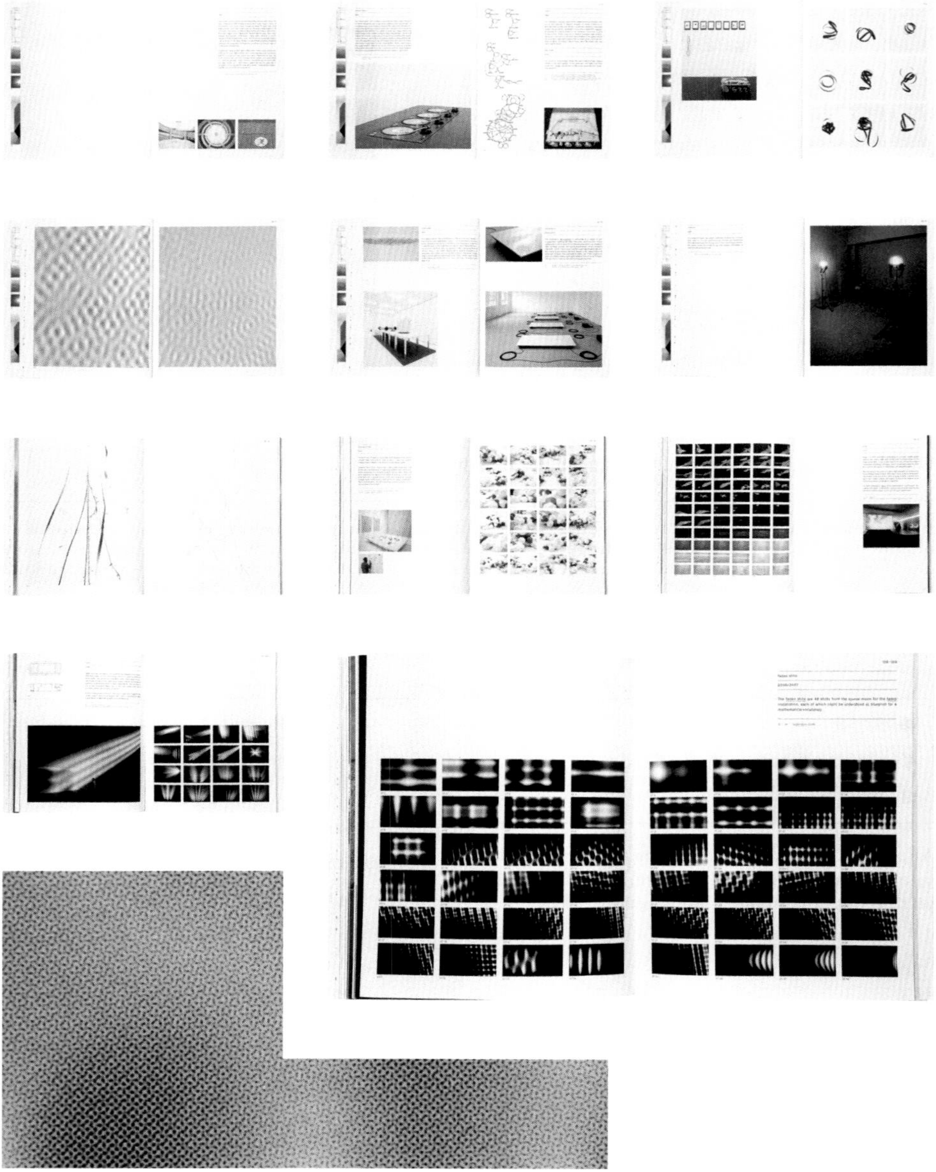

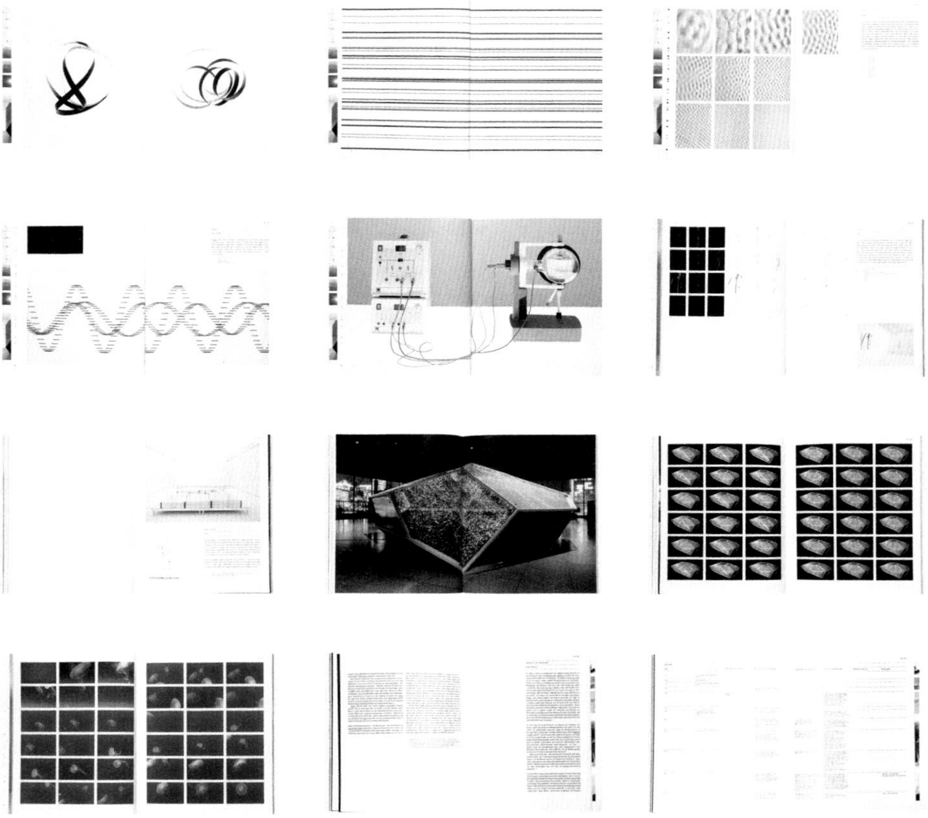

ms of
ion. Th

86 Josef Sudek
dokumentační fotografie
1937
77
008 Ha5

这是一本探讨捷克著名的文化和社会组织“Družstevní Práce”（以下简称 DP）及为其作出突出贡献的两位人物之间故事的论著。书籍讲述了 DP 产生的时代背景、摄影师约瑟夫·苏德克（Josef Sudek）的广告摄影作品、他与 DP 的关系、他与平面设计师拉吉斯拉夫·苏特纳尔（Ladislav Sutnar）配合工作等方面，展现了那个独特的年代。本书的装订形式为无线胶平装，外套无任何印刷和工艺的卡板书函。封面白卡红黑双色印刷覆哑膜。封面四块线条组成的留白暗示主要内容的四大块。内页前三分之二为捷克语主体部分，依照内容划分为四大板块，并由封面留白的四块花纹在下书口上的运用进行区隔，后附人物介绍、年表、资料等附文，书口花纹为单黑渐变。这部分采用偏暖的浅灰色哑面涂布纸四色印刷。内页的后三分之一为前文除彩图外的所有文字部分的英文版，胶版纸印单色黑。书籍设计严谨，版式考究，配合一定年代的彩图，具有很强的文献感。

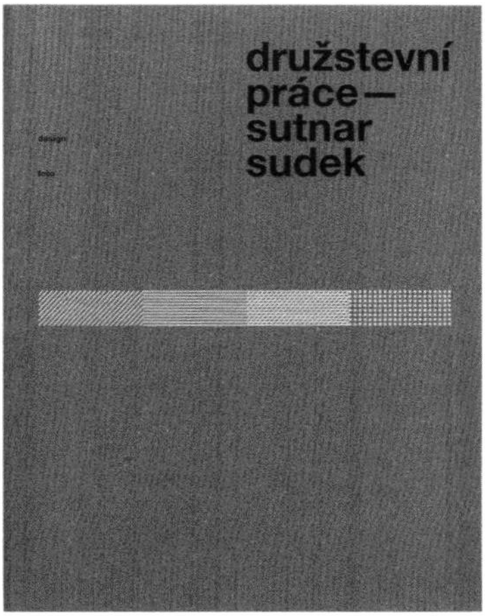
družstevní
práce—
sutnar
sudek
design
foto

This is a book about the famous Czech cultural and social organization 'Družstevní práce' (hereinafter referred to as DP) and the story between the two figures who made outstanding contributions to the organization. The book shows the unique era through explaining the background of DP, the advertising photography of photographer Josef Sudek, his relationship with DP, and his collaboration with graphic designer Ladislav Sutnar. This paperback book is prefect bound, enclosed by a slipcase which is made of graphic board without any printing or technologies. The laminated cover is made of white card printed in red and black. The blank space consisting of four lines on the cover implies that there are four parts of the main content. The first two-thirds of the book is written in Czech, which is divided into four parts according to the content. The introduction of characters, chronology, materials and other appendices are attached. This part is printed in four colors on warm light gray matte coated paper. The rest pages are printed on offset paper in monochrome black, containing the English version of the text without pictures. The book design is rigorous with the exquisite layout. Accompanied with aged colorful pictures, the book has a strong sense of document.

荣誉奖

合作工作：苏特纳尔 - 苏德克

捷克

This monograph documenting the creative partnership between Czech photographer Josef Sudek and art director Ladislav Sutner is a book as bold in its design as the works which it presents. Its decidedly black and white layout is sober and unpretentious. As such it presents a strong and individual contrast to the more overstated and colourful publications on similar themes.

Honorary Appreciation

Družstevní práce: Sutnar – Sudek

Czech Republic

Stěpán Malovec & Martin Odehnal

Lucie Vlčková (Ed.)

Uměleckoprůmyslové museum v Praze, Prague

270×205×20mm

1151g

250p

družstevní
práce —
sutnar
sudek

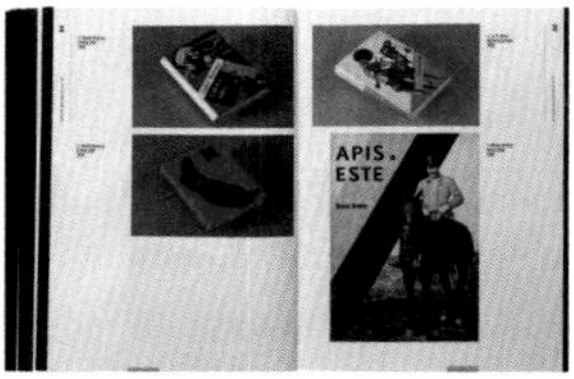
APIS a
ESTE

kalendář družstevní práce 1933

JAN MLČOCH — JOSEF SUDEK AND ADVERTISING PHOTOGRAPHY

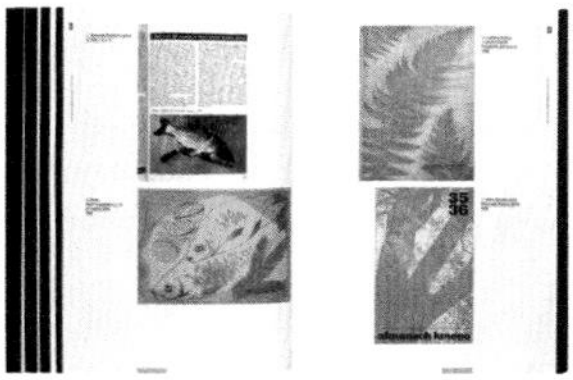
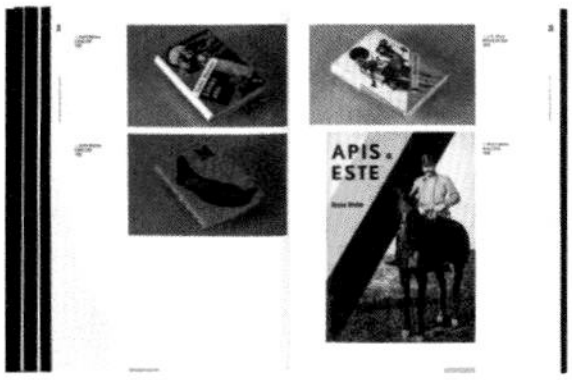

mysl
rár
nakladatelstvím

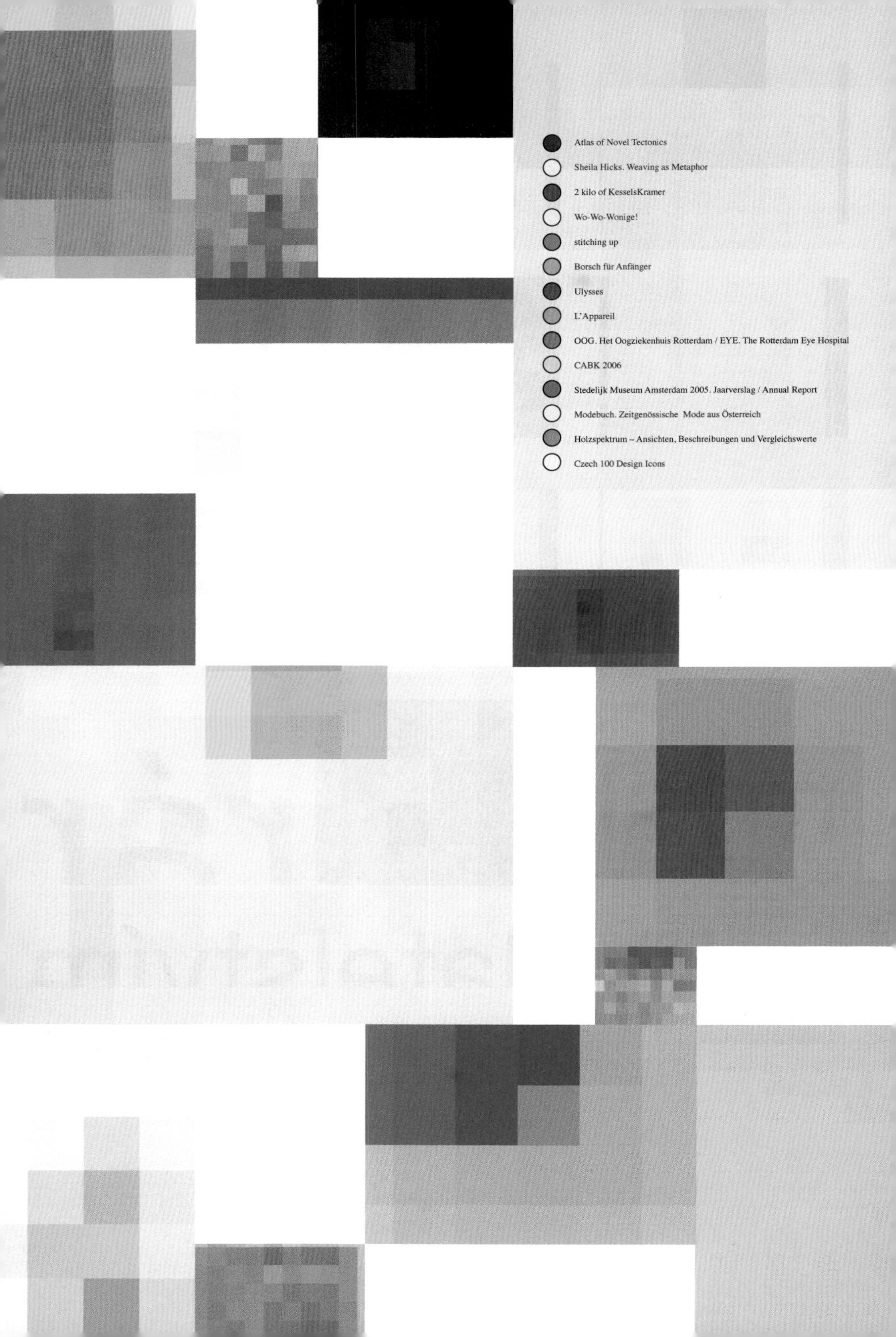

Atlas of Novel Tectonics
Sheila Hicks. Weaving as Metaphor
2 kilo of KesselsKramer
Wo-Wo-Wonige!
stitching up
Borsch für Anfänger
Ulysses
L'Appareil
OOG. Het Oogziekenhuis Rotterdam / EYE. The Rotterdam Eye Hospital
CABK 2006
Stedelijk Museum Amsterdam 2005. Jaarverslag / Annual Report
Modebuch. Zeitgenössische Mode aus Österreich
Holzspektrum – Ansichten, Beschreibungen und Vergleichswerte
Czech 100 Design Icons

2007

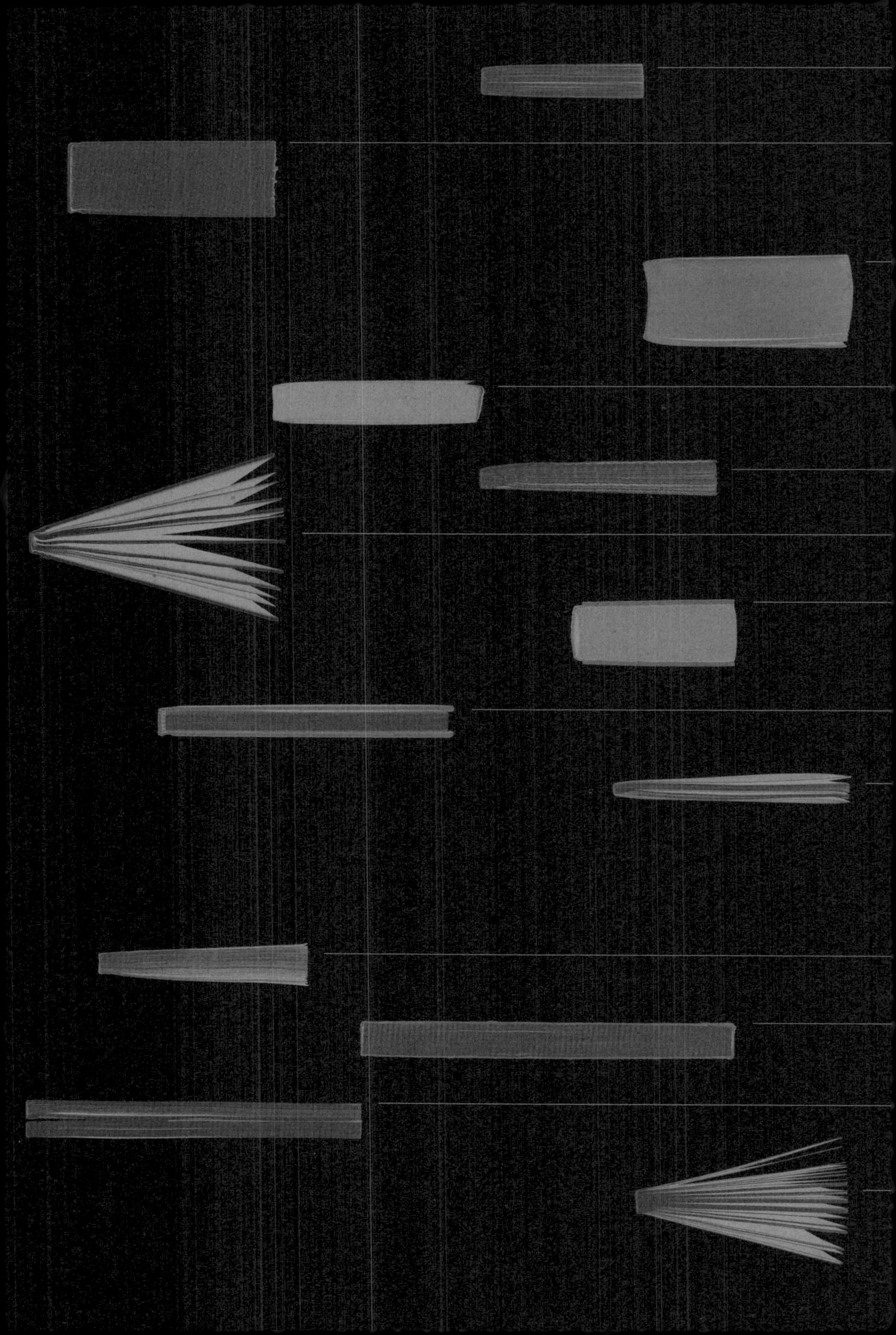

2007 Gl 1115 

2007 Gm 1123 

2007 Sm[1] 1129 

2007 Sm[2] 1135 

2007 Bm[1] 1141 

2007 Bm[2] 1147 

2007 Bm[3] 1153 
JAMES JOYCE ULYSSES

2007 Bm[4] 1159 

2007 Bm[5] 1165 

2007 Ha[2] 1171 

2007 Ha[3] 1175 

2007 Ha[4] 1179 

2007 Ha[5] 1183 

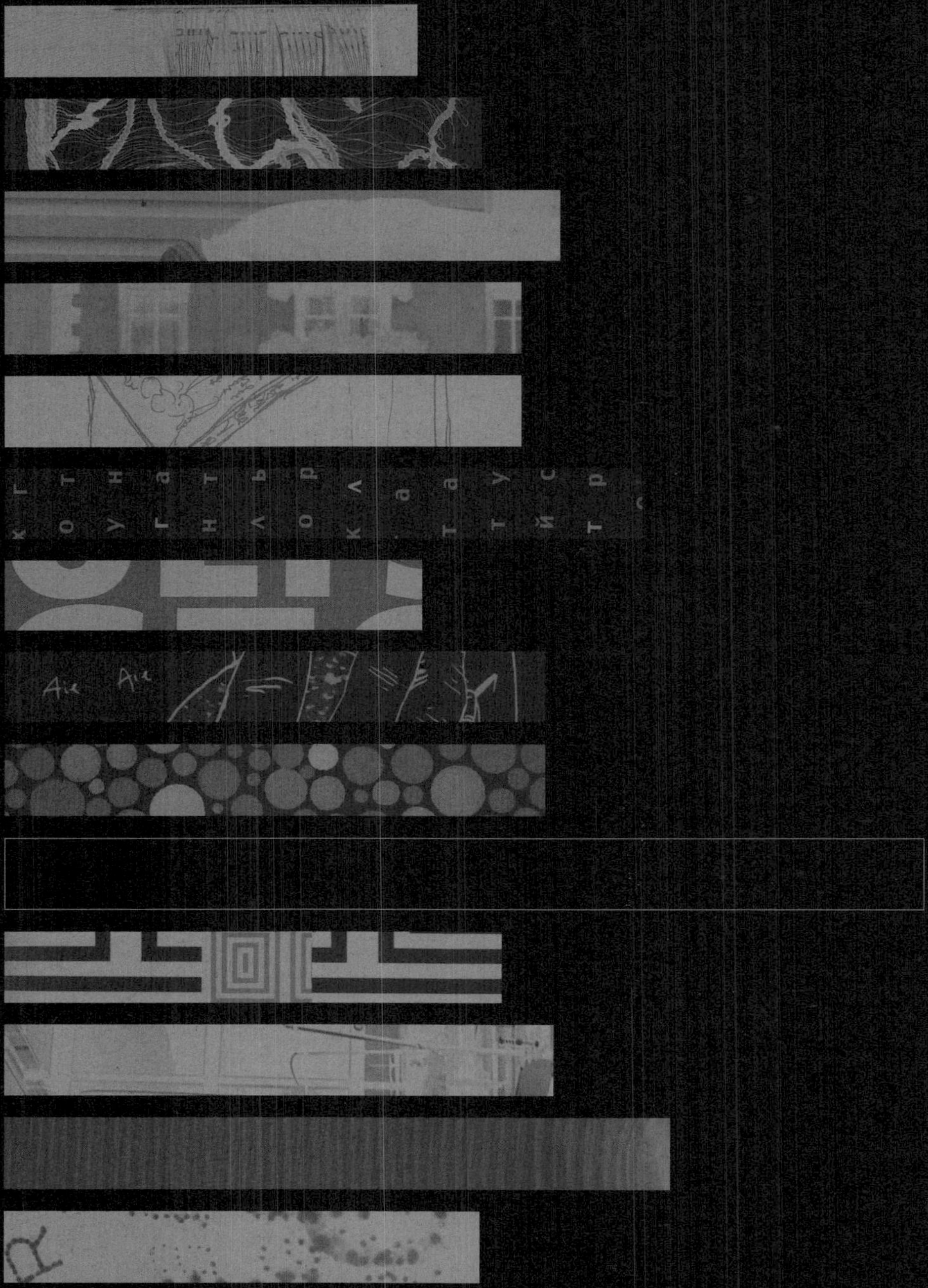
г т н а т ь р л а у с р
к о у г н л о к т т й т

by his alter ego to go out of his body in order to ge
One could parallel the desires in *Fight Club* with phe
desire to have everything grounded within the bo

and which serves him as the
root metaphor in his mature

日本と韓国でのワールドカップでオランダ代表チームが出場資格を得られなかったとき、ケッセルスクライマーは「負ける」ことの意味を深く考えるようになった。最上位に立つことに注目するより、負けるとはどういうことなのかを考えるときに来ていたのだろう。

weil man lediglich in den Tag hinein lebte oder aus Geda
kenlosigkeit. Die Unterlagen landeten in irgendwelchen
vaten Estrichen oder wurden weggeworfen. Besonders sc
gestaltete sich die Aufarbeitung der alternativen Wohn

“没想过。”

“他留长头发，衣服收腰，穿厚底鞋。就是那种看起来

Koreanischer Art

Besonderen Zuspruch bekam Salat Po-Korejsk
ren mit sehr heißem Knoblauchöl übergossen

Stattlich und feist erschien Buck Mulligan am Tre
tritt, ein Seifenbecken in Händen, auf dem gekreuzt ei
und ein Rasiermesser lagen. Ein gelber Schlafrock mit

moderne; rebelle et virtuose. Un bref séjour
en France a permis à ce jeune épicurien

Een patiënt meet de medische kwaliteit dan ook vaak af aan an
die hij wel kan beoordelen, zoals bejegening, voorlichting, wac
omgeving. Als specialisten zijn wij natuurlijk vooral bezig met

zelfstandige dienst
van de gemeente Am-
rdam was het museum

de oudste gemeentelijke
kunststichting. al verzelf-
standigd. De commissie

115

---WENDY&JIM Established in 1997 by Helga Schania and H
mann Fankhauser, the Wendy&Jim label is among the „shooting stars"
national fashion firmament. Particularly in Japan, the two former Helmut

diesem Nadelholz ist die Jahrringgrenze aufgrund deutli
Zellwandstärken von Spät- und Frühholz klar zu erkenne
dunkeln des Splints zu Honiggelb und des Kernbereichs
charakteristisches Bild, das im frischen Zustand noch nic

a professor of physics. **1912** ▶ Josef Goča

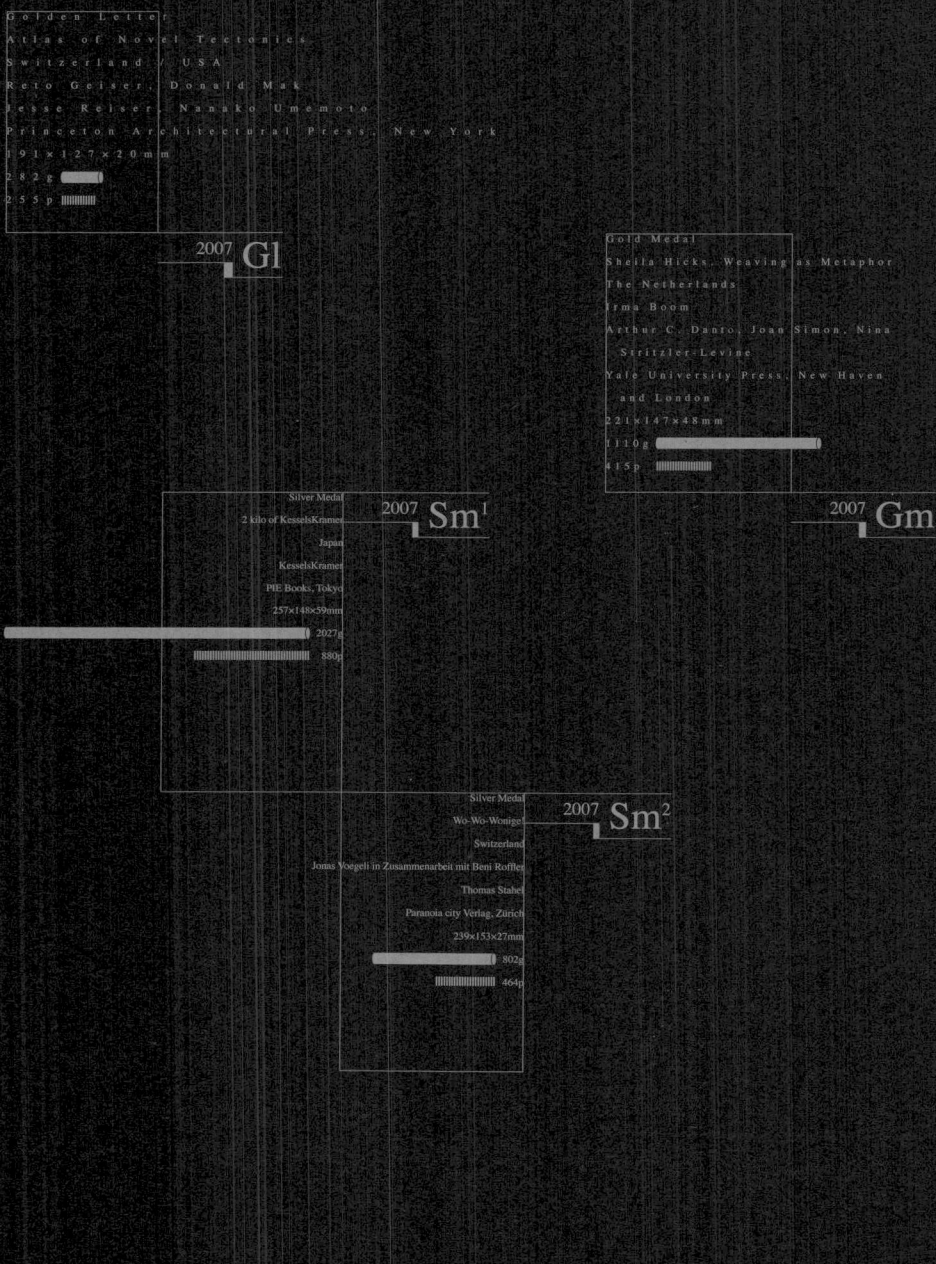

Golden Letter
Atlas of Novel Tectonics
Switzerland / USA
Reto Geiser, Donald Mak
Jesse Reiser, Nanako Umemoto
Princeton Architectural Press, New York
191×127×20mm
282g
255p

2007 Gl

Gold Medal
Sheila Hicks. Weaving as Metaphor
The Netherlands
Irma Boom
Arthur C. Danto, Joan Simon, Nina Stritzler-Levine
Yale University Press, New Haven and London
221×147×48mm
1110g
415p

2007 Gm

Silver Medal
2 kilo of KesselsKramer
Japan
KesselsKramer
PIE Books, Tokyo
257×148×59mm
2027g
880p

2007 Sm[1]

Silver Medal
Wo-Wo-Wonige!
Switzerland
Jonas Voegeli in Zusammenarbeit mit Beni Roffler
Thomas Stahel
Paranoia city Verlag, Zürich
239×153×27mm
802g
464p

2007 Sm[2]

2007 Ha[2]

Honorary Appreciation
Stedelijk Museum Amsterdam 2005. Jaarverslag / Annual Report
The Netherlands
Richard Niessen, Esther de Vries
Carolien de Bruijn (Ed.)
Stedelijk Museum Amsterdam
230×160×16mm
539g
176p

2007 Ha[3]

Honorary Appreciation
Modebuch. Zeitgenössische Mode aus Österreich
Austria
halle34 Albert Handler / Marcus Arige
Unit F büro für mode (Ed.)
Unit F büro für mode, Wien
254×178×23mm
895g
158p

2007 Ha[4]

Honorary Appreciation
Holzspektrum – Ansichten, Beschreibungen und Vergleichswerte
Austria
Reinhard Gassner, Marcel Bachmann,
Stefan Gassner (Atelier Reinhard Gassner)
Josef Fellner, Alfred Teischinger, Walter Zschokke
proHolz Austria
– Arbeits-gemeinschaft der österrei-chischen Holzwirtschaft, Wien
308×214×24mm
1328g
112p+96p

2007 Ha[5]

Honorary Appreciation
Czech 100 Design Icons
Czech Republic
Zuzana Lednická, Aleš Najbrt (Studio Najbrt)
Tereza Bruthansová, Jan Králíček
CzechMania, Prag
220×162×17mm
622g
100p

2007 Ha[1]

2007 Bm[1]

Bronze Medal
stitching up
China
Zhu Yingchun
Gu Shiju
Jiangsu Literature & Art Publishing House, Nanjing
239×163×14mm
417g
189p

2007 Bm[2]

Bronze Medal
Borsch für Anfänger
Germany
Tatjana Triebelhorn
Tatjana Triebelhorn, Roman Triebelhorn
Eigenverlag Tatjana Triebelhorn, Stuttgart
295×184×15mm
552g
103p

2007 Bm[3]

Bronze Medal
Ulysses
Germany
Werner Zegarzewski
James Joyce
Suhrkamp Verlag, Frankfurt am Main
193×122×47mm
676g
987p

2007 Bm[5]

Bronze Medal
OOG. Het Oogziekenhuis Rotterdam / EYE. The Rotterdam Eye Hospital
The Netherlands
Irma Boom
Ineke van Ginneke (Ed.)
Het Oogziekenhuis Rotterdam / The Rotterdam Eye Hospital
250×200×9mm
348g
96p

2007 Bm[4]

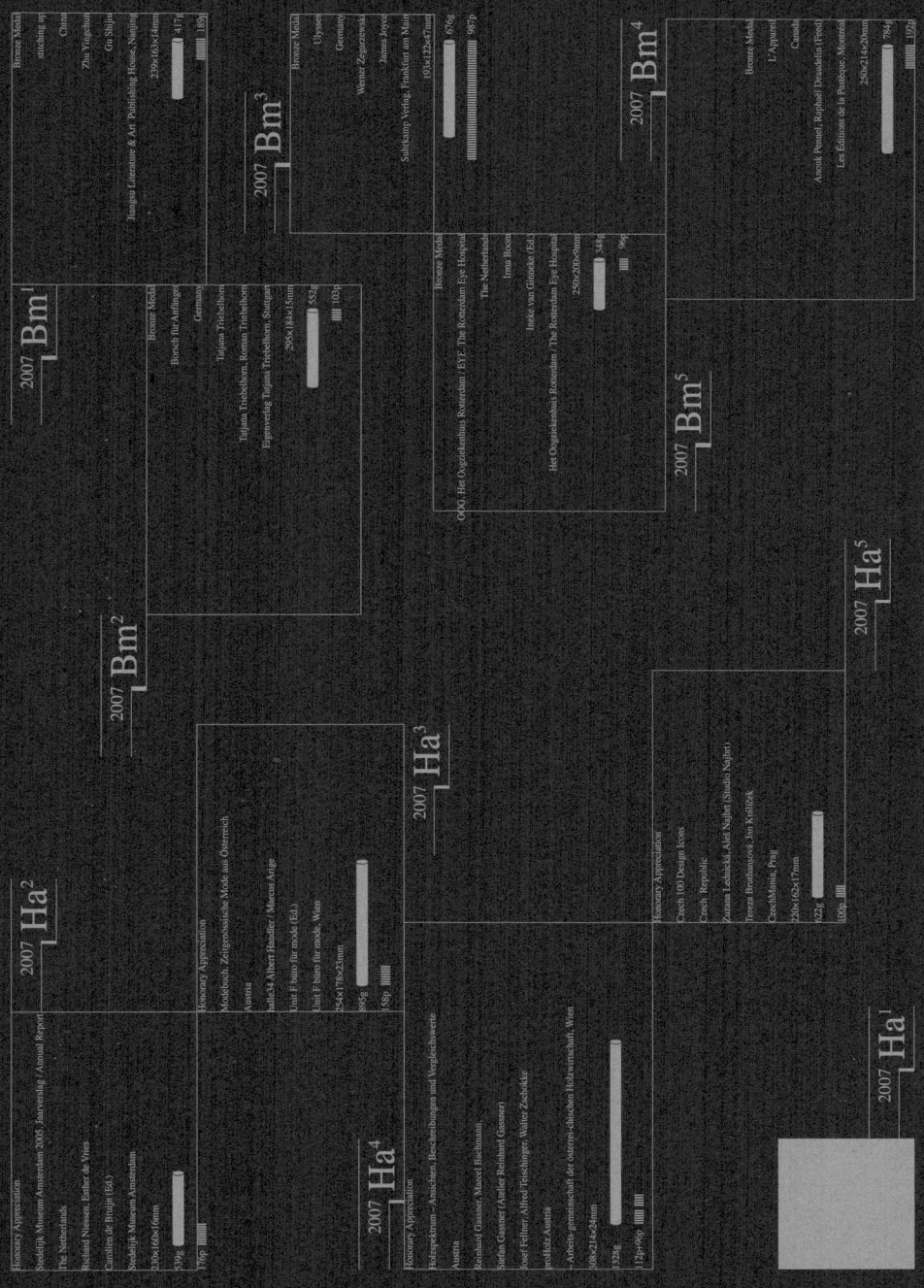

Bronze Medal
L'Appareil
Canada
Anouk Pennel, Raphaël Deaudelin (Feed)
Les Éditions de la Pastèque, Montréal
250×214×20mm
784g
192p

The International Jury

Rotraut Susanne Berner (Germany)

Markus Dreßen (Germany)

Reinhard Gassner (Austria)

Esther kit-lin Liu (China)

Péter Maczó (Hungary)

Thomas Narr (Germany)

Sophie Nicolas (France)

Designer

Reto Geiser, Donald Mak

Irma Boom

KesselsKramer

Jonas Voegeli in Zusammenarbeit mit Beni Roffler

Zhu Yingchun

Tatjana Triebelhorn

Werner Zegarzewski

Anouk Pennel, Raphaël Deaudelin (Feed)

Irma Boom

DesignArbeid

Richard Niessen, Esther de Vries

halle34 Albert Handler / Marcus Arige

Reinhard Gassner, Marcel Bachmann, Stefan Gassner (Atelier Reinhard Gassner)

Zuzana Lednická, Aleš Najbrt (Studio Najbrt)

2007

Country / Region

The Netherlands 4

Germany 2

Austria 2

Japan 1

Switzerland 1

China 1

Canada 1

Czech Republic 1

USA 0

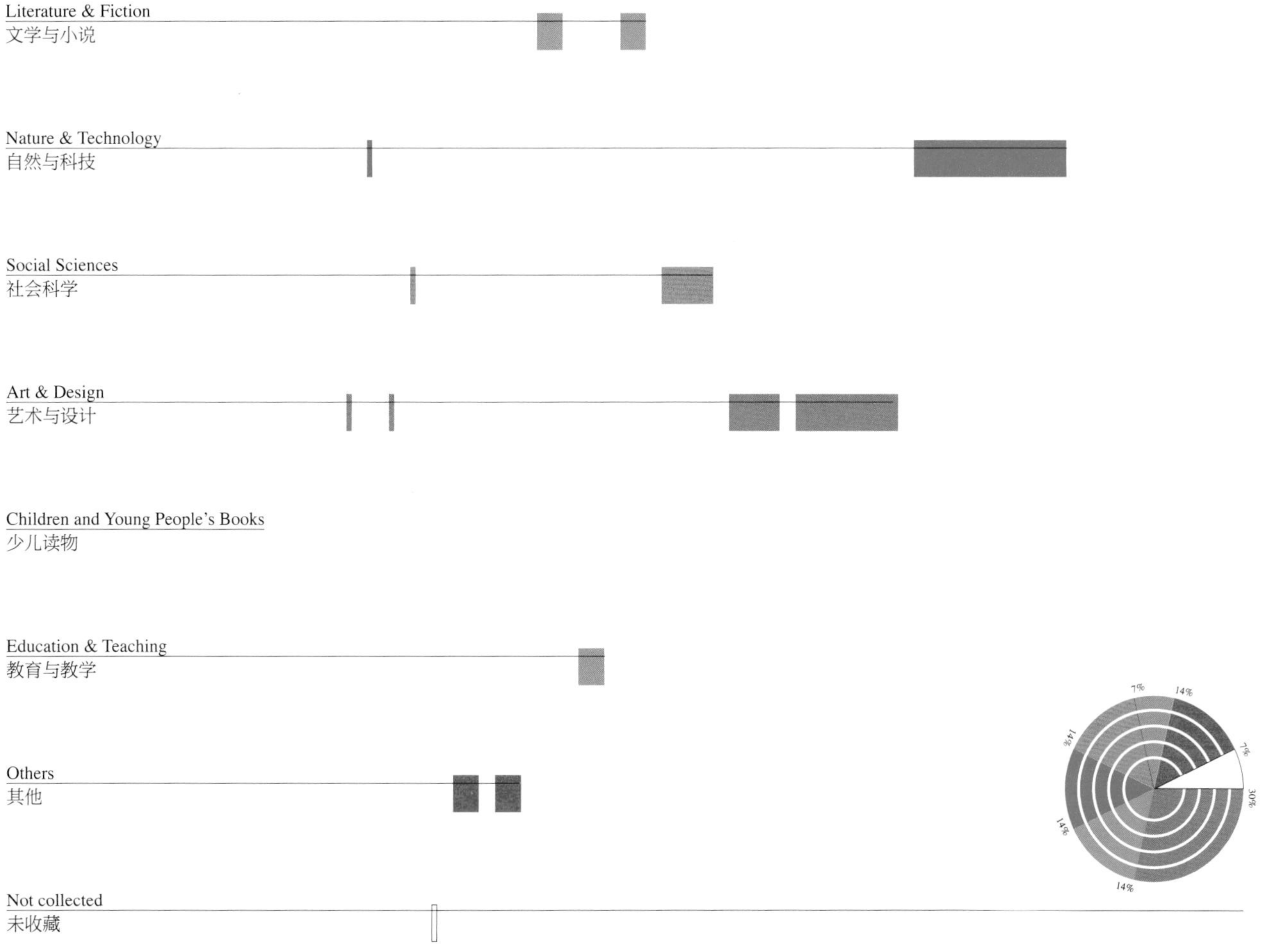
Literature & Fiction
文学与小说
Nature & Technology
自然与科技
Social Sciences
社会科学
Art & Design
艺术与设计
Children and Young People's Books
少儿读物
Education & Teaching
教育与教学
Others
其他
Not collected
未收藏
7%
14%
7%
30%
14%
14%
14%

007 G1

这是一本有关新型结构、材料和操作方面研究的专著，从建筑设计师、结构和材料工程师的研究出发，探讨了这些研究成果对他们设计的影响。这是一本设计精美的小书，锁线胶装，可以较好地翻开。黑色人造皮面压印书内结构线图，有如浮雕，体现内容的同时带来丰富的翻阅手感。封面和书脊烫银，居中排版，简洁明快，低调中透露出科技的光芒。内页双胶纸，黑色和墨绿色双色印刷。少量彩图在哑面铜版纸印刷后，以手工方式贴在各自页面。版面严谨而不缺变化，体现出一种存在于冷峻的技术中的浪漫情怀。

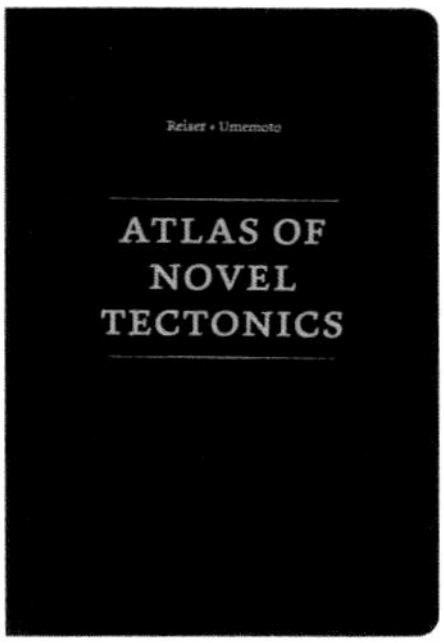
Reiser + Umemoto
ATLAS OF
NOVEL
TECTONICS

This is a monograph on the study of new structures, materials, and operations from the perspective of architects and material engineers. This beautifully designed little book is perfect-bound and can be fully opened. Black leather cover is embossed with structure diagrams, displaying the content while bringing rich meaning. Hot stamping silver is used for the cover and spine. The concise design highlights the technological elements in low key. The inside pages are made of adhesive paper, printed in black and dark green. A small number of colorful images are printed on matte coated paper and pasted on relevant pages by hand. The layout is rigorous without lack of changes, reflecting romantic feelings towards rational technology.

- 金字符奖
- 新构造图集
- 瑞士 / 美国

To hold this book, with its cover design, silver embossing on the matt laminated softcover and rounded-off corners, is like clasping a breviary. The initial darkgreen, pleasing endpapers are followed by a world of surprises waiting to be discovered. The book breathes thanks to bold contrasts as well as allusions and combinations: those of illustrations from contemporary architecture and classical, expertly set macro and microtypography (easy to navigate on account of recurring dark-green chapter beginnings) as well as picture pages on art-paper which are placed on the right-hand pages.

- Golden Letter
- Atlas of Novel Tectonics
- Switzerland / USA
- Reto Geiser, Donald Mak
- Jesse Reiser, Nanako Umemoto
- Princeton Architectural Press, New York
- 191×127×20mm
- 282g
- 255p

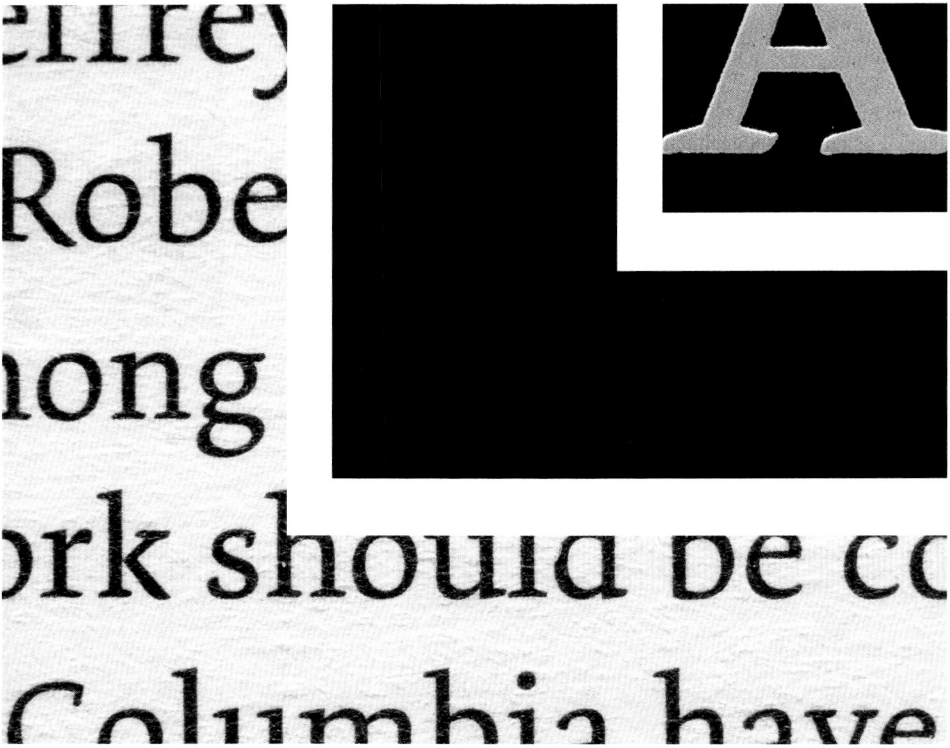

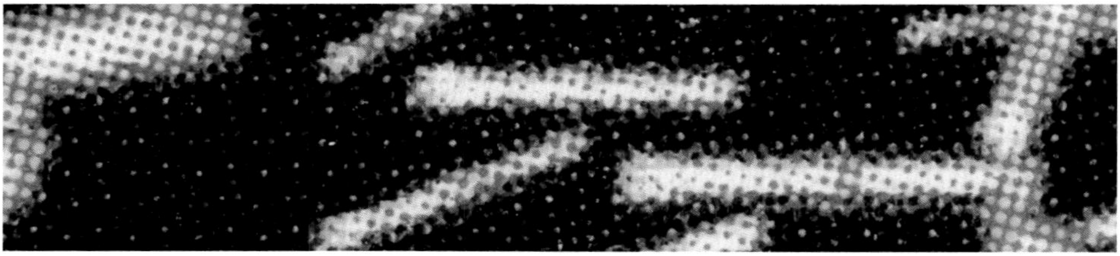

191×127×20

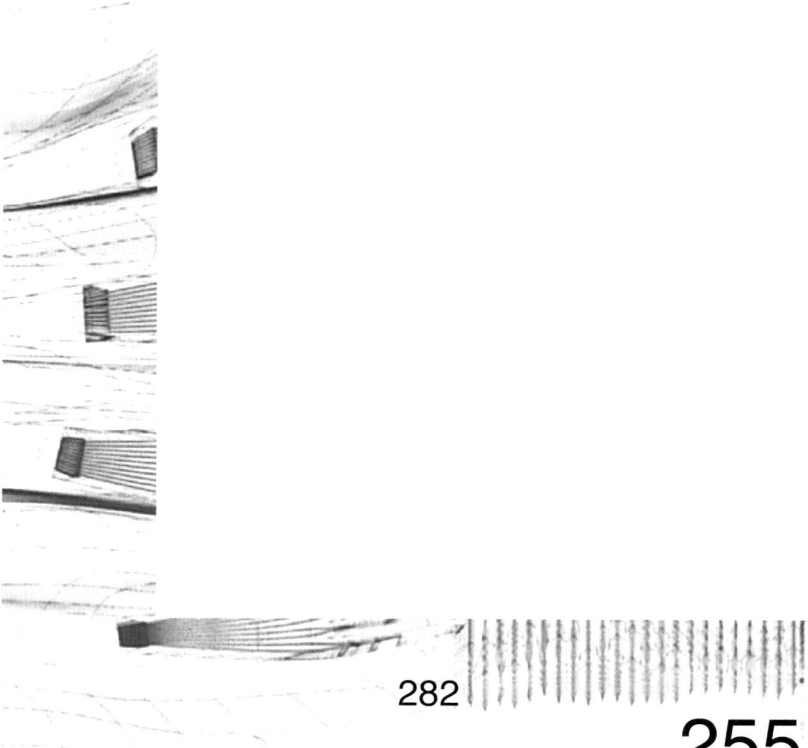

282

255

INTRODUCTION

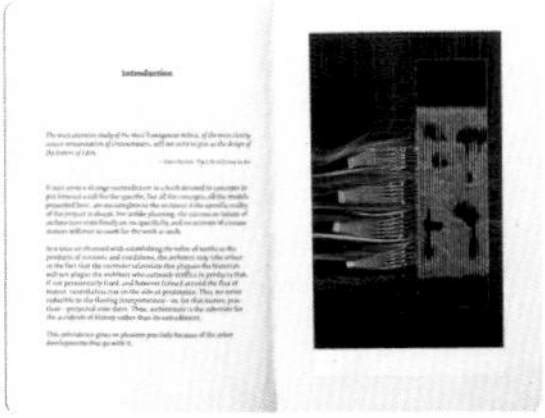

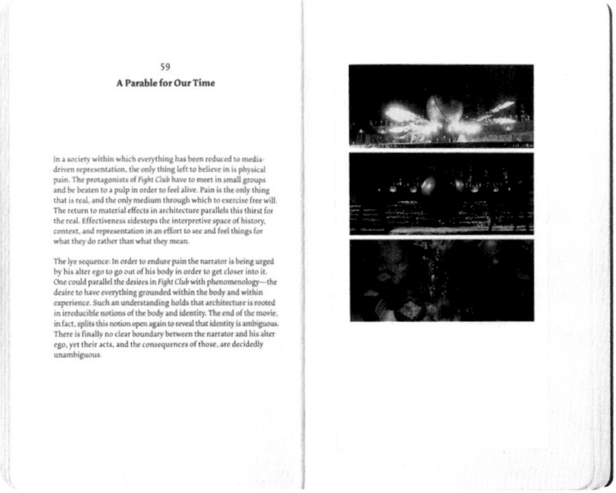

59

A Parable for Our Time

In a society within which everything has been reduced to media-driven representation, the only thing left to believe in is physical pain. The protagonists of *Fight Club* have to meet in small groups and be beaten to a pulp in order to feel alive. Pain is the only thing that is real, and the only medium through which to exercise free will. The return to material effects in architecture parallels this thirst for the real. Effectiveness sidesteps the interpretive space of history, context, and representation in an effort to see and feel things for what they do rather than what they mean.

The lye sequence: In order to endure pain the narrator is being urged by his alter ego to go out of his body in order to get closer into it. One could parallel the desires in *Fight Club* with phenomenology—the desire to have everything grounded within the body and within experience. Such an understanding holds that architecture is rooted in irreducible notions of the body and identity. The end of the movie, in fact, splits this notion open again to reveal that identity is ambiguous. There is finally no clear boundary between the narrator and his alter ego, yet their acts, and the consequences of those, are decidedly unambiguous.

ATLAS OF NOVEL TECTONICS
Contents
Acknowledgments

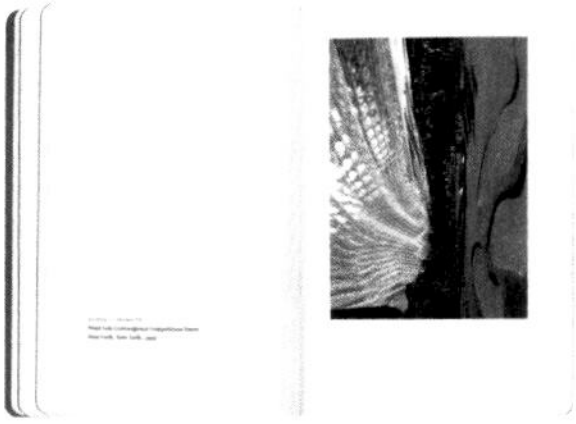

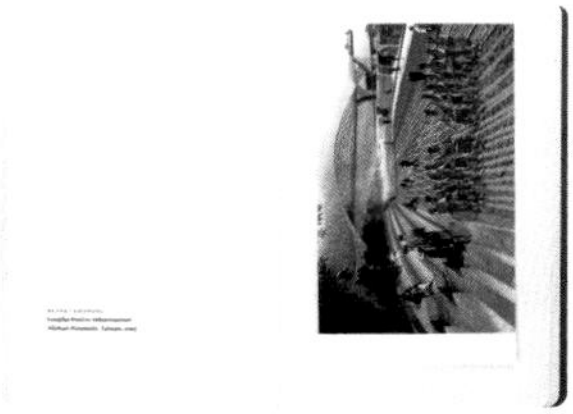

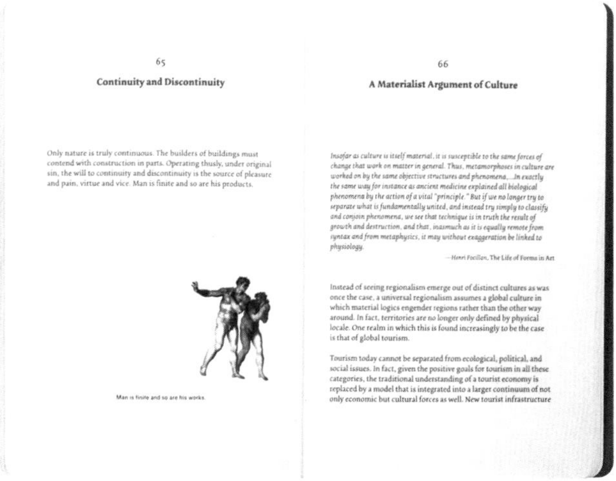

65

Continuity and Discontinuity

Only nature is truly continuous. The builders of buildings must contend with construction in parts. Operating thusly, under original sin, the will to continuity and discontinuity is the source of pleasure and pain, virtue and vice. Man is finite and so are his products.

Man is finite and so are his works.

66

A Materialist Argument of Culture

Insofar as culture is itself material, it is susceptible to the same forces of change that work on matter in general. Thus, metamorphoses in culture are worked on by the same objective structures and phenomena,...In exactly the same way for instance as ancient medicine explained all biological phenomena by the action of a vital "principle." But if we no longer try to separate what is fundamentally united, and instead try simply to classify and conjoin phenomena, we see that technique is in truth the result of growth and destruction, and that, inasmuch as it is equally remote from syntax and from metaphysics, it may without exaggeration be linked to physiology.

—Henri Focillon, The Life of Forms in Art

Instead of seeing regionalism emerge out of distinct cultures as was once the case, a universal regionalism assumes a global culture in which material logics engender regions rather than the other way around. In fact, territories are no longer only defined by physical locale. One realm in which this is found increasingly to be the case is that of global tourism.

Tourism today cannot be separated from ecological, political, and social issues. In fact, given the positive goals for tourism in all these categories, the traditional understanding of a tourist economy is replaced by a model that is integrated into a larger continuum of not only economic but cultural forces as well. New tourist infrastructure

FROM EXTENSIVE FIELD/INTENSIVE OBJECT TO INTENSIVE FIELD-OBJECT

In response to the perceived sterility and homogeneity of modern architecture, figures of the last generation as varied as Robert Venturi and John Hejduk selected and developed highly specific elements of the movement. This extraction of (generally figurative) motifs from the more systematic, Cartesian field they had occupied in high modernism was seen as a promulgation of uniqueness and variety in architecture. With equal ease, it could be defined through the development of a singular volume or figure or, in more discontinuous fashion, the collage technique. But this selective approach carried liabilities as well, for it dispensed with the grand systematic ambitions of modernism in favor of an idiosyncratic approach and concentrated on a revision of modernism that foregrounded the object divested of its field.

For architects, notions of space, until recently, remained trenchantly Cartesian, whether the field was recognized or the object premiated. The big shift, in which our work participates is the removal of the fixed background, of ordinates and coordinates, in favor of a notion of space and matter as being one. This shift is not simply one in concept or belief that would leave the architecture unchanged; at a fundamental level, it changes the way architecture is thought about and designed, and the way it emerges as a material fact.

Apologists for modernism—or those who simply want to extend the modernist project by updating their arguments while leaving the architecture unchanged—are in grave error. In their minds, the shifting paradigm is simply yet another shift in discourse, it doesn't affect the object, and the object has no effect on it. Discourse alone merely becomes a more fashionable view of the same universe, thus implying that the early model is but a failure of interpretation.

TOP **John Hejduk, Bye House (Wall House II)**
MIDDLE **Peter Eisenman, House II (Frank House)**
BOTTOM **Robert Venturi/VSBA, Gordon Wu Hall**

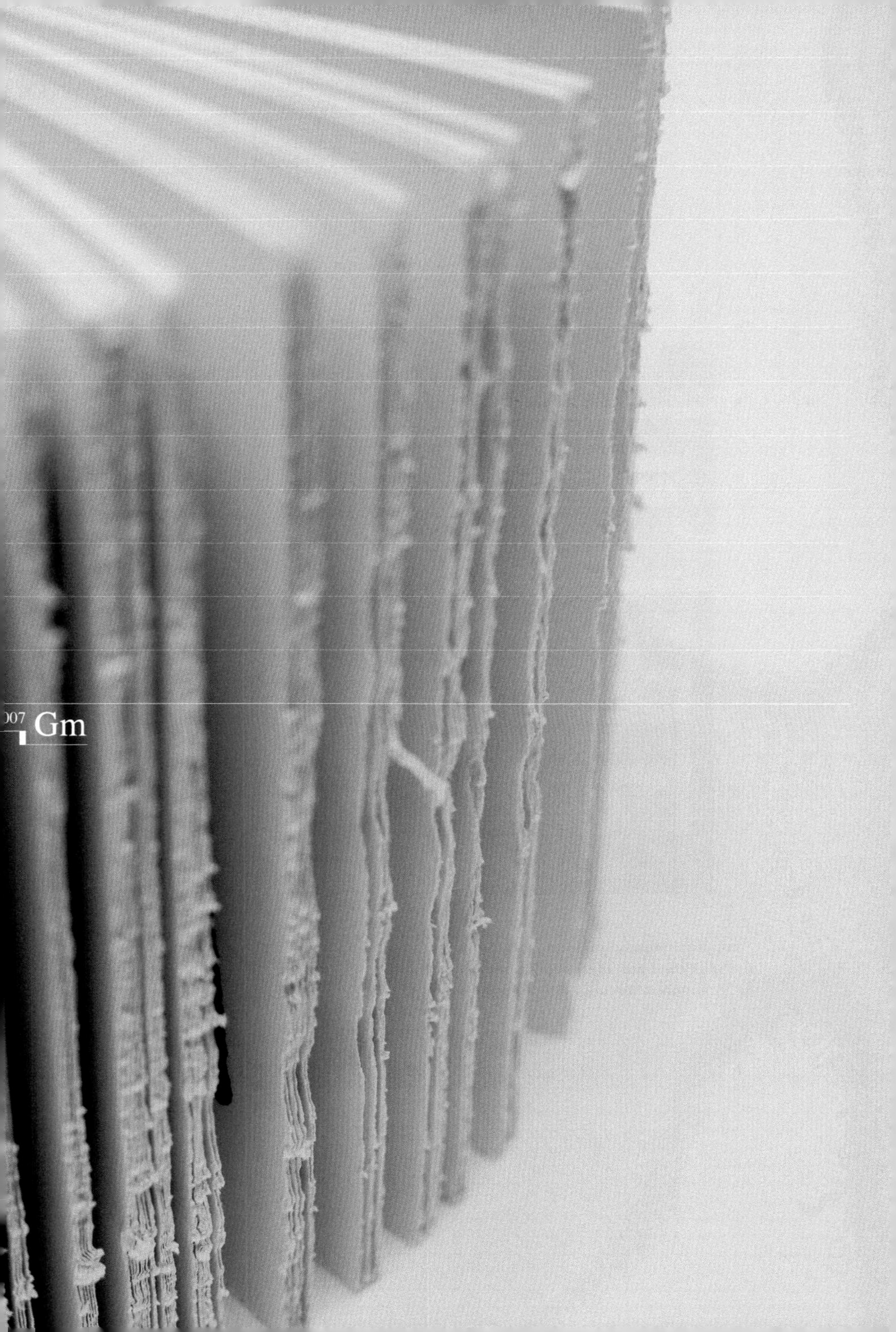
Gm

这本书是巴德研究生中心组织的一次展览作品集，展览内容为美国艺术家希拉·希克斯（Sheila Hicks）的作品阐释、工作方式和相关作品。本书为锁线硬精装，书籍的设计在方方面面体现着编织感。封面白色纸张裱覆在卡板上，封面压凹布纹编织肌理，与封底印刷的编织物形成呼应。内页纸张较厚，三面书口打毛处理，试图仿造层层织物叠压在一起的手感。内文按照主题及理念阐释、工艺和形式手法、展品图录等方面来全面展示这次展览。内文的排版很有意思，并没有像大多数书籍的正文那样统一字号和行距，而是通过各种整面大小变化的字号和行距，来形成不同的版面灰度，用以体现文字排版中类似织物的编织感。正文侧面安排放置了对应图片的编号，就像未完成的织物架在织物框上的线头，指引着读者找到相应的作品。

SHEILA HICKS

The book is a collection of works from an exhibition organized by Bard Graduate Center. The exhibition is about the interpretation of the American artist Sheila Hicks' woven works and the way she works. This hardcover book is bound by thread sewing, and the design idea reflects the sense of weaving in all aspects. White paper mounted on graphic board as the book cover, which is embossed weaving texture, and echoes the woven works printed on the back cover. Three deckled edges of the book endeavor to imitate layers of fabric stacked together. The exhibition is presented in terms of themes, concepts, techniques, forms and catalogue of exhibits. The layout of the text is very interesting – the font size and line spacing are not uniform as most books. Instead, different font size and line spacing are used to reflect the weaving sense of typesetting similar to fabric. The numbers of the corresponding pictures are arranged on the side of the text, just like the threads of unfinished fabric on the fabric frame, guiding readers to find the corresponding works.

金奖

希拉 · 希克斯——编织的隐喻

荷兰

Format, texture and weight of this book all speak for themselves. Its design and finish reflect its content – the metaphor of weaving – in a way which is both direct and honest. Even before opening the book the reader is given a foretaste of its content through the way it feels. The inner design interweaves text and image in a clear fashion. A book which commends itself in all its white beauty.

Gold Medal

Sheila Hicks. Weaving as Metaphor

The Netherlands

Irma Boom

Arthur C. Danto, Joan Simon, Nina Stritzler-Levine

Yale University Press, New Haven and London

221 × 147 × 48mm

1110g

415p

SHEILA
HICKS
WEAV
ING AS
META
PHOR

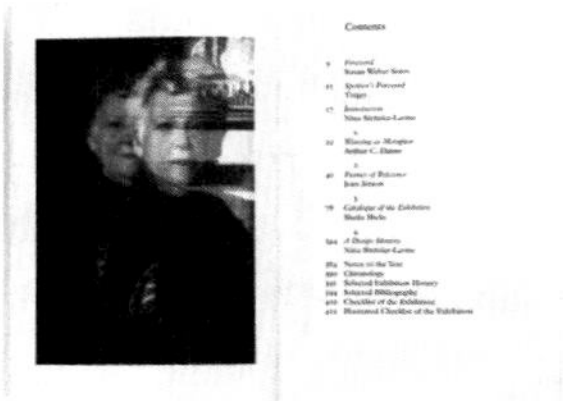
Contents

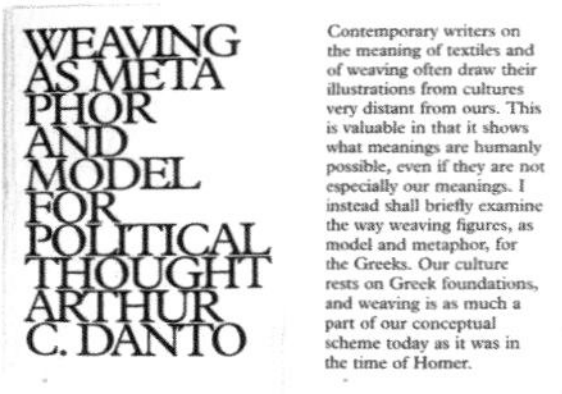
WEAVING
AS META
PHOR
AND
MODEL
FOR
POLITICAL
THOUGHT
ARTHUR
C. DANTO
Contemporary writers on the meaning of textiles and of weaving often draw their illustrations from cultures very distant from ours. This is valuable in that it shows what meanings are humanly possible, even if they are not especially our meanings. I instead shall briefly examine the way weaving figures, as model and metaphor, for the Greeks. Our culture rests on Greek foundations, and weaving is as much a part of our conceptual scheme today as it was in the time of Homer.

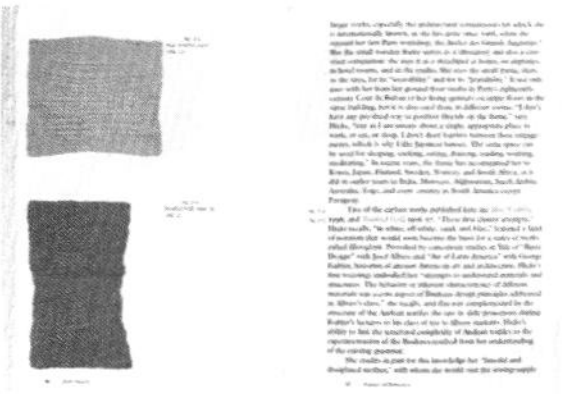

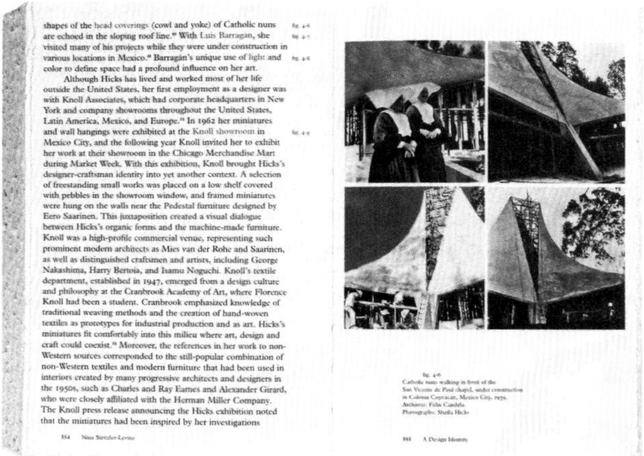
shapes of the head coverings (cowl and yoke) of Catholic nuns are echoed in the sloping roof line. With Luis Barragán, she visited many of his projects while they were under construction in various locations in Mexico. Barragán's unique use of light and color to define space had a profound influence on her art.
Although Hicks has lived and worked most of her life outside the United States, her first employment as a designer was with Knoll Associates, which had corporate headquarters in New York and company showrooms throughout the United States, Latin America, Mexico, and Europe. In 1962 her miniatures and wall hangings were exhibited at the Knoll showroom in Mexico City, and the following year Knoll invited her to exhibit her work at their showroom in the Chicago Merchandise Mart during Market Week. With this exhibition, Knoll brought Hicks's designer-craftsman identity into yet another context. A selection of freestanding small works was placed on a low shelf covered with pebbles in the showroom window, and framed miniatures were hung on the walls near the Pedestal furniture designed by Eero Saarinen. This juxtaposition created a visual dialogue between Hicks's organic forms and the machine-made furniture. Knoll was a high-profile commercial venue, representing such prominent modern architects as Mies van der Rohe and Saarinen, as well as distinguished craftsmen and artists, including George Nakashima, Harry Bertoia, and Isamu Noguchi. Knoll's textile department, established in 1947, emerged from a design culture and philosophy at the Cranbrook Academy of Art, where Florence Knoll had been a student. Cranbrook emphasized knowledge of traditional weaving methods and the creation of hand-woven textiles as prototypes for industrial production and as art. Hicks's miniatures fit comfortably into this milieu where art, design and craft could coexist. Moreover, the references in her work to non-Western sources corresponded to the still-popular combination of non-Western textiles and modern furniture that had been used in interiors created by many progressive architects and designers in the 1950s, such as Charles and Ray Eames and Alexander Girard, who were closely affiliated with the Herman Miller Company. The Knoll press release announcing the Hicks exhibition noted that the miniatures had been inspired by her investigations
354 Nina Stritzler-Levine
fig. 4-6
Catholic nuns walking in front of the San Vicente de Paul chapel, under construction in Colonia Coyoacán, Mexico City, 1959.
Architect: Félix Candela
Photographs: Sheila Hicks
355 A Design Identity

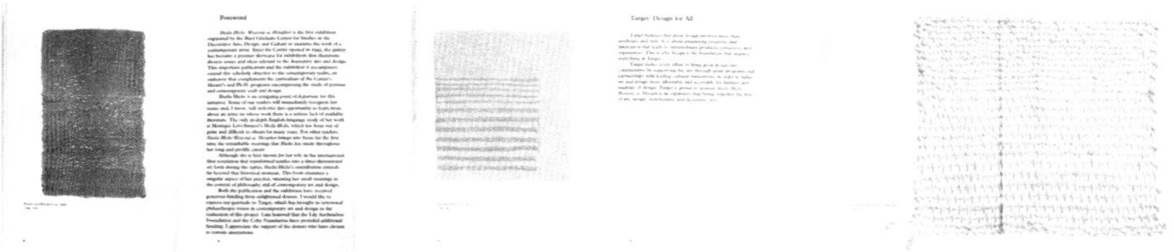
Foreword
Target: Design for All

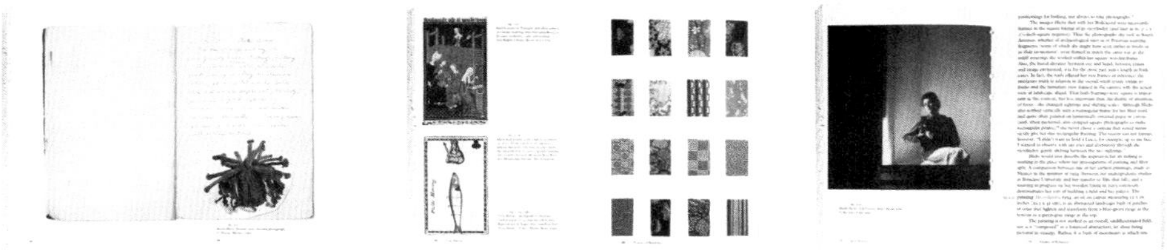

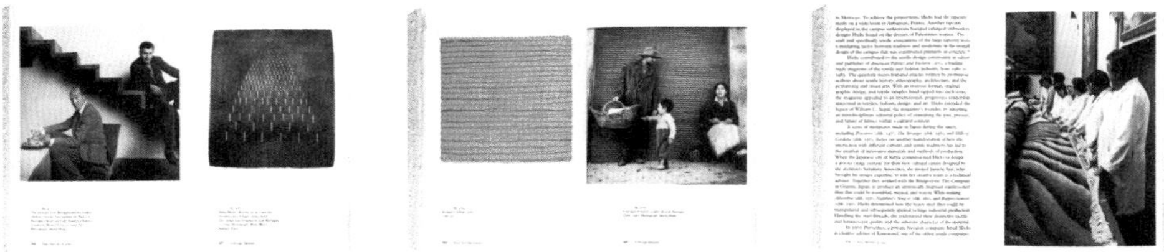

plex
ancy of

SHEILA HICKS

BGC
YALE

CATA
LOGUE
OF THE
EXHIB
ITION
SHEILA
HICKS

Each entry consists of title, date, materials (warp followed by weft), dimensions in centimeters and inches (height preceding width preceding depth), and place where the work was made, followed by a description that includes the work's principal colors, salient techniques, contextual references, and collection.

Sheila Hicks, Guerrero, Mexico, 1960
Photograph: Ferdinand Boesch

78

07 Sm1

这是荷兰的广告公司 KesselsKramer 的作品集，日英双语，为日本的出版社引进的书籍。全书以作品集为主，涵盖了 KesselsKramer 公司从 1996 年成立以来针对各种品牌、事件所做的市场、战略和传播方案，涉及品牌包括：Diesel、Channel 5 UK、喜力啤酒、绝对伏特加等知名品牌。本书的装订方式为锁线胶装，虽然尺寸偏长，但仍不影响完全展开。封面卡纸棕黑双色印刷，覆哑膜，十分朴实。本书的亮点正如书名“2 千克”所体现的，正是大约 2 千克的重量。内页的页码编排按照重量计算，并且在最后的统计页面计算出各服务品牌的“重量”（比例）。翻阅作品页的时候，页码“重量”逐渐减少，提示着读者还有多少“重量”的作品在后面等着翻阅。

2 kilo
OF KESSELSKRAMER

This is a collection of works by Dutch Design Studio KesselsKramer, written in both Japanese and English. Japanese publishers introduce the book to Japan. It is mainly composed of works, covering KesselsKramer Studio's marketing strategy and communication solutions for various brands and events since its establishment in 1996. The brands include Diesel, Channel 5 UK, Heineken Beer, Absolut Vodka and so on. This book is perfect bound with thread sewing and can be fully opened although the format is a bit too long. The matted cover in printed in brown and black. The title "2 Kg" highlights the weight of the book, which is about 2 Kilo. The pagination of the inside pages is calculated by weight, and the "weight" (proportion) of each serviced brand is calculated in the final statistics page. As you flip through the pages, the number "weight" gradually decreases, indicating how much more "weight" the reader has yet to flip through.

- 银奖
- 2 千克 Kessels Kramer
- 日本

- Silver Medal
- 2 kilo of KesselsKramer
- Japan
- KesselsKramer
- PIE Books, Tokyo
- 257×148×59mm
- 2027g
- 880p

»2Kilo« weighs two kilos, has pagination which records the weight of the pages up to any given point, and through its format and sturdiness gives the impression of being a brick. With its two-thousand pages, perfectly thread-bound and wrapped in a brick-coloured jacket – why pick up an object like this at all? Certainly not, as it says in the preamble, to throw through next door's window, but rather to learn more about KesselsKramer's fantastic work through its heavily compressed, pictorially rich and extremely well organised contents. A convincing archive of visual communication.

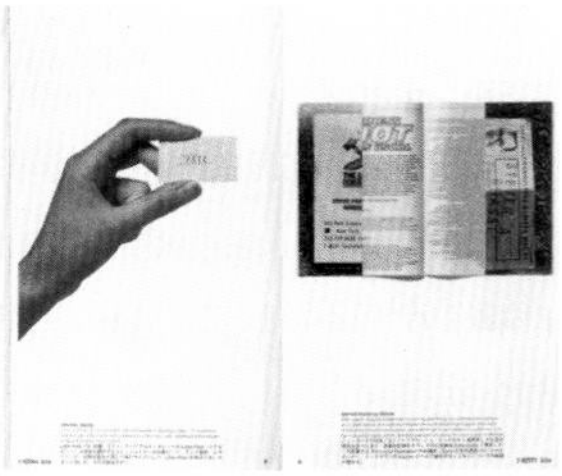

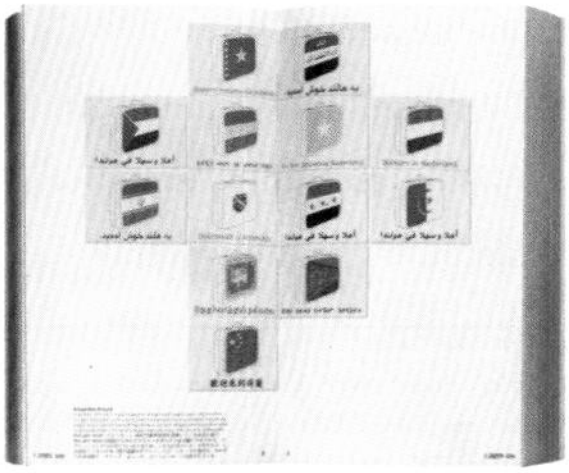

SMART
LAPPEN
FESTIVAL
20-21&22 NOV.
UTRECHT.

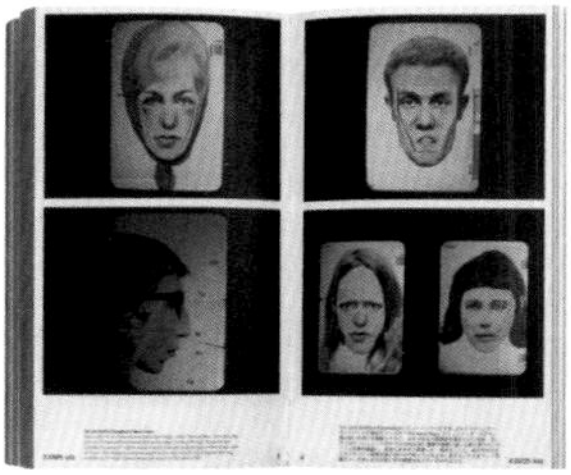

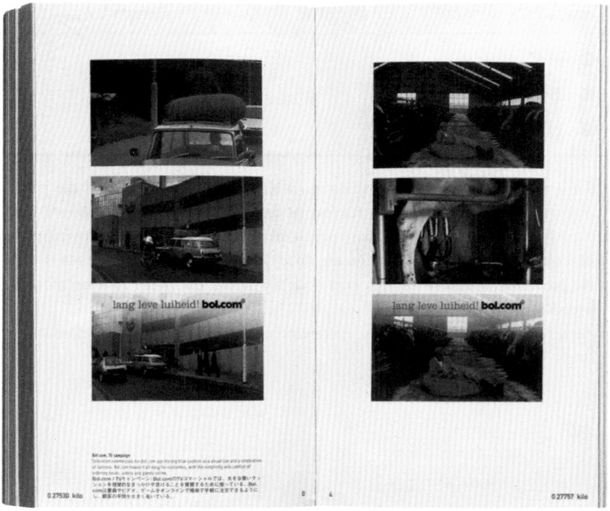

lang leve luiheid! bol.com
lang leve luiheid! bol.com
0.27530 kilo
0.27757 kilo

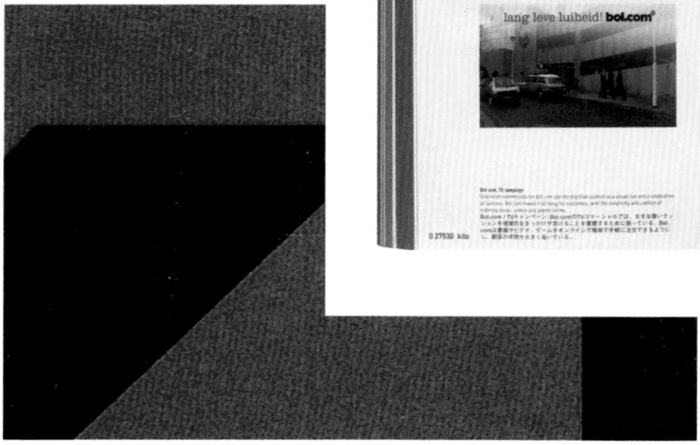

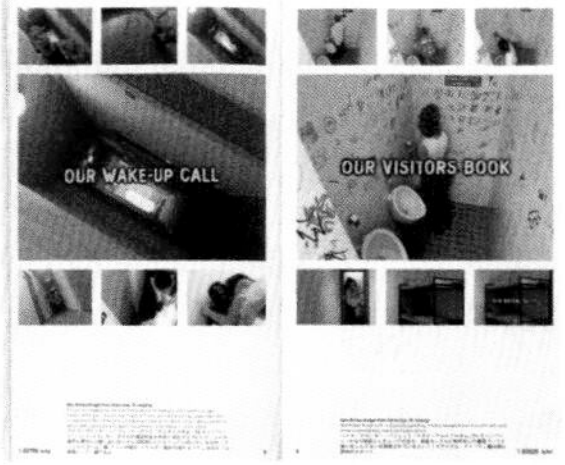
OUR WAKE-UP CALL
OUR VISITORS BOOK

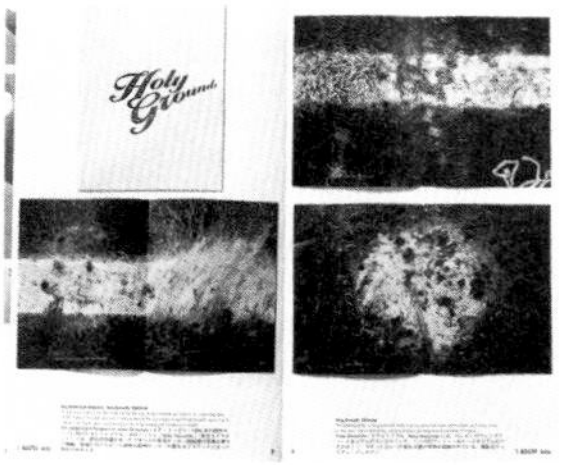

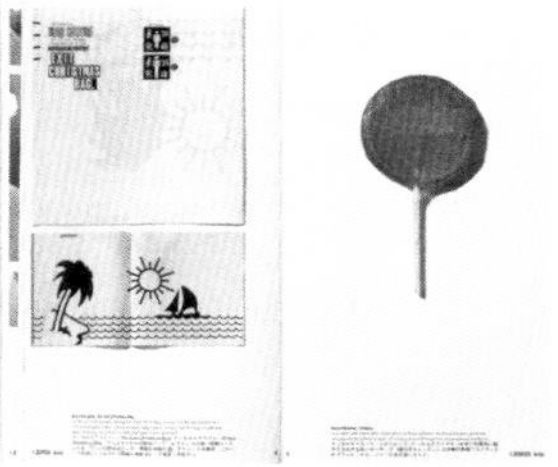

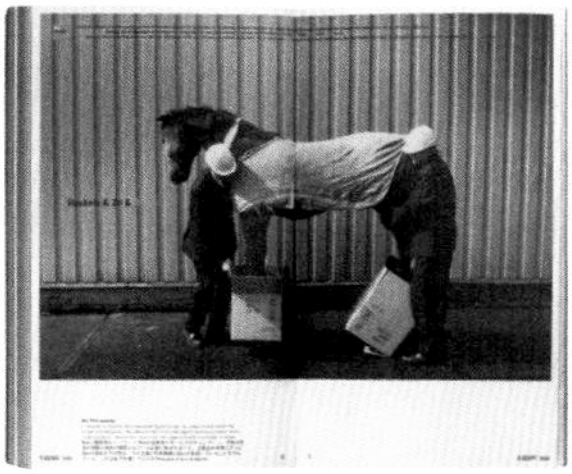

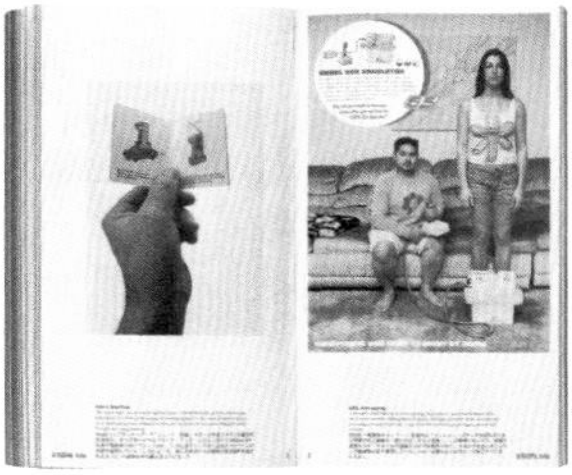

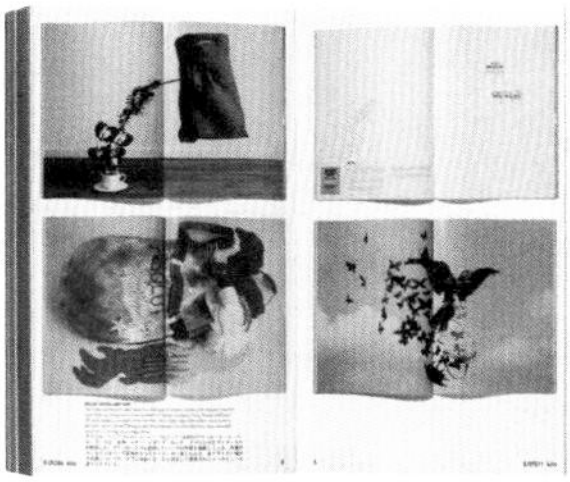

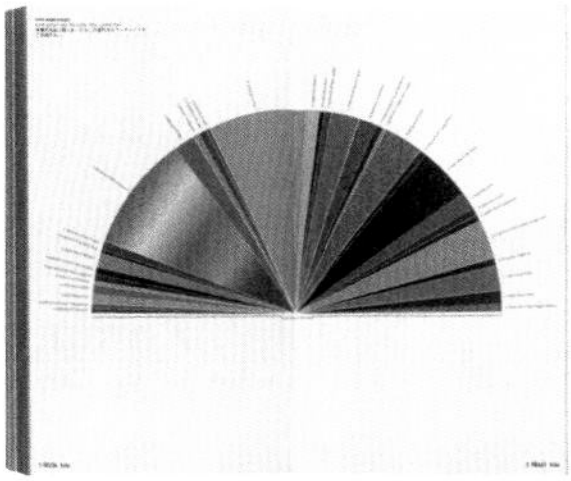

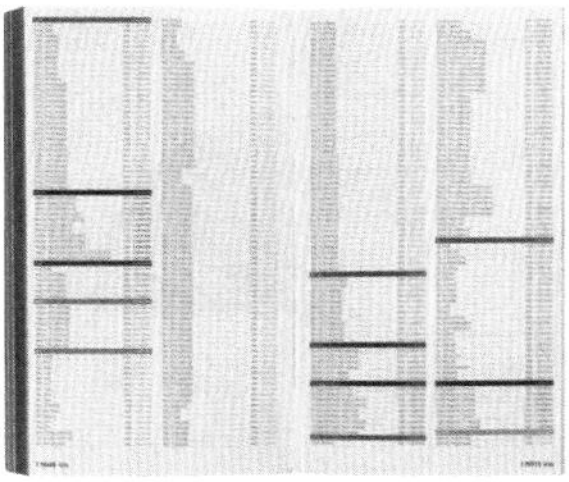

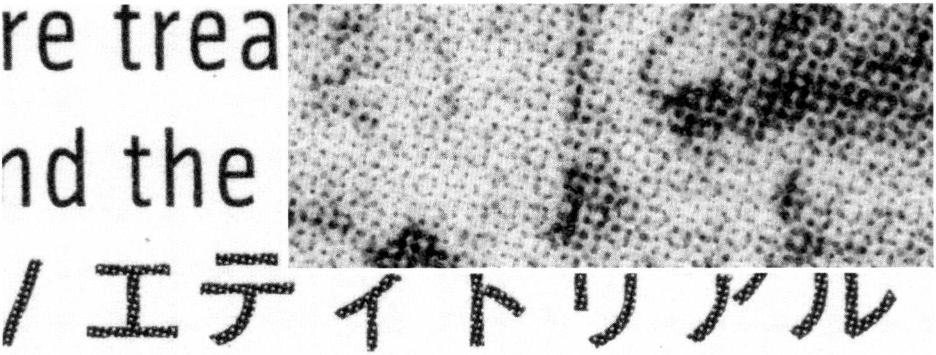
re trea
nd the
ノエディトリアル

2 kilo
OF KESSELSKRAMER
PIE
BOOKS

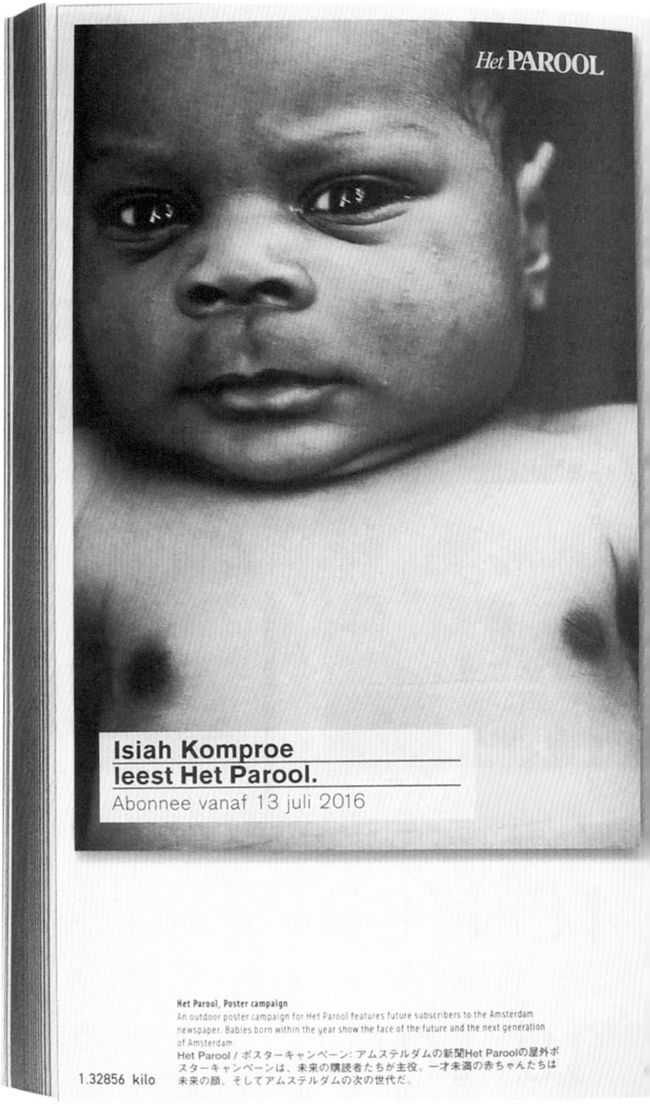
Het PAROOL
Isiah Komproe
leest Het Parool.
Abonnee vanaf 13 juli 2016
Het Parool, Poster campaign
An outdoor poster campaign for Het Parool features future subscribers to the Amsterdam newspaper. Babies born within the year show the face of the future and the next generation of Amsterdam.
Het Parool / ポスターキャンペーン: アムステルダムの新聞Het Paroolの屋外ポスターキャンペーンは、未来の購読者たちが主役。一才未満の赤ちゃんたちは未来の顔、そしてアムステルダムの次の世代だ。
1.32856 kilo

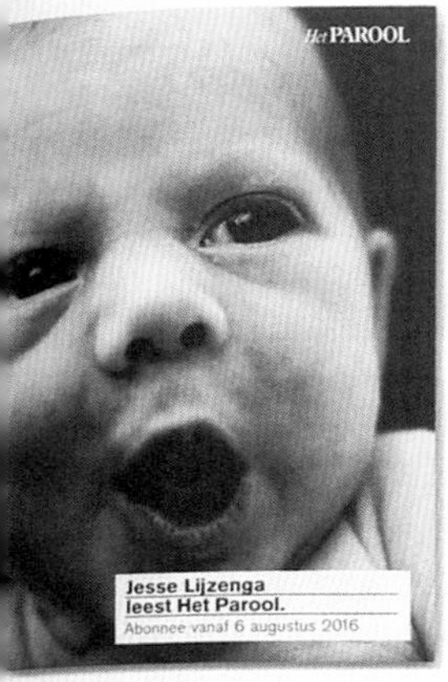
Het PAROOL
Jesse Lijzenga
leest Het Parool.
Abonnee vanaf 6 augustus 2016

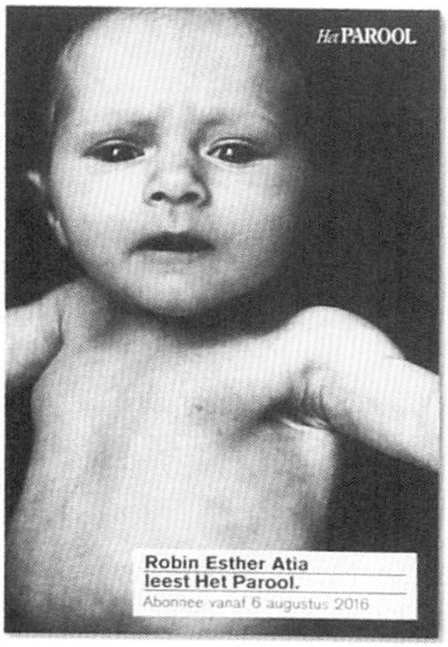
Het PAROOL
Robin Esther Atia
leest Het Parool.
Abonnee vanaf 6 augustus 2016

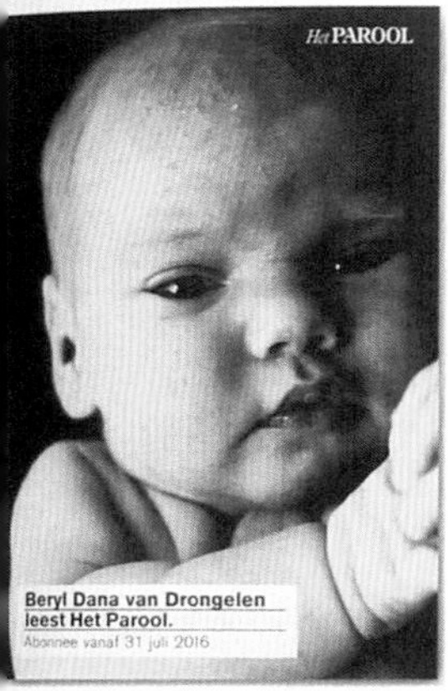
Het PAROOL
Beryl Dana van Drongelen
leest Het Parool.
Abonnee vanaf 31 juli 2016

1.33085 kilo

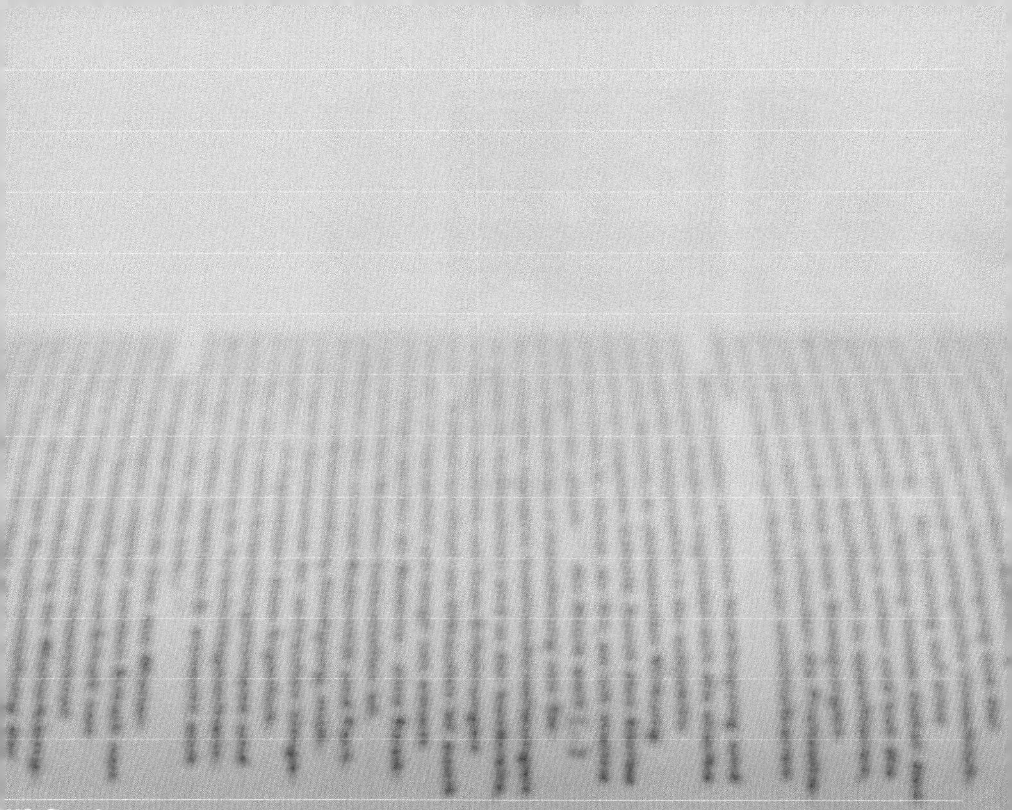

007 Sm²

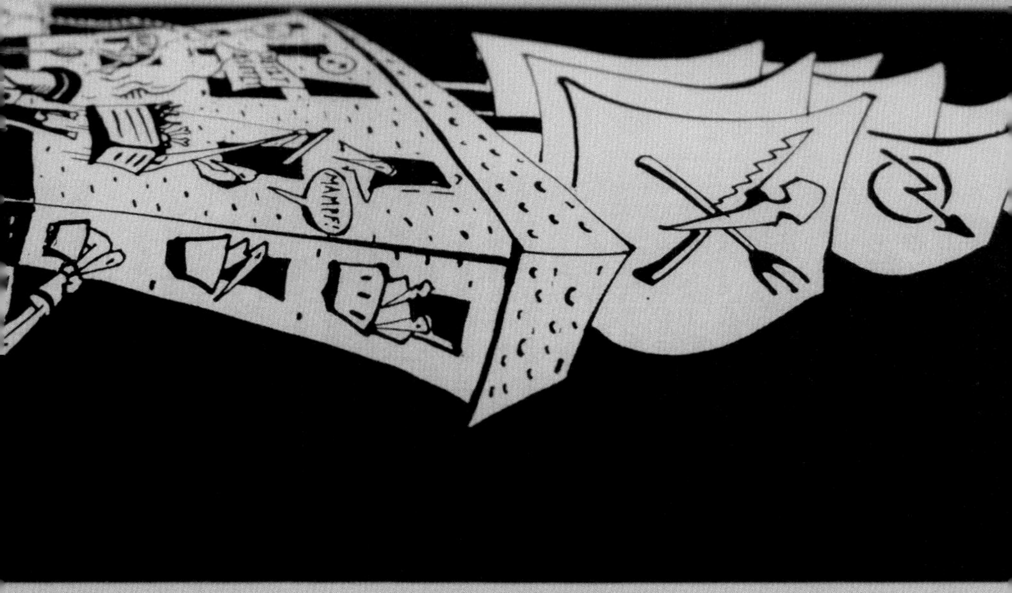
MAMPF!

这本书介绍了 1968 年以来与苏黎世城市发展和住房改革相关的政策和由此引发的社会运动。全书主要分为三个部分，分别介绍了 1945 年后瑞士的经济和城市住房的发展、居住形式与居住环境、城市住宅改革引发的民众抗议与后续措施。书籍的装订形式为无线胶平装。封面采用白卡纸印单黑，翻口部分通过前勒口的向后包裹折叠，需要完全展开才方便阅读。上下书口均刷黑色，与书脊和包裹在翻口上的部分形成连续的黑色四边，使得封面和封底白底黑字显得更加醒目。内页采用胶版纸张，除少量海报和照片采用四色印刷，其余以红黑两色为主，红色部分除了用于贯穿目录和内页的标题便于检索，也通过较为纤细的字体来体现页面内的注释。照片和海报通过字体、黑底反白等形式的另一套“检索”系统与全书红色建立的目录、注释的检索系统相区隔。文字部分有的专有名词采用加了下画线的打字机风格字体进行提示，丰富了版面节奏。

WO-
WO-
WONIGE!
STADT- UND
WOHN-
POLITISCHE
BEWEGUNGEN
IN ZÜRICH
NACH 1968
THOMAS STAHEL

This is a book about the policies and social movements related to urban development and housing reformation in Zurich since 1968. The book is divided into three parts. It introduces the development of Switzerland's economy and urban housing after 1945, the form and environment of residence, and the public protests and follow-up solutions caused by the urban housing reformation. The paperback book is perfect bound. The cover is made of white cardboard printed in black. The right edge is wrapped by the front flap and needs to be fully unfolded for reading. Both upper and lower edges are brushed black, forming a continuous black quadrilateral with the spine and the part wrapped by the flap, which makes the black characters on the white background of the front and back cover more striking. The inner pages are made of offset paper, except for a few posters and photos printed in four colors, the rest are mainly in red and black. The red color is used not only for the headings throughout the catalogue and the inner pages to facilitate retrieval, but also to display the annotations using relatively thin fonts. Photos and posters are distinguished from the red catalogue and annotation retrieval system by another set of 'retrieval' system with the different form of fonts and black-and-white color matching. The proper nouns in the text part are prompted by the underlined typewriter-style font, which enriches the layout rhythm.

银奖

Wo-Wo-Wonige!

瑞士

Silver Medal

Wo-Wo-Wonige!

Switzerland

Jonas Voegeli in Zusammenarbeit mit Beni Roffler

Thomas Stahel

Paranoia city Verlag, Zürich

239×153×27mm

802g

464p

This book discerningly manages to summarise political movements concern- ing urban and housing issues in Zurich since 1968. The combination of the bold monotype Grotesque with a typewriter-typeface successfully reflects the combative and activist nature of the movement. The typographic setting of picture references and footnotes is very meticulous and these make for optimum navigation between text, marginalia and reference illustrations throughout the book. Black double-pages divide the book into its three sections. Black edging on the top and the bottom plus a black flap along the front edge create the effect of the subject now being closed.

WIR
SUCHEN
WOHNUNG
ZIMMER
WOHNUNG

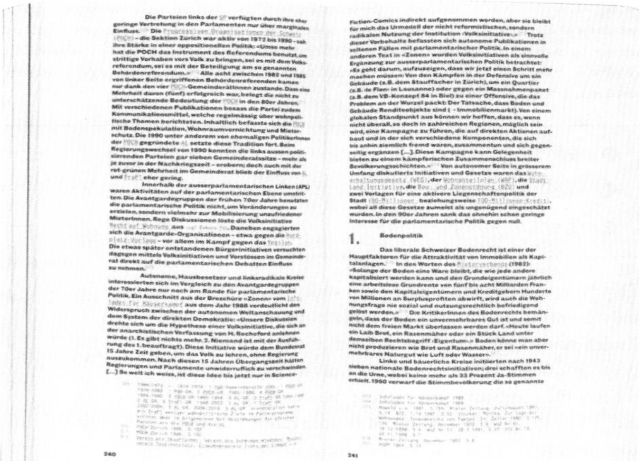

Die Parteien links der SP verfügten durch ihre eher geringe Vertretung in den Parlamenten nur über marginalen Einfluss. Die Progressiven Organisationen der Schweiz (POCH) – die Sektion Zürich war aktiv von 1972 bis 1990 – sah ihre Stärke in einer oppositionellen Politik: «Umso mehr hat die POCH das Instrument des Referendums benutzt, um strittige Vorhaben vors Volk zu bringen, sei es mit dem Volksreferendum, sei es mit der Beteiligung am so genannten Behördenreferendum.» Alle acht zwischen 1982 und 1985 von linker Seite ergriffenen Behördenreferenden kamen nur dank den vier POCH-GemeinderätInnen zustande. Dass eine Mehrheit davon (fünf) erfolgreich war, belegt die nicht zu unterschätzende Bedeutung der POCH in den 80er Jahren. Mit verschiedenen Publikationen besass die Partei zudem Kommunikationsmittel, welche regelmässig über wohnpolitische Themen berichteten. Inhaltlich befasste sich die POCH mit Bodenspekulation, Wohnraumvernichtung und Mieterschutz. Die 1990 unter anderem von ehemaligen PolitikerInnen der POCH gegründete AL setzte diese Tradition fort. Beim Regierungswechsel von 1990 konnten die links aussen politisierenden Parteien gar sieben Gemeinderatssitze – mehr als je zuvor in der Nachkriegszeit – erobern; doch auch mit der rot-grünen Mehrheit im Gemeinderat blieb der Einfluss von AL und FraP! eher gering.

Innerhalb der ausserparlamentarischen Linken (APL) waren Aktivitäten auf der parlamentarischen Ebene umstritten. Die Avantgardegruppen der frühen 70er Jahre benutzten die parlamentarische Politik nicht, um Veränderungen zu erzielen, sondern vielmehr zur Mobilisierung unzufriedener MieterInnen. Rege Diskussionen löste die Volksinitiative Recht auf Wohnung aus. Daneben engagierten sich die Avantgarde-Organisationen – etwa gegen die Marktplatz-Vorlage – vor allem im Kampf gegen das Ypsilon. Die etwas später entstandenen Bürgerinitiativen versuchten dagegen mittels Volksinitiativen und Vorstössen im Gemeinderat direkt auf die parlamentarischen Debatten Einfluss zu nehmen.

Autonome, Hausbesetzer und linksradikale Kreise interessierten sich im Vergleich zu den Avantgardegruppen der 70er Jahre nur noch am Rande für parlamentarische Politik. Ein Ausschnitt aus der Broschüre «Zonen» vom Infoladen für Häuserkampf aus dem Jahr 1988 verdeutlicht den Widerspruch zwischen der autonomen Weltanschauung und dem System der direkten Demokratie: «Unsere Diskussion drehte sich um die Hypothese einer Volksinitiative, die sich an der anarchistischen Verfassung von H. Rochefort anlehnen würde (1. Es gibt nichts mehr. 2. Niemand ist mit der Ausführung des 1. beauftragt). Diese Initiative würde dem Bundesrat 15 Jahre Zeit geben, um das Volk zu lehren, ohne Regierung auszukommen. Nach diesen 15 Jahren Übergangszeit hätten Regierungen und Parlamente unwiderruflich zu verschwinden. […] So weit ich weiss, ist diese Idee bis jetzt nur in Science-

240

Fiction-Comics indirekt aufgenommen worden, aber sie bleibt für mich das Urmodell der nicht reformistischen, sondern radikalen Nutzung der Institution ‹Volksinitiative›.» Trotz dieser Vorbehalte befassten sich autonome Publikationen in seltenen Fällen mit parlamentarischer Politik. In einem anderen Text in «Zonen» wurden Volksinitiativen als sinnvolle Ergänzung zur ausserparlamentarischen Politik betrachtet: «Es geht darum, aufzuzeigen, dass wir jetzt einen Schritt mehr machen müssen: Von den Kämpfen in der Defensive um ein Gebäude (z.B. dem Stauffacher in Zürich), um ein Quartier (z.B. ‹le Flon› in Lausanne) oder gegen ein Massnahmenpaket (z.B. dem VB-Konzept 84 in Biel) zu einer Offensive, die das Problem an der Wurzel packt: Der Tatsache, dass Boden und Gebäude Renditeobjekte sind (→ Immobilienmarkt). Von einem globalen Standpunkt aus können wir hoffen, dass es, wenn nicht überall, so doch in zahlreichen Regionen, möglich sein wird, eine Kampagne zu führen, die auf direkten Aktionen aufbaut und in der sich verschiedene Komponenten, die sich bis anhin ziemlich fremd waren, zusammentun und sich gegenseitig ergänzen […]. Diese Kampagne kann Gelegenheit bieten zu einem kämpferischen Zusammenschluss breiter Bevölkerungsschichten.» Von autonomer Seite in grösserem Umfang diskutierte Initiativen und Gesetze waren das Wohnerhaltungsgesetz (WEG), der Wohnanteilplan (WAP), die Stadt-Land-Initiative, die Bau- und Zonenordnung (BZO) und zwei Vorlagen für eine aktivere Liegenschaftenpolitik der Stadt (90-Millionen- beziehungsweise 100-Millionen-Kredit), wobei all diese Gesetze zumeist als ungenügend eingeschätzt wurden. In den 90er Jahren sank das ohnehin schon geringe Interesse für die parlamentarische Politik gegen null.

1. Bodenpolitik

Das liberale Schweizer Bodenrecht ist einer der Hauptfaktoren für die Attraktivität von Immobilien als Kapitalanlagen. In den Worten des Mieterverbands (1982): «Solange der Boden eine Ware bleibt, die wie jede andere kapitalisiert werden kann und den Grundeigentümern jährlich eine arbeitslose Grundrente von fünf bis acht Milliarden Franken sowie den Kapitaleigentümern und Kreditgebern Hunderte von Millionen an Surplusprofiten abwirft, wird auch die Wohnungsfrage nie sozial und nutzungsrechtlich befriedigend gelöst werden.» Die KritikerInnen des Bodenrechts bemängeln, dass der Boden ein unvermehrbares Gut ist und somit nicht dem freien Markt überlassen werden darf. «Heute laufen ein Laib Brot, ein Rasenmäher oder ein Stück Land unter demselben Rechtsbegriff ‹Eigentum›.» Boden könne man aber nicht produzieren wie Brot und Rasenmäher, er sei «ein unvermehrbares Naturgut wie Luft oder Wasser».

Linke und bäuerliche Kreise initiierten nach 1943 sieben nationale Bodenrechtsinitiativen; drei schafften es bis an die Urne, wobei keine mehr als 33 Prozent Ja-Stimmen erhielt. 1950 verwarf die Stimmbevölkerung die so genannte

241

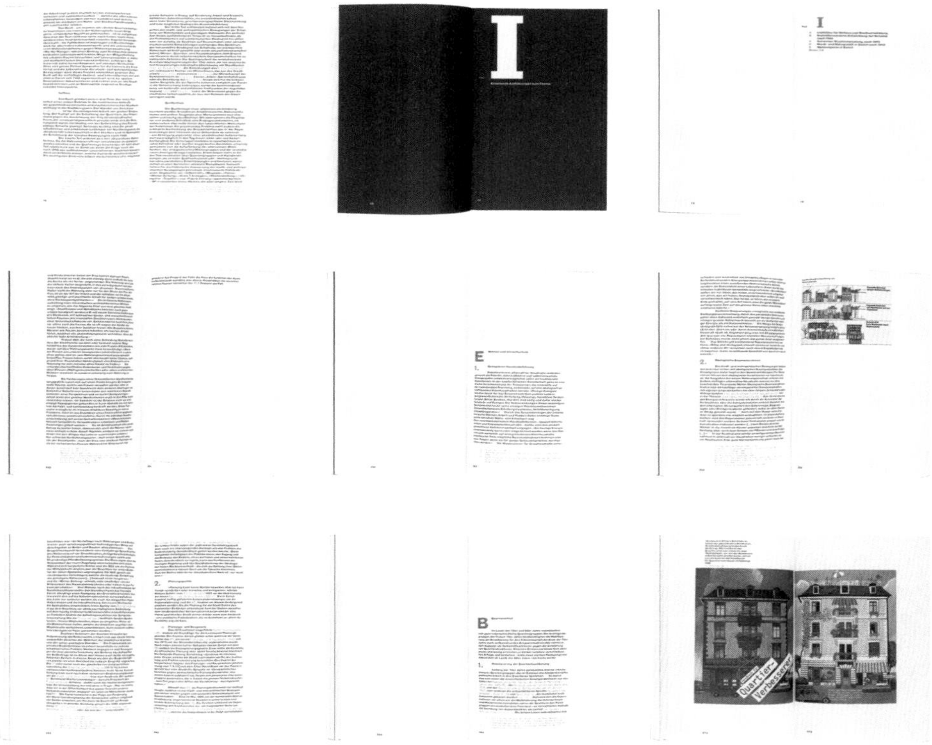
Quartier-
Verbesserer

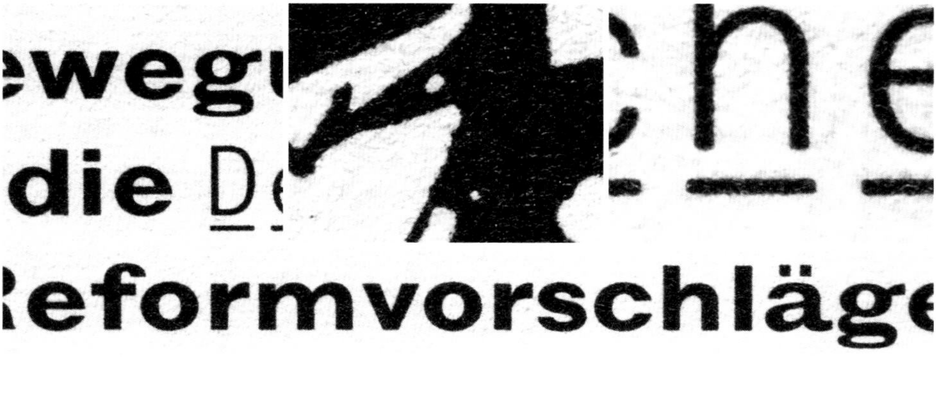
ewegu
die De
Reformvorschläge

Die Parteien links der SP verfügten durch ihre eher geringe Vertretung in den Parlamenten nur über marginalen Einfluss.[009] Die Progressiven Organisationen der Schweiz (POCH) – die Sektion Zürich war aktiv von 1972 bis 1990 – sah ihre Stärke in einer oppositionellen Politik: «Umso mehr hat die POCH das Instrument des Referendums benutzt, um strittige Vorhaben vors Volk zu bringen, sei es mit dem Volksreferendum, sei es mit der Beteiligung am so genannten Behördenreferendum.»[010] Alle acht zwischen 1982 und 1985 von linker Seite ergriffenen Behördenreferenden kamen nur dank den vier POCH-GemeinderätInnen zustande. Dass eine Mehrheit davon (fünf) erfolgreich war, belegt die nicht zu unterschätzende Bedeutung der POCH in den 80er Jahren.[011] Mit verschiedenen Publikationen besass die Partei zudem Kommunikationsmittel, welche regelmässig über wohnpolitische Themen berichteten. Inhaltlich befasste sich die POCH mit Bodenspekulation, Wohnraumvernichtung und Mieterschutz. Die 1990 unter anderem von ehemaligen PolitikerInnen der POCH gegründete AL setzte diese Tradition fort. Beim Regierungswechsel von 1990 konnten die links aussen politisierenden Parteien gar sieben Gemeinderatssitze – mehr als je zuvor in der Nachkriegszeit – erobern; doch auch mit der rot-grünen Mehrheit im Gemeinderat blieb der Einfluss von AL und FraP! eher gering.

Innerhalb der ausserparlamentarischen Linken (APL) waren Aktivitäten auf der parlamentarischen Ebene umstritten. Die Avantgardegruppen der frühen 70er Jahre benutzten die parlamentarische Politik nicht, um Veränderungen zu erzielen, sondern vielmehr zur Mobilisierung unzufriedener MieterInnen. Rege Diskussionen löste die Volksinitiative Recht auf Wohnung aus (vgl. Exkurs IV). Daneben engagierten sich die Avantgarde-Organisationen – etwa gegen die Hardplatz-Vorlage – vor allem im Kampf gegen das Ypsilon. Die etwas später entstandenen Bürgerinitiativen versuchten dagegen mittels Volksinitiativen und Vorstössen im Gemeinderat direkt auf die parlamentarischen Debatten Einfluss zu nehmen.[012]

Autonome, Hausbesetzer und linksradikale Kreise interessierten sich im Vergleich zu den Avantgardegruppen der 70er Jahre nur noch am Rande für parlamentarische Politik. Ein Ausschnitt aus der Broschüre «Zonen» vom Infoladen für Häuserkampf aus dem Jahr 1988 verdeutlicht den Widerspruch zwischen der autonomen Weltanschauung und dem System der direkten Demokratie: «Unsere Diskussion drehte sich um die Hypothese einer Volksinitiative, die sich an der anarchistischen Verfassung von H. Rocheford anlehnen würde (1. Es gibt nichts mehr. 2. Niemand ist mit der Ausführung des 1. beauftragt). Diese Initiative würde dem Bundesrat 15 Jahre Zeit geben, um das Volk zu lehren, ohne Regierung auszukommen. Nach diesen 15 Jahren Übergangszeit hätten Regierungen und Parlamente unwiderruflich zu verschwinden. [...] So weit ich weiss, ist diese Idee bis jetzt nur in Science-

009 1968–1972: –. 1974–1974: 1 PdA-GemeinderätIn (GR), 1 POCH-GR 1978–1982: 1 PdA-GR, 2 POCH-GR. 1982–1986: 4 POCH-GR 1986–1990: 3 POCH-GR 1990–1994: 4 AL-GR, 3 FraP!-GR 1994–1998: 2 AL-GR, 5 FraP!-GR. 1998–2002: 2 AL-GR, 1 FraP!-GR. 2002–2006: 3 AL-GR. 2006–2010: 5 AL-GR. Grundsätzlich hatte die FraP! weniger wohnpolitische Ziele im Parteiprogramm, vertrat aber im Allgemeinen bei Abstimmungen die gleichen Parolen wie die POCH und die AL.

010 POCH Zürich 1986, S. 102.

011 POCH Zürich 1986, S. 102.

012 Verein pro Stauffacher, Verein pro Schmiede Wiedikon, Mieterverein Tessinerplatz, Einwohnerverein links der Limmat u.a.

Fiction-Comics indirekt aufgenommen worden, aber sie bleibt für mich das Urmodell der nicht reformistischen, sondern radikalen Nutzung der Institution ‹Volksinitiative›.»[013] Trotz dieser Vorbehalte befassten sich autonome Publikationen in seltenen Fällen mit parlamentarischer Politik. In einem anderen Text in «Zonen» wurden Volksinitiativen als sinnvolle Ergänzung zur ausserparlamentarischen Politik betrachtet: «Es geht darum, aufzuzeigen, dass wir jetzt einen Schritt mehr machen müssen: Von den Kämpfen in der Defensive um ein Gebäude (z.B. dem Stauffacher in Zürich), um ein Quartier (z.B. ‹le Flon› in Lausanne) oder gegen ein Massnahmenpaket (z.B. dem VB-Konzept 84 in Biel) zu einer Offensive, die das Problem an der Wurzel packt: Der Tatsache, dass Boden und Gebäude Renditeobjekte sind (→ Immobilienmarkt). Von einem globalen Standpunkt aus können wir hoffen, dass es, wenn nicht überall, so doch in zahlreichen Regionen, möglich sein wird, eine Kampagne zu führen, die auf direkten Aktionen aufbaut und in der sich verschiedene Komponenten, die sich bis anhin ziemlich fremd waren, zusammentun und sich gegenseitig ergänzen [...]. Diese Kampagne kann Gelegenheit bieten zu einem kämpferischen Zusammenschluss breiter Bevölkerungsschichten.»[014] Von autonomer Seite in grösserem Umfang diskutierte Initiativen und Gesetze waren das Wohnerhaltungsgesetz (WEG), der Wohnanteilplan (WAP), die Stadt-Land-Initiative, die Bau- und Zonenordnung (BZO) und zwei Vorlagen für eine aktivere Liegenschaftenpolitik der Stadt (50-Millionen- beziehungsweise 100-Millionen-Kredit), wobei all diese Gesetze zumeist als ungenügend eingeschätzt wurden. In den 90er Jahren sank das ohnehin schon geringe Interesse für die parlamentarische Politik gegen null.

1. Bodenpolitik

Das liberale Schweizer Bodenrecht ist einer der Hauptfaktoren für die Attraktivität von Immobilien als Kapitalanlagen.[015] In den Worten des Mieterverbands (1982): «Solange der Boden eine Ware bleibt, die wie jede andere kapitalisiert werden kann und den Grundeigentümern jährlich eine arbeitslose Grundrente von fünf bis acht Milliarden Franken sowie den Kapitaleigentümern und Kreditgebern Hunderte von Millionen an Surplusprofiten abwirft, wird auch die Wohnungsfrage nie sozial und nutzungsrechtlich befriedigend gelöst werden.»[016] Die KritikerInnen des Bodenrechts bemängeln, dass der Boden ein unvermehrbares Gut ist und somit nicht dem freien Markt überlassen werden darf. «Heute laufen ein Laib Brot, ein Rasenmäher oder ein Stück Land unter demselben Rechtsbegriff ‹Eigentum›.» Boden könne man aber nicht produzieren wie Brot und Rasenmäher, er sei «ein unvermehrbares Naturgut wie Luft oder Wasser».[017]

Linke und bäuerliche Kreise initiierten nach 1943 sieben nationale Bodenrechtsinitiativen; drei schafften es bis an die Urne, wobei keine mehr als 33 Prozent Ja-Stimmen erhielt. 1950 verwarf die Stimmbevölkerung die so genannte

013 Infoladen für Häuserkampf 1988
014 Infoladen für Häuserkampf 1988.
015 Howald u.a. 1981, S.159. Mieter-Zeitung, Juli/August 1987, S.14. NZZ, 1.10.1981, S.52. Stocker, Monika. Zur Lage der Wohnenden. Innenansicht von Fakten. In: Zeller 1990, S.175, 178. Mieter-Zeitung, Dezember 1982, S.8. WoZ Nr.51, 19.12.1986, S.4. WoZ Nr.13, 28.3.1991, S.27. 21i Nr.13, 24.11.1988, S.1.
016 Mieter-Zeitung, Dezember 1982, S.8.
017 AGÖP 1984, S.33.

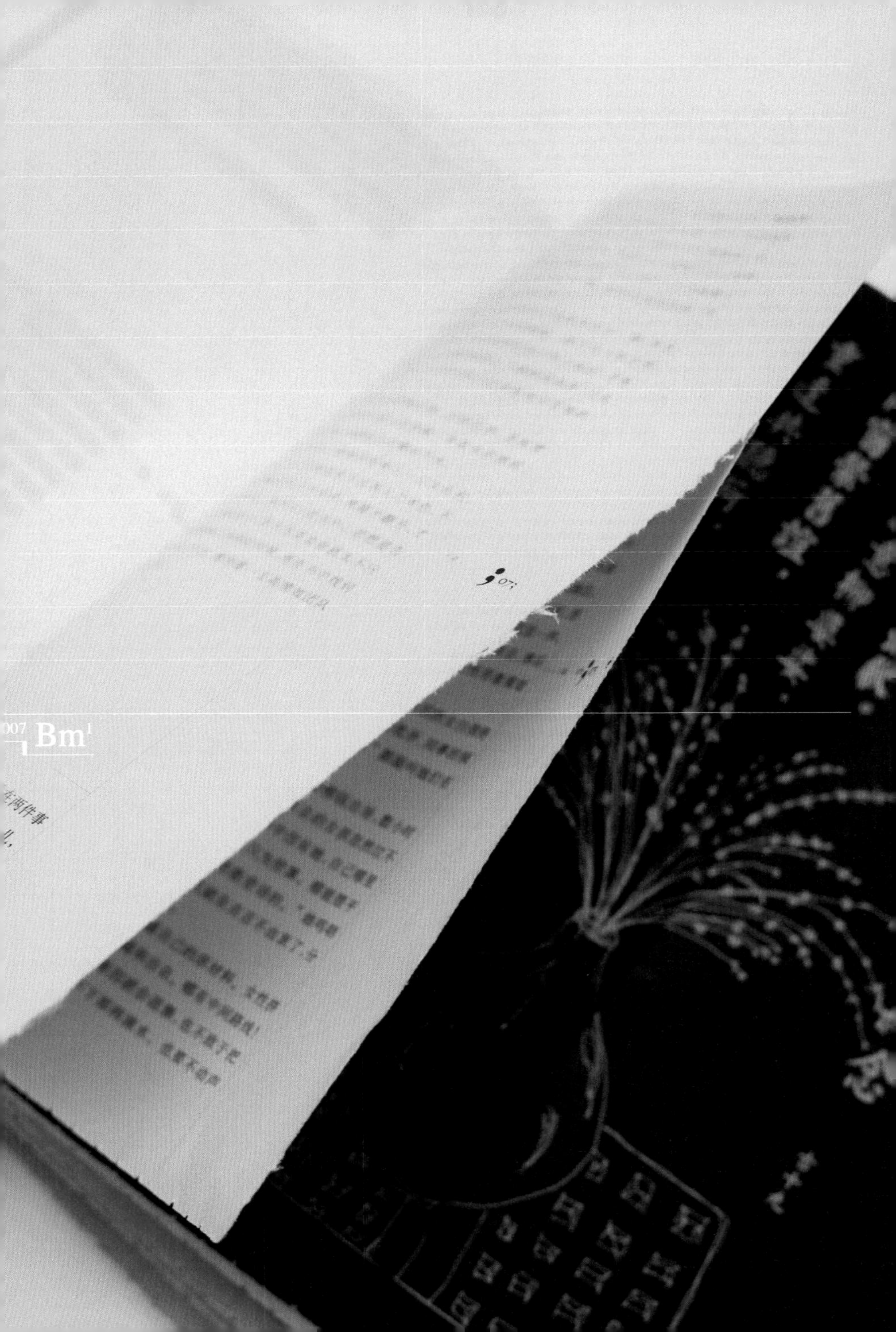

本书是广告人“古十九”的随笔集，主要为作者在报刊上发表过的各种评议杂谈、荒诞故事。全书锁线胶装，封面四色印刷并以红线贯穿，表现出想要将内容牢牢锁在里面的意味。内页单黑印刷，采用牛皮纸与黄色和白色两种双胶纸按照 1:4:1:4 的比例混合装订，上下书口打毛处理，而右侧书口部分故意不裁开，形成类似包背装的形式。但与传统包背装不同，这些未清边的页面需要读者自己裁开，才能看到里面的内容。读者的“裁”与作者的“不裁”，体现出作者的自谦，也为了让本书的阅读体验更加轻松随意，而减少过于精巧的书籍设计给人带来的疏离感。扉页部分，一把模切压出的纸刀在点题的同时，也在提醒着读者需要自己参与进来。

stitching up
不裁

This book is an essay collection by Gu Shijiu who works in advertising. All sorts of published reviews, miscellaneous talks and even absurd stories are collected in this book. It is glue-bound with thread sewing and the cover is printed in four colors with red lines running through, which means locking the content firmly inside. The inside pages are printed in black only, mixing kraft paper, yellow offset paper and white offset paper together. The upper and lower edges are rough, while the right edge is deliberately uncut. The book looks similar to a double-leaved one. Unlike traditional double-leaved books, these uncut pages need to be cut by the readers themselves. The ‘cutting’ of the readers and the ‘not cutting’ of the author, reflect the author’s self-humility. The design also makes the readers feel more relaxed and casual, and reduces the alienation brought by elaborate books. A die-cut paper knife in the title page reminds the readers of their own participation.

铜奖	Bronze Medal
不裁	stitching up
中国	China
朱赢椿	Zhu Yingchun
古十九	Gu Shijiu
江苏文艺出版社	Jiangsu Literature & Art Publishing House, Nanjing
	239×163×14mm
	417g
	189p

A red thread, a rough-textured bottom edge, the arrangement of typography, the surface and various colours of the paper – all make an excellent contribution to the sensitive and subtle transportation of the book’s content. A pre-perforated sewing utensil, worked into one of the endpapers and also suitable for opening the pages, can later be used as a bookmark. The design is in all its components well thought-through and technically well executed.

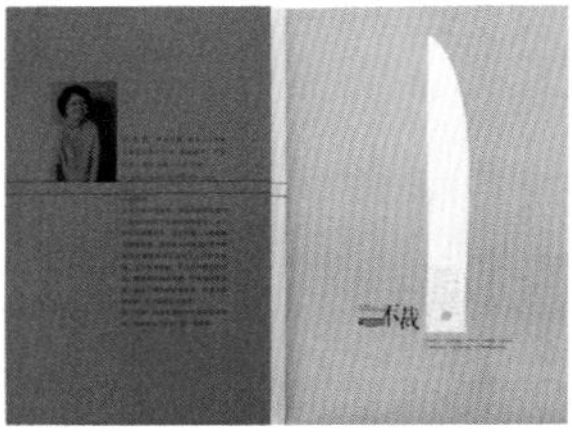
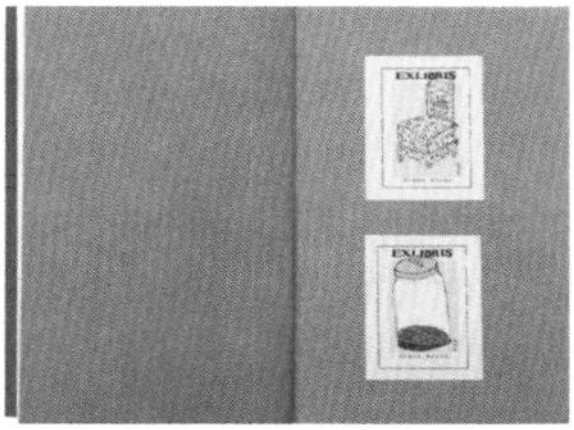

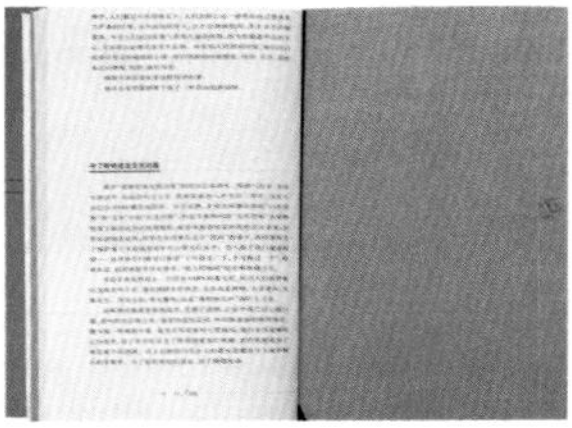

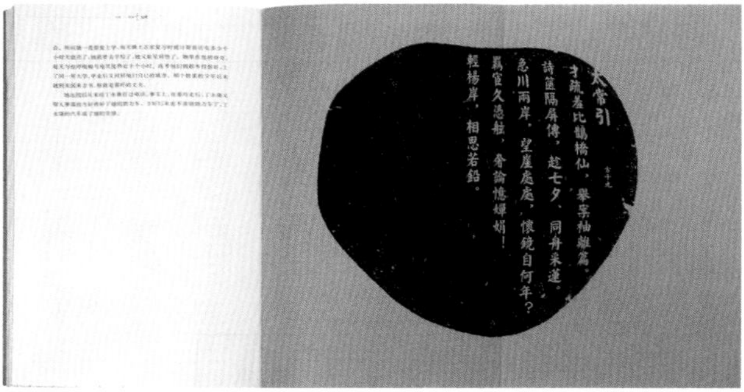
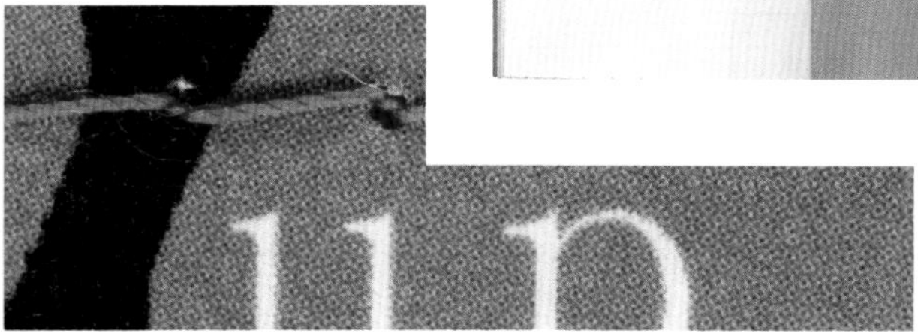

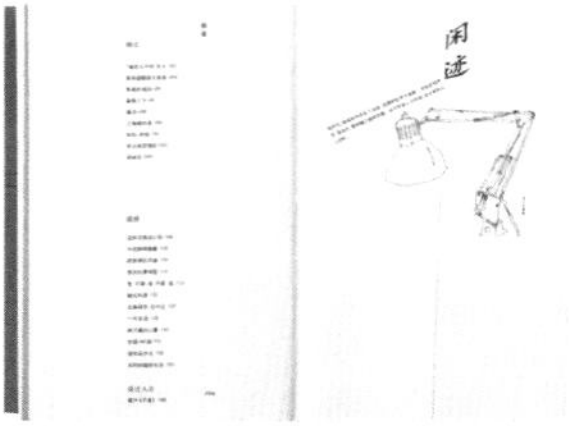

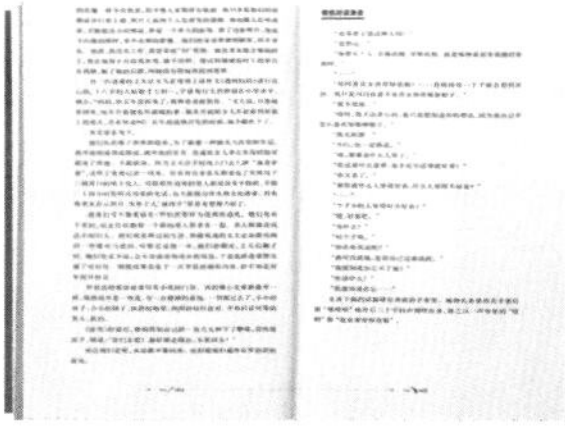

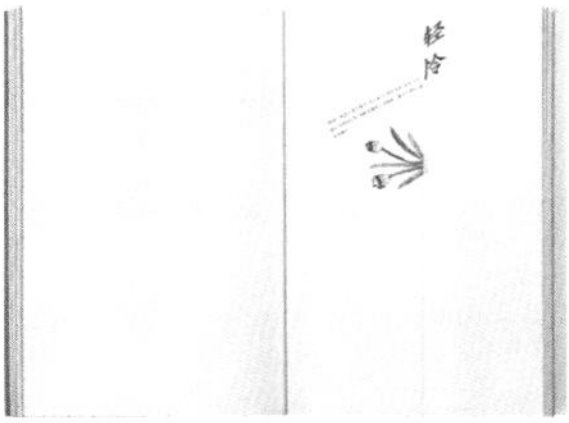

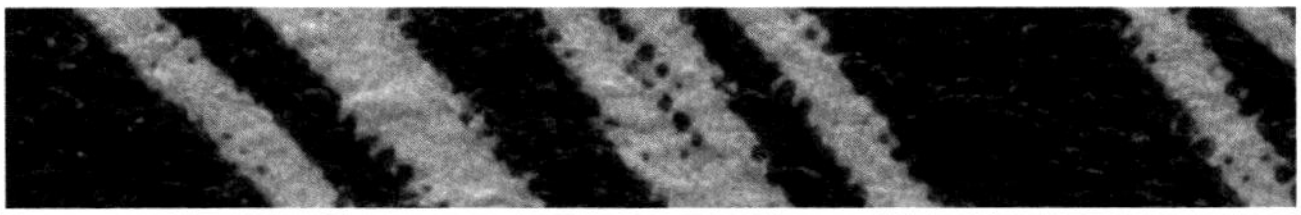

一把带刻度

stitching up
不裁
古十九 著
凤凰出版传媒集团
江苏文艺出版社

是没有公厕，僻静处老左的围墙拐角就成了他们面壁解决的上选之地。气得老左牙根发痒，但又抓不住现行，只得在家打印了一排“请勿随地小便，违者罚款”的字条贴在墙上，可惜收效甚微，所以每当潇涵来到他家时，远远就闻到一股异味。最终还是她出了个点子，漫画了只三脚着地、一脚踏墙的狗贴在墙上。现在的孩子再不接近自然，也在院中看过犬类方便的情态。教育意义显而易见，居然大见成效，老左家围墙再不必充当露天公厕了。他立即跑来拍潇涵的马屁，说，要不是你，我不知要受罪到几时呢。潇涵便笑道，那有什么。他马上又抛出一句现成的话：“你就是上帝派来救我的。” 她一听就知道是从冯玉祥向李德全求婚的传说中抄来的。

他做广告做得久了，渐渐混乱了什么是自己的原创，什么是借用。在他尚未成名前，当然那时的妻子也是“前妻”，前岳母在国家工厂任职，国庆前夕，单位要求她写一首赞美祖国的诗，前岳母当然想到这位女婿。老左那天只略略思索了几分钟就完稿了，结果事后被前岳母全家一顿痛骂，原来他写的竟然是“今天是你的生日，我的祖国，清晨我们放飞一群白鸽……”几乎和一首歌的歌词完全相同，显然是抄袭之作。不过老左当时写的时候一点都没有意识到，他还奇怪怎么这次灵感来得如此汹涌？

他就是这种毛病，动不动吊两句书袋。潇涵和晓世在一起，长了不少见识，人家整天和博物馆打交道，也没见他背些诗句嘛，不禁暗暗偷笑老左是半瓶醋。

老左常常跟潇涵回顾他一年前结束的那段婚姻，也不讳言那天跟他去开会的女秘书就是他现在的女朋友，也就是他所谓的“老婆”。他叹着气说：“唉，你别看我表面风光，其实一点劲都没有。按理说，我房子也买了，车子也买了，老婆……呃……也买了。”他说到这里，潇涵忍不住大笑起来，他自己也觉得可笑，两个人像傻瓜一样笑了半天。然后一起唱《THE SOUND OF SILENCE》，“And the people bowed and prayed/To the neon god they made. /And the sign flashed out its warming/In the words that it was forming……”

反正潇涵已经把他看作一般朋友了，直到有一个星期五。本来说好第二天去打彩弹珠的，结果他临时要去 W 城谈业务，只得改期。他在电话里说：“你要等我回来过节。”潇涵挂了电话，半天想不起来要过什么节。第二天早晨下了大雾，高速公路封了，她接到他的电话时

浣溪沙

古十九

躡足悄聲趕早歸，不期掩額遇斜暉。竊得閑意探薔薇。

任我山高連水遠，管它物是與人非。鳳臺虹影現城畿。

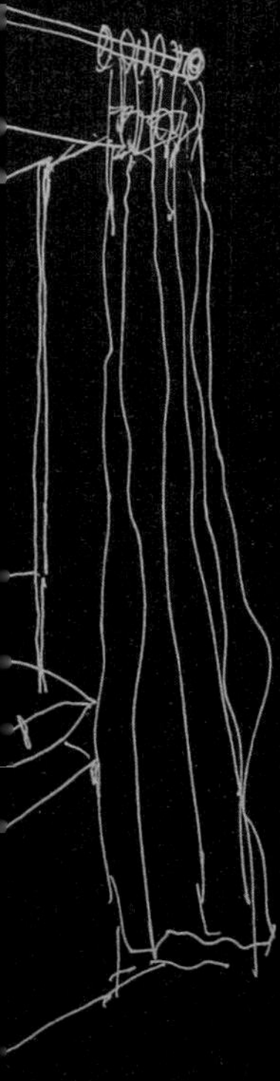

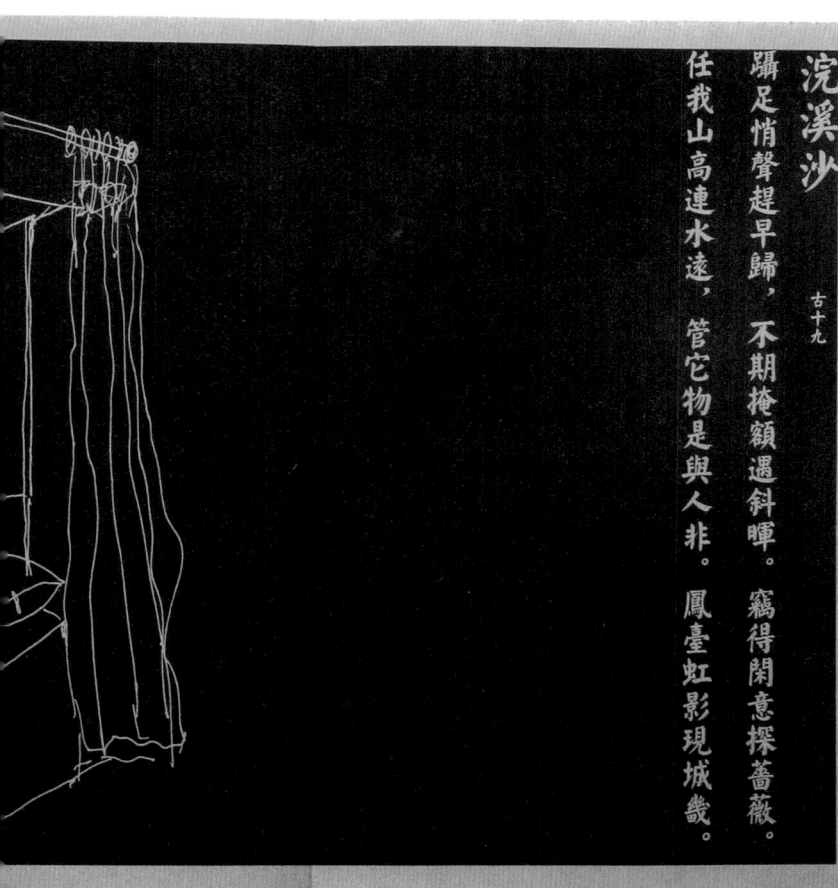

Die deutsche Sprache bildet bei Zahlen im Bezug auf die Reihenfolge von Zehner- und Einer-Namen eine Ausnahme. Wo im Russischen die Zehnereinheit zuerst kommt (**двадцать** пять), wird im Deutschen die Einereinheit zuerst genannt (**fünf**undzwanzig), ein Umstand, der vom arabischen Ursprung der Zahlen herrührt, es aber z B. beim Aufschreiben von Telefonnummern nicht einfacher macht.
Bm²

这是一本日记，也可以认为是作者在德国学习阶段的笔记。书籍的第一部分通过字母、单词和句子之对比来讲述德文与俄文之间语法结构的差异。第二部分则记录了在德国的见闻。书籍为硬精装，黑色与橙色荧光墨双色印刷，覆哑膜。环衬搭配橙色纸张以呼应。内页黑色正文，橙色荧光墨主要用于对照的俄文和需要突出的德文部分。目录被放在了最后，好让读者可以直接从最简单的德文和俄文字母的对比进入此书，而减弱目录划分内容的影响力。书内搭配黑橙双色书签条，将双色的主题用到极致。设计者通过黑橙双色来隐喻德俄两国及两国文化，罗宋汤作为俄国最被人熟知的美食，在这里也成了作者联系两国文化的锁链。

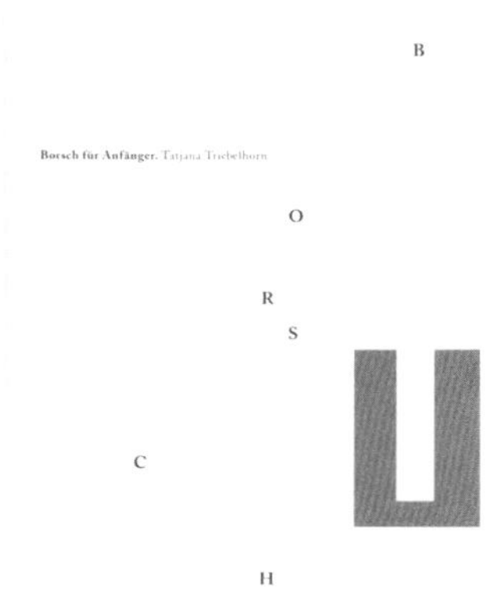
B
Borsch für Anfänger. Tatjana Triebelhorn
O
R
S
C
H

This is a diary, which can also be considered as the author's study notes in Germany. The first part of the book describes the differences in grammatical structure between German and Russian through the comparison of letters, words and sentences. The second part records what the author saw and heard in Germany. The hard cover is laminated and printed in black and orange, accompanied with orange end paper. Black is used for the text and fluorescent orange ink is mainly used to contrast Russian and German that need to be highlighted. The table of contents is placed in the final part, so that readers can start the book directly from the simplest comparison of German and Russian letters without distraction. Black and orange bookmarks are attached in the book, making the use of two-color theme to the extreme. The designer used black and orange as a metaphor for Germany and Russia and their cultures. Borscht, as the most well-known food in Russia, has also become the link between the two cultures.

铜奖

适合初学者的罗宋汤

德国

In this book the artist's method is very attractively illustrated and made easy to understand. She approaches her subject matter effectively from the outside through the script, and by means of typographical exercises demonstrates this with much temperament and emotion. Russian and German are juxtaposed with one another by means of their different alphabets. With much sensitivity to the total space available light, contoured and coloured letters are positioned on the pages. The materials used are appealing and the reducing of colours to orange and black is in keeping with everything else.

Bronze Medal

Borsch für Anfänger

Germany

Tatjana Triebelhorn

Tatjana Triebelhorn, Roman Triebelhorn

Eigenverlag Tatjana Triebelhorn, Stuttgart

295×184×15mm

552g

103p

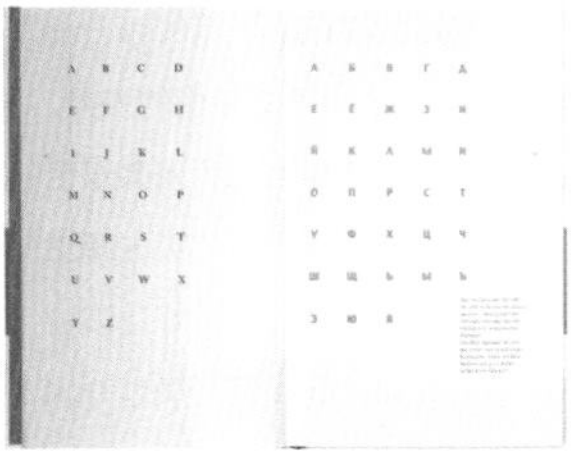
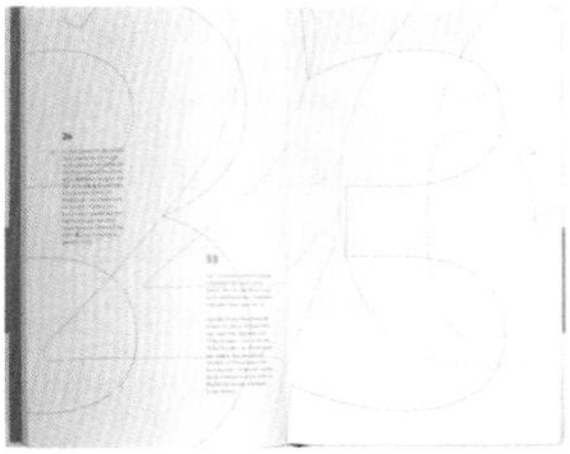
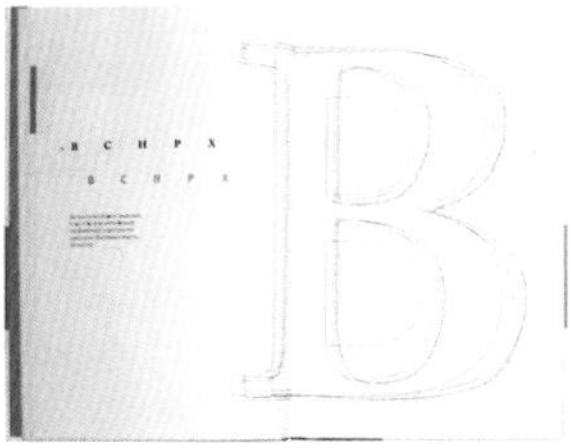
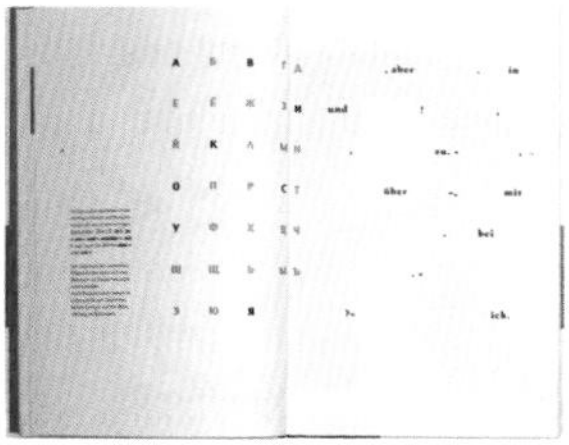
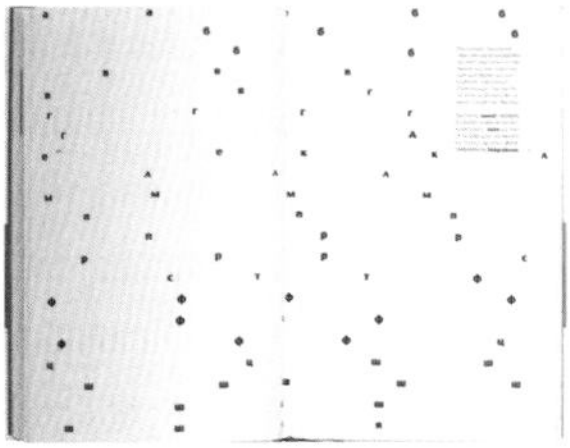
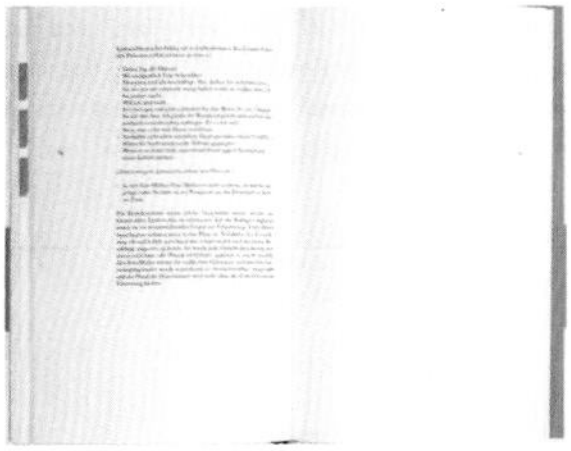
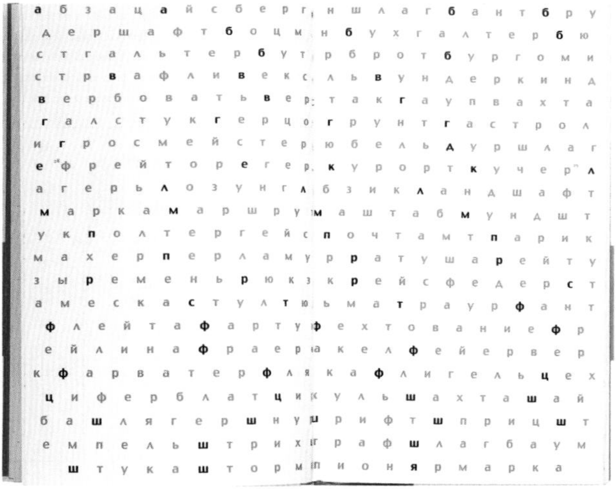
 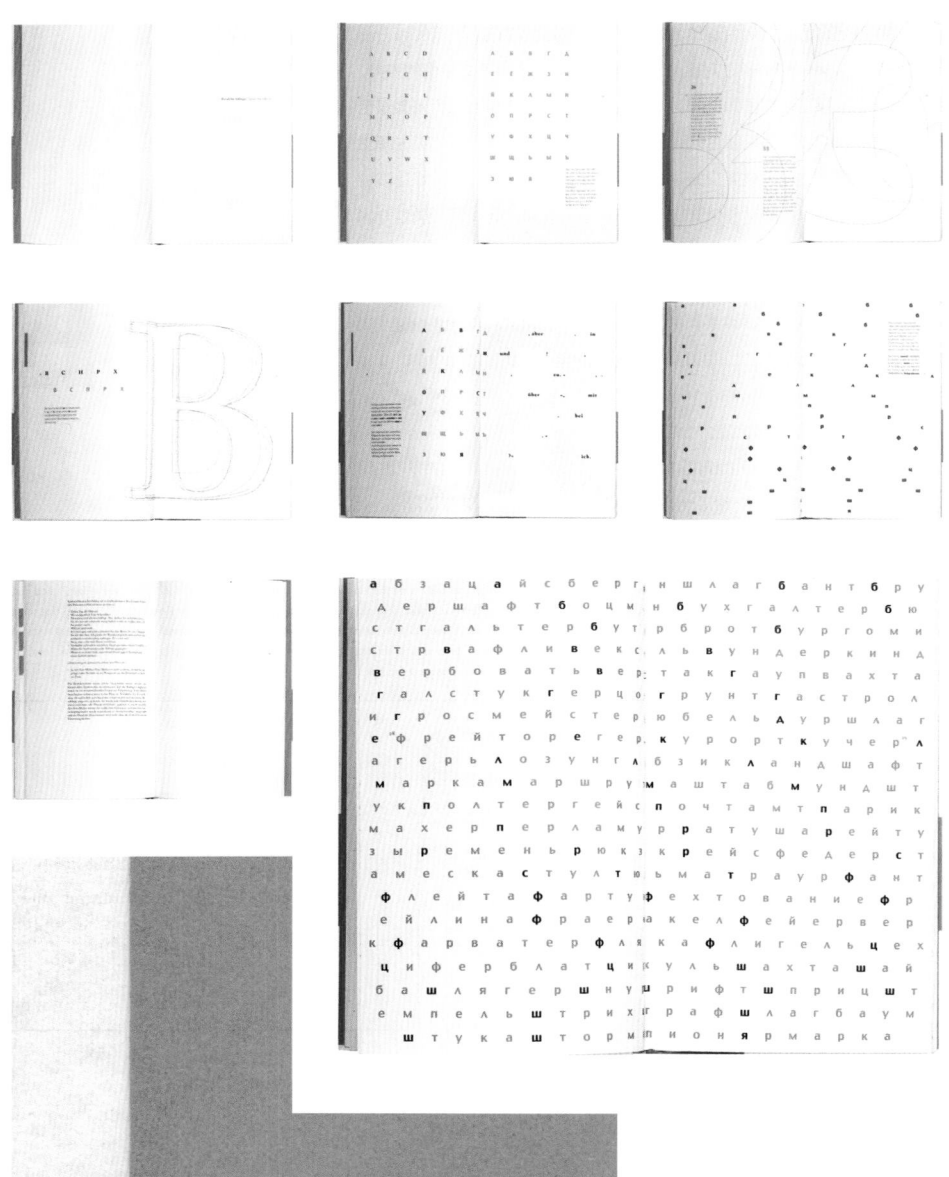

вод чка
RINDFLEISCHETIKETTIERUNGSUBERWACHUNGSAUFGABENUBERTRAGUNGSGESETZ

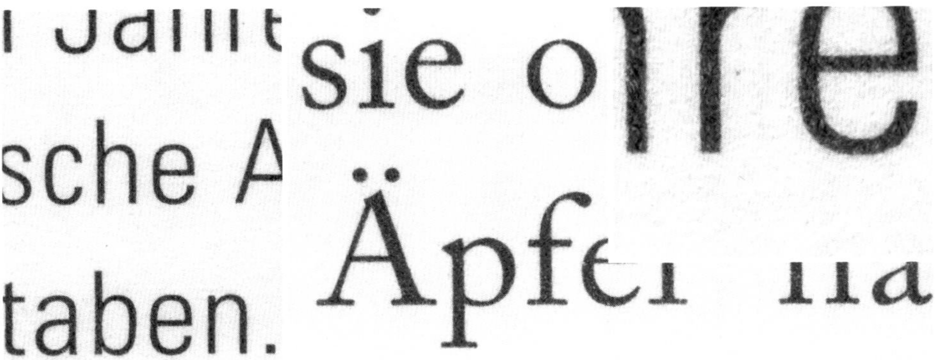
sche
taben.
sie o
re

Иван Иванович Иванов

Russische Personennamen bestehen aus drei Teilen: dem Vornamen, dem Vatersnamen und dem Familien- oder Nachnamen. Also zum Beispiel **Iwan Iwanowitsch Iwanow.**
Der Kundige erkennt, dass es sich um eine männliche Person namens **Iwan** handelt, dessen Vater auch **Iwan** heißt, und dass sein Nachname **Iwanow** lautet. Oder ein weibliches Beispiel: **Olga Iwanowna Iwanowa.** Sie könnte eine Schwester von Iwan Iwanowitsch sein und heißt mit dem Vornamen **Olga.** An den Endungen des Vatersnamens und Nachnamens erkennt man, dass es sich um ein weibliches Wesen handelt.

Die Vornamen werden häufig in diversen oft verwirrenden Verkleinerungsformen gebraucht. So wird der eigentliche **Iwan** auch **Wanja, Wan'ka, Wanjuscha** oder ebenfalls **Wanetschka** genannt.

Deutsche Eigennamen, die mit **H** anfangen, werden im Russischen meist mit Г (**G**) wiedergegeben und entsprechend ausgesprochen. Deshalb war einer der letzten deutschen Bundeskanzler hauptsächlich als **G**elmut bekannt.

ч

In Deutschland kann eine Person mehrere, muss aber mindestens einen Vornamen besitzen. Einem Neugeborenen dürfen maximal fünf Vornamen gegeben werden.
Bei Verwendung mehrerer Vornamen wird der Vorname, mit dem die Person »gerufen« wird, als Rufname bezeichnet. Dabei stellt die Reihenfolge der Vornamen keine Rangfolge dar. Nach höchstrichterlicher Rechtsprechung steht es in Deutschland dem Namensträger frei, zwischen seinen standesamtlich eingetragenen Namen zu wählen.

Ein »Rufname« ist also nicht unveränderlich festgelegt. Ein **Markus** kann in Vorlagen als **Florian** verzeichnet und in seinem Freundeskreis als **Veit** bekannt sein.

Ein Deutscher heißt im Russischen **Nemez**, was übersetzt »nicht eigener« oder auch »stumm« bedeutet. Ursprünglich bezeichnete der Name alle Ausländer, da sie der russischen Sprache nicht mächtig waren und dementsprechend mit dem einfachen Volk nicht kommunizieren konnten. Im Laufe der Geschichte kam es dazu, dass Fremde zum größten Teil aus Deutschland stammten, und die Bezeichnung hauptsächlich auf sie übertragen wurde.

45

alias Onkel Wanja

007 Bm³

本书是爱尔兰作家、诗人，20 世纪最重要的作家之一詹姆斯·乔伊斯（James Joyce）的英语现代主义文学里程碑《尤利西斯》。作者通过大量意识流技巧、揶揄的文笔以及其他创新的文学手法来刻画小说的主人公。本书采用锁线胶装，封面直接裱贴灰色卡板，标题文字巨大撑满封面，压印醒目的红色油墨。书脊则采用红色布面烫印白色，简单而具有强大的视觉冲击力。翻过鲜红的环衬，内文采用极薄的字典纸，单黑印刷。内文排版规整严谨，红色的书口刷边更将本书的整体气质推向了极致的经书风格，就像《尤利西斯》中将希腊神话引入现代文学的叙事结构，和大量篡改《荷马史诗》的标题，本书简洁有力的现代风格的封面与经书风格的内页也在此得到完美的融合。

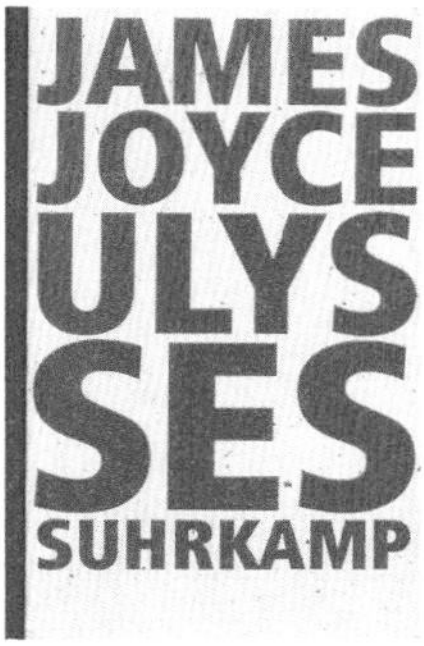
JAMES
JOYCE
ULYS
SES
SUHRKAMP

Ulysses is a milestone of English modernism literature by Irish writer James Joyce, one of the most important writers of the 20th century. The author portrays the protagonist of the novel through the use of stream-of-consciousness, wry expressions and other innovative literary techniques. This book is bound with sewing thread and the cover is directly mounted with gray graphic board. The huge title embossed with striking red ink occupies the whole cover. Followed by the bright red end paper, inside pages are printed in black on dictionary paper. The typesetting is neat and rigorous. Just as *Ulysses* introduced Greek mythology into the narrative structure of modern literature and modified many titles of *Homer's epics*, the book design harmonizes the modern cover with classic inside pages.

- 铜奖
- 尤利西斯
- 德国

A fine example of successful book-art. A bookblock comprising 990 pages with conventional interior finish is unadulterated before being presented in cardboard cut-flush binding with unfinished sheets of grey card. The book's cover is embossed in striking red Grotesque block capitals. The writing on the spine is in white on red cover material. The endpapers are also in red. Through red edging on all three sides the whole bookblock gains a striking new mantle, visible from every angle.

- Bronze Medal
- Ulysses
- Germany
- Werner Zegarzewski
- James Joyce
- Suhrkamp Verlag, Frankfurt am Main
- 193×122×47mm
- 676g
- 987p

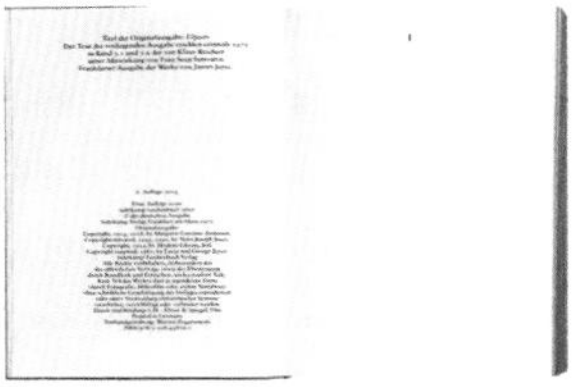

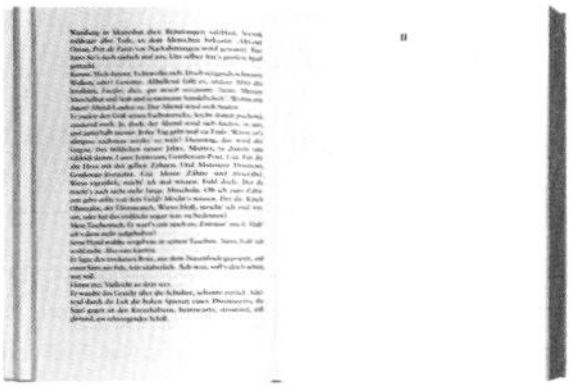

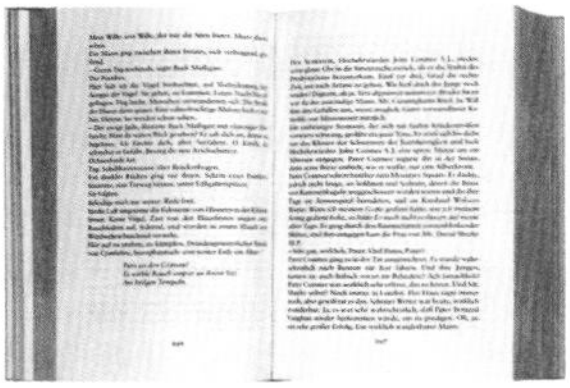

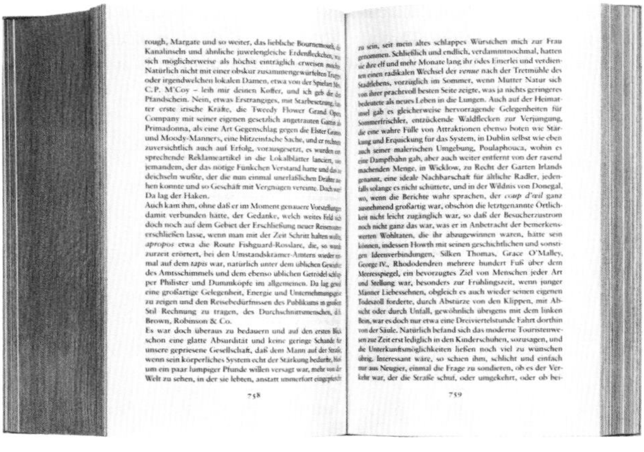

rough, Margate und so weiter, das liebliche Bournemouth, die Kanalinseln und ähnliche juwelengleiche Erdenfleckchen, was sich möglicherweise als höchst einträglich erweisen möchte. Natürlich nicht mit einer obskur zusammengewürfelten Truppe oder irgendwelchen lokalen Damen, etwa von der Spielart Mrs. C.P. M'Coy – leih mir deinen Koffer, und ich geb dir den Pfandschein. Nein, etwas Erstrangiges, mit Starbesetzung, unter erste irische Kräfte, die Tweedy Flower Grand Opera Company mit seiner eigenen gesetzlich angetrauten Gattin als Primadonna, als eine Art Gegenschlag gegen die Elster Grimes und Moody-Manners, eine blitzeinfache Sache, und er rechnete zuversichtlich auch auf Erfolg, vorausgesetzt, es würden entsprechende Reklameartikel in die Lokalblätter lanciert, von jemandem, der das nötige Fünkchen Verstand hatte und das zu deichseln wußte, der die nun einmal unerläßlichen Drähte ziehen konnte und so Geschäft mit Vergnügen vereinte. Doch wer? Da lag der Haken.

Auch kam ihm, ohne daß er im Moment genauere Vorstellungen damit verbunden hätte, der Gedanke, welch weites Feld sich doch noch auf dem Gebiet der Erschließung neuer Reiserouten erschließen lasse, wenn man mit der Zeit Schritt halten wolle, *apropos* etwa die Route Fishguard-Rosslare, die, so wurde zurzeit erörtert, bei den Umstandskrämer-Ämtern wieder einmal auf dem *tapis* war, natürlich unter dem üblichen Gewühle des Amtsschimmels und dem ebenso üblichen Getrödel schläfriger Philister und Dummköpfe im allgemeinen. Da lag gewiß eine großartige Gelegenheit, Energie und Unternehmungsgeist zu zeigen und den Reisebedürfnissen des Publikums im großen Stil Rechnung zu tragen, des Durchschnittsmenschen, d. h. Brown, Robinson & Co.

Es war doch überaus zu bedauern und auf den ersten Blick schon eine glatte Absurdität und keine geringe Schande für unsere gepriesene Gesellschaft, daß dem Mann auf der Straße, wenn sein körperliches System echt der Stärkung bedurfte, bloß um ein paar lumpiger Pfunde willen versagt war, mehr von der Welt zu sehen, in der sie lebten, anstatt immerfort eingepfercht

758

zu sein, seit mein altes schlappes Würstchen mich zur Frau genommen. Schließlich und endlich, verdammtnochmal, hatten sie ihre elf und mehr Monate lang ihr ödes Einerlei und verdienten einen radikalen Wechsel der *venue* nach der Tretmühle des Stadtlebens, vorzüglich im Sommer, wenn Mutter Natur sich von ihrer prachtvoll besten Seite zeigte, was ja nichts geringeres bedeutete als neues Leben in die Lungen. Auch auf der Heimatinsel gab es gleicherweise hervorragende Gelegenheiten für Sommerfrischler, entzückende Waldflecken zur Verjüngung, die eine wahre Fülle von Attraktionen ebenso boten wie Stärkung und Erquickung für das System, in Dublin selbst wie eben auch seiner malerischen Umgebung, Poulaphouca, wohin es eine Dampfbahn gab, aber auch weiter entfernt von der rasend machenden Menge, in Wicklow, zu Recht der Garten Irlands genannt, eine ideale Nachbarschaft für ältliche Radler, jedenfalls solange es nicht schüttete, und in der Wildnis von Donegal, wo, wenn die Berichte wahr sprachen, der *coup d'œil* ganz ausnehmend großartig war, obschon die letztgenannte Örtlichkeit nicht leicht zugänglich war, so daß der Besucherzustrom noch nicht ganz das war, was er in Anbetracht der bemerkenswerten Wohltaten, die ihr abzugewinnen waren, hätte sein können, indessen Howth mit seinen geschichtlichen und sonstigen Ideenverbindungen, Silken Thomas, Grace O'Malley, George IV., Rhododendren mehrere hundert Fuß über dem Meeresspiegel, ein bevorzugtes Ziel von Menschen jeder Art und Stellung war, besonders zur Frühlingszeit, wenn junger Männer Liebessehnen, obgleich es auch wieder seinen eigenen Todeszoll forderte, durch Abstürze von den Klippen, mit Absicht oder durch Unfall, gewöhnlich übrigens mit dem linken Bein, war es doch nur etwa eine Dreiviertelstunde Fahrt dorthin von der Säule. Natürlich befand sich das moderne Touristenwesen zur Zeit erst lediglich in den Kinderschuhen, sozusagen, und die Unterkunftsmöglichkeiten ließen noch viel zu wünschen übrig. Interessant wäre, so schien ihm, schlicht und einfach nur aus Neugier, einmal die Frage zu sondieren, ob es der Verkehr war, der die Straße schuf, oder umgekehrt, oder ob bei-

759

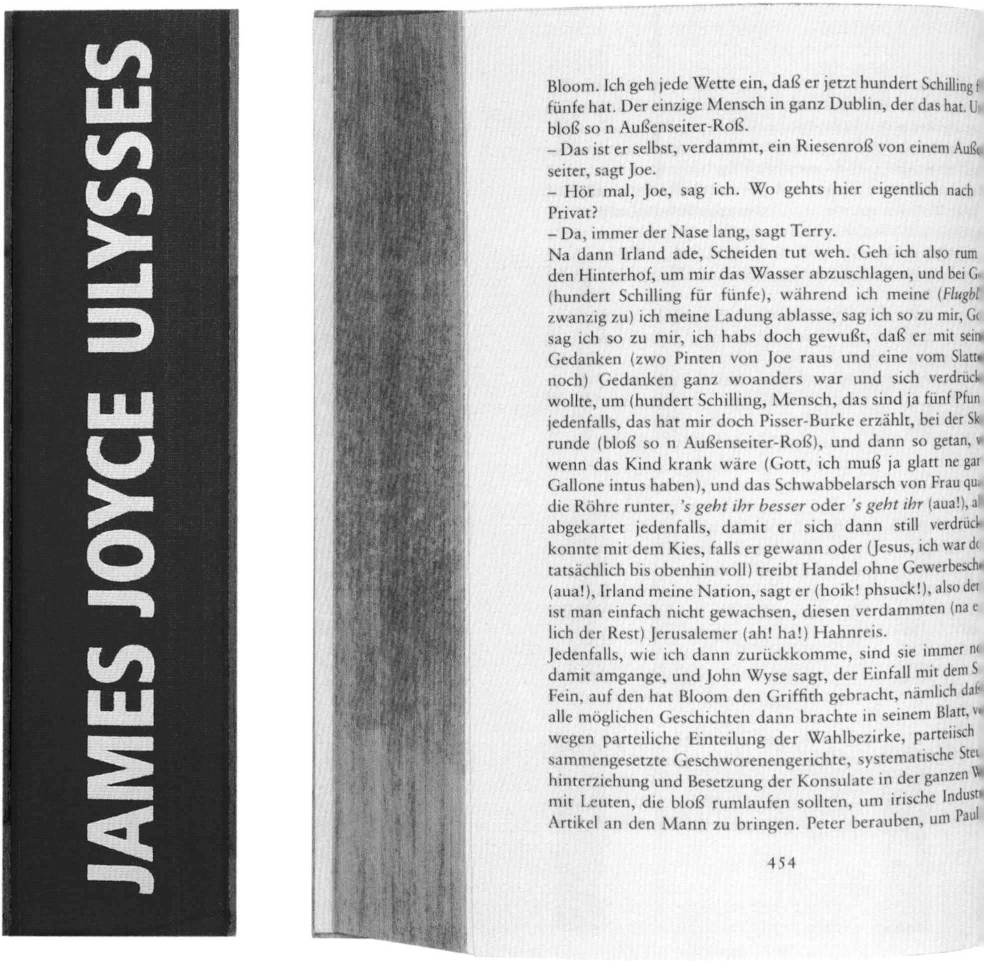
JAMES JOYCE ULYSSES

Bloom. Ich geh jede Wette ein, daß er jetzt hundert Schilling f
fünfe hat. Der einzige Mensch in ganz Dublin, der das hat. U
bloß so n Außenseiter-Roß.
– Das ist er selbst, verdammt, ein Riesenroß von einem Auße
seiter, sagt Joe.
– Hör mal, Joe, sag ich. Wo gehts hier eigentlich nach
Privat?
– Da, immer der Nase lang, sagt Terry.
Na dann Irland ade, Scheiden tut weh. Geh ich also rum
den Hinterhof, um mir das Wasser abzuschlagen, und bei G
(hundert Schilling für fünfe), während ich meine (*Flugbl*
zwanzig zu) ich meine Ladung ablasse, sag ich so zu mir, G
sag ich so zu mir, ich habs doch gewußt, daß er mit sein
Gedanken (zwo Pinten von Joe raus und eine vom Slatt
noch) Gedanken ganz woanders war und sich verdrück
wollte, um (hundert Schilling, Mensch, das sind ja fünf Pfun
jedenfalls, das hat mir doch Pisser-Burke erzählt, bei der Sk
runde (bloß so n Außenseiter-Roß), und dann so getan,
wenn das Kind krank wäre (Gott, ich muß ja glatt ne gar
Gallone intus haben), und das Schwabbelarsch von Frau qu
die Röhre runter, *'s geht ihr besser* oder *'s geht ihr* (aua!), a
abgekartet jedenfalls, damit er sich dann still verdrück
konnte mit dem Kies, falls er gewann oder (Jesus, ich war d
tatsächlich bis obenhin voll) treibt Handel ohne Gewerbesch
(aua!), Irland meine Nation, sagt er (hoik! phsuck!), also de
ist man einfach nicht gewachsen, diesen verdammten (na e
lich der Rest) Jerusalemer (ah! ha!) Hahnreis.
Jedenfalls, wie ich dann zurückkomme, sind sie immer n
damit amgange, und John Wyse sagt, der Einfall mit dem S
Fein, auf den hat Bloom den Griffith gebracht, nämlich da
alle möglichen Geschichten dann brachte in seinem Blatt, w
wegen parteiliche Einteilung der Wahlbezirke, parteiisch
sammengesetzte Geschworenengerichte, systematische Ste
hinterziehung und Besetzung der Konsulate in der ganzen W
mit Leuten, die bloß rumlaufen sollten, um irische Indust
Artikel an den Mann zu bringen. Peter berauben, um Paul

454

ezahlen. Gott, wenn das alte Schmuddelauge die Sache versaut, ist doch gleich alles im Eimer. Dann können wir uns verdammt gleich begraben lassen. Gott schütze Irland vor diesem verdammten Schnüffelkopp und allen seinesgleichen. Mr. Bloom mit seinem Larifari. Und sein Oller vor ihm, der hat genau solche Gaunereien begangen, der alte Methusalem Bloom, der räubernde Handlungsreisende, der sich dann mit Blausäure vergiftet hat, nachdem er das Land mit seinen Nippsachen und seinen Groschen-Diamanten überschwemmt hatte. Darlehen per Post zu leichten Bedingungen. Jeden Betrag als Vorschuß gegen einfache Unterschrift. Entfernung spielt keine Rolle. Sicherheiten nicht erforderlich. Gott, der ist doch genau so wie Lanty MacHale seine Ziege, die immer auf der Straße mitläuft, mit jedem ein Stückchen.

Aber bestimmt, das ist Tatsache, sagt John Wyse. Und da kommt der Mann, der euch Genaueres darüber erzählen wird, Martin Cunningham.

Und tatsächlich, da fährt doch der Schloßwagen vor, mit Martin drauf und dabei noch Jack Power und ein Bursche namens Crofter oder Crofton, irgendein Staatspensionär aus dem Oberfinanzamt, ein Orangist, den Blackburn auf der Liste hat und der sein Geld dafür bezieht, oder Crawford, daß er auf Kosten des Königs im ganzen Land herumscharwenzelt.

Unsere Reisenden erreichten die ländliche Herberge und stiegen von ihren Zeltern.

Ho, Bube! schrie derjenige, welcher seinem Gebaren nach der Anführer des Trupps zu sein schien. Frecher Schuft! Her zu mir!

Mit diesen Worten klopfte er laut mit seinem Schwertknauf an das offene Gitter.

Alsbald erschien der Herr Wirt auf die Mahnung, gegürtet mit seinem Wappenrock.

Einen guten Abend euch, ihr Herren, sprach er und verneigte sich unterwürfig.

Mach Beine, Kerl! schrie derjenige, welcher geklopft hatte. Kümmere dich um unsere Schlachtrosse. Und uns selber gib vom Besten, denn wahrlich, es tut uns not.

455

西瓜出版社（la Pastèque）致力于推广加拿大魁北克的漫画和文化，他们将漫画与魁北克独特的餐饮文化相结合，整理、绘制、设计和出版了这本独特的食谱。本书的装订形式为锁线硬精装。封面采用布纹纸，红色、亚金色与黑色三色印刷，裱附卡板。内页采用亚面艺术纸张，红黑双色印刷。书籍的内容将漫画、菜谱和厨师的介绍做了有趣的编排。全书共包含 10 位厨师与 10 位漫画家，并将其一一配对。查理斯·伊曼努尔·帕里索（Charles-Emmanuel Pariseau）为本书的主厨，在全书开篇和其他厨师之间的间隔部分提供一套包含前菜、主菜和甜点的西式菜单，而其他厨师各出三道菜品。每位厨师均由配对的漫画家为其绘制漫画形象，并配以简短的文字介绍，之后便是与饮食相关的漫画作品，风格迥异。菜单部分依照西餐菜单样式进行设计，双栏排版，信息清晰规整。||||||||||

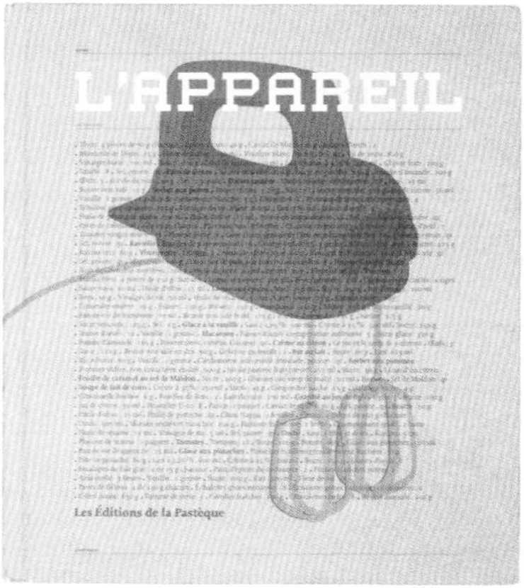
L'APPAREIL
Les Éditions de la Pastèque

Watermelon Publishing House (la Pastèque) is committed to promoting the cartoons and culture of Quebec, Canada. They combine the cartoons with Quebec's unique catering culture to organize, draw, design and publish this unique recipe book. The hardback book is bound with thread sewing. The cover is printed on wove paper in three colors – red, sub-gold and black, mounted with cardboard. The inner pages are printed on matte art paper in red and black. The book makes interesting arrangements for cartoons, recipes and chef's introductions. The book introduces 10 chefs and 10 cartoonists and pairs them one by one. Charles-Emmanuel Pariseau, the chef in the book, offers a Western-style menu of entrees, main courses and desserts at the interval between the book's opening and other chefs' introduction. The other chefs serve three dishes each. Every chef is given a cartoon image by his paired cartoonist, accompanied by a brief text introduction, followed by a variety of food-related cartoons in different styles. The menu is designed with double columns according to the menu style of Western food, so that the information is clear.

铜奖

炊具

加拿大

The convincing thing about this book is its bold idea of combining recipes with comic-style drawings. Each chapter of »L'Appareil« begins with the introduction of a chef and then continues as a comic strip, not showing how to cook but humorously getting the reader in the right mood. Then come the recipes, in a delicate typography which matches the book's refined colour-scheme very well. An unusual and successful format for a cookbook.

Bronze Medal

L'Appareil

Canada

Anouk Pennel, Raphaël Deaudelin (Feed)

Les Éditions de la Pastèque, Montréal

250 × 214 × 20mm

784g

192p

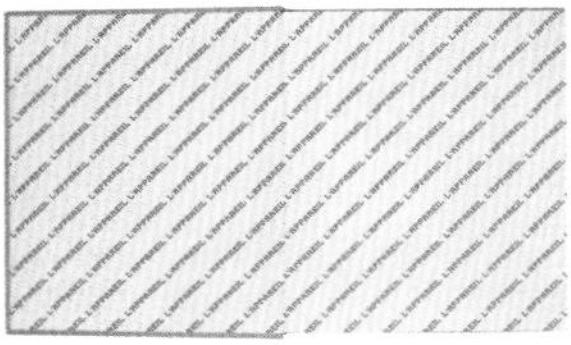

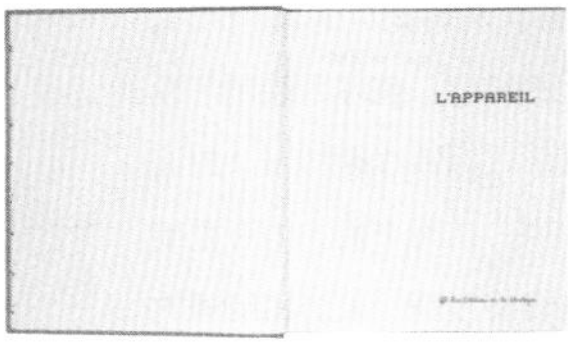
L'APPAREIL

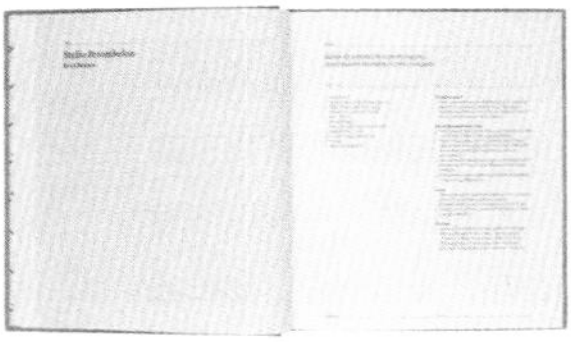

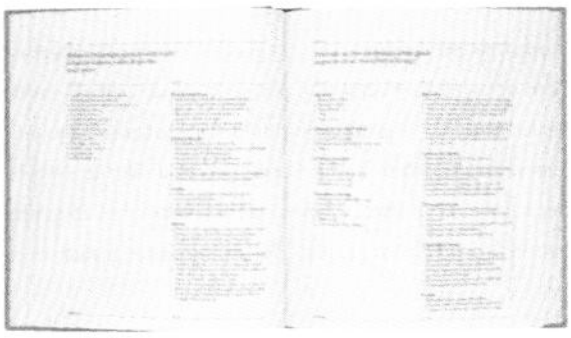

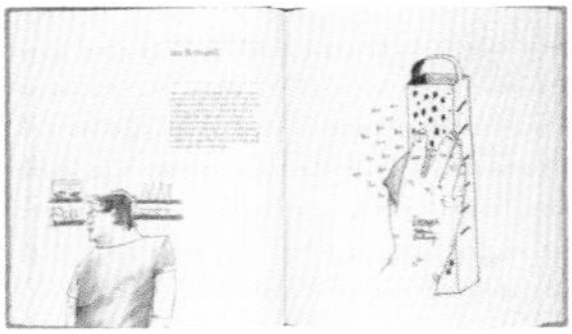

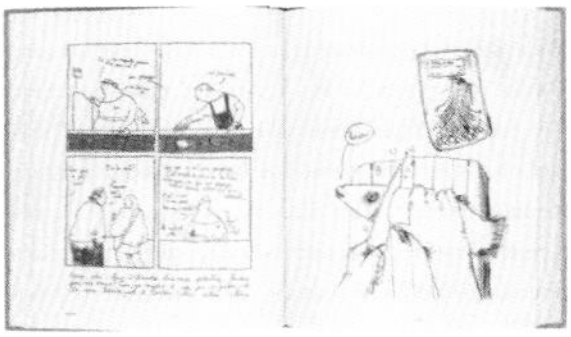

Crêpe vonnassienne de la mère Blanc,
pétoncles sautés, tombée de poireau,
sauce à la vanille de Madagascar
Sagà de framboise au parfum de réglisse,
sablé au sarrasin, mousse au chocolat blanc
citronné, roches à la framboise et au Campari

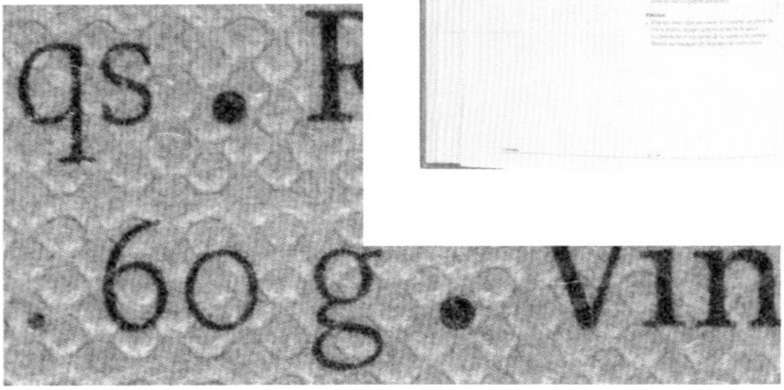
qs. R
. 60 g . Vin

THON HEURE SONNE.
DU RIFIFI DANS LES SUSHIS
L'hermaphrodite en a trop dit
le chef
PAINS PAS PERDUS POUR TOUT LE MONDE
du Verse
Perombe
Michel Ross de

Stelio Perombelon

La musique a perdu ce que l'art culinaire a gagné lorsque ce grand amateur de jazz a tout abandonné pour la cuisine. On a ainsi pu, après sa formation, remarquer son talent chez *Toqué!* auprès de Normand Laprise et aux *Caprices de Nicolas*, avec le regretté Nicolas Jongleux. On le retrouve chef de cuisine à l'ouverture du restaurant *Leméac* après un stage en France à *La Côte d'Or*, avec Bernard Loiseau, lui aussi disparu. Depuis avril 2003, il est chef et copropriétaire du restaurant *Les Chèvres*.

Les chèvres

CKAC
présente:
RADIO CUISINE
Avec
Lucie
et le grain de sel de Paul!
CHÈRES AUDITRICES, BONJOUR!
CETTE SEMAINE, NOUS PRÉPARERONS UN PLAT QUI FERA, À COUP SÛR, LE BONHEUR DE VOTRE MARI ET DE VOS ENFANTS...
EN ONDES
CKAC
IL S'AGIT DU CABILLAUD AU FOUR ACCOMPAGNÉ DE GYOZAS ET CONSOMMÉ DE TOMATES PIQUANT.
DES GYOZAS?! GUIZÉ ÇA?
CKAC
HA HA HA HA HA HA HA HA HA HA HA HA HA HA HA HA HA HA HA
CKAC
RADIO MONTRÉAL
RIRES
APPLAUDIS
HUÉES
DRUMROLL
CE SONT DES PÂTES CHINOISES FOURRÉES, MAIS JE VAIS EXPLIQUER ÇA TOUT À L'HEURE PAUL. D'ABORD, MESDAMES, VOICI LA LISTE DES INGRÉDIENTS. VOUS LA RETROUVEREZ ÉGALEMENT DANS MA CHRONIQUE CULINAIRE DU JOURNAL *LA PATRIE* DE SAMEDI PROCHAIN.
CKAC
Ingrédients
• PAVÉS DE CABILLAUD 4X130g
• PÂTE WONTON 8 FEUILLES
• CHAIR DE CRABE 200g
• CÉBETTES 2
• BLANC D'OEUF 1
• TOMATES MÛRES 12
• PIMENT SÉCHÉ 1
• GROS CHAMPIGNONS 16 TRANCHES
• HUILE DE CITRON
• SEL, POIVRE

007 Bm5

这是鹿特丹眼科医院的宣传资料。本书以医院建筑翻修为契机，向普通大众介绍了建筑翻修的目的、翻修后的室内设计，以及运用于建筑中的艺术，由此将视觉的重要性凸显出来。本书的装订为锁线胶平装。全书荷英双语排版，但完全按照人双眼的对称关系和色盲检查的网点图为元素进行设计，贯穿全书。封面封底分别采用红绿两色眼球图形为底印刷在白卡纸上，外包红绿双色网点图形的透明 PVC 材料。封面封底分别以荷英双语“眼睛”的单词进行设计，相同的字母并列，形成有如双眼圆睁的人脸意象。内页胶版纸、单面光铜版纸和双面光铜版纸混合装订，四色印刷，分别从前后开始进行荷英双语的内容排布，在目录页上可见其对称关系。书中包含色盲检查图、盲文、就诊和检查的照片、艺术家与建筑师的观点等内容，体现了医院的品质和服务标准，以及独立眼科医院的独特个性。||||||||||||||||||||||||||||||||||||

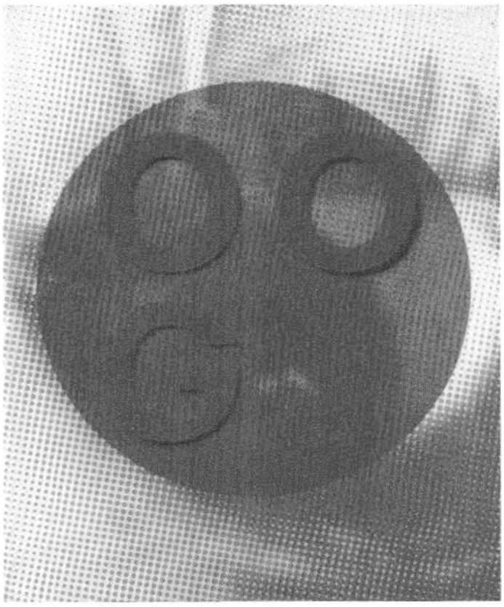

OO
G

This is a publicity material from Rotterdam Ophthalmological Hospital. Taking the renovation of hospital buildings as an opportunity, this book introduces the purpose of renovation, interior design after renovation and the art applied in architecture to the general public, thus highlights the importance of visual sense. The paperback book is glue bound. It is typeset in Dutch and English bilingually, and designed with the symmetry of human eyes and the dot map of color blindness examination as the elements. The front and back cover is printed on white cardboard with red or green eyeball pattern respectively, wrapped by the transparent PVC material with red-green dot pattern. There is the designed word 'eyes' in Dutch and English on the front and back cover, which looks like a face with open eyes. The offset page for the inside pages and the gloss coated art paper are bound mixed, printed in four colors. The Dutch text is typeset from the front while the English text is typeset from the back, and the symmetrical relationship can be seen from the catalogue page. The book contains dot map of color blindness examination, Braille, pictures of visits and examinations, and views of artists and architects, reflecting the quality and service standards of the hospital, as well as the unique characteristic of the independent ophthalmic hospital.

铜奖

眼睛——鹿特丹眼科医院

荷兰

Bronze Medal

OOG. Het Oogziekenhuis Rotterdam / EYE. The Rotterdam Eye Hospital

The Netherlands

Irma Boom

Ineke van Ginneke (Ed.)

Het Oogziekenhuis Rotterdam / The Rotterdam Eye Hospital

250×200×9mm

348g

96p

What can be more important to us than our eyes – our ability to see? The exterior of this intensely and excitingly composed book with its red matrix dots together with the Dutch word OOG creates the strong impression of a face. Its interior presents photographs of the building and of the patients documenting everyday life in the clinic. The oldest rendition of an eye, a drawing by Hunayn Ibn Ishag from the year 850, as well as portraits from today – magical pictures – join with the text to leave a convincing and powerful impression from beginning to end. Everything is placed within the format with sensitivity and a good sense of rhythm.

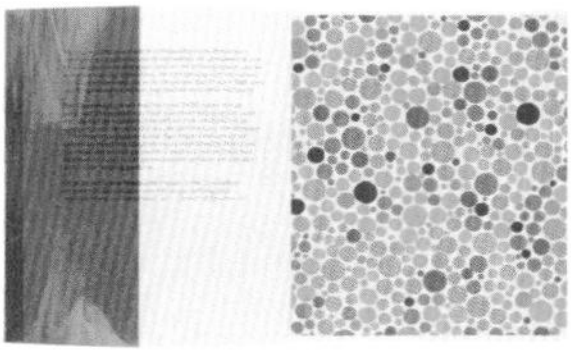

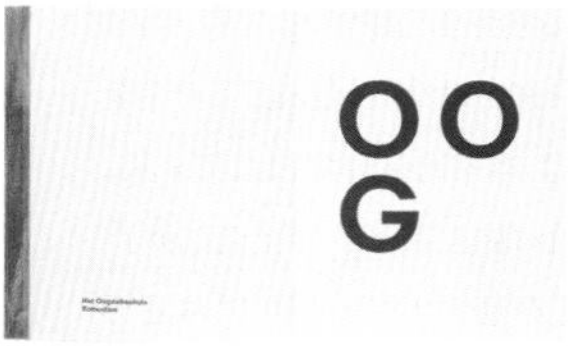
OO
G

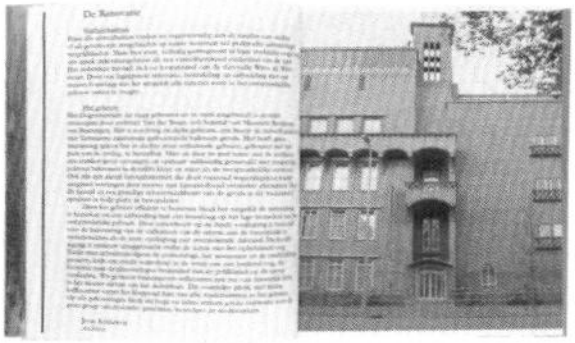

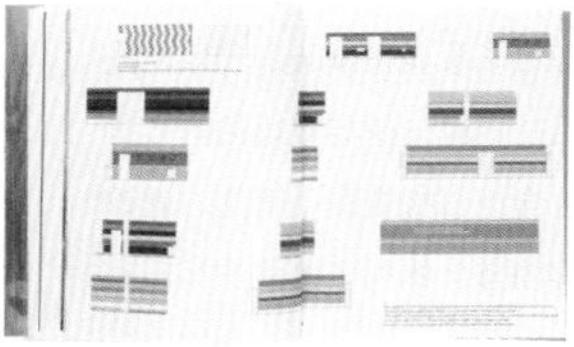

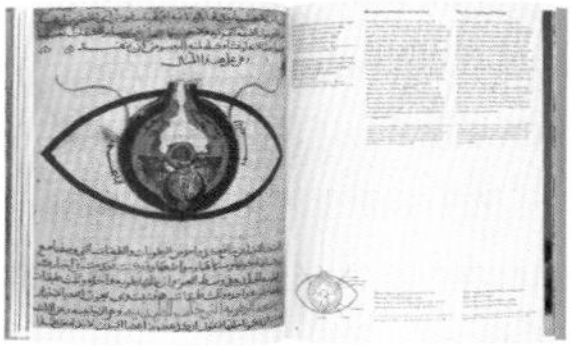

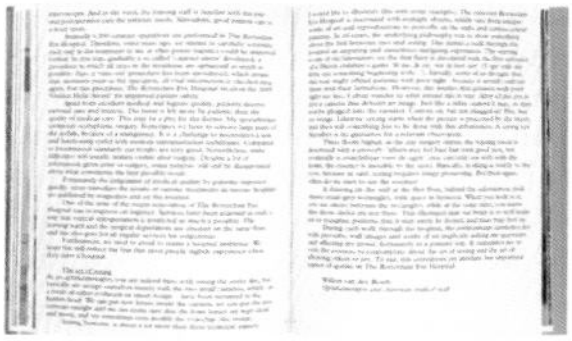

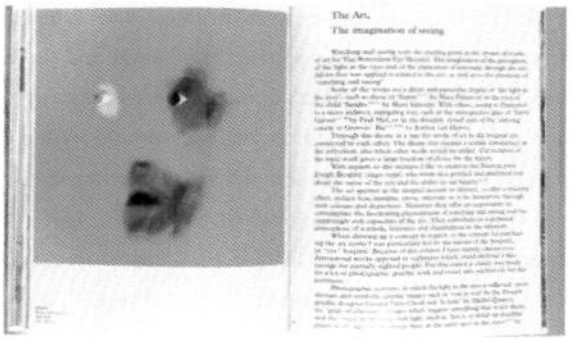
The Art,
The imagination of seeing

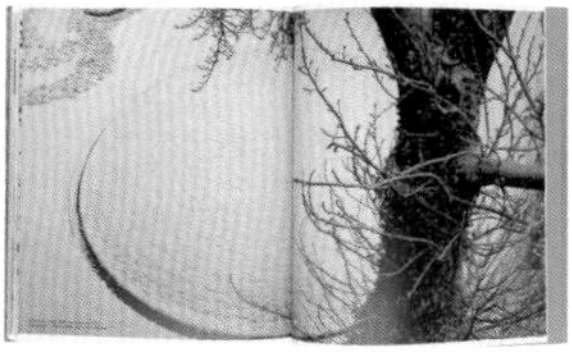

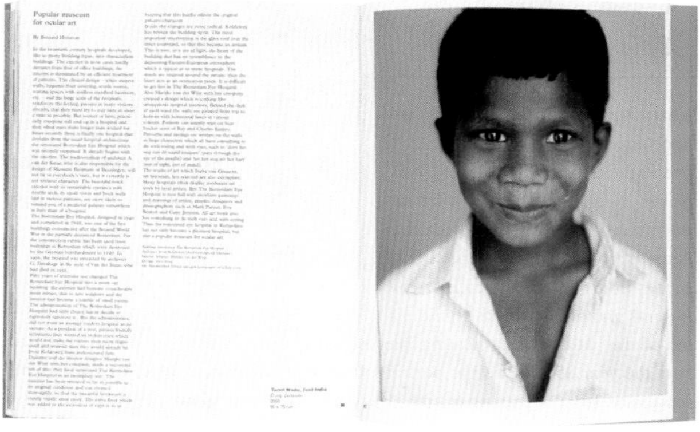
Popular museum
for ocular art

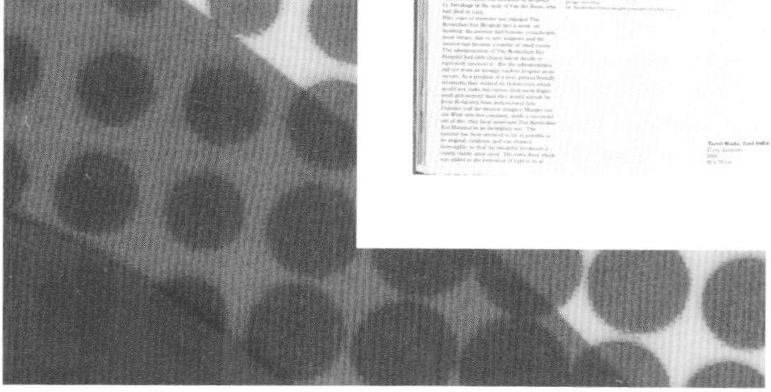

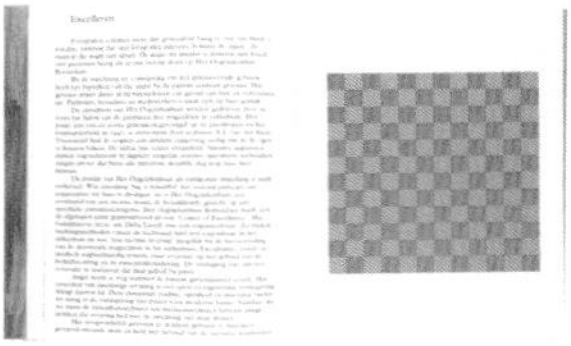

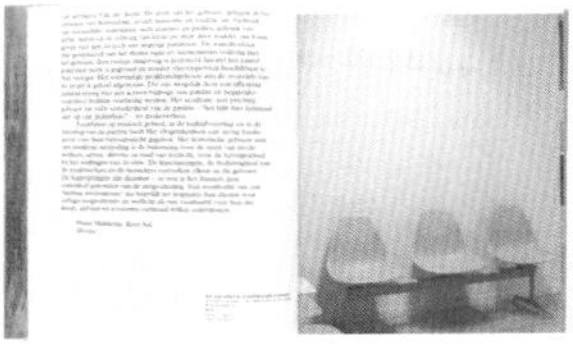

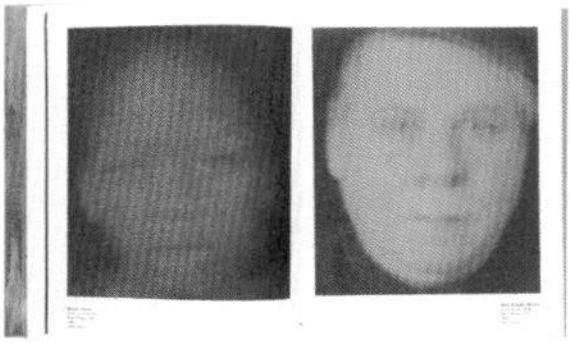

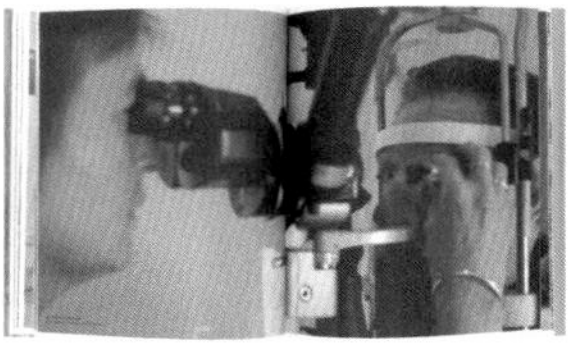

Kwaliteit in Het Oogziekenhuis

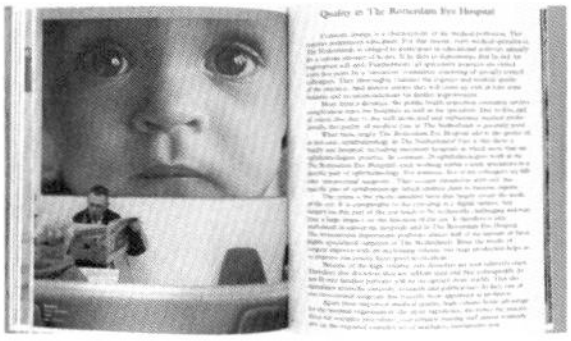

Quality in The Rotterdam Eye Hospital

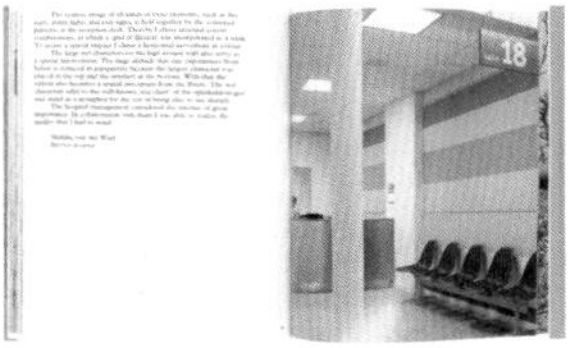

The Renovation

E
YE
The Rotterdam Eye Hospital

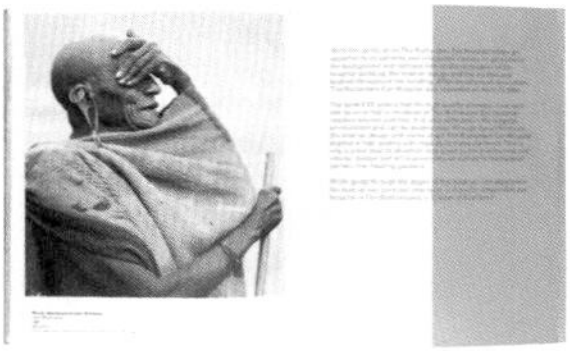

het zeke
He

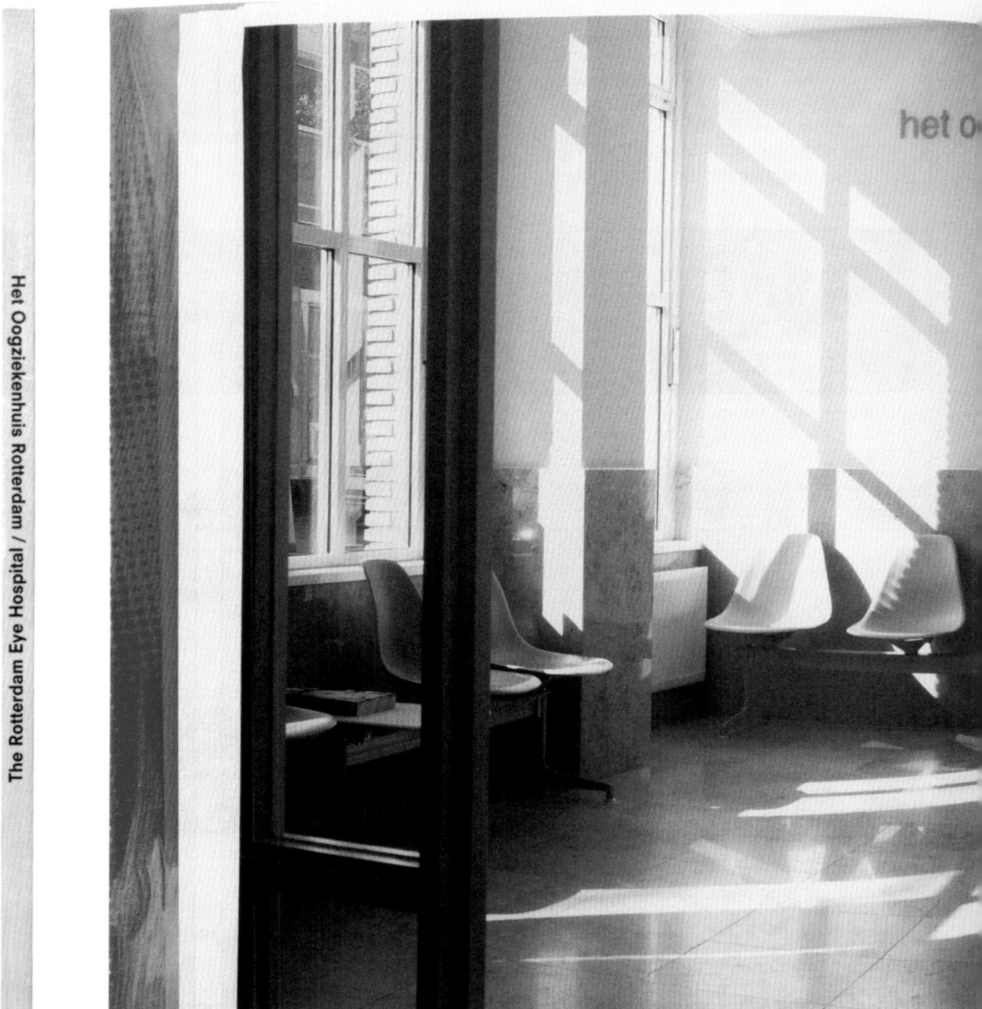
The Rotterdam Eye Hospital / Het Oogziekenhuis Rotterdam

Volksmuseum voor ogenkunst

Door Bernard Hulsman

In de twintigste eeuw hebben ziekenhuizen zich, zoals zoveel gebouwtypen, ontwikkeld tot karakterloze gebouwen. Het exterieur van wijkt meestal nauwelijks af van dat van kantoorgebouwen, het interieur staat in het teken van een efficiënte behandeling van de patiënten.
De klinische vormgeving – witte strakke muren, hygiënische vloerbedekking, steriele kamers, wachtruimtes met zielloos standaardmeubilair, enzovoorts – en de grote omvang van de ziekenhuizen versterken het toch al aanwezige gevoel van veel bezoekers dat ze hier zo kort mogelijk moeten zien te verblijven. Maar uiteindelijk komt bijna iedereen vroeg of laat wel in een ziekenhuis terecht en blijft er dan vaak langer dan gewenst.
Sinds kort is er in ieder geval één ziekenhuis dat afwijkt van de gebruikelijke ziekenhuisarchitectuur: het verbouwde Oogziekenhuis in Rotterdam dat onlangs werd heropend. Dat begint al met het exterieur. Het traditionalisme van architect A. van der Steur, die ook voor het Museum Boijmans Van Beuningen tekende, zal niet bij iedereen in de smaak vallen, maar karakterloos is het zeker niet. Het prachtige bakstenen exterieur doet met zijn opmerkelijke ingang met dubbele boog, zijn torentje en in verschillende verbanden gemetselde bakstenen muren eerder denken aan een middeleeuws palazzo ergens in Italië dan aan een ziekenhuis. Het Oogziekenhuis, in 1940 ontworpen en in 1948 voltooid, was een van de eerste gebouwen die na de Tweede Wereldoorlog werden neergezet in het deels verwoeste Rotterdam. Bij de bouw is nog gebruik gemaakt van het puin van Rotterdamse gebouwen die door het Duits bombardement in 1940 werden verwoest.
In 1956 werd het ziekenhuis door architect G. Drexhage in de stijl van de in 1953 overleden Van der Steur, uitgebreid.
Vijftig jaar intensief gebruik veranderde Het Oogziekenhuis in een versleten gebouw: het exterieur was door nieuwe ramen flink grover geworden en het interieur was een wirwar aan kleine ruimtes geworden. Het bestuur van Het Oogziekenhuis kon weinig anders dan besluiten het rigoureus te verbouwen. Maar het bestuur wilde geen doorsnee moderne ziekenhuisarchitectuur. Als pendant van een nieuwe, patiëntvriendelijke behandelwijze, wilde het bestuur een architectuur die de bezoekers niet nog banger en ongeruster maakten dan ze toch al zijn.
Daar zijn Joost Koldeweij van architectenbureau Duintjer en de interieurarchitect Marijke van der Wijst met haar bureau goed in geslaagd: ze hebben Het Oogziekenhuis voorbeeldig gerenoveerd. Het exterieur werd zoveel mogelijk in oude staat hersteld en grondig gereinigd, zodat het mooie metselwerk weer goed zichtbaar is. De extra verdieping die op de uitbreiding uit 1956 werd gezet is zo terughoudend dat deze nauwelijks afbreuk doet aan het oorspronkelijke palazzo-karakter.
Binnen zijn de veranderingen radicaler. Koldeweij heeft het gebouw opengebroken. Belangrijkste ingreep is het glazen dak op het binnenhof, zodat dit een atrium werd. Dit is nu, in een zee van licht, het hart van het gebouw dat niets heeft van de deprimerende Oost-Europese sfeer die eigen is aan zo veel ziekenhuizen. Om het atrium heen zijn de afdelingen gesitueerd: zo fungeert het hart als een oriëntatiepunt. Verdwalen in Het Oogziekenhuis is moeilijk.
Ook Marijke van der Wijst zorgde met haar bureau voor een inrichting die in bijna niets lijkt op anonieme ziekenhuis-interieurs. Achter de balie van elke afdeling zijn de muren van boven tot onder beschilderd met horizontale banen in verschillende kleuren. Wachten kunnen de patiënten veelal op blauwe kuipstoeltjes van Ray en Charles Eames. Op de wanden zijn in grote letters spreuken en gezegdes aangebracht die allemaal iets met zien en ogen hebben te maken, zoals 'door het oog van de naald kruipen' en 'uit het oog uit het hart'. Voorbeeldig zijn ook de kunstwerken die Ineke van Ginneke, kunsthistoricus, heeft uitgekozen. In veel ziekenhuizen hangen niet zelden matige kunstwerken van locale kunstenaars. Maar Het Oogziekenhuis hangt nu vol uitstekende schilderijen en tekeningen van kunstenaars, grafisch vormgevers en fotografen als Mark Pataut, Eva Besnyö en Cuny Janssen. Ook alle kunstwerken hebben iets te maken met ogen en zien. Zo is het verbouwde Oogziekenhuis in Rotterdam niet alleen een prettig ziekenhuis geworden, maar ook een volksmuseum voor ogenkunst.

Gebouw: renovatie Het Oogziekenhuis Rotterdam.
Architect: Joost Koldeweij (Architectengroep Duintjer).
Interieurarchitect: Marijke van der Wijst.
Ontwerp: 2002-2005.
NRC Handelsblad van 9 juli 2005

sociale proce
de maker.
Het werk van
gefotografeerd
De titels van de boe
holiday. Bilder von de
Faits divers.
Foto's van
eigenlijk? in een
gebracht.
78

这是阿姆斯特丹市立博物馆的年度报告，展示了博物馆2005年全年在各个职能方面的工作和统计。本书的装订形式为包背胶平装。封面采用中灰色布纹纸，蓝灰色满版压底，银色印刷图案。内页采用白、红、灰色纸张配合多色印刷形成的差异区分章节。书中包含博物馆在保护、收藏、捐赠、租借、活动、展览、组织、教育、出版等各方面的工作，也附加了为博物馆作出贡献的赞助者和捐助者。其中大事年表部分采用红色胶版纸黑、金双色印刷，捐助信息报表部分采用满版露白的专色绿外加黑、金三色印刷较为凸显，其余部分大多为白色纸张上印制黑、金双色，或者四色加金色的五色印刷。用于区别各章节且在整本书里形成统一风格的方式主要是依靠各式密排的连续纹样，通过纹样与标题字母、图片的排列组合形成丰富的版面形式，位于版面上中下三段的巨大页码也是本书特点。全书并未完全采用荷英双语对照，英文主要集中在书的最后，用以查阅。||||||||||||||||||

This is the annual report of the Municipal Museum of Amsterdam, which shows the work and statistics of the Museum in various functions throughout 2005. This double-leaved book is glue bound. The cover is made of medium gray wove paper with blue gray background printed in silver. The inner pages are differentiated by white, red and grey paper combined with multi-color printing. The book includes the museum's work in conservation, collection, donation, leasing, activities, exhibitions, organization, education, publishing and other aspects, as well as the sponsors and donors who contribute to the museum. The chronology of major events is printed on red offset paper in black and gold, while the donation report is printed in white, green and black gold. The rest is printed on white paper in black and gold, or in four colors plus gold. The way to distinguish chapters and form a unified style in the whole book mainly relies on the continuous patterns. Arranging and combining patterns with title letters and pictures form the rich layout. The huge page numbers on the pages are also the highlight of the book. The entire book is not completely Dutch-English bilingual comparison; English is mainly at the end of the book for reference.

- 荣誉奖
- 阿姆斯特丹市立博物馆2005年度报告
- 荷兰

- Honorary Appreciation
- Stedelijk Museum Amsterdam 2005. Jaarverslag / Annual Report
- The Netherlands
- Richard Niessen, Esther de Vries
- Carolien de Bruijn (Ed.)
- Stedelijk Museum Amsterdam
- 230×160×16mm
- 539g
- 176p

This book retrospectively presents the activities of the Amsterdam city museum in an annual report for 2005. Through the use of various papers and colours the chapters are respectively divided into clearly defined sections. The thin paper and the soft cover give the book a sleek exterior. Geometric patterns pervade the book creating a cross-cultural impression indebted to the Arabic world, to which the Japanese binding also contributes. In this way the museum's cultural activities underline its function as an institution promoting cultural dialogue.

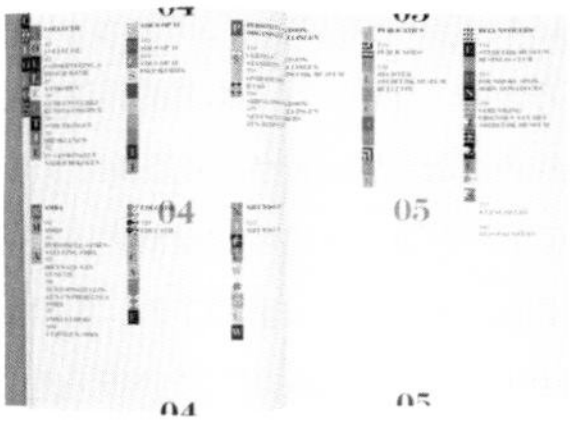

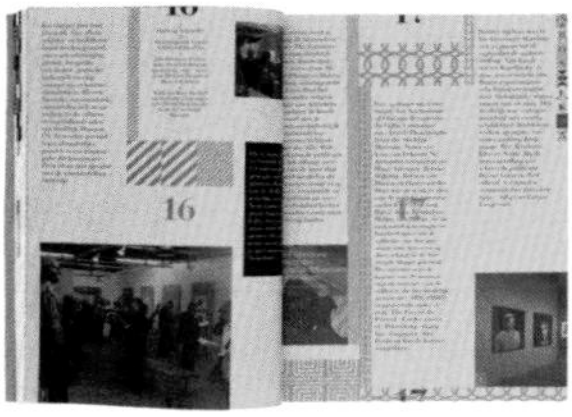

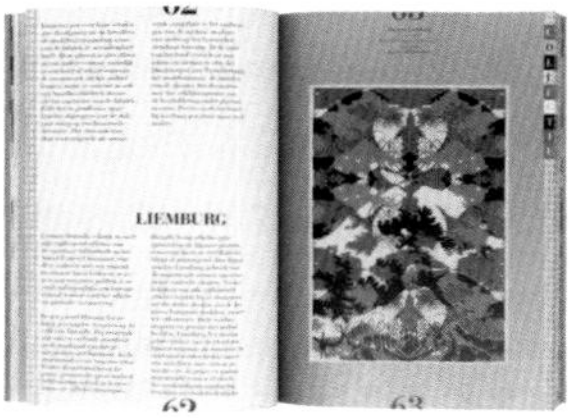

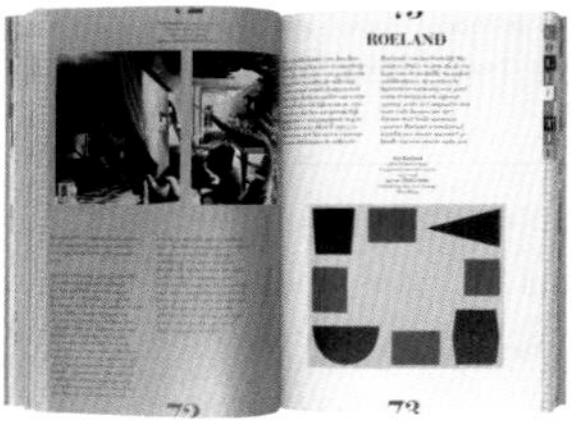

In zijn project liet Beltrán het publiek kennis maken met zijn intuïtieve, visuele ontdekkingsreis langs regels, punten en kalligrafische gebaren. Middels fotoreeksen, teksten, reclameschilderingen en een workshop onder leiding van de Chinese meester-kalligraaf Zou Le Sheng (Sjanghai), wilde hij greep krijgen op de raadsels van de letters. Bij de tentoonstelling verscheen SMBA Nieuwsbrief No. 84 met het essay De triomf van de Analfabeet van Matthijs van Boxsel, auteur van De encyclopedie van de domheid (1999).
In maart presenteerde SMBA de eerste Nederlandse solotentoonstelling van Lonnie van Brummelen: 'Frontyards and Backyards'. In deze presentatie waren twee nieuwe films te zien die situaties van gestremde mobiliteit toonden. Zowel in grootstedelijk Nederland als in een aantal Europese grensgebieden ging de kunstenaar op zoek naar knelpunten in de openbare ruimte; plekken waar een discontinuïteit bestaat
in de beweging van haar gebruikers. Zo filmde Van Brummelen voor Grossraum (2004/2005) (kleur, 35 mm) grensovergangen op twee lokaties in de periferie van het verenigde Europa: Ceuta, een Spaanse enclave in Noord-Marokko; en Hrebenne, een grenspost tussen Polen en Oekraïne. Onmiskenbaar politieker van aard dan haar eerdere werk laat deze film zien hoe de ogenschijnlijk lineaire structuur van landsgrenzen kan oplossen in chaotische opstoppingen, en hoe er naast officiële routes alternatieve paden en wegen ontstaan. Van Brummelen kreeg tijdens de presentatie in SMBA de eerste prijs van de Prix de Rome 2005 voor haar werk.
Gastconservator Xander Karskens introduceerde in mei het werk van Pablo Pijnappel in SMBA. Pijnappels soloproject Felicitas is een film- en dia-installatie waarin het levensverhaal wordt gereconstrueerd van Felicitas Baer (1910-2003), die na de Eerste Wereldoorlog Duitsland ontvluchtte en
92
93

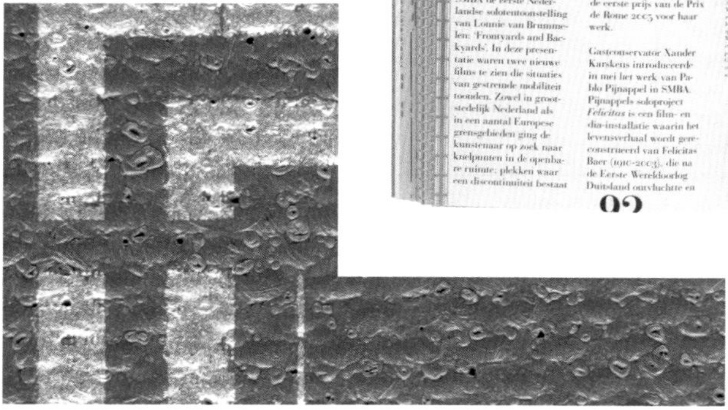

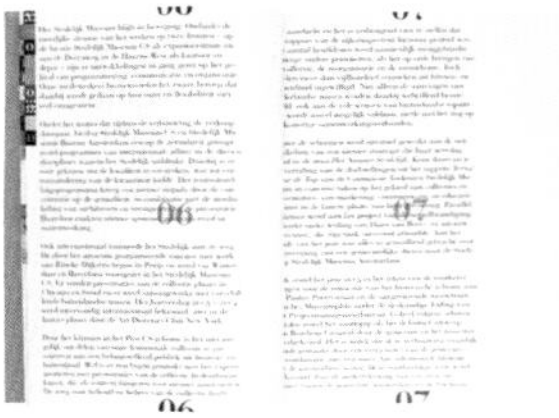

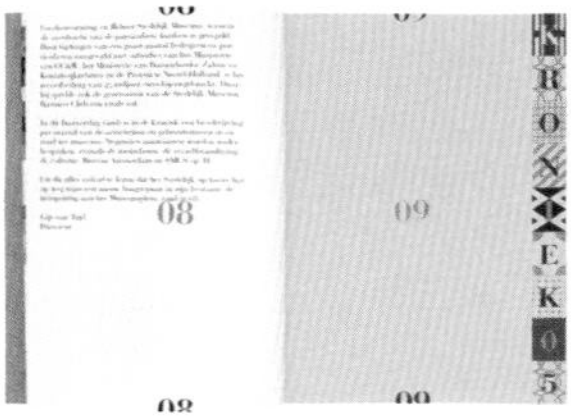

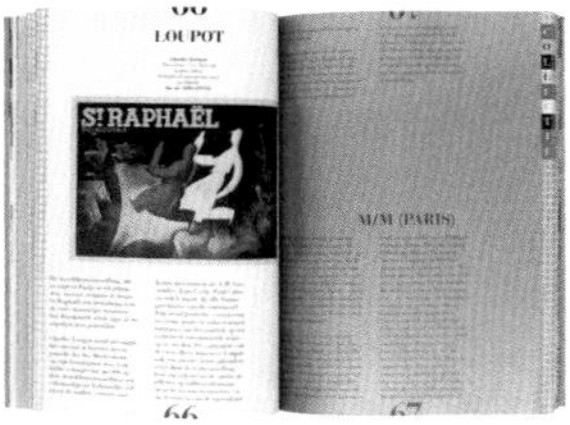
LOUPOT
M/M (PARIS)

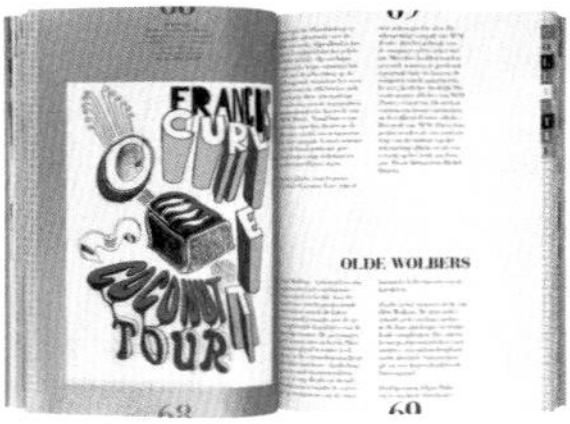
OLDE WOLBERS

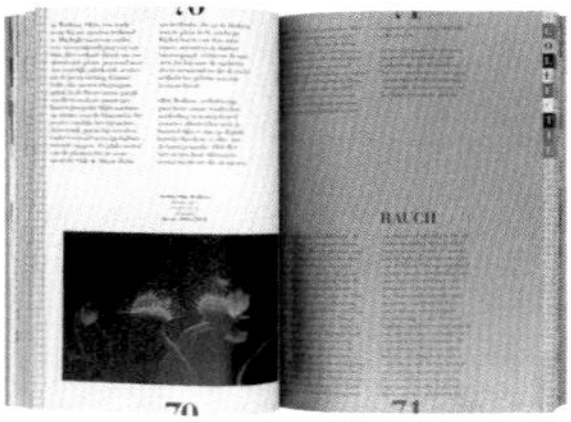
RAUCH

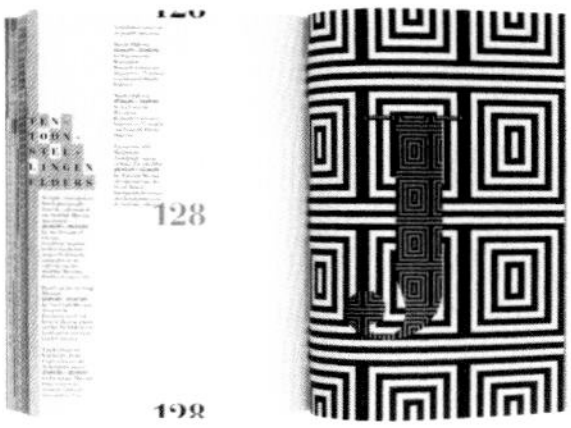

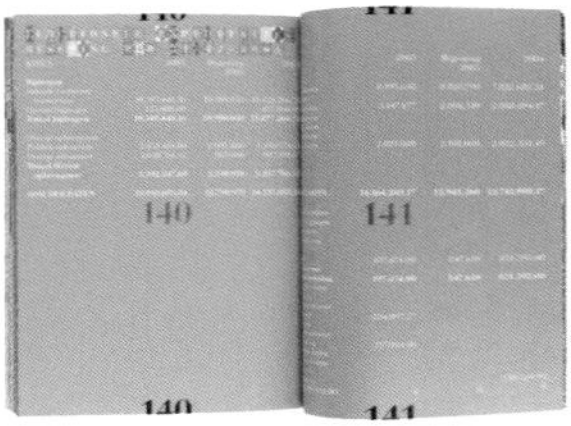

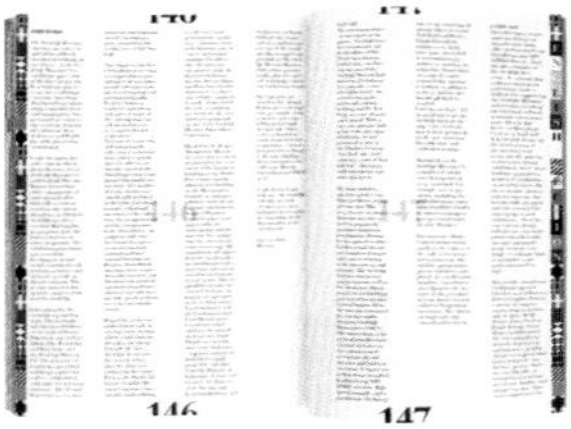

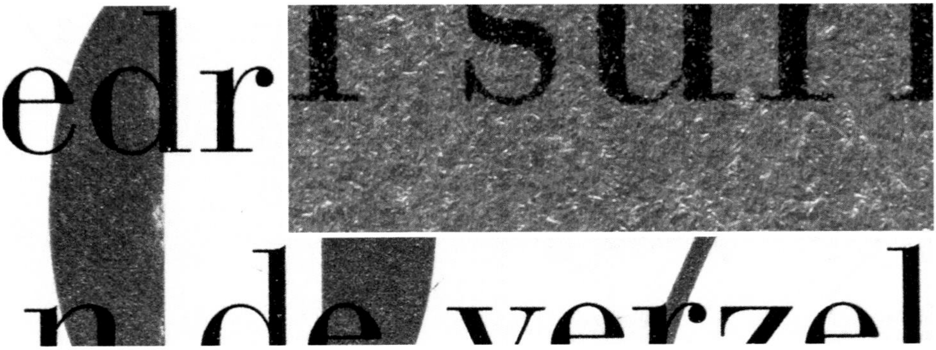

这是一本有关奥地利本土现代时尚产业的报告。本书从产业状况概述、大学专家访谈、商店活动、产业链、博物馆收藏、展览等方面，完整地介绍了当时奥地利时尚产业的发展现状。本书的装订形式很有特色，书函采用四边围合的卡纸单黑印刷，只呈现最少量的图书信息，书籍内容需要从书函上方抽取出来。书籍的装订形式类似于传统经折装，但上下均匀打孔，配上页面上压制的可撕虚线，更像是连续未拆开的多张快递单。书籍正反面分别为德英双语，标题和信息部分采用的打字机风格的等宽字体，也将当时的时尚风格演绎到了极致。

MODEBUCH

This is a report on Austria's local modern fashion industry. This book introduces the development of the fashion industry in Austria at that time through industry overview, interviews with university experts, store activities, industry chain, museum collection, exhibition and other aspects. The binding form of this book is distinctive – the slipcase is printed in single black with card paper, showing only a small number of book information. The top of the case is the opening part, from which the book should be taken out or put back. The binding form of this book is similar to the traditional accordion binding, but all the pages are perforated evenly with the dotted line that can be torn easily. It looks like a sequence of courier receipts. The typewriter-style typeface is used for the book title and other information, which also highlights the fashion style of the day.

- 荣誉奖
- 时尚图书——奥地利的当代时装
- 奥地利

- Honorary Appreciation
- Modebuch. Zeitgenössische Mode aus Österreich
- Austria
- halle34 Albert Handler / Marcus Arige
- Unit F büro für mode (Ed.)
- Unit F büro für mode, Wien
- 254×178×23mm
- 895g
- 158p

In its communication of contemporary fashion this book uses a completely new bookbinding concept. The first thing we all do is to look for the join where this seemingly endless leporello is bound together – only to find that there isn't one. The digital print as an endless form with its characteristic holes and perforations works perfectly and creates the impression of something which has been sewn, something textile. The strict double-page structure is skilfully broken up by interesting text/image combinations or pages with just pictures or text. The overview is maintained, however, thanks to the clear hierarchy conveyed by the chapter beginnings. An English translation is provided on the reverse side.

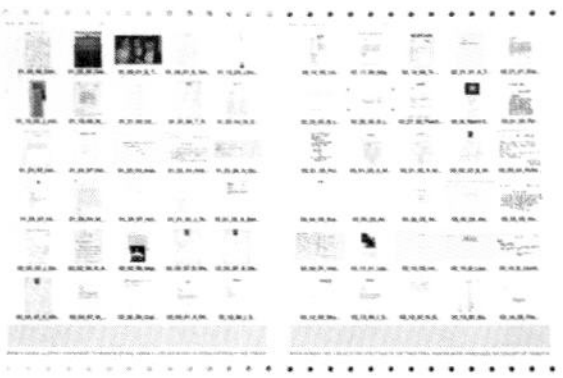
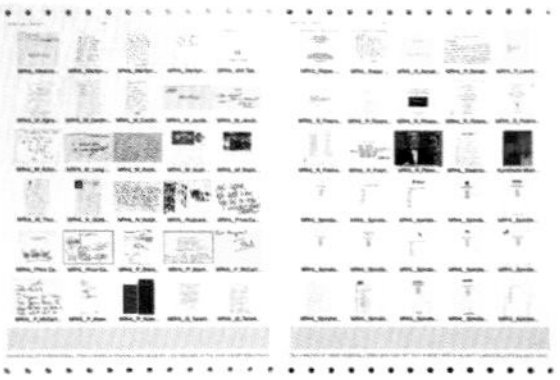

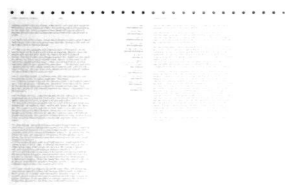

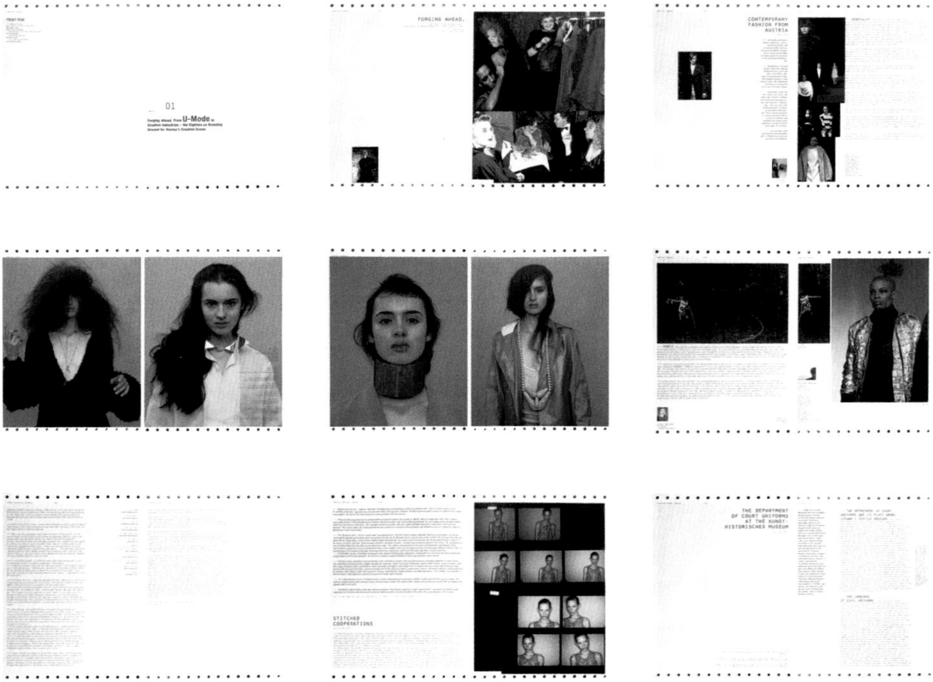
01
Forging Ahead. From U-Mode to Creative Industries – the Eighties as Breeding Ground for Vienna's Creative Scene
FORGING AHEAD.
CONTEMPORARY FASHION FROM AUSTRIA
STITCHED COOPERATIONS
THE DEPARTMENT OF COURT UNIFORMS AT THE KUNST-HISTORISCHES MUSEUM

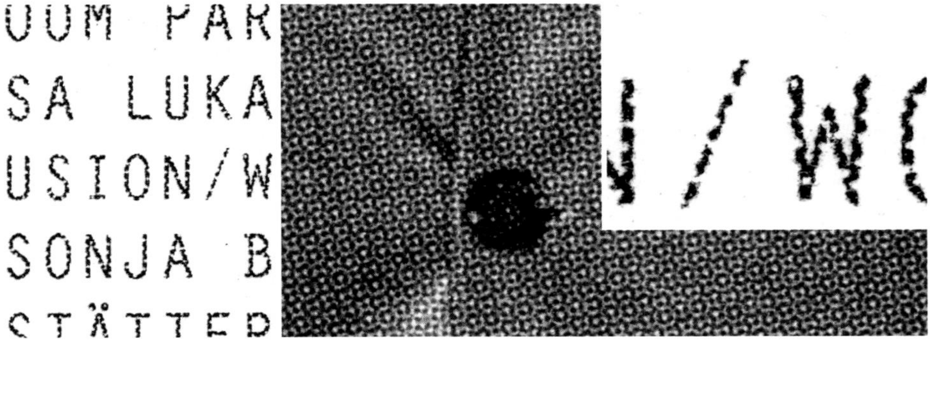
UUM PAR
SA LUKA
USION/W
SONJA B
STÄTTER

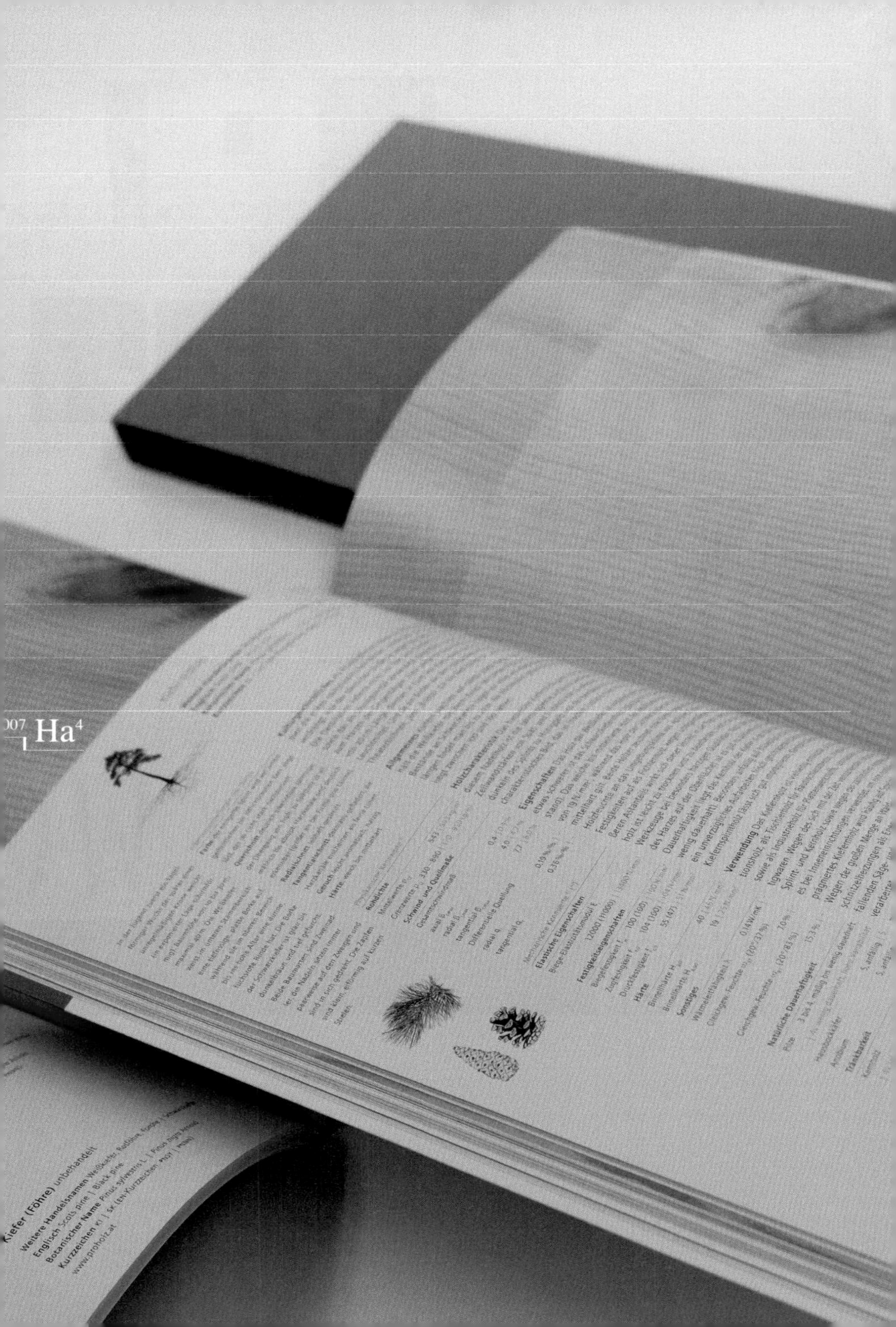
Kiefer (Föhre) unbehandelt
Englisch Scots pine | Black pine
Botanischer Name Pinus sylvestris L.
www.proholz.at

这是一套由奥地利木材工业协会发布的木材使用参考图谱。书中介绍了各类木材的信息、参数和样本，给木材使用机构予以专业的使用指导。本套书内含两本，均为无线胶平装。两本书由灰色布面裱覆卡板制作的书函进行收藏，书函的两面分别使用凹凸工艺加工出涵盖根部的树木剪影。第一本书为木材图谱手册，封面白卡纸四色印刷覆哑光膜，内页哑面铜版纸四色印刷。从木材与人类文化之间的关系开始阐述，列举了桦木、毛榉、橡木、红杉、樱桃木、梧桐等 24 种常用木材的树木信息、木材物理特性、加工参数、适用范围，以及横向、纵向剖面的原大图片，图片的天头部分采用过油工艺，模拟出木材在刷清漆后的效果。书的后部将这些木材的信息进行多方面的横向对比，为使用者提供更清晰的参考。第二本为这些木材的可撕式样本，封面哑面铜版纸四色印刷覆哑光膜，内页哑面铜版纸四色印刷，图片本身可与第一本的木材一一对应；每个木种提供 2 页的可撕式样本，每页按照 2×5 方格进行划分裁切，做可撕式虚线处理，每小块样本背后均有简要的木材信息。

This is a set of reference spectra for the use of wood published by the Austrian Wood Industry Association. The book introduces the information, parameters and samples of various types of wood, and gives professional guidance to the corresponding wood using agencies. This set of books contains two copies, both of which are perfect bound. The two books are put in a slipcase made of cardboard mounted by gray cloth. The two sides of the slipcase are processed with embossing technology to produce tree silhouettes including the roots. The first book is a wood atlas manual. The cover is made of white cardboard with four-color printing, laminated by matte film. The inside pages are printed on matte coated art paper in four colors. Starting from the relationship between wood and human culture, this book lists the tree information, physical characteristics, processing parameters and application scope of 24 common timbers, such as birch, beech, oak, sequoia, cherry, Chinese parasol and so on. The original graphs of transverse and longitudinal profiles of trees are displayed in the first book as well. The top margin of the page is varnished to simulate the effect of wood varnishing. In the latter part, the information of these timbers is compared in many ways to provide a clear reference for users. The second book contains samples of these timbers that can be torn off. Not only the cover but also the inside pages are made of matte coated art paper printed in four colors. The pictures correspond to the timbers introduced in the first book one by one. There are two pages of samples for each wood species and every page is die-cut according to 2 x 5. brief wood information is printed on the back of the samples.

荣誉奖

木材图谱——样本、信息和参数比较

奥地利

Honorary Appreciation

Holzspektrum – Ansichten, Beschreibungen und Vergleichswerte

Austria

Reinhard Gassner, Marcel Bachmann, Stefan Gassner (Atelier Reinhard Gassner)

Josef Fellner, Alfred Teischinger, Walter Zschokke

proHolz Austria – Arbeits-gemeinschaft der österrei-chischen Holzwirtschaft, Wien

308×214×24mm

1328g

112p+96p

A publication on the subject of wood, clearly conveying a wealth of technical information. The various broad columns and the entire typographical composition make for a high degree of lucidity with regard to this specialist subject. A brief introduction to the different types of trees and the shapes of their leaves by means of line drawings, alongside descriptive texts as well as wonderful reproductions of the structure of each wood, one per double-page, present the reader with a comprehensive view of the wood in question.

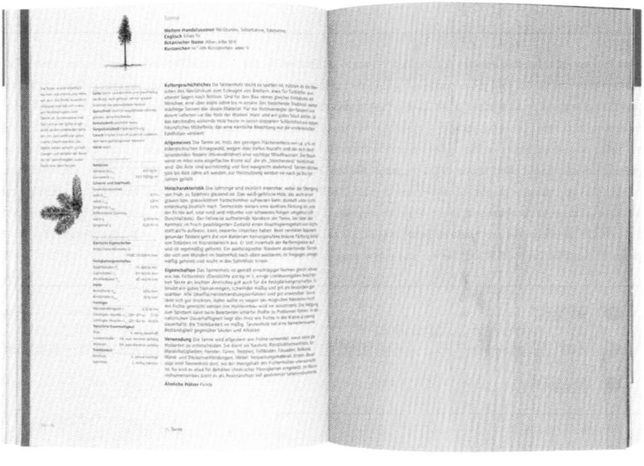
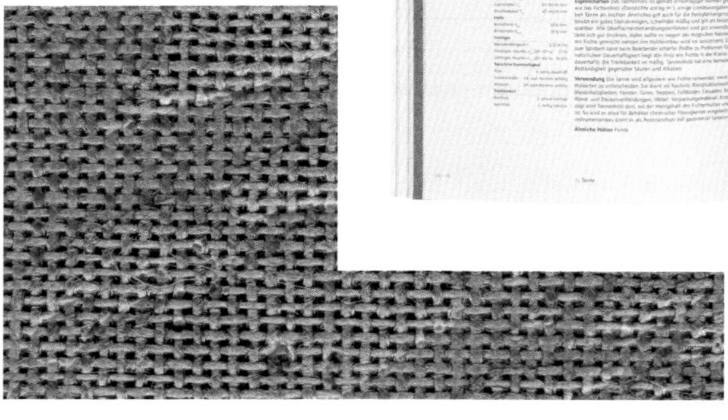

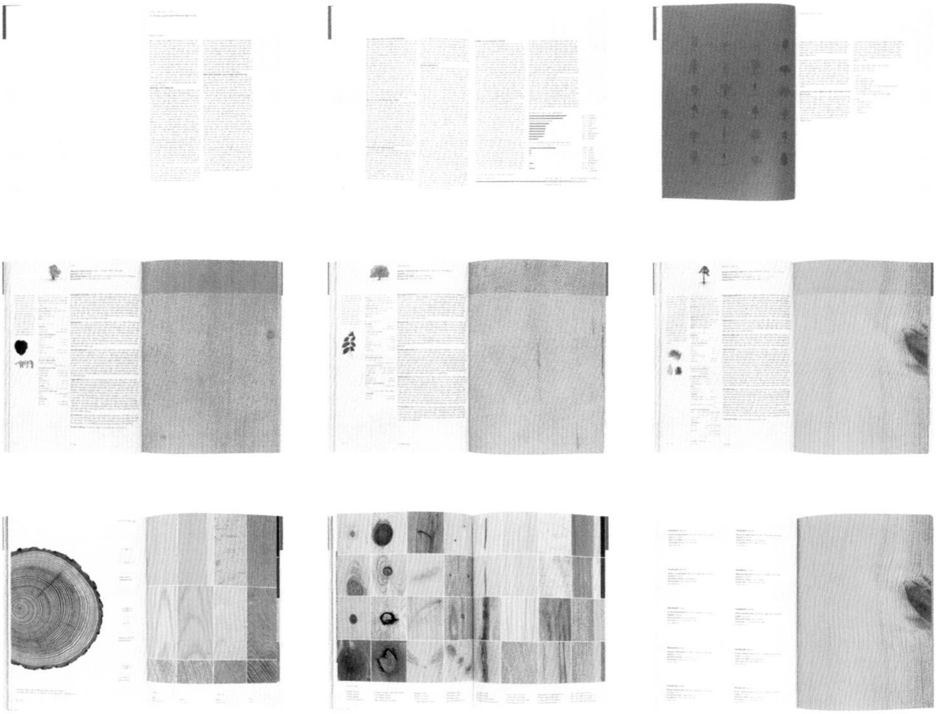

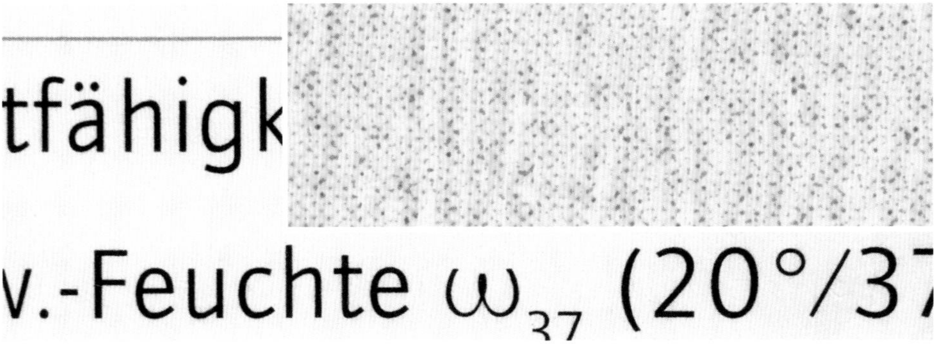
tfähigk
v.-Feuchte ω37 (20°/37

36
ANTONIN KYBAL
HANDGEWOBENER
ANTONÍN KYBAL
HAND-WOVEN TAPESTRY, 1948
In his work Antonín Kybal focused
almost solely on designing carpets.
He initially used abstract geometrical
patterns, but in the forties introduced
organic designs. In 1945 he became
Professor of Textile Art at the Academy
of Applied Arts. He designed a set
of textiles, including a curtain and
tapestries with motifs of castles and
chateaux, for the Czechoslovak pavilion
at the Brussels Expo.
OTAKAR DIBLÍK
AUTOKAR Š706 RTO, 1958
OTAKAR DIBLÍK
REISEBUS Š706 RTO, 1958
OTAKAR DIBLÍK
Š706 RTO COACH, 1958
007 Ha5

本书的主旨是通过展示和描述过去 100 年（出书年份往前）捷克各时期最有代表性的 100 件设计或项目，来描绘出捷克各时代的生活方式，并怀着自豪和自省的态度去看待历史和现在。书籍采用锁线胶装，封面卡纸四色印刷覆亮膜，采用书籍开篇所讲述的捷克国旗色彩演变史为封面的演绎元素。内页哑面铜版纸四色印刷，通过诗意的标题划分出章节，分别代表了 8 个不同的历史时期，并以不同的色彩进行区分。各时期均选取当时最有代表性的设计作品，包含建筑、海报、漫画、字体、家具、动画、汽车、电影等方方面面，一共 100 件，按照连续的编号在左、右上角清晰地标注。全书使用英、德、捷克三语混排。

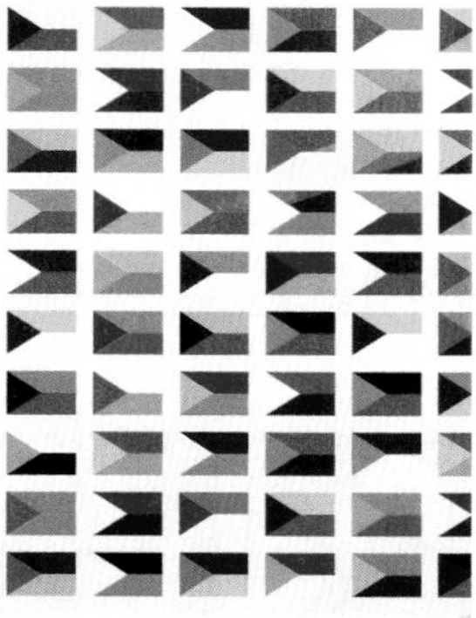

The book describes the Czech way of life by presenting 100 most representative designs or projects of the past 100 years in order to retrospect the history and the present with pride and introspection. The book is perfect bound with thread sewing. The laminated cover is printed in four colors. Colors used for the Czech national flag mentioned at the beginning of the book are adopted as the key elements of the cover design. The inside pages are printed in four colors on matte coated paper. The chapters are divided by poetic titles and distinguished by different colors to represent eight different historical periods. The most representative design works of the day were selected, including architecture, posters, cartoons, fonts, furniture, animation, automobile, films and other aspects. Total 100 pieces are arranged according to serial numbers clearly marked in the upper left or right corner. The book is written in English, German and Czech.

◎ 荣誉奖

《 捷克最有代表性的 100 个设计

◖ 捷克

◎ Honorary Appreciation

《 Czech 100 Design Icons

Czech Republic

Zuzana Lednická, Aleš Najbrt (Studio Najbrt)

Tereza Bruthansová, Jan Králíček

CzechMania, Prag

220 × 162 × 17mm

622g

100p

Content ideally matches design in this book. The jacket shows variations of the Czech flag modified in a range of different colours. The use of the national flag as a pictorial motif is testimony to Czech design's sense of its own identity, whilst the altering of its colours simultaneously shows openness to the wider world. The book's interior is divided into eight chapters, each being designated in a different colour. These colours return where the pages come together in the guise of narrow strips, creating a sort of illumination. The book is a magnificent reflection of the spirit of Czech design.

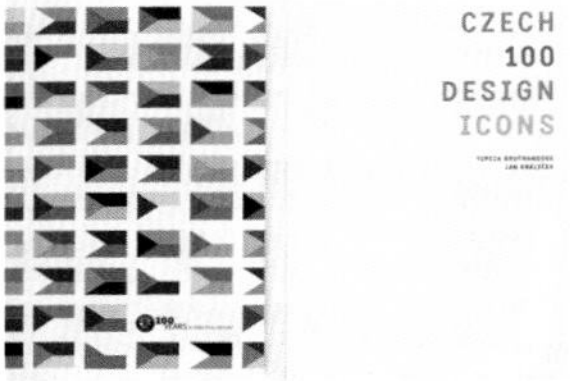
CZECH
100
DESIGN
ICONS

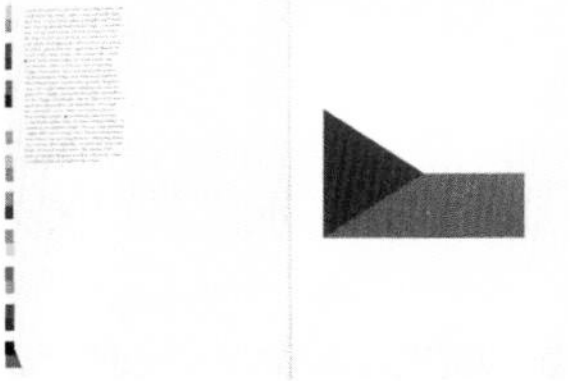

SCÉNA „ČERVENÁ SEDMA"
ARNOŠT DVOŘÁK MRTVÁ

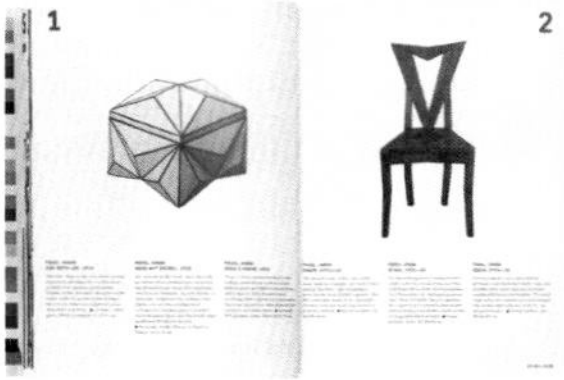
1
2

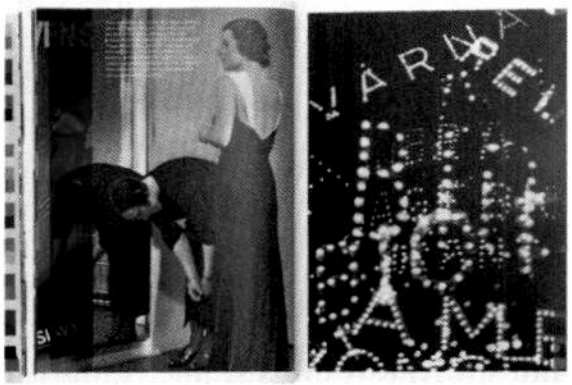

35
36

41

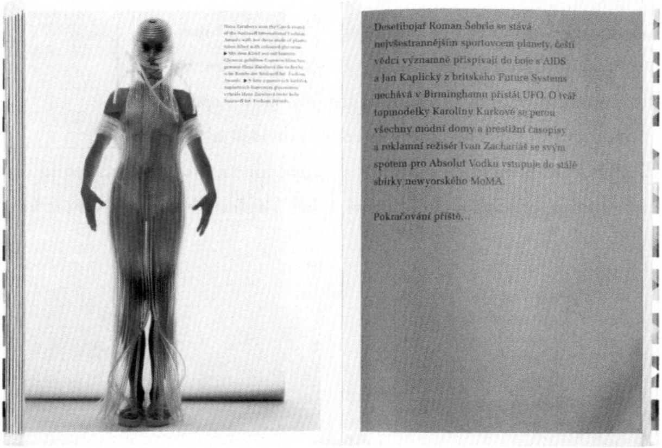
Desetibojař Roman Šebrle se stává nejvšestrannějším sportovcem planety, čeští vědci významně přispívají do boje s AIDS a Jan Kaplický z britského Future Systems nechává v Birminghamu přistát UFO. O tvář topmodelky Karolíny Kurkové se perou všechny módní domy a prestižní časopisy a reklamní režisér Ivan Zachariáš se svým spotem pro Absolut Vodku vstupuje do stálé sbírky newyorského MoMA.
Pokračování příště...

1–11
12–26
27–41
42–59
60–73
74–85
86–100

1910
1925

9
10

1920
1939

92
93

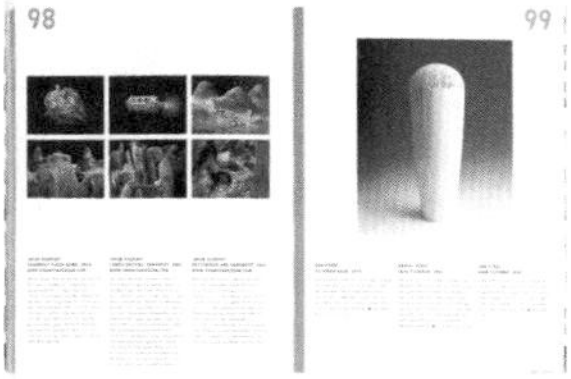
98
99

po šp  šr

rezentativní a při

- Jakob Demus. The Complete Graphic Work 1983–2005
- Metropolitan World Atlas
- Ellen Gallagher. Blizzard of White
- Strategien des Handwerks. Sieben Porträts au ergewöhnlicher Projekte in Europa
- Kekkonen
- Stedelijk Museum Amsterdam 2003/2004
- Opa
- Living Library: Wiel Arets
- Somewhere else is the same place
- Cao Xueqin's Art of Kite
- Max Bill. Maler, Bildhauer, Architekt, Designer
- Broedplaats Europoort v/h ›De Beer‹
- Russian Illustrated Magazine. 1703–1941
- Pohádky

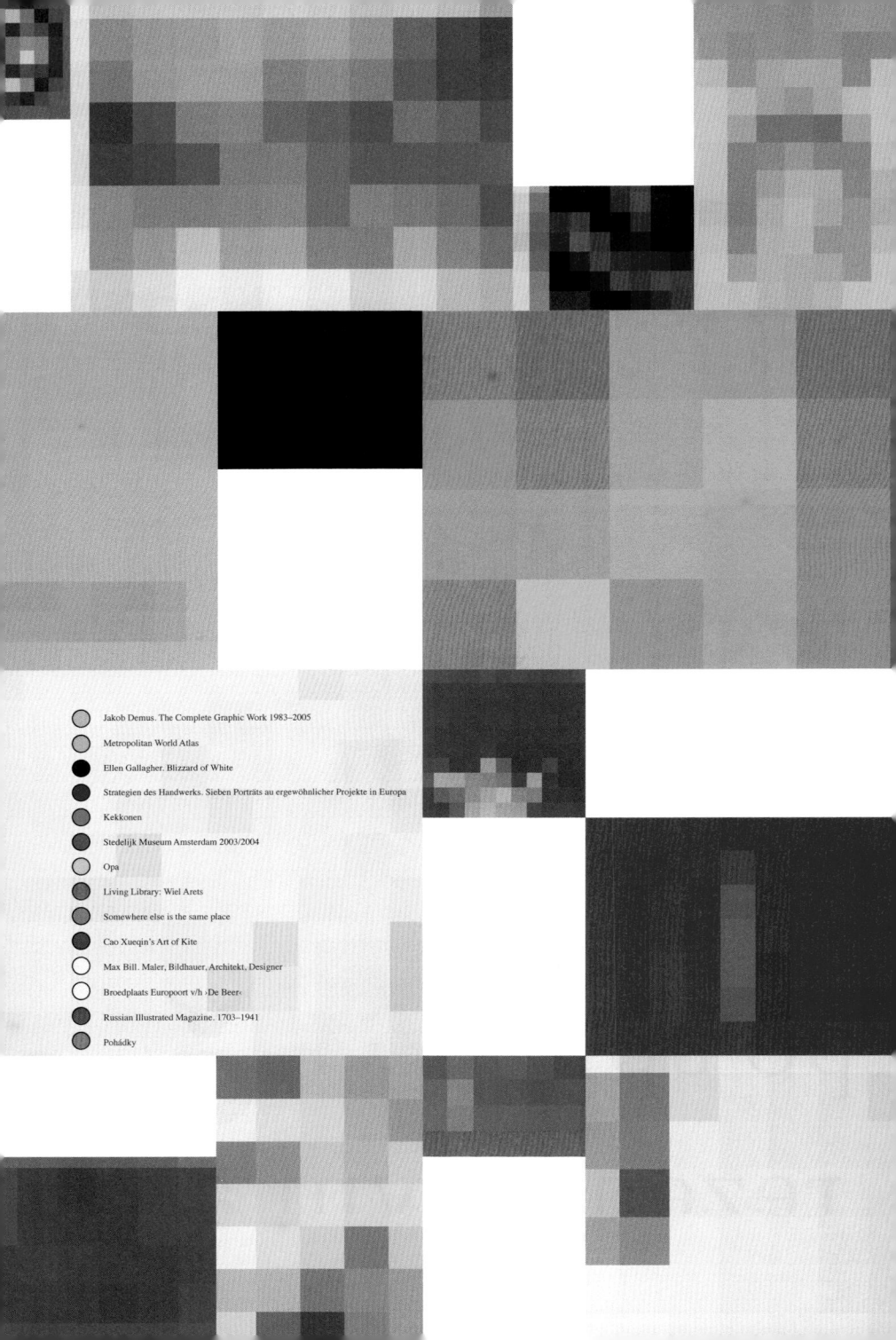

2006

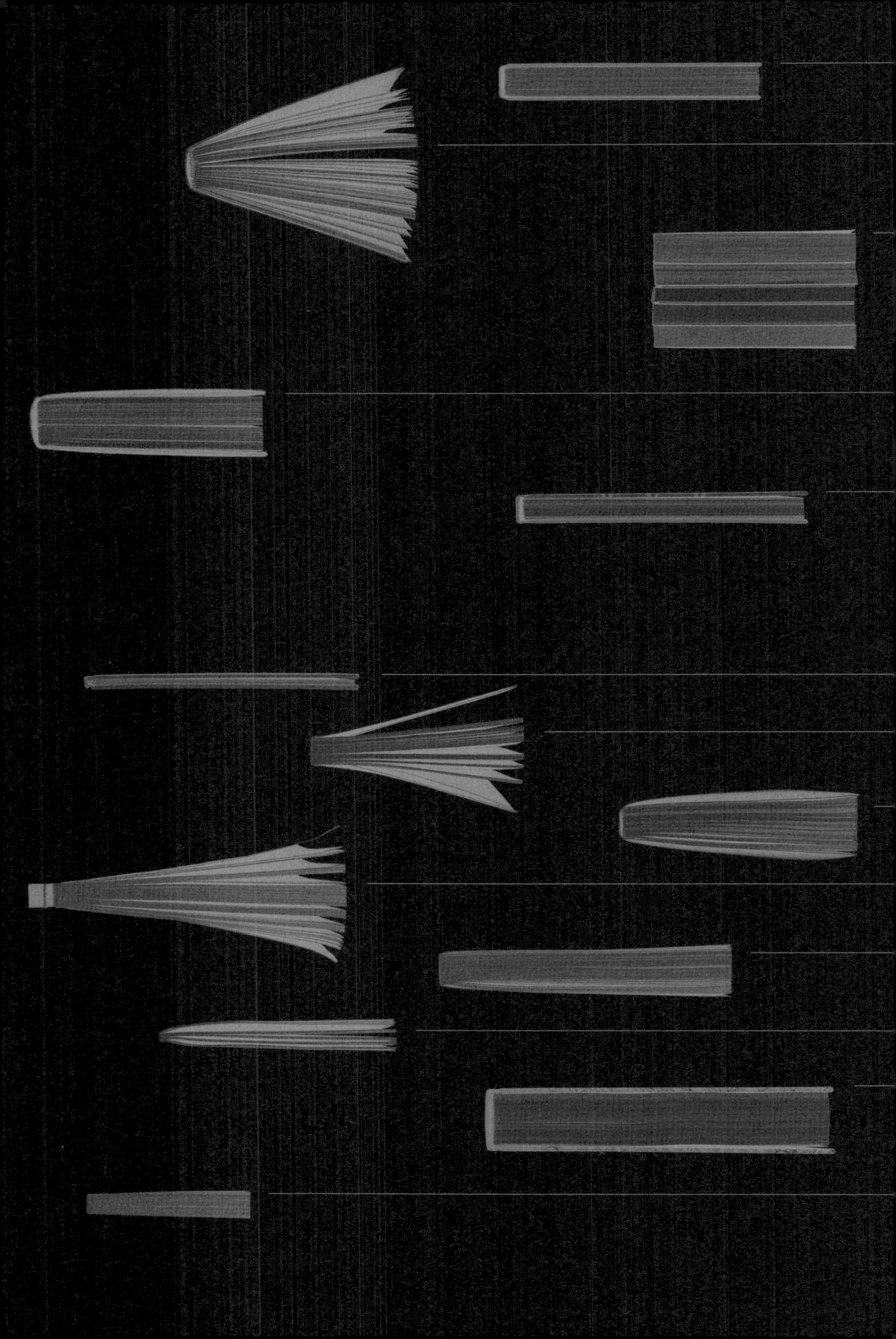

1197
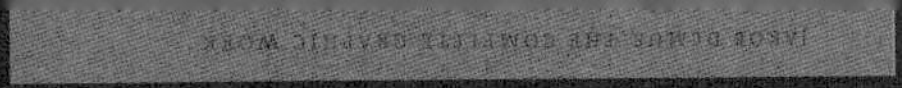
JAKOB DEMUS THE COMPLETE GRAPHIC WORK

1205
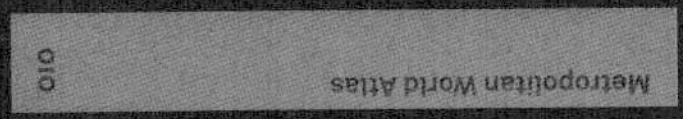
Metropolitan World Atlas
010

1211

1217

1223
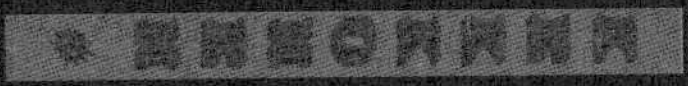

1229

1235

LIVING LIBRARY WIEL ARETS
PRESTEL

1241

1245
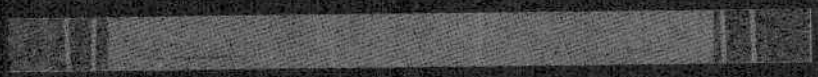

1249
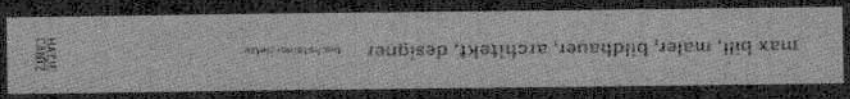
max bill, maler, bildhauer, architekt, designer

1253

1257

1261

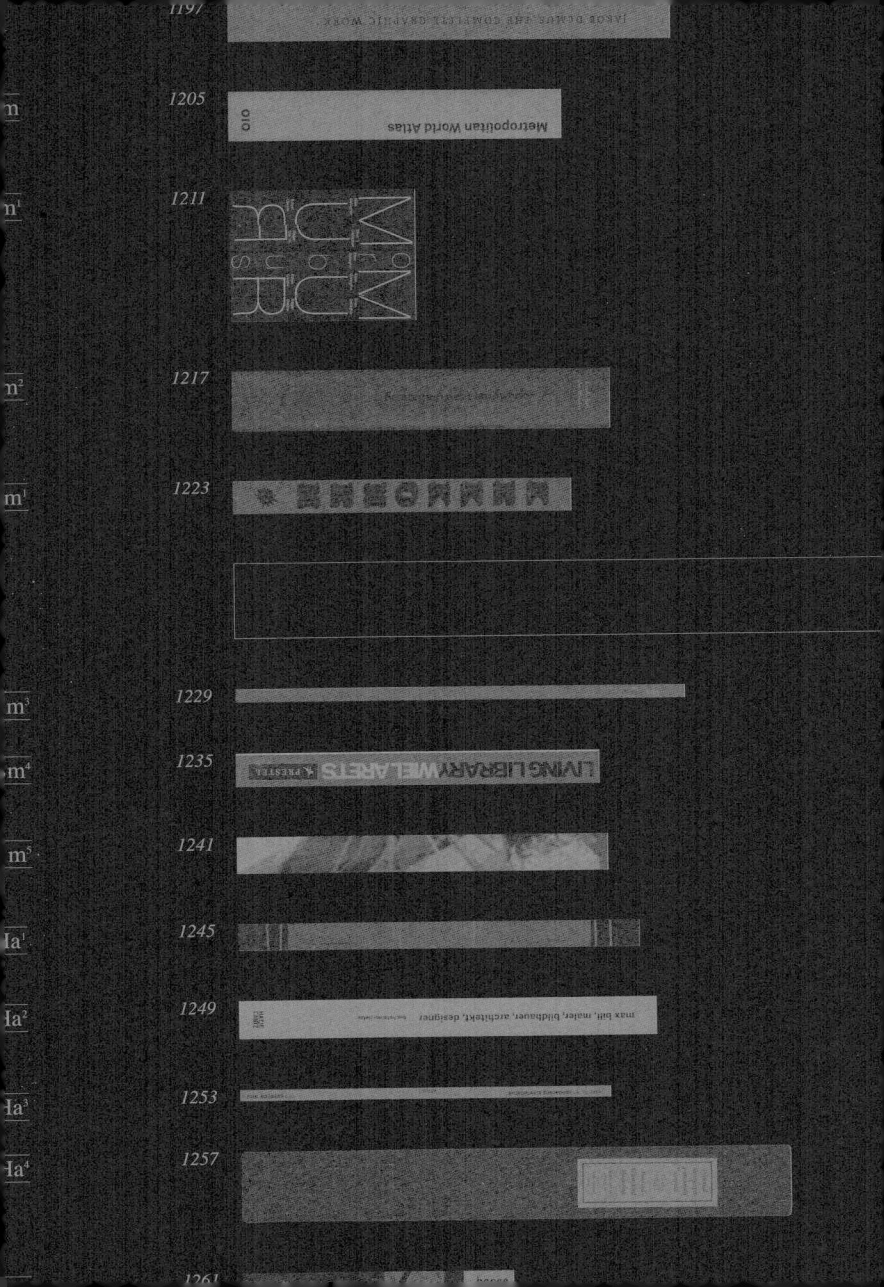

point technique directly into the exposed copper, that
acid, just engraving, then it is as if he writes on a blind

evel – the railway network. Among other things, thi
ccessibility both within regions and between citie
Cities that introduced an underground railway syste

Chief among these is her signat
lexicon of visual motifs gleaned
the debased signifiers of race or

Massschuh.de basiert auf zwei Grund-
gedanken: Handwerk soll kein teurer
Luxus sein, sondern ein Gut für die
Allgemeinheit. Diese Idee ist wiederum

Wichtige Projekterkenntni
•
Die Technologie muss sich n
Bedürfnissen des Handwerk

ПРЕЗИДЕНТ ФИНЛЯНДСКОЙ Р
ТРЕСКУ.

[redacted]

Volgens de vader van mijn opa had ze
twee gaten, in beide handen eén, maar d

is an outstanding example.
The physical aspects of the provision of information are clearly visible. The depots
many kilometres of information are not hidden away underground but hovezws the
the information. The library staff are also a distinct presence. They don't work in an
but in a real library. This is true both of the counter staff and those who work, metap
behind the scenes.

konstruktion handelt, die in einer Industriehalle in Japan steht, u
sowohl das Gestein wie der blaue Himmel mit Alpenpanorama auf
gedruckt sind. Dieses Spiel mit Schein und Sein erhält bei der

与此会而
述之手后
说明：
在《瓶湖梆
表演放风筝
文改编的文
一阵和风
鹰』的头
的余力未
右手顺势
一带步，
右下方做
只『苍鹰
，面向西
空一直钻
邦达说：
完，董邦

um 1950 entstand dann eine ganze anzahl von holzmöbeln für die wohnbedarf ag.[4] unter anderem auch eine version des damals populären nierentisches. die freie form dieses leitfossils jener zeit wurde mit der möglichkeit begründet, unterschiedlich grosse personengruppen an ein und demselben tisch versammeln zu können. bills dreirundtisch (abb. s. 109) wirkt vergleichsweise

dreirundtisch finden – erinnern sitz und rücken des
stuhls (1949, abb. s. 108), die für die firma junghans en
küchenuhr mit timer (abb. s. 114) sowie die bestrahlun
(abb. s. 113). weiter von den grundformen der geomet
sind die dreidimensional verformten sitz- und rücken
kreuzzargenstuhls (1951, abb. s. 110–111). sämtliche

Na de Tweede Wereldoorlog echter begon het
Rotterdam zich in snel tempo richting zee uit te br
vijftig was het eiland Rozenburg bereikt, maar de

Beketov's passion was collecting old Slavonic-Russian manuscripts, historical documents and various iconographic materials. Through collecting painted portraits, engravings and depictions of famous people for many years, he managed to amass a very fine collection. It was on the basis of his collection that Beketov decided to start a publication of "portraits of

Pantheon rossiis
avtorov
Page with illus
to the article "
by N. Karamz

po svatbě přišla při jednom neštěstí o manž
ní chudá a opuštěná vysedávala v malém p
a čekala na děťátko, které nebude mít otce.

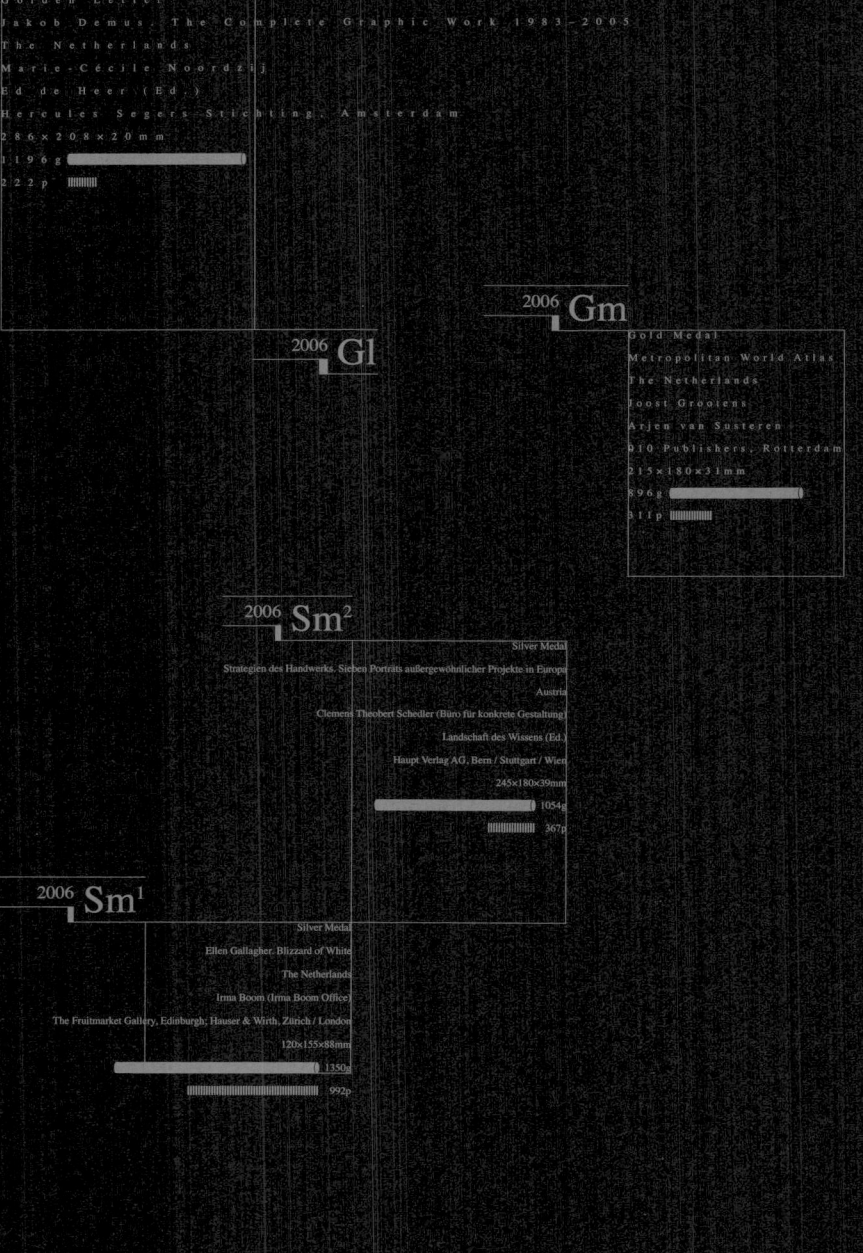

Golden Letter

Jakob Demus. The Complete Graphic Work 1983–2005

The Netherlands

Marie-Cécile Noordzij

Ed de Heer (Ed.)

Hercules Segers Stichting, Amsterdam

286×208×20mm

1196g

222p

2006 Gl

2006 Gm

Gold Medal

Metropolitan World Atlas

The Netherlands

Joost Grootens

Arjen van Susteren

010 Publishers, Rotterdam

215×180×31mm

896g

311p

2006 Sm[2]

Silver Medal

Strategien des Handwerks. Sieben Porträts außergewöhnlicher Projekte in Europa

Austria

Clemens Theobert Schedler (Büro für konkrete Gestaltung)

Landschaft des Wissens (Ed.)

Haupt Verlag AG, Bern / Stuttgart / Wien

245×180×39mm

1054g

367p

2006 Sm[1]

Silver Medal

Ellen Gallagher. Blizzard of White

The Netherlands

Irma Boom (Irma Boom Office)

The Fruitmarket Gallery, Edinburgh; Hauser & Wirth, Zürich / London

120×155×88mm

1350g

992p

2006 Ha[1]

Honorary Appreciation

Cao Xueqin's Art of Kite

China

Zhaojian Studio

Kong Xiangze, Kong Lingmin, Kong Bingzhang

Beijing Arts & Crafts Publishing House, Peking

260×231×19mm

608g

140p

2006 Ha[2]

Honorary Appreciation

Max Bill.

Maler, Bildhauer, Architekt, Designer

Germany

Sascha Lobe, Ina Bauer (L2M3 Kommunikationsdesign)

Thomas Buchsteiner, Otto Letze (Ed.)

Hatje Cantz Verlag, Ostfi Idern-Ruit

270×230×26mm

1367g

296p

2006 Ha[5]

Honorary Appreciation

Pohádky

Czech Republic

Luboš Drtina

Hermann Hesse

Argo, Prag

174×116×14mm

152g

165p

2006 Ha[3]

Honorary Appreciation

Broedplaats Europoort v/h ›De Beer‹

The Netherlands

René Put

Paul Bogaers u.a.

Artimo, Amsterdam

240×165×7mm

277g

104p

2006 Ha[4]

Honorary Appreciation

Russian Illustrated Magazine. 1703–1941

Russia

Masha Lyuledgian

Irina Mokhnacheva

Agey Tomesh, Moskau

356×218×45mm

2607g

368p

2006 Bm[1]

Bronze Medal

Kekkonen

Finland

Matti Hagelberg

Matti Hagelberg

Kustannusosakeyhtiö Otava, Helsinki

219×217×21mm

750g

200p

2006 Bm[4]

Bronze Medal

Living Library: Wiel Arets

The Netherlands

Irma Boom (Irma Boom Office)

Marijke Beek (Ed.)

Utrecht University Library; Prestel Verlag, München

235×170×22mm

897g

456p

2006 Bm[3]

Bronze Medal

Opa

The Netherlands

Irma Boom (Irma Boom Office)

Paul de Reus

Uitgeverij Van Waveren, Amsterdam

291×209×9mm

390g

44p

2006 Bm[2]

2006 Bm[3]

Bronze Medal

Somewhere else is the same place

Switzerland

Iza Hren, Georg Rutishauser

Monica Studer, Christoph van den Berg, Christoph Vögele (Ed.)

Kunstmuseum Solothurn; edition fink, Zürich

240×170×23mm

690g

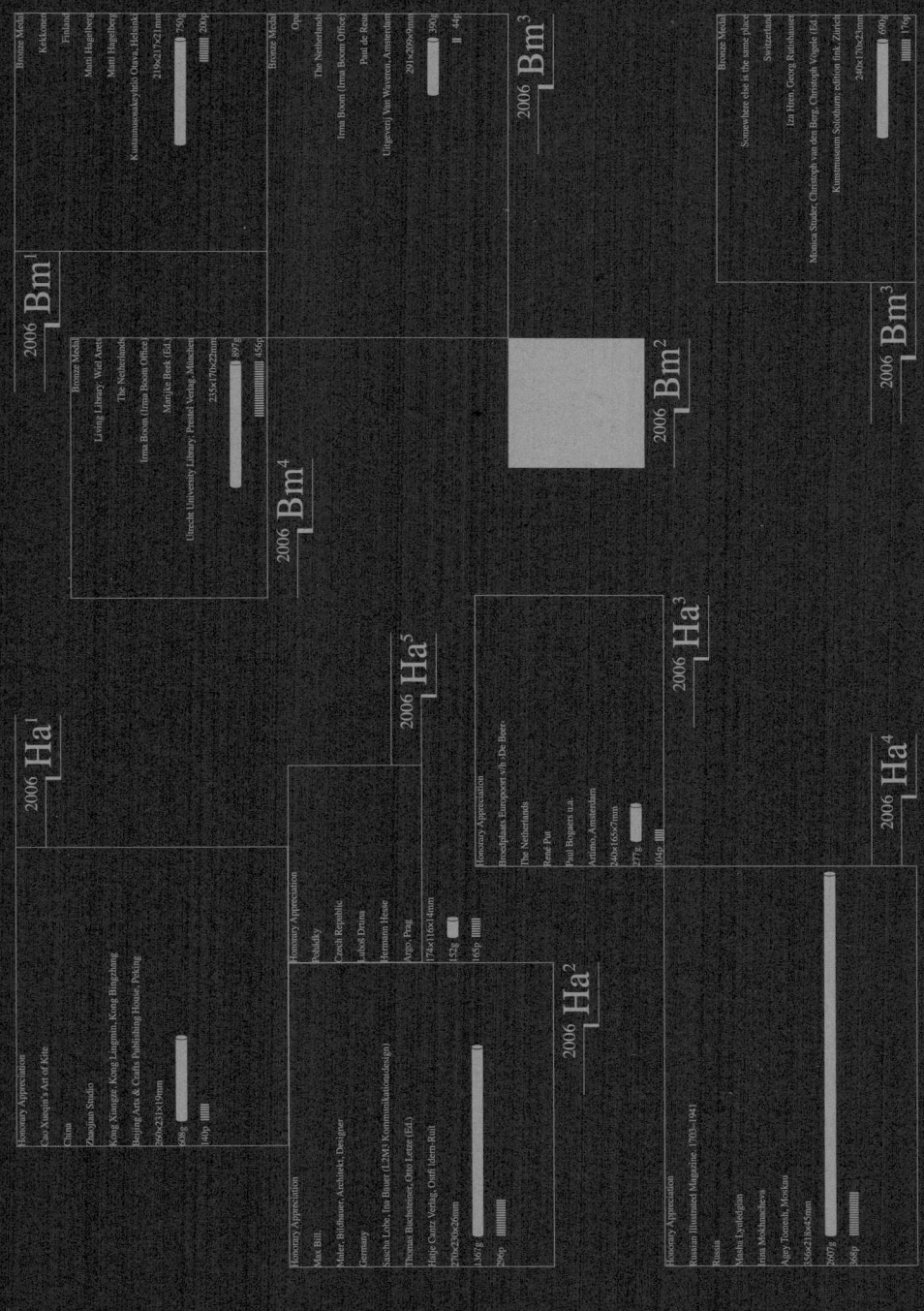

176p

The International Jury

Julia Blume (Germany)
Gabriela Fontanillas (Venezuela)
Juraj Horváth (Czech Republic)
Innokenty Keleinikov (Russia)
Rainer Leibbrand (Germany)
Christine Müller (Germany)
Ronald Widdershoven (The Netherlands)

Designer

Marie-Cécile Noordzij
Joost Grootens
Irma Boom (Irma Boom Office)
Clemens Theobert Schedler (Büro für konkrete Gestaltung)
Matti Hagelberg
Ben Laloua, Didier Pascal (mit Maaike Molenkamp)
Irma Boom (Irma Boom Office)
Irma Boom (Irma Boom Office)
Iza Hren, Georg Rutishauser
Zhaojian Studio
Sascha Lobe, Ina Bauer (L2M3 Kommunikationsdesign)
René Put
Masha Lyuledgian
Luboš Drtina

2006

Country / Region

The Netherlands 7
Austria 1
Finland 1
Switzerland 1
China 1
Germany 1
Russia 1
Czech Republic 1

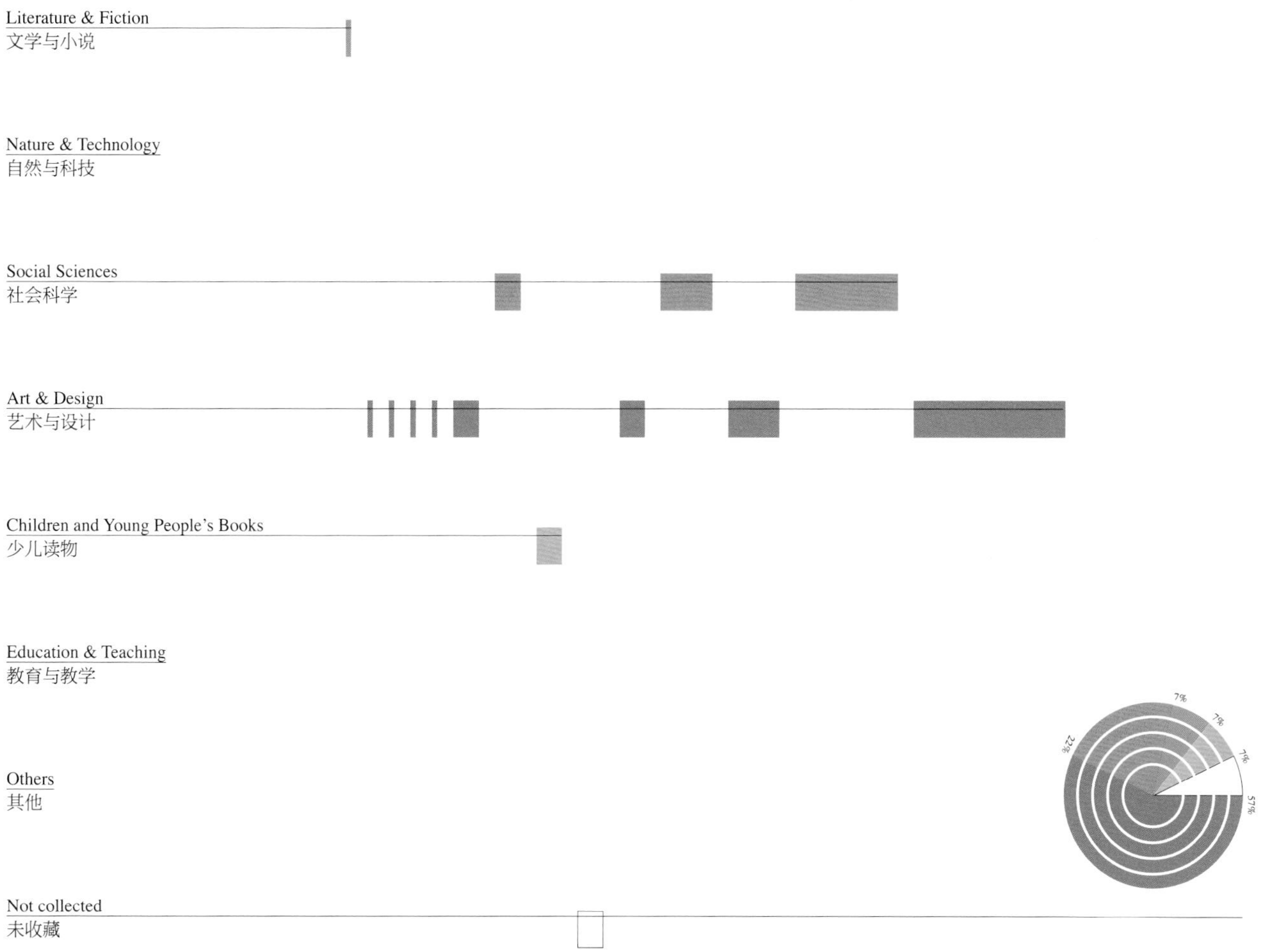

Literature & Fiction
文学与小说
Nature & Technology
自然与科技
Social Sciences
社会科学
Art & Design
艺术与设计
Children and Young People's Books
少儿读物
Education & Teaching
教育与教学
Others
其他
Not collected
未收藏
7%
7%
7%
22%
57%

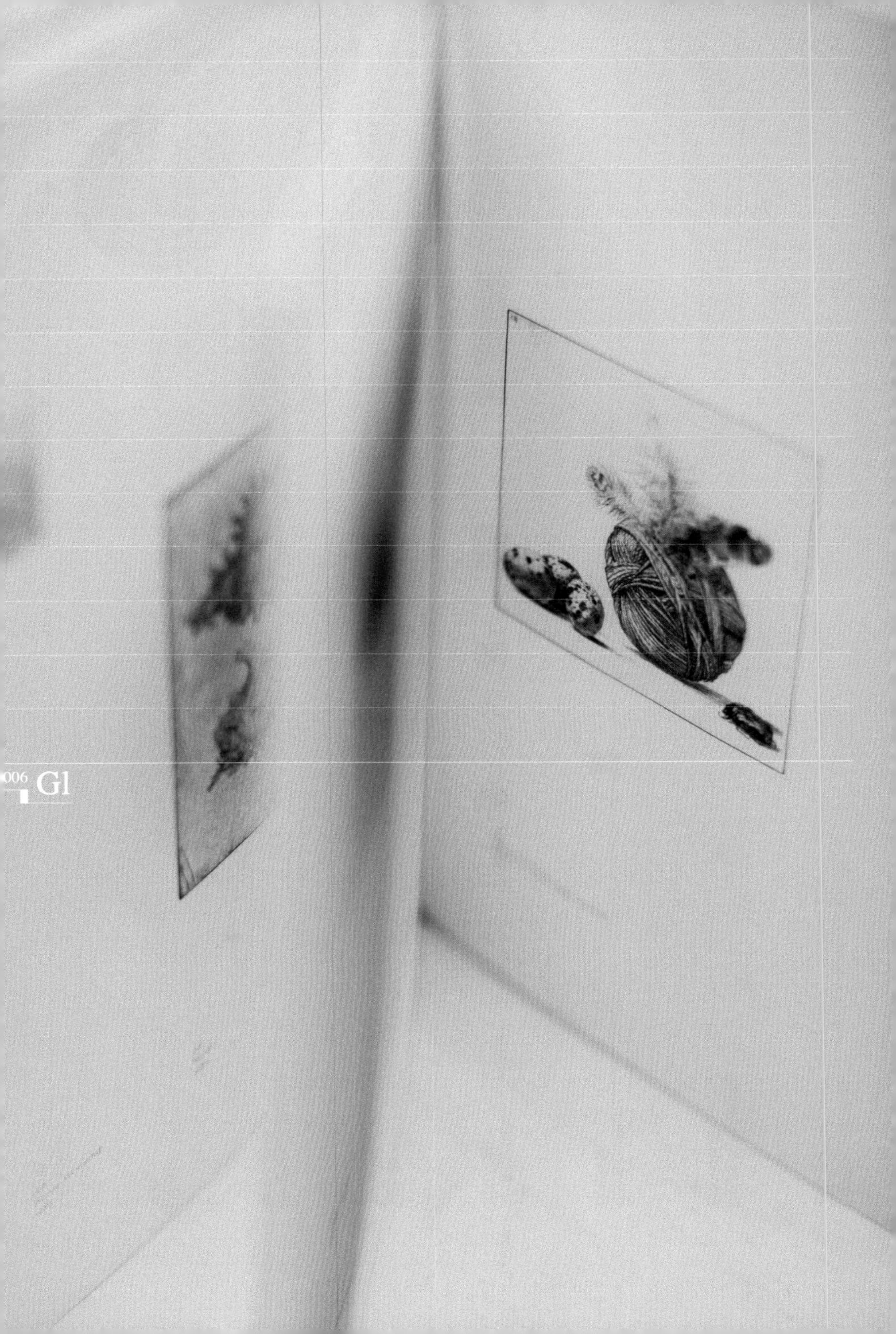

006 Gl

本书是奥地利艺术家雅各布·德穆斯（Jakob Demus）的作品集，呈现了他于 1983－2005 年创作的一系列作品，特别是他独有的钻石干点技术铜版画和一些混合技术的版画作品。本书采用褐黄色麻布面硬精装，封面和书籍只用了黑色烫印加凹版压印技术，沉稳简洁。内容大部分为哑光面涂布纸单黑色印刷，只有开篇部分页面使用四色印制少量非版画彩图作品。本书部分图片为原大尺寸，并配合详细文字重点介绍，其余部分均等比缩小至 40% 大小用以配合。作品页文字单栏设置，版心相对于本书开本来说较窄，这样留出大量空白，以主图加辅图相互搭配的方式来控制版面节奏，使得版面错落有致，充满诗意。

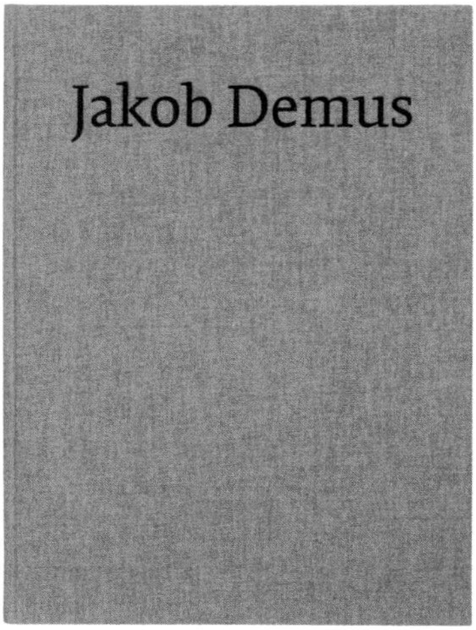

Jakob Demus

The book is a collection of works by the Austrian artist Jakob Demus, which presents a series of works he created from 1983 to 2005, especially his unique etchings engraved by diamond with mixed techniques. This hardback book, with the brown yellow linen surface, are printed using hot stamping and intaglio in black, thus the cover and the pages look composed and concise. Most of the contents are printed in black on matte coated paper sheets. Only the pages of the opening part print a small number of non-etching color works using four-color process. Some of the pictures in this book are of the original size with detailed text introduction, and the rest are reduced to 40% size to match. The text on the work page is set in a single column, and the layout heart is relatively narrow compared with the folio of this book. In this way, a large amount of blank is set aside, and the rhythm of the layout is controlled by the way of matching the main diagram with the auxiliary diagram, which makes the layout well-arranged and poetic.

- 金字符奖
- 雅各布·德穆斯 1983－2005 年的作品
- 荷兰

With its wonderful typography, easy to read and full of detail with excellent choice of typeface effectively set against the page, this book is a reader's delight. The pictures are superbly printed, its fold-out plates enhancing a perfectly bound book and rounding it off in style.

- Golden Letter
- Jakob Demus. The Complete Graphic Work 1983－2005
- The Netherlands
- Marie-Cécile Noordzij
- Ed de Heer (Ed.)
- Hercules Segers Stichting, Amsterdam
- 286×208×20mm
- 1196g
- 222p

begu

worl

only

or 'Bird Milk'. The

ple comparison. T

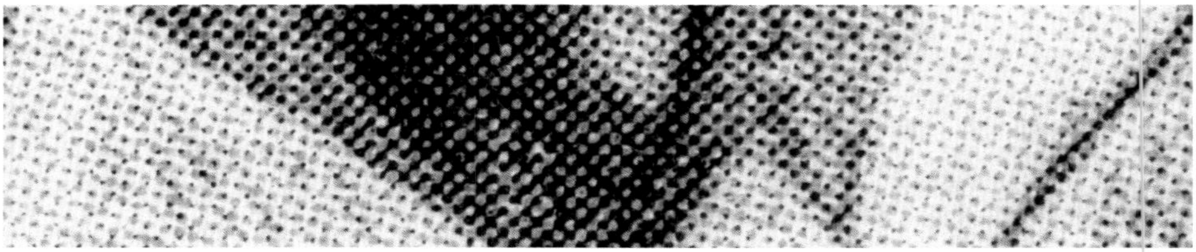

the Prate

186

The creation of this sheet
print. My eyes pierced the
ing coppery-blue leaves, w
dense protective case arou
restlessly climbs through
was driven with the same
som's honeycomb. Only ea
warded off by the azure bl

How often did the seduc
ingly similar to the devil i
the Devil. How often did he
me even further, he sent a
this continually growing c
directly above my head 'mi
tration.

But that he reminded me
For the knight assumed his
reflection of the thorny cr
ter night, defying the whis
of my diamond and I aime
tiful blossom surrounded
flame.

Though my longest and h
journey. My 'grail' shone th
And I sat firm in the saddle
temptation of death's hour

The *Eryngium* defeats the
night in the mountains, ju
name *Mannertreu* translate
to be true to myself.

It was only afterwards th
trait in the Louvre, Durer d
between his fingers – like a

By the way, the *Eryngium*
Its root is a great delicacy. I
eaten like asparagus, and li

And which root bears the

286×208×20

1196

222

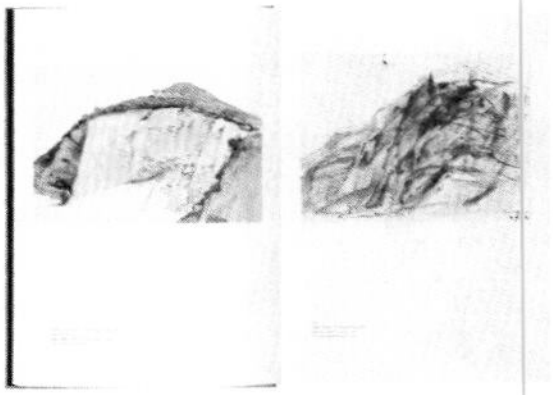

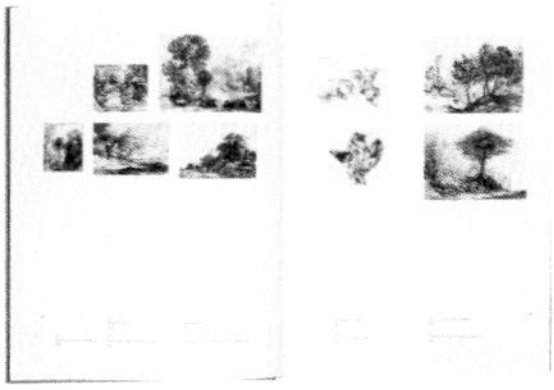

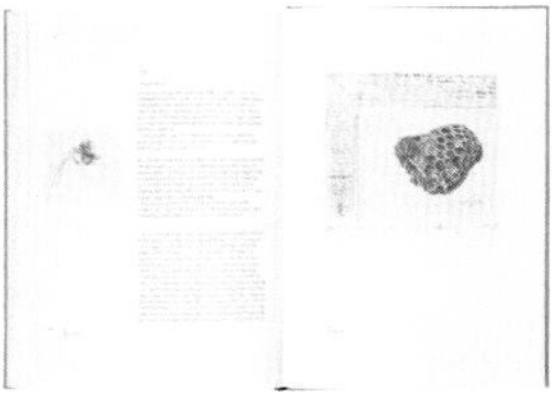

Jakob Demus
THE COMPLETE
GRAPHIC WORK
1983–2005

JAKOB DEMUS THE COMPLETE GRAPHIC WORK

170

What have tulips and dragons in common? Out of the bulb shoots the stem that bulges with the power of water and supports the swaying bud that has the same shaped head as the reptile. It glides through our hands, like a cool, glistening snake. Its flame colour – red and yellow – flickers out of the green, the colour of lizards.

The tulip is connected with the sea and its ships – which is why its triumphal entry into the Netherlands was unparalleled – and the ships of all peoples are dragon-headed. The curved petal of the tulip is like a full-blown sail or the extended wing of a dragon, and the strong central vein of a ship's keel. The edge of the petal is reminiscent of finely curled lines of foam created by waves breaking on the shore, or a wedding veil of the finest bobbin lace that flutters in a whispered breeze. The tulip expresses perfectly the marriage of water and earth, of land and sea, as well as that of fire and air. And this is exactly the meaning of the dragon of the four elements.

The tulip originated in Persia. And it is in Persia that the (European) dragon is presumed to have been created.

Dürer, who of course never knew the tulip, but was very familiar with the turban, was clearly drawn to dragons; their 'presence' marks his engravings, down to the finest line. It was such chased and finely wrought 'dragons' that I was thinking of when drawing these parrot tulips. It was the challenge of both Dürer and the tulip: to disentangle the twisted confusion of the unwinding petals.

Just as with the first version, cat. no. 163, I here followed the different stages of growth of the same bud. This time, a protruding dragon's head was the inspiration for the arrangement of the four views, which together form an open semi-circle that in turn expands to a spiral.

168
Altdorfer underneath a Pine Tree
1988

169
St John on Patmos (after Altdorfer)
1988

170
Four Parrot Tulips
1988

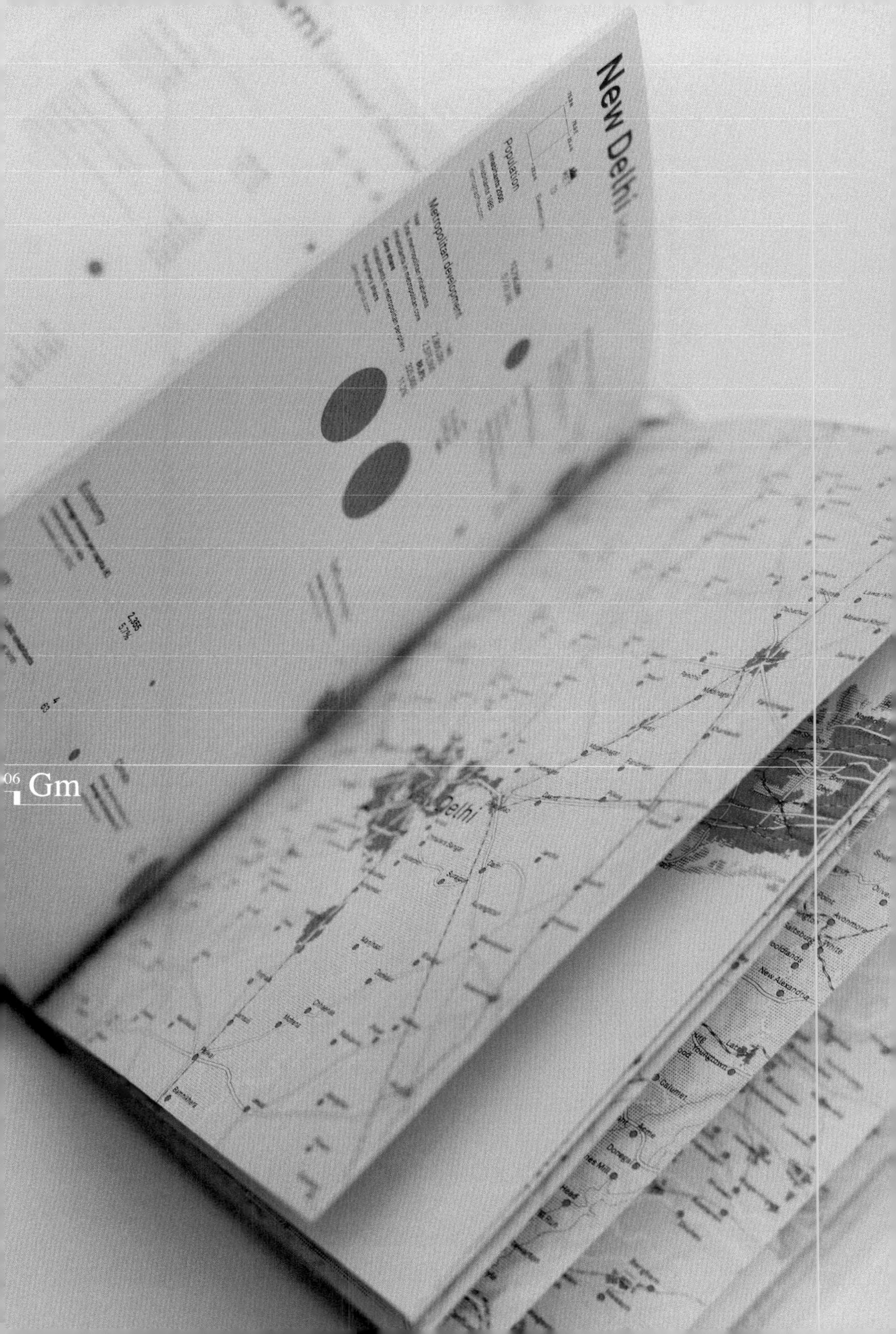
New Delhi
Population
Metropolitan development
2,355
5.7%
Delhi
New Alexandria
Calumet

这是一本对全球101个大都市（都市圈）进行大量数据统计和发展分析的专著。统计资料包含城市面积、人口、经济、机场、港口、医疗、犯罪、污染等多方面的详细数据。本书为软精装，封面黑色加荧光橙色双色印刷。内页为了体现数据的比较，采用黑色、荧光橙、银色、珠光黄四色印刷，呈现出独特的视觉效果。珠光黄色为参与统计的城市中数据的最大值，荧光橙则为当前城市的数据占比，如此可清晰地看出这个城市在此数据上的等级。最后的部分按照统计数据的分类，将这些城市的数据放在同一页内进行横向对比。书籍数据量巨大，但通过图式的方式得到有趣的呈现。书中包含白、银两条绢面书签，用于在数据的海洋中查阅时不至于迷路。

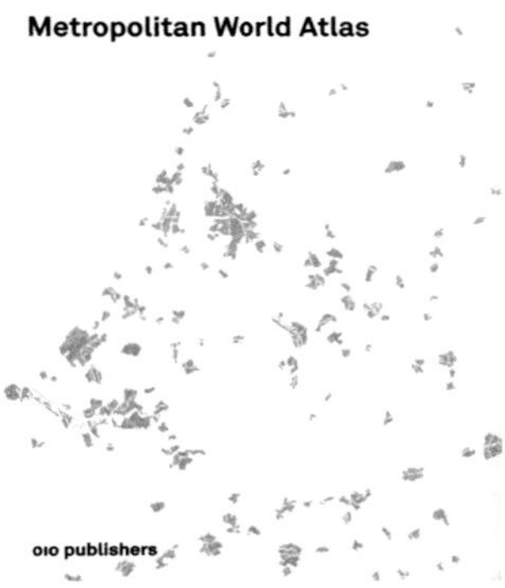
Metropolitan World Atlas
010 publishers

This is a monograph on the massive data statistics and analysis of 101 global metropolitan cities (areas). Statistics include detailed data on city size, population, economy, airports, ports, health care, crime, and pollution. The book is hardbound with soft cover, which is printed in black and fluorescent orange. In order to reflect the comparison of data, the inside pages are printed in black, fluorescent orange, silver and pearl yellow, presenting a unique visual effect. The pearl yellow represents the maximum value of the data of the cities involved in the statistics, while the fluorescent orange represents the data proportion of the current city, so it can be clearly seen that how the city ranks in this field. In the last part, the data of these cities are placed in the same page for synchronized comparison according to the classification of statistical data. A huge amount of contains in the book, but it is presented in an interesting way through the schema. The book is bound with two silk bookmarks, white and silver, so as not to get lost in the sea of data.

- 金奖
- 世界大都会地图集
- 荷兰

- Gold Medal
- Metropolitan World Atlas
- The Netherlands
- Joost Grootens
- Arjen van Susteren
- 010 Publishers, Rotterdam
- 215×180×31mm
- 896g
- 311p

Both in terms of content and design a compelling book. Its easy to understand and detailed codifi cation of complex material will make any reader want to reach for it. The map design is simple and balances well with the complexity of the material. Excellent and very clear typographical composition. In its stringency a most convincing atlas.

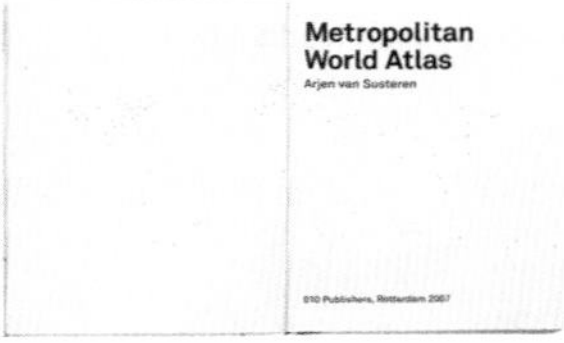
Metropolitan
World Atlas
Arjen van Susteren
010 Publishers, Rotterdam 2007

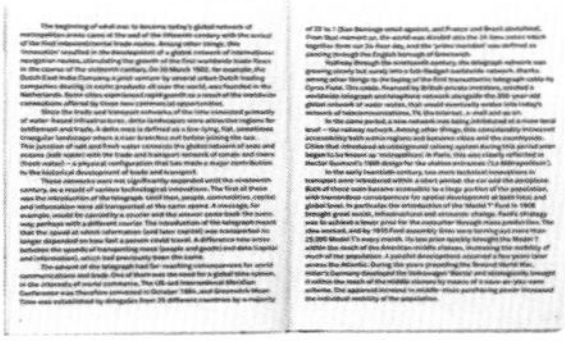

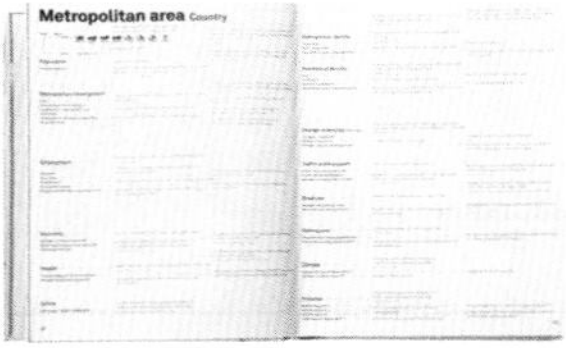
Metropolitan area Country

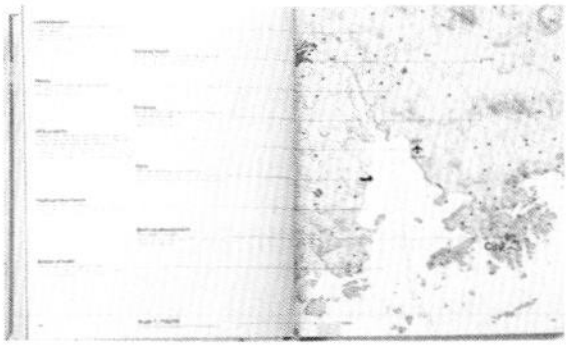

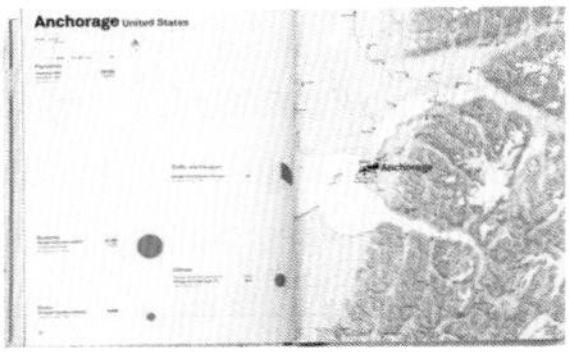
Anchorage United States
Anchorage

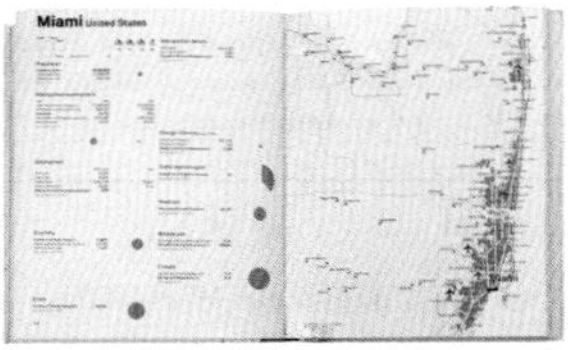
Miami United States

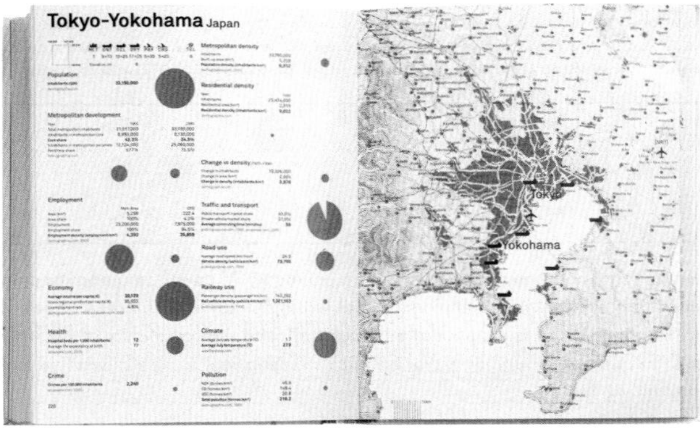
Tokyo-Yokohama Japan
Population
Metropolitan development
Employment
Economy
Health
Crime
Metropolitan density
Residential density
Change in density
Traffic and transport
Road use
Railway use
Climate
Pollution
Yokohama

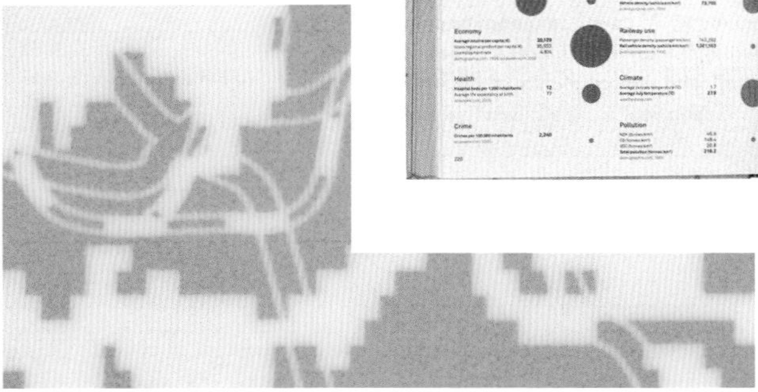

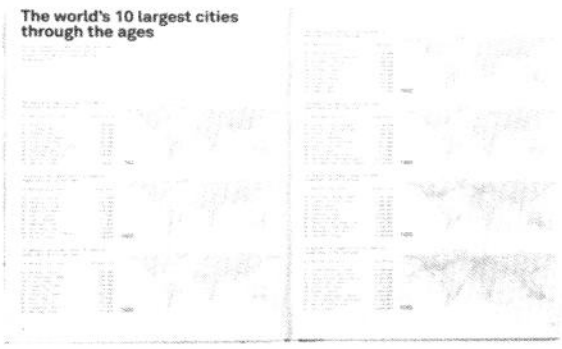
The world's 10 largest cities through the ages

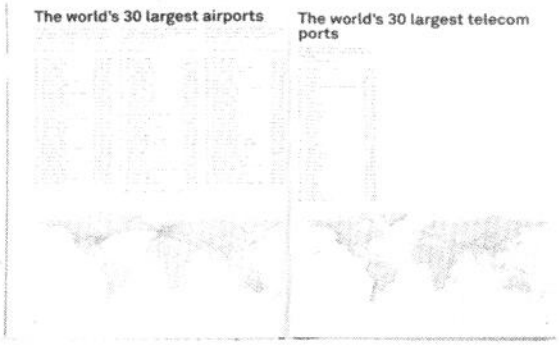
The world's 30 largest airports
The world's 30 largest telecom ports

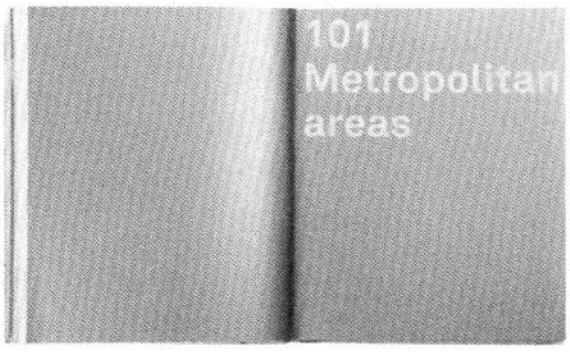
101
Metropolitan
areas

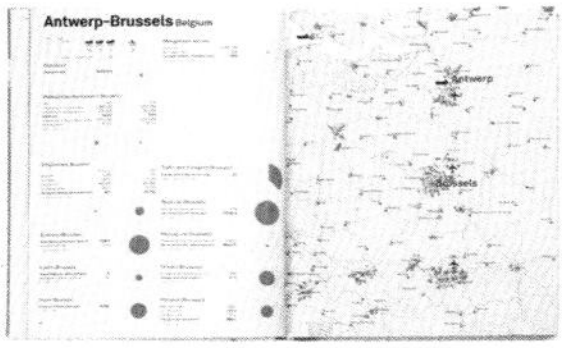
Antwerp-Brussels Belgium
Antwerp
Brussels

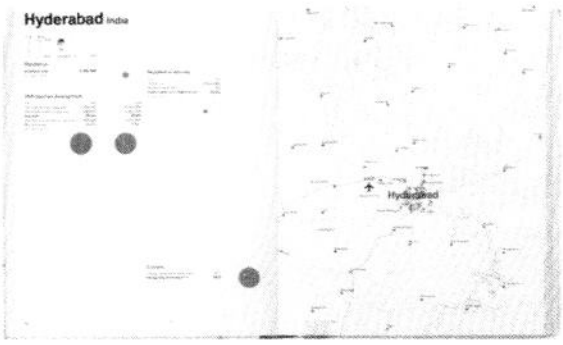
Hyderabad India
Hyderabad

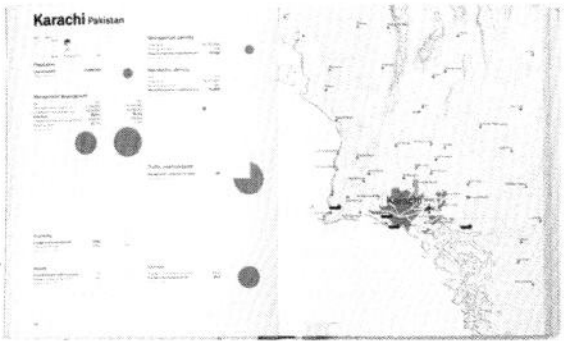
Karachi Pakistan

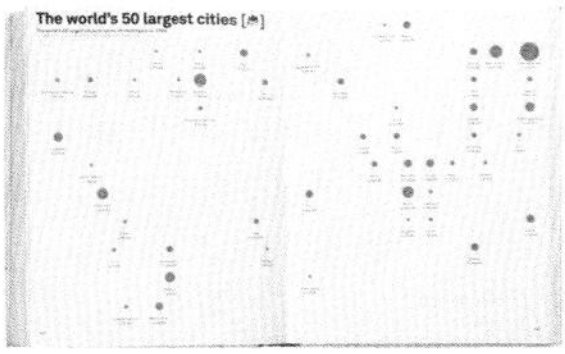
The world's 50 largest cities

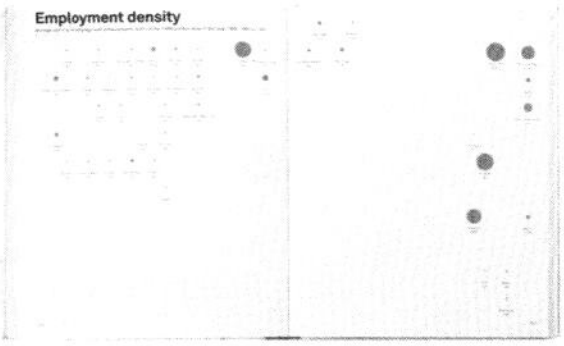
Employment density

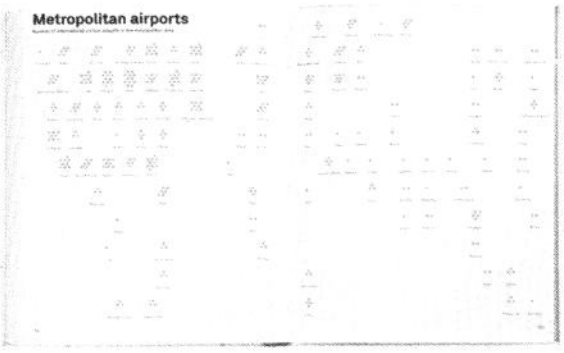
Metropolitan airports

Halle
share
and transportation in the

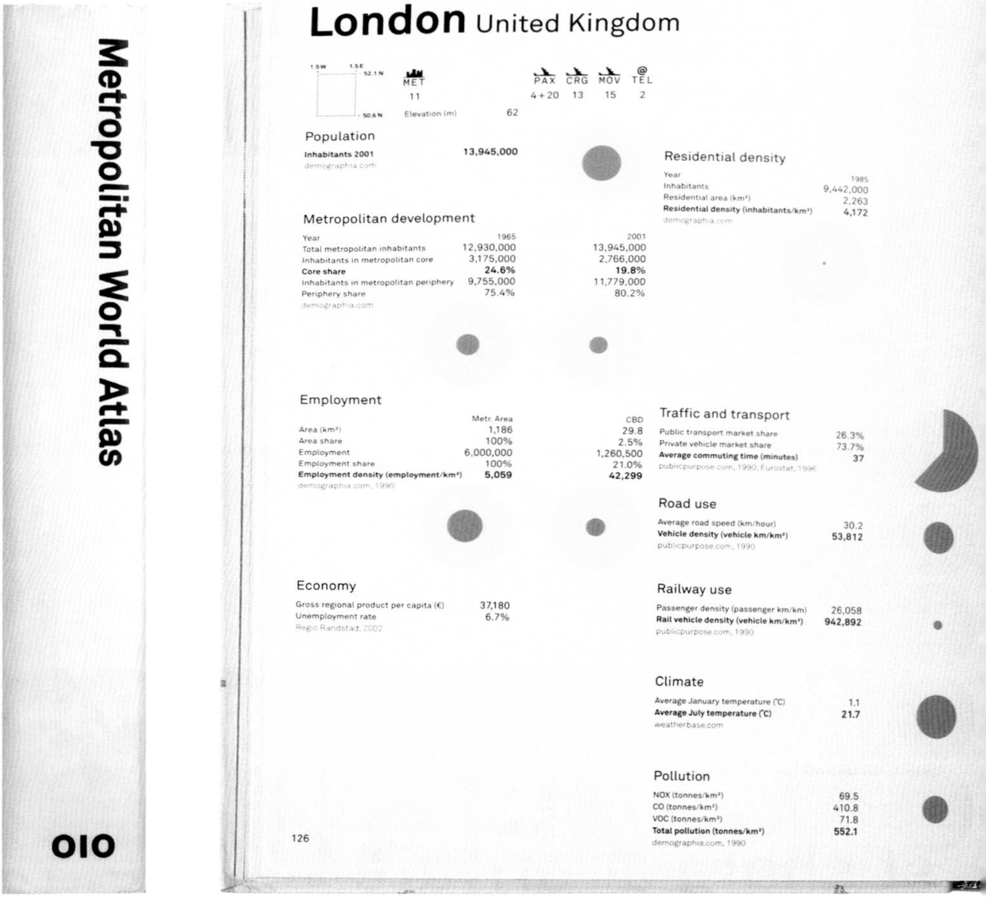

Metropolitan World Atlas

010

London United Kingdom

1.5W 1.5E 52.1 N 50.6 N

MET	PAX	CRG	MOV	TEL
11	4+20	13	15	2

Elevation (m) 62

Population

Inhabitants 2001	**13,945,000**

demographia.com

Residential density

Year	1985
Inhabitants	9,442,000
Residential area (km²)	2,263
Residential density (inhabitants/km²)	**4,172**

demographia.com

Metropolitan development

Year	1965	2001
Total metropolitan inhabitants	12,930,000	13,945,000
Inhabitants in metropolitan core	3,175,000	2,766,000
Core share	**24.6%**	**19.8%**
Inhabitants in metropolitan periphery	9,755,000	11,779,000
Periphery share	75.4%	80.2%

demographia.com

Employment

	Metr. Area	CBD
Area (km²)	1,186	29.8
Area share	100%	2.5%
Employment	6,000,000	1,260,500
Employment share	100%	21.0%
Employment density (employment/km²)	**5,059**	**42,299**

demographia.com, 1990

Traffic and transport

Public transport market share	26.3%
Private vehicle market share	73.7%
Average commuting time (minutes)	**37**

publicpurpose.com, 1990; Eurostat, 1996

Road use

Average road speed (km/hour)	30.2
Vehicle density (vehicle km/km²)	**53,812**

publicpurpose.com, 1990

Economy

Gross regional product per capita (€)	37,180
Unemployment rate	6.7%

Regio Randstad, 2002

Railway use

Passenger density (passenger km/km)	26,058
Rail vehicle density (vehicle km/km²)	**942,892**

publicpurpose.com, 1990

Climate

Average January temperature (°C)	1.1
Average July temperature (°C)	**21.7**

weatherbase.com

Pollution

NOX (tonnes/km²)	69.5
CO (tonnes/km²)	410.8
VOC (tonnes/km²)	71.8
Total pollution (tonnes/km²)	**552.1**

demographia.com, 1990

126

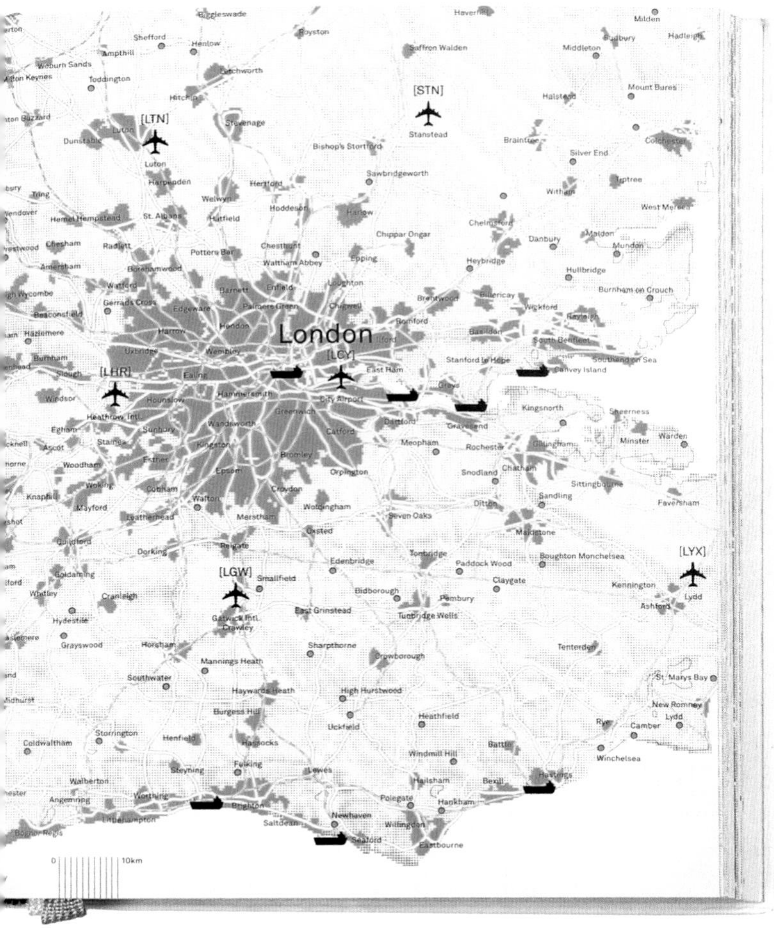
London
[STN]
Stanstead
[LTN]
Luton
[LHR]
Heathrow Intl.
[LCY]
City Airport
[LGW]
Gatwick Intl.
Crawley
[LYX]
Lydd
Biggleswade
Royston
Haverhill
Milden
Shefford
Henlow
Ampthill
Saffron Walden
Middleton
Sudbury
Hadleigh
Woburn Sands
Toddington
Letchworth
Hitchin
Halstead
Mount Bures
Stevenage
Dunstable
Bishop's Stortford
Braintree
Colchester
Silver End
Sawbridgeworth
Harpenden
Hertford
Tiptree
Witham
Welwyn
Hoddesdon
Harlow
West Mersea
Hemel Hempstead
St. Albans
Hatfield
Chelmsford
Chipping Ongar
Danbury
Maldon
Chesham
Radlett
Potters Bar
Cheshunt
Waltham Abbey
Epping
Mundon
Amersham
Borehamwood
Heybridge
Hullbridge
Watford
Barnett
Enfield
Loughton
Burnham on Crouch
Gerrards Cross
Edgeware
Palmers Green
Chigwell
Brentwood
Billericay
Wickford
Beaconsfield
Harrow
Hendon
Romford
Rayleigh
Hazlemere
Uxbridge
Wembley
Ilford
Basildon
South Benfleet
Burnham
Slough
Ealing
East Ham
Stanford le Hope
Southend on Sea
Canvey Island
Windsor
Hounslow
Hammersmith
Greenwich
Grays
Kingsnorth
Sheerness
Egham
Sunbury
Wandsworth
Catford
Dartford
Gravesend
Minster
Warden
Ascot
Staines
Kingston
Meopham
Rochester
Gillingham
Woodham
Esher
Bromley
Chatham
Epsom
Orpington
Snodland
Sittingbourne
Woking
Cobham
Croydon
Sandling
Knaphill
Mayford
Walton
Woldingham
Ditton
Faversham
Leatherhead
Merstham
Seven Oaks
Oxted
Maidstone
Guildford
Dorking
Reigate
Tonbridge
Boughton Monchelsea
Edenbridge
Paddock Wood
Godalming
Smallfield
Claygate
Kennington
Whitley
Cranleigh
Bidborough
Pembury
Ashford
Hydestile
East Grinstead
Tunbridge Wells
Grayswood
Horsham
Sharpthorne
Tenterden
Crowborough
Mannings Heath
St. Marys Bay
Southwater
Haywards Heath
High Hurstwood
New Romney
Burgess Hill
Heathfield
Lydd
Rye
Camber
Uckfield
Coldwaltham
Storrington
Henfield
Hassocks
Battle
Winchelsea
Windmill Hill
Steyning
Falmer
Lewes
Hailsham
Bexhill
Hastings
Walberton
Angmering
Worthing
Brighton
Polegate
Hankham
Littlehampton
Saltdean
Newhaven
Willingdon
Bognor Regis
Seaford
Eastbourne
0
10km

Super
Ellen Gallagher
2003, 4'51" loop, 16 mm.

06 Sm1

这是美国艺术家埃伦·加拉格尔（Ellen Gallagher）与德国艺术家埃德加·克莱恩（Edgar Cleijne）展览作品的展册。书籍为锁线胶装，分为五册，展示了混合材料绘画和视频作品，通过暗含在页面里的磁铁吸附成套。绘画作品的封面黑卡纸丝印白色，裱覆卡板。内页胶版纸四色印刷，每幅作品都需要翻开才能观看。视频作品的四本封面同样是黑卡纸丝印白色，无裱板，内页胶版纸四色印刷，为视频作品的静帧截图，封面上有视频长度和拍摄信息。||||||||||||||||

Blizzar
d of
White

This book is a collection of works by the American artist Ellen Gallagher and the German artist Edgar Cleijne. The collection, which is bound with thread sewing, is divided into five volumes showing mixed material paintings and video works. The volumes are attached together by magnets hidden in the pages. The cover of the painting volume is printed on black cardboard in white, covered with graphic board; inside pages are printed on offset paper in four colors. Black cardboards are also used as the cover of the four volumes of video works, without graphic board attached. Still frames of video works are printed on the inside pages in four colors while video length and shooting information are recorded on the cover.

银奖 — Silver Medal

埃伦·加拉格尔——暴雪的白色 — Ellen Gallagher. Blizzard of White

荷兰 — The Netherlands

Irma Boom (Irma Boom Office)

The Fruitmarket Gallery, Edinburgh; Hauser & Wirth, Zürich / London

120×155×88mm

1350g

992p

This finely designed book-object is an example of how through binding tradition and innovation can be brought together: the five brochures are held together by magnetic edges, optically underscored through the design of the spine. These sensitively designed volumes document Ellen Gallagher's artistic videos in the best possible way. The technical finish is also perfect, right down to the fold-out plates in the »orbus« volume.

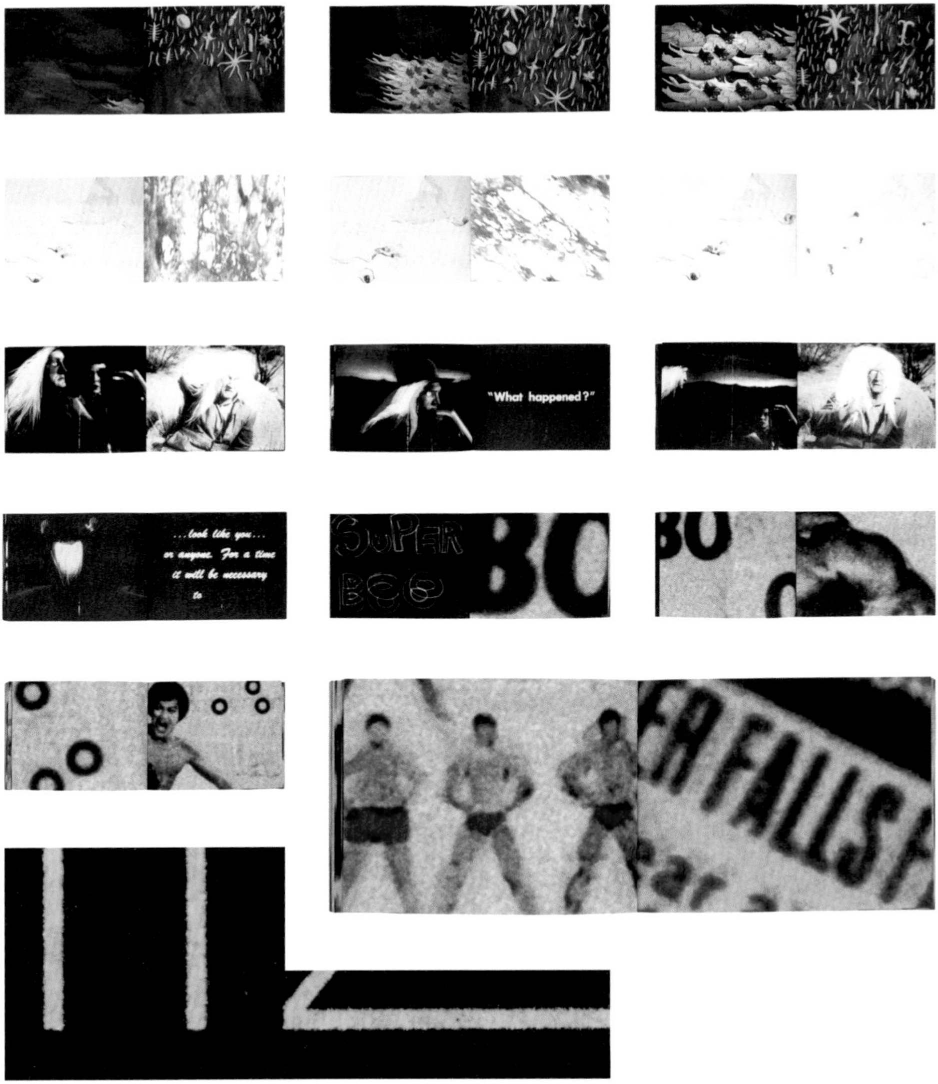
"What happened?"
...look like you...
or anyone. For a time
it will be necessary
to
SUPER
BO

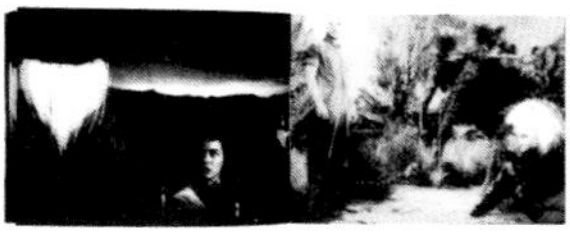

"Don't be afraid. It is within our power

GROW

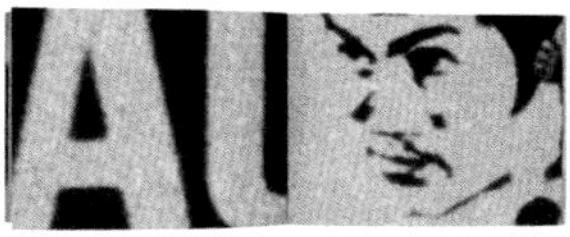
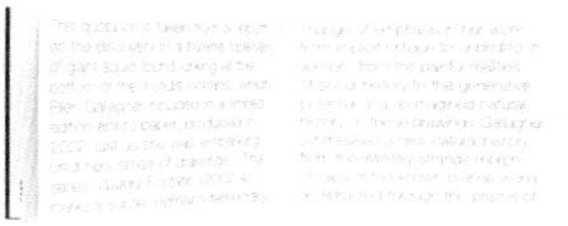

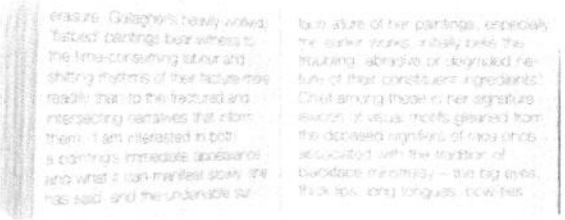
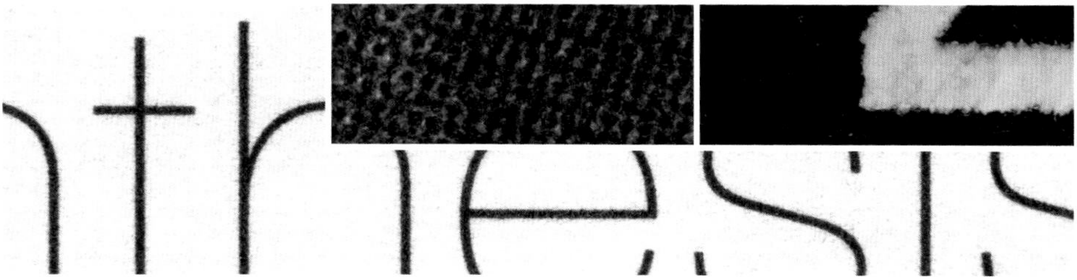

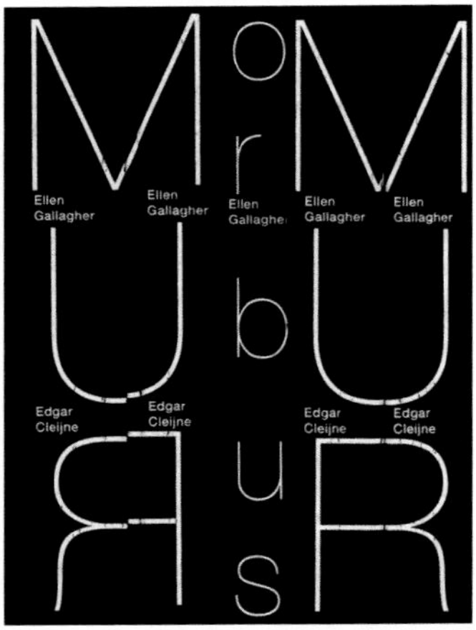
Ellen Gallagher
Ellen Gallagher
Ellen Gallagher
Ellen Gallagher
Ellen Gallagher
Edgar Cleijne
Edgar Cleijne
Edgar Cleijne
Edgar Cleijne
orbus

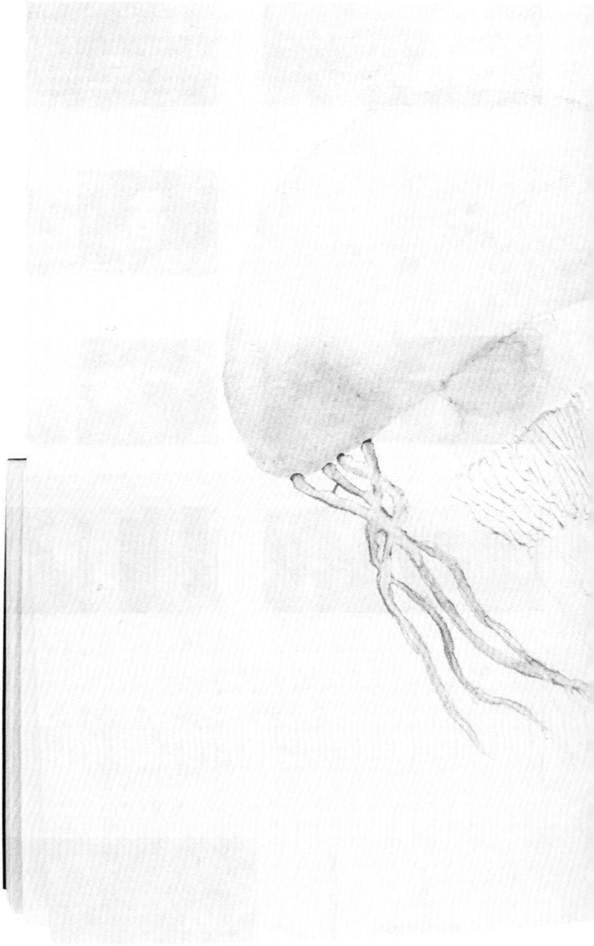

9
Deutsches Schuh-
museum (siehe
Anmerkung 2),
Kapitel 6.40.
Dazu passt,
dass in Manager-
Ratgebern der
neunziger Jahre
empfohlen wird,
die Fertigung zu
dezentralisieren
und auf »maßge-
schneiderte
Massenfertigung«
umzusteigen,
vgl. B. Joseph Pine:
Maßgeschneiderte
Massenfertigung.
Neue Dimensionen
im Wettbewerb.
Wien: Wirtschafts-
verlag Ueberreuter
1994.

这是一本有关传统手工艺发展现状的专著。通过对工艺、磨具、管理、旅游、工业化生产、知识、经营、培训、协会等领域的现状研究，试图找到欧洲各地传统手工业的发展之路。本书为硬精装，内封黑色纸张上印刷银色和红色，呈现丰富的标题层次，外包红色护封，灰色照片农田里风力发电机的倒影点明了本书主旨。内页为四色印刷，除少量图片保留真实色彩外，其余均为单色黑白照片。文本部分为黑色、红色和中灰色三色呈现，文本层次明晰，体现出质朴的文本感。

Landschaft des Wissens
Strategien des Handwerks
Sieben Porträts
außergewöhnlicher Projekte
in Europa
Haupt

This book is a monograph on the development of traditional handicraft. By studying the current situation of craft, abrasive tools, management, tourism, industrialized production, knowledge, operation, training, association and other fields, the author tries to find the development path of traditional handicraft industry in Europe. Printed in silver and red on hardcover black paper, this design presents a rich hierarchy of titles. The book is covered by red jacket, and uses gray photo of wind turbines in farmland to highlight the substance of the book. The inside pages are printed in four colors. Except a few pictures that retain the true colors, the rest are monochrome black-and-white photos. The text is presented in three colors of black, red and medium gray in order to show conciseness and simplicity.

银奖

手工艺战略

——欧洲非凡项目的 7 幅肖像

奥地利

Silver Medal

Strategien des Handwerks.

Sieben Porträts außergewöhnlicher Projekte in Europa

Austria

Clemens Theobert Schedler (Büro für konkrete Gestaltung)

Landschaft des Wissens (Ed.)

Haupt Verlag AG, Bern / Stuttgart / Wien

245 × 180 × 39mm

1054g

367p

Superb and very subtle typography. The page design is harmonious and very easy for the reader: the main body of text, annotations and secondary texts are clearly set out, the use of a second colour most delicate, and the still b/w photos create a warm atmosphere. The functionality of every element used – down to the choice of paper – make this non-fictional book into something really special.

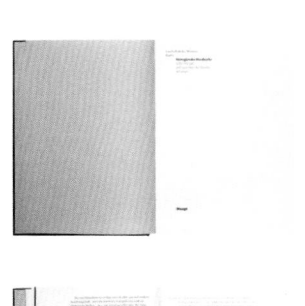

Landschaft des Wissens
Band 1
Strategien des Handwerks
Sieben Porträts
außergewöhnlicher Projekte
in Europa
Haupt

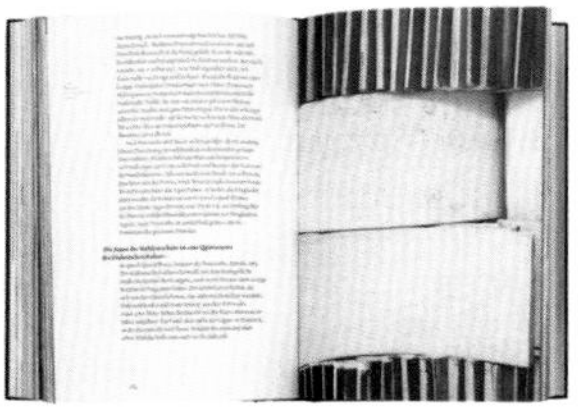

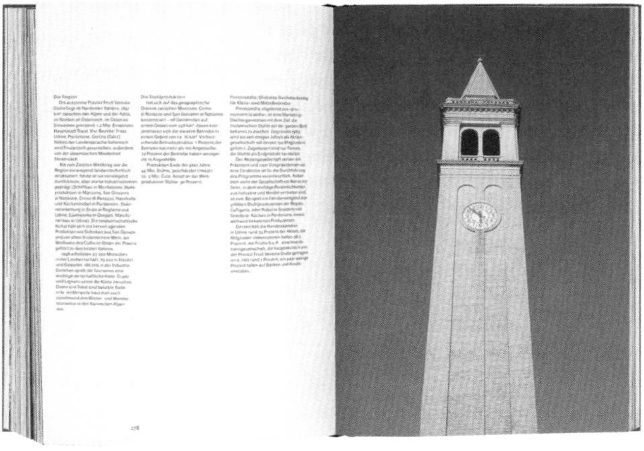
Die Region
Die Stuhlproduktion
278

erg

Das Handwerk der Zukunft
Kooperationsmodelle
für eine Ökonomie der Nähe

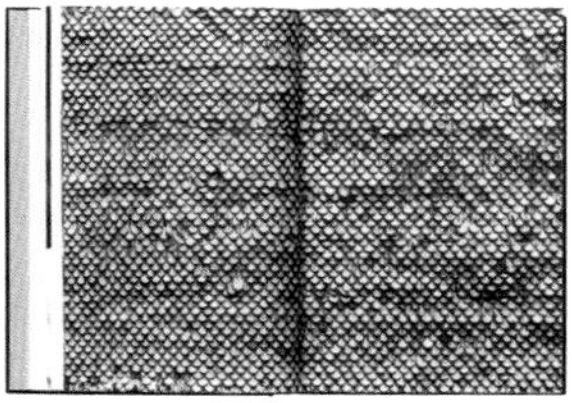

Autorinnen und Autoren
Selbstporträts

kor
von Ihnen ar

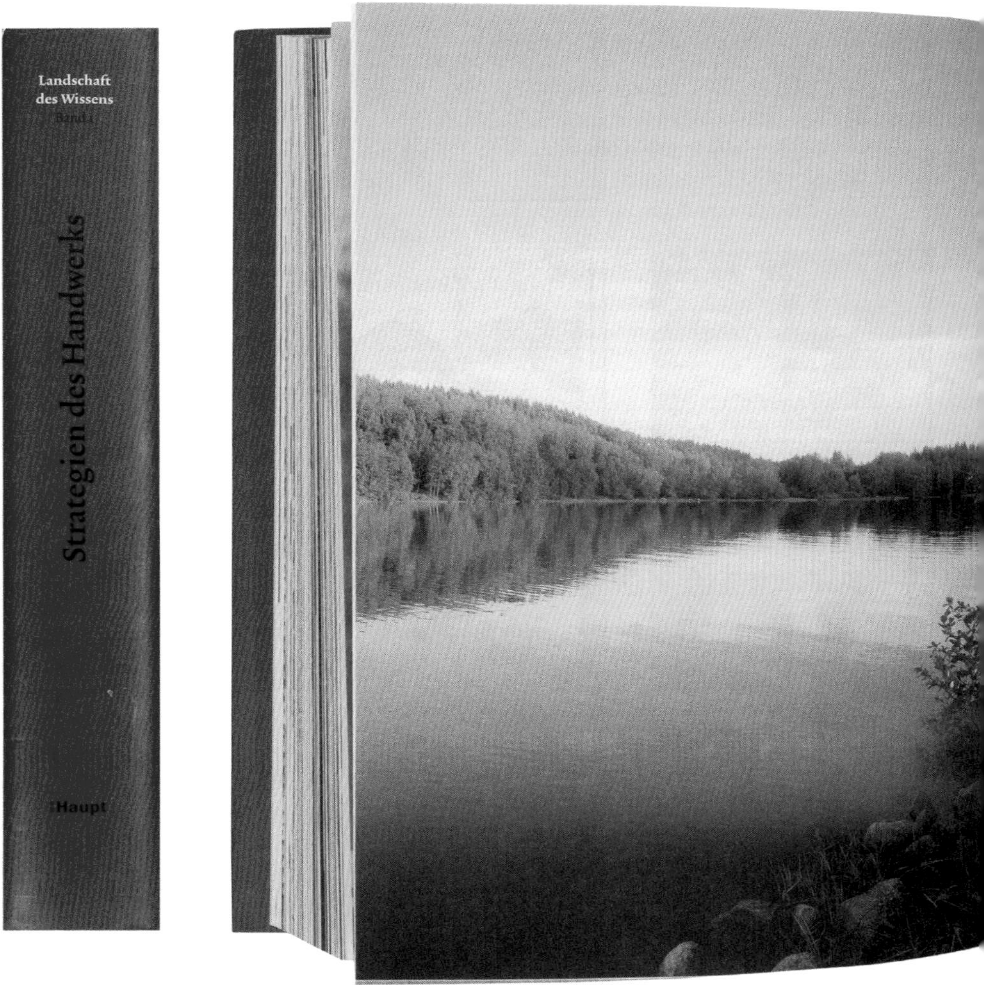
Landschaft
des Wissens
Strategien des Handwerks
Haupt

Es ist viel wärmer als wir es uns erwartet hätten. Eine Wärme, die ich mir zu Hause am Bodensee mitunter wünschen würde. Ein angenehmes Klima, wie es so auch südlich der Alpen, wo Christian, der Fotograf, lebt, in dieser Jahreszeit herrschen könnte. Als hätte sich die Natur, angeregt von der Europäischen Gemeinschaft und der einheitlichen Währung, auf eine europäische Einheitstemperatur für die Mitte des Juni geeinigt. Unter diesen Bedingungen drängt es uns, unter freiem Himmel zu leben. »Ein Himmel, der älter ist als die ganze restliche Welt.«* Wir gehen in einen Supermarkt, um uns mit Lebensmitteln und Werkzeugen für ein Picknick einzudecken. Zwischen den Regalen ein Gefühl von Überall. Tröstlich einerseits, weil die gewohnte Anordnung, auch die Produktauswahl die Annäherung erleichtert, und doch bedauerlich, weil wir plötzlich fürchten müssen, jede Sekunde sterbe irgendwo ein charakteristischer Wunsch, eine typische Geste, ein kleines, nur unter bestimmten geographischen Bedingungen nutzbares Talent für immer aus. Beim Obstregal stellen wir uns die Frage: Wohin retten sich regionale Handgriffe, Blicke, Gewohnheiten im Europa der Zukunft? Und äußern bei den Getränken schon wieder Zweifel: Brauchen wir diese Rustikalien überhaupt, haben wir wirklich ein Recht auf die Arterhaltung unseres Eigensinns?

* So schreibt Arto Paasilinna in *Der Sohn des Donnergottes*.

Der Ort

Petäjävesi, das Ziel unserer Reise, hat etwa 4000 Einwohner und liegt 35 Kilometer westlich von Jyväskylä, der Hauptstadt der Provinz Zentralfinnland. Der Ort sieht etwa so aus, wie die anderen Siedlungen, an denen wir im Laufe unserer Anreise vorbeigekommen sind. Eine Ansammlung von kleinen, großteils hinter Bäumen versteckten Giebelhäusern, meist aus Holz errichtet, oft ochsenblutfarben gestrichen, aber ohne erkennbare Tendenz zur Verdichtung angeordnet. Später erfahren wir, dass es hier wie überall der Supermarkt ist, auf den ganz selbstverständlich alles zuläuft.

Für uns liegt das Zentrum von Beginn an am Rand. Eine Esso-Tankstelle an der Hauptstraße. Dort sind wir verabredet, dort werden wir in den kommenden Tagen unser Frühstück zu uns nehmen, abends Bier holen und uns auch zwischendurch mit kaurismäkischen Finnland-Klischees versorgen.

这是一本由芬兰艺术家马蒂·哈格尔伯格（Matti Hagelberg）创作的黑色漫画集，书名“凯科宁”也正是本书漫画主人公的名字。本书为方形 20 开硬精装，封面并无特殊的印制工艺，但风格强烈且充满封面的漫画主人公头像和独特的绿色背景具有相当的视觉冲击力。内页胶版纸单黑印刷，除少数画面有灰层次表现，其余绝大多数为强烈的黑白双色漫画。漫画本身风格鲜明，视觉冲击强烈，画面分镜并非简单的四格排列，而是根据内容需要进行划分；内容涉及主人公日常生活中各种荒诞的想象，充满幽默与讽刺意味。

This is a dark humor comic book by the Finnish artist Matti Hagelberg. The title Kekkonen is named after the hero of this book. It is a square 20K hardcover book. The cover is not processed with any special printing technique, but the distinctive style, the cartoon hero's head that occupies the whole cover, and unique green background has a considerable visual impact. The inside pages are printed in black on offset paper, except for a few pictures with gray layers, the vast majority of the rest are strong black and white double-colored comics. The cartoon itself has distinct style and strong visual impact. The picture is not simply arranged in four grids, but divided according to the needs of the plot, which involves all kinds of absurd imagination of the protagonist's daily life, and is full of humor and irony.

铜奖
凯科宁
芬兰

Kekkonen – the book's title is also the name of its sad hero. The structure of this comic is both compact and simple. It is full of absurd stories and crazy details. The b/w wood engravings gel with the pseudo-techno typography. A very attractive picture book for adults which mixes »automated everyday boredom« with an expressive graphic idiom.

Bronze Medal
Kekkonen
Finland
Matti Hagelberg
Matti Hagelberg
Kustannusosakeyhtiö Otava, Helsinki
219×217×21mm
750g
200p

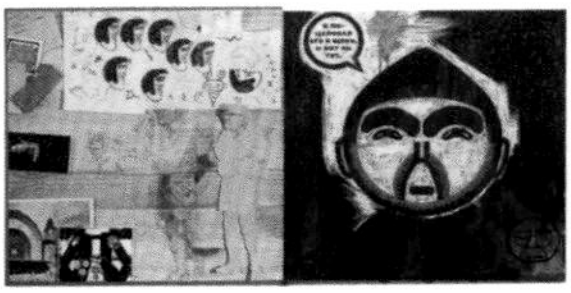

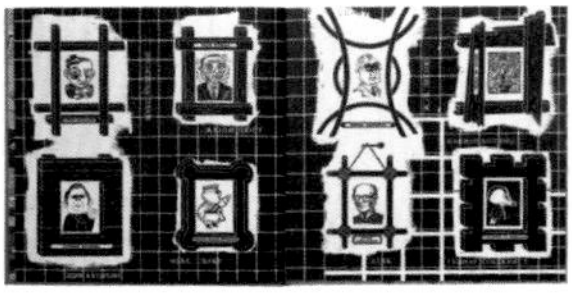

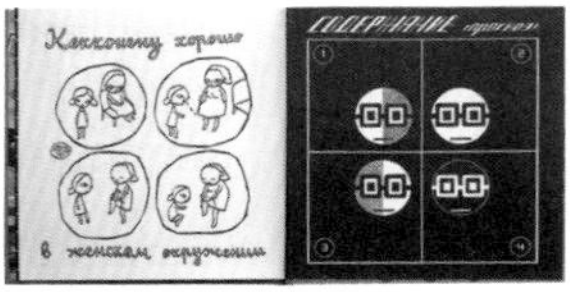

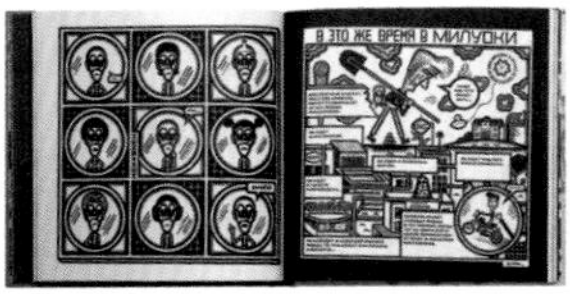
В ЭТО ЖЕ ВРЕМЯ В МИЛУОКИ

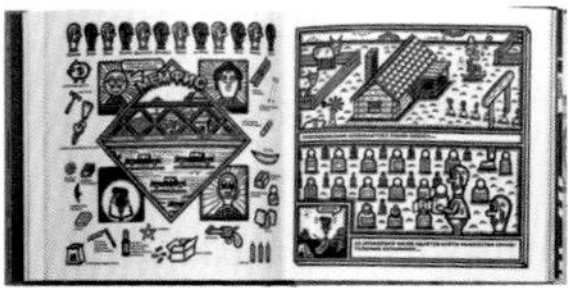

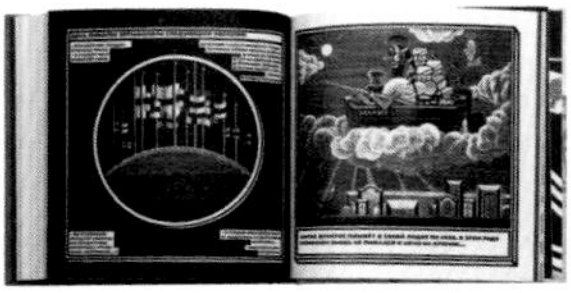

РОКОВЫЕ
ТРАМВАИ

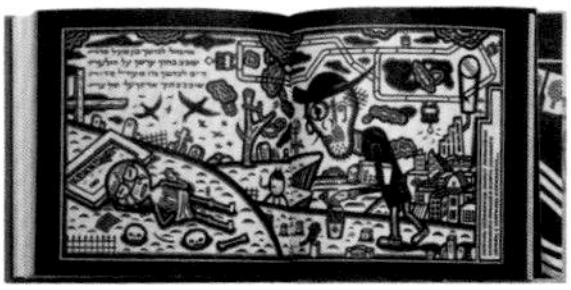

СЫН ЧЕЛОВЕЧЕСКИЙ ИЩЕТ ЗОЛОТО, А НАХОДИТ ЛИШЬ ЛАКРИЦУ.
СПОРТ
ЛОТО

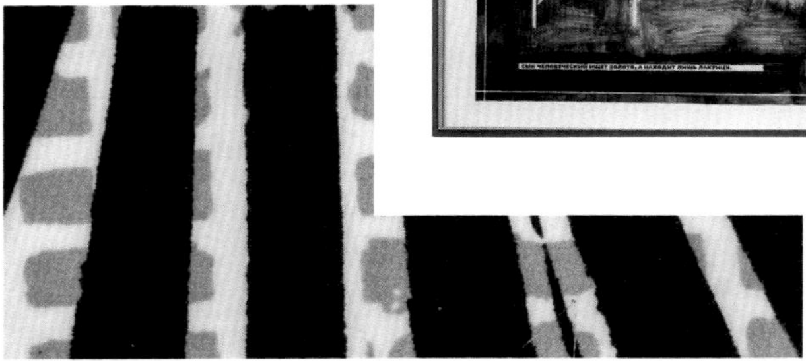

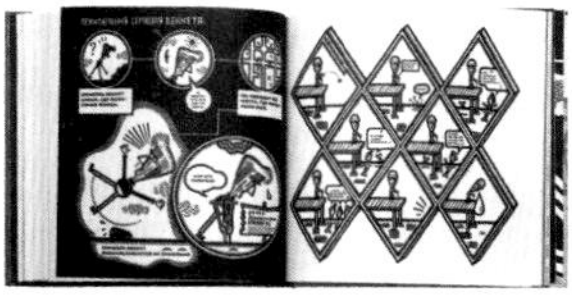
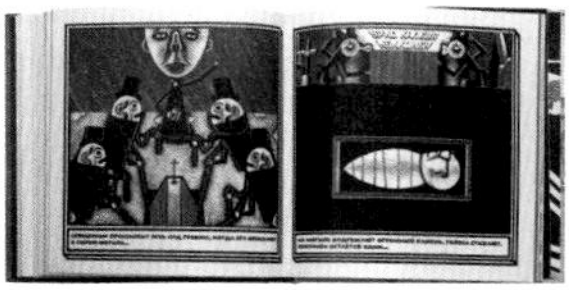
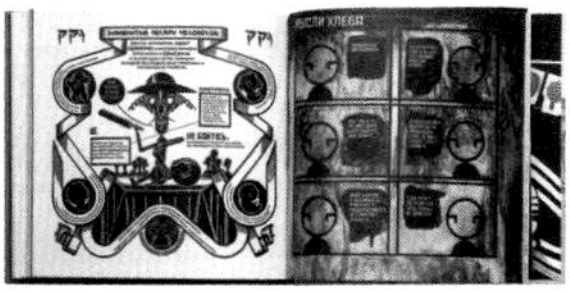

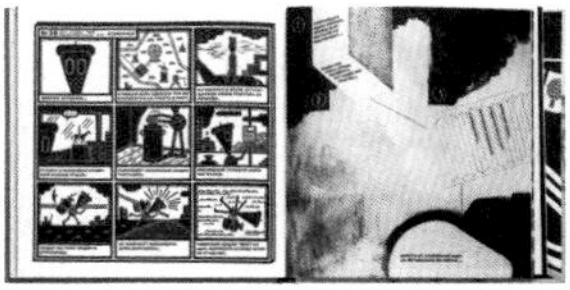
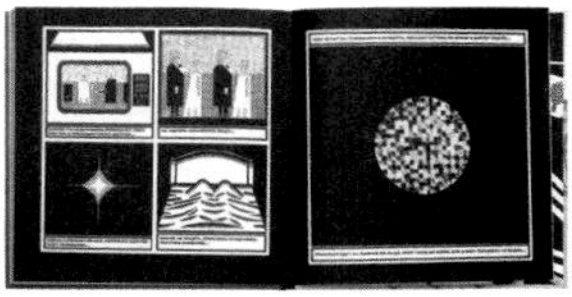
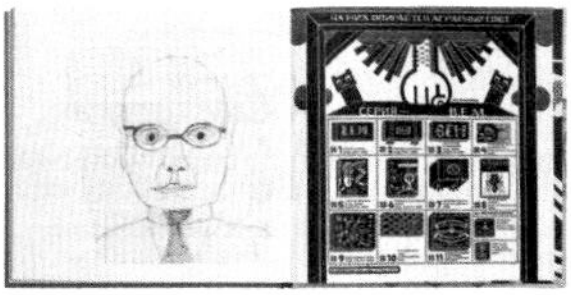

КЕККОНЕН

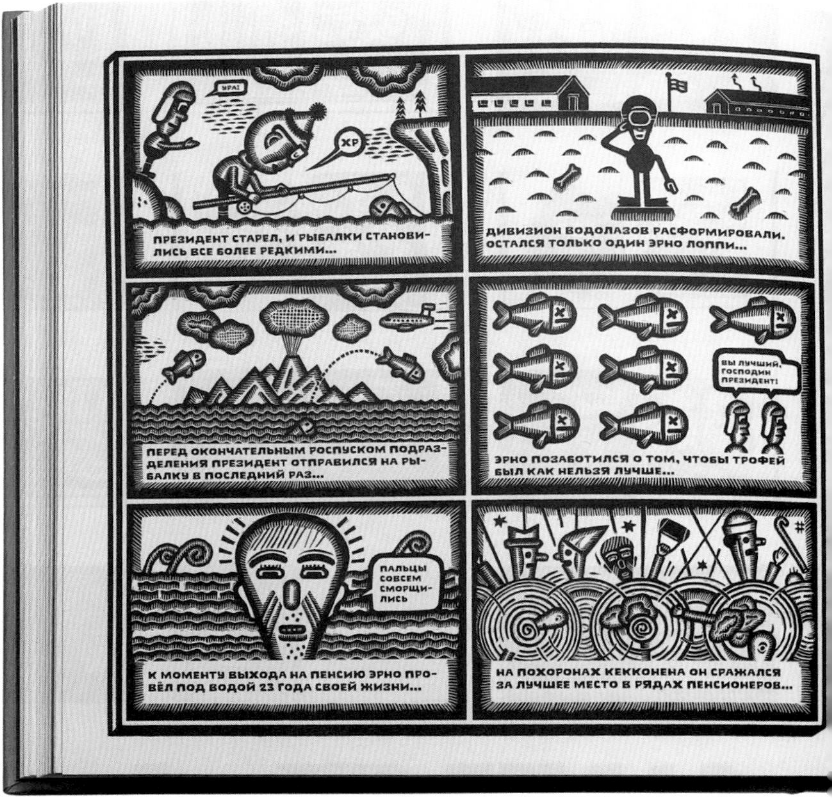
УРА!
ХР
ПРЕЗИДЕНТ СТАРЕЛ, И РЫБАЛКИ СТАНОВИ-ЛИСЬ ВСЕ БОЛЕЕ РЕДКИМИ...
ДИВИЗИОН ВОДОЛАЗОВ РАСФОРМИРОВАЛИ. ОСТАЛСЯ ТОЛЬКО ОДИН ЭРНО ЛОППИ...
ПЕРЕД ОКОНЧАТЕЛЬНЫМ РОСПУСКОМ ПОДРАЗ-ДЕЛЕНИЯ ПРЕЗИДЕНТ ОТПРАВИЛСЯ НА РЫ-БАЛКУ В ПОСЛЕДНИЙ РАЗ...
ВЫ ЛУЧШИЙ, ГОСПОДИН ПРЕЗИДЕНТ!
ЭРНО ПОЗАБОТИЛСЯ О ТОМ, ЧТОБЫ ТРОФЕЙ БЫЛ КАК НЕЛЬЗЯ ЛУЧШЕ...
ПАЛЬЦЫ СОВСЕМ СМОРЩИ-ЛИСЬ
К МОМЕНТУ ВЫХОДА НА ПЕНСИЮ ЭРНО ПРО-ВЁЛ ПОД ВОДОЙ 23 ГОДА СВОЕЙ ЖИЗНИ...
НА ПОХОРОНАХ КЕККОНЕНА ОН СРАЖАЛСЯ ЗА ЛУЧШЕЕ МЕСТО В РЯДАХ ПЕНСИОНЕРОВ...

КЕККОНЕН У ПАМЯТНИКА СЕМИ ГНОМАМ.
ВСЕХ ЖИВЫХ! ПРИХОДИТЕ И ПОПРОБУЙТЕ!
ПО РАДИО ПРЕРВАЛА УТРЕННИЙ МО-
ЭРНО ЛОППИ...
ГРР.
ЗООМАГАЗИН КОТ ШРЁДИНГЕРА
ОН НАДЕВАЕТ ВОДОЛАЗНЫЙ КОСТЮМ И СА-
ДИТСЯ В ТРАМВАЙ № 2...
ПРАВДА
ОН ЛЕЖИТ БЕЗ СОЗНАНИЯ У ЖЕЛЕЗНОДОРОЖНОГО ВОКЗАЛА...
НА МЕСТО ТРАГЕДИИ
НЕ ТРАТЯТ ВРЕМЯ НА
БЕЗ СОЗНАНИЯ...
ОНИ СПЕШАТ К ПАМЯТНИКУ СЕ-
МИ ГНОМАМ, ЧТОБЫ ВЗЯТЬ ИН-
ТЕРВЬЮ У КЕККОНЕНА И ПЕКАРЯ
ЧЕЛОВЕКОВ...

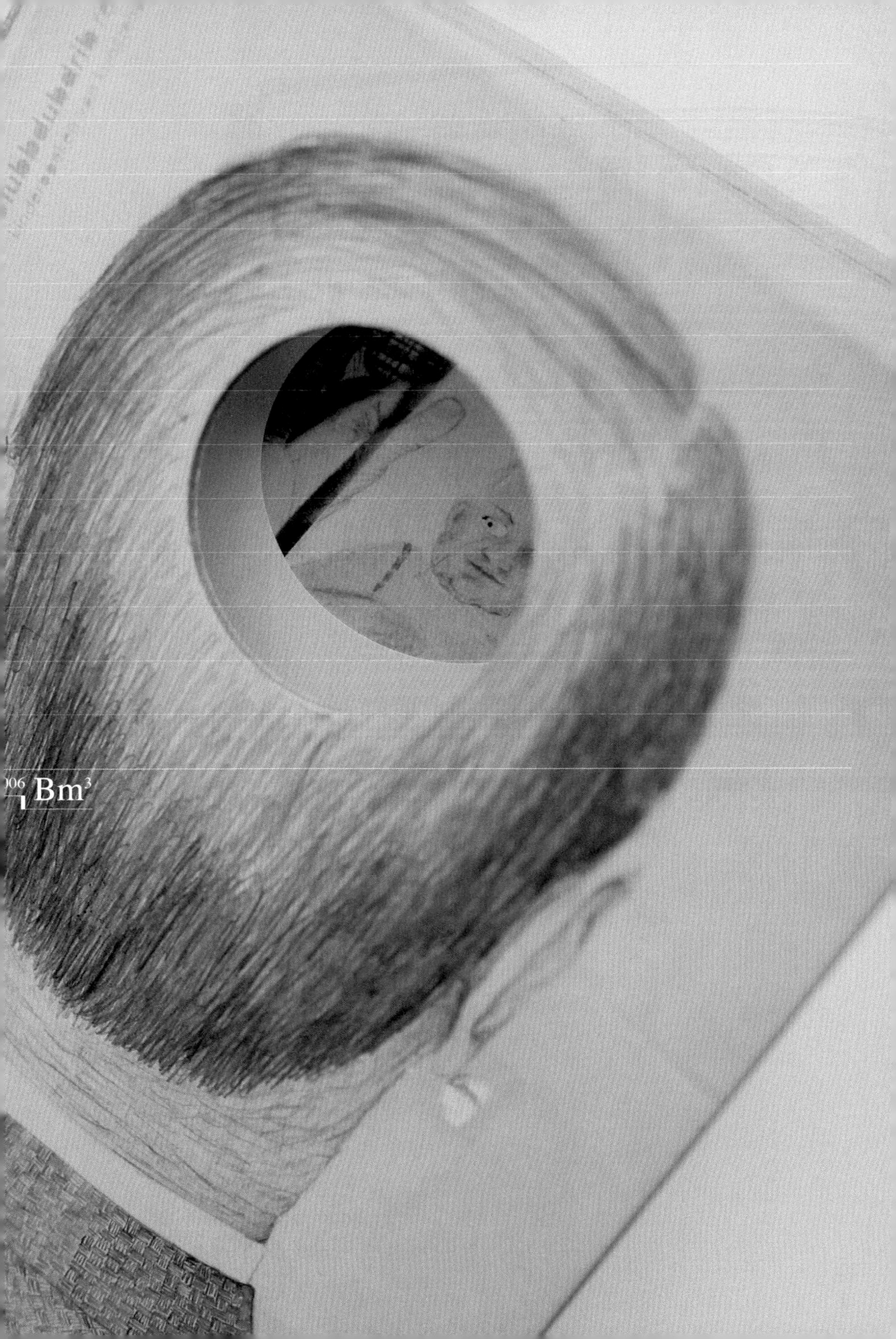
Bm³

本书为荷兰艺术家保罗·德雷乌斯（Paul de Reus）绘制的一本绘本，记录了他与祖父的一些遗传特征、祖父的遗物，并由此了解和叙述祖父的生平故事。这是一本典型的大16开硬精装欧式绘本，封面未覆膜，保留与内页一致的亲和的纸张手感。本书作者所记录的祖父的故事是通过一系列事物上大大小小的“圆洞”展开的，所以从封面开始到里面每一个内页都至少有一个洞，洞贯穿了故事，也成为观察祖父、了解祖父的特殊视角。很多圆洞相互透叠，形成了连续页之间丰富的联系。全书使用黑白及彩色铅笔绘制而成，文字部分也是手写，具有很好的亲和力。很多画面是主观视角，或者很强烈的透视视角，引人思考。

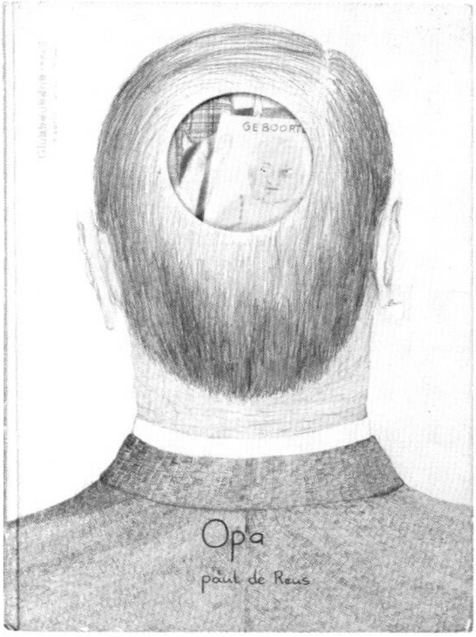
Opa
paul de Reus

This book is a picture book drawn by the Dutch artist Paul de Reus, which records some of his grandfather's genetic characteristics, his grandfather's relics, and thus learns and narrates the life story of his grandfather. This is a typical hardcover European picture book, the cover is not laminated in order to keep consistent with the inside paper. The story of the author's grandfather is unfolded through a series of "circular holes", big or small, on different things. Therefore, from the cover to each inside page, there is at least one hole, which runs through the story and becomes a special perspective to observe and understand the grandfather. Many circular holes overlap each other to form rich connections between successive pages. The whole book is drawn in black and colored pencils, and the text is also handwritten, which has a good affinity. Many of the images are strongly subjective or perspective, and thought provoking as well.

- 铜奖
- 祖父
- 荷兰

Opa by Paul de Reus is in more ways than one a delightful picture book, brimming with visual surprises. The punched holes provide a very particular kind of insight into the book, challenging the reader's expectations in a humorous fashion. The result is a superbly drawn book on the subject of perception with all its possible irritations, accompanied by an unpretentiously written text.

- Bronze Medal
- Opa
- The Netherlands
- Irma Boom (Irma Boom Office)
- Paul de Reus
- Uitgeverij Van Waveren, Amsterdam
- 291 × 209 × 9mm
- 390g
- 44p

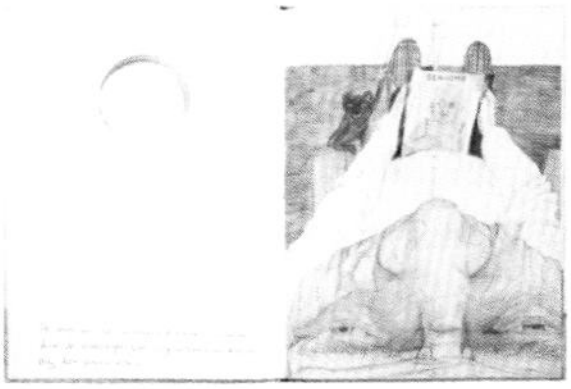

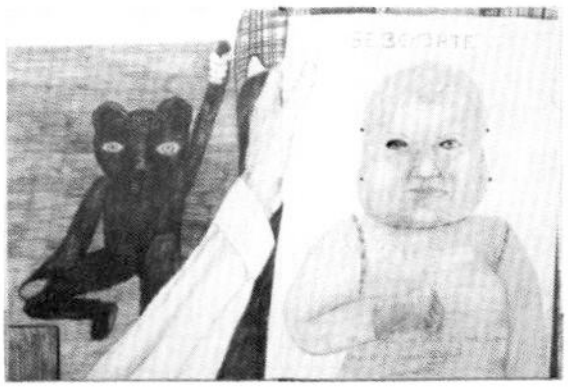

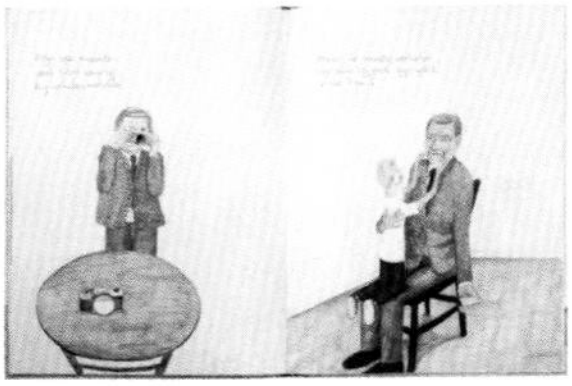

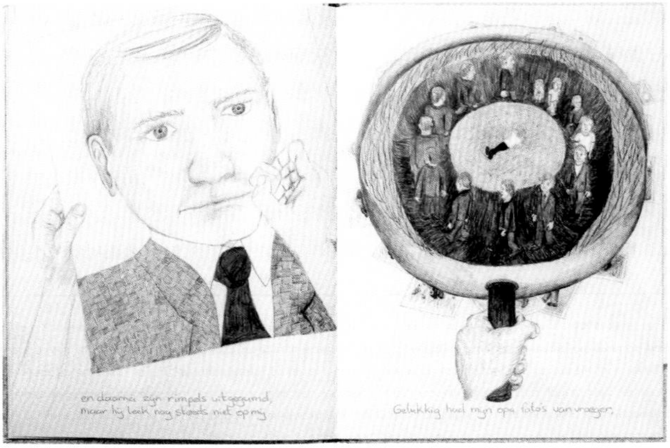
en daarna zijn rimpels uitgegumd,
maar hij leek nog steeds niet op mij
Gelukkig had mijn opa foto's van vroeger,

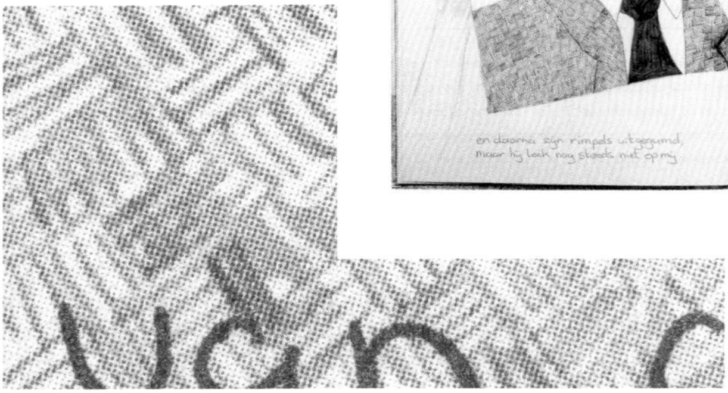

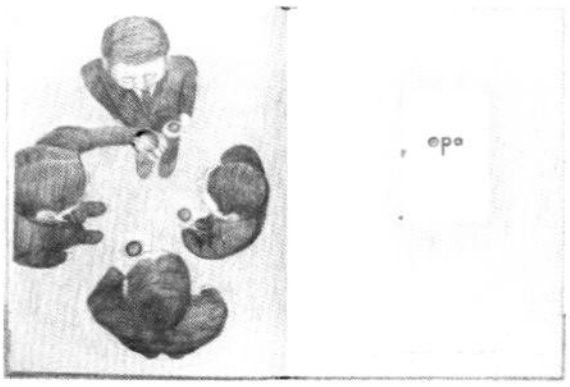
opa

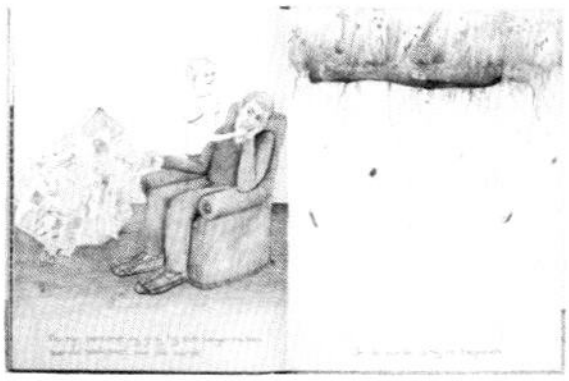

Opa Paul de Reus
Hij zou nog op vakantie, een reis naar de zon
die je in gedachten kan maken.

Nu zie ik mijn opa overal want veel
oude mensen lijken op hem,

The Library through History

这是一本介绍荷兰乌得勒支大学图书馆的书。这座图书馆由荷兰建筑师维尔·阿雷兹（Wiel Arets）设计，书中除了建筑类图书介绍建筑外观、内部流线、结构、亮点等常规内容，还加入了馆藏艺术项目、图书馆的使用状况、访谈、古今图书馆对比等生动的内容，在硬朗的建筑内暗藏柔软、丰富、值得思考的各方面文化内涵。书籍的装订形式为锁线胶装，封面白卡纸正反面均为黑红双色印刷，封底加印银，正面覆哑膜，过满版 UV，采用建筑玻璃上的植物图案挂粗网处理。勒口和封面的内侧均为此植物图案，把封面隐喻为进入建筑的一扇门窗。内页光面铜版纸四色印刷为主，局部夹插短于开本 3mm 的灰色胶版纸印单黑。铜版纸部分开始以游客视角游览建筑，之后通过建筑师间的对谈、建筑结构和细节图纸详细描述建筑。胶版纸部分则多为对现代图书馆的思考，以及相关作家的文章。

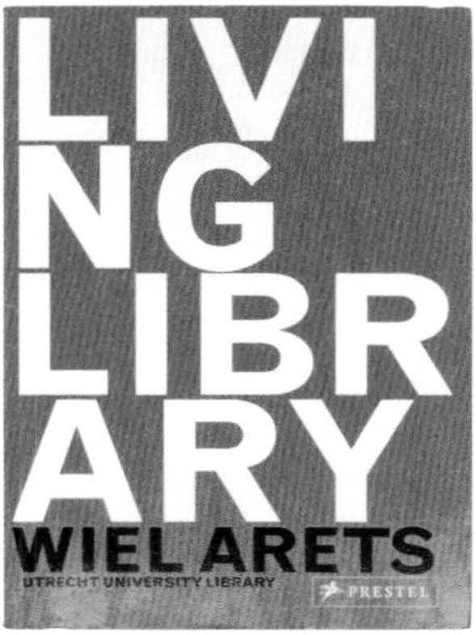
LIVING LIBRARY
WIEL ARETS
UTRECHT UNIVERSITY LIBRARY
PRESTEL

This is a book about the library of Utrecht University in the Netherlands, which was designed by the Dutch architect Wiel Arets. The book shows not only the exterior, internal path, structure and design highlights, but also the reserved art projects, the using condition of the library, deep interview, comparison of ancient and modern libraries. Inside the hale building, there hides the soft, rich and worth-thinking culture. The binding form of the book is French sewing, the front cover cardboard is printed in black and red on both sides, the back cover is printed with silver. The front cover is matt laminated with full UV. Both the flap and the inner side of the cover are decorated with plant patterns, which is a metaphor that the cover is a door of entering into the building. The inside pages are mainly printed in four colors on coated art paper, inserted by black printed gray offset paper which is 3mm shorter than the folio. The coated paper part begins with visiting the building from the perspective of tourists, and then describes the building in detail through the conversation between architects, architectural structure and detailed drawings. The offset paper part is mostly about the thinking of modern libraries and the articles of relevant authors.

铜奖
维尔·阿雷兹的“活的图书馆”
荷兰

The change of paper harmonises well with the pictures/photos reproduced on it. The printing is of a high quality and the typography – if not always perfectly legible – reflects the concentrated sequence of the pictures. Its well considered basic concept is enhanced by the subtle use of two different paper formats. An exceptional book on the subject of books and archives.

Bronze Medal
Living Library: Wiel Arets
The Netherlands
Irma Boom (Irma Boom Office)
Marijke Beek (Ed.)
Utrecht University Library; Prestel Verlag, München
235×170×22mm
897g
456p

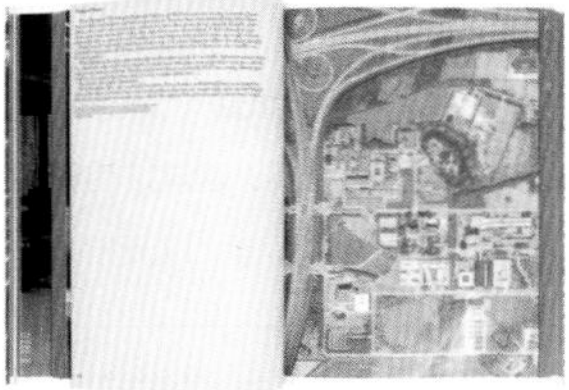

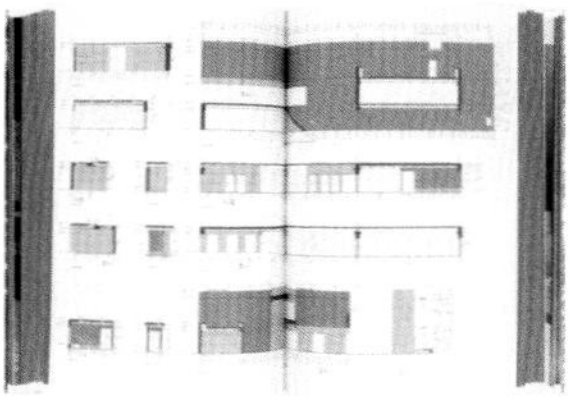

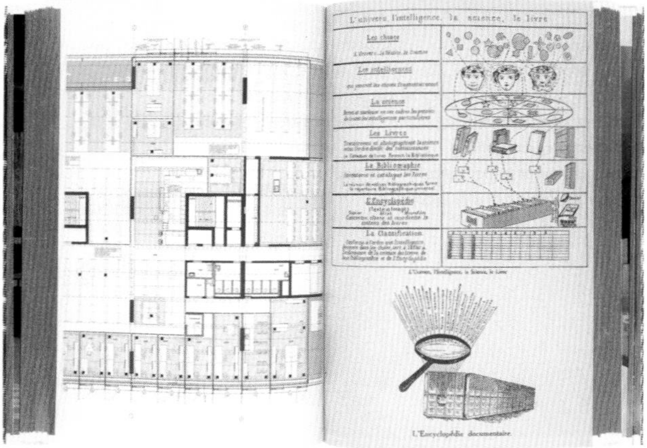
L'univers l'intelligence, la science, le livre
Les choses
Les intelligences
La science
Les Livres
La Bibliographie
L'Encyclopédie
La Classification
L'Encyclopédie documentaire.

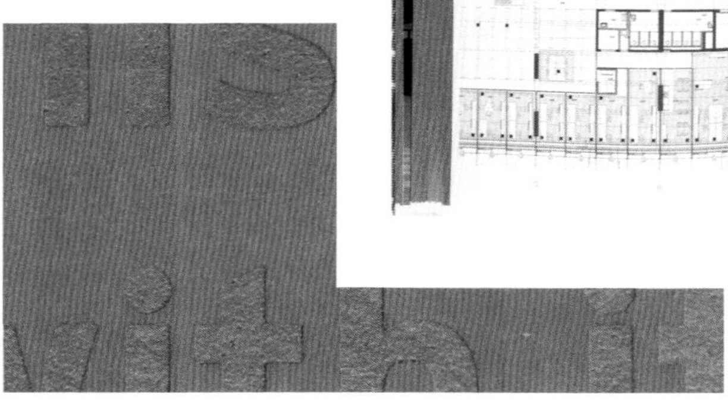

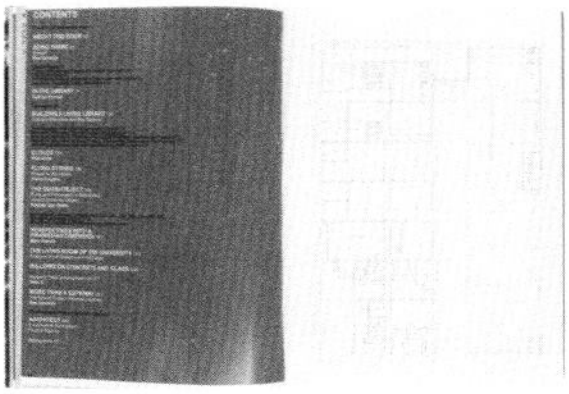

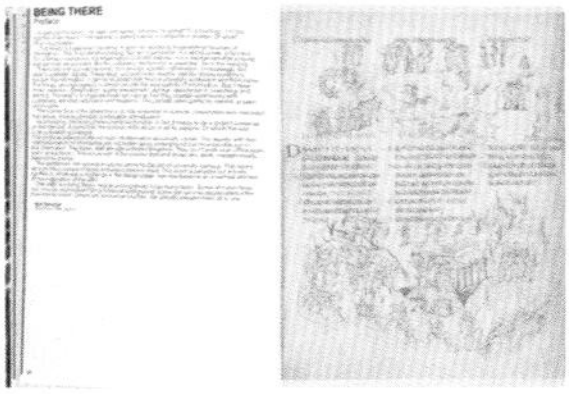

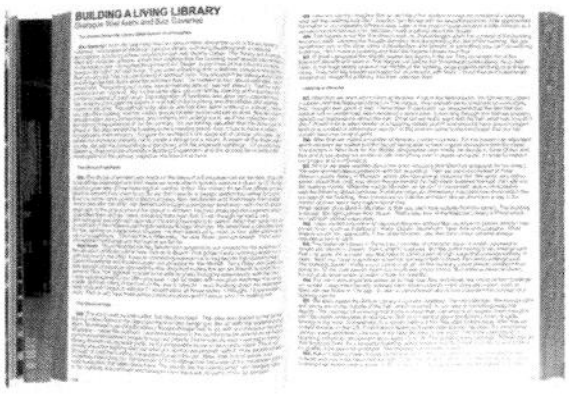
BUILDING A LIVING LIBRARY

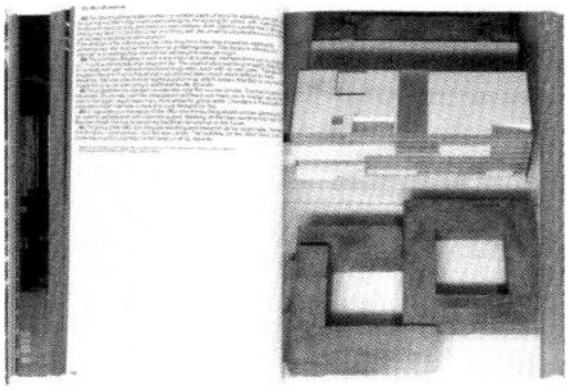

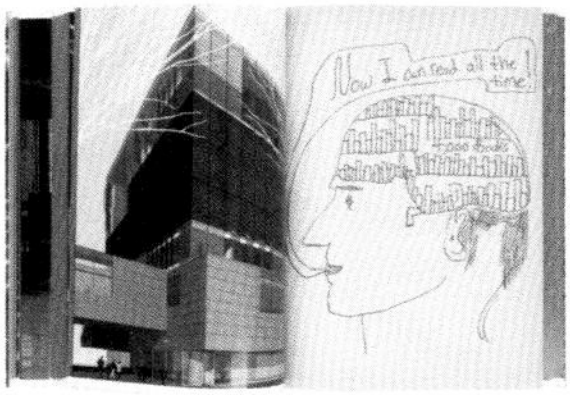

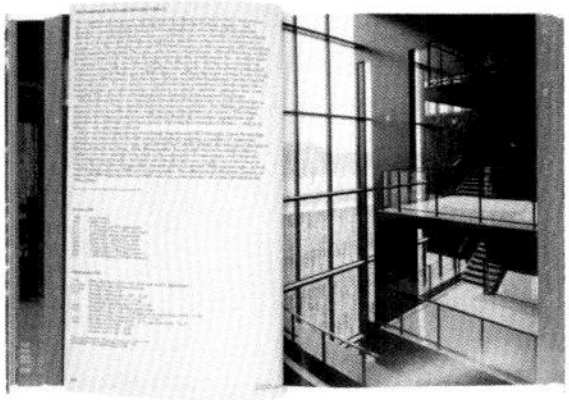

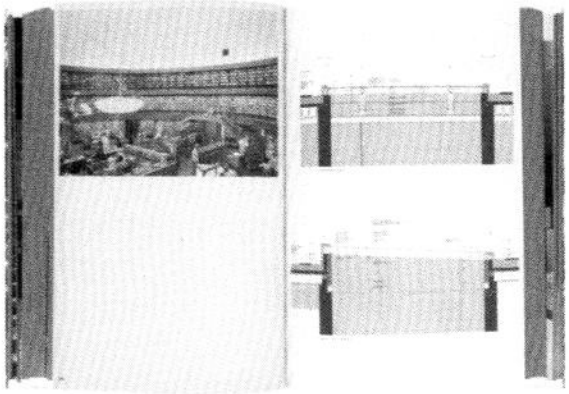

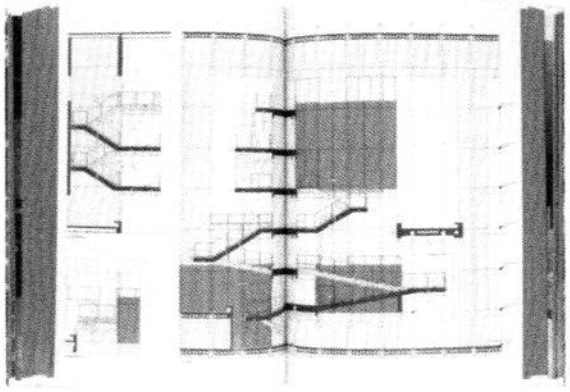

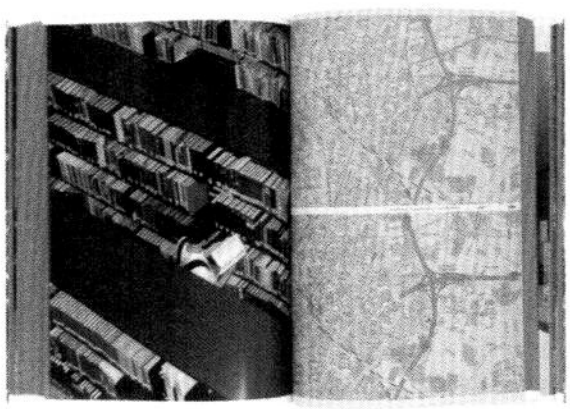

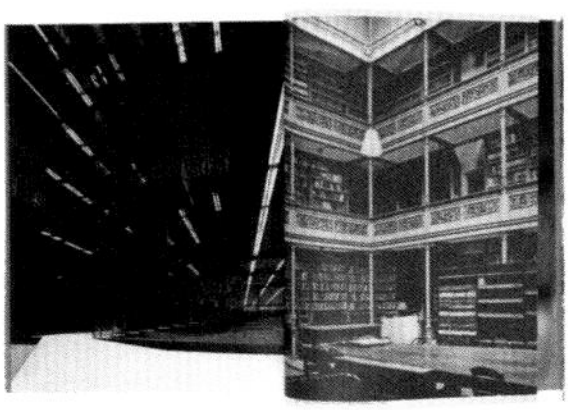

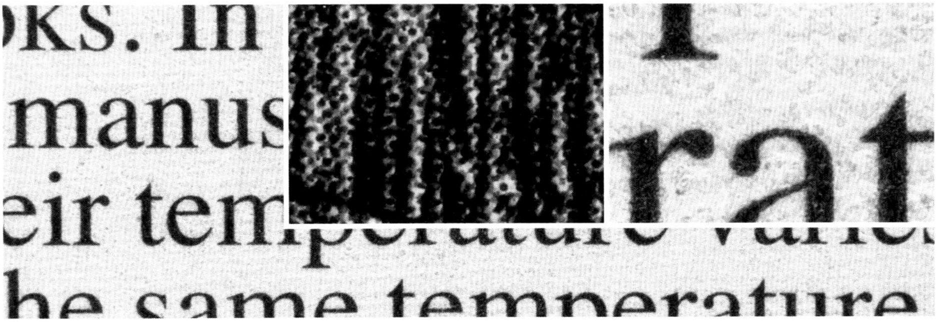

LIVING LIBRARY WIEL ARETS

PRESTEL

According to Louis Kahn in 1965:

'Exeter began with the periphery, where light is. I felt the reading room would be where a person is alone near a window, and I felt that would be a private carrel, a kind of discovered place in the folds of the construction. I made the outer depth of the building like a brick doughnut, independent of the books. I made the inner depth of the building like a concrete doughnut, where the books are stored away from the light. The center area is a result of these contiguous doughnuts; it's just the entrance where books are visible all around you through the big circular openings. So you feel the invitation of the books.'

Louis Kahn about Library and Dining Hall, Phillips Exeter Academy,
Exeter, New Hampshire, us, 1965-1972
Source: David B. Brownlee, David G. De Long, Kahn, Los Angeles 1992

Hans Scharoun on the National Library Berlin:

'Leading the visitor through the entrance hall to the visitor's catalogue, and from there over the platform-like, very flat and articulated stairs to the large reading room, provides spatial experiences which do justice to the high demands made of a national library. The reading room, whose design is impressive, lively and effective in several ways links well to the borrowing counter. The exit through a gallery and a second more intimate staircase creates new spatial impressions.'

'The aim is not the division into cabinets and rooms, nor even the central form of a large reading room crowned with a dome, but the equality and mutual interchange enhancing effect, and landscape-like structure of the reading room. There is sufficient opportunity for diversity and variation in this 150-metre-long, fluid space should the tasks of the library change over the course of time.'

'The "way of the book" starts with the underground delivery; proceeds to the postal department on the ground floor with the sorting of acquisitions, continues through the registration room and the adjoining large cataloguing hall. The newly acquired book is transported from here on a conveyor belt to the in-house bindery and finally for a last check to the store-rooms.'

From Hans Scharoun's text with his entry for the 'Staatsbibliothek' competition, Berlin 30-5-1964
Hans Scharoun, Bauten, Entwürfe, Texte, herausgegeben von Peter Pfankuch, Berlin 1974, pp.339-355

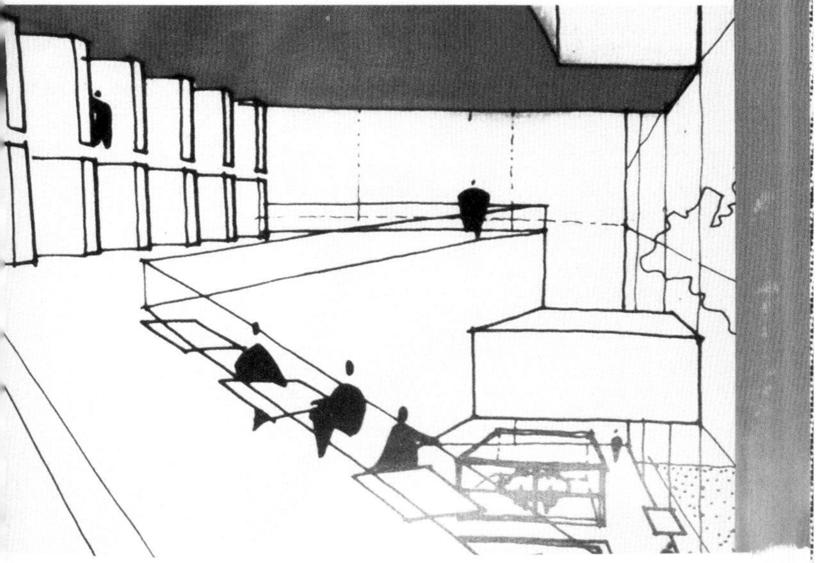

06 Bm5

of living life simultaneously in the real world and
ed one. It would seem all these projects are
pace of our perception. The digital image
ently a representation of a physical reality
can be sped up or slowed down. Nevertheless
inhabiting, a simulation is hard to do (as the
weather map suggest, or blue-screen special effects
With the Internet, however, some might argue this
boundary between physical and digital space is
web is awash with 'virtual worlds' (mostly graphical
boratively created for remote and telematic inhabit
Life23 to the artist-created 'palace/teepee' of the
aesthetics of these digital spaces is a main
amount of time we chose to spend in them.
This may be yet another reason to spend a moment
another) with the work of Studer/van den Berg which
ing and exploiting the oversaturated and networked
live – where all is interconnected and one can be in
once – also reminds us of where the ground is beneath
we really are

这是瑞士新媒体艺术家组合莫妮卡 · 施图德（Monica Studer）与克里斯托夫 · 范登贝尔赫（Christoph van den Berg）的艺术展览“别处都是同一处”（Somewhere else is the same place）的展览画册。艺术家长期以来致力于通过计算机对现实景观的“还原”，虚拟景观与现实的结合、对比和运用，来研究人、现实景观、虚拟景观之间的互动和影响。本书为无线包背胶平装。护封采用展览中搭建的“虚拟景观”的广告布裁切而成，每本的护封均不相同。内封白卡纸无印刷，内页胶版纸四色印刷。内容方面德英双语排版，在两种语言各自呈现的页面中间有相关作品展示的横向拉页，涉及虚拟场景、天气图景的运用、展览现场的搭建、全景视觉的研究，以及在现实场景中搭建虚拟感极强的“实体”模型。这些都基于两位艺术家在当时（2000 年左右）对于相关视觉艺术的研究，现在两人的网站上已在展示 VR 视觉方面的作品。

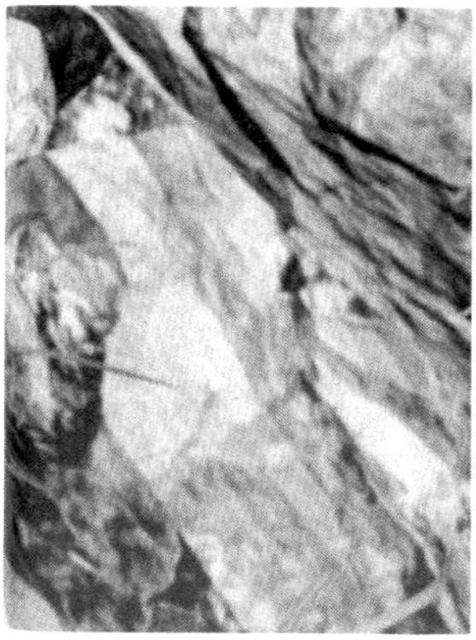

This book is an art album of an art exhibition "Somewhere else is the same place" by the Swiss new media artists Monica Studer and Christoph van den Berg. For a long time, artists have been committed to studying the interaction and influence between people, real landscape and virtual landscape through the "restoration" of real landscape by computer, the combination, comparison and application of virtual landscape and reality. This book is perfect-bound paperback. The jacket is cut from the advertising fabric of "virtual landscape" built in the exhibition, therefore the jacket of each book is different. Inside cover is using the white cardboard without any printing, while the inside pages are using the offset paper printed in four colors. The text is bilingually typeset in German and English. There are horizontal pull-ups of related works on the pages presented in both languages respectively, involving virtual scenes, the application of weather pictures, the construction of exhibition sites, the research of panoramic vision, and the construction of "entity" models with strong sense of virtual in real scenes. These are based on the two artists' research on the related visual art in 2000, and at this time their websites are showing their works in VR vision.

◉ 铜奖

《 别处也是同一处

◖ 瑞士

◎ Bronze Medal

《 Somewhere else is the same place

◖ Switzerland

D Iza Hren, Georg Rutishauser

△ Monica Studer, Christoph van den Berg, Christoph Vögele (Ed.)

|||| Kunstmuseum Solothurn; edition fink, Zürich

□ 240 × 170 × 23mm

690g

176p

This catalogue-book presents extracts from the artistic output of Monika Studer and Christoph van den Berg. A number of picture sequences create a further visual level between the texts which deal with the phenomenon of »landscape« or which grapple with the artistic work or rather its presentation in exhibition. This is a wonderful book to read, its unusual cover material making reference to the material used to display the pictures in the context of the exhibition.

Somewhe

re else is the same place

Inhalt / Content

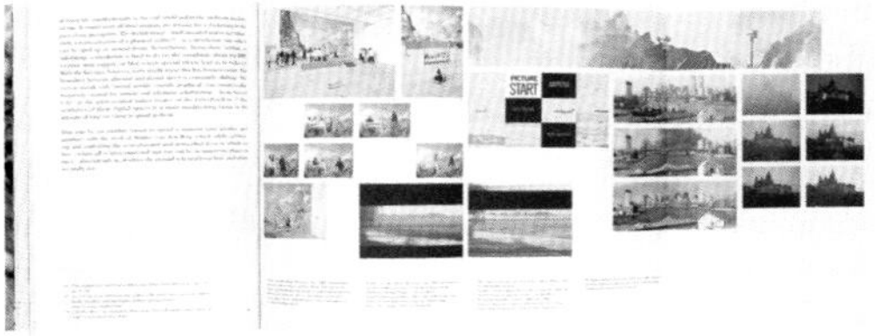
PICTURE
START

PREMIUM
BANANAS

Von der Illusion
des Überblicks

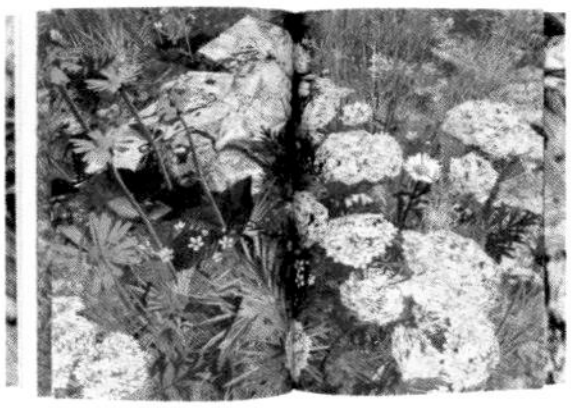

Bernadette Fülscher
Ausflug in die
Alpenwelt
Simulierte Gebirgslandschaften im
Werk von Studer / van den Berg
Von einer gelb gestrichenen Sitzbank aus geniessen Sie die
Aussicht auf die vor Ihnen liegende Berglandschaft; ein andermal
ren Sie bei strahlend blauem Himmel durch eine alpine Gegend
lassen sich von Ihrem (oder Ihrer) Liebsten vor
tergrund fotografieren; oder Sie sind zu Gast in Appenzell und
den reichhaltigen Schmuck der traditionellen Bauten im Zentrum
Städtchens. In allen drei Fällen haben Sie – vielleicht – eine
von Monica Studer und Christoph van den Berg besucht.
formatigen Installationen schaffen die beiden Künstler
Raumsituationen und versetzen uns in andere Welten – in
insbesondere in Alpenlandschaften (vgl. Abb. 1, 2, 3, 7).
jedoch nicht um die Darstellung des alpinen Kontexts: Im
künstlerischen Arbeiten von Studer / van den Berg stehen die Simulation
der begeh- und erfahrbaren Bergwelten inmitten urbaner und virtueller
Räume sowie der Bezug, den wir Städter zu solchen Orten haben.
Die hauptsächlich im musealen Rahmen präsentierten Werke zeichnen
sich einerseits durch einen stark inszenierten Auftritt sowie ein atmos-
phärisches Eigenleben aus; andererseits lassen sie die Künstlichkeit der
Anlage erkennen und entlarven damit die Simulation als Konstruktion.
Während also Der Berg – eine begehbare Alpenlandschaft der Gruppe
Panorama 2000 als Schweizer Beitrag an der Weltausstellung 2005 in
Aichi – beim Besucher Assoziationen an bereits gesehene und erlebte
Situationen (oder Bilder davon) weckt und ihn in entsprechende Stim-
mung versetzt, wird nicht vertuscht, dass es sich beim Berg um eine Holz-
konstruktion handelt, die in einer Industriehalle in Japan steht, und dass
sowohl das Gestein wie der blaue Himmel mit Alpenpanorama auf Planen
gedruckt sind. Dieses Spiel mit Schein und Sein erhält bei der 820 m²
grossen Anlage eine weitere Wendung: Wer von einem bestimmten Standort aus den Berg fotografiert – oder wer im BALTIC in die Kamera der
Installation Mountain Top blickt –, erhält ein Bild, das sich von der Aufnahme einer realen Alpensituation kaum unterscheidet. Dem Künstler-
paar geht es allerdings nicht darum, dem Besucher eine Ersatzwelt zu
bieten, sondern in einer für die Gegenwartskunst typischen Weise aus
der Metaebene heraus zu agieren und auf die Künstlichkeit simulierter
Welten ebenso anzuspielen wie auf die Mittel, die zu ihrer Herstellung
dienen. Bei der Installation Landschaft (2000) (Abb. 4) stehen sodann
nicht die gebaute Alphütte und der Aus- respektive Einblick in eine Berglandschaft mit See auf einem Panoramabild im Zentrum; indem die Aus-
senfassade der Hütte deren Innenansicht und das Innere der Hütte das
Aussenpanorama zeigen, werden Innen und Aussen vertauscht, und es
wird augenfällig, dass die fixen Regeln der Wirklichkeit im Bereich der
Simulation ihre Gültigkeit verloren haben.
Die Konstruktion imaginärer Welten war – nicht selten unter Einbezug
des Bergmotivs – bereits im 19. Jahrhundert sehr beliebt. In der Theater-
szenografie, bei Dioramen, beim «faux terrain» von Panoramen und auch
bei den grossen Welt- und Landesausstellungen kam es immer wieder zur
143

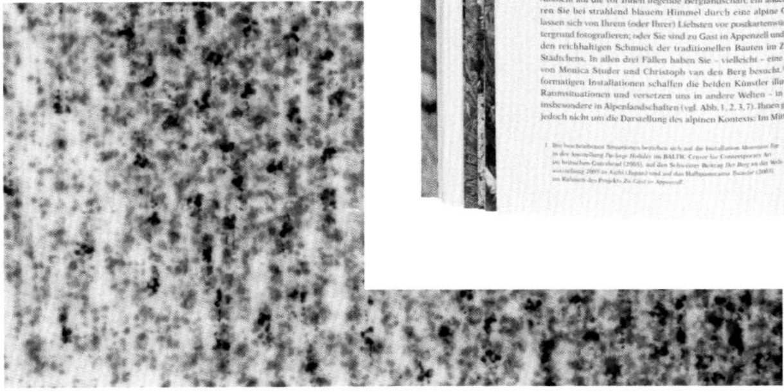

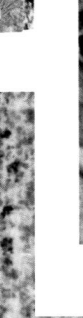

Vorwort zur zweiten Auflage
Digitale Heimatgefühle
Bildkommunikation und Mediatisierung im Werk von Monica Studer und Christoph van den Berg
Slow down
You move too fast
Try to make
the moment last
Überlegungen zur digitalen Verarbeitung von Kameradaten im Werk von Monica Studer und Christoph van den Berg

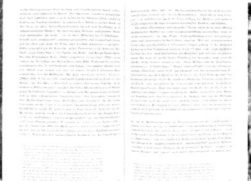

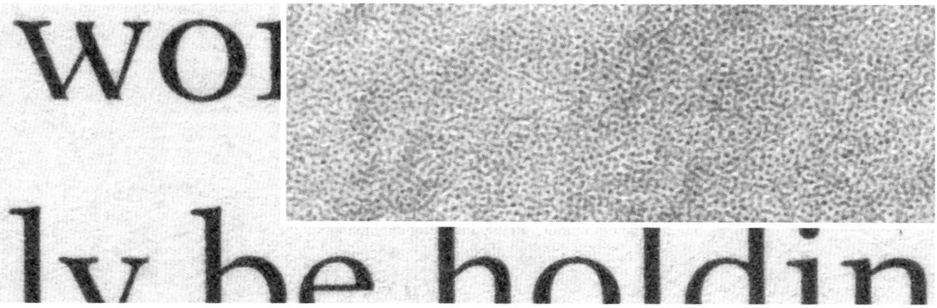
wo
ly be holdin

第叁章
06 Ha 1
第叁章
第贰章

本书是孔祥泽先生整理研究曹氏《南鹞北鸢考工志》以及相关史料的成果，介绍了曹氏风筝的发现、复制、分类、特点，风筝的制作工艺、艺术和独特历史，并配有大量图谱。本书为传统包背线装，方 12 开本。封面采用手揉纸印刷，手感舒适，适合翻阅展开。书脊处白色锁线与传统风筝工艺用线似暗暗呼应。内页版式疏朗，版面结构严谨而多变，大量留白的运用反映出水墨意象。书中印有大量精美的格式风筝图片，配以框架结构图和详细的文字描述，极具收藏和参考价值。

曹雪芹
风筝艺术

This book embodies the results of Mr. Kong xiangze's research on Cao Xueqin's *A Study of the Techniques for Kite-making in Both North and South China* and other relevant historical materials, introducing the discovery, reproduction, classification, characteristics, kite making technology and unique history of Cao's kite, illustrated with a large number of graphs. This book is double-leaved with Chinese traditional thread sewing and the format is 12mo. The cover is printed by hand kneading paper, which is comfortable and suitable for browsing. The white cord on the spine echoes the traditional kite craft line secretly. The layout of the inside pages is clear with rigorous and changeable structure, while the use of a lot of white space reflects the image of Chinese ink painting. There are a large number of exquisite kite pictures in the book, illustrated with frame structure graphs and detailed text description, which is valuable for collection and reference.

荣誉奖	Honorary Appreciation
曹雪芹风筝艺术	Cao Xueqin's Art of Kite
中国	China
赵健工作室	Zhaojian Studio
孔祥泽、孔令民、孔炳彰	Kong Xiangze, Kong Lingmin, Kong Bingzhang
中国友谊出版公司	Beijing Arts & Crafts Publishing House, Peking
	260×231×19mm
	608g
	140p

There are currently many books bound with Chinese binding, but this is an original. Its inner design has a very modern feel to it, not without small humorous additions. Distinguishing features are its clarity of structure and variety of layout. This is a book which is convincing on the strength of its fresh overall design, its choice of materials and high-quality binding.

风筝艺术

序

前言

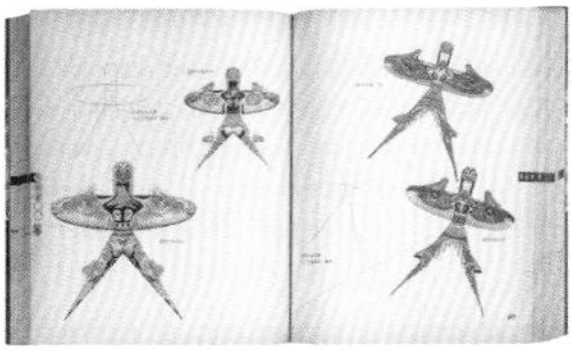

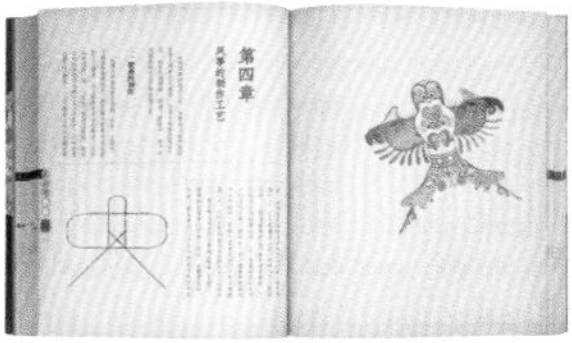
第四章

附录
一、《南鹞北鸢考工志》
部分原文
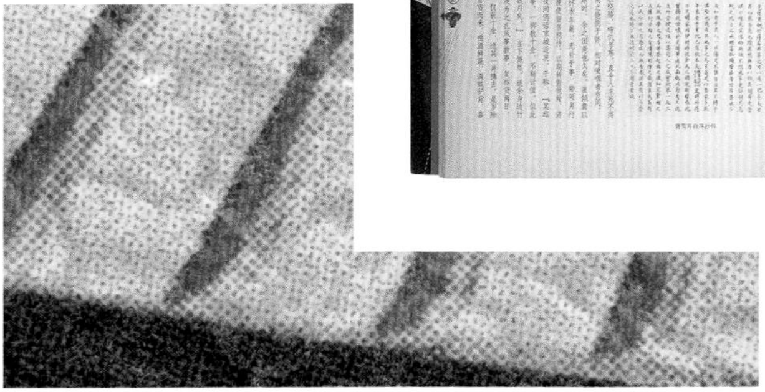

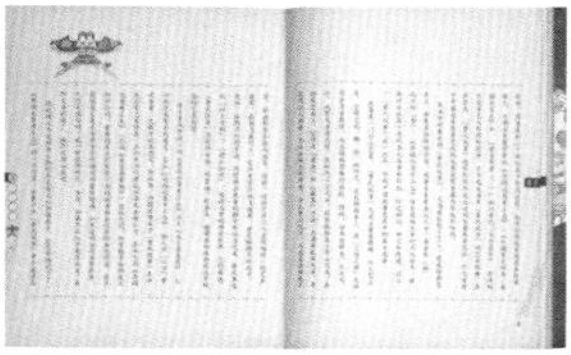

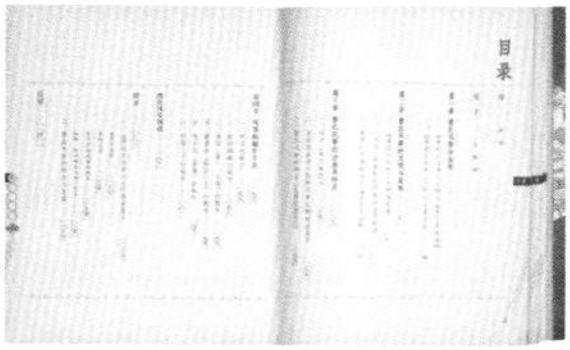
目录

大門
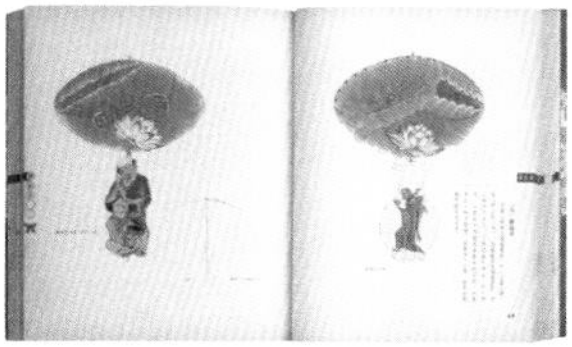
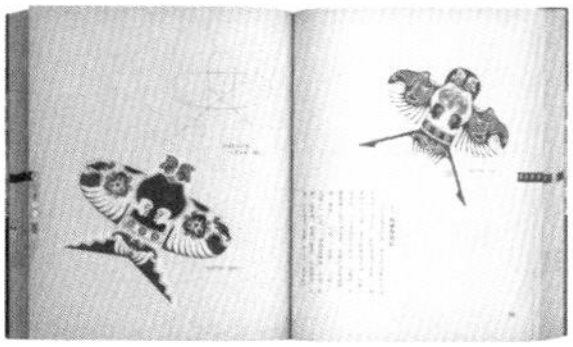
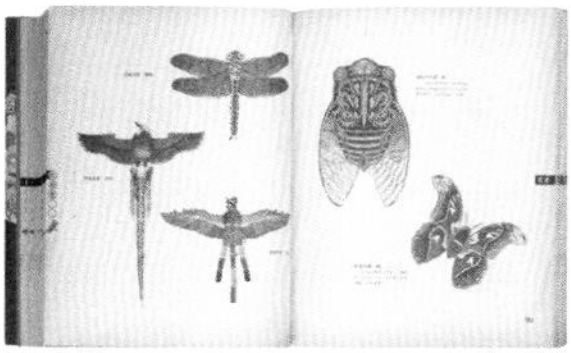
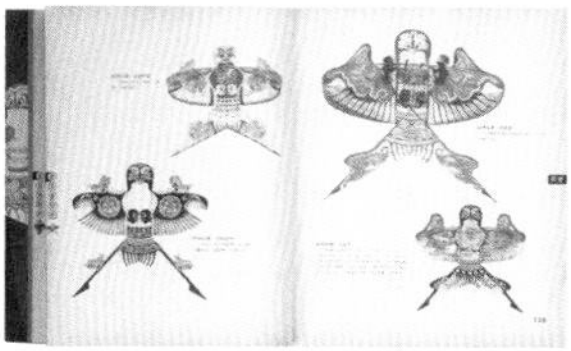
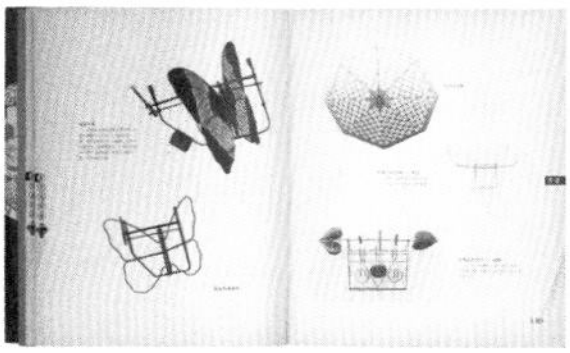

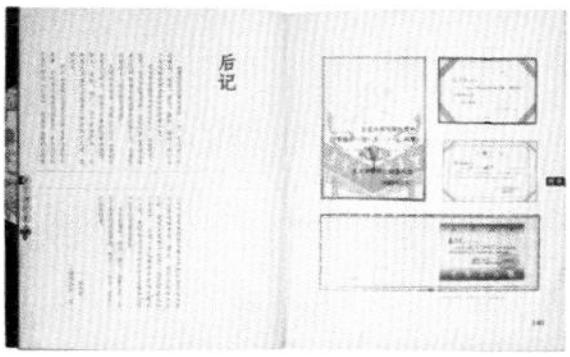
后记

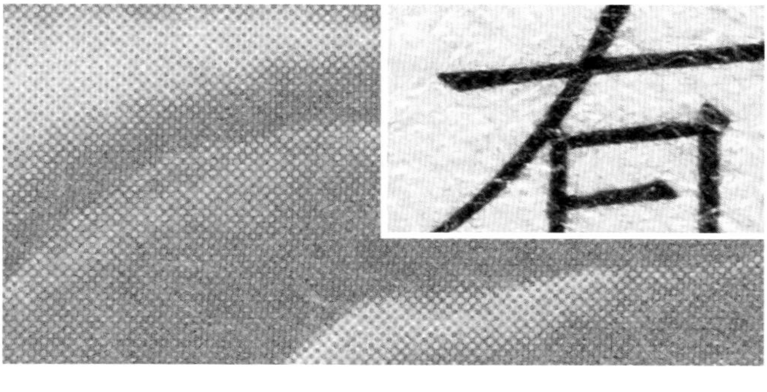
右

本书是瑞士著名设计师马克斯·比尔（Max Bill）的作品集。比尔早年毕业于德绍时期的包豪斯，师从康定斯基、保罗·克利以及奥斯卡·施莱默，之后在艺术、平面设计、广告、建筑等多项领域有所建树，对瑞士平面设计的建立和发展起到重要影响，是德国乌尔姆设计学院的创立者之一。本书的装订形式为锁线胶平装。封面采用白卡纸印单黑，表面覆哑光膜。内页微黄的胶版纸与哑面铜版纸混合装订，四色印刷。胶版纸部分为陈述文字，按照人物介绍、师徒关系、各领域的贡献、年表和摘录为顺序进行介绍。各领域部分将巨大的单词从中间断开，分别印在两页纸上，中间夹两份使用不同字号加以区分的介绍和讨论文字。铜版纸印制的相关作品跟随在各部分文字之后，用以浏览参考。建筑和工业设计作品大多采用黑白印刷，平面作品以彩色为主，印刷精美。

bill,

This book is a collection of works by Max Bill, a famous Swiss designer. Bill graduated from Bauhaus in Dessau in his early years. He was taught by Kandinsky, Paul Klein and Oscar Schleimer. Bill made achievements in many fields, such as art, graphic design, advertising, architecture and so on. He played an important role in the establishment and development of Swiss graphic design. He was one of the founders of the Ulm Institute of Design in Germany. This paperback book is perfect bound. The cover is made of white cardboard printed in black, and laminated with matte film. The yellowish offset paper and matte coated art paper are bound together with four-color printing. The text is printed on offset paper, including the introduction of characters, the relationship with Kandinsky and others, contributions in various fields, chronology and excerpts. In each section, huge words are separated from the middle and printed on two pages, on which introductions and discussions are typeset with different fonts. Most architectural and industrial design works are printed in black and white on coated art paper for browsing and reference; while graphic works are mainly printed in four colors exquisitely.

荣誉奖

马克斯·比尔

——画家、雕塑家、建筑师、设计师

德国

Honorary Appreciation

Max Bill.

Maler, Bildhauer, Architekt, Designer

Germany

Sascha Lobe, Ina Bauer (L2M3 Kommunikationsdesign)

Thomas Buchsteiner, Otto Letze (Ed.)

Hatje Cantz Verlag, Ostfi ldern-Ruit

270×230×26mm

1367g

296p

However many times one browses through, reads or just looks at this excellently presented book it still offers new and interesting features. The reader can enjoy a hint of the Bauhaus years through the clear and unfussy typography which characterises the book. In this respect its grandiose opening is particularly fascinating. Successful paper mix, contrasting typography, well-directed use of colour – a wonderfully designed homage.

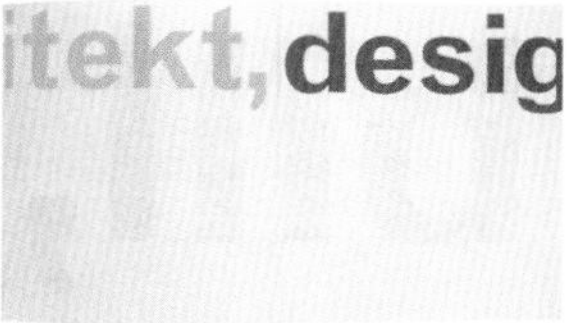
itekt, desig

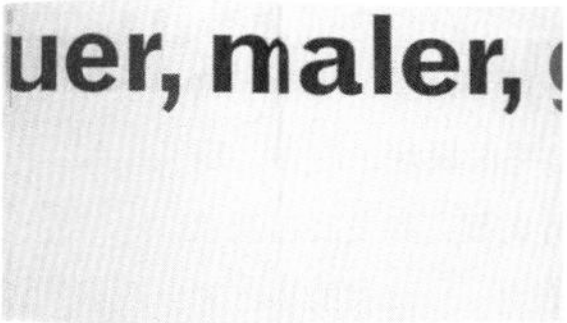
uer, maler,

grafiker, ty

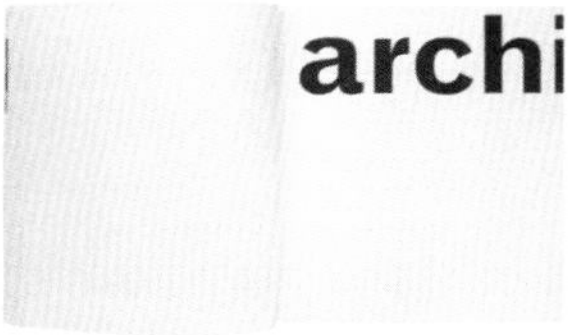
archi

itekt,

mal

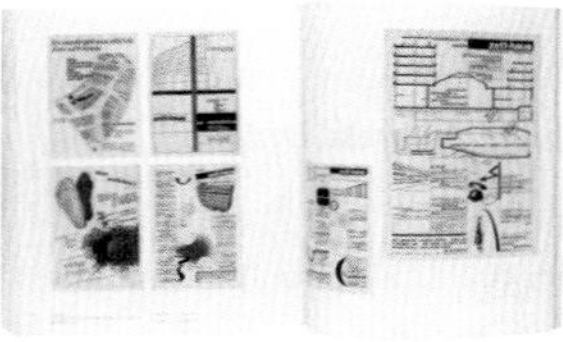

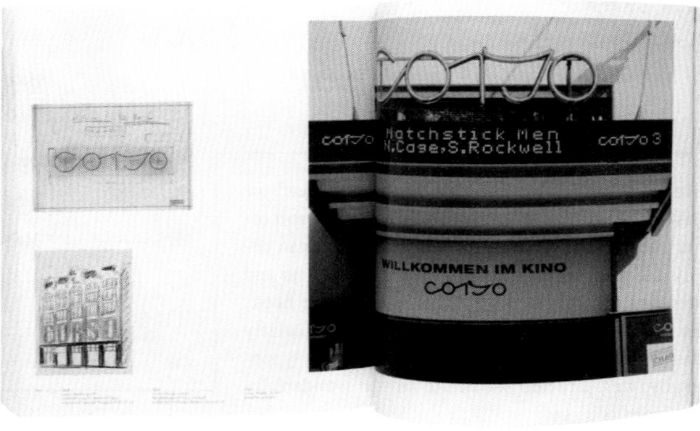
Matchstick Men
N.Case,S.Rockwell
WILLKOMMEN IM KINO

architekt,
maler,
designer,
grafiker,
bildhauer,
typograf
max bill
kandinsky als pädagoge und erzieher
desig
graf

dess
ammlungen ir

BROEDPLAATS EUROPOORT
v/h "DE BEER"
Ha3
om niet? En wel-
belang, natuurlijk.
eu zal in de toe-
naar belangrijker
wel eens een voor-
uitbreidingen en
te realiseren.
enorme schaal van
et plan voor zachte
(waarop riet kan
ng vindt, is dat een
olutie.
n is zeker zo ingrij-
oor het havenge-
n zogenoemde lei-
bebouwde stukken
le tientallen meters
eidingen met kero-
ater en stroom on-
Het havenbedrijf
ken angstvallig bij:
ar wordt gemaaid.
hier meer zijn gang
er binnen de kort-
zame orchideeën-
ipen er rugstreep-
Die zijn beschermd.
aal zijn, moeten ze
wet bepaald. Het
n dan niet bij zijn
l daarom generieke

这本书是荷兰艺术家保罗 · 博格尔斯（Paul Bogaers）的研究和展览项目。本展览记录并研究了已经被扩建为鹿特丹港口一部分的自然保护区 De Beer 的生态状况。本书由五部分组成，主书是一本记录当前状况的黑白摄影集，32 开锁线胶装，内部按照主体书的内容夹带四本骑马钉别册，用以补充尚未被开发破坏的地表植被、规划文本、露出地面的管线节点和逐步长出的苔藓绿植的相关资料。除规划文本为黑白扫描本外，其他均为彩色照片。图片根据横竖比例进行排版，由于大量横向图片，本书主体为书口朝下的上翻版面，文字部分为左翻横排，在空间利用率和文本阅读两者间找到了合适的平衡点，也增加了阅读的趣味。

PAUL BOGAERS
e.o.
BROEDPLAATS EUROPOORT
v/h "DE BEER"

The book is a research and exhibition project by the Dutch artist Paul Bogaers. The exhibition documents and studies the ecology of ‘De Beer’, a nature reserve that has been expanded as part of the port of Rotterdam. The book consists of five parts. The main book is a black and white album recording the current situation of ‘De beer’, perfect bound with thread sewing. Four saddle stitched supplements are attached to the main book to complement the following contents such as surface vegetation which has not been destroyed by the exploitation, planning text, pipeline nodes on the ground and growing bryophytes. Except the planning text is a black and white scanned copy, all other documents are full color photographs. The pictures are arranged according to the horizontal and vertical proportion. Due to a large number of horizontal pictures, the fore edge of main book is downward while the text part is horizontal layout with the right fore edge. The varied layouts keep a proper balance between space utilization and text reading, which also increase the reading interest.

- ◎ 荣誉奖
- « 欧罗波特 De Beer 自然保护区
- ◖ 荷兰

At first glance this book seems to be a documentary report dealing with the question of how industrial zones alter landscape. The book has a very simple structure and is printed in b/w. Coloured booklets are inserted between some of the book’s layers. The continuous pagination marks the respective position of these small-format booklets. The book is very functional but not devoid of poetry. In its production (binding and printing) its means are reduced but not in the same way in which its subject matter »environment« is reflected. A fascinating, quietly-spoken gem.

- ◎ ········· Honorary Appreciation
- « ··· Broedplaats Europoort v/h ›De Beer‹
- ◖ ········· The Netherlands
- D ········· René Put
- △ ········· Paul Bogaers u.a.
- ||||| ········· Artimo, Amsterdam
- □ ········· 240 × 165 × 7mm
- ⌷ ········· 277g
- ≣ ········· 104p

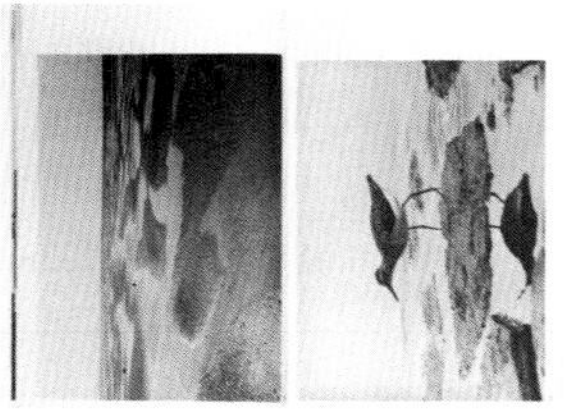

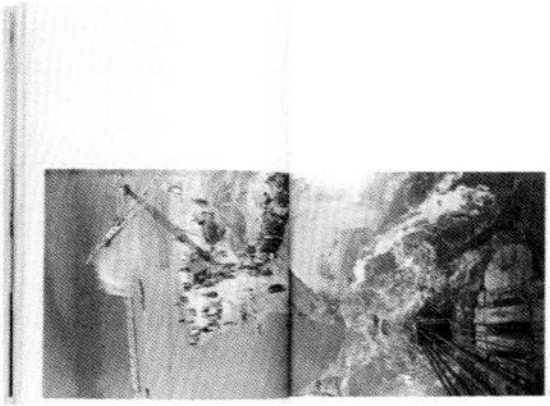
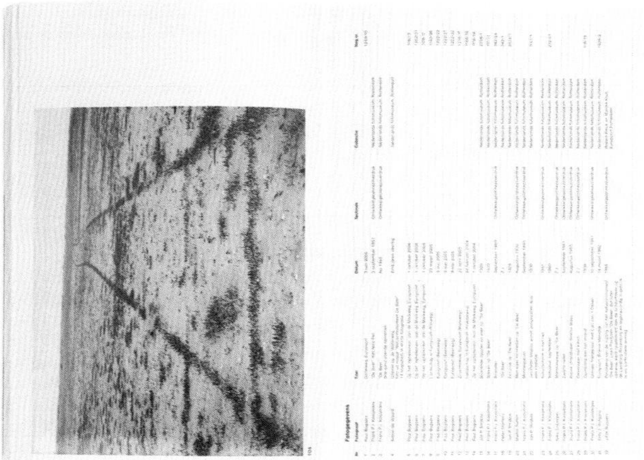
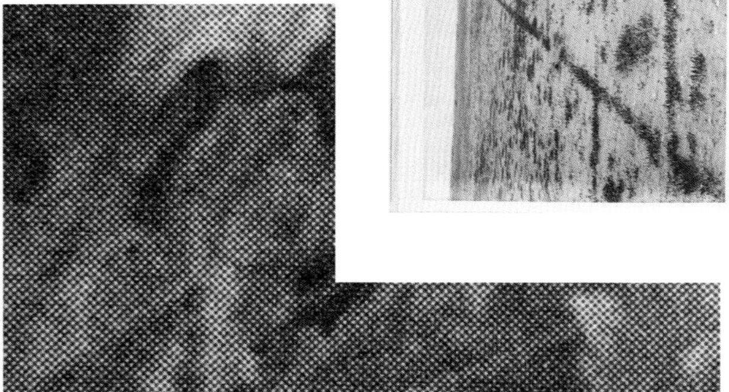

BROEDPLAATS EUROPOORT

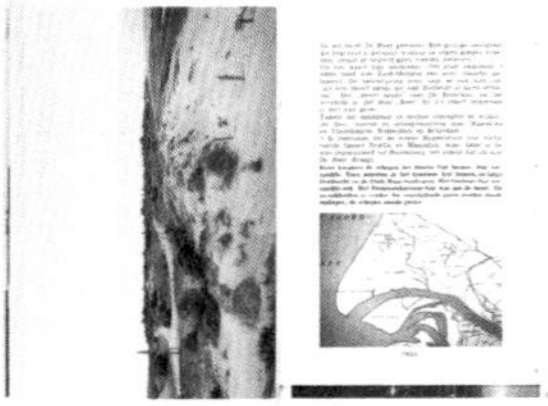

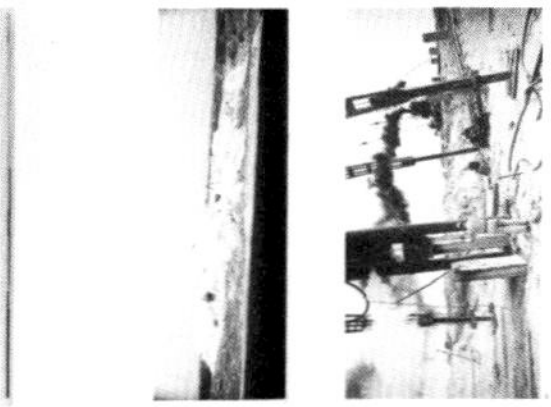

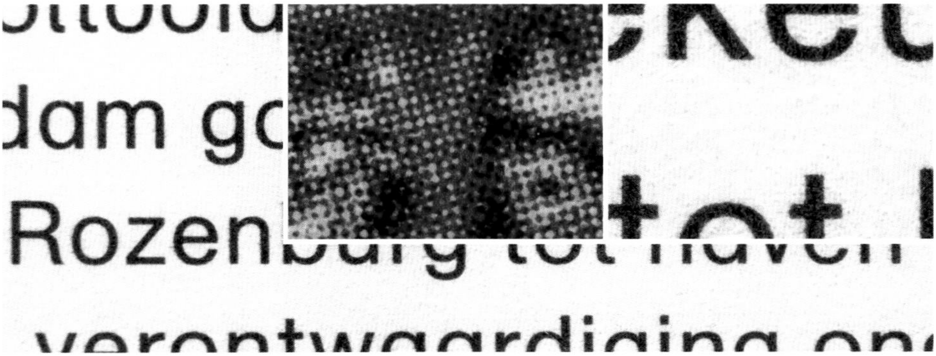
dam g
Rozen

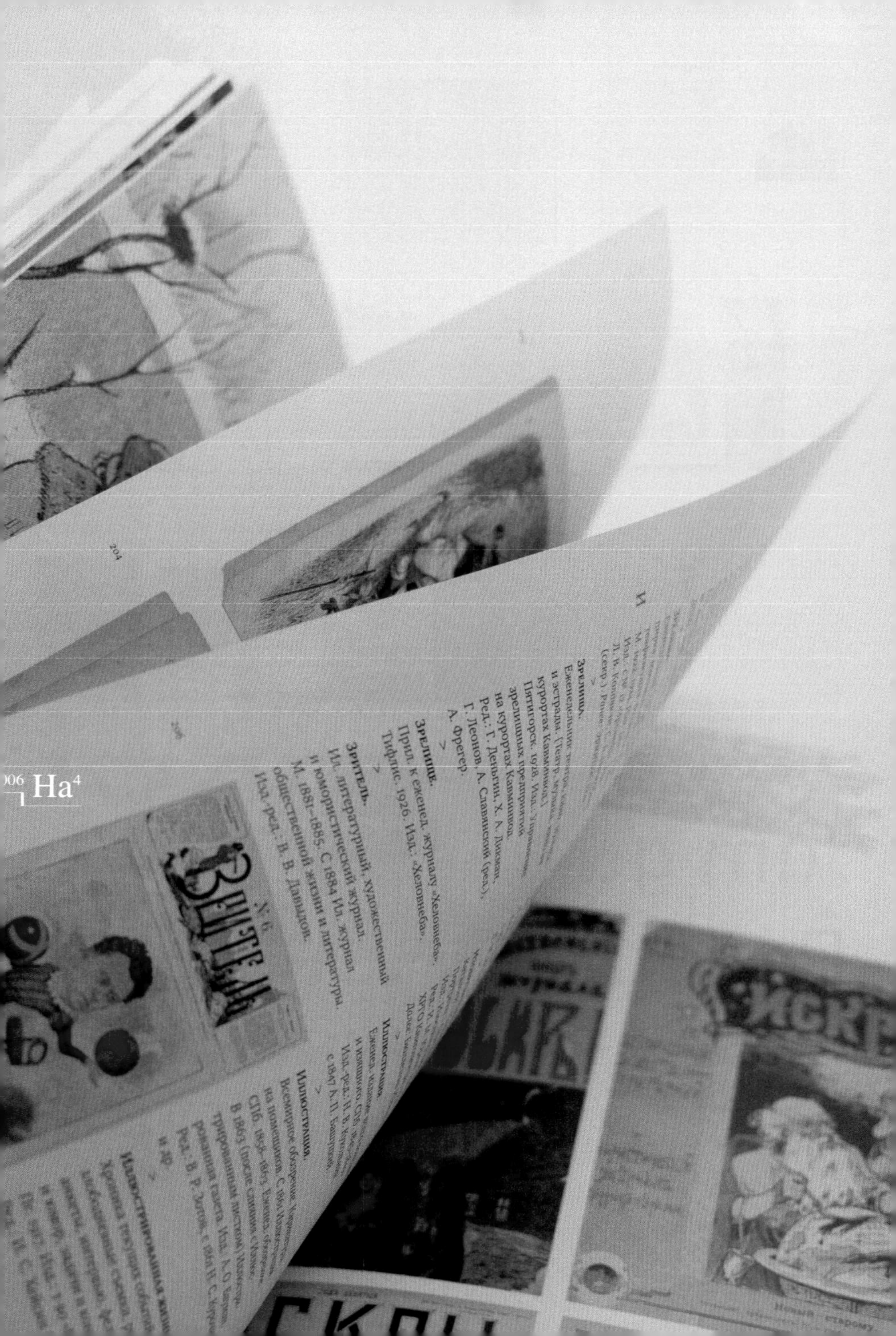
906 На4
Зрелища.
Зрелище.
Зритель.
Иллюстрация.

这是一本收录了俄罗斯及苏联时期775本插图杂志的合集，通过大量图片、版面的展示，以及出版商、编辑、插画家、工作人员的信息罗列，呈现出这段历史中社会各界对时代审美和社会文化的共同努力，也可以看出随着时代的改变，字体、印刷、版面、插图、摄影的不断进步。本书的装订形式为锁线硬精装，封面采用哑面铜版纸四色印刷，裱贴卡板，书脊处包覆红色布面，压凹裱贴书籍信息，外套同样装裱和印刷工艺的书函。暗红色艺术纸张的环衬更能衬托出本书的历史厚重感。内页哑面艺术纸四色印刷，全书俄英双语对照，图文精美。本书开篇对俄罗斯插图杂志的历史进行了综述，之后按照字母表顺序进行陈列，展示杂志的封面、内页和相关信息，前言部分也提到像这样按照字母顺序而不是时间顺序来陈列资料，会让本书的图片看起来缺乏连贯性，造成彩色和黑白交替的混乱感，但这是一种按照俄罗斯国家图书馆和具有相关藏品的博物馆陈列的原则，按照这一顺序也可以从另一个方面展现出历史长河中俄罗斯插画杂志发展的多样性。本书正文部分主要采用六栏设计，通过不同长度的栏宽组合（例如2+2+1+1、3+3、4+1+1等），实现丰富的版面排版。||||||||||||||||

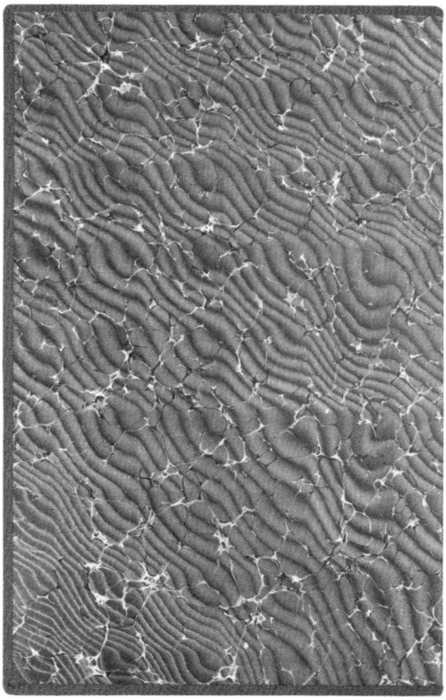

This is a collection of 775 illustrated magazines from Russia and the Soviet Union. Through displaying a large number of pictures and layouts, as well as listing information of publishers, editors, illustrators and staff, it shows the joint efforts of the society towards the aesthetics and social culture in that historical period. It also reflects the improvement of fonts, printing, layout, illustrations, and photography due to the changing times. The hardback book is bound by thread sewing. The cover is printed in four colors on matte coated art paper mounted with cardboard. The spine is covered with red cloth with embossed book information. The slipcase is using the same printing technology. The dark red endpaper gives the book historical appearance. Matte art paper with four-color printing is used for the inner pages. The book is typeset in Russian and English with exquisite pictures. This book begins with an overview of the history of Russian illustrated magazines, and then displays them in alphabetical order, showing the covers, inside pages and related information. The preface explains that presenting them in alphabetical order rather than chronological order will make this book lack of coherence and sequence, however, it follows the principle of exhibiting in accordance with the Russian National Library and museums with related collections. In this order, it can also show the diversity of the development of Russian illustrated magazines in the long history. The main body of the book is designed with six columns, and the enriched layout can be achieved through the combination of column widths of different lengths (e.g. 2+2+1+1, 3+3, 4+1+1 etc.).

- 荣誉奖
- 俄罗斯插图杂志 1703－1941
- 俄罗斯

This directory contains visual material of differing styles, sizes and proportions. The great challenge was to carefully present this very heterogeneous and not always compatible material in a logical and alphabetical order. Its designer has achieved a compelling visual performance beneath its marble-paper cover – from the interchange of the borderless double-pages right through to small illustrations inserted between the text.

- Honorary Appreciation
- Russian Illustrated Magazine. 1703–1941
- Russia
- Masha Lyuledgian
- Irina Mokhnacheva
- Agey Tomesh, Moskau
- 356×218×45mm
- 2607g
- 368p

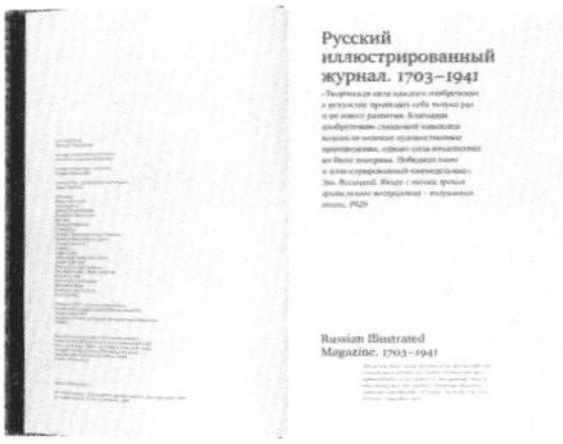
Русский иллюстрированный журнал. 1703–1941
Russian Illustrated Magazine. 1703–1941

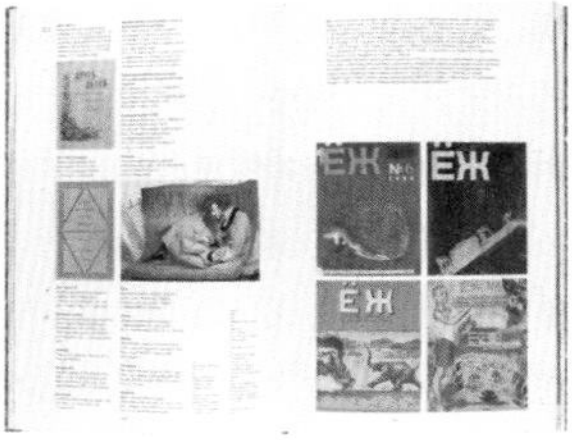
ЁЖ
ЁЖ
ЁЖ

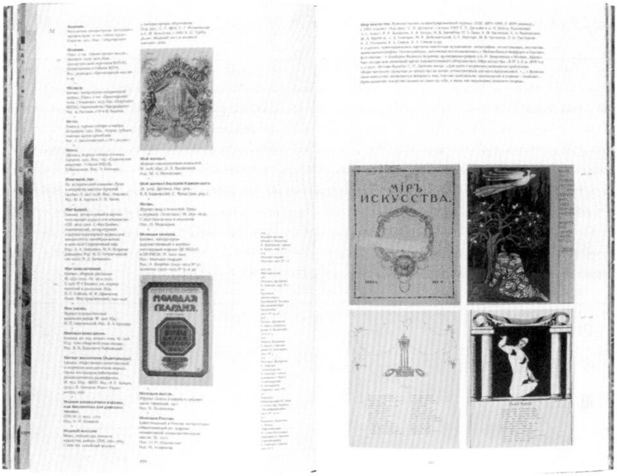
МІРЪ ИСКУССТВА.
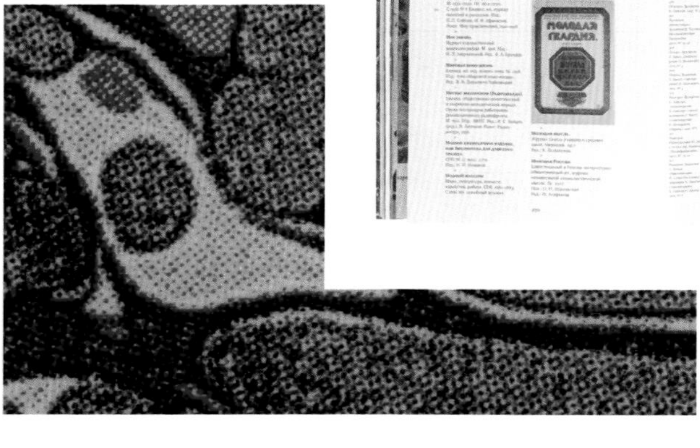

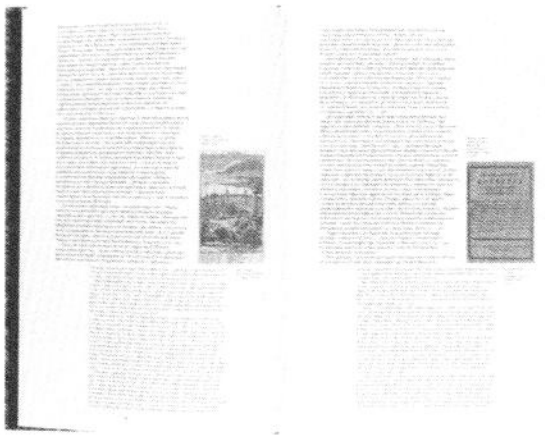

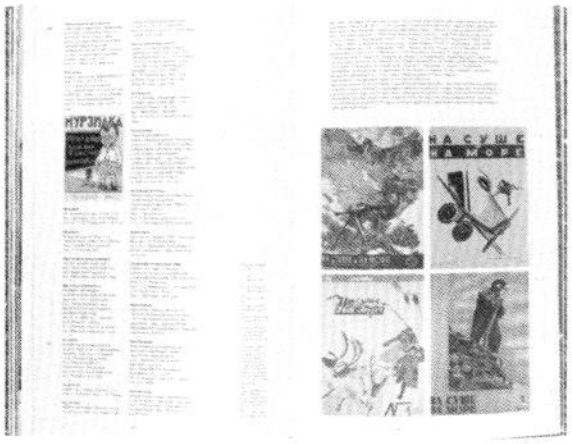
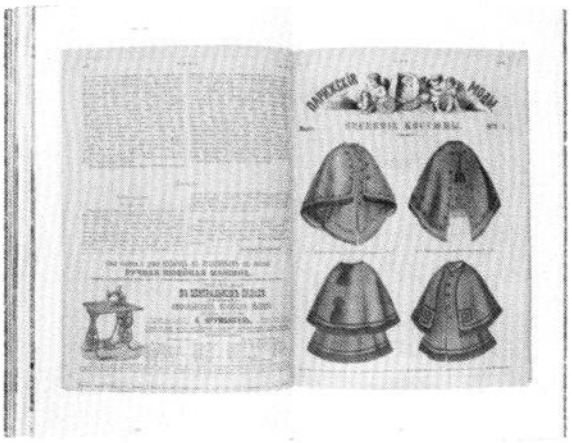
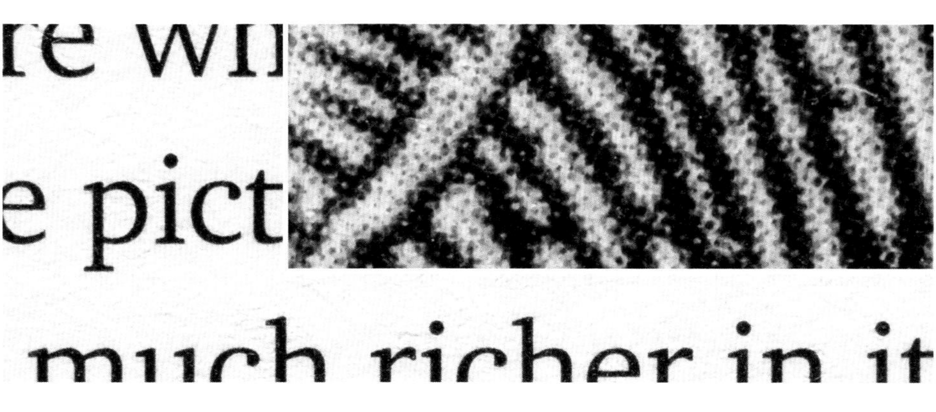
re wi
e pict
much richer in it

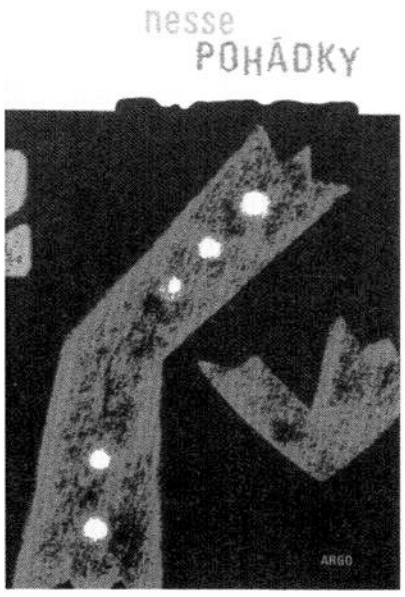
hesse
POHÁDKY
ARGO

这是一本由诺贝尔文学奖获得者，德国著名诗人、小说家赫尔曼·黑塞（Hermann Hesse）编写的童话故事集，内含八个独立的小故事。童话在这里并不是专门针对儿童的文体，而是一种写作手法，通过这种文风，创造出一个脱离现代世界的“虚拟”世界。本书是一本极小的口袋本，无线胶装，封面四色印刷覆哑膜，并无特殊工艺。但内页采用蓬松而轻型的纸张，整本书拿在手里非常轻巧，非常适合随身携带。内页采用黑橙双色印刷，橙色主要用于标题文字、插图细节和插图页背面满版印刷。目录被放在了全书的最后部分，可见作者和编者希望读者更专注于故事本身，希望阅读的过程是个线性连贯的过程。

This is a collection of fairy tales written by Hermann Hesse, who is a Nobel laureate in Literature, a German poet and novelist. The book contains eight individual stories. Fairy tale is not a literary style specifically aimed at children, but a writing method by which a "virtual" world separated from the modern world is created. This book is a very small pocket book with matt laminated and four-color printed cover, wireless binding. Though no special process is adopted, the inside pages are made of fluffy and lightweight paper, thus the book is light enough to carry around. The inside pages are printed in black and orange – the latter one is mainly used for title text, illustration details and full-page printing on the back of the illustration page. The table of contents is placed at the end of the book, indicating that the author and the editor want the reader to be more focused on the story itself and that the reading process should be linear and coherent.

荣誉奖

童话故事

捷克

At first glance this book seems to be a documentary report dealing with the question of how industrial zones alter landscape. The book has a very simple structure and is printed in b/w. Coloured booklets are inserted between some of the book's layers. The continuous pagination marks the respective position of these small-format booklets. The book is very functional but not devoid of poetry. In its production (binding and printing) its means are reduced but not in the same way in which its subject matter »environment« is reflected. A fascinating, quietly-spoken gem.

Honorary Appreciation

Pohádky

Czech Republic

Luboš Drtina

Hermann Hesse

Argo, Prag

174×116×14mm

152g

165p

HERMANN HESSE POHÁDKY

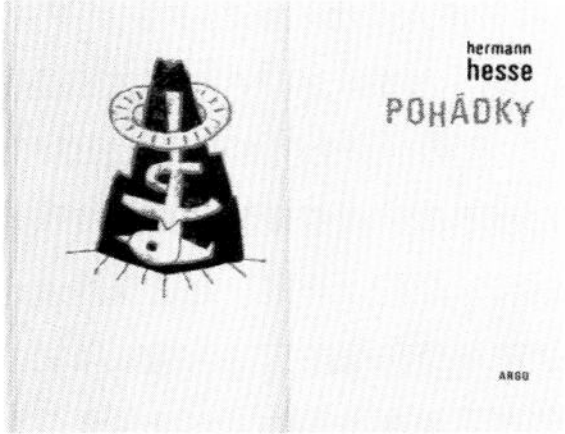
hermann
hesse
POHÁDKY
ARGO

AUGUSTUS

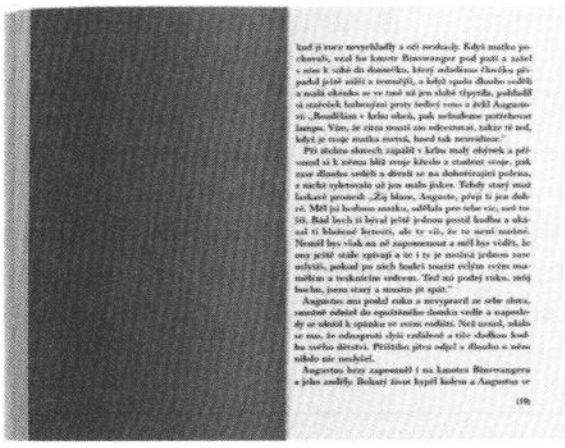

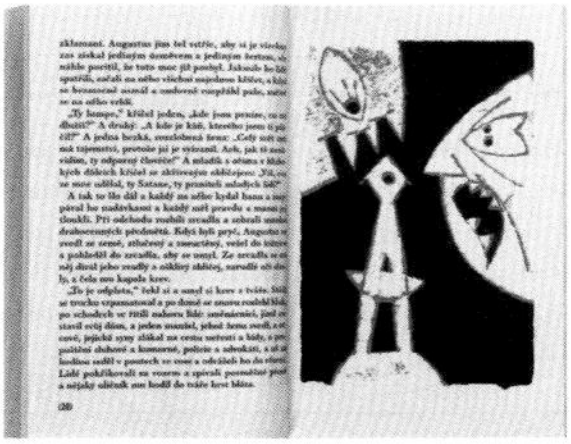

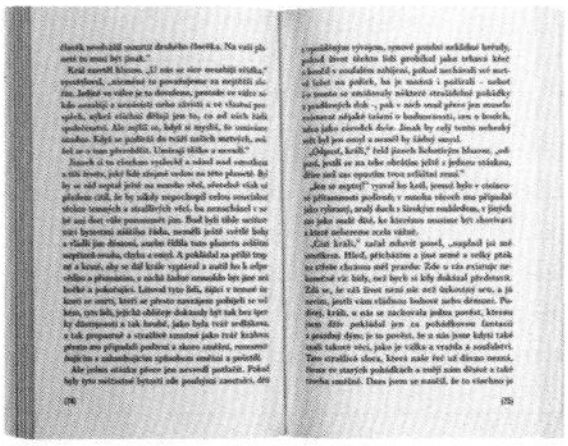

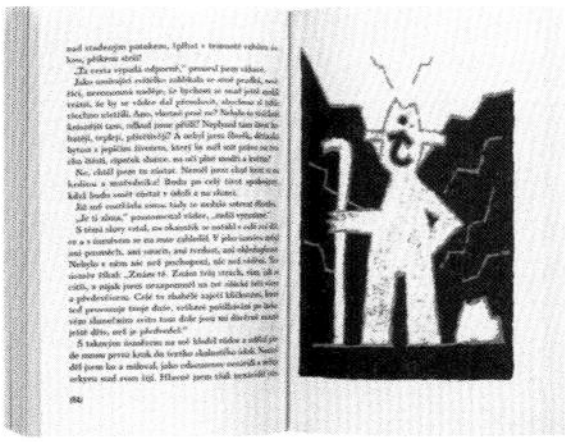

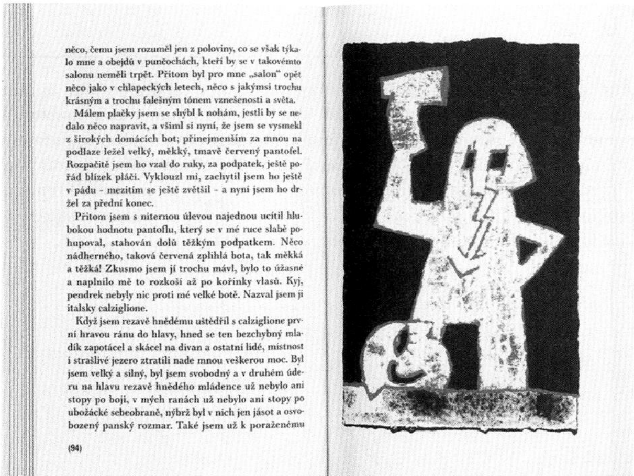
něco, čemu jsem rozuměl jen z poloviny, co se však týkalo mne a obejdů v punčochách, kteří by se v takovémto salonu neměli trpět. Přitom byl pro mne „salon" opět něco jako v chlapeckých letech, něco s jakýmsi trochu krásným a trochu falešným tónem vznešenosti a světa.

Málem plačky jsem se shýbl k nohám, jestli by se nedalo něco napravit, a všiml si nyní, že jsem se vysmekl z širokých domácích bot; přinejmenším za mnou na podlaze ležel velký, měkký, tmavě červený pantofel. Rozpačitě jsem ho vzal do ruky, za podpatek, ještě pořád blízek pláči. Vyklouzl mi, zachytil jsem ho ještě v pádu - mezitím se ještě zvětšil - a nyní jsem ho držel za přední konec.

Přitom jsem s niternou úlevou najednou ucítil hlubokou hodnotu pantoflu, který se v mé ruce slabě pohupoval, stahován dolů těžkým podpatkem. Něco nádherného, taková červená zplihlá bota, tak měkká a těžká! Zkusmo jsem jí trochu mávl, bylo to úžasné a naplnilo mě to rozkoší až po kořínky vlasů. Kyj, pendrek nebyly nic proti mé velké botě. Nazval jsem ji italsky calziglione.

Když jsem rezavě hnědému uštědřil s calziglione první hravou ránu do hlavy, hned se ten bezchybný mladík zapotácel a skácel na divan a ostatní lidé, místnost i strašlivé jezero ztratili nade mnou veškerou moc. Byl jsem velký a silný, byl jsem svobodný a v druhém úderu na hlavu rezavě hnědého mládence už nebylo ani stopy po boji, v mých ranách už nebylo ani stopy po ubožácké sebeobraně, nýbrž byl v nich jen jásot a osvobozený panský rozmar. Také jsem už k poraženému

(94)

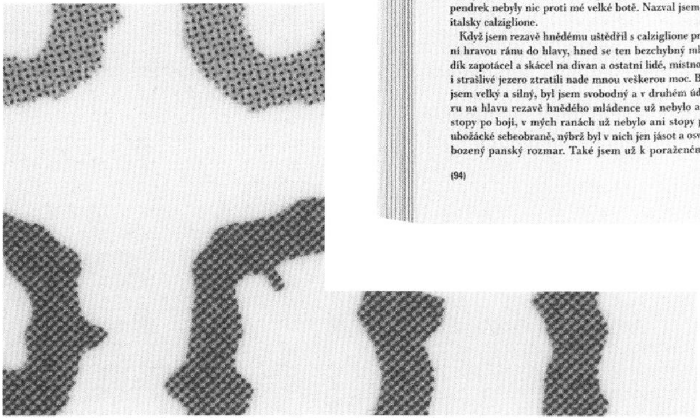

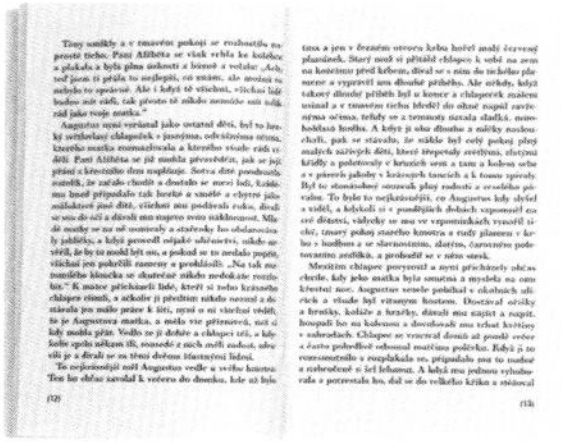
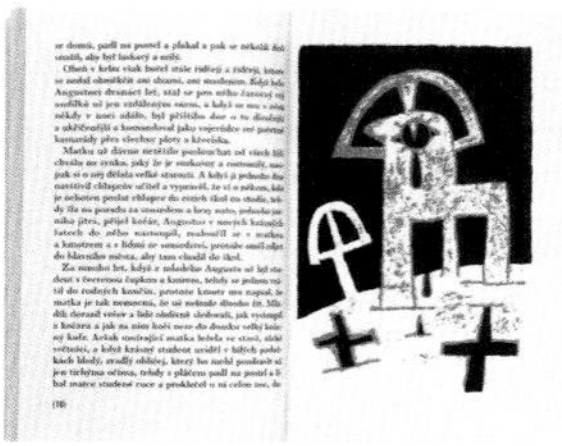

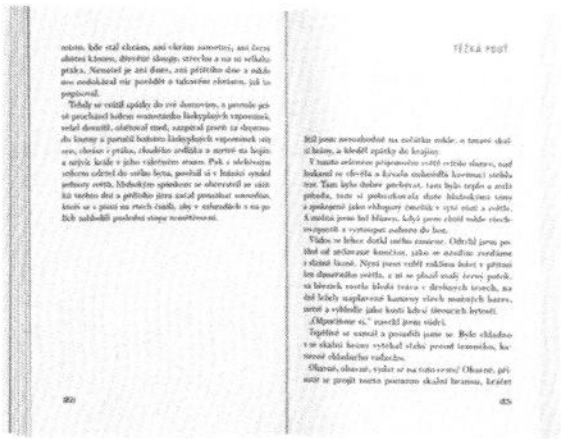

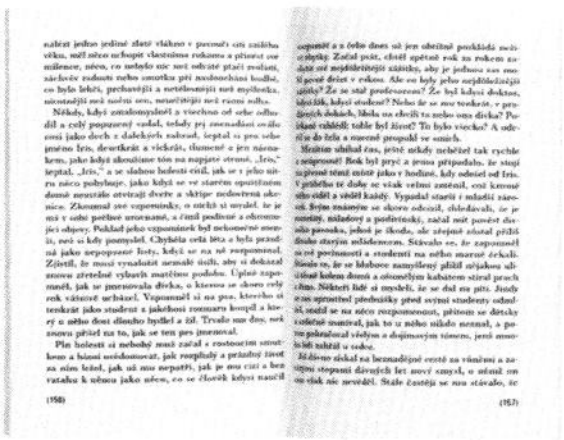
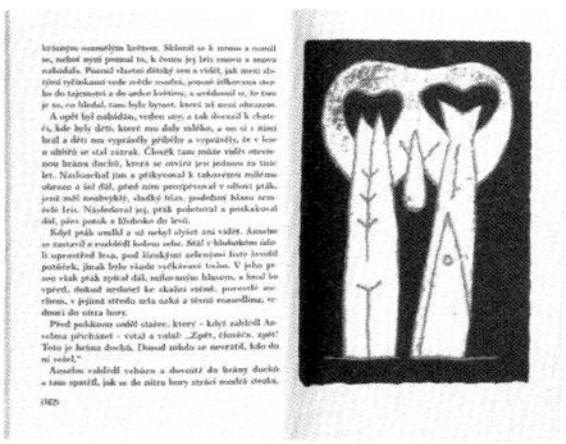
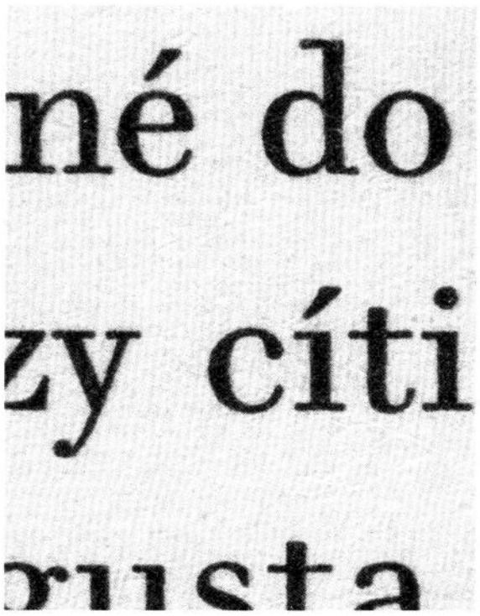
né do
zy cíti
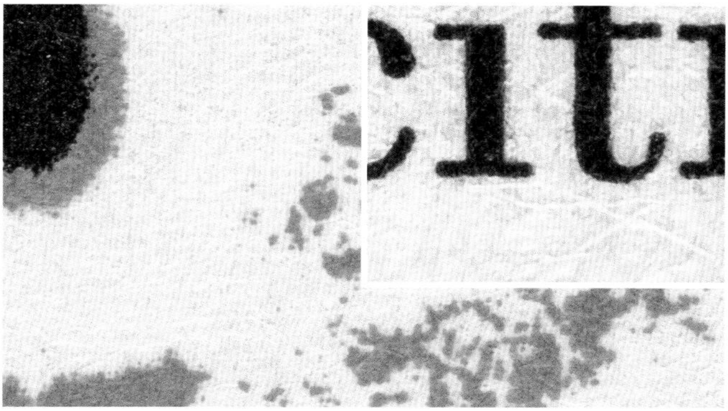

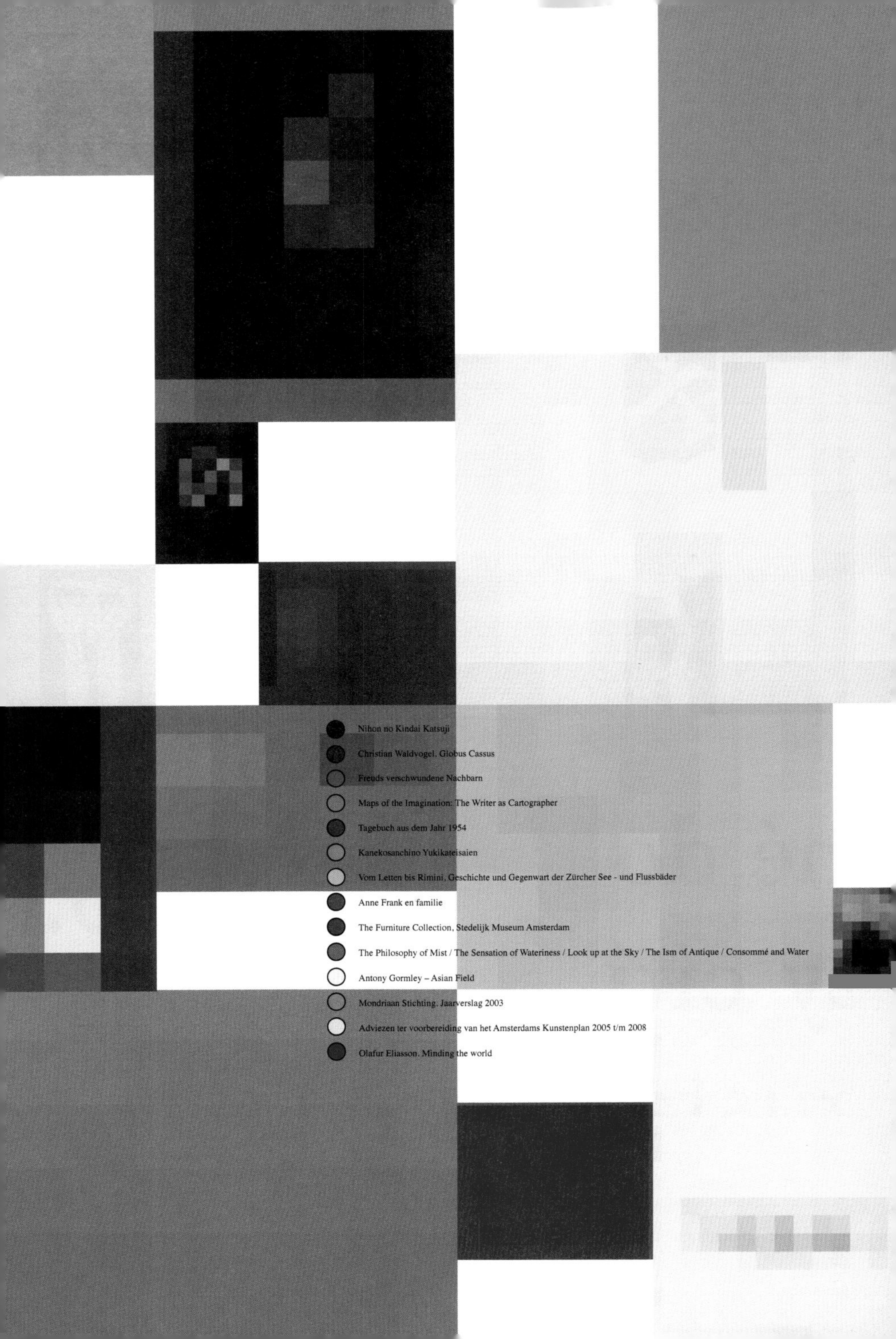
Nihon no Kindai Katsuji
Christian Waldvogel. Globus Cassus
Freuds verschwundene Nachbarn
Maps of the Imagination: The Writer as Cartographer
Tagebuch aus dem Jahr 1954
Kanekosanchino Yukikateisaien
Vom Letten bis Rimini. Geschichte und Gegenwart der Zürcher See - und Flussbäder
Anne Frank en familie
The Furniture Collection, Stedelijk Museum Amsterdam
The Philosophy of Mist / The Sensation of Wateriness / Look up at the Sky / The Ism of Antique / Consommé and Water
Antony Gormley – Asian Field
Mondriaan Stichting. Jaarverslag 2003
Adviezen ter voorbereiding van het Amsterdams Kunstenplan 2005 t/m 2008
Olafur Eliasson. Minding the world

2005

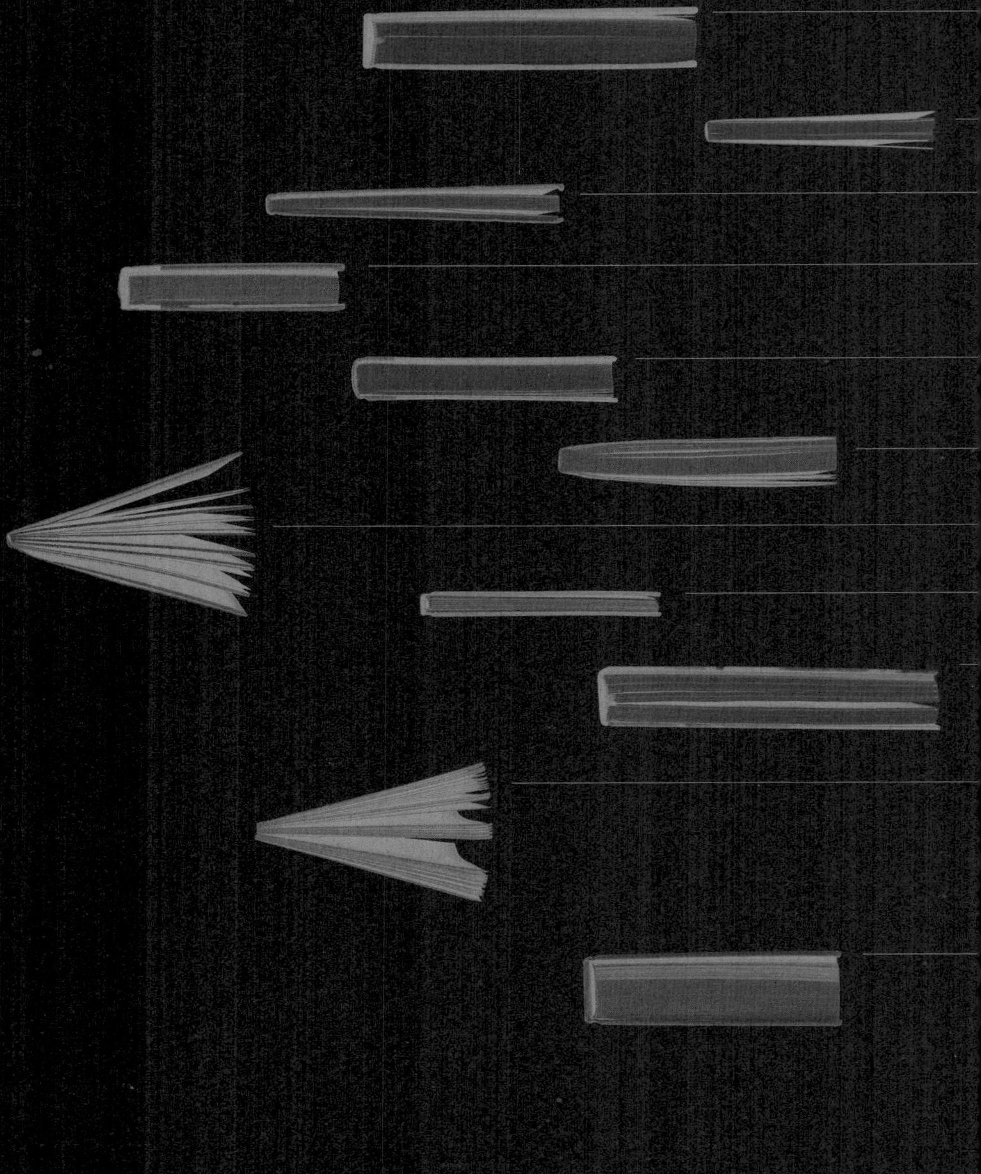

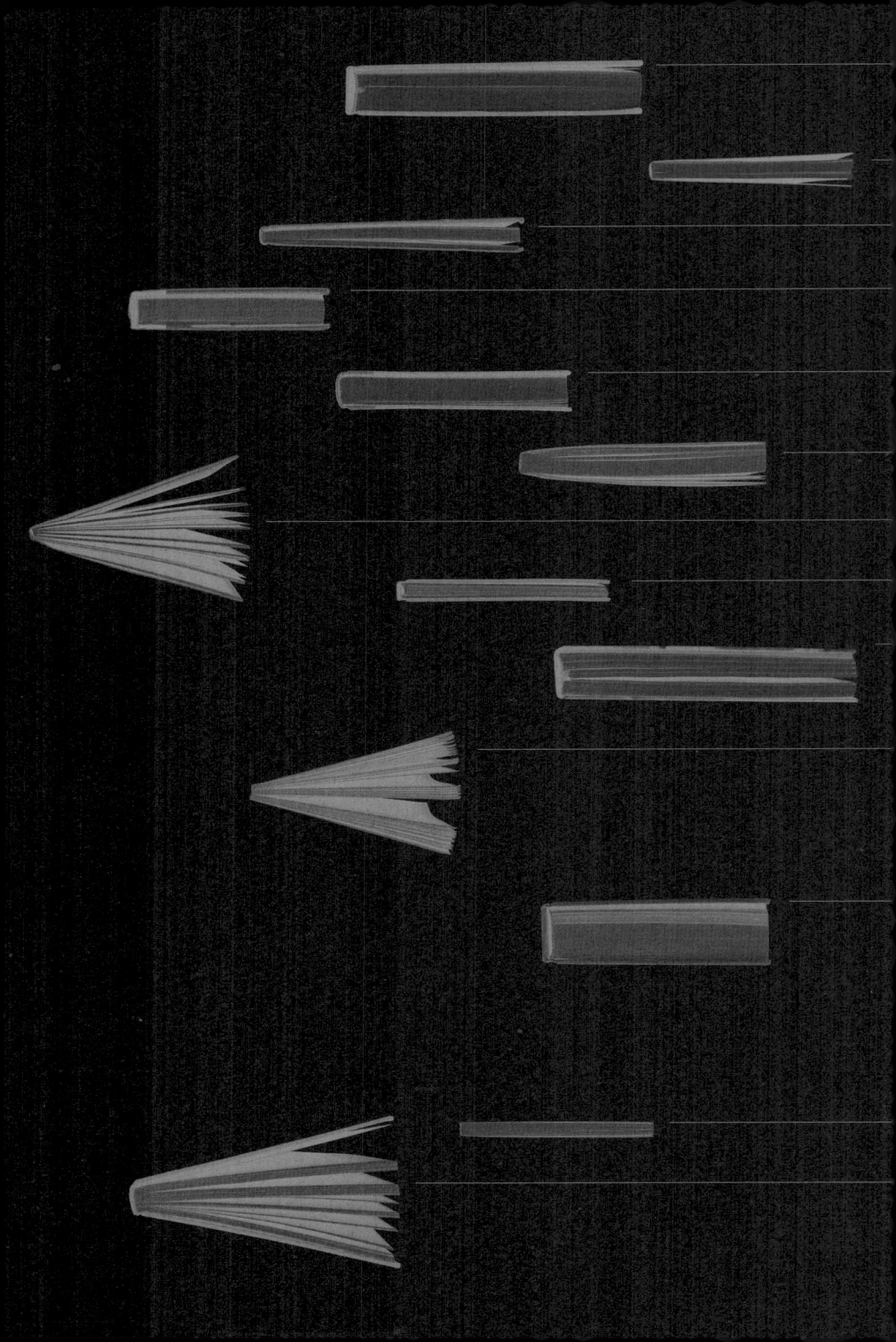

2005 Gl

1275

2005 Gm

1283

2005 Sm^1

1289

2005 Sm^2

1295

MAPS OF THE IMAGINATION: THE WRITER AS CARTOGRAPHER

2005 Bm^1

1301

2005 Bm^2

1307

2005 Bm^3

1313

2005 Bm^4

1319

2005 Bm^5

1325

The Furniture Collection
Stedelijk Museum Amsterdam

2005 Ha^1

1331

2005 Ha^2

1335

2005 Ha^4

1339

2005 Ha^5

1343

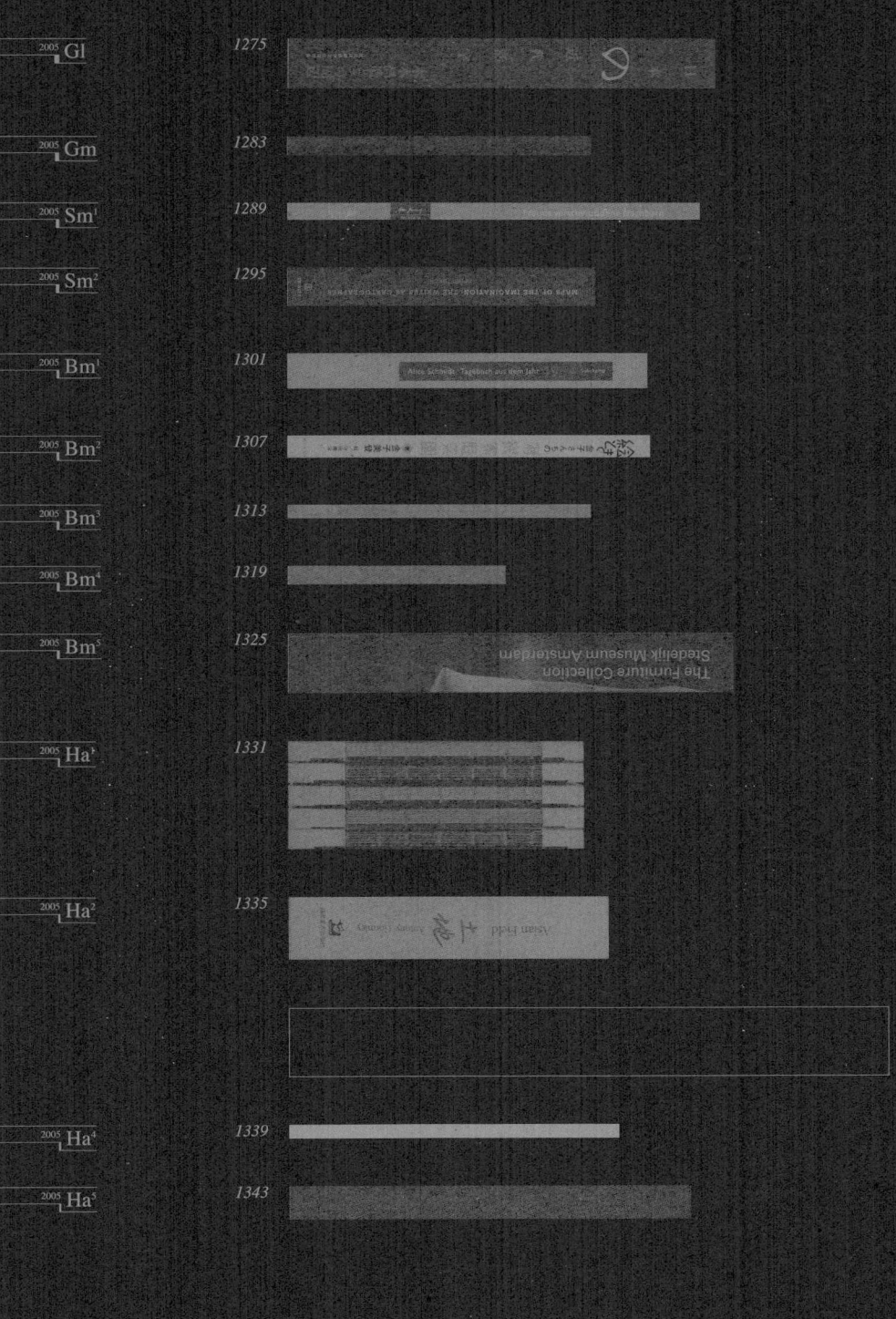

aanpassing
39
41
42
43
45
45
48
49

さらに譜のなか一六世紀されたこ技術への の目標にをもあわとした高。つまり、った。

われた活リシタン展をしる教信仰がして、妥年に死去ように説おもに社れなかっ

22.00 vue :

Les quatre câbles statiques ont tous attein
une longueur de 155'500 km. Ils touchent

inzwischen in Tel Aviv. Grete Humburger war seit 1935 mit dem Schriftsteller Leo Perutz verheiratet und im Juli 1938 mit ihm und seinen drei Kindern aus erster Ehe über Italien nach Palästina ausgewan-

Harvard Univ. Press, 1988.
———. *Why Read the Classics?* New York: Pantheo
Carson, Anne. "Beauty Prefers an Edge." Interview

:, geborene Murawski, im Jahre 1954 g

家庭菜園でてきたとこ終的に株間ます。畑やから移植すスト）が、キョウナはちんと畑をましょう。せないようす。ギザギザのくれ、病害長い収穫期キョウナはイのように下旬～三月できます。がおいしくい味になり冬は鍋物のおいしくい

n Wasser gelegenen Fabriken und einer ehemaligen lung. Die durch ein Kraftwerk erzeugte Strömung reisst en mit sich, die sich ihr getrost anvertrauen können, da Letten durch einen sicheren Gitterrost abgeschlossen ist. Letten wurde 1910 erstellt, als der Obere Letten nicht

mer worker's colony, this bath has a fascinatin Located downstream from a dam, visitors ca ing rush of water while braced securely by structed in 1910 because the Oberer Letten b unable to handle its growing number of visit

ekozen voor bepaalde

iomenten van fotografere

collections and scientific knowledge about it. That is also the reason it was decided to publish this catalogue in English. Thus an international public, both professionals

Stoffel Brouwers, for t the costs of this public panies its launch. That

我原先以为写文章用笔名，专门是为了暗箭伤人的。假面示人，便于躲
古则、藏头遮脸的干什么哪。藏头露尾似乎是鸵鸟的秉性，顾头不顾腚
从好汉做事好汉当的古训，既会暗箭伤人也要明火执仗，十八般武艺皆

姓名: 石川 年龄: 50 职业: 艺术家
地址:

[redacted]

gespecialiseerde kunstbibliotheek en een kunstenaarsarchief, die gebundeld zijn als kenniscentrum, en een projectbureau. De afdelingen ondersteunen

modernismens rum – og menneskeop
tnere, der i dag arbejder med rumlige problematikke

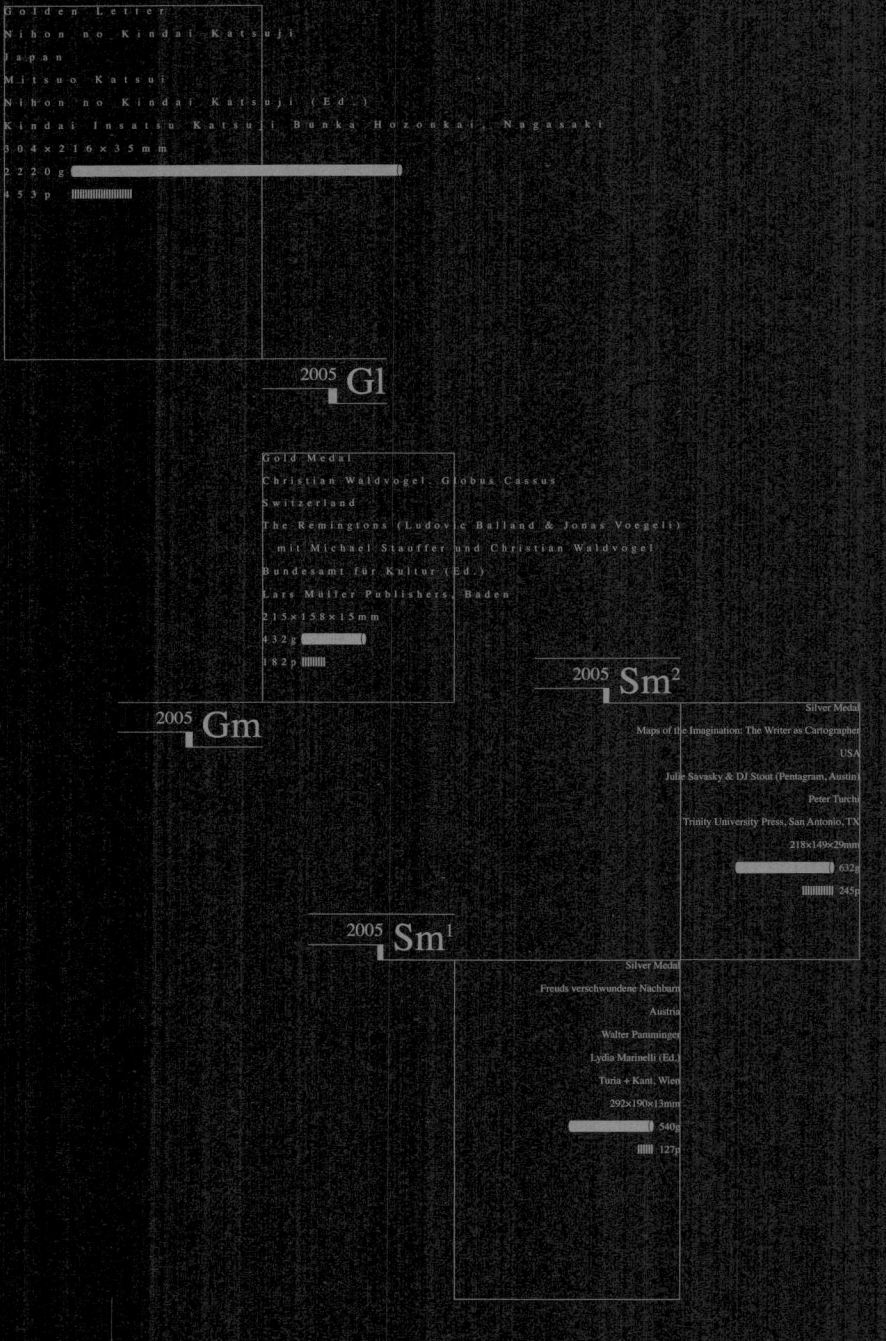

Golden Letter
Nihon no Kindai Katsuji
Japan
Mitsuo Katsui
Nihon no Kindai Katsuji (Ed.)
Kindai Insatsu Katsuji Bunka Hozonkai, Nagasaki
304×216×35mm
2220g
453p

2005 Gl

Gold Medal
Christian Waldvogel. Globus Cassus
Switzerland
The Remingtons (Ludovic Balland & Jonas Voegeli)
mit Michael Stauffer und Christian Waldvogel
Bundesamt für Kultur (Ed.)
Lars Müller Publishers, Baden
215×158×15mm
432g
182p

2005 Gm

2005 Sm2

Silver Medal
Maps of the Imagination: The Writer as Cartographer
USA
Julie Savasky & DJ Stout (Pentagram, Austin)
Peter Turchi
Trinity University Press, San Antonio, TX
218×149×29mm
632g
245p

2005 Sm1

Silver Medal
Freuds verschwundene Nachbarn
Austria
Walter Pamminger
Lydia Marinelli (Ed.)
Turia + Kant, Wien
292×190×13mm
540g
127p

2005 Ha[1]

Honorary Appreciation

The Philosophy of Mist / The Sensation of Wateriness / Look up at the Sky / The Ism of Antique / Consommé and Water

China

He Jun

Zhu Yeqing

China Friendship Publishing Co., Peking

210×152×57mm

949g

751p

2005 Ha[4]

Honorary Appreciation

Adviezen ter voorbereiding van het Amsterdams Kunstenplan 2005 t/m 2008

The Netherlands

Hansje van Halem & Gerard Unger

Amsterdamse Kunstraad, Amsterdam

234×165×9mm

316g

264p

2005 Ha[2]

Honorary Appreciation

Olafur Eliasson. Minding the world

Switzerland

groenland.berlin.basel (Dorothea Weishaupt & Michael Heimann)

ARoS Aarhus Kunstmuseum (Ed.)

Hatje Cantz Verlag, Ostfildern-Ruit

285×238×26mm

1238g

232p

2005 Ha[2]

Honorary Appreciation

Antony Gormley – Asian Field

China

Wang Xu & Associates

Hu Fang, Zhang Weiwei (Ed.) Zhang Wei (Ass. Ed.)

Hunan Fine Arts Publishing House, Hunan

227×177×45mm

1273g

488p

2005 Ha[3]

2005 Bm[1]

Bronze Medal

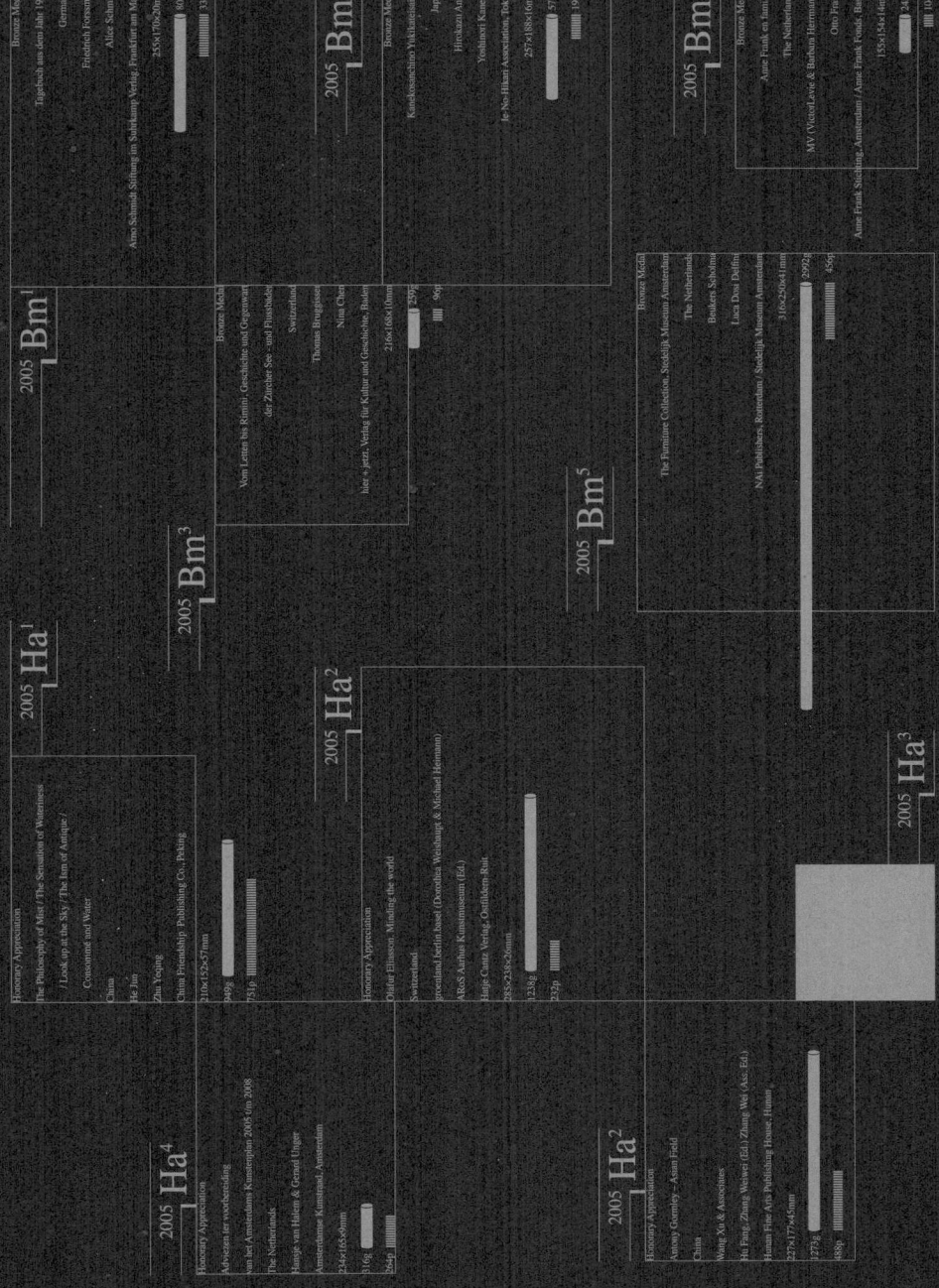

Tagebuch aus dem Jahr 1954

Germany

Friedrich Forssman

Alice Schmidt

Arno Schmidt Stiftung im Suhrkamp Verlag, Frankfurt am Main

255×170×20mm

803g

334p

2005 Bm[3]

Bronze Medal

Vom Letten bis Rimini, Geschichte und Gegenwart der Zürcher See - und Flussbäder

Switzerland

Thomas Bruggisser

Nina Chen

hier + jetzt, Verlag für Kultur und Geschichte, Baden

216×168×10mm

259g

96p

2005 Bm[2]

Bronze Medal

Kanekosanchino Yukikateisaien

Japan

Hirokazu Ando

Yoshinori Kaneko

Ie-No-Hikari Association, Tokyo

257×188×16mm

576g

192p

2005 Bm[5]

Bronze Medal

The Furniture Collection, Stedelijk Museum Amsterdam

The Netherlands

Beukers Scholma

Luca Dosi Delfini

NAi Publishers, Rotterdam / Stedelijk Museum Amsterdam

316×250×41mm

2992g

456p

2005 Bm[4]

Bronze Medal

Anne Frank en familie

The Netherlands

MV (VictorLevie & Barbara Herrmann)

Otto Frank

Anne Frank Stichting, Amsterdam / Anne Frank Fonds, Basel

155×154×14mm

245g

104p

The International Jury

Julia Blume (Germany)

Lotta Kühlhorn (Sweden)

Ed Marquand (USA)

Thomas M. Müller (Germany)

Birgit Tümmers (Germany)

Tatiana Wagenbach-Stephan (Switzerland)

Wim Westerveld (The Netherlands)

Designer

Mitsuo Katsui

The Remingtons (Ludovic Balland & Jonas Voegeli) mit Michael Stauffer und Christian Waldvogel

Walter Pamminger

Julie Savasky & DJ Stout (Pentagram, Austin)

Friedrich Forssman

Hirokazu Ando

Thomas Bruggisser

MV (VictorLevie & Barbara Herrmann)

Beukers Scholma

He Jun

Wang Xu & Associates

-SYB- (met dank aan Robin Uleman)

Hansje van Halem & Gerard Unger

groenland.berlin.basel (Dorothea Weishaupt & Michael Heimann)

2005

Country / Region

The Netherlands 4

Switzerland 3

Japan 2

China 2

Austria 1

USA 1

Germany 1

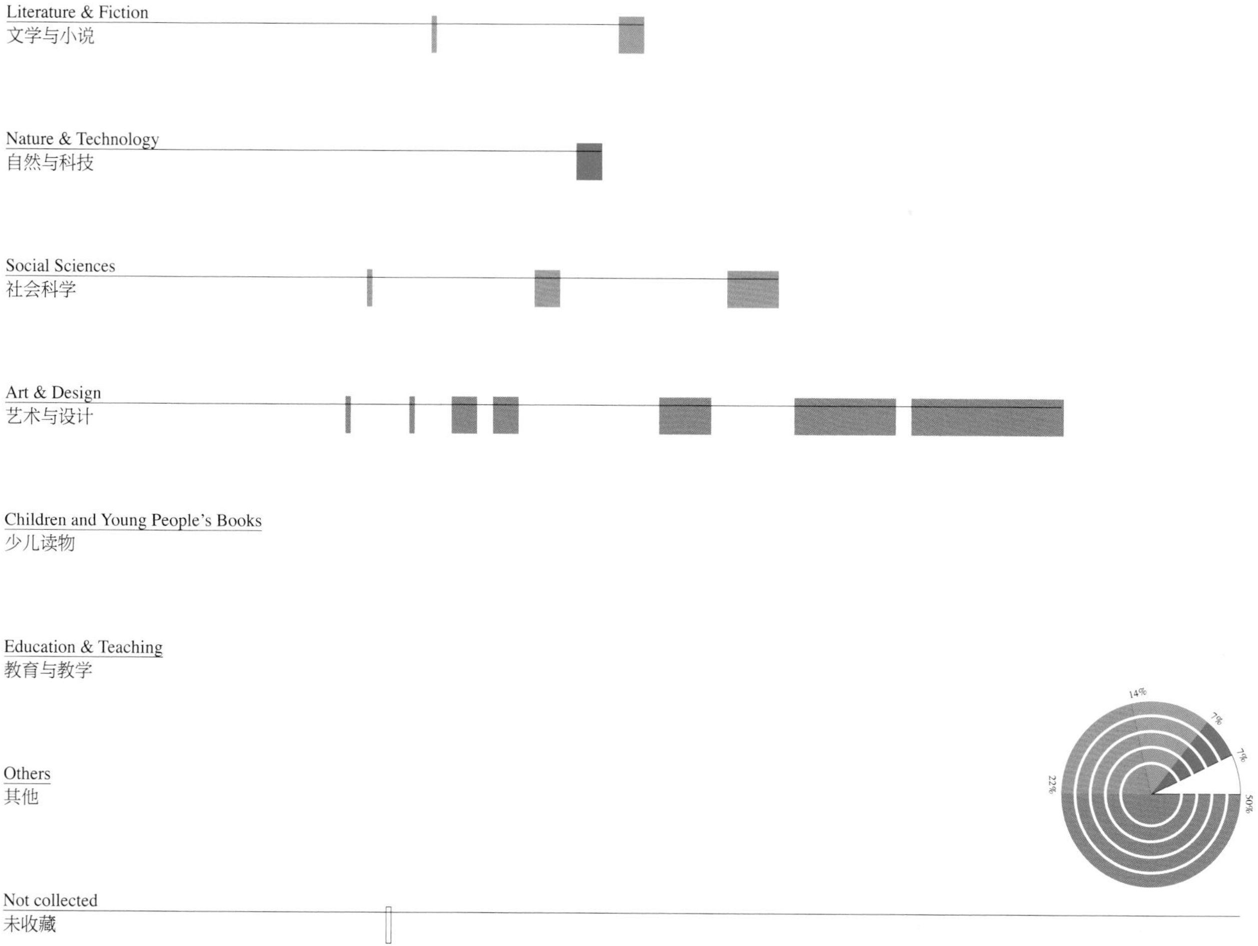

Literature & Fiction
文学与小说
Nature & Technology
自然与科技
Social Sciences
社会科学
Art & Design
艺术与设计
Children and Young People's Books
少儿读物
Education & Teaching
教育与教学
Others
其他
Not collected
未收藏
14%
7%
7%
22%
50%

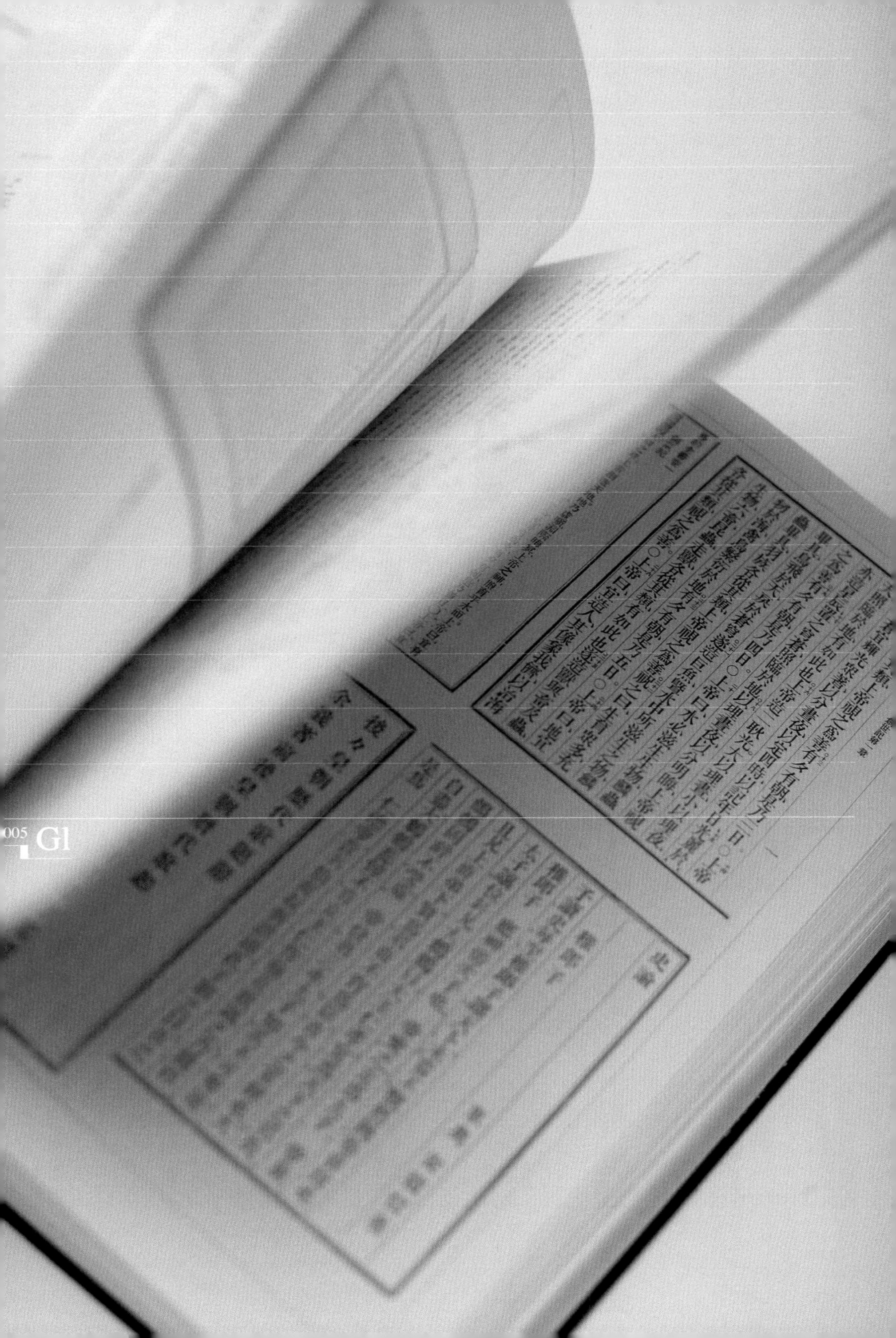

005 G1

这是一本有关日本近代活字历史的文献，由日本平面设计大师胜井三雄操刀设计。本书记录了近代活字印刷术从欧洲引入日本的历史、本木昌造的故事、活字印刷在日本的发展、活字印刷设备的改进等，内容详尽完满。本书的装订形式为锁线硬精装。书函为白色纸张裱覆，黑色印刷亮面 UV 工艺。封面为黑色绢面烫灰色和白色双色文字。内页图文混排，按照不同章节图文量和内容差异，分别采用纵向二、三、四栏的设计。书中包含大量手稿、图书封面、人物照片、印刷样张、印刷设备插图、拼版　结构示意、图表、活字字体、活字纹样、铸造技术等相关的图片，印刷精美，极具收藏价值。

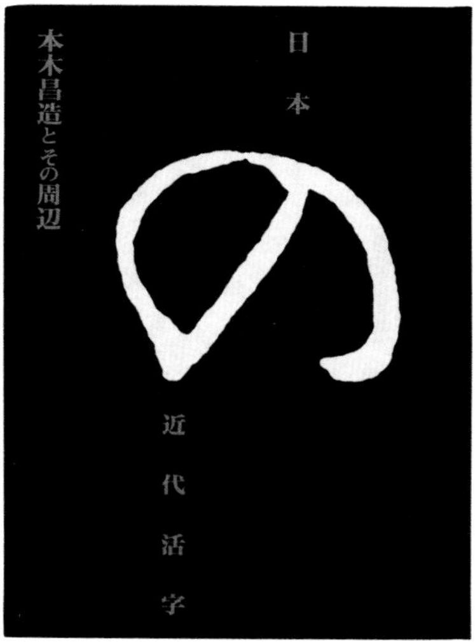
日本
の
近代活字
本木昌造とその周辺

This is a book about the history of modern movable type in Japan, designed by Japanese graphic design master Mitsuo Katsui. This book records the history of the introduction of modern movable-type printing from Europe to Japan, the story of Motoki Shōzō, the development of movable-type printing in Japan, and the improvement of movable-type printing equipment. The content is detailed and complete. The slipcase is mounted by white paper, printed in black and UV vanished. The cover is made of black silk with hot stamped text in grey and white. Graphics and text are combined on the side pages. According to the different amount and content of pictures and texts in each chapter, the pages are divided into 2 or 3 or 4 columns vertically. It contains a large number of manuscripts, book covers, portraits, printing samples, illustrations of printing equipment, diagrams, movable types, movable type patterns, casting technology and other related pictures, which are beautifully printed and highly valuable for collection.

- 金字符奖
- 日本近代活字
- 日本

An outstanding, opulent book with finely organised typography that remains *mobile*, never seeming rigid. The way in which pictures and text work together is compellingly and rhythmically composed. At the same time the book gives the impression of being accessible and informative. The overall finish is of the highest standard: simple, solid and harmonious, including attractively printed endpapers and a finely designed outer case.

- Golden Letter
- Nihon no Kindai Katsuji
- Japan
- Mitsuo Katsui
- Nihon no Kindai Katsuji (Ed.)
- Kindai Insatsu Katsuji Bunka Hozonkai, Nagasaki
- 304×216×35mm
- 2220g
- 453p

を得る
るに
是を銅板に打
を鋳込み以て
八米太人以
假亞中之近古利
心同
人之
而來
言者
擱入
ための
複製
造は
ノル鋳
のも

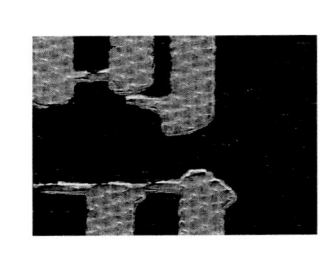

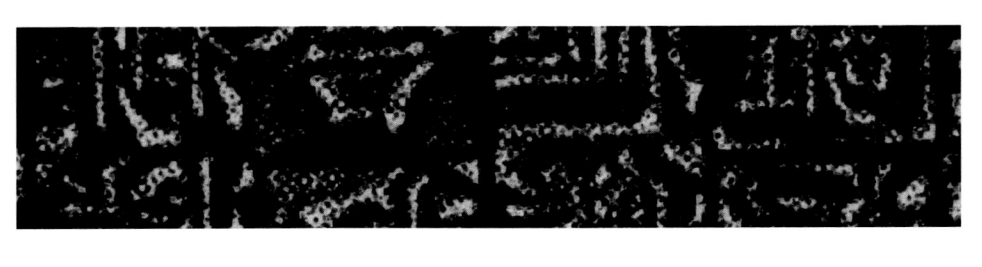

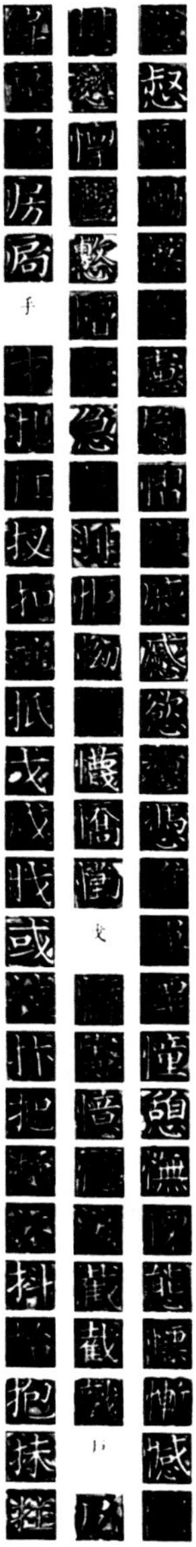

手

戈

戶

山

304×216×35

2220

453

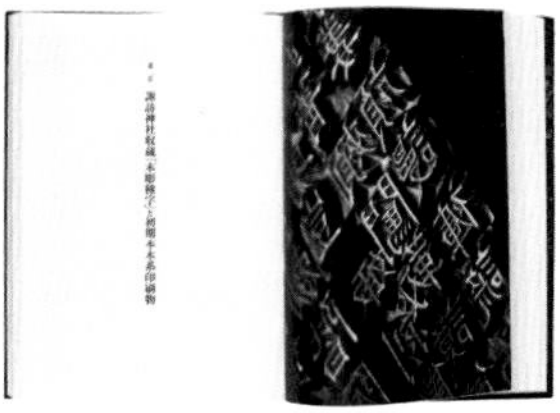

天下泰平國家安全

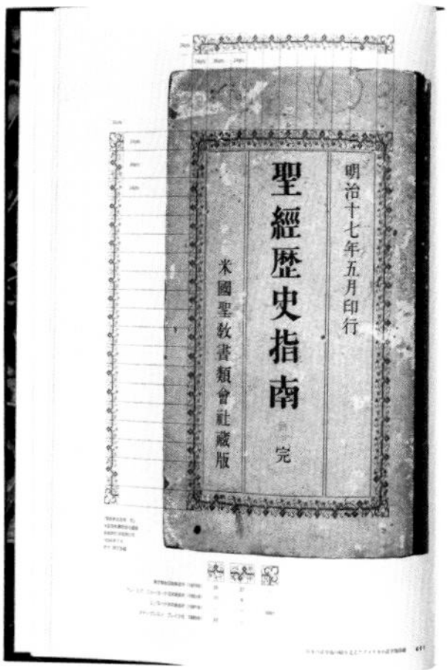
明治十七年五月印行
聖經歷史指南
完
米國聖教書類會社藏版
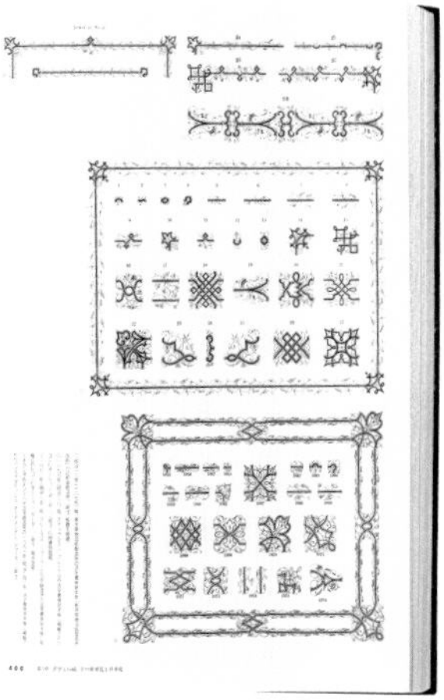
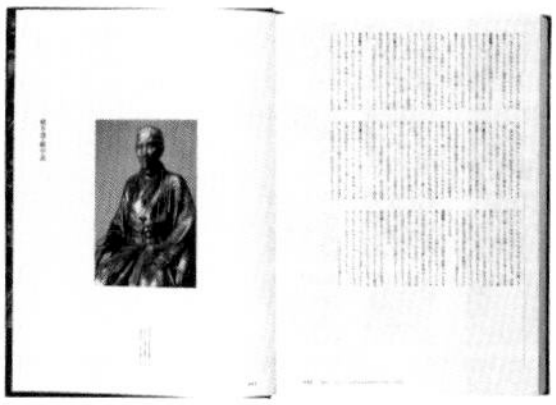
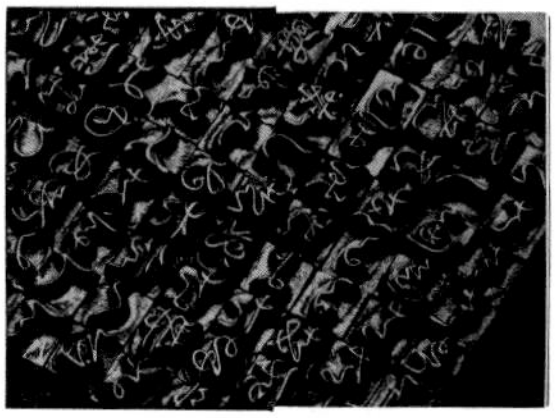

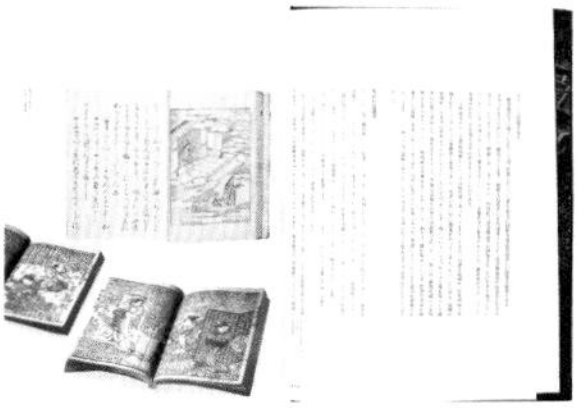

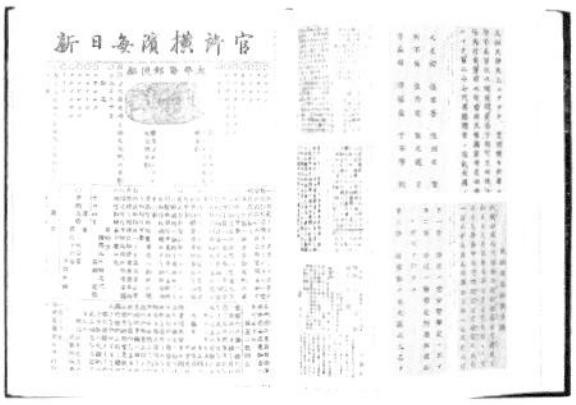
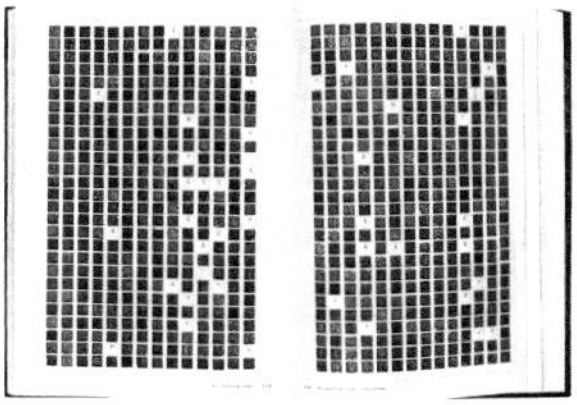
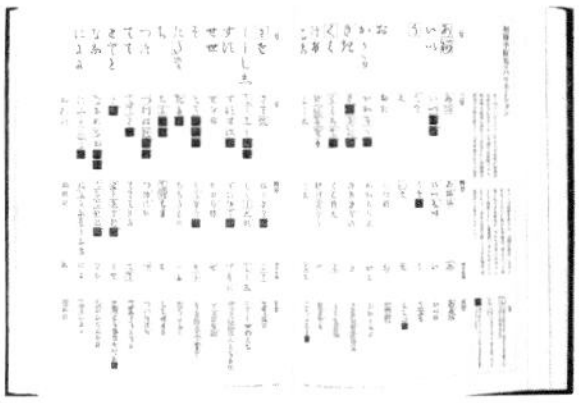
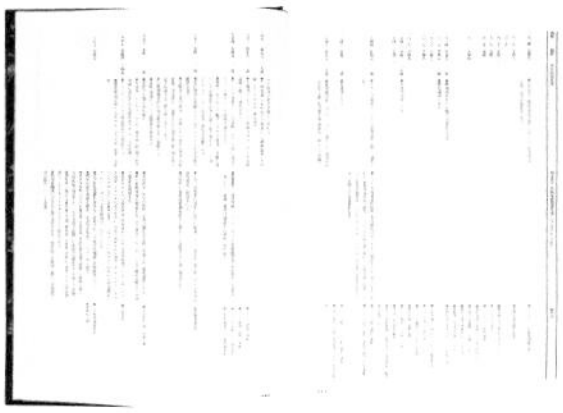
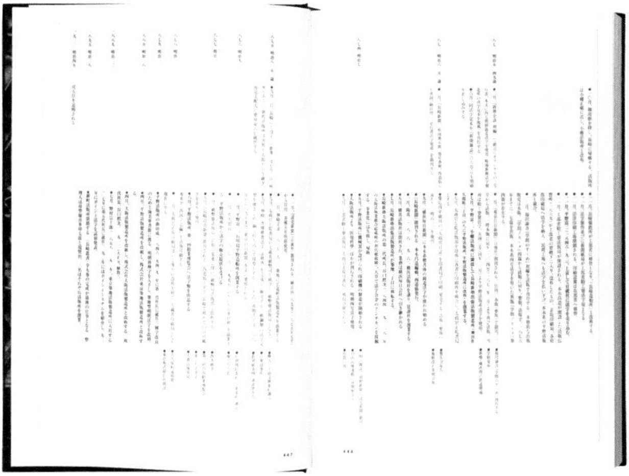

日本の近代活字

本木昌造とその周辺

近代印刷活字文化保存会

150%

三号漢字の本数は二、五〇〇本以上にのぼる。印刷物と参照した結果からも、相当の欠けがあることが分かるが、それでもセットとして保存する意図を感じさせるだけの本数である。一方で、日常的な印刷物には見出せないような字種も数多く含まれている。一体、オリジナルの漢字書体の製造に当たって、本木昌造らは何を手がかりに字種を選定したのだろうか。

ひとつ思いつくのが、前稿で鈴木広光が詳述しているW・ガンブルによる漢字使用頻度調査である。ガンブルは使用頻度調査によって得られた五、一五〇の字種にロンドン伝道会所有の活字見本帳から八五〇字を補充して、漢字一書体の必要字種を六千字と定めたが、明朝活字がガンブルのもたらした鉛活字を種字としてつくられたとすれば、本木らのオリジナル活字も明朝漢字の字種に準じたとしても不思議ではない。あるいは『漢字使用頻度調査書』そのものがガンブルによってもたらされたとも考えられる。

そこで、ガンブルの『漢字使用頻度調査書』(小宮山博史提供)に掲載されている六千字と、諏訪神社収蔵の三号楷書活字の字種を照合した。結果は、約二千字はガンブルの選定した六千字に含まれ、四〇〇字以上がリストから外れていた。

以下、三号漢字の全文字を原則的に『大漢和辞典』(諸橋轍次編、大修館書店刊行)の部首順にならって掲げた。誤字などについても同様に含めてある。ガンブルによる字種と同じとみられるものについてはGマークを、また重複字種にはWマークを付した。

グループA

グループB

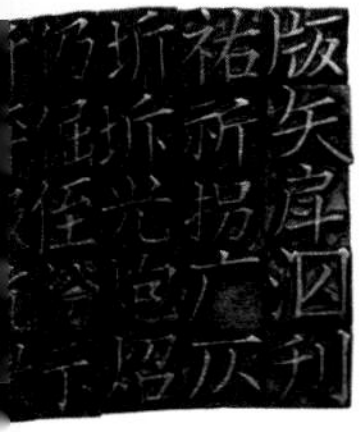

グループC

●三号楷書漢字の書風について

三号楷書体の種字を観察すると、全般に楷書特有のエレメントによって成り立っているといえるが、字面のとりかた、字画の構成、画線の抑揚などにばらつきがあり、必ずしも書風は一定していない。

木駒のサイズにもばらつきが見られるが、注目されるのは活字の高さと書風に相関関係が見られることである。例えば二一・五ミリ以上の活字高の高いグループ約二〇〇本には、やや小振りで右上がりの細い画線をもつもの（グループA）が集中し、最も低い活字高のグループ約三〇〇本には、字面が大きく筆画の抑揚や右上がりの度合いの小さいもの（グループB）が多いなど明確な偏りが見られた。書風の差と認められるものは五—六グループほどはあるとみられ、木駒のサイズと書風との関係も、版下や彫りを分業した結果であろうと考えられる。

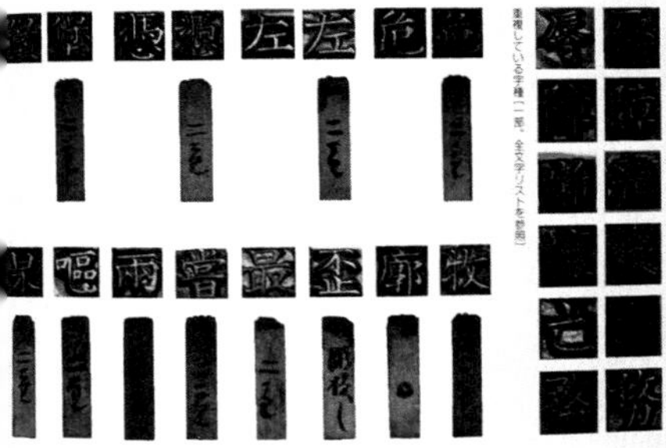

重複している字種（一部。全文字リストを参照）

●三号楷書漢字の重複字種とボディの書き込み

種字の中には、七〇組ほど重複した文字種が見つかっている。行書らしいものも含まれているが、これら重複する字種の種字の中には、ボディに筆による書き込みがあるものが四本あり、書き込みの内容から字種の重複と関係があるとも考えられる。しかし、書き込みのある種字であっても重複のないものが多く、ボディの書き込みと彫られた文字との関連は明らかではない。

●三号行書漢字

わずかに含まれる行書と思われる種字のうち、数文字が「西洋古史略」の印面に見られる文字の字形と一致している。ただし、行書のくずしの度合いに開きがあること、照合した印刷物の中にも楷書との混在が見られることなどから、楷書と行書を区別することは難しく、それぞれの本数は確定できていない。

諏訪神社収蔵種字

「西洋古史略」中の行書

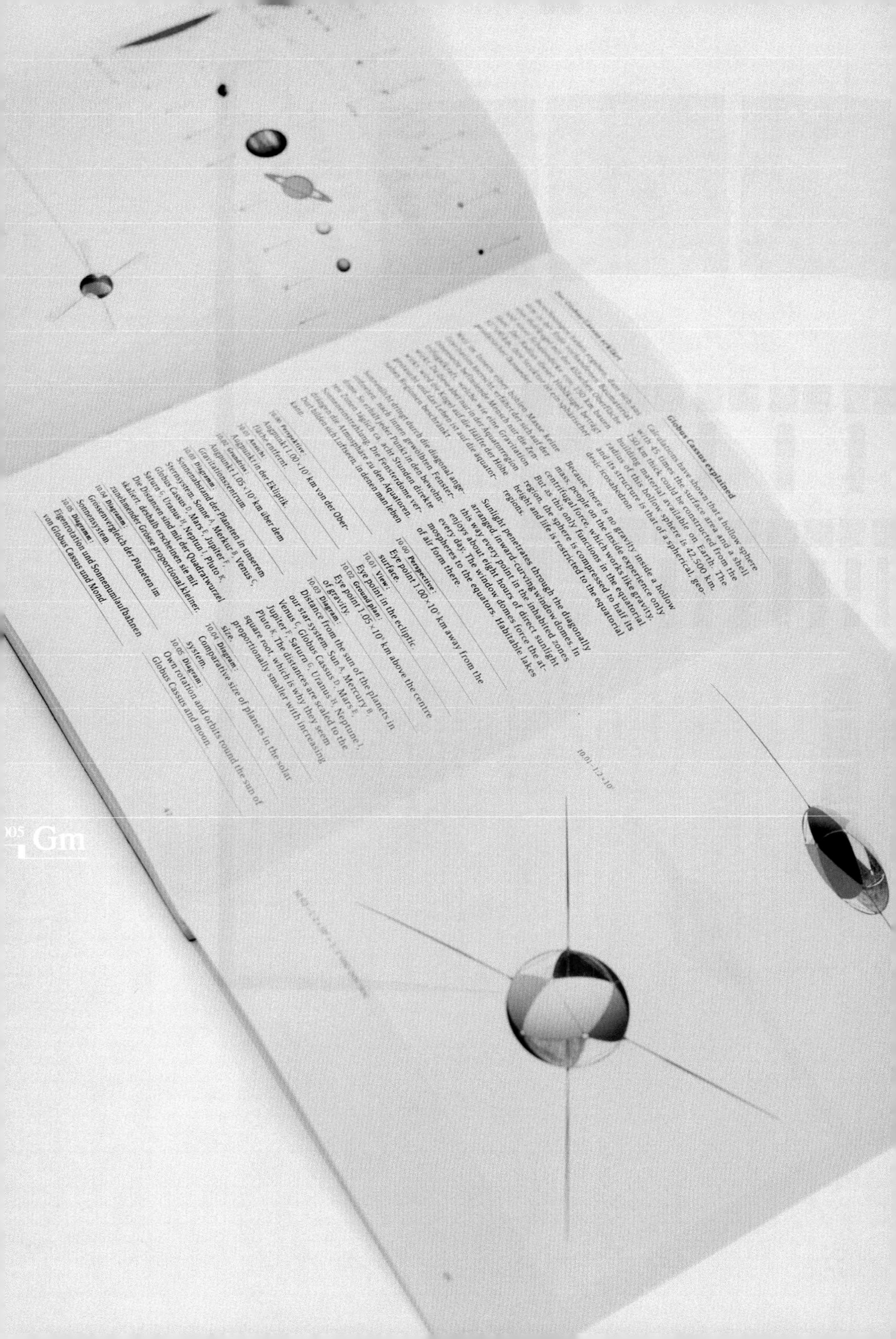

Globus Cassus explained
Calculations have shown that a hollow sphere with 45 times the surface area and a shell 150 km thick could be constructed from the building material available on Earth. The radius of this hollow sphere is 42,500 km, and its structure is that of a spherical, geodesic icosahedron.
Because there is no gravity inside a hollow mass, people on the inside experience only centrifugal force, which works like gravity. But as this only functions in the equatorial region, the sphere is compressed to half its height and life is restricted to the equatorial regions.
Sunlight penetrates through the diagonally arranged, inward-curving window domes. In this way every point in the inhabited zones enjoys about eight hours of direct sunlight every day. The window domes force the atmospheres to the equators. Habitable lakes of air form there.
10.00 Perspective: Eye point 1.00×10⁷ km away from the surface.
10.01 View: Eye point in the ecliptic.
10.02 Ground plan: Eye point 1.05×10⁷ km above the centre of gravity.
10.03 Diagram: Distance from the sun of the planets in our star system. Sun A, Mercury B, Venus C, Globus Cassus D, Mars E, Jupiter F, Saturn G, Uranus H, Neptune I, Pluto K. The distances are scaled to the square root, which is why they seem proportionally smaller with increasing size.
10.04 Diagram: Comparative size of planets in the solar system.
10.05 Diagram: Own rotation and orbits round the sun of Globus Cassus and moon.
10.00 Perspektive: Augpunkt 1.00×10⁷ km von der Oberfläche entfernt.
10.01 Ansicht: Augpunkt in der Ekliptik.
10.02 Grundriss: Augpunkt 1.05×10⁷ km über dem Gravitationszentrum.
10.03 Diagramm: Sonnenabstand der Planeten in unserem Sternsystem. Sonne A, Merkur B, Venus C, Globus Cassus D, Mars E, Jupiter F, Saturn G, Uranus H, Neptun I, Pluto K. Die Distanzen sind mit der Quadratwurzel skaliert, deshalb erscheinen sie mit zunehmender Grösse proportional kleiner.
10.04 Diagramm: Grössenvergleich der Planeten im Sonnensystem.
10.05 Diagramm: Eigenrotation und Sonnenumlaufbahnen von Globus Cassus und Mond.
10.01 – 1:2×10⁷
47
005 Gm

这本书记录了瑞士建筑师和艺术家克里斯蒂安·瓦尔德福格尔（Christian Waldvogel）的作品，为 2004 年威尼斯建筑双年展的参展项目。艺术家将地球假想为中空结构，并以此构建了人类在这个结构空间中的生存环境和建设过程。本书为锁线软精装，封面黑色皮面烫印银色，书脊烫印黑色，层次多元，质感丰富。本书以多语言排版加之大量图文混排，在内容的编排上独具一格。第一章节的每张图都有一独立编号，后跟德英双语详述，模拟用网络时代的交互特征，以下画线式的编号、页码“超链接”与图片建立关系。第二章则通过拉页的方式展现图表和德英双语旁述。以文本为主的章节则将德英双语分开呈现。最后偏黄的纸张用以印刷法文和意大利文的介绍和注解，配以相关“链接”以便查阅图片。整体设计严谨而独特，展现出浓重的学术艺术氛围。||||||||||||

GLOBUS
CASSUS

This book is an artistic project of the Swiss architect and artist Christian Waldvogel, which was exhibited in the 2004 Venice architecture biennale. The artist imagined the earth as a hollow structure and designed the living environment and construction process of human beings in this space. The book is bound by thread sewing with a soft hardcover. The cover is made of black leather with silver hot stamping, while the spine is black hot stamping with multiple layers and rich texture. This book is unique in content arrangement due to its multilingual layout and large amount of mixed pictures and texts. In the first chapter, each figure is numbered independently, followed by details in German and English. By simulating the interactive characteristics in the Internet era, numbers and page numbers are underlined as 'hyperlinks' to establish a relationship with the pictures. The ending yellow pages are used to print introductions and notes in French and Italian, with relevant 'links' for viewing images. The overall design is rigorous and unique, showing a strong academic and artistic atmosphere.

- 金奖
- 克里斯蒂安·瓦尔德福格尔的中空集群
- 瑞士

A mind game and hypothetical blueprint on a fictitious world taking the form of a scientific brochure whose contents are given a particularly objective and therefore all the more realistic appearance through well-balanced typographical features. From the choice of material for the cover through to the typography, the use of colour and choice of illustrations - a most compelling book both in terms of content and design.

- Gold Medal
- Christian Waldvogel. Globus Cassus
- Switzerland
- The Remingtons (Ludovic Balland & Jonas Voegeli) mit Michael Stauffer und Christian Waldvogel
- Bundesamt für Kultur (Ed.)
- Lars Müller Publishers, Baden
- 215 × 158 × 15mm
- 432g
- 182p

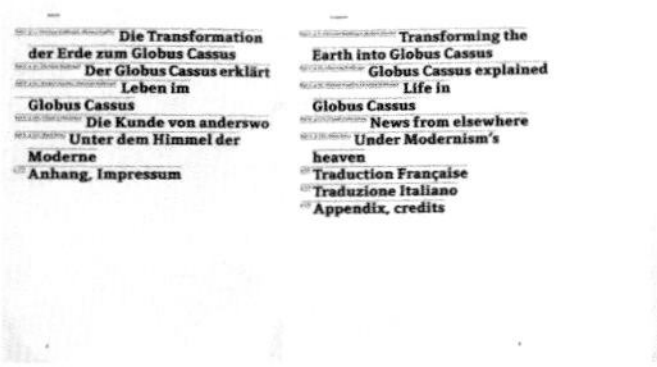
Die Transformation der Erde zum Globus Cassus
Der Globus Cassus erklärt
Leben im Globus Cassus
Die Kunde von anderswo
Unter dem Himmel der Moderne
Anhang, Impressum

Transforming the Earth into Globus Cassus
Globus Cassus explained
Life in Globus Cassus
News from elsewhere
Under Modernism's heaven
Traduction Française
Traduzione Italiano
Appendix, credits

Die Transformation der Erde zum Globus Cassus
Transforming the Earth into Globus Cassus

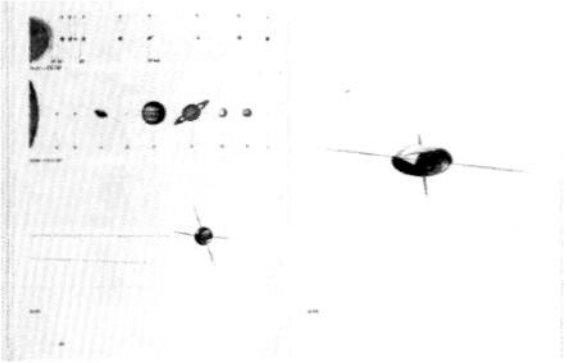

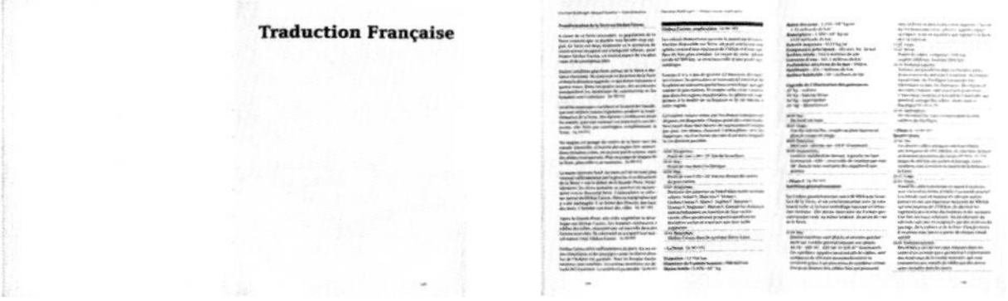

Traduction Française

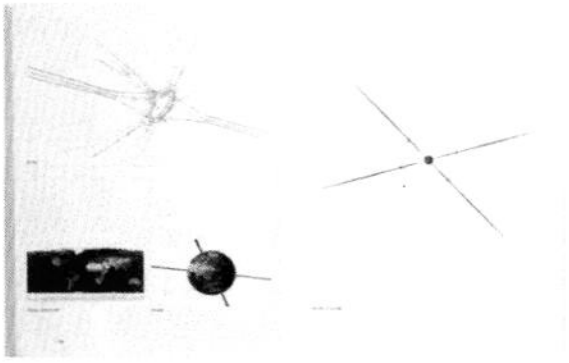

Die Erde
The Earth

Aus dem Erdinneren wird Magma zu den Knoten heraufgepumpt und aufgeschäumt. Das Magma wird zu einer sehr dünnen Schale geformt. In bestimmte Zonen werden nicht Schalen, sondern nach innen gewölbte, transparente Fensterdome gelegt. Je mehr Magma aus der Erde weggepumpt wird, desto mehr schrumpft sie.

Magma from inside the Earth is pumped up to the nodes and expanded. The magma is made into a very thin shell. In certain areas, no shells are made, but transparent, inward-curving window domes. The more magma is pumped away from the Earth, the more the Earth shrinks.

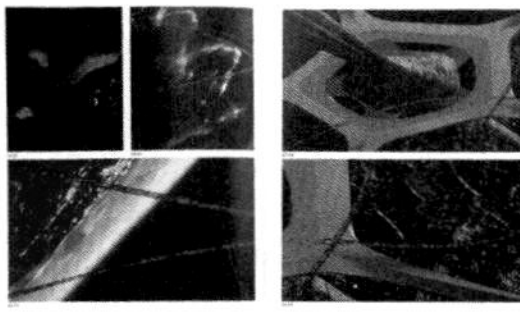

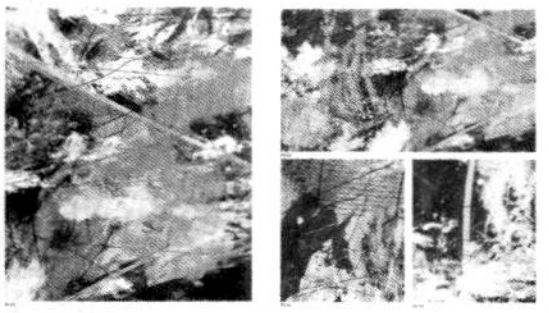

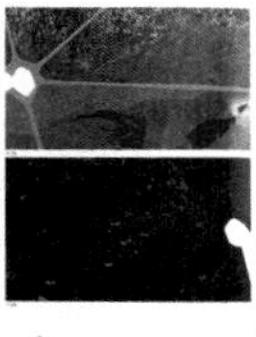

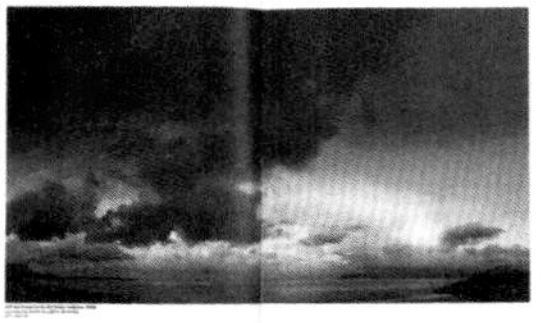

.02

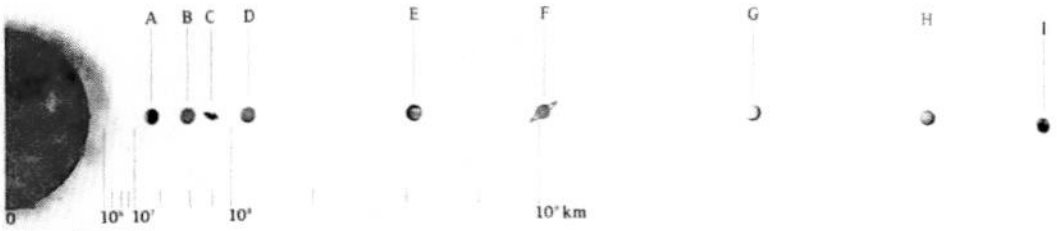
A B C D E F G H I
0 10⁶ 10⁷ 10⁸ 10⁹ km

10.03 – $\sqrt{1:6\times10^{9}}$

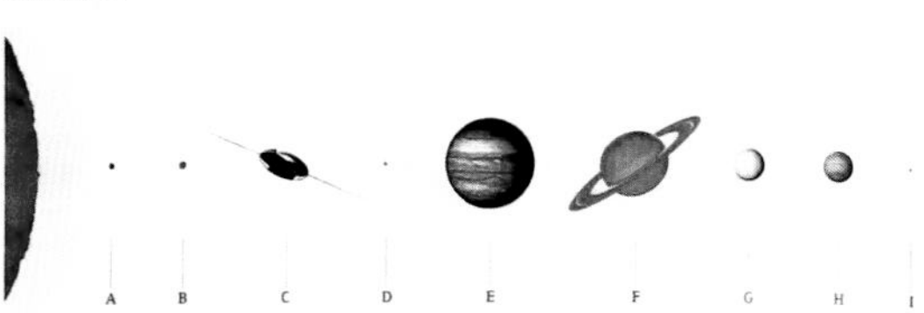
A B C D E F G H I

10.04 – $1:3.5\times10^{10}$

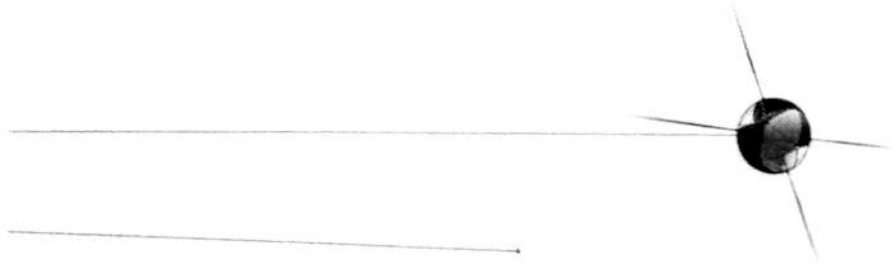

10.05

Der Globus Cassus erklärt

Berechnungen haben ergeben, dass sich dem in der Erde vorhandenen Baumate eine Hohlkugel mit der 45fachen Oberflä und einer Schalendicke von 150 km ba lässt. Der Radius dieser Hohlkugel bet 42'500 km, ihre Struktur ist ein sphärisc geodätischer Ikosaeder.

Weil im Innern einer hohlen Masse ke Gravitation herrscht, erfährt der sich auf Innenseite befindende Mensch nur die Z trifugalkraft, welche wie eine Gravita wirkt. Da diese aber nur in der Äquatorreg wirkt, wird die Kugel auf die Hälfte der H gestaucht und das Leben ist auf die äqua nahen Regionen beschränkt.

Sonnenlicht dringt durch die diagonal ar ordneten, nach innen gewölbten Fens dome. So erhält jeder Punkt in den bewo ten Zonen täglich ca. acht Stunden dire Sonneneinstrahlung. Die Fensterdome v drängen die Atmosphäre zu den Äquator Dort bilden sich Luftseen, in denen man le kann.

10.00 **Perspektive:**
Augpunkt 1.00×10^{5} km von der Ober- fläche entfernt.

10.01 **Ansicht:**
Augpunkt in der Ekliptik.

10.02 **Grundriss:**
Augpunkt 1.05×10^{5} km über dem Gravitationszentrum.

10.03 **Diagramm:**
Sonnenabstand der Planeten in unsere Sternsystem. Sonne A, Merkur B, Venu Globus Cassus D, Mars E, Jupiter F, Saturn G, Uranus H, Neptun I, Pluto K. Die Distanzen sind mit der Quadratwu skaliert, deshalb erscheinen sie mit zunehmender Grösse proportional klei

10.04 **Diagramm:**
Grössenvergleich der Planeten im Sonnensystem.

10.05 **Diagramm:**
Eigenrotation und Sonnenumlaufbah von Globus Cassus und Mond.

›bus Cassus explained

culations have shown that a hollow sphere :h 45 times the surface area and a shell) km thick could be constructed from the ilding material available on Earth. The lius of this hollow sphere is 42,500 km. l its structure is that of a spherical, geo-sic icosahedron.

:ause there is no gravity inside a hollow ss, people on the inside experience only trifugal force, which works like gravity. t as this only functions in the equatorial ion, the sphere is compressed to half its ght and life is restricted to the equatorial ions.

ılight penetrates through the diagonally anged, inward-curving window domes. In s way every point in the inhabited zones oys about eight hours of direct sunlight ery day. The window domes force the at-spheres to the equators. Habitable lakes air form there.

)0 **Perspective**:
:ye point 1.00 × 10' km away from the urface.

)1 **View**:
:ye point in the ecliptic.

)2 **Ground plan**:
:ye point 1.05 × 10' km above the centre of gravity.

)3 **Diagram**:
)istance from the sun of the planets in ur star system. Sun A, Mercury B, /enus C, Globus Cassus D, Mars E, upiter F, Saturn G, Uranus H, Neptune I, Pluto K. The distances are scaled to the quare root, which is why they seem roportionally smaller with increasing ize.

)4 **Diagram**:
:omparative size of planets in the solar ystem.

)5 **Diagram**:
)wn rotation and orbits round the sun of ilobus Cassus and moon.

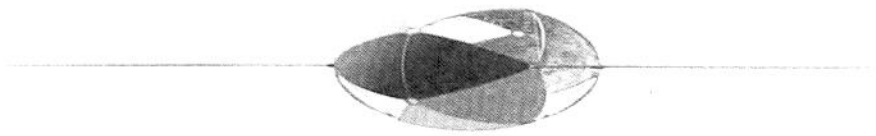

10.01 – 1:2 × 10'

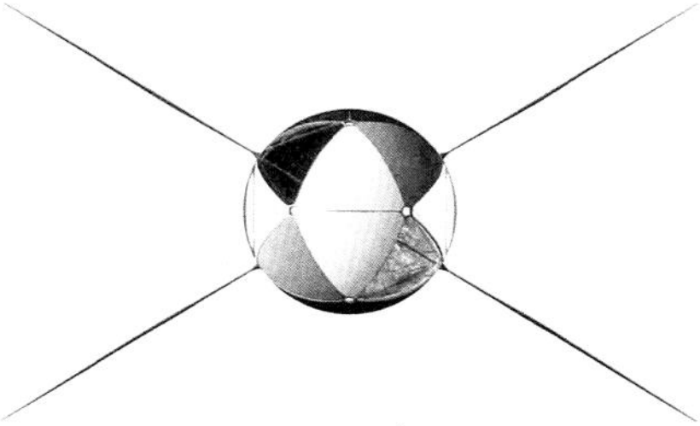

10.02 – 1:2 × 10^{9} = 1: 2'000'000'000

am selben Tag schrieben
an Erich Führer:
Freud, Adolfine Freud
werden, zeichnen wir
Erwartung,
weitere Anordnu
Entjudete
Entrechtung und
oder der Wohnung
Deportation
November 1941
Juden, die das
Vertreibung
Tür 9
Zelle 2
männliche Juden insgesamt
Wohnung auf Tür 9 aus 1 Zi
Juden, darunter
bei 17 Wohnung
Stadtteil Roßau zu den
gehörten in den
über die Situation
nung der Familie
ng, in die seit
delte sich

05 Sm1

本书是一本研究报告和相关展览的配套图书，主要记录了精神分析学家西格蒙德·弗洛伊德（Sigmund Freud）曾经的居所 Berggasse 大街 19 号住宅的相关研究。书中包含建筑结构、楼层分析图表、弗洛伊德曾经的邻居们的资料等，用以研究这些人住在这里时候的生活状态、犹太人受纳粹影响后的史实。书籍为硬精装灰纸表面烫黑加烫白工艺，封面仿照建筑楼层结构进行设计，在书的正文部分有更详细的楼层立面和平面图。全书黑白印刷，内页部分设置为双栏，主栏为正文，辅栏为注解，配合大量通栏、跨版心设置的图标和照片，使得内页排版呈现出超越题材本身的沉重感。书中大量资料表的影印，为读者还原出最真实的历史人文脉络。

Freuds
verschwundene
Nachbarn
3. Stock
2. Stock
1. Stock
Mezzanin
MUSEUM
Hochparterre
Parterre

The book is research report matched with a related exhibition, recording the research about 19 Berggasse Street, the former residence of the psychoanalyst Sigmund Freud. The book contains architectural structure, charts of each floor, and data of Freud's former neighbors, which are used to study the lives of these people and the history of Jews under the influence of the Nazis. The hardback book is using gray paper with black hot stamping and white hot stamping. The cover is designed according to the floor plan of the building. In the body part of the book, there are more detailed chart of building elevations and floor plans. The whole book is printed in black and white, and the inside pages are double columned – the main column is for the text and the auxiliary column is for the annotations. With a large number of full-column icons and photos, the layout of the inside pages represents the heaviness beyond the subject itself. The photocopied data sheets in the book manage to restore the most authentic historical reality for readers.

- 银奖
- 弗洛伊德那些消失的邻居们
- 奥地利

Successful combination of the original with its commentary. Beginning with an original text, without title-pages, it is immediately clear to the reader what the subjectmatter is. Clear typography is situated behind the illustrations, complementing and elucidating them. Superb choice of type, corresponding well with that used by typewriters.

- Silver Medal
- Freuds verschwundene Nachbarn
- Austria
- Walter Pamminger
- Lydia Marinelli (Ed.)
- Turia + Kant, Wien
- 292 × 190 × 13mm
- 540g
- 127p

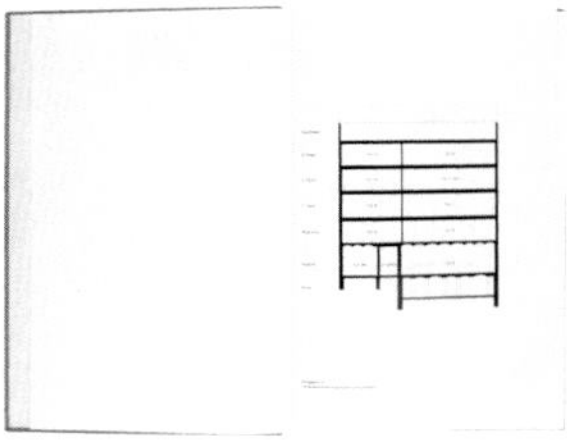

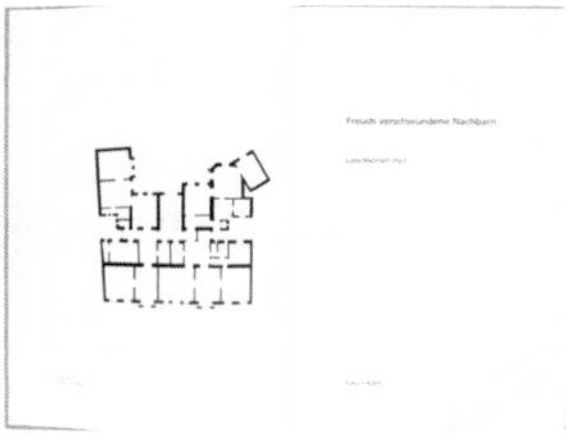

Freuds verschwundene Nachbarn

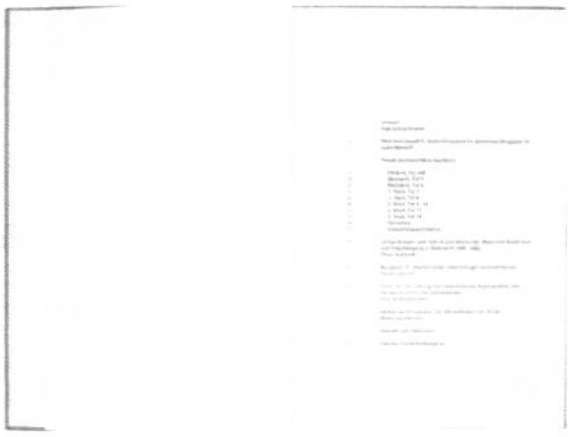

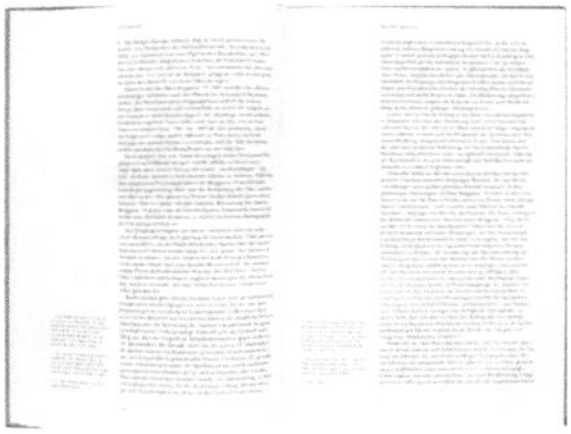

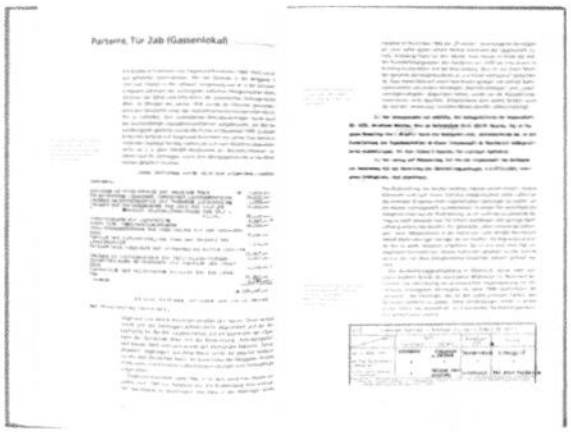

Parterre, Tür 2ab (Gassenlokal)

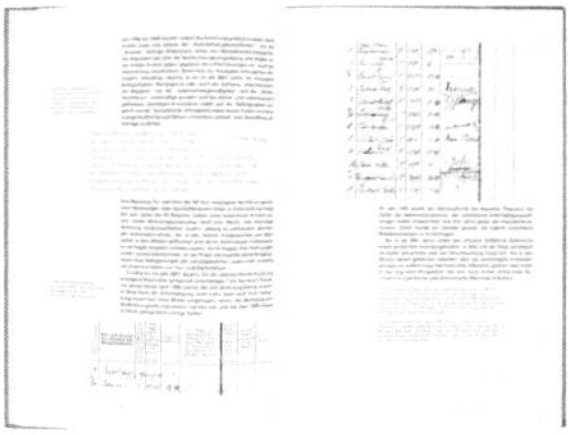

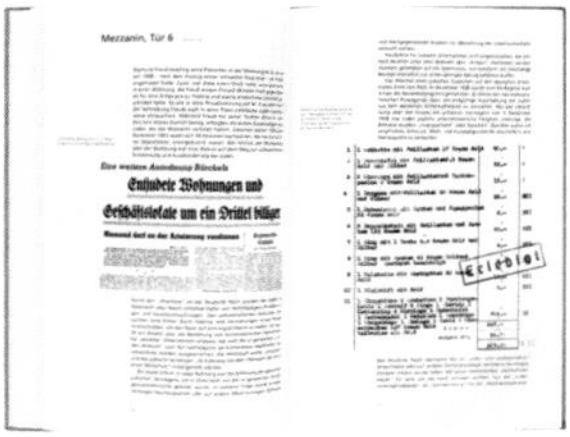

Mezzanin, Tür 6
Entjudete Wohnungen und
Geschäftslokale um ein Drittel billiger

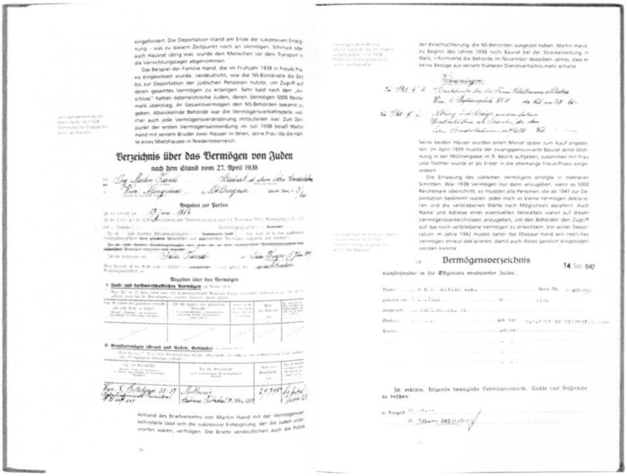

eingefordert. Die Deportation stand am Ende der sukzessiven Enteignung – was zu diesem Zeitpunkt noch an Vermögen, Schmuck oder auch Hausrat übrig war, wurde den Menschen vor dem Transport in die Vernichtungslager abgenommen.

Das Beispiel der Familie Hand, die im Frühjahr 1939 in Freuds Praxis eingewiesen wurde, verdeutlicht, wie die NS-Bürokratie die Zeit bis zur Deportation der jüdischen Personen nutzte, um Zugriff auf deren gesamtes Vermögen zu erlangen. Sehr bald nach dem „Anschluss" hatten österreichische Juden, deren Vermögen 5000 Reichsmark überstieg, ihr Gesamtvermögen den NS-Behörden bekannt zu geben. Abwickelnde Behörde war die Vermögensverkehrsstelle, welcher auch jede Vermögensveränderung mitzuteilen war. Zum Zeitpunkt der ersten Vermögensanmeldung im Juli 1938 besaß Martin Hand mit seinem Bruder zwei Häuser in Wien, seine Frau Ida die Hälfte eines Mietshauses in Niederösterreich.

Verzeichnis über das Vermögen von Juden
nach dem Stand vom 27. April 1938

Angaben zur Person

Angaben über das Vermögen

Anhand des Briefverkehrs von Martin Hand mit der Vermögensverkehrsstelle lässt sich die sukzessive Enteignung, der die Juden unterworfen waren, verfolgen. Die Briefe verdeutlichen auch die Politik

der Einschüchterung, die NS-Behörden ausgeübt haben. Martin Hand, zu Beginn des Jahres 1938 noch Baurat bei der Streckenleitung in Wels, informierte die Behörde im November desselben Jahres, dass er keine Bezüge aus seinem früheren Dienstverhältnis mehr erhalte.

Seine beiden Häuser wurden einen Monat später zum Kauf angeboten. Im April 1939 musste der zwangspensionierte Baurat seine Wohnung in der Müllnergasse im 9. Bezirk aufgeben, zusammen mit Frau und Tochter wurde er als Erster in die ehemalige Freud-Praxis eingewiesen.

Die Erfassung des jüdischen Vermögens erfolgte in mehreren Schritten. War 1938 Vermögen nur dann anzugeben, wenn es 5000 Reichsmark überschritt, so mussten alle Personen, die ab 1941 zur Deportation bestimmt waren, jedes noch so kleine Vermögen deklarieren und die verbliebenen Werte nach Möglichkeit abliefern. Auch Name und Adresse eines eventuellen Verwalters waren auf diesen Vermögensverzeichnissen anzugeben, um den Behörden den Zugriff auf das noch verbliebene Vermögen zu erleichtern. Vor seiner Deportation im Jahre 1942 musste daher das Ehepaar Hand sein restliches Vermögen erneut deklarieren, damit auch dieses gänzlich eingezogen werden konnte.

Vermögensverzeichnis

14 Sep 1942

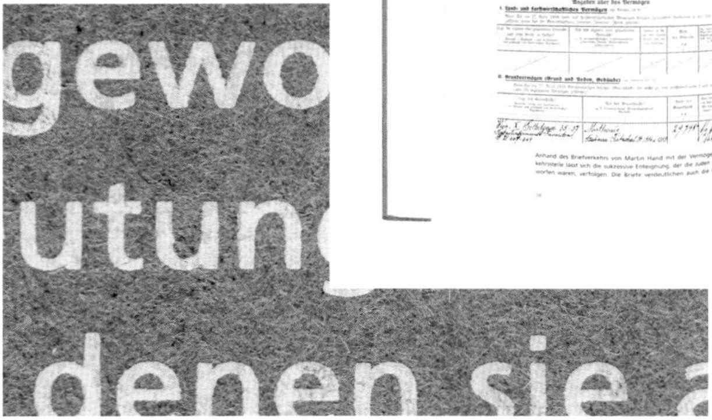

gewo
utung
denen sie a

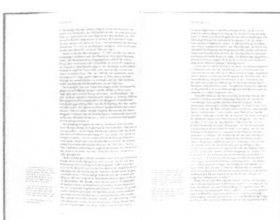

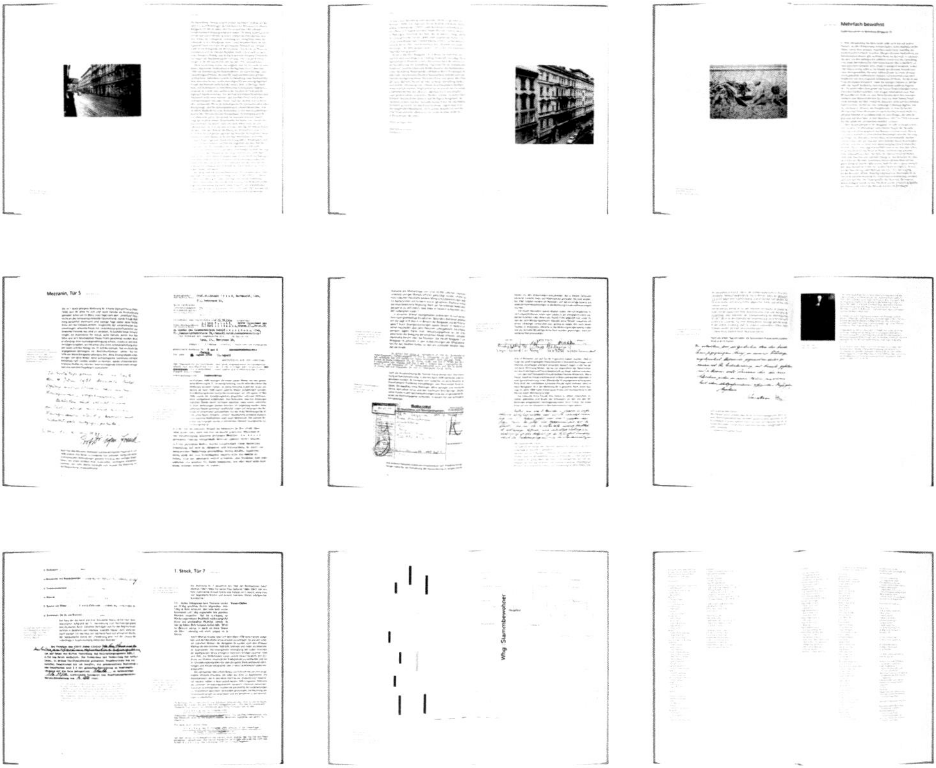
Mehrfach bewohnt
Mezzanin, Tür 5
1. Stock, Tür 7
Whg. Stammbewohner

er au
gewe
ns des Käufers

3. Stock, Tür 14

Überleben

Victor John (1870–1942) wohnte seit 1904 mit seiner Frau Antoinette (1879–1941) und seinen beiden Töchtern auf Tür Nr. 14 in der Berggasse 19. Mehr als 30 Jahre arbeitete er als Direktor der Versicherungsgesellschaft Riunione Adriatica di Sicurtà in Wien. Wenige Wochen nach dem März 1938 wurde er in den Ruhestand versetzt, mit ungewisser Aussicht auf eine Firmenpension. Auch er wurde ab 1939 gezwungen, ihm zugewiesene Juden in seine Wohnung aufzunehmen. Im Sommer 1939 teilten er und seine Frau sich die Wohnung mit einem älteren Ehepaar, bis 1941 stieg die Anzahl der zwangsweise einquartierten Personen in ihrer Wohnung auf zwölf. Victor John versuchte trotz seines fortgeschrittenen Alters zusammen mit seiner Frau im September 1941 nach Uruguay auszuwandern. Es gelang ihm zwar noch, sich eine Schiffspassage zu verschaffen, doch nicht mehr, die Reise anzutreten. Mit Jahresbeginn 1941 hatten bereits die Transporte in das besetzte Gebiet in Polen eingesetzt, und nur mehr wenige Juden schafften die Flucht aus Wien. Victor John musste im November 1941 seine Wohnung in der Berggasse aufgeben, in ein Sammellager in der Großen Sperlgasse ziehen und wurde von dort aus mit seiner Frau nach Lodz deportiert, von wo beide nicht mehr wiederkehrten.

Karteikarte, Victor John, 4.9.1941, Kartei der Devisenabteilung der Israelitischen Kultusgemeinde Wien Archiv der Israelitischen Kultusgemeinde Wien

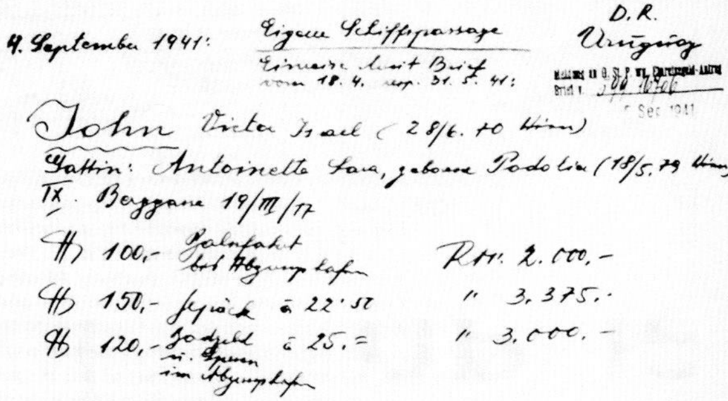
4. September 1941: Eigene Schiffspassage
Einreise laut Brief vom 18. 4. resp. 31. 5. 41:
D.R. Uruguay
5 Sep 1941

John Victor Israel (28/6. 70 Wien)
Gattin: Antoinette Sara, geborne Podolia (18/5. 79 Wien)
IX. Berggasse 19/III/14
$ 100,- Bahnfahrt bei Abgangshafen — RM. 2.000,-
$ 150,- Gepäck à 22.50 — " 3.375,-
$ 120,- Bordgeld u. Spesen im Abgangshafen à 25.= — " 3.000,-

Von den Bewohnern, mit denen das Ehepaar sich die Wohnung Nr. 14 zwischen 1939 und 1941 teilen musste, kehrten einige aus den Konzentrationslagern zurück: Elisabeth Winternitz wurde aus dem KZ Stutthof befreit, Emil und Ida Ehrenstein überlebten in Theresienstadt.

Emil Ehrenstein zählte zu jener Opfergruppe, die aufgrund ihres politischen Status einigermaßen rasch Anspruch auf Fürsorgeleistungen erhielt. Unmittelbar nach dem „Anschluss" wurde er für 80 Tage inhaftiert, weil er im Jänner 1938 einen Nationalsozialisten angezeigt hatte. Im September 1942 deportierte man ihn und seine Frau Ida nach Theresienstadt. Während ihr einziger Sohn in Auschwitz starb, wurden Emil und Ida Ehrenstein 1945 befreit und gingen nach dem Krieg wieder zurück nach Wien. Emil Ehrenstein bekam zunächst den Opferausweis ausgestellt, der im Wesentlichen nur geringfügige steuerliche Vorteile brachte. Unter Hinweis auf seine Inhaftierung im März

Antrag auf Ausstellung einer Amtsbescheinigung, Emil Ehrenstein an den Magistrat Wien, 12.6.1948
MA 12 – wien sozial

1938 und auf den „aktiven Einsatz" gegen den Nationalsozialismus, der ihm vom Bund der politisch Verfolgten bestätigt wurde, erhielt er schließlich die so genannte Amtsbescheinigung, die ihn zum Bezug einer Rente berechtigte.

Da ich nach Anzeige gegen einen Nazi im Jänner 1938 bei der Wr. Polizei 80 Tage hindurch (25/III – 15/VI 1938) in Haft war und 3 volle Jahre im K.Z. Theresienstadt zubringen mußte, wo ich körperlich misshandelt und dadurch eine schwere Herzerkrankung und gleich meiner Frau einen seelischen Zusammenbruch davontrug, woran ich noch heute leide, bin ich der Ansicht, dass mir auf Grund oben zitirten Gesetzes nach §1 Abs. d + e die Amtsbescheinigung zuzuerkennen ist.

Nach 1945 wurden zuerst nur jene als Opfer des Nationalsozialismus anerkannt, die aufgrund ihrer widerständischen Aktivitäten während des NS-Regimes verfolgt wurden. Die größte Opfergruppe, die aufgrund ihrer Herkunft Verfolgten, trat erst spät in das öffentliche Bewusstsein. Dementsprechend sah auch das erste Opferfürsorgegesetz von 1945 Leistungen nur für die „Opfer des Kampfes um ein freies, demokratisches Österreich" mit österreichischer Staatsbürgerschaft vor. Hingegen blieben zur Flucht gezwungene Österreicher, die mittlerweile eine andere Staatsbürgerschaft angenommen hatten, über weite Strecken von Ansprüchen auf Leistungen ausgeschlossen.

Bescheinigung des KZ-Verbandes für Emil Ehrenstein, 28.2.1947
MA 12 – wien sozial

Der KZ-Verband, Verband der österreichischen antifaschistischen Konzentrationslager-Schutzhäftlinge, hat im Sinne der Bestimmung des Abschnittes I, Abs. (4) 1. OFE., bei Behörden, sowie bei den vom Anspruchwerber

Name: Ehrenstein Emil geb.: 8.II.68.

wohnhaft: Wien 2, Malzgasse 7.

Mitglied des KZ-Verbandes Nr. 3473

ımhaft gemachten Zeugen und Mithäftlingen Erhebungen gepflogen, denen zufolge die Annahme begründet erscheint, daß der Inhaftnahme und Anhaltung des Anspruchswerbers in einem KZ (Anhaltelager) der politische Grund eines aktiven Einsatzes nach § 1, Abs. (1), II. OFV zugrunde lag.

Vor allem außenpolitische Motive bewogen die Politiker der Zweiten Republik, Österreich von jeglicher Mitschuld an den Verbrechen des Nationalsozialismus freizusprechen. Deshalb betonte man in den ersten Monaten nach Kriegsende auch die Rolle der „Kämpfer um ein freies, demokratisches Österreich". KZ-Überlebende hingegen, die keiner politischen Gruppierung zugeordnet werden konnten, stellten für die Regierungsvertreter der Zweiten Republik allzu oft eine Beunruhigung dar; die Geschichte ihrer Verfolgung belegte ja gerade die Beteiligung vieler Österreicher an den Verbrechen des Nationalsozialismus.

Die These von Österreich als erstem Opfer von Nazideutschland wirkte auch in den Regelungen der ersten Opferfürsorgegesetze weiter. Diese bevorzugten die Widerstandskämpfer gegenüber denen,

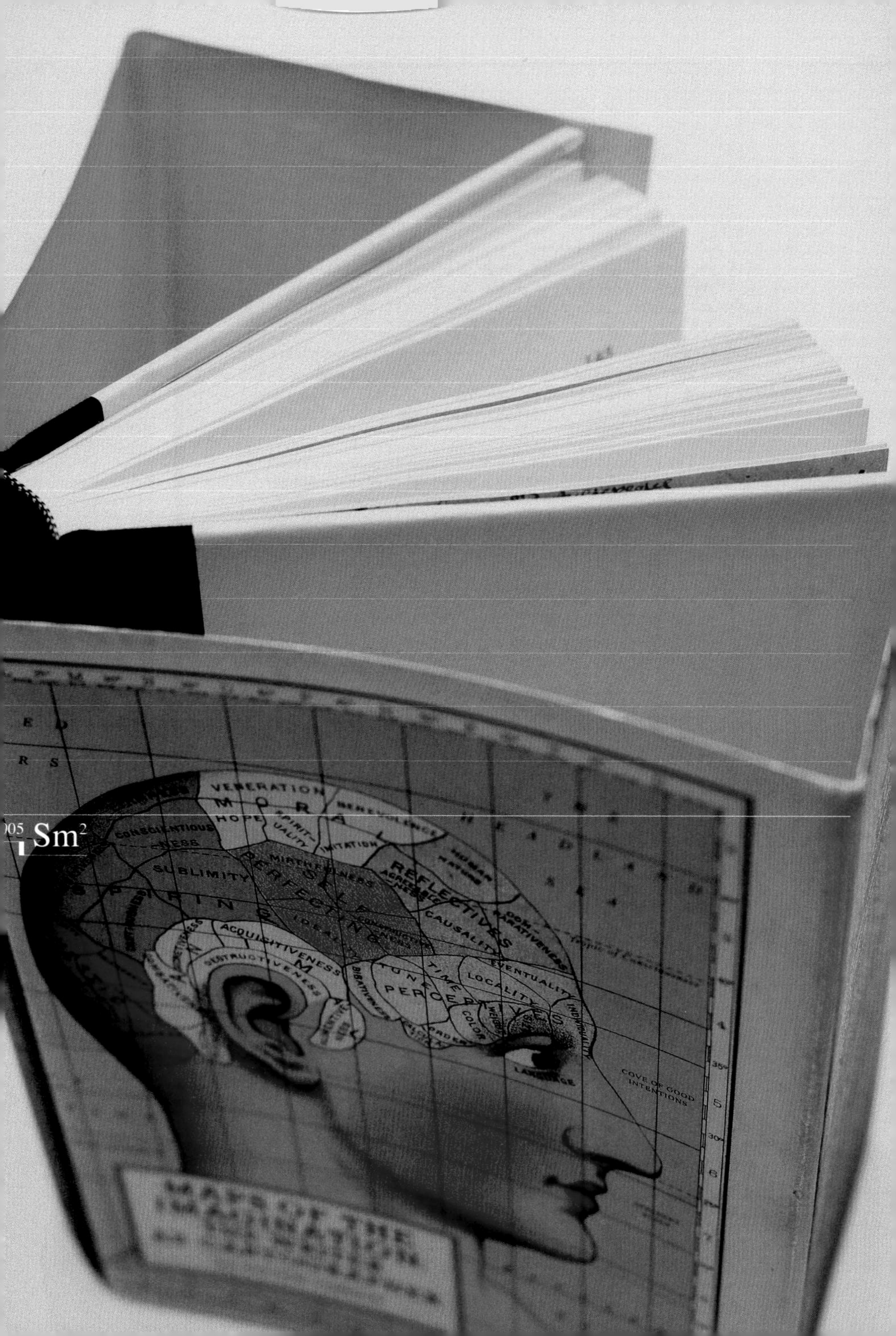
Sm²
VENERATION
MORAL
HOPE
SPIRIT-UALITY
BENEVOLENCE
IMITATION
CONSCIENTIOUSNESS
SUBLIMITY
MIRTHFULNESS
REFLECTIVES
CAUSALITY
ACQUISITIVENESS
DESTRUCTIVENESS
EVENTUALITY
LOCALITY
TIME
TUNE
PERCEPTIVES
COLOR
ORDER
LANGUAGE
INDIVIDUALITY
COVE OF GOOD INTENTIONS

这是一本有关大脑如何工作、抽象与具象思维如何协同工作绘制地图的书。作者通过大量蕴含在文字中的抽象隐喻与绘图家描绘物理世界的几何表现之间的联系，阐述地图绘制者创造的无限魅力。本书为布面圆脊硬精装，书脊烫银，精装封面外包内侧覆膜的四色印护封，地图画法与充满文字的大脑“地图”结合，具有强烈的视觉冲击力。内页四色印刷，书中段落之间使用地图的分界线来做分隔，多种类型的地图表现手法充满趣味，即使未必完全精准，却能通过出奇不意的想象力，呈现出精准地图所不具备的艺术价值和研究价值。

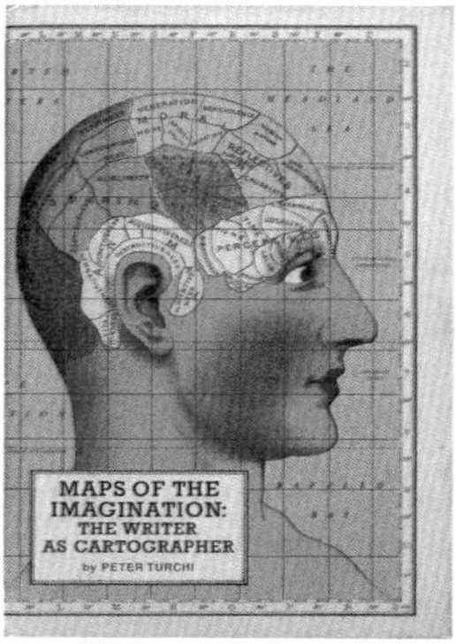
MAPS OF THE IMAGINATION:
THE WRITER AS CARTOGRAPHER
by PETER TURCHI

This is a book about how the brain works and how abstract thinking and concrete imaginary thinking coordinate to make maps. The author illustrates the infinite charm created by cartographers through the connection between the abstract metaphor hidden in words and the geometric representation of the physical world depicted by cartographers. The hardback book has a rounding flexiback and a cloth cover, and silver hot stamping is used on the spine. The four-color printed jacket is laminated inside, and the combination of map drawing and the brain 'map' has strong visual impact. The inside page is printed in four colors as well, and the borderlines used in maps are designed to separate the paragraphs. The expression methods of various types of maps are interesting – even though they may not be completely accurate, they are of artistic value and worthy of research by their amazing imagination.

- 银奖
- 想象的地图：作为制图师的作家
- 美国

◎	Silver Medal
«	Maps of the Imagination: The Writer as Cartographer
	USA
	Julie Savasky & DJ Stout (Pentagram, Austin)
△	Peter Turchi
	Trinity University Press, San Antonio, TX
□	218×149×29mm
	632g
	245p

One is already grabbed by the dust-jacket: a human head in profile portrayed as a continent on a historical maritime map. A label discloses the title. The style of an academic work pervades this small half-binding volume, the inside titles and chapter beginnings appear once again as labels, the typography is unfussy and shows great attention to detail: column titles in spaced capitals, scale lines used as an economical adornment. The richly coloured picture material is carefully incorporated into the type area or else placed on double-pages. The selected materials are harmonious. The overall appearance induces curiosity as to the book's content.

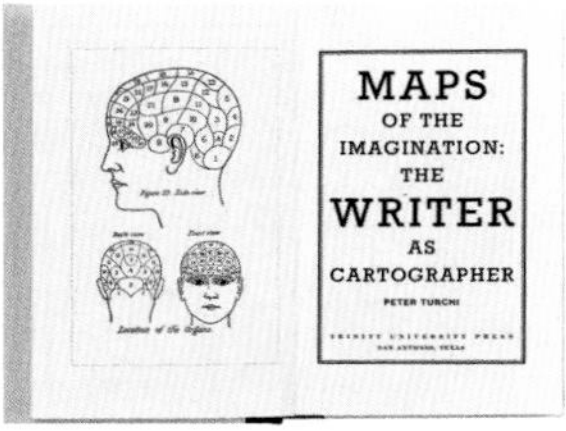

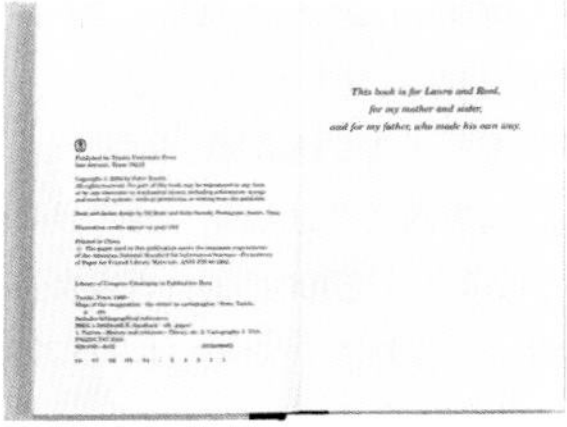

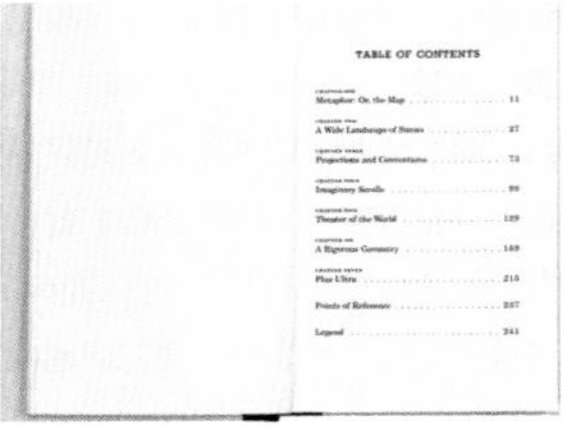

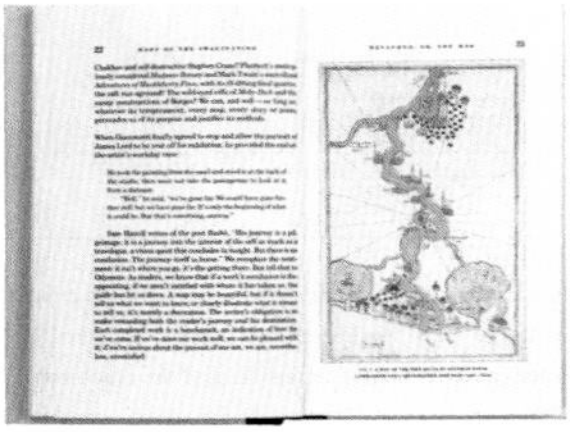

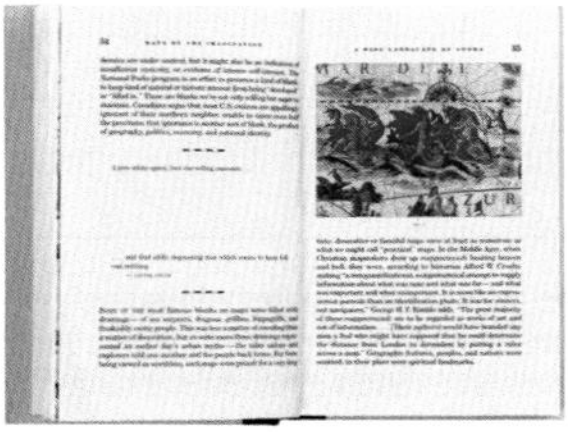

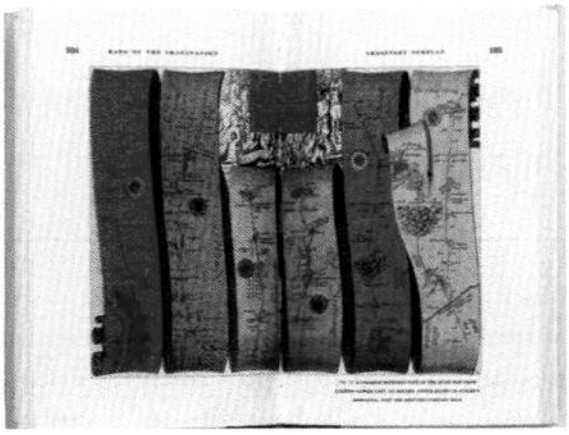

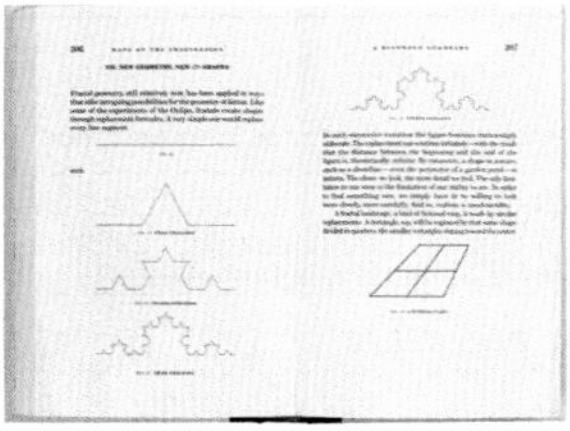

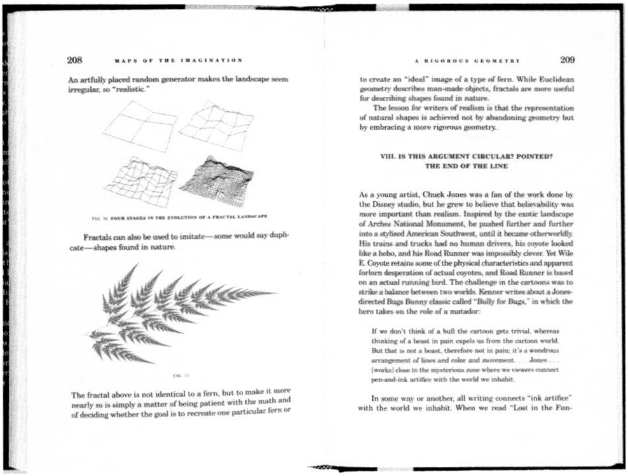

208 MAPS OF THE IMAGINATION

An artfully placed random generator makes the landscape seem irregular, so "realistic."

FIG. 70 FOUR STAGES IN THE EVOLUTION OF A FRACTAL LANDSCAPE

Fractals can also be used to imitate—some would say duplicate—shapes found in nature.

FIG. 71

The fractal above is not identical to a fern, but to make it more nearly so is simply a matter of being patient with the math and of deciding whether the goal is to recreate one particular fern or

A RIGOROUS GEOMETRY 209

to create an "ideal" image of a type of fern. While Euclidean geometry describes man-made objects, fractals are more useful for describing shapes found in nature.

The lesson for writers of realism is that the representation of natural shapes is achieved not by abandoning geometry but by embracing a more rigorous geometry.

VIII. IS THIS ARGUMENT CIRCULAR? POINTED? THE END OF THE LINE

As a young artist, Chuck Jones was a fan of the work done by the Disney studio, but he grew to believe that believability was more important than realism. Inspired by the exotic landscape of Arches National Monument, he pushed further and further into a stylized American Southwest, until it became otherworldly. His trains and trucks had no human drivers, his coyote looked like a hobo, and his Road Runner was impossibly clever. Yet Wile E. Coyote retains some of the physical characteristics and apparent forlorn desperation of actual coyotes, and Road Runner is based on an actual running bird. The challenge in the cartoons was to strike a balance between two worlds. Kenner writes about a Jones-directed Bugs Bunny classic called "Bully for Bugs," in which the hero takes on the role of a matador:

> If we don't think of a bull the cartoon gets trivial, whereas thinking of a beast in pain expels us from the cartoon world. But that is not a beast, therefore not in pain; it's a wondrous arrangement of lines and color and movement. . . . Jones . . . [works] close to the mysterious zone where we viewers connect pen-and-ink artifice with the world we inhabit.

In some way or another, all writing connects "ink artifice" with the world we inhabit. When we read "Lost in the Fun-

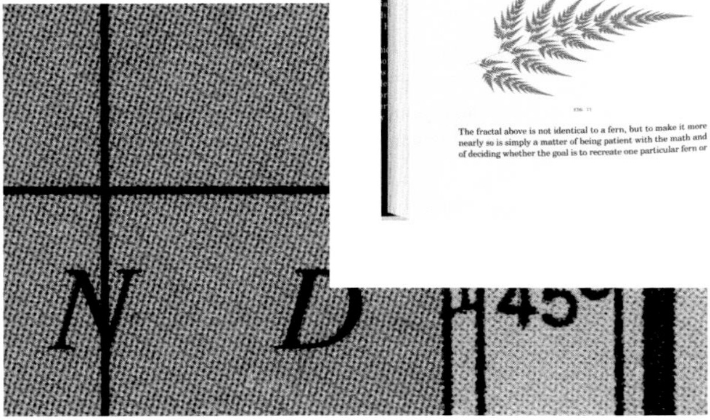

N
D
45°

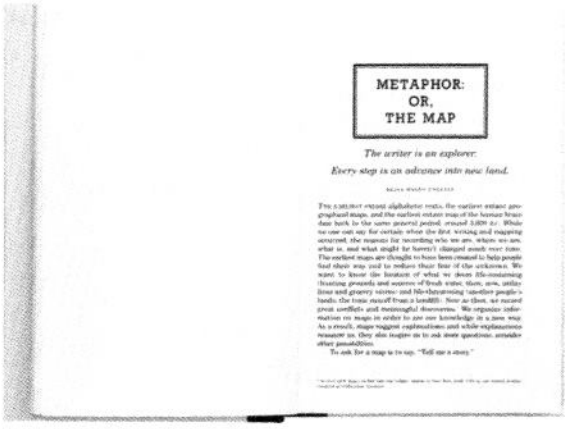
METAPHOR: OR, THE MAP
The writer is an explorer.
Every step is an advance into new land.
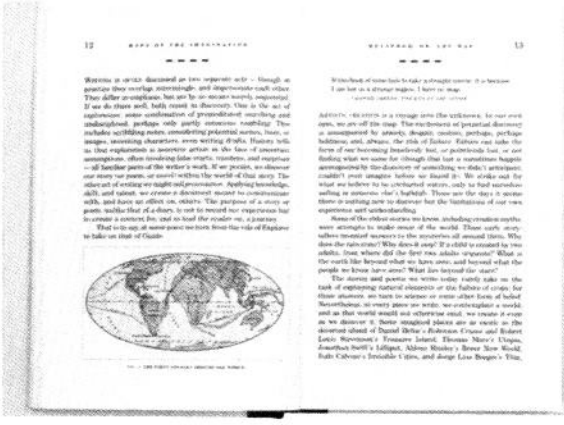
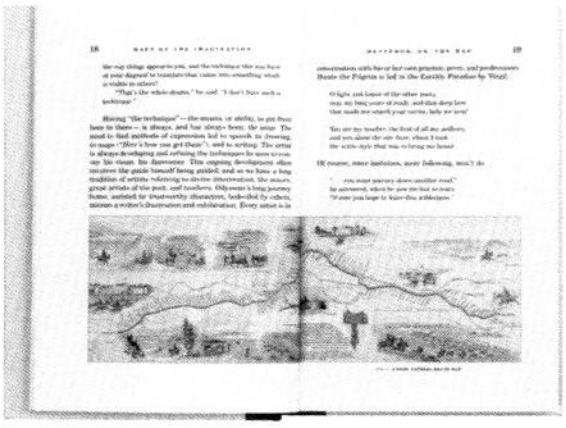
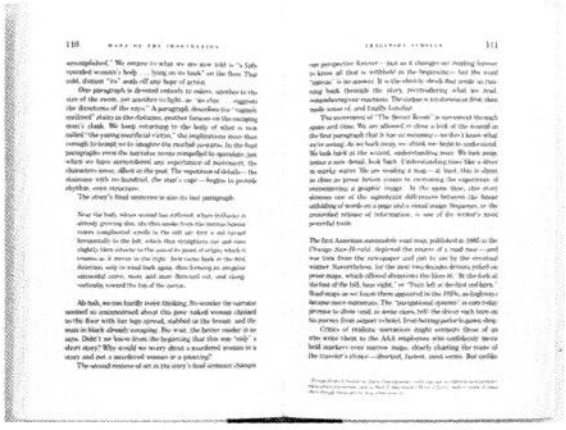
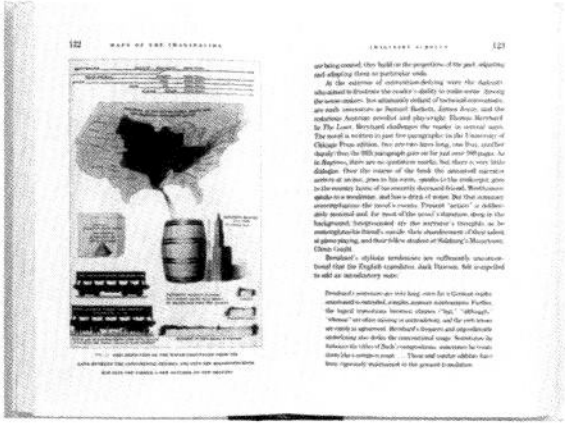
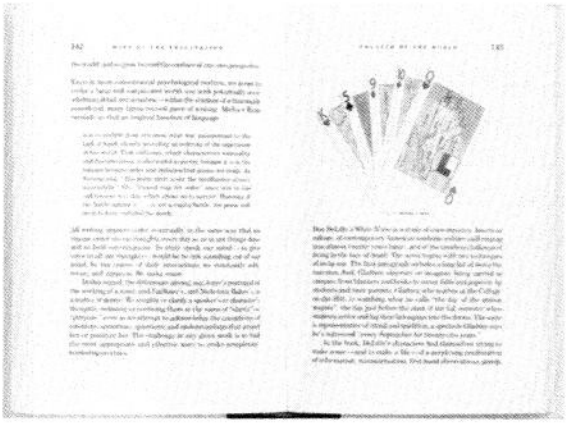
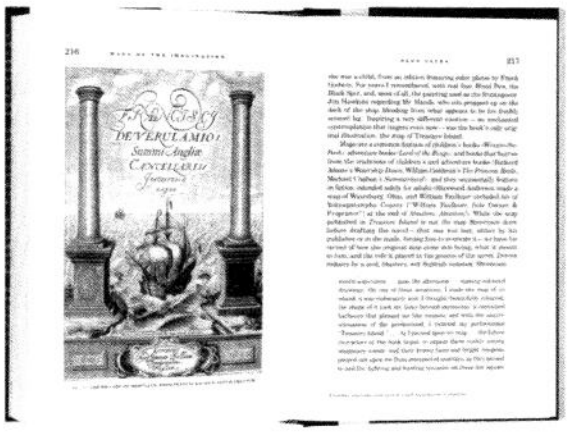
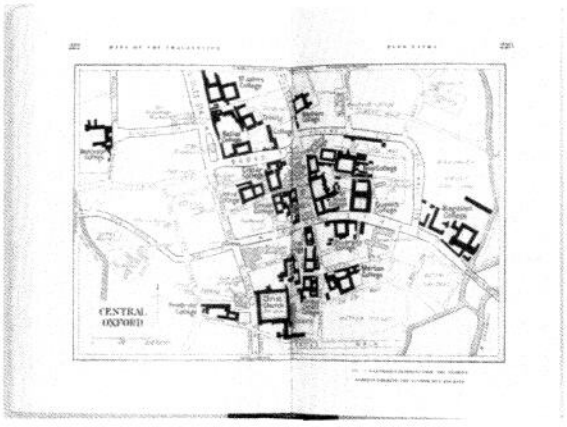
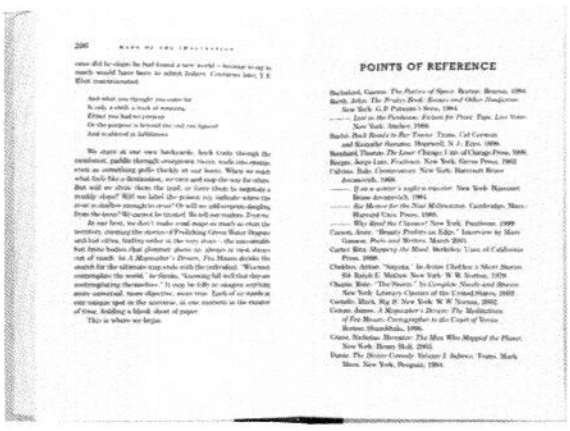
POINTS OF REFERENCE

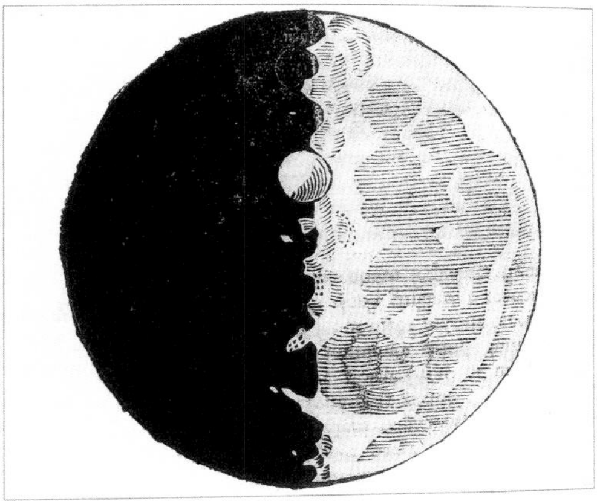

FIG. 72 ONE OF A SERIES OF COPPER ENGRAVINGS OF THE MOON BY GALILEO, ILLUSTRATING LUNAR SHADOWS

house," *The Unbearable Lightness of Being*, and *Lolita*, if we aren't persuaded to think of the characters in those narratives as people — if we aren't repulsed by Humbert Humbert, and occasionally amused—the fiction fails. But if we only feel impatient, in Barth's story, because it seems we'll never get to the beach, or feel exasperated, in Kundera's novel, because the narrator keeps "interrupting," if we only *feel*, we're missing the opportunity provided by fiction that means to engage our mind in other ways as well. Some of the work referred to as "postmodern" that is most difficult to access, even for dedicated readers, reads like integral calculus. Such work is more likely to be a tangent, in its extremity, than the next prevailing tradition. Nevertheless, a visit with the Oulipo-ets and other postrealists, via their work, is likely to be provocative and inspiring even to writers who have no intention of creating new methods of text generation.

John Barth used a phrase of Borges's to describe the blind librarian's stories; he said they combine "the algebra and the fire." Borges himself said, "I think that cleverness is a hindrance. I don't think a writer should be clever, or clever in a mechanical way." And while Nabokov called his characters galley slaves, the locations for scenes "sets" he "constructed," and his narrative structures "mechanisms," he also wrote:

> For me a work of fiction exists only insofar as it affords me what I shall bluntly call aesthetic bliss, that is a sense of being somehow, somewhere, connected with other states of being where art (curiosity, tenderness, kindness, ecstasy) is the norm. There are not many such books.

Tenderness. Kindness. Ecstasy. Nabokov's definition of art would surprise those who find him overly "intellectual." Elsewhere, he added, "you read an artist's book not with your heart (the heart is a remarkably stupid reader), and not with your brain alone, but with your brain and your spine. 'Ladies and gentleman, the tingle in your spine tells you what the author felt and wished you to feel.'"

Again and again, the best of the postrealists strike this note. Kundera writes that Cervantes and other early novelists "were not looking to simulate reality; they were looking to amaze, astonish, enchant. They were *playful*, and therein lay their virtuosity. . . . The great European novel started out as entertainment, and all real novelists are nostalgic for it!" This is a call for freedom, for play — but for serious play. "If I believed my writing were no more than . . . formal fun-and-games," Barth has told us, "I'd take up some other line of work. The subject of literature, says Aristotle, is 'human life, its happiness and its misery.' I agree with Aristotle. . . . What we want is passionate virtuosity. If these pieces aren't also *moving*, then the experiment is unsuccessful."

005 Bm1

这是一本特别的日记。作为德国著名作家、翻译家阿诺·施密特（Arno Schmidt）的妻子爱丽丝·施密特（Alice Schmidt），通过独特的方式，记录了 1954 年全年的生活，包括与友人的通信。本书为圆脊腔背硬精装，封面黑色纸张贴单色照片，只有红色“1954”的年份鲜明夺目。书脊布面包裹三色胶印。内页黑白印刷，包含大量照片、票据、手绘图、图表，记录了有关天气变化的详细资料。本书的特色在于全书并不都用文字来表述，而是对事件、人物进行大量简写，甚至图标化表现。整体设计清晰明了，以大量符号的堆砌与表达增加阅读的趣味性，排版生动，一目了然。

Alice Schmidt
Tagebuch aus dem Jahr
1954

This is a special diary. Alice Schmidt, the wife of the celebrated German author and translator Arno Schmidt, recorded her life throughout 1954 in a unique way, including correspondence with her friends. This hardback book has a rounding flexiback and a black-paper cover with monochrome photos. On the cover, only red ‘1954’ is bright and eye-catching. The spine is wrapped by cloth and offset printed in three colors. The inside pages, printed in black and white, contains a large number of photographs, notes, hand-drawn pictures and charts, recording detailed information about weather changes. What makes this book special is that it is not all written in words, but using simplified icon frequently to describe people and stories. The overall design is clear and simple, increasing the enjoyment of reading.

- 铜奖
- 1954 年的日记
- 德国

From the cover to the endpapers through to the paper used for the contents, a well-balanced choice of materials, very well finished. A pleasing, classical type area with typographical details which are incorporated perfectly from the first to the last page.

- ◎ Bronze Medal
- « Tagebuch aus dem Jahr 1954
- Germany
- Friedrich Forssman
- △ Alice Schmidt
- Arno Schmidt Stiftung im Suhrkamp Verlag, Frankfurt am Main
- □ 255×170×20mm
- 803g
- 334p

Arno Schmidt
Tagebuch aus dem Jahr
1954

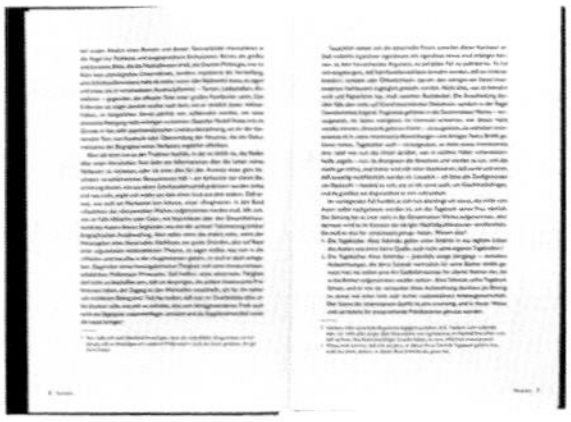

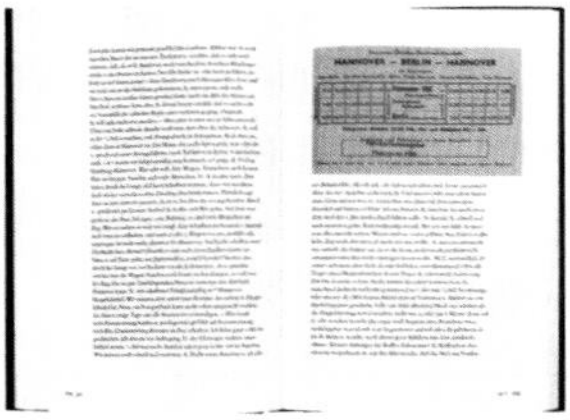

284 Fotografien
Fotografien 285

Edition

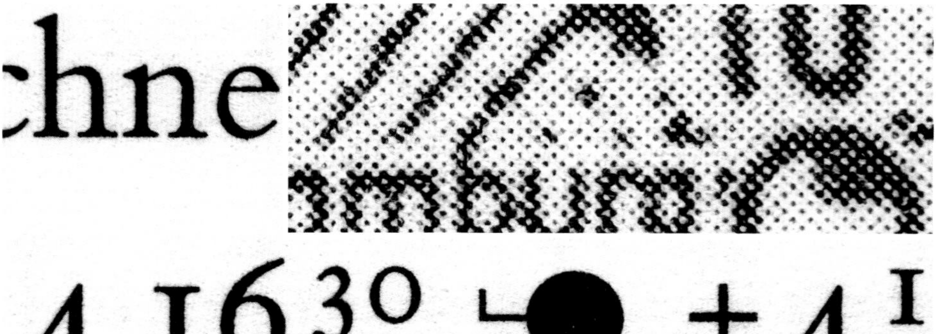
chne

So So. 5.9. 10^{20} ◕ +23; 13^{20} ◑ +26; Groß Uns-Wasch und ich: Arno Köpfel. – N – 18 ◑ +22 V: Goethe: Werther. – Unsere lieben Kätzel! Wären sie noch da!

Mo 6.9. 9 ● +16½; 12 ● +18; P; 15^{50} ● +23 •; N – 17^{50} ● +18 •; Kommentar des Senders der DDR. Herbert Geßner: Preissenkung in der Ostzone ab heute um ca 23%. Rias hätte ihnen vorgeworfen, sie hätten keine Gegenpartei! Brauchen wir nicht, denn das ist etwas für Demokratien wo es keine Preissenkungen gibt. Wie wäre denn eine Gegenpartei bei Preissenkungen von nöten. Wer würde denn eine solche wählen, die Preissteigerungen wollte, wie ja eine Gegenpartei müßte!? (hier murmelt Arno: dummes Schwein!) – Kongreß der Weltbevölkerung oder so ähnlich: glaub ein Chinese wars, der hätte Sterilisation und Geburtenbeschränkung vorgeschlagen. Dazu Geßner: »aber das ist schmutzig! Wir bejahen das Leben und lehnen dies somit ab!« – Na Arno, in so 'n Land willste?« A: »Ja wenn man so was wieder hört. Hast recht, da kann man ja ooch nich hin. So ein Blödling! Der machts ja genauso plump wie Fritzsche. –

Di 7.9. 9^{40} ● +18; 12^{40} ◕ +27; P; 14^{20} ● +26; N; 16^{40} ● +23; 17^{15} ϟ •; 19 ● +18 V: Lucian

Mi 8.9. 9 ● +15; 12 ◑ +24; P Celle: nichts v. Jansen da. – L – N – 18 ● +18] Arno fällt Gasthoftitel ein: »zum gelben Hintze.« – V. Lucian. A: ob sie meinen Cooper-Artikel nehmen? Davon hängt viel ab, weiß nicht was für ne Sorte ich ihnen sonst geben sollte. Solche könnte ich viel machen. –

Do 9.9. 8^{30} ♀ +17; 12^{10} ◑ +19; P; Arno schreibt »Undinenartikel« für Mai aber erst. – Sehr schön! – 17^{30} ◑ +19; 19 ◑ +18; Nachtregen. – Schweres Erdbeben in Orleansville/Algier. Stadt sieht wie zerbombt aus. 12 Sekunden. Verluste an Menschen noch unabsehbar. (Unsre Kätzel mußten auch sterben) – Radiomeldung eines W. Dtsch. Senders: der an der Erschießung v. Kanadiern beteiligt gewesene dtsch. General genannt Panzer-Meyer der v. allierten Gerichten erst zum Tode verurteilt, dann zu lebenslängl. Haft verurteilt wurde, ist heute nach 9jg. Haft entlassen worden. Zu

---

Fritzsche Hans Fritzsche (1900–1953), Rundfunkkommentator, ab 1942 Leiter der Rundfunkabteilung des Propagandaministeriums.
Undinenartikel *Undine* (BA III, 3, 112–116).
Panzer-Meyer Kurt Meyer (1910–1961) war der jüngste General der Waffen-SS. Das Urteil gegen ihn wurde 1954 aufgehoben, weil nicht nachgewiesen werden konnte, daß er den Befehl zur Erschießung der kanadischen Kriegsgefangenen gegeben hatte.

seinem Empfang (in einem Ort bei München-Gladbach) hatten sich mehrere 1000 Menschen eingefunden. Der Gesangverein und Turnverein holten ihn mit Musik ab. Zahlreiche Glückwunschtelegramme & Persönlichkeiten des Inn= und Auslandes. – Russen Sender bringt empörtes Kommentar darüber, sagen sogar ihm wäre v. Regierungsseite Scheck über 4.000 als Haftentschädigung überreicht worden. – Schön!

Fr 10.9. 10 ● +17: alte Frau Neises kommt u. bringt paar Birnen und Stückel Kuchen v. ihrem 70. Geburtstag. Sieht den leeren Jakobkäfig. Jammert um den Jakob wo der hin ist. Kommen auf die Kätzel. Erinnern uns gegenseitig wie schön und gut sie waren und einzelner Streiche. Kann mich nimmer halten und muß losheulen. Und auch der alten Fr. N. laufen die Tränen die Backen runter. Unsere geliebten Kleinen. Auch des Murrchen wird gedacht. – 12²⁰ ● +19; P: N 17 ● +19; 19 ◑ +17. V: aus dem lüneburger Heidebuch. –

Sa [11.9.] 8⁴⁰ ● +17: Hören, als wir noch im Bett liegen, Fr. N (Junge) Mies, Mies, komm mies mehrfach rufen. Neues Kätzel?! Als wir dann wach sind ruft Arno, soll mir's ansehn »Sieht aus wie 'n Satan« entfährts mir. »Ne wahr« sagt Fr. N. Hab mich selber so erschrocken, wie ich heute früh in die Küch kam, lags auf der Couch. Der Mann hats gestern abend

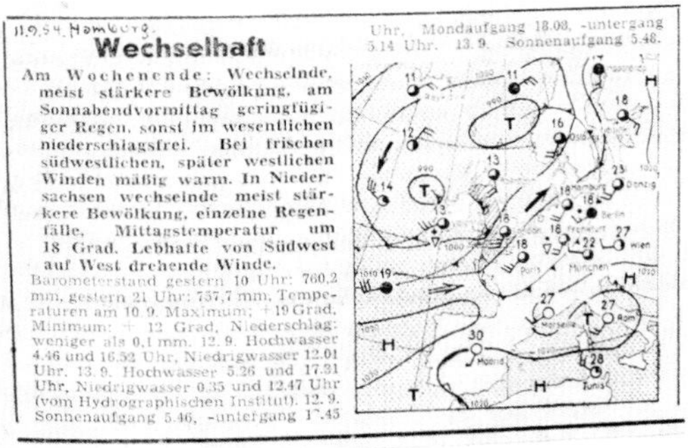
11.9.54. Hamburg.

Wechselhaft

Am Wochenende: Wechselnde, meist stärkere Bewölkung, am Sonnabendvormittag geringfügiger Regen, sonst im wesentlichen niederschlagsfrei. Bei frischen südwestlichen, später westlichen Winden mäßig warm. In Niedersachsen wechselnde meist stärkere Bewölkung, einzelne Regenfälle, Mittagstemperatur um 18 Grad. Lebhafte von Südwest auf West drehende Winde.

Barometerstand gestern 10 Uhr: 760,2 mm, gestern 21 Uhr: 757,7 mm. Temperaturen am 10.9. Maximum: +19 Grad, Minimum: + 12 Grad, Niederschlag: weniger als 0,1 mm. 12.9. Hochwasser 4.46 und 16.52 Uhr, Niedrigwasser 12.01 Uhr. 13.9. Hochwasser 5.26 und 17.31 Uhr, Niedrigwasser 0.35 und 12.47 Uhr (vom Hydrographischen Institut). 12.9. Sonnenaufgang 5.46, -untergang 17.45 Uhr. Mondaufgang 18.08, -untergang 5.14 Uhr. 13.9. Sonnenaufgang 5.48.

T
T
H
H
T
T
H
Paris
München
Marseille
Rom
Madrid
Tunis
Wien

---

luneburger Heidebuch Nicht mehr zu ermitteln.

小松菜は肥料好きな
ので、畑にたっぷり
完熟堆肥を鋤き込
み,よく耕しておく。
堆肥の量は3.3m²に
およそ10kg。
●緑肥を使ってみよう
4回ま
収穫を
Bm²

作为日本知名的农学家，金子美登除了在有机农业方面有所建树，也一直致力于有机种植的普及工作，本书便是这样一本全面介绍且让普通人可以实践的相关书籍。本书的装订形式为无线胶平装，右翻本。护封胶版纸黑绿双色印刷，内封艺术纸与环衬裱贴，书脊部分单印浅灰色文字。内页除前四面四色印刷有机蔬菜照片外，其余均采用胶版纸黑绿双色印刷。书籍的内容主要分为两大部分，第一部分为有机农业的全面介绍，包括场地、环境、工具、种植、品种、时节、肥料、病虫害防治等；第二部分则为具体的果蔬品种及详细的种植要点。最后配以所提果蔬种植时节及要点的完整表格，便于查阅。本书作为一本普及读物，为了使得普通读者在面对专业技术时不会觉得枯燥，由专业插画师进行大量配图，风格轻松谐趣，插图全部印刷为绿色。排版方面在维系统一体例的同时，也尽可能地多变，形成清晰且独特的阅读体验。||||||||||||||||||||||||||||||||

著 金子美登
絵／守田勝治
金子さんちの
有機家庭菜園
絵とき
家の光協会

As a well-known agronomist in Japan, Kaneko Yoshinori has been committed to the popularization of organic farming in addition to the achievements in organic agriculture. This book is a comprehensive introduction to organic farming and a guide to practice. The book is a perfect-bound paperback, rightward flipbook. The jacket is printed in black and green, while the texts on the spine are printed in light gray. The photos of organic vegetable are printed in four-colors on the first four pages of this book, while the other pages are printed on offset paper in black and green. The content of the book is mainly divided into two parts. The first part is a comprehensive introduction to organic agriculture, including site, environment, tools, planting, varieties, season, fertilizer, pest control and so on. The second part is concerning specific fruit and vegetable and detailed planting points. Finally, a complete table of planting season and key points of planting is provided for reference. As a popular reading material, the book is illustrated by experienced illustrators in order to not make general readers feel boring when facing the professional skills. All illustrations are printed in green.

◎ 铜奖

《 金子先生的有机家庭菜园

◖ 日本

A wonderful and inspiring book on the cultivation of bio-vegetables. The illustrations in green are amusing, very informative and natural on yellowish paper. The layout is diverse and open, playfully pleasing with many variations.

◎ ········ Bronze Medal

《 ···· Kanekosanchino Yukikateisaien

◖ ········ Japan

D ········ Hirokazu Ando

△ ········ Yoshinori Kaneko

|||| ···· Ie-No-Hikari Association, Tokyo

□ ········ 257×188×16mm

▯ ········ 576g

≣ ········ 192p

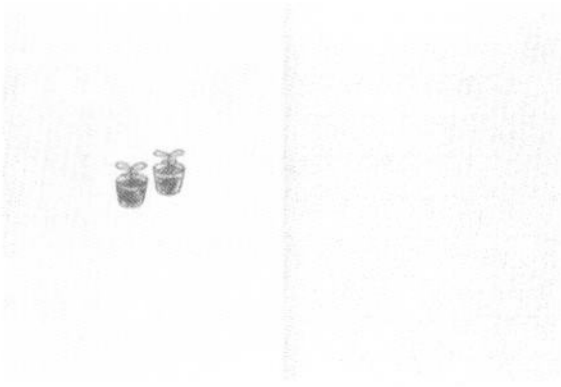

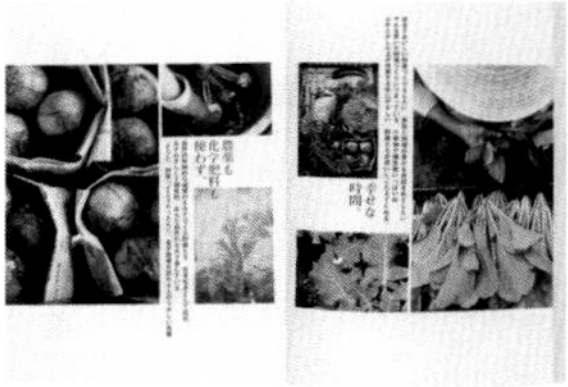

絵とき 金子さんちの
有機家庭菜園

有機農業の勘どころ

土づくり

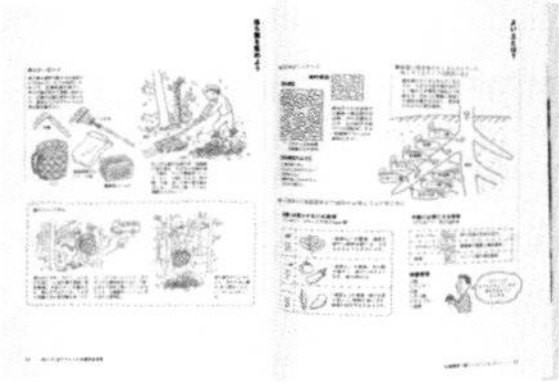

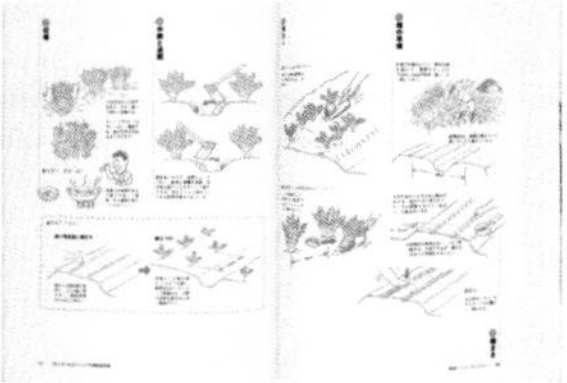

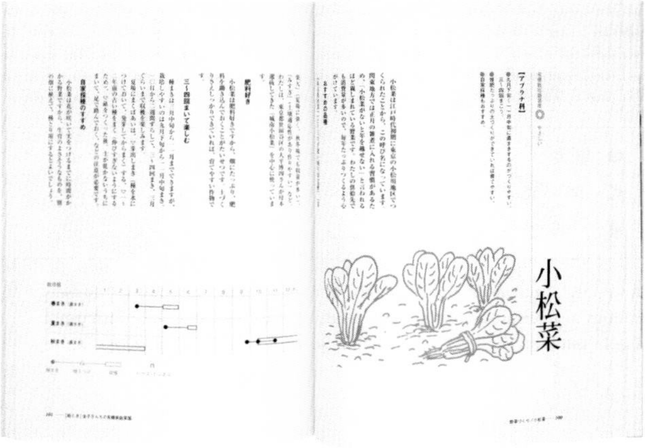
小松菜
【アブラナ科】
おすすめの品種
肥料好き
三〜四回まいて楽しむ
自家採種のすすめ

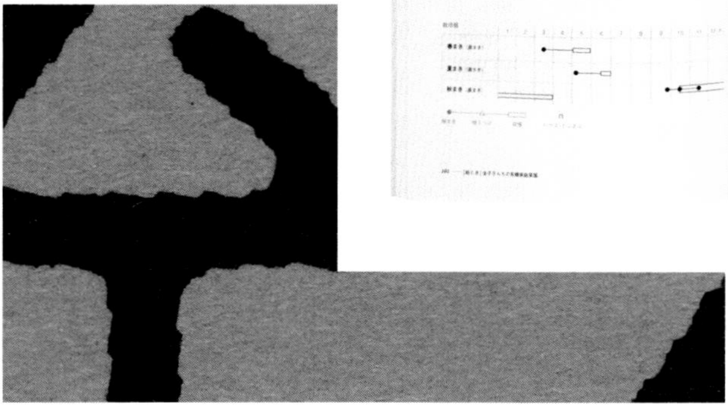

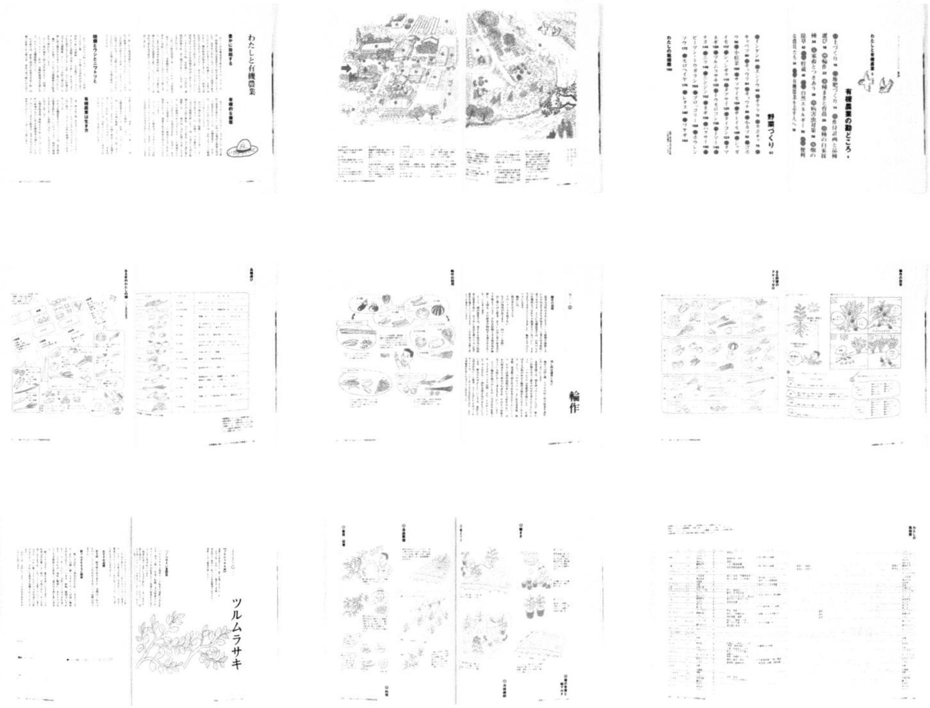
わたしと有機農業
野菜づくり
輪作
ツルムラサキ

絵とき
金子さんちの
有機家庭菜園
著 金子美登
絵／守田勝治
家の光協会

❶温床の枠をつくる

●ます枠をつくる

丸太に竹を渡し、わらを挟んだ枠をつくる。落ち葉、牛糞、切りわら、米糠、おからに水を補給しながら3～4層にして入れ、きつく踏み込むと、温度が上がり、温床になる。1月上旬にハウス内につくり、2月～3月上旬に種まきした苗箱を置く。

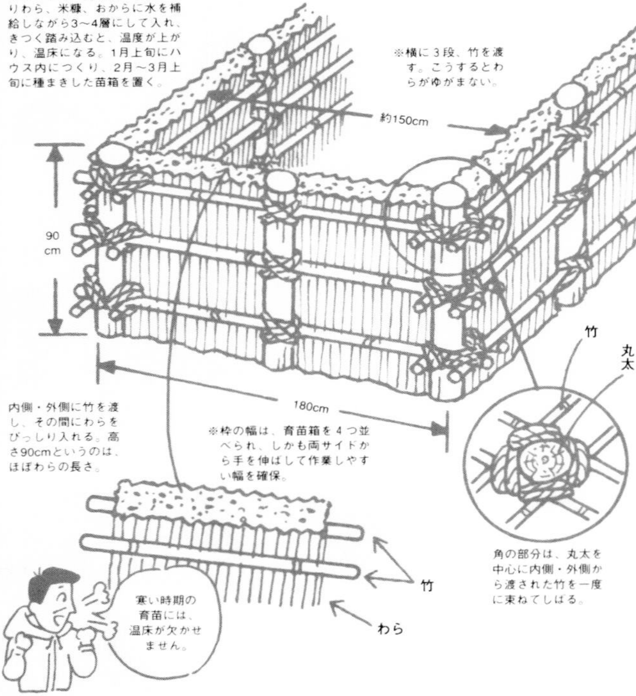

※横に3段、竹を渡す。こうするとわらがゆがまない。
約150cm
90cm
180cm
竹
丸太
内側・外側に竹を渡し、その間にわらをびっしり入れる。高さ90cmというのは、ほぼわらの長さ。
※枠の幅は、育苗箱を4つ並べられ、しかも両サイドから手を伸ばして作業しやすい幅を確保。
角の部分は、丸太を中心に内側・外側から渡された竹を一度に束ねてしばる。
竹
わら
寒い時期の育苗には、温床が欠かせません。

春まき

勘どころ 五

種まきと育苗

よい種を守り伝える

よい品種の種を農家自身が守っていくことは、とても重要です。とくに有機農業者は、市販の種がかならずしも有機農業向きというわけではないので、生産者自身が種を伝えていくことがたいせつになります。

また風土や環境によっても適する種は違いますから、五里四方（二〇キロ四方）の人たちが、自慢の種を交換しあうというのが理想です。わたしたちは毎年春に、有機農業の仲間と種苗交換会を行っています。

東京都世田谷区の大平博四さんの自慢の種は『どじょう』インゲンや『城南』小松菜。『どじょう』インゲンはおばあさんから「よそに出してはダメだよ」と言われていたほど、先祖伝来の貴重な在来種です。わたしのところの自慢はさやがおいしく、実が熟すと真っ白になる『白インゲン』と、繭玉のような形で秋口にうまい『金子トマト』と評判になり、地域おこしになると思います。

暑さ寒さに耐える苗

よい種が手に入ったら、次は種まきと育苗がポイントになります。春まき野菜は二～三月という寒い時期、夏まき野菜は七月～八月の暑い時期、秋まき野菜は九～一〇月で、少し遅くなると寒い時期になります。その暑さ寒さに耐える苗づくりをする必要があるのです。

春まきはハウスの中に踏み込み温床（二五～三〇度）をつくり、その上に育苗箱を置いて暖かくしてやります。電熱線など人工的なものは使わず、自然の材料でひと足早い春をつくってやります。

夏まきは、日当たりがよく、風通しのよいところで、冷床を工夫します。一般には高畝にして直まきにしますが、家庭菜園でそう多くまかない場合は、育苗箱を使ってもよいと思います。

秋まきは、早まきすると病害虫にやられてしまい、かといって、まきどきが遅くなると、今度は

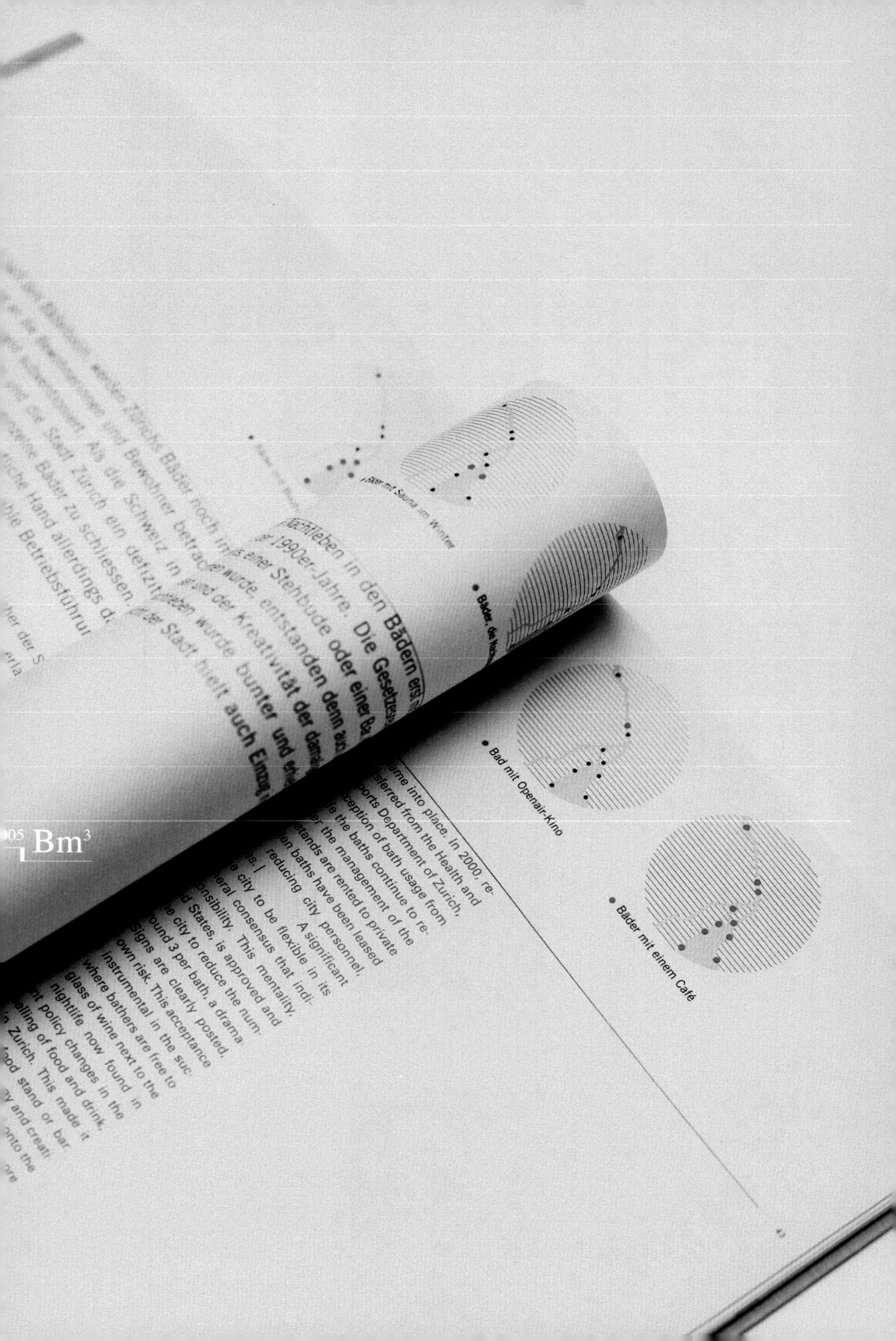
Bad mit Openair-Kino
Bäder mit einem Café

05 Bm³

这是一本美国作家的研究报告，研究了贯穿瑞士最大城市苏黎世的河流和湖泊上的浴场文化。书籍开篇是大量浴场照片，展现当地人热爱在自然水系里沐浴游泳的状态。之后通过水质分析、浴场文化历史、亲水平台和浴场设施建造、饮用水处理等方面来试图解读这一独特的文化现象。本书为锁线平装本，封面选用了半透明的浅蓝绿色纸，通过正面烫白、反面印深蓝色的方式呈现两级书名。封面纸张通过折叠的方式形成双层厚度，手感柔和，内嵌荧光黄色纸带与内文荧光色标记点形成呼应。内文照片部分光面铜版纸四色印刷，之后研究部分采用胶版纸，单黑印刷照片配合多色文字区分章节，荧光色用于标记最重要的几个浴场位置，并在书口上方便快速查找到详细介绍的页面。

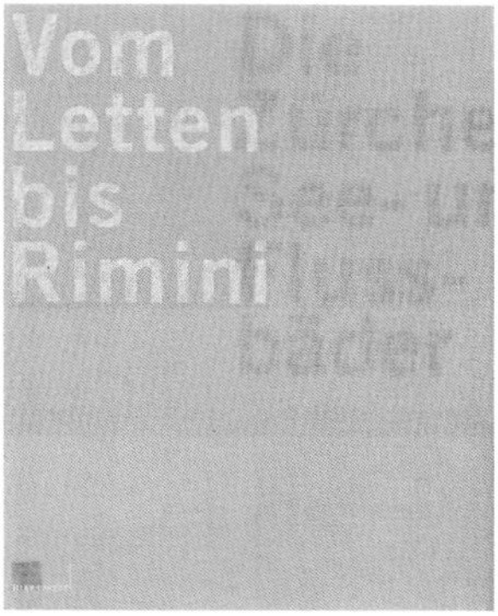
Vom
Letten
bis
Rimini

This is a research report by an American writer who examines the bathing culture on the rivers and lakes in Zurich, the largest city in Switzerland. The book begins with a large number of photos of bathing places, showing the local people's love for bathing and swimming in the natural water system. Later, the author tries to interpret this unique cultural phenomenon from different perspectives, such as water quality analysis, culture and history of bathing places, construction of bath facilities, and drinking water treatment. This book is a paperback with thread sewing. The cover is made of semi-transparent light blue-green paper, which is folded to form a double-layer thickness. The embedded fluorescent yellow paper tape echoes the fluorescent markers in the text. Part of the photos are printed on gloss artpaper in four colors, and the following report part is printed on offset paper. Single-black printed photos were combined with multi-color text to distinguish chapters. Fluorescent color is used to mark the positions of several important bathing places and help to locate the reference pages at the fore-edge.

铜奖

从拉脱维亚到里米尼
——苏黎世湖泊和河流浴场的历史和现状

瑞士

A book about the river and lake baths of Zurich. The fresh blue-green colours of the flexible cover create the mood of the subject matter as do the materials used. A sequence of photos extending over several pages conveys a bright overall mood and prepares the reader's senses for the factually informative second section, the texts of which receive a particular structuring through typography differentiated by colour. The book's format is easy to handle.

Bronze Medal

Vom Letten bis Rimini, Geschichte und Gegenwart der Zürcher See - und Flussbäder

Switzerland

Thomas Bruggisser

Nina Chen

hier + jetzt, Verlag für Kultur und Geschichte, Baden

216×168×10mm

259g

96p

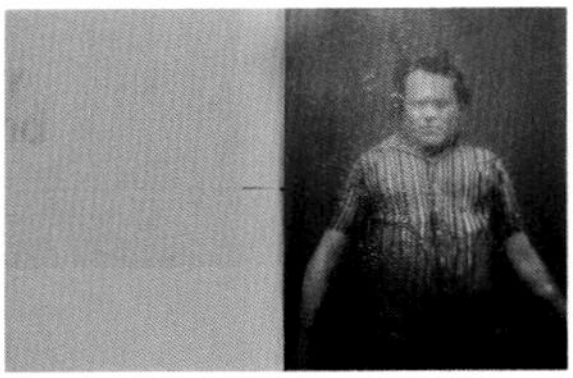

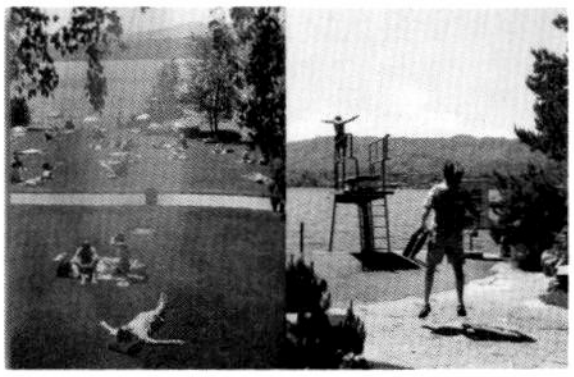

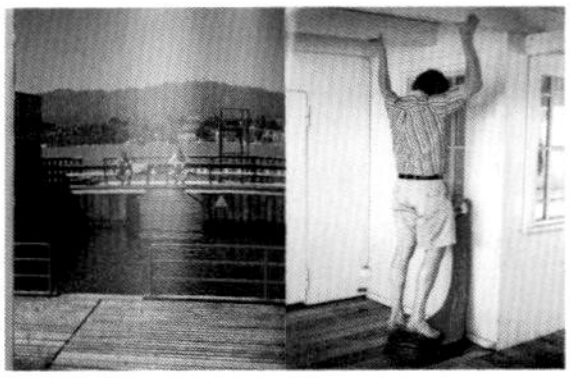

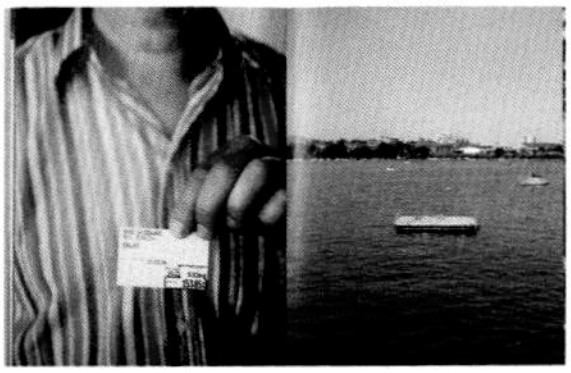

Vom
Letten
bis
Rimini

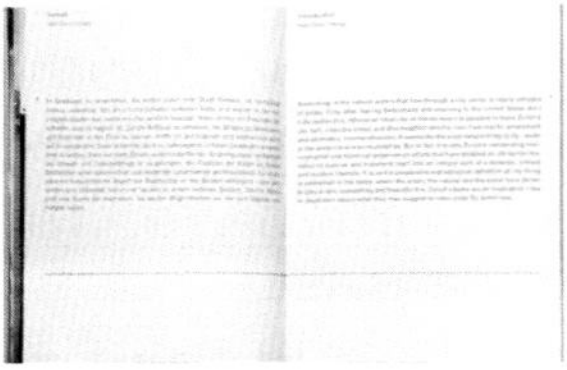

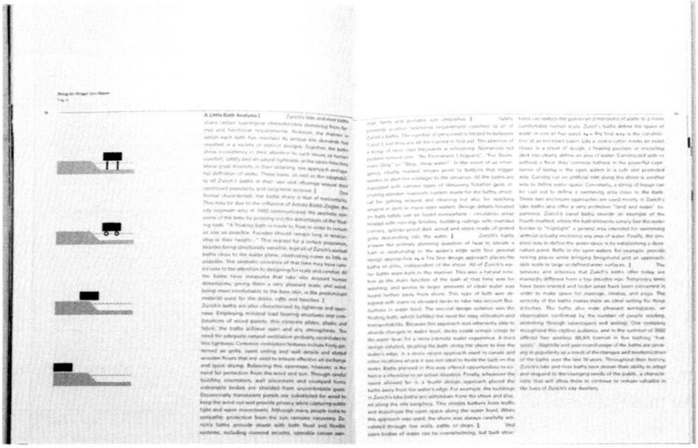

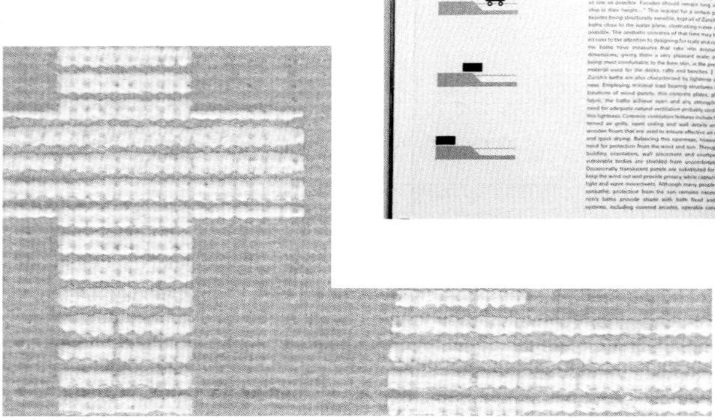

nking
f the
both
years the lake and

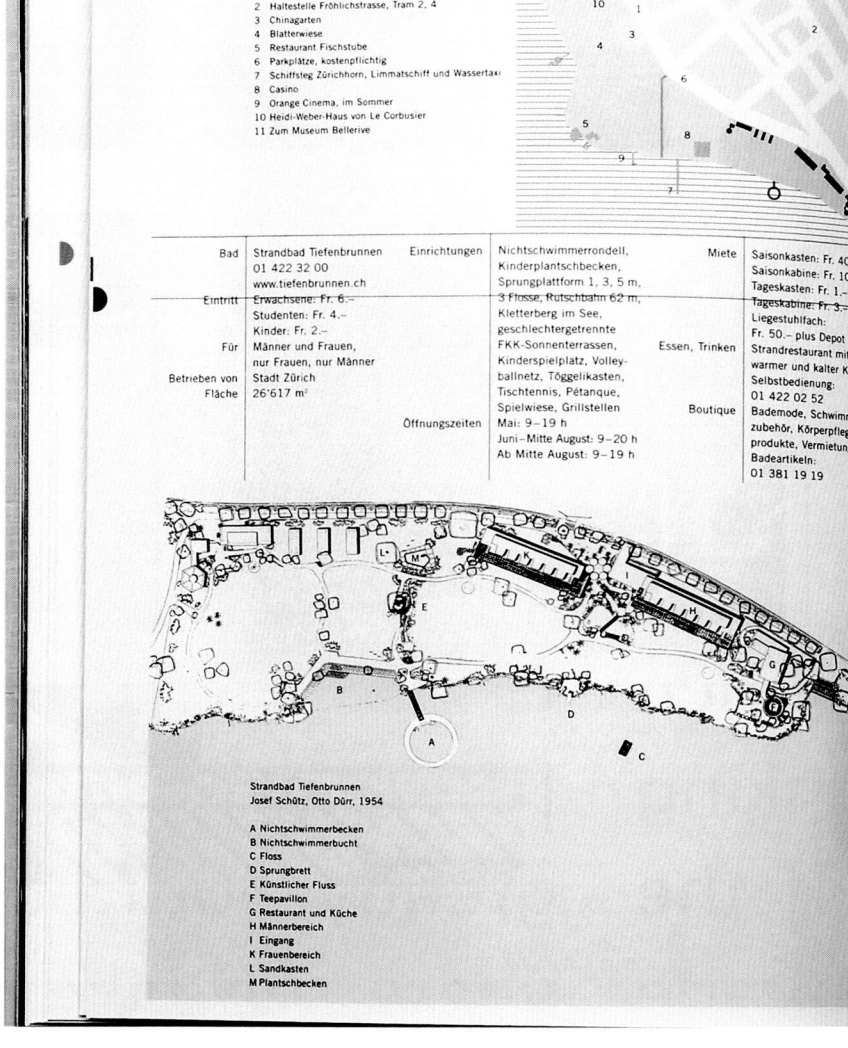

1 Haltestelle Chinagarten, Bus 912, 916
2 Haltestelle Fröhlichstrasse, Tram 2, 4
3 Chinagarten
4 Blatterwiese
5 Restaurant Fischstube
6 Parkplätze, kostenpflichtig
7 Schiffsteg Zürichhorn, Limmatschiff und Wassertaxi
8 Casino
9 Orange Cinema, im Sommer
10 Heidi-Weber-Haus von Le Corbusier
11 Zum Museum Bellerive

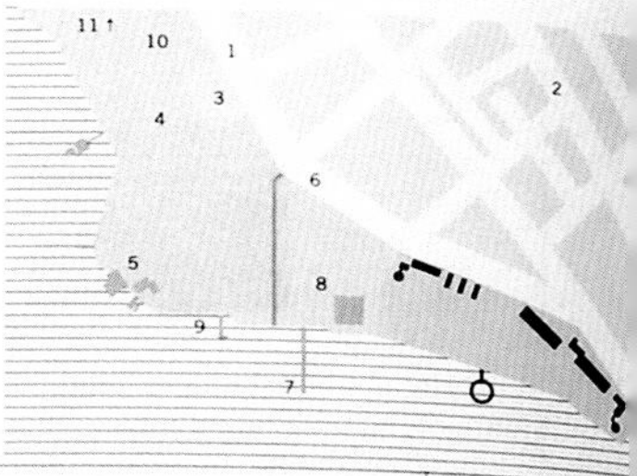
11 ↑
10
1
2
3
4
6
5
8
9
7

Bad	Strandbad Tiefenbrunnen 01 422 32 00 www.tiefenbrunnen.ch	Einrichtungen	Nichtschwimmerrondell, Kinderplantschbecken, Sprungplattform 1, 3, 5 m, 3 Flosse, Rutschbahn 62 m, Kletterberg im See, geschlechtergetrennte FKK-Sonnenterrassen, Kinderspielplatz, Volleyballnetz, Töggelikasten, Tischtennis, Pétanque, Spielwiese, Grillstellen	Miete	Saisonkasten: Fr. 40. Saisonkabine: Fr. 10 Tageskasten: Fr. 1.– Tageskabine: Fr. 3.– Liegestuhlfach: Fr. 50.– plus Depot
Eintritt	Erwachsene: Fr. 6.– Studenten: Fr. 4.– Kinder: Fr. 2.–				
Für	Männer und Frauen, nur Frauen, nur Männer			Essen, Trinken	Strandrestaurant mit warmer und kalter Kü Selbstbedienung: 01 422 02 52
Betrieben von	Stadt Zürich				
Fläche	26'617 m²				
		Öffnungszeiten	Mai: 9–19 h Juni–Mitte August: 9–20 h Ab Mitte August: 9–19 h	Boutique	Bademode, Schwimm zubehör, Körperpflege produkte, Vermietung Badeartikeln: 01 381 19 19

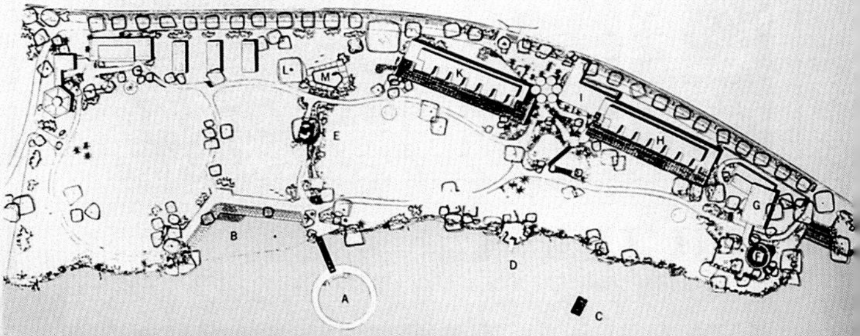
A
B
C
D
E
G
H
I
K
L
M

Strandbad Tiefenbrunnen
Josef Schütz, Otto Dürr, 1954

A Nichtschwimmerbecken
B Nichtschwimmerbucht
C Floss
D Sprungbrett
E Künstlicher Fluss
F Teepavillon
G Restaurant und Küche
H Männerbereich
I Eingang
K Frauenbereich
L Sandkasten
M Plantschbecken

Wellness	Massage und Open-Hair: 01 422 02 52
Am Abend	Grillspezialitäten bei Sonnenuntergang: 01 422 02 52
Im Winter	Park geöffnet
Ausserdem	Die Bäder der Stadt Zürich kann man für Anlässe mieten: 01 206 93 74
Besucher 2003	265'008

Strandbad Tiefenbrunnen

Das Strandbad Tiefenbrunnen ist sowohl ein öffentliches Bad als auch ein öffentlicher Park. Seine Tore stehen auch vor und nach der offiziellen Badesaison offen, der Park ist selbstverständliche Fortsetzung von Zürichs Seepromenade. Mit Mosaik ausgekleidete Springbrunnen, gepflegtes Grün, gelbe und rote Pavillons und einladende Becken schaffen einen idealen Rahmen für Familien. Die Bad- und Parkanlage wurde 1950 für 3,5 Millionen Franken auf dem ehemaligen Gelände der Schweizer Landesausstellung von 1939 angelegt, an Stelle eines älteren, 1886 erbauten Kastenbades. Felsbrocken aus einem nahe gelegenen Steinbruch sind dem Ufer entlang sorgfältig im Beton verlegt worden. Sie schaffen eine Landschaft, die natürlich gewachsen zu sein scheint, und laden die Besucher dazu ein, sich nahe des Wassers aufzuhalten. Nischen, lange und niedrige Mauern sowie ein ausgehobenes Becken mit Treppen ermöglichen eine vielfältige und abwechslungsreiche Erfahrung des Seeufers. Ein kreisförmiges, schwimmendes Becken, 27 Meter im Durchmesser, liegt auf Caissons aus Stahlbeton, während zwei Holzflosse, 60 Meter vom Ufer entfernt, den Schwimmbereich markieren. Die farbenfrohen Pavillons, die als Umkleidekabinen, Sonnenterrassen und Nebenräume dienen, sind wichtige Beispiele der Architektur der 1950er-Jahre. Im Bad Tiefenbrunnen breitet man sein Badetuch in einem botanischen Garten aus, unter hohen Föhren und seltenen Steineichen. 2004 feiert das Bad sein fünfzigjähriges Bestehen.

Tiefenbrunnen is interesting in that it is both a public bath and a public park. Before and after the official bathing season, the gates are opened and the park serves as a natural continuation of Zurich's lake promenade. Mosaic tiled fountains, vast areas of manicured greens, cheerful red and yellow pavilions and friendly pools create a favorite setting for families. Constructed in 1950 for 3.5 million CHF on a former site of the 1939 Swiss Expo, this park bath served as a replacement for an older courtyard bath from 1886. Boulders were delivered from a nearby quarry and carefully anchored in concrete along the shore to create a natural landscape and invite visitors to take a seat along the water's edge. Niches, long low walls and an excavated inlet pool with stairs create a diverse and changing shoreline experience from which to gaze upon the waters. A large circular floating pool, 27 meters across, floats serenely on concrete caissons, embracing the sky, while two wooden rafts 60 meters from shore define the swimming area. The colorful wood pavilions housing the lockers, sun terraces and services are important examples of 50s architecture. Here bathers spread their towels in a botanical garden underneath tall firs and rare species of stone oaks. This year the bath celebrates its 50th anniversary.

005 Bm4

这是二战中最著名的犹太受害者安妮·弗兰克（Anne Frank）及其家人的影集，拍摄者是安妮的父亲。本书的装订形式为锁线硬精装，采用棕色线进行装订。收藏版与第一版在封面材质上有所不同，第一版采用浅黄色布面，而收藏版则采用深蓝色乱纹艺术纸。封面纸张裱覆卡板，正中微微凹下的“相框”内裱贴照片。封面没有标题，呈现家庭相册的真实状态。内页最开始采用胶版纸印单黑，版面设置为三栏，荷、英、德三种文字的背景介绍，对照排版。中间照片部分采用哑面铜版纸黑棕双色印刷，呈现老照片的怀旧味道。内页后部采用胶版纸黑棕双色印刷，为照片注解。全书只有照片部分有页码，与后部注解形成索引关系。||||||

This is an album of Anne Frank, the most famous Jewish victim of World War II, and her family, filmed by Anne's father. The hardback book is bound by brown thread sewing. The cover material of the collector's edition is different from that of the first edition. The first edition uses light yellow cloth, while the collector's edition uses dark blue scribbled art paper. The cover is mounted with cardboard, and the photo is attached inside the slightly concaved 'photo frame'. There is no title on the cover, showing the true state of the family photo album. The first part of the inner pages is made of offset paper in black and typeset with three columns. The introduction written in Dutch, English and German are compared with each other. In the middle part, matte coated art paper printed in black and brown is used to present old photographs that are full of nostalgia. The last part is printed in black and brown on offset paper, containing photo annotations. Only the photo part of the book has page numbers, which index with the annotations.

◎ 铜奖

« 安妮·弗兰克与家人

◖ 荷兰

A quiet book which opens itself slowly to the reader and/or viewer. The black and brown used in its typography and pictures complement one another in exemplary fashion. The cover with no writing immediately arouses curiosity. Design and production are of the highest level, from the thread-binding with brown thread through to the partial glazing.

◎ ········ Bronze Medal

« ········ Anne Frank en familie

◖ ········ The Netherlands

D ········ MV (VictorLevie & Barbara Herrmann)

△ ········ Otto Frank

|||||| ··· Anne Frank Stichting, Amsterdam / Anne Frank Fonds, Basel

□ ········ 155×154×14mm

▯ ········ 245g

≡ ········ 104p

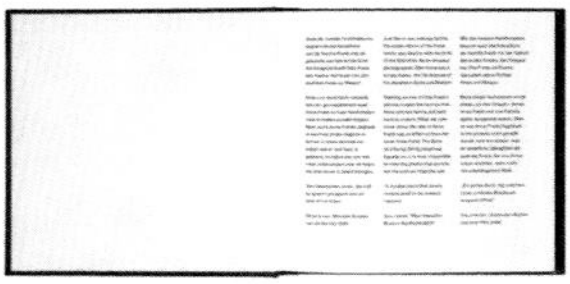

Anne Frank

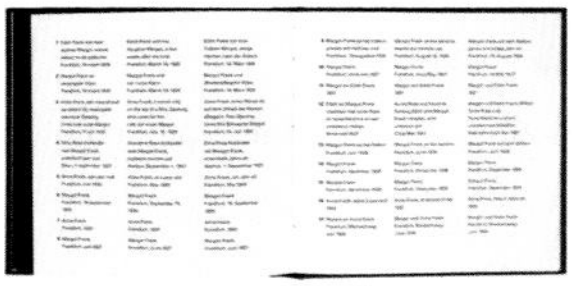

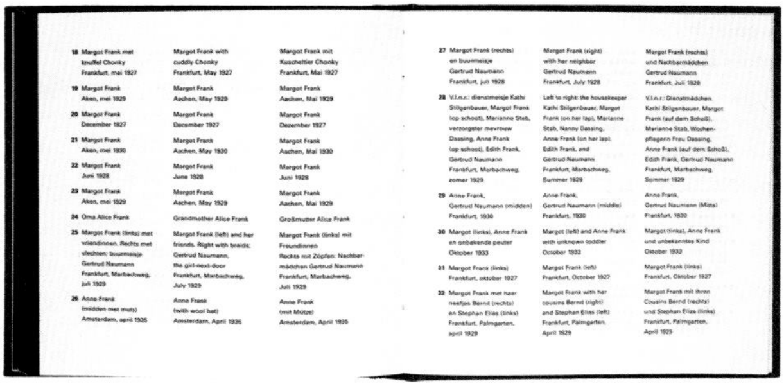
18 Margot Frank met knuffel Chonky Frankfurt, mei 1927
19 Margot Frank Aken, mei 1929
20 Margot Frank December 1927
21 Margot Frank Aken, mei 1930
22 Margot Frank Juni 1928
23 Margot Frank Aken, mei 1929
24 Oma Alice Frank
25 Margot Frank (links) met vriendinnen. Rechts met vlechten: buurmeisje Gertrud Naumann Frankfurt, Marbachweg, juli 1929
26 Anne Frank (midden met muts) Amsterdam, april 1935
Margot Frank with cuddly Chonky Frankfurt, May 1927
Margot Frank Aachen, May 1929
Margot Frank December 1927
Margot Frank Aachen, May 1930
Margot Frank June 1928
Margot Frank Aachen, May 1929
Grandmother Alice Frank
Margot Frank (left) and her friends. Right with braids: Gertrud Naumann, the girl-next-door Frankfurt, Marbachweg, July 1929
Anne Frank (with wool hat) Amsterdam, April 1935
Margot Frank mit Kuscheltier Chonky Frankfurt, Mai 1927
Margot Frank Aachen, Mai 1929
Margot Frank Dezember 1927
Margot Frank Aachen, Mai 1930
Margot Frank Juni 1928
Margot Frank Aachen, Mai 1929
Großmutter Alice Frank
Margot Frank (links) mit Freundinnen Rechts mit Zöpfen: Nachbarmädchen Gertrud Naumann Frankfurt, Marbachweg, Juli 1929
Anne Frank (mit Mütze) Amsterdam, April 1935
27 Margot Frank (rechts) en buurmeisje Gertrud Naumann Frankfurt, juli 1928
28 V.l.n.r.: dienstmeisje Kathi Stilgenbauer, Margot Frank (op schoot), Marianne Stab, verzorgster mevrouw Dassing, Anne Frank (op schoot), Edith Frank, Gertrud Naumann Frankfurt, Marbachweg, zomer 1929
29 Anne Frank, Gertrud Naumann (midden) Frankfurt, 1930
30 Margot (links), Anne Frank en onbekende peuter Oktober 1933
31 Margot Frank (links) Frankfurt, oktober 1927
32 Margot Frank met haar neefjes Bernd (rechts) en Stephan Elias (links) Frankfurt, Palmgarten, april 1929
Margot Frank (right) with her neighbor Gertrud Naumann Frankfurt, July 1928
Left to right: the housekeeper Kathi Stilgenbauer, Margot Frank (on her lap), Marianne Stab, Nanny Dassing, Anne Frank (on her lap), Edith Frank, and Gertrud Naumann Frankfurt, Marbachweg, Summer 1929
Anne Frank, Gertrud Naumann (middle) Frankfurt, 1930
Margot (left) and Anne Frank with unknown toddler October 1933
Margot Frank (left) Frankfurt, October 1927
Margot Frank with her cousins Bernd (right) and Stephan Elias (left) Frankfurt, Palmgarten, April 1929
Margot Frank (rechts) und Nachbarmädchen Gertrud Naumann Frankfurt, Juli 1928
V.l.n.r.: Dienstmädchen Kathi Stilgenbauer, Margot Frank (auf dem Schoß), Marianne Stab, Wochenpflegerin Frau Dassing, Anne Frank (auf dem Schoß), Edith Frank, Gertrud Naumann Frankfurt, Marbachweg, Sommer 1929
Anne Frank, Gertrud Naumann (Mitte) Frankfurt, 1930
Margot (links), Anne Frank und unbekanntes Kind Oktober 1933
Margot Frank (links) Frankfurt, Oktober 1927
Margot Frank mit ihren Cousins Bernd (rechts) und Stephan Elias (links) Frankfurt, Palmgarten, April 1929

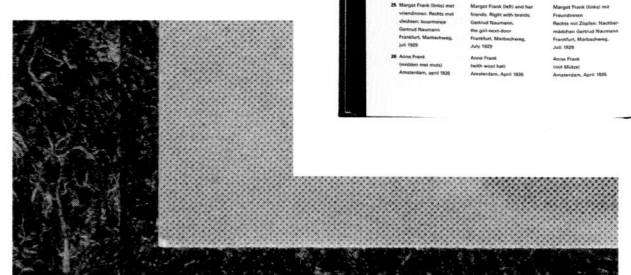

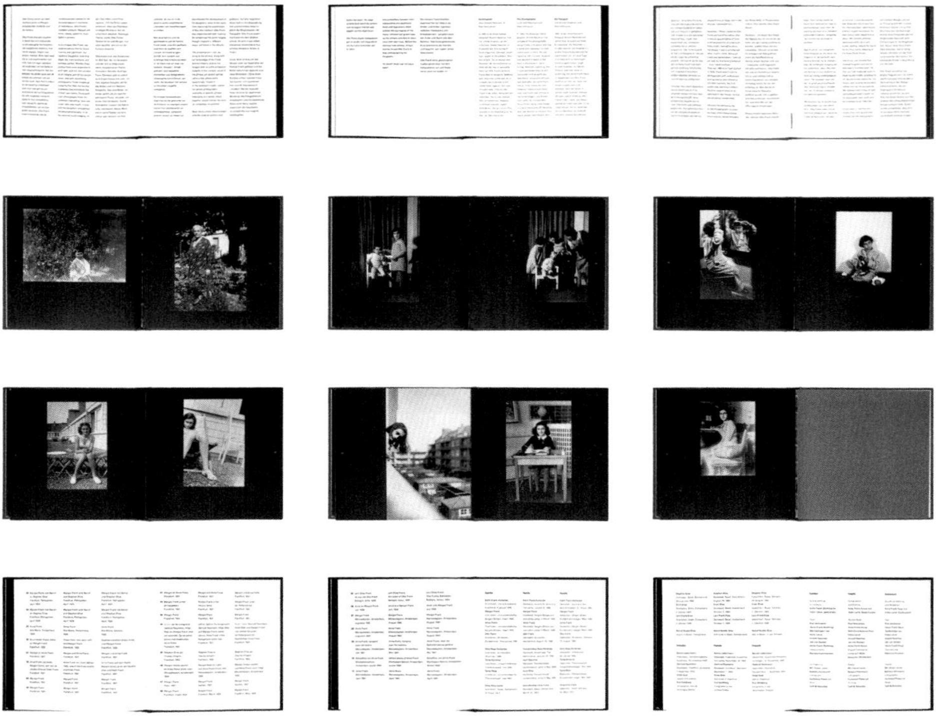

and. The ecre
ere in the Secret Annex.

53

54

05 Bm5

本书是荷兰阿姆斯特丹市立博物馆的家具收藏集。书籍最前面以时间线的方式展示了从 1840－2000 年的所有馆藏家具，后面的部分按照设计者的姓名字母排序，详细介绍了设计师和作品。两部分之间通过页码进行索引。书中还穿插了一些特殊主题的研究，包括这段时期内国际家具的代表作、荷兰家具的代表作、类似家具的对照研究、家具风格的发展延续、设计师和他们的作品，以及家具的维护与修复。本书为锁线硬精装，便于如此大的一本书彻底摊开。内文的历史介绍和具体藏品均为哑光纸黑白印刷，专题研究页面光面纸四色印刷，且纸张宽度较普通页面缩进 5mm，便于查找。书内设置黑、银双书签，增加了阅读节点记录和前后对照翻阅的可能性。

The Furniture Collection Stedelijk Museum Amsterdam
1850–2000 / From Michael Thonet to Marcel Wanders
NAi Publishers Rotterdam
Stedelijk Museum Amsterdam

The book is a collection of furniture of the Stedelijk Museum in Amsterdam, the Netherlands. At the beginning of the book, it presents a timeline of all the furniture from the museum collection from 1840 to 2000, and the later part elaborate the designers and their works in alphabetical order of the designer's name. The two sections are indexed by page numbers. The book is also interlaced with studies on some special subjects, including the master works of international furniture during this period, the representative works of Dutch furniture, comparative studies of similar furniture, the development of furniture style, designers and their works, as well as the maintenance and repair of furniture. This hardcover book is bound with thread sewing and available to spread out thoroughly. The introduction to the history and the specific collection are printed in black and white on matte paper, and the research part is four-color printed on glossy paper with the width indented 5mm compared with the ordinary page for quick reference. Black and silver bookmarks are attached in the book, which make cross-references possible when reading.

- 铜奖
- 阿姆斯特丹市立博物馆的家具收藏
- 荷兰

A well thought out basic concept, clear structuring, pleasing positioning of pictures, restrained use of colours, sophisticated overall finish. The colour illustrations are presented on a slightly narrower format and on satin-coated paper between black-and-white illustrations on matt paper. All of these wellbalanced aspects allow the book to make a very distinguished impression.

- Bronze Medal
- The Furniture Collection, Stedelijk Museum Amsterdam
- The Netherlands
- Beukers Scholma
- Luca Dosi Delfini
- NAi Publishers, Rotterdam / Stedelijk Museum Amsterdam
- 316×250×41mm
- 2992g
- 456p

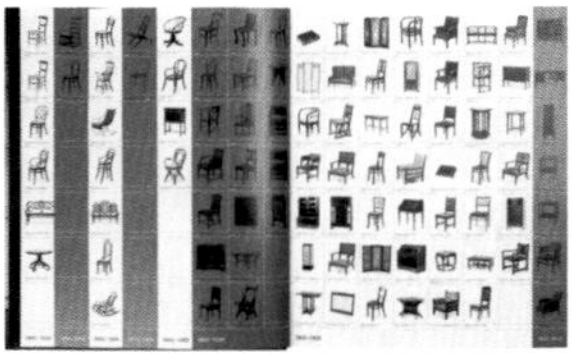
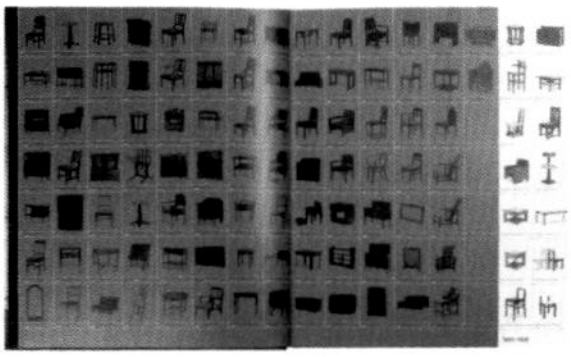
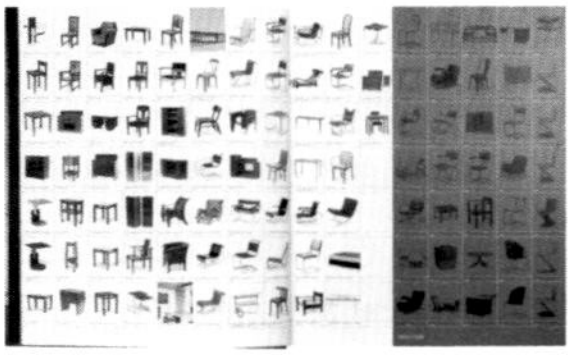
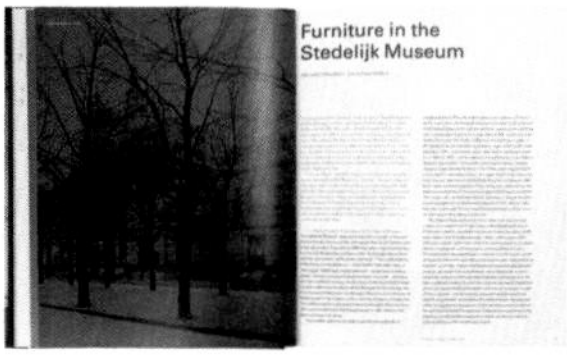
Furniture in the
Stedelijk Museum

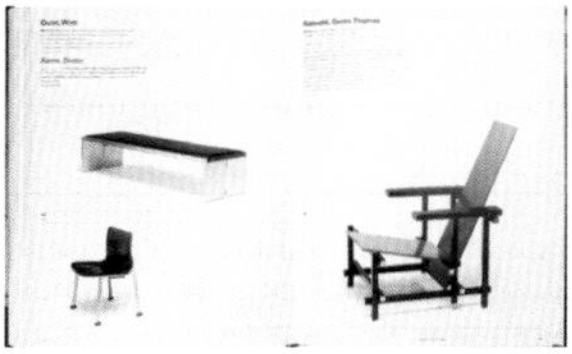
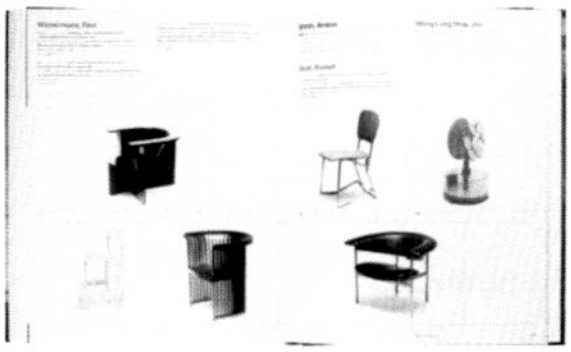

From Acquisition to Museum Collection Piece

The Furniture Collection Stedelijk Museum Amsterdam
Foreword

ng of th

The Furniture Collection
Stedelijk Museum Amsterdam

possibilities of new materials and techniques which creates a light yet indestructible chair. He uses epoxy to harden the chair's cord (Aramide rope with a carbon-core) so that the reinforced construction can support a person. The design is a veritable combination of old and new. Despite the playful association with furniture forms and conventions, the chair is clear evidence that Droog Design products are more restrained than the more exuberant ones from Memphis.

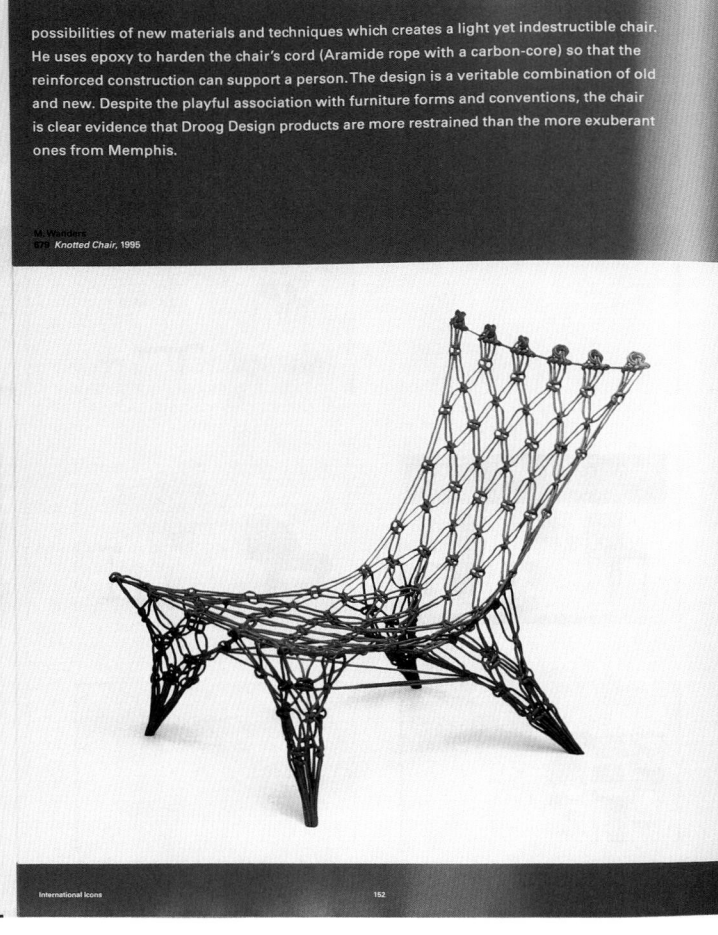

M. Wanders
679 *Knotted Chair*, 1995

International Icons 152

136 De Bazel, Karel P.C. Tea table, 1914–1917, produced by Meubelatelier De Ploeg, Amsterdam (NL)
76.5 × 72 × 72 cm; seat height 76.5 cm; solid mahogany and mahogany veneer construction; edged with ebony inlay; moulded legs with decorative carving; feet caps with gliding casters, four glass doors and two pull-out tops; gift from Mrs H. Magnus-Menko, Zeist, 1962
note: Part of the set of furniture cat. nos. 125 to 137
inv. no. KNA 2217

137 De Bazel, Karel P.C. Set of two small tables, 1914–1917, produced by Meubelatelier De Ploeg, Amsterdam (NL)
both tables: 51 × 54.5 × 35 cm; folding mahogany construction with moulded and carved legs; stored in matching storage cupboard KNA 2215; gift from Mrs H. Magnus-Menko, Zeist, 1962; note: Part of the set of furniture cat. nos. 125 to 137
inv. no. KNA 2255(1-2)

De Götzen, Giuseppe

138 De Götzen, Giuseppe Soubrette folding trolley, 1973–1974
produced by Guzzini, Recanati (I)
[illegible] cast polypropylene construction; moulded and painted steel and rubber strips; matt aluminium handle and wheels
[illegible]
inv. no. KNA 5583

De Klerk, Michel

139 De Klerk, Michel Lady's armchair, 1912–1913, produced under the designer's control
78 × 71 × 58.5 cm; seat height 42 cm; oak frame; chiselled decoration on legs, spindles and armrests; oak plywood (three-ply) backrest panel; deep sprung upholstered seat and flat upholstered backrest in moquette; gift from Mr J.H. Polenaar, Amsterdam, 1962; note: Designed for the home of Mr J.H. Polenaar, Frederiksplein 40, Amsterdam, on the occasion of his wedding in April 1913. The upholstery fabric was designed by T. Nieuwenhuis
inv. no. KNA 1359 and KNA 1360

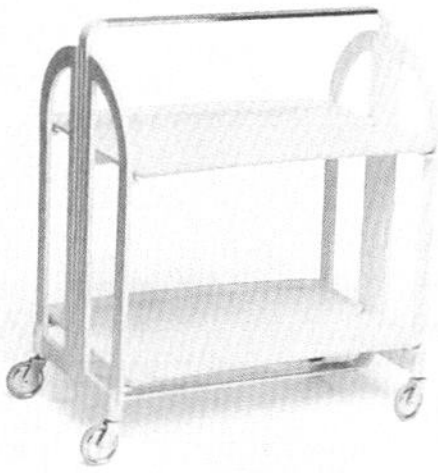

136
137

138
139

005 Ha 1

这是朱叶青的五本一套的杂文集，按照内容的细微差异，划分为五本。作者朱叶青为每本分别作序。书籍本身为裸脊锁线平装本，纸张柔软，便于翻阅，但为了长期保存，使用瓦楞纸制作了书函。书函的书脊部分为镂空状，露出裸脊锁线的装订特点。书函上下为开口部分，书籍需从中抽出或放回。封面上采用黑色加一套专色印刷，五本分设五个专色，除了书名、作者和出版社三个镂空位，通过专色油墨的镂空在黑色里呈现若隐若现的较大字号的书名，显得较为别致。内文胶版纸印单黑，排版较为质朴，着力在阅读本身上，穿插作者手绘小插图调整内容之间的节奏。||||||||||||||||||||||||||||||||||

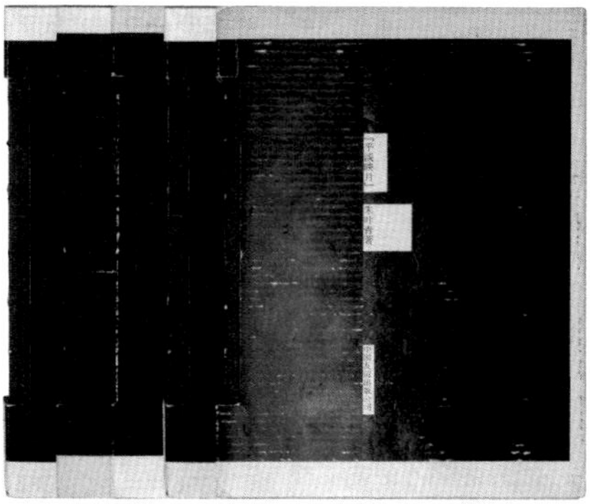

This is a collection of essays by Zhu Yeqing. The articles are divided into five volumes according to the nuances of the content. The author prefaces each book separately. The book is a thread sewing paperback with a bare spine and the paper is soft and easy to browse, but for long-term preservation, the slipcase is made of corrugated paper. The top and bottom of the case is the opening part, from which the book should be taken out or put back. The cover is printed in black plus one certain color. Five different colors are used for five books. The book title, name of the author and publishing house are presented in black on certain color. Inside pages are printed in black on offset paper. Typesetting is relatively simple, focusing on reading itself. The author's tiny hand-drawn illustrations are suitable to adjust the rhythm between the content.

◎	荣誉奖	Honorary Appreciation
«	朱叶青杂说系列	The Philosophy of Mist / The Sensation of Wateriness / Look up at the Sky / The Ism of Antique / Consommé and Water
◖	中国	China
D	何君	He Jun
△	朱叶青	Zhu Yeqing
▥	中国友谊出版公司	China Friendship Publishing Co., Peking
□		210×152×57mm
⌷		949g
≣		751p

The series demonstrates a convincing serial concept according to which each catalogue, printed on light, simple paper, is protected in a cardboard case. Catalogue and case correspond to one another in design and use of colour. Each volume in the series has a different, subtle colour scheme. The binding allows for excellent handling. The texts which in their typography are unpretentiously arranged are highlighted through poetic drawings and photographs.

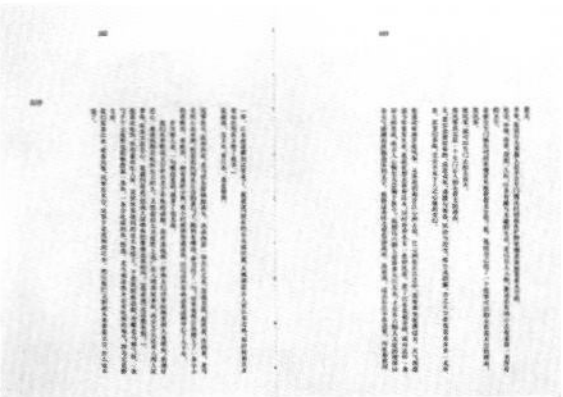

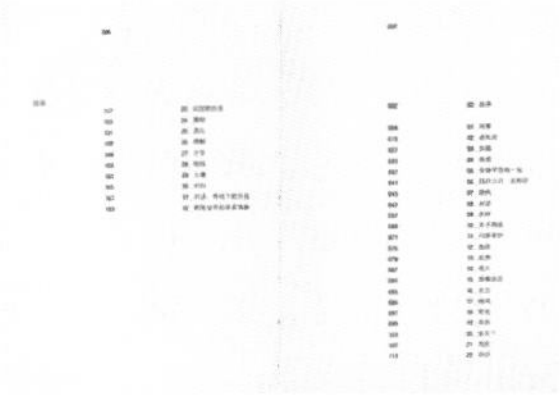

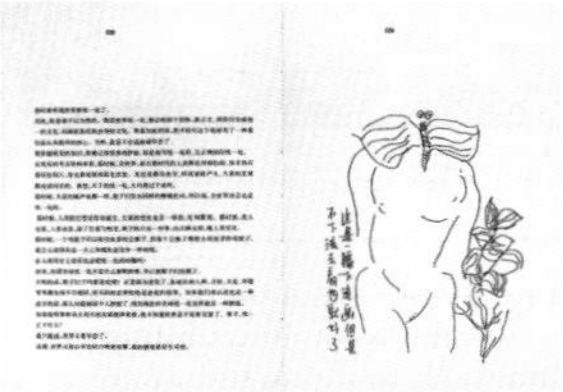

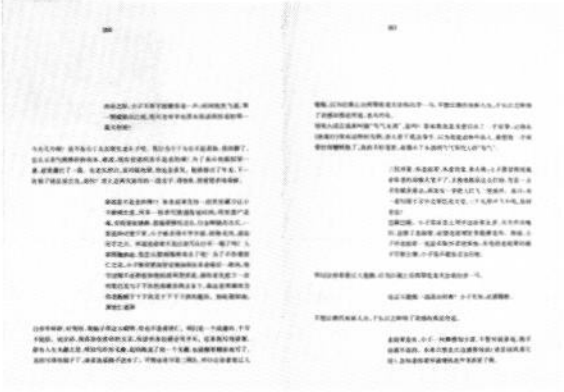

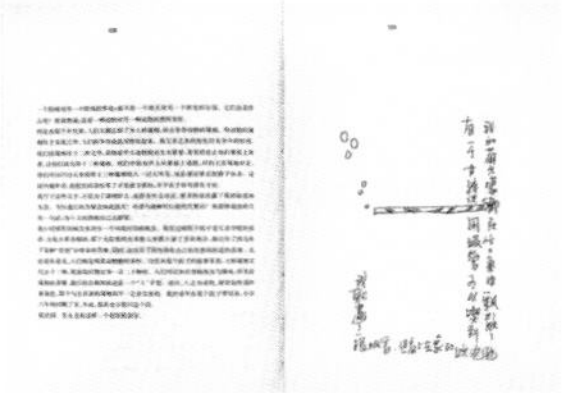

196

古，但是在艺术上究竟有多少是自己的创造，或者其艺术品味究竟有多高，依我看，皆是一些可以自有定论的话题。若是说得刻薄一点，可能是一位满族皇帝在追求中原文化时所表现的捉襟见肘的尴尬心态。我从个人的见解这样去看，乾隆皇帝是一位有福气之人，但却不是一个有艺术品味之人，即使相比其父其祖，康熙皇帝与雍正皇帝，无论是治国与治艺，也都是不可望其项背而企及的。康熙瓷器艺术独步本朝，晚清民初时期文物鉴赏家亦持有是论，皆目之康熙瓷器艺术为清朝之冠。至乎于清朝中晚期之后的瓷器，皆被视之为低劣什物。当时人之普遍鉴赏水平远甚于现在人之上，似乎也是一个很叫人感到无奈的事情。

乾隆皇帝之后的各代，也就是清代中晚期官窑瓷，更是一味对乾隆时期器物的拙劣模仿，而且，并非是艺术性地模仿却仅仅是工艺性地模仿，甚至在工艺上也是望其项背而不能企及的。于是就有这样一个公式，所谓晚清民国瓷器乃是"模仿之模仿"，二百余年来中国瓷器制造竟无多大的改变，各朝器物差异，仅仅是瓷器表面的或青花、或五彩、或斗彩，做着图案装饰与画风的变化。

总结乾隆官窑器物，应该说是有自己风格的，也可以说是富有创造性的，但是这个创造仅是局限在工艺方面，或谓之是仿生学的创造。譬如将瓷器的釉色模仿石头、木头、树皮、玉器、金属等等，林林总总，不一而足，总之在中国瓷器工艺制作史上达到了前所未有的地步，但同时，在中国瓷器艺术性上也达到了一个前所未有的地步，这一地步之审美性，是向着正极发展，还是向着负极发展，可能难以说得清楚。但是对于其后世的影响，却是很能够说得清楚的。

· · ·

197

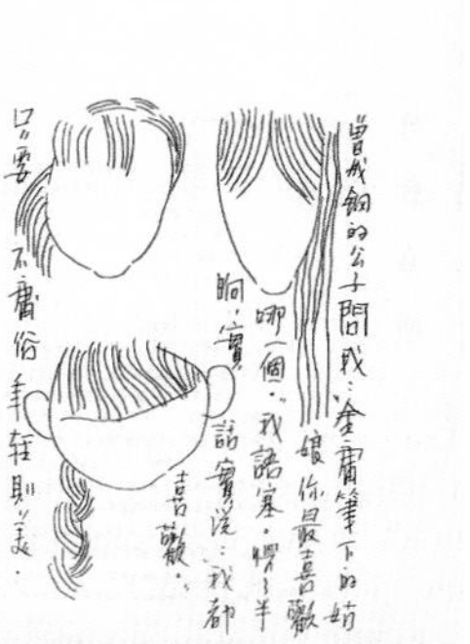

娘你最喜歡
我都
喜歡，
不庸俗手短則美。

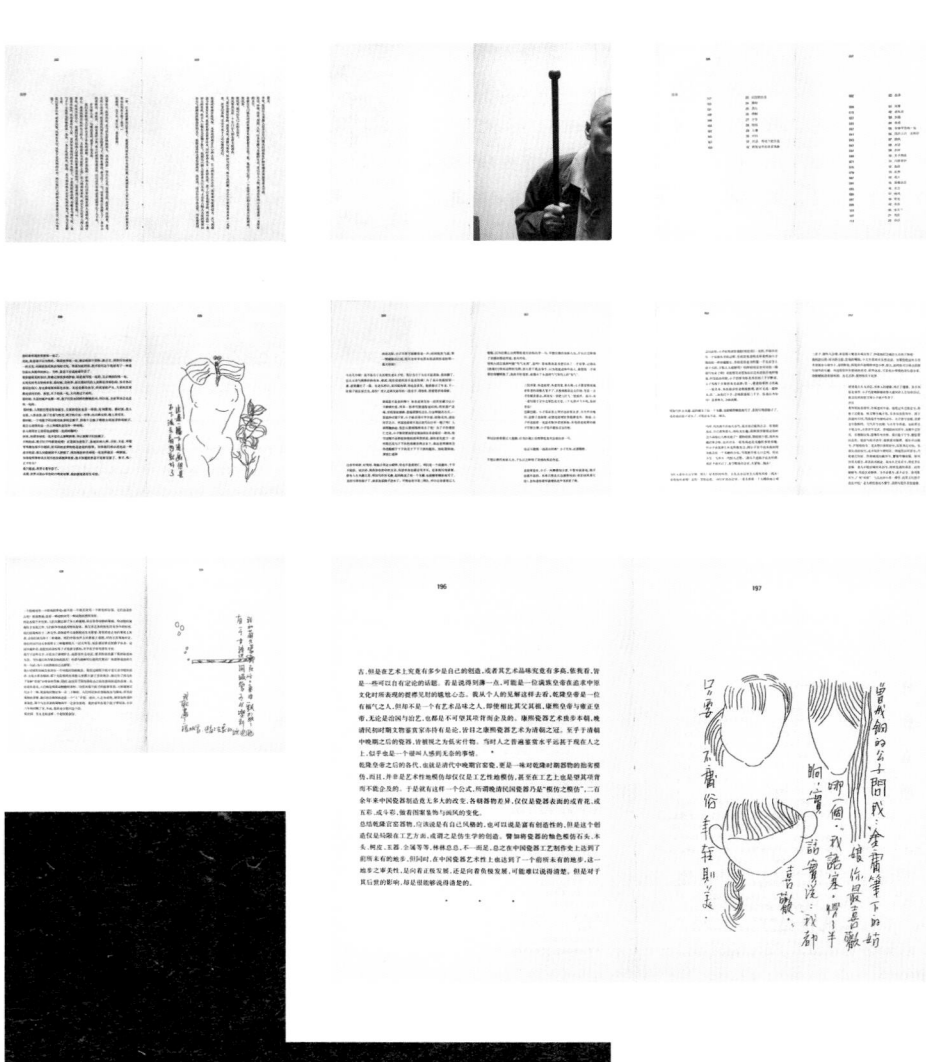

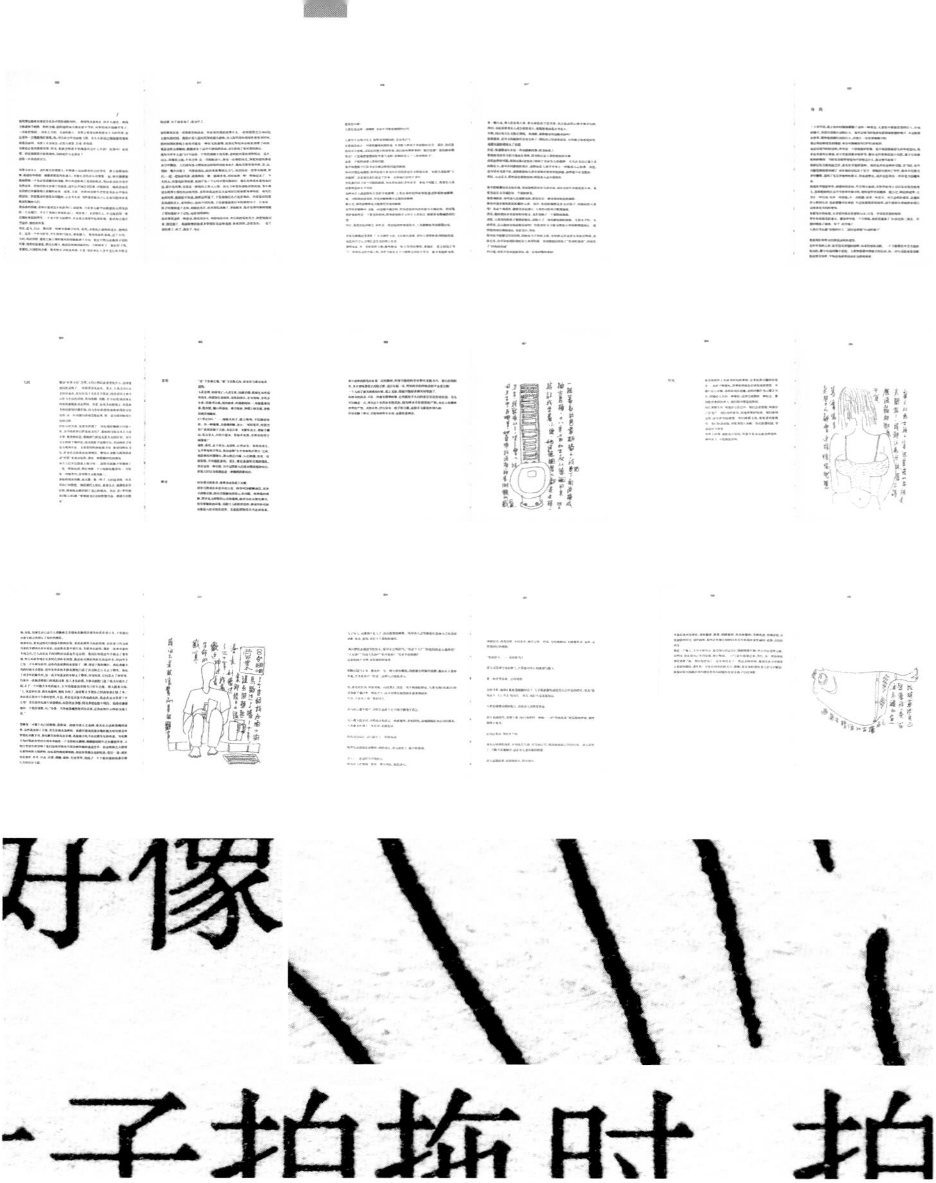

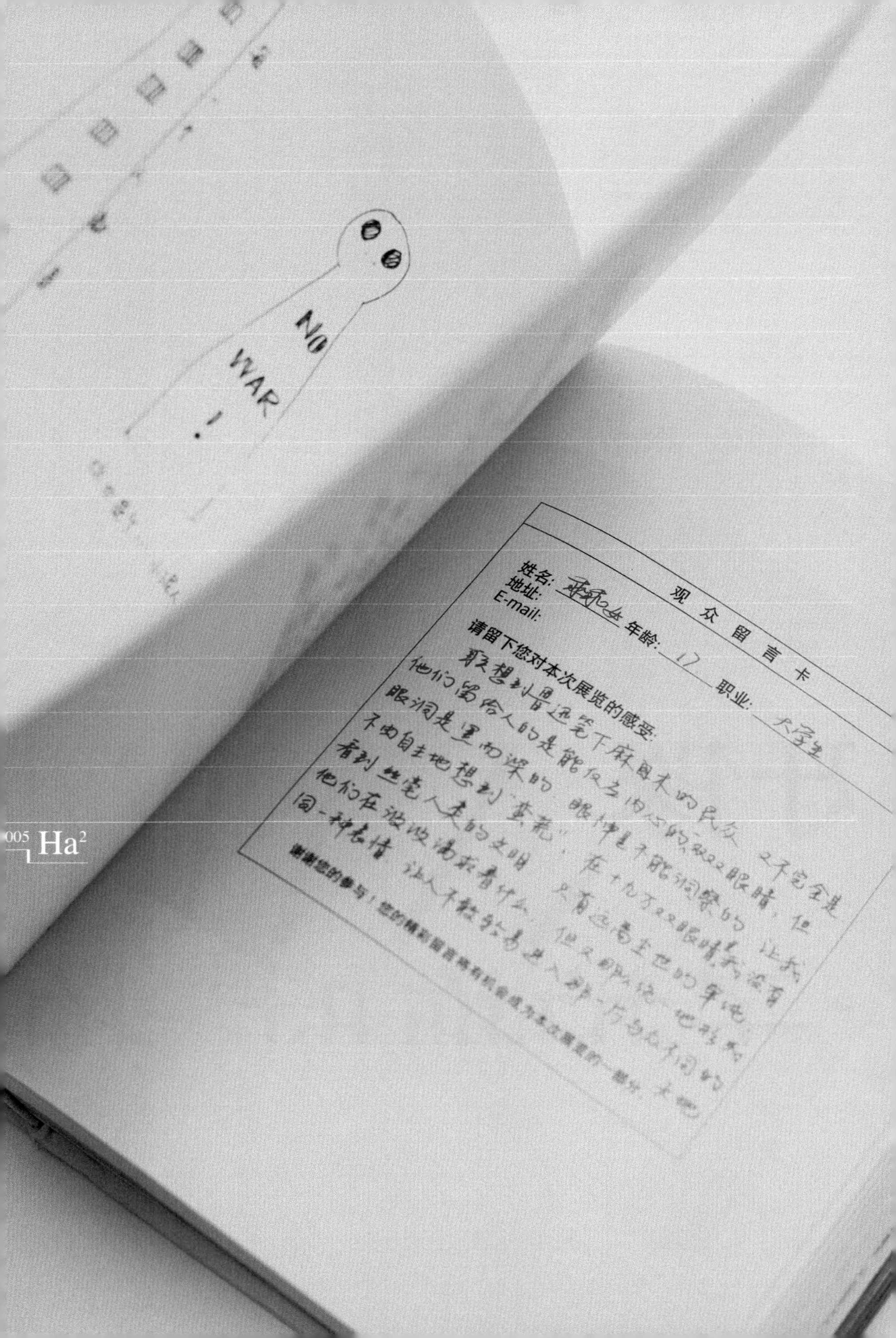
NO WAR !
观众留言卡
姓名:
地址:
E-mail:
年龄:
职业:
请留下您对本次展览的感受:
谢谢您的参与，您的精彩留言将有机会成为本次展览的一部分本书是英国著名雕塑家安东尼·葛姆雷（Antony Gormley）的大型展览“亚洲土地”（Asian Field）的综合记录，也是展览精神内核的延续。展览的作品是由安东尼·葛姆雷深入各地乡村指导人们用身边最熟悉的泥土创作的无数小雕塑。书籍记录了展览的现场、观众留言卡、网上留言、留言本上手绘的图画、创作过程和布展者名录等信息。作为书籍的阅读者，实际也在阅读和思考的过程中成为这件公众参与作品的一部分。本书20开大小，近500页，硬精装，体现出厚重的文本价值。封面裱贴一张空白的观众留言卡，意图将读者拉入作品公众参与的部分。内页胶版纸四色印刷，有如真实翻阅大量的留言簿，让读者具有身处现场的感觉，并逐渐参与其中。

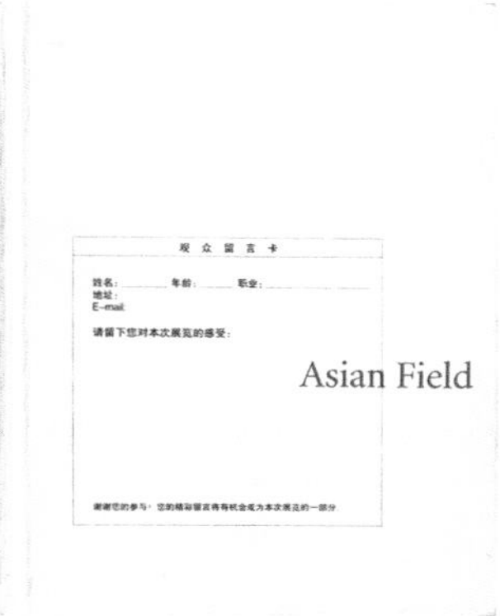
Asian Field

This book is a comprehensive record of the large-scale exhibition 'Asian Field' by the famous British sculptor Antony Gormley, and also a continuation of the core spirit of the exhibition. The works in the exhibition are countless small sculptures created by common people under the guidance of Anthony Gormley with the clay around us. The book records the information of the exhibition site, audience's message cards, online messages, hand-painted pictures on the message books, and the names of the creators and exhibitors. As readers of the book, they actually become part of this public work of public by participating in the process of reading and thinking. The hardcover book is nearly 500 pages, reflects the heaviness of the text. The cover is mounted with a blank comment card, intended to draw the reader into the public engagement of the work. The inside pages are printed on offset paper in four colors, just like reading a large number of guestbooks, so that the readers may have the feeling of being on the scene, and gradually participate into the exhibition.

◎ 荣誉奖	Honorary Appreciation
« 土地	Antony Gormley – Asian Field
中国	China
王序	Wang Xu & Associates
张卫	Hu Fang, Zhang Weiwei (Ed.) Zhang Wei (Ass. Ed.)
湖南美术出版社	Hunan Fine Arts Publishing House, Hunan
	227×177×45mm
	1273g
	488p

A classical book about a classical publisher. Interesting type-area using two columns and a marginal column for footnotes. The text elements seen together roughly form a square. A deliberately large space beneath the type-area is left free. Even the pagination is placed high up. The colour red is used as a structural text element. The perfection of the setting and lithography speaks for itself. Large illustrations are given a page to themselves, smaller ones are integrated into the text.

土
地

Asian Field

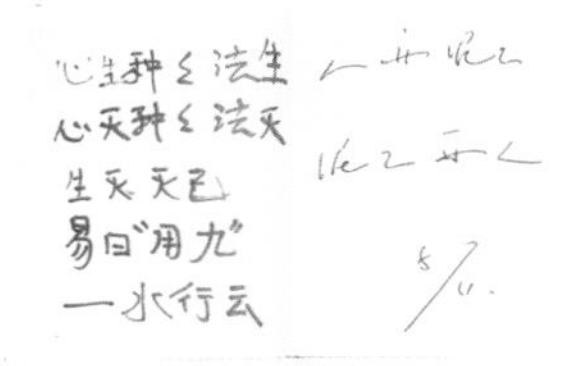

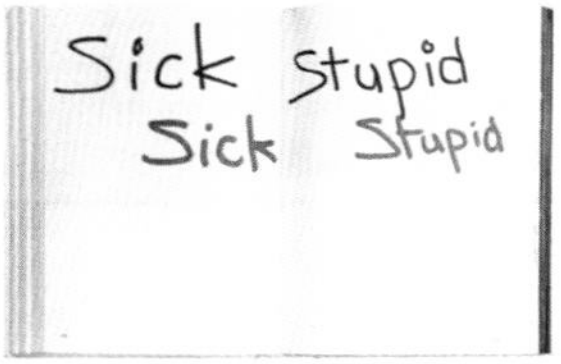
Sick Stupid
Sick Stupid

重庆
Chongqing

何处惹尘埃
本来无一物

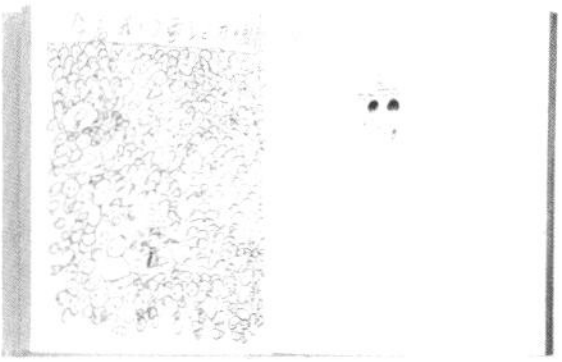

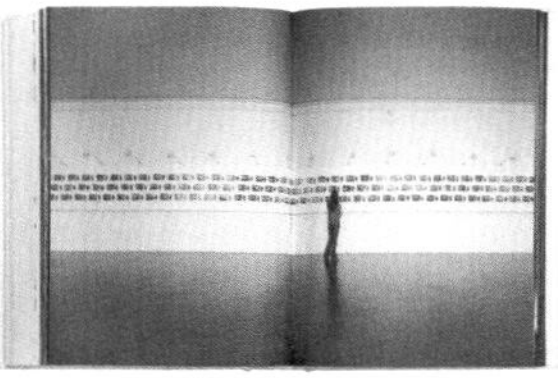

万众一心
众志成城

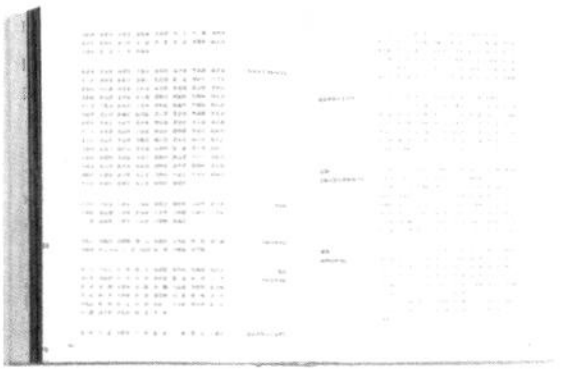

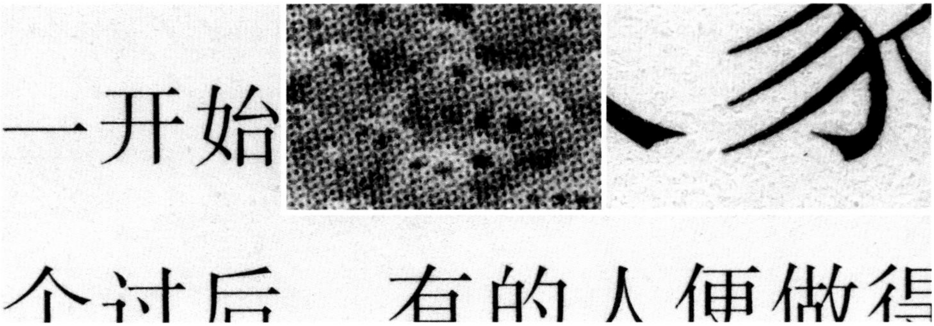
一开始

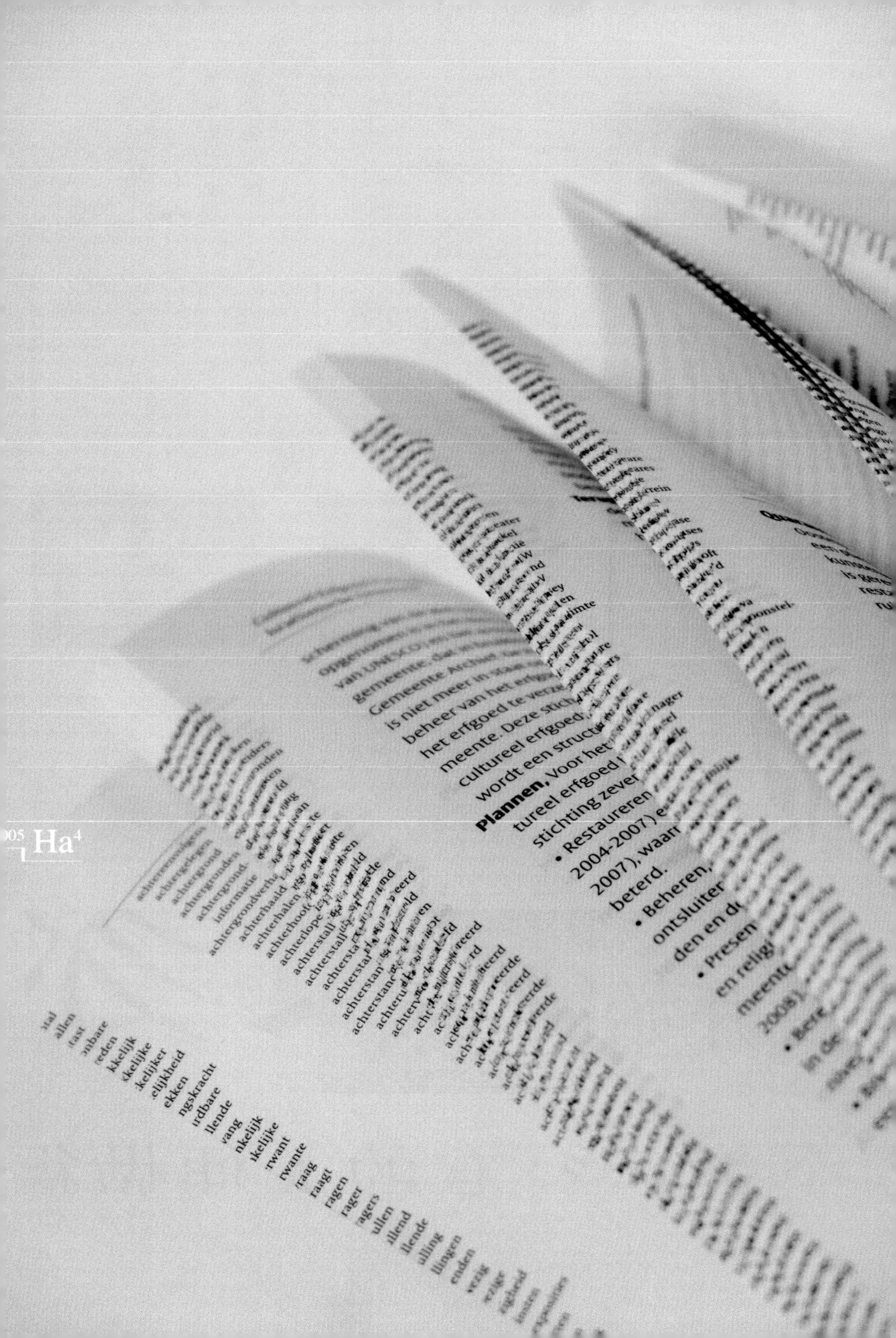

Ha
Plannen, Voor het
tureel erfgoed
stichting zeven
• Restaureren
2004-2007) e
2007), waar
beterd.
• Beheren,
ontsluiten
den en d
• Presen
en relig
meente

这是由阿姆斯特丹艺术委员会发布的文化展望建议书。此委员会是阿姆斯特丹市政委员会的常设咨询机构，主要职能为针对文化和艺术机构的发展情况、活动实施水平、拨款申请和未来预期作出判断，并为市政委员会最终确定四年一次的扶助计划提供参考建议。本书的装订形式为无线胶平装。封面采用白卡纸印单色黑，内页采用超薄字典纸印单色黑。内文的主要部分针对每个具体项目、机构和活动进行展开，包含项目或机构的简介、过去的发展情况、本期计划、艺术委员会的建议，以及对款项申请的建议。这部分内容按照字母表顺序进行梳理，项目或机构名称采用粗黑字体，比项目下的各段落标题字号略大，在提供了可分辨的差异用于查阅的情况下，版式具有高度的统一性。从封面贯穿至内页，遍布订口和翻口的单词是本书最大的设计特色，单词来自书的内容，按照字母表顺序进行罗列，在书口上形成整齐的肌理。

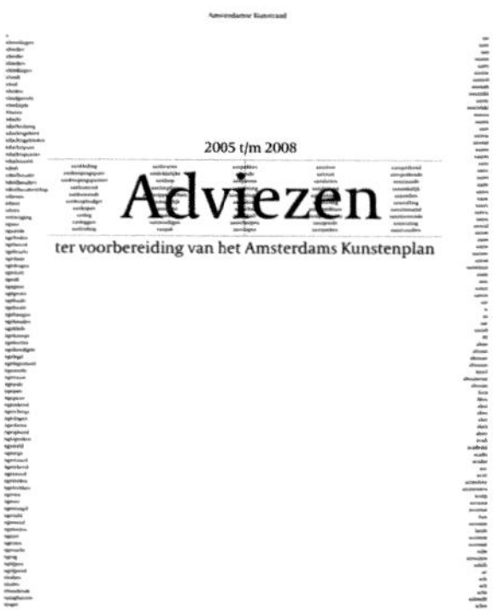
2005 t/m 2008
Adviezen
ter voorbereiding van het Amsterdams Kunstenplan

This is the proposal for cultural prospects issued by the Amsterdam Arts Committee. The committee is the permanent advisory organization of the Amsterdam Municipal Council. Its main functions are to judge the development of cultural and artistic institutions, the implementation of activities, applications for grants and future expectations, and to provide reference for the Municipal Council to finalize its quadrennial support plan. This paperback book is perfect bound. The cover is printed on white cardboard in monochrome black while the inner pages are printed on ultra-thin dictionary paper in black. The main part revolves around every specific project, institution and activity, including a brief introduction of the project or institution, past developments, current plan, recommendations of the Arts Committee, and proposals for the application of funds. This part is sorted out in alphabetical order. The names of projects or organizations are in boldface, which is slightly larger than the font size of paragraph headings. The format is highly uniform when distinguishable differences are provided for reference. From the cover to the inside pages, the most important design feature of the book is that many words are printed on the gutter and fore-edge. The words come from the book and are listed in alphabetical order, which form a neat texture on the edge of the book.

- 荣誉奖
- 关于编制 2005 年至 2008 年阿姆斯特丹艺术计划的建议
- 荷兰

- Honorary Appreciation
- Adviezen ter voorbereiding van het Amsterdams Kunstenplan 2005 t/m 2008
- The Netherlands
- Hansje van Halem & Gerard Unger
- Amsterdamse Kunstraad, Amsterdam
- 234×165×9mm
- 316g
- 264p

This understated book is designed solely on the basis of typographical means and distinguishes itself through its strong clarity. Particularly worthy of mention is the design of the book's outside edge and the binding edge using a 5-point type. The flexible volume has an outer envelope which makes use of the same design elements adding red as a distinguishing colour.

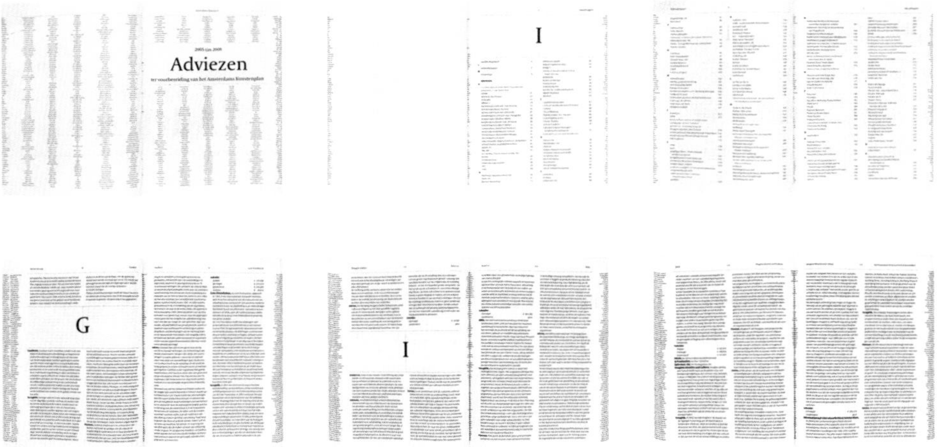

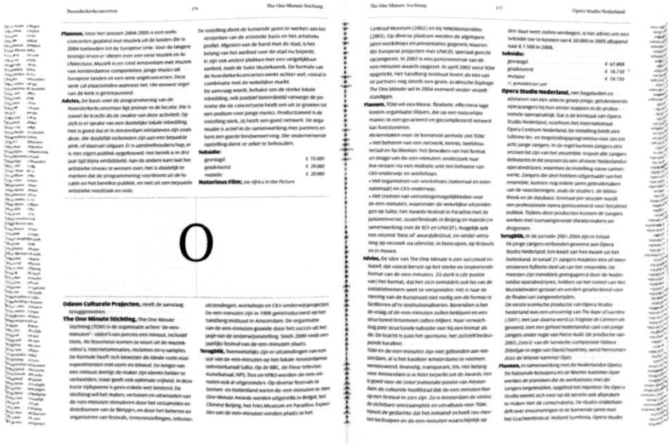

Noorderkerkconcerten 176 The One Minute Stichting

Plannen, Voor het seizoen 2004-2005 is een reeks concerten gepland met muziek uit de landen die in 2004 toetreden tot de Europese Unie. Voor de langere termijn leven er ideeën over een serie Muziek en Architectuur, Muziek in en rond Amsterdam met muziek van Amsterdamse componisten, jonge musici uit Europese landen en een serie orgelconcerten. Deze serie zal plaatsvinden wanneer het Hle-enowse orgel van de kerk is gerestaureerd.

Advies, De basis voor de programmering van de Noorderkerkconcerten ligt primair in de locatie. Die is zowel de kracht als de zwakte van deze activiteit. Op zich is er sprake van een duidelijke lokale inbedding. Het is goed dat er in Amsterdam initiatieven zijn zoals deze, die duidelijk verbonden zijn aan een bepaalde plek, of daarvan uitgaan. Er is aandeelhouderschap, er is een eigen publiek opgebouwd. Het bereik is in drie jaar tijd bijna verdubbeld. Aan de andere kant laat het artistieke niveau te wensen over. Het is duidelijk te merken dat de programmering voortkomt uit de locatie en het bereikte publiek, en niet uit een bepaalde artistieke noodzaak en visie.

De instelling dient de komende jaren te werken aan het versterken van de artistieke basis en het artistieke profiel. Afgezien van de band met de stad, is het belang van het aanbod voor de stad nu beperkt. Er zijn ook andere plekken met een vergelijkbaar aanbod, zoals de Suite Muziekweek. De formule van de Noorderkerkconcerten werkt echter wel, vooral in combinatie met de wekelijkse markt.

De aanvraag wordt, behalve om de sterke lokale inbedding, ook positief beoordeeld vanwege de potentie die de concertserie heeft om uit te groeien tot een podium voor jonge musici. Productioneel is de instelling sterk, zij heeft een goed netwerk. De organisatie is actief in de samenwerking met partners en kent een goede fondsenwerving. Die ondernemende opstelling dient ze zeker te behouden.

Subsidie:

gevraagd	€ 55.000
geadviseerd	€ 20.000
mutatie	€ 20.000

Notorious Film; *zie Africa in the Picture*

O

Odeon Culturele Projecten, Heeft de aanvraag teruggenomen.

The One Minute Stichting, The One Minute Stichting (TOM) is de organisatie achter 'de-een-minuten' – video's van precies een minuut, inclusief titels. Als fenomeen komen ze voort uit de muziekvideo's, internetanimaties, reclames en vj-samples. De formule heeft zich bewezen als ideale vorm voor experimenten met vorm en inhoud. De lengte van een minuut dwingt de maker zijn ideeën helder te verbeelden, maar geeft ook optimale vrijheid. In deze korte tijdspanne is geen enkele wet bindend. De stichting wil het maken, vertonen en uitwisselen van de-een-minuten stimuleren door het verzamelen en distribueren van de filmpjes, en door het beheren en organiseren van festivals, tentoonstellingen, televisie-uitzendingen, workshops en CKV-onderwijsprojecten.

De-een-minuten zijn in 1998 geïntroduceerd op het Sandberg Instituut in Amsterdam. De organisatie van de-een-minuten groeide door het succes uit het jaarlijks festival van de onderwijsinstelling. Sinds 2000 vindt een jaarlijks festival van de-een-minuten plaats.

Terugblik, Tweewekelijks zijn er uitzendingen van een uur van de-een-minuten op het lokale Amsterdamse televisiekanaal Salto. Op de BBC, de Finse televisie, Kunstkanaal, NPS, Teos en VPRO werden de-een-minuten ook al uitgezonden. Op diverse festivals in binnen- en buitenland waren de-een-minuten te zien. One-Minute Awards werden uitgereikt in België, het Chinese Beijing, het Fries Museum en Paradiso. Exposities van de-een-minuten vonden plaats in het

The One Minute Stichting 177 Opera Studio Nederland

Centraal Museum (2002) en bij NIMk/Montevideo (2003). Op diverse plaatsen werden de afgelopen jaren workshops en presentaties gegeven, waaronder Europese projecten met UNICEF, speciaal gericht op jongeren. In 2002 is een juniorenversie van de een-minuten-awards opgezet. In april 2002 werd TOM opgericht. Het Sandberg Instituut levert als één van de partners nog steeds een grote, praktische bijdrage. The One Minute wil in 2004 evenwel verder verzelfstandigen.

Plannen, TOM wil een kleine, flexibele, effectieve lage kosten organisatie blijven, die op een natuurlijke manier in een gevarieerd en gecompliceerd netwerk kan functioneren.

Als kerntaken voor de komende periode ziet TOM:

- Het beheren van een netwerk, kennis, beeldmateriaal en faciliteiten; het bewaken van het format en imago van de-een-minuten; onderzoek naar live-stream via een mobiele unit ten behoeve van CKV-onderwijs en workshops.
- Het organiseren van workshops (nationaal en internationaal) en CKV-onderwijs.
- Het creëren van vertoningsmogelijkheden voor de-een-minuten, waaronder de wekelijkse uitzendingen op Salto, het Awards-festival in Paradiso met de juniorenversie, zusterfestivals in Beijing en Nairobi (in samenwerking met de ECF en UNICEF), mogelijk ook een reizend 'best of'-awardsfestival, en verder vertoning op verzoek via televisie, in bioscopen, op festivals en in musea.

Advies, De idee van The One Minute is een succesvol initiatief, dat vooral berust op het sterke en inspirerende format van de-een-minuten. Zó sterk is (de positie van) het format, dat het zich inmiddels ook los van de initiatiefnemers weet te verspreiden. Het is naar de mening van de Kunstraad niet nodig om dit format te faciliteren of te institutionaliseren. Bovendien is het de vraag of de-een-minuten zullen beklijven en een structureel fenomeen zullen blijken. Naar verwachting past structurele subsidie niet bij een format als dit. De kracht is juist het spontane, het zichzelf bedruipende karakter.

TOM en de-een-minuten zijn niet gebonden aan Amsterdam, al is het karakter Amsterdams te noemen: vernieuwend, levendig, transparant, fris. Het belang voor Amsterdam is in feite beperkt tot de Awards. Het is goed voor de (inter)nationale positie van Amsterdam als culturele hoofdstad dat de-een-minuten hier op een festival te zien zijn. Zo is Amsterdam de vitrine, de zichtbare ontstaansplek en uitvalbasis voor TOM. Vanuit de gedachte dat het initiatief zichzelf zou moeten bedruipen en de-een-minuten waarschijnlijk op den duur weer zullen verdiegen, is het advies om een subsidie toe te kennen van € 30.000 in 2005 aflopend naar € 7.500 in 2008.

Subsidie:

gevraagd	€ 67.868
geadviseerd	€ 18.750 [1]
mutatie	€ 18.750

[1] *gemiddeld per jaar*

Opera Studio Nederland, Het begeleiden en adviseren van een selecte groep jonge, getalenteerde operazangers bij hun eerste stappen in de professionele operapraktijk. Dat is de kerntaak van Opera Studio Nederland, voorheen het Internationaal Opera Centrum Nederland. De instelling biedt een fulltime les- en begeleidingsprogramma voor zes tot acht jonge zangers. In de regel kunnen zangers één seizoen lid zijn van het ensemble. Vrijwel alle zangers debuteren in dat seizoen bij een of meer Nederlandse operabedrijven, waarmee de instelling nauw samenwerkt. Zangers die deel hebben uitgemaakt van het ensemble, kunnen nog enkele jaren gebruikmaken van de voorzieningen, zoals de studio's, de bibliotheek en de database. Eenmaal per seizoen wordt een professionele opera geënsceneerd voor betalend publiek. Tijdens deze producties kunnen de zangers werken met toonaangevende theatermakers en dirigenten.

Terugblik, In de periode 2001-2004 zijn in totaal 66 jonge zangers verbonden geweest aan Opera Studio Nederland. Een kwart van hen kwam uit het buitenland. In totaal 21 zangers maakten één of meer seizoenen fulltime deel uit van het ensemble. De meesten zijn inmiddels geëngageerd door de Nederlandse operabedrijven, hebben op het toneel van Het Muziektheater gestaan en werden geselecteerd voor de finales van zangwedstrijden.

De eerste scenische productie van Opera Studio Nederland was een uitvoering van *The Rape of Lucretia* (2001). Het jaar daarna werd *La Tragédie de Carmen* uitgevoerd, met een geheel Nederlandse cast van jonge zangers onder regie van Pierre Audi. De productie van 2003, *Zora D.* van de Servische componiste Isidora Žebeljan in regie van David Pountney, werd hernomen door de Wiener Kammer Oper.

Plannen, In samenwerking met de Nederlandse Opera, De Nationale Reisopera en de Wiener Kammer Oper werden de pianisten die de werksessies met de zangers begeleidden, opgeleid tot repetitor. De Opera Studio neemt zich voor op dit terrein ook afspraken te maken met de conservatoria. De studio onderhandelt over ensceneringen in de komende jaren met het Grachtenfestival, Holland Symfonia, Opera Studio

aanpass
aanpass...g
an...sing

II
A
J
K
S
V
evend e
kleurrijk
S, N
stellingsruimte

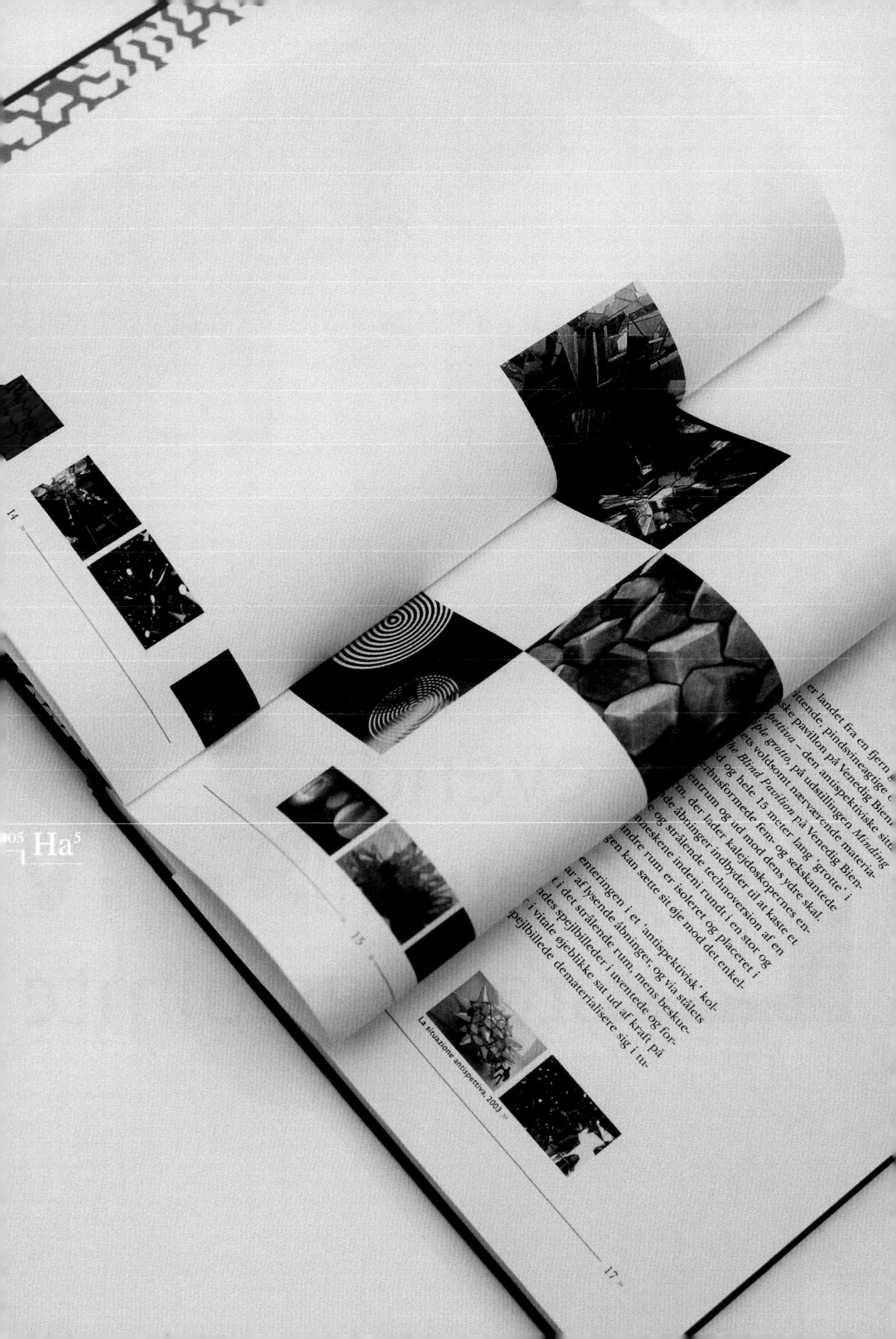

Ha5
14
15
er landet fra en fjern
sittende, pindsvineagtige
ske pavillon på Venedig Bien
pettiva – den antispektiviske sitt
ple grotto, på udstillingen Minding
ets voldsomt nærværende materia-
he Blind Pavilion på Venedig Bien-
ed og hele 15 meter lang 'grotte' i
erhusformede fem- og sekskantede
entrum og ud mod dens ydre skal.
m, der lader kalejdoskopernes en-
nde åbninger indbyder til at kaste et
e og strålende technoversion af en
nneskene indeni rundt i en stor og
indre rum er isoleret og placeret i
gen kan sætte sit øje mod det enkel-
enteringen i et 'antispektivisk' kol-
ar af lysende åbninger, og via stålets
t i det strålende rum, mens beskue-
ndes spejlbilleder i uventede og for-
i vitale øjeblikke sat ud af kraft på
pejlbillede dematerialisere sig i tu-
La situazione antispettiva, 2003 »
17 »

这是一本记录冰岛 - 丹麦艺术家奥拉维尔·埃利亚松（Olafur Eliasson）8 次展览和相关思考的合集。书中通过展览作品加作品分析、材料研究与友人的信件等展示，深度记录和探讨了艺术与创作艺术的过程。本书为圆脊布面精装，封面无印刷，通过压凹工艺呈现艺术家的代表图形和书名。内页四色印刷，通过鲜艳的绿色色块来区分 8 次展览，现场照片与结构图穿插，从构造、材料、光线、色彩等多角度呈现艺术家的作品。书籍的最后部分通过色块和数字标注的展览平面图来交代前文展品的展陈位置，并通过页码形成交互关系。

This collection records eight exhibitions of the artist Olafur Eliasson and his reflections. By presenting the works of exhibitions, analysis of the works, material research, and letters with friends, the book records and discusses art and the process of art creating in depth. This book has a rounding spine and a cloth cover that is not printed at all. The artist's representative graphics and the title are presented through embossing process. The inside pages are printed in four colors and bright green color blocks are used to distinguish 8 exhibitions. Site photos and structural drawings are interlaced to present artist's works from multiple perspectives, such as structure, material, light and color. In the last part of the book, the exhibition plans display the position of the works mentioned previously by color blocks and numbers, thus an interactive relationship is formed natually.

- 荣誉奖
- 奥拉维尔 · 埃利亚松——关注世界
- 瑞士

This book was the subject of much discussion in the jury and there was unanimous agreement on its inner design – the page-layout and typographical details create a very balanced effect. The photos and drawings are attractively placed on matt paper. A pleasing, calm and compelling inner layout, clearly legibly with attractive use of colours. Discussion centred merely on the use of colour in the cover and the band.

- Honorary Appreciation
- Olafur Eliasson. Minding the world
- Switzerland
- groenland.berlin.basel (Dorothea Weishaupt & Michael Heimann)
- ARoS Aarhus Kunstmuseum (Ed.)
- Hatje Cantz Verlag, Ostfildern-Ruit
- 285 × 238 × 26mm
- 1238g
- 232p

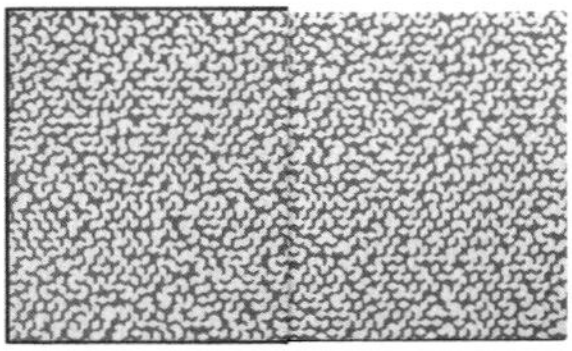

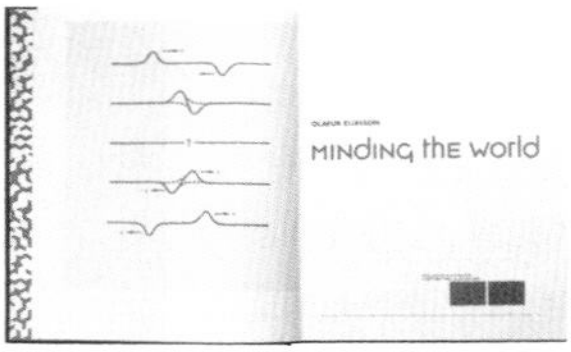
OLAFUR ELIASSON
MINDING the world

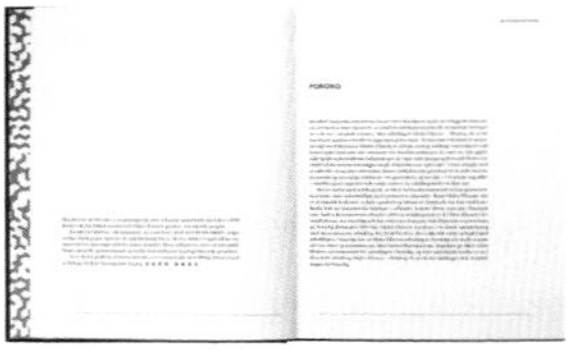

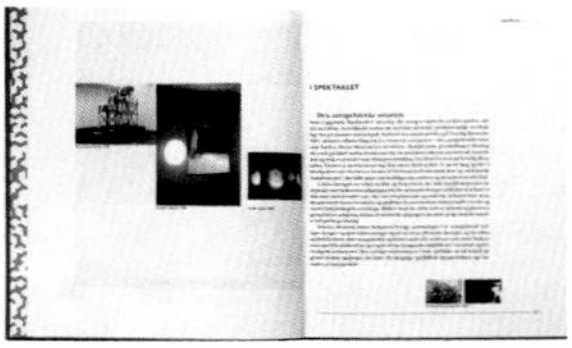

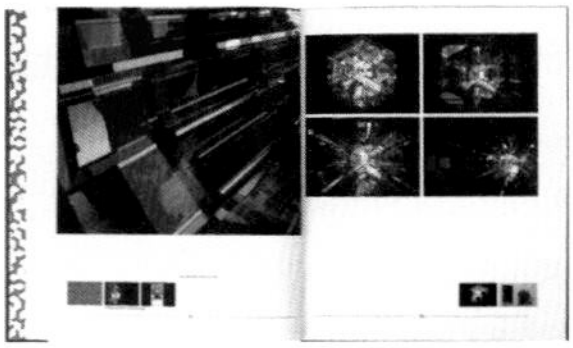

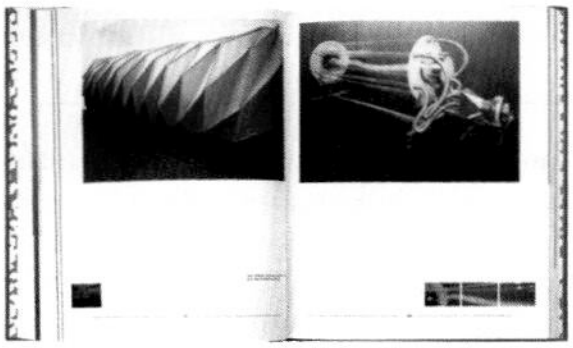

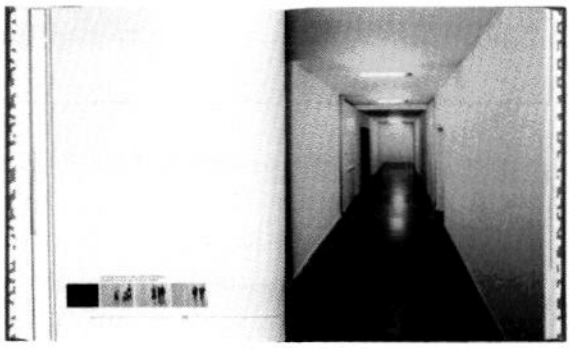

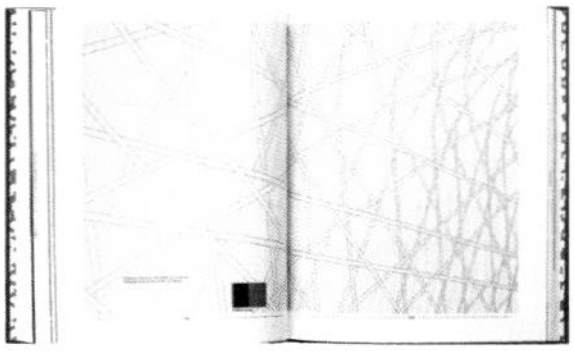

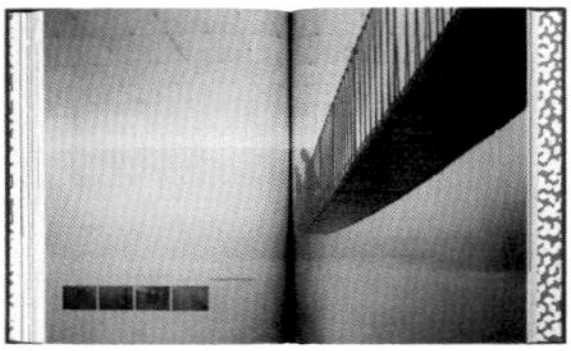

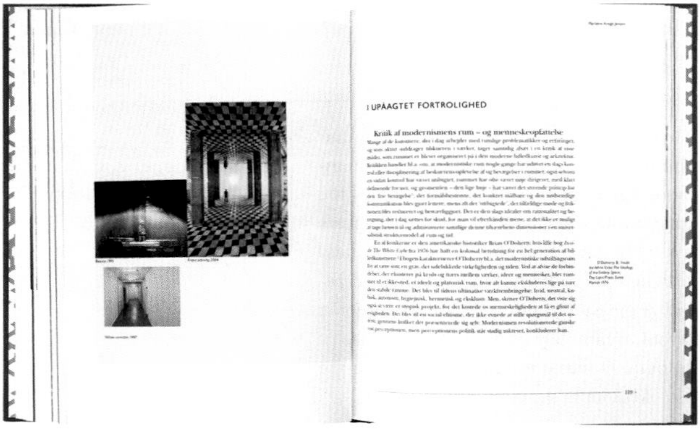
I UPÅAGTET FORTROLIGHED
Kritik af modernismens rum – og menneskeopfattelse

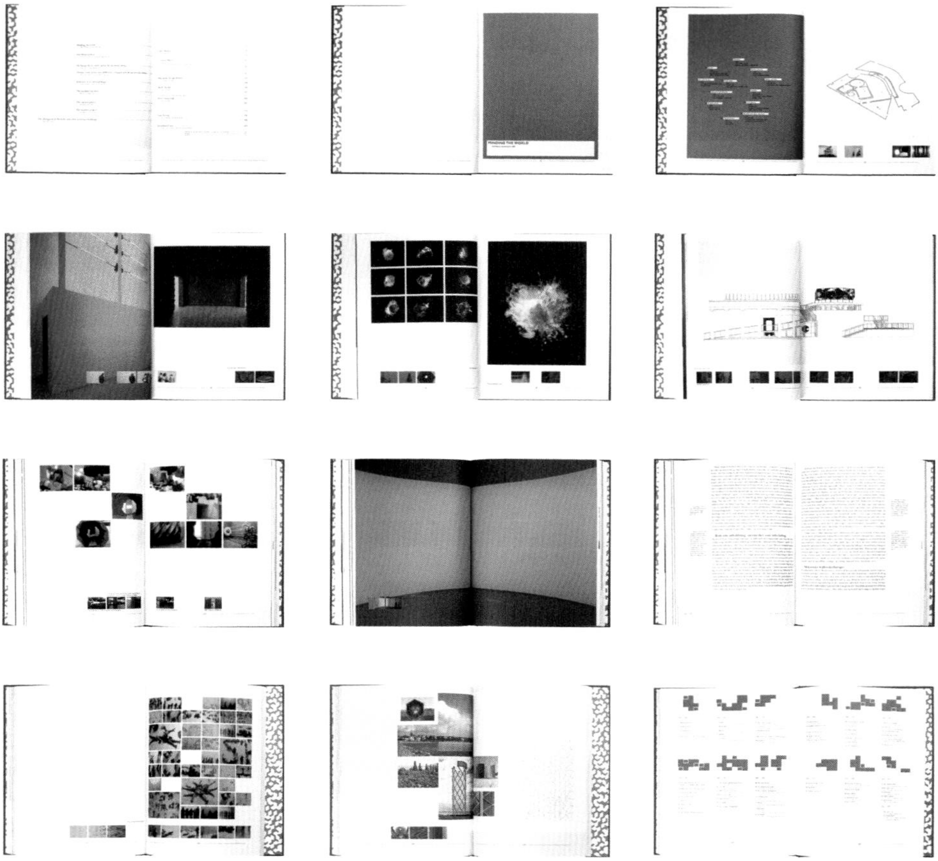

<<REWIND>>FORWARD

Materials of Mei Lanfang's Theatrical Performances

The Heart in Search of Beauty. An Exhibition Honoring the 100th Anniversary of Kobayashi Hideo's Birth

Paul Celan Verzamelde gedichten

Arbeiterlyrik 1842–1932

Caravaggios Rom. Annäherungen an ein dissonantes Milieu

Vermischungen. Interdisziplinäre Wochen an der Muthesius-Hochschule

Lebens-Mittel (Die Tollen Hefte 22)

Symbiont

nobody forever

Das Buch als Kunstwerk. Die Cranach Presse des Grafen Harry Kessler

Werktitel: Piet Gerards, grafisch ontwerper.

somnia. En bok om drømmer

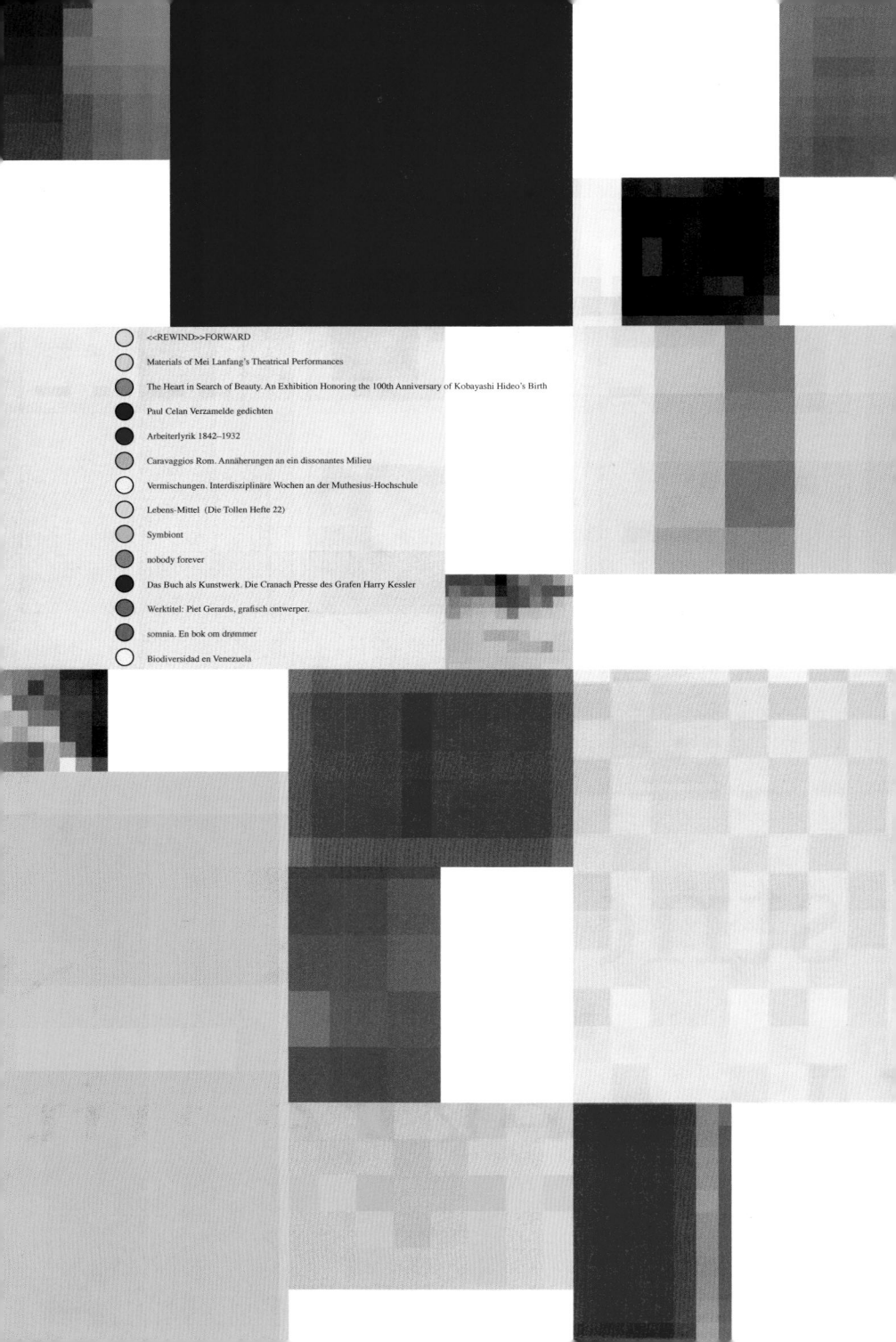

Biodiversidad en Venezuela

2004

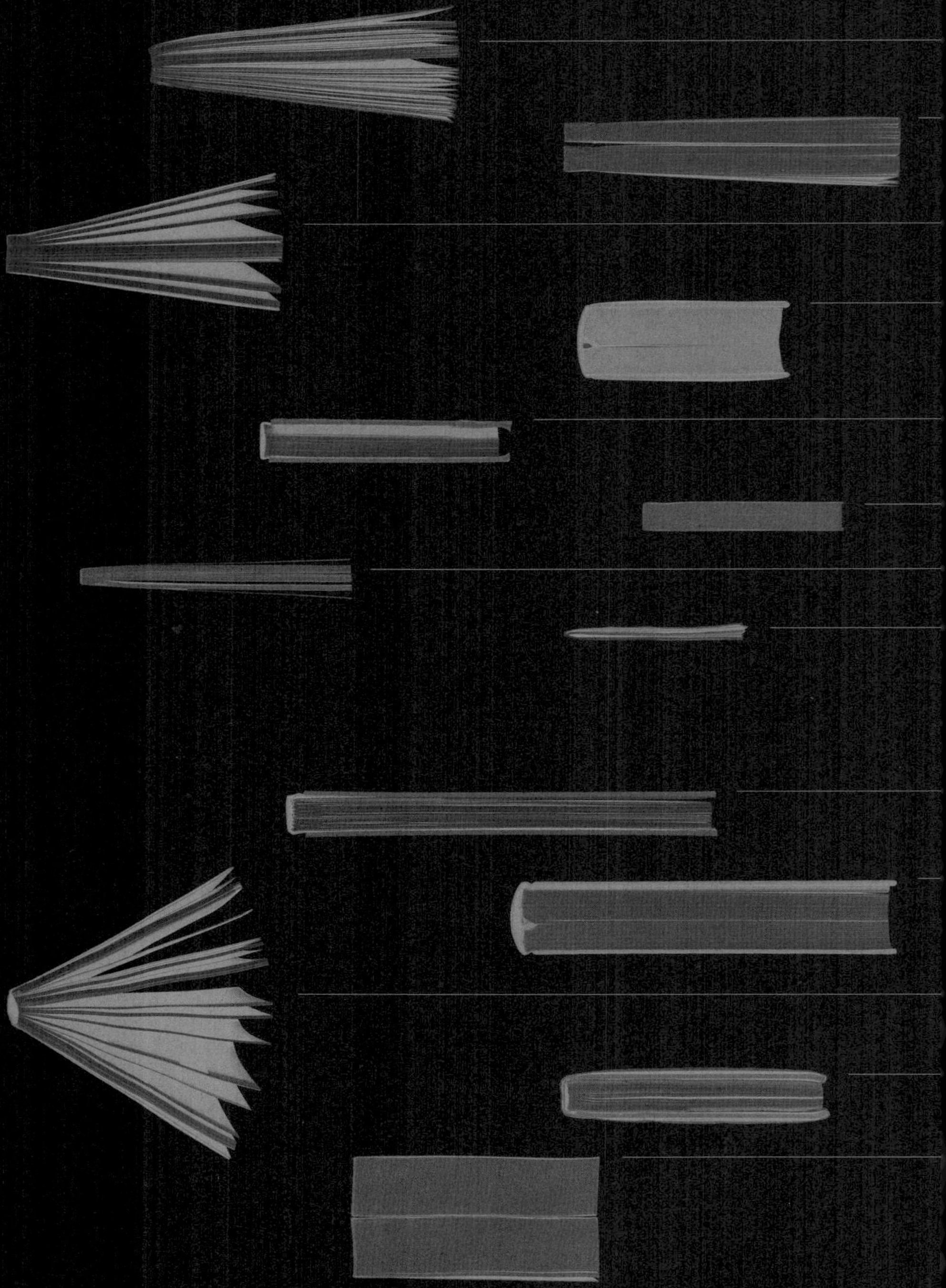

2004 Gl

1357

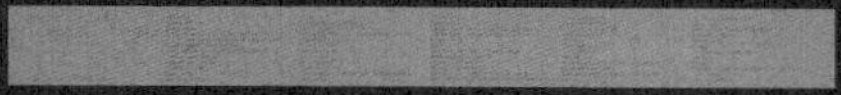

2004 Gm

1365

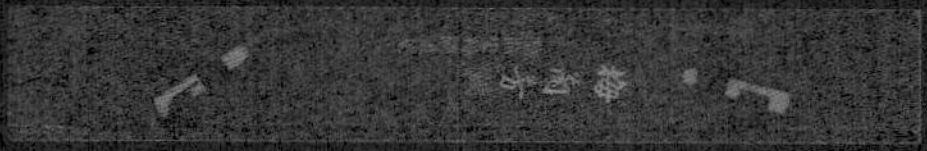

2004 Sm[1]

1371

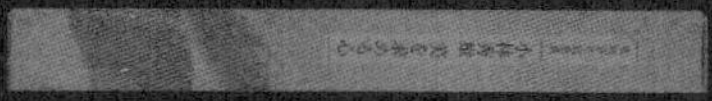

2004 Sm[2]

1377

CELAN

2004 Bm[1]

1383

Arbeiterlyrik 1842–1932

2004 Bm[2]

1389

CARAVAGGIOS ROM

2004 Bm[3]

1395

2004 Bm[4]

1401

2004 Ha[1]

1407

2004 Ha[2]

1413

2004 Ha[3]

1417

2004 Ha[4]

1421

2004 Ha[4]

1425

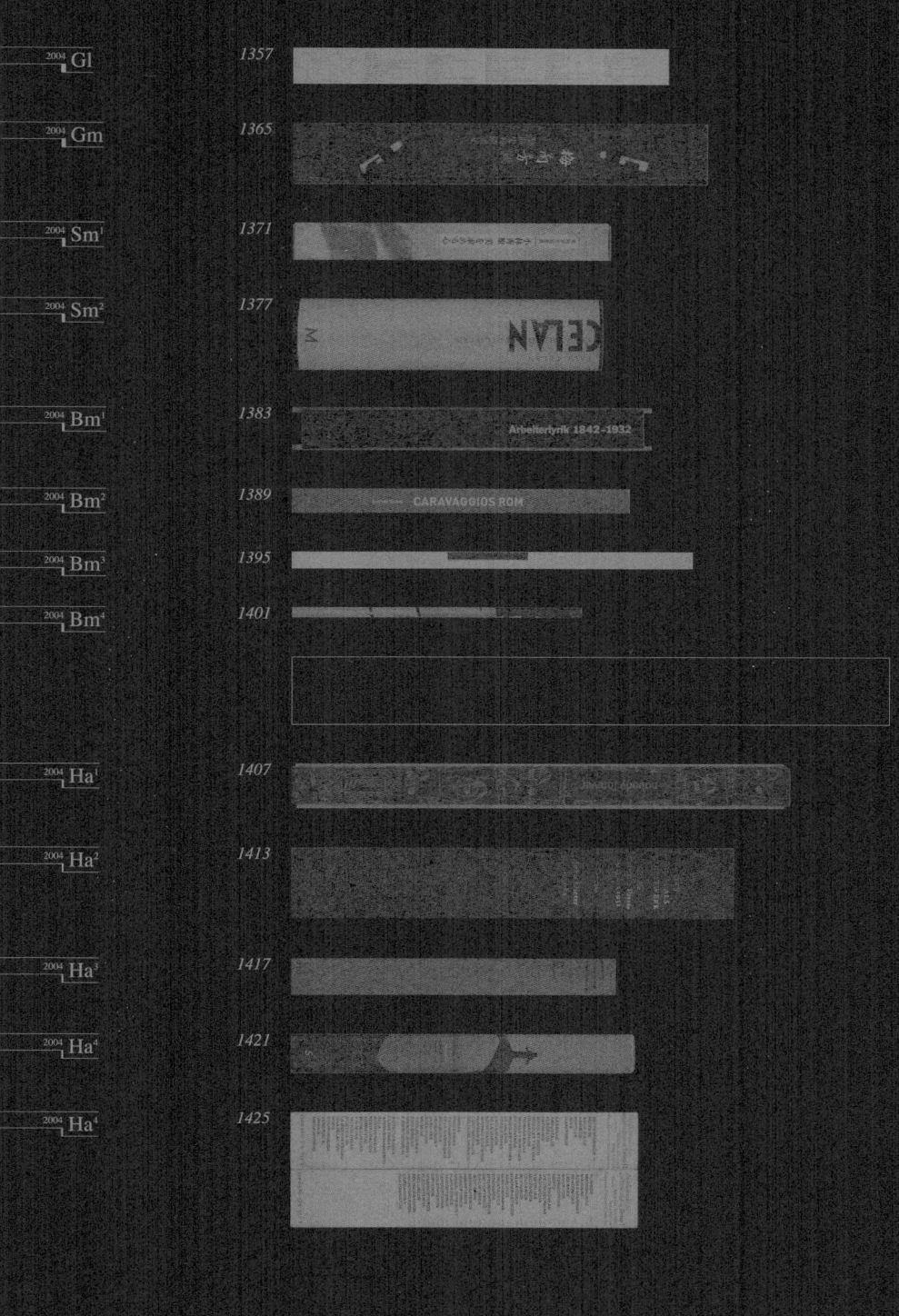

c.
d.

1992 erfolgt die Vergabe als eigenständiger
entliche Ausschreibung durch Jury-Ent
veiterten Kunstbeirats der Stadt Wolfsburg
sollen KünstlerInnen gefordert werden.

Olaf Nicolai has been honoured with the pr
For the exhibition associated with th
prize, Olaf Nicolai designed a new install
»Die Flamme der Revolution, liegend [in Wa

合いだとか、
を見る、そ
う。例えば、
。林檎は食
いをしている
していくるか、
ンヒルの店で
て買って来た
つけた訳だが
は一人もない
は何処のサイ
た人はない。
を一分間眺め
う。そしてラ
見ることは蝶
いていて一輪
。何だ、薔の
諸君は心の中
れば、諸君は
とです。薔の

ant wanneer de jood eraan komt en een andere joo
an is het snel gedaan met het zwijgen, ook in de bergen
n de natuur, dat zijn twee verschillende dingen, nog st

O nein, was sie den Wassern singt, ist nicht der Schmerz und nicht di
Ist Siegeslied, Triumpheslied, Lied von der Zukunft großem Tag!
Der Zukunft, die nicht fern mehr ist! Sie spricht mit dreistem Prophez

denn ihr Studium, unter Strafe stand.[93] Mög
vaggio sein Exemplar von einem befreundet

Gefangen von dieser Vorstellung trat Flaubert die Reise an, sah und empfand zunächst nach jenem Muster: »Eine Canja, einer Tartane ähnlich, gleitet über das Wasser: das ist der wahre Orient, melancholische und betäubende Wirkung: man

Mond oder einen Hasen malen. Manche Künstler nehmen
Leinwand und einen Rahmen aus Holz, aber das ist ziemli
Ehrlich gesagt, finde ich es auch zu umständlich. Ein Malka

wurde sie von einem Mann schwanger, der später kriminell wurde. Sie brachte ihr Kind in der geschlossenen Psychiatrie zur Welt.
Ihr zweiter Mann brachte ihr ebenso wenig Glück, dafür aber viele Beschimpfungen und Mißhandlungen. Ein dritter Mann schlug si
& blau, hatte eine Pistole und schoß auf sie und ihre Kinder. Ihr vierter Mann verließ sie mit einem Abonnement vom Scheißkermag

nd ihm kehrten die Sinne
len Schleier, und warf ihn
Wasser der Mündung.

noch weitere Bände über die fünfundzwanzig hinaus aufbinden zu lassen, um sie zu verkaufen.[91] Kessler aber blieb seiner noblen Geste, das Schöne verschenken zu wollen, treu.

Abb. 22 und
Die beiden
Bände mit
der vielleicht

ander werk. dienstbaar spel 264

documentatie: biografie **291** zuid-limburg **293** pan

våkne. Først og fremst inntreffer forandringer i det kjemiske milj
Signalstoffene serotonin og noradrenalin – som brukes til å sende meld
er mellom nervecellene – er til stede i store mengder i våken tilstand,

gos. Este es un número mayor de especies que el reportado para Colo
de ser Colombia mucho mejor conocido briológicamente. Más recien
et al. (1998) reportan dos nuevos registros de musgos para Venezuela p

2004 Gl

Golden Letter

<<REWIND>>FORWARD

Germany

Kristina Brusa, Markus Dreßen

Olaf Nicolai, Susanne Pfleger

Hatje Cantz Verlag, Ostfildern-Ruit

267×215×25mm

1126

168p

2004 Gm

Gold Medal

A Picture Album. Historical Materials of Mei Lanfang's Theatrical Performances

China

Zhang Zhiwei, Du Yuge, Gao Shaohong

Liu Cengfu, Zhu Jiajin

Hebei Education Publishing House, Shijiazhuan

295×220×43mm

1702g

229p

2004 Sm[1]

Silver Medal

The Heart in Search of Beauty. An Exhibition Honoring

the 100th Anniversary of Kobayashi Hideo's Birth

Japan

Kodama Yuuko, Ono Risa, Kawasina Hiroya

Hideo Kobayashi

Nihon Keizai Shimbun, Tokyo; Shinchosha, Tokyo

226×156×27mm

678g

256p

2004 Sm[2]

Silver Medal

Paul Celan Verzamelde gedichten

The Netherlands

Joost van de Woestijne; office of co

Paul Celan

Meulenhoff Amsterdam

221×150×53mm

1194g

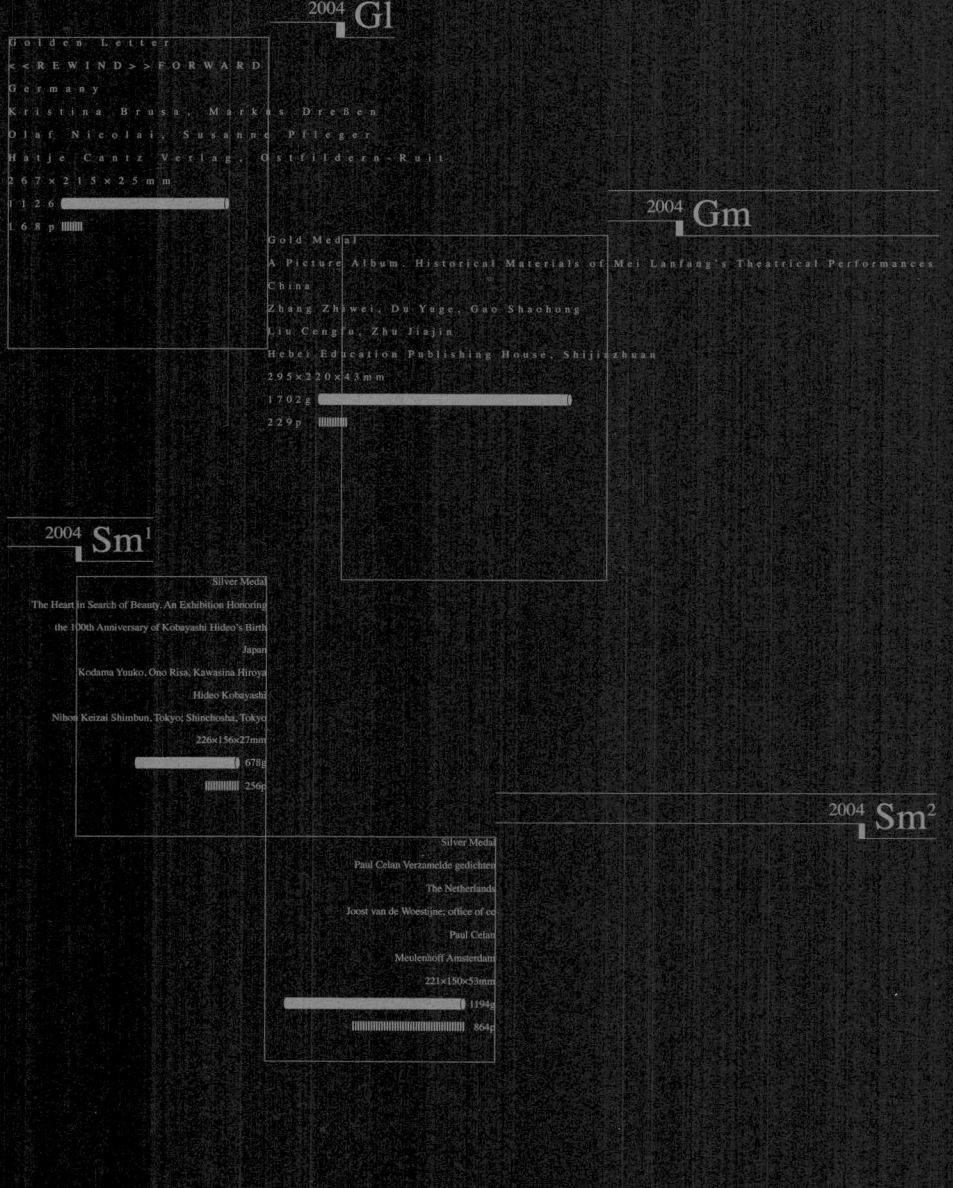

864p

2004 Ha1

Honorary Appreciation
nobody forever
Germany
Christine Lohmann, Walter Kerkhofs
Nick J. Swart
Edition Hamtil, Hamburg/Tilburg
355×288×29mm
1477g
72p

2004 Ha2

Honorary Appreciation
Das Buch als Kunstwerk. Die Cranach Presse des Grafen Harry Kessler
Germany
Sabine Golde
John Dieter Brinks (Ed.)
Triton Verlag, Laubach / Berlin
315×255×50mm
2923g
453p

2004 Ha3

Honorary Appreciation
Werktitel: Piet Gerards, grafisch ontwerper
The Netherlands
Bureau Piet Gerards
Ben van Melick
Uitgeverij 010 Publishers, Rotterdam
231×168×25mm
802g
366p

2004 Ha4

Honorary Appreciation
somnia. En bok om drømmer
Norway
Geir Henriksen
Julie Cathrine Knarvik, Anders Aabel
Christian Schibsteds Forlag, Oslo
245×176×28mm
926g
232p

2004 Ha5

Honorary Appreciation
Biodiversidad en Venezuela
Venezuela
Álvaro Sotillo, Luis Giraldo
Marisol Aguilera, Aura Azócar, Eduardo González Jiménez (Ed.)
Fundación Polar, Ministerio de Ciencia y Tecnología,
Fondo Nacional de Ciencia, Tecnología e Innovación (fonacit), Caracas
248×173×85mm
1663g+1695g
1080p

2004 Bm1

Bronze Medal
Arbeiterlyrik 1842–1932
Germany
Gaston Isoz
Heinz Ludwig Arnold (Ed.)
Parthas Verlag, Berlin
256×169×29mm
800g
255p

2004 Bm2

Bronze Medal
Caravaggios Rom. Annäherungen an ein dissonantes Milieu
Germany
Andreas Trogisch
Lothar Sickel
Edition Imorde, Emsdetten / Berlin
240×144×19mm
454g
268p

2004 Bm3

Bronze Medal
Vermischungen Interdisziplinäre Wochen an der Muthesius-Hochschule
Germany
Julia Hofmeister und Tim Albrecht (Muthesius-Hochschule)
Theresa Georgen (Ed.)
Muthesius-Hochschule, Kiel
285×188×11mm
359g
144p

2004 Bm4

Bronze Medal
Lebens-Mittel (Die Tollen Hefte 22)
Germany
Thomas M. Müller
Armin Abmeier, Rotraut Susanne Berner (Ed.)
Büchergilde Gutenberg, Frankfurt am Main / Wien / Zürich
207×137×7mm
104g
31p

2004 Bm5

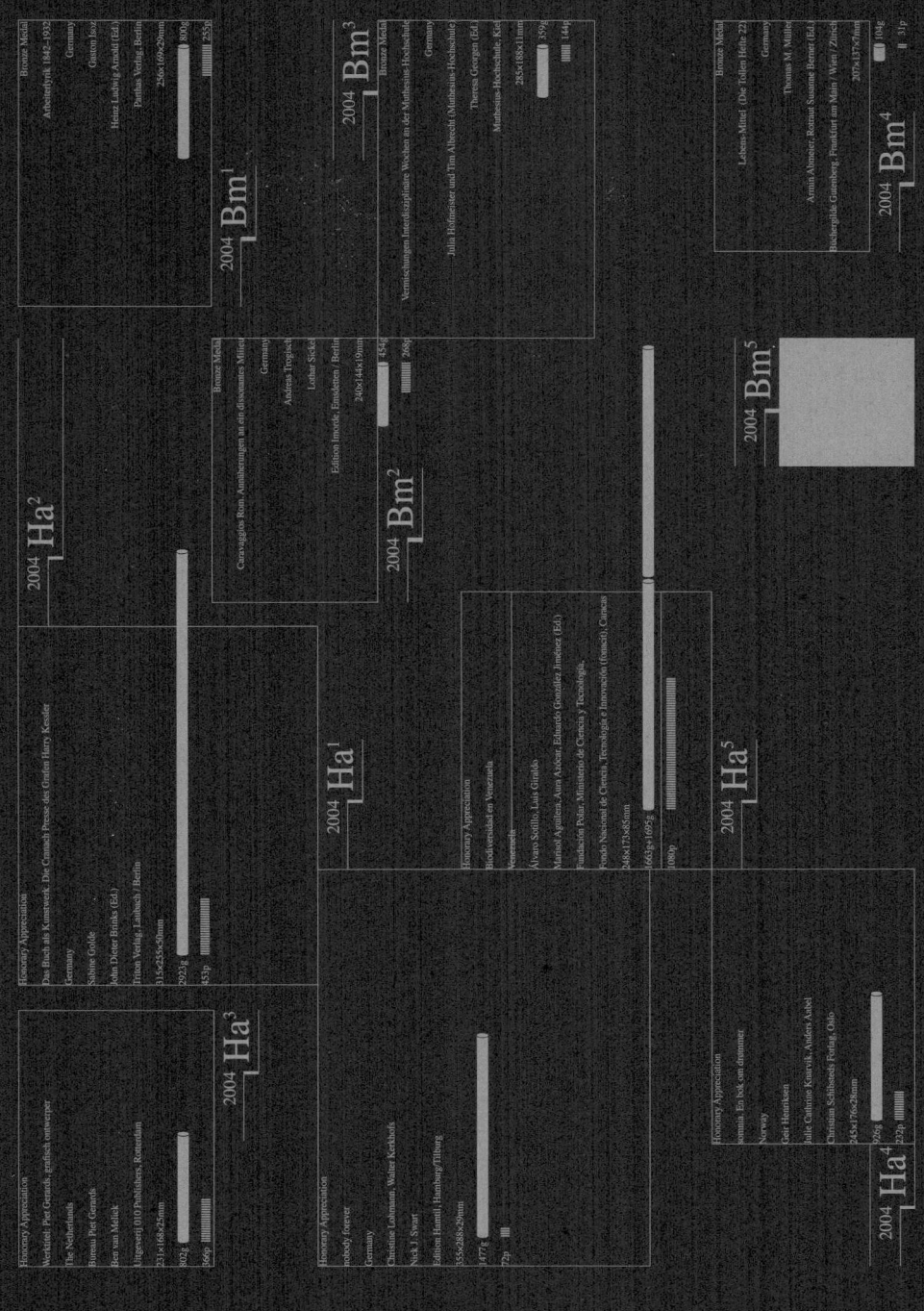

The International Jury

Hans Burkhardt (Switzerland)
Joachim Düster (Germany)
Dinu DumbrLavician (Romania)
Christine Fent (England)
Piet Gerards (The Netherlands)
Silke Nalbach (Germany)
Volker Pfüller (Germany)

Designer

Kristina Brusa, Markus Dreßen
Zhang Zhiwei, Du Yuge, Gao Shaohong
Kodama Yuuko, Ono Risa, Kawasina Hiroya
Joost van de Woestijne; office of cc
Gaston Isoz
Andreas Trogisch
Julia Hofmeister und Tim Albrecht (Muthesius-Hochschule)
Thomas M. Müller
Denisa Myšková
Christine Lohmann, Walter Kerkhofs
Sabine Golde
Bureau Piet Gerards
Geir Henriksen
Álvaro Sotillo, Luis Giraldo

2004

Country / Region

Germany 7
The Netherlands 2
China 1
Japan 1
Czech Republic 1
Norway 1
Venezuela 1

Literature & Fiction
文学与小说
Nature & Technology
自然与科技
Social Sciences
社会科学
Art & Design
艺术与设计
Children and Young People's Books
少儿读物
Education & Teaching
教育与教学
Others
其他
Not collected
未收藏
7%
7%
7%
7%
14%
58%

004 Gl

这是德国艺术家奥拉夫 · 尼科莱（Olaf Nicolai）的作品集，作品涉及工业产品、室内环境、书籍、海报、装置等方面。书中通过对每一件作品创作过程、细节、材料的展示来深度分析作品的内涵和创作意图。书籍的装订形式为锁线胶平装。封面以奥拉夫 · 尼科莱 2000 年创作的灯具作品为设计灵感来源，通过印银、UV、凹凸等不同的印刷工艺，在视觉与触觉感官上呈现出丰富的层次与肌理。书的内页采用老式卡带可撕式的便签虚线作为设计元素，对标题字体、页眉页码、版式分隔等局部进行设计，并采用老式打字机字体，体现出复古的文本感。本书试图还原出完整的创作过程与设计思路，仿佛一场“倒带”的创作回忆录。整本书通过多种纸张、开本、印刷和装订工艺的混合使用，有助于体现文本内容的不同方面，以多元化的视角与多层次的维度，在开拓阅读思路的同时，也增加了阅读的趣味性。||||||||||||||||||||||||||||||||||||||||||||||||||||||||

This is a collection of works by the German artist Olaf Nicolai, including the designs of industrial products, indoor environment, books, posters, installations and so on. By presenting the creation process, details and materials of each work, the book deeply analyzes the connotation and creative intention of his works. The paperback book is glue bound with thread sewing. The luminary created by Olaf Nicolai in 2000 inspires the book cover design. It presents rich layers and textures in the visual and tactile senses through different printing techniques, such as silver printing, UV and embossing. In the inside pages, the dotted lines from old style cassette tape are used as essential design element, and applied to the title font, page header, page number, layout and so on. Also the old style typewriter font reflects the retro quality in text. This book attempts to restore the complete creative process and design ideas, as if a “rewind” of the creative memoir. By using a variety of paper, formats, printing techniques and binding process, the book design helps to reflect the different aspects of the text content with a pluralistic perspective. It may increase the fun of reading while broadening the readers’ mind at the same time.

金字符奖

<< 倒带 >> 快进

德国

<<Rewind >>Forward is particularly impressive as a result of its uncompromising and clear typography set against a variety of special prints which are integrated in various sizes, shapes and materials. The cover is equally consistent in its elegantly structured arrangement of space and embossing. Innovative in terms of design and finish.

Golden Letter

<<REWIND>>FORWARD

Germany

Kristina Brusa, Markus Dreßen

Olaf Nicolai, Susanne Pfleger

Hatje Cantz Verlag, Ostfildern-Ruit

267×215×25mm

1126g

168p

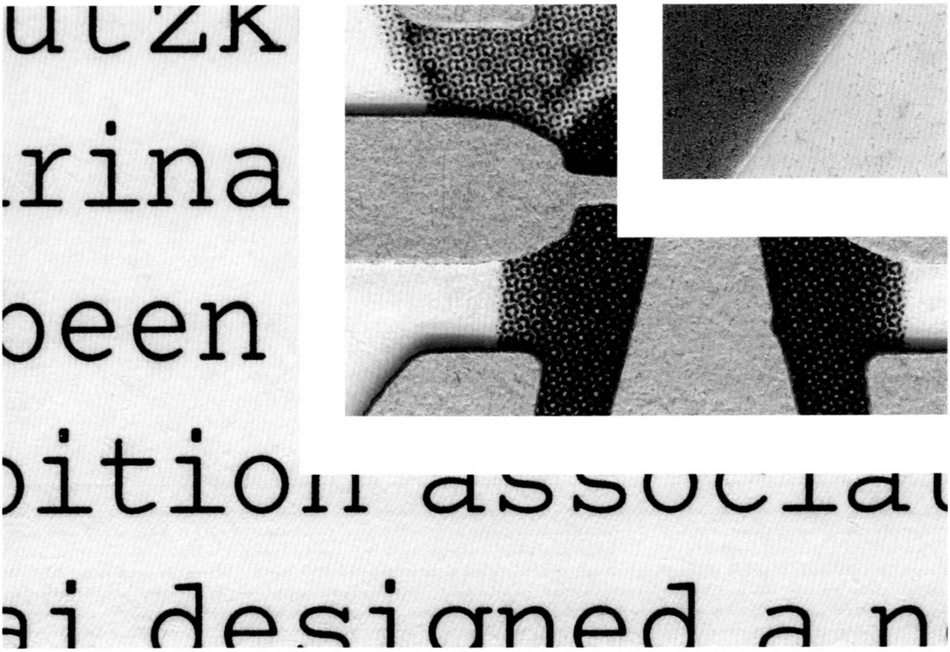

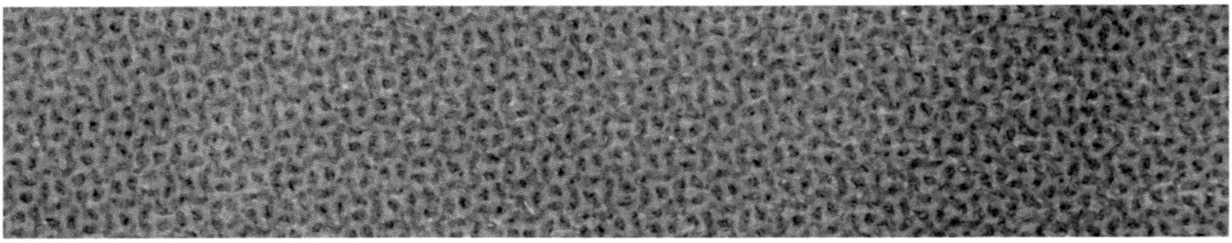

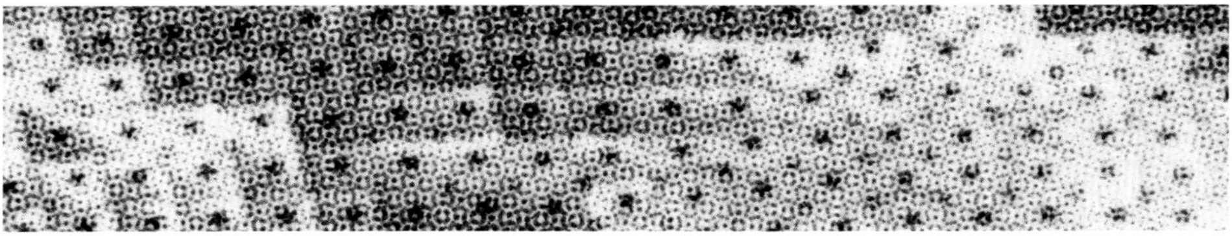

267 × 215 × 25

Depot/Landschaft. Ein Tableau
Die Flamme der Revolution, liegend...
Die Gabe. Eine Sammlung

Hello There/Hell Here
Hortus/Konstrukt
I.K. Star Course

1126

Mirror Cover [VOGUE]
modern dreams I [version dresden 1968]
modern dreams II

private mix, 1
Raumfenster
Reader

168

Und jedem Ende wohnt ein Anfang inne
Verglasung [Paar]
Viele, die eine Ahnung haben...

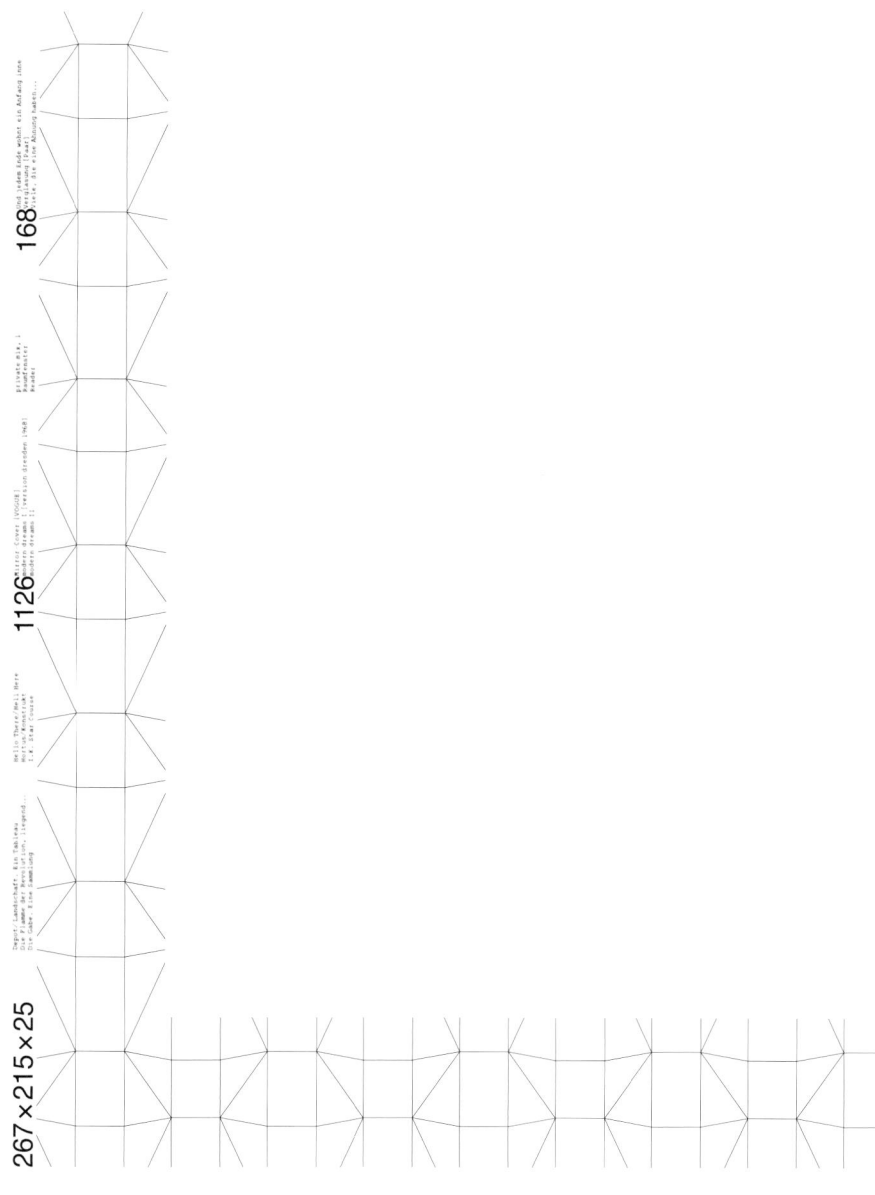

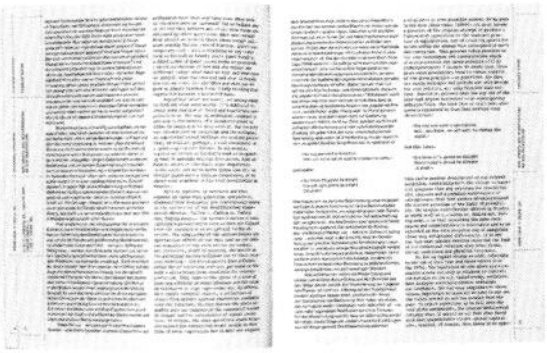

lucky thing. Of course, some of you ar
whole load of intrica

HELLO THERE
HELL THERE

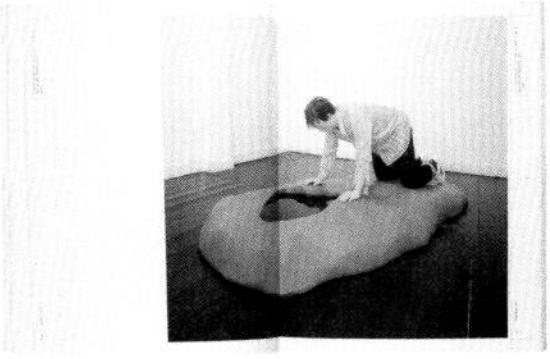

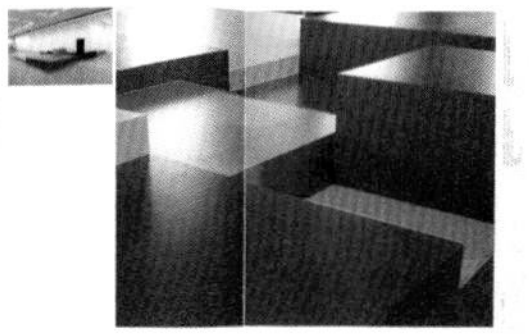

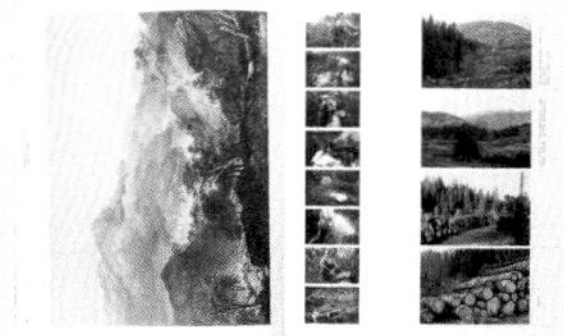
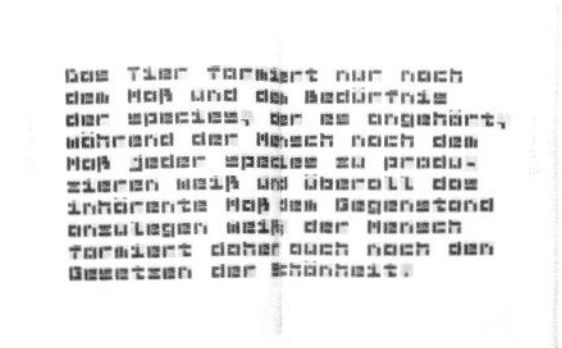
Das Tier formiert nur nach dem Maß und dem Bedürfnis der species, der es angehört, während der Mensch nach dem Maß jeder species zu produzieren weiß und überall das inhärente Maß dem Gegenstand anzulegen weiß; der Mensch formiert daher auch nach den Gesetzen der Schönheit.

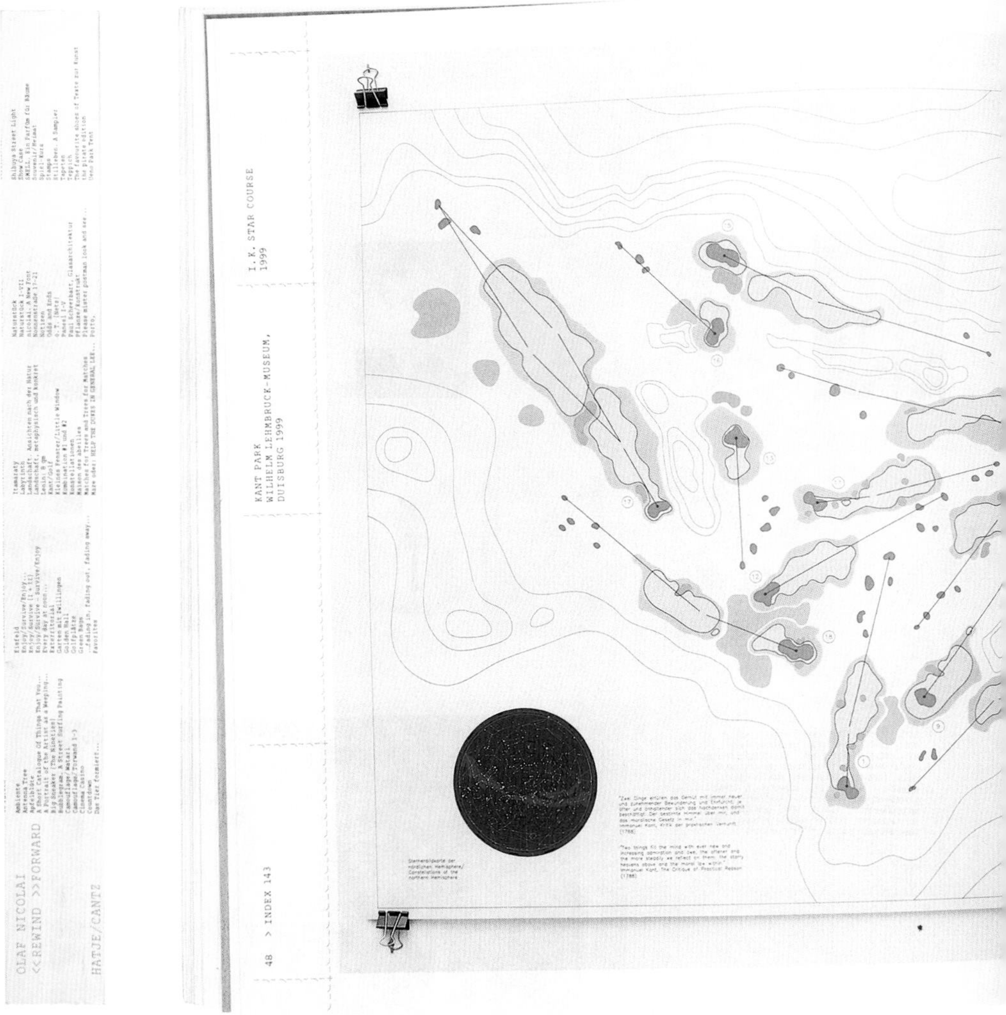
OLAF NICOLAI
<<REWIND >>FORWARD
HATJE/CANTZ
48 > INDEX 143
KANT PARK
WILHELM LEHMBRUCK-MUSEUM,
DUISBURG 1999
I.K. STAR COURSE
1999

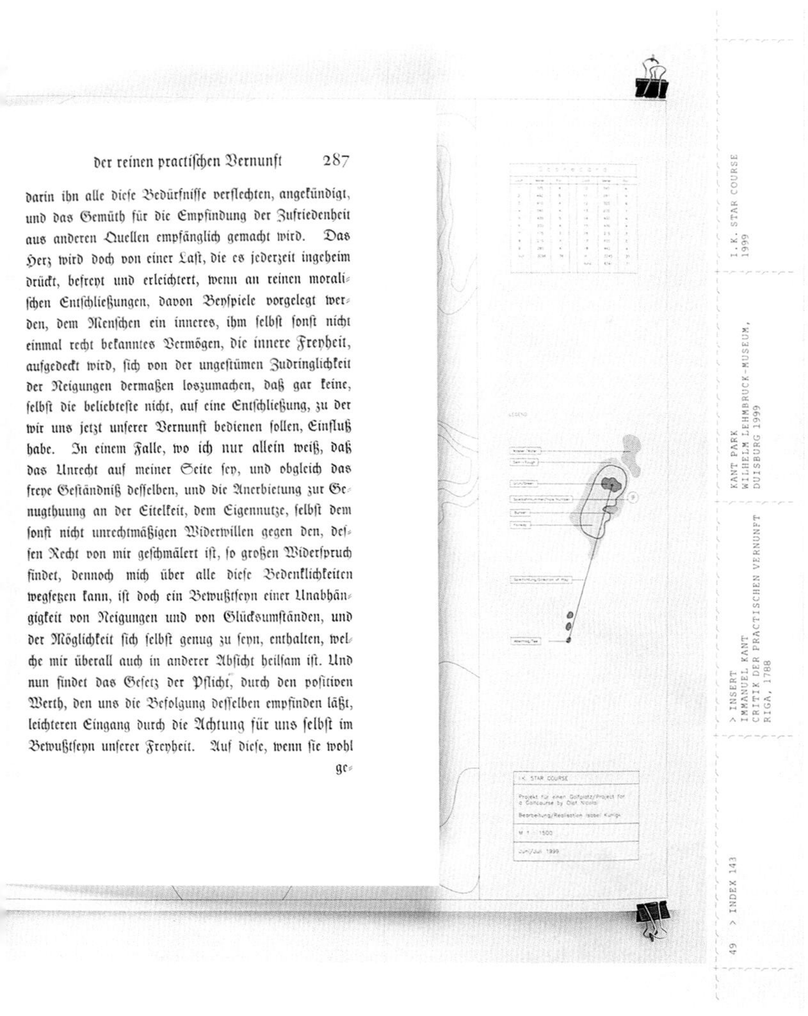

der reinen practischen Vernunft 287

darin ihn alle diese Bedürfnisse verflechten, angekündigt, und das Gemüth für die Empfindung der Zufriedenheit aus anderen Quellen empfänglich gemacht wird. Das Herz wird doch von einer Last, die es jederzeit ingeheim drückt, befreyt und erleichtert, wenn an reinen moralischen Entschließungen, davon Beyspiele vorgelegt werden, dem Menschen ein inneres, ihm selbst sonst nicht einmal recht bekanntes Vermögen, die innere Freyheit, aufgedeckt wird, sich von der ungestümen Zudringlichkeit der Neigungen dermaßen loszumachen, daß gar keine, selbst die beliebteste nicht, auf eine Entschließung, zu der wir uns jetzt unserer Vernunft bedienen sollen, Einfluß habe. In einem Falle, wo ich nur allein weiß, daß das Unrecht auf meiner Seite sey, und obgleich das freye Geständniß desselben, und die Anerbietung zur Genugthuung an der Eitelkeit, dem Eigennutze, selbst dem sonst nicht unrechtmäßigen Widerwillen gegen den, dessen Recht von mir geschmälert ist, so großen Widerspruch findet, dennoch mich über alle diese Bedenklichkeiten wegsetzen kann, ist doch ein Bewußtseyn einer Unabhängigkeit von Neigungen und von Glücksumständen, und der Möglichkeit sich selbst genug zu seyn, enthalten, welche mir überall auch in anderer Absicht heilsam ist. Und nun findet das Gesetz der Pflicht, durch den positiven Werth, den uns die Befolgung desselben empfinden läßt, leichteren Eingang durch die Achtung für uns selbst im Bewußtseyn unserer Freyheit. Auf diese, wenn sie wohl ge-

I.K. STAR COURSE
1999
KANT PARK
WILHELM LEHMBRUCK-MUSEUM,
DUISBURG 1999
> INSERT
IMMANUEL KANT
CRITIK DER PRACTISCHEN VERNUNFT
RIGA, 1788
49 > INDEX 143
I.K. STAR COURSE

梅蘭芳
戲曲史料圖畫集
上

该书收录了梅兰芳纪念馆收藏的一系列珍贵馆藏，包括戏曲脸谱、戏画、戏曲人物等珍贵资料，有明代遗珍、清代绝品、民初佳作等，是研究戏曲艺术与历史不可多得的文献图录。全书整体工艺考究，墨彩相映，古朴典雅。书函采用传统精装设计，象牙色骨签古朴清致。书函表面为玄色针织纹路纸张，压印戏曲人物脸谱，上下朱印点缀。内裱纸张为暗红色，印有黑色传统纹样。该书分为上下册，采用传统包背装，裸脊锁白线，虽未采用硬精装封面，但依然保留了上下堵头布，使得书脊部分易于长期保存。上下册的设计在开本、重量和翻阅手感等多方面找到了平衡。书封采用米色略带光泽的纸张，版式编排极具中国典籍神韵，天头地脚，空白留存得当。整体设计印刷精美，颜色优雅精致，纸质丰盈润和。

梅蘭芳
藏
戲曲史料圖畫集
上

The picture album records a series of valuable museum collections collected by Mei Lanfang Memorial hall, including precious materials such as facial make-ups in opera, opera paintings, and opera figures from Ming dynasty, Qing dynasty and early years of the Republic of China. It is a rare bibliography for the study of opera art and history. The book's production processes are sophisticated and the overall style is classical and elegant. The surface of the traditional slipcase is black knitting pattern paper embossed with opera facial make-ups, while the mounting page is dark red printed with black ancient patterns. The book is divided into two volumes, double-leaved binding with naked book back with white cord. Although it's not hardback, the book still retains the head bands which help to preserve the spine in a relatively long term. The choice of two volumes optimizes the feelings of book format, weight and browsing. Beige paper with slightly gloss is used for the cover, elegant in color and rich in paper. The layout of inside pages is typical Chinese style when white space is properly retained.

金奖	Gold Medal
梅兰芳（藏）戏曲史料图画集	A Picture Album. Historical Materials of Mei Lanfang's Theatrical Performances
中国	China
张志伟、蠹鱼阁、高绍红	Zhang Zhiwei, Du Yuge, Gao Shaohong
刘曾复、朱家溍	Liu Cengfu, Zhu Jiajin
河北教育出版社	Hebei Education Publishing House, Shijiazhuan
	295×220×43mm
	1702g
	229p

Two perfectly bound brochures in a sub-divided folding case with lock. The design for the contents is consistent and unfussy. The titles of the columns are printed opposite one another along facing edges of the fold. The drawings of masks are presented in a very clear and well-organised fashion. The paper used for the body of the brochures and the silk-structured dust-jacket, as well as the fabric covering, printed in more than one colour, with large areas of blindstamping are well matched. An extremely clean and exact degree of finish contributing to a very harmonious overall impression.

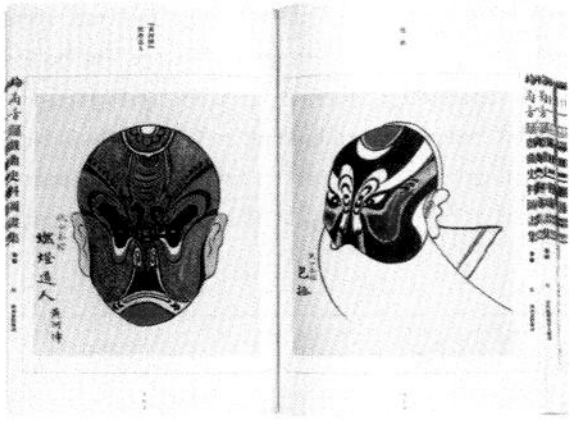
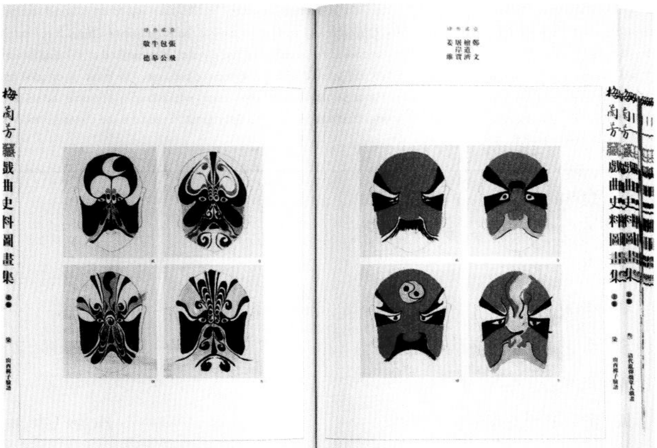
梅蘭芳藏戲曲史料圖畫集

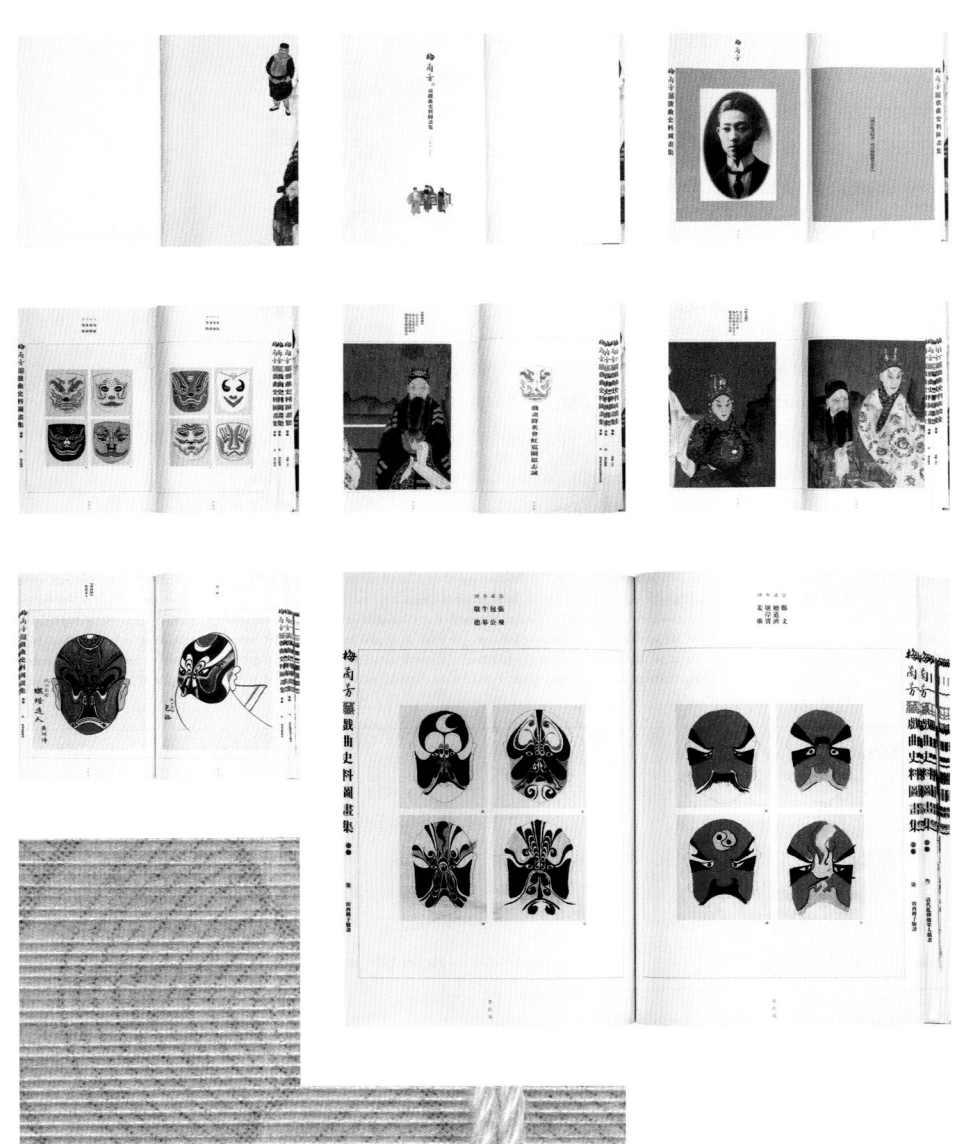

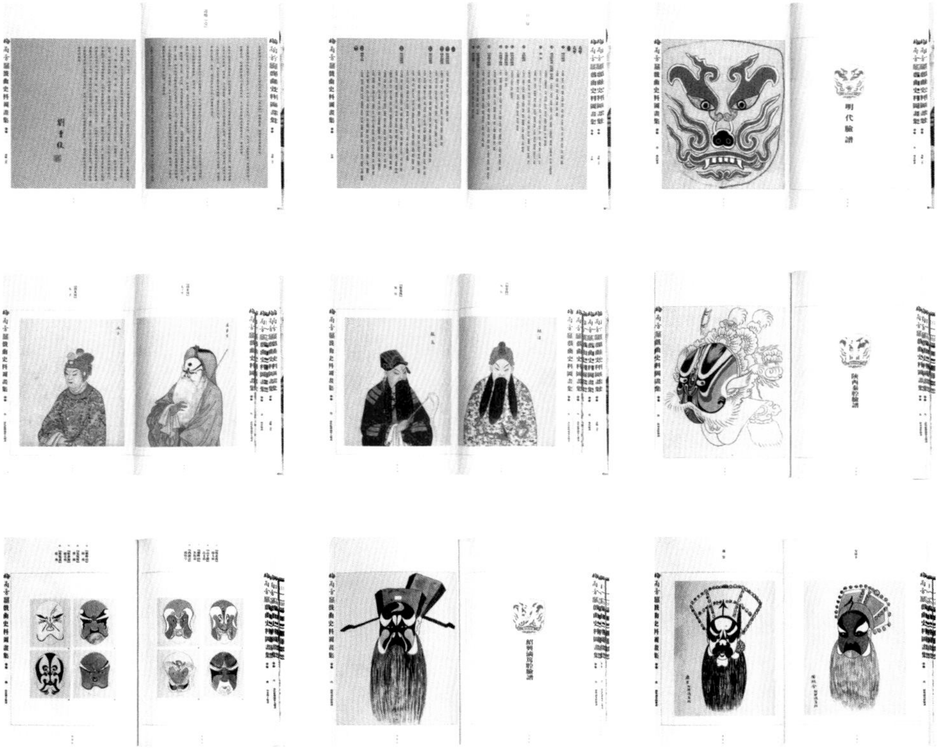
梅蘭芳藏戲曲史料圖畫集
明代臉譜
陝西秦腔臉譜
紹興調腔臉譜

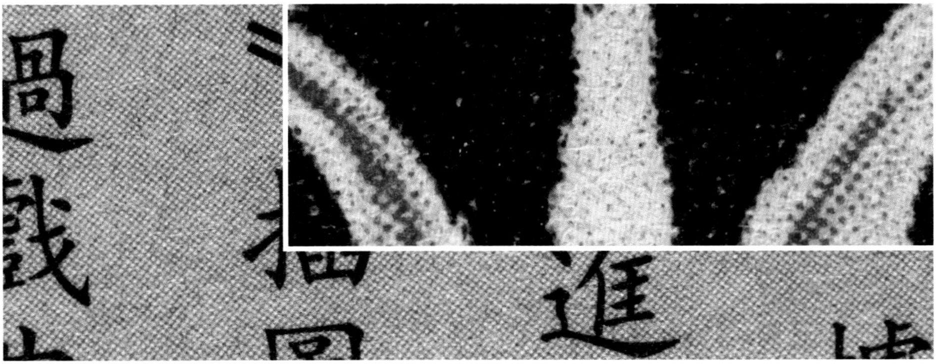

『思志誠』

自左向右
劉趕三
吴燕芳
（朱素雲之姐夫）
王彩琳
（王瑶卿之父）
朱蓮芬
（朱天祥之父）
方松齡
（小生錢俊仙之外祖父）
余紫雲
（余三勝之子、余叔岩之父）
小童董度雲（徐小香之徒）
葉中興
（葉春善之叔）
小童顧小農（朱雙喜之徒）
徐小香
（世所稱徐大老板）
小童曹福壽
楊鳴玉
（通呼楊三，所謂楊三已死無昆醜）
小童鄭多雲（徐小香之徒）
化虎
（羅百歲之師）
梅巧玲
（梅蘭芳之祖父）
時小福
（時慧寶之父）
楊朵仙
（楊小朵之父）
朱霞芬
（朱幼芬之父）
孫彩珠
黄三雄
孫元福

梅蘭芳藏戲曲史料圖畫集

上卷

說略「壹」

零貳肆

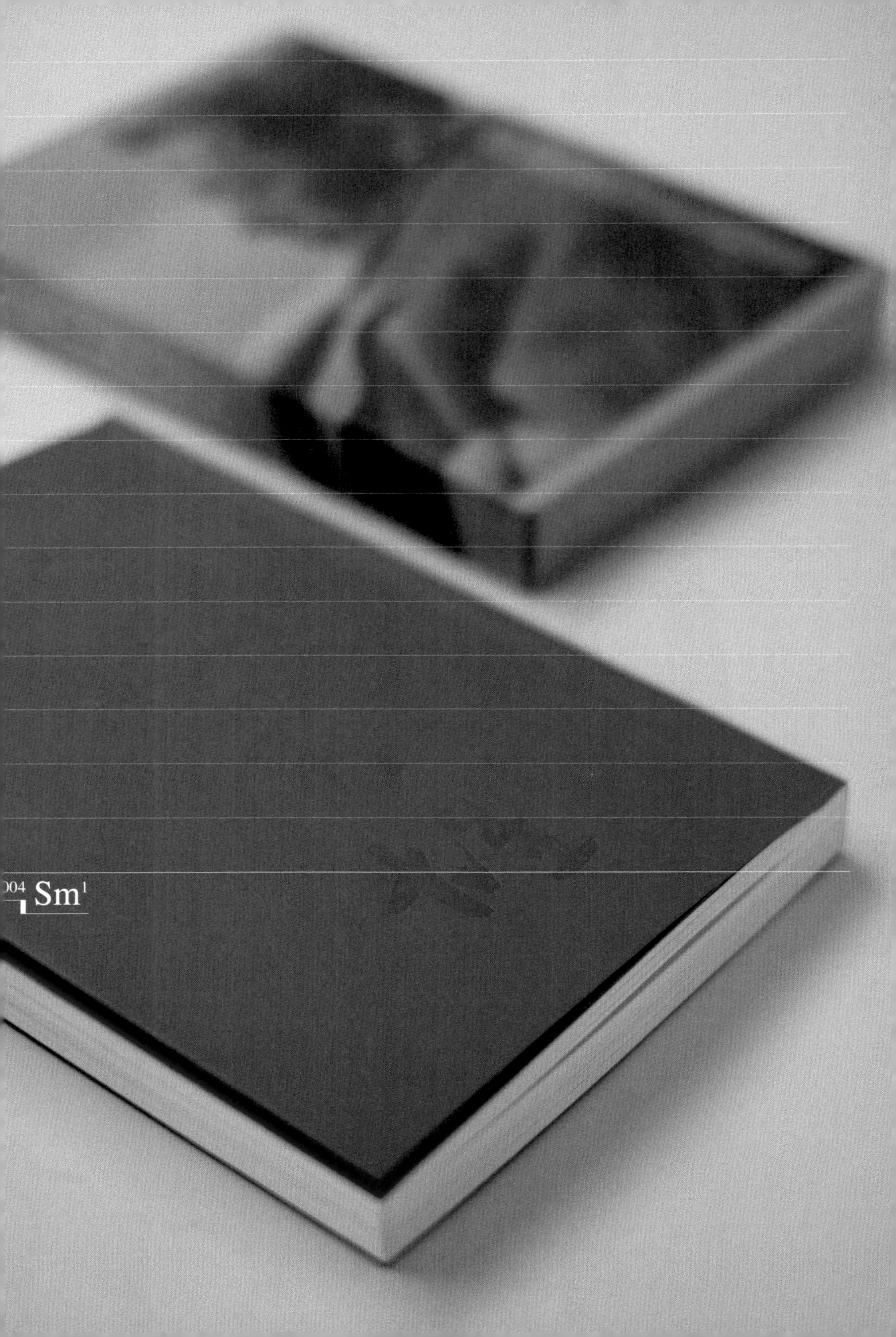

这是日本文艺评论界的灵魂人物小林秀雄诞辰100周年纪念展的展览画册。但画册并不局限于作为展陈信息的梳理和辅助作用，而是通过丰富细腻的资料整理，成为一本值得独立收藏的纪念资料集。本书装订为无线胶平装右翻本，书函部分为坚硬的黄褐色瓦楞纸壳内裱橙色特种纸张，外部裱贴胶版纸单黑印刷的照片和展览名文字。封面采用暖灰色仿皮面制作，封面封底文字主要采用压凹处理。内文的主体采用涂布艺术纸四色印刷，正文竖排。为小林秀雄针对艺术作品和艺术家发表的评论文字的节选，配合相关的西方近现代绘画、日本绘画，以及古董收藏共三部分。附录部分为著作照片、生平年表和图录作品的收藏信息。本书另外包含全英文翻译部分，这部分并不是直接与日文部分进行日英混排，而是遵照英文书籍的阅读方式，从书的封底开始采用左翻横排的设计，并通过标示页码的方法来对图片进行索引。||||||||||||||||||||||||||||||||||||||||||||||||

生誕百年記念展
小林秀雄 美を求める心

This is an album of the exhibition honoring the 100th anniversary of Kobayashi Hideo's birth. He is the soul of Japanese literary criticism. The album not only teases out the exhibition information, but also tries to become a collection of memorial materials. The book is a perfect-bound paperback. The slipcase is made of hard yellowish-brown corrugated paper mounted with orange special paper inside. The outer surface is covered with a black-printed photo and the title of the exhibition. The cover is made of warm gray imitation leather with embossed texts on it. The body text is vertically printed in four colors on coated art paper. According to Kobayashi Hideo's art review of various works and artists, the album is divided into three parts, such as modern western paintings, Japanese paintings, and antique collections. The appendix is the collection information of works' photos, life chronology and catalogue. In addition, the whole book is translated in English, which is not directly mixed with the Japanese narration. Instead, this part starts from the back cover of the book, following the typical style of English books, and uses the method of pagination to index the pictures.

◎ 银奖

« 寻美之心

——小林秀雄诞辰一百周年纪念展

◐ 日本

A thread-bound fold-out brochure in a case. Well-proportioned contents and very harmonious positioning of the texts with a refreshing degree of space left free. The distribution of the bilingual picture legends and texts intelligently follows the respective directions in which the two languages, Japanese and English, are read. The colours of the endpapers and the blind-stamped dust-jacket are well matched.

◎ Silver Medal

« The Heart in Search of Beauty. An Exhibition Honoring the 100th Anniversary of Kobayashi Hideo's Birth

◐ Japan

D Kodama Yuuko, Ono Risa, Kawasina Hiroya

△ Hideo Kobayashi

||||| Nihon Keizai Shimbun, Tokyo; Shinchosha, Tokyo

□ 226×156×27mm

⌷ 678g

≣ 256p

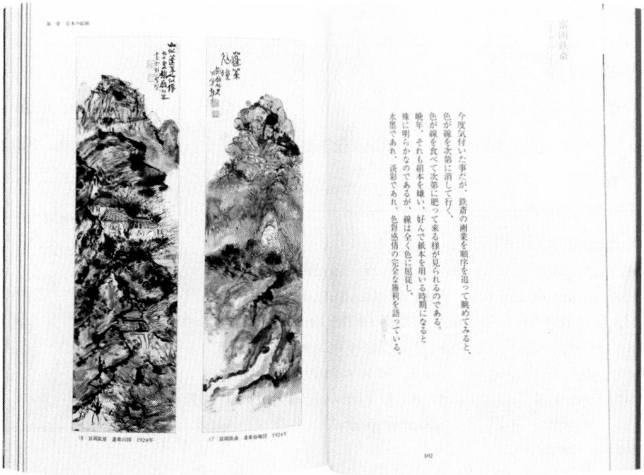
第二章　日本の絵画
富岡鉄斎
今度気付いた事だが、鉄斎の画業を順序を追って眺めてみると、
色が線を次第に消して行く、
色が線を食べて次第に肥って来る様が見られるのである。
晩年、それも絹本を嫌い、好んで紙本を用いる時期になると
殊に明らかなのであるが、線は全く色に屈従し、
水墨であれ、淡彩であれ、色彩感情の完全な勝利を語っている。
18　富岡鉄斎　蓬莱山図　1924年
17　富岡鉄斎　蓬莱仙境図　1924年
102
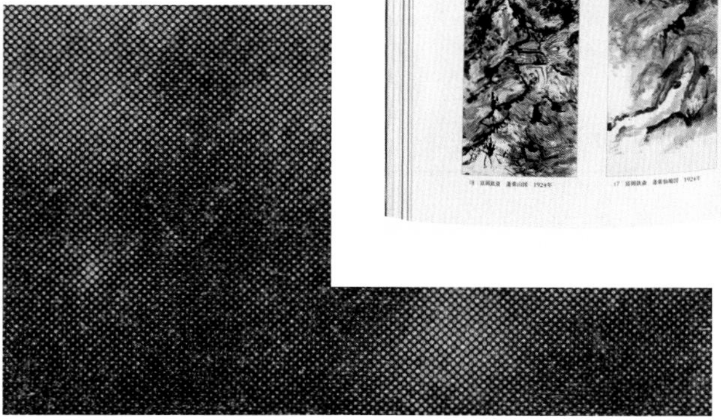

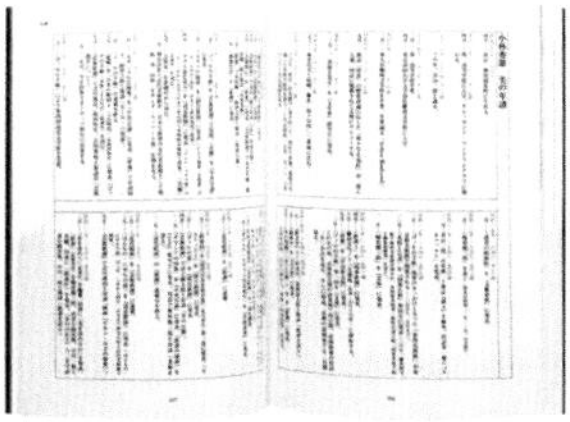

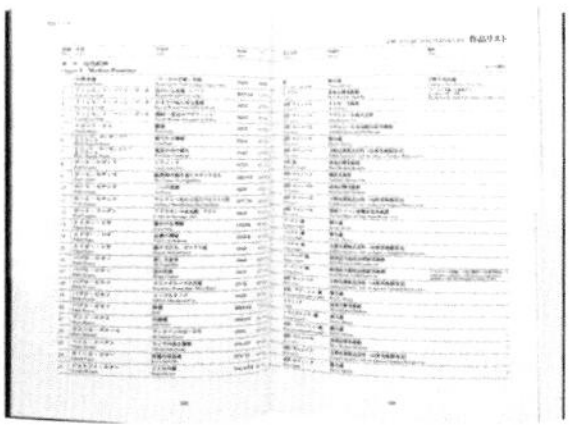

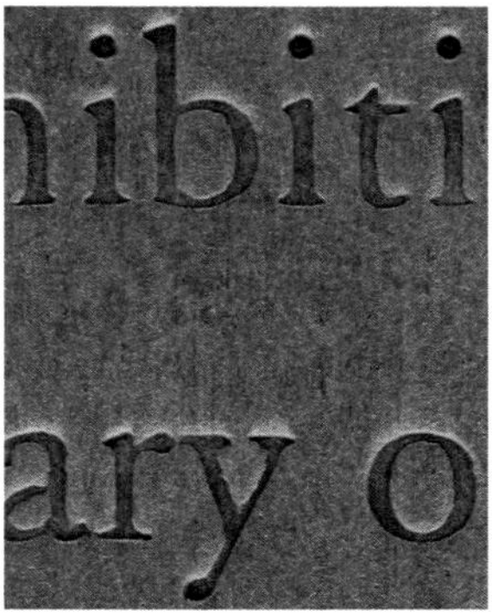

hibiti
ary o

生誕百年記念展
小林秀雄 美を求める心
49｜欅彫花文小箱 黒田辰秋作

黒田辰秋

Kuroda Tatsuaki 1904-1982

部厚な欅（けやき）の材を組合せた何気ない小箱なのだが、
漆を透して見事な木目が現れ、
蓋には非常に深い彫りの花模様があった。—中略—
今度、新宮殿を拝観し、黒田君の見事な作に、
たくさん接したが、やはり名人というものは、
個性は一貫したものだと、今更のように思った。

「『黒田辰秋　人と作品』序」

48｜拭漆栃彫花文小箱　黒田辰秋作

这是法籍犹太诗人保罗·策兰（Paul Celan）的诗歌全集的荷德双语版本。保罗·策兰是二战后最重要的德语诗人之一。本书为锁线圆脊硬精装。护封采用牛皮卡纸黑红金三色印刷，护封高度比精装封面略短，使得上下各露出大约 3mm 黑边。精装内封采用黑色布面裱卡板，封面正中和书脊部分文字压凹处理。环衬采用哑面艺术纸满版印黑，后叠印覆盖力较强的金色，只露出黑色小点。内页胶版纸印单黑，上下书口及右侧翻口处刷黑处理。由于书籍较厚，且不像小说类书籍的线性阅读方式，书籍搭配了黑金双色书签带，便于停顿在不同的位置。书籍整体黑红金三色搭配，与诗人的母语——德语暗合，但色彩并非完全的德国国旗配色，考虑到诗人的身份和经历，似有深意。书籍的内容以诗歌为主，按照创作年份进行组织，左页德文、右页荷兰文对照排版。中间穿插了三篇散文，按照上荷兰文、下德文的双区域对照排版。书籍的最后包含作者的笔记和年表信息。||||||||||||||||

Paul Celan

This is a bilingual edition (in Dutch and German) of the complete collection of Paul Celan's Poems, who is a French Jewish poet. Paul Celan is one of the most important German poets after World War II. This hardback book is bound by thread sewing with a rounded spine. The jacket is made of kraft cardboard printed in black, red and gold. The height of the jacket is slightly shorter than that of the hardcover, which exposes about 3 mm black edges. The inner cover is made of cardboard mounted with black cloth. The text in the middle of the cover and on the spine is embossed. The endpaper is made of matte art paper with full version black, and then printed in gold, only revealing black dots. The inside pages are made of offset paper in black, and three edges and brushed in black. Since the book is relatively thick and unlike fiction books that can be read in the way of linear reading, the book is equipped with black and gold double-color bookmarks, which make it easy to pause in different places. The book uses the color combination of black, red and gold, which coincides with the poet's mother tongue German, although the color combination is not completely matched with the German flag. Considering the poet's identity and experience, it seems meaningful. The content of the book is mainly composed of poems, organized according to the year of creation. The left page is typeset in German and the right page is in Dutch. There are three proses interspersed, which are typeset in German and Dutch as well. The final part of the book contains the author's notes and chronological information.

- 银奖
- 保罗 · 策兰诗集
- 荷兰

This book is inspiring on account of its very successful overall concept and colour composition made up of brown, black and gold. These colours permeate from the dust-jacket to the cover, from the endpapers and bookmarkers through to the all-edge coloring around the text block. The font (Garamond) is ideal for the interpretation of Celan's (bilingual) poems. The printing onto bookprinting paper is exemplary.

- ◎ ········ Silver Medal
- 《 ···· Paul Celan Verzamelde gedichten
- ········ The Netherlands
- ··· Joost van de Woestijne; office of cc
- △ ········ Paul Celan
- ········ Meulenhoff Amsterdam
- □ ········ 221×150×53mm
- ········ 1194g
- ········ 864p

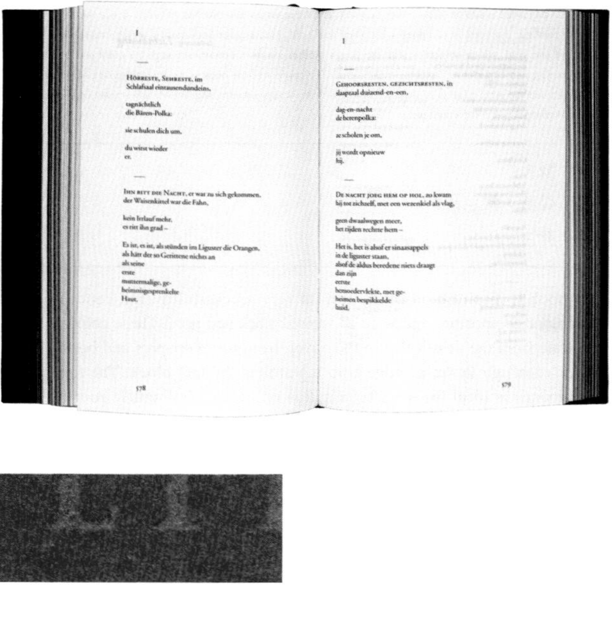
I
Hörreste, Sehreste, im
Schlafsaal eintausendundeins,
tagnächtlich
die Bären-Polka:
sie schulen dich um,
du wirst wieder
er.
Ihn ritt die Nacht, er war zu sich gekommen,
der Waisenkittel war die Fahn,
kein Irrlauf mehr,
es ritt ihn grad –
Es ist, es ist, als stünden im Liguster die Orangen,
als hätt der so Gerittene nichts an
als seine
erste
muttermalige, ge-
heimnisgesprenkelte
Haut.
378
I
Gehoorsresten, gezichtsresten, in
slaapzaal duizend-en-een,
dag-en-nacht
de berenpolka:
ze scholen je om,
jij wordt opnieuw
hij.
De nacht joeg hem op hol, zo kwam
hij tot zichzelf, met een wezenkiel als vlag,
geen dwaalwegen meer,
het rijden rechtte hem –
Het is, het is alsof er sinaasappels
in de liguster staan,
alsof de aldus beredene niets draagt
dan zijn
eerste
bemoedervlekte, met ge-
heimen bespikkelde
huid.
579

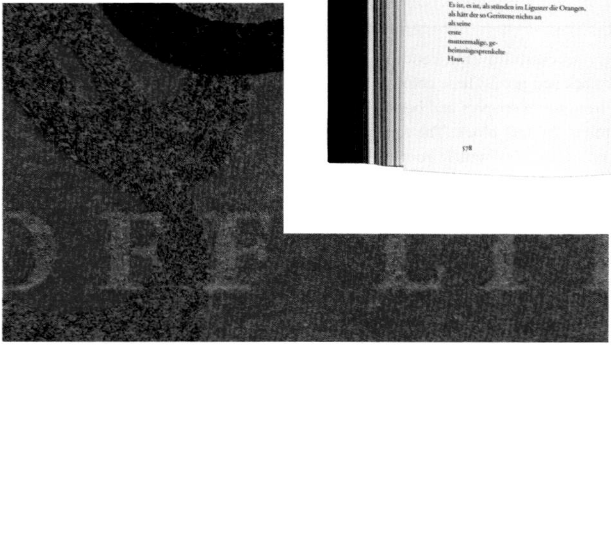

and v

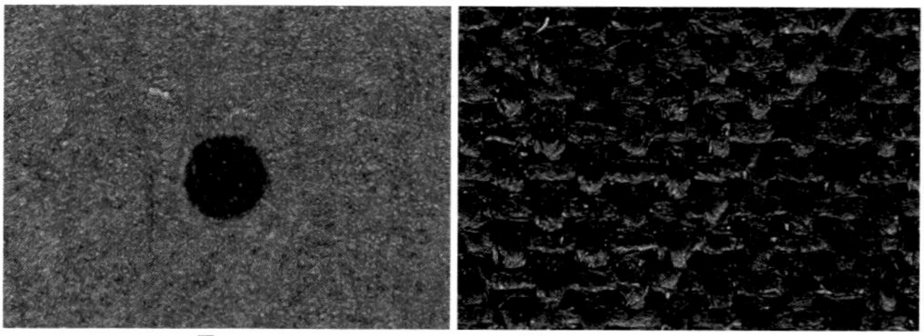

e haren beschorer

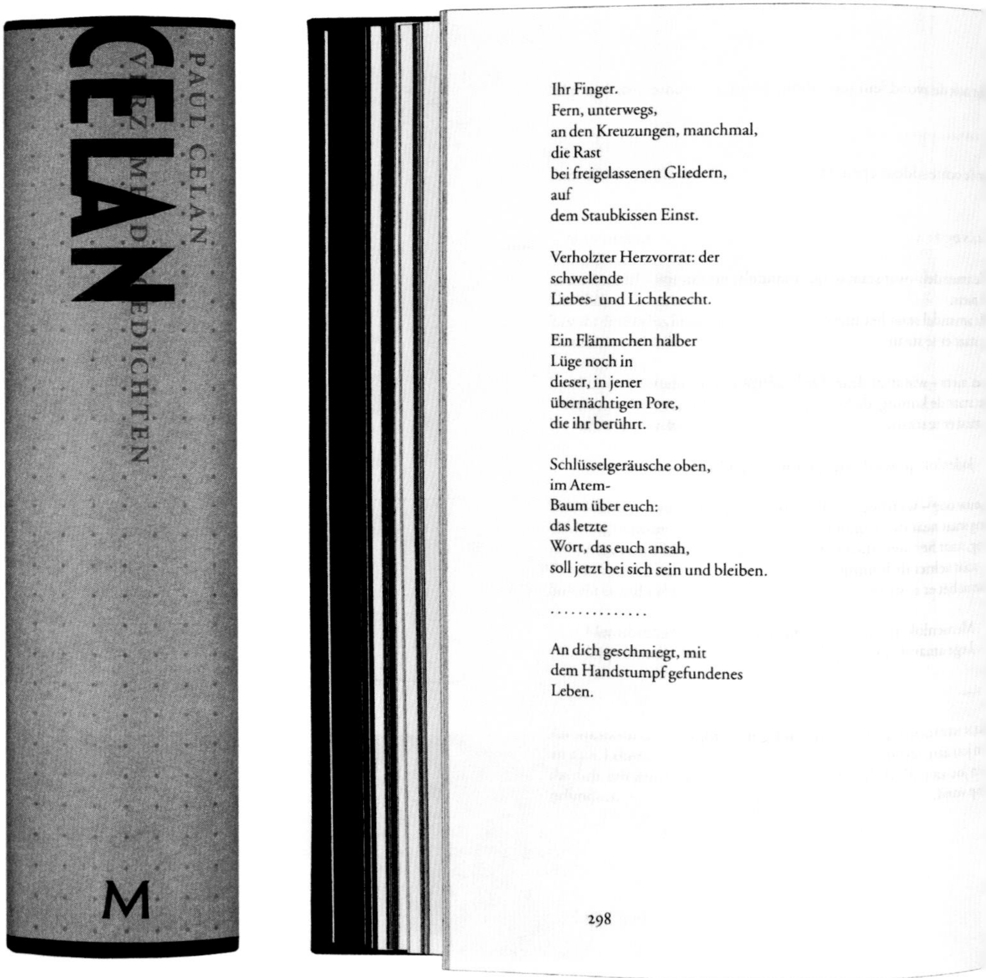
PAUL CELAN
CELAN
M

Ihr Finger.
Fern, unterwegs,
an den Kreuzungen, manchmal,
die Rast
bei freigelassenen Gliedern,
auf
dem Staubkissen Einst.

Verholzter Herzvorrat: der
schwelende
Liebes- und Lichtknecht.

Ein Flämmchen halber
Lüge noch in
dieser, in jener
übernächtigen Pore,
die ihr berührt.

Schlüsselgeräusche oben,
im Atem-
Baum über euch:
das letzte
Wort, das euch ansah,
soll jetzt bei sich sein und bleiben.

. . . .

An dich geschmiegt, mit
dem Handstumpf gefundenes
Leben.

298

O vingers.
Ver, onderweg,
op de kruisingen, soms,
de rustplaats
bij vrijgelaten ledematen
op
het stofkussen belofte.

Hout geworden hartvoorraad: de
smeulende
liefdes- en lichtknecht.

Een vlammetje halve
leugen nog in
deze, in gene
nachtbrakende porie,
die jullie aanraken.

Sleutelgeluiden boven,
in de adem-
boom boven jullie:
het laatste
woord dat jullie aanzag,
moet nu alleen zijn
en blijven.

....

Tegen jou aangevlijd, met
de handstomp gevonden
leven.

004 Bm1

本书为德国作家、记者、出版人海因茨·路德维希·阿诺尔德（Heinz Ludwig Arnold）采编整理的诗歌集。书籍收录了 1842－1932 年间德国工业迅猛发展至旧工人运动结束时期工人创作的诗歌。本书的装订形式为锁线胶装，封面采用 3mm 卡板印刷并直接裱贴在黑色布面书脊上，诗歌创作者的名字在封面左侧自成一列，封面右下角的文字“红色是炽热的余烬”反映了本诗歌集的整体风格导向。全书采用红黑双色设计，黑色环衬内包裹章节红色过门隔页。标题部分均采用粗黑有力的无衬线字体，诗歌正文部分则采用衬线字体，左对齐排版，版面布局疏朗匀称，整体体现出工人力量感的同时，又兼具包裹其间的浪漫情怀。书中插配大量工厂劳作与宣传主题的版画、宣传画、手稿作为插图，用以调剂节奏，使得整本书的层次丰富，井然有序。

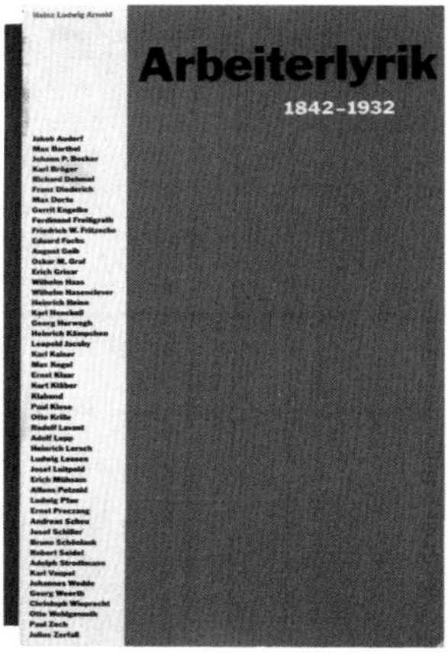
Heinz Ludwig Arnold
Arbeiterlyrik
1842–1932

The poetry anthology is collected and edited by the German writer, journalist and publisher Heinz Ludwig Arnold. The book records poems written by German workers between 1842 and 1932, from the period of rapid industrial development in German to the end of workers' movement. The book is glue bound with thread sewing. The cover is printed on 3mm graphic board and mounted directly on the black cloth spine. The names of the authors are arranged in a row on the left side of the cover, while the texts "red is fiery embers" are in the lower right corner of the cover, which reflects the overall style of the poetry collection. The book is designed in black and red – black end paper contrasts with red interleaves. The title adopts bold and strong sans serif font, and the body part of the poems uses serif font with left-aligned typesetting. The layout is clear and even, which embodies the workers' strength as well as the romantic feelings. A large number of engravings, propaganda pictures and manuscripts with the theme of factory work and publicity are inserted into the book as illustrations to adjust the rhythm and make the whole book rich and orderly.

- 铜奖
- 1842－1932 年工人的诗歌
- 德国

A book where one can positively smell *the working-class struggle*. The workshop character of this work is manifest in the materials used, the simple cardboard jacket as well as the colours red and black which permeate from the cover to the endpapers through to the typography. The overall aura is reinforced by the use of one sanserif and one serif font. A powerful, yet quietly spoken volume of poetry.

- Bronze Medal
- Arbeiterlyrik 1842–1932
- Germany
- Gaston Isoz
- Heinz Ludwig Arnold (Ed.)
- Parthas Verlag, Berlin
- 256×169×29mm
- 800g
- 255p

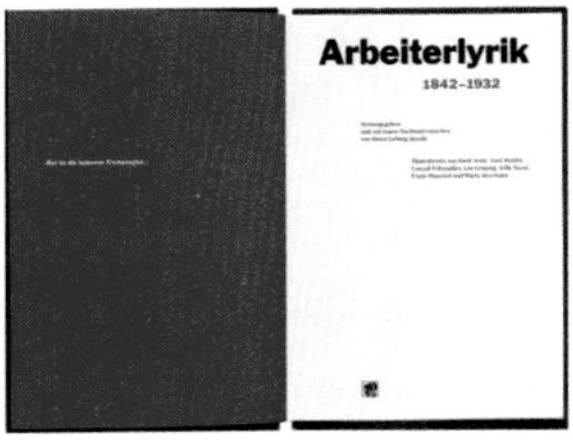
Arbeiterlyrik
1842–1932

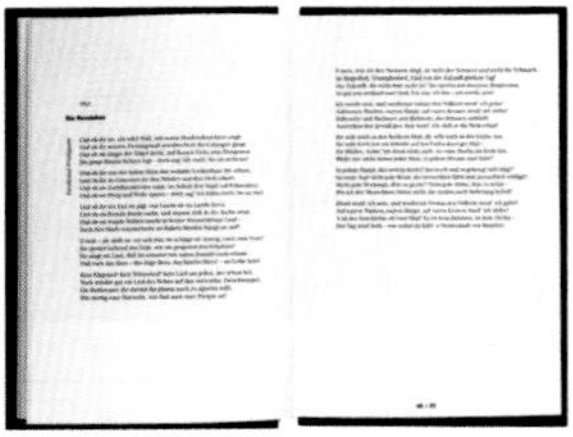

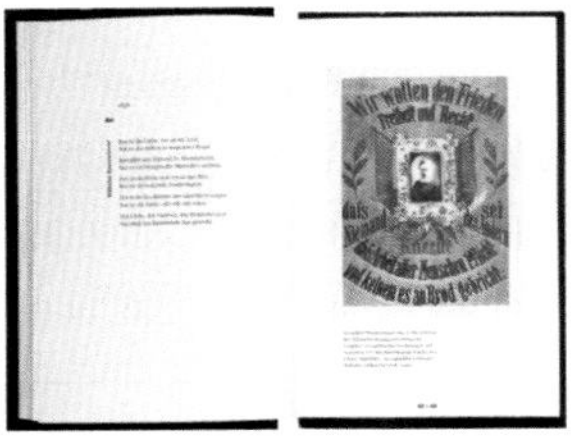

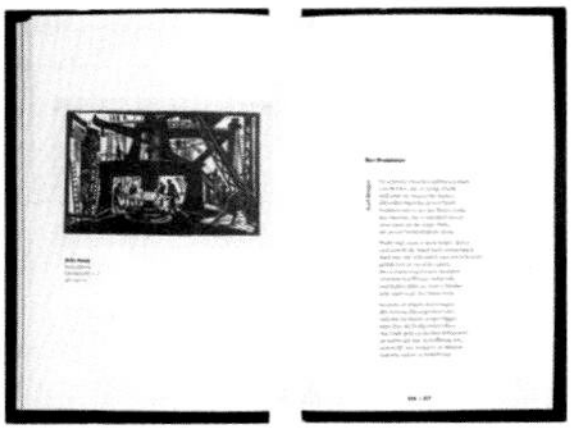

1919

Europa

Bruno Schönlank

Europa blutet Millionen Wunden
Haß säuft ihr Blut
Und schändet Heiligtümer.
Zerfetzte Brüder speit die Erde aus.
Europa,
Schlachtbank deiner Kinder.
Ekles Gewürm zehrt deinen Leib.
Ihr Brüder,
Eingegittert in die Käfige der Armeen.
Gepeitscht, gehetzt.
Die Erde dampft vom Blut der Gladiatoren.
Die Kreuze fluchen,
Lichtes Eiland stöhnt.
Fabriken ihr,
Und Schächte voller Erz,
Aus dunklen Nächten schweißgetränkte Kohlen.
Ihr Schlote qualmt,
Und brauset, ihr Maschinen,
So Tag wie Nacht der großen Schlacht zu dienen.
Nimm deine Hand, mit der du Kinder hegst,
Und drehe, Frau, todkreisende Granaten,
Füll Mord und Schmach und giftigschwere Schwaden ...
Ihr Reklamierten, bangend zwischen Gräben,
Vergittert euch mit neuen Eisenstäben.
Dein irres Lachen rast,
Europa,
Feil sind deine Glieder.
Die Augen brennen tolle Fieberglut.
Du schlägst dich selbst ans Kreuz,
Zeigst deinen Schwestern die zerrissenen Brüste
Und rufst sie alle, alle, dich zu foltern.
Millionen geistern über fahle Erde.

Fern flammt ein Licht –
Über dem Meer von Blut
Heilandsstimmen
Voll Liebe und Güte
Und seliger Opferglut.
Du aber würgst sie als Rebellen
Und wirfst sie in Gefängniszellen
Und Zuchthausqual.

Wo suchst du Gott?
Er wohnt in Kirchen nicht,
Wo falsche Priester seinen Namen schänden.
Er wohnt in Herzen,
Die sein Werk vollenden,
Die lauter
Und voll guten Willens sind.

Arbeitsbrüder,
Arbeitsschwestern,
Laßt eure Herzen Friedensglocken werden
Und eure Hände bahnen neuen Weg ...

Und die Erde trägt beseligt
Menschenbrüder,
Gottes liebe Kinder.

Die Arbeiter

Alfons Petzold

Sturm und Gewalt ist in unseren Händen,
stehn wir im räderdurchdonnerten Raum,
doch in dem keuchenden Beugen der Lenden
sind wir gar oftmals nur Andacht und Traum.

In dunkler Berge verlorner Kaverne
sind wir die Brüder der strahlenden Tage,
türmen wir Steine im Antlitz der Sterne,
lebt Gottes Sehnsucht in unserer Plage.

Unser Wille erschüttert die Erde,
und der heiligsten Unruhe voll
schenken wir ihr durch die stete Beschwerde
Ewigkeit, die unserm Schaffen entquoll.

138 | 139

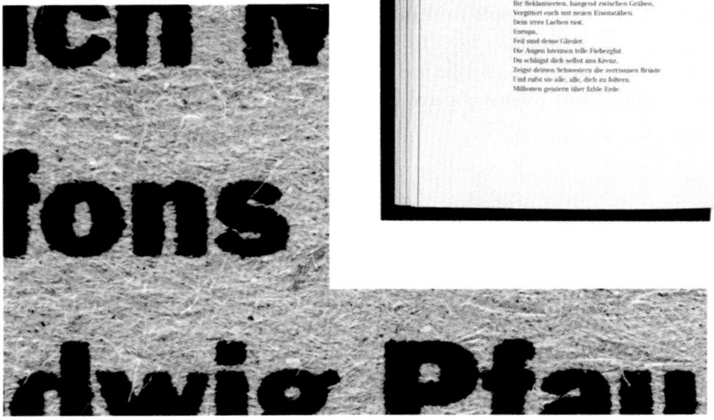

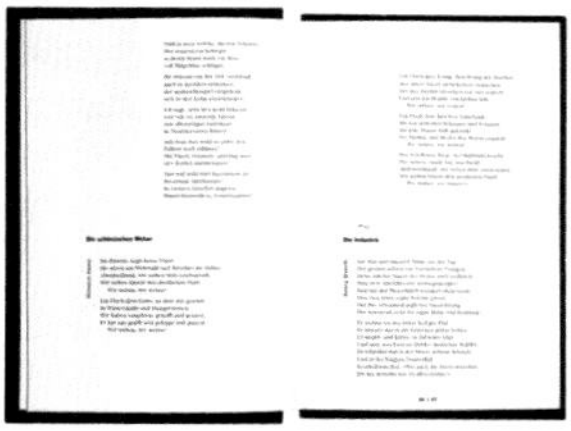

svon
en al
lager unsrer Zeit

Arbeiterlyrik 1842–1932

1924

Fabrikgang

Ludwig Lessen

Es stampft im Takt, es schlurrt im Schritt
des Morgens in der Frühe.
Sie gehn zu zwei'n, sie gehn zu dritt,
sie gehn den Gang der Mühe.

Sie schreiten ewig-gleichen Pfad
an jedem neuen Morgen.
Ein jeder Tag heischt neue Saat –
Saat gegen Not und Sorgen …
Die Frau'n, die Männer ziehn einher,
es schreitet Jugend und Alter, –
es stampft der Schritt so hart und schwer!
Und Nebel schwelt, ein kalter …

Im Nebel zieht die stumme Schar,
krumm sind die Rücken allen.
Der Frühwind zerrt an ihrem Haar …
und Schritte – Schritte hallen …

Leidensweg

Ludwig Lessen

Nicht beugen lassen! Den Rücken gestrafft
und hoch das Haupt getragen im Nacken!
In uns gärt Mut, in uns quillt Kraft, –
wir wollen das schwere Schicksal schon packen!

So leicht wird keiner zermalmt, der nicht
verloren sich gibt! Der Kampf macht zähe!
Und wo da funkelt ein helles Licht
sind immer auch Schatten in der Nähe …

Und ist unser Weg auch trümmerbestaubt,
wir schreiten ihn trotzig durch jegliches Grauen!
Hoch wollen wir tragen im Nacken das Haupt:
Wir wollen und müssen die Sonne schauen! …

Gerd Arntz
Fabrikbesetzung
Original-Holzschnitt, 1931
32,5 × 24 cm

004 Bm^2

本书讨论了意大利著名画家卡拉瓦乔（Caravaggio）在罗马时期的创作背后的故事。本书并没有将着眼点放在讨论画家的作品本身上，而是讨论了他在罗马如何逐渐变得出名，他的独特画法对巴洛克风格产生的深远影响，他与红衣主教的关系以及因此对艺术活动产生的影响，还有那些在他周边的人物以及为他的生活提供保障的赞助商和收藏家们。本书的装订形式为无线胶平装，封面采用卡纸荧光橙色和暗红色双色印刷，文字烫白。内页胶版纸荧光橙和黑色双色印刷。前后环衬分别为荧光橙色和暗红色底露出矩阵渐变白点的设计。书籍的外观将卡拉瓦乔最具代表性的戏剧化明暗对比法运用其间，书籍的封面、上书口和翻口为荧光橙色，而封底、书籍和下书口为暗红色。书籍被模拟成受单光源照射在一角上，从而“成为”卡拉瓦乔画作中强烈明暗对比的书本。内页部分只在章节标题和页码部分使用荧光橙色，烘托这种强烈的戏剧效果。注释文字被统一放置在每章节的后部，通过双栏设置与正文形成对比，调节整本书的阅读节奏。

Lothar Sickel
CARAVAGGIOS ROM
Annäherungen an ein dissonantes Milieu

This is an academic monograph on the story behind the creation of famous Italian painter Caravaggio in Roman times. Instead of focusing on the artist's works, the book discusses how he became famous in Rome, the far-reaching impact of his unique painting methods on the Baroque style, his relationship with the Cardinal and the consequent influence on artistic activities, as well as those sponsors and collectors who secure his life. The paperback book is perfect bound. The cover is printed on cardboard in two colors – fluorescent orange and dark red, with white hot-stamped text. The inside pages are printed on offset paper in fluorescent orange and black. The front and back endpapers are designed with gradient white dots that form a matrix. Caravaggio's most representative dramatic contrast of light and shade has been applied to the design of the book. The cover, upper edge and fore-edge are fluorescent orange, while the back cover and lower edge are dark red. The book seems to be illuminated in a corner by a single light source, thus forming a strong contrast between light and shade just like Caravaggio's paintings. Only the chapter titles and page numbers are printed in fluorescent orange on the inside pages, which shows a strong dramatic effect. The annotations are placed at the end of each chapter, and the reading rhythm is adjusted by the two-column layout comparing with that of the main body.

铜奖
卡拉瓦乔的罗马——走进一处失调之地
德国

Very contemporary design and interpretation of this academic subject matter. The use of two vibrant colours on the dust-jacket effectively reflects the book's title. The inner design is well proportioned to the unusually long and narrow format. Small touches such as orange pagination and chapter headings add to the book's atmosphere. Also the decision to set out the footnotes in two columns pleasantly breaks up the flow of the text.

Bronze Medal
Caravaggios Rom. Annäherungen an ein dissonantes Milieu
Germany
Andreas Trogisch
Lothar Sickel
Edition Imorde, Emsdetten / Berlin
240 × 144 × 19mm
454g
268p

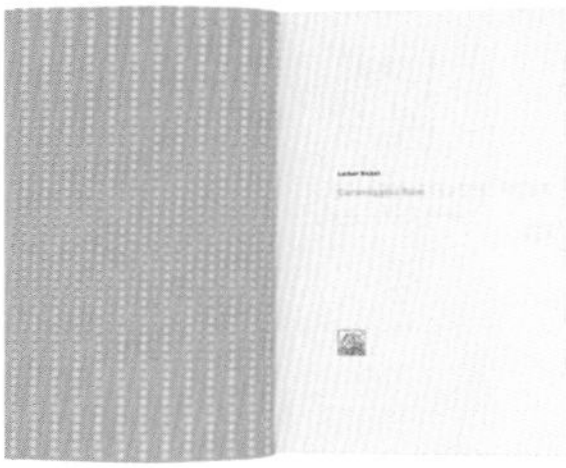

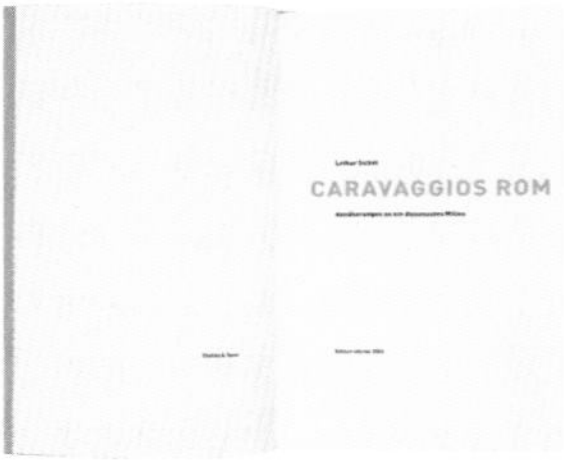
CARAVAGGIOS ROM

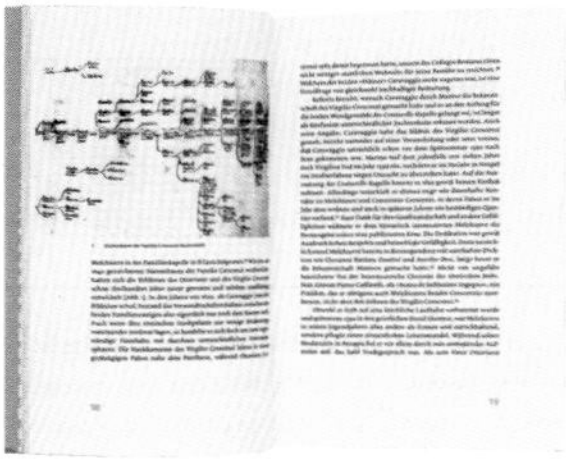

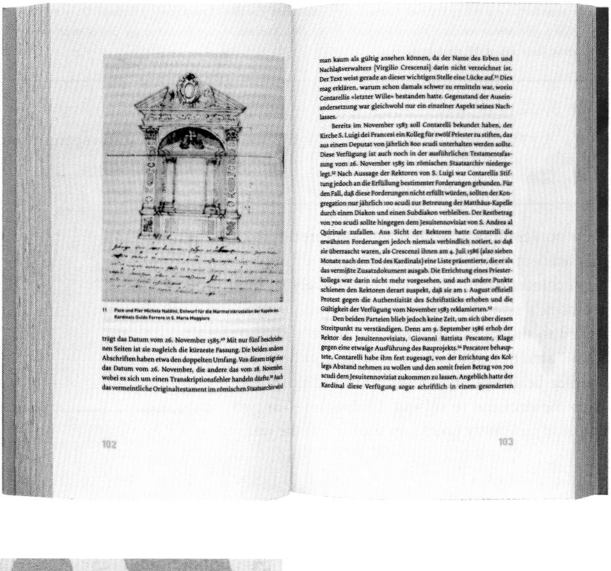
11 Pace und Pier Michele Naldini, Entwurf für die Marmorinkrustation der Kapelle des Kardinals Guido Ferrero in S. Maria Maggiore
trägt das Datum vom 26. November 1585.[29] Mit nur fünf beschriebenen Seiten ist sie zugleich die kürzeste Fassung. Die beiden anderen Abschriften haben etwa den doppelten Umfang. Von diesen trägt eine das Datum vom 26. November, die andere das vom 28. November, wobei es sich um einen Transkriptionsfehler handeln dürfte.[30] Auch das vermeintliche Originaltestament im römischen Staatsarchiv wird
102
man kaum als gültig ansehen können, da der Name des Erben und Nachlaßverwalters [Virgilio Crescenzi] darin nicht verzeichnet ist. Der Text weist gerade an dieser wichtigen Stelle eine Lücke auf.[31] Dies mag erklären, warum schon damals schwer zu ermitteln war, worin Contarellis «letzter Wille» bestanden hatte. Gegenstand der Auseinandersetzung war gleichwohl nur ein einzelner Aspekt seines Nachlasses.
Bereits im November 1583 soll Contarelli bekundet haben, der Kirche S. Luigi dei Francesi ein Kolleg für zwölf Priester zu stiften, das aus einem Deputat von jährlich 800 scudi unterhalten werden sollte. Diese Verfügung ist auch noch in der ausführlichen Testamentsfassung vom 26. November 1585 im römischen Staatsarchiv niedergelegt.[32] Nach Aussage der Rektoren von S. Luigi war Contarellis Stiftung jedoch an die Erfüllung bestimmter Forderungen gebunden. Für den Fall, daß diese Forderungen nicht erfüllt würden, sollten der Kongregation nur jährlich 100 scudi zur Betreuung der Matthäus-Kapelle durch einen Diakon und einen Subdiakon verbleiben. Der Restbetrag von 700 scudi sollte hingegen dem Jesuitennoviziat von S. Andrea al Quirinale zufallen. Aus Sicht der Rektoren hatte Contarelli die erwähnten Forderungen jedoch niemals verbindlich notiert, so daß sie überrascht waren, als Crescenzi ihnen am 4. Juli 1586 (also sieben Monate nach dem Tod des Kardinals) eine Liste präsentierte, die er als das vermißte Zusatzdokument ausgab. Die Errichtung eines Priesterkollegs war darin nicht mehr vorgesehen, und auch andere Punkte schienen den Rektoren derart suspekt, daß sie am 1. August offiziell Protest gegen die Authentizität des Schriftstücks erhoben und die Gültigkeit der Verfügung vom November 1583 reklamierten.[33]
Den beiden Parteien blieb jedoch keine Zeit, um sich über diesen Streitpunkt zu verständigen. Denn am 9. September 1586 erhob der Rektor des Jesuitennoviziats, Giovanni Battista Pescatore, Klage gegen eine etwaige Ausführung des Bauprojekts.[34] Pescatore behauptete, Contarelli habe ihm fest zugesagt, von der Errichtung des Kollegs Abstand nehmen zu wollen und den somit freien Betrag von 700 scudi dem Jesuitennoviziat zukommen zu lassen. Angeblich hatte der Kardinal diese Verfügung sogar schriftlich in einem gesonderten
103

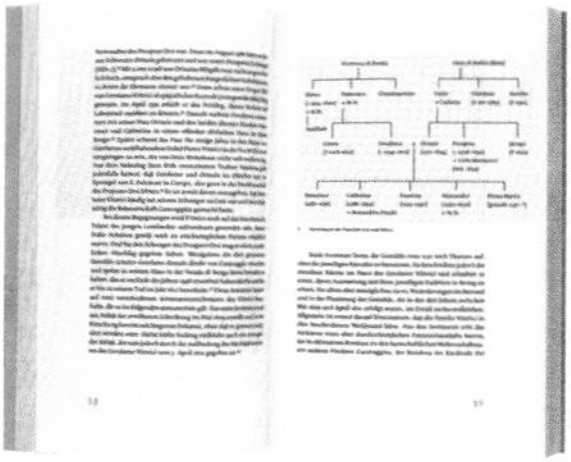

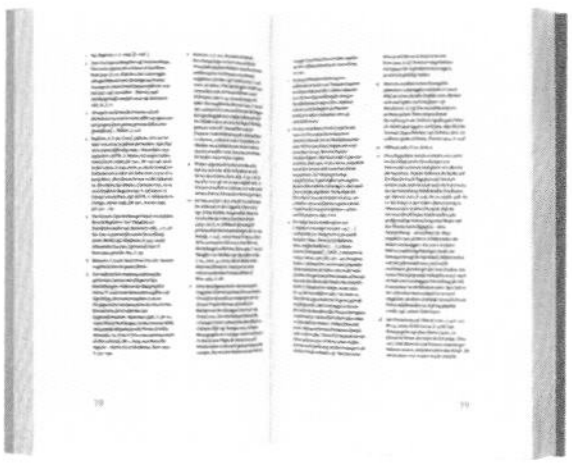

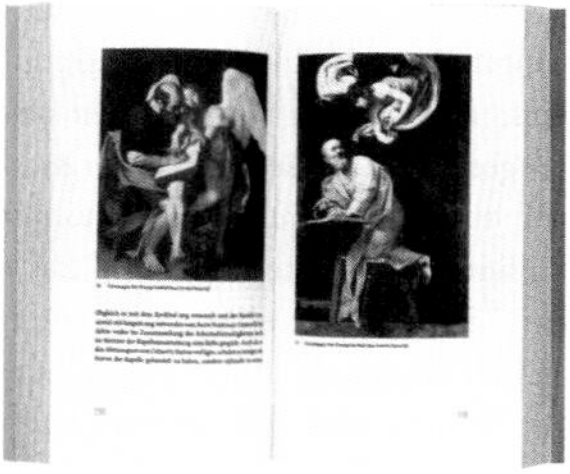

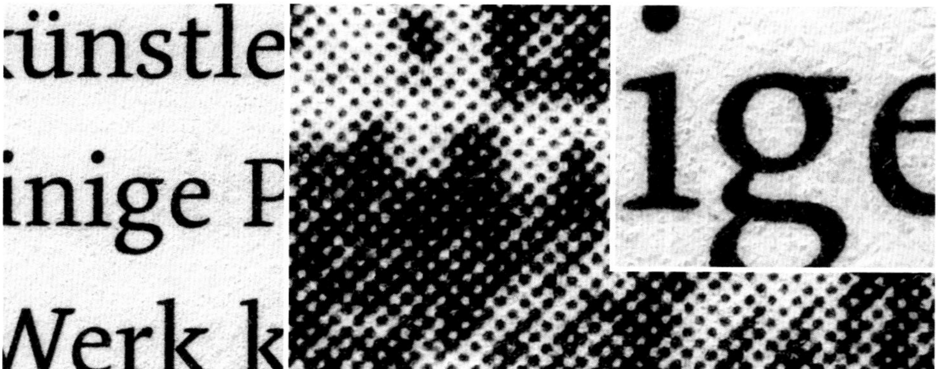

künstle
inige P
Werk k
ige

CARAVAGGIOS ROM

Lothar Sickel

«In atto di dispregiar' il mondo»

Caravaggios «Amor» und die Kultur Pasquinos

Der *Amor* in der Berliner Gemäldegalerie ist eines der bekanntesten Gemälde Caravaggios (Abb. 16). Es ist zugleich ein sehr untypisches Werk, denn es handelt sich um die einzige explizit allegorische Darstellung in seinem gesamten Œuvre. Die Umstände seiner Entstehung lassen sich recht genau rekonstruieren. Caravaggio schuf es in der ersten Hälfte des Jahres 1602 im Zuge seiner schon länger schwelenden Fehde mit Giovanni Baglione, die im Herbst 1603 kulminierte, als Baglione einen Strafprozeß wegen übler Nachrede gegen Caravaggio und dessen Freunde Onorio Longhi, Orazio Gentileschi und Filippo Trisegni anstrengte.[1] Wie Gentileschi in seiner Aussage vom 14. September 1603 berichtet, war der juristischen Auseinandersetzung ein regelrechter Bilderstreit vorausgegangen. Zunächst hatte Gentileschi in S. Giovanni dei Fiorentini eine großformatige Darstellung des Erzengels Michael ausgestellt, der Baglione einen *Himmlischen Amor* entgegengesetzt habe. Dabei handelte es sich sehr wahrscheinlich um die heute ebenfalls in Berlin bewahrte Fassung dieses Themas, das Baglione in zwei sehr ähnlichen Versionen gemalt hat.[2] Angeblich hatte Baglione dieses Bild in Konkurrenz zu Caravaggios «Amor terreno» gemalt, wie Gentileschi das Werk seines Freundes damals bezeichnete. Sein Bericht wird durch die Provenienz der beiden Gemälde bestätigt, denn sie waren etwa zur gleichen Zeit im Auftrag der Familie Giustiniani entstanden – allerdings für verschiedene Personen. Während Baglione seine beiden Fassungen des *Himmlischen Amor* dem Kardinal Benedetto Giustiniani zugeeignet hatte, war Caravaggio für dessen Bruder Vincenzo Giustiniani tätig gewesen. Angeblich hatte der Bankier 300 scudi, den Preis eines großen Altarbildes, für den *Amor* gezahlt.[3]

Es handelte sich also um eine Konkurrenzsituation in mehrfachem Sinn, denn anscheinend orientierte sich die Themenwahl an der jeweiligen Standeszugehörigkeit der Brüder Giustiniani. Für Baglione endete dieser Vergleich jedenfalls nicht eben glücklich. Nach eigenen Angaben machte ihm der Kardinal Giustiniani zwar eine goldene Kette zum Geschenk, von Seiten seiner Kollegen erntete er damals hingegen nur Spott. Gentileschi hielt den ersten *Himmlischen Amor* für völlig mißraten, und auch Caravaggio scheint sich derart in Häme ergangen zu sein, daß Baglione sich erdreistete, ihn wenig später, in seiner zweiten Version des Themas in der Gestalt eines Dämons zu porträtieren (Abb. 17). Im Palazzo Giustiniani hing dieses Gemälde der ersten Fassung Bagliones direkt gegenüber.[4] Seine Diffamierung Caravaggios ist vielleicht nicht sehr einfallsreich; sie erinnert aber daran, daß auch in Kunstwerken von augenscheinlich hoher Dignität mit satirischen Einlassungen zu rechnen ist. Dies gilt zumal für Caravaggio.

Sein *Amor* war bereits vielfach Gegenstand der Interpretation. Wohl kein Gemälde Caravaggios hat derart unterschiedliche Deutungen hervorgerufen, wie jene provozierende Darstellung eines lachenden Knaben, zu der ihm sein Schüler Francesco Buoneri, genannt Cecco del Caravaggio, Modell stand. Dessen ostentative Blöße galt als Ausweis homoerotischer Sinnlichkeit, schon bevor der um 1650 verfaßte Bericht des britischen Romreisenden Richard Symonds bekannt wurde, wonach jener Cecco mit Caravaggio einst auch das Lager geteilt habe.[5] An dieser Sichtweise stimmte jedoch skeptisch, daß sie dem Auftraggeber Vincenzo Giustiniani und dessen repräsentativen Interessen kaum entsprochen haben dürfte. Nach Aussage seines ehemaligen Kurators Joachim von Sandrart bildete das Gemälde den Schluß- und Höhepunkt einer jeden Führung durch die Galleria Giustiniani, und angeblich nur zur Steigerung des Präsentationseffekts war es – dreißig Jahre nach seiner Entstehung – mit einem Vorhang ausgestattet worden.[6] Im Hinblick auf jene repräsentative Funktion

Bm³

这是一本整理和出版穆特修斯大学（Muthesius-Hochschule）跨学科论坛的研究和会议结果的文献。论坛的内容涉及科学、历史、文化、艺术、心理、生理、材料、社会等多个领域，很多的议题都在多个学科领域里有所涉猎，将多领域的知识加以融合，得出新颖的结论。所以本书的设计也尽可能体现这种融合性，并在融合的同时注意对各领域内容加以标示和索引。书籍的装订形式为无线胶平装，封面通过深蓝和荧光橙双色字母的相互叠透来体现主题。内页胶版纸依然以这两种颜色为主，深蓝色为主要文字，荧光橙则用于注释、辅助图形、目录的文章分类索引等方面，图片部分采用深蓝色辅以少量黑色加重层次。内容的隔页部分采用薄透的描图纸正反印刷双色内容进行“融合”。书中的目录和交叉索引系统非常独特，在目录中将内容划分为“V”和“W”两大类，并在翻口部分的带有相应的线条符号进行内页索引，而不仅仅通过数字页码来进行指示。内页的隔页部分也会对相应内容的关键词进行分类，通过“V”和“W”构建的这一“超链接”指引到相应的章节。||||||||||||||||||||||||||

This is a document that collates the findings of the Muthesius-Hochschule interdisciplinary forum. The content of the forum covers many fields, such as science, history, culture, art, psychology, physiology, materials, society and so on. Many topics are discussed in many disciplines, integrating knowledge from many fields and drawing new conclusions. Therefore, the design of this book also reflects this integration as far as possible, and at the same time pays attention to labeling and indexing the content of various fields. The paperback books is perfect bound. The cover reflects the theme by overlapping dark blue and fluorescent orange letters. The inside pages are printed on offset paper in these two colors as well. Dark blue is used for the main text, while fluorescent orange is used for annotations, auxiliary graphics, classification index and so on. The separation pages are made of tracing paper, which is thin and translucent, printed on double sides to achieve ‘fusion’. The book’s catalogue and cross-index system are very unique. In the catalogue, the contents are divided into category ‘V’ and category ‘W’, and indexed with corresponding line symbols on the fore-edge, in addition to indexing with page numbers. The key words of the corresponding content are also categorized on the separation pages, are guided by the ‘hyperlinks’ constructed by ‘V’ and ‘W’.

铜奖

融合

——穆特修斯大学的跨学科周

德国

Bronze Medal

Vermischungen.

Interdisziplinäre Wochen an der Muthesius-Hochschule

Germany

Julia Hofmeister und Tim Albrecht (Muthesius-Hochschule)

Theresa Georgen (Ed.)

Muthesius-Hochschule, Kiel

285×188×11mm

359g

144p

A very pleasing, almost classical typography which makes use of a new font, Prokyron (combined with VandenKeere and ocr). The main text and photographs are printed in dark blue. Orange, the second colour, is used for graphics and footnotes, thus giving it a structural as well as aesthetic function. The large letters, printed in two colours on both sides of thin paper and inserted into the text, are very attractive. A most convincing documentation of an interdisciplinary forum.

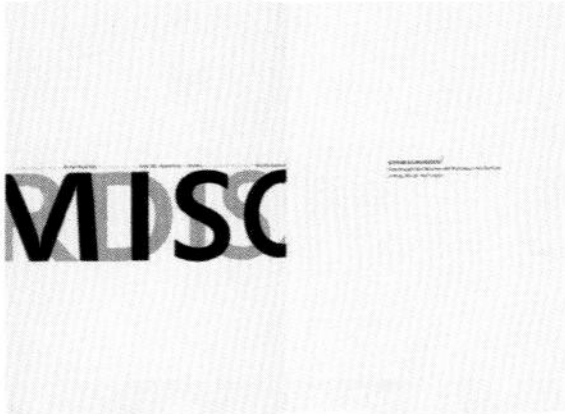
MISC

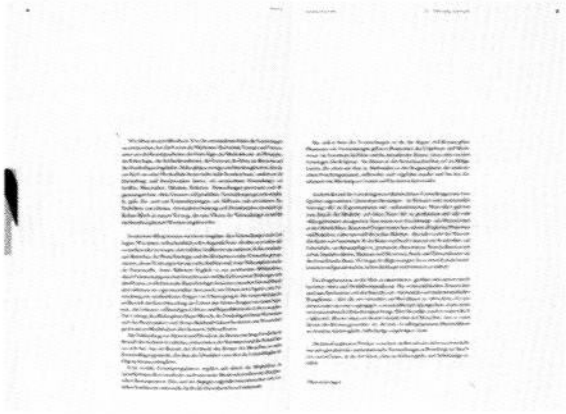

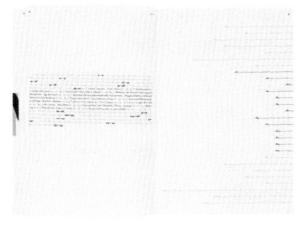

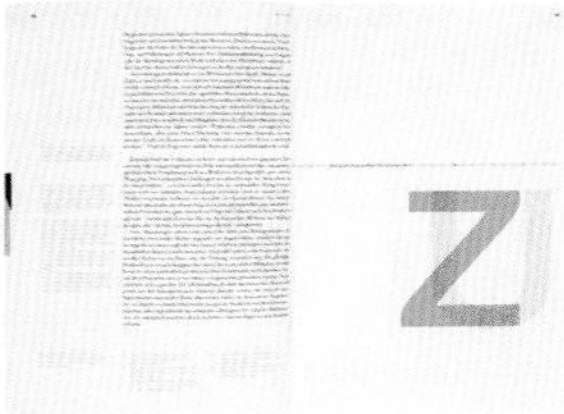
Z

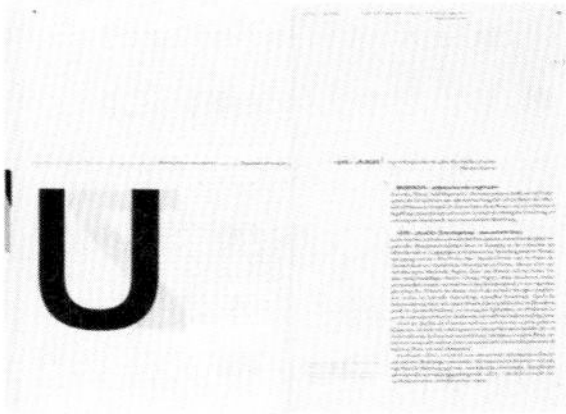
U

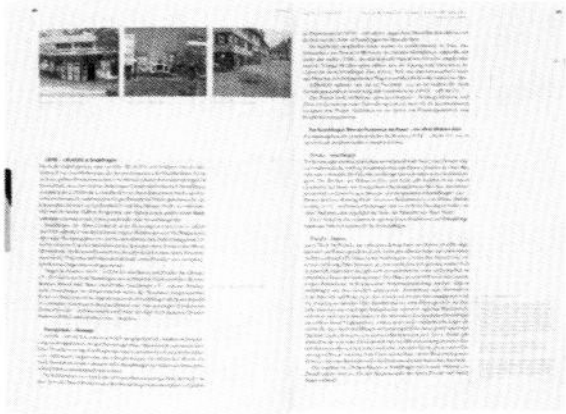

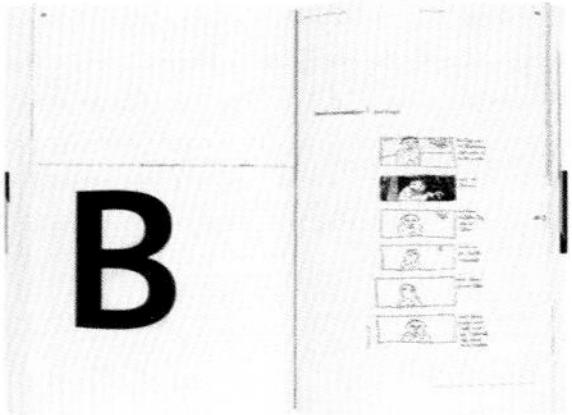
B

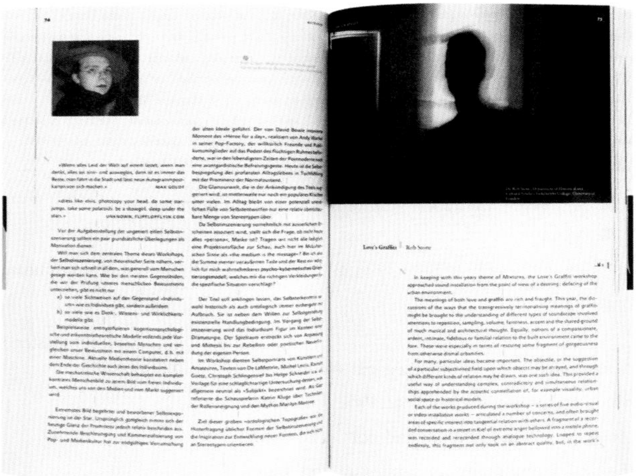
74
75
Love's Graffiti
Rob Storm

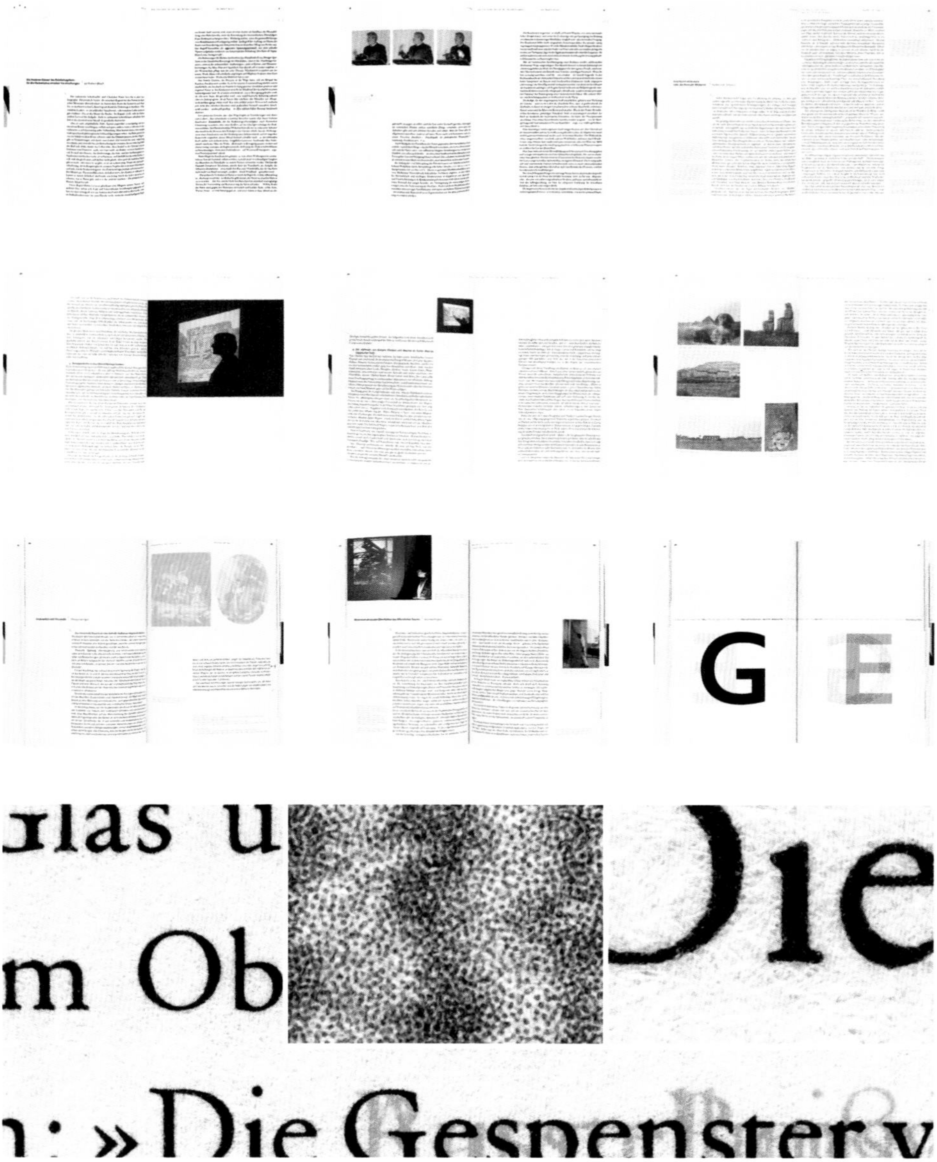
G
E
m Ob
»Die Gespenster

künsten der katholischen Kirche erzeugte sie vor allem Effekte zwischen bürgerlicher Selbstdarstellung und kirchlicher Glaubensvermittlung. Erst der Realismus im engeren Sinne, der als Stil nicht zufällig unmittelbar auf die Erfindung der Fotografie folgt, ließ Mitte des 19. Jahrhunderts eine Kunst der Augentäuschung entstehen, in der die Wahrheit der Abbildung ästhetischer Selbstzweck war. Doch dies ist schon ein erstes Kapitel der Avantgarde.

Charakteristisch für die klassische Kunst, wenigstens seit dem Barock, als sich gewissermaßen auf den verschiedensten Ebenen der Kultur die Richtung der strukturellen Dynamik der westlichen Zivilisation endgültig verfestigt hatte, war dagegen eine Verfeinerung der Fertigkeiten zur Augentäuscherei für manche profanen und sakralen Zwecke. Oliver Grau beschreibt beeindruckend den für die Zeitgenossen bis hin an die Grenze der Ununterscheidbarkeit perfektionierten dreidimensionalen Nachstellungen der Passion Christi, Inszenierungen, die ihre logische Fortsetzung im wohl häufigsten Panorama-Programm des 19. Jahrhunderts, der Kreuzigung, fanden. Technische Mittel zur Erregung von Gefühlen waren es, die schon damals virtuelle Welten erzeugten. Caravaggios Rosenkranzmadonna war eben keine Reflexion über den Status der Realität oder über die medialen Qualitäten der Ölmalerei und schon gar kein ›Stilwollen‹, sondern eine Auftragsarbeit für eine auf Seelenfang spezialisierte katholische Kirche der Gegenreformation. Und die Faszination der Simulationstechniken bestand nicht zu einem geringen Teil gerade darin, das Übersinnliche sinnlich fassbar zu machen.

Simulation – von Berninis hochbarocken Augentäuschereien und Lichtplastiken zur Überraschung der Gläubigen über die kommerziellen Panoramen zur Erhebung ›vaterländischer Gesinnung‹ bis zur schmelzenden Butter auf den Werbeplakaten der Lebensmittelindustrie – war so vor allem eine Praxis rhetorischer Überwältigung und niemals Selbstzweck. Auch das ›Wunder des Nordens‹, der ›Fotorealismus‹ des Jan van Eyk meinte ja keinen Realismus als Selbstzweck, sondern ausschließlich die präzise Wiedergabe der Natur als ›emanatio dei‹, also die getreue Wiedergabe der symbolischen Ordnung der Natur als göttlicher Erscheinung im Sinne der ›devotio moderna‹ und nicht des modernen Realismus. Realismus als Selbstzweck ist eben eine Erfindung des 19. Jahrhunderts, und Courbet wurde nicht von ungefähr als einer der ersten ›artiste moderne‹ gefeiert. Es ist wohl kein Zufall, dass eine der frühesten und bedeutendsten Abhandlungen über Simulationsapparate, die Ars magna lucis et umbrae des Athanasius Kircher aus dem Jahr 1643, eben das Werk eines Jesuiten darstellt, der gleich auch den Anwendungszweck solcher ›Jugentäuscherei‹ als Kupferstich vor Augen führt: ein kleines Teufelchen erscheint wundersam zur Bekehrung der armen Sünder.

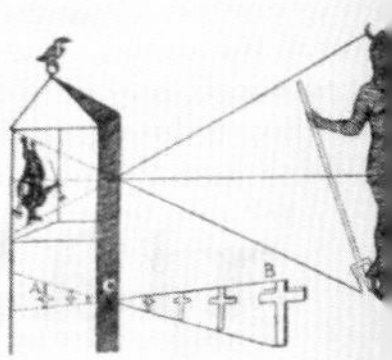

Das *originale Material*, seine Aura schließlich spielte gar keine Rolle, sieht man einmal von heiligen Reliquien und Ehrfurcht gebietenden Spolien ab. Nach Belieben haben Barockartisten wie Tiepolo Stuck und Stuccolo zur Täuschung der Sinne herangezogen. Adolf Loos präsentierte im Haus am Michaelerplatz statt solch ›verbrecherischer‹ Marmorsimulationen kostbares Originalmaterial. Noch ein Caspar David Friedrich – eigentlich Wegbereiter der Moderne – scheute sich nicht, bei günstiger Auftragslage auch einmal eine Replik eines Gemäldes herzustellen. Die Moderne schuf recht eigentlich den Kult um den wahren Gegenstand des Kunstwerks und begann die konkrete Dinglichkeit von Ölfarbe und Malgrund selbst zum Wert zu erklären. Erst die unendliche Reproduzierbarkeit verlieh also dem Bild als einmaligem materiellen Gegenstand seine einzigartige Bedeutung und ließ zuletzt das Bild, die raue Sinnlichkeit der Leinwand oder Leuchtkraft der Farbpigmente um ihrer selbst willen als Ort ›ästhetischer Transzendenz‹ erscheinen. Ein gut Teil der modernen Malerei gewinnt ihren Reiz eben erst durch die Differenz zu den Möglichkeiten der je aktuellen Reproduktionstechniken. Insofern ist auch eine solche alte Gattung, gerade da, wo sie scheinbar vom Stand der Technik vollkommen unberührt bleibt, immer nur aus ihrer Relation zu diesem zu verstehen und letztlich jede Kunst noch in der Negation eine Art Medienkunst.

Appropriation, also die freie Aneignung fremden Bildmaterials als postmoderne künstlerische Strategie war als Zitat geläufig und die Beherrschung überkommener Muster geradezu Essenz akademischer Ausbildung.[17] Wenngleich hier durchaus Unterschiede zu den klassischen Formen des bildlichen Zitats auszumachen sind, so ist doch die ›Ursprünglichkeit und Authentizität des individuellen Entwurfs‹ im Wesentlichen eine Forderung der Moderne gegenüber dem Konventionalismus akademischen Zitierens und Variierens. Erst recht das so ›unmoderne‹ Hollywood funktioniert fast gänzlich durch den Verweis auf die dem einzelnen Film übergeordneten Kontexte, eben vor allem die ausdifferenzierten Genresysteme innerhalb und außerhalb des Kinos, wodurch hochkomplexe und verdichtete Ausdrucksmöglichkeiten zur Verfügung stehen.[18] Ein John-Ford-Film erhält seine Dichte nicht zuletzt durch die Referenzen auf die amerikanischen Gründungsmythen und den Western und wäre als Schöpfung ex ovo nicht vorstellbar.[19]

Autorschaft und eine damit verbundene Arbeitsteilung war in der vormodernen Handwerkskunst vor allem eine Frage technischer Qualität und nur in den seltensten Fällen eigentlich ausschlaggebend. Ließ sich die Einmaligkeit einer künstlerischen Handschrift in den teilweise vorindustriell durchrationalisierten Bilderfabriken Hollands und Flanderns gelegentlich gar nicht mehr erkennen, so war sie – wenn sie denn zur Marktgröße wurde – allenfalls Synonym für technische Meisterschaft und Höhe intellektueller Konzeption, aber niemals Zeichen für die Selbstverwirklichung eines Künstlersubjekts. Rubens wäre ohne Werkstatt kaum imstande gewesen, die gewaltigen Illusionsinszenierungen etwa in White Hall oder der Antwerpener Jesuitenkirche zu vollenden.

Wenn der populäre Spielberg für seine Saurieranimation eine ganze Belegschaft von technischen Spezialisten rekrutierte, so ist das der Normalfall der westlichen Kunst. Auch der ›Filmkünstler‹ Greenaway suchte nicht etwa nach den Geheimnissen der Programmiersprache in der Paint-Box, sondern griff für *Prospero's Books* auf bewährte Techniker zurück. Man muss sich vergegenwärtigen, dass die Bildinszenierungen der Renaissance auf ihre Art häufig genug die fortgeschrittensten technologischen und organisatorischen Projekte ihrer Zeit darstellten. Sie waren nicht weniger Technologiezentren als die heute immer hart an der Grenze des

17 Zu dem von Craig Owens und Douglas Crimp Anfang der achtziger Jahre lancierten Begriff vgl. Groys, Boris: »Apropos Appropriation«, in: *Der Schnitt*, Nr. 18, H. [illegible], S. [illegible]

18 Vgl. Panofsky, Erwin: »*Stil und Medium im Film*« (1936–1947), in: *Die ideologischen Vorläufer des Rolls-Royce-Kühlers*. Frankfurt a. M. [illegible], S. [illegible]

19 Vgl. Baxter, John: *John Ford*. München [illegible], S. [illegible] ff.

LIEBE

14 LICHT

Zum Beispiel das Tageslicht, das auf diesen Buchseiten liegt, das sozusagen schläft mit offenen Augen. Hier, am Ende meines Zeigefingers.

Wenn ich das Buch zuklappe, stürzt kopfüber die Dunkelheit zwischen die Seiten hinein. Heißt das, es gibt eine Dunkelheit, die sich so dünn machen kann, dass sie beim Zuklappen nicht wieder herausgequetscht wird? Vielleicht ist es gar nicht dunkel in einem geschlossenen Buch. Ist es dann aber hell – oder was sonst?

Jürg Schubiger

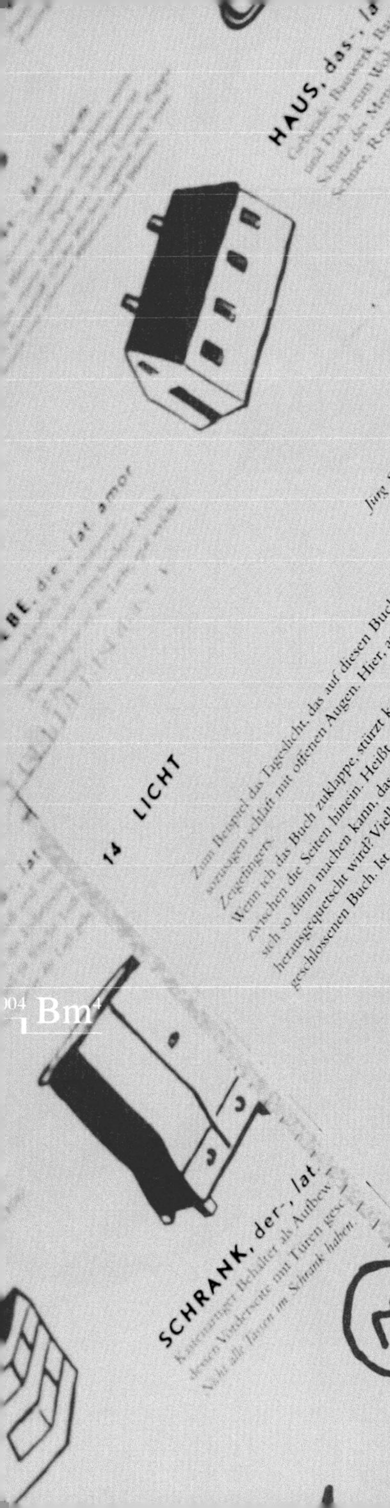

HAUS, das-, lat.
SCHRANK, der-, lat.
STIFT, der-

Keiner weiß, wie sie au
nicht mal.
Wenn Erwachse
Geld. Sie ma
wechselt
einen
L

本书是由藏书家阿尔敏·阿梅尔（Armin Abmeier）收集整理并出版的“伟大的书”（Die Tollen Hefte）书系里的一本，由他与 R. 苏珊·贝尔纳（R. Susanne Berner）整理。全书由德语作者的 26 篇文章合并而成，并由著名插画家托马斯·M. 穆勒（Thomas M. Muller）绘制插图。本书主要谈论了 26 种常见的生活用品和事物，涉及食品、家庭、娱乐、教育等多重领域。全书采用线装骑马钉外包护封的整体装订形式，32 开的非涂布纸张不仅增加了亲和力，更便于翻阅。内页部分含有一张 4 开海报，以双色印刷的方式阐述了全书所述的 26 种生活必需品，并配以解释文字，构成了一个饶有生趣的小型词汇表。整体设计风格诙谐有趣，细致精美。

Lebens-Mittel
WAS KINDER BRAUCHEN
Milch
Cacao
BÜCHERGILDE GUTENBERG

The book is collected and published by the bibliophile Armin Abmeier and it's one of the 'Die Tollen Hefte' (The Great Books) series, compiled by R. Susanne Berner and Armin Abmeier. The book concludes 26 articles by the German authors, illustrated by the celebrated illustrator Thomas M. Muller. This book mainly discusses 26 kinds of children's necessities, covering food, family, entertainment, education and other fields. The book is saddled stitched with thread sewing and protected with dust cover. Non-coated paper not only increases the affinity, but also makes it easier to read. A two-color printing poster is attached with the inside pages, illustrating the 26 daily necessities described in the book with explanatory text. The poster is like an easy and funny word list. The overall design style is humorous, interesting and delicate.

- 铜奖
- 生活用品（伟大的书 22）
- 德国

- Bronze Medal
- Lebens-Mittel (Die Tollen Hefte 22)
- Germany
- Thomas M. Müller
- Armin Abmeier, Rotraut Susanne Berner (Ed.)
- Büchergilde Gutenberg, Frankfurt am Main / Wien / Zürich
- 207 × 137 × 7mm
- 104g
- 31p

The witty and lively drawings are juxtaposed intelligently and with variety against the texts. Impressive use of the seven colours. The layout, in its calmness, gives pride of place to the illustrations whilst effectively offsetting them. The volume radiates an atmosphere thoroughly in keeping with the texts. Many original details such as an inserted poster, the glossary on the back of which refers back to words and topics from the volume itself. True to the title of the series, a really superb volume.

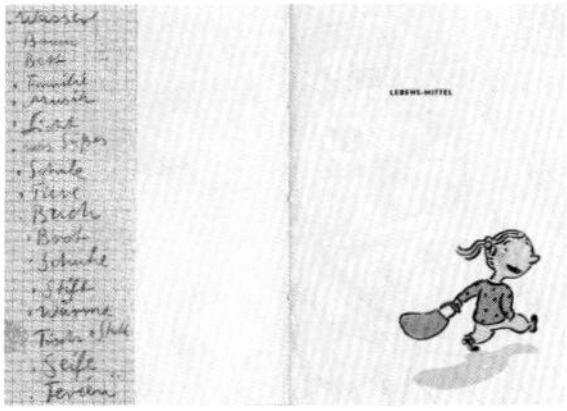
LEBENS-MITTEL

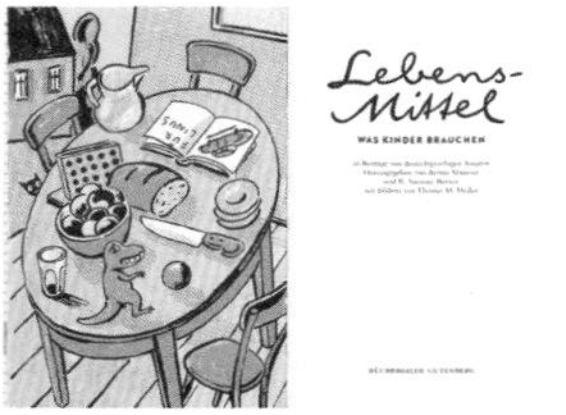
Lebens-Mittel
WAS KINDER BRAUCHEN

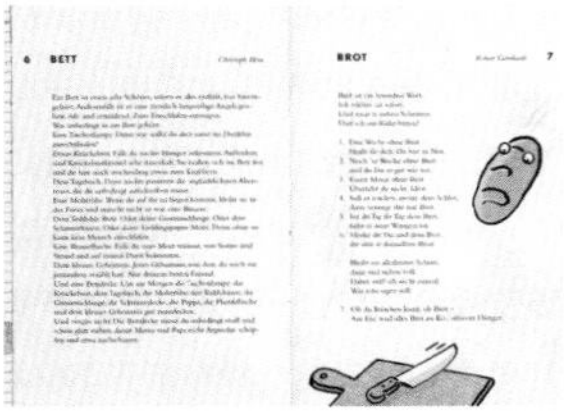
6 BETT
BROT 7

18 PAPIER
19

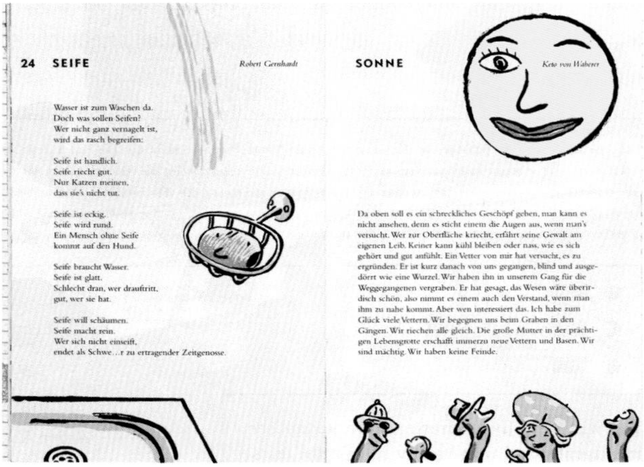
24 SEIFE *Robert Gernhardt*

Wasser ist zum Waschen da.
Doch was sollen Seifen?
Wer nicht ganz vernagelt ist,
wird das rasch begreifen:

Seife ist handlich.
Seife riecht gut.
Nur Katzen meinen,
dass sie's nicht tut.

Seife ist eckig.
Seife wird rund.
Ein Mensch ohne Seife
kommt auf den Hund.

Seife braucht Wasser.
Seife ist glatt.
Schlecht dran, wer drauftritt,
gut, wer sie hat.

Seife will schäumen.
Seife macht rein.
Wer sich nicht einseift,
endet als Schwe…r zu ertragender Zeitgenosse.

SONNE *Keto von Waberer*

Da oben soll es ein schreckliches Geschöpf geben, man kann es nicht ansehen, denn es sticht einem die Augen aus, wenn man's versucht. Wer zur Oberfläche kriecht, erfährt seine Gewalt am eigenen Leib. Keiner kann kühl bleiben oder nass, wie es sich gehört und gut anfühlt. Ein Vetter von mir hat versucht, es zu ergründen. Er ist kurz danach von uns gegangen, blind und ausgedörrt wie eine Wurzel. Wir haben ihn in unserem Gang für die Weggegangenen vergraben. Er hat gesagt, das Wesen wäre überirdisch schön, also nimmt es einem auch den Verstand, wenn man ihm zu nahe kommt. Aber wen interessiert das. Ich habe zum Glück viele Vettern. Wir begegnen uns beim Graben in den Gängen. Wir riechen alle gleich. Die große Mutter in der prächtigen Lebensgrotte erschafft immerzu neue Vettern und Basen. Wir sind mächtig. Wir haben keine Feinde.

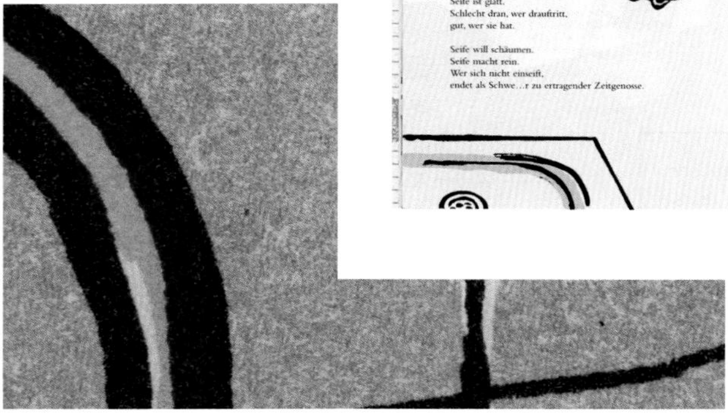

10 FERIEN
FREUNDE 11

HAUS
HEMD & HOSE 13

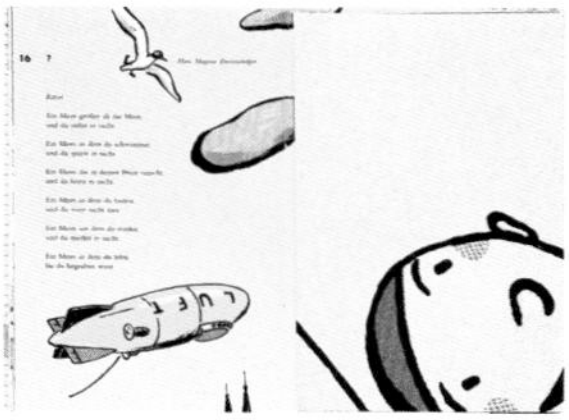
16 ?
LUFT

26 SPIEL
STIFT

TIERE
TISCH & STUHL 29

30 WÄRME
WASSER 31

st, aus
und die Lehrer un

22 SCHUHE

Brigitte Kronauer

Die Löwen wie die kleinen Schwarzfußkatzen,
auch meine mit den weiß gefleckten Fratzen,
pfeifen auf Schuhe, ob sie wachen oder ratzen.
Anders als dir und mir genügen ihnen: Vielzwecktatzen.

Musst du da nicht vor Neid fast platzen?

Bequem sind sie beim Fernsehen, die weichen
Pantoffeltatzen, und beim Jagen gut zum Schleichen
sind Stiefel, die bei strenger Kälte reichen,
beim Tanz jedoch für höchste Grazie das Zeichen.

Musst du da nicht vor Neid erbleichen?

SCHULE

Klaus Wagenbach **23**

Öh ist die langweilig, sagen alle.
Aber meine Lehrerin kann lesen und schreiben
und zeigt mir auch
wie's geht
und man ein Buch versteht
wie dieses.

Heimlich,

004 Ha[1]

这是一本汇集了来自德国、荷兰和英语区国家的丝网版画艺术家的作品集。作品风格差异巨大，套版数、创作手法也各不相同，但都在作品中极力展现着与生命、青春，以及面对生活的不幸表现出的不屈不挠的勇气。本书的装订形式为锁线硬精装。封面灰色卡板黄蓝双色套印，裱贴在印有浅蓝绿色图案的深蓝色布面书脊上，书脊内衬硬卡纸加固。封二、封三采用与书脊和扉页相同的由 Gummbah 创作的“nobody forever”人物图案，双色套印，内页选用厚实的版画用艺术纸张手工装订，按照每四页（八面）为一个折手，分为九个折手。除第一折手为扉页、版权信息和诗歌引言，其余页面展示了八位艺术家的作品，每人八件作品。每隔一折手，外包白卡纸灰绿色印刷的艺术家名字、作品名称“展签”一枚，用于区隔作品。考虑到艺术家来自不同的母语区或在不同地区的影响力，均将原母语语言信息放在首位，后跟德语（出版地），英语除外。

NOBODY

This is a collection of silkscreen woodcut paintings by artists from Germany, the Netherlands and other English speaking countries. The style of the works varies greatly, and the number of color plates and creative techniques are also different, but all of them are trying to show their indomitable courage to face the life, the youth and the misfortunes. The hardback book is bound by thread sewing. The cover is made of gray cardboard overprinted with yellow and blue, mounted on the spine wrapped by dark blue cloth and printed with light blue-green patterns. The spine is reinforced by hard cardboard. The inside front cover and the inside back cover are printed with ‘nobody forever’ figures created by Gummbah, the same as the spine and the title page. Two-color overprint is used. The inner pages are bound by hand with thick printmaking paper. Each four pages (eight sides) are bound into a fold. The book has nine folds in total. Except the first fold is used for the title page, copyright information and introduction, the remaining folds show the works of eight artists, and each of them shows eight works. Between every two folds, there is an ‘exhibition label’ made of white cardboard printed in gray-green with the name of artists and works. Considering that artists come from different mother tongue areas and considering their influence in different areas, this book gives priority to native language information, followed by German (the place of publication), with the exception of English.

◎ 荣誉奖
《 无人永生
◖ 德国

This book is a joint project from a number of artists from Germany and the Netherlands. Various silk-screen prints in strong colours, each by a single artist, are joined together and these are then united in an original, large-format hardcover volume. The various artistic thumbprints as well as the range of special colours lead to a visually highly attractive book, witty, modern and well-crafted.

- Honorary Appreciation
- nobody forever
- Germany
- Christine Lohmann, Walter Kerkhofs
- Nick J. Swart
- Edition Hamtil, Hamburg/Tilburg
- 355×288×29mm
- 1477g
- 72p

Avondland

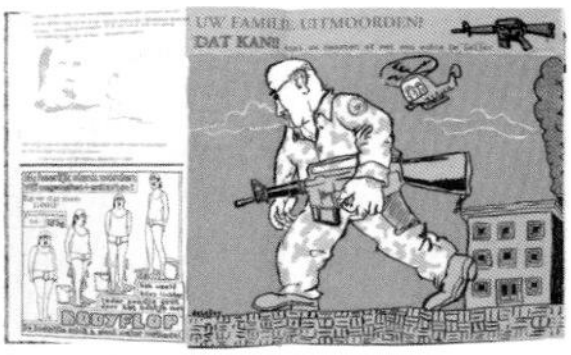
UW FAMILIE UITMOORDEN?
DAT KAN!

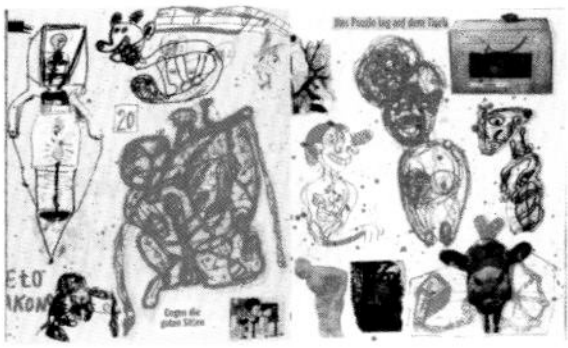

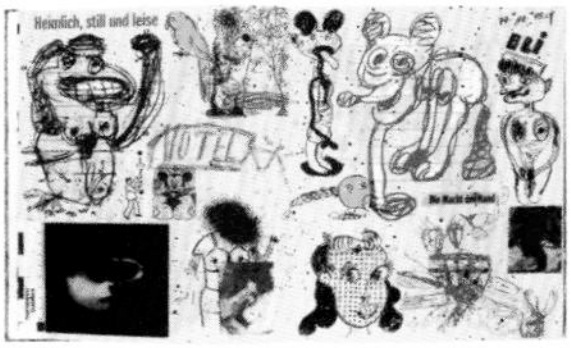
Heimlich, still und leise

DO NOT STAND ABOUT HERE
WHAT YOU SEE ON THE STREET DON'T BLAME ON THE PIGEONS
COURTESY WK

Abendland

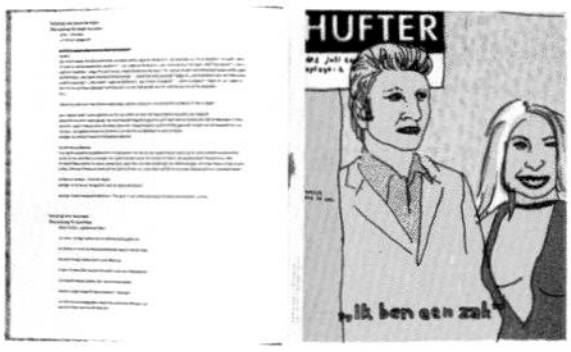
HUFTER
„Ik ben een zak

"Ik vind het gewoon lekker om in mijn broek te kakken!"

Geduld

HIJ WAS NIET WEG TE DENKEN MET ZIJN FLUIT OP TV

SCHITTERENDE BORSTEN OVERAL MAAR NAUWELIJKS ZWEMBADEN
VANDAAG IN 'OOG IN OOG MET VARKENS'...DE HYENA.

IK HAD HET AL OPGEGEVEN VOORDAT ER ZELFS MAAR SPRAKE WAS VAN HET ONZEDELIJK BETASTEN VAN WAT DAN OOK
KLES
NOBODY

SCHLUSS

et rol
e van

nobody forever

本书是一本有关图书的书，介绍了由哈里·凯斯勒伯爵（Grafen Harry Kessler）的 Cranach 出版社出版的书籍。书中探讨了字体的选用、插图、封面工艺等方面。大开本圆脊腔背硬精装的设计，体现了历史文献的厚重感，同时便于厚重的书籍可以完全打开。书函外壳为黑色，封面红色布面装裱，辅以金色烫纹和书名，内敛而稳重。环衬的中灰色与书签带的银色相得益彰。内文采用多栏设计，其中正文版心部分为双栏，并配以单栏 1/2 宽度的侧栏加注释。文中大量的花式风格的斜体设计构成了亮点所在。全书排版规整，印刷精美，装订考究，极具收藏价值。

It is a book about books that are published by Cranach press, founded by Grafen Harry Kessler. The book discusses the selection of fonts, illustrations, cover design and other aspects. Large format with rounding flexiback and hardcover, the design not only demonstrates the heaviness of historical documents, but also it is convenient for heavy book to be fully opened. The slipcase is black while the cover is mounted with red cloth with golden hot stamped title, introverted and stable. The neutral gray end paper and the silver bookmark band complement each other. The text is divided into multi-columns; for example, the body part is double columned with a sidebar 1/2 width of single column to add comments. A large number of fancy style italics in the text constitute the highlights. The book is well arranged, beautifully printed, elegantly bound and highly collectable.

荣誉奖

书籍的艺术

——哈里·凯斯勒伯爵的 Cranach 出版社

德国

A classical book about a classical publisher. Interesting type-area using two columns and a marginal column for footnotes. The text elements seen together roughly form a square. A deliberately large space beneath the type-area is left free. Even the pagination is placed high up. The colour red is used as a structural text element. The perfection of the setting and lithography speaks for itself. Large illustrations are given a page to themselves, smaller ones are integrated into the text.

Honorary Appreciation

Das Buch als Kunstwerk.

Die Cranach Presse des Grafen Harry Kessler

Germany

Sabine Golde

John Dieter Brinks (Ed.)

Triton Verlag, Laubach / Berlin

315×255×50mm

2923g

453p

DAS BUCH ALS KUNSTWERK

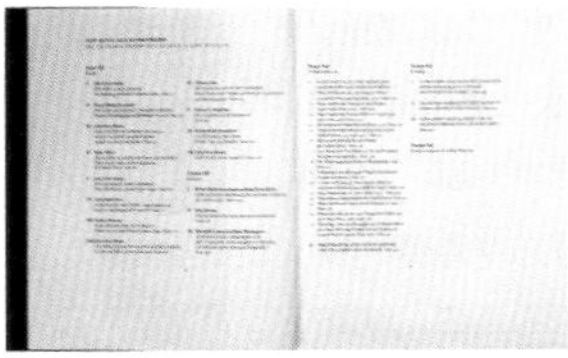

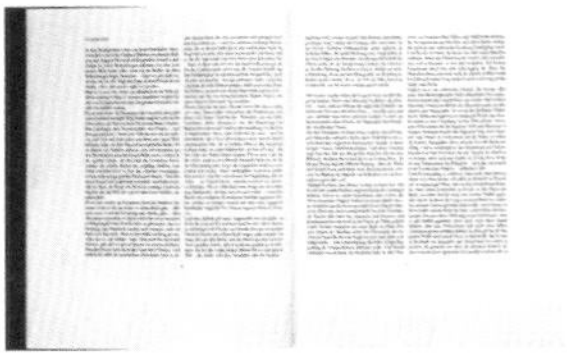

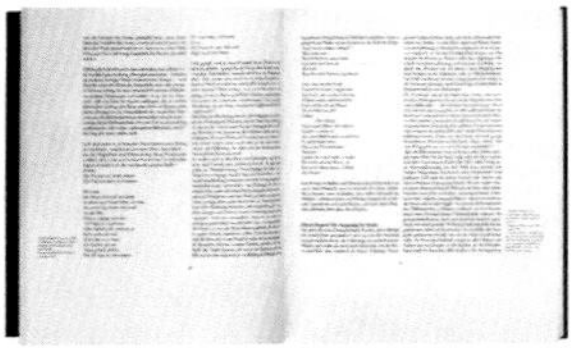

DIE ODYSSEE
DIE ODYSSEE

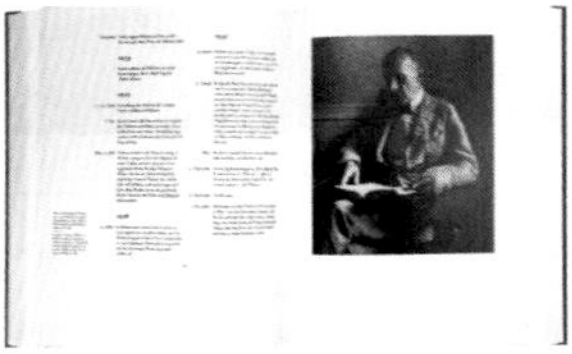

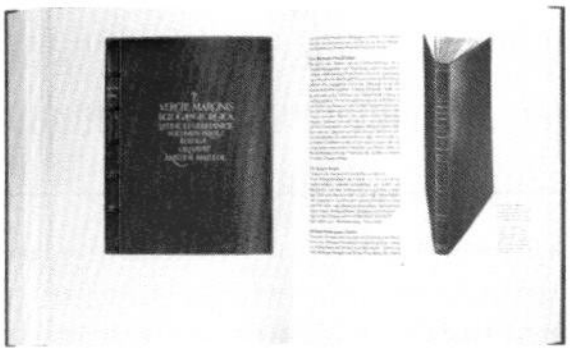

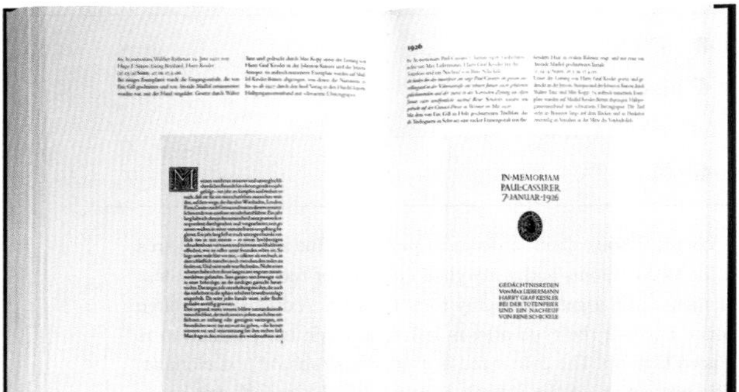
1926
IN·MEMORIAM
PAUL·CASSIRER
7·JANUAR·1926
GEDÄCHTNISREDEN
VON MAX LIEBERMANN
HARRY GRAF KESSLER
BEI DER TOTENFEIER
UND EIN NACHRUF
VON RENE SCHICKELE

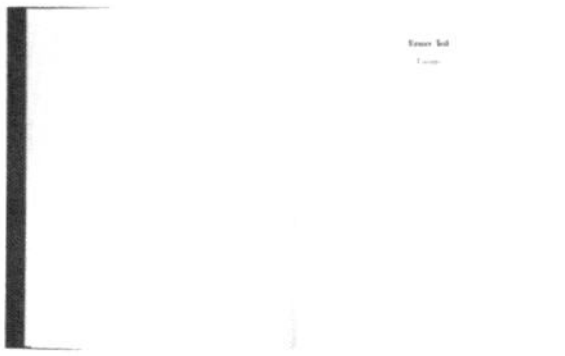
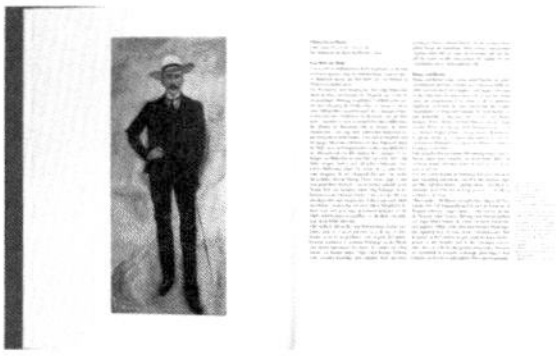

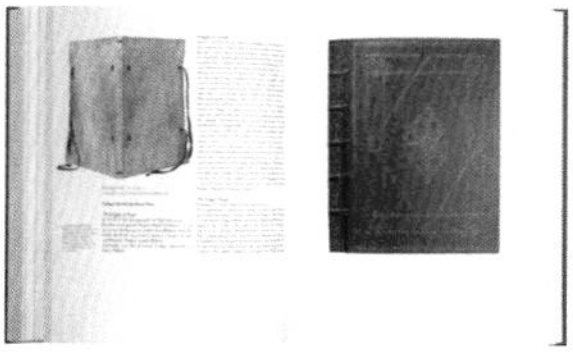

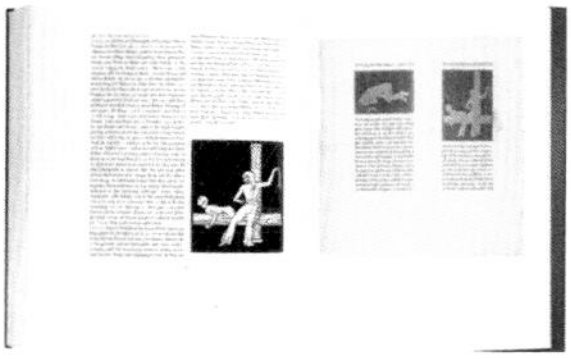

chal

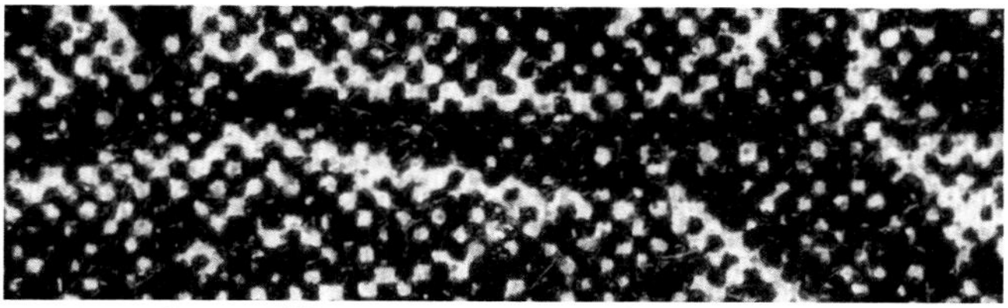

Rudolf Alexa

本书是一本介绍荷兰平面设计师皮特·吉拉德（Piet Gerards）的书籍。全书主要分为两部分，前一部分通过作品来分析其创作理念，后一部分则是个人成长、求学、思想和事业发展的生平传记。书籍采用锁线腔背软精装，便于展开翻阅。封面采用牛皮卡纸印黑，封面背面暗藏设计师书架的照片。全书以双语排版贯穿内外，正文部分通过大量引注页码，引证各个章节中的作品图片，由此而呈现出一种“链接式”的富文本特征，有如置身在信息通达的互联网时代。前一部分通过不同领域的平面设计作品，无衬线与衬线体的混用，来区分不同的内容部分，并通过鲜亮的大色块作为隔页来划分，此部分为涂布纸四色印刷。后一部分采用胶版纸印单黑，衬线体的使用只配合无衬线小标题，更适合大文字量阅读。

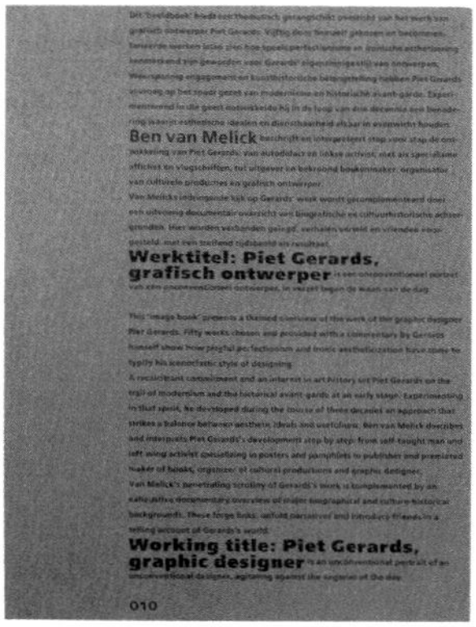
Ben van Melick
Werktitel: Piet Gerards, grafisch ontwerper
Working title: Piet Gerards, graphic designer
010

This is a book about Dutch graphic designer Piet Gerards. The book is divided into two parts: the first part analyzes his creative ideas through his works, and the second part is a biography of his personal growth, study, thoughts and career development. The book is bounded by thread sewing with flexiback and soft hardback. It is easy to unfold the pages. The cover is printed black with kraft paperboard and the photo of the designer's bookshelves is hidden on the reverse side of the cover. The text is composed bilingually; through quoting a mass of page numbers in the body text, the pictures of his works are cited in each chapter, thus showing a 'linked' text characteristic, as if in the accessible Internet era. In the first part, serif and sans serif fonts are used to distinguish different content parts, and pages with large bright color blocks are inserted to divide each chapter. This part is four-color printing on coated paper. The latter part is printed in black on offset paper, using serif body with no serif subtitle, which is more suitable for text reading.

- 荣誉奖
- 头衔：皮特·吉拉德，平面设计师
- 荷兰

This paperback is a monography of the well-known designer. It is particularly impressive on account of its clear, very finely detailed typography and excellent pictorial language. Well conceived font mixtures and a variety of papers determine the structure, resulting in an overall impression which is full of interest. The bilingual, typographical cover made from natural card the colour of wrapping-paper is innovative whilst remaining unassuming.

- Honorary Appreciation
- Werktitel: Piet Gerards, grafisch ontwerper
- The Netherlands
- Bureau Piet Gerards
- Ben van Melick
- Uitgeverij 010 Publishers, Rotterdam
- 231×168×25mm
- 802g
- 366p

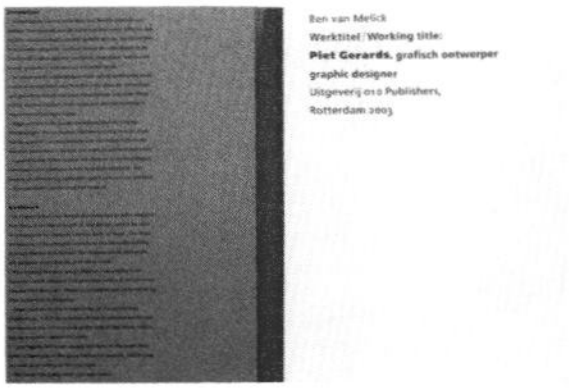
Ben van Melick
Werktitel / Working title:
Piet Gerards, grafisch ontwerper
graphic designer
Uitgeverij 010 Publishers,
Rotterdam 2003

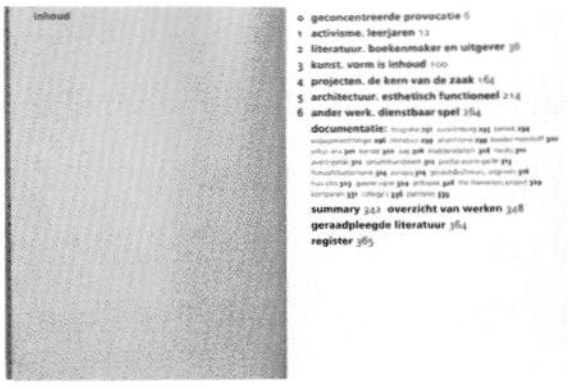
inhoud
0 geconcentreerde provocatie 6
1 activisme. leerjaren 12
2 literatuur. boekenmaker en uitgever 36
3 kunst. vorm is inhoud 100
4 projecten. de kern van de zaak 164
5 architectuur. esthetisch functioneel 214
6 ander werk. dienstbaar spel 264
documentatie:
summary 342 overzicht van werken 348
geraadpleegde literatuur 364
register 365

0 geconcentreerde provocatie

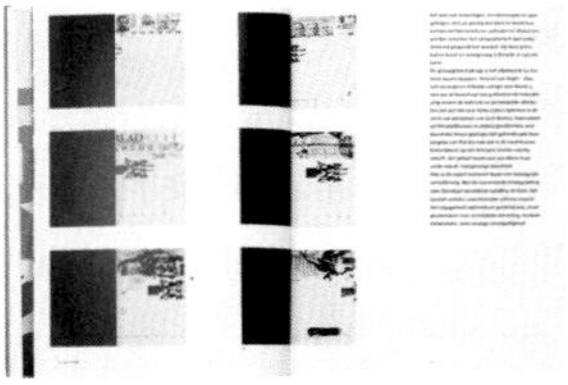

GALERIA ZACHĘTA
WARSZAWA WSPÓLNIE Z
KUNSTEN- MIN
CENTRUM SIONE FIRMY
HEERLEN (NL) POR
28/29/30 TRE
SIERPNIA TO
2004 WEJ
S.I. WITKIEWICZ

15.8-14.10.01
john
hejduk
fabriceren
in je hoofd

Protesterende
studenten
gaan vrij-uit
Engagement/religie
23/24
SEPTEMBER
VREDESWEEK
1972
ANTI
NATO
MARS
HEERLEN
BRUNSSUM
SCHINNEN
lijst 23
NEO
FASCISTEN
NEDERLAND
jozef luns
CHILI
CHILIDAG
ZONDAG 11 SEPTEMBER

effen

ANARCHISME/KULTUUR
4 NOVEMBER
BANAAL

Canetti

anatol stern
europa
mieczyslaw szczuka

Na
nú fragmenten 10
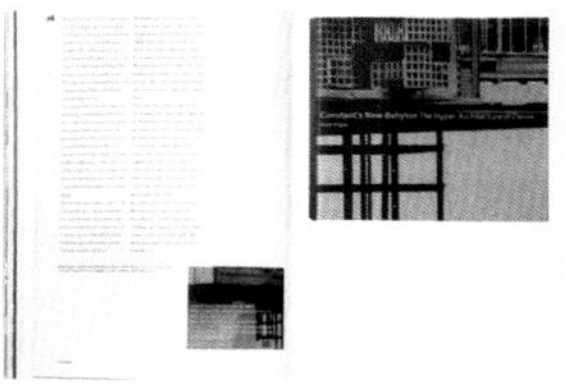

post-
ex-
sub-
dis-

Banaal: Anarcho-
cultureel project
in Kleine Comedie

alisme e
roep [300] De ironische

004 Ha4

这是一本有关梦境的书，撰稿人和摄影师合作，通过文字和摄影的方式来再现和研究梦境，在纸面上进行了一场关于梦的实验。本书的装订形式为锁线硬精装，护封铜版纸四色印刷覆哑光膜，内封铜版纸四色印刷覆哑光膜后裱覆卡板。护封与内封的照片是同一场景稍有变化的两张照片，通过微妙的人物变化和虚化差异，暗示亦实亦虚、无法参透的梦境世界。环衬采用大红色胶版纸，在冷色系封面和完全单黑印刷的内页间，仿佛一道屏障，提醒着读者接下来虚幻的世界。内页哑面涂布纸单黑印刷，使用照片与文字穿插的形式展现各类梦境的内容：美好的梦、梦中的动物、看不清的梦、噩梦、春梦、梦中的攻击、经常做的梦、飞翔的梦、坠落的梦、溺水的梦、清醒的梦，以及梦见死亡。文字部分单栏版心设置，口述梦境的文字采用斜体并在左侧设置巨大的缩进。摄影图片大多具有较宽的白边，通过大量连贯或同一场景的照片，极力再现不那么真实的梦境。||||||||||||||||||||||||||||

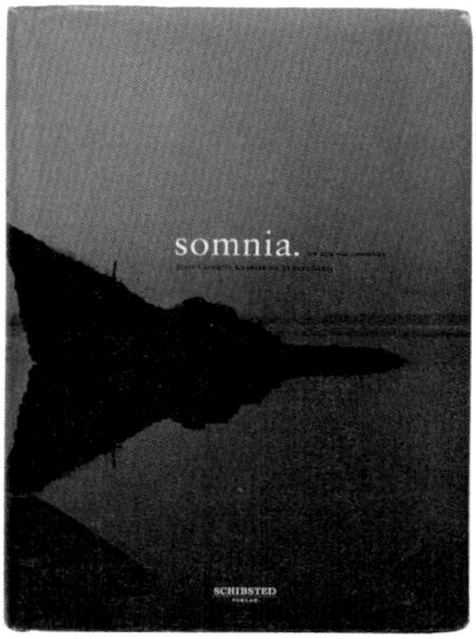
somnia.
SCHIBSTED

This is a book about dreams. The author and the photographer cooperated to reproduce and study dreams by means of words and photography, and carried out an experiment of dreams on paper. This hardback book is bound by thread sewing. The laminated jacket and inside cover are both made of coated art paper printed in four colors, while the latter one is mounted with cardboard. The pictures on the jacket and the cover are two photos with slight changes in the same scene. Through subtle changes in characters, it implies that a dream world is neither real nor visible. The endpapers are made of red offset paper, which acts as a barrier between the cold-colored cover and the completely black inner pages to remind the readers of the coming illusory world. The inside pages are made of coated paper printed in black. The contents of various dreams are presented with the mixture of photos and text: beautiful dreams, animals in dreams, blurry dreams, nightmares, dreams with sexual fantasy, attacks in dreams, frequent dreams, flying dreams, falling dreams, drowning dreams, lucid dreams, and dreams of death. The dictation of dream is italicized and has a huge indentation on the left. Most photographs have wide white edges. Through a large number of coherent pictures in the same scene, the book tries to reproduce less real dreams.

◎ 荣誉奖

《 梦境——一本有关梦的书

◐ 挪威

Particulary impressive about this book is its innovative and sensitive treatment of the photography. The positioning of pictures within the book's layout and the use of white frames and lines reinforces the character of the photography. The black and white inner design is impressively contrasted by the red end-paper. The typography is restrained and allows the pictures to speak for themselves. The calm design of the dust-jacket gives the book exactly the right tone.

◎ .... Honorary Appreciation

《 .... somnia. En bok om drømmer

◐ .... Norway

D .... Geir Henriksen

△ ... Julie Cathrine Knarvik, Anders Aabel

|||| ...... Christian Schibsteds Forlag, Oslo

□ .... 245×176×28mm

⌷ 926g

≣ 232p

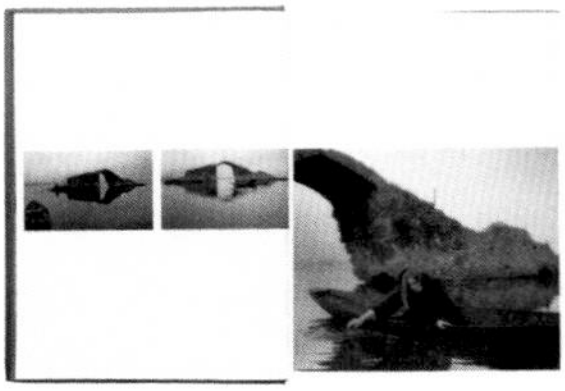

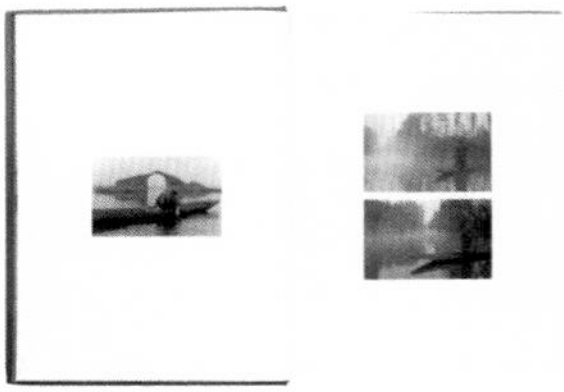

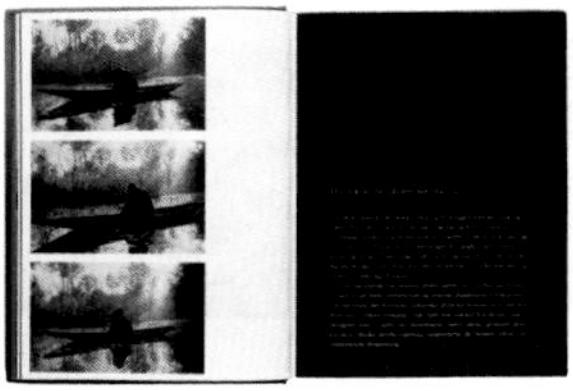

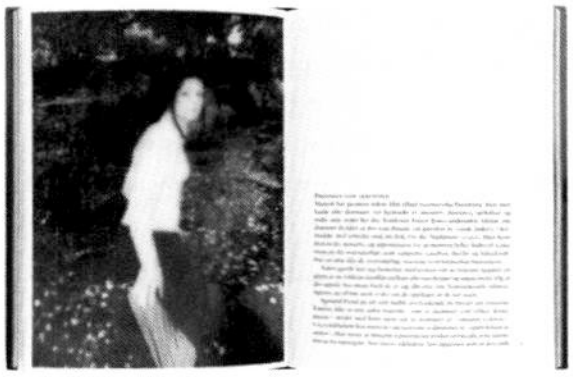

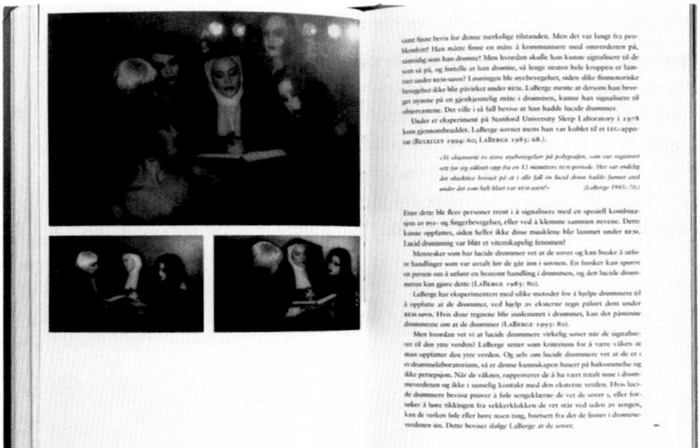

samt finne bevis for denne merkelige tilstanden. Men det var langt fra problemfritt! Han måtte finne en måte å kommunisere med omverdenen på, samtidig som han drømte! Men hvordan skulle han kunne signalisere til de som så på, og fortelle at han drømte, så lenge nesten hele kroppen er lammet under REM-søvn? Løsningen ble øyebevegelser, siden slike finmotoriske bevegelser ikke blir påvirket under REM. LaBerge mente at dersom han beveget øynene på en gjenkjennelig måte i drømmen, kunne han signalisere til observatørene. Det ville i så fall bevise at han hadde lucide drømmer.

Under et eksperiment på Stanford University Sleep Laboratory i 1978 kom gjennombruddet. LaBerge sovnet mens han var koblet til et EEG-apparat (Bulkeley 1994: 60; LaBerge 1985: 68.).

> *«Vi observerte to store øyebevegelser på polygrafen, som var registrert rett før jeg våknet opp fra en 13 minutters REM-periode. Her var endelig det objektive beviset på at i alle fall én lucid drøm hadde funnet sted under det som helt klart var REM-søvn!»* *(LaBerge 1985: 70.)*

Etter dette ble flere personer trent i å signalisere med en spesiell kombinasjon av øye- og fingerbevegelser, eller ved å klemme sammen nevene. Dette kunne oppfattes, siden heller ikke disse musklene blir lammet under REM. Lucid drømming var blitt et vitenskapelig fenomen!

Mennesker som har lucide drømmer vet at de sover og kan huske å utføre handlinger som var avtalt før de går inn i søvnen. En forsker kan spørre en person om å utføre en bestemt handling i drømmen, og den lucide drømmeren kan gjøre dette (LaBerge 1985: 80).

LaBerge har eksperimentert med ulike metoder for å hjelpe drømmere til å oppfatte at de drømmer, ved hjelp av eksterne tegn påført dem under REM-søvn. Hvis disse tegnene blir innlemmet i drømmen, kan det påminne drømmerne om at de drømmer (LaBerge 1995: 80).

Men hvordan vet vi at lucide drømmere virkelig sover når de signaliserer til den ytre verden? LaBerge setter som kriterium for å være våken at man oppfatter den ytre verden. Og selv om lucide drømmere vet at de er i et drømmelaboratorium, så er denne kunnskapen basert på hukommelse og ikke persepsjon. Når de våkner, rapporterer de å ha vært totalt inne i drømmeverdenen og ikke i sanselig kontakt med den eksterne verden. Hvis lucide drømmere bevisst prøver å føle sengeklærne de vet de sover i, eller forsøker å høre tikkingen fra vekkerklokken de vet står ved siden av sengen, kan de verken føle eller høre noen ting, bortsett fra det de finner i drømmeverdenen sin. Dette beviser ifølge LaBerge at de sover.

VIK OG AND

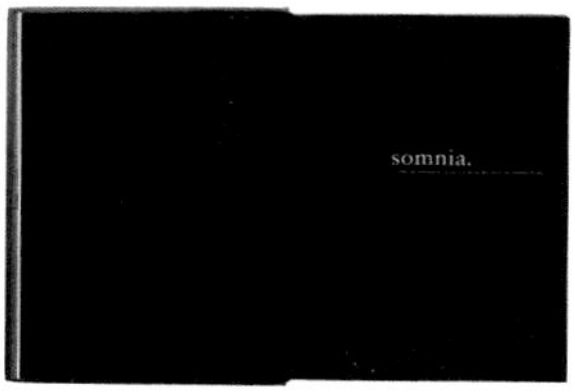
somnia.

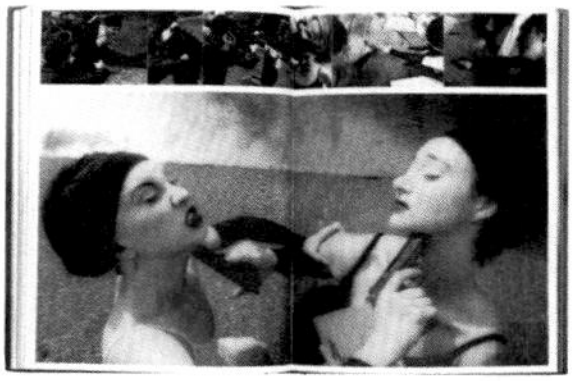

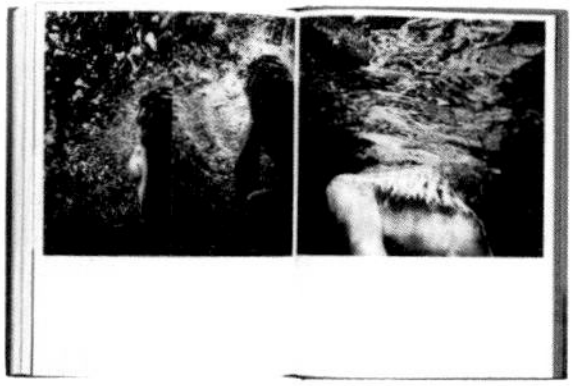

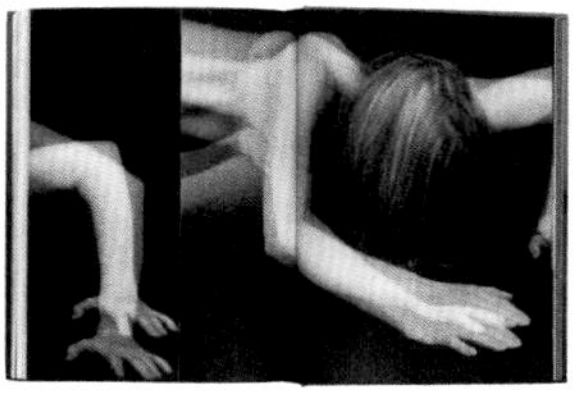

net c

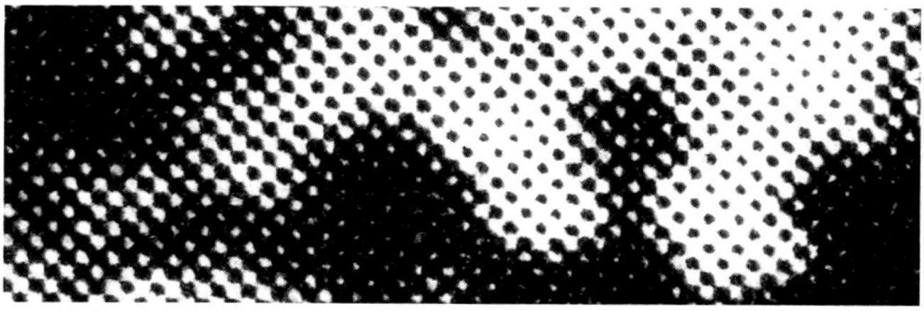

klar for natt

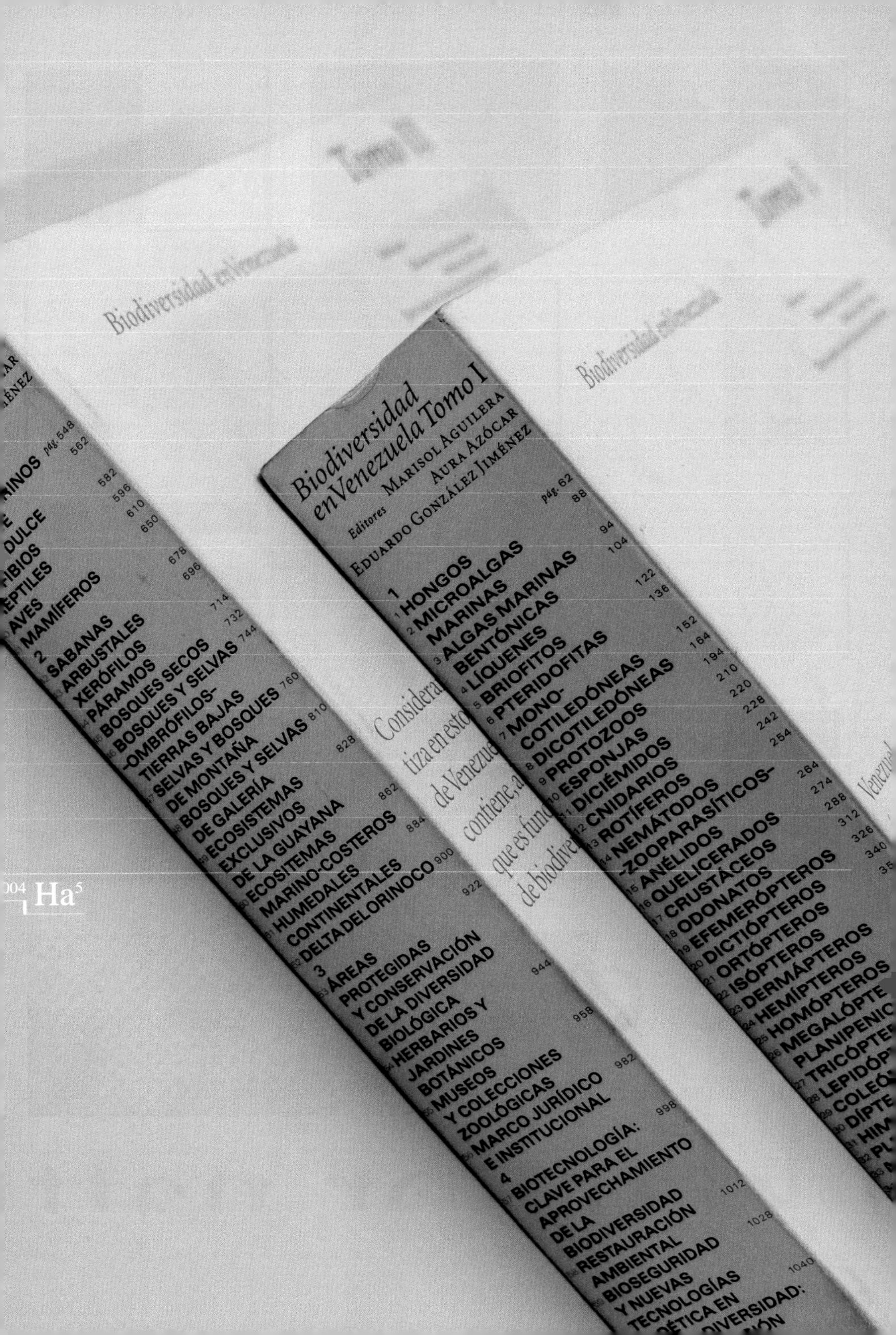

Biodiversidad en Venezuela
Biodiversidad en Venezuela Tomo I
Editores Marisol Aguilera Aura Azócar Eduardo González Jiménez
pág. 62
1 HONGOS
MICROALGAS MARINAS
ALGAS MARINAS BENTÓNICAS
LÍQUENES
BRIOFITOS
PTERIDOFITAS
MONO-COTILEDÓNEAS
DICOTILEDÓNEAS
PROTOZOOS
ESPONJAS
DICIÉMIDOS
CNIDARIOS
ROTÍFEROS
NEMATODOS
ZOOPARASÍTICOS
ANÉLIDOS
QUELICERADOS
CRUSTÁCEOS
ODONATOS
EFEMERÓPTEROS
DICTIÓPTEROS
ORTÓPTEROS
ISÓPTEROS
DERMÁPTEROS
HEMÍPTEROS
HOMÓPTEROS
MAMÍFEROS
2 SABANAS
ARBUSTALES XERÓFILOS
PÁRAMOS
BOSQUES SECOS
BOSQUES Y SELVAS OMBRÓFILOS-TIERRAS BAJAS
SELVAS Y BOSQUES DE MONTAÑA
BOSQUES Y SELVAS DE GALERÍA
ECOSISTEMAS EXCLUSIVOS DE LA GUAYANA
ECOSITEMAS MARINO-COSTEROS
HUMEDALES CONTINENTALES
DELTA DEL ORINOCO
3 ÁREAS PROTEGIDAS Y CONSERVACIÓN DE LA DIVERSIDAD BIOLÓGICA
HERBARIOS Y JARDINES BOTÁNICOS
MUSEOS Y COLECCIONES ZOOLÓGICAS
MARCO JURÍDICO E INSTITUCIONAL
4 BIOTECNOLOGÍA: CLAVE PARA EL APROVECHAMIENTO DE LA BIODIVERSIDAD
RESTAURACIÓN AMBIENTAL
BIOSEGURIDAD Y NUEVAS TECNOLOGÍAS

004 Ha5

南美的委内瑞拉地处南安第斯山脉、亚马孙平原和加勒比海之间，自然地貌丰富，具有极丰富的生物多样性。这套分为上下卷的大部头就是对委内瑞拉自然地理、生态保护和可持续利用方面的研究。本套书的装订形式为锁线胶平装。封面白卡纸蓝黑双色印刷，封面直接裱贴灰绿色的环衬，环衬上通过小圆孔来区分上下卷，同样的手法也用在了这位设计师获得的 2008 年金字符奖《委内瑞拉感性地理学史》的封面上（可在本书相关页面查看）。内页采用哑面涂布纸四色印刷，以单黑印刷的文本和插图为主，配以少量四色照片。书籍的内容共四个篇章，分别为：生物多样性、生态多样性、可持续利用的方式和保障，以及可持续利用上存在的技术和伦理挑战。本书章节较为繁复，并包含大量索引、参考文献、表格、技术插图等元素，但版面层次和细节分明。

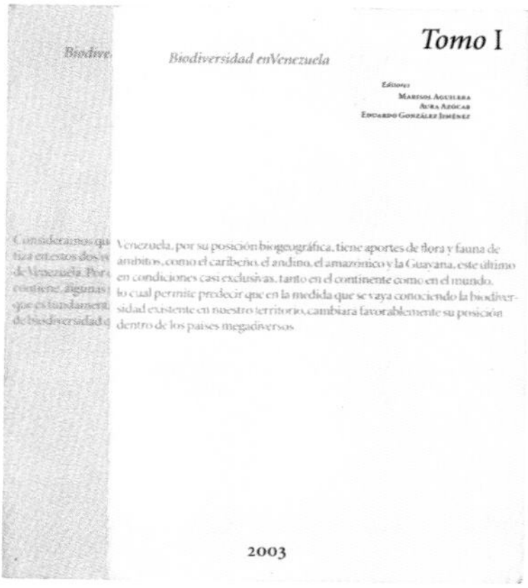
Tomo I
Biodiversidad enVenezuela
2003

Located between the South Andes Mountains, the Amazon Plain and the Caribbean Sea, Venezuela in South America is extremely rich in natural features and biological diversity. The major part of this tome including two volumes is the research of Venezuela's geography, ecological protection and sustainable use. This set of paperback books is glue-bound with thread sewing. The cover is printed on white cardboard in blue and black, directly mounted by gray-green endpapers. Small circular holes in the endpapers distinguish the first volume and the second volume. The same technique is also used on the cover of Venezuela's History of Sensitive Geography, the winner of Golden Letter in 2008, designed by the same designer. The inside pages are printed in four colors on matte coated paper, with text and illustrations printed in monochrome black and a small number of photographs in four colors. There are four chapters in the book: biodiversity, ecological diversity, ways and guarantees of sustainable use, and technological and ethical challenges in sustainable use. The chapters of this book are complex, containing a large number of elements such as index, references, tables, technical illustrations, but the layout and details are clear.

- 荣誉奖
- 委内瑞拉的生物多样性
- 委内瑞拉

- Honorary Appreciation
- Biodiversidad en Venezuela
- Venezuela
- Álvaro Sotillo, Luis Giraldo
- Marisol Aguilera, Aura Azócar, Eduardo González Jiménez (Ed.)
- Fundación Polar, Ministerio de Ciencia y Tecnología, Fondo Nacional de Ciencia, Tecnología e Innovación (fonacit), Caracas
- 248×173×85mm
- 1663g+1695g
- 1080p

The two book-designers have created an innovative typographical language in this two-volume specialised book. The information is well presented. Finely detailed typography, good quality of setting. Equally high is the quality of the graphics and illustrations which are well integrated into the layout. The outside design is restrained, purely typographical, yet innovative: the contents can already be viewed on the book's spine.

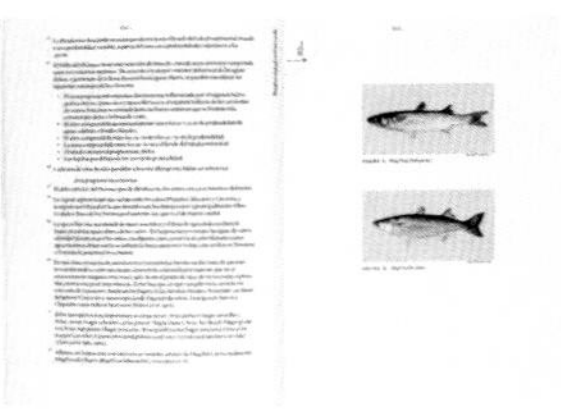

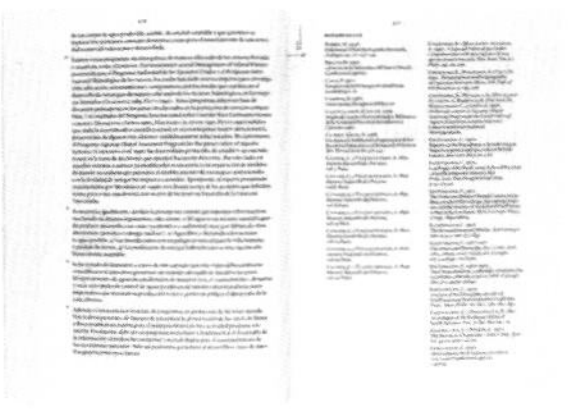

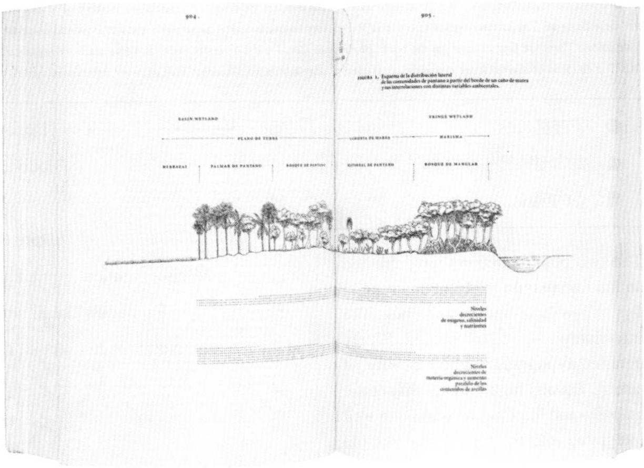

904.
905.
FIGURA 1. Esquema de la distribución lateral
de las comunidades de pantano a partir del borde de un caño de marea
y sus interrelaciones con distintas variables ambientales.
BASIN WETLAND
FRINGE WETLAND
PLANO DE TURBA
MARISMA
HERBAZAL
PALMAR DE PANTANO
BOSQUE DE MANGLAR
Niveles
decrecientes
de oxigeno, salinidad
y nutrientes
Niveles
decrecientes de
materia orgánica y aumento
paulatino de los
contenidos de arcillas

núme

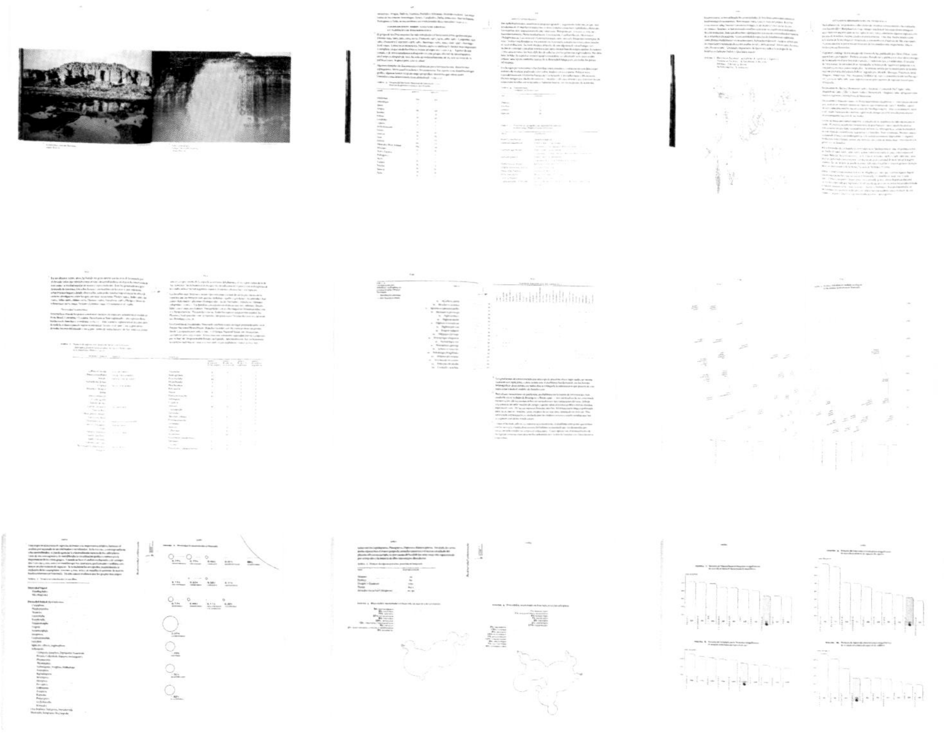

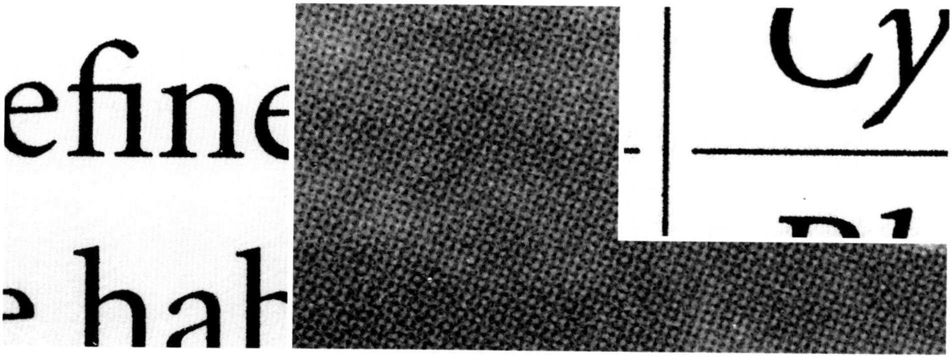
efine
Cy
e hab

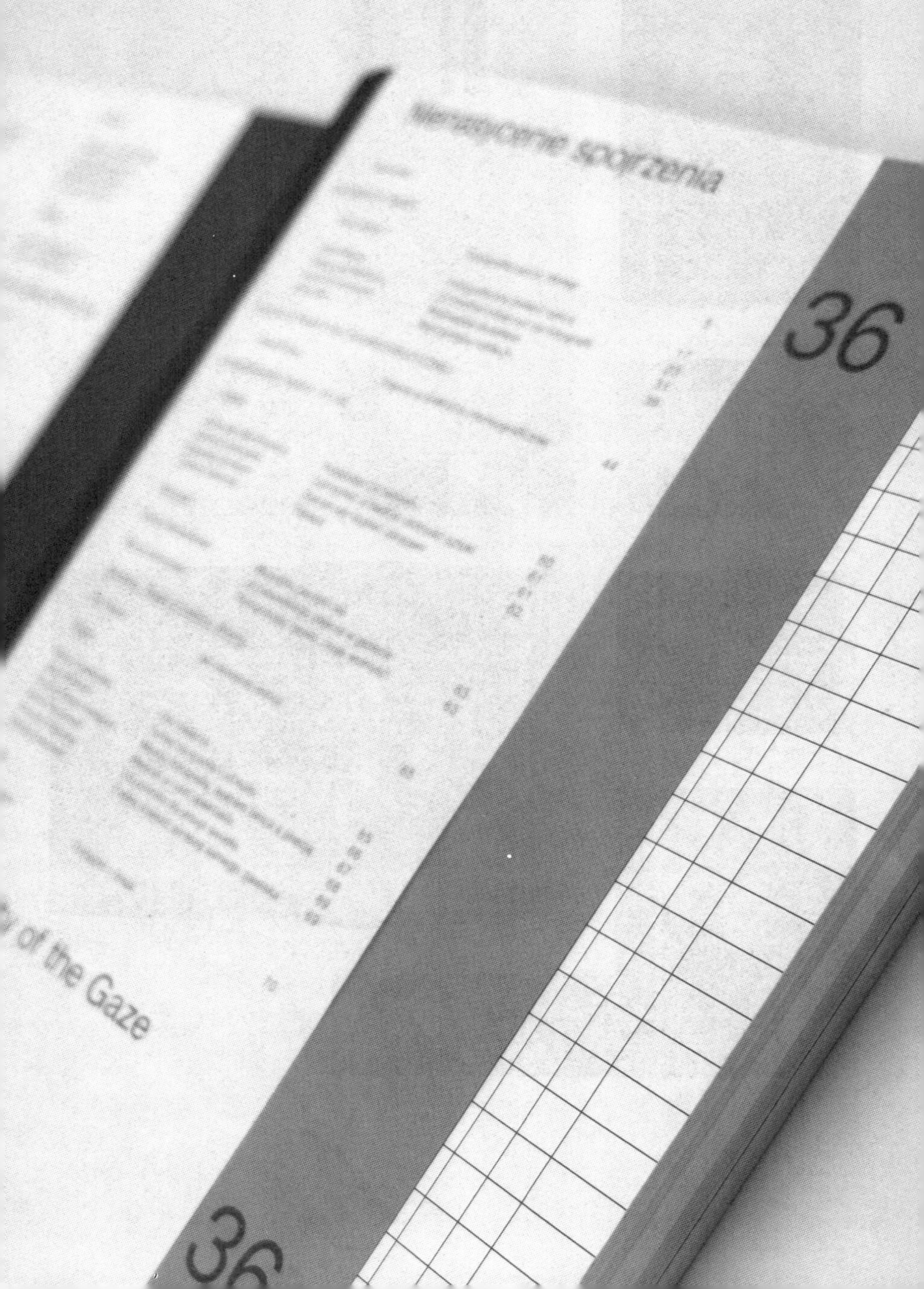
36
of the Gaze

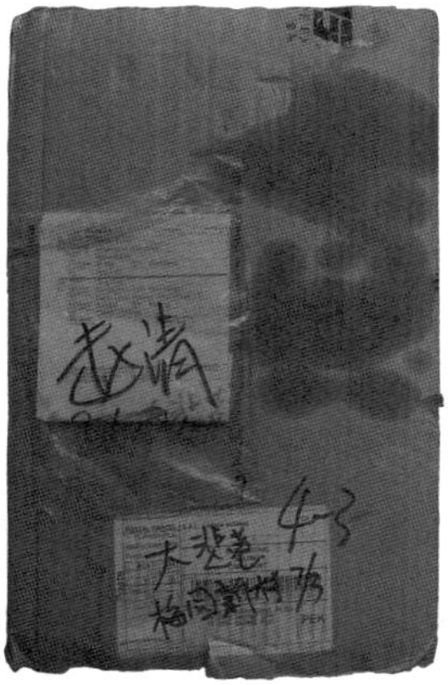

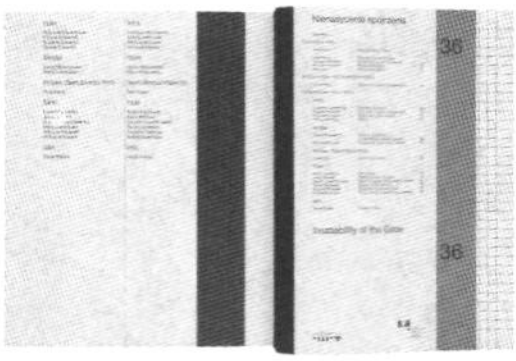
Nienasycenie spojrzenia
35
Insatiability of the Gaze
36

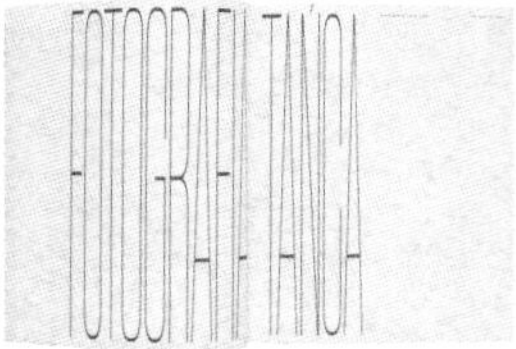
FOTOGRAFIA TAŃCA

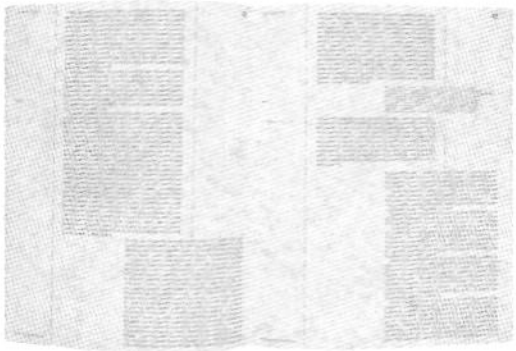

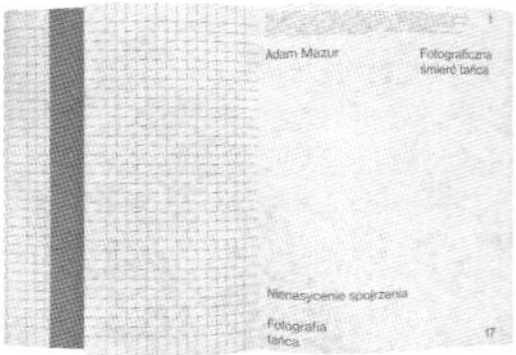
Adam Mazur
Fotograficzna śmierć tańca
Nienasycenie spojrzenia
Fotografia tańca
17

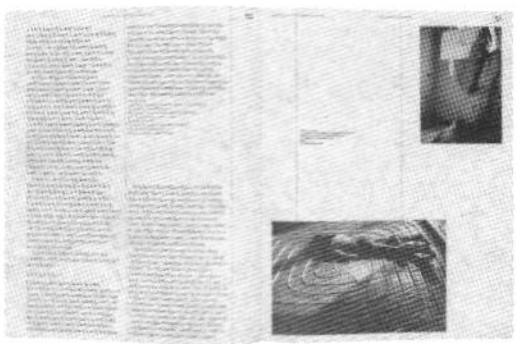

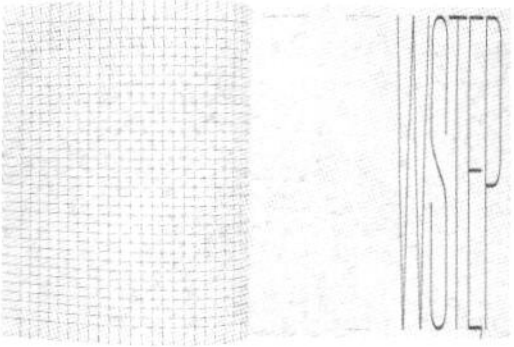
WSTĘP

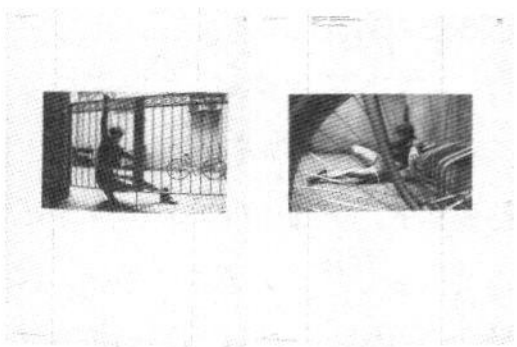

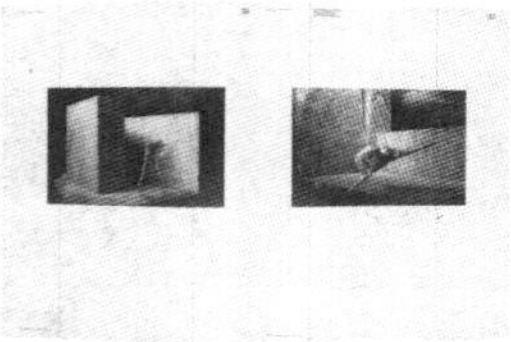
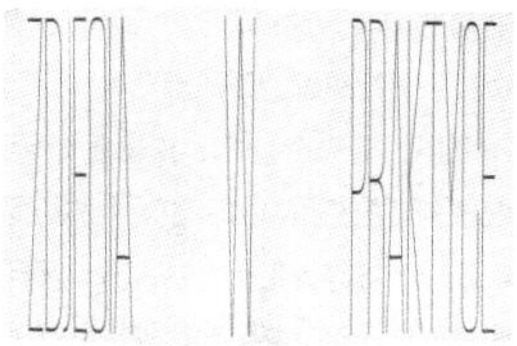
ZDJĘCIA W PRAKTYCE
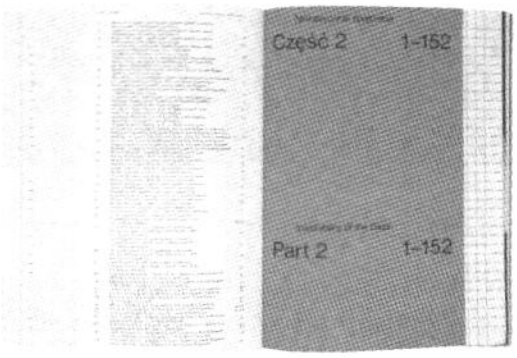
Część 2 1–152
Part 2 1–152

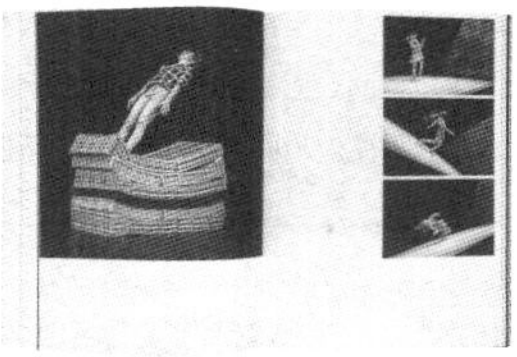
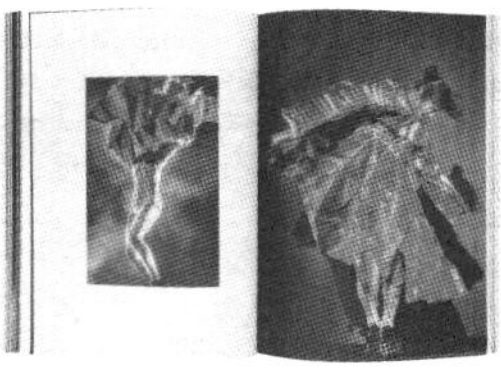
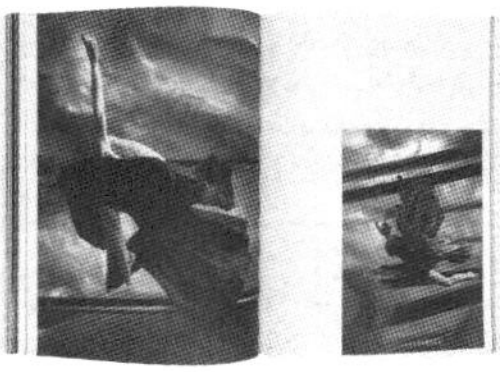

ania

这本书的编者安娜·克罗丽卡（Anna Królica）是一位波兰策展人和舞蹈理论家，她长期投身于波兰舞蹈的文化推广活动。在本书中她邀请了评论家、舞蹈家、演员和长期拍摄舞蹈的摄影师进行访谈，共同探讨舞蹈摄影相关的话题。本书的装订形式为锁线胶平装，封面采用特种纸印黑，标题烫金，内页混合装订了无涂布的胶版纸、带涂布的艺术纸、光面铜版纸和带纹理的艺术纸，用以呈现不同部分的内容。书籍的内容分为两部分，页码相互独立，采用艺术纸短夹页目录进行总领。第一部分为访谈，包括照片中能看到什么、舞蹈摄影的光影技术、摄影师如何进入表演的节奏、如何记录舞者的情绪、舞台和后台跳舞的异同、舞蹈摄影中“摄影眼”的培养等话题。这部分单黑印刷为主，文字混排黑白照片，夹插少量彩色照片控制阅读节奏。文中设定了一种四栏版式，两条竖线有如摄影中的三分法构图线，但排版并不拘泥于竖线划分的空间，而是通过整段缩进和照片的排列去打破空间。第二部分为舞蹈摄影作品，四色印刷，各位摄影师关注点不同，呈现出的作品也差异很大。||||||||||||||

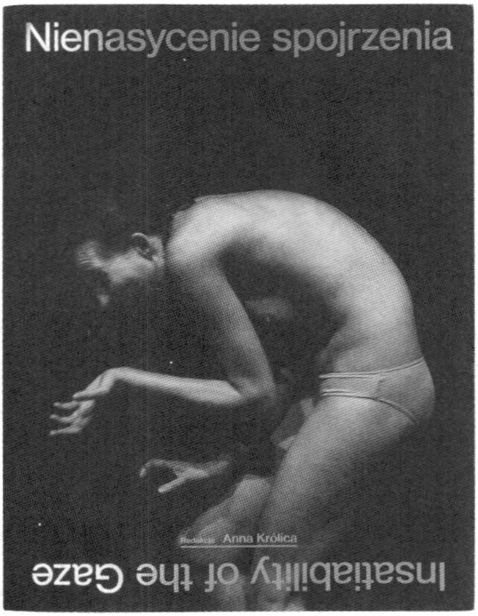
Nienasycenie spojrzenia
Anna Królica
Insatiability of the Gaze

Anna Królica, the editor of the book, is a Polish curator and dance theorist who has devoted herself to the cultural promotion of Polish dance for a long time. In this book, she invited critics, dancers, actors and photographers to talk about dance photography. The paperback book is glue bound with thread sewing. The cover is made of special paper printed in black with gold hot stamped title. The inside pages consist uncoated offset paper, coated art paper, glossy art paper and textured art paper to present different parts of the content. The book is divided into two parts with independent page numbers, and the art paper short folder catalogue is used to guide the readers. The first part contains interviews, including what can be seen in the photos, the light and shadow technology of dance photography, how the photographer enters the rhythm of performance, how to record the dancers' mood, the similarities and differences of stage performance and backstage dancing, and the cultivation of 'photographic eyes' in dance photography. This part is printed in single black and the text mixed with black-and-white photos, interspersed with a few color photos to control the reading rhythm. A four-column format is set up. The two vertical lines are like the three-component line in photography, but the layout is not restricted to the space divided by the vertical lines, but breaks the space by indenting the whole section and arranging the photos. The second part contains dance photography works with four-color printing. Since the photographers have different concerns, the works show great differences.

荣誉奖
贪婪的凝视
波兰

Honorary Appreciation
Nienasycenie spojrzenia / Insatiability of the Gaze
Poland
Ryszard Bienert
Anna Królica (Ed.)
Centrum Sztuki Mościce, Tarnów / Instytut Muzyki i Tańca, Warsaw
319×238×28mm
1607g
288p

Specialist articles on dance and photography are grouped in the first section; an atlas of illustrations showing dance photography in Poland forms the second half of this impressive Swiss brochure. The book opens with the picture of a dancer on the inside of the cover flap, the blurred motion shrouding her actions like a veil. Multiple exposure with phase shift captures the temporal sequence in a single picture. The spine of the innerbook has a black lining; the flat inner spine of the cover is white where the photo ends. The contents are listed on a narrower page, with grey structured paper projecting from underneath for the beginning of a chapter, itself narrower as well. Visible underneath, now across the full width of the format, there is an ornamental page with line grid. The titles of the articles are printed in ultra-light grotesque capitals whose height — despite the style being already narrowed — is reduced even further by several hundred per cent. It transforms the graphemes of a word into graceful, linear spatial investigations. Furthermore, there is alternation between matt white and satin gloss paper materials, the grey paper, and further series of narrower pages. In an unusual type area, the main column set in the middle is marked on both sides by vertical lines. These are broken by whole paragraphs; you could describe them as breaking rank. In the picture section, the positioning of the photographs on the pages creates such tension that it potentiates the dynamism of these choreographic depictions.

Ich möchte an dieser Stelle den Designern △

dieser schönen Bücher danken ,

Erst ihre Designs haben es möglich gemacht ,

« *das hier vorgelegte Buch herauszubringen* .

Thanks to the designers of these beautiful books, because of your design,

this book was born and presented .

在此要感谢这些美丽书的设计师们
因为你们的设计
才有了这本书的诞生与呈现

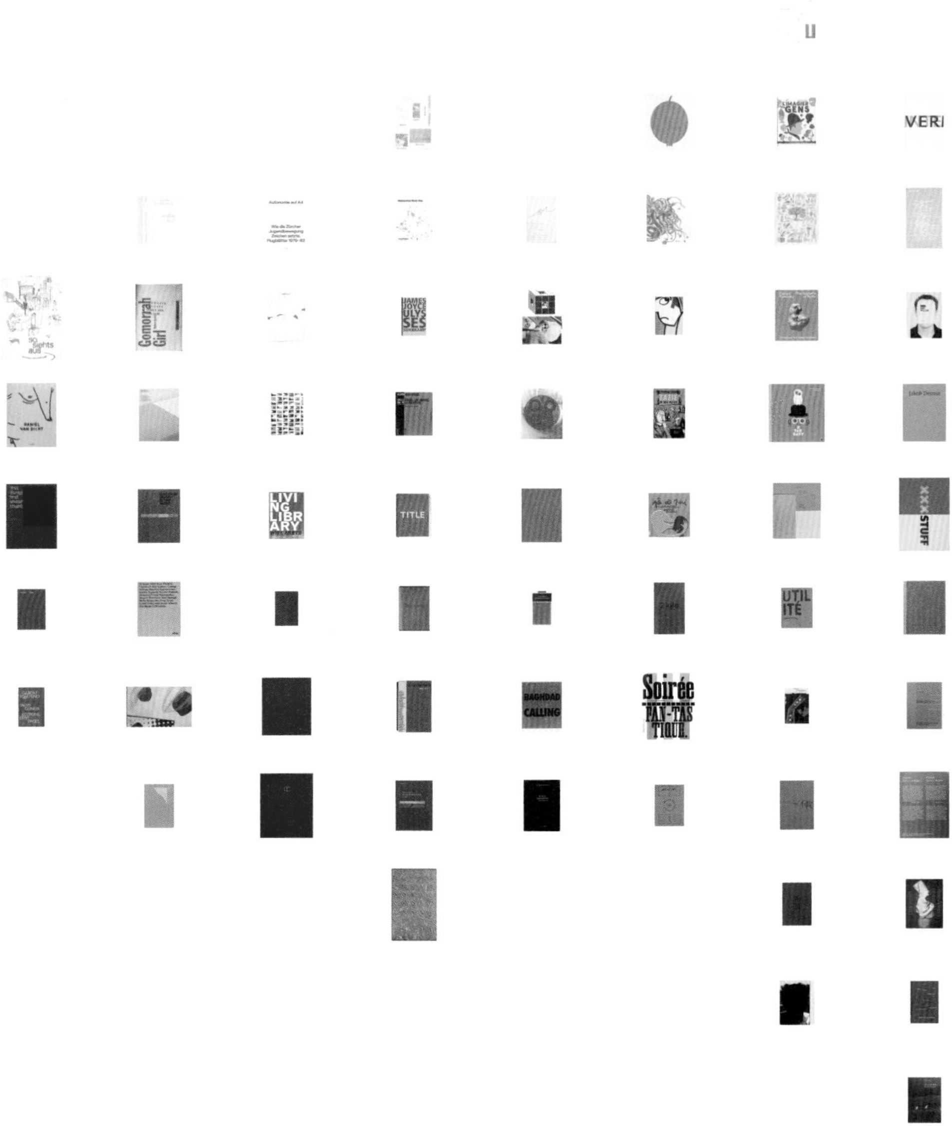
L'IMAGIER GENS
Autonomie auf A4
Wie die Zürcher Jugendbewegung Zeichen setzte. Flugblätter 1979–82
Gomorrah Girl
JAMES JOYCE ULYSSES
LIVING LIBRARY
TITLE
STUFF
UTIL ITÉ
BAGHDAD CALLING
Soirée FAN-TAS TIQUE.

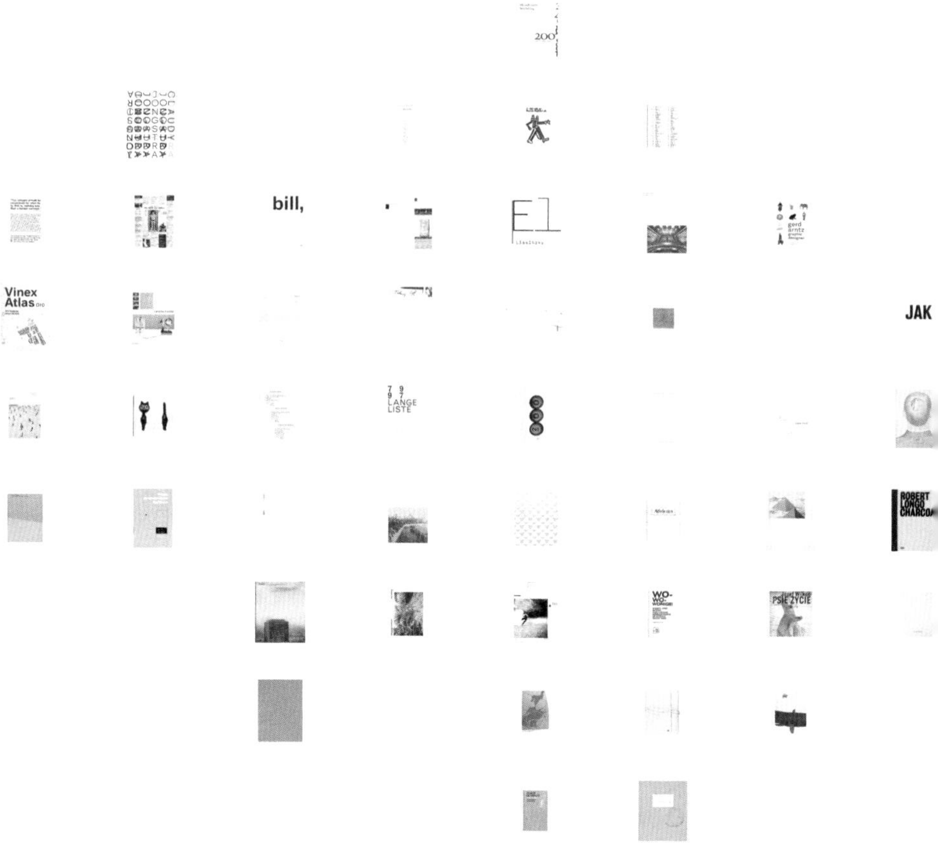
200
bill,
Vinex
Atlas 010
gerd
arntz
graphic
designer
JAK
7 9
9 7
LANGE
LISTE
ROBERT
LONGO
CHARCOA
PSIE ZYCIE

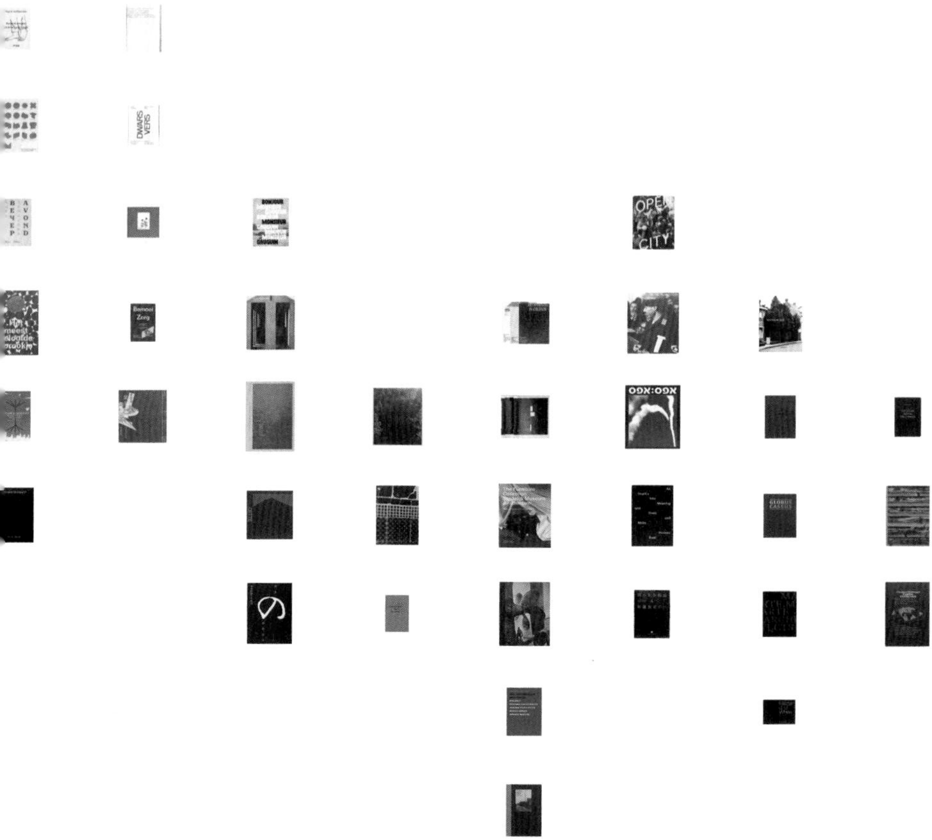
DWARS VERS
ВЕЧЕР
AVOND
BONJOUR
MONSIEUR
GAUGUIN
OPEN CITY
Bemoei Zorg
GLOBUS CASSUS

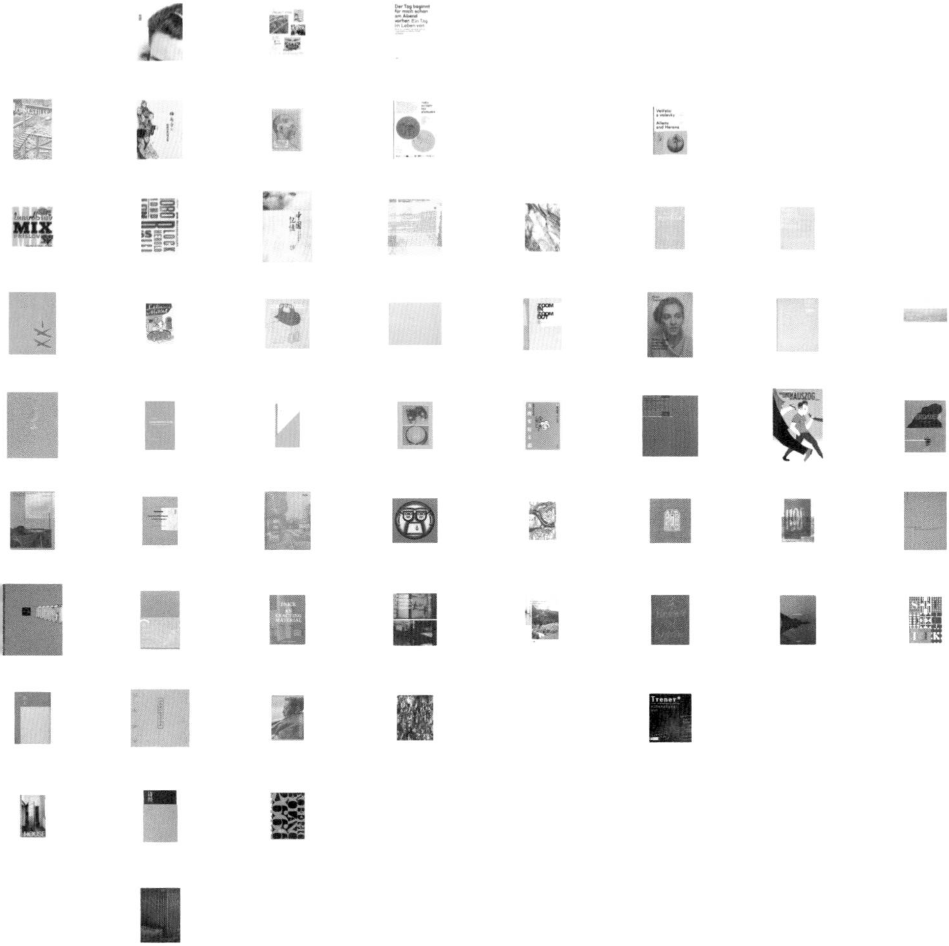
Der Tag beginnt für mich schon am Abend vorher. Ein Tag im Leben von
MIX
ZOOM IN ZOOM OUT
Trener*
HOUSE

Index *Index*

索引

Catalogue Archaeological Finds
Amsterdam's North/South Metro Line

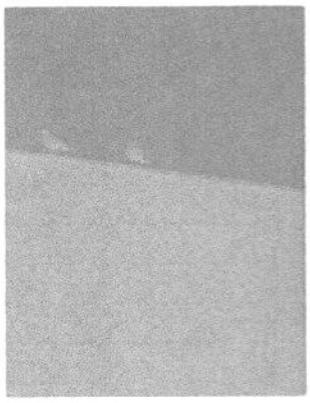
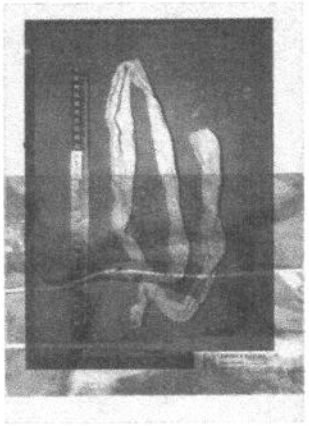

Scheidegger & Spiess
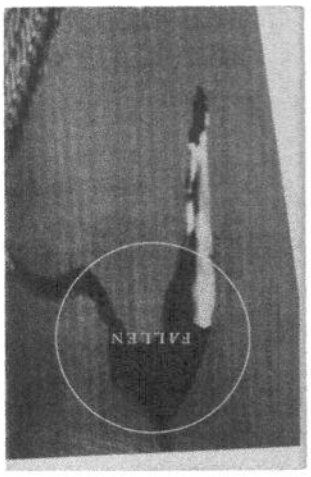
FALLEN

Atlas of the Conflict. Israel-Palestine
by Malkit Shoshan
design Joost Grootens
010 Publishers

XX-
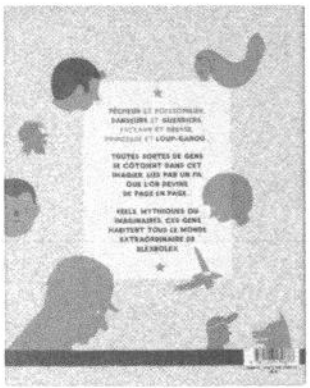

Princeton Architectural Press

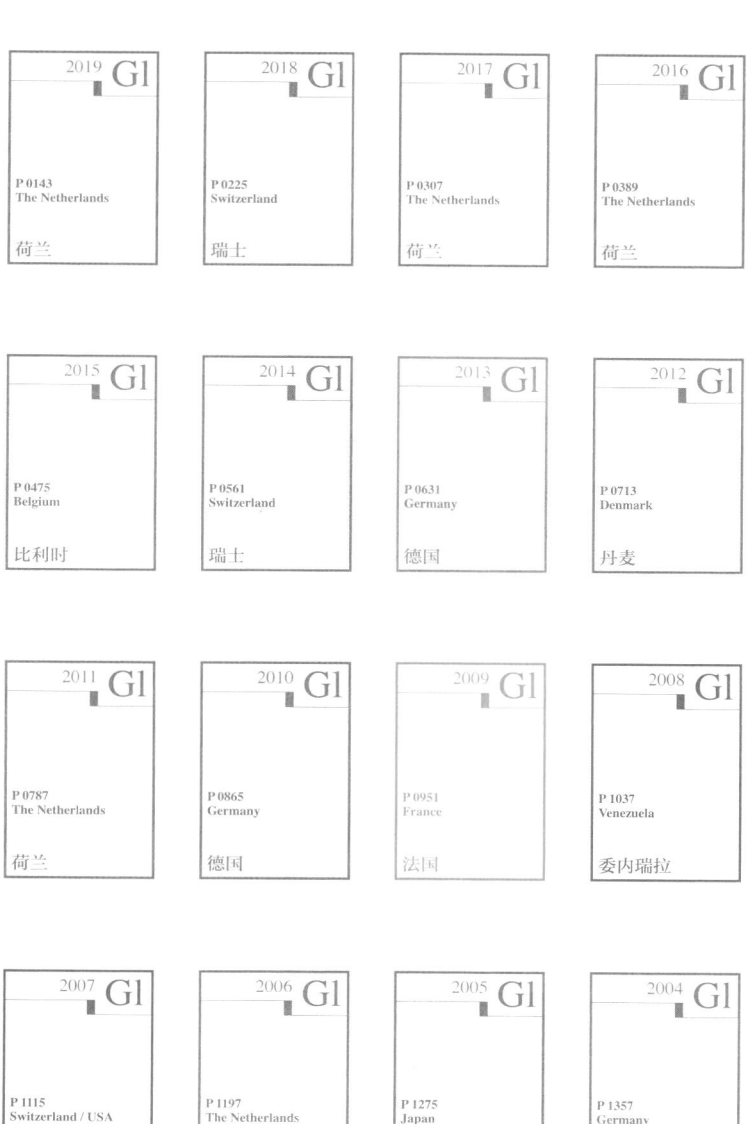

2019 Gl
P 0143
The Netherlands
荷兰

2018 Gl
P 0225
Switzerland
瑞士

2017 Gl
P 0307
The Netherlands
荷兰

2016 Gl
P 0389
The Netherlands
荷兰

2015 Gl
P 0475
Belgium
比利时

2014 Gl
P 0561
Switzerland
瑞士

2013 Gl
P 0631
Germany
德国

2012 Gl
P 0713
Denmark
丹麦

2011 Gl
P 0787
The Netherlands
荷兰

2010 Gl
P 0865
Germany
德国

2009 Gl
P 0951
France
法国

2008 Gl
P 1037
Venezuela
委内瑞拉

2007 Gl
P 1115
Switzerland / USA
瑞 士 / 美

2006 Gl
P 1197
The Netherlands
荷兰

2005 Gl
P 1275
Japan
日本

2004 Gl
P 1357
Germany
德国

Architecture from the Arab World
1914–2014 (a Selection)
مختارات من عمارة العالم العربي
١٩١٤–٢٠١٤
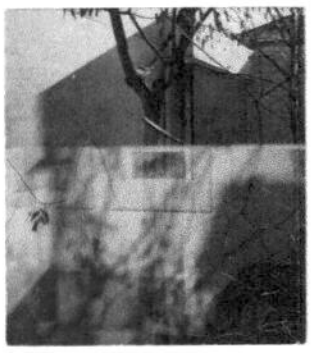
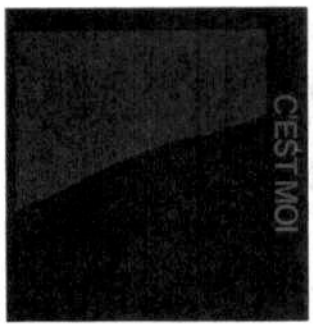
C'EST MOI

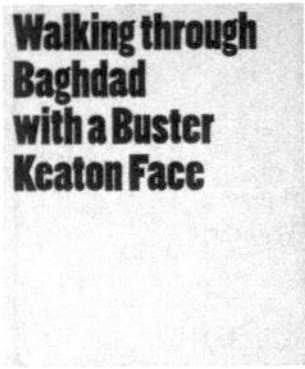
Walking through Baghdad with a Buster Keaton Face

EIN TAG

Metropolitan World Atlas

2019 Gm
P 0151
The Netherlands
荷兰

2018 Gm
P 0233
Germany
德国

2016 Gm
P 0397
China
中国

2015 Gm
P 0483
Switzerland
瑞士

2014 Gm
P 0569
Germany
德国

2013 Gm
P 0639
The Netherlands
荷兰

2012 Gm
P 0721
Austria
奥地利

2011 Gm
P 0795
Czech Republic
捷克

2010 Gm
P 0873
Switzerland
瑞士

2009 Gm
P 0959
The Netherlands
荷兰

2008 Gm
P 1045
Switzerland
瑞士

2007 Gm
P 1123
The Netherlands
荷兰

2006 Gm
P 1205
The Netherlands
荷兰

2005 Gm
P 1283
Switzerland
瑞士

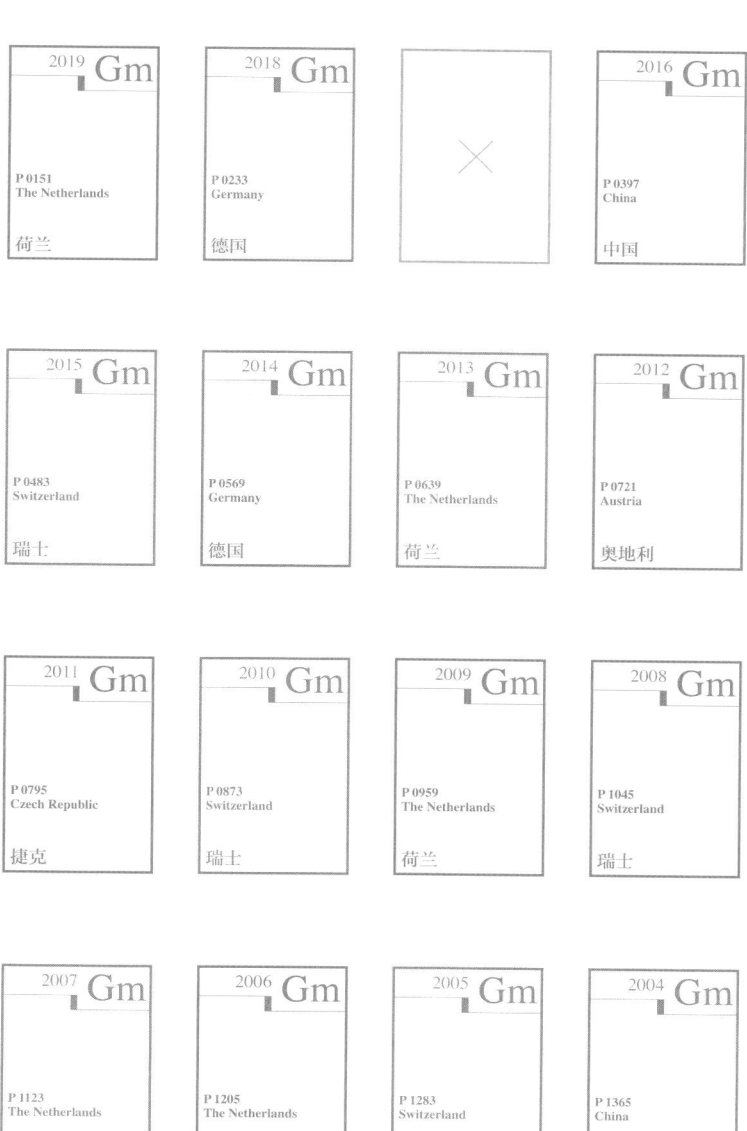

2004 Gm
P 1365
China
中国

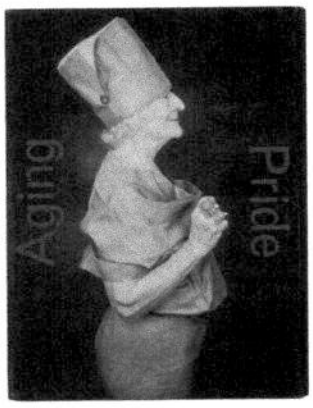
Aging
Pride

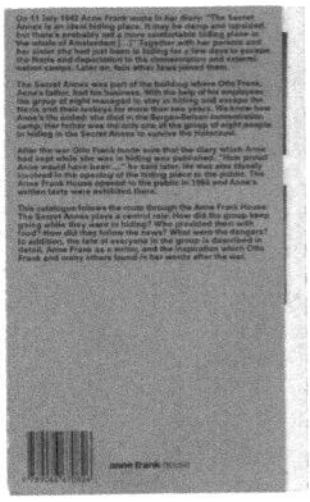

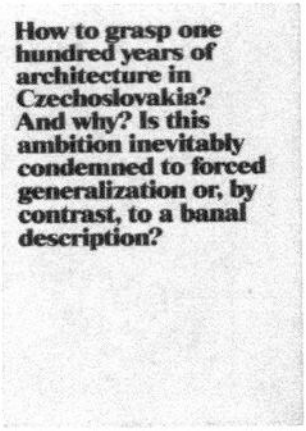
How to grasp one hundred years of architecture in Czechoslovakia? And why? Is this ambition inevitably condemned to forced generalization or, by contrast, to a banal description?

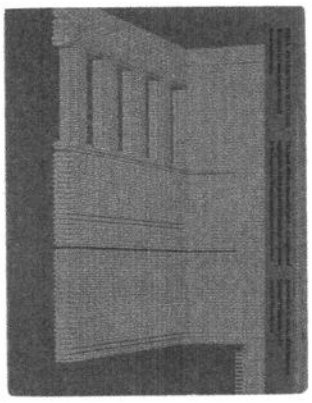

Sequester
Awoiska van der Molen

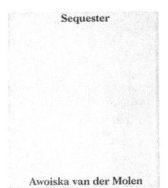

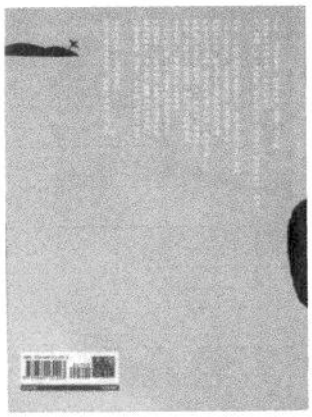

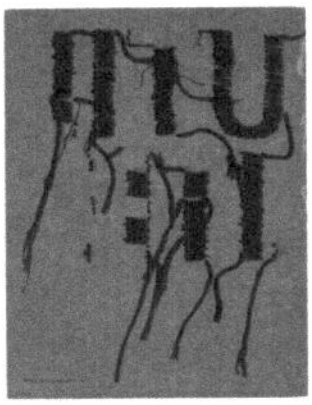

2019 Sm^1
P 0157
Austria
奥地利

2019 Sm^2
P 0163
The Netherlands
荷兰

2018 Sm^1
P 0239
China
中国

2018 Sm^2
P 0245
Japan
日本

2017 Sm^1
P 0315
China
中国

2017 Sm^2
P 0321
The Netherlands
荷兰

2016 Sm^1
P 0403
Czech Republic
捷克

2016 Sm^2
P 0409
The Netherlands and Flanders
荷兰及弗兰德斯地区

2015 Sm^1
P 0489
Denmark
丹麦

2015 Sm^2
P 0495
The Netherlands
荷兰

2014 Sm^1
P 0575
Austria
奥地利

×

2013 Sm^1
P 0645
Taiwan, China
中国(台湾地区)

2013 Sm^2
P 0651
Japan
日本

2012 Sm^1
P 0727
China
中国

2012 Sm^2
P 0733
The Netherlands
荷兰

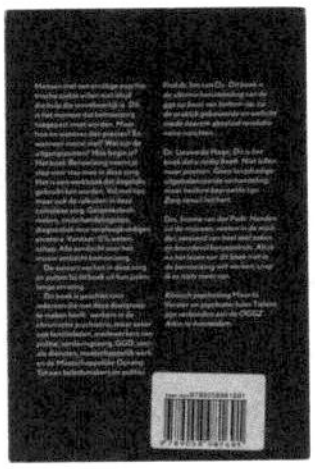
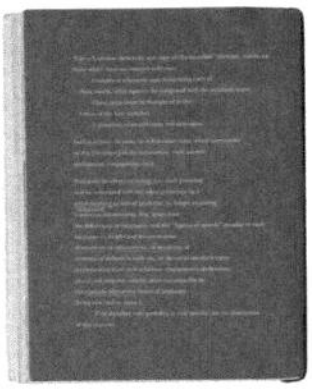

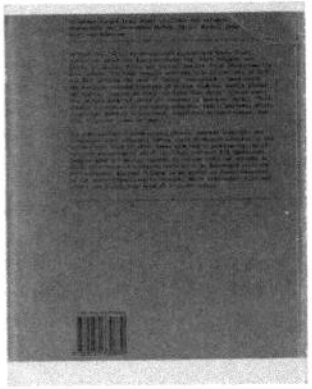

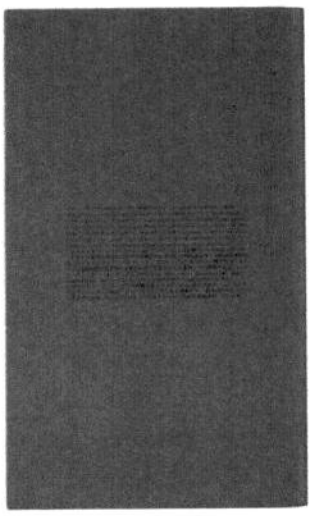
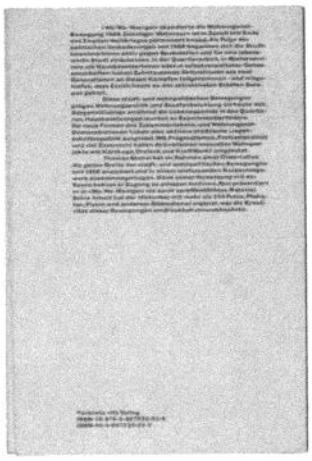

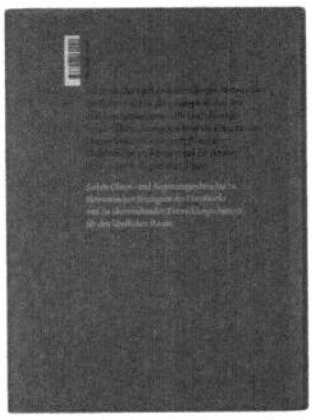
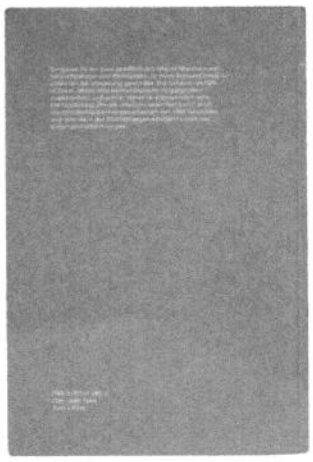

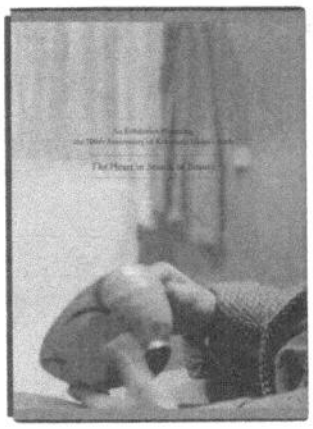

2011 Sm^1
P 0801
The Netherlands
荷兰

2011 Sm^2
P 0807
Switzerland
瑞士

2010 Sm^1
P 0879
Austria
奥地利

2010 Sm^2
P 0885
Switzerland
瑞士

2009 Sm^1
P 0965
The Netherlands
荷兰

2009 Sm^2
P 0971
Austria
奥地利

×

2008 Sm^2
P 1051
Japan
日本

2007 Sm^1
P 1129
Japan
日本

2007 Sm^2
P 1135
Switzerland
瑞士

2006 Sm^1
P 1211
The Netherlands
荷兰

2006 Sm^2
P 1217
Austria
奥地利

2005 Sm^1
P 1289
Austria
奥地利

2005 Sm^2
P 1295
USA
美国

2004 Sm^1
P 1371
Japan
日本

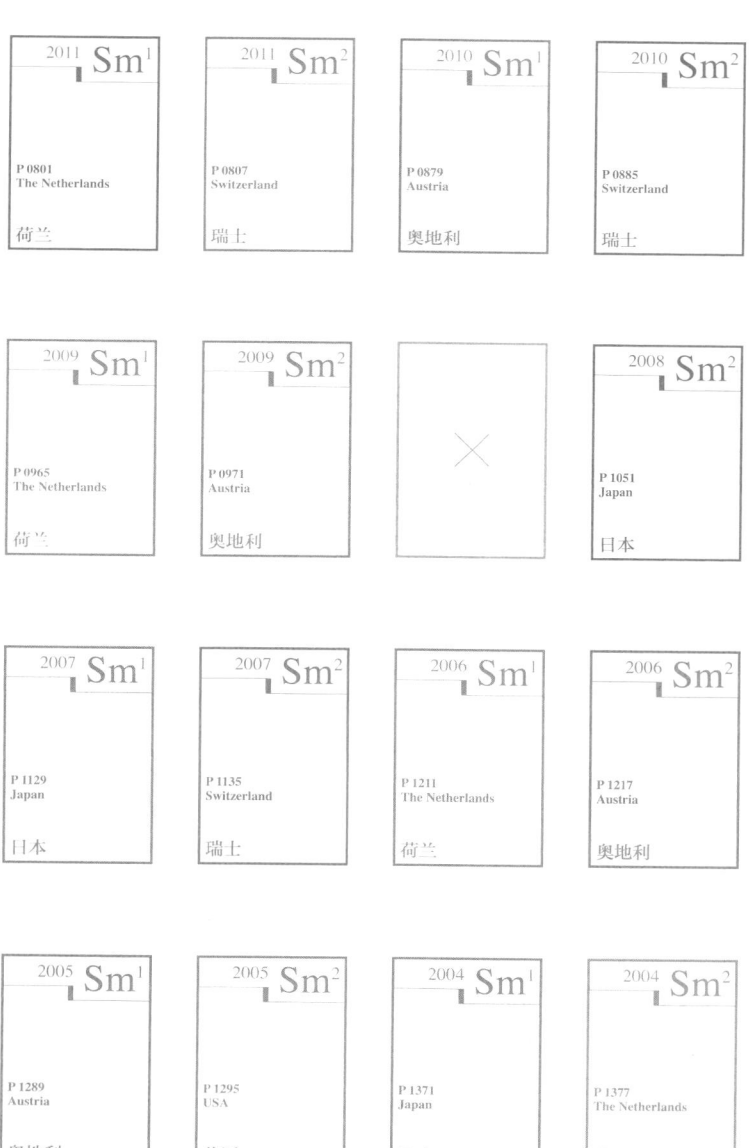

2004 Sm^2
P 1377
The Netherlands
荷兰

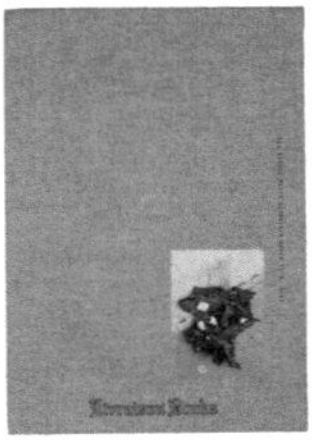

стихи
gedichten
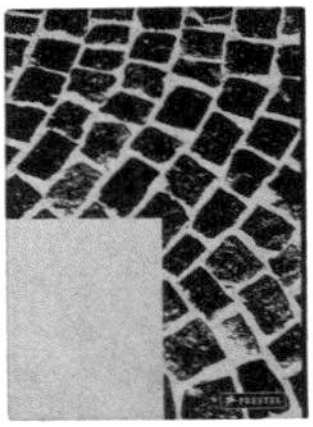
PRESTEL

UP UP: Stories of Johannesburg's Highrises presents a selection of buildings in the inner city of Johannesburg. The focus is on tall modernist towers that quickly became iconic in the architecture of the city. Witnesses to profound shifts in the political history of the metropolis, the skyscrapers have themselves endured immense changes. They are documented here in two ways: firstly, through historical facts, floor plans, archival records, and contemporary photographs of their formal architecture; and secondly, through reportage, interviews, or essays. The book offers new insights not only into these classic buildings, but also into contemporary urban life in South Africa.
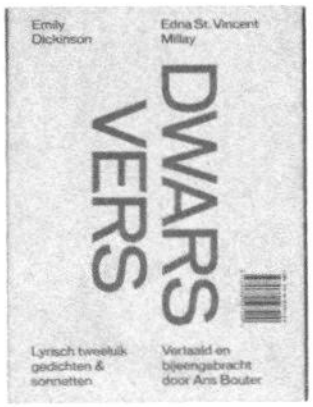
Emily
Dickinson
Edna St. Vincent
Millay
DWARS
VERS
Lyrisch tweeluik
gedichten &
sonnetten
Vertaald en
bijeengebracht
door Ans Bouter

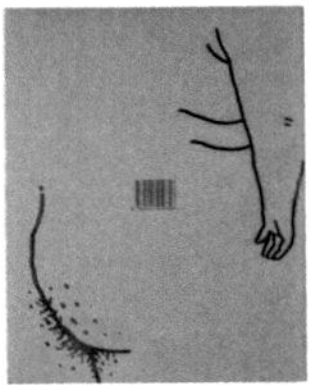

2019 Bm^{1}
P 0169
Czech Republic
捷克

2019 Bm^{2}
P 0175
Sweden
瑞典

2019 Bm^{3}
P 0181
The Netherlands
荷兰

2019 Bm^{4}
P 0187
Germany
德国

2019 Bm^{5}
P 0193
Ukraine
乌克兰

2018 Bm^{1}
P 0251
The Netherlands
荷兰

2018 Bm^{2}
P 0257
Switzerland
瑞士

2018 Bm^{3}
P 0263
Israel
以色列

2018 Bm^{4}
P 0269
Russia
俄罗斯

2018 Bm^{5}
P 0275
Switzerland
瑞士

2017 Bm^{1}
P 0327
Switzerland
瑞士

2017 Bm^{2}
P 0333
Germany
德国

2017 Bm^{3}
P 0339
The Netherlands
荷兰

2017 Bm^{4}
P 0345
Switzerland
瑞士

2017 Bm^{5}
P 0351
Switzerland
瑞士

2016 Bm^{1}
P 0415
Belgium
比利时

2016 Bm^{2}
P 0421
The Netherlands
荷兰

2016 Bm^{3}
P 0427
Austria
奥地利

2016 Bm^{4}
P 0433
China
中国

2016 Bm^{5}
P 0439
Norway
挪威

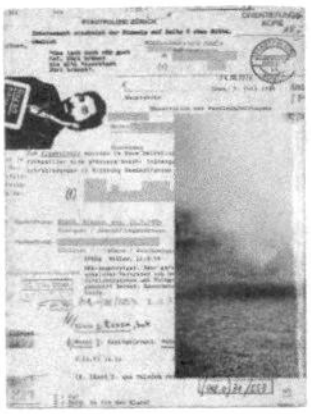

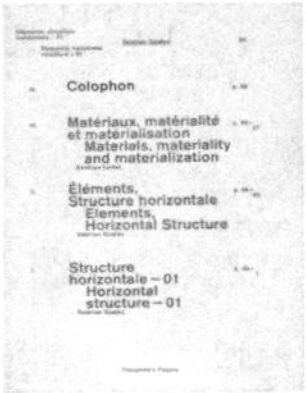
Colophon
Matériaux, matérialité et matérialisation
Materials, materiality and materialization
Éléments, Structure horizontale
Elements, Horizontal Structure
Structure horizontale – 01
Horizontal structure – 01

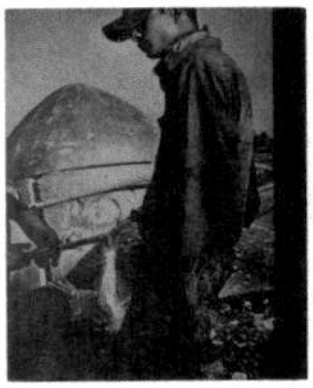

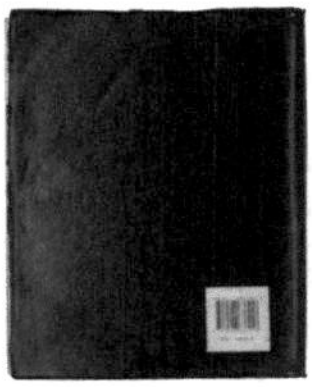

BASKERVILLE
SUPER GROTESK
COTTONWOOD
DAX
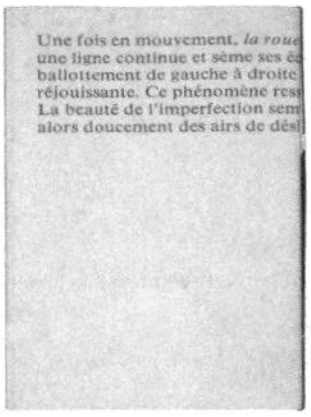
Une fois en mouvement, la roue
une ligne continue et sème ses é
ballottement de gauche à droite
réjouissante. Ce phénomène res
La beauté de l'imperfection sem
alors doucement des airs de dés

mowię do ciebie ja pies
twój
największy
przyjaciel

ZOOM
IN
ZOOM
OUT

2015 Bm1
P 0501
Japan
日本

2015 Bm2
P 0507
Germany
德国

2015 Bm3
P 0513
Switzerland
瑞士

2015 Bm4
P 0519
Belgium
比利时

2015 Bm5
P 0525
The Netherlands
荷兰

2014 Bm1
P 0581
Germany
德国

2014 Bm2
P 0587
Germany
德国

2014 Bm3
P 0593
Germany
德国

2014 Bm4
P 0599
China
中国

×

2013 Bm1
P 0657
Belgium
比利时

2013 Bm2
P 0663
Estonia
爱沙尼亚

2013 Bm3
P 0669
Belgium
比利时

2013 Bm4
P 0675
Poland
波兰

2013 Bm5
P 0681
Germany
德国

2012 Bm1
P 0739
The Netherlands
荷兰

×

2012 Bm3
P 0745
Austria
奥地利

2012 Bm4
P 0751
Taiwan, China
中国（台湾地区）

2012 Bm5
P 0757
Czech Republic
捷克

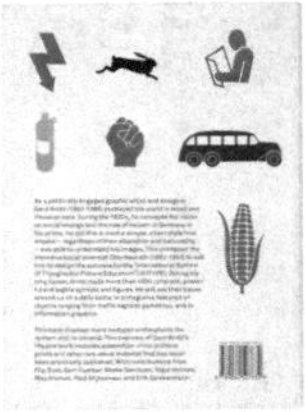

"The sectors of a city are to some extent decipherable."
Guy Debord
"These people go out from their countries driven by necessity that arises from famine, or war, and oppression, which in their own country is experienced by them, so that they are constrained to seek new land. Therefore they are most formidable, and if they are not confronted by good arms, can never be held back."
Machiavelli

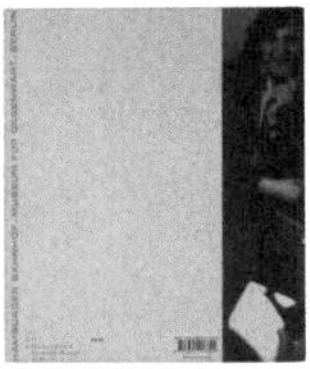

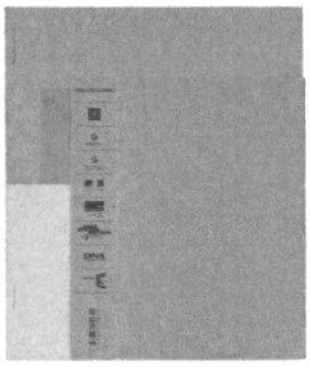

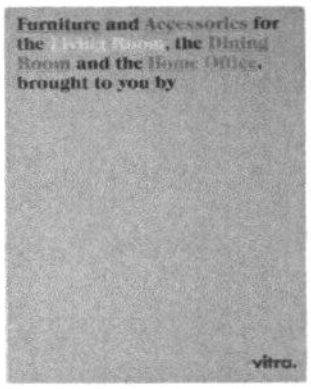
Furniture and Accessories for the Living Room, the Dining Room and the Home Office, brought to you by
vitra.

2011 Bm^{1} P 0813 The Netherlands 荷兰	2011 Bm^{2} P 0819 The Netherlands 荷兰	2011 Bm^{3} P 0825 Austria 奥地利	×	2011 Bm^{5} P 0831 Switzerland 瑞士
2010 Bm^{1} P 0891 Germany 德国	2010 Bm^{2} P 0897 Germany 德国	2010 Bm^{3} P 0903 Austria 奥地利	2010 Bm^{4} P 0909 Sweden 瑞典	2010 Bm^{5} P 0915 Switzerland 瑞士
2009 Bm^{1} P 0977 Germany 德国	2009 Bm^{2} P 0983 Germany 德国	2009 Bm^{3} P 0989 France 法国	2009 Bm^{4} P 0995 France 法国	2009 Bm^{5} P 1001 France 法国
2008 Bm^{1} P 1057 Germany 德国	2008 Bm^{2} P 1063 Germany 德国	2008 Bm^{3} P 1069 France 法国	2008 Bm^{4} P 1075 Switzerland 瑞士	2008 Bm^{5} P 1081 Switzerland 瑞士

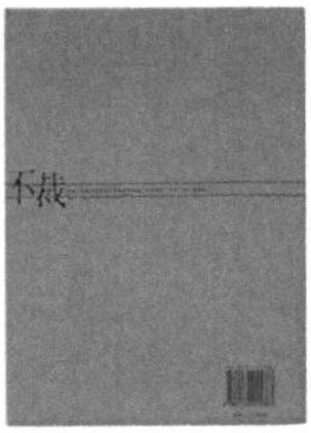
不裁

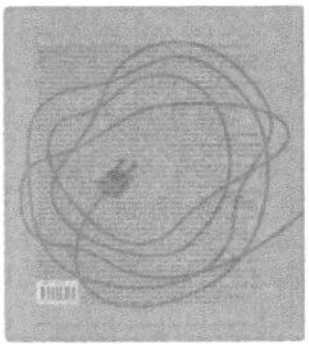

E
Y E

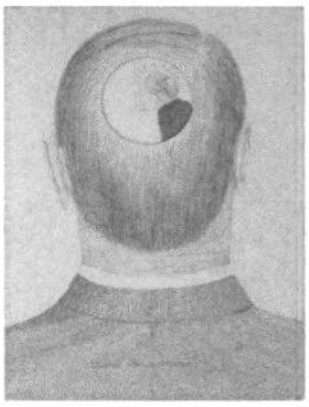

Living Library introduces the recently opened Utrecht University Library in the Netherlands designed by renowned Dutch architect Wiel Arets. In addition to illustrating its visual splendour and ingenious use of space, Living Library takes the reader on a journey through its inner world: the history of the library and its collection, its current and future use. The building by Arets is juxtaposed with a compilation of diverse and often surprising fragments made up of photographs, drawings, artworks, film stills, essays and interviews that address the library as an institution and metaphor in today's society. This intriguing combination, along with its innovative design, makes Living Library an exceptional book.

 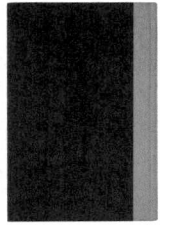
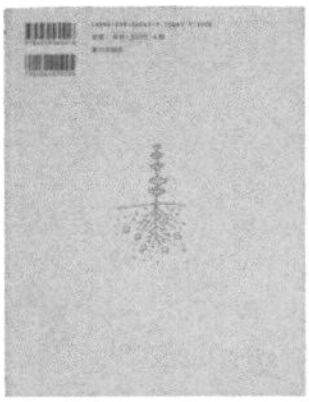

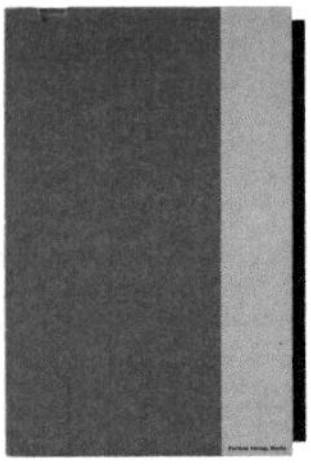

GEN

Die Tollen Hefte

	1	2	3	4	5
2007	Bm1 P 1141 China 中国	Bm2 P 1147 Germany 德国	Bm3 P 1153 Germany 德国	Bm4 P 1159 Canada 加拿大	Bm5 P 1165 The Netherlands 荷兰
2006	Bm1 P 1223 Finland 芬兰		Bm3 P 1229 The Netherlands 荷兰	Bm4 P 1235 The Netherlands 荷兰	Bm5 P 1241 Switzerland 瑞士
2005	Bm1 P 1301 Germany 德国	Bm2 P 1307 Japan 日本	Bm3 P 1313 Switzerland 瑞士	Bm4 P 1319 The Netherlands 荷兰	Bm5 P 1325 The Netherlands 荷兰
2004	Bm1 P 1383 Germany 德国	Bm2 P 1389 Germany 德国	Bm3 P 1395 Germany 德国	Bm4 P 1401 Germany 德国	

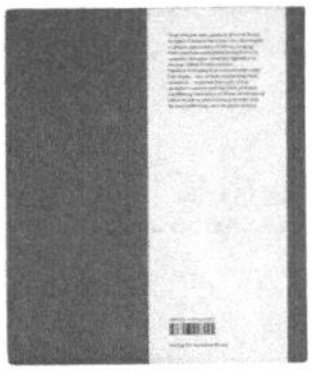

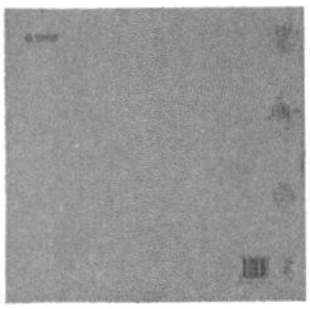

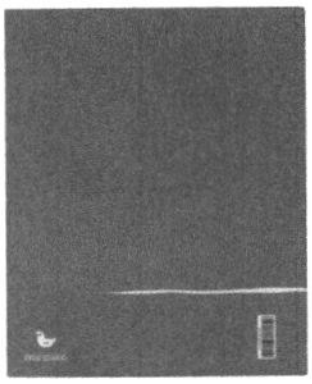

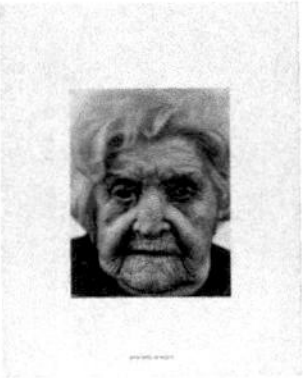
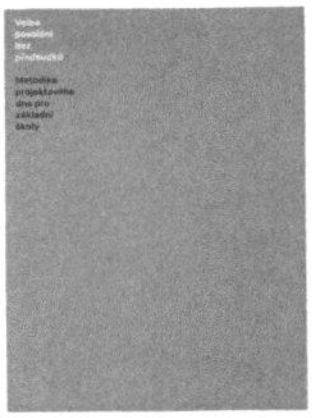
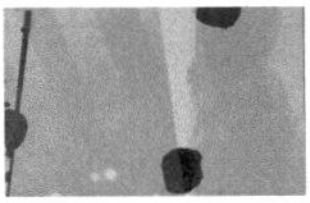

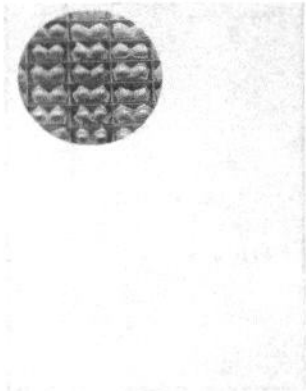

2019 Ha^{1} P 0199 Austria 奥地利	×	2019 Ha^{3} P 0203 Russian 俄罗斯	2019 Ha^{4} P 0207 Japan 日本	2019 Ha^{5} P 0211 China 中国
2018 Ha^{1} P 0281 Switzerland 瑞士	2018 Ha^{2} P 0285 The Netherlands 荷兰	2018 Ha^{3} P 0289 China 中国	2018 Ha^{4} P 0293 Venezuela 委内瑞拉	×
2017 Ha^{1} P 0357 Germany 德国	2017 Ha^{2} P 0363 China 中国	2017 Ha^{3} P 0367 Japan 日本	2017 Ha^{4} P 0371 Portugal 葡萄牙	2017 Ha^{5} P 0375 Germany 德国
2016 Ha^{1} P 0445 Norway 挪威	2016 Ha^{2} P 0449 Czech Republic 捷克	2016 Ha^{3} P 0453 Venezuela 委内瑞拉	2016 Ha^{4} P 0457 Germany 德国	2016 Ha^{5} P 0461 The Netherlands 荷兰

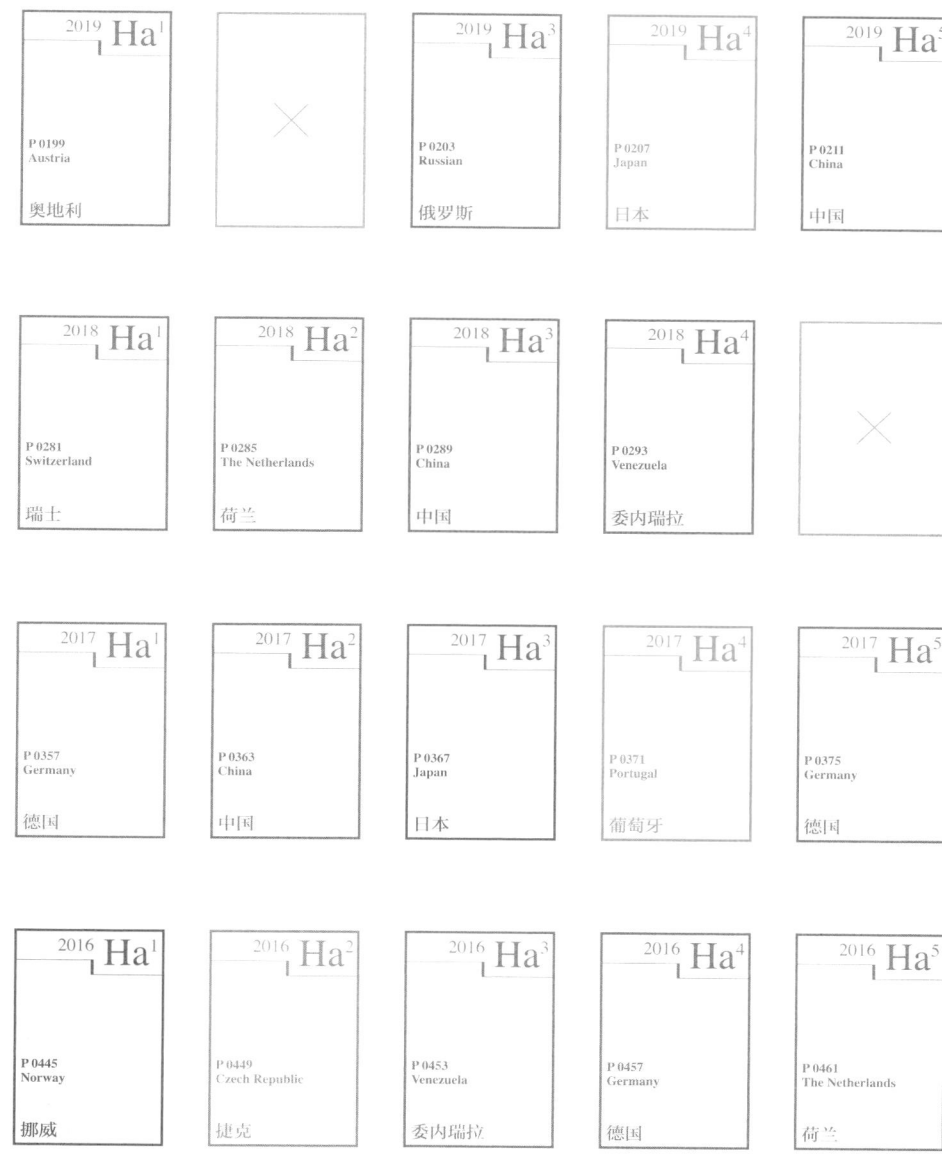

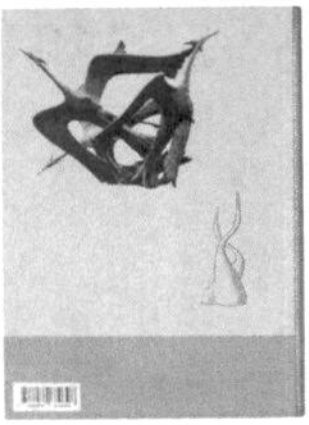

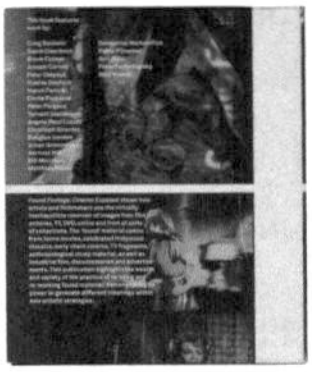
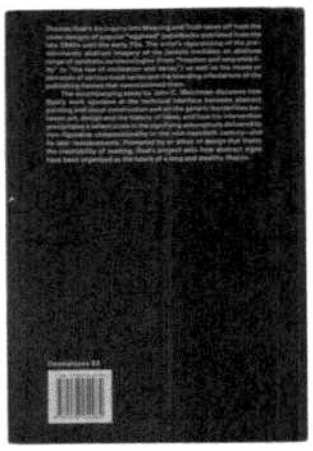

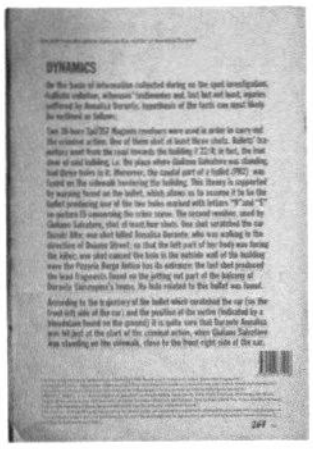

PAM
PI
LIO

2015 Ha1	P 0531	Romania	罗马尼亚
2015 Ha2	P 0535	Canada	加拿大
2015 Ha3	P 0539	Switzerland	瑞士
2015 Ha4	P 0543	Estonia	爱沙尼亚
2015 Ha5	P 0547	Czech Republic	捷克
2014 Ha1	P 0605	Iran	伊朗
×			
2014 Ha3	P 0609	Norway	挪威
2014 Ha4	P 0613	China	中国
2014 Ha5	P 0617	Japan	日本
2013 Ha1	P 0687	Germany	德国
2013 Ha2	P 0691	Germany	德国
2013 Ha3	P 0695	The Netherlands	荷兰
2013 Ha4	P 0699	The Netherlands	荷兰
×			
2012 Ha1	P 0761	China	中国
2012 Ha2	P 0765	The Netherlands	荷兰
2012 Ha3	P 0769	Poland	波兰
×			
2012 Ha5	P 0773	Sweden	瑞典

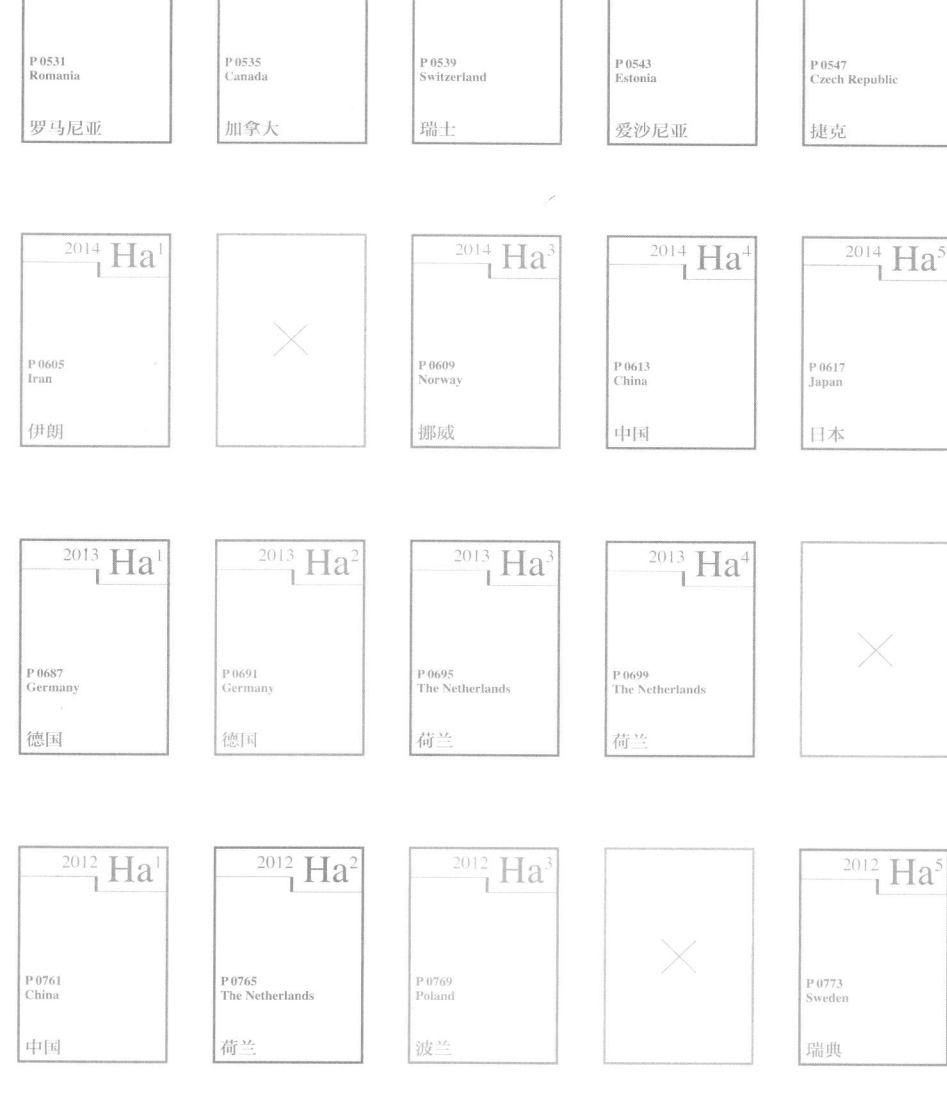

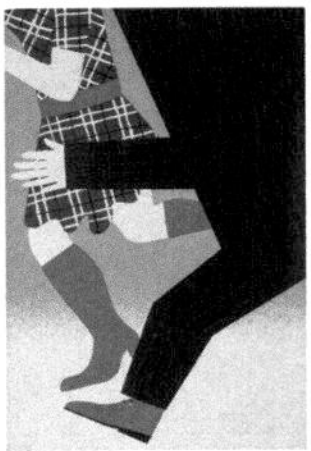
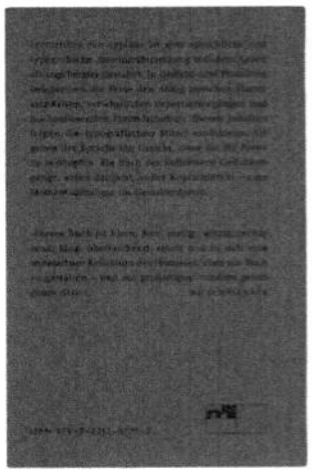

GÉVÉ
le vin qu

诗经

IABR
SUN
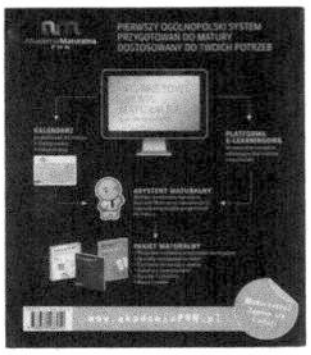

JULES
SPINATSCH

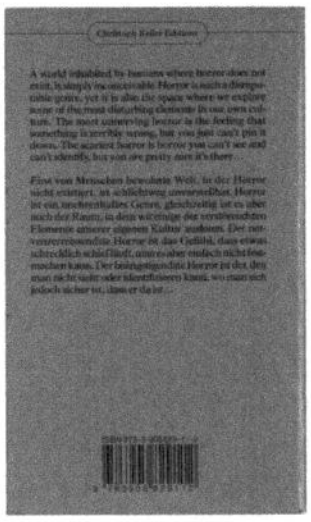

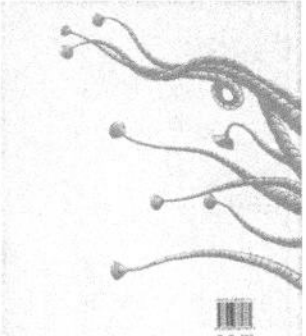

2011 Ha[1]	2011 Ha[2]	2011 Ha[3]	2011 Ha[4]	×
P 0837 China 中国	P 0843 Germany 德国	P 0847 Germany 德国	P 0851 Germany 德国	

2010 Ha[1]	2010 Ha[2]	2010 Ha[3]	2010 Ha[4]	2010 Ha[5]
P 0921 China 中国	P 0925 The Netherlands 荷兰	P 0929 Poland 波兰	P 0933 Switzerland 瑞士	P 0937 Czech Republic 捷克

2009 Ha[1]	2009 Ha[2]	2009 Ha[3]	2009 Ha[4]	2009 Ha[5]
P 1007 China 中国	P 1011 Germany 德国	P 1015 Germany / Switzerland 德国 / 瑞士	P 1019 The Netherlands 荷兰	P 1023 Switzerland 瑞士

2008 Ha[1]	2008 Ha[2]	2008 Ha[3]	2008 Ha[4]	2008 Ha[5]
P 1085 China 中国	P 1089 The Netherlands 荷兰	P 1093 Portugal 葡萄牙	P 1097 Switzerland 瑞士	P 1101 Czech Republic 捷克

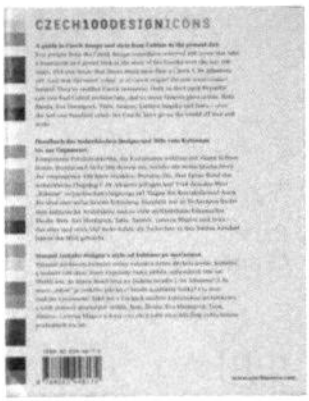
CZECH100DESIGNICONS

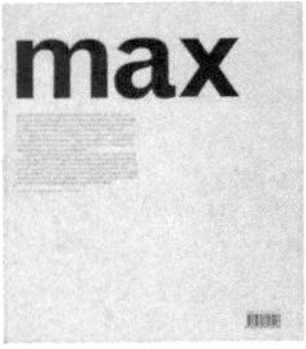
max

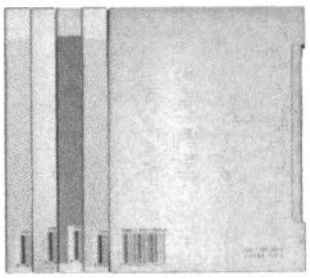

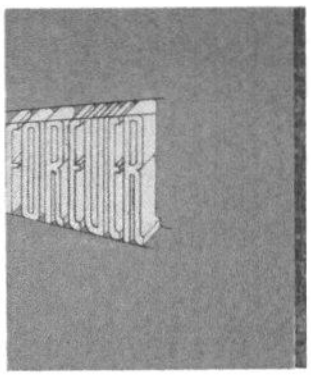

×	2007 Ha2 P 1171 The Netherlands 荷兰	2007 Ha3 P 1175 Austria 奥地利	2007 Ha4 P 1179 Austria 奥地利	2007 Ha5 P 1183 Czech Republic 捷克
2006 Ha1 P 1245 China 中国	2006 Ha2 P 1249 Germany 德国	2006 Ha3 P 1253 The Netherlands 荷兰	2006 Ha4 P 1257 Russia 俄罗斯	2006 Ha5 P 1261 Czech Republic 捷克
2005 Ha1 P 1331 China 中国	2005 Ha2 P 1335 China 中国	×	2005 Ha4 P 1339 The Netherlands 荷兰	2005 Ha5 P 1343 Switzerland 瑞士
2004 Ha1 P 1407 Germany 德国	2004 Ha2 P 1413 Germany 德国	2004 Ha3 P 1417 The Netherlands 荷兰	2004 Ha4 P 1421 Norway 挪威	2004 Ha5 P 1425 Venezuela 委内瑞拉

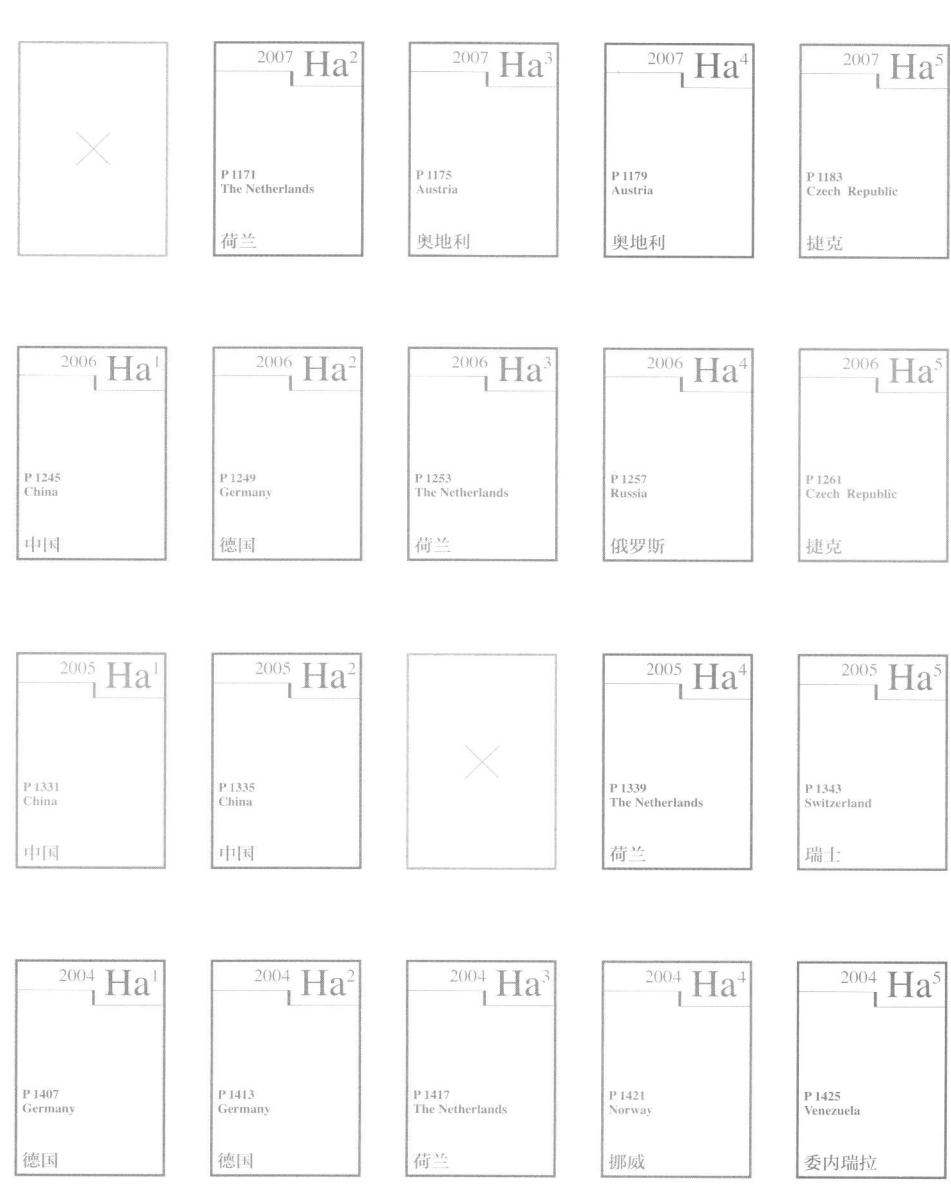

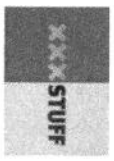
STUFF

P 0143

金字符奖 ||||| 阿姆斯特丹的"宝藏" || 荷兰

Golden Letter |||||| Amsterdam STUFF || The Netherlands || Willem van Zoetendaal || Jerzy Gawronski, Peter Kranendonk || Van Zoetendaal Publishers & De Harmonie, Monumenten & Archeologie Gemeente Amsterdam || 365×243×32mm || 2198g || 600p

P 0151

金奖 |||| 罗伯特·肯尼迪的葬礼火车——民众的视角 || 荷兰

Gold Medal ||||| Robert F. Kennedy Funeral Train – The People's View || The Netherlands // *Jeremy Jansen* // Rein Jelle Terpstra || Fw:Books, Amsterdam || 287×214×21mm || 922g || 140p

P 0157

银奖 |||| 年龄的力量 || 奥地利

Silver Medal |||| Die Kraft des Alters – Aging Pride || Austria // *Willi Schmid* // Stella Rollig, Sabine Fellner || Verlag für moderne Kunst, Belvedere, Vienna || 241×178×36mm || 1030g || 372p

HOUSE

P 0163

银奖 |||| 安妮·弗兰克之家 || 荷兰

Silver Medal |||| Anne Frank House || The Netherlands // *Irma Boom Office (Irma Boom, Eva van Bemmelen)* // Elias van der Plicht (Anne Frank Stichting) || Anne Frank Stichting, Amsterdam || 210×123×19mm || 452g || 336p

BONJOUR
MONSIEUR
GAUGUIN

P 0169

铜奖 ||| 你好，高更先生 || 捷克

Bronze Medal ||| Bonjour, Monsieur Gauguin || Czech Republic // *20YY Designers (Petr Bosák, Robert Jansa, Adam Macháček)* // Anna Pravdová || Národní galerie v Praze (National Gallery Prague) || 245×173×32mm || 879g || 308p

P 0175

铜奖 ||| 巴黎 || 瑞典

Bronze Medal ||| Paris || Sweden // *Sandberg&-Timonen* // Ola Rindal || Livraison Books, Stockholm || 347×235×18mm || 1095g || 140p

P 0181

铜奖 ||| 移民 || 荷兰

Bronze Medal ||| The Migrant || The Netherlands // *Teun van der Heijden* // Anaïs López || Self-Published || 325×246×20mm || 1048g+33g || 124p+16p

P 0187

铜奖 ||| 毁灭 || 德国

Bronze Medal ||| Nichtsein || Germany // *Katharina Schwarz* // Katharina Schwarz in co-operation with Ellen von den Driesch || Wissenschaftszentrum Berlin für Sozialforschung, UDK Berlin || 305×174×16mm || 553g || 160p

P 0193

铜奖 ||| 这就是我看到的 || 乌克兰

Bronze Medal ||| Я так бачу (This Is How I See) || Ukraine // *Art Studio Agrafka (Romana Romanyshyn, Andrii Lesiv)* // Art Studio Agrafka (Romana Romanyshyn, Andrii Lesiv) || Видавництво Старого Лева (Old Lion Publishing) || 287×267×10mm || 552g || 56p

P 0199

荣誉奖 || 切萨雷·费罗纳托——石头的解剖学 || 奥地利

Honorary Appreciation || Cesare Ferronato – Anatomie des Steins || Austria // *Raphael Drechsel* // Hannes Schüpbach || Verlag für moderne Kunst, Vienna || 255×207×27mm || 1042g || 224p

荣誉奖 ||

Honorary Appreciation || Nienasycenie spojrzenia – Insatiability of the Gaze || Poland // *Ryszard Bienert* // Anna Królica (Ed.) || Centrum Sztuki Mościce, Tarnów / Instytut Muzyki i Tańca, Warsaw

El
Lissitzky

P 0203

荣誉奖 || 埃尔·利西茨基 || 俄罗斯

Honorary Appreciation || El Lissitzky || Russian // *Evgeny Korneev* // Tatyana Goryacheva, Ruth Addison, Ekaterina Allenova || Tretyakov Gallery, Jewish Museum and Tolerance Center || 250×235×29mm || 1084g || 336p

P 0207

荣誉奖 || 第一件事 || 日本

Honorary Appreciation || The First || Japan // *Hideyuki Saito* || Kentarou Tanaka || Bonpoint Japon || 267×216×8mm || 418g || 46p

P 0211

荣誉奖 || 江苏老行当百业写真 || 中国 || 周晨 || 潘文龙、龚为（摄影）|| 江苏凤凰教育出版社

Honorary Appreciation || Old Trades of Jiangsu: A Glimpse || China // *Zhou Chen* // Pan Wenlong, Gong Wei (photographer) || Jiangsu Phoenix Education Publishing, Ltd || 288×284×38mm || 1817g || 646p

2018

P 0225

金字符奖 |||||| 家、工艺和日用品的乌托邦 || 瑞士
Golden Letter |||||| HEIMAT, HANDWERK UND DIE UTOPIE DES ALLTÄGLICHEN || Switzerland // *HUBERTUS, Jonas Voegeli, Scott Vander Zee, Kerstin Landis* // Uta Hassler || Hirmer || 265×202×48mm || 1897g || 568p

Soirée
PAN-TAS
TIQUE

P 0233

金奖 ||||| 迷幻的夜晚 || 德国
Gold Medal ||||| Soirée Fantastique || Germany // *Pierre Pané-Farré* // Pierre Pané-Farré || Institut für Buchkunst Leipzig || 328×275×12mm || 616g || 120p

P 0239

银奖 |||| 园冶注释 || 中国 || 张悟静 || 计成（原著）、陈植（注释） || 中国建筑工业出版社
Silver Medal |||| The Art of Gardening || China // *Zhang Wujing* // Ji Cheng (Ming Dynasty), Chen Zhi China || China Architecture & Building Press || 260×175×29mm || 645g || 424p

P 0245

银奖 |||| 成为一只小熊的过程 || 日本
Silver Medal |||| Process for becoming a little bear || Japan // *Akihiro Taketoshi (STUDIO BEAT)* // Nana Inoue || be Nice Inc. || 216×257×10mm || 383g || 34p

B A
E V
Ч O
E N
P D

P 0251

铜奖 ||| 黄昏 || 荷兰
Bronze Medal ||| Avond || The Netherlands // *Michael Snitker* || Anna Achmatova / Hans Boland || Stiching De Ross || 242×152×15mm || 388g || 128p

P 0257

铜奖 ||| 第 14 届卡塞尔文献展：日记账 || 瑞士
Bronze Medal ||| documenta 14: Daybook || Switzerland // *Julia Born & Laurenz Brunner, Zürich* // Quinn Latimer und Adam Szymczyk || Prestel, München || 293×204×20mm || 938g || 344p

P 0263

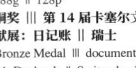

铜奖 ||| 0:0 || 以色列
Bronze Medal ||| EFES:EFES || Israel // *Dan Ozeri* // Dan Ozeri || Self expenditure || 310×264×20mm || 1044g || 192p

P 0269

铜奖 ||| 每分钟敲键 200 次 || 俄罗斯
Bronze Medal ||| 200 keystrokes per minute || Russia // *Igor Gurovich* // Narinskaya Anna || Moscow Polytechnic Museum || 297×211×20mm || 699g || 208p

P 0275

铜奖 ||| A4 上的自主权 || 瑞士
Bronze Medal ||| Autonomie auf A4 || Switzerland // *Atlas Studio (Martin Angereggen, Claudio Gasser, Jonas Wandeler)* // Peter Bichsel und Silvan Lerch || Limmat Verlag, Zürich || 318×241×24mm || 1156g || 288p

P 0281

荣誉奖 || 重中之重 || 瑞士
Honorary Appreciation || First things first || Switzerland // *NORM, Zürich* // Shirana Shahbazi || SternbergPress, Berlin || 326×245×9mm || 379g || 24p

P 0285

荣誉奖 || 克劳迪·扬斯特拉 || 荷兰
Honorary Appreciation || Claudy Jongstra || The Netherlands // *Irma Boom* // Louwrien Wijers, Lidweji Edelkoort, Laura M. Richard, Marietta de Vries, Pietje Tegenbosch || nai010 publishers || 350×260×20mm || 960g || 178p

P 0289

荣誉奖 || 茶典 || 中国 || 潘焰荣 || 陆羽（唐）等 || 商务印书馆
Honorary Appreciation || Tea Canon || China // *Pan Yanrong* // Lu Yu (Tang Dynasty) and others || The Commercial Press || 211×140×29mm || 665g || 772p

P 0293

荣誉奖 || 孟山都：摄影调查 || 委内瑞拉
Honorary Appreciation || Monsanto: A Photographic Investigation || Venezuela // *Ricardo Báez* // Mathieu Asselin || Verlag Kettler (English), Actes Sud (Français) || 300×270×21mm || 1266g || 148p

荣誉奖 ||
Honorary Appreciation || II Concurso Nacional de Poesía Joven Rafael Cadenas || Venezuela // *Juan Fernando Mercerón. Giorelis Niño (Assistent)* // AutoresVzlanos, Team Poetero Ediciones, Libros del Fuego (Editorial Production)

2017

P 0307

金字符奖 |||||| 鸟类学 || 荷兰
Golden Letter |||||| Ornithology || The Netherlands // *Jeremy Jansen* // Anne Geene, Arjan de Nooy || de HEF publishers, Rotterdam || 240×171×26mm || 666g || 336p

金奖 |||||
Gold Medal ||||| Palimpsest || Czech Republic // *Petr Jambor* // Petr Jambor || 4AM Fórum pro architekturu a média

P 0315

银奖 |||| 虫子书 || 中国 || 朱赢椿、皇甫珊珊 || 朱赢椿、皇甫珊珊 || 广西师范大学
Silver Medal |||| Bugs' Book || China // *Zhu Yingchun & Huang Fu Shanshan* // Zhu Yingchun & Huang Fu Shanshan || Guangxi Normal University Press || 200×141×25mm || 493g || 298p

P 0321

银奖 |||| （未）预料到的 || 荷兰
Silver Medal |||| (un) expected || The Netherlands // *Rob van HoeselPeter Dekens* // The Eriskay Connection, Amsterdam || 290×208×4mm || 357g || 144p

P 0327

铜奖 ||| 扣押原因: || 瑞士
Bronze Medal ||| Withheld due to: || Switzerland // *typosalon, Christof Nüssli* // Christof Nüssli || cpress || 224×167×19mm || 600g || 408p

P 0333

铜奖 ||| 拔地而起——约翰内斯堡的高楼故事 || 德国
Bronze Medal ||| UP UP – Stories of Johannesburg's Highrises || Germany // *Huber / Sterzinger, Zürich (CH)* // Nele Dechmann, Fabian Jaggi, Katrin Murbach, Nicola Ruffo || Hatje Cantz Verlag, Ostfildern || 270×192×24mm || 926g || 336p

P 0339

铜奖 ||| 交织的新鲜 || 荷兰
Bronze Medal ||| DWARS VERS || The Netherlands // *Team Thursday (Simone Trum & Loes van Esch)* // Emily Dickinson, Edna St. Vincent Millay || Ans Bouter en Benjamin Groothuyse || 210×150×20 mm || 469g || 272p

P 0345

铜奖 ||| 贝尔纳·夏德贝克——友好的警示 || 瑞士
Bronze Medal ||| Bernard Chadebec – Intrus Sympathiques || Switzerland // *Olivier Lebrun and Urs Lehni in collaboration with Simon Knebl, Phil Zumbruch, Saskia Reibel and Tatjana Stürmer (HfG Karlsruhe)* // Olivier Lebrun and Urs Lehni || Rollo Press, Zürich || 191×135×21 mm || 470g || 272p

P 0351

铜奖 ||| 伊娃·黑塞日记 || 瑞士
Bronze Medal ||| Eva Hesse – Diaries || Switzerland // *NORM, Zürich / Johannes Breyer, Berlin* // Eva Hesse / Barry Rosen || Hauser & Wirth Publishers / Yale University Press || 204×136×50 mm || 1190g || 900p

P 0357

荣誉奖 || 错误的线索 || 德国
Honorary Appreciation || Falsche Fährten || Germany // *Jonas Voegeli, Kerstin Landis, Scott Vander Zee – Hubertus Design, Zürich* // Peter Radelfinger || Edition Patrick Frey, Zürich || 296×213×42mm || 1756g || 584p

P 0363

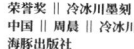

荣誉奖 || 冷冰川墨刻 || 中国 || 周晨 || 冷冰川 || 海豚出版社
Honorary Appreciation || Ink Rubbing by Leng Bingchuan || China // *Zhou Chen* // Leng Bingchuan || Dolphin Books || 260×184×39mm || 1622g || 642p

P 0367

荣誉奖 || 21世纪运动百科 || 日本
Honorary Appreciation || Encyclopedia of Modern Sport || Japan // *Shin Tanaka, Sozo Naito* // Toshio Nakamura, Takeo Takahashi, Tsuneo Sougawa, Hidenori Tomozoe || TAISHUKAN Publishing Co., Ltd. || 270×204×60mm || 2556g || 1378p

P 0371

荣誉奖 || 真的吗? || 葡萄牙
Honorary Appreciation || VERDADE?! || Portugal // *Pato Lógico* // Bernardo P. Carvalho || Pato Lógico || 256×200×9mm || 325g || 36p

P 0375

荣誉奖 || 信号突然中断 || 德国
Honorary Appreciation || Plötzlich Funkstille || Germany // *Benjamin Courtault* // Benjamin Courtault, Paris (FR) || kunstanstifter verlag, Mannheim || 305×187×9mm || 333g || 32p

1101

P 0389

金字符奖 |||||| 其他证据：蒙上眼睛 || 荷兰
Golden Letter |||||| Other Evidence: Blindfold || The Netherlands // *Titus Knegtel, Amsterdam* // Titus Knegtel || Titus Knegtel, Amsterdam || 239×160×40mm || 1117g || 335+335p

P 0397

金奖 ||||| 订单：方圆故事 || 中国 || 李瑾 || 吕重华 || 广西美术出版社
Gold Medal ||||| Order – The Story of Fangyuan Bookshop || China // *Li Jin* // Lü Chonghua || Guangxi Fine Arts Publishing House || 220×200×22mm || 557g || 259p

P 0403

银奖 |||| 2×100 mil. m2 || 捷克
Silver Medal |||| 2×100 mil. m2 || Czech Republic // *Mikuláš Macháček, Linda Dostálková* // Martin Hejl et. al. || Kolmo.eu, Prague || 200×135×32mm || 464g || 404p

BRICK
AN
EXACTING
MATERIAL

P 0409

银奖 |||| 砖块，一种精确的材料 || 荷兰及弗兰德斯地区
Silver Medal |||| Brick. An Exacting Material || The Netherlands and Flanders // *Studio Joost Grootens (Joost Grootens, Dimitri Jeannottat, Silke Koeck, Hanae Shimizu, Julie da Silva)* // Jan Peter Wingender || Architectura & Natura, Amsterdam || 245×172×32mm || 782g || 352p

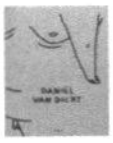
DANIEL
VAN DICHT

P 0415

铜奖 ||| 丹尼尔·范迪奇 || 比利时
Bronze Medal ||| Daniël van Dicht || Belgium // *Dear Reader, Matthias Phlips (Illustrations)* // Matthias Phlips || Lannoo || 320×240×18mm || 917g || 144p

Het
meest
geloofde
sprookje

P 0421

铜奖 ||| 最被人相信的童话 || 荷兰
Bronze Medal ||| Het Meest Geloofde Sprookje || The Netherlands // *De Vormforensen (Annelou van Griensven & Anne-Marie Geurink) & Lyanne Tonk* // De Vormforensen (Annelou van Griensven & Anne-Marie Geurink) || De Vormforensen (distribution De Vrije Uitgevers) || 326×222×8mm || 452g || 64p

P 0427

铜奖 ||| 漂泊者——全球化大潮中的非法移民 || 奥地利
Bronze Medal ||| Nowhere Men. Illegale Migranten im Strom der Globalisierung || Austria // *Christoph Miler, Zürich (CH)* // Christoph Miler || Luftschacht, Wien || 226×160×26mm || 669g || 336p

P 0433

铜奖 ||| 学而不厌 || 中国 || 曲闵民、蒋茜 || 周学 || 江苏凤凰美术出版社
Bronze Medal ||| Pleasure of Learning || China // *Qu Minmin, Jiang Qian* // Zhou Xue || Phoenix Fine Arts Publishing Ltd. || 232×152×26mm || 547g || 414p

Ingvar Ambjørnsen

P 0439

铜奖 ||| 小说再见：游离的 24 小时 || 挪威
Bronze Medal ||| Farvel til romanen. 24 timer i grenseland || Norway // *Aslak Gurholt Rønsen / Yokoland* // Ingvar Ambjørnsen, Editor: Bendik Wold og Nils-Øivind Haagensen / Flamme Forlag || Flamme Forlag || 212×134×14mm || 293g || 127p

P 0445

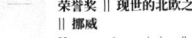

荣誉奖 || 现世的北欧之光 || 挪威
Honorary Appreciation || LIVING THE NORDIC LIGHT || Norway // *Snøhetta* // Åsne Seierstad, Po Tidholm, Lars Forsberg, Barbara Szybinska Matusiak, Vidje Hansen, Bruno Laeng Zumtobel AG || 271×213×37mm || 1394g || 400p

P 0449

荣誉奖 || 无偏见的职业选择 || 捷克
Honorary Appreciation || Career Choice Without Prejudice || Czech Republic // *Jan Šiller* // Anna Babanová, Jitka Hausenblasová, Jitka Kolářová, Tereza Krobová, Irena Smetáčková || Gender Studies, Prague || 297×211×10mm || 293g || 112p

P 0453

荣誉奖 || 照片的背后 || 委内瑞拉
Honorary Appreciation || Del reverso de las imágenes || Venezuela // *Álvaro Sotillo, Juan F. Mercerón (Assistent)* // Paolo Gasparini, Victoria de Stefano (Text), Ana Nuño (Translation) || Editorial mal de ojo || 201×315×15mm || 682g || 88p

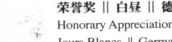

P 0457

荣誉奖 || 白昼 || 德国
Honorary Appreciation || Jours Blancs || Germany // *Nanni Goebel, Kehrer Design Heidelberg* // François Schaer (Photographs), Pauline Martin (Text) || Kehrer Verlag Heidelberg, Berlin || 330×300×14mm || 1058g || 100p

P 0461

荣誉奖 || 脱掉——我的邻居亨利 || 荷兰
Honorary Appreciation || Taking off. Henry my neighbor || The Netherlands // *Mariken Wessels, Jurgen Maelfeyt* // Mariken Wessels || Art Paper Editions, Gent || 330×242×25mm || 1617g || 326p

P 0475

金字符奖 |||||| 无题（九月杂志）保罗·艾利曼，2003 || 比利时

Golden Letter |||||| Untitled (September Magazine) Paul Elliman, 2003 || Belgium // *Paul Elliman, Julie Peeters* // Paul Elliman || Roma Publications and Vanity Press || 285×219×22mm || 1595g || 592p

P 0483

金奖 ||||| 原教旨主义者和其他阿拉伯现代主义——来自阿拉伯世界的建筑 1914-2014 || 瑞士

Gold Medal ||||| Fundamentalists and Other Arab Modernisms. Architecture from the Arab World 1914–2014 || Switzerland // *Jonathan Hares, Lausanne* // George Arbid, Kingdom of Bahrain National Participation, Biennale di Venezia 2014 || Bahrain Ministry of Culture, Bahrain; Arab Centre for Architecture, Beirut || 335×245×16mm || 713g || 160p

P 0489

银奖 |||| 丹麦艺术家的书 || 丹麦

Silver Medal |||| Danish Artists' Books / Danske Kunstnerbøger || Denmark // *Louise Hold Sidenius* // Thomas Hvid Kromann, Maria Kjær Themsen, Louise Sidenius, Marianne Vierø || Møller and Verlag der Buchhandlung Walther König || 325×238×30mm || 1448g || 302p

P 0495

银奖 |||| 与世隔绝 || 荷兰

Silver Medal |||| Sequester || The Netherlands // *Hans Gremmen* // Awoiska van der Molen || Fw:Books, Amsterdam || 290×245×14mm || 684g || 86p

MOTION SILHOUETTE

P 0501

铜奖 ||| 动态剪影 || 日本

Bronze Medal ||| Motion Silhouette || Japan // *Megumi KAJIWARA, Tatsuhiko NIIJIMA* // Megumi KAJIWARA, Tatsuhiko NIIJIMA || Megumi KAJIWARA, Tatsuhiko NIIJIMA || 250×149×8mm || 190g || 12p

P 0507

铜奖 ||| 米克洛什·克劳斯·罗饶 || 德国

Bronze Medal ||| Miklós Klaus Rózsa || Germany // *Christof Nüssli, Christoph Oeschger* // Christof Nüssli, Christoph Oeschger || Spector Books / cpress, Leipzig || 298×211×43mm || 2140g || 624p

P 0513

铜奖 ||| 陈佩之的新新约 || 瑞士

Bronze Medal ||| Paul Chan. New New Testament || Switzerland // *Kloepfer-Ramsey Studio, Brooklyn* // Laurenz Foundation, Schaulager, Basel; Badlands Unlimited, New York || Laurenz Foundation, Schaulager, Basel || 272×182×74mm || 2838g || 1092p

P 0519

铜奖 ||| 基础元素 01 || 比利时

Bronze Medal ||| Éléments Structure 01 || Belgium // *Alexis Jacob, Valérian Goalec* // Valérian Goalec, Béatrice Lortet || Théophile's Papers || 232×174×5mm || 109g || 60p

P 0525

铜奖 ||| 新地平线 || 荷兰

Bronze Medal ||| New Horizons || The Netherlands // *Rob van Hoesel* // Bruno van den Elshout || The Eriskay Connection, Breda || 340×235×68mm || 3993g || 212p

P 0531

荣誉奖 || 名片 || 罗马尼亚

Honorary Appreciation || Cartea de vizită || Romania // *72+87* // Fabrik-72+87 || Fabrik || 297×235×14mm || 680g || 64p

P 0535

荣誉奖 || 两个以上（让它们成为自己）|| 加拿大

Honorary Appreciation || More Than Two (Let It Make Itself) || Canada // *Jeff Khonsary (The Future)* // Micah Lexier || The Power Plant || 264×195×20mm || 877g || 224p

P 0539

荣誉奖 || 遗忘中的艺术处理 / 艾伯特宾馆 / 以巴斯特基顿的样子穿过巴格达 || 瑞士

Honorary Appreciation || Art Handing In Oblivion (fink twice 501) / Albert's Guesthouse (fink twice 502) / Walking through Baghdad with a Buster Keaton Face (fink twice 503) || Switzerland // *Rob van Leijsen, Georg Rutishauser, Sonja Zagermann, Petra Elena Köhle, Nicolas Vermot-Petit-Outhenin, Thomas Galler* // Rob van Leijsen / Petra Elena Köhle, Nicolas Vermot-Petit-Outhenin / Thomas Galler || edition fink, Verlag für zeitgenössische Kunst, Zürich || 200×145×21mm / 190×130×9mm / 210×168×10mm || 366g / 134g / 212g || 382p / 158p / 174p

P 0543

荣誉奖 || 塞托图书馆系列 || 爱沙尼亚

Honorary Appreciation || SERIES The Seto Library. Seto Kirävara || Estonia // *Agnes Ratas* // various || Seto Instituut || 166×114×23mm / 166×114×16mm / 247×166×23mm / 247×166×27mm / 247×166×22mm || 262g / 181g / 580g / 1038g / 522g || 288p / 176p / 256p / 396p / 200p

Vetřelci a volavky
Aliens and Herons

P 0547

荣誉奖 || 外星人和苍鹭 || 捷克

Honorary Appreciation || Aliens and Herons || Czech Republic // *Tereza Hejmová* // Pavel Karous (Ed.) || Arbor vitae, Academy of Arts, Architecture and Design, Prague || 240×167×38mm || 814g || 459p

P 0561

金字符奖 |||||| 梅雷特·奥本海姆——不要用有毒的字母包裹话语 || 瑞士
Golden Letter |||||| Meret Oppenheim. Worte nicht in giftige Buchstaben einwickeln || Switzerland // *Bonbon, Valeria Bonin und Diego Bontognali, Zurich* // Lisa Wenger, Martina Corgnati || Scheidegger & Spiess, Zurich || 330×220×38mm || 2018g || 474p

P 0569

金奖 ||||| Buchner Bründler——建筑物 || 德国
Gold Medal ||||| Buchner Bründler – Bauten || Germany // *Design Concept: Ludovic Balland, Andreas Bründler, Daniel Buchner – Basel (Switzerland) Design and Composition: Ludovic Balland und Gregor Schreiter / Ludovic Balland Typography Cabinet – Basel (Switzerland)* // gta D ARCH Ausstellungen, ETH Zürich (Ed.) || gta Verlag, Zurich || 273×232×36mm || 1643g || 338p

P 0575

银奖 |||| 无序的目录 || 奥地利
Silver Medal |||| Katalog der Unordnung || Austria // *Christoph Schorkhuber, Linz // Helmuth Lethen, IFK* // Internationales Forschungszentrum Kulturwissenschaften an der Kunstuniversität Linz, Linzt || 240×162×16mm || 478g || 191p

银奖 ||||
Silver Medal |||| Schwarze Hunde & Bunte Schafe || Austria // *Lisa Maria Matzi, Wien* // Lisa Maria Matzi

JAK

P 0581

铜奖 ||| JAK || 德国
Bronze Medal ||| JAK || Germany // *Demian Bern, Stuttgart* // JAK, Hamed Taheri || EXP.edition, Stuttgart || 245×166×23mm || 700g || 108p

P 0587

铜奖 ||| Keiko || 德国
Bronze Medal ||| Keiko || Germany // *Marek Mielnicki / veryniceworks, Warschau (Polen)* // Tomasz Gudzowaty (Hrsg.) || Hatje Cantz Verlag, Ostfildern || 319×243×20mm || 1531g || 144p

LANGE
LISTE

P 0593

铜奖 ||| 1979-1997 年朗格家的账单 || 德国
Bronze Medal ||| Lange Liste 79 – 97 || Germany // *Christian Lange, München* // Christian Lange || Spector Books, Leipzig || 340×241×19mm || 1038g || 186p

P 0599

铜奖 ||| 刘小东在和田 & 新疆新观察 || 中国 || 小马哥、橙子 || 侯瀚如、欧宁 || 中信出版社
Bronze Medal ||| Liu Xiaodong's Hotan Project & Xinjiang Research || China // *Xiao Mage & Cheng Zi* // Hou Hanru, Ou Ning || China CITIC Press || 239×175×25mm || 947g || 346p

铜奖 |||
Bronze Medal ||| Typografia niepokorna || Poland // *Monika Hanulak* // Monika Hanulak || Miasto Stołeczne Warszawa and Pracownia Ilustracji Akademii Sztuk Pięknych w Warszawie

P 0605

荣誉奖 || 你好石头 || 伊朗
Honorary Appreciation || Hello Stone (Sange Salam) || Iran // *Majid Zare* // Mohammad Reza Bayrami || Asr-e Dastan || 189×130×12mm || 235g || 240p

荣誉奖 ||
Honorary Appreciation || Ik ben een gemankeerde saxofonist. Lucebert & jazz || The Netherlands // *Piet Gerards Ontwerpers* // Ben IJpma, Ben van Melick || Huis Clos – Rimburg / Amsterdam

P 0609

荣誉奖 || 来自许多不同的世界——关于奥莱·罗伯特·桑德的写作 || 挪威
Honorary Appreciation || Som fra mange ulike verdener Om Ole Robert Sundes forfatterskap || Norway // *Andreas Töpfer* // Red. Audun Lindholm || Gyldendal Norsk Forlag AS || 211×135×26mm || 438g || 279p

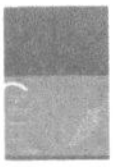

P 0613

荣誉奖 || 2010-2012 中国最美的书 || 中国 || 刘晓翔 || 上海市新闻出版局"中国最美的书"评委会 || 上海人民美术出版社
Honorary Appreciation || The Beauty of Books in China 2010–2012 || China // *Liu Xiaoxiang, Liu Xiaoxiang Studio* // Organizing Committee of "The Beauty of Books in China" of Shanghai Press and Publication Administration || Shanghai People's Fine Arts Publishing House || 291×192×40mm || 1709g || 294p

P 0617

荣誉奖 || Tottorich || 日本
Honorary Appreciation || Tottorich || Japan // *Masahiko Nagasawa* // Yuan Okada || Doyo Bijutsusha Shuppan Hanbai || 215×145×11mm || 273g || 93p

P 0631

金字符奖 |||||| 坠落 || 德国

Golden Letter |||||| Fallen || Germany // *Hans-Jörg Pochmann* // Gian-Philip Andreas, Gesine Palmer || Eigenverlag Hans-Jörg Pochmann, Leipzig || 205×130×12mm || 190g || 72p

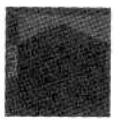

P 0639

金奖 ||||| 这座山是我 || 荷兰

Gold Medal ||||| Cette montagne, c'est moi || The Netherlands // *Hans Gremmen* // Witho Worms || Fw: Books, Amsterdam || 240×222×17mm || 813g || 175p

P 0645

银奖 |||| 坐火车的抹香鲸 || 中国(台湾地区) || NOBU || 王彦铠 || 大块文化出版股份有限公司

Silver Medal |||| A Cachalot on a Train || Taiwan, China // *NOBU* // Wang, Yen-Kai || Locus Publishing Company Limited || 210×150×13mm || 400g || 173p

P 0651

银奖 |||| 鲁迅箴言 || 日本

Silver Medal |||| The Words of Lu Xun || Japan // *Kenya Hara / Tei Rei* // Lu Xun || Heibonsha Limited Publishers, SDX Joint Publishing Company || 172×113×20mm || 285g || 259p

P 0657

铜奖 ||| 文化转型计划——鹿特丹、苏黎世、南特、兰斯塔德、波尔多 || 比利时

Bronze Medal ||| Changing Cultures of Planning– Rotterdam, Zürich, Nantes, Randstad, Bordeaux || Belgium // *Joris Kritis with Rustan Söderling* // Nathanaëlle Baës-Cantillon, Joachim Declerck, Michiel Dehaene, Sarah Levy || Architecture Workroom Brussels || 297×211×21mm || 1075g || 272p

P 0663

铜奖 ||| 从 Aa 到 Zz——字体排印简史 || 爱沙尼亚

Bronze Medal ||| Aa – Zz. CONCISE HISTORY OF TYPOGRAPHY || Estonia // *Ivar Sakk* // Ivar Sakk || Sakk&Sakk || 285×178×38mm || 1062g || 447p

P 0669

铜奖 ||| 隐秘的轮子 || 比利时

Bronze Medal ||| La roue voilée || Belgium // *PLMD (pleaseletmedesign) + interns: Audé Gravé, Rosalie Wagner* // David Widart || L'Amicale Books || 239×170×6mm || 174g || 84p

P 0675

铜奖 ||| 狗的生活 || 波兰

Bronze Medal ||| Psie życie || Poland // *Józef Wilkoń and Piotr Gil* // Józef Wilkoń || Wydawnictwo HOKUS-POKUS Marta Lipczyńs-ka-Gil || 226×205×7mm || 345g || 36p

P 0681

铜奖 ||| 云的研究 || 德国

Bronze Medal ||| Wolkenstudien / Cloud Studies / Études de Nuages || Germany // *Helmut Völter, Leipzig* // Marcel Beyer, Helmut Völter (Hrsg.) || Spector Books, Leipzig || 275×206×25mm || 1075g || 272p

P 0687

荣誉奖 || 罗伯特·隆戈——木炭画 || 德国

Honorary Appreciation || Robert Longo – Charcoal || Germany // *Stapelberg & Fritz, Stuttgart* // Hal Foster, Kate Fowle, Thomas Kellein || Hatje Cantz Verlag GmbH, Ostfildern || 304×259×33mm || 2114g || 251p

P 0691

荣誉奖 || 我们 || 德国

Honorary Appreciation || WIR || Germany // *Gaston Isoz, Berlin* // Jewgenij Samjatin || disadorno edition, Berlin || 217×120×22mm || 450g || 215+19p

P 0695

荣誉奖 || 寻找录像——电影曝光 || 荷兰

Honorary Appreciation || Found Footage. Cinema Exposed || The Netherlands // *Joseph Plateau, grafisch ontwerpers* // Marente Bloemheuvel, Giovanna Fossati, Jaap Guldemond (Hrsg.) || Amsterdam University Press, EYE Film Institute Netherlands, Amsterdam || 258×209×18mm || 438g || 255p

P 0699

荣誉奖 || 关于意义、真理及其他的探讨 || 荷兰

Honorary Appreciation || An Inquiry into Meaning and Truth and More… || The Netherlands // *Edwin van Gelder / Mainstudio* // Thomas Raat || Onomatopee, Eindhoven || 300×200×7mm || 330g || 80p

荣誉奖 ||

Honorary Appreciation || Nihil obstat. Lietuvos fotografija sovietmečiu || Lithuania // *Tomas Mrazauskas* // Margarita Matulytė || Vilniaus dailės akademijos leidykla / Vilnius Academy of Arts Press

2012

P 0713

金字符奖 |||||| 居所 || 丹麦
Golden Letter |||||| La Résidence || Denmark // *Greger Ulf Nilson* // JH Engström || Journal, Stockholm || 247×186×24mm || 823g || 180p

P 0721

金奖 ||||| 空间，随着时间扭转 || 奥地利
Gold Medal ||||| Raum, verschraubt mit der Zeit / Space, Twisted with Time || Austria // *Gabriele Lenz, Elena Henrich / büro für visuelle gestaltung* // Hubertus Adam || Birkhäuser, Basel || 285×207×20mm || 1066g || 186p

P 0727

银奖 |||| 剪纸的故事 || 中国 || 吕旻、杨婧 || 赵希岗 || 人民美术出版社
Silver Medal |||| The Story of Paper-cut || China // *Lü Min, Yang Jing* // Zhao Xigang || People's Fine Arts Publishing House, Beijing || 250×183×21mm || 434g || 342p

UTIL
ITÉ

P 0733

银奖 |||| 效用 || 荷兰
Silver Medal |||| Utilité || The Netherlands // *-SYB-(Sybren Kuiper)* // Ellen Korth || Ellen Korth, Deventer || 200×150×51mm || 1202g || 552p

P 0739

铜奖 ||| 我们穿着的 || 荷兰
Bronze Medal ||| What We Wear || The Netherlands // *Teun van der Heijden / Heijdens Karwei* // Pieter van den Boogert || Keff & Dessing Publishing, Amsterdam || 226×174×26mm || 691g || 140p

铜奖 |||
Bronze Medal ||| IM WALGAU GEMEINDEN gemeinsam || Austria // *Marcel Bachmann, Andrea Redolfi / Atelier Gassner-Redolfi*

P 0745

铜奖 ||| 画出一个假设 || 奥地利
Bronze Medal ||| Drawing a Hypothesis || Austria // *Simona Koch* // Nikolaus Gansterer || Springer Wien New York || 215×145×25mm || 574g || 348p

P 0751

铜奖 ||| 以有机为名 || 中国（台湾地区）|| 一元搥搥（方信元）|| 徐岩奇，徐岩奇建筑师事务所 || 田园城市出版社
Bronze Medal ||| Zoom in, Zoom out || Taiwan, China // *Fang Hsin-Yuan* // Victor HSU Architect & Partners; Zoom Design Atelier (Ed.) || Garden City Publishers Ltd., Taipei || 265×190×20mm || 661g || 228p

P 0757

铜奖 ||| 利布谢·尼克洛娃 || 捷克
Bronze Medal ||| Libuše Niklová || Czech Republic // *Zuzana Lednická / Studio Najbrt* // Tereza Bruthansová || Arbor vitae societas, Prag || 206×193×29mm || 975g || 299p

P 0761

荣誉奖 || 文爱艺诗集 || 中国 || 刘晓翔、高文 || 文爱艺 || 作家出版社
Honorary Appreciation || Poems by Wen Aiyi || China // *Liu Xiaoxiang, Gao Wen* // Wen Aiyi || Writers Press, Beijing || 235×144×23mm || 617g || 380p

P 0765

荣誉奖 || 蛾摩拉女孩 || 荷兰
Honorary Appreciation || Gomorrah Girl || The Netherlands // *-SYB-(Sybren Kuiper)* // Valerio Spada || Cross Editions, Paris || 335×227×3mm || 330g || 40p

P 0769

荣誉奖 || Pampilio || 波兰
Honorary Appreciation || Pampilio || Poland // *Monika Hanulak* // Irena Tuwim, Monika Hanulak || Wytwórnia, Warschau || 287×237×9mm || 455g || 32p

荣誉奖 ||
Honorary Appreciation || 100 % Ivanovo. Agitprop Fabrics and Designs of 1920–1930s || Russia // *Evgeny Korneev* // from the Collection of the Dmitry Burylin Museum. Vladimir Potanin Charity Foundation, Moskau

P 0773

荣誉奖 || 2666 || 瑞典
Honorary Appreciation || 2666 || Sweden // *Nina Ulmaja* // Roberto Bolaño || Albert Bonniers Förlag, Stockholm || 233×165×47mm || 1180g || 1056p

2011

P 0787

金字符奖 |||||| 以巴冲突地图集 || 荷兰
Golden Letter |||||| Atlas of the Conflict. Israel-Palestine || The Netherlands // *Studio Joost Grootens* // Malkit Shoshan || 010 Publishers, Rotterdam || 201×119×30mm || 597g || 479p

P 0795

金奖 ||||| 转型舆图 || 捷克
Gold Medal ||||| Atlas Transformace || Czech Republic // *Adéla Svobodová* // Zbyněk Baladrán, Vít Havránek, Věra Krejčcová (Ed.) || tranzit.cz, Prag || 220×163×40mm || 1271g || 834p

Bemoei
Zorg

P 0801

银奖 |||| 干预式护理 || 荷兰
Silver Medal |||| Bemoeizorg || The Netherlands // *René Put* // Jules Tielens, Maurits Verster || De Tijdstroom, Amsterdam || 190×120×25mm || 460g || 335p

TITLE

P 0807

银奖 |||| 标题 || 瑞士
Silver Medal |||| Title || Switzerland // *Ramaya Tegegne* // Ramaya Tegegne || Ramaya Tegegne, Genève || 219×174×17mm || 296g || 160p

P 0813

铜奖 ||| __ 与威廉——一个青年的档案 || 荷兰
Bronze Medal ||| — and Willem. Documentation of a Youth || The Netherlands // *Robin Uleman* // Willem Popelier || post editions, Rotterdam || 230×169×11mm || 393g || 170p

gerd
arntz
graphic
designer

P 0819

铜奖 ||| 格尔德·阿恩茨，平面设计师 || 荷兰
Bronze Medal ||| Gerd Arntz. Graphic Designer || The Netherlands // *Ontwerpwerk* // Ed Annink, Max Bruinsma (Ed.) || 010 Publishers, Rotterdam || 247×180×24mm || 763g || 288p

P 0825

铜奖 ||| 木质外墙 || 奥地利
Bronze Medal ||| Fassaden aus Holz || Austria // *Reinhard Gassner, Marcel Bachmann / Atelier Reinhard Gassner* // proHolz Austria (Ed.) || proHolz Austria, Wien || 296×211×15mm || 823g || 160p

铜奖 |||
Bronze Medal ||| Jak jsem potkal d'ábla || Poland // *Honza Zamojski* // Honza Zamojski || Galeria Miejska Arsenał, Wydawnictwo Morava, Poznań

P 0831

铜奖 ||| 克里斯蒂安·瓦尔德福格尔的极致地球 || 瑞士
Bronze Medal ||| Christian Waldvogel. Earth Extremes || Switzerland // *Jonas Voegeli, Christian Waldvogel, Benjamin Roffler* // Jacqueline Burckhardt, Christian Waldvogel, Jonas Voegeli (Ed.) || Verlag Scheidegger & Spiess AG, Zürich || 320×214×41mm || 2283g || 495p

P 0837

荣誉奖 || 漫游：建筑体验与文学想象 || 中国 || 小马哥、橙子 || 欧宁 || 中国青年出版社
Honorary Appreciation || Odyssey : Architecture and Literature || China // *Xiao Mage & Cheng Zi* // Ou Ning (Ed.) || China Youth Press, Beijing || 240×170×22mm || 792g || 448p

P 0843

荣誉奖 || 傻小子学害怕 || 德国
Honorary Appreciation || Von einem, der auszog das Fürchten zu lernen || Germany // *Doris Freigofas / Golden Cosmos* // Doris Freigofas || Golden Cosmos, Berlin || 370×243×4mm || 304g+24g || 24p+8p

P 0847

荣誉奖 || 留白 || 德国
Honorary Appreciation || Leerzeichen fur Applaus || Germany // *Jenna Gesse* // Jenna Gesse || Eigenverlag Jenna Gesse, Bielefeld || 184×114×12mm || 159g || 96p

P 0851

荣誉奖 || 地铁里的扎齐 || 德国
Honorary Appreciation || Zazie in die Metro || Germany // *Joe Villion* // Raymond Queneau || Büchergilde Gutenberg, Frankfurt am Main || 245×152×19mm || 615g || 221p

荣誉奖 ||
Honorary Appreciation || Serie Teaching Architecture. 3 Positions Made in Switzerland (Hong Kong Typology / Important Buildings / Radical Mix in Hanoi) || Switzerland // *Ludovic Balland, Ivan Weiss / Ludovic Balland / Ludovic Balland / Typography Cabinet, Basel* // gta Verlag, Zürich / Kaleidoscope Press, Milano / Kaleidoscope Press

2010

P 0865

金字符奖 |||||| XX-——打字机上的特殊字符 SS || 德国

Golden Letter |||||| XX-. Die SS-Rune als Sonderzeichen auf Schreibmaschinen || Germany // *Elisabeth Hinrichs, Aileen Ittner, Daniel Rother* // Elisabeth Hinrichs, Aileen Ittner, Daniel Rother || Institut für Buchkunst Leipzig an der HGB, Leipzig || 314×227×22mm || 1190g || 324p

P 0873

金奖 ||||| 托马斯·加勒——以巴斯特·基顿的样子穿过巴格达 || 瑞士

Gold Medal ||||| Thomas Galler. Walking through Baghdad with a Buster Keaton Face || Switzerland // *Georg Rutishauser* // Madeleine Schuppli, Aargauer, Kunsthaus, Aarau (Ed.) || edition fink, Verlag für zeitgenössische Kunst, Zürich || 300×240×15mm || 805g || 175p

P 0879

银奖 |||| 永恒的现代性——建筑界的五个代表 || 奥地利

Silver Medal |||| konstantmodern. Fünf Positionen zur Architektur || Austria // *Reinhard Gassner, Marcel Bachmann / Atelier Reinhard Gassner* // aut. architektur und tirol (Ed.) || Springer WienNewYork || 238×168×20mm || 648g || 256p

P 0885

银奖 |||| 大自然的声音——Pro Natura 100 周年 || 瑞士

Silver Medal |||| Die Stimme der Natur. 100 Jahre pro Natura || Switzerland // *Marco Müller* // Pro Natura (Ed.) Kontrast Verlag, Zürich; Edition Slatkine, Genève || 250×169×18mm || 670g || 196p

P 0891

铜奖 ||| 保留所有权 || 德国

Bronze Medal ||| Eigentumsvorbehalt || Germany // *Gaston Isoz* // Gaston Isoz (Ed.) || disadorno edition, Berlin || 305×224×11mm || 738g || 88p

P 0897

铜奖 ||| 威尼斯大都市 / 全球形势图集 || 德国

Bronze Medal ||| Migropolis. Venice / Atlas of a Global Situation || Germany // *Wolfgang Scheppe mit Katerina Dolejšová, Veronica Bellei, Miguel Cabanzo* // Wolfgang Scheppe, The IUAV Class on Politics of Representation (Ed.) || Hatje Cantz Verlag, Ostfildern || 240×170×73mm || 3252g || 1344p

P 0903

铜奖 ||| 1989 年是故事的结束，还是未来的开始? || 奥地利

Bronze Medal ||| 1989. Ende der Geschichte oder Beginn der Zukunft? || Austria // *Chris Goennawein* // Kunsthalle Wien, Gerald Matt, Cathérine Hug, Thomas Mießgang (Ed.) || Verlag für moderne Kunst Nürnberg || 240×190×33mm || 819g || 317p

P 0909

铜奖 ||| 一本戏剧家的日记本 || 瑞典

Bronze Medal ||| En dramatikers dagbok || Sweden // *Nina Ulmaja* // Lars Norén || Albert Bonniers Förlag, Stockholm || 209×148×52mm || 1354g || 1680p

P 0915

铜奖 ||| 虚空——回顾展 || 瑞士

Bronze Medal ||| Voids. A Retrospective || Switzerland // *Gilles Gavillet, David Rust mit Corinne Zellweger / Gavillet & Rust* // Mathieu Copeland mit John Armleder, Laurent Le Bon u. a. (Ed.) || JRP | Ringier Kunstverlag, Zürich || 286×222×37mm || 2134g || 544p

P 0921

荣誉奖 || 诗经 || 中国 || 刘晓翔 || 向熹（译注） || 高等教育出版社

Honorary Appreciation || The Book of Songs || China // *Liu Xiaoxiang* // Xiang Xi (Ed.) || Higher Education Press, Beijing || 257×167×27mm || 946g || 391p

OPEN
CITY

P 0925

荣誉奖 || 开放城市——设计共存 || 荷兰

Honorary Appreciation || Open City. Designing Coexistence || The Netherlands // *Mevis & Van Deursen mit Werkplaats Typografie* // Tim Rieniets, Jennifer Sigler, Kees Christiaanse (Ed.) || SUN architecture, Amsterdam || 270×202×29mm || 1051g || 464p

Trener*

P 0929

荣誉奖 || 教与练 * 数学 || 波兰

Honorary Appreciation || Trener* matematyka || Poland // *Grzegorz Podsiadlik* // Jan Górowski, Adam Łomnicki || Wydawnictwo Szkolne PWN, ParkEdukacja, Warszawa / Bielsko-Biała || 242×205×15mm || 671g || 328p

P 0933

荣誉奖 || 尤勒斯·施皮纳奇 || 瑞士

Honorary Appreciation || Jules Spinatsch || Switzerland // *Winfried Heininger / Kodoji Press* // Marco Obrist, Kunsthaus Zug (Ed.) || Kodoji Press, Baden || 235×170×25mm || 877g || 349p

MIX

P 0937

荣誉奖 || 赫鲁多斯·瓦劳谢克的谚语新解 || 捷克

Honorary Appreciation || Chrudošův mix přísloví || Czech Republic // *Juraj Horváth* // Chrudoš Valoušek || Baobab, Praha || 207×203×10mm || 320g || 46p

L'IMAGIER
GENS

P 0951

金字符奖 |||||| 人的意象 || 法国

Golden Letter |||||| L'Imagier des Gens || France // *Blexbolex* // Blexbolex || Albin Michel Jeunesse, Paris || 246×186×23mm || 632g || 196p

Vinex
Atlas

P 0959

金奖 ||||| Vinex 地图集 || 荷兰

Gold Medal ||||| Vinex Atlas || The Netherlands // *Joost Grootens mit Tine van Wel, Jim Biekmann und Anna Iwansson (Studio Joost Grootens)* // Jelte Boeijenga, Jeroen Mensink || Uitgeverij 010 Publishers, Rotterdam || 344×243×26mm || 1726g || 303p

BAGHDAD
CALLING

P 0965

银奖 |||| 巴格达的呼吁——关于土耳其、叙利亚、约旦和伊拉克的报告 || 荷兰

Silver Medal |||| Baghdad Calling. Reportages uit Turkije, Syrië, Jordanië en Irak || The Netherlands // *Mevis & Van Deursen* // Geert van Kesteren || episode publishers, Rotterdam || 254×192×14mm || 598g || 388p

P 0971

银奖 |||| Marte. Marte 建筑事务所 || 奥地利

Silver Medal |||| Marte. Marte Architects || Austria // *Reinhard Gassner & Andrea Redolfi (Atelier Reinhard Gassner)* // Bernhard Marte, Stefan Marte || Springer-Verlag Wien / New York || 225×162×41mm || 1059g || 415p

P 0977

铜奖 ||| 博伊斯——我们即革命 || 德国

Bronze Medal ||| BEUYS. Die Revolution sind wir || Germany // *Detlef Fiedler, Daniela Haufe, Nina Polumsky, Daniel Wiesmann (cyan)* // Eugen Blume, Catherine Nichols (Ed.) || Steidl Verlag, Göttingen || 305×247×41mm || 2458g || 407p

P 0983

铜奖 ||| 维特拉计划 || 德国

Bronze Medal ||| Projekt Vitra. || Germany // *Cornel Windlin* // Rolf Fehlbaum, Cornel Windlin (Ed.) || Birkhäuser Verlag, Basel / Boston / Berlin || 244×174×34mm || 1106g || 396p

P 0989

铜奖 ||| 以色列、阿拉伯人、巴勒斯坦 / 时间实验室 || 法国

Bronze Medal ||| Israël, Les Arabes, La Palestine / Les Laboratoires du Temps || France // *Julien Hourcade, Thomas Petitjean (Hey Ho)* // Série «Essais», Jean Daniel / Alain Fleischer || Galaade Éditions, Paris || 215×140×37mm / 215×140×28mm || 984g / 478g || 859p / 417p

P 0995

铜奖 ||| 阿尔普的艺术——图纸、拼贴、浮雕、雕塑、诗歌 || 法国

Bronze Medal ||| Art is Arp. Dessins, collages, reliefs, sculptures, poésie || France // *Sp Millot* // Editions des Musées de la Ville de Strasbourg, Strasbourg || 280×230×19mm || 1140g || 343p

P 1001

铜奖 ||| 方法 || 法国

Bronze Medal ||| Méthodes || France // *Manuela Dechamps Otamendi* // Cédric Libert & Atelier d'architecture Pierre Hebbelinck – Pierre Hebbelinck & Pierre de Wit Architectes (Ed.) || Wallonie-Bruxelles International, Bruxelles || 228×151×29mm || 807g || 360p

P 1007

荣誉奖 || 中国记忆——五千年文明瑰宝 || 中国 || 敬人设计工作室　吕敬人、吕旻 || 首都博物馆 || 文物出版社

Honorary Appreciation || The Chinese Memory: Treasures of the 5000-year Civilization || China // *Lü Jingren & Lü Min (Jingren Art Design Studio)* // Capital Museum (Ed.) || Cultural Relics Press, Beijing || 359×239×35mm || 2777g || 352p

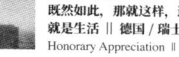

P 1011

荣誉奖 || 这就是它的样子 || 德国

Honorary Appreciation || So. siehts aus || Germany // *Kerstin Rupp* // Kerstin Rupp || Eigenverlag Kerstin Rupp, Leipzig || 426×294×14mm || 1149g || 96p

P 1015

荣誉奖 || Hardau——既然如此，那就这样，这就是生活 || 德国 / 瑞士

Honorary Appreciation || Hardau. Claro que si, c'est comme ça, c'est la vie || Germany / Switzerland // *Julia Ambroschütz, Jeannine Herrmann (Südpol)* // Julia Ambroschütz, Jeannine Herrmann || Salis Verlag, Zürich || 303×240×20mm || 911g || 120p

P 1019

荣誉奖 || 蒙德里安基金会 2007 年度报告 || 荷兰

Honorary Appreciation || Mondriaan Stichting. Jaarverslag 2007 || The Netherlands // *Ingeborg Scheffers* // Mondriaan Stichting, Amsterdam || 275×211×15mm || 723g || 220p

P 1023

荣誉奖 || 弗兰肯斯坦合集

Honorary Appreciation || Frankenstein Set || Switzerland // *Geoff Han* // Christian Jankowski || Christoph Keller Editions / JRP Ringier Kunstverlag, Zürich || 180×104×23mm || 345g || 440p

P 1037

金字符奖 ||||| 委内瑞拉感性地理学史 || 委内瑞拉
Golden Letter ||||| Geohistoria de la Sensibilidad en Venezuela || Venezuela // *Álvaro Sotillo mit Gabriela Fontanillas & Luis Giraldo* // Pedro Cunill Grau || Fundación Empresas Polar, Caracas || 285×220×71mm || 1618g+1638g || 528p

P 1045

金奖 |||| __ 生命中的一天 || 瑞士
Gold Medal |||| Ein Tag im Leben von || Switzerland // *Elektrosmog (Valentin Hindermann & Marco Walser & Simone Koller)* // Walter Keller (Ed.) || Salis Verlag AG, Zürich || 314×247×25mm || 1274g || 312p

银奖 ||||
Silver Medal |||| Handbuch der wildwachsenden Großstadtpflanzen || Germany // *Helmut Völter* // Helmut Völter || Institut für Buchkunst an der Hochschule für Grafik und Buchkunst, Leipzig

P 1051

银奖 |||| 江户川鸟类大词典 || 日本
Silver Medal |||| The Birds and Birdlore of Tokugawa Japan || Japan // *Omori Yuji* // Suzuki Michio (Ed.) || Heibonsha Ltd., Publishers, Tokyo || 269×192×64mm || 2355g || 762+51p

P 1057

铜奖 ||| 合一 || 德国
Bronze Medal ||| Oneness || Germany // *Hester Fell* // Mariko Mori || Hatje Cantz Verlag, Ostfildern || 261×209×25mm || 1337g || 304p

P 1063

铜奖 ||| 布雷姆的灭绝动物清单 || 德国
Bronze Medal ||| Brehms verlorenes Tierleben || Germany // *Hanna Zeckau & Carsten Aermes* // Hanna Zeckau, Carsten Aermes || Zweitausendeins, Frankfurt am Main || 250×175×20mm || 734g || 257p

GILBERT SORRENTINO
SALMIGONDIS
ÉDITIONS CENT PAGES
P 1069

铜奖 ||| Cent Pages 精选系列 || 法国
Bronze Medal ||| Collection cent pages || France // *sp Millot* // Éditions cent pages, Grenoble || 195×125×20mm || 392g || 512p

P 1075

铜奖 ||| 人民的中心 || 瑞士
Bronze Medal ||| Die Mitte des Volkes || Switzerland // *Prill & Vieceli* // Fabian Biasio, Margrit Sprecher || Edition Patrick Frey, Zürich || 285×211×10mm || 593g || 187p

P 1081

铜奖 ||| 维特拉家居集 2007–2008 || 瑞士
Bronze Medal ||| Vitra. The Home Collection 2007/2008 || Switzerland // *Cornel Windlin* // Vitra AG, Birsfelden || 274×210×15mm || 647g || 171p

P 1085

荣誉奖 || 之后 || 中国 || 耿耿、王成福 || 孙乃强、王磊、王成福、史长胜 || 天津杨柳青画社
Honorary Appreciation || The After Concept & Works Book || China // *Wang Chengfu & Geng Geng* // Sun Naiqiang, Wang Lei, Wang Chengfu, Shi Changsheng || Tianjin Yangliuqing Fine Arts Press, Tianjin || 256×216×40mm || 1710g || 456p

P 1089

荣誉奖 || 蒙德里安基金会2006年度报告 || 荷兰
Honorary Appreciation || Mondriaan Stichting. Jaarverslag 2006 || The Netherlands // *Joost Grootens* // Mondriaan Stichting, Amsterdam || 275×215×13mm || 587g || 192p

P 1093

荣誉奖 || 爸爸在身旁 || 葡萄牙
Honorary Appreciation || pe de pai || Portugal // *Bernardo P. Carvalho (Planeta Tangerina)* // Isabel Martins, Bernardo P. Carvalho || Planeta Tangerina, São Pedro do Estoril || 220×198×4mm || 145g || 32p

P 1097

荣誉奖 || 卡斯滕·尼科莱——渐显渐隐 || 瑞士
Honorary Appreciation || Carsten Nicolai. Static Fades || Switzerland // *Gavillet & Rust* // Dorothea Strauss, Haus Konstruktiv, Zürich (Ed.) || JRP Ringier Kunstverlag AG, Zürich || 286×238×13mm || 892g || 157p

P 1101

荣誉奖 || 合作工作：苏特纳尔－苏德克 || 捷克
Honorary Appreciation || Družstevní práce: Sutnar – Sudek || Czech Republic // *Štěpán Malovec & Martin Odehnal* // Lucie Vlčková (Ed.) || Uměleckoprůmyslové museum v Praze, Prague || 270×205×20mm || 1151g || 250p

P 1115

金字符奖 |||||| 新构造图集 || 瑞士 / 美国
Golden Letter |||||| Atlas of Novel Tectonics || Switzerland / USA // *Reto Geiser, Donald Mak* // Jesse Reiser, Nanako Umemoto || Princeton Architectural Press, New York || 191×127×20mm || 282g || 255p

P 1123

金奖 ||||| 希拉·希克斯——编织的隐喻 || 荷兰
Gold Medal ||||| Sheila Hicks. Weaving as Metaphor || The Netherlands // *Irma Boom* // Arthur C. Danto, Joan Simon, Nina Stritzler-Levine || Yale University Press, New Haven and London || 221×147×48mm || 1110g || 415p

P 1129

银奖 |||| 2 千克 KesselsKramer || 日本
Silver Medal |||| 2 kilo of KesselsKramer || Japan // *KesselsKramer* // PIE Books, Tokyo || 257×148×59mm || 2027g || 880p

P 1135

银奖 |||| Wo-Wo-Wonige! **|| 瑞士**
Silver Medal |||| Wo-Wo-Wonige! || Switzerland // *Jonas Voegeli in Zusammenarbeit mit Beni Roffler* // Thomas Stahel || Paranoia city Verlag, Zürich

P 1141

铜奖 ||| 不裁 || 中国 || 朱赢椿 || 古十九 || 江苏文艺出版社
Bronze Medal ||| stitching up || China // *Zhu Yingchun* // Gu Shijiu || Jiangsu Literature & Art Publishing House, Nanjing || 239×163×14mm || 417g || 189p

P 1147

铜奖 ||| 适合初学者的罗宋汤 || 德国
Bronze Medal ||| Borsch für Anfänger || Germany // *Tatjana Triebelhorn* // Tatjana Triebelhorn, Roman Triebelhorn || Eigenverlag Tatjana Triebelhorn, Stuttgart || 295×184×15mm || 552g || 103p

P 1153

铜奖 ||| 尤利西斯 || 德国
Bronze Medal ||| Ulysses || Germany // *Werner Zegarzewski* // James Joyce || Suhrkamp Verlag, Frankfurt am Main || 193×122×47mm || 676g || 987p

P 1159

铜奖 ||| 炊具 || 加拿大
Bronze Medal ||| L'Appareil || Canada // *Anouk Pennel, Raphaël Deaudelin (Feed)* // Les Éditions de la Pastèque, Montréal || 250×214×20mm || 784g || 192p

P 1165

铜奖 ||| 眼睛——鹿特丹眼科医院 || 荷兰
Bronze Medal ||| OOG. Het Oogziekenhuis Rotterdam / EYE. The Rotterdam Eye Hospital || The Netherlands // *Irma Boom* // Ineke van Ginneke (Ed.) || Het Oogziekenhuis Rotterdam / The Rotterdam Eye Hospital || 250×200×9mm || 348g || 96p

荣誉奖 ||
Honorary Appreciation || CABK 2006 || The Netherlands // *DesignArbeid* // CABK-ArtEZ, Zwolle

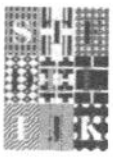
P 1171

荣誉奖 || 阿姆斯特丹市立博物馆 2005 年度报告 || 荷兰
Honorary Appreciation || Stedelijk Museum Amsterdam 2005. Jaarverslag / Annual Report || The Netherlands // *Richard Niessen, Esther de Vries* // Carolien de Bruijn (Ed.) || Stedelijk Museum Amsterdam || 230×160×16mm || 539g || 176p

P 1175

荣誉奖 || 时尚图书——奥地利的当代时装 || 奥地利
Honorary Appreciation || Modebuch. Zeitgenössische Mode aus Österreich || Austria // *halle34 Albert Handler / Marcus Arige* // Unit F büro für mode (Ed.) || Unit F büro für mode, Wien || 254×178×23mm || 895g || 158p

P 1179

荣誉奖 || 木材图谱——样本、信息和参数比较 || 奥地利
Honorary Appreciation || Holzspektrum – Ansichten, Beschreibungen und Vergleichswerte || Austria // *Reinhard Gassner, Marcel Bachmann, Stefan Gassner (Atelier Reinhard Gassner)* // Josef Fellner, Alfred Teischinger, Walter Zschokke || proHolz Austria – Arbeits-gemeinschaft der österrei-chischen Holzwirtschaft, Wien || 308×214×24mm || 1328g || 112p+96p

P 1183

荣誉奖 || 捷克最有代表性的 100 个设计 || 捷克
Honorary Appreciation || Czech 100 Design Icons || Czech Republic // *Zuzana Lednická, Aleš Najbrt (Studio Najbrt)* // Tereza Bruthansová, Jan Králíček || CzechMania, Prag || 220×162×17mm || 622g || 100p

P 1197

金字符奖 |||||| 雅各布·德穆斯 1983-2005 年的作品 || 荷兰

Golden Letter |||||| Jakob Demus. The Complete Graphic Work 1983–2005 || The Netherlands // *Marie-Cécile Noordzij* // Ed de Heer (Ed.) || Hercules Segers Stichting, Amsterdam || 286×208×20mm || 1196g || 222p

P 1205

金奖 ||||| 世界大都会地图集 || 荷兰

Gold Medal ||||| Metropolitan World Atlas || The Netherlands // *Joost Grootens* // Arjen van Susteren || 010 Publishers, Rotterdam || 215×180×31mm || 896g || 311p

P 1211

银奖 |||| 埃伦·加拉格尔——暴雪的白色 || 荷兰

Silver Medal |||| Ellen Gallagher. Blizzard of White || The Netherlands // *Irma Boom (Irma Boom Office)* // The Fruitmarket Gallery, Edinburgh; Hauser & Wirth, Zürich / London || 120×155×88mm || 1350g || 992p

P 1217

银奖 |||| 手工艺战略——欧洲非凡项目的 7 幅肖像 || 奥地利

Silver Medal |||| Strategien des Handwerks. Sieben Porträts au ergewöhnlicher Projekte in Europa || Austria // *Clemens Theobert Schedler (Büro für konkrete Gestaltung)* // Landschaft des Wissens (Ed.) || Haupt Verlag AG, Bern / Stuttgart / Wien || 245×180×39mm || 1054g || 367p

P 1223

铜奖 ||| 凯科宁 || 芬兰

Bronze Medal ||| Kekkonen || Finland // *Matti Hagelberg* // Matti Hagelberg || Kustannusosakeyhtiö Otava, Helsinki || 219×217×21mm || 750g || 200p

铜奖 |||

Bronze Medal ||| Stedelijk Museum Amsterdam 2003/2004 || The Netherlands // *Ben Laloua, Didier Pascal (mit Maaike Molenkamp)* // Stededelijk Museum, Amsterdam

P 1229

铜奖 ||| 祖父 || 荷兰

Bronze Medal ||| Opa || The Netherlands // *Irma Boom (Irma Boom Office)* // Paul de Reus || Uitgeverij Van Waveren, Amsterdam || 291×209×9mm || 390g || 44p

LIVI
NG
LIBR
ARY

P 1235

铜奖 ||| 维尔·阿雷兹的"活的图书馆" || 荷兰

Bronze Medal ||| Living Library: Wiel Arets || The Netherlands // *Irma Boom (Irma Boom Office)* // Marijke Beek (Ed.) || Utrecht University Library; Prestel Verlag, München || 235×170×22mm || 897g || 456p

P 1241

铜奖 ||| 别处也是同一处 || 瑞士

Bronze Medal ||| Somewhere else is the same place || Switzerland // *Iza Hren, Georg Rutishauser* // Monica Studer, Christoph van den Berg, Christoph Vögele (Ed.) || Kunstmuseum Solothurn; edition fink, Zürich || 240×170×23mm || 690g || 176p

P 1245

荣誉奖 || 曹雪芹风筝艺术 || 中国 || 赵健工作室 || 孔祥泽、孔令民、孔炳彰 || 中国友谊出版公司

Honorary Appreciation || Cao Xueqin's Art of Kite || China // *Zhaojian Studio* // Kong Xiangze, Kong Lingmin, Kong Bingzhang || Beijing Arts & Crafts Publishing House, Peking || 260×231×19mm || 608g || 140p

bill,

P 1249

荣誉奖 || 马克斯·比尔——画家、雕塑家、建筑师、设计师 || 德国

Honorary Appreciation || Max Bill. Maler, Bildhauer, Architekt, Designer || Germany // *Sascha Lobe, Ina Bauer (L2M3 Kommunikationsdesign)* // Thomas Buchsteiner, Otto Letze (Ed.) || Hatje Cantz Verlag, Ostfi ldern-Ruit || 270×230×26mm || 1367g || 296p

P 1253

荣誉奖 || 欧罗波特 De Beer 自然保护区 || 荷兰

Honorary Appreciation || Broedplaats Europoort v/h ›De Beer‹ || The Netherlands // *René Put* // Paul Bogaers u.a. || Artimo, Amsterdam || 240×165×7mm || 277g || 104p

P 1257

荣誉奖 || 俄罗斯插图杂志 1703-1941 || 俄罗斯

Honorary Appreciation || Russian Illustrated Magazine. 1703–1941 || Russia // *Masha Lyuledgian* // Irina Mokhnacheva || Agey Tomesh, Moskau || 356×218×45mm || 2607g || 368p

P 1261

荣誉奖 || 童话故事 || 捷克

Honorary Appreciation || Pohádky || Czech Republic // *Luboš Drtina* // Hermann Hesse || Argo, Prag || 174×116×14mm || 152g || 165p

P 1275

金字符奖 |||||| 日本近代活字 || 日本
Golden Letter |||||| Nihon no Kindai Katsuji || Japan // *Mitsuo Katsui* // Nihon no Kindai Katsuji (Ed.) || Kindai Insatsu Katsuji Bunka Hozonkai, Nagasaki || 304×216×35mm || 2220g || 453p

P 1283

金奖 ||||| 克里斯蒂安·瓦尔德福格尔的中空集群 || 瑞士
Gold Medal ||||| Christian Waldvogel. Globus Cassus || Switzerland // *The Remingtons (Ludovic Balland & Jonas Voegeli) mit Michael Stauffer und Christian Waldvogel* // Bundesamt für Kultur (Ed.) || Lars Müller Publishers, Baden || 215×158×15mm || 432g || 182p

P 1289

银奖 |||| 弗洛伊德那些消失的邻居们 || 奥地利
Silver Medal |||| Freuds verschwundene Nachbarn || Austria // *Walter Pamminger* // Lydia Marinelli (Ed.) || Turia + Kant, Wien || 292×190×13mm || 540g || 127p

P 1295

银奖 |||| 想象的地图：作为制图师的作家 || 美国
Silver Medal |||| Maps of the Imagination: The Writer as Cartographer || USA // *Julie Savasky & DJ Stout (Pentagram, Austin)* // Peter Turchi || Trinity University Press, San Antonio, TX || 218×149×29mm || 632g || 245p

P 1301

铜奖 ||| 1954年的日记 || 德国
Bronze Medal ||| Tagebuch aus dem Jahr 1954 || Germany // *Friedrich Forssman* // Alice Schmidt || Arno Schmidt Stiftung im Suhrkamp Verlag, Frankfurt am Main || 255×170×20mm || 803g || 334p

P 1307

铜奖 ||| 金子先生的有机家庭菜园 || 日本
Bronze Medal ||| Kaneko-sanchino Yukikateisaien || Japan // *Hirokazu Ando* // Yoshinori Kaneko || Ie-No-Hikari Association, Tokyo || 257×188×16mm || 576g || 192p

P 1313

铜奖 ||| 从拉脱维亚到里米尼——苏黎世湖泊和河流浴场的历史和现状 || 瑞士
Bronze Medal ||| Vom Letten bis Rimini || Switzerland // *Thomas Bruggisser* // Nina Chen || hier + jetzt, Verlag für Kultur und Geschichte, Baden || 216×168×10mm || 259g || 96p

P 1319

铜奖 ||| 安妮·弗兰克与家人 || 荷兰
Bronze Medal ||| Anne Frank en familie || The Netherlands // *MV (Victor Levie & Barbara Herrmann)* // Otto Frank || Anne Frank Stichting, Amsterdam / Anne Frank Fonds, Basel || 155×154×14mm || 245g || 104p

P 1325

铜奖 ||| 阿姆斯特丹市立博物馆的家具收藏 || 荷兰
Bronze Medal ||| The Furniture Collection, Stedelijk Museum Amsterdam || The Netherlands // *Beukers Scholma* // Luca Dosi Delfini || NAi Publishers, Rotterdam / Stedelijk Museum Amsterdam || 316×250×41mm || 2992g || 456p

P 1331

荣誉奖 || 朱叶青杂说系列 || 中国 || 何君 || 朱叶青 || 中国友谊出版公司
Honorary Appreciation || The Philosophy of Mist / The Sensation of Wateriness / Look up at the Sky / The Ism of Antique / Consommé and Water || China // *He Jun* // Zhu Yeqing || China Friendship Publishing Co., Peking || 210×152×57mm || 949g || 751p

P 1335

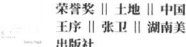

荣誉奖 || 土地 || 中国 || 王序 || 张卫 || 湖南美术出版社
Honorary Appreciation || Antony Gormley – Asian Field || China // *Wang Xu & Associates* // Hu Fang, Zhang Weiwei (Ed.) Zhang Wei (Ass. Ed.) || Hunan Fine Arts Publishing House, Hunan || 227×177×45mm || 1273g || 488p

荣誉奖 ||
Honorary Appreciation || Mondriaan Stichting. Jaarverslag 2003 || The Netherlands // *-SYB- (met dank aan Robin Uleman)* // Mondriaan Stichting, Amsterdam

P 1339

荣誉奖 || 关于编制2005年至2008年阿姆斯特丹艺术计划的建议 || 荷兰
Honorary Appreciation || Adviezen ter voorbereiding van het Amsterdams Kunstenplan 2005 t/m 2008 || The Netherlands // *Hansje van Halem & Gerard Unger* // Amsterdamse Kunstraad, Amsterdam || 234×165×9mm || 316g || 264p

P 1343

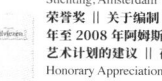

荣誉奖 || 奥拉维尔·埃利亚松——关注世界 || 瑞士
Honorary Appreciation || Olafur Eliasson. Minding the world || Switzerland // *groenland.berlin.basel (Dorothea Weishaupt & Michael Heimann)* // ARoS Aarhus Kunstmuseum (Ed.) || Hatje Cantz Verlag, Ostfildern-Ruit || 285×238×26mm || 1238g || 232p

P 1357

金字符奖 |||||| << 倒带 >> 快进 || 德国
Golden Letter |||||| <<RE-WIND>>FORWARD || Germany // *Kristina Brusa, Markus Dreßen* // Olaf Nicolai, Susanne Pfleger || Hatje Cantz Verlag, Ostfildern-Ruit || 267×215×25mm || 1126g || 168p

P 1365

金奖 ||||| 梅兰芳（藏）戏曲史料图画集 || 中国 || 张志伟、蠹鱼阁、高绍红 || 刘曾复、朱家溍 || 河北教育出版社
Gold Medal ||||| A Picture Album. Historical Materials of Mei Lanfang's Theatrical Performances || China // *Zhang Zhiwei, Du Yuge, Gao Shaohong* // Liu Cengfu, Zhu Jiajin || Hebei Education Publishing House, Shijiazhuan || 295×220×43mm || 1702g || 229p

P 1371

银奖 |||| 寻美之心——小林秀雄诞辰一百周年纪念展 || 日本
Silver Medal |||| The Heart in Search of Beauty. An Exhibition Honoring the 100th Anniversary of Kobayashi Hideo's Birth || Japan // *Kodama Yuuko, Ono Risa, Kawasina Hiroya* // Hideo Kobayashi || Nihon Keizai Shimbun, Tokyo; Shinchosha, Tokyo || 226×156×27mm || 678g || 256p

P 1377

银奖 |||| 保罗·策兰诗集 || 荷兰
Silver Medal |||| Paul Celan Verzamelde gedichten || The Netherlands // *Joost van de Woestijne; office of cc* // Paul Celan || Meulenhoff Amsterdam || 221×150×53mm || 1194g || 864p

P 1383

铜奖 ||| 1842–1932 年工人的诗歌 || 德国
Bronze Medal ||| Arbeiterlyrik 1842–1932 || Germany // *Gaston Isoz* // Heinz Ludwig Arnold (Ed.) || Parthas Verlag, Berlin || 256×169×29mm || 800g || 255p

P 1389

铜奖 ||| 卡拉瓦乔的罗马——走进一处失调之地 || 德国
Bronze Medal ||| Caravaggios Rom. Annäherungen an ein dissonantes Milieu || Germany // *Andreas Trogisch* // Lothar Sickel || Edition Imorde, Emsdetten / Berlin || 240×144×19mm || 454g || 268p

P 1395

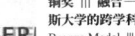

铜奖 ||| 融合——穆特修斯大学的跨学科周 || 德国
Bronze Medal ||| Vermischungen. Interdisziplinäre Wochen an der Muthesius-Hochschule || Germany // *Julia Hofmeister und Tim Albrecht (Muthesius-Hochschule)* // Theresa Georgen (Ed.) || Muthesius-Hochschule, Kiel || 285×188×11mm || 359g || 144p

P 1401

铜奖 ||| 生活用品（伟大的书 22） || 德国
Bronze Medal ||| Lebens-Mittel (Die Tollen Hefte 22) || Germany // *Thomas M. Müller* // Armin Abmeier, Rotraut Susanne Berner (Ed.) || Büchergilde Gutenberg, Frankfurt am Main / Wien / Zürich || 207×137×7mm || 104g || 31p

铜奖 |||
Bronze Medal ||| Symbiont || Czech Republic // *Denisa Myšková* // Denisa Myšková || Denisa Myšková an der Vysoká škola umělecko-průmyslová, (všup) Praha

P 1407

荣誉奖 || 无人永生 || 德国
Honorary Appreciation || nobody forever || Germany // *Christine Lohmann, Walter Kerkhofs* // Nick J. Swart || Edition Hamtil, Hamburg/ Tilburg || 355×288×29mm || 1477g || 72p

P 1413

荣誉奖 || 书籍的艺术——哈里·凯斯勒伯爵的 Cranach 出版社 || 德国
Honorary Appreciation || Das Buch als Kunstwerk. Die Cranach Presse des Grafen Harry Kessler || Germany // *Sabine Golde* // John Dieter Brinks (Ed.) || Triton Verlag, Laubach / Berlin || 315×255×50mm || 2923g || 453p

P 1417

荣誉奖 || 头衔：皮特·吉拉德，平面设计师 || 荷兰
Honorary Appreciation || Werktitel: Piet Gerards, grafisch ontwerper. || The Netherlands // *Bureau Piet Gerards* // Ben van Melick || Uitgeverij 010 Publishers, Rotterdam || 231×168×25mm || 802g || 366p

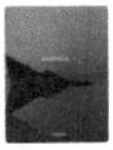

P 1421

荣誉奖 || 梦境——一本有关梦的书 || 挪威
somnia. En bok om drømmer || Norway // *Geir Henriksen* // Julie Cathrine Knarvik, Anders Aabel || Christian Schibsteds Forlag, Oslo || 245×176×28mm || 926g || 232p

P 1425

荣誉奖 || 委内瑞拉的生物多样性 || 委内瑞拉
Honorary Appreciation || Biodiversidad en Venezuela || Venezuela // *Álvaro Sotillo, Luis Giraldo* // Marisol Aguilera, Aura Azócar, Eduardo González Jiménez (Ed.) || Fundación Polar, Ministerio de Ciencia y Tecnología, Fondo Nacional de Ciencia, Tecnología e Innovación (fonacit), Caracas || 248×173×85mm || 1663g+1695g || 1080p

Nachbemerkungen — *Postscript*

跋

Bereits im 19. Jahrhundert sagte der deutsche Philosoph Hegel: „Schönheit ist die Manifestation von Ideen.“ Die Leipziger Auswahl „Schönste Bücher der Welt“ sollte man weniger als etwas zufällig Entstandenes ansehen als vielmehr ein ästhetisches Großereignis, auf das mühsam und unter großem Einsatz über lange Zeit hinweg hingearbeitet wurde. Es ist wie in Alexander Gottlieb Baumgartens erstem Buch, wo die Ästhetik als eine eigene Kategorie eingestuft wird. Wenn wir begreifen, dass Ästhetik eine emotionale Erkenntnis ist, kann die Schönheit, die durch dieses Verständnis geschaffen wird, endlich die Basis für eine angemessene Auslegung der Kunst schaffen.

Leipzig ist wie ein neues Berlin, voller Enthusiasmus in der Welt des künstlerischen Schaffens. In der Stadt ist es gelungen, Bücher aus der ganzen Welt zusammenzubringen und damit einen Raum zu öffnen für das nachhaltige Streben zum Verständnis der Buchgestaltung. Bei der Leipziger Auswahl kann keine Schönheit die andere ersetzen. Die Auswahl bei der „Schönheit“ ist vielfältig und reichhaltig, sie trägt verschiedenen Nationalitäten, unterschiedlichem Denken, verschiedenen künstlerischen Ausdrucksformen und kreativen Kerninhalten Rechnung. Es ist gerate diese bunte Vielfalt von Schönheit, die mich dazu veranlasst, ihre Ausdrucksmittel, die Bücher, mit anderen zu teilen, eine Hommage an das Buch darzubringen und eine Zeremonie für das Design zu vollführen.

In Shanghai wurden 2004 erstmalig mehr als ein Dutzend Bücher für die Sammlung „Chinas schönste Bücher“ ausgewählt, um im folgenden Jahr an der in Leipzig durchgeführten Wahl der weltweit schönsten Bücher teilzunehmen. China gewann eine Goldmedaille, und so ist 2004 das Ursprungsjahr für „Chinas schönste Bücher“ geworden. In den seitdem bis 2019 vergangenen anderthalb Jahrzehnten stammen 22 der weltweit schönsten Bücher aus China.

Dieses Buch beginnt seine Dokumentation im Jahre 2004. Es

In the nineteenth century, the German philosopher Hegel said, "Beauty is an emotional manifestation of an idea." Therefore, saying that the competition Best Book Designs from all over the World held in Leipzig started at an opportune moment does not mean that it has not accumulated knowledge about aesthetics over the years. Like him, Baumgarten – another German philosopher – gave aesthetics the position of a category in his first book. Only when we can also realize that aesthetics is a kind of cognitive perception, the beauty based on this understanding will be finally best interpreted in art.

The city of Leipzig seems a new Berlin whose art circles brim with enthusiasm. This city has collected books from many distant places all over the world and started a competition to understand and pursue the best book design. In the choices of Leipzig, there is not a kind of beauty that can replace another. In these choices, "beauty" is diverse and rich, represents different nationalities, ideas, art expressions, and creations. These extraordinarily varied forms of beauty have given me more reasons to share these books and to carry out a ceremony to pay my respects to them.

In 2004, in Shanghai, I decided for the first time to collect more than ten books that had been selected and awarded in the competition *Best Book Designs from all over the World* held in Leipzig. These books became the first edition of *China's Most Beautiful Books*. In the sixteen years after that, twenty weo Chinese books have been elected *Best Book Designs from all over the World*.

最美的选择

早在 19 世纪的时候，德国哲学家黑格尔就提出：“美是理念的感性显现。”与其说莱比锡“世界最美的书”是一场应运而生的比赛，倒不如说这是一场厚积薄发的美学盛宴。正如鲍姆加登出版的第一本书，它赋予了审美以范畴的地位。当我们也能同样领悟到审美是一种感性的认知时，基于这种理解所创造出来的美，才能最终在艺术中达到理想的最佳诠释。

莱比锡就像一座新柏林，在艺术创造圈里洋溢着澎湃的热情。这个城市从很远的地方回来，把世界各地的书都汇聚到一起，开启了一场对于书籍设计的理解与追求。在莱比锡的选择中，没有哪一种美是能够取代另一种的。“美”的选择是多元并且丰富的，它们携带着不同的国籍、不同的思考，不同的艺术表现与创作内核。这些异彩纷呈的美，让我更有理由要将这些书籍都分享出来，完成一场关于向书籍致敬，一场属于设计的仪式。

中国上海于 2004 年首次组织了十几本“中国最美的书”，参加了次年在莱比锡举办的“世界最美的书”评选并一举获得了

war unser Anliegen, auf mehr als tausend Seiten in diese Galerie für jedes Jahr der Zeit von 2014 bis 2019 jeweils 14 preisgekrönte Werke aufzunehmen. Die „Leipziger Auswahl" basiert auf eigenen Auswahlkriterien angesichts der zahlreichen internationalen Buchdesign-Wettbewerbe und ist zu etwas wie einem statistischen Nachschlagewerk der „schönsten Bücher" geworden. Zugleich handelt es sich bei der Sammlung auch um ein mobiles Kunstmuseum mit einem zeitlosen Ausstellungsthema – dem einzigartigen Charme der weltweiten Buchgestaltungskunst.

Aus verschiedenen Gründen waren wir nicht in der Lage, alle Bücher aus diesen anderthalb Jahrzehnten in die Sammlung aufzunehmen, mit insgesamt 208 Büchern kommen wir auf einen Anteil von 93%, es fehlen 16 Bücher noch.

Unser Dank geht an Frau Uta Schneider, die ehemalige Geschäftsführerin der Stiftung für Buchkunst. Die ersten Manuskriptentwürfe zu diesem Buch haben ihr gefallen, ihr spontaner Ausruf, das Buch in dieser Form zu veröffentlichen, war eine große Ermutigung für mich. Mein Dank geht auch an einen der Wegbereiter unseres Gewerbes der Buchgestaltung, Herrn Lü Jingren, der mir stets mit Rat und Tat zur Seite stand. Er lieferte nicht nur wichtige und immer wieder weiterführende Ideen für dieses Buch, vielmehr war es ihm auch stets ein Anliegen, aus dem ursprünglichen „Verpackungscharakter" chinesischer Buchumschläge überhaupt erst eine Designkunst zu machen.

Natürlich möchte ich auch der derzeitigen Geschäftsführerin der Stiftung Buchkunst, Frau Katharina Hesse, danken. Wir hatten zwar nur gelegentlich miteinander zu tun, doch war uns ihre Unterstützung für dieses Buch immer sicher. Nicht zuletzt geht mein Dank auch an meinen langjährigen Freund Liu Xiaoxiang, den Direktor des Book Design and Art Committee der China Publishing Association. Ihm verdanken wir es, dass China in der Vergangenheit so erfolgreich bei der „Leipziger Auswahl der weltweit schönsten Bücher" abgeschnitten hat. Zahlreiche wertvolle Kommentare von ihm haben zur Entstehung dieses Buches beigetragen.

Ich danke weiterhin Herrn Zhu Junbo, dem ehemaligen Direktor des Presse- und Publikationsbüros in Shanghai. Die ur-

This book started to be put together in 2004. Every year, in *Best Book Designs from all over the World* fourteen books were awarded. Since 2004 until 2019, sixteen years have passed and the awarded works have reached more than one thousand pages. Among many international book design competitions and exhibitions, *The Choice of Leipzig* presents the criteria and features for its selections. It is a systematic reference work about "the most beautiful books" while, at the same time, it is also a moving art museum. Its subject is one that will never be outdated – the singular beauty of the art of international book design.

Of course, because of numerous different reasons, we have not been able to collect all the books of these sixteen years. At this point, we have collected 93%, 208 books in total. Only sixteen books are missing.

I would like to thank Uta Schneider, the managing director of the Stiftung Buchkunst, for her joyful words when she saw the first draft. With them, she gave me a great deal of encouragement. I would also like to express my gratitude to Lu Jingren, a member of the previous generation of book designers who has always guided and helped me along the road of this profession. He has not only been an important guide for the concepts in the book, but has also provided numerous indications on the evolution of China's ancient book binding into book design.

Of course, I also want to thank Katharina Hesse, current president the Stiftung Buchkunst. Even though we have only met a few times, she leaves a deep impression on people. For sure, she has also provided an important foundation for this book. In addition, I would also like to thank my old friend Liu Xiaoxiang, who has been the winner of the most prizes of

金奖，所以说 2004 年成为“中国最美的书”的元年，自此至 2019 年的 16 年间，来自中国的书籍设计产生了 22 本“世界最美的书”。

这本书便由 2004 年开始收录。每年 14 本获奖作品，从 2004 年直至 2019 年，前后跨越 16 年，最终在纸上幻化成 1000 余页的陈列。《莱比锡的选择》是在众多国际书籍设计比赛展览中呈现出它的选择标准和面貌，是一本关于“最美的书”的统计辞典。与此同时，它也是一座移动艺术馆，有着永不过期的展览主题——关于世界书籍设计艺术的独特魅力。

当然由于种种原因，我们未能收藏到这 16 年所有的书，现达到了 93%，即 208 本，离完全收录相差 16 本。

感谢德国图书艺术基金会的前任主席乌塔女士，当看到呈上的初步样稿时，她欣喜的一句“就这样出吧”给了我极大的鼓励。也感谢一直在书籍设计之路上给予我指导提携帮助的书籍设计前辈吕敬人先生，他不仅给本书提供了重要的思想指导，更是一直持续发力指引着中国书籍从“装帧”到“书籍设计”的观念更新。

当然也要感谢德国图书艺术基金会的现任主席、美丽的黑塞女士，虽然只是几面之交，但给人印象深刻，她的肯定也为本书

sprüngliche Initiative zur Anfertigung einer Auswahl „Chinas schönste Bücher" geht auf Herrn Zhu zurück, ihm ist es zu verdanken, dass Chinas Buchdesign den Weg auf die Bühne in Leipzig fand.

Meiner Meinung nach ist es gut möglich, dass Bücher durch eine künstlerische Veränderung der sprachlichen Form eine neue Ära einleiten und wir zu einem ganz neuen Verständnis und einer neuen Definition von Papiermedien kommen. „Die schönsten Bücher der Welt" zeigt, was möglich ist: Exquisites, Alternatives, Blutiges, Dokumentarisches und Fantastisches begegnen einander lebendig und frisch auf dem Papier. Es geht nicht nur um Papier als Material, das man berühren kann und als Gegenstand eines Druckprozesses, sondern viel mehr noch um das Lesen des Textes und die Wertschätzung des Visuellen. Alles soll zu einer harmonischen Schönheit zusammengefügt werden, erst so ergibt sich das ganze Bild des Buches.

Der Verlauf zur Herstellung dieses Buches war nicht einfach, allein die Sammlung der prämierten Werke war außerordentlich mühsam und mit vielen Schwierigkeiten verbunden. Für eine langfristig angelegte Suche mussten verschiedene Kanäle eröffnet werden. Es konnte sein, dass andernorts ganz plötzlich ein Wandel in den Entwicklungen eingetreten war, der Empfang lange ersehnter Pakete aus der ganzen Welt löste stets Freude aus, es gab so manches Anekdotenreifes, aber auch geradezu Lächerliches. Zum Beispiel eine mit der „Goldenen Letter" ausgezeichnete Arbeit, die zwar erfolgreich alle nationalen Grenzen und Abstände zwischen den Kontinenten überwunden hatte, am Ende aber den Launen eines Haustiers zum Opfer fiel und geschreddert wurde.

Viele Bücher sind aufgrund der fremden Sprache, in der sie abgefasst sind, geraume Zeit nicht jedem Leser zugänglich. Daher ist das Übersetzen eine der vorrangigen Aufgaben. Für die Lektüre eines Buches und die Erarbeitung eines Konzeptes, das ihm zugrunde liegt, benötigt man durchaus ein paar Tage. Ich will nicht weiterhin viele Worte verlieren über die Sichtung Tausender von Fotos, die Analyse der unablässig eingehenden Informationen, die Freuden und Qualen dieser ganzen Arbeit. Aufrichtig danken möchte ich

Leipzig's *Best Book Designs from all over the World* in China. As the director of the Publisher's Association of China, he has suggested a great amount of valuable insight for this book.

I would also like to thank Zhu Junbo, from the Shanghai News Department, who initiated the contest *China's Most Beautiful Books*, opening the road to Leipzig for Chinese book design and allowing it to flourish in the competition of this German city.

As I see it, through the changing artistic language of their form, books have the possibility to start a new era, leading to a new interpretation and definition of the printed media. *Best Book Designs from all over the World* shows that these possibilities – exquisite, alternative, bloody, documentary, and fantastic – all appear in the lively and vivid paper. Paper not only consists in a material or a printing process, but it is rather part of reading characters and visual appreciation. All this is comes together creating a harmonic beauty and finally composing a complete vision of the book.

The process of book making is really difficult, but collecting awarded books means undergoing all kinds of obstacles. Carrying out a long-term search means to use every means in one's hand and to ship goods from different places. Receiving parcels from everywhere around the world is indeed a joy, and one goes through experiences when one is between laughter and tears. For example, an awarded book that has crossed borders and continents marked with the symbol of the *Best Book Designs from all over the World* was destroyed by a naughty pet who chewed it. Many books often give a lot of trouble because of the problem of language barriers – translation is one of the biggest difficulties. It usually takes at least a few days to read a book and

的出版提供了重要的依据。还要感谢至今我国获“莱比锡世界最美的书”奖次数最多，现任中国出版协会书籍设计艺术委员会主任的多年老友刘晓翔先生也多次给本书提出了许多宝贵意见。也要感谢来自上海新闻出版局的老局长祝君波先生，是祝先生开创促成举办了“中国最美的书”的评选活动，打开了中国书籍设计通向莱比锡的通路，让中国书籍设计在莱比锡的舞台上绽放。

在我看来，书籍经由一种艺术转换的形式语言，很有可能开启一个全新的时代，引发一种关于纸媒的全新理解与定义。“世界最美的书”展示的就是这样一种可能，精致、另类、血腥、纪实、幻想，都是最生动与鲜活的纸上相遇。不仅在于材料触感与印刷工艺，更在于文字阅读与视觉欣赏，所有的一切，综合而成的和谐之美，才最终构成了这本书的全貌。

做书的过程实属不易，单是搜集书中获奖作品，就历经了各种艰辛。长期的搜寻，打通各种渠道，异地的转运，收到时时期盼的世界各地来的邮包的欣喜。当然插曲时时有，也曾有过令人啼笑皆非的经历，比如一本穿越洲际与板块的边界而来的“金字符”奖作品，却因宠物的顽皮嚼成碎纸落地。不少书籍由于语言不通的问题常常大费周章，翻译是头等难事，通读完一本

dem ganzen Team von Mitarbeitern, das sich um die Erstellung der „Leipziger Auswahl" verdient gemacht hat. Von Beginn an mit dabei war meine Studentin Zhou Chao, deren Forschungen für die über die Jahre hinweg anfallenden Arbeiten eine hervorragende Grundlage boten. Danken möchte ich auch meinem Kollegen Zhu Tao für sein sorgfältiges Mitwirken. Qing Yun und Wang Jia haben unendlich viel Mühe auf die Beschaffung von Büchern und Materialien verwandt, ihnen meinen Dank. Ich möchte auch dem Jiangsu Phoenix Fine Arts Publishing House für diese Buchveröffentlichung danken. Der Verlag hat keine Mühen gescheut, den Lesern dieses wunderbare Buch zugänglich zu machen.

Die „Leipziger Auswahl" ist eine offene Plattform der Schönheit. Die enorme Anziehungskraft von Büchern hat mit den in ihnen vorhandenen Räumen zu tun, in denen sich Emotionen entwickeln können und Fenster zur kognitiven Welt angelegt sind. Nur wenn es gelingt, mit wohl temperiertem Design das Temperament eines Textes zu vermitteln, lassen sich mit der Kraft des Papiers Zeitgrenzen überwinden.

Als Buchgestalter war und bin ich Anhänger der Idee, dass der Reiz der Schönheit aus der Vitalität kommt.

Nur Schönheit, die der Vitalität entstammt, kann zu einer echten Schönheit werden, die frei, mild, dauerhaft und grenzenlos ist.

Zhao Qing
am 14.März 2019 im Meiyuan, Nanjing

understand its concept. All the laughter and tears that we have undergone cannot fully describe everything we have gone through making near ten thousand photographs, analyzing and producing a great number of information charts. Therefore it was necessary to write an article to thank all the team of people who have put effort into *The Choice of Leipzig*.

First I want to thank my master-degree classmate Zhou Chao because his research gave me a very solid foundation for the following work. Then I would also like to express my deep gratitude to my colleague Zhu Tao, who carried out a complicated work sparing no effort. Also I would like to extend my thanks to assistant Qing Yun and master degree student Wang Jia for their perseverance in searching and buying the books. Finally, I would like to express my appreciation to Jiangsu Phoenix Fine Arts Publishing for publishing this book for all readers who are fond of beautiful books.

On the stage of the competition *Best Book Designs from all over the World, The Choice of Leipzig* is always open. The reason why this book is so fascinating is the inner space that it contains. It is not only a tranquil tree hole, but also a window to get to know the world. The strength of paper can only overcome the barriers of time by using the warmth of design to transmit the character of a work.

As a well-known book designer, I completely agree with this opinion: the appeal of beauty should stem from vitality. Only the beauty that comes from vitality can only become the most graceful and, from the beginning to the end, be always free, warm, long lasting, and unlimited.

Zhao Qing
on March 14th, 2019 at Meiyuan, Nanjing

书再梳理出概念至少要花费几天时间。近万张照片的整理拍摄，大量的信息图表的分析制作，许多笑与泪就不再赘述了。值此付梓之际，由衷地感谢所有为《莱比锡的选择》付出心力的团队和朋友们，从一开始参与的我的研究生周超同学，她的研究给后续工作提供了一个强有力的基础。感谢后续繁复工作的执行中同仁朱涛尽力细心的付出。还有助理卿云、研究生王嘉在收购书籍中所付出的坚持。也要感谢江苏凤凰美术出版社把此书出版奉献给爱美书的读者们。

在“世界最美”的舞台上，“莱比锡的选择”永远是开放的。书籍之所以能够存在巨大的魅力，是因为它所包含的内置空间，不仅是情绪安放的树洞，也是认知世界的窗口 。用设计的温度去传递文本的气质，纸张的力量才能穿透时间的限制。

作为一名书籍设计师，我始终认同一个观点：美的感染力，应当源自生命力。

缘由生命力而播种的美，才能成长为最美，自始至终，永远自由，温热，恒久，无界。

赵清

2019 年 3 月 14 日　于梅园

Mitglied der Internationalen Vereinigung der Grafiker (AGI); Stellvertretender Vorsitzender der Kommission für Grafikkunst innerhalb der Chinesischen Vereinigung der Mitarbeiter des Publikationswesens; Mitglied der Shenzhen Graphic Designers Association (GDC) und Gründer der Nanjing Graphic Designers Alliance. Im Jahr 2000 gründete Zhao Qing die Firma Han Qingtang Design Co., Ltd. und übernahm den Posten als Direktor für Design. Zhao ist stellvertretender Herausgeber des Verlags Phoenix Jiangsu Science and Technology Press und Betreuer von Masterstudenten der Designer-School der Kunstakademie von Nanjing. Seit mehr als 30 Jahren setzt sich Zhao für das Grafikdesign in allen Bereichen der Praxis und Forschung ein. Zahlreiche seiner eigenen Designerarbeiten gewannen Preise oder wurden weltweit im Rahmen fast aller wichtigen Grafikdesign-Wettbewerbe und Ausstellungen für Preise nominiert. Unter den internationalen Designerpreisen, mit denen Zhaos Werk ausgezeichnet wurde, sind u.a. anzuführen ADC, TDC, One Show Design (New York); D-AD (Großbritannien); Red dot, IF (Deutschland); Golden Bee (Russland); JTA (Japan); TDC (Tokyo); GDC (Shenzhen); GDA (Hong Kong), DFA (Hong Kong); Golden (Taiwan) usw.

The member of Alliance Graphique Internationale (AGI), Deputy Dean of The Committe of Book Design and Art of The Publishers Association of China, the member of Shenzhen Graphic Design Association (GDC), and the founder of Nanjing Graphic Design Association. In 2000, Zhao Qing established "Han Qing Tang Design Co., Ltd." and hosted the post of Chief Design Director. As the Associate Senior Editor of Jiangsu Phoenix Science Press and postgraduate supervisor of School of Design of Nanjing University of the Arts, Zhao Qing has devoted to practice, research and development of graphic design in various fields for more than three decades. His personal works are either rewarded or selected among almost all important worldwide graphic design contests and expositions. Moreover, he earned numerous international design rewards such as ADC (New York), TDC (New York), One Show Design (New York), D&AD (the UK), Red dot (Germany), IF (Germany), Golden Bee (Russia), JTA (Japan) , TDC (Tokyo), GDC (Shenzhen), GDA (Hong Kong), DFA (Hong Kong), Golden (Taiwan), etc.

赵清
Zhao Qing

国际平面设计联盟(AGI)会员,中国出版工作者协会书籍设计艺术委员会副主任,深圳平面设计师协会(GDC)会员,南京平面设计师联盟创始人。2000年创办瀚清堂设计有限公司并任设计总监。江苏凤凰科学技术出版社副编审,南京艺术学院设计学院硕士生导师。三十几年来坚持致力于平面设计各个领域的实践与研究推广,个人设计作品获奖或入选世界范围内几乎所有重要的平面设计竞赛和展览,并获得了美国纽约ADC、TDC、One Show Design、英国D&AD、德国Red dot、IF、俄罗斯Golden Bee、日本JTA、东京TDC,以及中国香港GDA、DFA、台湾Golden、深圳GDC等众多设计奖项。

SHEILA HICKS
2 kilo
JAMES JOYCE ULYSSES
Ein Tag im Leben von
CZECH 100 DESIGN ICONS
诗经
SUN
Found Footage
Cinema Exposed
Changing Cultures of Planning
PUMP
DANIËL VAN DICHT
BRICK
CHAQUE CHOSE A SA PLACE
The Migrant
The Anatomy of Stone
CARAVAGGIOS ROM
Stedelijk Museum Amsterdam